standard handbook of plant engineering

OTHER McGRAW-HILL HANDBOOKS OF INTEREST

standard handbook of plant engineering

Robert C. Rosaler, P.E. **Editor in Chief**

James O. Rice Associates
New York

James O. Rice **Associate Editor**

James O. Rice Associates
New York

McGraw-Hill Book Company

New York St. Louis San Francisco Auckland
Bogotá Hamburg Johannesburg London Madrid
Mexico Montreal New Delhi Panama Paris
São Paulo Singapore Sydney Tokyo Toronto

This Handbook Dedicated to Saul Poliak

Library of Congress Cataloging in Publication Data

Main entry under title:
Standard handbook of plant engineering.

Includes index.
1.Plant engineering—Handbooks, manuals, etc. I.Rosaler, Robert C. II.Rice, James O'Neill, date
TS184.S7 1983 690'.54 82-9988
ISBN 0-07-052160-3 AACR2

1234567890 KPKP 898765432

ISBN 0-07-052160-3

The editors for this book were Patricia Allen-Browne and Ruth L. Weine, the designer was Mark E. Safran, and the production supervisor was Thomas G. Kowalczyk. It was set in Century Schoolbook by University Graphics, Inc.

Printed and bound by The Kingsport Press

contents

board of advisors

contributors

Willard C. Allphin
Lighting Products Group
GTE Sylvania

Curtis W. Allshouse
Project Engineer
General Battery Corporation

Charles H. Artus
Product Application Engineer
Gates Rubber Co.

David R. Bender
Lexington, Ky

Anne Bernhardt
Staff Engineer
Gulf Research & Development Co.

Richard Best
Sr. Fire Analysis Specialist, Research and Fire Information Services
National Fire Protection Association

Raymond J. Birck
The Falk Corporation

Michael M. Calistrat
Manager, Mechanical Development,
Metal Products Division
Koppers Company, Inc.

Marvin Campen
Product Manager, Synthetic Fluids
Gulf Oil Chemicals Company

William Carter, Standards Engineer
The Torrington Company

Howard Cary
Vice President, Welding Systems
Hobart Brothers Co.

Kao Chen, P.E.
Westinghouse Electric Corp.

Karl J. Danz, P.E.
Principal Engineer
Roy F. Weston, Inc.

Rodger C. Dishington
Sr. Engineer
Imperial Oil & Grease Company

Martin C. Doherty
Consulting Applications Engineer
Thermal Power Systems Engineering
General Electric Co.

Richard J. DuMola
Materials Engineer
METCO, Inc.

John K. Edmonds
Executive Vice President
American Iron and Steel Institute

Wally Erickson
Chief Engineer, Product Application Dept.
Gates Rubber Co.

J. Mark Gilstrap
Technical Manager, Mechanical Engineering Services
Bently Nevada Corp.

Bruce Glidden
Assistant Vice President of Engineering
United States Steel Corp.

Nils R. Grimm, P.E.
Syska & Hennessy, Inc., Engineers

A. J. Grosso
Vice President, Engineering
American District Telegraph Corporation

Edward R. Guerdan
Senior Engineer
Westinghouse Electric Corp.

James E. Gutzwiller
Boston Gear Division,
INCOM International, Inc.

Michael R. Harrison
Manager, Engineering and Technical Services
Insulation Systems Dept.
Johns-Manville Sales Corp.

William S. Harrison, P.E.
Director of Technical Services
Engelhard Industries Division
Engelhard Corp.

Ronald E. Heil
Manager, Structural Section
Buildings Division
Sverdrup & Parcel & Associates

Leo J. Horvath
Director, Technical Affairs
Association of Asbestos Cement Pipe Producers

Joseph C. Jackson
President
Association of Asbestos Cement Pipe Producers

Walter D. Janssens
Director, Solid Research
Imperial Oil & Grease Company

Paul Jensen
Manager, Industrial Services
Bolt Beranek and Newman Inc.

Charles Albert Johnson, Ph.D.
Technical Director
National Solid Wastes Management Association

Paul A. Johnson
Manager, Civil Engineering Section
Transportation and Public Works Division
Sverdrup & Parcel & Associates

Robert J. Klepser
Sr. Development Chemist
Heavy Duty Maintenance Coatings
PPG Industries, Inc.

E. G. Kominek
Program Director, Advanced Water and Waste Treatment
Envirotech Corporation

Harold O. Kron
Philadelphia Gear Corp.

William J. Landman
Manager, Applications Engineering
The Trane Company

James A. Larson
Sr. Project Manager, Transportation and Public Works Division, and Sr. Geotechnical Engineer
Sverdrup & Parcel & Associates

Gordon F. Leitner
Vice Chairman,
Water Services of America, Inc.

Richard T. Lohr
President, International Chimney Corp.

Chris S. Louskos
O-Ring Division
Parker-Hannifin Corp.

Duane E. Lyon
Associate Professor
Mississippi State U.

C. J. McCann
Chairman
Tower Performance, Inc.

Fred A. Mañuele
Sr. Vice President
M & M Protection Consultants
Marsh & McLennan, Inc.

Philip H. Maslow, P.E., FCSI
Consultant
Chemical Materials for Construction

Arthur R. Meenen
Manager, Special Structures Section
Industrial Division
Sverdrup & Parcel & Associates

Frank H. Miller, R.A.
Specialist: Architectural Systems
Giffels Associates, Inc.

Russel N. Mosher
Assistant Executive Director
American Boiler Manufacturers Association

Stanley A. Mruk
Technical Director
Plastics Pipe Institute

Francis A. Murad, Ph.D., P.E.
Assistant Director of Industrial Engineering
Giffels Associates, Inc.

John F. Murray
Lexington, Kentucky

Herbert T. Nock
Application Engineer
General Electric Co.

John B. Painter
Gasket Division
Parker-Hannifin Corp.

Ralph E. Peterson
Parker Packing
Parker-Hannifin Corp.

Donald H. Pratt
Executive Vice President and
General Manager
Butler Manufacturing Company

P. Eric Ralston
Manager, Manufacturing Services
Babcock & Wilcox Co.

Richard G. Ramsdell
Seal Group
Parker-Hannifin Corp.

Richard D. Ramsey
Section Manager
Technical Services, Buildings Division
Sverdrup & Parcel & Associates

Ranjit S. Randhawa
The Foxboro Company

Thomas C. Roberts
Product Engineering Manager
S.S. White Industrial Products Division
Pennwalt Corporation

Robert C. Rosaler, Vice President
James O. Rice Associates, Inc.

Thomas J. Rost
Project Engineer
Chain Division
FMC Corporation

Richard B. Ruch, Jr.
Project Manager
Roy F. Weston, Inc.

Richard Ryan, P.E.
Manager, Plant Engineering
Hamilton Standard Division
United Technologies Corp.

Robert William Ryan
Assistant Director
Department of Environmental Safety
University of Maryland

Ken S. Satija, Ph.D., P.E.
United Engineers & Constructors
Inc.

John B. Scannell
Parker Packing
Parker-Hannifin Corp.

William M. Scardino, P.E.
Consulting Engineer

Benjamin A. Schranze, P.E.

Jaswant Singh, Ph.D., CIH
Vice President and Technical Director
Clayton Environmental
Consultants, Inc.

Robert L. Smith, Jr.
General Electric Co.

William H. Snyder
Director, Corporate Marketing
Research
Johns-Manville Sales Corp.

Arthur Spiegelman, P.E.
M & M Protection Consultants
Marsh & McLennan, Inc.

Georg Stromme, P.E.
Product Manager
Electrical Power Systems
Onan Corporation

William M. Throop, P.E.
Manager, Market Development
Envirex Inc.

Eric E. Ungar
Principal Engineer
Bolt Beranek and Newman Inc.

Jack P. Waite
Manager, Engineering Field Services
Imperial Oil & Grease Company

George W. Walsh
General Electric Co.

Dale F. Willcox
Sr. Vice President
and Chief Engineer
Furnas Electric Co.

B. J. (Jack) Williams, Jr.
President,Twin City
Roofing Inc.

Don Williams
Technical Service Director
Huntington Laboratories, Inc.

Edward Willoughby, P.E.
Director of Quality Assurance
Giffels Associates, Inc.

D. G. Wilson
Vice President and
General Manager
Facility Plans Engineering and
Construction
United States Steel Corp.

Terrence E. Wilson, P.E.
Project Engineer, Industrial
Engineering
Giffels Associates, Inc.

James V. Yu
Manager, New Market and Product Planning
Electric Motor Division
Gould Inc.

Aluminum Association
Washington, D.C.

American Gear Manufacturers Association
Arlington, Virginia

American Iron and Steel Institute
Washington, D.C.

Executone, Inc.
Long Island City, N.Y.

General Battery Corporation
Reading, Pennsylvania

Machine Design **A Penton/IPC Publication**

National Electrical Contractors Association, Inc.
Washington, D.C.

K. W. Tunnell Company, Inc.
King of Prussia, Pennsylvania

The Valve Manufacturers Association (Communications Committee)
McLean, Virginia

Waukesha Engine Division
Dresser Industries Inc.
Waukesha, Wisconsin

Weston Instruments Division
Sangamo-Weston Inc.
Newark, NJ

preface

Virtually every industrial activity has been affected, often in revolutionary ways, by the surge of technology. Because it is so central to virtually all manufacturing and service facilities, plant engineering is *uniquely* affected. It is "in the middle" in the sense that the plant engineer must, increasingly, have a broader knowledge of an ever-widening universe.

Events of the past decade have further served to accent the importance of the plant engineer's role in corporate operations, notably the demands for energy conservation and pollution control.

This Handbook is a response to these changing conditions and needs.

Arranging a logical structure and index to meet the needs of all engineers required considerable thought. The structure finally developed here is a reflection of the procedural sequences that occur in the plant facility itself: Planning and Construction, Plant Equipment Procurement and Operation, Maintenance. Individual equipment is covered broadly with descriptions of operational features, installation, and maintenance. Managerial aspects are included only where they interface closely with technical matter.

The objective of the book is to provide the reader with sufficient data on any specific equipment to permit judgment on choices and an insight into "how it works" and how to maintain it.

We want to express our appreciation to the authors and their organizations for their generous contributions and prompt execution of their tasks, and to the Board of Advisors for their guidance, particularly Leo Spector, Chairman of the Board and Editor of *Plant Engineering* magazine. For initial suggestions on the outline, we wish to thank Stewart Burkland. We also received excellent guidance and encouragement from Harold Crawford, Ruth Weine, and M. Joseph Dooher of McGraw-Hill. Our thanks go, too, to Dorothy Smith and Betsy Watson for helping to keep the project moving.

Saul Poliak, to whom this Handbook is dedicated, founded the National Plant Engineering Exposition and Conference which has been held both nationally and regionally since 1950. He has also pioneered similar expositions and conferences in the United Kingdom, Europe, Central America, and the Far East. It is widely recognized that he has been a major force in advancing the awareness of the critical plant engineering function in the industrial societies.

Robert C. Rosaler
James O. Rice

part A

The Basic Plant Facility: Construction, Equipment, and Maintenance

section 1

Planning the New or Remodeled Facility

Prepared by
Giffels Associates, Inc.
Southfield, Michigan

Authors

Frank H. Miller, R.A.
Specialist: Architectural Systems

Francis A. Murad, P.E., Ph.D.
Assistant Director of Industrial Engineering

Edward Willoughby, P.E.
Director of Quality Assurance

Terrence E. Wilson, P.E.
Project Engineer, Industrial Engineering

Glossary

These terms are excerpted from the *Architect's Handbook of Professional Practice*, Glossary of Contruction Industry Terms: AIA Document M101,* 1970 Edition, Copyright © 1970, by the American Institute of Architects with their permission under Application Number 81056. Usage of AIA language does not imply endorsement or approval of this Glossary. Further reproduction of AIA copyrighted materials is prohibited. Because AIA Documents are revised from time to time, users should ascertain from the AIA the current edition of the Document reproduced herein. Italicized type indicates language provided by author.

Addendum A written or graphic instrument issued prior to the execution of the contract which modifies or interprets the bidding documents, including drawings and specifications, by additions, deletions, clarifications, or corrections. Addenda become part of the contract documents when the construction contract is executed.

Area and volume method (of estimating cost) Method of estimating probable total construction cost by multiplying the adjusted gross project floor areas or volumes by a predetermined cost per unit.

Bid A complete and properly signed proposal to do the work or designated portion thereof for the sums stipulated therein, supported by data called for by the bidding requirements.

Bidding documents The advertisement or invitation to bid, instructions to bidders, bid form, and proposed contract documents including any addenda issued prior to receipt of bids.

Bidding or negotiation phase The fourth phase of an architect's or engineer's basic services, during which competitive bids or negotiated proposals are sought as the basis for awarding a contract for construction.

Budget, construction (1) The sum established by an owner as available for construction of the project; (2) the stipulated highest acceptable bid price or, in the case of a project involving multiple construction contracts, the stipulated aggregate total of the highest acceptable bid prices.

Budget, project The sum established by an owner as available for the entire project, including the construction budget, land costs, equipment costs, financing costs, compensation for professional services, contingency allowance, and other similar established or estimated costs.

Change order A written order to the contractor signed by the owner and the architect or engineer, issued after the execution of the contract, authorizing a change in the work or an adjustment in the contract sum or the contract time. A change order may be signed by the architect or engineer alone, provided there is written authority from the owner for such procedure and that a copy of such written authority is furnished to the contractor upon request. A change order may also be signed by the contractor when agreeing to the adjustment in the contract sum or the contract time. The contract sum and the contract time may be changed only by change order.

Construction cost The cost of all of the construction portions of a project, generally based upon the sum of the construction contract(s) and other direct construction costs. Construction cost does not include the compensation paid to an architect or engineer and consultants or the price of the land, rights-of-way, or other items defined in the contract documents as being the responsibility of the owner.

Construction documents phase The third phase of an architect's or engineer's basic services. In this phase an architect or engineer prepares from the approved design devel-

*AIA Document M101, Glossary of Construction Industry Terms, 1970 Edition, Copyright © 1970 by the American Institute of Architects. All Rights Reserved and further reproduction prohibited AIA ®.

opment documents, for approval by the owner, the working drawings and specifications and the necessary bidding information. In this phase an architect or engineer also assists the owner in the preparation of bidding forms, the conditions of the contract, and the form of agreement between the owner and the contractor.

Construction phase—administration of the construction contract The fifth and final phase of an architect's or engineer's basic services, which includes their general administration of the construction contract(s).

Contingency allowance A sum designated to cover unpredictable or unforeseen items of work or changes subsequently required by the owner.

Contract administration The duties and responsibilities of an architect or engineer during the construction phase.

Contract documents The owner-contractor agreement, the conditions of the contract (general, supplementary, and other), the construction documents (drawings and specifications), all addenda issued prior to execution of the contract, all modifications thereto, and any other items specifically stipulated as being included in the contract documents.

Cost-plus fee agreement An agreement under which a contractor (in an owner-contractor agreement) or an architect or engineer (in an owner-architect or owner-engineer agreement) is reimbursed for direct and indirect costs and, in addition, is paid a fee for services. The fee is usually stated as a stipulated sum or as a percentage of cost.

Critical path method (CPM) A charting of all events and operations to be encountered in completing a given process, rendered in a form permitting determination of the relative significance of each event, and establishing the optimum sequence and duration of operations.

Design development phase The second phase of an architect's or engineer's basic services. In this phase an architect or engineer prepares from the approved schematic design studies, for approval by the owner, the design development documents consisting of drawings and other documents to fix and describe the size and character of the entire project as to structural, mechanical, and electrical systems, materials, and such other essentials as may be appropriate. The architect or engineer also submits to the owner a further statement of probable construction cost.

Drawings The portion of the contract documents showing in graphic or pictorial form the design, location, and dimensions of the elements of a project.

Estimate of construction cost, detailed A forecast of construction cost prepared on the basis of a detailed analysis of materials and labor for all items of work, as contrasted with an estimate based on design area, volume, or similar unit costs.

Facility design program *A statement of an architectural or engineering design problem and the requirements to be met by proposed design solutions.*

Feasibility study A detailed investigation and analysis conducted to determine the financial, economic, technical, or other advisability of a proposed project.

Field order A written order effecting a minor change in the work, not involving an adjustment in the contract sum or an extension of the contract time, issued by the architect or engineer to the contractor during the construction phase.

Final acceptance An owner's acceptance of the project from the contractor upon certification by the architect or engineer that it is complete and in accordance with the contract requirements. Final acceptance is confirmed by the making of final payment, unless otherwise stipulated at the time of making such payment.

General conditions (of the contract for construction) That part of the contract documents which sets forth many of the rights, responsibilities, and relationships of the parties involved.

General contract (1) Under the single contract system, the contract between an owner and the contractor for construction of the entire work. (2) Under the separate contract system, the contract between an owner and a contractor for construction of architectural and structural work. *In some cases the bias of work may be mechanical or some other work, and in this case the mechanical or another contractor may be awarded the general contract.*

Joint venture A collaborative undertaking by two or more parties or organizations for a specific project or projects, having the legal characteristics of a partnership.

Letter of intent A letter signifying an intention to enter into a formal agreement, usually setting forth the general terms of such agreement.

Low bid Bid stating the lowest bid price, including selected alternates, and complying with all bidding requirements.

Percentage fee Compensation based upon a percentage of construction cost.

PERT schedule An acronym for *P*roject *E*valuation *R*eview *T*echnique. The PERT schedule charts the activities and events anticipated in a work process.

Preliminary drawings Drawings prepared during the early stages of the design of a project.

Prequalification of prospective bidders The process of investigating the qualifications of prospective bidders on the basis of their competence, integrity, financial status, and responsibility relative to the contemplated project.

Prime contractor Any contractor on a project having a contract directly with the owner. *On occasion, several prime contractors may be assigned by an owner to the general contractor for purposes of supervision, administration, and coordination.*

Progress schedule A diagram, graph, or other pictorial or written schedule showing proposed and actual times of starting and completion of the various elements of the work.

Record drawings (as-built drawings) Construction drawings revised to show significant changes made during the construction process, usually based on marked-up prints, drawings, and other data furnished by the contractor to the architect or engineer.

Schematic design phase The first phase of an architect's or engineer's basic services. In this phase, an architect or engineer consults with the owner to ascertain the programmed requirements of the project and prepares schematic design studies consisting of drawings and other documents illustrating the scale and relationship of the project components for approval by the owner. The architect or engineer also submits to the owner a statement of probable construction cost.

Shop drawings Drawings, diagrams, illustrations, schedules, performance charts, brochures, and other data, prepared by the contractor or any subcontractor, manufacturer, supplier, or distributor, which illustrate how specific portions of the work shall be fabricated and/or installed.

Specifications A part of the contract documents contained in the project manual consisting of written descriptions of a technical nature of materials, equipment construction systems, standards, and workmanship. Under the *Masterformat system,* the specifications comprise *17* divisions.

Statement of probable construction cost Cost forecasts prepared by an architect or engineer during the schematic design, design development, and construction documents phases of basic services for the guidance of the owner.

Study Preliminary sketch, drawing, *or equipment description and comparison to facilitate or evaluate the development of a project.*

Supplementary conditions A part of the contract documents which supplements and may also modify the provisions of the general conditions.

Unit prices Amounts stated in a contract as prices per unit of measurement for materials or services as described in the contract documents.

Introductory Statement

In theory, a product may be designed, a means of production may be developed, the facility for this production may be defined, a site for it may be selected and then master-planned; the actual facility may also be designed, constructed, equipped, and finally put into operation.

In practice, this series of events is not nearly so linear and involves multiple conditions of feedback, cross-checking, and redefinition. The complexity is due to a not unreasonable conflict of local optimums and a desire for overall balance of quality and economy. It is compounded by such factors as manufacturing policy, product markets, contractual concerns, labor characteristics, government regulations, technology, time and financial limitations, etc. In summary, all factors of industry, commerce, and economy become involved in this planning process.

The facility may be considered similar to a living entity with a cycle of birth, life, and death, all of which are affected by the factors mentioned. Some of the influences can be readily recognized and quantified; others may not even be known until they appear. Those readily recognized may include corporate manufacturing policies, product markets, etc. The process of designing, constructing, and later operating a new or modernized facility involves the never-ending identification, quantification, and utilization of a seemingly endless list of such socioeconomic influences. Profits, both tangible and intangible, may often be considered the most potent of those influences which determine change.

Section 1 deals with planning the new or remodeled facility and is the core of the birth phase of that facility's life cycle.

Since the planning process itself can be infinite with respect to time and content, one could plan forever and accomplish nothing. Planning must therefore conclude with a decision to implement the results, based on the planning to date. In no way does this end all planning. Because of its very nature, planning continues whether directed or not, on both an informal or formal basis, and results in decisions that may or may not modify prior decisions. This never-ending interaction between planning and decision-making processes gives rise to the general qualification of any plan: It will change!

To discuss planning the new or remodeled facility, it is necessary to abstract certain tasks and explain them as if they existed in a pure state. The logic of these uncomplicated procedures is reinforced by the long-run economies of proceeding with a classic design approach and is conversely undermined by the often-critical immediacy of need.

The planning process, in any event, presumes a beginning, and, as in this case where the facility alone is of prime concern, the product and means of production are thought of as existing. Often the development of a facility design is interrupted by major product or production revisions. Revisions may be caused by realizing the facility and facility cost implications of product or of production means, and may cause aspects of the facility to be reconsidered at the most basic program level, even when design may have gone quite far. To some extent, this cannot be avoided, but in the great majority of instances, adherence to the described planning processes should point out these implications at an early stage and minimize the pain and expense of revision. The processes, if followed in order, should also eliminate revision due to facility concerns alone. Although some conditions calling for design revision may yet exist, as progress resumes it should do so in the original logical sequence. The premise stating the existence of product and means of production remains valid.

Overall company or corporate concerns, such as manufacturing policies, product markets, etc., show some influence on the facility. They are also held outside the planning process by presuming that company or corporate directions, strategies, and policies have addressed and resolved such issues. These particular topics cover such a wide spectrum that facility concerns are normally of less importance and hence should rightfully support broader objectives.

Some planning tasks are on such a small scale that only handbook perusal is in order.

Other planning tasks are nearly monumental, and can easily reduce the handbook explanations to outline-level discussion. More complex tasks demand education and experience commensurate with the undertaking.

In displaying the planning process, the details of design are kept outside the text, and the processes themselves (developing a facility design program, a site selection, a long-range master plan, and a responsive facility design solution) are given emphasis.

The intent of this section cannot be tutorial. It can only lay down the framework of design and design management which is necessary for construction projects, define terms commonly used, and provide sources whence additional information may be obtained. It should be recognized that not all projects will require all the steps described.

Chapter 1-1. "Establishing Requirements," deals with determining physical needs and design objectives. Methods and materials used in making these determinations and developing a facility design program are discussed.

Chapter 1-2. "Site Selection," presents a logical process and important criteria that may be used to investigate, evaluate, and compare alternative project sites. Problems concerned with cost and analytic modeling are discussed.

Chapter 1-3. "Master-Planning," presents systematic steps and important issues that may be used to produce and continuously update a long-range master plan for the facility. Problems associated with long-term usability, future alterations, and expansions are discussed.

Chapter 1-4. "Facility Design," explains the use and meaning of the schematic, design development, and construction documents phases which lead to the bidding or negotiation and construction phases. Follow-up procedures are also discussed.

Chapter 1-5. "Scheduling, Estimating, and Contracting," explains the development and limitations of project planning activities. Methods and procedures used in establishing and developing project functions are discussed.

These five topics represent the major elements included in "Planning the New or Remodeled Facility."

Facilities are the buildings, the site, and the equipment within them. Because so much is "building," a great deal of the text relies not on the language of engineering but on the language of architecture. This is important not only to present things as they are in the building industry but also to avoid creating a confusing list of parallel terminology and procedures. For this reason, this section started with a glossary which should prove helpful.

chapter 1-1

Establishing Requirements

Establishing requirements for a facility is an analytic process that includes collecting and processing information and presenting the results necessary for stating an architectural or engineering design problem and the requirements to be met by proposed design solutions.

INTRODUCTION

Before the design effort may begin, a clear definition of goals, functional concepts, criteria, and problems is essential. A listing of these design requirements, to be referenced as design criteria, must be developed, and this listing is termed a facility design program.

The publication of the American Institute of Architects, *Architect's Handbook of Professional Practice,* in discussing project procedures, begins with the first, or project program, stage by stating: "Before any real progress can be made in the design of a building it is necessary to have a definite design program giving conditions precedent to construction and requirements to be met by the proposed project."

A fairly competent technician can design/specify a building to enclose space and exclude weather, but postoccupancy rearrangement and a plethora of subsequently applied appendages often bear evidence to the fact that the wrong building was designed in the first place. The fault here does not necessarily lie with the designer. Most probably the building or facility was not, in these instances, correctly programmed to address the proper requirements. Programming, the process by which facility requirements are established in sufficient detail to permit later project implementation, occurs prior to expending time and effort on design. Later stages involve only confirmation of this programming effort and not the actual formulation of it.

The appropriateness of many design solutions must often be answered by the designer's experience and intuition. A far greater portion of assurance results if the designer has followed a carefully researched, analyzed, and documented facility design program. The level of detail of the program, as well as the type of programming methods used, is determined by project conditions and the designer's needs for sufficient and accurate information. Some judgment is called for to make this determination.

The decisions made during this initial programming become the foundation for a multitude of subsequent decisions. For this reason, the care, expertise, and depth of research and analysis involved in formulating the program should not be underestimated. Projects incorporating programming errors require great sums to rectify those errors in the later stages of design and even greater sums to rectify them in the stages of construction or operation.

PURPOSE

The explicit purpose of programming is to develop design requirements detailing what is to be designed.

Implicit to this task are major adjunct purposes, without which the essential assuredness of the program is not possible. These include the gathering of information, the resolution of preferences, and the presentation for required approvals.

Because the information is rarely close at hand, the gathering of the necessary material, data, plans, observations, measurements, comments, and opinions is a long and tedious task. Yet, the results of the effort must be comprehensive, exact, and, most of all, appropriately addressed to the real issues. The programmer should usually be familiar with the objectives of the activities that are to be involved in the design procedure. If this is not the case, then during the programming process the programmer must develop an association with those who are familiar with the objectives. Consequently, a percentage of the programming effort may appear trivial, obvious, or naturally assumable. Nevertheless, this portion must be tabulated and supported since approvals by management and major crucial decisions by other persons emanate from their understanding of the information presented in the program.

Much of the information, when structured, can be considered to produce self-evident facility design requirements. Some information produces only general and undefined awareness of problems. When such a vagueness is present, there remains the process of selecting an optimum or priority which provides direction to the subsequent design effort. This selection is of an abstract functional concept, not a physical design response, which describes how the operations or activities should perform.

The idea of feasibility studies originates at this stage. A feasibility study, while certainly not limited to programming, must often be done to determine the desirability and viability of chosen functional concepts. Without such early resolution of intent, later design stages may become burdened with the examination of endless random solutions.

To communicate programming results to the designer without having obtained required approvals not only improperly bypasses many authority structures but also does not represent the best way to obtain long-run correct implementation. Management must be made aware, in as much detail as is prudent, of the implications of the proposed actions. In cases where consultants are to be retained for design, an approved facility design program is mandatory and may become contractually binding. Examples of design that was carried to great lengths, on the basis of unofficial or divergent program premises, are numerous and unfortunate. The basic step of acquiring management approval reduces these risks. It also lends important credibility to the program subject matter, and subjects the program to another level of expert analysis.

The ultimate purpose of programming is to assemble all results and relevant support information into a detailed report describing what is to be designed; this is the facility design program. This document should rightfully be regarded as a project milepost and key to project success. In no case should it be treated lightly.

TIMING

Adequate time for programming is essential. Facility requirements should be established as early as possible to permit studies concerning such things as remodeling vs. new construction to be an integral part of the programming activity. The urgency of some projects often precludes the setting aside of a distinct programming time slot, in which case the task is nonetheless performed, often informally, as the initial step of design. In this latter instance the semantics may vary, but the activity must be carried out, albeit briefly.

At times programming may be performed to provide data for just site selection or master-planning. In these cases the level of program detail will naturally vary, since facility design is not the end product. Very often the level of detail is deceptively low. The needed program material is at the summary or conclusion level; this demands a considerable amount of investigation, calculation, deliberation, and resolution not always apparent in the brief final statement.

It is not uncommon for facilities programming efforts to be included in feasibility analyses regarding products, production methods, product markets, etc., since analysts or management are often concerned about the facilities implications of these issues.

PERSONNEL

Establishing facilities requirements (programming) is accomplished by a team comprising an owner's representatives, consultants, or a combination of these two groups. Some specific programming efforts are considered a specialty and as a result have sometimes been delegated to consultants (architects, engineers, management or manufacturing consultants, etc.) with highly effective results. When dealing with such consultants, the input of the owner's representatives, because of their intimacy with operations, is essential. The programming team may require individuals with education, experience, and above-average analytical and communications skills in the following disciplines:

- Architectural planning and design
- Civil engineering
- Construction engineering and cost estimating
- Construction specification
- Electrical engineering
- Environmental protection
- Financial analysis
- Industrial engineering—specifically plant layout, process and storage systems, process control systems, and materials handling
- Mechanical engineering
- Personnel and labor management
- Plant engineering and maintenance
- Plant management
- Product engineering

- Production engineering
- Purchasing
- Tax law
- Transportation and traffic control

It is important to provide the individuals responsible for programming with an appropriate central authority for information gathering and reporting. This can be a single individual or a committee. If it is a committee, there should be a coordinator or director, with responsibility and authority, who can ensure against delays and deviation from the established objectives. Such an individual becomes the project manager during programming, and may continue in this capacity during subsequent design and construction phases.

In a situation where all the required expertise is not present among the persons preparing the program, members of the committee may be used to provide additional input. This may take the form of establishing program criteria and providing evaluation for consultants or may involve the more demanding task of providing full segments of the actual written facility design program. An example of this would occur in an industrial firm which has extensive in-house capability in plant layout, process and storage systems, process control systems, and materials handling, yet wishes to retain consultant architects and engineers to develop a complete facility design program. In this case, the consultants would likely incorporate committee inputs directly into the final program, placing on the respective committee members the dual tasks of management responsibility and study execution.

A corporation may have individuals with all these needed skills on its staff and yet elect to obtain additional expertise through consultants, as would a much smaller company. This would be done for many reasons, including excessive work load, maintenance of confidentiality, the desire for special expertise, fresh ideas, etc. In the final analysis, all areas of concern identified during programming require appropriate skills to achieve satisfactory results.

METHODOLOGY

The methodology for establishing facility requirements is that of investigating, tabulating, evaluating, acquiring approvals, and composing a facility design program in a logical and sequentially proper fashion. The end product should be descriptive of what is to be designed (and should *not* include how it is to be designed) in sufficient detail that the designer may productively work along a defined path to develop responsive solutions.

A reasonable order of events for this procedure is as follows:

- Existing facility investigation
- Future facility evaluation
- Constraints evaluation
- Requirements determination
- Program presentation and approval
- Program composition

Investigation of Existing Facility (Generally applicable to addition or remodeling)

A precise inventory of current facilities is essential. This should include not only listings of materials, equipment, etc., but also relevant notes concerning capacities, condition, and other subjective or intangible characteristics. Even operating methods, where they have facility implications, should be noted. To confirm the actual physical status of an existing facility, there is no substitute for on-the-spot personal observation. The assembled material should be collected, indexed, and either bound or filed in preparation for the anticipated referencing; it should include the following:

- Site information
- Plans, sections, elevations, and details of construction
- Area and volume tabulations
- Equipment (type, capacities, condition, etc.) data
- Energy and utilities usage and location data
- Personnel and personnel distribution data
- Flow/organization/operation charts
- Operations and maintenance practices data
- Photographs (including aerial, if possible)
- Interviews with individuals
- Personal observations

The organization of raw information, such as personnel or accounting data, is often not in the desired format, and this is the proper time to organize this material. Care should be taken to rearrange data within the available level of knowledge concerning the history and structure of the information, but this should be done only when necessary to maintain project programming applicability.

Evaluation of Future Facility

Any facility expansion must be examined in terms of the motivating need and analyzed in terms of the possibilities available without involving new building work. It should be remembered that construction has the potentially negative aspects of permanence, expense, and disruption, all of which should be balanced by the accrued benefit of the proposed project.

Initially, one should ask the question: Is any construction genuinely required? It is often possible to solve many needs without new construction and for less capital than construction would entail. In other cases the opposite is true. Much of this may require feasibility studies as a part of the programming effort.

To illustrate this point, the following example may be helpful:

> Consider a parts receiving department where additional loading docks appear to be necessary to reduce extensive truck waiting time outside the delivery entrance. A simple addition to or rescheduling of lift truck operations might allow present personnel to accommodate peak loading in a shorter turnaround time. This may alleviate the problem at minimal increase in operating expense and avoid the need for new construction. On the other hand, the installation of additional truck docks may allow rescheduling of existing personnel and lift trucks to other areas during nonpeak periods, permitting a reduction of operating expense. This may permit a short payback period for the construction investment.

If a satisfactory resolution to the problem can be found without involving construction, such a determination should be made at this earliest of stages. Few actual conditions are as elementary as the example, and often the basic question of construction desirability must be asked over and over again at all stages of project development. As the designer should continuously question the programmer, so should the programmer question the production and operations personnel who are the eventual facility users.

Evaluation of Constraints

Every aspect of facility design and construction contains constraints, both the type which limit options and the type which direct options. The effects of such constraints can vary from barely discernible to extensive.

The constraining influences affecting facility design and construction may include the following:

- Corporate policy
- Product markets
- Long-range plans
- Operations and manufacturing policies
- Contractual concerns
- Labor requirements (including contracts, agreements, etc.)
- Governing regulations (Office of Safety and Health Administration, Environmental Protection Agency, zoning, codes, etc.)
- Utility, energy, and transportation requirements
- Financial limitations
- Physical limitations

Each of these should be examined, tabulated, and evaluated on the basis of the source of the constraint, the impact of the constraint, and the desirability and/or possibility of circumventing or eliminating the constraint.

Those constraints which emanate from important and powerful sources, such as upper management and the government, are generally regarded as fixed, especially when they involve interrelationships with vital corporate and social infrastructures. Where the impact is minimal, or so universal as to apply to virtually all project design and construction, the constraint is in all probability also fixed. When the measures necessary to evade the influence of constraint are extremely circuitous or time-intensive, again the constraint is probably fixed. Finally, when, in the professional opinion of responsible project personnel, for legal or other reasons, the wisdom of some constraints is sufficient to accept them, these also may be considered as fixed.

Conditions, however, may exist in which it is desirable to examine the potentials of constraint elimination or modification. Some might include situations where significant economic impact arises or where there is chance inclusion of certain constraints, arbitrary regulation, etc. Programming should be prepared under the premise of constraint evasion only when both the possible resultant benefits are significant and the likelihood of successful constraint evasion is high; otherwise a significant portion of subsequent activities may be rendered worthless. In any case, the evasion of constraints is a tenuous venture and should be initiated only with proper managerial concurrence.

Determination of Requirements

A comprehensive description, categorically comparable to that developed during the investigation of existing facilities, must be compiled for the proposed new facility. Since the new or remodeled facility does not yet exist, the level of detail may be different or more general in nature.

The established means of production and other basic activities must be described, tabulated, ordered, sized, and arranged to satisfy operational needs. Support, circulation, logistic factors, and constraints are included. If this may be completed on a numeric and written basis alone, no two-dimensional programming effort (the preparation of graphics) is required, but in many cases some basic diagrams must be made.

Where existing facilities are involved, revised block layouts and occasional equipment layouts are often assembled to examine the interface of proposed new activities with existing conditions. These diagrams should show variations in the present activities, enlargements of the present activities, additions of new activities, and alterations to other activities.

Where new construction alone is envisioned, similar diagramming may be required but without the influence of existing conditions.

As with most such situations, there are generally as many ways to approach a problem as there are planners. In most cases not all functional concepts are acceptable and those which are not should be eliminated to allow later efforts to be concentrated on viable options. Some functional concepts can be eliminated at the programming stage without developing schematic designs or making feasibility studies. Such arbitrary elimination is certainly not an easy task and relies on the programmer's understanding of the project objectives.

Where concepts quickly come to light they should be evaluated first in terms of acceptability, second in terms of known constraints, and last to see if they have a possibility of satisfying the program goals. If these criteria are met, the question of economic feasibility should be addressed. Some functional concepts are such that an arrangement may be quickly specified in sufficient detail that a cost opinion can be generated, and hence the feasibility, or at least the desirability, can be examined. Others may be such that they cannot be subjected to a cost estimate without extensive design development. In this instance, the feasibility study may have to be relegated to a later design stage, when details are properly included.

In order to generate a cost opinion, it is necessary to establish both a requisite area or volume and a level of quality. The sizes must be either calculated on the basis of known experiences and from published standards or guessed. When this information is available, a cost may be established, as described in Chap. 1-5 under "Estimating," and integrated with the project budget. If the costs are found to be too great, then perhaps the area, volume, quality level, or some other program requirement may have to be questioned.

The determination of facility requirements is not designing. The programming should be carried only to the extent of describing what is to be achieved by the design and should incorporate feasibility studies only if they are required to understand and evaluate functional concepts. The results should rely greatly on experience, expertise, established standards, safety factors, multipliers, contingencies, etc.

The question constantly posed during programming should be: What must the design solution address with respect to project needs or requirements? It is a quantifying and analytical function.

Program Presentation and Approval

Not every functional concept need be presented for management review in the final facility design program; nor is it necessary, in cases where a single option is possible, to generate multiple choices in the hope of being perceived as comprehensive. Although in some instances management persons reviewing a program elect to examine the document minutely, more than likely the persons responsible for investigation and analysis should do some screening to reduce the number of choices to a recommended few.

Each functional concept qualifying for final consideration should be selected on the basis of a distinct, positive, and unique feature which is both the reason for its inclusion and for the rejection of others. Those not included should be appended. Those preferred should be elaborated with respect to at least the following characteristics:

- Spaces, volumes, and relationships
- Personnel and personnel distribution
- Equipment (type, capacities, grouping, and segregation)
- Service and service distribution
- Security
- Priorities
- Movement (flow and interaction of process, materials, vehicles, and people)
- Flexibility
- Energy and utilities usage

- Analysis of probable implementation time
- Analysis of probable implementation costs
- Analysis of probable economy (life-cycle and operating costs and profitability)

Emphasis should be given to variance. Where there is an insignificant differential between options, possibly in the categories of equipment, energy and utilities usage, or personnel and personnel distribution, then repetitive display documentation is of limited value; cross references may be used.

The objective of the program presentation is to gain approvals. As a consequence, the submission of material for review prior to the final presentation, especially material involving radically different or unexpected approaches, is extremely important. There should be no big surprises in a presentation.

It is not uncommon for management to prefer that program data be presented by means of visual or graphic comparison to familiar existing facilities or to alternative options. The resultant graphics are often referred to as *schematic* drawings though they are really preschematic or program drawings. They are not design drawings and are made for programming purposes during the programming phase. They should be labeled program proposals to avoid later confusion.

Program Composition

Because the programming effort yields a compiled facilities design program, ultimately carrying the endorsement of management, and because it may be incorporated as a component in contracts with design consultants, it should be formally drafted. When the document becomes a formal description of scope of services, then the specific wording should be reviewed by persons familiar with design contracts.

The composite results of the programming effort become the facility requirements. Although persons executing the subsequent design may be given strict design criteria, it is often valuable to supplement these criteria with raw information from the programming work. The designer is thus made more aware of important factors underlying certain program decisions and is often permitted to proceed without redeveloping a basis for design.

If the programming and resultant facilities design program are indeed comprehensive and accurate, then it is reasonable to assume that conformance with criteria, schedules, and budgets contained within them will follow. When this is not the case, the possibility that subsequent design and construction decisions may diverge from project objectives must be accounted for by contingency allowances. Contingency allowances in design features, time allotments, and cost allocations are, historically, a pragmatic component. The lack of contingency considerations increases the likelihood of program abbreviations, time extensions, or cost escalations.

ISSUES

A comprehensive and interrelated set of nine issues must be addressed in the facility design program. These issues must be resolved in the context of the particular conditions of a given industrial situation or given location. The first issue is goals.

Goals

Often it is management who establishes the goals for a new project. Such goals can be a major document or a one-line statement; the statement can address many issues or only a few, and it can be absolutely specific on some points and all but overlook others.

Some typical brief statements of goals encountered during programming activities might be: "The safety methods right now are inadequate; in the new facility it would be desirable to make it next to impossible for an accident to occur," or "Maintenance costs are out of hand; with this remodeling there should be a vast reduction in that kind of expense, even if it means additional investment," or "The employees in the old buildings shouldn't have to feel like second-class citizens compared to those in the new one." These statements, while they may be unquantified and general, can tell the designers a great deal; they can affect many subsequent decisions, and it may be wise to document their source.

There should also be a more specific goal, which is the actual reason for the project itself, and this, too, should be stated. This specific goal should not be a public relations promise but rather a statement of what the project is to achieve. An example might be: "This warehouse addition is intended to increase our storage capacity of purchased parts supply from 30 days to 60 days." Care should be taken to make sure the goal is one which facility design and construction can achieve, and not one which requires management or operational change.

Functional Concepts

This issue is often an extension of manufacturing engineering data, as compiled by the programmer, and is descriptive of the activity which takes place in the completed facility. This material should follow the format for functional concepts developed in programming presentation and approval.

Criteria for Design

This information conveys the established performance criteria which the designer is expected to meet and usually pertains to types of systems, not details of systems. Such design criteria may include the following:

1. Master Plan (if executed)
2. Plan Requirements
   - Areas, dimensions, volumes, and capacities
   - Shapes
   - Adjacency
   - Sequence relationships
   - Segregation (acoustical, thermal, hazards, contaminants, etc.)
   - Screening of unsightly activities and industrial hygiene requirements
3. Function Requirements
   - Access and control
   - Control and communications systems
   - Energy, utilities, and waste-treatment systems
   - Service/maintenance systems and methods
   - Security system and methods
   - Personnel (number, distribution, etc.)
4. Building Type Requirements
   - Architecture and planning
   - Building systems/construction methods
   - Image

Criteria for Construction

These data are used to convey standards, preferences, and requirements for the materials, related details, or methods of assembly to be used.

1. Drawing and Specification Format
2. Materials and Construction Methods

Detail or material preferences are often in evidence as a result of individual owner experiences. These should be communicated to the designer at this very early stage to assure that the resultant design solution uses the desired components.

Site Conditions

If a site has been selected prior to programming, the following information should appear in the facilities design program.

1. Location and Property Description (surveys, photographs, drawings of existing structures, etc.)
2. Utilities (location, specifications, restrictions, etc.)
3. Zoning Codes and Permits (agencies, contacts, and restrictions of contact)

If a site has not been selected, feasibility studies or site selection could become a part of the programming effort. More about this subject appears in Chap. 1-2, "Site Selection."

Schedule for Design and Construction

Schedules developed during programming can vary from an elaborate network plan to merely a list of critical dates; a bar chart is sometimes adequate.

Budget for Design and Construction

1. Financing Concerns (source of funds, interest rates, etc.) Programming may disclose severe financial constraints or cash flow considerations. Such matters may interact with scheduling and estimating market forecasts.
2. Budget (construction, equipment, fees, permits, contingencies, etc.)
3. Cost Estimating Requirements

Contracting Methods

This issue is briefly discussed here, and additional information on the subject is found in Chap. 1-5 under "Contracting."

The Design Work

One of the most complete and well-written descriptions of design contract concerns is AIA Document B141a, "Standard Form of Agreement between Owner and Architect." This distributes the design contracting concerns among 14 articles. No matter what method is used, the following points must be considered:

- Architect's or engineer's services and responsibilities
- Owner's responsibilities
- Construction cost
- Direct personnel expenses
- Reimbursable expenses
- Payments to the architect or engineer
- Architect's or engineer's accounting records
- Ownership and use of the documents
- Arbitration

- Termination of agreement
- Miscellaneous provisions
- Successors and assigns
- Extent of agreement
- Basics of compensation

The Construction Work

It may be possible during programming to determine the number of construction contracts required and the distribution of work between them. It may also be necessary to determine how the construction work will be contracted, a subject which is discussed in detail in Chap. 1-5 under "Contracting." There are several ways in which this contracting may be accomplished, as listed below:

- Integral with the design as a "design/build" package
- With a single general contractor, with multiple independent contracts, or through a construction manager
- Through competitive bidding or negotiated price
- For a fixed sum or the cost of the work plus a fee, the latter with or without guaranteed maximum cost limits or contractor participation in cost savings

Problem Statement

There should be a summary statement which describes the design intent. This is a brief premise for design and should cover four topics:

- Function
- Physical character
- Level of quality
- Life expectancy

In summary, establishing facility requirements for a new or modernized plant is most successfully accomplished by a programming activity leading to a facility design program, which describes the design problem. Each programming effort must be tailored to the proposed project scope. (Two worker-years of programming prior to the construction of the proverbial birdhouse is irrational.)

chapter 1-2

Site Selection

Site selection is an analytic process of narrowing the facility location options to a comprehensively evaluated final selection on the basis of the company or corporate directions described in the facility design program or other design documents.

INTRODUCTION

From the initial question of locating additional or new facilities, the process of site selection becomes one of elimination.

This logical process should be a multilevel effort, examining first major regions, then individual states, SMSAs (standard metropolitan statistical areas), counties, municipalities, and finally individual proposed sites. The SMSA data are perhaps the most useful, because of the generous amount of information available under the title from both state and federal agencies, and are more indicative of real conditions than any politically defined area.

Great care should be taken to make comparisons or evaluations of equivalent levels in order to ensure the validity of each decision. There is often a tendency to orient quickly to a specific site when it is encountered, even while broad regions, states, etc., are still being evaluated. This is a risky situation, and can result in significant distortion of regional, state, or SMSA data with site-specific characteristics.

The possibility that any one site may eventually prove difficult or impossible to obtain, together with the hazards of an evaluation bias caused by use of site-specific data, makes it prudent to conclude the site selection process with two or three reasonably equivalent selections.

The nature of searching for a site assures encounters with persons who have vested interests. Certainly persons in state or municipal development authorities or members of chambers of commerce have vested interests, as do representatives of industrial parks, utilities, railroads, etc. This is to be expected, but it should be recognized that such people speak from prejudiced positions. Care should be taken to exclude from the search consultants, or even in-house personnel, who can be identified as having potential conflicts of interest. Examples of such individuals might be an architect, engineer, or construction manager whose operating range includes only a portion of the study area and who stands to gain further commissions only through selections within a specific region or state; an employee whose career future may be affected by relocation; a consultant who provides services to both the communities seeking industry as well as to the industry seeking a community in which to locate (a case of dual allegiance); and local enthusiasts who do not have the power or authority to deliver promised location inducements.

Site selection is accomplished by applying comprehensive evaluation to a tabulation of issues and by integrating these conclusions into an array for which there is, hopefully, a quantifiable optimum. Though a location decision is not always concluded in terms of single site response, the in-depth evaluative process, conducted as described, should provide the basis for an intelligent decision.

PURPOSE

The purpose of the site selection process is ultimately to provide the best location for a specific facility. This has a significantly broader scope than simply locating the best possible property, though the latter task is certainly an integral part of a comprehensive site selection process. An important early facet is indeed that of regional analysis or, in the case of some projects, possibly national or even international analysis. In any case, it remains the process of narrowing the choices until eventually the specific property is selected and acquired.

TIMING

Site selection generally follows a determination that new facilities are in order, though it may also be carried out in varying degrees to examine the feasibilities of relocation. On occasion, it has been done years in advance of anticipated future needs in order to acquire a site and, in effect, put that site on the shelf. Desirable industrial properties are becoming harder to find and are escalating in price, and, even if the site is ultimately not needed, it may at least develop as a good investment.

The selection process may proceed consecutively with or subsequent to programming, prior to facility master-planning, or at any time prior to final design. Most analysts would probably concur that final site selection is most appropriately made after a conceptual master plan has been produced, as described in Chap. 1-3, "Master-Planning."

PERSONNEL

Site selection is usually effected by a team comprising owner's representatives, consultants, or a combination of these two groups. The entire activity is often considered a specialty and, as a result, is now almost always handled by specialized consultants (architects, engineers, management and accounting consultants, site selection consultants, etc.). The input of the owner's representatives must not, however, be minimized, since the final decision is their responsibility. The team may be required to have specialists with education and experience in the following areas:

- Architectural planning and design
- Civil engineering
- Construction engineering and cost estimating
- Demographics
- Economics
- Electrical engineering
- Environmental protection
- Industrial development financing
- Industrial engineering—specifically plant layout
- Mechanical engineering
- Personnel and labor management
- Plant management
- Political science
- Real estate
- Tax law
- Transportation

The activity of individual specialists naturally becomes greater or less as the selection process proceeds from a regional scale to the evaluation of specific parcels of land. Each nonetheless has valuable and often overlapping roles.

It is important to provide those persons making the actual site selection with appropriate investigative and reporting mechanisms. These mechanisms should be similar to those described in Chap. 1-1, "Establishing Requirements."

METHODOLOGY

The methodology of site selection is that of analytically evaluating site data in terms of the criteria which describe corporate concern for location or relocation. The process may be used for any facility where location is a question, and presumes sufficient definition of program requirements to assure applicability. An effective series of steps for doing this is as follows:

1. **Criteria Development** Formulate site selection criteria.
2. **Weighting** Individually weight each element of the criteria.
3. **Investigating** Obtain regional, local, or actual site information related to each of the criteria for each of the compared locations.
4. **Rating** Individually analyze and rate the information gathered, with respect to the criteria, for each of the compared locations.
5. **Scoring** Factor the rating and the weighting for each criterion at each location, or, in the case of only one location, compare this with anticipated norms.
6. **Selecting** Resolve a locational optimum and obtain purchase options for the best site(s).

An example of this kind of tabular comparison is shown in Table 2-1, "Regional Analysis Summary."

TABLE 2-1 Regional Analysis Summary

Criterion category (weight)	West coast	Midwest location	Southeast location	Border state location	Present location
	Region: (rate) and scoring				
1. Climate (4)	(9) 36	(4) 16	(3) 12	(6) 24	(5) 20
2. Land (2)	(2) 4	(7) 14	(7) 14	(8) 16	(6) 12
3. Proximity (8)	(3) 24	(7) 56	(6) 48	(6) 48	(5) 40
4. Economy (6)	(6) 36	(4) 24	(8) 48	(8) 48	(5) 30
5. Labor (9)	(4) 36	(3) 27	(6) 54	(8) 72	(5) 45
6. Transportation (6)	(8) 48	(3) 18	(6) 36	(8) 48	(5) 30
7. Energy/utilities (4)	(3) 12	(8) 32	(7) 28	(6) 24	(7) 28
8. Construction (3)	(4) 12	(4) 12	(5) 15	(9) 27	(3) 9
9. Government (6)	(5) 30	(7) 42	(7) 42	(9) 54	(5) 30
10. Social (5)	(7) 35	(2) 10	(5) 25	(8) 40	(3) 15
Total (53)	273 (5.1)	251 (4.9)	322 (6.0)	401 (7.6)	259 (4.9)

The formulating and weighting of the site selection criteria comprise a matter in which all parties, especially the responsible management staff, should achieve full initial concurrence. This step is essential to establishing the validity of the process. The research, analysis, and rating of gathered information are the responsibility of persons with related expertise. Even when a final tabulation provides undesirable results, only the ratings and their supportive analysis, not the criteria or their weightings, should generally be the subject of a critique.

It should be noted that an absolute summated score in excess of another absolute summated score does not necessarily denote a "winner." The absence of extremely low ratings and the balance of specific lows with specific highs may also be considered important.

Results generally indicate simple groupings into unacceptable, acceptable, and preferred selections. Only on rare occasions are absolute answers apparent. The emphasis should be on developing appropriate, adequate, rational criteria and on providing research sufficient to rate the site in terms of each criterion at a consistent level of detail. If these steps are taken, the resulting comparative analysis will not suffer from inappropriateness or incompleteness. Where information is not available to analyze a given site in terms of a given criterion, this should, of course, be noted, but the rating should not be reduced. To reduce the rating might artificially reward completeness, whereas the favorable or unfavorable nature of the content should be the only factors taken into account.

Additional information concerned with this methodology may be found under the heading "Notes" and comprises the last paragraphs of this chapter.

An effective tool for providing location comparison and presentation of that comparison is a cost model. This is presented as no more than a tool because of the many nonpecuniary factors intrinsic to any business decision. The cost model might be used to compare either the one-time costs of establishing a new facility in certain locations, the ongoing annual operating costs at these contemplated locations, or both.

Comparison of initial costs might include such factors as:

- Land and land acquisition
- Existing plant acquisition
- Permits
- Roadway, railway, and utility connections
- Building(s), services, and site development
- Furniture and equipment and installation
- Off-site construction requirements
- Operations start-up
- Fees (real estate, legal, architectural, engineering, etc.)
- Facilities relocation
- Personnel relocation
- Personnel acquisition and training
- Administration
- Financing
- Contingency

The impact of these costs can be greatly increased or decreased by the level of assistance provided by economic development authorities. A well-coordinated effort by these persons, coupled with service incentives, may not only reduce costs but also provide schedule improvements, with attendant cash-flow benefits and minimized interest costs.

The time element is worthy of note, as it can vary significantly, and only a portion of the impact of schedule delays may be reflected in the costs. Some examples of schedule impediments are permit requirements, Environmental Impact Statement requirements, site conditions, training demands, etc. It is entirely possible to encounter schedule extensions that may delay a project for years, and this may offset otherwise notable cost advantages. Again, state and local development agencies are well aware of how discouraging these and other factors might become and are usually prepared to offer assistance.

Comparison of annual operating costs can include:

- Proximity (this includes service and support functions, raw materials, manufacturing support, and product shipping)
- Labor (this includes payroll, fringe benefits, and local practices costs)
- Energy and utilities (this includes electricity, gas, water, wastewater, etc.)
- Government (taxes)

These costs must be compared with initial costs and other fiscal concerns in order to determine the composite financial picture at a studied location.

An example of this kind of tabular comparison is shown in Table 2-2, "Annual Operating Cost Model."

A model of annual operating costs involves the interrelationship of cost categories and is valid only if an equivalent level of detail and accuracy is present throughout all categories. Individual categories such as labor or proximity costs may easily be developed in comparative model form for different locations by use of hypothetical criteria. A comprehensive cost model, however, depends on the hypothetical criteria being very representative of the actual conditions.

The following example should be helpful in explaining this relationship:

> When comparative labor costs are developed for all locations, on the basis of a hypothetical working staff of 250 persons, the benefits of one area over another may be observed. When proximity costs developed for the locations are based on a hypothetical shipping cost of 1500 $\times 10^6$ $ft^3 \cdot mi$, the benefits of one area over another may also be observed. But to consider both

TABLE 2-2 Annual Operating Cost Model

	Present facility		Present location		Southeast location		Border state location	
3. Proximity								
Raw materials shipping	.229		.229		.450		.415	
Service and support	.217		.120		.085		.069	
Manufacturing support	2.616		2.616		2.100		1.900	
Manufacturing support shipping	.030		.030		.059		.029	
Product shipping	1.654		1.654		2.742		2.268	
		4.746		4,649		5.436		4.681
5. Labor								
Payroll	4.990		4.230		3.390		3.860	
Fringe benefits	1.650		1.590		.950		.850	
Productivity factor	—		—		.434		.165	
		6.640		5.820		4.774		4.875
7. Energy and utilities								
Electricity	.612		.485		.292		.351	
Gas	.161		.108		.135		.114	
Water and wastewater	.008		.006		.007		.100	
Other	.440		.195		.221		.202	
		1.181		.794		.655		.767
9. Government (Taxes)								
Land	.110		.100		.017		.030	
Property (capital)	.470		.540		.369		.405	
Property (inventory)	.135		.130		.104		.099	
Corporate income	1.000		1.000		.790		.670	
Sales	.700		.700		.322		.355	
Franchise	—		—		.095		.035	
Workers' compensation	.078		.072		.028		.030	
		2.493		2.542		1.725		1.624
Total annual cost		15.060		13.805		12.590		11.947
Variation		—		—(8%)		—(16%)		—(20%)

labor and proximity costs together may be valid only when the 250 persons and 1500×10^6 $ft^3 \cdot mi$ are genuine. New, and possibly differing, results would occur if the figures became 290 persons and 1200×10^6 $ft^3 \cdot mi$ in locations where labor and shipping costs do not necessarily rise or fall in tandem, which is often the case. The need for careful programming in advance of site selection becomes evident.

Another noteworthy characteristic of an annual operating cost model is that many important factors that may have a cost effect cannot be evaluated in this comparison. Such factors are climate, economy, transportation, and social conditions. For example:

- The climate may include particularly harsh winters or tornado seasons which may result in unplanned and irregular work stoppages.
- The local economy may have an unstable cost-of-living history which, in varied cycles, may create wage-increase pressures.
- Transportation facilities may not accommodate some growth conditions, which may result in overutilization, delay, or disruption.
- The social conditions may not be attractive, which may impair recruitment and retention of key employees.

None of these can be inserted into an annual operating cost model unless at a factored rate, which produces only a cost-oriented weighting and rating array, with money as the common denominator. This is not an incorrect procedure, but it no longer reveals a true annual operating cost. However, if an acceptable level of detail and accuracy is known for the categories of proximity, labor, energy and utilities, and government, an annual operating cost model is valuable. A comprehensive regional analysis summary, though not directly a measure of costs, includes a great deal of information with nontabulatable or nonquantifiable cost implications.

ISSUES

Site selection on any scale must address a comprehensive and interrelated set of issues. There may be no absolute or objective optimal resolution for many of these issues, but they must be addressed in the light of specific industrial and location factors even when they contradict each other.

Barring situations in which the intention is rapid improvement and resale, the purchase of land involves the concept of long-term ownership. To properly evaluate land or site acquisition, the duration of ownership (or the facility life expectancy) should be defined. This is important because the issues to be addressed are often modified by generally intangible "trends." Quantifying the time element limits the modifying effect of these trends, just as the absence of an established time frame can result in giving an undue weighting to some other factor.

A comprehensive range of issues to be considered might include the following.

Climate

1. General Description
2. Temperature Range
3. Rainfall/Snowfall
4. Wind Range
5. Solar Range

Emphasis is placed on cost implications, quality-of-life implications, and potential work disruptions. Description includes macro- and microcharacteristics.

Land

1. Geography
   - General description of countryside
   - Topography
   - Seismology
   - Hydrology, soils, and subsoils
   - Drainage and waterways
   - Vegetation
   - Existing usage and structures
   - Environs
   - Economic implications
2. Availability
   - Description of land market
   - Description of land available
   - Range of property costs (not including taxes)
   - Trends

Proximity

1. Major Raw Material Components and Sources
2. Primary Service and Support Facilities
3. Manufacturing Support Facilities
4. Product Distribution
5. Competitor Profiles

Economy

1. Industrial/Commercial/Agricultural
   - General description
   - SIC (*S*tandard *I*ndustrial *C*lassification) distribution
   - Sales volume, value added, expenditures
   - Characteristics
   - Business climate data
   - Trends
2. Demography
   - General description
   - Income distribution
   - Cost of living and standard of living
   - Population (distribution and growth)
   - Trends

Labor

1. History and Characteristics
   - Unionization
2. Availability
   - Description of labor market

- Tabular profile of labor distribution
- Trends

3. Costs
   - Direct costs/fringe-benefit costs (not including taxes)
   - Productivity, absenteeism, turnover
   - Trends

Transportation

1. Air
2. Rail
3. Roadway
4. Waterway and Ports
5. Mass Transit

For each factor, areas of comparison are:

- Available facilities
- Utilization
- Limitations, regulations, etc.
- Costs (not including taxes)
- Proximity, access, and access requirements
- Trends

Energy and Utilities

1. Electricity
2. Natural Gas
3. Bottled Gases
4. Water
5. Wastewater Treatment
6. Telephone
7. Other (coal, fuel oil, etc.—if applicable)

For each element, areas of comparison are:

- Available facilities and capacities
- Utilization
- Limitations, regulations, etc.
- Costs (not including taxes)
- Proximity, access, and access requirements
- Easements
- Trends

Construction

1. Materials
   - Availability or proximity
   - Service availability or proximity
   - Trends

2. Construction Trades
   - Availability or proximity
   - Trends
3. Costs
   - Regional modifying characteristics and indices
   - Site conditions and access
   - Cost range
   - Trends

Government

1. Governing Building/Operating Regulations
   - Permits required
   - Agencies involved
   - Zoning and planning agencies' concepts
   - Codes, regulations, and restrictions
2. Environmental Impact Statement (EIS) or Prevention of Significant Deterioration (PSD) Regulations
3. Taxes
   - Description of taxation
   - Agencies involved
   - Tabulation of taxes and tax rate
   - Tax incentives
   - Trends
4. Financial Incentives
5. Service Incentives

Social

1. Residential Characteristics
   - Housing—types and distribution
   - Purchases housing—costs range and availability
   - Rental housing—costs range and availability
   - Trends
2. Education
   - Facilities, demographic characteristics, and utilization
   - Vocational programs
3. Recreational and Cultural Characteristics
4. Public Safety (fire and police)
5. Health Services
6. Community Services
7. Religion
8. Communications and Media
9. Character and Attitudes of Population
10. Quality-of-Life Description (Will the area attract and retain employees?)
11. Trends

NOTES

The use of a weighted and rated evaluation system can result in a rational evaluation of dissimilar data, but a number of safeguards should be observed to avoid numerical distortion of the results.

Presuming each alternative region, locality, or site is rated in each category on a scale of 1 (low) through 10 (high), then:

A rating of either 1 or 10 should seldom be used, the assumption being that something worse or better may exist in some other mythical area. These ratings often become subjective and should be avoided.

A rating of 2 or 3 is one which should be looked at with care and depth, as it is indicative of potential problems. Likewise, a rating of 8 or 9 should be found only when the potential for benefit is assuredly excellent. The middle range of ratings, from 4 through 7, indicates average values or only a trend toward negative or positive effects. These last ratings denote a general lack of both problems and benefits, within the depth of the analysis.

Since a region or locality is rated as objectively as possible, judgment is often limited to the availability of facilities or opportunities. The rating should reflect the degree of availability, and only in some cases should the apparent cost differential be sufficient to offset variances in availability.

As a result, the more urbanized regions usually receive higher scores. This situation seems prejudicial in favor of metropolitan areas. This is because, in rating an area, the subjective characteristics are minimized and the availability of resources and opportunities is emphasized; this, as a rule, occurs in urban areas.

The site selection study should be well referenced and indexed since it is possible to accumulate too much resource material to include in the final site selection study documents.

Since a good deal of the information used to develop the study may originate from promotional material, the information cannot always be taken at face value. Even when presented with seemingly impressive statistical data, such information must be treated with caution. For the same reason, a great deal of the information may not be directly comparable to information from other areas. This problem can be somewhat avoided by generous use of federal documents, but no matter what the source, comparisons may be complicated by time-frame discrepancies, since a common year is not always available for comparison of all categories.

In spite of the fact that data may be promotional, dissimilar, or discordant, no attempt should be made to bring material arbitrarily or mathematically onto a common ground. Without extensive background knowledge concerning the history and compilation of such data, such an attempt would risk distortion of the content.

It is wise to pay close attention to data definition and dates.

chapter 1-3

Master Planning

Master-planning is the development of a comprehensive long-range plan, based on the design program or other design documents, into which future facility requirements, alterations, and expansions fit as orderly additive increments.

INTRODUCTION

A master plan directly identifies and addresses specific growth-related problems and should be recognized as just that by all parties concerned. A management commitment to master-planning may add greatly to the impact and the implementation of the results, whereas a lack of such commitment may generate disinterest and a general feeling that the undertaking is nonessential, or a frill.

Important to the worth of master-planning is the concept of keeping the master plan current. Just as the planned facility is dynamic and probably undergoing continuous change, the long-range plans must be amended and updated, lest they become obsolete. Updating is usually not complex, yet is often neglected, with unsatisfactory results.

Since the people who prepare a master plan often include consultants, there is a noticeable undercurrent to the discussion concerning how one might best deal with the specialist who is retained. The plant engineer's role in this interaction often consists of providing criteria, objectives, restraints, and evaluation, which are basic owner activities. These tasks contribute immeasurably to producing valuable results.

The master-planning methodology consists of discussing how persons intimately familiar with operations and operational needs may develop the critical long-range facility requirements. These requirements become the complete long-range facility description only when they are applied to the expansive tabulation of issues in a fashion similar to that described in Chap. 1-2, "Site Selection." The result is a master-plan program. How a responsive design is developed for an individual facility becomes a master-plan design solution. Discussion of the details of creating master-plan design solutions would likely result in divergence to synthesizing processes and hence is not included.

PURPOSE

The purpose of master-planning is to make provision for orderly future expansions. In addition and more importantly, the purpose is to assure that present facilities and equipment remain usable and viable within the expanded facility's framework. Despite the seemingly ample capacity of most initial construction undertakings, time often produces a pattern of growth greatly in excess of initial estimates. The eventual size of a matured facility may become so large as to completely subvert the initial facility. The effort of master-planning is to develop means and provisions by which the initial segments of a total facility will remain contributory to the ultimate expanded facility and certainly not become an impediment to it.

This can be true for the largest undertaking; it can also be true for the most minor of plant alterations. In either case, a master plan should provide the framework for decision making with regard to the long-term effects of construction solutions responsive to short-term needs.

TIMING

Master-planning may be done at any point in the life of a facility. The earliest would be prior to site selection in order to develop a conceptual plan (as only a conceptual plan is possible without a specific site) as criterion for the actual selection of a location. In some ways, the most appropriate time is prior to facility design, when comprehensive expansion requiremeents may be used to establish master-plan directions for the entire design solution. Unfortunately, the most common time for master-planning is prior to an expansion increment, when existing and not always durable facility elements are present. The use of a master plan is no greater or less at any project development stage, though its use and scope are certainly different. Most planners would probably concur that "the sooner, the better" is an appropriate adage with respect to master-plan timing.

PERSONNEL

Master-planning is accomplished by a team comprising an owner's representatives or consultant(s), or a combination of these two groups. This activity is considered a specialty and, like site selection, has come to be handled almost exclusively by specialized consultants (architects, engineers, land planners, etc.). As with all consultants, the input of the owner's staff should be given its fair weight. This team should have individuals with education and experience in the following areas:

- Architectural planning and design
- Civil engineering
- Construction engineering and cost estimating
- Ecology
- Electrical engineering
- Environmental protection

- Industrial engineering—specifically plant layout, process and storage systems, process control systems, and materials handling
- Landscape planning and architecture
- Mechanical engineering
- Personnel or labor management
- Plant engineering and maintenance
- Plant management
- Transportation and traffic analysis

It is important to provide those persons doing the master-planning with an appropriate information gathering and reporting mechanism. This mechanism should be similar to that described in Chap. 1-1, "Establishing Requirements."

METHODOLOGY

The methodology of master-planning is that of projecting various expansions, evaluating these expansions, and providing program requirements (not designs) to accommodate the change. The results may be applied to any subsequent facility design solution. The tasks include:

- Projecting expansions
- Evaluating expansions

Projecting Expansions

An appropriate master plan must be founded on the process or activity which is the intent of the facility. A plan simply showing a site fully covered with future buildings is a land-use study, not a master plan. Hence, an examination of the potential future expansions of the initial activity is of the first order of importance.

Changes in the activity and in the resultant facility requirements may be brought about by:

- Variation in the activity
- Enlargement of the activity
- Addition of related activities
- Addition of unrelated activities
- Alteration to another activity

A number of changes can occur without significant alteration of facilities—a change of management, means of control, equipment, number of employees, level of automation, etc. The considerations pertinent to master-planning are those which result in facility modification.

Evaluating Expansions

Each proposed facility modification must be evaluated in order to determine the degree of attention necessary in master-planning. Some of the criteria for this evaluation may be:

- Probability of occurrence
- Importance of occurrence
- Impact
- Cost
- Time of occurrence

Evaluative criteria might well be weighted and rated for each of the specific potential activity changes to be considered. Discussions relative to this weighting and rating process are in Chap. 1-2, "Site Selection." The result of this evaluation should be some concrete definition of expansion criteria ranked by importance. Where expansion of facilities appears probable, certain characteristics of the expansion should then be developed. Some of these may be:

- Type
- Degree
- Sequence
- Flexibility
- Function of initial components in the expanded facility
- Relocation requirements
- Demolition requirements
- Disruption during construction
- Costs

The expansion characteristics may be applied specifically to most physical components of the facility, including land, process operations equipment, buildings, outdoor spaces, utilities, materials handling equipment, roadways, parking areas, railways, and individual building systems. This procedure may also be used in the development of performance criteria for design solutions or for equipment and material selections.

Though master-plan design solutions are significantly more subjective than the master-plan criteria, they may be first logically and then empirically refined until a final approved master plan evolves. The results should satisfy activity and facility concerns of the ultimate expanded facility. Initial construction may be planned to conform with projected goals and hence to obtain maximum long-range usability of the initial construction.

The acceptability of the master-plan design solution is based on the presence of flexibility, adaptability, and a degree of functional viability and expansion alternatives at each planned increment of expansion.

ISSUES

Master-planning of the facility must address a complex and interrelated set of issues. Issues must be resolved in the context of specific company or corporate goals and growth projections, as well as in the sometimes contradictory context of each other.

Not all instances of master-planning involve the development of total facility plans so significant interaction with all issues may not occur. Take, for example, the master-planning for a single system (e.g., systems for electrical power distribution or trucking access, holding, loading, and unloading). A single system may well be analyzed, programmed, planned, and implemented by a limited number of individuals for a very specific scope. As long as this effort relies on the established expansion requirements, is respectful of the fact that it is only a portion of a much greater whole, and pays heed to interfacing with related issues, then the limited-scope master plan is entirely reasonable. For such an endeavor, the full complement of issues serves as a master list from which appropriate items are selected.

This comprehensive range of issues includes:

Climate

1. General Description
2. Temperature Range
3. Rainfall and Snowfall (Precipitation)

4. Wind Range
5. Solar Range

Emphasis is placed on construction implications, maintenance implications, operating-cost implications, and potential work disruptions. Description includes activity and building program requirements.

Land Use

1. Geography
   - General description of countryside
   - Topography
   - Seismology
   - Hydrology, soils, and subsoils
   - Drainage and waterways
   - Vegetation
   - Existing usage and structures
   - Environs
   - Economic implications
2. Utilization
   - Description of use distribution
   - Areas, setbacks, and easements
   - Coverage areas and ratios
   - Visibilities/views/image
   - Social impact
   - Expansion

Government

1. Governing Building/Operating Regulations
   - Permits required
   - Agencies involved
   - Zoning and planning agencies concepts
   - Codes, regulations, and restrictions
   - Trends
2. *E*nvironmental *I*mpact *S*tatement (EIS) or *P*revention of *S*ignificant *D*eterioration (PDS) Regulations
3. Taxes
   - Description of taxation
   - Agencies involved
   - Tabulation of taxes and tax rate
   - Tax incentives
   - Trends
4. Financial Incentives
5. Service Incentive

Emphasis is placed on design and operational requirements, from the standpoint of life-cycle investment planning. Also important are permit costs and time requirements, as they affect project implementation.

Arrangement

1. Activities
   - Description of process operations
   - Adjacency
   - Sequence relationships
   - Segregation (hazards, contaminants, etc.)
   - Material flow interface
   - Expansion
2. Buildings
   - Areas/dimensions/volumes/capacities
   - Adjacency
   - Sequence relationships
   - Segregation (acoustical, thermal, etc.)
   - Screening of unsightly activities
   - Expansion
3. Outdoor Spaces
   - Function
   - Access and control
   - Landscaping (type, function, and appearance)
   - Visibilities/views/image
   - Expansion

Movement

1. Materials Flow
   - Loading and unloading (quantity, location, size, type, and control means)
   - Storage (quantity, location, type, access, density, and control means)
   - Staging
   - Aisles/ramps/doors
   - Containerization (quantity, size, type, and transfer points)
   - Inspections interface
   - Use interface
   - Flow (pattern, quantity, vehicle type, and control means)
   - Expansion
2. Vehicular
   - Site entry
   - Staging
   - Paved areas (quantity, location, type, access, and control means)
   - Flow (pattern, quantity, vehicle type, and control means)
   - Special vehicles
   - Service vehicles
   - Emergency vehicles
   - Expansion
3. Parking
   - Paved areas (pattern, quantity, location, entry access, building access, and control means)
   - Segregation (administration/operations/visitors)

- Personnel interface
- Expansion

4. Rail
   - Site entry
   - Staging
   - Flow (pattern, quantity, switching sequence, and control means)
   - Expansion
5. Aircraft and Flight Patterns
6. Watercraft and Port Facilities
7. Mass Transit
8. Personnel
   - Flow (pattern, quantity, and control means)
   - Segregation (administration/operations/service/visitors)
   - Expansion

Systems

1. Controls/Communications
2. Energy and Utilities
   - Site entry
   - Components (quantity, locations, size, and type)
   - Distribution (trestles, tunnels, etc.)
   - Expansion
3. Services and Maintenance
   - Facilities (location, type, and access)
   - Expansion
4. Security
   - Facilities (pattern, location, type, entry points, and operating methods)
   - Automated surveillance and control systems
   - Integration with plant operation norms
   - Personnel interface
   - Expansion

Construction

1. Design
   - Description of architecture and planning
   - Construction methods/building systems
   - Image
   - Social impact
   - Expansion suitability
   - Trends
2. Materials
   - Availability/proximity
   - Durability and service availability or proximity
   - Expansion suitability
   - Trends

3. Construction Trades
   - Availability/proximity
   - Jurisdictional situation during expansion
   - Trends
4. Costs
   - Regional and local modifying characteristics and indices
   - Site conditions/access
   - Costs range
   - Trends

NOTES

Expansion is a potentially vague aspect of any master plan. Though a certain amount of variation, enlargement, addition, and alteration to the original activity may be predicted, the influence of time is generally unsettling to the best of plans.

The answer to the dilemma is to include only a minimum of restrictions in the initial facility and allow for a maximum of flexibility. Consultants who deal in expansions could point out many examples of inadequate provisions for expansion in an initial design which was done without sufficient master-planning. This is unfortunate, particularly in cases where otherwise equivalent materials and methods could have been used. Not every component of a facility can be master-planned for later expansion in every way, since the cost of constructing to prepare for so many possibilities would be self-defeating. The point is that certain options exist during facility programming and design, and these options should be weighed not only for their primary function—quality and cost—but also for their compatibility with the expanded facility as defined by the master plan.

It is noteworthy to mention the previously listed expansion criteria ranked in order of importance. It is through reference to these criteria that the details of programming and design may be evaluated as to their compatibility with anticipated expansion. Where expansion requirements are determined to be of significance, the initial facility design may well include such things as oversizing and overspacing to allow for probable additions. Conversely, when expansion criteria are adjudged to be little more than speculative, the initial facility design may do no more than provide for some additional undeveloped land.

chapter 1-4

Facility Design

Facility design is a synthesizing process that includes the formulation of schematic design solutions, the development of system and detail definition, and the production of documents necessary to obtain construction proposals and accomplish the work described in the design program.

INTRODUCTION

Facility design is a creative process, in which the finite requirements of the facility design program are systematically filled by design solutions. At its beginning the design may be originated by one individual, an architect or engineer, who can envision a means to solve the problem. But facility design is also a technical process, and the design solution must be developed in all the detail necessary to obtain construction proposals and to execute the work. For any major project this procedure will probably involve the efforts of many persons, including specialists. The particulars of a project dictate the composition of the team and the extent of each person's involvement. The nature of the design process dictates beginning with the creative process and progressing toward the technical development and description necessary for construction.

No designer, either a member of the owner's staff or a consultant, can be expected to continuously respond to design problems without some process for doing so. Although intuitive responses are important, it is simply not wise or reliable to react to all facility requirements with experience and intuition alone. As a consequence, overall facility design, as well as its component parts, is discussed both as a logical process and as a procedure that requires systematic empirical methods.

The process is distributed into five phases, each of which terminates with a management decision approving the development of the project up to that point. The process of approvals is vital to management as it confirms that the development of the project is along lines which it, as owner, requires and prefers. The process of approvals is equally vital to the designer as it confirms the efforts to be productively applied to the anticipated result.

PURPOSE

The purpose of facility design is to logically develop a design solution responsive to the requirements of the approved facility design program. Some review of the program may be called for, followed by a trial-and-error run through multiple schematic design solutions and a degree of balancing between quality, time, and cost issues. The product must be a single design solution which meets all the requirements of the program.

The final purpose is to create and develop graphic and written construction documents which clearly and totally define the design solution for a building construction project, an equipment procurement project, or any combination thereof, in terms understood by construction contractors and equipment suppliers.

TIMING

Facility design has a number of prerequisites. A facility design program must have been compiled to obviate an empirical search through many alternatives; a site selection must have been made to avoid discussions of hypothetical codes, soils, environs, etc.; and a master plan must have been drawn up to ensure proper attention to future expansion requirements.

Facility design, however, must be completed prior to construction or equipment procurement. When the demand for completion of the construction is great, it is sometimes necessary to compress the sequence of events. This may be done by separating the construction into a number of activities, each with individual beginning and ending dates. In such a case the design might also be separated into comparable activities, each with individual beginning and ending dates. This is termed *fast track* and can be so condensed that some design activities have not yet begun when certain construction activities have been completed. Of course, every design activity must be completed prior to its corresponding construction activity. In addition, the project must pass as a single cohesive solution through some level of design detail (either schematic design or design development) prior to the time when activities may be broken up.

PERSONNEL

The time and effort spent on each phase is determined by the nature and complexity of the project, as well as by the time and resources available. Where the project is of such magnitude or complexity that the design effort may tax the available in-house capabilities, many owners retain the services of architects, engineers, or other consultants to supplement their staff or provide expertise not otherwise available.

Facility design is accomplished by a team which, depending upon the project scope, may have individuals with education and experience in the following areas:

- Architectural planning and design
- Architectural detailing
- Civil engineering
- Construction engineering, specifications, and cost estimating
- Control systems engineering
- Ecology
- Electrical engineering
- Energy conservation
- Environmental protection
- Industrial engineering—specifically plant layout
- Landscape planning and architecture
- Process engineering
- Mechanical engineering
- Structural engineering
- Traffic analysis and materials handling

Recent progress has rendered the computer a powerful aid, and computer-aided design (CAD) is having a marked effect on many aspects of design. With CAD it is possible to analyze a greater number of available potential solutions to levels of depth and detail that would require prohibitive time and cost if carried out manually. The advances in computer graphic systems enable the development of three-dimensional models and the viewing and analysis of these models while varying many of the parameters or dimensions involved. Such systems are also used for building design, process layout development, and equipment design, and are often used to produce finished drawings.

In addition, computers are often employed to prepare specifications, cost estimates, and design and construction schedules, and to provide management assistance in the administration of contracts. As in other fields which have been influenced by computer technology, *use of the computer within the design process can be expected to further expand as the technology becomes more versatile and less expensive.*

The commercially available CAD systems should be thoroughly evaluated before a firm invests in them. *To be economical, these systems must be structurally adaptable to the specifics of the project, and all involved personnel must have a working knowledge of the CAD system selected for use.*

METHODOLOGY

Facility design is normally segmented into separate, dependent phases, which together form a logical progression to the completed facility. It should begin with the goals, functional concepts, criteria, and problems analyzed during the programming stage where the requirements are established. The process initially proposes schematic design solutions and becomes more specific as time and effort progress. The preferred solution is subsequently synthesized into detailed building and equipment drawings and specifications.

The division of the design effort into phases provides a proper mechanism for continued evaluation of the viability of the project and allows for a systematic decision-making process. The most universally accepted terminology for the various phases is that promulgated by the American Institute of Architects (AIA). The Glossary at the beginning of this section gives specific meanings to terms which are not fully covered by the dictionary meaning of the word.

Depending on the size of the project, the type of the project (whether only construction work is involved or equipment procurement is also involved), and the contracting methods, certain design phases may be combined and others may be expanded and given more importance. It is also not unreasonable to expect that, with time constraints, the various phases may sometimes overlap.

In either all new facilities or remodeling projects involving building work or equipment the design tasks to be performed are:

1. Schematic, or Feasibility Study Phase
2. Design Development Phase, during which preliminary designs are prepared
3. Construction Documents, or Final Design Phase during which construction, fabrication, and procurement documents are prepared
4. Bidding or Negotiating Phase when bids are solicited and contracts awarded
5. Construction Phase during which building construction and equipment fabrication are carried out

A time for review should be reserved after each design phase is complete. At the review a decision must be made to proceed with the design, to retreat to the previous phase and rework the design, to delay the project, or to cancel it. The various phases of design and their interaction with each other are shown in Fig. 4-1.

Most companies have an established decision-making framework or process, and it is important that this framework be used to arrive at prompt decisions during each review and to keep the project on schedule. Orderly decision making should not be sacrificed to save time, as doing so may lead to future setbacks.

SCHEMATIC PHASE

This phase, sometimes called Feasibility or Preliminary Design, commences after the conclusion of programming and uses the requirements and functional concepts formulated during that stage. The requirements often include rearrangement of activities, the addition of new activities or equipment, new building construction, or a combination of all. The schematic phase should consist of investigating all feasible methods of satisfying the requirements and therefore quite often requires participation of an interdisciplinary group to perform the design tasks.

The schematic design phase is an appropriate time for investigating solutions which involve new techniques and technologies; it is also a time for general "brainstorming." This involves intuition and experience, and is the creative essence of architecture and engineering—the formulation of design solutions. As many alternative ideas or solutions as possible should be conceived and evaluated, though not all of the alternatives need be developed completely. Some may be discarded due to prohibitive costs or other considerations; however, care should be taken to not reject any idea too early.

Alternative design solutions may be ranked in several ways: from traditional or current methods through totally new techniques, or combinations in between; from the least

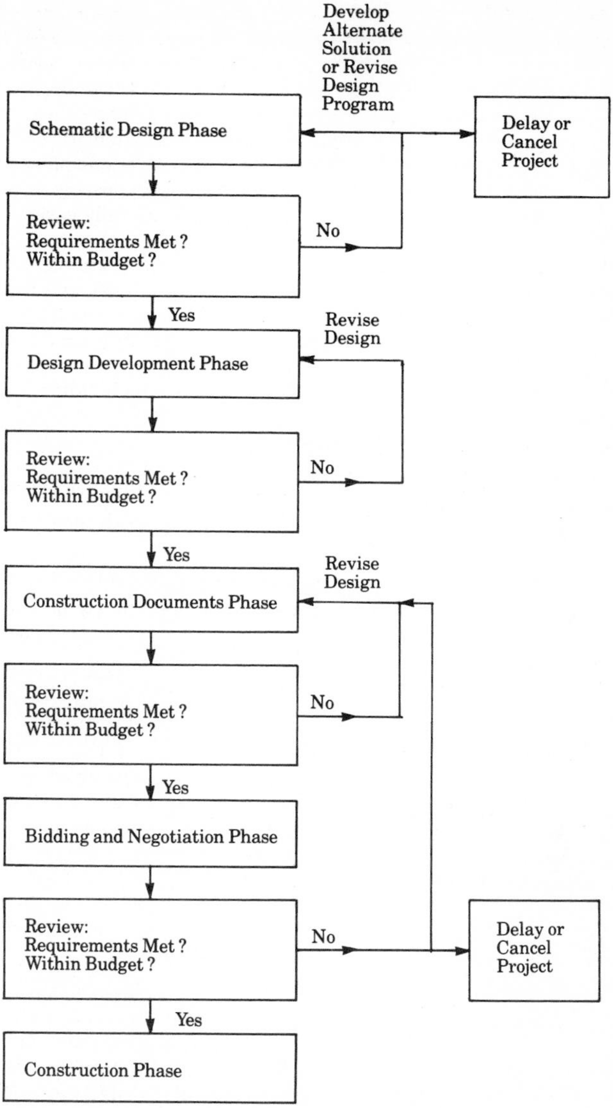

Figure 4-1 Design phases.

amount of cost to the most; from purely manual to highly automated; or any other logical sequence that fits the task at hand.

In this phase, one should look at the overall facility and how a particular solution fits in with the operation of adjacent departments or facilities, as well as how it fits into overall future requirements of the master plan. Tendencies to optimize a subsystem at the expense of the overall operations, particularly in support activities such as materials handling, warehousing, and the like, should be avoided.

No hard-line drawings need normally be prepared in this phase. Sketches which provide enough detail to differentiate between alternative design solutions and which can be used to develop comparative costs are sufficient. These sketches may take the form of block diagrams and layouts, which define space requirements and overall equipment arrangements for each of the operations involved, and flow diagrams, which establish departmental and functional relationships and interdependencies.

To develop and justify the proposed design solutions, the following activities are usually performed during the schematic phase, some of which may be repeated, in more detail, later in the design process:

- Analysis of constraints
- Analysis of codes
- Analysis of standards
- Analysis of future requirements
- Site analysis
- Analysis of energy and environmental requirements
- Analysis of impact on production
- Development of design and construction schedules
- Analyses of investment and annual operating cost
- Composition of schematic design report

Analysis of Constraints

Though "real" constraints must be acknowledged, they should not be overemphasized during the schematic phase. A real constraint could be one which is cost-prohibitive, technologically impossible, time-prohibitive, or dictated by government regulations. Although few constraints cannot be eliminated given enough time or money, both time and money are often constraints. The point is, constraints limit the options available, and, at this stage of the design process, only design solutions which are viable should be retained.

Analysis of Codes

Local, state, and national code requirements need to be investigated, but only to the extent that they may restrict design or limit the options available. This could be in such areas as dimensional limitations, zoning setback requirements, and environmental considerations.

Analysis of Standards

Standards of design should be analyzed with respect to their impact on available design solutions. Most large organizations have compiled corporate standards for materials, minimum quality control and assurance levels, installation procedures, maintainability requirements, and the like. These standards have in many cases evolved over a long period of time and reflect the company's experiences and preferences. Industrywide standards are also published: American National Standards Institute (ANSI) publishes minimum requirements for most construction materials and installation procedures; national associations and institutes, such as the Joint Industrial Council (JIC), publish standards expressing minimum requirements for electrical, pneumatic, hydraulic, and other systems; vendor associations, such as the Crane Manufacturers Association of America (CMAA) and the National Machine Tool Builders Association (NMTBA), publish standards for machine tools and other industrial equipment. A full list of such organizations would be extensive; but some are listed after Chap. 1-5.

TABLE 4-1 Comparison of Initial Costs and Annual Operating Costs of Conventional Storage vs. Automated High-Rise Storage

Costs	Conventional storage	High-rise storage, single-depth	High-rise storage, double-depth
Initial Costs			
1. Land	$ 63	$ 21	$ 18
2. Site preparation	32	10	9
3. Building	1,575	820	720
4. Racks	300	480	480
5. Captive pallets	—	98	98
6. Storage retrieval machine	—	600	330
7. Computer control	—	150	150
8. Lift trucks	240	160	160
9. Tractor/trailer train	100	100	100
10. Fire protection	100	100	100
11. Investment tax credit	(64)	(159)	(132)
Total	$2,346	$2,380	$2,033
Annual Operating Cost			
1. Labor	$ 800	$ 560	$ 560
2. Depreciation	95	140	128
3. Interest	265	279	238
4. Maintenance	50	35	35
5. Utilities	70	40	40
6. Damage	50	10	10
7. Pilferage	50	10	10
8. Insurance	40	50	50
9. Housekeeping	20	5	5
Total	$1,440	$1,129	$1,076
Payback period			

$$\text{Single-depth storage vs. double-depth storage} = \frac{\text{Investment cost difference}}{\text{Annual operating cost difference}}$$

$$= \frac{\$2{,}380 - \$2{,}033}{\$1{,}129 - \$1{,}076} = \frac{347}{53} = 6.5 \text{ yr}$$

effort required for design development depends on the size of the project, its complexity, and the time available.

The schematic design solution must be developed in such detail that all basic components are described by type and principal dimension, sufficient to allow the subsequent composition of construction documents. To accomplish this, the design development phase consists of the following main activities:

- Analysis of codes
- Site analysis
- Development of preliminary drawings
- Development of outline specifications
- Development of design and construction schedules
- Analysis of preliminary costs
- Composition of design development report

Analysis of Codes

All land in the United States is governed by at least one building code, and many areas are covered by multiple codes. Codes investigation should include local, state, and federal codes, though many cities or counties have adopted the state or federal codes with only minor modifications. When multiple codes are in force, the strictest code governs the design. Environmental protection regulations must also be considered.

Site Analysis

Unless current applicable surveys and soil borings are available from previous work on the site, the services of a competent surveyor and testing laboratory should be procured. The surveys should include utilities and services below ground as well as above ground.

Development of Preliminary Drawings

Sketches developed in the schematic phase should be expanded. Building configurations in the form of sketches or block layouts should be converted to dimensioned drawings which, in addition to showing the areas allocated to operations, should show aisles, rest rooms, cafeterias, etc. Adequate space for such items as maintenance access should be provided around operating machinery. Basic decisions, such as the type of HVAC system, primary power distribution, method of conveying, etc., need to be made and systems should be shown in single-line fashion.

If possible, drawings may be planned in a manner that allows their use in later design phases; however, detailed design should be kept to a minimum. An example of a preliminary design drawing is shown in Fig. 4-2.

Development of Outline Specifications

Materials and systems must be determined and described in the outline specifications. These documents include description of the materials and the systems which are the basis for further statements of probable construction cost and criteria for later development of construction documents. As can be seen from Table 4-2, outline specifications are usually segmented into sections following the Masterformat–Master List of Section Titles and Numbers, which is discussed later.

TABLE 4-2 Example of Outline Specifications

B. BUILDING STRUCTURE AND SYSTEMS

- **1.** Substructure (Alternative no. 1 and no. 2):
  - **a.** First floor: reinforced concrete slab on grade.
  - **b.** Grade beam: 1′-0″ x 3′-0″ reinforced concrete beams supported at footings.
  - **c.** Foundation: reinforced concrete footings on timber piles.
- **2.** Superstructure (Alternative no. 1 and no. 2):
  - **a.** Roof: metal deck over steel purlins and trusses.
  - **b.** Mezzanine floor: 6″ reinforced concrete slab on metal deck, steel beams, and trusses.
  - **c.** Underhung cranes: supported by crane runway spanning between roof and mezzanine trusses.
  - **d.** Top running cranes (Alternative no. 2 only): supported by crane girders on crane columns at 20′-0″ spacing.
  - **e.** Columns: steel, spacing as shown.
  - **f.** Bracing: at roof and along perimeter and expansion joint columns.
- **3.** Roofing:
  - **a.** Existing building no. 369 (Main building, monitor and sloped roof):
    - **(1)** 3½″ rigid fiber/urethane foam insulation.
    - **(2)** 4-ply built-up roofing.
    - **(3)** 4″ fiber cants.
    - **(4)** Felt-base flashing.
    - **(5)** Metal counter and cap flashing.

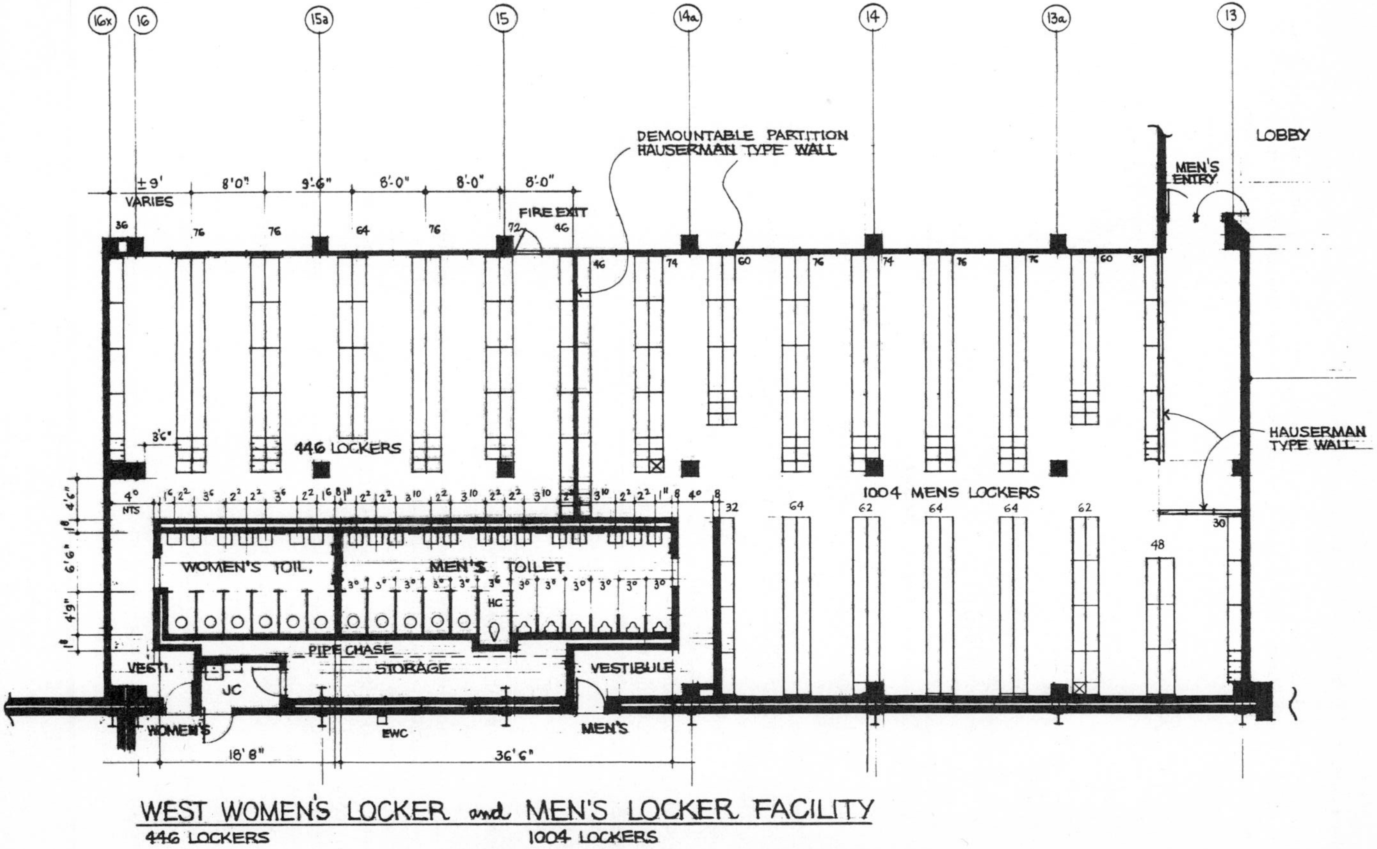

Figure 4-2 Design development (preliminary) drawing.

Development of Design and Construction Schedules

The schedules developed during the schematic phase should be updated and refined. Decisions made during the design development phase and any additional developed detail should be incorporated, and production impact plans should be reviewed for changes.

Analysis of Preliminary Costs

These are called Statements of Probable Construction Cost, and are derived from more definitive data and drawings than were available during the schematic design phase. For instance, since building materials should be selected, building costs can be calculated in terms of "in-place" building units (type of wall, type of roof, type of floor, etc.), instead of on an area- or volume-cost basis. Cost contingencies must still be included since not all details are known; however these most recent estimates are normally of better accuracy than those developed during the schematic design phase.

Composition of Design Development Report

All of the above information should be presented in a report. This is used as a presentation aid and as support for decisions affecting the project; it also serves as a record of the design activity.

Submission of the report should again be followed by a review period. During this time the design solution should be scrutinized to ensure conformance with program requirements.

If the current cost opinion differs appreciably from previous ones, the causes of the differences should be determined. Differences often result from changes in the parameters of the program, such as the scope of the project or the quality of materials and systems. If budget allocations cannot be changed, reduction of quantity or quality levels, deferment of certain portions of the project, or outright abandonment of the project may be required. If extensive design changes are proposed, it may be necessary to revert to an earlier point in the process and repeat some of the activities involved.

CONSTRUCTION DOCUMENTS PHASE

This phase, sometimes called *final design,* commences with approval of the design development documents and takes into consideration any changes in scope, size, or character resulting from the approval process. The selected schematic design, having been developed during the design development phase, is now further defined in detail to become the construction documents, or working drawings and specifications. These documents are in turn the basis for taking bids, entering into construction or procurement contracts, and carrying out the work.

The following activities are normally performed during this phase:

- Preparation of construction drawings
- Preparation of construction specifications
- Preparation of construction schedule
- Cost estimating
- Preparation of bidding documents

The drawings graphically depict the project, the specifications verbally describe the work, and both must accurately establish all that is to be accomplished. The drawings and specifications provide information from which contractors can determine the extent of the project, the materials and methods to be used, special conditions, and any other information required to render a competitive bid and carry out the work. The construction documents must do this precisely because they eventually become a portion of the construction contract documents. Should the documents be incomplete or ambiguous, then the accuracy by which bidders can determine their prices is affected and in turn both increased costs and contractual problems may result.

Preparation of Construction Drawings

At this point the designers must render all details of design necessary to construct the project. The working drawings include plans, sections, elevations, and many special details. They must be complete and accurate and must convey the intent of design. The framework for this work may exist in the design development drawings and drawing lists.

Most companies have established drafting standards, which specify sheet sizes, lettering styles and sizes, standard details of design, etc. Since drawings are often developed by various groups, each representing a different discipline, it is necessary to coordinate all, in order to ensure accuracy and compatibility.

An example of final drawing detail is provided in Fig. 4-3.

Preparation of Construction Specifications

The construction specifications are a detailed elaboration, in the language of the construction industry, of the materials and components in the outline specifications portion of the approved design development report. Whereas the drawings provide a graphic representation of the project, the specifications define the materials to be used and set forth the requirements for construction.

Specifications are written in either of two ways: (1) Listing all materials, the specific quantities or extent of coverage for each, their quality requirements, and the method of carrying out the work so that the desired result is achieved. This type is defined as "procedure" specifications. (2) Listing only the performance required of the components and delegating to the contractor both the choice of materials and the installation methods necessary to bring about the specified results. This type is commonly referred to as "performance" specifications.

Project specifications may contain a combination of the two forms. An example of both types is provided in Table 4-3.

TABLE 4-3 Types of Construction Specifications

PERFORMANCE SPECIFICATION
1.03.2 Wastewater Pumps:
Pump impeller diameter shall not exceed 90 percent of the maximum published diameter for the casing, or one which is smaller than the smallest published diameter for the casing.
Pumps shall be suited to operate at indicated temperature without vapor binding and without cavitation under any system operating condition.
Pumps of the same duty condition classification, and same accessories or with specified accessory deviation, shall be identical and from the same manufacturing source.
PROCEDURE SPECIFICATION
1.03.2 Wastewater Pumps:
Provide as indicated on drawings. Pumps shall have an RPM of 1200 or less; shall be able to pass 3″ diameter spherical solids; and shall be self-priming, and suction, horizontal, centrifugal. They shall be equipped with spool flanged fittings for suction and discharge, a priming fill port, a removable access cover and a pressure relief valve. Gorman-Rupp Model T6A3-B, or other approved.

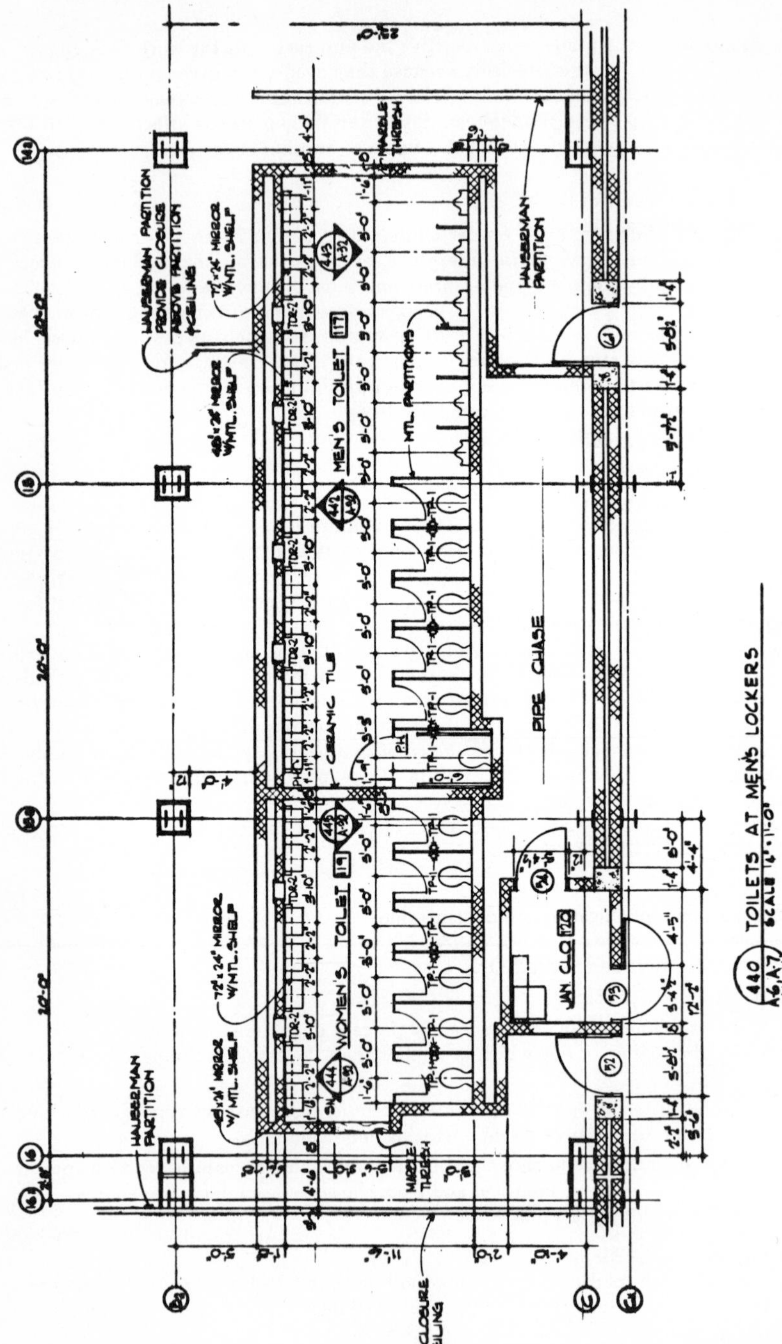

Figure 4-3 Construction (final) drawing.

Coordination between drawings and specifications is imperative, since neither of these two documents is mutually exclusive. It should be mentioned that most construction contract general conditions specify that in cases of disparity between drawings and specifications, the latter shall govern.

The industry standard for organizing the descriptions of construction materials (i.e., specifications) is the *C*onstruction *S*pecifications *I*nstitute (CSI)/Construction Specifications Canada: Masterformat–Master List of Section Titles and Numbers. This established format arranges detail specification into 17 divisions:

- Bidding and contract requirements
- General requirements
- Site work
- Concrete (structural, architectural, and paving)
- Masonry
- Metal (structural, architectural, and miscellaneous)
- Wood and plastics
- Thermal and moisture protection
- Doors and windows
- Finishes
- Specialties
- Equipment
- Furnishings
- Special construction
- Conveying systems
- Mechanical
- Electrical

National professional organizations, including the AIA and the CSI have standard specifications which are widely accepted and are often used as a model. Publications of the AIA and CSI, listed at the end of Chap. 1-5, contain further coverage of this subject.

Computerized specification systems have gained wide acceptance. With the systems, project specifications are prepared by choosing applicable standard sections from an all-inclusive master text and copying these sections onto word processing storage devices. Changes, reflecting individual project variances from the master, are made to the copy, after which the word processor prints the entire specification. Caution must once more be advised, since to be economical these commercially available systems must be structured in a manner which is suited to the project at hand, and those persons interfacing the system must have a working familiarity with it.

Preparation of Construction Schedule

Refinement of the construction schedule should be made during this phase. More project detail is available, and it is likely that anticipated construction processes may be segmented into a greater detail of activities than was previously possible.

Cost Estimating

A detailed cost estimate may now be prepared using the complete drawings and specifications. This is covered in depth in Chap. 1-5 under "Estimating."

Preparation of Bidding Documents

The final step in this phase is to assemble the construction documents (drawings and specifications) that are enclosed as a part of the bidding documents. Where multiple construction or procurement contracts are involved, exact lines of demarcation between contracts must be established, such that the scope of work under each is clearly defined, all-inclusive, and not overlapping.

BIDDING OR NEGOTIATING PHASE

The activities discussed include those normally performed by design personnel. Because this chapter provides a complete coverage of all the design tasks, mention of construction-related activities is included; however, Chap. 1-5, under "Contracting," contains more complete detail.

During this phase the following activity is normally performed.

Bid Review and Analysis

If the bidding documents include procedure-type specifications and drawings, if the bidders are deemed equally qualified, and if all bidders have complied completely with the bidding documents, the construction contract award is usually made on the basis of price.

If performance-type specifications and drawings are used for the bidding documents, a comparative analysis of the bids is required. The bidders may not be proposing identical or equal materials, systems, or levels of quality, and the final award is often to the most cost-effective proposal, not necessarily the lowest price bid. All components of the project should be incorporated in this analysis. Major systems and subsystems, quantities and quality of materials, project implementation schedule, proposed subcontractors, and bidder exceptions should be examined. If possible, each of these areas should be compared with the base requirements outlined in the bidding documents, and with the schedule and estimate prepared during the construction documents phase. The purpose of the evaluation is to ascertain whether any deviations from the baseline are balanced by increased or decreased values, time, or costs.

CONSTRUCTION PHASE

Once a construction or procurement contract has been executed, the design process does not necessarily come to an end. Those post-award activities that are normally performed by the design staff vary according to the nature of the contract and the project; however, the effort required can be significant and may sometimes consume between 15 and 25 percent of the overall design budget. This is often called *follow-up work* and requires both an office staff to perform consulting services and a field staff for on-site inspections. Both staffs are involved throughout the construction period.

Activities performed during the construction phase may be generally categorized under the following headings:

- Contract administration
- Quality assurance

Contract Administration

Even with explicitly written specifications and well-executed drawings, questions and misinterpretations of the construction documents may occur. This is especially true when performance contracts are awarded. To avoid delays and clarify issues, a specific group of design personnel should have the task of resolving the problems. The liaison group should be totally familiar with the construction documents, construction procedures, and the jobsite. Bulletin–change orders or field orders may be needed because of owner-initiated changes in the scope of the work and field conditions that occurred after execution of the contract. Design revisions which alter the contract should be tightly controlled, since the advantages of the competitive bidding process are no longer present.

Major changes in the construction documents are often best made by members of the design staff. Once the necessary design change has been performed, the drawings and portions of the specifications that have been altered are reissued as bulletins, with the revised sections clearly indicated to the contractor. No action is permitted by the contractor until a price for the change has been quoted and submitted to the owner. Once approval is given, a formal change order, authorizing the contractor to proceed with the work, is issued. This alters the contract accordingly.

A field order, which is not considered major and does not alter the contract, is generally authored by a field representative.

These administrative procedures are historically necessary, despite the best of attempts for quality and completeness in the design process, and should not be thought of as brought about by error. To plan for them is wise, and to not plan for them is to misinterpret the realities of design and construction.

Quality Assurance

Quality assurance during the construction phase essentially consists of ensuring that materials and workmanship comply with the construction documents. This is normally accomplished by checking the proposed materials and installation methods, and inspecting the work during construction.

The construction documents normally include provision for the submission of shop drawings, samples, brochures, and similar representation of materials from the contractor before such materials are ordered and installed. This is done even when procedure specifications are used, and materials are explicitly called for, as choices are made available to the contractor. When performance specifications are used, and materials are not specifically identified, it becomes even more important for the design staff to confirm that proposed materials meet the specification requirements.

Contractors often submit materials or equipment which differ from those specified, but which are maintained to be substantially equivalent. The difference may come about due to an unavailability of the specified product, preferences, or other reasons, and these substitutions must be carefully analyzed before approval is granted.

Independent testing consultants are often employed to analyze and report on various construction materials and practices, such as concrete mixtures and welded joints. In many cases inspections by regulatory agencies must also have design personnel in attendance to provide any necessary clarification.

Some equipment or systems must be checked for acceptance by testing their operation. The purpose of acceptance testing is to ensure that the fabrication and installation has been performed in accordance with the provisions of the contract. When such testing is necessary, general procedures for these tests should be included in the specifications, though details may be added during the bidding or negotiating phase. Once the work has been completed, the design staff uses the acceptance testing procedures to ascertain both conformance and completion.

As construction nears the end, the design staff should maintain a list of incomplete or unacceptable work, called a "punch list." Before the final payment to the contractor is approved, an inspection of the entire project should be carried out, and any work which has not been completed according to contract documents should be recorded. Notice of this should be relayed to the contractor for completion and to the owner.

Continuous attention to construction quality may appear to place design and construction parties in adversary roles. This can happen but is not necessary. Both are in fact bound to the owner by contracts involving quality, time, and cost. The contractor's low bid is given with the confidence that the requirements of the contract documents may be met as a minimum, and the work is undertaken with the responsibility for doing so. The monitoring of quality assures that such is the case.

In a broad sense, the responsibilities for quality, time, and cost are shared, and in general it is only an individual party's concern for profitability which can endanger these features. For the owner, the contractor, and the designer, the greatest loss of profitability would in most cases come about from constructing the project without the contracted level of quality. From this standpoint, the designer's assurance protects not only the project but each concerned party.

chapter 1-5

Scheduling, Estimating, and Contracting

INTRODUCTION

As the realization of a project calls for certain design and construction activities, it also requires some parallel functions to assure timely, efficient, and orderly completion. The functions include scheduling, estimating, and contracting.

Time and cost are important factors in our economy and require due attention as a part of planning. There are some important differences, however, between the design/construction sequence and both scheduling and estimating. First, scheduling and estimating take place not once but on many occasions throughout the project development and require administration after they are accomplished. Second, they tend to quantify or qualify the content of the design/construction tasks, and as such are often viewed as constraints. Both of these characteristics have obvious negative overtones, but such is not the case when the project is viewed from a broad company or corporate vantage point, where time and cost are usually critical.

Because scheduling and estimating deal with the future and because their accuracy is often critical to success, they are frequently criticized for any inaccuracy. The condition of inaccuracy may be the result of error or of luck, but is most often due to the inherent fact that all prediction is a bit risky. The resolution of this dilemma is the incorporation of contingency allowances throughout schedules and estimates to cover a percentage of the unforeseen future.

Few owners have the capacity to undertake more than a limited number of design and construction tasks with in-house personnel. Engaging others to accomplish tasks (contracting) involves a knowledge of both facilities and business administration. It is important to understand the particulars of construction work to be able to select the proper contracting methods and to engage in the often technical details of managing the contract. However, the efficient and orderly execution of the work is largely a result of proper business practices.

These subjects are highly structured and exacting—features that can be a benefit to the project when followed and a detriment when violated. Their effects should be realized and respected.

SCHEDULING

Scheduling is the development of a comprehensive time plan, based on the facility design program or other design documents, for the orderly accomplishment of design and construction activities.

Scheduling is an integral part of project planning and allows the many decisions which have time-frame implications to be resolved with foresight about the impact of the project schedule.

PURPOSE

The prime purpose of scheduling is to provide a forecast of the time required to accomplish the projected work so that interfacing or resultant manufacturing operations may be planned. Most construction work impedes ongoing activities somewhat and, with completion, creates some alteration to the activity. Both this ongoing activity and the construction work require adequate interdependent planning efforts for propriety and efficiency.

The supplementary purposes of scheduling relate to the internal efficiency of design and construction. The construction sequence depends on long lead times for the purchase of equipment or materials. Design and construction tasks must be defined, their duration estimated, and their vital relationships examined to ensure efficient execution of the work. The feasibility of accomplishing some construction while maintaining ongoing activities must be confirmed. The design and construction of systems must be evaluated in terms of many varied criteria, including the time factor, in order to sequence the work logically and balance time concerns with other important factors. Cash flow must be projected.

Scheduling is undertaken at varied levels of detail, which relate to the overall project levels of detail. Efforts to generate complex project schedules are useless until both the time requirements and project detail of the integral activities are known. Such schedules serve only as conceptual "time models," and cannot become a genuine calendar of events until there is a detailed project activity identification, description, and time duration established for each task. On the other hand, certain projects are so basic or simple in scope that simple (uncomplex) schedule planning systems are irreducible without becoming an exercise in overmanagement. There is no standard or rule of thumb to govern the extent of scheduling detail, except those of experience and common sense.

TIMING

The earliest point at which a schedule might be formulated would be at project inception, when perhaps only an approximate scope or rough description of activities may be available. At this point a scheduler can generate little beyond a gross time estimate for probable overall project duration, based mostly on previous experience or just an intuitive guess. This type of time estimate might be typified by the example: "For an engine plant that will produce 160,000 units per year, design and construction require 30 months." These estimates, made by experienced schedulers, can sometimes be surprisingly accurate.

When a facility design program is complete, describing the areas, basic system types, and some minimal degree of site information, then individual categories of work may be scheduled using Gantt chart, or bar chart, format. Should the project area requirement be subdivided into area types, the bar chart may also be subdivided. Some chronology high points become evident from this type of schedule. This clarity makes the bar chart appropriate for project presentations and evaluations. If the project is on a relatively small scale, or comprises a minimum of tasks or interdependent activities, this method may be sufficient to preclude further scheduling.

Once a schematic design is complete, then the individual construction contracts, if there are several, may be identified. At this point a network plan of the work is conceivable. The network, using either the CPM (critical path method) or PERT (project evaluation and review technique) procedures, can begin as a general time model. As the design continues through the design development phase, the construction documents phase, and finally through construction, the network may simulatneously be expanded in detail (task subdivision). This expansion is essential to all major projects.

A valuable step toward a well-developed network is a schedule-planning session. At this meeting experts from various areas of design and construction can "talk through" the project. It is very likely that such a gathering would develop most of the major project schedule relationships, and do so in a rather short time.

PERSONNEL

Scheduling is usually done by a company's own staff or by consultants such as architects, contruction managers, CPM consultants, engineers, etc. The choice of a scheduler depends on the type and scope of the project, as well as on the personnel available and their expertise. If in-house personnel, who are not normally engaged in design or construction scheduling, are called on to accomplish the task, there are several points of caution. First, special care should be taken in dealing with long-lead procurement items and resource availability issues, as both involve continuous attention to market fluctuations, with which a specialist would normally be familiar. In these and some other areas it is often of value to request information from suppliers or contractors. Second, commercially available computerized scheduling services should be carefully evaluated before a company invests in them. These reservations are detailed in Chap. 1-4, "Facility Design."

METHODOLOGY

The methodology of scheduling consists of determining the activities of project design and construction, their durations, and their relationship to each other. The procedure can begin with a fixed project duration, in which case the scheduling becomes an examination of the feasibility of completing everything within the established period. The process can also begin with an open-ended time scale. When this most common condition prevails, there will likely result an infinite combination of time-vs.-cost conflicts, which management and not the scheduler may be called upon to resolve. For either condition, the tasks of scheduling include:

- Activity determination
- Determination of activity time
- Determination of activity relationship

Activity Determination

The very first concern of scheduling is that of how the project implementation should be broken into activities. It may be found that a hierarchy of activity subdivisions is in order, in which case it is valuable to check that the several sublevels consist of parallel and exclusive categories which, when totaled, represent the entire activity. Design and construction of a project may have activities identified under the following general classification headings:

- Contracts (where there are several)
- Areas of the project site
- Engineering discipline (primarily for design work)
- Construction systems
- Material or facility component (Masterformat specification divisions)
- Trade or crew (primarily for construction work)

Consultants for design and construction contractors usually schedule their efforts in a different manner than that used to determine the overall project schedule. Their schedules are not necessarily for the same purpose but do bear a relationship to the total project plan.

Determination of Activity Time

When individual activities are established, each must be assigned a duration. Schedules are structured to use either deterministic or probabilistic times, and are covered from this standpoint in the network planning discussion.

To define the activity time requirements one must deal directly with experienced persons, each of whom can describe the length of time necessary to accomplish the task about which he or she is knowledgeable. For example:

A structural engineer can best provide the time requirement to design and specify a certain structural framing system; an excavation contractor can best provide the time requirement to grade a certain area; a masonry contractor can best provide the time requirement to lay a certain wall; etc. These time requirements may be calculated on the basis of equipment capacities, crew size, work rates, etc., or they may be estimated.

Persons supplying such information would probably be assisted by knowing the scale and scope of the whole project in order to approximate or compute the number and size of work crews to best undertake the task. When certain task durations are incorporated into network arrangements and found to be either excessively long or on the critical path, the personnel component, on which the task duration was based, may have to be re-evaluated. Knowledge of the individual tasks and of the entire scale and scope of the project is required to permit adjustment of the schedule to be made within practicable limits.

Determination of Activity Relationship

Individual activities can be dependent upon certain other activities, which govern when any one activity may begin, and which other activities are restrained until the task is complete. When these connections are graphically portrayed, the result is a network. Even when a bar chart is employed and the connections are not indicated, the network relationships exist because the concept of "restraints" is an actual condition.

The relationship of activities involves the logic of construction. The basic elements are obvious: Excavation must precede substructure work, which must precede superstructure work, which must precede roof-metalwork, which must precede roofing work, etc. A complete activity network is definitely more complex and more sophisticated, but the theory remains the same.

METHODS

The project schedule is usually displayed by following one of three types of procedures: gross time estimating, bar chart planning, or network planning. Because the full tasks of design and construction include a multitude of intricately related activities, any schedule should become concerned with all of the complexities of the most involved network planning. Time availability, information availability, project scope, or proposed schedule use may amend this concern and limit the schedule type actually produced to a system that does not portray the total project in detail.

Gross Time Estimate

As has been mentioned, this idea of scheduling relies on experience and intuition, especially when concerned with broad-ranging tasks. Scheduling by this method is sufficient and reliable for individual and small component tasks. The following example may illustrate this: A hardware supplier may quickly answer a client inquiry, "Installing new locks on those 16 doors will be completed in 15 working days." This is a gross time estimate, but it includes the knowledge and experience that a certain period is required to place an order, ship the order, schedule an installation crew, and accomplish the job–and it includes some contingency time. When this is a part of new construction, involving only the installation effort, the answer would naturally be a shorter period of time (such as 2 days for the actual installation).

When assembling major schedules, the separate portions of a bar chart or network are often composed by making such gross time estimates. These estimates are based on the scope of the work and are sensitive to relationships with other associated activities. Any inaccuracy of an individual task duration estimate is usually balanced by the multiplicity of tasks in the entire project.

Whether the time element is the result of a gross time estimate or of a calculated rate of activity accomplishment, some contingency is necessary to the overall schedule. A full discussion of these contingencies is covered under "Notes" near the end of this chapter.

Bar-Chart Planning

A fairly simple design or construction schedule is the Gantt chart or, as it is more commonly called, the bar chart. This chart normally lists the activities (items of work) down the left-hand side of the page and the time scale across the top. A line or bar is drawn opposite each activity delineating the interval between the beginning and ending dates of the activity. This chart may be quite adequate when relatively few activities are involved and their relationships are either weak or few.

For summary or presentation purposes this format is very good. The complexity of a full network diagram is such that full comprehension is often limited to those experienced in such systems. Fine points are often understood only through an extensive review of activity number and schedule codes. In contrast, the bar chart is clear and may be visually grasped in a short time. For this reason the bar chart is suitable for display purposes, where verbal descriptions can develop understanding of the relationships, and for summary purposes, where reference to an accompanying text can do the same.

Network Planning

Major projects demand a more complex, but more useful and comprehensive scheduling method, network planning. Networks include CPM, or deterministic, diagramming and PERT, or probabilistic, diagramming methods. The application of one rather than the other to a certain project is based on experience, special characteristics of that project, and the preference of the scheduler.

CPM schedule planning is considered deterministic because for each activity a time is allocated, based on the known time requirement for undertaking and accomplishing the activity. Each activity may be shown as a box, or node, containing several word or letter codes and connected to other activities by arrows which denote the order of event precedence. Each activity may alternatively be shown as an arrow, beginning and ending in a circle, or node, each of which is in turn given a number or letter code. The nodes and arrows are usually drawn from left to right across a horizontal time scale, but are normally not to scale.

PERT schedule planning is considered probabilistic because each activity is assigned a minimum, probable, and maximum time duration. Either precedence or arrow diagramming may be used.

Originally the two systems had many differences, but with development and refinement they have come together on most points other than duration assignment. The arrow diagram can also prove to be a bit more difficult to edit, and this lessens its applicability to projects requiring frequent schedule changes. It should be noted that a critical path may be found through the PERT network as well as through a CPM network. In fact, with computerized scheduling it is possible for a user of the schedule to engage in several types of interactive situations with either system and never use a graphic network display.

The purpose of the network is to present the project activity in a logical arrangement that is indicative of the real order of events. To make the logic complete, "dummy" activities may be inserted and notated by the use of dashed lines. This latter step is a working tool for the scheduler and does not represent an error in project procedure.

The full subject of network planning is complicated and can only be explained here with respect to principles. The major principle of any network is the *critical path.* This is defined as the longest logically connected sequence of activities in the network, which finally dictates the total project design and construction period.

Not every activity is on the critical path. Assume that two activities, A and B, must begin on the same day and also end on the same day, because of other related activities.

If activity A requires 30 days and activity B requires 20 days, A will control the duration and hence be on the critical path. Activity B, on the other hand, has 10 days of "float" or "slack"; that is, it can begin 10 days later or end 10 days sooner, or any combination thereof. From this illustration the concepts of *early start, late start, early finish,* and *late finish* can be understood. Should there be a desire to shorten overall project time, perhaps personnel could be doubled on activity A, thus effectively cutting the time required in half. From this illustration the concept of "resource management" can be understood. (This situation would now place activity B, with its 20-day duration, on the critical path.) Should the staff be at full employment, perhaps overtime could be considered, even though it represents additional expense. From this illustration the concept of *time vs. cost* can be understood. Should both activities actually begin on the same day and each encounter a 5-day delay, obviously activity A at 30 days is experiencing the harmful delay, which requires remedy. From this illustration the concept of *expediting* can be understood.

In these ways, the schedule is planned and managed. The real world naturally produces situations less pure, but the concepts remain. Figure 5-1 is an example of a working planning chart.

ESTIMATING

Estimating is an analytic process of determining costs of the component materials, labor, and other expenses necessary to accomplish the work described in the facility design program or other design documents. Like scheduling, estimating is an integral part of overall project planning and provides means to settle multiple decisions which have financial implications with some idea of the effect on cost.

PURPOSE

The central purpose is to present an estimate of the money required to undertake projected work and hence allow fiscal evaluation. Rarely is facility need so imperative that it precludes cost concern, and only via the development of an estimate is it possible to generate a cost analysis. On most projects the relationship of program necessity to budget limitations can bring about an intense management interest. The product may be numerous program and design determinants. Indeed, inquiry as to whether the project is in any way realistic, within certain budget restraints, is often critical to continued design development. When any estimate indicates that a project has grown beyond established budget limits, it may be necessary to revise the design, and time should be reserved for such work. These judgments may be made only if cost estimating parallels other project efforts.

The supplementary purposes of estimating orient principally to the internal efficiency of facility design and construction. Many design and construction methods must be evaluated in terms of multiple varied factors, including cost, in order to balance the facility requirements with cash flow and the cost criteria.

Estimating is undertaken at varied levels of detail, commensurate with the overall project development of detail. Previous to the time when specific project requirements are identified, any efforts to generate intricate cost estimating systems are merely exercise. They serve only as conceptual cost models and cannot become genuine estimates of construction cost until there is commensurate elaboration of project detail. On the other hand, some projects are so basic or limited in scope that the most simple estimating methods are completely adequate. There is no standard or rule of the road to govern applicability of cost estimating methods, except those of experience and common sense.

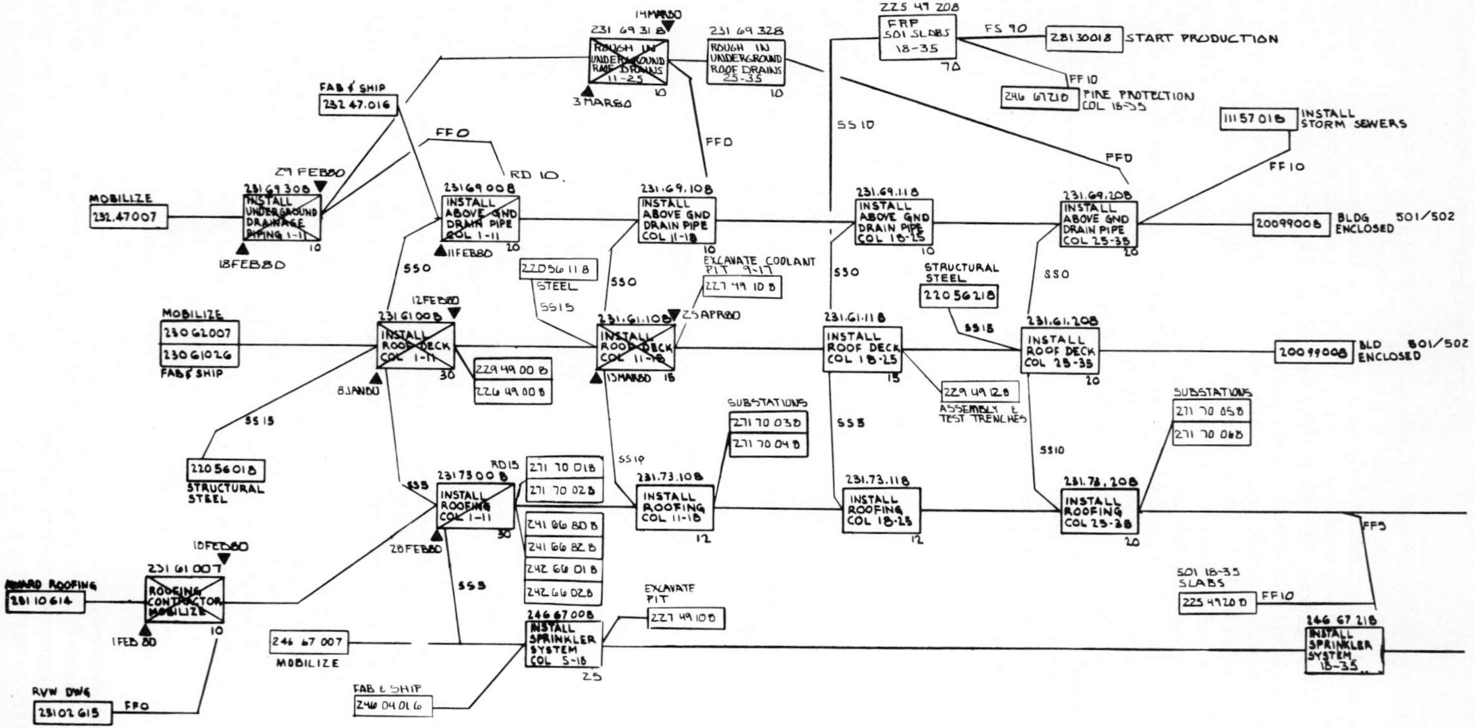

Figure 5-1 Scheduling diagram.

TIMING

The earliest point at which a cost opinion might be formulated would be at project initiation, when little more than a general scope or basic description of project needs is available. At this point an estimator can develop no more than a gross opinion of probable construction cost, based mostly on previous experience or just an intuitive guess. This type of cost opinion might be exemplified by the following: "An assembly plant intended to produce 60 trucks a day, including equipment but not tooling, will cost 50 to 60 million dollars." These opinions are vague, should be formulated to cover a range which is never low, and can be negated by design or site specifics which are particular to any one project.

When a facility design program is complete, describing the areas, basic system types, level of quality, and some acceptable degree of site information; a cost opinion based on the overall cost per unit of area or volume is possible. Should the project area requirement be subdivided into area types, the opinion may also be subdivided in like fashion, the refinement improving the accuracy. This opinion may be further upgraded when schematic design is complete, and the areas may be abstracted from the actual design drawings.

When design development is complete, including some outline specifications for materials, methods of construction, basic systems, equipment, and site data, then a cost opinion based on the in-place building units is possible. From the design plans and the outline specifications, it should be reasonable to expect that approximate areas or volumes of in-place units (brick walls, concrete slabs, steel deck, doors, etc.) can be tabulated. Units may be priced on the basis of published figures to provide the "in-place" cost, and an overall opinion formulated.

When construction documents are complete, a detailed estimate of construction cost, based on extensive material tabulation, is possible. Though this activity is not always included in a consultant's basic services, it is common to request that it be added. Only a material quantity and installation labor summary results in a true "estimate" of probable construction cost, since before this time a complete listing of components has not been known. The earlier cost opinions are just that, because they do not have the necessary detail elaborated within their referenced design documents. The level of detail required for a project estimate may vary, but it is mandatory that it be related to the level of detail in the design documents used as the basis for the estimate.

PERSONNEL

Estimating is accomplished by a company's own staff or by consultants such as architects, construction managers, cost-estimating consultants, engineers, etc. This choice naturally depends on the type and scope of the project, as well as the personnel available and their expertise. When in-house employees, who are not normally engaged in design/construction estimating, are called on to perform the task, there are several points of caution. Such points of caution are the same as those observed when preparing construction schedules and have been listed previously. A special problem is that the production of many construction materials and their distribution currently occurs on a worldwide basis, which produces susceptibility to international relations, finances, and politics; and changes in these can come about with alarming suddenness.

METHODOLOGY

The methodology of estimating consists of determining the quantities of the materials required to construct a project of a particular design, the cost to supply and install those quantities, and the market or project particulars which influence the final price. The process can begin with a fixed budget, in which case the estimating is somewhat an examination of the feasibility of completing everything within the established limit. The process can also begin with an open-ended cost limit. When the project is judged to be feasible, there likely follows an infinite combination of cost -vs.- time or-quality conflicts, which management personnel and not the estimator may be called upon to resolve.

For any condition, the tasks of estimating include:

- Quantity determination
- Quantity pricing
- Location, market, and project conditions

Quantity Determination

The very first concern of estimating is: What quantities of component materials and labor are necessary to complete the construction? Each project is unique in this regard, no matter how much standardization is employed. At some levels of design there does not exist enough detail to consider the actual tabulation of materials, and other units of measurement must be used as a basis for the estimate. The following categories of measurement units may be used:

- Square footage (SF) or cubic footage (CF) of the building, and acreage of the site
- Square footage or cubic footage of portions of the building and site (e.g., manufacturing area, warehouse area, office area, etc., until all areas are included)
- In-place units (or construction systems, e.g., concrete slabs, masonry walls, steel deck, etc., until all systems are included)
- Materials (e.g., CF of concrete, SF of formwork, number and size of reinforcing bars, anchor bolts, etc., until all materials are included) and labor
- Contracts (where there are several)
- Areas of the project site

Square footage and cubic footage are normally measured by the formulas described in AIA Document D101. For leasing or other purposes, the Building Owners and Managers Association International (BOMA) standard method of floor measurement might be used. The latter method is primarily for another purpose and may produce different quantities.

Design consultants and construction contractors each estimate their projects and may do so using different methods than the overall project budget estimate. Their estimates are not necessarily for the same purpose, but procedural divergence alone does not relieve the need to conform to the total project budget.

When quantities are not known, as in the earlier stages of design, they must sometimes be generated in order to develop an estimate (or cost opinion). This is termed a *conceptual estimate* and is not only an estimator's task, but also involves programmers, designers, etc. This type of estimate employs certain individuals' ideas of what might be involved to complete a project. The result can be expanded into a *cost model* against which the eventual design and construction may be measured in terms of cost performance. This conceptual estimate, if made by persons skilled and experienced in the task and if properly structured to suit the project, can prove to be very accurate.

Any estimate depends on the quantities upon which it is based. This obvious condition underscores the need for accuracy and completeness. Completeness involves adding not only the material, labor, profit, and overhead required for construction but also the other real costs. For a total project budget these may include:

- Land and land acquisition
- Permits
- Roadway, railway, and utility connections
- Building(s), services, and site development

- Furniture and equipment and installation
- Off-site construction costs
- Operations and start-up
- Fees (real estate, legal, architectural engineering, etc.)
- Facilities relocation
- Personnel relocation
- Personnel acquisition and training
- Administration
- Financing
- Contingencies

Quantity Pricing

When project quantities are established, each must be assigned a unit and total price. The total price differs from the quantity times unit price in that it includes a cushion factor, or contingency fund, to complete the work. This factor covers those details or specialties not indicated in the unit price, but still a part of accomplishing the activity. A full discussion of contingencies is covered later in this chapter under "Notes."

To define the unit prices one must refer to published cost data, previous experience, or supplier information. Two well-known cost data books are: McGraw-Hill's *Dodge Construction Systems Costs* (indexing data by building type, building system, material, and labor, including some cost ranging and locality adjustment) and *Means Construction Cost Index* (indexing data by Masterformat–Master List of section titles and numbers, material, and labor, including crew, daily output, and contractor overhead and profit information). Both are published annually.

Cost-estimating consultants obtain price information not only from literature but also from experience. To do so requires some attention to current markets, as cost changes can be very rapid and frequent.

In many cases it is necessary to consult supplier catalogs and price lists, or even to request specific price quotations. This is necessary when there is no other way to gain the information. Assistance from suppliers is generally forthcoming through sales representatives, whose job is to help both owners and consultants. It would not be unusual for representatives to request preliminary drawings, specifications, or requirements and to provide a fairly detailed cost opinion, though not a contractually binding one. The opinion is very useful for special equipment such as escalators and elevators, generators, pumps, sterilizers, substations, etc.

The final step in pricing is to add a percentage for contractor or subcontractor overhead and profit. This percentage is not fixed and is heavily dependent on market conditions.

Location, Market, and Project Conditions

Every project cost is eventually influenced by its location, the market, and special conditions of the project. Some may be calculated and follow predictable multipliers, while others cannot be accurately forecast. An estimate deals with things to come, and the further into the future the estimate must project the greater the possibility of eventual deviation from that estimate.

Location involves macro and micro adjustments. When prices are abstracted from literature which uses a national average, a regional multiplier must be used to convert such a price to the labor costs of the area where the project is to be constructed. This is a macro adjustment. A good source of regional comparative wage rates for various trades is *Means Construction Cost Index,* Appendix 1. Where certain local conditions create a supply problem for any material or labor, which is special and at variance with regional averages, additional multipliers must be used for those components. This is a micro adjustment. An example would be a rural location where specialized machinery installation personnel would require travel expenses to commute from a distant urban center.

The market can be a vague entity, in which resolution of many concerns cannot follow any prescribed formula. The overall economy of the country and, to some extent, the world may have some bearing. Subjects such as inflation, escalation, recession, taxation, capital investments, Consumer Price Index, interest rates, etc., combine to influence eventual construction costs. A bidding climate in which contractors are busy generally does not produce the low bids found when work is scarce. When an estimate directly precedes bidding on a 10-month construction phase, the market might be fairly well forecast. When an estimate is formulated 2 years ahead of a 4-year construction effort, virtually no corrective factor can take everything into account. This could be emphasized by recollection of certain product prices having risen as much as 100 percent in as little as 1 month. The only possible way of dealing with such intangibles is to make note of them.

Despite any such market concern, it may always be possible to receive a very good (low) bid due to other circumstances. Perhaps one contractor may exist, in the busiest construction times, who simply does not have work. It may also be the case that a contractor is situated and mobilized with a good proximity to the project site and, hence, may also give a low bid. Such situations are very satisfactory ones. On the other hand, a contractor may inadvertently submit a low bid due to an estimating error. This is not a good situation, because the contractor will make subsequent prejudiced decisions in an effort to recoup an impending loss, many of which may be to the detriment of project quality. The error can be recognized only through vigilance, for which a detailed construction cost estimate is a good baseline reference.

All these factors make estimating a less-than-exact science.

There also exist project conditions which can influence the unit price. Some examples of these include schedule or other time limitations which impose overtime or premium time costs, construction-method restrictions such as area limitations (which might impede steel shakeout activities), insurance limitations (which may preclude the use of helicopters to set roof-mounted air-handling units), utility limitations (which may require the contractor to use temporary electrical generators), payment methods (which may require the contractor to include financing charges), etc.

METHODS

Estimates or earlier cost opinions are produced by following one of several procedures. These rely on varied data inputs and may all be done for a major project, each at selected phases in the project development. Not all are likely to be used on a small project.

Gross Cost Opinion

As has been mentioned, this type of opinion relies on experience and intuition, especially when used for major projects. On large-scale projects with budgets in the millions of dollars, an estimate or cost opinion can become bogged down in the detail tabulation of certain small items, many of which may not be specified until late in the design process. In such a case it may be well, by using this method of estimating, to develop an "allowance." Although the project may have progressed significantly far into the construction documents phase, certain components may not yet be specific enough to allow any form of estimate other than the allowance. Typical of such items are platforms, guardrails, kitchen or laboratory equipment, first-aid equipment, hardware, landscaping, etc. The components should obviously not be excluded nor should other prices be arbitrarily increased to cover the unspecified component. The allowance is a proper way of including costs until the time when a more exact component estimate can be developed.

Area and Volume

For this type of cost opinion the area or volume of the building is multiplied by a determined cost per square or cubic unit. This area or volume may be either a calculated total, as would be found in the facility design program, or a scaled amount, as would be taken from a schematic design drawing. The area or volume may be for the entire facility or may be subdivided into smaller, distinct space types.

This cost opinion cannot include site improvements, utilities, etc., within the above figures, and additional figures must be separately added in order that the cost opinion be complete. The costs which are used must be adjusted from general averages to accommodate the determined quality level and project specifics. Contingencies must be added to cover site conditions, escalation, market conditions, etc. If the design has not proceeded sufficiently far, then a contingency allowance must also be added to cover final design development.

The overall unit cost can also be subdivided by design discipline and a unit cost assigned to each. It might be done as follows:

- Architectural and structural work
- Mechanical work
- Electrical work
- Equipment and installation work
- Site work

In-Place Unit

For this cost opinion, the basic construction systems must be identified and tabulated and a price assigned to each. The level of detail must again follow the level of design completeness, for example: When a simple "massing study" is complete, an estimator can do little more than use an average height times the perimeter length, possibly subtracting a percentage for doors, windows, and other openings, to develop a cost opinion for the exterior wall; when design development drawings are complete, the same exterior wall might be exactly scaled and tabulated. In either case, the wall area could be assigned a price per area unit, to supply and install the wall material.

Basic subdivisions of construction, on which this type of cost opinion might be assembled, may follow this categorization:

- Site preparation
- Foundations and subgrade work
- Slabs on grade
- Superstructure
- Roof
- Exterior walls
- Interior walls
- Vertical circulation
- Finishes (by floors, walls, and ceilings)
- Miscellaneous finish items
- Heating, ventilating, and air conditioning
- Plumbing
- Fire protection
- Process (services, drainage, and exhaust)
- Power systems (primary and secondary)
- Temporary power systems

- Auxiliary power systems
- Lighting
- Process control
- Equipment (by subsystem)
- Allowances and contingencies

Quantity and Cost

This is also referred to as a *time and materials* estimate. For the cost opinion or detail estimate a complete tabulation of the materials of construction is abstracted from the design documents. The materials are then priced. The labor to install the materials is calculated and it is then priced. General site and contract conditions are added, as is the contractor's overhead and profit. The last item, as has been mentioned, can vary significantly.

An estimate produced in this manner can remain an opinion, or, if developed in proper detail and with proper safeguards, can become an estimate. Construction contractors would most likely follow this format.

CONTRACTING

Contracting is the process of engaging with others in a legally enforceable agreement to provide services and materials necessary to accomplish the work described in the contract.

PURPOSE

The purpose of contracting is to engage others to provide services and materials that are generally beyond the capabilities of an owner's staff. When a task might be accomplished to the requisite level of detail and excellence by in-house personnel, or even by a limited and temporary extension of personnel, then perhaps contracting is unnecessary. If the latter conditions prove to be the case, such a determination should be made early and no contracts entered. However, when distinct reasons for contracting are in evidence (scope of work, personnel availabilities, specialized tasks, etc.), the process of contracting for services in required.

In these cases an owner may contract for design or for construction. Considerable detail concerning the acquisition of design services has been developed in Chap. 1-1, "Establishing Requirements," and may be undertaken at any point after which some degree of design programming has been completed. Other consulting services (feasibility studies, site selections, master plans, etc.) may be acquired in a like fashion, after their respective program requirements have been established. Though any contracting process involves a great commonality of events, the remainder of this discussion orients to buildings and equipment procurement and installation.

TIMING

Contracting should only be done after contract documents are completed. These may include drawings and/or specifications, or, in the case of certain design-build or design-purchase-install agreements, only a facility design program. Although there are certain steps which may be initiated in the absence of document completion, most contracts cannot likely be executed without some supporting documents. For legal purposes and for permit approvals, as well as for contract definition, the construction documents must be completed prior to execution. When multiple contracts, involving several parties, are involved, the completion and certification of construction documents is the foundation of basic liability definition.

PERSONNEL

The actual contract execution is handled by an owner's authorized employee, probably assisted or possibly represented by an attorney, and no other person. There are numerous valuable and productive capacities in which a consultant, especially a construction manager, may assist, advise, manage, administrate, etc. In the end, with all services provided, the owner is the individual entering into the contract. An architect, construction manager, engineer, etc., does not contract for an owner except in unique circumstances. Purchasing department personnel are frequent participants. Since purchasing agents are often unfamiliar with either design or construction contracting, they may sometimes incorporate other contracting conditions with standardized contracts.

It is of value to refer to the AIA *Architect's Handbook of Professional Practice,* vol. 2. The included documents, all definitions, etc., are widely used and accepted by not only the AIA, but by AGC (Associated General Contractors of America), CEC (American Consulting Engineers Council), CSI (Construction Specifications Institute), NSPE (National Society of Professional Engineers), and, to a great degree, the courts of the land.

It may be strongly emphasized that it is next to impossible to properly manage a poorly formulated contract, while little more than basic business administration is required to manage a good one.

METHODOLOGY

There are a number of procedures which must be carried out as part of contracting for construction work. These may be classified as those which occur before execution of the contract and those which occur afterward.

Activities Prior to Contract Execution

Certain activities must occur in order to enter into a contract for construction. Some arrangements with consultants provide that the consultant accomplish a portion of the services, though such arrangements relate not to the process but to the delegation of responsibility. Variations may also be found where company or corporate policy has determined that other events should be included.

It is also presumed that the number of contracts, content of each, and schedule for the contract activities have been established prior to entering this phase of the project.

First, determine if the method of awarding the contract is to be by direct selection and negotiation or by competitive bidding. If a contract is to be negotiated, then the remaining steps, up to the execution of the contract, may be carried out in a somewhat less formal manner. The contractor's qualifications must be confirmed, but the documents need not be carried to the same level of completion, and a contractor's price must still be ascertained. In some ways, the contractor must undertake more effort than when bidding, because a complete understanding of the construction cost must be communicated to the owner. This is essential to negotiating a fair and adequate price.

Second, the contractors (or bidders) must be selected and examined. This is called *prequalification.* Bidding must be advertised or contractors may be chosen from known parties whose work is respected. In both cases (negotiated or bid contracts) the contractor(s) should be prequalified by requesting letters of reference (trade, financial, bonding, etc.), examples of recent work, organization and personnel data, financial statements, etc. A Dun and Bradstreet statement might be obtained and, finally, the contractor(s) might be interviewed.

Third, information must be distributed to the contractor(s). All documents must be assembled, printed, and distributed either by the owner, the designer, through open plan rooms, or in some other way. Each bidder should be furnished a number of sets of the drawings and specifications for all portions of the work involved in the bid. When bidding time is limited, or many subtrades are involved, increasing the number of sets of documents distributed to each bidder can save time. Sets of documents may also be provided at central trade offices or open plan rooms in the area in which the project is located, so as to increase interest in bidding and heighten the level of competitiveness. Persons receiving the documents may be asked to submit a deposit to ensure their return after bidding or negotiation is complete. Bidding documents should include:

- Invitation or advertisement to bid
- Notice to the bidders, describing the requirements for submitting bids
- Instructions to the bidders, giving definitions, representations, bidding procedures, examination of documents, etc.
- Bid forms
- Proposed construction contract
- Bid security or bond requirements
- Performance and labor/material payment bond requirements
- General conditions of the contract, describing the contract documents; owner, designer, contractor, and subcontractor responsibilities; other provisions; time; payments; safety; insurance; changes in the work; correction of the work; and termination of the contract
- Supplementary conditions—specifics of the site, project, etc.
- Construction documents, including drawings and technical specifications

Any changes and additional materials, referred to as addenda, must be issued to all persons who have copies of the documents. On most occasions it is of value to have a bidder(s) meeting, prior to the submission of bids. The meeting is usually held to allow the prospective bidders to visit the work site, acquaint themselves with the scope of the project, and voice any questions they may have regarding the work involved. This is especially useful both if the project involves renovation of facilities or addition to existing facilities, and if it involves equipment furnishing and installation contracts.

Fourth, the bids must be received, held, and, at the appointed time, opened. The opening may be done either in public or private and may have any number of concerned parties present. Bids should be tabulated, including all alternatives (variations in the work which the contractors have been asked to price or which are volunteered).

Fifth, the bid(s) must be reviewed and recommendations made. Basic issues of completeness, accuracy, etc., must be settled, as there is no real benefit in entering a contract the price of which is based on error. Purchasing, legal, design, etc., personnel should examine the bid(s) and, if there are variations, alternates, etc., the contents must be analyzed. It may be valuable to have meetings with one or more of the bidding contractors. When the owner and a selected contractor have finally reached and committed themselves to an agreement, the contract is considered awarded. In the interest of expediting progress, a letter of intent may be issued to allow work to commence before appointed representatives of the two parties can meet and sign contracts.

Sixth, at an arranged meeting, the contract is executed.

One of the most complete and well-written descriptions of construction contractual components is the AIA Document Series A (101a through 701a). It describes important construction contracting considerations, including the following matters:

- Bidding
- Contract document
- The work
- Owner's responsibilities

- Contractor's responsibilities
- Subcontractors
- Work by the owner or by separate contractors
- Time of commencement and (substantial) completion
- Contract sum
- Progress payments
- Final payment
- Protection of persons and property
- Insurance, bonds, and surety
- Changes in the work
- Correction of work
- Termination of the contract
- Temporary facilities and services

Activities Subsequent to Contract Execution

The activities after execution of the contract are each delineated to some extent in the bidding and construction documents and in the contract. All parties are concerned, though the activities may on different occasions be carried out by varied combinations of the parties.

First, details of communication must be established and administered. In addition to naming the persons who hold positions and explaining the internal methods of operation particular to each party, interface procedures between the owner, designer, and contractor must be established. These are the lines of project communication. The following characteristics must be established:

- Channels, procedures, format, intervals, and responsibilities
- Correspondence and documentation requirements
- Meeting schedules, locations, and participants
- Discrepancy resolution procedures

Second, responsibilities between the contractor(s) and the owner's ongoing operations must be administered. A great deal of this is naturally included in the contract, but when other contractors or the owner are supplying equipment, a coordinated effort is essential. Details should be worked out at the beginning in order to avoid problems.

Third, responsibilities existing between the contractor(s) and the designer must be administered. Because of shop drawing and submissions review, the design effort often extends beyond the bidding or negotiating phase. If this effort is not integrated into construction operations, progress may likely be impeded. A complete schedule of activities involving the two parties must be arranged.

Fourth, construction progress reporting and payment procedures must be administered. As increments of the work are completed, inspections and approvals must take place and payments must follow. The contractor(s) must have progress payments as defined in the contract documents and they must be made in a timely fashion to avoid financing charges or halting progress. A contractor cannot profitably operate if unanticipated finance charges are incurred; the costs may be passed to the owner as specified in the contract. Efficient contract administration can keep such events from occurring.

Fifth, mechanisms for change to the contract must be administered. No matter how excellent the design or the construction documents, changes may come about during construction. Formal modification to the contract may be required and includes:

- Revisions: Major changes submitted in bulletin form to the contractor, and on which a price must be quoted. Changes to the contract only occur and may only be implemented when a change order is issued by the owner

- Field orders: Minor work changes, given directly by the owner, architect, or engineer to the contractor, and not altering the contract price or time
- Discrepancies in the construction documents or the field conditions: Problems noted and analyzed, which result in changes initiated by bulletin–change order procedures or field orders
- Claims: Extra compensation or other consideration demanded by the contractor(s) or owner and submitted to negotiations, arbitration, or legal action

Sixth, inspection and completion procedures must be administered. Inspection and reporting for conformance with the construction documents must be carried out, normally by the designer. Inspection and approval for completeness of the work must also be done. At the termination of the construction work, records, shop drawings, records of "as-built" conditions, guarantees, operating manuals, inspection reports, waivers of lien, etc., must be compiled and presented to the owner.

METHODS

The varied types of contracting arrangements involve the relationship of the construction contractor to the design contractor, types of contractors, methods of awarding contracts, and contract price definition.

It is probably helpful to describe the "traditional" approach to contracting, termed *design-bid.* In this process the work is designed and specified by consultants (architects, engineers, etc.) who produce construction and contract documents. The documents describe what is to be built or purchased and installed and are released to contractors who submit a bid, or price, to accomplish the described work. If the contractor's bid is acceptable, the owner contracts the party to do the work. In summary, the project is designed first, then bid, and finally constructed. For many reasons there is now a wide spectrum of variation from the traditional approach.

Relationship of Design and Construction Contractors

Although the designer, when retained, is normally termed a consultant, in this context it is not incorrect to refer to the designer as a contractor. The designer and the construction contractor(s) may have one of four relationships: the designer may retain the construction contractor, they may be the same party, the construction contractor may retain the designer, or they may be separately contracted.

For industrial projects, which are the central theme of this text, it is uncommon for the designer to retain the construction contractor. In other types of construction this does occur, though in only a small percentage of cases. It is not uncommon for the designer to also act as a construction manager and such a relationship is becoming popular, but, even here, the construction contractor is not under contract to the designer.

When the designer is the owner, as is probably the case when the design work is done by the plant engineering staff, the construction contractor is retained by the designer. An important aspect is that the designer possesses potentially conflicting responsibility for both time-cost performance control and quality control.

When the designer and the construction contractor are the same party, or when the designer is retained by the construction contractor, then the term *design-build* is often employed. In this relationship there is generally one prime contract, usually with a fixed budget and time frame, and a necessarily strict description of project requirements. Without a very detailed facility design program (project requirements) the budget or quality level would have to be considered open-ended. Construction and design quality must be well-controlled by this program, lest the contractor make multiple detail and material decisions which consider only the contractor's prime interest: project profitability. In some cases the requirements include so much detail that the facility design program actually becomes a design development set of drawings and specifications.

Some building work is nonetheless obtained in this fashion since the condition of single responsibility is attractive to many owners. Much operations equipment is obtained by design-build, or (design) furnish and install contracts. It is important to note that in all situations design is necessary and is included in the price; no one provides the service without charge.

When both the designer and the construction contractor are the owner, or when the designer is retained as a consultant and construction is performed by the owner's staff, similar problems of ensuring quality arise. The design and construction must be closely monitored.

When the designer and construction contractor are under separate contract, the design-bid or other variation of the traditional process is the case.

Types of Contractors

It has become common for a single contractor to enter many types of contracts and hence to become, by experience, a multipurpose contractor. This is done to suit project conditions, client requirements, etc.

General contractors undertake some work with their own crews, and subcontract the remaining work to other specialized subcontractors. The owner enters into only one prime contract, with the general contractor (GC)—all others being subcontractors to the latter. The "general," as this party is often referred to, manages the project in the field; it prepares schedules and budgets, orders materials, supervises and coordinates, expedites, etc. The designer is still often required to perform construction inspection and other follow-up services as may be specified in the design contract.

A variation of the general contracting method is that of *multiple contracting.* This may include a general contractor, depending upon the scope of the work and other factors. The owner enters into a number of independent prime contracts, usually delegating the field management and other items to specific contractors; this is quite common where equipment is involved. Multiple contracting might be typified by the following example:

> A general contract is awarded for all building and site work, but does not include certain process-exhaust and waste-treatment systems. The general contract includes supervision and scheduling in the field. The process-exhaust system is a separate contract in its entirety. The waste-treatment system is also a separate contract, but the GC is to be retained, as subcontractor to the waste-treatment contractor, to install certain control and piping interfaces through its mechanical and electrical subcontractors. Industrial equipment is purchased under six contracts: one furnish and install, four furnish, and one install.
>
> The rationale is as follows: A general is desired to manage the entire project and provide all building and site work. The process-exhaust and waste-treatment systems have a long lead time and cannot be installed or debugged until the equipment is in and operations ready to begin—and the general is off the site. Certain plumbing and control systems of the waste-treatment contract are integral with the building work, and the general can best oversee their integration with other building systems. For example, one piece of industrial equipment, a high-density stacker storage system, needs a very long lead time, must be ordered very early, and is best ordered to include both fabrication and installation. The other four contracts for industrial equipment also need very long lead times and the equipment involved should also be ordered very early (before the general contract is effective), but a single installation contract for all remaining equipment is desired and obtained at the same time.
>
> Some contract management is necessary, and has to be accomplished by the owner. This is because there is no true hierarchy of contracts (prime contracts being executed with the general, process-exhaust, and waste-treatment contractors, and six industrial equipment contractors), and there are specified arrangements between contractors (the general is in charge of supervision and scheduling in the field, the waste-treatment contractor is to use the general and its mechanical and electrical subcontractors for some work, and four industrial equipment suppliers are to use one installer).

When multiple prime contracts are used, the timing of each is probably staggered to result in a fairly compressed schedule and the term "fast track" is often employed to describe the method of contracting. It can be a time-efficient method, but requires effective contract administration.

When a single contract is used, responsibility is easily affixed. The responsibility is a portion of the risk, undertaken by a contractor, for which the owner must pay.

Construction managers do not normally undertake work with their own crews. They manage. They do not contract; the owner contracts. The owner is required to enter into multiple prime contracts for varying portions of the work and relies on the guidance of a construction manager (CM). The CM manages the project in the field, prepares schedules and budgets, supervises and coordinates, expedites, etc. Even here the designer is required to exercise construction inspection functions and other follow-up services of the design contract.

The general contractor normally executes the work for a *percentage profit* and the construction manager normally works for a *percentage fee.* The fee is generally lower, as are the risks, but on any venture there are risks and someone must take them. With the CM, it is the owner.

Method of Awarding Contracts

The price may be arrived at by competitive bidding or by direct selection and negotiated price. In a competitive situation the owner has a certain edge, which can keep the price to a minimum, although, during periods of intense construction industry activity, this edge can disappear. A bid contract also usually gives the contractor, and not the owner or designer, the choice of selecting many specific items or materials, usually with the project's profitability as a guidepost. Very complete bidding or construction documents must be prepared, to ensure certain specified minimums, and the effort can affect the project schedule. Competitive bidding is generally used to arrive at a fixed sum.

A direct selection and negotiated price method does not use competition to obtain minimum cost. In contrast to the bid condition, the owner or designer has more of an option in selecting specific items and materials, because in the negotiation a contractor is not faced with a loss of profitability. The documents need not be developed to such a complete condition, also for the same reason. Negotiated pricing does not always result in a fixed sum.

Defining the Price of the Contract

The price may be described as a fixed sum (fee) or as the cost of the work plus a fee (time and materials). When a fixed sum is requested, the contractor is assuming risk, maintains a choice in selecting specific items and materials, requires very complete bidding or construction documents, and retains its interest in the project's profitability. The owner is, of course, guaranteed a fixed cost of the construction.

When time and materials (plus a markup) are the basis of the price, the owner, not the contractor, assumes the risk. As would be anticipated, the owner or designer retains a greater say in the selection of specific items and materials, and completed documents are not a prerequisite. A variation would occur when this method is used and a guaranteed maximum or NTE (not to exceed) price is established. Here, the owner has some assurance of the maximum cost and the benefit of paying only for what is in the job. The last factor is very helpful in some bidding climates. It should be emphasized that, when a maximum is guaranteed, the contractor is once again likely to make many decisions that ensure this cost to be, in the end, no more than the maximum. Not all of the decisions may be in the interest of quality.

As an incentive to keep the cost down, contractors are often given a percentage of any savings below a guaranteed maximum. The situation creates a very complex, if not messy, possibility of conflict of interest and has lost popularity.

The selection of methods by which to contract depends greatly upon the specifics of the project and upon company or corporate policy. Regardless of the contracting method used, the owner can only get what is paid for; true bargains are rarely obtainable.

NOTES

A part of planning for time or cost is to provide some buffer which allows for those things not known at the time of the planning. This buffer is called a contingency or contingency allowance, and consists of a general contingency and a construction contingency. The contingency should be applied to both the schedule and the estimate, and should be consumed only as required.

An important point to recall is that a contingency allowance has the prime intent of providing for unseen problems, not error or major catastrophe. Each contingency allowance is therefore qualified as to just what it does and does not provide for.

General Contingency

This contingency allowance covers what is required to complete the work, yet cannot accurately be predicted at the outset. Some typical conditions:

A schedule may include a certain number of owner reviews of the project, prior to release of documents for bidding purposes. Each review is supposed to include all concerned parties. Experience with certain owner-consultant relationships may indicate that some time contingency be provided for a practical inability to assemble all persons on all occasions and to resolve all points within all time limits. One key individual's unplanned absence, if it precedes an event such as the Christmas-to-New Year shutdown, could possibly result in a 2-week delay. This could be very harmful to a 14-week design schedule.

A detailed project budget may include a certain cost for platforms and walkways, each of which has been developed in good detail. Experience with certain machinery and equipment types may indicate that some cost contingency be provided for a practical inability for a standard design to access all necessary areas on all necessary occasions for all possible equipment which may be provided. It is possible that key equipment may finally be supplied by an unplanned source, resulting in an alteration and increase in the platform and walkway systems designs, and in their costs.

These are changes to the detail of design and construction, but not to the scope of the project. Additional time or cost may be incurred in the *normal* execution of the project, the responsibility for which is the owner's. Contingency should be reserved.

Where errors are made in design or construction they must be corrected, but to a great extent this is a burden covered by contracts and insurance. The contingency is not intended to absorb these types of problems whether they are the owner's responsibility or not. Where serious difficulties arise and produce radical project alteration, the contingency cannot be expected to handle such excessive problems. The latter is more likely an instance where project redefinition may be in order. The following examples may typify these cases:

A schedule could be seriously extended by the inability to resolve process-emissions legalities which become design criteria. Although similar agency rulings may not normally have evolved into EIS efforts, certain officials may in one case decide to request such data—adding months to the schedule.

A budget could be seriously altered by the discovery of very poor soil conditions, causing extensive use of piling where it was not thought to be necessary. Although similar local construction indicates that preliminary budgets should not include money for such measures, test borings may later show serious problems, for which different and more costly measures must be taken.

These changes are the owner's responsibility, cannot be predicted, and could only be covered by contingency were the allowance extremely large. To retain a general contingency allowance of such magnitude is generally not prudent.

Construction Contingency

Some changes are necessitated during construction and are no one's fault, but they are still the owner's responsibility. Consider, if during excavation, an unknown and yet in-use sanitary-sewer line were discovered directly below the proposed building location. The controlling government agency cannot be held at fault for the maps not being complete; the surveyor cannot be held at fault for not finding what was hidden; the designer cannot be held at fault for what was not known about; and the contractor cannot be asked to absorb the expense for rerouting the line (work not covered in the contract). In this case, a construction contingency allowance would be necessary to avoid any schedule or budget increase.

There are projects which are completed without using more than a fraction of such a contingency allowance, and that is partly due to good luck. There are those which greatly exceed the contingency allowance, and that can be partly bad luck. The construction contingency allowance is usually a percentage of the construction cost and only that. It is based on experience, and is reserved for resolving problems encountered after execution of the contract.

BIBLIOGRAPHY

American Institute of Architects: *Architects Handbook of Professional Practice,* vols. 1 and 2, American Institute of Architects, Washington, D.C., 1980.

Means Construction Cost Index, Robert S. Means Company, Inc., Duxbury, Mass., Annual publication.

Bureau of Economic Analysis, Regional Economics Information System, U.S. Department of Commerce, Washington, D.C., Annual publication.

Callender, John H. (ed. in chief): *Time-Saver Standards* for *Architectural Design Data,* 6th ed., McGraw-Hill, New York. 1982.

Clough, Richard H.: *Construction Project Management,* Wiley, New York, 1972.

CSI Manual of Practice, Construction Specifications Institute, Washington, D.C., 1970.

Fabrycky, W. J., and G. J. Thuesen: *Economic Decision Analysis,* 2d ed.,Prentice-Hall, Englewood Cliffs, N.J., 1980.

Hunt, William Dudley, Jr. (ed.): *Creative Control of Building Costs,* McGraw-Hill, New York, 1967.

Meier, Hans W.: *Construction Specifications Handbook,* Prentice-Hall, Englewood Cliffs, N.J., 1978.

Muther, Richard, and Lee Hales: *Systematic Planning of Industrial Facilities,* Management and Industrial Research Publications, Kansas City, Mo., 1979.

Pena, William: *Problem Seeking, An Architectural Programming Primer,* CBI Publishing Company, Boston, Mass., 1977

Preiser, Wolfgang F. E. (ed.): *Facility Programming,* Dowden, Hutchinson & Ross, Stroudsburg, Pa., 1978.

Ramsey, Charles G., and Harold R. Sleeper: *Architectural Graphic Standards,* 7th ed.,Wiley, New York, 1980.

Taylor, George A.: *Managerial and Engineering Economy,* 3d ed., Van Nostrand, New York, 1975.

PROFESSIONAL ASSOCIATIONS

The following professional organizations publish handbooks, standards, etc., which may be of value in the administration or execution of many planning activities. The addresses given are for the national headquarters of these organizations, though it should be noted that many have state and local chapters through which information may be obtained.

American Association of Cost Engineers (AACE)
308 Monongahela Building, Morgantown, WV 26505

Associated General Contractors of America (AGC)
1957 E Street, N.W., Washington, DC 20006

American Institute of Architects (AIA)
1735 New York Avenue, Washington, DC 20006

American Institute of Industrial Engineers (AIIE)
25 Technology Park/Atlanta, Norcross, GA 30092

American Institute of Plant Engineers (AIPE)
3975 Erie Avenue, Cincinnati, OH 45208

American National Standards Institute (ANSI)
1430 Broadway, New York, NY 10018

American Society of Civil Engineers (ASCE)
345 East 47th Street, New York, NY 10017

American Society of Landscape Architects (ASLA)
1900 M Street, N.W., No. 750, Washington, DC 20036

American Society of Mechanical Engineers (ASME)
345 East 47th Street, New York, NY 10017

American Society of Professional Estimators (ASPE)
7789 Othello Avenue, Suite A, San Diego, CA 92111

Construction Specifications Institute (CSI)
1150 17 Street, N.W., Washington, DC 20036

Institute of Electrical and Electronics Engineers (IEEE)
345 East 47th Street, New York, NY 10017

National Society of Professional Engineers (NSPE)
2029 K Street, N.W., Washington, DC 20006

Individual State Economic Development Agencies

section 2

Building Construction and Maintenance

chapter 2-1

Soils, Rock, and Drainage

by
James A. Larson
Senior Project Manager, Transportation and Public Works Division
and Senior Geotechnical Engineer
Sverdrup & Parcel and Associates, Inc.
St. Louis, Missouri

INTRODUCTION

Soils and rocks are important materials to be considered in the design, construction, and performance of all facilities which they support or with which they interface. They may also be the sole construction material for dams, levees, and embankments. Soil and rock may be used as they are found in their natural state, or they may be remolded or chemically treated under controlled procedures to provide uniform design performance values and characteristics. When rock is found as a continuous consolidated layer, it is referred to as bedrock.

Since these materials have a great influence on the design and performance of any structure or utility, it is essential that a competent and experienced geotechnical engineer determine the design parameters and develop proper construction procedures and methods.

The geotechnical engineer will determine the nature and condition of the soil and rock to be encountered or influenced by the proposed facility by reviewing the stratigraphic column of the materials. This will be done by taking borings, digging test pits, and performing other investigative procedures.

Borings are a basic and universal method used to penetrate and sample the soil and rock and to study water conditions below the ground surface. Samples are obtained according to established procedures as the borings are advanced. Standard penetration tests involve driving a split spoon (split pipe), which results in a measurement of driving resistance in blow counts and provides a sample for viewing and possible testing. Thin-walled tubes are also pushed into the soil to obtain relatively undisturbed samples for laboratory testing or visual inspection. More sophisticated sampling and testing equipment and procedures may be used as dictated by the particular engineer and the materials and conditions encountered. Rock samples are usually obtained with core barrels which cut a core of rock and retain it for withdrawal.

A boring log is a record of each boring which includes date, location, elevations of the various soil strata, the equipment used, personnel involved, and other pertinent general information. Information derived from the boring is usually recorded beside a scaled vertical column to graphically represent the materials and conditions encountered. The geotechnical engineer prepares this record, which appropriately describes essential features at the correct location on the column. Standard penetration, pocket penetrometer, and other test determinations will also be recorded. These tests provide a measure of the strength and density parameters which may be used for preliminary evaluations.

For engineering purposes, soil is composed of mineral particles that can be separated by mechanical means. Air, water, or organic matter (the decayed remains of vegetable materials) may also be present. The solid particles are materials derived from the physical and chemical weathering of rocks and the organic remains of plant and animal organisms. Rock is a natural aggregate of minerals bonded together by strong permanent cohesive forces which may require great force or power to disaggregate. However, irregularities and discontinuities appear in the rocks, as erosion, weathering, and other forces deteriorate the rock mass, eventually creating soil.

Soils are described as residual or transported, according to the origin of their constituents. Residual soil is soil that has remained at its place of origin and may exhibit remanent rock structure. These soils are generally strong and stable, but may be surrounded by or contain blocks, pinnacles, or slabs of unweathered rock. Should the soil be compressible, the combination with relatively unweathered rocks may create foundation problems. Transported soils are the product of rock weathering and/or alteration; water, wind, or other means have carried them to a new location. Several deposits of transported soil, and most soils of organic origin, are soft or loose and unstable and may cause significant foundation problems.

SOIL PROPERTIES AND CLASSIFICATION

The physical attributes of the soil particles and soil aggregates are the index properties on which the distinction between different kinds of soil are based. Such properties can readily be determined by simple classification tests made according to standard procedures of the American Society for Testing and Materials (ASTM). These can be correlated with more complex engineering properties such as compressibility, permeability, strength, and swelling.

For engineering purposes, classification may be in accordance with the Unified Classification System (Table 1-1). Boring logs should include descriptions of the basic soil group to which the materials belong (as sand, silt, clay, organic soil, or gravel) and include adequate descriptions pertaining to relative density, consistency, color, odor, grain shape, stratification, interbedding, varved layering, and minor constituents. The method of deposition or origin should preferable be given.

Particle Size and Shape

Particle size has the most influence on the physical characteristics; it is evaluated by determining the size and the distribution of sizes with sieve and/or hydrometer analyses. The results are conveniently presented on the semilogarithmic particle-size curve shown in Fig. 1-1, which provides information from which the permeability of the soil can be estimated and shows whether it is poorly or well graded. Particle shape depends to some extent upon the mineralogical constituents, origin, and geologic history. Silts, sands, and gravels composed of hard minerals like quartz may be less rounded than those derived from softer minerals under similar weathering conditions. These hard or soft particles may be angular or subangular and, if well-worn by abrasion or attrition during transportation, rounded. Clay particles are flat and elongated, or lamellar.

Soil Structure

Soil structure refers to the geometric arrangement of soil components. Clean sands, gravels, and silts have single-grain structure which may be loose, honeycombed, or compact. Clay soils have dispersed or flocculent arrangements. A matrix binder which holds soil particles may create a framework of coarse grains either held in contact or held apart to form a *void-bound* structure with large void (or empty) spaces. The weight-volume relationship and other engineering properties described in the following paragraphs are frequently used to judge a soil structure.

Weight-Volume Relationships

In practice the unit weight υ, moisture content w, and specific gravity of solids G_s are readily determined, and from these the volume of air, water, and solids can be computed. Some of these material characteristics are not only related to each other, but they can be correlated to many soil-engineering properties. For instance, the void ratio and the density are influenced by particle gradation as well as arrangement, and they are inversely related to each other; a soil having a high density is likely to reduce potential settlement, permeability, and the detrimental effects of water absorption, and increase both the bearing capacity and the shearing resistance.

Relative Density

The compactness or looseness of a granular soil, such as gravel, sand, inorganic silt, or a combination of these, can range from a minimum for that particular soil to a maximum to be compared to the existing void ratio. The relative density is expressed as a percentage between the loosest and densest. Thus, a soil will have a relative density of 0 in its loosest possible condition and 100 percent in its densest condition (Fig. 1-2).

Consistency

The consistency of undisturbed cohesive soil depends on its unconfined compressive strength and is described as very soft, soft, medium, stiff, very stiff, and hard (Table 1-2). The effect of remolding on the consistency of a clay is the degree of sensitivity S_t

TABLE 1-1 Soil Classification Chart

MAJOR DIVISIONS			GROUP SYMBOLS	TYPICAL NAMES
COARSE-GRAINED SOILS More than 50% retained on No. 200 sieve*	GRAVELS 50% or more of coarse fraction retained on No. 4 sieve	CLEAN GRAVELS	GW	Well-graded gravels and gravel-sand mixtures, little or no fines
			GP	Poorly graded gravels and gravel-sand mixtures, little or no fines
		GRAVELS WITH FINES	GM	Silty gravels, gravel-sand-silt mixtures
			GC	Clayey gravels, gravel-sand-clay mixtures
	SANDS More than 50% of coarse fraction passes No. 4 sieve	CLEAN SANDS	SW	Well-graded sands and gravelly sands, little or no fines
			SP	Poorly graded sands and gravelly sands, little or no fines
		SANDS WITH FINES	SM	Silty sands, sand-silt mixtures
			SC	Clayey sands, sand-clay mixtures
FINE-GRAINED SOILS 50% or more passes No. 200 sieve*	SILTS AND CLAYS Liquid limit 50% or less		ML	Inorganic silts, very fine sands, rock flour, silty or clayey fine sands
			CL	Inorganic clays of low to medium plasticity, gravelly clays, sandy clays, silty clays, lean clays
			OL	Organic silts and organic silty clays of low plasticity
	SILTS AND CLAYS Liquid limit greater than 50%		MH	Inorganic silts, micaceous or diatomaceous fine sands or silts, elastic silts
			CH	Inorganic clays of high plasticity, fat clays
			OH	Organic clays of medium to high plasticity
Highly Organic Soils			PT	Peat, muck, and other highly organic soils

*Based on the material passing the 3-in. (75-mm) sieve.

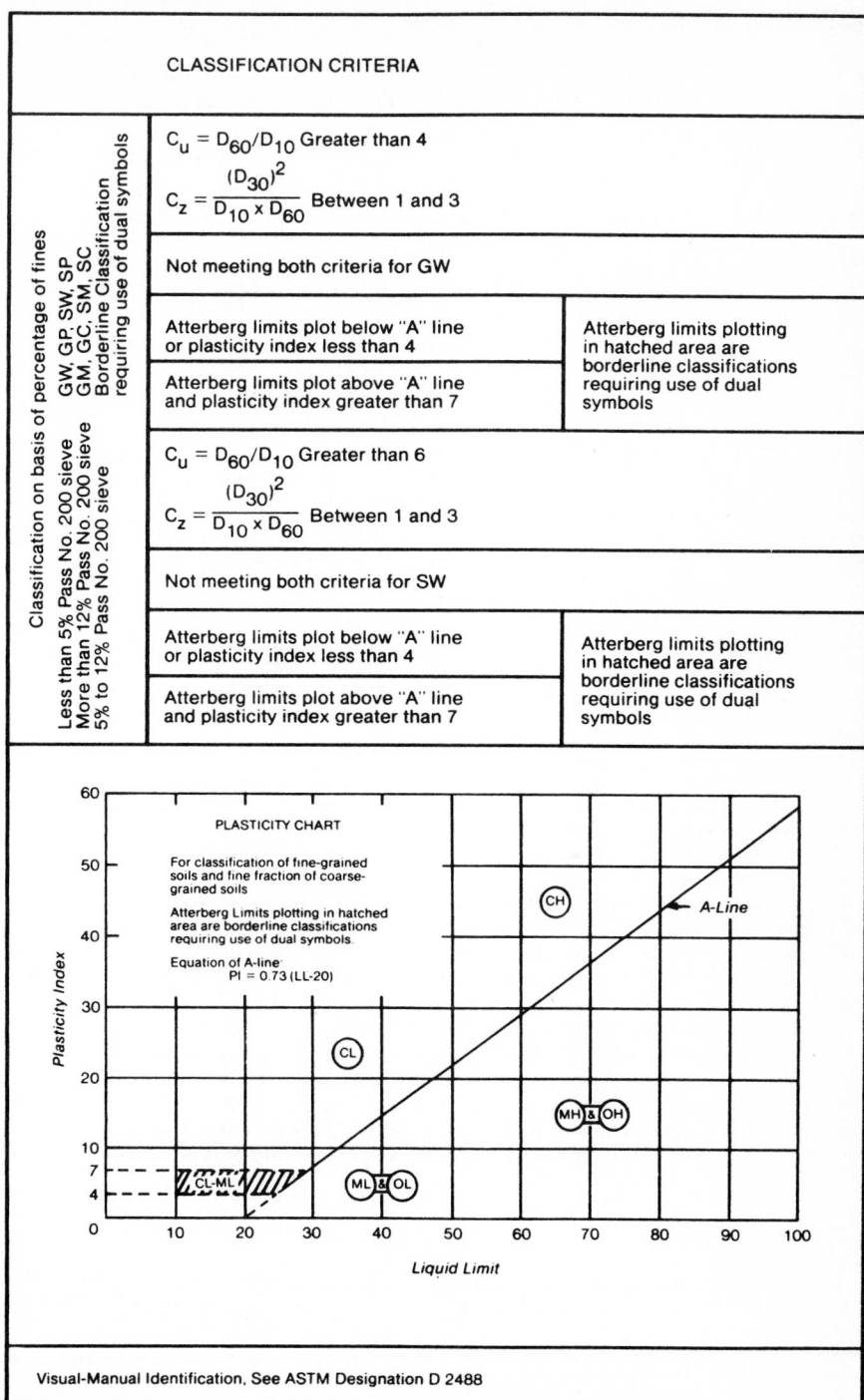

CLASSIFICATION CRITERIA		
Classification on basis of percentage of fines Less than 5% Pass No. 200 sieve GW, GP, SW, SP More than 12% Pass No. 200 sieve GM, GC, SM, SC 5% to 12% Pass No. 200 sieve Borderline Classification requiring use of dual symbols	$C_u = D_{60}/D_{10}$ Greater than 4 $C_z = \frac{(D_{30})^2}{D_{10} \times D_{60}}$ Between 1 and 3	
	Not meeting both criteria for GW	
	Atterberg limits plot below "A" line or plasticity index less than 4	Atterberg limits plotting in hatched area are borderline classifications requiring use of dual symbols
	Atterberg limits plot above "A" line and plasticity index greater than 7	
	$C_u = D_{60}/D_{10}$ Greater than 6 $C_z = \frac{(D_{30})^2}{D_{10} \times D_{60}}$ Between 1 and 3	
	Not meeting both criteria for SW	
	Atterberg limits plot below "A" line or plasticity index less than 4	Atterberg limits plotting in hatched area are borderline classifications requiring use of dual symbols
	Atterberg limits plot above "A" line and plasticity index greater than 7	

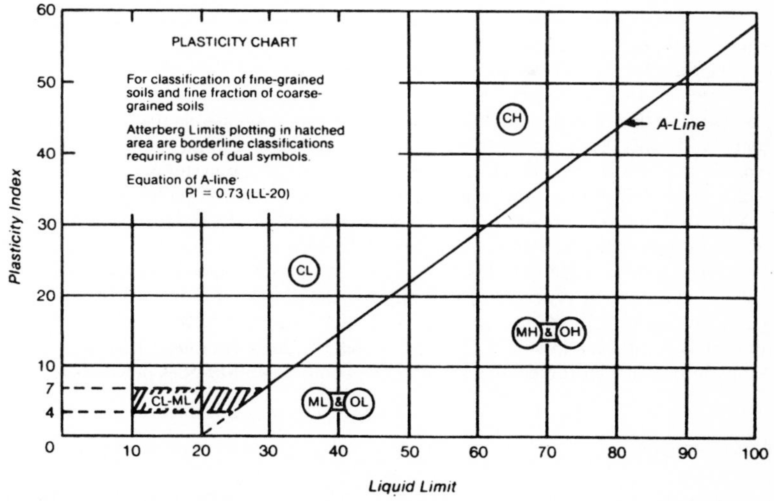

Visual-Manual Identification, See ASTM Designation D 2488

(From 1980 Annual Book of ASTM Standards, Part 19, "Soil and Rock; Building Stones," Standard D 2487, pp 377-388. Used by permission.)

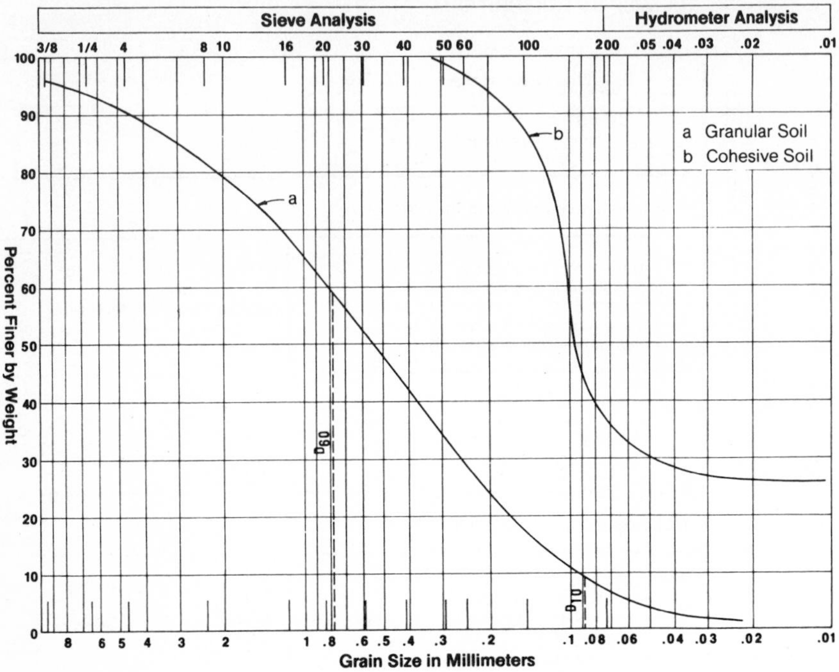

Figure 1-1 Grain-size distribution curve.

expressed as a ratio of the unconfined compressive strength of the undisturbed soil to that of remolded soil at the same moisture content. The values of S_t vary between 2 and 4 for most clays, but for "quick" clays they may exceed 16.

The consistency of remolded soil changes with an increase or decrease in moisture content. With very high moisture content the soil may be fluid, and as the moisture decreases the soil will range down through a viscous fluid, a plastic solid, and a semisolid to a solid. Atterberg limits show the moisture content between the liquid, plastic, semisolid, and solid states of cohesive soil. These are known as the liquid (LL), plastic (PL), and the shrinkage (SL) limits. The numerical difference between the liquid and plastic limits is termed the plasticity index (PI or I_p). Liquid and plastic limits are both dependent on the amount and type of clay in a soil, but the plasticity index is ordinarily dependent only on the amount of clay present. With increasing LL, both the permeability and compressibility increase in soils having identical PI. Soils of equal LL show a decrease in permeability with an increase in PI, and the compressibility does not change.

SOIL HYDRAULICS

The design, construction, and performance of facilities are influenced by the location of the groundwater table, permeability, capillarity, and pore water, as well as by effective stresses.

Groundwater Table

The upper surface of the zone of saturation is called the groundwater table (GWT), or the phreatic surface. Below this, the spaces between the soil particles are filled with

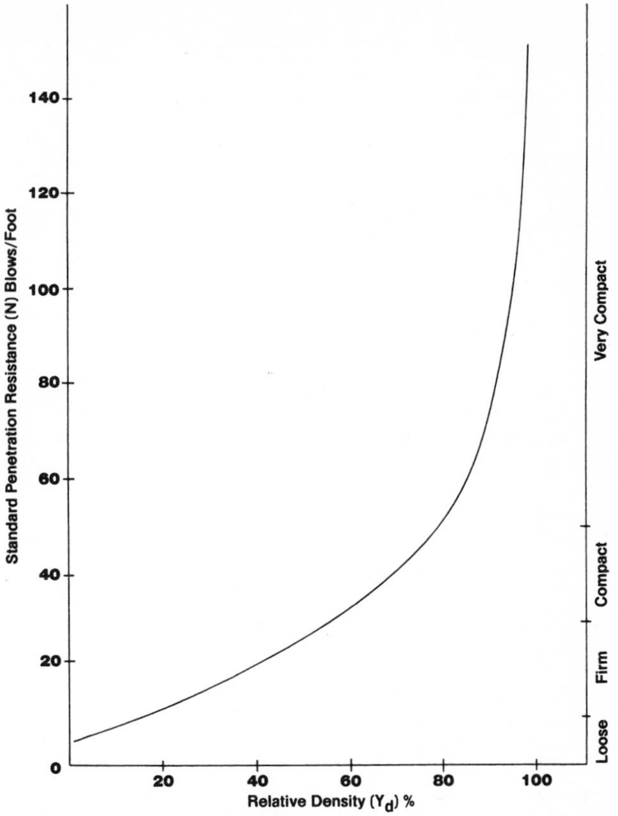

Figure 1-2 Standard penetration—relative density comparison.

TABLE 1-2 Consistency and Strength of Cohesive Soil

Consistency	Field identification	Unconfined compressive strength q_u, tons/ft²*
Very soft	Easily penetrated and deeply penetrated by fist or comparable blunt object	Less than 0.25
Soft	Easily and moderately penetrated a few inches by thumb or short blunt stick	0.25–0.5
Medium	Penetrated by thumb or stick, with effort up to 1½ in	0.5–1.0
Stiff	Can be indented by thumb and penetrated about ¼ in by ⅜-in-diam stick with much effort	1.0–2.0
Very stiff	Easily penetrated by fingernail; stick makes slight indentation	2.0–4.0
Hard	No penetration with stick; fingernail slightly penetrates	>4.0

*1 ton/ft² = 0.9765 kg/cm²; the units are therefore used interchangeably. Cohesive strength c and shear strength s are equal, and each is equal to one-half the unconfined compressive strength.

water. If the ground water is interrupted by dikes or a stratum of impervious soil above the groundwater table causing ground water to accumulate, a perched water table results, which must be considered as a groundwater situation in any foundation design.

Permeability

Permeability depends on the size of the voids between particles, which in turn depends on the size, shape, and the state of packing. Under identical voids and densities, the coarse-grained soils (sands and gravels) are more permeable than the fine-grained silts and clays. Where soils have similar textural characteristics, the permeability declines as the density rises. The permeability of the soil influences its drainage, compressibility, and susceptibility to frost action.

Capillarity

Water flows into unsaturated soils located above the groundwater table because of surface tension and attraction to the water at the phreatic surface. The height to which the water will rise depends on both the affinity of the soil for water and the size of the voids. For equal affinities, both the height and rate of rise are inversely related to the size of the voids. The height of capillary movement is also affected by evaporation and by changes in the groundwater level. For example, open gravel or rock roads which have given satisfactory performance for years may, when paved, develop problems from subgrade weakening due to capillary saturation.

Porewater Pressure and Effective Stress

The water in voids in saturated soil exerts a pressure (as it does in any vessel) that is called porewater pressure. The pressure is calculated by determining the height of the vertical unit column of water below the point of phreatic surface. The effective stress in any direction is the difference between all stresses (total stress) in that direction and the porewater pressure.

Both the deformability and the strength of a soil are dependent on effective stress. For a layer of fine-grained soil rapidly loaded locally, the viscous retardation of porewater flow builds up excess porewater pressure. The excess porewater pressure is not only a function of the loading change, but also a function of soil properties. Excess porewater pressures can cause soil movements and failures.

Shear Strength

The shear strength s of a soil in any direction is the maximum shear stress that can be developed in the soil structure in that direction, c is a measure of cohesion, ϕ is the angle of internal friction, and σ is the normal stress on the shear plane. The values of c and ϕ for any soil at a specified initial moisture content and unit weight depend on the conditions under which the loading is applied to the soil. Saturated clays tested under undrained conditions exhibit a constant shearing resistance. The angle of internal friction for the same clay tested under drained conditions may be as much as 30°. Clean sands and gravels, dry or saturated, do not display any cohesion. The angle of internal friction for dense sand varies between 33 and 46°, for loose sand between 28 and 34°, for dense inorganic silt between 25 and 35°, and for loose inorganic silt between 20 and 30° For an extreme, dense, well-graded gravel, ϕ may be as high as 50°.

Shear-strength properties can be determined in the laboratory with triaxial compression (Fig. 1-3) or direct shear tests. In the field, standard penetration, pocket penetrometer, vane shear, or plate-loading tests are some that are used. In all cases, evaluation must be made with judgment. Factors affecting the results of soil strength tests are type of soil and test, size and shape of sample, dry density and moisture content, method and rate of loading, drainage conditions during test, permeability and structure, climatic conditions, sample disturbance, and time between sampling and testing.

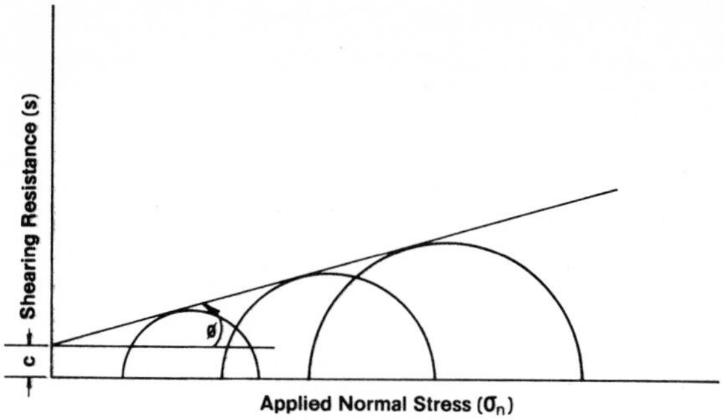

Figure 1-3 Strength envelope for soil.

Volume Change

Certain clay soils have large volume change with a change in moisture, swelling with the addition of water and shrinking with dehydration. When swelling, they produce pressure which may seriously damage buildings and other facilities. These *volume-change soils* are difficult to recognize in borings; their potential for swelling and the pressures they will exert by swelling must be determined by testing.

Soil deformation may be defined as volume change attributed to rearrangement of solid particles, a change in shape of solid particles, and deformation by extrusion of pore water or air. In cohesionless soils and somewhat in clays, the distortion of grains is mainly responsible for a volume change which is largely elastic and consequently reversible. Particle slipping and bending, and extrusion of pore water and air are major mechanisms in cohesive soils which exhibit slow time-dependent deformations which may or may not be reversible depending on the interaction of many factors.

To determine time-dependent compressibility or swelling characteristics, tests are made on undisturbed or remolded specimens as a uniaxial test with restraint of lateral deformation. The results, the pressure-to-void ratio and time-to-compression ratio or swell, are graphically presented in Fig. 1-4.

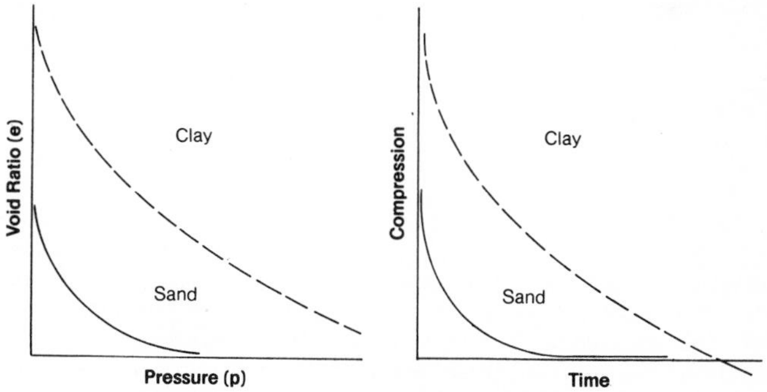

Figure 1-4 Diagrams of compressibility of soil.

Clays and sands undergo reduction in void ratio with pressure increase. For sand, the rate of reduction decreases with load increases. The major part of the compression in sands under an increment of load takes place instantaneously. Clays require considerable time to attain final compression under a load increment.

Depending upon geologic history, soils are divided into normally consolidated or preconsolidated (overconsolidated) categories. The first has never been subjected to a pressure greater than the present overburden and, in the case of clay, has a liquidity index

(I_L) that varies from about 0.6 to 1.0. The second has had a pressure greater than the present overburden, and has an I_L varying from 0.0 to 0.6 for clay. Generally, normally consolidated rocks are more compressible than preconsolidated ones.

Unlike adding load, which causes settlement, removing the load can cause the soil to rise, or heave. Heaving can occur in the bottom of an excavation because of stress release, excess hydrostatic pressure, and lateral displacement of soil from below.

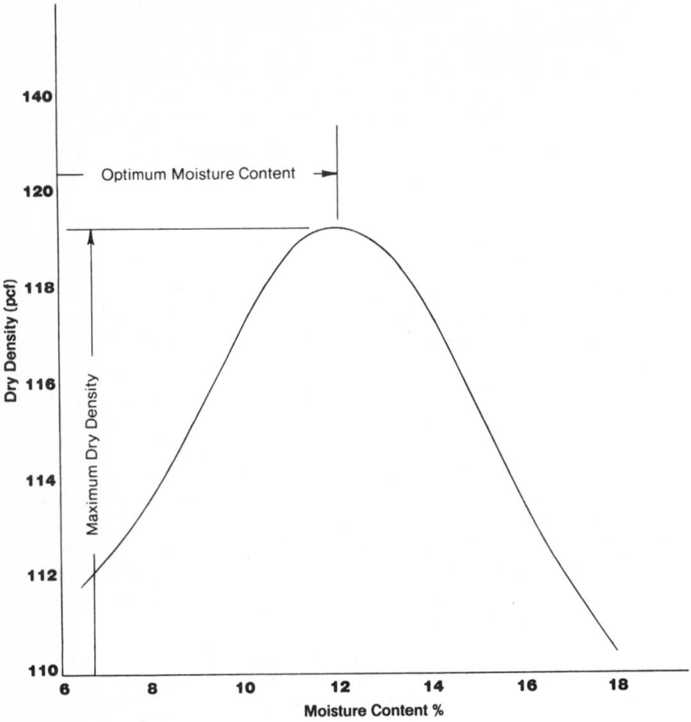

Figure 1-5 Relationship between dry density and moisture content for constant amount of compaction.

Compaction

Compaction is the process in which the particles are constrained and packed closer together by a reduction in voids, and is quantitatively expressed in terms of soil dry density. For each soil with a given amount of compaction there is an optimum moisture content at which the dry density is maximum (Fig. 1-5).

Soil type, moisture content, and the amount and method of densification affect the compaction. Increased compaction will result in an increase in maximum dry density. Soil compaction improves strength, reduces compressibility, and decreases the ability to change the moisture content (which causes swelling or shrinkage).

CLASSIFICATION AND PROPERTIES OF ROCK

The behavior of a rock mass is governed by mechanical properties and also by the number and nature of discontinuities in the rock. Lithology, weathering, quality, structure, ground water, permeability, and engineering properties are significant factors to be considered during design and construction.

Lithology

Rock materials are classified by such lithologic terms as shale, granite, schist, and limestone which reflect mineralogy, texture, and fabric of the rock. Lithology suggests the mechanical properties and helps associate *in situ* features. For example, schists have foliated textures with directional differences in both strength and deformation moduli.

Weathering

Weathering is the action of altering rock properties of color, texture, composition, or form. The effects extend below the earth's surface by groundwater movement through and around grains or along discontinuities or fissures. The variations in depth of weathering are often dictated by the differences in lithology. The strength of the rock is reduced by an increasing degree of weathering. This condition will usually be described in boring logs in relative terms.

Rock Quality

Rock is characterized by discontinuities and the degree of weathering or alteration. Rock quality is used to define the approximate bounds of behavior and to locate zones of poor quality. Rock quality designation (RQD) is a procedure based on modified core recovery that is most frequently used in estimating average rock quality. An RQD of less than 25 percent represents *in situ* rocks of poor quality, while over 90 percent is indicative of excellent quality. If applied, RQD should be noted in the boring logs.

Rock Structure

Discontinuities such as shears, shear zones, foliations, and faults influence the behavior of rock through interactions between rock blocks and discontinuities. Several engineering properties of discontinuities including attitude, frequency, location, continuity, shape, roughness, tightness, and coating and/or filling materials play significant roles in assessing behaviors of both the rock mass and individual blocks of rock contained in them.

Ground Water and Permeability

Ground water in rocks usually moves through joints, fissures, or other openings. High hydrostatic pressures can cause instability along planes of discontinuity, loss of strength, and excessive water discharge.

Strength

For intact rock, the uniaxial compressive strength and the modulus of elasticity at a stress level equal to one-half the ultimate strength of the intact rock are used to classify the rock materials and to verify the degree of weathering or alteration in the rock. The compressive strength is also used in ascertaining bearing capacities in certain jointed rocks.

The strength along discontinuities is generally derived from the basic shear strength and the resistance against dilation due to interlocking surface projections. Plane discontinuities may have a very low angle of inclination, whereas very irregular surfaces may display a high inclination.

Shearing resistance along the discontinuity depends also on the roughness and character of filling materials. A smooth, slickensided joint may display a low resistance, while the very rough surfaces will show much higher resistance. The shear strength of fill in discontinuities is largely dictated by the thickness and properties of the filling materials such as composition, grain size, moisture content, and plasticity. Highly plastic clayey fills of significant thickness generally have low frictional resistances. However, in discontinuities with very thin filling or coating materials, the initial strength properties may be dictated by the filling or coating, but the final strength may be controlled by the rock surface properties as the filling or coating is ruptured through.

Swelling

Free swell or excessive swelling pressure is ordinarily encountered in fill materials and in weathered or altered zones. The extent of these behaviors is dictated by the amount, composition, and properties of the material. Clay-mineral and particle-size determinations, Atterberg limit tests, and free-swell tests may be used in ascertaining the swelling index properties of rock minerals.

DRAINAGE

Surface and subsurface water must be considered as essential elements of design and construction. The subsurface investigation program includes obtaining information on subsurface water conditions and reliable location of the groundwater table at the time of the study. Historical data that may be available can be very useful to the designers.

Local, state, or national agencies may have requirements to be met in the drainage design and may specify certain design procedures. These requirements will assist in determining runoff and sizing the drainage system.

Surface Drainage

All parts of the facility require good performance from the surface drainage system. The areal drainage pattern and surface topography will affect the site grading plan, which will include disposing or rearranging surface materials to eliminate surface water problems both during and after construction. The plan must also consider environmental protection regulations for the site. Topographic maps prepared by the U.S. Geological Survey are useful for analyzing both topography and drainage, and show elevation contours, bodies of water, streams, and other features from which the rates of water runoff can be established. Aerial photographs can also provide useful information for site evaluation. For final designs, however, an on-site survey is needed for more accurate detail.

Surface topography, precipitation, rate of runoff, and the nature of the soil must be considered in preparing a good drainage design. These parameters should also receive careful consideration when establishing both the floor elevations and grading plan elevations, since building distress has often been caused by inadequate drainage of surface water.

Ditches, swales, sewers, and earth slopes are the common ways of removing surface water to avoid ponding or water flowing into the excavation. Slope stability and erosibility should be analyzed to predict the possibility of slope failure and blocking of ditches and sewers. Silt fences or retention ponds may be required to retain eroded soil on the site. If space permits, ditches may be deepened so as to lower the ground water, but this is usually limited to a shallow depth.

Silty and fine sand soils require flat slopes, while soils with clay can be relatively steeper. Water seeping out of slopes exerts a seepage pressure which leaves the slopes less stable. To prevent erosion and scour, all slopes should have a protective covering compatible with both the soil type and the water velocities. During embankment construction and before each rain, embankments should be sloped for drainage and compacted to a smooth surface for quick runoff. Permanent slope protection should be installed as soon as a slope is finished.

Slope protection may be obtained from many different materials. For example, sodding or seeding works well on many slopes, while gravel or rock blankets may be selected for others. The availability of materials in an area may influence the type of material to be used. The higher water velocities in ditches may require using riprap, concrete, or asphaltic pavement for slope protection. Many specialty products will provide good protection, and these may be economically feasible if local materials are limited. Wire baskets or "sausages" filled with smaller rocks than those required for riprap can contain a mass as effectively as riprap. Soil additives that serve as a cementing material to soil may increase the soil's erosion resistance sufficiently to withstand the design velocities.

Strength, workability, trafficability, and other soil characteristics will change with soil moisture changes. The best soil condition for most construction situations may be a dense soil at or very near its optimum moisture content. When a soil used for embankment construction becomes wet, the moisture content must be adjusted downward by either drying out the soil, blending it with other dryer material, or using additives (lime, cement, or silica grout) to gain stability.

Subsurface Drainage

Soil-boring programs obtain information on ground water; however, the design must consider the changes in groundwater depth that may come with seasonal change, the amount of precipitation, and other natural phenomena. Water wells in the area, their effects on the groundwater level, and the effects of their retirement from service may be very important. Piezometers installed in bore holes will provide both short- and long-term information on groundwater elevations.

When the depth of excavation is greater than the distance to free water in a soil, drainage will be required during construction and may be required for the finished facility. In relatively impervious soils like clays and silty clays, the water inflow may be on the order of a seep; but pervious soils may provide an abundant flow. Drainage will permit construction in-the-dry and maintain the stability of the slopes and bottom.

A sump with ditches in the bottom of an excavation will collect water for pumping from the excavation, but well points, wells, or other extensive measures may be necessary for more severe water conditions. Berms and well points in stages on the berms may be the dewatering method for saturated slopes. Deep wells will withdraw large volumes of water to lower the ground water below the excavation depth in very pervious soils.

Drainage systems are used to keep water from contacting completed structures. An impervious blanket is placed at the surface to slope away from the building and carry surface water away. Perforated pipe sloped to drain in pervious material at or below the foundation level will remove moderate amounts of water. Such a system may be necessary under floor slabs to keep them dry and prevent uplift. Figure 1-6 shows a subsurface drainage system.

Drainage systems must be kept clean to function as intended, and an occasional cleanout opening should be considered for long runs of pipe. Most systems are closed, and the only infiltration may be from the particular soil being drained. A filter of graded sand and rock layers or filter fabric will prevent soil migration (see Fig. 1-7). The filter fabric is more easily installed and permits various arrangements.

An alternative to drainage is waterproofing building components that may be in contact with free water. Waterproofing must be carefully installed and may be expensive. Since waterproofing will not relieve water pressure against walls and slabs, this pressure must be included in the wall and slab design. Figure 1-8 shows a waterproofing design method.

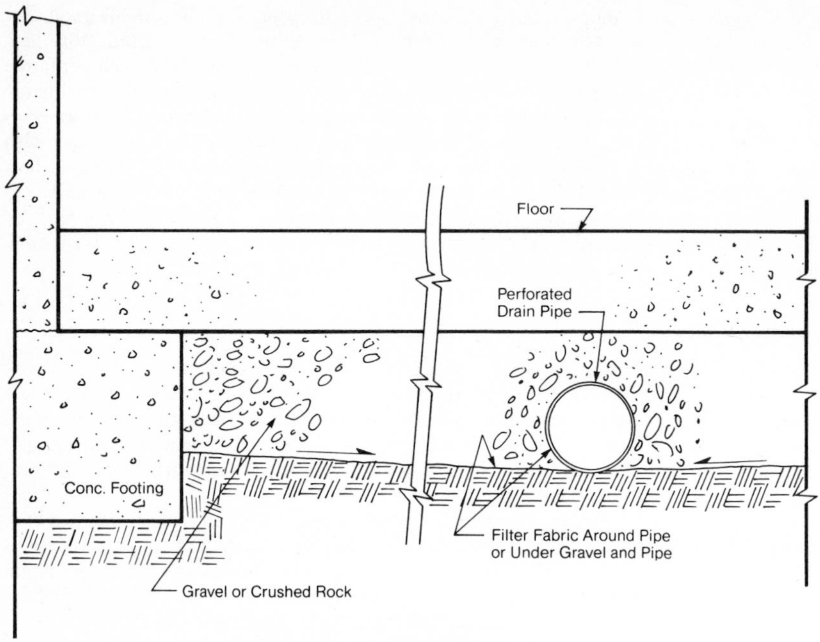

Figure 1-6 Subsurface drainage system.

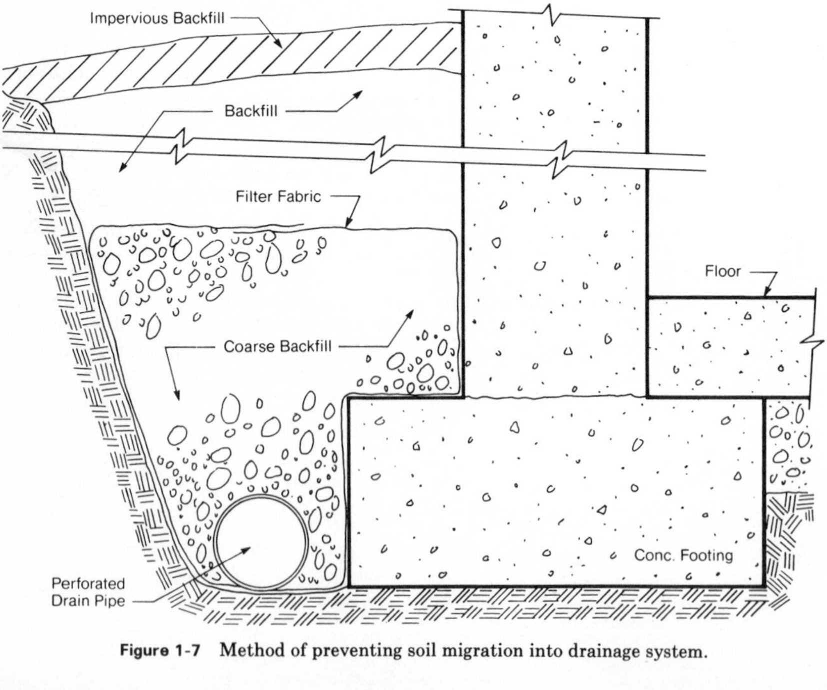

Figure 1-7 Method of preventing soil migration into drainage system.

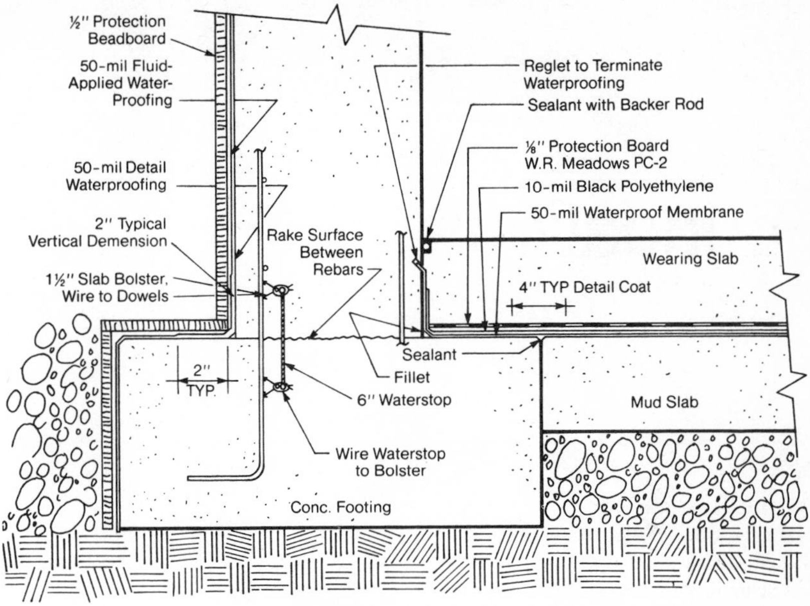
½" Protection Beadboard
50-mil Fluid-Applied Water-Proofing
50-mil Detail Waterproofing
2" Typical Vertical Demension
1½" Slab Bolster, Wire to Dowels
Reglet to Terminate Waterproofing
Sealant with Backer Rod
⅛" Protection Board W.R. Meadows PC-2
10-mil Black Polyethylene
50-mil Waterproof Membrane
Rake Surface Between Rebars
Wearing Slab
4" TYP Detail Coat
2" TYP
Sealant
Fillet
6" Waterstop
Mud Slab
Wire Waterstop to Bolster
Conc. Footing

Figure 1-8 Waterproofing design method.

chapter 2-2

Foundations

by
Arthur R. Meenen
Manager, Special Structures Section,
Industrial Division,
Sverdrup & Parcel and Associates, Inc.
St. Louis, Missouri

James A. Larson
Senior Project Manager, Transportation and Public Works Division
and Senior Geotechnical Engineer,
Sverdrup & Parcel and Associates, Inc.
St. Louis, Missouri

REVIEW

The foundation is the part of the structure that distributes the weight of the structure and its supported loads to the underlying earth or rock substrata. The distribution may occur by direct bearing of a footing on soil or rock or by transmitting the loads to a deeper strata through driven piles or drilled piers.

Many materials have been used for constructing soil-bearing footings, including timber and steel grillages, clay masonry, cut rock slabs, and formed concrete. Except for formed concrete, most of these systems are generally not used today. Foundation concrete is usually reinforced, especially where heavy concentrated loads are to be supported.

Piling and drilled piers are of the same types and materials which have been in use for many years. Their length and capacity have increased considerably, however.

SOILS INVESTIGATIONS REQUIRED

Knowledge of the subsurface materials and conditions that will interface with a structure is a critical requirement. Procedures and methods for obtaining this information are given in Chap. 2-1, "Soils, Rocks, and Drainage." This information will reveal the best site location for a building, the possible need for dewatering, the data needed to establish the allowable foundation support, the anticipated settlement, shear strengths, swell potential, and the need for bracing and shoring for foundations and pipe trenches. A subsurface investigation will provide information and samples from which allowable soil bearing, consolidation characteristics, shear strengths, swell potential, and other design parameters are determined.

The investigative program must be extensive and go beyond the limits of vertical influence for both shallow or deep foundations. Where there is flexibility in locating the structure, or when the exact location, size, and shape of the building are not firm, the program must extend to the horizontal limits of the possible building footprint. An experienced geotechnical engineer can plan an effective program, but, since earth materials vary in consistency and stratification, there must be flexibility in the program as the work progresses.

If the subsurface program is planned after conceptual studies of the new facility have been made, the anticipated loading conditions, floor and grade elevations, and any special features should be given to the geotechnical engineer who will then emphasize and intensify the investigation in the critical areas.

The subsurface program includes determining the groundwater conditions at the time of the investigation. Piezometers may be installed to study porewater pressures in isolated soil zones or observation wells to observe static water levels for the period of the program or for extended periods of time.

All possibilities of water and its influence must be considered in establishing floor and foundation elevations, the need for drainage, and the problems and additional cost inherent in constructing each type of foundation.

Unsatisfactory soil at a site may be removed and replaced under controlled-density procedures, removed and replaced with other material (also under controlled-density procedures), or altered by grouting to gain strength, increase density, or replace ground water.

DESIGN CONSIDERATIONS IN SELECTING THE FOUNDATION

After the soils investigation has been completed, a foundation type must be selected.

The selection is based on many factors relating to the soil and the structure. Among these are magnitude of the load; type of load—dead, live, dynamic; application of load—vertical, lateral; moments; depth to supporting soil strata; supporting capacity of the soil; settlement anticipated under load; and the type of structure to be supported.

Types of Foundations

The three basic types of footings used for building foundations are soil-bearing, pile, and drilled piers. Any of these may be used for individual column support, either as combined or strapped foundations or as large mats. The decision on the footing type and the interrelation of column supports must be based on the factors given in the preceding paragraph.

The nearness of a column to a property line may require strapping a wall footing to an interior column footing. The imposition of large moments may require combining the footings in strips or mats.

Soil-Bearing Footings

With soil-bearing footings the maintenance of equal settlement under dead loads may require proportioning the footings for reduced bearing pressures as the column load and the footing size increase.

Generally, the exact amount of settlement that can occur is not critical. The differential amount of settlement between footings causes the problems in a structure. More settlement, of course, increases the possibility of large differential amounts.

The amount of settlement (total or differential) will have different significance depending on the type and use of the structure. In a structure designed with continuous beams or frames, differential settlement between columns can induce large stresses. Unless these stresses are added to the design parameters, structural distress or failure may occur. Even when a structure is designed to provide simple span beams between columns, this settlement can cause problems in the function of the structure. Elevator guides may require realignment, doors may bind in the jambs, pressure may crack the glass in sashes, and the walls may crack, allowing moisture to penetrate and start corrosion of the supporting members.

When the strata which are relatively close to the surface are not adequate to support the anticipated loads, the loads must be carried to more solid material at a greater depth by drilled piers or driven piles.

Drilled Piers

Drilled piers may be cased or uncased, with a straight shaft or with belled bottoms. The choice depends on the material and the depth of the pier. A material with sandy seams or saturated soils may require casing to be lowered into the hole as drilling progresses. To be economical, this casing must be pulled as the concrete is placed. Because of the instability of the soil and casing required, these piers will not be belled. Piers drilled in stiff firm clays can be belled to provide increased bearing area on the supporting soil layer. When the piers are relatively short, drilling larger-diameter shafts may be more economical than changing the auger to a belling device. When the pier is to be drilled to rock, a belling device will generally not be usable and the straight shaft must be used. Piers will not require reinforcing unless they are in an area of high seismic activity or are subject to lateral or tension forces. When reinforcing is required it is commonly only needed in the top one-fourth to one-third of the pier.

Concrete to be placed in the piers should be a stiff but flowable mix placed using a tube called an "elephant trunk" to prevent segregation of materials. Vibrating the mix will prevent honeycomb, or voids, in the shaft. Having all the operations taking place within the small working area available in smaller piers is difficult, and vibration may be impractical.

Pile Footings

Driven piles are another method of obtaining support from deep strata. The most commonly used piles are steel H, concrete-filled pipe, concrete-filled steel thin-shell, reinforced or prestressed-precast concrete, and timber piles.

Friction timber piles are generally used in lengths up to 60 ft and with loads to 40 tons [36.3 t (metric tons)]. Steel H piles are used for long high-capacity piling bearing either on rock or into a hard soil layer. Tapered precast-concrete piling is generally used for conditions similar to those for which timber piles are used, but with longer lengths and higher capacity. Pipe, shell, and large, uniform-diameter precast piling are used for displacement friction or for end-bearing piling. The common range of loads now carried on piling is from 20 to 200 tons (18.1 to 181 t) per pile.

When driving high-displacement piling, some preboring may be required to prevent disturbing either adjacent structures or previously driven piles. All field-poured piles in a cluster should be driven before filling with concrete. Piles should not be driven closer than 4½ pile diameters from concrete which is less than 24 h old.

The selection of the pile type depends on the capacity and length required, the soil encountered and, of course, the cost of the piling. All these factors must be studied for each job since they will vary considerably with site, time, and availability of materials.

The total economy of piling and cap must be studied when selecting the piling system. The piling capacity should be selected to fully load the minimum number of piles at the columns with the lightest loads.

While the settlement of spread footings is usually recognized, piling settlement is often forgotten. The total movement at the pile butt will result from the tip movement caused by both loading the soil and elastic shortening of the pile, which can be considerable in long piling. In addition, concrete piling can experience long-term shortening due to creep in the concrete.

Generally, clusters of vertical piling will resist the amount of lateral forces to be expected in a building by bending of the piles and passive resistance of the soil against the cap. Where large lateral forces are expected, the piling may have to be driven on a batter, or angle. A slope of 1 horizontal to 2 vertical is a practical limit for driving batter piles. A pile cluster with batter piles must be checked for stability for all loading conditions, which may include a condition in which there is no lateral load.

Although the batter piles may be adequate for all lateral loads, the effects of pile settlement should also be considered. For instance, lateral movement of the pile caps from pile settlement, while limited, may cause distress in a rigid-frame structure.

Installation Problems and Construction Effects on Nearby Structures

Finally, in selecting the type of foundation, consideration must be given to installation problems and construction effects on nearby structures. This is not limited to vibrations caused by pile driving. Consider the case in which a lightweight building may be supported on a relatively weak soil layer, and a heavily loaded structure is to be built nearby. Say two options can be considered: (1) Piles or drilled piers to carry the loads to a deeper, firmer soil layer or (2) excavating to a medium-firm layer and constructing a mat foundation. Removing the soil weight during excavation for the mat could result in movement of the soil mass supporting the nearby building, resulting in cracking or collapse of parts or all of the older structure.

GROUNDWATER EFFECTS ON FOUNDATION DESIGN AND CONSTRUCTION

Ground water is discussed under "Soil Hydraulics and Drainage" in Chap. 2-1, "Soils, Rocks, and Drainage." In many cases, the groundwater table is subject to variations throughout the year. The table is usually not a straight line, and will show undulations when profiled across the plant site. The subsurface investigation program will show a reliable elevation for the table at the time of the study. The variations may be determined by installing piezometers and observing them for a period of time. This information, however, may not be part of the subsurface investigation report. The rate of variation depends on soil permeability, while the amount of water is influenced by streams, tides, water wells, nearby impoundments, and precipitation. Rain or snow melt may saturate the soil as it filters down to the groundwater table.

Soil moisture content is a most important soil characteristic. Consistency, stability, strength, and sometimes the unit weight are functions of moisture content. Foundation design procedures allow for moisture changes, and these procedures usually include the worst moisture condition for the *in situ* soils. Maintaining the integrity of the foundation support material during construction is important.

Removing water and lowering the groundwater table are very common construction problems, and may be necessary during construction both to improve the working conditions and to ensure construction that meets the design intent. The soil trafficability may be very poor in the presence of free water, for example, and the workers and equipment may disturb and remold the soil in undesirable ways.

Water should be removed as it enters an excavation. In relatively impervious soils, the ground water may be a small amount that decreases as the slopes dry out. Small amounts may be diverted into sumps and pumped out. Where the excavation goes below the groundwater table in soil whose permeability may be on the order of 5×10^{-6} cm/s, special dewatering procedures will be required with well points or wells. The designer may specify the method, or leave it to the contractor's option. Conditions at some sites may require special protection for foundations because of flowing water, while precautions may be required at others to prevent settlement due to loss of ground water. Seasonal fluctuations may be of significant magnitude to schedule construction so as to eliminate the need for a dewatering system.

Free water exerts a hydrostatic force against foundations, floor slabs, and walls. The force, independent of the soil, is equal to the unit weight of the water times the height to which it can accumulate. For this condition, either the structure must be designed to resist the force, or the water must be removed with a permanent dewatering system. Other methods such as grouting the soil may be used to alleviate or remove the problem. All dewatering in the construction sequence must be continued until the facility has been completed to the point that raising the groundwater level will not cause damage to the structure or interfere with its use.

STRUCTURAL DESIGN

The structural design of all foundations involves two considerations: first, the function of the foundation as a supporting unit and, second, the function of the materials themselves used in constructing the foundations.

The function of the foundation involves the stiffness of the unit and the balance of the loads against the supporting pressure. The service load condition should be used for establishing the stability of the foundation to applied loads.

The stiffness becomes critical when large footings, mats, or strapped foundations are considered. While lack of stiffness may not cause complete failure, it can lead to settlement or twisting of the superstructure. A foundation that is not designed to balance supporting pressures uniformly against gravity loads will create eccentricity, again resulting in settlement or twisting in the superstructure.

Particular care must be taken in the structural design of large footings or mats by the present design (ultimate) procedures to assure that enough stiffness is built into the footing. If concrete is the controlling material and requires large amounts of reinforcing, the design is probably too flexible.

As in all parts of the structure, the idea is to provide a design for the least cost which will adequately support the structure. However, it should be remembered that the results of a foundation failure are not only very serious, but repairs are difficult and costly.

Spread footings may be constructed in three basic ways: a uniform-thickness slab, sloped top, and stepped top. Because of the high cost of field labor involved in the last two, footings are generally limited to uniform-thickness slabs.

Soil-Bearing Footings

Soil-bearing isolated footings are designed as flat plates with a vertical load and/or moment transmitted to the footing by the column, pedestal, or wall that it supports. Design should be according to the requirements of the chapter on footings in ACI 318. The footing depth must be adequate to provide anchorage of column bars or dowels in either tension or compression as the design conditions require.

For two-way rectangular footings, the ACI code specifies that the area of reinforcing calculated from the bending moment requirements shall be placed with a large portion located in a band equal to the narrow width of the footing. This can lead to placing mistakes in the field. Another solution is to place the same concentration of bars uniformly across the entire footing. This can be done by using the following equation:

$$A_s \text{ to be placed} = A_s \text{ calc. by mom.} \frac{2 \times \text{dim. of long side}}{\text{dim. of long side} + \text{dim. of short side}}$$

Most concrete design textbooks follow through the design calculations for isolated soil-bearing footings. The Concrete Reinforcing Steel Institute (CRSI) *Handbook* and Portland Cement Association (PCA) *Notes on ACI 318* are readily available and provide sample calculations.

The bar size selected should always be checked to see that the distance from the critical section for moment to the edge of the footing at least equals the development length of the reinforcing plus 3 in.

If a footing is subjected to large external moments, reinforcing may be required in the top of the footing.

Although the code is not too clear on the requirement for minimum reinforcing, it is generally accepted that the requirement for $A_s = 0.0018bt$ (where b = breadth of slab and t = total thickness) be used each way. If the footing is rectangular with basically moment in one direction only, it may be prudent to provide $A_s = 200/f_y\, bd$ (where f_y = stress at yield point and d = effective depth to the reinforcing) in that direction to prevent sudden failure.

Pile Footings

The design of footings supported on piles follows the same basic structural design as for spread footings on soil. The difference is in the distribution and application of load. Also, the load carried by a pile footing is generally much greater than that carried by a spread footing of similar size. As a result shear is usually a controlling factor in establishing the thickness of the footing.

When one- or two-pile footings are used, they must be tied together with beams to prevent bending caused by eccentricity of the load when the pilings are driven off-center. It is generally considered adequate to design these beams for a moment of the column load times the radial tolerances specified for the pile group.

In addition to the reqirements for shear at distances of d or $d/2$ from the face of the column, as outlined in the ACI code, shear should be checked around the perimeter of an individual pile or across the corner of a cap. In two-pile footings in which the piles are located inside the critical sections for shear, the code commentary recommends checking the footing for shear in deep flexural members. Some design handbooks also check for an extension of the Corbel shear provisions to this case. It could also be critical to ensure that adequate bond of the tension reinforcing is provided beyond the outside face of the pile.

When pile footings are used in seismic zones, they must be tied together with tension-compression struts to prevent lateral moment and subsequent failure of the piles or structure caused by the eccentricity of the load on the piles.

Drilled Piers

Piers of 18-in or greater diameter are called drilled piers. When smaller-diameter piles are drilled, they are usually considered drilled-in-place piles. A lining must be lowered into the hole when access into the pier must be provided. This lining cannot be of less than 30 in diameter, which requires a hole of approximately 36-in diameter.

While piers may be used as single units under column loadings, they can be clustered and tied together with a cap to provide more resistance in cases with large applied moments.

Some codes require the pier length to be limited to 30 times the diameter of the pier.

Strap or Combined Footings

A column or other load source will often be located near an existing structure foundation or property line. An adequate isolated footing could not be used without experiencing eccentricity. In this case a common solution is to tie two foundations together to eliminate the eccentric moment. This can take the form of a combined foundation, or two foundations strapped together with a wall or grade beam. (See Fig. 2-1.) The figures show a few of the configurations that can be used.

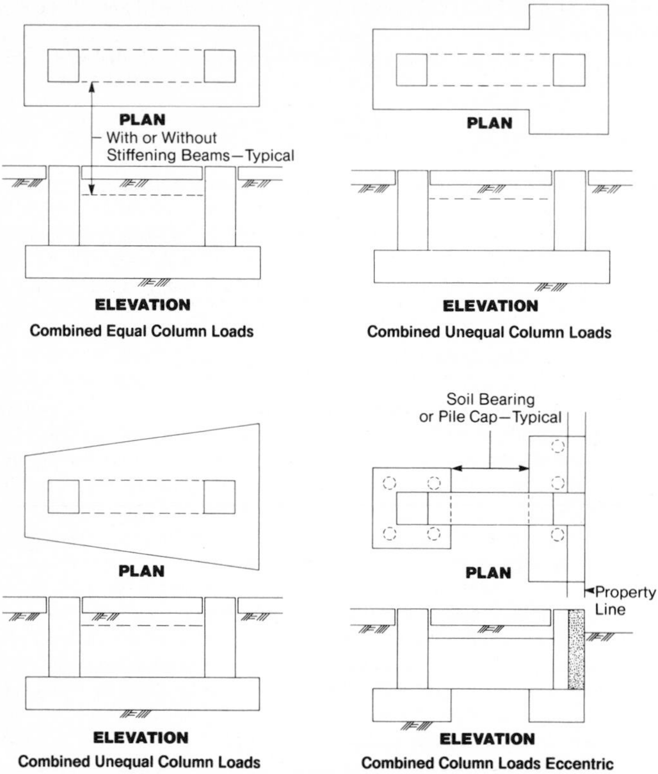

Figure 2-1 Strapped or combined footings.

Any solution that is stable and structurally sound is acceptable and can be used for soil-bearing, pile-supported, or drilled-pier foundations.

Mats

When loads are to be supported on a relatively weak soil, the footing sizes or pile spacing may almost result in an overlap of foundations. In this case, the best solution may be to provide a continuous mat foundation under the entire structure. Even when piles are used, a mat with a pattern of piles may be necessary to provide adequate support. An analysis based on the interaction between soil settlement and foundation deflections will give the best design. Care should be taken in selecting the thickness of the mat since one that is too flexible will not provide the necessary distribution of applied loads.

The mat should be extended beyond the perimeter columns if possible, again to provide for better load distribution. Stiffening provided by a perimeter grade beam will aid in the distribution along an edge.

Posttensioning

The use of posttensioning tendons in lieu of regular reinforcing may have some advantages in foundation design. While regular reinforced foundations will deflect under load and create some unequal soil pressure distribution, load-balancing posttensioning may result in more uniform soil pressures with less materials. In addition, the posttensioning tendons can be stressed in stages to increase the prestressing as continuing construction operations increase the load.

Normal prestressed concrete design concepts must be followed. This may require a review of the foundation for various load stages during construction as well as the final service load combinations and final design (ultimate load) load considerations.

RETAINING WALLS

When usage and site conditions require excavating the existing grade to provide two distinct levels, a retaining wall may be required to hold the earth without a long natural ground slope.

Retaining walls may be built using many different construction methods. Some of the more common methods are concrete, sheet piling, concrete or piling cribs, soil anchoring, reinforced earth, slurry walls, and gabions. At most plants, concrete walls are probably adjacent to, or are used as part of, the building foundations. Sheet piling will often be driven first to permit excavation, with a concrete wall-surfacing placed afterward to provide a more attractive surface. Concrete cribs are frequently used to provide an extended change in site grade to permit constructing the buildings or access roads. Figures 2-2 to 2-8 show typical retaining wall designs.

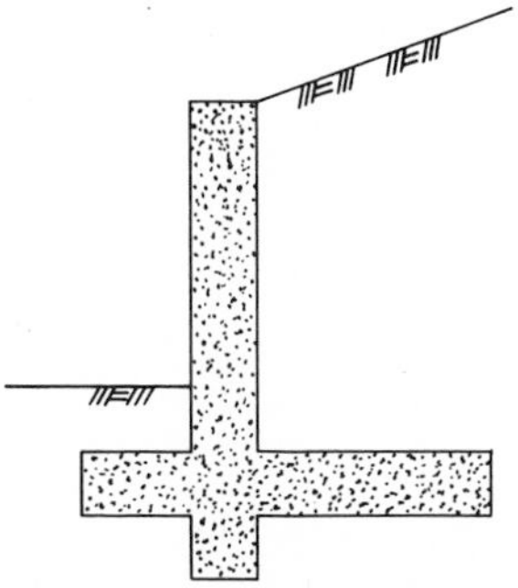

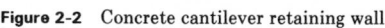

Figure 2-2 Concrete cantilever retaining wall.

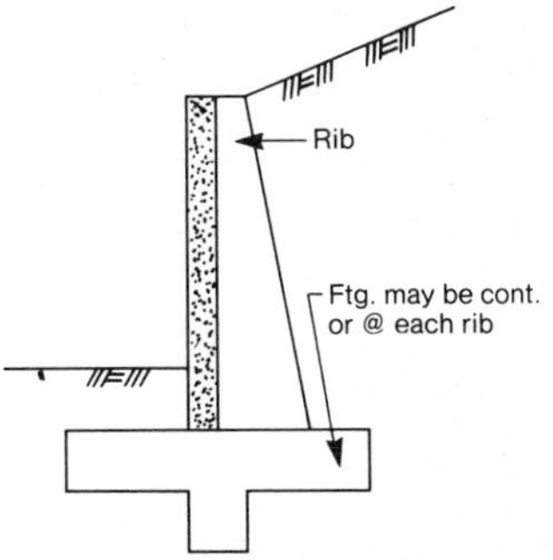

Figure 2-3 Concrete counterfort retaining wall.

Concrete retaining walls may be designed to provide lateral stability and structural capability in various ways. The wall may be analyzed as a unit wall: each foot of wall is expected to retain a foot of earth behind the wall.

Another method is to span the wall horizontally between counterforts. This is usually more economical for very high walls, but the additional forming costs do not warrant its use for relatively low walls. Of course, the wall may be soil-bearing or pile-supported. For soil-bearing walls a shear key may be required to provide the necessary passive soil pressure to prevent sliding. When the wall is used as part of a building foundation, however, the floor slab may provide the necessary base restraint. The design should be

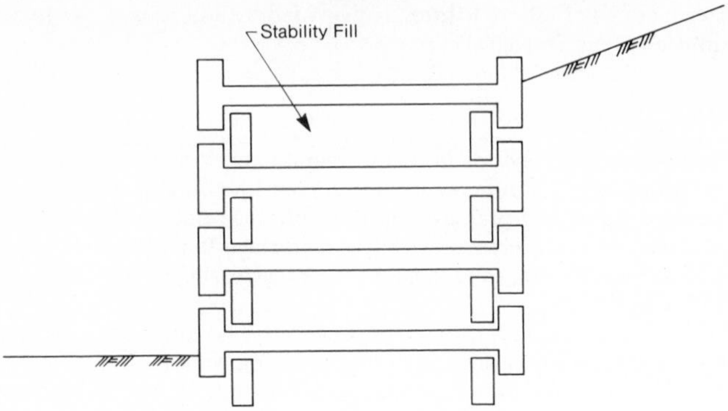

Figure 2-4 Concrete crib wall.

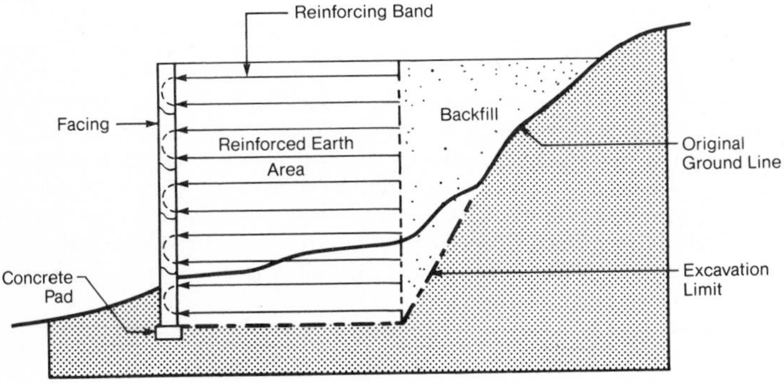

Figure 2-5 Reinforced earth structure.

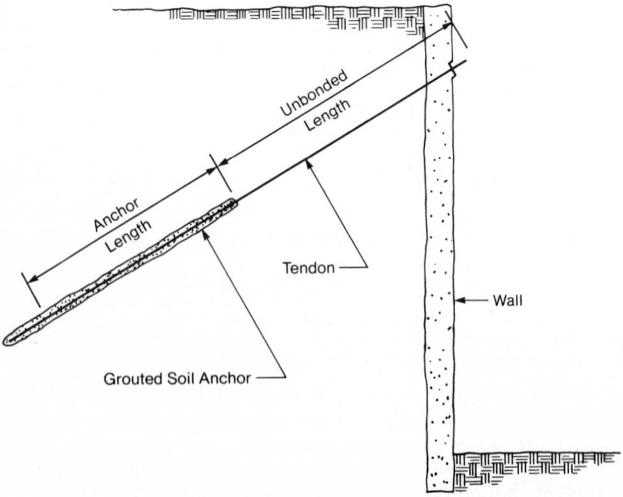

Figure 2-6 Soil-anchored wall.

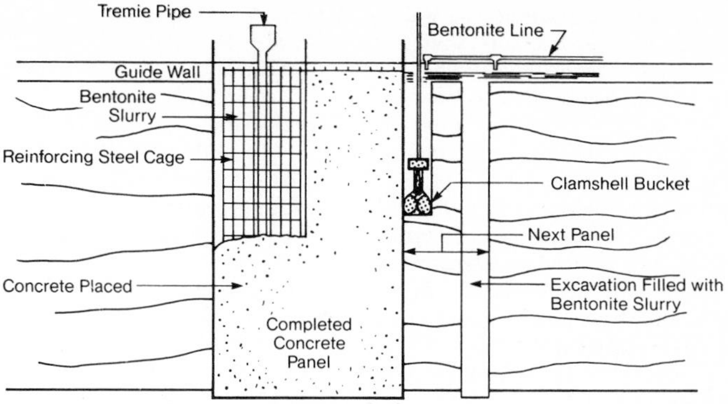

Figure 2-7 Slurry wall.

checked for a temporary condition during construction, but with a reduced safety factor to assure stability during all phases. A reduced safety factor is warranted since sliding failures are usually gradual movements, and the short period during construction will not permit this to happen.

When the wall is pile-supported, the lateral resistance of the piles coupled with a shear key may provide adequate lateral restraint for short walls. Again, the floor slab may provide support for the long-term condition. If the wall is high and pile-supported, the use of batter piles may be necessary.

When a wall has its face exposed to the atmosphere, a sand-gravel drainage area is

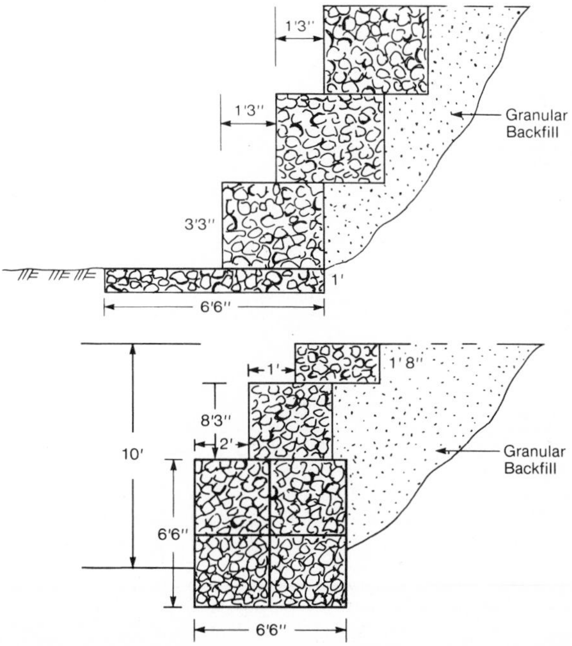

Figure 2-8 Gabion walls.

commonly provided just behind the wall, with drain holes spaced through the wall to relieve the groundwater pressure that could otherwise be created. If local conditions indicate groundwater pressure could be a problem and the wall is part of a building foundation, a perforated drainpipe may have to be provided at the bottom of the sand-gravel area to divert the water into a storm-water drainage system.

As is the case for all walls, a retaining wall should be built with control joints spaced 20 to 30 ft (6 to 9 m) on-center to isolate the shrinkage cracking that will occur. This is even more critical for an exterior wall subject to extreme temperature changes.

Good practice dictates placing the control joints at the same location in both the continuous footing and the wall. Although the footing is not subject to much temperature change, shrinkage will occur and the footing will crack indiscriminately if there are no control joints, and these cracks will trace up through the wall at locations other than the wall joints. Drain-hole pipes in the wall should be centered on the control joints, or wall cracking may also occur at the drain location.

When a retaining wall is also part of a building foundation, it must be designed for all conditions of loading which can occur. This can often result in a difficult design problem in establishing a footing that will provide the proper bearing and stability under varying loading conditions.

One solution that can be considered when this occurs is to separate the retaining wall from the building columns and footings. The effect of this solution on the support of the wall system must be studied, but no great problems will result if the support and connection details are studied and designed to agree with the anticipated movements.

chapter 2-3

Concrete

by
Ronald E. Heil
Manager, Structural Section,
Buildings Division,
Sverdrup & Parcel and Associates, Inc.
St. Louis, Missouri

GLOSSARY

Many of the terms used to describe concrete, its use, and the materials from which it is made are found in reference material printed and updated at intervals by the American Concrete Institute (ACI)[2] and the American Society for Testing and Materials (ASTM). The reader is urged to become familiar with the latest issue at the time of use.

Structural concrete Concrete used to form elements which are intended to support the loads generated by external forces and by their own weight. The term also identifies concrete of a quality specified for structural use. Materials most often used for structural concrete consist of a binder made of portland cement and water, and a variety of inert coarse and fine aggregates. Structural concrete may be plain or reinforced, but the most common type is reinforced.

Plain concrete Concrete that is unreinforced or contains less reinforcement than the minimum amount specified for reinforced concrete in ACI Standard 318.[3] ACI 322, "Standard Building Code Requirements for Structural Plain Concrete," presents the standard specification for using plain concrete for structural purposes. These uses should be limited to structures having continuous contact with the ground or to structures capable of providing continuous vertical support. Arch-type structures in which the members are in compression under all loading conditions are an exception to this limitation. Footings, pedestals, walls, and slabs supported on the ground, or equipment pads supported by a structural system capable of providing continuous vertical support to the pad are common structural elements that meet the above requirements. In general, plain concrete is used only when the actual computed internal tensions are less than 75 lb/in^2 (518 kPa). Plain concrete used for structural purposes should have a minimum compressive strength of 2500 lb/in^2 (17.24 MPa) as required by ACI Standard 322.

Reinforced concrete Contains steel bars or wires that act together with the concrete to resist imposed forces. Reinforced concrete design in building construction is usually required to conform with ACI Standard 318, "Building Code Requirements for Reinforced Concrete."[3] Reinforcement is used to overcome the weak tensile strength of concrete as well as add to its compressive load-carrying capacity. In members subject to bending or tension forces, the concrete is assumed to have no tensile strength; therefore steel reinforcement is provided at locations subject to tensile stress. This assumption is made since concrete will often develop small cracks as it hardens, leaving no predictable capability to resist tension stress. Reinforced concrete can be used for members supported above the ground on columns, walls, beams, or other isolated supports, in addition to foundations and slabs supported on the ground that are subject to large tension stresses caused by bending.

Normal-weight concrete This has a unit weight of about 150 lb/ft^2 (2400 kg/m^3), and is made with normal-weight aggregates commonly and naturally available in the construction locality. Natural rock having weight, strength, and durability properties similar to limestone is most often used for normal-weight concrete. This type concrete is frequently used because of its low cost and the local availability of the aggregates.

Lightweight concrete This has a substantially lower unit weight than normal-weight concrete. The unit weight of structural lightweight concrete is between 90 and 120 lb/ft^3 (1440 and 1920 kg/m^3) in contrast with the 150 lb/ft^3 (2400 kg/m^3) unit weight of normal-weight concrete. It is usually made with normal-weight fine aggregate (sand) and lightweight coarse aggregate [⅜ in (1 cm) or greater in size]. Concrete made with this combination usually weighs about 115 lb/ft^3 (1840 kg/m^3). When both the fine and coarse aggregates are lightweight material, the unit weight can be as low as 90 lb/ft^3 (1440 kg/m^3). The selection of structural lightweight concrete is usually based on economic advantages inherent in supporting a lighter-weight structure. These advantages may not always exist, however, because of predicted higher cost for lightweight aggregates. Most lightweight aggregate is made by heat processes requiring a relatively high energy input and its price is, accordingly, escalating with energy costs. The availability of lightweight aggregate may therefore significantly diminish because of reduced demand. Nonstructural insulating lightweight concrete, primarily used for thermal insulation, has a unit weight of 15 to 90 lb/ft^3 (240 to 1440 kg/m^3).

Heavyweight concrete This is made with high-specific-gravity aggregates such as barite, ilmenite, iron, limonite, magnetite, or steel, and is often used for radiation shielding. Its unit weight varies from 180 to 350 lb/ft^3 (2884 to 5607 kg/m^3).

Admixtures Materials or substances other than cement, water, and aggregate which are added to the concrete immediately before or during mixing. They are used to modify the properties of the concrete before or after it has hardened to make it either more economical or more suitable to conditions during and after placement. Admixtures can modify the properties before the concrete hardens by increasing workability, retarding or accelerating initial setting, and improving pumpability. The effects after hardening include increasing the strength without increasing the amount of cement; increasing the durability or resistance to freezing temperatures; increasing the bond between old and new concrete; inhibiting the corrosion of embedded corrodible metal; producing fungicidal, germicidal, and insecticidal properties; and producing colored concrete. More information on admixtures can be found in "The Guide for Use of Admixtures in Concrete" and "Admixtures for Concrete" reported by ACI Committee 212 in the *ACI Journal of the American Cement Institute,* September 1971 and November 1963, respectively, and also in the *ACI Manual of Concrete Practice,* part 1.[4]

Prestressed concrete Reinforced concrete in which internal compressive stresses have been introduced to reduce potential tensile stresses resulting from internal or external loads. The internal stress is generally supplied by tensioned steel bars or wires, called tendons, that are anchored in the concrete before the load is applied. Steel tensioning done before the concrete is placed is called pretensioning; when done after placement it is called posttensioning. The advantages of prestressing include providing an almost zero-deflected structure when subjected to its own weight and providing nearly crack-free elements.

Pretensioning methods are normally used at plants manufacturing precast concrete structural components such as slabs, beams, walls, and flanged members having a T or double-T (TT) configuration. Pretensioning requires anchoring and tensioning the tendons before the concrete is cast. The tendons are released from their independent anchorage when the concrete is sufficiently hardened, compressing the concrete since the force in the tendons is then transferred to each end of the member through the anchorage created by the bond between the tendons and the hardened surrounding concrete. Information on designing pretensioned structures can be obtained from the Prestressed Concrete Institute (PCI), *PCI Design Handbook,* 2d ed., and other PCI materials.

Posttensioning is most often done at the construction site. Tendons are preplaced in ducts or tubes slightly larger than the tendons themselves and have the capability of being anchored at each end of the member to be stressed. Concrete is cast around the ducts complete with tendon anchorage devices. The tendons are tensioned with a portable jack when the concrete has reached about 75 percent of its specified design strength, and are anchored when the desired prestressing force is obtained. Two types of duct systems are available—bonded and unbonded. The bonded system is one in which a cement grout is injected into the duct after the tendon has been anchored. The grout fills the space between the tendon and the inside of the oversized duct, effectively bonds the tendon to the duct, and protects the tendon against corrosion. In the unbonded system (see Fig. 3-1), the tendons are pregreased with a corrosion-resistant material, and

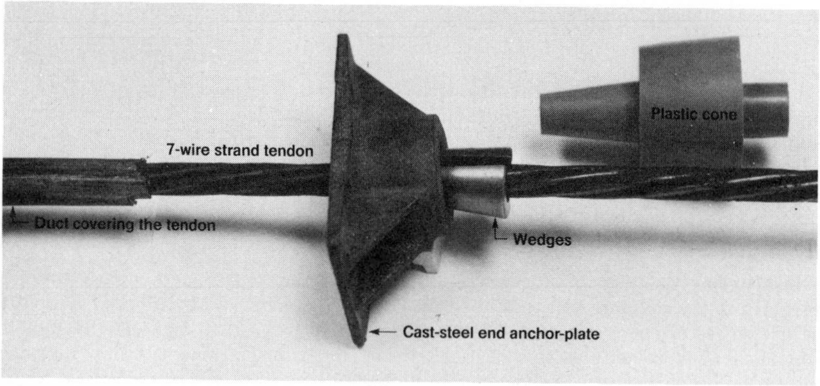

Figure 3-1 Unbonded posttensioning system.

grout is not used. Information on designing posttensioned structures and associated specifications can be obtained from the Post-Tensioning Institute's (PTI) *Post-Tensioning Manual* and other PTI and ACI technical data.

Precast concrete A term used to identify concrete members that are cast and cured in other than their final position, either in precast concrete manufacturing plants or at the construction site. Some of the advantages of these members include eliminating the formwork and shoring required at the construction site; economies from mass production and reducing forming requirements; reducing concrete placing problems resulting from extreme weather conditions; and accommodating fast construction schedules since units can be made while or before the foundations are installed. Products available from manufacturers generally include wall systems of solid or ribbed construction; hollow-core and solid- and ribbed-slab systems as well as beams and girders in a variety of I and L shapes. They are available in conventionally reinforced concrete and pretensioned, prestressed reinforced concrete.

Site-cast precast concrete is often used for tilt-up wall panels or lift-slab construction. In tilt-up construction the wall slabs are usually cast against the building floor slabs and are tilted-up into their final position and anchored to the structural frame after they have hardened and cured. Lift-slab construction is a procedure in which the floor slabs of a multistory building are cast against the ground-floor slab and are jacked into final position on the building columns.

USE AS A BUILDING MATERIAL

Concrete is a composite material basically consisting of a binding medium within which are embedded various sizes of aggregate. The information in this chapter is largely directed to structural concrete in which the binding medium consists of hydraulic cement mixed with water. Portland cement and/or blends of portland cement are normally used for building construction. The cement and water combine chemically to harden to a rock-like material.

Concrete of this type is used plain, but more often it is used in combination with steel reinforcing bars to form a variety of structural systems and building components. See Chap. 2-10 for a description of building systems utilizing concrete.

The proper use of concrete requires an understanding of its constituent materials, its structural properties, and its behavior in concrete building systems. This chapter seeks to equip plant engineers with a broad knowledge of concrete terminology and its uses to permit them to deal intelligently with consultants they may hire to do structural concrete design and specification work. It is also designed to help the plant engineers assure themselves that design work they buy conforms to generally accepted professional standards. Consequently, this chapter discusses the data required to specify the performance and properties of concrete and its components for use as a building material. The reader is also directed to various references for information concerning the analysis of structural properties and building systems, and to appropriate specifications and standards.

CONCRETE MATERIAL SPECIFICATIONS AND USAGE

Cements

Portland Cement

ASTM C150 is the governing specification for this cement, which is available in five basic types. The normal color is light to medium gray, depending on the manufacturer, but white cements are available. White cement is often used for terrazzo-concrete floors where architecturally desirable, and when color admixtures are used to provide a colored concrete. Portland cement meets requirements for use in making structural concrete in accordance with ACI 318. The uses for the five types are:

Type I. For normal use when no special properties are required.

Type II. For general use, especially when moderate sulfate resistance or moderate heat of hydration is desired. It should be used for concrete either in direct contact with sewage or exposed to a moderate sulfate environment containing 150 to 1000 ppm (parts per million) of sulfate.

Type III. For use when high early strength is required. The normally specified 28-day strength can be obtained in 7 days. It is often used to speed form removal, or when service loads need to be applied sooner than 28 days.

Type IV. For use when low heat of hydration is desired. Heat is produced by the chemical reaction that occurs between the water and the cement. Very thick concrete sections lose the heat of hydration at a slow rate, allowing destructive temperatures to develop. The slower rate of temperature rise allows more time for the heat to dissipate, thus reducing the potential harm to the structure. This type of cement is rarely used and its availability is decreasing.

Type V. Used when high sulfate resistance is desired. Concrete exposed to substances with a sulfate content of over 1000 ppm should be made with this type of cement. Common uses include foundations in direct contact with high-sulfate soils, and structures associated with high-sulfate-content wastewater.

Types IA, IIA, and IIIA. Air-entraining cements with the same uses as types I, II, and III, respectively. These are used to provide greater workability of fresh concrete, but are more often used to provide greater capacity to resist the effects of freezing and thawing after the concrete is cured.

Blended Hydraulic Cements

These are available in three basic types: portland blast-furnace slag cement, portland pozzolan cement, and slag cement. The governing specification is ASTM C575. Concretes made with these cements result in varieties of gray color and are usually darker than concrete made with portland cement. A tan-buff cement is also available. Blended hydraulic cements, except for slag cements, may be used for making structural concrete according to ACI 318. The availability of these cements is not as great as portland cement; portland pozzolan cements may become more available, however, with the increasing use of coal as a fuel, making more fly ash available as a pozzolan material. In general, blended hydraulic cements have a lower heat of hydration and gain strength at a slower rate than portland cement; they usually need a longer curing time, especially in cold weather.

ASTM designations for these types of cements are: type 1S, portland blast-furnace slag cement; types 1P and P, portland pozzolan cements; and type S, slag cement. Types P and S are not recommended for structural concrete.

Calcium Aluminate Cement

This is not covered by an ASTM specification, nor is it included in ACI 318 as an acceptable material for structural concrete. Its primary use is for concrete that is subjected to high temperatures [>800 to 3000°F (427 to 1649°C)]. Lightweight aggregate is often used with calcium aluminate cement to provide greater heat resistance and insulation. Typical uses include lining steel chimneys and stacks, and other vessels and equipment for process industries; lining gas washers, ducts, sintering, or drying equipment; and insulating furnace walls. It has also been used in jet-engine testing structures. Manufacturers' literature often indicates using this cement as a structural material; however, thorough investigation and testing are advisable to establish its properties, especially those related to long-term strength and durability under in-service moisture, heat, and weather conditions, before it is used as a structural material. This cement hardens very rapidly and has a high heat of hydration, making placing and curing procedures very critical in producing a satisfactory material. Information needed for using this cement is contained in a special 1978 ACI publication SP57, "Refractory Concrete."

A summary of the state-of-the-art of using refractory concrete including calcium aluminate cement has been reported by ACI Committee 547 in *Concrete International,* vol. 1, no. 5, May 1979.

Expansive Cement

This is a hydraulic cement which when mixed with water forms a paste that, after setting, tends to increase in volume to a significantly greater degree than portland cement paste.

It is used primarily to compensate for volume decrease due to shrinkage that normally occurs as concrete hardens. The governing specification is ASTM C845. The desirability of using these cements should be checked when sulfates will be present in the environment. Three types of expansive cements, identified as K, M, and S, differ in the principal constituents and reactive aluminates which form the expansive qualities. These cements are often used to counteract the normal tendency for cement to shrink as it hardens. Common uses have been for grade floor slabs in an effort to reduce or eliminate joints placed in slabs to control shrinkage and to reduce or eliminate cracks that form because of drying shrinkage. This material has been extensively studied and continues to be investigated to determine the long-term results. ACI Standard 223, "Recommended Practice for the Use of Shrinkage-Compensating Concrete," provides recommendations for using these cements. A high degree of quality control is needed in the proportioning, mixing, and placing to produce the desired shrinkage compensating and expansive effects.

Water

Water is required in combination with cement to form the binding substance in concrete. The water should be fresh (not salt water or ocean water), clean, and drinkable. Minerals and bacteria in the water are not usually harmful to the concrete if their concentration does not render the water unpotable. The Portland Cement Association (PCA) Booklet EB001.12T, "Design and Control of Concrete Mixtures," gives suggested limits for alkalis, chlorides, sulfates, and solids in mixing water. Water which is not drinkable may be used if it can produce mortar cubes having 7- and 28-day strengths equal to at least 90 percent of the strength of similar specimens made with drinkable water. The test method for mortar cubes is established by ASTM C109, "Method of Test for Compressive Strength of Hydraulic Cement Mortars."

Aggregates

These are generally classified in terms of size (fine and coarse) and are identified as two separate ingredients. Fine aggregate material is either natural or manufactured sand (or both) graded and having a maximum size of ⅜ in (9.5 mm) and having only 2 to 10 percent passing a No. 100 sieve (150-μm mesh). Coarse aggregate is graded material normally having a maximum size not over 1½ in (38 mm), and with no more than 5 percent passing a No. 16 sieve (1.18 mm mesh). To permit concrete to completely surround the concrete reinforcement, coarse aggregate is usually specified to have maximum sizes of 1, ¾, ½, or ⅜ in, (or 25.4, 19, 13, or 9.5 mm). The corresponding ASTM C33 grading designations for these coarse aggregate maximum sizes are 57, 67, 7, and 8, respectively. Aggregates are additionally classified in terms of weight—heavy, normal, and light—and apply to both fine and coarse materials. To provide uniformity, only one source of aggregate should be used. The aggregate comprises most of the concrete mix and has a significant influence on concrete properties and durability. Detailed information can be obtained from ACI 221 R-61 "Selection and Use of Aggregates for Concrete" reported by ACI Committee 621 in the *ACI Manual of Concrete Practice,* part 1.

Normal-Weight Aggregates

Normal-weight aggregates consisting of hard, stable, durable types of rock available in the construction area are most commonly used and should conform to ASTM Specification C33. Natural sand and manufactured sands, gravel, crushed gravel, crushed stone, or air-cooled blast furnace slag having a bulk dry unit weight of about 100 lb/ft^3 (1600 kg/m^3) are included in this classification.

Lightweight Aggregates

These are covered by ASTM Specification C330. Lightweight aggregate is not as strong as normal-weight aggregate and has a bulk dry-unit weight between 30 and 70 lb/ft^3 (480 and 1120 kg/m^3). Additional cement may be needed to produce concrete having the same strength as normal-weight concrete due to the lower aggregate strength. The primary usefulness of lightweight aggregate is to produce concrete weighing about 35 lb/ft^3 (560 kg/m^3) less than normal-weight concrete, and to provide concrete with insulating properties greater than those of normal-weight aggregate. Most lightweight aggregate is manufactured from clay, shale, slate, industrial cinders, and fly ash. Natural lightweight aggregates consist of pumice, scoria, volcanic cinders, tuff, and diatomite, but these materials are not widely available.

Reinforcement

Reinforcement may consist of deformed steel bars, welded wire fabric, and prestressing tendons, or combinations of each. Also, smooth undeformed wire may be used for the spiral reinforcement for concrete columns, and rolled structural steel sections may be used in composite, concrete-and-steel compression members. Nonmetallic materials such as fiberglass have been used, but are not widely accepted or tested at this time. Fiberglass bars can be obtained by special order.

Deformed Reinforcement

Deformed bars should be specified to meet the following ASTM standards as well as standards in the ACI Standard 318, "Building Code Requirements for Reinforced Concrete."[3] These bars have deformations, or ridges, around their circumference. Bar diameters range from $^{3}/_{8}$ to $^{18}/_{8}$ in. Bar size-designations are related to the bar diameter given in multiples of $^{1}/_{8}$ in. For example, a No. 4 bar has a diameter of $^{4}/_{8}$ in. Grade 40 and grade 60 steel, having yield points of 40,000 and 60,000 lb/in^2 (276 and 414 MPa), respectively, are commonly used. Steel with greater yield points may be used as prescribed by ACI 318. Welding of reinforcement is not generally recommended because of problems that may result from creating a brittle condition in the steel. Welding, if required, should be done in accordance with the "Reinforcing Steel Welding Code" (AWS D12.1) of the American Welding Society. Steel used for deformed bars includes billet steel conforming to ASTM A615 (the type of steel most often used); rail steel, axle steel, and low-alloy steel bars conforming to ASTM A616, A617, and A706, respectively, are all acceptable per ACI 318.

Welded Wire Fabric (WWF)

This is available with either smooth or deformed wires. Smooth WWF and the wire material are covered by ASTM Specifications A185 and A82, respectively. Deformed WWF and its wire material are covered by ASTM Specifications A497 and ASTM 496, respectively. Additional WWF information is available in the *Manual of Standard Practice* and *Designing and Detailing Manual for Structural Concrete Slabs* published by the Wire Reinforcement Institute. Smooth WWF is most commonly used. WWF is commonly used to reinforce slabs on-grade, elevated slabs supported by columns, walls, etc., and lightly reinforced walls.

The wire sizes vary in diameter from 0.135 to 0.628 in. The sizes were formerly given in U.S. steel wire gages, usually 4, 6, 8, or 10 gage. Present practice identifies the wire sizes by their cross-sectional area in hundredths of an inch; i.e., wire size 5.0 has a cross-sectional area of 0.05 in^2. "W" or "D" preceding the wire size number denotes smooth or deformed wire, respectively. The new wire-size designations of 4.0, 2.9, and 1.4 are the same as the commonly used old wire gage numbers of 4, 6, and 10.

The method used to designate WWF includes the spacing of the longitudinal and transverse wires. A typical designation is 6 × 8—W2.9 × W1.4, which means the longitudinal wires are spaced at 6 in and the transverse wires at 8 in, the longitudinal smooth wire size is 2.9, and the transverse smooth wire size is 1.4.

WWF is available in roll or sheet form. To permit handling, the sheet width is usually limited to 90 in (2.3 m) or less. The maximum length is usually limited to 40 ft (12 m) for flat sheets and 200 ft (61 m) for rolls. WWF in rolls having longitudinal wire sizes of W2.9 or less will lie sufficiently flat to permit proper placement. Use of flat sheets is advisable where longitudinal wire sizes exceed W2.9 to allow satisfactory placement.

Prestressing and Posttensioning Tendons

Three types of material are usually used for prestressing tendons: uncoated stress-relieved wire, ASTM Specification A421; uncoated seven-wire stress-relieved strand, ASTM Specification A416; and uncoated high-strength steel bar, ASTM Specification A722. Although tendons used for posttensioning can be obtained as a separate item, they are more often purchased as an integral part of a system complete with surrounding ducts or sheaths and end anchorages. Anchorages for posttensioned-type systems should be required to develop the ultimate capacity of the tendon without exceeding the anticipated anchor set, and should be able to transmit the tendon force under both static and cyclic loading conditions. For more detailed descriptions of both bonded and unbonded posttensioned materials refer to the Prestressed Concrete Institute's "Guide Specification for Post-Tensioning Materials," and Fig. 3-1, which illustrates an unbonded system.

Admixtures

Commonly used admixtures are air-entraining types conforming to ASTM C260 specifications; water-reducing types conforming to ASTM C494 Type A; set-control types conforming to ASTM C494, Types B and D which are retarders and a combination of water-reducing and retarder, respectively; and fly ash conforming to ASTM C618 Class F. Air-entraining admixtures are used to provide better workability of fresh concrete in addition to increasing the durability of hardened concrete when subjected to freezing conditions. Fly ash is used to increase the workability and plasticity of fresh concrete and can be used to partially replace some of the cement in concrete without significantly affecting the strength of hardened concrete. Fly ash is also helpful in reducing the aggressive destructive action of seawater on hardened concrete. Admixtures containing chlorides should be used only after a thorough investigation is made of the possible adverse effects they may have on embedded metallic materials, reinforcing, conduit, and the concrete itself. Using admixtures in concrete is covered in depth by an ACI Committee 212 report in the *ACI Journal* of September 1971 and also in the *ACI Manual of Concrete Practice,* part 1.[4] Ensuring the compatibility of different admixtures in concrete is most important, preferably during the preparation of trial-mix designs.

PROPORTIONING CONCRETE MATERIALS

General Requirements and Procedures

Concrete materials are proportioned to provide necessary placeability, strength, durability, and density. These are obtained by altering the quantities of the five major ingredients—cement, water, fine and coarse aggregates, and admixtures.

The basic objective in proportioning the ingredients is to produce, for the least possible cost, concrete with the specified physical, mechanical, and chemical properties needed in both its plastic and hardened states.

Selecting the proportions is basically a trial-and-error procedure. Each trial mixture is tested for its various plastic-state properties (such as slump, air content, and volume of concrete produced from the established amounts of each ingredient) and then placed in containers and allowed to harden into specimens suitable for testing the properties of the hardened concrete. The process is repeated until the specified properties are produced.

While the qualities and properties of each ingredient will vary, each has sufficient uniformity and identifiable properties to permit using preevaluated proportions and techniques to reduce the number of trials necessary to establish the desired results. Sufficient test data are also usually available for a variety of concrete mixes that can be used to demonstrate ability to meet the specified properties. ACI 318 recognizes this by providing criteria for establishing proportions on the basis of field experience. If this experience is not available, the proportions must be established by making trial batches in a testing laboratory.

ACI Standards 211.1 and 211.2 are often used for establishing proportions for normal and lightweight concrete, respectively. Additional information and requirements concerning trial batches are provided in the PCA Bulletin EB 001.12T, *Design and Control of Concrete Mixtures,* and ACI 318. Methods for testing various properties of concrete are provided in part 14 of the Book of ASTM Standards. Whether proportions are established by field experience methods or by trial batches, the services of a professional materials testing laboratory should be used wherever concrete is used for structural purposes.

Proportioning for Wet Concrete Properties

The properties of wet, or "fresh," concrete must be controlled to produce the required properties and appearances of hardened concrete. Most fresh concrete properties relate to the ease and homogeneity with which the materials can be mixed, the ability of the concrete to be properly placed in forms which establish its final hardened shape, and the ease of finishing the unformed surfaces. Concrete proportions should provide fresh concrete having a plastic cohesive mass in which the aggregates are thoroughly coated with cement paste and are held in suspension in a well-distributed array. Mixes containing poorly graded coarse aggregate that lacks the smaller size ranges or insufficient fine aggregate can produce a harsh texture that is difficult to consolidate, finish, or make into smooth, formed surfaces. Such a mixture is considered to have poor workability. Mixes with excessive fines create a very smooth texture which usually require greater amounts of cement paste and water to provide the required wetness for proper placement. This results in concrete that is more costly and which tends to have more cracks from drying shrinkage.

Concrete must be capable of both filling all corners in the forms and completely surrounding the reinforcing or other items embedded in the concrete without creating voids in the member being cast. The slump test, made in accordance with ASTM C143, establishes a measurement for evaluating the stiffness or consistency of fresh plastic concrete. The test is made by placing fresh concrete into a 12-in-high truncated cone mold, removing the mold immediately after it is filled, and measuring the distance the concrete slumps [12 in (305 mm) minus the height of the concrete mound left after removing the cone form]. See Fig. 3-2. The commonly specified slumps are between 3 and 5 in (76 and 127 mm). If a mechanical vibrator is used to consolidate the fresh concrete, a 4-in (102-mm) slump is satisfactory for most work. Water-reducing and air-entraining admixtures are often used to improve workability and consistency.

Proportioning for Hardened Concrete Performance

Most of the properties of hardened concrete are controlled by the amount and quality of its constituent materials. The material quality, although not constant, is generally required to conform to governing ASTM standards, as discussed earlier in this chapter. Given a certain source of acceptable materials, the amounts of each material are established to meet certain desired properties of hardened concrete. Various types of strength (usually compressive strength), durability, and weight are the most commonly specified properties. Proportions selected for one property will usually affect the other properties, which are interrelated.

Compressive Strength

Compressive strength is usually specified relative to its age. Concrete gains strength with time when subjected to a moist environment. Building codes such as ACI 318 require the strength evaluation to be made after a 28-day time period in a specified moist environment. Compressive strength is determined by compressive tests made on samples of a specified size and shape. The most commonly used test methods are as follows:

ASTM C192. "Method of Making and Curing Concrete Test Specimens in the Laboratory"

Figure 3-2 Slump test.

ASTM C39. "Method of Test for Compressive Strength of Cylindrical Concrete Specimens"

ASTM C172. "Method of Sampling Fresh Concrete"

ASTM C31. "Method of Making and Curing Concrete Test Specimens in the Field"

For a mix to be acceptable according to ACI 318, test specimens cured and tested in the laboratory must always have a compressive strength that is 400 to 1200 lb/in^2 (2.76 to 8.27 MPa) greater than the specified design strength, depending upon the test history of strength for a particular concrete mix. A mix without a history of at least 30 tests made within the preceding 12 months is required to have a strength that is 1200 lb/in^2 (8.27 MPa) greater than the specified design strength. With a history of at least 30 consecutive tests, the additional strength required varies from 400 to 1200 lb/in^2 (2.76 to 8.27 MPa), according to the computed standard deviation. Refer to ACI 214, "Recommended Practice for Evaluation of Strength Test Results of Concrete," and to ACI 318.

With the aggregate in a mix being constant, strength is basically a function of the cement content and the water-cement ratio for a unit volume of concrete. Given a constant cement content, strength will decrease as the water-cement ratio increases. When strength data are not available from tests made on trial batches or from field experience, water-cement ratios and minimum cement contents can be used as the basis for selecting proportions to meet strength requirements (see ACI 318 and ACI 211.1) Such methods should be used only for normal-weight concrete with a specified strength of 4000 lb/in^2 (27.58 MPa) or less, and the concrete should contain no admixtures other than those provided for air entrainment. Although this method of proportioning without prior strength tests may be used, acceptance of concrete in place must be based on acceptable test results of the concrete placed in a structure (see "Concrete Quality Control" in this chapter).

Flexural Strength

This is usually specified in relation to concrete for slabs on-grade, or roads or driveways subjected to high concentrated loads. Flexural strengths are not normally specified as criteria for proportioning concrete materials except where the concrete will primarily be used for paving, a large amount of which will be subjected to heavy concentrated loads. When flexural strength is specified, it is usually required to be determined after a specimen has cured 28 days in a laboratory. The method for determining the flexural strength is specified by ASTM C78. A standard specimen is 6 in (152 mm) square in cross section and about 24 in (610 mm) long (refer to ASTM C31 and ASTM C192). Concrete used for slabs subjected to heavy concentrated wheel loads [18,000 lb (3629 kg) or greater] should normally have a minimum flexural strength of 650 lb/in^2 (4.49 MPa) at 28 days. With a consistent source of aggregate, the cement content and the water-cement ratio have the greatest effects on this property. Strength decreases with decreasing amounts of cement and increasing amounts of water.

Splitting Tensile Strength

This is usually specified in connection with lightweight aggregate concrete. Because lightweight aggregate is not usually as strong as normal-weight aggregate and has potentially a greater variation in strength, this property is often specified where shearing forces are critical. Tests for this property are specified by ASTM C330, "Standard Specification for Lightweight Aggregates for Structural Concrete," and ASTM C496, "Test for Splitting Tensile Strength of Cylindrical Concrete Specimens." With a given source of aggregate, the cement content and the water-cement ratio have the greatest effect on this property. Strength decreases with decreasing amounts of cement and increasing amounts of water.

Proportioning for Durability

The satisfactory performance of hardened concrete can be threatened by chemical, climatic, abrasive, and erosive action. Resistance against such action can be achieved through material selections, as well as through varying proportions of materials. "Guide to Durable Concrete" (ACI 201.2R), "The Recommended Practice for Selecting Proportions for Normal and Heavyweight Concrete" (ACI 211.1), and "Erosion of Concrete in Hydraulic Structures" (ACI 210), all included in part 1 of the *ACI Manual of Concrete Practice,* provide practical information for producing concrete that resists the various types of deterioration. Durability against most forms of deterioration is improved as the water-cement ratio is decreased to about 0.45. The type of cement and aggregate used greatly influence resistance to various types of chemical attack (see the section "Concrete Material Specification and Usage"). Aggregates should be selected on the basis of their inertness and reactance to the particular chemical environment they are to be subjected to. Resistance to abrasion is enhanced by proportioning concrete for a strength of at least 4000 lb/in^2 (27.58 MPa) and using hard and tough fine and coarse aggregates. Resistance to the climatic conditions of freezing and thawing is increased with the entrainment of air in the cement paste and using aggregates that do not absorb large amounts of water. Frost-resistant concrete should have 4½ to 6½ percent entrained air, by volume, depending on the size of the aggregate. Air entrainment is usually achieved by using an air-entraining cement or admixture (see the section "Concrete Material Specifications and Usage").

Proportioning for Density

Concrete density is largely changed by the type of aggregate that is used. Heavyweight concrete is often used in connection with radiation shielding. A guide to proportioning for heavyweight concrete is provided in Appendix 4 of ACI 211, "Recommended Practice for Selecting Proportions for Normal and Heavyweight Concrete." Unit weights from 180 to 350 lb/ft^3 (2880 to 5600 kg/m^3) can be obtained depending on the type of aggregate used. Iron and steel pellets are often used when attempting to achieve 350 lb/ft^3 (5600 kg/m^3) concrete.

Lightweight concrete is often used for its insulating and fire-protection qualities in addition to providing a lightweight structure. ACI 211.2, "Recommended Practice for Selecting Proportions for Structural Lightweight Concrete," provides data for this type of concrete. Unit weights from 90 to 115 lb/ft^3 (1440 to 1840 kg/m^3) can be obtained depending on the type of lightweight aggregate and the amount of natural sand that is used in the mix.

Minor changes in density can occur without changing the aggregate. These changes usually amount to about 15 lb/ft^3 (240 kg/m^3) and are related to both the amount of entrained and entrapped air, and the gradation and type of aggregate.

Proportioning Materials for Small Jobs

When concrete design strengths above 3000 lb/in^2 are not required, and when only a small amount of concrete is required for a job (making field-mixed concrete possible), it may be impractical to determine the proportions in a testing laboratory in accordance with standard recommended procedures. Under such conditions, proportions established in accordance with Table 3.6.1 in "Concrete Mixes for Small Jobs" in AC1 211.1–77, "Recommended Practice for Selecting Proportions for Normal and Heavyweight Concrete," will usually provide concrete of satisfactory strength and durability provided the amount of water added will not produce a slump greater than 5 in (127 mm). Concrete having a slump of 5 in (127 mm) or less will tend to stick to the shovel and slide off as a unit, but will not run off. If it runs off it is too wet. Water should be added gradually and be thoroughly mixed with the other materials to avoid overwatering. Material proportions for 1 ft^3 (0.02832 m^3) of concrete consisting of 23 lb (104 N) cement, 47 lb (212 N) sand, and 64 lb (288 N) of coarse aggregate [stone or gravel of ¾ in (19 mm) maximum size] will usually result in a satisfactory mix for non-air-entrained concrete. Depending on the gradation of coarse aggregate, the sand and cement may have to be adjusted to produce the desired workability.

BATCHING AND MIXING CONCRETE

Batching consists of weighing, or volumetrically measuring, the ingredients for a batch or single quantity of concrete and placing them in a mixer. Mixing consists of combining the ingredients into a uniform plastic mass. The various ingredients need to be accurately measured and thoroughly mixed into a uniform mass within a specified time before placement. Mixing should be scheduled so that no more than 1½ h elapses between the time cement is charged into the mixer and the time it is discharged for placement. If concrete is not satisfactorily measured and mixed there can be adverse affects on placeability, strength, durability, and permeability, in addition to the creation of internal voids and surface defects. The temperature of mixed concrete should be controlled so that when delivered to the site for placement it is between 50 and 90°F (10 and 32°C). When fresh concrete is hotter, it may set too fast. Special treatment is often necessary during extreme cold or hot weather (see the section "Environmental Effects on Concrete" in this chapter). A guide for batching and mixing concrete is provided in ACI Standard 304, "Recommended Practice for Measuring, Mixing, Transporting and Placing Concrete."

It is important that the total quantity of concrete needed for any one element be batched, mixed, delivered to the point of placement, and placed before initial set occurs as well as that concrete be provided at such a rate that the element can be completely cast in a continuous operation without incurring cold joints between successively placed masses. A cold joint is a joint or discontinuity formed when a concrete surface hardens before the next batch is placed against it, and it is characterized by poor bond to the hardened surface. Concrete that has lost its flowability (determined by its slump) as a result of partial initial set should not be retempered to the desired slump by adding water to the mix. Adding water under this condition will result in reduced strength and durability. If it can be determined that the loss of slump is not the result of initial set caused by delays in delivery or placement, additional water can be added as long as the designed water-cement ratio or designed slump is not exceeded.

Mechanically Mixed Concrete

Concrete in amounts of 2 ft^3 (0.057 m^3) or more is most economically mixed in mechanically powered mixing machines. Semiportable and stationary type mixers are normally available in capacities ranging from about 2 ft^3 to 1 yd^3 (0.057 to 0.76 m^3). Truck-mounted mixers range in capacity between 3 and 15 yd^3 (2.29 and 11.47 m^3). Mixers with a rated capacity of 1 yd^3 (0.076 m^3) or larger should conform to the requirements of the Plant Mixer Manufacturers Division of the Concrete Plant Manufacturers Bureau. When mixers larger than 1 yd^3 (0.76 m^3) are used, some means of mechanical batching is usually required, especially when the element being cast requires a greater volume of material than can be mixed in a single batch. The requirements for batching and mixing equipment should meet the recommendations established by the Concrete Plant Manufacturers Bureau. In urban and neighboring areas, mechanically mixed concrete can usually be purchased from a ready-mix concrete plant.

Specifications governing batching, mixing, and transporting ready-mix concrete are established by ASTM C94, "Specification for Ready-Mixed Concrete." This specification includes a requirement for the purchaser to specify coarse aggregate size, slump, air content when air-entrained concrete is desired, the unit weight when lightweight concrete is used, and the 28-day design compressive strength. Two other alternatives for establishing material proportions can be utilized depending upon the amount of responsibility the purchaser wishes to accept. Plant equipment and facilities should conform to the "Check List for Certification of Ready-Mixed Concrete Production Facilities" of the National Ready-Mixed Concrete Association. For reasons of quality control and economy, ready-mix concrete is used more often than site-batched and -mixed concrete. When it becomes necessary to set up a site batching and -mixing plant, its equipment and operation should meet ASTM C94 specifications and the recommended practice presented in ACI Standard 304. Materials should be placed in the mixers in a manner to obtain preblending of cement, aggregate, and admixtures. Water should enter the mixer first, but should continue to flow while other ingredients are entering. About 10 percent of the aggregate should enter the mixer before the cement begins to flow. Admixtures that are liquid should enter with the water and those that are powdered should enter with the cement. After these starting procedures, the materials should be charged into the mixer simultaneously, and charging should be complete within the first 25 percent of the mixing time. Mixing for batches of 1 yd^3 (0.076 m^3) or less should continue for not less than 1 min, and 15 s of mixing time should be added for each additional cubic yard (0.076 m^3) or fraction thereof being mixed.

Manually Mixed Concrete

When concrete is manually mixed, the aggregates and cement should be thoroughly mixed and blended before the water is added. Water should be gradually added to and mixed with the preblended dry materials. Mixing should be completed within 5 min.

PLACING CONCRETE

Concrete placement should be scheduled so that discharge from the container which delivered the concrete and placement of the batch can be completed within about ½ h after delivery to the site. Placing the concrete involves transporting the wet concrete from the point of delivery to the location where it is required and consolidating it so it is free of voids and completely surrounds all embedded reinforcement and other miscellaneous items.

Equipment used to place concrete should be clean and made of materials that will not cause adverse chemical reactions with fresh concrete. For this reason such equipment should not be made of aluminum. Metallic equipment should be and normally is made of steel. Recommended practices for transporting and placing concrete are included in a paper Title No. 68-33 reported by ACI 304, included in the *ACI Manual of Concrete Practice.*

Methods of Transporting Concrete

In many cases, where concrete is being placed at grade level it can be dispensed directly from a truck-mounted mixer into the desired location by chutes which are normally carried on the truck. Other methods are often required, however, because of height restrictions or inaccessibility caused by formwork or reinforcement. Such methods consist of crane-hoisted buckets, manual or motor-propelled buggies, conveyor belts, and pumps and hoses. Concrete that is to be pumped or carried on conveyors often requires special consideration when aggregate size and slump are being established. Equipment should not travel directly over, or be supported on, reinforcement or other items embedded in the concrete in order to prevent moving them from their designated location. Runways or other supports independent of reinforcement and embedded items should be provided.

Final Placement and Consolidation

The location for placing concrete and reinforcement should be clean of debris and free of ice and water. Concrete can be placed under water if the special procedures described later in this paragraph are followed. Placement for a member should proceed at a rate such that fresh concrete is placed over underlying layers or adjacent to masses that are in a plastic state. This is required to prevent the formation of cold joints in a member. Concrete should be deposited in a manner to prevent segregation of aggregate. Deep members should be placed in horizontal layers not exceeding 2 ft (0.61 m), and concrete should not be allowed to drop freely more than 5 ft (1.52 m). In walls or columns over 6 ft (1.83 m) deep, concrete should be placed by using a tube, called an elephant trunk, sufficiently long so that the concrete will not fall freely more than 5 ft (1.52 m). Each layer should be compacted by a mechanical vibrator in addition to rodding, to work out air pockets or voids and drive the concrete into all corners and around reinforcement and embedded items. Each layer should be worked at least 6 in (152 mm) into a preceding layer. Additional information for consolidating concrete is included in ACI Standard 309, "Recommended Practice for Consolidation of Concrete."

Shotcrete is a process by which concrete can be placed in thicknesses of about 6 in (152 mm) without using forms. The concrete is conveyed at high velocity through a hose and projected onto a surface. The special considerations for use of shotcrete are presented in the "Recommended Practice for Shotcreting," ACI 506.

Concrete can be placed under water if the water is not flowing at the time of placement. It must be deposited without allowing the concrete to drop freely through the water. A 10- to 12-in (254- to 305-mm-) diameter pipe called a tremie is used to carry the concrete from a point above the water surface to its desired location beneath the water. The bottom of the tremie should remain in fresh concrete at all times during placement until the work is complete. Since the concrete cannot be rodded, vibrated, or otherwise disturbed for consolidating after it is deposited, a 6- to 9-in (152- to 229-mm) slump must be used. The recommended water-cement ratio for tremie concrete is 0.44 by weight. More detailed information concerning the equipment, methods, and mixtures used for tremie concrete operations are included in ACI 304, "Recommended Practice for Measuring, Mixing, Transporting and Placing Concrete."

FORMWORK FOR CONCRETE

Fresh plastic concrete is made to conform to desired shapes by placing it in a mold. For structures supported on earth, the mold can be formed by shaping the earth. This method is usually restricted to members having a shallow depth, such as slabs, footings, and shallow beams. A temporary structure, commonly termed formwork, is normally used to form the required shape of the molds or to support prefabricated molds. The term formwork commonly includes the total system of support for freshly placed concrete, including all hardware and necessary bracing. ACI Standard 347, "Recommended Practice for Concrete Formwork," presents guidelines relative to types of forming systems as well as to their design.

Selection of Materials

Materials in contact with the concrete should be selected to produce the desired finishes and should not cause adverse reactions with fresh concrete. Common materials often used in contact with concrete are wood (plywood, or boards dressed on at least two edges and one face), wood products, steel, plastic, or fiberglass. Many manufactured forming systems are available for walls, columns, beams, joists (including two-way ribbed slabs), and necessary supports, scaffolding, and accessories. Form ties (which are rods used to hold opposite vertical form surfaces in position), should be factory-fabricated, be adjustable in length, and be capable of being removed so as to leave no metal closer than 1½ in (38 mm) to concrete surfaces. Snap-off-type ties are normally used and meet these requirements. Ties for below-grade construction should incorporate a water-seal washer. Wood or other porous materials used as forms should be coated with a material that will keep the forms from absorbing moisture from fresh concrete. Form-release agents should be used to prevent concrete bonding to the forms. Form-release agents should be types that will not impair or stain the concrete surface or prevent bonding of subsequent surface treatments.

Design and Construction

Formwork has to be designed to adequately support the lateral fluid pressures created by fresh concrete, as well as the weight of fresh concrete and construction and wind loads. Forms should be fabricated to be easily removed without damaging concrete surfaces. Formwork should be positioned and braced to produce concrete work conforming to the tolerances required by ACI Standard 347.

Form Removal

Formwork should not be removed until the concrete has sufficient strength to support both itself and the weights it will have to support during construction. Refer to the paragraphs "Curing Temperature" and "Duration of Curing" for information about the rate at which concrete gains strength and the procedures for determining concrete strength for terminating curing and removing forms.

CURING

The strength and durability of concrete is dependent upon its age and the temperature and moisture conditions to which it is subjected as it ages. Curing concrete relates to maintaining the moisture and temperature of freshly placed concrete during some definite period following placing, or finishing to assure satisfactory hydration of the cementitious materials and proper hardening of the concrete. Detailed recommendations for curing are given in ACI 308, "Recommended Practice for Curing Concrete."

Curing Temperature

The temperature of the concrete during curing should be kept above freezing and below 180°F (82°C), but the strength gain is very slow below about 40°F (4.4°C). Reasonable curing periods can be obtained when the temperature of the concrete is kept above 50°F (10°C). Concrete temperatures over 110°F (43°C) are not usually reached except during steam curing procedures. Refer to ACI 517, "Recommended Practice for Atmospheric Pressure Steam Curing of Concrete." Special provision for hot- and cold-weather conditions should be followed, as described in the section in this chapter titled "Environmental Effects on Concrete." See Figs. 3-3 and 3-4. for the effects of low and high curing temperatures on concrete strength.

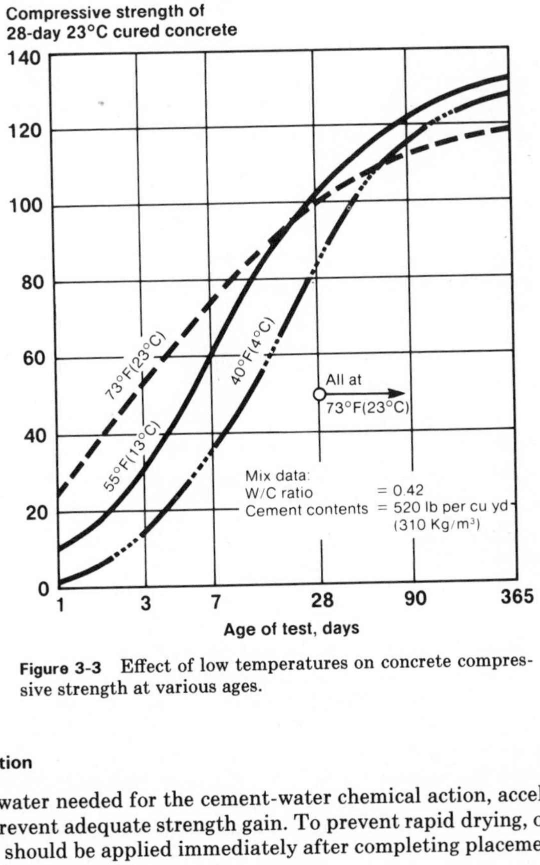

Figure 3-3 Effect of low temperatures on concrete compressive strength at various ages.

Moisture Retention

Due to loss of water needed for the cement-water chemical action, accelerated drying of concrete will prevent adequate strength gain. To prevent rapid drying, one of the following procedures should be applied immediately after completing placement and finishing.

1. Keep absorptive mats or fabric continuously wet.
2. Use waterproof sheet materials conforming to ASTM C171, "Specifications for Waterproof Sheet Materials for Curing Concrete"
3. Use compounds conforming to ASTM C309 "Specifications for Liquid Membrane-Forming Compounds for Curing Concrete." Some of these compounds may prevent the later satisfactory bonding of concrete coatings, adhesives for floor coverings, or other materials requiring bonding to the concrete. These compounds should be used in accordance with the manufacturer's recommendations and be applied immediately

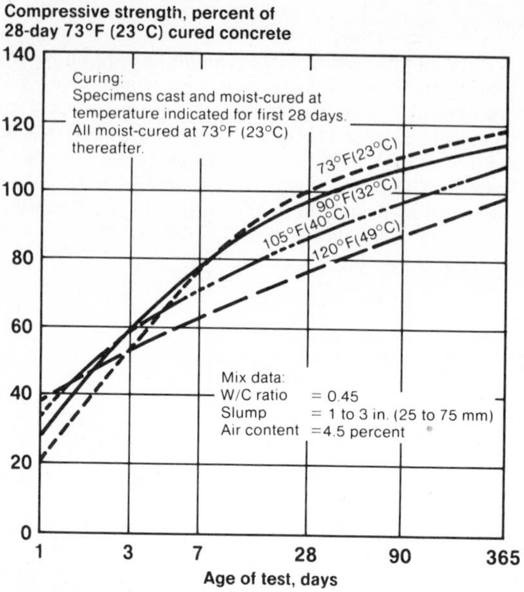

Figure 3-4 Effect of high temperatures on concrete compressive strength at various ages.

after the disappearance of any water sheen that may appear after finishing or after the use of other curing methods.

4. Keep nonabsorptive forms or absorptive forms moist and leave them in place.
5. Use a continuous steam or mist spray on the concrete surface.
6. Effect ponding or continuous sprinkling with water or application of sand kept continuously wet. These methods may adversely affect finished surfaces.

Duration of Curing

Concrete should be cured for the period of time necessary to reach a compressive strength of at least 70 percent of the specified 28-day design strength. The length of time will vary depending upon the temperature and effectiveness of the moisture-retaining procedures used during the curing period. When the temperature during curing is above 50°F (10°C), a period of 7 days is usually satisfactory and is generally considered to be the minimum desirable time, except 3 days is usually satisfactory when high early strength cements are used, and as little as 18 to 24 h will suffice if steam curing is used. Test specimens used to establish strength criteria for terminating curing or removing forms should be cured under the same conditions as those used for the concrete work under consideration, and should conform to ASTM standard methods C31, C39, and C78. Forms used to support concrete elements should not be removed until the concrete has gained sufficient strength for the member to safely support its own weight and any additional superimposed loads to which it may be subjected. In some cases strengths above 70 percent of the specified 28-day design strength may be required.

QUALITY CONTROL

Because concrete is made of many different materials produced by a variety of processes and trades of workers, there is considerable opportunity for errors that can result in producing an inferior product in which life safety is jeopardized. To provide a satisfactory concrete product, quality control is needed in the areas of design, specification, production, transportation, placement, finishing, curing, and maintenance of concrete structures.

Structural engineering consultants should normally be retained for producing designs and specifications, for reviewing shop fabrication drawings, and for field observation during construction. They should be selected on the basis of their experience with the particular type and size of project and their quality control and checking procedures. This is the first step in obtaining quality concrete work. The second is similar, having to do with selecting contractors experienced in the type and size of the project and their ability to provide a similarly qualified construction superintendent. The contractor is basically responsible for carrying out the work described by the designs and specifications, preparing shop fabrication drawings, and supplying and installing the proper materials. The third step has to do with selecting a materials testing laboratory that is independent of the engineer or contractors to perform the normally specified inspection and testing of concrete ingredients, batching and mixing equipment, and fresh and hardened concrete. This laboratory should be other than the one used to design the proportions for the types of concrete specified.

Concrete Specifications

The project specifications establish the basic quality of materials and construction, and the requirements for their testing. ACI Standard 301, "Specifications for Structural Concrete for Buildings,"[5] includes specifications that can be used by referral in the specific project specifications with addition of supplemental requirements. Mandatory supplemental requirements include specifying the required 28-day design compressive strength of concrete used in each portion of the structure, and the various types and grades of reinforcing steel by ASTM specification number and yield strengths as applicable. Other supplemental requirements are also usually needed and are listed as options in these specifications. In addition to ACI Specification 301, commercially produced guide specifications are available, some of which are adaptable to computerized reproduction and editing. Both publications need to be thoroughly reviewed and supplemented or edited for the specific project. Information about the availability of guide specifications can be obtained from the Construction Specifications Institute (CSI).

Shop Drawings

Shop drawings, catalog abstracts, and certificates of compliance are normally required to be submitted by the fabricator or manufacturer through the general contractor for review and acceptance by the owner or owner's representative to verify that the design drawings and specifications have been correctly interpreted. Having the consulting engineer who prepared the designs review such submittals is a good practice since they relate to design considerations. Commonly required submittals are as follows:

1. Reinforcing steel shop drawings prepared in accordance with the Concrete Reinforcing Steel Institute's *Manual of Standard Practice* MSP-1; the ACI Committee 315 publication, *Manual of Standard Practice for Detailing Reinforced Concrete Structures* for deformed bars; and the Wire Reinforcement Institute, Inc. publication, *Welded Wire Fabric Manual of Standard Practice* for welded wire fabric.
2. Manufacturer's specifications and installation instructions for reinforcing proprietary materials, including reinforcement accessories and bar couplings.
3. Mill certificates for reinforcing steel indicating the mill analysis, and the tensile and bend tests.
4. Manufacturer's specifications and installation instructions for admixtures, bonding agents, waterstops, joint systems, sealants, chemical floor hardeners, and other such proprietary materials.

5. Shop drawings of embedded steel items such as curb angles, connection plates, or other anchoring devices.
6. Material certificates for cement and aggregates signed by the contractor and manufacturer certifying that each material complies with or exceeds the specified requirements.
7. Laboratory reports on design mixes and results of concrete tests made during construction.
8. Samples of concrete finishes, layouts of form work showing form-tie locations and types of form liners and construction-joint locations are required when architectural considerations are of importance; many of these considerations can be met by requiring a mock-up sample to be built on the site to demonstrate the desired finishes for the project.

Quality Control Tests During Construction

Fresh concrete should be sampled and tested to verify that it is in accordance with the construction specifications and the design mix, except in small jobs of little structural importance [50 yd^3 (38 m^3) or less]. Sampling should be done in accordance with ASTM C172, "Standard Method of Sampling Fresh Concrete." ACI Standard 318 requires that "samples for strength tests of each class of concrete placed each day shall be taken not less than once a day, nor less than once for each 150 yd^3 (115 m^3) of concrete, nor less than once for each 5000 ft^2 (465 m^2) of surface area for slabs or walls." On a given project, if the total volume of concrete is such that the frequency of testing required by that previously stipulated "would provide less than five strength tests for a given class of concrete, tests shall be made from at least five randomly selected batches or from each batch if fewer than five batches are used." In variance to the preceding requirements, a greater frequency of testing is often specified. Instead of "once for each 150 yd^3 (115 m^3)," once for 100 yd^3 or portion thereof (76.5 m^3) is specified. Additionally, three to four specimens are made for each test. One specimen is tested at 7 days and two at 28 days. The fourth may be reserved for later testing if required.

Slump tests, in accordance with ASTM C143, should be made for at least one in every five concrete loads at the point of discharge; and one test should be made for each set of compressive-strength test specimens. (See Fig. 3.2.)

Air-content tests should be made in accordance with ASTM C173, volumetric method: one test for every five loads of concrete at point of discharge when air entrainment is specified.

Concrete temperature should be tested hourly when the air temperature is 40°F (4.4°C) and below, when 80°F (27°C) and above, and each time a set of compression test specimens is made.

When concrete is pumped, sampling should be done on the concrete discharged at the effluent end of the hose as well as at the point of discharge from the container used for delivery.

Observation of Work During Construction

During construction the work should be observed by an engineer familiar with the project plans, specifications, and construction techniques for the particular class of work being accomplished. This observer should be someone other than the construction contractor's representatives, preferably the engineer responsible for the design, or a qualified representative of the owner. The primary function of the observer is to see that the installation is proceeding in compliance with the general intent of the plans and specifications. A guide for observation or inspection of the work is provided by ACI Standard 311, "Recommended Practice for Concrete Inspection," and the *ACI Manual of Concrete Inspection* (ACI 311.1R). Although it is basically the responsibility of the contractor to perform the work in accordance with the plans and specifications, it is in the best interest of the owner to make provisions for independent observation and inspection, considering that once the work is in place it is extremely costly and time consuming to repair defective work and defend or prosecute liability claims. Items that should be observed are as follows:

1. The basic building or structure layout
2. Formwork conformance to the required shapes
3. Conformance of the strength, size, and spacing, as well as the location, support, and anchorage of reinforcing to that shown on the shop fabrication and design drawings and described in the project specifications
4. Removal of debris in forms where concrete is to be placed
5. Conformance of fresh concrete to the specified properties and the performance of sampling and testing required
6. The suitability of weather conditions and the institution of protective measures for cold or hot weather
7. Procedures used for placement and consolidation
8. Prompt institution of curing methods and maintenance thereof for the required time periods
9. Review of laboratory test results

Conditions not fulfilling the intent of the design and specifications should be reported to the contractor immediately upon discovery so corrective measures can be initiated in a timely manner and not delay the work. Defective work that is left uncorrected should also be reported to the owner so corrective measures can be worked out with the contractor.

TESTING HARDENED CONCRETE MEMBERS

Testing hardened concrete, by methods other than using concrete test specimens made at the time of placing concrete, is done to determine the adequacy of in-place concrete that is suspected of being inferior, to determine the strength of concrete before testing specimens made at the time of concrete placement, and to help evaluate the structural capacity of existing buildings. The test methods used are basically of two types, destructive and nondestructive. Detailed information on available testing methods can usually be obtained from a testing laboratory or from ACI. Testing should always be done with the cooperation of, or under the direction of a qualified engineering consultant.

Destructive Testing

This is usually accomplished by removing a core of concrete from the structure and testing it in a laboratory. Diamond-tipped core bits are usually used to obtain undamaged specimens. Core samples are usually 3 to 4 in (76 to 102 mm) in diameter and have a length about twice the diameter. Such testing should be conducted in accordance with ASTM C823, "Standard Recommended Practice for Examination and Sampling of Hardened Concrete in Constructions," ASTM C42, "Standard Method of Obtaining and Testing Drilled Cores and Sawed Beams of Concrete," and ACI paper Title No. 64-61, "Strength Evaluation of Existing Concrete Buildings," reported by ACI Committee 437. The amount of destruction associated with this method is only at the point of core removal. Surrounding areas are not usually adversely affected. Metal-detecting devices can be used to locate reinforcement so cores can be taken without damaging reinforcement.

Nondestructive Testing

This usually consists of one of the following methods:

1. Measuring the flow or travel of pulse or train of waves through a measured path length through the concrete. Refer to ASTM C597, "Standard Test Method for Pulse Velocity Through Concrete."
2. Measuring the absorption or scatter of x-rays or gamma radiation.
3. Measuring surface hardness by rebound or identation-measuring devices, respectively, per ASTM C805, "Test Method for Rebound Number of Hardened Concrete," and ASTM C803, "Penetration Resistance of Hardened Concrete."

New methods are constantly being developed, and information on recent developments can be obtained from ACI and testing laboratories. For any of the nondestructive methods, it is usually necessary to calibrate the equipment against compression tests of concrete core specimens representative of the concrete being tested.

ENVIRONMENTAL EFFECTS ON CONCRETE

The exposure of concrete to certain temperature extremes, wind, precipitation, and various chemicals can have detrimental effects on concrete during both its plastic and hardened states. In most cases, procedures can be adopted to minimize or eliminate detrimental effects if the adverse environment is known before the design and specifications are prepared. A description of commonly experienced adverse environmental conditions and recommended procedures to prevent or minimize their detrimental effects on concrete is presented in the following paragraphs. Information in addition to that presented can be obtained from ACI paper Title No. 59-57, "Durability of Concrete in Service," reported by ACI Committee 201.

Temperature Extremes

Low Temperatures

When concrete is placed during periods when the mean daily temperature falls below 40°F (4.4°C), the procedures described in ACI Committee 306 Report, "Cold Weather Concreting," should be followed to prevent fresh concrete from freezing and to provide proper curing conditions. These provisions make recommendations concerning the following: the temperature of fresh concrete delivered to the site; using insulation on the forms; covering concrete with insulating blankets during curing and heating the surrounding air; the curing interval and the interval before form removal, depending on maintaining the concrete temperature; and the requirements for maintaining above-freezing temperatures of forms or other surfaces against which concrete is placed. For the effects of low temperatures during curing on concrete compressive strength, see Fig. 3-3.

Hardened concrete exposed to freezing and thawing is subject to deterioration which is believed to be largely caused by hydraulic pressures created by an expanding ice-water system during freezing of the water that is absorbed in the concrete. To provide resistance against freezing and thawing, air entrainment is used (4.5 to 6.5 percent by volume depending on the aggregate size), aggregates adequate for the exposure are selected, and a low water-cement ratio is used [about 5 gal (19 L) water per 94 lb (423 N) of cement]. Other items which should be considered are providing sloped surfaces to prevent water puddles and providing joints in the work at close enough intervals to minimize development of tensile stresses and resulting cracking caused by thermal contraction and expansion.

High Temperatures

During hot weather, usually above 80°F (27°C), there is a potential for the quality of both fresh and hardened concrete to be impaired by premature set and by loss of the moisture needed for proper hydration of the cement. The potential is highest during dry and windy conditions. During these conditions, the recommendations in ACI Committee 305 Report, "Hot Weather Concreting," should be followed. The basic requirements consist of delivering and placing concrete at the lowest practical temperature, erecting wind shields and sunshades, cooling forms and dampening subgrades before concrete placement, completing finishing as rapidly as possible, and starting curing operations immediately after finishing. The use of set-retarding admixtures is helpful. Usually a combination water-reducing and retarding admixture is used to increase workability while maintaining the lowest possible water-cement ratio. For the effects high temperatures during curing have on concrete compressive strength, see Fig. 3-4.

Exposure to high temperatures after concrete has hardened can also have adverse affects. When members are subjected to a temperature gradient between opposite faces, such as that which occurs when the top of a slab or beam receives greater exposure to heat from the sun in comparison to a bottom shaded surface, a bending stress is induced in the member that can cause cracking. To control such cracking it may be necessary to provide additional reinforcement and/or to extend reinforcement beyond points at which bars are normally terminated or greatly reduced. Concrete exposed to extremely high temperatures above 400°F (204°C) experiences a loss of compressive strength as the temperature continues to rise. Strength reduction will be in the range of 15 to 40 percent at 600°F (316°C).[6] The use of concrete in temperatures above 600°F (316°C) is not recommended without a detailed investigation as to the actual conditions of exposure and the types of cements and aggregates that are are to be used.

Wind

Since it relates to premature loss of moisture in fresh concrete or the rate of cooling during cold weather, wind is detrimental only during placing and curing concrete. Placing concrete during windy periods may require provision of windshields, fog sprays, or enclosures to prevent detrimental effects such as plastic shrinkage cracking.

Precipitation

In the form of rain, precipitation can be detrimental during concrete placement. The greatest problem is caused by erosion of the cement paste before hardening occurs. Damage is usually limited to the exposed concrete surface finish and can often be repaired without having to demolish and replace the structure. If the concrete has not completely set after the rain ends, work can continue with additional concrete placement or filling in eroded areas. Puddles of water should be removed before placing additional concrete. When concrete is hardened and additional concrete cannot be worked into previous placements, patching can often be accomplished by using bonding agents before the concrete repairs. When erosion is deep and the concrete has hardened, repairs should be made only after consultation with and the approval of a professional structural engineer. See also Technical Bulletin 17, "Concrete Pavements Exposed to Rain During Construction," published by American Concrete Paving Association, 2625 Clearbrook Drive, Arlington Heights, IL 60005.

Chemical Attack

This largely affects hardened concrete and is most often produced by chemicals in solution. Chemicals that attack concrete are inorganic and organic acids, alkaline solutions, salt solutions, bromine (vapor), sulfite liquor, chlorine (gas), and seawater. Sulfates often occur in soils and can cause serious deterioration when in contact with concrete, depending upon the concentrations. Groundwater in earth fills containing blast furnace slag or cinders is also likely to contain damaging amounts of sulfates. Inorganic acids may occur in peat soils as well as in process plants using various acids. Organic acids are often present from spillage in food processing plants or wood pulp mills.

Salts may be present in soils, and are often in areas where ice removal is required. Salt and similar deicing chemicals are especially destructive if applied to concrete less than 4 months old. Aggressive chemicals should preferably never be used, but if they are, they should not be used during the first year. Many of the chemicals that adversely affect the concrete also promote corrosion of steel reinforcement. Protection against sulfate attack is obtained by using sulfate-resistant cements and a dense, high-quality concrete. Protection against high concentrations of acid is often provided by coatings or linings that prevent contact with the concrete. This method could be used to provide resistance for most chemical attack; however, the coatings or linings are usually very expensive and require frequent maintenance. This and additional information concerning chemical attack is provided in ACI 201.2R, "Guide to Durable Concrete."

REPAIRING CONCRETE

The repairs used for concrete structures are usually of two basic types—cosmetic (relating to appearance) or structural. Materials and methods used are related to the type of repairs being made. The selection of materials that will form a strong bond to concrete surfaces and be at least equal in strength to the concrete is basic to either type of repair. Oil, dust, dirt, or other coatings that will prevent materials from bonding to concrete must be removed from concrete surfaces receiving a patch. Removing rust from reinforcing bars in the patch areas is equally important. The concrete to which patches are applied should be sound material. Loose or soft concrete should be removed before patching is done. In cases requiring the removal of deteriorated materials, the structural element may have to be temporarily shored before any materials are removed. Except for very minor defects, repairs should not be made without consultation with and direction from a professional structural engineer.

Repairing "Green" Concrete

"Green" concrete is concrete which has set but has not appreciably hardened. Minor defects less than about 1½ in (38 mm) deep that offer no threat to structural capacity can often be patched by coating the defective surface with a neat cement grout and patching with a 2½:1 sand/cement mixture. For deeper defects, similar to those caused by rock pockets or honeycombing, all material should be removed down to sound concrete. A bonding agent should be applied to the sound surface before patching with a concrete mixture having a strength equal to that specified for the member being patched. Patching of new work should be accomplished within a day after discovery and before the concrete has dried. Patches should be moist-cured for at least 3 days. For additional information on repairing concrete surfaces, refer to Chap. VII on the *Concrete Manual,* 7th ed., U.S. Department of the Interior, Bureau of Reclamation.

Repairing Hardened Concrete

Flowable pressure-injected expoxies are often used for repairing cracks for structural considerations. Holes and other such defects are often patched by using epoxy grouts. Epoxies used for repairing concrete work usually provide a very strong bond to the concrete. Epoxies that can be applied to damp surfaces are often desirable, since creating or maintaining dry surfaces is often a problem. Detailed information concerning the use of epoxies with concrete is available in the *Journal of the American Concrete Institute* for September 1973, Title No. 70-56, "Use of Epoxy Compounds With Concrete," reported by ACI Committee 503; it is included in the *Manual of Concrete Practice,* part 3. Material is presented in "A Guide to Repair of Concrete" which includes information for cosmetic and structural repair (*Concrete Construction,* vol. 22, no. 3, March, 1977).

PHYSICAL AND MECHANICAL PROPERTIES OF HARDENED CONCRETE

The properties of hardened concrete vary with different types of aggregates, the water-cement ratio, the amount of entrained air, the 28-day compressive strength, curing methods, and the amount of moisture retained in the concrete. If accurate values of the various properties are desired, they should be established from laboratory tests made on specific design mixes. For most designs, however, approximate values are satisfactory, and approximate values are given for the properties described in the following text.

Compressive Strength

The values of compressive strength (f'_c) used in design are based on concrete specimens tested at 28 days; however, concrete will continue to gain strength after 28 days. See the section "Proportioning Concrete Materials" in this chapter. Strengths vary to the greatest extent with cement content and water-cement ratios, and to a minimum extent with types of aggregate. Strength is also affected by the temperature of the concrete during curing (see Fig. 3-4). The strengths most often specified are 3000, 4000, and 5000 lb/in^2 (20.68, 27.58, and 34.47 MPa). Strengths as high as 8000 lb/in^2 can be obtained without great dificulty.

Common Uses of Concretes of Various Strengths

3000 lb/in^2 (20.68 MPa). Grade beams; footings; slabs on-grade not exposed to heavy traffic or metal wheels; beams; lightly loaded columns; and elevated slab systems consisting of two-way or one-way slabs supported on beams, joists, or waffle slabs.

4000 lb/in^2 (27.58 MPa). Pile caps; drilled piers; spread footings on soils having a bearing capacity over 6000 lb/ft^2 (287,000 Pa); slabs on-grade for warehouses and light-to-medium industries; columns supporting heavy loads, flat slabs, flat plates, and elevated floor systems supporting live loads over 150 lb/ft^2 (7185 Pa); truck dock slabs; pavement for roads; and prestressed and precast concrete construction.

5000 lb/in^2 (34.47 MPa). Pile caps for 100-t (900,000-N) or greater-capacity piles, heavy industrial floors, columns supporting heavy loads, precast concrete construction, and prestressed concrete supporting heavy loads, or used on long spans.

6000 to 8000 lb/in^2 (41.34 to 55.12 MPa). Industrial floors subject to steel wheels, columns, and precast concrete.

Tensile and Flexural Strength

Concrete used in a structural member is not considered to have any ability to resist axial tension forces. Concrete actually does have some tensile strength, but it is very low (about 7 percent of the compressive strength) and is unpredictable. The only types of tensile stress resistance considered in the design of concrete structures is that associated with flexural strength and diagonal tension as it relates to shear strength. Both splitting tensile strength (used in evaluating the shear strength of lightweight concrete) and flexural strength should be determined by tests. Refer to the sections "Proportioning for Performance of Hardened Concrete," "Flexural Strength," and "Splitting Tensile Strength" in this chapter. The approximate values of flexural strength, often referred to as the modulus of rupture, are as follows:

$$9\sqrt{f'_c}\ \text{lb/in}^2 \qquad (0.72\sqrt{f'_c}\ \text{MPa})$$

as used in design of pavement[7] and

$$7.5\ \sqrt{f'_c}\ \text{lb/in}^2 \qquad (0.62\ \sqrt{f'_c}\ \text{MPa})$$

as used in determining stiffness of concrete members according to ACI 318 (f'_c is the compressive strength).

Shear Strength

There are basically two ways in which the shear capacity of concrete is evaluated. One is a measure of diagonal tension, and the other is a measure of interface shear friction. Shear friction is the resistance to relative displacement of a common point on each side of a common plane in a direction parallel to the direction of the applied load or torque. As a measure of diagonal tension resistance, shear strength varies with applied axial tension and compression, flexural stress, and torsional stress. A conservative estimate of ultimate shear strength as a measure of diagonal tension, not subject to axial tension, is $2\ \sqrt{f'_c}$ lb/in² ($0.17\ \sqrt{f'_c}$ MPa), where f'_c is the compressive strength. Interface shear friction resistance for unreinforced uncracked concrete is approximately 12 percent of the compressive strength. Both interface shear friction resistance and shear as a measure of diagonal tension resistance are greatly increased by the use of steel reinforcement bars. For detailed evaluations of shear capacity, the requirements of ACI 318, "Building Code Requirements for Reinforced Concrete," should be followed. Additional information concerning concrete shear strength is available in ACI 426R, "The Shear Strength of Reinforced Concrete Members," reported by ACI-ASCE Committee 426.

Modulus of Elasticity

Unlike steel, the stress/strain relationship of concrete is not exactly a direct proportion, and primarily varies with the weight and strength of concrete. For calculations involving the modulus of elasticity, E_c, a value equal to $W_c^{1.5}33\sqrt{f'_c}$ lb/in² ($W_c^{1.5}0.043\sqrt{f'_c}$ MPa) has been widely accepted. Refer to ACI 318, "Building Code for Reinforced Concrete Buildings" and its commentary.[8] W_c is unit weight of concrete in pounds per cubic foot, and f'_c is the compressive strength of concrete. A standard test method for determining the modulus of elasticity is presented in ASTM C469.

Poisson's Ratio

This ratio of lateral strain to axial strain for concrete is variable. Values as low as 0.10 and as high as 0.30 have been observed.[9] Values between 0.15 and 0.20 are often used in analysis. A test method for determining Poisson's ratio is presented in ASTM C469, "Standard Test Method for Static Modulus of Elasticity and Poisson's Ratio of Concrete in Compression."

Weight (Density)

The density of structural concrete primarily varies with the type of aggregate as follows:

Normal or regular weight: 150 lb/ft³ (2400 kg/m³)

All lightweight: 90 to 100 lb/ft³ (1400 to 1600 kg/m³)

Sand lightweight: 110 to 120 lb/ft² (1760 to 1920 kg/m³)

Heavyweight: 180 to 350 lb/ft³ (2880 to 5600 kg/m³)

(See ACI Standard 211, Appendix 4.)

Shrinkage

As concrete dries, it decreases in volume. This type of volumetric change is called concrete shrinkage. The amount of shrinkage increases with increasing amounts of cement used in the concrete mix and decreases with the addition of bonded reinforcement. Shrinkage is also affected by ambient relative humidity and its volume/surface ratio.

Considering relative humidity (RH) and volume-to-surface (V/S) ratios, effective shrinkage (SH) can be determined approximately by the following relationship[10]:

$$\text{SH} = 550 \times 10^{-6}\left(1 - 0.06\frac{V}{S}\right)(1.5 - 0.015\text{RH}) \qquad \text{in/in}$$

Creep

As defined by ASTM, creep is the time-dependent deformation which continues after the application of a load that is maintained on a solid material. Consideration of creep in concrete is important when evaluating deflections from bending or axial compressive loads. The presence of deformed reinforcing steel in the direction and at the location of applied compressive stress reduces the amount of creep that would occur in unreinforced concrete. Creep is also related to concrete strength, the relative humidity of the surrounding air, volume-to-surface ratios, and the age of the concrete when the compressive load is applied. The ultimate creep strain of unreinforced concrete is approximately equal to $2f_c/E_c$, where f_c is the stress in concrete due to applied compressive load and E_c is the modulus of elasticity of concrete.

Thermal Expansion Coefficient

The coefficient of thermal expansion varies primarily with the type of aggregate, but is about the same as steel. The approximate coefficient of thermal expansion for concrete commonly used in design is 0.00055 per unit of length per 100°F (0.00099 per 100°C).

Thermal Conductivity

The thermal conductivity of concrete varies primarily with the type of aggregate used. The conductivity in British thermal units per square foot per hour per degree Fahrenheit per inch thickness is about 12 for normal-weight concrete and 4 for lightweight concrete.

Fire Resistance

Concrete is often used to protect steel against loss of strength when subjected to high temperatures. Structures made entirely of reinforced concrete have very good fire ratings. Lightweight concrete provides a better fire resistance rating than normal weight concrete. An analytic procedure for determining the fire resistance of a concrete structure is presented in Armand H. Gustaferro and Leslie D. Martin's *Design for Fire Resistance of Precast Prestressed Concrete,* published by the PCI. Thicknesses of concrete to be used as insulation to provide desired fire ratings for steel construction are provided by various insurance agencies and in Louis Przetak's *Standard Details for Fire-Resistive Building Construction,* McGraw-Hill, New York, 1977.

ANALYSIS OF CONCRETE STRUCTURES

Analysis consists of determining the various types and maximum magnitudes of forces a member is required to support for prescribed loading conditions and proportioning members to provide sufficient strength to resist the applied forces without exceeding prescribed factors of safety. Proportioning members generally consists of selecting the reinforcement as well as the cross-section dimensions of the member. Loads consist of dead and live loads. Dead loads are static loads which include the mass of the members, the supported structure, and permanent attachments or accessories. Live loads may be static loads that can be moved from one part of the structure to another and/or moving or dynamic loads not permanently attached to the structure. Equipment or machinery can be considered a dead or live load depending on whether it is permanently a part of the building.

People, furnishings, traveling equipment, wind, snow, rain, the effects of earthquakes and temperature changes, moving water, ice pressure, debris, and lateral soil pressure are typical examples of live loads. Live loads are established by considering the actual loads associated with the use of the structure, environmental conditions to which the structure will be subjected, and various types and minimum magnitudes of loads established by local governing authorities for various types of structures. Local governing authorities often adopt basic building codes such as *The Uniform Building Code* (UBC), the *Building Officials and Code Administrators Code* (BOCA), the American Insurance Association *National Building Code,* or the American National Standards Institute "Building Code Requirements for Minimum Design Loads in Buildings and Other Structures," ANSI A58.1. These and other similar codes establish minimum loadings and requirements for designing and building structures.

The commonly referenced requirements for designing and constructing concrete are those established by the *A*merican *C*oncrete *I*nstitute (ACI). The often-used ACI codes and standards are as follows:

1. ACI Standard 30, "Recommended Practice for Concrete Floor and Slab Construction."
2. ACI Standard 313, "Recommended Practice for Design and Construction of Concrete Bins, Silos, and Bunkers for Storing Granular Materials."
3. ACI Standard 318, "Building Code Requirements for Reinforced Concrete," and commentary.
4. ACI Standard 322, "Building Code Requirements for Structural Plain Concrete."
5. ACI Standard 350, "Concrete Sanitary Engineering Structures."
6. ACI Standard 443, "Analysis and Design of Reinforced Concrete Bridge Structures."

Strength of Concrete Members

Two basic methods for evaluating the strength of concrete members are widely accepted. One method [commonly called *strength design method* (SDM)] is based on analytic procedures that predict the maximum forces that can be sustained just before failure of concrete in compression or yielding of reinforcement in tension, or both simultaneously. The maximum forces predicted are compared to actual loads multiplied by a factor greater than 1. Assigning a larger factor to live loads than to dead loads has been customary. Recent practice has been to use load factors of 1.7 and 1.4, respectively, times the actual live and dead loads. Additionally, the strength provided by steel reinforcement is always made to be less than that which will allow sudden failure of the concrete in compression, so the yielding of steel and the associated large deflections will give a visual warning of excessive load without the threat of sudden collapse. This is the most widely used method at the time of this writing.

The second method, called the alternative method, limits the applied loads to those established by a theory based on a straight-line relationship of stress to strain and by limiting internal member stresses to values well within the elastic range of the materials. Designs based on this method usually result in using greater amounts of concrete or reinforcement or both, than is required when using the SDM.

The theories used for both design methods are described in the edition of ACI 318 current at the time of this writing. Design aids published at the time of this writing for both types of design are as follows:

1. *Design Handbook in Accordance with the Strength Design Method of ACI 318-71,* ACI 340.IR-73.
2. *Design Handbook in Accordance with the Strength Design Method of ACI 318-77,* vol. 2, "Columns," ACI 340.2R-78.
3. "PCA Notes on ACI 318-77 Building Code Requirements for Reinforced Concrete with Design Applications," The Portland Cement Association. In the past PCA has updated this reference to be consistent with revisions made in the ACI Standard 318.
4. Concrete Reinforcing Steel Institute's *Handbook.*

Considerations for Proportioning Concrete Members Based on the Strength Design Method (SDM)

Flexural Members

Members proportioned by the SDM can be quite flexible if stressed to the maximum compressive capacity of the concrete and if span-to-depth ratios are greater than 18 for members supported at both ends and greater than 8 for cantilever spans. If span-to-depth ratios are kept below these values, M_u 12,000/bd^2 is less than about 500 for concrete strengths below 4000 lb/in^2, and the reinforcing steel has a yield point no greater than 60,000 lb/in^2 (413.4 MPa), the deflections and flexural stresses will not normally be critical and the proportions of the member will be reasonably economical. In the formula, M_u is the moment, in foot-kips, at critical sections that results from factored dead or live loads (1.4*D*, 1.7*L*); *d* is the depth, in inches, from the compressive face to the center of tension reinforcement; *b* is the width, in inches, of the beam or the loaded width of slab. *Deflections* should always be evaluated. Some conditions that are especially sensitive to deflections are as follows:

1. First interior spans
2. Members supporting masonry
3. Members over and under windows and doors
4. Long spans [30 ft (9.14 m) or greater] supporting drywall partitions or other fragile materials
5. Members supporting equipment sensitive to deflection
6. Heavily loaded spans [greater than 100 lb/ft^2 (4790 Pa]
7. Locations where clearances are critical to the function of the structure
8. Lightly loaded floors without dampening members such as walls
9. Cantilevers

Additional precautions when proportioning members are as follows:

1. Select member widths so the concrete can be easily consolidated around the reinforcement.
2. Use higher load factors when members are subject to cyclic loading that may fatigue concrete. Refer to *ACI Symposium,* vol. SP-41, "Fatigue of Concrete."
3. Maintain low tensile stress in the reinforcement to minimize cracking in concrete structures exposed to corrosive environments or liquid-containing structures.

Compression Members

The size of columns should be proportioned so the concrete can be thoroughly compacted around the vertical and lateral reinforcement. This can usually be done by carefully selecting the size of reinforcement and sizing the column cross section to keep the required reinforcement percentage below six percent of the column cross-section area. Where columns are located in areas subject to damage from the impact of materials or mobile equipment, special consideration should be given to protecting the column against loss of material. Also, in such cases, providing columns with sufficient stability to withstand such abuses is important.

REFERENCES*

1. ACI Publication SP-19, "Committee 116 Report on Cement and Concrete Terminology."
2. American Society for Testing and Materials: ASTM Standard C125, "Definition of Terms Relating to Concrete and Concrete Aggregates, Philadelphia, Pa., latest edition.
3. ACI Standard 318, "Building Code Requirements for Reinforced Concrete."
4. American Concrete Institute, *ACI Manual of Concrete Practice,* pts. 1, 2, and 3, Detroit, Mich., latest edition. This manual includes the four following references, along with many other ACI standards and reports relating to numerous phases of concrete technology.
5. ACI Standard 301, "Specifications for Structural Concrete for Buildings."
6. "Effect of High Temperature in Hardened Concrete," *Concrete Construction,* November 1971.
7. "Slab Thickness Design for Industrial Concrete Floors On-Grade," The Portland Cement Association, Philadelphia, Pa., latest edition.
8. ACI Standard 318, Commentary on "Building Code Requirements for Reinforced Concrete."
9. Troxell, George E. and Harmer E. Davis: *Composition and Properties of Concrete,* 1st ed., McGraw-Hill, New York, 1956.
10. Zia, Paul, H. Kent Preston, Norman L. Scott, and Edwin B. Workman: "Estimating Prestress Losses," *Concrete International,* June 1979.

*References to appropriate books, specifications, and standards are also included in the text discussions for the particular subjects.

chapter 2-4

Floors

by
Richard D. Ramsey
Section Manager, Technical Services
Buildings Division
Sverdrup & Parcel and Associates, Inc.
St. Louis, Missouri

INTRODUCTION

Plant floors are generally subjected to extreme abuse, and receive minimum maintenance. Not only foot traffic, but vehicles with steel wheels, spilled chemicals, dropped products, and equipment combine with age to deteriorate floors of all kinds. Floors must be considered as both working surfaces and traffic ways for moving people and products. Well-designed and well-maintained floor surfaces are safe and, in the long run, economical—the most expensive maintenance is usually replacement.

Although plant engineers may seldom have a voice in floor selection for a building, they must live with the results. A poorly designed or selected floor can occasionally be altered or replaced to the engineers' satisfaction, but more frequently they make the best of it. Floor product developments can sometimes correct problem floors. Floor product availability changes rapidly, with new developments happening almost weekly. Other products fail and disappear.

Although many floor products have come and gone, concrete remains the basic material. As a finished working surface, or with a wide variety of both integral and applied surface treatments, concrete continues to serve as the workhorse of floor construction. Service usage may generally be classified as light, medium, or heavy-duty. Some floors are considered special-purpose and require special treatments, for example, floors in food and beverage preparation areas.

Floor finishes and treatments have improved greatly in both durability and appearance. Products are available for almost any type of service. This is good and bad–good in that many problems are more easily solved, but bad in that the plant engineer must be constantly aware of various product limitations and suitability for specific use. It was simple to say "paint it"; now we must ask "with what, how much, what preparation?" and other questions. When using a floor finish, coating, or treatment, suitability for the job must always be the concern. The best source for this kind of information is the product manufacturer, who is usually expected to guarantee the product's performance.

In addition to suitability of use, safety considerations are also important. The restrictive fire- and life-safety regulations that are so prevalent affect not only structures but finishes as well. The inappropriate application of a coating, for instance, may change a complying nonslip surface into a hazardous, slippery surface—or into a surface that spreads fire, smoke, or toxic fumes. When a new plant is occupied, it is presumably in compliance with all applicable regulations and it should be maintained in the same way.

BASIC FLOOR MATERIALS

Concrete

Concrete, the basis for most plant floors, has the attributes of durability and strength. Without surface treatment it is absorbent, allowing soiling, buildup of grease and other chemicals, and surface dusting. Products and procedures are available to control these problems, however, making concrete an economical flooring material.

Proper concrete floor design includes numerous structural and nonstructural considerations, and should be done by professional consultants. This chapter concentrates on finishes, maintenance, and repairs.

The basic concrete floor is an on- or above-grade monolithic slab that is placed and finished in continuous sequence and cured after finishing. Finishing usually consists of troweling, with possibly special-purpose aggregates troweled into the uncured surface.

Finishing

Troweling the concrete surface consolidates (densifies) the surface, and imparts a smooth or textured surface. Since the durability of the concrete surface depends greatly on the aggregate (not the cement), care must be taken to prevent compressing the aggregate too far below the surface. If this happens, additional aggregate should be tamped into the surface, and troweling continued.

Different surface characteristics are obtained by using trowels of various materials—smooth surfaces with a steel trowel and various textures and nonslip finishes with wood, magnesium, or aluminum trowels. The uncured surface can also be textured with burlap or stiff bristle brushes for rougher surfaces.

Special aggregates other than the mix aggregate may be tamped into the surface while finishing. Metallic aggregates, such as processed iron borings, are frequently used to create highly durable surfaces in hard-use areas. By saturating the surface with well-compacted aggregate, the exposed cement can be reduced to as little as 5 percent of the exposed surface (normal concrete exposes 20 to 35 percent cement). Similarly, natural aggregates such as crushed granite, trap rock, and hard river gravel are used to increase the surface durability; they can also be decorative.

Color is occasionally applied to concrete during the finishing process. This can be useful in identifying particular plant areas and hazardous areas. Color shakes (sprinkling the surface with dry color) should be kept to a minimum so as not to disturb the cement-to-aggregate balance, and color uniformity is difficult to control with this application method. Better uniformity can be obtained, but at higher cost, by purchasing the concrete with the color premixed.

Multislab Construction

As an alternative to monolithic slabs, floors may be of multislab construction. Multislab construction usually consists of a basic concrete slab, as in monolithic construction, but with an applied topping varying from ½ to 3 in (1.27 to 7.62 cm) in thickness, depending upon the type of topping. Additionally, the topping may be in more than one layer and of differing materials, such as a metallic aggregate waterproofing layer plus the wearing surface.

Multislabs are generally safer and easier to repair than monolithic, since there is no need for concern about maintaining the structural integrity of the basic slab. Also, totally changing the topping to permit a different use for the floor is possible without having to disturb the structural slab. The system, however, is usually more expensive and involves more construction time than the monolithic construction.

Waterproofing is frequently placed in the multislab construction, when on-grade. It may be a built-up fabric and asphalt membrane system, sheet- or liquid-applied membrane, or troweled metallic aggregate. The waterproofing should be carefully restored when repairs are made.

Multislab Toppings

Multislab toppings may be ordinary cement and aggregate or one of many special formulations designed to provide a specific service. Ordinary toppings usually do not serve any better than monolithic slabs, but can be easier to maintain or replace. Special toppings commonly contain a metallic or hard natural aggregate mixed with cement or a plastic matrix. There are too many proprietary formulations to consider here, but most are variations of the following types:

Metallic Aggregates

Metallic aggregate toppings are usually iron particles mixed with binders and a cement matrix. Their purpose is to increase the surface durability and occasionally to increase conductivity of the floor, preventing buildup of static electricity. These toppings are usually mixed with a minimum of water, placed on the rough slab, tamped to a dense mass, and troweled to the desired finish. Surface rusting in little-used areas is not unusual. Emery, aluminum oxide, and ceramic granules may be similarly used in some systems.

Natural Aggregates

As in monolithic slabs, the natural aggregate toppings may be granite, quartz, or gravel. These toppings are sometimes ground and polished to a very smooth surface to form a typical terrazzo finish. Although durable and easy to clean, terrazzo is a form of concrete, and is subject to abrasive wear and staining. Maintenance and sealing are important if its inherent durability is to be preserved.

Some terrazzo toppings are placed over a sand bed or slip sheet. These should be preserved when repairs are made since they isolate the topping from base slab movement. Likewise, most terrazzo has brass or aluminum strips at frequent intervals which should also be preserved to help control cracking. Newer terrazzo systems use a synthetic matrix with rather small aggregate, and are applied as thinly as ½ in (1.27 cm) and without metal strips; repair materials should be compatible with the original system.

Asphaltic Concrete

Asphaltic concrete toppings over rough slabs are declining in use but are still considered when easily replaceable toppings at relatively low cost are needed. These toppings are not usually used where excessive oils, gasoline, or other solvents are present, unless special surface sealers are used. These sealers usually have a limited durability and must be reviewed periodically. See Chap. 2-9 for further reference to asphaltic concrete.

Clay Products

Flooring products of a basic clay (earthen) composition are brick, quarry tile, and ceramic tile. These products are most frequently laid over a rough concrete slab, and are set in mortar or in sand (for flexibility). Products vary in size from 1 × 1 in (2.54 × 2.54 cm) face to 12 × 12 in (30.48 × 30.48 cm) and larger, with thickness from ¼ to 2¾ in (0.635 to 6.98 cm). This type of flooring is characterized by numerous mortar joints and a hard, nonresilient surface which is highly resistant to chemical attack. The joints are the most vulnerable part of the system and should be selected with care.

Brick

Paving bricks were widely used as plant flooring, but they are not so frequently found in modern construction. Improvements in concrete surfacing plus other product developments have reduced their uses. Bricks are molded clays, burned at high temperatures, to form a dense, hard surface that can have various textures and glazes. Good-quality paving brick is limited in both availability and the number of sizes, and the plant engineer may have problems in obtaining bricks that match older installations. Bricks that are too large can be trimmed with a masonry saw if only small quantities are needed.

Quarry and Ceramic Tile

Quarry and ceramic tile are available as both glazed and unglazed units. Unglazed tiles are usually used for flooring and may have abrasive units mixed with regular units for increased slip resistance—about 7½ percent abrasive tile is normal. Because of their higher cost, these products are usually confined to areas where sanitation or appearance is important. They are also used as decorative floor treatment. These products are treated further in Chap. 2-5.

Wood

Once a commonly used plant flooring, wood is now infrequently used because of improvements and development of other types of flooring and increasing emphasis on reducing the fire load of buildings. Few types of wood flooring possess the durability and low maintenance qualities demanded for modern plant usage. Only end-grain treated wood blocks and some radiation-treated forms possess these qualities, the latter at very high cost.

Wood Blocks

End-grain wood blocks of oak or yellow pine are pressure-treated with penta or creosote to form a basic flooring block. Finishes varying from pitch to urethane are applied according to the final usage. Pitch finishes are serviceable for hard-use industrial purposes, where appearance and foot tracking of oils to adjacent areas are not important. Urethane finishes over penta-treated blocks provide nontracking surfaces that are easily maintained. With either treatment, wood-block flooring is resilient and has impact resistance superior to normal concrete, requires a minimum of maintenance, and can take more abuse than most other flooring materials.

Wood Flooring

For finished plant areas such as showrooms, display areas, and lobbies, there are a number of prefinished and unfinished wood floor products. The products previously mentioned plus parquet and strip flooring of oak, pine, maple, and teak are all suitable for this use. These flooring materials require daily maintenance to preserve the exceptional finishes.

Metal

Metal is most frequently used for flooring in the form of gratings and plate, usually applied over a structural steel framework. Both stairs and platform (or walkway) surfaces are made of these products. Strength, durability, low maintenance, and ease of cleaning are features of metal flooring of this type. Metal plate or grating systems can also be made portable for increased flexibility of use.

The metal plates may be steel, galvanized steel, or aluminum, with plain or "checkered" surfaces, and gratings may be had of the same materials. The material size varies and should be engineered to support the loads imposed. Plate for walking surfaces is usually selected as checkered, with a raised-pattern nonslip surface. A liquid- or sheet-applied abrasive can be used on smooth-surface plate to provide a nonslip surface; these surfaces must be occasionally renewed.

Another form of metal flooring formerly in general use is the embedded floor grid. Cast-iron or steel grids of varying sizes are embedded in concrete and the grid filled with concrete and troweled flush with the grid surface. Many of these grids have disappeared from the market. Some forms of cast-iron grids are still available through foundries, and these are excellent for steel-wheeled vehicle traffic and heavy abuse areas.

APPLIED FINISHES

In addition to integral finishes, there are hundreds of applied finishes. These consist of liquid coatings and applied solid sheet materials. Their purposes vary from simply hardening and reducing dusting to decorative application.

Coatings

Surface treatments which harden and reduce dusting, such as sodium silicate crystals and magnesium fluorosilicate crystals, both mixed with water, are basic to almost all concrete installations. Liquid curing agents to promote uniform curing are similarly applied, but final finish should be considered because curing agents can have an adverse effect on some adhesives used for applying resilient flooring.

Surface hardeners should be applied as soon after finishing the surface as the manufacturer's recommendations permit. The treatment should be repeated when dusting becomes noticeable.

Paint coatings are sometimes applied to improve cleanability, to provide identification of areas, and to identify hazards. Paint coatings on floors were traditionally considered to be limited-life, high-maintenance finishes. This is still somewhat true, but these materials are now greatly improved and the market offers a greater variety to suit special needs. Such coatings as urethane-based and epoxy-based products offer more durability and cleanability than most of the old pigmented oils. Somewhat more difficult to apply, they will cure (rather than dry) to form a tough film that will withstand hard usage on light-duty and some medium-duty floors. Surface preparation is highly critical in accomplishing this durability, and the manufacturer's recommendations should be carefully followed in preparing new or old floors to preserve the product guarantees. Always verify the suitability of specific use with the manufacturer of special coatings before proceeding.

Concrete Stains

A number of concrete stains are available, some of which both color and harden the surface. In addition, some wood-stain products can serve adequately as concrete stains.

Seamless Flooring

A class of applied floor finishes referred to as *seamless flooring,* although not new, has greatly expanded in both quantity of systems and in quality of performance. Traditional seamless flooring, such as cupric oxychloride and magnesium oxychloride, is still available but has in many instances been replaced by newer products. Seamless floor formulations of epoxy, urethane, vinyl, or rubber are easy to apply and afford greater water and chemical resistance than traditional systems, which were primarily troweled coatings. Like the traditional systems, the newer products are also very critical in base preparation and application techniques. Installing and repairing these flooring products are best left to the professional.

Most of these systems are proprietary and vary in formulations and application among the manufacturers. Generally, a system will consist of a primer, a tack coat into which a granule or plastic flake may be broadcast, and one or more clear or colored topcoats. Failure of these systems usually occurs as delamination from the base (a result of faulty preparation or moisture drive through slabs on-grade) or wearing of the topcoat. Most of the systems can be recoated in high-traffic areas, but delamination should be repaired by professionals to prevent recurrence of the failure.

Many of the liquid-applied systems are acceptable for food and beverage preparation areas, laboratories, clean rooms, toilet facilities, and wherever a low-maintenance, sanitary finish is needed. The liquid systems, however, are not generally recommended for hard-use industrial service. Some spray- or trowel-applied systems are available for harder-use areas, but must be carefully selected. Always verify the application with the manufacturer. Compatibility must be checked when recoating or applying a different flooring over an existing seamless floor.

Liquid treatments for sealing, waxing, cleaning, sanitizing, and other purposes are available from janitorial supply companies. Most such products are proprietary, and claims of performance must be carefully scrutinized. Linseed oil can effectively seal floor surfaces and prevent deterioration caused by some chemicals, and is particularly effective against deicing salts.

Resilient Flooring

Resilient flooring, in the form of tile and sheet goods, is primarily for light- to medium-duty finished areas where cleanability or appearance is important but a seamless application is not necessary. Tile is easy to install and repair, but has numerous joints. Sheet-goods flooring is more difficult to install and requires more fitting; however, it is not difficult to repair, and has fewer seams. Rolls of the goods are usually a minimum of 6 ft (1.83 m) wide and are made in long roll lengths.

Resilient Tile

Resilient tile flooring is usually $12 \times 12 \times \frac{1}{8}$ in ($30.48 \times 30.48 \times 0.318$ cm) thick; $\frac{1}{16}$-in (0.159-cm) (service gauge) and $\frac{3}{32}$-, $\frac{3}{16}$- and $\frac{1}{4}$-in (0.238-, 0.476-, and 0.635-cm) thicknesses are available from some manufacturers. Some 9×9 in (22.86×22.86 cm) tile is still available, but in limited materials and colors. Vinyl asbestos (VA) tile is the most common. Asphalt, vinyl, and rubber tiles are also available.

Asphalt Tile

Asphalt tile (sometimes referred to as *mastic*) is a durable tile product that has gradually been phased out by VA tile, which is far superior to asphalt tile in resiliency, impact resistance, and resistance to chemical attack. A better selection of colors is also available in VA tile products. Vinyl and rubber tiles are not often used for plant work because they are rather expensive; they are used primarily for decorative purposes.

Vinyl Composition Tile

Vinyl composition floor tile, a new product, is a modified VA tile which eliminates asbestos from the formulation and ends a potential health hazard. Even without asbestos, these products have the best characteristics of VA tile plus apparently better maintenance qualities.

Resilient tile and sheet-goods products, as well as accessories such as rubber and vinyl base, are all adhesive-applied to concrete or wood subfloors. Adhesive must be used as recommended by the tile manufacturer for the particular installation—above grade, below grade, wet area, alkali resistance, etc.

Linoleum

A long-time standard for sheet-goods products and tile, linoleum has virtually disappeared from the flooring market. Almost all linoleum is now imported, and in limited quantity. Other sheet goods have replaced linoleum and have superior performance characteristics. As existing linoleum installations deteriorate, it is better to replace them with one of the newer flooring products than to attempt repairs.

Carpet

Very little carpet finds its way into the industrial plant. Usually confined to offices and showrooms, carpet has excellent properties for this service. Conceivably, as fiber developments and construction techniques improve, carpet may eventually be used in the plant.

MAINTENANCE

Floor maintenance is a continuing process, beginning immediately with first use. Floor surface deterioration is a direct result of debris accumulation. Where abrasive debris is present, sweeping one or more times a day will be necessary to prevent rapid damage. Daily sweeping is advisable for all floors, with more frequent attention to trouble spots. Whether manual or power sweeping is used is unimportant; the choice is a function of the size of the area requiring maintenance.

Continuing attention to spills and leaks of liquids of all kinds, from water to acids and alkalies, is essential to long floor life. This is probably the most difficult type of cleaning to control, until a walking hazard is developed. Some types of spills can be controlled by dry absorbent floor granules if a supply is maintained in the area of frequent problems.

Finished Floors

Floors of a more finished nature, such as resilient flooring, seamless flooring, terrazzo, and some clay tiles, require more than just sweeping. Mopping, anywhere from daily to weekly as the area dictates, is essential. Latent abrasive dust left after sweeping is very destructive to these surfaces and can only be controlled by mopping (or power scrubbing). Water from this type of cleaning should be quickly removed to prevent deterioration of the adhesives.

Concrete and Wood Floors

Maintenance of concrete floors (topped or monolithic) and of pitch-coated wood block is by sweeping as noted before. Accumulations of deleterious materials from concrete surfaces are removed by scrubbing with an appropriate cleaner (sometimes requiring soaking and scraping) followed by thorough flushing with clean water. Done at frequent intervals, this task remains relatively easy. Scrubbing wood block is discouraged, except when urethane finishes are applied—and then only a minimum of water is advisable. Wood block floors, however, can be sanded to expose a new surface, and the proper surface sealer (pitch or urethane) reapplied.

Other Floors

Other types of flooring should be maintained strictly as recommended by the manufacturer. If the manufacturer is not known, most good janitorial supply companies can evaluate and recommend acceptable products. Product compatibility is important—both between adjacent systems and with maintenance products. For additional information refer to Chap. 14-6, "Sanitation Control and Housekeeping."

REPAIRS

Repair work can be classified as *emergency* or *permanent*. Permanent repairs are those which restore a floor to its original state; they usually require professional help. Too frequently emergency repairs become permanent if successfully accomplished! But the plant engineer should proceed with permanent repairs as soon as feasible after the emergency—delay usually results in an increasingly difficult repair job.

Emergency Repairs

Emergency repairs are necessary when floor damage results in a hazardous condition for walking or vehicular traffic, or structurally endangers the building. The emergency repair should relieve rather than accentuate the condition. Expedients such as placing boards or plywood over a hole, for instance, may constitute a hazard equal to that of the hole, unless properly secured.

One method of emergency patching holes of moderate size in concrete floors is to clean out the hole and line it with polyethylene film or similar plastic, then fill the hole with a fairly dry concrete mix (1:3:4 or even 1:4, with a minimum of water added), screen off and tamp smooth with a board; add mix and tamp again until the surface stays flush with the adjacent flooring. Protect for about 24 h; then use with care. This is a safe patch for walkways, but permanent repairs should be made as soon as possible. The patch will break out easily and leave a clean hole for permanent repairs.

Most applied flooring materials do not require emergency repairs unless they are loose. When sheet goods, tile, or other filmlike finishes come loose, temporary repairs can sometimes be made with a typical water-soluble white glue, with the area weighted for a few hours. When the permanent repair is made, the white glue can be removed.

Permanent Repairs

Permanent repairs require the services of experts who are knowledgeable about the various systems. Ordinary concrete, wood block, and resilient tile repairs are frequently handled by plant maintenance personnel. Resilient sheet goods, seamless flooring, terrazzo, and other special concrete toppings and integral finishes require skillful repairs to restore the floor properly for its intended purpose. The use of professional finishers for this work is advisable.

Concrete

Repairing ordinary concrete floors can be routine, requiring only a few preparations. Materials needed will be cement (portland cement type I or type III), clean sand, and coarse aggregate [clean rock or gravel ½ to 1 in (1.27 to 2.54 cm), approximating the aggregate used in the floor]. A good concrete bonding agent and a water supplement (such as liquid latex formulations to improve bond) can be very helpful but are not necessary. All patching materials should be clean and the cement should be fresh.

Patching procedures should approximate the following:

1. Clean area to be patched; remove all loose material and dust; wire-brush reinforcing steel, if any (do *not* remove); repair vapor barrier, waterproofing, or insulation, if any.

2. Undercut edges of holes about ½ in (1.27 cm); cut out cracks to about ½- in (1.27 cm) width and 1- in (2.54 cm) depth; clean.
3. Prepare patching material in clean mixer; for larger holes a 1:2:4 or 1:2½:5 mix by volume of cement, sand, and coarse aggregate; use cement and sand only for small patches and cracks, about 1:2½. Mix with only enough clean, potable water to make a stiff mix (should form a cohesive ball when squeezed, without excess water). Water supplement may replace some mixing water to improve the bond to old concrete.
4. Dampen surfaces to be patched, particularly earth under slabs. Apply bonding agent to old concrete and reinforcing steel, or brush with a creamlike slurry of cement and water.
5. Place patch-mix in hole or crack; compact it. Be sure mixture fills undercuts and has good contact with reinforcing steel. Tamp in mixture until slightly higher than adjacent surface.
6. Trowel-finish, compacting flush with adjacent surfaces; use trowel to match existing finish.
7. Apply surface hardener and/or sealer.

Again, control of water is important. Excessive water contributes to shrinkage, which only weakens the patch. If a patch occurs at an expansion or control joint, the joint should be restored.

Where severe structural cracking may occur in a concrete floor, repairs must be engineered to restore the design strength of the slab. This is best left to professionals. Systems involving pressure-injected adhesives are also available for such restoration.

Refer to Chap. 2-3 for additional information on repairing hardened concrete.

Resilient Tile and Wood Block

Repairs to resilient tile and wood block are simpler if replacement units to match the original are available. Repair consists of removing and replacing the tile or block with a new unit. Most tile can be removed easily, but if a problem occurs, placing dry ice on the broken tile for a short time will make removal easier. Color and pattern matching is always a problem with tile, so obtaining extra tile at the time of construction is advisable, if possible, to store for future repairs.

Wood-block floor repairs are frequently needed because of disruption of the base surface or because of buckling due to excess water. Generally, the blocks can be removed to permit repairing the base, and the good block can be reused. Water-swollen block should be thoroughly dried and, if not deformed, may be reused.

SUMMARY

Floor-finish treatments, coatings, systems, and additives have multiplied to the point where it is impossible to categorize and comment on all. Even when restricted to those suitable for plant use, the number of products is awesome. This chapter has only briefly touched on some of the more commonplace floors. There is a floor for virtually every plant condition, with some form of concrete being the closest thing to an all-purpose floor. Obtaining and following the manufacturer's assistance and directions in maintaining and repairing proprietary products and systems is again stressed. *Remember, neglect is the worst enemy of any floor.*

Table 4-1 summarizes the material discussed in this chapter.

TABLE 4-1 Summary

Floor type	Service	Maintenance factor	Repair	Appearance	Durability	Uses
Concrete, hardened	Light & medium	Frequent, absorbent	Simple but messy*	Gray + color	Fair to good	Utility areas; moderate-traffic plant areas; not for sanitary areas
Concrete, special surface treatments	Medium & heavy	Frequent, may be absorbent	Difficult to match original*	Range of color	Good to excellent	Heavy-duty factory floors and warehouses
Clay products (tile, brick)	Light & medium	Seal-joint	Difficult material, obsolete	Can be decorative	Good to excellent	Areas requiring frequent cleaning, sanitary, no impact loads; acid areas
Seamless flooring (applied)	Light & medium	Easy to clean	Difficult, requires professional help	Utility to decorative	Fair to good	Highly chemical-resistant & sanitary, can be used for floors & walls in laboratories, food & beverage areas, locker rooms, & toilets
Resilient flooring	Light & medium	Good	Simple, replace	Large selection	Good	Offices, corridors, washroom & locker areas, etc.
Terrazzo	Light to medium-heavy	Excellent	Difficult, requires professional help	Large selection	Excellent	Where long service life & low maintenance are requirements, can be decorative; not usual for factory areas
Impregnated wood block	Medium & heavy	Minimum	Average difficulty	Utility dark	Excellent	Machinery areas; not for chemical processing
Untreated wood	Light	Frequent	Simple		Poor	Not recommended for plant use

*Requires protection until sufficiently cured.

BIBLIOGRAPHY

ASTM Standards in Building Codes, latest ed., American Society For Testing and Materials, 1916 Race St., Philadelphia, PA 19103.

Callender, John Hancock: *Time-Saver Standards for Architectural Design Data,* 4th ed., McGraw-Hill, New York, 1966.

Concrete Floors on Ground, EBO75, 010, 1978, Portland Cement Association, 5420 Old Orchard Rd., Skokie, IL 60077.

Handbook for Ceramic Tile Installation, latest ed., Tile Council of America, Inc., 800 Second Ave., New York, NY 10017.

Hornbostel, Caleb, and William J. Hornung: *Materials and Methods for Contemporary Construction,* 2d ed., Prentice-Hall, Englewood Cliffs, N.J., 1982.

Ramsey, C. G., and H. R. Sleeper: *Architectural Graphic Standards,* 7th ed., Wiley, New York, 1980.

Watson, Don A.: *Construction Materials and Processes*, 2d ed., McGraw-Hill, New York 1978.

chapter 2-5

Walls, Windows, and Entrances

by
Richard D. Ramsey
Section Manager, Technical Services
Buildings Division
Sverdrup & Parcel and Associates, Inc.
St. Louis, Missouri

INTRODUCTION

Walls, windows, and entrances are those parts of the plant which provide weather protection, visual and sound control, access to spaces and, frequently, fire control. Most common building materials may be used as part of a system of walls, windows, and entrances. This chapter considers only those most prevalent in plant structures: wood, masonry, concrete, and metal walls, and metal windows and entrances.*

It is important for the plant engineer to recognize that certain walls, windows, and entrances in modern plants are part of a fire-control system established by building codes

*A more detailed description of materials in Don A. Watson, *Construction Materials and Processes,* 2d ed., McGraw-Hill, New York, 1978, is recommended for reference.

and insurance requirements. In the case of doors and some windows, attached labels identify them as such. But ceilings, walls, or floors are usually not so marked. Before modifying these components, or before repairing, establish whether they are fire-rated and, if so, protect the integrity of the system. Reference to local code authorities or your professional consultant may be necessary to determine whether a fire rating is required.

WALLS

Masonry

Wall construction of brick, stone, and other earthen or cementitious building blocks is referred to as *unit masonry.* Units of various sizes are stacked, and the whole is unified with a cementitious jointing material, or mortar. Many types of units and their joints are water-absorptive. Masonry walls are constructed of one or more wythes (rows front to back), depending on the expected service. For exterior walls, the wythe exposed to the weather should be of durable quality with low absorption; the remaining wythes are basically for strength. Weathering ability is not important for interior walls and partitions; however, many plant situations demand high-performance masonry for chemical resistance, sanitation, or other special conditions.

Some of the traditional masonry products, such as gypsum block and structural clay tile (both glazed and unglazed), have limited availability. Glazed structural clay tile is still available from a limited number of manufacturers, but matching the color and size of older units may be difficult.

Modern masonry construction frequently consists of brick and concrete masonry because the materials are readily available, durable, and relatively economical. Concrete masonry units (blocks and brick made of portland cement and aggregate) are a common backup material, and are widely used for interior and exterior wall facing. Various coatings may be applied to provide specific weathering or service characteristics. Coatings, of course, must occasionally be renewed. Where brick, a glazed unit, or stone is the exposed face, a coating is generally applied only to reduce absorption of water or soiling.

Standards

The quality of most masonry products is controlled by industry and national standards too numerous to list here. The most definitive of these is the American Society for Testing and Materials (ASTM). Most of the material standards that will be needed by a plant engineer are included in the publication *ASTM Standards in Building Codes,* available from ASTM. This publication also contains standards for many other building materials, as well as masonry.

Classifications

Modern masonry product lines have expanded to include a number of grades (quality) and finishes for the old traditional materials—particularly brick, concrete masonry, and mortar. Although this complicates the selection process, it does permit selecting an economical product for a particular use.

Brick is classified by ASTM as SW (severe weather), MW (moderate or normal weather), or NW (nonweathering). This allows the plant engineer to select an exposed brick suitable for a particular climate or use, and yet select a less-expensive product where weathering is not critical.

Concrete masonry units are classified as *load-bearing* or *non-load-bearing* and are also controlled by ASTM by grade—Grade N, (above- or below-grade, exposed to moisture), and Grade S (limited to above-grade use). Concrete masonry may also be *standard* or *lightweight,* depending on the type of aggregate used in the concrete mix.

Mortar, however, is graded by types according to its strength (see Table 5-1). Masonry mortar is frequently site-mixed from bulk ingredients, but the prepared masonry mortar mixes are convenient for maintenance and minor repairs. This eliminates stockpiling cement, sand, and lime. Such prepared mixes are usually locally avail-

TABLE 5-1 Masonry Mortar*

Mortar type	Average compressive strength at 28 days, lb/in² (MPa)	Parts by volume of portland cement, cement, or portland blast-furnace slag cement	Parts by volume of masonry cement	Parts by volume of hydrated lime or lime putty	Aggregate, measured in a damp, loose condition
M	2500(17.2)	1	1	—	Not less than 2¼ and not more than 3 times the sum of the volumes of the cements and lime used
		1	—	¼	
S	1800(12.4)	½	1	—	
		1	—	over ¼ to ½	
N	750(5.2)	—	1	—	
		1	—	over ½ to 1¼	
O	350(2.4)	—	1	—	
		1	—	over 1¼ to 2½	
K	75(0.5)	1	—	over 2½ to 4	

*After ASTM C270.

able in 50- or 90-lb bags and, occasionally, in ready-to-use caulking-type cartridges. The cartridges can be very handy for resetting loose brick or stone, tuckpointing eroded joints, and resetting loosened quarry tile or brick flooring.

Loading Characteristics

Masonry walls can be designed to support heavy loads applied vertically, but are weak resisting lateral (horizontal) loads, as from high winds or vehicle impact. Masonry walls subject to lateral loads are usually reinforced with steel ladder-type or truss-type joint reinforcing and are also designed with deformed reinforcing steel rods. Designing such structures is best left to professional consultants. However, it is important that the plant engineer, in restoring damaged walls, should replace reinforcing where affected by the patchwork.

Wood

A minimum amount of wood is used for modern plant construction. Although it is one of the least expensive construction products, fire codes prohibit such construction in larger plants, or require special fire treatment. Smaller new plants and older plants may still contain considerable amounts of wood.

Metal

Durability, strength, low maintenance costs, and reasonable first costs are factors that characterize metal products in plant construction. Walls, roof systems, doors, windows, and many other components are primarily metal, usually some form of steel or aluminum. The availability of many alloys, sizes, and shapes of both steel and aluminum contribute to their great versatility. The many forms of sheet and strip stock, bars, rods, tubes, and pipe, as well as structural shapes and custom-designed shapes, are evidence of this versatility.

A large consumer of metal products for construction is the building systems industry. Pre-engineered metal structures have steel supporting members, metal roofing, and metal wall panels. These systems are discussed further in Chap. 2-10, "Building Systems."

Many of the building systems components, such as wall panels and roofing sheets, can be used for additions and alterations in other types of construction. Also, many products of a similar type are available in various strengths, configurations, and finishes. Insulated sandwich and plain metal panels for roofs and walls are frequently used with conventional steel and masonry construction, and have become a common form of large plant design. Some shapes, such as common corrugated panels, have become industry standards and do not change from year to year. These are good selections where the possibility of frequent replacement because of damage is expected and where future expansion is a consideration.

Metals are used in plants in many other miscellaneous ways: as light framing members of wall and partition systems and window and entrance units (discussed further in this chapter), as floor components (Chap. 2-4), and for security and safety applications too numerous to list here.

Other Construction Materials

Cementitious Products

These are usually some form of cement mixed with an aggregate and water, with the most common products used in plants being concrete or some form of plaster. The structural aspects of concrete are discussed in other chapters.

Cementitious products have a major advantage of providing excellent fireproofing qualities. Usually applied or erected in a plastic state (precast or sitecast), cementitious products have great flexibility, which allows them to achieve special shapes and to conform to unique substrates. The disadvantages are difficulty of repair and a lack of resiliency when cured. Because of its greater durability, concrete is the main cementitious product used in plant work.

Plaster is primarily a finish coating rather than a structural material. Most problems with plaster come from physical damage, and plaster use is therefore usually restricted to various fireproofing tasks and must be protected in contact areas. Some hard plasters, such as Keene's cement, are available for locations needing extra durability.

Plastics

Current use is limited to insulation, glazing, and a few wall-protective products. Some attempts have been made to develop plastic windows, doors, and structural shapes, but success has so far been limited to residential applications. As protective coatings and sheets, however, plastics find an increasing use in industrial applications. Tank linings, machinery finishes, wall and floor coatings, pipe, and food processing area finishes are a few of these uses.

WINDOW AND ENTRANCE UNITS

Metal window and entrance units are common in modern plant construction. For many years, steel windows dominated the industry. Currently, there are fewer manufacturers of steel windows, but aluminum windows are plentiful. There is little difference in overall performance between the materials, but there is some variation in individual characteristics. For instance, steel is usually stronger, but aluminum is more weathertight because of better seals and requires no painting. In addition, because of the greater number of aluminum window manufacturers, there are more different window types available in aluminum. The major window problems are centered around glass breakage and hardware failure.

Steel entrance units continue their long-time popularity for plant installations. Many manufacturers produce units (doors and frames), and a wide selection of type and quality is available. Aluminum and stainless steel are most frequently used as major entrances to public and office areas and at special locations where greater corrosion resistance is needed. For openings where fire ratings are necessary, steel frames with wood or steel doors are used. There are presently no fire-rated aluminum entrance assemblies.

MAINTENANCE AND REPAIRS

The materials previously discussed are combined in many ways with many other products to form wall, window, and entrance systems. "Systems" here is the key word, for all these elements must work together to provide a particular performance—for fire protection, for weatherability, or for service. In maintenance and repair work, the integrity of the system must be preserved.

Walls

Walls are most commonly built of masonry, concrete, or steel. Both masonry and concrete have general, widespread usage because of their durability and strength. Separately or in many combinations, the resulting walls and partitions offer the widest range of finished surfaces and a broad range of performance.

Plaster and Concrete Repairs

Routine wall and partition maintenance consists of cleaning and prompt patching of damage. Plaster and concrete repairs are not greatly different from concrete floor patching (see Chap. 2-4). It is essential to cut back damaged areas to solid, undamaged material; thoroughly clean the area; use a bonding agent or cement-water paste; and patch with a material matching the original. It is advisable to occasionally dampen the patch to promote curing and reduce shrinkage. Small breaks in concrete masonry can be patched in the same way. A piece of cardboard larger than the hole with a piece of string through the center can be used to keep patching materials from falling into the cells.

Masonry Repairs

Repairing masonry walls should be done with care to preserve the integrity of the original wall. When feasible, broken units should preferably be replaced with whole units of the same type. This usually means cutting out the broken pieces with a chisel, including all of the old exposed mortar, exposing the adjacent units. A bonding additive can be added to the mortar for improved bond. To insert the masonry unit, butter the edges of the unit and the sides of the hole with a fairly stiff mortar, insert and position the unit, and compress the mortar into the joint. Finish by tooling the joint to match the existing joints, adding mortar as needed. If the old units are very absorbent, dampening the new units before insertion is advisable.

Cracks in concrete, masonry, and plaster occur frequently; they require prompt attention. Static (nonmoving) cracks resulting from shrinkage or physical impact should be repaired with the same base material after cleaning the crack and cutting back to sound, tight material. It is helpful to lightly spray the crack with water to aid bonding of cementitious materials. Dynamic cracks resulting from expansion and contraction, or continuing building movement, should always be repaired with flexible sealants. This procedure prevents a buildup of rigid material in the crack, which will cause greater separation with the next movement (see Chap. 2-8).

Wall Panel Products

Wall panel products, such as drywall, hardboard, cement-asbestos board, and metal panels are usually difficult to repair. Most frequently the quickest, most economical method is whole-panel replacement. Drywall repairs are possible but are time-consuming. A procedure similar to that for repairing holes in concrete block can be effectively used for small drywall patches.

Wood

The most important aspect of maintaining wood is maintaining the wood finish. Untreated wood, other than redwood, some cedars, and cypress must be finished to protect against early deterioration. Any disruption of the finish can lead to rotting and vermin infestation. Frequent inspection and touch-up of finishes are the best preventative.

Wood repairs usually involve replacing the damaged member, but occasionally minor damage can be repaired with wood fillers applied in thin layers to reduce shrinkage. If the structural integrity of the wood is reduced by damage, then replacement or "scabbing" (if not visually critical) is best to restore strength. Treated woods should always be replaced or repaired with like treated woods.

Windows

Windows are usually identified or described by function, for example, single-hung, double-hung, center-pivoted, projected, or sliding. Windows that operate have a frame and a sash, with the frame being the fixed portion that is built in or anchored to the wall. The sash is the fixed or operating frame containing the glazing, stops (or putty), and, occasionally, muntins (glass dividers). A fixed window has only the frame and glazing (maybe muntins) and is frequently referred to as a *fixed sash.*

Steel Windows

Plant buildings usually have windows of the steel pivoting or projected type, which are the simplest in operation and the easiest to install for remote operation. There is little that can go wrong with the functioning of these types of windows. Glass breakage is the major problem, and will be discussed under "Glazing." Rusted-out sills and hardware breakage (usually latches) are frequent problems with older windows. Although rusted-out sills can be repaired by removing the damaged part and welding in a new piece, it is sometimes difficult to find new sections to match the old. In this case replacement is advised. The cost difference between replacement and major repairs may be insignificant, depending on plant location. Replacement hardware may be unavailable locally for older windows, but companies such as Blain Window Hardware, Inc., stock many obsolete parts and can frequently supply replacement parts. It is advisable to send a sample for matching purposes.

Glazing

Replacing broken glass is a major maintenance cost for most plants. If breakage is in random locations, replacing with glass of the same kind is practical and generally the most economical. However, if breakage occurs frequently in one particular area, the plant engineer should consider using plastic glazing sheets for replacement.

Translucent fiberglass-reinforced plastic (FRP), acrylic, and polycarbonate sheets (clear, tinted, and opaque) make excellent replacements for glass in problem areas, but a few precautions are in order. If the existing sash has a glazing leg (recess) less than ½ in (1.27 cm) deep, only small panes of plastic glazing should be used since these products expand more than glass and require added leg space. Also, large pieces are flexible enough to blow out of frames with small glazing legs in high wind pressures; this feature may also be a security problem. The manufacturers can provide tables of recommended leg depths which can be used for pane selection or modifying the sash. In corrosive atmospheres, the plastic should be checked for chemical resistance to the agents present in the space.

Flat FRP panels are usually the most economical of the plastics, but are not available as clear sheets. Acrylic and polycarbonate sheets are closest to exact replacements for glass. Both have surfaces somewhat softer than glass and require care when cleaning to prevent scratching. In an area of blowing sand or other windblown abrasive, these products may become cloudy from the abrasion. The plant engineer must determine the problems this characteristic will cause in the specific locality.

Painting and Caulking

The most important aspects of routine window maintenance are painting and caulking (see Chap. 2-8). Steel and wood windows must be kept painted to prevent deterioration. All rust, dirt, and loose paint should be removed before painting, and all bare steel or wood should be primed. A good-quality exterior enamel gives good performance. Some of the water-based paints are formulated for this use and can help reduce peeling paint on wood surfaces.

Entrances

Entrances include doors, door frames, hardware and, sometimes, glazing. Entrances are necessary for circulation of people and materials throughout spaces requiring fire, security, privacy, or environmental separation. Entrance units are usually described by the function and material of the door, such as single-swing wood, double-acting steel, and sliding aluminum. Steel doors of many functions are widely used in modern plants. Older buildings may have wood units. Aluminum and stainless steel are usually used as decorative entrances, except in areas where their superior corrosion resistance is warranted.

Hardware

Entrance units are second only to floors when it comes to abuse. Hardware failure is a major source of problems because the hardware is the working part of the entrance system. The door leaf and frame are subjected to impact damage and dents and punctures. Hardware, however, wears out, gets out of adjustment, is overstressed, and is otherwise abused. Thus, hardware maintenance is important to the service life of the entrance unit. For the larger plant, a stock of items that need frequent replacement or repair, such as closers and locksets, is recommended. Obtaining replacement parts for older door hardware is often difficult. Frequent tightening of bolts and screws helps reduce hardware breakage. Modern heavy-duty hardware is durable and reliable if reasonably maintained.

When considering hardware replacement, it is important that new hardware functions match the existing units as nearly as possible. Butts (hinges) may be plain or ball-bearing type. Ball-bearing butts are usually selected for heavy-duty service and should be replaced with the same type. Panic devices, if UL-rated, should be replaced with UL-rated devices. Avoid installing door-stop devices any closer than 18 in (45.72 cm) (preferably 24 in, or 60.96 cm) to the butt edge of the door to prevent overstressing the butts.

Hardware on fire doors is particularly important. Only the whole entrance assembly is fire-rated, and not the individual item, such as door, frame, or lockset. Therefore, fire-door hardware must be properly maintained if it is to function as originally rated and installed.

Unit Replacements

When severe damage occurs at an entrance unit, doors and hardware are rather easily replaced. Frames, however, can be difficult. In plant construction, door frames are often steel, either rolled sheet-steel (hollow metal) or structural-steel channel sections with bar stops welded to the face. Both types of frames are often built in as walls are erected, and are almost impossible to remove without destroying the anchors (removing channel frames usually involves removing part of the wall). If the frames are clipped to the floor, the task is more difficult.

There is usually enough space at the jambs of hollow metal frames to slip in a hacksaw and cut the jamb anchors, generally three on each side of the opening. Frames are not usually anchored at the head in one- or two-door openings. If there are no floor clips, the frames can then be removed and repaired. If there are floor clips, the frame is usually cut out and replaced with a new one. The plant engineer should consider the replacement costs of severely damaged steel doors and frames as opposed to repairs. For standard sizes and configurations, replacement may be less costly than major repairs.

Miscellaneous Doors

Wood doors and frames are simple to repair, but are often replaced with steel which is more durable but also more difficult to repair. More complex types of doors, such as rolling steel slat doors, overhead acting doors, and most types of ornamental doors, are usually best repaired by service companies, particularly when such doors are equipped with power-actuated operators. Such operators may require sensitive adjustments for proper operation, and are easily damaged by improper maintenance. Problems related to these doors include malfunctioning relays, motor burnout, and broken springs.

Fire Doors

Older types of fire doors are often equipped with closing devices controlled by fusible links, which permit the door to be held in the open position and still be self-closing in a fire. If a link breaks for some reason, as they do on occasion, the closer functions and the door closes even though there is no fire. Too often, a wood block or other device is then used to hold the door open, preventing its usefulness as a fire door. Preferably, the broken link should be immediately replaced with a new one of like kind.

BIBLIOGRAPHY

Architectural and Engineering Concrete Masonry Details for Building Construction, Natural Concrete Masonry Association, 6845 Elin St., McLean, VA 22101, 1967.

Blain Window Hardware, Inc., catalog, 1919 Blain Dr., Hagerstown, MD 21740.

Callender, John Hancock: *Time-Saver Standards for Architectural Design Data,* 6th ed., McGraw-Hill, New York, 1982.

Dalzell, J. Ralph: *Simplified Masonry Planning and Building,* McGraw-Hill, New York, 1955.

——— and G. Townsend: *Masonry Simplified,* 2 vols., American Technical Society, 848 East 58th Street, Chicago, 1957.

——— and ———: *Bricklaying Skill and Practice,* American Technical Society, Chicago, 1954.

Hornbostel, Caleb, and William J. Hornung: *Materials and Methods for Contemporary Construction,* 2d ed., Prentice-Hall, Englewood Cliffs, N.J., 1982.

Ramsey, C. G., and H. R. Sleeper: *Architectural Graphic Standards,* 7th ed., Wiley, New York, 1980.

Structural Clay Facing Tile Handbook, Facing Tile Institute, 333 North Michigan Avenue, Chicago, 1959.

Watson, Don A.: *Construction Materials and Processes,* 2d ed., McGraw-Hill, New York, 1978.

SPECIFICATIONS AND STANDARDS

"Abbreviations and Symbols Used in Builders' Hardware Schedules and Specifications," Builders' Hardware Manufacturers Association, 60 East 42nd Street, New York, NY 10017, 1961.

ANSI A42.1, "American Standard Specifications for Gypsum Plastering," American National Standards Institute, 1430 Broadway, New York, NY 10018, 1955.

ANSI A42.4, "American Standard Specifications for Interior Lathing and Furring," American National Standards Institute, New York, 1955.

ANSI A42.2 and A42.3, "American Standard Specifications for Portland Cement Stucco and Portland Cement Plastering," American National Standards Institute, New York, 1946.

"Guide to Portland Cement Plastering," Committee 524, American Concrete Institute, P.O. Box 4754, Bedford Station, Detroit, MI 48217.

ASTM Standards in Building Codes, latest ed., American Society for Testing and Materials, 1916 Race St., Philadelphia, PA 19103.

"Basic Builders' Hardware," Builders' Hardware Manufacturers Association, New York, 1969.

"Hardware for Hospitals," Builders' Hardware Manufacturers Association, New York, 1965.

"Hardware for Labeled Fire Doors," Builders' Hardware Manufacturers Association, New York, 1970.

"Hardware for Schools," Builders' Hardware Manufacturers Association, New York, 1966.

"Nomenclature for Steel Doors and Frames," A123.1, American National Standards Institute, New York, 1967.

"Recommended Practice for Portland Cement Plastering," Committee 624, American Concrete Institute, Detroit.

"Specifications for Aluminum Windows," Architectural Aluminum Manufacturers' Association, 35 East Wacker Drive, Chicago, IL 60601, 1970.

"Standardization of Terms and Nomenclature of Keying," Builders' Hardware Manufacturers Association, New York, 1969.

chapter 2-6

Roofing

by
B. J. (Jack) Williams, Jr.,
President
Twin City Roofing Inc.
Wahpeton, North Dakota
Chairman, NRCA Specifications Review Committee
National Roofing Contractors Association
Oak Park, Illinois

GLOSSARY

Base felt Heavyweight roofing felt, generally coated with bitumen, and used as the first layer in the roofing membrane.

Base flashing Also called composition or membrane flashing. The waterproofing components of the roof system, where termination or projections occur, composed of roof membrane materials.

Blowing point (BP) The temperature of the residual bitumen at which air is introduced into the material during manufacturing. The blowing point controls softening temperature.

Built-up roofing (BUR) A roofing membrane composed of bitumens and felt, in which the membrane construction is "built-up" by the felt plies and interspersing bitumen layers. A viscoelastic membrane.

Cant strip Manufactured from wood or fibrous insulation and beveled at 45° angles, they reduce the bending radius of the base flashing and offer support to the flashing.

Counterflashing A formed sheet-metal strip flashing installed over the base flashing of the roofing membrane.

Elastomeric membrane Roof membranes manufactured from plastics, synthetic rubbers, or modified bitumens installed as a single-layer membrane and either attached or unattached to the roof insulation or deck. Unattached systems are ballasted with aggregates. These membranes exhibit elastic properties.

Equiviscous temperature (EVT) The temperature at which heated asphalt bitumen has a viscosity of 125 centistokes (cSt). This temperature [±25°F (14°C)] is the optimum spreading temperature of the particular bitumen.

Finishing felt A lighter-weight roofing felt, generally saturated with bitumen, manufactured from (1) asbestos paper, (2) organic wood fibers, or (3) glass fibers. Used in the construction of the top layers of the BUR membrane. The reinforcing component of the membrane. May be used with or without base felt.

Flash point (FP) The temperature at which bitumen ignites.

Inverted roof assemblies (IRA) Roof systems where the roof membrane is installed below the roof insulation.

k factor Insulative value. The time rate of heat flow through a homogeneous material under controlled temperature differences. Value is expressed in British thermal units per hour per square foot per degree difference per inch thickness.

Roofing assembly Those components of the roof structure including roof deck, vapor retarder, (if required) roof insulation, roof membrane, surfacing, and flashings.

Roofing bitumen A mineral substance derived from the residual fraction of petroleum or coal products, used as the cementing and waterproofing agent in the construction of BUR membranes.

Roof insulation Usually a board type, manufactured from various materials to be installed above the roof deck as a thermal barrier.

Roofing membrane In a BUR roof system, the layers of felt, interspersed with bitumen that compose the waterproofing film. In elastomeric systems, the film is composed of vinyl or synthetic rubbers.

Roofing system All components of the roofing assembly above the roof deck.

Roof surfacing In BUR roofing, a layer of waterproofing bitumen installed on the top surface of the membrane as a wearing surface. Additional protection may be provided by the addition of aggregates. In elastomeric systems, the ballasting aggregate.

Softening point (SP) The temperature, determined by laboratory test, at which bitumen starts to soften.

Vapor retarder A semi-impermeable membrane placed between roof insulation and roof deck to inhibit the flow of water vapor into the roofing system from within the building.

INTRODUCTION

The vast majority of nonresidential construction projects in the United States are constructed with low, sloped roof structures. These roofing assemblies provide the thermal barrier at the roof line and have the essential purpose of forming a waterproofing shelter for the occupants, the contents, and the other components of the building. The practice of architecture could be described as the "creation of an artificial environment within a harsh, natural physical environment." The geographical vastness of the United States creates a vast spectrum of physical environments with which the designer must deal. The artificial environment factors, fewer in number, also create degradation factors.

DEGRADATION FACTORS

Exterior

Sunlight

Portions of the sun's spectrum (infrared, ultraviolet) degrade roofing materials.

Water

The primary purpose of the roof structure is to repel water. Yet water is a universal solvent and when combined with other degradation factors such as sunlight, the effects are magnified.

Ice

Ice formation produces additional loading on the structure and increased deflections. In contact with the roofing system, it can produce thermally induced tensile loading of the roof system.

Snow

Snow produces additional loading on the structure which may exceed design load. Removal of snow can create roof damage.

Temperature Change

The roof, always exposed, is subjected to continuous temperature cycling. The components of the roofing system undergo varying thermal cycling with varying movement potential, and stress is created in the roofing membrane.

Wind

Uplift wind vectors created by the aerodynamic shape of the structure can exceed normal wind velocities. The roof structure must resist these forces.

Hail

Hail can create physical damage by puncture of the membrane.

Fire

The roof must resist the spread of fire. Some components of roofing systems are flammable.

Other

Earthquake and sonic forces can create stress in the roofing system. Airborne chemicals may affect roofing materials. Physical damage.

Interior

Water Vapor

Certain occupancies and processes create high relative humidity levels within the building. Water-vapor pressures can distort and damage roof assembly components, and reduce insulating values of the roof insulation.

Temperature Change

Negligible in most buildings, interior temperature change, when it occurs, can have drastic effects on the roof structure from thermal movements.

Fire

The roof structure must not contribute to the effects of an interior fire in the building.

DECK DESIGN AND CONSTRUCTION (Fig. 6-1)

Loading and Deflection

Roofing systems are dependent upon the structural integrity of the roof deck—the foundation of the roof assembly. Consideration must be given to normal design live and dead loads. In addition, the designer must consider the concentrated load of roof installation equipment. Deflection should be limited to $\frac{1}{240}$th of the span, considering either concentrated or uniform loading.

Expansion Joints

Expansion joints should always be provided through the roofing and roof deck for the following conditions:

1. Where steel framing, structural steel, or deck materials change direction
2. Where separate wings of L, U, T, or similar configurations exist
3. Where types of deck materials change—i.e., where precast concrete and steel decks abut
4. Wherever control, expansion, or contraction joints are provided in the structural system
5. Whenever additions are connected to existing buildings
6. At junctions where interior heating conditions change, i.e., a heated office abutting an unheated warehouse, etc.
7. Where movement between vertical walls and roof deck may occur

In the design and placement of expansion joints, the designer must consider the thermal movement characteristics of the structural roof deck, the roof insulation, the roofing membrane, and the climatic conditions to be encountered. Joints must be continuous and not be terminated short of the edge of the roof. Water drainage should never be attempted through or over the joint, and it must be so detailed and constructed to provide a minimum height of 8 in (20 cm) above the finished roof line.

Slope and Drainage

All roofs must be designed and constructed to drain properly. Ponding water is detrimental to roofing membranes and results in (1) deterioration of the surface and membrane, (2) debris accumulation, (3) deck deflections, (4) tensile failure due to freezing water, and (5) difficulties in repair should leaks occur.

Slope requirements must be considered according to type of deck, deflection design, drainage system, and building layout characteristics. For example, the accepted structural limitation for deflection of roof decks is $\frac{1}{240}$th of the span. Allowable deflection of

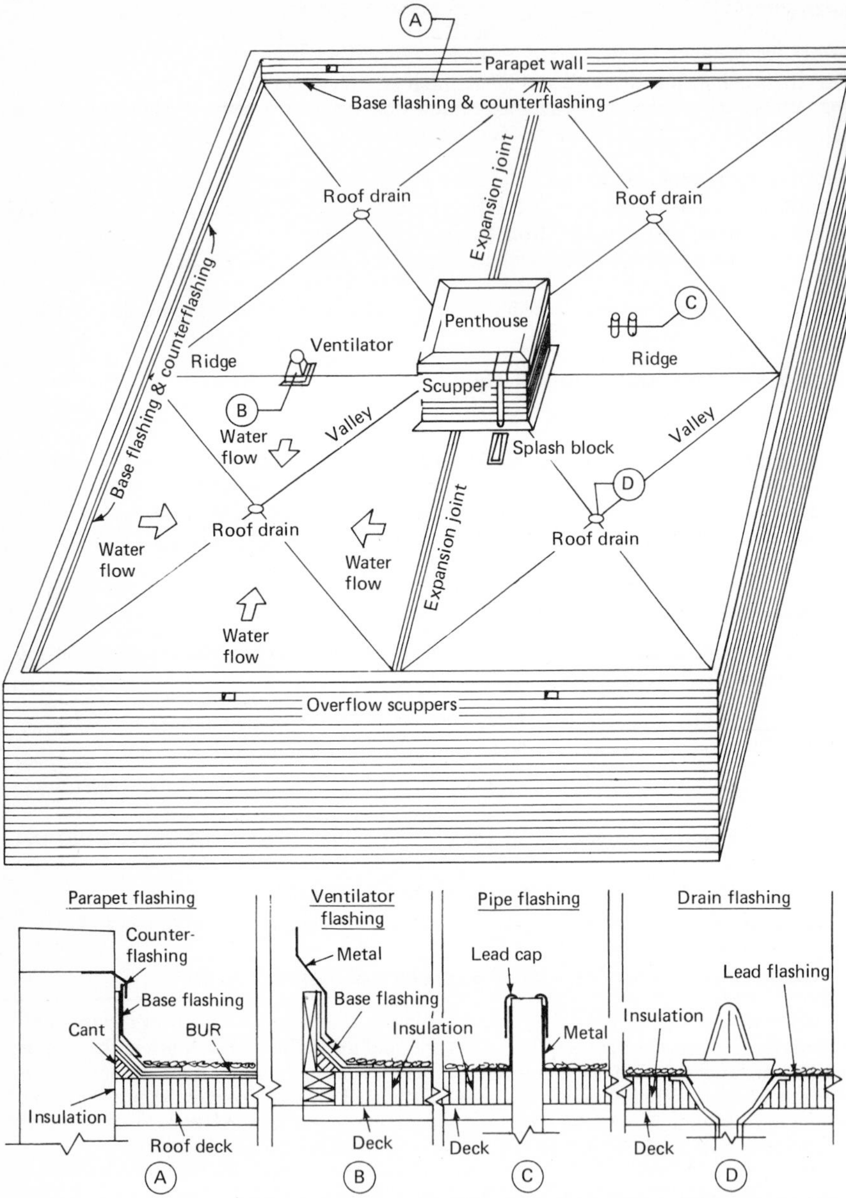

Figure 6-1 Typical roof arrangement (schematic).

a 50-ft (16-m) span would be 2.5 in (6 cm). Thus, the designer would have to provide 2.5 in (6 cm) of slope just to keep this deck level. After the deflection slopes are computed, the designer must provide additional slope for positive drainage, which will require ⅛ in (3 mm) per foot. Thus to drain the example 50-ft (16-m) span, the designer would have to provide 6.25 in (15 cm) positive drainage and 2.5 in (6 cm) deflection drainage for a total slope of 8.75 in (24 cm). The designer may provide these slopes in the structural deck or provide for them by use of tapered insulations or poured slope systems. Certain decks, such as precast concrete, have camber built into the member, which must be considered in the design of the slope system.

Deck-Mounted Equipment

All curbs, holes, and projections through the deck must be in place prior to roofing. All required wood blocking must be in place. Electrical conduit, other piping, or bolts must *not* be placed on the roof deck. All equipment must be mounted on curbs raised above the roof surface with sufficient room for roof installation beneath. Pitch boxes or pans are prohibited.

Types of Decks

Steel Decks

Deck material must not be lighter than 22 gage. Deck placement must be straight and securely welded. Insulation must be applied to steel decks prior to installation of membranes and should be applied in two layers. For the first layer, insulation thicknesses which will span the flutes of the metal deck should be selected. Mechanical fasteners should be used for attachment of the first layer of insulation. The second layer should be laid in solid moppings of hot asphalt with all joints offset from the bottom layer.

Wood and Plywood Decks

Wood decks must be dry when installed, and protection from moisture must be provided immediately. Lumber or plywood treated with oil-based preservatives should not be used. Attachment of the membrane directly to wood decks should be by approved mechanical fasteners only.

Lightweight Concrete Decks

These decks must have a minimum thickness of 2 in (5 cm), a minimum dry density of 22 lb/ft^3 ($35kg/m^3$), and a minimum compressive strength of 125 lb/in^2 (850,000 N/m^2). A venting provision, allowing latent moisture to escape, must be provided in the deck system. A coated base ply should be used as the first ply of a BUR membrane and attached by approved mechanical fasteners.

Thermosetting Insulating Fill Decks

Minimum compacted density should be no less than 18 (700,000) and no more than 22 lb/ft^3 (850,000 N/m^2). These insulative fills must be protected from moisture immediately, and roofing should be applied each day. The base ply is installed in a solid mopping of steep asphalt.

Gypsum Concrete (Poured)

The gypsum must be reinforced with wire mesh and have a minimum thickness of 2 in (5 cm). A minimum of ½ in (12 mm) of gypsum should be above the top of bulb tees. If insulations are used, they must be protected by vapor barriers. In direct application of the membrane, a coated base ply is mechanically attached with approved mechanical fasteners. Venting of the fill must be provided in the deck system.

Precast Plank

Precast plank is constructed from gypsum, lightweight concrete, or cementitious wood fiber and used as structural decking. Voids and joints over bulb tees must be grouted. Certain decks have restrictions on maximum interior moisture conditions; care is also required in storage and handling. In direct application of the membrane a base ply is mechanically attached with approved mechanical fasteners.

Precast or Prestressed Concrete

Surfaces must be smooth, level, and free of moisture for proper adhesion of the roofing system. Variations between units must be leveled prior to roofing. Wood blocking at edges, wood nailers for flashing attachment, and metal flashing reglets are required.

Reinforced Concrete

Poured surfaces must be smooth, level, and free of moisture for proper adhesion of the roofing system. Wood blocking and nailers must be provided at all perimeters and penetrations. The deck must be primed. If roof insulation is to be installed, either provision must be made for venting of latent moisture or a vapor barrier must be provided.

Insurance Requirements

A series of loss prevention data publications are issued by Underwriters Laboratories, Inc., and the Factory Mutual System. Consult current issues for the latest information on installation methods and approved materials.

Considerable impact has been made on roofing systems by these requirements; they are heavily weighted toward the prevention of insurance losses, but do not always result in quality roofing systems.

ROOFING SYSTEM DESIGN AND CONSTRUCTION

Vapor Retarders

Single-ply membranes, used to inhibit water-vapor infiltration of roof insulation and roofing systems, present extreme difficulties in lap and penetration sealing. They should be used in conjunction with insulation relief vents.

Vapor barriers are multi-ply, constructed similarly to roofing membranes, allowing no penetration of vapor. Vapor barriers should always be used over high-moisture-content areas such as swimming pools, showers, etc. They may be required over any structural deck containing moisture, if insulation is used, unless adequate venting provisions are made.

Vapor retarders or barriers should be compatible with other components of the roofing system. Materials such as vinyls, which can be melted by hot bitumen, must be avoided.

Insulation Selection

Roof insulation, the thermal barrier, also provides the surface substrate to which the roof membrane is applied. It must be compatible with, and provide support to, the membrane. A brief description of currently marketed roof insulation types and characteristics will aid the selection process:

Wood Fiber

Board type, manufactured from wood fiber. Organic, and subject to moisture deterioration. Flammable. A strong, dimensionally stable board with low coefficient of expansion. Average k factor.

Perlite

Board type, manufactured from inorganic expanded perlite. An organic binder is used which somewhat negates the inorganic advantages. Dimensionally stable with low coefficient of expansion. The board is fragile and easily damaged. Nonflammable. Average k factor.

Fibrous Glass

Board type, manufactured from inorganic fibrous glass with organic binder and facing sheet. Less dimensionally stable and higher coefficient of expansion than wood fiber or perlite. Strong. Above-average k factor. Flammable.

Urethane

Board type, manufactured by various processes from foamed polyurethane and skinned with organic sheet. Sometimes dimensionally unstable. Moisture-absorbent. Inorganic. Good strength properties. Very high coefficient of expansion. Flammable. Excellent k factor. Light in weight.

Combination Board

Board type, manufactured by laminating perlite and urethane boards. Dimensionally stable. Inorganic but skinned with organic sheet. Good strength properties. Although frequently specified in one-layer systems, the two-layer system should always be used with perlite layers on top and bottom. Flammable. Excellent k factor. Stabilized coefficient of expansion.

Beadboard, Polystyrene

Board type, manufactured from pressed polystyrene beads. Material has very low melting temperature and is not compatible with hot bitumen which melts it. Inorganic. Very high coefficient of expansion. Flammable. Excellent k factor. This board must always be used in conjunction with other insulation types and not with hot bitumen systems.

Styrofoam

Board type, manufactured from foamed polystyrene. Closed cell. Good strength properties. Flammable. High coefficient of expansion. Ultraviolet sensitive. Used primarily in construction of inverted membrane roofs. Excellent k factor.

Foamglass

Board type, manufactured from foamed glass. Average k factor. Less dimensionally stable than perlite or wood fiber and higher coefficient of expansion. Fragile and easily damaged. Closed cell. Excellent compressive strength. Nonflammable.

Poured-in-Place

Lightweight concretes produced at jobsite from expanded perlite or vermiculite ore. Insertion of beadboard polystyrene with venting slots increases insulative factors. May be sloped to drain easily. Nonflammable. Others are thermosetting fills produced at jobsite of expanded perlite ore and mixed with hot asphalt. Flammable.

Insulation Installation

Board-type thermal insulations should always be installed in two layers with all joints offset between the upper and lower layers. All roof insulations must be protected from the elements before, during, and after installation to prevent moisture entrapment. Roof insulation and roofing membrane should be installed perpendicular to the flow of water on the roof. Roofing membranes should not be installed directly to any foam type insulation. A separation layer of wood fiber, perlite, glass fiber insulation, or a venting base sheet should be attached to the foam insulation to which the membrane is installed.

Roof Membrane Selection

Low-slope roofing systems are each designed for a specific set of conditions of substrate, desired life, roof slope, and climatic conditions. The designer must weigh these conditions in relationship to the field conditions in determining the proper specification. The geographical location and climatic conditions often dictate the most suitable roof membrane.

In the past, the designer worked under the handicap of a lack of engineering data on existing and proposed membranes. Some recent technical studies have provided much-needed technical information for use in the selection process. In 1974, the National Bureau of Standards published the Building Science Series 55, *Preliminary Performance Criteria for Bituminous Membrane Roofing.* It proposed performance criteria for BUR membranes and selected these performance attributes: (1) tensile strength, (2) thermal expansion, (3) thermal shock factor, (4) flexural strength, (5) tensile fatigue, (6) flexural fatigue, (7) punching shear strength, (8) impact resistance, (9) wind resistance, (10) fire resistance. The designer is urged to study this information and understand the relationship to the membrane selection process. Other studies are being made with proposed performance criteria for elastomeric systems.

Roof Membrane Installation

BUR roofing felts should be laid in bitumen at the correct application temperature (EVT). The rates of bitumen application will vary according to EVT temperature, felt weights, brooming intensity, and felt materials, and will vary considerably. The application of bitumens at excessive application rates is detrimental to the membrane. It results in (1) slippage problems, (2) high coefficient of expansion properties, (3) *lower* tensile strengths, and (4) poor wetting and adhesion. The application of all plies, in shingle fashion, each day, is recommended. If not possible, all felts should be protected from moisture by glaze coating with bitumen. The membrane application should commence at the low points and be perpendicular to the flow of water. All felts should be broomed.

Roofing Bitumen Heating

Recently, an operating and application temperature concept for asphalt bitumens was introduced. Called equiviscous temperature (EVT), it is that application temperature (± 25°F or ±14°C tolerance) which results in optimum bonding, waterproofing, and spreading qualities. The EVT is variable for different asphalt grades and fluxes and is to be provided by each manufacturer of asphalt. Asphalt should not be heated at or above the flash point (FP) or the blowing point (BP). These high temperatures must not be maintained for more than 4 h. The EVT concept is endorsed by the Roofing Systems Technical Committee, a joint committee of the Asphalt Roofing Manufacturers Association and the National Roofing Contractors Association.

Coal-tar roofing bitumens are produced by fewer manufacturers and have fewer material variations. Manufacturers recommend heating temperatures of 425°F (218°C) with application temperatures from 325 to 400°F (163 to 204°C).

Roofing Bitumen Application

Interply moppings of bitumen should be thin and continuous and at the correct application temperature.

Surfacing and Aggregates

Roofing membranes require some type of wear surfacing material. Surfacing should be applied as soon as practicable after the membrane is installed. Gravel, slag, marble chips, etc., are used for aggregate-surfaced roofs. Liquid surface coatings are used for smooth-surfaced roofs.

Flashing and Cants

All vertical surfaces to which flashing is applied must have cant strips at the angle of the roof and the surface. A wood nailer must be provided at the top of the base flashing. The minimum height of the base flashing should be no less than 8 in (20 cm).

Metal flashing must be isolated from the roofing membrane and not connected. Flanged metal flashing such as gravel stops should be raised above the waterline by tapered cants and wood blocking. All walls and projections that receive base flashing should have metal counterflashing installed in the wall above and should be of two-piece design.

Inverted Roof Assembly (IRA)

See Glossary at the beginning of this chapter. Its advantage is the reduction of thermal movement of the BUR. The problems with IRA result from reduction of the insulative factors due to moisture infiltration at joints, restriction of water flow, ultraviolet sensitivity of the insulation, and floating and displacement of the insulation.

Elastomeric Systems

Unattached systems have the advantage that stress transfer between components cannot occur. The elastic properties of membranes are also beneficial. Degradation factors appear to be ultraviolet rays or ozone.

Weather Considerations

Roofing materials cannot be applied in cold weather unless correct bitumen-application temperatures can be maintained. The lack of cooling of the applied bitumen in warm weather can induce foaming, sticking, and susceptibility to tearing by equipment and foot traffic. Hot materials may be blown about by wind, creating a safety hazard; and "wind chill" must be considered in maintaining proper application temperatures. No roofing should be installed if precipitation is occurring or if moisture is present on the roof surface. When weather conditions preclude installation of a finished membrane, a temporary roof should be installed.

REROOFING

Each reroofing project has specific conditions that require individual assessment. The designer must study the failure and degradation factors of the existing system to properly prepare reroofing specifications. The imperative rule is that no moisture be trapped within the existing roofing system after installation of the new roofing. New roofing should never be adhered to old roofing with bitumen.

The upgrading of insulative factors should be considered when reroofing specifications are prepared. Attic or air-space fill insulations, added to low-slope structures, present extreme difficulties in design if condensation is to be avoided. When adding roof insulation, the designer must provide for additional wood blocking heights at perimeters and projections. The drainage provisions of the system must be carefully considered.

MAINTENANCE

All roofs require periodic maintenance for maximum roof life. Routine maintenance procedures that should be the responsibility of the owner are as follows.

1. Periodic removal of debris from the roof and roof drains
2. Immediate notification of the roofing contractor should leaks occur
3. Filing of all job records for future reference
4. Restriction of foot traffic on the roof and logging of all access to roof
5. In general, no attempt at roof repairs without consultation with roofing contractor or specialist

6. Periodic inspection from a qualified roofing contractor who should be asked for suggestions about in-house maintenance (generally, this will be cost-effective in the long term)
7. Performance of major maintenance and repairs by a qualified roofing contractor

Roofs do not age uniformly and certain high-wear areas of the roof may require repair. The equalization of these wear areas by repairs affords maximum roof life. The following guidelines apply.

1. Smooth-surfaced roofs may require periodic resurfacing with roof coatings. The coatings must be chemically compatible with the existing roofing materials.
2. Aggregate-surfaced roofs generally do not require resurfacing. Wind and water erosion may occur in certain areas, and resurfacing of these areas may be required to equalize wear.
3. The highest stress concentrations in the roofing system occur at the base flashing where connection is made to exterior walls or projections. Most roof leaks originate at these points, and periodic inspections should be made to detect the stress points. Reinforcement of these points will equalize wear.
4. Penetrations and projections through the roofing system are high-stress points. Inspections should include these areas, and repair may be required.
5. The roofing membrane is inherently fragile and susceptible to damage by nonroofers working on the roof surface. The detection of damage, especially on aggregate-surfaced roofs, is most difficult. The roofing contractor can suggest methods for avoiding damage when such work is required. Repairs must be made immediately should such damage occur.
6. In snow and ice areas, removal of deposits may create roof damage. A roofing contractor should be consulted before snow removal is attempted.
7. Any changes in building usage, occupancy, or relative humidity conditions should be reported to the roofing contractor.

BIBLIOGRAPHY

Mathey, Robert, and Cullen, William C.: *Preliminary Performance Criteria for Bituminous Membrane Roofing.* National Bureau of Standards, Building Science Series, No. 55, U.S. Department of Commerce, Washington, D.C., 1974.

Griffin, C. W., Jr. (for The American Institute of Architects): *Manual of Built-Up Roof Systems,* McGraw-Hill, New York, 1970.

National Roofing Contractors Association: *Roofing Manual,* Oak Park, Ill., 1981.

—— *The Built-Up Roof,* Oak Park, Ill., 1978.

—— *Energy Manual,* Oak Park, Ill., 1977.

—— *Proceedings of Symposium on Roofing Technology 1977,* Oak Park, Ill., 1978.

Sources of Additional Information

Factory Mutual Research Corp., 1151 Boston-Providence Turnpike, Norwood, MA 02062
Underwriters Laboratories, Inc., 207 E. Ohio St., Chicago, IL 60611.

chapter 2-7

Thermal Building Insulation

by
William H. Snyder
Director, Corporate Marketing Research
Johns-Manville Corp.
Denver, Colorado

GLOSSARY

Building insulation A product or system of products designed and installed in or at the building envelope to retard the flow of heat either out of the building (in winter) or into the building (in summer).

Thermal properties

***k* (thermal conductivity).** Amount of heat in British thermal units (kilojoules) transmitted per hour through a square foot of material one inch thick when the temperature difference between the two surfaces of the material is maintained at 1°F. Units: $(\text{Btu})(\text{in})/(\text{h})(\text{ft}^2)(°\text{F})$ $[\text{W}/(\text{m})(\text{K})]$.*

***C* (thermal conductance).** k/thickness Units: $(\text{Btu})/(\text{h})(\text{ft}^2)(°\text{F})$ $[\text{W}/(\text{m}^2)(\text{K})]$.

***R* (thermal resistance).** $1/C$ Units: $(\text{h})(\text{ft}^2)(°\text{F})/\text{Btu}$ $[(\text{K})(\text{m}^2)/\text{W}]$.

***U* (thermal transmittance).** The overall coefficient of heat transmission through a system of materials with possibly different cross sections (e.g., studs in walls) which considers the insulating value of air films at each surface:

$$U = 1/\Sigma R \qquad \text{Btu}/(\text{h})(\text{ft})^2(°\text{F})\ [\text{W}/(\text{m}^2)(\text{K})]$$

where ΣR is the sum of the R values of each material, cavities, and air surfaces, through the cross section.†

INTRODUCTION

Basic Theory

Heat is transmitted from a heat source to a colder surface or area in three modes:

1. **Radiation** This, the primary mode of heat transfer, operates via electromagnetic waves even in a vacuum (e.g., the sun heating the earth). The rate of heat flow is proportional to the fourth power of absolute temperature (R^4 or K^4) of the hot surface (K_H^4) minus K^4 for the cold surface (K_C^4); it is inversely proportional to the reflectivity of the surfaces and can be controlled by placing absorbing or reflective materials between the hot and cold surfaces.
2. **Conduction** This mode operates as atomic or molecular activity: the more densely packed the molecules the higher the conductivity (e.g., steel conducts heat more readily than air). Conduction can be controlled by the use of less conductive materials and a discontinuous structure (e.g., plastic or glass instead of steel and foam or fiber instead of solids).
3. **Convection** This mode operates by the natural flow of gases or liquids caused by the changes in density occurring with temperature differences (e.g., smoke and heated air from the fireplace are forced up the chimney by the denser cold air which displaces the lighter products). It can be controlled by limiting the size of spaces and hence the temperature differentials which would promote convection.

Therefore the total apparent conductivity k of a material as measured in tests is actually a complex function of the different forms of heat transmission described above. A graph of the components of thermal conductivity vs. density is shown in Fig. 7-1. Both the shape and position of the thermal conductivity curve will vary for different insulation materials, for different designs (e.g., fiber diameter, pore size, Freon, etc.) and for different mean test temperatures.

*SI conversions based on Standard for Metric Practice, ASTM E380-76 (also IEEE Standard 268-1976 and American National Standard Z210.1-1976).

†For systems of more than one type of cross section, weight each $1/\Sigma R$ by its proportion of the total area.

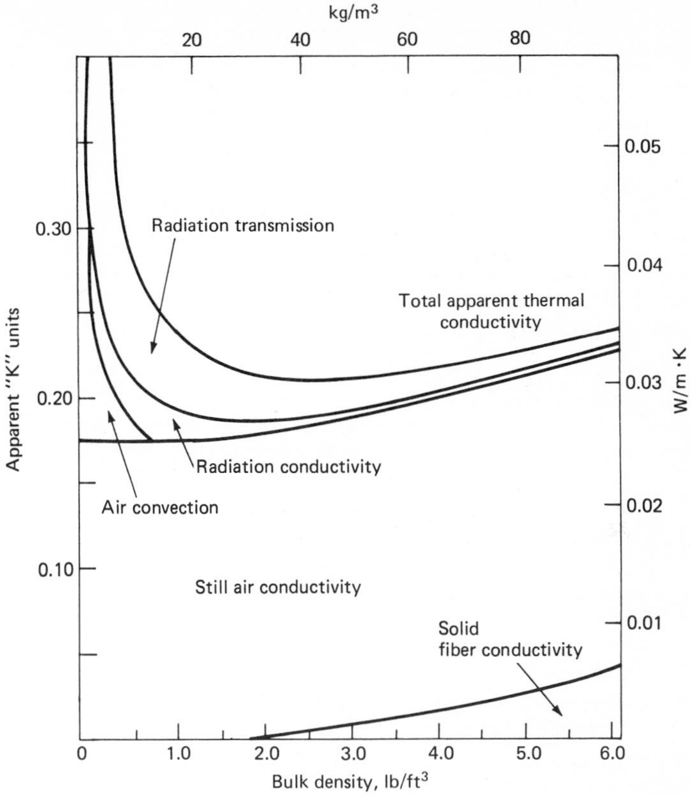

Figure 7-1 Graph of components of thermal conductivity vs. density for a typical glass fiber insulation. (*From C. M. Pelanne, "Heat Flow Principles in Thermal Insulation," Journal of Thermal Insulation, vol. 1, July 1971, Techonomic Publishing Co. Westport, Conn.*)

Amount of Insulation Required

The "right" amount of insulation can be defined as either the amount which meets standards and building codes or the economical amount. Since the oil embargo of 1973, the prices of energy have escalated rapidly, so the economical amount of insulation has become considerably more than previously accepted standards. Basically the right amount of insulation is that which minimizes the total costs over the life cycle on a present-value basis. Total costs are the sum of investment in insulation plus the energy cost to heat and cool the building.

Since insulation can be added in discrete increments (e.g., *R*-11, *R*-19, *R*-22, *R*-30, etc.) the minimum total cost can be achieved by going to successively higher levels of insulation as long as the ratio of discounted life-cycle *a*dded *s*avings to *a*dded *c*ost, AS/AC, exceeds 1.0. AC is the added cost in dollars per square foot to go to the next higher level of insulation. *Note:* In some cases the HVAC system can be down-sized because of the additional insulation, and this reduces the added effective cost of the insulation.

PWF is the present-worth factor for annual energy savings over the life cycle (no units) as shown in Table 7-1. (Note that this variable has a wide range.)

TABLE 7-1 Present-Worth Factors*†

D (discount rate) minus EI (energy inflation rate)	Life cycle L, yr 15	30	50
0.10 (or 10%)	8.0	10.2	10.9
0.05 (or 5%)	10.7	16.2	19.6
0 (equal)	15.0	30.0	50.0

*PWF = $[(1 + \text{EI})/(D + \text{EI})]\{1 - [(1 + \text{EI})/(1 + D)]^L\}$

†Table calculated based on EI = 0.10 (table accurate within 8 percent for EIs from 0.05 to 0.15).

In insulation calculations:

HES = heating energy savings for incremental insulation [dollars per square foot (square meter) per year]

EH = seasonal efficiency of heating system typically from 0.60 to 0.75 for gas and oil, approximately 1.0 for electric resistance and around 1.8 for heat pumps

CES = cooling energy savings for incremental insulation [dollars per square foot (square meter) per year]

EC = seasonal efficiency of cooling system (no units), typically around 2.0

Degree days = the sum of the daily average temperature differences from 65°F (60 or 55°F are sometimes used) (18.3°C, 15.6 or 12.8°C), counting only those days which are below the base, for an average year. These data are widely published.

\$/Btu = local energy cost per unit per British thermal unit of content per unit (dollars per joule).

The HES, heating energy savings, can be fairly easily calculated by using the equation:

$$\text{HES} = \Delta U \times \frac{\text{degree days}}{\text{year}} \times 24 \frac{\text{h}}{\text{day}} \times \frac{\$}{\text{Btu}} \qquad \text{(SI: J)}$$

where ΔU = the change in thermal transmittance from one level of insulation (or none) to the next higher level.*

The CES (cooling energy savings) is more complicated to calculate, and the reader is referred to more detailed references such as Ref. 2. Generally the cooling calculations do not contribute significantly to increased insulation unless the climatic conditions have low heating and high air-conditioning loads.

Example On a financial basis, what is the right amount of roof insulation for a metal roof deck without a ceiling (underside of roof exposed) considering heating only? (See Table 7-2.)

Solution†

Assumed for this example:

Degree days = 5500

\$/Btu = \$4.00/10^6 Btu (\$3.79/10^9 J)

*Handbooks are available which list thermal design values for various building materials and describe methods for calculating U's. (See Ref. 1.)

†Conversion factors to SI units:

C and U: watt per meter² per kelvin = 5.68 Btu/(h)(ft²)(°F)

R: kelvin·meter per watt = 0.176 (h)(ft²)(°F)/Btu

Savings: \$/meter²·year = 10.76 \$/(ft²)(year)

TABLE 7-2 Factors for Determining Amount of Roof Insulation

Calculating U's	Case 1	2	3	4	5	6	7	8
Amount of insulation $C = R$	None	0.15	0.13	0.10	0.08	0.07	0.06	0.05
R (resistance) values								
Outside air (15 mi/h)	0.17	0.17	0.17	0.17	0.17	0.17	0.17	0.17
Built-up roofing	0.33	0.33	0.33	0.33	0.33	0.33	0.33	0.33
Insulation	0	6.67	7.69	10.00	12.50	14.29	16.67	20.00
Metal deck	0	0	0	0	0	0	0	0
Structural beam	0	0	0	0	0	0	0	0
Inside surface (still air)	0.61	0.61	0.61	0.61	0.61	0.61	0.61	0.61
Total resistance (ΣR)	1.11	7.78	8.80	11.11	13.61	15.40	17.78	21.11
U Values								
Transmittance $U = 1/\Sigma R$	0.901	0.129	0.114	0.0900	0.0735	0.0649	0.0562	0.0474
ΔU	—	0.772	0.015	0.024	0.0165	0.0086	0.0087	0.0088
Calculating savings/cost ratio:								
HES = $\Delta U \times$ d.d. $\times$ 24 $\times$ \$/Btu [\$/(ft²)(yr)]	—	0.408	0.0079	0.0127	0.0087	0.0045	0.0046	0.0046
HES/EN, \$/(ft²)(yr)	—	0.583	0.0113	0.0181	0.0124	0.0064	0.0066	0.0066
AS = PWF(HES/EN), \$/(ft²)(yr)	—	9.45	0.183	0.293	0.201	0.104	0.107	0.107
AC, \$/ft²	—	0.834	0.051	0.116	0.125	0.090	0.119	0.167
Ratio AS/AC	—	11.3	3.6	2.5	1.6	1.2*	0.9	0.6

EH = 0.70

PWF = 16.2 (for discount rate = 15 percent, energy inflation = 10 percent and life = 30 years)

AC = \$0.50 installation cost plus \$0.05 per R of insulation per square foot (\$5.38 + \$3.05 per R per square meter)

Therefore the level of insulation which is the "best" from the financial standpoint is Case 6 from Table 7-2: $C = 0.07$ or $R = 14.29$ (SI: C = 0.40 or $R = 2.52$).

INSULATING MATERIALS

Uses

Insulating materials commonly used for the building envelope of industrial and commercial buildings differ by area of application shown in Table 7-3.

TABLE 7-3 Insulation for Building Envelopes

Area	Products used: New buildings	Retrofit
Roof-ceiling	Cellular plastics Perlite boards Fiberglass bds. Insulating concrete Cellulose (spray-on)	Cellular plastics Perlite boards Fiberglass bds. Insulating concrete Cellulose (spray-on)
Walls	Cellular plastics Fiberglass batts Rock wool batts Perlite Vermiculite	Cellular plastics Fiberglass batts Rock wool batts
Floors		
Wood-framed	Fiberglass batts Rock wool batts	Fiberglass batts Rock wool batts
Masonry	Cellular plastics	

TYPES OF INSULATING MATERIALS

Rock and Slag Wool

These are terms which denote a fibrous-type insulation produced by melting and fiberizing rock or the slag obtained as a by-product of the smelting of metallic ores. These products are available both in batt and loose-fill forms, which are principally used in residential applications.

Fiberglass

Fiberglass insulation wool is in many ways similar to rock and slag wool but utilizes a more refined process to produce glass, which is then fiberized. These products are available in batts, loose fill, and boards for both residential and industrial/commercial building envelopes.

Mineral Wool

This is a generic term which includes both fiberglass, rock, and slag wool.

Cellulose

Cellulose insulation is typically produced by converting used newsprint with the incorporation of flame-retardant chemicals to produce loose fill primarily for residential applications and spray-on forms for industrial and commercial applications.

Cellular Plastics

These are available in different compositions and forms. Polystyrene foam is used in the form of boards for industrial/commercial and residential applications. Polyurethane (and polyisocyanurate) foams are either boards or foamed-in-place primarily for industrial and commercial as well as some residential applications. Urea-formaldehyde foam is foamed in place and used almost entirely for residential applications.

Perlite

Perlite insulation is produced by expanding the naturally occurring perlite, a siliceous volcanic glass, into lightweight beads by rapid heating. The product is used to some extent as a loose fill but primarily manufactured with fibers to form roof insulation boards for industrial and commercial building.

Vermiculite

This type of insulation is produced by expanding the naturally occurring mica-like hydrated laminar mineral by heating it to a high temperature. It is used as loose fill, primarily in residential applications.

Insulating Concrete

This product is made by combining expanded perlite or expanded vermiculite with concrete or by adding a foaming agent to concrete. It is primarily used for industrial and commercial roof decks.

Comparative Properties

The properties of the various insulation products listed above are detailed in Table 7-4 (see also Ref. 2). Product information is available from individual manufacturers, and general information may be obtained from the following associations: Thermal Insulation Manufacturers Association, National Insulation Contractors Association, Mineral Insulation Manufacturers Association, National Cellulose Insulation Manufacturers Association, Perlite Institute, Vermiculite Association, Society of the Plastics Industry.

DESIGN AND INSTALLATION

Roof and Ceiling Insulation Assemblies

Insulation of roofs is normally the most cost-effective area of low-rise industrial/commercial buildings. (See also Chap. 2-6, "Roofing.")

Conventional Overdeck Insulation

This uses a board-type insulation (fiberglass, perlite, foam, or a combination product) which is applied between the roof deck and the built-up roofing. To obtain hourly fire-resistance ratings, it is necessary to use perlite, gypsum board, or insulating concrete. Attaining Class A, B, or C burning brand ratings on the exterior surface depends on the exterior surfacing materials used.

Insulation above Roof Membrane

This uses the same elements as the conventional system but with a moisture-resistant insulation (cellular plastic board) applied above the built-up roof and covered with a layer of crushed stone. Major advantages are improved membrane life through reduced temperature fluctuations and ability to add insulation to existing roofs of good condition (ability of roof to sustain added roof load must be checked).

Underdeck Insulation

This type is applied below the roof deck. Mineral fiber batts or board insulation can be used. Care must be taken to avoid condensation problems at the deck (either a good vapor barrier on warm side or insulation above deck combined with the underdeck insulation will be needed). Spray-on foams or cellulose systems or mineral wool systems could also be used, depending on fire-code and vapor-barrier requirements.

Wall Insulation

Cavity Walls

These consist of a structural masonry wall with rigid board insulation attached to the exterior before adding the face veneer (or brick, stone, block, stucco, etc.). Alternatively, the two masonry portions can be erected with a space for pourable insulation. Normally an air space is left between the board insulation and the face brick for passage of any wind-driven rain. Vapor barriers may or may not be necessary. Advantages of the cavity wall system are its high fire safety and thermal mass in the interior of the building. A disadvantage is the inability to add more insulation at a later time.

Interior Wall Insulation

These systems involve attaching a layer of board-type insulation to the interior of the load-bearing masonry wall and then adding an interior finish such as gypsum board. Alternatively, studs and batt-type insulation with gypsum board can be used. Advantages of this system include ease of construction and ability to use as a retrofit method on existing buildings. A disadvantage is the loss of the thermal mass wall on the interior side.

Insulated Frame Wall Insulation

Used on practically all residential buildings, it is also found on a significant portion of the small industrial and commercial buildings. Typically mineral-fiber batts are used to

TABLE 7-4 Properties of Insulating Materials*

Property	Fiberglass	Rock wool	Cellulose
Density, lb/ft^3	0.6–1.0	1.5–2.5	2.2–3.0
Conductivity, k	Varies by density	0.27–0.34	0.27–0.31
Thermal resistance, R	3.16R/in (batts) 2.2R/in (loose)	3.7–3.2R/in (batts) 2.9R/in (loose)	3.7–3.2 R/in
Water vapor permeability	> 100 perm-in	> 100 perm-in	High
Water absorption	< 1% by wt	2% by wt	5–20% by wt Not known
Capillarity	None	None	
Fire resistance	Noncombustible	Noncombustible	Combustible
Flame spread	15–20	15	15–40
Fuel contributed	5–15	0	0–40
Smoke developed	0–20	0	0–45
Toxicity	Some	None	Develops CO when burned
Aging effect on			
Dimensional stability	None (batt)	None (batt)	Settles 0–20%
Thermal performance	Settling (loose)	Settling (loose)	Not known
Fire resistance	None	None	Inconsistent data
Degradation due to temperature	None below 180°F	None	None
Cycling	None	None	Not known
Animal activity	None	None	Not known
Moisture	None	Transient	Not severe
Fungal and bacterial action	Does not promote growth	Does not support growth	May support growth
Weathering	None	None	Not known
Corrosiveness	Noncorrosive	None	May corrode steel, aluminum, copper
Odor	None	None	None

*(from Ref. 1): *SI conversion factors Density: kilogram per meter^3 = 1.602 pounds per foot^3; Conductivity: watt per meter kelvin = 0.144 × k^2; Thermal resistance: kelvin meter^2 per watt per meter = 6.94 × R/in; Permeability: kilogram per pascal second meter = 1.46 E − 12 × perm-in; Temperature: kelvin = (°F + 459.7)/1.8 = °C + 273.2

Polystyrene foam	Polyurethane foam	Perlite (loose fill)	Vermiculite	Insulating cement
0.8–2.0	2.0	2–11	4–10	12–88
0.20 (extruded) 0.23–0.26 (molded)	0.16–0.17 (unfaced & aged) 0.13–0.14 (impermeable skin)	0.27–0.40	0.33–0.41	1.17 @ 40 lb/ft^3 0.83 @ 25 lb/ft^3
5 R/in (extruded) 3.85–4.35 R/in (molded)	5.8–6.2R/in (unfaced & aged) 7.1–7.7 (impermeable skin)	2.5–3.7R/in	2.4–3.0R/in	0.85R/in @ 40 lb/ft^3 1.2R/in @ 25 lb/ft^3
0.6 perm-in (extruded) 1.2–3.0 perm-in (molded)	2–3 perm-in	High	High	Varies with density
0.02–4% by vol	Negligible	Low	None	NA†
None	None	NA†	None	None
Combustible	Combustible	Noncombustible	Noncombustible	Noncombustible
5–25	25–50	0	0	0
5–80	5–25	0	0	0
10–100	55–500	0	0	0
Develops CO when burned	Develops CO when burned	Not toxic	None	None
None	0–12% change	None	None	None
None	0.11 new to 0.17 aged	None	None	None
None	None	None	None	None
None below 165°F	None below 250°F	None below 1200°F	None below 1000°F	None below 1000°C
None	Not known	None	None	None
None	None	None	None	None
None	Limited data	None	None	None
Does not support growth	Does not promote growth	Does not promote growth	Does not promote growth	Does not support growth
Exposure to uv Light causes degradation	None	None	None	Frost damage < 30 lb/ft^3
None	None	None	None	None
None	None	None	None	None

insulate the wall cavities. Alternatively, the cavities can be filled with foam-in-place polyurethane or a blown-in loose-fill fiber insulation. Additionally, cellular plastic foam boards can be added as exterior sheathing for added insulation. Advantages of this type of wall are its simple construction and ability to reach very low heat transmittance values (*U*'s) with the combination of cavity and sheathing insulation. Disadvantages include limited fire resistance.

Sandwich Panels

These are prefabricated building components used as walls with the insulation either foamed in place or laminated between the facings of the panel. Usually, panels are of steel and aluminum and are ½ to 6 in (0.025 to 0.15 m) thick. Advantages include high thermal performance per unit of wall thickness, high strength-to-weight ratio, and elimination of thermal short circuits through framing members. Disadvantages include poor acoustics and limited fire resistance unless cores include gypsum or other mineral boards.

Foundation and Floor Insulation

Most industrial and commercial buildings utilize a slab on or below grade. Insulation is accomplished with cellular plastic foam boards either on the interior or exterior of the foundation extending at least 2 ft below grade. Insulation beneath the slab is not used unless the slab is heated directly. Retrofit insulation is sometimes possible on the exterior foundation walls. The foam must be covered by the earth or a barrier coating to prevent degradation by ultraviolet light. Calculations of heat transmission through floors and foundations are quite involved. (See Ref. 2.)

Insulation for Metal Buildings

These represent a significant special case, at least for their walls and roofs. Because of their unique construction, many of the major insulation manufacturers have designed and marketed special insulation systems for metal buildings. Special computer analyses are generally available from manufacturers to compute energy savings and optimal levels of insulation for specific applications.

STANDARDS AND CODES

ASHRAE 90-75

American Society of Heating, Refrigeration, and Air Conditioning Engineers Standard 90-75 has provided the basis for energy conservation requirements in many building codes. This standard is a component performance-type specification for new buildings based on steady-state conditions with empirical adjustments for air infiltration, solar gains, and shading. At the time of this writing ASHRAE had a series of proposed standards in the 100.X series covering energy conservation in existing buildings.

Model Codes

The U.S. Department of Energy sponsored development of energy conservation codes through the National Conference of States on Building Codes and Standards (NCSBCS) for the three model codes:

1. The Uniform Building Code, administered by the International Conference of Building Officials in Whittier, California.
2. The Basic Building Code, administered by the Building Officials and Code Administrators International, Inc., in Chicago, Illinois.

3. The Standard Building Code published by the Southern Building Code Congress International, Inc., in Birmingham, Alabama. Most of the states have adopted one of the model code's energy conservation requirements and a few states have written their own requirements.

In addition to the energy conservation requirements, the building codes influence the use or manner of use of many insulation materials through their fire and smoke requirements.

REFERENCES

1. For technical information: *ASHRAE Handbook—1977 Fundamentals,* American Society of Heating, Refrigeration and Air-Conditioning Engineers, Inc., New York, 1977.
2. For information on insulation materials and applications: *An Assessment of Thermal Insulation Materials and Systems for Building Applications,* prepared by Brookhaven National Laboratory for the U.S. Department of Energy, June 1978 (available from the Superintendent of Documents, U.S. Government Printing Office, Washington, D.C. 20420, as stock number 061-000-00094-1).

chapter 2-8

Construction Sealants

by
Richard D. Ramsey
Section Manager, Technical Services
Buildings Division
Sverdrup & Parcel and Associates, Inc.
St. Louis, Missouri

GLOSSARY

Adhesive (adhesion) The ability of the material to remain adhered to the substrate.

Caulking (calking) A puttylike material having little or no flexibility after drying, little adhesion to adjacent surfaces, and a low percentage of solids. For small, nonmoving joints (static) only.

Cohesive (cohesion) The ability of the material to "hang together" without a splitting or tearing failure of the material itself.

Gun A mechanical device, either manually or power operated, used to extrude construction sealant into a joint.

Joint filler A class of materials used to fill space in joints not occupied by construction sealants.

Primer A sealer for porous materials that is used before applying certain sealants, usually as recommended by the manufacturer.

Sealant Commonly referred to as *elastomeric,* a medium- to heavy-bodied material, poured or extruded into a joint, which cures to a flexible, tightly adhered seal, with a high percentage of solids, and little, if any, shrinkage; for moving joints (dynamic).

INTRODUCTION

Construction sealants is a generic term describing materials and products having the primary function of sealing a building against weather intrusion. These materials are used wherever construction tolerances require a space to permit installing glazing, door frames, and window frames, and at other miscellaneous penetrations through floors, walls, and roofs. A frequent location is at expansion and control joints, which are designed to permit movement of building components.

The sealant industry has developed numerous products which vary widely in composition and performance. The quality in each class of products discussed in this chapter varies from poor to excellent. The traditional caulking products (asphalt, tar, and oleoresinous compounds) are still available and in continuous use, but their application has been limited by newer products that perform better.

One of the results of having many products available is that compatibility between products as well as between products and adjacent materials becomes critical. Ignoring compatibility can lead to staining, splitting, and premature failure of joint seals. Compatibility is also critical in new joints as well as in resealing old joints. The manufacturer's advice is often necessary if intelligent repair work is to be done.

MATERIALS

Construction sealants are classified by generic type, with different grades available in some types. Since there is currently no industry standard controlling the manufacture and classification of sealants, manufacturers' terminology varies considerably, as does the performance of similar generic classes of sealants. The classification and performance data in this chapter should therefore be considered rule-of-thumb guidelines only.

The construction sealant used in many older structures was oleoresinous-type caulking (oil base), with comparatively poor performance characteristics; it was too often installed and forgotten, and was usually repaired or replaced only during painting. Reasonable performance was obtained from such caulking materials only if they were kept well-painted, but their usefulness was frequently gone after 2 or 3 years. Now the selection of caulking products includes not only good-quality oleoresinous compounds, but vinyl, butyl, and any of the one-part acrylic, urethane, and polysulfide products. If maintenance is expected to be infrequent (as in inaccessible locations), the better products should be used.

Although the Introduction covers a major part of construction-sealant use, other uses are of equal importance. Basically, these are in joints designed to control cracking from building movement, or dynamic joints.

All joints in a building are dynamic to a point, of course, but joints for construction tolerance (such as around doors and windows) move so little they are considered static for sealing purposes. Building expansion joints, however, may move from ⅛ to 1 in (0.318 to 2.54 cm) or more between seasons. Therefore, an elastomeric sealant must be used that has the elasticity, cohesion, and adhesion to keep the joint sealed during these cycles. Silicone and two-part polysulfides and urethane products are typical sealants of this type.

The successful use of elastomeric sealants is dependent upon proper joint design and cleaning, joint priming (when required), proper adhesion of the sealant to the substrate, and effective cohesion of the sealant. Failure of any of the foregoing will result in joint failure.

MAINTENANCE

Successful maintenance and repair of sealed joints is a simple three-step procedure—clean out the joint, correct the cause of failure, and reseal. It is never practical to either reseal over old material or to reseal without first determining why the joint failed.

Poor joint preparation is the most common cause of joint failure. With elastomeric sealants, a properly primed clean joint is necessary; otherwise failure will occur in the form of loss of adhesion to the substrate or as a bleeding stain from a contaminant. A

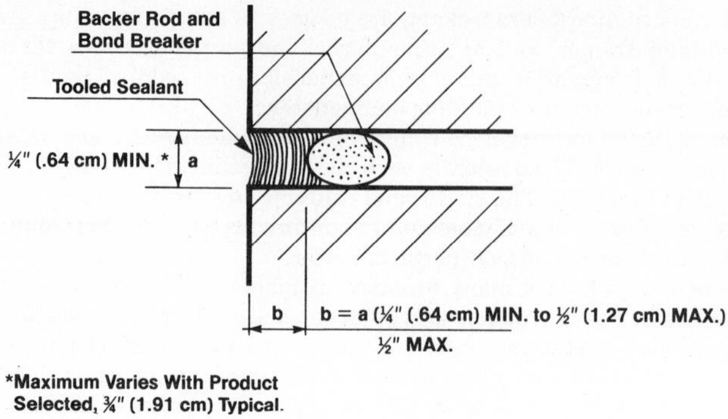

Figure 8-1 Typical sealant joint for polysulfide.

cohesion failure is recognizable by a split in the sealant and is usually the result of poor-quality material, poor joint design, or incorrect selection of material. Good joint design usually includes a bond breaker at the joint between the sealant and joint backup, so that adhesion is only to the sides of the joint.

Proper joint design varies with the type of sealant and must be verified with the manufacturer. Figure 8-1 is typical of the elements of good joint design for polysulfide sealants. The bond breaker and joint filler may be one material in some cases. The purpose of a joint filler is mainly to control the amount of sealant in the joint. In the case of sealants, *more* is ***not*** *better.*

Joints caulked with oleoresinous sealants are best repaired with the same or a similar material because it is difficult to clean such joints sufficiently to get good adhesion with other products.

BIBLIOGRAPHY

Callender, John Hancock: *Time-Saver Standards for Architectural Design,* 6th ed., McGraw-Hill, New York, 1982.

Hornbostel, Caleb, and William J. Hornung: *Materials and Methods for Contemporary Construction,* 2nd ed., Prentice-Hall, Englewood Cliffs, N.J., 1982.

Ramsey, C. G., and H. R. Sleeper: *Architectural Graphic Standards,* 7th ed., Wiley, New York, 1980.

Watson, Don A.: *Construction Materials and Processes,* 2d ed., McGraw-Hill, New York, 1978.

ASTM Standards in Building Codes, latest ed., American Society for Testing and Materials, published annually by ASTM, 1916 Race St., Philadelphia, PA 19103.

chapter 2-9

Roads and Parking Lots

by

Paul A. Johnson
Manager, Civil Engineering Section
Transportation and Public Works Division
Sverdrup & Parcel and Associates, Inc.
St. Louis, Missouri

DESIGN

Design Coordination

Roads and parking areas are service facilities which should not in any way inhibit plant production or detract from or interfere with the surrounding community. Design coordination is therefore the key to successfully designing new plant roads and parking areas or expanding existing facilities.

Design Data Required

A broad range of data is needed to design roads and parking lots so they both serve their intended functions efficiently and require minimal maintenance to keep them in good condition.

TABLE 9-1 Site and Plant Data

Data required	Possible sources
Topographic surveys of site and surrounding roads or streets showing contours, all surface improvements, and property boundaries	Plant manager, developer; obtain by field surveys or aerial mapping.
Location of existing utilities	Utility companies
Location of springs or swamps	Site inspection
Location of rock outcrops	Site inspection
Present legal access to property	Deeds, plats, tax assessor's office
Condition and general capacity of existing roads and streets surrounding property	Site inspection; state, county or municipal agency
Location and dimensions of plant building(s)	Plant layout, plant manager, plant designer
Location of shipping and receiving facilities. Rail? Truck? Number and type of trucks each day	Plant manager, plant designer, corporate planner
Plant employment and hours by shifts	Plant manager, corporate planner
Location of employee entrances	Plant layout
Visitor requirements, plant tours? Will buses be involved?	Plant manager, corporate planner
Plant security requirements	Plant manager
Aesthetic requirements	Plant manager

Site and Plant Data

Table 9-1 shows the site and plant data that should be acquired, and the appropriate information sources. Be sure to carefully check your sources to be certain the data are accurate since a small variation at this level can make a big difference in either the usefulness and efficiency of the resulting facilities or the costs to construct and maintain them.

Do not accept shortcut instructions which simply order you to place a road or parking lot in a given location. Table 9-1 is designed to provide information which confirms or denies the advisability of placing the facility on a given site, and may avoid considerable later expense and inconvenience. Carefully follow through on the items in Table 9-1 to assure yourself the points have been considered and all instructions are valid.

Regulation and Permit Data

Investigate federal, state, county, and municipal regulations to determine the requirements, what approvals and permits are required, and when submittals must be made. Agencies having jurisdiction might include (1) planning commissions, (2) local drainage districts, (3) environmental protection agencies, (4) departments of public works, (5) highway and street departments, and (6) zoning commissions.

Soils Data

The amount and type of soils data required will vary widely between sites. An appropriate sampling and testing program is required to develop the data upon which the foundation designs can be based.

Running the sampling and testing program, analyzing the test data, making recommendations for slopes and special subdrainage, and designing the base, pavement, and surfacing are geotechnical functions which should be done by an experienced geotechnical engineer. The novice should never attempt to develop this important and critical information.

Traffic Data

Traffic data are the determining factors in selecting road width, turning lane length, and total parking area, and in locating parking lot entrances and connections. The data given in Table 9-2 must be obtained or developed.

TABLE 9-2 Traffic Data

Data required	Possible source
Starting and quitting time for each shift at the plant	Plant manager, plant planner
Total employees each shift (maximum)	Plant manager, plant planner
Predicted employee traffic each shift	Plant manager, plant planner
Traffic counts on surrounding roads or streets and predicted future volumes	City, county or state traffic agency, actual field count
Locations of signals on surrounding roads and their probable capacity	City, county or state traffic agency, field check and field count
Predicted truck traffic to the plant, expected peak hours, probable loads and probable configurations (single unit, semitrailer, double trailer, etc.)	Plant manager, plant planner
Are mass transit facilities available and convenient to the plant, or are they planned?	Local transit agency, field inspection
Does the plant have or plan to have special programs to reduce employee traffic (plant bus service, or incentives for van car pools) and, if so, how many employees are expected to use these programs?	Plant manager, plant planner
Do railroad lines cross the roadways and, if so, what is the anticipated train schedule?	The railroad, plant manager, plant planner

Road Geometry

Site Location

The roadway location will be controlled by several factors, as follows:

1. Limits within which a permit can be obtained to connect to existing roads
2. Preferred intersection location within above limits
3. Location of employee, truck, and bus parking
4. Location of existing facilities which are to remain
5. Location of future planned facilities

These items show where the road must be connected and the areas to avoid. The remainder of the site offers possible road locations and must be further examined to determine the most feasible one. Further examination should include the location of springs, swamps, sinkholes, ravines, steep hillsides, rock outcrops, areas difficult to drain, and refuse disposal areas. Although roads can be built in locations with features like these, they will certainly increase costs, and such areas should be avoided if possible.

Widths

Roadway width design is a function of two factors, traffic volume and interferences.

Traffic volume for design is generally expressed in terms of design hour volume (DHV). The peak volumes for most plant roads will be of short duration and will desirably occur for about 30 min for incoming and for about 15 min for outgoing flows.

Interferences have a direct bearing on the capacity and service level of any given roadway width, and in many instances become the controlling factor in establishing width. Interferences include road intersections, at-grade railroad crossings, security gates, parking lot entrances and exits, and pedestrian crossings. Each interference must be considered in establishing the roadway width. Desirably, the design of each interference should be such that it does not control roadway capacity, but many times this is not possible or practical and the roadway width must be adjusted to minimize the interference. This can often be done by adding lanes near the interference without extending those lanes for the entire length of roadway.

For final design, the traffic analysis for roadway widths should be carefully completed, and an experienced traffic engineer should be used. The engineer will probably rely on procedures and analysical data contained in the *Highway Capacity Manual*[1] or the *Transportation and Traffic Engineering Handbook.*[2]

For preliminary approximations, the following rules of thumb may be appropriate, but they should never be used for final analyses:

1. On long stretches of road [over 1000 ft (305 m)] without interferences, each lane can carry 1000 to 1200 passenger cars per hour.
2. A single lane making an uninterrupted right-angle right turn will carry about 600 cars per hour.
3. Through lanes operating through a signal system with approximately equal cross traffic will carry about 600 cars per lane per hour.

Turning Lane Requirements

At intersections, the roadway capacity can often be improved by adding lanes reserved solely for turning vehicles. The required lengths of these added lanes are completely dependent upon the amount of cross traffic, whether the intersection is signalized, and (in the case of left turns) the amount of opposing traffic. The *Highway Capacity Manual*[1] or the *Transportation and Traffic Engineering Handbook*[2] should be used to determine these requirements.

Horizontal Alignment

The preferred horizontal alignment is, of course, a straight line; but when that is not possible the radius used for any given curve is dependent upon superelevation (banking) and design speed. Maximum superelevation for a plant road will normally be 0.06 ft/ft (6 cm/m), although 0.08 ft/ft (8 cm/m) might be appropriate on very long roads where icing conditions do not occur.

The minimum permissible radius can be calculated from the formula:

$$R = \frac{V^2}{15(e + f)}$$

where R is the minimum radius, V is the design speed, e is the maximum superelevation rate [normally 0.06 ft/ft (6 cm/m) for plant roads], and f is the friction factor (for paved roads use 0.17 for $V = 20$, 0.16 for $V = 30$, 0.15 for $V = 40$, 0.14 for $V = 50$, and 0.13 for $V = 60$).

Using the previously established design controls discussed in connection with "Site Location" and the above curve criteria, the designer can quickly establish a suitable horizontal alignment by trial and error.

Vertical Alignment

The road should be designed with a smooth grade line, with gradual changes conforming to the existing terrain. A line with numerous breaks and short lengths of grade should be avoided. Grades over 6 percent should be avoided, and the flattest grades consistent with the terrain should be used.

Vertical curves should be used to connect tangent grade lines, and the curves should be long enough to provide minimum stopping sight distance, comfort, and appearance. The curve connecting two grades should not be shorter than

$$L = KA$$

where L is the length of vertical curve in feet, A is the algebraic difference of the intersecting grades in percent, and the values of K are as follows:

	Minimum K value	
Design speed, mph (km/h)	Crest vertical curves	Sag vertical curves
30 (48)	28	35
40 (64)	55	55
50 (80)	85	75
60 (97)	160	105

Figure 9-1 illustrates both crest and sag vertical curves.

Where the roadway has curbs or curb and gutter, a minimum grade of 0.5 percent should be used to assure adequate drainage.

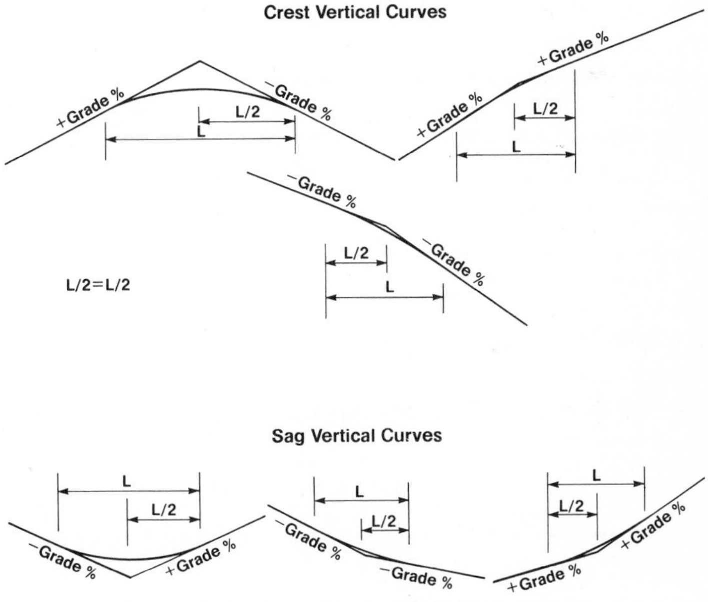

Figure 9-1 Typical vertical curves.

Parking Lot Geometry

Determining the Number of Lots and Location.

The number of parking lots to be constructed and their locations are a commonsense decision based on considering the following:

1. Available space
2. Proximity to facility served (employee entrance, truck dock, visitors' entrance)
3. Separation of cars, trucks, and buses if possible
4. Convenient entrances and exits

5. Number of each type of vehicle to be parked
6. Good drainage
7. Easy snow removal

Geometric Limitations

Parking lot geometry choices will be limited at most locations by the size and shape of the plant site, the location and layout of plant facilities, and the location of the plant road.

Efficient Layout

Capacity is the prime consideration for plant parking regardless of whether the lot will be used by employees, visitors, trucks, or buses. Circulation is also a very important factor for employee parking lots.

Each site must be considered on its own merits, and several preliminary layouts should be prepared to determine the most efficient mix of capacity and circulation.

The following general rules will usually result in the most efficient use of space.

1. Make the lot rectangular if possible.
2. Make the long sides of the area parallel.
3. Avoid irregular shapes.
4. Align traffic aisles parallel to the longest side of the lot.
5. The perimeter of the lot should be used for parking and not for traffic aisles.
6. Each traffic aisle should serve parking stalls on both sides of the aisle.

Determining Spaces

The following terms are generally used in describing the dimensions of parking lot units:

1. **Parking angle** The angle between the center line of the traffic aisle and the center line of the parking stall.
2. **Stall width** The width at a right angle to the parking stall.
3. **Curb length** The width of the stall parallel to the traffic aisle.
4. **Stall depth** The stall dimension at right angles to the traffic aisle.

The selected parking angle will affect the capacity and circulation within the lot. Maximum capacity will result from a 90° parking angle, but reduced angles are easier to enter and exit and, therefore, generally result in better circulation. Parking angles of 45, 60, and 90° are the most commonly used, and the angle should be selected on the basis of the best mix of capacity and circulation, determined by evaluating trial layouts. A common stall width is 8.5 ft (2.59 m), and the other design dimensions for that width for the 45, 60, and 90° parking angles are as follows:

Parking Angle	Curb Length	Stall Depth	Aisle Width
45°	12′0″ (3.66 m)	19′5″ (5.92 m)	13′6″ (4.11 m)
60°	9′10″ (3 m)	20′9″ (6.32 m)	18′6″ (5.64 m)
90°	8′6″ (2.6 m)	19′0″ (5.79 m)	25′0″ (7.62 m)

Some agencies having jurisdiction over parking lot designs require wider stalls, thus proportionately altering the above numbers. Design work should not be started until these width requirements are determined.

More detailed information on space design can be obtained from *Parking Principles*, Special Report 125,[3] or from the *Parking Design Manual*.[4]

The dimensions quoted and those in the suggested references are for passenger cars only and are still in common use, but the present trend to smaller, more maneuverable cars will doubtless result in smaller future space requirements. The same basic principles also apply for trucks and buses, but the dimensions for both stall and aisle widths must be adjusted to fit the equipment to be parked.

Entrances and Exits

Entrances and exits should be designed for free traffic flow and, therefore, the number of entrances and exits will be determined by a study of the lot size and the amount of interference by street traffic. Entrances and exits should not be located close to other intersections, and separate entrances and exits should be considered if the maximum width of driveway opening cannot exceed 30 ft (9.1 m). In general, more entrances and exits will increase the circulation and permit rapid loading and unloading of the lot.

Base, Pavement, and Surfacing

Data Required

Good soils sampling, testing, and analysis are essential in designing base, pavement, and surfacing, and a geotechnical engineer should be retained for this work.

Samples must be taken below the subgrade elevation to determine the soil-bearing capacity and water content.

In addition, data on the qualities of local aggregates, the projected traffic, and expected wheel loads are required.

Type and Design Selection

There are many types of base materials and asphalt and concrete pavements, all of which are suitable if properly designed for the soil and load conditions. Selection of type of base, pavement, and surfacing is therefore usually based on the availability and quality of local materials, the personal preference of the designer, cost or, sometimes, upon the appearance desired.

The design begins at subgrade, and the analysis and design of the load-carrying surface should be made by an experienced geotechnical engineer.

If sufficient soil data are known, preliminary "shotgun" design can be accomplished by the novice using design manuals published by the Asphalt Institute or the Portland Cement Association. These are available on request from local offices of either organization.

Asphalt pavements generally require seal coats, and it is important for parking lots to be sealed with materials that are resistant to gasoline and oil drippings. The Asphalt Institute manuals also cover seal coats.

Storm Drainage

Evaluating the Existing Drain System

The existing drain system will ultimately dictate the required discharge points for road and parking lot drainage, and it is therefore essential that complete data on the existing system be obtained.

Where there is no storm sewer system, the following data must be obtained:

1. Location and direction of flow of all ditches, gulleys, or streams.
2. Does the existing system pond or serve as a retention basin, and should it?
3. Do all parts of the existing system discharge from the plant property (are discharge points in some areas into sinkholes)?

Where there is an existing system, obtain the following data:

1. System location
2. The size and type of pipe in all segments of the system
3. Location, size, flow-line elevation, and top elevation of all manholes and inlets
4. The unused system capacity

Drainage Layout

A system of properly designed ditches and culverts (open system) will be the least expensive if sufficient area is available and the roads and parking lots do not have curbs. Make sure that all runoff is uninterrupted or is picked up by a culvert or ditch.

If a closed pipe system is to be used, some areas may require minor ditches to direct the flow to planned gutters or inlets which discharge to a pipe system. Inlets will be located at all gutter low points and at any planned low points in parking areas. Additional inlets will be located along the roadway to prevent excessive width of spread in the gutter flow. This spacing depends on the inlet capacity, the gutter grade, and the roadway cross slope.

Determining Runoff and Sizing the System

Almost every site is now under the jurisdiction of some local, state, or national agency with respect to drainage design, and these agencies specify the procedures to be used in determining runoff, inlet spacing, and system sizing. The procedures vary from agency to agency, and it is imperative the designer contact the appropriate agency and follow its procedures. In the absence of any jurisdictional agency, the procedures of the county or state highway department are suggested for use so consistency is developed within the area.

Signing

Signing on plant roads is often designed for aesthetics as well as to deliver messages and, therefore, may be very individual in character. General guidelines for signing principles are contained in The Federal Highway Administration's *Manual on Uniform Traffic Control Devices.*[5]

Signals

Signals for plant roads usually occur at the intersections with existing roads and are therefore usually designed, constructed, and maintained by the local or state road agency. The agency should be contacted in the early phases of the project to confirm this and to furnish design data. Signal designs are based on the procedures described in *Highway Capacity Manual,* Spec. Rep. 87.[1]

Lighting

There are many light posts and standards of varying designs to achieve certain aesthetic requirements, and these can only be selected by searching supplier catalogs.

Certain lighting levels should be attained by spacing the selected light standards correctly. Rely on a competent lighting engineer to provide this design. Basic lighting information is found in Ref. 6 and is included in Chap. 3-5, "Lighting."

MAINTENANCE AND REPAIR

Many maintenance and repair procedures are weather-sensitive and should not be scheduled during adverse conditions. Cold and, particularly, freezing weather adversely affect the placing of asphalts and concrete, embankment construction, and beds for pipe laying. In emergencies, special measures will permit this construction without reducing quality, but the costs will be high. All work should be suspended when it rains or the work is saturated or flooded. Use the standard specifications of the state highway agency as a guide to permissible working conditions.

Maintenance

Roadway and parking lot maintenance will consist of cleaning out the inlets on a regular schedule, sealing pavement cracks, sealing asphalt pavements, cleaning out ditches at regular intervals, removing snow, restriping, and replacing the luminaires in lighting standards.

Parking lot sizes and road length will dictate whether purchasing extensive maintenance equipment is warranted, whether the maintenance should be contracted for, or whether parts of the maintenance will be done with hand equipment. Each maintenance phase should be evaluated to determine the most cost-effective procedure.

Repairs

Repairs result from neglected maintenance and their extent will vary widely. Each repair should be evaluated separately to determine if it should be contracted or if it can be handled by plant forces.

REFERENCES

1. *Highway Capacity Manual,* Spec. Rep. 87, Highway Research Board of the National Academy of Sciences-National Research Council, publication 1328, 1965.
2. *Transportation and Traffic Engineering Handbook,* The Institute of Traffic Engineers, Prentice-Hall, Englewood Cliffs, N.J., 1976.
3. *Parking Principles,* Spec. Rep. 125, Highway Research Board of the National Academy of Sciences–National Academy of Engineering, 1971.
4. *Parking Design Manual,* Education Fund of the Parking and Highway Improvement Contractors Association, Inc., 1968.
5. *Manual on Uniform Traffic Control Devices for Streets and Highways,* U.S. Department of Transportation, Federal Highway Administration, 1978.
6. Illuminating Engineering Society: *IES Lighting Handbook,* 5th ed., IES, New York, 1972.

chapter 2-10

Building Systems

by
Donald H. Pratt
Executive Vice President and General Manager
Butler Manufacturing Company
Buildings Division
Kansas City, Missouri

GLOSSARY

Anchor bolt plan Shows size, location, and projection of all anchor bolts for the foundations of the metal building system's components. Sometimes includes column reactions and minimum base plate dimensions.

Base angle An angle secured to the foundation perimeter for support and closure of wall panels.

Bonded roof A written warranty with respect to weathertightness of a roof for a specified number of years.

Closure strip Resilient strip, formed by the contour of ribbed panels used to close openings created by joining metal panels and flashings.

Collateral loads All specified additional dead loads other than the metal building systems framing. Examples are sprinklers, mechanical and electrical systems, and ceilings.

Expansion joint A break or space in construction to accept thermal expansion and contraction of the structure's materials.

Fixed base A column base designed to resist rotation and horizontal or vertical movement.

Fascia A decorative trim or panel projecting from the face of a wall.

Filler strip See "Closure Strip."

Finial Gable closure at ridge.

Flashing A sheet-metal closure which functions primarily to provide weathertightness in a structure and secondarily to enhance appearance.

Gable A triangular portion of the end wall of a building directly under the sloping roof and above the eave line.

Gable roof A ridged roof that terminates in gables.

Girt A secondary horizontal structural member attached to sidewall or end-wall columns to which wall covering is attached and supported horizontally.

Haunch The deepened portion of a column or rafter, designed to accommodate the higher bending moments at such points. (Usually occurs at connection of column and rafter.)

Header A horizontal framing structural member over a door, window, or other framed opening.

Hip roof A roof which rises by inclined planes from all four sides of building. The line where two adjacent sloping sides of a roof meet is called the *hip*.

Knee The connecting area of a column and rafter of a structural frame such as a rigid frame.

Lean-to A structure such as a shed, having only one slope or pitch and depending upon another structure for partial support.

Liner panel A panel applied as an interior finish.

Mastic A type of caulking or sealant; normally used in sealing roof panel laps.

MBDA Metal Building Dealers Association.

MBMA Metal Building Manufacturers Association.

Peak The uppermost point of a gable.

Peak sign A sign attached to the peak of the building at the end wall showing the building manufacturer.

Prepainted coil Coil steel which received a paint coating prior to the forming operation.

Primary members The main load-carrying members of a structural system, including the columns, end-wall posts, rafters, or other main support members.

Purlin A secondary horizontal structural member attached to the primary frame which transfers the roof loads from the roof covering to the primary members.

Rake The intersection of the plane of the roof and the plane of the gable (as opposed to end walls meeting hip roofs).

Rake angle Angle fastened to purlins at rake for attachment of end-wall panels.

Rake trim A flashing designed to close the opening between the roof and end-wall panels.

Ridge Highest point on the roof of the building which describes a horizontal line running the length of the building.

Sandwich panel A panel assembly used as covering; consists of an insulating core material with inner and outer skins.

Secondary members Members which carry loads to the primary members. In metal building systems, this term includes purlins, girts, struts, diagonal bracing, wind bents, flange, and knee braces, headers, jambs, sag members, and other miscellaneous framing.

Side lap fastener A fastener used to connect panels together at the side lap.

Single slope A sloping roof with one surface. The slope is from one wall to the opposite wall of a rectangular building.

Siphon break A small groove to arrest the capillary action of two adjacent surfaces.

Soffit The underside covering of any exterior portion of a metal building system.

Tapered member A built-up plate member consisting of flanges welded to a variable depth web.

Thermal block A spacer of low-thermal-conductance material.

Truss A structure made up of three or more members, with each member designed to carry a tension or compression force. The entire structure in turn acts as a beam.

Valley gutter A channel used to carry off water from the "V" of roofs or multigabled buildings.

Wainscot Wall material used in the lower portion of a wall that is different from the material in the rest of the wall.

Wall covering The exterior wall skin consisting of panels or sheets and their attachments, trim fascia, and weather sealants.

INTRODUCTION

Their ability to control the cost and timetable of a typical building has earned pre-engineered metal building systems a prominent role in nonresidential construction. Applications commonly include factories, warehouses, utility structures, distribution centers, and related one- and two-story office buildings.

The term *building system* applies to a variety of construction products. However, *pre-engineered* building system refers to an entire method of building, a true alternative to more labor-intensive traditional materials and site-built techniques. A series of factory-produced structural members, wall panels, metal roof systems, related window and door units, and trim components constitute the pre-engineered approach. These components are predesigned and prefabricated for assembly into a building unit, an approach that permits full-scale testing and more definable quality control for the in-place construction.

Although the industry's basis is standardized components and modules, virtually any design can be accommodated by custom component combinations. Custom-fabricated components are produced to accomodate unusual loadings or dimensions. These normally incorporate standard components to preserve the pre-engineered benefits.

Pre-engineered building systems lose some cost effectiveness compared to conventional construction when (1) the building is designed with very short spans, (2) the building would house corrosive processes or other atmospheric conditions damaging to steel construction, or (3) the layout totally compromises the building system's framing modularity. Pre-engineered building systems deserve investigation in the preliminary planning of most projects.

GENERAL DESCRIPTION

Pre-engineered building systems include structural framing, roof, and walls. Some manufacturers also offer architectural fascia, vents, skylights and related accessories, paneling, and even interior components.

The main structural framing members, or *primary frames,* span the building's width.

They attach with anchor bolts to concrete footings to resist both vertical and horizontal thrusts produced by rigid-frame connections. The frames are generally welded construction using plate and bar stock having yield strengths up to 50,000 lb/in^2.

Types of Primary Frames

Primary frames exist in several types: single-span rigid frames, tapered beams, continuous-beam frames, single-span and continuous trusses, and lean-to, width extensions.

The specific application depends on an analysis of the intended layout, building end use, internally mounted equipment, and other related factors.

A single-span rigid frame (Fig. 10-1) eliminates interior columns for a clear-span interior. The earlier roof slopes of 3 to 12 in or 4 to 12 in are still used, but slopes of 1 to 12 in or ½ to 12 in are more common today. Standard offerings provide spans from 40 to 120 ft in 10-ft increments, although frames have been supplied over 200 ft of clear span on a custom-produced basis. Column designs vary from tapered to straight.

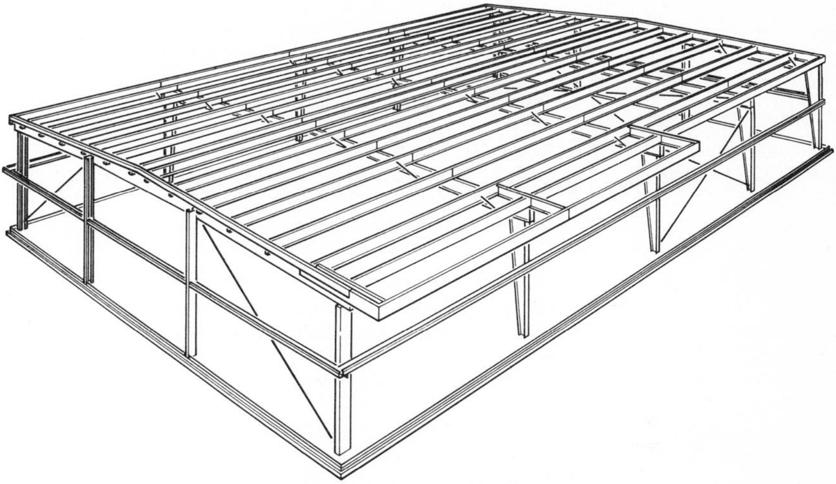

Figure 10-1 Single-span rigid frame.

Continuous-beam, or modular, rigid frames (Fig. 10-2) are used when interior columns are not objectionable occupancy factors. This frame employs column modules across the building's width in increments such as 40, 50, 60, or 75 ft. Standard buildings may be up to 300 ft wide with nonstandard modules and dimensions also available. The use of interior columns reduces the girder size and makes this frame more economical

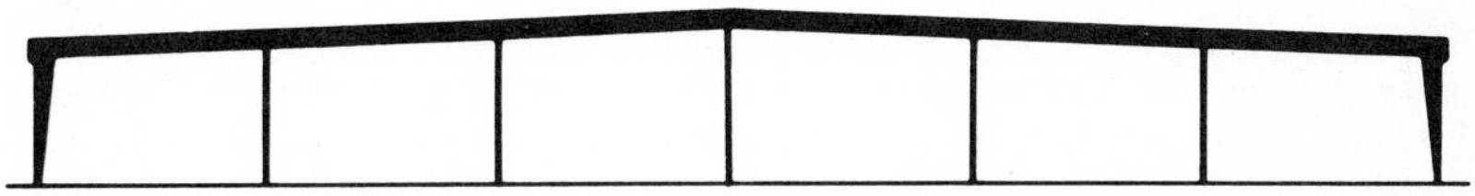

MODULAR RIGID FRAME

Roof Slope: ½" in 12"
Bay Lengths: 20', 25', 30'
Frame Modules: 40', 50', 60', 75'

Figure 10-2 Continuous beam/modular rigid frame.

than large, clear-span rigid frames. The exterior columns are again either tapered or straight.

Several variations of post-and-beam primary framing are offered. Generally these are used for buildings where deep-tapered columns would interfere with interior space utilization. They may be single- or double-sloped and can have clear spans up to 60 ft before encountering cost penalties. Bays between these frames are usually 20, 25, or 30 ft. Larger bays up to 50 ft are available when light trusses are used instead of cold-formed "Z" purlins.

Width extensions (Fig. 10-3) of lean-to design are advantageous for additions where vertical loads are the only consideration, since they derive lateral support from the adjacent structure. These adjoin either at the existing building's roofline or at a point below the eave height.

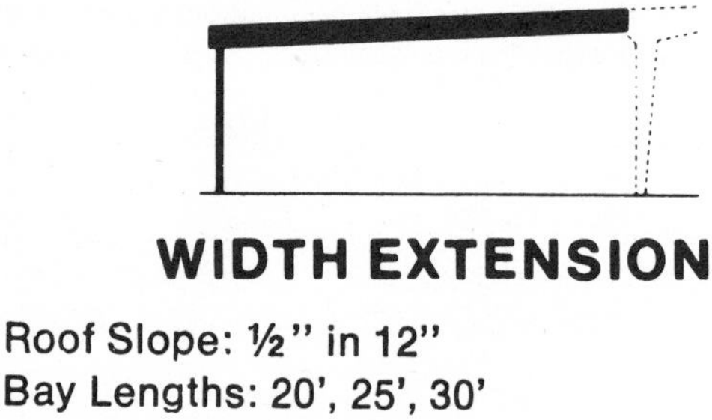

Figure 10-3 Width extension.

End-wall framing is independent of the interior primary frames. Some building systems eliminate the need for wall rod bracing by having end-wall posts fixed to the concrete foundation. This is especially true in buildings having large glass expanses, rail-car doors, or multiple sidewall openings.

If future expansion is anticipated, an interior primary frame can be substituted for the end-wall framing. Otherwise, diagonal brace rods normally are employed in end walls to resist wind loads on the roof and the sidewalls of the end bays. Diagonal brace rods are also installed in at least one interior bay, extending the building's width in the roof and on opposite sidewalls. This transfers any roof or end-wall wind load into the building's foundation. Additional flexibility is obtained by using fixed-base wind posts and/or portal frames.

Several manufacturers offer two-story building systems (Fig. 10-4) in their line. These utilize the same structural concepts as single-story post-and-beam systems and are applied to free-standing office buildings or buildings adjacent to plant or warehouse buildings.

Secondary Structural Members

These components (Fig. 10-1) span the bay between primary frames and support the roof and wall assemblies. The preferred roof and wall girts are cold-roll-formed sections having factory-punched fastener holes to assure proper framing alignment. The factory punching matches design dimensions and is an important element to ensure quality control during the erection process.

ROOF SYSTEMS

There are three types of roof panels: simple lapped panels, foam-insulated sandwich panels, and standing-seam panels.

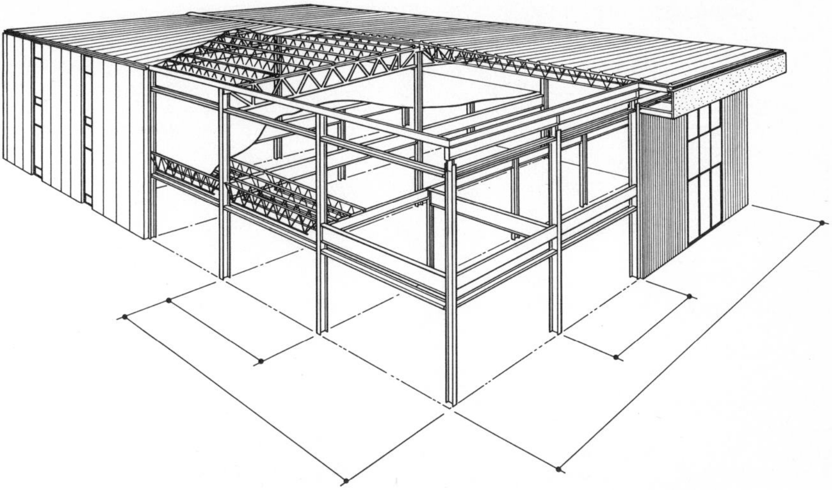

Figure 10-4 Two-story building system with component assemblies.

The simple panel (Fig. 10-5) lapped at sides and ends is also the most commonly used wall panel. Major corrugations are 12 in on center on a panel providing 36-in coverage in width and as much as 40 ft in length. Panels are often factory-punched to assure alignment and positive end-lap assembly.

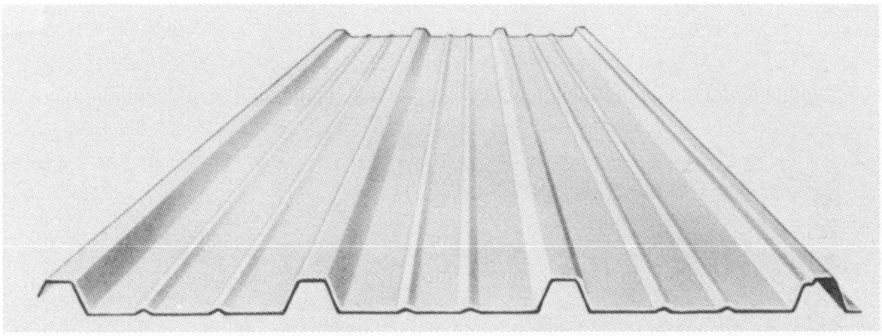

Figure 10-5 Simple lapped panel.

Slotted holes (Fig. 10-6) are available at end laps and allow panels to respond to temperature-induced expansion and contraction without building up stress. Steel panels can move ¾ in in 100 ft where a ±100°F (55°C) temperature differential occurs. Panels are attached to purlins with specialized sheet-metal screws or machine-driven blind rivets. Blanket insulation is installed between the purlins and roof panels.

Factory-insulated sandwich panels (Fig. 10-7) can consist of metal surfaces sandwiching a foam core with U values as low as 0.05. The liner panel, providing a finished interior appearance, ranks as a unique asset. Insulation performance is excellent, but core material may require sprinkler installation.

Standing-seam roofs (Fig. 10-8) minimize the use of fasteners across the roof surface. Most standing-seam systems employ concealed clips at the seams. This method permits using insulation blocks to separate the panel from the purlins, thus greatly diminishing the thermal loss experienced with simple panel roofs. The clip attachment method also permits some designs to accept thermal expansion and contraction. With this provision for thermal movement, buildings may be very wide. The watertightness of standing-seam roofs allows roof slopes as low as ¼ in in 12 in.

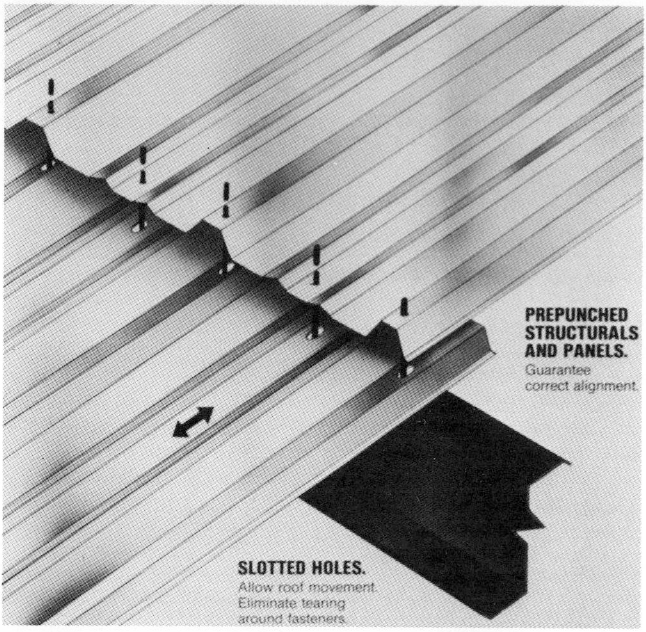

Figure 10-6 Slotted holes to allow panels to accept expansion and contraction at structural-to-panel connections.

Figure 10-7 Foam-insulated sandwich panel.

Roof panel materials include galvanized steel, painted galvanized steel, aluminum–zinc-alloy–coated steel, aluminum-coated steel, or aluminum. Choice of material depends on the particular panel type and its application; however, it should be matched to the environmental conditions.

Many roof systems have an Underwriters Laboratories Class 90 Wind Uplift rating, which is the highest rating available for any roof. The predictable integrity of metal roofs, especially certain standing-seam types, is a key reason why metal building systems are widely used.

Roof systems also include accessories such as ventilators and other provisions for roof openings. Roof-mounted mechanical equipment units are handily accommodated, although ground mounting is preferable engineering practice. Just as with other roofing materials, the proper installation of roof openings is a vital ingredient for roof weathertightness.

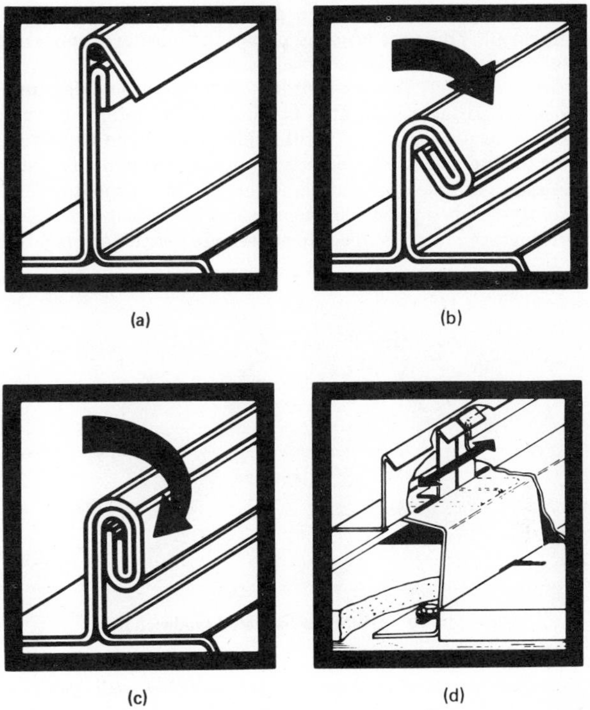

Figure 10-8 Standing-seam roof system.

WALL SYSTEMS

The most typical wall-system is the simple lapped panel (Fig. 10-5) with corrugations 12 in on-center and usually 1½ in deep. A girt at 7 or 8 ft provides support. Blanket insulation is installed between panels and girts.

Interior liner panels are installed on the inside flanges of girts for appearance, abuse resistance, and/or added thermal protection.

Some single-skin wall panels 16 in wide employ only inside fastening for a cleanly executed exterior appearance (Fig. 10-9). These panels also provide support and attachment surfaces for interior lining with Sheetrock gypsum board or other conventional

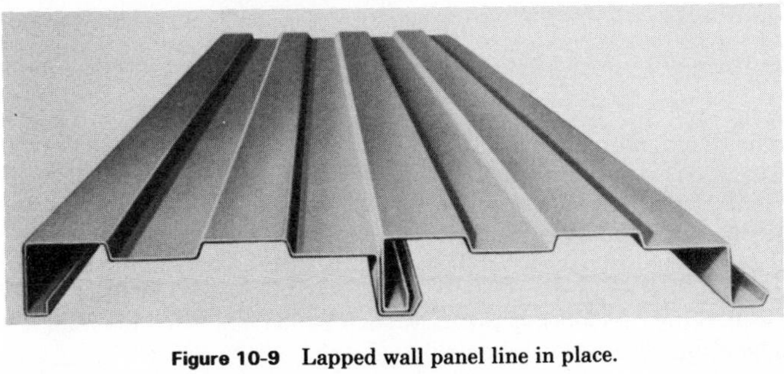

Figure 10-9 Lapped wall panel line in place.

materials. Insulation is placed inside each panel and, if desired, behind panels when using a furred liner wall. This type of panel proves the most economical for many applications with interior wallboard finishes.

Panels that are completely factory-finished, inside and outside, are also available. Some consist of metal surfaces separated by thermal insulators at vertical joints and filled with glass fiber insulation (Fig. 10-10). These panels can be installed quickly on buildings much as tongue-and-groove boards would be and eliminate exterior fasteners. Improved appearance and ease of maintenance are distinct benefits over simple lapped panels. Other factory-finished panels are foam-sandwich assemblies. These have excellent thermal performance but are subject to some code restrictions regarding sprinklers.

Figure 10-10 Fiberglass-insulated sandwich panel.

Some building systems manufacturers produce a window-wall system which includes extruded aluminum transitions specially designed for exact fit with a variety of wall-panel systems. The window-wall units are completely factory-assembled, including glazing, and are attached to building structural parts in the same manner as wall panels.

At least one manufacturer has a fully compatible prestressed-concrete wall panel for use on otherwise pre-engineered steel building systems (Fig. 10-11). This panel is available in a choice of textured finishes, and with varying cores of foam insulation in a sandwich construction. Structural framing members are specially sized and detailed for the panel's connection requirements. Concrete and masonry walls of various types find frequent use on metal buildings. Such a wall guards against abuse and helps meet covenants in certain industrial parks. The building girt is commonly used for wind-load transfer on these low masonry walls.

Figure 10-11 Factory-insulated prestressed-concrete wall panel component.

Each wall system includes accessories such as louvers, windows, and doors predesigned to be installed exactly and quickly.

FASCIA SYSTEMS

Fascia paneling systems are frequently used for a plant's adjacent offices. A fascia frame attaches to the building structure or exterior wall, depending on the manufacturer, and various paneling systems then attach to the frame. Most fascia systems extend above the roof in a manner similar to a parapet wall.

Regions of heavy snow require that any space behind the fascia be left open. This design avoids accumulations on the roof of ice and snow that could otherwise overload the building. Some fascia systems are made to be installed directly below the eave to avoid the snow-related risks of parapet types. This approach works well if the end walls are flat, as is the case of one structural system that uses a standing-seam roof exclusively.

Fascia panel materials range from simple metal wall panels to molded plastic shapes, stone aggregate panels, batten and panel systems, or conventional materials such as wood, plaster, etc.

SPECIAL REQUIREMENTS

Most manufacturers regularly produce specially dimensioned parts and structural frames for unusual loads, such as traveling cranes. Even crane rails have become part of the pre-engineered product line.

Large projects become likely candidates for treatment as custom-designed buildings because material use can be additionally optimized with computer techniques. Custom-designed buildings—such as single-slope buildings—that have been repeated frequently, often evolve into standard product offerings.

WHERE TO START

Any building is essentially designed around its owner's functional operations. It is therefore essential to identify both present and projected operational requirements before initiating the design process.

A good starting point is the floor plan. Nearly all manufacturers use a 5-ft planning module. The floor plan should outline the entire area to be covered by the proposed structure. Elements within it should identify the location of offices, machinery, pits, truck wells, loading docks, access points, traffic aisles, and the flow of raw materials being transformed into finished goods.

Equipment elevations should be indicated in a similar manner, along with the required access clearances. These considerations will then enable the contractor or designer to determine the size, type, and shape of the required building.

As the design process progresses, the engineer should determine what pre-engineered building systems provide the best cost-benefit value for the investment dollar. In many instances, insurance, operating cost, and maintenance savings may justify the specification of systems or additional insulation requiring a modest initial premium.

The directions for future expansion should be determined so that steps can be taken initially to economically provide this flexibility. Also, any planned adjustments in crane population or location, bus-duct routing, mezzanines, or other load-imposing criteria should be engineered at the outset, if possible, to avoid incurring modification costs later.

BIBLIOGRAPHY

"Metal Building Systems," Metal Building Dealers Association, Dayton, Ohio, and Metal Building Manufacturers Association, Cleveland, Ohio, 1980.

section 3

Using Electric Power

chapter 3-1

Power Distribution Systems

prepared by

National Electrical Contractors Association, Inc.
Washington, D.C.

GLOSSARY

Frequently used terms are defined here. More complete definitions are available in the latest edition of the ***National Electrical Code®***, published by the National Fire Protection Association.*

Ampacity Current-carrying capacity of electric conductors, expressed in amperes.

Ampere The unit of measure of electric current. Electric current is measured by the number of electrons that flow past a given point in a circuit in 1 s.

Branch Circuit The conductors between the final overcurrent device protecting the circuit and the outlets or utilization equipment.

Bus(es) Metal conductors, usually copper or aluminum, of large size utilized to transmit large blocks of power.

Capacitor A device capable of storing electric energy. It is basically constructed of two conductor materials separated by an insulator.

Circuit breaker A device designed to open and close a circuit manually, and to open the circuit automatically (trip) on a predetermined overcurrent without injury to itself when properly applied within its rating.

Connected load The sum of the continuous loads of the connected power-consuming apparatus.

Controller A device, or group of devices, that serves to govern, in some predetermined manner, the electric power delivered to the apparatus to which it is connected.

Current The movement of electrons through a conductor material.

Demand The peak rate at which energy is consumed, specified usually in kilowatts.

Electromagnet A magnet in which the magnetic field is produced by an electric current. A common form of electromagnet is a coil of wire wound on a laminated iron core, such as the potential element of a watthour meter.

Electromotive force (emf) The force which tends to produce an electric current in a circuit. The common unit of electromotive force is the volt.

Electron A negatively charged particle which revolves about the nucleus of an atom.

Equipment A general term including material, fittings, devices, appliances, fixtures, apparatus, and the like used as a part of, or in connection with, an electric installation.

Fault Any system problem, but usually a short between phase conductors or a short to ground.

Frequency The number of cycles of an alternating current completed in a certain period of time, usually 1 s.

Fuse A protective device made up of a conductor which melts and opens when the current through it is more than the ampere rating of the fuse.

Ground A conducting connection, either intentional or accidental, permitting current to flow between an electric circuit or equipment and the earth.

Ground-fault circuit interrupter A device whose function is to interrupt the electric circuit to the load when a fault current to ground exceeds some predetermined value that is less than that required to operate the overcurrent protective device of the supply circuit.

Impedance The total vector sum of resistance and reactance opposing current flow in an ac system.

Inductance The property of a coil or any part of a circuit which causes it to oppose any change in the value of the current flowing through it. The unit of measure of inductance is the henry (H).

**National Electical Code®* is a Registered Trademark of the National Fire Protection Association, Inc., Quincy, MA 02269.

Kilo A prefix meaning thousand, or 10^3.

Load The equipment or appliance which is operated by electric current. Also, the current drawn by such a device.

Load factor The ratio of average current to the maximum demanded.

Mega A prefix meaning million, or 10^6.

Ohm The unit of electric resistance. A circuit has a resistance of 1 Ω when 1 V applied to it produces a current of 1 A in the circuit (Ohm's law).

Outlet A point on the wiring system at which current is taken to supply utilization equipment.

Overcurrent Any current in excess of the ampacity of equipment or conductor. It may result from overload, short circuit, or ground fault.

Overload Operation of equipment in excess of normal, full-load rating or of a conductor in excess of rated ampacity.

Panelboard A single integral enclosed unit including cabinet buses and automatic overcurrent protective devices, with or without manual or automatic control devices, for the control of electric circuits; designed to be accessible only from the front.

Peak load The maximum demand on an electric system during any particular period. Units may be kilowatts or megawatts.

Power factor The relationship between the active power (watts) and the voltamperes in any particular ac circuit. It is defined as the ratio of the total active power to the total voltamperes. It is also numerically equal to the cosine of the angle of phase difference between the total circuit voltage and current.

Relay A switch operated by means of electromagnetism.

Resistance The tendency of a device or a circuit to oppose the movement of current through it. The unit of resistance is the ohm.

Switchboard An integrated, factory-coordinated combination of circuit protective devices, control devices, meters, relays, busbars, and wireways enclosed in a single, preplanned unit designed to be a self-contained center. Protective devices may be individually compartment-mounted (switchgear) or group-mounted in barriered compartments. The entire structure is designed and built to operate as a coordinated unit.

Switching device A device designed to close and/or open electric circuits either manually or automatically.

Transformer A device which transfers electric energy from one coil to another by means of electromagnetic induction.

Utilization equipment Equipment which converts electric energy to mechanical work, chemical energy, heat, or light, or performs similar conversions.

VAR The term commonly used for voltamperes reactive.

Volt The practical unit of electromotive force, or potential difference. One volt will cause 1 A to flow when impressed across a 1-Ω resistance.

Voltampere Voltamperes are the product of volts and the total current which flows because of the voltage. In dc circuits and ac circuits with unity power factor, the voltamperes and the watts are equal. In ac circuits at other than unity power factor, the voltamperes equal the square root of (watts squared plus reactive voltamperes squared).

Voltage (of a circuit) The measured potential difference between any two circuit conductors or any conductor and ground.

Watt The practical unit of active power which is defined as the rate at which energy is delivered to a circuit. It is the power expended when a current of 1 A flows through a resistance of 1 Ω. $P_w = I^2R$

Watthour The unit volume of electric energy which is expended in 1 h when the power is 1 W.

INTRODUCTION

Approximately 36 percent of all energy consumed in the United States each year is used by industry. About one-third of the energy is used in the form of electricity.

Some 80 percent of the electricity used by industry is applied for electric drives which are elements of electromechanical systems. These systems are used to *form* (extrude, roll, cast, press, and spin), *shape* (mill, ream, drill, hone, and tap), and *transport* (conveyors, elevators, brakes, fans, pumps, and compressors). The remaining 20 percent of industry's electrical use is applied for electrolytic processes, process heating, lighting, and comfort conditioning.

The larger a plant is, the more important is its electric distribution system. It must be capable of meeting the needs of all electric equipment, from the point of entrance of the power company's service (or the plant generating powerhouse) to the terminals of the utilization equipment. If electric power is not available when and where needed, the owner's investment in both plant and inventory becomes idled.

The electric distribution system which is best for a given plant depends on the value assigned to dependability and flexibility. For example, if electricity is needed to manufacture a product whose design is frequently changed, a flexible, easily changed system is best. When continuity of service is essential, as in some chemical processes when a batch could be ruined by a power failure, an extremely reliable system is best.

Flexibility and reliability are only two concerns affecting electric distribution systems. There are several others, not the least of which is cost, including initial cost and life-cycle cost.

The way in which an industrial power distribution system is designed, installed, and maintained has considerable influence on virtually all aspects of system performance. For optimum performance, those involved with operation of the electric distribution system require at least a basic understanding of the factors involved in the generation, transformation, distribution, and utilization of electricity.

Basics of Electricity

Direct Current

Direct current is current which flows through the circuit in the same direction at all times. A dc flow in which the level is always constant is often called a *continuous* current.

Direct current is used in some industrial electrolytic processes, in almost all vehicles, and for certain motors. However, most of the electric energy used in America is generated as alternating current.

Alternating Current

An *alternating current* is one which passes through a regular succession of changing positive and negative values by periodically reversing its direction of flow. Total positive and negative values of current are equal.

If a typical alternating voltage is plotted against time, it will resemble the curve shown in Fig. 1-1.

The curve is called a sine wave because it has the same shape as the curve described by the equation

$$e = E_{max} \sin \theta$$

where θ is an angle and e is the instantaneous voltage.

Figure 1-1 shows the variation in ac voltage through two cycles. Voltage is zero at the beginning of the cycle, rises to a maximum value, and then falls back to zero halfway through the cycle. In the second half of the cycle, the voltage achieves a maximum negative value and then returns to zero at the end of the cycle. The number of cycles which the voltage goes through in 1 s is called *frequency*. Frequency is expressed in *cycles per second*, also known as *hertz*. Most common ac power supplies in the United States have a frequency of 60 cycles per second (cps), or 60 hertz (Hz).

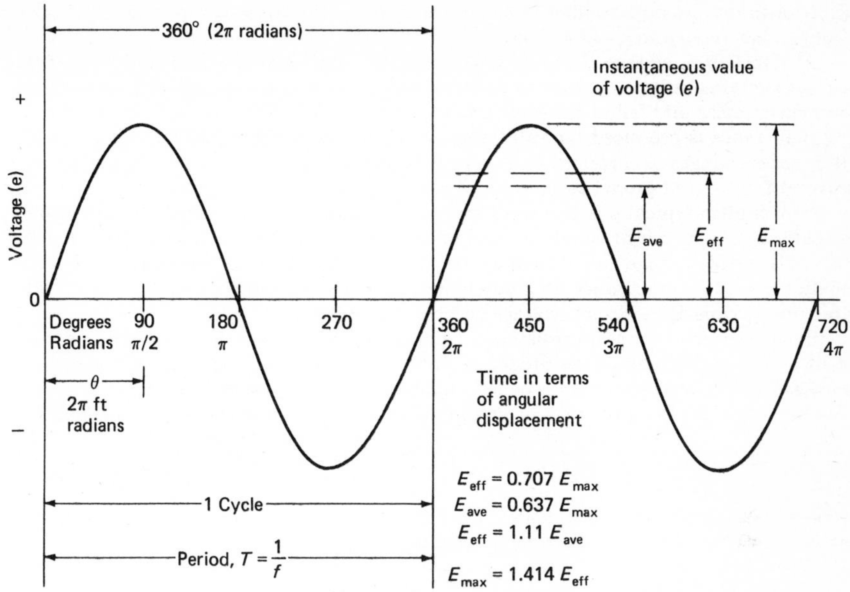

Figure 1-1 Sine-wave relationships.

Figure 1-1 shows that a cycle covers a definite period of time and is completed in 360° of the rotation of the armature of the generator. Accordingly, time may be expressed in terms of an angle of rotation:

$$\text{Period } T \text{ in seconds} = 360° \text{ or } 2\pi \text{ rad}$$

To express time, angle θ becomes

$$\theta = 2\pi ft \text{ rad}$$

where t is the time in seconds and f is frequency in hertz.

As such, the equation for the sine wave shown in Fig. 1-1 may be expressed as

$$e = E_{max} \sin 2\pi ft$$

or

$$e = E_{max} \sin \omega t$$

where $\omega = 2\pi f$ or

$$e = E_{max} \sin \theta$$

A voltage sine wave can be expressed in three values by:

1. Maximum or peak value E_{max}.
2. Average value E_{av}, which is equal to average value of e for the positive half or negative half of the cycle.
3. The root mean square (rms) or effective value (E_{eff}), a value of current which gives the same heating effect in a given resistor as the same value of direct current. Unless another description is specified, mention of alternating currents or voltages refer to the effective (rms) value.

Resistance is the only factor which opposes flow of current in a dc circuit.

Resistance also opposes flow in an ac circuit, but so do two other qualities: reactance from circuit *inductance* and *capacitance.*

As alternating voltage rises and falls, ac circuit amperage also rises and falls. If the circuit contains resistance only, voltage and current cycles are "in phase"; both voltage and current rise and fall at the same time, as shown in Fig. 1-2.

Inductance is produced typically when a coil of wire is connected into an ac circuit. It opposes changes in current flow as voltage changes during a cycle, causing voltage and current to be out of phase, as shown in Fig. 1-2.

Capacitance typically is produced when a capacitor or condenser is inserted into an ac circuit. A capacitor or condenser also opposes voltage changes produced by a generator. As a result, voltage and current go out of phase, but in a direction which is opposite to that caused by inductance. In a purely inductive circuit, voltage leads current by 90°. In a purely capacitive circuit, voltage lags current by 90°.

Since all circuits contain resistance, total opposition to flow of ac current is dependent on the vector sum of the resistance, inductive reactance, and capacitive reactance in the circuit. This vector sum is called *impedance,* and is measured in ohms.

Ohm's law can be applied to ac circuits by substituting impedance for resistance:

$$I = \frac{E}{Z}$$

where I is the current, E is in volts, and Z is in impedance in ohms. In dc circuits, the power in watts is simply voltage times current, $P = EI$.

When ac current is used to operate a magnetic device, current lags applied voltage creating an out-of-phase relationship.

In Fig. 1-3 for example, the current wave is said to be θ degrees out of phase with the voltage wave, because it lags the voltage wave by the angle θ. It reaches its peak value θ degrees of angular displacement after the voltage wave reaches its peak. Thus, $P = EI \cos \theta$.

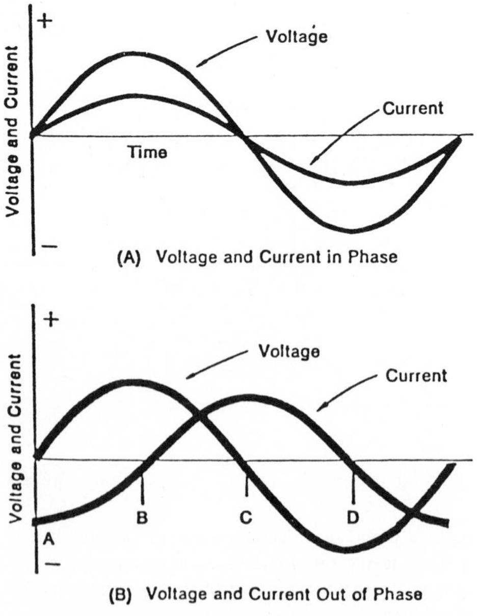

Figure 1-2 Voltage and current graphs.

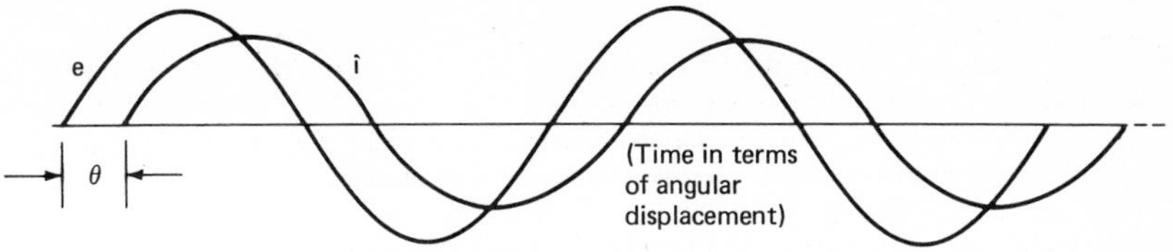

Figure 1-3 Current wave lagging voltage wave.

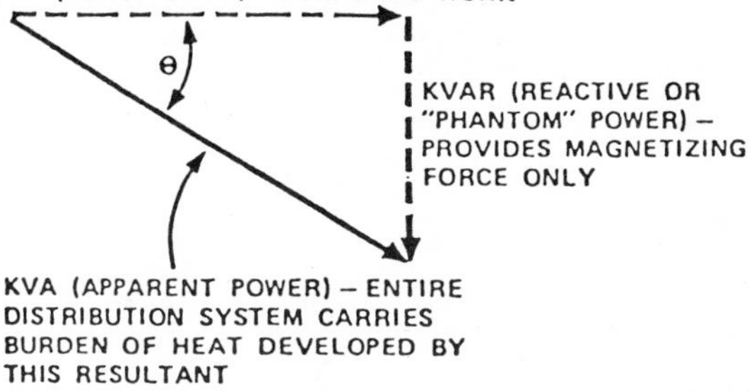

Figure 1-4 Power factor relationships.

The cosine of the angle between the current and the voltage is known as the *power factor*. It is a measure of reactive power which produces additional heat in electric system components, *but performs no real work*. This relationship is illustrated mathematically in Fig. 1-4.

Buying and Conserving Electric Energy

Purchased energy represents an increasing share of manufacturing costs. Most utilities use somewhat complicated methods to charge their customers for energy use. The basic charge is for actual kilowatthours used in a billing period. This charge is computed on a scale that recognizes blocks of energy use at different rates. The more energy used, the less is paid for each unit. This reflects the fact that overall utility company costs go down in proportion to the quantity of constant energy supplied by the system.

Most utilities also impose a charge for demand, expressed in kilowatts. This is a charge for maximum power requirements. It is determined by dividing energy consumed in a short (15- or 30-min) time period by the length of the time period. Thus, if 100 kWh is consumed in a 15-min demand interval, demand is 400 kW (100 kWh ÷ 0.25 h). Modern computer technology makes it possible to select the one demand interval of the billing period during which maximum wattage demand occurs.

A third charge sometimes imposed is a penalty for poor power factor. This charge is made to cover the cost of generating the unusable portion of electricity consumed in reactive power losses.

A fourth component of most utility bills, over which users have no control, is a *fuel adjustment charge*. This is added to cover the cost of fuel not included in the basic rate structure.

Careful examination of plant electric systems, utilizing a one-line diagram, can identify many steps to reduce energy consumption and electrical costs. Manual or automatic shutoff devices can be placed on short-term or rarely used equipment. Load-limiting and demand-limiting control devices can be installed. Power factor can be improved by the installation of capacitors or synchronous motors. Management can also investigate the feasibility of installing multifunction programmable controllers or centralized automated, computerized control to limit power and energy within the facility.

One-Line Diagram

There is no industrywide standard for the graphic description of electric systems, but most designers use a one-line diagram.

A one-line diagram of a plant's basic electric system should be developed and maintained. It can be used for system maintenance and as a management tool which can provide "instant" information on system loading and capacity.

Commonly used one-line diagram symbols are shown in Fig. 1-5; they are defined in IEEE Standard 315-1975, "Graphic Symbols for Electrical and Electronics Diagrams," ANSI Y32.2-1975.

The devices in switching equipment are referred to by numbers, according to the functions they perform. These numbers are based on a system which has been adopted

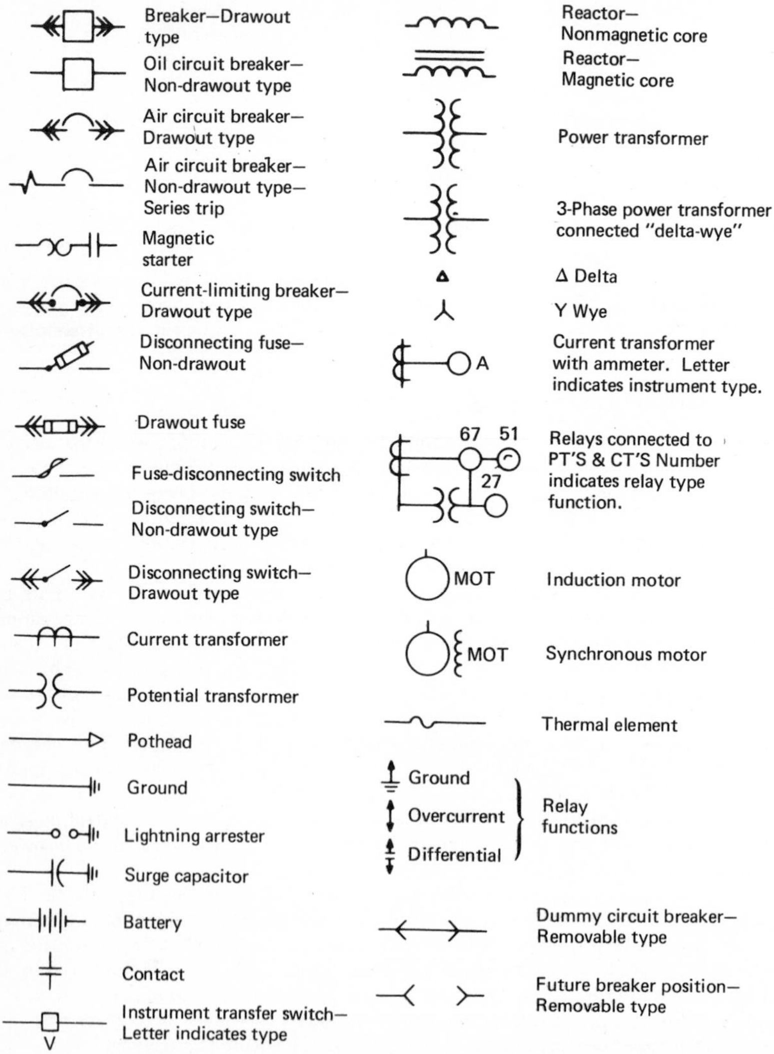

Figure 1-5 Commonly used symbols for one-line electrical diagrams.

as standard for automatic switchgear by IEEE. Numbers and their meanings are shown in Table 1-1.

A typical power-distribution diagram for an industrial facility is shown in Fig. 1-6. Note that the diagram is schematic. Locations shown refer to system relationships and not geography within the plant. One-line diagrams should be kept up to date as changes occur.

TABLE 1-1 IEEE Device Numbers and Functions

Device no.	Function	Device no.	Function
1	Master element	37	Undercurrent or underpower relay
2	Time-delay starting or closing relay	38	Bearing protective device
3	Checking or interlocking relay	39	(Reserved for future application)
4	Master contactor	40	Field relay
5	Stopping device	41	Field circuit breaker
6	Starting circuit breaker	42	Running circuit breaker
7	Anode circuit breaker	43	Manual transfer or selector device
8	Control power disconnecting device	44	Unit sequence starting relay
9	Reversing device	45	(Reserved for future application)
10	Unit sequence switch	46	Reverse-phase-balance current relay
11	(Reserved for future application)	47	Phase-sequence voltage relay
12	Overspeed device	48	Incomplete sequence relay
13	Synchronous-speed device	49	Machine or transformer thermal relay
14	Underspeed device	50	Instantaneous overcurrent or rate-of-rise relay
15	Speed or frequency matching device	51	AC time overcurrent relay
16	(Reserved for future application)	52	AC circuit breaker
17	Shunting or discharge switch	53	Exciter or dc generator relay
18	Accelerating to running transition contactor	54	High-speed dc circuit breaker
19	Starting to running transition contactor	55	Power-factor relay
20	Electrically operated valve	56	Field application relay
21	Distance relay	57	Short-circuiting or grounding device
22	Equalizer circuit breaker	58	Power rectifier misfire relay
23	Temperature control device	59	Overvoltage relay
24	(Reserved for future application)	60	Voltage balance relay
25	Synchronizing or synchronism check	61	Current balance relay
26	Apparatus thermal device	62	Time delay stopping or opening relay
27	Undervoltage relay	63	Liquid or gas pressure level or flow relay
28	(Reserved for future application)	64	Ground protective relay
29	Isolating contactor	65	Governor
30	Annunciator relay	66	Notching or jogging device
31	Separate excitation device	67	AC directional overcurrent relay
32	Directional power relay	68	Blocking relay
33	Position switch	69	Permissive control device
34	Motor-operated sequence switch		
35	Brush-operating or slip-ring short-circuiting device		
36	Polarity device		

TABLE 1-1 IEEE Device Numbers and Functions (*Continued*)

Device no.	Function	Device	Function
70	Electrically operated rheostat	85	Carrier or pilot-wire receiver relay
71	(Reserved for future application)		
72	DC circuit breaker	86	Locking-out relay
		87	Differential protective relay
73	Load resistor contactor		
74	Alarm relay	88	Auxiliary motor or motor generator
75	Position-changing mechanism		
		89	Line switch
76	DC overcurrent relay	90	Regulating device
77	Pulse transmitter		
78	Phase-angle measuring or out-of-step protective relay	91	Voltage directional relay
		92	Voltage and power directional relay
79	AC reclosing relay	93	Field-changing contactor
80	(Reserved for future application)		
81	Frequency relay	94	Tripping or trip-free relay
82	DC reclosing relay	95	(Reserved for special applications)
83	Automatic selective control or transfer relay	96	(Reserved for special applications)
		97	(Reserved for special applications)
84	Operating mechanism	98	(Reserved for special applications)
		99	(Reserved for special applications)

Source: ANSI C37.2-1970.

Standards

Standards relate to requirements, including agreed-upon definitions of terms, measurement and test procedures, and equipment dimensions and ratings. Some of the more commonly used standards relating to electric distribution systems follow.

National Electrical Manufacturers Association Standards

National Electrical Manufacturers Association (NEMA) standards establish dimensions, ratings, and performance requirements for electric equipment, regardless of manufacturer. NEMA standards are used widely in specifications.

National Fire Protection Association Standards Documents

National Fire Protection Association (NFPA) standards specify requirements for fire protection and safety.

*NFPA's National Electrical Code®**

The NFPA's ***National Electrical Code®*** **(NEC)*** is recognized as the minimum standard for the "practical safeguarding of persons and property from hazards arising from use of electricity." The **NEC®** may be supplemented by local requirements and ordinances pertaining to electric systems requiring more stringent safety measures. In general, the **NEC®** is nationally recognized. This Code is under constant review and is updated every 3 years. All plant engineers should have working knowledge of this Code.

Underwriters Laboratories, Inc. Standards

Underwriters Laboratories (UL), Inc., develops safety testing standards for electric equipment. Only equipment which complies with those requirements may be listed or labeled.

American National Standards Institute

The American National Standards Institute (ANSI) promotes, coordinates, and approves as American National Standards documents which have been prepared in

****National Electrical Code®*** is a Registered Trademark of the National Fire Protection Association, Inc., Quincy, MA 02269.

accord with ANSI regulations. American National Standards carry the identification numbers both of ANSI and originating organizations. The sponsoring organization is responsible for keeping a standard up-to-date.

Occupational Safety and Health Act

The Occupational Safety and Health Act (OSHA) requires that employers provide a safe and healthful work place for all employees.

Local Codes

Some large cities develop their own codes, but these generally are based on the ***National Electrical Code®.****

SYSTEM COMPONENTS

From point of service to point of use, electric power is directed, protected, and modified by segments of the system whose function, performance, and efficiency are vital to the utilizing equipment. To operate and maintain an industrial electric distribution system properly and to understand system planning fully, it is necessary to comprehend the function of each system component and its place in the overall system. Continued reference to Fig. 1-6, the one-line diagram, will aid in putting the various elements of the system into proper perspective.

Note that some system components have no normal operation and maintenance requirements. For such a device, it is necessary to have only a basic understanding of its appearance, construction, and function.

One such device is the *pothead*, used for connecting medium- and high-voltage cable systems to the internal distribution system. It is a terminator which permits termination of a complex cable at one end and attachment of cable lugs at the other. Its body usually is cast and can be filled with an insulating compound like petroleum jelly.

Outdoor high- and medium-voltage substations usually employ *oil-immersed circuit breakers* as a system protective device. These devices are equipped with direct-acting internal current transformers and trip coils enclosed in a jacket filled with a mineral oil. The oil jacket provides rapid cooling for contacts which become heated due to arcing action created by each switching operation. The oil-filled jacket is a better heat dissipator than air, so the device can be smaller than one which is air-cooled.

An oil-immersed circuit breaker must be located with care. The oil is flammable; a serious explosion or fire could result from a spark coming into contact with the cooling medium. In addition, in case of a leak a means for disposing oil, usually in the form of a gravel-filled drain, must be provided.

The oil must be examined periodically. Inspection ports are provided for this purpose. If the oil becomes contaminated, it must be replaced or drained and filtered.

Transformers

Transformers make possible the use of the high distribution and utilization voltages found in industrial electric systems. They are used to transform one primary voltage to a second primary level, to step from primary down to secondary voltage (not more than 600 V), and to step a secondary distribution voltage to a secondary utilization level.

The names used to categorize transformers relate to their applications:

General-Purpose Transformers. Dry-type units rated 600 V or less. They are used for local step-down from a secondary distribution voltage to a utilization level, serving

**National Electrical Code®* is a Registered Trademark of the National Fire Protection Association, Inc., Quincy, MA 02269.

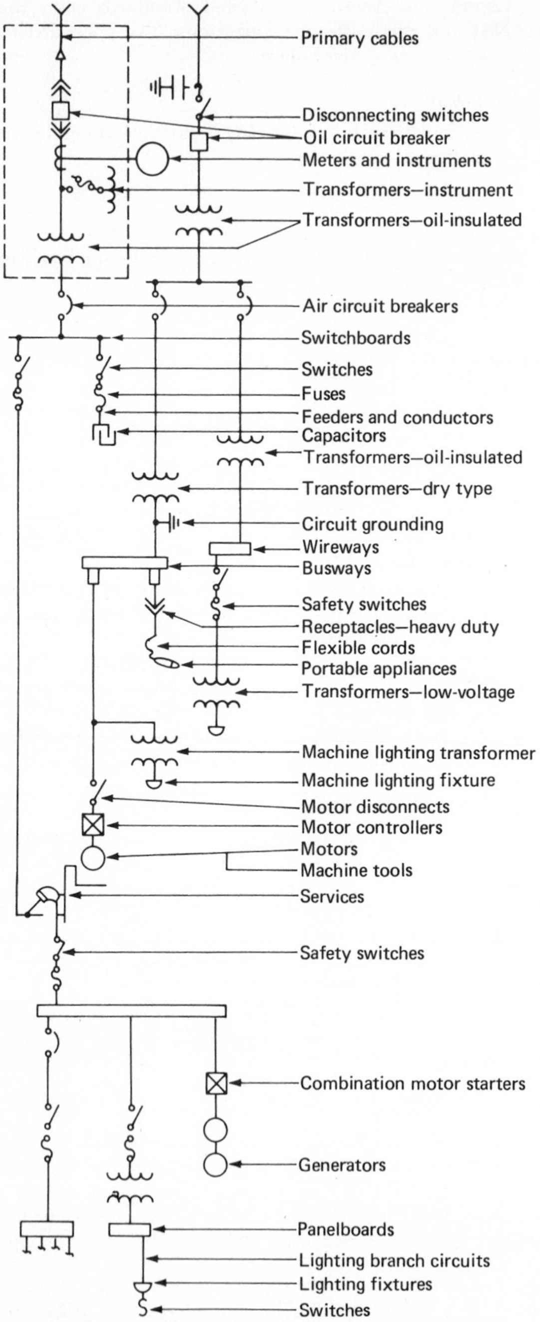
Primary cables
Disconnecting switches
Oil circuit breaker
Meters and instruments
Transformers—instrument
Transformers—oil-insulated
Air circuit breakers
Switchboards
Switches
Fuses
Feeders and conductors
Capacitors
Transformers—oil-insulated
Transformers—dry type
Circuit grounding
Wireways
Busways
Safety switches
Receptacles—heavy duty
Flexible cords
Portable appliances
Transformers—low-voltage
Machine lighting transformer
Machine lighting fixture
Motor disconnects
Motor controllers
Motors
Machine tools
Services
Safety switches
Combination motor starters
Generators
Panelboards
Lighting branch circuits
Lighting fixtures
Switches

Figure 1-6 Typical electrical distribution one-line diagram.

lighting and appliance loads. They also are called *general power* and *light* transformers or *lighting transformers*. Ceiling-suspended and floor-standing units are available.

Load-Center Transformers. Either dry-type or liquid-filled units, primary rated from 2400 to 15,000 V. They are used for both indoor and outdoor applications to step down to a voltage of 600 V or less. Load-center transformer units may be used separately, in combination with separate protective and switching devices and secondary distribution switchboards, or in combination with primary and secondary switching and protection in a packaged unit called a substation or load-center substation. They are base-mounted, free-standing units.

Distribution Transformers. Single-phase and three-phase oil-immersed, pole-mounted, or platform-mounted units. They are primary rated from 480 to 15,000 V, with step-down to a secondary level (or to a lower high-voltage level for units over 10-kV primary). Distribution transformers in capacities up to 167 kVA are used for pole-line distribution. Other outdoor wiring systems use platform units rated up to 500 kVA.

Substation Transformers. Oil-immersed units which are primary rated from 2400 to 67,000 V, with secondary ratings ranging from less than 600 to 15,000 V. Substation transformers are used in utility distribution and industrial substations. Power transformers are available for over 67 kV.

A transformer's nameplate indicates its kVA rating.

Transformer Ratings

Insulation classifications include liquid and dry types. Liquid insulation can be subclassified by type of liquid. Dry-type transformers can be subclassified as ventilated or sealed gas-filled.

kVA rating includes both self-cooling and forced-draft (fan-cooled) ratings for a specified temperature rise. The self-cooled rating should not exceed the peak demand by more than 25 percent to achieve the most efficient operation. A significant increase in transformer capacity can be obtained by the application of proper fan cooling. This permits transformers to be applied to present load conditions with an optimum load factor while still being able to serve expanded loads.

Transformers are designed to withstand short-duration overloads without any damage except a shortening of the useful life. The permissible overload and its duration vary with ambient temperature, preloading condition, and duration.

Transformer voltage ratings include primary and secondary continuous duty levels at specified frequencies, along with each winding's basic impulse level (BIL). The primary winding's continuous rating is the nominal line voltage of the system. The secondary voltage rating is the value under no-load conditions. Secondary voltage change experienced under load is termed *regulation*. It is a function of the system including the transformer impedance and the load power factor. Good regulation can often be achieved by tap adjustments on the primary side.

The *basic impulse level* rating for a transformer winding identifies the transient overvoltage withstand capability of its insulation.

Voltage Taps

Voltage taps are used either to compensate for small changes in primary supply to the transformer or to vary secondary voltage level with changed load requirements. A manually adjustable no-load tap changer is a common standard arrangement since tap changing under load is a highly complex problem.

Connections

Delta primary and wye secondary connections such as those shown in Fig. 1-7 are the most common transformer connections utilized today. In older plants, such variations as the grounded wye secondary and delta-delta connections to provide power to three-phase equipment are frequently encountered.

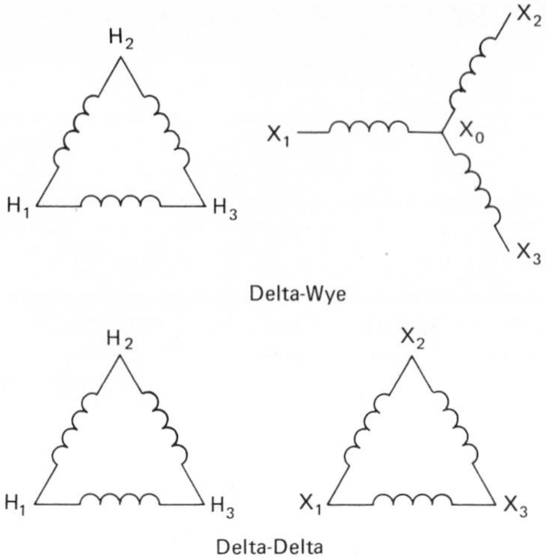

Figure 1-7 Three-phase transformer configurations.

Impedance

A transformer's impedance is its opposition to current flowing through it with the secondary short-circuited. Impedance is expressed as the percentage of normal-rated primary voltage which must be applied to cause full-rated load current to flow in the short-circuited secondary. For example, when a 480- to 120-V transformer is impedance-rated at 5 percent, it means that 5 percent of 480 V, or 24 V, must be applied to the primary to cause rated load current to flow in the shorted secondary. When 5 percent of primary voltage causes such current, 100 percent of primary voltage will cause 20 times full-load-rated secondary current to flow through a solid short-circuit on the secondary terminals.

The lower a transformer's impedance, the higher the short-circuit current it can deliver. The impedance values of general-purpose transformers generally range between 3 and 6 percent.

Circuit Breakers (600 V and Below)

Two types of circuit breakers are commonly used for applications at 600 V and less: *power circuit breakers* and *molded-case circuit breakers*.

Power circuit breakers are open-construction assemblies on metal frames for service in switchgear compartments and similar enclosures. Molded-case circuit breakers also are used in switchgear. While their protective value is high, they offer few control options.

Circuit breakers can be subcategorized in a variety of ways, based on different characteristics, as follows:

Adjustable. Indicates that the overcurrent device of the circuit breaker can be set to trip at different values of current and/or time within a predetermined range.

Instantaneous Trip. Indicates that no delay is purposely introduced in the tripping action of the circuit breaker. Automatic circuit breakers may have both time overcurrent and instantaneous tripping devices.

Inverse Time. Indicates that a delay in the tripping action of the circuit breaker is purposely introduced. The delay decreases as the magnitude of the current increases.

Nonadjustable. Indicates that the circuit breaker does not have any adjustment to alter the value of current at which it will trip or the time required for its operation.

Generally speaking, circuit breakers are rated by a frame size which determines the maximum overcurrent trip setting and an interrupting capacity indicating the maximum short-circuit current which can be interrupted safely without damage to the circuit breaker.

Circuit breakers are sometimes used with integral current-limiting fuses to provide for interrupting current requirements up to 200,000 A, symmetrical rms.

Switchboards and Switchgear

Switchboards and switchgear include buses, conductors, control devices, protective devices, circuit switching, interrupting devices, interconnecting wiring, accessories, supporting structures, and enclosures.

Switchgears can serve as main secondary service equipment, as main primary service equipment, and as load center equipment when located near load concentrations. The assembly and its devices provide for the control and distribution of electricity to utilization or subdistribution equipment. Most switchgear manufactured since 1955 has been metal-enclosed, dead-front, free-standing type, with its circuit protective devices enclosed each in its own compartment. For certain types of load applications, the protective devices are group-mounted in separate cubicles instead of individual compartments.

Older switchboards may be live-front type with the devices wired and mounted to slate panels with the wiring exposed on the rear of these panels.

A preventive maintenance plan should be conducted for all plant switchgear. Normal maintenance includes checking and tightening cable connections, testing air circuit-breaker operations, and checking relay and meter calibrations.

Types and Functions of Switchgear

The three most common types of switchgear used in industrial plants are:

1. Medium- and high-voltage metal-clad circuit breaker
2. Low-voltage power circuit breaker, either drawout or stationary (Molded-case breakers are being used more frequently for switchboard applications.)
3. Load interrupter switch and fused switchgear

Metal-Clad Switchgear. This is metal-enclosed power switchgear characterized by removable circuit switching and interrupting devices. Major parts of the primary circuit are enclosed by grounded metal barriers. Instruments, secondary control devices, meters, relays, and their wiring also utilize grounded metal barriers to isolate them from primary circuit elements. All live parts are enclosed in grounded metal compartments, bus conductors are insulated, and mechanical interlocks are provided.

Low-Voltage Power Circuit Breaker Switchgear. This is metal-enclosed and generally consists of: low-voltage power circuit breakers; bare or insulated bus and connections; control instruments, meters, and relays; and control wiring and accessories, together with cable and busway termination facilities. Circuit breakers may be drawout type, stationary air circuit breakers, or molded-case circuit breakers, usually of the solid-state, adjustable trip variety.

Metal-Enclosed Load Interrupter Switch and Fuse Switchgear. This consists of fused load interrupter switches, bare or insulated bus and connections, power fuses, instruments for control, and control wiring and accessories. Both stationary and removable interrupter switches and power fuses are used.

Protective Relays

Protective relays consist of an operating element and a set of movable contacts. In today's solid-state relays it is difficult to distinguish each of these elements visually. The operating element receives power from a control power source within the switchgear It obtains input from a sensing element in the circuit. This input can be in the form of either a voltage or current variation. The relay measures the variation against its standard settings and, when a preset limit has been exceeded, it initiates an action by causing its contacts to assume a new configuration. This action is used to perform some other control function such as sounding an alarm or tripping a circuit breaker.

It is essential for all protective relays to be maintained in proper working order well within prescribed tolerances. Relay manufacturers provide complete maintenance and setting instructions including test procedures for all their relays. Such instructions should be kept in an overall operating manual in the plant engineer's office.

Directional Relays. Directional relays are used to initiate an action when current flows in a direction other than normal. Such relays will initiate actions preventing the reverse flow of current through equipment.

Differential Relays. Differential relays are so named due to their current-balance characteristic. If the current entering the protected section is not balanced by the current leaving it, the relay initiates an action. These relays are used most frequently for protection of transformers and substation buses where they permit or prevent casing of the breakers.

Voltage Relays. Voltage relays are fundamentally similar to overcurrent relays, except voltage is used to activate the operating element. Some voltage relays are activated by overvoltage, others by overcurrent, and still others by a combination of the two conditions.

Ground Relays. Ground relays are instantaneous relays that respond to unwanted ground current. As such, they are unaffected by load currents and so may be set to operate for single-phase-to-ground currents that are smaller than full-load currents.

Control power for relay use is supplied generally by the use of current and potential transformers. Low-voltage equipment operating at less than 1000 V to ground generally requires no potential transformers because relays and instruments are designed to function at normal low voltages.

Current Transformers

Current transformers are transformers in which the switchgear bus usually forms the single turn of the primary winding while the secondary is a combination of many turns. As such, the ratio of the current flowing in the main bus to that which flows in the secondary of the transformer is directly proportional to the bus rating. All relays and instruments are calibrated to read directly based on this ratio. The full secondary current is usually 5 A. Maintenance personnel must be careful not to open-circuit the secondary side of a current transformer. The resulting voltage across the open terminals is the transformer turn ratio times the primary voltage and is very dangerous.

Safety Switches and Disconnecting Means

Each piece of utilization equipment located remote from its source of power or overcurrent protective device requires a disconnecting means. In most cases, a safety switch is provided for plant machinery which is not built with its own internal on-off switch designed to completely disconnect the device from the electrical system. A safety switch may be either fused or unfused. The disconnecting means is provided to permit equipment maintenance without the possibility of some remote-control device functioning to turn the machine on while the maintenance person is in contact with some dangerous part. Some of the commonly used switch types include safety switches and toggle, or snap, switches. A *safety switch* is a spring-loaded, quick-make, quick-break, device enclosed in a metal housing. Also known as service switches or disconnect switches, they are rated in amperes and/or horsepower. *Toggle, or snap, switches* are also used for disconnect purposes. They are designed for outlet box mounting and are rated by voltage and current. Another such device is the thermal motor switch or manual motor starter, intended as the overcurrent protective device for small motors. It is a combination of motor circuit protector circuit breaker and thermal overload device, built as an integrated unit designed for outlet box mounting.

Fuses

Many types of fuses are used in switches and other protective devices. The National Electrical Manufacturers Association lists standardized fuse types.

A fuse protects a circuit by means of a "link" or internal element which melts because of the heat caused by excessive current, thereby opening the circuit.

Fuses are selected on the basis of voltage, current-carrying capacity, and interrupting rating, in accordance with NEMA standards.

Spare part planning should include provisions to have the plant stock at least one complete set of each fuse in use in the plant.

Power Fuses (Over 600 V)

The NEMA E rating for power fuses requires that all fuses must carry their current rating continuously. Two types of power fuses are in common use: *current-limiting* and *expulsion-type*.

Current-Limiting Type. These power fuses are designed in such a way that the melting introduces high arc resistance into the circuit during the first half-cycle's peak current, thus restricting short-circuit current.

Their current-forcing action produces transient overvoltages. Surge-protective apparatus may be needed to compensate.

Expulsion Type. These fuses are generally used in distribution system cutouts or disconnect switches. An arc-confining tube with a deionizing fiber liner and fusible element provide fault current interruption. Production of pressurized gases inside the fuse tube extinguishes the arc by expelling it from the open end of the fuse.

Low-Voltage Fuses (600 V and below)

There are two styles of low-voltage fuse design, *plug* and *cartridge*.

Plug. These fuses are of three basic types, all rated 125 V or less to ground and up to 30 A maximum. Although they have no interrupting rating, they are subjected to short-circuit test with an available current of 10,000 A. The three types are: Edison base without time delay and all ratings interchangeable, Edison base with time delay and interchangeable ratings, and type S base available in three noninterchangeable current ranges: 0 to 15, 16 to 20, and 21 to 30 A. These last two types normally have a time-delay characteristic.

Cartridge. These fuses are either renewable or nonrenewable. Nonrenewable fuses are factory-assembled and must be replaced after operating. Renewable fuses can be disassembled and the fusible element replaced. Consult the NEMA standards and the code for fuse class definitions.

Wires and Cable

Wire and cable used for the distribution of electric power consists of a conducting medium usually enclosed within an insulating sheath and sometimes further protected by an outer jacket. The conducting medium is generally either copper or aluminum. The insulating medium can be made of any one of several materials depending on the ambient characteristics in which the wire is to be applied. Modern building wire is usually insulated with some form of plastic insulation. The various insulating materials are described in and defined by the ***National Electrical Code,***® Tables 310 to 313.*

Outer cable jacket materials range from cross-linked polyethylene to copper armor. Each jacket type has its own unique set of rules for application.

The cable protection method, conduit, cable tray, or bare exposed, depends on many design considerations. Cables require no basic maintenance except to assure that they are not applied beyond their ampacity. Cable and wire ampacities are published in the ***National Electrical Code***®* and by the manufacturers who make cables. All cables used for normal electrical usage are sized in the United States in accordance with the American Wire Gauge (AWG) or circular mils.

Terminations of aluminum cables require periodic examination to ensure that "cold flow" has not loosened the connection. Lugs connecting aluminum cable must be tightened in exact accordance with the recommendations of the equipment manufacturer. Maintenance personnel should possess a set of torque wrenches for aluminum cable maintenance. A program for terminal examination should be included in maintenance manuals.

Busways and Busbars

Busways and busbars perform the same functions as wires and cables, but their construction features and applications are far different.

Busbars can be cut, welded, bent, punched, drilled, plated and insulated to meet specifications. They are most practical when configurations must be precise, terminal points inflexible, and installation locations confined.

Standard busway types, ampacities, components, and accessories are so numerous that almost any routing plan can be created. They can be enclosed or ventilated, installed indoors or out, and used either as point-to-point power feeders, as multipoint power sources, or as continuous power takeoff routes.

Motor Controllers and Motor Control Centers

All motor-operated devices require some form of motor controller to provide start-stop, reversing, or other functions. Sometimes these control elements are built into the machine and the controls are machine-mounted. More often than not the control of a building utility or process system motor is from a remote motor controller.

Motor controllers—commonly called *starters*—are usually magnetically operated devices with thermal overload protection built-in through the application of melting alloy links. A preventive maintenance program is essential for motor starters if they are to continue to function properly. Remote controls with operating lights indicate only that the controls functioned. Therefore, it is necessary to make periodic inspections of these devices.

It is often more economical to group-mount motor controllers rather than individually mount them. A group of starters collected together into one integrated unit, is called a *motor control center.* NEMA has well-defined standards for type and class of wiring and the degree of interconnecting and interwiring present in any given center. This data can be obtained from the manufacturer and should be filed in the plant maintenance manuals together with complete detailed instructions for a periodic maintenance program. For further discussion, see Chap. 3-4.

Panelboards

Panelboards are assemblies of switching and overcurrent devices enclosed in a cabinet, usually wall-mounted, and protected by a cover or trim which can have a hinged door.

****National Electrical Code***® is a Registered Trademark of the National Fire Protection Association, Inc., Quincy, MA 02269.

Panelboards provide for local protection and control of apparatus and lighting. They are usually composed of integrated groups of circuit breakers and fused switches providing protection for circuits terminating in their immediate area.

All of the maintenance and operating considerations previously discussed for circuit breakers and switches apply to panelboards. Unless the circuit breakers in a panelboard bear a *switching duty rating* and the panel nameplate or breaker is so marked, they are not designed for such service on a constant basis. Constant use for switching will shorten the life of a standard circuit breaker. Occasional use will not have a negative effect.

Panelboards serving lighting and appliance circuits which must be switched on at fixed times and off at other times can well be equipped with magnetic contactors. These devices operate in a manner similar to motor controllers except they contain no overload devices. They permit remote switching of panelboards without using the breakers as a switching device.

PRIMARY SERVICE METHODS

There are a number of different distribution methods used in modern industrial systems. In all of these, electric power is received from a public utility or in-house generating plant at some convenient primary voltage and passed through a control and distribution system to the point of utilization. There is no one standard system of industrial plant distribution. Each system is tailored to meet conditions specified at time of design. Prime factors in the type of system chosen are the required utilization voltage and the distances involved in the distribution. Because the number of plants in many areas has increased and moved further from the utility generators, utilities have been using higher voltages to transmit electricity to the plants. This technique also minimizes line losses and provides good voltage regulation. As a result, many utilities are distributing at up to 230 kV and some at 350 kV. In many cases the rates paid for power at higher voltages are low enough to generate savings which repay higher installation costs for a main substation within several years.

Systems utilizing primary power as their source require some point at which this power is received and transformed to a useful form. The point of connection is normally the plant substation. If the primary voltage is in the high voltage classification (34,500 V and above), chances are that the service is aerial. In these cases, service enters the substation from an overhead structure containing a series of disconnect switches and lightning arresters. The substation requires oil-filled circuit breakers on each feeder. Liquid-filled transformers change the incoming voltage to some more usable level and provide a method (either cable or bus) for distribution to the main secondary switchboard.

Main Substations

If the purchased power voltage can be used for the plant primary system without transformation, a plant main bus can serve the same purpose as a main substation.

The principal functions of a main substation are indicated in Fig. 1-8. This is a simple arrangement which meets the requirements of many smaller plants. More complex arrangements are needed when there is more than one incoming line, or more than one power transformer, or one of a number of other bus arrangements. For large plants with heavy loads in widely separated areas, substations may require transmission voltage feeders connected to the incoming-line bus.

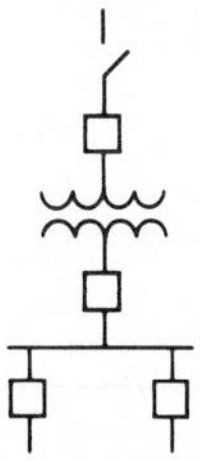

Figure 1-8 A typical main substation arrangement used by an industrial plant.

SECONDARY SERVICE METHODS

Many plants are switching to higher in-plant voltages to feed power over long distances throughout the plant without excessive line losses or loss of regulation. The most commonly used utilization voltages in new plants are 480 and 240 V, with 480 V becoming more popular. The best overall secondary utilization voltage is 480 V. It costs less and provides fewer line losses and less voltage drop.

Most new plants use load-center distribution, with unit substations being close to the loads and primary distribution being made at 2400, 4160, or 13,800 V. Primary-distribution voltage between master substation and load center should be selected based on size of the load and the distance to be covered. In general, 4160 V is used for loads under 10,000 VA and 13.8 kV for loads over 20,000 VA. For loads between 10,000 and 20,000 VA, 4160 V is used when plant layout is compact and 13.8 kV for long, rambling layouts.

The standard method of receiving and distributing secondary power is through use of radial systems.

Radial Systems

Conventional Simple Radial System

A conventional simple radial system (Fig. 1-9) receives power at the utility supply voltage. Voltage is stepped down to the utilization level by a transformer.

Because the full building load is served from a single incoming substation, diversity among loads can be used to full advantage; installed transformer capacity can be minimized. The system's drawbacks include poor voltage regulation and poor service reliability.

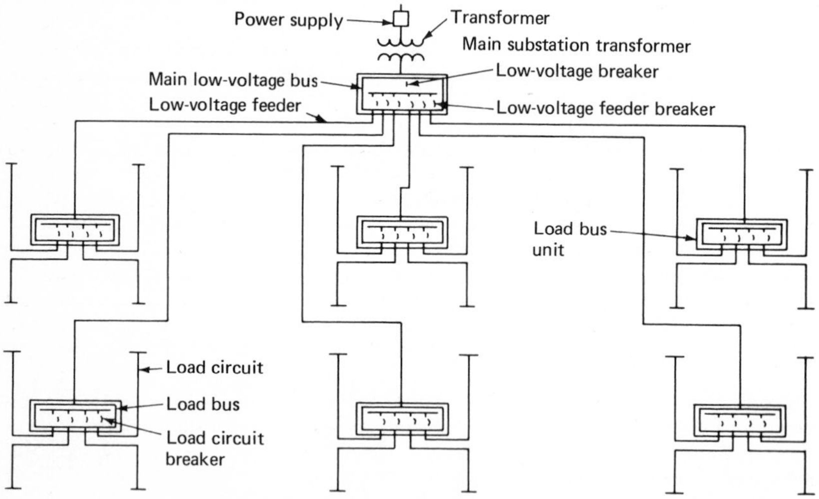

Figure 1-9 Conventional simple radial system.

Modern Load-Center Simple Radial System

The modern load-center simple radial system (Fig. 1-10) distributes incoming power at the primary voltage to power-center transformers located in building load areas. These transformers step the voltage down to utilization levels.

Each transformer must have enough capacity to handle the peak load of its specific load area. Combined transformer capacity requirements, therefore, may exceed those of a conventional simple radial system. This approach results in reduced losses, improved

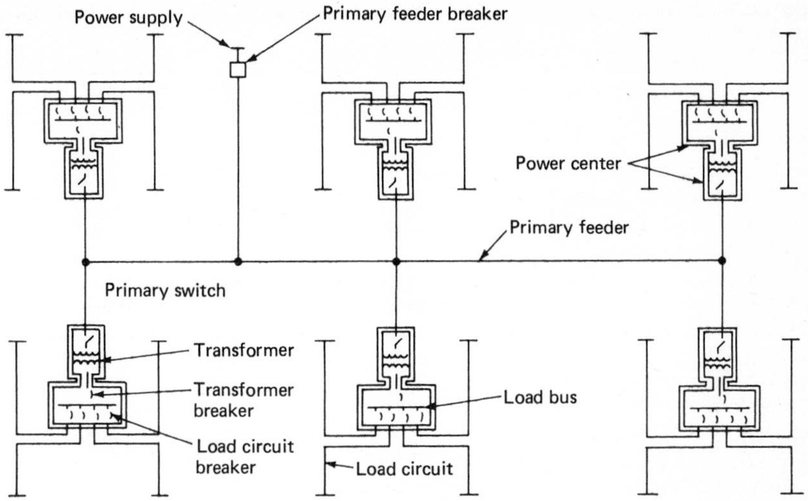

Figure 1-10 Modern simple radial system.

voltage regulation, reduced cost of feeder circuits, and no need for large low-voltage feeder circuit breakers.

Unit Substations

A unit substation contains one or more sections of each of three main components.

A *primary section* provides for connection of incoming medium- or high-voltage circuits, usually with disconnecting and circuit protective devices such as switches and circuit breakers. A *transformer section* includes one or more transformers. A *secondary switchboard section* provides for connection of secondary distribution feeders, each with a circuit protective switching and interrupting device.

Unit substation sections usually are subassemblies designed for field connection into an integral single unit. Numerous types are available for both indoor and outdoor applications and are described fully in manufacturers' catalogs.

Unit substations can be either single-ended, fed by a single primary feed, or double-ended, fed from either end by a separate primary feed. Double-ended substations are usually designed so that either transformer can assume two thirds of the load in the event of a failure of one primary feeder. This is accomplished by the inclusion of a normally open tie circuit breaker in the switchgear lineup. The application of double-ended substations to industrial plants increases the reliability of the system and allows partial operation in the event of partial power failures. Plant operators must have a plan for immediate dumping of nonessential loads before closing the tie circuit breaker. This plan will keep essential processes functioning while power is down.

System Voltage

The voltage class of both primary and secondary distributions are spoken as *nominal system voltage.* This term identifies the basic voltage normally utilized, such as 120/208 or 277/480 V. The actual voltage of each nominal system may be a slight variation such as 125/216 or 265/460 V. Each utility company uses its own selected secondary system. When a plant buys or generates primary power, the secondary voltage can be set very close to the nominal system rating.

Ranges for standard nominal system voltages are covered in ANSI C84.1-1977. Nominal system voltages most often found in the United States are indicated below.

120 V. A single-phase, two-wire system. Used for convenience outlets and incandescent lighting.

120/240 V. A single-phase, three-wire system. Nominal voltage between the two-phase conductors is 240 V. Nominal voltage from each phase conductor to ground is 120 V; used for power equipment, power outlets, electric heating processes, and in some cases high-intensity discharge lighting.

240 V. A three-phase, three-wire system, delta-connected, with 240 V between phase conductors and no ground or neutral conductor. Used for motor and three-phase power loads. This system is gradually being replaced by the more modern 120/208 V system with grounded neutral.

120/208 V. A three-phase, four-wire system, wye-connected with 208 V between phase wires and 120 V between phase and ground. This system permits single-phase, three-wire circuits to be taken from the system as well as 120 V single-phase circuits. The system is in general use for all types of loads. Recently, larger buildings have been designed to use a higher utilization voltage, but it often is converted to 120/480 V for convenience outlets and incandescent lighting.

277/480 V. A system similar in use and characteristics to 120/208 V for direct operation of motors, process equipment, and all forms of discharge lighting including fluorescent. Transformers are required to convert 277/480 to 120/208 V for uses described under that system.

4160 or 2400/4160 V. Sometimes used in large plants for internal distribution or for direct power operation on motors in excess of 250 hp.

Medium- and high-voltage systems are rarely used for distribution within plant buildings. In multibuilding sites higher distribution voltages may be found running between plant areas and between substations. A detailed discussion of medium- and high-voltage distribution or transmission systems is beyond the scope of this chapter.

Conductor Sizing and Load Growth

Wire and cables are usually sized exactly for the loads they serve. The sizes of feeders running to panelboards or switchboards usually allow for all spare circuits built into these devices, computed as if they were half-loaded.

Feeders serving grouped loads can usually have some additional load added to them because they are protected from overload by the natural diversity of equipment operation. Some study must be made of the actual use pattern of the equipment to determine the extent of the excess capacity available through this diversity.

All feeders serving either lighting, receptacle, or motor loads are sized to include a 25 percent factor to account for heating due to continuous loading. This factor must be maintained even when maximum loading is desired.

Most modern designers utilize only half a circuit's normal capacity during initial design; thus a 20-A receptacle circuit is loaded to only about 10 A initially. This approach permits addition of another 5 A of continuous load.

GROUNDING

The subject of electric system grounding is broad and complex, and is discussed here in brief overview.

Contemporary approaches hold that all power systems should have a grounded neutral included in the system. It is imperative for life safety that all metal elements of electric systems remain at ground potential at all times. Grounding should also take into account that buildings and equipment can build up a hazardous static charge of far greater magnitude than that encountered in winter when walking across a carpet and touching a doorknob. For tall structures or buildings located in isolated surroundings without other construction around, lightning can pose a potential hazard, which must be considered.

Ungrounded Systems

For many years industrial plants relied on an *ungrounded system,* essentially a delta-connected system without a grounded neutral. In this system, a single line-to-ground fault does not cause automatic circuit tripping. However, a second undetected ground on such a system can cause continuous nuisance tripping and even equipment burnout on devices not connected to the affected circuits. This is especially true on higher voltage systems. Ungrounded systems also pose the problem of transient overvoltages caused by grounded circuits. For all of these reasons there are now relatively few of these systems being installed, and existing ungrounded systems are being converted.

System Grounding

There are several types of grounded systems generally used in today's industrial plants. These are described briefly in the following paragraphs.

Resistance-Grounded Systems

Resistance-grounded systems are characterized by a resistance connection between system neutral and ground. This system introduces impedance in the ground path which tends to limit the current flow to ground. This technique also limits overvoltages caused by an intermittent-contact line-to-ground short circuit. Resistance is used to provide a ground-fault current which can be used for protective relaying operation.

Solidly Grounded Systems

Solidly grounded systems permit better control of overvoltages than any other scheme, but ground-fault currents can be higher.

These systems are used at operating voltages up to 600 V. The low line-to-neutral voltage reduces the risk of dangerous voltage gradients. A large magnitude ground-fault current helps attain optimum performance of phase-overcurrent protective devices.

Equipment Grounding

Equipment grounding is provided by a system of conductors utilized to maintain the metallic housings of electric system devices at ground potential. By grounding equipment in this manner, the system provides life safety and severely limits the fire hazard of short-circuit currents by providing a simple path to ground. This system should be periodically inspected and tested to maintain it in proper working order.

In general, it is an essential of a safe electric system that everything which might come into contact with a live system be maintained at ground potential. Further safety is gained by providing the system with a grounded neutral leg in the system.

System grounds can be derived from a ground attachment to a cold-water pipe ahead of the water meter or by a system of ground electrodes or some combination of these. A single ground point for the system should be established and periodically checked for continuity. Provision should be made to disconnect the system neutral during tests.

FAULT PROTECTION AND SYSTEM COORDINATION

Fault protection also is a complex subject. The principal types of faults which plague electric systems are three-phase short circuits, line-to-ground faults, and intermittent ground faults.

A three-phase bolted short circuit can be caused by any accident. The instant voltage fluctuations and the large overcurrent flowing in the system, coupled with the rapid decay of the system voltage, cause the circuit breaker to trip or the fuses to blow at the nearest point. The elapsed time for a circuit breaker to clear such a fault is from three to eight cycles depending on the size of the breaker. It is here that the design of the system is tested, because the circuit breaker is forced to open a circuit of far greater current than its trip rating. If breaker interrupting ratings are selected properly, clearing of three-phase bolted faults is simple. A fuse operates in the first one-half cycle. Fig. 1-11 illustrates a typical operating characteristic of current-limiting during a high-fault current interruption.

A solid line-to-ground fault causes system protective devices to react as they do for a three-phase short circuit.

The intermittent ground fault is the most difficult and therefore the most dangerous

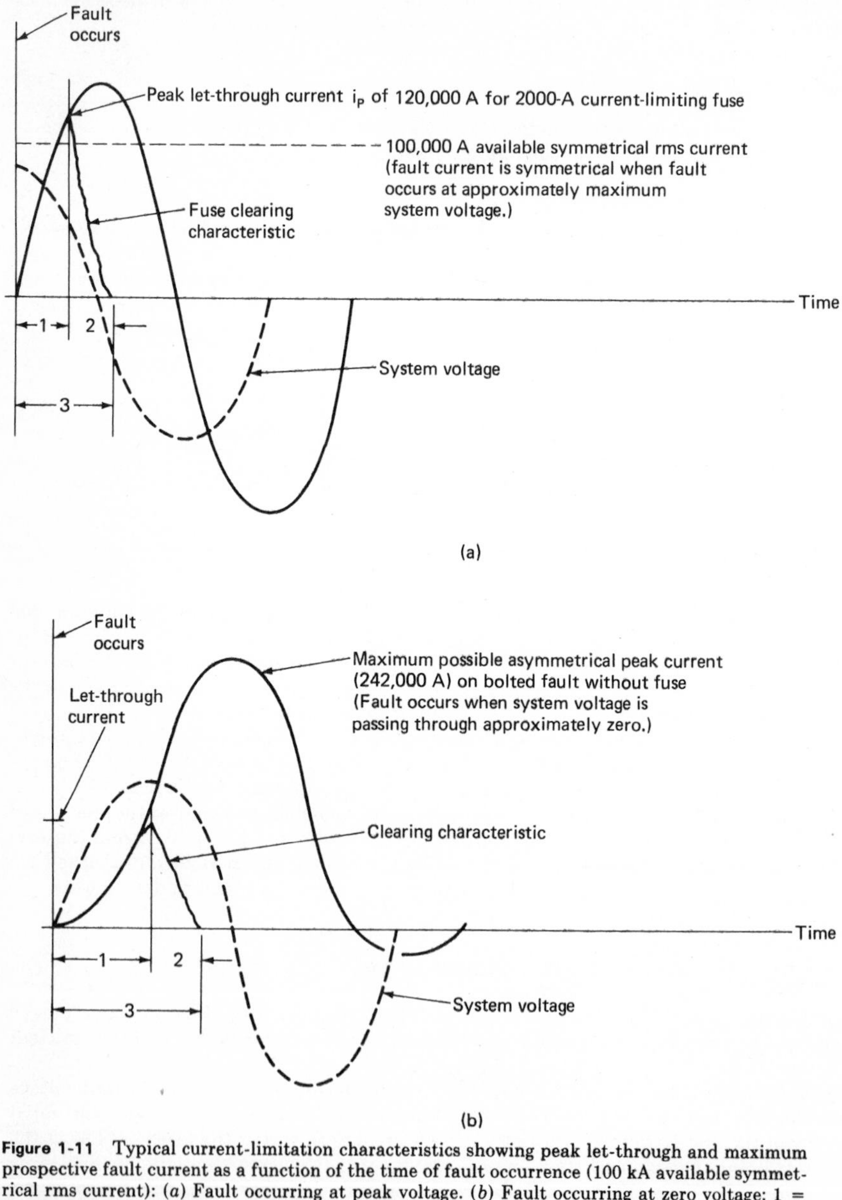

Figure 1-11 Typical current-limitation characteristics showing peak let-through and maximum prospective fault current as a function of the time of fault occurrence (100 kA available symmetrical rms current): (*a*) Fault occurring at peak voltage. (*b*) Fault occurring at zero voltage: 1 = melting time, 2 = arcing time, and 3 = total clearing time. *(Source: IEEE Standard 141-1976.)*

type of system fault. At no time does an overcurrent flow for a period long enough for the protective device to detect the problem and react. Intermittent ground faults are detected best by a *zero sequence* protection system. This system measures any current flow in the ground path and uses this current to operate a relay to cause the system protective device to operate and clear the fault.

Protective devices are usually coordinated so that the unit closest to the fault opens first. If the first unit in the system fails to clear the fault, the next one acts, and so on until the main opens, shutting the whole system down.

All circuit breakers, fuses, and most relays come with, or have available from the manufacturer, time-current operating curves. See Fig. 1-12 for comparative time-current

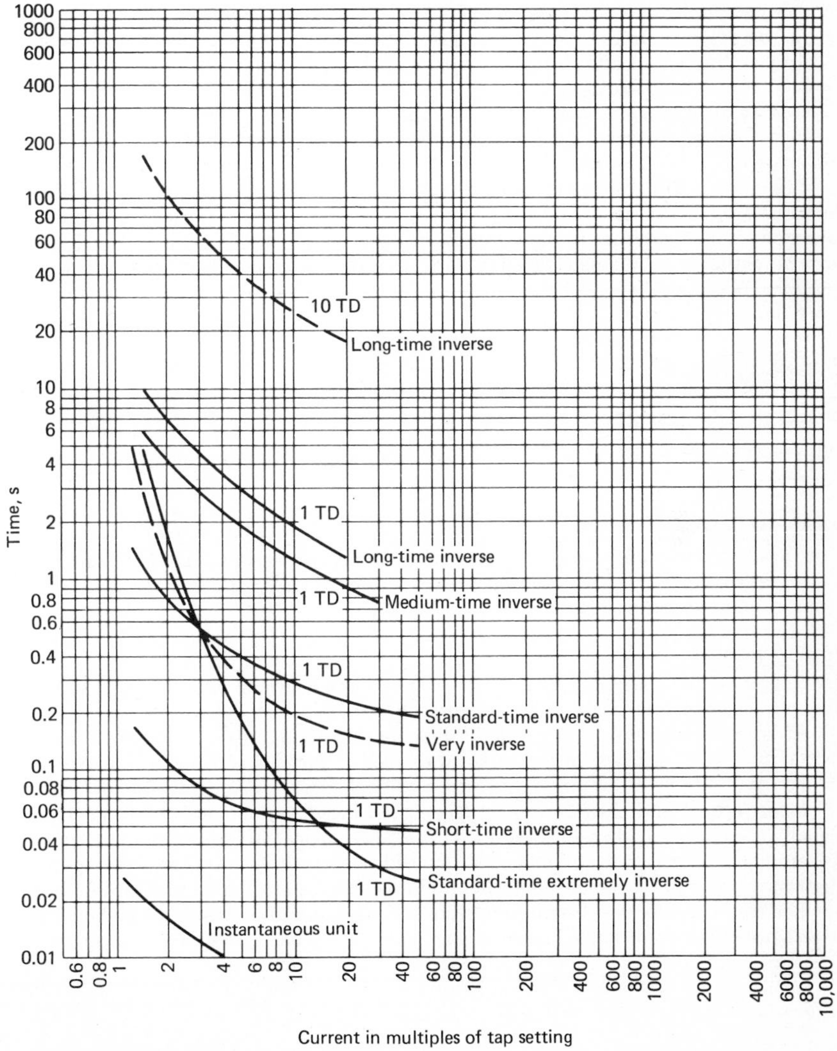

Figure 1-12 Comparative time-current curves of a typical induction overcurrent relay plus instantaneous trip.

curves of a typical induction overcurrent relay, plus instantaneous trip. By overlaying these curves and selecting services with coordinated operating times, it is possible to achieve the sequential operation needed.

Primary-system and even secondary-system protective relays are set to achieve this same system of sequential tripping.

It is incumbent on the operating personnel to never simply reclose a protective device that has opened on fault without first ascertaining the cause of the fault and either correcting it or removing the offending piece of equipment from the circuit.

SWITCHGEAR LOCATION AND WIRING

The best location for secondary switchboards is near the center of load. Frequently these are located on mezzanines above the plant floor, to isolate them from unauthorized personnel as well as to have them located near the load without taking valuable floor space.

In most industrial applications, a special enclosure is provided for a switchboard to isolate it from unauthorized personnel.

Wiring to equipment should be kept as short as possible, making load-center location of switchgear advantageous. Power switchboards are passive devices and require almost no operation. They are reset only if they trip under fault conditions.

Metering of utility company services can be done best at the single point of entry of service to the plant site. Metering of individual circuits for the purpose of monitoring energy use or some other function can be done anywhere in the plant. All such meters can be gathered in a single location with appropriate transmitters and telemetering equipment.

WIRING METHODS

Primary and Secondary Feeders

Two methods of running primary and secondary feeders in industrial plants are used: *overhead* and *underground.*

Overhead Distribution

Overhead distribution is used to interconnect buildings of the same facility which are separate from one another, when plants are operating on a temporary basis, or when a plant is more or less in a state of continual change.

Underground Distribution

Underground distribution is used for short distances, congested plant areas, or when the importance of high reliability cost justifies the extra expense.

A variety of raceways is available for distribution wiring. Rules governing the application and installation of raceways are discussed in various sections of the ***National Electrical Code.***®* Adhering to these rules will help assure a safe, reliable installation.

Main feeders and *distribution feeders* for localized switchboards and panelboards usually are run in rigid metal conduit. Large power blocks are distributed in busways. Where plant floor use or equipment is changed frequently, it often is advantageous to distribute all main feeders in the form of interlocked armored cable run in cable trays.

The method of power distribution which is best for a given plant depends on characteristics of the plant itself. In some, work stations are fixed. In others, the production system is fluid and electrical demands change from product to product.

In the case of a fixed production system, electricity can be brought to the point of utilization in rigid conduit or by any of the other means already discussed. Plants which

****National Electrical Code***® is a Registered Trademark of the National Fire Protection Association, Inc., Quincy, Massachusetts.

require more flexible power distribution often rely on a grid of busways which permits circuit-breaker-protected power takeoffs at any point.

Busways

Busway systems are widely used for feeders, plug-in subfeeders, and plug-in and trolley-type branch circuits.

They permit frequent changes in load layout, because they can be easily adapted to production-line rearrangement without seriously disrupting production.

Busway systems also have the benefit of controlled voltage drop characteristics, permitting them to meet different voltage stability requirements.

The types of busway commonly employed for industrial applications follow.

Feeder Busway

Low-impedance-type busway is frequently used for all types of high-capacity feeders and risers. Typical applications include transformer-vault-to-switchboard service runs, switchboard-to-load-center panel feeders, and welder feeders.

Plug-in Busway

Plug-in distribution busway is extremely flexible because it has easily accessible plug-in openings along its length, permitting tap-ins. It can be run from a switchboard or tapped from a run of feeder busway to carry power to closely spaced machines and other loads. Branch circuits to motors also can be tapped in.

UTILIZATION EQUIPMENT CONTROLS

Small equipment and hand tools are run from plug-in-type receptacles. These receptacles should be grounded and, for safety, equipped with ground-fault interrupter devices. Locking-type receptacles should be used for tools or equipment which are subject to extensive movement during use.

All equipment and machinery should be supplied with a disconnect device within sight of the machine, preferably within easy reach of the operator. This is a code requirement because it is mandatory for maintenance of a safe work space.

Utilization equipment, such as fans, pumps, oil burners, and lighting fixtures on branch circuits, can be controlled by contactors such as push button switches, toggle or tumbler switches, and rotary snap-action switches.

Magnetic switches are widely used for motor control and where it is necessary to operate the switch contacts from remote pushbutton stations.

Small pieces of equipment and hand tools usually have controls built into them.

A variety of controls can be added to an existing plant to increase safety and/or reduce energy consumption and expense.

Microprocessor-based programmable controllers are available in different sizes, offering much diversity. Among the most popular is the so-called demand-based multifunction controller. Typically, this device provides demand control, remote start-stop, duty cycling, and optimal start-stop. The number of loads which can be controlled depends on the nature of the controller involved. In many cases they are modular, enabling addition of loads as needed. Devices such as these can be used for machinery, lighting, comfort cooling and heating, and other purposes. Each of the functions provided by a multifunction device generally can be performed by a smaller device which performs one function only, such as remote start-stop.

Timers and time-clock controls have been used for many years. Their function is generally to start and stop a certain operation at a predetermined time, for example, outdoor lighting.

Photocell controls are applied to lighting. As ambient lighting conditions fall to a certain predetermined point, the photocell activates lighting. When ambient lighting rises above that point, lighting is deactivated.

Dimming controls are available for fluorescent as well as mercury vapor lamps. They reduce light output to what is required for the tasks involved. The percentage of light reduction is close to the percentage of energy consumption reduction. In no case should lighting levels be reduced to less than what is required. Worker productivity will suffer; safety and security hazards may be created.

HAZARDOUS LOCATIONS

The dust and fumes put into the air by many industrial processes create an environment which is classified as electrically hazardous. Although the airborne substances may not be toxic, they can create an atmosphere which is subject to explosion and fire when heated by an electrical switching operation.

Electrical hazards are classified into three groups by the ***National Electrical Code,®*** in the chapter titled "Hazardous Locations."* Each group is further subdivided into divisions by degree of hazard and type of substance present.

The basic classifications are as follows:

Class I. Air is contaminated by hazardous fumes or vapors including gases and airborne chemicals. This is subdivided into two divisions.

Division 1 Locations where hazardous vapors and gases can or do exist under normal operations.

Division 2 Areas where flammable vapors or gases are handled in proper containers but where hazardous concentrations are normally prevented by forced ventilation.

Class II. Similar to class I except it includes atmospheres containing or likely to contain combustible dust. The divisions of this class resemble those of class I.

Class III. This class of hazard deals with the presence of combustible materials in the manufacturing process and the presence of airborne particles created by the process. The divisions deal with the manufacturing site and the storage sites.

In addition to class and division of hazard, electrical hazards are further defined in each segment by group and designated by a letter such as "A," "B," etc.

The ***National Electrical Code®*** accurately defines each class, division, and type of hazard. The methods of wiring required are also defined in the same Code.*

The plant operator must be able to identify those areas of the plant presenting identifiable hazards and must maintain wiring systems which conform to the probable hazard present. Great skill is needed to keep to a minimum the electric devices actually located in the highest hazard areas. Careful arrangement keeps control and switching devices in the area with the lowest classification. Explosion-proof wiring is expensive. It is often less expensive to remove or reduce the hazard than it is to provide the appropriate electric system.

It is essential for the plant engineer to become familiar with the rules of the Code and also with other publications of the National Fire Protection Association that further define the steps which can reduce the hazards.

BIBLIOGRAPHY

1. Beeman, D.L. (ed.): *Industrial Power Systems Handbook,* McGraw-Hill, New York, 1955.
2. "Constructing Electrical Systems," (Special Report), *Electrical Construction and Maintenance,* May 1979, pp. 59–90, 164–170.
3. *G.E. Specifier's Guide,* General Electric Co., Schenectady, New York

****National Electrical Code®*** is a Registered Trademark of the National Fire Protection Association, Inc., Quincy, MA 02269.

4. *IEEE Recommended Practice for Electric Power Distribution for Industrial Plants,* The Institute of Electrical and Electronics Engineers, Inc., New York, 1976.
5. *Industrial and Commercial Power Distribution Course,* The Electrification Council, New York, 1975.
6. McPartland, Joseph F., and William J. Novak: *Electrical Equipment Manual,* 3d ed., McGraw-Hill, New York, 1965.
7. ***National Electrical Code®,*** National Fire Protection Association, Quincy, Massachusetts, 1981.
8. NECA Electrical Design Library: *Fault Detecting and Disconnecting Devices,* National Electrical Contractors Association, Washington, D.C., June 1977.
9. NECA Electrical Design Library: *Power Distribution Systems,* National Electrical Contractors Association, Washington, D.C., March 1977.
10. NECA Electrical Design Library: *Specifying Electrical Conductors,* National Electrical Contractors Association, Washington, D.C., June 1978.
11. *NECA Wiring Symbols Standard,* National Electrical Contractors Association, Washington, D.C., May 1976.
12. *Recommended Practice For Electrical Equipment Maintenance,* NFPA 70-B, National Fire Protection Association, Quincy, Massachusetts, 1977.
13. Smeaton, Robert W.: *Switchgear and Control Handbook,* McGraw-Hill, New York, 1976.
14. *System Neutral Grounding and Ground Fault Protection,* Westinghouse Relay Instrument Division, Westinghouse Electric Corporation, Coral Springs, Florida, January 1978.
15. Thumann, Albert: *Electrical Design, Safety, and Energy Conservation,* Fairmont Press, Atlanta, Georgia, 1976.
16. *Westinghouse Construction Specification,* 5th ed., Westinghouse Electric Corporation, Pittsburgh, Pennslyvania, 1978.

chapter 3-2

Electric System Management

by

Edward R. Guerdan, P.E.
Pittsburgh, Pennsylvania

with contributions by

Kao Chen, P.E.
Westinghouse Electric Corp.
Bloomfield, New Jersey

GLOSSARY

Kilowatt (kW) The basic unit of electric power delivery or consumption, i.e., the *rate* of energy delivery.

Kilowatthour (kWh) The basic unit of electric energy delivered.

Kilovoltampere (kVA) A measure of the total capacity of electric service delivery.

Power factor A measure of the effectiveness of energy utilization, the ratio kW/kVA.

Demand The rate of energy draw from the electric system during a prescribed period.

Peak demand The maximum rate of energy draw during the prescribed period.

Time-of-day (TOD) Reference to provisions in utility rate structures for discount allowances or penalties on energy consumption, depending on period of energy usage.

Rate schedule (structure) The document defining utility billing parameters, costs, and usage stipulations.

On-peak A period during the day defined by the utility to imply that its system is in high demand.

Off-peak The remainder from the on-peak period when utility system loading is low.

Blackout The complete loss of power.

Brownout An intentional reduction of power generation by the utility in order to effect rationing, generally in the form of a voltage reduction.

INTRODUCTION

Electric system management is the systematic control of electric energy utilization in a plant or building; it is a management effort that integrates and accounts for the effects of the utility rate structure, the user's electric distribution system, geographical location, activity, management style, and objectives of the user's operations. A well-controlled electric management system provides significant savings and increased reliability through reductions in energy utilization and utility costs, optimization of capital equipment requirements, improvements in the efficiency of plant operations, and improvements in the availability of critical equipment during emergencies.

Electric system management is not the same as *energy management systems.* Energy management systems are usually defined as the application of computer- or microprocessor-based intelligence systems specifically designed to control the sequential application and utilization of energy-consuming equipment with the goal of optimizing overall energy requirements and operating conditions.

However, energy management systems (whether computer-based or organizational) can be an integral part of electric systems management, providing an additional technique to knit together the many elements involved in managing an electric system. This chapter addresses those elements of electric systems that provide opportunities for saving through proper planning and control.

UTILITY RATE STRUCTURES

The utility rate structure is the basic reference, the basis of utility remuneration for services and energy delivered; it is a major cornerstone of plant electric system managerial strategy.

Utility rate structures (or rate schedules) are typically complex contracts individually tailored to each utility's load profile, type of power generating equipment, and energy source. These rates are also influenced by the distance between generating stations and the user, which affects power transmission costs. It is therefore impossible to define and describe a general rate schedule. *This fact places the responsibility on the plant engineer to become familiar with the applicable rate structures imposed by the electric utility serving each facility.*

Although the responsibility of final interpretation and adaptation of the rate structure to the electric system management of the facility rests with the plant engineer, utility representatives are usually willing to assist in interpretation and implementation of the schedule. Common elements of rate structures as well as special areas to be considered are described below.

Demand

The electric utility rate schedule basically addresses two portions of electric energy delivery. First is the charge for energy consumed, measured in kilowatt-hours (kWh). Second is the *demand* or *capacity* charge for kilowatts (kW) drawn from the electric utility during the prescribed billing period (usually monthly). It is designed to compensate the utility for its investment in equipment and overhead required to deliver power to the customer. The kilowatt demand is defined as the average rate of energy draw from the utility system during a specified time segment (usually 15 min, sometimes 30 or 60 min).

On-Peak / Off-Peak

Since energy drawn from the utility's generating facilities is normally greatest during the daylight hours, the utility often provides a penalty factor on the demand portion of the electric bill for demand incurred during the daylight periods (*on-peak*). On the other hand, it may provide incentives (such as reduced demand charges) to the plant to maximize the load during the nondaylight periods (*off-peak*) in an effort to have the customer draw a more uniform overall load on the system. This is a billing strategy designed to maximize the use of the utility's equipment and to minimize its investment requirements. This element of billing is often referred to as *time-of-day* (TOD) billing, although the meaning of TOD billing can often be narrowed to a particular time target of only a few hours per day.

By judicious rescheduling of certain plant operations, the plant engineer can bring about substantial savings in electric energy bills. The nature of most plant operations causes peak demand to occur usually during the daylight hours and generally for only a few hours a day. Whether the utility rate structure provides direct incentives with TOD rate differentials, there is nevertheless a real incentive for the plant to spread electric load throughout the day in order to reduce the peak demand charge attendant with compressed operations schedules.

Seasonal Demand

In an effort to impose uniformity of loads throughout the year, utilities may also assess a demand penalty which is related to the time of the year it occurs. Some utilities are summer-peaking (air-conditioning loads), others are winter-peaking (heating loads). The penalty may be in the form of carrying the demand charge established in the defined summer months as a minimum demand charge during the defined winter months, or vice versa. This billing characteristic is commonly referred to as a *ratchet*.

Power Factor

The power in kilowatts that the utility delivers represents that portion of power which does real work (motors) and provides heat and light. In addition, there is a component of current which is necessary to provide the magnetic field in transformers and motors; it is known as the *reactive component*. This current does not produce real work; in electric theory it is considered to be vectorially in quadrature with the power components of current. (See Chap. 3-1.) The vector sum of the two currents is the amount of current actually drawn from the utility's system, and as such it determines the size of transmission facilities that the utility must have to furnish power to the customer. *Power factor* is the ratio kW/kVA, or

$$\frac{\text{Power component of current} \times \text{voltage}}{\text{Total current} \times \text{voltage}}$$

The utility will often assess a billing penalty for excess use of reactive current. This may be done in several ways as described in the following paragraphs.

kVA Demand

In lieu of measuring peak demand by kW, the utility may measure and charge for the peak kVA incurred during the billing period. Hence, the energy demand is effectively augmented by the reciprocal of the power factor.

Power Factor

The utility may measure and charge for the power factor directly. The power factor may be measured coincident with the peak kW during the billing period, or the utility may measure the reactive kilovoltampere-hours (kvarh) and relate this factor to the kWh, to assess an average power factor during the billing period. Although these differences seem subtle, they can affect the methods and economic benefits that the plant engineer must consider when designing the electric management system.

Kilowatthour

The kilowatthour is the basic energy unit used to measure electric power consumption, and the kilowatthour charge is therefore unassailable as a billing element. Energy conservation opportunities that may be addressed within the facility's operation—such as high-efficiency lighting and motors—are discussed more fully elsewhere in this handbook.

Miscellaneous

Several additional elements of the rate structure may provide opportunities for savings and should be investigated.

Ownership

The main-plant transformation equipment may be wholly owned by the customer or may be leased from the utility. In lieu of a direct rental agreement, the utility may assess a charge on the demand or energy (or both) portions of the bill reflecting rental costs; alternatively it may define credits within the rate structure on demand or energy if the customer provides the main substation equipment. The utility is usually amenable to negotiations for sale or purchase of the main substation equipment, subject to the satisfaction of economic interests on both sides.

Schedule Transfer

The utility normally has several rate schedules which bracket the customer's usage profile. Due to changes in usage patterns, the plant engineer may choose another rate schedule that is more economical. The utility, upon request, will usually provide an analysis based on recent billings to confirm the applicability of the most favorable rate schedule.

Taxes

Many states allow for discounting of state taxes for utilities or energy used for specific activities such as manufacturing, health care, etc. The utility representative will advise the customer of the entitlements in this area.

LOAD CURVE SURVEY

In order to institute an electric management system in the facility, it is important that the plant engineer thoroughly understand how the facility's electricity consumption patterns vary and their dependence on time of day, day of week, and season of year.

Analysis of the characteristics of the electricity consumption patterns is aided by use of the *demand chart.* The utility company will often maintain recorded kW (or kVA) demand (15 min, 30 min, or an alternative interval period) on a chart very similar to that shown in Fig. 2-1 to permit identification and subsequent billing for the peak demand established during the billing period. Alternatively, the utility's metering system may incorporate electronic pulse recording which transmits and records the equivalent infor-

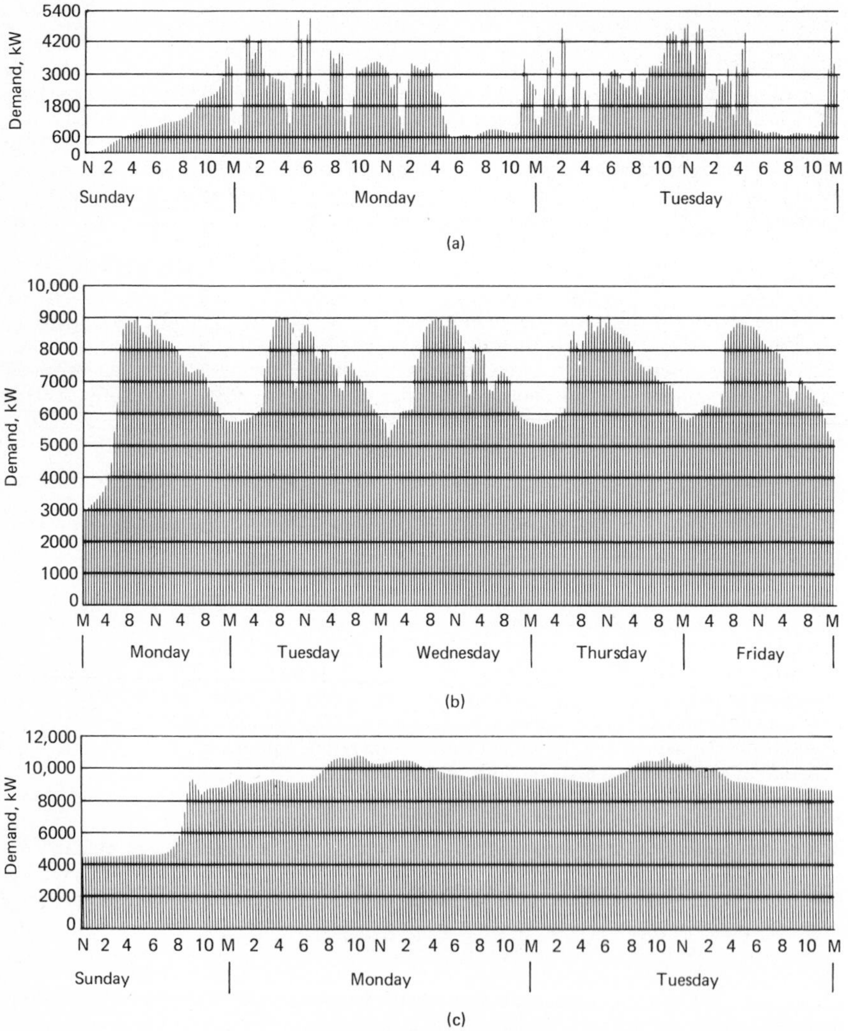

Figure 2-1 Demand charts.

mation on magnetic tape. These data may be retrieved to permit manually generating an equivalent demand chart which aids in the visual interpretation of the plant electric consumption characteristics. When no continuous recording is maintained, the utility should be requested to temporarily install demand recording instrumentation to enable the attendant analysis.

Daily Load Pattern

Each plant or facility has a unique demand situation. Fig. 2-1 represents three disparate load patterns that could be encountered. In Fig. 2-1*a*, dramatic spike peaks of usage are caused by uncontrolled simultaneous application of large loads. Significant savings may be realized by controlled alternate application of the identified loads. This situation may provide the best opportunity for computer-based demand control. In Fig. 2-1*b*, the chart shows a fairly normal industrial characteristic. Modest nighttime activity, followed by rapid increase in the morning, tapering off toward late afternoon, is indicative of shift work activity. The uniformity of the load curve in Fig. 2-1*c* appears to offer little opportunity for demand reduction. However, the relatively high weekend load suggests that opportunities for energy reduction through greater equipment turnoffs may be available.

Daily Load–Weather Dependence

Superposition of weather data on daily load curves for various seasonal loading conditions permits identification of the facility's weather dependence and opportunities for improved energy performance, as well as tailoring of the load curve to utility rate structures.

Figure 2-2 shows three load profile curves of different seasonally dependent characteristics. This curve represents a facility which is mostly or totally dependent on electricity.

- During the summer months, air-conditioning load dependence is evident, as power demand peaks during the midday high cooling demand period.
- In a typical winter season, the power demand may actually be greater during the night and early morning due to higher heating loads.
- In the spring and fall, when heating and/or cooling requirements are slight, the influence of the normal plant activity dominates the load curve, with power demand rising and falling according to plant or facility business activity.

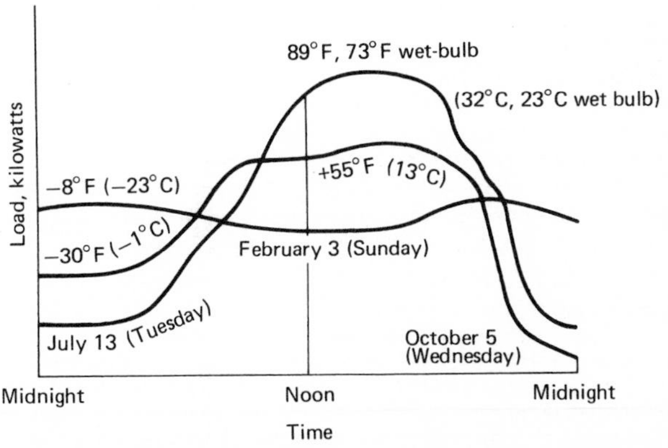

Figure 2-2 Electrical load profile curves.

PLANNING CONSUMPTION CONTROL

The previous sections have identified those elements of electric energy consumption which must be considered and accounted for in the development of the electric management system. Following are some of the remedial actions that the plant engineer should consider to maximize the cost effectiveness of the electric energy service to a facility, particularly as influenced by the utility rate structure.

Power-Factor Correction (Improvement)

Unless the utility rate structure specifically includes penalty charges for low power factor, there is usually minimal economic incentive to correct power factor. Although low power factor implies higher current loading which increases the distribution power losses, the cost of correcting individual offending loads is generally much greater than the energy savings to be realized from reduced distribution losses. However, when power-factor improvement to mitigate the utility's penalties is desired, then correcting the power factor of the offending load will provide the added benefit of increased bus capacity and reduced voltage drop with an attendant energy savings and improvement in performance of the user load.

Individual load correction procedures include:

- Use of high-power-factor utilization equipment such as high-power-factor lighting ballasts and distribution transformers.
- Operating induction motors at near full load to improve the power factor of the motor. Shutting down idling motors reduces kvar draw and eliminates wasted energy consumption.
- For larger motor applications, it is often economically attractive to apply synchronous motors in lieu of induction motors. Synchronous motors can operate at the leading power factor, therefore *supplying* kvar to the inductive (lagging) power factor loads.
- Application of power-factor correcting capacitors is one of the most common methods of improving power factor. When applied to an individual load at the load's terminals, the capacitors supply the required leading kvar to the inductive load, and are removed upon disconnection of the load. This procedure has the added benefit of inhibiting a voltage rise associated with kvar fed back through distribution-system transformers during periods of low activity; if excessive, this can be injurious to certain voltage sensitive loads. This voltage rise is defined by the relationship

$$\%\text{ voltage rise through transformer} = \frac{\text{capacitor kvar} \times \text{transformer }\%\text{ reactance}}{\text{kVA of transformer}}$$

When the intent of power factor improvement is primarily to satisfy utility requirements and to offset utility billing penalties, it is usually most economical to apply the power-factor-correcting capacitors at the main incoming power substation. Fewer switching and protective devices, less distribution wiring, and lower unit cost of capacitors at the higher voltages contribute to this economy.

Plant engineers may obtain more detailed information on calculation of capacitors required to correct the power factor in the referenced articles, in capacitor manufacturer's catalogs, and from local electric utility representatives.

Demand Factor Reduction

Analysis of the facility's demand-load (as typified in Fig. 2-1) and its relationship to the utility rate structure, identify to the plant engineer the strategies to be incorporated into electric system management.

In all cases, regardless of load curve characteristics, a major tenet of demand reduction, (load management) includes energy conservation efforts. It is evident that the demand curve, which represents the *rate* at which electric energy is consumed in the facility, is composed of two portions. One, a *base load* which is fairly constant, represents the inherent electric service required to maintain the facility's environment and comfort conditions. The other, *variable load,* is superimposed on this constant load and is dependent upon the business activity of the facility, weather conditions, and other variables.

There are several stategies that the plant engineer may incorporate as elements of management of the electric system to effect reductions in the demand curve and hence the cost of the electric energy service to the facility:

- **Base Load Reduction.** As previously implied, this goal will be realized by implementation of an energy-conservation program which addresses the optimization of utilization of lighting, HVAC, motors, etc. These areas are discussed more fully in the appropriate sections of this Handbook.
- **Variable Load Reduction.** There are several tactics which can be employed. Each approach must consider the effects of the utility rate structure, the facility's electric system, and its relation to the operations. Although the following examples imply manual or organizational actions, where a great number of the loads involved are to be identified, the appropriate decision making may be best implemented through the application of a computer- or microprocessor-based demand-energy controller.

  Inhibit— Interlocking of large loads to prevent their simultaneous application during the utility demand period (15 min, 30 min, etc.), such as large chillers or heat treatment furnaces.

  Substitute— A particular process may be better served with an alternative fuel or the attendant demand reduction may justify the replacement of old, inefficient equipment.

  Reschedule— Certain operations may be shifted to other periods of the day to minimize their impact on the peak demand. Other once-a-week operations might be performed on the weekend or at night if the attendant savings justifies overtime premiums. Likewise, when the utility rate structure contains TOD provisions, it may be justifiable to transfer entire operations (with little or no concern for additional controls) to nighttime.

  Generate— When emergency generating equipment is available, the plant engineer may consider utilizing this equipment (in lieu of the utility service) to serve a particular load or group of loads during periods of high activity which may threaten establishing a new peak demand. Such a consideration must be carefully studied regarding energy cost effectiveness and potential overutilization of the critical emergency equipment.

  Storage— Storing energy during inactive periods in anticipation of retrieval in high demand periods can take on many forms, e.g., precooling or preheating domestic water or other process fluids, pumping these fluids to high storage tanks, or battery storage.

When the utility employs a seasonal ratchet, precooling or heating of the facility during the night (which in effect permits reduction of the base load during the day) may be advantageous. In considering this tactic, it is important to verify that the demand-factor savings has not been exchanged for an energy consumption penalty.

ELECTRIC SYSTEM MANAGEMENT REEVALUATION

The preceding section emphasized the controllable aspects of kW (or kVA) demand, i.e., the rate of electric energy consumption. It should be noted that the relative magnitude of the demand portion of the electric utility bill vs. the energy (kWh) portion has varied from time to time. During periods of high economic growth, with higher capital investment by the utilities to maintain generating and distribution capability, the demand portion of the bill is relatively large. Conversely, during periods of economic stagnation with rapidly escalating energy costs, the energy portion of the bill dominates. These perturbations can greatly influence the plant engineer's economic justification to address either factor.

However, the demand portion has been stressed primarily due to its controllability and the significant spin-off benefits associated with an effective demand reduction program. These benefits are discussed in the next three paragraphs.

Energy Savings

It has been demonstrated that with the implementation of an effective demand reduction program, there is an accompanying *sympathetic* energy conservation effect, usually in the order of 20 to 30 percent of the demand reduction.

Released Capacity

By reducing the demand on the electric distribution system, the plant engineer will achieve a proportional increase in plant electric system capacity. This can defray the cost and often *enable* plant or operation expansion with little or no electrical capacity investment.

Reliability

Alternatively, redistribution of the electric system may be accomplished, permitting the reduction of loading on severely overloaded portions of the circuit. In the extreme, *total* sections of the system may be able to be disconnected, supplying additional savings with the elimination of no-load transformer losses and spare capacity and equipment for the remainder of the plant electric system.

Note: Shutting down surplus transformers *can* provide significant savings through the elimination of the no-load losses. However, it is recommended that the transformer manufacturer be contacted to determine appropriate mothballing techniques, in order to preserve the integrity and availibility of the transformer in the event of an emergency.

PROVIDING FOR EMERGENCIES

The occurrence of emergency conditions with a utility system must be anticipated, and appropriate action should be incorporated in electric system management. Emergency conditions can take on two basic forms:

- Total and unexpected *blackout* or loss of power due to system failure or "acts of God" (utility or facility system).
- Phased power or voltage reductions (*brownout*) by the utility, designed to preserve the ability to serve the system load. This condition can be caused by unusually high system loads in excess of the utility's capacity, loss of generating or transmission capacity, or energy source interuption or reduction.

Blackout

When this occurs, plants must have security provisions in the order of *personnel safety first* and *facilities loss prevention second.* Other considerations include protection of critical processes and process equipment and continuity of production operations.

Personnel Safety

The primary consideration is to provide for safe egress from the facility. This usually means providing *adequate* emergency lighting to identify and gain access to exits. Depending on the size of the facility, these requirements can be satisfied either with battery-operated unit lighting systems, central battery systems, or engine-generator sets. (See Chap. 3-3.)

Prevention of Facilities Loss

Under emergency conditions the possibility of concurrent power failure with local fire or other major catastrophe must be presumed. Hence, other considerations include the uninterrupted availability of fire protection or communications signaling systems and emergency power to operate fire water pumps.

Generally, fire codes do *not* encourage the utilization of elevator systems for exiting, but may require that elevators automatically "home" to a preassigned landing with one car preserved for official use.

Production-Process Equipment

These systems must be factored in after the foregoing considerations and are included relative to their loss-prevention exposure.

Brownout

The objectives in the occurrence of brownout are to prevent the destruction of sensitive equipment due to undervoltage and to preserve the continuity of the production operation. In areas where brownouts are known to be frequent, it is often advisable to provide voltage regulation equipment at the facility's service entrance. Alternatively, certain loads may be transferred to engine-generator or battery supply, or the plan may include the shedding of major loads in order to compensate for the reduced capacity on the system.

BIBLIOGRAPHY

Bello, Louis A.: "Blackouts Last Longer for Some Buildings than Others," *Buildings*, November 1977.

Chen, Kao, and Ed Palko: "An Update on Rate Reform and Power Demand Control," *IAS Transactions*, vol. IA-15 no. 2, March/April 1979.

Conlon, Joseph: "Selecting an Effective Energy Management System," *Electrical Construction and Maintenance*, June 1980.

Daley, James M. and Rene Castenschiold: "Demand Peak Shaving with On-Site Generators," *Electrical Construction and Maintenance*, February 1981.

"Electric Utility Rate Schedules," Federal Power Commission, U.S. Department of Energy, 1980–81.

Fenster, Larry C. and A. Jay Granties: "How a Contracting Firm Looks at Energy Management," *Heating/Piping/Air Conditioning*, October 1979.

Knisely, J. R.: "Power Reliability for Telephone Switching Center," *Electrical Construction and Maintenance*, March 1979.

Lawrie, Robert J. and Dennis O'Connell: "Emergency Power System Performance," *Electrical Construction and Maintenance*, October 1977.

"Maintaining Emergency Systems," *Electrical Construction and Maintenance*, March 1979.

Manhal, Harvey C.: "Understanding the Controllable Factors that Affect Your Electric Bill", *Plant Engineering*, August 17, 1978.

Mathews, L. Turner: "Proposed Electric Power Rate Reform", *Plant Engineering*, December 13, 1979.

Palko, Ed: "Cutting Costs with Power Capacitors," *Plant Engineering*, February 24, 1972.

Palko, Ed: "Installing Power Factor Improvement Capacitors," *Plant Engineering*, January 11, 1979.

"Power Factor Correction," *Electrified Industry*, January 1975.

"Recommended Practice for Emergency and Standby Power Systems," IEEE Standard 446, 1974.

"Trends for the Electrical Eighties," *Electrical Construction and Maintenance*, October 1979.

Turrell, Donald L.: "Utility Rate Factors," *Building Operating Management*, March 1979.

chapter 3-3

Standby and Emergency Power

PART 1

Rotating Equipment Systems

by

Georg Stromme, P.E.
Product Manager, Electrical Power Systems
U.S. Power Products Division
Onan Corporation
Minneapolis, Minnesota

GLOSSARY

Automatic sequential paralleling system (ASPS) An emergency power supply system that operates multiple engine-generator sets in parallel. It allows the first set to reach acceptable speed after receiving the start command to be immediately connected to critical loads. It automatically synchronizes and connects remaining sets. It is often applied in health-care facilities with a 10-s transfer requirement.

Automatic transfer switch (ATS) An electric switching device that alternately connects the load to the normal or emergency power source as required without operator involvement. It may also include controls that start and stop the engine-generator set, monitor voltage, and frequency of both power sources and other timing and logic functions. See also nonautomatic transfer switch.

Nonautomatic transfer switch An electric switching device that alternately connects the load to the normal or emergency source as determined and initiated by an operator.

Prime mover The engine that drives the generator through permanent coupling. It may be either spark ignited, diesel cycle, or gas turbine.

Paralleling switchgear Switchboards dedicated to control of engine-generators. Contains start sensor, synchronizer, reverse-power relay, and other controls. It closes generator output to the emergency bus through a fast-acting circuit breaker.

INTRODUCTION

The emergency power supply system (EPSS) encompasses a wide variety of equipment. Equipment under the EPSS heading ranges from a simple, self-contained battery light to a complex, highly engineered, multiple engine-generator set system with a capacity of several megawatts. These two extreme examples suggest a division of these systems based on the source of emergency power. The first example uses a stored energy device, a battery, as the power source while the second example uses rotating equipment, engine-generator sets, as the power source. Stored-energy systems and rotating-equipment systems are the two categories of emergency power supply systems. The subject of this part is rotating-equipment systems.

SYSTEM DESCRIPTION

When the normal power source fails, the EPSS functions to supply electric power to specific, select loads. The equipment that comprises the system is largely determined by the characteristics and requirements of the loads being served. Two equipment groups, the emergency power source and the electric switching equipment, subdivide the system based on function. While the two equipment groups have independent functions, the groups are interrelated and both serve the common purpose of the complete system.

Emergency Power Source

The function of this equipment is to generate electric power. The engine-generator set is the principal part of the group. The generator is permanently coupled to and driven by a prime mover which may be a diesel, gasoline, or gaseous engine or a gas turbine. Figure 3-1 illustrates a typical diesel engine-generator set. Included in this group are an independent fuel supply with storage and transfer equipment and the engine-generator set(s) with supporting equipment such as the governor, voltage regulator, exciter, cooling system, ventilation equipment, exhaust system, and engine control with meters and alarms.

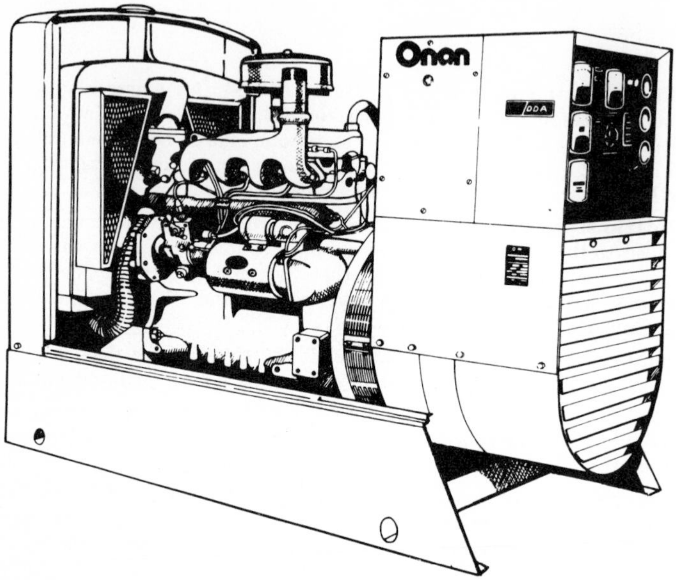

Figure 3-1 Engine-generator set. *(Onan Corporation.)*

Electric Switching Equipment

The function of the electric switching equipment is to interconnect the generator power with the utilization equipment. Included in this group are transfer switches, illustrated in Fig. 3-2, either automatic or nonautomatic. The transfer switch is interlocked to prevent simultaneous closing to both the normal and emergency power source. Automatic transfer switches also monitor both sources and initiate starting of the system. Other electric switching equipment includes bypass switches if needed, overcurrent protection, and in the case of parallel operation of multiple engine-generators, paralleling and totalizing switchboards.

SYSTEM CLASSIFICATION

Recognizing the diversity of applications for emergency power supply systems, the National Fire Protection Association (NFPA) Committee on Emergency Power Supplies is considering the following system definitions based on type, class, category, and level.[1]

Type

Response time is the criterion for determining the system type. Types range from uninterruptible power supply systems that "float" on-line to power systems with no response time requirement. Systems with response times of 60 s or shorter generally are automatic systems, while systems with no response time limit are often manual-starting or portable.

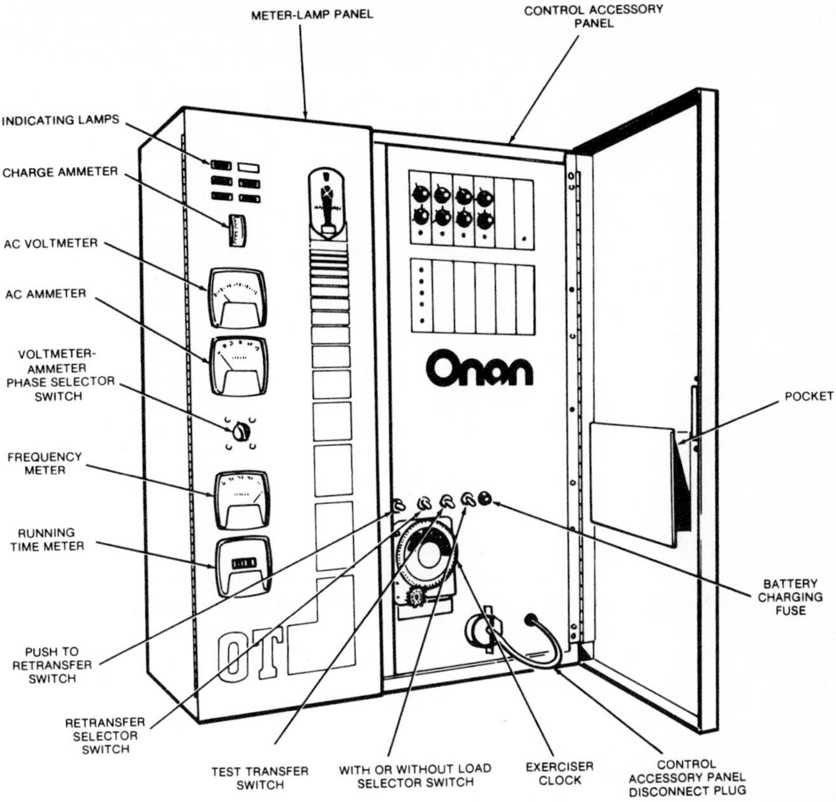

Figure 3-2 Transfer switch. *(Onan Corporation.)*

Class

Systems are placed in a class based on the length of full-load operation possible without refueling or recharging. The extremes of class are those of very short duration, 5 min, to those systems with indefinite duration based on the user's needs. Properly designed rotating-equipment systems have the capability of unlimited operation.

Category

The NFPA Committee on Emergency Power Supplies is considering two categories. One category is stored-energy systems that receive energy from the normal power source. The second category is rotating-equipment systems that use engine-generator sets as the power source.

Level

The critical nature of the load being served by the EPSS determines the system level. For example, a system that supplies loads which support human life and safety is the most critical level. The lowest level defines a system that supplies loads which would result in economic loss when without power. The level of a legally required system greatly influences the requirements the equipment must meet.

SYSTEM JUSTIFICATION

Loss prevention, safety, and legal requirements are three common justifications for an EPSS. A fourth justification, energy economics, is becoming more common as the cost of energy rises. The significant savings on insurance also justifies the cost.

Loss Prevention

An EPSS in an industrial plant can prevent several kinds of measurable loss. Some examples are the loss of wages during production downtime, the loss of in-process product, the loss of data processing, and the loss of refrigerated storage. These examples, of course, are not exhaustive.

Safety

Loss of electric power can directly threaten personnel safety. Industrial processes that present a hazard when without power are one example. Emergency power is also needed to operate elevators, fire pumps, fire alarms, communication networks, and other safety-related equipment.

Legal Requirement

Most states and some major cities have adopted codes that require EPSSs in specific buildings. The classification of the required system is determined by the building occupancy. Because legal requirements change often and are different from state to state, check the most current local regulations. A good source of additional information is the local inspector.

Energy Economics

Peak shaving, cogeneration, and heat reclamation are potentials for saving energy costs that are becoming feasible as the cost of energy rises. Peak shaving uses the emergency system to reduce utility demand charges by assuming loads during periods of peak demand. Cogeneration uses heat from the engines to make steam or hot water for indus-

TABLE 3-1 Emergency Electric Power Supplies—Load Tabulations for Example Location or Building (Onan-Fridley); Nominal Voltage 120/208, Frequency 60 Hz

Load (blocks) functional description	Load				Phase	Circuit no.	Phases used	Static loads		
	Type	Class	Category	Level				kW	PF	A
Incadescent lights exit-evacuation	10	2	A	1	1	5	A–N	2	1.0	−17
Fluorescent lights office-general	60	8	B	2	1	7	A–B–C	5.5	0.95	16
Fire pump, 1	10	2	B	1	3	2	A–B–C	—	—	—
Pump chiller, 1	60	8	B	3	3	7	A–B–C	—	—	—
Elevators, 3 each	10	8	B	1	3	4	A–B–C	—	—	—
Air compressor-1	60	8	B	2	3	6	A–B–C	—	—	—

trial processes. Heat reclamation captures waste heat from the engine through heat exchangers for hot water or space heating or for cooling with absorption chillers.

SYSTEM CONSIDERATIONS

Most system considerations can be grouped under three headings: reliability, capacity, and quality.

Reliability

System reliability is the capability of the EPSS to fulfill its intended purpose predictably without problems. The degree of reliability necessary for an emergency system is closely related to the critical nature of the load being served. Many factors contribute to system reliability; those given here are preinstallation considerations. After installation, effective maintenance, exercise, and experienced operators contribute to reliability.

1. **Compatibility** Since all the equipment in the system is interrelated and must function toward a common purpose, it is important that each piece of equipment is coordinated with the rest of the system. A single design source for the complete system helps to ensure compability.
2. **Testing** Prototype testing during the development process of the engine-generator set ensures the fitness of the equipment for its intended purpose. Prototype testing helps to ensure performance under a variety of conditions such as overloads, surges, short circuits, and others that may occur in actual service. Testing of the complete system for compatability and performance as specified before installation also increases reliability.

Capacity

One of the most imporatnt considerations for an EPSS is capacity, or the amount of power the system can generate. Determination of capacity requires careful study of the load equipment. List the voltage, current, and power requirements of each piece of load equipment. Include pertinent information about the load such as inertia, power factor, starting method, and other data which would affect capacity. Table 3-1 shows a load tabulations example.

Running loads				Motor starting data					Notes: Reduced-voltage starting, SCR controls, wound rotor motor, compression relief, etc.
				NEMA motor code	Locked rotor				
hp Output	kW Input	PF	A		kVA	PF	kW	Starting sequence	
——	——	——	——	——	——	——	——	1	
——	——	——	——	——	——	——	——	3	
40	35	0.85	114	F	212	0.4	84.8	1	Across line starting
40	35	0.85	114	F	212	0.4	84.8	3	Reduced-voltage starting, series reactor, 80%, closed transition
20	17.5	0.85	57	F	106	0.4	42.4	1	SCR controlled—one elevator at a time
20	17.5	0.85	57	F	106	0.4	42.4	2	Compression relief during starting, low inertia load

Choose between full protection or selected protection. Full protection means a larger system, and selected protection means less capacity and special wiring costs. If needs increase in the future, consideration given to expansion can pay off.

Quality

The EPSS should generate power that is essentially equal in quality to the power of the normal source. Requirements such as frequency regulation, voltage regulation, waveform deviation, harmonic content, and noise interference define the quality of the emergency power.

REPRESENTATIVE SYSTEM

Illustrated in the simplified block diagram, Fig. 3-3, is a typical automatic sequential paralleling emergency system. This system, designed to meet NFPA-76A requirements, consists of engine-generator sets, paralleling switchboards, totalizing board, and automatic transfer switches. In this system, three 400-kW engine generators furnish 1200 kW of power. The loads, 1200 kW total, are divided into three priorities of 400 kW each, matching the capacity of one engine-generator. If utility power is interrupted, an automatic transfer switch initiates starting of all three engine-generator sets simultaneously. When the first of the three engine-generators reaches operating voltage and frequency, the corresponding paralleling switchboard closes its circuit breaker connecting the generator power to the emergency bus. With one generator on-line, the priority control enables the first priority transfer switches and the most critical loads to receive power. The shaded line in Fig. 3-3 represents this power distribution. The second and third engine-generators are automatically synchronized with and closed to the emergency bus. As each successive generator is closed to the bus, the priority control enables the second- and third-priority transfer switches to operate.

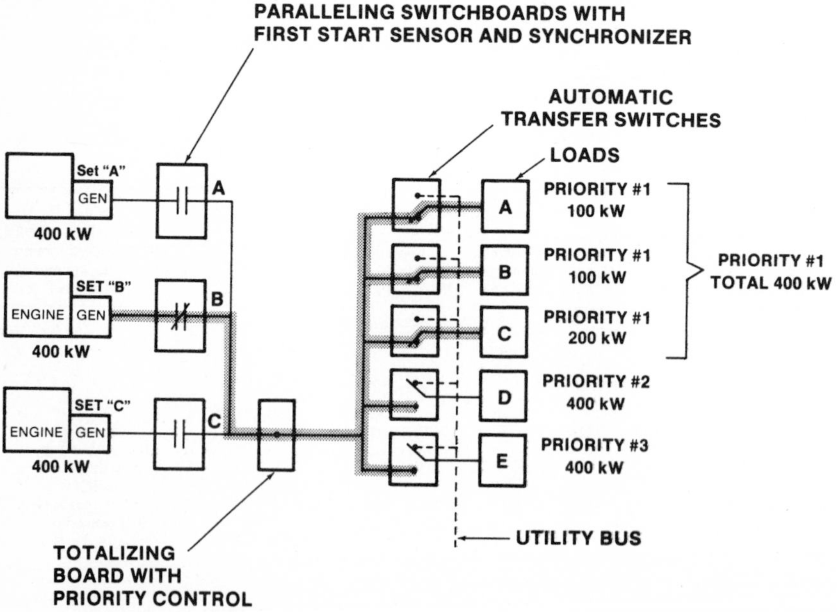

Figure 3-3 Automatic sequential paralleling system. *(Onan Corporation.)*

Comparison of Prime Movers

The following table reviews the advantages and disadvantages of various types of systems.

Advantages	Disadvantages
Gasoline reciprocating	
Lower initial cost than diesel Quick starting, especially in low ambients Lightweight	Fuel storage Gasoline deteriorates over time Lower thermal efficiency than diesel
Gaseous reciprocating	
Lower initial cost than diesel Fuel does not deteriorate over time More efficient combustion than gasoline Lower maintenance needs. Easy starting	Requires high Btu content (1100 Btu/ft^3) or derating necessary May not be permitted in seismic risk areas without backup fuel storage
Diesel reciprocating	
Low maintenance requirements Easy fuel storage Low operation costs Good thermal efficiency Availability in wide range of kW ratings	Higher initial cost than gasoline, gaseous
Gas turbine	
Lightweight Smaller than equivalent diesel Low vibration, low noise No cooling water required Adaptable for cogeneration Meets pollution requirements easily Multiple fuel capability Low maintenance	Higher initial cost

Standard Units

Major components of an EPSS include a generator set, transfer switch control, and if applicable, paralleling switchgear. Units discussed are representative only of the sizes and features generally available from EPSS manufacturers.

Table 3-2 lists representative generator sets for EPSSs. For this type of service, a generator set should have at least a battery-charging alternator, battery-charge-rate ammeter, oil-pressure gauge, coolant temperature gauge, and a run-stop swtich on the engine control (Fig. 3-4). Desirable features, some of which may be required by codes, include an engine-cranking limiter, low-oil-pressure shutdown, high-coolant-temperature shutdown (water cooling), and running-time meter. Water-cooled engines will generally have available as options heat exchanger, city water, or remote radiator systems, and water-cooled exhaust manifolds.

Common sizes of transfer switch controls range from 30 to 2000 A. An automatic transfer switch offers completely automatic, unattended operation and can include time delays for engine starting, load transfer to the emergency power source, retransfer of load to the normal power source, and stopping of the engine (Fig. 3-5). It often has voltage sensors for sensing either undervoltage or overvoltage conditions of the normal source voltage and usually for sensing undervoltage conditions of only the emergency source voltage. Operation indicator lamps and meters are also available, if not standard features. An exerciser which automatically exercises the generator set on a regularly scheduled basis is a popular option.

TABLE 3-2 Representative Generator Sets*

Fuel	Generator set continuous standby rating, kW	Phase	Power factor	r/min	Standard cooling: Air	Standard cooling: Radiator	Approx. weight: lb	Approx. weight: kg
Gasoline†	2.5	1	1.0	1800	X		200	90
	5.0	1	1.0	1800	X		350	160
	7.5	3	0.8	1800	X		475	215
	12.5	3	0.8	1800	X		650	295
	15.0	3	0.8	1800		X	900	410
	30.0	3	0.8	1800		X	1500	680
	55.0	3	0.8	1800		X	1900	860
	85.0	3	0.8	1800		X	2500	1135
Diesel	6.0	3	0.8	1800	X		475	215
	12.0	3	0.8	1800	X		700	315
	15.0	3	0.8	1800		X	900	410
	30.0	3	0.8	1800		X	1600	725
	75.0	3	0.8	1800		X	2600	1180
	100.0	3	0.8	1800		X	2750	1250
	150.0	3	0.8	1800		X	5900	2680
	200.0	3	0.8	1800		X	6200	2815
	300.0	3	0.8	1800		X	7800	3540
	500.0	3	0.8	1800		X	11,500	5220
	750.0	3	0.8	1800		X	16,500	7490

*Generator sets listed have reciprocating engines. Generator sets with gas, diesel, or combination fuel turbine engines are available and can range in size from several hundred to several thousand kilowatts.

†Gaseous fuel operation optional (some as combination gas-gasoline carburetor systems).

Paralleling switchgear includes, for each generator set of the paralleling system, an ac ammeter, ac voltmeter, frequency meter, wattmeter, synchronizing lights, a circuit breaker for connecting generator output to the bus, and voltage and frequency adjustment controls. An ac ammeter, an ac voltmeter, and a wattmeter are also connected to the bus for readings of the total paralleling system output. While manual paralleling or automatic paralleling systems are available, the use of automatic systems is more prevalent. (Automatic paralleling switchgear has provisions for manual paralleling if necessary.)

SYSTEM PREPAREDNESS

Once an EPSS is correctly installed and correctly interconnected, system failure results usually from lack of maintenance support. Whether manual or automatic system startup, preparedness depends on a maintenance program, system exercise, and competent maintenance personnel.

Maintenance Program

Personnel establishing the maintenance program must prepare a maintenance performance schedule for the complete EPSS. Next, they must establish a means to document the system history of maintenance and service performed.

Although maintenance personnel prepare the maintenance schedule, they must include manufacturer's recommendations of maintenance items and schedules. Due to the infrequent nature of operation for these systems, some maintenance items require service intervals of operational hours while other items require time intervals of days, weeks, months, or years. Some items require both. (The EPSS should have a running time meter to indicate hours of operation.)

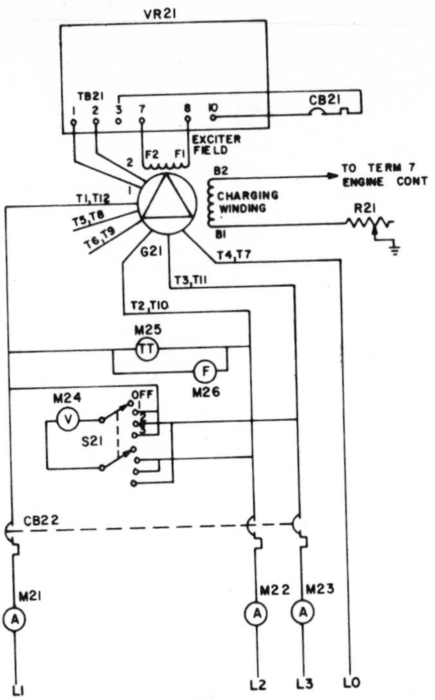

(a)

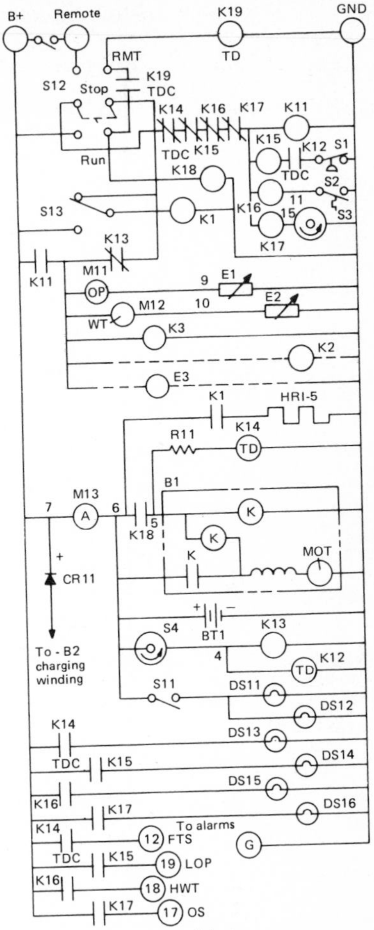

(b)

Figure 3-4 Schematic diagram of generator set. (*a*) AC generator set. (*b*) DC generator set control. All components are shown in the de-energized position. FTS—failed to start; LOP—low oil pressure; HWT—high water temperature; OS—overspeed. *(Onan Corporation)*

Key

B1—Engine starter & solenoid
BT1—Battery

CB21,22—Circuit breaker
CR11—Charger rectifier

DS11,12—Control panel lamp
DS13—Generator set failed to start lamp
DS14—Low oil pressure lamp
DS15—High water temperature lamp
DS16—Overspeed lamp

E1—Oil pressure sender
E2—Water temperature sender
E3—Fuel pump

G21—Generator

HR1—Manifold heater
HR2,3,4,5—Glow plug heater

K1—Heater relay
K2—Water solenoid
K3—Fuel solenoid
K11—Fuel relay
K12—Oil pressure relay
K13—Start disconnect relay
K14—Cranking limiter relay
K15—Low oil pressure shutdown relay
K16—High water temperature shutdown relay
K17—Overspeed shutdown relay
K18—Starter pilot relay
K19—Preheat time delay relay

M11—Oil pressure gauge
M12—Water temperature gauge
M13—DC charge ammeter
M21,22,23—AC ammeter
M24—AC voltmeter
M25—Running time meter
M26—Frequency meter

R11,21—Resistor

S1—Low oil pressure switch
S2—High water temperature switch
S3—Overspeed switch
S4—Centrifugal switch
S11—Panel light switch
S12—Operation mode selector switch
S13—Manifold heater switch
S21—AC voltage selector switch

VR21—Voltage regulator

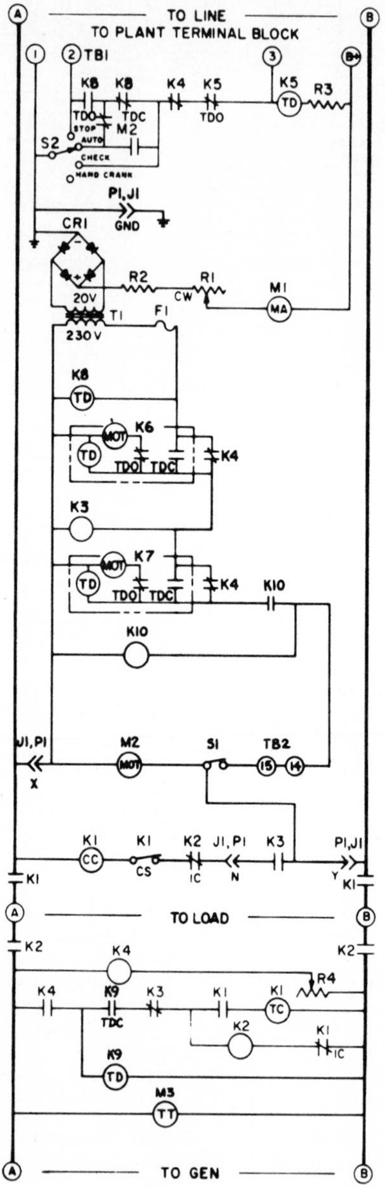

Figure 3-5 Schematic diagram of transfer switch. *(Onan Corporation.)*

The following gives a representative listing of maintenance items and maintenance intervals. Use only as a guideline for recognizing particular system maintenance needs and for establishing a schedule. A health-care facility, for example, might require additional items.

Every 8 Operational Hours

1. Check coolant level (water cooling).*
2. Check crankcase oil level.* Wait 15 min after shutdown for accuracy.
3. Check oil sump level (turbine).
4. Visually inspect generator set. Look for fuel, oil, or coolant leaks. Check exhaust if possible with generator set running. Note security of hardware and fittings.
5. Check fuel level.

Every 50 Operational Hours

1. Check air cleaner.* Perform more often in extremely dusty conditions. Replace if necessary.
2. Inspect governor and carburetor–injector-pump linkage. Clean if necessary. Perform more often in extremely dusty conditions.
3. Drain fuel filter sediment.

Every 100 Operational Hours

1. Clean and inspect crankcase breather.*
2. Change engine crankcase oil.* Change oil at least every 3 months, more often in extremely dusty conditions.
3. Replace engine oil filter element.* Coincide with engine oil changes.
4. Clean engine-cooling fins (air-cooling).*

Every 250 Operational Hours

1. Replace fuel filter element.* For diesel fuel systems with two filters, the second filter from main fuel tank usually needs replacement after several thousand hours.
2. Inspect battery charging alternator.*
3. Replace the ignition points and spark plugs; time ignition (spark ignition).*
4. Check water filter (if equipped).*

Every 500 Operational Hours (Turbine Engine Only)

1. Replace fuel filter.
2. Clean and inspect fuel drain valve.
3. Replace oil filter element.
4. Clean and inspect combustion chamber liner assembly (liquid fuel).
5. Clean and inspect fuel nozzle assembly (liquid fuel).
6. Change oil. Turbine engine manufacturer may allow longer oil change periods if oil sample tests are performed and oil meets engine manufacturer's specifications.

Every 1000 Operational Hours

1. Check generator brushes (if applicable). Brushes must not stick in brush holders.
2. Clean generator. Blow out with low-pressure, filtered, compressed air.

*Reciprocating engine only.

Every 2500 Operational Hours (Turbine Engine Only)

1. Clean and inspect combustion chamber liner assembly (gaseous fuel).
2. Clean and inspect fuel nozzle assembly (gaseous fuel).
3. Clean and insepct igniter plug.

In contrast to operational hours, the following items usually require inspection or maintenance on a regular-time-interval basis. Time intervals used are week, month, half-year, and year.

Every Week

1. Main fuel tank level—Keep full as much as possible.
2. Day tank fuel level.
3. Coolant level (water cooling)—Coolant should have rust inhibitor and antifreeze, if applicable.
4. Fan and alternator belts.
5. Hoses and connections.
6. Coolant heater operation (if applicable).
7. Oil heater operation (if applicable).
8. Batteries—Check cleanliness, electrolyte level, and cable connections.
9. Battery charger—Note charge rate.
10. Exhaust condensation trap—Drain out water.
11. Emergency power supply area—Note general cleanliness. Wipe down entire system. For an exceptionally clean area, longer intervals could be used.
12. Running tests.—Start the generator set and note following (load is preferable):
    - *a.* Fuel system—Check operation of fuel solenoid auxiliary fuel pump (if applicable), and general fuel system operation.
    - *b.* Lubrication system—Note engine oil pressure and record.
    - *c.* Exhaust system—Inspect for tight connections and leaks. Note condition of muffler, exhaust line, and exhaust supports.
    - *d.* Cooling system—Note operating temperature and record (engine must run long enough to warm up).
    - *e.* Battery charging—Note charge rate of generator set.
    - *f.* Meters—Note general operation.
13. System documentation—Check that operation manual, wiring diagram, maintenance schedule, and log are accessible to maintenance personnel.

Every Month

1. Cooling system (water cooling)—Inspect for adequate water flow. Remove any material which interferes with radiator airflow, etc.
2. Ventilation—Air inlets and outlets should have unrestricted airflow. Check security of duct work. Check operation of any motor-operated louvres.
3. Fuel system—Drain water from main fuel tank and day tanks if applicable. Check fuel tank vents.
4. Battery—Check specific gravity of electrolyte. Clean battery terminals.
5. System operation indicator lamps—Test lamps with test switch, if equipped.
6. Transfer switch and paralleling switchgear (if applicable)—Inside cabinets should be clean and free of foreign objects. Check appearance of wiring insulation and color of terminals.

Every Six Months

1. Cooling system (water cooling)—Check for rust and scale. If necessary, flush out system and replace coolant.
2. Engine alarm shutdown devices
3. Transfer switch control—Inspect components and check settings of time delays, voltage sensors, and exerciser, if applicable. Clean cabinet with low-pressure, filtered, compressed air.
4. Generator set control—Clean interior with low-pressure, filtered, compressed air.
5. Paralleling switchgear (if applicable)—Inspect components, busbars, and feeder connections. Clean cabinet with low-pressure, filtered, compressed air.

Every Year

1. Generator—Measure insulation resistances of windings with a megger. Record readings.
2. Paralleling switchgear (if applicable)—Perform insulation tests and record.

Records of maintenance and service performed on the emergency power supply system have two main benefits. They help to ensure that maintenance procedures were performed, and they provide an excellent system history. A maintenance record form should have entry provisions for the date, maintenance work performed, personnel involved, and general comments (form could also include provisions for labor and parts costs). It will show if schedules were met and if authorized personnel performed the maintenance.

Maintenance history can point out repeating problems or symptoms of a problem in the system. It can become a communications reference with a manufacturer for troubleshooting, for repair, or for warranty purposes.

System Exercise

Most engines left idle for long periods of time will have difficulty starting. For this same reason, the EPSS needs a regularly scheduled exercise program to promote operation readiness. The program might utilize either an automatic exercise feature of a transfer switch control or manually initiated testing.

Frequent system use especially benefits the generator set. It causes water to evaporate from the lubrication system and generator windings and causes the engine to coat internal moving parts with a film of oil.

Exercise with load if possible for at least 30 min should cause water evaporation in the lubrication system and minimize engine carbon buildup and exhaust system fouling. System exercise should occur at least once a week.

An automatic exerciser can have settings for the length and number of unattended exercise periods. Manually initiated testing gives the same system benefits as automatic exercising, except it does require the presence of maintenance personnel. However, personnel can use this time to increase their own system familiarity, to train other personnel, or to perform system inspection.

Competent Maintenance Personnel

One popular method to obtain competent maintenance personnel is through maintenance contracts. The manufacturer or manufacturer's representative usually offering the contract service ensures that personnel performing the maintenance are trained for the equipment and that maintenance procedures meet agreed schedules. If plant personnel perform maintenance procedures, they should receive prior equipment training. (Some manufacturers of emergency power supply systems offer such service schools or training sessions.)

REFERENCE

1. National Fire Protection Association, *NFPA Technical Committee Reports,* Boston, Mass., 1979, pp. 22–54.

STANDARDS

"Electrical Power Distribution for Industrial Plants," IEEE Standard 141-1969.
"Recommended Practice for Emergency and Standby Power Systems," IEEE Standard 446-1974.
"Motors and Generators," ANSI/NEMA Standards Publication MG1-1978.
"Centrifugal Fire Pumps," NFPA 20.
"Stationary Combustion Engines and Gas Turbines," NFPA 37.
"National Electrical Code®," NFPA 70.*
"Essential Electrical Systems for Health Care Facilities," NFPA 76A.
"Life Safety Code®," NFPA 101.*

BIBLIOGRAPHY

Stromme, Georg: "Coordination Procedures for On-Site Power Generators," *Specifying Engineer,* May 1977, pp. 116–119.
Stromme, Georg: "Emergency and Standby Power Systems," *Building Operation Management*, July 1977.

* ***National Electrical Code***® and ***Life Safety Code***® are Registered Trademarks of the National Fire Protection Association, Inc., Quincy, MA 02269.

PART 2

Batteries

prepared by

General Battery Corporation
Reading, Pennsylvania

GLOSSARY

Alkaline cell Primary cell with excellent leakage protection capable of higher energy output than carbon-zinc cells.

Ampere-hour A unit of electricity (symbol A·h) equal to the current flowing past any point in a circuit for 1 h at a constant A.

Ampere-hour capacity The number of ampere-hours which a cell delivers under specified conditions of discharge rate, temperature, initial specific gravity, and final voltage.

Carbon-zinc cell Low-cost primary cell with moderate leakage protection and low energy output.

Cycle A discharge and its subsequent recharge.

Cycle service A type of battery operation in which a battery is repeatedly discharged and recharged during the life of the battery.

Equalizing charge An extended charge given to a cell or battery to ensure the complete restoration of the active materials within a cell to the fully charged condition.

Final voltage A prescribed voltage at which a discharge is to be terminated, usually chosen to realize optimum useful capacity without overly discharging the battery.

Finish rate The maximum value of current at which a charged or nearly charged cell or battery may be charged without causing excessive gassing or heating equal to 5 A per 100 A·h of 6- or 8-h rating.

Float service A method of battery operation in which a battery is continuously connected to a bus whose voltage is set slightly higher than the open-circuit voltage of the battery. Under these conditions the battery will either charge or discharge into the load, according to the fluctuations in the bus voltage occasioned by varying load conditions. The bus voltage is set to maintain the battery during normal operation in a fully charged condition with a minimum of overcharging.

Gassing The evolution of gases from either the positive or negative plate of a cell.

Level indicators A float or visible reference mark used to indicate the electrolyte level within a cell.

Nickel-cadmium cell A secondary cell commonly used in portable power applications.

Nominal voltage The voltage rating of a cell or battery arbitrarily assigned to a cell type for the purpose of establishing the operating voltage range. For example, the nominal voltage of a lead-acid cell is 2 V and the nominal voltage of a nickel-cadmium cell is 1.2 V.

Specific gravity The ratio of the density of the electrolyte to the density of water at standard conditions. The specific gravity (spgr) of battery electrolyte is usually measured with a hydrometer. Temperature, water loss, and state of charge will all effect the specific gravity of a cell.

INTRODUCTION

A *battery* is a device consisting of one or more cells that store chemical energy which can be converted into electric energy on demand. The unit of measure of this electric energy is the kilowatthour (kWh), but battery output is often rated in ampere-hour capacity (A · h) because it is an easily measured quantity used to indicate the work capability. A *cell* is the smallest unit a battery can consist of. The minimum components of a cell are two dissimilar electrodes, an electrolyte, a means of conducting the electric power from the cell, and a container. Other components such as separators, covers, and vents are added to improve performance, life, or usage of the cell. The capacity of a battery depends on the internal construction of the cells. The voltage of the battery is the sum of the voltage of each cell connected in series. Cell voltage is a function of the electrode and electrolyte material.

A *primary cell* is a cell that cannot be easily recharged because the electrochemical reaction is nonreversible. The major types of primary cells are carbon-zinc (CZn) and alkaline. These batteries are best used in applications where long shelf life, low current draw, infrequent use, and low initial cost are important. The disadvantages of these cells include low output current, high voltage drops at high current, and the inability to be recharged. The selection of the proper type will depend upon cost, energy output, current draw, frequency of use, and amount of leakage protection required. Primary cells can be found in flashlights, instrumentation, alarm systems, cameras, and many portable low-power devices.

Secondary cells are fully reversible. The chemical energy can readily be restored by supplying electric energy to the cell in a process called *recharging*. Cells can store only a limited amount of energy. No amount of overcharging will store additional energy. The lead-acid battery is the most common form of secondary cell; it provides most of the traction, stationary, and engine-starting requirements of industry. Advances in technology have resulted in sealed lead-acid batteries being used more frequently in portable power applications. The other type of secondary cell used in industrial plants is the nickel-cadmium (NiCd) battery. It is occasionally used in traction, stationary, and engine-starting applications when cost is not a major factor. The nickel-cadmium battery is frequently used in portable power applications because of its ability to provide lightweight, high-current output in repeated-cycle service for a reasonable cost.

BATTERY CLASSIFICATION BY USAGE

An industrial plant usually has many different types of batteries in a great variety of sizes and shapes. Batteries can range from a single cell weighing a few ounces to a large battery that fills a room and weighs many tons. The electric output of batteries can range from a few milliwatts to hundreds of kilowatts. Industrial plants frequently use batteries for portable power, motive power, engine cranking, and/or stationary applications.

Portable Power

These cells and batteries are designed to be easily carried and supply energy requirements for portable lighting, power tools, instrumentations, communications, and alarm signals. They are sealed to prevent leakage and can be either nonrechargeable (primary cell) or rechargeable (secondary cell) depending on load and usage. These batteries usually provide intermittent or steady low-current power for long periods of time. The life can range from a few hours to many years depending on load and cycle conditions. Table 3-3 lists typical ratings used in the selection of portable power cells.

TABLE 3-3 Typical Range of Ratings of Portable Power Cells

	Primary	Secondary
Weight, lb	0.1–5.0	0.1–5.0
Volume, in^3	0.5–200	1–200
Voltage per cell	1.3–1.5	1.2–2.0
5-h capacity, A·h	0.1–6.0	0.1–10.0
No. of cycles	None	100–1000
Cost, dollars per cell*	½–10	1–20
General types	Carbon	Nickel-cadmium
	Mercury	Lead-acid (GEL)
	Alkaline	Lead-acid (sealed)

*1980 dollars.

Motive Power

These batteries are designed for repeated-cycle service supplying the energy to propel and operate electrically powered industrial trucks, sweepers, scrubbers, personnel carriers, mine equipment, and over-the-road electric vehicles. Lead-acid industrial, golf cart, and automotive batteries are the most frequently used types for this service. These

TABLE 3-4 Typical Range of Ratings of Motive Power Batteries

	Minimum	Maximum
Weight, lb	30	7000
Length, in	10	60
Width, in	7	45
Height, in	9	36
Voltage per battery, V	6	96
No. cells	3	48
6-h capacity, A·h	50	2000
Cycles	1000	2000
Life, years	1	7
Cost, dollars per battery*	50	8000
General types: Industrial truck, mine, golf cart, automotive, electric vehicle		

*1980 dollars.

batteries usually provide intermittent moderate and high-current power for 3 to 10 h to and 80 percent depth of discharge between recharges. They are discharged 3 to 10 times a week with a life ranging from 1 to 7 years. Table 3-4 lists data on and ratings of typical motive power batteries. Figure 3-6 shows the internal construction of a typical motive power cell. This diagram points out some of the important design criteria used to achieve good performance and long life in motive power service.

Engine Cranking

These batteries are designed to furnish the electric energy requirements of internal combustion engines used in vehicular and stationary applications. These requirements include starting, lighting, and ignition (SLI). Some nickel-cadmium batteries are used in this type of service but most are either automotive or industrial lead-acid storage batteries. These batteries usually provide high-current power for a very short period of time during the engine-starting process. This results in a very shallow depth of discharge and very many cycles. Table 3-5 list information and ratings of typical engine-starting batteries.

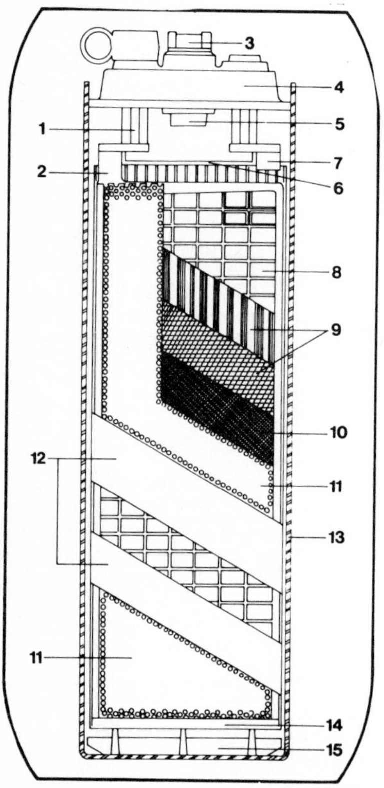

Figure 3-6 Internal construction of typical motive power cell. Key: 1, terminal post; 2, positive grid; 3, vent cap; 4, cover; 5, acid level indicator; 6, separator protector; 7, negative grid; 8, active material; 9, positive active material retention material; 10, positive active material retention material; 11, positive active material retention material; 12, separator; 13, container or jar; 14, positive active material retention material; 15, plate rest and sediment area.

Stationary

Batteries designed to be permanently installed on supporting racks and operated in float service as the backup or emergency energy source for communication systems, switchgear and control equipment, emergency dc power for lighting and essential equipment, and uninterrupted power supplies (UPS). The most common types of stationary batteries are antimony and calcium lead-acid. A few nickel-cadmium and Edison batteries are also found in this service. The power requirements of stationary batteries vary depending on the service. Carefully selecting a battery to match the load and cycle requirements of a particular operation is very important in achieving the desired life. Stationary batteries are often custom-designed to match the usage requirements. Table 3-6 lists information about and ratings of typical stationary cells.

BATTERY CHARGING

There are many ways to charge a battery. Modern chargers are becoming more and more automated reducing the maintenance needs and increasing the life of the battery. The charging methods used will depend on the type of service and battery. Battery recharge should always be conducted to return the correct amount of charge. Both *overcharging* and *undercharging* are detrimental to battery life.

TABLE 3-5 Typical Range of Ratings of Engine-Cranking Batteries

	Minimum	Maximum
Weight, lb	5	1600
Length, in	4	45
Width, in	3	30
Height, in	3	20
Voltage per battery, V	6	64
No. cells	3	32
Cranking rate, A	50	3500
Life, years	2	8
Cost, dollars per battery*	15	6000
General types: Motorcycles, automotive, truck, marine, aircraft, diesel locomotive, and stationary diesel engine		

*1980 dollars.

TABLE 3-6 Typical Range of Ratings of Stationary Cells

	Minimum	Maximum
Weight per cell, lb	10	1800
Length per cell, in	3	19
Width per cell, in	3	18
Height per cell, in	6	60
No. cells	3	120
Voltage per battery, V	6	250
8-h capacity, A·h	10	8000
Life, years	5	20
Cost, dollars per cell*	50	2500
General types: Communication, utility, emergency lighting, uninterrupted power supplies		

*1980 dollars.

Constant Current

A charge conducted at a constant rate. The rate is usually at or below the finish rate. A *trickle charge* is a low-rate constant-current charge given to maintain the battery at a fully charged condition. This should be used only when the charging rate is matched to the battery. Some portable power and a few small-size stationary batteries use the trickle-charge method.

Two-Step Charging

This is a motive power charge technique commonly used on motor generators. The type of charge consists of a high rate charge followed by a lower finishing rate charge. The finish rate charge is often initiated by a temperature voltage relay (TVR) which detects the gassing voltage. A timer to limit the length of the finish rate charge is often started by the TVR in order to limit overcharging.

Modified Constant Potential

This is a charging method frequently used in motive power service to automatically regulate the charging rate throughout a recharge. Ferromagnetic circuits are commonly used to control the initial rate, taper, and finishing rate of the recharge over a broad range of conditions.

Constant Voltage (Potential)

This is charging method frequently used with engine cranking and stationary batteries employing a charger with a voltage which is maintained at a constant value. The charging current is dependent on battery needs once the battery reaches the fixed potential. Stationary batteries frequently use a *float charge* which is a constant-voltage charge in which the voltage is set at a value slightly greater than the open-circuit potential to maintain all internal losses without overcharging the battery.

Charger Selection

Battery-charger selection is an important part of achieving good battery life and recharge efficiency. The following items should be considered when selecting a charger: automatic charging rate control; provisions for equalizing charge; fail-safe design (component failure will not harm battery); low electrolyte temperature rise during recharge; automatic over- and undercharge protection; automatic charger termination; charger protected from shorted and open-circuit output; charger protected from reverse polarity; polarity- and voltage-keyed connectors; high electrical efficiency and power factor; and low ac line draw.

BATTERY MAINTENANCE

It is a good practice to keep batteries clean and dry. Cleaning reduces losses due to contact resistance and prevents shorting or grounding through conductive dirt films. The heat dissipation is improved on clean batteries which helps reduce operating temperature. Batteries are commonly operated between 40 and 120°F (6 and 71°C). Freezing should be avoided because it can permanently damage a battery. High-temperature storage or operation will effectively reduce life. For optimum performance and life, the manufacturers recharge and maintenance instructions should be closely followed.

Portable power batteries require little maintenance beyond being kept clean, dry, and cool. Primary cell maintenance is limited to replacing when leaking or discharged. Secondary portable power cells require charging when discharged and replacement when leaking. Since most secondary portable power cells are nickel-cadmium, recharging should be conducted after a full discharge. Repeated charging after partial discharge will reduce the capacity.

Motive power, engine cranking, and stationary batteries require cleaning, watering, and charging. When cleaning, electrolyte on the cell covers or connectors should be neutralized and rinsed to prevent corrosion and shorting. Water that is approved for batteries or distilled water should be added as required to keep the electrolyte between the high and low level. The electrolyte level should always be above the plates. Gas bubbles created during charging displace volume causing the electrolyte level to increase. Therefore, water should be added when the battery is on charge and gassing. If water is added after the gas has dissipated from the electrolyte, room should be left for expansion.

Battery connections and charging equipment should be checked at least once a month. A loose or dirty connection can reduce performance or cause an explosion. Malfunctions in charging equipment can result in over- or undercharging which will reduce the life of the battery.

Maintenance records of batteries are useful in scheduling periodic maintenance functions such as checking charging equipment, keeping the battery clean, and maintaining levels. Records can also be used to locate problem areas.

BATTERY SAFETY

Caution should be used when storing, operating, or repairing a battery because of the chemical, explosive, and electrical safety hazards associated with all batteries.

Chemical

Batteries contain corrosive liquids which can be harmful on contact. Always wear protective clothing when exposed to corrosive liquids.

Explosive

Some batteries present an explosive safety hazard because of hydrogen gas released during charging. This hazard is controlled by ventilation and preventing ignition by sparks or open flame in charging areas.

Electrical

High-voltage batteries should be treated like any other-high-voltage source for protection from shock hazards. Precaution should be taken to keep metal objects away from battery connectors and terminals to prevent shorting. A battery stores a large amount of energy which can be released rapidly when shorted.

Batteries are safe when proper safety practices are followed. All personnel working with batteries should be trained in the operation and safety practices provided by the battery manufacturer to prevent injury and damage to equipment.

SOURCES OF INFORMATION

Federal Specifications

Superintendent of Documents, U.S. Government Printing Office, Washington, DC 20402.

NEMA Standards

National Electrical Manufacturers Association, Suite 300, 2101 L Street, N.W., Washington, DC 20037.

BCI Standards

Battery Council International, 111 East Wacker Drive, Chicago, IL 60601.

IEEE Standards

Institute of Electrical and Electronic Engineers, 345 East 47th Street, New York, NY 10017.

Book

Vinal, George Wood: *Storage Batteries,* 4th ed., Wiley, New York, 1955.

chapter 3-4

Motors and Motor Controls

PART 1

Electric Motors

by

James V. Yu

Manager, New Market and Product Planning
Electric Motor Division, Gould Inc.
St. Louis, Missouri

GLOSSARY OF COMMON MOTOR TERMS

Ambient temperature the temperature of the surrounding cooling environment.

Amperes The unit of intensity of electric current flowing in a conductor produced by the applied voltage.

Full-load amperes (FLA): The current drawn by the motor when the motor is delivering its rated horsepower at its rated speed (full-load speed) and with rated voltage and frequency applied to the motor.
Locked-rotor amperes (LRA) (starting current per ampere): The current drawn by the motor with the rotor locked (zero speed) and with rated voltage and frequency applied to the motor.
Service-factor amperes (SFA): The current drawn by the motor when the motor is delivering its service-factor horsepower with rated voltage and frequency applied to the motor.

Efficiency How well a motor turns electric energy into mechanical energy, expressed as the ratio of mechanical power output (watts) to electric power input (watts).

$$\text{Efficiency } \% = (\text{hp} \times 746)/(\text{input watts}) \times 100$$

Frequency The number of complete cycles of current per second made by alternating current. While sometimes called cycles per second, the preferred terminology is hertz. The standard frequency in the United States is 60 Hz; however 50 and 25 Hz can be found in the United States as well as in other industrial nations.

NEMA The National Electrical Manufacturers Association (NEMA), an organization which establishes voluntary standards and represents general practice in industry. NEMA has defined motor standards that include nomenclature, construction, dimensions, tolerances, operating characteristics, performance, quality, rating, and testing.

Phase The number of circuits over which electric power is supplied. In a single-phase motor, power is supplied in a single circuit or winding. In a three-phase system, power is provided over three circuits, each circuit reaching corresponding cyclic values at 120° intervals. AC motors are typically qualified as single-phase or polyphase.

Poles The number of magnetic poles in a motor, determined by the location and connection of the windings:

$$\text{Synchronous speed} = \frac{\text{frequency} \times 120}{\text{number of poles}}$$

Power *Mechanical power* is the rate of doing work, usually expressed in terms of horsepower.

$$\text{Power (ft}\cdot\text{lb/min)} = \frac{\text{work (ft}\cdot\text{lb)}}{\text{time (min)}}$$

$$\text{Horsepower (hp)} = \frac{\text{ft}\cdot\text{lb/min}}{33{,}000} = \frac{\text{Watts}}{746}$$

Electric power is measured and expressed in watts, kilowatts (1 kW = 1000 W), or megawatts (1 MW = 10^6 W).

In *dc circuits*,

$$P = VI$$

where P = power in watts, V = line voltage in volts, and I = line current in amperes.
In *single-phase ac circuits,*

$$P = VI \times \mathrm{PF}$$

where PF is the power factor.
For three-phase ac circuits,

$$P = \sqrt{3}\, VI \times \mathrm{PF}$$

If the voltage and current are not in phase (as in most magnetic circuits), the volt-amp product is not actual power but apparent power (voltage times total line current) measured in voltamperes (VA) or kilovoltamperes (kVA).

Power Factor The cosine of the phase angle between line voltage and current in ac circuits. The phase angle is determined by the electrical characteristics of the load. See pp. 3-8 and 3-9 for a detailed discussion of power factor.

Rotor The rotating part of an electric motor.

Rated temperature rise The NEMA allowable rise in temperature for a given insulation system above ambient, when operating under maximum allowable load (i.e., service-factor load).

Stator The stationary part of an electric motor.

Service factor A multiplier which indicates what percent higher than the nameplate horsepower can be accommodated continuously at rated voltage and frequency without injurious overheating (i.e., exceeding NEMA allowable temperature rise for given insulation systems).

Slip The percentage reduction in speed from synchronous speed to full-load speed (known as *percent slip*). All ac induction squirrel-cage motors have slip.

Speed The rotational velocity of the motor shaft, measured in terms of revolutions per minute (r/min).

Full-load speed: Motor speed at which rated horsepower is developed;
No-load speed: Motor speed when allowed to run freely with no load coupled.
Synchronous speed: The synchronous speed of an ac motor is that speed at which the motor would operate if the rotor turned at the exact speed of the rotating magnetic field. However, in ac induction motors, the rotor actually turns slightly slower. This difference is the slip and is expressed in per cent of synchronous speed. Most induction motors normally have a slip of 1 to 3 percent.

Squirrel cage A term used to describe the construction of one type of induction motor. The rotor is made of an iron core mounted on a concentric shaft. Copper, brass, or aluminum bars run the entire length of the core in slots on the core. These bars act as conductors and are fastened on each end of the rotor to end rings in order to form a complete short circuit within the rotor.

Torque The turning effort of a motor, normally expressed in ounce-feet (for fractional horsepower motors) or pound-feet (for integral horsepower motors). A motor developing 15 lb·ft of torque develops a force of 15 lb at the end of a 1-ft-radius lever arm.

Accelerating torque: The difference between the torque developed by the motor and torque required by the load at any given speed. This excess torque accelerates the motor and load.
Breakdown torque: The maximum torque developed by the motor at rated voltage and frequency, without an abrupt drop in speed.
Locked-rotor torque (also called *starting torque, static torque, breakaway torque*): The minimum torque developed at rest for all angular positions of the rotor with rated voltage applied at rated frequency.

Pull-in torque: The torque of a *synchronous motor* that brings the driven load into synchronous speed. (There is no corresponding term for induction motors.)

Pull-out torque: The maximum torque of a *synchronous motor* developed at synchronous speed with rated frequency and excitation.

Pull-up torque: The minimum torque developed by the motor during the period of acceleration from zero speed (rest) to the speed at which breakdown occurs.

Voltage A unit of electromotive force. One volt applied to a conductor offering 1 Ω of resistance will produce a current in that conductor of 1 A.

Wound rotor A term used to describe the construction of a type of induction motor. Similar in configuration to the squirrel cage, a wound rotor motor has windings or coils in the slots on the rotor core.

INTRODUCTION

Electric-motor application for the plant engineer is the common-sense matching of load requirements with motor characteristics. Motor types, styles, sizes, mountings, and enclosures vary greatly. So the first step in using electric motors correctly is to understand them and the terminology the motor industry uses to describe them.

To help achieve such an understanding, the first part of this section presents a glossary of common motor terms essential to motor use and application. After that, motors and the many variables by which they are classified are discussed in terms of National Electrical Manufacturers Association (NEMA) standards, the unifying doctrine within the motor industry.

There are many ways to classify motors. But whichever one might choose, a familiarity with NEMA classifications and standards will at some point be necessary. Thus NEMA standards are as good a basis as any on which to organize a discussion of motors for the plant engineer.

NEMA is a nonprofit trade organization whose voluntary standards have been widely adopted by motor manufacturers and users alike. The NEMA standard "Motors and Generators" (MG1-1967) is designed to eliminate misunderstandings between manufacturer and purchaser and to assist the purchaser in selecting and obtaining the proper product for his or her particular needs.

Motors are classified by size, application, electrical type, NEMA design letter, and environmental protection and cooling methods. They are rated for special standard environmental and operating service conditions by performance and mechanical configuration: voltage and frequency, locked-rotor kVA, service factor, horsepower, speed, torque, locked-rotor current, performance, temperature rise, duty cycle, and frame size.

Because of the large number of motor types and configurations, this section is limited to motors of most interest to plant engineers. Subfractional horsepower and special-use motors (such as small-instrument, small-fan, stepping, and timing motors for example) are only rarely specified for plant-engineering applications.

Motor types and classifications are followed by a brief guide to motor selection, the essentials of picking the right motor for either a new or replacement application. Following this application guide are discussions of two of the more practical concerns that face the electric motor user: troubleshooting and energy efficiency.

The section ends with a list of references and sources of additional information for each of the areas discussed.

MOTOR CLASSIFICATION

NEMA standards for electric motors cover frame sizes and dimensions, horsepower ratings, service factors, temperature rises, and performance characteristics. Such standards provide greater availability, more convenience in use, a basis for accurate comparison, faster repair service, shorter delivery times, and maximum mechanical and electrical interchangeability from motor to motor.

Motors are classified by size, application, electrical type, design letter, environmental protection, and cooling methods.

Classifying by Size

Virtually all electric motors used by the plant engineer can be classified as either fractional or integral horsepower. Despite the obvious distinction of horsepower, frame size (discussed later in this section) actually determines to which category a motor belongs.

A fractional horsepower (FHP) motor is either a motor built in a frame which is designated by a two-digit frame number or in a three-digit frame smaller than the 140 series. An integral horsepower (IHP) motor is one built in a frame which has a three-digit frame number from 140 to 680.

Classifying by Application

NEMA also classifies electric motors by application as general-purpose, definite-purpose, or special-purpose. A general-purpose motor is an induction motor which has a continuous rating, service factor, and temperature rise in accordance with NEMA standards. General-purpose motors are built in quantity in standard ratings with standard operating characteristics and mechanical construction for a wide variety of common applications.

A definite-purpose motor, on the other hand, is designed for specific service conditions and applications. It differs from the general-purpose motor with respect to rating, serving factor, and temperature rise, one or all of which have limits much narrower than those of general-purpose motors. Definite-purpose motors conform to established NEMA standards, are produced in high volume, and are often low in cost compared with general-purpose motors of the same ratings. However, use of a definite-purpose motor for a duty other than that for which it was intended must be carefully considered.

Special-purpose motors incorporate specialized operating characteristics and/or mechanical construction to serve one-of-a-kind applications not satisfied by general- and definite-purpose motors.

Because of both the limited scope of this discussion and the fact that the great majority of plant applications require general-purpose motors, only general-purpose motors are discussed here.

Classifying by Electrical Type

Figure 4-1 illustrates the family of ac and dc general-purpose motors which together can serve virtually all needs of the plant engineer. This family is organized according to the characteristics of the electric power driving the motor and the variations in motor winding and rotor configuration.

AC Motors

Single-Phase Induction Motors. Alternating current motors fall into three major categories: single-phase, polyphase, and universal (ac-dc). Single-phase induction motors are inherently unable to start themselves. Thus these motors are classified by their means of starting as well as by their basic design, either induction or synchronous. Induction motors fall into two further categories: squirrel-cage and wound-rotor.

The squirrel-cage motor consists of a wound stator and laminated, cylindrical iron-core rotor. Cast-aluminum conductors imbedded within the rotor and short-circuiting end rings form a "squirrel-cage" configuration.

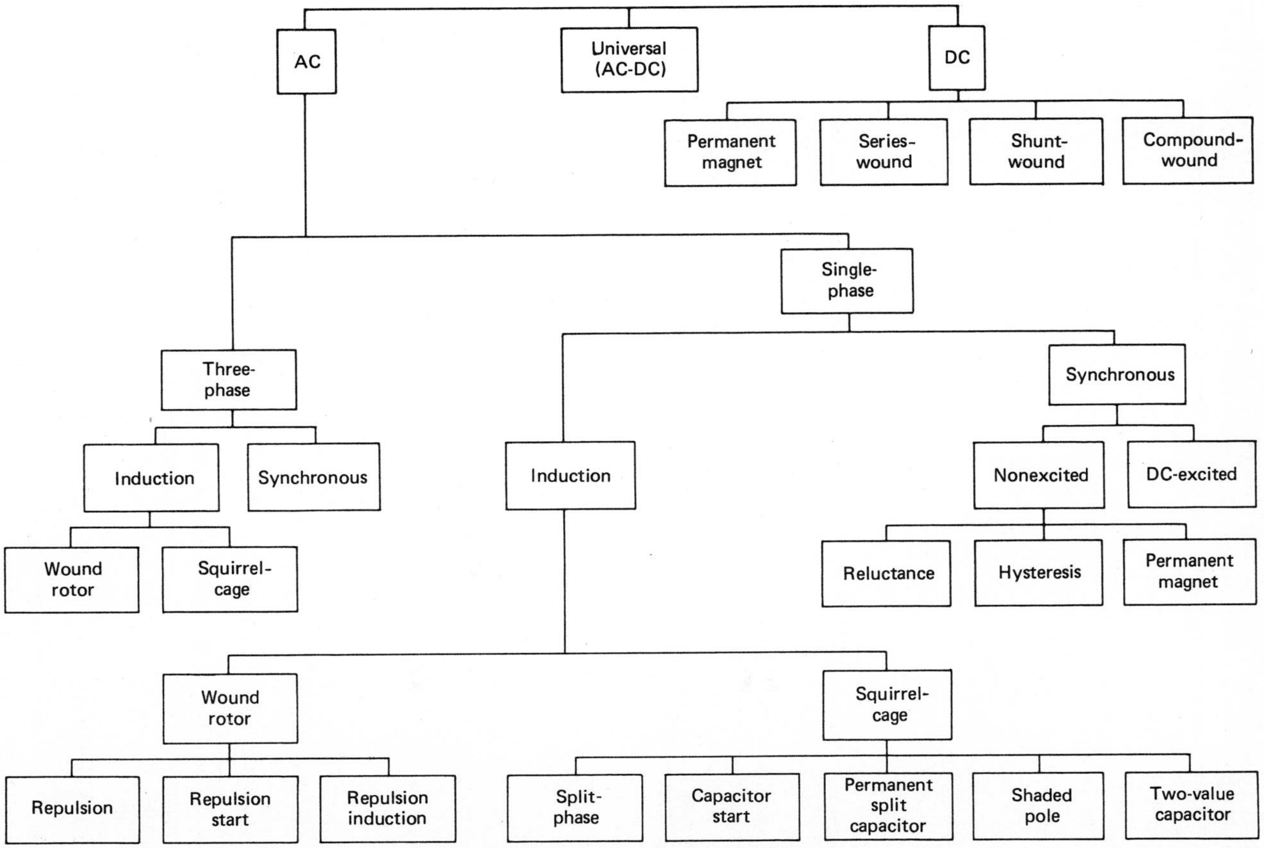

Figure 4-1 Electric motor family for plant engineers.

Split-Phase Motors. Split-phase motors use two windings—a main winding and a start winding (Fig. 4-2). The high resistance of the start winding creates a phase shift and induces a torque that causes initial motor rotation and acceleration. When a predetermined speed (the cutout point) is attained, a centrifugal mechanism opens the start-winding circuit. The motor then accelerates with only the main winding energized and runs as an induction motor.

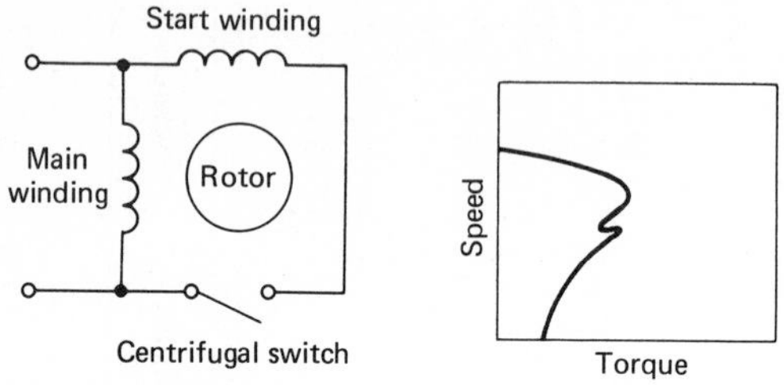

Figure 4-2 Schematic diagram and speed vs. torque diagram—ac split-phase motor.

Capacitor-Start Motors. Capacitor-start motors are similar to split-phase motors except that a capacitor is placed in series with the start winding to produce greater starting and accelerating torque. (Fig. 4-3). After the start winding is removed from the circuit by a centrifugal or electronic switch, performance is identical to that of split-phase motors.

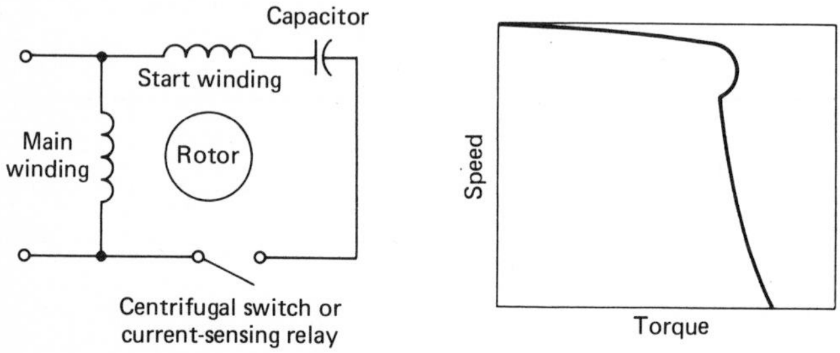

Figure 4-3 Schematic diagram and speed vs. torque diagram—ac capacitor-start motor.

Permanent Split-Capacitor Motors. Permanent split-capacitor motors also have a start winding with a capacitor (Fig. 4-4). Because the capacitor and start windings are continuously energized, these motors operate at a higher power factor than other designs, although at the expense of a lower locked-rotor torque. Since no centrifugal switch is needed, the motor is usually shorter and often more reliable than other single-phase designs.

Two-Value-Capacitor Motors. Two-value-capacitor motors have both a switched-start capacitor and a run capacitor to improve full-load current, starting torque, and power factor (Fig. 4-5). Both are connected in parallel to the start winding, with the start capacitor disconnecting as the motor accelerates. These motors provide good overall torque characteristics and are quiet-running.

Shaded-Pole Motors. Instead of a start winding, shaded-pole motors have a continuous solid-copper loop around a small portion of each salient pole (Fig. 4-6). This shading coil causes the reaction necessary to start the motor, but produces rather low starting

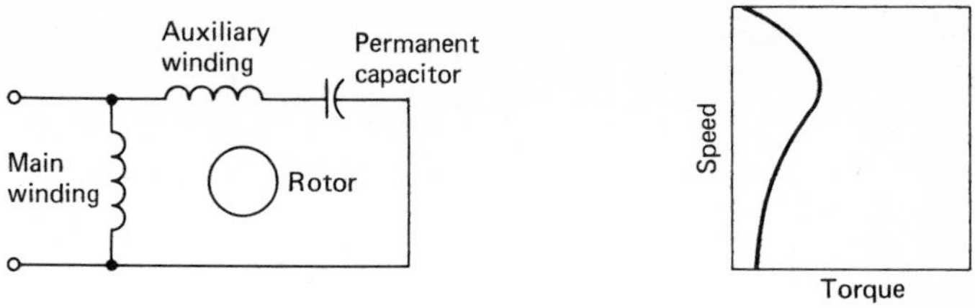

Figure 4-4 Schematic and speed vs. torque diagram—ac permanent split-capacitor motor.

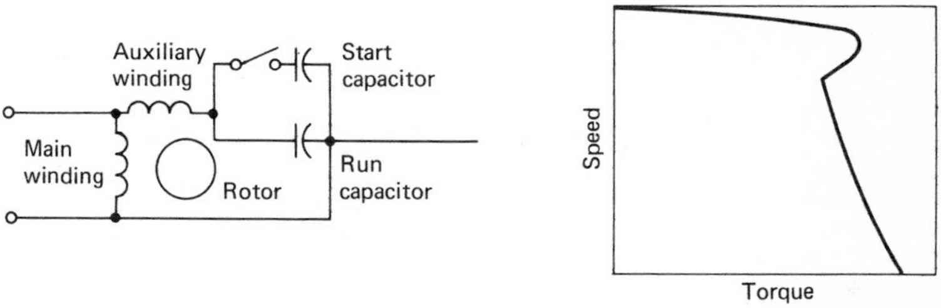

Figure 4-5 Schematic and speed vs. torque diagram—ac two-value-capacitor motor.

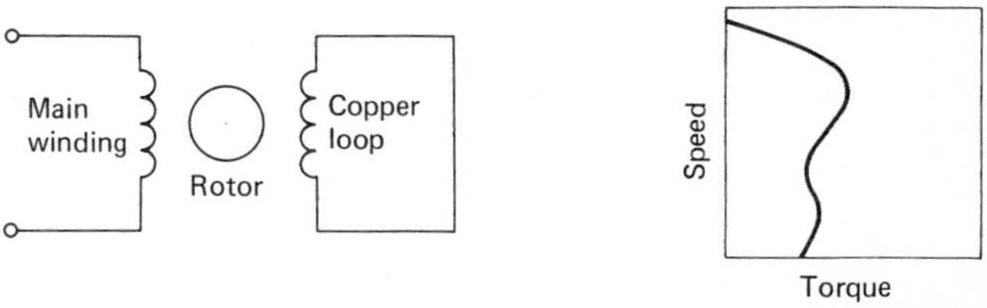

Figure 4-6 Schematic and speed vs. torque diagram—ac shaded-pole motor.

and accelerating torque. Because of their low starting torque, shaded-pole motors are best suited to light-duty applications such as direct-drive fans and blowers. Efficiency and power factor are also lower than that of other single-phase motors.

Wound-Rotor Motors. Wound-rotor motors have a stator winding connected to the power source and a rotor winding connected to a commutator.

Unlike the squirrel-cage induction motor, the wound-rotor motor has controllable speed and torque. Their application is considerably different from squirrel-cage motors because of the accessibility of the rotor circuit. Various performance characteristics can be obtained by inserting different resistances in the rotor circuit.

Wound-rotor motors may be used as constant-speed or as adjustable-speed motors. They are frequently used where high locked-rotor and accelerating torque with low starting current are required.

Repulsion Motors. Repulsion motors have an armature winding, commutator, and brushes (Fig. 4-7). Brushes are short-circuited and shifted to give the effect of two stator windings: a field winding at right angles to the brush axis and an induction winding along the brush axis. The induction winding induces current in the armature winding that reacts with the magnetic field set up by the field winding to produce starting torque. Repulsion motors feature good starting characteristics and are often used for heavy hard-to-start loads.

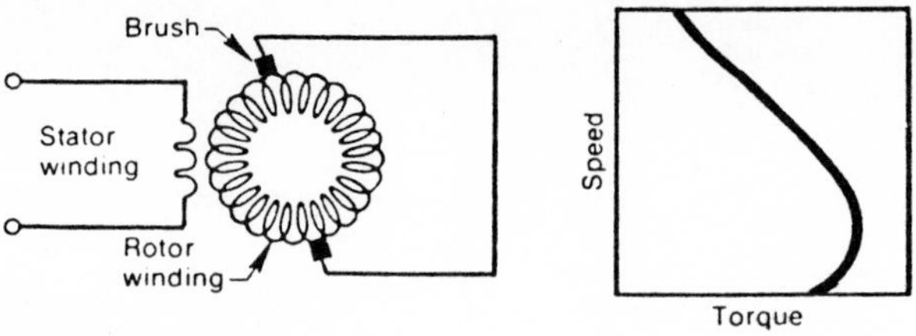

Figure 4-7 Schematic and speed vs. torque diagram—ac repulsion motor.

Repulsion-Start Motors. Repulsion-start motors are repulsion motors with a centrifugal switch (Fig. 4-8). At about 75 percent of synchronous speed, the switch short-circuits the commutator bars and the motor performs like a squirrel-cage motor.

Repulsion-start motors are expensive and no longer widely used in industry.

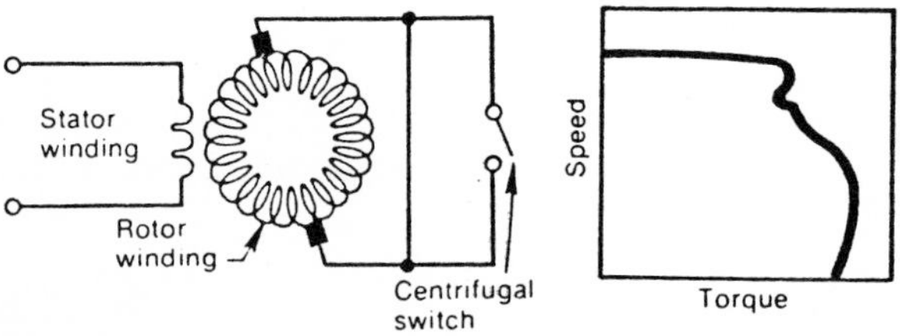

Figure 4-8 Schematic and speed vs. torque diagram—ac repulsion-start motor.

Repulsion-Start Induction-Run Motors. Repulsion-start induction-run motors are simply repulsion-start motors with the addition of a squirrel-cage rotor winding to improve speed regulation (Fig. 4-9). At a predetermined speed, a centrifugal switch shorts the commutator and the motor operates as a squirrel-cage induction motor.

These motors are ideal for applications requiring high starting torque and low starting current.

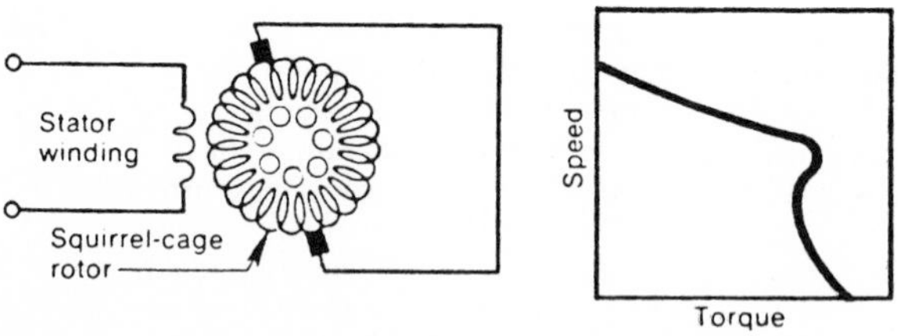

Figure 4-9 Schematic and speed vs. torque diagram—ac repulsion-induction motor.

Single-Phase Synchronous Motors. Single-phase synchronous motors are constant-speed motors that operate in synchronism with line frequency. As with squirrel-cage induction motors, speed is determined by the number of pairs of poles and is always a ratio of the line frequency.

Synchronous motors range from subfractional self-excited units to large horsepower, dc excited motors for industrial drives. In the single-phase fractional horsepower range, synchronous motors are primarily used where precise constant speed is required.

Like single-phase induction motors, synchronous motors cannot start themselves. They employ self-starting circuits.

Nonexcited Reluctance Motors. Reluctance synchronous motors have squirrel-cage construction with salient poles (Fig. 4-10). The rotor has one cutout for each pole, which together cause magnetic reluctance to be greater between poles than along the axis. The motor locks into synchronism in less than one cycle of applied voltage.

Efficiency and power factor are lower than for dc-excited synchronous or squirrel-cage induction motors. However, the motor is inexpensive, simple, and suitable for light loads.

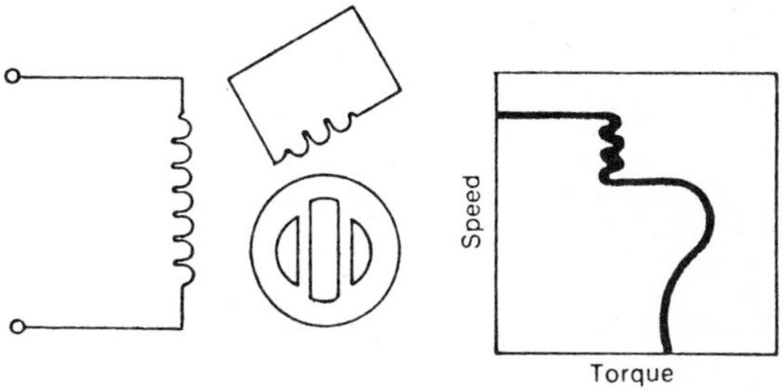

Figure 4-10 Schematic and speed vs. torque diagram—ac reluctance motor.

Nonexcited Hysteresis. Hysteresis motors have no physical pole arrangement on their rotors but develop fixed magnetic poles in some random angular position as they reach synchronous speed (Fig. 4-11).

Used as timing motors or for applications requiring precise constant speed, they are of secondary interest to plant engineers.

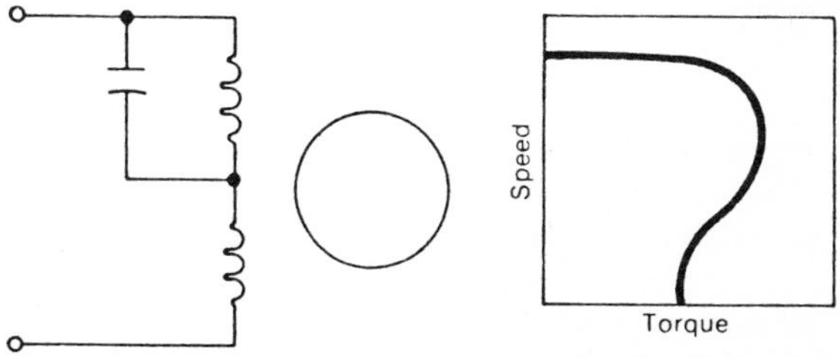

Figure 4-11 Schematic and speed vs. torque diagram—ac hysteresis motor.

Nonexcited Permanent Magnet Motors. Permanent magnet motors have permanent magnets embedded in a squirrel-cage-type rotor (Fig. 4-12). They produce fixed poles that lock into step with the armature field. Because of its relatively high efficiency and power factor, this motor is popular in the fractional and lower integral horsepower range.

Three-Phase Motors. The rotating magnetic field provided by three-phase ac power permits a simple and low-cost means of constructing an electric motor. In general-pur-

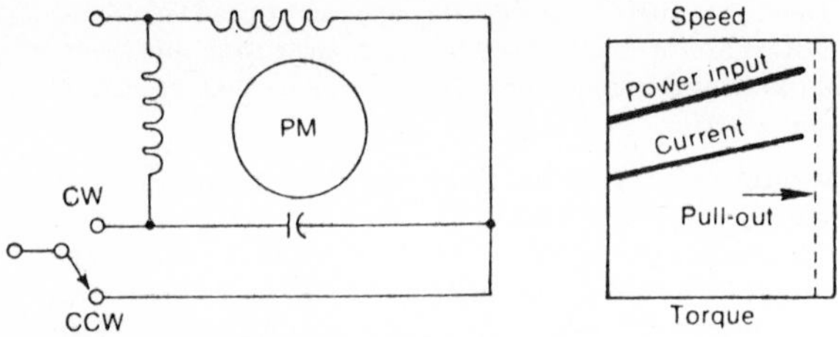

Figure 4-12 Schematic and speed vs. torque diagram—ac permanent-magnet motor.

pose use, three-phase motors require no start windings, switches, or starting or running capacitors, thereby eliminating some major sources of failures in single-phase motors.

The horsepower of three-phase motors ranges from ½ to 2500 or more. Starting current required is low to medium, about 5 to 7 times full-load current.

Three-phase motors can be easily reversed electrically, making them useful for applications involving control of direction of rotation or remote positioning. Different combinations of speed-torque characteristics are also available so that motor performance can be specifically matched to an application.

Three-Phase Induction Squirrel-Cage Motors. Three-phase squirrel-cage motors are basically constant-speed machines, although operating characteristics can be varied to some degree by modifying the rotor design. These variations produce predictable changes in torque, current, and full-load speed.

Evolution and standardization within the motor industry have resulted in five fundamental types of three-phase induction motors known by NEMA design letters (discussed later in this section).

Three-Phase Induction Wound-Rotor Motors. Wound-rotor motors offer more speed and torque control than squirrel-cage induction motors, as well as the major advantage of accessibility of the rotor circuit. Performance characteristics can be varied merely by inserting different values of resistance in the rotor circuit.

Wound-rotor motors may be used as either constant-speed or adjustable-speed motors. With full load, the speed may be reduced by as much as 50 percent of synchronous speed for certain fixed loads such as fans or compressors. These motors are frequently used when high locked-rotor and accelerating torque with low starting current are required. They are also used where heavy or delicate loads must be accelerated gradually and smoothly, as in hoists and elevators. A variety of solid-state control systems are available for use in the rotor circuit of wound-rotor motors.

AC Three-Phase Synchronous Motors. Like single-phase motors, three-phase synchronous motors cannot start by themselves. One of two starting methods is used in most motors: dc excitation and reluctance.

Synchronous motors employing dc excitation, although offered in sizes as small as 20 hp, are primarily used in applications requiring 50 to several thousand horsepower. High-speed synchronous motors (from 514 to 1800 r/min) are normally used for the same applications as NEMA design A, B, or F squirrel-cage motors. Low-speed synchronous motors (below 450 r/min) are usually used as direct-connected drives for compressors and pumps, where they are more economical than induction motors with gear, chain, or belt drives.

Reluctance-type synchronous motors are usually limited to 30 hp and smaller sizes. They are self-contained, self-excited units that use special rotor laminations to get the synchronous speed characteristics. Reluctance types produce higher locked-rotor, pull-up, and pullout torques than comparable induction motors, but also require relatively high locked-rotor current.

DC Motors

DC motors see a wide variety of industrial applications because their speed-torque relationships can be varied to almost any useful form—for both motor and regeneration applications and in either direction of rotation. Many dc motors can be operated continuously over a speed range of 8:1. Speed control down to zero for short durations or for driving reduced loads is also common.

AC motors lose speed rapidly and sometimes stall at loads above twice their rated torque. DC motors, by contrast, are often applied where they momentarily deliver three or more times their rated torque. And in emergency situations, dc motors can supply over 5 times rated torque for a limited time without stalling if the required power is available.

DC motor speed can be regulated smoothly down to zero, immediately followed by acceleration in the opposite direction without power circuit switching. DC motors also respond quickly to changes in control signals due to their high torque-to-inertia ratio.

Wound-field dc motors are classified by the type of motor field: shunt-wound, series-wound, and compound-wound. Permanent magnet types are also popular, normally as fractional horsepower motors.

Shunt-Wound Motors. Shunt-wound and stabilized shunt-wound dc motors can supply both constant speed at any control setting and a wide speed range that is field-controllable (Fig. 4-13). Most shunt motors are operated from adjustable voltage power supplies and, therefore, do not need auxiliary starting provisions.

A stabilizing winding helps prevent speed increases as the load increases at weak field settings. This winding has disadvantages in reversing applications, however, because it must be reversed with respect to the shunt winding when the armature voltage is reversed. Reversing contactors are normally used.

The shunt winding can either be connected to the same power supply as the armature (self-excited) or be separately excited. Care must be taken never to open the field of a shunt-wound motor that is running unloaded. The loss of field flux causes motor speed to increase to dangerously high levels.

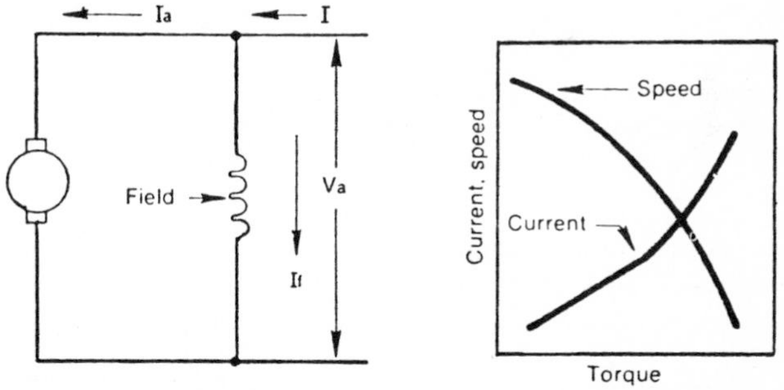

Figure 4-13 Schematic and speed vs. torque diagram—dc shunt-wound motor.

Series-Wound Motors. In series-wound motors, the field flux is created by coils that are electrically in series with the armature (Fig. 4-14). When the motor starts, the current, and consequently the magnetic field are at maximum values, producing a large starting torque. As the motor speeds up and the current is reduced, the field flux also becomes smaller. With no external load on the shaft, the field flux drops nearly to zero and motor speed becomes dangerously high. For this reason, series-wound motors should be used only where the load is directly connected or geared to the shaft.

Compound-Wound Motors. Compound-wound motors combine both series and shunt fields (Fig. 4-15). The disadvantage of series-motor overspeeding at light loads is

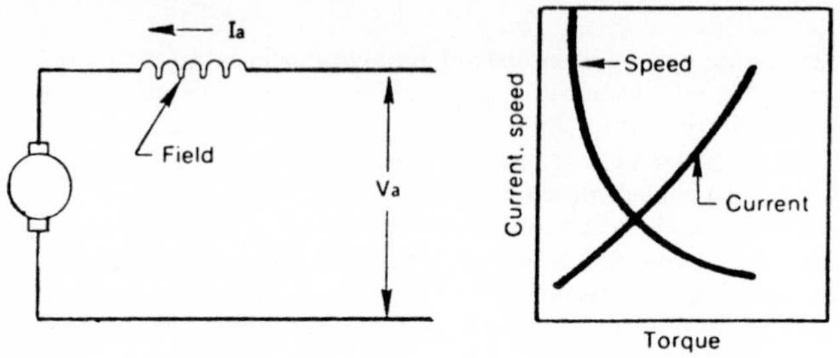

Figure 4-14 Schematic and speed vs. torque diagram—dc series-wound motor.

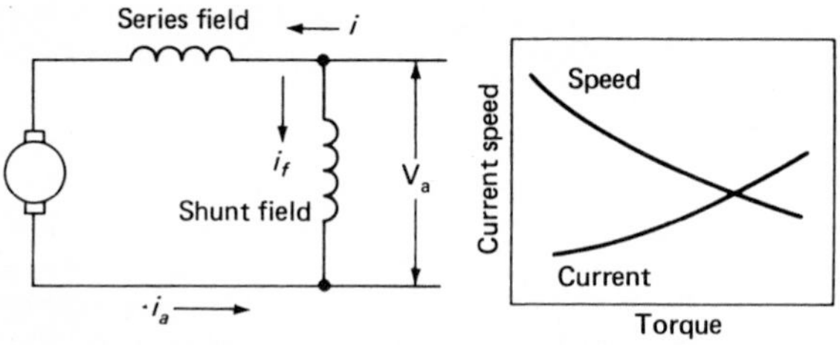

Figure 4-15 Schematic and speed vs. torque diagram—dc compound-wound motor

avoided since there is so little current in the series field at no load that speed is determined by the shunt field alone. At higher loads, speed depends on the sum of the two fields, making speed reduction similar to that of a series motor.

Compound motors have high starting torques and fairly flat speed-torque characteristics at rated load. Because of the elaborate circuits needed to control compound motors, however, only large bidirectional types are built.

Permanent-Magnet Motors. Permanent-magnet motors have fields supplied by permanent magnets (Fig. 4-16). Those fields create two or more poles in the armature by passing magnetic flux through it. The magnetic flux causes the current-carrying armature conductors to move, creating a torque. This flux remains basically constant at all motor speeds; speed-torque and current-torque curves are linear.

Permanent-magnet motors, available in fractional and low-integral horsepower sizes, have several advantages over field-wound types. Excitation power supplies and associ-

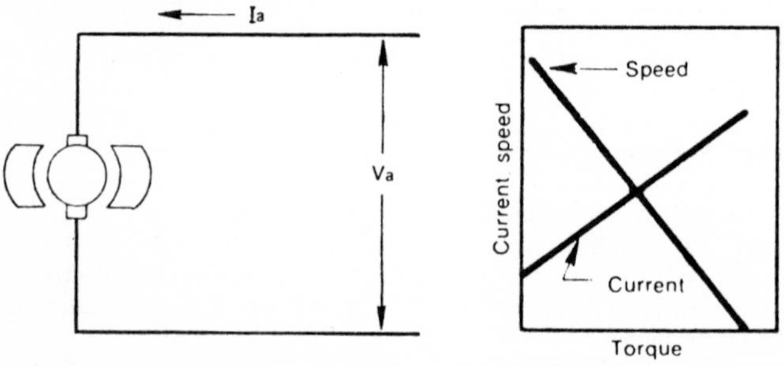

Figure 4-16 Schematic and speed vs. torque diagram—dc permanent-magnet motor.

ated wiring are not required, and reliability is improved. Efficiency and cooling are also improved by elimination of the power loss associated with an excited field.

Disadvantages are the absence of field control and special speed-torque characteristics. Overloads may also cause partial demagnetization that changes motor speed and torque characteristics until magnetization is fully restored.

Permanent-magnet motor performance is actually a compromise between compound-wound and series-wound motors. It has better starting torque but approximately half the no-load speed of a series motor. In applications where compound motors are traditionally used, the permanent-magnet motor can offer slightly higher efficiency and greater overload capacity. In series-motor applications, permanent-magnet motors provide a cost advantage.

Excited Motors. DC-excited synchronous motors, practical for most cases in sizes larger than 1 hp, use direct current supplied through slip rings for starting. The direct current may be supplied from either a separate source or a dc generator connected directly to the motor shaft.

DC-excited motors in fractional-horsepower sizes use starting methods common to induction motors (split-phase, capacitor-start, repulsion-start, and shaded-pole). After acceleration, these motors automatically switch to synchronous operation.

Although the dc-excited motor has a squirrel-cage winding for starting, the inherent low starting torque and the need for a dc power source requires a starting system that gives a full motor protection while starting, applies dc field excitation at the proper time, removes field excitation at rotor pullout (synchronization), and protects the squirrel-cage winding against thermal damage under out-of-step conditions.

Universal Motors

Universal motors are essentially series motors that operate with nearly equivalent performance on direct current or alternating current up to 60 Hz. They differ from dc series motors in that they have different winding ratios and thinner iron laminations. A dc series motor runs on alternating current, but inefficiently. A universal motor, however, runs on direct current with essentially equivalent ac performance, but with poorer commutation and brush life than an equivalent dc series motor.

A universal motor has the highest horsepower per pound ratio of any ac motor because of its ability to operate at much greater speeds than any other 60-Hz motor. It is ideally suited for operation at a rated output, where an occasional overload or intermittent heavy load occurs. Stall torque may be as much as 10 times the continuous rated torque. The motor may even be operated in a stalled condition for short periods of time.

High starting torque, adjustable-speed characteristics, small size, and economy are all advantages of the universal motor. Universal motors are not more widely used, however, because their operating life is shorter, their size range is limited (to about 2 hp), and their very high speeds limit their applications.

Universal motors can be built to deliver speeds ranging from 4000 to 24,000 r/min and rated power from 0.1 to 1 hp. Efficiency varies from 30 percent for small sizes to 75 percent for large sizes.

Classifying By Design Letter

Three-phase, squirrel-cage, IHP motors are the most widely used ac induction motors in industry. Classification of performance requirements has results in NEMA standardized designs that satisfy torque, horsepower, speed, and current requirements for a large number of applications. The classifications are distinguished by a NEMA design letter (A, B, C, and D).

Single-phase motors also have NEMA design letters, but they do not specify the performance characteristics to the same extent that the three-phase design letter does. The letters N and O are used for FHP motors and L and M for IHP motors.

Design A

NEMA design A motors are general-purpose motors with high starting (locked-rotor) currents and normal locked-rotor torques (Fig. 4-17). Slip is less than 5 percent, except for motors with 10 or more poles which have a slightly greater slip. These motors are suitable in applications where load inertia is small and starts are infrequent. They are normally used where high breakdown torque (compared to NEMA design B motors) is required.

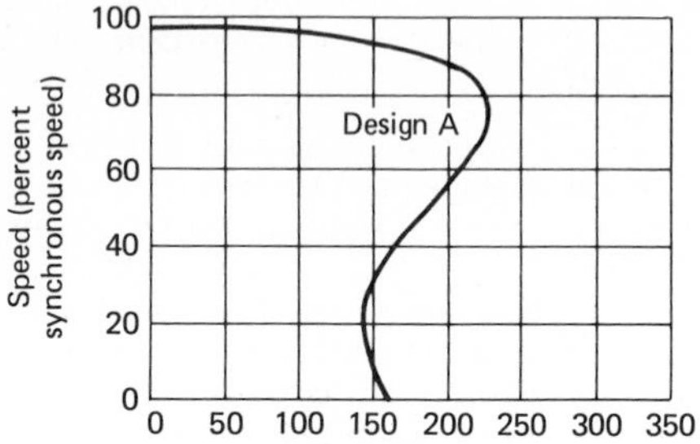

Figure 4-17 NEMA Design A.

Design B

NEMA design B motors are general-purpose motors with normal starting torques and currents and relatively high breakdown torques (Fig. 4-18). Pull-up torque normally available allows rapid acceleration to full load speed. Slip is the same as design A. Design B motors are the most popular in industry.

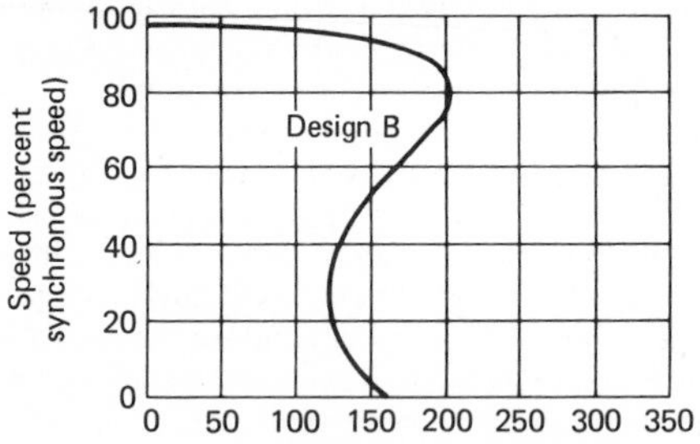

Figure 4-18 NEMA Design B.

Design C

NEMA design C motors have high starting torques with normal starting currents (Fig. 4-19). Breakdown torques are normal, though slightly less than design B motors. Although slip is less than 5 percent, it is higher than the slip of a design B motor. These motors are typically used in applications where breakaway loads are high and where starting torques higher than those available from design B motors are required. The motors have good running characteristics, although efficiency is somewhat poorer than that of design B motors.

Design D

NEMA design D motors have very high starting torques, with moderate starting currents and high breakdown torques (Fig. 4-20). Slip is high, 5 percent or more at rated load.

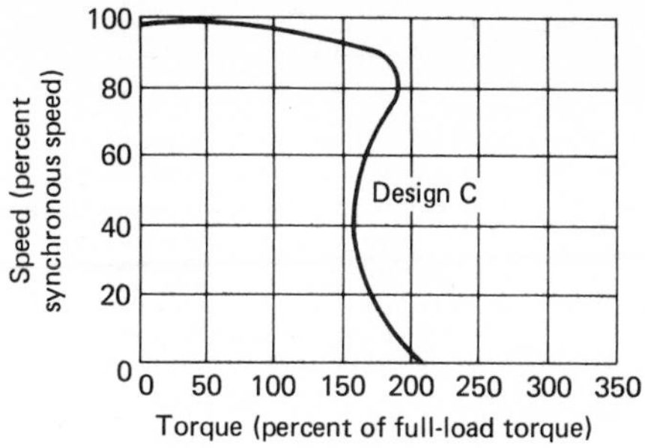

Figure 4-19 NEMA Design C.

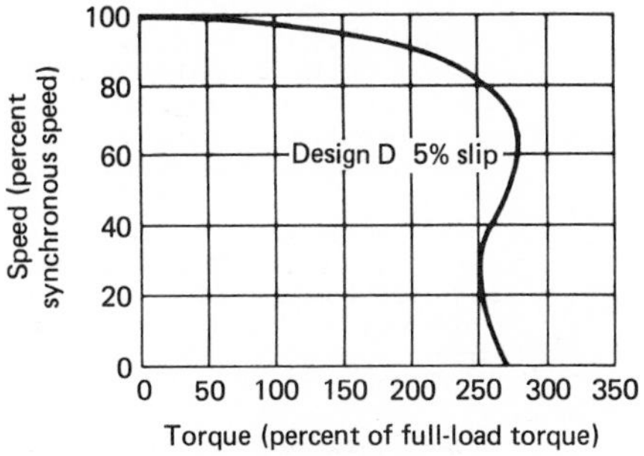

Figure 4-20 NEMA Design D.

Thus speed can fluctuate significantly with changing loads. For practicality these designs have been subdivided into several groups in terms of slip (5 to 8 percent, 8 to 13 percent, etc.). These motors are typically found in applications where heavy loads are suddenly applied or removed at frequent intervals.

Classifying By Environmental Protection and Cooling Methods

The two general NEMA classifications for motor enclosures are open (O) and totally-enclosed (TE). An open machine is one having ventilating openings which permit passage of external cooling air over and around the windings of the motor. A totally enclosed machine is constructed so as to prevent the free exchange of air inside and outside the motor, but not sufficiently enclosed to be termed airtight. These enclosures are further designated by the degree of protection they provide (Table 4-1).

MOTOR RATINGS

The NEMA rating of a motor consists of the output of that machine along with any other characteristics assigned to it by the manufacturer. These characteristics include, but are not limited to speed, voltage, current, and service factor.

Voltage and Frequency

The standard voltages for FHP dc motors, universal motors, and 60-Hz, single-phase ac motors are 115 and 230 V. Three-phase ac motors at 60 Hz have standard voltages of 115 (15 hp and smaller), 230, 460, 575, 2300, 4000, 4600, and 6600 V.

Motors must operate successfully under running conditions within the limits speci-

TABLE 4-1 Motor Enclosures

Types	Characteristics
	Open
Dripproof (ODP)	Will not allow dripping liquids or solids to enter the motor when falling on the motor at an angle not greater than 15 degrees from vertical.
Splash-proof	A splash-proof motor is similar to a dripproof motor, except it is constructed to exclude liquids and solids falling on it any angle not more than 100° from the vertical.
Guarded	Motor openings are limited in size. Openings giving access to live or rotating parts do not permit the passage of a rod ¾ in diam and are at least 4 in away from those parts.
Semiguarded	Top half of the motor with limited-size openings as defined under Guarded motor.
Externally ventilated	Motor cooled by air circulated by a separate motor-driven blower. Normally used on large low-speed motors where rotor fans cannot move sufficient air to properly cool the motor.
Pipe ventilated	Motor end shields are constructed to accept pipe or ducts to supply ventilating air. Air can be supplied from a source remote from the motor, if necessary, to supply clean air for cooling.
Weather protected, type 1	Ventilating passages minimize entrance of rain, snow, and airborne particles. Passages are less than ¾ in diam
Weather protected, type 2	Motors have, in addition to type 1, passages to discharge high-velocity particles blown into the motor.
	Totally enclosed
Nonventilated (TENV)	Enclosed motor not equipped for cooling by external means.
Fan-cooled (TEFC)	Cooled by external integral fan mounted on the motor shaft.
Explosion-proof	Enclosed motor which withstands internal gas explosion and prevents ignition of external gas.
Dust-ignition-proof	Excludes ignitable amounts of dust and amounts of dust that would degrade performance.
Waterproof	Excludes liquids and airborne solids except around shaft.
Pipe-ventilated	Openings accept air inlet and/or exit ducts or pipe for air cooling.
Water-cooled	Cooled by circulating water.
Water–air-cooled	Cooled by water-cooled air.
Air-to-air cooled	Cooled by air-cooled air.
TEFC guarded	Fan cooled and guarded by limited-size openings.
There are two other classes of protection for both open and totally enclosed motors:	
Encapsulated windings	Machine having random windings filled with resin.
Sealed windings	Machine with form wound coils with windings and connections sealed against contaminants.

fied for voltage and frequency variation by the NEMA operating-service conditions listed below. Successful operation within the variations specified, however, does not necessarily mean that the motor will be able to start and accelerate the load to which it is applied.

Service Conditions

Usual environmental service conditions are defined by NEMA as:

1. Ambient temperatures in the range of 32 to 105°F (0 to 40°C) or, when water cooling is used, in the range of 50 to 105°F (10 to 40°C).
2. Barometric pressure corresponding to an altitude not exceeding 3300 ft (1000 m)
3. Installation on a rigid mounting surface
4. Installation in areas or supplementary enclosures which do not seriously interfere with ventilation

Usual operating service conditions are:

1. Voltage variation up to 6 percent of rated voltage for universal motors and 10 percent of rated voltage for ac and dc motors.
2. Frequency variation not more than 5 percent above or below rated frequency.
3. Combined voltage and frequency variation not more than 10 percent above or below the rated voltage and frequency.
4. V-belt drive, flat-belt, chain, and gear drives, in accordance with NEMA standards.

Locked-Rotor kVA

Every ac motor (except three-phase wound-rotor types) rated 1/20 hp and larger has a letter designation for locked-rotor kVA per horsepower (Table 4-2). It is calculated as follows:

$$\text{kVA/hp} = \frac{IE}{\text{hp} \times 1000} \quad \text{for single-phase motors}$$

$$\text{kVA/hp} = \frac{IE\sqrt{3}}{\text{hp} \times 1000} \quad \text{for three-phase motors}$$

where I is the locked-rotor amperage at rated voltage E.
For motors with dual ratings, the following rules determine the code letter to be used:

Motor Type	*Code letter corresponds to kVA/hp for*
Multispeed	
Variable torque	Highest speed
Constant torque	Highest speed
Constant horsepower	Highest kVA/hp
Wye delta, starting on wye	Wye connection
Dual voltage	Highest kVA/hp
Dual frequency, 60/50 Hz	60-Hz kVA/hp
Part-winding start	Full-winding kVA/hp

Service Factor

NEMA service factor is a multiplier which indicates what percent higher than the nameplate horsepower can be accommodated continuously at rated voltage and frequency without injurious overheating (i.e., exceeding NEMA allowable temperature rise for given insulation systems). Service factors for general-purpose ac motors are shown in Table 4-3.

TABLE 4-2 Locked-Rotor kVA/hp Code Designations

Code letter	Locked-rotor kVA/hp	Code letter	Locked-rotor kVA/hp
A	0–3.15	L	9.0–10.0
B	3.15–3.55	M	10.0–11.2
C	3.44–4.0	N	11.2–12.5
D	4.0 –4.5	P	12.5–14.0
E	4.5 –5.0	R	14.0–16.0
F	5.0 –5.6	S	16.0–18.0
G	5.6 –6.3	T	18.0–20.0
H	6.3 –7.1	U	20.0–22.4
J	7.1 –8.0	V	22.4 and up
K	8.0 –9.0		

TABLE 4-3 Service Factors

hp	Synchronous speed, r/min 3600	1800	1200	900	720	600	514	
1/20	1.4	1.4	1.4	1.4	—	—	—	Fractional horsepower motors
1/12	1.4	1.4	1.4	1.4	—	—	—	
1/8	1.4	1.4	1.4	1.4	—	—	—	
1/6	1.35	1.35	1.35	1.35	—	—	—	
1/4	1.35	1.35	1.35	1.35	—	—	—	
1/3	1.35	1.35	1.35	1.35	—	—	—	
1/2	1.25	1.25	1.25	1.15*	—	—	—	
3/4	1.25	1.25	1.15*	1.15*	—	—	—	
1	1.25	1.15*	1.15*	1.15*	—	—	—	Integral horsepower motors
1½–125	1.15*	1.15*	1.15*	1.15*	1.15*	1.15*	1.15*	
150	1.15*	1.15*	1.15*	1.15*	1.15*	1.15*	—	
200	1.15*	1.15*	1.15*	1.15*	1.15*	—	—	

*In the case of three-phase squirrel-cage integral-horsepower motors, these service factors apply only to Design A, B, and C motors.

Level of Performance

NEMA also rates motors by level of performance, specifically horsepower, speed, torque, and locked-rotor current for each category of motor type and size. Tables specifying acceptable limits for each are too detailed to include here, but may be found in NEMA standard MG1-1967.

Temperature Rise

All standard motors, unless otherwise stated, are designed for use in a maximum ambient temperature no greater than 40°C (104°F). Allowable temperature rise for motors without special insulation is 40°C for motors with service factors greater than 1.0 and 50°C for motors with service factors of 1.0.

Special insulation provisions, organized into four NEMA classes, increase allowable temperature rise, according to Table 4-4.

Duty Rating

The duty or time rating of an electric motor is determined by the electrical design, the enclosure, the method of cooling, and the class of insulation system used. This rating is

TABLE 4-4 Allowable Temperature Rise

Class of insulation	Temperature, °C A	B	F	H
1. Windings—fractional horsepower motors				
a. Open motors other than those given in parts 1*b* and 1*d*	60	80	105	125
b. Open motors with 1.15 or higher service factor	70	90	115	—
c. Totally enclosed nonventilated and fan-cooled motors, including variations	65	85	110	135
d. Any motor in a frame smaller than the 42 frame	65	85	110	135
2. Windings—integral horsepower motors				
a. Motors other than those given in parts 2*b*, 2*c*, 2*d*, and 2*e*	60	80	105	125
b. All motors with 1.15 or higher service factor	70	90	115	—
c. Totally enclosed fan-cooled motors, including variations	60	80	105	125
d. Totally enclosed nonventilated motors, including variations	65	85	110	135
e. Encapsulated motors with 1.0 service factor, all enclosures	65	85	110	—

the length of time the motor may operate without causing overheating or reducing the normal life of the motor.

An intermittent duty motor is intended to operate for short periods of time totalling no more than 1 or 2 h/day. Motors with short-time ratings are designed to operate for no longer than the period shown on the nameplate. They must be shut off and allowed to cool to room temperature before being reactivated. Typical short-time duty ratings are 5, 15, 30, and/or 60 min. A motor with a continuous duty rating can be run indefinitely at rated load and voltage without injurious overheating or reduction in motor life.

Frame Designations and Assignments

The NEMA system for designating frame dimensions of motors consists of a series of numbers (frame number) in combination with letters. For convenience, manufacturers are allowed to use letters of the alphabet preceding the frame number for their own identification. However, such letters have no reference to the standards in Table 4-5 and vary in meaning from manufacturer to manufacturer.

The NEMA frame numbering system provides a useful relationship between motor horsepower and speed ratings and motor size. Although no such relationship currently exists for FHP motors, IHP motor horsepower and speed ratings and their assigned motor frames are supplied in MG1-1967.

APPLICATION AND SELECTION

Motor application begins by matching load requirements with motor characteristics. A correctly applied motor must be able to start the load, bring it up to operating speed, and run as long as necessary through all expected variations in the load.

The first step is determining the load characteristics to which the motor will be matched: power, torque, speed, and duty cycle. Starting torque, as well as running torque, must be considered. It may vary from a small percentage to a value several times the full-load torque. The greater the excess torque than that required to start the load, the more rapid the acceleration.

Power is the product of torque times speed. The term *load* usually refers to the horsepower required to drive a machine:

$$\text{Horsepower (hp)} \cong (\text{torque} \times \text{r/min})/5250$$

Duty cycle (how much is asked from a motor how often and for how long) is the final parameter. The more that is asked the tougher the motor should be.

Having established what load the motor must drive, the correct motor can be selected by considering the following factors. The speed-vs.-torque curve tells much about the performance characteristics of a motor (Fig. 4-21).

Power Supply

Power supply will be either single- or three-phase. A single-phase motor of the proper voltage can be used on a three-phase system if properly connected. But a three-phase motor cannot be used on single-phase supply. Three-phase motors generally cost less, perform better, and last longer than single-phase motors of the same size.

Voltage

Motor rating must match the nominal voltage and frequency of electricity supplied. Most motors are available in several standard voltages, but generally, the highest available voltage gives the lowest installation cost.

TABLE 4-5 Frame Designations

FHP Motors	
The following letters immediately follow the frame number and denote specific variations.	
O	Face mounting
G	Gasoline-pump motors
H	A frame having an F dimension larger than that of the same frame without the suffix letter H
J	Jet-pump motors
K	Sump-pump motors
M	Oil-burner motors
N	Oil-burner motors
Y	Special mounting dimensions (consult manufacturer)
A	All mounting dimensions are standard except the shaft extension

IHP Motors	
The following letters immediately follow the frame number and denote specific variations.	
C	Face mounting on drive end (When the face mounting is at the end opposite the drive, the prefix F is used, making the suffix letters FC.)
CH	Face mounting dimensions different from those for the frame designation having the suffix letter C (The letters CH are considered to be one suffix and should not be separated.)
D	Flange mounting on drive end (When the flange mounting is at the end opposite the drive, the prefix F is used, making the suffix letters FD.)
E	Shaft extension dimensions for elevator motors in frames larger than the 326U frames
HP and HPH	Vertical solid-shaft motors having dimensions in accordance with MG1-18.625 (The letters HP and HPH are to be considered as one suffix and should not be separated.)
JM	Face-mounted close-coupled pump motor having antifriction bearings and dimensions in accordance with NEMA standards (The letters JM are to be considered as one suffix and should not be separated.)
JP	Face-mounted close-coupled pump motor having antifriction bearings and dimensions in accordance with NEMA standards (The letters JP are to be considered as one suffix and should not be separated.)
LP and LPH	Vertical solid-shaft motors having dimensions in accordance with NEMA standards (The letters LP and LPH are to be considered as one suffix and should not be separated.)
P and PH	Vertical solid-shaft motors having dimensions in accordance with NEMA standards
R	Drive-end tapered-shaft extension having dimensions in accordance with NEMA standards
S	Standard short shaft for direct connection
T	Included as part of a frame designation for which standard dimensions have been established
U	Previously used as part of a frame designation for which standard dimensions had been established
V	Vertical mounting only
VP	Vertical solid-shaft motors having dimensions in accordance with NEMA standards (The letters VP are to be considered as one suffix and should not be separated.)
X	Wound-rotor crane motors with double shaft extension.
Y	Special mounting dimensions (Consult manufacturer.)
Z	All mounting dimensions standard except the shaft extension(s) (Also used to designate machine with double shaft extension.)

Suffix letters are added to the frame number in the following sequence.

Suffix letters	Sequence
A	1
T, U, HP, HPH, JM, JP, LP, LPH, and VP	2
R and S	3
C, D, P, and PH	4
FC and FD	5
V	6
E, X, Y, and Z	7

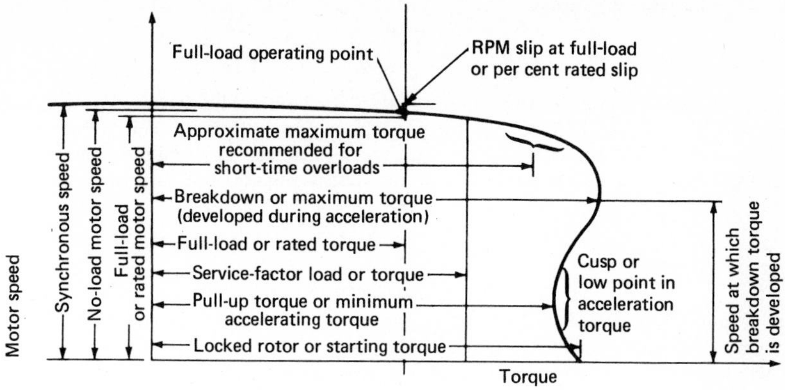

Figure 4-21 Speed-torque characteristics.

TABLE 4-6 Single-Phase Motors—Selection by Characteristics

Type	Horsepower ranges	Load-starting requirement	Starting current	Characteristics	Electrically reversible
Split-phase	1/20 to 1/2	Easy	High	Small, inexpensive, simple construction; nearly constant speed	Yes
Capacitor-start	1/8 to 10	Hard	Medium	Simple construction, long service; nearly constant speed	Yes
Two-value capacitor	1/4 to 20	Hard	Medium	Simple construction, long service; nearly constant speed	Yes
Permanent-split capacitor	1/20 to 1	Easy	Low	Inexpensive, simple construction; speed reduced by lowering voltage	Yes
Shaded-pole	1/400 to 1/2	Easy	Medium	Inexpensive for light duty	No
Wound-rotor (repulsion) types	1/6 to 10	Very hard	Low	Larger than other equivalent single-phase motors	No
Universal or series	1/150 to 2	Hard	High	High speed, small size; speed changes with load variations	Yes, some types
Synchronous	Very small fractional	——	——	Constant speed	——

Horsepower

Mechanical power available at the motor shaft is the nominal horsepower rating at rated revolutions per minute. When installing a motor on new equipment, match the motor to the required horsepower computed when determining the load characteristics. When replacing a worn-out motor, provide a motor that duplicates the nameplate data from the old motor.

Type

Type of motor best suited to a particular application is a major question because of the wide variety of motor types available. Tables 4-6 and 4-7 provide some guidelines, the first one by motor characteristics, the second by application.

Coupling

If motor speed matches the input shaft speed, a simple mechanical coupling can be used. But if it turns at a speed different from that recommended or calculated for the equipment, a speed conversion drive is needed. It includes pulley and belt, gear, or chain and sprocket.

TABLE 4-7 Motor Selection by Application*

	Single-phase		Three-phase	
Application	Small, hp	Large, hp	Small, hp	Large, hp
Compressors				
Air	CS	CS,CP	B	B
Refrigeration	CS	CS,CP	B	C
Centrifugal	SP,CS	CS,CP	B	B
Reciprocating				
Loaded	CS	CS,CP	B	C
Unloaded	SP,CS	CS,CP	B	B
Conveyors and elevators				
Unloaded	CS,WR	CS,WR	B	B
Loaded	CS,WR	CS,WR	B	C
Cooling towers	CS	CS	B	B
Dryers	CS	CS,CP	B	B
Fans and blowers				
Centrifugal	SP,C	CS,CP	B	B
Propeller	SP,C,P	CS,CP	B	B
Unit heaters	SP,C,P	CS,CP	B	B
Machine tools				
Lathes	CS	CS	B	B
Milling machines	CS	CS	B	B
Drill presses	CS	CS	B	B
Grinders	CS	CS	B	B
Oil burners	SP	CS	B	B
Pumps				
Reciprocating	SP,CS	CS,CP,WR	B	B
Centrifugal	CP,CS	CS,CP,WR	B	B
Heavy oil	WR	CP,WR	B	D
Saws				
Metal, band saw	CS,U	CS	B	B
Wood, circular	CS,U	CS	B	B

*SP, split phase; CS, Capacitor start; CP, two-value capacitor; C, permanent split capacitor; P, shaded pole; WR, wound rotor; U, universal; B, NEMA Design B; C, NEMA Design C; D, NEMA Design D.

Enclosures

The drip-proof, general-purpose motor is suitable for dry, clean, and ventilated locations. Wet, dirty, or explosive conditions require other enclosures, such as totally enclosed fan-cooled or totally enclosed nonventilated motors in standard, severe duty, or explosion-proof construction. Table 4-8 matches enclosure to environment.

TROUBLESHOOTING

Table 4-9, a chart of motor problems and possible causes, can serve as a guide in identifying and correcting motor-system malfunctions.

ENERGY EFFICIENCY

Energy efficiency is a growing issue in the motor industry, and for good reasons. In addition to scarce energy resources, a great potential for energy conservation and reduced operating costs exists with the electric motor.

TABLE 4-8 Motor Enclosures

		Standard		
Condition or application	Drip-proof	Totally enclosed	Severe-duty	Explosion-proof
Atmospheric				
Dry	x	—	—	—
Humid	x*	—	—	—
Outdoor, mild	x*	—	—	—
Outdoor, severe	x*	x*	x	—
Chips				
Metal or plastic	—	x	x	—
Wood	x	x	x	—
Dust				
Abrasive, nonexplosive	—	x	x	—
Abrasive, explosive	—	—	—	x
Carbon, coal, or coke	—	—	—	x
Flour	—	—	—	x
Metal, nonexplosive	—	x	x	—
Metal, explosive	—	—	—	x
Sand	x	x	x	—
Sawdust	x	x	x	—
Textile fibers	—	x	—	—
Fumes				
Explosive	—	—	—	x
Nonexplosive	—	x	x	—
Corrosive	—	x*	x	—
Liquids				
Acid or alkali	—	x*	x	—
Dripping water	x*	—	—	—
Explosive	—	—	—	x
Nonexplosive	—	x	x	—
Paint	—	—	—	x
Petroleum, oil	—	—	—	x
Splashing water	x*	x*	x	—
Solvents				
Corrosive, nonexplosive	—	x*	x	—
Noncorrosive, nonexplosive	—	x	x	—
Noncorrosive, explosive	—	—	—	x

*Depending on concentration and/or severity, additional protection may be necessary.

TABLE 4-9 Troubleshooting Guide

Trouble	Cause	What to do
Motor fails to start	Blown fuses	Replace with time-delay fuses matched to nameplate amperes. If motor has service factor greater than 1.0, use fuse size up to 12.5% of motor amperes. Check for grounded winding.
	Improper power supply	Check to see that power supplied (voltage-frequency phases) agrees with motor nameplate.
	Less than 208 V on a 208 V system	Use a 200-V motor.
	Low voltage	Inadequate wiring. Long or inadequate extension cords.
	Improper line connections	Check connections against diagram supplied with motor.
	Overload (thermal protector) tripped	Check and reset overload relay in starter. Check heater rating against motor nameplate current rating. Check motor load. If motor has a manual reset thermal protector, check whether tripped.
	Open circuit in winding or starting switch	Indicated by humming sound when switch is closed. Check for loose wiring connections, whether starting switch inside motor is closed, or if capacitor is defective.
	Mechanical failure	Check to see if motor and drive turn freely. Check belts, bearings, and lubrication.
	Short-circuited stator	Indicated by blown fuses and/or high no-load line current. Motor must be rewound or replaced.
	Poor stator coil connection	Repair or replace.
	Rotor defective	Look for broken bars or end rings.
	Motor may be overloaded	Reduce load; increase motor size.
	If three-phase, one phase may be open	Indicated by humming sound. Check lines for open phase. Check voltage with motor disconnected from line; one fuse may be blown.
	Defective capacitor	Check for short-circuited grounded, open or low value capacitor. Replace if necessary.
Motor stalls	Wrong application	Change type or size of motor. Consult motor service firm.
	Overloaded motor	Reduce load or increase motor size. Belt may be too tight.
	Low motor voltage	See that nameplate voltage is maintained.
Motor runs and then dies down	Power failure	Check for loose connections to line, to fuses, and to control. Check if thermal protector tripped, fuses blown, check overload relay, starter, and pushbuttons.
Motor does not come up to speed	Not applied properly	Consult motor service firm for proper type and size of motor. Use larger motor.

TABLE 4-9 Troubleshooting Guide (*Continued*)

Trouble	Cause	What to do
	Voltage too low at motor terminals (because of line voltage drop or voltage drop in wiring to motor)	Use higher-voltage tap on transformer terminals; increase wire size.
	Starting load too high	Check load capability of motor.
	Shorted or weak capacitor	Replace capacitor.
	Low temperature (below 0°F)	Replace capacitor with one of higher value. Check ball bearing lubricant—use low-temperature grease.
	Broken rotor bars	Look for cracks near the end rings or broken rotor bars. A new rotor may be required as repairs are usually temporary.
Motor takes too long to accelerate	Excess loading; tight belts; high inertia load	Reduce load; increase motor size. Loosen belts.
	Inadequate wiring	Check for high resistance; increase wire size.
	Defective rotor	Replace with new rotor.
	Applied voltage too low	Check incoming voltage drop in wiring to motor. Power may have to supply higher voltage.
	Weak or shorted capacitor	Replace capacitor.
	Low starting torque	Replace with larger motor.
Motor overheats while running under load	Overload	Reduce load; increase motor size; belts may be too tight.
	Insufficient airflow over shell of "air-over" motor	Modify installation or change to a self-cooled motor.
	May be clogged with dirt to prevent proper ventilation of motor.	Good ventilation is apparent when a continuous stream of air leaves the motor. If it does not after cleaning, check with manufacturer.
	Motor may have one phase open (three-phase motors)	Check to make sure that all leads are well-connected and a fuse is now blown in one line.
	Grounded or shorted coil.	Repair or replace the motor.
	Unbalanced terminal voltage (three-phase motors)	Check for faulty leads, connections, and transformers. Excessive single-phase loads on one circuit.
	Faulty connection	Clean, tighten, or replace.
	High or low voltage	Check voltage at motor with voltmeter; should not be more than 10% above or below rated.
	Rotor rubs stator bore	If not poor machining, replace worn bearings.

TABLE 4-9 Troubleshooting Guide (*Continued*)

Trouble	Cause	What to do
Motor vibrates after corrections have been made	Motor misaligned	Realign.
	Coupling out of balance	Balance coupling.
	Driven equipment unbalanced	Rebalance driven equipment.
	Defective ball bearing	Replace bearing.
	Balancing weights shifted	Rebalance rotor.
	Polyphase motor running single phase	Check for open circuit or blown fuses.
	Excessive end play	Adjust bearing or add shims to eliminate excess end play.
	High voltage	Correct.
Unbalanced line current on polyphase motors during normal operation	Unequal terminal volts	Check leads and connections. Check transformers.
	Single-phase operation	Check for open contacts, blown fuses.
	Unbalanced supply	Check line-to-line voltage.
Scraping noise	Fan rubbing air shield	Repair.
	Fan striking insulation	Repair.
	Bent shaft	Straighten shaft or replace rotor.
Noisy operation	Air gap not uniform	Check and correct end shield fits or bearing.
	Rotor unbalance	Rebalance.
	High voltage	Reduce line voltage by changing power taps.
Hot bearings, general	Loose on mounting surface	Tighten holding bolts.
	Bent shaft	Straighten or replace shaft or rotor.
	Excessive belt pull	Decrease belt tension.
	Pulleys too far away from bearing nose	Move pulley closer to motor bearing.
	Pulley diameter too small; slipping	Use larger pulleys (both motor and load). Check belt tension.
	Misalignment	Correct by realignment of drive.
Hot bearings, sleeve	Oil grooving in bearing obstructed by dirt	Remove end shield, clean bearing housing, and oil grooves; renew oil.
	Oil too heavy	Use recommended oil.

TABLE 4-9 Troubleshooting Guide (*Continued*)

Trouble	Cause	What to do
	Oil too light	Use recommended oil.
	Too much end thrust	Reduce thrust induced by drive, or supply external means to carry thrust.
	Dry bearing	Add oil.
	Badly worn bearing	Replace bearing.
	Feeder wick not touching shaft	Repair or replace wicking.
Hot bearings, ball	Insufficient grease	Maintain proper quality of grease in bearing. Replace sealed bearing.
	Deterioration of grease or lubricant contaminated Water in bearing	Remove old grease, wash bearings thoroughly in clean kerosene, and replace with new grease. Replace bearing if sealed type. Eliminate source of moisture.
	Excess lubricant	Reduce quantity of grease. Bearing should not be more than ½ filled.
	Overloaded bearing	Check alignment, side, and end thrust.
	Broken ball or rough races	Replace bearing. First clean end shield housing thoroughly.

The greatest conservation potential begins with motors rated from 1 to 20 hp. Although there is a large population of motors rated under 1 hp, their conservation potential is limited.

The first energy-efficient motors therefore consisted of four-pole units from 1 to 25 hp. New energy-efficient motors have been made available in ratings to 250 hp in two-, four-, and six-pole speeds in drip-proof, totally-enclosed fan-cooled (TEFC) and TEFC severe-duty construction. The rationale for extending energy-efficient design to larger motors is that even a small improvement in efficiency will have a large impact on total energy consumed.

Efficiency and Power Factor

Motor efficiency is simply a measure of mechanical work output over electric power input. A motor's power factor is a measure of how well the motor uses the current it draws.

Total line current is made up of two components: real and reactive. Power factor is the ratio of *real power* to *apparent power*, or the cosine of the angle between the two. Technically, it is input kilowatts divided by the input kilovoltamperes.

Figure 4-22 shows efficiency and power factor as functions of loading for a typical small-integral horsepower polyphase motor. Efficiency is relatively stable over a wide range of loading conditions. But power factor drops off fast as the motor is unloaded—about a 15-percent spread between half- and full load. Proper application and sizing are thus important to maintain good power factor. A high power factor means that the motor requires less total current, and the resulting lower line current means that less energy is wasted in all feeder circuits serving the motor.

Any inefficiency in the process of converting electric into mechanical energy occurs as heat, often referred to as watt losses, or simply losses.

To increase efficiency, it is necessary to minimize these losses. The rotors of energy-

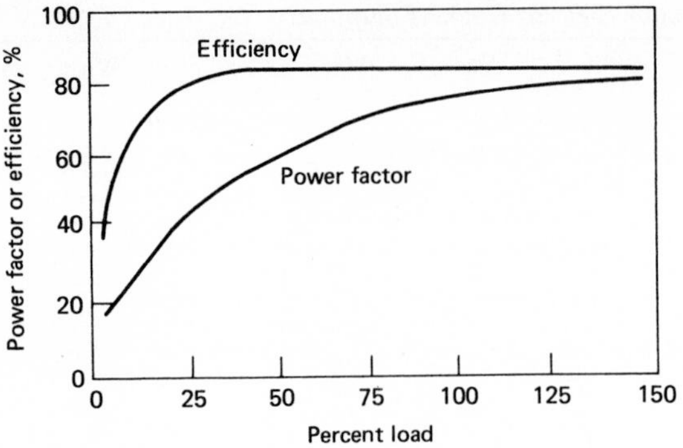

Figure 4-22 Efficiency and power factor vs. load.

efficient motors are built with additional aluminum to reduce losses resulting from current flowing in the aluminum rotor bars. Additional copper is used in the stators to reduce losses in the motor. Most motors have 100 percent copper stator windings; few are aluminum-wire-wound. More steel, together with special processing, is also used in the motors to reduce the stator and rotor losses. This specially processed steel includes thinner laminations as well as silicon electric steels.

Windings in the most energy-efficient motors are designed for optimum winding distribution. The air gap is also optimized for the best power factor and efficiency performance. Stator and rotor slots are usually designed for optimum performance and are not compromised to serve other uses.

Power factor is improved by decreasing line current. The most direct way is to add

TABLE 4-10 Efficiency and Power Factor of Energy-Efficient Motors vs. Industry Average Standard Motors (Nominal Values—Drip-proof, 4-Pole)

	Efficiency		Power factor	
Motor hp	Energy-efficient	Industry* average	Energy-efficient	Industry* Average
1.0	0.825	0.759	0.840	0.709
1.5	0.840	0.777	0.857	0.752
2	0.840	0.785	0.857	0.764
3	0.865	0.815	0.864	0.794
5	0.875	0.840	0.870	0.817
7.5	0.885	0.843	0.867	0.799
10	0.895	0.855	0.855	0.831
15	0.917	0.869	0.856	0.815
20	0.917	0.878	0.870	0.841
25	0.917	0.882	0.898	0.827
30	0.924	0.894	0.889	0.852
40	0.924	0.891	0.857	0.846
50	0.930	0.904	0.859	0.850
60	0.930	0.909	0.888	0.829
75	0.936	0.912	0.880	0.847
100	0.924	0.921	0.858	0.861
125	0.936	0.924	0.869	0.868
150	0.936	0.929	0.858	0.884
200	0.941	0.935	0.850	0.879

*Average of 10 manufacturers.

Source Gould Inc., Electric Motor Division

magnetic materials to the rotor and to the stator core. Decreasing the air gap on several ratings and decreasing flux density also decreases line current.

Table 4-10 compares efficiency and power factor of typical energy-efficient motors with the standard industry averages. The values for the new motors, although only 1 to 6 percentage points higher than the industry average, represent a reduction of as much as 27 percent of the losses over a standard industry average motor.

Energy-efficient motors all carry a price premium, one which based on current electricity rates is often paid off in a matter of months. The payback formula for energy-efficient motors is

$$\text{Payback (months)} = \frac{\text{price premium (dollars)}}{\text{kW saved} \times \text{cost per kWh} \times \text{annual running time (h)}} \times 12$$

To realize maximum energy savings, energy-efficient motors must be properly applied. In fact, a standard motor properly applied can save more than an energy-efficient motor improperly applied. Thus, if in doubt about a particular application, consult the motor manufacturer.

BIBLIOGRAPHY

Alger, Philip L.: *Induction Machines,* 2d ed., Gordon Breach, New York, 1970.
Bodine, Clay: *Small Motor, Gearmotor, and Control Handbook,* 4th ed., Bodine Electric Co., Chicago, 1978.
Fitzgerald, A. E., and Charles Kingsley, Jr.: *Electric Machinery,* McGraw-Hill, New York, 1952.
Lamkey, F. R.: *Alternating Current Electric Motors,* 2d ed., Gould Inc., Electric Motor Division, St. Louis, Missouri, 1971.
Lloyd, Tom C.: *Electric Motors and Their Applications,* Wiley Interscience, New York, 1969.
Puchstein, A. F., and T. C. Lloyd: *Alternating Current Machines, 2d ed.,* Wiley, New York, 1949.
Schweitzer, Gerald: *Basics of Fractional Horsepower Motors and Repair,* John F. Rider, New York, 1972.

ACKNOWLEDGMENTS

The following publications are acknowledged as partial sources for this section.
Form 5S1248, "Answers to Common Electric Motor Problems," Dayton Electric Manufacturing Company, Chicago.
"Motor Standard," ELC-1/Nat, National Electrical Manufacturers Association, 2101 L St. N.W., Washington, DC 20037
Machine Design, Electrical and Electronics Reference Issue, sec. 7, "Motors," May 17, 1979, pp. 7–42.

PART 2

Motor Controls

by

Dale F. Willcox

Senior Vice President
and Chief Engineer,
and
The Engineering Staff
Furnas Electric Co.
Batavia, Illinois

GLOSSARY

Definitions essential to an understanding of ac and dc motor control are provided. Most definitions are in accordance with National Electrical Manufacturers Association (NEMA) and American National Standards Institute (ANSI).

Accelerating relay A relay used to aid in motor starting or accelerating from one speed to another. It may function by: current-limit (armature-current) acceleration; counter-emf (armature-voltage) acceleration; or definite-time acceleration.

Across-line starting A method that connects the motor directly to the supply line on starting.

Actuator The cam, arm, or similar mechanical piece used to actuate a device.

Ambient conditions Conditions of the atmosphere adjacent to the electric apparatus. The specific reference may apply to temperature, contamination, humidity, etc.

Antiplugging protection A control function that prevents application of counter-torque until the motor speed has been reduced to an acceptable value.

Brake An electromechanical friction device employed to stop and hold a load. When the brake is set, a spring pulls the braking surface into contact with a braking wheel which is directly coupled to the motor shaft.

Branch circuit The portion of a wiring system extending beyond the final overcurrent device protecting the circuit.

Blowout coil Electromagnetic coil used in contactors and starters to deflect an arc when a circuit is interrupted.

Circuit breaker A device designed to open and close a circuit by nonautomatic means and to open the circuit automatically at a predetermined overload of current without injury to itself when properly applied within its rating (ampere-, volt-, and horsepower-rated).

Closed-circuit transition (as applied to reduced-voltage controllers) A method of motor starting in which power to the motor is not interrupted during the normal starting sequence.

Combination starter A magnetic starter having a manually operated disconnecting means built into the same enclosure. The disconnect may be a motor-circuit switch (with or without fuses) or a circuit breaker.

Contactor A device for repeatedly establishing and interrupting an electric power circuit. Definite-purpose contractors are those designed for specific applications such as air-conditioning, heating, and refrigeration equipment control.

Controller service The specific application of the controller. General purpose: standard or usual service. Definite purpose: specific application other than usual.

Control, three-wire Control function incorporating a momentary contact pilot device and a holding circuit contact to provide undervoltage protection.

Control, two-wire Control function utilizing a maintained contact pilot device to provide undervoltage release.

Controller A device, or group of devices, that governs in a predetermined manner the electric power delivered to the apparatus to which it is connected.

Controller function To regulate, accelerate, decelerate, start, stop, reverse, or protect devices connected to an electric controller.

Controller, drum A manual switching device having stationary contacts connected to a circuit by the rotation of a group of movable contacts.

Current-responsive protector Devices that include time-lag fuses, magnetic relays, and thermal relays (normally located in the motor or between the motor and controller) that provide a degree of protection to the motor, motor control apparatus and branch-circuit conductors from overloads or failures to start.

Dash pot Device employed to create a time delay. It consists of a piston moving inside a cylinder filled with a liquid or gas which is allowed to escape through a small orifice in the piston. Moving contacts actuated by the piston close the electric circuit.

Drop-out (voltage or current) The voltage or current at which a device will return to its de-energized position.

Duty Specific controller functions. Continuous: constant load, indefinitely long time period. Short time: constant load, short or specified time period. Intermittent: varying load, alternate intervals, specified time periods. Periodic: intermittent duty with recurring load conditions. Varying: varying loads, varying time intervals, wide variations.

Dynamic braking The process of disconnecting the armature from the power source and either short-circuiting it or adding a current-limiting resistor across the armature terminals while the field coils remain energized.

Electromechanical Term applied to any device which uses electric energy to magnetically cause mechanical movement.

Electronic control Term applied to define electronic, static, precision, and associated electronic control equipment.

Electronic protector A monitoring system in which current in the motor leads is sensed and the thermal characteristic of the motor reproduced during both heating and cooling cycles. Other circuits compensate for copper and iron losses and detect a lost phase.

Electropneumatic controller An electric controller having its basic functions performed by air pressure.

Faceplate controller Controller having multiple switching contacts mounted near a selector arm on the front of an insulated plate. Additional resistors are mounted on the rear to form a complete unit.

Feeder The circuit conductors between the service equipment and the branch-circuit overcurrent device.

Float switch A switch responsive to the level of liquid.

Foot switch A switch designed for operation by an operator's foot.

Frequency The number of complete variations made by an alternating current per second, expressed in hertz.

Fuse An overcurrent protective device with a circuit-opening fusible member that is severed by the heat developed from the passage of overcurrent through it.

Field Weakening A method of increasing the speed of a wound field motor by reducing field current to reduce field strength.

Instantaneous A qualified term applied to the closing of a circuit in which no delay is purposely introduced.

Interlock An electric or mechanical device, actuated by an external source, and used to govern the operation of another device.

Interrupting capacity The highest current at rated voltage that a device can interrupt.

Inverse time A qualifying term indicating that a delayed action has been introduced. This delay decreases as the operating force increases.

Jogging (inching) The intermittent operation of a motor at low speeds. Speed may be limited by armature series resistance or reduced armature voltage.

Latching relay Relay that can be mechanically latched in a given position when operated by one element and released manually or by the operation of a second element.

Limit switch A device that translates a mechanical motion or a physical position into an electric control signal.

Locked-rotor current Steady-state current taken from the line with the rotor locked and with rated voltage (rated frequency in the case of ac motors) applied to the motor.

Locked-rotor torque The minimum torque a motor will develop at rest for all angular positions of the rotor with rated voltage applied at the rated frequency.

Low-voltage protection (magnetic control only) The opening of the motor circuit by the controller upon a reduction or loss of voltage. Manual restarting is required when the voltage is restored.

Low-voltage release (magnetic control only) The effect of a device, operative on the reduction or failure of voltage, to cause the interruption of power supply to the equipment, but not preventing the re-establishment of the power supply on return of voltage.

Microprocessor controller Motor controller that utilizes feedback control and a computer complete with processor, memory, I/O, keyboard, and display.

Motor circuit switch A switch, rated in horsepower, capable of interrupting maximum operating overload current of a motor of the same horsepower rating at rated voltage.

Motor control circuit The circuit that carries the electric signals directing controller performance but does not carry the main power current. Control circuits tapped from the load side of motor branch circuits' short-circuit protective devices are not considered to be branch circuits and are permitted to be protected by either supplementary or branch-circuit overcurrent protective devices.

Nonreversing A control function that provides for motor operation in one direction only.

Normally open or closed Terms used to signify the position of a device's contacts when the operating magnet is de-energized (applies only to nonlatching-type devices).

Off-delay timer A device whose output is discontinued following a preset time delay after the input is de-energized.

Open-circuit transition A method of reduced-voltage starting in which power to the motor is interrupted during normal starting sequence.

Operating overload The overcurrent to which electric apparatus is subjected in the course of normal operating conditions that it may encounter. *Note:* Maximum operating overload is considered to be 6 times normal full-load current for ac industrial motors and control apparatus and 4 to 10 times normal full-load current, respectively, for dc industrial motors and control apparatus used for reduced- or full-voltage starting. It should be understood that these overloads are currents that may persist for a very short time only, usually a matter of seconds.

Operator's control (pushbutton) station A unitized assembly of one or more externally operable pushbutton switches, sometimes including other pilot devices.

Overcurrent Any current in excess of equipment or conductor rating. It may result from overload, short circuits, or ground faults.

Overload Operation of equipment in excess of normal full-load rating. A fault such as a short circuit or ground fault is not an overload.

Phase-failure protection Protection provided when power fails in one wire of a polyphase circuit to cause and maintain the interruption of power in all wires of the circuit.

Phase-lock servo A digital control system in which the output of an optical tachometer is compared to a reference square wave to generate a system error signal proportional to both shaft velocity and position.

Phase-reversal protection The prevention of motor energizing under conditions of phase sequence reversal in a polyphase circuit.

Pilot device A low-current status indicating or initiating device such as a pilot light, pushbutton, and limit or float switch.

Plugging Motor braking by reversal of the line voltage or phase sequence in order to develop a counter-torque which exerts a retarding force.

Programmed control A control system in which operations are directed by a predetermined input program consisting of cards, tape, plug boards, cams, etc.

Proximity switch A device that reacts to the presence of an actuating means without physical contact or connection.

Pull-up torque (ac motors) The minimum torque developed by the motor during the period of acceleration from rest to the speed at which breakdown occurs.

Pushbutton A master switch having a manually operable plunger or button for actuating the switch.

Rating, continuous The substantially constant load that can be carried for an indefinite time.

Rating of a controller Designation of operating limits based on power governed and the duty and service required.

Rating, eight-hour The rating of a magnetic contactor based on its current-carrying capacity for 8 h without exceeding established limitations. Rating considerations include new, clean contact surfaces, free ventilation, and full-rated voltage on the operating coil.

Rating, make or break The value of current for which a contact assembly is rated for closing or opening a circuit repeatedly under specified operating conditions.

Reactor, saturable An inductor having the means to change the degree of magnetic saturation of its core(s) to control the magnitude of alternating current supplied to a load.

Regenerative braking In ac motors, it results from the motor's inherent tendency (through a negative slip) to resist being driven above synchronous speed by an overhauling load. In shunt-wound dc motors, it occurs when driven by an overhauling load, when shunt field strength is increased, or when armature voltage is decreased (in adjustable-voltage drives).

Relay A device operated by a variation in the conditions in one electric circuit to effect the operation of other devices in the same or other circuits. Examples include: current, latching, magnetic control, magnetic overload, open-phase, low or undervoltage, and overload.

Reset A manual or automatic operation that restores a mechanism or device to its prescribed state.

Resistance starting A form of reduced-voltage starting employing resistances that are short-circuited in one or more steps to complete the starting cycle.

Reversing Changing the operation of a drive from one direction to the other.

Rod-and-tube (rate-of-rise) sensor A thermostat consisting of an external metal tube and an internal metal rod that operates as a differential-expansion element. This element actuates a self-contained snap switch.

Service of a controller The specific application in which the controller is to be used—either general-purpose or definite-purpose (crane and hoist, elevator, machine tool, etc.).

Starter An electric controller for accelerating a motor from rest to normal speed. *Note:* A device designed for starting a motor in either direction includes the additional function of reversing and should be designated a reversing controller.

Starting, slow-speed A control function that provides for starting an electric drive only at the minimum speed setting.

Static control A system that may contain electronic components that do not depend on electronic conduction in a vacuum or gas. The electrical function is performed by semiconductors or the use of otherwise completely static components such as resistors, capacitors, etc.

Static controller A controller in which the major portion of all of the basic functions are performed through the control of electric or magnetic phenomena in solids such as transistors, etc.

Synchronous motor controller A controller consisting of a three-pole starter for the ac stator circuit, a contactor for the dc field circuit, an automatic synchronizing device to control the dc field contactor, and a cage-winding protective relay to open the ac circuit without synchronizing, in order to start a synchronous motor, accelerate it to synchronous speed, and synchronize it to supply frequency.

Switch A device for making, breaking, or changing the connections in an electric circuit. In controller practice, a switch is considered to be a device operated by other than magnetic means.

Switch selector A manually operated multiposition switch for selecting alternative control circuits.

Temperature-responsive protector A protective device for assembly as an integral part of a motor that provides a degree of protection to the motor against dangerous overheating due to overload and failure to start.

Tests, application Tests performed by a manufacturer to determine those operating characteristics not necessarily established by standards but which have application interest.

Test, dielectric The application of a voltage higher than the rated voltage for a specified time to determine the adequacy of insulating materials and spacings against breakdown under normal conditions.

Thermal cutout An overcurrent protective device that contains a heater element and renewable fusible member which open the motor circuit. It is not designed to interrupt short-circuit currents.

Thermistors Devices that sense temperature through changes in resistance. Signals from a thermistor may be amplified to interrupt the contactor holding coil to provide a degree of protection against motor locked-rotor conditions and running overloads.

Threading Signifies low-speed operation similar to jogging, but for longer periods with interlocked control.

Time accelerating The time to change from one specified speed to a higher or lower speed while operating under specified conditions.

Time delay A time interval purposely introduced into the performance of a function.

Time response An output, expressed as a function of time, resulting from the application of a specified input under specified operating conditions.

Torque A turning or twisting force that tends to produce rotation.

GENERAL CONSIDERATIONS

Speed Control of Electric Motors

The term *speed control,* as applied to electric motors, covers a wide range of control functions (see Figs. 4-23 to 4-34) including motor starting, control of motor speed during normal operation, and reversing and stopping of motors.

Operating requirements for specific motor applications must also be considered and generally include constant, variable, adjustable, and multispeed operation.

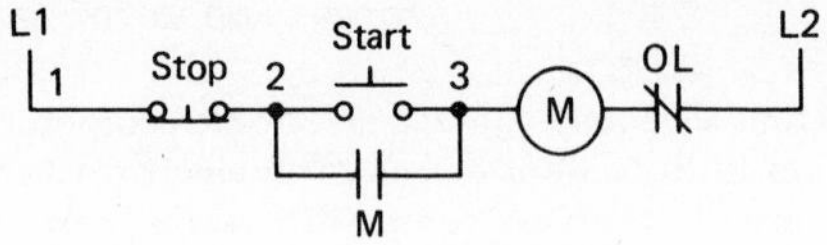

Figure 4-23 Low-voltage protection with single pushbutton station.

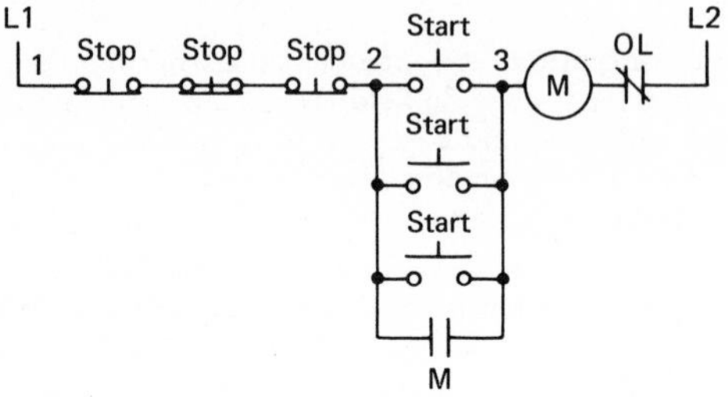

Figure 4-24 Low-voltage protection with multiple pushbutton stations.

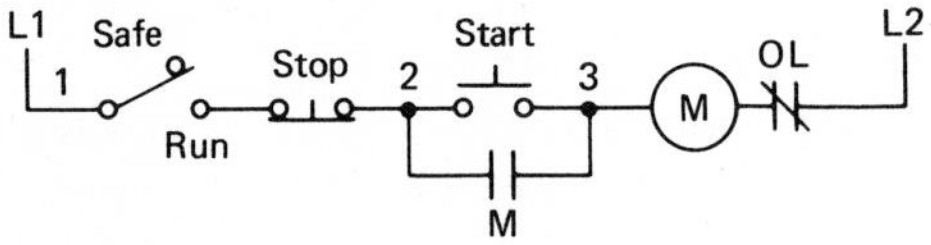

Figure 4-25 Low-voltage protection with safe-run switch.

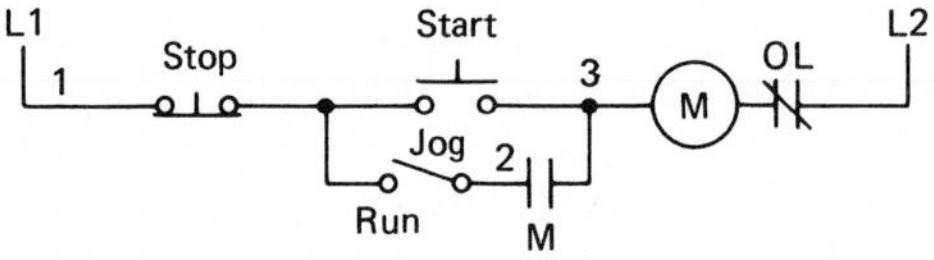

Figure 4-26 Control for jog or run using pushbuttons and selector switch.

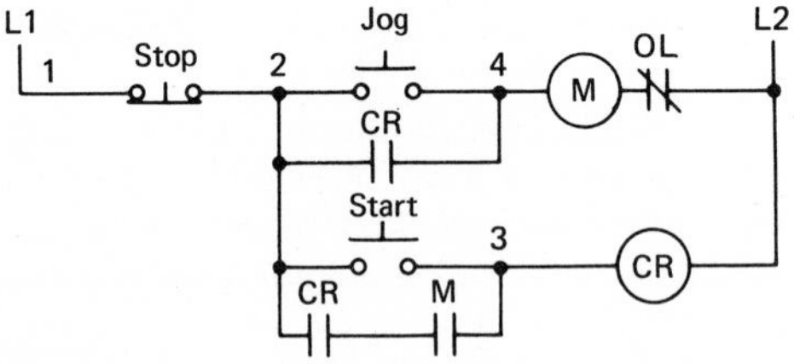

Figure 4-27 Control for jog-start-stop using pushbuttons.

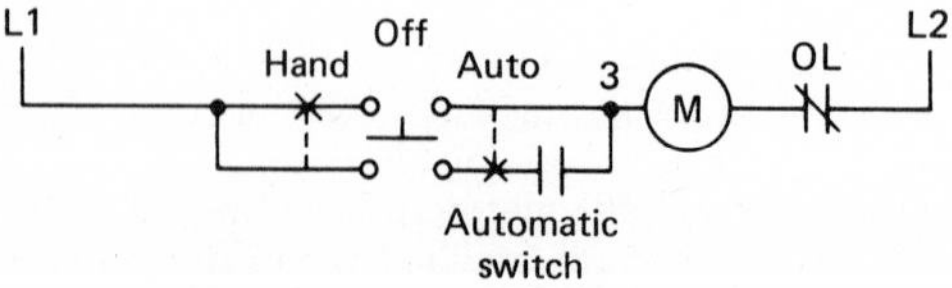

Figure 4-28 Low-voltage release.

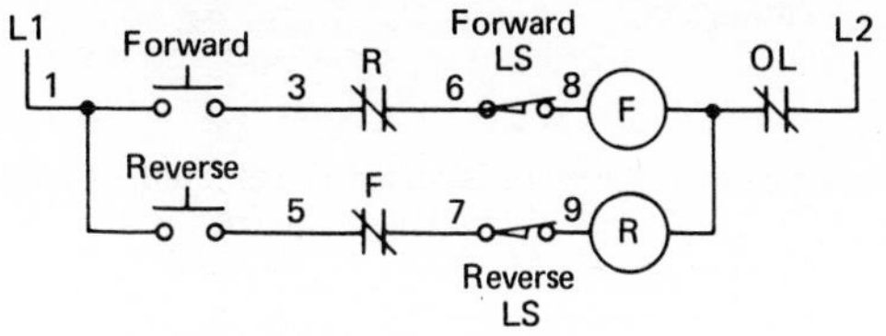

Figure 4-29 Two-wire control for reversing jogging.

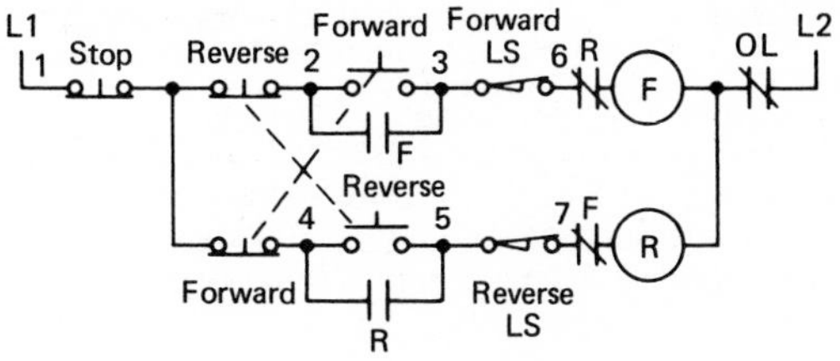

Figure 4-30 Control for instant reversing.

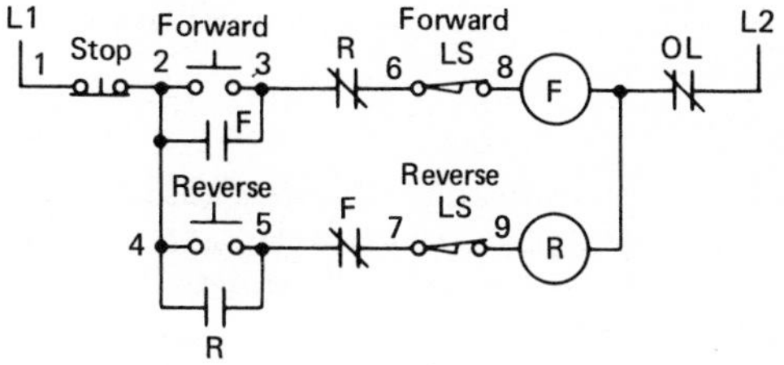

Figure 4-31 Control for reversing.

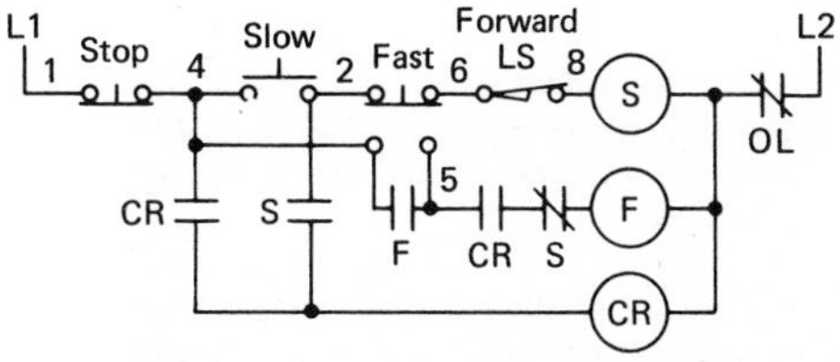

Figure 4-32 Control for two-speed starter.

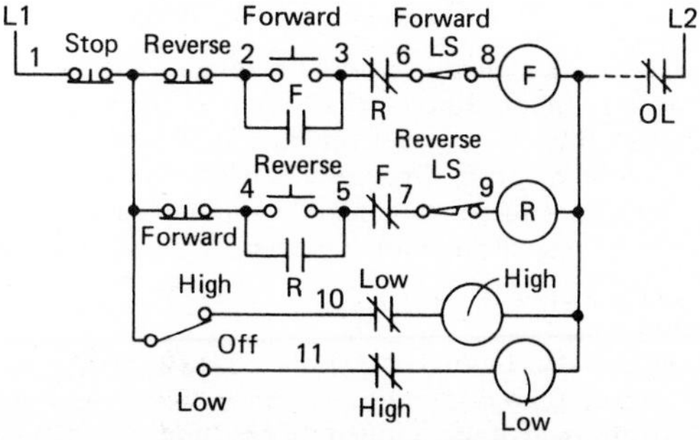

Figure 4-33 Control for two-speed reversing starter.

Constant Speed

Motors of this type may be designed with speed ratings from 80 r/min and horsepower ratings ranging up to 5000 hp. A typical application is with water pumps.

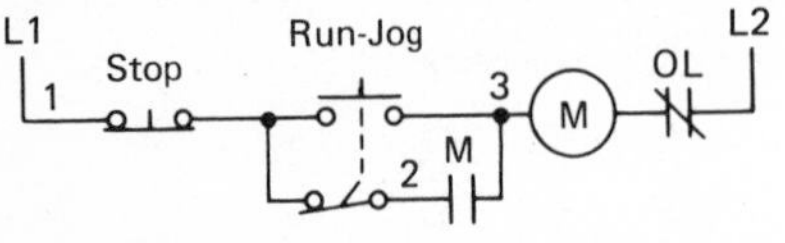

Figure 4-34 Selector push contacts for three- or two-wire operation.

Variable Speed

Variable-speed motors slow as the applied load is increased and speed up as the load is decreased. Applications include use with equipment such as cranes and hoists.

Adjustable Speed

Adjustable-speed motors can be varied over a wide speed range while the motor is running. Motor speed remains almost constant once set, even with load applied. A typical application is with machine tools.

Multispeed

Multispeed motors are designed to operate at two or more definite speeds. However, once adjusted to a particular speed, the motor speed remains nearly constant regardless of applied load changes. A typical application is with turret lathes.

Control Selection Considerations

The selection of a specific motor speed control system also requires the consideration of a number of factors. Depending upon the particular type, size, and application of the motor to be controlled and the particular characteristics of the driven load, motor control selection may be simple or complex.

For example, the motor controller (Fig. 4-35) may be manually operated, magnetically operated (which includes both manual and automatic types), static, or some combination of static and electromagnetic.

Manual Speed Control

One of the first forms of motor speed control employed rheostat (variable-resistance) controls, which have a limited effective range—approximately 4:1. Other disadvantages of rheostat-type controls include relatively high power consumption and poor speed regulation with load torque changes.

Another manual control device is the faceplate controller. Connection is made from the line to a lever that carries a contact brush or shoe. The brush rides on a series of stationary contact buttons to which resistance is connected. As the lever is moved, it shorts out resistance in steps until the motor is connected directly to the line. The circuit is maintained as long as the voltage remains on-line. If the voltage drops out, the lever is released and returns to its full resistance position to prevent the possibility of starting the motor at full voltage.

A drum-type controller is a manual controller having its electric contacts mounted on the surface of a rotating cylinder. Stationary contacts (fingers) are riveted to a support channel, with the supply line, motor, and resistors connected to one end and held in contact at the other end by a spring. Generally, a drum controller can withstand more abuse than multiple-switch or faceplate controllers, offers easier serviceabilty, and can be equipped with blowout magnets and arc barriers.

Automatic Speed Control

For speed-control ranges of up to 10:1, a variable transformer (Variac) with rectifiers can be used with dc motors. The more efficient Variac speed controls use an autotransformer starter to step down voltage applied to the motor terminals which, in turn, reduces the motor starting current.

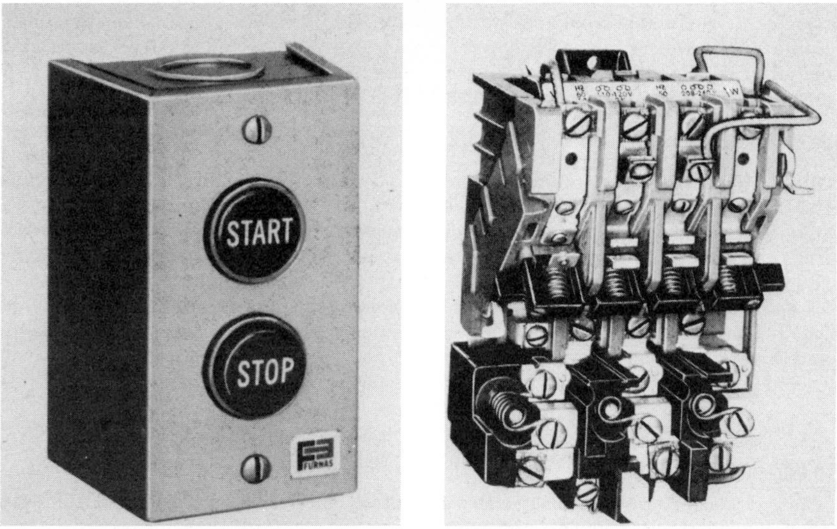

Figure 4-35 Typical motor controllers. (*Furnas Electric Co.*)

A second method of automatic speed control is accomplished with silicon-controlled rectifier (SCR) controls. With full-wave rectification, SCR systems can control motor speed over ranges up to 20:1.

Solid-State Speed Control

One method of solid-state speed control for permanent magnet and shunt motors uses a transistor in place of a variable resistor for input voltage control. More complex solid-state starters use semiconductors and trigger circuits instead of mechanical contacts to accelerate the motor. Motor circuit voltage is varied from zero to maximum on a stepless ramp while producing negligible inrush current. Using solid-state starters, mechanical starting shock to the driven load is minimized. The ramped-down stopping feature also minimizes mechanical shock to low-inertia loads that could be damaged by abrupt stops. Solid-state starters generally provide thermal-overload, phase-loss, and phase-reversal protection.

When precise speed control is required, a tachometer can be used to measure actual speed which is then compared with a reference, and the output difference is fed to a power conversion unit that supplies power to the motor. This type of system is known as an analog speed regulator.

Digital speed regulators are used to control ac or dc drives and generally consist of a pulse generator that produces pulses in proportion to motor speed and an oscillator used as a reference. Pulse generator and oscillator counts are compared, and a corresponding error signal is fed to an analog regulator.

Special Considerations

Selection of the proper motor control system also involves several other key factors. These include: operator-vs.-automatic-machine starting, expected starting requirements, continuous or intermittent machine operation, and special functions, if any, required during operation.

Separate from these special functions are requirements that specify the need to reverse direction or stop the motor, and the types and number of protective devices necessary to assure proper and continued operation.

In order to simplify this apparently detailed process of motor control selection, the above considerations will be discussed in terms of general control requirements (types of starters, drives, and general applications of speed control) and specific control requirements (application of controls to specific motor types, starting and braking, speed-torque characteristics, and protective devices). General motor control requirements are considered here; specific requirements are discussed in subsequent sections.

Motor Drive Circuits

Motor drive circuits vary in design from ac to dc motors. Generally, however, they take the form of simple ON-OFF switches, speed controls, or position controls. Controlling the frequency of the supply power to ac motors is one of the most practical ways of controlling ac motor speed. In dc motors, however, speed is controlled either by adjusting the shunt field or the armature voltage.

Solid-state drive circuits used to control ac and dc motors are static versions of their electromechanical counterparts. A basic three-phase static drive consists of three diodes and three controlled rectifiers (thyristors). Two such convertor-inverter units can be connected in a back-to-back arrangement to provide a reversing, regenerative drive.

AC Drive Characteristics

Cycloconverter

The cycloconverter is essentially a step-down frequency drive, consisting of a number of thyristor switches set between the power source and the load. Dual converters are used in all three phases to produce three single-phase outputs. Spaced sine-wave references turn on the switches in proper sequence, changing 60-Hz ac power to adjustable-frequency ac power. It is best suited for low-speed drives and has a normal power range from 500 to 10,000 hp.

Inverter Drives

Inverter drives can be used with industrial ac induction or synchronous motors for adjustable speed control where a source of dc power is available. For three-phase motor application, three single-phase inverters are arranged for 120° phase displacement and applied individually to the motor windings. A three-phase inverter basically consists of six sequentially operated switches, with only three closed at any given time.

Pulse-Width Modulation Drive

Pulse-width modulation (PWM) inverter drives rectify ac power and then invert it to adjustable-frequency ac power. Some designs use rectifier switching to force commutation of the inverter bridge. The drive controls voltage by varying pulse width and frequency. It is designed for use with squirrel-cage motors within a range of 5 to 150 hp.

Adjustable-Voltage Constant-Frequency Drive

By adjusting the point on the ac wave at which the thyristors are turned on, ac source voltage can be varied to control the speed of squirrel-cage motors (motor torque also varies as the square of the voltage). Disadvantages of this type of drive include excessive motor heating at low speeds, losses which are proportional to motor slip, and applications limited to soft-start, constant-torque loads. The drive can also be used with wound-rotor induction motors equipped with permanent secondary resistance and a tachometer feedback. Power ratings range from 5 to 50 hp; applications include pump and fan loads.

Slip-Power Reclamation Drive

The speed of wound-rotor motors is controlled with this drive by changing the values of resistance connected in the secondary circuit. AC slip power from the rotor is rectified to direct current, inverted, and fed back to the line. Using a current error signal to turn on the thyristors and control the rate at which the motor secondary power is returned to the line, the system performs as a torque-regulated drive. A speed error reference signal is used to adjust torque and maintain a preset speed. Speed range is from zero to near synchronous speed, and the normal power range is from 300 to 5000 hp.

DC Drive Characteristics

Adjustable-Voltage Drive

DC motor armature voltage can be varied to obtain wider ranges of speed control than are available with standard dc motors. An ac-to-dc motor-generator set controls the dc motor speed by increasing the generator voltage while holding the motor field at full strength. When generator voltage reaches full value, the motor runs at base speed. If the speed control is increased further, the motor field is weakened while generator voltage is held constant, causing the motor speed to increase above its base.

Single-Phase Dual Convertors

Control of speed and torque is accomplished by controlling voltage and current in both directions of motor rotation. Gating and control circuitry permit smooth transition from forward to reverse speeds and prevent circulating currents between thyristors.

Single-Phase Semiconvertors

Thyristors are pulsed alternately with the changing polarity of the ac input current. Output current is regulated by adding a phase shift that changes the level of the trigger pulse. A commutating diode bypasses inductive motor current that would keep the thyristor on.

Single-phase semiconvertors are normally used for fractional to 5-hp motors and can be used with options such as dynamic braking, jogging, threading, acceleration-deceleration, and speed regulation.

Three-Phase Dual Convertors

Although similar to single-phase dual convertors, they have additional diodes and thyristors to handle a three-phase input. Power range is from 20 to 7500 hp.

Three-Phase Semiconvertors

While similar to single-phase semiconverters, they have additional diodes and thyristors to handle the three-phase input. Power range is from 5 to 200 hp.

Three-Phase Single-Way Convertors

Voltage and current can be controlled in either direction with dual convertors and require only six thyristors to achieve a full-reversing, regenerative system. A transformer is necessary to connect one end of the motor armature to neutral. Power range is from 5 to 100 hp.

Three-Phase Single Convertors

DC output voltage is controlled by introducing a phase shift that alters the reference level at which the rectifiers operate. By reversing the motor field with contactors, a reversing regenerative drive may be achieved with these inherently nonreversing and nonregenerative drives. The usual range of single converters is from 20 to 7500 hp.

Chopper Drives

The high-speed switch (chopper) characteristics of phase-controlled rectifiers can be used to create adjustable-speed dc drives. A constant-voltage rectifier or a dc bus is used as the incoming supply. When the first thyristor is turned on by a reference-level signal, the dc motor is connected to the constant-voltage bus. When motor speed rises above the reference level, a second thyristor is turned on, shutting off the first, and the motor is driven from power stored in an inductor. When the motor speed drops below the reference level, the first thyristor is turned on, turning off the second which again supplies power from the bus. Current transfer between thyristors is regulated by capacitor discharge through a transformer. Dynamic braking can be added by turning on a third thyristor that dissipates energy into a dynamic braking resistor; it is impossible to return current to the line through the rectifier bridge.

This type of drive is often used in applications such as adjustable-voltage hoists because power supplies can be subjected to wide voltage deviations.

Mechanical Drives

Directly Coupled Drives

Synchronous motors can be adapted for direct-coupling machines operating at speeds from 3600 r/min down to 80 r/min, with horsepower ratings ranging from 20 to 5000 hp and larger. Low-speed, directly coupled induction motors are rarely used at speeds below 500 r/min because of low power factors and low efficiency.

When using directly coupled drives, the alignment of the devices must be checked from four positions spaced 90° around the motor shaft.

Lead Screw Drive

The lead screw drive consists of a motor turning a lead screw which moves a mass.

Belt-Pulley Drive

Flat belts, V belts, chains, or gears are driven by the motor through a system of pulleys to accomplish smooth speed changes at a constant revolutions per minute. Offset belt-pulley drives offer the advantage of being easier to align than direct drives.

In order to select proper pulley sizes, the drive revolutions per minute, the driven revolutions per minute, the diameter of the drive pulley, and the diameter of the driven pulley must be known.

Gear Drive

The gear motor is a speed-reducing motor giving direct power from a gear or series of gears linked to the motor output shaft. Economical operating ranges are from 1 r/min to approximately 780 r/min and may run at constant or adjustable speed.

Magnetic Drive

This drive couples the motor to its load magnetically, without mechanical contact between rotating members. Torque is transmitted between two rotating units electromagnetically by energizing a coil winding.

Motor Starters

Whether operated manually, electromagnetically, or statically, starters designed for ac or dc electric motors can be divided into two categories based on the motor starting-voltage conditions. The two basic categories are full-voltage (across-the-line) and reduced-voltage starters.

Proper size selection of starters can result in cost and space savings. Matching the motor to its starter eliminates overcapacity purchasing. NEMA standard and custom starter sizes are available (see Table 4-11) for the various voltage- and horsepower-rated motors.

Full-Voltage Starting (Across-the-Line)

Under ideal conditions, any size ac motor operating at any voltage can be started at full voltage. In actual application, however, when full voltage is applied to the motor terminals, its locked-rotor current may range anywhere from 6 to 8 times the value of normal running current. By design, this current may not harm the motor; however, it may damage the driven load due to the motor's starting torque. Under these conditions, the application of a reduced-voltage starter could eliminate such potential problems.

Full-voltage manual starters and contactors provide direct control for applications not requiring remote control and which permit automatic restarting. Overload relays are generally provided to protect against excessive current damage which may result from sustained overloads, low line voltages, etc.

TABLE 4-11 Motor Starter Sizes

Single-phase horsepower		Three-phase horsepower			
115 V	230 V	200 V	230 V	460–575 V	Starter size*
⅓	1	1½	1½	2	00
1	2	3	3	5	0
2	3	7½	7½	10	1
3	5	10	10	15	1P, 1¾
3	7½	10	15	25	2
—	—	15	20	30	2½
—	—	25	30	50	3
—	—	30	40	75	3½
—	—	40	50	100	4
—	—	50	75	150	4½
—	—	75	100	200	5
—	—	150	200	400	6

*NEMA Standard and custom sizes included.

Full-voltage starting of dc motors is limited to motors rated 2 hp or less. In fractional-horsepower ratings, double-pole switches of the toggle, key, and lever type can be used. For integral-horsepower dc motors up to 2 hp, integral pushbutton starting may be used. Magnetic contactors also can be used for starting, stopping, and reversing dc motors of this rating range.

Reduced-Voltage Starting—AC

A number of design approaches are employed for reduced-voltage starting, including the following.

Primary Resistor Starting. Using this method, series resistance is added in each conductor to the motor. A rheostat control is used to gradually short out the resistance as the motor comes up to speed, until the motor is connected to the full line voltage.

Primary Reactor Starting. This motor-starting method is similar to the primary resistor method except that reactors are substituted for the resistors.

Autotransformer Starting. At starting, wye-connected autotransformer coils reduce motor terminal voltage in each conducter to 50, 65, or 80 percent of line voltage. After a timed interval, a manual switch or a contactor connects the motor across the line and bypasses the autotransformer coils, employing either an open or closed transition.

Part-Winding (Increment) Starting. Though technically not reduced-voltage starting, part-winding starters apply starting current in timed steps to part-winding motors. Two-step controllers apply voltage through one starter to one motor winding followed by the second starter which connects voltage to the second winding. Timed automatic closing of the second controller can minimize voltage fluctuations. Three-step starters incorporate a series resistance with one motor winding to increase the motor terminal voltage gradually as the voltage drops across the resistor.

Wye-Delta Starting. Although technically not reduced-voltage starting, a wye-delta starter energizes the motor windings through electric contacts that form a wye connection giving about 33 percent of full line voltage across each winding. After a set time delay, the motor windings are connected in a delta configuration. Wye-delta starting can be used in applications requiring a low starting torque when supplying full starting current would cause significant voltage drops.

Reduced Voltage Starting—DC

To minimize high starting currents in dc motors rated 2 hp and above, series resistance is added to the motor winding. As the motor begins to rotate, it generates a counter emf which increases the internal resistance of the motor and lowers the effective voltage across the motor. Either manual or automatic starters are used to decrease the value of the connected resistance gradually as the motor goes from standstill to full speed.

Reduced-Voltage Starting—Solid-State

Solid-state starters accelerate motors using semiconductor devices in the power and trigger (reference level) circuits rather than mechanical contacts. Acceleration times are adjusted through the trigger circuit to maximum delays of approximately 10 s. Power ratings range from 2 to 400 hp. Smooth starting and gradual stopping capability are used to protect the driven load from mechanical shock.

Open- vs. Closed-Loop Transition

In applications where reduced-voltage starting is required to limit starting current or torque, motor acceleration is accomplished in a series of steps. Depending on the torque limitations of the driven load and the capacity of the power source, the plant engineer will have to select between open-transition (interrupted) or closed-transition (continuous) starting.

Open-Loop Transition

Transition from reduced-voltage starting to full-voltage operation is accomplished (in automatic autotransformer starters) by magnetic contactors in combination with a timing device. In this method, the motor is allowed to accelerate for a preset time at a reduced voltage. The timing device then opens the starting coil circuit, breaking the starting contacts, and recloses the motor circuit to full line voltage.

Closed-Loop Transition Starting

In contrast, closed-loop transition starting introduces an additional step that prevents line transients during starting and eliminates secondary inrush currents normally experienced when connecting the motor to full line voltage. Power to the motor is not interrupted during the normal starting sequence, thereby providing a smoother acceleration than is possible with open-circuit transition starting.

Motor Acceleration

Once started, a motor will continue to accelerate if the load torque requirements do not exceed the motor torque. An increase in motor counter-voltage, however, accompanies the acceleration process and tends to reduce output torque. Since resistance frequently is added to limit initial inrush current, a gradual shorting out of this resistance as motor speed increases will maintain motor torque above load torque, allowing the motor to continue accelerating to the desired operating speed.

The resistance steps can be calculated using the following equations:

$$R_t = \frac{V_r}{I_s}$$

where

R_t = total circuit resistance, including motor and leads
V_r = rated motor voltage
I_s = maximum inrush current (a percentage of full load current)

As the motor accelerates, inrush current drops to a normal full-load value. The voltage that can be measured across the connected resistance is

$$V_1 = R_t \times I_{fl}$$

where I_{fl} = motor full load current.

At this point, a portion of the connected resistance must be shorted out for the motor to continue accelerating to its full-load speed. The generally accepted method of calculating the value of resistance to be cut is to determine the value that will result in a second inrush current equal to the first, or

$$R_1 = \frac{V_1}{I_s}$$

The motor will again accelerate, generating a counter-voltage and, in turn, drop the motor current to its normal full-load value. The motor voltage at this point will be

$$V_2 = R_1 \times I_{fl}$$

The value of subsequent resistance step reductions can then be calculated in the same manner as R_1. When the total of the motor resistance and each of the resistance step reductions is equal to R_t, a sufficient number of accelerating steps will be provided to bring the motor to its full-load speed. These values can also be determined graphically (as in Fig. 4-36) using a plot of the motor characteristic curve as a starting point.

Motor Braking

Electrically operated mechanical brakes are used for two purposes: (1) to provide the means to stop a driven load quickly and accurately and (2) to hold the load in place after stopping.

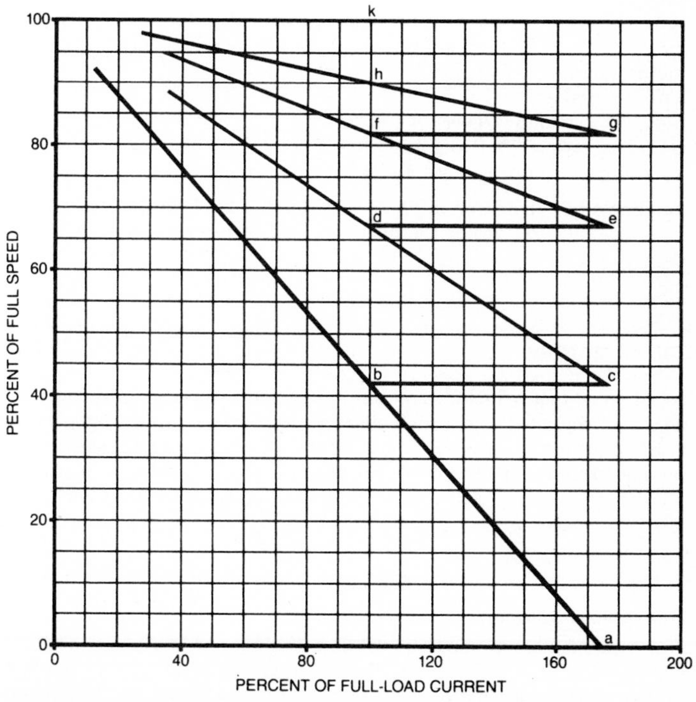

Figure 4-36 Motor characteristic curves: $R_1 = (db/kb)R_t$; $R_2 = (fd/kb)\,R_t$; $R_3 = (hf/kb)R_t$.

To select the proper size brake for a particular application, the required brake torque can be calculated by the following equation:

$$T = \frac{5250\ \text{hp}}{N} \qquad \text{in lb·ft}$$

or

$$T = \frac{726\ \text{hp}}{N} \qquad \text{in kg·m}$$

where

T = torque, lb·ft
hp = rated motor horsepower
N = full-load motor speed, r/min

The brake selected should have torque rating at least equal to the value calculated from the above equation.

The time required to change from one operating speed to another is

$$t = \frac{WR^2(N_2 - N_1)}{308T}$$

The time required for stopping is

$$t = t_a + \frac{WR^2N_b}{308T}$$

where

t = time, s, required to stop moving mass after brake is de-energized
t_a = time, s, for brake shoe to apply against wheel after brake is de-energized
N_2 = motor r/min at highest speed
N_1 = motor r/min at lowest speed
N_b = r/min of brake wheel
R = radius of gyration of mass, ft (m)
T = retarding torque, lb·ft (kg·m)
W = weight of the rotating mass, lb (kg)

Other factors to be considered in brake selection are the heat-dissipating capacity of the brake and the duty ratings of the operating coil or mechanism used to release the brake shoes.

CONTROL OF AC MOTORS

The control of an ac motor includes motor starting and stopping; governing the motor speed, torque, horsepower, and other characteristics; and protecting personnel and equipment. Simple motor controllers are called starters; more complex systems are called drives. Each has been designed to cover the wide range of available ac motor types and sizes and to meet the varying characteristics of specific motor applications.

Starters

Fractional-Horsepower Starting

The simplest type of manual starting switch is the one- or two-pole fractional-horsepower toggle switch consisting of an on-off snap-action mechanism. This method is generally applied to single-phase motors with ratings up to a maximum of 1 hp at 120 or 240 V, where only infrequent starting and stopping are required.

Integral-Horsepower Starting

These manual starting switches consist of full-voltage starters and mechanically operated electric contacts. General applications are with single-phase and polyphase integral-horsepower motors operating at up to 600 V ac and having maximum ratings of 5 and 7.5 hp, respectively. The manual switch operation may be initiated by pushbuttons or a toggle handle.

Variations include a reversing manual starter, designed for use with polyphase ac motors, and a two-speed manual starter, for two-speed separate-winding wye-connected motors.

Polyphase Magnetic Starting

Polyphase magnetic starters are designed for full-voltage starting of squirrel-cage induction motors when full starting torque and current surge are permitted. They are also used for primary circuit control for wound-rotor (slip-ring) motors that have provision for manual starting and speed control in their secondary circuits.

Wound-Rotor Controls

To control starting, accelerating, and regulation, a variable resistance is added to the rotor circuit. Full rotor resistance is used during motor starting. As the motor begins to accelerate, the resistance is reduced in steps. When the motor is connected to full line voltage (all resistance shorted out), it acts as a squirrel-cage motor.

The basic control circuit consists of a full-voltage starter and a balanced, adjustable three-phase resistor, wye-connected in the rotor circuit. Speed can be established for a given load by adjusting rotor resistance; once set, speed will vary with load conditions.

A static controller may be used to control the operating speeds of wound-rotor motors. In one method, a controlled saturable reactor is placed in the rotor circuit with the accelerating resistance. For fixed operating speeds, reactor saturation (which controls the motor speed) can be varied using a control resistor. Static controllers can also be used to reverse the direction of motor rotation by placing saturable reactors in the motor primary circuit rather than reversing contactors. Controlled reactor saturation directs the reversal of the motor rotation.

Synchronous Motor Starting

This method is used for power-factor correction of heavy concentrations of induction motors. It is also used for constant-speed, slow-speed industrial drive applications and for maximum efficiency on continuous heavy loads in excess of 75 hp. Three-phase ac power is connected to the stator and dc to the rotor (which has both a field and a squirrel-cage winding).

A full-voltage magnetic contactor connects the ac motor winding to the line, and the rotor winding is closed through a starting and discharge resistor. The motor starts and comes up to speed similar to a squirrel-cage motor. At the correct rotor speed, a polarized-field frequency relay and reactor automatically apply dc excitation to the field to synchronize the motor with maximum synchronizing torque, while drawing minimum line current.

Speed-Torque Characteristics

Polyphase ac motors are designed to operate at speeds directly proportional to the frequency of the voltage applied to the stator field. However, while the motor's synchronous speed is directly proportional to the applied frequency, it is inversely proportional to the number of motor poles.

Since induction motors rely upon rotor bars or windings to cut the flux of the rotating field to turn the rotor, they will operate at a speed slightly less than synchronous speed.

In constant-horsepower designs, output torque can vary inversely with motor speed. For applications requiring constant output torque, however, the air-gap flux must be held constant over the entire speed range of the motor.

Multispeed motors may be divided into three basic groups, based on their speed-vs.-torque response:

Constant-Torque. These motors produce a horsepower output that varies with the speed of the motor. For example, a constant-torque motor with a 10 hp rating at a speed of 1800 r/min will deliver 5 hp at 900 r/min.

Variable-Torque. Motors of this type have a torque output that varies as the square of motor speed. The term is generally applied to drives that have fan or centrifugal pump loads.

Constant-Horsepower. This type of motor can deliver rated horsepower at any of the rated operating speeds. Motor torque, however, varies inversely with the speed. Motors of this type are used with machine tools where the motor speed governs the speed of the cutting tool.

Speed Control

Adjustable-Frequency Drives

By design, ac motor speed is regulated by a rotating field impressed upon the stator. Controlling the frequency of the supply power, then, is an obvious method to control the speed of an ac motor. The more efficient adjustable-frequency systems control both supply voltage and frequency in order to maintain the required torque characteristics as well.

The use of frequency adjustment to control motor speed permits operation of motors in applications such as pump and compressor drives at values closer to the rated capacity. It also has been suggested that increasing motor speeds beyond 3600 r/min will allow the design of smaller, lighter weight motors without changing the power output. Such systems may use microprocessors to obtain accurate digital control of power-transistor-type switches.

DC Source Control

Where a dc power supply is available, the speed of ac motors may be controlled using dc-powered inverter systems that incorporate phase shifting and sequentially operated switching to convert dc source power to adjustable-frequency ac power. Motor speed control is then accomplished in a manner similar to other adjustable-frequency control systems.

Selecting Speed Controls

AC Motor Starting and Braking

The frequency of motor starts and stops will determine the length of satisfactory controller service because magnetic switches (such as motor starters, relays, and contactors) will eventually self-destruct through repeated opening and closing. Solid-state controllers eliminate much of this problem since switching of the motor is accomplished electronically instead of electromagnetically.

The potential for successfully selecting the proper controller can also be increased by choosing the controller design best suited to handle the particular requirements of the installation.

Polyphase Magnetic Starters. These starters are designed for the full-voltage starting of squirrel-cage induction motors. They also can be applied for primary circuit control of wound-rotor motors that incorporate provisions for manual starting and speed control in their secondary circuits.

Manual Primary Resistor Starters. These starters are used for reduced-voltage starting of polyphase squirrel-cage motors in applications where full starting current or torque cannot be tolerated.

Primary Reactor Starters. These starters offer heavy-duty service for induction motors rated for 2200- to 5000-V operation, and can also be combined with jogging controls.

Autotransformer (Compensator) Starters. These starters can also be used for reduced-voltage starting polyphase squirrel-cage motors, and offer the additional advantages of current-limiting and higher starting torque without the energy loss of resistor-type starters.

Part-Winding Starters. Automatic-type starters are designed for use with squirrel-cage motors that have two separate parallel windings on the stator.

Wye-Delta Starters. Automatic-type starters have use limited to reduced-voltage starting of motors whose windings can be reconnected in delta. As the motor reaches full speed, the windings are reconnected in a delta configuration for running.

Motor Braking

Changing the speed of, or stopping, an electric motor can be accomplished using electrically operated mechanical brakes, electrically through dynamic or regenerative braking or plugging, or by a combination of the two. Electrically operated mechanical brakes are used for two purposes: (1) to provide the means of stopping a driven load quickly and accurately and (2) to hold the load in place after stopping is accomplished.

Dynamic Braking. When very quick or accurate stopping is not a necessity, the motor may be used to stop itself. AC motors can be dynamically braked (Fig. 4-37*a*, *b*, and *c*) by removing the ac power source and reconnecting a dc power source (supplied either from batteries or rectified ac power). The motor then acts like a dc generator with a short-circuited armature. Energy is dissipated in the form of rotor heat.

Regenerative braking. In ac motors, regenerative braking (Fig. 4-38*a* and *b*) is developed by the motor's natural tendency to resist being driven above synchronous speed by an overhauling load. As this situation occurs, a negative slip is developed. The energy absorbed in slowing down is put back into the power supply. Regenerative braking generally is not used with rectified power supplies because this process requires the reversal of armature current.

CONTROL OF DC MOTORS

The control of dc motors, as with ac motors, includes motor starting and stopping; governing motor speed, torque, horsepower, and other characteristics; and ensuring the safety of personnel and equipment. The methods of controlling dc motors, however, vary considerably.

DC motor speed is directly proportional to the applied armature voltage and inversely proportional to the field flux. Shunt field control and armature voltage control are, then, the two basic methods used to accomplish speed adjustment.

Speed Control

Shunt Field Control

With the addition of a rheostat to the shunt field circuit, control is obtained by adjusting the rheostat to weaken the shunt field current. This, in turn, increases the motor speed and reduces the output torque. Tests indicate that this type of system can be used only to obtain speeds above the motor's base speed (defined as motor speed with full-rated voltage applied to the armature and field circuits).

For short periods of time only, speeds below the base may be obtained by overexciting the field. Extended operation in this manner generally results in excessive motor heating.

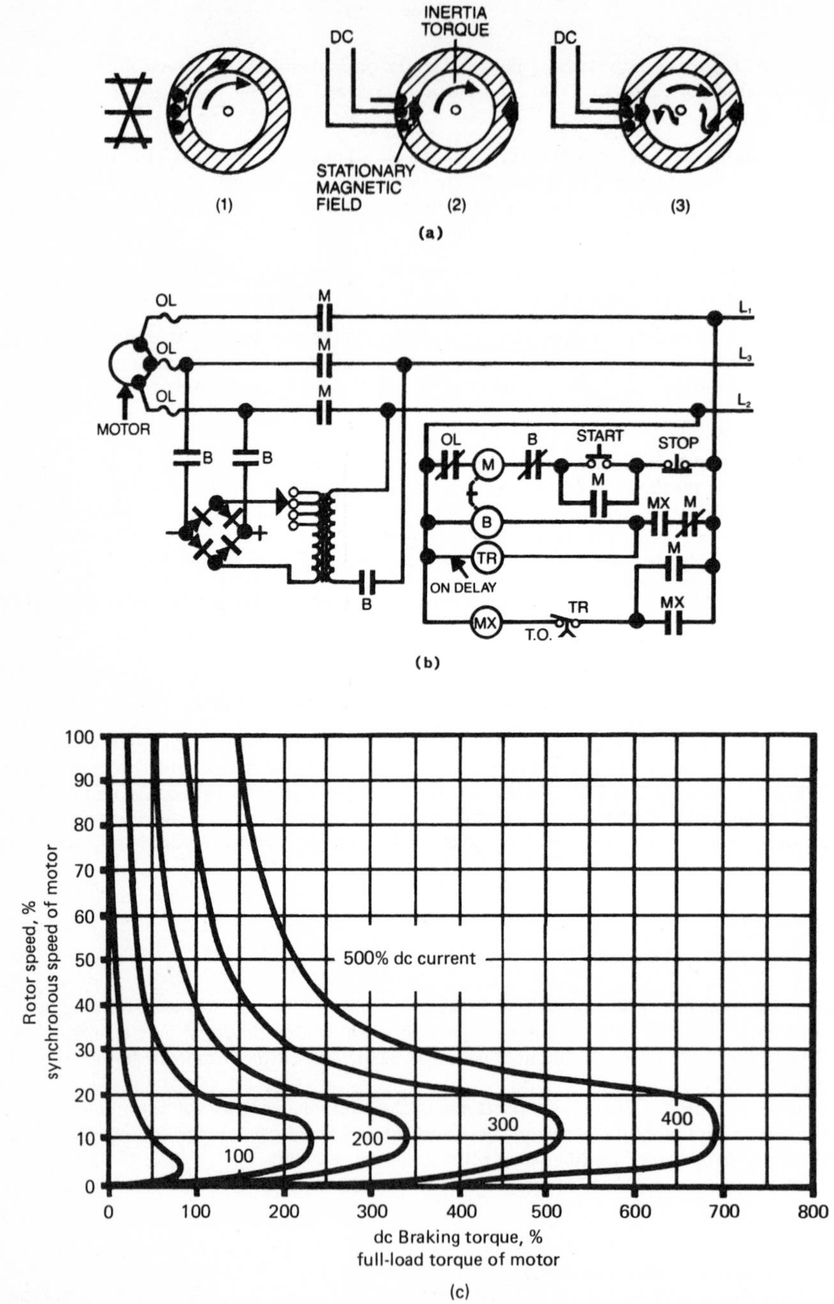

Figure 4-37 (*a*) Dynamic braking in ac motors. This is accomplished by disconnecting the motor from its ac source and then reconnecting it to a dc source. The ac motor then functions as a dc generator. (*b*) Dynamic braking circuitry for ac motors. The circuitry required to achieve dynamic braking of ac motors includes a dc source (either a battery or a rectified ac source), an auxiliary relay (to transfer between ac and dc sources), and a timer (to disconnect the dc source after the motor has been brought to a stop). (*c*) DC braking torque. This increases slowing as the motor slows down. The braking torque is controlled by the magnitude of the applied dc voltage.

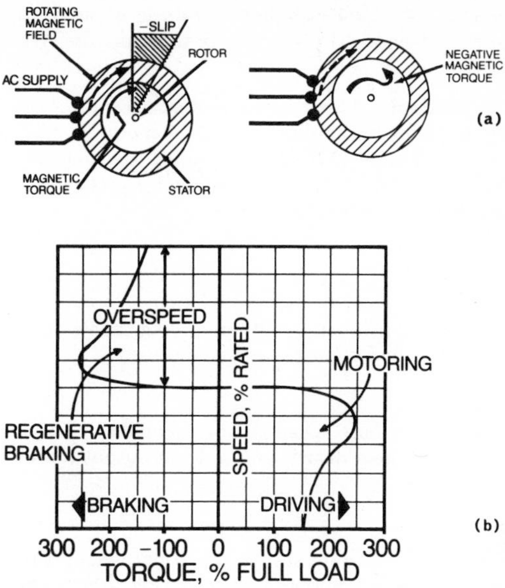

Figure 4-38 In ac motors, regenerative braking is developed by the motor's inherent tendency to resist being driven above its synchronous speed. A negative slip is developed that results in a negative torque (*a*), producing the braking action. This is seen graphically in (*b*).

The maximum standard speed range using field control is 3:1; specially designed motors can achieve speed ranges of 4:1 or more. For a specific motor, the nameplate rating should be checked or the manufacturer consulted if the allowable range of motor speed is not known.

Armature Voltage Control

This method requires the application of a variable voltage power supply with a capacity equal to that of the motor, while holding the shunt field current constant. The resulting speed is proportional to the motor counter-emf (which is equal to applied voltage minus armature circuit *IR* drop). The torque remains constant at rated current, regardless of motor speed. In applications where a speed range of greater than 3:1 is required, armature voltage control should be used.

Basic Variations

Combinations and variations of shunt field control and armature voltage control can be applied to fractional-horsepower wound field motors and generally include the following types.

Straight-Series Motor Control. Under no-load conditions, a straight-series (wound field) motor could theoretically run away as motor speed increases. Reduced motor current and stator field flux causes a further increase in motor speed. Internal friction and winding losses, however, provide sufficient load to keep the motor speed within safe operating limits.

Split-Series Motor Control. This method is performed similar to straight-series control with the exception that the motor's two oppositely wound field coils can be used for polarity switching to produce rapid changes in the direction of motor rotation.

Shunt Motor Control. Control of shunt motors is obtained through parallel-connected armature and field coils. A line current in the armature creates a field in opposition to the main field.

Compound Motor Control. This type of motor has both a series and shunt field. When the series winding aids the shunt winding, the motor is called a "cumulative compound" motor, and when it opposes the shunt winding, the motor is called a "differential compound" motor. If reversing of the compound motor is required, the polarity of both fields, or the armature, must be switched.

Selecting Speed Controls

The speed of dc motors is controlled by either varying the voltage across the armature, the field, or both. Each method offers certain advantages.

Field Series Resistance

Adjustment of up to 4 times base speed can be provided by varying shunt-motor field resistance. Output power is held nearly constant but output torque is reduced with the reduction in field strength.

Armature Shunt Resistance

By reducing the effective line current seen at the armature, a braking action can be provided in either shunt or series motors. The resulting drop in armature voltage also reduces the motor speed.

Armature Series Resistance

Varying a resistance placed in series with shunt or series motors can lower base motor speed up to 50 percent by reducing armature voltage.

Series-Parallel Combinations

For speed control in multiple series or shunt motor installations, two identical motors may be connected in series or parallel. In parallel connections, each motor receives full voltage and operates at base speed. In series connections, the armature voltage and speed of each motor is cut in half.

AC Source Control

Since most electric distribution systems supply three-phase, 60-Hz ac power, the power must be converted from ac to adjustable-voltage dc power.

Three methods to convert ac to dc include: (1) a fully controllable motor-generator set, (2) static rectifiers, and (3) semiconductor-type devices such as thyristors. The operating principles of these devices are covered in "General Considerations," earlier in this part.

Starting and Braking

Starting Considerations

Starting of dc motors rated 2 hp or less can generally be accomplished by manually operated, full-voltage starters. For dc motors rated above 2 hp, reduced-voltage starting is usually recommended to avoid commutator damage. A common method of limiting starting current is by the addition of resistance (which is decreased in steps). The final step (larger motors or those with smooth starting requirements use several steps) is to connect the motor to full line voltage.

Dynamic Braking

As in the case of ac motors, dc motors may be brought to a stop quickly using electromechanical brakes. DC motors can also be braked dynamically to eliminate the need for or reduce the size of mechanical brakes.

Dynamic braking is achieved by disconnecting the armature circuit supply and connecting a resistive load (Fig. 4-39*a* and *b*) while the field remains energized. The net effect is to make the motor act as a generator connected to a resistive load. As the armature current is reversed, a negative torque develops which acts to slow and stop the machine. As the motor slows, the negative torque automatically decreases in equal proportion.

Regenerative Braking

In dc motors, regenerative braking occurs any time a shunt-wound motor operates in the generator mode (see Fig. 4-40). This occurs when the motor is driven by an overhauling

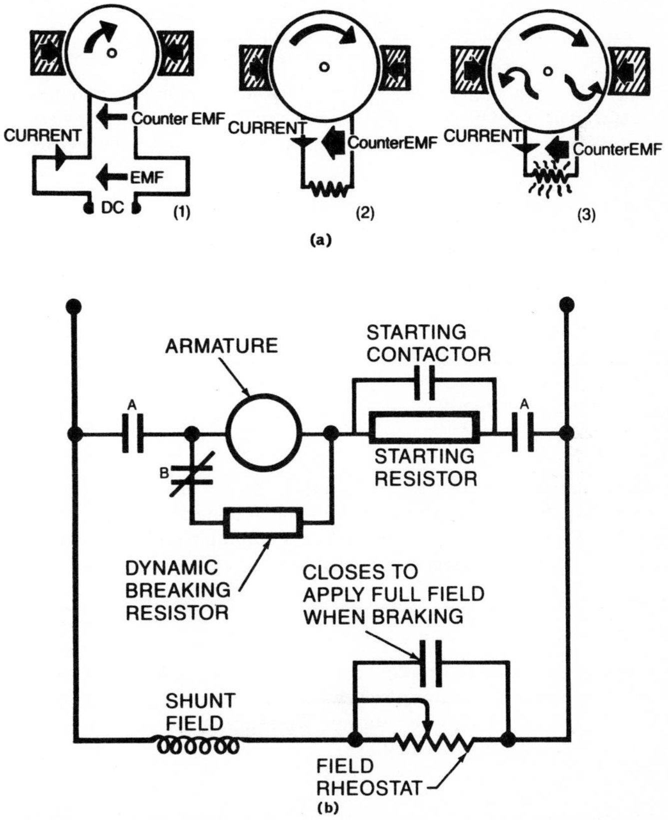

Figure 4-39 (*a*) Dynamic braking in dc motors. This is accomplished by disconnecting the power supply from the armature and reconnecting a resistive load. Operating as a generator feeding a load, the motor develops a negative torque which stops it. (*b*) Dynamic braking circuitry for dc motors. The circuitry required to achieve dynamic braking of dc motors includes a braking resistor and an auxiliary relay which disconnects power from the armature and connects the resistor. The auxiliary relay can also be used to short out the field rheostat when full field braking is desired. A and B are braking contactors.

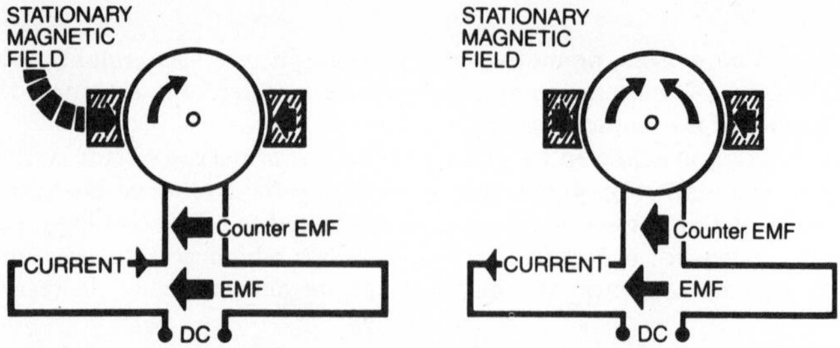

Figure 4-40 Regenerative braking in dc motors. Regenerative braking occurs in dc motors whenever the motor is driven by an overhauling load, when the shunt field strength is increased, or when the armature voltage is decreased. Operating in the generator mode, the dc motor produces a counter-emf and a resulting negative torque that stops the motor.

load, when the shunt field strength is increased, or when the armature voltage is decreased (in adjustable-voltage drives).

Regenerative braking is generally not applicable to highly compounded machines unless the series field is removed or shorted out during the braking process. Also, since this process involves the reversal of armature current, it is not normally used with rectified power supplies.

CONTROL-SYSTEM PROTECTION

Part of the process of selecting motor control equipment involves providing protection of the controller, all connected equipment, and personnel from abnormal operating conditions. Protective techniques may employ physical shielding of the controller from its environment and/or tampering, or devices that sense the presence of abnormal conditions.

Mechanical Protection

In certain environments, enclosures can increase the life span and ensure trouble-free controller operation. Special-rated enclosures such as general-purpose, watertight, dustproof, explosion-proof, and corrosion-resistant are available for specific applications that must meet national and local electrical and building codes.

Open-Field Protection

DC shunt and compound wound motors can be protected against the loss of field excitation by installing field loss relays in the shunt field circuit. Larger dc motors may race dangerously with the loss of field excitation, while other motors may not race due to friction and the fact that they are small.

Open-Phase Protection

Phase failure relays are used to prevent motor starting if one phase in a three-phase circuit has been opened or reversed or if a preset degree of voltage imbalance is present. Phase failure may be caused by a blown fuse, an open connection, or a broken line. If phase failure occurs while the motor is stopped, stator currents will rise to a very high value while the motor remains stationary. Since motor windings cannot be properly ventilated while the motor is stationary, excessive heating may damage the windings.

Overload (Running) Protection

A controller, with overload protection provided by thermal bimetallic overload relays, magnetic–thermal-overload relays, and thermal-solder–ratchet devices, which monitors motor current (see Fig. 4-41) will protect a motor while allowing it to reach maximum available power. Motors may also contain inherent protection, built into their windings, which is used to monitor motor temperature. Overload and high-temperature conditions can be caused by overloaded machinery, by low line voltage, or by single-phase operation in a polyphase system.

Overspeed Protection

To prevent damage to a driven machine, materials in the industrial process, or motor, overspeed protection is provided by controlling the power supply frequency in ac motors and by limiting the maximum shunt field resistance or armature voltage in dc motors. Typical applications include paper and printing plants, steel mills, processing plants, and the textile industry.

Overtravel Protection

In applications where precise positioning is required, control devices (such as limit switches, cams, photoelectrics, etc.) are used to govern the starting, stopping, and reversal of electric motors. These devices can be used to control regular operation or as emergency switches to prevent the improper functioning of machinery.

Overvoltage Protection

Devices such as overvoltage relays provide a reduction in the level of applied voltage or a maintained interruption of power to the motor circuit when excessive voltage levels are experienced. High induced voltages may occur as a result of the interruption of inductive motor control circuits.

Reversed-Current Protection

Rectifiers are used to protect dc controllers used with dc or three-phase ac systems that can be subject to damage when experiencing phase failures and phase reversals. It is also important to provide reversed-current protection for battery-charging equipment.

Reversed-Phase Protection

Phase-failure and phase-reversal relays are used to prevent the operation of elevators and industrial machinery to protect motors, machines, and personnel from potential hazards when phase reversal or loss occurs. Interchanging two phases of the supply of a three-phase induction motor will reverse its direction of rotation.

Short-Circuit Protection

For motors with greater than fractional horsepower ratings, devices such as fuses and circuit breakers may be installed in the same enclosure as the motor-disconnecting means to protect branch-circuit conductors, motor control apparatus, and the motor itself against fault conditions that may be the result of short circuits or grounds, and prolonged or excessive starting currents.

Overload Relay		Current Rating (Amps)			Auxiliary Contacts: Form		Auxiliary Contacts: Rating					Compensation		Heater								Reset		U.L.		
		25, 60, 100, 180	30, 60	20, 30, 45, 65, 95, 140	S P S T	S P D T	Standard Duty	Heavy Duty	300V. AC Max.	600V. AC Max.	Continuous Carrying Amps	Non-Comp	Comp.	Std. Trip	Quick Trip	5000 Amps Fused Short Circuit Capacity	Trip Current Adjustment	In Line Wiring	Trip Free	Manual Trip	Aluminum Wire Terminations	Manual	Automatic	Recognized	Listed	Cost (5)
I	Melting alloy																									
	1. Single element	✓			✓		✓			✓	10	✓		✓		✓			✓			✓			✓	3
	2. Tri-element (1)		✓		✓			✓		✓	10	✓		✓		✓		✓	✓		✓	✓			✓	2
II	Bimetal																									
	1. Single element	✓			✓		✓			✓	10	✓	✓	✓	✓	✓		✓	✓			✓	✓		✓	3
	2. Tri-element (1)		✓		✓	✓		✓		✓	10	✓	✓	✓	✓	✓	✓	✓	✓	✓	✓	✓	✓		✓	2
III	Electronic																									
	1. Three phase (1)			✓	✓		✓		✓		5		✓	(2)	(2)	✓	✓	(3)	✓		(3) ✓	(4)	(4)	✓		1

Notes:

1. These units can also be supplied for two phase or single phase operation.
2. Heater elements are not required (a number of standard trip curves are available).
3. No power terminal wiring is required (motor leads pass through current loops).
4. Requires additional circuitry to obtain manual reset.
5. Cost is listed with No. 1 being highest.

Figure 4-41 Overload relay selection chart.

Undervoltage Protection

Undervoltage relays and three-wire control are employed to initiate the opening and maintaining open of a motor circuit upon a reduction or failure of voltage. Undervoltage protection (Fig. 4-42) generally is applied where uncontrolled motor starting could result in potential hazards.

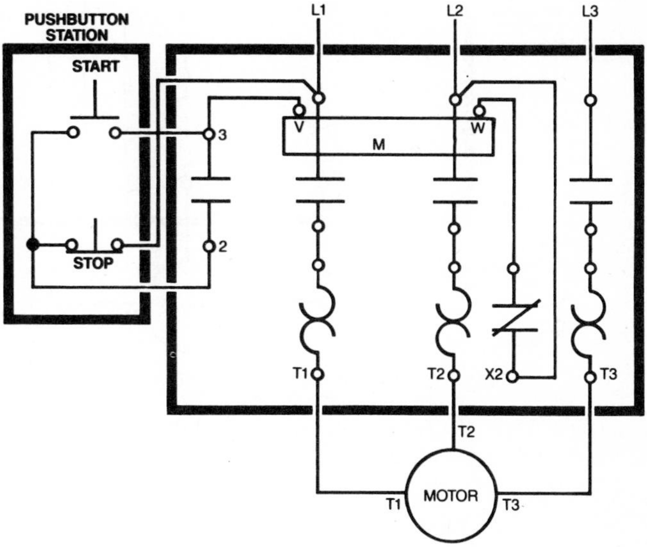

Figure 4-42 Basic three-wire control circuit.

Undervoltage Release

Undervoltage relays and two-wire control are employed to initiate the temporary opening of a motor circuit (Fig. 4-43) upon the reduction or failure of voltage. Upon restoration of full voltage, the motor is automatically restarted.

TROUBLESHOOTING ELECTRIC MOTOR CONTROLS

Varying factors such as temperature, humidity, and atmospheric contamination may adversely affect the performance of motor controls. Misapplication of a control may also lead to serious trouble and is often regarded as the major cause of motor control problems. Visual inspection every 6 months or so, and less-frequent electrical checks with the proper instruments, will help to ensure that production will not be interrupted because of a starter failure that could have been prevented.

It is important to make a complete mechanical check of motor control equipment before and after installation. Damaged or broken parts can usually be found easily and quickly, and replaced if necessary. Visual checks should be made with the aid of a flashlight, air hose, and a small brush. Debris and dirt can be brushed from contacts and other areas of the switch; light rust and dirt on pole faces can be removed with compressed air and brush. Never use a file or abrasive of any kind on pole faces since this can upset the precise fit between core components. A simple tightening of terminal screws should be sufficient to correct many motor controller problems.

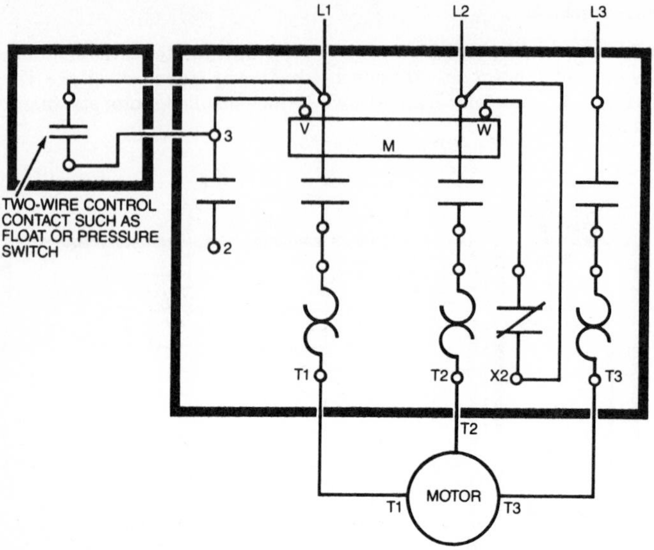

Figure 4-43 Basic two-wire control circuit.

It is recommended that the following general procedures be observed by qualified personnel in the inspection and repair of motor controller involved in a fault. Manufacturer's service instructions should be consulted for additional details.

CAUTION*— It must be understood that all inspections and tests are to be made on controllers and equipment which are de-energized, disconnected, and isolated so that accidental contact cannot be made with live parts and so that all plant-safety procedures will be observed.*

Procedures

Enclosure

Substantial damage to the enclosure, such as deformation, displacement of parts, or burning, requires replacement of the entire controller.

Circuit Breakers

Examine the enclosure interior and the circuit breaker for evidence of possible damage. If evidence of damage is not present, the breaker may be reset and turned on. If it is suspected that the circuit breaker has opened several short-circuit faults or if signs of possible deterioration appear within the enclosure, the test described in paragraph AB 1-2.38 of the NEMA standards publication, "Molded Case Circuit Breakers," Publication AB 1, should be performed before restoring the breaker to service.

Disconnect Switch

The external operating handle must be capable of opening the switch. If it fails or if visual inspection after opening indicates deterioration beyond normal, such as overheating, contact blade or jaw pitting, insulation breakage, or charring, the switch must be replaced.

Fuse Holders

Deterioration of fuse holders or their insulating mounts requires their replacement.

Terminals and Internal Conductors

Indications of arcing damage and/or overheating, such as discoloration and melting of insulation, require the replacement of damaged parts.

Contactor

Contacts showing heat damage, displacement of metal, or loss of adequate wear allowance require replacement of the contacts and, where applicable, the contact springs. If deterioration extends beyond the contacts, such as binding in the guides or evidence of insulation damage, the damaged parts of the entire contactor must be replaced.

Overload Relays

If burnout of the current element of an overload relay has occured, the complete overload relay must be replaced. Any indication that an arc has struck and/or any indication of burning of the insulation of the overload relay also requires replacement of the overload relay. If there is no visual indication of damage, the relay must be electrically or mechanically tripped to verify proper functioning of the overload-relay contacts.

Final Check

Before returning the controller to service, checks must be made for the tightness of electric connections and for the absence of short circuits, grounds, and leakage. All equipment enclosures must be closed and secured before the branch circuit is energized.

For these and other complex problems, the manufacturer's wiring and schematic diagrams should be reviewed before attempting repairs. A listing of standard wiring diagram symbols is shown in Fig. 4-44. Also, to aid in troubleshooting, a listing of many of the possible motor control problems and their probable causes and solutions are presented in Table 4-12.

MOTOR CONTROL CENTERS

Motor control centers are floor-mounted assemblies of one or more enclosed vertical sections having a horizontal common power bus and principally containing combination motor control units. The units are mounted one above the other in the vertical sections. These sections may incorporate vertical buses connected to the common power bus, thus extending the common power supply to the individual units. Units may also connect directly to the common power bus by suitable wiring (definition as stated by NEMA).

The common enclosure design (Fig. 4-45) and the use of combination starters offer both economy and ease of installation in multiple motor control installations. In addition, motor control centers (MCCs) provide proper coordination between short-circuit protective devices and the controller. Since MCCs are engineered systems, the components are closely coordinated to work together, and the unit is rated for a particular value of short-circuit interrupting duty at the point of its installation. MCCs may contain a molded-case circuit breaker and starter, or a fused switch and a starter.

MCCs centralize all the electric control apparatus for a given installation in housings which are convenient for easy field installation and maintenance. Use of MCCs can minimize the total amount of floor and wall space required by isolated motor control apparatus. The individual units generally are interchangeable and easily removable as well. With proper planning, MCCs can be designed with provisions to allow for future system expansion.

Vertical sections generally are 20 in (50.8 cm) wide by 90 in (228.6 cm) high, and most designs can accommodate up to six motor starter units per section.

The guide on pp. 3-129–3-133 is provided to assist the plant engineer in specifying MCCs.

ELECTRICAL SYMBOLS

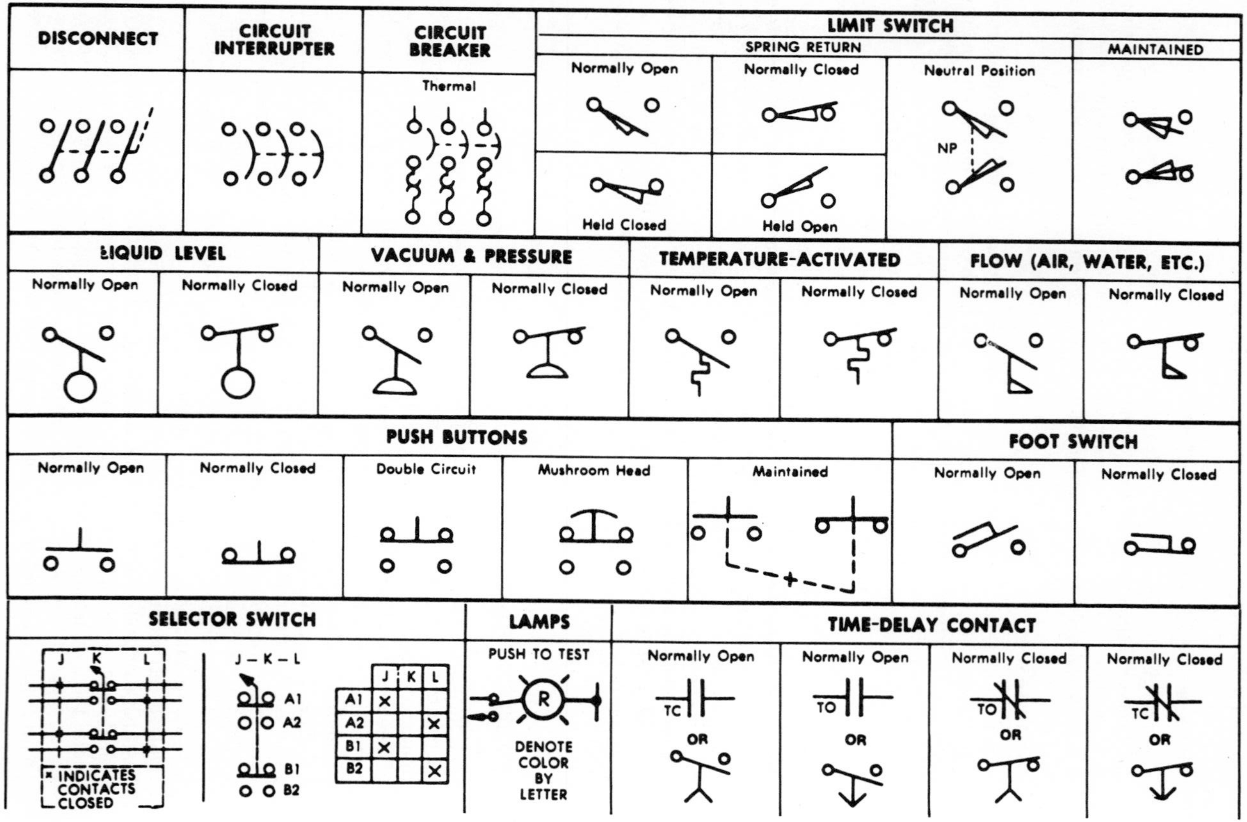

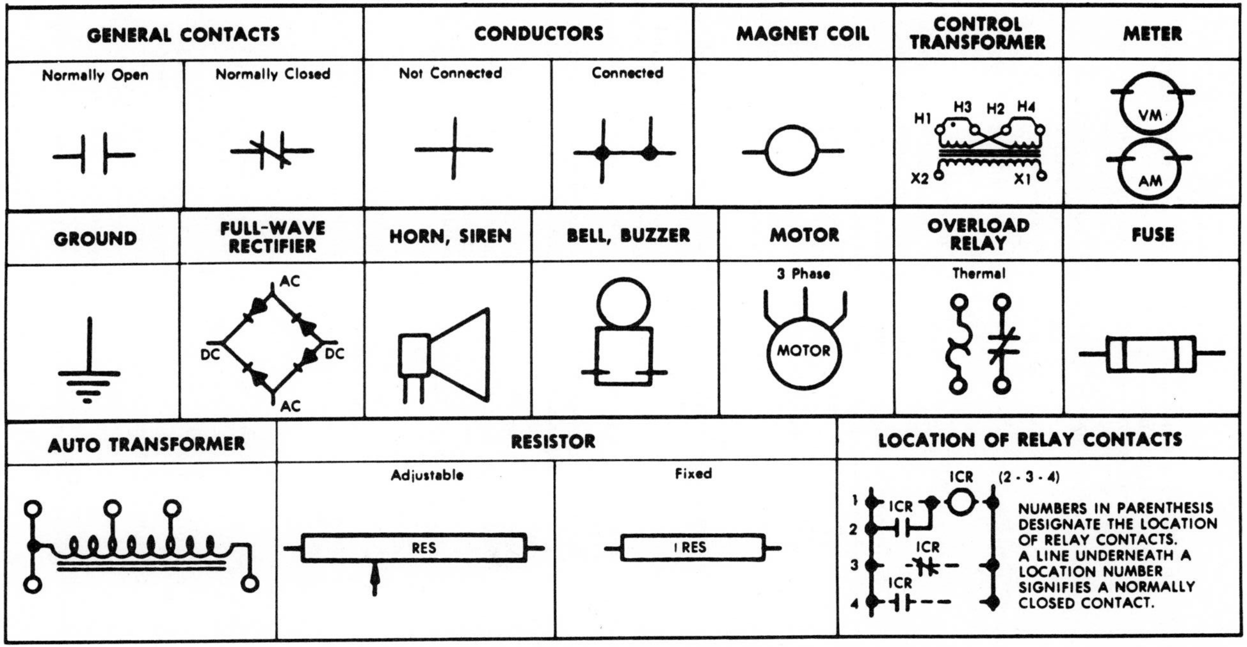

Figure 4-44 Wiring diagram symbols.

TABLE 4-12 Troubleshooting Motor Controls

Problem	Possible cause	Solution
I. Magnetic and mechanical parts		
Noisy magnet (humming)	1. Misalignment or mismating of magnet pole faces	1. Realign or replace magnet assembly.
	2. Foreign matter on pole face (dirt, lint, rust, etc.)	2. Clean (do not file) pole faces; realign if necessary.
	3. Low voltage applied to coil	3. Check system and coil voltage. Observe voltage variations during start-up time.
Noisy magnet (loud buzz)	4. Broken shading coil	4. Replace shading coil and/or magnet assembly.
Failure to pick up and seal in	1. Low voltage	1. Check system and coil voltage; watch for voltage variations during starting.
	2. Wrong magnet coil or wrong connection	2. Check wiring, coil nomenclature, etc.
	3. Coil open or shorted	3. Check with an ohmmeter, and when in doubt replace.
	4. Mechanical obstruction	4. Disconnect power and check for free movement of magnet and contact assembly.
Failure to drop out or slow dropout	1. "Gummy" substance on pole faces or magnet slides	1. Clean with nonvolatile solvent or degreasing fluid.
	2. Voltage to coil not removed	2. Shorted seal-in contact (exact cause found by checking coil circuit).
	3. Worn or rusted parts causing binding	3. Clean or replace worn parts.
	4. Residual magnetism due to lack of air gap in magnet path	4. Replace any worn magnet parts or accessories.
	5. Mechanical interlock binding (reversing starters)	5. Check interlocks for free pivoting. New bushing or light lubrication may be required.
II. Contacts		
Contact chatter (source probably from magnet assembly)	1. Broken shading coil	1. Replace assembly.
	2. Poor contact continuity in control circuit	2. Improve contact continuity or use three-wire control.
	3. Low voltage	3. Correct voltage condition. Check momentary voltage dip on starting.
Welding	1. Abnormal inrush of current	1. Use larger contactor; check for grounds, shorts, or excessive load current.
	2. Rapid jogging	2. Install larger jogging-rated device or caution operator.
	3. Insufficient tip pressure	3. Replace contact springs; check contact carrier for deformation or damage.
	4. Low voltage preventing magnet from sealing	4. Correct voltage condition. Check momentary voltage dip on starting.
	5. Foreign matter preventing contacts from closing	5. Clean contacts with nonvolatile solvent. Low-current or -voltage contactors, starters, and control accessories should be cleaned with solvent and acetone to remove solvent residue.

TABLE 4-12 Troubleshooting Motor Controls (*Continued*)

Problem	Possible cause	Solution
	6. Short circuit	6. Remove short fault and check for correct fuse or breaker size.
Short contact life or overheating	1. Filing or dressing	1. Do not file silver contacts. Rough spots or discoloration will not harm or impair their efficiency.
	2. Interrupting excessively high currents	2. Install larger device or check for grounds, shorts or excessive motor currents.
	3. Excessive jogging	3. Install larger device rated for jogging or caution operator.
	4. Weak contact pressure	4. Replace contact springs; check contact carrier for deformation or damage.
	5. Dirt or foreign matter on contact surface	5. Clean contacts with nonvolatile solvent.
	6. Short circuits	6. Remove short fault and check for correct fuse or breaker size.
	7. Loose connection	7. Clean and tighten.
	8. Sustained overload	8. Install larger device; check for excessive load current.
	9. Excessive wear	9. Higher than normal voltage may result in mechanical wear and bounce.
Contacts, supports, discoloring	1. Loose connections	1. Tighten hardware or replace.*
III. Coils		
Open circuit	1. Mechanical damage	1. Handle and store coils carefully.
Cooked coil (overheated)	1. Overvoltage or high ambient temperature	1. Check application and circuit. Coils will operate satisfactorily over a range of 85 to 110% of rated voltage.
	2. Incorrect coil	2. Check rating; replace with proper coil if incorrect.
	3. Shorted turns caused by mechanical damage or corrosion	3. Replace coil.
	4. Undervoltage, failure of magnet to seal in	4. Correct system voltage.
	5. Dirt or rust on pole faces increasing air gap	5. Clean pole faces.
	6. Sustained low voltage	6. Remedy according to local code requirements, low-voltage system protection, etc.
IV. Overload relays		
Nuisance tripping	1. Sustained overload	1. Check for equipment grounds and shorts and and excessive motor currents due to overload. Check motor winding resistance to ground.
	2. Loose connections	2. Clean connections and tighten. This includes load wires and heater-element mounting screws.
	3. Incorrect heater	3. Check heater sizing and also check ambient temperature.

TABLE 4-12 Troubleshooting Motor Controls (*Continued*)

	Problem	Possible cause	Solution
	Failure to trip out (causing motor burnout)	1. Mechanical binding, dirt, corrosion, etc.	1. Clean or replace.
		2. Incorrect heater or jumper wires used or heaters omitted	2. Recheck ratings and heater size. Correct if necessary.
		3. Wrong calibration adjustment	3. Consult factory. Calibration adjustment is normally not recommended unless under factory supervision. It is customary to return units to factory for check and calibration.
V.	Manual starters		
	Failure to operate (mechanically)	1. Mechanical parts, including springs, worn or broken	1. Replace parts as needed.
		2. Welded contacts due to application or other abnormal cause	2. Replace contacts and recheck operation.
	Trips out prematurely	1. Motor overload, incorrect heaters, or misapplication	1. Check conditions; replace or adjust as needed.
VI.	Timers		
	A. Pneumatic		
	Erratic timing	1. Foreign matter in valve	1. Clean if at all possible, or replace timing head completely and exchange unit with factory.
	Contacts do not operate	1. Adjustment incorrect on time-actuating screw	1. Follow service bulletin instructions for desired adjustment.
		2. Worn or broken parts in switch assembly	2. Replace defective parts.
	B. Electronic relay		
	Erratic timing	1. Loose connections	1. Check over unit visually.
		2. Timing relay worn out	2. Plug in new tested relay.
		3. Defective components	3. Check and replace if necessary.
	Timer stops operating	1. Mechanical relay	1. Substitute good relay.
		2. Defective circuit components	2. Check over unit visually and with a vacuum tube voltmeter (VTVM). Replacement of circuit board probably better than repair if relay normal.
	C. Electronic (solid-state)		
	Erratic timing	1. Loose connections	1. Visually inspect all connections.
		2. Check external connections	2. Check functions with VTVM in accordance with prescribed service instructions.
	Timer will not time out	1. External connections	1. Systematically check system.
		2. Check power supply to timer	2. Fuses, etc.
		3. Initiation circuit open	3. See step 1.
		4. Contacts dirty	4. See step 1; clean if necessary.
VII.	Limit switches		
	Broken parts	1. Excessive overtravel of actuator	1. Use resilient actuator or operate within device tolerance.

TABLE 4-12 Troubleshooting Motor Controls (*Continued*)

Problem	Possible cause	Solution
Inoperative	1. Switch actuator out of position or broken 2. Lack of contact continuity	1. Inspect, repair, or replace. 2. Clean contacts; replace contact block if necessary.
VIII. Drum controls		
Poor contact	1. Dirty rotor contacts and fingers	1. Inspect contacts; if copper, burnish with 4-0 sandpaper until clean; if silver, use a suitable solvent. Check for approximately 3/64-in finger movement.
	2. Dirt or other foreign matter on horizontally mounted units	2. Clean systematically with carbon tetrachloride and air.
Sluggish or hard operating switch	1. Dry bearings 2. Worn parts	1. Lubricate bearings sparingly. 2. Inspect carefully; replace worn parts.
IX. Pressure switches		
Pressure switch inoperative	1. Foreign matter in pressure-sensing area 2. Contacts burned	1. Remove switch and clean opening. 2. Clean contacts; replace if necessary.
Erratic operation	1. Worn parts 2. Diaphragm faulty	1. Inspect, adjust, or replace. 2. Replace diaphragm.
Very frequent operation	1. Likely due to a waterlogged system	1. Drain part of water from pressure tank and, if possible, pump in about 4 lb air.
X. Pushbuttons		
Button inoperative (mechanical)	1. Shaft has dirt or residue binding 2. Contact board spring broken	1. Check, clean, and clear. 2. Replace contact board.
(electrical)	3. Contaminated contacts and corrosion	3. Clean.
XI. Pushbutton, pilot lamp		
No light	1. Bulb out or burned out 2. Broken parts, wire, or transformer 3. Short life of bulb due to excessive high voltage	1. Replace with proper unit. 2. Inspect, repair, or replace. 3. Replace with next higher voltage pilot lamp (brillance may be reduced slightly).

*Any contact replacement should include a complete set of replacement including the support springs, screws, etc.

Motor Control Center Specifications

Usually the following information is required to process an MCC order properly.

1. Service voltage
2. Configuration
3. Size of main horizontal bus
4. Bus bracing
5. Incoming service
6. Main protective device
7. Wiring NEMA class

8. Wiring NEMA type
9. Metering
10. Branch protective devices

1. Service Voltage

Low-voltage ratings for industrial control apparatus are based on utilization voltage per NEMA ICSI-112.22 voltage rating and are as follows for 60-Hz alternating current, multiphase: 115, 200, 230, 460 and 575 V. Corresponding system voltages are 120, 208, 240, 480, and 600 V.

Enter the proper values for hertz, phase, and number of wires used in system.

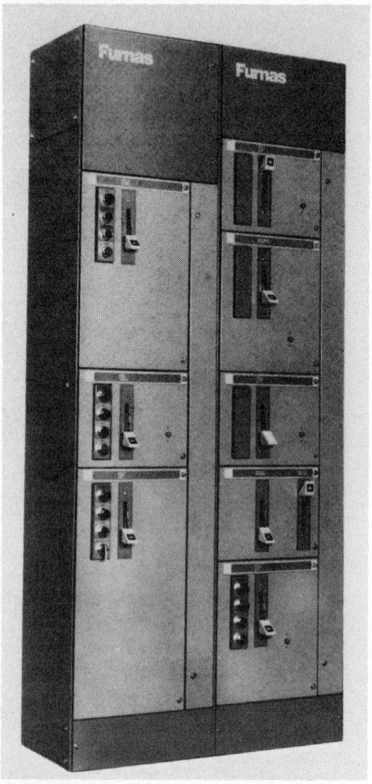

Figure 4-45 Motor control center. (*Furnas Electric Co.*)

2. Configuration

The MCC should be of dead-front, indoor design and fabricated from code-gauge steel, with all sections joined to form a single assembly. All side, front, and rear cover plates should be field-removable.

The MCC enclosure should be (unless otherwise noted):

- NEMA 1 indoor construction
- Non-walk-in—front accessible

Main and branch units should be front-connected. The MCC should have space or provisions for future expansion, as noted on the plans.

The MCCs are to be constructed in accordance with the latest NEMA-ICS and UL 845 standards.

Individual sections are to be front-accessible, not less than 15 (38.1 cm) deep, and the rear of all sections should align. All bolts used to join current-carrying parts should be installed so as to permit servicing from the front only so that no rear access is required. An unobstructed conduit entry area is to be provided at the top and bottom of each standard control section.

Paint. If any color other than USAS 61 or USAS 49 is required, a paint chip or complete Munsel number should be supplied.

3. Size of Main Horizontal Bus

Mark as required. Note there are two ratings. The higher ratings are per NEMA, the smaller per Underwriters Laboratory. UL ratings must be used if UL labels are a requirement on structures.

4. Bus Bracing

Main Bus. The MCC should be bused with rectangular busbars made of ___ aluminum ___ copper, and braced for ___ 50,000 ___ 75,000 A symmetrical.

The through bus on the end section should be extended and predrilled to allow the addition of future sections with standard busbar splice plates.

Neutral Bus. Indicate size of neutral bus required in amps or physical size; specify location required (top or bottom of sections). Specify lug size required for neutral cable or actual neutral cable size.

Ground Bus. If required, specify amperage or size. Specify location if specific location is desired. Recommend always mounting in bottom front. Specify lug size required or ground cable size.

5. Incoming Service

Underground Service. To isolate incoming underground service conductors, an underground pull, or auxiliary, section should be used. This section should be of the ___ nonbused ___ top-bused type, and sealable per local utility requirements. ___ Lugs ___ compression connectors to terminate ___ copper ___ aluminum cable, should be furnished as detailed on plans.

Overhead Service.

Cable entry
___ Lugs ___ compression connectors to terminate ___ copper ___ aluminum cable, should be furnished as detailed on plans.
Bus duct entry
The MCC is to be fed by ___ copper ___ aluminum, _______ A, _______ bus duct, as detailed on plans, ___ and other sections of the specification. The MCC manufacturer is to be responsible for coordination, proper phasing, and internal busing to the incoming bus duct.

6. Main Protective Device

The main protective device, to be installed in the service section, is as indicated below.

Molded-Case Circuit Breaker of the quick-make, quick-break, trip-free, ___ heavy duty, ___ extra heavy duty, ___ energy limiting, ___ solid-state type, It should be an _______ frame ___ two-pole ___ three-pole, 600-V breaker with a trip current rating of:

_ 400 A	_ 1000 A	_ 2000 A
_ 600 A	_ 1200 A	_ 2500 A
_ 800 A	_ 1600 A	_ 3000 A

and an interrupting capacity of not less than _______ A rms symmetrical at the system voltage.

The following accessory features are to be included: ___ shunt trip, ___ electrical operator, ___ ground-fault relaying, ____________________ (other).

Fusible Switch of the quick-make, quick-break type. It should be a ___ two-pole ___ three-pole, ___ 240-V ___ 600-V unit with a continuous current rating of ___ 400 ___ 600 ___ 800 ___ 1200 A, and with _______ A class _______ fuses, suitable for application on a system with _______ A symmetrical available fault current.

Bolted Pressure Switch of the quick-make, quick-break type. It should be a ___ two-pole ___ three-pole ___ 240-V ___ 480-V unit with a continuous current rating of:

_ 800 A	_ 1600 A	_ 2500 A
_ 1200 A	_ 2000 A	_ 3000 A

and with _______ A class L fuses, suitable for application on a system with _______ A symmetrical available fault current.

The following accessory features are to be included: ___ shunt trip, ___ electrical operator, ___ ground-fault relaying, ____________________ (other).

Power Circuit Breaker with a ___ stationary ___ drawout frame, and a current rating of:

_ 600 A	_ 2000 A
_ 800 A	_ 3000 A
_ 1600 A	

It is to be ___ manually ___ electrically operated with a(n) ___ electromechanical ___ solid-state trip device, and an interrupting capacity of________ A rms symmetrical at the system voltage.

The following accessory features are included: ___ short time delay, ___ ground-fault relay (trip), ___ shunt trip (M.O.C/B only), ___ control power transformer, ________________ (other).

7. Wiring NEMA Class

Indicate class I or class II as required. If class II, elementary or schematic drawings should be supplied with the order.

8. Wiring NEMA Type

Mark as required. If C terminals are mounted in bottom wiring space section, additional unit space will be required for more than one row of terminals.

Unit Terminals (B or C Wiring). If B or C wiring, NEMA class is required indicate type of terminals desired.

Pull-Apart Terminals. Indicate where required.

9. Metering

The following customer metering equipment should be furnished as shown on the plans.
Main bus

____ Voltmeter, with ___ -phase transfer switch
____ Ammeter with ___ -phase transfer switch
____ Watthour meter(s) (two) (three)-element
(with) (without) demand attachment
____ ________________ Current transformer(s), ________________ /5 or suitable rating
____ ________________ Potential transformer(s), suitable rating

Branch circuits

____ Ammeter(s), with ___ -phase transfer switch
____ ________ Current transformers(s), ________ /5 or suitable rating

10. Branch Protective Devices

All molded-case circuit breakers, fusible switches, and/or motor starter units used as a protective device in a branch circuit should meet the requirements of the appropriate paragraph below.

Each protective device should have an individual door over the front, equipped with a voidable interlock that prevents the door from being opened when the switch is in the "on" position, unless the interlock is purposely defeated by activation of the defeater mechanism.

Molded-Case Circuit Breakers should be of the quick-make, quick-break, trip-free, ____ motor circuit protector (MCP), ___ thermal-magnetic type ___ solid-state, with frame, trip, and voltage ratings, either two-pole or three-pole, as indicated. All breakers should have an interrupting capacity of not less than ___ A rms symmetrical at the system voltage. All breakers should be removable from the front of the MCC without disturbing adjacent units. The MCC should have space or provision for future units.

Fusible Switches should be quick-make, quick-break units and conform to the ratings shown on the plans.

All switches should have externally operated handles. Switches should be equipped with ________ fuse holders and class ________ fuses of ampere rating and type as indicated. Suitable for application on a system with ________ A symmetrical available fault current.

NEMA or Custom-Rated Magnetic Starters are to be furnished of the type and horsepower ratings as indicated. Thermal overload relays on starters should be ___ non-compensated melting-alloy type (unless otherwise noted). Three overload elements should be furnished on each starter. The overload heater elements should be sized from the actual motor nameplate data.

The following accessory features should be furnished on each starter:

___ Individual control power transformers
___ Pilot light(s)
___ Auxiliary interlocks
______________ NO ______________ NC

Pushbuttons, selector switches, and other pilot devices shall be furnished as indicated.

APPLICABLE CODES AND STANDARDS

American National Standards Institute (ANSI). Standards on a wide variety of electrical equipment, published by the American National Standards Institute.

Canadian Standards Association. Canadian standards on a wide variety of electrical equipment, published by the Canadian Standards Association, 178 Rexdale Boulevard, Rexdale, Ontario M9W1R3.

Machine Tool Electrical Standards. Standards of the National Machine Tool Builders Association, 7901 Westpark Drive, McLean, VA 22101.

National Electrical Code®. Standard of the National Fire Protection Association, Quincy, MA 02269.*

National Electrical Manufacturers' Association (NEMA). Standards on motors and control published by the National Electrical Manufacturers' Association, 2101 L Street N.W., Washington, D.C. 20037.

Underwriters Laboratories Inc. (UL). An independent organization that tests and makes recommendations based on safety and fire hazard conditions relating to the tested equipment. Standards are published by Underwriters Laboratories Inc., 333 Pfingsten Rd., Northbrook, IL 60062.

BIBLIOGRAPHY

Alerich, Walter N.: *Electric Motor Control,* Van Nostrand Reinhold, New York, 1975.
Chestnut, Harold: *Systems Engineering Tools,* Wiley, New York, 1965.
Chestnut, Harold, and Robert W. Mayer: *Servomechanism and Regulating System Design,* 2d ed., vol. I, and vol. II, Wiley, New York, 1959.
Cockrell, Wm. D.: *Industrial Electronics Handbook,* McGraw-Hill, New York, 1958.
DC Motors-Speed Controls-Servo Systems, Electro-Craft Corp., Hopkins, Minnesota, 1972.
Heumann, G. W.: *Magnetic Control of Industrial Motors,* Wiley, New York, 1961.
Keucken, John A.: *Solid-State Motor Controls,* Tab Books, Blue Ridge Summit, Pennslyvania, 1978.
Kintner, Paul M.: *Electronic Control Systems in Industry,* McGraw-Hill, New York, 1968.
McPartland, J. F.: *Motor and Control Circuits,* McGraw-Hill, New York, 1975.
Millermaster, Ralph A.: *Harwood's Control of Electric Motors,* 4th ed., Wiley, New York, 1970.
American National Standards Institute: ANS/C2-1977, ***National Electrical Code,*** New York, 1977.
Siskind, Charles S.: *Electrical Control Systems in Industry,* McGraw-Hill, New York, 1963.

*National Electrical Code® is a Registered Trademark of the National Fire Protection Association, Inc., Quincy, MA 02269.

chapter 3-5

Lighting

by

Willard C. Allphin
Lighting Products Group
GTE Sylvania, Inc.
Danvers, Massachusetts

GLOSSARY

Adaptation A lighting condition to which the eye is accustomed.

Angstrom A unit of wavelength equal to one-ten billionth of a meter.

Arc discharge An electric discharge characterized by high cathode current density and a low voltage drop at the cathodes.

Ballast A device used with an electric discharge lamp to obtain the necessary circuit conditions (voltage, current, and waveform) for starting and operation.

Blackbody The theoretical body which radiates the maximum at every wavelength.

Brightness Perceived luminance.

Candlepower In practical terms, the luminous intensity in all directions from an international standard candle.

Cavity ratio A number depending upon the length, width, and height of a cavity.

Ceiling cavity The cavity formed by the ceiling, the upper walls, and the plane of the fixtures.

Coefficient of utilization (CU) Ratio of the lumens reaching the work plane to those leaving the lamps.

Color adaptation The condition of the eyes when they are "used to" a particular color of illumination.

Color rendering index The percentage by which color emitted by a lamp approaches that of a blackbody emitting light of the same correlated color temperature.

Color temperature (of a light source) The temperature in kelvins at which a blackbody must operate to emit the same color of light as the light source.

Correlated color temperature (of a light source) The temperature of a blackbody, the color of which most nearly matches that of the source.

Cosine law The law that illumination on any surface varies as the cosine of the angle of incidence.

Cutoff angle (of a fixture) The angle between a horizontal line and the line of sight at which a bare lamp first becomes visible.

Diffuse reflection Light reflected at various angles from a matte (nonglossy) surface.

Diffuse transmission Light emitted in various directions after passing through a translucent material.

Disability glare Light causing reduced visibility or visual performance.

Discomfort glare Glare which produces discomfort. It is not necessarily associated with disability glare.

Efficacy Lumens per watt of a light source (lm/W).

Electroluminescence Light emitted by a phosphor excited by an electromagnetic field.

Electromagnetic spectrum Electric and magnetic radiation including all wavelengths.

Equivalent sphere illumination (ESI) The level of sphere illumination equivalent to the measured illumination on a task.

Fixture The common term for a *luminaire* which is fixed in place.

Fluorescent lamp A low-pressure mercury discharge lamp with fluorescent phosphors which shift the invisible shortwave radiation to longer, visible wavelengths.

Footcandle (fc) The illumination on one square foot of surface with one lumen evenly distributed over it; or the illumination on one square foot of curved surface all points of which are at a distance of one foot from a standard candle.

Footlambert (fL) The luminance of a uniformly transmitting or reflecting surface emitting one lumen per square foot (1 lm/ft^2).

Glare Luminance in the visual field which produces discomfort or reduces the ability to see.

High-intensity discharge lamp (HID). One of a group, including mercury, metal halide, and high-pressure sodium lamps.

High-pressure sodium lamp Has a sodium vapor discharge in a ceramic tube surrounded by an outer bulb.

Illuminance The amount of light per unit area falling on a surface.

Indirect lighting Illumination from fixtures sending almost all of their light upward.

Lamp A synthetic source of light. Unfortunately, the term is also used for a portable lamp bulb, housing, and shade plugged into an outlet.

Light-loss factor (LLF) In aggregate, the illumination just before cleaning and relamping, divided by the initial (100 h for fluorescent) illumination.

Lumen The amount of luminous flux on one square foot of surface, all points of which are one foot from a standard candle.

Lumen method Method used to determine the number of lamps or fixtures needed to provide uniform illumination on a horizontal work surface.

Luminance Number of lumens per square foot leaving a surface toward the eye or a measuring instrument.

Lux (lx) The illumination on one square meter of surface with one lumen evenly distributed over it, or the illumination on one square meter of curved surface all parts of which are one meter from a standard candle.

Maintenance factor Average illumination just before cleaning and relamping, divided by the initial (100 h for fluorescent) illumination. (Now called light-loss factor.)

Mercury lamp A lamp whose light comes from a mercury arc in a tube surrounded by an outer bulb.

Metameric pair Two colors which look alike under one color of light but different under another color.

Nanometer Unit of wavelength equal to one-billionth of a meter (1×10^{-9} m).

Nit The luminance of a uniformly transmitting or reflecting surface emitting one lumen per square meter.

Photopic vision Vision when luminance of visual field is above about one footlambert.

Point-by-point method Determining illumination at a point by adding the contributions of all light sources in the vicinity.

Polarized light Light whose vibrations are confined to one plane.

Radiant energy Energy traveling in the form of electromagnetic waves.

Scotopic vision Vision where luminance of the visual field is below about 0.01 fL.

Shielding angle (of a fixture) Angle between a horizontal line and the line of sight at which a bare lamp first become visible.

Starter A device used with fluorescent lamps other than rapid-start or instant-start, which first closes a circuit through the cathodes and then opens it, causing the arc to strike.

Troffer A lighting unit, usually longer than it is wide, which is installed in the ceiling.

Tungsten-halogen lamp A gas-filled tungsten lamp containing certain halogens.

Ultraviolet radiation (uv) Radiation of shorter wavelengths than the visual range.

Veiling reflections Light reflected from a visual task which partially or entirely obscures the task.

Zonal cavity method Method for determining coefficients of utilization by taking into account the shapes and reflectances of three cavities for suspended fixtures and two cavities for ceiling-mounted fixtures.

LIGHT AND THE EYE

We are surrounded by radiation, most of it invisible: radio waves, microwaves, and a host of others, as indicated in Fig. 5-1. Only a small band of this enormous array of radiation is visible to the normal human eye, a band including from about 400 to 700 nm. (One nanometer equals ten angstroms.) As shown in Fig. 5-2, the band of visible radiation ranges through the colors of the spectrum from violet to red. Beyond the red is infrared (radiant heat) and beyond violet in the other direction is ultraviolet. We often say "ultraviolet light" but, strictly speaking, it is not light since it is invisible.

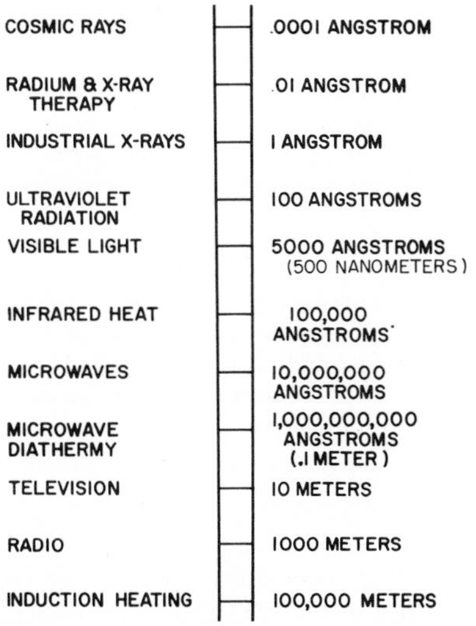

Figure 5-1 The electromagnetic spectrum, with a few types of radiation indicated. Each value actually represents a range.

As shown by the curve in Fig. 5-2, the eye is not equally sensitive to all parts of the visible spectrum, its peak of sensitivity being in the yellow-green. This is the *photopic* curve for luminance above about 1 fL. For night vision of luminances below 0.01 fL, the *scotopic* curve is displaced to the left by about 50 nm. Other curves fall between these two for luminances between 0.01 and 1 fL.

Near-ultraviolet from the sun is responsible for sunburn and tanning. Produced from electric sources, it is used to excite fluorescent paints and dyes, making them glow. Far uv has a sterilizing effect, but can be very harmful to eyesight. As will be seen later, the inner tube of a mercury lamp produces harmful uv which is filtered out by the outer bulb.

"White" light usually includes all the colors in the visible spectrum, but the propor-

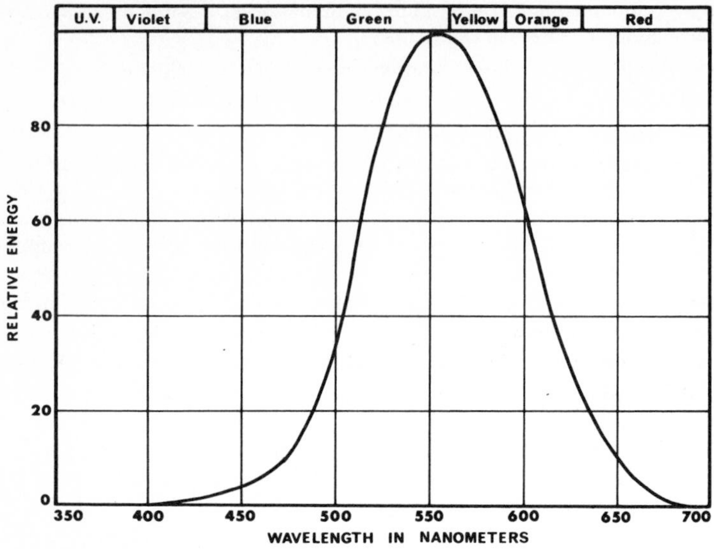

Figure 5-2 Photopic sensitivity curve of an average human eye.

tions can vary. The light from Cool White fluorescent lamps mixes well with daylight coming through windows, and the light from Warm White fluorescent lamps mixes well with the light from incandescent lamps.

Illumination Levels

The eye adapts itself to an enormous range of illuminations from moonlight to sunlight, a range of almost a million to one, but neither condition would be desirable for close visual tasks. Working levels from electric lighting commonly range from 10 to 200 fc (108 to 2150 lx) but higher levels are used in special cases, particularly over small areas.

Speed and accuracy of seeing depend upon the size of the detail and its contrast with its background, as well as the illumination.

It should be noted that older eyes require more light than young eyes for the same visual performance.

Changes in adaptation take time, more in going from high to low than from low to high levels, as when one goes from daylight into a darkened movie theater. At first one can see scarcely anything, but as the eyes adapt, people and objects gradually become visible. Under industrial conditions, visual performance is reduced when a worker has to look back and forth between areas of widely different luminance.

Luminance

An important distinction must be made between brightness and luminance, since the two are often confused. Luminance has to do with an object and is a measure of the number of lumens per unit area going in a particular direction. *Brightness* concerns the effect on the eye and the brain produced by the luminance. For example, consider the headlights of an approaching car at night. If they are on high beam the brightness is so great as to be almost blinding, yet the same lighted headlamps seen in daylight are scarcely noticeable. The luminance toward the eye is the same in both cases, but the adaptation of the eye makes the headlights seem much less bright in daylight. Just as footcandles are lumens per square foot falling on a surface, footlamberts are lumens per square foot leaving a surface toward the eye or a measuring instrument.

Color Adaptation

The eye also adapts to colors. Consider, for example, adjacent rooms, one lighted by Cool White and the other by Warm White fluorescent lamps. Strictly speaking, the rendition of colors is not perfect in either case, but color adaptation of the eye makes both seem "normal." However, if someone who had been in the Cool White room for some time steps into the Warm White room, things will seem rather yellow at first. Soon this feeling disappears and things seem normal. In the same way, going into the Cool White room after being adapted to the Warm White room will make things seem bluish-green at first. Accurate matching of colors is not possible in either case, as will be seen in the "Inspection Lighting" section.

Glare

Glare is an important consideration in all lighting. There are two types: *discomfort glare* and *disability glare*. Discomfort glare is subjective, depending upon how the observer feels about it, and it varies greatly from one person to another in the same situation. *Disability glare* is objective and interferes, to a greater or lesser degree, with seeing the visual task. It is often assumed that discomfort glare and disability glare are merely different degrees of the same thing, but this is not the case. Either can be present without the other for a particular individual in a particular situation.

Veiling Reflections

Veiling reflections are a special case of disability glare. Often when looking at a glossy photograph or magazine page the reflection of a window or a lighting fixture prevents seeing details unless we tilt the material. What is seldom realized is that subtle reflections appear on most visual tasks in business and industry, and reduce ability to see detail easily and rapidly. Veiling reflections can usually be reduced by changing the relationship between the light sources and the work, but whatever veiling reflections remain must be compensated for by increasing the illumination.

Behavior of Light

When a ray of light strikes a smooth reflecting surface, most of it is reflected with the angle of reflection equal to the angle of incidence. This is *specular* reflection, as in Fig. 5-3. A ray striking an entirely matte surface is reflected equally in all directions. This is *diffuse* reflection, as in Fig. 5-4. Most surfaces give a combination of specular and diffuse reflection. See Fig. 5-5. No reflecting surface is perfect, so some of the light is absorbed by the material and goes off in heat.

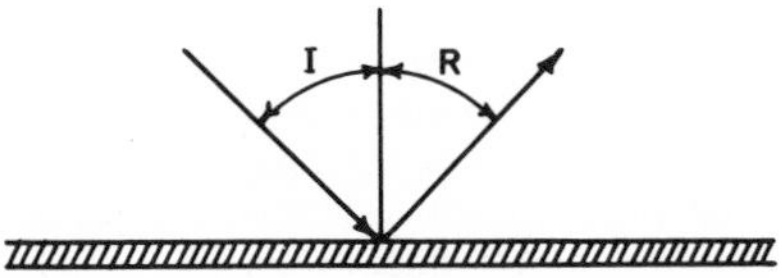

Figure 5-3 Specular reflection from a surface.

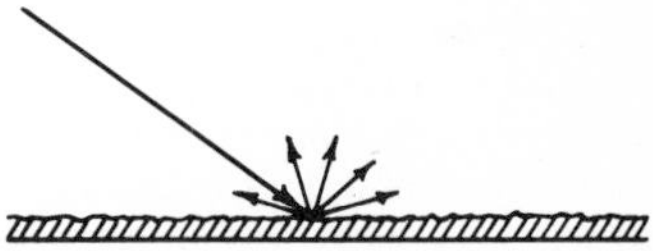

Figure 5-4 Diffuse reflection from a surface.

Light passing through a translucsent material leaves it as diffused light, as in Fig. 5-6.

A ray of light striking a transparent material is bent when entering the material and bent again when leaving it. See Fig. 5-7.

Light passing through a transparent object whose sides are not parallel is bent as in the prism of Fig. 5-8. This effect is used in some lighting fixtures, sometimes to reflect

Figure 5-5 Combination of specular and diffuse reflection.

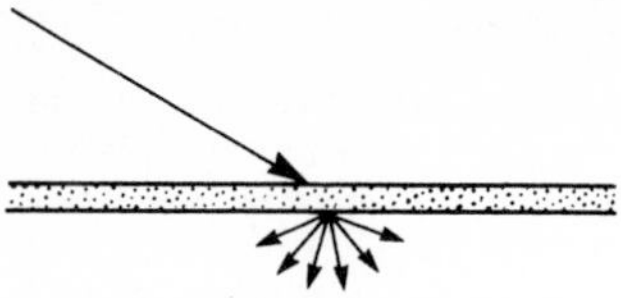

Figure 5-6 Transmission through a translucent substance.

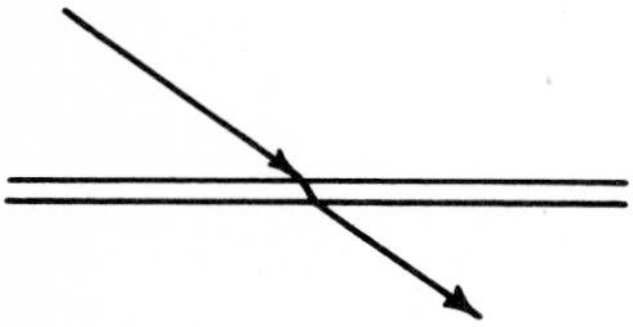

Figure 5-7 Transmission through a transparent substance.

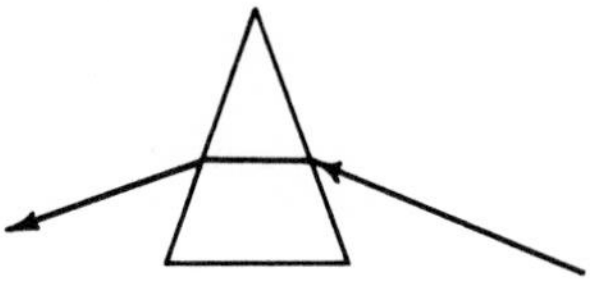

Figure 5-8 Path of light ray through a prism.

light and sometimes to refract it (change its direction). Transparent materials absorb some of the light just as reflecting surfaces do.

Polarized Light

Polarized light has been used to some extent in illumination and may become more widely used in the future. Polarizing materials can be used in lighting fixtures to reduce reflected glare spots and veiling reflections, but for each fixture there is only one angle from the vertical (Brewster's angle) at which the greatest reduction occurs.

INCANDESCENT LAMPS

Incandescent lamps are too familiar to require any discussion of their operating principles, but certain characteristics are important to note. For example, Fig. 5-9 shows the effects of changes in the applied voltage on lumen output, wattage, and life. This chart offers an answer to claims for lamps with tremendously long life. For example, a lamp designed to last 3 times as long as normal would give only about 76 percent of the light. This may be desirable when small numbers of lamps are in difficult-to-reach locations, but is not economical for general lighting applications. The need for longer life in these cases is met by extended service lamps. See lamp manufacturers' catalogs.

There are also changes through life due to gradual wasting away of filament material and to interior bulb blackening. Figure 5-10 indicates these changes.

The *rated average life* (not average rated life) of an incandescent lamp is the point at which 50 percent of a large batch will have failed under normal operating conditions. As seen in Fig. 5-11, some will fail early and some will last longer than average life. Operating lamps much beyond the rated point is not economical, however, because of the reduced output.

Reflector Lamps

R and PAR lamps have internal reflectors. PAR lamps are made of thicker glass and can be used outdoors without protecting them. Both are made in spot and flood types, the floodlamps having a partly diffusing face to spread the light over wider areas.

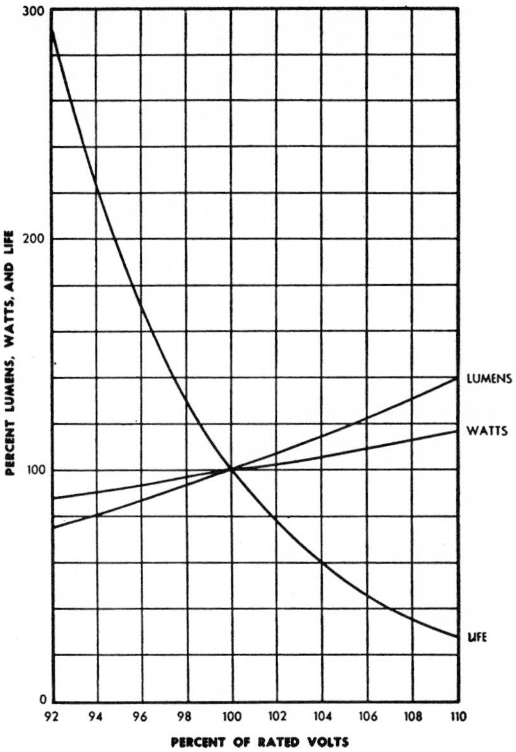

Figure 5-9 Incandescent lamp characteristics as affected by voltage.

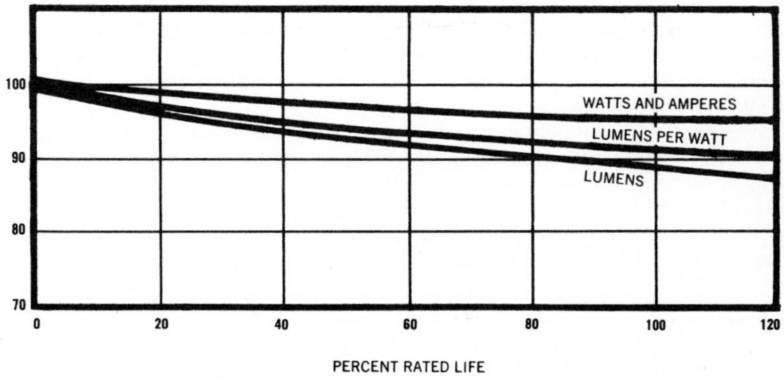

Figure 5-10 Lamp characteristics as they change throughout lamp life.

Dichroic Reflector Lamps

Dichroic reflector lamps have coatings thinner than the wavelength of light involved. They can reflect the light without most of the heat, which passes back through the reflector. The same principle is used on the faces of some PAR lamps to pass a selected color without the absorption of color filters.

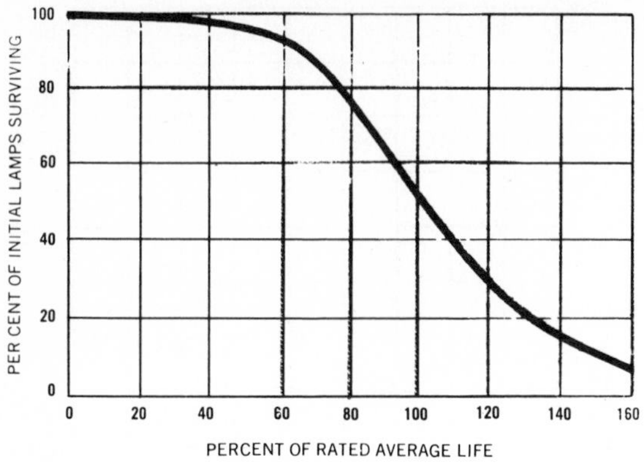

Figure 5-11 Typical life-expectancy curve for incandescent lamps.

Tungsten-Halogen Lamps

Tungsten-halogen lamps provide longer life and much better lumen maintenance than do conventional lamps. Their light output at end of life is down only about 3 percent from initial. This is accomplished by the iodine, bromine, or other vapors inside the bulb (actually a tube) which causes the tungsten boiling off the filament to be redeposited there.

ARC DISCHARGES

With the exception of exotic types such as luminescent panels and LEDs (light-emitting diodes), all electric light sources depend upon either an incandescent filament or an arc.

When a mercury arc, assisted by a small quantity of argon, is formed between two electrodes, it emits most of its energy in narrow bands or "lines" as shown in Fig. 5-12. The lines are drawn to the same height for general illustration, but in actual lamps the heights will vary depending upon gas pressure. Note that most of the lines lie in the invisible uv range. Only two of them are in the visible range, and these account for the bluish-green color of clear mercury lamps.

Whereas incandescent lamps are self-limiting devices, a mercury arc would "run away with itself" if placed directly on the line. This is because the more current passing through an arc, the less its resistance. Thus a limiting device, usually a choke coil, is required. With all but the smallest sizes, more than line voltage is needed to start and operate the lamp, so a *ballast* both steps up the voltage and limits the current. See the section "High-Intensity Discharge Lamps" for more on mercury lamps.

Fluorescent Lamps

In a fluorescent lamp, a mercury arc, assisted by a small amount of argon, is formed between two coated-filament electrodes in a long tube. The arc emits most of its energy in invisible uv radiation which does not pass through the glass, but the coating of phosphors inside the tube converts this to visible light by shifting the short wavelengths to longer wavelengths.

The arc must be started either by applying a high voltage to ionize the gas, or by putting a current through each electrode, or *cathode*, to boil off electrons. Small lamps such as those used in desk lamps are sometimes started manually. With lamp and ballast in series across the line, a button is pressed momentarily to put a current through the cathodes. When the button is released, the arc strikes between the cathodes. Most of the

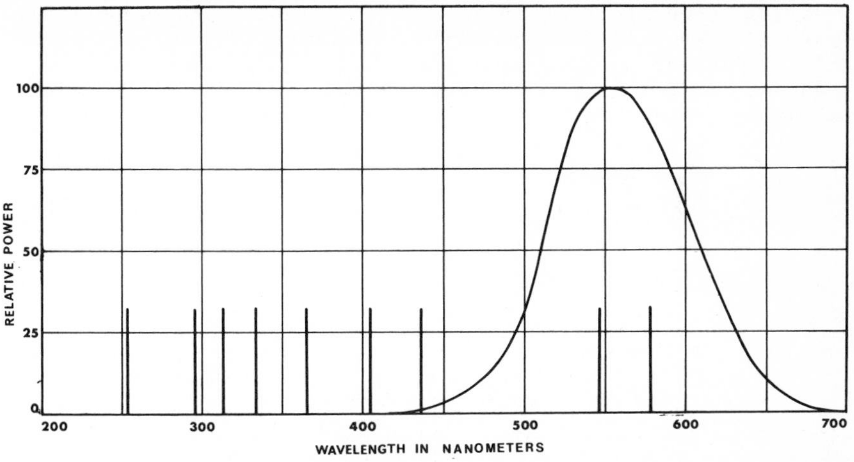

Figure 5-12 Energy-emission lines and the visibility curve for an average human eye. Mercury lines are shown with same length. Actual lamps produce different proportions of energy in different lines, the amount depending on the internal gas pressure.

early 40-W lamps had starters which performed this operation automatically, and some of these systems are still in use.

Instant-Start Lamps

Instant-start lamps use a much higher voltage to start the lamps without preheating the electrodes; the ballasts are larger and more expensive, and lamp life is somewhat less. Long instant-start lamps are also called *slimlines.*

Rapid-Start Lamps

Rapid-start circuits such as in Fig. 5-13 are used for nearly all 40-W fluorescent lamps and for high-output and very-high-output lamps. Current through the cathodes drops when the arc strikes.

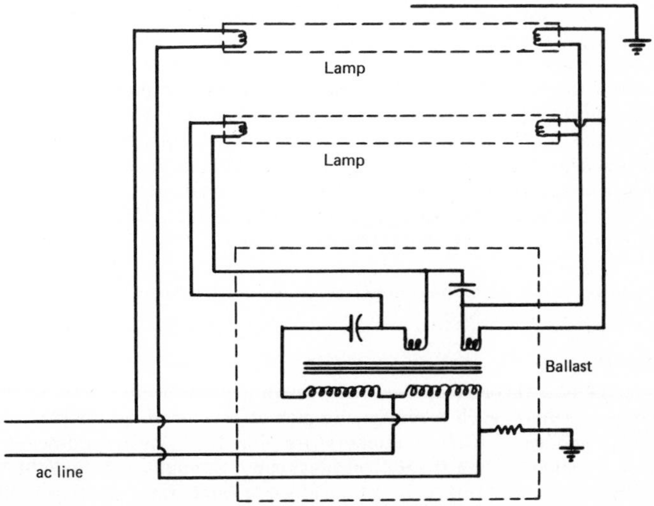

Figure 5-13 Series rapid-start circuit.

Ballasts are available for distribution systems from 120 to 277 V, and high-frequency operation has been tried. This permits much smaller ballasts. Also, most ballasts for 40-W lamps and larger are power-factor-corrected.

Efficient operation requires that a ballast provides good wave form, approximating a sine wave. Figure 5-14 shows a poor waveform from an inferior ballast.

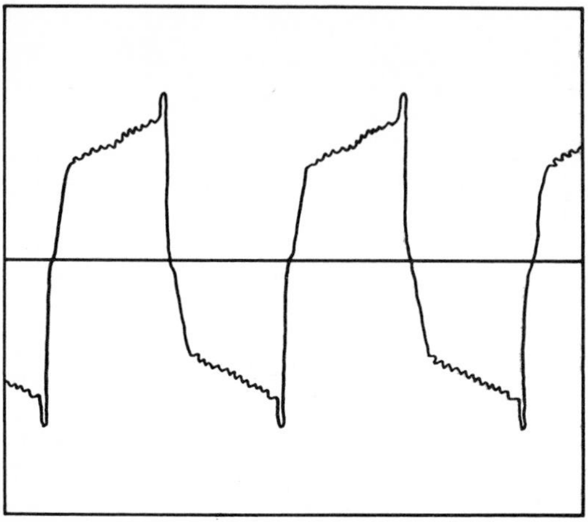

Figure 5-14 Poor waveform. Voltage across a fluorescent lamp operated on an inferior ballast.

Fluorescent Lamp Colors

The use of differing phosphors permits a considerable range of colors. The daylight lamp was first, but it has been largely supplanted by the Cool White, the spectral power curve for which is shown in Fig. 5-15. The mercury lines get through the phosphors to some extent and contribute to the color. They have been drawn wider in Fig. 5-15 to keep them on the chart.

In Fig. 5-16 for the Warm White lamp, it is evident that there is less blue and green, more yellow, and slightly more red than from the Cool White. The total lumen output is the same. Although usually not noticed by nonartists, both of these lamps distort colors of objects somewhat.

It is possible to achieve better color balance by adding red phosphors, at the sacrifice of about one-third of the lumen output. Figures 5-17 and 5-18 are the spectral power curves for Deluxe Cool White and Deluxe Warm White lamps. Note the greater proportion of red radiation. There are other types which sacrifice efficacy in an attempt to get closer matches to incandescent lamps or to some sort of daylight. *Efficiency* is sometimes used in place of efficacy, but this is technically incorrect, since efficiency is output divided by input, in the same units whereas efficacy is measured in lumens per watt (lm/W).

There is also a lamp which has three phosphors, each producing a different color, the curves of which do not overlap.

Color Temperature

Physicists speak of a theoretical *blackbody* which absorbs all light falling on it. When heated to incandescence, such a body would give off a color of light depending upon its temperature in kelvins, K (Celsius temperature plus 273°). An incandescent lamp filament is close in behavior to a theoretical blackbody. Sunlight and daylight, since they originate from an incandescent body, are given color temperature ratings in kelvins.

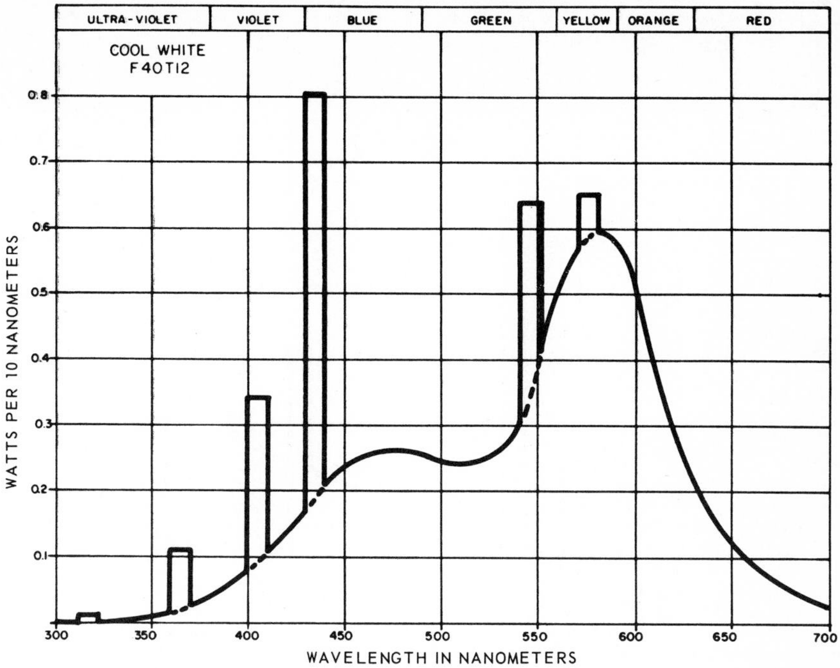

Figure 5-15 Spectral power curve for Cool White fluorescent lamp.

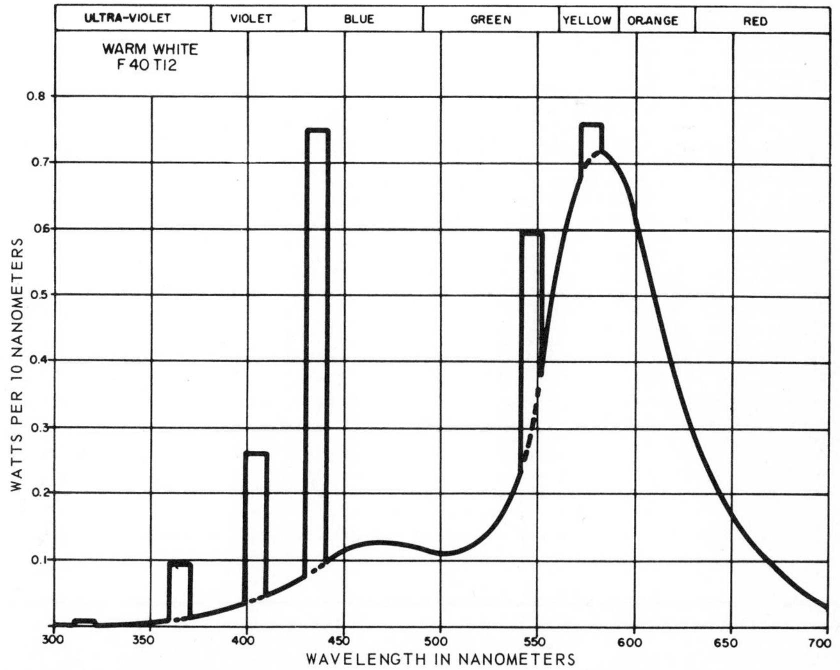

Figure 5-16 Spectral power curve for Warm White fluorescent lamp.

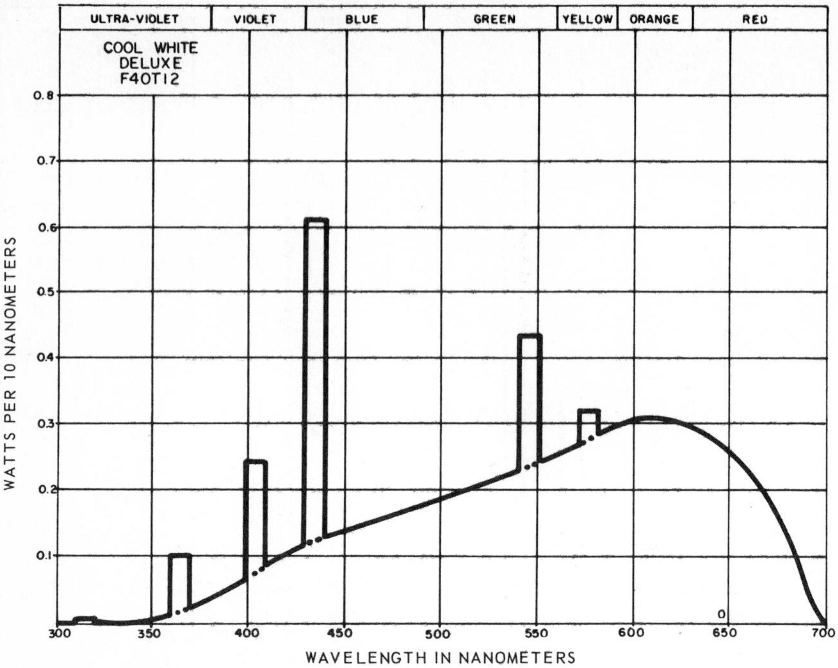

Figure 5-17 Spectral power curve for Deluxe Cool White fluorescent lamp.

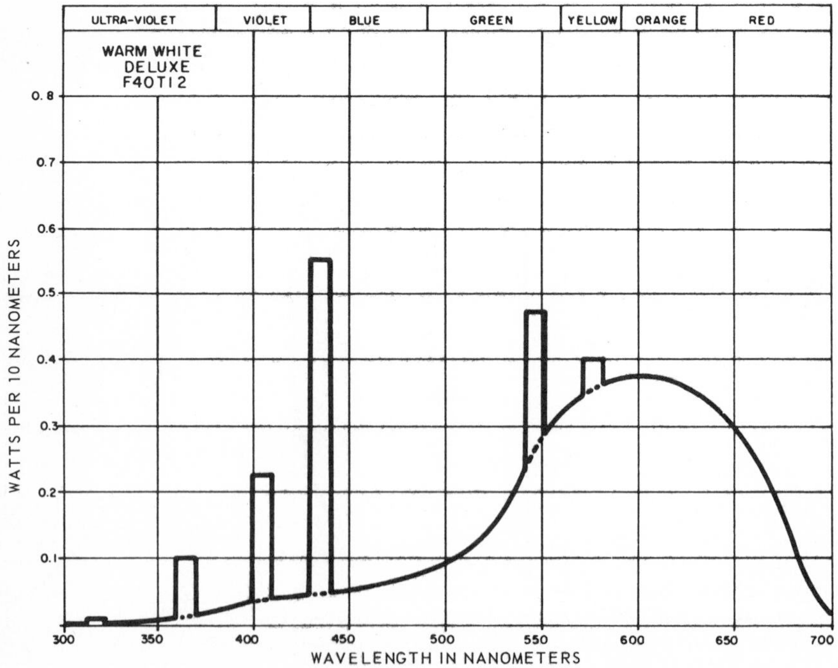

Figure 5-18 Spectral power curve for Deluxe Warm White fluorescent lamp.

Correlated Color Temperature

Fluorescent lamps and high-intensity discharge (HID) lamps, not having smooth spectrums, do not resemble a blackbody enough to have a true color temperature rating. However, for any such lamp there is a color temperature which it resembles. For example, a Cool White lamp having a correlated color temperature of 4500 K is closer to an incandescent lamp of that color temperature than it is to a lamp of any other color temperature.

Color Rendering Index

Color rendering index (CRI) is a measure, in percent, of how closely the correlated color temperature of a lamp approaches a true color temperature. In other words, the correlated color temperature of a lamp is the color the lamp is trying to produce. The CRI tells how well it is succeeding.

Voltage Variations

Voltage variations have less effect on fluorescent lamps than they do on incandescent lamps, but the voltage applied to a fluorescent lamp cannot be ignored. Standard ballasts are designed to operate at 118 V, but lamps on these ballasts will operate satisfactorily over a range from 110 to 125 V. When the voltage is too low, it may be difficult to start the lamp, particularly under humid conditions. When the voltage to a rapid-start lamp is too high, it may operate as an instant-start lamp, causing the cathode coating to be used up more rapidly.

Effect of Starts on Lamp Life

The number of starts does affect the life of a fluorescent lamp because a little of the cathode coating is lost at each start. Figure 5-19 shows mortality curves for 40-W rapid-start lamps for different operating periods. It used to be considered economical to leave lamps burning rather than shut them off for short periods. Now, with the emphasis on saving energy, and with the higher cost of electricity, the practice is to turn them off whenever not in use.

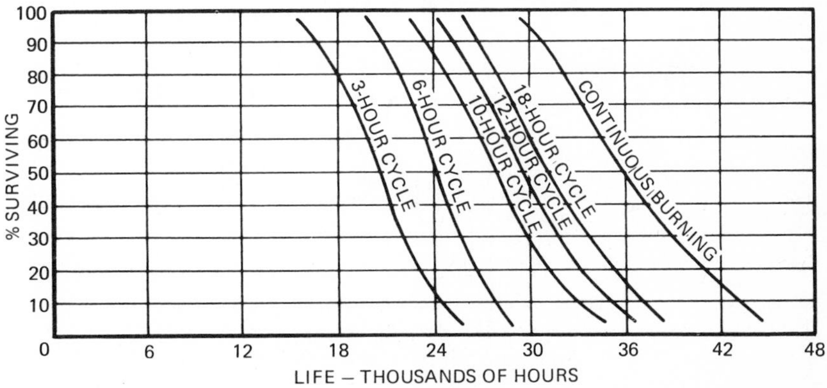

Figure 5-19 Typical mortality curves as a function of burning cycles for 40-W rapid start lamps with a rated life of 20,000+ hours.

Lumen Maintenance

Lumen maintenance for fluorescent lamps is better than for incandescent lamps. Some manufacturers, in addition to giving initial lumens, give approximate lumens at 40 percent of rated average life. These run about 88 percent of initial for 40-W lamps, 87 percent for 48-in rapid start, and 75 percent for very-high-output lamps.

Low-Temperature Operation

Low-temperature operation of standard lamps may not be satisfactory at ambient temperatures below 50°F (10°C). However, jacketed lamps are available and can be operated on low-temperature ballasts.

Special Types

A reflector lamp has a reflective coating over part of the circumference inside the tube, between the phosphor and the glass. This increases the output at certain angles, with some loss in overall efficacy. An aperture lamp has a reflector over a large part of the circumference, and the remainder is clear of phosphor, so the light comes through a slit. For display purposes there are colored lamps, most with filter coatings with the red having a red glass tube. Lamps designed to conserve energy are discussed later in this chapter, under "Energy Conservation," as well as in Sec. 15.

High-Intensity Discharge Lamps

High-Pressure Mercury

High-pressure mercury, the basic (HID) lamp, is still used in some industrial and street-lighting installations. Construction details are shown in Fig. 5-20. When the lamp is turned on, full starting voltage is applied across the main electrodes and between the lower electrode and the probe near it, causing a glow discharge between the electrode and the probe. This discharge enables an arc to strike in the argon gas between the main

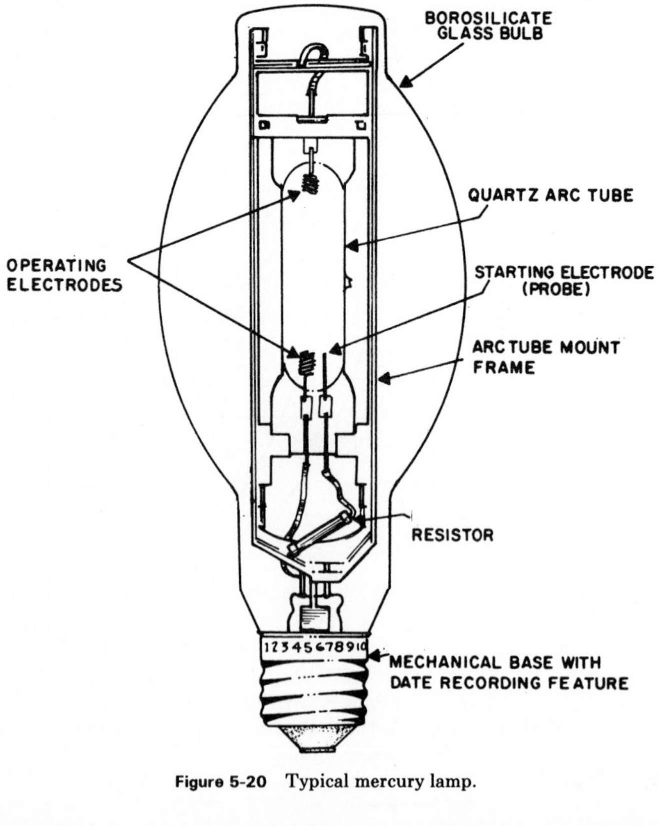

Figure 5-20 Typical mercury lamp.

electrodes. At first the arc gives off very little light, but its heat gradually vaporizes most of the mercury in the tube. This change builds up pressure and the arc becomes a true mercury-vapor arc. The quartz arc tube passes far uv, but the outer bulb absorbs this. Not shown in Fig. 5-20 is a mechanical-electric device for shutting off the arc if the outer bulb is broken. This is for the protection of eyesight. The characteristic blue-green color of the light from a clear mercury lamp is explained by the spectral power curve of Fig. 5-21; but there are various color-improved mercury lamps which have phosphor coatings inside the outer bulbs. One of these is shown in Fig. 5-22. Note the considerable addition of red radiation.

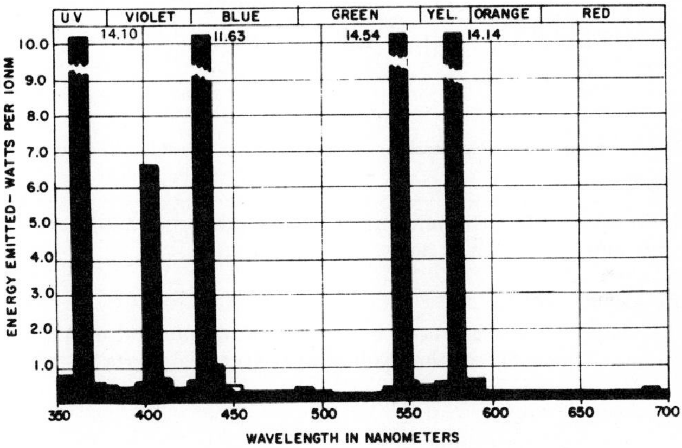

Figure 5-21 Spectral power curve of clear mercury lamp.

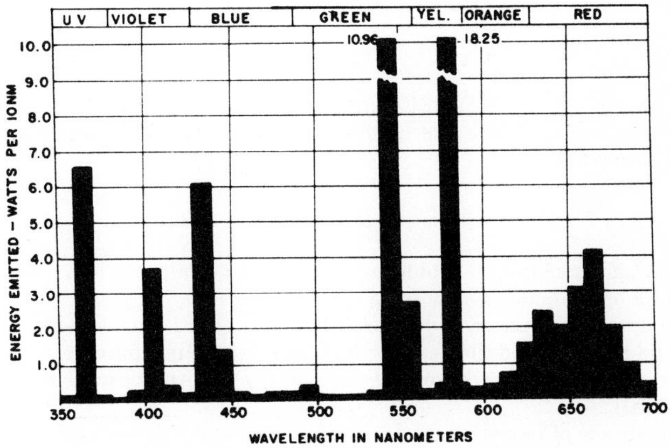

Figure 5-22 Spectral power curve of color-improved mercury lamp.

Metal Halide Lamps

Metal halide lamps are similar in general principles to mercury lamps but the arc tube is smaller and contains halides such as thorium iodide, sodium iodide, and scandium iodide in addition to the argon and mercury. The lamp produces a "whiter" light than does color-improved mercury, as shown by the spectral power curve for a 400-W clear lamp in Fig. 5-23. Further color improvement can be achieved with phosphor coatings

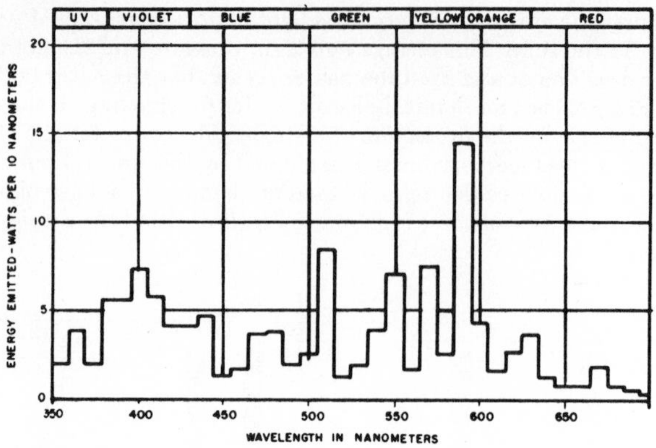

Figure 5-23 Spectral power curve of clear metal halide lamp.

inside the bulb. Operating position is important, and different lamps are made for base-down, base-up, and base-down or horizontal operation.

Low-Pressure Sodium Lamps

Low-pressure sodium lamps are used in some parking lots and at highway ramps for contrast with other lighting, but their yellow light greatly distorts the colors of objects.

High-Pressure Sodium Lamps

High-pressure sodium lamps differ from other HID lamps in that the sodium arc operates inside a translucent ceramic tube. The lamp gives a somewhat yellow light, but its high efficacy leads to some uses in industrial and outdoor areas.

ILLUMINATION

Illumination deals with both quantity and quality of light. Quantity is concerned with footcandles on the work plane and quality with such elements as color of light, glare, contrast, and shadows.

Quantity

For task lighting, enough light should be supplied to permit rapid and accurate seeing; and in nontask areas, enough for safety.

The Illuminating Engineering Research Institute (IERI) sponsors research on light and vision and has provided the basis for footcandle recommendations in the *IES Lighting Handbook*. Some of these have been adopted by the American National Standards Institute (ANSI).

The high efficacy of modern light sources and the relatively low cost of electricity led to the general practice of providing uniform lighting over an entire area, based on the most difficult visual tasks performed there. Now, with higher electricity costs and demands to conserve energy, emphasis has turned to the treatment of individual work spaces. However, many areas must still be provided with general lighting, and this will be discussed further later in this chapter under "Lighting Layout."

The fifth edition of the *IES Lighting Handbook* (1972) carried detailed tables of footcandle recommendations, some samples of which are: corridors, 20 (216 lx); machine work, 50 (538 lx); fine bench and machine work, 100 (1076 lx); office work, 70 to 150 (753 to 1614 lx); drafting, 200 (2160 lx); inspection, 50 to 1000 (538 to 10,760 lx). The 1981 edition of the *IES Lighting Handbook* employs a different system of criteria which should be referred to. The latest government, ANSI, and IES publications should also be consulted when a layout is to be made.

Equivalent Sphere Illumination

Equivalent sphere illumination (ESI) is a concept which has been used by some people and challenged by others. It attempts to compensate for loss of visibility from veiling reflections. Considering a horizontal task, for example, the smallest loss from veiling reflections would occur if a direct light came over the shoulder of the subject. On the other hand, the loss would be nearly as small if the subject were inside a sphere with the light coming to the task uniformly from all directions. This situation can be approximated by measuring instruments which can be standardized. Using this method might show, for example, that where a footcandle meter placed on the work reads 100 fc (1076 lx), the ESI might be only 70 fc (753 lx).

Quality

Quality includes freedom from glare and excessive contrasts and light of suitable color quality.

Discomfort Glare

Discomfort glare depends upon the average luminance of the lamp or fixture, or in cases of very uneven luminance, the brightest portion. It also depends upon the size and position in the field of view and the overall luminance to which the eye is adapted.

A subjective measure called *borderline between comfort and discomfort* (BCD) has been used in research to rate on a horizontal line of sight from a seated position near the rear wall of the room. Researches using BCD have led to the development of *visual comfort probability* (VCP). VCP means the percentage of a large group of people who would find a lighting installation comfortable.

For office work a value of 70 percent or higher is usually sought. Some fixture manufacturers publish VCP ratings for particular fixtures in various situations.

Color quality should be sufficient to meet the requirements of the work, and in offices lighting "pleasantness" should be considered. This depends on color of finishes as well as on color of light.

Shadows

Shadows can be either bad or good, depending upon their degree. Sharp, black shadows of objects not only have an unpleasant appearance, they can cause accidents. At the other extreme, an installation without shadows has a "bland" appearance and objects lose their roundness or depth. There should be shadows which are illuminated, but to a lesser degree than the objects casting them.

LIGHTING CALCULATIONS

The two basic ways of designing for footcandles are the *lumen method*, sometimes called the flux method, and the *point-by-point method.*

Lumen Method

The lumen method is for uniformly distributed light on a horizontal work plane, usually 30 in (1 m) above the floor. In a large area without tall obstructions, it can be assumed that illumination on vertical surfaces will be roughly half that on horizontal surfaces.

The basic lumen formula is:

$$\text{footcandles} = \frac{\text{total lamp lumens} \times \text{CU} \times \text{LLF}}{\text{area of work plane in square feet}}$$

$$\text{lux} = \frac{\text{total lamp lumens} \times \text{CU} \times \text{LLF}}{\text{area of work plane in square meters}}$$

where CU is the coefficient of utilization and LLF is the light-loss factor (formerly called maintenance factor).

Coefficient of Utilization

The coefficient of utilization is obtained from a fixture catalog or data sheet for the fixture which has been chosen. The information which must be fed into the CU table includes ceiling reflectance, wall reflectances, and a factor depending on fixture mounting height and shape of room.

Zonal Cavity Method. In the zonal cavity method, this factor is broken down into three cavity ratios based on the volume between floor and work plane, that between work plane and fixtures, and that between fixtures and ceiling. Each cavity ratio depends upon the dimensions and reflectances of the cavity. Tables of cavity ratios can be found in the *IES Lighting Handbook* and in the *Primer of Lamps and Lighting.*

Determining Reflectances. A white ceiling in good condition can be assumed to have a reflectance of about 80 percent. For walls which are to be newly painted, some paint manufacturers furnish reflectances with their color chips. Existing walls can be measured with a footcandle meter and something of known reflectance. A convenient standard is the Kodak Neutral Test Card sold in photographic supply stores. One side has a reflectance of 18 percent and the other a reflectance of 90 percent. A clean piece of white blotting paper, if used as a standard, can be assumed to have a reflectance of about 85 percent.

Hold the standard against the wall and point the footcandle meter *toward* it from a distance of about 6 in, taking care not to cast a shadow on the card. Read the meter, then remove the card and take another reading where the card had been. The unknown reflectance will be:

$$\frac{\text{Reading from card}}{\text{Reading from wall}} \times \text{reflectance of card}$$

Light-Loss Factor

Light-loss factor theoretically includes all of the following:

Light-loss factors not to be recovered:
- Fixture ambient temperature
- Voltage to fixture
- Ballast factor
- Fixture surface depreciation
- Room surface depreciation

Light-loss factors to be recovered:
- Room surface's dirt depreciation
- Lamp burnouts
- Lamp lumen depreciation
- Fixture dirt depreciation

In practice, some of these are difficult to predict. Where a new area is to be lighted in an existing plant, the engineer can arrive at an overall LLF by taking footcandle readings in other areas just before fixtures are going to be cleaned and relamped and comparing them with the original values. It should be mentioned that initial ratings for fluorescent lamps are for lumens after 100 h operation, since there is a sizable drop in the output of a new lamp in the first few hours. In calculations for lighting layouts the lumen formula is rearranged to the form:

$$\text{Number of fixtures} = \frac{\text{designed fc} \times \text{floor area}}{\text{no. of lamps per fixture} \times \text{lumens per lamp} \times \text{CU} \times \text{LLF}}$$

where floor area is in square feet; if the lux is used in the calculation, floor area is in square meters.

Point-by-Point Method

When illumination is not to be uniform, the lumen method does not work. It is necessary to take a given point on the work plane and compute how much each fixture in the vicinity contributes to that point. For outdoor lighting, this gives a reliable value for the illumination at the point. For indoor lighting, the value will be a little low because the lumens reflected and re-reflected around the room are neglected. This is negligible in most cases because the fixtures are likely to be of the direct type.

The basic formula is:

$$\text{Illumination} = \frac{I \times \cos\theta}{d^2}$$

where illumination can be in either footcandles or lux and

I = intensity in candlepower from the light source in the direction toward the measurement point

θ = the angle between a line from the center of the fixture and the work plane at the point

d = distance, ft (m), from center of light source to the point

The intensity is obtained from the candlepower distribution curve of the fixture. Figure 5-24 is an example of such a curve for an incandescent fixture. The values are plotted on a vertical plane passed through the center of the fixture. Since the fixture is symmetrical about its own vertical axis, the curve shows the intensity in all directions. For a rectangular fixture, curves are commonly shown for two vertical planes, one through the long axis of the fixture and the other at right angles to it as in Fig. 5-25. For more accurate results, a 45° plane can usually be obtained from the fixture manufacturer, and values between the planes can be estimated. Such work can be very laborious, and the method was seldom used until the advent of computers. Now there are computer programs which can make such computations very rapidly, once candlepower and positional data are fed in.

LIGHTING LAYOUTS

Steps in making a lighting layout can be summarized as follows:

1. Analyze lighting needs.
2. Establish illumination level.
3. Decide lamp type and size.
4. Choose lamp color.
5. Select fixture.
6. Decide mounting height.
7. Estimate maintenance conditions.

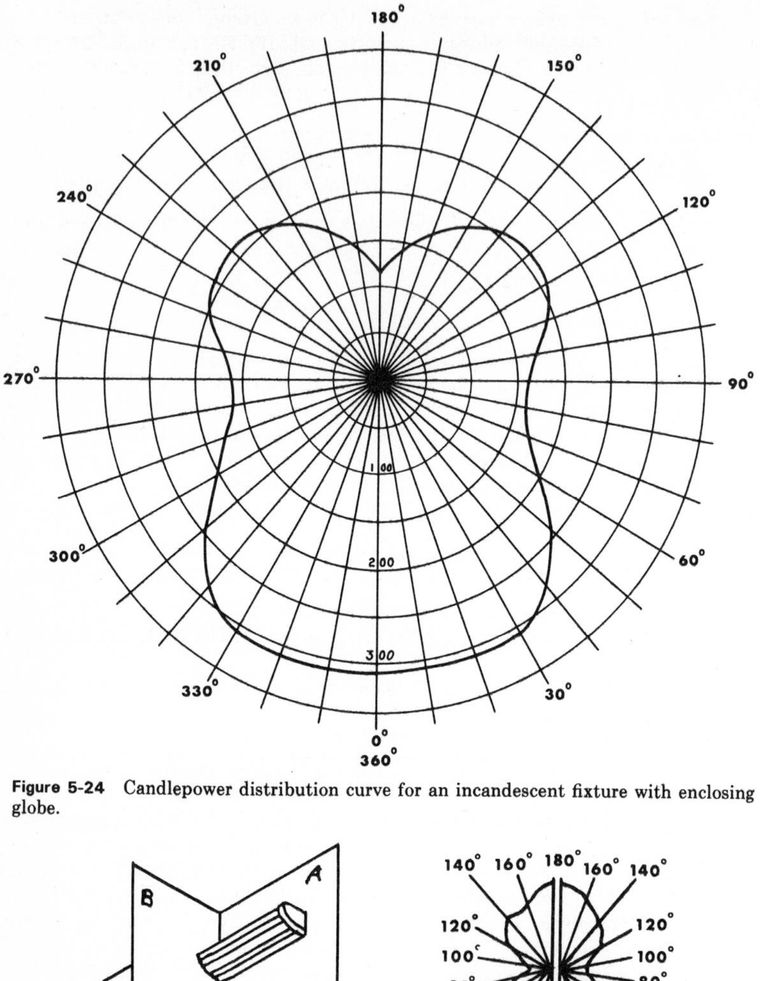

Figure 5-24 Candlepower distribution curve for an incandescent fixture with enclosing globe.

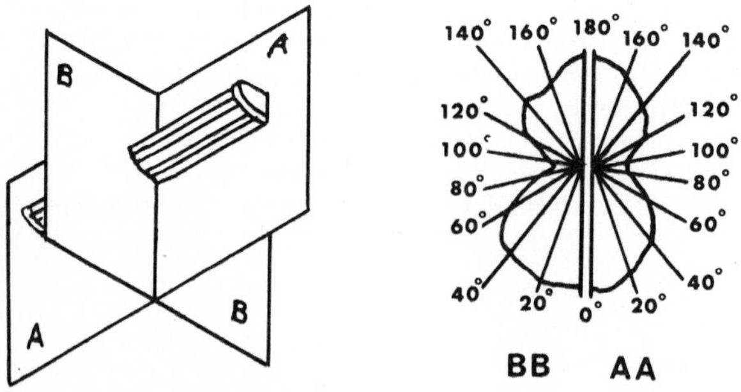

Figure 5-25 Candlepower distribution curves for a fluorescent fixture.

And, for uniform illumination by lumen method:

8. Measure or estimate reflectances.
9. Use method such as zonal cavity to find coefficient of utilization.
10. Determine or estimate light-loss factor.
11. Compute number of fixtures required.
12. Determine maximum spacing.
13. Make fixture layout.

Some of the steps have already been discussed. Steps 3 and 4 involve the suitability of the lamp for the work. For example, a 400-W HID lamp would not be mounted 8 ft above the floor because its high luminance would cause direct glare. Again, a clear mercury lamp would not be used in an office on account of the color of its light.

Initial cost of lamp and equipment are important. Table 5-1 gives partial data for a few typical lamps used in industry. Others will be found in large lamp ordering guides or catalogs. With fluorescent and HID lamps, the actual efficacies are about 5 percent lower than those in Table 5-1, on account of ballast losses.

TABLE 5-1 Some Typical Lamps*

Watts	Type	Color	Length, in	Initial lumens	Lumens per watt	Life, hours	Corr. color temp., K	CRI
40	Fl	CW	48	3,150	78.7	20,000	4300	67
40	Fl	CWX	48	2,200	55.0	20,000	4100	85
60	Fl HO	CW	48	4,300	71.7	12,000	4300	67
75	Fl	CW	96	6,300	84.0	12,000	4300	67
115	Fl VHO	CW	48	6,750	58.7	10,000	4300	67
100	Inc	IF	—	1,740	17.4	750	2900	90
110	Fl HO	CW	96	9,050	82.3	12,000	4300	67
215	Fl VHO	CW	96	15,000	69.8	10,000	4300	67
500	Inc.	IF	—	10,600	21.2	1000	3050	90
400	Hg	Clr	—	20,500	51.3	24,000	5900	22
400	Hg	Imp	—	23,000	57.5	24,000	3800	51
400	MH	Clr	—	34,000	85.0	15,000	4500	65
400	MH	Imp	—	34,000	85.0	15,000	3800	70
400	S	Clr	—	50,000	125	24,000	2100	21

*Inc, Incandescent; Fl, Flurescent; IF, inside frost; CW, Cool White; CWX, Cool White Deluxe; HO, High Output; Clr, clear; VHO, very high output; Hg, mercury; MW, metal halide; Imp, color improved; S, high-pressure sodium; CRI, Color Rendering Index.

For brevity and because they are the most widely used, only Cool White and Deluxe Cool White fluorescent lamps have been listed in Table 5-1. Actually, there are fluorescent lamps with correlated color temperatures ranging from 3000 to 7000 K. The lower numbers indicate "warm" light and the upper ones "cool" light. In wall colors, the reds are considered to be warm and the blues and greens cool. Warm light enhances warm colors and grays cool ones. Likewise, cool light enhances cool colors and grays warm ones. ing in mind that glare should not reach the worker's eyes from normal viewing angles and that light from shiny surfaces should not be reflected into the worker's eyes.

Fluorescent lamps have low luminance, so they can be mounted lower without excessive glare. They offer a choice of colors.

HID lamps give a large amount of light from a small package, so they can be mounted higher on wider spacings to reduce first costs.

In selecting an open-bottom fixture, cutoff angle is important. See Fig. 5-26. Its significance is illustrated in Fig. 5-27. Shielded fixtures have translucent plastic or prismatic enclosures which partially direct the light.

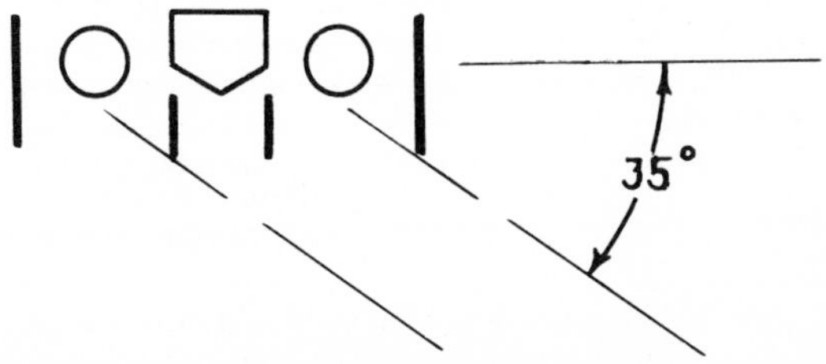

Figure 5-26 Meaning of cutoff angle.

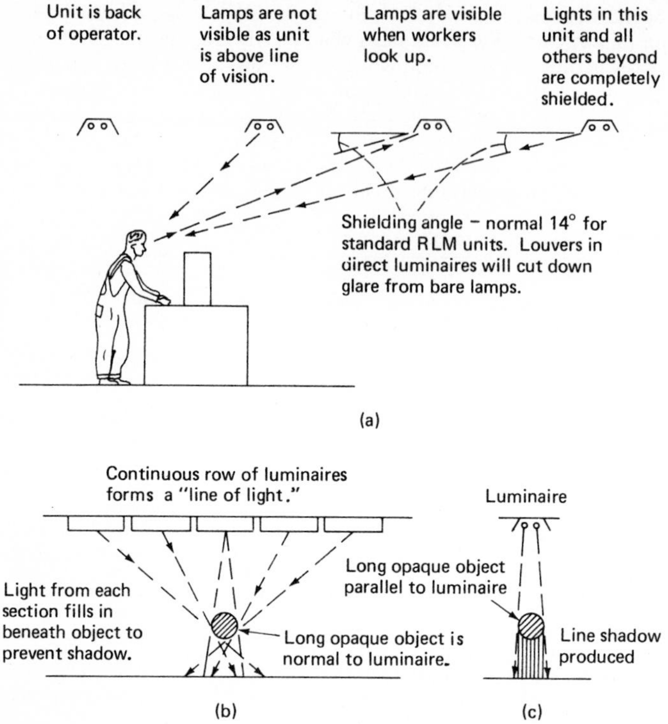

Unit is back of operator.
Lamps are not visible as unit is above line of vision.
Lamps are visible when workers look up.
Lights in this unit and all others beyond are completely shielded.
Shielding angle – normal 14° for standard RLM units. Louvers in direct luminaires will cut down glare from bare lamps.
(a)
Continuous row of luminaires forms a "line of light."
Light from each section fills in beneath object to prevent shadow.
Long opaque object is normal to luminaire.
Luminaire
Long opaque object parallel to luminaire
Line shadow produced
(b)
(c)

Figure 5-27 Shielding angles for direct fixtures. *(From: W. Staniar (ed.),* Plant Engineering Handbook *(2d ed.), McGraw-Hill, New York, 1959, p. 10-18. Copyright © McGraw-Hill Book Company, Inc. Reproduced by permission of McGraw-Hill Publishing Co.)*

Indirect fixtures, which send 60 percent or more of their light upward, are often used in offices where the ceilings are not dropped. With dropped ceilings troffers are usually set in flush, each replacing an acoustic panel. One advantage of this method is that fixtures can be relocated with relative ease.

Another advantage of enclosed, flush-mounted troffers is that they can be connected to ducts for removing much of the heat given off by lamps and ballasts. In summer the heat can be discharged outdoors and in winter it can be used to help heat the building.

Downward distribution from fixtures varies from fairly narrow to wide, depending on the shape of their reflectors and the presence of control elements such as prisms. In a very-high-ceilinged industrial area, a narrow distribution may be used to avoid sending too much light to the walls. On the other hand, if there are tall machines with important details on vertical surfaces, wide-angle fixtures will help to illuminate the vertical surfaces.

In some general offices where people face up and down the room but not crosswise, fixtures with "bat-wing" distribution have been used. They emit most of their light sideways at high angles, so that light reaching a worker's desk comes from the sides. This minimizes veiling reflections.

Recommended spacing-to-mounting-height ratios are given in fixture catalogs and these are intended as maximums, not minimums. Illumination will be quite uniform if the ratios are not exceeded. It is customary to make the spacing from wall to first row half of that between rows.

For nonuniform lighting, fixtures are located with reference to the work spaces, bearing in mind that glare should not reach the worker's eyes from normal viewing angles and that light from shiny surfaces should not be reflected into the worker's eyes.

Daylight

Although daylight from windows, skylights, and clerestories is useful in permitting lights to be turned off at times, no allowance should be made for it in lighting layouts because on a dark day in winter its effect is negligible.

APPLICATIONS

Industrial Lighting

Industrial lighting tasks vary from the rough work of a foundry to the delicate tasks of assembling microcomputers.

For rough work, a general lighting installation of moderate footcandle level should suffice. For general machine work, the lighting can be from continuous rows if the machines are aligned in rows. Rows should be spaced with the machines, closer to the working sides.

For work such as on lathes and milling machines, a local lighting unit is usually mounted on the machine with its light directed toward the cutting point. Textile machinery is lighted by long rows parallel to the machines on the side from which the operators work. On automobile assembly lines, in addition to overall general lighting, continuous rows of fluorescent fixtures about 10 ft above the floor are tilted so that they provide vertical lighting as well as some additional horizontal lighting. Assembly lines where workers are seated at benches are usually lighted from continuous rows of fluorescent fixtures over the benches.

The most critical visual work such as the assembly of delicate instruments is usually done in "white rooms," the "white" referring to both the interior finishes and the costumes of the workers. In these rooms, lighting is customarily from enclosed troffers mounted flush in the ceiling for cleanliness.

Inspection Lighting

Inspection lighting may involve color, direction, and amount of diffusion of the light. In many cases, color is not critical and one of the deluxe types of fluorescent lamps will serve. When color is critical, particularly when colors must be matched to standards, special color-matching fixtures may be required. These are usually designed to provide "north sky daylight," but one or two other colors may be available—incandescent for example.

Metamerism is a phenomenon in which two colors match under one light source but fail to match under another. Such colors are called *metameric pairs*. The existence of the phenomenon explains the need for special care in inspecting colors.

Scratches are best revealed by sharp light as from a clear incandescent lamp, directed at a low angle perpendicular to the direction of the scratches. This also applies to small dents and "dings."

Transparent objects are often inspected on a table top of translucent glass or plastic lighted from below. Another example is the "perch" in a textile mill, where the cloth is pulled across a frame and seen against a window or a lighted panel.

Very small objects such as filaments or electronic grids can be inspected by silhouetting them on a translighted table.

Waves in shiny surfaces can be detected by viewing the reflected image of a fluorescent fixture having a diffusing panel on which narrow cross lines have been painted. Any waves will distort the reflected grid pattern.

Fine cracks in castings can be revealed by coating them with a fluorescent liquid and viewing them in a darkened room with uv produced by "black light" lamps. Any cracks will glow brightly.

One of the most subtle flaws to detect is the slight loss of sheen in a piece of plastic counter-top material. An effective method is to suspend over the work a fluorescent fixture with a diffusing bottom, bordered on each side with a black panel as large as the fixture. Looking down at the material, the inspector sees a reflection of alternating dark and light areas. The important thing is the border between a light and dark area. When the product moves along, or when the inspector moves, and a reflected border enters a flawed area, it "leaps to view."

Warehouses

Fixtures are mounted as high as possible, and over aisles when stacks are permanently located. HID lamps are the usual choice but in dead storage areas where lights are seldom turned on, incandescents may be used to save first costs. However, the effect of higher wattage on a demand rate should be considered.

Offices

Much of what has been said about lamps, fixtures, and general lighting applies to office lighting, but there is a development which reduces or eliminates general lighting in favor of fluorescent lighting units fastened to desks. Some of these are located lengthwise above the back edge of the desk, but they give some trouble with veiling reflections. Others are mounted over side edges of the desk, one on each side, to reduce reflections toward the eye.

In open offices, where individual work areas are set off by low partitions, fluorescent units are often fastened to the partitions. Some of these send light both downward to the desk and upward for a little general lighting.

Drafting rooms are usually lighted from continuous rows of fluorescent fixtures, to higher levels than needed for general office work. Sometimes the rows are run at a 45° angle to the walls in order to reduce shadows from T squares and triangles.

Private offices are usually lighted from surface or flush-mounted fixtures. One good arrangement is a U-shaped array of fixtures with the bottom of the U over the occupant's head and the open side of the U pointed away from her or him. The office of a top-level executive can be quite elaborate, with remote-control buttons on the desk to provide more than one level of general lighting and to turn on spotlights aimed at pictures and art objects.

Daylight should be used whenever feasible to conserve energy, but direct sunlight should be kept from work areas by some form of shielding such as venetian blinds. Adjustable vertical louvers are more expensive but are used in some private offices.

Switching

Too many units should not be controlled from one switch. Where daylight is useful, units should be switched in rows parallel to the windows so that some rows can be turned off on bright days. Another reason for switching in smaller groups is so that if someone works late in a general office he or she need not light a large area of the room. Also, some units can be switched so as to provide a path of light for the watchman.

Painting

When decorating a new office or redecorating an old one, light colors should be used for large areas because of their higher reflectance and to give a psychological "lift" to the occupants. Three harmonizing colors, a slightly darker one for the lower walls, and a dark trim color for accents make a good arrangement.

OUTDOOR LIGHTING

The principal applications of outdoor lighting for industrial plants are building floodlighting and area lighting. Floodlighting is primarily for advertising purposes, but it does have a safety element if operated all night. It can be done from cornices, from units extending out on brackets, or from poles. Parking areas are lighted by clusters of floodlights on tall poles. HID lamps are indicated, and high-pressure sodium is often used on account of its high efficacy. Floodlighting and area-lighting calculations were originally made by cut-and-try or by the point-by-point method, but now the footcandles can be obtained from manufacturers' tables.

SAFETY LIGHTING

Both indoors and outdoors, lighting should be such that people will not trip over obstacles. Indoors this is not much of a problem because suitable general lighting will reveal any hazards. However, stairwells require particular attention. The edge of a step should not cast a shadow on the step below it, nor should a person cast a strong shadow across the steps. A continuous row of fluorescent strip fixtures, mounted overhead and shielded to prevent direct glare, is a good answer. Also, steps and risers can be given different colors.

Areas having inflammable or explosive vapors or dusts should be lighted with approved, enclosed safety fixtures.

Emergency Lighting

Emergency lighting requirements are defined by the ***National Electrical Code***® and the ***Life Safety Code***®,* as well as by state codes.* Methods include generators which start automatically and serve emergency circuits when electric service fails, central battery systems, and individually battery-operated lights at strategic points.

Discussions of methods and equipment are found in the Codes.

MAINTENANCE

Cleaning

The long life of fluorescent and HID lamps necessitates cleaning between relampings if efficent delivery of light is to be maintained. A posted schedule on which the cleanings of different areas can be checked off is highly recommended.

Different reflecting and transmitting materials require different cleaning agents. Glass and porcelain enamel can be cleaned by most nonabrasive cleaners such as detergents or automobile glass cleaners. Synthetic enamels should be cleaned only with detergents. Plastics have a tendency to collect dust due to static charges. Detergents containing a destaticizing agent will prevent this, providing the plastic is merely rinsed after washing, and not wiped dry.

Relamping

In office areas, lamps are usually replaced as they burn out. With high-mounted industrial fixtures, however, scattered outages are sometimes allowed to accumulate until a noticeable reduction in light occurs at some area.

Large installations of fluorescent lamps where all lamps are turned on and off at the same time are sometimes handled by group replacement. At some point such as 80 percent of rated average life as predicted by the lamp mortality curve, a trained crew using equipment designed for the job, replaces all lamps. Under this plan, when a lamp fails very early, its replacement is marked with a grease pencil so that when group replacement occurs, it can be saved and used to replace a future early failure.

Instead of replacing all lamps at 80 percent of rated life, another method is to keep

*_**National Electrical Code**_® and _**Life Safety Code**_® are Registered Trademarks of the National Fire Protection Association Inc., Quincy, Massachusetts.

count of early replacements and make the group replacement when early replacements have reached 20 percent of the lamps in service. This method automatically makes allowance for any deviation from rated average life caused by more or less frequent turning on and off. However, conservation pressures to switch in smaller groups and only light areas in actual use make group relamping benefits less likely.

Any HID lamp which fails early should be replaced at once because each lamp covers a wider area than does a fluorescent lamp.

Disposal of Burned-Out Lamps

Disposal of burned-out fluorescent and HID lamps must be handled with care. The amount of mercury in a lamp is small, but if it falls into a crevice on the floor or elsewhere, the mercury vapor can be dangerous. It is many years since fluorescent powders contained beryllium, but a lamp could have been used very little in some out-of-the-way place, or could have been mislaid for a long time before being used. Beryllium is highly toxic. For this reason and to reduce the buildup of scrap, fluorescent lamp breakers are available.

TROUBLESHOOTING

Incandescent Lamps

Incandescent lamps present few problems in troubleshooting. The most common problem is short life, and this usually means that the socket voltage is higher than the rated lamp voltage. If some of the early failures do not show the normal blackening expected at full life, it may be that the lamps are subjected to shocks or blows, and rough-service lamps should be used. Or, if they are subjected to vibration, vibration lamps are recommended.

Sometimes small clearances, or venting arrangements in fixtures, cause the lamps to run hotter than they should. An indication of this is often a dark brown discoloration of the base. Another evidence is a loose base.

Fluorescent Lamps

A complete guide to troubleshooting fluorescent and HID lamps would take more space than is available here. Lamp manufacturers publish bulletins listing pages of troubles and remedies—not that the troubles are necessarily frequent. There are also testing instruments and test lamps for such things as checking the current delivered through the cathodes of a rapid-start lamp.

A few of the more obvious points for fluorescent lamps are:

Lamp improperly seated in lampholders

Corroded contacts in lampholders

Lamp at end of life, as evidenced by dense, black areas at ends

Ballast at end of life

HID Lamps

Similar points for HID lamps are:

Poor contact in socket

Voltage too high or low

Ballast not grounded

Lamp at end of life

Ballast at end of life

COMPLAINTS

Industrial workers seldom complain about their eyes in regard to the work situation. In fact, it is remarkable what some inspectors put up with in the way of glare. Office workers, on the other hand, are more likely to say, "The lights are bothering my eyes," often without any perceptible reason. The fact is that people tend to use their eyes as "scapegoats." If something is bothering them, often subconsciously, they may tend to think, "My eyes are bothering me. It must be those new lights." A good first approach to this problem is to perform a little experiment. If the installation has Cool White lamps, change them to Warm White, but if it already has Warm White, change to Cool White. Chances are that the change will eliminate complaints. Other complaints, such as flicker, etc., are usually caused by the items listed for checking under "Troubleshooting" above.

ENERGY CONSERVATION

There are many considerations involved in energy conservation. On one hand, lighting represents only a small part of today's energy consumption. It is estimated that of the energy used for all purposes, electricity accounts for only 25 percent. Of this, lighting is only about 20 percent. Thus light represents only about 5 percent of total energy use.

On the other hand, lighting makes the most conspicuous use of electricity. Office buildings and factories which are fully lighted after working hours make a bad impression on people.

Regulations

The federal government and some state governments have proposed setting limits on lighting loads, unfortunately often on a basis of footcandles rather than on consumption of electricity. This would mean that there would be no penalty for using lamps of low efficacy. Instead, a regulation should reward those who use efficient systems and turn them off when not in use.

The 10-30-100 rule is frequently mentioned. This was promulgated by the Federal Energy Administration in 1974, and actually means 100, 75, or 50 fc on the task, depending on its difficulty, 30 fc for surround lighting, and 10 fc for remote areas and corridors.

The energy bill of 1975 requires that states adopt standards based on wattage for the total building. With the situation changing from year to year and from month to month, a handbook cannot provide a working guide to current regulations, but it can offer hints on reducing wattage and on conservation.

Methods of Conservation

Sections on lamps, switching, and use of daylight have already mentioned energy-saving aspects. For new installations, select the lamp of highest efficacy whose color is acceptable, even though fixed charges may be higher than for other types. For existing installations, the problem is to reduce the wattage somewhat without greatly reducing the quantity or quality of the illumination.

Merely removing some of the fluorescent lamps in an installation is not a good solution. Depending upon the type of circuit, removing one lamp from a two-lamp fixture may cause the other lamp to go out or to operate improperly. If both lamps are removed from a two-lamp fixture, a gap is left in the lighting pattern. If it is felt necessary to remove two lamps from a four-lamp fixture, be sure that both lamps are served from the same ballast. Some power will still be consumed unless the ballast is disconnected. A much better way is to substitute one of the new energy-saving lamps.

Energy-Saving Lamps

The energy shortage has prompted lamp manufacturers to design and produce lamps which have the same physical sizes as the lamps they replace and operate on the same ballasts but give somewhat less light and use less wattage. Thus, a plant can keep all fixtures properly lamped and save wattage without greatly reducing the illumination level.

Various lamps are available to replace one fluorescent lamp on a two-lamp ballast. Reductions in wattage and lumens range from 12 to 50 percent, and some types lose less in lumens than in watts. Of course, a 30 percent reduction in wattage means only a 15 percent reduction for a two-lamp fixture.

With HID lamps, a dramatic saving can be made if mercury lamps are being operated on lag-type ballasts and a change in color of light does not matter. There is a high-pressure sodium lamp which can be substituted for a mercury lamp in the same socket, on the same ballast. In the 400-W size, it saves 10 percent in wattage and gives 80 percent more light.

Switching

A large installation which is controlled by one switch can be subdivided by a system of remote-controlled switches operated by carrier currents sent along the regular wiring to individual parts of the circuit. These make expensive changes in wiring unnecessary.

Summary

Existing Systems

Consider energy-saving lamps in fluorescent systems.

Consider substituting high-pressure sodium for mercury lamps when they are on lag-type ballasts.

Add switches where circuits cover too much area.

Use daylight when available.

Turn off lights when not in use.

New Installations

Specify lamps with highest efficacy consistent with color requirements.

Lay out switching in rows parallel to windows and switch for small areas.

Plan to turn lights off when not in use.

BIBLIOGRAPHY

Allphin, Willard: *Primer of Lamps and Lighting*, 3d ed., Addison-Wesley, Reading, Massachusetts, 1973.

IES Lighting Handbook, 5th ed., Illuminating Engineering Society, New York, 1972.

IES Lighting Handbook, Illuminating Engineering Society, New York, 1981.

Lamp Engineering Bulletins, GTE Sylvania Inc., Danvers, Massachusetts.

chapter 3-6

Plant Communication Systems

prepared by

Executone, Inc.
Long Island City, New York

GLOSSARY

Ambient noise The level of background sound that is present in a room or area.

Amplification An increase in signal level from one point of a circuit to another.

Amplifier A device which generates an increase in signal level.

Axial sensitivity An imaginary line through the center of a speaker or trumpet on which the acoustic power output levels are measured.

Attenuation A decrease in signal level from one point of a circuit to another.

Attenuator A variable network which reduces the signal level from one circuit to another.

Audio-frequency spectrum A range of frequencies from 20 to 20,000 Hz. Telephone spectrum is between 300 and 3000 Hz.

Balanced line In communication, a pair of wires—from one circuit device to another—which carries an electric signal which is symmetrical with respect to a common reference point, normally ground.

Bridging amplifier An amplifier with an input impedance which is higher than the circuit to which it is connected. As a result, the amplifier does not substantially affect the signal level of the circuit.

Channel An audio path used for communication.

Compressor A device which generates a smaller output signal for a given input amplitude range.

Crosstalk Unwanted signals appearing in a given circuit due to inductive or capacitive coupling with other circuits.

Decibel (dB) A measurement for power levels defined as follows:

$$\text{dB} = 10 \log \frac{P_{\text{out}}}{P_{\text{in}}}$$

Distortion The unwanted change of a signal. Four types are common: amplitude, harmonic, intermodulation, and phase.

Dynamic range The minimum and maximum frequency that a device will accept.

Echo A signal which is returned to its generating point and can be detected at this point because of its magnitude and phase difference.

Gain Increase in signal level from one point of a circuit to another.

Hybrid transformer A transformer so designed to allow circuits of different impedance to be interconnected. In communications, the transformer is normally used to interconnect four-wire lines to two-wire lines.

Line transformer A transformer used to electrically isolate and match two individual circuits.

Mixer A device having two or more inputs and one output. The device is designed so that the input signals do not affect each other significantly.

Motorboating An unwanted low-frequency oscillation in an audio system. Usually caused by an overloaded power supply.

Octave The band of frequency between two frequencies having a 2:1 ratio.

Preamplifier A device used to amplify low-level signals so as to maintain a good signal-to-noise ratio.

Program Audio signal normally used for entertainment.

Public address (PA) system An audio system consisting of distributed speakers, trumpets, amplifiers, and microphones used for paging, announcements, music, etc.

Reverberation The lingering of sound in a room due to repeated reflections. The harder the surfaces of a room, the longer the sound will remain. Rooms or areas with high reverberation can cause difficulty in understanding speech.

Roll-off A decrease in amplitude with respect to an increase or decrease of frequency.

Unbalanced line Usually a circuit in which one side is grounded.

INTRODUCTION

A communication system is a virtual necessity to increase efficiency and safety in plant operations. Plant systems have two basic requirements: first, that the system be capable of locating roving personnel; second, that a communications link be provided for instructions or conversation.

Location of personnel can be accomplished by utilizing either a wired system (PA system) or a wireless system (wireless paging system).

WIRED VOICE SYSTEMS

In most plant operations, the maintenance and service personnel are distributed in various parts of the plant. Therefore, it becomes extremely important, in plant communication systems, that a personnel-locating system be integrated with the communication system. A basic plant system consists of a single telephone channel system interfaced to a sound or PA system and is shown in Fig. 6-1.

The operation of the system allows the office staff to page and converse with roving maintenance personnel in the following manner:

1. Office person lifts handset, depresses a page button on handset, and announces the name of the person desired. The page is broadcast throughout the plant via the plant speakers.
2. Upon hearing his or her name, the paged person goes to the nearest station and lifts the handset for instant communication with the office.
3. Telephone handsets are normally provided with a "busy" light to indicate the system is already in use, or busy.

In large plants a *multichannel* system is preferable. In mulitchannel systems, the telephone handset is designed to allow more than one conversation to take place at the same time. A selector switch is provided on the telephone handset unit to allow any number of channels to be selected. Standard commercial units are available in one-, two-, or six-channel models. A typical system is shown in Fig. 6-2. These models allow for greater flexibility in paging. Multichannel systems may be set up so that a number of channels can cover individual plant zones and one channel can be used to page all plant zones simultaneously. For instance, channel 1 can be used to page in the office area, channel 2 to page into the security area, channel 3 to page the plant, and channel 4 to page all areas simultaneously. For such a system the operation will be as follows:

1. Office person selects, say, channel 3, lifts handset, depresses page button and pages as follows: "Mr. Jones, channel 3 please."
2. Mr. Jones goes to the nearest telephone handset, turns the channel selector to channel 3, and lifts the handset to his ear.
3. Mr. Jones and the caller can now hold a conversation.
4. When conversation is terminated each party replaces the handset on its cradle.

Standard commercial handsets are available in different models to meet the local ambient conditions. These models include explosion-proof telephone handsets, weatherproof telephone handsets, desk-mounted units, and wall-mounted units. Examples of these are shown in Fig. 6-3.

Some manufacturers offer, as options, telephone units with built-in speakers so that external speakers need not be used. This unit is especially designed for use in areas where high-level paging would be burdensome to the personnel (Fig. 6-4).

In most systems, an unlimited number of telephone handset stations can be accommodated and multichannel units can be mixed with single-channel units.

Cable used to interconnect the system is usually run from handset station to handset station and finally to a central control box. The control box contains circuitry to provide "busy" indication and interfaces the telephonic channel to the PA amplifier and pickup points for the power supply.

Telephone handset cabling should always be twisted and shielded, and, wherever possible, run in conduit. Such precautions will protect the system from picking up unwanted hum, crosstalk, and other extraneous electrical noises. The telephone cabling, however,

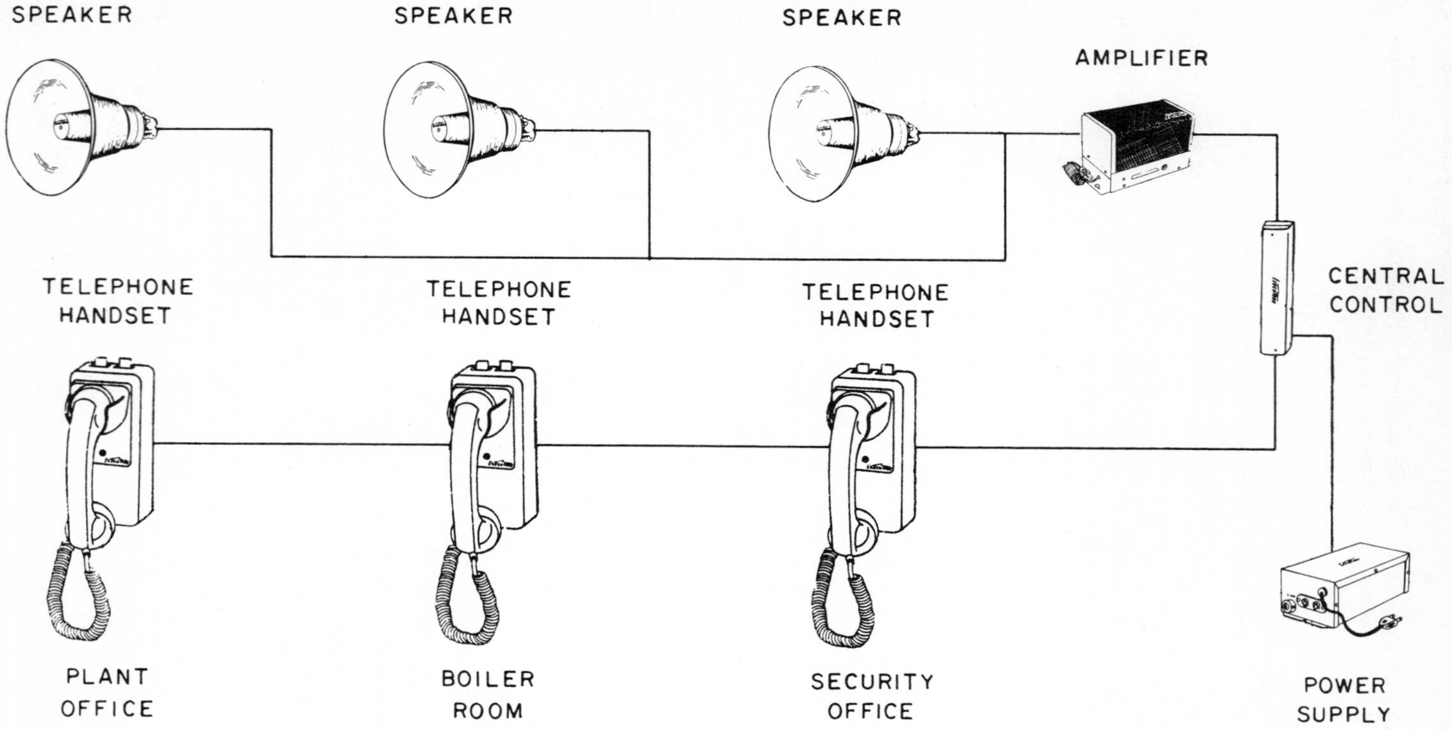

Figure 6-1 Basic single-channel telephone paging system.

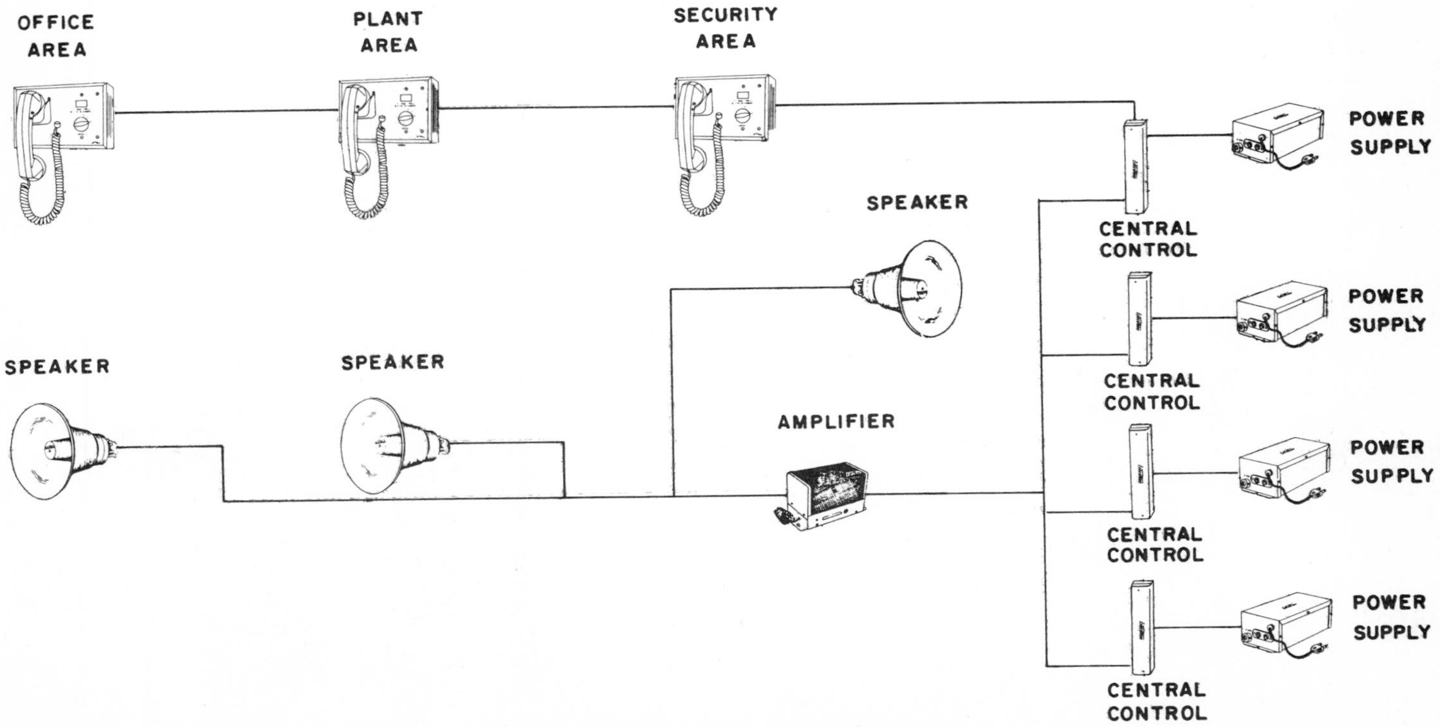

Figure 6-2 Multichannel telephone paging system.

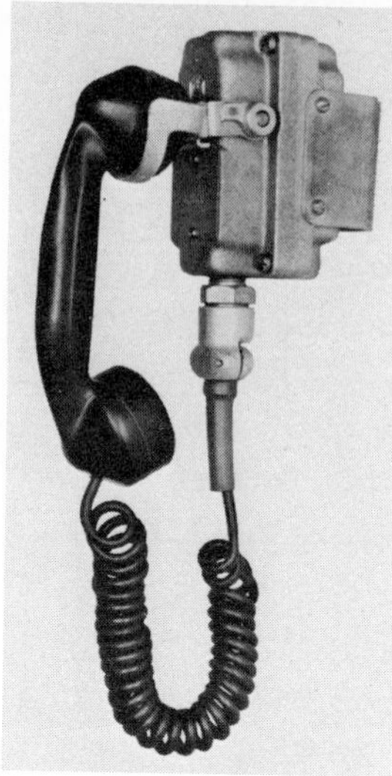

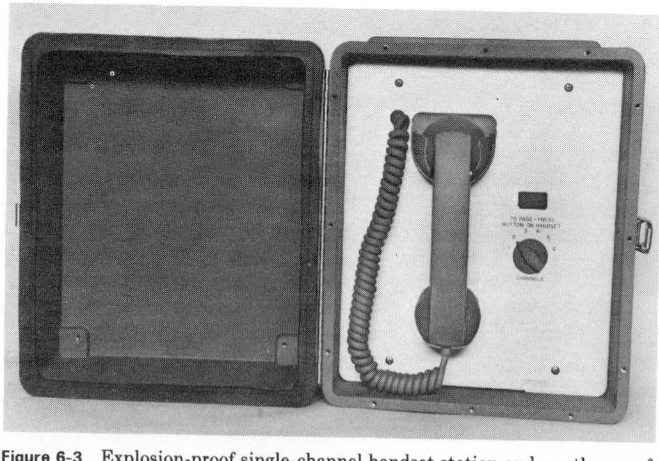

Figure 6-3 Explosion-proof single-channel handset station and weatherproof multichannel handset. (*Executone*)

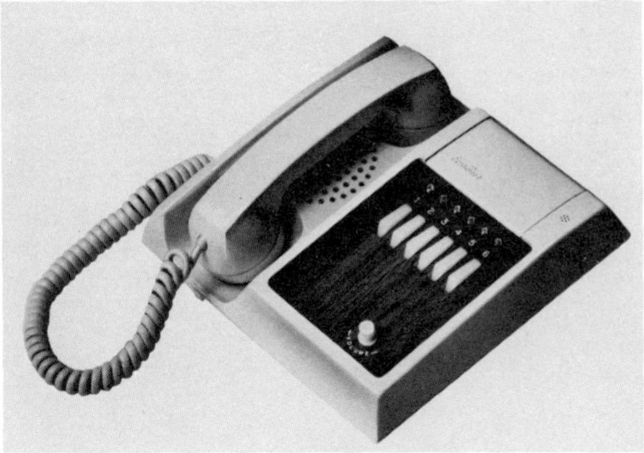

Figure 6-4 Telephone with built-in speaker. (*Executone*)

should never be run in conduit containing other conductors carrying voltages greater than 1 V rms.

The most critical part of the plant communication or PA system is the proper selection and location of the speakers. Three factors affect a speaker's performance:

1. Dispersion angle
2. Frequency response
3. Power level

Trumpet reproducers rather than speakers are commonly used in plant areas because their dispersion angles, frequency response, and power levels are better suited to the conditions found in plants.

Dispersion Angle

Plants by their very nature consist of areas with very hard walls and surfaces. The sound from a paging speaker has the tendency to bounce from surface to surface, resulting in a lingering of the sound (reverberation), and making it difficult to understand speech. It is therefore important to choose speakers that aim the sound in the general area of the room but at the same time prevent the major part of the sound from hitting hard surfaces such as ceilings, floors, and walls. Hard ceilings are the major obstacles to good speech comprehension. They normally reflect a major part of the sound introduced by a paging speaker, so sound should be kept away from the ceiling as much as possible. A trumpet-type speaker is used for this purpose (Fig. 6-5). This type of trumpet is flat at top and bottom to allow the sound to be diverted into a middle area and away from the ceiling and floor. It has a wide dispersion angle in the horizontal direction and a small dispersion angle in the vertical plane.

In areas where the surfaces are not very hard, trumpets with uniform horizontal and vertical dispersion angles can be used. These trumpets are circular in construction, a typical unit is shown in Fig. 6-6. Trumpets are also available in explosion-proof models for location in hazardous areas.

Frequency Response

Since low-frequency sound (below 500 Hz) has a tendency to have a longer reverberation time than high-frequency sound (above 500 Hz), the understanding (intelligibility) can be increased by decreasing the low-frequency sound components in the paging circuits. This is normally done by introducing a speech filter at the input of the paging amplifier. Many commercial amplifiers are provided with this feature.

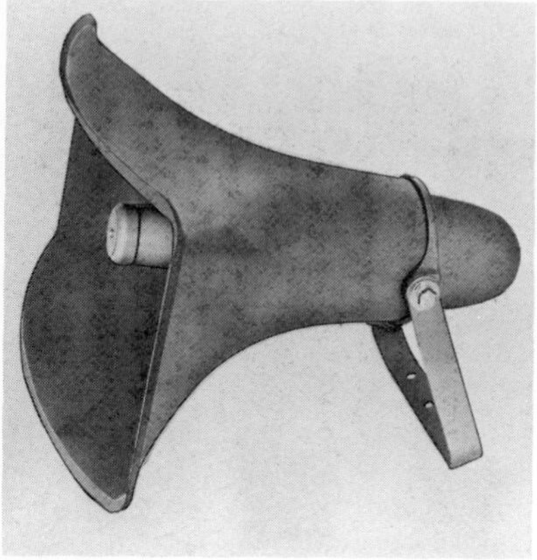

Figure 6-5 Trumpet-type paging speaker. (*Executone.*)

Figure 6-6 Circular-trumpet paging speaker. (*Executone.*)

In amplifiers with tone controls, the bass (low-frequency) and treble (high-frequency) controls should be adjusted to allow for minimum bass and maximum treble.

Power Levels

The power output of a speaker or trumpet is normally stated in decibels (dB). The decibel is an acoustic power level.

The speaker or trumpet manufacturer will normally specify the acoustic power level (dB) in relation to the electric power provided to the speaker. The measurement is normally stated at particular distances. A typical specification by the manufacturer will be:

Axial sensitivity: 93dB at 4 ft (1.2 m), with 1 W input to speaker coil.

The axial sensitivity means that the measurement was made on the axis of the speaker or trumpet, an imaginary line through the center of the speaker or trumpet, as shown in Fig. 6-7.

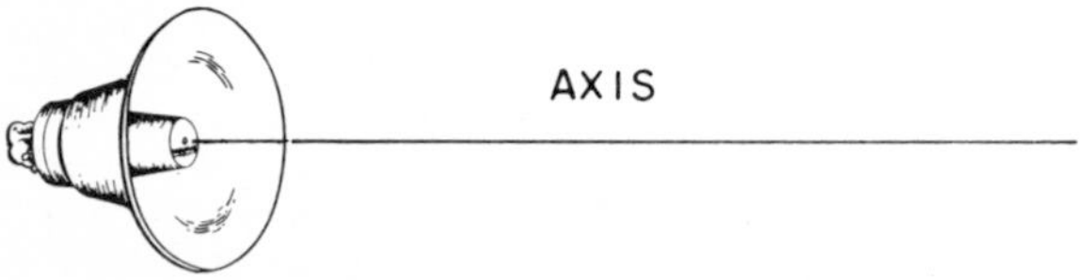

Figure 6-7 Speaker axial sensitivity.

This statement indicates that with 1 W of power (normally at 1000 Hz) put into the voice coil of the speaker, a level of 95 dB was measured 4 ft (1.2 m) away from the speaker along its axis.

Utilizing this information, it is easy to calculate the power required for coverage. The key items to remember are: 6 dB is *lost* when the distance is doubled; 3 dB is *gained* when the power is doubled. Figure 6-8 indicates output levels of a typical speaker for various voltage inputs at various distances.

Note that with 1 W at 4 ft (1.2 m) the sound level is 93 dB; 8 ft (2.4 m) feet away from the speaker we lose 6 dB and the level is 87 dB. If we now put 2 W (double the power) into the speaker at 4 ft (1.2 m), the speaker will produce a sound level of 96 dB

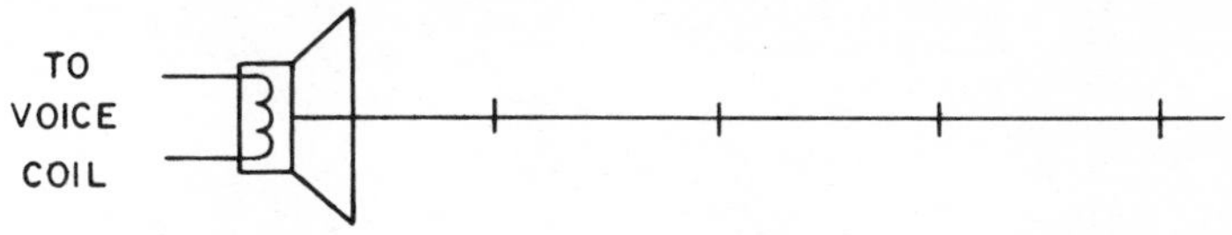

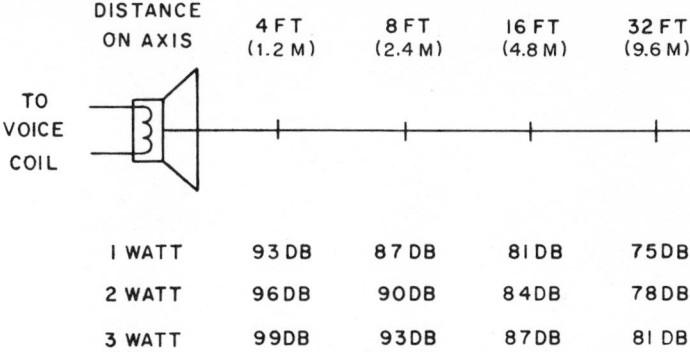

DISTANCE ON AXIS	4 FT (1.2 M)	8 FT (2.4 M)	16 FT (4.8 M)	32 FT (9.6 M)
1 WATT	93 DB	87 DB	81 DB	75 DB
2 WATT	96 DB	90 DB	84 DB	78 DB
3 WATT	99 DB	93 DB	87 DB	81 DB

Figure 6-8 Speaker power vs. distance at various power inputs.

(a gain of 3 dB). At a distance of 8 ft (2.4 m), the level will be 90 dB with 2 W in, etc. If a page is to be heard, the level of the page must be higher than the ambient (background) noise. In areas where the background noise contains high frequencies (frequencies higher than 800 Hz), the paging level must be at least 6 dB higher than the ambient noise. Areas where ambient noise is below 800 Hz require that the paging level be only 3 dB higher than the ambient noise.

Measurement of the ambient noise can be made with a sound-level meter much as the one shown in Fig. 6-9.

EXAMPLE To calculate the power requirement for the trumpet shown in Fig. 6-8 whose axial sensitivity is 93 dB at 4 ft (1.2 m) with 1 W. How much power will we require to page into an area with ambient noise (containing high frequencies) of 82 dB at a distance of 32 ft (9.6 m) from the trumpet?

We know that at 4 ft (1.2 m) the sound level is 93 dB with 1 W into the speaker; 32 ft (9.6 m) away from the speaker the level will be 75 dB. To overcome the ambient noise, the level of page must be at least 6 dB higher, meaning that we will require a paging level of 81 dB. Therefore, we must provide a trumpet with 4 W of power.

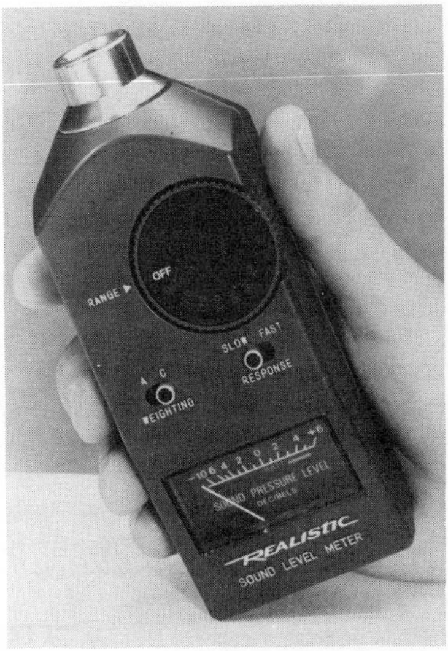

Figure 6-9 Typical audio sound-level meter. (*Tandy Corp.*)

Along with telephone handsets, it is possible to interface the paging or handset system with a privately owned telephone system.

Telephone systems are divided into two major categories: key systems and PBXs. Figures 6-10 and 6-11 illustrate a typical phone for each of the latter two systems. If a key system is provided in the facility, it can be tied into the PA or intercom system via a matching interface unit. These units are manufactured by makers of the key system to allow the system to match the levels and signals required by the paging or intercom system. With this interface, the key system can access the PA system to page and then converse on the telephone channel with the handsets in the plant area.

Plants that utilize a PBX require that a PA or intercom system be accessed by dialing

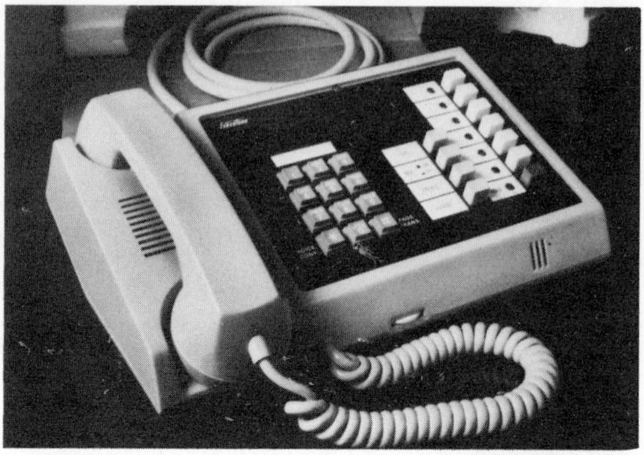

Figure 6-10 Typical key-system telephone. (*Executone.*)

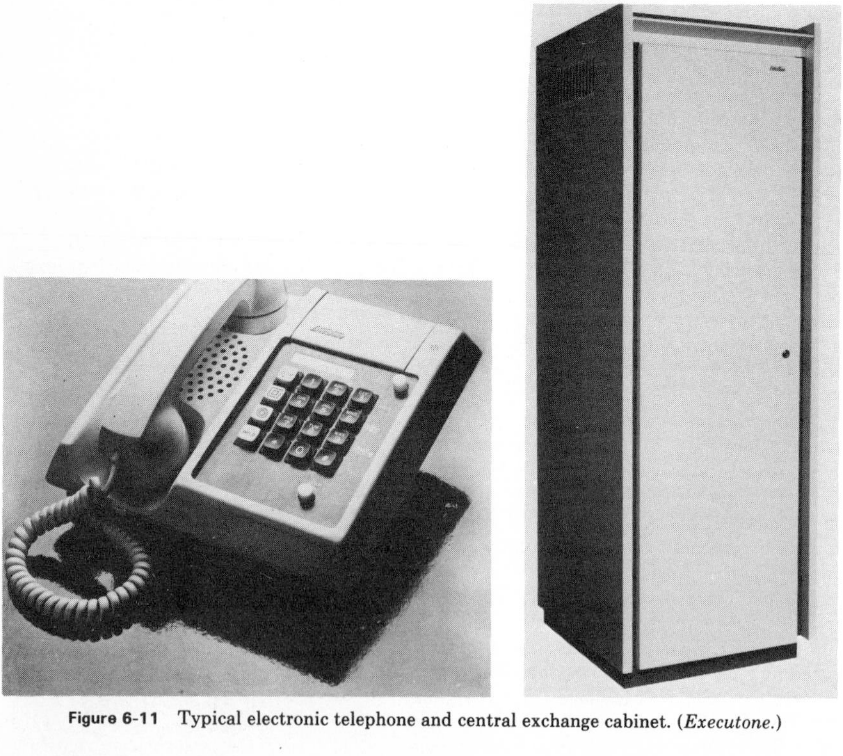

Figure 6-11 Typical electronic telephone and central exchange cabinet. (*Executone.*)

a preassigned number(s). Once the preassigned number is dialed, the calling party may make a page and receive a reply via the intercom system.

An option offered by many manufacturers is to allow the PA system to be tied into an alarm panel. These alarm panels will allow fire, civil defense, and time alarms to be transmitted throughout the plants.

Figure 6-12 Basic wireless paging system. (*Executone.*)

WIRELESS SYSTEMS

Wireless systems provide an alternative solution to the problem of locating roving personnel in very large plants which cannot easily accommodate conduit for the PA or intercom wiring, or in plants that have extremely high ambient noise levels.

Wireless transmitter systems consist of an encoder, a transmitter, and an antenna. The system transmits a tone to a receiver carried by roving plant personnel. A typical system is shown in Fig. 6-12.

In most commercially available systems, in addition to a keyboard encoder a pocket page-to-PBX tie-in is available. Operation of the wireless system is as follows: Person at keyboard selects the number of the receiver to page and activates a transmitter, which in turn sends out a tone signal. The roving person, hearing the "beep" tone, calls a preassigned extension or location for his or her message. If the system is equipped with a telephone tie-in, any person may dial a preassigned extension number, then, upon hearing a second tone, dial the receiver number. The transmitter will send out a beep tone to roving employee. Upon hearing the beep the paged person goes to the nearest phone, dials a preassigned number(s) and is in immediate contact with the paging party.

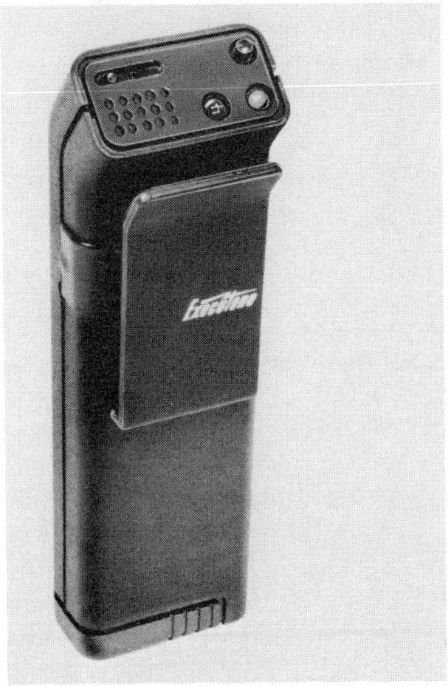

Figure 6-13 Digital-readout VHF receiver. (*Executone.*)

One option available with the wireless system is a beep with one-way voice. In this system both the receiver and the transmitter are equipped with a voice channel. The paging party utilizes the keyboard encoder in the same manner as to make a tone call. However, after the tone is sent out, the calling party may give a voice message. The voice message is also available if the pocket page system is tied to a PBX.

Other options currently available are group call receivers, digital readout receivers, and vibrating receivers. The *group call receiver* allows the keyboard operator to page more than one receiver at the same time. This page may also be followed by a voice message. Digital readout receivers (see Fig. 6-13) are provided with a solid-state readout. When the receiver is paged, the operator at the encoder depresses another digit on the keyboard. This digit corresponds to a preassigned message. The digit will then appear on the receiver readout. The person receiving the page looks at the readout and receives the preassigned message. Such receivers are a great help in high-noise areas where it may be difficult to hear the voice page from the receiver or to converse over a telephone.

Receivers are also available which do not emit a tone but *vibrate* when they are called. Again, such receivers are useful in high-noise areas or when a tone from the receiver would be disturbing.

In the past, wireless paging systems utilized a two-tone arrangement to alert a receiver. The two audio tones would modulate an RF carrier. Both AM and FM systems have been available, the FM system being slightly superior since it was less susceptible to false calling and to interference. Current digital systems broadcast a digital code superimposed on the RF carrier. The receiver generates its own code. When the receiver "looks" at the RF carrier, it compares the digital code on the carrier to the one it generates. If the two match, the receiver generates a tone. If the codes do not match, the receiver resets. The advantage of the digital arrangement is that it prevents the receiver from receiving false calls from other RF interference.

In addition to telephone interfacing system options that are available with wireless systems, there are interfaces to boiler gauges, call-in buttons, fire-alarm systems, etc. The interface in the wireless radio system normally will signal a receiver or group of receivers if a set of contact closures at a boiler gauge indicates that the boiler is going into an overheat condition. A signal can also be sent to respond to a doorbell (call-in button) or a fire-alarm signal.

INSTALLATION, MAINTENANCE, AND REPAIR

Installation Practices

Cables

It is most important that wires carrying different voltage potentials (line levels) not be run together. Three types of line levels are normally found in communication and sound systems. These levels are:

1. Microphonic line: 10 to 100 mV (rms)
2. Zero-level line: 0.25 to 1 V (rms)
3. High-level line: 20 to 70 V (rms)

The following rules should be observed in wiring:

1. The three types of lines should *never* be run in the same cable or conduit.
2. If these lines are run in cable trays, they should be separated by a minimum of 6 in (15 cm)
3. *Never* run these lines near high-voltage lines, power transformers, fluorescent lamps, etc.
4. All three different types of lines should *always* be twisted. The microphonic lines should be both shielded *and* twisted.
5. When using shielded lines, the shield should be grounded at *one end* of the line only. Grounding the shield at both ends will cause ground loops which will make the shield useless.

Housings

Housing or racks used to house electronic equipment should at least have top, bottom, and side perforations to allow for convection venting. In some cases, a fan should be added to prevent heat buildup. The manufacturer's specifications and guidelines should be consulted in this area.

Large power supplies should be mounted at the top of the housing or racks. Since hot air rises, heat from large power supplies placed at the bottom of a housing or rack will add additional unwanted heat to the electronic circuits mounted in the rack. Small power supplies supplying less than 200 W may normally be placed at the bottom of the housing or racks. Again, the equipment manufacturer's specifications and guidelines should be observed in these matters.

As with conduits, cables carrying different voltage levels should be run on opposite sides of the housing or at least 6 in (15 cm) away from each other. Cables run in housings should be harnessed or run in wire rings to prevent accidental breakage of the equipment.

Cable Termination

Every installation should be made with maintenance and service in mind. Units such as telephone stations, amplifiers, speakers, etc. should be provided with plug-in-type connectors. This allows a unit to be replaced easily. Units which have screw terminals make it difficult for service personnel to replace a unit without closing down the whole system.

In many cases, junction boxes will be required. These junction boxes should be set up to allow the wiring for different areas of a plant to be isolated. For example, let us say that we have six areas in a plant to which we have run separate wiring. If a speaker wire is shorted in one area, it could prevent the whole paging system from operating until the short is found. In large plants this may take some time. However, if the runs can be easily disconnected at a junction box, the shorted run can easily be identified, permitting operation of the remainder of the system while the short is repaired.

Telephonic systems often have common wiring running from station to station in contrast with individual wiring to each station. In systems with such common wiring, station wiring should be zoned to allow for the disconnection of a zone in the event of a wiring defect in that zone.

Documentation

To facilitate maintenance and repair, system documentation is vital. Medium to large installations should be provided with the minimum documentation described in the following paragraphs.

Riser Diagrams. These diagrams contain a simple line drawing showing how the cables are run throughout the structure to the different units (speakers, telephone stations, etc).

Installation Diagrams. These diagrams show how each of the units is interconnected. *It is very important that color coding of the cables is shown* in these diagrams. The terminals of each unit should be labeled with a functional designation such as "control line," "tip," "ring," "+24," etc.

Schematics. These are the electrical drawings of the internal wiring of the components without regard for physical location.

Load Factor Notations. It is important that power supply loads and power amplifier loads be noted on a per-zone or -area basis. If additions are to be made to a system, it will then be easy to see if more power supplies or power amplifiers will be required to power the new units.

Maintenance

A systematic schedule of maintenance should be maintained on all communication and PA systems. Depending on personnel available, a biannual check should be made on all mechanical or electromechanical components of a system.

A music or tone source should be continuously fed to the PA system and each speaker should be checked for performance. All electromechanical devices, such as relays and stepper switches, should be checked annually for signs of wear or pitting. Dirty relay contacts should be burnished. Under no circumstances should relay contacts be sprayed with contact cleaners. These cleaners put a thin coat of film on the contacts which prevent them from making good contacts. If a relay burnishing tool is not available, an emery board can be used.

Some newer electronic communication systems utilize voltage-regulated power supplies to minimize problems due to power-line voltage fluctuations. These power supplies should be periodically checked to see if the output voltage is within manufacturer's recommendations.

A supply of lamps and fuses used in the communication and PA systems should always be kept for immediate replacement. It is good practice to list all instruments and units in the communication and PA systems and have each unit checked out at least once a year.

Service

Many manufacturers hold factory training seminars on the equipment they produce. It is always advantageous to have one or more of the in-plant personnel enrolled in these seminars. Armed with a knowledge of the system, in-plant personnel can correct minor problems, which usually are the most frequent.

It is helpful to stock major subassemblies of the communication and PA systems in the plant so that, when a problem is isolated to a particular unit, the subassembly of that unit can be immediately replaced. The subassembly can then be repaired in the service shop. This approach to servicing will minimize the system *downtime*.

SPECIFICATIONS AND STANDARDS

These can be obtained from:
Acoustical Society of America (ASA), 335 East 45 Street, New York, NY 10017.
American Society for Testing and Materials (ASTM), 1916 Race Street, Philadelphia, PA 19103.
American National Standards Institute (ANSI), 1430 Broadway, New York, NY 10018.
Institute of Electrical and Electronic Engineering (IEEE), 345 East 47 Street, New York, NY 10017.

BIBLIOGRAPHY

Audio

Bruel and J. A. Kjaer: *Acoustic Noise Measurements,* 2d ed., Naerum, Denmark, 1971.
Davis, Don, and Carolyn Davis: *Sound System Engineering,* Sams, Indianapolis, 1975.
Peterson, A. P. G., and E. E. Gross: *Handbook of Noise Measurements,* 7th ed., General Radio Company, Cambridge, Massachusestts, 1972.

Telephony

Carlson, A.: *Communication System: An Introduction to Signals and Noise in Electrical Communication,* 2d ed., McGraw-Hill, New York, 1975.
"Fundamentals of Telephony," TM 11-6-78, U.S. Government Printing Office, 1978.
Hamsher, D.: *Communication System Engineering Handbook,* McGraw-Hill, New York, 1967.
ITT Reference Data for Radio Engineers, 5th ed., Sams, Indianapolis, 1968.
Smith, Emerson C.: *Glossary of Communications,* Telephony Publishing Co., Chicago, 1971.

chapter 3-7

Electric Measuring Instruments

prepared by
Weston Instruments Division
Sangamo-Weston, Inc.
Newark, New Jersey

GLOSSARY

Accuracy The limit of the error of an instrument, usually expressed as a percent of full-scale value.

Average value 0.636 of the peak value of a sine wave, normal shape of power-based electricity.

Full scale (FS) The largest value of the actuating electrical quantity that can be indicated on a meter scale.

Multimeter Alternative designation for VOM.

Sine wave Waveform corresponding to a pure single-frequency oscillation.

Root mean square (rms) 0.707 of the peak value of a sine wave. Typically, the value sensed and displayed by ac instruments because it is most relevant to power analysis and comparisons.

Transducer Device for converting a physical quantity into an electric signal that can be displayed on an electric instrument.

Varmeter Measures reactive power in a circuit. Voltamperes reactive (var) do not provide any useful power.

VOM Volt-ohm-milliammeter measurement capabilities provided in one instrument.

INTRODUCTION

Electric measuring instruments play a key role in many areas of plant maintenance: repair, preventive testing, load balancing and total demand reduction. Some of these tasks require only simple "yes or no" checks, e.g., continuity, presence or absence of line power, or current flow. Tools for these tests may be rugged and inexpensive. Yet they must be safe and reliable, as lethal voltages must be checked. Other measurements, such as power factor and demand, can yield cost savings proportional to instrument accuracy.

Plant loads may be divided into five broad categories requiring significant electrical or fuel utility costs: air distribution, heating, air conditioning, lighting, and process power. Load requirements vary with geographical location, type of industry and building, and working-shift schedules. Typical maintenance instruments for each load category are indicated in Table 7-1.

TABLE 7-1

Plant load	Air distribution	Heating	Air conditioning	Lighting	Process power
Types of loads	Motors Control relays Gas monitors	Motors Circuit breakers Controllers	Motors Low-voltage thermostats Valves	Incandescent, flourescent, HPM,HPS lamps Controllers Generators	Motors Welding equip. Ovens Switchgear Cables
Electrical measuring instruments	Analog VOM	(AC line voltage, low AC voltage, low resistances)			
	Clamp-on ammeter	Clamp-on ammeter Power factor meter Wattmeter Recorder	Clamp-on Ammeter Power factor meter Wattmeter Recorder	Clamp-on ammeter Light meter	Clamp-on ammeter Digital multimeter Power factor meter Wattmeter Recorder Ground Fault detector

In the well-designed plant, electric power requirements are calculated with some allowance for load changes. Even though the circuits are initially "in balance" they tend to get out of balance as a result of machines being added to the system, changes in ovens, pumps, etc. Clamp-on ammeters and voltmeters can be used to determine the loading and to check that current densities are not being exceeded for the cables and wiring involved. These individual readings may also be taken when there is a bus duct distribution system. The electric meters are used to observe out-of-balance conditions which can then be corrected by a redistribution of the load.

Ground-fault detection can prevent massive destruction to electric equipment. In general, ground-fault detection is located at the main substation but may also be incorporated at power panels, load centers, etc. Ground-fault equipment generally has its own meter to indicate defects in the distribution system and is usually programmed to trip out key circuit breakers when the condition becomes unsafe.

In virtually any plant a considerable number of individual machines will be equipped with their own load meters, wattmeters, or similar monitors for the purpose of indicating speeds, feeds, welding current, or motor loading.

It is considered good practice to keep a record of the current loads in the various plant feeders and branches as a form of preventive maintenance. In the event that the manufacturing processes necessitate shifting around equipment, the road map or network distribution system and its readings, previously recorded by electric indicating instruments, will prove invaluable.

ANALOG DC AMMETERS AND VOLTMETERS

Analog dc measurement is accomplished by use of the D'Arsonval mechanism. (See Figure 7-1.) The current being measured passes through a coil of wire suspended in the field of a permanent magnet. The resultant motor action causes the pointer, attached to the coil, to move across a calibrated scale to indicate the quantity being measured.

Two types of coil suspension are used. Taut-band suspension can provide higher sensitivity and better repeatability. Pivot-and-jewel suspension can provide better repeatability under high vibration conditions and permits the smaller meter height desirable for edgewise panel meters.

Ranges of up to 50 A or 500 V can be measured by incorporating shunts or series resistors inside the dc meter case. Higher-range measurement is possible with external range extenders.

DC meters are typically available in the following types:

Type	*Typical accuracy (of full scale)*
Panel	
(Round, rectangular, edgewise)	±2%
Switchboard	±1%
Portable (lab or shop)	±¼% to ±1%

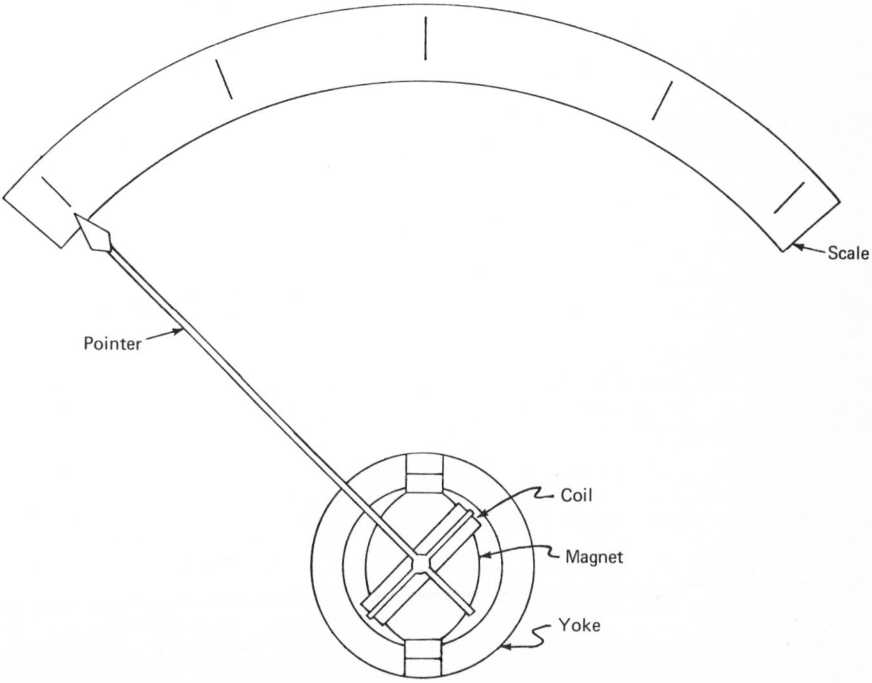

Figure 7-1 D'Arsonval mechanism.

Most dc meters have 90° or 100° FS angles. Edgewise types, which takes up less space than other panel types, have 50 to 60° angles. Angles of 250° are available in switchboard or round-panel types.

DC meters can be used with transducers to measure variables other than current or voltage. One example is a pyrometer that uses a thermocouple connected to a dc meter to measure temperature.

Another electromechanical mechanism is the moving-magnet type, where the coil is stationary and the magnet moves in response to the signal. This mechanism is less sensitive and less accurate (±3 to 5 percent FS rating) than the D'Arsonval type.

Electronically actuated bar-graph displays are becoming available to display analog signals. External power, other than the quantity being measured, is required for operation.

ANALOG AC AMMETERS AND VOLTMETERS

The flow of current in an ac circuit can be measured with indicating instruments operating on (1) electrodynamometer, (2) iron-vane, (3) thermal, and (4) rectifier principles. All *electrodynamometer, iron-vane, and thermal instruments* read true rms values of ac current or voltage, while rectifier instruments sense average values but are calibrated in rms values. In circuits where a small amount of energy is available, rectifier-type instruments are quite effective. They can measure the low-level current; however, their main limitation is a sensitivity to waveform errors, and they should be used on applications where waveform is sinusoidal. Standard panel-type instrument accuracies are typically 3 percent FS, and portable type accuracies are 1½ to 3 percent FS. A typical full bridge rectifier circuit is shown in Fig. 7-2.

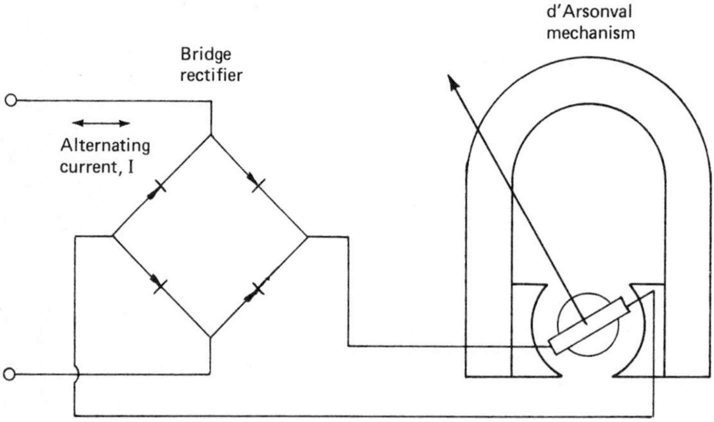

Figure 7-2 Rectifier meter for ac measurements.

Thermocouple instruments (Fig. 7-3) are used on dc, ac and audio and radio frequencies, particularly where other waveforms or frequencies would cause errors in other types of indicators. Since they are of low resistance and draw very little power they are useful in low-energy circuits. These instruments, however, are very sensitive to overloads so they should be limited to circuits where overloads seldom occur. Standard panel-type instrument accuracies are typically ±2 percent FS and portable type accuracies are ½ to 1 percent FS.

There are several moving-iron-vane designs, but the most widely used is the repulsion type (Fig. 7-4). This type incorporates two magnetic iron vanes located within a field coil. One vane, with a pointer attached, is free to rotate. As current flows in the field coil,

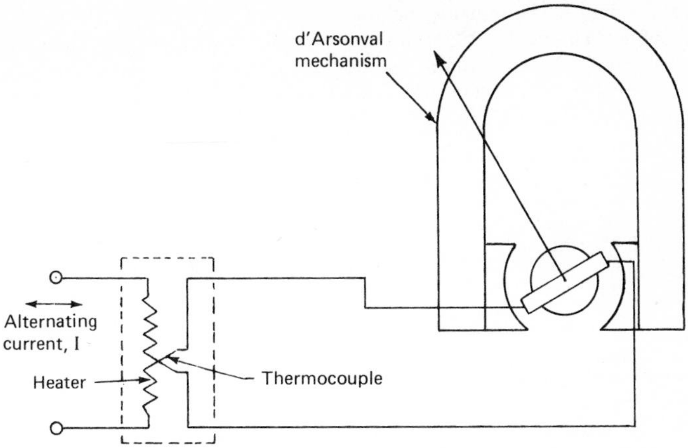

Figure 7-3 Thermocouple meter for ac measurements.

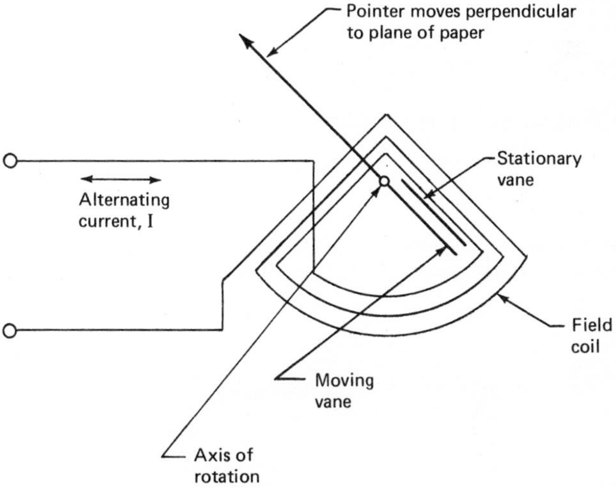

Figure 7-4 Iron-vane mechanism for ac measurements.

the two vanes are inductively magnetized with like polarity and produce a repelling force proportional to that of the current. This repelling force varies inversely as the square of the distance between the vanes. Standard panel-type instrument accuracies are typically 2 percent FS and portable-type accuracies are ½ to 1 percent FS.

The electrodynamometer movements, which usually have one fixed and one movable coil, measure true rms and are used in the high-accuracy portable type. An important feature of these instruments is their ability to measure direct or alternating current with equal accuracy. Standard portable accuracies are typically ±¼ to ±¾ percent FS.

ANALOG VOM

The analog VOM (volt-ohm-milliammeter) is a versatile instrument that combines measurement capabilities normally requiring several different instruments. The VOM can be used as a portable servicing instrument as well as for bench-top use.

Typical in-plant measurements include ac line voltage and continuity (resistance) of fuses and cables. The analog VOM, although generally less accurate than a digital version, is adequate for most in-plant use, has the advantages of being self-powered, and can provide quick trend information. The following table shows the measurement capabilities of a typical analog VOM:

Function	Range span	Accuracy
DC volts	250 mV–1000 V	±2% FS
AC volts	2.5–1000 V	±3% FS
DC current	50 μA–10 A	±2% FS
Resistance	0–2000 Ω	±2% of scale length
	0– to 20 MΩ	

The dB scale can be used in conjunction with the ac voltage measurement to give an indication of the power gain of an amplifier. A zero (dB) power level must be chosen, typically 1 mW in 600 Ω.

The power gain is defined as:

$$\text{dB} = 10 \log \frac{P_2}{P_1}$$

The dB scale shown corresponds to the lowest ac voltage range on the meter. The chart provided on the dial is used to determine the add-on dB value for the higher-voltage ranges.

Figure 7-5 shows a typical analog VOM.

CLAMP-ON AMMETER AND VOLTMETER

The ac clamp-on ammeter is actually a special application of current transformer. It is used for measuring very high currents or when the circuit under test cannot be inter-

Figure 7-5 Analog VOM. (*Sangamo–Weston.*)

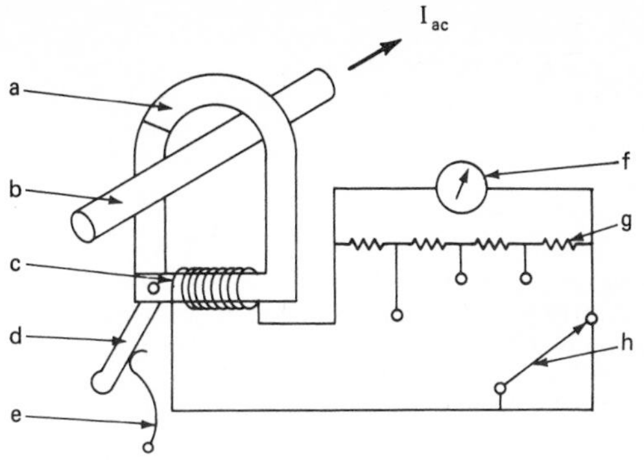

Figure 7-6 Schematic diagram of a clamp-on ammeter. *Key:* a, iron core; b, circuit conductor carrying current to be measured (serving as primary of transformer); c, wire coil (serving as secondary of transformer); d, lever arm; e, spring; f, ammeter; g, shunts; h, selector switch.

rupted. A schematic diagram of a typical clamp-on is shown in Fig. 7-6. The core is made up of two parts. The smaller is provided with a lever arm and is movably attached to the larger part. The primary winding is the conductor of the circuit of which the current is to be measured. The secondary winding is on the larger core part located inside the instrument housing. The secondary winding is connected to an ammeter, permanently mounted on the instrument housing. The ammeter is usually of the D'Arsonval type with a rectifier circuit. Multiple ranges may be provided by a switch selecting different shunt resistors. The internal hook-up wiring is such that the secondary windings are always closed.

For operation, the lever arm of the smaller core piece located outside the instrument housing is depressed so that the larger core piece can be hooked around the current-carrying conductor. After release of the lever arm, a built-in spring will keep the core closed. The range-selector switch should be set at the highest range and if necessary turned to lower ranges in order to avoid overloading the internal circuitry.

The standard accuracy of a clamp-on ammeter is ±3 percent FS. The accuracy may be adversely affected by circuit frequencies other than for which the clamp-on is calibrated or by other current-carrying conductors in close vicinity of the core.

Depending on the size of the instrument, ranges covered may be as high as 2000 A. More common ranges are: 500, 250, 100, 50, and 10 A ac. Some clamp-on meters also provide ac voltage measurement capability through use of separate plug-in leads. DC clamp-on meters are also available, but they utilize a different operating principle.

ELECTRIC POWER MEASUREMENT

Wattmeter

Wattage is instantaneous power. Energy (watthours) is power integrated over a time period. By measuring wattage, circuit conditions can be observed on a continuous basis. Money can be saved by reducing peak power demands by shifting loads to off-hours. Motor damage due to overloads can be controlled if power to variable loads such as mixing machines is monitored.

A true power measurement requires that the wattmeter respond to the magnitude of both voltage and current as well as the phase angle between them. In addition, the meter must have a frequency span wide enough to handle harmonics of a distorted wave. A wattmeter will measure either alternating or direct current and will reverse should the flow of power reverse.

Wattmeter Configuration

Dynamometer. The dynamometer consists of a fixed coil around a moving coil with pointer and a scale. As a mechanism which provides a true watt response, the dynamometer was used in the first wattmeters and is still widely used today.

A listing may be set up in terms of accuracy. Meter cost increases rapidly with better accuracy.

1. 1 percent accuracy, fixed installation of panel and switchboard types; 1, 2 and 2½ elements available. A 1-element wattmeter contains one voltage and one current circuit; a 2-element wattmeter contains two independent sets of voltage and current circuits, the outputs of which can be separate or combined; and a 2½-element wattmeter is the same as the 2-element type except for the addition of a common current circuit between the two circuits. The 1-element type is used on single-phase circuits, and the other types are used on polyphase circuits.
2. ½ percent accuracy, small general-purpose portable meters, moderate pricing, 1 or 2 elements.
3. ¼ percent accuracy, some laboratory types ⅒ percent accuracy; high-priced, restricted to one element.

Watt Transducers. These are devices which produce a dc output in direct proportion to input wattage. Readout can be an analog or digital meter or computing equipment. There are no moving parts, circuits being solid-state, Hall effect, or thermocouple.

Advantages compared to analog dynamometer wattmeters are:

1. A strong dc signal to drive readout devices.
2. Good accuracy in the ±1 to ±¼ percent range at moderate cost.
3. Available in 1, 2, 2½, and 3-element types (also var meters).
4. Several outputs can be added and sent via telemetry.
5. Some transducers are powered from test circuit but will only function over a narrow voltage span.
6. Protection against lightning strikes on power lines is available.

Disadvantages compared to analog dynamometer wattmeters:

1. Some form of readout is required.
2. Certain units have very narrow frequency span; may not be accurate on chopped waveforms.
3. Most are restricted to ac operation only.
4. 120-V operating power may be needed.
5. Require fixed installation.

Watthour Meters*

Operating Principles

The great majority of ac watthour meters in present-day usage are of the induction type. An induction watthour meter has a voltage measuring component which consists of many turns of small wire wound on a laminated member. A second current-sensing component has few turns of large-size wire wound on a laminated steel member. The flux due to these two components operates on an aluminum disk to produce a torque proportional to power. The disk rotates at a speed proportional to watts, and the number of revolutions of the disk is proportional to energy. Revolutions are counted by a geared register

*This material was written by Dale Becker, Sangamo Energy Management Division of Sangamo-Weston, Inc., Newark, New Jersey.

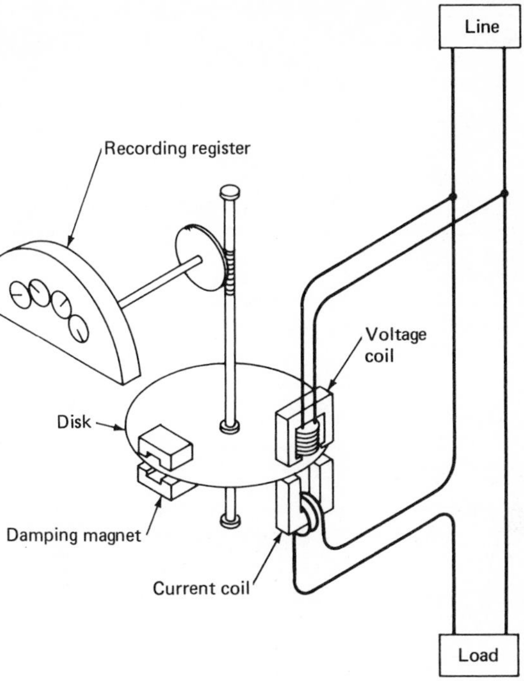

Figure 7-7 Operating principle of the watthour meter.

and displayed as watthours. Figure 7-7 shows the relative arrangement of voltage coil, current coil, disk, magnet, and recording register in a typical single-phase meter.

Polyphase meters employ two or more single-phase meters either operating on separate disks connected by a common shaft or on the same disk.

Meter Types

Meters can be classified by application, current-carrying capacity, and construction details. The broad classification as to application and current-carrying capacity is the so-called "transformer-rated" types as opposed to the self-contained types. Transformer-rated meters are provided as 2½- or 5-A nominally rated meters. Self-contained meters may be nominally rated at 15, 30, or 50 A. The maximum current rating is usually 400 or 667 percent of nominal for modern meters. The transformer type is used in conjunction with instrument transformers external to the meter, and the self-contained type permits current to pass directly through the meter. Such meters are made for single-phase, three-phase, three-wire, wye, and delta networks.

Another classification involves actual construction. For practical purposes, no more than six basic constructions will be encountered, and these are shown in Fig. 7-8.

Accuracy

Modern meters are compensated to operate over a wide range of voltage, current, and power factor. They also operate well over a wide range of temperature and resist lightning surges, transient voltage, and harsh environments. The accuracy curves of a typical modern single-phase meter are shown in Fig. 7-9. These curves apply to a perfectly calibrated meter of most recent manufacture. An installed meter may have an accuracy that is better or worse than that shown. Meters do, however, perform well for long periods of time. Frequent inspection and recalibration are not required.

Figure 7-8 Construction types of watthour meters: (*a*) Four-terminal, single-phase, bottom-connected construction. (*b*) Four-terminal, single-phase, socket-type construction. (*c*) Four-terminal, polyphase, bottom-connected construction. (*d*) Six- or eight-terminal, polyphase, bottom-connected construction. (*e*) Seven-, eight-, or thirteen-terminal, polyphase, socket-type construction. (*f*) Switchboard-type, single-phase or polyphase construction. (*Weston–Sangamo.*)

DIGITAL MULTIMETERS

Digital multimeters (DMM) are instruments that display a measurement as a discrete number rather than as a pointer deflection as on an analog meter. Scale resolution for this type of meter may vary from 3½ digits (up to 1999 in increments of 1 count) all the way up to 6 digits (up to 999.999). Scale ranges (actually the decimal point location) normally range from milli units to mega units. Digital multimeters are normally configured to measure ac and dc volts, ac and dc current, and ohms. Options on some units include ratio and temperature measurements. For dc measurement, polarity is automatic and displayed to the left of the numeric readout. Digital multimeters designed for system applications are provided with a digital output capable of driving a printer. Many digital multimeters have a "hold reading" feature that can be operated remotely by means of a special hold probe. Thus one can use hands (and eyes) to make a measurement in a difficult area and then read the frozen value when convenient.

Specifications

Sensitivity and Resolution

Sensitivity and resolution are two voltmeter parameters that are often confused. They are related but define different capabilities. *Resolution* is a pure number. Resolution of a DMM is the ratio of the maximum number of counts that can be displayed to the least number of counts. For example, a 5½-digit DMM with 20 percent overranging can display

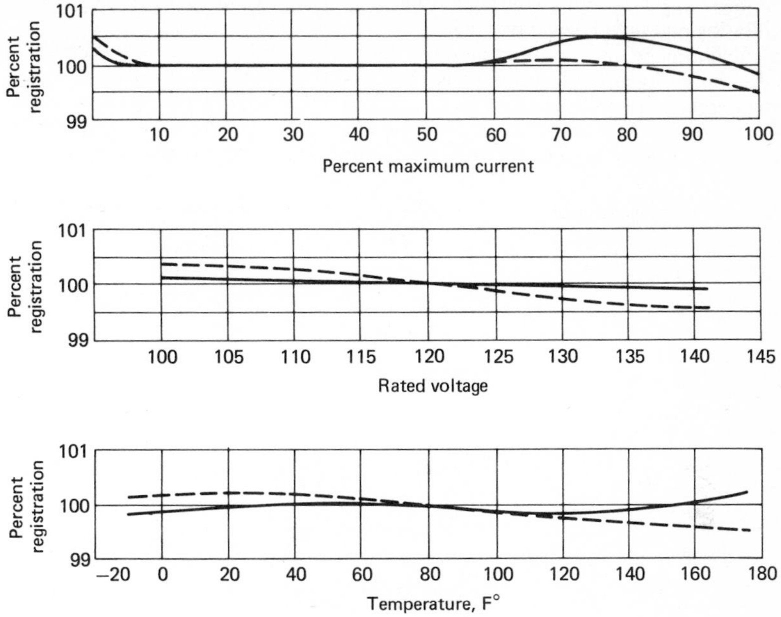

Figure 7-9 Accuracy curves for a watthour meter.

120,000 counts. Its resolution is then 120,000 to 1. Generally, overranging is ignored and the full-scale resolution of a 5½ digit DMM is 100,000 to 1, or 0.001 percent. *Sensitivity* is the ability of a DMM to respond to small voltage changes. For example, a 5½-digit DMM with a 100-mV-FS range has a 1-μV sensitivity. The least-significant digit is 1 μV.

Accuracy

Accuracy is expressed as a percent of reading plus a percent of range (or full scale). The percent of range may alternately be expressed as $\pm X$ digits or $\pm X$ counts.

The accuracy of a DMM may be considerably less than the resolution. For example, a 4½-digit DMM having a resolution of $\pm$ 0.05 to $\pm$ 0.08 percent is sufficient. If the DMM has good linearity, short-term stability, and low noise, it is well-suited to most applications. In short, if the DMM can make repeatable measurements, its absolute accuracy need not match its resolution.

To be meaningful, accuracy must be stated along with the conditions under which it will hold. These conditions should include time, temperature, line variations, and relative humidity. These conditions should be realistic relative to the DMM's intended application. For example, most manufacturers specify a temperature range of 23 $\pm$ 5°C which covers the majority of environments. Other DMMs are specified for $\pm$1°C which is fine in a lab but hardly suitable for on-site testing or production use. Time indicates calibration cycle. DMMs are specified to hold their accuracy for 90 days, 6 months, or 1 year. Many times, several accuracy specifications are included for various times.

Noise Rejection

There are two types of noise which may affect accuracy and sensitivity of a DMM: normal mode and common mode. Normal-mode noise enters the DMM with the signal and is superimposed on it. Filtering is the simplest way to cut down on noise but it slows measurement speed. Integration "calculates" noise out of the measurement by looking at the input signal over a period of time equal to the period of expected noise. Filtering is advantageous for rejecting line-related noise. Figure 7-10 shows typical noise rejection for filtering and integrating methods.

Common-mode noise appears between the DMM's input terminals and ground. It is usually caused by grounding differences between the DMM and the device being measured. Errors caused by common-mode noise may be reduced by a passive technique called *guarding*. Guarding shunts the noise to ground away from input terminals. By proper connection of the guard (Fig. 7-11), a remarkable improvement can be seen in a

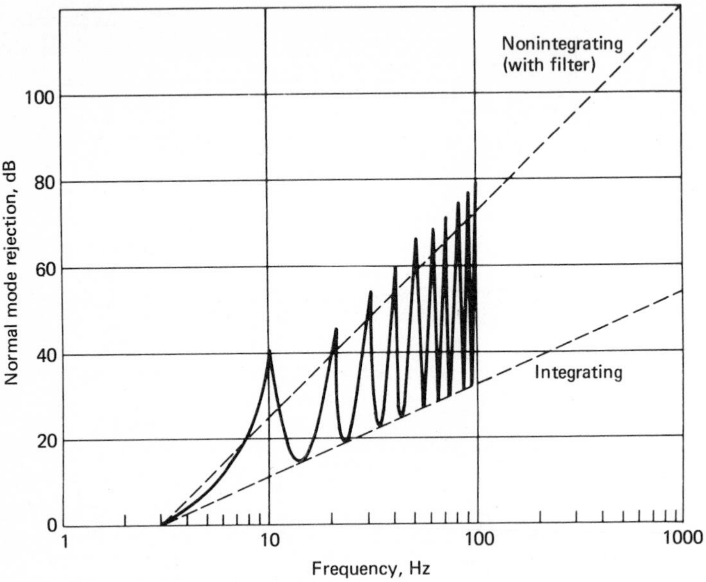

Figure 7-10 Digital meter normal-mode rejection.

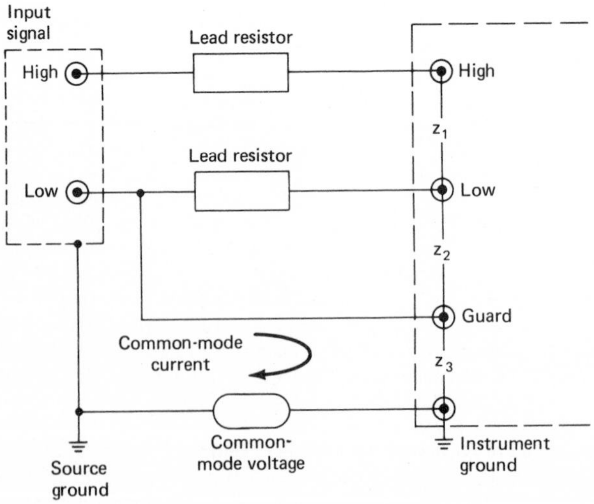

Figure 7-11 Common-mode measurement.

DMM's ability to reject common-mode noise. *Effective* common-mode rejection (CMR) is the specification that usually appears in data sheets. Effective refers to the final reading. Effective CMR is the combined result of "pure" CMR due to guarding plus normal-mode rejection by the instrument.

Measurement Capabilities

DMMs are generally designed to have five dc voltage ranges, five ac voltage ranges, and four or five ac/dc current ranges. The ac/dc voltage ranges generally extend from 200 mV to 100 V and have 100-μV sensitivity for 3½-digit units and 10.5-μV sensitivity for 4½-digit units. In reviewing specifications for ac volts, care must be taken to determine the high-voltage range specified for peak value or an average value. The six resistance ranges for digital instruments normally extend from 200 Ω to 20 MΩ and have a sensitivity on the 200-Ω range of 0.1 Ω for 3½ digits and 0.01 Ω for 4½-digit units.

In reviewing specifications for DMMs, a determination must be made as to whether the resistance measurement is made using a two-wire or four-wire technique. Many DMMs offer a feature called "Hi/Lo" ohms. The Lo power ohm ranges will not turn on silicon junctions, hence allowing in-circuit resistance measurements. Hi power ohms supply enough voltage to turn on junction which allows a *diode test*.

Current ranges for DMMs generally go from the 200-μV range to the 2-A range, although some instruments offer 10 A internal as a standard feature. Sensitivity on the 2-mA range is 1 μA for 3½-digit units and 0.1 μA for 4½-digit units.

Accessories

Most DMMs are offered with a line of accessories for making measurements that go beyond the basic multimeter capabilities.

A high-voltage probe is normally offered, extending the range up to 15 or 40 kV depending upon the particular probe. Guaranteed probe accuracy is normally about 1 percent, making the probe a useful tool in TV series for CRT anode voltage measurements.

A clamp-on current transformer probe can extend the ac measurement capability of the basic DMM up to hundreds of amperes. The clamp-on transformer design allows accurate measurements to be made without breaking the circuit under test.

A variety of radio frequency (RF) probes are offered for converting the DMM into a wide-range RF voltage meter. RF probes are available with bandwidths from 50 kHz to 500 MHz. This type of probe is available for signal tracing and checking levels in RF circuits.

A variety of plug-in current shunts are normally offered extending the DMM capability up to 10/20 A.

Temperature probes are available that convert DMMs into an accurate thermometer with the range of temperature from −60 to +300°F (−50 to 150°C). These probes are normally designed with a probe tip having a low thermal mass to provide fast response.

Surface measurements are particularly applicable since no surface presentation is required as with thermocouples.

BIBLIOGRAPHY

Handbook of Electrical Measurements, Instruments Publishing Company, Inc., Pittsburgh, 1965.

Harris, Forest K.: *Electrical Measurements*, Wiley, New York, 1952.

"Requirements for Electrical Analog Indicating Instruments," ANSI C39.1-1972, American National Standards Institute, New York.

Ross, Edward A.: "Digital Panel Meters," *Digital Design*, July 1977.

"Safety Requirements for Electrical and Electronic Measuring and Controlling Instrumentation," ANSI C39.5-1974, American National Standards Institute, New York.

section 4

In-Plant Prime Power Generation and Cogeneration

chapter 4-1

Applied Thermodynamics*

by

P. Eric Ralston
Manager, Manufacturing Services
Industrial and Marine Division
Babcock & Wilcox Co.
Wilmington, North Carolina

GLOSSARY

Thermodynamics The science concerned with conservation of energy and the rules governing energy changes into transient forms, namely, heat and work.

Energy The capacity to do work.

Heat Q [British thermal units per pound (joules per kilogram)] is a form of energy that is transferred across a boundary because of a temperature difference.

*A portion of this material was adapted by permission from *Steam/It's Generation and Use,* published by Babcock & Wilcox Co. in 1980.

Work *W* [British thermal units per pound (joules per kilogram)] is a force acting through a distance in the direction of the force.

System In the case of expansion work, a system is a fluid capable of expansion or contraction under the influence of pressure changes, temperature changes, chemical reactions, or all three.

Process This defines how these changes are constrained to take place.

Cycle A system which passes through a sequential arrangement of processes, returning to its original state. A thermodynamic power cycle includes a number of processes for converting heat energy into mechanical or electrical work.

Saturation temperature [degrees Fahrenheit (degrees Celsius)] or the boiling point of a liquid is the temperature at which its vapor pressure is equal to the total pressure above its free surface. Steam at saturation is known as saturated steam.

Quality (x, percent) The ratio of the mass of vapor to the total mass of liquid and vapor.

Superheated steam This results when heat is added to saturated steam out of contact with the liquid.

Specific volume v [cubic feet per pound (meter^3 per kilogram)] of a substance is defined as the volume per unit mass, the reciprocal of density.

Internal energy u [British thermal units per pound (joules per kilogram)] is energy stored within a substance associated with molecular and atomic motions and forces and accounts for all forms of energy other than the kinetic and potential energies of the collective masses of the molecules.

Specific enthalpy h [British thermal units per pound (joules per kilogram)] is a thermodynamic property defined as the sum of the internal energy u and the product of the pressure P and specific volume v:

$$h = u + Pv \qquad \text{Btu/lb (J/kg)}$$

Specific entropy s [British thermal units per pound per degree Rankine (joules per kilogram per kelvin)] is an abstract concept and a thermodynamic property defined as the quotient of a quantity of heat [British thermal units per pound (joules per kilogram)] divided by its absolute temperature [degrees Rankine (kelvins)]. Any thermodynamic change which takes place results in a net entropy increase when both the system and its surroundings are considered.

Adiabatic process A process in which there is no heat transfer between a system and its surroundings.

Isentropic (ideal) process A process in which entropy remains constant.

INTRODUCTION

Thermodynamics is the science concerned with the conservation of energy and especially with the conversion of heat into work.

Applications of this science are governed by two basic principles, called the first and second laws, which are firmly established in the engineering field. These laws originally evolved from the development of the steam engine and its demonstration of the conversion of heat to mechanical work.

The terms heat, work, and energy have little practical significance unless the concepts of systems, processes, and cycles are included.

Steam may be considered a thermodynamic system and, because of its availability and advantageous characteristics, has inevitably become a favored system in power generation and heat transfer.

Steam generated in power boilers is used in prime movers for the development of mechanical work and is a source of heat for various processes and for a combination of these purposes.

PROPERTIES OF WATER AND STEAM

Reliable information on steam properties is important in the design of steam generating equipment and cycle application, for both power and heat transfer.

Data selected from the ASME Steam Tables are given in Tables 1-1 to 1-3.

The first two columns of Tables 1-1 and 1-2 define the pressure-temperature correspondence for saturation, the equilibrium between liquid and vapor phases.

The steam tables also include the intensive properties specific volume, specific enthalpy, and specific entropy for specified temperatures, pressures, and states (liquid or gas). Intensive properties are those which are independent of mass. They are also independent of the type of process or any past history. These are the state and thermodynamic properties required for numerical solutions to design and performance problems involving steam for heat transfer and power.

If either the temperature or pressure is known, the enthalpy, entropy, and specific volume for saturated water (h_f, s_f, v_f) and saturated steam (h_g, s_g, v_g) can be found. The increases in these properties during evaporation are also given (h_{fg}, s_{fg}, v_{fg}.)

Knowing the quality (x) of wet steam, the properties can be computed from data from the steam tables using the following relationship illustrated for enthalpy:

$$h = h_f + xh_{fg}$$

Properties of superheated steam and compressed water are tabulated in Table 1-3. If the temperature and pressure are known, the value of the properties can be read from the table.

Data on steam are often illustrated in chart form such as the Mollier chart or enthalpy-entropy relationship (Fig. 1-1, page 4-14). This chart contains lines of constant temperature, constant pressure, and quality of steam. Enthalpy and entropy for a specific steam condition are found at the intersection of the appropriate lines. The properties of superheated steam are found at the intersection of the appropriate pressure and temperature lines, saturated steam at the intersection of the pressure and saturated steam lines, and wet steam at the intersection of the appropriate pressure and quality lines.

THE FIRST LAW OF THERMODYNAMICS

According to the law of the conservation of energy, or the first law of thermodynamics, energy can be neither created nor destroyed. This principle means that an exact balance between input and output of energy must prevail whenever energy is put into or removed from a system. There can be no excess or deficiency. Within the system, or device, energy may be converted from one form to another, such as heat to work, but there must be complete conservation of the aggregate.

Expressed as an equation, the restricted energy conservation law is

$$E_2 - E_1 + E(t) = Q - W \tag{1}$$

$E_2 - E_1 = \Delta E$ is the change in stored energy at boundary states 1 and 2 of the system, and $E(t)$ accounts for energy changes due to unsteady-state performance [for steady-state systems $E(t) = 0$]. Q is heat added to, and W is work done by the system. If the system is open in the sense that mass enters and leaves, the term ΔE reflects stored energy entering and leaving with the mass.

Problems in the power field are generally concerned with open systems (Fig. 1-2, page 4-15). For open systems, an alternative form of Eq. (1) breaks down the ΔE terms into internally and externally stored energy entering and leaving the process on a unit mass basis. This alternative of Eq. (1) is sometimes called the general energy equation, and for the flowing fluid it is

$$\Delta u + \Delta(Pv) + \Delta\frac{V^2}{2g_c} + \Delta Z = Q - W \tag{2}$$

TABLE 1-1 Properties of Saturated Steam and Saturated Water (Temperature)*

Temp °F	Press. psia	Volume, ft³/lb			Enthalpy, Btu/lb			Entropy, Btu/lb, F			Temp °F
		Water v_f	Evap v_{fg}	Steam v_g	Water h_f	Evap h_{fg}	Steam h_g	Water s_f	Evap s_{fg}	Steam s_g	
32	0.08859	0.01602	3305	3305	−0.02	1075.5	1075.5	0.0000	2.1873	2.1873	**32**
35	0.09991	0.01602	2948	2948	3.00	1073.8	1076.8	0.0061	2.1706	2.1767	**35**
40	0.12163	0.01602	2446	2446	8.03	1071.0	1079.0	0.0162	2.1432	2.1594	**40**
45	0.14744	0.01602	2037.7	2037.8	13.04	1068.1	1081.2	0.0262	2.1164	2.1426	**45**
50	0.17796	0.01602	1704.8	1704.8	18.05	1065.3	1083.4	0.0361	2.0901	2.1262	**50**
60	0.2561	0.01603	1207.6	1207.6	28.06	1059.7	1087.7	0.0555	2.0391	2.0946	**60**
70	0.3629	0.01605	868.3	868.4	38.05	1054.0	1092.1	0.0745	1.9900	2.0645	**70**
80	0.5068	0.01607	633.3	633.3	48.04	1048.4	1096.4	0.0932	1.9426	2.0359	**80**
90	0.6981	0.01610	468.1	468.1	58.02	1042.7	1100.8	0.1115	1.8970	2.0086	**90**
100	0.9492	0.01613	350.4	350.4	68.00	1037.1	1105.1	0.1295	1.8530	1.9825	**100**
110	1.2750	0.01617	265.4	265.4	77.98	1031.4	1109.3	0.1472	1.8105	1.9577	**110**
120	1.6927	0.01620	203.25	203.26	87.97	1025.6	1113.6	0.1646	1.7693	1.9339	**120**
130	2.2230	0.01625	157.32	157.33	97.96	1019.8	1117.8	0.1817	1.7295	1.9112	**130**
140	2.8892	0.01629	122.98	123.00	107.95	1014.0	1122.0	0.1985	1.6910	1.8895	**140**
150	3.718	0.01634	97.05	97.07	117.95	1008.2	1126.1	0.2150	1.6536	1.8686	**150**
160	4.741	0.01640	77.27	77.29	127.96	1002.2	1130.2	0.2313	1.6174	1.8487	**160**
170	5.993	0.01645	62.04	62.06	137.97	996.2	1134.2	0.2473	1.5822	1.8295	**170**
180	7.511	0.01651	50.21	50.22	148.00	990.2	1138.2	0.2631	1.5480	1.8111	**180**
190	9.340	0.01657	40.94	40.96	158.04	984.1	1142.1	0.2787	1.5148	1.7934	**190**
200	11.526	0.01664	33.62	33.64	168.09	977.9	1146.0	0.2940	1.4824	1.7764	**200**
210	14.123	0.01671	27.80	27.82	178.15	971.6	1149.7	0.3091	1.4509	1.7600	**210**
212	14.696	0.01672	26.78	26.80	180.17	970.3	1150.5	0.3121	1.4447	1.7568	**212**
220	17.186	0.01678	23.13	23.15	188.23	965.2	1153.4	0.3241	1.4201	1.7442	**220**
230	20.779	0.01685	19.364	19.381	198.33	958.7	1157.1	0.3388	1.3902	1.7290	**230**
240	24.968	0.01693	16.304	16.321	208.45	952.1	1160.6	0.3533	1.3609	1.7142	**240**
250	29.825	0.01701	13.802	13.819	218.59	945.4	1164.0	0.3677	1.3323	1.7000	**250**

260	35.427	0.01709	11.745	11.762	228.76	938.6	1167.4	0.3819	1.3043	1.6862	**260**
270	41.856	0.01718	10.042	10.060	238.95	931.7	1170.6	0.3960	1.2769	1.6729	**270**
280	49.200	0.01726	8.627	8.644	249.17	924.6	1173.8	0.4098	1.2501	1.6599	**280**
290	57.550	0.01736	7.443	7.460	259.4	917.4	1176.8	0.4236	1.2238	1.6473	**290**
300	67.005	0.01745	6.448	6.466	269.7	910.0	1179.7	0.4372	1.1979	1.6351	**300**
310	77.67	0.01755	5.609	5.626	280.0	902.5	1182.5	0.4506	1.1726	1.6232	**310**
320	89.64	0.01766	4.896	4.914	290.4	894.8	1185.2	0.4640	1.1477	1.6116	**320**
340	117.99	0.01787	3.770	3.788	311.3	878.8	1190.1	0.4902	1.0990	1.5892	**340**
360	153.01	0.01811	2.939	2.957	332.3	862.1	1194.4	0.5161	1.0517	1.5678	**360**
380	195.73	0.01836	2.317	2.335	353.6	844.5	1198.0	0.5416	1.0057	1.5473	**380**
400	247.26	0.01864	1.8444	1.8630	375.1	825.9	1201.0	0.5667	0.9607	1.5274	**400**
420	308.78	0.01894	1.4808	1.4997	396.9	806.2	1203.1	0.5915	0.9165	1.5080	**420**
440	381.54	0.01926	1.1976	1.2169	419.0	785.4	1204.4	0.6161	0.8729	1.4890	**440**
460	466.9	0.0196	0.9746	0.9942	441.5	763.2	1204.8	0.6405	0.8299	1.4704	**460**
480	566.2	0.0200	0.7972	0.8172	464.5	739.6	1204.1	0.6648	0.7871	1.4518	**480**
500	680.9	0.0204	0.6545	0.6749	487.9	714.3	1202.2	0.6890	0.7443	1.4333	**500**
520	812.5	0.0209	0.5386	0.5596	512.0	687.0	1199.0	0.7133	0.7013	1.4146	**520**
540	962.8	0.0215	0.4437	0.4651	536.8	657.5	1194.3	0.7378	0.6577	1.3954	**540**
560	1133.4	0.0221	0.3651	0.3871	562.4	625.3	1187.7	0.7625	0.6132	1.3757	**560**
580	1326.2	0.0228	0.2994	0.3222	589.1	589.9	1179.0	0.7876	0.5673	1.3550	**580**
600	1543.2	0.0236	0.2438	0.2675	617.1	550.6	1167.7	0.8134	0.5196	1.3330	**600**
620	1786.9	0.0247	0.1962	0.2208	646.9	506.3	1153.2	0.8403	0.4689	1.3092	**620**
640	2059.9	0.0260	0.1543	0.1802	679.1	454.6	1133.7	0.8686	0.4134	1.2821	**640**
660	2365.7	0.0277	0.1166	0.1443	714.9	392.1	1107.0	0.8995	0.3502	1.2498	**660**
680	2708.6	0.0304	0.0808	0.1112	758.5	310.1	1068.5	0.9365	0.2720	1.2086	**680**
700	3094.3	0.0366	0.0386	0.0752	822.4	172.7	995.2	0.9901	0.1490	1.1390	**700**
705.5	3208.2	0.0508	0	0.0508	906.0	0	906.0	1.0612	0	1.0612	**705.5**

*Abstracted from *Thermodynamic and Transport Properties of Steam*, Copyright 1967, The American Society of Mechanical Engineers.

TABLE 1-2 Properties of Saturated Steam and Saturated Water (Pressure)

Press. psia	Temp °F	Volume, ft³/lb			Enthalpy, Btu/lb			Entropy, Btu/lb, F			Energy, Btu/lb		Press. psia
		Water	Evap	Steam	Water	Evap	Steam	Water	Evap	Steam	Water	Steam	
		v_f	v_{fg}	v_g	h_f	h_{fg}	h_g	s_f	s_{fg}	s_g	u_f	u_g	
0.0886	32.018	0.01602	3302.4	3302.4	0.00	1075.5	1075.5	0	2.1872	2.1872	0	1021.3	**0.0886**
0.10	35.023	0.01602	2945.5	2945.5	3.03	1073.8	1076.8	0.0061	2.1705	2.1766	3.03	1022.3	**0.10**
0.15	45.453	0.01602	2004.7	2004.7	13.50	1067.9	1081.4	0.0271	2.1140	2.1411	13.50	1025.7	**0.15**
0.20	53.160	0.01603	1526.3	1526.3	21.22	1063.5	1084.7	0.0422	2.0738	2.1160	21.22	1028.3	**0.20**
0.30	64.484	0.01604	1039.7	1039.7	32.54	1057.1	1089.7	0.0641	2.0168	2.0809	32.54	1032.0	**0.30**
0.40	72.869	0.01606	792.0	792.1	40.92	1052.4	1093.3	0.0799	1.9762	2.0562	40.92	1034.7	**0.40**
0.5	79.586	0.01607	641.5	641.5	47.62	1048.6	1096.3	0.0925	1.9446	2.0370	47.62	1036.9	**0.5**
0.6	85.218	0.01609	540.0	540.1	53.25	1045.5	1098.7	0.1028	1.9186	2.0215	53.24	1038.7	**0.6**
0.7	90.09	0.01610	466.93	466.94	58.10	1042.7	1100.8	0.3	1.8966	2.0083	58.10	1040.3	**0.7**
0.8	94.38	0.01611	411.67	411.69	62.39	1040.3	1102.6	0.1117	1.8775	1.9970	62.39	1041.7	**0.8**
0.9	98.24	0.01612	368.41	368.43	66.24	1038.1	1104.3	0.1264	1.8606	1.9870	66.24	1042.9	**0.9**
1.0	101.74	0.01614	333.59	333.60	69.73	1036.1	1105.8	0.1326	1.8455	1.9781	69.73	1044.1	**1.0**
2.0	126.07	0.01623	173.74	173.76	94.03	1022.1	1116.2	0.1750	1.7450	1.9200	94.03	1051.8	**2.0**
3.0	141.47	0.01630	118.71	118.73	109.42	1013.2	1122.6	0.2009	1.6854	1.8864	109.41	1056.7	**3.0**
4.0	152.96	0.01636	90.63	90.64	120.92	1006.4	1127.3	0.2199	1.6428	1.8626	120.90	1060.2	**4.0**
5.0	162.24	0.01641	73.515	73.53	130.20	1000.9	1131.1	0.2349	1.6094	1.8443	130.18	1063.1	**5.0**
6.0	170.05	0.01645	61.967	61.98	138.03	996.2	1134.2	0.2474	1.5820	1.8294	138.01	1065.4	**6.0**
7.0	176.84	0.01649	53.634	53.65	144.83	992.1	1136.9	0.2581	1.5587	1.8168	144.81	1067.4	**7.0**
8.0	182.86	0.01653	47.328	47.35	150.87	988.5	1139.3	0.2676	1.5384	1.8060	150.84	1069.2	**8.0**
9.0	188.27	0.01656	42.385	42.40	156.30	985.1	1141.4	0.2760	1.5204	1.7964	156.28	1070.8	**9.0**
10	193.21	0.01659	38.404	38.42	161.26	982.1	1143.3	0.2836	1.5043	1.7879	161.23	1072.3	**10**
14.696	212.00	0.01672	26.782	26.80	180.17	970.3	1150.5	0.3121	1.4447	1.7568	180.12	1077.6	**14.696**
15	213.03	0.01673	26.274	26.29	181.21	969.7	1150.9	0.3137	1.4415	1.7552	181.16	1077.9	**15**
20	227.96	0.01683	20.070	20.087	196.27	960.1	1156.3	0.3358	1.3962	1.7320	196.21	1082.0	**20**
30	250.34	0.01701	13.7266	13.744	218.9	945.2	1164.1	0.3682	1.3313	1.6995	218.8	1087.9	**30**
40	267.25	0.01715	10.4794	10.497	236.1	933.6	1169.8	0.3921	1.2844	1.6765	236.0	1092.1	**40**
50	281.02	0.01727	8.4967	8.514	250.2	923.9	1174.1	0.4112	1.2474	1.6586	250.1	1095.3	**50**

60	292.71	0.01738	7.1562	7.174	262.2	915.4	1177.6	0.4273	1.2167	1.6440	262.0	1098.0	**60**
70	302.93	0.01748	6.1875	6.205	272.7	907.8	1180.6	0.4411	1.1905	1.6316	272.5	1100.2	**70**
80	312.04	0.01757	5.4536	5.471	282.1	900.9	1183.1	0.4534	1.1675	1.6208	281.9	1102.1	**80**
90	320.28	0.01766	4.8777	4.895	290.7	894.6	1185.3	0.4643	1.1470	1.6113	290.4	1103.7	**90**
100	327.82	0.01774	4.4133	4.431	298.5	888.6	1187.2	0.4743	1.1284	1.6027	298.2	1105.2	**100**
120	341.27	0.01789	3.7097	3.728	312.6	877.8	1190.4	0.4919	1.0960	1.5879	312.2	1107.6	**120**
140	353.04	0.01803	3.2010	3.219	325.0	868.0	1193.0	0.5071	1.0681	1.5752	324.5	1109.6	**140**
160	363.55	0.01815	2.8155	2.834	336.1	859.0	1195.1	0.5206	1.0435	1.5641	335.5	1111.2	**160**
180	373.08	0.01827	2.5129	2.531	346.2	850.7	1196.9	0.5328	1.0215	1.5543	345.6	1112.5	**180**
200	381.80	0.01839	2.2689	2.287	355.5	842.8	1198.3	0.5438	1.0016	1.5454	354.8	1113.7	**200**
250	400.97	0.01865	1.8245	1.8432	376.1	825.0	1201.1	0.5679	0.9585	1.5264	375.3	1115.8	**250**
300	417.35	0.01889	1.5238	1.5427	394.0	808.9	1202.9	0.5882	0.9223	1.5105	392.9	1117.2	**300**
350	431.73	0.01913	1.3064	1.3255	409.8	794.2	1204.0	0.6059	0.8909	1.4968	408.6	1118.1	**350**
400	444.60	0.0193	1.14162	1.1610	424.2	780.4	1204.6	0.6217	0.8630	1.4847	422.7	1118.7	**400**
450	456.28	0.0195	1.01224	1.0318	437.3	767.5	1204.8	0.6360	0.8378	1.4738	435.7	1118.9	**450**
500	467.01	0.0198	0.90787	0.9276	449.5	755.1	1204.7	0.6490	0.8148	1.4639	447.7	1118.8	**500**
550	476.94	0.0199	0.82183	0.8418	460.9	743.3	1204.3	0.6611	0.7936	1.4547	458.9	1118.6	**550**
600	486.20	0.0201	0.74962	0.7698	471.7	732.0	1203.7	0.6723	0.7738	1.4461	469.5	1118.2	**600**
700	503.08	0.0205	0.63505	0.6556	491.6	710.2	1201.8	0.6928	0.7377	1.4304	488.9	1116.9	**700**
800	518.21	0.0209	0.54809	0.5690	509.8	689.6	1199.4	0.7111	0.7051	1.4163	506.7	1115.2	**800**
900	531.95	0.0212	0.47968	0.5009	526.7	669.7	1196.4	0.7279	0.6753	1.4032	523.2	1113.0	**900**
1000	544.58	0.0216	0.42436	0.4460	542.6	650.4	1192.9	0.7434	0.6476	1.3910	538.6	1110.4	**1000**
1100	556.28	0.0220	0.37863	0.4006	557.5	631.5	1189.1	0.7578	0.6216	1.3794	553.1	1107.5	**1100**
1200	567.19	0.0223	0.34013	0.3625	571.9	613.0	1184.8	0.7714	0.5969	1.3683	566.9	1104.3	**1200**
1300	577.42	0.0227	0.30722	0.3299	585.6	594.6	1180.2	0.7843	0.5733	1.3577	580.1	1100.9	**1300**
1400	587.07	0.0231	0.27871	0.3018	598.8	576.5	1175.3	0.7966	0.5507	1.3474	592.9	1097.1	**1400**
1500	596.20	0.0235	0.25372	0.2772	611.7	558.4	1170.1	0.8085	0.5288	1.3373	605.2	1093.1	**1500**
2000	635.80	0.0257	0.16266	0.1883	672.1	466.2	1138.3	0.8625	0.4256	1.2881	662.6	1068.6	**2000**
2500	668.11	0.0286	0.10209	0.1307	731.7	361.6	1093.3	0.9139	0.3206	1.2345	718.5	1032.9	**2500**
3000	695.33	0.0343	0.05073	0.0850	801.8	218.4	1020.3	0.9728	0.1891	1.1619	782.8	973.1	**3000**
3208.2	705.47	0.0508	0	0.0508	906.0	0	906.0	1.0612	0	1.0612	875.9	875.9	**3208.2**

TABLE 1-3 Properties of Superheated Steam and Compressed Water (Temperature and Pressure)

Abs press. lb/sq in (sat. temp)		Temperature, °F 100	200	300	400	500	600	700	800	900	1000	1100	1200	1300	1400	1500
1	v	0.0161	392.5	452.3	511.9	571.5	631.1	690.7								
	h	68.00	1150.2	1195.7	1241.8	1288.6	1336.1	1384.5								
(101.74)	s	0.1295	2.0509	2.1152	2.1722	2.2237	2.2708	2.3144								
5	v	0.0161	78.14	90.24	102.24	114.21	126.15	138.08	150.01	161.94	173.86	185.78	197.70	209.62	221.53	233.45
	h	68.01	1148.6	1194.8	1241.3	1288.2	1335.9	1384.3	1433.6	1483.7	1534.7	1586.7	1639.6	1693.3	1748.0	1803.5
(162.24)	s	0.1295	1.8716	1.9369	1.9943	2.0460	2.0932	2.1369	2.1776	2.2159	2.2521	2.2866	2.3194	2.3509	2.3811	2.4101
10	v	0.0161	38.84	44.98	51.03	57.04	63.03	69.00	74.98	80.94	86.91	92.87	98.84	104.80	110.76	116.72
	h	68.02	1146.6	1193.7	1240.6	1287.8	1335.5	1384.0	1433.4	1483.5	1534.6	1586.6	1639.5	1693.3	1747.9	1803.4
(193.21)	s	0.1295	1.7928	1.8593	1.9173	1.9692	2.0166	2.0603	2.1011	2.1394	2.1757	2.2101	2.2430	2.2744	2.3046	2.3337
15	v	0.0161	0.0166	29.899	33.963	37.985	41.986	45.978	49.964	53.946	57.926	61.905	65.882	69.858	73.833	77.807
	h	68.04	168.09	1192.5	1239.9	1287.3	1335.2	1383.8	1433.2	1483.4	1534.5	1586.5	1639.4	1693.2	1747.8	1803.4
(213.03)	s	0.1295	0.2940	1.8134	1.8720	1.9242	1.9717	2.0155	2.0563	2.0946	2.1309	2.1653	2.1982	2.2297	2.2599	2.2890
20	v	0.0161	0.0166	22.356	25.428	28.457	31.466	34.465	37.458	40.447	43.435	46.420	49.405	52.388	55.370	58.352
	h	68.05	168.11	1191.4	1239.2	1286.9	1334.9	1383.5	1432.9	1483.2	1534.3	1586.3	1639.3	1693.1	1747.8	1803.3
(227.96)	s	0.1295	0.2940	1.7805	1.8397	1.8921	1.9397	1.9836	2.0244	2.0628	2.0991	2.1336	2.1665	2.1979	2.2282	2.2572
40	v	0.0161	0.0166	11.036	12.624	14.165	15.685	17.195	18.699	20.199	21.697	23.194	24.689	26.183	27.676	29.168
	h	68.10	168.15	1186.6	1236.4	1285.0	1333.6	1382.5	1432.1	1482.5	1533.7	1585.8	1638.8	1992.7	1747.5	1803.0
(267.25)	s	0.1295	0.2940	1.6992	1.7608	1.8143	1.8624	1.9065	1.9476	1.9860	2.0224	2.0569	2.0899	2.1224	2.1516	2.1807
60	v	0.0161	0.0166	7.257	8.354	9.400	10.425	11.438	12.446	13.450	14.452	15.452	16.450	17.448	18.445	19.441
	h	68.15	168.20	1181.6	1233.5	1283.2	1332.3	1381.5	1431.3	1481.8	1533.2	1585.3	1638.4	1692.4	1747.1	1802.8
(292.71)	s	0.1295	0.2939	1.6492	1.7134	1.7681	1.8168	1.8612	1.9024	1.9410	1.9774	2.0120	2.0450	2.0765	2.1068	2.1359
80	v	0.0161	0.0166	0.0175	6.218	7.018	7.794	8.560	9.319	10.075	10.829	11.581	12.331	13.081	13.829	14.577
	h	68.21	168.24	269.74	1230.5	1281.3	1330.9	1380.5	1430.5	1481.1	1532.6	1584.9	1638.0	1692.0	1746.8	1802.5
(312.04)	s	0.1295	0.2939	0.4371	1.6790	1.7349	1.7842	1.8289	1.8702	1.9089	1.9454	1.9800	2.0131	2.0446	2.0750	2.1041
100	v	0.0161	0.0166	0.0175	4.935	5.588	6.216	6.833	7.443	8.050	8.655	9.258	9.860	10.460	11.060	11.659
	h	68.26	168.29	269.77	1227.4	1279.3	1329.6	1379.5	1429.7	1480.4	1532.0	1584.4	1637.6	1691.6	1746.5	1802.2
(327.82)	s	0.1295	0.2939	0.4371	1.6516	1.7088	1.7586	1.8036	1.8451	1.8839	1.9205	1.9552	1.9883	2.0199	2.0502	2.0794

	v	0.0161	0.0166	0.0175	4.0786	4.6341	5.1637	5.6831	6.1928	6.7006	7.2060	7.7096	8.2119	8.7130	9.2134	9.7130
120	h	68.31	168.33	269.81	1224.1	1277.4	1328.1	1378.4	1428.8	1479.8	1531.4	1583.9	1637.1	1691.3	1746.2	1802.0
(341.27)	s	0.1295	0.2939	0.4371	1.6286	1.6872	1.7376	1.7829	1.8246	1.8635	1.9001	1.9349	1.9680	1.9996	2.0300	2.0592
	v	0.0161	0.0166	0.0175	3.4661	3.9526	4.4119	4.8585	5.2995	5.7364	6.1709	6.6036	7.0349	7.4652	7.8946	8.3233
140	h	68.37	168.38	269.85	1220.8	1275.3	1326.8	1377.4	1428.0	1479.1	1530.8	1583.4	1636.7	1690.9	1745.9	1801.7
(353.04)	s	0.1295	0.2939	0.4370	1.6085	1.6686	1.7196	1.7652	1.8071	1.8461	1.8828	1.9176	1.9508	1.9825	2.0129	2.0421
	v	0.0161	0.0166	0.0175	3.0060	3.4413	3.8480	4.2420	4.6295	5.0132	5.3945	5.7741	6.1522	6.5293	6.9055	7.2811
160	h	68.42	168.42	269.89	1217.4	1273.3	1325.4	1376.4	1427.2	1478.4	1530.3	1582.9	1636.3	1690.5	1745.6	1801.4
(363.55)	s	0.1294	0.2938	0.4370	1.5906	1.6522	1.7039	1.7499	1.7919	1.8310	1.8678	1.9027	1.9359	1.9676	1.9980	2.0273
	v	0.0161	0.0166	0.0174	2.6474	3.0433	3.4093	3.7621	4.1084	4.4505	4.7907	5.1289	5.4657	5.8014	6.1363	6.4704
180	h	68.47	168.47	269.92	1213.8	1271.2	1324.0	1375.3	1426.3	1477.7	1529.7	1582.4	1635.9	1690.2	1745.3	1801.2
(373.08)	s	0.1294	0.2938	0.4370	1.5743	1.6376	1.6900	1.7362	1.7784	1.8176	1.8545	1.8894	1.9227	1.9545	1.9849	2.0142
	v	0.0161	0.0166	0.0174	2.3598	2.7247	3.0583	3.3783	3.6915	4.0008	4.3077	4.6128	4.9165	5.2191	5.5209	5.8219
200	h	68.52	168.51	269.96	1210.1	1269.0	1322.6	1374.3	1425.5	1477.0	1529.1	1581.9	1635.4	1689.8	1745.0	1800.9
(381.80)	s	0.1294	0.2938	0.4369	1.5593	1.6242	1.6776	1.7239	1.7663	1.8057	1.8426	1.8776	1.9109	1.9427	1.9732	2.0025
	v	0.0161	0.0166	0.0174	0.0186	2.1504	2.4662	2.6872	2.9410	3.1909	3.4382	3.6837	3.9278	4.1709	4.4131	4.6546
250	h	68.66	168.63	270.05	375.10	1263.5	1319.0	1371.6	1423.4	1475.3	1527.6	1580.6	1634.4	1688.9	1744.2	1800.2
(400.97)	s	0.1294	0.2937	0.4368	0.5667	1.5951	1.6502	1.6976	1.7405	1.7801	1.8173	1.8524	1.8858	1.9177	1.9482	1.9776
	v	0.0161	0.0166	0.0174	0.0186	1.7665	2.0044	2.2263	2.4407	2.6509	2.8585	3.0643	3.2688	3.4721	3.6746	3.8764
300	h	68.79	168.74	270.14	375.15	1257.7	1315.2	1368.9	1421.3	1473.6	1526.2	1579.4	1633.3	1688.0	1743.4	1799.6
(417.35)	s	0.1294	0.2937	0.4307	0.5665	1.5703	1.6274	1.6758	1.7192	1.7591	1.7964	1.8317	1.8652	1.8972	1.9278	1.9572
	v	0.0161	0.0166	0.0174	0.0186	1.4913	1.7028	1.8970	2.0832	2.2652	2.4445	2.6219	2.7980	2.9730	3.1471	3.3205
350	h	68.92	168.85	270.24	375.21	1251.5	1311.4	1366.2	1419.2	1471.8	1524.7	1578.2	1632.3	1687.1	1742.6	1798.9
(431.73)	s	0.1293	0.2936	0.4367	0.5664	1.5483	1.6077	1.6571	1.7009	1.7411	1.7787	1.8141	1.8477	1.8798	1.9105	1.9400
	v	0.0161	0.0166	0.0174	0.0162	1.2841	1.4763	1.6499	1.8151	1.9759	2.1339	2.2901	2.4450	2.5987	2.7515	2.9037
400	h	69.05	168.97	270.33	375.27	1245.1	1307.4	1363.4	1417.0	1470.1	1523.3	1576.9	1631.2	1686.2	1741.9	1798.2
(444.60)	s	0.1293	0.2935	0.4366	0.5663	1.5282	1.5901	1.6406	1.6850	1.7255	1.7632	1.7988	1.8325	1.8647	1.8955	1.9250
	v	0.0161	0.0166	0.0174	0.0186	0.9919	1.1584	1.3037	1.4397	1.5708	1.6992	1.8256	1.9507	2.0746	2.1977	2.3200
500	h	69.32	169.19	270.51	375.38	1231.2	1299.1	1357.7	1412.7	1466.6	1520.3	1574.4	1629.1	1684.4	1740.3	1796.9
(467.01)	s	0.1292	0.2934	0.4364	0.5660	1.4921	1.5595	1.6123	1.6578	1.6990	1.7371	1.7730	1.8069	1.8393	1.8702	1.8998

TABLE 1-3 Properties of Superheated Steam and Compressed Water (Temperature and Pressure) (*Continued*)

Abs press. lb/sq in (sat. temp)		Temperature, F														
		100	200	300	400	500	600	700	800	900	1000	1100	1200	1300	1400	1500
600	v	0.0161	0.0166	0.0174	0.0186	0.7944	0.9456	1.0726	1.1892	1.3008	1.4093	1.5160	1.6211	1.7252	1.8284	1.9309
	h	69.58	169.42	270.70	375.49	1215.9	1290.3	1351.8	1408.3	1463.0	1517.4	1571.9	1627.0	1682.6	1738.8	1795.6
(486.20)	s	0.1292	0.2933	0.4362	0.5657	1.4590	1.5329	1.5844	1.6351	1.6769	1.7155	1.7517	1.7859	1.8184	1.8494	1.8792
700	v	0.0161	0.0166	0.0174	0.0186	0.0204	0.7928	0.9072	1.0102	1.1078	1.2023	1.2948	1.3858	1.4757	1.5647	1.6530
	h	69.84	169.65	270.89	375.61	487.93	1281.0	1345.6	1403.7	1459.4	1514.4	1569.4	1624.8	1680.7	1737.2	1794.3
(503.08)	s	0.1291	0.2932	0.4360	0.5655	0.6889	1.5090	1.5673	1.6154	1.6580	1.6970	1.7335	1.7679	1.8006	1.8318	1.8617
800	v	0.0161	0.0166	0.0174	0.0186	0.0204	0.6774	0.7828	0.8759	0.9631	1.0470	1.1289	1.2093	1.2885	1.3669	1.4446
	h	70.11	169.88	271.07	375.73	487.88	1271.1	1339.2	1399.1	1455.8	1511.4	1566.9	1622.7	1678.9	1735.0	1792.9
(518.21)	s	0.1290	0.2930	0.4358	0.5652	0.6885	1.4869	1.5484	1.5980	1.6413	1.6807	1.7175	1.7522	1.7851	1.8164	1.8464
900	v	0.0161	0.0166	0.0174	0.0186	0.0204	0.5869	0.6858	0.7713	0.8504	0.9262	0.9998	1.0720	1.1430	1.2131	1.2825
	h	70.37	170.10	271.26	375.84	487.83	1260.6	1332.7	1394.4	1452.2	1508.5	1564.4	1620.6	1677.1	1734.1	1791.6
(531.95)	s	0.1290	0.2929	0.4357	0.5649	0.6881	1.4659	1.5311	1.5822	1.6263	1.6662	1.7033	1.7382	1.7713	1.8028	1.8329
1000	v	0.0161	0.0166	0.0174	0.0186	0.0204	0.5137	0.6080	0.6875	0.7603	0.8295	0.8966	0.9622	1.0266	1.0901	1.1529
	h	70.63	170.33	271.44	375.96	487.79	1249.3	1325.9	1389.6	1448.5	1504.4	1561.9	1618.4	1675.3	1732.5	1790.3
(544.58)	s	0.1289	0.2928	0.4355	0.5647	0.6876	1.4457	1.5149	1.5677	1.6126	1.6530	1.6905	1.7256	1.7589	1.7905	1.8207
1100	v	0.0161	0.0166	0.0174	0.0185	0.0203	0.4531	0.5440	0.6188	0.6865	0.7505	0.8121	0.8723	0.9313	0.9894	1.0468
	h	70.90	170.56	271.63	376.08	487.75	1237.3	1318.8	1384.7	1444.7	1502.4	1559.4	1616.3	1673.5	1731.0	1789.0
(556.28)	s	0.1289	0.2927	0.4353	0.5644	0.6872	1.4259	1.4996	1.5542	1.6000	1.6410	1.6787	1.7141	1.7475	1.7793	1.8097
1200	v	0.0161	0.0166	0.0174	0.0185	0.0203	0.4016	0.4905	0.5615	0.6250	0.6845	0.7418	0.7974	0.8519	0.9055	0.9584
	h	71.16	170.78	271.82	376.20	487.72	1224.2	1311.5	1379.7	1440.9	1499.4	1556.9	1614.2	1671.6	1729.4	1787.6
(567.19)	s	0.1288	0.2926	0.4351	0.5642	0.6868	1.4061	1.4851	1.5415	1.5883	1.6298	1.6679	1.7035	1.7371	1.7691	1.7996
1400	v	0.0161	0.0166	0.0174	0.0185	0.0203	0.3176	0.4059	0.4712	0.5282	0.5809	0.6311	0.6798	0.7272	0.7737	0.8195
	h	71.68	171.24	272.19	376.44	487.65	1194.1	1296.1	1369.3	1433.2	1493.2	1551.8	1609.9	1668.0	1726.3	1785.0
(587.07)	s	0.1287	0.2923	0.4348	0.5636	0.6859	1.3652	1.4575	1.5182	1.5670	1.6096	1.6484	1.6845	1.7185	1.7508	1.7815
1600	v	0.0161	0.0166	0.0173	0.0185	0.0202	0.0236	0.3415	0.4032	0.4555	0.5031	0.5482	0.5915	0.6336	0.6748	0.7153
	h	72.21	171.69	272.57	376.69	487.60	616.77	1279.4	1358.5	1425.2	1486.9	1546.6	1605.6	1664.3	1723.2	1782.3
(604.87)	s	0.1286	0.2921	0.4344	0.5631	0.6851	0.8129	1.4312	1.4968	1.5478	1.5916	1.6312	1.6678	1.7022	1.7344	1.7657

1800 (621.02)	v	0.0160	0.0165	0.0173	0.0185	0.0202	0.0235	0.2906	0.3500	0.3988	0.4426	0.4836	0.5229	0.5609	0.5980	0.6343
	h	72.73	172.15	272.95	376.93	487.56	615.58	1261.1	1347.2	1417.1	1480.6	1541.1	1601.2	1660.7	1720.1	1779.7
	s	0.1284	0.2918	0.4341	0.5626	0.6843	0.8109	1.4054	1.4768	1.5302	1.5753	1.6156	1.6528	1.6876	1.7204	1.7516
2000 (635.80)	v	0.0160	0.0165	0.0173	0.0184	0.0201	0.0233	0.2488	0.3072	0.3534	0.3942	0.4320	0.4680	0.5027	0.5365	0.5695
	h	73.26	172.60	273.32	377.19	487.53	614.48	1240.9	1353.4	1408.7	1474.1	1536.2	1596.9	1657.0	1717.0	1777.1
	s	0.1283	0.2916	0.4337	0.5621	0.6834	0.8091	1.3794	1.4578	1.5138	1.5603	1.6014	1.6391	1.6743	1.7075	1.7389
2500 (668.11)	v	0.0160	0.0165	0.0173	0.0184	0.0200	0.0230	0.1681	0.2293	0.2712	0.3068	0.3390	0.3692	0.3980	0.4259	0.4529
	h	74.57	173.74	274.27	377.82	487.50	612.08	1176.7	1303.4	1386.7	1457.5	1522.9	1585.9	1647.8	1709.2	1770.4
	s	0.1280	0.2910	0.4329	0.5609	0.6815	0.8048	1.3076	1.4129	1.4766	1.5269	1.5703	1.6094	1.6456	1.6796	1.7116
3000 (695.33)	v	0.0160	0.0165	0.0172	0.0183	0.0200	0.0228	0.0982	0.1759	0.2161	0.2484	0.2770	0.3033	0.3282	0.3522	0.3753
	h	75.88	174.88	275.22	378.47	487.52	610.08	1060.5	1267.0	1363.2	1440.2	1509.4	1574.8	1638.5	1701.4	1761.8
	s	0.1277	0.2904	0.4320	0.5597	0.6796	0.8009	1.1966	1.3692	1.4429	1.4976	1.5434	1.5841	1.6214	1.6561	1.6888
3200 (705.08)	v	0.0160	0.0165	0.0172	0.0183	0.0199	0.0227	0.0335	0.1588	0.1987	0.2301	0.2576	0.2827	0.3065	0.3291	0.3510
	h	76.4	175.3	275.6	378.7	487.5	609.4	800.8	1250.9	1353.4	1433.1	1503.8	1570.3	1634.8	1698.3	1761.2
	s	0.1276	0.2902	0.4317	0.5592	0.6788	0.7994	0.9708	1.3515	1.4300	1.4866	1.5335	1.5749	1.6126	1.6477	1.6806
3500	v	0.0160	0.0164	0.0172	0.0183	0.0199	0.0225	0.0307	0.1364	0.1764	0.2066	0.2326	0.2563	0.2784	0.2995	0.3198
	h	77.2	176.0	276.2	379.1	487.6	608.4	779.4	1224.6	1338.2	1422.2	1495.5	1563.3	1629.2	1693.6	1757.2
	s	0.1274	0.2899	0.4312	0.5585	0.6777	0.7973	0.9508	1.3242	1.4112	1.4709	1.5194	1.5618	1.6002	1.6358	1.6691
4000	v	0.0159	0.0164	0.0172	0.0182	0.0198	0.0223	0.0287	0.1052	0.1463	0.1752	0.1994	0.2210	0.2411	0.2601	0.2783
	h	78.5	177.2	277.1	379.8	487.7	606.9	763.0	1174.3	1311.6	1403.6	1481.3	1552.2	1619.8	1685.7	1750.6
	s	0.1271	0.2893	0.4304	0.5573	0.6760	0.7940	0.9343	1.2754	1.3807	1.4461	1.4976	1.5417	1.5812	1.6177	1.6516
5000	v	0.0159	0.0164	0.0171	0.0181	0.0196	0.0219	0.0268	0.0591	0.1038	0.1312	0.1529	0.1718	0.1890	0.2050	0.2203
	h	81.1	179.5	279.1	381.2	488.1	604.6	746.0	1042.9	1252.9	1364.6	1452.1	1529.1	1600.9	1670.0	1737.4
	s	0.1265	0.2881	0.4287	0.5550	0.6726	0.7880	0.9153	1.1593	1.3207	1.4001	1.4582	1.5061	1.5481	1.5863	1.6216
6000	v	0.0159	0.0163	0.0170	0.0180	0.0195	0.0216	0.0256	0.0397	0.0757	0.1020	0.1221	0.1391	0.1544	0.1684	0.1817
	h	83.7	181.7	281.0	382.7	488.6	602.9	736.1	945.1	1188.8	1323.6	1422.3	1505.9	1582.0	1654.2	1724.2
	s	0.1258	0.2870	0.4271	0.5528	0.6693	0.7826	0.9026	1.0176	1.2615	1.3574	1.4229	1.4748	1.5194	1.5593	1.5962
7000	v	0.0158	0.0163	0.0170	0.0180	0.0193	0.0213	0.0248	0.0334	0.0573	0.0816	0.1004	0.1160	0.1298	0.1424	0.1542
	h	86.2	184.4	283.0	384.2	489.3	601.7	729.3	901.8	1124.9	1281.7	1392.2	1482.6	1563.1	1638.6	1711.1
	s	0.1252	0.2859	0.4256	0.5507	0.6663	0.7777	0.8926	1.0350	1.2055	1.3171	1.3904	1.4466	1.4938	1.5355	1.5735

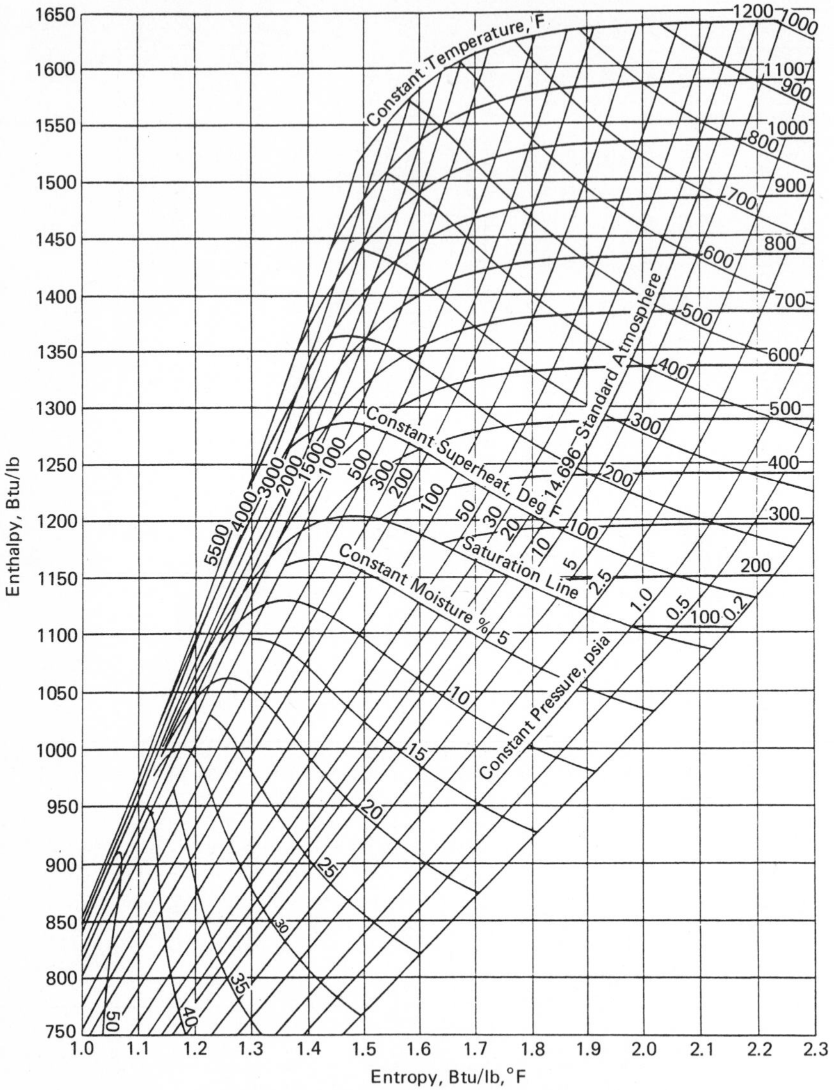

Figure 1-1 Mollier diagram (h-s) for steam.

This equation is represented diagrammatically in Fig. 1-2, where the enclosure represents any one or some combination of devices. Δu represents the difference in internally stored energy associated with molecular and atomic motions and forces. Internally stored energy, or simply internal energy, accounts for all forms of energy other than the kinetic and potential energies of the collective masses of the molecules.

$\Delta(Pv)$ is externally stored energy in that it reflects the difference in work required to move a unit mass into and out of the system. Both $\Delta(Pv)$ and Δu as used here differ from the other forms of stored energy in that they are dependent only on the state variables of P, v, and T through an equation of state.

The remaining terms of externally stored energy, $\Delta V^2/2g_c$ and ΔZ, depend on other physical aspects of the system. $\Delta V^2/2g_c$ represents a difference in the total kinetic energy

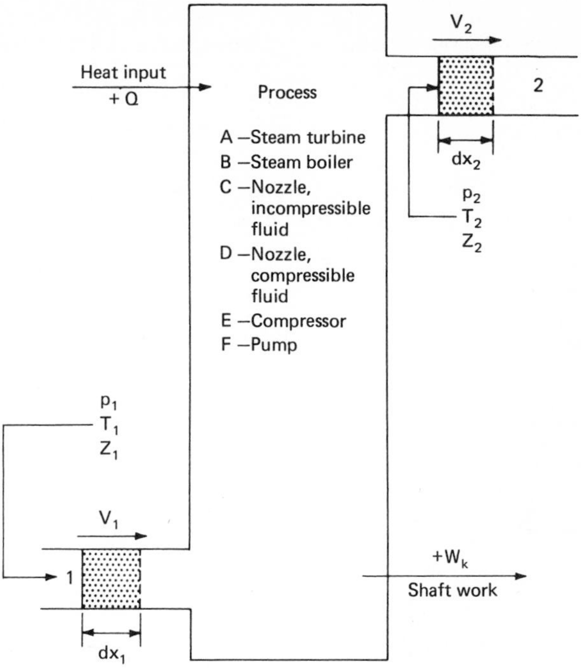

Figure 1-2 Diagram illustrating thermodynamic processes.

of the fluid between two reference points. ΔZ represents any differences in energy as a result of changes in elevation in a gravitational force field.

In keeping with the concept that changes in stored energy are the manifestation of differences in heat and work effects [Eq. (1)], heat quantities are positive when entering the system and shaft work is positive when leaving the system.

Since both u and Pv of Eq. (2) are stored energy, they are system properties and their sum is also a property. Moreover they are both functions of the state variables and cannot be changed independently like the other two terms of stored energy. For this reason it is customary and convenient to consider the sum $u + Pv$ as a single property h, called enthalpy:

$$h = u + Pv$$

In most practical problems ΔZ and $\Delta(V^2/2g_c)$ are small and negligible. The energy equation then becomes

$$h_2 - h_1 = Q - W \qquad \text{Btu/lb (J/kg)} \tag{3}$$

The following examples illustrate the application of the steady-state open-system energy equation (2) and the usefulness of enthalpy in the energy balance of specific equipment.

Steam Turbine

In most practical cases ΔZ, ΔV, and Q from throttle, 1, to exhaust, 2 (Fig. 1-2), are small compared with $h_1 - h_2$. This reduces the energy equation (2) to

$$u_2 + P_2v_2 - u_1 - P_1v_1 = -W$$

or

$$h_1 - h_2 = W \qquad \text{Btu/lb (J/kg)}$$

From this equation it is evident that the work done W for the steam turbine is equal to the difference between the enthalpy of the steam entering (h_1) and the enthalpy of the steam leaving (h_2). However, it seldom occurs that both h_1 and h_2 are known, and further description of the process is required for a numerical solution of most problems.

Steam Boiler

Since the boiler does no work, $W_k = 0$; and since ΔZ and ΔV from feedwater inlet, 1, to steam outlet, 2, are small compared to $h_1 - h_2$, Eq. (2) becomes

$$u_2 + P_2v_2 - u_1 - P_1v_1 = Q$$

or

$$Q = h_2 - h_1 \qquad \text{Btu/lb (J/kg)}$$

Based on this equation the heat added Q (positive), in the boiler per pound of feed, is equal to the difference between h_2 of the steam leaving and h_1 of the feedwater entering.

Water Flow through a Nozzle

The equation is not derived here, but for water flowing through a nozzle, if the velocity of approach (V_1) is zero, the velocity of the jet (V_2) becomes

$$V_2 = \sqrt{2gH} = 8.02\sqrt{H} \qquad \text{ft/s}$$

where H is the static head.

Flow of a Compressible Fluid through a Nozzle

When steam, air, or any other compressible fluid flows through a nozzle, it can be shown that the energy equation reduces to

$$V_2 = 223.9\sqrt{h_1 - h_2} \qquad \text{ft/s}$$

Compressor

The work done by the ideal compressor is equal to the difference between the enthalpies of the fluid leaving and entering the unit

$$W = h_2 - h_1 \qquad \text{Btu/lb (J/kg)}$$

Pump

In pumping a truly noncompressible fluid, all the energy added as work goes ideally to raising the static head H

$$W = H \qquad \text{ft}$$

THE SECOND LAW OF THERMODYNAMICS

Although the first law treats heat and work as interchangeable, it is also a matter of experience that certain qualifications apply. All forms of energy including the transient form, work, can be wholly converted to heat, but the converse is not generally true. Given a source of heat coupled with a heat-work cycle, such as heat released by high-temperature combustion in a steam power plant, only a portion of this heat can be converted to work. The rest must be rejected as heat to the stored energy of a sink at a lower temperature, such as the atmosphere. This is in essence the Kelvin statement of the second law of thermodynamics. It can also be shown that it is equivalent to the Clausius statement: Heat, in the absence of some form of external assistance, can flow only from a hotter to a colder body.

Entropy

Since it is a form of energy in transition, heat, like work, is a function of potential difference. That potential is easily measured as temperature. If a quantity of heat is divided by its absolute temperature, the quotient can be considered a type of distribution property or factor complementing the intensity factor of temperature. Such a property, proposed and named entropy by Clausius, is widely used in all branches of thermodynamics because of its close relationship to the second law.

Entropy can be more generally defined as the property which measures that portion of the heat added which cannot be converted into work, no matter how nearly perfect the operation may be.

Reversible and Irreversible Processes

Reversible thermodynamic processes exist in theory only, but serve the important function of limiting cases for heat flow and work processes which may be represented as total differentials. Reversible thermodynamic processes are confined to paths that describe continuous functional relationships on coordinate systems of thermodynamic properties. These properties, in turn, are homogeneous in the sense that there are no variations between any subregions of the system. Moreover, during interchanges of heat or work between a system and its surroundings, only corresponding potential gradients of infinitesimal magnitude may exist.

All actual processes are classified as irreversible. To occur, they must be under the influence of a finite potential difference measurably different from zero. A temperature difference supplies this drive and direction of heat flow. The work term, on the other hand, is more complicated since there are as many different potentials (generalized forces) as there are forms of work. However, the main concern here is expansion work for which the potential is clearly a pressure difference.

The substitution of reversible processes for real processes is illustrated in Fig. 1-3, which represents the adiabatic expansion of steam in a steam turbine, or any gas expanded from P_1 to P_2 in order to produce shaft work. T_1, P_1, and P_2 are given. The value of h_1 is fixed by T_1 and P_1 for the single-phase problem (vapor) and may be found from the steam tables, an h-s diagram, or Mollier Chart (see Fig. 1-1). From the combined first and second laws it is established that the maximum energy available for work in an adiabatic open system is $h_1 - h_3$ (Fig. 1-3), where h_3 is found by the adiabatic isentropic expansion (expansion at constant entropy so that $\Delta s = 0$) from P_1 to P_2. It is also a matter of experience that a portion of this available energy, usually about 10 to 15 percent, will represent lost work (W_L) of friction and shock, limiting the Δh for shaft work to $h_1 - h_2$. The two reversible paths used to arrive at point b are ac at constant entropy and cb at constant pressure:

$$(h_1 - h_3) + (h_3 - h_2) = h_1 - h_2$$

Point b now fixes T_2 as well as h_2, and both v_1 and v_2 are available from tabulated values of v, not shown in the figure.

Increases in Entropy

Increases in entropy are a measure of that portion of heat involved in a process which is unavailable for conversion to work.

Even though the net entropy change in any system executing a cycle of processes is always zero (because the cycle requires restoration of all properties to some designated starting point), the sum of all entropy increases has a special significance. These

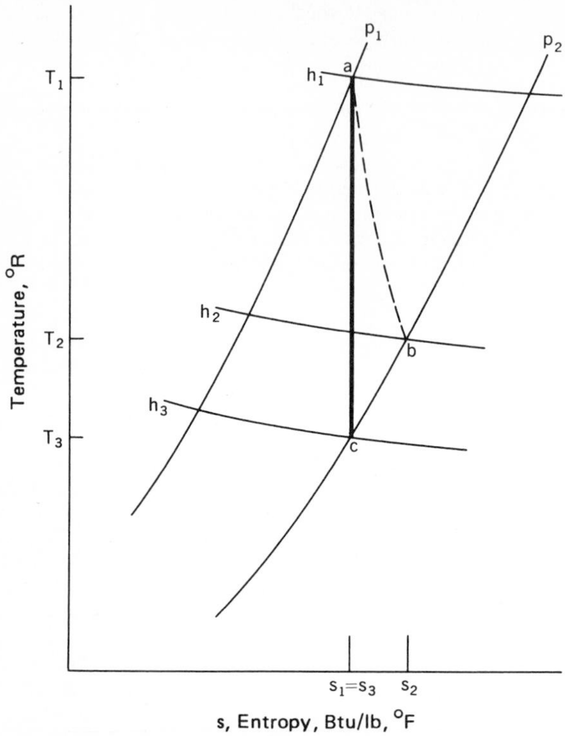

Figure 1-3 Irreversible expansion, state *a* to state *b*.

increases in entropy, less any decreases due to recycled heat within a regenerator, multiplied by the appropriate sink temperature (R), are equal to the heat flow to the sink. In this case the net change of entropy of the system is zero, but there is an increase in entropy of the surroundings. Any thermodynamic change which takes place, whether it is a one-time process or cycle of processes, results in a net entropy increase when both the system and its surroundings are considered.

POWER-PLANT CYCLES

Up to this point only thermodynamic processes have been discussed. The next step is to couple processes in some special way so that heat may be converted to useful work on a continuous basis. This is done by selectively arranging a series of thermodynamic processes in a cycle forming a closed curve on any system of thermodynamic coordinates. Since the main interest here is steam, the following discussion emphasizes expansion or Pdv work.

Carnot Cycle

This cycle, on a temperature-entropy diagram, is shown in Fig. 1-4 for a two-phase saturated vapor.

The Carnot cycle consists of the following processes:

1. Heat added to the working medium at constant temperature ($dT = 0$) from an appropriate heat source, resulting in expansion work and changes in enthalpy

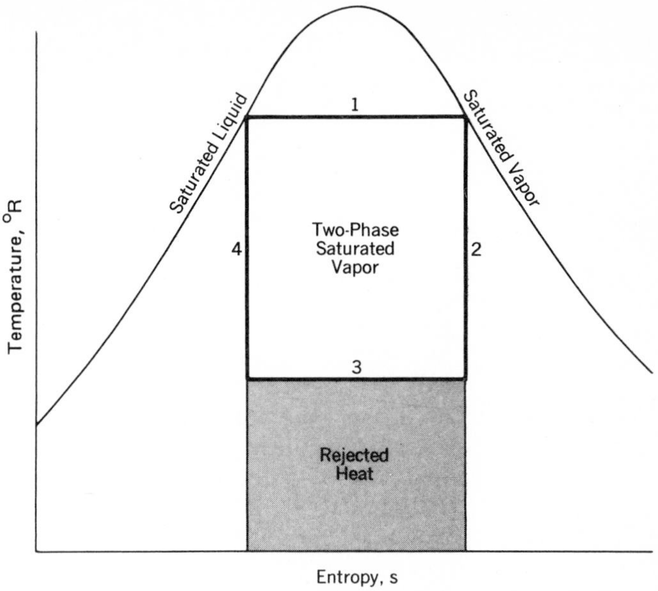

Figure 1-4 Temperature-entropy diagram, saturated vapor.

2. Adiabatic isentropic expansion ($ds = 0$) with expansion work and an equivalent decrease in enthalpy
3. Constant-temperature heat rejection to the surroundings equivalent to the compression work and any changes in enthalpy
4. Adiabatic isentropic compression back to the starting temperature with compression work and an equivalent increase in enthalpy

This cycle has no counterpart in practice. Nevertheless, the Carnot cycle clearly illustrates the basic principles of thermodynamics. Since the processes are reversible, the Carnot cycle offers maximum thermal efficiency attainable between any given temperatures of heat source and sink. Moreover, this thermal efficiency depends only on these temperatures:

$$\eta = \frac{T_1 - T_2}{T_1} = 1 - \frac{T_2}{T_1} \tag{4}$$

where

η = thermal efficiency of heat-to-work conversion

T_1 = absolute temperature of heat source, °R

T_2 = absolute temperature of heat sink, °R

Rankine Cycle

Early thermodynamic developments were centered around the performance of steam engines and, for comparison purposes, it was natural to select a reversible cycle which more nearly approximated the processes related to its operation. The Rankine cycle shown in Fig. 1-5, proposed independently by Rankine and Clausius, meets this objective. All steps are specified for the system only (working medium) and carried out revers-

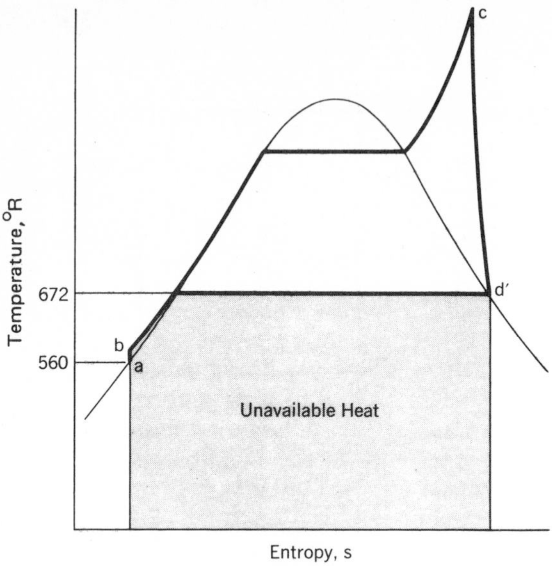

Figure 1-5 Open (noncondensing) cycle, 100°F feed.

ibly in the vapor, liquid, and two-phase states as indicated in the figure. Liquid is compressed isentropically from *a* to *b*. From *b* to *c* heat is added reversibly in the compressed liquid two-phase and superheated states. Isentropic expansion with shaft work output takes place from *c* to *d′*, and unavailable heat is rejected to the atmospheric sink from *d′* to *a*.

The main feature of the Rankine cycle is compression confined to the liquid phase only, avoiding the high compression work and mechanical problems of a corresponding Carnot cycle with two-phase compression. This part of the cycle from *a* to *b* in Fig. 1-5 is greatly exaggerated since the difference between the saturated liquid line and reversible heat addition to compressed liquid is too small to show in proper scale. For example, the temperature rise with isentropic compression of water from a saturation temperature of 212°F and 1 atm to 1000 psia is less than 1.0°F.

Open steam cycles are still found in small unit sizes and some special process and heating load applications coupled with power. Usually the condensate from process and heating loads is returned to the power cycle for economic reasons.

If the Rankine cycle is closed in the sense that the same fluid repeatedly executes the various processes, it is termed a condensing cycle.

The higher efficiency of the condensing steam cycle is a result of the particular pressure-temperature relationship between water and its vapor state, steam. The lowest temperature at which an open or noncondensing steam cycle may reject heat is approximately the saturation temperature of 672°R (212°F). This corresponds to normal atmospheric pressure of 14.7 psia. The closed, or condensing, cycle takes advantage of the much lower sink temperature for heat rejection available in natural bodies of water and the atmosphere. Being closed, the back pressure is no longer limited to normal atmospheric pressure but rather to saturation pressure corresponding to a condensing temperature of approximately 100°F and lower. Because the maximum possible $P\,dv$ work per pound of a compressible fluid is directly related to a function of the pressure ratio available for expansion, as well as the initial absolute temperature, an increase in this pressure ratio means an increase in the available work.

The arrangement of equipment in a rudimentary condensing steam power plant is shown schematically in Fig. 1-6. In addition to the prime mover, there are the boiler, the condenser, and the feed pump. The functional elements of boiler, prime mover, condenser, and feed pump are essential for a workable steam plant that complies with the basic thermodynamic cycle.

Figure 1-7 illustrates the difference between a closed Rankine cycle and the open cycle illustrated in Fig. 1-5. Liquid compression takes place from *a* to *b*, and heat is added from *b* to *c*. The work and heat quantities involved in each of these processes are the same for both cycles. Expansion, conversion of stored energy to work, takes place from *c* to *d′* for the open cycle and *c* to *d* for the closed cycle. Since this process is shown for the irreversible case, there is internal heat flow and an increase in entropy. From *d′* to *a* and *d* to *a*, heat is rejected. Because this last portion of the two cycles is shown as reversible, the shaded areas are proportional to rejected heat. The larger amount of rejected heat for the open cycle is clearly indicated.

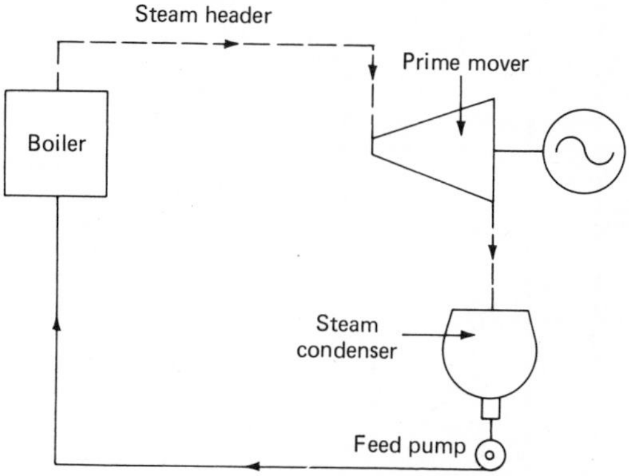

Figure 1-6 Diagram of rudimentary steam power plant.

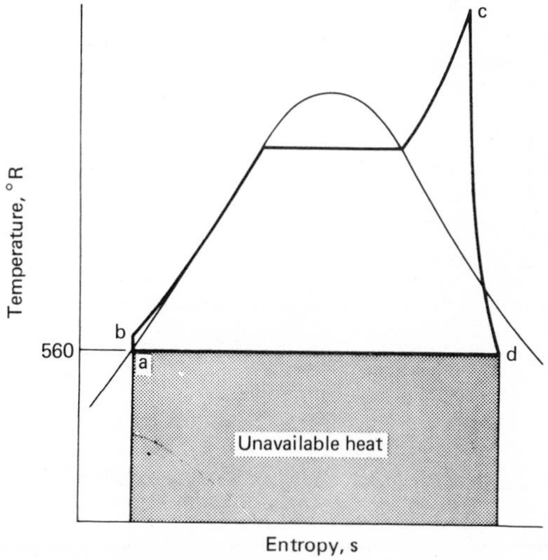

Figure 1-7 Closed (condensing) cycle, 100°F, 0.95 psia condensing conditions.

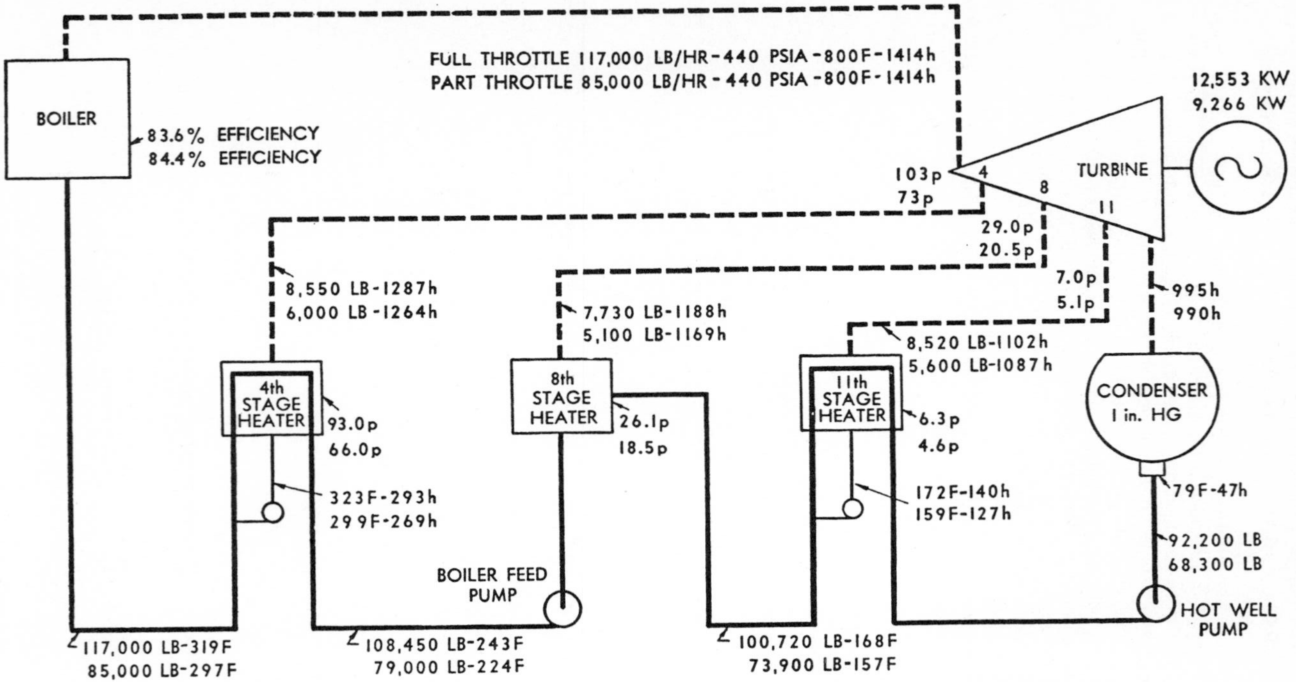

CASE	GROSS GENERATION, KW	NET PLANT SENDOUT, KW	HEAT SUPPLIED IN FUEL, BTU/HR	OVERALL PLANT HEAT, RATE, BTU/KW HR
FULL THROTTLE	12,553	12,050	165.7×10^6	13,750
PART THROTTLE	9,266	8,849	122.9×10^6	13,900

Figure 1-8 Heat balance diagram for 10,000-kW steam power plant.

Regenerative Rankine Cycle

The reversible cycle efficiency given by Eq. (4), where T_2 and T_1 are mean absolute temperatures for rejecting and adding heat, respectively, indicates only three choices for improving ideal cycle efficiency: decreasing T_2, increasing T_1, or both. Not much can be done to reduce T_2 in the Rankine cycle because of the limitations imposed by the normal temperatures of available sinks for rejected heat. Some leeway becomes available by selecting variable condenser pressures for the very large units with two or more exhaust hoods, since the lowest temperature in a condenser is set by the exit temperature of the cooling water. On the other hand there are many ways to increase T_1 even though the maximum steam temperature may be limited by the materials problems of high-temperature corrosion and allowable stress at elevated temperatures.

One early improvement to the Rankine cycle was the adoption of regenerative feedwater heating. This is done by extracting steam at various stages in the turbine to heat the feedwater as it is pumped from the hot well to the boiler economizer.

Figure 1-8 is a cycle diagram that shows the schematic arrangement of various components including the feedwater heaters. Regardless of whether the cycle is high temperature or high pressure, regeneration is used in all modern condensing steam power plants. It not only improves cycle efficiency but has other advantages, among which are lower volume flow in the final turbine stages and a convenient means of deaerating the feedwater.

Multipurpose Steam Power Plants

In fossil-fuel steam power plants operated solely for the generation of electric power, thermal efficiencies that are economically justifiable range up to about 40 percent, including the optimum use of steam pressure and temperature and bleed preheating of feedwater. This means that more than half of the heat released from the fuel is wasted and must be transferred to the environment in some way. This is usually done through a condenser, resulting in the heating of some body of water.

Higher-efficiency steam plants can be obtained by the use of high steam temperatures but, after many years of research on high-temperature metals, steam temperature is limited economically to about 1000°F (540°C).

One practical means available for improving the utilization of energy in steam is the use of multipurpose steam plants, where steam is exhausted or extracted from the turbines at a proper pressure level for use in an industrial process. If the power and process heat loads can be coordinated, it is entirely practical to generate steam in a boiler at an elevated pressure, pass it through turbines or engines, and then exhaust or extract the steam at a proper pressure level for use in the process. With such arrangements it is possible to obtain an overall thermal utilization between 65 and 70 percent. Figure 1-9 is an example of a multipurpose steam power plant.

Detailed example calculations for the numerous possible steam-plant cycles may be found in any of the many textbooks on thermodynamics.

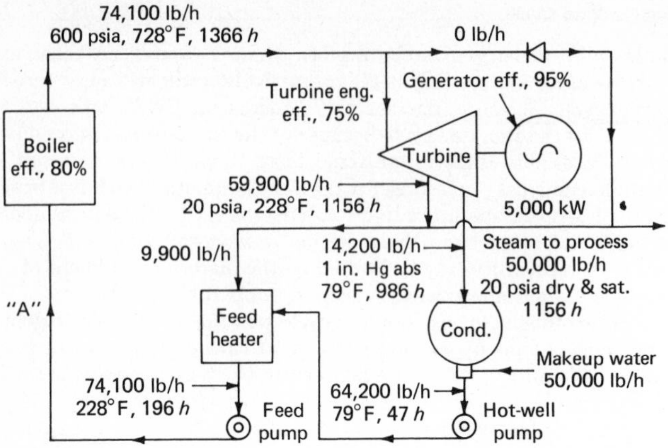

Figure 1-9 By-product generation of electric power and process steam.

chapter 4-2

Boilers

by
Russell N. Mosher
Assistant Executive Director
American Boiler Manufacturers Association
Arlington, Virginia

TERMINOLOGY

Since the term *boiler* alone does not adequately describe the complete system for providing the motive force or energy, it is necessary to know and understand the component functions that go into making the boiler a complete unit. Comprehending the definitions in the terminology used by the industry is the first step in sorting out the perceived mysteries of understanding.

A *boiler* is a closed vessel in which water is heated, steam is generated, or steam is superheated (or any combination of these) under pressure or vacuum by the application of heat from combustible fuels, electricity, or nuclear energy. Boilers are generally subdivided into four classic types—residential, commercial, industrial, and utility.

Residential boilers produce low-pressure steam or hot water primarily for heating applications in private residences.

Commercial boilers produce steam or hot water primarily for heating applications in commercial use, with incidental use in process operations.

Industrial boilers produce steam or hot water primarily for process applications, with incidental use as heating.

Utility boilers produce steam primarily for the production of electricity.

Within these four generic types of boilers, specific types of boilers emerge, with their classification based on their use. An example would be a *heat-recovery boiler* (Fig. 2-1) which recovers normally unused energy and converts it into usable heat. Likewise, a *fluidized-bed boiler* is one which utilizes a fluidized-bed combustion process. In this type of process, fuel is burned in a bed of granulated particles which are maintained in a mobile suspension by an upward flow of air and combustion products. This technology, in use for over 40 years, is currently being adapted as a combustion process to be installed within a boiler.

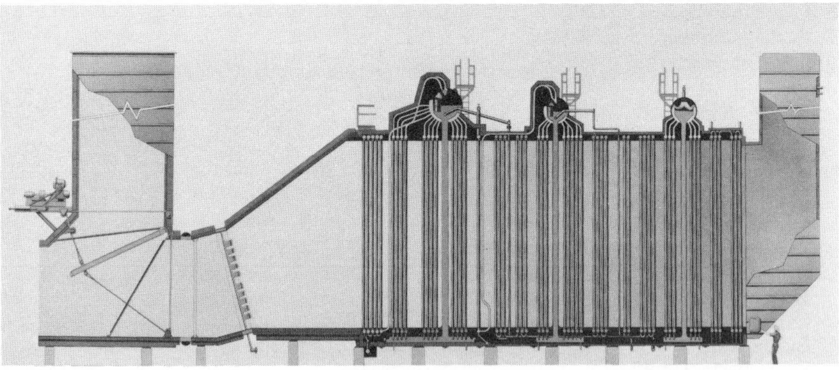

Figure 2-1 Heat-recovery boiler.

Generally, boilers are furnished either *packaged* or *field-erected*. A *packaged boiler* is one which is equipped and shipped complete with fuel-burning equipment, mechanical-draft equipment, automatic controls, and accessories. It is usually shipped in one or more major sections. *Field-erected boilers* are those that are shipped from the factory as tubes, casings, drums, fittings, etc., and completely assembled in the field.

The generally accepted reference for the *capacity* of a boiler is the manufacturer's stated output rate for which the boiler is designed to operate over a period of time. The *maximum continuous rating* is the maximum load in pounds (kilograms) of steam per hour for a specific period of time for which the boiler is designed. Likewise, then, the *capacity factor* is the total output over a period of time divided by the product of the boiler capacity and the time period.

INTRODUCTION

Modern boilers provide most of the motive force in the world and are probably the least understood of all mechanical pieces of equipment. They are subjects for engineering, congressional legislation, agency legislation, and physical laws, all of which shape their destiny. Despite all this attention this age-old energy source is still shrouded in mystery.

This chapter explains the theoretical and practical aspects of modern boiler equipment. The objective is to enlighten the reader on the principles of boiler design, to characterize the available equipment, and to explain the need to achieve maintenance for maximum and sustained utilization of equipment.

BOILER APPLICATION

Modern boilers range in size from those required to provide steam or hot water to heat homes, through mid-size units which provide energy to drive presses, to very large units used as the primary motive force in producing electric power. Boilers can be arranged for firing almost any type of fuel available, provided the designer is cognizant of the fuel to be employed prior to making the initial calculations for sizing.

The primary purpose of a boiler is to generate steam or hot water at pressures and/or temperatures above that of the atmosphere. Steam or hot water is produced by the transfer of heat from the combustion process taking place within the boiler, thereby elevating its pressure and temperature.

With this higher pressure and temperature, it follows that the containment vessel or pressure vessel must be designed in such a way as to encompass the desired design limits with a reasonable factor of safety. For the sake of economy, the capacity of the unit must be generated and delivered with minimum losses.

In smaller boilers used in home heating applications, the maximum operating pressure for steam is usually 15 psig (104 000 N/m^2). In the case of hot water, this is equal to 450°F (232°C).

Larger boilers are designed for various pressures and temperatures, depending upon the application within the heat cycle for which the unit is being designed. A boiler designed to heat a large college campus may require a certain capacity at an elevated pressure and a superheated temperature which provides the force to transmit the steam to its final use point. In other cases, very high pressures and temperatures are required in order to implement chemical reactions, to provide drying steam in a paper cycle, or to provide the needed energy to drive a large piece of mechanical equipment.

The dependability and safety record displayed by today's modern boilers is the product of almost 100 years of design experience, control fabrication, and monitored operation of boilers. The properties of steam and water have been accurately graphed for use by the engineer.

With the use of computers, the boiler design engineer has gained a new understanding of boiler thermal dynamics and heat transfer and has expanded the understanding of the burnability of fuels in a safe and efficient manner, thus developing units to produce the large amounts of steam required today. Advances in metallurgical fields have yielded better-quality steels and alloys, which allow use of the high pressures and temperatures required.

A large central-station boiler is designed on the basis of the output cost of electricity produced. Its operation is under close control, and the load cycle follows a very well defined and predictable pattern based on area power requirements.

The industrial boiler however, is usually a single unit installed primarily as an important step in the production of a product. Its use is merely a means whereby the product can be fabricated in the shortest amount of time, with the lowest materials cost, and shipped. Hence, it is frequently called upon to perform a difficult task, often under unfavorable conditions of steam load, water, and fuel. Plant load cycles are usually highly unpredictable, and the boiler must be ready at any time to achieve the required capacity in the shortest amount of time without hesitation.

Many factors must be taken into consideration when one is designing a boiler. After the decision has been made as to what fuel must be burned, it is necessary to determine how much input steam is necessary to satisfy the requirements or demands upon the boiler. Operating parameters include minimum, maximum, and normal load range; length of time in a cycle; and type of load, whether constant or fluctuating. All of these parameters must be analyzed for proper size selection.

The basic components of a boiler are the *furnace* and the *convection* sections. In the *furnace* section (Fig. 2-2), the products of combustion are consumed and heat is released and transferred into the water, thereby producing steam or heating the water. This space must be designed for the three "T's" of combustion—time, turbulence, and temperature. *For complete combustion, it is necessary for the fuel to have sufficient time to be completely consumed;* there must be sufficient turbulence for complete mixing of fuel and air for efficient burning; and there must be a high enough temperature to allow the products to be ignited.

The shape of the furnace is controlled by the type of fuel and burning method. Adequate provisions must be made for instituting and maintaining ignition and combustion of the fuel. For those units with solid-fuel firing, adequate provision must be made to allow for the removal of unburned combustibles and/or ash.

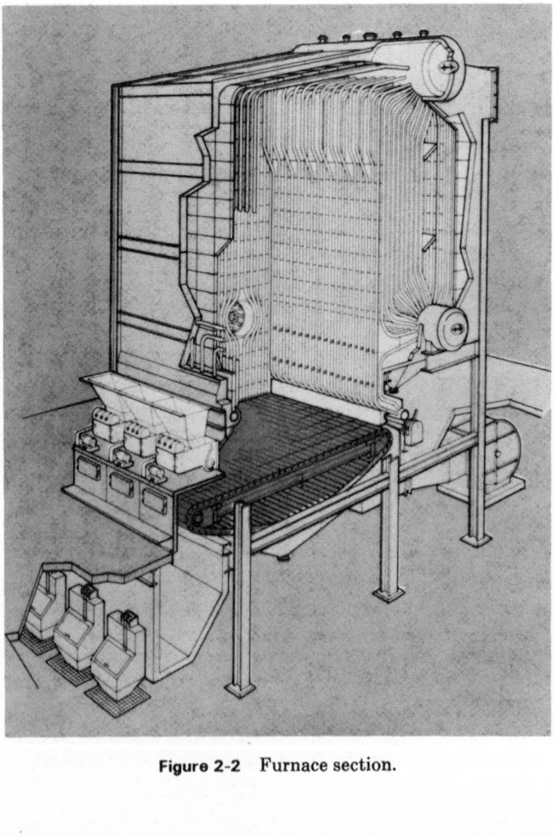

Figure 2-2 Furnace section.

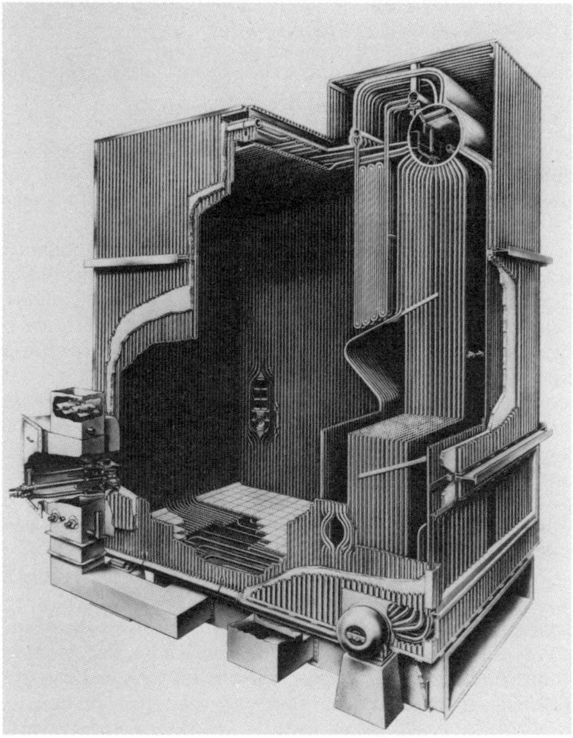

Figure 2-3 Convection section.

The *convection* section (Fig. 2-3) of a boiler is that portion in which heat contained in the combustion gases is transferred to the water for the production of steam. The selection of heating surface and tube spacing depends completely upon the type of fuel producing the flue gas with its entrained particles. Adequate provision must be made in this section to allow any unburned particles to pass through and be collected in downstream separators. Pressure drop and volumetric flow are influencing factors which dictate the overall design of the convection section. Generally, a higher gas volume through a fixed flow area produces a greater pressure drop. A greater pressure drop enhances the transfer to a point where economic decisions must be made on inputs of auxiliary equipment to produce the necessary flow of these gases.

The steam and water circulation rate within the pressure vessel decides the effectiveness of the heat-transfer surface. When new feedwater is added to the system, precipitates fall out which might be removed as blowoff. Provisions are usually made in the lower portion of the convection section whereby a boiler operator may remove these particular precipitates by opening boiler blowoff valves.

Many applications utilizing boiler equipment require steam at a very high rate of purity. Boiler designers therefore install steam-separating equipment in the boiler drum to remove entrained moisture and solids before the steam is taken from the boiler to the system. These steam separators come in a variety of types including cyclones, mist deflectors, and baffle plates.

In some applications, the heating surface or tubes are of the bare-tube type. In other cases, the heating surface will be of the extended surface or fin-tube type. Utilizing fin tubes allows greater tube surface within the convection section. Greater heat-transfer levels can be achieved with the use of this type of tubing.

BOILER CLASSIFICATIONS

Characteristically, boiler types are generally classified as either *firetube* or *watertube.*

Firetube Boilers

In the *firetube* boiler (Fig. 2-4), the flame and products of combustion pass through the tubes. The heated water or other medium surrounds the internal furnace and the tube bundles.

Various types of furnaces are used in conjunction with firetube boilers. Some are long, cylindrical tubes while others are firebox (Fig. 2-5) arrangements allowing the burning of solid fuels. In most cases, the firetube boiler includes a shell to contain the water and steam space. Within the shell will be tube sheets and tubes which are portions of the pressure vessel containment. The furnace or firebox provides space for the combustion process from the heat source.

Many types of firetube boilers are being supplied to industry. One type is the *horizontal return tubular boiler* (Fig. 2-6). In this unit, the products of combustion travel across the shell and back through the tubes within the pressure vessel. These units are usually horizontally brick-set.

Another type of firetube unit is the *Scotch marine boiler.* This design was developed originally for shipboard installation. This type of boiler can be fired with either solid, liquid, or gaseous fuels.

Because the Scotch marine boiler is a very compact type of unit, it has become readily adaptable for stationary service. When modifications to the basic type are made in adapting it to process and heating use, it is called a *modified Scotch marine boiler.*

Another type of firetube unit currently being marketed is the *vertical-type boiler.* In this particular unit the fuel or heat source is in the bottom, and the products of combustion rise up through tubes and are emitted from the top of the unit.

Figure 2-4 Firetube boiler.

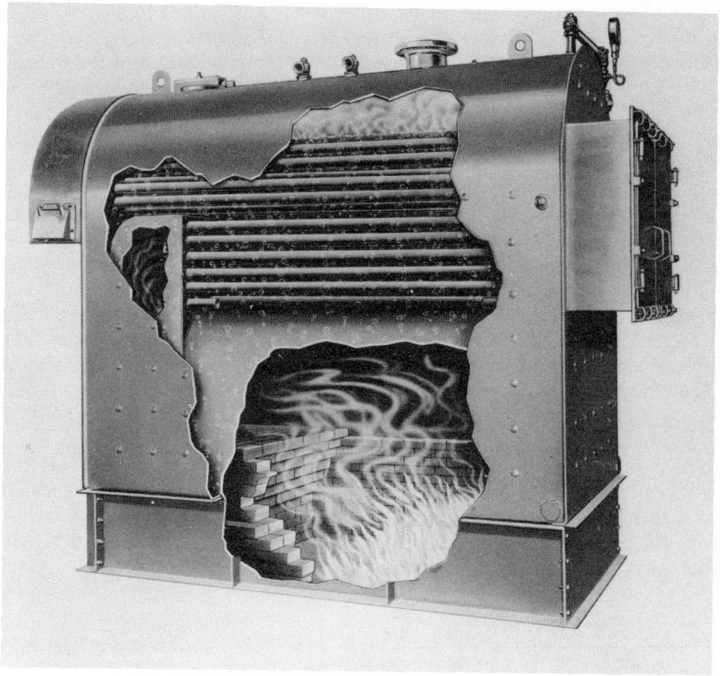

Figure 2-5 Firebox boiler.

Figure 2-6 Horizontal-return tubular boiler.

Watertube Boilers

Watertube boilers come in a variety of arrangements and designs. In this type of unit, the products of combustion usually surround the tubes (Fig. 2-7), and the water is inside the tubes which are inclined upward toward a vessel or drum at the highest point of the boiler. The configuration of these tubes generally describes the type of boiler. Some manufacturers offer straight-tube units while others offer units with bent tubes (Fig. 2-8). Other configurations of watertube boilers describe the various types in terms of the variations of pressure vessel arrangement.

In a *box-header watertube* boiler, the watertubes are connected to rectangular headers which are arranged so that the circulating water and steam mixture will rise toward a collection drum. The box headers are usually on either end of the tube bundles, and the products of combustion pass between the headers and around tube bundles.

Some boilers are of the *long-drum* type; that is, when viewed from the front of the boiler (Fig. 2-9), the drum is the length of the boiler. Its corollary is the *cross-drum* boiler. When viewed from the front of the unit, the drums are installed perpendicular to the long centerline or across the the boiler.

Firetube units are generally furnished in applications up to approximately 30,000 lb (13 500 kg) of steam per hour. They are furnished for low-pressure operation [15 psig (104 kN/m^2) and under] and as power boilers [up to approximately 300 psig (2100 kN/m^2) of steam pressure]. Watertube boilers for use in industrial applications are furnished in capacities up to almost 1 million lb (450 000 kg) of steam per hour. Design pressures vary from 100 psig (700 kN/m^2) up through 1200 or 1400 psig (8.3 or 9.6 MN/m^2) with steam temperatures ranging from saturated to 1000°F (540°C).

Figure 2-7 Commercial watertube boiler.

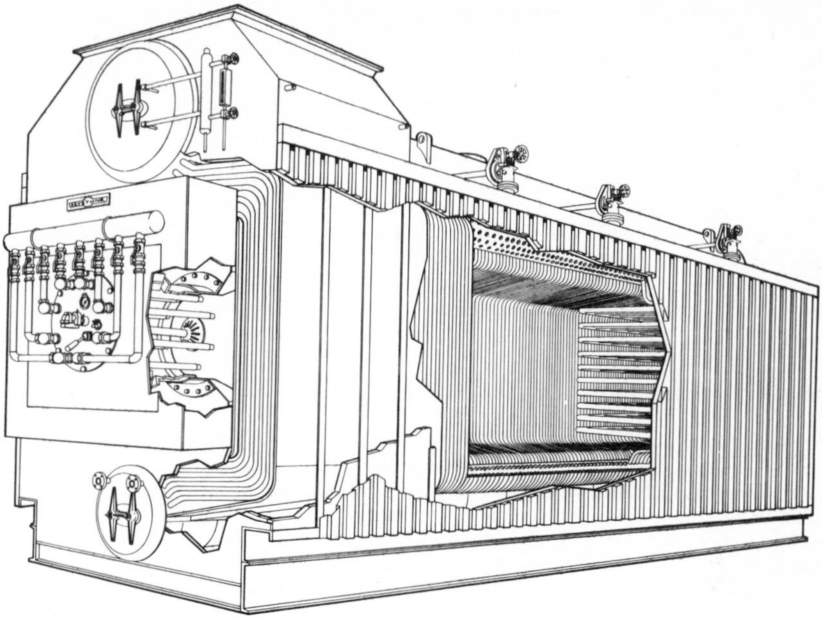

Figure 2-8 Bent-tube boiler.

Figure 2-9 Long-drum boiler.

Packaged Boilers

Many manufacturers supply both watertube and firetube units already packaged, or shop-assembled. A *packaged* steam or hot-water boiler is one which is generally shop assembled and includes all major components: burner, draft equipment, pressure vessel, trim, and controls. The main limitation to packaging boilers is the capability of their being handled by trucking or railroad equipment. Most manufacturers have designed complete lines of firetube and watertube boilers which can be shipped by either truck or railroad. Some manufacturers have utilized shop-assembling procedures to fabricate a boiler which can be assembled easily in the field. These components may be a single portion of the vessel or multiple major parts which can be brought together in the field.

BOILER COMPONENTS

To understand the operation of a boiler, it is necessary to observe what happens from input to output of the unit. Several cycles are involved in the complete operation of the unit. The heat cycle, the water and steam cycle, and the boiler-water circulation cycle all interact to produce the output of the boiler. Fuel and water are brought to the unit; water is heated to its final predesignated condition (water and/or steam) and transported to the point of its end use. When the heat has been taken out of the water, the remaining steam and water mixture (or condensate), if usable, is returned to the unit and recycled.

Furnace*

In the fuel cycle, the solid, liquid, or gaseous fuel is delivered to the boiler where it is mixed with air and burned. This liberation of heat is usually achieved in the *furnace* portion of the boiler (Fig. 2-1). Furnaces can be of either the *refractory* or the *water-cooled* type.

In the *refractory-type* furnace, refractory brick forms the envelope of the furnace. These refractory furnaces are usually backed with insulation and a casing material. For the *water-cooled* wall-type furnace, the envelope consists of tubes placed close to each other, which thereby absorb heat and help in the production of steam. These water-cooled furnaces can have either tube and tile, tangent-tube, or welded-membrane walls.

The basic function of the furnace is to allow for the combustion of the fuel. It is necessary that the furnace size be sufficient to allow for adequate combustion of the fuel, time for its combustion, and for enough turbulence to permit efficient combustion.

Boiler Section

This is usually referred to as the boiler or convection section of the unit. Closely spaced tubes are arranged to allow passing of the products of combustion around the tubes or through the tubes, depending on the type of unit. Most of the steam is generated in the boiler portion of the unit. In watertube units, if additional steam temperature is required by the process, the steam is then routed to a superheater.

Superheater

In a superheater unit (Fig. 2-10), the steam is directed back through the products of combustion to take on additional heat. This additional heat results in considerable energy gain by the steam which will be liberated in end use. This end use can be a steam turbine or other type of equipment requiring considerable energy release for its operation.

Superheaters are either of the radiant or convection type. In a radiant superheater,

*See also Chap. 4-3.

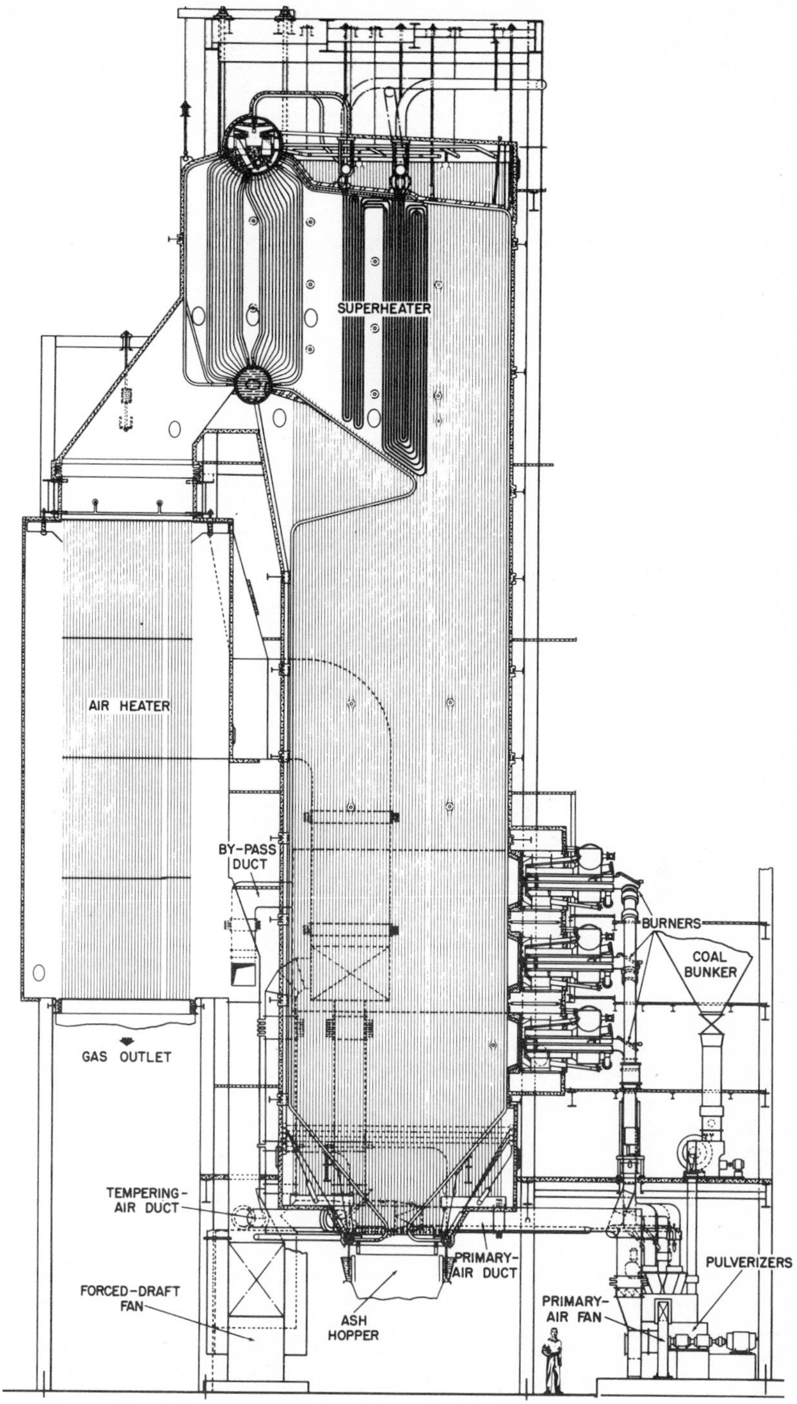

Figure 2-10 Boiler with superheater unit.

the tubes are usually located in the furnace section of the boiler. Convection-type superheaters are usually located behind the screenwall of the convection section. Radiant-type superheaters receive their heat by direct radiation from the flame, while convection superheaters receive their heat primarily from the passage of the products of combustion around the tubes.

Air Heaters

It is often desirable to preheat the air for combustion prior to bringing it in contact with the fuel (Fig. 2-11). This is necessary when burning fuels of very high moisture content. In an air heater, the ambient air volume is brought in and preheated by utilizing sensible heat from the boiler flue gas being discharged from the unit. This increases overall efficiency, eliminating the use of extra fuel for this purpose. This is one type of *heat recovery unit.*

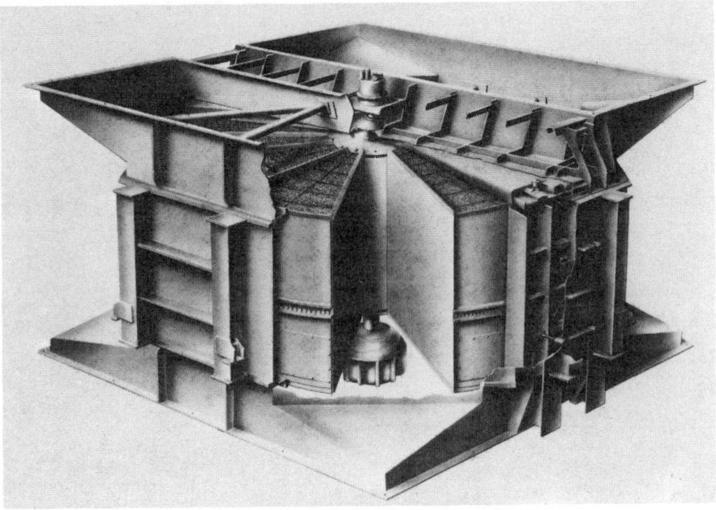

Figure 2-11 Air heater.

Economizers

An economizer (another type of heat recovery unit) is a boiler component which preheats incoming feedwater from its supplied temperature, utilizing sensible heat from the boiler outlet flue gas being exhausted from the unit. As in the air-heater principle, raising this inlet feedwater temperature (Fig. 2-12) increases the efficiency of the unit by eliminating the use of additional fuel for this operation.

OPERATION AND MAINTENANCE

Start-Up

Before any preparation can be made to start up a boiler, new or otherwise, the operator's manual furnished by the boiler manufacturer for the particular make and model unit must be available. It is important that operating personnel carefully follow the procedures in the manual, particularly the safety precautions, before attempting to activate the equipment.

When a new boiler is prepared for its initial operation, procedures should be followed

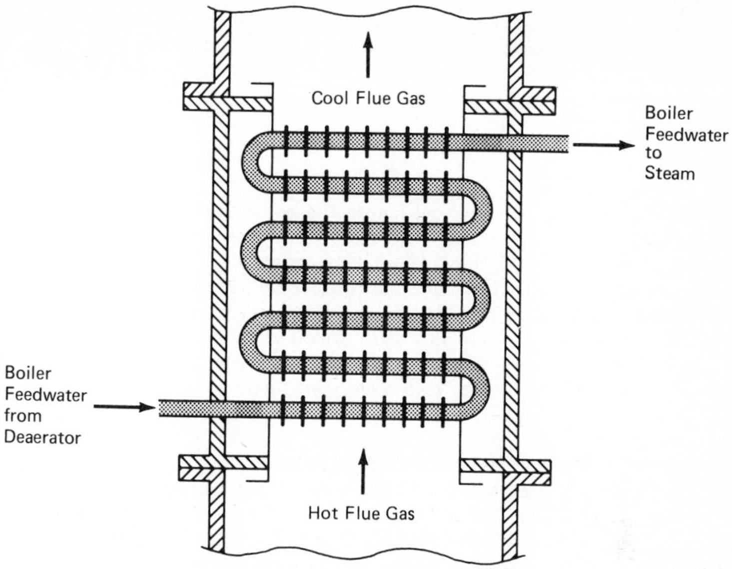

Figure 2-12 Principle of economizer.

to ensure high efficiency at which the unit can operate a long life, and the economies to be expected from an engineered piece of equipment. Even though the unit has been checked by the manufacturer, the following precautions should be observed:

1. The unit should be thoroughly examined on both the water side and the fire side to make sure that no foreign material is present.
2. All piping such as blowdown piping, steam piping, and feedwater piping should be checked to ensure that the piping has been properly installed so that there will be no danger to any individuals. Items such as the gauge, gauge glasses, and controls should be checked for any evidence of damage or breakage either incurred during transportation or caused by installation personnel working around the equipment after it was placed.
3. Electrical equipment such as motors, pumps, blowers, and compressors should be operated whenever possible to assure proper rotation. Items such as control valves, interlocks, motorized valves, and limit switches should be checked wherever possible to ensure their proper operation.
4. All fuel lines should be checked per the installation instructions in the manual.

After a thorough inspection of the unit has been completed, the next steps to start-up follow.

Caution*—All manufacturers of boilers usually supply operating instruction manuals with their equipment. Before starting the unit, be sure that the manual of instructions has been thoroughly read and understood.*

Drying Out and Boiling Out

Since the refractories and insulation of the unit may contain absorbed moisture and since (at initial start-up) the boiler is filled with water that is at the supplied temperature (which is colder than normal), the initial firing of the unit should be maintained at as low a level as possible. The boiling-out period should be continued at approximately 50 percent of the unit operating pressure for a long enough time to ensure that all oils and materials to be removed by the boiling-out process have been dislodged. Experience has indicated that a 12-h minimum boil-out period is generally sufficient to complete the cleaning. However, factors such as chemical concentration, amount of material to be removed, and pressure may modify boil-out time.

Cleaning

The system must be carefully cleaned before the boiler is connected into the system. Many clean boilers have been ruined with system contaminants such as pipe dope, cutting oil, and metal shavings or chips. Many contractors will use a new boiler for heating and curing a building under construction. Special care must be taken to assure that *adequate water treatment is provided by the contractor during this initial use of the boiler.* Succeeding owners can receive a badly scaled or damaged boiler through contractor misuse. Moreover, as new zones are cut into a system, *cleaning of these zones is required to prevent damage to the boiler. Only one boiler should be used to boil out a system.*

Cleaning Improves a Steam or Hot-Water Heating System

One important phase in completing boiler installation is too often neglected in the specifications. *No provision usually has been made for cleaning the system.* It is sometimes drained for changes and adjustments but never actually cleaned. The architect, engineer, or contractor selects boilers for applicable installations. The selection may represent the best system; but it will be better if it is a clean system.

How to Tell If a System Needs Cleaning

There are definite symptoms of an unclean system. A typical checklist follows. If any of the items are positive, the system needs cleaning.

1. Obviously discolored, murky, dirty water
2. Gases vented at high points in the heating area that ignite and burn with an almost invisible bluish flame
3. A pH alkalinity test that gives a pH test reading below 7 (A pH lower than 7 indicates the water in the system is acid.)

No matter how carefully a system is installed, certain extraneous materials do find their way accidentally into the system during construction. Pipe dope, thread-cutting oil, soldering flux, rust preventives or slushing compounds, coarse sand, welding slag, and dirt, sand, or clays from the jobsite are usually found. Fortunately the amounts of these are usually small and do not cause trouble. However, in some instances there may be sufficient quantities to break down chemically during the operation of the system, causing gas formation and acid in the water system. Hot-water systems, in most cases, naturally operate with a pH of 7 or higher. The condition of the water can be quickly tested with Hydrion paper, which is used in the same manner as litmus paper except that it gives specific pH readings. A color chart on the side of the small Hydrion dispenser gives the readings in pH units. Hydrion paper is inexpensive and readily obtainable through appropriate wholesale and retail channels.

A system that tests acid (below 7 on the scale, sometimes as low as 4) will usually have the following symptoms:

1. Gas formation (air trouble)
2. Pump seal and gland problems
3. Air vent sticking and leaking
4. Frequent operation of relief valves
5. Piping leaks at joints

Once this condition exists, the symptoms continue until the situation is corrected by cleaning the system. Many times, because of gas formation, automatic air vents are added throughout the system to attempt a cure. The excessive use of automatic air vents can defeat the function of the air-elimination system since the small quantities of entering air must be returned to the expansion tank to maintain the balance between the air cushion and the water volume.

If a system is permitted to deteriorate with resultant leaks and increased water losses, serious boiler damage can occur. Therefore, the chief consideration is to maintain a closed system that is clean, neutral, and watertight.

How to Clean a Heating System

Cleaning a system (either steel or copper piping) is neither difficult nor expensive. The materials for cleaning are readily available. Trisodium phosphate, sodium carbonate, and sodium hydroxide (lye) are the materials most commonly used for cleaning. They are available at paint and hardware stores.

The preference is in the order named, and the substances should be used in the following proportions; use a solution of *only one type* in the system.

Trisodium phosphate, 1 lb for each 50 gal (1 kg for 420 L) in the system

Sodium carbonate, 1 lb for each 30 gal (1kg for 240 L) in the system

Sodium hydroxide (lye), 1 lb for each 50 gal (1 kg for 420 L) in the system

Fill and vent the system and circulate the cleaning solution throughout, allowing the system to reach design or operating temperatures if possible. After the solution has been circulated for a few hours, the system should be drained completely and refilled with fresh water. Usually, enough of the cleaner will adhere to the piping to give an alkaline solution satisfactory for operation. A pH reading between 7 and 8 is preferred, and a small amount of cleaner can be added if necessary.

A clean, neutral system should *never* be drained except for an emergency or for such servicing of equipment as may be necessary after years of operation. Antifreeze solution in the system should be tested from year to year as recommended by the manufacturers of the antifreeze used. Without a doubt, the clean system is the better system.

Arrangements for Cleaning Heating Systems

Much of the dirt and contamination in a new system can be flushed out prior to boil-out of the system. This is accomplished by first flushing the system to waste with clear water and then using a chemical wash.

The boiler and circulating pump are isolated with valves, and city water is flushed through the successive zones of the system, carrying chips, dirt, pipe joint compound, etc., to waste with it. This is followed by a chemical flush. Removal of pipe chips and other debris before operating the isolation valves of the boiler and pump will help to protect this equipment from damage by such debris. After this flushing process is complete, the usual boil-out procedure is accomplished.

CAUTION*—If one zone is flushed and boiled out before other zones are completed or connected, this flushing process should be repeated on completion of additional zones, loops, or sections of the piping.*

When a boiler is fired for the first time (or started again after repairs or inspection), vapor and water may be observed as a white plume in stack discharge or as condensate on the boiler fire sides and services. Generally, this condition is temporary and it will disappear after the unit reaches operating temperature. This condensation should not be confused with the stack plume that occurs when the boiler is operating during extremely cold weather.

When cool-down of a boiler is required, the unit should be permitted to cool over a period of 12 h, losing its heat to the atmosphere. Forced cooling is not recommended; it will possibly loosen tubes in the tube sheets or cause other damage to the pressure parts.

Water Treatment*

Water treatment is required for satisfactory operation of a boiler at the initial start-up to prevent any deposition of scale and to prevent any corrosion from acids, oxygen, and other harmful substances that may be in the water supply. A qualified water-treatment specialist should be consulted and the water should be appropriately treated.

The basic aims and objectives of boiler-water conditioning are to:

1. Prevent the accumulation of scale and deposits in the boiler
2. Remove dissolved gases from the water
3. Protect the boiler against corrosion
4. Eliminate carryover and/or timing (steam)
5. Maintain the highest possible boiler efficiency
6. Decrease the amount of boiler downtime for cleaning

Water treatment should be checked and maintained whenever the boiler is fired.

CAUTION—The purchaser should be sure that the boiler is not operating for approval tests or any other operation of firing without water treatment.

It should also be noted that water boilers may well need chemical treatment for the first filling of water as well as additional periodic chemical treatment depending on the system's losses and the make-up requirements. Water treatment may vary from season to season or over a period of time and, therefore, there should be a requirement that the water-treatment procedure be checked no fewer than four times a year and possibly more frequently if the local water conditions require it. When the system is drained and then refilled, chemical treatment is required inasmuch as raw water has been put into the boiler system.

There are two major methods of boiler feedwater treatment, external and internal.

External Feedwater Treatment

This type of treatment is performed in separate tanks, containers, or other necessary devices for the removal of oxygen and other detrimental gases, and the removal of magnesium carbonate, calcium carbonate, silica, iron, etc. There are also filters available for the removal of foreign matter. A common method of removing gas from the boiler water is to use deaerating feedwater heaters. One method of removing the magnesium and calcium carbonates is to use sodium zeolite softeners. There are filters and other equipment presently manufactured that will cover virtually every requirement for water treatment.

Internal Feedwater Treatment

Internal feedwater treatment is generally nothing more than the addition of the proper chemicals to prevent the deposition of scaling materials on the hot surfaces of the boiler. A sludge formed by the chemicals with calcium and/or magnesium carbonates drops to the bottom of the boiler or remains in suspension. In steam boilers, this sludge can be removed by proper blowdown procedures. The chemicals that are to be added to the boiler water, the blowdown procedure, and the analysis and maintenance of the feedwater conditioning should be handled by a water-treatment consultant.

Care of Idle Boilers

Boilers that are used on a seasonal basis and will be idle for a long period of time (more than 30 days) should be laid up by using either a dry or a wet method of protection during the periods of inactivity.

*Also see Chap. 6-2.

Boilers Laid Up Dry

If a boiler is subject to freezing temperatures or if it is to be idle for an excessive period of time, the following procedures should be carried out so that the boiler is not damaged during its period of inactivity:

1. Drain and clean the boiler thoroughly (both fire and water sides) and dry the boiler out.
2. Place lime or another water-absorbing substance in open trays inside the boiler and close the unit tightly to exclude all moisture and air.
3. All allied equipment such as tanks, pumps, etc., should be thoroughly drained.

Boilers Laid Up Wet

In order to protect the boiler during short periods of idleness, it should be laid up wet and in the following manner:

1. Fill the boiler to overflowing with hot water. The water should be at approximately 120°F (45°C) to help drive out the free oxygen. Add enough caustic soda to the hot water to maintain approximately 350 parts per million (ppm) of alkalinity and also add enough sodium sulfite to produce a residue of 50 to 60 ppm of this chemical.
2. Check all boiler connections for leaks and take a weekly water sample to make sure that alkalinity and sulfite content are stable.

Restarting Boilers

Upon restarting a boiler that has been laid up dry, laid up wet, or has been cooled down for repairs, be sure to follow the recommended start-up procedure as defined in the operating manual provided by the manufacturer.

Preparation for Lay-Up

When a boiler is being cleaned in preparation for lay-up, the water side of the unit should be cleaned and then the unit should be fired to drive off gases. The fire side should then be cleaned. An oil coating on the fire-side metal surfaces is beneficial when the boiler is not used for extended periods of time. Another helpful treatment would consist of completely filling the boiler with an inert gas and sealing it tightly to prevent any leakage of the inert gas. This will help prevent oxidation of the metal. Fuel-oil lines should be drained and flushed of residual oil and refilled with distillate fuel. If all boilers are to be laid up, care of oil tanks, lines, pumps, and heaters is similarly required.

Burner Care

A planned preventive maintenance program is a direct route to safe, dependable boiler unit operation. Boilers are supplied with engineering fuel-burning equipment that must be maintained through a regular, conscientious maintenance program to keep it in satisfactory operating condition. Oil nozzles, igniters, electrodes, and internal burner parts should be checked as part of a regular monthly maintenance program. The settings of spark gaps and nozzle openings as well as their general dimensions should be checked for both wear and cleanliness. Specific instructions as to the method of cleaning, methods of adjustment, and particular dimensions are contained in the instruction manual furnished by the boiler manufacturer.

The best method of keeping a planned preventive maintenance program in effect is to keep a daily log of pressure, temperature, and other gauge data as well as of water-treatment data. In the event that a variation appears from the normal readings, the trouble can be quickly analyzed and corrected to avoid serious problems. For example, an oil-fired unit showing a drop in oil pressure can indicate a faulty regulating valve, a plug strainer, an air leak in the suction line, or a change in the operation of some other piece of equipment in the oil line. A decrease in oil temperature can indicate malfunction of the temperature controls or a malfunction or fouling of the heating element.

For example, in a gas-fired unit, a decrease in the gas pressure can indicate a malfunction of the regulator, a drop in the gas supply pressure, or some restriction in the gas flow possibly caused by one or more controls or valves not operating properly.

Items to Be Checked Periodically

Such items as linkages and other mechanical fastenings and stops should be periodically checked for tightness and visually checked for any movements or vibration. Any items that are loose or that have changed in position should be thoroughly checked and readjusted as necessary.

Stacks should be checked daily for haze or smoke conditions. A cloudy, hazy, or smoky stack indicates a possible need for burner adjustment. The fire may not be receiving enough air; there may be improper control of air/fuel ratios; there could be a change of fuel delivered, etc.

Stack temperatures should be checked and noted on the log; however, one must bear in mind that a rise in stack temperature does not always mean poor combustion or a fouled water site or fire site. Stack temperatures will vary proportionally as the low point changes. Stack temperatures should be observed in relation to the firing rate and a comparison should be made with previous records of the same firing rates. Stack temperatures can vary as much as 100°F (55°C) from high fire levels to low fire levels, and therefore caution must be exercised before interpreting the stack temperature reading.

Fire-Side Maintenance

Periodic inspection and fire-side cleaning should be performed when the exit fuel gas temperature is more than 100 to 175°F (55 to 97°C) above normal operating temperature. Cleaning should be performed immediately upon shutdown. The boiler manufacturer's recommended procedures for all fire-side maintenance should be followed.

Burner, access, or head gaskets should be inspected and replaced as required. All refractories should be inspected for excessive cracking, chipping, erosion, or loose sections. This inspection should be carried out when the boiler is open for cleaning, or at least once a year.

Fuel solenoid valves and motorized valves should be visually checked by observing the fire when the unit shuts down. If the fire does not cut off immediately, the valve could be fouling or showing wear. If this occurs, the valve should be repaired or replaced immediately to avoid any serious problems.

All switches, controls, safety devices, and other equipment associated with the boiler should be periodically checked. Do not assume that all safety devices, switches, controls, etc., are operating properly. They should all be checked periodically on a planned maintenance schedule and any malfunctions noted and repaired.

Spare Parts

Be sure spare parts for your equipment are readily available. Consult the boiler manufacturer for a suggested parts list.

Manual

The operator's manual supplied with your boiler is an excellent guide to the control functions, control care, and control adjustments. It is important to know and use the instruction manual. The manual should be kept in a place where it will be readily available for the operator's use.

Water-Level Controls

The purpose of water-level controls is to maintain the water inside the boiler at the proper operating level. All water-level controls have a range of operations—not one set point. Water-level controls with gauge glasses should be so set that the water level is never out of sight, either low or high.

Water columns, gauge glasses, and low-water cutoffs on a steam boiler should be flushed at least once every shift. The purpose of this flushing is to prevent any accumulation of sludge or dirt that could possibly cause a control failure. When flushing the water column, it is advisable to test the operation of the low-water cutoff.

The water-level control that is most frequently found on boilers is a combination level indicator and low-water cutoff that is often incorporated in a water-column arrangement. This combination control allows for visual inspection of the water level, and in addition it functions to interrupt the electric current to a burner circuit in the event that an unsafe water level should develop.

Local water conditions and the introduction of treatment chemicals to a boiler will vary the amount of sediment accumulation in a control float bowl or a water column. For heating boilers and power boilers it is recommended that the boiler safety control be blown down regularly at least once a week when the boiler is in operation; however, power boilers may require a more frequent blowdown depending on operating and water conditions. When blowing down a control, it is advisable to check the operation of the low-water cutoff at a low-fire burner setting.

Monthly, the low-water cutoff should be tested under actual operating conditions. With the burner operating and the boiler steaming at the proper water level, close all the valves in the feedwater and condensate return lines for the duration of the test and shut off the feedwater pump, if required, so that the boiler will not receive any placement water. Then carefully observe the waterline to determine where the cutoff switch stops the burner in relation to the lowest permissible waterline established by the boiler manufacturers. The boiler water level should never be allowed to drop below the lowest visible part of the water-gauge glass. If the cutoff does not function during this test, immediately stop the operation of the burner. Then determine the cause of the failure and remedy it. The slow steaming evaporation test should then be repeated to verify that the control does function correctly.

If the burner cutoff level is not at, or slightly above, the lowest permissible water level, the low-water cutoff should be moved to the proper elevation, or should be serviced, repaired, or even replaced if necessary.

The low-water cutoff should be dismantled and checked at yearly intervals by a qualified technician to the extent necessary to ensure freedom from obstructions and proper functioning of the working parts. Inspect connecting lines to the boiler for any accumulation of sediment or scale, and clean as required. Examine all visible wiring for brittle or worn insulation, make sure electric contacts are clean, and where applicable, check mercury switches for any discoloration or mercury separation. Normally, operating mechanisms should not be repaired in the field. Replacement parts and complete replacement mechanisms, including necessary gaskets and installation instructions, are available from the manufacturer.

On boilers, the low-water cutoff may be checked periodically by manually tripping the control. Instructions on the method of tripping the specific control are found in the operator's manual supplied by the manufacturer.

Allied Boiler Equipment

It is recommended that a thorough check be made of all grease fittings, oil fittings, and other lubrication points and that a maintenance program be instituted to assure proper lubrication of all moving parts as required by the manufacturer. Such items as air compressors, blower bearings, motors, and other mechanically operated equipment do require lubrication from time to time. Consult the instruction manual for the detailed points and instructions; do not fail to establish a continuing maintenance program.

Depending upon the particular conditions of installation, such items as oil strainers, air filters, screens, etc., will accumulate foreign matter. All filters, screens, and strainers should be periodically cleaned to avoid any obstruction or malfunction. A smoke condition in the burner can be caused by an obstructed air inlet to the blower or by an obstructed compressor air filter just as well as by excess fuel input. A reduction in oil flow can be caused by a dirty oil strainer. A regular maintenance program should be set up according to the particular installation conditions to prevent any accumulation of foreign material.

Fuel-Oil System

Tanks should be checked annually for the presence of water and sludge. Fill should be checked for tightness of covers and proper gaskets after each delivery. Make certain the fill box is above grade to prevent water seepage into the tanks. Check vent pipes for obstructions.

Heaters

Heaters should be checked annually for the presence of water or sludge and for "coking" of heat-exchanger surfaces.

Pumps

Pumps should be checked at least monthly for leaky shaft sills and worn or loose drive mechanisms.

Piping

Oil lines should be checked at least monthly for external leaks and damage.

General Precautions

The oil-line vacuum gauge should be checked daily. Gauge readings that increase or are erratic indicate potential trouble. In this case, strainers should be checked and cleaned, and oil lines checked for obstructions or internal sludge buildup.

When taking an oil circulation and heating system out of service, precautions should be taken so that start-up will not be obstructed by congealed fuel oil. This may require flushing the lines with a lighter-grade oil and/or shutting down the system on the lighter-grade fuel.

Electric Contacts

Electric contacts on such items as starters, contactors, and controls should be periodically checked for cleanliness and arc burns. Dust should not be permitted to accumulate on any contact. Covers on controls must be in place and control cabinet doors should be closed except during the period when access is actually required for service. ***CAUTION****—Power must be shut off before any control cover or cabinet is opened.*

Steam Systems

It is recommended that a periodic check of the steam system be conducted to prevent boiler malfunction due to external problems.

- Check to assure that cold makeup water is not being fed into a hot boiler.
- Check the feedwater-treatment equipment to make sure that it is operating properly.
- Check items such as feed pumps, valves, and other equipment to assure proper performance.
- Check safety valves in accordance with instructions by ASME Boiler & Pressure Vessel Code, Sec. VI, "Recommended Rules for Care and Operation of Heating Boilers."

Hot-Water Systems

The following checks should be made periodically to prevent boiler problems:

- Expansion tanks should be checked to ensure proper performance.
- Water circulators should be checked to make sure that circulation is maintained in all operating conditions.
- Extreme fluctuations of pressure gauges can be indicators of system problems. Check the system for a possible lock in the expansion tank. The air-removal device or connection on top of the boiler should be checked to ensure that it is functioning properly.
- Piping should be checked to make sure that there are no stoppages due to the piping being air-bound.
- Air vents at the high points of the systems should be checked to ensure that they are not bleeding air into the system during pump starts.
- Weeping safety valves not only cause undesirable watermarks and steam losses but may indicate valve malfunction. A check for excessive amounts of system makeup water should be made by means of a water meter on the inlet line.
- A check should be maintained whenever the equipment is switching from a cooling to a heating cycle to ensure that the boiler is not shocked by extreme system changes. When a boiler is used with a cooling/heating system, switching to the chilling cycle should be handled with care. Proper controls should be installed in the system to prevent chilled water from entering the boiler.

When a hot-water system is in use, it is recommended that circulation be maintained. If it is necessary to shut down the circulators, the system will cool down to ambient temperature and can then cause damage at start-up. This is referred to as *thermal shock*. Particular instructions and recommendations made in the instruction manual should be observed to assure the long life of the boiler.

When a boiler is being drained for inspection, it is recommended that a flow of water be maintained in the boiler with a high-pressure hose to keep any sediment thoroughly agitated and in suspension to prevent caking of the sludge that can be extremely difficult to remove at a later date.

Recommended Periodic Testing and Verification

Once equipment has been placed in service, it becomes the owner's responsibility to maintain it. Maintenance should include periodic testing and verification of controls and safety devices. Records or logs of such maintenance should be kept by the owner and/or boiler operator.

The maintenance and testing is in addition to those inspections required by the various governmental agencies or insurance companies. A list showing the recommended frequency of periodic testing and verification is found in Table 2-1.

BIBLIOGRAPHY

"Steam/Its Generation and Use," Babcock & Wilcox Co., Wilmington, N.C., 1980.
"Combustion Engineering," Combustion Engineering, Inc., Stamford, Conn., 1967.
Packaged Firetube Engineering Manual, American Boiler Manufacturers Association, 1971.
Boiler & Pressure Vessel Code, American Society of Mechanical Engineers (ASME), New York, 1980.
Carl D. Shields: *Boilers*, McGraw-Hill, New York, 1961.

TABLE 2-1 Recommended Periodic Testing and Verification Checklist

Item	Frequency	Accomplished by	Remarks
Gauges, monitors, & indicators	Daily	Operator	Make visual inspection and record readings in log
Instrument & equipment settings	Daily	Operator	Make visual check against factory-recommended specifications
Firing rate control	Weekly	Operator	Visual inspection
	Semi-annually	Service technician	Verify factory settings; check with combustion test instruments
Fuel valves			
Pilot valves, Main gas valves, Main oil valves	Weekly	Operator	Open limit switch; make audible and visual check; check valve position indicators; check fuel meters
	Annually	Service technician	Perform leakage tests; refer to manufacturer's instructions
Combustion safety controls			
Flame failure	Weekly	Operator	Close manual fuel supply for (1) pilot, (2) main fuel cock and/or valve(s); check safety shutdown timing; log.
Flame signal strength	Weekly	Operator	If flame signal meter has been installed, read and log; for both pilot and main flames, notify service organization if readings are very high, very low, or fluctuating. Refer to manufacturer's instructions.
Pilot turn down tests	As required/annually	Service technician	Required after any adjustments to flame scanner mount or pilot burner; verify annually.
Refractory hold in	As required/annually	Service technician	See pilot turn down test.
Low-water cutoff	Monthly	Operator	
High limit safety control	Annually	Service technician	Refer to manufacturer's instructions.
Operating control	Annually	Service technician	Refer to manufacturer's instructions.
Low draft interlock	Annually	Service technician	Refer to manufacturer's instructions.
Atomizing air steam interlock	Annually	Service technician	Refer to manufacturer's instructions.
High- & low-gas-pressure interlock	Annually	Service technician	Refer to manufacturer's instructions.
High- & low-oil-pressure interlock	Annually	Service technician	Refer to manufacturer's instructions.
High- & low-oil-temperature interlock	Annually	Service technician	Refer to manufacturer's instructions.
Fuel valve interlock switch	Annually	Service technician	Refer to manufacturer's instructions.
Purge switch	Annually	Service technician	Refer to manufacturer's instructions.
Burner position interlock	Annually	Service technician	Refer to manufacturer's instructions.
Rotary cup interlock	Annually	Service technician	Refer to manufacturer's instructions.
Low fire start interlock	Annually	Service technician	Refer to manufacturer's instructions.
Automatic changeover control (dual fuel)	At least annually	Service technician	Under supervision of gas utility.
Safety valves	As required	Operator	In accordance with procedure in ASME boiler code.

chapter 4-3

Fuels and Combustion Equipment*

by
P. ERIC RALSTON
Manager, Engineering Services
Industrial and Marine Division
Babcock & Wilcox Co.
Wilmington, North Carolina

*A portion of this material was adapted by permission from *Steam/It's Generation and Use,* published by Babcock & Wilcox Co. in 1980.

GLOSSARY

Grindability A term used to measure the ease of pulverizing a coal in comparison with a standard coal chosen as 100 grindability.

Gross (higher) heating value The heat released from the combustion of a unit of fuel quantity (mass) with the products in the form of ash, gaseous CO_2, SO_2, N_2, and liquid water exclusive of any water added directly as vapor.

Net (lower) heating value Calculated from the gross heating value as the heat produced by a unit quantity of fuel when all water in the products remains as vapor. This calculation (ASTM Standard D 407) is made by deducting 1030 lb of water derived from the fuel, including both the water originally present as moisture and that formed by combustion.

Proximate analysis The determination of moisture, volatile matter, and ash and the calculation of fixed carbon by difference.

Ultimate analysis The determination (using a dried sample) of carbon, hydrogen, sulfur, nitrogen, and ash and the estimation of oxygen by difference.

SOLID FUELS

CHARACTERISTICS

Coal

Coal Analysis

Customary practice in reporting the components of a coal is to use proximate and ultimate analyses (see Glossary).

The scope of each is indicated in the analyses of a West Virginia coal (Table 3-1) in which the ultimate analysis has been converted to the as-received basis. The analysis on the as-received basis includes the total moisture content of the coal received at the plant. Similarly, the as-fired basis includes the total moisture content of the coal as it enters the boiler furnace or pulverizer. Standard laboratory procedures for making these analyses appear in ASTM D 271, "Sampling & Analysis, Laboratory, Coal & Coke."

A list of other testing standards is given in Table 3-2.

TABLE 3-1 Coal Analyses on As-Received Basis
(Pittsburgh Seam Coal, West Virginia)

Proximate Analysis		Ultimate Analysis	
Component	Weight, %	Component	Weight, %
Moisture	2.5	Moisture	2.5
Volatile matter	37.6	Carbon	75.0
Fixed carbon	52.9	Hydrogen	5.0
Ash	7.0	Sulfur	2.3
Total	100.0	Nitrogen	1.5
		Oxygen	6.7
		Ash	7.0
Heating value, Btu/lb	13,000	Total	100.0

ASTM Classification by Rank

Coals are classified in order to identify their end use and also to provide data useful in specifying and selecting burning and handling equipment and in the design and arrangement of heat-transfer surfaces.

One classification of coal is by rank, i.e., according to the degree of metamorphism, or progressive alteration, in the natural series from lignite to anthracite. In the ASTM classification, the basic criteria are the fixed-carbon content and the calorific values (in British thermal units) calculated on a mineral-matter-free basis.

In establishing the rank of coals, it is necessary to use information showing an appreciable and systematic variation with age. For the older coals, a good criterion is the "dry, mineral-matter-free fixed carbon or volatile [matter]." However, this value is not suitable for designating the rank of the younger coals. A dependable means of classifying the latter is the "moist, mineral-matter-free Btu," or calorific value, which varies little for the older coals but appreciably and systematically for younger coals. Table 3-3, ASTM D 388, is used for classification according to rank or age.

The basis for the two ASTM criteria (the fixed-carbon content and the calorific value calculated on a moist, mineral-matter-free basis) are shown in Fig. 3-1 for over 300 typical coals of the United States. The classes and groups of Table 3-3 are indicated in Fig. 3-1. For the anthracitic and low- and medium-volatile bituminous coals, the moist, mineral-matter-free calorific value changes very little; hence the fixed-carbon criterion is used. Conversely, in the case of the high-volatile bituminous, subbituminous, and lignitic

TABLE 3-2 ASTM Standards for Testing Coal, Specifications, and Definitions of Terms

ASTM Standards for Testing Coal	
*D 1756	Carbon Dioxide in Coal
*D 2361	Chlorine in Coal
*D 291	Cubic Foot Weight of Crushed Bituminous Coal
*D 440	Drop Shatter Test for Coal
*D 547	Dustiness, Index of, of Coal and Coke
*D 1857	Fusibility of Coal Ash
*D 1412	Equilibrium Moisture of Coal at 96 to 97% Relative Humidity and 30°C
*D 2014	Expansion or Contraction of Coal by the Sole-Heated Oven
*D 720	Free-Swelling Index of Coal
D 409	Grindability of Coal by the Hardgrove-Machine Method
*D 2015	Gross Calorific Value of Solid Fuel by the Adiabatic Bomb Calorimeter
D 1812	Plastic Properties of Coal by the Gieseler Plastometer
D 2639	Plastic Properties of Coal by the Automatic Gieseler Plastometer
D 197	Sampling and Fineness Test of Powdered Coal
*D 271	Sampling and Analysis, Laboratory, of Coal and Coke
D 492	Sampling Coals Classified According to Ash Content
*D 2234	Sampling, Mechanical, of Coal
*D 2013	Samples, Coal, Preparing of Analysis
*D 410	Screen, Analysis of Coal
D 311	Sieve Analysis of Crushed Bituminous Coal
D 310	Size of Anthracite
*D 431	Size of Coal, Designating from its Screen Analysis
*D 1757	Sulfur in Coal Ash
*D 2492	Sulfur, Forms of, in Coal
*D 441	Tumbler Test for Coal
Specifications	
*D 388	Classification of Coals by Rank
*E 11	Wire-Cloth Sieves for Testing Purposes
E 323	Perforated-Plate Sieves for Testing Purposes
Definitions of Terms	
*D 121	Coal and Coke
D 2796	Lithological Classes and Physical Components of Coal
*D 407	Gross Calorific Value and Net Calorific Value of Solid and Liquid Fuels

*Approved as American National Standard by the American National Standards Institute.

coals, the moist, mineral-matter-free calorific value is used, since the fixed-carbon value is almost the same for all classifications.

Other Classifications of Coal by Rank

There are other classifications of coal by rank (or type) which are currently in limited use on the European continent. These are the International Classification of Hard Coals by Type, and the International Classification of Brown Coals. Other criteria for the classification of coal by rank have been proposed by various authorities.

Volatile Matter and Heating Value

The composition of the fixed carbon in all types of coal is substantially the same. The variable constituents of coals can, therefore, be considered as concentrated in the volatile matter. One index of the quality of the volatile matter, its heating value, is perhaps the most important property as far as combustion is concerned, and this bears a direct relation to the properties of the pure coals (dry, mineral-matter-free). The volatile matter in coals of lower rank is relatively high in water and CO_2 and consequently low in heating value. The volatile matter in coals of higher rank is relatively high in hydrocarbons, such as methane (CH_4), and consequently is relatively high in heating value.

The relationship of the heating value of the volatile matter to the heating value of the pure coal is shown in Fig. 3-2 for a large number of coals.

TABLE 3-3 Classification of Coals by Rank* (ASTM D 388)

Class	Group	Fixed carbon limits, % (dry, mineral-matter-free basis)		Volatile matter limits, % (dry, mineral-matter-free basis)		Calorific value limits, Btu/lb (moist,† mineral-matter-free basis)		Agglomerating character
		Equal or greater than	Less than	Greater than	Equal or less than	Equal or greater than	Less than	
I. Anthracitic	1. Meta-anthracite	98	—	—	2	—	—	Nonagglomerating
	2. Anthracite	92	98	2	8	—	—	Nonagglomerating
	3. Semianthracite‡	86	92	8	14	—	—	Nonagglomerating
II. Bituminous	1. Low-volatile bituminous coal	78	86	14	22	—	—	Commonly agglomerating¶
	2. Medium-volatile bituminous coal	69	78	22	31	—	—	Commonly agglomerating¶
	3. High-volatile *A* bituminous coal	—	69	31	—	14,000§	—	Commonly agglomerating¶
	4. High-volatile *B* bituminous coal	—	—	—	—	13,000§	14,000	Commonly agglomerating¶
	5. High-volatile *C* bituminous coal	—	—	—	—	11,500	13,000	Commonly agglomerating¶
						10,500	11,500	Agglomerating
III. Subbituminous	1. Subbituminous *A* coal	—	—	—	—	10,500	¶11,500	Nonagglomerating
	2. Subbituminous *B* coal	—	—	—	—	9,500	¶10,500	Nonagglomerating
	3. Subbituminous *C* coal	—	—	—	—	8,300	¶ 9,500	Nonagglomerating
IV. Lignitic	1. Lignite *A*	—	—	—	—	6,300	8,300	Nonagglomerating
	2. Lignite *B*	—	—	—	—	—	6,300	Nonagglomerating

*This classification does not include a few coals, principally nonbanded varieties, which have unusual physical and chemical properties and which come within the limits of fixed-carbon or calorific value of the high-volatile bituminous and subbituminous ranks. All of these coals either contain less than 48 percent dry, mineral-matter-free fixed carbon or have more than 15,500 moist, mineral-matter-free British thermal units per pounds.

†Moist refers to coal containing its natural inherent moisture but not including visible water on the surface of the coal.

‡If agglomerating, classify in low-volatile group of the bituminous class.

§Coals having 69 percent or more fixed carbon on the dry, mineral-matter-free basis shall be classified according to fixed carbon, regardless of calorific value.

¶It is recognized that there may be nonagglomerating varieties in these groups of the bituminous class, and there are notable exceptions in high-volatile *C* bituminous group.

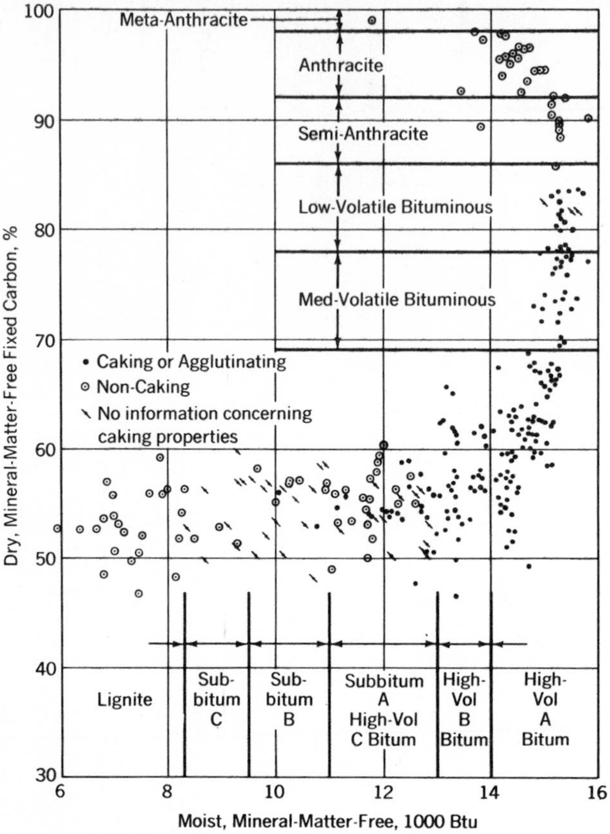

Figure 3-1 Distribution plot for over 300 coals of the United States, illustrating ASTM classification by rank as defined in Table 3-3.

Commercial Sizes of Coal

Bituminous. Sizes of bituminous coal are not well standardized, but the following sizings are common:

Run of Mine. This is coal shipped as it comes from the mine without screening.

Run of Mine (8 in). This is run of mine with oversized lumps broken up.

Lump (5 in). This size will not go through a 5-in round hole.

Egg (5 × 2 in). This size goes through 5-in holes and is retained on 2-in round-hole screens.

Nut (2 × 1¼ in). This size is used for small industrial stokers and for hand firing.

Stoker Coal (1¼ × ¾ in). This is used largely for small industrial stokers and for domestic firing.

Slack (¾ in and under). This is used for pulverizers and industrial stokers.

Anthracite. Definite sizes of anthracite are standardized in Table 3-4.

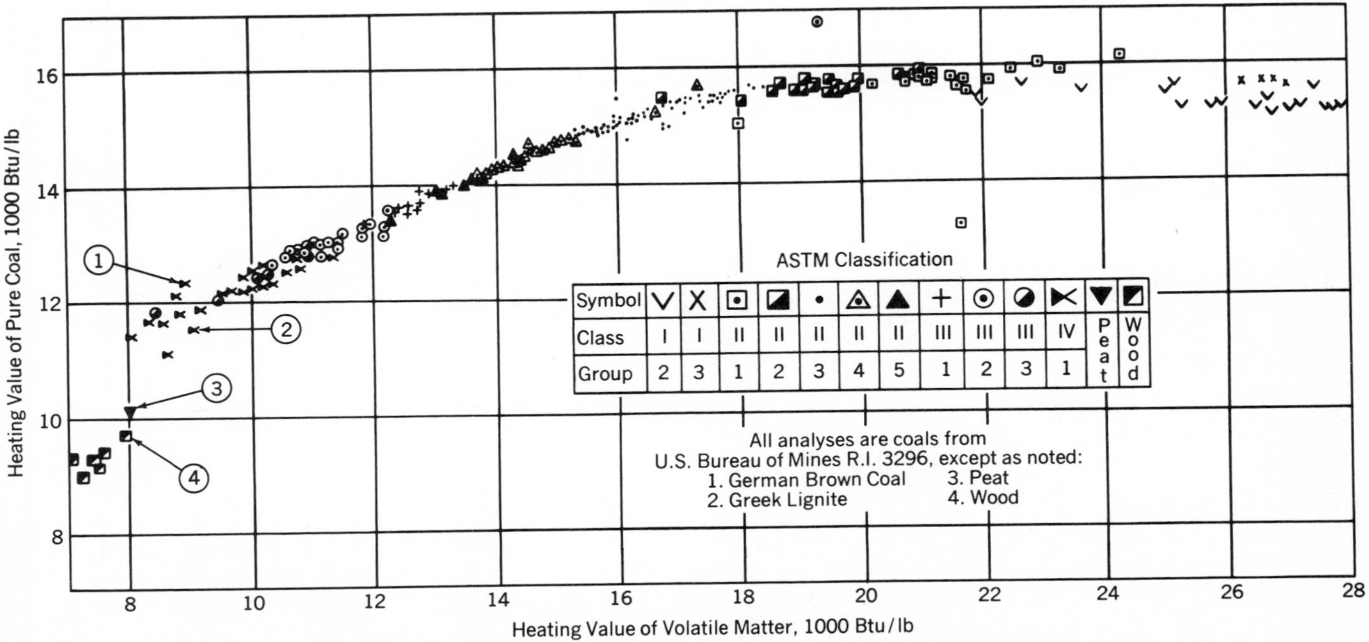

Figure 3-2 To illustrate a suggested coal classification using the relationship of the respective heating values of "pure coal" and the volatile matter.

TABLE 3-4 Commercial Sizes of Anthracite (ASTM D 310)
(Graded on round-hole screens)

	Diameter of Holes, in	
Trade name	Through	Retained on
Broken	4 3/8	3 1/4 to 3
Egg	3 1/4 to 3	2 7/16
Stove	2 7/16	1 5/8
Nut	1 5/8	13/16
Pea	13/16	9/16
Buckwheat	9/16	5/16
Rice	5/16	3/16

Moisture Determination

Moisture in coal is determined quantitatively by definite prescribed methods. It is preferable to determine the moisture in two steps: (1) prescribed air drying to equilibrium at 10 to 15°C above room temperature, and (2) prescribed oven drying for 1 h at 104 to 110°C, after pulverizing (see ASTM Standard D 271).

Since it is the surface moisture that must be evaporated prior to efficient pulverizing of coal, the air-dried component of the total moisture value should be reported separately.

ASTM Standard D 1412 provides a means of estimating the bed moisture of either wet coal showing visible surface moisture or coal that may have lost some moisture. It may be used for estimating the surface moisture of wet coal, i.e., the difference between total moisture, as determined by ASTM Standard D 271, and equilibrium moisture.

Grindability Index

A coal is harder or easier to grind if its grindability index is less or greater, respectively, than 100. The capacity of a pulverizer is related to the grindability index of the coal.

Coal Ash

The presence of ash is accounted for by minerals associated with initial vegetal growth or by those which entered the coal seam from external sources during or after the period of coal formation.

The composition of the coal ash is customarily determined by chemical analysis of the residue produced by burning a sample of coal at a slow rate and at moderate temperature (1350°F) under oxidizing conditions in a laboratory furnace. It is thus found to be composed chiefly of compounds of silicon, aluminum, iron, and calcium, with smaller amounts of magnesium, titanium, sodium, and potassium. The analyses of coal ash in Table 3-5 indicate what may be expected of selected coals from various areas of the United States.

Coals may be classified into two groups on the basis of the nature of their ash constituents. One is the bituminous-type ash and the other is the lignite-type ash. *Lignite-type ash* is ash having more CaO plus MgO than Fe_2O_3. By contrast, *bituminous-type ash* has more Fe_2O_3 than CaO plus MgO.

Free-Swelling Index. The ASTM Standard method D 720 is used for obtaining information regarding the free-swelling property of a coal. Since it is a measure of the behavior of rapidly heated coal, it may be used as an indication of the caking characteristics of coal burned as a fuel.

Ash Fusibility. The determination of ash fusion temperatures (initial deformation, softening, hemispherical, and fluid) is a laboratory procedure, developed in standardized form (ASTM Standard D 1857).

TABLE 3-5 Ash Content and Ash Fusion Temperatures of Some U.S. Coals and Lignite

	Rank						
	Low-volatile bituminous	High-volatile bituminous				Subbituminous	Lignite
Seam	Pocahontas No. 3	No. 9	Pittsburgh	No. 6			
Location	West Virginia	Ohio	West Virginia	Illinois	Utah	Wyoming	Texas
Ash, dry basis, %	12.3	14.10	10.87	17.36	6.6	6.6	12.8
Sulfur, dry basis, %	0.7	3.30	3.53	4.17	0.5	1.0	1.1
Analysis of ash, % by wt							
SiO_2	60.0	47.27	37.64	47.52	48.0	24.0	41.8
Al_2O_3	30.0	22.96	20.11	17.87	11.5	20.0	13.6
TiO_2	1.6	1.00	0.81	0.78	0.6	0.7	1.5
Fe_2O_3	4.0	22.81	29.28	20.13	7.0	11.0	6.6
CaO	0.6	1.30	4.25	5.75	25.0	26.0	17.6
MgO	0.6	0.85	1.25	1.02	4.0	4.0	2.5
Na_2O	0.5	0.28	0.80	0.36	1.2	0.2	0.6
K_2O	1.5	1.97	1.60	1.77	0.2	0.5	0.1
Total	98.8	98.44	95.74	95.20	97.5	86.4	84.3
Ash fusibility							
Initial deformation temperature, °F							
Reducing	2900+	2030	2030	2000	2060	1990	1975
Oxidizing	2900+	2420	2265	2300	2120	2190	2070
Softening temperature, °F							
Reducing		2450	2175	2160		2180	2130
Oxidizing		2605	2385	2430		2220	2190
Hemispherical temperature, °F							
Reducing		2480	2225	2180	2140	2250	2150
Oxidizing		2620	2450	2450	2220	2240	2210
Fluid temperature, °F							
Reducing		2620	2370	2310	2250	2290	2240
Oxidizing		2670	2540	2610	2460	2300	2290

Ash melts when heated to a sufficiently high temperature. If insufficiently cooled, the ash particles remain molten or sticky and tend to coalesce into large masses in the boiler furnace or other heat-absorption surfaces. This problem is dealt with by adequate design of burners and furnace arrangement, knowing the ash fusion temperatures for the fuels to be burned.

Viscosity of Coal-Ash Slag. Measurement of viscosity of coal-ash slags provides reliable data that can be used for determining suitability of coals for use in slag-tap-type boilers.

Other Solid Fuels

Coke

When coal is heated with a large deficiency of air, the lighter constituents are volatilized and the heavier hydrocarbons crack, liberating hydrogen and leaving a residue of carbon. The carbonaceous residue containing the ash and a part of the sulfur of the original coal is called *coke*.

Undersized coke, called *coke breeze*, usually passing a $\frac{3}{8}$-in screen, is unsuited for charging blast furnaces and is available for steam generation. A typical analysis of coke breeze appears in Table 3-6.

Coke from Petroleum

The heavy residuals from the various petroleum cracking processes are presently utilized to produce a higher yield of lighter hydrocarbons and a solid residue suitable for fuel. Solid fuels from oil include delayed coke, fluid coke, and petroleum pitch. Some selected analyses are given in Table 3-7.

Some of these cokes are easy to pulverize and burn, while others are quite difficult.

TABLE 3-6 Selected Analyses of Bagasse, Coke Breeze, and Fluidized-Bed Char

Analyses (as fired), wt %	Bagasse	Coke breeze	Fluidized-bed char
Proximate			
Moisture	52.0	7.3	0.7
Volatile matter	40.2	2.3	14.7
Fixed carbon	6.1	79.4	70.4
Ash	1.7	11.0	14.2
Ultimate			
H_2 Hydrogen	2.8	0.3	—
C Carbon	23.4	80.0	—
S Sulfur	Trace	0.6	4.1
N_2 Nitrogen	0.1	0.3	—
O_2 Oxygen	20.0	0.5	—
H_2O Moisture	52.0	7.3	—
A Ash	1.7	11.0	—
Heating value, Btu/lb	4000	11,670	12,100

TABLE 3-7 Selected Analyses of Solid Fuels Derived from Oil

Analyses (dry basis), wt %	Delayed	Coke	Fluid	Coke
Proximate				
Volatile matter	10.8	9.0	6.0	6.7
Fixed carbon	88.5	90.9	93.7	93.2
Ash	0.7	0.1	0.3	0.1
Ultimate				
Sulfur	9.9	1.5	4.7	5.7
Heating value, Btu/lb	14,700	15,700	14,160	14,290

The low-melting-point pitches may be heated and burned like heavy oil, while those with higher melting points may be pulverized and burned.

Wood

Analyses of wood ash and selected analyses and heating values of several types of wood are given in Table 3-8.

Wood or bark with a moisture content of 50 percent or less burns quite well; however, as the moisture content increases above this level, combustion becomes more difficult. With moisture content above 65 percent, a large part of the heat in the wood is required to evaporate the moisture, and little remains for steam generation.

Bagasse

Bagasse is the refuse from the milling of sugar cane. It consists of matted cellulose fibers and fine particles. Mills grinding sugar cane commonly use bagasse for steam production. A selected analysis of bagasse is given in Table 3-6.

Other Vegetable Wastes

Food and related industries produce numerous vegetable wastes that are usable as fuels. They include such materials as grain hulls, the residue from the production of furfural from corncobs and grain hulls, coffee grounds from the production of instant coffee, and tobacco stems.

Solvent-Refined Coal

Solvent-refined coal is made by dissolving the organic material in coal in a coal-derived solvent. The finished product is a solid with a melting point of 284 to 293°F (140 to 145°C). It can be burned as an oil or a solid.

TABLE 3-8 Analyses of Wood and Wood Ash

Analyses	Pine bark	Oak bark	Spruce bark*	Redwood bark*
Wood analyses (dry basis), wt %				
Proximate				
Volatile matter	72.9	76.0	69.6	72.6
Fixed carbon	24.2	18.7	26.6	27.0
Ash	2.9	5.3	3.8	0.4
Ultimate				
Hydrogen	5.6	5.4	5.7	5.1
Carbon	53.4	49.7	51.8	51.9
Sulfur	0.1	0.1	0.1	0.1
Nitrogen	0.1	0.2	0.2	0.1
Oxygen	37.9	39.3	38.4	42.4
Ash	2.9	5.3	3.8	0.4
Heating value, Btu/lb	9030	8370	8740	8350
Ash analyses, wt %				
SiO_2	39.0	11.1	32.0	14.3
Fe_2O_3	3.0	3.3	6.4	3.5
TiO_2	0.2	0.1	0.8	0.3
Al_2O_3	14.0	0.1	11.0	4.0
Mn_3O_4	Trace	Trace	1.5	0.1
CaO	25.5	64.5	25.3	6.0
MgO	6.5	1.2	4.1	6.6
Na_2O	1.3	8.9	8.0	18.0
K_2O	6.0	0.2	2.4	10.6
SO_3	0.3	2.0	2.1	7.4
Cl	Trace	Trace	Trace	18.4
Ash fusibility, °F				
Reducing				
Initial deformation	2180	2690		
Softening	2240	2720		
Fluid	2310	2740		
Oxidizing				
Initial deformation	2210	2680		
Softening	2280	2730		
Fluid	2350	2750		

*Saltwater stored.

Municipal Solid Waste

Municipal solid waste (MSW) can be burned as received or by preparing a refuse-derived fuel (RDF) by shredding the MSW and removing ferrous and nonferrous inorganic materials. The heating value of MSW can range from 3500 to 6500 Btu/lb, and the moisture and ash can vary just as widely in similar or opposite directions.

COAL PROCESSING AND TRANSPORTING

The purpose of coal preparation is to improve the quality of the coal or to make it suitable for a specific purpose by (1) cleaning to remove inorganic impurities, (2) special treatment (such as dedusting), or (3) sizing by crushing and screening.

Economic Factors

All coals have certain properties which place limitations on their most advantageous use.

To define the limitations of various types of coal-burning equipment in service, specifications covering several important properties of coal are necessary. For certain types of stokers, for instance, a minimum ash-softening temperature, maximum ash content, and maximum sulfur content of the coal are specified to prevent excessive clinkering. For pulverized-coal firing, it is necessary to specify ash-slagging and ash-fouling parameters for a dry-ash installation. Within these and other equipment limitations there is usually a wide range of coals that can be satisfactorily burned in a specific steam boiler, and the choice depends primarily on economics—which coal will produce steam at the lowest overall cost, including cost at the mine, shipment, storage, handling, operating, and maintenance costs. To burn a wide range of coals often requires a larger initial investment than would otherwise be necessary. However, fuel cost represents a large part of the overall operating cost.

Nature of Impurities in Coal

Impurities can be divided into two general classifications—inherent and removable. The inherent impurities are inseparably combined with the coal. The removable impurities are segregrated and can be eliminated, by available cleaning methods, to the extent that is economically justified.

Mineral matter is always present in raw coal and forms ash when the coal is burned. The ash-forming mineral matter is usually classified as either inherent or extraneous. Ash-forming material organically combined with the coal is considered as inherent mineral matter. Generally the inherent mineral matter contained in coal is about 2 percent or less of the total ash. Extraneous mineral matter is ash-forming material that is foreign to the plant material from which the coal was formed. It consists usually of slate, shale, sandstone, or limestone.

Sulfur is always present in raw coal in amounts ranging from traces to as high as 8 percent or more. This results in the emission of sulfur oxides in the stack gases when the coal is burned.

Moisture, which is inherent in coal, may be considered to be an impurity. It varies with different ranks of coal, increasing for lower-rank coals from 1 to 2 percent in anthracite to 45 percent or more in lignite. Moisture that collects on the exposed surfaces of coal is commonly called *surface moisture* and is removable.

Cleaning Methods

The principal benefit of cleaning is the reduction in ash content. With ash content reduced, shipping costs and the requirements for storage and handling decrease because of the smaller quantity of coal necessary per unit of heating value.

Cleaning at Mine Face

Efforts to reduce the impurities loaded in the mine usually result in increased mining cost; hence economics plays an important part in determining the amount of cleaning done.

Mechanical Picking

Many mechanical picking devices employ differences in physical picking dimensions of coal and impurities as a means of separation. Bituminous coal fractures into rough cubes, whereas slate and shale normally fracture as thin slabs. A slot-type flat picker can be installed on shaker screens, shaking conveyors, and chutes.

Froth Flotation

The incoming coal feed may be agitated in a controlled amount of water, air, and reagents that cause a surface froth to form, the bubbles of which selectively attach themselves to coal particles and keep them buoyant while the heavier particles of pyrite, slate, and shale remain dispersed in the water. This method can be used in cleaning smaller particles, $\frac{1}{10}$ in by 0 (no minimum), particularly particles smaller than 48 mesh.

Gravity Concentration

Removal of segregated impurities in coal by gravity concentration is based on the principle that heavier particles separate from the lighter ones when settling in a fluid. This principle is applicable because most common solid impurities are heavier than coal (Table 3-9).

The commercial processes used in gravity concentration can be divided into two main classifications—wet and dry (or pneumatic).

TABLE 3-9 Specific Gravities of Coal and Impurities

Material	Specific gravity
Bituminous coal	1.12–1.35
Bone coal	1.35–1.7
Carbonaceous shale	1.6–2.2
Shale	2.0–2.6
Clay	1.8–2.2
Pyrite	4.8–5.2

Cleaning by Gravity Concentration

Wet Processes

Most cleaning of bituminous coal and lignite is accomplished by wet processes. Table 3-10 lists the methods and types of equipment generally used and the extent of application of each method, expressed in percent.

TABLE 3-10 Types of Wet Processes Used for Cleaning of Bituminous Coals and Lignite

Equipment or method used	Percent of cleaning done by wet processes
Jigs	46.7
Dense-media method	29.2
Concentrating tables	13.9
Flotation method	2.6
Classifiers	1.4
Launderers	1.3
Total	95.1

Dry Processes

The principal pneumatic cleaning process uses tables equipped with riffles, with air as the separating medium. Admitted through holes in the tables, the air is blown up through the bed of coal. The motion of the tables plus the airflow segregates the coal and impurities.

Washability Characteristics of Coal

There is a general correlation between specific gravity and ash content, although the relation differs for various coals. Ash contents corresponding to various specific gravities for bituminous coal are shown in Table 3-11.

Float-and-sink tests run in a laboratory provide data useful for rating and controlling cleaning equipment and evaluating efficiencies obtained, both quantitative and qualitative.

TABLE 3-11 Ash Content of Various Gravity Fractions in Bituminous Coal

Specific gravity	Ash content, %
1.3–1.4	1–5
1.4–1.5	5–10
1.5–1.6	10–35
1.6–1.8	35–60
1.8–1.9	60–75
1.9 and above	75–90

Special Treatment

Dedusting

Dedusting is a type of air separation with classification according to size. Air passed through the coal entrains a large percentage of the "fines." The fine coal is recovered from the air with cyclone separators and bag filters. This process is often employed to remove fines prior to wet cleaning.

Mechanical Dewatering and Heat Drying

The larger sizes of coal (above ⅜ in) can be easily dewatered, and natural drainage is sometimes provided by special hoppers (or bins), screen conveyors, perforated-bucket elevators, and fixed screens. However, when the fine sizes must be dried or when lower moisture content is required for the large coal, mechanical dewatering or thermal drying is necessary.

Mechanical dewatering devices may be shaker screens, vibrating screens, filters, centrifuges, and thickening or desliming equipment. Thermal drying is used to obtain low-moisture-content coal, especially for the finer sizes. Various types of thermal dryers are rotary, cascade, reciprocating-screen, conveyor, suspension (including flash and venturi dryers), and fluidized-bed.

Dustproofing

Oil and calcium chloride are commonly used for dustproofing coal. When coal is sprayed with oil, the film causes some dust particles to adhere to the larger pieces of coal and others to agglomerate into larger lumps not easily airborne. Calcium chloride absorbs moisture from the air, providing a wet surface to which the dust adheres.

Freezeproofing

To prevent freezing of coal during transit or storage, spray oil may be used, and it is applied to the coal in the same manner as dustproofing. As an alternative and less costly method, the car hoppers may be heavily sprayed with oil. Another freezeproofing method is thermal drying of the fine coal.

Sizing by Crushing and Screening

Sizing Requirements

Generally acceptable coal sizings for various types of fuel-burning equipment are given in the section covering the equipment. The degradation in coal sizing resulting from handling is an important consideration in establishing sizing specifications for steam plants, where the maximum permissible quantity of fines in the coal is set by firing-equipment limitations.

Crushers and Breakers

Many types of crushing and screening equipment are used commercially. Representative types are the Bradford breaker, single-roll crusher, double-roll crusher, and hammer mill.

Screens

Many types of screens are used for sizing, including the gravity bar screen or grizzly, revolving screen, shaker screen, and vibrating screen.

Transportation of Coal

Transportation costs may represent a large portion of the delivered cost of coal. Transportation may affect the condition of the received coal if freezing occurs in transit or if there is a change in moisture content or a degradation of size.

When freezing in transit is anticipated, its effect may be ameliorated by special treatment of the coal at the point of loading.

The moisture content of the coal will depend on sizing, condition at time of loading, and weather conditions during transit.

Size degradation depends in a large measure on the friability of the coal but it is also affected by the amount of handling and shaking in transit.

HANDLING AND STORAGE AT CONSUMER'S PLANT

Unloading

When coal is received by rail, if the surface moisture of the coal is high, it may be necessary to rap the car sides with a sledge or to use a slice bar from above in order to start the coal flowing. If this high-moisture coal has been in transit several days at freezing temperatures, it may be frozen into a solid mass, and unloading is a serious problem. In hot, dry weather the coal, on arrival, may be so dry that a high wind can blow the fines away as dust.

Frozen Coal

Successful equipment for unloading frozen coal includes steam-heated thawing sheds, oil-fired thawing pits, and the radiant-electric railroad-car thawing system. Another device is a car shakeout, in which a motor-driven eccentric shaker clamped to the top flanges of the car transmits a vibratory motion to the car body.

When expensive equipment is not economically justified, the methods used depend primarily on manual labor and slice bars, sledges, portable oil torches, and steam or hot water.

Mechanical Handling

Extensive equipment, covering the range of plant requirements from a few tons per day to the largest, is available for transferring coal from the railroad cars either to outside storage or to inside bunkers or hoppers. When pulverizers are used, it is desirable to include a magnetic separator somewhere in the system. Frequently a coal crusher will make it economically possible to use coal that is not screened or sized.

Outside Storage at or near Plant

Outside storage of coal at or near the plant site is often necessary, and in some cases required, to assure continuous plant operation.

The changes that may affect the value of stored coal are loss in heating value, reduction of coking power, reduction of average particle size (weathering or slacking), and, most important, losses from self-ignition or spontaneous combustion. Direct loss of stored coal by wind or water erosion is possible and may become a serious nuisance.

Oxidation of Coal

Many constituents of pure coal begin to oxidize when exposed to air. Oxidation of coal is affected by length of time exposed to air, coal rank, size or surface area, temperature, amount and size of pyrites particles, moisture, and amount of mineral matter or ash.

All coals may be safely stored, provided proper procedures are used. Coal can be successfully stored by following methods outlined by the Bureau of Mines.

Selection of Storage Site

The site should be level solid ground, free of loose fill, properly graded for drainage, and compacted by a bulldozer. Consideration should be given to access and provision for coal delivery as well as to protection from prevailing winds, tides, flooded rivers, and spray from salt water.

Storage in Small Amounts

No special precautions are required for anthracite. Sized (double-screened) bituminous coals may be stored in small conical piles from 5 to 15 ft high. Slack sizes of bituminous coal may be compacted with a bulldozer. Subbituminous coal and lignite should be stored in small piles and thoroughly packed.

Storage in Large Amounts

Anthracite coal is very safe to store in large high piles that permit drainage.

Bituminous coal should be stockpiled in multiple horizontal layers. Each layer should be spread out to vary from 1 to 2 ft in thickness and thoroughly packed to eliminate air spaces. The top should be slightly crowned and symmetrical to permit even runoff of water. All sides and the top should be covered with a 1-ft compacted layer of fines and then capped with a 1-ft layer of sized lump coal.

Subbituminous coal and lignite should be stockpiled by using the same system of layering and compacting, but the coal layers should be thinner, not more than 1 ft thick, to ensure good compaction.

Inside Storage and Handling

Dry anthracite coal flows easily; however, other fine, wet coals feeding from a bottom-outlet bunker may have a tendency to "rat-hole" or "pipe" all the way to the top surface. When this occurs, coal flow to the feeders is intermittent and spasmodic. From studies of actual plant operations it has been found that hang-ups of coal may begin when there is a surface-moisture level of 5 to 6 percent.

Bunker Design

The purpose of the bunker is not only to store a given capacity of coal but also to function efficiently as part of the system in maintaining a continuous supply of fuel to the pulverizer or stoker. In the design of bunkers, careful consideration should be given to capacity, shape, material, and location.

Bunker Fires

A fire in the coal bunker should be recognized at once as a serious danger to personnel and equipment. Steam or carbon dioxide may be piped in directly to covered bunkers and to the affected areas in open-top bunkers. If the hot coal is run through the fuel-burning equipment, special care should be taken to feed uniformly, without interruption.

Mechanical Stokers

Mechanical stokers are designed to feed fuel onto a grate within the furnace and to remove the ash residue. The grate area required for a stoker of given type and capacity is determined from allowable rates established by experience. Table 3-12 lists recom-

TABLE 3-12 Maximum Allowable Fuel-Burning Rates

Type of stoker	Btu/(ft²)(h)
Spreader—stationary and dumping grate	450,000
Spreader—traveling grate	750,000
Spreader—vibrating grate	400,000
Underfeed—single or double retort	425,000
Underfeed—multiple retort	600,000
Water-cooled vibrating grate	400,000
Chain grate and traveling grate	500,000

mended fuel-burning rates (in British thermal units per square foot per hour) for various types of stokers, based on using coals suited to the stoker type in each case. The practical steam-output limit of boilers equipped with mechanical stokers is about 400,000 lb/h.

Almost any coal can be burned successfully on some type of stoker. In addition, many by-products and waste fuels, such as coke breeze, wood wastes, pulpwood bark, and bagasse, can be used either as a base or auxiliary fuel.

Mechanical stokers can be classified into four main groups, based on the method of introducing the fuel to the furnace:

1. Spreader stokers
2. Underfeed stokers
3. Water-cooled vibrating-grate stokers
4. Chain-grate and traveling-grate stokers

Spreader Stokers

The spreader stoker is the one most commonly used in the capacity range from 75,000 to 400,000 lb of steam per hour because it responds rapidly to load swings and can burn a variety of fuels. The spreader stoker is capable of burning a wide range of coals, from high-rank Eastern bituminous to lignite or brown coal, as well as a variety of by-product waste fuels.

As the name implies, the spreader stoker projects fuel into the furnace over the fire with a uniform spreading action, permitting suspension burning of the fine fuel particles (Fig. 3-3). The heavier pieces, that cannot be supported in the gas flow, fall to the grate for combustion in a thin, fast-burning bed.

Grates for Spreader Stokers. Spreader-stoker firing is old in principle, and the first grate design was a stationary type, with the ash removed manually through the front doors.

Stationary grates were soon followed by dumping-grate designs, in which grate sections are provided for each feeder and the undergrate air plenum chambers are correspondingly divided. This permits the temporary discontinuance of the fuel and air supply to a grate section for ash removal without affecting other sections of the stoker.

The continuous-ash-discharge traveling grate has no interruptions for removing ashes and because of the thin, fast-burning fuel bed, this design increased average burning rates approximately 70 percent over the stationary and dumping-grate types. Continuous-cleaning grates of reciprocating and vibrating designs have also been developed.

The normal practice of all continuous-ash-discharge spreader stokers is to remove the ashes at the front or feeding end of the stoker. This permits the most satisfactory fuel-distribution pattern and provides maximum residence time on the grates for complete combustion of the fuel.

Overfire Air. An overfire air system, with pressures from 27 to 30 in of water, is essential to successful suspension burning. It is customary to provide at least two rows of evenly spaced high-pressure-air jets in the rear wall of the furnace and one in the front wall. This air mixes with the furnace gases and creates the turbulence required for complete combustion.

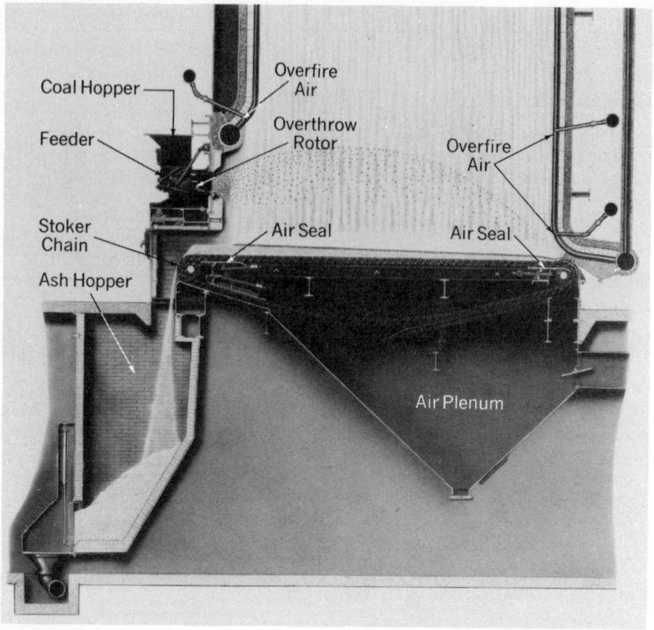

Figure 3-3 Traveling-grate spreader stoker with front-ash discharge.

Fly-Carbon Reinjection. Partial suspension burning results in a greater carryover of particulate matter in the flue gas than occurs with other types of stokers. In general the arrangement of the collection equipment is such that the coarse carbon-bearing particles can be returned to the furnace for further burning and the fine material discharged to the ash removal system.

Reintroducing the fly carbon into the furnace results in an increase in boiler efficiency of 2 to 3 percent.

Fuels and Fuel Bed. All spreader stokers, and in particular the traveling-grate spreader type, have an extraordinary ability to burn fuels with a wide range of burning characteristics, including coals with caking tendencies. High-moisture, free-burning bituminous and lignite coals are commonly used, and some low-volatile fuels, such as coke breeze, have been burned in a mixture with higher-volatile coal. Anthracite coal, however, is not a satisfactory fuel for spreader-stoker firing.

Coal size segregation is a problem with any type of stoker, but the spreader stoker can tolerate a small amount of segregation because the feeding rate of the individual feeder-distributors can be varied. Size segregation, where fine and coarse coal are not distributed evenly over the grate, produces a ragged fire and poor efficiency.

Firing of By-Product Waste Fuels. By-product wastes having considerable calorific value can be used as a base fuel or as a supplementary fuel in the generation of steam for power, heating, or industrial processes. Bark from wood-pulping operations and bagasse from sugar refineries are good examples. Others include coffee-ground residue from instant-coffee manufacture, corncobs, coconut and peanut hulls from furfural manufacture, bark and sawdust from woodworking plants, and municipal solid waste. Spreader-stoker firing provides an excellent way to burn these wastes.

Waste fuels with high moisture content may present problems in maintaining combustion unless there is enough auxiliary fuel to maintain the average moisture of the total fuel input at a maximum of about 50 percent. Preheated air, at temperatures dependent upon the fuel moisture content, aids in drying and igniting the fuel as it is fed into the furnace. Air temperatures up to 450°F are common in bark-fired units. High air temperatures may require the use of alloy grate materials in order to reduce maintenance.

Underfeed Stokers

Underfeed stokers are used principally for heating and for small industrial units with a capacity of less than 30,000 lb of steam per hour. Underfeed stokers are generally of two types: the horizontal-feed, side-ash-discharge type, Fig. 3-4; and the gravity-feed, rear-ash-discharge type, Fig. 3-5.

In the side-ash-discharge underfeed stoker (Fig. 3-4), fuel is fed from the hopper by means of a reciprocating ram to a central trough called the retort. On very small heating stokers, a screw conveys the coal from the hopper to the retort. A series of small auxiliary pushers in the bottom of the retort assist in moving the fuel rearward, and as the retort is filled, the fuel is moved upward to spread to each side over the air-admitting tuyères and side grates. The fuel rises in the retort and burns, and the ash is intermittently discharged to shallow pits, quenched, and removed through doors at the front of the stoker.

The single-retort and double-retort, horizontal-type stokers are generally limited to 25,000 to 30,000 lb of steam per hour with burning rates of 425,000 Btu/(ft²)(h) in furnaces with water-cooled walls. The multiple-retort, rear-end-cleaning type (Fig. 3-5), has

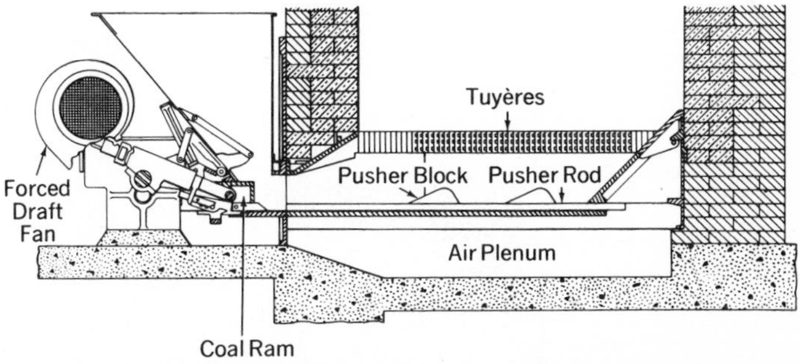

Figure 3-4 Single-retort, horizontal-feed, side-ash-discharge underfeed stoker.

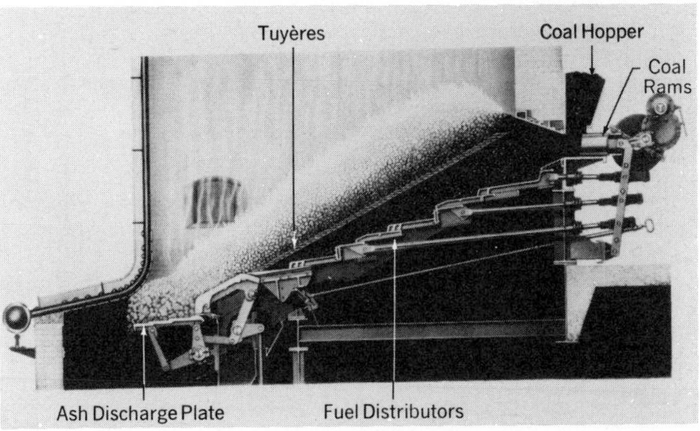

Figure 3-5 Multiple-retort gravity-feed type, rear-ash-discharge underfeed stoker.

a retort and grate inclination of 20 to 25°. This type of stoker can be designed for boiler units generating up to 500,000 lb of steam per hour. Burning rates up to 600,000 Btu/(ft^2)(h) are practicable.

The burning rates for underfeed stokers are directly related to the ash-softening temperature. For coals with ash-softening temperature below 2400°F, the burning rates are progressively reduced.

With multiple-retort stokers, overfire-air systems are also frequently provided.

The size of the coal has a marked effect on the relative capacity and efficiency of an underfeed stoker. The most desirable size consists of 1¼ in × 0 nut, pea, and slack in equal proportions. A reduction in the percentage of fines helps to keep the fuel bed more porous and extends the range of coals with a high coking index.

In general, underfeed stokers are able to burn caking coals. The range of agitation imparted to the fuel bed in different stoker designs permits the use of coals with varying degrees of caking properties. The ash-fusion temperature is an important factor in the selection of the coals. Usually, the lower the ash-fusion temperature, the greater the possibility of clinker trouble.

Water-Cooled Vibrating-Grate Stokers

An entirely different design of stoker is the water-cooled vibrating-grate hopper-feed type, Fig. 3-6. This stoker consists of a tuyère grate surface mounted on, and in intimate contact with, a grid of water tubes interconnected with the boiler circulation system for positive cooling. The entire structure is supported by a number of flexing plates allowing the grid and its grate to move freely in a vibrating action that conveys coal from the feeding hopper onto the grate and gradually to the rear of the stoker. Ashes are automatically discharged to an ash pit.

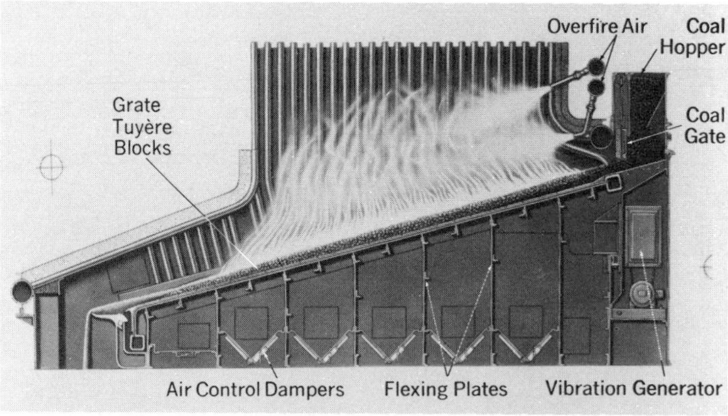

Figure 3-6 Water-cooled vibrating-grate stoker.

Vibration of the grates is intermittent, and the frequency of the vibration periods is regulated by a timing device to conform to load variations, synchronizing the fuel feeding rate with the air supply.

Water cooling of the grates makes this stoker especially adaptable to multiple-fuel firing since a shift to oil or gas does not require special provision for protection of the grates. A normal bed of ash left as a cover gives adequate protection from furnace radiation.

Chain-Grate and Traveling-Grate Stokers

Traveling-grate stokers, including the specific type known as the chain-grate stoker (Fig. 3-7), have assembled links, grates, or keys joined together in endless belt arrangements

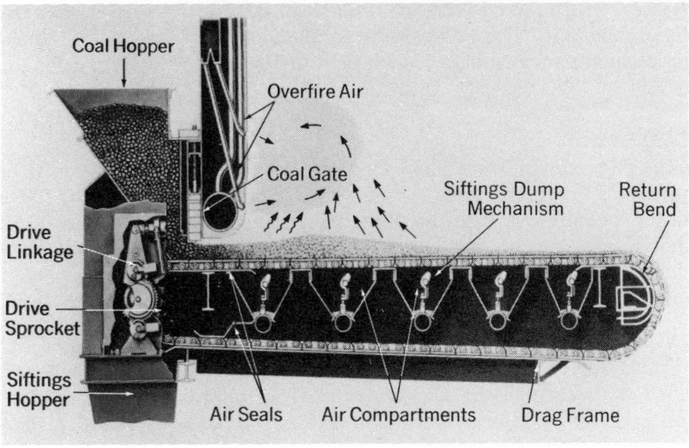

Figure 3-7 Chain-grate stoker. *(Laclede Stoker Co.)*

that pass over the sprockets or return bends located at the front and the rear of the furnace. Coal enters the furnace after passing under an adjustable gate to regulate the thickness of the fuel bed, is heated by radiation from the furnace gases, and is ignited. The fuel bed continues to burn as it moves along, and ash is discharged from the end of the grate into the ash pit. Generally these stokers use furnace arches (front and/or rear) to improve combustion by reflecting heat into the fuel bed.

Chain-and traveling-grate stokers can burn a wide variety of fuels.

PREPARATION AND UTILIZATION OF PULVERIZED COAL

The capacity limitations imposed by stokers are overcome by the pulverized-coal system. This method of burning coal also provides:

1. Ability to use coal from fines up to 2-in maximum size
2. Improved response to load changes
3. An increase in thermal efficiency because of lower excess air used for combustion and lower carbon loss than with stoker firing
4. Improved ability to efficiently burn coal in combination with oil and gas

Pulverized-Coal Systems

The function of a pulverized-coal system is to pulverize the coal, deliver it to the fuel-burning equipment, and accomplish complete combustion in the furnace with a minimum of excess air.

A small portion of the air required for combustion (15 to 20 percent) is used to transport the coal to the burner. This is known as primary air and is also used to dry the coal in the pulverizer. The remainder of the combustion air (80 to 85 percent) is introduced at the burner and is known as secondary air.

Two principal systems—the bin system and the direct-firing system—have been used for processing, distributing, and burning pulverized coal. The direct-firing system is almost exclusively the one being installed today.

Bin System

The bin system is primarily of historical interest. In this system the coal is processed at a location apart from the furnace, and the end product is pneumatically conveyed to cyclone collectors which recover the fines and clean the moisture-laden air before returning it to the atmosphere. The pulverized coal is discharged into storage bins and later conveyed by pneumatic transport through pipelines to utilization bins and from there to the furnace.

Direct-Firing System

The pulverizing equipment developed for the direct-firing system permits continuous utilization of raw coal directly from the bunkers. This is accomplished by feeding coal of a maximum top size directly into the pulverizer, where it is dried as well as pulverized, and then delivering it to the burners in a single continuous operation. Components of the direct-firing system are illustrated in Fig. 3-8.

There are two direct-firing methods in use—the pressure type and the suction type.

Pressure Firing. In the pressure method, the primary-air fan, located on the inlet side of the pulverizer, forces the hot primary air through the pulverizer where it picks up the pulverized coal and delivers the proper coal-air mixture to the burners. On large installations, primary-air fans operating on cold air force the air through the air heater first and then through the pulverizer.

Suction Firing. In the suction method, the air and entrained coal are drawn through the pulverizer under negative pressure by an exhauster located on the outlet side of the pulverizer. With this arrangement the fan handles a mixture of coal and air, and distribution of the mixture to more than one burner is attained by a distributor beyond the fan discharge.

In the direct-firing system the operating range of a pulverizer is usually not more than 3 to 1 (without change in the number of burners in service) because the air velocities in lines and other parts of the system must be maintained above the minimum levels to

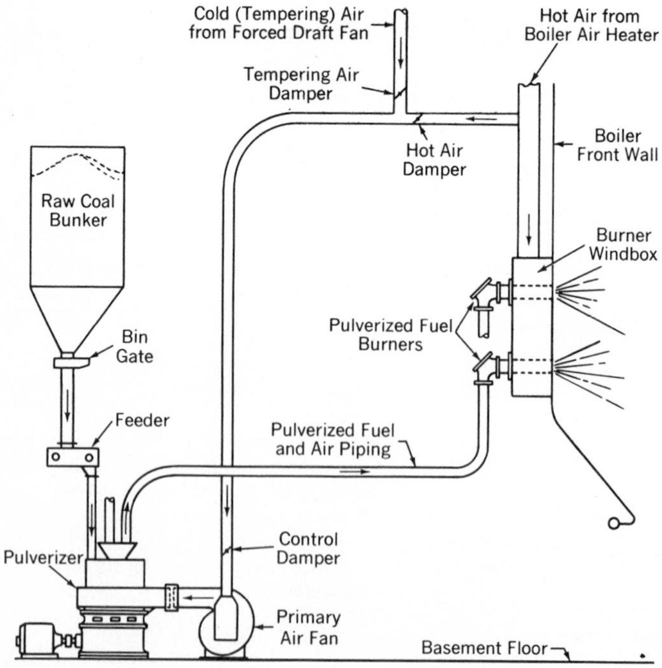

Figure 3-8 Direct-firing system for pulverized coal.

keep the coals in suspension. Most boiler units are provided with more than one pulverizer, each feeding multiple burners. Load variations beyond 3 to 1 are accommodated by shutting down (or starting up) a pulverizer and the burners it supplies.

Types of Pulverizers

All pulverizing machinery operates to grind by impact, attrition, compression, or a combination of two or more of these.

Medium-Speed Pulverizers

There are two groups of medium-speed (75 to 225 r/min) pulverizers, classified as the ball-and-race and roller types. The principle of pulverizing by a combination of crushing under pressure, impact, and attrition between grinding surfaces and material is used in each group, but the method is different.

Medium-Speed Ball-and-Race Pulverizers. The ball-and-race pulverizer works on the ball-bearing principle. The Type EL pulverizer, illustrated in Fig. 3-9, has one stationary top ring, one rotating bottom ring, and one set of balls that make up the grinding elements. The pressure required for efficient grinding is obtained from externally adjustable dual-purpose springs. The bottom ring is driven by a yoke which is attached to the

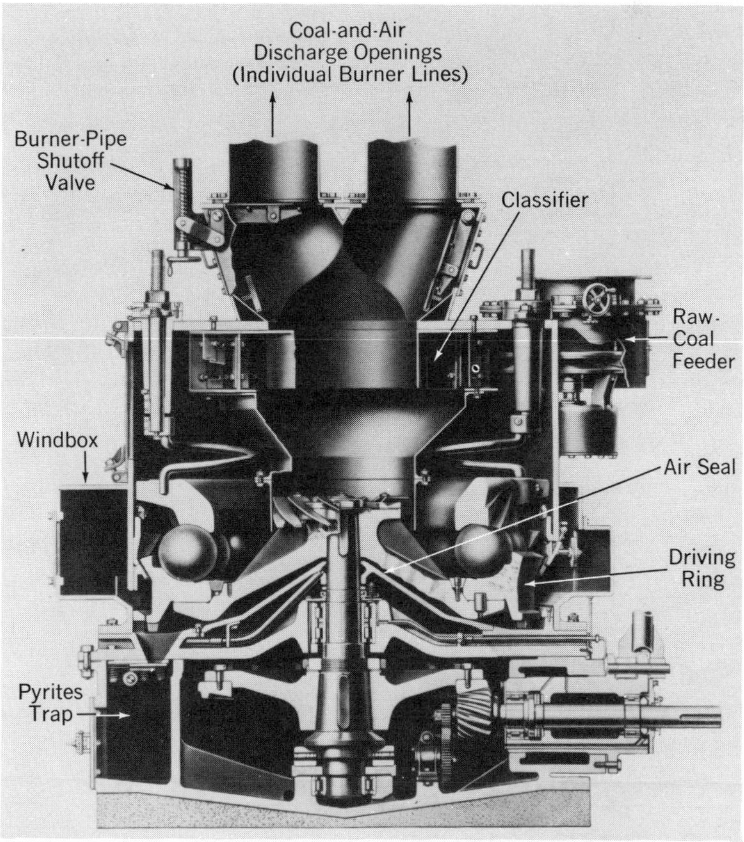

Figure 3-9 Babcock & Wilcox Type EL single-row ball-and-race pulverizer.

vertical main shaft of the pulverizer. The top ring is held stationary by the dual-purpose springs.

Medium-Speed Roller-Type Pulverizers. This type of pulverizer can be of a design in which the ring is stationary and the rolls rotate or one in which the rolls are mounted off the mill housing and the ring rotates. In the first type, grinding elements consisting of three or more cylindrical rolls, suspended from driving arms, revolve in a horizontally positioned replaceable race. The principal components of the second type, the bowl mill, are a rotating bowl equipped with a replaceable grinding ring, two or more tapered rolls in stationary journals, a classifier, and a main drive.

Tube Mill. One of the oldest practical pulverizers is the tube mill, in which a charge of mixed-size forged-steel balls in a horizontally supported grinding cylinder is activated by gravity as the cylinder is rotated. The coal is pulverized by attrition and impact as the ball charge ascends and falls within the coal.

High-Speed Pulverizers

High-speed pulverizers use impact as a primary means of grinding through the use of hammerlike beaters, wear-resistant pegs rotating within a cage, and fan blades integral with the pulverizer shaft.

Selecting Pulverizer Equipment

A number of factors must be considered in the selection of pulverizer equipment. If selection anticipates the use of a variety of coals, the pulverizer should be sized for the coal that gives the highest base capacity. *Base capacity* is the desired capacity divided

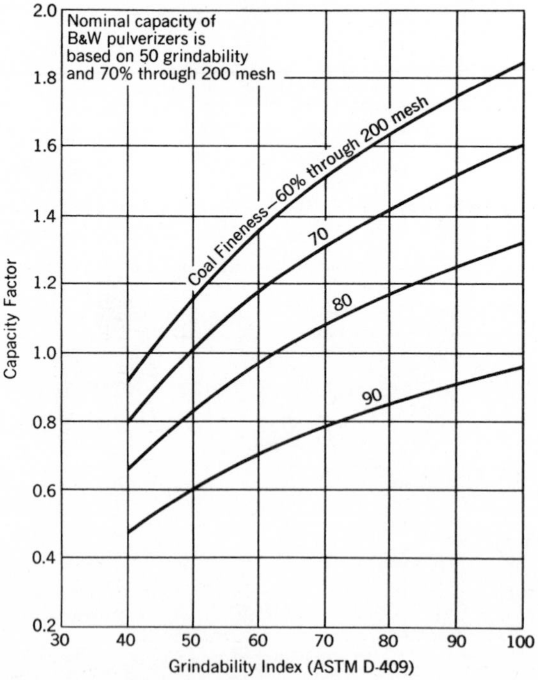

Figure 3-10 Effect of grindability and fineness on pulverizer capacity.

by the capacity factor. The latter is a function of the grindability of the coal and the fineness required (see Fig. 3-10).

The percentage of volatile matter in the fuel has a direct bearing on the recommended temperature for combustion of the mixture of primary air and fuel. The generally accepted safe values for pulverizer exit temperatures for fuel-air mixtures are given in Table 3-13. The temperature of the primary air entering the pulverizer may run 650°F or higher, depending on the amount of moisture in the coal.

Fine grinding of coal is necessary to assure complete combustion of the carbon for maximum efficiency. The required pulverized fuel fineness is expressed as the percentage of the product passing through various sizes of sieves. Coal classification by rank and end use of product determine the fineness to which coal should be ground (Table 3-14).

TABLE 3-13 Prevalent Pulverizer Exit Temperatures for Mixtures of Primary Air and Fuel

Fuel	Exit temperature, °F
Lignite	120–140
High-volatile bituminous	150
Low-volatile bituminous	150–175
Anthracite	175–212

TABLE 3-14 Required Pulverized Fuel Fineness
Percent through 200 U.S. sieve*

	ASTM Classification of Coals by Rank					
	Fixed Carbon, %			Calorific Value, Btu/lb†		
Type of furnace	97.9–86 (Petroleum coke)	85.9–78	77.9–69	Above 13,000	12,900–11,000	Below 11,000
Water-cooled	80	75	70	70	65‡	60‡
Cement kiln	90	85	80	80	80	——
Metallurgical	As determined by process, generally from 80 to 90%					

*The 200-mesh screen (sieve) has 200 openings per linear inch or 40,000 openings per square inch. For U.S. and ASTM sieve series, the nominal aperture for 200 mesh is 0.0029 in. (0.074 mm). The ASTM designation for 200 mesh is 74 μ

†Coal with fixed carbon value below 69%.

‡Extremely high ash content coals will require higher fineness than indicated.

Burning Equipment for Pulverized Coal

The burner is the principal component of equipment for firing pulverized coal. Coal must be pulverized to the point where particles are small enough to assure proper combustion (Table 3-14). In the direct-firing system the coal is dried and delivered to the burner in suspension in the primary air, and this mixture must be adequately mixed with the secondary air at the burner.

Piping and Nozzle Sizing Requirements

Size selection of nozzles for pulverized-coal piping and burners requires flow velocities that are high enough to keep the coal particles in suspension in the primary air stream. This generally requires 40 to 70 percent of the pulverizer's full-load airflow requirement at zero output. Horizontally arranged burner nozzles should be sized for no less than 3000 ft/min at the minimum pulverizer capacity.

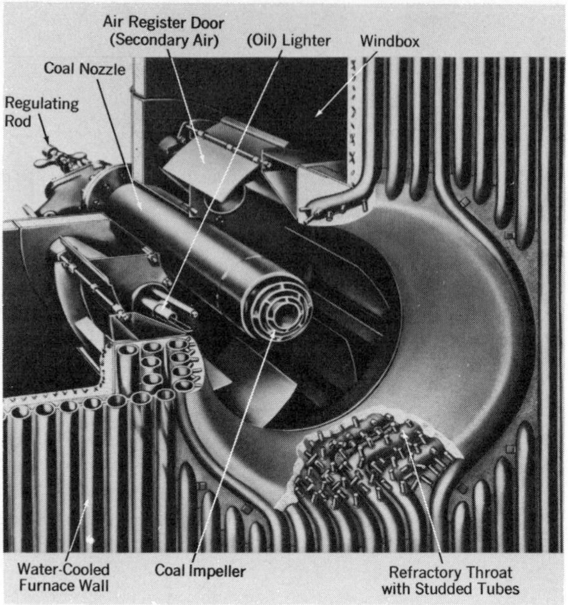

Figure 3-11 Circular register burner for pulverized-coal firing.

Standards of Burner Performance

Operators of pulverized-coal equipment should expect ignition of the pulverized coal to be stable, without the use of supporting fuel, over a load range of approximately 3 to 1. Most boilers are equipped with many burners so that a wider capacity range is readily obtained by varying the number of burners and pulverizers in use. The loss of unburned combustibles should be less than 2 percent with excess air in the range of 15 to 22 percent, measured at the furnace outlet. The design should avoid the formation of deposits that may interfere with the continued efficient and reliable performance of the burner.

Burners. The most frequently used burners are of the circular single-register type designed for firing pulverized coal only (Fig. 3-11). This type can be equipped to fire any combination of the three principal fuels. However, combination firing of pulverized-coal with oil in the same burner should be restricted to short emergency periods because of the possibility of coke formation on the pulverized-coal element.

Lighters (Igniters) and Pilots. In starting up the burner on pulverized coal, it is necessary to keep the igniters in operation until the temperature in the combustion zone becomes high enough to assure self-sustaining ignition of the main fuel. The self-igniting characteristics of pulverized coal vary from one fuel to another. In some instances completely reliable ignition is obtained down to a quarter load. When firing pulverized coal with a volatile matter content less than 25 percent, it may be necessary to activate the igniters even at high loads.

Excess Air

Pulverized coal requires more excess air for satisfactory combustion than either oil or natural gas. An acceptable quantity of unburned combustible coal is usually obtained with 15 percent excess air at high loads. This allows for the normal maldistribution of primary-air, coal, and secondary air.

LIQUID FUELS

CHARACTERISTICS

Fuel Oil

It is common practice in refining petroleum to produce fuel oils complying with several specifications prepared by the ASTM and adopted as a commercial standard by the National Bureau of Standards (Table 3-15).

Fuel oils are graded according to gravity and viscosity, the lightest being No. 1 and the heaviest No. 6. Grades 5 and 6 generally require heating for satisfactory pumping and burning. The range of analyses and heating values of the several grades of fuel oils are given in Table 3-16.

The gross heating value, density, and specific gravity of various fuel oils for a range of API gravities are shown in Fig. 3-12. The abscissa on this figure is the API (American Petroleum Institute) gravity and sp gr at 60–60°F represents the ratio of oil density at 60°F to water density also at 60°F.

Fuel oils are generally sold on a volume basis, with 60°F as the base temperature. Correction factors are given in Fig. 3-13 for converting known volumes at other temperatures to the 60°F standard base. This correction is also dependent on the API gravity range as illustrated by the three parametric curves of Fig. 3-12.

Since equipment for handling and, especially, burning of fuel oil is usually designed for a maximum oil viscosity, it is necessary to know the viscosity characteristics of the fuel oil to be used. If the viscosities of heavy oils are known at two temperatures, viscosities at other temperatures can be closely predicted by a linear interpolation between these two values located on the standard ASTM chart of Fig. 3-14. Viscosities of light oils at various temperatures within the region designated as No. 2 fuel oil can be found by drawing a line parallel to the No. 2 boundary lines through the point of only one known viscosity and temperature.

Shale Oil

Oil shale is not actually a shale nor does it contain oil. It is generally defined as a fine-grained, compact, sedimentary rock containing an organic material called kerogen. Heating the oil shale to about 875°F decomposes this material to produce shale oil.

Pitch and Tar

The liquid and semiliquid residues from the distillation of petroleum and coal are known as pitch and tar. Most of these residues are suitable for use as boiler fuels. Some handle as easily and burn as readily as does kerosene, whereas others give considerable trouble.

Coal Oil Mixture (COM)

Pulverized coal can be mixed with oil and kept in suspension by agitation, recirculation, or by use of additives. Transportation by pipeline is possible, but truck, train, or barge shipment may be more economical. COM is burned with equipment similar to that used for oil firing.

PREPARATION AND UTILIZATION OF OIL

Preparation

Most petroleum is refined to some extent before use, although small amounts are burned without processing. The refining of crude oil yields a number of products having many different applications. Those used as fuel include gasoline, distillate fuel, residual fuel, jet fuels, still gas, liquefied gases, kerosene, and petroleum coke.

TABLE 3-15 ASTM Standard Specifications for Fuel Oils[a,b]

No. 1: A distillate oil intended for vaporizing pot-type burners and other burners requiring this grade of fuel
No. 2: A distillate oil for general purpose domestic heating for use in burners not requiring No. 1 fuel oil
No. 4: Preheating not usually required for handling or burning

Grade of fuel oil	Flash point, °F (°C) Min	Pour point, °F (°C) Max	Water and sediment, % by volume Max	Carbon residue on 10% bottoms, % Max	Ash, % by weight Max	Distillation temperatures, °F (°C) 10% point Max	90% point Min	90% point Max
No. 1	100 or legal (38)	0	trace	0.15	—	420 (215)	—	550 (288)
No. 2	100 or legal (38)	20[c] (−7)	0.10	0.35	—	[d]	540[c] (282)	640 (338)
No. 4	130 or legal (55)	20 (−7)	0.50	—	0.10	—	—	—
No. 5 (light)	130 or legal (55)	—	1.00	—	0.10	—	—	—
No. 5 (heavy)	130 or legal (55)	—	1.00	—	0.10	—	—	—
No. 6	150 (65)	—	2.00[g]	—	—	—	—	—

[a]Recognizing the necessity for low-sulfur fuel oils used in connection with heat-treatment, nonferrous metal, glass, and ceramic furnaces and other special uses, a sulfur requirement may be specified in accordance with the following table:

Grade of Fuel Oil	Sulfur, max, %
No. 1	0.5
No. 2	0.7
No. 4	no limit
No. 5	no limit
No. 6	no limit

Other sulfur limits may be specified only by mutual agreement between the purchaser and the seller.

Transportation, Handling, and Storage

A worldwide system for distributing petroleum (and its products) has been developed because petroleum has a high calorific value per unit volume, is in easily handled liquid form, and has varied applications.

The serious hazard inherent in possible oil-storage-tank failure is overcome by storing oil in underground tanks or by protecting surface tanks by surrounding them with cofferdams of sufficient capacity to hold the entire contents of any tank so protected. The National Fire Protection Association has prepared a standard set of rules for the storage and handling of oils.

To facilitate pumping heavy fuel oil, heating equipment is usually provided in storage and transportation facilities. Storage tanks, piping, and heaters for heavy oils must be cleaned at intervals because of fouling or sludge formation. Various commercial compounds are helpful in reducing sludge.

No. 5 (Light): Preheating may be required depending on climate and equipment
No. 5 (Heavy): Preheating may be required for burning and, in cold climates, may be required for handling
No. 6: Preheating required for burning and handling

Saybolt viscosity, s				Kinematic viscosity, centistokes				Gravity, deg API	Copper strip corrosion
Universal at 100°F (38°C)		Furol at 122°F (50°C)		At 100°F (38°C)		At 122°F (50°C)			
Min	Max	Min	Max	Min	Max	Min	Max	Min	Max
—	—	—	—	1.4	2.2	—	—	35	No. 3
(32.6)[f]	(37.93)	—	—	2.0[e]	3.6	—	—	30	—
45	125	—	—	(5.8)	(26.4)	—	—	—	—
150	300	—	—	(32)	(65)	—	—	—	—
350	750	(23)	(40)	(75)	(162)	(42)	(81)	—	—
(900)	(9000)	45	300	—	—	(92)	(638)	—	—

[b]It is the intent of these classifications that failure to meet any requirement of a given grade does not automatically place an oil in the next lower grade unless in fact it meets all requirements of the lower grade.

[c]Lower or higher pour points may be specified whenever required by conditions of storage or use.

[d]The 10% distillation temperature point may be specified at 440°F (226°C) maximum for use in other than atomizing burners.

[e]When pour point less than 0°F is specified, the minimum viscosity shall be 1.8 cs (32.0 s, Saybolt universal) and the minimum 90% point shall be waived.

[f]Viscosity values in parentheses are for information only and not necessarily limiting.

[g]The amount of water by distillation plus the sediment by extraction shall not exceed 2.00%. The amount of sediment by extraction shall not exceed 0.50%. A deduction in quantity shall be made for all water and sediment in excess of 1.0%.

Source ASTM D 396.

OIL-BURNING EQUIPMENT

The burner is the principal component of equipment for firing oil. Burners are normally located in the vertical walls of the furnaces.

Oil Burners

The most frequently used burners are the circular type. Figure 3-15 shows a single circular-register burner for gas and oil firing.

The maximum capacity of the individual circular burner ranges up to 300×10^6 Btu/h. In circular burners the tangential "doors" built into the air register provide the turbulence necessary to mix the fuel and air and control flame shape. Although the fuel mixture as introduced to the burner is fairly dense in the center, the direction and velocity of the air, plus dispersion of the fuel, completely and thoroughly mix it with the combustion air.

TABLE 3-16 Range of Analyses of Fuel Oils

Characteristic	Grade of fuel oil No. 1	No. 2	No. 4	No. 5	No. 6
Weight, %					
Sulfur	0.01–0.5	0.05–1.0	0.2–2.0	0.5–3.0	0.7–3.5
Hydrogen	13.3–14.1	11.8–13.9	(10.6–13.0)*	(10.5–12.0)*	(9.5–12.0)*
Carbon	85.9–86.7	86.1–88.2	(86.5–89.2)*	(86.5–89.2)*	(86.5–90.2)*
Nitrogen	Nil–0.1	Nil–0.1	—	—	—
Oxygen	—	—	—	—	—
Ash	—	—	0–0.1	0–0.1	0.01–0.05
Gravity					
Deg API	40–44	28–40	15–30	14–22	7–22
Specific,	0.825–0.806	0.887–0.825	0.966–0.876	0.972–0.922	1.022–0.922
lb/gal	6.87–6.71	7.39–6.87	8.04–7.30	8.10–7.68	8.51–7.68
Pour point, °F	0 to −50	0 to −40	−10 to +50	−10 to +80	+15 to +85
Viscosity					
Centistokes @ 100°F	1.4–2.2	1.9–3.0	10.5–65	65–200	260–750
SSU @ 100°F	—	32–38	60–300	—	—
SSF @ 122°F	—	—	—	20–40	45–300
Water & sediment, vol %	—	0–0.1	Tr to 1.0	0.05–1.0	0.05–2.0
Heating value, Btu/lb, gross (calculated)	19,670–19,860	19,170–19,750	18,280–19,400	18,100–19,020	17,410–18,990

*Estimated.

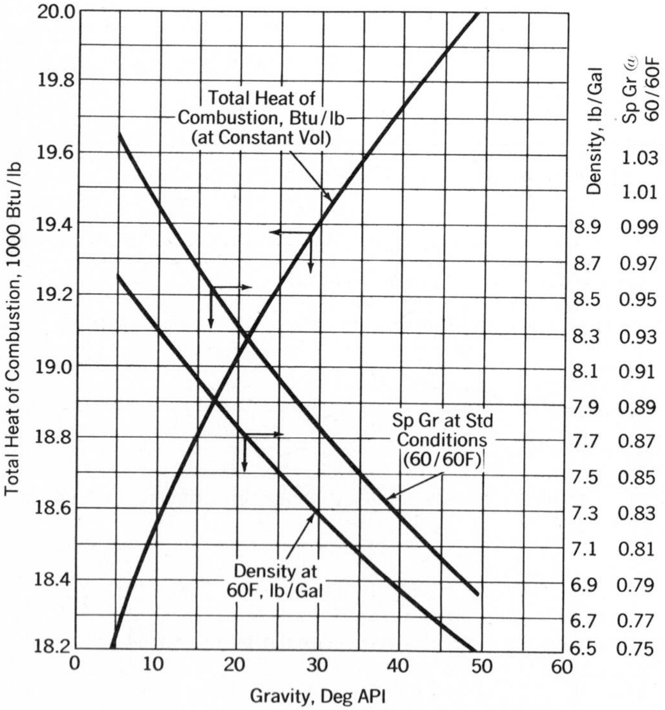

$$\text{Deg API} = \frac{141.5}{\text{Sp Gr @ 60/60F}} - 131.5$$

Figure 3-12 Heating value, weight (pounds per gallon), and specific gravity of fuel oil for a range of API gravities.

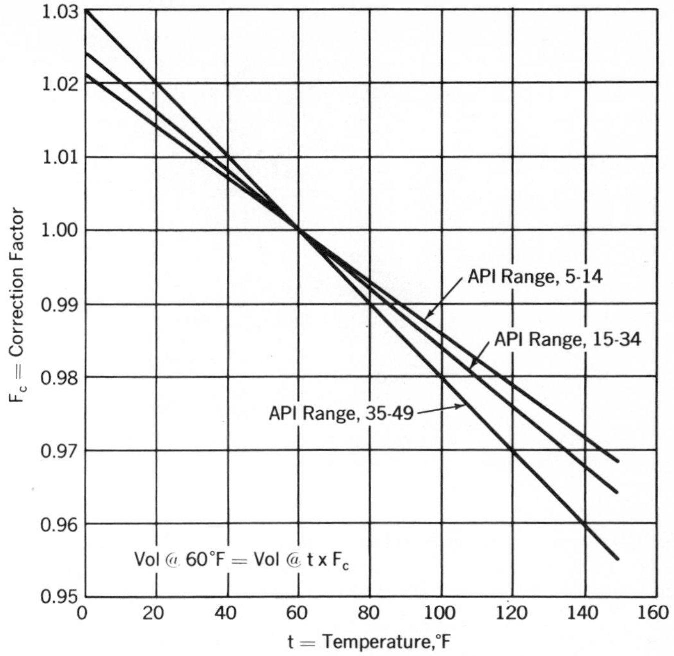

Figure 3-13 Temperature-volume correction factor for fuel oil.

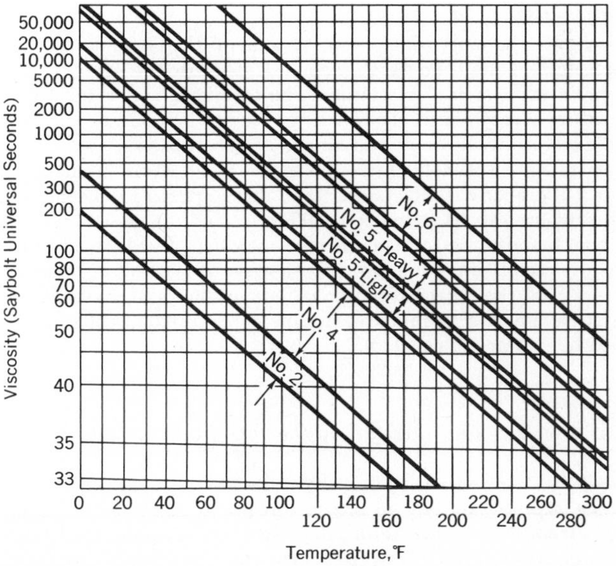

Figure 3-14 Approximate viscosity of fuel oil at various temperatures. *(Source: ASTM.)*

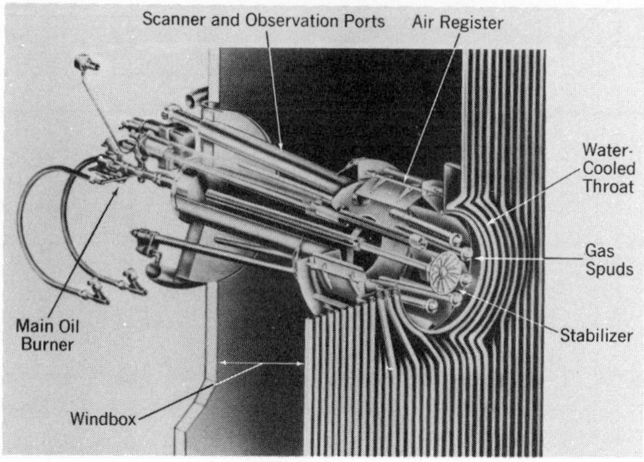

Figure 3-15 Circular register burner with water-cooled throat for oil and gas firing.

In order to burn fuel oil at the high rates demanded of modern boiler units it is necessary that the oil be *atomized,* i.e., dispersed into the furnace as a fine mist, to expose a large amount of oil particle surface to the air and assure prompt ignition and rapid combustion.

For proper atomization, oil of grades heavier than No. 2 must be heated to reduce viscosity to 135 to 150 SSU (Saybolt seconds universal). Steam or electric heaters are required to raise the oil temperature to the required level, i.e., approximately 135°F (57°C) for No. 4 oil, 185°F (74°C) for No. 5 oil, and 200 to 220°F (93 to 104°C) for No. 6 oil.

Steam or Air Atomizers

Steam atomizers are the most widely used. In general they operate on the principle of producing a steam-fuel emulsion which, when released into the furnace, atomizes the oil through the rapid expansion of the steam. The atomizing steam must be dry because entrained moisture causes pulsations which can lead to loss of ignition. Where steam is not available, moisture-free compressed air can be substituted.

Steam atomizers are available in sizes up to 300 $\times$ 10^6 Btu/h input—about 16,500 lb of oil per hour. Oil pressure is much lower than that required for mechanical atomizers. Maximum oil pressure can be as much as 300 lb/in^2 and maximum steam pressure 150 lb/in^2. The steam atomizer performs more efficiently over a wider load range than other types. It normally atomizes the fuel properly down to 20 percent of rated capacity.

A disadvantage of the steam atomizer is its consumption of steam. A good steam atomizer can operate with a steam consumption as low as 0.02 lb of steam per pound of fuel oil at maximum atomizer capacity.

Mechanical Atomizers

In mechanical atomizers the pressure of the fuel itself is used as the means for atomization.

The return-flow atomizer is used in many units where the use of atomizing steam is objectionable or impractical. The oil pressure required at the atomizer for maximum capacity ranges from 600 to 1000 lb/in^2, depending on capacity, load range, and fuel. Mechanical atomizers are available in sizes up to 180 $\times$ 10^6 Btu/h input— about 10,000 lb of oil per hour.

Excess Air

It is necessary to supply more than the theoretical quantity of air to assure complete combustion of the fuel in the furnace. The amount of excess air provided should be just enough to burn the fuel completely in order to minimize the sensible heat loss in the stack gases. The excess air normally required for oil firing, expressed as percent of theoretical air, is generally between 5 and 7 percent.

GASEOUS FUELS

CHARACTERISTICS

Natural Gas

Of all chemical fuels, natural gas is considered the least troublesome for steam generation. It is piped directly to the consumer, eliminating the need for storage at the consumer's plant. It is substantially free of ash and mixes intimately with air to provide complete combustion at low excess air without smoke.

The high hydrogen content of natural gas compared with that of oil or coal results in the production of more water vapor in the combustion gases, thus causing a correspondingly lower efficiency of the steam-generating equipment. This can be taken into account in the design of the equipment and evaluated in comparing the cost of gas with other fuels.

Analyses of natural gas from several United States fields are given in Table 3-17.

Gaseous Fuels from Coal

A number of gaseous fuels are derived from coal either as by-products or from coal gasification processes. Table 3-18 lists selected analyses of these gases according to the various types described in the following paragraphs.

TABLE 3-17 Characteristics of Selected Samples of Natural Gas from United States Fields

Characteristic	Sample no. and source of gas				
	1 Pa.	2 So. Calif.	3 Ohio	4 La.	5 Okla.
Analyses					
Proximate constituents, vol %					
H_2 Hydrogen	—	—	1.82	—	—
CH_4 Methane	83.40	84.00	93.33	90.00	84.10
C_2H_4 Ethylene	—	—	0.25	—	—
C_2H_6 Ethane	15.80	14.80	—	5.00	6.70
CO Carbon monoxide	—	—	0.45	—	—
CO_2 Carbon dioxide	—	0.70	0.22	—	0.80
N_2 Nitrogen	0.80	0.50	3.40	5.00	8.40
O_2 Oxygen	—	—	0.35	—	—
H_2S Hydrogen sulfide	—	—	0.18	—	—
Ultimate constituents, wt %					
S Sulfur	—	—	0.34	—	—
H_2 Hydrogen	23.53	23.30	23.20	22.68	20.85
C Carbon	75.25	74.72	69.12	69.26	64.84
N_2 Nitrogen	1.22	0.76	5.76	8.06	12.90
O_2 Oxygen	—	1.22	1.58	—	1.41
Specific gravity (rel to air)	0.636	0.636	0.567	0.600	0.630
Higher heat value					
Btu/ft^3 @ 60°F & 30 inHg	1,129	1,116	964	1,002	974
Btu/lb of fuel	23,170	22,904	22,077	21,824	20,160

TABLE 3-18 Selected Analyses of Gaseous Fuels Derived from Coal

Analytical result	Coke-oven gas	Blast-furnace gas (lean)	Carburetted water gas	Producer gas
	Fuel analyzed			
Constituents, vol %				
H_2 Hydrogen	47.9	2.4	34.0	14.0
CH_4 Methane	33.9	0.1	15.5	3.0
C_2H_4 Ethylene	5.2	—	4.7	—
CO Carbon monoxide	6.1	23.3	32.0	27.0
CO_2 Carbon dioxide	2.5	14.4	4.3	4.5
N_2 Nitrogen	3.7	56.4	6.5	50.9
O_2 Oxygen	0.6	—	0.7	0.6
C_6H_6 Benzene	—	—	2.3	—
H_2O Water	—	3.4	—	—
Specific gravity (relative to air)	0.413	1.015	0.666	0.857
Higher heat value, Btu/ft^3				
@ 60°F & 30 inHg	590	—	534	163
@ 80°F & 30 inHg	—	83.8	—	—

Coke-Oven Gas

A considerable portion of coal is converted to gases or vapors in the production of coke. The noncondensable portion is called *coke-oven gas*. Constituents depend on the nature of the coal and coking process used (Table 3-18).

Blast-Furnace Gas

The gas discharged from steel-mill blast furnaces is used at the mills in heating furnaces, in gas engines, and for steam generation. This gas is quite variable in quality but generally has a high carbon monoxide content and low heating value (Table 3-18).

Water Gas

The gas produced by passing steam through a bed of hot coke is known as water gas. Carbon in the coke combines with the steam to form hydrogen and carbon monoxide.

Water gas is often enriched by passing the gas through a checkerwork of hot bricks sprayed with oil, which in turn is cracked to a gas by the heat. Such enriched water gas is called *carburetted water gas* (Table 3-18).

Producer Gas

When coal or coke is burned with a deficiency of air and a controlled amount of moisture (steam), a product known as *producer gas* is obtained. This gas, after removal of entrained ash and sulfur compounds, is used near its source because of its low heating value (Table 3-18).

"Synthetic Natural Gas"

"Synthetic natural gas" is made from coal by one of the many gasification processes resulting in a low-Btu gas which is then methanated.

Carbon Monoxide

In the petroleum industry, the efficient operation of a fluid-catalytic-cracking unit produces gases rich in carbon monoxide. To reclaim the thermal energy represented by these gases, the fluid-catalytic-cracking unit can be designed to include a CO boiler that uses the CO as fuel to generate steam for use in the process.

PROCESSING AND UTILIZATION OF NATURAL GAS

Processing

Propane and butane are often separated from the lighter gases and are widely used as bottled gas. They are distributed and stored liquefied under pressure. Natural gas containing excessive amounts of hydrogen sulfide is commonly known as "sour" gas. The sulfur is removed before distribution.

Transportation, Handling, and Storage

A pipeline is an economical means for transporting natural gas overland. Tankers are employed for overseas transportation of natural gas. The gas is liquefied under pressure (liquefied natural gas, LNG) for ease of transportation.

GAS-BURNING EQUIPMENT

Natural-Gas Burners

An example is the variable-mix multispud gas element (Fig. 3-15) for use with circular-type burners. Simultaneous firing of natural gas and oil in the same burner is possible.

To provide safe operation, ignition of a gas burner should remain close to the burner wall throughout the full range of allowable gas pressures, not only with normal airflows, but also with much more airflow through the burner than is theoretically required.

Burners for Other Gases

Many industrial applications utilize coke-oven gas, blast-furnace gas, refinery gas, or other industrial by-product gases. With these gases, the heat release per unit volume of fuel gas may be very different from that of natural gas. Hence, gas elements must be designed to accommodate the particular characteristics of the gas to be burned. Other special problems may be introduced by the presence of impurities in industrial gases, such as sulfur in coke-oven gas and entrained dust in blast-furnace gas.

Lighters (Igniters) and Pilots

Usually the ignition device is a spark device energized only long enough to assure that the main flame is self-sustaining. Although ignition should be self-sustaining within 1 or 2 s after the fuel reaches the combustion air, in a fully automated burner it is customary to allow 10 to 15 s "trial for ignition" so that the fuel can reach the burner after the fuel shutoff valve on the burner is opened. There are applications where a continuously burning lighter or pilot is needed. This is particularly true in the use of a by-product fuel, such as gas from a chemical process.

Excess Air

It is possible to operate most units with as little as 5 to 7 percent excess air at the furnace outlet at full load, and some boilers have operated with less than 2.5 percent excess air without excessive loss of unburned combustibles.

At partial loads on all units, regardless of the fuel fired, it is necessary to increase the excess air as the load is reduced. Burner dampers are designed not to close tightly in order to permit the air to protect the idle burner(s) from overheating by radiant heat from nearby operating burners.

EFFECT OF COAL AND MULTIFUEL FIRING ON INDUSTRIAL BOILER DESIGN*

Coal is a complex fuel, and it is necessary to establish its source as well as its physical and chemical characteristics before considering boiler size or selecting equipment. Adding wood, oil, or gas to obtain combined fuel firing further affects the design criteria employed to properly design the boiler furnace.

A properly designed furnace performs two functions: (1) burning the fuel completely and (2) cooling the products of combustion sufficiently so that the convection passes of the boiler unit are maintained in a satisfactory condition of cleanliness with a reasonable amount of soot blowing. When the average temperature of gas leaving a coal-fired furnace is too high, the ash particles are molten or sticky, a condition which leads to an excessive need for cleaning the ash deposits from the upper furnace and the high-temperature zones of the convection passes.

The fouling and slagging classification of the coal, as characterized by the ash analysis, establishes furnace sizing, spacing and arrangement of tubes in convection passes, and placement of soot blowers for the furnace walls and convection passes. The slagging characteristic governs burner clearances, heat input per unit of furnace cross-sectional area, and the number of furnace-wall soot blowers provided to control buildup of slag on the furnace walls. The fouling characteristic sets a relationship of gas temperature entering the tube bank for a given side spacing of tubes and tube bank depth according to soot-blower cleaning radius.

EFFECT OF DESIGN PARAMETERS ON BOILER SIZING

The substantial effect on boiler sizing of the coal-ash classification is made clear by comparison of units sized for three types of coal. In general, an increased fouling tendency requires an increase in the furnace surface to lower the gas temperature entering the superheater, and an increased slagging potential of a coal causes furnaces designed for Western coals to be substantially more conservative than those designed to fire Eastern bituminous coals.

Figure 3-16 shows three boilers sized for a steam capacity of 500,000 lb/h. Boiler A is designed to fire a West Virginia bituminous coal classified as having a low slagging and low fouling potential. The slightly larger boiler B is designed to fire an Alabama bituminous coal classified as having a medium slagging and a medium fouling potential. Boiler C is the largest, designed to fire a Wyoming subbituminous lignitic ash coal classified as having a severe slagging and medium fouling potential.

The difference in the size of the unit for the West Virginia coal and the unit for the Alabama coal can be attributed primarily to the difference in fouling potential. The furnace height has been increased substantially to reduce the temperature of the gas entering the superheater, thereby reducing the fouling problem. The increased furnace depth and height of the Wyoming coal unit over the dimensions of the unit sized for the Alabama coal are primarily attributed to the difference in the slagging potential of the ash. The lignitic ash results in considerably more slag buildups on the furnace walls, thereby decreasing heat transfer and requiring considerably more furnace surface. The furnace depth has been increased to control slagging by reducing the input per plan area.

OIL/GAS CAPABILITY

A boiler designed to use coal and/or wood as the primary fuel is frequently required to provide steam-generating capability when fired with oil and/or natural gas.

*A portion of the material in this section has been adapted with permission from the paper of the same title by J. D. Blue, J. L. Clement, and V. L. Smith that was presented at the TAPPI Engineering Meeting, October, 1974.

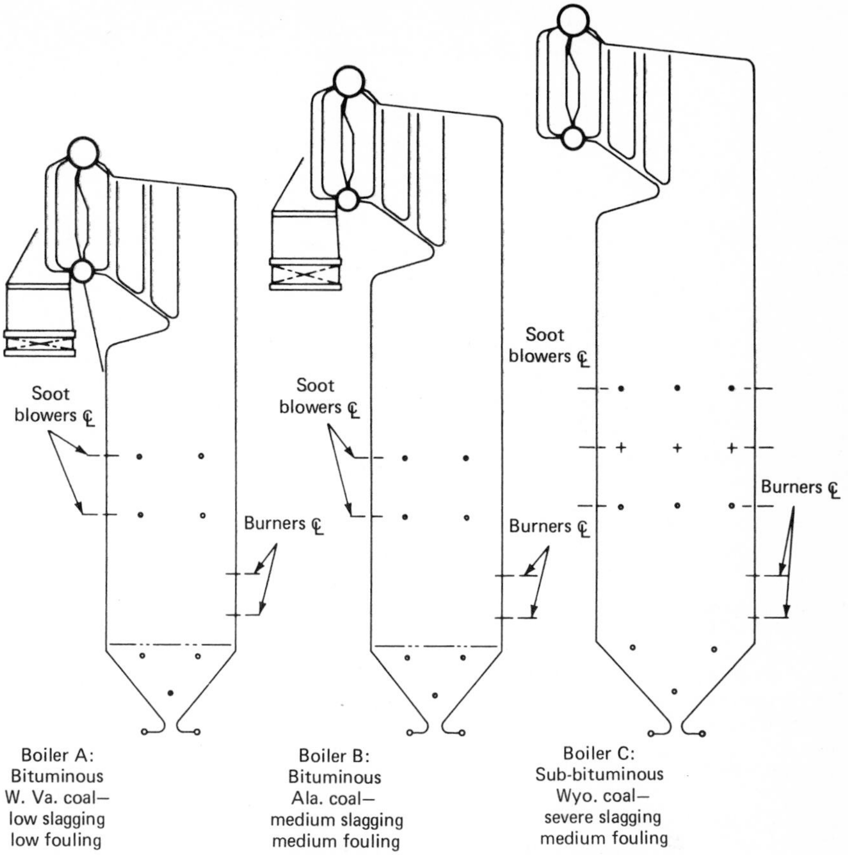

Figure 3-16 Influence of ash characteristics on furnace size.

The design of a boiler for alternative fuels generally requires a compromise. For example, the steam temperature resulting from the effects of a fixed furnace and superheater surface on various fuels is illustrated in Fig. 3-17. The superheater surface in this illustration is designed to provide full-rated steam temperature on oil. The operation with gas or bark as a fuel produces the desired steam temperature at 45 and 55 percent of full load rating, respectively. The steam temperatures on these fuels would have to be moderated at higher loads to control the terminal temperature. The extent of alloy tubing used in fabrication of the superheater is governed by gas firing where the highest potential steam temperatures exist.

STOKER OR PULVERIZED COAL

A decision to design for stoker or pulverized-coal firing is generally one of economics. A comparison of boilers designed to generate equal steam rates is characterized as follows:

1. The furnace volume is less for stoker firing because burning lower in the furnace gives more effective use of total furnace surface.
2. The furnace width is less for pulverized-coal firing.
3. Pulverized-coal-fired boilers are more responsive to load change.

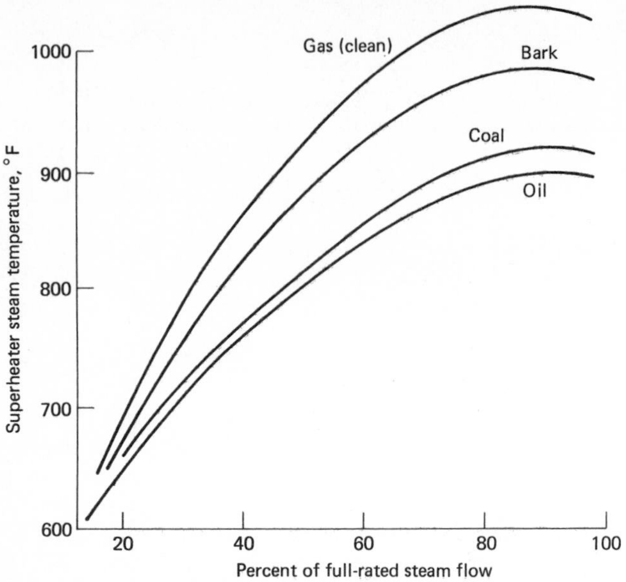

Figure 3-17 Effect of fuel on uncontrolled superheater steam temperature.

4. Stoker firing requires a lower operating horsepower.
5. Thermal efficiency is higher for pulverized coal (PC) by virtue of the lower unburned-carbon loss (UCL); the efficiency for stoker firing is dependent upon reinjection of ash as shown in Table 3-19.

The heat input to the stoker is decreased with increased reinjection rate to accommodate the reinjection of ash across the width of the furnace with reasonable carryover. With zero reinjection, heat inputs would generally be 10 percent higher than those on a stoker with 70 percent reinjection of the total ash.

TABLE 3-19 Thermal Efficiency

	Pulverized coal	Stoker coal		
Reinjection, %	0	0	30	70
Unburned-carbon loss, % efficiency	0.5	6.0	4.6	1.6
UCL difference, %	Base	5.5	4.1	1.1
Efficiency, % of heat input	86.0	80.5	81.9	84.9
Increase in heat input for constant steam flow, % ratio to PC	1.0	1.068	1.050	1.013

FURNACE MAINTENANCE

Preventive maintenance includes a policy of operating equipment properly and within its range of capability and keeping the equipment clean and in prime operating condition. This is verified by instrumentation and in-service observations. Preventive maintenance also includes regularly scheduled outages to make those inspections which cannot be made during operation and to perform necessary repairs.

In-Service Maintenance

Safety

Primary emphasis should be on safe operation, the avoidance of conditions that could result in explosive mixtures of fuel in the furnace or other parts of the setting, and the protection of furnace pressure parts to prevent excessive thermal stresses or overheating that would result in failure.

Furnace and Setting

The prevention of furnace explosions deserves top priority because of potential personnel injury, the high cost of repairs, and the effect of outage time on the industrial processes. Four items are of major importance in the prevention of furnace explosions:

1. Optimum operating procedures and operator training
2. Optimum burner observation with prompt detection of flame failure
3. Detection of unburned combustibles in the flue gas
4. Positive, immediate indication of fuel-air relation at the burners

The majority of furnace explosions result from failure to detect a loss of ignition, even though other indications such as dropping boiler pressure, steam temperature, and exit-gas temperature show that fuel is either not being burned or is being burned incompletely. This emphasizes the fact that *nothing takes the place of seeing the fire*. Therefore, a reliable remote indication of positive ignition on all burners must be relayed to the operator at the central control point. A *combustibles alarm* to indicate the presence of unburned combustibles in the flue gas is considered a good backup for flame observation.

The measurement of furnace pressure on both pressurized and suction-type units is required to assure that design pressures of the casing or containment are not exceeded. Some form of differential gas- or air-pressure measurement, in conjunction with time lapse, is required as an indication that the setting has been adequately purged prior to firing. This is a necessary operating guide to assure a proper fuel-air ratio at the burners.

Variables in Efficiency Losses

There are two variables in dry gas loss: stack temperature of the gas and weight of the gas leaving the unit. Stack temperature varies with the degree of deposit on the heat-absorbing surfaces throughout the unit, and it varies with the amount of excess combustion air. The effect of excess air is twofold: (1) It increases the gas weight, and (2) it raises the exit-gas temperature. Both effects increase dry gas loss and thereby reduce efficiency. A rough rule of thumb gives an approximate 1 percent reduction in efficiency for about a 40°F (22°C) increase in stack-gas temperature on coal-fired installations.

In coal-fired units, unburned-combustible loss includes the unburned constituents in the ash-pit refuse and in the flue dust.

Monitoring Efficiency

Continuous monitoring of flue-gas temperatures and flue-gas oxygen content by a regularly calibrated recorder or indicator and by periodic checks on combustibles in the refuse will indicate if original efficiencies are being maintained. If conditions vary from the established performance base, corrective adjustments or maintenance steps should be taken.

High exit-gas temperatures and high draft losses with normal excess air indicate dirty heat-absorbing surfaces and the need for soot blowing.

High excess air normally increases exit-gas temperatures and draft losses and indicates the need for an adjustment to the fuel/air ratio. The high excess air may, however, be caused by excessive casing leaks or by cooling air, sealing air, or air-heater leaks.

High combustibles in the refuse indicates a need for adjustments or maintenance of fuel-preparation and -burning equipment.

Water-Side Cleanliness

One of the best preventive steps that can be taken to assure safe, dependable operation is the maintenance of boiler-water conditions that will insure against any internal tube deposits that could cause overheating and failure of furnace tubes.

Outage Maintenance

Outages for preventive maintenance should be scheduled as required to prevent equipment failures.

Water-Side Cleaning

Chemical or acid cleaning is the quickest and most satisfactory method for the removal of water-side deposits. It is, however, extremely important to use a procedure of known reliability under careful control.

Gas-Side Inspection

During outage periods, the furnace should be thoroughly inspected; the objectives should be:

1. To detect any possible signs of overheating of tubes; furnace wall tubes should be examined for swelling, blistering, or warping
2. To discover any possible signs of erosion or corrosion
3. To detect any misalignment of tubes from warpage
4. To locate any deposits of ash or slag, not removed by sootblowing, that interfere with heat transfer or free gas flow in the furnace
5. To determine the condition of fuel-preparation and -burning equipment, particularly if routine sampling of ash has indicated the presence of increasing amounts of unburned combustible material
6. To determine the condition of any refractory exposed to furnace gases, such as burner throats or furnace walls

Gas-Side Cleaning

When ash deposits contain an appreciable amount of sulfur, they should be removed prior to any extended outage, since these deposits can absorb ambient moisture to form sulfuric acid that will corrode furnace pressure parts.

chapter 4-4

Electric Generators

by
Martin C. Doherty
Consulting Application Engineer
Thermal Power Systems Engineering
General Electric Co.
Schenectady, New York

Herbert T. Nock
Application Engineer
Medium Steam Turbine Generators
General Electric Co.
Lynn, Massachusetts

GLOSSARY

Armature The member of a rotating electric machine in which an alternating voltage is generated by virtue of relative motion with respect to a magnetic-flux field.

Core An element of magnetic material, serving as part of a path for magnetic flux. In rotating machines, this is frequently part of the stator, a hollow cylinder of laminated magnetic steel, slotted on the inner surface for the purpose of containing the armature windings.

Exciter The source of all or part of the dc field current for the excitation of an electric machine.

Field An insulated winding in rotating synchronous electric machinery whose purpose is the production of the main electromagnetic field of the machine.

Permeability A property of materials which expresses the relationship between magnetic induction and magnetizing force. An indication of the relative ability to conduct magnetic flux.

Power factor The ratio of the active power in watts to the apparent power in voltamperes. It is also the cosine of the phase angle between the voltage and current.

Reactance The imaginary part of impedance.

Reactive power The imaginary power required to magnetize the air gaps in motors and transformers in ac electric circuits.

Short-circuit ratio In synchronous electric machines, the ratio of the field current for rated open-circuit armature voltage and rated frequency to the field current for rated armature current on sustained symmetrical short circuit at rated frequency.

Slip Difference between synchronous and actual speed, usually expressed as a percentage of synchronous speed.

Synchronous speed Speed in revolutions per minute relating to ac frequency. It equals 120 × frequency in hertz per number of poles.

INTRODUCTION

There are three basic types of rotating electric generators:

Synchronous ac
Induction ac
Rotating dc

Virtually all of the power generated by electric utilities and industrial turbine generators is supplied by synchronous ac generators. This type of generator includes an excitation system which is used to regulate the output voltage and power factor. The emphasis in this chapter will therefore be on synchronous ac generators.

Induction generators are squirrel-cage induction motors which are driven above synchronous speed. They do not have an excitation system and hence cannot control voltage or power factor. The system must supply the excitation. Induction generators are generally applied where relatively small waste energy or hydro potential exists; they are driven by a steam turbine, a gas expander, or a hydraulic turbine to recover the power in the energy stream. In these cases it is economical to adjust power factor and voltage on other larger synchronous generators in the system.

Rotating dc generators have been replaced almost entirely by static silicon rectifiers. The demand for rotating dc generators is limited to a few very special applications such as elevators and large excavators. No practical method has been developed for reducing the high maintenance associated with the commutators and brushes of dc generators.

SYNCHRONOUS GENERATORS

The fundamental principle of operation of synchronous ac generators is that relative motion between a conductor and a magnetic field induces a voltage in the conductor. The magnitude of the voltage is proportional to the rate at which the conductor cuts lines of flux. The most common arrangement is with a cylindrical electromagnet rotating inside a stationary conductor assembly. The electromagnet is called the *field* and it is shown in simplified schematic form in Fig. 4-1. The conductors constitute the *armature* and are illustrated in Fig. 4-2. An external source of dc power is applied through the collector rings on the rotor. The flux strength and hence the induced voltage in the armature are regulated by the dc current and voltage supplied to the field. Alternating current is produced in the armature by the reversal of the magnetic field as north and south poles pass the individual conductors.

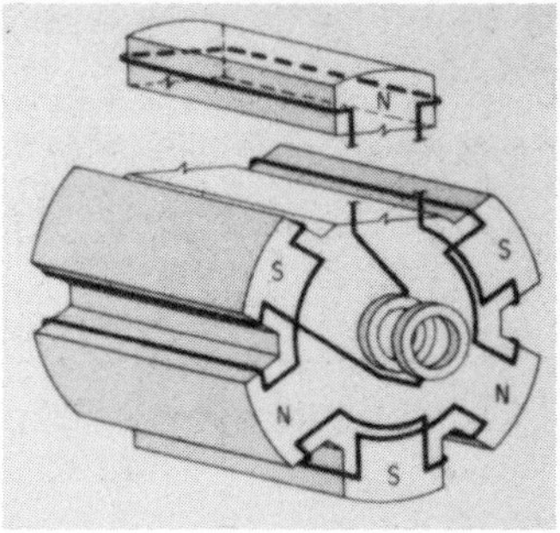

Figure 4-1 Simplified six-pole generator field.

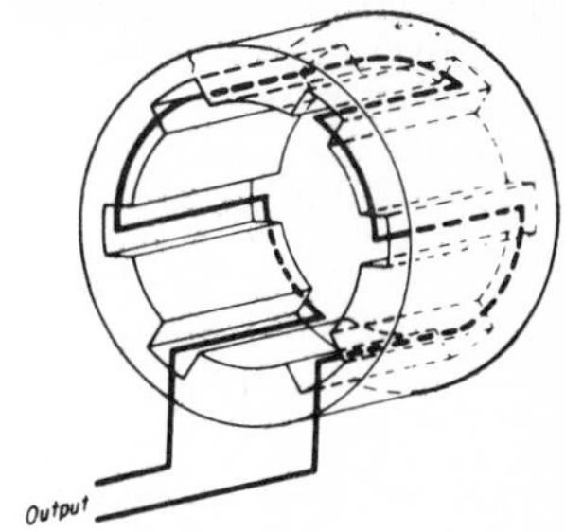

Figure 4-2 Simplified generator armature.

Lines of magnetic flux always form a closed circuit as shown in Fig. 4-3. Confining the flux field in materials with high permeability (low resistance to magnetic flux) intensifies the flux density. The permeability of certain steels is thousands of times greater than air. The flux density at the pole faces is proportional to the ampere turns on the poles and the combined permeability of all the materials in the circuit including the rotor core, the stator core, and the air gap.

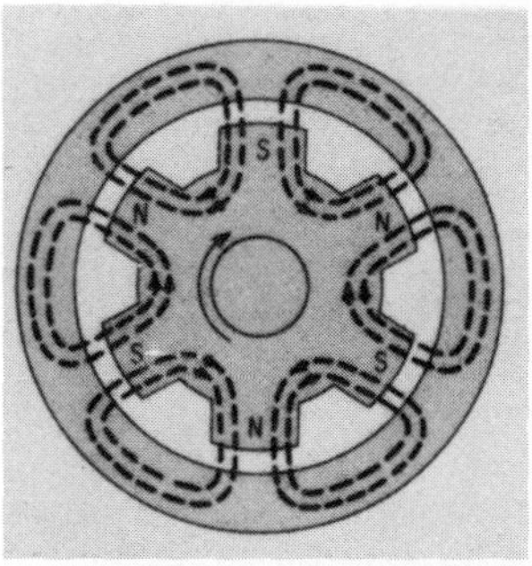

Figure 4-3 Generator magnetic circuit.

The stator core is built up with steel laminations to provide both the high permeability magnetic path and a high-resistance electric path to minimize induced voltage and inherent heat generation.

The simplified drawing of an armature winding in Fig. 4-2 shows only a single phase. All generators except those with very small ratings have three phases, each phase consisting of several conductors.

There are two parameters which limit the output of a generator:

1. **Flux density saturation** As field-exciting current is increased, a point is reached where the flux density no longer increases because of iron saturation in the core. Normally the generator rating in kilovoltamperes (kVA) is near this flux saturation point.
2. **Temperature rise in the windings and insulation due to losses** This includes losses due to excitation current in the field windings, ac current in the armature windings, the magnetic circuit, and any stray currents or magnetic fields which are generated.

Synchronous ac generators are classified by their construction, method of cooling, and excitation system. The design chosen is determined by the type of prime mover driving the generator, the power required, and the operating duty (continuous versus intermittent operation, clean versus dirty environment).

Construction

Generator rotors are made of forgings of high-permeability magnetic steel. Conducting coils are assembled on the rotor to produce the desired number of electromagnetic poles which rotate with the rotor. The number of poles and the rotational speed of the rotor will determine the frequency of the power generated by the following relationship:

$$f\ (\text{Hz}) = \frac{\text{no. of poles}}{2} \times \frac{\text{r/min}}{60}$$

There are two methods of rotor fabrication. In the four-pole generator shown in Fig. 4-4, separate poles and windings are attached to the rotor. This *salient-pole* type of construction is distinguished by the protusion of the poles from the shaft. Turbine- and engine-driven salient-pole generators generally run at 900 to 1800 r/min and are available up to 40,000 kVA. Slower-speed (lower than 500 r/min) salient-pole generators are used with hydraulic turbines in sizes up to 800,000 kVA.

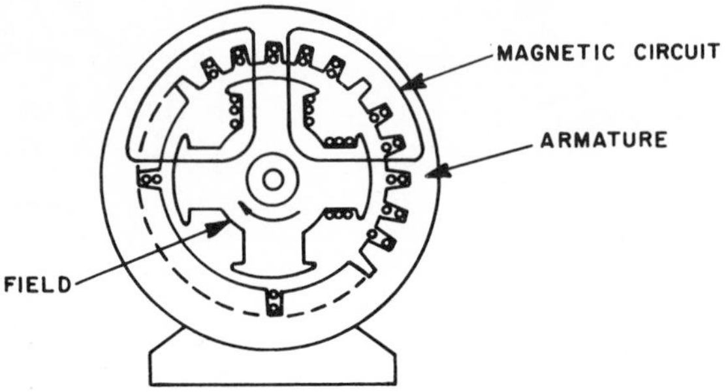

Figure 4-4 Salient-pole generator.

Figure 4-5 illustrates a cylindrical or integral slot rotor. This type of construction is used for speeds of 1800 or 3600 r/min for 60-Hz applications. Most steam- and gas-turbine generators in ratings greater than 15,000 kVA have generators with cylindrical rotors.

Figure 4-5 Cylindrical generator rotor. *(General Electric Co.)*

The stator of the generator consists of the frame, core, windings, and cooling system. It must be capable of supporting the static weight of the entire generator plus securing the stator core against the torsional forces due to load. The stator core is constructed of grain-oriented, silicon steel laminations, selected for high permeability, which are used to hold the copper windings of the armature. The windings are made of individually insulated strands of copper and are arranged so that each strand takes a different depth in the slot. This creates the same voltage in each strand and eliminates current loops that would form within the windings if unequal strand voltages existed.

Excitation System

The excitation system provides magnetizing power (about 1 percent of generator output power) for the rotating field winding and accurately controls the amount of magnetizing power to maintain close regulation of the generator output voltage and power factor.

Several excitation systems presently exist; these are classified according to the exciter power source:

- DC generator with commutator
- AC generator and stationary rectifiers
- AC generator and rotating rectifiers (brushless)
- Transformers on the main generator and rectifiers (static excitation)

A schematic diagram of the dc generator with commutator connected to the main shaft is shown in Fig. 4-6. (They can be driven by separate motors or steam turbines.) The excitation power is taken from the commutator on the dc-generator rotor and applied to the main-generator rotating field through collector rings. The main-generator output voltage is controlled by using a voltage regulator to vary the excitation of the dc-generator stator.

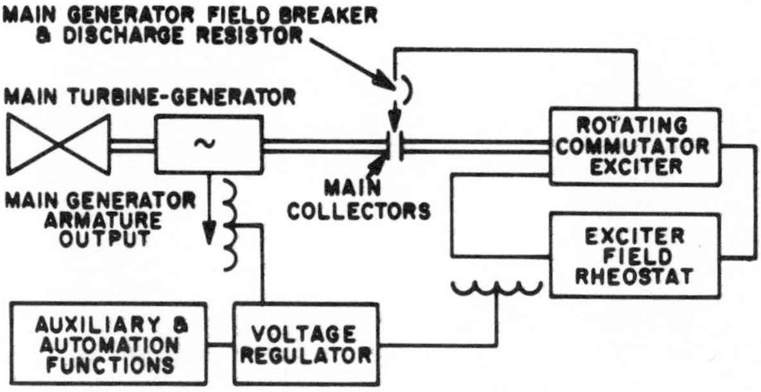

Figure 4-6 DC commutator excitation system.

Since commutator-type dc-generator excitation systems are inherently high in maintenance due to the commutator and brushes, the invention of solid-state rectifiers has reduced the usage of this equipment in favor of ac generators rectified to dc using silicon diode rectifiers. There are two methods of implementing these systems: stationary rectifiers and rotating rectifiers. The stationary rectifier system is shown schematically in Fig. 4-7. The ac exciter has a rotating field, as does the main generator. The exciter out-

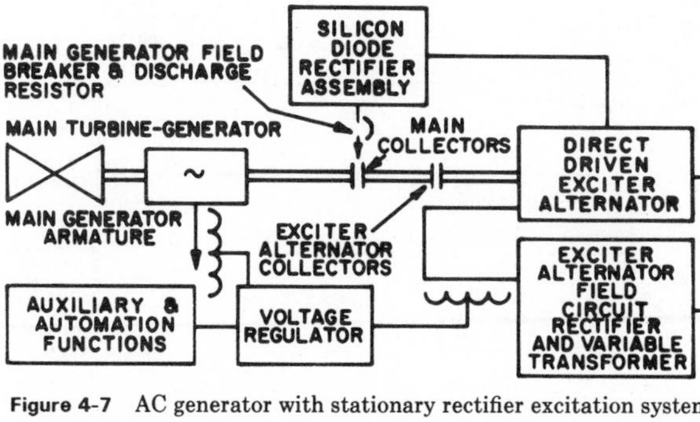

Figure 4-7 AC generator with stationary rectifier excitation system.

put is taken from its stationary armature windings, converted to dc by silicon diode rectifiers, and applied to the main-generator rotating field through collector rings. The control system is similar to the dc-generator system, except that the excitation to the exciter rotating field is transferred by collector rings. This type of system is used for generators larger than 400,000 kVA where the excitation power can be as high as 7000 kW.

An alternative ac exciter system known as *brushless* or *rotating rectifier* is shown in Fig. 4-8. It reverses the exciter field and armature and thereby eliminates both sets of collector rings. The main generator output voltage is controlled through the exciter field in the stator. The exciter armature and the silicon diode rectifiers are on the main shaft, directly connected to the main-generator field, and generator control is affected through the air gap of the exciter by varying the stationary exciter field current. This system eliminates all collector rings, hence the name *brushless exciter.*

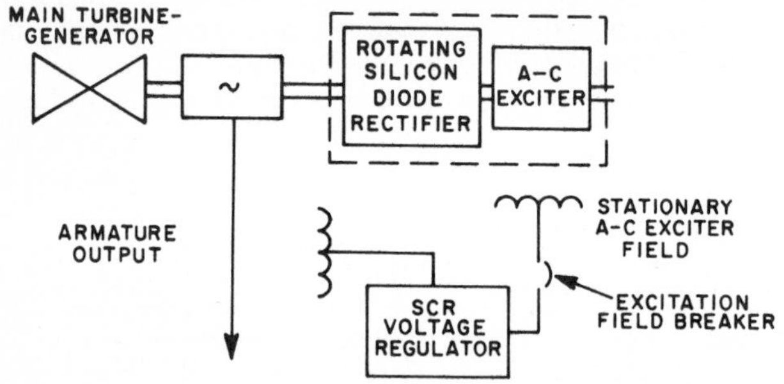

Figure 4-8 AC generator with rotating rectifier (brushless) excitation system.

A static excitation system (see Fig. 4-9) eliminates the need for a separate generator for excitation. The excitation power is provided by the main-generator terminals through excitation transformers. The controlled ac output of the transformers is converted to dc by silicon diode rectifiers and applied to the main-generator field through collector rings.

When comparing excitation systems, each has its advantages: While the brushless system eliminates collector rings, a failure of the rotating rectifier can cause a shutdown. In contrast, the static excitation system normally provides parallel sets of stationary rectifiers, so a full load can be carried with one bank out of service. However, this system requires periodic brush maintenance which can be done while operating.

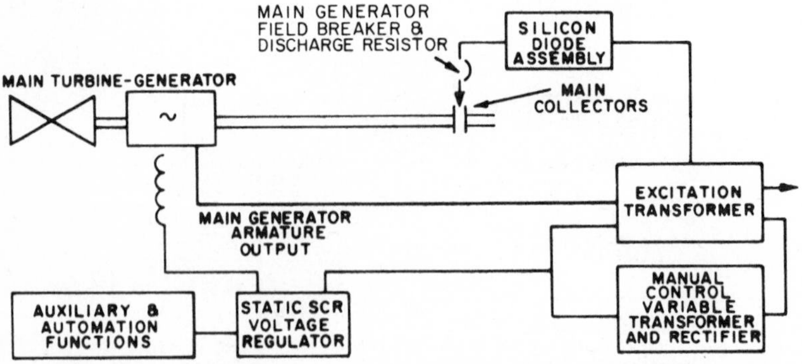

Figure 4-9 Static excitation system.

Cooling System

Owing to the flow of current in the field and armature, heat will be generated which must be removed by a cooling system. These systems can be defined several ways:

1. **Cooling Medium** Generally air or hydrogen is used.
2. **Direct vs. Indirect Cooling** In direct cooling, the cooling medium comes in direct contact with the windings being cooled, while in indirect cooling the heat must pass through the insulation prior to reaching the cooling medium. Most generators of less than 250 MVA are indirectly cooled.
3. **Self- vs. Separately Ventilated** Self-ventilated units use fans on the generator rotor to circulate the cooling media while separately ventilated generators use external fans.
4. **Open vs. Enclosed Ventilation** An open ventilated unit allows ambient air to enter, normally through a filtration system, and leave the generator freely, while an enclosed generator will recirculate the air inside the generator frame. Enclosed units allow no interface of the cooling media with the ambient air and must use heat exchangers (generally with water) to remove the heat from the generator. A typical coolant flow diagram for an enclosed self-ventilated generator is shown in Fig. 4-10.

Figure 4-11 is a photo of a totally enclosed air-to-water–cooled generator with the water coolers mounted in the corners of the generator casing.

In air-cooled generators, the generator circulates its own weight in air every 45 min. In larger generators, the power required to circulate the air can amount to 1 to 2 percent

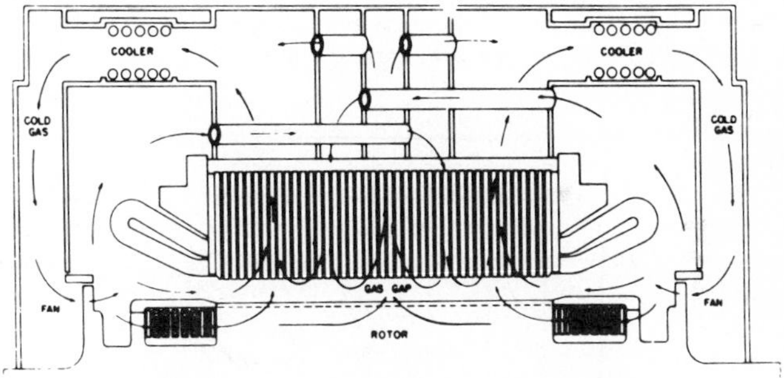

Figure 4-10 Typical generator cooling system.

Figure 4-11 Photo of a totally enclosed air-to-water–cooled generator. *(General Electric Co.)*

of the kilowatt rating which tends to limit the economics of air cooling to ratings of less than 50 MVA. Above this rating, hydrogen-cooled generators dominate because the superior conductivity and lower density of hydrogen make these generators smaller and ½ to 1 percent more efficient at full load than air-cooled generators, with up to 2 or 3 percent improvement at partial loads.

While hydrogen does require a sealing system for the generator, hydrogen-cooled units are more efficient than enclosed air-cooled generators. Hydrogen is not combustible above a purity of 75 percent, so it can be used safely at the 96 to 98 percent purity levels of today's generators. The internal components of hydrogen-cooled units also stay cleaner since they are isolated from the ambient environment.

Operation

When a generator is operating alone with an isolated load, the power factor and reactive power are determined by the load. When it operates as part of a system of generators and loads (power grid), however, the reactive power supplied by each generator can vary depending on the level of excitation of each machine. Therefore, changing the excitation of one generator in a system will change the power factor of that unit, while the voltage and frequency of the power grid will remain constant by the collective action of the other generators.

The effect of varying excitation or field current on generator performance is typically shown in excitation *V* curves and reactive capability curves. Typical curves of this type are shown as Figs. 4-12 and 4-13, respectively, for a hydrogen-cooled generator. The excitation *V* curve shows the amount of excitation required to produce a desired power factor at any power output of the generator, while the reactive capability curve defines the limits of operation. Using these curves an operator can cause a generator to produce reactive power by increasing the field current (excitation) or absorb reactive power by decreasing the field current. This would correspond to moving along the vertical axis on

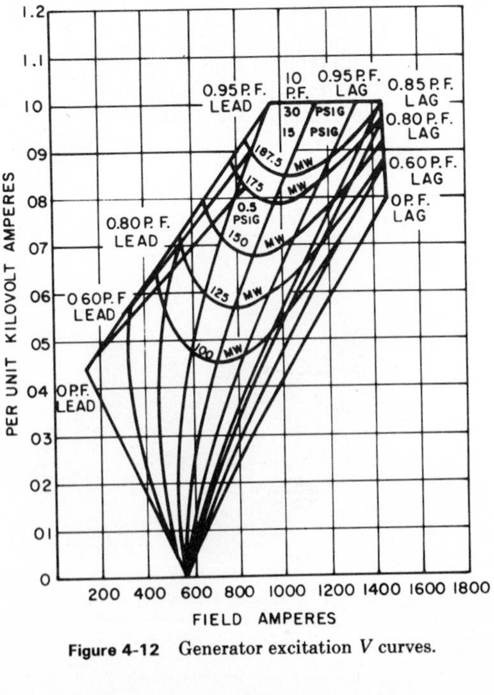

Figure 4-12 Generator excitation *V* curves.

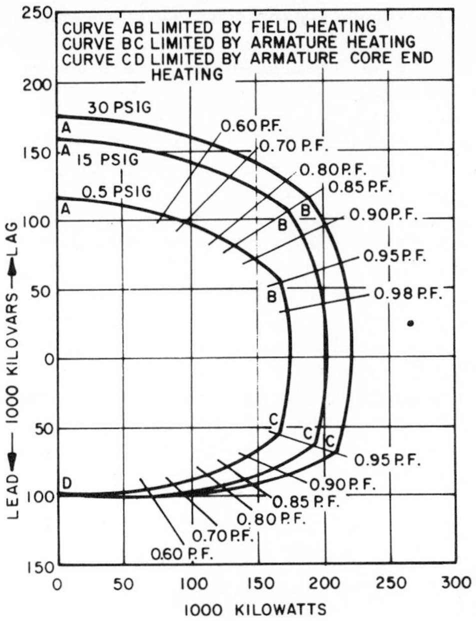

Figure 4-13 Reactive capability curve.

the reactive capability curve. To move horizontally along this curve, the power supplied to the generator by the driving machinery would have to be increased. This highlights two important facts of generator operation in a power system:

1. The reactive power (power factor) of the power delivered by the generator is controlled by the generator field current (excitation).
2. The amount of power delivered by the generator is controlled by the machinery driving the generator.

The stability of a generator is a function of its short-circuit ratio and the response time of its excitation system. The short-circuit ratio is the ratio of field current for open-circuited armature voltage at rated frequency to the field current for rated armature current on a sustained symmetrical short circuit at rated frequency. The higher the short-circuit ratio of a generator, the more stable the design.

A typical type of upset condition examined when determining generator stability is a short circuit. In examining the transient conditions during upsets, the term *reactance* is used. It is defined here as the ratio of open-circuit armature voltage at some field current to the short-circuit armature current at the same field current and is expressed in ohms (Ω). Reactances are usually given as a ratio to a base reactance in either percent or per unit. The base is generally taken as the ratio of phase voltage to phase current. Following a short circuit, the reactance can decrease to about 0.15 per unit and the armature current can increase to 10 or 15 times rated value. This condition will occur for a few cycles and is called the subtransient reactance (X_d''). The reactance increases after a few cycles to the transient reactance (X_d') and continues to increase to a steady-state value known as the synchronous reactance (X_d). The synchronous reactance is typically 1.5 per unit, and the armature current will be about twice the rated current assuming a three-phase short circuit. Since the synchronous reactance is inversely proportional to the short-circuit ratio when the saturation effects of the magnetic circuit are neglected and since it is desirable to minimize the changes in generator operation (changes in armature current) due to upset conditions, a short-circuit ratio is a good measure of generator stability.

When a generator is started, it is first brought to rated speed by the driving equipment and the field current is adjusted to produce rated generator voltage on open circuit. The value of field current is shown as the extreme lower end of the excitation *V* curve. The frequency, voltage, and phase angle of the generator output are checked to ensure they are consistent with the power system , and the circuit breakers are closed to connect the generator to the system.

During operation a number of protective devices are used to detect abnormal conditions and promptly isolate the troubled area to prevent damage to the generator. Relays can be provided to trip the generator upon the occurrence of any of the following conditions: unbalanced differential phase currents, phase-to-phase or ground faults, external phase-to-phase or ground faults, loss of excitation, out-of-step reverse power flow, unbalanced loading, lightning, and overvoltage. Alarms also indicate a field ground fault or excessive stator winding temperatures which are measured by resistance temperature detectors (RTDs). The latter two conditions may not require an instantaneous trip, and the operator can plan the maintenance shutdown.

INDUCTION GENERATORS

The stator of an induction generator is similar to a synchronous generator. The rotor differs from the synchronous generator rotor in that there is no excitation and the conductors are shorted together at the rotor ends by an annular ring. This arrangement resembles a squirrel cage, which lends its name to the type of winding.

The induction generator supplies real power in *kilowatts,* which displaces high-cost energy from the system. The imaginary power, *kilovars,* is drawn by the induction generator; it requires installed capability by some other device on the system, but consumes only a negligible amount of energy.

An induction machine operates at synchronous speed at zero load. The rotor turns at the same speed as the rotating flux field in the stator, and no lines of flux are cut. When a load torque is applied, the rotor speed drops off or "slips" until full torque is reached at 2 to 5 percent slip. As a generator, the driver must overspeed the generator by 2 to 5 percent to achieve full electric output.

Induction generators cannot operate independently in an isolated system. They can only function in parallel with synchronous generators which regulate voltage and supply the kilovars necessary to overcome the lagging power of the induction generation.

Induction generators are simple and lower in initial cost than synchronous generators. They have been applied to recover power by expanding waste-gas streams and low-pressure steam. In some applications an energy-recovery turbine or expander drives an induction generator-motor and another pump or compressor on the same shaft. The generator-motor can either supply or absorb torque when the power of the other two devices is out of balance.

DC GENERATORS

The operating principle of the dc generator is very similar to that of the ac generator. In the dc generator the field is located in the stator while the armature rotates, generating alternating current in the armature windings. The commutator and brushes provide a means of transferring the output from the rotor to the stator, as well as of mechanically rectifying the alternating current. Figure 4-14 illustrates a dc-generator brush rigging and commutator. The commutator is a wearing surface for the brushes. It consists of individual copper segments insulated from each other by mica and connected to the armature windings. The armature-winding connections to the commutator and the brush

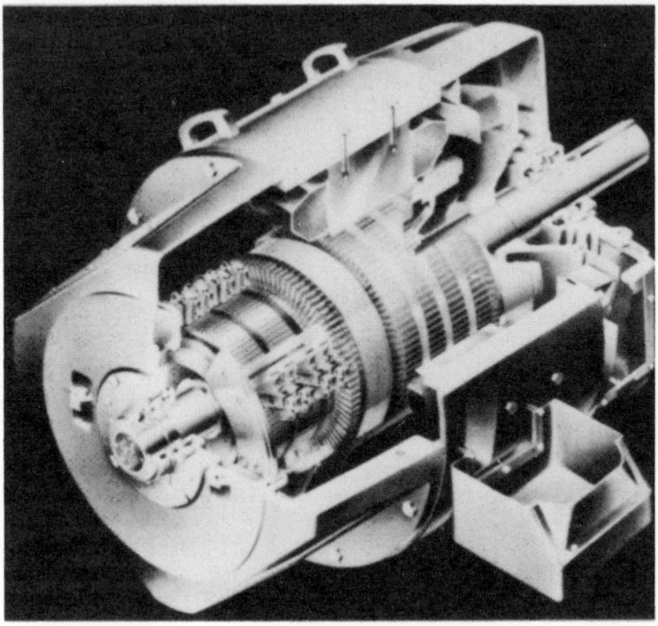

Figure 4-14 DC generator. *(General Electric Co.)*

spacing have to be carefully arranged so that brushes of opposite polarity contact windings which are 180 electrical degrees out of phase.

Many dc generators driven by motors have been installed in industrial plants, such as steel mills, to provide power for variable-speed drives. However, the advances in static silicon-rectifier dc power sources have reduced the market for dc generators primarily to replacement and repair parts, with very few new installations.

BIBLIOGRAPHY

"American Standard Requirements for Cylindrical Rotor Synchronous Generators," C50.13-1965, American Standards Association (now ANSI) New York, 1965.

"Electric Generators," special report, *Power,* McGraw-Hill, New York, March 1966.

Fink, Donald G. (ed.): *Standard Handbook for Electrical Engineers,* 11th ed. McGraw-Hill, New York, 1978.

"Guide for Operation and Maintenance of Turbine-Generators," IEEE Standard 67-1972 (ANSI C50.30-1972), The Institute of Electrical and Electronics Engineers, Inc., New York, 1972.

Jay, Frank (ed.); *IEEE Standard Dictionary of Electrical and Electronics Terms,* 2d ed., The Institute of Electrical and Electronics Engineers, Inc., New York., 1977. Distributed in cooperation with Wiley-Interscience, New York.

chapter 4-5

Stationary Turbines

PART 1

Gas Turbines

by
Martin C. Doherty
Consulting Application Engineer
Thermal Power Systems Engineering
General Electric Co.
Schenectady, New York

GLOSSARY

Aircraft derivative gas turbine An aircraft jet engine modified for ground applications to produce shaft power instead of thrust.

Base rating The designed rating point of a gas turbine at which it is suitable for continuous operation. Referenced to standard ISO conditions. (See "ISO Rating.")

Combined cycle A combined steam and gas turbine arrangement in which the gas turbine exhaust is ducted to a heat-recovery steam generator which supplies steam to the steam turbine.

Compression ratio The ratio of the compressor discharge pressure to the suction pressure.

Firing temperature The mass-flow mean total temperature of the working fluid measured in a plane immediately upstream of the first-stage turbine buckets.

Fuel consumption The input fuel heating value per unit of time to a gas turbine, generally measured in British thermal units per hour (kilojoules per hour). Also called heat consumption. Generally stated in terms of the lower heating value (LHV) of the fuel by gas turbine manufacturers. Can also be in terms of pounds per hour, where typical heating values are 18,500 Btu/lb for liquid fuels and 21,500 Btu/lb (LHV) for natural gas. Also see "Specific Fuel Consumption."

Heat rate The fuel consumption of a gas turbine divided by the output. For mechanical drive gas turbines this is the net output including the on-base auxiliary power losses. For generator-drive gas turbines this includes these auxiliaries plus the generator losses. Does not include the power requirements for off-base lubrication oil cooling or heavy fuel treatment, unless specified as in certain totally packaged designs. Expressed as British thermal units or kilojoules per kilowatthour or horsepower hour. Generally in terms of the LHV of the fuel.

Heavy-duty industrial gas turbine A type of gas turbine designed specifically for ground applications using a design philosophy similar to that of the steam-turbine industry. Casings are split on the horizontal centerline, with onsite maintenance planned after long periods of operation.

Heavy fuel Liquid petroleum fuels which are ash bearing and not true distillates. Can be crude oil or residuals (No. 5 or 6), or a blend of a distillate and residuals.

High heating value The gross heating value of the fuel. Includes the latent heat required to vaporize the water in the products of combustion, which is not truly available to a combustion device having an exhaust temperature higher than 212°F (100°C).

Hot gas path The path of the working fluid of a gas turbine during and after combustion. Includes the fuel nozzles, combustion chamber and liner (if required), transition pieces to the turbine, stationary and rotating airfoils (nozzles and buckets), and exhaust plenum and ducting.

ISO rating The rated output of a gas turbine at the standard site conditions specified by the International Standards Organization: sea-level altitude, standard atmospheric pressure of 14.7 psia (101.4 kPa) at the turbine inlet and exhaust, 59°F (15°C) ambient temperature, and 60 percent relative humidity.

NO_x Oxides of nitrogen include both NO and NO_2. Emission limits are generally based on parts per million by volume (ppmv) of the total of NO and NO_2 emitted by combustion devices.

Peak rating The designed rating point for gas turbines for peak-load service generally operated at less than 1000 hours per year. Not generally used for industrial applications which are base-loaded.

Regenerative cycle A gas turbine which includes a gas-to-gas heat exchanger which transfers heat from the exhaust to the compressor discharge air to reduce fuel consumption.

Simple cycle A gas turbine which exhausts to the atmosphere without heat recovery.

Specific fuel consumption (SFC) The gas turbine fuel consumption per unit of output. Generally in terms of pounds per kilowatthour or pounds per horsepower hour using LHV.

Specific work The output of a gas turbine per unit of air flow. Can be horsepower seconds per pound or British thermal units per pound (kilojoules per kilogram).

Thermodynamic efficiency The net output of a gas turbine divided by the input. It is the reciprocal of heat rate after normalizing units. For example:

$$\text{Thermal efficiency} = \frac{1}{\text{heat rate (Btu/kWh)}} \times 3412 \text{ Btu/kWh}$$

PRINCIPLES

Thermodynamic Fundamentals

A schematic diagram of a simple-cycle gas turbine is shown in Fig. 5-1. It consists of an air compressor, a combustion section, and an expansion turbine. The Brayton cycle is the thermodynamic process which is the basis of the gas turbine. Classical pressure-volume (*P-V*) and temperature-entropy (*T-S*) diagrams for this cycle are shown in Fig. 5-2. The numbers on these diagrams correspond to the numbers on Fig. 5-1. For the ideal

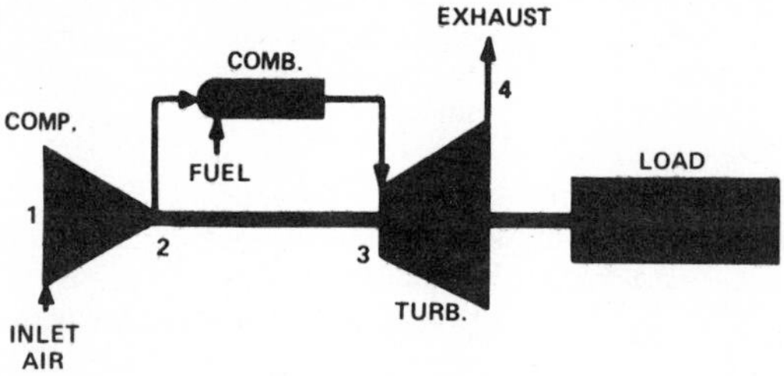

Figure 5-1 Schematic diagram of a simple-cycle, single-shaft gas turbine.

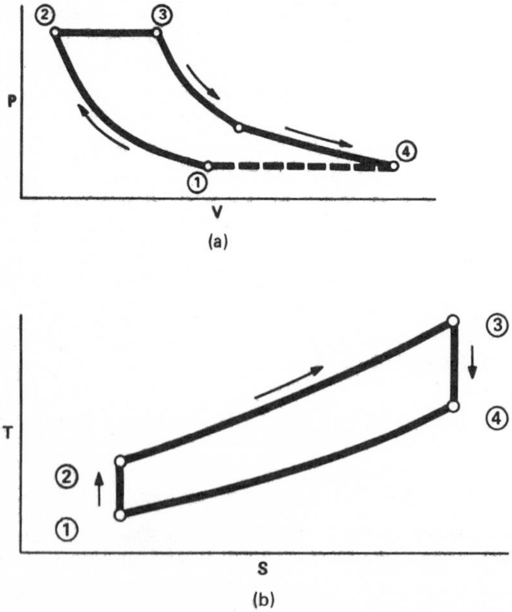

Figure 5-2 Brayton cycle: (*a*) pressure-volume and (*b*) temperature-entropy diagrams.

cycle, path 1 to 2 is an isentropic compression. No heat is added; however there is an increase in temperature due to the work of compression. Path 2 to 3 is the addition of heat at constant pressure. The isentropic expansion along path 3 to 4 produces net output, after subtracting the compression work, which drives the load.

In closed cycles, heat is rejected to the surroundings at constant pressure along path 4 to 1. In an open cycle the exhaust gas at point 4 is rejected to the atmosphere and fresh ambient air is drawn in at point 1.

A limited number of closed-cycle gas turbines have been developed primarily in Europe. In these, the heat addition along path 2 to 3 is by a heat exchanger with the heat supplied by an external source. In actual open cycles the heat addition results from the injection of fuel and its combustion along path 2 to 3.

The thermal efficiency of the Brayton cycle can be calculated using classical thermodynamic analysis. The compression ratio of the working fluid and the temperatures of heat addition and heat rejection are very important parameters. The results of such an analysis are shown in Fig. 5-3. These calculations are based on an ambient temperature of 59°F (15°C), actual component efficiencies, and real gas relationships. The results are plotted as thermal efficiency vs. specific work for two different firing temperatures.

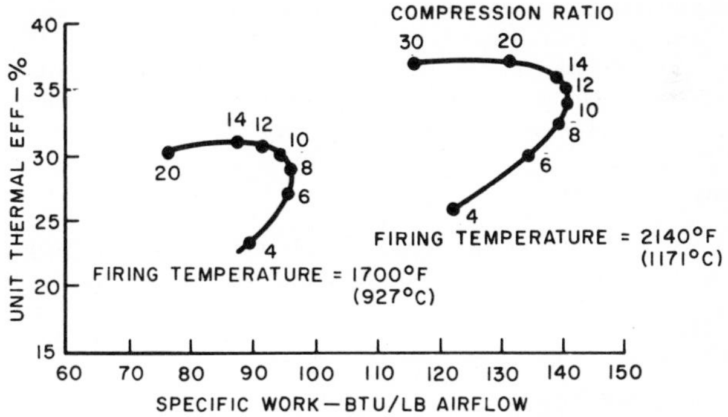

Figure 5-3 Efficiency vs. specific work of gas turbine cycles.

The observations which can be made from these curves are:

- Thermal efficiency increases as heat is added at higher temperature.
- For a given firing temperature there is an optimum pressure ratio for achieving maximum thermal efficiency.
- For a given firing temperature there is an optimum pressure ratio for achieving the maximum specific work which is different from the optimum thermal-efficiency pressure ratio.

A typical value of exhaust temperature for simple-cycle gas turbines is 950°F (510°C). Obviously, the thermal efficiency of the cycle can be improved if the heat ejected to the surroundings is at a lower temperature. One method of recovering some of the exhaust heat is to add a regenerator to the cycle such as shown in Fig. 5-4. The regenerator is a gas-to-gas heat exchanger which uses exhaust gas to heat

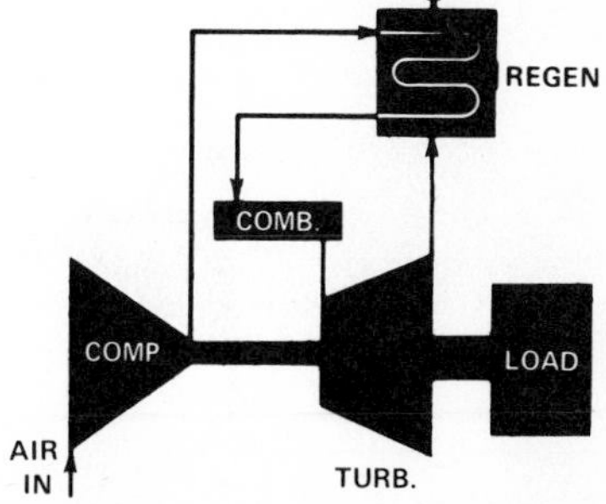

Figure 5-4 Schematic diagram of a regenerative cycle gas turbine.

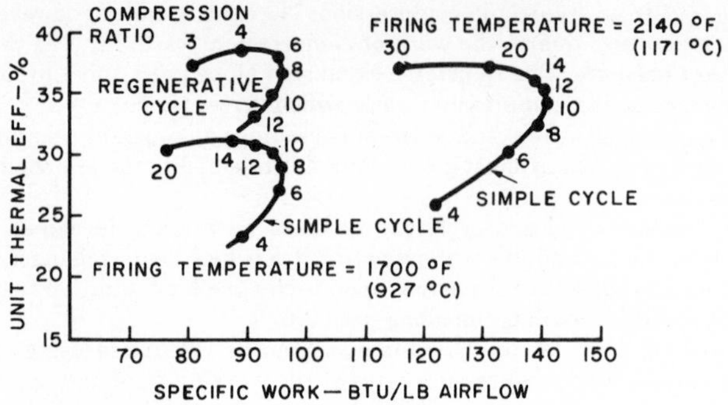

Figure 5-5 Efficiency vs. specific work of simple and regenerative cycle gas turbine.

the compressor discharge air prior to the combustion chamber. Fuel consumption can be reduced by as much as 25 percent by a regenerator. However, the gains due to the regenerator are sensitive to compression ratios. This is shown in Fig. 5-5. Higher values of compression ratio increase the compressor discharge temperature which reduces the amount of heat which can be picked up from the exhaust gas. Actual regenerative cycle units typically have compression ratios of less than 10:1.

Design Features

There are many different design features among the gas turbines available for industrial plant applications. Some of the more important characteristics are:

- One or two shafts
- Heavy-duty industrial type or aircraft derivative
- Combustion-chamber design

Figure 5-1 illustrates a single-shaft gas turbine in which all stages of the air compressor and the expansion turbine are mounted on the same shaft, which is also coupled to the driven load. Figure 5-6 shows a two-shaft gas turbine. A high-pressure turbine is mounted on the same shaft as the air compressor. The low-pressure turbine is mounted on a different shaft which is coupled to the driven load. Both types of designs are in widespread use.

The single-shaft type of gas turbine is ideally suited for electric-generator drive applications, while the two-shaft designs are more suited to mechanical-drive service. However, applications have not been exclusively along these lines. Single-shaft gas tur-

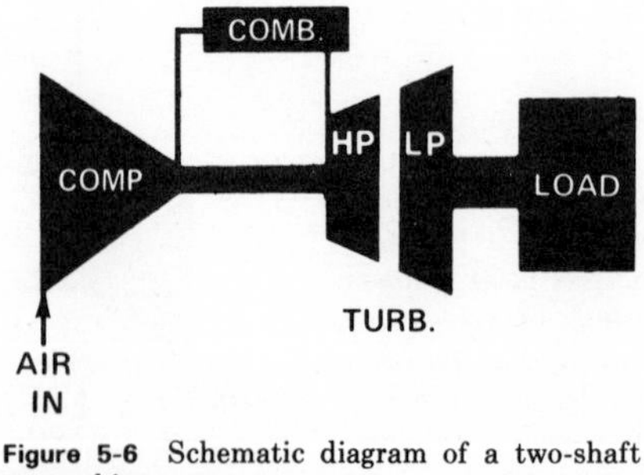

Figure 5-6 Schematic diagram of a two-shaft gas turbine.

bines also drive pumps and compressors, while two-shaft units have been applied to generator drives. Single-shaft units are suited to generator drives because of the compatibility of operating speed range and starting torque requirements. The high-efficiency axial-flow air compressor of many modern gas turbines is limited to a narrow speed range to stay within stall margins. This is compatible with electric-generator drive service since synchronous generators operate essentially at constant speed. Electric generators can also be easily unloaded during start-up to minimize the torque requirements on the starting device. This can be very important with single-shaft gas turbines. The starting motor or engine must accelerate the gas turbine and the driven load until a speed is reached at which the gas turbine can be fired; it then becomes self-sustaining.

Pumps and compressors are more difficult to unload during start-up. The starting torque can be reduced by adding a recycle piping system, but these can be complex and expensive. With a two-shaft gas turbine, the starting device need only accelerate the gas turbine's air compressor and high-pressure turbine, and not the low-pressure turbine and driven device. During normal operation, the gas turbine's low-pressure turbine typically has a speed range of 50 to 105 percent of rated speed, while the high-pressure turbine remains at constant speed or within a very moderate speed range (± 5 percent).

The designs of gas turbines have evolved from two distinct philosophies. Heavy-duty or industrial-type units have been based on the technology developed in the steam-turbine industry for large central stations. Of robust construction, with casings split along the horizontal centerline, these units are designed for long periods of continuous operation, generally have the capability to burn a variety of fuels, and are maintained on site. Aircraft-derivative gas turbines are jet engines modified to produce shaft power instead of thrust. Of lightweight construction, aircraft-derivative gas turbines are generally derated from flight-takeoff firing temperatures to allow long periods of continuous operation; they can usually be maintained on site, or are suitable for quick change-out and replacement with a spare engine. Aircraft-derivative units generally do not have the fuel flexibility of heavy-duty units.

Another distinguishing characteristic of gas turbine designs is the type of combustion section. There are three general types: a series of small cylindrical chambers or cans, an annular chamber surrounding the shaft, or a large single off-base combustor. The series of small cylindrical combustors is best suited to full-scale combustion development testing. New materials and designs can be developed without going to the expense of prototype testing. Investigations can also be simply made of unusual fuels and methods of reducing objectionable emissions such as NO_x. The annular combustion chamber has minimum ducting, weight, and length and therefore is best suited to aircraft-type units.

Figures 5-7 and 5-8 illustrate many of the characteristics of different gas turbine designs mentioned above. Figure 5-7 is the General Electric Model Series 6001A single-shaft heavy-duty gas turbine, and Fig. 5-8 is the General Electric LM2500 two-shaft aircraft-derivative unit. Notice the 10 different cylindrical combustors in the MS6001 and the single annular combustor of the LM2500. The MS6001 has two pressure-lubricated journal bearings. The two-shaft LM2500 has four rolling-contact (ball) bearings.

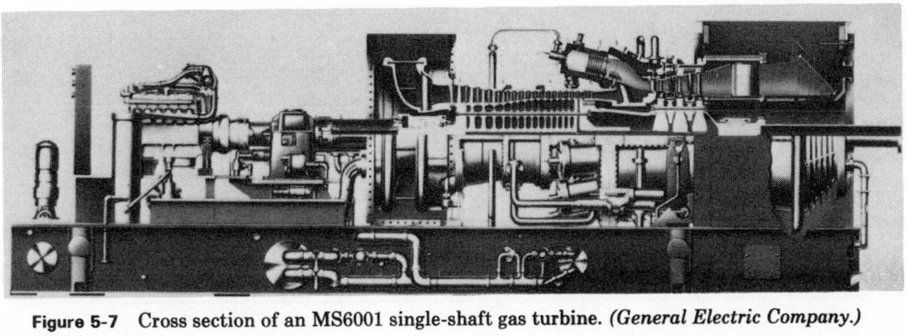

Figure 5-7 Cross section of an MS6001 single-shaft gas turbine. *(General Electric Company.)*

Figure 5-8 Cross section of an LM2500 two-shaft gas turbine. *(General Electric Co.)*

Some aircraft-derivative-type units consist of a jet engine or gas generator supplied by one manufacturer and a low-pressure turbine supplied by another manufacturer, who generally takes overall packaging responsibility.

PERFORMANCE CHARACTERISTICS

Gas Turbine Ratings

Since the introduction of the first industrial gas turbines in the 1950s there has been a continuous growth in performance. During this period there have been significant developments in the metallurgy of hot-gas path parts and in coatings, cooling techniques, instruments and control systems, and component efficiencies. Ratings for specific frame sizes have grown threefold. Indications are that ratings and efficiency values will con-

TABLE 5-1 Performance Data, Typical Gas Turbine–Generator, Distillate Oil Fuel

	Model			
Characteristic	LM2500	MS5001P	MS6001A	MS7001E
ISO rating base, kW	20,100	24,700	31,800	75,000
Heat rate, Btu/kWh (LHV)	9,800	12,230	11,250	10,590
kJ/Wh	10,340	12,900	11,870	11,170
Pressure ratio	18.0	10.2	11.5	11.5
Airflow, lb/s	144	269	304	604
kg/s	65.3	122.0	137.9	274.0
Turbine inlet temp, °F	2167	1730	1850	1985
°C	1186	943	1010	1085
Exhaust temp, °F	922	901	901	977
°C	504	483	483	525
Dry weight, lb	220,000	398,000	440,000	587,000
kg	99,800	180,500	199,500	266,200

Source: General Electric Company.

TABLE 5-2 Performance Data, Typical Mechanical-Drive Gas Turbine, Natural Gas Fuel

Characteristic	Model			
	MS3002	LM2500	MS5002(B)	MS5002R(B)
ISO rating, hp	14,600	27,500	35,000	32,000
kW	10,890	20,500	26,100	23,850
Heat rate, Btu/hph (LHV)	9,530	7,110	8,830	7,070
kJ/Wh	13,480	10,060	12,490	10,010
Pressure ratio	6.0	18.0	8.2	8.3
Airflow, lb/s	113	144	257	257
kg/s	52.6	65.3	116.6	116.6
Turbine inlet temp, °F	1730	2139	1700	1710
°C	943	1171	927	932
Exhaust temp, °F	979	922	915	666
°C	526	494	491	352
Dry weight, lb	120,000	52,000	257,000	457,000
kg	54,400	23,600	116,600	207,300
Output shaft speed, r/min	6500	3600	4670	4670

Source: General Electric Company.

tinue to increase as new techniques are developed for increased air and water cooling of hot-gas parts.

Therefore, any table of specifications for gas turbines can only represent a "snapshot" in time of what is a dynamic, ever-changing picture. Nevertheless, Tables 5-1 and 5-2 are offered to represent the state of the art of gas turbine technology in 1980.

In Table 5-1 the models MS5001, MS6001, and MS7001 are heavy-duty gas turbine–generator sets. The MS5001 has a firing temperature of 1730°F (943°C), which is typical for turbine buckets of advanced superalloys without internal bucket air cooling. Only the first-stage nozzles (stationary airfoils) are cooled by compressor discharge air in this unit. Figure 5-9 is a photo of a segment of an MS5001 first-stage nozzle illustrating the cooling-air holes on the trailing edge of the nozzles.

The MS6001 and MS7001 incorporate internal air cooling of the turbine bucket (rotating airfoil) which allows the firing temperature to rise to 1985°F (1085°C) while maintaining the same metal temperatures. Figure 5-10 illustrates a cutaway view of the first-stage air-cooled bucket of the MS7001. Notice that the inlet and exit holes are not directly in the hot-gas path. This allows burning heavy ash-bearing fuels without external plugging of the cooling holes. Figure 5-11 illustrates the flow path of the cooling air from the compressor discharge down into the rotor and up through the buckets.

The modern heavy-duty gas turbines listed in Table 5-1 have moderate compression ratios and values of specific work, which enhance their heat recovery capability. They

Figure 5-9 First-stage air-cooled nozzle of an MS5001 gas turbine. *(General Electric Company.)*

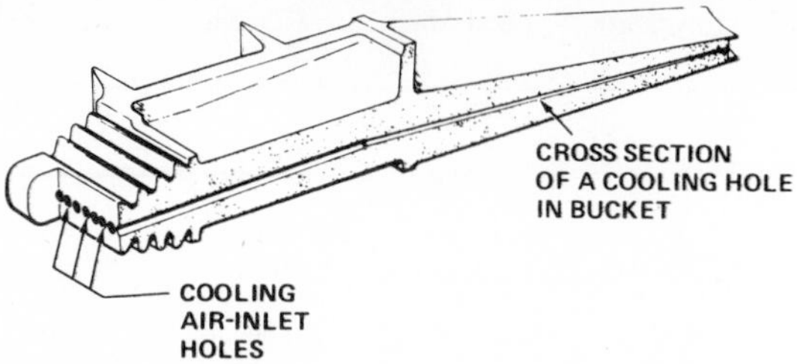

Figure 5-10 Air cooling of the first-stage bucket of an MS7001 gas turbine. *(General Electric Co.)*

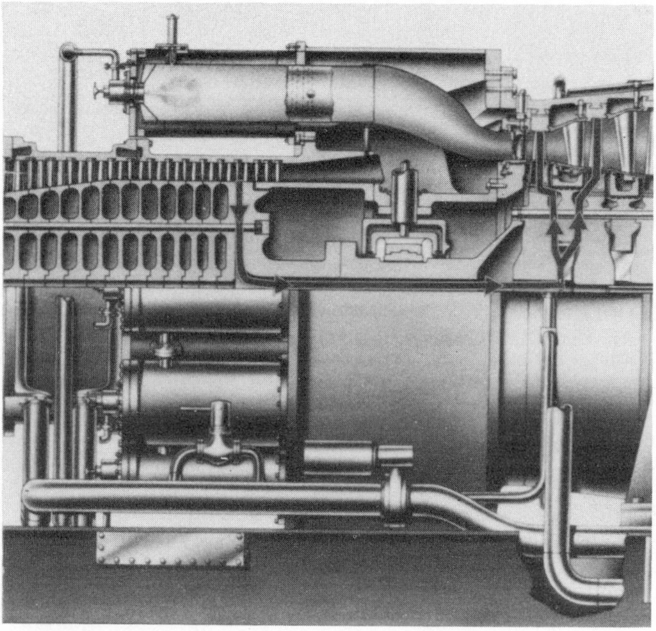

Figure 5-11 MS7001 gas turbine bucket cooling airflow path. *(General Electric Co.)*

can be incorporated into high-efficiency combined steam and gas turbine cycles, or gas turbine cycles with heat recovery for process.

The LM2500 is known as a second-generation aircraft-derivative gas turbine. The high firing temperature of 2139°F (1171°C) is accomplished while maintaining similar metal temperatures as heavy-duty units by additional air cooling. Air is bled from three different locations on the air compressor and cools the two stages of the high-pressure turbine blades and nozzles. (In the aircraft engine industry, the term *blade* is used instead of *bucket.*) Figure 5-12 illustrates the extensive cooling of the first-stage blade of the LM2500 which allows the relatively high firing temperature. The high compression ratio and firing temperature result in high efficiency and specific work. Therefore, this type of gas turbine is ideally suited for simple-cycle applications and on offshore platforms where weight and "footprint" are important.

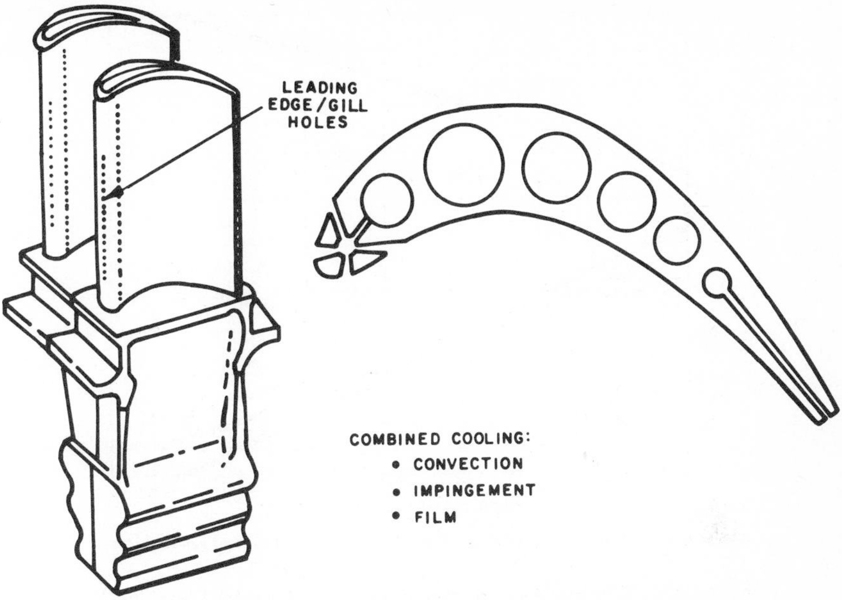

Figure 5-12 LM2500 gas turbine first-stage blade *(General Electric Co.)*

The performance characteristics of the gas turbine generators shown in Table 5-1 are the output at the generator terminals and account for generator losses and gear losses, where required. Generally, large gas turbine–generator sets of 60 MW and up are direct drive. That is, the turbine speed is either 3600 r/min for 60 Hz or 3000 r/min for 50 Hz. Most gas turbines rated from 60 MW down to 15 MW operate at 4000 to 7000 r/min and drive the generator through a reduction gear.

Table 5-2 lists typical heavy-duty mechanical-drive gas turbines as well as the LM2500. The MS5002R(B) is the regenerative-cycle version of the MS5002(B). Use of the regenerator results in a 25 percent improvement in heat rate. However, the exhaust temperature beyond the regenerator is reduced to 666°F (352°C), which decreases the usefulness of the exhaust gas as a source of waste heat. All the units listed in Table 5-2 are of the two-shaft design and can be operated over a range of 50 to 105 percent of rated speed.

Site Performance

The performance characteristics listed in Tables 5-1 and 5-2 are based on the standard ISO conditions: 59°F (15°C) ambient temperature and 14.7 psia (101.4 kPa) pressure at the inlet and exhaust. The performance characteristics must be corrected to reflect the actual conditions at the site caused by the following: ambient temperature, elevation above sea level, inlet pressure losses due to filters and silencers, and exhaust losses due to silencing and/or waste-heat recovery equipment. Figure 5-13 is a curve of correction factors of typical temperature effects for a gas turbine. Table 5-3 lists typical correction factors for inlet and exhaust pressure losses.

OPERATION AND MAINTENANCE

Starting Procedures

In order to start up a gas turbine, another small prime mover is required to accelerate the unit to a preselected speed until firing occurs and the unit becomes self-sustaining.

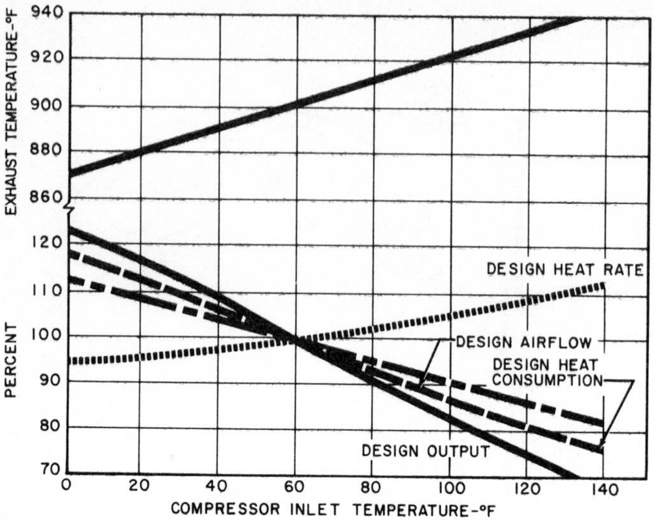

Figure 5-13 Temperature effects correction factors for a typical gas turbine.

TABLE 5-3 Typical Effects of Inlet and Exhaust Pressure Losses on Gas Turbine Performance

	Output, %	Heat rate, %	Exhaust temperature, °F
4 in H_2O inlet	−1.7	+0.7	+1.0
4 in H_2O exhaust	−0.7	+0.7	+1.0

The starting device is then uncoupled from the gas turbine by a clutch. Starting devices can be:

Motors
Diesel engines
Expansion turbines
Steam turbines

Frequently the starter turbine simply expands a small quantity of the natural-gas fuel supply for start-up. When steam turbines are used, these can also continue to operate to boost the gas turbine output. These are called "helper" turbines. Diesel engines and turbines can be used to provide the gas turbine with "black start" capability. That is, the unit can come on-line without requiring an external electric power source.

Gas turbine control systems are sophisticated electrical devices which contain many control functions. On start-up, fuel flow is scheduled as a function of time and speed. Figure 5-14 indicates a typical start-up control sequence. In this figure VCE is the electronic voltage control signal which modulates fuel flow. Modern control systems for gas turbine–generator sets include automatic synchronizing capability which closes the generator breaker when voltage and phase relationships are within proper limits.

Normal Operation

Most gas turbine–generators normally operate at synchronous speed at full capability. Fuel flow is governed to maintain the firing temperature at its design limit. Therefore, the output of the unit will vary with ambient temperature. Units that are synchronized

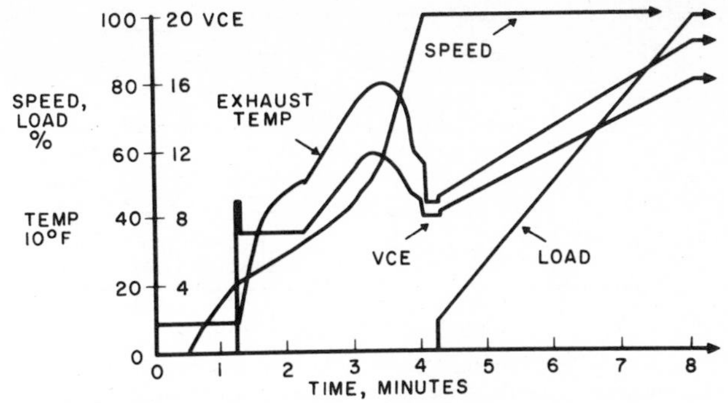

Figure 5-14 Gas turbine start-up control characteristics.

in a grid can also operate at part load using a droop or speed/load control characteristic. This is illustrated in Fig. 5-15. Because the unit is synchronized to the system, the speed is essentially constant. Therefore, varying the speed set point effectively varies the load. The family of diagonal lines represents different settings of the speed/load control knob. Isolated gas turbine–generators can be furnished with an isochronous control mode. Load changes result in transient speed excursions which are instantaneously corrected by modulating fuel flow. Whether a gas turbine is on droop or on isochronous control, the maximum firing temperature control will always provide an upper limit to prevent overfiring. Many other backup and protective controls and alarms are also provided.

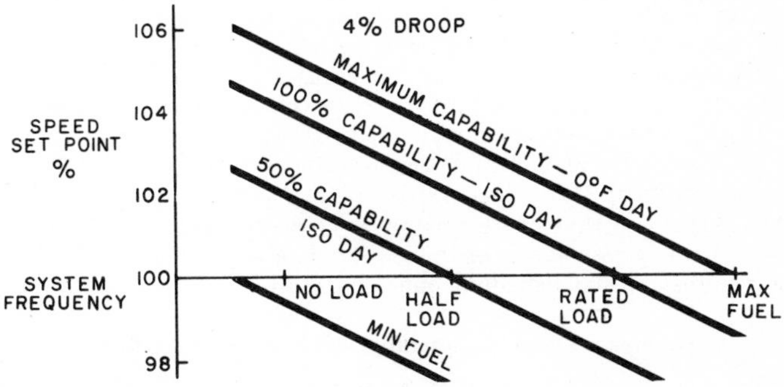

Figure 5-15 Typical droop speed control characteristics.

Mechanical-drive gas turbines normally operate on speed/load control with the set-point provided by the process control system. Figure 5-16 depicts a typical performance curve for a two-shaft mechanical-drive gas turbine, with the load characteristic of a process compressor system superimposed. A process controller might receive the suction or discharge pressure signal of the driven compressor and generate the appropriate speed/load set-point of the gas turbine. Again the fuel flow is still limited by the maximum-firing-temperature control.

Maintenance

Periodic inspection, repair, and replacement of parts are required to maintain gas turbines. The frequency of maintenance is heavily dependent on the type of fuel, the start-up frequency, and the environment. Although control systems carefully sequence start-up, there is an inherent thermal cycle which reduces parts life if frequently repeated.

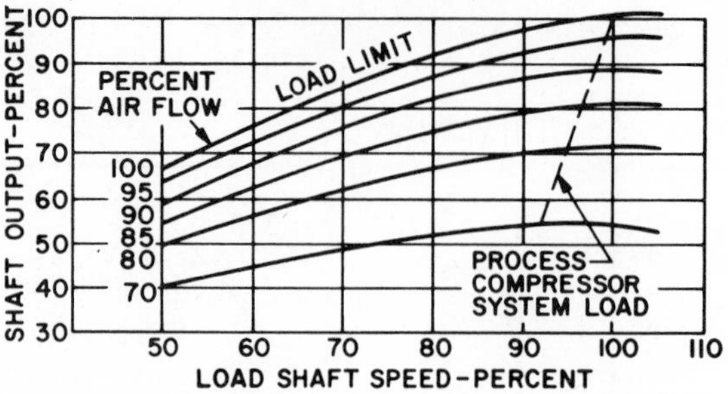

Figure 5-16 Two-shaft mechanical drive gas turbine performance curve with a process compressor system load curve.

The parts life of peaking gas turbines which run for 4 h/day is lower than that of continuous-duty units. However, most industrial plants operate for many more than 100 fired hours per start and therefore do not have this problem.

The general environment can also affect the parts life of gas turbines. Many plants are located in areas with corrosive or abrasive matter in the atmosphere. Desert sandstorms, saltwater mist, chemical fumes, and airborne fertilizers are examples. However, the effects of these types of environments can be minimized by multistage high-efficiency inlet-air filters and mist eliminators as well as the use of correct materials and protective coatings in the compressor and turbine.

The most important factor in gas turbine maintenance is the type of fuel burned. Natural gas is the cleanest fuel and incurs minimum maintenance costs and downtime. It is common for gas turbines in base-load industrial service to operate at full load on maximum-exhaust-temperature control continuously for 3 years. Not many industrial plants or processes can operate for such long periods, hence gas turbines are generally maintained at shorter intervals during process outages.

Normally, No. 2 distillate oil contains very little contamination, but it does burn with greater radiation, or luminosity, than natural gas. This decreases the life of hot-gas-path parts. The low lubricity of distillate oil decreases the life of parts of fuel forwarding and metering systems as well. Heavy fuel oils, both crude and residual, generally burn with additional radiation and have contaminants which accelerate corrosion and deposition on the hot-gas-path parts. Sodium and potassium must be removed from these fuels to prevent hot corrosion, and vanadium must be inhibited by the use of magnesium additives.

Preventive maintenance practices generally consist of several different types of maintenance procedures:

- Running inspection
- Combustion inspection
- Hot-path inspection
- Major inspection

Running inspections include load vs. exhaust temperature measurements, vibration monitoring, and fuel-flow and -pressure measurements. Sophisticated electronic equipment is planned to enhance trend monitoring and on-line diagnostics.

In the combustion inspection the unit is shut down, and some disassembly is required to repair or replace combustion parts such as fuel nozzles and liners. Visual or boroscope inspections can also be made of turbine nozzles and buckets during these inspections.

A hot-gas-path inspection includes disassembly of the turbine casing. A major inspection includes a disassembly of the compressor casing as well as the turbine casing. A major inspection essentially returns the gas turbine to its new, or *zero time,* condition.

Many gas turbine parts are fabricated from expensive superalloys. Minimum maintenance costs can be achieved by repairing these parts during an inspection to extend their life. Spare sets of parts can be used as replacements to minimize downtime. In some critical continuous-process plants, it is more economical to maintain production without outages and not extend parts life by repairs.

Typical maintenance requirements for an MS5001 gas turbine for various fuels are shown in Table 5-4. The worker-hour requirements are based on the assumption that replacement parts are available.

For purposes of feasibility studies, engineers sometimes estimate life-cycle maintenance costs at an average rate of 2.5 percent of the total installed cost per year for natural gas fuel and 5 percent per year for heavy fuel oil.

TABLE 5-4 Typical Maintenance Requirements of an MS5001 Gas Turbine

Type of inspection	Fuel type	Inspection interval, 1000 h	Inspection worker-hour requirements		
			Worker hours	Crew size	8-h shifts
Combustion	Natural gas	8–10	160	4	5
	Distillate oil	6–8			
	Heavy	1–5			
Hot-gas path	Natural gas	16–20	480	6	10
	Distillate oil	12–16			
	Heavy oil	6			
Major	Natural gas	32–40	1280	8	20
	Distillate oil	24–32			
	Heavy oil	18–23.5			

APPLICATIONS IN PLANTS

General Discussion

There are several different application categories for stationary gas turbines. These include:

Industrial plants
Pipeline pumping stations
Offshore platforms
Electric utility stations
  Base-load
  Midrange (1500 to 3000 h/year)
  Peaking duty

The emphasis of this chapter is on industrial plants, which are generally base-loaded (330 to 365 days/year). However, a few comments on the other application categories are in order.

Pipeline pumping stations are generally base-loaded around the year, or through all except the summer months. There are many applications of regenerative cycles and of simple-cycle gas turbines in remote areas. There have also been a very small number of combined steam and gas turbine cycles in this category.

Most offshore-platform applications have been simple cycles due to weight and "footprint" constraints, but there have also been some regenerative-cycle applications.

In electric utility service, thousands of gas turbines around the world have been applied to serve peak loads (up to 1500 h/year) in the simple-cycle mode. To meet these loads, fuel consumption is not as significant a factor as are capital costs, operating labor, and maintenance. Most gas turbines which are applied in midrange or base-load electric utility service combine steam and gas turbine cycles, but a small number have also used regenerative cycles.

Many of the gas turbines applied in electric utility combined-cycle service are supplied as part of a complete package by the gas turbine manufacturer. The manufacturer supplies or specifies all the major equipment, such as heat-recovery steam generators (HRSGs), steam turbines, and condensers, and then assumes responsibility for the overall heat rate of the plant. One specific gas turbine design might be offered in several different combined-cycle configurations. Table 5-5 lists typical performance specifications for three versions of combined cycles based on the MS7001E gas turbine.

TABLE 5-5 Typical Performance of a Combined Cycle, Based on 59°F (15°C) Sea-Level Site, with No. 2 Distillate Fuel Oil

Plant designation	Output kW	Heat rate (LHV) Btu/kWh (kJ/kWh)	Gas turbine configuration
STAG* 107E	99,300	7700 (8120)	One MS7001E
STAG* 407E	396,700	7650 (8070)	Four MS7001E
STAG* 607E	595,600	7630 (8050)	Six MS7001E

*Registered trademark of General Electric Co.

Heat Recovery and Efficiency

Most gas turbines applied in industrial plants are in base-load service. There are many simple-cycle gas turbines applied throughout the world in industrial plants where fuel supplies are abundant and are located either in underdeveloped nations or in remote or harsh environments. Examples are Indonesia, the Sahara, and the Alaskan North Slope. However, generally all gas turbines applied in industrial plants in developed nations are equipped with some type of heat recovery. Figure 5-17 illustrates some of the ways in which the high-temperature exhaust of gas turbines has been recovered in industrial plants. In Fig. 5-17*a* the exhaust gases are used to generate low-pressure process steam. The HRSGs can be unfired or have supplementary firing to increase steam output. In Fig. 5-17*b* higher-pressure steam is generated for a steam turbine. Typical upper limits for steam conditions of unfired HRSGs are 850 psig, 825°F (5964 kPa, 441°C). Fired HRSGs have been applied with steam conditions as high as 1450 psig, 950°F (10,100 kPa, 510°C). In Fig. 5-17*c* a two-pressure HRSG is shown. When high-pressure turbine inlet steam is generated in an unfired HRSG, typical stack temperatures are a relatively high 400 to 450°F (204 to 232°C). Additional heat can be recovered when a 25- to 150-psig (276- to 1138-kPa) saturated steam generation section is included.

In Fig. 5-17*d* a regenerative-cycle gas turbine is followed by a low-pressure process steam generator. One of the consequences of the low fuel consumption of the regenerative-cycle gas turbine is a reduction of the regenerator exhaust gas temperature to approximately 600°F (316°C). This arrangement is selected when only a relatively small amount of process steam is required.

Finally, in Fig. 5-17*e*, the heat in the exhaust gas is used directly in the process or as preheated combustion air for a fired process heater.

In all these cycles the fuel-utilization effectiveness is improved by recovering heat from the gas turbine exhaust. In order to evaluate the relative efficiency of these cycles which supply process heat and power, a yardstick called fuel chargeable to power (FCP) has been developed. Figure 5-18 illustrates the method of calculating FCP. Performance data are required for a plant which supplies heat and power as well as for a plant which supplies the same amount of heat only. These performance data should include all energy-related factors, such as boiler efficiency, boiler blowdown, and auxiliary power

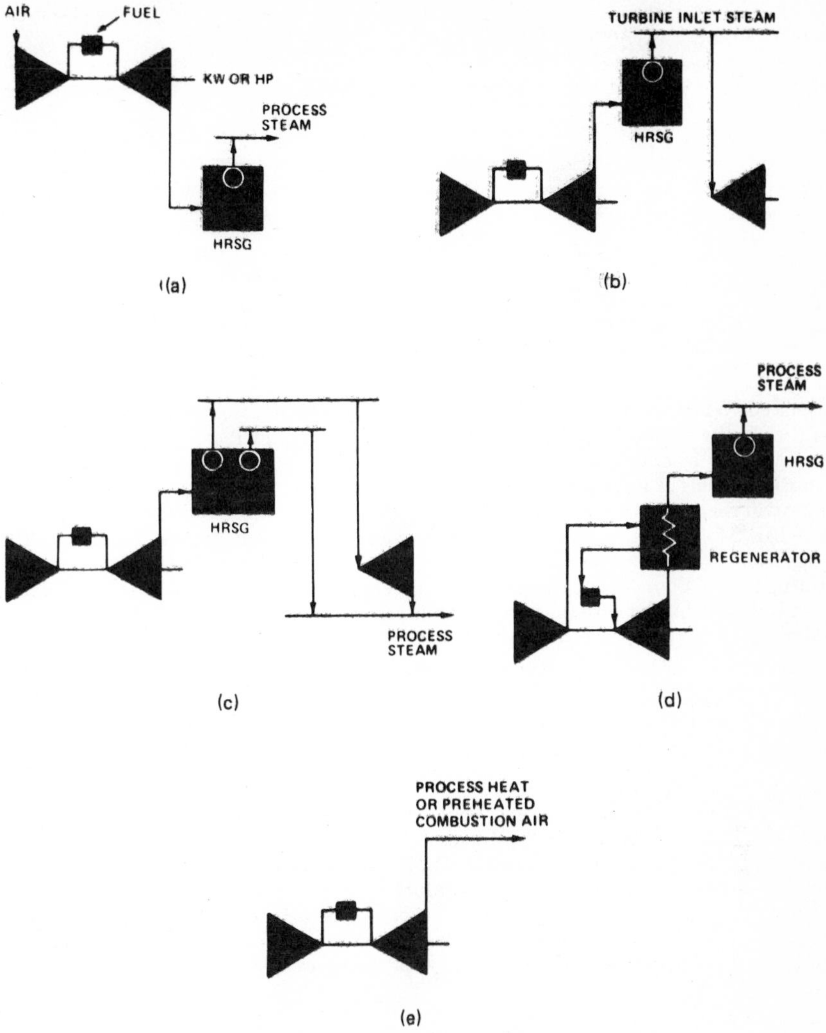

Figure 5-17 Industrial gas turbine heat recovery cycles.

requirements. Fuel chargeable to power (FCP) is defined as:

$$\text{FCP} = \frac{\text{fuel}(2) - \text{fuel}(1)}{\text{gross power}(2) - \text{aux. power}(2) + \text{aux. power}(1)}$$

where

fuel(2) = the fuel consumption for plant supplying heat and power
fuel(1) = the fuel consumption for plant supplying heat only
gross power(2) = power generated
aux. power(2) = auxiliary power for plant supplying heat and power
aux. power(1) = the auxiliary power for plant supplying heat only

Stated in words, FCP is the incremental fuel consumption for power generation, divided by the net power generation, with a credit for the auxiliary power which has been

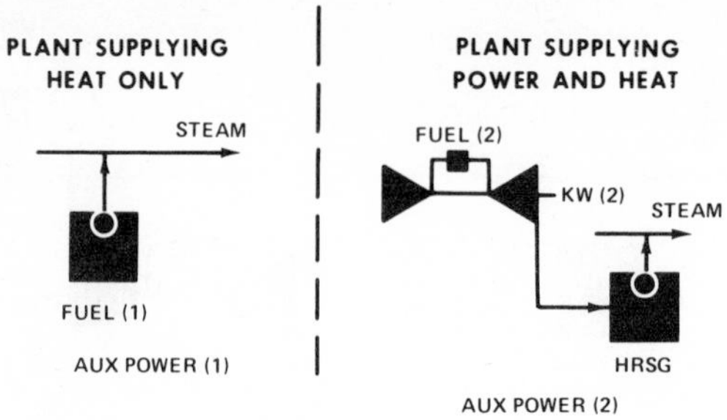

$$FCP = \frac{\text{FUEL (2)} - \text{FUEL (1)}}{\text{KW (2)} - \text{AUX POWER (2)} + \text{AUX POWER (1)}}$$

Figure 5-18 Diagram illustrating FCP calculation.

displaced. When the amount of process heat is reduced to zero, the definition of FCP reduces to the accepted definition of net station heat rate. FCP can be used to measure the efficiency of any heat engine which supplies both heat and power, and it can be defined in terms of British thermal units per kilowatthour (kilojoules per kilowatthour) or British thermal units per horsepower hour for mechanical-drive systems.

When gas turbines operate with heat-recovery equipment, typical values of FCP range from 7600 Btu/kWh (8018 kJ/kWh) to 4500 Btu/kWh (4748 kJ/kWh) HHV (higher heating value). Equivalent thermal efficiencies are in the range of 50 to 75 percent HHV.

It is worthwhile to note that simple-cycle gas turbine performance is normally specified in terms of the lower heating value (LHV) of the fuel. However, when prime movers also provide process heat, the overall system performance is more frequently expressed, in the United States, in terms of higher heating value of the fuel. Typical ratios of HHV to LHV are 1.11 for natural gas and 1.06 for distillate and residual oils. When a gas turbine fuel consumption in terms of LHV is to be combined with boiler fuel consumption in terms of HHV, the gas turbine fuel consumption rate should be increased by the appropriate HHV/LHV ratio.

The heat-recovery capability and fuel chargeable to power values of typical gas turbines are shown in Table 5-6. Notice that the FCP values for the heavy-duty units approach those of the aircraft-derivative LM2500. This indicates that there is an optimum compression ratio and firing temperature for each HRSG cycle, as was shown earlier in Fig. 5-4 for the regenerative cycle. A specific gas turbine design is often applied in simple, regenerative, and combined cycles. Therefore, the designer often must select a compromise configuration, suitable to all types of applications.

Cogeneration

Steam turbines are often used in cogeneration systems which produce heat for industrial processes as well as power. A typical application is shown in Fig. 5-19. In this case an automatic-extraction noncondensing unit supplies steam at two different pressure levels to the process. A typical value of fuel chargeable to power for noncondensing steam turbine cycles is 4200 Btu/kWh (4431 kJ/kWh) HHV. This is an equivalent thermal efficiency of 80 percent, which is far higher than that of most other types of prime movers. The high efficiency of the noncondensing steam turbine cycle is due to the fact that heat losses to the surroundings are minimized. The only losses are the boiler inefficiency (stack losses), generator, seals, bearing friction, radiation, and additional auxiliary power requirements.

However, the amount of high-efficiency power which can be produced in this manner is limited by the process steam demand. Most industrial plants require far more power than can be generated by noncondensing steam turbine cycles. The additional power can be furnished by one of the following methods: adding a condensing section to the indus-

TABLE 5-6 Typical Performance of Gas Turbines with Unfired HRSGs*

Performance characteristics	Gas Turbine Model							
	LM2500		MS5001(P)		MS6001(A)		MS7001(E)	
ISO Rating, kW	20,100		24,700		31,050		75,000	
Site output, kW	16,410		21,630		27,550		67,230	
Gas turbine fuel (HHV), 10^6 Btu/h	177.9		296.7		343.6		783.3	
(Gj/h)	(187.7)		(313.0)		(362.5)		(826.4)	
	Steam flow, 1000 lb/h (t/h)	GT & HRSG FCP Btu/kWh (kJ/kWh)	Steam flow, 1000 lb/h (t/h)	GT & HRSG FCP Btu/kWh (kJ/kWh)	Steam flow, 1000 lb/h (t/h)	GT & HRSG FCP Btu/kWh (kJ/kWh)	Steam flow, 1000 lb/h (t/h)	GT & HRSG FCP Btu/kWh (kJ/kWh)
Steam conditions								
250 psig, sat.	75.7	5490	133.2	6560	140.6	6540	348.0	5650
(1827 kPa, sat.)	(34.3)	(5790)	(60.4)	(6921)	(63.8)	(6900)	(157.9)	(5961)
600 psig, 750°F	58.5	5960	100.6	7340	105.2	7230	271.0	6160
(4240 kPa, 399°C)	(26.5)	(6288)	(45.6)	(7744)	(47.7)	(7628)	(122.9)	(6499)
850 psig, 825°F	54.8	6140	93.1	7650	96.9	7520	255.0	6340
(5964 kPa, 441°C)	(24.9)	(6478)	(42.2)	(8071)	(44.0)	(7934)	(115.7)	(6689)

Basis: No. 2 distillate fuel; pressure losses: inlet 4 inH_2O (1 kPa), exhaust 10 inH_2O (2.5 kPa); 3 percent blowdown; 228°F (109°C) feedwater; 92 percent effectiveness; steam generated at 5 percent higher pressure, 5°F higher temperature; FCP based on HHV, fuel displaced from 88 percent efficiency boilers.

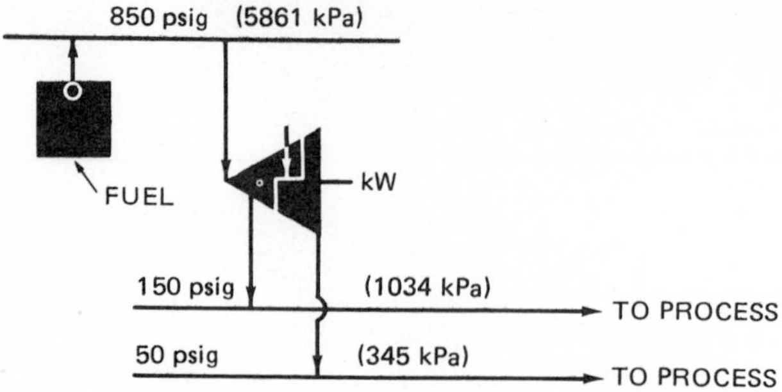

Figure 5-19 Typical noncondensing steam turbine application.

trial steam turbine, importing electricity from an electric utility system, or adding gas turbines. Typical values of FCP for industrial condensing steam turbine sections are 12,000 to 14,000 Btu/kWh (12 660 to 14 770 kJ/kWh) HHV, and there is an additional requirement for circulating water to condense the steam with its associated thermal emissions. The average heat rate for electric utilities is 10,500 Btu/kWh (11 078 kJ/kWh) HHV at the central station, with associated transmission losses to the industrial plant. But the gas turbine heat-recovery systems shown in Fig. 5-17 can cogenerate power with FCP values of 4500–8000 Btu/kWh (4 748–8 440 kJ/kWh), without requiring circulating water. From an energy-efficiency point of view, large blocks of power can best be generated by combined gas and steam turbine cogeneration systems.

Figure 5-20 illustrates how gas turbines have been integrated into a modern industrial combined cycle. Steam turbines expand high-pressure steam down to process pressure and cogenerate by-product power. Gas turbines cogenerate additional by-product and supply steam for the steam turbines and thus reduce the fuel required for the power boilers.

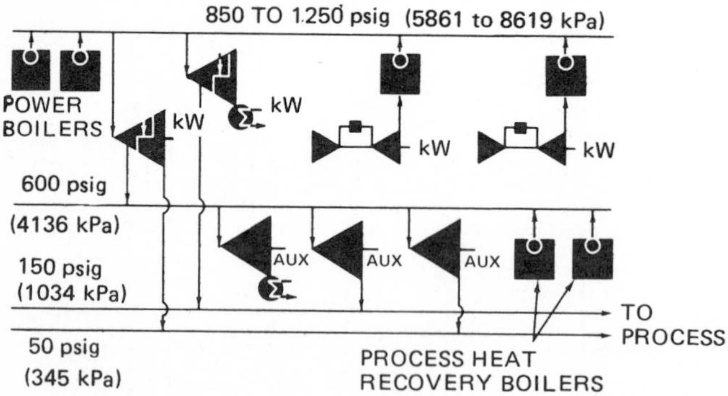

Figure 5-20 Modern industrial combined-cycle cogeneration plant.

Fuel chargeable to power is not the only criterion used for justifying cogeneration plants. As compared with the industrial plant, economy of scale gives electric utility generation equipment an advantage in capital costs. This equipment can utilize a mix of nuclear power, coal at unit train rates, and hydropower. At present, the gas turbine cogeneration plant requires natural gas or petroleum fuels, and these are considered premium fuels. The final decision on cogeneration equipment generally depends on an economic evaluation to determine the rate of return on the incremental capital costs of the system.

Gas Turbine Emissions

The gas turbine is one of many types of combustion devices which have been subject to strict environmental codes in recent years. Limits have been placed on the following types of objectionable emissions:

- Oxides of nitrogen NO_x
- Oxides of sulfur, SO_x
- Particulates
- Unburned hydrocarbons
- Carbon monoxide, CO

The development of combustion systems has progressed to the point where typical gas turbine emissions of particulates, unburned hydrocarbons, and carbon monoxide fall well below the environmental limits. Figure 5-21 shows the reverse-flow cannular combustion system which has been the object of some of the most intensive development programs. Liquid fuels are atomized by high-pressure air as they are injected through the fuel nozzle. This has been very effective in reducing emissions of unburned hydrocarbon and particulates. Most of the compressor discharge air flows through the combustion liner downstream of the reaction zone. This cools the products of combustion and reduces the formation of NO_x. Injection of water or steam into the reaction zone further reduces NO_x formation.

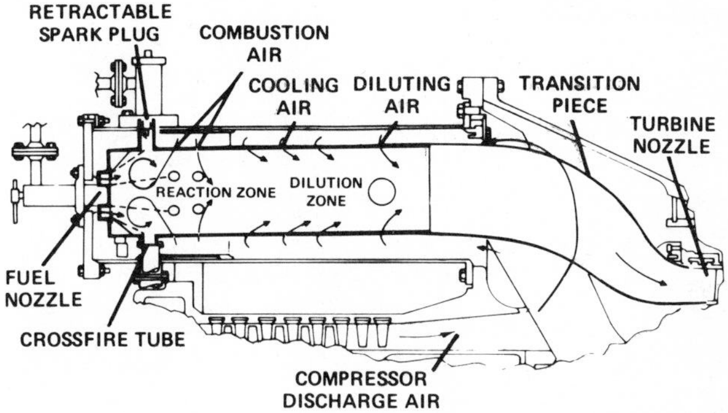

Figure 5-21 Gas turbine reverse-flow combustion system. *(Courtesy of the General Electric Company.)*

The federal New Source Performance Standards (NSPS) for stationary gas turbines can generally be attained by modern gas turbines. However, the injection of water or steam for NO_x abatement, an additional operating cost for the user, is required. To protect the turbine parts, the steam or water must be treated to obtain boiler feedwater quality, and the water is not recovered. A great deal of combustion-system development is being conducted to meet NO_x emission standards without the use of water or steam, i.e., to develop the so-called "dry" combustion system.

The NSPS limits NO_x emissions to 75 parts per million by volume (ppmv), normalized to 15 percent excess O_2. That is, if the exhaust O_2 is 14 percent, then the limit is increased by the ratio of 15:14. Additional increases in the NO_x limit are allowed for increased turbine efficiency, and if the fuel has a high content of fuel-bound nitrogen.

Stationary gas turbines are also subject to federal Prevention of Significant Deterioration Standards (PDS). The Environmental Protection Agency has mathematical models for determining ground-level concentrations of pollutants from combustion devices. The aggregate of all contributing combustion devices in a given area is not permitted to cause significant increase of ground-level pollutants. Therefore, new gas turbines can be located only in areas which have attained acceptable environmental conditions, and to prevent deterioration in a specific area, only a limited amount of incremental emissions are permitted.

BIBLIOGRAPHY

1. "Performance Specifications 1980," *Gas Turbine World,* December 1979.
2. In *Power Engineering,* a series of seven articles:
   (a) Carlstrom, L. A., H. F. Heissenbuttel, and A. H. Perugi: "Gas Turbine Combustion System: Key to Improved Availability," May 1978.
   (b) Patterson, J. R., and C. M. Grant: "Operating Gas Turbines for Extended Component Life," June 1978.
   (c) DuBois, M. R., and R. J. Fresneda: "Inspection and Maintenance of Gas Turbine Nozzles, Buckets and Rotors," July 1978.
   (d) Scheper, G. W., A. J. Mayoral, and E. J. Hipp: "Maintaining Gas Turbine Compressors for High Efficiency," August 1978.
   (e) Bingham, P. J., P. H. Huhtanen, and H. G. Starnes: "Maintenance of Gas Turbine Accessory Equipment," Sept. 1978.
   (f) Kiernan, J. G., A. D. Foster, and D. T. Harden: "Gas Turbine Fuels and Fuel Systems," October 1978.
   (g) Stretch, R. H., J. N. Shinn, and D. B. Brudos: "Calibration and Troubleshooting of Gas Turbine Controls," November 1978.
3. Doherty, M. C., and D. R. Wright: "Application of Aircraft Derivative and Heavy Duty Gas Turbines in the Process Industries," ASME Paper 79-GT-12, *ASME Gas Turbine Conference,* San Diego, Calif., March 1979.

PART 2

Steam Turbines

by

Martin C. Doherty
Consulting Application Engineer
Thermal Power Systems Engineering
General Electric Co.
Schenectady, New York

GLOSSARY

Automatic-admission turbine A steam turbine with the capacity to admit steam at two or more pressures. Valve gear at the low-pressure opening can automatically control the pressure in that header.

Automatic extraction turbine A steam turbine with the capacity to extract steam. The pressure of the extracted steam is controlled by a valve gear at that opening. *Note:* Steam turbines can be furnished with automatic extraction and admission capability at the same opening.

Available energy The difference in enthalpy between an inlet steam condition (a specific pressure and temperature) and an exhaust pressure along a path of constant entropy.

Back-pressure turbine A steam turbine which exhausts at a pressure equal to or greater than atmospheric pressure. Also a noncondensing steam turbine.

Bottoming cycle An energy recovery cycle which uses waste heat from another source to generate power. A steam turbine bottoming cycle uses steam generated by a waste-heat exhaust stream.

By-product power Power generated coincidentally when supplying useful heat. See also "Noncondensing power" and "Cogeneration."

Cogeneration The simultaneous production of power and other forms of useful energy—such as heat or process steam. U.S. government agencies restrict the definition of cogeneration to electric power generation only, while various industrial associations extend it to include mechanical-drive power and electric power.

Feedwater heater A steam-to-water heat exchanger which heats the boiler feedwater with steam extracted from a steam turbine. In a closed feedwater heater the two fluids are separated by the use of shell and tube construction. In an open feedwater heater the fluids are mixed. A deaerator is an open feedwater heater which separates entrained gases from the feedwater by vigorous agitation with steam.

Mechanical-drive steam turbine A turbine used to drive devices other than electric generators such as pumps or compressors. Generally designed to operate over a wide speed range.

Mollier diagram A plot of enthalpy vs. entropy of a fluid which includes lines of constant pressure and temperature.

Noncondensing power Power generated by steam which is expanded through a turbine and either exhausts or extracts at a pressure equal to or greater than atmospheric pressure.

Pressure control The ability of steam turbine governor systems to maintain constant pressure in a steam line by the action of a valve gear, as the turbine supplies steam to the header or draws steam from it.

Pressure rise point In an automatic-extraction steam turbine the maximum exhaust flow with zero extraction flow.

Steam rate The weight flow rate of steam required to produce a unit of output; pounds per kilowatthour (kilograms per kilowatthour) or pounds per horsepower hour. The *theoretical steam rate* (TSR) defines a perfect expansion process between two conditions. An *actual steam rate* defines the actual expansion, including the inefficiency of the turbine and generator.

Uncontrolled extraction An opening in a steam turbine casing between two stages. The pressure of the extraction steam is uncontrolled and is a function of the steam flow to the following stage.

TYPES OF TURBINES

There are many different types of steam turbines in industrial-plant service. These can be broadly classified as either *condensing* or *noncondensing*. Condensing steam turbines have a subatmospheric exhaust pressure, while noncondensing units exhaust at atmospheric or higher pressure. When steam is expanded to subatmospheric pressure in a condensing steam turbine its temperature is generally reduced to less than 130°F (54°C). This low-temperature energy is not often useful and is generally classified as waste heat. Its disposal is more of a liability than an asset. On the other hand, steam exhausted at positive pressure from noncondensing steam turbines is much higher in temperature and is useful in many industrial processes or heating applications.

Figure 5-22 shows schematic diagrams of three different configurations of noncon-

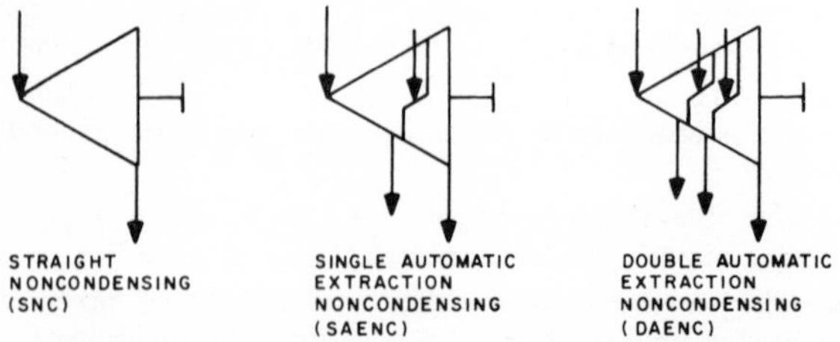

Figure 5-22 Schematic diagram of types of noncondensing steam turbines.

densing steam turbines. All three exhaust into a low-pressure header. The *s*ingle *a*uto-matic-*e*xtraction *n*on*c*ondensing (SAENC) unit also extracts steam at another, higher-pressure, header. The *d*ouble *a*utomatic-*e*xtraction *n*on*c*ondensing (DAENC) unit extracts steam at two additional pressure levels. The term *automatic extraction* implies that the steam flow is automatically controlled (governed) to maintain constant pressure in the header independent of the extraction flow. Figure 5-23 is a cross-sectional view of a single automatic-extraction noncondensing steam turbine. The high-pressure section consists of the turbine inlet valve gear, a high-pressure turbine control stage, five additional turbine stages, and the extraction opening. The exhaust section includes the extraction valve gear, three additional turbine stages, and the exhaust opening. The turbine governor operates both sets of valve gear in concert to control the pressure in the extraction steam line and one other variable, such as exhaust pressure or the power output. Figure 5-24 is an axial cross-sectional view of the upper-inlet valve gear. It consists of six poppets which are individually operated through cam action lifts. This arrangement minimizes throttling losses for high-efficiency operation over a wide range of steam flow. Units can be furnished with only the upper-inlet gear or an upper- and lower-inlet valve gear for increased flow capacity. Figure 5-25 is an axial cross-sectional view of a spool-type extraction valve. It is a variable restriction which is mounted downstream of the extraction opening. Notice that the steam flow to the exhaust passes through the extraction valve and not the extracted steam.

Condensing steam turbines in industrial plants can also include automatic-extraction capability. Figure 5-26 is a schematic illustration of four types of condensing steam turbines; straight condensing, single automatic extraction condensing (SAEC), double automatic extraction condensing (DAEC), and triple automatic extraction condensing (TAEC). A cross-sectional view of a DAEC steam turbine is shown in Fig. 5-27. Notice

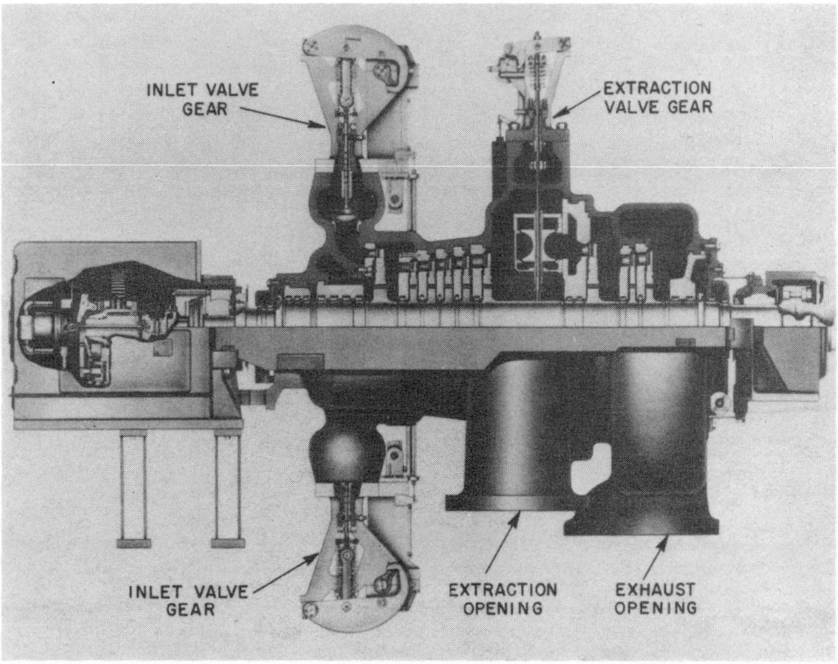

Figure 5-23 Cross-sectional view of a single automatic-extraction noncondensing steam turbine. *(General Electric Co.)*

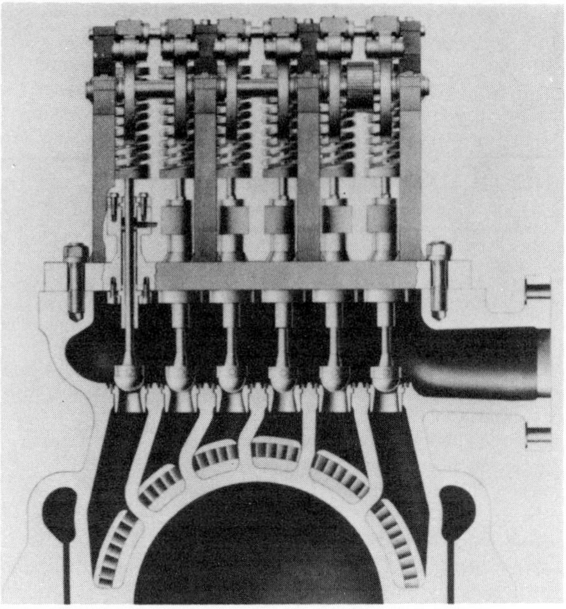

Figure 5-24 Axial cross-sectional view of a steam turbine inlet valve gear. *(General Electric Co.)*

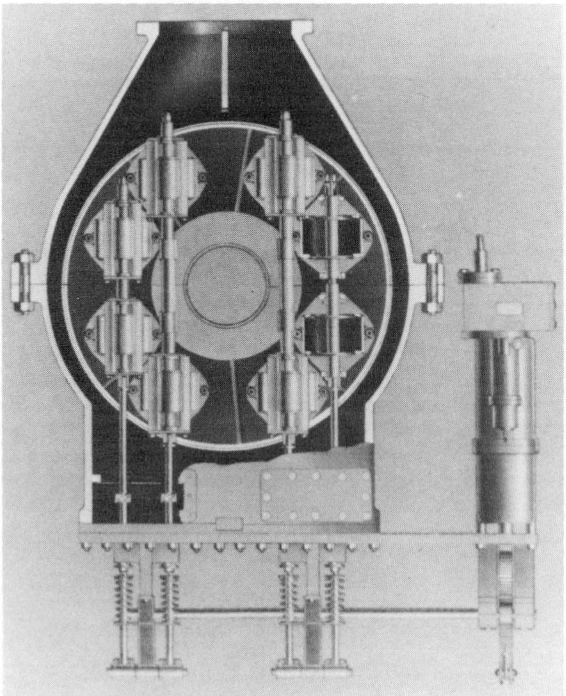

Figure 5-25 Axial cross-sectional view of a spool-type extraction valve. *(General Electric Co.)*

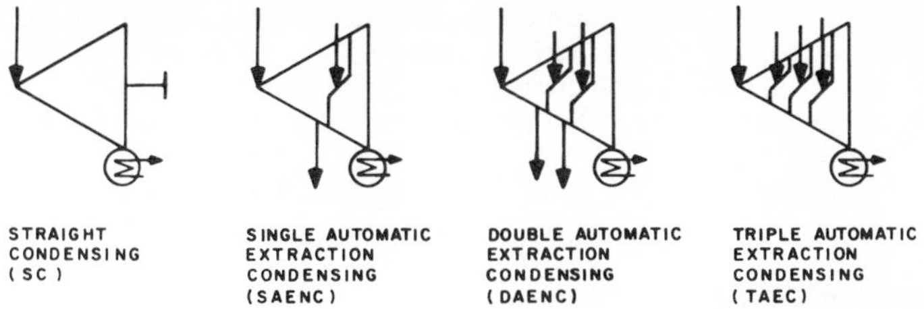

Figure 5-26 Schematic diagram of types of condensing steam turbines.

the relatively large exhaust casing required to pass the low-density subatmospheric exhaust steam flow. A DAEC steam turbine has the capability to control two extraction-pressure levels and also independently control the amount of power produced by varying steam flow to the condensing exhaust.

Automatic admission capability is another feature which can be specified for industrial steam turbines. Some industrial plants have an excess of low-pressure steam which can be admitted through an extraction admission opening and then expanded through the low-pressure section. In some cases, the turbines extract steam during normal operation and only admit steam during process upsets or outages.

Steam turbines can also be furnished with uncontrolled extraction openings. However, the pressure at an uncontrolled extraction varies with steam flow. The variation from normal conditions in absolute pressure is approximately proportional to the variation in flow through the following stage. Therefore, uncontrolled extractions are not generally suitable to supply process steam headers. On the other hand, uncontrolled extractions are suitable to provide feedwater heating in many cases. When steam is extracted from a turbine to heat only the feedwater for that turbine, then the requirements for extraction steam will also be proportional to steam flow. Large steam turbines in central stations can have as many as six uncontrolled extractions to supply different stages of feedwater heaters.

The steam turbines illustrated in Figs. 5-23 to 5-27 are of multivalve, multistage type

Figure 5-27 Cross-sectional view of a double automatic-extraction condensing steam turbine. *(General Electric Co.)*

of construction. Smaller steam turbines are available with a single throttling valve on the inlet and with one (or more) turbine stage(s). These units are classified as single-valve, multistage, and single-valve, single-stage. This type of construction reduces the initial cost of the steam turbine, but at a penalty in efficiency, which can be less than half of that of the multistage, multivalve type. Single-valve, single-stage units are generally applied in small (fewer than 1000 hp) mechanical-drive service.

THERMODYNAMIC CONSIDERATIONS

The properties of steam have been well documented in steam tables.[1] These tables are useful in the calculation of detailed steam turbine cycle performance. A familiar graphical representation of steam properties is the Mollier diagram shown in simplified form in Fig. 5-28. The ordinate is *enthalpy* in British thermal units per pound (kilojoules per kilogram), which is a measure of the internal energy plus the flow energy of the steam. It is used to make most of the energy calculations in steam turbine cycles. The abscissa is *entropy* in British thermal units per pound per degree Rankine (kilojoules per kilogram per kelvin). Processes which are reversible and adiabatic (no losses or heat transfer) are isentropic and occur along vertical paths on the Mollier diagram.

Point 1 represents typical inlet steam conditions for industrial steam turbines: 1250 psig, 900°F (8720 kPa, 482°C). For an isentropic expansion to a 4-in Hg (13.55 kPa) pressure, the path of the steam would be from point 1 to point 2. Notice point 1 is above the saturation line in the superheated steam region and point 2 is below the line in the two-phase or wet steam region. The difference in enthalpy from point 1 to point 2 is called *available energy*. In this case it would be:

Available energy = 1438 − 917 = 521 Btu/lb (1212 kJ/kg)

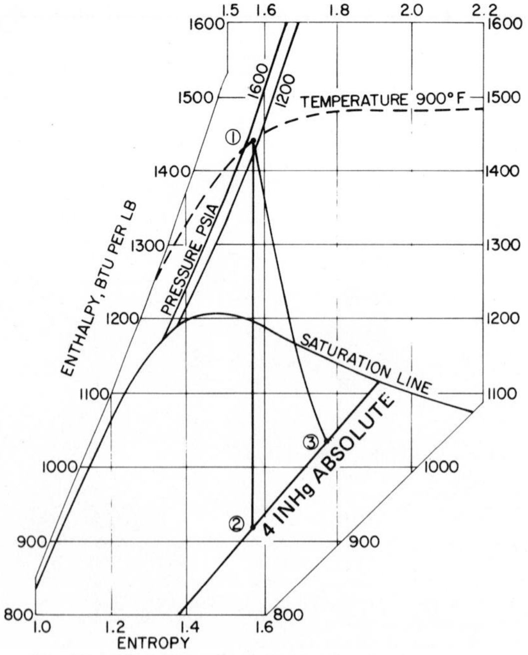

Figure 5-28 Simplified Mollier diagram.

For each pound of steam expanding through this process, 521 Btu (550 kJ) of output power can be produced.

Reading the Mollier chart can be difficult, and the chart is of limited accuracy. To facilitate these types of calculations the data along isentropic expansions have been compiled in a table of *Theoretical Steam Rates*.[2] The units of the theoretical steam rates (TSRs) have been converted to the more common measure of steam turbine output of kilowatts instead of British thermal units (kilojoules). Figure 5-29 is a sample TSR tabulation. Notice that the value for the expansion along path 1 to 2 of Fig. 5-28 is 6.541 lb/kWh (2.97 kg/kWh). This is calculated from

$$\text{TSR (lb/kWh)} = \frac{3412 \text{ Btu/kWh}}{521 \text{ Btu/lb}} = 6.54 \text{ lb/kWh}$$

THEORETICAL STEAM RATE IN LB PER KWHR

1250 LB PER SQ. IN. GAGE, INITIAL PRESSURE

INITIAL TEMPERATURE, DEGREES FAHRENHEIT									
775	800	825	850	900	950	1000	1050	1100	
INITIAL SUPERHEAT, DEGREES FAHRENHEIT									
201.1	226.1	251.1	276.1	326.1	376.1	426.1	476.1	526.1	EXHAUST PRESSURE IN. HG ABS.
INITIAL ENGHALPY, BTU/LB									
1360.0	1376.4	1392.4	1408.0	1438.4	1468.1	1497.4	1526.4	1555.2	
5.9624	5.8641	5.7702	5.6801	5.5095	5.3490	5.1969	5.0517	4.9128	0.5
6.284	6.178	6.076	5.9791	5.7948	5.6216	5.4573	5.3008	5.1510	1.0
6.499	6.387	6.281	6.179	5.9853	5.8033	5.6309	5.4665	5.3094	1.5
6.666	6.550	6.440	6.334	6.133	5.9442	5.7653	5.5948	5.4318	2.0
6.805	6.686	6.572	6.463	6.256	6.061	5.8767	5.7011	5.5332	2.5
6.925	6.803	6.686	6.574	6.362	6.162	5.9730	5.7928	5.6207	3.0
7.032	6.907	6.787	6.673	6.456	6.252	6.058	5.8741	5.6981	3.5
7.128	7.001	6.879	6.762	6.541	6.332	6.135	5.9474	5.7680	4.0
7.217	7.087	6.963	6.844	6.618	6.406	6.206	6.014	5.8319	4.5
7.299	7.167	7.041	6.920	6.690	6.475	6.271	6.076	5.8909	5.0
7.913	7.765	7.623	7.487	7.229	6.986	6.757	6.539	6.330	10.0
8.348	8.188	8.034	7.887	7.608	7.346	7.098	6.862	6.638	15.0
8.700	8.530	8.367	8.210	7.914	7.636	7.372	7.122	6.884	20.0
9.003	8.823	8.652	8.488	8.176	7.884	7.607	7.344	7.094	25.0
									PSIG
9.268	9.081	8.902	8.731	8.406	8.100	7.812	7.538	7.277	0
9.748	9.547	9.355	9.170	8.820	8.492	8.181	7.886	7.606	5
10.168	9.953	9.749	9.552	9.180	8.830	8.501	8.188	7.890	10
10.546	10.320	10.104	9.896	9.504	9.135	8.787	8.458	8.145	15
10.895	10.658	10.431	10.213	9.801	9.415	9.050	8.705	8.377	20
11.221	10.974	10.737	10.510	10.079	9.676	9.295	8.935	8.593	25
11.530	11.272	11.026	10.789	10.341	9.922	9.526	9.151	8.796	30
11.825	11.557	11.302	11.056	10.591	10.156	9.745	9.357	8.989	35
12.108	11.831	11.566	11.312	10.831	10.380	9.955	9.553	9.174	40
12.382	12.095	11.822	11.559	11.061	10.595	10.156	9.742	9.354	45
12.647	12.352	12.069	11.798	11.284	10.804	10.351	9.924	9.530	50
13.158	12.844	12.545	12.257	11.713	11.203	10.724	10.277	9.869	60
13.647	13.316	13.000	12.696	12.121	11.584	11.080	10.618	10.197	70
14.120	13.772	13.439	13.120	12.515	11.950	11.426	10.950	10.516	80
14.580	14.215	13.866	13.531	12.897	12.305	11.765	11275	10.828	90
15.031	14.649	14.283	13.932	13.269	12.654	12.098	11.595	11.135	100
16.126	15.702	15.295	14.905	14.170	13.509	12.916	12.379	11.888	125
17.196	16.727	16.280	15.850	15.056	14.354	13.724	13.153	12.632	150
18.253	17.741	17.251	16.785	15.943	15.198	14.531	13.927	13.376	175
19.310	18.752	18.221	17.727	16.828	16.051	15.347	14.709	14.127	200

Figure 5-29 Sample *Theoretical Steam Rate* table. *(Courtesy of The American Society of Mechanical Engineers.)*

The performance of actual steam turbines includes several different types of losses and irreversibilities. The efficiency of a steam turbine–generator (TG) is defined as the output at the generator terminals divided by the *available energy*. Typical industrial steam turbine efficiencies are 70 to 80 percent. The *actual steam rate* (ASR) is defined as the TSR divided by the turbine-generator efficiency. In this case the ASR would be

$$\text{ASR} = \frac{\text{TSR}}{\text{TG efficiency}} = \frac{6.541}{0.75} = 8.72 \text{ lb/kWh (3.96 kg/kWh)}$$

If an output of 20,000 kW was required in this case, the steam flow would be

$$\text{Steam flow (lb/h)} = \text{ASR (lb/kWh)} \times \text{output (kW)}$$
$$= (8.72)(20{,}000) = 174{,}400 \text{ lb/h (79,100 kg/h)}$$

The expansion through an actual steam turbine would be approximately along path 1 to 3 on the Mollier diagram. Point 3 is at the same exhaust pressure as point 2, but at a different value of enthalpy.

When steam turbines exhaust or extract into a process heater, the value of enthalpy at that point is also of interest. Process heating loads are effectively expressed in British thermal units per hour (kilojoules per hour), even though they are frequently specified as weight flow per hour at a specific pressure. In order to determine exhaust or extraction enthalpy, the factors which contribute to the overall efficiency of a turbine generator should be considered. These include:

Mechanical Losses. Bearings, lube oil pump, and radiation

Generator losses. Electrical and windage

Packing leakage.

Wheel Efficiency. Including inlet valve loss

All these losses except wheel efficiency result in heat rejection from the cycle. The inefficiency of the steam expansion, or the adiabatic losses, result in heat remaining in the exhaust steam. Exact calculations of all these losses can be very detailed. For approximate calculations, a common assumption is that the losses other than the wheel efficiency are 2.5 percent of the generator output. The first law of thermodynamics can be used to derive an expression for exhaust or extraction enthalpy

$$\Delta \text{ Energy of steam} = \text{power output} + \text{losses} \tag{1}$$

Substituting,

$$\Delta \text{ Energy of steam} = \text{steam flow } (H_{in} - H_{out}) \tag{2}$$

$$\text{Power output} = \frac{\text{steam flow}}{\text{ASR}} \tag{3}$$

$$\text{Losses} = 0.025 \times \text{power output} \tag{4}$$

results in

$$\text{Steam flow } (H_{in} - H_{out}) = \frac{\text{steam flow}}{\text{ASR}}(1.025)$$

Rearranging,

$$H_{out} = H_{in} - \frac{1.025}{\text{ASR}}$$

and converting units results in

$$H_{out} \text{ (Btu/lb)} = H_{in} \text{ (Btu/lb)} - \frac{3500 \text{ (Btu/kWh)}}{\text{ASR (lb/kWh)}} \tag{5}$$

or

$$H_{out} \text{ (kJ/kg)} = H_{in} \text{ (kJ/kg)} - \frac{3693 \text{ (kg/kWh)}}{\text{ASR (kg/kWh)}}$$

Equation (5) is very useful for approximate calculations. For example, returning to the case with initial steam conditions of 1250 psig, 900°F (8720 kPa, 482°C), the exhaust or extraction enthalpy could be found at 100 psig (793 kPa) as follows:

$$\text{ASR} = \frac{13.269}{0.75} = 17.69 \text{ lb/kWh } (8.024 \text{ kg/kWh})$$

The enthalpy at 100 psig would be

$$H_{out} = 1438.4 - \frac{3500}{17.69} = 1240.5 \text{ Btu/lb}$$

or

$$H_{out} = 334.5 - \frac{3693}{8.024} = 2885 \text{ kJ/kg}$$

Referring to the steam tables,[1] the temperature of the steam at 100 psig (793 kPa) and 1240.5 Btu/lb (2885 kJ/kg) would be 429°F (221°C).

When more exact values are required, particularly at partial load operation, extraction and/or exhaust enthalpy can generally be obtained from the manufacturer.

PERFORMANCE CHARACTERISTICS

Performance Curve of Steam Turbine–Generators

Large (greater than 5000 kW) industrial steam turbines of multivalve, multistage construction generally have peak efficiency values in the range of 70 to 80 percent. The use of automatic-extraction valve gear can reduce these efficiency values by 2 to 5 percent.

Figure 5-30 is a typical performance curve of a straight condensing or noncondensing steam turbine. The plot of throttle flow vs. generator output approximates a straight line. The plot is known as the *Willans line.* The intercept of this line at zero load represents the steam flow required to supply the no-load losses of the set.

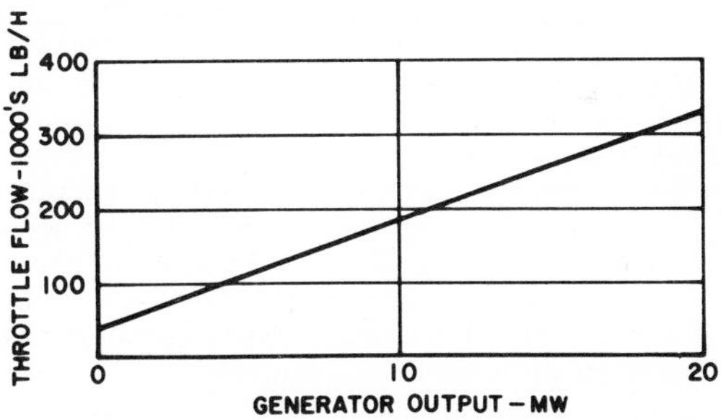

Figure 5-30 Performance curve of a straight condensing or straight noncondensing steam turbine.

The performance curve of a typical single automatic-extraction turbine is shown in Fig. 5-31. The zero extraction line is similar to the performance of a straight condensing or noncondensing unit. The family of parallel lines indicate the performance for various values of extraction flow. A certain minimum amount of flow is required at the exhaust end to cool the turbine buckets. This limit is indicated on the left-hand side of the plot. Also, the geometry of the section will limit it to a maximum flow capability which is shown on the right-hand side. The intercept of the maximum exhaust flow line and the zero extraction line is called the *pressure rise* point. Additional flow can be put through the exhaust end, but only if the pressure ahead of the extraction valve gear increases. This also causes an increase in the extraction pressure. This is not a normal mode of operation.

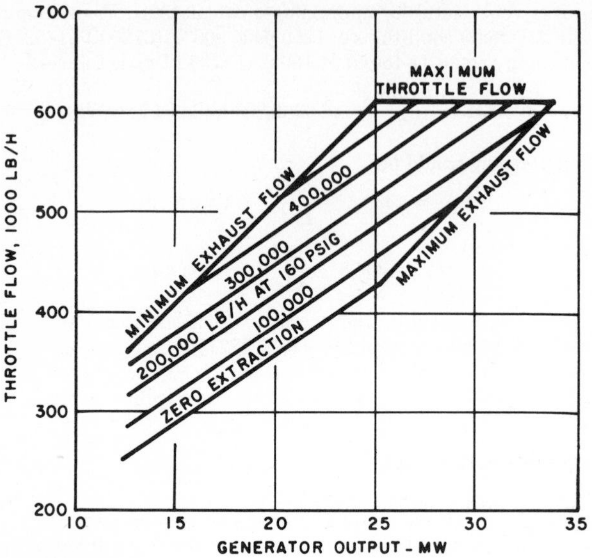

Figure 5-31 Performance curve of a single automatic-extraction steam turbine.

Performance of Mechanical-Drive Steam Turbines

Steam turbines are also selected as drivers for pumps and compressors which generally require variable speed. Process compressors as large as 60,000 hp (44 800 kW) are in service with variable-speed steam turbine drives. Special robust buckets are required for these turbines to withstand the vibratory stresses inherent with continuous operation over a wide speed range. In other respects large multistage, multivalve mechanical-drive steam turbines are similar to corresponding turbine-generator sets.

Single-valve, single-stage mechanical-drive steam turbines are very common in small sizes (less than 1000 hp) in industrial plants. This type of construction has a low cost, but efficiency drops to a range of 40 to 50 percent. These small units are often used in parallel or as a backup to motor drives to provide added reliability against a loss of either electric power or steam supply. Small mechanical-drive steam turbines are also selected as drives for hazardous locations. However, the low efficiency of these types of turbines is an increasingly important disadvantage with high energy costs.

Control Systems

Many industrial steam turbines are furnished with *e*lectro*h*ydraulic *c*ontrol (EHC) systems. The components in this type are illustrated in Fig. 5-32. Electronic control signals generated in the console are amplified by two stages of relays to control the steam valves: an electrohydraulic servo valve, and a 1500-psig (10 450-kPa) hydraulic ram. The console control action can respond to operators' commands, steam pressure signals, turbine speed (permanent magnet generator), and position feedback signals from the steam-valve actuators.

Earlier turbines have been furnished with mechanical hydraulic control (MHC) systems. These generally have lower-pressure hydraulic actuators [200 psig (1480 kPa)], a flyball speed sensor, and three bar linkages with springs and dashpots.

Most industrial steam turbine generators are synchronized with an electric grid. The speed/load control mode illustrated in Fig. 5-15 for gas turbine generators (Chap. 4-5,

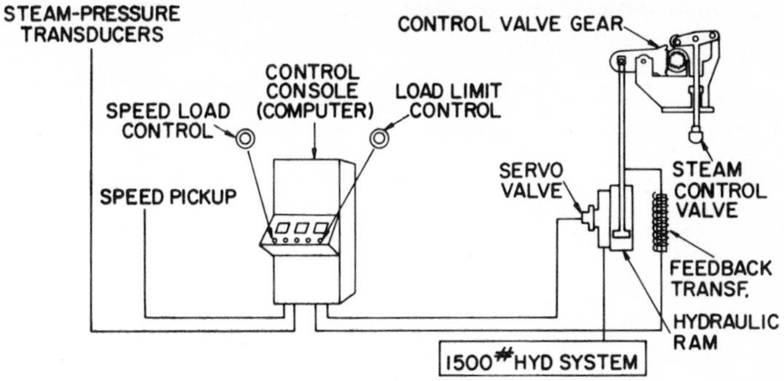

Figure 5-32 Electrohydraulic control system.

Part 1) is also used for steam turbines in these applications. Similarly, isochronous governing can also be used with steam turbines operating in isolated electric systems. However, steam turbine governing systems have many other capabilities to control steam pressures. The most common feature is steam pressure control of straight noncondensing or automatic-extraction noncondensing units. The synchronized generator is "locked into" the grid frequency, and the turbine inlet and extraction control valves maintain flow through the sections of the turbine in response to extraction and exhaust steam pressure signals. This type of system follows heat demands. The electricity which is generated in this manner is called *by-product power.*

When automatic-extraction steam turbines have a condensing exhaust, the control system can govern extraction pressure as well as power generation by varying flow to the condenser. In some cases the amount of power called for by the steam turbine governor is continuously modulated by an external control system such as a utility tie-line controller or by a master control system designed to parallel two or more turbine-generator sets.

Another control mode is initial pressure control. When a steam turbine is supplied by a heat recovery boiler or a by-product fuel boiler, the amount of steam generated can vary independently. Initial pressure governing allows the steam turbine to draw all the available steam out of the header while maintaining constant pressure.

OPERATION AND MAINTENANCE

Starting Procedures

Before a steam turbine–generator can be started, proper operation of a number of auxiliary systems must be assured, including:

Bearing oil pump and seal oil pump, when applicable

Hydraulic control fluid pump

Condenser vacuum pump, when applicable

Generator cooling system (air or hydrogen pressure and circulating water)

Emergency trip system

Drains in all steam lines must be opened to prevent slugs of water from entering the turbine. Units of 10 MW or larger are generally furnished with a turning gear. This is then engaged to rotate the unit slowly. The steam lines are preheated and the gland seal exhaust system is put into operation. Steam is very gradually admitted to the turbine, and the turbine rolls off the turning gear and is gradually accelerated. The rate of acceleration selected will depend on the casing metal temperature, which depends on the duration of the previous outage. Acceleration time can vary from 10 to 30 min. During this period vibration and turbine shell temperature are monitored.

As the unit approaches synchronous speed, the generator breaker is in the open position and the excitation system regulates voltage to match the bus voltage. The operator then matches the frequency and phase angle before the generator breaker is closed. The unit then picks up load at a rate again depending on the temperature of the unit before startup. Typical loading times are 30 to 60 min, depending on whether the unit was hot or cold before starting.

Normal Operation

During periods of normal operation several parameters are monitored, including turbine shell temperature and pressure, exhaust hood temperature, bearing oil temperature and pressure, condenser vacuum, shaft vibration, hydraulic oil pressure, and generator gas and winding temperatures.

Control systems generally contain many automatic protection and alarm systems. A typical protection system includes a main stop valve which is spring-loaded to close and is mounted ahead of the turbine inlet valve. Hydraulic oil keeps this valve open if several protection devices all indicate safe operation. These include an overspeed governor, bearing oil pressure relay, a manual trip relay, and when applicable, a low-vacuum relay and nonreturn valve relays.

Maintenance

A limited amount of running maintenance is required in addition to data logging. This includes periodic lubrication of valve gear (monthly or quarterly) and H_2 replenishment for hydrogen-cooled generators (one bottle two or three times a month).

The turbine shell is generally removed for a warranty inspection after a year's operation. Shutdowns thereafter can be at intervals of 3, 4, or 5 years. Long periods of operation with minimal maintenance require steam of high purity. Carryover of certain contaminants in the steam can cause deposits, erosion, and stress cracking. A list of common deposit- and corrosion-causing contaminants is given in Table 5-7.

TABLE 5-7 Common Water Contaminants

Deposit-forming	Corrosion-causing
Calcium salts	Ammonia
Magnesium salts	Oxygen
Silica	Chlorides
Iron salts	Sulfides
Organic matter	Carbonates
Copper salts	Bicarbonates
Sulfates	
Nitrates	

These contaminants can enter the steam supply system with the makeup water or in process heat exchange equipment. These species can exist in the boiler drum in relatively high concentrations without causing problems. It is only when they are carried over into the exiting steam that they enter the turbine. Efficient boiler drum separators can limit total dissolved solids to as little as 0.5 to 1.0 ppm. Silica is difficult to separate from the steam and must be controlled in the boiler feedwater. Care must also be taken with the fluid used to attemperate the steam exiting the superheater. Contaminated process returns can bypass the steam separation in the drum as attemperator fluid. This should be maintained at less than 1.0 ppm total dissolved solids.

Primary water treatment conditions the makeup water before it enters the cycle. Secondary water treatment is the addition of chemicals to the cycle to "polish" the feedwater. Boiler blowdown is the removal of solids from the cycle to prevent high concentrations in the drum.

Monitoring steam purity is a very important part of steam turbine maintenance. Deposits can be detected by abnormal shell pressure readings. There is a trend toward increased use of boroscopes to detect erosion and stress cracking during a turbine outage without requiring disassembly of the turbine casing. Many types of deposits can be removed by washing the turbine at reduced speed with wet steam.

Persistent deposits are removed by opening the turbine. Water washing and rinsing remove most soluble deposits. Steam cleaning and/or blasting with a fine-grade abrasive may be required for hard deposits like silica or iron oxide.

APPLICATIONS IN PLANTS

Cogeneration

Cogeneration is the simultaneous generation of power and useful heat. Steam turbines are the most common prime movers used in cogeneration systems. The fuel chargeable to power in noncondensing steam turbines typically ranges from 4000 to 4400 Btu/kWh HHV (4220 to 4640 kJ/kWh). In order to achieve the optimum economic gain from a cogeneration system several factors should be considered in the plant design stages. These include:

Throttle steam pressure

Extraction and exhaust steam pressure

Turbine efficiency

Feedwater heating cycle

High-cost energy has made initial steam conditions for industrial plants economical up to 1250, 1450, and even 1800 psig (8 720, 10 100, and 12 514 kPa). Higher boiler and turbine equipment prices are justified by the additional by-product power available from a given process heat load. Similarly, to further increase by-product power, steam should be expanded down to the lowest pressure consistent with the process heating temperature. For example, many industrial plants distribute steam with a single header of, say, 250 psig (1 827 kPa), when most of the demand is at lower pressures. Two or three pressure levels of distribution can result in the production of many additional kilowatthours of low-cost by-product power.

Figure 5-33 illustrates the effects of turbine inlet and exhaust pressure on by-product power generation. Past practice might have resulted in steam conditions of 850 psig, 825°F, to 250 psig (5 964 kPa, 454°C, to 1 827 kPa) to meet a 350°F heating demand. Each 100×10^6 Btu (105.5 GJ) of heat supplied to process would generate 3600 kW. A more energy-conscious design would be 1450 psig, 950°F, to 150 psig (10100 kPa, 510°C, to 1138 kPa). The by-product power would increase to 6000 kW per 100×10^6 Btu/h (105.5 GJ/h) of heat supplied. Assuming a power cost of 5 cents/kWh and a fuel cost of \$4 per 10^6 Btu (\$3.79/GJ), and 350 days/year of operation, the preferred steam conditions would save \$670,000 per year in energy costs.[5]

Steam turbine efficiency is another important parameter in cogeneration plants. In noncondensing steam turbine applications, the turbine efficiency has only a minor effect on fuel chargeable to power. However, the amount of high-efficiency by-product power generated is directly proportional to the turbine efficiency. Converting fuel energy into low-temperature heat is a simple process, and high efficiencies of 85 to 90 percent can be achieved. However, converting fuel energy into power is much more difficult, and efficiencies are only 33 to 35 percent in the largest, most modern central stations. Therefore, high-efficiency multistage, multivalve noncondensing steam turbines should be preferred over inefficient single-valve, single-stage designs because more of the fuel energy is converted to the more valuable quantity—power.

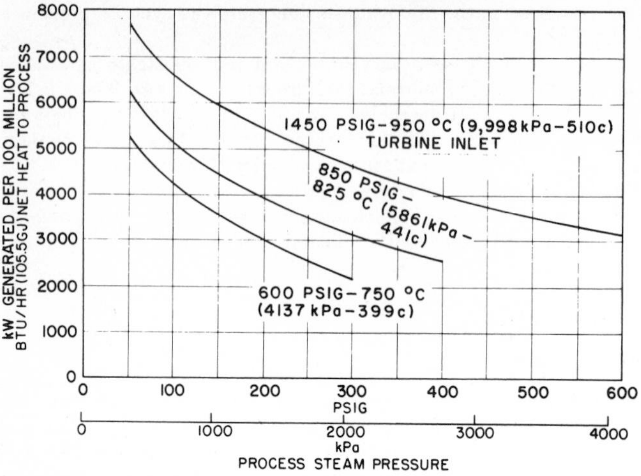

Figure 5-33 Effects of steam turbine inlet and exhaust pressure on by-product power generation.

Feedwater Heating

This is another method of enhancing the amount of by-product power generated for a given process heat load. Figure 5-34 illustrates a single automatic-extraction noncondensing steam turbine with a single heater and with a second higher-pressure feedwater heater. The second heater increases the amount of by-product power by 2550 kW. Using the same economic analysis as the previous example, it can be calculated that this heater would save $710,000 per year in energy costs.

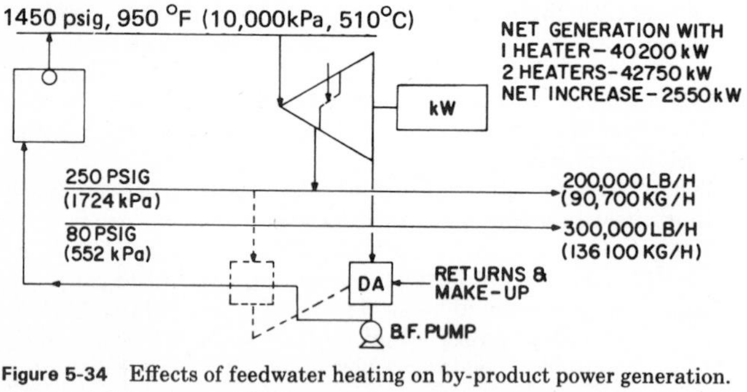

Figure 5-34 Effects of feedwater heating on by-product power generation.

As was the case with gas turbines or any other type of heat engine applied to a cogeneration system, fuel chargeable to power, or the quantity of by-product power, is not the only factor to be considered. The investment in this equipment must be justified in terms of rate of return or a similar criterion.

In the case of large variable-speed drivers, mechanical-drive steam turbines often are the only viable solution. For example, a schematic diagram of a large ethylene plant driver is shown in Fig. 5-35. The charge gas compressor has a speed and power beyond that of suitable gas turbines or direct-drive motors. Process heat recovery produces steam at 1500 psig, 950°F (10 446 kPa, 510°C) and there is a large process heat demand at 300 psig (2070 kPa). A single automatic-extraction condensing steam turbine is ideally suited as a driver in these types of plants.

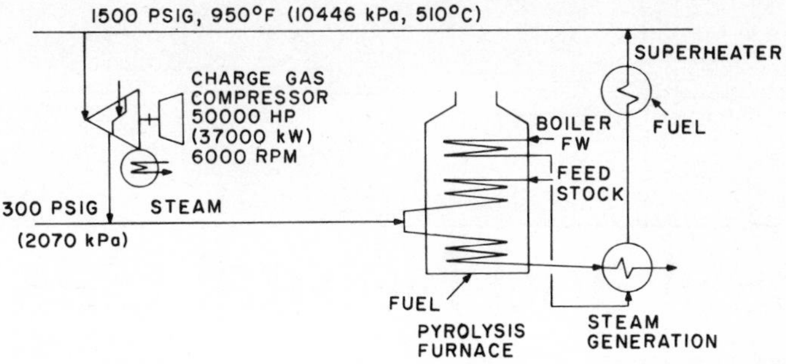

Figure 5-35 Schematic diagram of a large mechanical-drive steam turbine in an ethylene plant.

REFERENCES

1. *967 ASME Steam Tables,* The American Society of Mechanical Engineers, New York, 1967.
2. *Theoretical Steam Rate Tables,* The American Society of Mechanical Engineers, New York, 1969.
3. Salisbury, J. K.: *Steam Turbines and Their Cycles,* Wiley, New York, 1950.
4. Newman, L. E.: *Modern Turbines,* Wiley, New York, 1944.
5. Kovacik, J. M., and W. B. Wilson: "Turbine Systems to Reduce Petroleum Refining and Petrochemical Plant Energy Costs," *Petroleum Division Conference,* ASME Paper 76-PET-62, Mexico City, Mexico, Sept. 1976.

chapter 4-6

Diesel and Natural Gas Engines

prepared by

Waukesha Engine Division

Dresser Industries, Inc.

Waukesha, Wisconsin

GLOSSARY

Brake horsepower (bhp) The horsepower delivered by the engine shaft at the output end.[1] The name is derived from the fact that it originally was determined by a braking device on the engine flywheel.

Brake mean effective pressure (BMEP) The average cylinder pressure to give a resultant torque at the flywheel:[2]

$$\text{BMEP (lb/in}^2) = \frac{792{,}000 \times \text{bhp}}{(\text{r/min}) \times \text{displacement (CID)}} \text{ (four-stroke cycle)}$$

$$\text{BMEP (lb/in}^2) = \frac{396{,}000 \times \text{bhp}}{(\text{r/min}) \times \text{displacement (CID)}} \text{ (two-stroke cycle)}$$

$$\text{BMEP (kPa)} = \frac{600{,}000 \times \text{kW}}{(\text{r/min}) \times \text{displacement (L)}} \text{ (two-stroke cycle)}$$

$$\text{BMEP (kPa)} = \frac{1{,}200{,}000 \times \text{kW}}{(\text{r/min}) \times \text{displacement (L)}} \text{ (four-stroke cycle)}$$

Compression ratio The ratio of the volume of cylinder space above the top piston ring with the piston at bottom dead center to the volume in the cylinder above the top ring when the piston is at top dead center.

Cycle The complete series of events in each cylinder, including introduction and compression of air, burning of fuel, expansion and expulsion of the working medium in the engine.[1]

Diesel engine An engine in which the fuel is ignited entirely by the heat resulting from the compression of the air supplied for combustion.[1]

Displacement The volume of an engine's cylinder swept by the piston. It is equal to the area of each piston multiplied by the stroke multiplied by the number of cylinders and is expressed in cubic inches displacement (CID) or liters.

Duty cycle A term used to describe the load pattern imposed on the engine. Continuous, or heavy-duty, service is generally considered to be 24 h/day with little variation in load or speed.[2] Intermittent, or standby, service is classed as duty where an engine is called upon to operate only in emergencies or at infrequent intervals.

Four-stroke cycle engine An engine completing one cycle in four strokes of the piston or two shaft revolutions. The cyclic events are designated by the following strokes: (1) induction or suction stroke, (2) compression stroke, (3) expansion stroke, (4) exhaust stroke.[1]

Gas engine An engine in which the fuel in its natural state is a gas and the air-fuel mixture is ignited by a spark within the combustion chamber.

Intercooler A device used to cool the compressed air after it leaves a turbocharger but before it enters the engine cylinders. This device is sometimes called an aftercooler.

Naturally aspirated A term used to describe an engine which used no device to increase the pressure on the intake air of the engine before it enters the cylinder.

Opposed piston engine One that uses the working medium simultaneously between two pistons in the same cylinder.[1]

Power The rate of doing work. Engine power output is expressed in units of horsepower, equivalent to 550 ft·lb/s, or kilowatts, equivalent to 1000 J/s. Here are some conversion formulas:

Useful Equivalents

1 hp (U.S.) = 550 ft · lb/s
1 hp (metric) = 0.9864 hp (U.S.)
1 hp (U.S.) = 1.0138 hp (metric)
1 kW = 1 000 J/s
= 1 000 N·m/s
1 kW = 1.341 hp (U.S.)

Engine power expressed by various engine manufacturers throughout the world should be specified as to the conditions under which the engine was run or the conditions to which the horsepower is corrected. Rating societies such as the Diesel Engine Manufacturers Association (DEMA), the Society of Automotive Engineers (SAE), German Industry Standards (DIN), British Standards (BS), and International Standards Association (ISO) all have their ratings for temperature and barometric conditions to which engine manufacturers relate their data. Manufacturers usually specify their engine performance and rating standards as well as methods of correcting the power available to expected site conditions of temperature and barometric pressure.

Speed regulation The incremental difference between no-load (NL) speed and full-load (FL) speed divided by the full-load speed. This is sometimes referred to as speed droop.

$$\frac{(\text{NL speed}) - (\text{FL speed})}{(\text{FL speed})} \times 100 = \text{percent speed regulation}$$

Torque Twisting effort of an engine described in pound·feet:[2]

$$T\ (\text{U.S.}) = 5252 \times \frac{\text{bhp}}{\text{r/min}}$$

Turbocharger A rotary air compressor driven by a turbine using exhaust gas as the driving fluid and compressing intake air into the engine.

Two-stroke cycle engine An engine completing one cycle in two strokes of the piston or one shaft revolution. The cyclic events are designated by the following strokes: (1) induction and compression stroke, (2) expansion and exhaust stroke.[1]

INTRODUCTION

Stationary internal combustion engines are employed in plant applications primarily for emergency and prime electric power generation as well as for compressors, blowers, and pumping equipment.

Diesel and gas engines have been widely used for fire and flood pumps as well as for in plant engine–generator sets and they have found acceptance in some areas as prime movers for compressor equipment used for air conditioning or refrigeration.

In sewage treatment plants, where gas is produced as a by-product and captured for use as engine fuel, gas engines have become popular for driving blowers required for the treatment process, pumping effluent, and for prime power generation.

DIESEL ENGINES

Principles of Diesel Engines

The diesel engine is an engine in which the cylinder is charged with air which is then compressed until it is hot enough to ignite fuel injected into the combustion space. The fuel is ignited by the hot air, and the expanding gases drive the piston down on the power stroke.

The compression ratios of diesel engines range from 12:1 to as high as 23:1.

Engine types fall into two categories: two-stroke cycle and four-stroke cycle. Typical two-stroke-cycle engine arrangements are shown in Figs. 6-1 and 6-2. A typical four-stroke-cycle engine arrangement is shown in Fig. 6-3.

There are many types of combustion chamber designs which manufacturers use to mix fuel and air. There are claimed advantages for each type. These may be fuel economy, ability to make use of simpler fuel systems, wide speed range coverage, or firing pressure minimization. These systems generally fall into an *open-chamber* (Fig. 6-4) or *divided-chamber* (Fig. 6-5) category.

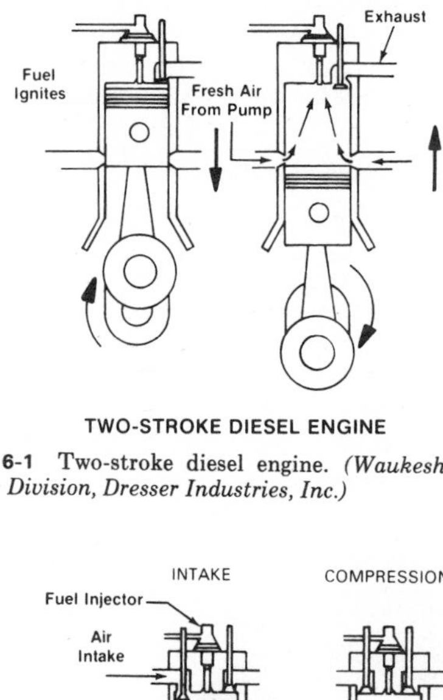

Figure 6-1 Two-stroke diesel engine. *(Waukesha Engine Division, Dresser Industries, Inc.)*

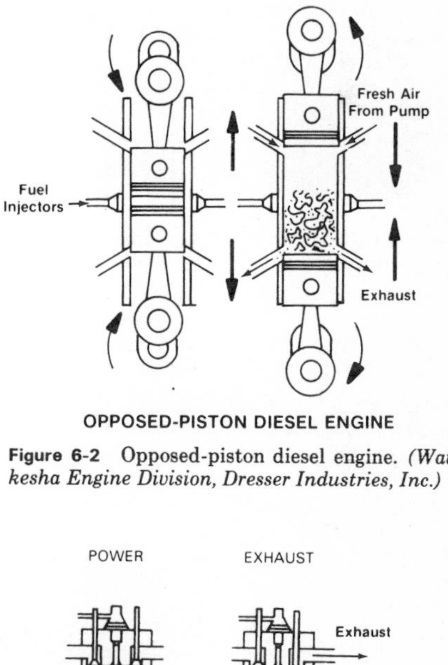

Figure 6-2 Opposed-piston diesel engine. *(Waukesha Engine Division, Dresser Industries, Inc.)*

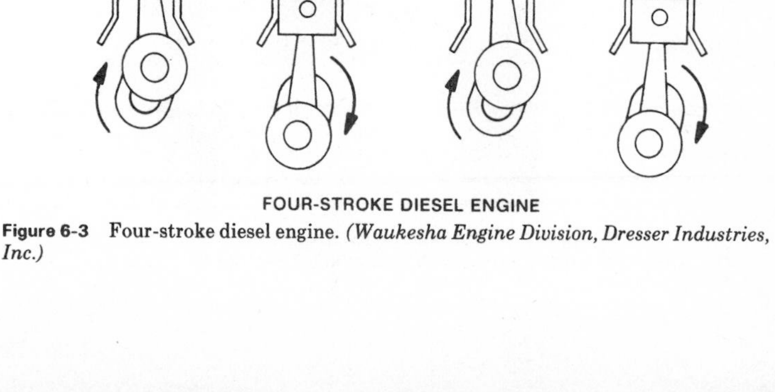

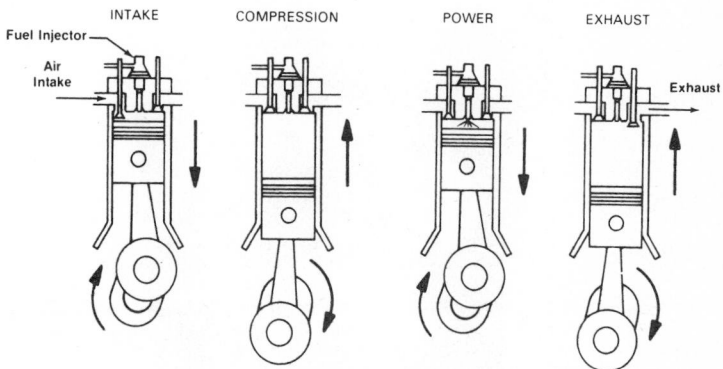

Figure 6-3 Four-stroke diesel engine. *(Waukesha Engine Division, Dresser Industries, Inc.)*

The open chamber has less heat rejected to the cooling system but has greater demands on the injection system. The divided-chamber engines can operate with less sophisticated injection equipment since the air and fuel mixing is aided by more rapid air motion. Better fuel economy can be attained with an open-chamber design, while a divided chamber permits lower emissions.

Figure 6-4 Open combustion chamber. *(Waukesha Engine Division, Dresser Industries, Inc.)*

Engines are frequently turbocharged to increase the horsepower taken from a given displacement engine. Turbocharging utilizes some of the waste heat energy and velocity of engine exhaust gas to drive a turbine connected to a high-speed centrifugal compressor. The power from a given package size can be more than doubled provided the engine components are strong enough to withstand the higher cylinder pressures. The highly turbocharged engines usually have a means of reducing the air temperature after the air leaves the compressor by means of an air-to-air cooler or an air-to-water cooler. These devices are known as intercoolers.

Performance Characteristics

The performance of a diesel engine is affected by air temperature, air pressure, and humidity. Correction factors are employed to assure that the power specified takes into account the losses that are expected with altitude or temperature conditions at the site. These correction factors are usually specified by the Diesel Engine Manufacturers Association (DEMA), the Society of Automotive Engineers (SAE), German Industry Standards (DIN), British Standards (BS), or the International Standards Organization (ISO).

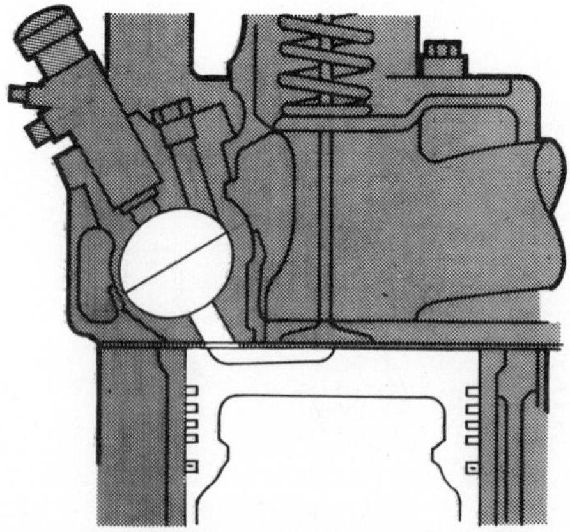

Figure 6-5 Divided combustion chamber. *(Waukesha Engine Division, Dresser Industries, Inc.)*

Fuels

Introduction

Diesel engines can be designed to use a wide variety of fuels; however, if fuel other than diesel grade 1 or 2 is used, the manufacturer should be consulted. Jet A fuel can be used in diesel engines if it has a 40 octane minimum.

Fuel consumption rates on diesel engines are generally specified as brake specific fuel consumption and its units are in pounds per brake horsepower hour, grams per brake horsepower hour, or grams per kilowatthour. (See Fig. 6-6.)

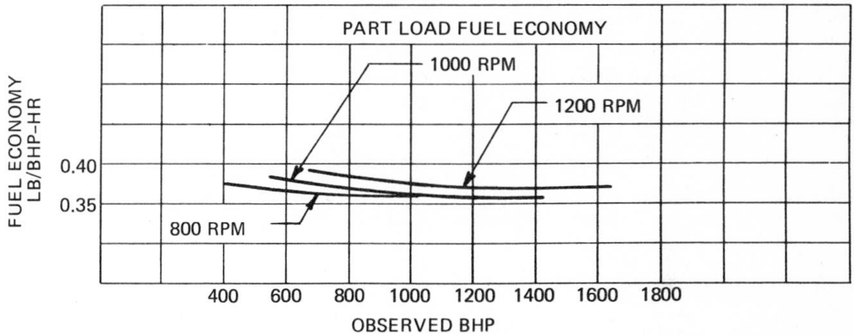

Figure 6-6 Fuel consumption rates for the diesel engine. *(Waukesha Engine Division, Dresser Industries, Inc.)*

Diesel Fuel Oil Specifications

It is important that the fuel oil purchased for use in an engine be as clean and water-free as possible. Dirt in the fuel can clog injector outlets and ruin the finely machined precision parts in the system. Water in the fuel will accelerate corrosion of these parts. Reputable fuel suppliers deliver clean, moisture-free fuel. Most of the dirt and water in the fuel is introduced through careless handling, inadequate filtration, dirty storage tanks or lines, and poorly fitted tank covers.

There are fuel composition requirements that must be met when purchasing diesel fuel. Table 6-1 lists fuel properties for No. 2 diesel fuel and their limits. Definitions of the more critical properties follow.

Diesel Fuel Properties

API Gravity. The specific gravity, or density in pounds per gallon.

Ash. The mineral residue in fuel. High ash content leads to excessive oxide buildup in the cylinder and/or injector.

Cetane Number. Ignitability of the fuel. The lower the certane number, the harder it is to start and run the engine. Low-cetane fuels ignite later and burn more slowly. Explosive detonation could be caused by having excessive fuel in the chamber at the time of ignition.

Cloud and Pour Points. The pour point is the temperature at which the fuel will not flow. The cloud point is the temperature at which the wax crystals separate from the fuel. The cloud point must be no more than 10°F (5.5°C) above the pour point so the wax crystals will not settle out of the fuel and plug the filtration system. The cloud point should also be at least 10°F (5.5°C) below the ambient temperature to allow the fuel to move through the lines.

Distillation Point. Temperature at which certain portions of the fuel will evaporate. The distillation point will vary with the grade of fuel used.

TABLE 6–1 Fuel Oil Recommendations Chart

Fuel oil physical properties	Limits	ASTM test method
Fuel grade	Diesel no. 2	
API gravity	30 min	D 287
Cetane number	40 min*	D 613
Sulfur, %	0.7 max	D 129
SSU viscosity, at 100°F (37.7°C)	30–50	D 88
Water and sediment, %	0.1	D 96
Pour point, min.	10°F 5.5°C below ambient air	D 97
Carbon residue	0.25%	D 189
Ash, % max.	0.02	D 482
Aklali or mineral acid	Neutral (pH 7)	D 974
Distillation point		D 158
10% min	450°F 232°C	
50%	475–550°F 246–288°C	
90% max	675°F 357°C	
End point max	725°F 385°C	
Cloud point	†	D 97

*For automatic starting units, a fuel with 50 cetane minimum is recommended.

†Cloud point should not be more than 10°F (5.5°C) above pour point.

Source Waukesha Engine Division installation manual.

Sulfur. Amount of sulfur residue in the fuel. The sulfur combines both with any moisture in the fuel and the water vapor formed during combustion to form sulfuric acid. This acid can quickly corrode engine parts. The lower the sulfur content of the fuel, the better.

Viscosity. Influences the size of the atomized droplets during injection. Improper viscosity will lead to detonation, power loss, and excessive smoke.

Fuel Systems

Installation of a diesel engine requires special planning to handle fuel delivery, storage, and piping (Fig. 6-7).

Fuel Tanks

Most diesel engine installations utilize a two-tank system, with both a main storage tank and a day tank.

Day Tanks

A day tank is designed to keep a clean supply of fuel oil close to the engine and to provide an immediate supply of fuel when the engine is started. By locating the day tank close to the engine, the fuel-transfer pump will be able to draw fuel more easily, without having to develop high suction pressures.

The day tank is generally sized to hold enough fuel for several hours of operation or whatever local fire codes allow. It is often a 275-gal standard commercial tank installed at or above floor level.

If the day tank is positioned above the level of the engine fuel-injection pump, the flow of the fuel will maintain a constant pressure at the fuel-transfer pump inlet.

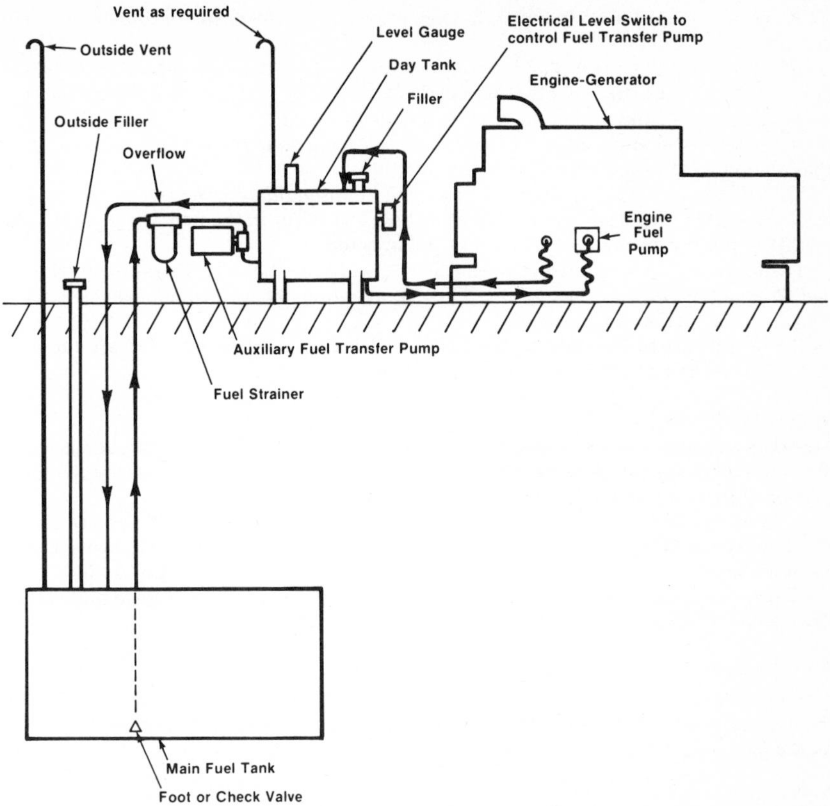

Figure 6-7 Typical diesel fuel-oil system. *(Waukesha Engine Division, Dresser Industries, Inc.)*

A positive shutoff should be added just beyond the day tank whenever the tank is above the level of the fuel-injection pump. If the day tank is mounted on a structure that is subject to vibration, a flexible connector should be added between the tank and the fuel piping. (Actually, flexible connections are recommended between an engine and any support system, whether it involves fuel, air, water, or oil.) The weight of both the tank and the fuel must be considered when designing and mounting the tank.

Main Storage Tanks

Spherical or cylindrical tanks should be used for added strength. Avoid square tanks. Main storage tanks are generally large enough to hold a 10-day fuel supply. (Local codes may dictate tank sizes.) If fuel delivery is uncertain due to weather, traffic, or any other reason, the tank size should be increased.

The location of the main storage tank is influenced by the method of delivery. If the fuel oil will be delivered by railroad tank cars, the tank and filler opening should be near the railroad siding. If a truck will be delivering the fuel, the tank and filler opening should be close to the road. Tanks should always be located so as to minimize the length of the fuel lines.

Ideally, exterior main storage tanks should be located underground, below the frost line. This will insulate the tank against temperature changes, and water condensation in the tank will be held to a minimum. For underground storage tanks the filler pipe should be located high enough above grade level to prevent fuel contamination from ground water and dirt.

As with day tanks, the fuel pickup should never draw from the bottom of the main storage tank.

The tank should be installed lower at one end to allow the dirt and water that settle out of the fuel to be drained off or pumped out.

Underground tanks should be bottom pumped at least twice each year to remove all accumulated water and sludge. Above-ground tanks are more subject to condensation, so they should be bottom pumped more frequently.

CAUTION*—Diesel fuel tanks, fittings, and lines should never be made of galvanized steel nor should they be of a zinc alloy material.* The sulfur in the fuel will corrode these metals, gumming up the injection pump and injectors.

Heating lines can be added to warm the fuel and keep it at a temperature at which it can be easily pumped to the engines. The manufacturers of such components should be consulted for further information.

The vent pipe to the outside must have at least a 1-in diameter, and an approved flame arrester must be incorporated into the vent.

Strainers and Filters

Strainers and filters are an important part of any diesel fuel system. Without the cleaning action of these components, the dirt and grime in the fuel would destroy the finely machined parts in the injectors and the injection pump.

A filter should be placed just before each meter and each pump. To assure proper maintenance, it should be located in an easy-to-reach position. Also, try to leave enough room under each filter for a catch basin to avoid messy, dangerous fuel oil spills.

Separators should be added to a system, particularly at the main storage tank outlet, to remove sediment.

Shutoff Valves

A shutoff valve should be incorporated in the fuel system at the fuel tank outlet and at the point where the fuel line enters the building or engine room and wherever applicable local codes so dictate.

Fuel-Transfer Pumps

Transfer pumps are used to supply fuel to the injection pump, or to raise fuel to a tank or engine at a higher level. Centrifugal pumps cannot be used as transfer pumps because they are not self-priming. Positive-displacement pumps must be used.

Fuel Return Lines

Fuel return lines take the hot excess fuel not used in the engine cycle away from the injector and back to either the fuel storage tank or the day tank. The heat from the excess fuel is dissipated in the tank. ***CAUTION****—Never run a fuel return line directly back to the engine fuel supply lines. The fuel will overheat and break down.*

The fuel return lines should always enter the storage or day tank above the highest fuel level expected. This will prevent fuel in the storage tank from running back into the fuel return line.

NATURAL GAS ENGINES

Principles

Gas engines can be two-cycle or four-cycle and either naturally aspirated or turbocharged. The fuel can be introduced into the air by a carburetor or injected into the intake port just ahead of the inlet valve or directly into the combustion chamber.

The combustion chamber can be either the open type or the divided type, as discussed in the section on diesel engines.

Performance Characteristics

Like diesel performance, the performance of a gas engine is affected by air temperature, air pressure, and humidity. Correction factors are used to ensure that the power specified will take into account the losses expected with the altitude and temperature conditions anticipated at the site. The rating societies mentioned previously, in the discussion of diesel engine performance, supply methods of correcting power on spark-ignited engines.

The fuel consumption of a gas engine usually is expressed in British thermal units per brake horsepower (kilocalories per metric horsepower) (see Fig. 6-8).

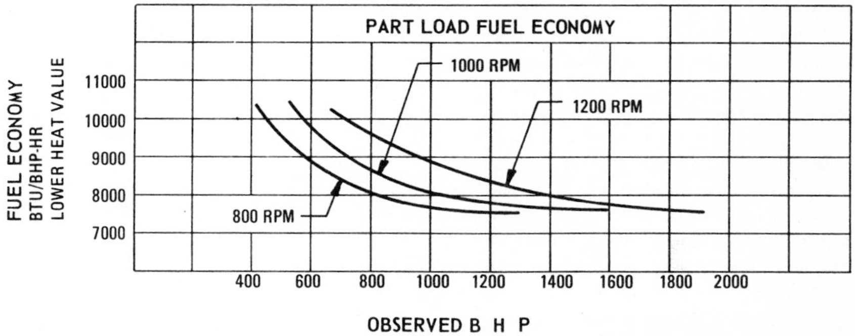

Figure 6-8 Fuel consumption rates for the natural gas engine. *(Waukesha Engine Division, Dresser Industries, Inc.)*

Fuels

A gas engine can be adjusted to accept a wide variety of fuels. Among these are:

1. Pipeline quality natural gas
2. Digester gas from sewage treatment plants
3. Methane from sanitary landfills
4. Propane and LP gas
5. Pyrolytic gas from hydrocarbon sources
6. Field gas (wellhead gas)

The octane number for various gaseous fuels available for gas engines can be calculated from their known constituents. The heating value of engine fuel can also be obtained with a calorimeter and is nearly always expressed as lower heating value (LHV).

Fuel System

The gas engine can be naturally aspirated or turbocharged. Since the manifold pressure of the turbocharged units can be as high as 15 psig, the pressure of the natural gas supply should be in the 20- to 25-psig range in order to have the engine operate properly. Conversely, the gas supply to the naturally aspirated engine can be at a pressure as low as 0.5 psig (but is normally preferred at a 5- to 10-psig span) since the engine manifold operates at atmospheric pressure or less.

The gas pressure regulators to reduce the supply pressures to the required engine carburetor inlet pressures are usually mounted on the engine. Experience has proved that it is best to supply one regulator per carburetor (on dual-carburetor engines), mounted as close to the carburetor as possible. This minimizes line drops and governing instability.

The type and number of gas shutoff devices and pressure-sensing switches are determined by applicable safety codes and control circuit requirements. A low gas pressure switch is commonly specified.

SYSTEMS COMMON TO BOTH DIESEL AND NATURAL GAS ENGINES

Lubrication System

Although the lubrication system is one of the simplest systems of the engine, its importance should not be underestimated. It is the most important support system of the engine and cannot be neglected except at the expense of premature engine failure.

The lubrication system of an industrial engine is almost completely assembled before it leaves the factory. On large-displacement engines, the free-standing lube oil filter (and in some cases the oil cooler) is the only major lubrication component usually shipped free of the engine. (Smaller engines have the oil filter mounted directly on the engine.)

Lubricating Oil Filter Installation

Position the lube oil filter as close to the engine as possible.

CAUTION*—Do not put the filter near the exhaust outlet or other places where the temperature could become excessively warm. Excessive heat will speed oil deterioration and will also create a fire hazard in the event of an oil spill or line rupture.*

It is important to use pipe of adequate size between the engine and lube oil filter in order to maintain the proper oil pressure to the engine. Consult the engine supplier for recommendations. Black iron or steel pipes should be used to carry oil. ***CAUTION****—After welding, flush pipes with muriatic acid to remove all welding scale, and rinse thoroughly to neutralize the acid and ensure clean piping.*

Flexible Connections

Flexible connections designed to handle hot lubricating oil at pressures up to 100 lb/in^2 should be used between the engine and the free-standing oil filter. Position the connections as close to the engine as possible. Supports should be added under the oil filter lines to minimize vibration and prevent breakage.

Lubricating Oil Recommendations

Lube oil selection is the responsibility of the engine operator and the oil supplier. The refiner is responsible for the performance of the lubricant.

Most engine warranties are limited to the repair or replacement of parts that fail due to defective material or workmanship during the warranty period. These warranties do not include satisfactory performance of lube oil. That is considered the responsibility of the oil supplier.

Most engine manufacturers do not recommend lubricants by name or brand. Assistance in lubricant selection can be obtained from a publication of the Engine Manufacturer's Association, One Illinois Center, 111 E. Wacker Drive, Chicago, IL 60601. This book, *EMA Lubrication Oils Data Book for Heavy-Duty Automotive and Industrial Engines*, has a table of lubricants and their performance grades. Check with the engine manufacturers concerning oil recommendations for their various engines. It is common to find that different oils are used for gas and diesel engines, as well as for different models of each type.

Accessories available for lube oil systems are such items as flowmeters and oil-level regulators. These can be unit-mounted and automatically add oil as it is consumed as well as measure and record the quantity of oil consumed. It is also possible to obtain engine-mounted switches for low and high engine lube oil levels to signal a warning and/or cause engine shutdown.

On larger engines, use an air-motor- or electric-motor-driven prelube pump to fill and pressurize the lube oil system before cranking and operating the engine. In addition, use a lube oil heater to keep the oil warm and in condition for positive lubrication of vital areas on start-up wherever low ambient temperature may be encountered.

Cooling Systems

Cooling systems in liquid-cooled engines are affected by the mineral content and corrosiveness of water put into the system—be it cooled by radiator, heat exchanger, or standpipe. In all cases, recommended practice is to use treated water so as to minimize long-term effects on the engine.

Radiator Cooling

The most common cooling arrangement is that of the stationary engine with a unit-mounted radiator. In this case, the cooling fan is belt-driven from the front of the engine. Usually a pusher fan is used to prevent hot air from being drawn over the set and its operator. The radiator has to be sized for the maximum expected operating temperature, the maximum rated engine horsepower or kilowatt load, and the type of coolant used (ethylene glycol). The radiator has to cool the heat rejected to the engine-jacket water, the heat rejected from the engine lube oil, and (if applicable) the heat rejected from the intercooler circuit. Some radiators are dual-core units, with one core for the jacket-water circuit and a second core for the intercooler oil-cooler circuit. This is required on natural gas engines to keep the water entering the intercooler at 130°F (54°C) maximum.

A radiator-cooled unit (Fig. 6-9) is self-contained and, therefore, adaptable to mobile installations and/or relocation. Since the radiator is sealed, it does not waste water and does not contaminate drain water. However, the horsepower required to drive the fan is lost for other uses, and the relatively large radiator airflow both in and out of the engine room has to be handled judiciously. Other installation items are louvers (fixed or variable), vertical discharge air scoops, and air discharge duct adaptors.

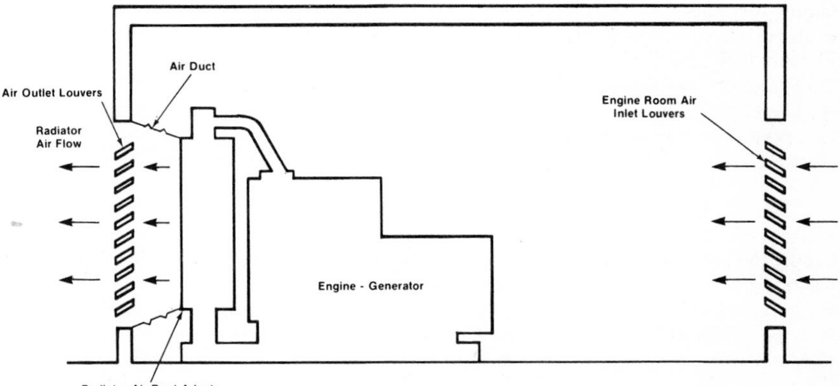

Figure 6-9 Radiator cooling. *(Waukesha Engine Division, Dresser Industries, Inc.)*

Cooling by use of a remote-mounted radiator is also common. This system is quite flexible since the radiator can be mounted outside, which reduces engine-room airflow. Also, the core can be horizontal, which makes the radiator insensitive to wind direction. Roof mounting of the radiator is quite common; if the roof height causes an increase over the recommended jacket-water pressure, a hot well (Fig. 6-10) and auxiliary pump can be added. Other installation considerations are automatic louvers, vertical vs. horizontal core, remote surge tank, piping sizes to control line-pressure drops, and sound-limit requirements.

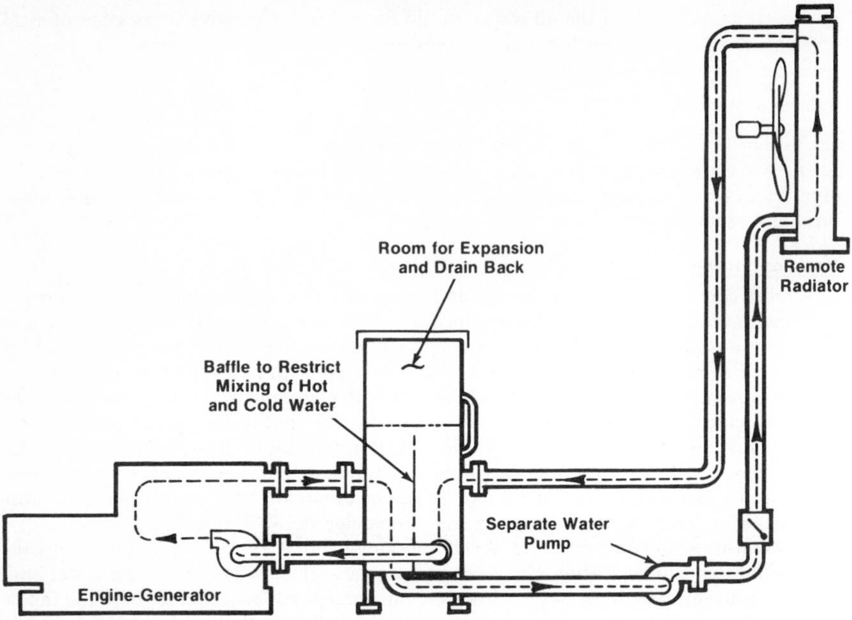

Figure 6-10 Hot well cooling. *(Waukesha Engine Division, Dresser Industries, Inc.)*

Heat-Exchanger Cooling

In applications where cool water is plentiful, or where it is desirable to preheat a water supply for other processes, engines using heat-exchanger (Fig. 6-11) cooling are used. If conservation of raw water is important, raw water can be controlled thermostatically to limit its flow to the minimum required.

The raw-water pressures are not imposed on the engine jacket; engine cooling is not greatly affected by ambient temperatures; engine-room airflow requirements are much lower than unit-mounted radiator applications; and engine-jacket-water heat can be used constructively.

City-Water Cooling or Standpipe Cooling

Occasionally used with standby-service engine–generator systems, city-water cooling is simply the blending of cold raw water into the jacket water during operation through the use of water-pressure-regulating and thermostatic-control valves and diverting the excess to waste (Fig. 6-12). ***CAUTION****—Although inexpensive to install, this method introduces minerals and corrosive elements into the engine water jacket and is dependent on a municipal water supply which could become inoperative in time of disaster.*

Ebullition (Ebullient) Cooling

Engines and engine-generator systems, intended for applications where recovery of the heat in jacket water (for utility heating, air conditioning through absorption chillers, or processing needs) may be of genuine economic value, are sometimes equipped with ebullition cooling (Fig. 6-13).

Ebullient cooling is a process whereby jacket water is circulated through the engine at near-boiling temperature, vaporized at the top of the engine, and then condensed and recirculated. Cooling is accomplished by capturing the heat of vaporization for process or other uses.

By omitting the jacket-water pump and taking advantage of the natural circulation of water with temperature differences, the engine may be operated at jacket-water tem-

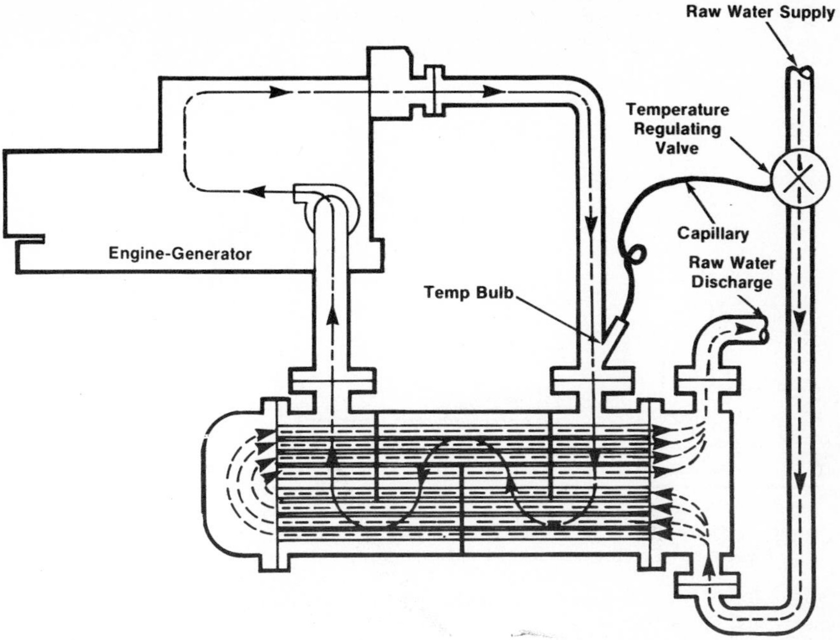

Figure 6-11 Heat-exchanger cooling. *(Waukesha Engine Division, Dresser Industries, Inc.)*

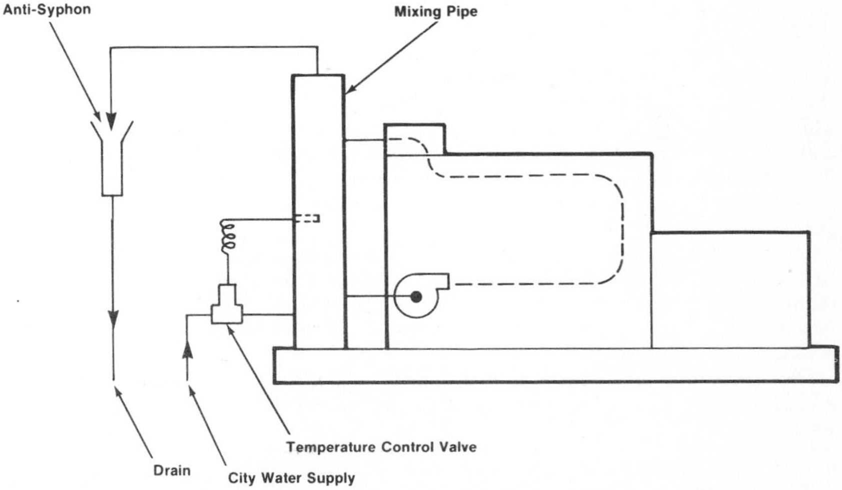

Figure 6-12 City-water cooling. *(Waukesha Engine Division, Dresser Industries, Inc.)*

peratures well into the boiling range (15 psig, 250°F max). A considerable amount of heat in the form of low-pressure steam may be recovered. When ebullition cooling is being considered, however, the engine manufacturer should be consulted since some engine designs are not suitable for this.

Two big advantages of ebullition cooling are the recovery of normally wasted heat

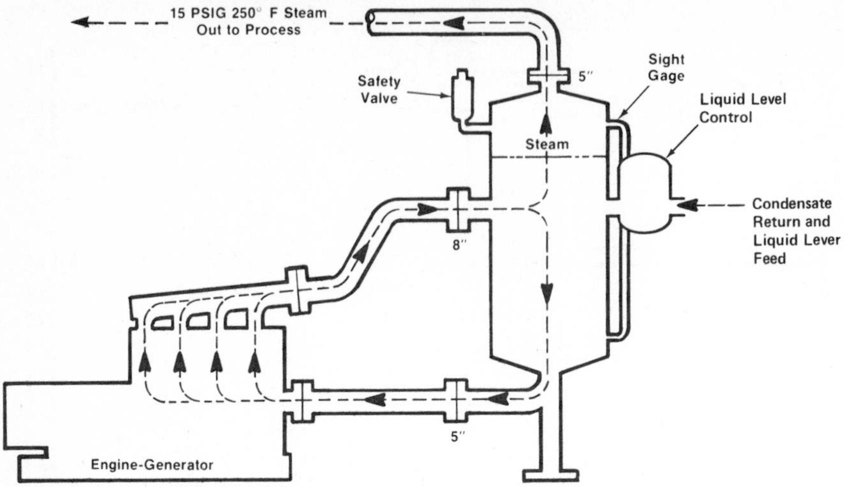

EBULLITION COOLING SYSTEM

Figure 6-13 Ebullition cooling. *(Waukesha Engine Division, Dresser Industries, Inc.)*

energy and optimum use of water. The comparative cost of the installation requires a certain minimum number of operating hours per year to make ebullition cooling economical.

Waste-Heat Recovery Systems

Where the waste heat of an engine can be used, electric power generation with waste-heat recovery has pronounced advantages. With a conventional engine generator, something on the order of 32 percent of the fuel energy is converted to useful electric power, while the remaining 68 percent is wasted to the atmosphere. With an engine jacket-water and exhaust-heat recovery system, the recovered heat energy is about 50 percent of the fuel burned, reducing waste heat by about 18 percent.

A typical heat balance on engines of identical size for both gas and diesel engine-generators is shown in Table 6-2.

Figure 6-14 shows a simple representative arrangement for engine heat recovery. Engine jacket water and exhaust are passed through a first heat exchanger where the water temperature is increased as the exhaust gas is cooled. It then passes through a second heat exchanger where the building or process water is heated and the jacket water is returned to the engine for cooling.

The amount of waste heat readily recoverable in a representative system of this kind is approximately 71 Btu/(bhp)(min) for a gas engine and 56 Btu/(bhp)(min) for a diesel engine.

Exhaust System

Every engine manufacturer specifies a maximum exhaust back pressure limit at the exhaust outlet of the engine (Fig. 6-15). Typical values can range from 12 to 20 inH_2O depending on the engine models. To exceed these values could result in engine damage or interfere with good long-life engine operation.

Since sound levels are becoming more critical daily, silencers are selected to give adequate noise attenuation. Unfortunately, this usually increases the exhaust flow pressure or drop through the silencer.

TABLE 6–2 Typical Heat Balance

	Conventional cooling system		Cooling system with engine jacket and exhaust heat recovery	
500-kW natural gas engine generator*				
Electric power	30%		30%	
Jacket water heat	38%	70% wasted	38%	54% recoverable
Exhaust heat	24%		Exh recoverable 16%	
			Exh lost 8%	16% wasted
Radiated heat lost to atmosphere	8%		8%	
	100%		100%	
500-kW diesel engine generator†				
Electric power	35%		35%	
Jacket water	32%	65% wasted	32%	48% recoverable
Exhaust heat	24%		Exh recoverable 16%	
			Exh lost 8%	17% wasted
Radiated heat lost to atmosphere	9%		9%	
	100%		100%	

*Based on 7900 Btu (bhp)(h) at rated load (LHV) and 95 percent generator efficiency.
†Based on 0.380 lb/(bhp)(h) at rated load [18,200 Btu/(lb)(LHV)] and 95 percent generator efficiency.

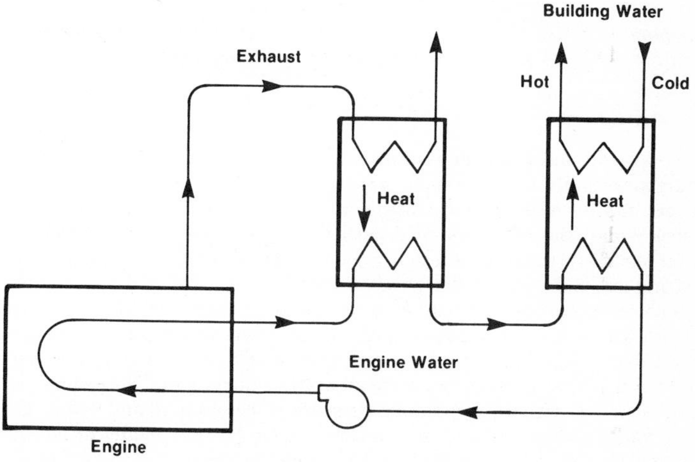

Figure 6-14 Representative heat-recovery arrangement. *(Waukesha Engine Division, Dresser Industries, Inc.)*

Silencers and exhaust pipes in engine rooms are usually insulated to reduce heat radiation and noise. A flexible exhaust connection is mounted at the engine to isolate engine vibration; also, the exhuast line is sloped away from the engine with a drain to avoid the accumulation of condensation.

Emissions

Exhaust products of either diesel or natural gas engines contain oxides of nitrogen (NO and NO_2), carbon monoxide (CO), hydrocarbons (HC), and sulfur oxides (SO_2 and SO_3) if the fuel contains sulfur.

Emissions of these products are subject to regulation under federal, state, and local

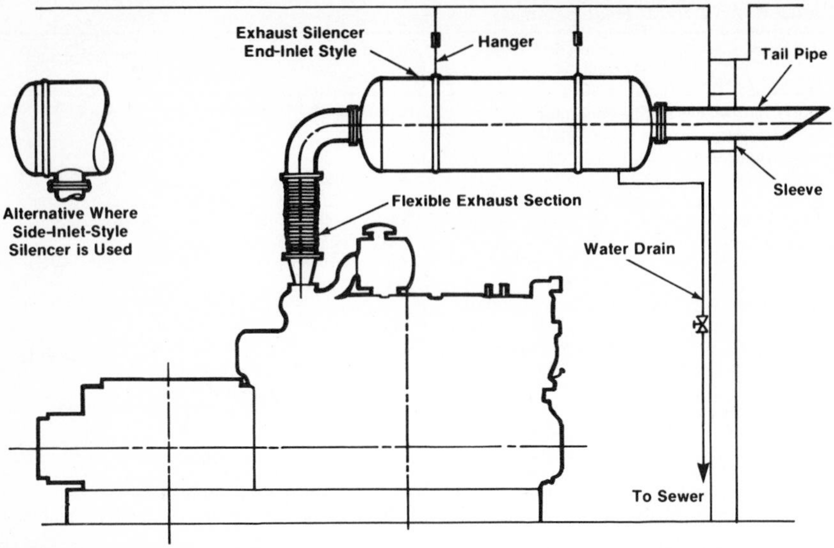

Figure 6-15 Horizontal exhaust schematic. *(Waukesha Engine Division, Dresser Industries, Inc.)*

laws. The plant engineer should be aware of the regulations covering a specific site and, if necessary, consult the engine manufacturer for emission information.

Starting Systems

The most frequently used starter systems are electric motors and air starters. The electric starters are usually 12- or 24-V dc cranking motors. The higher voltage is preferred on larger engines for annunciator controls and/or power circuit-breaker tripping. Adequately sized and maintained lead-acid batteries are the most practical and economical solution for the battery source today. However, nickel-cadmium and other types of battery requiring less maintenance are often specified.

All the batteries require an adequately sized battery charger to keep the battery fully charged both with the engine operating and when down. In low ambient temperatures, both batteries and engines must be kept warm to ensure good starting and operation.

Many starting systems use one or more air starters per engine. The common air motor will crank an engine with an applied air pressure of between 90 and 150 psig. Starters can be controlled through a hand or a solenoid valve both manually or automatically. They also have an inline oiling system to ensure motor life.

The air starting system usually includes a remote high-pressure air receiver (approximately 250 psig) with an air regulator to drop pressure to the cranking motor. Again the pipe losses and line sizes have to be considered to ensure good cranking. The compressors to pressurize the receivers are normally driven by electric motors and are automatically started and stopped by a pressure-switch sensor at the air receiver.

Engine-Speed Governing System

The engine governor is the most important device with respect to engine performance. On smaller units, engine-mounted mechanical governors provide adequate speed and frequency control with approximately 3 to 10 percent steady-state speed regulation from no load to full load.

Hydraulic isochronous governors are also available for medium to large engines, and these provide excellent speed control with adjustable speed regulation of 0 to 5 percent.

In engine-generator applications, this permits running isochronously as a single unit or with adjustable regulation for multiunit manual parallel operation and load sharing.

More precise governing is available on both small and large engines with electric governors. This type depends on engine speed with a magnetic pulse pickup on the flywheel ring gear teeth, and provides speed and load control with an electric or electrohydraulic actuator connected to the engine throttle or fuel injection pump.

The electronic control of the govenor is solid-state and is usually mounted in the control panel or generator switch gear.

Electric governing permits operation of multiengine generator units in parallel isochronously, that is, at constant frequency under steady-state operation regardless of load, with each engine taking an equal share of the overall load.

Additionally, the electric governor is extremely fast with respect to transient load response and will handle up to 50 percent sudden load changes with a speed deviation of approximately 3 percent and recovery within approximately 2s for diesel engines, and with a speed deviation of approximately 4 to 5 percent and recovery within approximately 3s for carburetted gas engines.

Where automatic operation includes coming on-line in parallel with other engine generators or the utility, the electric governor with isochronous load-sharing control is desirable. In conjunction with the proper switchboard relaying, units can be added or removed on a load-sharing basis or on load control basis as the situation may demand. A good 24-V dc battery system is required as a power supply to the electric governor.

INSTALLATION AND MAINTENANCE OF DIESEL AND NATURAL GAS ENGINES

The engine installation should be designed with maintenance requirements in mind. Serviceable components such as filters, fittings, and connections should be readily accessible to the engine operator. Routine engine maintenance will not be neglected if the operator has easy access to the engine.

Sufficient service space must be present on all sides of the engine to allow for removal of even the largest engine components. An overhead crane should be incorporated into the engine-room design to assist the mechanic-operator in removing heavy assemblies.

Ventilation

In engine-generator installations sufficient airflow must be provided into the engine room for ventilation and combustion air. It is also good practice to calculate the amount of heat transferred to the room air (i.e., engine and generator radiator heat, plus any other heat sources) to determine the temperature rise of the engine-room air. In many cases it is necessary to increase the engine-room airflow to maintain reasonable operating temperatures.

The following are *general rule of thumb values* that assume the only radiating heat source in the engine room is the engine-generator set. For greater accuracy, an independent engineering study should be made covering the following points:

Cubic feet per minute of air required to limit air room temperature rise to 18°F, over normal ambient = 45 × kilowatt rating

Cubic feet per minute of combustion air required = 3.5 × kilowatt rating for diesel engines

Cubic feet per minute of combustion air required = 2.4 × kilowatt rating for gas engines

The total air requirement equals the sum of the cubic feet per minute of combustion air plus the cubic feet per minute required to limit the room temperature rise.

Other ventilation considerations are filters for sandy or dusty areas and louvered openings at both inlet and outlet air openings. The louvers can be motor-operated and temperature-controlled.

Cooling System

Potential problems with the engine cooling system can be avoided if the following considerations are incorporated into the design and installation of the cooling system.

Excessive fittings, elbows, and connectors in the system piping will impede coolant flow. Use of fittings should be kept to a minimum.

An expansion-tank balance line should be incorporated into the cooling system, running to the suction side of the water pump. This balance line will maintain a net positive suction at the inlet of the pump and reduce the possibility of air locks and cavitational erosion.

All filters, fill points, and bleed cocks should be installed in an easy-to-reach location.

Place the radiator away from a wall or any other obstruction that causes air recirculation or restricts airflow. These obstructions would also include any dirt source, vehicle travel path, air-conditioning units, or exhaust stacks and chimneys. Remember that the radiator must be in a location where it can be cleaned and serviced.

In installations where gaseous or LP fuels are used, keep all floor drains and service trenches out of the engine enclosure. LP and some constituents of natural gas can be heavier than air and will quickly flow into such low spots, creating a fire hazard.

Exhaust System

Plan the exhaust system so that the gases are expelled to a safe outside area, consistent with all local building and environmental codes. Do not discharge gases near windows, ventilation shafts, or air inlets. The exhaust outlet must be designed to keep out water, dust, and dirt.

To avoid metal stress and turbocharger damage, support the exhaust system independently, keeping the weight of the piping off the engine. Roller-type supports and flexible exhaust connections should be used to absorb thermal expansion. (If overhead cranes and hoists are used in the engine room, the exhaust-system piping may have to be supported from below.)

A condensate trap and drain should be designed into the exhaust system. The drain should be in an easy-to-reach location.

If the exhaust systems of more than one engine are to be connected to a common exhaust, the engine manufacturer should be consulted beforehand. Exhaust-system back flow (common in such connections) could result in an engine that is not running.

Exhaust-system back pressure should be checked periodically. The back pressure must fall within the limits established by the engine manufacturer.

Air Induction

As with other engine systems, accessibility is the key to air-induction system maintenance. The filter element should be positioned so that it can be easily removed and replaced. The filter should always be positioned at the entrance to the air induction system; when combustion air is ducted in from outside the engine room, the filter should be at the opening to the piping. All systems should be equipped with a restriction indicator to show excess pressure drop due to filter-element plugging.

Always locate the air inlet away from concentrations of dirt, exhaust stacks, fuel tanks, tank vents, and stockpiles of chemicals and industrial wastes. Try to duct air to the engine from a cool, dry, dirt-free area. The ambient temperature at the air inlet location should ideally be 60 to 90°F (15 to 32°C).

Run all air ducts away from engine exhaust pipes, heating lines, or other hot areas. Remember to allow clearance for overhead lifts and cranes when air ducts run through the engine room.

Air ducts should be thoroughly sealed to avoid drawing dirty air in behind the filter. The ducting must be checked periodically for leaks.

Air-system ducting should be seamless or welded-seam piping. Flanged fittings with gaskets, not threaded connections, should be used between pipe sections to avoid restrictions in the system. The best ducting system is as short and straight as possible, using long-radius bends and low-restriction fittings. Never allow air-duct restriction to exceed 2 in (50.8 mm) of water column.

Engine Alignment

The alignment of the engine mount and the alignment between the engine and the driven equipment is critical to long engine life. Alignment should be checked periodically according to the manufacturer's recommendations.

SELECTION OF ENGINES

There are several factors to consider in the selection of an engine.

Type of Fuel

The type of fuel chosen will depend to a large degree on availability, price, local building codes, and pollution restrictions. For instance, in some communities, natural gas is not available for industrial use, while in others, local building codes will prohibit storage of diesel fuel. The price of one fuel may preclude its use in comparison with another type of fuel. Moreover, emissions restrictions may have a bearing on what type of fuel may be used. All of these are factors in choosing an engine.

Horsepower Load and Speed

The load and speed of the equipment to which the engine will be coupled are key considerations in engine selection. The aim is to match the prime mover to the power required at a speed which will be compatible with the equipment to be driven, while maintaining optimum efficiency.

Duty Cycle

The duty cycle should be examined to determine whether continuous or intermittent operation of the equipment is required, because this affects engine selection.

Brake Mean Effective Pressure (BMEP)

This is important because the higher the figure, the greater the chance for higher stress on working parts of the engine. This could mean higher maintenance costs and earlier need for rebuilding.

Fuel and Oil Consumption

With the cost of fuel rising ever higher, this is becoming an increasingly significant factor in the selection of an engine.

Torsional Compatibility

A torsional analysis should be conducted to determine whether the engine and driven equipment are compatible with respect to operating stress in the shaft system.

SELECTION OF ENGINE-GENERATORS

Both diesel and gas engines perform very satisfactorily for in-plant power generation when properly applied and installed. The varied nature of these applications depends on factors such as:

1. The specific characteristics of the plant's product and its production process as well as its sensitivity to power failure
2. The plant power needs in terms of availability of purchased power and the reliability of normal power sources
3. Economic considerations in terms of location, cost of purchased power, demand charges, and equipment
4. Economic considerations of in-plant generated power in terms of capital outlay, operation, maintenance, and fuel cost

Once the decision has been made that in-plant engine-driven power generation will be used to provide electric power, whether on the basis of prime, cogeneration, peaking service, emergency standby, or a combination of any of these, there are a number of considerations to be investigated.

Purpose

The purpose of the installation should be stated and careful thought given to definition of the electrical load requirements.

- Will the power generators supply all of the plant load or only a part of the total?
- What is the power factor characteristic of the load? (Engine-generators are normally rated at 0.8 power factor lagging.)
- Will there be any attempt to control or improve power factor?
- What is the largest block kilowatt loading anticipated and what large motors are to be started and on what basis—across the line, reduced voltage?
- Are computer loads, SCRs, inverters, x-ray, or heavy welding equipment involved?

Amount of Equipment

Selection of the size and number of units to handle the load demands thoughtful consideration.

Emergency standby protection (see also Chap. 3-3.) may be provided by a single unit rated to protect a known segment of load that can be isolated with a two-way transfer (normal emergency) switch (Fig. 6-16). The size of this type of unit may range from 50 kW or under to approximately 1500 kW (Fig. 6-17).

For loads well over the 1000-kW range, it is customary to provide multiple engine-generator units, sometimes as many as eight units operating in parallel on a common or split bus for total load capability to 10,000 kW and beyond.

Generally available emergency standby units are rated at 1800 r/min synchronous speed for 60-Hz service up to approximately 1000 kW and at 1200 r/min up to 2000 kW, while prime power units are available in both 1200 and 900 r/min synchronous speed for 60-Hz service up to 3000 kW rating per unit.

The same ratings are available for 50-Hz service at synchronous speeds of 1500, 1000, and 750 r/min. Much larger units and units with slower speeds are also available.

When selecting size and number of units, it is most helpful to study the plant electrical load profile over the course of a year. If this is not available, a profile based on connected loads with anticipated load and diversity factors can be developed. Single-unit emergency standby generators are often applied with load factors of only 50 to 75 percent of rating. Multiunit prime power installations favor load factors in the 75 to 90 percent range, as this selection results in optimum fuel economy and overall operating efficiencies. Future growth of plant electrical loads must be anticipated.

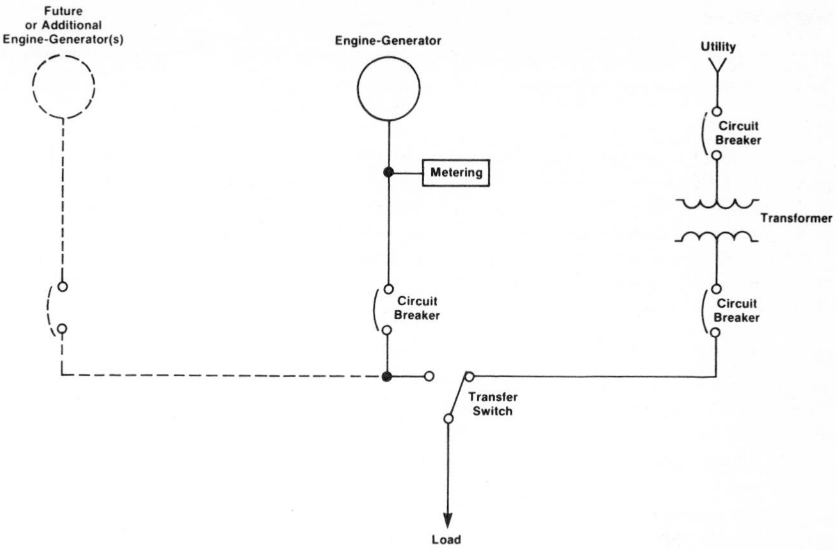

Figure 6-16 Normal emergency standby without utility parallel. *(Waukesha Engine Division, Dresser Industries, Inc.)*

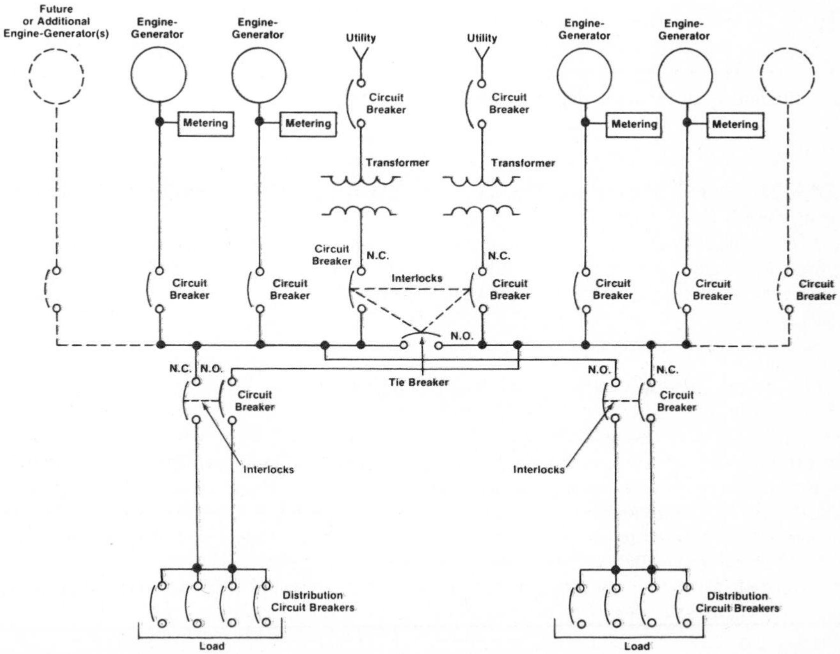

Figure 6-17 Normal emergency standby with peaking capability or selection of critical loads for limited kilowatt generation (without utility parallel). *(Waukesha Engine Division, Dresser Industries, Inc.)*

Fuel Selection

As with engines in direct-drive applications, this may be determined by availability and cost.

Natural Gas

This is an excellent engine fuel and is often available on an uninterrupted basis. Where necessary, LPG or propane can be stored as backup or secondary fuel supply.

Sewage treatment plants are increasingly turning to power generation, utilizing the process-waste gas from digesters to fuel the engine. Coal gasification, wood-chip processing, and numerous other process-waste gases may be suitable for burning in a gas engine but should never be used without consulting the engine manufacturer for specific approval.

No. 2D Diesel Fuel Oil

This is commonly used in small and medium-size engines. Large diesel engines have the capability of burning heavier fuels such as No. 4D and heavier distillates; these may require special handling, heating, centrifuging, and filtering. Diesel fuel has the advantage of large-volume onsite storage capability.

Location of Equipment

Locating the engine-generator(s) in the plant may involve many practical considerations. Often it is advantageous to locate them near other heavy plant equipment such as boilers, large air conditioners, or compressors. Standby power generation equipment is normally automatically controlled and, by locating it close to other equipment requiring periodic operator attention, it may also get the attention it deserves.

Figure 6-18 shows typical envelope dimensions for both gas and diesel generator sets. Minimum clearance on all sides is also charted and can be used for preliminary space estimates. For final determinations, always work from the equipment manufacturer's specific space requirements for optimum clearances both around the unit and overhead for efficient maintenance, major overhaul, and repair.

Cooling

Adequate cooling of both gas and diesel engines is essential to good performance and equipment life.

On emergency standby units, it is common to waste the recoverable heat. On prime power applications, there is increasing attention given to heat recovery from the engine jacket water and exhaust for process use in the plant.

The Electric Control System

This system, together with its attendant complexities, requires determination of the power-generation voltage and control voltages.

Decisions as to the mode of operation, such as manual, semiautomatic, atuomatic, or unattended operation, must be made. Adequate metering, monitoring, readout, and display decisions must be made. Independent engine generation on isolated plant loads or multiunit engine generation in parallel operation against plant loads must be determined. Will the engine generators always operate isolated from the utility or will there be occasions to parallel with the utility's power?

Both peaking service and cogeneration may call for parallel operation with the utility, and this requires review with the power company at the planning stage.

Electric Controls

After the kilowatt size of unit(s) is selected, the generated output voltage should be considered.

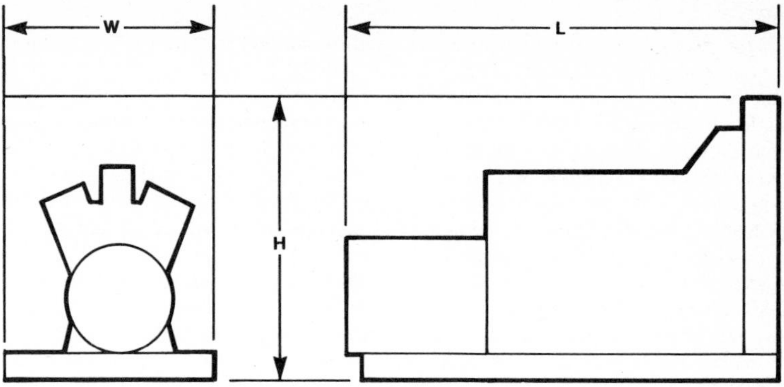

(Length, Width, and Height Are In Inches)

kW	RPM	L	W	H	WT, LB
50	1800	88	30	50	2,300
100	1800	110	36	55	4,200
150	1800	114	36	60	4,500
200	1800	114	42	60	6,000
250	1800	114	42	60	7,000
300	1800	140	56	90	13,000
400	1800	160	69	90	16,000
500	1800	180	72	90	19,000
600	1200	210	74	121	32,000
700	1200	216	84	124	32,500
800	1200	222	84	124	33,000
900	1200	222	96	124	35,000
1000	1200	222	96	127	36,000
*1200	1200	186	80	98	38,000
*1500	1200	262	78	113	42,000
*2000	900	330	72	139	45,000
*2500	900	363	72	139	52,000
*3000	900	363	72	139	52,000

* - Indicates less cooling radiator

Figure 6-18 Approximate space envelope for gas and diesel engine generators. *(Waukesha Engine Division, Dresser Industries, Inc.)*

1. Voltages of 208/120 wye or 240/139 wye can be economically utilized in the lower kilowatt ranges since no transformers should be required in plants (facilities) using only these voltages.
2. Voltage of 480/277 wye is the most commonly generated voltage with minimum capital cost in both generator and switchgear.
3. Voltage of 4160/2400 wye is the common generated voltage for larger plants and more extensive power distribution systems. This voltage level increases the cost of the engine-generator unit and its associated generator switchgear, but may effect savings in the plant distribution system.

A further consideration is the engine-generator and panel control voltage. Generally, this is supplied from a dc battery source which allows control and operation of the engine and breakers without ac voltage from either utility or generators. Engine-generator sets in the 150- to 1500-kW range require a 24-V dc battery source to start and run. An alternate is 100 to 150 psig air pressure to start and 24, 48 or 125 V dc to run and control. For large, multiunit engine-generator sets and their associated switchgear, 48 or 125 V dc is recommended for the most reliable circuit-breaker operations.

Metering

When 24-V dc engine starting batteries are used for control purposes, the control should be run from the set of batteries with the highest voltage to minimize the effects of voltage dip when starting engines or tripping circuit breakers.

Single-unit metering would include at least a frequency meter, voltmeter, ammeter and hour meter; for emergency standby service, 3½-in ±2 percent accuracy panel meters are both generally acceptable and adequate. Prime power service with long-term continuous operation, often with multiple units in parallel, demands 4½-in switchboard meters of 1 percent accuracy. These sets should also include an indicating wattmeter for each engine generator. Additional meters that may be considered are: those reading kilowatthours, kilovoltamperes reactive (kvar), and power factor. Of increasing interest on prime power multiunit applications are recording meters on the output bus for voltage, frequency, and kilowatts and/or kilowatthours. More recently available are digital readout meters providing greater accuracy, but with limitations with respect to indication of transients, which is an important characteristic of engine-generator sets.

Protection

The electrical protection of the generator system can vary from simple molded-case circuit breakers to insulated-case breakers and to metal-frame air circuit breakers. Both insulated-case and air breakers are available with either fixed mounting or drawout provisions. For optimum flexibility in long-term use, the drawout breaker is superior and recommended. Since engine generators have a limited short-circuit capability, care should be taken in breaker selection for both generator and distribution to achieve proper selective trip coordination. Where selective trip coordination is required, the generator must be supplied with three per unit short-circuit sustain capability for 10 s.

Additional protection to consider is generator differential and ground fault. Usually, generator-differential protection is used on 4160-V systems where generator internal damage requires early detection to protect capital investment.

Ground-fault protection also protects equipment and personnel and is less expensive to install. Most circuit-breaker manufacturers provide optional ground-fault tripping integral with the breaker. Many different types of ground-fault protection are available, and consideration must be given to select and specify protection consistent with the existing plant grounding system and specific application. More sophisticated methods of fault detection are available, but are usually associated with megawatt generator sizes since protection cost is very high.

When engine generators are operated in parallel with the utility, certain cautions must be observed. The utility, being an "infinite" source, can cause severe damage to the engine generator should the utility's protection system open and then reclose (out of phase). Conversely, should the upstream utility protection system open and then not reclose, the generators can feed a fault from the reverse direction, which can be hazardous to both personnel and equipment. Whenever parallel operation with the utility is being considered, review the relaying used to protect against the occurrences mentioned with the local utility company.

Many breaker and/or transfer switch configurations are possible, each with its own advantages and disadvantages dependent upon application (Figs. 6-19 to 6-21).

Engine Control

After the power distribution format is known, the engine-control (starting and stopping) mode(s) must be considered. For single emergency standby units the unit should start, come up to rated frequency and voltage, and provide power whenever the utility source fails. Normal engine protection would include shutdown on engine low oil pressure, high water temperature, overspeed, and failure to start. Additional features available and actually included for hospital duty are warnings before shutdown on low oil pressure and high water temperature. On prime power units, additional considerations are warning and/or shutdown on high oil temperature, low oil level, high vibration, and engine overload.

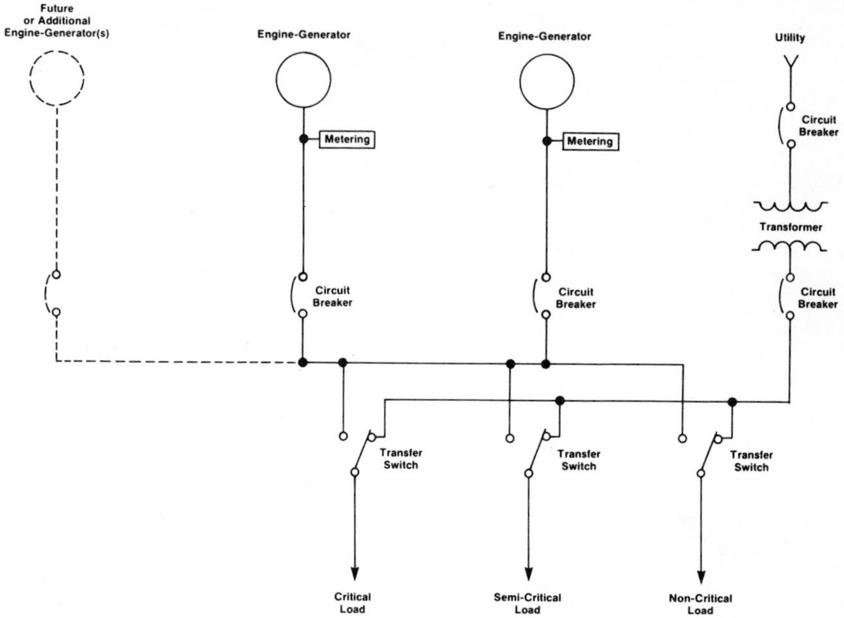

Figure 6-19 Multiunit standby or cogeneration in parallel with utility or in independent operation. *(Waukesha Engine Division, Dresser Industries, Inc.)*

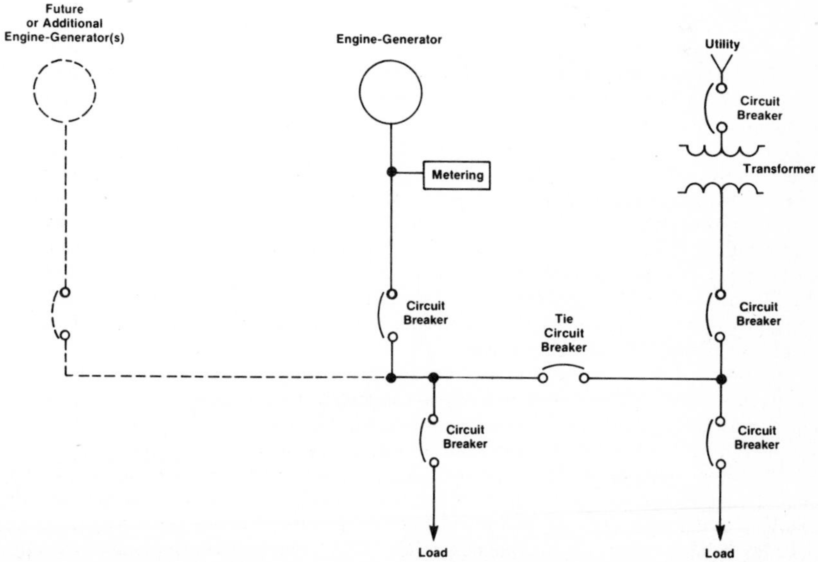

Figure 6-20 Isolated loading for peak shaving without parallel utility (breakers interlocked) or peak shaving and parallel operation with utility. *(Waukesha Engine Division, Dresser Industries, Inc.)*

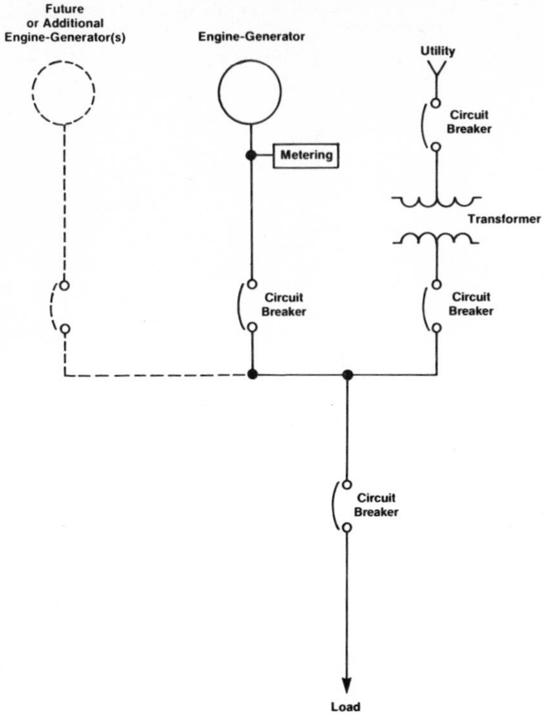

Figure 6-21 Isolated loading for standby (breakers interlocked) or peak shaving and cogeneration with parallel operation with utility. *(Waukesha Engine Division, Dresser Industries, Inc.)*

Some extremely critical applications may call for the generator set not to stop for any reason, as nuisance shutdown cannot be tolerated and the possibility of generator set failure or destruction may be less costly than the process it controls.

Engine control is only a part of the overall system control. System control is offered in varying degrees of complexity. Single standby units have very simple system controls, while prime power multiunit applications must determine the quantity of untis required to satisfy power demand while also controlling plant loads to prevent high peak loading or overloading. When operating in parallel with the utility, controls can be programmed to accept: (1) fixed utility supply with load variations picked up by the engine-generator sets; (2) variable utility supply with peak demands picked up by the generator sets; (3) fixed generator set output with load variations being picked up by the utility (this mode includes capability of supplying power to the utility over low plant load conditions).

Generally speaking, the control logic can do almost anything the engineer may require in a given situation; however, simplicity is the key to both cost and long-term performance. Simpler systems are easier to maintain, whereas complex systems may require a wider variety of engineering skills to maintain and troubleshoot problems.

The system control can start and stop generator sets on the basis of actual kilowatt demand; an alternative simple approach would be to know the load profile and manually operate or utilize real-time devices to start and stop units.

Today solid-state programmable controllers, which can be used in place of the more conventional relay logic in control circuits, are available. Some of these controllers are versatile and allow control logic to be changed simply. Many allow computer input-output links for record-keeping purposes.

These versatile programmable controllers provide flexibility and a ready means of tailoring sequence, time, and load functions to a particular plant's needs. Selected performance indicators can be linked to provide readouts and printouts as may be required for efficient operation and recording purposes.

REFERENCES

1. *Standard Practices for Low and Medium Speed Stationary Diesel and Gas Engines,* Diesel Engine Manufacturers Association, Cleveland, Ohio, 1972.
2. Gunther, F. J.: "Engines," in *Pump Handbook,* Igor J., Karassik, William C. Krutzsch, Warren H. Fraser, and Joseph P. Messina (eds.), sec. 6.1.3, McGraw-Hill, New York, 1976.

chapter 4-7

Smokestacks and Chimneys

by
Richard T. Lohr
President, International Chimney Corp.
Buffalo, New York

INTRODUCTION

The word *chimney* generally refers to masonry and concrete chimneys and *stack* refers to steel or fiber-reinforced polyester (FRP) construction. However, many chimneys or stacks include dual material for inside and outside walls. This chapter calls all types "chimneys."

The purpose of a chimney is to emit a waste gas at a desired elevation. There are many emission sources of waste gases: boilers, high-temperature processing, offensive fumes, processes, etc. Gas travels from the emission source through a breeching to the chimney.

A chimney is a major part of the plant facility. It is a structure that also functions as a piece of equipment, like any other essential machinery for process or mechanical use.

Readers of this section most likely are choosing a new chimney, suspect problems with an existing chimney, or are considering a retrofit (new fuel, boiler conversion, elimination of incineration, etc.). In every case the reader should start with basic design considerations. For a retrofit, it would be wise to determine first what chimney design would best perform the job and then determine adaptability of the existing chimney.

NEW CONSTRUCTION

Design considerations are size, material, initial cost, and lifetime cost.

Size

The first design consideration is size of the *bore* (the smallest inside diameter, which is usually at the top of the chimney); next is *height*.

Bore capacity depends on the type and volume of gas to be emitted and how the gas is fed to the chimney (i.e., forced, induced, or natural draft). The right bore (internal diameter) helps maximize efficiency of the system, and minimize corrosion of the chimney interior. Size of bore affects flow of gas and can help keep temperatures hot enough (or cool enough, depending on the gas) to minimize corrosion.

Flow to the chimney, as well as size of bore, determines how fast gas moves through the chimney. Flow is *forced-draft, induced-draft,* or *natural-draft.* Forced-draft flow means that a fan at the emission source pushes gas through ductwork and breeching; induced-draft flow means that a fan in the ductwork, chimney, or breeching pulls gas from the emission source and pushes it through the chimney; natural-draft flow means that the gravitational action of hot gas rising up the chimney pulls other gas from the emission source through the breeching. (In forced- or induced-draft flow, specifications from the fan manufacturer and/or boiler manufacturer will help determine stack size. A recognized chimney designer and/or contractor can advise how much gas of a given temperature moves per minute in a chimney of a given size and location by natural draft.)

Chimney height is the second design consideration. The taller the chimney, the more widely the gas is disbursed and the better the natural draft characteristics. Local air-pollution codes, nature of the gas, surrounding structures, topography, and prevailing climatic conditions determine the desired height. In a natural-draft system the height must also be that which creates the desired draft.

A chimney designer or contractor actively involved in this work is the best source for determining the optimum bore and height, after considering the temperature and corrosive nature of gas from the emission source.

Materials

A chimney must have structural integrity against windloading, climate, earthquake, fire, etc. Radial brick, reinforced concrete, steel (carbon and certain alloys), and more recently FRP are commonly used materials, singly or in combination. The chimney must also resist any destructive heat and chemical effects of the gas it handles; an interior lining is often mandatory to provide reasonable service life.

If moderate heat is the only destructive effect of the gas, then brick, concrete, or steel are suitable in single-wall, single-material configuration. Size might determine which material should be used. For an unguyed self-supported chimney beginning at grade elevation on a foundation in a good soil bearing area, steel is probably most economical for a chimney under 10 ft (3.1 m) in diameter up to 150 ft (45.8 m) high where steel sections are shippable as full cylinders. For larger diameters or heights over 200 ft (61 m), radial brick or reinforced concrete are usually more economical. Between 100 ft (31 m) and 375 ft (114 m), radial brick is usually most reasonable. Over 375 ft (114 m), reinforced concrete may be more reasonable compared with other materials. If the chimney must be building-supported or if soil conditions are poor, steel (as the lightest material) is more economical in a wider range than mentioned above. For maintenance and repair costs, radial brick is generally superior, concrete next, carbon steel poorest.

For a corrosive gas of constantly low temperature with no chance of a boiler or process "runaway," lower-cost FRP is feasible. FRP also offers resistance against the destructive effects of acid-laden gases. FRP is very light, however, and there are structural limitations connected with its use. FRP chimneys must be designed carefully for static and dynamic stability. FRP can also serve as a lining material in a stack constructed of other materials.

In most new installations, a protective lining for the chimney interior is necessary. The most common destructive effects are extreme heat, wide temperature fluctuations, and corrosive gas. Pneumatically applied refractory cement, acid-resistant cement, abrasion-resistant cement, or a combination can be applied onto the structural material as a lining. Some protective linings are "hung" or are partly attached by support shelves to the external structure, for example, corbelled brick linings, hung steel linings, brick linings on shelf angles, and FRP on corbels. Some linings are independent inner columns of steel, concrete, radial brick, or solid brick (firebrick for heat resistance, or ASTM Type H, M, or L double-solid brick for corrosion resistance).

A relatively new design is an independent insulated steel interior to a steel chimney. The assumption is that—for marginal operating temperatures—the insulating value of the dual walls will keep gases above dew points. This mandates that flue gas temperature be kept hot enough at all times; otherwise the interior steel will be corroded by acid attack. Also in dual-wall steel construction it is conceivable that total failure of the structure can result before anyone looking at the chimney exterior notices that corrosive gases have leaked through the inner wall to attack the outer wall. (On the other hand, a failure in a pneumatically applied technical cement lining will produce a localized failure in the outer steel wall that will be noticed and repaired before structural integrity is jeopardized.) *This stack type must be checked for dynamic stability.*

Another modern trend in linings is a liner designed to be disposable, or easily replaceable. Sometimes this is a design compromise that is the only answer to a difficult set of operating conditions. In fact, there are always trade-offs in selection of structural material, decisions to use linings, lining materials, and costs. The lined chimney will always cost more than the same structure without lining, but will usually extend service life. Sometimes the lining can enhance structural strength, so the structure can be constructed more economically anticipating the added lining strength. But when initial cost is reduced this way, repairs are usually more expensive.

Chimney designers and contractors are a good source of the best designs for your needs. However, remember each may specialize in one type of chimney. Table 7-1 suggests, very roughly, the options for various operating conditions. It is to be considered only a general guide, subject to modification for specific situation; it is not to be used as an unconditional recommendation by the author.

TABLE 7-1

	Operating conditions					Cost	
Type of construction	Extremely high temp.*	Moderately high temp.†	Low to high temp.	Constantly low temp.‡	Corrosion resistance	Initial cost	Maintenance and repair
Radial brick	—	OK	—	OK	—	Med	Best
Radial brick lined	OK	OK	OK	OK	OK	High	Best
Reinforced concrete	—	OK	—	OK	—	Med	Good
Reinforced concrete lined	OK	OK	OK	OK	OK	High	Good
Single-wall steel	—	OK	—	OK	—	Low	Fair
Single steel lined	OK	OK	OK	OK	OK	Med	Fair
Dual-wall steel	—	—	—	OK	OK for some uses	High	Fair
FRP	—	—	—	OK	OK	Med	Good

*Typically 800°F (430°C).
†Typically 250–800°F (120–430°C).
‡Typically less than 250°F (120°C).

Cost Trade-Offs

A chimney designer and contractor is also in the best position to recommend trade-offs between cost and optimum chimney size. Cost is a factor in chimney size, since tall chimneys cost more per foot than shorter chimneys of the same construction material (the bottom 100 ft of a 200 ft chimney must be made stronger since it has to support both itself *and* another 100-ft chimney on top of it.) Chimneys with small bores cost absolutely less, but proportionately more, than those with large bores. To double the surface area at the top of the chimney to increase capacity requires approximately 41 percent more materials (bore capacity increases as $A = \pi r^2$, materials increase as $C = \pi d$).

Other recommendations might change how the gas enters the breeching (with stronger fans, different flue temperatures, etc.) to allow less expensive construction to perform well.

EXISTING INSTALLATIONS

Causes of Deterioration

Chimneys deteriorate with age due to weathering as well as through normal usage under original design considerations. Also, operating conditions, and hence the nature of the emitted gas, may change over the years. Many common changes produce gases that are

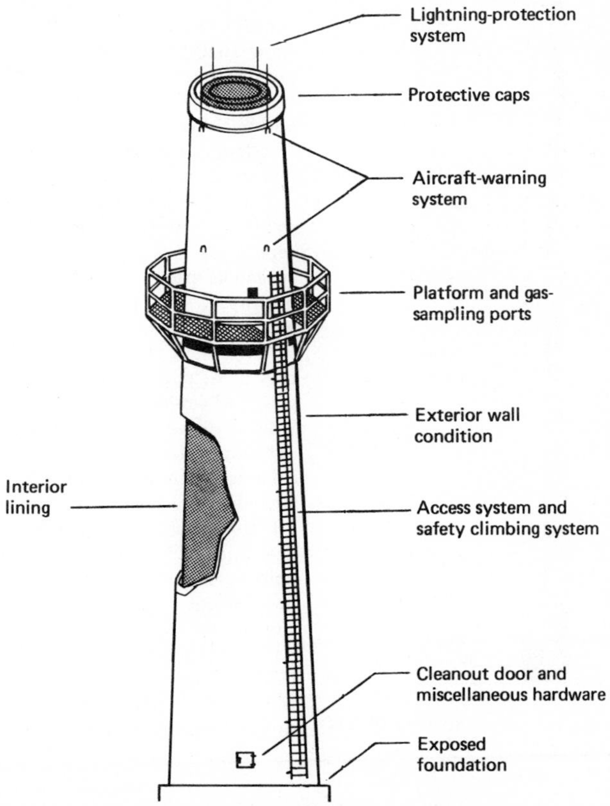

Figure 7-1 Inspection and maintenance areas to consider.

more corrosive, or lower the dew point of already corrosive gases (causing condensation on the interior wall that the chimney's original interior cannot handle).

Examples of changes that invite corrosion are use of new fuels, boiler conversion, elimination of incineration, scrubber installation, cyclone installation, use of economizers and other energy-saving devices, use of other pollution-reducing devices, and capacity adjustments. A chimney design firm can analyze an existing structure in terms of current operating conditions and easily advise if there is a potential corrosion problem.

Inspection and Repair

Chimneys must be inspected and maintained so that minor repairs will not escalate into major repairs (Fig. 7-1). Most reputable chimney contractors provide a visual inspection of the exterior, on a regular schedule or on demand, for a nominal fee. Chimneys of all materials could have problems such as deteriorated or corroded hardware (concrete and cast iron caps, pollution test platforms, access ladders to testing platforms, lightning protection systems, warning lights, etc.). Radial brick exteriors may show weathering of mortar joints, spalling of brick, and stress cracking—either horizontal or vertical. Concrete exteriors may show spalling, cracks, delamination, and porous pour seams. Steel exteriors may show corrosion, fatigue, splits, hot spots, and buckling.

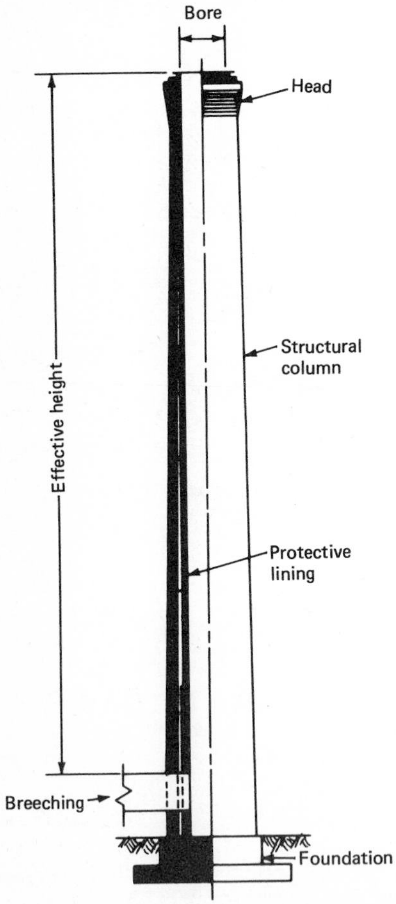

Figure 7-2 View of elevation (half external, half sectional).

Figure 7-2 shows an elevated view (external and sectional) of a typical chimney. Repairs to chimney exteriors include welding, tuck-pointing, cutting out and repointing cracks, banding to prevent spread of cracks, repair or replacement of hardware, corsetting to spread load in weak areas, etc.

Chimney engineers and contractors may do a "climbing" inspection, using rigging and sophisticated testing equipment. This involves taking a closer look at the exterior, but its major value is that it includes an inspection of the interior. Routine climbing inspections are advisable on a 1- to 5-year cycle, depending on chimney age and corrosion potential. (Concrete chimneys especially are subject to deterioration not spottable by a visual inspection of the exterior.) Preventive maintenance and timely repair of small problems in the chimney interior will minimize repair costs. Perhaps most important, with proper inspections, anticipated repairs can be on a scheduled—not emergency—basis.

Routine climbing inspections are scheduled when the chimney can best be taken out of service. Operating chimneys can be inspected with heat-resistant photographic equipment if necessary.

The climbing inspection report should:

- Identify deterioration and cause
- Recommend extent of any repairs needed and procedures
- Estimate cost of repairs
- Contain a development drawing and photographs (so any deterioration can be plotted, and repairs scheduled and budgeted)
- Discuss (if pertinent) relative cost benefits of *retrofitting* (major changes to match current or anticipated operating conditions) vs. making temporary repairs that will have to be repeated

Repairs to interiors include the installation of pneumatically applied protective cements, lining replacement, and lining resurfacing. *Retrofits* are major repairs or rebuilding of the existing chimney so it can handle new operating conditions. A retrofit chimney is usually far superior in performance to a replacement installation of minimum quality and will cost less over the lifetime of the structure. Very often the retrofit will be lower in initial cost, too, and will avoid demolition and new foundation expense as well.

An existing chimney should be taken out of service permanently only when (1) repair and/or retrofitting would be prohibitively expensive or (2) the repaired chimney still would be only marginally able to handle existing or anticipated conditions.

Climbing inspections, repairs, retrofitting, and demolition are all specialized work, best reserved for chimney contractors. Technical information is best obtained from chimney design and construction firms, which are listed in industrial directories.

To help ensure impartial recommendations, it is best to deal with a firm that has licensed professional engineers for design and a good record of reliable chimney construction. It is also wise to use a firm that deals with all types of chimneys rather than with one specialized product.

section 5

Heating, Ventilating, and Air Conditioning

chapter 5-1

Heating and Ventilating

by
Nils R. Grimm, P.E.*
Syska & Hennessy, Inc., Engineers
New York, New York

*The material presented in this chapter is based on the author's notes, technical experience, and data from: The American Society of Heating, Refrigerating and Air Conditioning Engineers (ASHRAE): *Industrial Ventilation;*[1] the American Conference of Industrial Ventilation Design Manual: *Mechanical Engineering*; NAVDUCKS DM-3; Buffalo Forge: *Fan Engineering;* Carrier Corporation's *System Design Manual;*[2] and *Woods Practical Guide to Fan Engineering.*[3]

INTRODUCTION

The purpose of this chapter is to provide the plant engineer with a ready reference to practical solutions of day-to-day heating and ventilating problems.

One of the cardinal rules of a good, economical, energy-efficient design is to avoid designing a system (be it heating, ventilating, exhaust, humidification, dehumidification, etc.,) to meet only the stringent (critical) requirements of a small or minor portion of the total area served. If possible, the critical area should be isolated and treated separately.

The formulas and data presented in this chapter are for sea-level locations. Correction factors are discussed in the section of this chapter entitled "Altitude Corrections for Heating and Ventilating Systems."

CLIMATIC CONDITIONS

When possible, the outdoor design conditions should be obtained from official weather stations such as those of the U.S. Weather Service, U.S. Air Force, U.S. Navy, Canadian Atmospheric Environment Service, etc., or from the latest edition of ASHRAE's *Handbook of Fundamentals* in the chapter "Weather Data and Design Conditions."

Data from many weather stations at specific locations and elevations furnish a network from which, by interpolation, good estimates can be made of the expected conditions at locations without weather stations.

ADJUSTMENTS FOR CLIMATE

When design requirements are extremely important, it may be advisable to retain a competent applied meteorologist to develop data that can show a comparative relationship with the nearest official station having a record for a long period. If this is not feasible, the following general rules will apply in adjusting the design data supplied for the weather stations listed to fit some other location.

Adjustment for Elevation

For a lower elevation, the design values should be increased, while for a higher elevation, the values should be decreased. The increments used in these adjustments are generally

Dry-bulb (db) temperature	1°F per 200 ft (1°C per 100 m)
Wet-bulb (wb) temperature	1°F per 500 ft (1°C per 300 m)

In the winter, where cold-air drainage (principally in hilly or mountainous areas) or considerable radiation cooling occurs at a site, these adjustments do not apply.

Adjustment for Mass Modification

Short-distance variations are most extreme near large bodies of water where air moves from the water over the land in summer. Along the west coast of the United States, both dry-bulb and wet-bulb temperatures increase with distance from the ocean. In the region north of the Gulf of Mexico, dry-bulb temperatures increase for the first 200 to 300 mi (322 to 482 km) with a slight decrease in wet-bulb temperatures caused by mixing with drier air inland. Beyond this 200- to 300-mi (322- to 482-km) belt, both dry-bulb and wet-bulb values tend to decrease at a somewhat regular rate.

Adjustment for Vegetation

The difference between large areas of dry surfaces and large areas of dense foliage upwind from the site can account for variations of up to 2°F (1.1°C) wb and 5°F (2.8°C) db. The warmer temperatures are associated with the dry surfaces. Adjustments for vegetation require the assistance of a consulting applied meteorologist.

SEASONAL CONSIDERATIONS

Preferably the winter outdoor design temperature should be based on a temperature minimum that will exist 99 percent of the time. The 99 percent value represents a specific temperature that will be equaled or exceeded in a negative directioin by 1 percent of the approximately 2200 h of total annual heating. Then, in a normal winter, the outdoor design temperature will be equal to or below this temperature for approximately 22 h.

It is recommended that, if economically possible, the summer outdoor design (dry-bulb and wet-bulb temperatures) conditions for ventilation systems be based on conditions not exceeded more than 5 percent of the time. The 5 percent value represents specific dry-bulb and wet-bulb temperatures that will not be equaled or exceeded more than 5 percent of the approximately 2900 h of total annual cooling. Thus, in a normal summer, the outdoor design conditions will be equal to or above these conditions a total of approximately 145 h.

HEATING

GENERAL

In general there are three types of heating systems:

Steam, hot-water, and hot-air. These, in turn, are further categorized below.

Steam

Low-Pressure Steam

Less than 15 psig (1 bar) operating pressure. This is the most common steam system for comfort heating.

High-Pressure Steam

Greater than 15 psig (1 bar) operating pressure. This system is used mainly in process heating and in distribution systems.

Vacuum Steam

This system operates below atmospheric pressure. Less than 1 psig (0.07 bar) operating pressure. It is rarely used today for comfort heating.

Hot-Water

Though there are no standards on the precise values that define "low," "medium," and "high" temperatures, the following are typical.

Low-Temperature, ≤200°F (93°C)

This is one of the most common heating systems.

Medium-Temperature, >200°F (93°C) but <260°F (127°C)

The application of this system is the same as high-temperature water systems.

High-Temperature, 260 to 450°F (127 to 232°C)

This system is rarely if ever used for space heating because of the potential for scalding personnel as a result of a leak. It is mainly used for heat-distribution systems and, in some processes, heating where water is preferred to steam in the same temperature range.

Hot-Air

Comfort Heating

The temperature of the discharge supply air emanating from the heating device is generally not greater than 200°F (93°C).

Process Heating

The temperature of the discharge supply air of the heating device can be considerably higher than 200°F (93°C), depending on the process requirements.

Infrared

Infrared heating systems are used primarily for industrial heating requirements where (1) it is impractical to heat the total volume of space, (2) it is required to bring the total area only to minimum standards, or (3) it is necessary to heat a small area or areas (such as a work station) to an acceptable level. It is also used in some dyeing and baking ovens.

In designing heating systems using infrared heaters for industrial spaces (especially with high ceilings and cold climates), the relative humidity at the roof level must be maintained below the point where condensation will occur. Figure 1-4 in the section, "Humidity Control," later in this chapter, shows maximum relative humidity values that can be maintained without condensation occurring.

ENERGY SOURCES

The following table lists common energy sources for the heating systems listed above:

Energy source	Comment
Coal	Various grades of hard and soft coal.
Oil	No. 2, 4, 5, or 6.
Gas	Natural, liquid natural, propane or butane.
Electricity	60 Hz, 120-V, single-phase; 240-V three-phase. Industrial process heaters frequently operate at higher voltages.
Wood	Depending on the moisture content and type, the heating value can generally range from 4000 to 7500 Btu/lb (9 to 17 kJ/kg).
Solar	Applicable with low-temperature water and steam systems. The capability of generating medium-temperature water and high-pressure steam [125 psig (8.6 bars)] has still not been developed for commercial purposes.

SYSTEM COMPONENTS

Steam-Heating System

The principal components are: boilers or generators; the boiler feed system; the deaerator; makeup water; chemical treatment; the fuel system; distribution supply and return (condensate), piping, supports, expansion, and insulation; end-use distributors such as radiators, coils, and radiation panels (radiant heat); process equipment; valves, vents, traps, and drains; condensate pumps and receivers; controls and safety devices.

Vacuum System

In general this system has the same components as a steam system except for the following: the condensate pump is replaced by a vacuum pump; all air is vented from the system by the vacuum pump and there are no other automatic air vents; the traps used are specifically designed for vacuum service.

Hot-Water System

The basic components here are: generator (boiler); feedwater, fuel, and makeup water systems; chemical treatment; expansion or compression system; pumping system; supply and return piping systems and their supports, expansion, and insulation; end-use distributors such as convectors (radiators), coils, radiation panels (radiant heat); process equipment; valves, vents, and drains; controls and safety devices.

EQUIPMENT SELECTION

In this chapter we are concerned more with the end use than with the distribution and pumping systems. The reader is referred to Sec. 9 of this book for a discussion of distribution and pumping systems.

Suggested Procedure

Environmental Conditions

Outdoor Design Considerations. The procedure to determine the impact of outdoor design conditions was discussed previously in this chapter, under "Climatic Conditions."

Indoor Design Selection. In order to avoid overdesigning the heating system so as to conserve energy and to minimize construction costs, each space or area should be analyzed separately to determine the minimum temperature that can be maintained and whether humidity control is required or desirable. For a discussion on humidity control see "Humidity Control" later in this chapter.

The U.S. government has set 68°F (20°C) as the maximum design indoor temperature for personnel comfort during the heating season in areas where *employees* work. In *manufacturing areas* the process requirements govern the actual temperature. From an energy-conservation point of view, if a process requires a space temperature greater than 5°F (2.8°C) above or below 68°F (20°C), it should, if possible, be treated separately and operate independently from the general personnel comfort areas. The staff members working in such areas should be provided with supplementary spot (localized) heating and/or ventilating systems as the conditions require in order to maintain personnel comfort.

If the stored products permit, garages and warehouses should be designed for the lowest temperature that is needed to prevent freeze-ups. In such buildings, locations to which personnel are assigned could have supplemental electric heaters, or hot-air or infrared heating systems. The process engineering department or quality control group should determine the manufacturing process space temperature. The manufacturer of the particular process equipment can be an alternative source for the recommended space temperature. The air temperature at the ceiling may vary beyond the comfort range and should be considered in calculating the overall heat transmission to the outdoors. A normal 0.75°F (0.42°C) increase in air temperature per foot (0.3 m) of elevation above the breathing level [5 ft (1.5 m) above finish floor] would be expected in normal applications, with an approximately 75°F (42°C) temperature difference between indoors and outdoors.

DETERMINING THE DESIGN HEATING LOAD

General

From a physiological point of view, the ideal heating system would minimize the radiation and convection losses from the worker's body to the surrounding surfaces (to the point where the body will not detect a cooling sensation), while maintaining warm feet and a cool head for the workers. From an industrial or process point of view, the ideal heating system is one that satisfies the equipment and product's requirements at minimum energy expenditure and cost. Most heating systems must be designed as a compromise between these ideal goals and minimizing the life-cycle (sum of capital, energy, and maintenance) costs.

In colder climates, having floors (above unheated areas or on grade) and exterior walls well insulated is no guarantee that downdrafts from windows or exposed walls will not create a pool or pools of chilly air over large areas of the floor. When the floor slabs are on grade or above unheated spaces, it is especially important to provide a heating system that will deliver sufficient heat near the floor to counteract the downdrafts at the exterior walls and windows and the heat loss through the floor slabs.

It is reasonable to expect a normal commercial space-heating control system to maintain a specified space temperature within ±2°F (±1°C) of the thermostat set point. This is within the tolerance of personnel comfort heating. However, some industrial processes may be adversely affected if the ambient temperature varies more than ±0.5°F (±0.3°C) from the set point. If closer temperature control is required, that is, less than ±1°F (±0.6°C) from the set point, more sophisticated (and expensive) industrial controls must be considered.

Design

With the outdoor and inside design temperatures for each space determined, the heating load for each space can be calculated. When time does not permit (or the project requirements do not warrant the more detailed method for calculating heating loads as found in standard references such as the latest edition of ASHRAE *Handbook of Fundamentals*[4]), the following procedure can be used: The heating load from each space consists of the heat loss through the walls, floors, roof, and glass by conduction and convection plus the infiltration and ventilation losses. Radiation losses from the space to the outside are neglected.

Heat Loss

Refer to "Infiltration" and "Ventilation" in this chapter for a method of determining the ventilation and infiltration heating losses.

Heat Loss by Conduction and Convection Heat Transfer through Any Surface. The basic formula is

$$g = AU(t_i - t_o) \tag{1}$$

where

- g = heat transfer through the walls, roof, ceiling, floor, or glass, Btu/h (metric conversion: Btu/h × 0.0293 = W)
- A = area of wall, glass, roof, ceiling, floor, or other surface area through which heat is flowing, ft². If accurate plans, sections, and elevations of the spaces to be heated are available, the appropriate areas can be taken from them. However, if plans, sections, and elevations are not available, these areas must be field-measured.
- U = air-to-air heat-transfer coefficient, Btu/(h)(ft²)(°F). For a discussion on U values, see the last part of this section.
- t_i = indoor air temperature near surface involved, °F
- t_o = outdoor air temperature, or temperature of adjacent unheated space or ground, °F

Heat Loss through Below-Grade Walls and Floors. U values can be determined from Tables 1-1 and 1-2. When using Table 1-1 to calculate the heat loss through the below-grade wall sections, it may be easier to calculate each foot of vertical wall separately and add the British thermal units per hour for each section to determine the total heat loss through the below-grade portion of the wall. For the case where a wall is more than 7 ft (2 m) below grade, it is suggested that the value shown for 7 ft (2 m) should be used. This will result in a conservative heat loss.

t_g is the equivalent temperature of the ground in degrees Fahrenheit. Selecting the appropriate ground temperature can be a problem, since the ground surface temperature is known to fluctuate about a mean value which varies with the geographical location and

TABLE 1-1 *U* Values for Heat Loss Below Grade in Basement Walls,* Btu/(h)(ft²)(°F)

Below grade, ft (m)	Uninsulated	Heat loss through walls with insulation of various thicknesses: 1 in (2.5 cm)	2 in (5 cm)	3 in (7.6 cm)
0–1 (1st) (0–0.3)	0.410	0.152	0.093	0.067
1–2 (2d) (0.3–0.6)	0.222	0.116	0.079	0.059
2–3 (3d) (0.6–0.9)	0.155	0.094	0.068	0.053
3–4 (4th) (0.9–1.2)	0.119	0.079	0.060	0.048
4–5 (5th) (1.2–1.5)	0.096	0.069	0.053	0.044
5–6 (6th) (1.5–1.8)	0.079	0.060	0.048	0.040
6–7 (7th) (1.8–2.1)	0.069	0.054	0.044	0.037

*K_{soil} = 9.6 Btu/(h)(in)/(ft²)(°F)(in thickness); $K_{insulation}$ = 0.24 Btu/(h)(in)/(ft²)(°F)(in thickness).

TABLE 1-2 *U* Values for Average Heat Loss Through Basement Floors, Btu/(h)(ft²)(°F)

Depth of foundation (basement floor) below grade, ft (m)	Width of wall, ft (m): 20(6.0)	24(7.3)	28(8.5)	32(10)*
5 (1.5)	0.032	0.029	0.026	0.023
6 (1.8)	0.030	0.027	0.025	0.022
7 (2.2)	0.029	0.026	0.023	0.021

*Or more

the surface cover. The ground temperature t_g can be estimated in various ways. ASHRAE recommends subtracting the mean value of the amplitude A of ground surface temperature obtained from Fig. 1-1 from the mean annual air temperature t_a, obtained from meteorological records:

$$t_g = t_a - A$$

Figure 1-1 shows annual ranges in ground temperature at a depth of 4 in. Additional constant-amplitude curves can be obtained from Chang's *Ground Temperature.*[5]

Another method the author has used to determine the ground temperature (t_g) is based on Table 1-3.

Heat Loss through Slab on Grade. The heat loss through a concrete slab on grade is more nearly proportional to the perimeter than to the floor area. It can be estimated with the equation:

$$g = F_2 P(t_i - t_o) \quad (2)$$

where

g = heat loss through the slab, Btu/h (metric conversion: Btu/h × 0.0293 = W)
F_2 = heat loss coefficient, Btu/(h)(lf)(°F). The values of F_2 range from 0.81 for a floor with no edge insulation to 0.55 for a floor with 1 in of edge insulation.
P = perimeter of exposed edge of slab, lf (linear foot)
t_i = indoor air temperature, °F
t_o = outdoor air temperature, °F

An alternative method for calculating the heat loss through a slab on grade is to use Eq. (3) and Table 1-4:

$$g = F_1 P \quad (3)$$

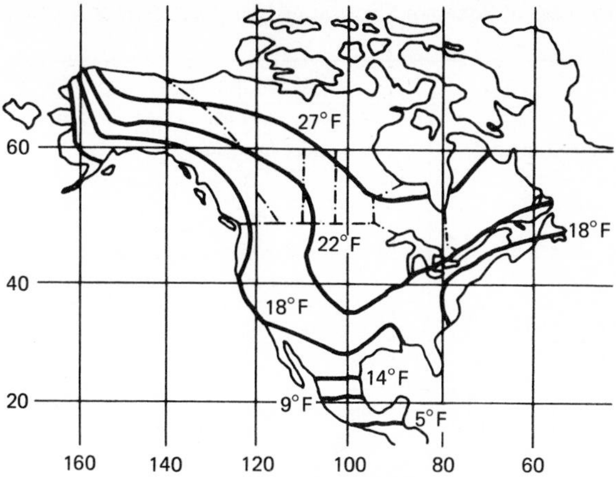

Figure 1-1 Lines of constant amplitude for Canada, the United States, and Mexico.

TABLE 1-3 Ground Temperature t_g

Outdoor design temperature		Ground temperature, t_g	
°F	°C	°F	°C
−30	−34.4	40	4.4
−20	−28.9	45	7.2
−10	−23.3	50	10.0
0	−17.8	55	12.8
+10	−12.2	60	15.6
+20	−6.70	65	18.3

Source: Carrier Air Conditioning Company, *System Design Manual*, Part 1, "System Load Estimating," Syracuse, N.Y., 1960.

where

g = heat loss through the slab, Btu/h (metric conversion: Btu/h × 0.0293 = W)
F_1 = heat loss coefficient from Table 1-4, Btu/(h)(lf)
P = perimeter of exposed edge of slab, lf

The insulation should extend horizontally for 2 ft (0.6 m) under the slab and can also be located along the vertical foundation wall with equal effectiveness if it extends 2 ft (0.6 m) below the floor level.

Adjacent Unheated Spaces. The heat loss from heated rooms to unheated rooms or spaces must be based on the estimated or assumed temperature in such unheated spaces. This temperature will lie in the range between the indoor and outdoor temperatures. Assuming that the respective surface areas adjacent to the heated room and exposed to the outdoors are the same and the heat transfer coefficients are equal, the temperature in the unheated space may be assumed to be almost equal to the average of the indoor and outdoor design temperatures. If, however, the surface areas and coefficients are unequal, the temperature in the unheated space may be estimated from the latest ASHRAE *Handbook of Fundamentals*[4] in the chapter "Heating Loads."

TABLE 1-4 Heat Loss of Concrete Floors at or Near Grade Level per Linear Foot of Exposed Edge

Outdoor design temperature, °F	Heat loss per linear foot of exposed edge, Btu/h	
	Recommended 2 in (5 cm) edge insulation	1 in (2.5 cm) edge insulation
−20 to −30	50	55
−10 to −20	45	50
0 to −10	40	45
	1 in (2.5 cm) edge insulation	No edge insulation*
−20 to −30	60	75
−10 to −20	55	65
0 to −10	55	60

*This construction not recommended; shown for comparison only.

Unheated Attic. The reduction in temperature difference between attic air and outside air is approximately linear, with attic ventilation (natural and/or mechanical) between 0 and 0.5 $ft^3/(min)(ft^2)$ [0 and 0.0025 $m^3/(s)(m^2)$] of attic floor area.

When relatively large louvers are installed (customary in the warmer climates), the attic temperature is often assumed to be an average between the indoor and outdoor temperatures. For a short approximate method of calculating heat losses through attics, the combined ceiling and roof heat-transfer coefficient may be used.

Unheated Crawl Space. The temperature in the crawl spaces below floors will vary widely, depending on the number and size of wall vents (if any), amount of warm pipes and ducts running through the crawl space and the type and thickness of the insulation, if any. It is necessary therefore to evaluate temperatures by judgment. As an expedient it may be satisfactory to average the ground temperature with the basement or space temperature.

Load Tabulation

When the areas requiring heat are numerous, a systematic approach to keeping track of the individual areas is necessary. Though there are many ways to accomplish this, the author has found that the format provided in Table 1-5 is helpful.

Table 1-5 can be repeated several times on a standard sheet of paper, thus enabling many areas to be tabulated on the same sheet. Each area to be heated is identified by a room number or name. If this is not practical and a layout or floor plan is available, the spaces to be heated can be sequentially numbered on the layout or plan and the corresponding identification number placed on the table. The floor area and space volume are filled in at the top.

The appropriate U is multiplied by the difference in temperature t and is placed in the factor column. The respective wall, roof, etc., factors need not be recalculated unless the corresponding U or Δt changes.

This table allows the designer to add convection, conduction, infiltration, and ventilating heating loads for each heating zone if a hot-air system is selected. Alternatively, if a steam or hot-water system is selected, the convection, conduction, and infiltration loads for each area are listed to select the space radiation, unit heater, etc., requirements. The makeup air heater can be sized by totaling the ventilation loads. If the ventilation system will pressurize the area, the infiltration load can be deleted.

Determining (Selecting) U Values

If possible the U value should be obtained from standard heating textbooks, such as ASHRAE *Handbook of Fundamentals,* in the chapter "Heating Transmission Coefficients."

TABLE 1-5 Summary of Heating Loads

Space Identification

Area__________ft² Volume__________ft³

Item	Factor, $U \times t$	Area, ft²	Btu/h	Remarks
Roof				
Skylight				
Wall (below grade)				
Wall (above grade)				
Glass				
Ceiling				
Floor				
Floor (on grade)				
Partition				
Subtotal				
Convection and radiation load				
Infiltration	$1.08 \times \Delta t$	ft³/min		
Subtotal				
Convection, radiation, and infiltration load				
Ventilation	$1.08 \times \Delta t$	ft³/min		
Subtotal				
Process heating load				
Grand total				

In the event a standard reference source is not available, the values in Tables 1-6 to 1-11 will, in general, result in a conservative design. These tables represent weighted averages based on the author's judgment from data printed in Chap. 5 of Carrier Air Conditioning's *System Design Manual,* part 1.[2]

TABLE 1-6 U Values for Glass, Btu/(h)(ft²)(°F temp diff)

Material	Installed vertically	Installed horizontally
Single pane	1.13	1.4
Thermal (double pane)	0.6	0.7
Triple pane	0.4	—
Glass block		
normal 4″ thick	0.6	—

TABLE 1-7 *U* Values for Flat Roofs with Built-Up Roofing, Btu/(h)(ft²)(°F temp diff)*

Material	Ceilings			
	None	None or plastered	Suspended plaster	Suspended acoustical tile
Flat metal	0.74	—	0.35	0.25
2″ thick preformed slab wood fiber and cement binder	—	0.22	0.17	0.14
4″ thick poured concrete slab	—	0.56	0.31	0.23
2″ thick wood	—	0.31	0.25	0.21

*If insulation is used, enter Table 1-11 with the appropriate value from this table (without insulation) to obtain the *U* value with insulation.

TABLE 1-8 *U* Values for Pitched Roofs, Btu/(h)(ft²)(°F temp diff)*

Material	Ceilings		
	None	Gypsum board or plaster	Acoustical tile on furring
Asphalt shingles paper on plywood sheathing	0.56	0.33	0.25
Asbestos cement shingles or asphalt roll roofing			
Building paper on plywood sheathing	0.65	0.38	0.28
Sheet metal building paper on plywood	0.7	0.4	0.26

*If insulation is used, enter Table 1-11 with the appropriate value from this table (without insulation) to obtain the *U* value with insulation.

INFILTRATION

The *infiltration heating load* is the amount of heat that must be provided to raise the temperature of the outside air leaking into the heated space through cracks and/or spaces in and around doors, windows, skylights, floors, walls, shafts, stairwells, etc., to the inside design air temperature.

Outside air infiltrates the building structure as a result of the pressure difference between inside and outside of the structure (inside is negative with respect to outside). The pressure difference may be due to wind or to the density difference of outside and inside air, often called the "chimney" or "stack" effect. It also can be caused mechanically by exhausting more air from the structure than is supplied to the structure.

Outside air infiltration is usually a significant portion of the heating load, especially in the colder climates. It is also important in determining the humidification requirements if the relative humidity of the space must be maintained above 10 percent during the heating season in the colder climates.

In buildings that have adequate makeup air, the cracks around windows and doors are usually the major source of air leaking into the structures.

For a detailed discussion of infiltration, wind effect on structures, air density difference, and methods of calculating the quantity of air infiltrating a building, the reader is referred to the chapter "Infiltration and Ventilation" in the latest ASHRAE *Handbook of Fundamentals.*[4]

Methods for Determining Infiltration Heating Load

Air-Change Method

A quick method of determining the infiltration heating load is the *air-change method.* Though this method is not as accurate as the crack method, if used properly it will produce acceptable results.

TABLE 1-9 *U* Values for Walls, Btu/(h)(ft^2)(°F temp diff)*

Material	Thickness, in	Interior Finish: None	Interior Finish: Gypsum board or plaster	
Brick	6	0.48	0.38	
	8	0.41	0.32	
Concrete	6	0.75	0.5	
	8	0.67	0.48	
Hollow concrete block	8	0.52	0.4	
	12	0.47	0.36	
4-in face brick with 8-in concrete block backing	——	0.41	0.33	
4-in face brick veneer on wood sheathing		0.42	0.28	
Wood siding or shingles on wood sheathing		0.36	0.25	
			Flat iron	3/4″ wood
3/8-in corrugated transit, no sheathing		1.16	0.55	0.35
24-gauge corrugated iron, no sheathing		1.4	0.6	0.38

Partition	3/4″ wood panel	Gypsum board or plaster
Finish one side	0.43	0.61
Finish both sides	0.24	0.35

*If insulation is used, enter Table 1-11 with the appropriate value from this table (without insulation) to obtain the *U* value with insulation.

TABLE 1-10 *U* Values for Ceilings and Floors, Btu/(h)(ft^2)(°F temp diff)*

Material	Ceilings: None	Ceilings: Acoustical tile glued	Ceilings: Gypsum board or plastered	Ceilings: Acoustical tile on furring
Concrete slab with linoleum or tile floor				
4″ (10 cm) thick slab	0.63	0.36	0.41	0.26
8″ (20 cm) thick slab	0.52	0.32	0.37	0.24
Concrete slab with linoleum or tile on plywood on sleepers				
4″ (10 cm) thick slab	0.31	0.23	0.3	0.18
8″ (20 cm) thick slab	0.28	0.21	0.27	0.17
Wood subfloor only	0.45	——	0.3	0.8
Wood subfloor and hardwood floor or linoleum on plywood	0.33	——	2.5	0.19

*If insulation is used, enter Table 1-11 with the appropriate value from this table (without insulation) to obtain the *U* value with insulation.

TABLE 1-11 *U* Values for Walls, Ceilings, Roofs, and Floors, Btu/(h)(ft^2)(°F temp diff), with insulation

U value before adding insulation wall, ceiling, roof, floor	Addition of fibrous insulation thickness, in (cm)		
	1 (2.5)	2 (5)	3 (7.6)
0.60	0.19	0.11	0.08
0.58	0.19	0.11	0.08
0.56	0.18	0.11	0.08
0.54	0.18	0.11	0.08
0.52	0.18	0.11	0.08
0.50	0.18	0.11	0.08
0.48	0.17	0.11	0.08
0.46	0.17	0.10	0.08
0.44	0.17	0.10	0.07
0.42	0.16	0.10	0.07
0.40	0.16	0.10	0.07
0.38	0.16	0.10	0.07
0.36	0.15	0.10	0.07
0.34	0.15	0.10	0.07
0.32	0.15	0.10	0.07
0.30	0.14	0.09	0.07
0.28	0.14	0.09	0.07
0.26	0.13	0.09	0.07
0.24	0.13	0.09	0.07
0.22	0.12	0.08	0.06
0.20	0.12	0.08	0.06
0.18	0.11	0.08	0.06
0.16	0.10	0.07	0.06
0.14	0.09	0.07	0.05
0.12	0.08	0.06	0.05
0.10	0.07	0.06	0.05

The air-change method assumes the total air volume within a space will be completely exchanged with (displaced by) outside air a certain number of times each hour. In using this method it is difficult to define the proper space volume, particularly for high-ceiling areas. There is a tendency to assume a high rate of change per hour, especially in high-volume areas. This will result in overestimation of the heating load. As a result, in order to obtain a good estimate of the infiltration load, a greater amount of experience and judgment is required with this method than with the crack method. Nevertheless, the author has obtained satisfactory results with this method, using Table 1-12 in conjunc-

TABLE 1-12 Air Changes Occurring under Average Conditions Exclusive of Air Provided for Ventilation*

Kind of room	Number of air changes per hour
Spaces with no windows or exterior doors	0.5
Spaces with windows or exterior doors on one side	1
Spaces with windows or exterior doors on two sides	1.5
Spaces with windows or exterior doors on three sides	2

*Spaces with weatherstripped windows, or with storm sash, use two-thirds these values.

TABLE 1-13 Infiltration through Doors in Winter With a 15 mi/h Wind Velocity and Doors on One or Adjacent Windward Sides

	ft³/(min) per (ft²) area	
Description	Infrequent use	Average use 1 & 2 story building.
Revolving door	1.6	10.5
Glass door, 3/16″ crack	9.0	30.0
Wood door, 3′ × 7′	2.0	13.0
Small factory door	1.5	13.0
Garage and shipping room door	4.0	9.0
Ramp garage door	4.0	13.5

tion with Table 1-13, when time and/or the degree of accuracy did not permit or warrant the use of the crack method.

The air-change-per-hour infiltration-load formula is

$$g = \frac{\text{CH}}{\text{h}} \times \frac{\text{space}}{\text{volume}}\, 0.018(t_o - t_i) \tag{4}$$

where

g = space-infiltration load, Btu/h (metric conversion: Btu/h × 0.0293 = W)

CH/h = the number of air changes per hour selected from Table 1-12 or as specified by the engineer

space volume = the volume of the space corresponding to the selected CH/h, ft^3

0.018 = a constant equal to the specific heat of moist air at 70°F db and 50 percent relative humidity (RH) (0.244 Btu/(lb)(°F) divided by the specific volume of moist air at 70°F db and 50 percent RH (13.5 ft^3/lb)

$$\frac{0.244\ \text{Btu/(lb)(°F)}}{13.5\ \text{ft}^3/\text{lb}} = 0.018\ \text{Btu/(ft)}^3\text{(°F)}$$

t_o = outside design air temperature, °F

t_i = inside (space) design air temperature, °F

Door Infiltration Method

The door infiltration-load formula is

$$g = \text{outside airflow}\ (1.08)(t_o - t_i) \tag{5}$$

where

g = door infiltration load, Btu/h (metric conversion: Btu/h × 0.0293 = W)

outside airflow = cubic feet of air per minute selected from Table 1-13 times the number of square feet of door openings, ft^3/min

1.08 = a constant equal to the specific heat of moist air at 70°F db and 50 percent RH [0.244 Btu/(lb)(°F)], divided by the specific volume of moist air at 70°F db and 50 percent RH (13.5 ft^3/lb) multiplied by 60 to convert minutes into hours, Btu/(ft^3/min)(h)(°F)

t_o = outside air temperature, °F

t_i = inside (space) air temperature, °F

To obtain the total building-infiltration heating load in British thermal units per hour, one must add all the individual space-infiltration loads calculated by the air-change method [Eq. (4)] to the door-infiltration loads calculated with Eq. (5).

Though Table 1-12 was originally developed for residential buildings, here it is used in conjunction with Table 1-13 to estimate the infiltration loads of industrial office and manufacturing spaces with the following guidelines:

1. Table 1-12 can be used as shown for offices, laboratories, cafeterias, and other spaces having hung ceilings no higher than 10 ft (3 m).
2. For manufacturing plants, warehouses, and other areas with operable windows that occupy almost the entire exterior wall area, the number of changes per hour shown in this table should be increased by approximately 50 percent.
3. For manufacturing plants, warehouses, and other areas that have no exterior windows or a moderate number of them, with a height from the floor to the underside of the floor or roof above of 15 to 25 ft (4.6 to 7.6 m), the number of changes per hour shown in this table should be decreased by approximately 50 percent.

All values in Table 1-13 are based on the wind blowing directly at the door. When the prevailing wind direction is oblique to the doors, multiply the above values by 0.60 and use the total door area on the windward side(s). Table 1-13 is based on a wind velocity of 15 mi/h. For design wind velocities different from the base, multiply the table values by the ratio of velocities. Multiply the table values by ($V_e - V/15$) for doors on the leeward side of the building. V_e is the equivalent velocity in miles per hour. V is the design or actual velocity in miles per hour.

Doors on opposite sides increase the above values 25 percent. Vestibules may decrease the infiltration as much as 30 percent when door usage is light. If door usage is heavy, the vestibule is of little value in reducing infiltration. Heat added to the vestibule will help maintain room temperature near the door.

The installation of air curtains can, if properly sized and installed, reduce the infiltration on heavily used doors from 60 to 80 percent. However, installation of air curtains will not significantly reduce the heating energy cost; in fact, it may increase that cost. Therefore, the use of air curtains should be evaluated from a comfort, hygienic (if food and drugs are manufactured or processed), and life-cycle-cost standpoint.

VENTILATION

GENERAL

The optimum design goals of a general ventilation system for any space are to promote the health, comfort, and well-being of the occupants and the quality of the manufacturing process therein at minimum life-cycle and energy costs. These objectives can be accomplished by controlling the thermal conditions or the amounts of contaminants, or both, in the space environment.

Due to the high ambient dry- and wet-bulb temperatures prevalent during the summer months in many parts of the world, it is virtually impossible to achieve comfortable indoor environmental conditions with ventilation air only. Where ventilation systems are designed, it is recommended that they be designed for the greatest change of air that can be economically provided and that, where possible, the discharge distribution be designed to increase the evaporation rate from the bodies of personnel occupying the area.

The ventilation requirements for providing a safe working environment for personnel exposed to toxic, noxious, or hazardous gases or liquids are beyond the scope of this chapter. The reader is referred to Chap. 14-5 and to standard references such as: *Industrial Ventilation, A Manual of Recommended Practice;*[1] *Handbook of Ventilation for Contaminant Control;*[6] *Plant and Process Ventilation;*[7] and OSHA's requirements.

Likewise, determining the ventilation needed to maintain the thermal equilibrium of the body is beyond the scope of this chapter. The reader is referred to standard references such as the latest editions of *Industrial Ventilation, A Manual of Recommended Practice*[1] and ASHRAE's *Handbook of Fundamentals.*[4]

The two basic types of ventilation systems are *natural* and *mechanical.*

NATURAL VENTILATION

Natural, or gravity, ventilation has a limited application since its effectiveness is directly dependent on prevailing winds outside the building and temperatures (stack effect) inside the building. It should be considered only for locations that have a reliable prevailing wind and where the personnel, manufacturing process, or product stored can tolerate temperature and/or humidity conditions above or below the design space values for prolonged periods.

In considering the feasibility of designing a natural ventilating system, the following guidelines are suggested:

1. Systems should be designed for wind velocities of half the average seasonal prevailing velocity.
2. In order to take maximum advantage of the stack effect (density difference), the supply air should enter through openings at or near the floor level of the space to be ventilated and leave through openings high in the wall and/or through gravity roof ventilators.
3. Locate air inlet openings on the side of the building facing directly into the prevailing wind.
4. Locate air outlets where prevailing wind movements will create low-pressure areas, i.e., on the side directly opposite the direction of the prevailing wind. Outlets may be placed on a roof in the form of individual gravity ventilators, continuous monitors, or ridge ventilators.
5. Inlet openings should not be obstructed by buildings, trees, signboards, indoor partitions, etc.
6. Greatest flow per unit area of total opening is obtained by using inlet and outlet openings of nearly equal areas.
7. Direct short circuits between openings on two sides at a high level may clear the air at that level without producing any appreciable ventilation at the lower level of occupancy.
8. The vertical distance between the inlets and outlets should be as great as possible in order to develop the greatest ventilation benefit from the temperature difference.
9. In multistory structures, openings in the neutral pressure zone are least effective for ventilation.

In general, natural ventilation for spaces is inadequate for the following cases:

1. Offices having an open window area that is less than 5 percent of the floor area
2. Offices more than 24 ft (7.3 m) deep and lacking cross ventilation
3. Offices with cross ventilation but having occupied space that is more than 35 ft (10.7 m) from a window or air inlet
4. Toilet rooms having window area that is less than 9 ft^2 (0.8 m^2) or less than 0.2 ft^2 (0.02 m^2) for each foot (0.3 m) of height or 5 percent of the floor area.
5. Cafeteria or assembly areas having a window area that is less than 6 percent of the floor area.

An estimate of the quantity of ventilation air required for natural ventilation can be obtained by applying the following rule of thumb (originally derived for residences, it can be applied to offices and other light-work areas): The ventilation air quantity should be based on changing the air within the space (with outside air) at least 30 times per hour in locations above latitude 37°N or below 37°S and 60 times per hour in locations between those latitudes and the equator.

For a detailed discussion of wind velocity and stack effect on buildings, the reader is referred to the latest edition of ASHRAE's *Handbook of Fundamentals*.[4]

Calculations for Natural Ventilation

The minimum airflow quantities for ventilation can be calculated from:

$$Q = \frac{H}{60 C_p \rho (t_i - t_o)} \tag{6}$$

where

- Q = ventilation airflow required, ft³/min (metric conversion: ft³/min × 0.000472 = m³/s)
- H = quantity of heat required to be removed, Btu/h. See discussion under "Mechanical Ventilation."
- 1/60 = constant to convert hours to minutes
- C_p = specific heat of air at constant pressure at 70°F db and 50 percent RH = 0.244 Btu/(lb)(°F)
- ρ = density of standard air (at 70°F db and 50 percent RH) = 0.0741 lb/ft³
- t_i = inside (space) design air temperature, °F db
- t_o = outside design air temperature °F db (See discussion under "Climatic Conditions" to select values for t_o and t_i)

Note that the constant 1.1 can generally be substituted for the factor $60C_p\rho$ since it equals 60(0.244) × 0.0741 which closely approximates 1.1.

The minimum size required for the inlet air openings to provide the design airflow rates at the prevailing wind velocity can be calculated from:

$$A = \frac{Q_a}{EV} \tag{7}$$

where

- A = free area of inlet opening, ft² (metric conversion: ft² × 0.0929 = m²)
- Q_a = required ventilation airflow, ft³/min, through the opening of area A. The number of openings should, where feasible, be selected to obtain a velocity as uniform as possible from the ventilation air across the space. Therefore, Q_a is equal to Q from Eq. (6) divided by the number of openings; i.e., Q_a = Q/(no. of openings).
- E = effectiveness of opening (E should be taken as 0.50 to 0.60 for perpendicular winds, and 0.25 to 0.35 for diagonal winds.)
- V = wind velocity, ft/min (mi/h × 88 = ft/min)

It is recommended that one-half the average velocity of the seasonal prevailing wind be used.

The flow of air due to the thermal forces within a building or space that has minimum internal resistance (to airflow) can be calculated using the following formula:

$$Q_t = 9.4A\sqrt{h(t_i - t_o)} \tag{8}$$

where

- Q_t = air flow, ft³/min, due to thermal forces only (metric conversion: ft³/min × 0.000472 = m³/s)
- A = free area, ft², of inlets or outlets if they are equal. If they are not equal use the smaller value.
- h = vertical height, ft, between inlet and outlet
- t_i = average temperature of indoor air, °F db, at height h above the floor

t_o = outdoor temperature, °F db. For this formula it is assumed that both the indoor temperature at the floor and the outdoor temperature are close to 80°F.

9.4 = constant of proportionality, including a value of 65 percent for effectiveness of openings. This should be reduced to 50 percent (constant = 7.2) if conditions are not favorable.

The increase in airflow from unequal inlet and outlet areas can be approximated from Fig. 1-2. The value C_t obtained from Eq. (8) can be increased by increasing the percent obtained from Fig. 1-2.

Ratio of Outlet to Inlet or Vice Versa

The combined flow due to wind and thermal forces (stack effect) is not equal to the flows estimated separately [Q_a from Eq. (7) and Q_t from Eq. (8)]. The flow can be approximated from Fig. 1-3 by entering at the bottom the ratio $Q_t/(Q_a + Q_t)$ and reading the factor that must be multiplied by the flow due to thermal effect Q_t to obtain the combined total flow. The combined flow is Q_t (factor from Fig. 1-3).

When the flows Q_a and Q_t are equal, the actual combined flow will be about 30 percent greater than either Q_a or Q_t.

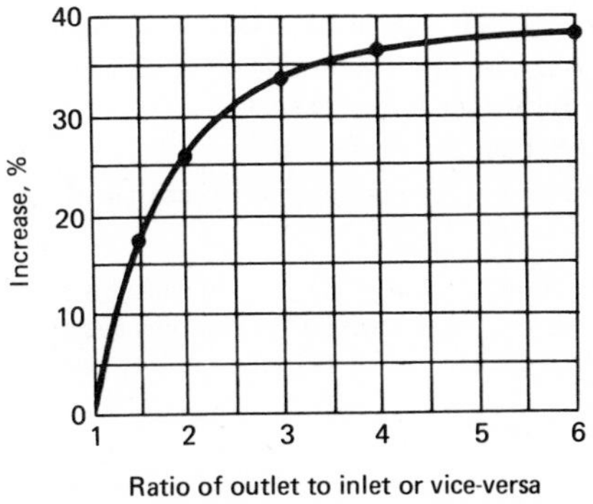

Figure 1-2 Increase in flow caused by excess of one opening over another.

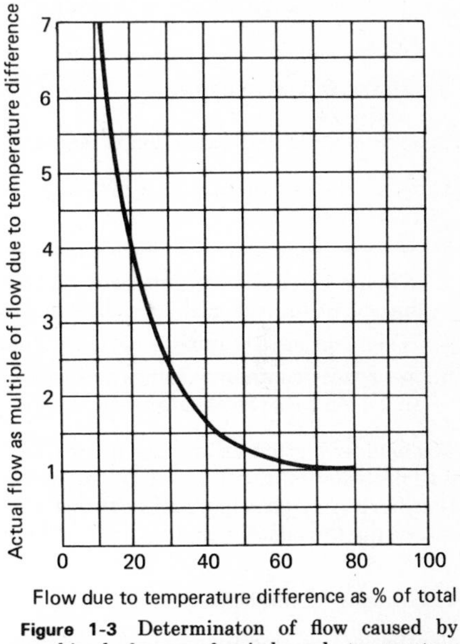

Figure 1-3 Determinaton of flow caused by combined forces of wind and temperature difference.

Selection of Natural Ventilation Equipment

Once the number of cubic feet per minute of air required for natural ventilation and the inlet and outlet areas have been determined, the following items of equipment can be selected:

1. Operable windows
2. Weatherproof louvers
3. Doors

4. Operable skylights
5. Roof ventilators (gravity type)
6. Specially designed inlets and outlets

The location and size of the equipment (inlets and outlets), as previously discussed, are critical. The possible requirement that all or part of the ventilating equipment may have to be operating during severe precipitation storms, without incurring water and wind damage to the space, must not be overlooked when selecting equipment.

Gravity Roof Ventilator

In general, all equipment should be selected for high flow-coefficient values. One of the most efficient units is the gravity roof ventilator. A roof ventilator should be selected for its:

1. Ability to utilize wind energy to induce flow by centrifugal or ejector action and the chimney effect.
2. Ruggedness.
3. Corrosion resistance.
4. Stormproof features (weatherproof inlets and outlet). Ventilators can be stationary, pivoting, oscillating, or rotating. They may also have manual dampers or dampers controlled automatically by a space thermostat or the wind velocity.

MECHANICAL VENTILATION

General mechanical ventilation is required:

1. When the design quantities of outside ventilation and exhaust air cannot be supplied continuously by natural forces
2. When it is mandatory to have a positive supply and/or exhaust ventilating system
3. When it is required to pressurize an area by supplying substantially more ventilation supply air (outside air) than the exhaust or return air from the space
4. When a process requires a specified quantity of supply or exhaust air
5. For spaces containing fumes and vapor with specific gravity greater than that of air (in which case the exhaust intakes must be located at the floor level)

Mechanical ventilation systems range from the simplest type (consisting of through-the-wall propeller fans with roof-type exhaust fans and manual controls) to complex systems that have multiple supply and exhaust fans, distribution ducts, registers and/or grilles, filters, duct insulation, and automatic controls.

Calculations for Mechanical Ventilation

Ventilation systems are designed to perform one or both of the following functions:

1. To control odors, maintain acceptable O_2 and CO levels, and to provide the quantities of supply and exhaust air required by the processes within an area
2. To maintain space temperatures (as far as possible without air conditioning) at a specified design temperature

Pressure Requirements

The ventilation-air quantity calculated from the appropriate factors listed in Table 1-14 should be considered the exhaust-air quantity. Therefore, in order to determine the quantity of ventilation supply air, the designer has to determine whether the space in question should be under a negative, neutral, or positive pressure.

Positive Pressure. When there is no code or process requirement, it is suggested that the ventilated space be under a positive pressure. That is, the supply air quantity to the space is greater than the exhaust air quantity from the space. Under these conditions the quantity of supply air can be up to 10 percent greater than the quantity of exhaust air.

Neutral Pressure. The supply air quantity equals the exhaust air quantity.

Negative Pressure. The supply air quantity is less than the exhaust air quantity. If no specific value is stipulated, then the quantity of supply air should not be less than 90 percent of the quantity of exhaust air.

Controlling Odors

The quantity of ventilation exhaust air required to control odors and maintain acceptable O_2 and CO levels within a space can be calculated with the following formula:

$$Q_p = f_{vp}N_p \tag{9}$$

where

Q_p = ventilation exhaust air required, ft³/min, because of number of workers within the space (metric conversion: $\text{ft}^3/\text{min} \times 0.000472 = \text{m}^3/\text{s}$)

f_{vp} = ventilation airflow, ft³/min, per person based on activity level of those working. This value can be obtained from Table 1-14.

N_p = number of people working in or assigned to the space

Note: If the physical activity levels differ markedly among the workers within a space, it is a matter of engineering judgment whether the ventilation rates should be calculated separately to determine the total airflow or whether one activity level can be selected to determine the required rate.

The quantity of ventilation exhaust air from spaces where people are not permanently assigned (transient), or from spaces where the number of people assigned per square foot of floor area is so small that another factor (e.g., area, number of workers) will govern, can be calculated with the following formula:

$$Q_s = f_sN_u \tag{10}$$

where

Q_s = ventilation exhaust air required, ft³/min, to flow from the space (metric conversion: $\text{ft}^3/\text{min} \times 0.000472\ \text{m}^3/\text{s}$)

f_s = ventilation airflow per unit, such as $\text{ft}^3/(\text{min})(\text{ft}^2)$ of floor area or ft³/min per locker. This value is obtained from Table 1-14.

N_u = number of units corresponding to the space and f_s

Process Requirements

The process exhaust requirements should be obtained from the process engineering department. As a second choice, the manufacturer of the process equipment should be consulted.

The quantity of ventilation air calculated by Eq. (9) or (10) for the same space must be compared with the process supply and exhaust rates. If it is greater than the required process exhaust, then the quantity of ventilation exhaust air calculated by Eq. (9) or (10) will govern and be used to determine the quantity of space supply air. If, however, the process exhaust air requirements are greater than the ventilation exhaust calculated by Eq. (9) or (10), then the quantity of process exhaust air will govern and will be used to determine the quantity of ventilation supply air.

The ventilation supply and exhaust rates that satisfy the space odor, O_2, CO, and process requirements must be checked to see if they have sufficient cooling capacity to remove the internal space heating load during the summer season.

Internal Space Heat Gain

The internal space heat gain is the sum total of the sensible heat from people, lights, process, and solar radiation. For our ventilation-load calculations only, the space heat

TABLE 1-14 Ventilation Requirements for Occupants[a]

Typical application	Estimated persons per 100 ft^2 (98.9 m^2) of floor area	Required ventilation air, per human occupant: Min ft^3/min per person (m^3/s)	Required ventilation air, per human occupant: Recommendations ft^3/min (m^3/s) per person
Commercial public rest rooms	100	15 (0.007)	20–25 (0.0094–0.012)
General requirements, merchandising (apply to all forms unless specially noted)	30	7	10–15 (0.0033) (0.0047–0.0071)
Sales floors (upper floors)	20	7 (0.0033)	10–15 (0.0047–0.0071)
Storage areas	5	5 (0.0024)	7–10 (0.0033–0.0047)
Dressing rooms	—	7 (0.0033)	10–15 (0.0033–0.0071)
Malls and arcades	40	7 (0.0033)	10–15 (0.0033–0.0071)
Shipping and receiving	10	15 (0.007)	15–20 (0.0033–0.0094)
Warehouses	6	7 (0.0033)	10–15 (0.0033–0.0071)
Elevators	—	7 (0.0033)	10–15 (0.0047–0.0071)
Meat processing[b]	10	5 (0.0024)	5 (0.0024)
Pharmacists, workrooms	10	20 (0.0094)	25–30 (0.012–0.014)
Greenhouses[c,d]	1	5 (0.0024)	7–10 (0.0033–0.0047)
Dining rooms	70	10 (0.0047)	15–20 (0.0071–0.0094)
Kitchens[m]	20	30 (0.014)	35 (0.017)
Photo studios			
Camera rooms	10	5 (0.0024)	7–10 (0.0033–0.0047)
Stages[f]			
Darkrooms	10	10 (0.0047)	15–20 (0.0071–0.0094)

Shoe repair shops (combined workrooms and trade areas)	10	10 (0.0047)	15–20 (0.0071–0.0094)
Garages, auto repair shops, service stations			
Parking garages[e]	——	1.5[e] (0.0076)[e]	2–3[e] (0.010–0.015)[e]
Auto repair workroom[e]	——	1.5[e] (0.0076)[e]	2–3[e] (0.010–0.015)[e]
Service station offices	——	20 (0.0033)	7–15 (0.0033–0.0071)
Transportation			
Waiting room	50	15 (0.007)	20–25 (0.0094–0.012)
Ticket & baggage areas, corridors, and gate areas	——	50	20–25 (0.0094–0.012)
Control towers	50	25 (0.012)	30–35 (0.014–0.017)
Hangars[g]	2	10 (0.0047)	15–20 (0.0071–0.0094)
Platform	150	10 (0.0047)	15–20 (0.0071–0.0094)
Concourses	150	10 (0.0047)	15–20 (0.0071–0.0094)
Repair shops	——	10 (0.0047)	15–20 (0.0071–0.0094)
Offices			
General office space	10	15 (0.0071)	15–25 (0.0071–0.012)
Conference rooms	60	25 (0.012)	30–40 (0.014–0.019)
Drafting rooms, art room	——	20 (0.0033)	7–15 (0.033–0.0071)
Doctors' consultation room	——	10 (0.0047)	10–15 (0.0033–0.00471)
Waiting rooms	30	10 (0.0047)	15–20 (0.0071–0.0094)
Lithographing rooms[h]	20	7 (0.0033)	10–15 (0.0047–0.0071)
Diazo printing rooms[h]	20	7 (0.0033)	10–15 (0.0047–0.0071)
Computer rooms	20	5 (0.0024)	7–10 (0.0033–0.0047)

TABLE 1-14 Ventilation Requirements for Occupants[a] *(Continued)*

Typical application	Estimated persons per 1000 ft^2 (98.9 m^2) of floor area	Required ventilation air, per human occupant: Min ft^3/min per person (m^3/s)	Required ventilation air, per human occupant: Recommendations ft^3/min (m^3/s) per person
Keypunching rooms	30	7 (0.0033)	10–15 (0.0047–0.0071)
Communication			
Press rooms	100	15 (0.0071)	20–25 (0.0094–0.012)
Composing rooms	30	7 (0.0033)	10–15 (0.0047–0.0071)
Engraving shops	30	7 (0.0033)	10–15 (0.0047–0.0071)
Telephone switchboard rooms (manual)	50	7 (0.0033)	10–15 (0.0047–0.0071)
Telephone switchboard rooms (automatic)	—	7 (0.0033)	10–15 (0.0047–0.0071)
Teletype writer and facsimile rooms	—	5 (0.0024)	7–10 (0.0033–0.0047)
Research institutes	50	15 (0.0071)	20–25 (0.0094–0.012)
Laboratories[i]	50	15 (0.0071)	20–25 (0.0094–0.012)
Machine shops	50	15 (0.0071)	20–25 (0.0094–0.012)
Darkrooms, spectroscopy rooms	50	10 (0.0047)	15–20 (0.0071–0.0094)

Animal rooms[j]	20	40 (0.019)	45–50 (0.021–0.024)
Miscellaneous			
Classrooms	50	10 (0.0047)	10–15 (0.0047–0.0071)
Libraries	20	7 (0.0033)	10–12 (0.0047–0.0057)
Locker rooms[k]	20	30 (0.014)	40–50 (0.019–0.024)
Incinerator service areas[l]	—	5 (0.0024)	7–10 (0.0033–0.0047)

[a]Industrial (including agricultural processing) occupational safety laws in the various states in the United States usually regulate the ventilation requirements, which are almost always far in excess of the ventilation requirements for the occupants. ASHRAE Standard 62-73 lists requirements for occupants only. In general, 25 ft^3/min (0.012 m^3/s) per occupant is recommended, except for mining or metalworking, where 40 ft^3/min (0.019 m^3/s) per occupant is recommended.

[b]Spaces maintained below 50°F (10°C) are not covered by these requirements unless the occupancy is continuous. Ventilation from adjoining space is permissible. When occupancy is intermittent, infiltration will normally exceed ventilation requirements.

[c]Ventilation to optimize plant growth, temperature, and humidity will almost always be greater than shown.

[d]Maximum alllowable concentration (MAC) for sulfur dioxide is 30 $\mu g/m^3$.

[e]Cubic feet per minute per square foot (cubic meters per second per square meter) of floor area.

[f]Thermal effects probably determine requirements.

[g]Special solvent and exhaust problems handled separately.

[h]Installed equipment must incorporate positive exhaust and control (as required) of undesirable contaminants (toxic or otherwise).

[i]Special contaminant control system may be required.

[j]Special requirements or codes may determine requirements.

[k]Cubic feet per minute per locker.

[l]Special exhaust systems required.

[m]Exhaust to outside; source control as required.

loss through the exterior walls, doors, windows, and roof can be neglected. Note that since we are not considering air conditioning, the space design temperature must be higher than the outdoor design temperature.

Personnel. The heat gain from the personnel working within the space can be calculated from

$$g_p = f_{shp} N_p \tag{11}$$

where

g_p = heat gain from the people within the space, Btu/h (metric conversion: Btu/h × 0.0293 = W)

f_{shp} = sensible heat rate per person, Btu/h. This value depends on the activity level of the workers. It is obtained from Table 1-15.

N_p = number of people working or assigned to the space

Lights. The heat gain from the lights is equal to the total wattage installed in the space multiplied by 3.4.

$$g_l = 3.4 t_l \tag{12}$$

where

g_l = heat gain from electric lights, Btu/h (metric conversion: Btu/h × 0.0293 = W)

t_l = total number of watts from the lights that are energized during the time the design outdoor air temperature is anticipated. When fluorescent lights are used, the heat generated in the ballast must be added to the tube wattage to obtain the total lamp wattage. It is usually satisfactory to increase the tube wattage by 25 percent to include the ballast wattage.

Process Load. Information concerning the amount of heat from the process should come from the process engineering department and the quantity should be expressed in British thermal units per hour. This load, like the lighting load, must coincide in the time with the maximum design outdoor air temperature.

If process-load data cannot be obtained from the process engineering department, then an estimate can be made by the following methods:

1. If the process is electrically powered with all the process heaters, motors, etc., located within the space, and if it is possible to measure the average power demand in watts or watthours for 1 h coincident with the time of the design outdoor temperature, then the process internal load will equal this value converted directly into British thermal units per hour.
2. If this is not possible, list the total nameplate horsepower and heater loads of all the process equipment that will be energized during the time of the design outdoor temperature. The values in this list should all be converted to one set of units, then multiplied by a diversity factor to obtain the simultaneous demand. Generally, multiplying the total connected nameplate load by 50 percent will give a reasonable approximation.
3. If the process equipment is supplied with other energy sources, such as steam, hot water, gas, and oil, then the manufacturer of the equipment should be contacted for the internal heat load to the space.
4. The internal power load is generally constant throughout the year. If this is the case, then the internal load may be measured at any season of the year.

Total Internal Space Heat Load. The total internal space heat load is equal to the heat gain from personnel plus heat gain from lighting plus the total process load:

$$g_{ti} = g_p + g_l + g_{\text{process}} \tag{13}$$

where

g_{ti} = total internal heat gain to the space (people + lights + process), Btu/h (metric conversion: Btu/h × 0.0293 = W)

To check whether the ventilation flow, in cubic feet per minute (to control odor, CO, and O_2), is adequate to maintain the design space temperature in the summer, the following formula should be used:

$$t_i = t_o + \frac{g_{ti}}{1.08Q_v} \tag{14}$$

where

t_i = inside space dry-bulb temperature, °F [metric conversion: (°F − 32)5/9 = °C]

t_o = outside design dry-bulb temperature in summer, °F (See "Climatic Conditions.")

Q_v = ventilation airflow determined in this section, ft³/min

If t_i from Eq. (14) is less than the inside design space dry-bulb temperature, then the ventilation air Q_v is adequate to maintain space dry-bulb temperatures in the summer. The space humidification should be checked at this point (see the discussion on dehumidification under "Humidity Control"). If t_i from Eq. (14) is greater than the inside

TABLE 1-15 Rates of Heat Gain from Occupants of Conditioned Spaces*

Degree of activity	Typical application	Total heat† adjusted		Sensible heat		Latent heat	
		Btu/h	kcal/h	Btu/h	kcal/h	Btu/h	kcal/h
Seated, very light work, writing	Offices, hotels, apts.	420	105	230	55	190	50
Seated, eating	Restaurant‡	580‡	145	255	60	325	80
Seated, light work, typing	Offices, hotels, apts.	510	130	255	60	255	65
Standing, light work or walking slowly	Retail stores, banks	640	160	315	80	325	80
Light bench work	Factory	780	195	345	90	435	110
Walking, 3 mi/h, light machine work	Factory	1040	260	345	90	695	170
Heavy work, heavy machine work, lifting	Factory	1600	400	565	140	1035	260

*Tabulated values are based on 78°F room dry-bulb temperature. For 80°F, the total heat remains the same, but the sensible heat value should be decreased by approximately 8 percent and the latent heat values increased accordingly.

†Adjusted total heat gain is based on normal percentage of men, women, and children for the application listed, with the postulate that the gain from an adult female is 85 percent of that for an adult male, and the gain from a child is 75 percent of that for an adult male.

‡Adjusted total heat value for eating in a restaurant includes 60 Btu/h for food per individual (30 Btu sensible and 30 Btu latent).

All values rounded to nearest 5 W or kilocalories per hour or to nearest 10 Btu/h.

design space temperature, then the ventilation air Q_v is inadequate and must be increased or a higher design temperature must be accepted.

Ventilation Air Quantity

To calculate the quantity of ventilation air required to maintain the original space design dry-bulb temperature, the following formula should be used:

$$Q = \frac{g_{ti}}{1.08(t_i - t_o)} \tag{15}$$

where

Q = summer ventilation air quantity, ft³/min (metric conversion: ft³/min × 0.000472 = m³/s)

g_{ti} = total internal heat gain to the space, Btu/h (people + lights + process)

t_i = inside dry-bulb design air temperature, °F

t_o = outside dry-bulb design air temperature, °F

The preceding procedure is repeated for each space within the structure.

For energy conservation the additional ventilation required during the summer months should be provided by two-speed fans or additional exhaust fans that will be used only during warm weather. The reader is referred to Sec. 15 of this book for a comprehensive discussion of energy conservation technology.

As the temperature differential between the inside space and outside design diminishes, the required ventilation airflow increases rapidly. There is a point where it will be more economical to air-condition the space, rather than to try to ventilate it. To determine whether it is more economical to air-condition or to ventilate, a detailed study should be performed. Although such a study is beyond the scope of this chapter, the following guidelines are suggested:

1. If the design space temperature is no more than 5°F (2.8°C) higher than the outdoor design temperature, air conditioning is usually more economical than ventilation to maintain the space temperature.
2. If the design space temperature is between 5 and 10°F (2.8 and 5.6°C) higher than the outside design temperature, a ventilation system may be more economical than an air-conditioning system to maintain space dry-bulb temperature.
3. When the space temperature is more than 10°F (5.6°C) higher than the outdoor design temperature, a properly designed ventilating system will usually be capable of maintaining the space dry-bulb conditions. However, space dehumidification may be required to control the humidity.

Selection of System Size

After determining the ventilation supply and exhaust quantities for each space within the structure, the designer must decide on the type and size of mechanical ventilating system that will most effectively meet the project's needs.

Some of the factors to be considered are:

1. Should the total building be ventilated with one system or with more than one independent system?
2. Is a supply fan system required to provide the makeup air for the exhaust system?
3. Although we are only considering ventilation in this section, space for heating equipment, depending upon the geographic location, may also be required.
4. The fewer the number of pieces of mechanical equipment installed, the less maintenance is required.

If the structure is one-story, with spaces having at least one external wall, it may be more economical to install rooftop exhaust fans with through-the-wall supply fans, thereby keeping the ductwork and automatic controls to a minimum.

On the other hand, in multistory structures or single-story structures with numerous spaces per floor, a ducted supply and exhaust system is more common. The designer should combine compatible respective supply and exhaust systems where governing codes or process requirements permit. This will usually require more sophisticated automatic controls, especially if the process equipment in all the spaces is not operated at the same time.

Good engineering practice as well as energy conservation considerations mandate that all spaces operating at the same time be served by the same system. This does not preclude the use of many fans or systems. It does mean that areas that operate at different times or processes that are operated intermittently should be on separate systems, so that they can be shut off when not required. Although duct design is beyond the scope of this chapter, the reader is referred to standard references such as the latest edition of *Industrial Ventilation, A Manual of Recommended Practice*[1] and ASHRAE's *Handbook of Fundamentals*.[4] As a guide, Table 1-16 lists representative velocities in ducted systems.

The reader is referred to the following sections of this book for additional related information:

- Section 11 for instrumentation and automatic controls
- Section 12 for noise and vibration control
- Section 15 for energy conservation techniques

In order to achieve the maximum cooling effect (air motion in the occupied zone), registers or grilles are preferred to diffusers for distributing the ventilation supply air.

Depending on the geographic location, different quantities of supply air may be highly desirable to maximize the cooling effect during the warmer months and minimize the energy cost during the heating season. The capacity of the exhaust fan system need not be increased during the summer ventilation season, if additional, strategically located relief air capability is provided. Methods of providing increased supply air (depending on the building interior space arrangement, construction, and proposed ven-

TABLE 1-16 Suggested and Maximum Duct Velocities

	Recommended velocities, ft/min (m/s)		Maximum velocities, ft/min (m/s)	
Designation	Schools, theaters, public buildings	Industrial buildings	Schools, theaters, public buildings	Industrial buildings
Outside air intakes	500 (2.5)	500 (2.5)	900 (4.6)	1200 (6.1)
Filters*	300 (1.5)	350 (1.8)	350 (1.8)	350 (1.8)
Heating coils	500 (2.5)	600 (3)	600 (3)	700 (3.6)
Air washers	500 (2.5)	500 (2.5)	500 (2.5)	500 (2.5)
Suction connections	800 (4.1)	1000 (5.1)	1000 (5.1)	1400 (7.1)
Fan outlets	1300–2000 (6.6–10.2)	1600–2400 (8.1–12.2)	1500–2200 (7.6–11.2)	1700–2800 (8.6–11.2)
Main duct	1000–1300 (5.1–6.5)	1200–1800 (6.1–9.1)	1100–1600 (5.6–8.1)	1300–2200 (6.6–14.2)
Branch ducts	600–900 (3–4.6)	800–1000 (4.1–5.1)	800–1300 (4.1–6.6)	1000–1800 (5.1–9.1)
Branch risers	600–700 (3–3.6)	800 (4.1)	800–1200 (4.1–6.1)	1000–1600 (5.1–8.1)

*These velocities are for total face area, not the net free area; other velocities in this table are for net free area.

TABLE 1-17 High Environmental Dry- and Wet-Bulb Temperatures* That Can Be Tolerated in Daily Work by Healthy, Acclimatized Employees Wearing Warm-Weather Clothing

Activity	Relative humidity, %	Air velocity: 15–25 ft/min (0.08–0.13 m/s) Dry-bulb, °F(°C)	15–25 ft/min (0.08–0.13 m/s) Wet-bulb, °F(°C)	100 ft/min (0.5 m/s) Dry-bulb, °F(°C)	100 ft/min (0.5 m/s) Wet-bulb, °F(°C)	300 ft/min (1.5 m/s) Dry-bulb, °F(°C)	300 ft/min (1.5 m/s) Wet-bulb, °F(°C)
Summer season,	80	89 (31.7)	84 (28.9)	91 (32.8)	85 (29.4)	93 (33.9)	87 (30.6)
Light sedentary activities	60	94 (34.4)	82 (27.8)	96 (35.6)	84 (28.9)	98 (36.7)	85 (29.4)
	40	100 (37.8)	79 (26.1)	101 (38.3)	81 (27.2)	103 (39.4)	82 (27.8)
	20	109 (42.8)	75 (23.9)	110 (43.3)	75 (23.9)	110 (43.3)	75 (23.9)
	5	119 (48.3)	69 (20.6)	118 (47.8)	69 (20.6)	117 (47.2)	68 (20.0)
Summer season, heavy work	80	83 (28.3)	78 (25.6)	86 (30.0)	81 (27.2)	89 (31.7)	83 (28.3)
	60	88 (31.1)	76 (24.4)	90 (32.2)	78 (25.6)	93 (33.9)	80 (26.7)
	40	93 (33.9)	73 (22.8)	95 (35)	75 (23.9)	97 (36.1)	76 (24.4)
	20	100 (37.8)	69 (20.6)	101 (38.3)	70 (21.1)	102 (38.9)	70 (21.1)
	5	107 (41.7)	64 (17.8)	107 (41.7)	54 (17.8)	106 (41.1)	63 (17.2)
Winter season, light or heavy	80	78 (25.6)	73 (22.8)	81 (27.2)	77 (25.0)	85 (29.4)	79 (26.1)
	60	81 (27.2)	71 (21.7)	85 (29.4)	74 (23.3)	88 (31.1)	76 (24.4)
	40	86 (30.0)	68 (20.0)	89 (31.7)	70 (21.1)	91 (32.8)	72 (22.2)
(75 ET)	20	91 (32.8)	63 (17.2)	93 (33.9)	65 (18.3)	94 (34.4)	66 (18.9)
	5	97 (36.1)	58 (14.4)	97 (36.1)	58 (14.4)	97 (36.1)	59 (15)

*Including radiation effect.

TABLE 1-18 Suggested Terminal Velocity for Supply Registers and Grilles

Area	Terminal Velocity ft/min	m/s
Private office	50–100	0.25–0.50
General office	100–150	0.50–0.75
Industrial plants and process areas	150–200	0.75–1
Corridors	300	1.5

tilating system) can vary from opening windows to installing manual or automatic relief dampers (connected to exterior relief louvers), to increasing the proposed exhaust system's capacity proportionally.

The supply registers or grilles should be selected so that they will deliver the required air quantities in cubic feet per minute, have terminal velocities comparable to those shown in Table 1-18, and have a maximum resistance to airflow of 0.25 in H_2O (62.25 Pa).

If, during the summer, the cooling effect from the ventilation system is to be maximized, higher terminal velocities should be considered (see Table 1-18). During the heating season the lower velocities will result in a more comfortable environment for the staff.

Supply registers and grilles should be the double-deflecting type. Greater care is required for the selection of systems using supply grilles than for registers to assure proper air-conditioning capability. In lieu of air conditioning, the degree of cooling comfort can be maximized if the supply air is capable of absorbing heat. If the dry-bulb temperature of the air is reasonably lower than human body temperature, or if its humidity is low enough to allow evaporation of sweat, a blast of air with a velocity up to 400 or 500 ft/min (2 or 2.5 m/s) directed at the workers will be effective. Velocities higher than 500 ft/min (2.5 m/s) should not be used. If the air is at or above body temperature, or nearly saturated, this method of blast or velocity cooling should not be considered since its use will actually decrease the workers' comfort level.

Table 1-17 can be used as a guide to the maximum environmental dry- and wet-bulb temperatures that healthy workers can tolerate in an 8-h day.

In general, in office space or equivalent, return registers or grilles, when located at or adjacent to the floor, should be sized for low air volumes per grille or register, and the velocities should be between 300 and 500 ft/min (1.5 and 2.5 m/s). If return registers and grilles are mounted high on the sidewall or in the ceiling, they may be selected to carry air at 600 to 1200 ft/min (3 to 6 m/s) with no objectionable noise generated. The return

TABLE 1-19 Suggested Gross Velocities for Return Air Registers or Grilles

	Gross velocity access face, ft/min (m/s)
Office area (location of returns)	
Above occupied zone	800–1250 (4.1–6.4)
Within occupied zone not near personnel	600–800 (3–4.1)
Within occupied zone near personnel	300–600 (1.5–3)
Door or wall louvers	500–1000 (2.5–5)
Industrial plants Process areas—Corridors	800–1200 (4.1–6)

registers or grilles should be selected so that they will deliver the required air quantities, have velocities over the gross area as shown in Table 1-19, and have a maximum resistance to airflow of 0.2 in of water (50 Pa).

Exterior intake (supply) storm louvers should be designed to minimize the entrance of rain and/or snow. A plenum should be provided immediately behind the louver, with its bottom pitched so that any moisture entering the louver will drain out the bottom portion of the louver. All joints in the lower portion of this plenum must be waterproof. Louvers should be provided with bird screen [0.25 in (0.006 m) grid minimum]. In cold climates there should be a means to stop the airflow so as to minimize heat loss when the system is not operating. Dampers are commonly used for this purpose; they can be manual or automatic depending on the complexity of the system.

HUMIDITY CONTROL

GENERAL

Depending on the activity or functions taking place within a space, humidity control may or may not be required.

For comfort and prevention of material deterioration, the relative humidity generally should not exceed 60 percent at any point in occupied spaces. A notable exception is in textile mills. Normally the relative humidity should not fall below 20 percent to prevent human throats and nostrils from becoming dry and furniture from drying excessively.

COMFORT

During the heating season it is beneficial to the health and comfort of the people within a space to maintain the relative humidity above 25 to 30 percent. In colder climates the possibility of excessive condensation on cold surfaces such as exterior glass windows

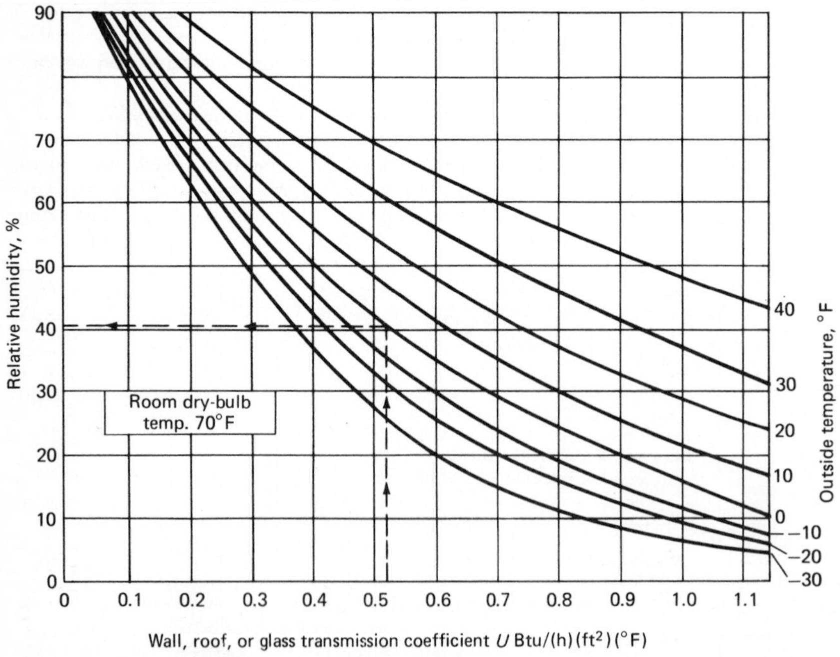

Figure 1-4 Maximum room RH without condensation.

must be avoided. See Fig. 1-4 for maximum space relative humidity without condensation on surfaces.

For a comfort heating application it is possible to supply an acceptable degree of comfort to the average person with a relatively low space dry-bulb temperature and a higher relative humidity in lieu of a relatively high dry-bulb temperature and lower relative humidity. *Under certain conditions this can result in lower total energy costs.*

Humidity control during the summer months may be justified from both a personal comfort and productivity point of view. The locations in the world where it is more economical (or for that matter, physically possible) to control the humidity (dehumidify) for personnel comfort (space design conditions around 80°F db and 50 percent RH) without air conditioning are extremely few.

Table 1-20 is to be used with Fig. 1-4 to correct relative humidities where condensation will occur for space or room temperatures of 60 and 80°F (15.6 and 26.7°C).

TABLE 1-20 Correction in Room RH, Percent, for Wall, Roof, or Glass Transmission Coefficient U

	Transmission Coefficient, Btu/h(ft²)(°F)					
	$U = 1.1$		$U = 0.65$		$U = 0.35$	
	Room Temp, °F db					
Outdoor temp, °F db	60	80	60	80	60	80
−30	+1.0	−1.0	+1.5	−2.0	+2.5	−2.0
−20	+1.0	−1.5	+2.5	−2.5	+3.0	−2.0
−10	+2.0	−2.0	+3.5	−3.0	+3.0	−2.0
0	+3.5	−2.5	+4.0	−4.0	+3.5	−2.5
10	+5.0	−3.5	+5.0	−4.5	+4.0	−3.0
20	+7.0	−4.0	+6.5	−5.0	+4.5	−3.5
30	+9.0	−7.5	+8.5	−6.0	+5.0	−4.0
40	+12.0	−9.5	+9.5	−7.5	+6.0	−4.5

PROCESS

It is not uncommon for a particular manufacturing process or product storage area to require that the space humidity and temperature be controlled year-round within a specified range. Whenever possible the manufacturer's required relative humidity and room temperature should be maintained. When the manufacturer's or process humidity range is not available, Table 1-21 can be used as a guide.

The industries and products in Table 1-21 will derive great benefit when protected against humidities lower than the levels shown. Humidifiers, with the capacity to maintain at least the relative humidities indicated, should be provided where required.

STATIC ELECTRICITY CONTROL

In explosive environments, humidity control that maintains the space relative humidity at or above 50 percent is strongly recommended to minimize the possibility of static electricity causing an explosion. Some codes and insurance carriers specify minimum space humidity levels in explosive environments.

In situations where a person receiving a static electricity shock could sustain an injury or where a product could be damaged during manufacturing or processing, the economic merits of humidity control should be evaluated. There are other processes, such as computer applications, where humidity control is required not only to minimize the static electricity problem but also to stabilize the physical size of the computer key punch cards. Computer manufacturers' specifications should be carefully followed.

TABLE 1-21 Recommended Relative Humidities

Industries or products	RH, %	Industries or products	RH, %	Industries or products	RH, %
Abrasives	50–60	Cordage	60–70	Labels	40–50
Agronomy	60–70	Cotton	60–70	Laboratories	50–*
Air conditioning	30–40	Decals	50–60	Lace	50–60
Animal rearing	50–60	Egg storage	70–80	Leather	45–55
Antiques	30–40	Elastic yarns	50–60	Letterpress printing	40–50
Apple storage	70–80	Electronic computers	40–50	Lithography	45–55
Art galleries	30–40	Environmental chambers	*	Meats	75–85
Bag making	40–50	Film processing	50–60	Mullers	80–90
Bag storage	50–60	Film storage	40–50	Museums	40–50
Bakeries	60–70	Florists	50–60	Pharmaceuticals	*
Belting	50–60	Food storage	60–70	Photography	40–50
Bowling alleys	30–40	Fruit storage	70–80	Pipe organs	40–50
Braiding	45–55	Furniture	40–50	Printing	40–50
Breweries	65–75	Glass (lenses)	50–60	Radium	40–50
Cabinet making	30–40	Gloves	50–60	Rayon	45–50
Candy	40–50	Gluing	50–60	Silks	50–60
Carpet	50–60	Greenhouses	*	Synthetics	45–55
Cartons	40–50	Hatcheries	60–70	Tapes	40–50
Cellophane	40–50	Hats (fur felt)	50–60	Textiles	45–55
Ceramics	40–50	Horticulture	40–50	Tobacco	50–60
Cereals	35–45	Hosiery	50–60	Wood	40–50
Cigarettes	50–60	Hospitals	40–50	Wool	50–60
Cigars	60–70	Incubators	60–70	X-ray	45–55
Containers, paper	40–50	Knitting	50–60	Yarn	50–60

*For these applications the range can be so great that consultation with specialists in these areas is recommended.

DEHUMIDIFICATION

Space dehumidification is generally accomplished by supplying air to the space to be dehumidified at a sufficiently lower moisture content that the supply (ventilating) air can absorb the space latent load without exceeding the design condition. The three most common methods used to dehumidify a space are:

1. Passing the total supply (ventilation) air through a dehumidifier prior to discharging it within the space
2. Passing part of the supply (ventilation) air through a dehumidifier, then mixing the dehumidified airstream with the mainstream prior to discharging the total supply (ventilation) air within the space
3. Recirculating space air through one or more dehumidifiers located within or adjacent to the space to be dehumidified

Moisture can be removed from an airstream by passing it through a *mechanical* or *chemical* dehumidifier.

Mechanical

With *mechanical dehumidification* the airstream to be dehumidified is passed over the fins of a cooling coil where it is cooled below its dew point, thereby condensing moisture out of the incoming airstream. The amount of moisture condensed depends on how much lower the average coil temperature is than the dew point of the entering airstream. The greater the differential, the more water is condensed, and the drier the outgoing air. The desired cooling-coil temperature is usually controlled by circulating either a refrigerant (e.g., one of the fluorocarbons such as R11 or R22) or chilled water through the coils.

Chemical

With *chemical dehumidifiers* the airstream is brought in contact with a substance which absorbs moisture out of the airstream that is passed over it. The amount of moisture absorbed depends on various factors such as dew point of the entering air, strength of the sorbent, surface area of sorbent, efficiency of contact of air molecules with the sorbent, duration of contact with sorbent, etc. Since the sorbent becomes saturated with moisture and must be replaced or regenerated (usually heat-dried) to enable it to again absorb moisture, a duplex system is desirable. This will assure continuous operation with one of the units in operation while the other is in a regeneration cycle.

Sorbents

Sorbents are solid or liquid materials which have the property of extracting and holding other substances (usually gases or vapors, e.g., water vapor) brought into contact with them. The sorption process always generates heat, the major part of which is the result of the condensation of water vapor. For commercial dehumidifying systems and equipment, a sorbent should have the following characteristics:

1. Suitable vapor-pressure characteristics, including high absorptive capacity.
2. Stability, i.e., it should not break down structurally or chemically in its operation and application, and should resist contamination by impurities.
3. Relative chemical inertness, i.e., it should be noncorrosive, odorless, nontoxic, and nonflammable.
4. Low viscosity and good heat transfer characteristics (liquids); relatively high density to avoid excessive bulk (solids)
5. Capability of regeneration or reactivation with methods and temperatures generally available.
6. Ready availability at moderate cost.

Common absorbents are lithium chloride, calcium chloride, lithium bromide, silica gel, and alumina gel.

Load Calculations

General

The actual net internal moisture (latent) load of a space is the algebraic summation of the following loads:

1. Gain in latent load from the occupants (See Table 1-15.)
2. Net gain or loss in latent load from the process or products (gain if process or products give up moisture, loss if they absorb moisture)
3. Gain in latent load from aqueous (water) surfaces within the space
4. Gain in latent load from open flames within the space
5. Net gain or loss from water vapor migration through cracks, around doors, windows, and other openings*
6. Net gain or loss from water vapor transmitted through the building surfaces*
7. Net gain or loss from the moisture load of the ventilation (outside) air*

Generally, in a reasonably constructed building the internal space moisture load due to moisture migration through cracks and openings and its transmission through the building surfaces can be neglected.

Latent Loads

Latent Load from Personnel. The space latent load due to the personnel working within the space can generally be neglected if there is an average >300 ft^2 (27.9 m^2) of

*The load will be a gain in space latent load if the design space moisture content in grains of moisture per pound of dry air is less than the moisture content at the adjacent area or of the outside air, in grains of moisture per pound of dry air.

floor area per person. If necessary, however, it can be calculated using the following formula.

$$\text{PLL} = \text{LL} \times \text{NP} \times \frac{1}{h_{fg}} \tag{16}$$

where

PLL = personnel latent load per hour, lb/h (metric conversion: lb/h × 0.4536 = kg/h)

LL = latent load per hour, Btu/h per person. This value is obtained from Table 1-15. Enter at appropriate degree of activity of the personnel within the space.

NP = average number of personnel working, or in the space, for more than 1 h

h_{fg} = difference in enthalpy, Btu/lb between saturated vapor and liquid. This value is obtained from Table 1-22. Enter the space dry-bulb temperature in degrees Fahrenheit.

Latent Load from Process or Product. This value must be obtained from the process engineering department or its equivalent. It should be given in units of pounds (or kilograms) of moisture dissipated into the space or absorbed from the space per hour of production. Average hourly values are generally adequate.

Evaporation Load from Open Tanks. If this load is not available from the process engineering department and it can reasonably be assumed that the fluid within the tank or tanks is similar to water as far as evaporation is concerned, the following formulas can be used to estimate this load:

$$w_v = S\left(\frac{95 + 0.425V}{\lambda_v}\right)(e_w - e_a) \quad \text{For air flow parallel to the long axis of the tank} \tag{17}$$

$$w_v = S\left(\frac{201 + 0.88V}{\lambda_v}\right)(e_w - e_a) \quad \text{For air flow transverse to the long axis of the tank} \tag{18}$$

where

w_v = pounds of water evaporated into the airstream (space), lb/h (metric conversion: lb/h × 0.4536 = kg/h)

S = area of exposed water surface, ft²

V = velocity of air flowing across the tank, ft/min. This velocity can be the design velocity across the tank, or it can be movement across the tank. In existing systems it may be possible to measure the actual velocity across tanks.

λ_v = latent heat of evaporation, Btu/lb. This value is obtained from Fig. 1-5 by entering the graph with the temperature of the fluid in the tank.

e_w = vapor pressure of the liquid in the tank, inHg. This value is obtained from Table 1-23 by entering the table with the temperature of the fluid in the tank.

e_a = vapor pressure of the air (for our purposes) within the space, inHg $e_a = e_w h$, where h is space relative humidity. This is the design space relative humidity.

Open Flames. Generally, the amount of moisture added to a space as a result of burning a hydrocarbon or hydrogen fuel in an open flame within the space can be neglected.*

Latent Load from Migration Through Cracks and Openings. The infiltration latent load will increase the space design dehumidification load when the moisture content of the design outside air conditions exceeds the space design conditions and reduce

*If there is a need to calculate this value, refer to one of the many references on the combustion reactions of common fuels.

TABLE 1-22 Saturated Steam Table*

	Absolute Pressure, p		Enthalpy				Absolute Pressure, p		Enthalpy		
Temp. t, °F	lb/in²	inHg @ 32°F	Sat. liquid h_f	Evap. h_{fg}	Sat. vapor h_g	Temp. t, °F	lb/in²	inHg @ 32°F	Sat. liquid h_f	Evap. h_{fg}	Sat. vapor h_g
32	0.0886	0.1806	0	1075.1	1075.1	90	0.6980	1.421	58.00	1042.1	1100.1
34	0.0961	0.1957	2.01	1074.0	1076.0	92	0.7429	1.513	59.99	1040.9	1100.9
36	0.1041	0.2120	4.03	1072.9	1076.9	94	0.7902	1.609	61.98	1039.8	1101.8
38	0.1126	0.2292	6.04	1071.7	1077.7	96	0.8403	1.711	63.98	1038.7	1102.7
40	0.1217	0.2478	8.05	1070.5	1078.6	98	0.8930	1.818	65.98	1037.5	1103.5
42	0.1315	0.2677	10.06	1069.3	1079.4	100	0.9487	1.932	67.97	1036.4	1104.4
44	0.1420	0.2891	12.06	1068.2	1080.3	102	1.0072	2.051	69.96	1035.2	1105.2
46	0.1532	0.3119	14.07	1067.1	1081.2	104	1.0689	2.176	71.96	1034.1	1106.1
48	0.1652	0.3364	16.07	1065.9	1082.0	106	1.1338	2.308	73.95	1033.0	1107.0
50	0.1780	0.3624	18.07	1064.8	1082.9	108	1.2020	2.447	75.94	1032.0	1107.9
52	0.1918	0.3905	20.07	1063.6	1083.7	110	1.274	2.594	77.94	1030.9	1108.8
54	0.2063	0.4200	22.07	1062.5	1084.6	112	1.350	2.749	79.93	1029.7	1109.6
56	0.2219	0.4518	24.07	1061.4	1085.5	114	1.429	2.909	81.93	1028.6	1110.5
58	0.2384	0.4854	26.07	1060.2	1086.3	116	1.512	3.078	83.92	1027.5	1111.4
60	0.2561	0.5214	28.07	1059.1	1087.2	118	1.600	3.258	85.92	1026.4	1112.3
62	0.2749	0.5597	30.06	1057.9	1088.0	120	1.692	3.445	87.91	1025.3	1113.2
64	0.2949	0.6004	32.06	1056.8	1088.9	122	1.788	3.640	89.91	1024.1	1114.0
66	0.3162	0.6438	34.06	1055.7	1089.8	124	1.889	3.846	91.90	1023.0	1114.9
68	0.3388	0.6898	36.05	1054.5	1090.6	126	1.995	4.062	93.90	1021.8	1115.7
70	0.3628	0.7387	38.05	1053.4	1091.5	128	2.105	4.286	95.90	1020.7	1116.6
72	0.3883	0.7906	40.04	1052.3	1092.3	130	2.221	4.522	97.89	1019.5	1117.4
74	0.4153	0.8456	42.04	1051.2	1093.2	132	2.343	4.770	99.89	1018.3	1118.2
76	0.4440	0.9040	44.03	1050.1	1094.1	134	2.470	4.029	101.89	1017.2	1119.1
78	0.4744	0.9659	46.03	1048.9	1094.9	136	2.603	5.300	103.88	1016.0	1119.9
80	0.5067	1.032	48.02	1047.8	1095.8	138	2.742	5.583	105.88	1014.9	1120.8
82	0.5409	1.101	50.02	1046.6	1096.6	140	2.887	5.878	107.88	1013.7	1121.6
84	0.5772	1.175	52.01	1045.5	1097.5	142	3.039	6.187	109.88	1012.5	1122.4
86	0.6153	1.253	54.01	1044.4	1098.4	144	3.198	6.511	111.88	1011.3	1123.2
88	0.6555	1.335	56.00	1043.2	1099.2	146	3.363	6.847	113.88	1010.2	1124.1
						148	3.536	7.199	115.87	1009.0	1124.9

*t = temperature, °F
P = pressure absolute = gauge pressure plus 14.7 lb/in²
h_f = enthalpy of saturated liquid, Btu/lb
h_{fg} = Δ difference in enthalpy, Btu/lb, vapor to liquid or liquid to vapor
h_g = enthalpy of saturated vapor, Btu/lb

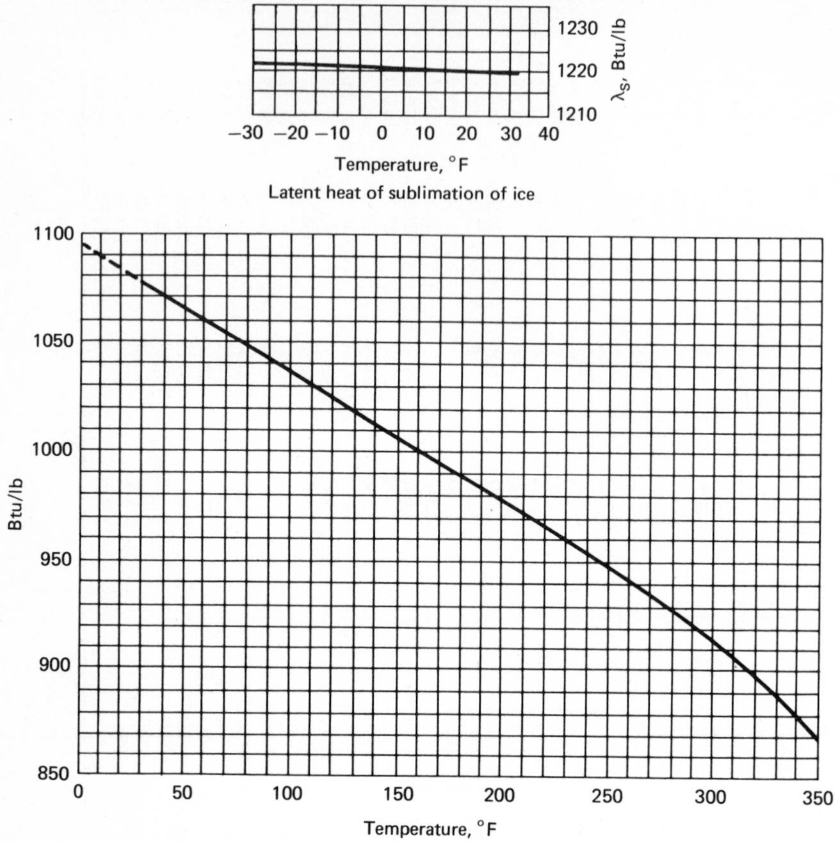

Figure 1-5 Latent heats of water and ice.

the load when it is less. From the psychrometric chart shown in Fig. 1-6, one can compare the design outside air moisture content with the space design conditions to determine if this load will increase or decrease the space dehumidification load.

When the outside air has less moisture than the space (which is usually the case during the heating season), this load can be omitted, since it will result in a conservative design. There are three general categories of this load:

Category One. The space is mechanically ventilated. Furthermore, the space is not under a negative pressure (that is exhausted and/or returned from it or does not exceed the air supplied to it). Under these conditions, the latent load due to infiltration for all practical reasons will be zero and therefore can be omitted.

Category Two. The space is mechanically ventilated. Furthermore, the space is under a negative pressure. Under these conditions the latent load from the infiltration must be accounted for. Assuming that the difference between the total air quantity exhausted from the space and that supplied to it will come from the outside air infiltrating the building, one can reasonably estimate this load by using the following formula.

$$\text{ILL} = \frac{(\text{exhaust} - \text{supply})}{100{,}000} \times 60\,\frac{\text{min}}{\text{h}}\,(f) \qquad (19)$$

TABLE 1-23 Vapor Pressures e_w of Ice* and Water,† inHg

t, °F	0	1	2	3	4	5	6	7	8	9
−20	.0126	.0119	.0112	.0106	.0100	.0095	.0089	.0084	.0080	.0075
−10	.0222	.0209	.0199	.0187	.0176	.0168	.0158	.0150	.0142	.0134
−0	.0376	.0359	.0339	.0324	.0306	.0289	.0275	.0259	.0247	.0233
+0	.0376	.0398	.0417	.0441	.0463	.0489	.0517	.0541	.0571	.0598
10	.0631	.0660	.0696	.0728	.0768	.0810	.0846	.0892	.0932	.0982
20	.1025	.1080	.1127	.1186	.1248	.1302	.1370	.1429	.1502	.1567
30	.1647	.1716	.1803	.1878	.1955	.2035	.2118	.2203	.2292	.2383
40	.2478	.2576	.2677	.2782	.2891	.3004	.3120	.3240	.3364	.3493
50	.3626	.3764	.3906	.4052	.4203	.4359	.4520	.4686	.4858	.5035
60	.5218	.5407	.5601	.5802	.6009	.6222	.6442	.6669	.6903	.7144
70	.7392	.7648	.7912	.8183	.8462	.8750	.9046	.9352	.9666	.9989
80	1.032	1.066	1.102	1.138	1.175	1.213	1.253	1.293	1.335	1.378
90	1.422	1.467	1.513	1.561	1.610	1.660	1.712	1.765	1.819	1.875
100	1.932	1.992	2.052	2.114	2.178	2.243	2.310	2.379	2.449	2.521
110	2.596	2.672	2.749	2.829	2.911	2.995	3.081	3.169	3.259	3.351
120	3.446	3.543	3.642	3.744	3.848	3.954	4.063	4.174	4.289	4.406
130	4.525	4.647	4.772	4.900	5.031	5.165	5.302	5.442	5.585	5.732
140	5.881	6.034	6.190	6.350	6.513	6.680	6.850	7.024	7.202	7.384
150	7.569	7.759	7.952	8.150	8.351	8.557	8.767	8.981	9.200	9.424
160	9.652	9.885	10.12	10.36	10.61	10.86	11.12	11.38	11.65	11.92
170	12.20	12.48	12.77	13.07	13.37	13.67	13.98	14.30	14.62	14.96
180	15.29	15.63	15.98	16.34	16.70	17.07	17.44	17.82	18.21	18.61
190	19.01	19.42	19.84	20.27	20.70	21.14	21.59	22.05	22.52	22.99
200	23.47	23.96	24.46	24.97	25.48	26.00	26.53	27.07	27.62	28.18
210	28.75	29.33	29.92	30.52	31.13	31.75	32.38	33.02	33.67	34.33
220	35.00	35.68	36.37	37.07	37.78	38.50	39.24	39.99	40.75	41.52
230	42.31	43.11	43.92	44.74	45.57	46.41	47.27	48.14	49.03	49.93
240	50.84	51.76	52.70	53.65	54.62	55.60	56.60	57.61	58.63	59.67
250	60.72	61.79	62.88	63.98	65.10	66.23	67.38	68.54	69.72	70.92
260	72.13	74.36	74.61	75.88	77.16	78.46	79.78	81.11	82.46	83.83
270	85.22	86.63	88.06	89.51	90.97	92.45	93.96	95.49	97.03	98.61
280	100.2	101.8	103.4	105.0	106.7	108.4	110.1	111.8	113.6	115.4
290	117.2	119.0	120.8	122.7	124.6	126.5	128.4	130.4	132.4	134.4
300	136.4	138.5	140.6	142.7	144.8	147.0	149.2	151.4	153.6	155.9
310	158.2	160.5	162.8	165.2	167.6	170.0	172.5	175.0	177.5	180.0
320	182.6	185.2	187.8	190.4	193.1	195.8	198.5	201.3	204.1	206.9
330	209.8	212.7	215.6	218.6	221.6	224.6	227.7	230.8	233.9	237.1
340	240.3	243.5	246.8	250.1	253.4	256.7	260.1	263.6	267.1	270.6
350	274.1	277.7	281.3	284.9	288.6	292.3	296.1	299.9	303.8	307.7
360	311.6	315.5	319.5	323.5	327.6	331.7	335.9	340.1	344.4	348.7
370	353.0	357.4	361.8	366.2	370.7	375.2	379.8	384.4	389.1	393.8
380	398.6	403.4	408.2	413.1	418.1	423.1	428.1	433.1	438.2	443.4
390	448.6	453.9	459.2	464.6	470.0	475.5	481.0	486.6	492.2	497.9
400	503.6	509.3	515.1	521.0	526.9	532.9	538.9	545.0	551.1	557.3

*Adapted from data of *International Critical Tables,* vol. 3, National Research Council, by McGraw-Hill, New York, 1928, p. 210.

†Adapted from data of J. H. Keenan and F. G. Keyes, *Thermodynamic Properties of Steam,* Wiley, New York, 1936. These data differ but slightly from the data of J. A. Goff and S. Gratch, "Thermodynamic Properties of Moist Air," *Trans. ASHVE,* vol. 51, pp. 125–164, 1945, and corrections thereto by J. A. Goff, "Saturation Pressure of Water on the New Kelvin Temperature Scale," *Trans. ASHVE,* vol. 63, pp. 347–354, 1957.

Source: This table was reproduced with permission from R. Jorgensen (ed.): *Fan Enineering,* 7th ed. Chap. 1, p. 8. copyright © 1970 by Buffalo Fuge Company.

where

ILL = infiltration latent load through cracks and openings, lb/h (metric conversion: lb/h × 0.4536 = kg/h)

exhaust = total exhaust plus return air from the space, ft^3/min

supply = total supply to the space, ft^3/min

Project________________ Date__________
System________________ By__________
Coil________________________________

HEAT GAIN DATA

OA______	db______	wb______	cfm	
RA______	db______	wb______	cfm	
		Total ______	cfm	
Ent. Air______	db______	wb______	h_{ea}	
Lv. Air______	db______	wb______	h_{1a}	
		______	Δh	

GTH = Total cfm × 4.45 × Δh = Btu/hr.

GTH = __________ × 4.45 × ________

= ______________ Btu/hr.

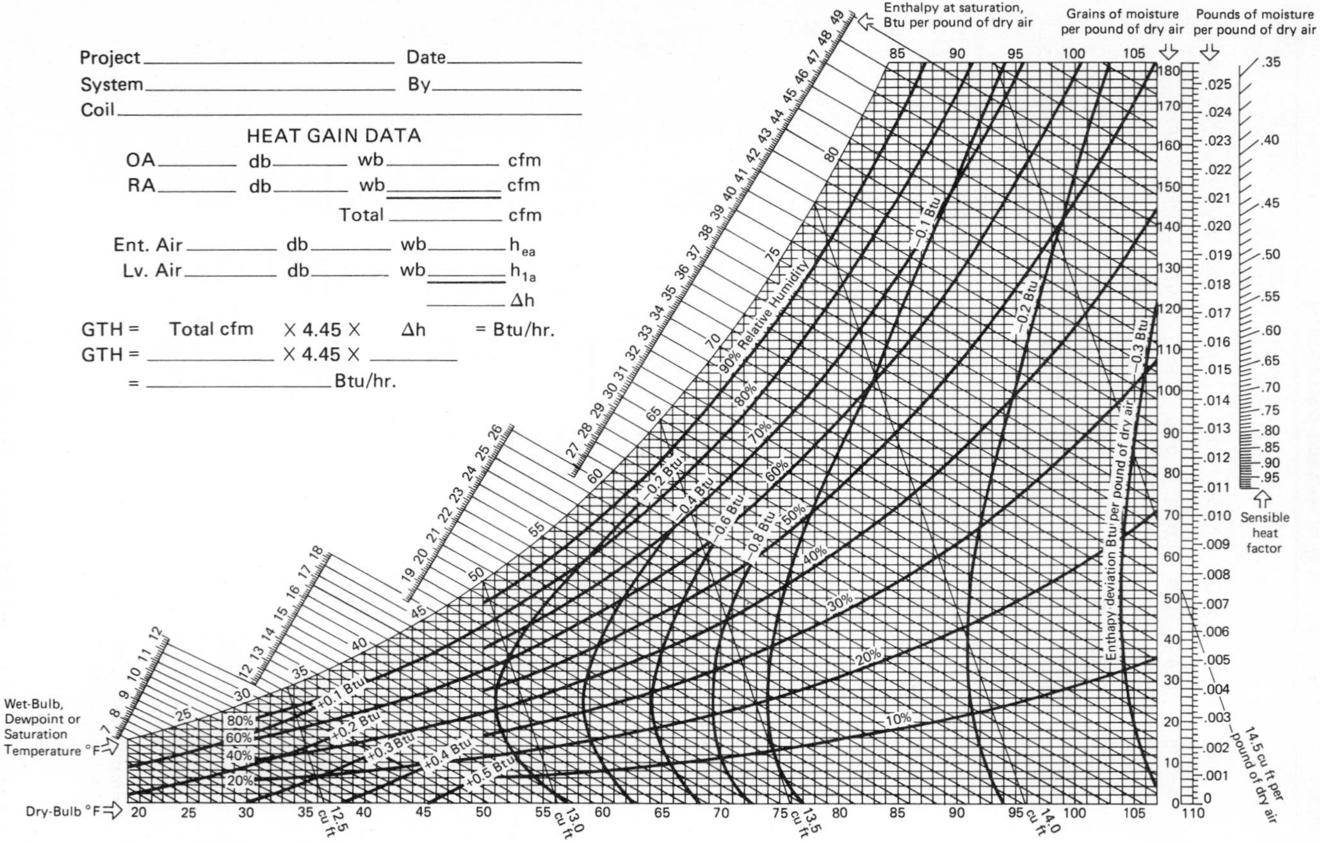

Below 32°F, properties and enthalpy deviation lines are for ice.

Figure 1-6 Psychrometric chart.

TABLE 1-24 Space Temperature, °F db*

RH, %	50	55	60	65	70	72	75	80	85	90
30	0	4	8	12	17	19	24	31	39	48
35	4	7	12	17	21	25	30	37	47	59
40	7	11	16	22	29	30	37	48	59	72
45	9	14	20	27	35	36	44	55	68	83
50	12	16	24	32	40	42	50	63	78	95
55	15	19	28	36	45	47	57	69	87	105
60	17	22	32	41	51	53	64	79	96	116
65	20	26	36	45	57	59	70	87	106	128
70	23	30	40	50	62	65	77	95	115	140
75	25	33	43	54	67	70	83	100	121	152
80	28	36	47	59	73	76	89	108	130	164

*Figures in this table are grains of moisture per pound of dry air. This table assumes that entering air will always provide 15 gr.

f = factor taken directly from Table 1-24 when the number of grains of moisture per pound of outside air of the space design is greater than that of the outside air. (This can be verified by comparing the design space conditions with the design outside air conditions on the appropriate psychrometric chart. See Fig. 1-6.) If the number of grains of moisture per pound of air at the design outside conditions is greater than that of the design space (inside) conditions, then the factor f must be obtained from a psychrometric chart. Enter Fig. 1-6 with the design outside dry-bulb and wet-bulb temperatures or the relative humidity and read the corresponding grains of moisture per pound of air. Repeat the same procedure with the space design conditions. Factor f is the difference between the outside and the inside grains per pound of air.

Category Three. The space is ventilated with a natural (gravity) ventilation system. Under this condition, the latent load due to infiltration must be accounted for. This load can be estimated by using the air-change method (described under "Infiltration" previously in this chapter) in conjunction with the following formula:

$$\text{ILL} = \left(\frac{\text{NCH} \times S}{100{,}000}\,\frac{1}{60}\right) f \qquad (20)$$

where

ILL = infiltration latent load through cracks and openings, lb/h (metric conversion: lb/h × 0.45359 = kg/h)

NCH = number of changes of air within the space per hour. See "Infiltration" to determine this value.

S = space volume, ft^3

f = factor. See notes under Eq. (19).

Latent Load from Outside Air. The discussion in the first paragraph, "Latent Loads from Migration through Cracks and Openings," is equally applicable to this load. In this case we are concerned only with the latent load that results in the introduction of outside air (via the mechanical ventilation system) into the space. Therefore, this load can be calculated using the following formula:

$$\text{OLL} = \frac{(\text{OA} \times 60/1)f}{100{,}000} \qquad (21)$$

where

OLL = outside air latent load, lb/h (metric conversion: lb/h × 0.45359 = kg/h)

OA = outside air (quantity), ft^3/min

f = factor. See notes under Eq. (19).

Total Space Dehumidification—Design Latent Load

The total dehumidification latent load (TDLL) for a space is equal to the algebraic summation of the personnel latent load (PLL), process latent load, evaporation open tank load, latent load from open flame(s), infiltration latent load from moisture migration through cracks and openings (ILL), and the outside air latent load (OLL):

$$\text{TDLL} = \text{PLL} \pm \text{process load } w_v + \text{open flame load} \pm \text{ILL} \pm \text{OLL} \qquad (22)$$

Units for this load are pounds per hour (metric conversion: lb/h × 0.45359 = kg/h).

Selection of Dehumidifying Unit

The decision to use a mechanical or chemical dehumidifier depends on the designer's experience and the application, availability, and serviceability of the manufacturer's units. Generally, mechanical dehumidifiers are used when the design space relative humidity is 45 percent or higher and the dry-bulb temperature is approximately 75°F (23.9°C). Chemical dehumidifiers are used when lower space moisture values are required.

After calculating the design space dehumidification load, the designer should review the manufacturer's catalog selection data and determine which types and capacities of units are most appropriate.

HUMIDIFICATION

Space humidification is generally accomplished by either adding moisture to the supply (ventilation) air prior to its discharge into the space or adding moisture directly to the space. Sufficient moisture must be added to the normal (nonhumidified) space moisture content to satisfy the design (space) relative humidity at the design dry-bulb temperature.

Table 1-25 lists the various types of humidifiers commercially available in order of increasing maintenance requirements.

The five types of humidifiers are available in both the duct type (adds moisture to the supply airstream) and the space recirculating type (adds moisture directly to the space).

TABLE 1-25 Humidifier Types

1. Steam
2. Water spray (atomizing)
3. Centrifugal (atomizing)
4. Pan evaporation
5. Wick evaporation

Load Calculations

The procedure for calculating the design humidification load is as outlined in the dehumidification "Load Calculations," except for the following modifications:

Evaporation Load from Open Tanks

Since the evaporation from open tanks located within the space will always add moisture to the surrounding area, this load may safely be omitted if the exposed tank surface area is a small fraction of the space area. On the other hand, if the tank surface area is large in comparison with the space, omitting this load can result in seriously overdesigning the humidification system.

Latent Load from Migration through Cracks and Openings

The moisture migrating through cracks and openings will increase the space design humidification load whenever the outside or adjacent area's moisture content is greater than the space moisture content and decrease it when it is less. This load calculation generally can be omitted when time or the degree of accuracy of the humidification design load does not warrant it. It can also be safely omitted when the outside air moisture content is greater than the space moisture content (which is not the usual case during the heating season), since it will result in a conservative design load.

Latent Load from Outside Air

This load must be accounted for since it is generally the major portion of the space humidification design load.

Total Space Humidification—Design Latent Load

The total humidification latent load (THLL) for a space is equal to the algebraic summation of the personnel latent load (PLL), process latent load, evaporation open tank load, latent load from open flame(s), infiltration latent load from moisture migration through cracks and openings (ILL), and the outside air latent load (OLL).

$$\text{THLL} = -\text{PLL} \pm \text{process load} - w_v - \text{open-flame load} \pm \text{ILL} \pm \text{OLL} \qquad (23)$$

THLL is expressed in pounds per hour (metric conversion: lb/h × 0.45359 = kg/h).

Selection of Humidifying Unit

The decision as to which type of humidifying unit should be selected depends on the application, availability, and serviceability as well as the designer's experience. After calculating the design space dehumidification load, the designer should review the manufacturer's catalog selection data and determine which types and capacities of units are the most appropriate.

The manufacturer's sizing and installation requirements should be adhered to in order to avoid the annoying problems of condensation of water, caused by impingement of the discharge from the humidifier on adjacent surfaces (especially on colder surfaces), and of supersaturation of the air in the vicinity of the humidifier discharge to the point where condensation commences.

ALTITUDE CORRECTIONS FOR HEATING AND VENTILATING SYSTEMS

GENERAL

The following material is a simplified discussion of the general altitude corrections required for heating and ventilating systems. The formulas and data presented in this chapter are for sea-level locations and generally can be used without correction for elevations up to 1500 ft (460 m). The effect of altitude at elevations up to 10,000 ft (3050 m) on the thermal properties of air (such as viscosity, thermal conductivity, and specific heat) is very small and can be neglected.

CORRECTIONS

When designing heating and ventilating systems for higher elevations [above 1500 ft (460 m)], corrections for air density, airflow (in cubic feet per minute or cubic meters per second), duct air friction, specific humidity, steam gauge pressure, and boiling temperature should be made. The correction factors should be obtained from the manufacturer of the equipment to be selected. The procedure presented here is to design a system as if it were at sea level, then (before selecting the equipment) to adjust the manufacturer's

TABLE 1-26 Airflow Correction Factor, ft³/min (m³/s)

Elevation, ft (m)	Correction factor at 70°F (21.1°C)
2,500 (760)	1.1
5,000 (1500)	1.2
7,500 (2290)	1.33
10,000 (3050)	1.46

TABLE 1-27 Duct Air Friction Correction Factors*

Elevation, ft (m)	Temp corr factor F_1	Temp, °F (°C)	Temp corr factor F_2
2,000 (610)	0.944	0 (18)	1.120
4,000 (1220)	0.890	50 (10)	1.031
6,000 (1830)	0.838	100 (38)	0.957
8,000 (2440)	0.788	150 (66)	0.894
10,000 (3050)	0.742	200 (93)	0.840
		250 (121)	0.792
		300 (149)	0.749

*Duct air friction (at altitude and temperature) = duct air friction (standard air) $\times F_1 \times F_2$.

sea-level capacities by using the appropriate correction factor corresponding to the altitude of installation. In the event manufacturer's elevation (altitude) correction factors are not available, the factors from Tables 1-26 to 1-30 can be used.

Airflow Correction

Multiply the airflow at sea level by the appropriate correction factor to maintain the same mass flow rate (heating or cooling capability) at the actual altitude as exists at sea level.

Air Friction Correction

The duct air friction (standard air) is obtained from a standard duct friction chart, with the corrected airflow (in cubic feet per minute or cubic meters per second) for altitude from Table 1-27 and the duct sizes.

Heating Medium Corrections

Hot Water

No correction is needed.

Steam

Although altitude causes no change in the saturated temperature-pressure relationship, there is a change in the temperature corresponding to a certain gauge pressure (Tables 1-28 and 1-29).

Gas

The recommendation of the American Gas Institute for operating gas-fired heating units for altitude operation is as follows: Ratings need not be corrected for elevations up to 2000 ft (610 m). For elevations above 2000 ft (610 m), ratings should be reduced 4 percent for each 1000 ft (305 m) above sea level.

TABLE 1-28 Steam Gauge Pressure Reductions from Sea Level*

Elevation, ft (m)	Gauge pressure correction, psig (bars)
2,500 (760)	−1.3(− 0.09)
5,000 (1500)	−2.6(−0.182)
7,500 (2290)	−3.6(−0.252)
10,000 (3050)	−4.6(−0.323)

*Altitude gauge pressure = sea-level gauge pressure − gauge pressure correction.

TABLE 1-29 Corresponding Saturated Steam Temperature at Various Altitudes and Gauge Pressures*

			Saturated temperatures, °F (°C)			
			Steam pressure, psig (bars)			
Elevation, ft (m)	Barometric pressure, psia (bars)	Saturated temperature at barometric pressure, °F (°C)	2 (0.14)	10 (0.69)	50 (3.45)	100 (6.9)
0 (0)	14.7 (1.01)	212 (100)	218.5 (103.6)	239.4 (115.2)	253.6 (123.1)	288.6 (142.6)
2,500 (760)	13.41 (0.92)	207.3 (97.4)	214.4 (101.3)	236.4 (113.6)	296.3 (146.8)	287.9 (142.2)
5,000 (1500)	12.23 (0.84)	202.9 (94.9)	210.4 (99.1)	233.6 (112)	251.4 (121.9)	287.2 (141.8)
7,500 (2290)	11.12 (0.77)	198.3 (92.4)	206.3 (96.8)	230.9 (110.5)	293.9 (145.5)	286.6 (141.4)
10,000 (3050)	10.10 (0.7)	193.7 (89.8)	202.4 (94.7)	228.2 (109)	292.8 (144.9)	285 (141.1)

*The heating equipment must be selected at the reduced temperatures.

TABLE 1-30 Minimum Required Primary Air (ft^3/min and m^3/s) for Electric Resistance Heaters at High Altitudes

	Elevation							
Heater wattage	2500 ft	760 m	5000 ft	1500 m	7500 ft	2290 m	10,000 ft	3050 m
500	39	0.018	43	0.020	47	0.022	51	0.024
1000	77	0.036	85	0.040	93	0.043	102	0.048
1500	110	0.052	121	0.057	132	0.062	146	0.069
2000	148	0.070	162	0.076	178	0.084	197	0.093
2500	181	0.085	198	0.093	218	0.102	240	0.113
3000	219	0.103	241	0.114	264	0.125	291	0.137
3500	252	0.119	277	0.131	304	0.143	335	0.158
4000	290	0.137	319	0.151	350	0.165	385	0.182
4500	323	0.152	354	0.167	382	0.180	429	0.202
5000	362	0.171	396	0.187	437	0.206	480	0.227

Electricity

Altitude does not affect the capacity output of electric resistance heaters. However, it is necessary to increase the minimum required actual primary air volume over that published in order to maintain the same minimum air weight flow (Table 1-30). Failure to compensate for the reduction in air weight flow may trip the heater element's thermal overload.

Pump Correction

Altitude affects the operation of pumps installed in open systems because it reduces the available net positive suction head (NPSH). The available NPSH must always be equal to or greater than the required NPSH in order to produce flow through the pump.

Motor Correction

Since the effectiveness of cooling air depends on its density, motor cooling decreases with altitude. To compensate for this decrease, it is necessary to provide additional margin for the increase in motor temperature. Contact the motor manufacturer for recommendations and requirements.

Relative Humidity Correction

The following adjustments are required to calculate loads at high altitude:

1. The design outside and room air moisture content must be adjusted to the new elevation by one of the following methods:
   a. If the dry-bulb temperature and percent relative humidity are given, divide the specific humidity at sea level by the air-density ratio.
   b. If the dry-bulb and wet-bulb temperatures are given, obtain the specific humidity at the altitude by using the following formula:

$$W_1 = W_0 + \frac{(P_0 - P_1)}{P_1} \times W_s \tag{24}$$

   where
   W_1 = specific humidity
   W_0 = specific humidity at altitude for specified db and wb, lb/lb (kg/kg) of dry air
   W_s = specific humidity at sea level and saturated wb temperature, lb/lb (kg/kg) of dry air
   P_0 = barometric pressure, psia (bars)
   P_1 = altitude pressure, psia (bars)

2. The values of specific humidity can be obtained from the National Weather Service or from standard design references, such as ASHRAE's *Handbook of Fundamentals*.[4]

EQUIPMENT SELECTION

GENERAL

When selecting equipment, check with the equipment manufacturer to be sure that the catalog used to select the equipment is current; also be sure to follow recommended selection procedures. If there is any doubt, consult the manufacturer's representative to confirm that the unit and/or equipment selection is appropriate. Guidelines in the selection of dehumidification and humidification units are discussed under "Humidity Control."

Generally systems with the lowest installed cost are less efficient and may have the highest operating costs. *The total life-cycle cost should be determined,* since only with proper evaluation of initial, operating, and maintenance costs over the expected life of alternative equipment selections, can the most energy- and cost-efficient unit be selected.

VARIABLE-AIRFLOW UNITS

From an energy-conservation standpoint, when requirements permit the supply and/or return air quantities to be varied, variable-airflow units should be evaluated. There are four principal ways to achieve variable airflow. They are listed below in order of electric power saved:

1. Variable-pitch axial-flow fans
2. Fan-speed control
3. Inlet-vane (vortex dampers) control
4. Discharge-damper control

Variable-pitch axial-flow fans are not generally available in commercial heating and ventilating units. However, where large-capacity [above 75,000 ft^3/min (35.4 m^3/s)] supply and exhaust volumes are required, this method of varying the fan's capacity should be evaluated.

Fan speed control is commercially available in two forms:

1. **Electric control of speed of induction motors** by one of the following methods:
   a. Voltage control
   b. Frequency control
   c. Multiple-speed (winding) motors
   d. Wound rotor motors
2. **Mechanical control of speed of fans** by one of the following methods:
   a. Fluid couplings
   b. Eddy current couplings
   c. Torque converters

Control of fans' speed is usually economical when large fans are required.

Though variable-voltage direct-current motors give excellent speed control, their high cost and limited availability for heating and ventilating applications make this option extremely remote. Since most manufacturers of heating and ventilating equipment do not offer all the options noted above, the designer usually has limited options available unless the unit is custom-made. However, *current trends are in the direction of lower cost and availability of such controls,* spurred by the development of lower-cost electronic controls and increased emphasis on energy conservation.

OUTDOOR INSTALLATION

If units are to be installed outdoors, it is recommended that they be specifically designed by the manufacturer and factory-assembled for outdoor installation. If such units are not available, then the following weatherproofing should be provided.

1. Gaskets for all access doors and panels
2. Sealing washers under all removable panel screws
3. Two coats of epoxy paint, minimum total thickness 10 mils (0.000254 m)
4. Cadmium-plated or stainless-steel damper shafts
5. Stainless-steel or nonferrous-metal damper linkages
6. Stainless-steel or anodized aluminum dampers
7. Stainless-steel panel fasteners
8. Totally enclosed removable belt guards, if the fan drive is external to the unit
9. Totally enclosed motors

Heating and ventilating air-handling units generally are not designed as roof-mounted units (that is, for mounting directly on roof curbs). They should be mounted on a steel dunnage, fabricated to the particular unit's dimensions with appropriate pitch pockets at all roof penetrations. If ductwork and piping installation does not require more height, a minimum of 18 in (0.457 m) clearance under the dunnage beams should be provided for inspection and painting of the unit and the steel.

UNIT SELECTION CRITERIA

The engineer must determine whether a single-zone heating and ventilating unit or a multizone heating and ventilating unit is more appropriate. See Figs. 1-7 and 1-8. Though the capacity and types of fans commercially available in heating and ventilating units vary with each manufacturer, the following are fairly typical.

1. **Single-zone draw-through units** These are nominally available in the following capacity ranges:

   With forward-curved low- and medium-pressure fans (see discussion, "Fan Selection Procedures," for pressure ranges): 800 to 60,000 ft^3/min (0.378 to 28.32 m^3/s). With low-pressure fans, capacities up to 75,000 ft^3/min (35.4 m^3/s) are available.
   With backward-inclined or airfoil low- and medium-pressure fans: 1500 to 50,000 ft^3/min (0.708 to 28.32 m^3/s).
   With low-pressure fans, capacities up to 75,000 ft^3/min (35.4 m^3/s) are available.
   With airfoil high-pressure fans: 2500 to 50,000 ft^3/min (1.18 to 23.6 m^3/s).

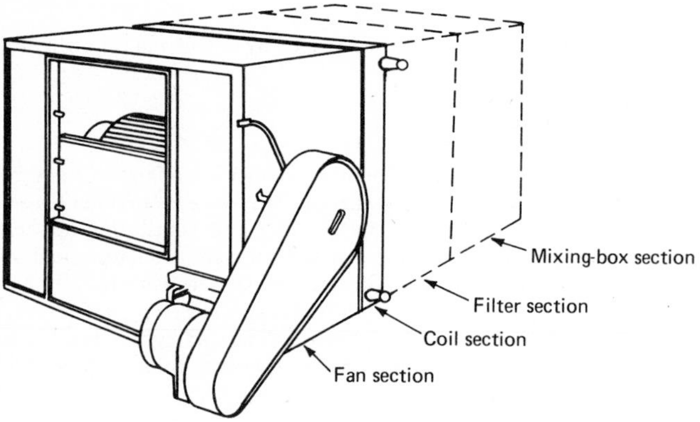

(a)

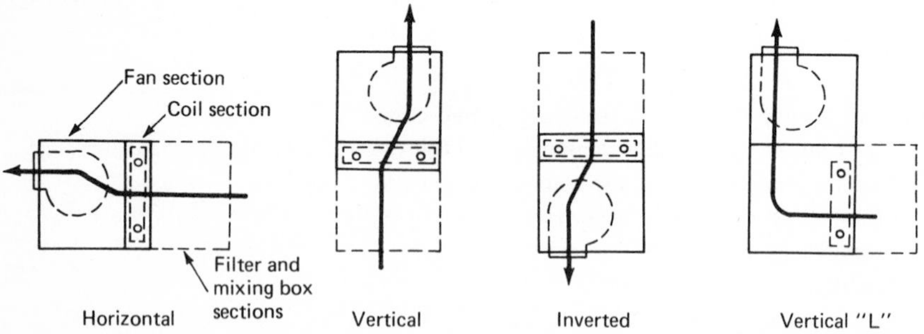

(b)

Figure 1-7 (*a*) Typical single-zone draw-through heating and ventilating unit. (*b*) Typical arrangements for a single-zone draw-through heating and ventilating unit; the dotted line indicates an accessory section. *Note:* Blow-through units are available from some manufacturers. (*American Air Filter Co., Inc.*)

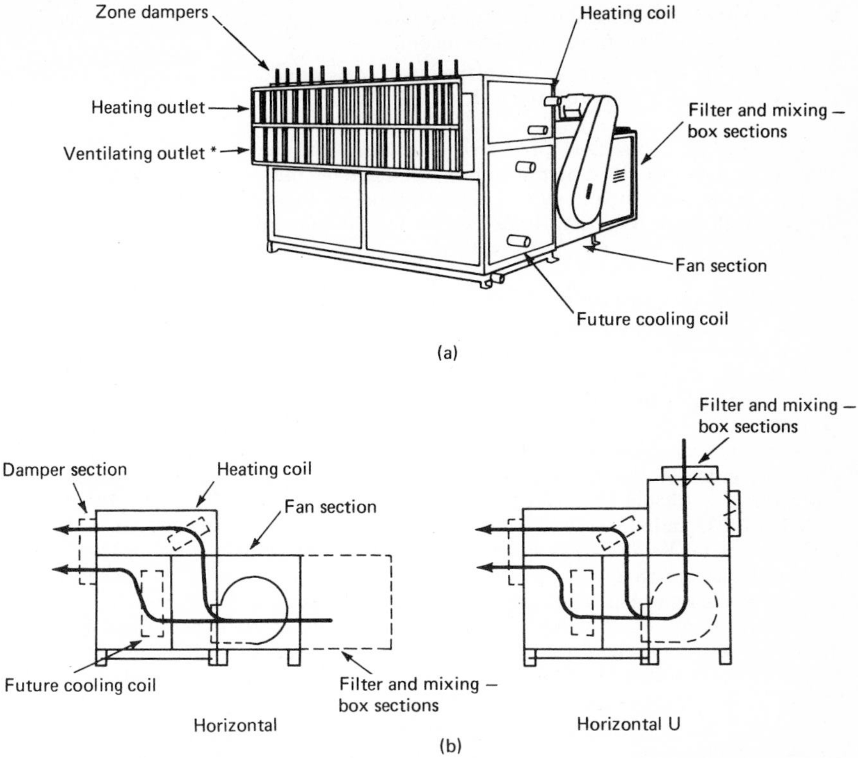

Figure 1-8 (*a*) Typical multizone heating and ventilating unit. Such units have the ability to simultaneously heat and ventilate with provision for future cooling. For heating and ventilating service, a cooling coil would not be provided; a balancing plate would be provided instead of the cooling coil. *If a cooling coil is installed, this will be the cold air outlet. (*b*) Typical arrangements for a multizone unit; dotted lines indicate the accessory section. (*American Air Filter Co., Inc.*)

2. **Multizone blow-through units** These are nominally available in the following capacity ranges:

   With forward-curved low- and medium-pressure fans: 1200 to 40,000 ft^3/min (0.556 to 18.88 m^3/s).
   With backward-inclined or airfoil low- and medium-pressure fans: 1500 to 40,000 ft^3/min (0.708 to 18.88 m^3/s).
   With airfoil high-pressure fans: 2500 to 35,000 ft^3/min (1.18 to 16.52 m^3/s).

If the unit is required to serve areas that will be satisfied with the same supply-air temperature, then a single-zone heating and ventilating unit should be selected. If various supply-air temperatures are required for the areas to be served, then the engineer must choose between a multizone unit and a single-zone unit with reheat coils in the zone ducts. This decision should be based on evaluating initial, operating, and maintenance costs. It may be stated that:

1. Long duct runs and a readily available heating medium in the areas to be served favor a single zone with reheats.

2. From a maintenance and operating standpoint, it is desirable to serve all equipment from one area. This favors a multizone unit.
3. If there is a potential for air conditioning in some or all of the areas to be served, a multizone unit would probably be a wise choice.

With the required airflow (previously calculated from the appropriate design section of this chapter) and the design engineer's decision on the type of unit to be installed (single-zone or multizone), one can determine the unit size (unit number) by entering the "Typical Quick Selection Guide" tables for the selected type of unit (single-zone, Fig. 1-9, or multizone, Fig. 1-10), with the required airflow and reading off the unit size.

An example is indicated in Figs. 1-9 to 1-11. The assumption is that (1) an 11,500 ft^3/min (5.428 m^3/s) single-zone unit is required and (2) an 8500 ft^3/min (4.01 m^3/s) multizone unit is required.

It should be noted that Fig. 1-10 is based on the assumption that all multizone units shown are going to be used for air conditioning. Therefore, if the unit being selected has no future air-conditioning requirement, it should be selected at a higher coil velocity, 800 to 1000 ft/min (4.06 to 5.08 m/s), not at 400 to 600 ft/min (2.03 to 3.05 m/s) as indicated. Likewise, if the single-zone unit being selected has a future air-conditioning requirement, it should be selected at a coil velocity of about 550 ft/min (2.79 m/s), not at 600 to 1000 ft/min (3.05 to 5.08 m/s) as indicated. This is the principal reason why the single-zone size 18 unit in this example has a capacity of 11,500 ft^3/min (5.43 m^3/s) and the multizone size 18 unit has a capacity of 8500 ft^3/min (4.01 m^3/s). If they were both selected at the same coil velocity they would have the same airflow capacity.

In the example, a size 18 unit would be a satisfactory choice in both cases. However, in the case of a single-zone unit (Fig. 1-9), a size 15 unit with a 14.5-ft^2 (1.35-m^2) coil area or a size 22 unit with a 12- or 16-ft^2 (1.11- or 1.45-m^2) coil area could provide alternative selections. Likewise, in the case of a multizone unit (Fig. 1-10), a size 15 unit with a 15-ft^2 (1.39-m^2) coil area or a size 22 unit with an 18.5-ft^2 (1.72-m^2) coil area could provide alternative selections.

Once the unit size has been chosen, the following items can be read from Fig. 1-9 or 1-10:

1. Number and size of fans that will fit within the unit
2. Areas of coils that will fit within the unit
3. Size of the unit

From Fig. 1-11, the following component items that can be provided in the unit selected can be determined by reading down, under the appropriate unit number column:

1. Type and size of coils
2. Type and size of fans and maximum horsepower
3. Type and size of filters

From Fig. 1-12, the outlet velocity, fan speed (revolutions per minute), and brake horsepower at the corresponding static pressure (across the fan) can be determined. Figure 1-12 is for backward-inclined fans; similar data are available for forward-curved and airfoil fans.

To determine the required fan speed and the minimum brake horsepower to deliver the design airflow, one must first calculate the total system static pressure against which the fan must operate. The total system static pressure is the summation of the following losses: duct, diffusers and/or registers, unit casing, coils, filters, etc.

From Fig. 1-13, the physical dimensions of the selected unit can be determined. Also, the physical dimensions of a selected multizone unit can be obtained from Fig. 1-14. Similar data are available for high-pressure and vertical single- and multizone units.

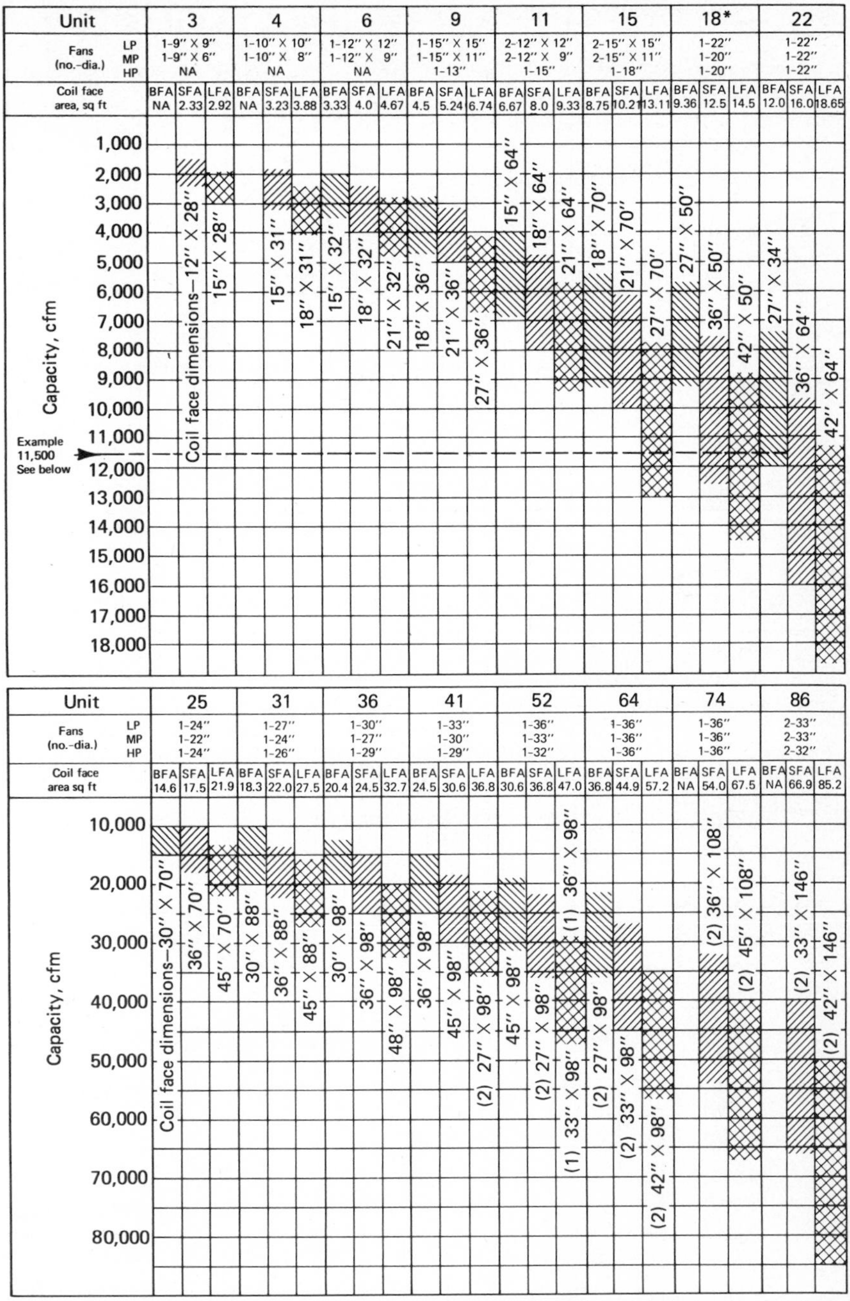

600 fpm / 1000 fpm — Balanced flow coil. Used only on make-up air applications.

600 fpm / 1000 fpm — Small face area coil. Normally used only with internal face and bypass dampers.

600 fpm / 1000 fpm — Large face area coil. Most commonly used.

*Example: 11,500 cfm

1. Select size 15 LFA (most economical).
2. Select size 18 SFA (if internal bypass required).
3. Select size BFA (if balance flow required).

Figure 1-9 Typical single-zone quick-selection guide. (*American Air Filter Co., Inc.*)

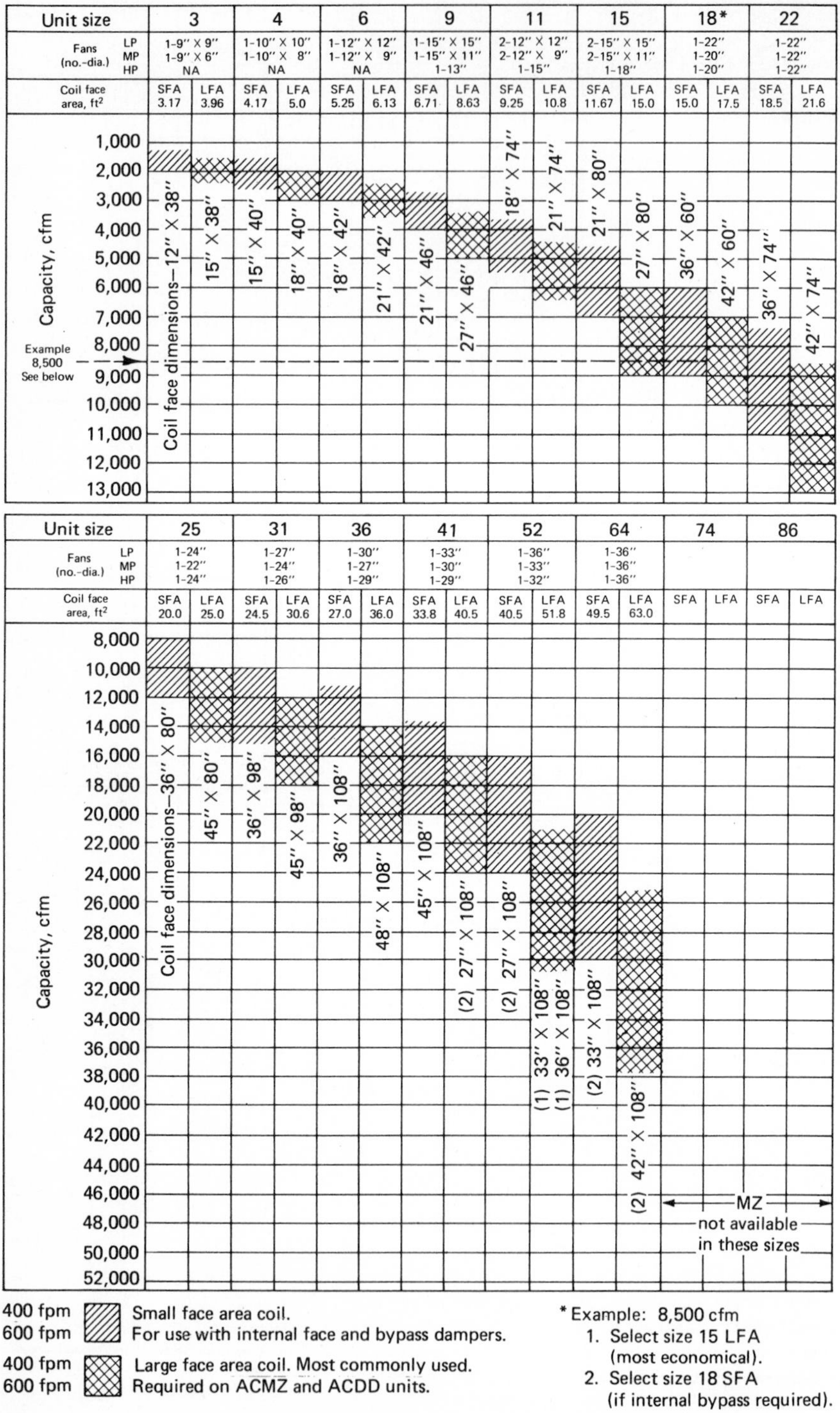

Figure 1-10 Typical multizone quick-selection guide. (*American Air Filter Co., Inc.*)

COIL SELECTION PARAMETERS

General

Heating and ventilating units in general can be supplied with *steam, hot-water, or electric heating coils* (see Fig. 1-15).

Steam Coils

Steam coils are available in two basic types, standard and distributing (nonfreeze). *Standard* coils are less expensive than distributing coils and should be used whenever the supply-air temperature to the coil is above the freezing temperature. *Distributing* coils are designed to prevent, or at least to minimize, the possibility of freezing the steam condensate within the coil, when the supply-air temperature is below freezing. Therefore, distributing (nonfreeze) coils should be used when the air to the coil could be below freezing temperature.

Hot-Water Coils

Hot-water coils are available in two basic types, standard and cleanable. *Standard* coils are less expensive and should be used whenever the heating water is clean. That is, the hot water will not form scale or precipitate sludge deposits (within the tubes) that will cause a measurable loss in the coil's ability to transfer heat from the water to the coil fins.

Cleanable coils are designed to be used when the heating water has the tendency to form scale on, or to precipitate sludge deposits within, the tubes. With cleanable coils, the cleaning or cleanability referred to deals with the *inside* of the tubes. There is no difference between the air-side design of cleanable coils and that of standard coils having the same number of fins per inch. When cleanable coils are specified, removable plugs on both ends of every tube in the coil should be specified so that each tube can be thoroughly cleaned (punched through).

Electric Coils

Electric resistance coils are best employed where the electric energy rates are lower than the prevailing fossil fuel rates for the same amount of heat (British thermal units per hour or watts) transferred from the coil to the airstream, or where the required service life of the heating system in question is so short (such as a temporary installation) that the higher capital and operating costs of a heating plant cannot be economically justified.

Coil Construction

The coil frames (casings) are generally of galvanized steel. If a hygienic environment is necessary or a corrosive environment is encountered, special frames of aluminum, stainless steel, etc., can be furnished. Heating coils are usually one- or two-row coils. Four- and six-row coils can be provided on special order. Water and steam coils are normally furnished with copper tubes.

Fins

Aluminum or copper coil fins are standard on water and steam coils. Since the copper fins are of significantly greater cost than aluminum, one should evaluate the corrosiveness of the airstream with respect to the application of aluminum fins on a life-cycle cost basis, before selecting the fin material. Water and steam coils are available with fin spacings of 6, 8, 11, and 14 fins per inch (2, 3, 4, and 6 fins per centimeter). In general, the more fins per inch, the greater the ability to transfer heat from the heating medium to the airstream for the same coil velocity, material, number of rows, and mean effective

UNIT SIZE				3	4	6	9	11	15	18*	22
FANS	Low Pressure	Forward Curved Steel Wheels	No. and Dia. of Wheels Tot. Outlet Area (Sq. Ft.)	1–9 X 9 .93	1-10 X 10 1.225	1-12 X 12 1.84	1-15 X 15 2.70	2.12 X 12 3.67	2-15 X 15 5.40	1-22 1/4 6.69	1-22 6.69
			Max. RPM Std. Shaft Shaft Size & Bearing Size	2050 1 7/16	1850 1 7/16	1730 1 7/16-1 3/16	1331 1 7/16	1694 1 7/16	1587 1 7/16	811 1 15/16-1 11/16	811 1 11/16-1 7/16
			Max. RPM Low Sp. Shaft Bearing Dia. (In.)	1735 1 3/16	1530 1 3/16	1303 1 3/16	—	1271 1 3/16	—	700 1 11/16	641 1 7/16
		Backward Inclined Steel Wheels	No. and Dia. of Wheels Tot. Outlet Area (Sq. Ft.)	NA	NA	1-12 1.86	1-15 2.77	2-12 3.72	2-15 5.54	1-22 6.69	1-22 6.69
			Max. RPM Shaft Size & Bearing Size	NA	NA	2668 2 3/16-1 7/16	2231 2 7/16-1 7/16	2539 1 15/16-1 7/14	2242 2 3/16-1 7/16	1553 2 15/16-1 11/16	1553 2 7/16-1 7/16
	Medium Press.	Forward Curved Steel Wheels	No. and Dia. of Wheels Tot. Outlet Area (Sq. Ft.)	1–9 X 6 .65	1-10 X 8 .96	1-12 X 9 1.435	1-15 X 11 2.135	2-12 X 9 2.870	2-15 X 11 4.27	1-20 5.45	1-22 6.69
			Max. RPM Shaft Size & Bearing Size	2860 1 11/16-1 7/16	2460 1 11/16-1 7/16	1725 1 7/16-1 7/16	1722 1 11/16-1 7/16	1750 1 7/16 X 1 7/14	1676 1 7/16-1 7/16	1286 2 7/16-1 11/16	1161 1 15/16-1 11/16
		Backward Inclined Steel Wheels	No. and Dia. of Wheels Tot. Outlet Area (Sq. Ft.)	NA	NA	1-12 1.86	1-15 2.77	2-12 3.72	2-15 5.54	1-20 5.45	1-22 6.69
			Max. RPM Shaft Size & Bearing Size	NA	NA	3403 2 15/16-1 7/16	2600 2 15/16-1 7/16	3433 2 11/14-1 7/16	2733 2 11/16-1 7/16	2370 3 15/16-1 11/16	2150 3 3/16-1 11/16
	High Press.	Airfoil Steel Wheels	No. and Dia. of Wheels Tot. Outlet Area (Sq. Ft.)	NA	NA	NA	1-13 7/32 1.55	1-14 9/16 1.89	1.17 13/16 2.81	1-19 11/16 3.42	1-21 3/16 4.18
			Max. RPM Shaft Size & Bearing Size	NA	NA	NA	4046 2 3/16-1 7/16	3675 3 3/16-1 7/16	2994 3 3/16-1 7/16	2716 2 11/16-1 7/16	2481 3 3/16-1 11/16
Max. Motor Frame Size	On Discharge Face in Cowl		Open Drip Proof	215	215	215	215	215	215	254	254
	In Plenum		T.E., Multi Speed	213	215	213	213	184	184	215	254
AC COILS	Large Face Area ②		Dims.	15 X 38	18 X 40	21 X 42	27 X 46	21 X 74	27 X 80	42 X 60	42 X 74
			Sq. Ft.	3.96	5.0	6.12	8.63	10.8	15.0	17.5	21.6
	Small Face Area ③		Dims.	12 X 38	15 X 40	18 X 42	21 X 46	18 X 74	21 X 80	36 X 60	36 X 74
			Sq. Ft.	3.17	4.17	5.3	6.71	9.3	11.65	15.0	18.52
	Max. No. of Rows AC Units ① ⑤	Hori. Units, Standard Depth	Single Coil	10	10	10	10	10	10	10	10
			Combination of Two Coils	8	8	8	8	8	8	8	8
		Hori. Units, Extra Depth	Single Coil	13	13	13	15	13	15	21	21
			Combination of Two Coils	10	10	10	12	10	12	18	18
		Verti. Units, Standard Depth	Single Coil	8	8	8	8	8	8	14	14
			Combination of Two Coils	5	5	5	5	5	5	10	10
HV COILS	Large Face Area ②		Dims.	15 X 28	18 X 31	21 X 32	27 X 36	21 X 64	27 X 70	42 X 50	42 X 64
			Sq. Ft.	2.92	3.88	4.67	6.74	9.33	13.11	14.5	18.65
	Small Face Area ③		Dims.	12 X 28	15 X 31	18 X 32	21 X 36	18 X 64	21 X 70	36 X 50	36 X 64
			Sq. Ft.	2.33	3.23	4.0	5.24	8.0	10.21		16.0

Group	Description		Item								
HV COILS	Balanced Flow [1]		Dims.	NA	NA	15 X 32	18 X 36	15 X 64	18 X 70	27 X 50	27 X 64
			Sq. Ft.			3.33	4.5	6.67	8.75	9.36	12.0
	Max. No. of Rows HV Units ⑤ — Hori. Vert. Inv. Units	Standard Depth	Single Coil	4	4	4	4	4	4	4	4
			Combination of Two Coils	—	—	—	—	—	—	—	—
		Extra Depth	Single Coil	8	8	8	8	8	8	8	8
			Combination of Two Coils	4	4	4	4	4	4	4	4
	Vert. L Shape	Standard Depth	Single Coil	9	9	9	9	9	9	14	14
			Combination of Two Coils	6	6	6	6	6	6	9	9
ACMZ-ACDD COILS	Cooling Coil ⑤		Dims.		18 X 40	21 X 42	27 X 46	21 X 74	27 X 80	42 X 60	42 X 74
			Sq. Ft.		5.0	6.12	8.63	10.8	15.0	17.5	21.6
			Max. No. of Rows		8	8	8	8	8	8	8
	Heating Coil ⑤		Dims.		9 X 40	12 X 42	15 X 46	12 X 74	15 X 80	21 X 60	21 X 74
			Sq. Ft.	NA	2.5	3.5	4.8	6.2	8.4	8.8	10.8
			Max. No. of Rows		4	4	4	4	4	4	4
FLAT BANK UNIT FILTER	2" Thick Throwaway, HV2, Renew., Polyurethane		No. & Size	2-16 X 20	2-20 X 20	2-20 X 20	2-20 X 25	2-16 X 20 2-20 X 20	4-20 X 25	3-20 X 25 3-20 X 20	2-20 X 25 2-20 X 20 2-16 X 20 2-16 X 25
			Total Area (Sq. Ft.)	4.44	5.56	5.56	6.95	10.0	13.89	18.7	22.5
ANGLE BANK UNIT FILTER	2" Thick Throwaway, HV2, Renew., Polyurethane		No. & Size	4-16 X 20	4-16 X 20	4-16 X 20	4-20 X 25	6-20 X 20	2-16 X 25 6-20 X 25	12-16 X 20	12-16 X 25
			Std. Arr.	VEE	VEE	VEE	VEE	VEE	VEE	W	W
			Total Area (Sw. Ft.)	8.9	8.9	8.9	13.9	16.7	26.4	26.6	33.3
HIGH VOLUME UNIT FILTER	2" Thick Throwaway, HV2, Renew., Polyurethane		No. & Size	NA	NA	4-20 X 20	6-20 X 20	6.20 X 25	9-20 X 25	12-20 X 20	12-20 X 25
			Std. Arr.	NA	NA	VEE	ZEE	VEE	ZEE	W	W
			Total Area (Sw. Ft.)	NA	NA	11.1	16.7	20.8	31.2	33.3	41.6
AUTOMATIC FILTERS	TYPE H ROLL-O-MATIC MODEL G		Size Standard Capacity	NA	25-38	25-38	33.50	25-70	33-70	45-54	45-70
			Size High Capacity	NA	25-48	25-68	33-68	25-88	33-88	45-78	45-88
SIDE ACCESS FILTERS	ServiSide Electro-Pak ROLLOTRON ROLL-O-PAK		Size Standard Capacity	NA	3H28	3H28	4H34	6H24	7H34	5H48	6H48
			Size High Capacity	NA	4H24	6H24	6H34	8H24	8H34	7H48	8H48
										↑ Example	

① No. of Rows without Access Door
② Used with External Face and By Pass Dampers
③ Used with Internal Face and By Pass Dampers
④ Used with Balanced Flow Face and By Pass Dampers
⑤ Maximum number of rows shown is approximate and is based on use of Types W, S & DE Coils
* Example

Figure 1-11 Typical physical data. (*American Air Filter Co., Inc.*)

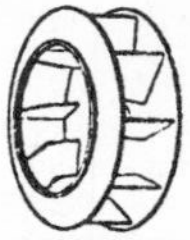

Size 6 LP (1 - 12¼" BI Fan)

f factor = .131

CFM	OUTLET VEL.	STATIC PRESSURE — INCHES OF WATER																													
		.2		.4		.6		.8		1.0		1.2		1.4		1.6		1.8		2.0		2.2		2.4		2.6		2.8		3.0	
		RPM	BHP	RPM	BHP	RPM	BHP	RPM	BHP	RPM	BHP	RPM	BHP	RPM	BHP	RPM	BHP	RPM	BHP	RPM	BHP	RPM	BHP	RPM	BHP	RPM	BHP	RPM	BHP	RPM	BHP
1500	804	1026	0.1	1197	0.2	1338	0.3	1456	0.4	1564	0.4	1663	0.5	1756	0.6	1846	0.7	1933	0.8	2017	0.9	2098	1.0	2177	1.1	2254	1.2	2329	1.4	2401	1.5
2000	1072	1261	0.2	1399	0.3	1529	0.4	1647	0.5	1752	0.6	1845	0.7	1932	0.8	2015	1.0	2093	1.1	2169	1.2	2241	1.3	2311	1.4	2380	1.5	2447	1.6	2513	1.8
2500	1340	1503	0.4	1627	0.5	1735	0.6	1843	0.8	1943	0.9	2037	1.0	2124	1.1	2204	1.3	2279	1.4	2351	1.5	2420	1.7	2486	1.8	2551	1.9	2613	2.1		
3000	1608	1756	0.7	1865	0.8	1964	0.9	2053	1.1	2145	1.2	2232	1.3	2315	1.5	2395	1.6	2471	1.8	2543	2.0	2611	2.1								
3500	1876	2017	1.1	2109	1.2	2202	1.3	2285	1.5	2360	1.6	2439	1.8	2517	1.9	2592	2.1	2663	2.3												
4000	2144	2282	1.5	2360	1.7	2442	1.8	2522	2.0	2596	2.1	2661	2.3																		
4500	2413	2541	2.1	2616	2.3																										

Size 9 LP (1 - 15" BI Fan)

f factor = .362

CFM	OUTLET VEL.	STATIC PRESSURE — INCHES OF WATER																													
		.2		.4		.6		.8		1.0		1.2		1.4		1.6		1.8		2.0		2.2		2.4		2.6		2.8		3.0	
		RPM	BHP	RPM	BHP	RPM	BHP	RPM	BHP	RPM	BHP	RPM	BHP	RPM	BHP	RPM	BHP	RPM	BHP	RPM	BHP	RPM	BHP	RPM	BHP	RPM	BHP	RPM	BHP	RPM	BHP
2500	901	900	0.2	1031	0.4	1145	0.5	1241	0.6	1327	0.8	1406	0.9	1480	1.0	1550	1.2	1618	1.3	1683	1.5	1747	1.6	1809	1.8	1869	2.0	1928	2.2	1986	2.4
3000	1082	1030	0.4	1143	0.5	1249	0.6	1345	0.8	1431	1.0	1507	1.1	1578	1.3	1646	1.4	1710	1.6	1771	1.8	1830	1.9	1888	2.1	1944	2.3	1999	2.5	2052	2.7
3500	1262	1161	0.5	1264	0.7	1360	0.8	1450	1.0	1534	1.2	1612	1.4	1683	1.6	1748	1.7	1810	1.9	1869	2.1	1926	2.3	1981	2.5	2035	2.7	2087	2.9	2137	3.0
4000	1442	1296	0.8	1394	0.9	1477	1.1	1561	1.3	1640	1.5	1715	1.7	1786	1.9	1853	2.1	1915	2.3	1972	2.5	2028	2.7	2081	2.9	2133	3.1	2183	3.3		
4500	1623	1435	1.0	1524	1.2	1604	1.4	1677	1.6	1752	1.8	1823	2.0	1891	2.2	1956	2.5	2018	2.7	2077	2.9	2133	3.2	2186	3.4						
5000	1803	1576	1.4	1656	1.6	1734	1.8	1801	2.0	1867	2.2	1935	2.4	2000	2.6	2062	2.9	2122	3.1	2181	3.4										
5500	1983	1720	1.8	1790	2.0	1863	2.2	1931	2.4	1990	2.7	2051	2.9	2112	3.2	2173	3.4	2230	3.7												
6000	2164	1864	2.3	1928	2.5	1995	2.8	2061	3.0	2120	3.2	2174	3.5	2230	3.7																
6500	2344	2006	2.9	2067	3.1	2129	3.4	2191	3.6																						
7000	2524	2146	3.6	2208	3.8																										

Size 11 LP (2 - 12¼" BI Fans)

f factor = .262

CFM	OUTLET VEL.	STATIC PRESSURE — INCHES OF WATER																													
		.2		.4		.6		.8		1.0		1.2		1.4		1.6		1.8		2.0		2.2		2.4		2.6		2.8		3.0	
		RPM	BHP	RPM	BHP	RPM	BHP	RPM	BHP	RPM	BHP	RPM	BHP	RPM	BHP	RPM	BHP	RPM	BHP	RPM	BHP	RPM	BHP	RPM	BHP	RPM	BHP	RPM	BHP	RPM	BHP
3000	804	1026	0.3	1197	0.4	1338	0.6	1456	0.7	1564	0.9	1663	1.0	1756	1.2	1846	1.4	1933	1.6	2017	1.8	2098	2.0	2177	2.3	2254	2.5	2329	2.7	2401	2.9
3500	938	1142	0.4	1296	0.5	1433	0.7	1552	0.9	1655	1.1	1751	1.3	1841	1.4	1926	1.6	2008	1.8	2086	2.0	2162	2.3	2237	2.5	2309	2.7	2380	3.0	2449	3.2
4000	1072	1261	0.5	1399	0.7	1529	0.9	1647	1.1	1752	1.3	1845	1.5	1932	1.7	2015	1.9	2093	2.1	2169	2.3	2241	2.6	2311	2.8	2380	3.0	2447	3.3	2513	3.5
4500	1206	1381	0.7	1509	0.8	1631	1.0	1743	1.3	1847	1.5	1942	1.7	2028	2.0	2108	2.2	2184	2.4	2258	2.7	2328	2.9	2396	3.2	2462	3.4	2526	3.7		
5000	1340	1503	0.9	1627	1.1	1735	1.3	1843	1.5	1943	1.8	2037	2.0	2124	2.3	2204	2.5	2279	2.8	2351	3.1	2420	3.3	2486	3.6						
5500	1474	1628	1.1	1746	1.3	1846	1.5	1946	1.8	2042	2.0	2132	2.3	2219	2.6	2300	2.9	2376	3.2	2447	3.5	2515	3.8								
6000	1608	1756	1.4	1865	1.6	1964	1.9	2053	2.1	2145	2.4	2232	2.7	2315	3.0	2395	3.3	2471	3.6												
6500	1742	1885	1.7	1986	2.0	2083	2.2	2166	2.5	2250	2.8	2334	3.1	2414	3.4	2491	3.7														
7000	1876	2017	2.1	2109	2.4	2202	2.6	2285	2.9	2360	3.2	2439	3.5	2517	3.9																
7500	2010	2150	2.6	2233	2.8	2321	3.1	2404	3.4	2476	3.7																				
8000	2144	2282	3.1	2360	3.4	2442	3.7	2522	4.0																						
8500	2278	2413	3.7	2487	3.9																										

Size 15 LP (2 - 15" BI Fans) **f factor = .724**

CFM	OUTLET VEL.	.2 RPM	.2 BHP	.4 RPM	.4 BHP	.6 RPM	.6 BHP	.8 RPM	.8 BHP	1.0 RPM	1.0 BHP	1.2 RPM	1.2 BHP	1.4 RPM	1.4 BHP	1.6 RPM	1.6 BHP	1.8 RPM	1.8 BHP	2.0 RPM	2.0 BHP	2.2 RPM	2.2 BHP	2.4 RPM	2.4 BHP	2.6 RPM	2.6 BHP	2.8 RPM	2.8 BHP	3.0 RPM	3.0 BHP
4000	721	780	0.3	926	0.5	1040	0.7	1140	0.9	1230	1.2	1314	1.4	1394	1.7	1471	1.9	1545	2.2	1617	2.5	1686	2.8	1753	3.1	1816	3.5	1878	3.8	1937	4.1
5000	901	900	0.5	1031	0.7	1145	1.0	1241	1.2	1327	1.5	1406	1.8	1480	2.0	1550	2.3	1618	2.6	1683	2.9	1747	3.3	1809	3.6	1869	4.0	1928	4.3	1986	4.7
6000	1082	1030	0.7	1143	1.0	1249	1.3	1345	1.6	1431	1.9	1507	2.2	1578	2.5	1646	2.9	1710	3.2	1771	3.5	1830	3.8	1888	4.2	1944	4.5	1999	4.9	2052	5.3
7000	1262	1161	1.1	1264	1.4	1360	1.7	1450	2.0	1534	2.4	1612	2.7	1683	3.1	1748	3.5	1810	3.8	1869	4.2	1926	4.6	1981	5.0	2035	5.3	2087	5.7	2137	6.1
8000	1442	1296	1.5	1394	1.8	1477	2.2	1561	2.5	1640	2.9	1715	3.3	1786	3.7	1853	4.2	1915	4.6	1972	5.0	2028	5.4	2081	5.8	2133	6.3	2183	6.7	2232	7.1
9000	1623	1435	2.1	1524	2.4	1604	2.8	1677	3.2	1752	3.6	1823	4.0	1891	4.5	1956	4.9	2018	5.4	2077	5.9	2133	6.4	2186	6.8	2236	7.3				
10000	1803	1576	2.8	1656	3.2	1734	3.5	1801	4.0	1867	4.4	1935	4.8	2000	5.3	2062	5.8	2122	6.3	2181	6.8	2237	7.4								
11000	1983	1720	3.7	1790	4.0	1863	4.4	1931	4.9	1990	5.3	2051	5.8	2113	6.3	2173	6.8	2230	7.3												
12000	2164	1864	4.7	1928	5.0	1995	5.5	2061	6.0	2120	6.4	2174	7.0	2230	7.5																
13000	2344	2006	5.8	2067	6.2	2129	6.8	2191	7.3																						
14000	2524	2146	7.1	2208	7.7																										

Example → **Size 18 LP (1 - 22¼" BI Fan)** **f factor = 2.77**

CFM	OUTLET VEL.	.2 RPM	.2 BHP	.4 RPM	.4 BHP	.6 RPM	.6 BHP	.8 RPM	.8 BHP	1.0 RPM	1.0 BHP	1.2 RPM	1.2 BHP	1.4 RPM	1.4 BHP	1.6 RPM	1.6 BHP	1.8 RPM	1.8 BHP	2.0 RPM	2.0 BHP	2.2 RPM	2.2 BHP	2.4 RPM	2.4 BHP	2.6 RPM	2.6 BHP	2.8 RPM	2.8 BHP	3.0 RPM	3.0 BHP
4000	598	464	0.3	565	0.5	643	0.6	713	0.8	776	1.1	836	1.3	891	1.6	945	1.9	999	2.2	1051	2.5	1101	2.8	1149	3.1	1194	3.5	1238	3.8	1280	4.1
5000	747	534	0.4	620	0.7	698	0.9	763	1.1	822	1.3	878	1.6	930	1.9	979	2.2	1027	2.5	1073	2.8	1117	3.1	1160	3.5	1203	3.8	1247	4.2	1289	4.6
6000	897	606	0.6	679	0.9	753	1.2	818	1.4	876	1.7	928	1.9	977	2.2	1024	2.5	1069	2.8	1112	3.2	1154	3.5	1195	3.9	1234	4.3	1272	4.7	1310	5.1
7000	1046	681	0.8	751	1.1	809	1.5	874	1.8	930	2.1	983	2.4	1031	2.7	1075	3.0	1118	3.3	1159	3.7	1199	4.0	1238	4.4	1275	4.8	1312	5.2	1347	5.6
8000	1196	760	1.2	822	1.5	879	1.8	929	2.2	986	2.6	1038	3.0	1085	3.3	1131	3.7	1173	4.0	1212	4.3	1250	4.7	1287	5.1	1323	5.5	1358	5.9	1393	6.3
9000	1346	834	1.5	894	1.9	950	2.3	998	2.7	1042	3.1	1093	3.6	1141	4.0	1185	4.4	1227	4.8	1267	5.2	1305	5.5	1341	5.9	1375	6.3	1408	6.7	1441	7.1
10000	1495	917	2.0	971	2.4	1022	2.9	1069	3.3	1111	3.7	1151	4.2	1195	4.7	1241	5.2	1282	5.7	1321	6.1	1359	6.5	1396	6.9	1431	7.3	1464	7.8	1495	8.2
11000	1645	1000	2.6	1050	3.1	1094	3.5	1141	4.0	1183	4.5	1220	4.9	1256	5.5	1295	6.0	1338	6.5	1378	7.1	1415	7.6	1450	8.1	1484	8.5	1518	9.0	1551	9.4
12000	1794	1084	3.3	1120	3.7	1170	4.2	1212	4.8	1254	5.3	1292	5.8	1327	6.3	1359	6.9	1394	7.5	1431	8.1	1470	8.7	1506	9.3	1540	9.8				
13000	1944	1169	4.1	1202	4.5	1248	5.1	1286	5.7	1326	6.3	1363	6.8	1398	7.4	1431	7.9	1461	8.5	1492	9.2	1525	9.8								
14000	2093	1254	5.1	1284	5.4	1328	6.2	1362	6.7	1398	7.3	1435	8.0	1470	8.6	1502	9.1	1532	9.7												
15000	2243	1337	6.1	1368	6.6	1396	7.1	1440	7.9	1473	8.5	1507	9.2	1541	9.9																
16000	2392	1417	7.3	1452	7.9	1478	8.3	1521	9.2	1550	9.9																				
17000	2542	1498	8.6	1536	9.3																										

MAXIMUM COOLING COIL FACE VELOCITY LIMITED TO 600 FPM.

$$\text{Max. BHP} = f \times \left(\frac{\text{RPM}}{1000}\right)^3$$

Figure 1-12 Typical low-pressure draw-through units—backward-inclined fan wheels. (*American Air Filter Co., Inc.*)

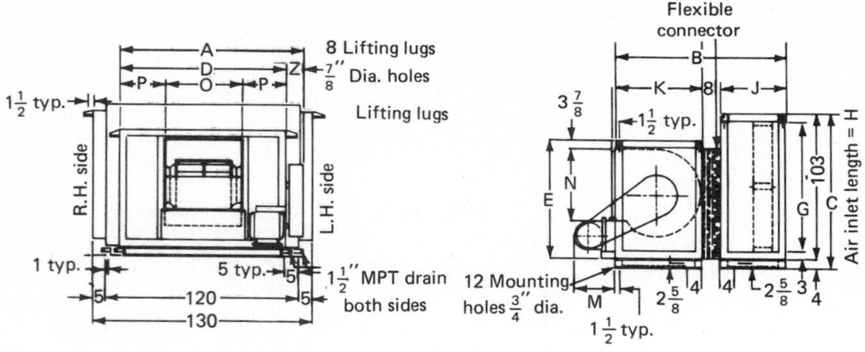

Front View

Unit Size 74
(Must be deck-mounted)

Left Side View

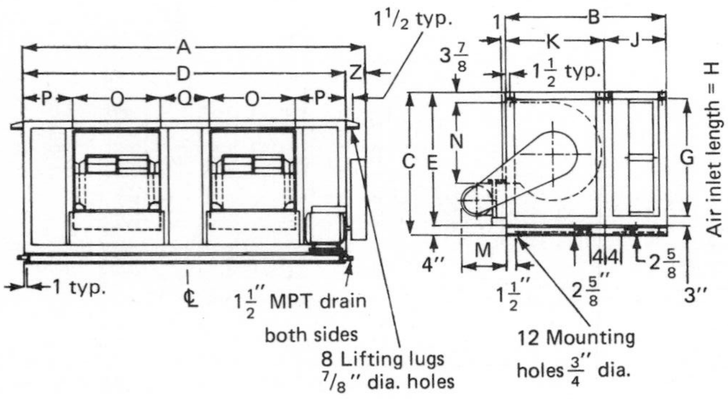

Unit Size 86
(Must be deck-mounted)

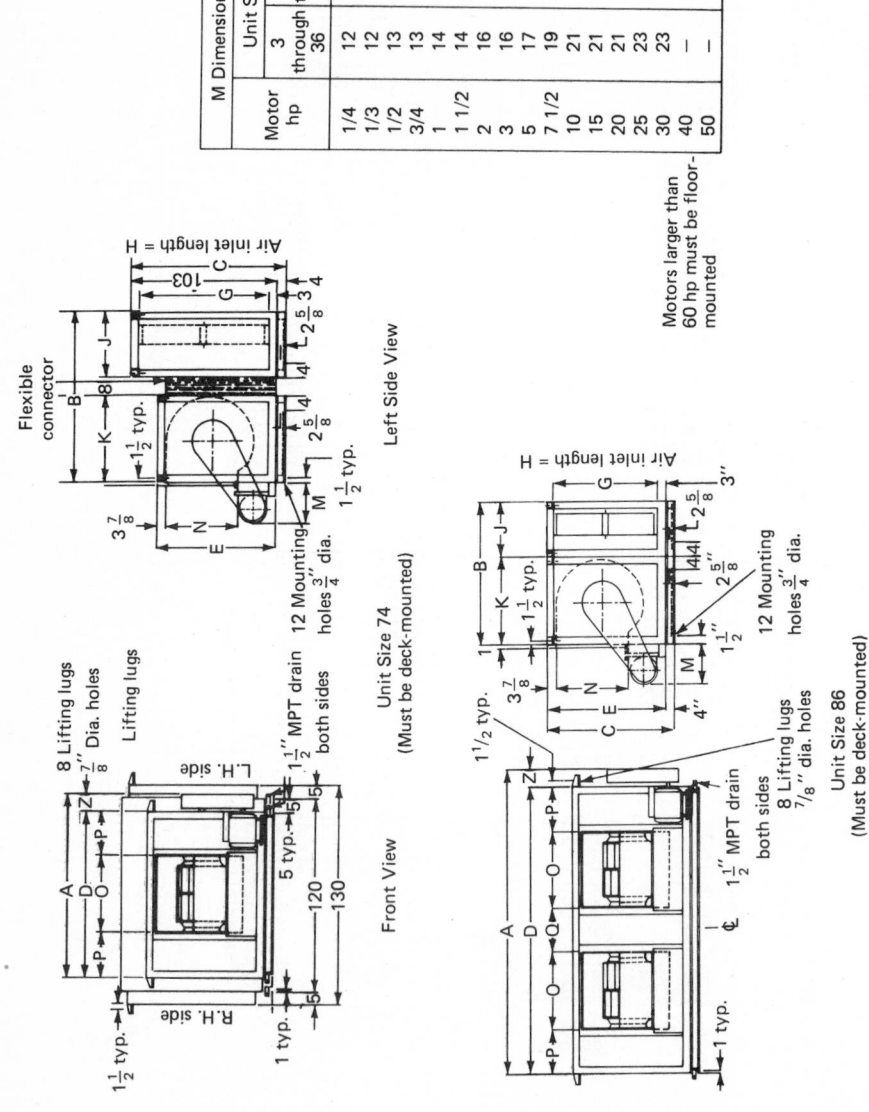

M Dimensions		
	Unit Size	
Motor hp	3 through 36	41 through 86
1/4	12	–
1/3	12	–
1/2	13	–
3/4	13	–
1	14	–
1 1/2	14	–
2	16	–
3	16	–
5	17	22
7 1/2	19	27
10	21	27
15	21	28
20	21	28
25	23	30
30	23	30
40	–	30
50	–	30

Motors larger than 60 hp must be floor-mounted

Figure 1-13 Typical dimensions* of horizontal draw-through, heating and ventilating low- and medium-pressure units. (*American Air Filter Co., Inc.*)

Unit size	A	B				C	D	E	G	H	J		K		Low pressure								Medium pressure								Z
		1		2									1	2	FC				BI				FC				BI				
		Std. depth coil section	Extra depth coil section	Std. depth coil section	Extra depth coil section						Std.	Extra depth			N	O	P	Q	N	O	P	Q	N	O	P	Q	N	O	P	Q	
3	51	46½	54	46½	54	25½	40	22½	18½	36	24	31½	22½	22½	11¼	11⅞	14	—	—	—	—	—	11¼	8⅜	15¾	—	—	—	—	—	6
4	53	49½	57	49½	57	28½	42	25½	21½	38	24	31½	25½	25½	13¼	13¼	14⅜	—	—	—	—	—	13¼	10⅝	15⅝	—	—	—	—	—	6
6	55	46	53½	52½	60	31½	44	28½	24½	40	24	31½	22	28½	17	15¾	14⅛	—	15⅝	17½	13¼	—	17	12½	15¾	—	15⅝	17½	13¼	—	6
9	59	50	60½	58½	69	37½	48	34½	30½	44	24	34½	26	34½	21	18¾	14⅝	—	19⅛	21	13½	—	21	14¾	16⅝	—	19⅛	21	13½	—	6
11	88	46	53½	52½	60	31½	76	28½	24½	72	24	31½	22	28½	17	15¾	12	20½	15⅝	17½	11⅞	17¼	17	12½	13⅝	23¾	15⅝	17½	11⅞	17¼	7
15	94	50	60½	58½	69	37½	82	34½	30½	78	24	34½	26	34½	21	18¾	12	20½	19⅛	21	12	16	21	14¾	14	24½	19⅛	21	12	16	7
18	74	61½	81	74½	94	53½	62	50½	46½	58	24	43½	37½	50½	30⅞	31½	15¼	—	30⅞	31½	15¼	—	27¾	28¾	16⅝	—	27¾	28¾	16⅝	—	7
22	88	61½	81	74½	94	53½	76	50½	46½	72	24	43½	37½	50½	30⅞	31½	22¼	—	30⅞	31½	22¼	—	30⅞	31½	22¼	—	30⅞	31½	22¼	—	7
25	94	65	84½	79	98½	58	82	55	51	78	24	43½	41	55	34	35	23½	—	34	35	23½	—	30⅞	31½	25¼	—	30⅞	31½	25¼	—	7
31	112	69½	89	79	98½	58	100	55	51	96	24	43½	45½	55	37½	38½	30¾	—	37½	38½	30¾	—	34	35	32½	—	34	35	32½	—	7
36	123	74	93½	85	104½	64	110	61	57	106	24	43½	50	61	41¾	42¼	33⅞	—	41¾	42¼	33⅞	—	37½	38½	35¾	—	37½	38½	35¾	—	8
41	123	88½	99½	100	111	70	110	66	62	106	34	45	54½	66	45¾	47	31½	—	45¾	47	31½	—	41¾	42¼	33⅞	—	41¾	42¼	33⅞	—	8
52	123	96	107	113	124	83	110	79	74	104	34	45	62	79	46	51¾	29⅛	—	46	51¾	29½	—	45¾	47	31½	—	45¾	47	31½	—	8
64	123	96	107	128	139	98	110	94	89	104	34	45	62	94	50⅝	51¾	29⅛	—	50⅝	51¾	29⅛	—	50⅝	51¾	29⅛	—	50⅝	51¾	29⅛	—	8
74	118	118	—	150	—	107	110	94	97	114	48	—	62	94	50⅝	51¾	29⅛	—	50⅝	51¾	29⅛	—	50⅝	51¾	29⅛	—	50⅝	51¾	29⅛	—	8
86	166	110	—	—	—	101	158	97	91	152	48	—	62	—	45¾	47	16	32	—	—	—	—	—	—	—	—	45¾	47	16	32	8

*All dimensions in inches. Dimensions subject to change.

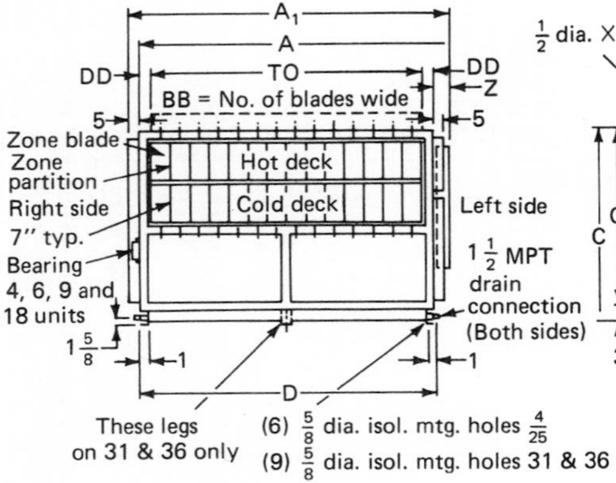

Horizontal Multi-zone Front View

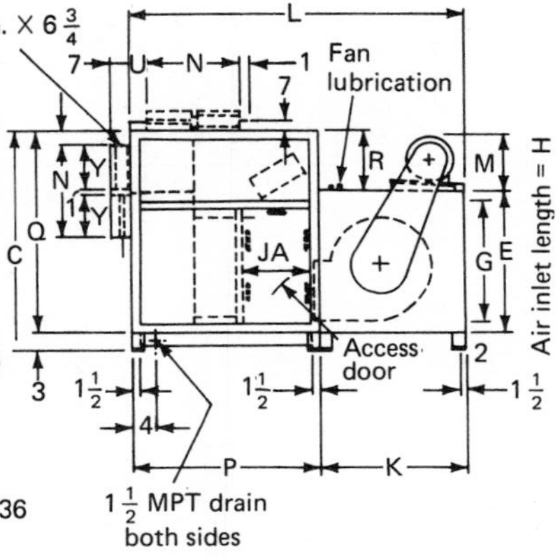

Left Side View Horizontal Unit

Unit Sizes 4, 6, 9, 11, 15, 18, 22, 25, 31, and 36

M Dimensions

Motor hp	Unit Size 1016 through 36	Unit Size 41 through 64
1/4	12	—
1/3	12	—
1/2	13	—
3/4	13	—
1	14	—
1 1/2	14	—
2	16	—
3	16	—
5	17	22
7 1/2	19	27
10	21	27
15	21	28
20	21	28
25	23	30
30	23	30
40	—	30
50	—	30

Motors larger than 60 hp must be floor-mounted.

Figure 1-14 Typical dimensions* of horizontal blow-through, multizone heating and ventilating, low- and medium-pressure units. (*American Air Filter Co., Inc.*)

Unit size	A	A_1	C	D	E	G	H	Horiz. K	Horiz. L	M	N	TO	P	Q	R	U	Y	Z	BB	DD	JA
4	48	53	44¾	42	25½	21½	38	25½	57	Varies with size of motor see engine M dimension table	15½	35	31½	41¾	16¼	2½	7¼	6	5	3½	5¼
6	50	55	48¾	44	28½	24½	40	22	56½		20¾	35	34½	45¾	17¼	2	9⅞	6	5	4½	3⅞
9	54	59	58	48	34½	30½	44	26	69½		24¾	42	43½	55	20½	2	11⅞	6	6	3	9⅞
11	83	88	48¾	76	28½	24½	72	22	56½		20¾	70	34½	45¾	17¼	2	9⅞	7	10	3	3⅞
15	89	94	58	82	34½	30½	78	26	69½		24¾	77	43½	55	20½	2	11⅞	7	11	2½	9⅞
18	69	74	76	62	50½	46½	58	37½	88		38	56	50½	73	22½	2	18½	7	8	3	12⅜
22	83	88	76	76	50½	46½	72	37½	88		38	70	50½	73	22½	2	18½	7	10	3	12⅜
25	89	94	79½	82	55	51	78	41	96		38	77	55	76½	21½	2	18½	7	11	2½	16⅞
31	107	112	79½	100	55	51	96	45½	100½		38	91	55	76½	21½	2	18½	7	13	4½	16⅞
36	118	123	82½	110	61	57	106	50	111		38	105	61	79½	18½	2	18½	8	15	2½	18⅞

*All dimensions in inches. Dimensions subject to change.

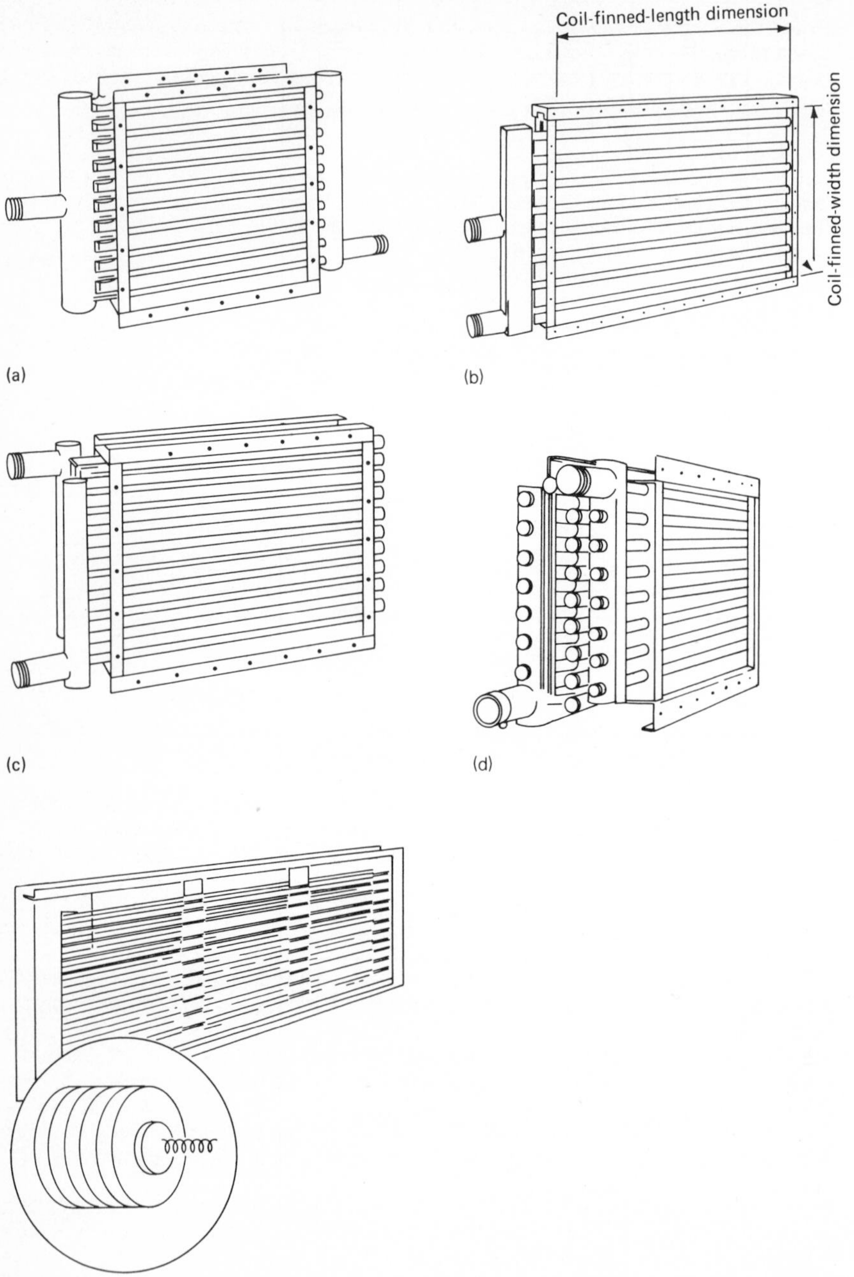

(a) (b) (c) (d)

Figure 1-15 Typical heating coils: (*a*) Standard steam coil. (*b*) Distribution coil. (*c*) Standard water coil. (*d*) Cleanable-tube water coil. (*e*) Electric resistance coil.

temperature difference. From the manufacturer's point of view, for the same heating requirement and coil velocity, it is cheaper to produce coils with more fins per inch and fewer rows. However, from the operation performance and maintenance standpoint (though there may be little difference between the air resistance of coils with more fins per inch—and fewer rows—and coils with fewer fins per inch—and more rows), if the airstream is dirty, the installation with more fins per inch will have a greater potential for plugging (filling the space between the fins). This not only will significantly increase the coil's air resistance but can create a fire hazard if the particles retained undergo spontaneous combustion due to the heat they will absorb from the hot coil. Furthermore, both the time required for cleaning and the frequency of cleaning will increase as the space between adjacent fins is decreased.

Here is a guide to determine the optimum number of fins per inch:

1. Select no more than 8 fins per inch (3 fins per centimeter), provided that the heating load can be met with no more than two rows.
2. If the air is dirty (has a tendency to plug the coil), select fewer than 8 fins per inch (3 fins per centimeter) and no more than 2 rows. This may require increasing the coil area or temperature or both.
3. If the air is clean (does not have a tendency to plug the coil) and increasing the number of fins per inch will result in reducing the number of rows, then the greater fin spacing may be desirable.

Though heating coils can be selected with velocities ranging from as low as 200 ft/min (1.02 m/s) to as high as 1500 ft/min (7.62 m/s), the optimum range for overall economy and performance is 800 to 1000 ft/min (4.06 to 5.08 m/s). Exceptions to this are:

1. If the air is dirty, it may be desirable to decrease the coil velocity in order to satisfy the heating load with no more than 8 fins per inch and 2 rows (greater coil area).
2. If a single-zone unit is selected and there is a requirement for future air conditioning, the air velocity of the coil selected should be 500 to 550 ft/min (2.54 to 2.79 m/s).

Hot-water coil ratings are based on counterflow of air and water, i.e., the hot-water supply to the coils must enter the coil tubes on the leaving-air side and the water must leave the coil tubes on the entering-air side. The optimum water velocity for economical heat-transfer rate and water head loss will normally be in the range of 2.5 to 4 ft/s (0.0127 to 0.0203 m/s). Water velocities above 8 ft/s (0.041 m/s) or below 0.5 ft/s (0.0025 m/s) should not be used. The water velocity within the tubes is a function of the number of coil circuits for coils of the same tube diameter.

Most manufacturers provide three standard circuiting arrangements to produce the optimum velocity for each application: single serpentine, half-serpentine, double serpentine. Single serpentine has the greatest applicability and is the most commonly used. Half-serpentine increases water velocity; double serpentine decreases water velocity.

The availability of the heating medium (steam or hot water) usually determines whether a steam- or water-heated coil will be selected. Likewise, the choice between standard and steam distribution type (if steam coils are selected) or standard and cleanable type (if water coils are selected) depends on the installation in question (see under "Coil Selection Parameters").

The plant engineer should determine the preferable coil fin spacing on the basis of previous discussion and past experiences. This data of the discussion may have to be modified, depending on actual fin spacing available from coil manufacturers.

Since the actual procedure in selecting a heating coil varies with each manufacturer and there are many parameters in the selection procedure, the coil manufacturer's procedure must be strictly followed in order to select a coil with a capacity equal to or slightly greater than the required heating load.

FILTER SELECTION

General

The selection of the type of filter should be based on the following parameters:

1. **Efficiency** This parameter is defined as the ability to retain airborne particles. What degree of air cleanliness is required? *The greater the particle retention* (which could be viewed technically as obstructions in the airstream), *the higher the operating and maintenance costs.*
2. **Air volume** How critical are fluctuations in air volume to the performance of the heating and ventilation system as a result of increasing system resistance (static pressure) caused by dust and/or the retention of dust on the filter medium? The Roll-O-Matic® (American Air Filter Co., Inc.) and electrostatic filters maintain a relatively constant static pressure drop across the filter medium for the life of the filter. In contrast, as cartridge-type filters go from the clean to dirty state, the static pressure across them undergoes the greatest increase.

Replacement

As a rule of thumb, filters should be replaced or cleaned when the initial air resistance has doubled. However, if the system's fans can deliver the required air quantities against the higher system static pressure, then the point at which the filters should be replaced

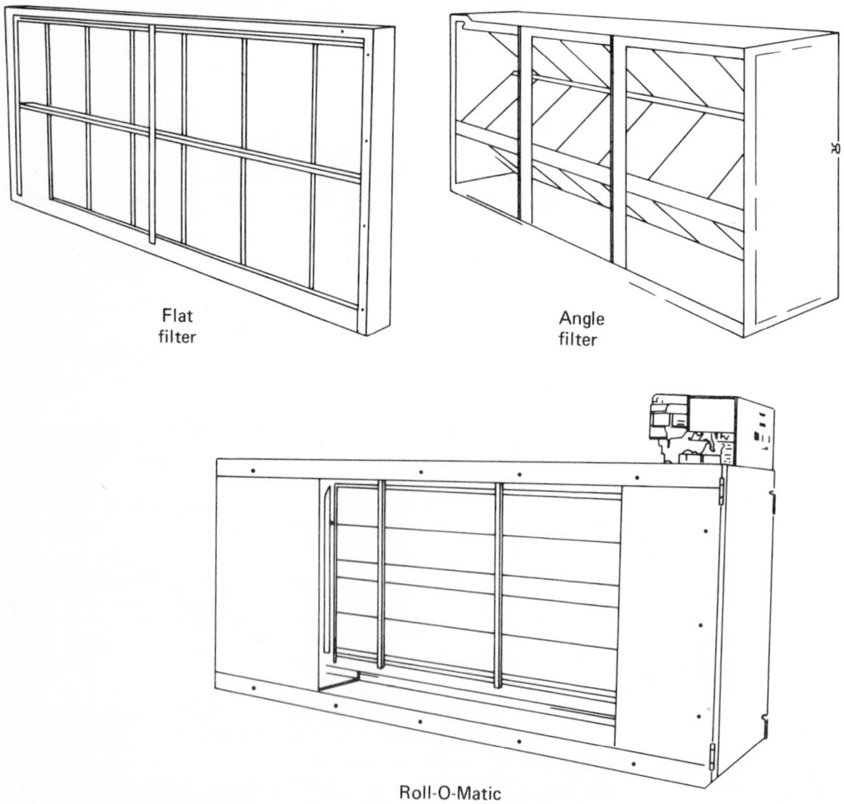

Figure 1-16 Typical filters and filter boxes. Generally throwaway, renewable, or cleanable filter types are available in flat or angle-unit filter boxes. (*American Air Filter Co., Inc.*)

or cleaned can be determined by economics. That is, filters should be changed when the replacement or cleaning costs are equal to the cost of the additional energy (due to the increase in air resistance above the initial resistance) in the time span it takes to go from clean to dirty filters.

The types of filters and filter boxes commercially available are shown in Fig. 1-16. Typical data on filter media, efficiency, maximum air volumes, air velocity, and air resistance are shown in Fig. 1-17 for different types of filters commercially available.

FAN SELECTION

General

Heating and ventilating fans are available in three pressure classes. These are determined by the total static pressure the unit's fan (or fans) is designed to develop. The pressure classes are:

1. Low-pressure (class 1) total static pressure up to 3 inH_2O (747 Pa)
2. Medium-pressure (class 2) total static pressure 3 to 5 inH_2O (747 to 1245 Pa)
3. High-pressure (class 3) total static pressure 5 to 10 inH_2O (1245 to 2490 Pa)

Heating and ventilating unit fans are typically centrifugal. They are, depending on the unit's capacity, available in forward-curved, backward-inclined, and airfoil wheel

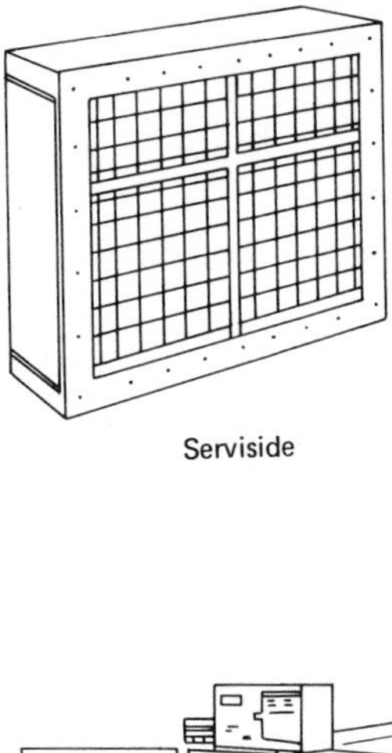

Serviside

Cartridge

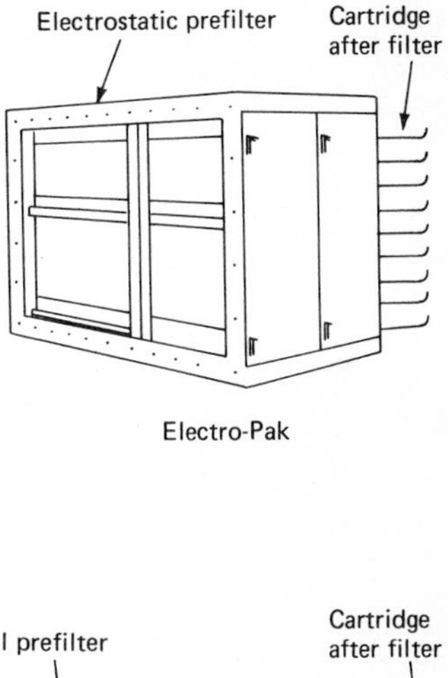

Electro-Pak

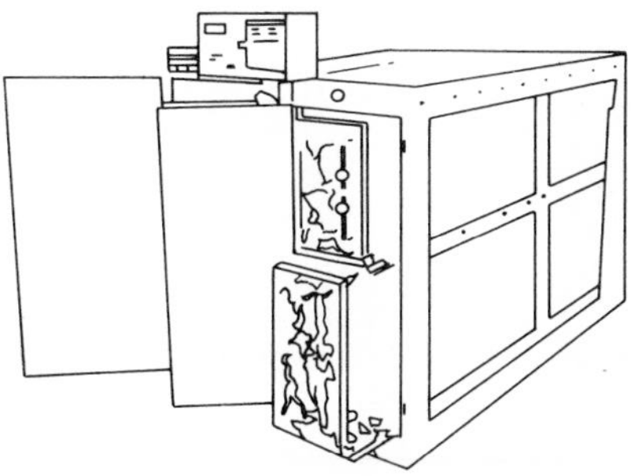

Roll-O-Tron

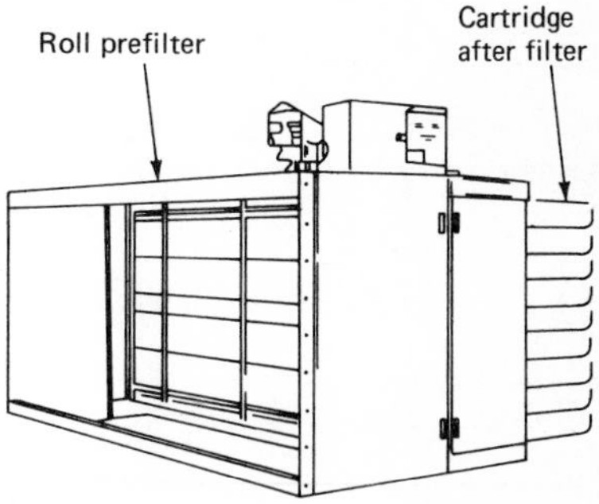

Roll-O-Pak

Electrostatic

Figure 1-16 *(Continued)*

(*a*) Typical Comparative Filter Performance

Type		Maintenance cost	Maintenance required	Constant air volume	Media number	Approx. efficiency		ServiSide® housing extension required & size	Maximum face volume, ft/min	Recommended face volume, ft/min
						AFI (weight)	NBS dust spot (discoloration)			
5700 [Throwaway]		High	Replace filters every 4–10 weeks	No	—	70–75%	—	—	500	300
Amer-Frame [Renewable]	Glass	Low	Change media every 6–12 weeks	No	—	75–80%	—	—	500	300
	Polyurethane	Moderate	Clean filters every 6–12 weeks	No	—	65–75%	—	—	450	350
HV-2 [Cleanable]		Moderate	Clean filters every 6–12 weeks	No	—	70–80%	—	—	625	500
Roll-O-Matic®		Low	Change media once a year	Yes	—	75–80%	—	—	600	500
ServiSide® [Cartridge]	Dri-Pak®	Low	Change media once a year	No	30 40 60 90 100 or 10	— — — — —	30% 38–40% 50–55% 80–85% 93–97%	No Yes—10″ Yes—26″ Yes—26″ Yes—26″	625 625 500 or 625 500 or 625 500 or 625	625 625 500 or 625 500 or 625 500 or 625
	Varicel®	Low	Change every 6–12 months	No	6 9 10	— — —	55–60% 85–90% 90–95%	No No No	625 500 500	625 500 500

*2000 Series rated at 500 ft/min, 2500 Series rated at 625 ft/min.

(*b*) Typical Maximum Air Volumes—Standard-Capacity Filters

Air-handling unit size	Unit filters						Roll-O-Matic®			Roll-O-Tron®, ServiSide®, Electro-Pak®, & Roll-O-Pak®							
	Flat bank			Std.-angle bank								ServiSide®				Roll-O-Pak®	
	5700 & glass	Poly.	HV-2	5700 & glass	Poly.	HV-2	Model no.	@ 500 ft/min	@ 600 ft/min	Model no.	Roll-O-Tron®	Quantity and Size of Filter Cells	@ 500 ft/min	@ 625 ft/min	Electro-Pak®	@ 500 ft/min	@ 600 ft/min
3	2220	2000	2780	4460	4000	5560	NA	—	—	NA	—	—	—	—	—	—	—
4	2780	2500	3480	4460	4000	5560	25-38	2765	3320	3H28	NA	1A & 1C	3000	3750	NA	NA	NA
6	2780	2500	3480	4460	4000	5560	25-38	2765	3320	3H28	NA	1A & 1C	3000	3750	NA	NA	NA
9	3480	3120	4350	6950	6250	8700	33-50	5415	6495	4H34	4660	2B & 2C	5300	6500	4960	5300	6000
11	5000	4500	6250	8350	7500	10400	25-70	5840	7010	6H24	4580	3B	4950	6000	4860	4950	6000
15	6950	6250	8700	13200	11900	16500	33-70	7915	9495	7H34	8320	3B & 3C	7950	9750	8860	7950	9750
18	9350	8400	11700	13300	12000	16700	45-54	8170	9790	5H48	8990	4A & 2C	10000	12500	9570	10000	10500
22	11300	10150	14100	16700	15000	20800	45-70	11085	13295	6H48	10980	6A	12000	15000	11700	12000	12600
25	13300	12000	16700	20800	18700	26000	45-78	12250	14700	7H48	12490	6A & 2C	14000	17500	13250	14000	14700
31	16600	15000	20800	26600	24000	33300	45-88	14000	16800	8H48	14480	8A	16000	20000	15370	16000	16800
36	18000	16300	22600	27800	25000	34700	57-94	19500	23385	9H50	17850	8B & 4C	17200	21000	18990	17200	21000
41	20400	18400	25500	34800	31300	43500	57-94	19500	23385	9H58	20610	12B	20000	25000	21870	20000	24300
52	28100	25300	35100	41700	37500	52100	69-94	23830	28580	9H68	24720	12A & 3C	27000	33750	26190	27000	29700
64 & 74	30700	27600	38300	52000	46800	65000	4545-100	32670	39190	9H4040	27480	16B	26400	32000	29160	26400	32000
86	NA	NA	NA	NA	NA	NA	4545-134	44340	53180	13H4040	39160	24B	39600	48000	38880	39600	48000

†Applicable to Size 64 only.
A = 24″ × 24″; B = 20″ × 24″; C = 12″ × 24″

Figure 1-17 Typical filter performance data. (*American Air Filter Co., Inc.*)

(*c*) Typical Filter Air Resistance

Reference velocity, ft/min	Filters* (Inches of water)								
	Flat bank		Standard angle						
	TA ren. poly	HV-2	TA ren. poly.	HV-2	Servi-Side®	Roll-O-Matic®	Roll-O-Pak®	Electronic	Electro-Pak®
300	0.12	0.06	0.09	0.04	—	0.40	—	—	—
400	0.16	0.11	0.12	0.07	—	0.40	—	—	—
500	0.21	0.16	0.14	0.09	0.5	0.40	0.6	0.15	0.55
600	0.28	0.24	0.18	0.13	—	0.40	—	—	—
700	—	0.32	0.23	0.17	—	0.40	—	—	—

*Resistance shown based on clean filters.

Figure 1-17 (*Continued*)

designs (see Fig. 1-18). *Forward-curved* fans are most common in heating and ventilating, especially in the smaller units. *Backward-inclined* fans are used in the larger units where their nonoverloading characteristics and greater stable operating range (compared to forward-curved fans) are desired. Because of their greater efficiency, nonoverloading characteristics, and quieter operation, *airfoil* fans are frequently used in high-pressure applications where greater motor horsepower is required.

Fan noise is a function of the fan design, air-volume flow rate, total pressure, and efficiency. The quietest fan operation will be obtained when a fan is selected at its most efficient operating point. Since it is not realistic to assume the system air volume and resistance will be constant or will match the fan's performance at maximum efficiency, fans should be selected, if possible, in the range from maximum efficiency to 90 percent maximum efficiency. Sound power level specifications for the fans being selected are not always available. Since low outlet velocities do not necessarily assure quiet operation, Fig. 1-19 can be used to determine the appropriate fan outlet velocity range. By entering Fig. 1-19 with the total static pressure, one can read the outlet velocity range corresponding to the maximum efficiency and 90 percent maximum efficiency curves.

Adjustable motor-drive sheaves should be used on V-belt-driven fans to provide the capability for minor adjustment of airflow and static pressure in the field.

The types and capacity ranges of typical commercially available ventilation fans are shown in Fig. 1-20. The fan section of a heating and ventilating unit (see Fig. 1-7) can also be used as a ventilation fan with or without ductwork.

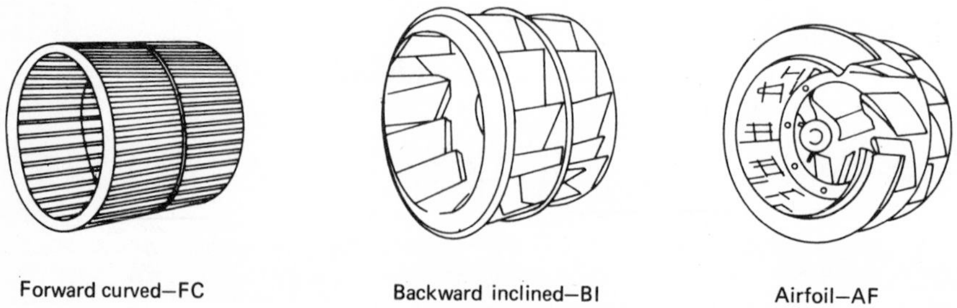

Figure 1-18 Typical fan wheels.

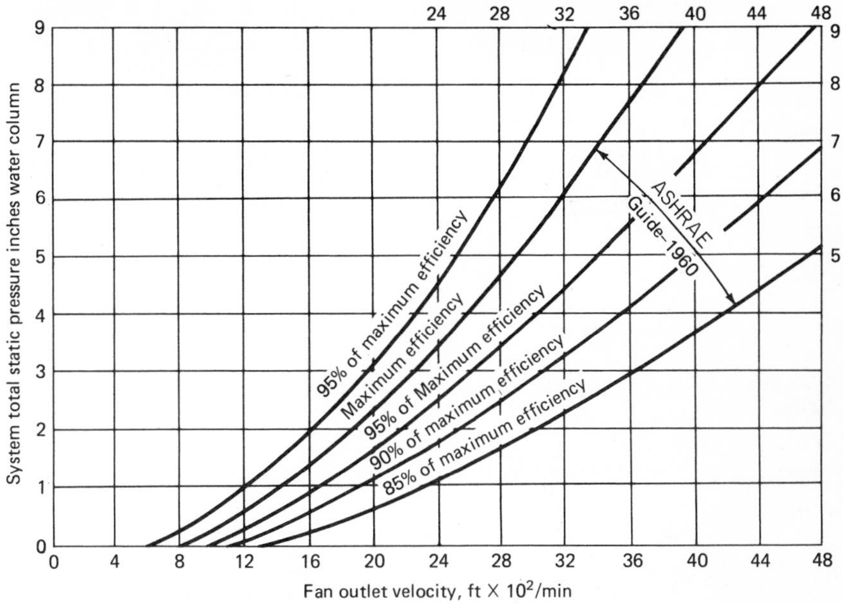

Figure 1-19 Typical fan performance. The suggested range of outlet velocity approximates the curves plotted for maximum efficiency and 85 percent of maximum efficiency as indicated by arrows.

SELECTING UNIT HEATERS

General

The revolving unit heater is ideally suited for large industrial spaces. Some of the advantages are:

1. More heat coverage: up to 145 × 145 ft (44.2 × 42.2 m) with a standard heater
2. Uniform heat: minimizes hot and cold spots, no steady hot blasts
3. Low floor-to-ceiling temperature differentials
4. Steam, hot water, high-temperature hot water, and gas can be used as heating media
5. Discharge designs to 65-ft (19.81-m) mounting heights
6. Effective summer air circulation
7. Low installed and operating costs

Unit heaters can be powered by gas, propane, or electricity. Data on them can be obtained from manufacturers.

When selecting electric unit heaters care must be taken to ensure electrical compatibility with other electric systems of the unit. Adequate power and wiring should also be available.

Controls

Summer ventilation can be provided, with fixed or revolving units mounted within the roof truss spaces or adjacent to the exterior walls, by installing a fresh-air intake duct from the roof or exterior wall. This provides a source of fresh outside air for some cooling and maintains positive pressure within the building, thus increasing the efficiency of the exhaust system.

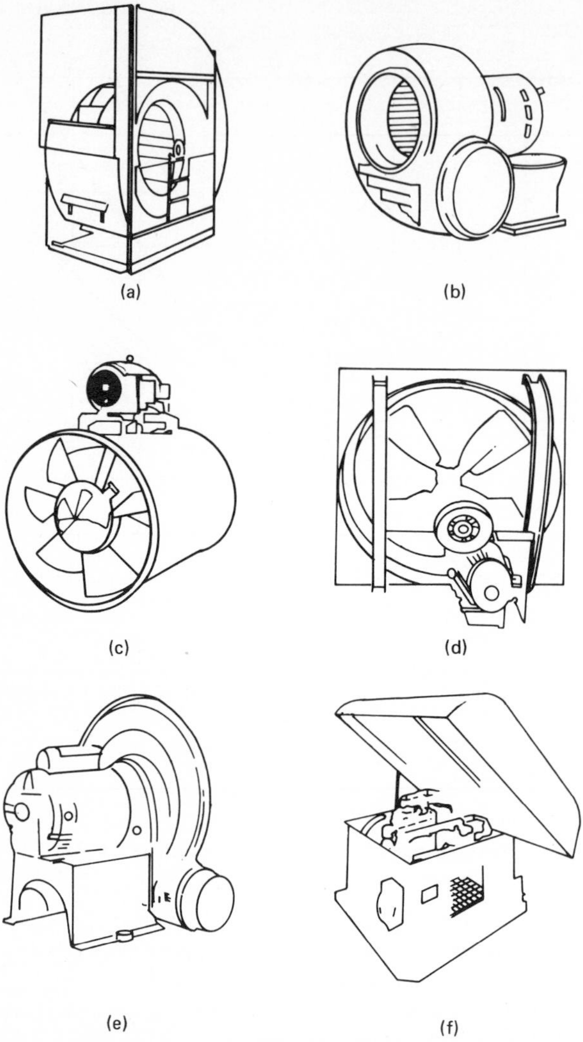

Figure 1-20 Typical ventilating fans. (*a*) Aerofoil®, a high-efficiency, low-noise-level fan for quality, optimum design air-conditioning and ventilation systems. Construction features include wheel blades that combine the best of backward-curved and airfoil designs. Capacities vary from 687 to 661,000 ft^3/min and pressures up to 12 inHg. (*b*) Baby vent sets—a quiet, reliable fan with a sturdy corrosion-resistant cast-iron housing that is rotatable to obtain any 45° angular discharge. Widely used as a component on many industrial products. Features include all-weather cover, drain, nonspark aluminum wheel, heat slinger, antivibration pads, special corrosion-resistant materials and/or protective coatings. Capacities range from 47 to 1765 ft^3/min with pressures to 1 inHg. (*c*) Axial-flow fan. These efficient, low-noise vane-axial fans with limited load horsepower characteristics are available with either fixed-pitch steel or aluminum, or adjustable-pitch aluminum wheels, enabling the user to match pressure/volume requirements exactly. Capacities greater than 300,000 in^3/min are available. (*d*) Propeller fan. Such fans may include belt or direct drive, penthouses, wall or ceiling shutters, wall boxes, filters, wire guards, and spark-resistant construction. Wheel sizes from 8 to 120 in. Capacities from 500 to 240,000 ft^3/min and up to 1 in SP or higher. (*e*) Electric blower/exhauster. Small package units like this are used for small furnace draft, handling particles of wood, metal, and abrasives in collecting systems, and many more jobs requiring moderate volume against low pressures. Their capacities range from 30 to 700 ft^3/min. (*f*) Power roof ventilators are engineered and built to meet the demand for high-capacity exhaust and supply air. Will withstand hurricane-force winds, heavy snow loads, and corrosive atmospheres. Filters, dampers, screens, heating coils available. Capacities from 1000 to 250,000 ft/in. Pressures from free delivery to over 1 in SP. (*Buffalo Forge Co.*)

A damper box mounted above the motor and fan provides for admission of air from above the roof for summer ventilation or from the truss area for winter heating. This is controlled within the box by the functioning of the two sets of dampers which are manually or automatically operated, as desired. Fresh-air roof intakes and damper boxes are available from manufacturers.

Automatic temperature controls should be provided to cycle the units on and off in order to maintain space temperatures during the heating season and provide outside ventilation air during the summer.

Selection Procedure

To minimize installation, maintenance, and energy costs, it is desirable to select the minimum number of units that will provide a uniform distribution of heat (minimum temperature differential) throughout the space. For exterior areas, unit heaters should be positioned so that their discharge "wipes" the exterior walls and windows. It is more practical to treat exterior door heating requirements separately, especially loading-dock areas. For interior areas, the unit heaters should be positioned so that their discharge provides uniform air motion (heat distribution).

The major exception to these guidelines occurs when spot heating is the primary requirement. In this case, the units should be positioned to provide a uniform distribution over the area where localized heating is required and to position units in the remaining area so as to maintain a specified minimum temperature determined by the process or product stored or to prevent freeze-ups.

Since there are various combinations of unit heater capacities and distribution patterns, the selection is a trial-and-error procedure. It is recommended that the units be selected from manufacturers' data, and that the selection be based on appropriate distribution patterns and the available mounting height for the layout of the space to be heated before checking for adequate capacity. This is preferable to selecting the units on the basis of capacity and *then* checking for adequate distribution.

If the total capacity of the unit heaters tentatively selected is less than the required design space heating load, proceed as follows.

1. If unit heaters with greater capacity and approximately the same distribution pattern are available, they should be selected, and the new total heating capacity should then be checked against the required design load.
2. If units of larger capacity are not available, more unit heaters must be added. Care must be exercised to add the required units where the greatest heat loss occurs so as to minimize the possibility of overheating some areas.
3. In some cases it may be required to combine steps 1 and 2 to obtain the best solution.

If the total capacity of the selected units is greater than the space heating load, proceed as follows.

1. If possible, select units with less capacity but with approximately the same distribution pattern.
2. Select automatic controls that will cycle the unit heaters without causing excessive temperature variation.
3. Combining steps 1 and 2 may be the most practical solution.

The final selection should result in the installed heating capacity of the units being equal to or slightly more than the design heating load.

Figure 1-21 illustrates typical revolving and fixed vertical-discharge steam or hot-water unit heaters. These units are available in increment sizes and typically have a capacity of 30,000 to 60,000 Btu/h (879 to 1758 W) in the smaller sizes and 750,000 to 1,200,000 Btu/h (21,975 to 35,160 W) in the large sizes, depending on whether the heating medium used is steam or hot water. Steam units are generally rated at 5 lb/in^2 (0.345 bar) steam pressure and 60°F (15.6°C) entering air to the heating coil. Hot-water units

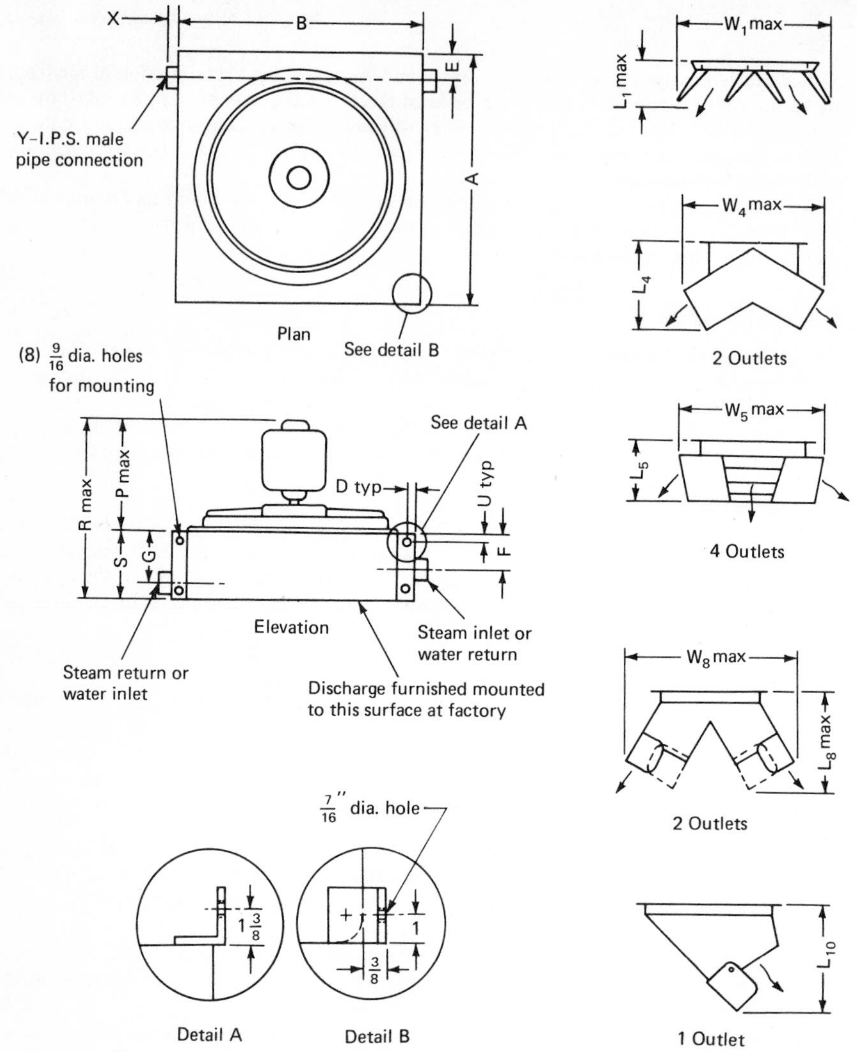

Figure 1-21 Typical revolving and fixed vertical-discharge steam or hot-water heaters. (*Wing Industries Inc.*)

are generally rated at 200°F (93.3°C) entering water and 60°F (15.6°C) entering air to the heating coil.

Whenever the steam pressure, entering hot-water temperature, or entering air temperature is greater or less than the values that the manufacturer's published capacities were based upon, the published capacities must be corrected in order to obtain the heating capacity of the unit selected under the actual operating conditions: the temperature of the entering heating medium (steam or hot water) and the entering-air.

The physical sizes of the basic fixed- and revolving-discharge unit heaters are the same. Typically the smaller sizes are about 16 × 16 in by 18 in (0.406 × 0.406 m by 0.457 m) high, with the larger sizes about 50 × 60 in by 50 in (1.27 × 1.52 by 1.27 m) high. The fixed discharge models will add about 6 in (0.152 m) to the height of the smaller units and as much as about 10 in (0.254 m) to the larger units. The revolving discharge models will add about 12 to 24 in (0.301 to 0.610 m) to the height of the smaller units and as much as 24 to 50 in (0.610 to 1.27 m) to the height of the larger units, depending on the type of discharge selected.

In general, the larger the unit, the higher the (maximum) mounting height and the greater the area covered per unit. Also revolving discharge unit heaters (especially in the larger sizes) can cover greater areas than fixed discharge units. Typically, fixed and revolving discharge heaters have a maximum mounting height in the smaller sizes of 8 to 12 ft (2.44 to 3.66 m), depending on the model selected. The larger sizes have a maximum mounting height of 25 to 50 ft (7.62 to 15.24 m), depending on the model selected for fixed and revolving discharge heaters. However, the largest revolving unit heater has a minimum mounting height of about 25 ft (7.62 m) and covers an area of about 90 × 90 ft (27.4 × 27.4 m) to a maximum mounting height of about 65 ft (19.81 m) and covers an area of about 140 × 140 ft (42.67 × 42.67 m).

Figure 1-22 illustrates a typical horizontal discharge steam or hot-water unit heater. Such units are used when it is desirable to obtain heat distribution by using a number of small heaters, in areas such as offices, showers, stockrooms, etc., where it is neither economically nor physically practical to use large heaters. These units are available in increment sizes and typically have a capacity of about 15,000 Btu/h (439.5 W) in the smallest size to about 10^6 Btu/h (29,300 W) in the largest size.

Whenever the steam pressure, entering hot-water temperature, or entering air temperature is greater or less than the values that the manufacturer's published capacities were based upon, the published capacities must be corrected in order to obtain the heating capacity of the unit selected under the actual conditions: entering heating medium (steam or hot water) and entering air temperature.

The smaller units are typically 12 in (0.305 m) wide by 12 in (0.305 m) high by 12 in (0.305 m) deep (face of coil to back of motor). The smallest unit has a maximum mounting height of about 8 ft (2.44 m) and a horizontal effective heating throw of about 20 ft (6.10 m). The largest unit has a maximum mounting height of about 20 ft (6.10 m) and a horizontal effective heating throw of about 100 ft (30.5 m).

Figure 1-23 illustrates a typical electric horizontal-discharge unit heater. These units have the same application as horizontal steam or hot-water unit heaters. They are available in increment sizes and typically have a capacity of about 4 kW in the smallest size to about 45 kW in the largest. The electrical characteristics of the units are typically 208 and 240 V, single- or three-phase; 277 V single-phase; and 480 V three-phase up to about 10-kW capacity. Above 10-kW capacity, only three-phase is available at 208, 240, and 480 V, depending on the size. The smallest unit has a maximum mounting height of about 8 ft (2.44 m) and a horizontal effective heating throw of about 12 ft (3.66 m). The largest unit has a maximum mounting height of about 18 ft (5.49 m) and a horizontal effective heating throw of about 40 ft (12.19 m).

Whenever the entering air temperature to the heater is greater or less than the value that the manufacturer's published capacities were based upon, the published capacities must be corrected in order to obtain the heating capacity of the unit selected under the actual entering air temperatures.

SELECTION OF INFRARED RADIANT HEATERS

Since infrared radiant energy produces heat only when it is absorbed by a body whose temperature then rises (thereby producing heat), the manufacturer's selection procedures should be strictly followed. Following is a list of general limitations on the application of infrared heaters;

1. Do not mount in an explosive environment.
2. Observe minimum spacing to combustible materials [approximately 24 in (0.61 m) minimum, end or side spacing from fixture to combustible materials and at least 5 ft above them].
3. Do not mount in a recessed position, unless approved by the manufacturer.
4. For efficient operation, do not direct radiation onto window glass.

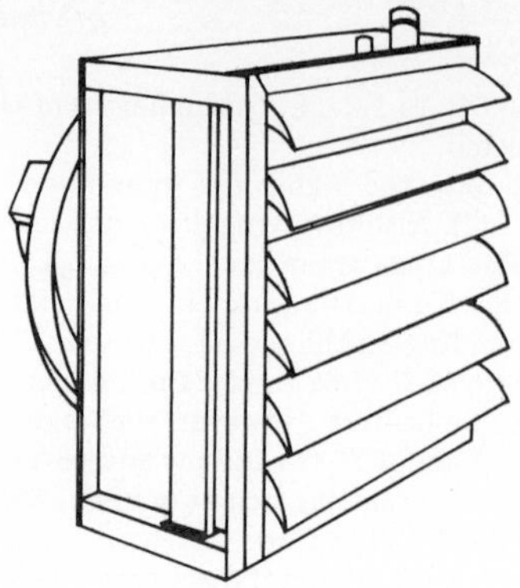

(a)

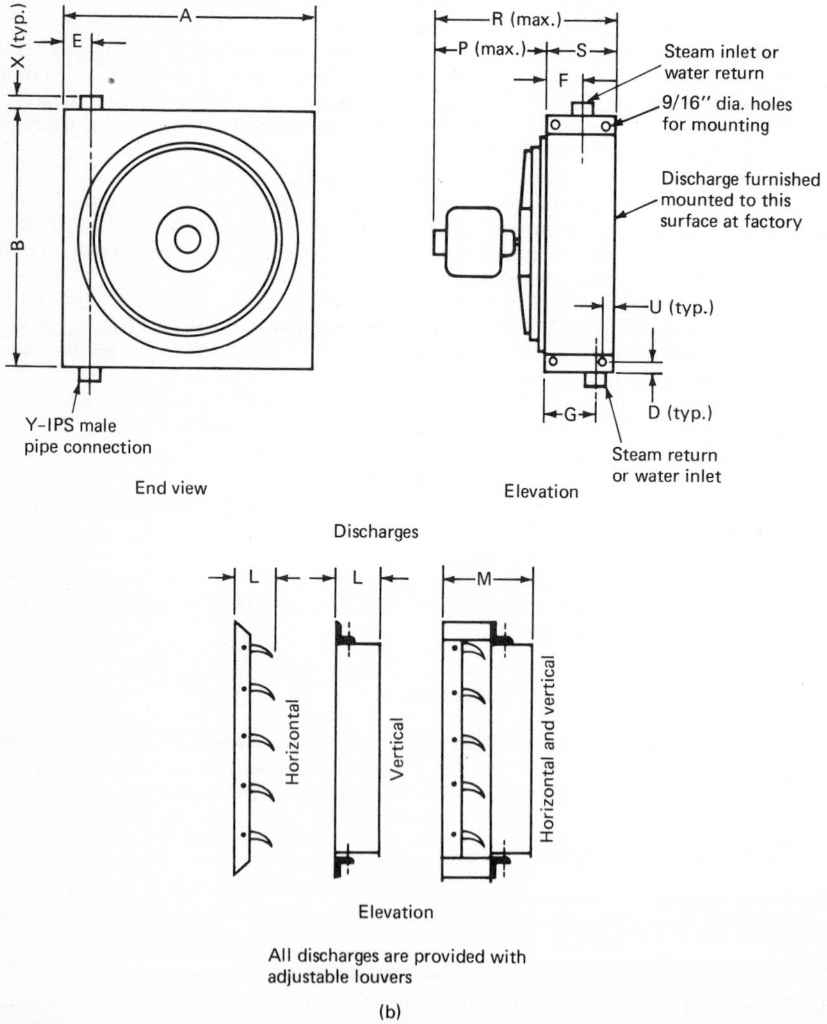

A
E
X (typ.)
B
Y-IPS male
pipe connection
End view
R (max.)
P (max.)
S
F
Steam inlet or
water return
9/16" dia. holes
for mounting
Discharge furnished
mounted to this
surface at factory
U (typ.)
D (typ.)
G
Steam return
or water inlet
Elevation
Discharges
L
L
M
Horizontal
Vertical
Horizontal and vertical
Elevation
All discharges are provided with
adjustable louvers

(b)

5. Do not use in high humidity or in corrosive environments.
6. Do not mount closer to personnel than 4 ft (1.22 m).

From the point of view of both energy conservation and personnel comfort it is desirable not only to cycle the panels on and off automatically from a space thermostat, but also to de-energize selective panels as the heating load decreases. In determining the panels that are to be de-energized in milder weather, care must be taken so that the remaining active panels will uniformly heat the space without overheating the area directly below the (active) panels.

Electric and gas-fired infrared unit heaters are commercially available, generally with three distribution patterns:

1. Narrow distribution pattern (45°)
2. Medium distribution pattern (60°)
3. Broad distribution pattern (90°)

One of the significant parameters in determining the distribution pattern (narrow, medium, or broad) is the required mounting height of any unit. In general:

Narrow (beam) distribution pattern is suitable at mounting heights of 40 to 45 ft (12.2 to 13.7 m).

Figure 1-23 Typical electric horizontal unit heater. (*Emerson Electric Co.*)

Figure 1-22 Typical horizontal steam or hot-water unit heater. (*a*) Sketch, (*b*) End and elevation views. (*Wing Industries Inc.*)

Mounting Heights and Throws for Horizontal Unit Heaters

Heater size	Max mtg height, ft	Effective heating throw, ft
1½-U	8	20
2-U	9	25
2½-U	9	30
13-U	10	35
15-U	12	40
17-U	12	45
18-U	13	50
19-U	13	55
20-U	13	60
22-U	14	65
23-U	14	70
25-U	15	75
26-U	17	80
28-U	18	88
30-U	18	92
33-U	19	102
36-U	19	110
38-U	20	122
40-U	22	135
43-U	25	155
44-U	28	180

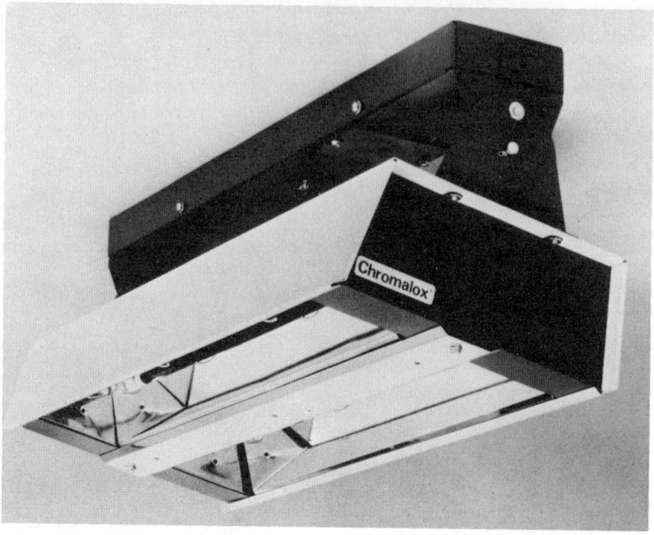

Figure 1-24 Typical electric narrow-beam infrared heater. (*Chromalox®—E. L. Wiegand Div., Emerson Electric Co.*)

Medium (beam) distribution pattern is suitable at mounting heights up to 35 ft (10.7 m). Broad (beam) distribution pattern is suitable at mounting heights up to 25 ft (7.6 m).

Figure 1-24 illustrates a typical electric narrow-beam infrared heater. Such a heater has the following characteristics:

1. This type is typically a deep-well reflector with precise optical design that confines energy within a 45° beam angle with little spill or scatter. It is a good choice for high mounting, since it generally does not require heaters with 5 percent additional wattage for each foot (0.33 m) above a 10 ft (3.1 m) mounting height.
2. The housing on some models can be swiveled up to 45° in either direction from the vertical.
3. They are available for indoor or outdoor exposed applications.

Three different heating elements are available: metal sheath, quartz tube, and quartz lamp. When used outdoors, the *metal sheath* must be shielded from the wind for maximum heating efficiency. Units with metal-sheath elements are available in increments from 600 to 2000 W, with electrical characteristics of 120, 208, and 240 V single-phase for the smaller units and 208, 240, 277, and 480 V single-phase for the larger ones. Units with *quartz-tube* elements are available, in increments, from 550 W to 15.4 kW, with electrical characteristics of 120, 208, and 240 V single-phase for the smaller units and 208, 240, and 277 V single-phase for the larger ones. Units with *quartz-lamp* elements are available, in increments, from 800 to 3800 W, with electrical characteristics of 120, 208, and 240 V single-phase for the smaller units and 480 and 600 V single-phase for the larger ones.

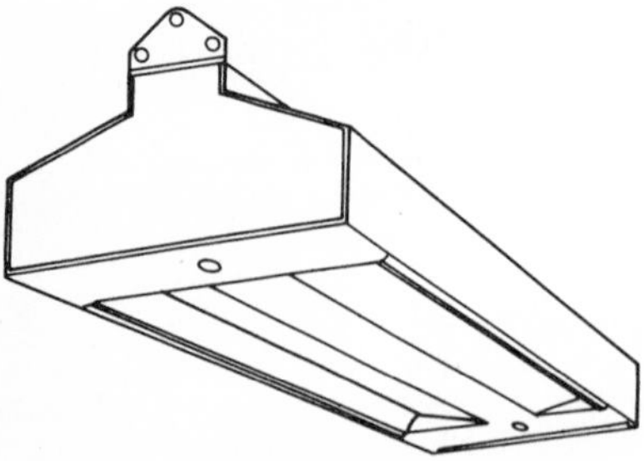

Figure 1-25 Typical electric medium- or broad-beam infrared heater.

Figure 1-25 illustrates a typical electric medium- or broad-beam infrared heater.

Double-element fixtures are available which provide double heating capacity in a single infrared radiant heater. They are applicable to indoor and outdoor protected use. They can be mounted directly to a ceiling or hung from chains. *Broad-beam* units have a 90° distribution pattern, with a symmetrical reflector for low to medium mounting height. *Medium-beam* units have a 60° symmetrical distribution pattern for medium mounting height. Units are available for medium mounting height; these have a 60° asymmetrical perimeter distribution pattern. Units are available with metal-sheath, quartz-tube, or quartz-lamp elements. Nominal capacities range from 1100 to 7600 W. Though all units require single-phase voltage, their voltage requirements vary, depending on the model and type of element. In general, the smaller units are available in 120, 208, and 240 V, though some units are also available in 277 and 480 V. The medium-size units generally are available in 208 and 240 V, though some units are also available in 277 and 480 V. The larger units generally are available in 480 and 600 V, except for the largest unit which may be restricted to 600 V.

Figure 1-26 illustrates a typical gas infrared heater. Code requirements may restrict the use of, or require special installation procedures for, gas heaters. These heaters are available with natural gas or liquid propane (LP) gas fuels. In general, units are available in increments from 30,000 to 160,000 Btu/h (879 to 4588 W).

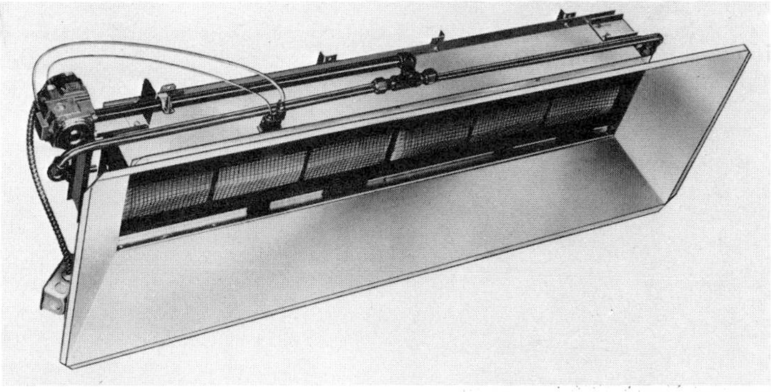

Figure 1-26 Typical gas infrared heater. (*Chromalox®—E. L. Wiegand Div., Emerson Electric Co.*)

SELECTION OF RADIANT CEILING PANELS

The type of radiant ceiling panel shown in Fig. 1-27 is most effective for spot or local heating of finished or semifinished areas with moderate mounting heights. The usual mounting height is about 9 to 10 ft. The number of panels required is obtained by dividing the heating capacity per panel into the required heating load for the area or space. The electric service must be checked to ensure that there is adequate capacity for these heaters. From the point of view of both energy conservation and personnel comfort, it is desirable not only to cycle the panels on and off automatically from a space thermostat, but also to de-energize selective panels as the heating load decreases. In determining the

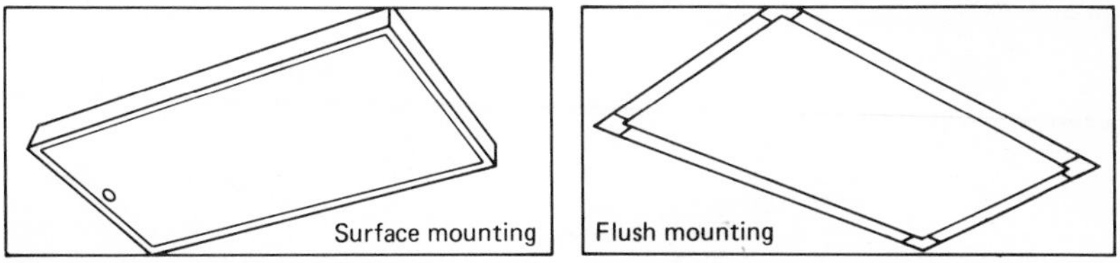

Figure 1-27 Typical radiant ceiling panels.

panels that are to be de-energized in milder weather, care must be taken so that the remaining active panels will uniformly heat the space without overheating the area directly below the (active) panels.

Figure 1-28 illustrates a typical electric surface- and flush-mounted radiant ceiling panel. Such units are prewired with 40-in (1.02-m) flexible cable [standard 2 × 4 ft (0.61 × 1.22 m) module] and are designed to produce uniform heat over the entire panel surface. They are available in capacities of 500 and 750 W at 120/240, 208, and 277 V single-phase. The units can be controlled by line or low-voltage thermostats, manual switches, or automatic time control switches.

REFERENCES

1. *Industrial Ventilation, A Manual of Recommended Practice,* latest ed., American Conference of Industrial Ventilation (Committee on Industrial Ventilation), Lansing, Michigan, 1980.
2. Carrier Air Conditioning Company: *System Design Manual,* part 1, "System Load Estimating," Syracuse, N.Y., 1960.
3. Daly, B. B. (ed.): *Woods Practical Guide to Fan Engineering,* 3d ed., International Publications Service, New York, 1978.
4. ASHRAE: *Handbook of Fundamentals,* latest ed., American Society of Heating, Refrigeration and Air Conditioning Engineers, New York.
5. Chang, Jen-Hu: *Ground Temperature,* vols. I and II, Bluehill Meteorological Observatory, Harvard University, Cambridge, Massachusetts, 1958.
6. McDermott, Henry J.: *Handbook of Ventilation for Contaminant Control,* Ann Arbor Science, Ann Arbor, Michigan, 1976.
7. Hemeon, W. C. L.: *Plant and Process Ventilation,* Industrial Press, New York, 1963.

chapter 5-2

Air-Conditioning the Basic Facility

by
William J. Landman
Manager, Applications Engineering
The Trane Co.
LaCrosse, Wisconsin

GLOSSARY

Absorbent A material or solution which extracts one or more substances from a liquid or gaseous medium with which it comes in contact and changes physically or chemically (or both) during the process. Calcium chloride (material), lithium bromide, lithium chloride and ethylene glycol in solution are examples.

Air, ambient The air surrounding an object.

Air, outside Air in the atmosphere which surrounds a conditioned space.

Air, recirculated Air returned from the conditioned space to the air-conditioning unit, reconditioned, and then supplied to the conditioned space.

Air, reheated Air which is heated after passing through the air-conditioning unit because the temperature is too low.

Air, saturated A mixture of dry air and saturated water vapor at the same dry-bulb temperature, and which, if cooled to a lower dry-bulb temperature, will cause the saturated vapor to condense out as liquid water.

Air, changes A method used to express the number of times the air volume in a conditioned space is completely changed. Usually given in terms of the number of times it is changed per hour.

Air cleaner or filter A device used to remove airborne impurities from the air to be conditioned.

Air conditioning, comfort The process of simultaneously controlling the temperature, humidity, air movement, and the quality of air in a space to meet the comfort requirements of the occupants.

Air conditioning, process The conditioning of air for purposes other than comfort.

Air washer A device consisting of a casing and a set of sprays, through which air is passed for the purpose of conditioning it relative to its cleanliness and moisture content (humidity).

Anemometer A device for measuring the velocity of a fluid.

Barometer An instrument for measuring atmospheric pressure.

British thermal unit (Btu) Approximately the heat required to raise one pound of water from 59 to 60°F.

Capacitor A device used to introduce capacitance into an electric circuit.

Capacity The usable net output of a system or a system component.

Coil A cooling or heating apparatus made of pipe or tubing and fins.

Coil, direct expansion A coil in which the refrigerant is directly expanded.

Comfort chart A chart showing effective temperatures with dry-bulb temperatures and humidities by which the effects of various air conditions on human comfort may be compared.

Comfort zone The range of effective temperatures over which the majority (50 percent or more) of adults feel comfortable.

Compressor, centrifugal A nonpositive displacement compressor which depends upon centrifugal force to produce pressure rise.

Compressor, reciprocating A positive displacement compressor which changes internal volume of the compression chamber(s) by reciprocating the motion of one or more pistons.

Condensate The liquid formed by condensing a vapor, i.e., water extracted from a mixture of air and water vapor.

Convection Heat transfer by the movement of a fluid (air, water, brine, or refrigerant).

Cooling medium Any substance whose temperature is such that it can be used, with or without a change of state, to cool other bodies or substances.

Decibel A unit used in acoustics for expressing the relation between two amounts of power. Used in determining noise levels.

Dehumidification (1) Condensation of water vapor from the air by cooling below the dew point. (2) Removal of water vapor from the air by a chemical or a physical method.

Dew-point temperature The temperature corresponding to saturation (100 percent RH) for a given absolute humidity at constant pressure.

Diffuser, air A circular, square, rectangular, or linear air distribution outlet, generally located in the ceiling and arranged to promote mixing of supply air with secondary room air.

Duct A passageway made of sheet metal, or other suitable material, used for conveying air or other gases at low pressures.

Enthalpy A thermodynamic property of a substance defined as the sum of its internal energy plus the quantity Pv/j, where P is pressure of the substance, v is its volume, and j is the mechanical equivalent of heat. In air conditioning this was formerly called total heat or heat content.

Evaporation A change of state from liquid to vapor.

Fan An air-moving device consisting of a wheel or blade, and a housing or orifice plate.

Fan, centrifugal A fan rotor or wheel within a scroll-type housing which surrounds the wheel.

Fluid A gas, vapor, or liquid.

Gauge An instrument used for measuring pressure or liquid level.

Heat A form of energy that is transferred by a difference in temperature.

Heat, latent In air conditioning—the heat associated with the evaporation of body moisture, or other sources of water vapor, that must be removed from the air to induce moisture condensation and, therefore, humidity control within the conditioned space.

Heat, sensible In air conditioning—the heat, imparted to the air by occupants and other heat-producing sources, that must be removed by the supply air to maintain the desired dry-bulb temperature.

Heat pump, cooling and heating A refrigeration system designed to use alternately or simultaneously the heat extracted at a low temperature and the heat rejected at a higher temperature to produce cooling and heating, respectively.

Heat transmission The time rate of heat flow; usually refers to conduction, convection, and radiation combined.

Heat-transmission coefficient Refers to any number of values of coefficients used in calculating heat transmission by conduction, convection, and radiation through various materials and structures.

Humidifier A device used to add moisture to the air.

Humidify To add water vapor to the air.

Humidistat A control device, actuated by changes in humidity in the air.

Humidity Water vapor in the air of a given space.

Humidity, absolute The weight of water vapor per unit volume.

Humidity, percentage The ratio of the specific humidity of air to that of saturated air at the same temperature and pressure.

Humidity, relative (RH) The approximate ratio of the partial pressure or density of the water vapor in the air to the saturation pressure or density, respectively, of water vapor at the same temperature.

Humidity, specific The weight of water vapor, expressed in pounds or grains, associated with each pound of dry air.

Hygrometer An instrument responsive to humidity conditions of the atmosphere.

Inch of water A unit of pressure equal to the pressure exerted by a column of liquid water 1 in high and at 39.2°F (4°C).

Infiltration Air flowing into a space through a wall or through leaks in the building.

Load factor The ratio of the actual mean load to the maximum load that the system can produce in a given period.

Main A pipe or duct for distributing or collecting from various branches.

Mass The quantity of matter in a body or the measure of the inertia in a body.

Meter An instrument for measuring rates or integrating rates over a period of time.

Output Capacity; duty; performance.

Power The rate of performing work.

Pressure, gauge The pressure above atmospheric pressure.

Pressure, static Practically, it is the normal force per unit area at a small hole in a wall of a duct or pipe through which the fluid is flowing.

Pressure, absolute The sum of the gauge pressure and the barometric pressure.

Pressure, vapor The pressure exerted by a vapor, e.g., the pressure of water vapor in a mixture of air and water vapor.

Psychrometer An instrument for ascertaining the humidity or hygrometric state of the atmosphere.

Psychrometric chart A graphical representation of the thermodynamic properties of moist air.

Pyrometer An instrument for measuring high temperatures.

Radiation The transmission of energy by means of electromagnetic waves.

Radiation, thermal The passage of heat from one object to another without warming the space between.

Refrigerant The fluid used in a refrigeration system, which absorbs heat at a low temperature and a low pressure of the fluid and rejects the heat at a higher temperature and a higher pressure.

Regain The amount of moisture absorbed by a material, expressed in percent of weight of the material.

Return, dry A return pipe in a steam heating system which carries both condensate and air. It is usually above the level of the water line in the boiler.

Return, wet A return main in a steam heating system which is filled with water of condensation. It is usually below the level of the water line in the boiler.

Saturation The condition for stable equilibrium of a vapor and liquid or a vapor and solid phase of the same substance, i.e., steam over water from which it is being generated.

Saturation, degree of The ratio of the specific humidity of humid air to that of saturated air at the same pressure and temperature.

Sensible heat ratio The ratio of the sensible cooling to the total cooling required.

Steam, dry saturated Steam at the saturation temperature corresponding to the pressure and containing no water in suspension.

Steam, superheated Steam at a higher temperature than the saturation temperature corresponding to the pressure.

Steam, wet saturated Steam at the saturation temperature corresponding to the pressure and containing suspended water particles.

System A heating or refrigerating scheme or machine usually confined to those parts in contact with a heating or refrigerating medium.

System, central fan A mechanical indirect system of heating, ventilating, or air conditioning in which the air is treated by equipment located outside of the room served and conveyed to and from the rooms by a system of supply and return ducts.

System, duct A series of ducts, elbows, and connectors to convey air from one location to another.

System, refrigerating Any system which, in operation between a heat source and a heat sink (in the thermodynamic sense) at different temperatures, can absorb heat from the heat source at the lower temperature and reject it to the heat sink at a higher temperature.

Temperature, absolute The temperature expressed in kelvins (Fahrenheit temperature plus 459.6 or Celsius temperature plus 273).

Temperature, dry-bulb (db) The temperature of a gas or mixture of gases as indicated by an accurate thermometer after correction for radiation.

Temperature, effective The dry-bulb temperature of a black enclosure at 50 percent RH and sea level, in which a solid body or occupant would exchange the same amount of heat by radiation, convection, and evaporation as in the existing nonuniform environment.

Temperature, wet-bulb (wb) The temperature indicated by a wet-bulb psychrometer constructed and used according to specifications.

Thermostat An automatic control device actuated by temperature and designed to be responsive to temperature.

Ton of refrigeration A useful refrigerating effect equal to 12,000 Btu/h (3516 W).

Total heat (See Enthalpy).

Tower, water-cooling An enclosed device for evaporatively cooling water by contact with air.

Transmission In thermodynamics, a general term for heat travel.

Transmittance, thermal (*U* factor) The time rate of heat flow per unit area under steady conditions from a fluid on the warm side of a barrier to a fluid on the cold side, per unit temperature difference between the two fluids.

Vapor, water Commonly used in air-conditioning parlance to refer to steam in the atmosphere.

Ventilation The process of supplying or removing air by natural or mechanical means to or from a space.

Volume, specific The volume of a substance per unit of mass: the reciprocal of density.

Water, cooling The water used in a condenser for condensing a refrigerant.

INTRODUCTION

Air conditioning is the simultaneous control of temperature, humidity, air movement, and quality of air in a space. The use of the conditioned space determines the temperature, humidity, air movement, and quality of the air to be maintained. Air conditioning is able to provide widely varying atmospheric conditions.

The range of temperatures and humidities used in comfort air conditioning fall within a small band. The location of this band on the psychrometric chart depends on the season of the year and the type of application.

Cleanliness and air movement must always be considered. Air should be clean, that is, free of dust and soot particles. Air cleanliness is important from the standpoints of both health and building maintenance.

Air should circulate freely in the room to which it is delivered. This allows it to absorb heat and moisture uniformly throughout the entire room. At the same time, the air movement should be gentle, or it will cause objectionable drafts.

Comfort can be defined as any condition which, when changed, will make a person uncomfortable. Though this sounds paradoxical, it means that a comfortable person is

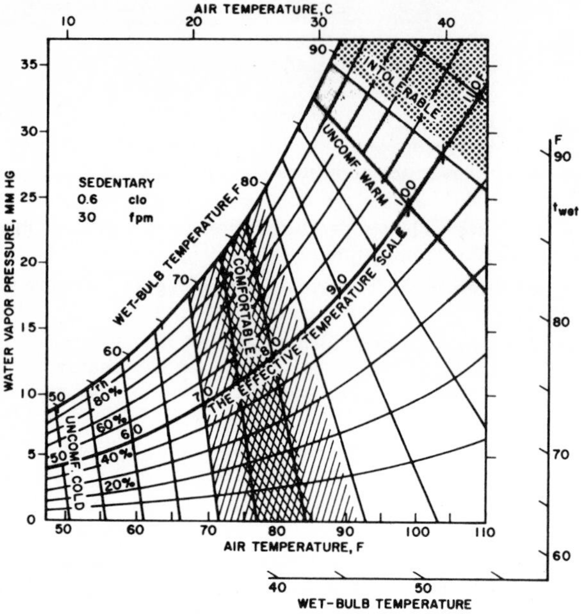

Figure 2-1 Comfort zone temperatures and humidity. (Reproduced with permission from ASHRAE: *Handbook of Fundamentals* © 1981, ASHRAE, New York)

not aware of the air-conditioning system. If a person is uncomfortably warm, the temperature or humidity (or both) are too high. In a first-class air-conditioning system, people are not really conscious of the temperature or humidity because they are comfortable. Neither are they aware of equipment noise or air movement.

Figure 2-1 (ASHRAE effective temperature chart) shows the comfort zone resulting from tests made on selected individuals performing sedentary work and wearing standard clothing. While this chart shows the ideal conditions arrived at in tests, some leeway is usually taken, in actual practice, to reduce the tonnage and energy required. For example, in commercial air conditioning, inside design conditions are usually selected at 78° db (25°C) and 50 percent RH.

In industrial air conditioning, even more leeway can be taken. A slightly higher design dry-bulb temperature may be used [say 80°F (26°C), to reduce the tonnage and air quantities. Higher diffuser air outlet velocities may be used because they are usually located at a higher level than in commercial systems. This results in a smaller number of diffusers and less-complicated ductwork, resulting in lower first costs.

To offset the slightly higher dry-bulb design temperature, a lower relative humidity is selected (say 45 percent). This, together with slightly higher air velocities, results in proper evaporative cooling of the workers. Maintaining a lower relative humidity may first appear to offset the savings resulting from using a slightly higher dry-bulb temperature, but plotting the comparative results on the psychrometric chart will illustrate the savings. See Fig. 2-2.

The environmental objective in most factory general shop "air-conditioning" applications is to produce a condition of average comfort for the type of work involved. Unlike most commercial applications, the system required to accomplish this can be simple and inexpensive.

Simplicity and low cost do not mean reduced quality standards for industrial applications. Rather, low cost is a natural consequence of system simplicity. Problems that are typical of large commercial designs are often eliminated. Such problem areas as complex air-supply systems resulting from areas of variable load, modular control of systems, stringent operating-noise levels, and tight space limitations seldom occur.

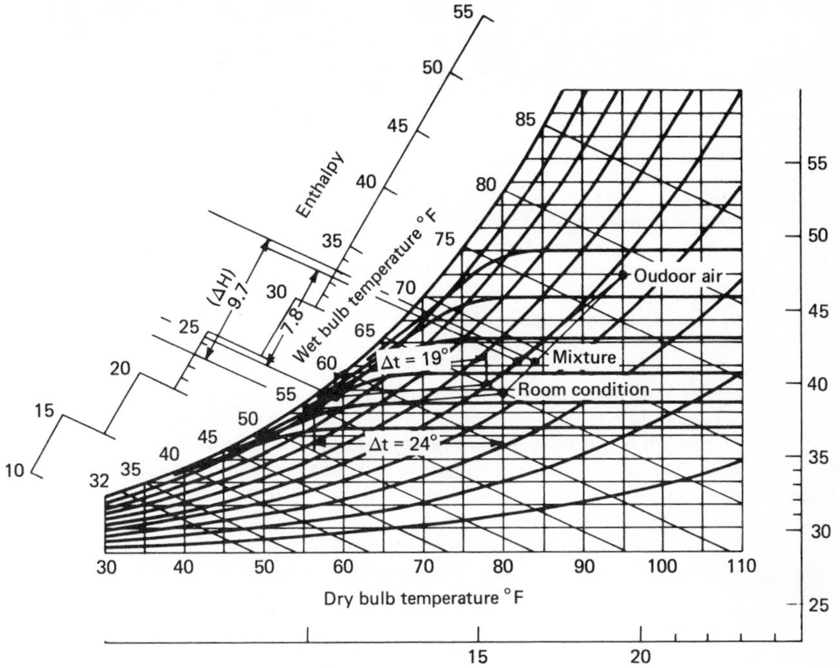

Figure 2-2 Calculations of savings using lower relative humidity.

Unfortunately, the basic differences between commercial and industrial air conditioning are not always recognized. Quite often, analysis, design, and cost estimates are influenced too heavily by commercial design experience. The resulting overcomplication and high cost needlessly reduce net returns and sometimes discourage projects entirely.

Industrial facilities must, by nature, be highly flexible. Continual layout revision and expansion, not common to most commercial installations, must be considered in the initial system choice. For this reason, the owner must define the eventual plant size and layout flexibility required to provide the designer with the basis for a system selection.

INSTRUMENTATION

Contrary to the usual impression, controls and instrumentation are not identical. Controls adjust the system operation to meet the changing load conditions. Control application and design are subjects which require extensive study and explanation and, therefore, cannot be adequately covered in a short discussion. Consequently, this discussion is limited to the instrumentation used for the measurement of temperature, humidity, pressure, power, air and water flow, and noise in industrial air-conditioning systems.

Temperature Measurement

Temperature is measured by a thermometer. Usually the term *thermometer* refers to the ordinary liquid-in-glass temperature-indicating device. The liquid may be either mercury (range −40 to about 150°F) or alcohol (range −94 to 148°F). Most thermometers have Fahrenheit and/or Celsius scales etched in the glass stems.

Industrial thermometers are fitted with metal guards to prevent breakage and are installed in ducts or pipes to indicate air, gas, or liquid temperatures. When thermometers are inserted into ducts or pipes which carry high velocities or high pressures, a well, made of thin metal, is used to prevent breakage of the instrument. The well is filled with a heat conductive liquid such as oil, water, or mercury.

When an hourly record of temperatures is desired, a recording thermometer is used. This instrument consists of a time-clock arrangement with a chart on which a continuous temperature record is drawn. The temperatures are measured by means of a remote bulb.

Humidity Measurement

Humidity can be measured in a number of different ways. The most common method of measurement is with a *psychrometer*. This consists of two thermometers mounted side by side in a frame and fitted with a handle for rotating the instrument in the air. One thermometer bulb is fitted with a cloth wick or sock. This wick is wetted with distilled water which evaporates when the instrument is rotated. The plain thermometer registers the dry-bulb temperature of the air, while the one with the wet wick (because of evaporation) indicates the thermodynamic wet-bulb temperature. The difference in readings is called the wet-bulb depression. Then, by referring to a psychrometric chart or tables, the relative humidity can be determined.

Other *hygrometers* measure humidity by sensing the flow of an electric current through a thin coating of a material which changes its electrical resistance with changes in the moisture content of the air which surrounds it.

Some other hygrometers make use of the process of condensing the moisture out of the air and determining its weight to ascertain the dew point of the moist air. Then by referring to psychrometric tables, the relative humidity can be obtained.

Many organic materials such as nylon, Dacron, human hair, wood, and paper change dimension with changes in the moisture content of the air. These materials, when connected to a suitable mechanical linkage can be used to construct recording hygrometers. In many cases they are combined with a recording thermometer to give an hourly record of temperature and humidity on one chart.

Measurement of Air Pressure

In industrial air-conditioning systems, it is essential to know air pressures at various points. Knowing the air pressures on the inlet and outlet side of the fan, which supplies the air for the system, and knowing the speed (in revolutions per minute) of the fan, the volume of air supplied by the fan can be determined. Likewise, knowing the air-pressure drop across the filters and coils can indicate the need for cleaning or the replacement of the filters. Also, knowing the pressure drop in sections of the distributing ductwork indicates whether the correct amount of air is flowing through the duct.

All these air-pressure drops can be determined by various forms of manometers or by a Pitot tube in conjunction with a manometer.

Pressure gauges for measuring the pressures in water, steam, or refrigerant pipes are important in the determination of proper system operation. These gauges are usually constructed with a power element which may take the form of a bellows, a diaphragm, or a Bourdon tube. Gauges are manufactured for specific applications. Therefore, when selecting gauges, the type, pressure range, and application should be clearly specified.

Power Measurement

Energy conservation is important in these days of high fuel costs and fuel shortages. Therefore, instruments such as recording wattmeters and other electric energy-recording devices are important to any system. By use of these instruments, the demand can be checked frequently and, if possible, reduced by alternate use and starting of equipment. This not only conserves the energy that the power plant must be capable of supplying but also reduces the system operating cost.

Another consideration is a periodic check of the power factor of the electric system. The lower the power factor, the more energy is used to handle the actual load. If the power factor is much below 1, it would be well to consider the installation of a synchronous motor or condenser to bring the power factor as close to 1 as possible.

Measuring Air and Water Flow

Establishing the required flow of air in the distributing ducts and the water flow in the piping system is important in obtaining the required capacity. This can be accomplished by using the pressure-drop method and applying the proper formulas or by using flowmeters, venturi nozzles, rotameters, or other direct-reading instruments. The latter method results in additional initial expense which in most cases is not justified. Therefore, the pressure-drop method is usually used.

ESSENTIALS OF AN AIR-CONDITIONING SYSTEM

Air conditioning is defined as the simultaneous control of temperature, humidity, quality, and movement of air in a conditioned space or building.

An air-conditioning system is, therefore, defined as an arrangement of equipment which will air-condition a space or a building. Thus a complete air-conditioning system includes a means of refrigeration, one or more heat-transfer units, air filters, a means of air distribution, an arrangement for piping the refrigerant and heating medium, and controls to regulate the capacity of these components. The components must also be arranged and controlled in the proper sequence to provide the required conditions.

Refrigeration Methods

The most common types of refrigeration machines, classified according to their type of operation are (1) mechanical compression, (2) absorption, and (3) vacuum. In addition, well water may be used to supply the refrigeration in some cases.

Mechanical compression machines may be divided into reciprocating, centrifugal, and rotary types.

The term *heat pump* is occasionally used to describe a refrigeration machine. However, a heat pump is a refrigeration cycle—either reciprocating or centrifugal—in which the cooling effect as well as the rejected heat are used to furnish cooling or heating to the air-conditioning units, either simultaneously or separately.

Reciprocating compressors can be used in a system that circulates the refrigerant through remote, direct-expansion heat-transfer surfaces. Alternatively, they can be used in conjunction with water-chilling heat exchangers to produce chilled water for circulation through remote heat-transfer surfaces which cool and dehumidify the air.

Centrifugal refrigeration machines usually are not suitable for circulating and expanding the liquid refrigerant in remote heat-exchange surfaces. Centrifugal machines are therefore primarily used to chill water or brine for circulation through remote heat-exchange surfaces.

Absorption machine cycles are similar to mechanical-compression machine cycles only to the extent that both cycles evaporate and condense a refrigerant liquid. They differ in that the mechanical-compression cycle uses a compressor to obtain the necessary pressure differential, whereas the absorption cycle uses heat. The mechanical-compression cycle uses purely mechanical processes, while the absorption cycle uses physiochemical processes to produce the refrigeration effect.

Vacuum refrigeration machines are seldom used in comfort air-conditioning systems. The steam and water vapor necessary to produce the refrigeration effect make these machines more applicable to process work.

Heat-Transfer Methods

In an air-conditioning system, there are two general methods of transferring heat to or from air. These are fin-and-tube coils and air washers.

Fin-and-tube coils are widely used for cooling air with cold water or refrigerants and for heating air with steam or hot water. They owe their popularity to their relative compactness and to the ease with which they can be installed.

When the water or refrigerant in a coil is at a lower temperature than the dew point of the entering air, moisture will be condensed out of the air. Thus, coils can be used for dehumidification as well as for cooling.

For many years prior to the advent of the finned coil, water sprays were used almost exclusively for conditioning air. This was accomplished by bringing the air and water into contact with each other within a spray chamber. Air was drawn through the chamber by means of a fan. Such spray chambers, complete with water tank, eliminator plates, spray nozzles, and other auxiliary equipment are known as air washers. Air washers can be used not only to dehumidify and cool the air with chilled water, but also to humidify the air with warm water.

Air washers are usually constructed in four different lengths: 5, 7, 9, and 14 ft long. Five-foot-long washers generally have a single bank of sprays and are used mainly for humidifying purposes. Seven- and nine-foot-long washers are equipped with two banks of sprays and are used mainly for cooling and dehumidifying purposes.

Fourteen-foot-long washers are usually made up of two 7-ft-long units and are of the two-stage type. They are generally used in process applications where high saturation and counterflow of air and water are desired.

Filters

All air used in an air-conditioning system must be filtered to maintain a clean atmosphere in the conditioned space. Outside air always contains a number of contaminants, such as bacteria, pollens, insects, soot, ash, dust, and dirt. Return air has such contaminants as dust, lint, soot, and ash. The concentration of these contaminants in the air, and the degree of cleanliness required in the conditioned space, will determine the type of filter or filters which should be used.

Filters can be divided into three classifications: (1) viscous impingement type, (2) dry type and, (3) electronic air cleaners.

Viscous impingement–type filters can be further subdivided into (1) throwaway, (2) cleanable, and (3) roller or awning curtain types.

The throwaway type can be taken out and discarded when loaded with dirt or contaminants. The frames of this type of filter are usually made of cardboard with an expanded metal or plastic screen on the front and back. The filter medium between the two screens may be spun glass, copper or aluminum wool, animal hair, hemp fibers, felt, or other materials.

The viscous cleanable type is constructed in much the same manner as the throwaway, except the frames and screens are all metal. The filtering medium is usually composed of a number of zigzag screens. When the filter becomes full of dirt, it is replaced by another section. The dirty filter is then washed in a cleaning solution, dried, and given a coat of viscous oil.

Roller- or curtain-type filters consist of two rollers, one at the top of the air passage and one at the bottom. The filtering medium may be disposable or cleanable. The disposable type consists of a roll of medium placed on the top roller and extended to the bottom roller. As the dirt accumulates, it is periodically rolled up on the bottom roller and, when fully wound on the bottom roller, is taken out, discarded and replaced with a new roll.

The cleanable type of curtain filter consists of a number of metal plates fastened to a chain that runs over sprockets on the top and bottom rollers. An oil bath under the bottom roller provides cleaning and recoating of the plates with a viscous oil.

Dry filters all have a disposable medium and may be stationary or rotating. The stationary type is made in two versions. One uses a dense mat of glass, cellulose, or cotton fibers, which is held in place in the airstream by V-shaped frames. The other uses a cotton-fiber medium in the shape of bags with the air circulating through the bags from the inside. Both types of medium are disposable. The rotating type is similar to the rotating impingement type, using rolls of cotton fibers which move from one roller to the other. The main difference is that the medium is dry.

Electronic air cleaners are of two types: (1) those which give the dust particles a charge by passing them through an ionizing zone, (2) those which use a charged medium to charge the particles.

In the first type, closely spaced metal plates are arranged parallel to the airflow, with adjacent plates having opposite polarity. This is accomplished by charging alternate plates with approximately 12,000 V dc and grounding the other plates. Preceding the bank of plates by a few inches, there are vertical wires having a high positive charge. Any particle going through the electronic air filter receives a positive charge and is attracted to the negatively charged plates. To clean the plates, the power supply is disconnected, and the plates are washed down with hot water or a solvent. After the plates have been washed down, they are given a coat of adhesive before they are put back into service.

In the charged-medium filter, fiberglass is held in nonconductive frames and installed in a V-shaped bank across the inlet to the conditioning unit. The frames are held between a positively charged grid and the grounded metal support frame. The charge placed on the medium is 10,000 to 12,000 V dc and causes the approaching particles to be attracted to the filter medium. This type has the advantage that, if the rectifying unit fails, the filter medium still is efficient as an impingement filter.

Air Distribution

The satisfactory distribution of conditioned air requires a well-designed duct system and a properly chosen fan.

Ducts

Three duct design methods are commonly used:

1. Velocity
2. Equal friction
3. Static regain

The choice of design method depends almost entirely upon the size of the duct arrangement. Small duct systems (homes, small shops, or a few office rooms), are commonly designed by the velocity method. Large high-pressure systems are most frequently designed by the static regain method. Duct arrangements between these two extremes are nearly always laid out by the equal friction method. Sometimes a duct arrangement is designed by a combination of two methods. For instance, the trunk duct is laid out by the static regain method and the branch duct runs by the equal friction method.

There are, in general, two methods of supplying air to the conditioned space: single-duct and dual-duct. With the single-duct arrangement, all the supply air, either heated or cooled, is furnished to the conditioned space through a single duct. With the dual-duct arrangement, cold air is supplied through one duct and warm air through the second duct. By mixing some air from each duct at the outlets, the proper air temperature is supplied to the space.

Fans

Fans for supplying air to the space through the ductwork can be placed in one of two general classifications: centrifugal fans or axial fans.

Centrifugal fans induce airflow within the fan wheel that is substantially radial to the shaft, while axial fans induce flow within the wheel that is substantially parallel to the shaft. Centrifugal fans operate in a scroll-type housing. Axial fans, on the other hand, operate within a cylindrical or ring-type housing.

Fans for heating, ventilating, and air-conditioning systems are used for three general purposes:

1. To move the conditioned air from the conditioning apparatus to the conditioned space.
2. To return the air from the conditioned space to the conditioning apparatus.
3. To exhaust the air from the conditioned space to the outside or to supply air to the conditioning apparatus.

Centrifugal fans may be further classified as backward-inclined (BI), forward-curved (FC), and airfoil.

In the BI fan, the tip of the fan blade is sloped backward, opposite to the direction of rotation. In the FC fan, the tip of the blade is sloped forward in the direction of wheel rotation. As with the BI fan, the blade of the airfoil fan is sloped backward, opposite to the direction of rotation. However, the design of the blade is such that it results in a higher efficiency, a wider range of high efficiency, and a lower noise level.

With the forward-curved blade wheel, a forward movement is imparted to the air by the blade. This motion causes the air to leave at a higher resulting velocity than that produced by either the backward-inclined blade or airfoil blade.

Because the pressure produced by a fan is a function of the forward motion of the air at the tip of the blade, a fan with forward-curved blades operates at a lower speed for a given duty than a backward-inclined fan. In other words, where large volumes of air are required at a comparatively low rotative speed, the fan wheel with forward-curved blades is usually applied.

Centrufugal fans with forward-curved, backward-inclined, and airfoil blades all have been used successfully in ordinary air-conditioning and ventilating work. Generally, size-for-size for a given duty, the backward-inclined or airfoil fan uses less horsepower than the forward-curved fan. In spite of the higher speed of the backward-inclined or airfoil fans, they are, when properly selected, as quiet as or quieter than the forward-curved fan, making them a logical selection in the larger sizes. In small sizes (generally less than 24-in wheel diameter) the high speed of the backward-inclined or airfoil fans may result in excessive belt speeds. Here the low rotating speed characteristic of the forward-curved fan is desirable.

Where high static pressures along with medium air volumes are required, the airfoil fan is usually the one that is selected.

There are three general types of axial flow fans: (1) tube-axial, (2) vane-axial, and (3) propeller.

Tube-axial fans may be considered to be heavy-duty propeller fans. They are generally arranged for duct connection, while propeller fans are usually arranged for wall mounting. Actually, it is poor practice to connect either a propeller fan or a tube-axial fan to discharge through ductwork unless special precautions are taken. All axial fan wheels discharge air with a spiral motion. This spiral flow produces higher system losses than would be produced by straight airflow. Tube-axial fans are built for pressures up to 2.5 to 3 inH_2O. Generally, their application is limited to industrial duties where noise considerations are not important.

Vane-axial fans are tube-axial fans plus vanes. The vane is located behind the wheel to straighten the spiral flow, increase the static efficiency, and permit diversified duct applications. The vane-axial and tube-axial fans are often used where space considerations are important. For this reason, they are often the logical choice in transportation and marine applications. Improved vane-axial fans can be both as quiet and as efficient as centrifugal fans.

Propeller fans are low-pressure, high-capacity fans which are seldom used in applications that require more than 0.75 in (20 mm) of static pressure. They consist basically of a stamped or cast wheel located in an orifice ring. Propeller fans are built either for direct connection to an electric motor or for V-belt drive.

The horsepower required by a propeller fan is lowest at maximum air volumes. This characteristic is opposed to centrifugal fans that require minimum horsepower at no air delivery. Consequently, for delivering large volumes of air at low pressures, the propeller fan is better suited than the centrifugal fan. Propeller fans will deliver large air volumes with less horsepower than centrifugal fans when selected to occupy the same amount of space or to have the same first cost.

There is one other type of centrifugal fan which is more or less a centrifugal fan within a tube-axial fan housing. It is known as a tubular centrifugal fan. This fan design, while using a centrifugal fan wheel, provides straight-line airflow. This is done by placing a centrifugal fan wheel within a tubular housing which has conversion vanes before and after the fan to convert radial airflow to axial airflow.

In some applications, such as within a straight run of circular duct, a tubular centrifugal fan may save space. However, a scroll centrifugal fan is generally more efficient and quieter than a tubular centrifugal fan.

Piping Arrangements

Water piping arrangements for heating and cooling can be divided into three general classifications:

1. One-pipe
2. Two-pipe
3. Four-pipe

The first two are arranged so that either hot or cold water can be circulated through the piping. However, when hot water is being circulated in one-pipe and two-pipe arrangements, only hot water is available to each coil connected to that circuit. Conversely, when cold water is being circulated, only cold water is available to each coil.

The four-pipe arrangement is actually two 2-pipe arrangements. One of the two supply pipes carries cold water and the other supply pipe carries hot water to the coils connected to the circuit. There are also two return lines connected to each coil—one return carries the hot water back to the boiler or heat exchanger and the other carries the cold water back to the chiller. Thus, hot and cold water do not mix.

One-Pipe Arrangement

In the one-pipe arrangement, a single pipe loops the building and constitutes both the supply and return main. The same pipe size is maintained throughout its length because all of the water that is circulated goes through this one main.

At each branch takeoff to the unit, a flow fitting is installed. These flow fittings cause some of the water from the main to flow up and through the coil in the unit. The pressure drop through the flow fitting must be equal to or greater than the pressure drop of the riser, plus the pressure drop through the unit, plus the pressure drop of the return piping which connects the unit to the main.

The water returning to the main is at a different temperature from the water flowing in the main. The mixing of the water from the unit and the water flowing in the main changes the water temperature in the main so that the water entering the next unit is at a different temperature. However, the temperature change is small, since the quantity of water flowing through the unit is small when compared with the total amount of water flowing in the main. In any event, it is good practice to check the temperature drop because it may be advisable to select the unit at the end of the main for a different water temperature than those at the start of the main.

The coils in the units must be designed for a low-pressure drop, since the flow fittings are designed for a small pressure drop in order to keep the pumping head of the system within reasonable limits. The maximum loss through the coil in the unit and the runouts is fixed, once a specific flow fitting is selected. The only way this maximum pressure loss can be altered is by changing the flow fitting or by changing the water flow in the main. The water flow through the takeoffs and through the unit can be changed by increasing or decreasing the size of piping to and from the unit, by selecting a unit with greater or lesser pressure drop through the coil, or by installing flow control valves in the runout piping.

Some use has been made of the one-pipe arrangement for cooling purposes, although this arrangement is not used as extensively for cooling as it is for heating. However, by proper design and application, a one-pipe arrangement for cooling represents a considerable savings in first cost. Actually, the arrangement used for one-pipe cooling is not a true one-pipe arrangement.

Two-Pipe Arrangement

There are two general methods of arranging piping for two-pipe water circulation. One is called a direct-return, two-pipe arrangement and the other is called a reverse-return arrangement.

In the direct-return arrangement, the length of travel of the water is different for each unit. The water flowing through the first unit has a much shorter distance to travel than the water flowing through the last unit. Therefore, balancing this arrangement is a problem because the pressure drop from the supply main, through the first unit, and back to the return must be increased by means of some restriction. Usually, a valve or orifice is used; otherwise, most of the water will flow through the first unit.

In the reverse return arrangement, the length of travel of the water from the chiller or boiler, through any of the units and back to the chiller or boiler is essentially the same. This is accomplished by making the unit nearest the chiller or boiler on the supply main the farthest away on the return main. In other words, the water flowing from the supply main to the first unit must flow through the full length of the return main before it returns to the chiller or boiler.

From the description of this system, it appears that this arrangement is self-balancing. Actually, in practice, it is necessary to install balancing valves or orifices in the piping to obtain the exact water flow to each unit.

Obviously, the two-pipe arrangement is more costly than the one-pipe arrangement, but its flexibility and simplicity of design make it more widely used.

Four-Pipe Water Circulation

The four-pipe water distribution arrangement is essentially two 2-pipe circuits with one 2-pipe circuit circulating cold water and the other circulating hot water. There is one distinct difference between the four- and the one- and two-pipe systems. Since the circuit through which the hot water is flowing handles only hot water, the water temperature can be adjusted upward to reduce the pipe size in this circuit. In this arrangement, no water is mixed, since the cold water returns to the chiller, and the hot water returns to the boiler or heat exchanger.

By using separate coils for cooling and heating, it is possible to dehumidify and reheat with this arrangement (see Fig. 2-3). Thus, cooling and heating can be made available to all units simultaneously.

When only one coil is used for both cooling and heating, the same type of valve is used on the supply lines to each coil. In addition, another valve is used on the return line from each unit to divert the water to the proper return line (see Fig. 2-4).

Another method of using a single coil on the four-pipe arrangement is to use a three-way valve to allow either cold or hot water to be circulated through the coil. Two swing check valves are used to prevent the circulation from the line not being used. In this method the flow through the coil is reversed when the valve changes its position (see Fig. 2-5).

In many large buildings it is customary to use two hot-water circuits, one for cooling and heating units and one at a higher temperature for the convectors or fin radiation in toilets, stairwells, storage rooms, etc. With the four-pipe arrangement, it is unnecessary to use two heating circuits, since the hot-water circuit is entirely separate from the

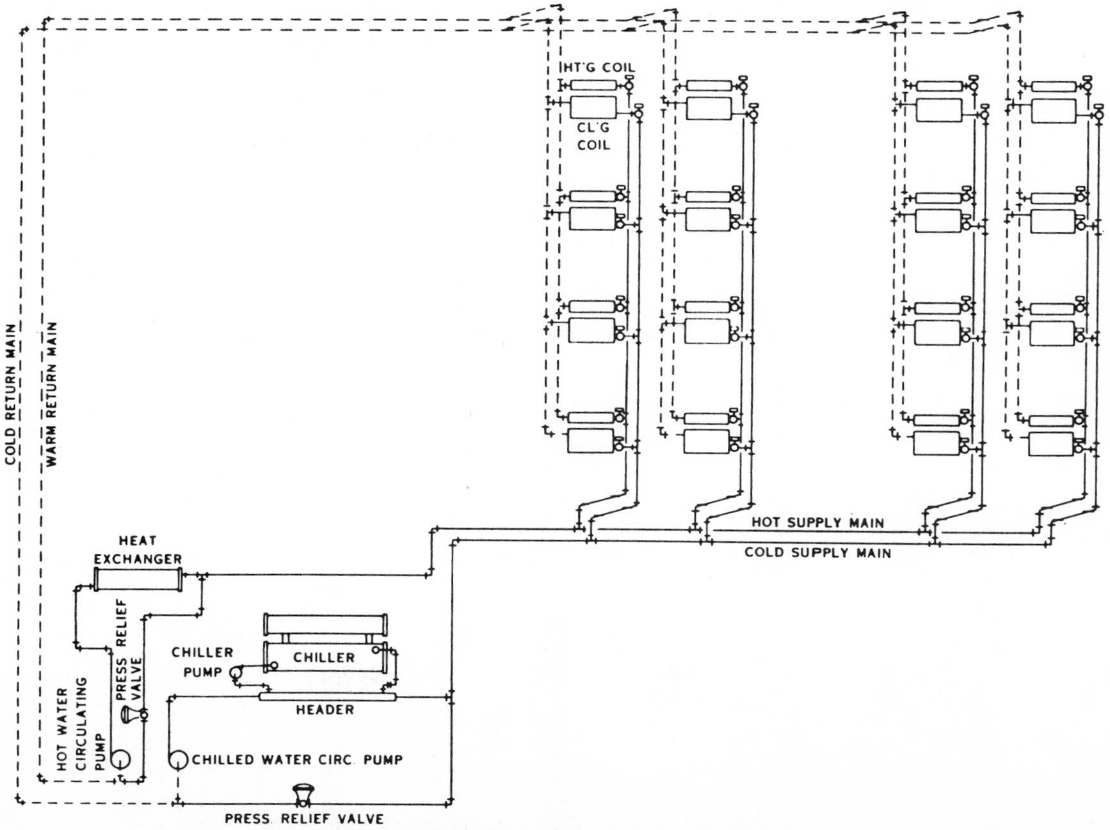

Figure 2-3 Four-pipe distribution with two coils. (*The Trane Co.*)

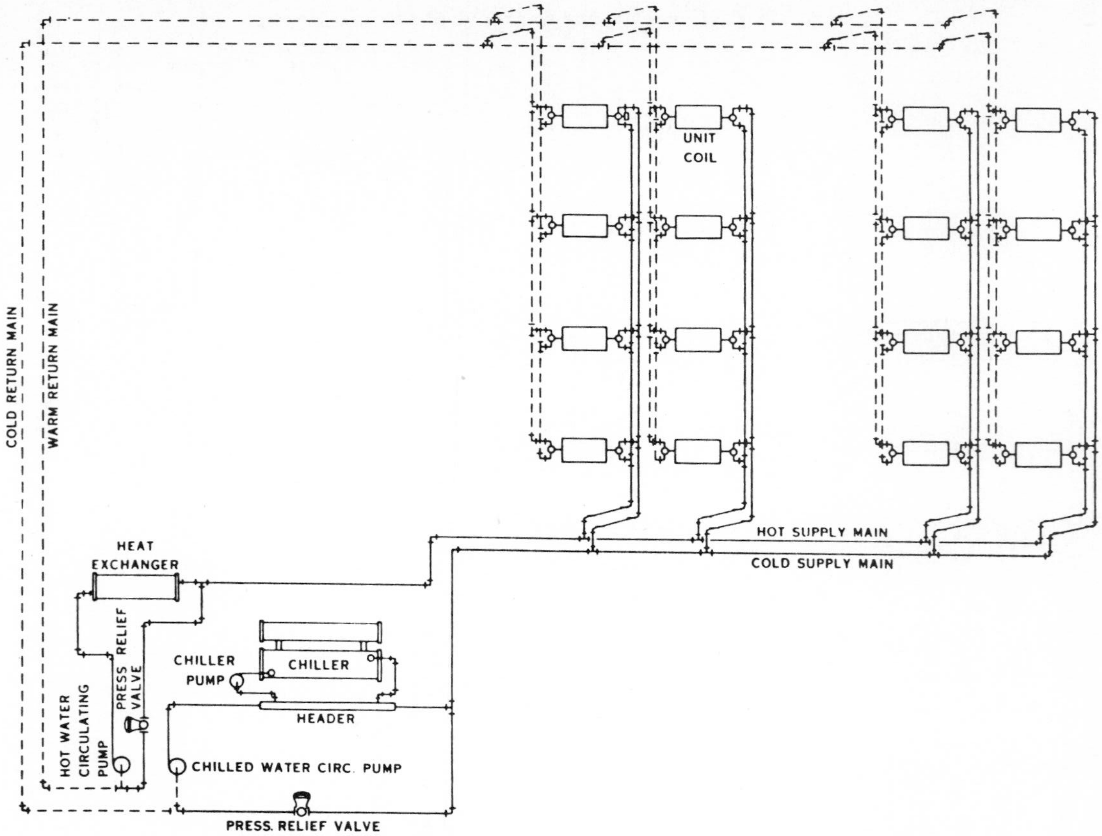

Figure 2-4 Four-pipe distribution with one coil. (*The Trane Co.*)

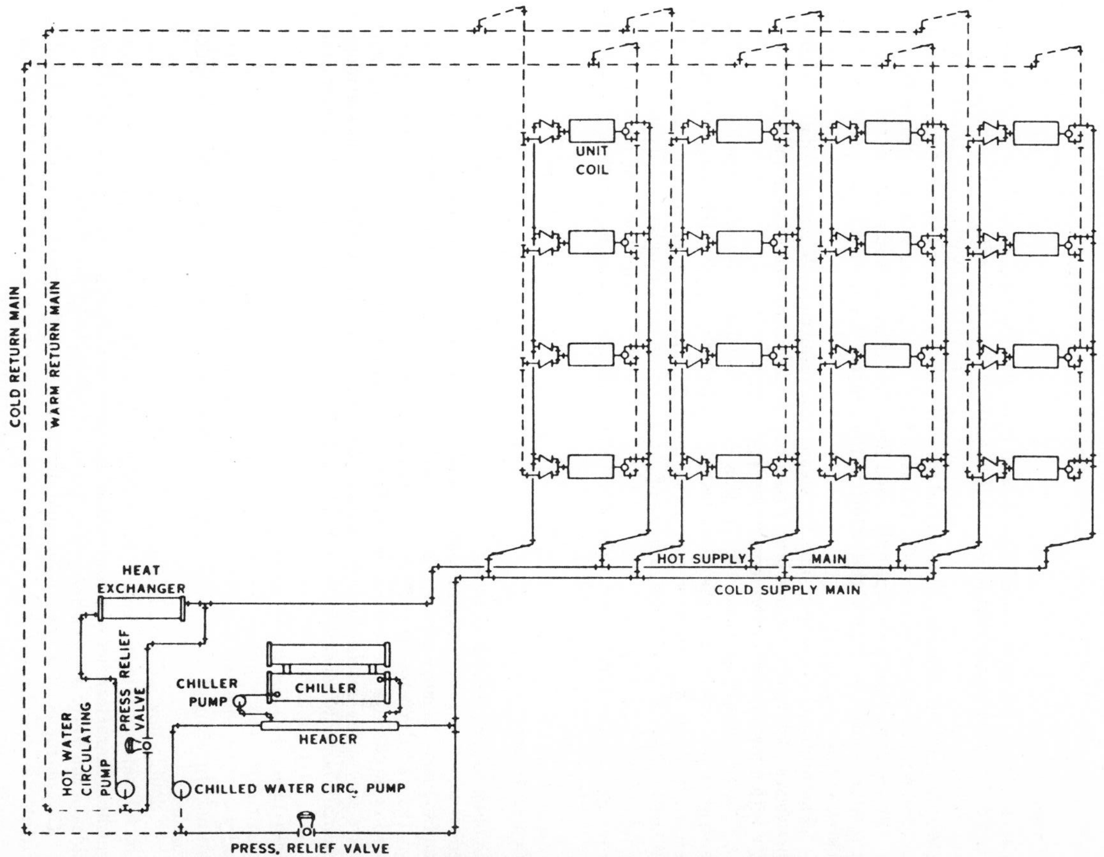

Figure 2-5 Four-pipe distribution with reverse flow through coil. (*The Trane Co.*)

chilled-water circuit. Neither is it necessary to zone the piping, because each unit has become a zone by itself, furnising either heating or cooling as required. If a unit, such as a convector, is not to be used for cooling, it is necessary only to omit the cold-water connections to it and merely connect it to the hot-water supply and return lines.

With the four-pipe arrangement, the ultimate in simplicity of design, room control, and economy of operation are accomplished with only a slight increase in first cost.

Steam Piping

Steam piping, for steam-heating coils used in air-conditioning systems, is usually based on low-pressure steam(that is, 0 to 15 lb/in^2 gauge). Steam-heating coils are those which use steam within the tubes, while the air to be heated is circulated around the outside of the tubes and fins. Steam coils are used primarily for preheating and reheating the air in central fan or zone units and, in some cases, for remote heating and reheating coils.

Steam flows through a pipe only when the pressure at one end is greater than that at the other end. The higher the velocity, the greater the pressure difference must be between inlet and outlet.

The problem in sizing a pipe for steam flow resolves itself into striking a balance between a high velocity with a large pressure drop requiring a small pipe and a low velocity with a low pressure drop requiring a large pipe.

The pressure drop caused by friction does not cause a loss in energy because this energy appears as heat. If the steam leaving the boiler is wet or saturated, the heat generated by friction tends to evaporate the moisture into steam or even to superheat it. Whether the heat gain at the expense of pressure drop is used or wasted depends on the equipment that is using the steam. Naturally, if the heat is not used, the pressure used to overcome the friction is a total loss. In the average heating application, the loss of pressure is used and not wasted.

Cooling Towers

Cooling towers (see also Chap. 6-4) are one of the most common methods used to cool condenser water for refrigeration systems. They are usually classified according to the method used to move air through the tower as induced draft or forced draft.

Induced-draft towers are provided with top-mounted fans that induce atmospheric airflow up through the tower as warm water falls downward. An induced-draft tower may have only spray nozzles for water breakup or it may use various slat and deck arrangements. There are several types of induced-draft cooling towers. In the counterflow draft tower, a top-mounted fan induces air to enter all four sides of the tower and flow vertically upward as the water cascades down through the tower. The counterflow tower is particularly well adapted to a restricted space, since the discharge air is directed vertically upward, and the four sides require only a minimum clearance for air intake. The primary breakup of water may be either by pressure spray or by gravity from water-filled flumes.

A double-flow induced-draft tower, has a top-mounted fan to induce air to flow across the fill material. The air is then turned vertically in the center of the tower. The distinguishing characteristics of a double-flow induced draft tower are two intakes on opposite sides of the tower and the horizontal flow of air through the fill sections.

Comparing counterflow and double-flow induced-draft towers of equal capacity, the double-flow tower would be somewhat wider but the height is much less. Cooling towers must be braced against the wind. From a structural standpoint, therefore, it is much easier to design a double-flow than a counterflow tower because the silhouette of the double-flow type offers much less resistance to wind force.

Mechanical equipment for counterflow and double-flow towers is mounted on top of the towers and is readily accessible for inspection of maintenance. Water distributing systems are completely open on top of the tower and can be inspected during operation. This makes it posisble to adjust the float valves and clean stopped-up nozzles while the towers are operating.

A cross-flow, induced-draft tower is a modified version of the double-flow induced-draft tower. The fan in the cross-flow cooling tower draws air through a single horizontal opening at one end and discharges the air at the opposite end.

A forced-draft cooling tower, on the other hand, uses a fan to force air into the tower. In the usual installation, the fan shaft is in the horizontal plane. The air is forced horizontally through the fill and upward to be discharged at the top of the tower.

Under-flow cooling towers are an improvement on the forced-draft tower that retain all of the advantages of the efficient double-flow design. Air is forced into the center of the tower at the bottom. The air is then turned horizontally, both right and left, through filled chambers and is discharged vertically at both ends. By forcing the air to flow upward and outward through the fill, and leave at the ends, operating noise is baffled and a very desirable reduction of sound level is achieved. All sides of the under-flow tower are smoothly encased with no louver openings. This blends with modern architecture and eliminates the necessity of the masonry walls or other screening devices often necessary to conceal cooling towers of other types.

Water treatment is an important part of the operation of a cooling tower. The evaporation of water from the cooling tower leaves certain solids behind, the same sort as the lime found in a tea kettle. The recirculation of water in a condenser–cooling-tower circuit, and the accompanying evaporation, causes the concentration of solids to increase. The concentration must be controlled or scale and corrosion will result. It is recommended that a water-treatment specialist be consulted, for overtreatment can materially reduce the performance or life of a condenser circuit.

Rooftop Units

Rooftop units usually consist of all the essentials for an air-conditioning system: fans, coils, compressor, filters, controls, structural supports, and vibration isolators. These units are mounted on relatively flat roofs of one- or two-story industrial plants.

Refrigeration is accomplished by expansion of the refrigerant within the cooling coils. Heating can be by either gas- or oil-fired heat exchangers, steam, hot water, or electric resistance heaters; in some cases the unit is constructed to operate as a heat pump. Air-distribution ducts must run through the roof into the space to be conditioned.

LOAD CALCULATIONS

Human Factors

Before selecting the temperature and humidity conditions to be maintained in the conditioned area, it is necessary to analyze further the effects of temperature and humidity on the human body.

The metabolic rate of individuals is given in Table 2-1. The amount of heat given in this table does not consist entirely of sensible heat. The percentages of sensible and latent heat depend on the dry-bulb temperature of the ambient air and the degree of worker activity. Total heat (sensible plus latent), however, remains the same over a range of dry-bulb temperatures in combination with a given rate of activity.

TABLE 2-1 Average Metabolic Rate* of Adult Males

Degree of activity	Heat release, Btu/h
Sitting quietly	400
Seated, light work (typing)	640
Standing, light bench work	880
Light machine work, some walking	1040
Heavy work, heavy machine work, lifting	1600
Sustained heavy work	2000

*Also the amount of heat produced by the body.

TABLE 2-2 Rate of Heat Release M from Adult Males, Btu/h*

Activity	Average heat release	Ambient dry-bulb temp, °F			
		78		80	
		Sensible	Latent	Sensible	Latent
Seated, light work	640	320	320	295	345
Seated, light bench work	880	390	490	360	520
Light machine work, some walking	1040	345	695	315	725
Heavy work, heavy machine work, lifting	1600	565	1035	520	1080
Heavy sustained work	2000	705	1295	650	1350

*Heat release M from an adult female is approximately 85 percent of that for an adult male.

Table 2-2 illustrates this. As the dry-bulb temperature is increased from 78 to 80°F, the percentage of sensible heat is reduced, and the latent heat percentage is increased. For the same rate of activity, the total heat remains the same in all cases.

These facts, in addition to the knowledge that high temperature and high humidity retard both the sensible heat loss and the evaporation of moisture from the skin, can be used to advantage in reducing the cost of industrial air conditioning.

For instances, increasing the air velocity within limits increases the evaporation of moisture from the skin, thereby increasing the pickup of latent heat. Using higher air velocities also means that fewer outlets are required, and the outlets used can be smaller.

Lowering the dew-point temperature of the room air increases its ability to pick up latent heat. In addition, to achieve a low dew-point temperature, the dry-bulb temperature of the supply air must be similarly reduced. This results in a greater differential between the supply- and room-air dry-bulb temperatures, reducing the air quantity needed, as well as the size of the fans and ducts. Although lowering the room dry-bulb and dew-point temperatures increases the sensible- and latent-heat pickup from the occupants, the lower dry-bulb temperature increases the transmission load of the building. Therefore, to balance the transmission load, the supply-air quantity, plus the duct and fan sizes, must be adjusted upward accordingly.

The amount of heat produced by the body varies with race, sex, age, size, and weight. Since it is impossible to predict these variables for each employee in advance, a general average is used for all workers. Average values for adult males engaged in various activities, are given in Table 2-1.

Dew points below 40°F are difficult to maintain unless the building is specially constructed for this type of operation. This usually means extremely tight construction and, in most cases, the existence of a vapor seal.

Velocities higher than the maximum shown in Table 2-3 may cause discomfort if they are maintained for any length of time.

TABLE 2-3 Velocity Chart

Type of work	Conditions*		Recommended velocity, ft/min
	M	t_g	
Desk or bench work	Low	Low	25–50
	Low	High	Up to 75 (max)
Medium work (machine operator)	Medium	Low	Up to 100 (max)
	Medium	High	Up to 150 (max)
Heavy work (heavy lifting)	High	Low	Up to 200 (max)
	High	High	200 and up

*M is the metabolic rate; t_g is the globe temperature.

Insulaton and Shielding

Manufacturing processes differ quite widely in the amount of heat and humidity produced. Some tend to create high temperatures and high humidities in the occupied area. In addition, the building itself, if not properly insulated and enclosed, may allow large quantities of heat and humidity to enter. It is only logical, therefore, to recommend that when air conditioning is considered, a study be made to determine the economics of installing proper insulation which would make the building as tight as possible before the installation is made.

It may appear to be out of place to discuss insulation and building construction, but it is just as important to reduce the radiant effect of hot roofs and walls as it is to reduce the radiant effect of ovens and furnaces.

The use of shields to absorb radiant heat cannot be overemphasized. In installations where there is a great amount of radiant heat impinging on workers, lowering the temperatures and dew point and increasing the air velocity cannot, by themselves, sufficiently improve the working atmosphere. In addition, using heat shields to absorb radiant heat means that refrigeration machines, air-handling equipment, and heat-transfer surfaces can be of a smaller size.

Design Factors

Factories are heated by warm walls, ceilings, windows, people, lights, and motors and other process equipment. Each source that contributes heat to the conditioned space must be identified.

The design of an air-conditioning system depends upon the calculation of these heat gains. The calculations must be made with care and accuracy to produce a satisfactory installation. Undersized air-conditioning systems are never satisfactory. Grossly oversized systems, on the other hand, are an unnecessary economic drain.

Heat always flows from a warm environment to a colder environment. Thus, the term "heat loss" refers to heating loads and "heat gains" refers to cooling loads. If the temperature inside a building is lower than the outdoor temperature, there will be a continuous flow of heat inward (heat gain) through the exterior walls. Sensible heat applied to any substance will cause the temperature of that substance to rise. Therefore, heat flowing into a building will cause a rise in the inside temperature, unless heat is removed as fast as it flows in. If only part of the heat coming in through the walls is removed, the indoor-outdoor temperature balance will take place at a higher indoor temperature. The heat flowing through the walls, roofs, glass, and other exposures depends upon the indoor-outdoor temperature difference as well as the amount of energy that is directed on the wall, roof, or glass by the sun. To design an air-conditioning system, the indoor-outdoor temperature difference as well as the geographic location must be determined. The temperature difference and the amount of solar heat depend almost entirely upon the geographical location of the factory. While the designer has little control over the geographical location of the plant, the climate in the selected location should be carefully analyzed to determine the most economical and practical system of air conditioning. To do this means that the range of outside dry-bulb temperatures and humidity values must be studied to arrive at proper outdoor design conditions. This does not mean that the highest dry-bulb values and the highest humidity values in the area are selected as design conditions. Obviously, the system would be oversized, because the highest dry-bulb temperature does not occur simultaneously with the highest humidity value, and usually both are of relatively short duration.

The customary procedure is to base the computations on those outdoor dry-bulb and coincident wet-bulb temperatures that are not exceeded more than 2.5 percent of the days during the summer.

The recommended design conditions for many localities have been worked out and appear in the ASHRAE *Handbook of Fundamentals*. But it is not reasonable to assume that the design conditions for a city will also apply to the area in which a plant is located, because the plant is within a few miles of the city. Many factors such as prevailing winds, average wind velocity, elevation, and proximity of water may mean other design temperatures should be used.

It is therefore advisable, if no design data are available for the immediate area, to conduct a survey and establish accurate design conditions for the plant.

When environmental design conditions have been determined in accordance with the previous discussion and the outside design conditions have been determined from a survey, a complete analysis of the building should be made. This analysis should include such items as wall and roof construction, type of glass, ceiling heights, truss construction, floor areas, and orientation. In addition to the types of construction, the area(s) of the walls, glass, roof, etc., should be calculated.

If building plans and details are not available, a scale drawing should be made of the building. On this drawing, the process loads, lights, motors, and other load-producing equipment should be noted. It is also advisable to divide the plant into areas which have approximately the same type of manufacturing processes and the same loading from the standpoint of lights, people, and ventilation requirements.

The following schedule will be helpful when plant and equipment selections are to be made.

Procedure for Load Calculation

1. Divide the plant into areas which have approximately the same type of manufacturing process and the same load from the standpoint of lights, people, and ventilation requirements per square foot.
2. Calculate the heat gains on the basis of typical areas.
   a. Transmission
      (1) Wall
      (2) Glass
      (3) Roof
   b. Solar
      (1) Wall
      (2) Glass
      (3) Roof
   c. People
      (1) Sensible heat
      (2) Latent heat
   d. Lights
   e. Process
      (1) Sensible heat
      (2) Latent heat
   f. Ventilation
      (1) Sensible heat
      (2) Latent heat
3. Calculate the square feet per ton (each area).
   a. Internal sensible heat
   b. Internal latent heat
   c. Internal total heat
4. Calculate the airflow (in cubic feet per minute) required for each area
   a. Airflow (ft^3/min) is equal to internal British thermal units (sensible) divided by $1.1 \times \Delta T$.

$$\text{ft}^3/\text{min} = \frac{\text{Btu (sensible)}}{1.1 \times \Delta T}$$ (See Table 2-4 for ΔT values.)

$$\text{Sensible heat ratio} = \frac{\text{sensible internal}}{\text{total internal}}$$

TABLE 2-4 Room Sensible Heat Ratio

T,°F	100	98	96	94	92	90	88	86	84	82	80	78	76	74	72	70
ΔT	18.0	18.4	18.7	19.0	19.5	20.0	20.5	21.0	21.5	22.0	22.5	23.0	23.5	24.0	24.5	25.0

5. Calculate building volume in cubic feet for each area.
   a. Length × width × average height
   b. If stratification is used, average height equals 15 ft.
6. Calculate the number of air changes for each area.
   a.
   $$\text{Air changes} = \frac{(\text{ft}^3/\text{min}) \times 60}{\text{volume}}$$
   b. If stratification is used, volume from item **5b** applies
7. If air changes are fewer than 6, increase airflow to obtain at least 6 air changes per hour.

The location and selection of equipment will be discussed in the next subsection.

EQUIPMENT

Fans

Among the centrifugal fans there are three principal types, each distinguished by the type of fan wheel used.

Forward-Curved Fans

The first of these centrifugal fan wheels to be considered has blades that are curved in the direction of wheel rotation (Fig. 2-6). These are called *forward-curved* or FC blades.

FC wheels are operated at relatively low speeds and are used to deliver large air volumes against relatively low static pressures. The inherently light construction of the forward-curved blade does not permit this wheel to be operated at the speeds needed to generate high static pressures.

The maximum static efficiency of the FC fan is approximately 60 to 68 percent, and it occurs just to the right of the maximum static pressure points on the fan curves.

The curved shape of the FC blade imparts a forward motion to the air as it leaves the blade tip. This, together with the wheel speed, causes the air to leave at a relatively high velocity.

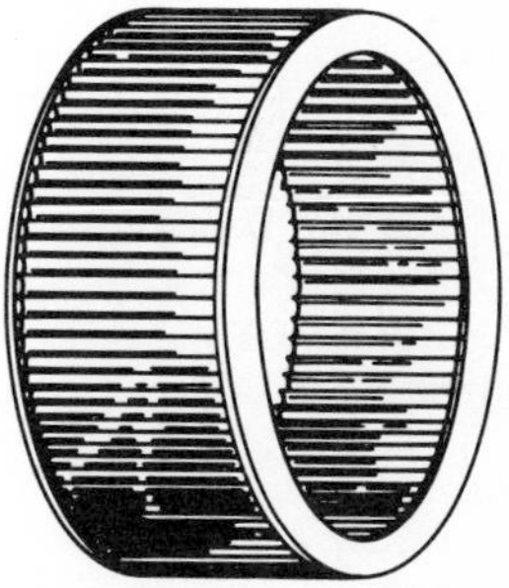

Figure 2-6 Forward-curved or FC fan.

Because the pressures produced by a fan are a function of the forward motion of the air at the blade tip, the FC fan can perform a given air-moving task, which is within its airflow and static performance range, at lower speeds than other fan types.

The application range of the FC fan is from approximately 45 to 80 percent of wide-open airflow (Fig. 2-7). Selection of an operating point on any of the performance curves that places the air delivery rate below approximately 45 percent for that curve may place the fan in an area of instability. Similarly, an operating point that places the air delivery rate beyond 80 percent of wide-open airflow typically produces noise and inefficiency (Fig. 2-8).

Notice, on the typical FC fan curve of Fig. 2-8, how the brake horsepower lines cross the fan performance curves. Therefore, if the system resistance were to drop from 2 to

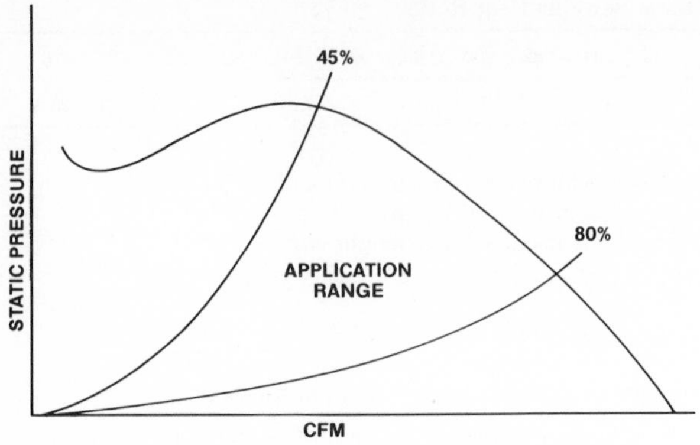

Figure 2-7 Application range of FC fan. (*The Trane Co.*)

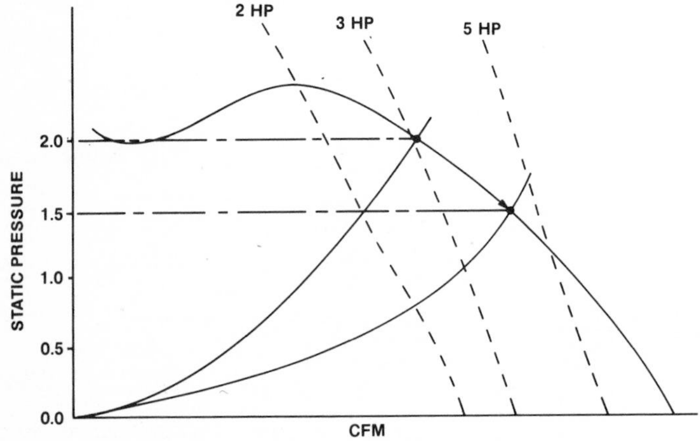

Figure 2-8 Operating points on FC fan curve. (*The Trane Co.*)

1.5 in of static resistance, in this example, the fan brake horsepower requirement would rise from 3 bhp to something over 4 bhp, possibly overloading the motor. Consequently, the FC fan is referred to as an overloading-type fan.

Like all fan types, the FC fan can exhibit unstable operation, or surge. However, under certain conditions, an FC fan can surge without producing any noticeable disturbance. Since FC fans are used typically in low-static-pressure applications, many small FC fans can operate in surge without noticeable noise and vibration.

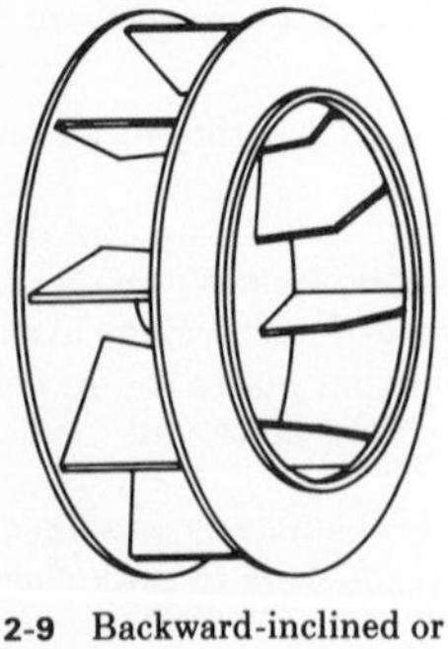

Figure 2-9 Backward-inclined or BI fan.

Backward-Inclined Fans

The second wheel design (Fig. 2-9) has blades that are slanted away from the direction of wheel travel. The term applied to this type of blading is *backward-inclined* (BI). The performance of this wheel is characterized by high efficiency and high airflow, and its rugged construction makes it suitable for high-static-pressure applications.

The maximum static efficiency of the BI

wheel is approximately 75 to 80 percent, and it occurs at approximately 50 percent of wide-open airflow on the fan curves.

The angle of the backward-inclined blades causes the air leaving the wheel to be bent back against the direction of rotation. However, combined with the forward movement of the wheel, the air assumes a vector sum velocity at a higher rotative speed.

The application range of the BI fan is from approximately 40 to 85 percent of wide open airflow (Fig. 2-10). As before, an operating point above 85 percent typically produces noise and inefficiency.

Since the magnitude of surge is pressure-related, the surge characteristic exhibited by the BI fan is greater than that of the FC fan.

Unlike those of the FC fan, the brake horsepower lines of a BI fan are, for the most part, parallel to the fan performance curves (Fig. 2-11). Therefore, if the system resistance were to drop from 4 to 2 in of static pressure in this example, the fan brake horsepower would change only slightly. For this reason, BI fans are referred to as nonoverloading fans.

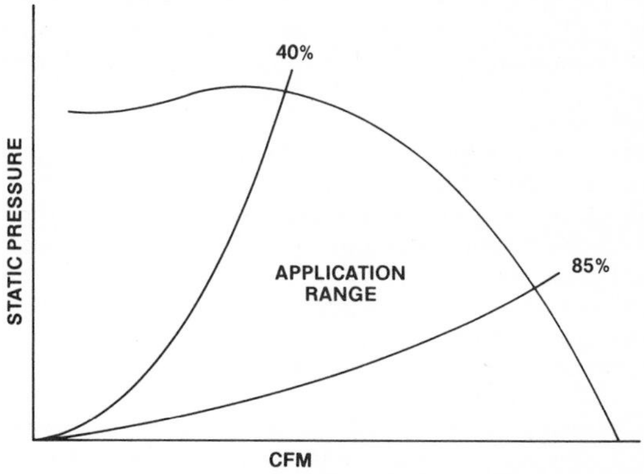

Figure 2-10 Application range of BI fan. (*The Trane Co.*)

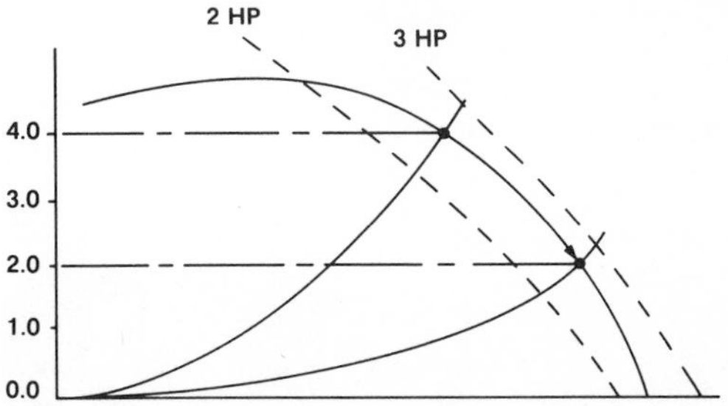

Figure 2-11 Operating points on BI fan curve. (*The Trane Co.*)

Airfoil Fans

A refinement of the BI wheel blade changes its shape from a flat plate to that of an airfoil (Fig. 2-12). The airfoil blade induces a smooth airflow across the blade surface, reducing eddy currents which produce turbulence and noise within the wheel. This results in increased fan static efficiency and reduced overall sound level. Static efficiencies as high as 86 percent are achieved with airfoil fans. In general, airfoil (AF) fans exhibit characteristics that are essentially the same as those of the plate-bladed BI fan.

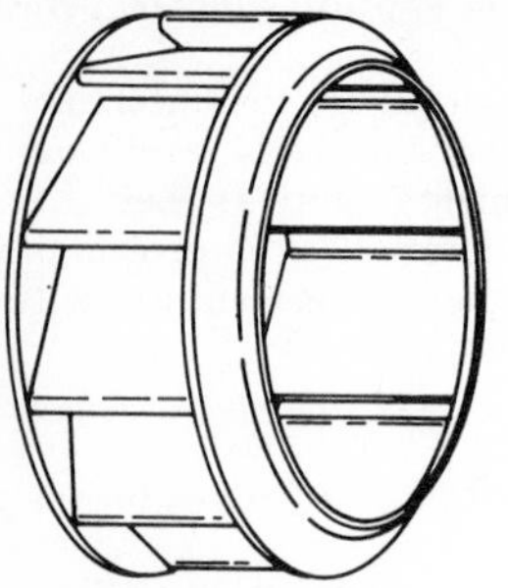

Figure 2-12 Airfoil fan.

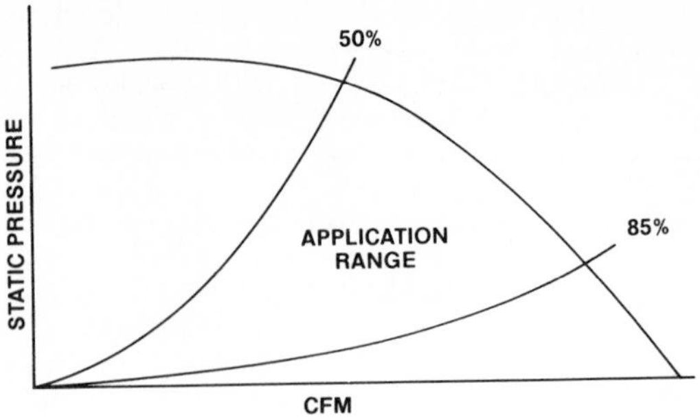

Figure 2-13 Application range of airfoil fan. (*The Trane Co.*)

The application range of the AF fan is from approximately 50 to 85 percent of wide-open airflow (Fig. 2-13). This is a narrower application range than that of either the FC or BI fan. This is because the AF fan surges at a greater percentage of wide-open airflow, placing the surge line farther to the right on the fan curve. This, in turn, reduces the application area of the fan.

Because surge occurs at a higher airflow, the magnitude of the surge characteristics of the AF fan is greater than those of the flat-bladed BI and FC fans.

The other fan commonly used in air-conditioning work is of the vane-axial design (Fig. 2-14). This fan is fundamentally a propeller fan mounted in a cylindrical housing. Since a propeller fan inherently produces a spiral airstream, vanes are installed at the leaving side of the fan to convert this spiraling stream into a straight airflow path. As might be expected, a spiraling airstream produces greater duct friction losses than straight airflow; therefore the straightening of the air path improves the fan's efficiency. Static efficiencies of 70 to 72 percent are achieved with vane-axial fans. The airflow and static performance range of the vane-axial fan are similar to those of the BI and AF fans.

The application range of the vane-axial fan is approximately 60 to 90 percent of wide-open airflow (Fig. 2-15). As for the BI fan, the horsepower lines are essentially parallel to the fan curves; therefore the vane-axial fan is considered a nonoverloading design.

The type of fan to be applied in a particular constant-volume system is a decision based on system size and space availability.

The FC fan is well applied in the small system requiring air volumes to 20,000 ft^3/min or less and static pressures to 4 in. The FC fan is the least costly and is generally compatible with the budgetary considerations of the job.

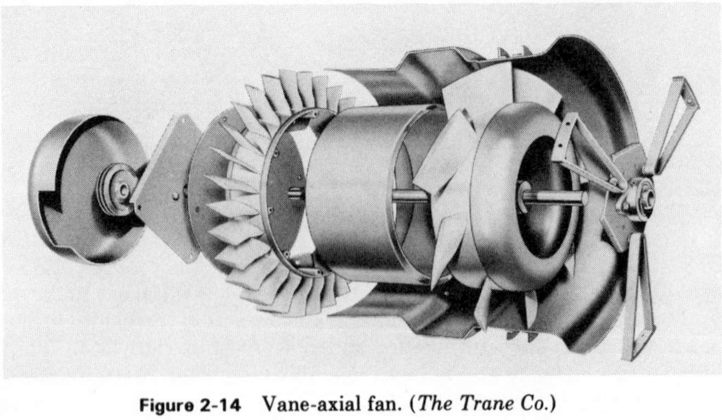

Figure 2-14 Vane-axial fan. (*The Trane Co.*)

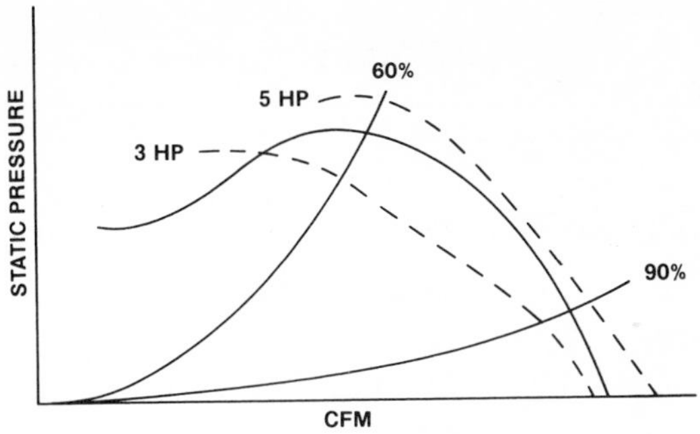

Figure 2-15 Application range of vane-axial fan. (*The Trane Co.*)

On the other hand, systems requiring air volumes in excess of 20,000 ft^3/min and 3 in of static pressure are usually better served by the more efficient BI or AF fans. With the larger fan sizes requiring larger motors, the higher fan efficiencies result in significant brake horsepower savings.

When space is a prime consideration, the vane-axial fan provides a solution. Straight-through airflow permits this fan to be installed in limited space.

Heating and Cooling Coils

A wide variety of extended surface coils are available to transfer heat to or from the conditioned air. Common heat-transfer fluids are water, steam, refrigerant, or antifreeze solution. Coil equipment is used singly, as multiples in built-up banks, and as components of central station air handlers, terminal units, and factory-assembled unitary air conditioners. Most coils used in HVAC systems are of the fin-type or spiral-fin type.

To meet the wide variety of applications and performance criteria, several coil parameters can be altered.

Coil Variables

Fin material and size	Number of rows of tubes
Tube material and size	Fin configuration
Tube pitch	Header construction
Tube spacing	Length and height dimensions
Fin spacing	Special coatings
Tube circuiting	

Performance Criteria

Heat flux (heat-transfer quantity)	Condensate carryover
Effectiveness (heat-transfer efficiency)	Corrosion protection
Air-side pressure loss	Water, steam, or refrigerant pressure
Water-side pressure loss	Cleanability
Refrigerant distribution	Drainability

Coil maintenance is confined to cleaning, corrosion protection, and freeze protection. Since coils have no moving mechanisms, maintenance is fairly simple. Failure to perform simple maintenance, however, can result in coil failure or unnecessary power and energy consumption in addition to inadequate performance (see Fig. 2-16).

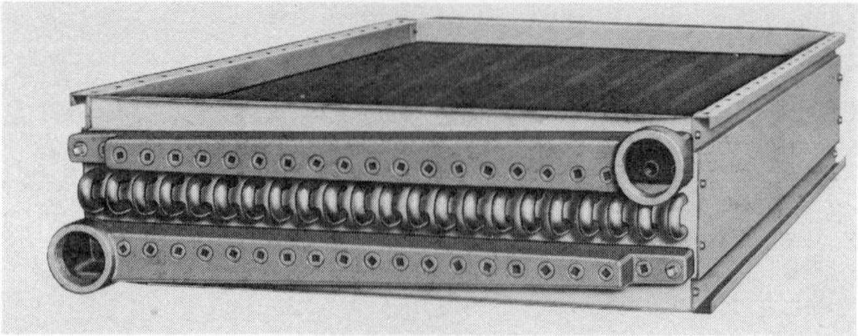

Figure 2-16 Heating-cooling coils. (*The Trane Co.*)

Refrigeration Machinery

Direct-expansion and chilled-water systems employ compressors of various types for the cooling function (see Figs. 2-17 to 2-22). Major compressor classifications are:

1. Positive displacement
2. Nonpositive displacement (dynamic)

Table 2-5 lists the more popular compressors used today.

Heating Equipment

Directly and indirectly fired heaters are described in Chap. 5-1. Various forms of electric heating devices are also explained in that section. A third type of heating equipment involves the application of a reverse compression-expansion refrigerant cycle, sometimes referred to as a heat pump; this concept uses equipment similar to that described as "refrigeration machinery."

Figure 2-17 Typical reciprocating compressor employing halogenated hydrocarbon refrigerants. (*The Trane Co.*)

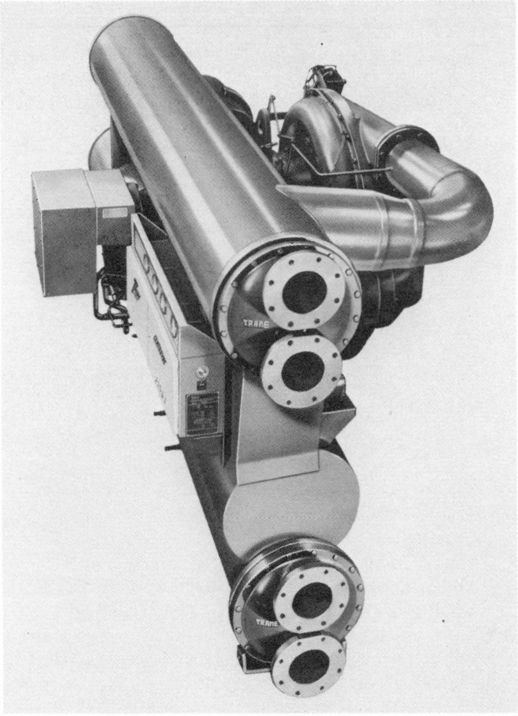

Figure 2-18 Typical packaged water chiller using a centrifugal-type compressor and water-cooled condenser. (*The Trane Co.*)

Figure 2-19 Typical air-cooled water chiller employing reciprocating compressor(s). (*The Trane Co.*)

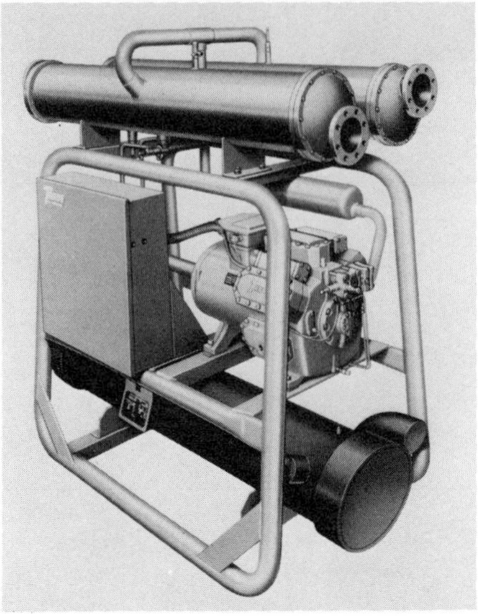

Figure 2-20 Typical packaged water-cooled chiller arranged to recover condenser heat for separate water-heating service. (*The Trane Co.*)

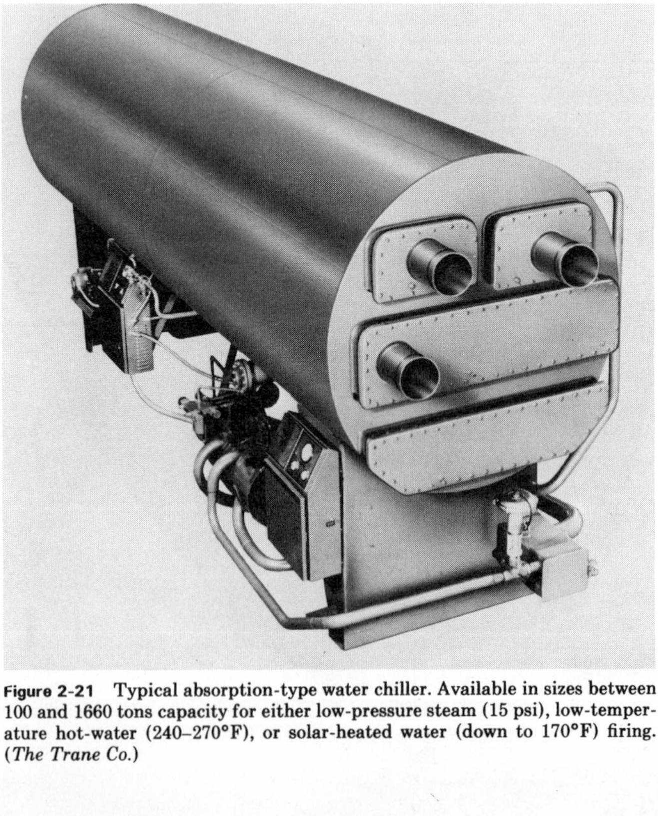

Figure 2-21 Typical absorption-type water chiller. Available in sizes between 100 and 1660 tons capacity for either low-pressure steam (15 psi), low-temperature hot-water (240–270°F), or solar-heated water (down to 170°F) firing. (*The Trane Co.*)

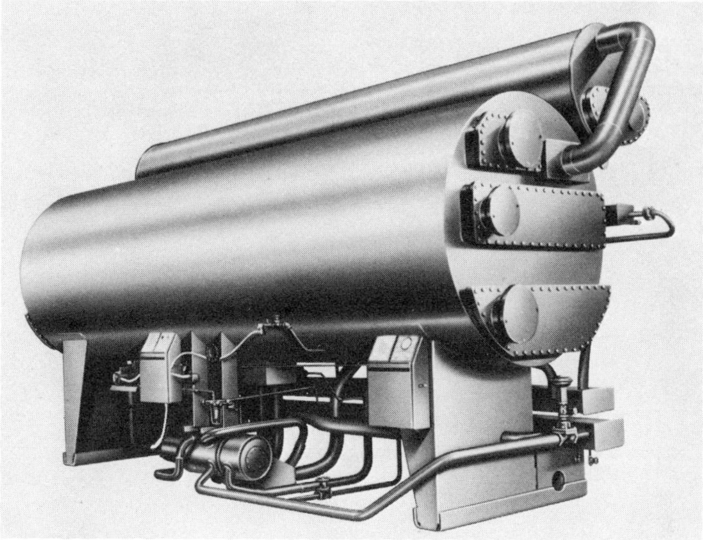

Figure 2-22 Two-stage absorption chiller available in sizes from 500 to 1100 tons. Utilizes medium pressure steam (125–150 psi) or high-temperature hot water (up to 370°F). (*The Trane Co.*)

TABLE 2-5 Compressors Commercially Available For Comfort Air Conditioning

Compressor type	Classification	Size range, tons
Reciprocating	Positive displacement	½–200
Rotary	Positive displacement	½–5
Helirotor (screw)	Positive displacement	30 –800
Centrifugal	Dynamic	100–10,000

Specific examples

Fig. 2-17	Typical reciprocating compressor employing halogenated hydrocarbon refrigerants.
Fig. 2-18	Typical packaged water chiller using a centrifugal-type compressor and water-cooled condenser. This type of equipment is also available with air-cooled condensers.
Fig. 2-19	Typical air-cooled water chiller employing reciprocating compressor(s).
Fig. 2-20	Typical packaged water-cooled chiller arranged to recover condenser heat for separate water heating service. Heated water may be used for space heating, domestic water heating, or process.
Fig. 2-21	Typical absorption-type water chiller. Available in sizes between 100 and 1660 tons capacity for either low-pressure steam (15 lb/in^2), low-temperature hot water (240–270°F), or solar-heated water (down to 170°F) firing.
Fig. 2-22	Two-stage absorption chiller available in sizes from 500 to 1100 tons. Utilizes medium pressure steam (125–150 lb/in^2) or high-temperature hot water (up to 370°F).

Heating equipment is often a combination of heating coils, fans and a suitable enclosure. Figure 2-23 shows a factory-built central-station air handler. It consists of an air mover, air-filter section, heating coil, cooling coil, and inlet air mixing section.

Heating is often accomplished by "radiation" equipment which, despite its name, heats mostly by convection. Finned tube elements in combination with various enclosures are used for this purpose (Fig. 2-24).

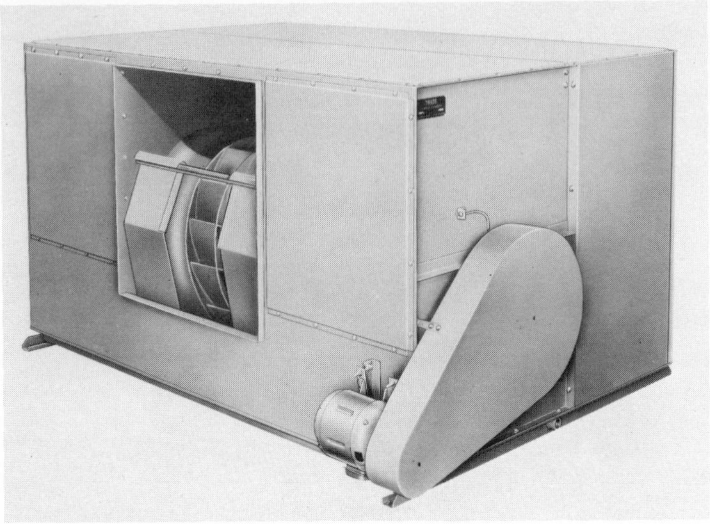

Figure 2-23 Factory-built central station air handler. (*The Trane Co.*)

Figure 2-24 Finned-tube "radiation" enclosures (*The Trane Co.*)

Air-Distribution Devices

Grilles and registers of various configurations are used to distribute conditioned air. Recent advances in variable air-volume systems have been accompanied by special air-distribution devices. Figure 2-25 shows an example of a variable air-volume diffuser device.

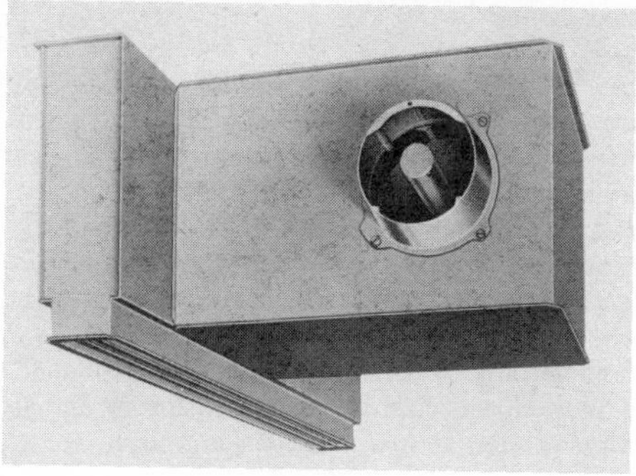

Figure 2-25 Variable air volume diffuser. (*The Trane Co.*)

Heat-Rejection Equipment

All mechanical refrigeration devices absorb heat from cooling loads. This heat, plus the energy to drive the process, must be rejected to the environment. Since water has the ability to absorb great amounts of heat, it is commonly used as a vehicle to reject heat. Water-cooled condensers of the shell-and-tube type are popular and efficient (Fig. 2-26).

Where water is scarce, costly, or difficult to manage, air is used to reject refrigeration-cycle heat. The most common type of air-cooled condenser employs fin-and-tube construction (Fig. 2-27).

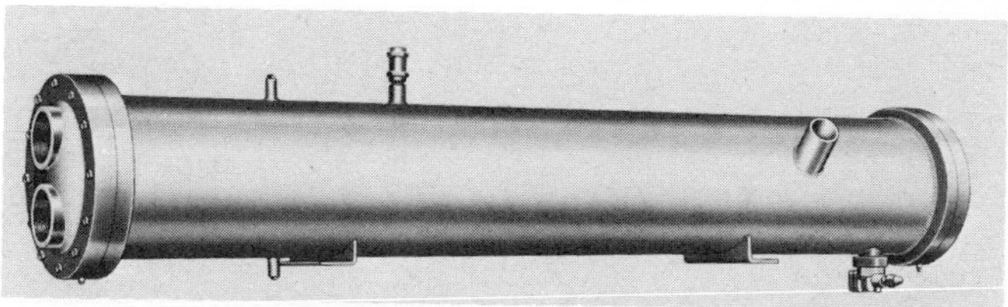

Figure 2-26 Water-cooled condenser—shell-and-tube type. (*The Trane Co.*)

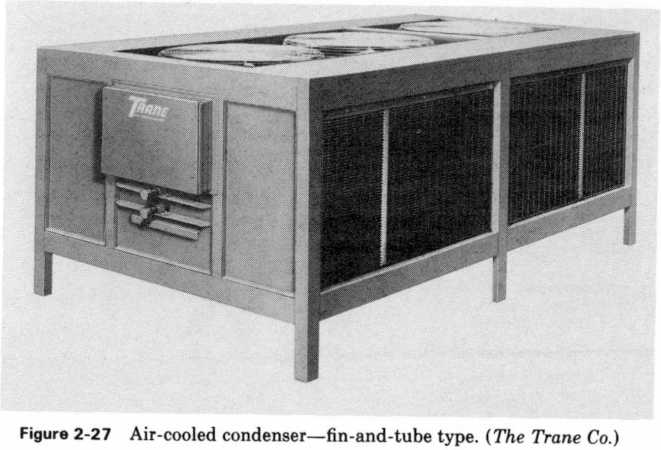

Figure 2-27 Air-cooled condenser—fin-and-tube type. (*The Trane Co.*)

In many areas, water cannot be used for heat rejection unless it is conserved by recycling through an evaporative process. A variety of devices perform this function, the simplest being an open-circuit cooling tower. Heat is rejected by evaporating (boiling) water at the partial pressure corresponding to the wet-bulb temperature of the ambient air. Throughout most of the United States, this method provides a very adequate heat sink at a maximum temperature of 78 to 80°F (25° to 27°C).

A more complicated evaporative device is a closed-circuit cooling tower, or evaporative cooler. A heat-transfer surface (usually copper tubing) is placed between the recirculated condenser water and the evaporative water system, thus preventing contaminants from entering the condenser water circuit.

In a similar way, refrigerant condensing can be accomplished directly by using the heat-transfer surface in a closed-circuit cooling tower to act as the condenser. This arrangement is sometimes called an *evaporative condenser.*

chapter 5-3

Air-Conditioning the Manufacturing Facility

by
William J. Landman
Manager, Applications Engineering
The Trane Co.
LaCrosse, Wisconsin

GLOSSARY

Air, makeup Air that is introduced into a space to replace air that has been exhausted.

Air, outside Air that is taken from outside the conditioned space or process area and introduced into the conditioned space or process area.

Air, exhaust Air that is removed from a space, either naturally or by mechanical means, and discharged to the atmosphere.

British thermal unit (Btu) See Glossary, Chap. 5-2.

Combustion The burning of any substance (solid, liquid, or gas) by a chemical union with oxygen.

Contaminants Any substance (solid, liquid, or gas), added to a mixture of air and water vapor, that is undesirable because it is injurious to health or to the condition of a product or because it has an objectionable smell.

Dehumidification See Glossary, Chap. 5-2.

Dew-point temperature See Glossary, Chap. 5-2.

Drying To separate or remove a liquid or vapor from another substance.

Heat, convection Usually referred to as heated air or vapor that rises because of a difference in temperature.

Heat, radiant See Radiation, thermal, Glossary, Chap. 5-2.

Heat shield A barrier to prevent the passage of radiant heat.

Heat of vaporization Usually referred to as the heat required to change a liquid to a vapor at the same temperature and pressure.

Heat, latent See Glossary, Chap. 5-2.

Load factor See Glossary, Chap. 5-2.

Mean radiant temperature The temperature of a uniform black enclosure in which a solid body or occupant would exchange the same amount of heat as in the existing nonuniform environment.

Molecular weight The weight of a substance equal to the weight of one mole.

Pressure, negative A pressure below atmospheric pressure.

Reheat Usually refers to adding heat to air which has been previously cooled or cooled and dehumidified or cooled and humidified.

Stratification The process of warm air rising to accumulate near the top of a structure and the cold air remaining near the bottom.

Ton of refrigeration See Glossary, Chap. 5-2.

Transformer An electric device for increasing or decreasing electric voltage.

Vapor pressure See Pressure, vapor, Glossary, Chap. 5-2.

Velocity The time rate at which a substance is moving.

Viscosity The property of semifluids, fluids, and gases to resist an instantaneous change of shape or arrangements of parts.

INTRODUCTION

Some designers are in the habit of adding process heat and humidity to the building air-conditioning load and sizing the system accordingly. This method usually increases both the initial cost of the air-conditioning system and the future owning and operating costs. Heat and humidity liberated by manufacturing processes should be handled independently of the air-conditioning system whenever possible.

It is possible to reduce, and in some cases eliminate entirely, loads generated by manufacturing processes. This greatly simplifies the design of the air-conditioning system.

Some of the many types of process loads, and a method of reducing or eliminating them, are discussed later.

Process loads can be classified as follows:

1. **Process Heat and Humidity** These include such process areas as steam-cleaning areas, open heated-water tanks, dryers, gas burners, and chemical bath tanks.
2. **Equipment and Machinery** These include such items as motor heat, oven heat, forges, molding machines, plastic machines, welding machines, heat-treating ovens, process piping, transformers, and process materials.
3. **Process Ventilation** This includes loads due to makeup air that must be supplied for such processes as spray-paint booths, fumes from welding operations, odor-producing processes, dust-producing operations such as grinding of castings, lint-producing operations, and liquid mist- or fog-producing operations.

The methods of reducing or eliminating these loads will depend upon the type and size of the load, the area in which it is located, and the comparative cost of absorbing these loads with the air-conditioning system.

MINIMIZING PROCESS LOSSES

Process Heat and Humidity

Steam-Cleaning Areas

Areas where steam is used for cleaning purposes, or where steam is escaping or being generated from open tanks of boiling water, usually require special treatment. Consideration should always be given to isolating these rooms as much as possible from other manufacturing areas.

Steam that is liberated into the conditioned area raises the dew point of the air in the space. Therefore, the steam must be removed by supplying air at a dew point which is low enough to absorb this moisture without exceeding the dew point to be maintained in the conditioned space. Areas liberating enough steam to cause fogging (i.e., supersaturated air) require supply air that is less than saturated. The introduction of cool, saturated air, will only aggravate the fogging condition.

The quantity of steam from cleaning nozzles can usually be obtained from the manufacturer's ratings, or by condensing the steam and weighing the condensate. The heat load can then be obtained from the following formula:

$$Q_c = G \times h_s \tag{1}$$

where

Q_c = latent heat load, Btu/h
G = pounds of steam or pounds of condensate per hour
h_s = heat of vaporization at the temperature and pressure of the steam

The amount of water evaporated from open tanks can be approximated by means of the following formulas.

For V_p/u less than 30,

$$E = \left(0.46 + 0.0693\frac{V_p}{u}\right) M_v D_v^{2/3} \left(\frac{u}{p}\right)^{1/3} \log\left(\frac{30 - P_a}{30 - P_v}\right) \tag{2}$$

For V_p/u greater than 30

$$E = 16.2\left(\frac{V_p}{u}\right)^{0.78} M_v D_v^{2/3} \left(\frac{u}{p}\right)^{1/3} \log\left(\frac{30 - P_a}{30 - P_v}\right) \tag{3}$$

where

E = evaporation rate, lb/(h)(ft^2)
V_p= velocity of air across the surface, ft/s
p = density of the air flowing over the surface, lb/ft^3
u = viscosity of the air flowing over the surface, lb/(ft)(h) (0.042 at 50°F, 0.044 at 100°F, 0.049 at 200°F, and 0.060 at 400°F)
M_v = molecular weight of the vapor (18.01 for water vapor)
D_v = diffusion coefficient of the water vapor at the average of the air and liquid temperature, ft^2/h (0.76 at 0°F, 1.07 at 100°F, and 1.42 at 200°F)
P_v = vapor pressure of the water at the water temperature, inHg
P_a = vapor pressure of the water vapor in the air, inHg

As stated previously, it is an important economic move to isolate steam rooms and other areas generating a large amount of water vapor. Once isolated, several different methods can be used to maintain satisfactory working conditions inside the room.

One method is to bring in outside air, through a ventilating unit, in sufficient quantity to pick up the excess moisture in the space. This method works satisfactorily, provided the outside air is not saturated and the temperature is low enough to give some cooling effect. In the winter, and during particularly humid days, the air must be heated in order to prevent cold drafts and fogging. Heating the air also provides the necessary differential between dry-bulb and dew-point temperatures to pick up moisture. This arrangement may, however, cause difficulty on days of high temperature and high humidity—or on extremely cold days when humidity exhaust causes ice to form on the exterior of the building or in exhaust ducts.

A better method, and one which maintains conditions year-round regardless of outside conditions, incorporates a dehumidifier and reheat cycle. With this arrangement, a quantity of outside air is introduced to supply the necessary ventilation. The outside air is mixed with recirculated air, dehumidified, and then reheated with condenser water, hot refrigerant gas, or some external medium such as steam or hot water. When outside air is at the right conditions to provide dehumidification by itself, 100 percent outside air is used.

A diagram of the dehumidifier-reheat arrangement is shown in Fig. 3-1.

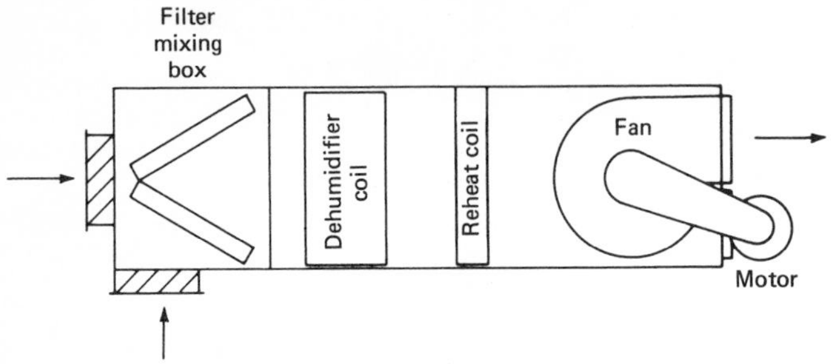

Figure 3-1 Dehumidifier-reheat arrangement.

Usually it is most economical to introduce the conditioned air as close to the source of moisture as possible. This picks up and exhausts the moisture before it can disperse into the room. Thus, the necessity of introducing large quantities of air into the room, at low dew points, is eliminated. Figures 3-2 to 3-4 illustrate some of the methods which can be used.

Open Flames

Open gas flames such as Bunsen burners, soldering torches, welding and cutting torches, gas burners for drying ovens, and other heat- and moisture-producing flames are com-

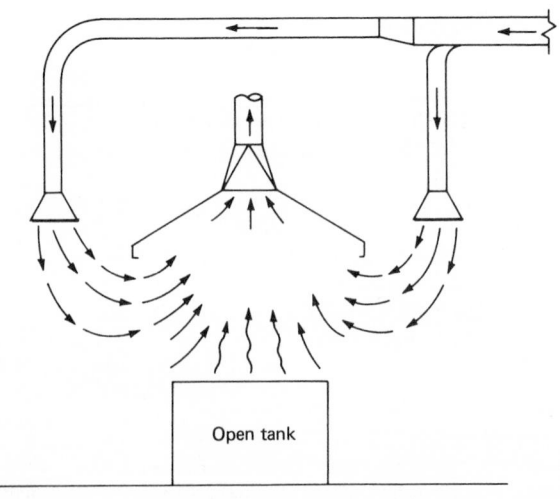

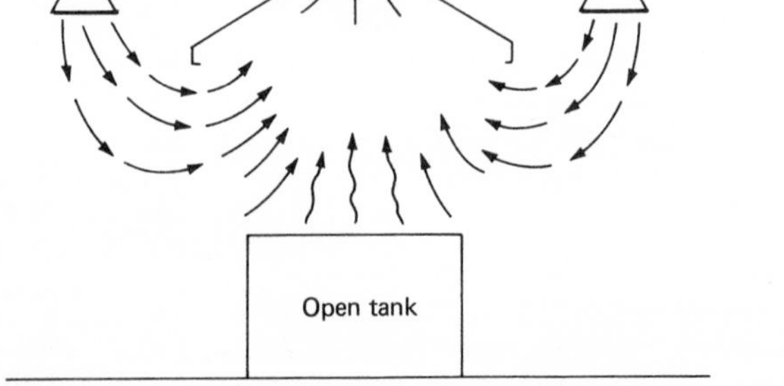

Figure 3-2

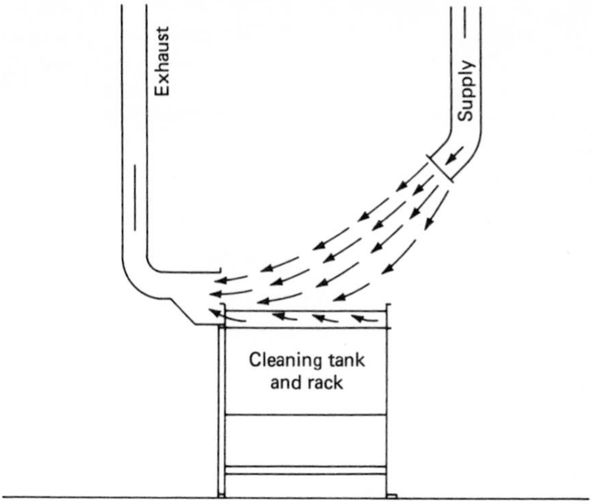

Figure 3-3

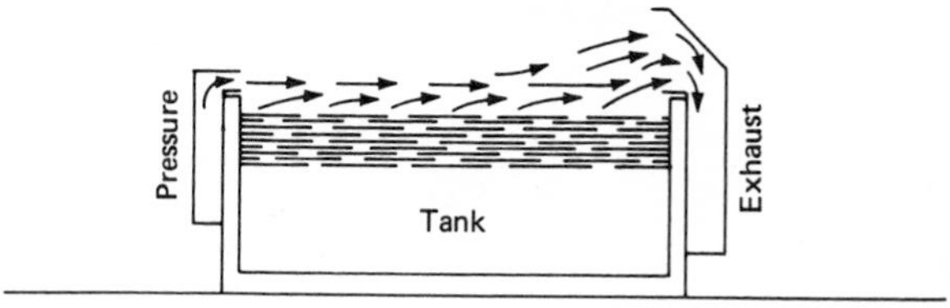

Figure 3-4

mon in industrial plants. These devices produce sensible heat as well as latent heat or water vapor, which is produced by the combination of oxygen and the gases used.

Many approximations have been made regarding the amount of sensible and latent heat given off by these devices. The most accurate method is to obtain the gas consumption of each device and then calculate the heat output.

For new plants, the consumption can be obtained from the manufacturers of the heat-producing equipment. Then the rate of usage must be estimated for the type of operation to be performed.

For plants already in operation, natural or manufactured gas consumption can be obtained from the gas meters. For acetylene, propane, butane, and other gases, the consumption may be obtained by referring to the number of drums or containers emptied during the period of heaviest usage. Then, with this information, and the data in Table 3-1, the heat load can be calculated.

TABLE 3-1 Heat Liberated by Combustion of Gases, Btu/ft^3

Type of gas	Density,[a] lb/ft^3	Sensible heat	Latent heat	Total
Natural gas	0.049	800	200	1000
Manufactured gas	0.025	440	110	550
Butane	0.157	2663	450	3113
Propane	0.117	2031	354	2385
Methane	0.043	747	166	913
Ethane	0.080	1382	259	1641
Hydrogen	0.0053	181	94	275
Acetylene	0.070	1341	112	1453

[a]For 70°F and atmospheric pressure.

Air used in combustion must be replaced with ventilation air. The analysis for removal of steam can also be applied to gas burners and gas-burning equipment. That is, air supplied to the room must be at a low enough dew point to pick up moisture from the burners. Yet the dew point, after the moisture pickup, should not exceed design conditions. At the same time, the air must be at a dry-bulb temperature which will permit it to pick up the sensible heat, without exceeding the room-design dry-bulb temperature.

Introducing the air at a point as close as possible to the source of heat and humidity and then exhausting it will, of course, eliminate handling this air through the regular air-conditioning system. However, this must be done carefully in the case of gas burners and open-flame operations.

Dryers

Dryers may be of various types: open, closed, vented, unvented, steam-heated, gas-fired, or electrically heated. In order to determine the amount of heat and moisture added to the conditioned space, the type of installation must be analyzed and the amount of heat and humidity calculated. As an example, consider an application where heat is liberated to a space from an electrically heated, closed dryer with the waste air vented to the outside. The air supply is also taken directly from the outside. For this example, the only heat liberated to the conditioned space would be the radiant and convective heat from the outside surface of the dryer. There would be no latent heat liberated unless there was leakage from the exhaust ducts.

Table 3-2 is included to assist in analyzing the heat and humidity problems concerning dryers.

Chemical Bath Tanks

Many vapors other than water vapor must be exhausted to maintain a safe atmosphere. Some of these vapors may be explosive in certain concentrations. Others, although not explosive, may be injurious to the health of the workers, may be unpleasant to breathe, or may be only a nuisance. In any case, the most effective and usually the most economical way to remove them is to ventilate.

Air that is exhausted must be replaced. Replacement of the exhaust air can be accomplished in two ways. The replacement air can be drawn into the space by allowing it to "leak in" around doors and windows and other openings, or it can be supplied by means of a fan.

TABLE 3-2 Heat Liberated by Dryers

Type of dryer	Heating medium	Heat liberated
Open, unvented	Steam	All heat supplied by the steam to be figured as room sensible load, except heat required to evaporate the moisture from the product which is to be figured as latent heat
Open, unvented	Electricity	Same as steam-heated
Open, unvented	Open gas flame	Heat supplied by the gas to be divided into sensible and latent. Heat required to evaporate moisture from product to be figured as latent and subtracted from sensible heat of gas
Closed, unvented	Steam	Figure same as "open, unvented" because all the heat will eventually be liberated into the space
Closed, unvented	Electricity	Same as steam-heated
Closed, unvented	Open gas flame	Figure same as "open, unvented" with gas
Closed, vented to atmosphere	Steam	Figure only radiant and convective heat from outside surface of the drier
Closed, vented to atmosphere	Electricity	Figure same as steam-heated for "closed, vented" drier
Closed, vented to atmosphere	Gas flame, vented	Figure same as steam-heated for "closed, vented" drier

TABLE 3-3 Maximum Allowable Concentration of Toxic Fumes*

Gases, $ft^3/10^6 ft^3$ air		Solvents, $ft^3/10^6 ft^3$ air		Metallic fumes, mg/m^3 air	
Arsine	1	Amyl acetate, butyl acetate, ether	400	Cadmium	0.1
Ammonia	100	Aniline, nitrobenzene	5	Chromic acid	0.1
Carbon monoxide	100	Benzene	75	Lead	0.15
Chlorine	1	Carbon bisulfide	15	Manganese	30.0
Formaldehyde	20	Carbon tetrachloride	30	Mercury	0.1
Hydrogen cyanide	20	Dichlorbenzene, monochlorbenzene	75	Zinc oxide (fumes)	15.0
Hydrochloric acid	10	Dichlorethyl ether	15		
Hydrogen fluoride	3	Ethylene dichloride	100		
Hydrogen sulfide	20	Methanol, toluene, turpentine	200		
Nitrogen dioxide	10	Naphtha, gasoline	1000		
Ozone	1	Tetrachlorethane	10		
Phosphine	2	Tetrachlorethylene	200		
Sulfur dioxide	10	Trichlorethylene	200		
		Xylene, coal-tar naphtha	200		

*From "Toxic Fumes in Massachusetts Industry," published by Massachusetts Division of Occupational Hygiene. These values are constantly under revision. Consult the applicable state agency for current limits.

When the makeup air is allowed to leak in around windows and doors, it is difficult to maintain the quality and the quantity of air. Any opening of doors or windows allows a greater quantity of unfiltered air to be drawn in because of reduced resistance. This larger quantity adds to the conditioning load and upsets the temperature control in the conditioned space.

If the makeup air is introduced by means of a supply-air fan, the conditioned space can be maintained at a zero pressure difference or at a slightly negative pressure. The result is a low variation in room load when doors are open or closed. A slightly negative pressure is more desirable in the case of fume exhaust since this prevents contamination of adjacent areas.

The quantity of air to be exhausted depends upon the allowable concentration of the vapor, the area of the exposed tank surface, the volume of the space from which the fumes are to be exhausted, and requirements set up by local or state codes. Some of the allowable concentrations of various toxic fumes and the recommended ventilation rates are given in Tables 3-3 and 3-4.

Equipment and Machinery

Motor Heat

Electric motors add sensible heat to any space in which they are located. The amount of heat added depends upon the size of the motor, motor efficiency, and whether the entire energy of the motor is expended in doing work in the conditioned space. Neglecting the motor efficiency, each horsepower of motor load gives off 2545 Btu/h.

Motor efficiency varies with motor size. Thus, a small fractional-horsepower motor may have an efficiency as low as 50 percent. A 10-hp motor may have an efficiency as high as 88 percent. Motor manufacturers usually supply efficiency curves for their motors, provided the motor serial number and type are known. A curve showing approximate motor efficiencies for various horsepower ratings is shown in Fig. 3-5. See also Chap. 3-4 for additional discussion of motors.

The equation for calculating the heat load (in British thermal units per hour) from a motor is as follows:

$$Q_m = \frac{\text{hp rating}}{\text{motor eff}} \times 2545 \times F_3 \quad (4)$$

TABLE 3-4 Ventilation Rates For Open Surface Tanks*

Process	Minimum ventilation rate, ft³/min per hood opening: Enclosing hood, One open side	Enclosing hood, Two open sides	Canopy hood, Three open sides	Canopy hood, Four open sides	Minimum ventilation rate, ft³/(min) per ft² tank area, lateral exhaust, $\frac{W}{L} = \frac{\text{tank width}}{\text{tank length}}$ ratio: W/L 0–0.24, A	W/L 0–0.24, B	W/L 0.25–0.49, A	W/L 0.25–0.49, B	W/L 0.50–1.0, A	W/L 0.50–1.0, B
Plating										
Chromium (chromic acid mist)	75	100	125	175	125	175	150	200	175	225
Arsenic (arsine)	65	90	100	150	90	130	110	150	130	170
Hydrogen cyanide	75	100	125	175	125	175	150	200	175	225
Cadmium	75	100	125	175	125	175	150	200	175	225
Anodizing	75	100	125	175	125	175	150	200	175	225
Metal cleaning (pickling)										
Cold acid	65	90	100	150	90	130	110	150	130	170
Hot acid	75	100	125	175	125	175	150	200	175	225
Nitric & sulfuric acid	75	100	125	175	125	175	150	200	175	225
Nitric & hydrofluoric acid	75	100	125	175	125	175	150	200	175	225
Metal cleaning (degreasing)										
Metal cleaning (caustic or electrolytic)										
Not boiling	65	90	100	150	90	130	110	150	130	170
Boiling	75	100	125	175	125	175	150	200	175	225
Bright dip (nitric acid)	75	100	125	175	125	175	150	200	175	225
Stripping										
Conc. nitric acid	75	100	125	175	125	175	150	200	175	225
Conc. nitric & sulfuric acid	75	100	125	175	125	175	150	200	175	225
Salt baths (molten salt)	50	75	75	125	60	90	75	100	90	110
Salt solution (parkerize, bonderize, etc.)										
Not boiling	75	90	100	150	90	130	110	150	130	170
Boiling	90	100	125	175	125	175	150	200	175	225
Hot water (if ventilated)										
Not boiling	50	75	75	125	60	90	75	100	90	110
Boiling	75	100	125	175	125	175	150	200	175	225

**Source* ASHRAE Guide, 1958.

where

Q_m = heat load, Btu/h

F_3 = usage or load factor

As explained previously, the motor efficiency can be found by referring to the manufacturer's curves or can be approximated from Fig. 3-5. The factor F_3 is the fractional part of the motor load being used at the time the air-conditioning load is calculated. This can be approximated by estimating usage if the motor is not installed, and by using Eq. (4).

If the motor is installed and in use, the heat load can be calculated from the following formulas:

For single-phase

$$Q_m = \text{amps} \times \text{volts} \times \text{PF} \times 3.4 \tag{5}$$

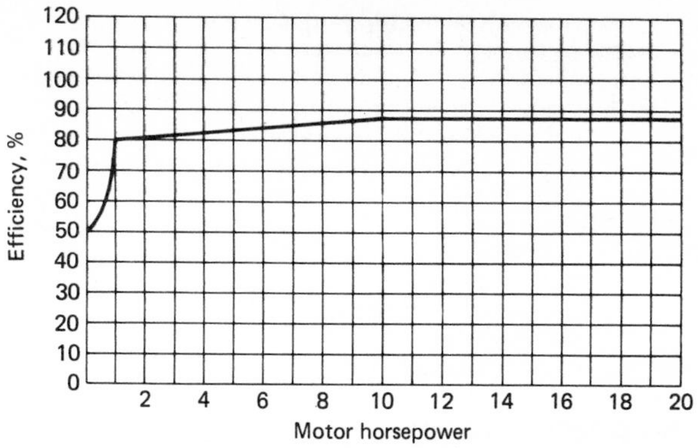

Figure 3-5 Motor efficiencies vs. power rating.

and for three-phase

$$Q_m = \text{amps} \times \text{volts} \times 1.73 \times \text{PF} \times 3.4 \tag{6}$$

where

Q_m = heat load, Btu/h
amps = amperes going to motor
volts = voltage across the lines to the motor
PF = Power factor of the motor

It is relatively easy to calculate the amount of motor heat that should be charged to the conditioned space. If the following simple rules are followed, there will be little chance for error.

1. If both the motor and the work done by the motor are in the conditioned space, consider that all of that work, plus the motor heat, will be dissipated into the conditioned space. Use Eq. (4). Examples: Punch press, drill press, lathe, etc.
2. If the motor is located in the conditioned space and the heat is dissipated outside the conditioned space, consider that only the heat due to the inefficiency of the motor will be dissipated into the conditioned space. Use

$$Q_m = \left[\left(\frac{\text{hp rating}}{\text{motor eff}}\right) - (\text{hp rating})\right] \times 2545 \times F_3 \tag{7}$$

Examples: Centrifugal exhaust fans and pumps discharging outside the conditioned space, refrigeration compressors, etc.
3. If the motor is located outside the conditioned space and the heat is dissipated in the conditioned space, then only the motor load will be dissipated into the conditioned space. Use

$$Q_m = \text{hp rating} \times 2545 \times F_3 \tag{8}$$

Examples: Ventilating fans and motors located outside of the space which they ventilate.
4. The work done by a fan motor in an air-conditioning unit within the air-conditioned space is considered a part of the coil load. Thus, the only time the motor would add heat to the conditioned space is when it is located in the space; and then only the heat due to the inefficiency of the motor would be added to the space. Use Eq. (7). Example: An air-conditioning unit with an exposed motor located within the conditioned space.

An analysis of the foregoing statements will show that any efforts to absorb the heat from anything besides very large motors by a separate ventilation and exhaust system would be economically unfeasible. Motors of 50 hp and larger usually fall into the large motor classification. Some methods of providing ventilation and absorbing the heat from large motors are illustrated in Figs. 3-6 to 3-8. As indicated in the illustrations, air for motor ventilation should always be filtered to prevent both clogging of motor air passages and formation of deposits on the windings of open motors.

When water-cooled motors are used, it is, of course, unnecessary to calculate any heat pickup in the room except when the work done by the motor is dissipated in the conditioned space.

Oven Heat

Industrial ovens of all types, sizes, and shapes usually constitute a large source of heat load.

The heat given off by ovens is largely radiant. It is absorbed by the objects surrounding the oven, which in turn, will give off some radiant and some convective heat. Radiant heat shares many of the properties of light. For example, radiant heat needs no actual

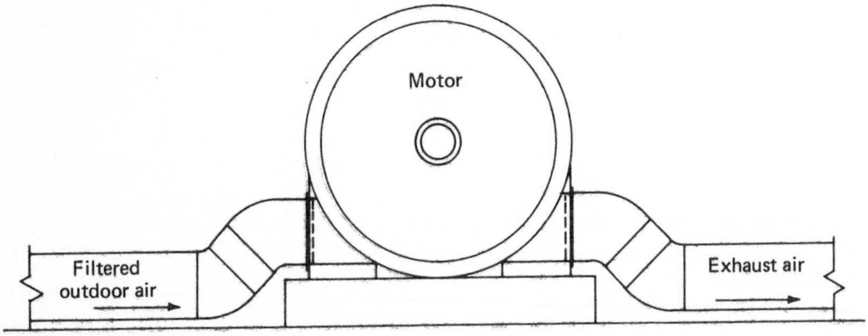

Figure 3-6

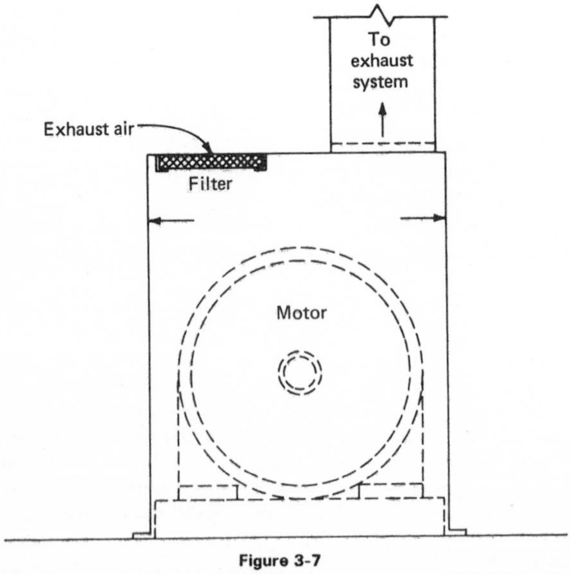

Figure 3-7

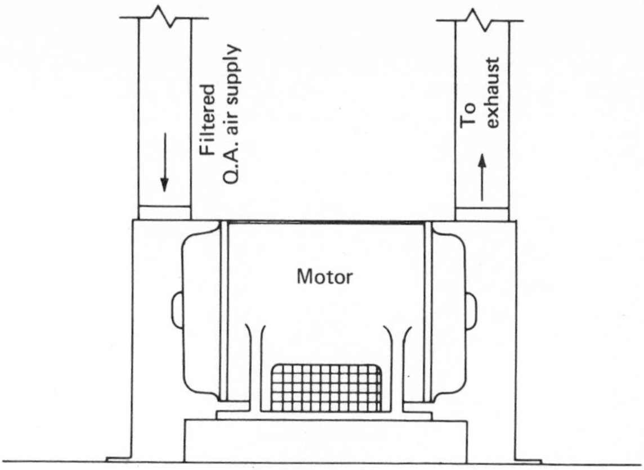

Figure 3-8

substance to carry it from one object to another. For this reason, it is difficult to absorb radiant heat by the circulation of air. Other methods, such as reflecting shields, air-cooled shields, or water-cooled shields, should be used.

The amount of heat given off by ovens can be determined only by test or by contacting the oven manufacturer.

To determine the amount of radiant heat by test, use an accurate mercury thermometer shielded from the radiant heat, a globe thermometer to measure the "blackbody" or globe temperature, and a thermal anemometer or Kata thermometer for measuring convection currents. With these instruments it is possible to obtain information which, when substituted in the following equation,* will yield the mean radiant temperature (MRT):

$$T_w^4 = T_g^4 + 0.103 \times 10^9 \sqrt{V}\,(t_g - t_a) \tag{9}$$

where

T_w = mean radiant temperature, °F absolute
T_g = globe temperature, °F absolute
V = air velocity, ft/min
t_g = globe temperature, °F
t_a = ambient temperature, °F

When the mean radiant temperature has been determined, the surface temperature of the oven must be determined, preferably by means of a thermocouple. These values can be substituted in the following equation* to obtain the amount of radiant heat transmitted, in British thermal units per square foot per hour:

$$Q_r = A \times \frac{1}{(1/E_1) + (1/E_2) - 1} \times (1073 \times 10^{-12}) \times (T_1^4 - T_2^4) \tag{10}$$

where

Q_r = radiant heat, Btu/h
A = area, ft^2 of radiating surface
E_1 = emissivity of radiating surface
E_2 = emissivity of receiving surface
T_1 = radiating surface temperature, °F absolute
T_2 = receiving surface temperature, °F absolute, assumed the same as globe temperature, °F absolute

*ASHRAE Guide 1958.

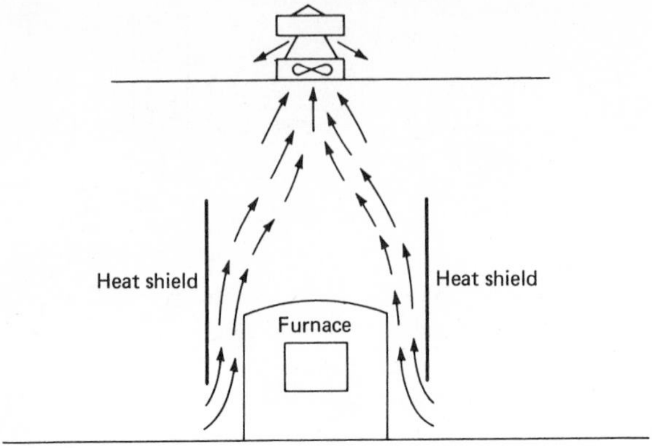

Figure 3-9

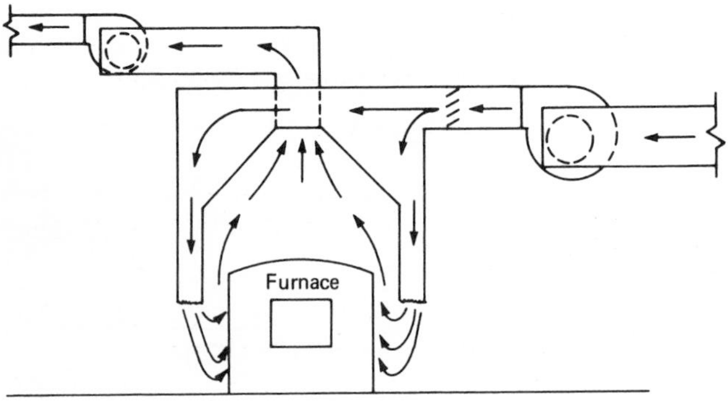

Figure 3-10

When Eq. (10) is used for estimating the heat absorbed by a radiant shield, the value of A should be taken as the radiation area which actually faces the radiation shield.

The convective heat that must be offset by the circulation of air can be estimated by means of the following formula:*

$$Q_c = A \times C \left(\frac{1}{D}\right)^{0.2} \left(\frac{1}{T_{av}}\right)^{0.181} (t_s - t_f)^{1.27} \qquad (11)$$

where

Q_c = convective heat, Btu/h
A = area of surface, ft^2
C = a constant = 1.39 for vertical surfaces
D = height of vertical wall, in (at 24 in this value becomes constant)
T_{av} = average of surface and surrounding air temperature, °F absolute
$(t_s - t_f)$ = temperature excess between warm surface and surrounding air, °F

Some methods of heat pickup from ovens are shown in Figs. 3-9 to 3-11.

Forge Shops

One of the most challenging types of manufacturing processes to condition is forging. The large amount of radiant heat and the complicated nature of the operation may at

*ASHRAE Guide 1958.

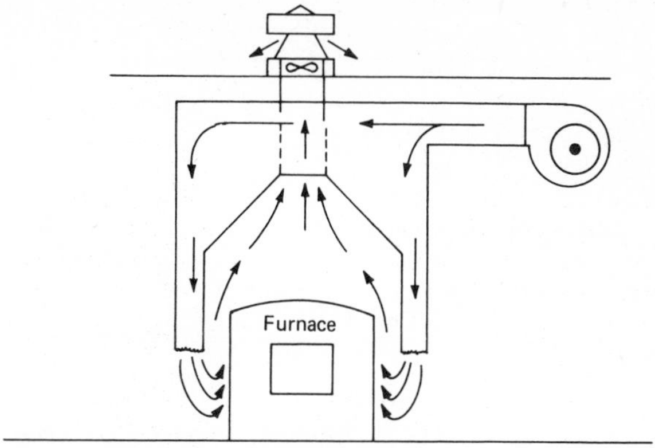

Figure 3-11

first lead one to believe that air conditioning is impossible. This is not true. By proper design of heat shields, installation of exhaust air hoods to exhaust convective heat, and proper introduction of makeup air, a forge shop can be conditioned to maintain reasonable working conditions.

It is possible that some rearrangement of the equipment and storage of finished material may be necessary, but the increased efficiency and the reduction in absenteeism usually makes it economically feasible to provide more acceptable working conditions.

The radiant and convective heat in a forge shop must be estimated in much the same manner as for ovens. The design of suitable reflective shields and exhaust hoods requires ingenuity and careful thought because of the complex nature and variable aspects of the process.

Molding Machines and Foundries

Sensible heat liberated by molding machines and castings in a foundry is usually secondary in importance to the contaminants released. Silica dust, smoke, and fumes are more serious than heat, since their inhalation can be far more injurious to the health of the workers.

Heat shields can be used to absorb the heat of the metal castings. However, the fumes and dusts must be carried away by ventilation to improve the overall working conditions most. This category will be discussed further under "Process Ventilation."

Plastic Molding and Extrusion

Plastic molding and extruding machines are either electrically or steam heated. The heat load from these machines can be obtained from manufacturers' ratings or estimated by means of the following equations.

For electrically heated machines:

$$Q_e = W \times F_1 \times 3.41 \tag{12}$$

where

Q_e = sensible heat, Btu/h
W = electricity, W
F_1 = usage factor = wattage in use/installed wattage
3.41 = a constant to convert W to Btu/h

For steam-heated machines, it is necessary to measure the amount of condensate and its temperature and then convert to British thermal units per hour using the following equation:

$$Q_c = (G \times h_s) + G(t_s - t_c) \tag{13}$$

where

Q_c = heat load, Btu/h
G = pounds of condensate per hour
h_s = heat of vaporization of steam at the temperature and pressure being used
t_s = saturation temperature of the steam, °F
t_c = temperature of the condensate, °F

Since a large amount of the heat in this process is radiant heat, ways and means to absorb it with heat shields and heat-absorbing panels should be thoroughly investigated. Much the same type of thinking and analysis must be used as in the case of forge shops. Seldom is it possible to use the same type of shields in all plants, since the end product may be of a variety of shapes and sizes.

Arc- and Spot-Welding Machines

Arc-welding, spot-welding, and resistance-welding operations all contribute heat to the conditioned space. Although some of the heat from the welded parts may be in the form of radiant heat, it is the usual practice to calculate the load as being the equivalent of the power input to the machines. These machines are usually rated in kilovoltamperes so the load in British thermal units per hour is

$$Q_w = \text{kVA} \times \text{PF} \times 3415 \times F_1 \tag{14}$$

where

Q_w = sensible heat, Btu/h
kVA = kilovoltamperes of the machine
PF = power factor
3415 = constant to convert kVA to Btu/h
F_1 = usage factor, percent of operating time

The British thermal units per hour from the above formula, being all sensible heat, must be absorbed by the conditioned air unless a local supply and exhaust system can be used for each machine (see Fig. 3-12).

In addition to the sensible heat, there are fumes which are given off by the arc-welding process. These fumes must be removed by ventilation air introduced either through

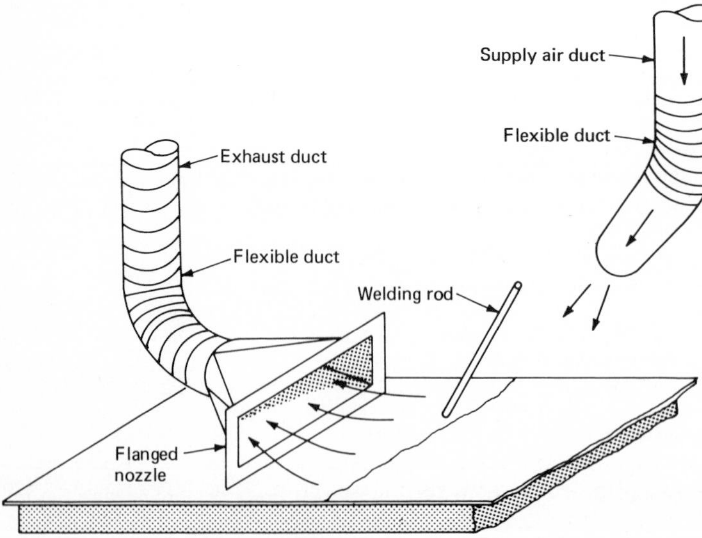

Figure 3-12

the air-conditioning system or locally as shown in Fig. 3-12. Local supply and exhaust will be the most economical from the standpoint of operating cost, since the full load of the ventilating air is not added to the air-conditioning load.

Heat-Treating Ovens

Heat-treating ovens can be considered in the same classification as industrial ovens. As such, they can be analyzed and the load calculated in the same way as in industrial ovens. In many cases, it is possible to isolate the heat-treating area or room so that little or no heat from this area penetrates other departments. This can be accomplished by proper insulation of partition walls, the use of automatically controlled, quick-opening and -closing doors, and proper cooling of the heat-treated parts before they are transported to other departments.

In the heat-treating area itself, the use of heat shields plus proper ventilation can be used to reduce the conditioning load.

Process Piping

Steam piping, hot-water piping, or any type of process piping that conveys hot liquids or gases can be a source of heat unless properly insulated. In some cases these pipes may be concealed in the walls or in furred spaces. Therefore, when making an analysis of an industrial plant for air conditioning, it is particularly important that a thorough examination be made of the plans and premises.

The amount of heat given off by process piping, the variable conductivity of insulation, and the variable temperatures encountered preclude a simplified method of determining the most economical type of insulation to use in every plant.

A more practical method is to determine the savings in owning and operating cost for comparative types of insulation and then to determine the length of time that is needed to pay for the insulation with the savings.

Transformers

The efficiency of a transformer bank is usually high. Since the only heat released to the air-conditioned space is the heat due to inefficiency, the amount is rather small compared to the total capacity of the transformer.

The load from transformers is usually assumed to be all sensible convective heat and can be calculated from the formula

$$Q_T = \text{kVA} \times (100 - \text{eff}) \times 3415 \tag{15}$$

where

Q_T = sensible heat, Btu/h
kVA = kilovoltampere rating
eff = efficiency of the transformer
3415 = a constant to convert kVA to Btu/h

The efficiency of transformers usually runs from 95 to 99 percent. Therefore, a 95 percent efficiency is a safe assumption.

Where the transformer bank is isolated in a separate vault or room, forced outside air for ventilation is usually sufficient to maintain the ambient temperature at an acceptably low level.

If the transformers are located in the air-conditioned space, the heat load must be picked up by the conditioned air. One method of conserving on this load is to use exhaust air to pick up the heat before the exhaust air is discharged from the building.

Processed Material

Materials which have been processed and brought into the space at a temperature higher than the room dry-bulb temperature add heat to the room. The amount of heat added will depend upon the weight, type, temperature, and the length of time the material is held in the conditioned space.

When buildings already constructed and in operation are to be conditioned, it is an easy matter to measure the temperature of the material as it enters and leaves the air-conditioned space. Then, knowing the weight of the material, the specific heat, and the time elapsed between entering and leaving the air-conditioned room, it is possible to obtain the heat given up to the space.

In new plants it is necessary to estimate the temperature of the materials from previous or similar processes. The basic equations for radiation and convection can then be used to obtain the heat transferred to the conditioned space.

Process Ventilation

Spray-Paint Booths

The ventilation rate required for large spray-paint booths is usually 100 to 150 ft³/min per square foot of cross-sectional area, and 75 to 100 ft³/min per square foot of cross-sectional area for small spray booths. Since all of this air must be exhausted to the outside, it requires an equal amount of makeup air introduced through the air-conditioning system or by some other route.

To add this load to the air-conditioning system means an increase in refrigeration tonnage as well as an increase in first costs and owning and operating costs.

For example, a 10-by-20-ft cross-sectional area spray-paint booth requires a minimum of 20,000 ft³/min of exhaust air. If the outside air–design conditions are 95°F db and 78°F wb, and the inside conditions are to be 80°F db and 68°F wb, the load in tons is

$$\frac{\text{ft}^3/\text{min} \times 4.5 \times (h_{oa} - h_{ra})}{12{,}000} = \text{tons} \tag{16}$$

or

$$\frac{20{,}000 \times 4.5 \times (41.2 - 32.3)}{12{,}000} = 66.75 \text{ tons}$$

where

$41.2(h_{oa})$ = enthalpy of the outside air
$32.3(h_{ra})$ = enthalpy of the room air

On the basis of $500 per ton for air conditioning, this adds $33,375 to the initial cost of the air-conditioning installation.

By introducing outside air directly in front of the spray booth and allowing this air to be exhausted, the necessity of introducing outside air through the air-conditioning equipment is eliminated (see Fig. 3-13). The air thus introduced is not conditioned for summer operation but must be tempered in the winter to prevent cold-air circulation over workers in the spray booth.

In some cases, outside summer temperatures may reach levels that are uncomfortable if outside air is brought directly into the spray booth. In these cases, consideration should be given to precooling the outside air with condenser water or taking a portion of the air in through the air-conditioning system and a portion directly from the outside.

Arc-Welding Fumes

The exhaust air necessary to carry away fumes from arc welding can be replaced by introducing outside air through the air-conditioning system or by local exhaust. The choice depends upon the particular application.

When circumstances are such that makeup air must be supplied through the air-conditioning system, the amount of air exhausted and, consequently, the amount introduced can be estimated from the number of welders aand the size of welding rod used. Table 3-5 will assist the designer in estimating the required quantity of exhaust air and makeup air.

For local exhaust, the quantity of exhaust air depends mainly on how close the local exhaust is to the welding operation. Good design dictates that a local exhaust hood, with 1500 ft/min entrance velocity approximately 6 in from the point of welding, should exhaust at least 250 ft³/min. For distances greater than 6 in, the air quantity should be

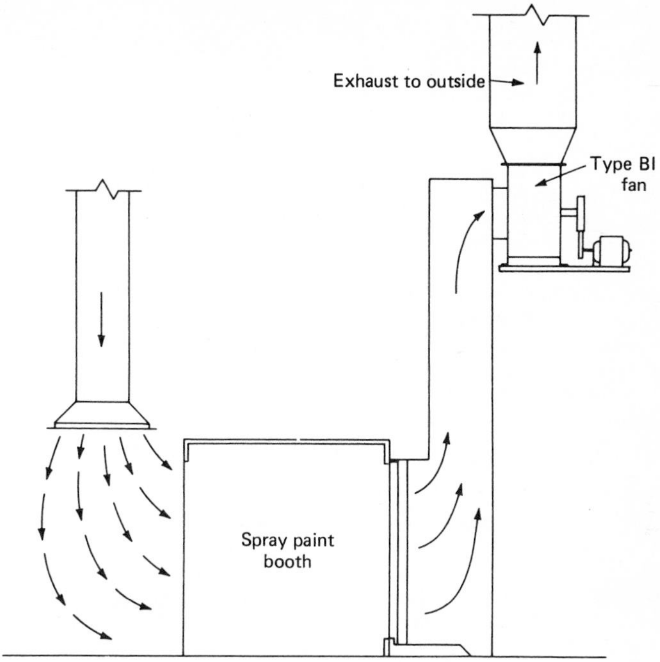

Figure 3-13 Direct air exhaust.

TABLE 3-5 Exhaust Air Requirements

Welding rod diameter, in	Airflow, ft³/min per welder*
5/32	1000
3/16	1500
1/4	3500
3/8	4500

*American Conference of Governmental Industrial Hygienists.

increased according to the ratio of the squares of the distances—up to a total of 1000 ft^3/min per welder or approximately 12 in from the point of welding. At greater distances, Table 3-5 should be used.

When the exhaust air must be replaced by the introduction of outside air through the air-conditioning system, the load on the air conditioner can be calculated using Eq. (16).

Odor-Producing Processes

In manufacturing plants processes which produce offensive odors have long been a nuisance to workers and residents in the vicinity. Although control of temperature and humidity for air conditioning may seem remote from odor-producing processes, there is a definite connection between the two—since the definition of true air conditioning includes control of the purity or cleanliness of air, as well as temperature, humidity, and air motion. In addition, test data indicate some odors are more offensive in an atmosphere of high humidity than in one of low humidity.

One of the most common methods of removing odors in manufacturing plants is to exhaust the odors to the outside along with enough air to dilute them—thereby making the odors less offensive and preventing them from pervading the balance of the manufacturing area. In some cases, the mixture is discharged at a height that will dilute the odors and prevent their being offensive. Unfortunately, this does not always work. Atmospheric conditions and downdrafts may bring the odors down before they have been properly diluted. Also, this is quite costly, since all the air exhausted must first be brought in from the outside and conditioned, especially in the winter. In some cases, local exhaust can also be used, but is not always effective with respect to odors.

Substances such as activated charcoal or chemical sprays are often used to absorb odors. With this arrangement, only the outside air required for ventilation purposes is introduced. The return and exhaust air is circulated through the activated charcoal filters or chemical spray. Thus, the outside air load is held to a minimum.

Still another method recently adopted uses *reodorants.* This consists of introducing a masking agent into the odor-containing atmosphere, resulting in a nonoffensive scent. The masking agent must not be so intense that it, in itself, creates objectionable odors. Odor masking is a process by which one odor is cancelled by superimposing another odor. A great amount of research is still required in this area to make it generally applicable.

It is apparent that with this arrangement, it is necessary to introduce only the amount of air required for ventilation purposes. The load on the conditioning equipment is, therefore, much lower than when large quantities of outside air are used to dilute odors.

Dust- and Lint-Producing Processes

The basic difference between dust- and lint-producing processes and processes in which fumes or odors are produced is that enough air must be echausted not only to eliminate the dust and lint but also to carry the particles through the exhaust system and prevent them from accumulating. The amount of air required depends upon the size, weight, quantity, and the allowable concentration of the particles. The selection of the air quantity and the design of the exhaust system are too extensive to cover in detail. The primary purpose of mentioning them is to point out the importance of local supply and exhaust in dust- and lint-producing processes. This cannot be overemphasized. Not only does it reduce load, and thereby reduce initial and operating costs, but the maintenance of return air filtering equipment is significantly reduced because of lower dust and lint accumulation on filters.

Some of the operations that can be listed as dirt- and lint-producing processes are woodworking, grinding, drilling and cutting of castings, and foundry operations. Of these, the foundry operation is the most intricate because heat, gas, and dust are liberated in the molding, pouring, and shakeout operations. Where the foundry is completely

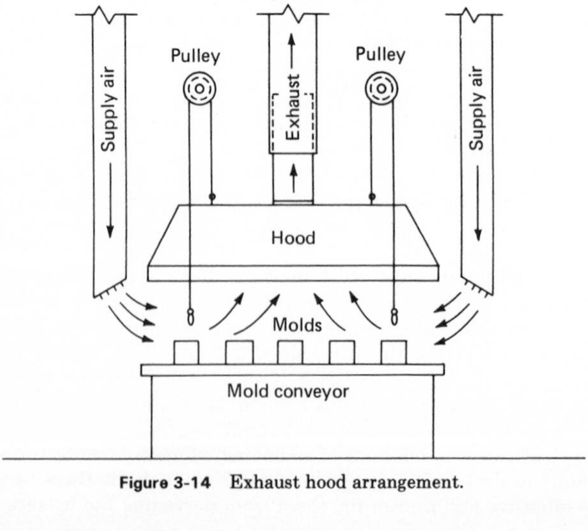

Figure 3-14 Exhaust hood arrangement.

mechanized, with automatic molding machines, conveyers, etc., it is not a difficult task to provide the necessary supply and exhaust. With unmechanized foundries, it is a matter of providing flexible exhaust hoods and supply ducts to the area to be isolated (see Fig. 3-14).

Liquid Mist- or Fog-Producing Processes

Operations that require spraying of liquids (other than water), that produce a mist or fog not absorbed by the air, can be treated in exactly the same manner as spray-painting booths. Most of the information under that heading applies here as well.

One interesting, though specialized, method of preventing misting has been developed for application on newspaper presses. The ink spray from the rollers of these presses has been a difficult problem. This method uses a charged wire located adjacent to each roller. The wire's charge is opposite to that put on the rollers. Ink, sprayed from rollers at high speeds, is given a charge opposite to that of the rollers. This causes the ink to be immediately attracted to the rollers, preventing it from entering the air surrounding the press. All indications are that it performs satisfactorily.

Stratification

Most of the heat generated in a building, or entering a building from external sources, will collect, or stratify, near the ceiling. This fact has been used to advantage in buildings which have high ceilings.

By making use of the stratified area, the required supply-air quantity can usually be reduced. The reason for the reduction in air quantity is that the transmission through the roof can be assumed to be absorbed in the stratified area and eliminated. Also, the solar load through the roof can be reduced by approximately 40 percent. However, when stratification is used, the supply air must be introduced approximately 10 to 12 ft above the floor level, and the return air withdrawn at a height not to exceed 10 to 12 ft from the floor level. A building with ceiling heights of less than 15 ft does not lend itself readily to stratification because of the truss construction and the height required to suspend the air conditioning. If the lights are installed above the stratification zone, approximately 40 percent of the heat generated by the lights can also be eliminated. It has been demonstrated that with proper distribution of the supply air, some of the heat from process work can also be eliminated; however, the approximate amount has yet to be determined through further testing.

Stratification is also an important consideration in installations which require large quantities of outside air to replace the air which must be exhausted. In the design of a system of this type, the supply air is introduced at about the 10-ft level. This air then picks up the heat generated in the occupied space. A small amount of return air is picked up at the same level as the supply. Some of the air to be exhausted is allowed to circulate up to the roof, where it is exhausted after picking up some of the heat from the lights, some solar heat, and heat transmitted through the roof.

The fact that stratification does occur in industrial plants can also be used to good advantage during the heating season. By the use of a runaround cycle, using one coil in the exhaust air to pick up the heat and another coil in the incoming outside air to transfer the heat to this air, with a circulating fluid between the two coils, a large portion of the heat can be recovered for use in heating the plant.

CONCLUSION

Although the list of process loads discussed is not completely exhausted, an attempt has been made to cover areas that are most frequently dealt with. From the examples given, it should be obvious that process loads differ widely, as does the method of handling them. One fact, however, seems to hold true in most situations:

Whenever possible, it is usually more efficient and less expensive to handle process heat and humidity separately from the regular air-conditioning system.

section 6

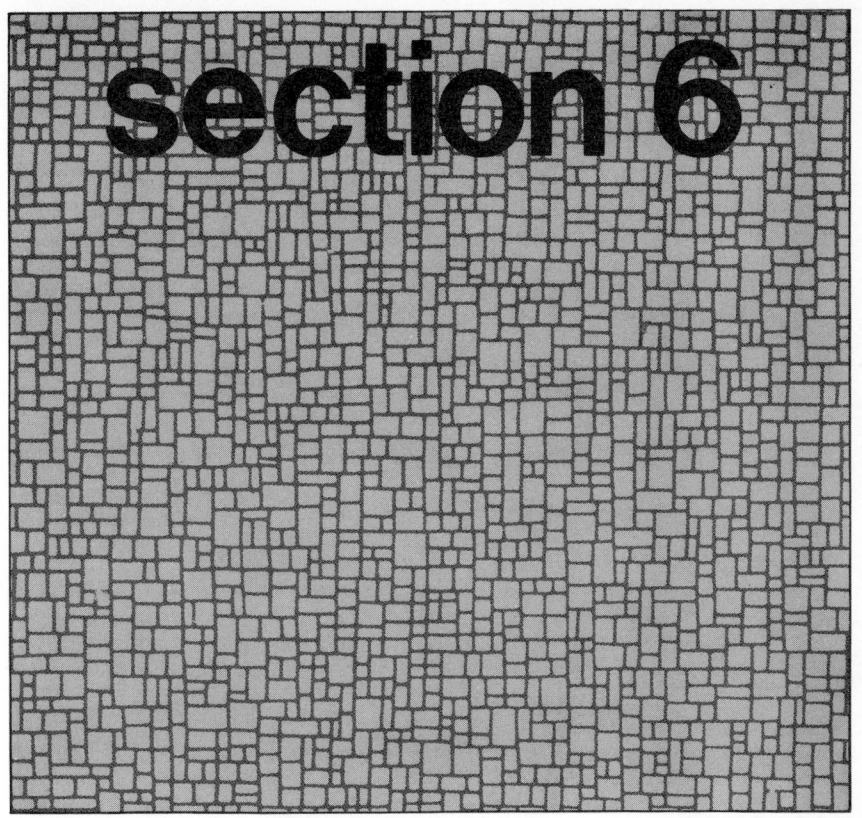

Water: Sources, Use, and Disposition

chapter 6-1

Sources of Supply

by
Gordon F. Leitner
Vice Chairman
Water Services of America, Inc.
Milwaukee, Wisconsin

USES OF WATER IN THE PLANT

Water is essential to enable any industrial plant to fulfill its mission. The amount of water required for a commercial enterprise or industrial plant can vary from as little as 25 gal (95 L) per person per day to many thousands of gallons per employee per day. The functions or uses of water in the plant will usually fall into one or more of the following categories:

- Personal use by employees or customers, including drinking, and use as a medium of transport for waste disposal
- Preparation of food or other manufacturing processes
- Comfort and process cooling: In this application it can be used directly for cooling via a suitable heat exchanger, or as makeup water for an open recirculating stream in which the heat absorbed by the water in comfort or process cooling is dissipated to the atmosphere via a cooling tower or evaporative condenser.
- Comfort heating: The heat from a given water source is raised to usable levels by a heat pump.
- Makeup for steam generators or plant processes: In this case the water may require pretreatment, ranging from basic filtration and/or chlorination or ozonization to elaborate decontamination and demineralization or distillation processes.
- Plant waste disposal: In this case water is usually used as a medium of transport, both as a rinsing medium and as a medium to transport the material removed by rinsing to a processing system or to a disposal sewer line.

SOURCES OF WATER

The selection of a source of water to meet the needs of a particular plant will depend upon several factors:

- Quantity of water required
- Quality of water required; often several different water qualities are required, and this may suggest consideration of more than one source
- Cost of water from available sources
- Quantity available from each source

The first source to consider is the local municipal supply. If the available quantity, quality, or the cost of this source is unacceptable, then these additional potential sources should be investigated:

AUDIT FOR
WATER SUPPLY REQUIREMENT

Plant Location:

Date of Audit:

Source:* (1) ________ (2) ________ (3) ________

Requirement	Source	Gallons requirement				Gallons waste-water effluent
		Quantity	Avg. daily	Total daily	Peak consump-tion	
Personal						
Fountains						
Urinals						
Toilets						
Showers						
Cleaning						
Kitchen or Commissary						
Food Preparation						
Cleanup						
Dishwasher						
Ice Production						
Drinking Water						
Laundry						
Washing Machines						
Heating & Cooling Plant						
Boiler Makeup						
Cooling Tower Makeup						
Humidifier Makeup						
Equipment Cooling						
Process Requirement						
Water Addition to Product						
Rinse						
Cleanup						

**NOTE:* Source (2) or (3) may be separate supply sources or the basic source—with treatment.

Figure 1-1

- Rivers
- Wells
- Lakes or ponds, natural or constructed
- Seawater (including seawater desalination if required)
- Effluent from wastewater-treatment plants
- Ponds: Pond development requires consideration of loss due to infiltration and evaporation. Infiltration losses can be controlled by plastic liners; several types are commercially available. Infiltration can also be controlled by the addition of certain additives to the water to seal the ground. Evaporation, too, can be controlled but at substantial additional cost.
- Rainwater-collecting areas: in certain parts of the world, particularly in some island communities, water-collecting reservoirs are used. These are sometimes built with concrete liners and cover the entire side of a hill. In other cases, rainwater from roofs is collected in ground-level cisterns.

CALCULATION OF REQUIREMENTS

The increasing requirement for conservation of existing water supplies as well as economic factors dictates the necessity of a periodic water audit. In recent years, water costs have escalated faster than inflation. This trend can be expected to continue in the future for several reasons: increasing costs for development of new water sources to meet increasing demand, increased cost of energy for pumping, increased treatment costs, and the growing practice of relating charges for wastewater discharge to the amount of fresh water supplied.

In preparing a water audit for an existing plant or in the design of a new plant, the completion of a summary sheet similar to Fig. 1-1 will provide the basic planning information necessary.

BIBLIOGRAPHY

The Water Encyclopedia, Water Information Center, Port Washington, N.Y., 1970.

chapter 6-2

Purification and Treatment

by

E. G. Kominek
Fluor Engineers & Constructors, Inc.
Irvine, California

Robert W. Okey
S. D. Heden
Eimco Processing Machinery Division
Envirotech Corporation
Salt Lake City, Utah

GLOSSARY*

Aeration Intimate contact between air and water.

Chemical coagulation Destabilization of colloidal and finely divided suspended matter with floc-forming chemicals.

Chlorine demand The amount of added chlorine chemically reduced by organic substances in water at the end of a specified contact period.

Clarification The processes used to reduce suspended solids in water.

Deaeration The removal of oxygen from water.

Dechlorination The reduction of residual chlorine by chemical or physical processes.

Filtration Passing water through a filtering medium to remove suspended solids.

Flash mixer A device for dispersing chemicals uniformly.

Flocculation The agglomeration of colloidal and finely dispersed suspended matter after coagulation by gentle stirring.

Ion exchange The process by which ions from two different molecules are exchanged through the use of ion-exchange materials.

Lime-soda softening A process whereby calcium and magnesium ions are precipitated from water by reaction with lime and soda ash.

Plain sedimentation Separation of suspended solids unaided by chemicals.

Pressure filter A closed, granular-medium filter inserted in a pressure line.

Recarbonation The introduction of carbon dioxide into water for pH adjustment.

Sludge dewatering Concentration of sludge by draining, evaporation, pressing, vacuum filtration, centrifuging, settling, or flotation.

Stabilization Any process that will minimize or eliminate scale-forming tendencies of water.

INTRODUCTION

The unit operations for water purification and treatment include:

Sedimentation	Filtration
Disinfection	Coagulation
Ion exchange	Activated-carbon treatment
Lime-soda softening	Reverse osmosis

The industries that are major users of water are:

Power generation	Petroleum
Beverage	Pulp and paper
Aluminum	Chemical
Iron and steel	Food
Textile	

Major water uses are for drinking, boiler feed, industrial processes, and cooling-tower makeup.

Figure 2-1 illustrates the unit operations most commonly used for treating raw water.

*Definitions from *Glossary of Water and Waste Control Engineering*—APHA, ASCE, AWWA and WPCF, 1969. Reprinted by permission.

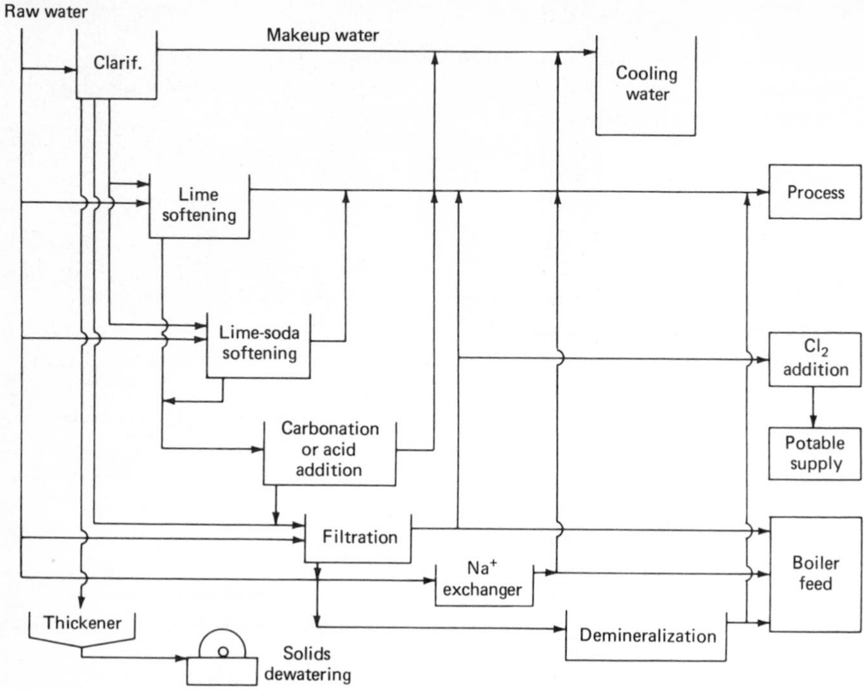

Figure 2-1 Water-treatment diagram.

It also illustrates what can be done to treat water for reuse and to approach a zero effluent for compliance with proposed U.S. Environmental Protection Agency Standards.

IMPURITIES IN WATER

All natural waters contain suspended or dissolved inorganic or organic chemicals to some degree. Whether they are present in high enough concentrations to be considered impurities depends on the water use(s).

The impurities can be classified as:

Inorganic	*Organic*	*Biologically Active*
Suspended	Suspended	Bacteria
Colloidal	Immiscible	Viruses
Dissolved	Miscible	Algae
	Soluble	Protozoa

In turbulent streams, suspended solids range from small pebbles down to colloidal clay particles 10^{-5} to 10^{-7} cm in diameter. The water may also contain organic solids, algae, and bacteria. Dissolved inorganic solids are usually bicarbonates, sulfates, and chlorides of calcium, magnesium, and sodium as well as compounds of silica, iron, and manganese. Nonferrous metals and organic compounds may be present in low concentrations which nevertheless exceed the EPA's proposed limits.

The biggest problem caused by dissolved solids in water is *hardness*, i.e., the presence of calcium and magnesium compounds. Other dissolved solids may be considered as impurities, depending upon their concentrations and the intended water use.

SEDIMENTATION PROCESSES

Sedimentation processes are used to clarify turbid and/or colored waters and to remove dissolved impurities such as iron, manganese, calcium, and magnesium compounds as well as silica and fluorides.

Removal of Suspended Solids

On a few occasions plain sedimentation will be employed to remove suspended solids, but usually the process will include the addition of chemicals to improve removal of solids. The particles suspended in water result in a turbid, or colored, appearance that is objectionable, and they have a static charge (usually negative) that causes the particles to repel each other and remain suspended. By adding certain chemicals it is possible to neutralize these charges, permitting the particles to agglomerate and settle from the liquid more effectively. The current practice is to refer to this neutralization or destabilization step as *coagulation* and the subsequent gathering together of the particles into larger, more separating "flocs" as *flocculation.* Inorganic chemicals such as aluminum sulfate (alum), ferrous sulfate (copperas), ferric chloride, and sodium aluminate, as well as a long list of organic polymers, are used for coagulating and flocculating suspended matter from water. Coagulation takes place very quickly—a few seconds to one minute—and the chemicals should be added with intense mixing in order to obtain maximum efficiency from their use. Flocculation, on the other hand, normally requires detention periods of from 20 to 45 min, and, once the chemicals have been added, should be accomplished by relatively gentle mixing. The mixing is carried out in a flocculation basin, and its purpose is to bring about the maximum collisions between the suspended particles without shearing or breaking apart the particles that have already been formed.

This will limit the particle velocities to the range of 1 to 6 ft/s (0.305 to 1.83 m/s), depending upon the "toughness" of the floc that is produced. Equipment should be provided for varying the intensity of the mixing so that the optimum velocities can be achieved.

Organic polymers (polyelectrolytes) are also frequently used as flocculant aids. They facilitate the gathering of the already coagulated or destabilized particles into larger and less fragile floc particles that have better settling characteristics. The type and amount of chemicals required to treat a given supply of water are best determined by laboratory jar tests. These tests should be made on fresh samples at the same temperatures and other conditions that will be present in the full-scale plant. No other reliable method has been found for predicting the best chemicals and optimum dosages for clarifying a water on the basis of a study of the physical and chemical water analyses. Typically, inorganic coagulants (alum, ferric chloride, etc.) might range from 10 to 100 mg/L, cationic organic coagulants from 1 to 5 mg/L, and anionic organic flocculants from 0.1 to 1.0 mg/L. Many suppliers of chemicals and equipment manufacturers will perform jar-test studies at reasonable fees and make recommendations as to the best chemicals, equipment, and sizing for treating a water supply.

Sedimentation, with or without chemical treatment, is usually carried out in continuous, flow-through settling units with horizontal flow patterns (Fig. 2-2). Important exceptions to this are the solids contact clarifiers and reactors which are discussed later. Settling units are normally rectangular, with the flow along the length of the basin, or circular, with the flow radially outward from a central inlet compartment. Mechanical scraping mechanisms are employed to move the settled sludge along the bottom to hoppers or sumps from which it is discharged. The mechanical design of the scraper mechanism should be carefully evaluated when considering such equipment, as should the inlet distribution and effluent collection system. The settling zone of the unit must provide sufficient area and volume so the bulk of the solids will settle before reaching the effluent collector. Proper design of the inlet distributor and effluent collector will result in the most effective operation of the settling zone. Typical settling-area designs would

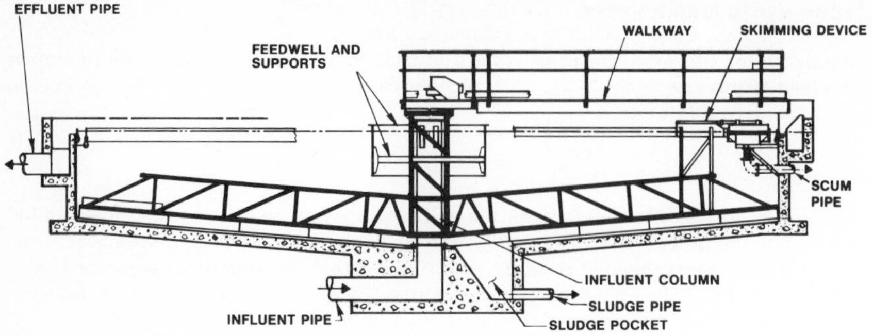

Figure 2-2 Illustrative section of settling unit.

allow liquid flow to range from 0.35 to 0.75 gal/(min) (ft^2) (0.85 to 1.80 m^3/h) and settling-unit detention times from 2.5 to 4.0 h.

Temperature and changes in temperature of the liquid being treated are extremely important considerations in the design and operation of sedimentation units. The rate at which a particle settles in water is inversely proportional to the kinematic viscosity, a property that varies with the temperature. Thus, the settling rate of a given particle at 40°F (4.4°C) is only 63 percent of what it would be at 70°F (21.1°C). Rapid changes in inlet water temperatures to settling units will cause thermal currents, which at best are disruptive to the settling of particles and at worst are totally upsetting. Manufacturers typically limit changes to 2°F (1°C) per hour in their performance warranties. The rate at which chemical reactions proceed in water is higher at higher water temperatures. A common rule of thumb is a doubling of the rate for each 10°F (4°C) increase in temperature. In general, warm but constant water temperature is desirable in treating water.

Solids contact units combine the coagulation and flocculation function within the settling unit together with the ability to internally recirculate solids that have been formed by earlier reactions (Fig. 2-3). Besides the economies attainable by combining these functions within a single unit, improvements in settling characteristics and reaction rates permit higher design ratings for this type of unit. Most industrial water-treatment applications now use some form of solids contact or combination treatment equipment. A variety of units are available from manufacturers of such equipment, and a careful comparison of the process and the mechanical features of each is advisable. The American

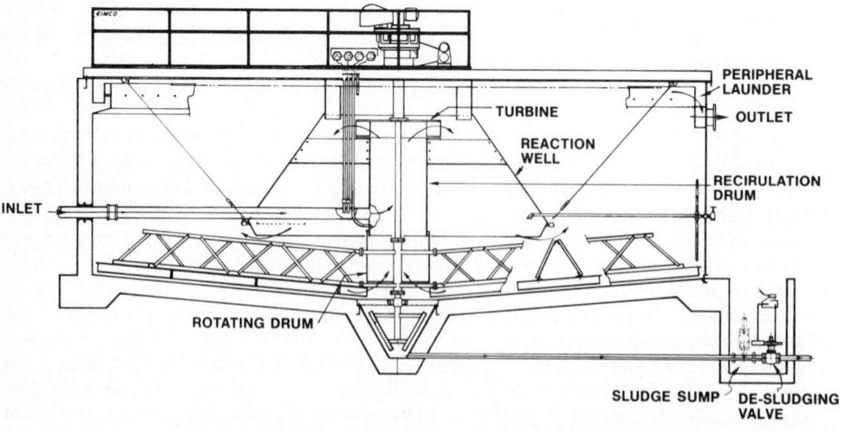

Figure 2-3 Illustrative section of solids contact unit.

Water Works Association (AWWA), the American Society of Civil Engineers (ASCE), and the Conference of State Sanitary Engineers (CSSE) have provided guidelines for the design of solids contact units as well as for other sedimentation and water-treatment processes.

Removal of Dissolved Impurities

Up to this point we have discussed sedimentation processes primarily from the standpoint of removing suspended materials from water. The removal of dissolved mineral impurities such as in the lime or lime–soda ash softening process is an equally important aspect of sedimentation in the treatment of water. In the softening process, hydrated lime or lime and soda ash are added to react with the dissolved CO_2 and the calcium and magnesium salts that commonly cause the hardness of water. The following equations describe some of the reactions that take place in the formation of the calcium carbonate and magnesium hydroxide precipitates:

$$CO_2 + Ca(OH)_2 \longrightarrow CaCO_3 + H_2O$$
$$Ca(HCO_3)_2 + Ca(OH)_2 \longrightarrow 2CaCO_3 + 2H_2O$$
$$Mg(HCO_3)_2 + 2Ca(OH)_2 \longrightarrow 2CaCO_3 + Mg(OH)_2 + 2H_2O$$
$$CaSO_4 + Na_2CO_3 \longrightarrow CaCO_3 + Na_2SO_4$$

A coagulant is normally added along with the lime and soda ash to improve clarity of the product water. When the lime requirements are high, economies can usually be realized by feeding quicklime (CaO) and using a lime slaker to convert it to the hydrated lime [$CA(OH)_2$]. When requirements reach approximately 200 lb/h (90 kg/h), the economies of quicklime should be investigated.

Neither calcium carbonate nor magnesium hydroxide is completely insoluble, so some amount, depending upon the type of treatment, temperature, and other conditions will remain in solution. Unlike most compounds, $Mg(OH)_2$ and $CaCO_3$ are less soluble at higher temperatures, rather than more soluble. In addition to the viscosity and reaction rates mentioned earlier, this is a further advantage to treating warm water. The hot-process softener, sometimes used for treating boiler feedwater, uses steam to heat the water to more than 200°F (93°C) and takes advantage of these factors to produce a water of lower hardness in smaller-sized units than could be used in a cold process. These high-temperature units also accomplish silica (SiO_2) removal, which is often necessary when treating boiler feedwater. This is described in further detail later.

As already mentioned, solids contact units are particularly well suited to lime-soda softening applications. Calcium carbonate and, to some degree, magnesium hydroxide have a tendency to supersaturate (remain in solution at considerably higher than theoretical concentrations), and solids contact operation reduces this tendency. By mixing the chemicals and the untreated water in the presence of recycled solids, a large surface area is provided on which the precipitate will form. This "seeding" not only reduces the supersaturation, but also results in the growth of larger particles that settle more rapidly. Overall, chemical usage efficiency is improved because the lower calcium and magnesium values attained through solids contact require no additional chemicals.

Hot-process lime treatment has been used to remove silica from boiler feedwater where extremely low concentrations requiring ion exchange are not necessary. The silica is removed by adsorption on freshly precipitated magnesium hydroxide. The removal is therefore dependent upon the amount of $Mg(OH)_2$ precipitated. The effectiveness of the process is enhanced by the high temperatures and can be further improved by solids contact. Often there is insufficient natural magnesium available in the water, so it is supplemented by the addition of dolomitic lime, which contains a high percentage of magnesium oxide. This is feasible in a high-temperature process where the magnesium hydroxide will hydrate, something it will not usually do at normal water temperatures.

Cold lime-treatment processes are now being employed to reduce silica for cooling-water systems in specially designed solids contact units. By providing intense mixing and pumpage for high solids contact concentration and long detention times, the silica can be reduced at cold-water temperatures to levels formerly attainable only in hot-process treatment. Some of these systems use a two-stage sedimentation process where lime treatment at a high pH is used in the first stage. Soda ash and CO_2 (carbonic acid) are added in the final stage for stabilization.

Tube and Plate Settlers

Tube and plate settlers have been used for about 20 years as sedimentation units. There are a variety of designs, but basically all of them use inclined surfaces with relatively close spacing. The water to be treated is passed between the surfaces at velocities which permit suspended solids to settle and to coalesce on the lower tube or plate surface. The angle of inclination in excess of 45° causes the settled solids to slide downward into a sludge-concentration compartment located at the bottom of the treatment unit.

These units permit efficient clarification at detention times substantially less than those used in conventional clarifiers. Total settling detention times may be as low as 10 min, greatly reducing the plant size.

The units are cost-effective for many applications, even though the inclined tubes or plates add appreciably to the equipment cost. These settling units are best for discrete solids settling where detention time is not a significant factor.

GRANULAR-MEDIA FILTRATION

Granular-media filters are used in water treatment to remove relatively small amounts of suspended matter where a high degree of clarity is needed. The solids are collected in the filter until such time as the pressure drop across the unit becomes excessive or the unit begins to "break through" and pass excessive solids into the effluent. When this point is reached, the unit is backwashed at high flow rates to flush out the collected solids and the filter is returned to service. Normally the water volume required for washing is less then 5 percent of the filtered water produced. Usually the water will have been chemically treated with coagulant and/or flocculating chemicals, but this is not always necessary. Frequently the filters will be downstream in the process from a sedimentation unit in which the chemical conditioning has taken place.

There are a limited number of filters that operate in the upflow mode, but the majority operate downflow during the filtering cycle and upflow during the backwash cleaning cycle. A further general differentiation is between the gravity and enclosed-pressure filters. The choice between these two is largely dependent upon where the filters fit into the process flow diagram, and pumping considerations are of particular importance.

Some state boards of health will require that any potable water filtration, particularly if it is a surface water supply, be done in open-top gravity filters. However, for most industrial applications either type of filter will produce effluents of similar quality if they have the same filter medium configuration.

The internals of a filter, whether it be steel or concrete, pressure or gravity, open or enclosed, will consist of an inlet distributor and backwash collection system, the granular-media beds, and the underdrain system. Most water filters now have two and sometimes more layers of granular media for removing the suspended solids, and in many cases a layer of gravel is provided to support the filter media, prevent them from entering the underdrain system, and properly distribute the backwash water. Current practice is to so select the size and chemical conditioning of the various media that solids are collected through the entire depth of the filter rather than on the top few inches. For that reason, a coarse layer of lighter material (1-mm coal) is usually installed over a fine layer of heavier material (0.5-mm sand). The coarser upper layer collects most of the suspended solids and is usually about 2 ft (0.61 m) thick. The finer layer acts as a barrier to prevent solids breakthrough into the effluent and is normally on the order of 1 ft deep. There is considerable latitude in the selection of sizes of media, depending upon the nature of the solids being removed and the results desired. Generally, the objective is to make use of the full depth of the top layer for collecting solids without going through to the second barrier layer which will seal off rather quickly. As has been mentioned, underdrain systems that have large openings through which the media could pass require a graded gravel supporting bed. Many filters now use underdrain nozzles with openings designed to retain the media particles; this considerably simplifies installation of the beds and avoids any potential of disrupting the supporting gravel beds.

Filter designs that incorporate air-wash cleaning have special provisions for distributing the air evenly across the bed area during the period of air wash. This is accomplished either by an air piping grid or by specially designed underdrain nozzles with tail pipes and orifices that meter the air into each nozzle.

Filters can be backwashed with only water, with air followed by water, or with air and water together followed by water alone. If water alone is used, it should be supplemented with a surface water wash. The equipment used for this is usually a rotary jet employing a separate high-pressure water supply. When air is employed, the water level in the filter must first be lowered to the point that no medium is lost through the backwash collection system during the vigorous air agitation. When air and water wash are used simultaneously, the water level is usually first lowered to the top of the medium and the air is applied along with the water at a reduced backwash rate of backwash water. A very popular style of filter incorporates a backwash storage compartment above the filtering compartment (Fig. 2-4). This is an economical as well as easily maintained and automated system for many industrial filter applications.

Flow rates through filters are controlled either at the upstream side or on the effluent side. Another type, less frequently used, operates without any constant flow control and actually operates as a declining-rate filter as the pressure drop across the filter gradually increases. All types if properly designed can be equally effective.

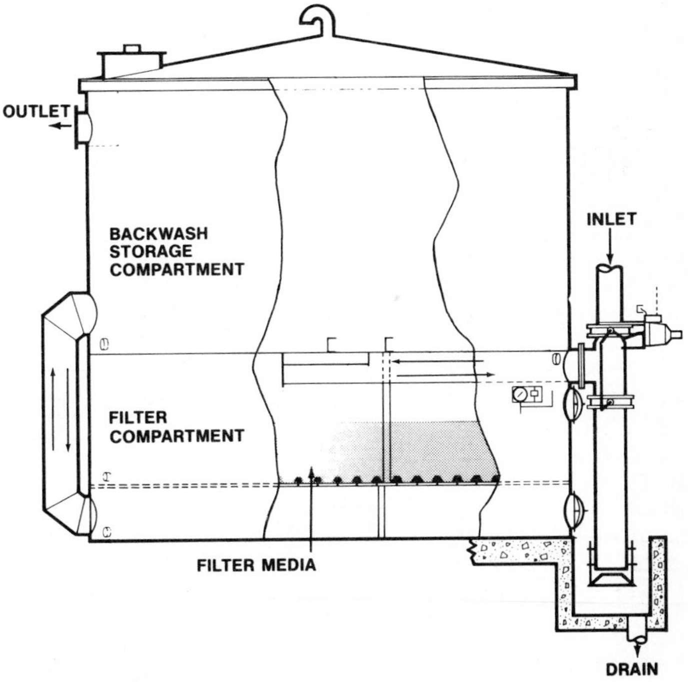

Figure 2-4 Granular-media filter.

APPLICATIONS (FLOWSHEETS)

Figure 2-5 depicts many treatment systems that can be developed by applying the various treatment methods that have been discussed. Many options are shown, and the most suitable can be selected on the basis of raw-water characteristics and the treated-water requirements. The diagram is by no means complete as to methods or combinations possible.

Obviously, raw-water supplies vary widely in quality and, although not indicated on the diagram, can be used without any treatment in some cases. Potable supplies will all require disinfection with chlorine or ozone as a minimum. When selecting any system to be used for potable purposes, the local governing health departments should be consulted so all their requirements are met.

The minimum water quality required at each point of use must be established and the method or series of treatments selected that can conservatively meet these requirements under any and all raw-water conditions, including temperature. Consideration must be given to the degree of automation desired and the quality and training of operators and maintenance personnel.

Chemical feeders, controllers, and instrumentation are invariably a part of water-treatment installations. These accessories are probably the source of most of the operational and maintenance problems, and the same care should be given to their selection as to the major items of equipment.

Water treatment is a continuous process operation that requires some chemical control testing; the degree of complexity depends upon the process. Qualified personnel must be assigned to this task—preferably laboratory technicians, if they are available, since they are skilled in running similar tests.

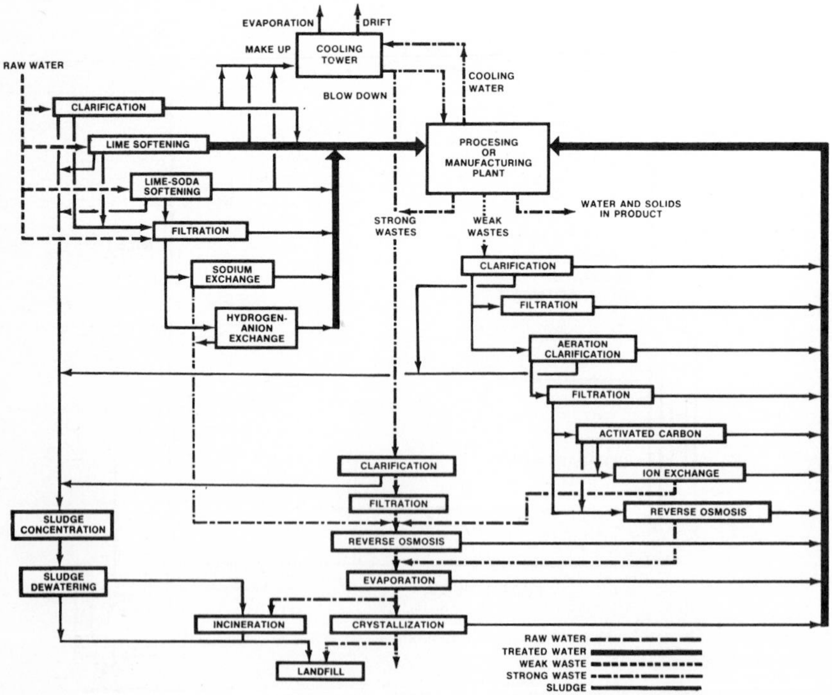

Figure 2-5 Diagram of water and waste treatment for zero effluent.

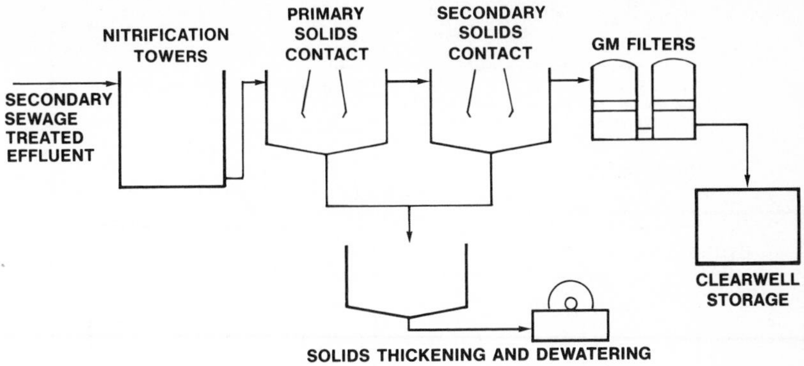

Figure 2-6 Flow diagram for reuse treatment.

WATER RECYCLING AND REUSE

A reuse-treatment plant is usually nothing more than a water-treatment plant using as a source of supply the discharge from either a domestic or an industrial wastewater treatment plant. Figure 2-6 is a flow diagram of such a plant using treated domestic sewage as a source and treating it further for use in an electrical generating station. Except for the nitrification towers which are included because of a considerable savings in operating chemicals, the balance of the plant is a typical water-treatment plant.

Waste-treatment plants, particularly those using biological treatment, are subject to upsets on occasion, and this can be troublesome to the operation of the reuse-treatment plant. The design of a reuse treatment plant should include adequate storage, recycle, or bypass facilities needed for the inevitable upsets that will occur upstream.

ION EXCHANGE

Applications of ion exchange for softening, dealkalizing or demineralizing water include municipal water supply, boiler feedwater, and industrial process water as well as treatment of industrial process water and boiler feedwater.

Sodium exchange substitutes sodium ions for all cations having two or more positive charges. When water containing salts of calcium and magnesium is passed through a sodium exchanger, the sodium ions of the bed replace them to produce an effluent of close to "zero" hardness. When the exchange resin approaches exhaustion, a salt solution is used for regenerating it. The theoretical salt requirement is about 0.17 lb (0.08 kg) per 1000 grains of hardness removed.

It is possible to obtain higher exchange capacities by increasing the amounts of salt used per regeneration; however, for average industrial applications, the amount of salt used with the high-capacity resin cation exchanger is usually 0.4 lb/1000 gr (0.18 kg/1000 gr).

When regenerated with acid, cation-exchange resin will exchange hydrogen for metallic cations such as calcium, magnesium, and sodium. As the cations are exchanged for hydrogen ions, the bicarbonate, sulfate, nitrate, and chloride anions are converted to their respective acids. When the exchanger approaches exhaustion, it is backwashed and regenerated with acid. After rinsing, the bed is again ready for use.

For hydrogen exchange, either sulfuric or hydrochloric acid is used. Sulfuric acid is generally used because of its low cost, but hydrochloric acid is employed when the high-calcium water would result in the precipitation of calcium sulfate in the exchanger bed.

Depending upon the raw-water composition and the desired effluent composition, the

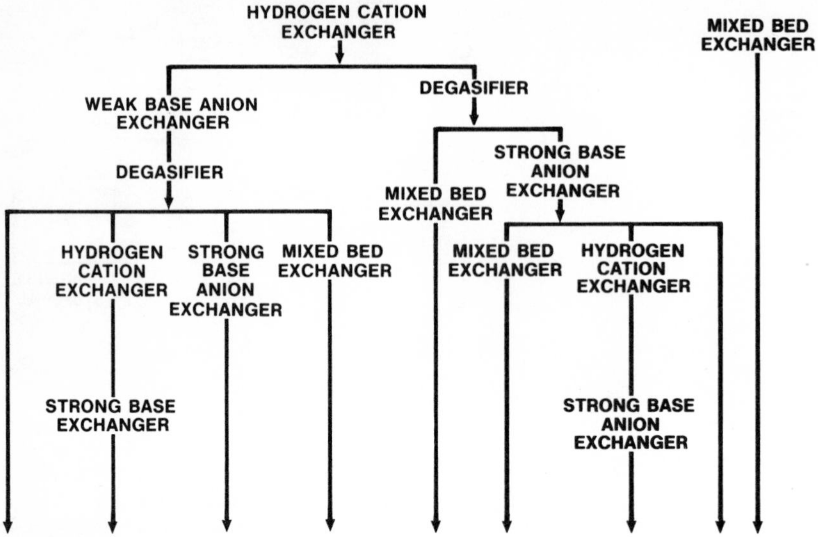

Figure 2-7 Alternatives for a demineralization system.

operation of a sodium exchanger in parallel with a hydrogen exchanger will produce an effluent which approaches zero hardness and also has a low alkalinity.

The terms *demineralization* or *deionization* refer to the removal of cations and anions from water by cation-anion exchange. The cations are removed in hydrogen exchanger(s). The effluent from the hydrogen cation exchanger contains acids equivalent to the anions present in the untreated water.

Anion-exchange resins may be weakly or strongly basic. The weakly basic resins will remove the strongly ionized acids but not the weakly ionized ones. The effluent contains the same amount of silica as the influent water plus carbon dioxide equivalent to the bicarbonate alkalinity and free carbon dioxide. The concentration of carbon dioxide can be reduced to 5 to 10 mg/L in a degasifier or a vacuum deaerator. Strongly basic resins remove both strongly ionized and weakly ionized acids. At the end of each operating run, the weakly basic anion exchanger is backwashed, regenerated with a solution of sodium carbonate or caustic soda, rinsed, and returned to service. Caustic soda is used to regenerate the strongly basic resins.

Where a minimum amount of dissolved solids is required in the treated water, mixed cation-anion beds are used. Regeneration of the mixed bed consists of separating the anion resin from the cation resin and regenerating each separately.

Figure 2-7 illustrates the arrangement of equipment used in nine basic types of demineralizing systems. Selection of ion-exchange processes is based upon:

- Raw-water characteristics
- Treated-water characteristics required
- Operating costs: chemicals, energy, waste disposal, qualified operating labor, and maintenance
- Capital costs
- Available space

In most cases, the services of a qualified consulting engineer should be used to evaluate alternatives and recommend the optimum method of treatment.

It is essential that the influent water to ion exchangers be free of suspended solids.

Silt and organic matter are particularly objectionable, since these materials can deposit in the exchanger beds and reduce the capacity of the units by coating the exchanger resins with films which either prevent or retard the movement of anions and cations through the resins. The influent water may require a preliminary treatment of sedimentation and/or filtration.

REVERSE OSMOSIS

Systems

A reverse-osmosis (RO) system consists of the following essential parts:

1. A pretreatment section (which usually includes chemical feed systems for the injection of a coagulant), acid for the control of pH and hence calcium carbonate precipitation, a sequestering agent (often polyphosphate compounds) for the control of iron and calcium sulfate precipitation, and a granular-media filter for the removal of coagulated solids. When the suspended solids level is consistently high (>50 mg/L), the overall system may be best served by including a clarifier in the flowsheet, in effect, to increase the length of the filter runs. Also, in many cases, it will be necessary to remove chemically some hardness and perhaps silica from the system in a presoftening step. Raw-water silica values which will concentrate to 150 mg/L will require SiO_2 control. However, the economics of pH adjustment, the danger of membrane damage, and the possible precipitation of silica should also be considered.
2. The reverse-osmosis system itself consists of the high-pressure pump or pumps, the pressure vessels, and the membranes themselves, along with the necessary interconnecting piping, valves, and fittings.
3. A posttreatment system which consists of chemical feed systems for the feeding of corrosion control chemicals and chlorine for control of biological growths in the distribution system.

Any RO system that does not receive water from a highly controlled source will require pretreatment for the removal of suspended solids and colloids.

In general, some provision should be made for cleaning the membranes even though the pretreatment system is nominally satisfactory for the reduction of solids and colloidal material in the raw water. Reverse-osmosis membranes make a separation between the dissolved solids phase and the liquid phase by reversing the flow that normally occurs through a semipermeable membrane when the concentration of a given salt is different on the two sides of the membrane. Normally, the flow will proceed in such a fashion as to equalize the concentration, and hence pressure equal to the osmotic pressure plus an additional driving force is provided to permit flow from the more concentrated to the less concentrated side—thus the term *reverse osmosis*.

The flow proceeds through the membrane, passing between the atoms in the polymer lattice. Flow is accompanied by a loose association between the transported species and the membrane driven by the pressure differential. Most membrane systems operate at system pressures between 20 and 68 atm. Typically, brackish water systems operate at 20 to 40 atm and seawater systems at 48 to 68 atm.

Figure 2-8 is a schematic drawing of a typical RO system. The illustration shows all the essential components identified in the preceding discussion together with a cleaning system. The cleaning system is employed simply by opening the brine valve and reducing the pressure to zero in the membranes themselves and flushing detergent-carrying water through the membranes at relatively high velocity. After the membranes have been scrubbed this way for 15 to 30 min, the flux will be restored to normal or near normal values unless the fouling is extraordinary.

The quality of the water with respect to its suitability for introduction into a reverse-osmosis system may be measured in a crude fashion through the use of a test yielding

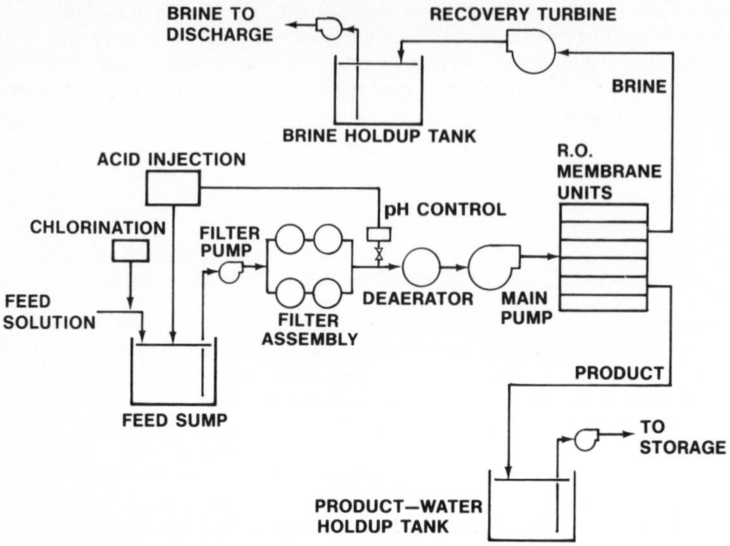

Figure 2-8 Basic flow diagram for a single-state reverse-osmosis plant.

what is called the *silting or fouling index*. The silting index (S.I.) is described in the following equation:

$$\text{S.I.} = 1 - \frac{T_1}{T_2} 100$$

In this case, T_1 represents the time to collect 500 mL of filtrate through an 0.45-μm filter at the initiation of the test. T_2 represents the time required to collect 500 mL of filtrate after 15 min from the time of test initiation. The filter employed is a conventional 0.45 Micropore filter. The filter assembly is normally constructed of light metal and is designed to be operated at 30 lb/in^2 (2 atm).

In general, waters that have a silting index of less than 30 are considered highly satisfactory for RO treatment. Waters with a silting index of 30 to 50 are considered marginal. Those waters with a silting index greater than 50 are considered unsatisfactory and may require some form of additional treatment.

Applications

Reverse-osmosis systems find their application in a number of areas. However the most significant are:

1. Inclusion in systems for providing low-conductivity water for boiler makeup and related purposes
2. The production of potable water where such supplies are not conveniently available
3. Industrial waste treatment and water recovery

Of the preceding, perhaps the most significant application of reverse osmosis is in the deionization systems for the production of low-conductivity boiler feedwater.

In most instances where the raw-water supply contains more than 300 to 400 mg/L total dissolved solids (TDS), an overall system economy may be shown by utilizing reverse osmosis for removing the bulk of the dissolved solids at a high recovery rate prior to discharge of the product to the ion-exchange polishing system. The approach is to replace the cationic-exchange, degasification, and anionic-exchange subsystems with the reverse-osmosis system and to employ the mixed-bed polishing subsystem for the removal of residual solids. In some instances, it can be shown that this puts a somewhat higher load on the mixed-bed polisher. However, if this load is less than 50 mg/L TDS, it does not constitute an improper or unreasonable burden in terms of cycle time or regeneration frequency. This type of system usually produces a less expensive water at roughly the same recoveries as would be experienced with the full deionizing (DI) system and often does not produce the quantity of conservative solids in the backwash and regenerating stream that the DI system will produce. The preceding is a generalization only, and a careful cost comparison should be made in any specific situation.

Sometimes only high-salinity waters are available for process and other purposes, and the supply of potable water is either limited or nonexistent. In these cases, a reverse-osmosis facility can be installed to produce water for the plant at comparatively low cost.

Many industrial wastes are highly amenable to treatment by reverse-osmosis systems. The most commonly encountered RO application is in the area of metal contamination or the presence of excessive dissolved materials in the effluent. Often when RO systems are employed for treatment, the water is suitable or can easily be made suitable for reuse and can constitute a supply for processes or sometimes even boiler feedwater.

Cost of RO Systems

Table 2-1 contains a brief summary of capital and operating costs of RO systems as a function of salinity. The actual cost will vary as a function of the overall recovery employed in the system. However, the data presented in Table 2-1 represent a reasonable generalization for recoveries of between 75 and 80 percent on brackish water and 15 to 20 percent on seawater. Membrane life is expected to be in excess of 3 years.

EVAPORATIVE SYSTEMS

Evaporative systems are used in some applications for the production of very high quality water from saline waters or wastewater. With high energy costs, evaporative systems (except under very special circumstances) appear to offer a less satisfactory solution than

TABLE 2-1 Membrane Systems Costs—1979 Dollars

Capacity, gal/day	Pressure, lb/in²	Feed salinity, mg/L NaCl	Capital cost, $	O & M cost, $/1000 gal*
50,000	420	5,000	65,000	0.80†
150,000	420	5,000	130,000	0.76†
500,000	420	5,000	420,000	0.73†
1,000,000	420	5,000	800,000	0.71†
2,500,000	420	5,000	1,800,000	0.68
10,000	850	35,000	120,000	4.80‡
50,000	850	35,000	200,000	3.80‡
100,000	850	35,000	380,000	3.55‡
500,000	850	35,000	1,500,000	3.30‡

*Data supplied by Fluid Systems division, UOP Inc. San Diego, Calif. To convert gallons to cubic meters divide by 264 gal/m^3.

†Based on power at $0.04 per kWh, $0.10 to $0.18 per 1000 gal membrane replacement, $0.04/per 1000 gal for chemicals, $0.01 to $0.09 per 1000 gal for attendence, $0.04 per 1000 gal for maintenance, all at 75 percent recovery.

‡Same but with 10 to 15 percent recovery and $0.60 per 1000 gal membrane replacement.

alternative systems to many desalination problems. Nevertheless, there undoubtedly are many instances where the recovery of a very high quality water through the use of waste energy in an evaporative system is indicated. Where waste energy is available, evaporation procedures should be considered as a possible candidate in any water recovery system analysis.

The two most commonly encountered evaporator types are the *m*ulti*s*tage *f*lash (MSF) and the falling-film vertical tube evaporator (VTE). Rising-film vertical-tube systems and single-stage units with vapor recompression systems are also employed. Generally speaking, economy ratios of 5 to 12 lb of water per 1000 Btu (2 to 5 kg H_2O per 10^6 J) are possible, and systems are available which can provide the recovery ratios cited. Corrosion and special maintenance problems such as tube fouling should be considered in examining this option. The substantial advantage that an evaporator offers is that even if the source has variable salinity or extremely high salinity, the quality of the product will be essentially unchanged. Furthermore, the amount of energy required to run the system is independent of the salinity.

System costs for some evaporator systems are shown in Table 2-2. The operating cost is estimated on the basis of \$2.50 per 10^6 Btu (1.055×10^9 J).

ULTRAFILTRATION SYSTEMS

Ultrafiltration (UF) is a pressure-driven membrane process similar in many ways to reverse osmosis. However, in this case, as opposed to the situation encountered in RO systems, the flow of water through the membrane is generally through pores and not through the space between the lattices in the polymer. Furthermore, there is little or no chemical interaction between the transported species and the membrane itself. UF membranes may be tailor-made to meet virtually any type of removal specification.

UF systems are often used to remove very fine particles from water streams, e.g., in preconditioning water prior to RO treatment and sometimes for the removal of large organic molecules. UF systems have been used in this regard to remove colored colloids as well as fine suspended solids. In water-reuse applications, UF has been employed to remove submicron-size particles of activated carbon from the treated waste stream.

In recent years one of the more important uses of UF systems has been in the removal of tramp oils from various wastewater streams. In some instances it has been possible to concentrate the oil in the UF system up to as much as 50 percent, thereby making the recovery of that material comparatively easy.

A schematic diagram of a typical UF system is shown in Fig. 2-9. As can be seen, UF systems usually employ a fairly high volume of recycle. The quantity of UF on each pass is comparatively small relative to the material which is passed by the membrane.

TABLE 2-2 Evaporator Capital and Operating Costs—1979 Dollars

Capacity, 10^6 gal/day	Capital costs,*,† $\$ \times 10^6$	Operating costs,‡ \$/1000 gal§
0.2	1.6	3.07
1.0	6.5	3.89
5.0	25.0	2.73

*In-place price including civil works.

†Prices supplied by Envirogenics Systems Company, El Monte, Calif.

‡Based on \$2.50 per 10^6 Btu, economy ratio of 10, \$0.26 per 1000 gal for chemicals, \$0.03 to 0.20 per 1000 gal for attendance, 2 percent of first cost for annual replacements.

§To convert 1000 gal to cubic meters, multiply by 3.8 m^3 per 1000 gal, 10^6 Btu = 1.005×10^9 J.

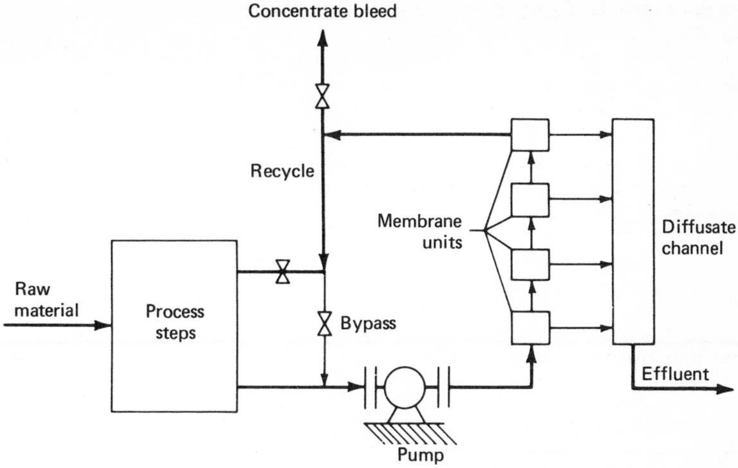

Figure 2-9 Typical flowsheet for an ultrafiltration system.

SPECIAL-PURPOSE TREATMENTS

Special-purpose treatments are classified as treatment methods used to remove impurities not removed by other methods, except for reverse osmosis.

Chlorination

Chlorine compounds are strong oxidizing agents and also have potent bactericidal properties. Chlorine reacts with ammonia or amines to form chloramines, which are weaker disinfectants than chlorine but are useful for maintaining a residual chlorine content in water mains. The disadvantage of chlorination is its reaction with many organic chemicals to form trihalomethanes and other potential carcinogens. The maximum contaminant level (MCL) for trihalomethanes for drinking water has been tentatively set at 0.10 mg/L by the EPA. Chlorination of natural colored water forms chlorine addition compounds, commonly called haloforms. Chloraromatics may also be formed. Chlorine will also oxidize ferrous iron, manganese, and sulfide ions. In the oxidation process, chlorine is reduced to the chloride ion.

Chlorine dioxide is also used as an oxidant and disinfectant to a limited extent. When it is used for the treatment of potable water, trihalomethanes are not produced.

Ozonization

Ozone, produced from air or pure oxygen, is a powerful oxidizing agent. It is used for color, taste, and odor removal and for organic chemical oxidation, bacterial disinfection, and virus inactivation. Based upon European experience, ozonization followed by activated carbon treatment appears to be the most effective means for meeting the EPA's proposed drinking-water standards. Ozone is extremely toxic, however, and the off-gas from ozone contactors must be passed through a catalytic reactor before being vented.

Activated Carbon Treatment

Granular activated carbon has been used for many years to remove tastes and odors from municipal drinking-water supplies. The EPA's *Interim Primary Drinking Water Regu-*

lations recommends the use of granular activated carbon for the removal of synthetic organic chemicals from potable water supplies contaminated with industrial pollution.

When carbon approaches its maximum effective loading with organics, it requires reactivation. The spent carbon can be removed from the absorption columns and processed in regeneration furnaces by controlled oxidation.

If water is ozonized, the byproduct oxygen in the water will promote the growth of bacteria in the carbon columns. These oxidized biodegradable organic compounds increase the permissible carbon loadings and reduce the required frequency for thermal reactivation.

RECOMMENDED REFERENCE BOOKS

Applebaum, Samuel P.: *Demineralization by Ion Exchange,* Academic, New York, 1968.

The Betz Handbook of Industrial Water Conditioning, 8th ed. Betz Laboratories, Trevose, PA 19047, 1980.

Nordell, Eskell: *Water Treatment for Industrial and Other Uses,* 2d ed., Reinhold, New York, 1961.

Drew Principles of Industrial Water Treatment, 4th ed. Drew Chemical Corporation, Boonton, N.J., 1981.

Standard Methods for the Examination of Water and Waste Water, 15th ed., American Public Health Association, Washington, D.C., 1981.

Water: The Universal Solvent, Nalco Chemical Company, Oak Brook, IL 60521, 1977.

Water Treatment Plant Design, American Water Works Association, Denver, 1969.

chapter 6-3

Water-Cooling Systems

by
C. J. McCann
Chairman
Tower Performance, Inc.
Fairfield, New Jersey

GLOSSARY

Familiarity with some of the more commonly used terms is a useful preliminary in considering cooling-tower operations.

Approach The difference between the cold-water temperature in degrees Fahrenheit (degrees Celsius) and the ambient or inlet wet-bulb temperature.

Blowdown Water discharged from the system to control concentration of salts or other impurities in the circulating water.

Cell The smallest tower subdivision which can function as an independent unit with regard to air and water flow; it is bounded on exterior walls or partitions. Each cell may have one or more fans or stacks and one or more distribution systems.

Cold-water temperature (CWT) Temperature of the water entering the cold-water basin before addition of makeup or removal of blowdown.

Counterflow tower One in which air, drawn in through the louvers (induced draft) or forced in (forced draft) at the base by the fan, flows upward through the fill material and interfaces countercurrently with the falling hot water.

Crossflow tower One in which air, drawn or forced in through the air intakes by the fan, flows horizontally across the fill section and interfaces perpendicularly with the falling hot water.

Design conditions In mechanical draft towers: the hot-water temperature (HWT), cold-water temperature (CWT), flow in gallons per minute (gpm) or cubic meters per hour (m^3/h), and wet-bulb temperature (wbt). In natural draft towers: HWT, CWT, flow, wbt plus either dry-bulb temperature (dbt) or relative humidity (RH).

Distribution system Those parts of a tower, beginning with the inlet connection, which distribute the hot circulating water within the tower to the points where it contacts the air. In a counterflow tower, this includes the header, laterals, and distribution nozzles. In a crossflow tower, the system includes the header or manifold, valves, distribution box, basin pan, and nozzles.

Drift Water lost from the tower as liquid droplets entrained in the exhaust air. It is independent of water lost by evaporation. Units may be in pounds per hour (kilograms per hour) or percentage of circulating water flow. Drift eliminators control this loss from the tower.

Drift eliminator An assembly constructed of wood, plastic, cement asbestos board, steel, or other material which serves to remove entrained moisture from the discharged air.

Fan stack Cylindrical or modified cylindrical structure in which the fan operates. Fan stacks are used on both induced-draft and forced-draft axial flow propeller fans. Also known as *cylinder*.

Hot-water temperature (HWT) Temperature in degrees Fahrenheit (degrees Celsius) of circulating water entering the distribution system.

Makeup Water added to the circulating water system to replace water lost from the system by evaporation, drift, blowdown, and leakage.

Plenum The enclosed space between the eliminators and the fan stack in induced-draft towers or the enclosed space between the fan and the filling in forced-draft towers.

Pumping head Minimum pressure required to lift the water from basin curb to the top of the system. Pumping head is equal to static head plus friction loss through the distribution system.

Range Difference between the hot-water and cold-water temperatures. Units: degrees Fahrenheit (degrees Celsius). Also known as *cooling range*.

Recirculation A condition in which a portion of the discharge air enters the tower along with the fresh air. The amount of recirculation is determined by tower design, tower placement, and atmospheric conditions. The effect is generally evaluated on the basis of the increase in the entering wet-bulb temperature compared to the ambient.

Water loading Circulating water flow, expressed in gallons per minute per square foot (cubic meters per hour per square meter), of effective horizontal wetted area of the tower.

Wet-bulb temperature (wbt) Temperature indicated by a psychrometer. Also known as the thermodynamic wet-bulb temperature or the temperature of adiabatic saturation. Units: degrees Fahrenheit (degrees Celsius).

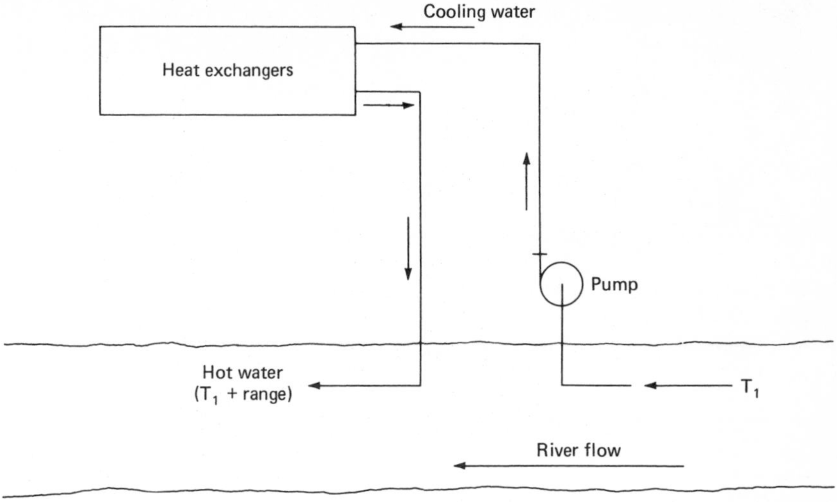

Figure 3-1 Once-through cooling.

INTRODUCTION

With the growth in the number and sizes of manufacturing plants of all types and the attendant higher heat rejection rates, cooling-tower requirements have increased dramatically. These trends are coupled with environmental aspects, including water conservation and limitations on thermal and chemical discharges. As a result, the plant engineer has witnessed an upsurge in the specification and use of cooling towers.

COOLING SYSTEM OPTIONS

Once-Through Cooling Systems

Many plants operating today are on once-through cooling systems, as shown in Fig. 3-1. They utilize water from a lake or river to supply cooling water to the heat exchangers. The heated water is then returned to the body of water.

As a result of all the heat being discharged to rivers, lakes, etc., by plants operating with once-through cooling systems, the term "thermal pollution" has assumed significance and has resulted in the enactment of environmental-related legislation. Consequently, once-through cooling is not available as an option in many cases.

Closed-Cycle Cooling Systems

Closed-cycle cooling refers to the water side of the system and generally favors the use of a cooling tower. Figure 3-2 shows the relationship of the cooling tower to the cooling system. The cooling water is continuously recirculated through the plant. The cooling tower is used to remove the heat added to the circulating cooling water by the heat exchangers. Water withdrawn from the natural source would be used only for makeup of losses.

Cooling Towers

Cooling-tower designs currently available to plant designers fall into natural-draft and mechanical-draft designs. The natural-draft design utilizes large-dimension concrete

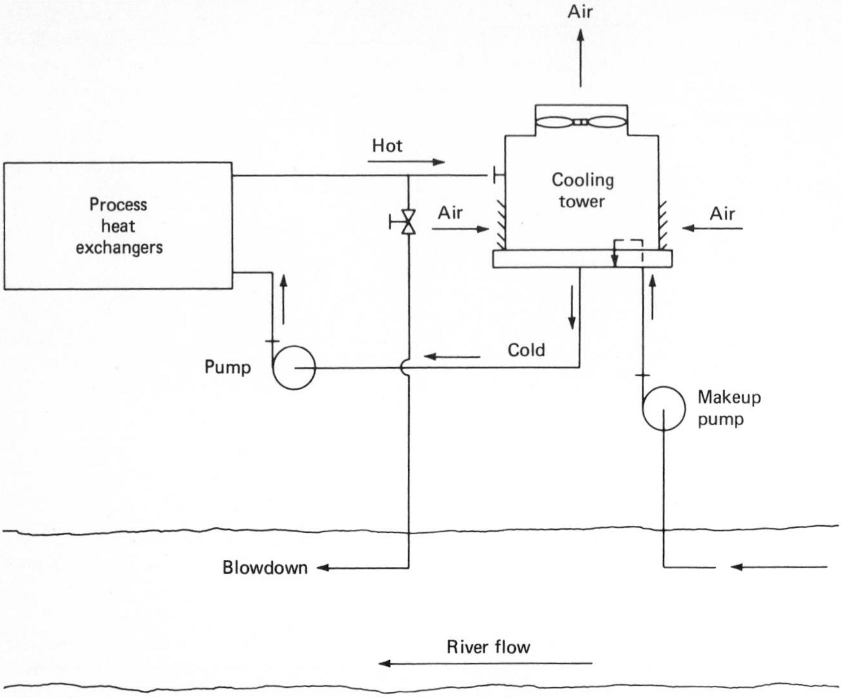

Figure 3-2 Closed-cycle cooling-tower system.

chimneys to induce air through the cooling medium. In the mechanical-draft design, large-diameter fans driven by electric motors induce or force the air through the circulated water which flows over fill surface provided to interrupt the flow of water and increase the time of contact between air and water. This permits the efficient transfer of heat from the water to air.

Spray Ponds

An alternative to cooling towers in closed-cycle cooling systems is a spray pond, where warm water is pumped through pipes from the heat exchangers and then out of the spray nozzles. The nozzles atomize the warm water into fine droplets. The basic arrangement of a spray pond is shown in Fig. 3-3. The spray nozzles are usually located about 5 ft

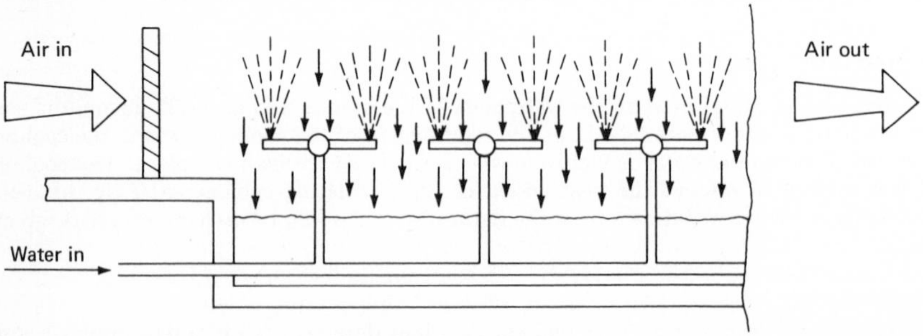

Figure 3-3 Spray pond.

above the pond surface. Height of the sprays is about 6 ft. A nominal water loading rate of 1 gal/min/ft^2 (2.44 m^3/h/m^2) of pond area and wind speed of 5 mi/h (8.05 km/h) would be typical design parameters for such a pond. Performance is strongly dependent on wind speed and direction, and is limited by the relatively short contact time between the air and water spray.

Spray ponds cause excessive water losses due to drift, and this in turn may cause localized icing and fogging and relatively high pumping costs—all disadvantages to the use of spray ponds. A spray-pond system uses eight times more land than a cooling-tower installation.

Atmospheric Cooling Towers

When there is a need for larger cooling ranges (the gap between hot- and cold-water temperatures) and closer approaches (difference between cold-water and inlet wet-bulb temperatures), the natural draft or atmospheric spray and filled-deck towers might be considered. Wooden or plastic filled decks are installed on spray coolers to increase the time of contact between water and air. Different types of fill packing and spacing are utilized, and heights of towers may vary with the extent of cooling needed. The cooling is dependent on the efficiency of the filled deck and the wind velocity passing through the tower as the water descends through the deck.

The advantages are: (1) no electric power required except for pumping head, (2) no mechanical equipment necessary, reducing maintenance requirements.

The disadvantages are: (1) Atmospheric towers have limited capacities as they are dependent solely on ambient atmospheric conditions. (2) At low- or no-wind conditions they are inefficient. (3) Water loss as a result of high wind velocities can be appreciable. (4) A rather high pumping head is required to allow for maximum air-water contact time.

Large natural-draft hyperbolic cooling towers are found only in utility power station service in the United States. The economics of plant design will favor the mechanical-draft type because of its rather short amortization period. Natural-draft towers work better when wet-bulb temperatures are low and relative humidity is high or if heavier demand is in the winter. A combination of low wet-bulb with high inlet and exit water temperatures enhances the operation of a hyperbolic tower. Because of the tremendous size of these units [500 ft (150 m) high and 400 ft (120 m) in diameter at the base], they are more practical when the circulating cooling water flow rate is about 200,000 gal/min (45,400 m^3/h) and higher.

Mechanical-draft towers have positive control of the air delivery through the packing with the use of large-diameter fans. Therefore, they can be designed for close control of cold-water temperature.

Counterflow and crossflow designs are illustrated in Figs. 3-4 and 3-5.

COOLING-TOWER OPERATION

Theoretical Concepts

The basic equations covering combined mass- and heat-transfer phenomena have been covered in the literature.[1] The analysis combines the sensible- and latent-heat transfer into an overall process based on enthalpy potential as the driving force.

The process is shown schematically in Fig. 3-6, where each particle of bulk water in the tower is assumed to be surrounded by an interfacial film to which heat is transferred from the water. This heat is then transferred from the interface to the main air mass by (1) a transfer of sensible heat and (2) mass-heat transfer (latent) resulting from the evaporation of a portion of the bulk water. This can be represented by

$$\frac{KaV}{L} = \int_{T_2}^{T_1} \frac{dT}{h_w - h_a} \tag{1}$$

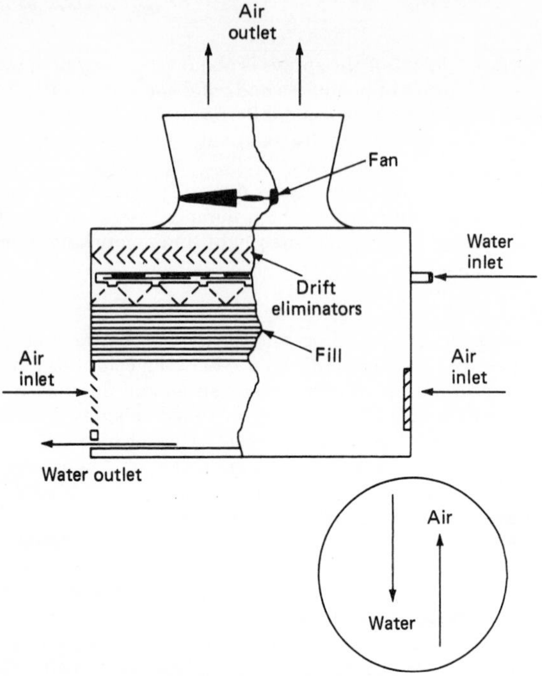

Figure 3-4 Mechanical-draft counterflow tower.

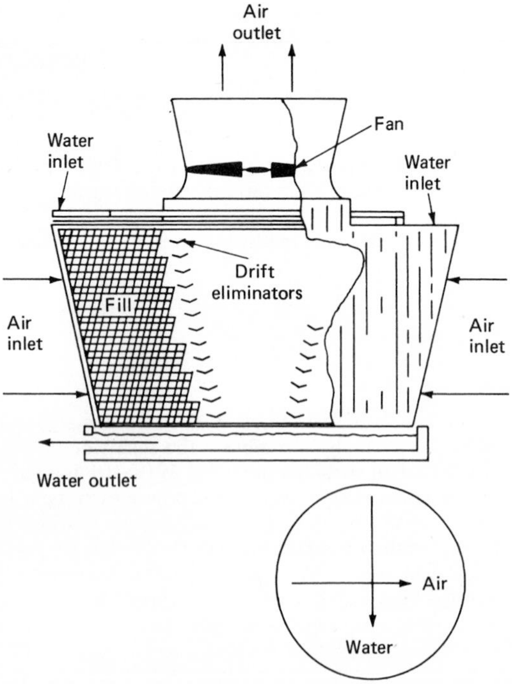

Figure 3-5 Mechanical-draft crossflow tower.

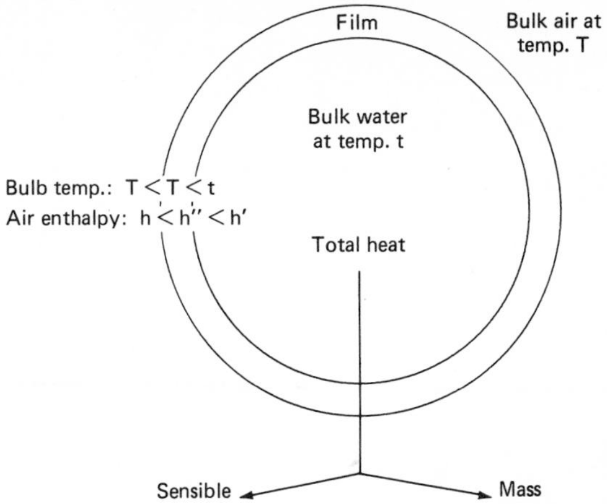

Figure 3-6 Schematic of a water droplet with interface film.

where

KaV/L = tower characteristic
T_1 = hot-water temperature, °F (°C)
T_2 = cold-water temperature, °F (°C)
T = bulk water temperature, °F (°C)
h_w = enthalpy of air–water vapor mixture at bulk water temperature, Btu/lb dry air (J/kg)
h_a = enthalpy of air–water vapor mixture at wet-bulb temperature, Btu/lb dry air (J/kg)

This equation is commonly referred to as the Merkel equation. The derivation can be found in Ref. 2.

The left side of the equation is called the *tower characteristic*. The laws of thermodynamics demand that the heat discharged by the water descending through the cooling tower must equal the heat absorbed by the air rising upward through the tower, or

$$L(T_1 - T_2) = G(h_2 - h_1)$$
$$\frac{L}{G} = \frac{h_2 - h_1}{T_1 - T_2} \tag{2}$$

where

L = mass water flow, lb/(h)(ft²) [kg/(h)(m²)] plan area
T_1 = hot-water temperature, °F (°C)
T_2 = cold-water temperature, °F (°C)
G = mass airflow, lb dry air/(h)(ft²) [kg/(h)(m²)]
h_2 = enthalpy of air–water vapor mixture at exhaust wet-bulb temperature, Btu/lb dry air (J/kg)
h_1 = enthalpy of air–water vapor mixture at inlet wet-bulb temperature, Btu/lb dry air (J/kg)
L/G = liquid-to-gas ratio, lb water/lb dry air (kg/kg)

Equations (1) and (2), or the tower characteristic, can be represented graphically by the diagram in Fig. 3-7. The interfacial film is assumed to be saturated with water vapor at the bulk water temperature T_1 (A in Fig. 3-7). As the water is cooled to temperature T_2, the film enthalpy follows the saturation curve to B.

Air entering the tower at wet-bulb temperature T_{wb} has an enthalpy C'. The origin of the air-operating line, point C, is vertically below B and is positioned to have an enthalpy corresponding to that of the entering wet-bulb temperature. The heat removed from the

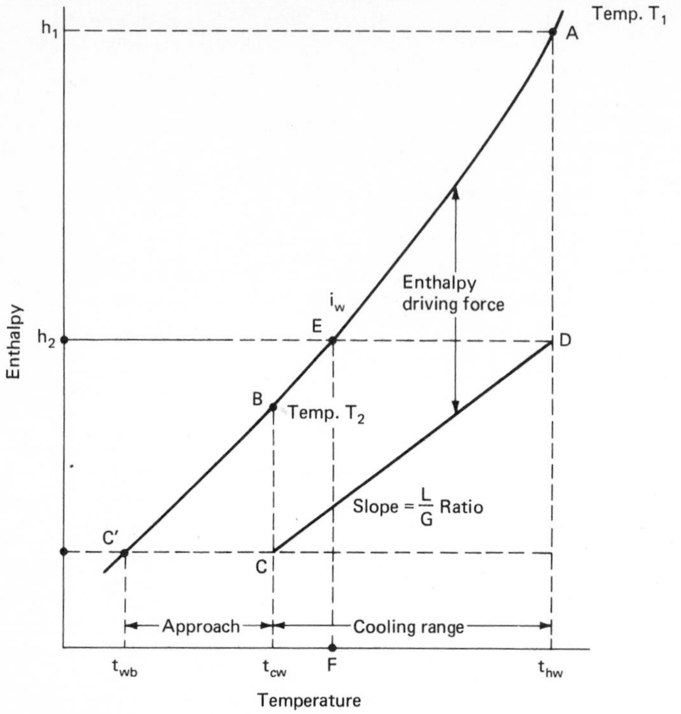

Figure 3-7 Graphical representation of tower characteristic.

water is added to the air so its enthalpy increases along line CD having a slope equaling the L/G ratio. The vertical distance BC represents the initial driving force.

Point D represents the air leaving the cooling tower. It is the point on the air-operating line vertically below A. The projected length CD (or AB) is the cooling range.

The coordinates refer directly to the temperatures and enthalpy of the water-operating line AB, but refer directly only to the enthalpy of a point on the air-operating line CD. The corresponding wet-bulb temperature of any point on CD is found by projecting the point horizontally to the saturation curve, then vertically down to the temperature coordinate. DEF shows this projection for the outlet air wet-bulb temperature of point D. Point F is the outlet air wet-bulb temperature.

The following integral is represented by the area ABCD:

$$\int_{T_2}^{T_1} \frac{dT}{h_w - h_a}$$

where

T_1 = hot-water temperature, °F (°C)
T_2 = cold-water temperature, °F (°C)
T = bulk water temperature, °F (°C)
h_w = enthalpy of air–water vapor mixture at bulk water temperature, Btu/lb dry air (J/kg)
h_a = enthalpy of air–water vapor mixture at wet-bulb temperature, Btu/lb dry air (J/kg)

This value is characteristic of the tower, varying with the rates of water and airflow. An increase in the entering air wet-bulb temperature moves the origin C upward, and the line CD shifts to the right to establish equilibrium. Both the inlet and outlet water temperatures increase, while the approach decreases. The curvature of the saturation line is such that the approach decreases at a progressively slower rate as the wet-bulb temperature increases.

An increase in the heat load increases the cooling range and increases the length of line CD. To maintain equilibrium, the line shifts to the right, increasing hot- and cold-water temperatures and the approach.

The increase causes the hot-water temperature to increase considerably faster than does the cold-water temperature.

In both these cases, the area ABCD should remain constant—actually it decreases about 2 percent for every 10°F (5.6°C) increase in hot-water temperature. The cooling-tower designers take this into consideration in their initial design by applying a hot-water temperature correction to design figures when the design hot-water temperature exceeds 110°F (43.5°C) (see Fig. 3-8).

However, a change in L/G will change this area. It has been found that a logarithmic plot of L/G vs. KaV/L, the tower characteristic, at a constant airflow results in straight line. This line or curve, when plotted on the demand curve for the design conditions, is the tower characteristic curve. The slope of the curve depends upon the tower packing. In the absence of more specific data, splash-type packing will have a slope of -0.6.

Knowing the wet-bulb temperature, the range, the approach, and the L/G ratio, KaV/L can be determined by reference to the charts in the Cooling Tower Institute Blue Book.[3] A typical tower characteristic curve which would be submitted by a manufacturer

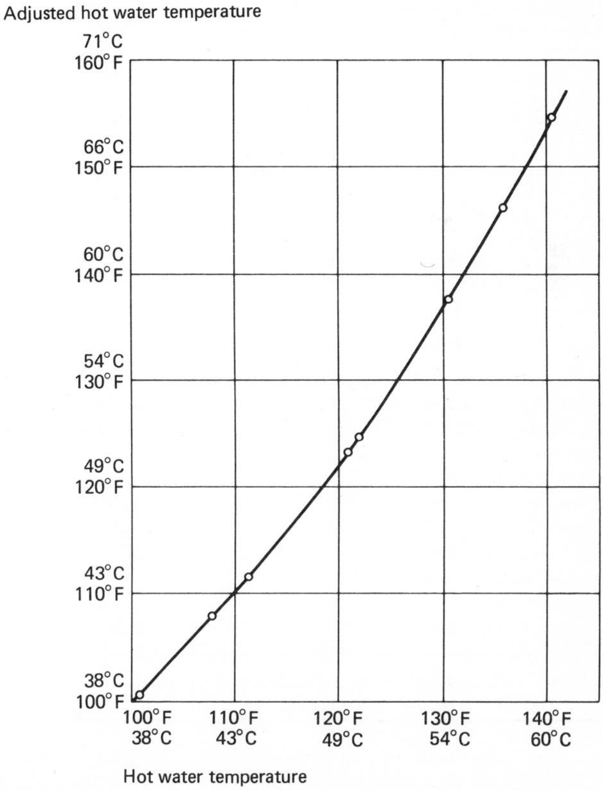

Figure 3-8 Plot of hot-water temperature adjustment.

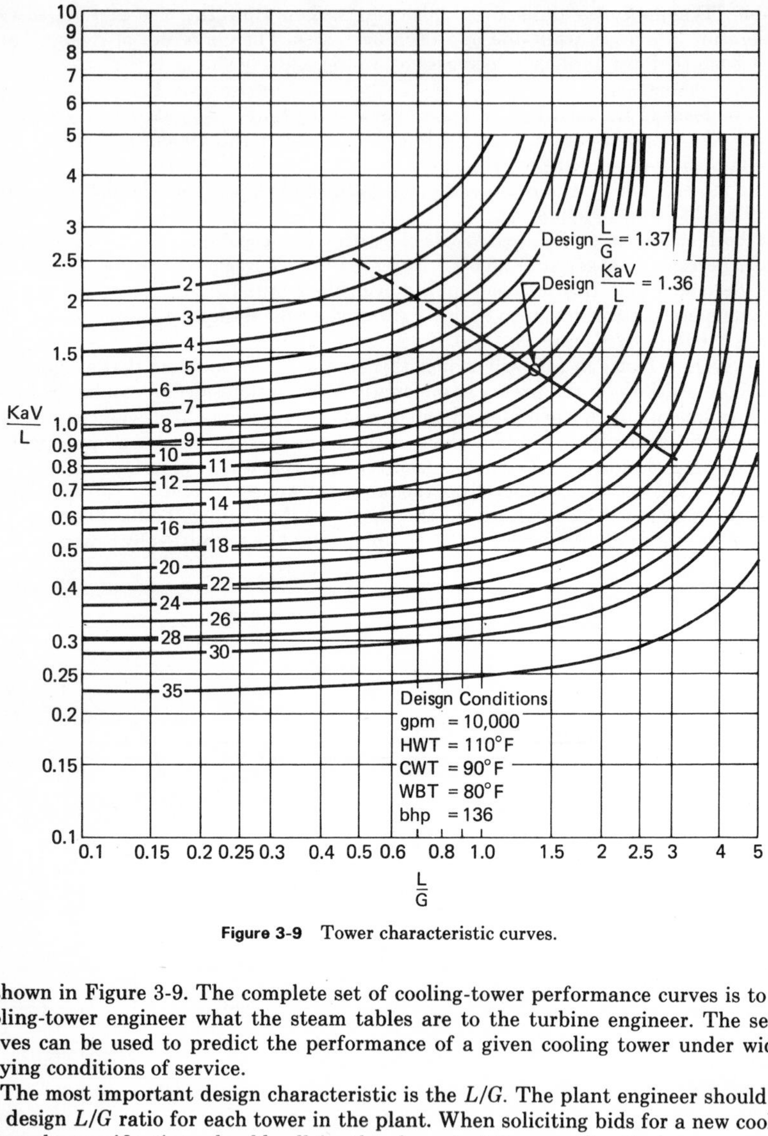

Figure 3-9 Tower characteristic curves.

is shown in Figure 3-9. The complete set of cooling-tower performance curves is to the cooling-tower engineer what the steam tables are to the turbine engineer. The set of curves can be used to predict the performance of a given cooling tower under widely varying conditions of service.

The most important design characteristic is the *L/G*. The plant engineer should file the design *L/G* ratio for each tower in the plant. When soliciting bids for a new cooling tower, the specifications should call for the characteristic curve for the tower proposed to be submitted with the bid package.

Design Parameters

Cooling towers are designed to meet a condition of operation specified by the plant engineer. This condition requires the removal of a heat load of a specified magnitude. When the cooling-water flow rate is selected, the specification can be set. The cooling tower is specified to cool a water quantity (gallons per minute) through a definite temperature gradient (range) to a final temperature which is a certain number of degrees above the design wet-bulb temperature (approach). Only infrequently will the tower operate at this point, since the plant will normally level out at a slightly different requirement and/or the wet-bulb temperature will be other than design. For this reason the tower characteristic curve should be supplied by the manufacturer.

The design wet-bulb temperature is usually based on National Weather Service records and is chosen as the temperature which will not be experienced more than, say 1, 2½, or 5 percent of the time during the summer months of June to September. The design wet-bulb temperature should be selected only after some reference has been made to the economics of the plant being served, the seasonal requirements of cooling, and the tabulated Weather Service record for the locality. Choosing the 1 percent frequency level would be judicious when the cooling tower is serving a temperature-sensitive process or in production of a high-profit product.

The three design levels shown in Table 3-1 are 1, 2½, and 5 percent of the 2928 summer hours, June to September, rounded off to the nearest whole degree. For selecting design temperatures at locations between the cities shown in the table, use data from the city or town nearest the locality of the plant installation. For major installations involving large expenditures or critical temperature balance, make further detailed studies with the assistance of a meteorologist.

TABLE 3-1 Summer, June–September, Design Wet-Bulb Temperatures[4]

	1%	2½%	5%
Alabama			
Birmingham	79	78	77
Huntsville	78	79	76
Mobile	80	79	79
Alaska			
Anchorage	63	61	59
Fairbanks	64	63	61
Arizona			
Flagstaff	61	60	59
Phoenix	77	76	75
Tucson	74	73	72
Arkansas			
El Dorado	81	80	79
Fayetteville	77	76	75
Little Rock	80	79	78
California			
Arcata/Eureka	60	59	58
Bakersfield	72	71	70
Fresno	73	72	71
Los Angeles	69	68	67
Sacramento	72	70	69
San Diego	71	70	68
San Francisco	65	63	62
Colorado			
Denver	65	64	63
Grand Junction	64	63	62
Pueblo	68	67	66
Connecticut			
Hartford	77	76	74
New Haven	77	76	75
Delaware			
Wilmington	79	77	76
District of Columbia			
Washington	78	77	76
Florida			
Jacksonville	80	79	79
Miami	80	79	79
Orlando	80	79	78
Pensacola	82	81	80
Tampa	81	80	79
Georgia			
Atlanta	78	77	76
Macon	80	79	78
Valdosta	80	79	78
Hawaii			
Honolulu	75	74	73
Idaho			
Boise	68	66	65
Idaho Falls	65	64	62
Pocatello	65	63	62
Illinois			
Chicago	78	76	75
Peoria	78	77	76
Springfield	79	78	77
Indiana			
Evansville	79	78	77
Indianapolis	78	77	76
South Bend	77	76	74
Iowa			
Des Moines	79	77	76
Mason City	77	75	74
Sioux City	79	77	76
Kansas			
Dodge City	74	73	72
Goodland	71	70	69
Topeka	79	78	77
Wichita	77	76	74
Kentucky			
Lexington	78	77	76
Louisville	79	78	77
Paducah	80	79	78
Louisiana			
Baton Rouge	81	80	79
New Orleans	81	80	79
Shreveport	81	80	79
Maine			
Augusta	74	73	71
Portland	75	73	71
Maryland			
Baltimore	79	78	77

TABLE 3-1 (*Continued*)

	1%	2½%	5%
Massachusetts			
Boston	76	74	73
Worcester	75	73	71
Michigan			
Battle Creek	76	74	73
Detroit	76	75	74
Saginaw	76	75	73
Minnesota			
Alexandria	76	74	72
Duluth	73	71	69
Minneapolis/St. Paul	77	75	74
Rochester	77	75	74
Mississippi			
Greenwood	81	80	79
Meridian	80	79	78
McComb	80	79	79
Missouri			
Joplin	79	78	77
Kansas City	79	77	76
Springfield	78	77	76
Montana			
Billings	68	66	65
Butte	60	59	57
Great Falls	64	63	61
Nebraska			
Grand Island	76	75	74
North Platte	74	73	72
Omaha	79	78	76
Scotts Bluff	70	69	67
Nevada			
Elko	64	62	61
Las Vegas	72	71	70
Reno	64	62	61
New Hampshire			
Concord	75	73	72
Manchester	76	74	73
New Jersey			
Newark	77	76	75
Trenton	78	77	76
New Mexico			
Albuquerque	66	65	64
Carlsbad	72	71	70
Roswell	71	70	69
New York			
Albany	76	74	73
Binghamton	74	72	71
Buffalo	75	73	72
New York	77	76	75
North Carolina			
Asheville	75	74	73
Charlotte	78	77	76
Raleigh	79	78	77
North Dakota			
Bismarck	74	72	70
Fargo	76	74	72
Minot	72	70	68
Ohio			
Cincinnati	78	77	76
Cleveland	76	75	74
Columbus	77	76	75
Oklahoma			
Ponca City	78	77	76
Oklahoma City	78	77	76
Tulsa	79	78	77
Oregon			
Medford	70	68	66
Pendleton	66	65	63
Portland	69	67	66
Pennsylvania			
Altoona	74	73	72
Harrisburg	76	75	74
Philadelphia	78	77	76
Pittsburgh	75	74	73
Rhode Island			
Providence	76	75	74
South Carolina			
Charleston	81	80	79
Columbia	79	79	78
Spartanburg	77	76	75
South Dakota			
Pierre	76	74	73
Rapid City	72	71	69
Sioux Falls	77	75	74
Tennessee			
Chattanooga	78	78	77
Knoxville	77	76	75
Memphis	80	79	78
Nashville	79	78	77
Texas			
Abilene	76	75	74
Amarillo	72	71	70
Austin	79	78	77
Big Spring	75	73	72
Corpus Christi	81	80	80
Dallas	79	78	78
El Paso	70	69	68
Houston	80	80	79
Utah			
Richfield	66	65	64
St. George	71	70	69
Salt Lake City	67	66	65
Vermont			
Burlington	74	73	71
Virginia			
Norfolk	79	78	78
Richmond	79	78	77
Roanoke	76	75	74
Washington			
Ellensburg	67	65	63
Seattle/Tacoma	66	64	63
Spokane	66	64	63
West Virginia			
Charleston	76	75	74
Morgantown	76	74	73
Parkersburg	77	76	75
Wisconsin			
Green Bay	75	73	72
Madison	77	75	73
Milwaukee	77	75	73
Wyoming			
Casper	63	62	60
Cheyenne	63	62	61
Rock Springs	58	57	56

STANDARDS AND SPECIFICATIONS

The Cooling Tower Institute has developed over the years several standards which are important to industrial cooling-tower design. When writing a specification for a new tower, if you write, "This is to be a CTI code tower," immediately the CTI standards become a part of your specifications and your contract with that manufacturer; for example:

1. "Redwood Lumber Specification," Standard 103
2. "Gear Speed Reducers," Standard 111
3. "Pressure Preservative Treatment of Lumber," Standard WMS-112
4. "Douglas Fir Lumber Specifications," Standard 114
5. "Timber Fastener Specification," Standard 127
6. "Asbestos Cement Materials," Standard 127
7. "Acceptance Test Code," Bulletin ATC-105

Manufacturers are protected since all bidding will be on the same basis; the buyer is protected with the assurance of getting a quality product. A sample set of specifications follows.

SUGGESTED COOLING TOWER SPECIFICATIONS

I. General

This specification covers the construction of an induced-draft, counterflow water cooling tower at (location) for (company) (hereafter referred to as Owner).

Each cell of the cooling tower is to be capable of individual operation with its own water supply and mechanical equipment. The design and construction of the cooling tower shall conform to the latest applicable provisions of the Cooling Tower Institute Standards and shall be a CTI Code tower.

Bids are to be submitted on CTI Bid Forms STD-118, pages 1 through 4, with all items completed.

Attached Plant Safety Requirements are a part of this specification and will become a part of the issued contract.

II. Facilities Furnished by Others

a. Power wiring, motor controls, and all electrical labor

b. Materials and installation labor for external piping to and from the tower, including valves

c. Necessary concrete cold-water basin

d. 110-V, 60-cycle, 1-kW, single-phase power to contractor at one location at the tower site (additional facilities which others are willing to supply to contractor)

III. Design Data

Circulation rate	________gal/min (m^3/h)
Water temperature to tower (hot-water temperature)	________°F (°C)
Water temperature from tower (cold-water temperature)	________°F (°C)
Inlet wet-bulb temperature	________°F (°C)
Range	________°F (°C)

Approach	________°F (°C)
Wind velocity maximum	________mi/h (m/s)
Wind loading maximum (standard 30 lb/ft^2)	________lb/ft^2 (kg/m^2)
Basin depth	________ft (m)
Tower location	________ft above grade (m)
Drift loss (0.2 percent of circulation rate standard)	________percent

Bidder to include tower characteristic curve with bid and state design L/G ratio.

IV. Evaluation

a. Fan horsepower evaluation will be added to the base price of the tower at $________ per horsepower (or $________ per horsepower per year for ________ years).

b. Pumping head evaluation will be added to the base price of the tower at $________ per foot (meter) of head.

c. Concrete cold-water basin evaluation will be added to the base price of the tower at $________ per square foot (square meter) of plan area at the base of the tower.

V. Materials

A. General

1. Lumber

All lumber used shall be heart redwood as graded and specified in CTI Standard 103, Grades II and III. No plywood is allowed in any portion of the tower. (Alternate: Douglas Fir)

2. Preservation Treatment

All lumber shall be treated. Lumber shall be cut to dimensions, notched, and drilled prior to preservative treatment. Treatment is to be with chromated copper arsenate, Type B (CCA-b) and in accordance with CTI Standard WMS-112.

3. Hardware

(See CTI Standard TPR-126 for charting of materials.)

All bolts, nuts, and washers shall be ____________. All nails shall be ________. Other hardware, such as connectors and base anchors, shall be galvanized or cast iron.

B. Component Parts

1. Framework

a. All tower columns shall not be less than 4 × 4 in (10 × 10 cm) nominal.

b. All connections and joints are to be carefully fitted and bolted. Nailing or notching of structural members will not be permitted. Nonframework members such as fill, sheathing, and louvers shall not be called upon to furnish part of the structural strength of the tower.

2. Fan Deck

a. Fan deck shall be designed for a live load of 60 lb/ft^2 (kg/ m^2) and shall be reinforced for any concentrated or distributed dead loads. Fan decking shall be tongue-and-groove with nominal 2 in (5-cm) thickness.

b. On counterflow selections one access door per fan cell shall be furnished through the fan deck. A ladder is to be supplied for access from the fan deck to the drift eliminators.

3. Fill

Tower fill shall be pressure-treated, clear heart redwood, treated Douglas fir, cement asbestos, or PVC. On crossflow selections fill supports are to be PVC-coated steel hangers. Bidder to specify vertical and horizontal spacing of fill bar, fill bar size, and fill depth.

4. Drift Eliminators

Shall be treated redwood, treated Douglas fir, PVC, or cement asbestos.

5. Hot-Water Distribution System

Counterflow The tower shall be provided with a complete water distribution system fabricated from Schedule 80 PVC pipe header and laterals with self-draining, nonclogging, full-pattern spray nozzles. Piping shall terminate with one flanged connection for each cell approximately 1 ft outside the tower casing to permit shutdown of any cell without affecting operation of other cells.

Crossflow The hot-water distribution basin floor shall be constructed of tongue-and-groove redwood (or Douglas fir) with downtake orifice nozzles constructed of polypropylene with integral splash surface diffusers. Each basin shall have a flow-control valve capable of full shutoff.

6. Fans and Drives

a. Each fan shall be of propeller type having at least six adjustable-pitch blades of glass-reinforced epoxy. The fans shall be properly rated P.F.M.A. and shall be statically and dynamically balanced prior to shipment.

b. All motors shall be suitable for across-the-line starting and shall be designed for cooling-tower service. Motors shall be installed outside the exit airstream, and nameplate rating shall not be exceeded when tower is operating within the limits of design conditions specified. Motors shall be wired for three-phase, 60-cycle, _________-V power.

c. The fans shall be driven through right-angle, heavy-duty, cooling-tower reduction-gear assemblies having a minimum service factor of 2.0. Reduction gears shall be provided with vent line and oil-fill line extending to outside of fan stack. Oil-fill line shall include oil level sight gauge. Reducers shall conform to CTI Standard 111.

d. The drive-shaft assembly connecting the motor and gear reducer shall be of nonlubricated design, Thomas SN, or equal. Two drive-shaft guards shall be supplied, one at each end of each shaft.

e. Supports for motor and reducer assembly shall be of unitized steel construction. Minimum thickness of steel employed shall be ¼ in (6.4 mm).

f. One vibration cutout switch shall be provided with each fan.

7. Fan Stack

Fan stacks shall be venturi entrance type, not less than 6 ft (1.83 m) high, constructed of glass-reinforced polyester.

8. Partitions

Towers consisting of two or more cells shall have a solid transverse partition wall between all cells. It will extend from louver face to louver face and from basin-curb level to fan-deck and distribution-deck level.

All partition walls will be constructed of treated redwood, Douglas fir, or corrugated fiberglass-reinforced polyester plastic.

9. Casing

Tower casing shall be single-wall, 8-oz, 4.2 corrugated fiberglass-reinforced polyester plastic. Casing sheets shall have a minimum of one corrugation overlap at all seams and shall be of watertight design.

10. Air Inlet Louvers

Louvers shall be of treated redwood, Douglas fir, or fiberglass-reinforced polyester plastic supported on spans of 4 ft (1.22 m) or less.

11. Access

a. At least one ladder and one stairway at opposite ends of the tower shall be provided, extending from the ground level to the top of the fan deck.

b. Stairways and ladders shall be in accordance with OSHA requirements.

c. Handrails around the top of the tower shall be provided in accordance with OSHA requirements.

d. Access to plenum chamber on crossflow towers shall be through the end-wall casing at basin-curb elevation. Access doors will be provided through each partition wall. A walkway shall be provided at basin-curb level from end wall to end wall.

VI. Drawings

The cooling-tower manufacturer shall submit three copies of complete drawings for approval. Catalog drawings will not be considered acceptable as approval drawings. Approval drawings shall clearly show exact dimensions and all construction details.

VII. Testing

PERFORMANCE TEST The cooling-tower manufacturer shall conduct a performance test in accordance with the "Cooling Tower Institute Acceptance Test Procedure for Industrial Water Cooling Towers," CTI Bulletin ATC-105, latest revision. The cooling-tower manufacturer shall quote a separate price for conducting the performance test.

VIII. Guarantee

The cooling tower shall be guaranteed for a period of 1 year after structural completion or 18 months after shipment, whichever occurs first. Any defective parts or workmanship shall be repaired at the cooling-tower manufacturer's expense.

OPERATION AND MAINTENANCE

Evaporation Loss

In the usual cooling-tower operation the water evaporation rate is essentially fixed by the rate of removal of sensible heat from the water, and the evaporation loss can be roughly estimated as 0.1 percent of the circulating water flow for each Fahrenheit degree of cooling range.

Drift Loss

Cooling-tower drift loss is the entrained liquid water droplets being discharged with the exit air. The function of the drift eliminators is to limit the number of escaping droplets to an acceptable level. Most design specifications state the permissible drift loss as a percentage of the circulating water flow. Most modern cooling towers are designed with drift eliminator face velocities below 650 ft/min (198.12 m/min) and entrainment losses of less than 0.1 percent of the water circulation rate.

Blowdown

Cooling-tower blowdown is a portion of the circulating water that is discharged from the system to prevent excessive buildup of solids. The maximum concentration of solids that can be tolerated is usually determined by the effects on the various components of the cooling system, such as piping, pumps, heat exchangers, and the cooling tower itself. The required blowdown rate is determined from a material balance, yielding

$$b = \frac{e}{r - 1} - d$$

where
- b = blowdown rate, gal/min (m^3/h)
- e = evaporation loss rate, gal/min (m^3/h)
- r = ratio of solids in blowdown to solids in makeup, cycles of concentration
- d = drift loss, gal/min (m^3/h)

Makeup-Water Rate

To hold a given solids concentration ratio, sufficient water must be added to the recirculating water system to make up for evaporation, blowdown, drift, and other losses. The required makeup rate may be computed by either of the equations

$$M = b + e + d$$

or

$$M = \frac{r}{r - 1}(e)$$

where
- M = makeup rate, gal/min (m^3/h)
- b = blowdown rate, gal/min (m^3/h)
- e = evaporation loss rate, gal/min (m^3/h)
- d = drift loss, gal/min (m^3/h)
- r = ratio of solids in blowdown to solids in makeup (cycles of concentration)

Proper maintenance of the mechanical equipment and distribution system will ensure optimum operation from the cooling tower over a long period of time. Motors must remain properly lubricated, gearbox oil should be maintained at the proper level, and drive-shaft alignment should be checked on a regular basis. Uniform distribution of hot water within the tower is essential in order to maintain optimum tower performance. Refer to Table 3-2 for a suggested preventive maintenance schedule for key tower components.

COLD-WEATHER OPERATION

General

The successful operation of induced-draft cooling towers during extremely cold weather very often presents a problem to the plant operators. Ice tends to form on the air-intake louvers and the filling immediately adjacent to the louvers. This is because the water in contact with the airstream will intermittently splash on the louver boards, where it freezes, and ice will eventually build up to the point where the flow of air is restricted. Also, subcooling will be obtained in the area of the filling immediately adjacent to the louvers, where ice will build up during periods of light load or high air velocity. Particular care must be taken in starting up fans that have been shut down for an appreciable length of time because of the possibility of unequal ice loading on the blades. If a fan is started up in an unbalanced condition, the gear unit could be torn from its mounting,

TABLE 3-2 Master Periodic Inspection Chart

	What to look for	What to do
Weekly		
Fan	Noise, vibration, tower sway	Visually check for damage; check weep holes for water in fan blades.
Speed reducer	Noise, rapid vibration, oil leaks	Check oil level, check oil for water or other contamination; check breather pipe for clogging, shaft for misalignment. Run idle units for 10 min.
Drive shaft	Vibration, broken disks	Replace broken disks; tighten bolts.
Suction pit screen	Debris	Remove debris.
Monthly		
Speed reducers	Routine check	Check oil level and contamination. Inspect oil on high- and low-speed shafts for lubrication.
Drive shaft	Routine check	Check alignment.
Nozzles	Scale, corrosion, debris	Clean; replace if damaged.
Headers and laterals	Scale or clogging	Spot-check nozzles on side of tower opposite from risers. Clean.
Distribution decks	Algae, debris, channeling, deposits of lime, scales, etc.	Clean with steam, high-pressure hose, or stiff brush.
Yearly		
Tower	Routine check	Shut down; clean thoroughly from top to bottom, including basin, with steam or fire hose.
Grid decks	Warping, water channeling	Replace individual bars as needed.
Structure	Decay, excessive delignification	Use ice pick. Replace structural members if necessary.
Bolts	Looseness, corrosion	Tighten all bolts; replace those corroded.
Wall sheathing	Leaks	Caulk as necessary. Keep tower wet.
Mechanical equipment	Fan-blade damage, unbalanced pitch, fan-shaft looseness, speed-reducer wear, shaft alignment	Make checks and corrections as necessary.
Basin	Dirt, debris, signs of oil	Clean thoroughly.

causing permanent damage to the mechanical equipment and surrounding tower structure.

Control of Ice on Air-Intake Louvers and Filling

Every effort should be made to maintain the design water quantity and heat load per cell. In the event of a reduction in the plant heat load, it is extremely important that the water quantity be reduced proportionately and that cells be shut down to maintain the design quantity per cell; that is, riser valves shut on idle cells. In addition, the water temperature in the basin should be maintained at a reasonable level, such as 60 to 70°F (15.6 to 21.1°C) by reducing the volume of air entering the tower.

The greater the air volume handled by the fans, the greater the amount of ice formation. It is our recommendation that the fan motors on a multicell cooling tower be reduced from full speed to half speed as required to maintain the temperature of the water in the basin at 60° to 70°F (15.6°C to 21.1°C). In the event that all fans are running at half speed and the basin temperature falls below 60°F (15.6°C), it will then be necessary to shut down the fans as required to maintain this temperature.

Usually the water concentration in gallons per square foot (liters per square meter) of cell area is sufficient to cause a reverse of airflow when the fans are not operating. This will tend to melt any ice that is formed on the filling and the louvers. In a multicell tower, the fans should remain turned off for approximately 12 h and then turned back to low speed. The adjoining cell should then have its fan turned off for the same period of time, and this operation should be repeated on the other cells during cold weather.

Should this procedure prove ineffective in controlling the ice formation, one of the following recommendations should be followed:

1. Remove ice from the louvers with a steam hose. Be careful not to allow the ice load on the fill to exceed its design, causing the filling to collapse while thawing.
2. Remove louver boards. Some counterflow and straight-sided crossflow towers have the top louver board and every fifth louver beneath of double width. The intermediate louver boards may be removed, thereby reducing the amount of ice forming between these louver blades and choking off the air supply.
3. Install reversing switches on low-speed fan motor terminals. This will reverse the flow of air through the tower, which will quickly melt any ice that is formed on the louvers and filling. The fan motors should not be operated in reverse for more than about 30 min at a time. During extremely cold weather, each of the fans operating at half speed should be reversed once a day to remove ice from the air-intake louvers.

REFERENCES

1. Sherwood, T. K., and R. L. Pigford: *Absorption and Extraction,* 2d ed., McGraw-Hill, New York 1952, pp. 102–104.
2. Kern, D. Q.: *Process Heat Transfer,* McGraw-Hill, New York, 1950.
3. *Cooling Tower Performance Curves* Blue Book, Cooling Tower Institute, Houston, Texas, 1967.
4. "Evaluated Weather Data for Cooling Equipment Design," Addendum No. 1, Fluor Products Co., Inc., 1964.
5. Baker, D. R., and L. T. Mart: "Cooling Tower Characteristics as Determined by the Unit Volume Coefficient," *Refrigerating Engineering,* vol. 60, pp. 965–971 (1952).

part B

Plant Operation Equipment: Selection and Maintenance

section 7

Mechanical Power Transmission

chapter 7-1

Gearing and Enclosed Gear Drives

prepared by

American Gear Manufacturers Association*
Arlington, Virginia

GEARING

Functions of Gear Drives

The major functions of gears and gear drives are: reduction of speed, multiplication of torque, and positioning of shafts.

Speed Reducer

Economically, it is normally better to use a small, high-speed prime mover and gear-reducer combination than a larger, low-speed power source.

*Individual contributors:
James E. Gutzwiller, Boston Gear Division of INCOM International, Inc.
Harold O. Kron, Philadelphia Gear Corporation
Raymond J. Birck, The Falk Corporation

Speed Increaser

In some instances, it is impractical to operate a prime mover at a speed high enough to suit requirements of the driven equipment. For such applications, gears may be used as a speed increaser.

Shaft Orientation

Gears may provide desired orientation and relative rotation of shafts. Some common arrangements available are: in-line, parallel shaft, and right angle. Miter gears (1:1 ratio bevel gears), for example, serve the specific purpose of providing a 90° shaft orientation. Other angles can be supplied by specially designed gears of several types.

Gear Types

The most common types of gears are illustrated in Figs. 1-1 to 1-11. Other available types are generally modifications of the basic gears shown.

Spur Gears. These are cylindrical in form and operate on parallel axes. The teeth are straight and parallel to the axes (Fig. 1-1).

Spur Rack. A spur gear has straight teeth that are at right angles to the direction of motion (Fig. 1-1).

Helical Gears. A helical gear is cylindrical in form and has helical teeth. Parallel helical gears operate on parallel axes and, where both are external, the helices are of opposite hand (Figs. 1-2, 1-3, 1-12, and 1-13).

Double Helical Gears. Each of these has both right-hand and left-hand helical teeth, and operates on parallel axes. These also are known as *herringbone gears* (Fig. 1-4).

Crossed Helical Gears. These gears operate on crossed axes and may have teeth of the same or opposite hand. The term *crossed helical gears* has superseded the old term *spiral gears* (Fig. 1-5).

Worm Gear. This gear is the mate to a worm. A worm gear that mates with a cylindrical worm is said to be single-enveloping. Worm gearing operates on nonintersecting axes (Figs. 1-6, 1-7, 1-14, and 1-15).

Cylindrical Worm Gearing. This is a form of helical gearing that mates with a worm gear (Fig. 1-6).

Double-Enveloping Worm Gearing. This comprises hourglass worms mated with a worm gear (Fig. 1-7).

Bevel Gears. These are conical in form and operate on intersecting axes that usually are at right angles (Figs. 1-8 to 1-10).

Straight Bevel Gears. These gears have straight-tooth elements which, if extended, would pass through the point of intersection of their axes (Fig. 1-9).

Spiral Bevel Gears. These have teeth that are curved and oblique (Figs. 1-10 and 1-16).

Hypoid Gears. Similar in general form to bevel gears, hypoid gears operate on nonintersecting axes (Fig. 1-11).

Gear Geometry

Gear-Tooth Action

Gears of all types have the common characteristics of theoretically smooth transmission of motion through engagement of successive teeth. Certain design parameters must be met to ensure that one pair of teeth starts engagement before the preceding pair leaves off.

Standard Tooth Forms

The involute form is almost universally used for spur and helical gears and has some application for worm and bevel gears. The involute provides for accuracy of motion transmission, even when there is some change in center distance between gears. It also offers a number of manufacturing advantages. In worm and bevel gearing, conjugate forms are used to suit the manufacturing processes employed.

Modified and Special Tooth Forms

Common modifications to involute gear teeth are: long and short addenda, stub teeth, and tip and root relief. Special tooth forms are sometimes used to provide higher capacity. Normally, accurate mounting of the gears must be maintained to realize this advantage.

Gear Materials and Heat Treatment

Common Materials

Gears may be made up of a wide range of materials that may be ferrous, nonferrous, and nonmetallic. For industrial applications, steel and iron are most commonly used for spur, helical, and bevel gears. Worm-gear pairs generally are made up of bronze or iron for the gear and steel for the worm. Use of plastics is generally limited to light applications, particularly those in which minimal lubrication is available.

Hardened vs. Unhardened Material

Since the gear set rating is governed to a great extent by hardness of the teeth, heat treating often is used to provide higher capacity in the same space. Pinions, which endure more load cycles, generally are made slightly harder than their mating gears. Hardening adds to the cost, particularly where an additional finishing operation is required to correct distortion. For some applications, untreated gears may prove more economical. Obviously, this will be true when the equipment arrangement will not permit use of smaller hardened gears.

Material Selection

Information regarding material selection and properties is contained in AGMA 390, "Gear Handbook" (part II, sec. 2) and Ref. 1.* The selection of specific material and treatment combinations should be based on an analysis of the overall requirements and conditions. Some of the fundamental factors to be considered when making material and treatment selections for gearing are as follows:

- Factor of safety, loading, duty cycle, mounting, gearing enclosure, lubrication, and ambient atmospheric conditions.
- For replacement gearing, the life obtained from the previous gearing should be evaluated. If satisfactory, replace with similar material. If longer life is required, selection of a heat-treatment specification yielding a higher hardness and, if necessary, a better material, may provide the desired improvement.
- Annealed carbon steels, bar stock, forgings, or castings, are usually satisfactory for pinions and gears for uniform or moderate shock loads when the size of the gearing is not an important factor.
- Annealed carbon-steel pinions with cast-iron gears are sometimes used for the same reason mentioned above for annealed carbon steels.

*References are listed at the end of the chapter.

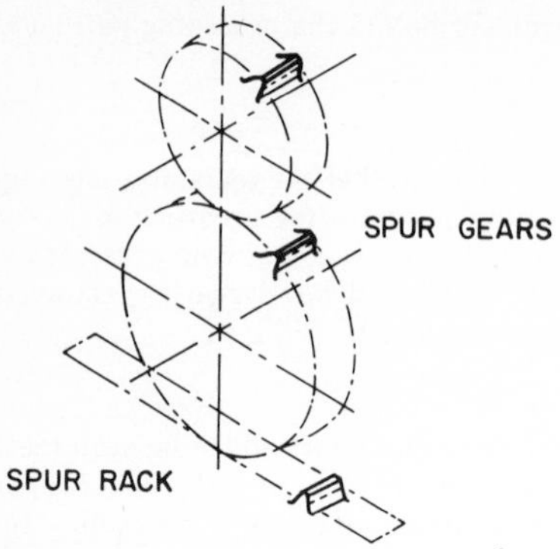

Figure 1-1 Spur gears and spur rack. *(Extracted from Ref. 2.)*

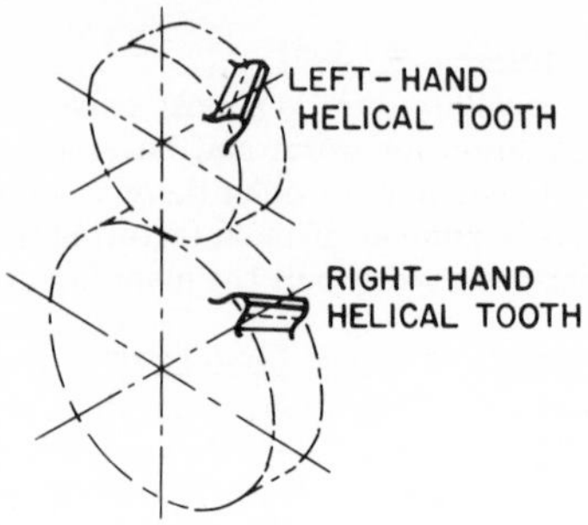

Figure 1-2 Parallel helical gears. *(Extracted from Ref. 2.)*

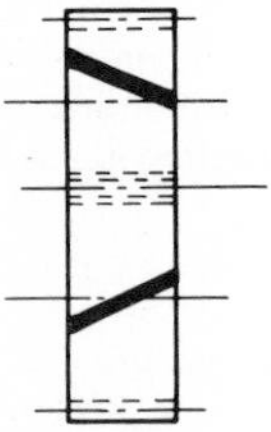

Figure 1-3 Single helical gears. *(Extracted from Ref. 2.)*

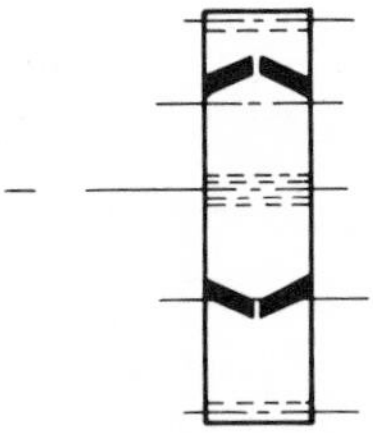

Figure 1-4 Double helical (herringbone) gears. *(Extracted from Ref. 2.)*

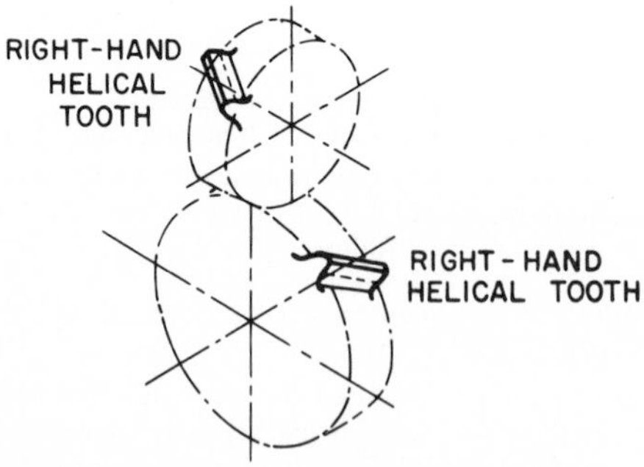

Figure 1-5 Crossed helical gears. *(Extracted from Ref. 2.)*

- Alloy-steel pinions are used when there are increased loads or greater life is desired. They may be used with cast iron or annealed (forged or cast) steel gears, usually when the ratio is about 6:1 or higher.
- Alloy-steel pinions and gears, heat-treated, should be used with the higher hardness ranges when space limitation is a factor, i.e., where smaller center distances and face widths may be necessary.

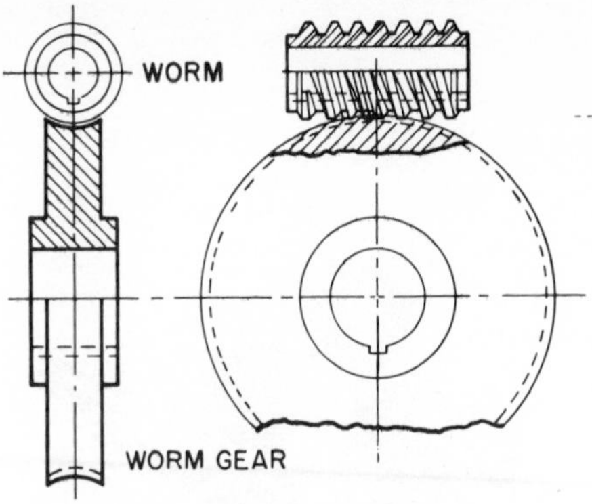

Figure 1-6 Cylindrical wormgearing, single-enveloping. *(Extracted from Ref. 2.)*

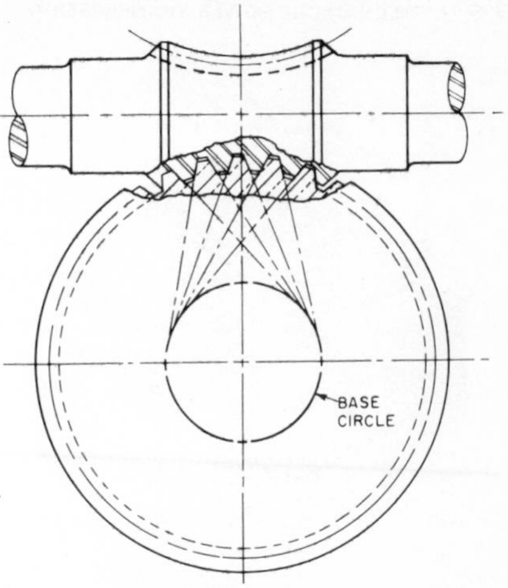

Figure 1-7 Double-enveloping Cone® worm gearing. *(Extracted from Ref. 2.)*

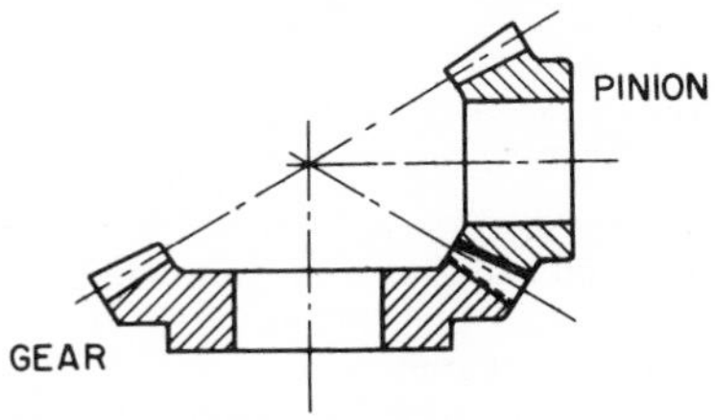

Figure 1-8 Bevel gears. *(Extracted from Ref. 2.)*

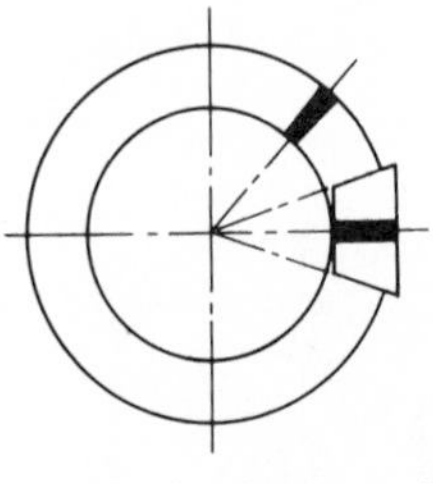

Figure 1-9 Straight bevel gears. *(Extracted from Ref. 2.)*

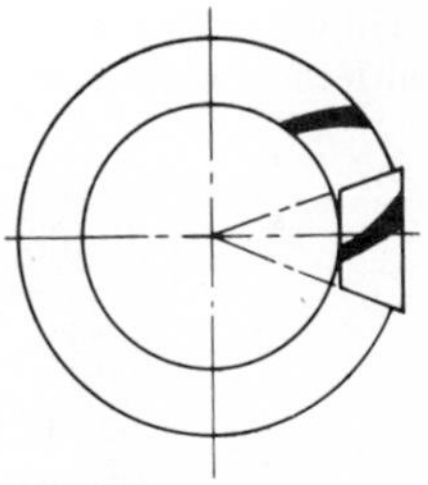

Figure 1-10 Spiral bevel gears. *(Extracted from Ref. 2.)*

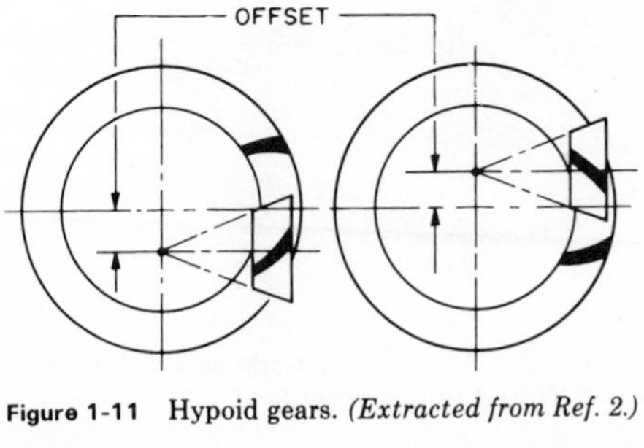

Figure 1-11 Hypoid gears. *(Extracted from Ref. 2.)*

Figure 1-12 High-speed single-reduction helical gear drive. *(Philadelphia Gear Corporation.)*

Figure 1-13 Double-reduction helical gear drive. *(Philadelphia Gear Corporation.)*

- Steel pinions and gears which are to be machined and cut after heat treatment should have the pinion hardness specified as follows:
  Ratios up to 2:1; pinion and gear to be same minimum hardness
  Ratios from 2:1 to 8:1; minimum hardness of pinion to be 40 Bhn (Brinell hardness number) higher than minimum gear hardness
  Ratios of 8:1 and higher; pinion to be more than 40 Bhn harder at minimum than gear
- Steel pinions and gears hardened to 400 Bhn or higher after cutting are generally specified with the same hardness, unless extremely high hardness is desired for the pinion.

Figure 1-14 Double-reduction worm-gear drive. *(Boston Gear Division of INCOM International Inc.)*

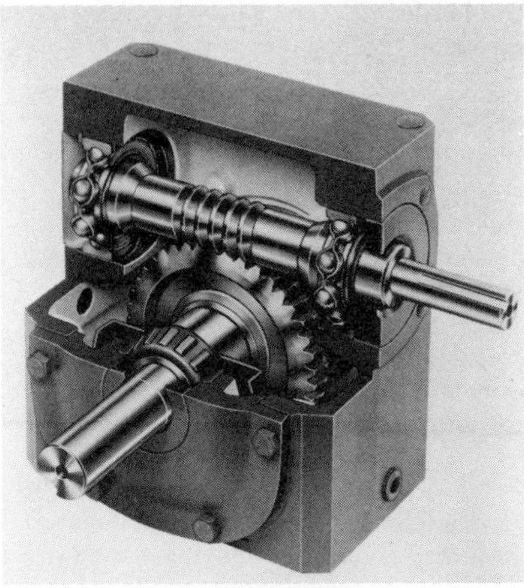

Figure 1-15 Single-reduction worm-gear drive. *(Boston Gear Division of INCOM International Inc.)*

- A range of 40 Bhn should be specified for minimum hardnesses up to 285 Bhn. A range of 50 Bhn should be specified for minimum hardnesses over 302 Bhn.[1] For example, a pinion might be specified with a range from 285 to 321 Bhn and the mating gear from 223 to 262 Bhn.
- When core hardness is a requirement, consideration should be given to the size and shape of the cross sections involved.
- Where impact loads exist, the use of alloys and the lowering of hardness are recommended for carburized gears and pinions.
- When accelerated wear is encountered in service, heat-treated gearing providing greater hardness will, in most cases, help to alleviate this condition. Consult with an AGMA company member for appropriate recommendations.

Figure 1-16 Single-reduction spiral bevel gear drive. *(The Falk Corporation.)*

- On flame- and induction-hardened gears, where hardness in the root area is a requirement, it should be specified.

Manufacturing Processes

Generating

Most gear teeth are produced by a generating process that takes into account the shape of the tool and relative motions between the tool and workpiece. Examples are: hobbing, shaping, rolling, grinding, and shaving.

Forming

The forming process produces gear teeth by a direct replication method. Examples are: casting, molding, and broaching.

Finishing

Finishing operations remove a relatively small amount of material from gear teeth. These may be used to improve accuracy and finish. Shaving is used to improve profile and finish on relatively soft gears. Grinding and honing are employed for harder gears.

Size Limitations

General limitations on gear diameters are listed below. A specific company may have the capability of producing larger gears of a particular type. Spur gears of very large diameter for low-speed application have been cut in arc sections and also fabricated from rack sections, bent to the proper radius of curvature.

	Approximate Maximum Diameter	
Gear Type	*in*	*m*
Spur	400	10
Helical	400	10
Straight bevel	100	2.5
Spiral bevel	90	2.3
Worm gear	150	4

Accuracy Requirements

Requirements for gear accuracy may be governed by several factors. Operating speed and gear size are probably the most important. From an application standpoint, specifications on sound level and smoothness of motion transmission are often controlling.

AGMA 390, "Gear Handbook," contains a classification system covering tolerances for common types of gears. In this system tolerances are tabulated for a series of *quality numbers* that range from Q3 to Q16. The higher the quality number the more precise are the tolerances. The handbook covers the gears themselves and does not apply to gears in enclosed drives.

By far the majority of applications can be satisfied by gears in the range of Q5 to Q8. Rarely does a job require a gear more precise than class Q14. AGMA 390 tabulates a range of suggested quality numbers for various applications. This is only a guide. Judgment should be used in making a final selection to meet the specific conditions involved.

Precise Motion Transmission

Many applications use gears as a means for precise transmission of motion. Usually, high AGMA quality numbers (Q12 to Q14) are required for these gears. Examples of such applications are: navigational tracking, telescopes, and index devices for machine tools.

High-Speed Considerations

Gear inaccuracies become more critical as operating speed increases. The most obvious problem may be a high noise level. In addition, the dynamic loads on the teeth caused by these errors may be a substantial part of the total transmitted load.

Pitch line velocity, which takes into account size as well as rotational speed, is the usual measure of gear speed. The unit normally used to measure this velocity at tooth mesh is feet per minute.

AGMA 390 suggests the following quality numbers for power drives in the machine tool industry.

Gear Pitch Line Velocity, ft/min	*Quality Number*
0–800	Q6–Q8
800–2000	Q8–Q10
2000–4000	Q10–Q12
Over 4000	Q12 and up

Accuracy vs. Cost

Cost of manufacturing and inspection escalates rapidly with increasing quality number. Therefore, the user should use care to avoid specifying higher classes than required for the application.

Mounting of Gears

Close attention must be given to the mounting of gears and the maintenance of alignment under operating conditions. Obviously, precision of mounting must match that of the gears.

Enclosed Gear Drives

Advantages

Enclosed gear drives marketed by gear manufacturers offer several advantages over open power-transmission devices.

- Safety—protection from moving parts
- Retention of lubricant
- Protection from environment
- Economics of quantity manufacture

Types and Features

Enclosed gear drives generally are classified by the principal type of gearing used. They may have a single set of gears or additional gears of either the same or different types to form multiple reductions. Figures 1-12 to 1-17 provide illustrations of common types. Table 1-1 covers important features of each.

Mounting

Gear drives may be designed for base, flange, or shaft mounting. The last type makes use of a hollow output shaft for direct mounting on the driven shaft (refer to Fig. 1-18). A reaction arm or similar device is required to secure the unit against rotation.

Gear Motors

A gear motor is an integral drive unit incorporating an electric motor and gear reducer, with the frame of one supporting the other. Some designs use motors with special shaft ends and/or mountings, while others adapt standard motors (refer to Fig. 1-19).

Figure 1-17 Double-reduction spiral bevel–helical gear drive. *(The Falk Corporation.)*

Normal Speed vs. High Speed

AGMA standards for enclosed gear drives, used in general industrial service, limit input speed to 3600 r/min. An additional limitation is imposed: 5000 ft/min pitch line velocity for helical and bevel units and 6000 ft/min sliding velocity for cylindrical worm gears. Above these limits, special consideration should be given to such items as gear quality, lubrication, cooling, bearings, etc. See Refs. 3 and 4.

TABLE 1-1 Enclosed Gear-Drive Units—Features (Refer to Figs. 1-12 to 1-15.)

Type of unit	Ratio range*	Horsepower range*	Efficiency %†	Fig. no.
Helical & double helical				
Single reduction	to 7:1	to 20,000	96–98	1-12
Double reduction	5:1–40:1	10,000	95–97	1-13
Triple reduction	20:1–300:1	3,000	93–95	——
Quadruple reduction	150:1–1000:1	1,000	91–93	——
Bevel, helical	5:1–40:1	3,000	95–97	——
Bevel helical, helical	20:1–300:1	2,000	93–95	——
Bevel				
Single bevel	to 9:1	2,000	96–98	
Worm gear				
Single reduction	5:1–70:1	300	50–96	1-15
Double reduction	25:1–4900:1	100	20–92	1-14
Helical, worm	20:1–300:1	150	50–94	——

*The information on ratio and horsepower ranges is approximate and is for the product usually offered.

†Efficiency is based on transmission of full rated power.

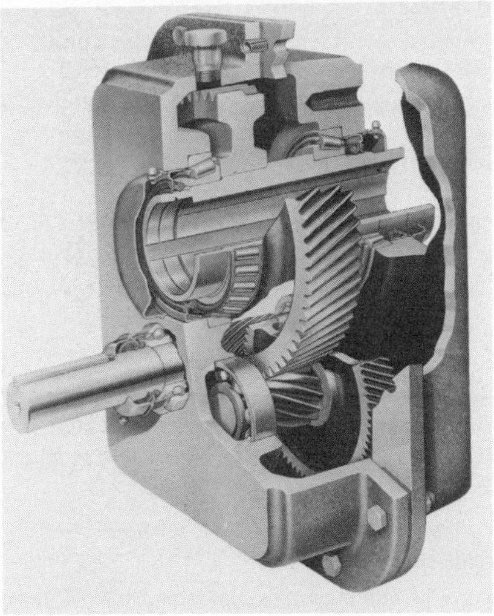

Figure 1-18 Shaft-mounted double-reduction helical gear drive. *(The Falk Corporation.)*

INSTALLATION AND MAINTENANCE

The variety of types and sizes of gears and enclosed gear drives makes it impractical to cover installation and maintenance in specific detail. The user should refer to the manufacturer's literature, nameplate data, and warning tags. Such information should take precedence over the generalized comments that follow.

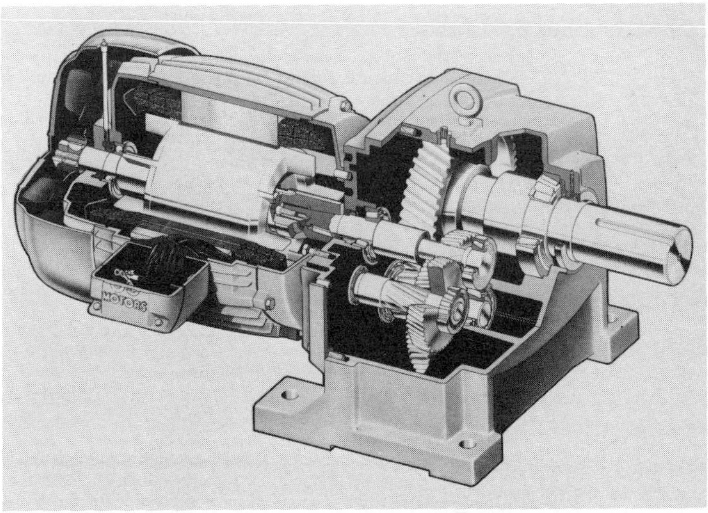

Figure 1-19 Triple-reduction helical gear motor. *(U.S. Electrical Motors.)*

Mounting and Installation of Gears

Accuracy of mounting must be commensurate with the quality of the gears themselves to obtain optimum results. Some types of gears require close endwise positioning of either or both members of a pair to obtain proper operation. Examples are bevel gears (both), cylindrical worm gears (gear only), double-enveloping worm gears (both). This positioning must be provided by bearings of suitable capacity to accommodate the thrust loads involved.

Provision must be made for adequate lubrication of open or semi-enclosed gears and guarding for safety. See Chap. 17-1, "Lubricants," for information on gear lubrication. Reference 5 also provides information on lubricant types and methods of lubrication.

Installation and Start-Up of Enclosed Gear Drives

The handling, installation, and servicing of a new enclosed gear drive deserves close attention to avoid damage and to assure proper operation. A checklist of important items is provided in Table 1-2.

Lubrication of Enclosed Gear Drives

Improper lubrication is one of the major causes of failure of gear drives. The gear manufacturer's instructions must be followed to assure proper operation.

Reference 6 provides detailed information on recommended lubricants and maintenance procedures. Gear type, size, and speed, along with ambient temperature range, are major influencing factors on lubricant selection.

The gear unit should be drained and cleaned with a flushing oil after 4 weeks of initial operation. For refilling, either the filtered original lubricant or new lubricant may be used. For normal operation, oil changes should be made after every 2500 h of service.

Periodic checks must be made on oil levels, oil cups, and grease fittings. When pressure lubrication is used, proper functioning of pump, filter, and cooler should be frequently audited.

TABLE 1-2 Installation and Start Up Checklist

Step	Instruction
Storage	If necessary to store or maintain the gear unit in an inactive condition for more than a month, contact the manufacturer to determine need for protective action.
Handling	Observe manufacturer's instructions for unpacking and handling of gear drive.
Foundation	Provide adequate foundation commensurate with size and type of unit. Surface is to be level unless drive has been specifically designed for other positioning.
Accessibility	Provide adequate space to permit future maintenance.
Auxiliary parts	Assemble components, such as coupling hubs, sprockets, etc., to shafts with shrink or slip fits in accordance with instructions. Do not force fit.
Alignment	Align shafts and auxiliary drives accurately. Most couplings are designed to accommodate only minor misalignment.
Guards	Install suitable guards for safety in accordance with OSHA standards.
Lubrication	Observe manufacturer's instructions, using the specified lubricants. *Note: Most gear drives are shipped without lubricant.*
Rotation	Check for proper rotation and freedom from obstructions before start of full operation. For pressure lube systems, make certain pump is delivering oil.
Operation	Inspect for oil leaks and unusual noise or vibration immediately after start-up. Check oil temperature after several hours. A temperature of 180 to 200°F is not unusual for most gear drives. After first week, recheck alignment and tightness of all fasteners, fittings, and pipe plugs.

TABLE 1-3 Trouble Chart

Trouble	What to inspect
Heating	Is unit and fan assembly covered with dirt?
	Is unit overloaded?
	Has recommended oil level been exceeded or is level too low?
	Are couplings in alignment?
	Have bearings been properly adjusted?
	Are oil seals or stuffing boxes the cause?
	Is oil clean or is sludge content high?
	Has oil filter been cleaned?
	Is oil pump functioning?
Shaft failure	Check alignment; most shafts fail owing to misalignment. Some troubles are caused by use of rigid couplings. Is overhung load beyond capacity of unit?
	Is unit subject to high energy loads or extreme repetitive shocks not previously considered?
Bearing Failure	Rust formation caused by high humidity or the entrance of water.
	Unsuitable lubricant.
	Abnormal loading causing excessive deflection results in flaking, cracks, and fractures. Improper adjustment causes abnormal loading if bearings are pinched or abnormal gear wear if bearings are too free. This is dependent upon type of bearing and the possible lack of lubrication.
Oil leakage	Check oil seals and replace if worn.
	Check stuffing boxes and adjust or replace packing.
	Check tightness of drain, level, and other plugs or fittings.
Wear	Backlash may be insufficient.
	Misalignment due to worn bearing.
	Incorrect lubrication.
	Insufficient lubrication.
	Lubricant carrying foreign matter, viz., abrasive dirt or particles of worn metal teeth.[7]
	Excessive temperature.
	Excessive speeds.
	Excessive loads.
Noise or vibration	Bad alignment.
	Loose or worn bearings.
	Insufficient lubrication.
	Excessive lubrication.

Troubleshooting

Alertness to changes in operating characteristics, such as increased temperature rise over ambient noise and vibration and oil leakage, can prevent costly shutdowns. Table 1-3, "Trouble Chart," provides a checklist to diagnose various problems in operation.

APPLICATION OF GEARING AND ENCLOSED GEAR DRIVES

Gear Ratings

AGMA has developed rating formulas for most types of gearing and enclosed gear drives. The ratings determined from these formulas are intended for applications where loads of a uniform nature are applied for no more than 10 h/day. It is these ratings that normally are tabulated in manufacturers' catalogs.

Service Factors

In selecting gearing or an enclosed gear drive for an application, the horsepower to be transmitted is multiplied by a service factor to determine an *equivalent horsepower*. Service factors have been developed from the experience of manufacturers and users to allow for the nature and duration of the transmitted load. Table 1-4 provides a tabulation of service factors extracted from Ref. 3 for enclosed speed reducers or increasers using spur,

TABLE 1-4 AGMA Standard Practice for Enclosed Speed Reducers or Increasers Using Spur, Helical, Herringbone, and Spiral Bevel Gears

		Driven machine load classifications		
Prime mover	Duration of service, h/day	Uniform	Moderate shock	Heavy shock
Electric motor, steam turbine, or hydraulic motor	Occasional, ½	0.50	0.80	1.25
	Intermittent, 3	0.80	1.00	1.50
	Over 3 up to and incl. 10	1.00	1.25	1.75
	Over 10	1.25	1.50	2.00
Multicylinder internal combustion engine	Occasional, ½	0.80	1.00	1.50
	Intermittent, 3	1.00	1.25	1.75
	Over 3 up to and incl. 10	1.25	1.50	2.00
	Over 10	1.50	1.75	2.25
Single-cylinder internal combustion engine	Occasional, ½	1.00	1.25	1.75
	Intermittent, 3	1.25	1.50	2.00
	Over 3 incl. 10	1.50	1.75	2.25
	Over 10	1.75	2.00	2.50

helical, herringbone, and spiral bevel gears. The factors for other types of gears vary slightly from those shown.

Application Classification

Most AGMA standards for enclosed gear drives provide tables for various applications as a guide for selecting service factors. This information is usually also contained in manufacturers' catalogs.

Product Selection

After the equivalent horsepower has been determined, selection of gearing or enclosed gear drives can be made by comparing this figure with the basic rating. It is necessary that the product selected have a rated load capacity equal to or in excess of the equivalent horsepower. An enclosed gear drive usually must also be checked for thermal rating. This is the horsepower that can be transmitted continuously for 3 h or more without causing a temperature of more than 100°F above ambient temperature. Should this limitation prevail, several alternatives are available, such as auxiliary cooling systems, oil pans to reduce churning, or selection of a larger unit.

Systems Considerations

An essential phase in the design of a system of rotating machinery is the analysis of the dynamic (vibration) response of a system to excitation forces.

The dynamic response of a system results in additional loads imposed on the system and relative motion between adjacent elements in the system. The vibratory loads are superimposed upon the mean running load in the system and, depending upon the dynamic behavior of the system, could lead to failure of the system components.

In a gear unit, these failures could occur as tooth breakage or pitting of the gear element, shaft breakage, or bearing failure.

Any vibration analysis must consider the complete system, including prime mover, gear unit, driven equipment, couplings, and foundations. The dynamic loads imposed upon a gear unit are the result of the dynamic behavior of the total system and not of the gear unit alone.

For further information, see Ref. 8.

Sound and Vibration

The greatest concern regarding sound and vibration of gear drives is the contribution to the industrial noise level. A second concern is that these may be sypmtomatic of abnormal wear and impending failure.[9–12] Refer to Sec. 12 of this Handbook, "Noise and Vibration Control," for additional information.

REFERENCES: AGMA STANDARDS APPLICABLE TO ENCLOSED GEAR DRIVES

1. "Gear Materials Manual," AGMA 240.01, 1972.
2. "Terms, Definitions, Symbols, and Abbreviations," ANSI/AGMA 112.05, 1976.
3. "Practice for Enclosed Speed Reducers or Increasers Using Spur, Helical, Herringbone and Spiral Bevel Gears," AGMA 420.04, 1975.
4. "Practice for High Speed Helical and Herringbone Gear Units," AGMA 421.06, 1969.
5. "Lubrication of Industrial Open Gearing," AGMA Specification 251.02, 1974.
6. "Lubrication of Industrial Enclosed Gear Drives," AGMA Specification 250.03, 1972.
7. "Gear-Tooth Wear and Failure," ANSI/AGMA 110.03, 1962.
8. "Systems Considerations for Critical Service Gear Drives," AGMA Information Sheet 427.01, 1976.
9. "Measurement of Sound on High Speed Helical and Herringbone Gear Units," AGMA Specification 295.04, 1977.
10. "Sound for Enclosed Helical, Herringbone and Spiral Bevel Gear Drives," AGMA 297.01, 1973.
11. "Sound for Gearmotors and In-Line Reducers and Increasers," AGMA 298.01, 1975.
12. "Fundamentals of Sound as Related to Gears," AGMA 299.01, sec. 1, 1978.
13. "Manual for Assembling Bevel and Hypoid Gears," AGMA 331.01 (R1976), 1969.
14. "Manual for Machine Tool Gearing," AGMA 360.02, 1971.
15. "Gear Classification, Materials and Measuring Methods for Unassembled Gears," "AGMA Gear Handbook," AGMA 390.03, Vol. I, 1973.
16. "Practice for Single- and Double-Reduction Cylindrical-Worm and Helical-Worm Speed Reducers," AGMA 440.04, 1971.
17. "Practice for Single, Double and Triple-Reduction, Double-Enveloping Worm and Helical-Worm Speed Reducers," AGMA 441.04, 1978.
18. "Practice for Worm Hollow Output Shaft Speed Reducers," AGMA 442.01 (R1974), 1965.
19. "Practice for Gearmotors Using Spur, Helical, Herringbone and Spiral Bevel Gears," AGMA 460.05, 1971.
20. "Practice for Worm Gearmotors," AGMA 461.01 (R1974), 1966.
21. "Practice for Spur, Helical and Herringbone Gear Shaft-Mounted Speed Reducers," AGMA Information Sheet 480.06, 1977.
22. "Spiral Bevel, Helical and Herringbone Gear Units for Water Cooling Tower Fans," AGMA Information Sheet 490.02, 1972.

chapter 7-2

Bearings

PART 1

Rolling Element Bearings

by
William H. Carter
Standards Engineer,
The Torrington Company
South Bend, Indiana

GLOSSARY*

Aligning bearing A bearing which, by virtue of its shape, is capable of considerable misalignment.

*Extracted with permission from Anti-Friction Bearing Manufacturers Association Standards.

Antifriction bearing A term commonly given to ball and roller bearings.

Average life The summation of all bearing lives in a series of life tests, divided by the number of life tests.

Axial clearance See end play.

Basic load rating The calculated, constant, radial, or centric thrust load which a group of apparently identical bearings can theoretically endure for 1 million (10^6) revolutions.

Basic static-load rating The static load which corresponds to a total permanent deformation of ball and race, or roller and race, at the most heavily stressed contact of 0.0001 in (0.00254 mm) of the ball or roller diameter.

Bearing series A graduated dimensional listing of a specific type of bearing.

Bore The area of the bearing making contact with the bearing seat on the shaft.

Boundary dimensions Dimensions for bore, outside diameter, width, and corners.

Cage A device which partly surrounds the rolling elements and travels with them. The main purpose of the cage is to space the rolling elements in ball bearings and space and guide the elements in roller bearings.

Combined load A combination of all radial and axial forces on a bearing.

Cone The inner ring of a tapered roller bearing.

Conrad ball bearing A nonfilling-slot type of ball bearing.

Cup The outer ring of a tapered roller bearing.

Diametral clearance See radial internal clearance.

Double-row bearing A bearing with two rows of rolling elements.

End play The measured maximum possible movement parallel to the bearing axis of the inner ring in relation to the outer ring.

Equivalent radial load The calculated, constant, stationary, radial load, which, if applied to a radial bearing, would give the same life as attained under actual conditions of load and rotation.

Equivalent thrust load The calculated, constant, centric axial load which, if applied to a thrust bearing, would give the same life as attained under actual conditions of load and rotation.

Fixed bearing A bearing which positions a part of the shaft against axial movement.

Floating bearing A bearing so designed or mounted as to permit axial displacement between shaft and housing.

Full-complement bearing A cageless bearing with a maximum number of rolling elements.

Housing bearing seat The part of the housing bore which contacts the outside diameter of the bearing.

Housing fit The amount of interference or clearance between the bearing outside diameter and the housing bore seat.

Inch bearing A bearing designed in inch dimensions.

Internal clearance See radial internal clearance.

Lateral travel See end play.

Life The life of an individual ball or roller bearing is the number of revolutions (or hours at some given constant speed) which the bearing runs before the first evidence of fatigue develops in the material of either the ring (or washer) or of any of the rolling elements.

Load rating Load ratings for specific speeds are based on a rating life of 500 h (AFBMA).

Loading groove A slot in the raceway shoulder that permits assembly of a maximum number of rolling elements.

Lubrication groove A continuous recess in a bearing for conveying lubricant.

Lubrication hole A hole in the rings to pass lubricant to rolling elements.

Metric bearing A bearing designed to metric dimensions.

Multirow bearing A bearing with more than two rows of rolling elements.

Needle roller A load-carrying rolling element generally understood to be long in relation to its diameter.

Outside diameter The area of the bearing making contact with the housing bearing seat.

Pitch diameter, rolling elements The diameter of the pitch circle generated by the center of a rolling element as it traverses the bearing's axis of rotation.

Pocket (cage) The portion of cage which is shaped to receive the rolling element.

Preload An internal loading characteristic in a bearing which is independent of any external radial and/or axial load carried by the bearing.

Races The inner ring or outer ring of a bearing.

Raceway The path of the rolling element on either ring of a bearing.

Radial bearing An antifriction bearing primarily designed to support a load perpendicular to the shaft axis.

Radial internal clearance For a single-row radial contact bearing, this is the average outer-ring raceway diameter, minus the average inner-ring raceway diameter, minus twice the rolling-element diameter.

Radial load Radial load is that load which may result from a single force or the *resultant* of several forces acting in a direction at right angles to the bearing axis.

Radial play See radial internal clearance

Rating life For a group of apparently identical bearings the rating life (L_{10}) is the life in millions of revolutions that 90 percent of the group will complete or exceed.

Retainer See cage.

Sealed bearing A ball or roller bearing protected against loss of lubricant and from outside contamination.

Self-aligning bearing A bearing with built-in compensation for shaft or housing deflection or misalignment.

Self-contained bearing Unit bearing assembly (nonseparable).

Separable bearing A bearing assembly that may be separated completely or partially into its component parts.

Separator See cage.

Shaft bearing seat The portion of the shaft upon which the bearing is mounted.

Shaft fit The amount of interference or clearance between the bearing inside diameter and the shaft bearing seat diameter.

Shield A circular part affixed to one bearing ring to cover the interspace, but not to run in contact with the other ring.

Single-row bearing A bearing having only one row of rolling elements.

Special bearing A bearing not meeting the requirements of standard or established line bearings.

Spherical roller bearing (radial) A bearing which, by virtue of the raceway or outer-ring construction, is capable of considerable misalignment.

Spherical roller thrust bearing An antifriction thrust bearing using spherical rollers as rolling elements.

Standard bearing An antifriction bearing conforming to AFBMA, "General Boundary Plans of Metric and Inch Dimensions."

Static equivalent load The calculated, static radial or static centric thrust load, which if applied to a bearing would cause the same total permanent deformation at the most heavily stressed rolling element and raceway contact as that which occurs under an actual condition of loading.

Static load Static load is a load acting on a nonrotating bearing.

Tapered roller A roller with one end smaller than the other (the frustum of a cone).

Thrust bearing A bearing designed primarily to support a load parallel to shaft axis.

Thrust load The load which results from a single force or the resultant of several forces acting in a direction parallel to the bearing axis.

Washer (thrust bearing) An annular ring upon which thrust bearing raceways are ground.

INTRODUCTION

Antifriction (rolling-element) bearings continue to find increased use wherever the reduction of friction is required at the interface of dynamic and static components in machinery. A principal advantage of these bearings is the ability to operate at friction levels considerably lower than those of plain or oil-film bearings, while maintaining coefficients of friction at start-up that are close to those in normal operation.

The complement of rolling elements in antifriction bearings comprises balls or rollers (or even a combination of both in some special designs). In concept, the balls or rollers are arranged within a bearing primarily to support either pure radial or pure thrust loading, but they sometimes have a capability for accomplishing both.

Standard (off-the-shelf) bearings usually consist of four essential components. The inner ring and outer ring with their raceways, and a complement of rolling elements (balls or rollers) contained and separated by the cage, retainer, or separator. In operation the hardened and ground surfaces of the raceways form the track for the rolling elements (load-supporting members) to follow while transmitting load from the dynamic members of an assembly to the stationary members.

APPLICATION INFORMATION

Bearing Life

The life of a bearing is expressed as the number of revolutions or the number of hours, at a given speed, for which the bearing will operate before any evidence of fatigue develops on the rolling elements or the raceways. Life may vary from one bearing to another, but stabilizes into a predictable pattern when considering a large group of the same size and type of bearings. The *rating life* of a group of such bearings is the number of hours or revolutions (at a given constant speed and load) that 90 percent of the tested bearings will exceed before the first evidence of fatigue develops. This is called L_{10} *life* or *minimum life.*

Average Life

The results of testing a large group of ball or roller bearings may be graphically illustrated. The distribution curve shown is obtained by plotting relative life vs. percent of bearings tested. See Fig. 2-1.

From the curve in Fig. 2-1 it is apparent that the average life is approximately five times the minimum life. About 50 percent of the bearings will exceed the average life. Since it is not possible to predict the exact life of a single bearing, a safety factor must

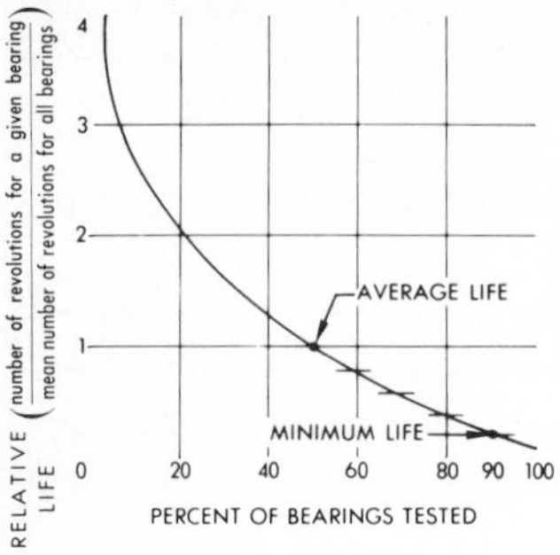

Figure 2-1 Average-life curve.

be allowed to minimize the chances of early failure. The cost of replacing a bearing plus the expense of machine downtime may greatly exceed the relatively low cost of the bearing. Therefore, most designers prefer to use minimum life as a basis for design. In some applications where safety or maintenance economy is not critical and low initial bearing cost is desirable, the average-life value may be used.

Life-and-Load Relationship

Empirical calculations and experimental data point to a predictable relationship between bearing load and life. This relationship may be expressed by formulas. In these empirical formulas the bearing life is found to vary inversely as the applied load to an exponential power. The assigned value of the exponent depends upon the basic type of rolling element.

For all types of roller bearings the formula is

$$\text{Life} = \left(\frac{\text{basic dynamic capacity}}{\text{load}}\right)^{3.33} \quad 10^6 \text{ revolutions}$$

For all types of ball bearings the formula is

$$\text{Life} = \left(\frac{\text{basic dynamic capacity}}{\text{load}}\right)^{3} \quad 10^6 \text{ revolutions}$$

Ring Rotation Factors (RF)

The basic dynamic capacity of a radial bearing is based on the inner ring rotating with respect to the applied load. Recent findings indicate that the bearing capacity does not have to be downgraded when the outer ring rotates. If the inner ring or outer ring rotates with respect to load, the rotation factor is 1.0.

Effect of Load

It is also evident from the exponential character of the basic life–load relationships, that for any given speed, a change in load may have a substantial effect on the life. For a roller bearing, if the load is doubled, the life is reduced to one-tenth its former calculated duration. Similarly, if the load is halved, the life is increased tenfold. For a ball bearing, if the load is doubled, the life is reduced to one-eighth its former value. Likewise, if the load is reduced one-half, the life is increased eightfold.

Effect of Speed

The preceding expressions are independent of the speed of the bearing and are valid for speeds ranging from 10 to 10,000 r/min. For speeds below 10 r/min, consult the engineering department of a bearing manufacturer.

If bearing life is measured in hours, an increase in speed results in a decrease in hours of life. The number of revolutions per unit time determines the hours of life available before the fatigue limit of the bearing is reached. If the speed is doubled, the hours of life are halved. Conversely, if the speed is reduced by 50 percent the hours of life will be doubled.

Selection of Bearing Type

Selection of bearing type is made after the general design concept of the machine has been established and the magnitude of the loads and speeds estimated. Special conditions can directly affect bearing operation and must be considered. These include ambient or localized temperatures, shock or vibration, dirt or abrasive contamination, difficulty in obtaining accurate alignment, space limitations, need for shaft rigidity, etc.

Selection of the proper type of bearing is not an exact science. The fields of application for many types of bearings overlap, and the value of experience in bearing applications cannot be overemphasized. Each type of bearing, however, has inherent features, which determine its relative suitability for a specific application. Careful analysis of these features and familiarization with the fundamental characteristics of each type of bearing will help in selecting the proper bearing.

As an aid to experienced designers and inexperienced bearing users alike, the similarities and differences of ball and roller bearings are outlined. Table 2-1 shows the relative operating characteristics of each cataloged bearing type.

TABLE 2-1 Relative Operating Characteristics

Bearing type	Radial capacity	Thrust capacity	Limiting speed	Resistance to elastic deformation	
				Radial	Axial
Ball radial					
Conrad	Moderate	Moderate in both directions	High	Moderate	Low
Maximum-capacity	Moderate +	Moderate in one direction	High	Moderate +	Low+
Angular-contact	Moderate	Moderate + in one direction	High −	Moderate	Moderate
Roller radial					
Cylindrical roller	High	None	Moderate +	High	None
	High	Light in one direction	Moderate +	High	Not recommended
	High	Light in both directions	Moderate +	High	Not recommended
Journal	High +	None	Low	High	None
Spherical roller	High	Moderate in both directions	Moderate	High −	Moderate
Tapered roller					
Single-row	High −	Moderate + in one direction	Moderate	High −	Moderate
Single-row, steep angle	Moderate +	High in one direction	Moderate −	Moderate	High
Double-row	High	Moderate + in both directions	Moderate	High	Moderate
Double-row, steep angle	Moderate +	High in both directions	Moderate −	Moderate	High
Four-row	High +	High in both directions	Moderate −	High +	High
Thrust					
Angular-contact ball thrust	Low +	High − one direction	Moderate	Low	High −
Ball thrust	None	High one direction	Moderate −	None	High
Roller thrust	None	High + one direction	Low	None	High +
Self-aligning roller thrust	None	High + one direction	Low	None	High +
Tapered roller thrust	Locational only	High + one direction	Low	None	High +

Ball Bearings vs. Roller Bearings

Using balls as the rolling elements in bearings offers certain performance advantages. Most of the advantages of ball bearings are derived from the small area of contact between ball and raceway.

Ball bearings may be operated at higher speeds, with less internal friction and with less heat generation. They have a greater inherent ability to accommodate slight misalignment. Under certain conditions of combined loading, ball bearings occupy less space than required for roller bearings of the same bore size.

Rollers are not limited to a single geometric shape. There are several types, such as tapered rollers, spherical rollers, and cylindrical rollers. For a given load and diameter of rolling element, rollers transmit load through a larger contact area than do balls. This allows roller bearings to support greater loads and accommodate far more shock than ball bearings of equivalent size. For a given applied load, contact area stresses for roller bearings are lower than for ball bearings, and, therefore, they produce lower elastic deformation. Since the larger contact areas create more friction, permissible operating speeds for roller bearings are lower than those for ball bearings.

Selection of Bearing Size

Once the designer selects a suitable type of bearing for a set of specific conditions, the size of the bearing needed to provide adequate service life is determined. Many cases exist in which more than one bearing type will satisfy the operating conditions. In these instances the designer should determine the most suitable size of each type considered and make the final selection on the basis of mounting simplicity, space considerations, and overall economy.

The basic parameters affecting the choice of bearing size are radial load, thrust load, speed, required life, ring rotation, and shock or vibration conditions. Other factors such as misalignment, abnormal temperature, contamination, and poor lubrication will seriously reduce service life, but their exact effect cannot be determined. These factors should be eliminated by proper mounting design rather than by attempting to estimate their effect on bearing life.

Limiting Speeds

The ability of a bearing to operate at high speeds is dependent upon the rate at which generated heat is dissipated. Maximum speed is governed by bearing type, size, bearing load, ambient temperature, and type of lubricant.

The geometric design of a bearing and the method of positioning the rolling elements basically determine the coefficient of friction. Since frictional losses are proportional to the peripheral speed, it follows that the smaller the bearing the greater the speed at which it may operate.

Elastic deformation of the raceways and rolling elements is increased by heavy loads, and this creates additional heat, thereby limiting the allowable speed.

Ambient temperature may affect the rate of heat dissipation. Applications having a high ambient temperature require careful selection of the method of lubrication.

The type of lubricant is a basic criterion for establishing limiting speeds. Lubricants with high viscosity offer more frictional resistance, therefore oil is preferable to grease for higher speeds. Even with the proper frequency and amount of lubrication, limiting speeds for grease are approximately 50 percent of the given values. The use of circulating oil or oil mist will allow higher speed limits than does oil-bath lubrication.

Fits of Shaft and Housing

To ensure the full utilization of bearing capacity under operating conditions, it is important to have the proper fit between inner ring and shaft, and outer ring and housing. The tolerances to which the bearing is made are standardized, so desired fits may be obtained by controlling the dimensions and tolerances for the shaft and housing.

Normally, the problem in fit determination is to make the rotating ring of the bearing and its associated shaft or housing rotate as a single unit by using an interference fit. The fit of the nonrotating ring should be loose, with minimum clearance, for ease of assembly and axial movement in the housing.

The amount of interference fit employed should not create in the bearing rings excessive stress that might result in early fatigue failure. Under conditions of light load, the interference fit can be small. As the loads increase or shock loading is introduced, the interference must be increased so that no clearance exists and none can be induced by the load. This is the only effective means of preventing "creep." As a rule, axial clamping cannot be relied on to prevent creep since the clamping force must be excessively high. Thus, the heavier the load, the tighter the fit.

The degree of fit is labeled in the Anti-Friction Bearing Manufacturers Association (AFBMA) tolerance system which has been adopted where applicable. This tolerance system is in accordance with that adopted by the American National Standards Institute (ANSI). This system applies to all ball thrust bearings and radial bearings (except tapered roller bearings). For tapered roller bearings, an adaptation of the recommended AFBMA fitting practice has been made for the convenience of designers.

Special Materials for Bearings

Industrial demands for special bearings to meet abnormal service requirements spur the continual search for new and improved bearing materials. High temperatures, corrosive atmospheres, massive size, marginal lubrication, complex design, and space and weight limitations are typical abnormal requirements.

Conventional bearing steels are often inadequate when these problems are present. Sustained high operating temperatures reduce hardness, wear resistance, yield strength, and, therefore, bearing life. Conventional bearing steels also lack resistance to the oxidation which takes place at elevated temperatures.

Several materials such as 440-C stainless; 18-4-1 high-speed steel; M-1, M-2, M-10, and MV-1 high-speed steels; and high-cobalt alloys are used for high-temperature applications. For extremely high temperatures, materials such as metallic carbides and ceramics are used.

The combination of bearing size, complexity of design, and space and weight limitations can be a governing factor in the selection of bearing material. For example, a large-diameter, thin-section bearing with integral gear teeth and bolt holes would require a material which could be selectively hardened.

Corrosion is also a factor to be considered. The corrosive agent, load, temperature, and lubrication all affect the final choice of material. Some high-temperature materials with capacities equivalent to conventional bearing steel are used in both the normal- and high-operating-temperature ranges because of their good corrosion resistance. Other corrosion-resistant materials such as Monel, 18-8 stainless steel, and beryllium-copper alloys have sharply limited capacity, but excellent resistance to specific kinds of corrosion.

Monel and beryllium copper are not as hardenable as bearing steels and are nonmagnetic and resist saltwater corrosion. These qualities make them excellent materials for marine applications.

A variety of materials is also used for cages. Synthetic resin–impregnated fabrics, aluminum, and nylon products are becoming popular in the normal temperature ranges. Ductile iron, certain stainless steels, and iron-silicon bronze are used for higher temperatures.

Mounting Design

Mounting design varies widely, depending upon the type of bearing used and the requirements of the application. Selection of bearing type and mounting design are closely related since many cases exist where selection of bearing type is influenced by mounting design considerations.

Most applications require the use of more than one bearing on a shaft. Two identical bearings or a combination of different types and sizes may be used on a common shaft. The advantages of each combination should be evaluated by the designer in selecting bearings.

A *fixed*-bearing mounting locates the shaft and carries any thrust loads which exist in the application. A *float*-bearing mounting accommodates relative axial movement between the shaft and housing. Various combinations of these two basic mountings are used:

1. Fixed-float mounting
2. Fixed-fixed (or opposed) mounting
3. Float-float mounting

The fixed and float arrangement is necessary when: (1) a long shaft is used, (2) thermal expansion or contraction of the shaft with respect to the housing occurs, or (3) separate housings are required for two or more bearings on a common shaft.

The fixed and float combination offers many desirable features for heavy industrial equipment. The upper half of Fig. 2-2 shows a typical arrangement. The axial location

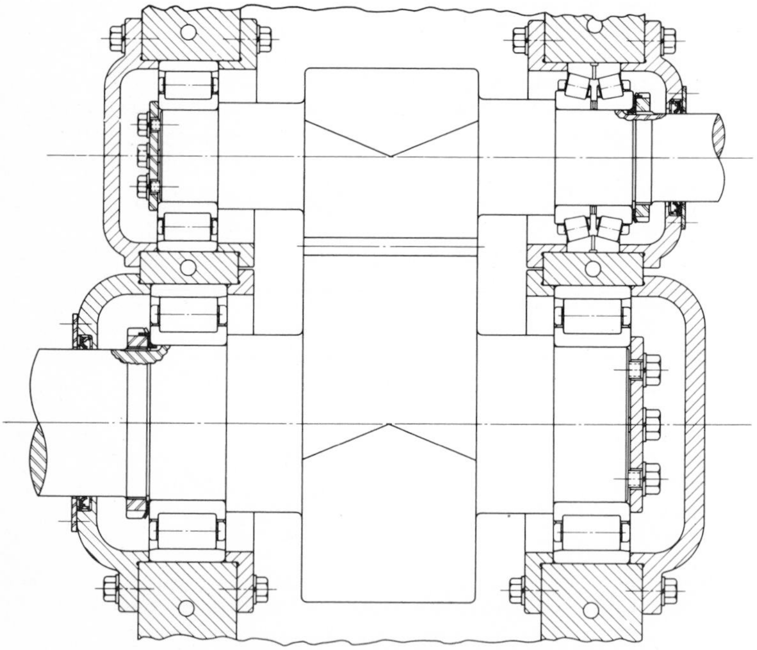

Figure 2-2 Typical fixed- and float-bearing arrangements.

is accomplished through the fixed bearing since it is clamped rigidly to the shaft and the housing. A two-row tapered roller bearing is shown in the fixed position. However, other radial bearings capable of taking thrust loads in both directions and applied radial loads may be used. For the float position the cylindrical roller bearing shown supports relatively heavy loads and permits free axial displacement. A type TDO or TNA tapered roller bearing may be used if the sliding pressures between the outer ring and the housing bore are not excessive.

The lower half of Fig. 2-2 demonstrates the float-float mounting. A pair of cylindrical roller bearings permits the entire shaft to float. A herringbone gear is used and the floating shaft allows the gears to mesh properly. The arrangement shown permits manufacturing economies since gear axial alignment is assured without very accurate machining or shimming of the bottom housing cover and end plate. A pair of type TDO or TNA tapered roller bearings, not fixed axially in their housings, may be used if the sliding pressures are not excessive.

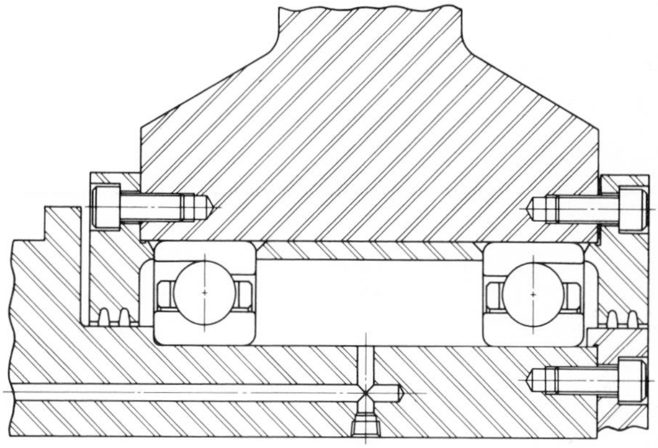

Figure 2-3 Typical fixed-fixed bearing arrangement.

Figure 2-3 illustrates a typical mounting arrangement where the bearings will be subjected to a radial load with some locational thrust load. A gap is left between the clamping ring and the face of the inner ring to provide for tolerance accumulation. Use of a spacer between the faces of the outer rings allows through-boring of the housing. Although maximum-capacity (types BH or BIH) ball bearings are shown, Conrad (types BC or BIC) ball radial bearings could also be used when there are moderate loads. Some applications, such as flywheels or spur gear hubs, require a tight fit of the outer ring in the housing with a loose fit of the inner ring on the shaft.

Seals and Closures

Seals are used to protect the bearing from contamination as well as to retain the lubricant. There are three basic types of seals:

1. **Lip Contact, Commercial Seals** These are usually standard components made by several manufacturers.
2. **Annulus and Labyrinth Seals** These are noncontact seals with slight clearance between stationary and rotating members depending on the lubricant to effect a frictionless closure.
3. **Slinger Seals** External types depend on centrifugal force to fling foreign matter away from the shaft. Internal slinger-type seals are used to distribute lubricant within the housing and to shield the bearing.

Basic types of sealing arrangements are shown in Fig. 2-4.

Type A. Commercial seal. The contact lip may be synthetic rubber or leather, spring-backed for more positive sealing. Consult manufacturers' literature for shaft finish and limiting speeds. May be used for either grease or oil.

Type B. Annular grooves. Shown here with drain slot at bottom. These may be in either the shaft or the housing, or in both. The effectiveness is increased by keeping the running clearance small and by using multiple grooves. Used for oil or grease lubrication.

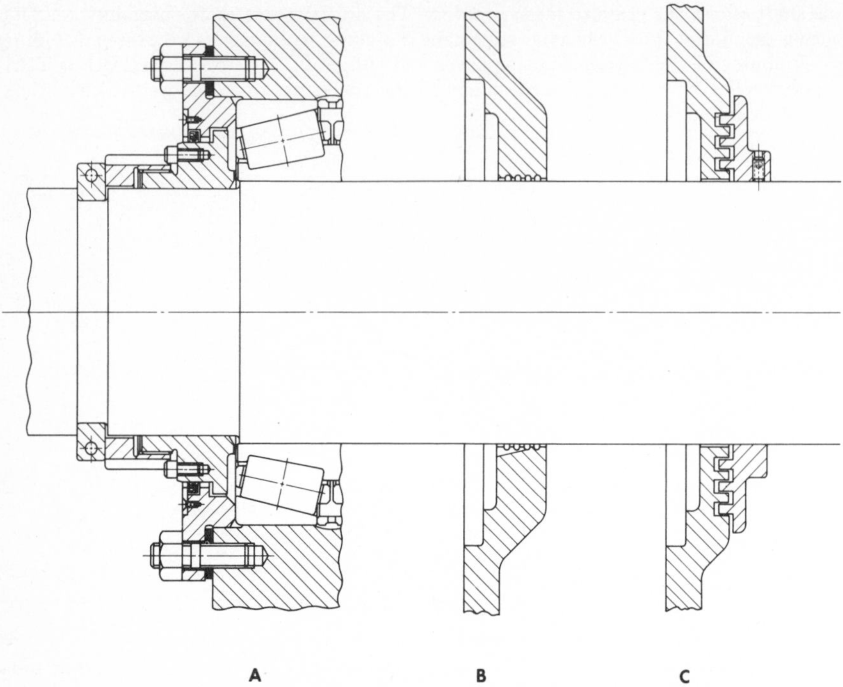

Figure 2-4 Basic sealing arrangements.

Type C. Axial labyrinth seal. Does not require a split housing. Clearance must be allowed for axial movement. Effective for abrasive environment. Use for oil or grease.

Type D. Radial labyrinth seal. Used with a split housing or end cap. Bore of grooved sealing ring is slightly larger than shaft, allowing it to float axially. Angular surface of housing groove reduces pumping action. Suitable for oil or grease.

Type E. Felt seal. Provides medium effectiveness at low speeds but it loses its effectiveness at high speeds and high temperatures. In most cases, it functions as a contact seal, but after "wearing in" often functions as a simple close-clearance annulus seal. Not suitable for an abrasive environment. Use for grease lubrication.

Type F. The piston ring seal is a modification of the labyrinth seal. The piston rings are stationary and are mounted under radial compression in the housing. The split rings have rabbetted joints. Clearances between the rings and grooves are slight. This seal accommodates axial displacement. It is easy to install or remove, and can be used with a spacer or in shaft grooves as shown. Accepting deflection and misalignment, it offers a relatively positive closure augmented by a grease annulus. A particularly effective seal for an abrasive environment, it is suitable for grease and oil.

Type G. This is a combination annular groove-axial labyrinth seal. Annular grooves retain lubricant in the housing. The external flinger serves as a shield and flings contaminants away from the seal. It is suitable for grease or oil.

Type H. This is a triple combination of lip contact seals. Two commercial seals are mounted and opposed with a spacer in between to allow relubrication of the contact surfaces. A face contact seal is also used to prevent the entrance of contaminants. It is used primarily for grease.

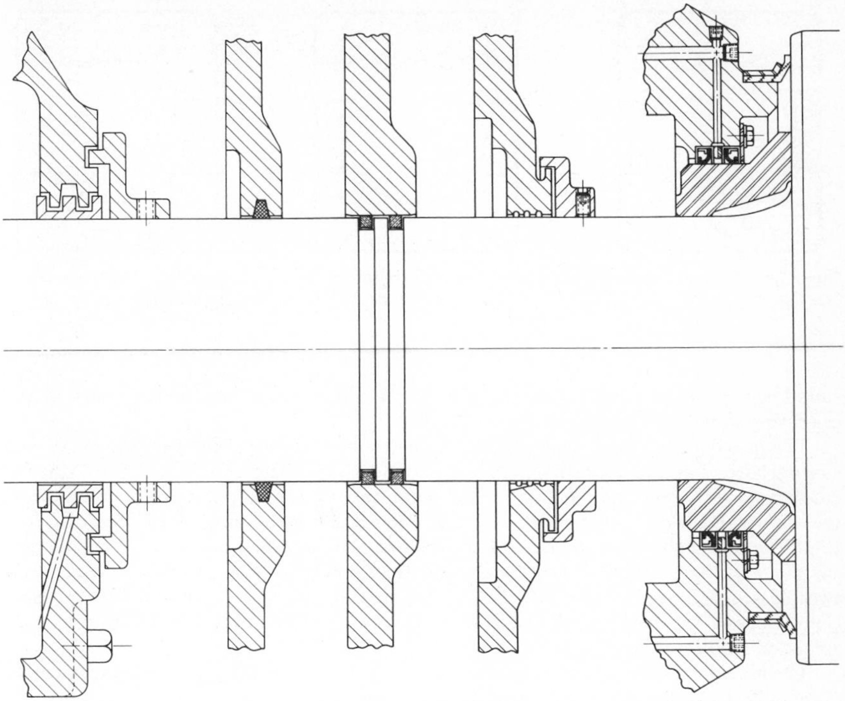

Figure 2-4 Basic sealing arrangements. (*Continued*)

PHYSICAL DESCRIPTION

Ball Bearings

Ball Radial Bearings

There are three major types of ball radial bearings with metric bore range of 100 to 320 mm and inch bore range of 4 to 40 in normally listed in bearing catalogs.

Conrad. See Fig. 2-5.
Type BC—metric sizes
Type BIC—inch sizes

Maximum Capacity. See Fig. 2-6.
Type BH—metric sizes
Type BIH—inch sizes

Angular Contact. See Fig. 2-7.
Type BA—metric sizes
Type BIA—inch sizes

For the convenience of the designer, metric and inch sizes of the three ball-bearing types are tabulated in order of increasing bore. Selection of the proper type of ball bearing is determined by consideration of the direction and magnitude of the bearing load. Ball bearings are particularly suited to high-speed operation because they have a lower coefficient of friction than roller bearings.

Figure 2-5 shows the most widely used ball-bearing type. Although it is primarily a

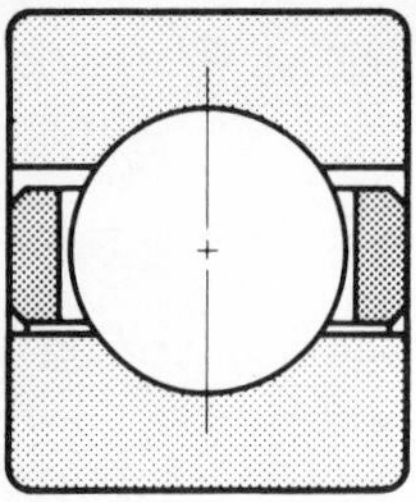

Figure 2-5 Conrad-type ball bearing.

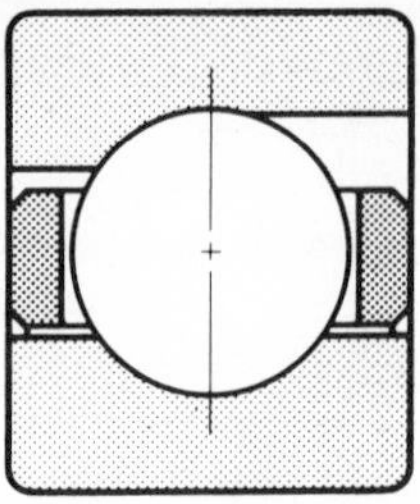

Figure 2-6 Maximum-capacity-type ball bearing. BH, BIH

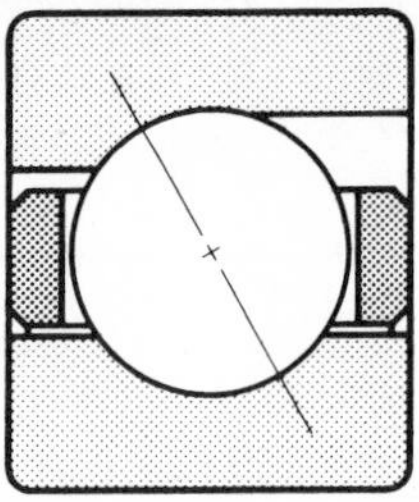

Figure 2-7 Angular-contact-type ball bearing.

radial bearing, it is capable of handling moderate thrust loads from either direction and operating at relatively high speeds.

The bearing rings have symmetrical, deep-grooved raceways without filling slots or a counterbore. The raceways are precision-ground to conform closely to ball curvature, consistent with minimum friction, maximum capacity, and practical manufacturing techniques. Balls are selected for uniformity to ensure optimum internal load distribution. The bearing utilizes the maximum number of balls which can be inserted between the raceways by eccentrically displacing the inner and outer rings. Balls are spaced by a two-piece, machined bronze cage. For higher speeds, and when other operating conditions warrant, the cage may be made of other materials. The nonseparable bearing construction facilitates handling.

The maximum-capacity ball bearing shown in Fig. 2-6 has the greatest radial capacity obtainable in a single-row ball bearing with cage, but it can take thrust in only one direction.

This type of bearing has an inner ring with a deep groove, such as the Conrad bearing, but the outer ring is counterbored; this substantially reduces the raceway shoulder on one side. The counterbore allows a maximum complement of balls in a one-piece, machined bronze cage to be assembled in the bearing. The outer ring is thermally expanded and slipped over the cage, ball, and inner-ring assembly. After cooling, the bearing is nonseparable.

Types BH and BIH are suited for higher radial loads than Conrad bearings of equivalent size as well as for combined radial and moderate unidirectional thrust loads. Since thrust capacity is unidirectional, the preferred mounting is in opposed pairs. A maximum-capacity bearing may also be opposed by a Conrad or other type of axially locating bearing. The center distance between bearings should be held to a minimum. Otherwise, axial shaft expansion may impose thrust on the shallow shoulder of the maximum-capacity bearing or cause preloading.

The design of angular-contact ball bearings, as shown in Fig. 2-7, is the same as that of the maximum-capacity type except the contact angle is 30°. This design feature allows the bearing to resist heavier thrust loads and minimizes axial deflection under load. The limiting speed of an angular-contact ball bearing is less than that of a Conrad or maximum-capacity type.

Angular-contact ball bearings are suited for applications where the thrust load is of the same order as or greater than the radial load. The preferred mounting arrangement is two bearings per shaft, with their contact angles opposed. As in the case of maximum-capacity bearings, the center distance between opposed bearings should be held to a minimum.

Angular-contact bearings can be furnished in matched pairs for duplex mounting (Fig. 2-8). When the bearings supplied for opposed mounting are placed together, there will be a small gap between the inner rings (type DB) or outer rings (type DF) before clamping. After clamping together, an internal preload is introduced in the set which

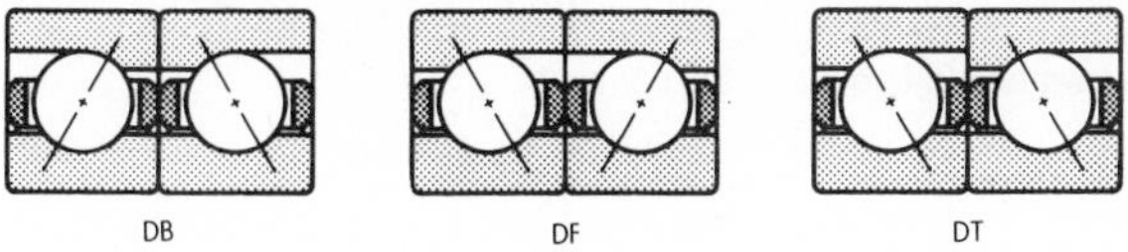

Figure 2-8 Duplex mounting of matched-pair angular-contact ball bearings.

increases the axial and radial rigidity of the duplexed pair. When supplied for tandem mounting (type DT), very heavy unidirectional thrust loads may be almost equally distributed among the bearings of the set. Details of application for DB and DF pairs should be submitted to the engineering department of the bearing manufacturer for preload recommendations.

Ball Thrust Bearings

Ball thrust bearings are used for lighter loads and higher speeds than roller thrust bearings.

The type TVB ball thrust bearing (Fig. 2-9) is separable, and consists of two hardened and ground steel washers with grooved raceways, and a cage which separates and retains precision-ground and -lapped balls. The standard cage material is bronze, but this may be varied according to the requirements of the application.

The type TVB bearing provides axial rigidity in one direction and its use to support radial loads is not recommended. It is very easily mounted. Usually the rotating washer is shaft-mounted. The stationary washer should be housed with sufficient outside diameter clearance to allow the bearing to assume its proper operating position. In most sizes both washers have the same bore and outside diameter. The housing must be designed to clear the outside diameter of the rotating washer, and it is necessary to step the shaft to clear the bore of the stationary washer.

Type TVL (Fig. 2-10) is a separable angular-contact ball bearing designed primarily for unidirectional-thrust loads. The angular contact design, however, will accommodate

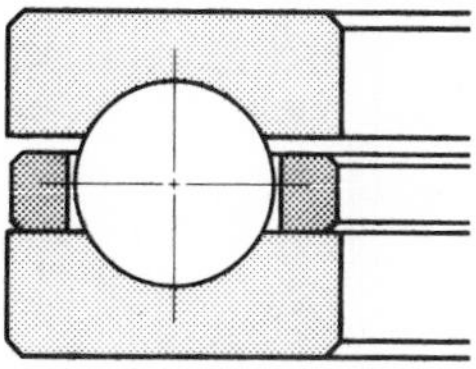

Figure 2-9 Ball thrust bearing.

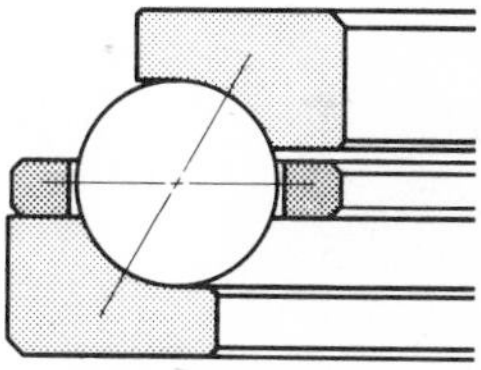

Figure 2-10 Angular-contact ball thrust bearing.

combined radial and thrust loads since the loads are transmitted angularly through the balls.

Tbe bearing has two hardened and ground steel rings with ball grooves and a one-piece bronze cage which spaces the ball complement. Although not strictly an annular ball bearing, the larger ring is called the outer ring, and the smaller the inner ring.

Usually the inner ring is the rotating member and is shaft-mounted. The outer ring is normally stationary and should be mounted with outside diameter clearance to allow the bearing to assume its proper operating position. If combined loads exist, the outer ring must be radially located in the housing.

The type TVL bearing should always be operated under thrust load. Normally, this presents no problem as the bearing is usually applied on vertical shafts in oil-field rotary tables and machine-tool indexing tables. If a constant-thrust load is not present, it should be imposed by springs or other built-in devices.

Low friction, cool running, and quiet operation are advantages of the type TVL bearing, which may be operated at relatively high speeds. The bearing is also less sensitive to misalignment than other types of rigid thrust bearings.

Roller Bearings

Tapered Roller Bearings

Tapered roller bearings are generally considered to offer the best support for combinations of heavy radial and thrust loads at moderate speeds.

A single-row tapered roller bearing consists of an inner ring (called a *cone*), an outer ring (called a *cup*), a bronze or steel cage, and a complement of controlled-contour rollers. In multiple-row tapered roller bearings one or more cones, cups, and cage assemblies may be used.

Tapered rollers and raceways are designed on the geometric principle of a cone (Fig. 2-11). Extensions of the lines of contact between the rollers and the raceways all meet at

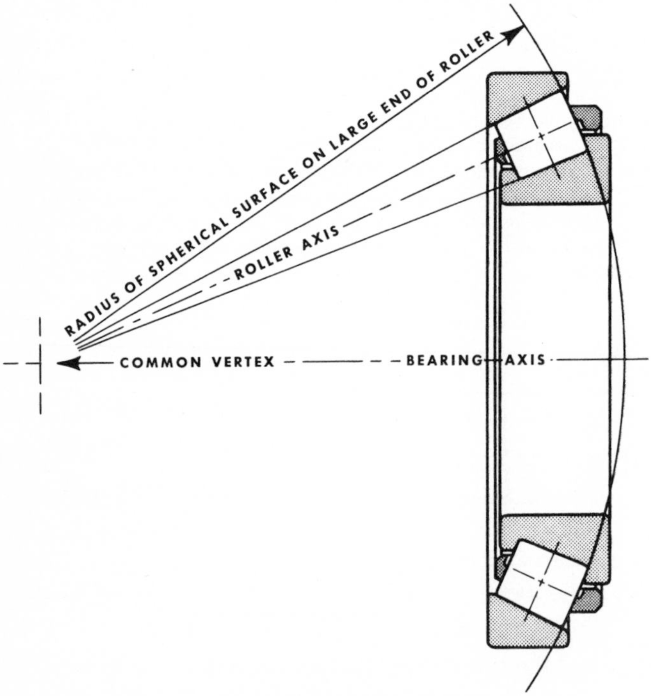

Figure 2-11 Geometric principle of tapered roller bearing.

a common point on the axis of the bearing. The design assures true geometric rolling. The large ends of the tapered rollers are spherically ground to match the spherically ground face on the guiding cone rib. Under load, the nominal pressure exerted between these two ground surfaces accurately positions the rollers within the load zone.

Single-Row Tapered Roller Bearings. Three types (TS, TSF, and TSS) of single-row tapered roller bearings are offered (Fig. 2-12). Each has a cup and a cone with a cage and roller assembly. Type TS serves as the basic design for the others.

Since a single-row tapered roller bearing supports thrust loads from only one direction, the preferred mounting is in opposed pairs. The proper internal clearance for the two bearings may be obtained by axial adjustment at the time of assembly.

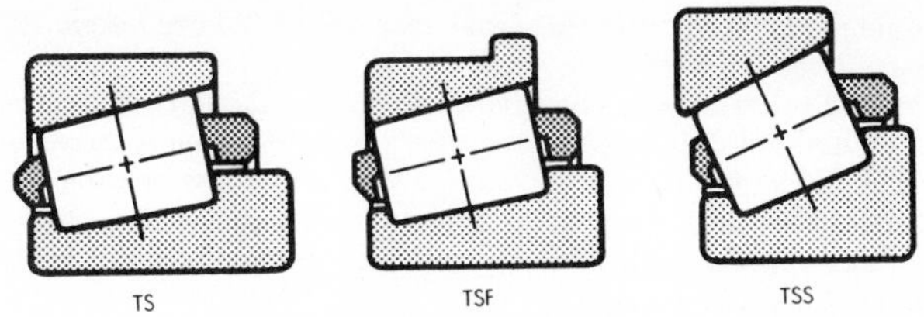

Figure 2-12 Single-row tapered roller bearings.

Type TSF bearings are identical with type TS except that the cup of the TSF type incorporates an external flange which, in some mountings, facilitates location and permits economies in design. When through-boring of the housing is advantageous, the use of the type TSF bearing is suggested.

Type TSS bearings are similar to type TS, but have a steeper angle of contact. These bearings are recommended for applications where the thrust load is predominant.

Two-Row Tapered Roller Bearings. Three basic types (TDI, TDO, and TNA) of two-row tapered roller bearings are available (Figs. 2-13 and 2-14). Types TDIS, TDIE,

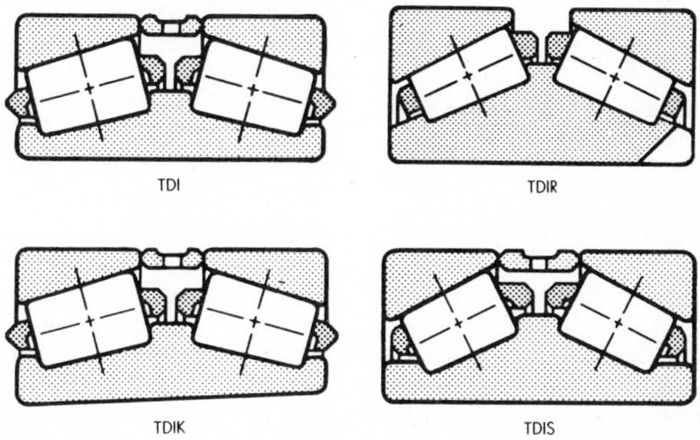

Figure 2-13 Two-row tapered roller bearings with converging angles of contact.

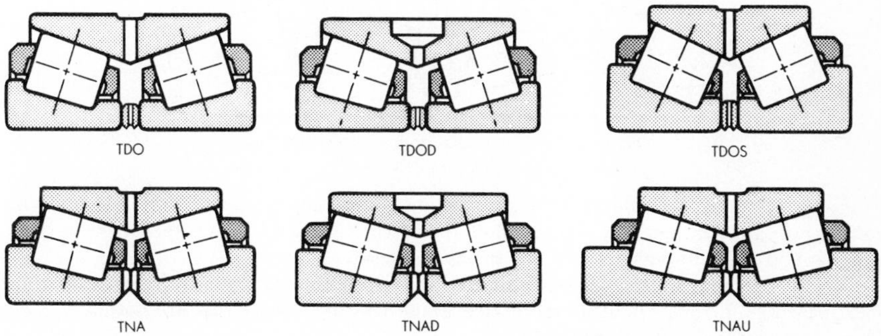

Figure 2-14 Two-row tapered roller bearings with diverging angles of contact.

and TDOS are steep-angle versions of the basic types. Type TDIE is supplied with a face keyway and without a cup spacer. Type TDIK is identical with type TDI except it has a tapered bore. Type TNAU is an extended-cone rib version of type TNA. Types TDOD and TNAD are identical, respectively, to types TDO and TNA, except they are supplied with a dowel hole in the outer ring. *All types (except TDIE) are normally furnished as preadjusted assemblies.*

Two-row tapered roller bearings have twice the radial capacity of single-row bearings of the same series and are used in positions where radial loading is too severe for single-row bearings. They have the further advantage that a two-row bearing can take thrust loads in both directions, thus allowing all applied thrust and shaft location to be taken at one position. This often simplifies design and reduces the danger of bearing clearance changes due to axial shaft expansion.

The steep-angle versions (TDIS and TDIE) offer greater thrust capacity with reduced radial capacity. These are widely used as backup thrust bearings in rolling mills and other applications where heavy thrust loads are encountered.

The design of types TDI, TDIS, and TDIE is such that the contact angles converge as they approach the axis of rotation (Fig. 2-15). Consequently, the use of these bearings will not appreciably increase the rigidity of the shaft mounting, and they should not be used singly on a shaft since they will not resist overturning moments.

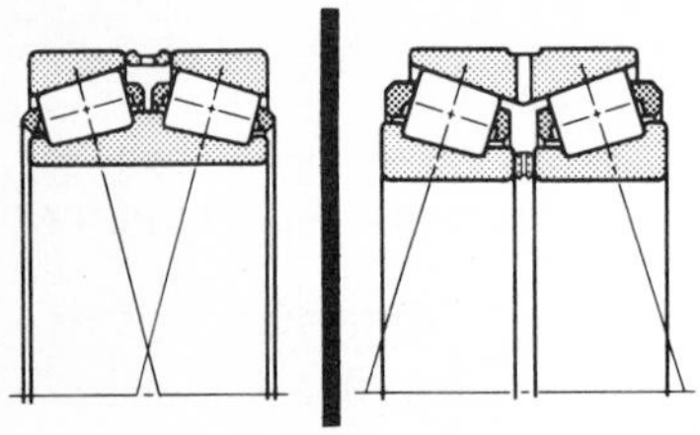

Figure 2-15 Comparison of contact angles for two-row tapered roller bearings.

In types TDO, TDOS, and TDOD, the contact angle lines diverge as they approach the axis of rotation, thus increasing the rigidity of the shaft mounting (Fig. 2-15). Therefore, these bearings are suited for resisting overturning moments. Due to the increased bearing ridigity, housing bore alignment is somewhat more critical than with types TDI, TDIS, and TDIE.

Four-Row Tapered Roller Bearings. The four-row tapered roller bearing (type TQO) (Fig. 2-16) is an extremely high capacity bearing designed primarily for heavy-duty roll-neck applications in metal rolling mills. The bearing is composed of two double-cone assemblies, one double cup, two single cups, and factory-adjusted cup and cone spacers. Spacers for each bearing are face-ground, after accurate measurement of the distance between adjacent cups and cones, to obtain the required initial internal clearance. Because spacers are ground to specific dimensions for each bearing, they are not to be interchanged and they are individually marked for proper assembly. It is important to submit sufficient information for proper internal clearance to be established.

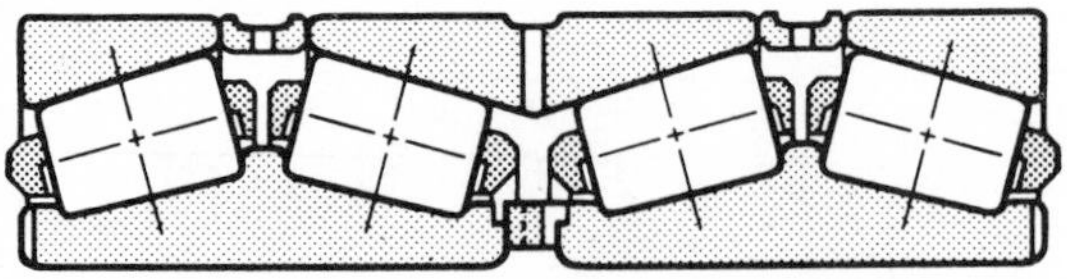

Figure 2-16 Four-row tapered roller bearing.

Lubrication grooves and oil holes are provided in the cup spacers and double cup. For rolling-mill applications, lubrication slots in the cone faces and cone spacer permit the lubricant to pass through the bearing to the roll neck.

Since wear- and shock-resistance under heavy rolling loads are requirements of roll-neck bearings, highest quality carburizing-grade bearing alloy steels are normally used in all four-row tapered roller bearings.

The type TQOK bearing (Fig. 2-17), with tapered bore for roll-neck mounting, is a development for modern high-speed rolling-mill requirements. The TQOK bearing features high capacity, compactness, and a mounting system that guarantees a positive interference fit on roll necks. The tightly-fitted twin-cone mounting of type TQOK provides extra mill rigidity and rolling accuracy by eliminating extraneous clearance. Roll-

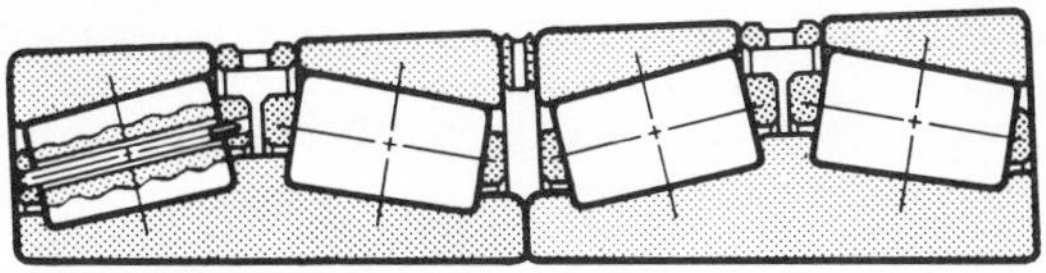

Figure 2-17 Four-row tapered roller bearing with tapered bore.

ing-mill speeds in excess of 5000 ft/min are permissible. Gauge uniformity is assured during the acceleration cycles of feeding the strip through the mill. The mill can be stopped and started without downtime for screw adjustment.

The advantages of compactness of the TQOK design are numerous, but the chief advantage is the short lever arm required on the roll neck with reduction in neck bending stress. Conversely, more capacity can be obtained for any given space limit than with other bearing designs. These and other design features greatly improve rolling-mill production and make the bearing ideal for backup and work-roll mountings in four-high and two-high mills.

Cylindrical Roller Bearings

Six standard types of cylindrical roller bearings are shown in Fig. 2-18. All six types have the same roller complements and, consequently, the same capacity for a given size. All

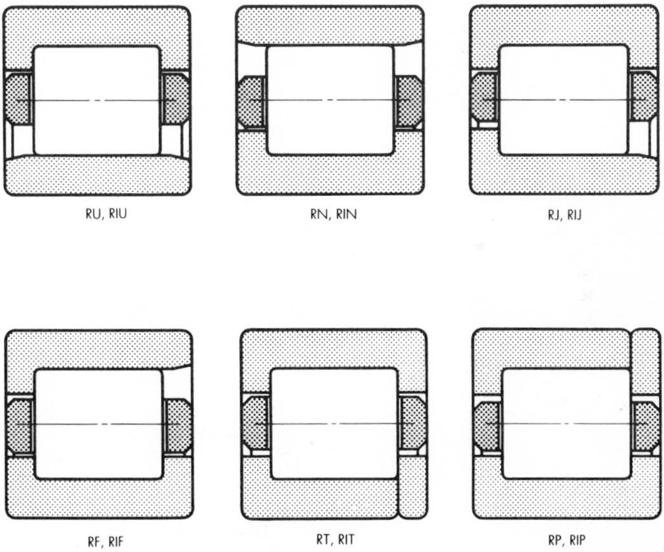

Figure 2-18 Standard types of cylindrical roller bearings.

types can be mounted with interference fits on either the inner or the outer ring, or both. In the latter case, a bearing with increased internal clearance must be specified to provide proper running clearance.

For convenience, bearings are listed according to bore, with both metric and inch bearings in the same figure. Inch bearings are identified by the letter "I" in the type code of the bearing number; thus where RN denotes a particular type of metric bearing, RIN denotes an inch bearing of the same type.

Types RU and RIU have double-ribbed outer and straight inner rings. Types RN and RIN have double-ribbed inner and straight outer rings. The use of either type at one position on a shaft is ideal for accommodating nominal expansion or contraction. The relative axial displacement of one ring to the other occurs with minimum friction while the bearing is rotating. These bearings may be used in two positions for shaft support if other means of axial location are provided.

Types RJ and RIJ have doubled-ribbed outer and single-ribbed inner rings. Types RF and RIF have double-ribbed inner and single-ribbed outer rings. Both types can support heavy radial loads, as well as light unidirectional thrust loads up to 10 percent of the radial load. The thrust load is transmitted between the diagonally opposed rib faces in a sliding action rather than a rolling action. Thus, when limiting thrust conditions are approached, lubrication can become critical. When thrust loads are very light, these bearings may be used in an opposed mounting to locate the shaft. In such cases, shaft end play should be adjusted at time of assembly.

Types RT and RIT have a doubled-ribbed outer ring and a single-ribbed inner ring with a loose rib which allows the bearing to provide axial location in both directions. Types RP and RIP have a double-ribbed inner ring and a single-ribbed outer ring with a loose rib.

Types RT and RP (as well as RIT and RIP) provide heavy radial capacity and light thrust capacity in both directions. Factors governing the thrust capacity are the same as for types RF and RJ bearings.

A type RT or RP bearing may be used in conjunction with a type RN or RU bearing for applications where axial shaft expansion is anticipated. In such cases the fixed bearing is usually placed nearest the drive end of the shaft to minimize alignment variations in the drive. Shaft end play (or float) is determined by the axial clearance in the bearing.

Spherical Roller Bearings

The self-aligning spherical roller bearing (Fig. 2-19) is a combination radial and thrust bearing designed for taking misalignment under load. When loads are heavy, alignment of housings difficult, and shaft deflections excessive, the use of spherical roller bearings assures best service life results.

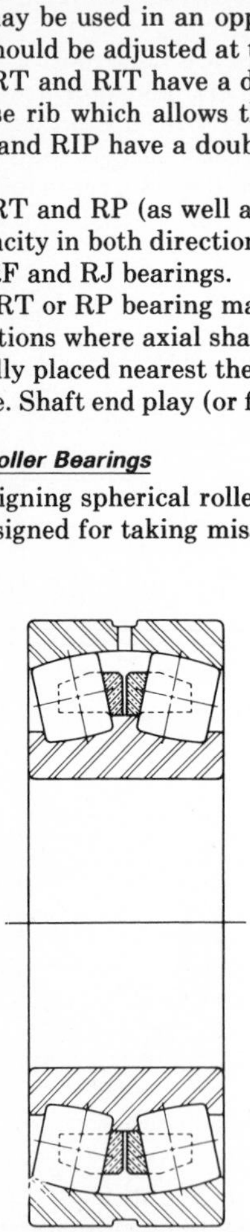

Figure 2-19 Self-aligning spherical roller bearing.

With the spherical rollers operating on the spherically shaped outer race, the assembly of the inner ring, retainers, and rollers may take up to $\pm 1\frac{1}{2}°$ of misalignment and continue to function with full capacity. High radial load capacity is secured by a large area of roller-to-race contact. Double-direction thrust capacity results from angular location of rollers relative to bearing axis.

Shaft deflections and housing distortions caused by shock loads are compensated for by the internal self-alignment of the free-rolling bearing elements. The binding stresses that limit service life of non-self-aligning bearings cannot develop in spherical roller bearings. The retainers ride on the center flange of the inner ring rather than on the rollers. Unit design and construction make the spherical roller bearing simple to handle at assembly point or during removal for maintenance.

Cylindrical Roller Thrust Bearings

Cylindrical roller thrust bearings withstand heavy loads at relatively moderate speeds. Standard bearings can be operated at bearing outside-diameter peripheral speeds of 3000 ft/min (1000 m/min). Special design features can be incorporated into the bearing and mounting to attain higher operating speeds.

Because loads are usually high, extreme pressure lubricants should be used with roller thrust bearings. Preferably, the lubricant should be introduced at the bearing bore and distributed by centrifugal force.

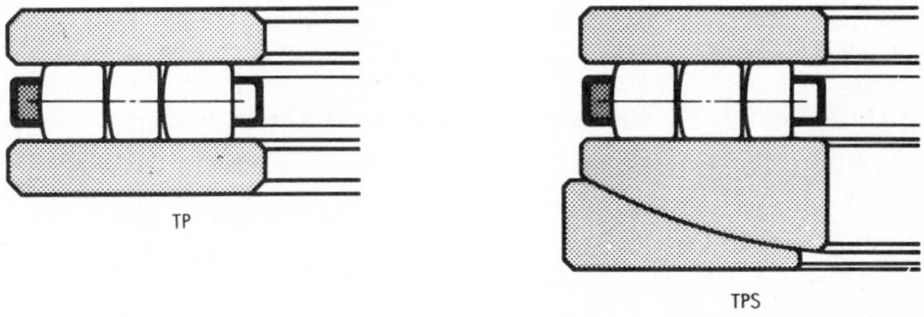

Figure 2-20 Cylindrical roller thrust bearings.

The type TP cylindrical roller thrust bearing (Fig. 2-20) has two hardened and ground steel washers, with a cage retaining one or more rollers in each pocket. When two or more rollers are used in a pocket they are of different lengths and are placed in staggered position in adjacent cage pockets to create overlapping roller paths. This prevents wearing grooves in the raceways and prolongs bearing life.

Because of the simplicity of their design, type TP bearings are economical. Since minor radial displacement of the raceways does not affect the operation of the bearing, its application is relatively simple and often results in manufacturing economies for the user. Shaft and housing seats, however, must be square to the axis of rotation to prevent initial misalignment problems.

Type TPS bearing (Fig. 2-20) is the same as type TP bearing, except one washer is spherically ground to seat against an aligning washer, thus making the bearing adaptable to initial misalignment. Its use is not recommended for operating conditions where alignment is continuously changing (dynamic misalignment).

The type TTHD tapered roller thrust bearing (Fig. 2-21) has an identical pair of hardened and ground steel washers with conical raceways, and a complement of tapered rollers equally spaced by a cage.

In the design of type TTHD, the raceways of both washers and the tapered rollers have a common vertex at the bearing center. This assures true rolling motion. The large end of each tapered roller is spherically ground to match the concave faces of both washer ribs. The pressure exerted under load by the roller ends at the rib surfaces accurately guides the rollers. The center of the large end of each roller is counterbored to improve lubrication at the guiding rib surfaces.

TTHD bearings are well suited for applications such as crane hooks, where extremely high thrust loads and heavy shock must be resisted and some measure of radial location obtained. For very low speed, extremely heavily loaded applications, these bearings are supplied with a full complement of rollers for maximum capacity. The TSR spherical roller thrust bearing design (Fig. 2-22) achieves a high thrust capacity with low friction and continuous roller alignment. The bearings can accommodate pure thrust loads as

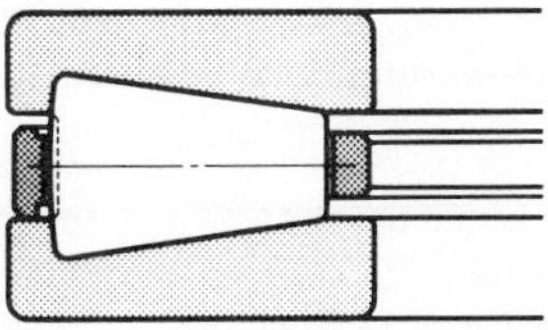

Figure 2-21 Tapered roller thrust bearing.

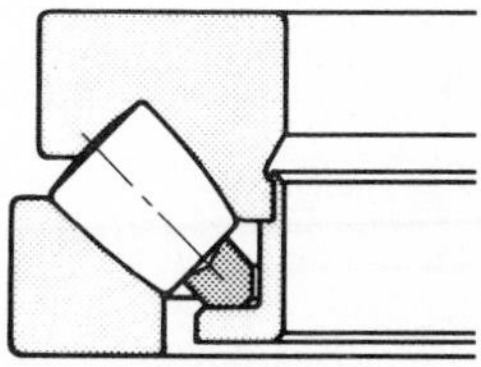

Figure 2-22 Spherical roller thrust bearing.

well as combined radial and thrust loads. Higher speeds are obtainable with this design than with any of the other roller thrust bearings. Typical applications are air regenerators, centrifugal pumps, and deep-well pumps.

Because the spherical roller thrust bearing must always carry some thrust load, it cannot be used as a float bearing. Most applications require the use of two or more bearings on a shaft: a fixed bearing and a float bearing.

As a rule, coil springs must be applied against the stationary ring of the bearing. Among the more common conditions requiring the use of springs are:

1. Applied thrust load is too low to overcome induced thrust load.
2. Load is temporarily reversed, and excessive lateral movement would be detrimental.
3. Thrust load varies from zero to a maximum and back to zero, and roller-to-raceway contact must be maintained.

Needle Roller Bearings

Size for size, needle bearings (Fig. 2-23) have more rollers and more lines of contact, and thus, generally, have higher capacities than roller bearings. This is particularly true

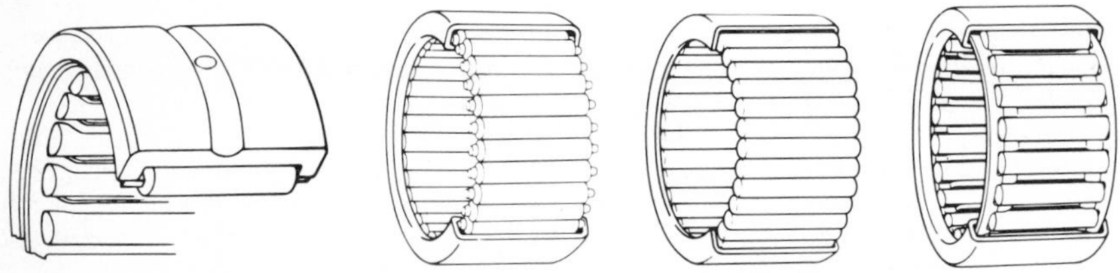

Figure 2-23 Needle roller bearings.

under static, slow-rotating, or oscillating conditions. Both types have far higher capacity in less radial space than other antifriction bearings.

The heavy-duty bearing has an outer race which is deeply hardened, while the race of the drawn-cup bearing has a thin, hardened case over a relatively ductile core. Thus, the heavy-duty bearing can withstand heavier shock or more continuous loads than the drawn-cup bearing, which should never be dynamically loaded, even momentarily, beyond the load limit given in the manufacturer's tabular data.

The smaller, more compact cross section is provided by the drawn-cup bearing with its smaller roller diameter and thin-steel outer race, as opposed to the larger roller diameter and thick outer race of the heavy-duty bearing.

For moderate speeds, both types are satisfactory. However, the cage in the roller bearing allows higher speeds for a given shaft size.

MOUNTING AND MAINTENANCE

General Installation Rules

Depending on the size of bearing and the application, there are different methods for mounting bearings. In all methods, however, certain basic rules must be observed.

Maintain Cleanliness

Choose a clean environment. Work in an atmosphere free from dust or moisture. If this is not obtainable, and sometimes in the field it isn't, the installer should make every effort to ensure cleanliness by use of protective screens, clean cloths, etc.

Plan the Work

Know in advance what you are going to do and have all necessary tools at hand. This reduces the amount of time for the job and lessens the chance for dirt to get into the bearing.

Preparation and Inspection

All component parts of the machine should be on hand and thoroughly cleaned before proceeding. Housings should be cleaned, even to blowing out the oil holes. Do not use an air hose on bearings. If blind holes are used, insert a magnetic rod to remove metal chips that might have become lodged during fabrication.

Shaft shoulders and spacer rings contacting the bearing should be square with the shaft axis. The shaft fillet must be small enough to clear the radius of the bearing.

On original installations, all component parts should be checked against the detailed specification prints for dimensional accuracy. Shaft and housing should be carefully checked for size and roundness.

Mounting Straight-Bore Bearings

Heat-Expansion Method

Most applications require a tight interference fit on the shaft. Mounting is simplified by heating the bearing to expand it sufficiently to slide easily onto the shaft. Two methods of heating are in common use:

1. Immersion in a tank of heated oil (Fig. 2-24)
2. Induction heating

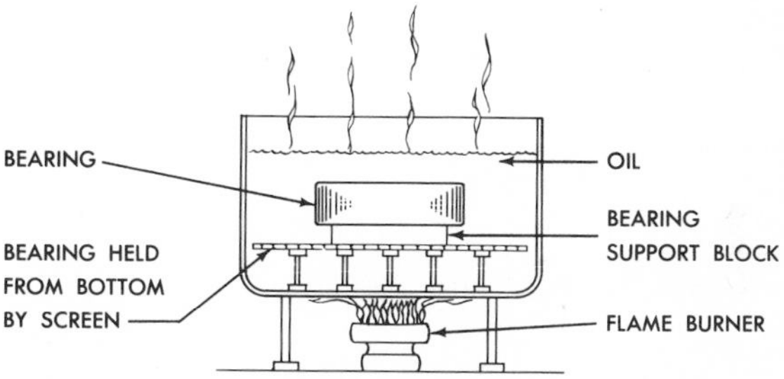

Figure 2-24 Typical arrangement for heat-expansion method.

The first is accomplished by heating the bearing in a tank of oil having a high flash point. The oil temperature should not exceed 250°F (120°C). A temperature of 200°F (93°C) is sufficient for most applications. The bearing should be heated at this temperature, generally for 20 or 30 min, until it is expanded sufficiently to slide onto the shaft very easily.

The induction-heating method is particularly suited for mounting small bearings in production-line assembly. Induction heating is rapid, and care must be taken to prevent bearing temperature from exceeding 200°F (93°C). Trial runs with the unit are usually necessary to obtain the proper timing. Thermal crayons which melt at predetermined temperatures can be used to check the bearing temperature.

While the bearing is still hot, it should be positioned squarely against the shoulder. Lock washers and lock nuts, or clamping plates, are then installed to hold the bearing against the shoulder of the shaft. As the bearing cools, the lock nut or clamping plate should be tightened.

The oil bath is shown in Fig. 2-24. The bearing should not be in direct contact with the heat source. The usual arrangement is to have a screen several inches off the bottom of the tank. Small support blocks separate the bearing from the screen. It is important to keep the bearing away from any localized high-heat source that may raise its temperature excessively, resulting in race hardness reduction.

Flame-type burners are commonly used, but may have to be replaced with some other heat source to comply with local safety regulations. An automatic device for temperature control is desirable. If safety regulations prevent the use of an open heated-oil bath, a mixture of 15 percent soluble oil in water may be used. This mixture may be heated to a maximum temperature of about 200°F (93°C) without being flammable. The bath should be checked from time to time to ensure its proper composition as the water evaporates. The bath leaves a thin film of oil on the bearing sufficient for temporary rust prevention, but normal lubrication should be supplied to the bearing as soon as possible after installation. Be sure all of the oil-in-water solution has been drained away from the bearing.

Arbor Press Method

The alternative method of mounting, generally used only on smaller sizes (Fig. 2-25) is to press the bearing onto the shaft or into the housing. This can be done by using an arbor press and a mounting tube. The tube can be of soft steel with inside diameter slightly larger than the shaft. The outside diameter of the tube should not exceed the maximum shoulder height given in the tables of dimensions. The tube should be faced square at both ends, thoroughly clean inside and out, and long enough to clear the end of the shaft after the bearing is mounted.

If the outer ring is being pressed into the housing, the outside diameter of the mounting tube should be slightly smaller than the housing bore, and the inside diameter should not be less than the recommended housing-shoulder diameter in the tables of dimensions.

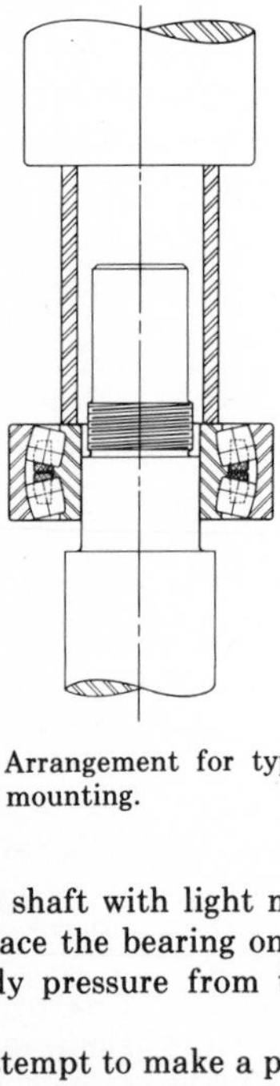

Figure 2-25 Arrangement for typical Arbor press bearing mounting.

Coat the shaft with light machine oil to reduce the force needed for the press fit. Carefully place the bearing on the shaft, making sure it is square with the shaft axis. Apply steady pressure from the arbor ram to drive the bearing firmly against the shoulder.

Never attempt to make a press fit on a shaft by applying pressure to the outer ring, or to make a press fit in a housing by applying pressure to the inner ring.

Straight-Bore-Bearing Removal

Bearing pullers of various types and designs are available from several manufacturers. These pullers are useful in removing bearings up to about 10-in bore. The preferred method utilizes a split ring which is placed behind the inner ring. The pulling device is assembled so as to cause the split ring to push the bearing off the shaft. If machine elements interfere with this method, a prong-type puller may be used with a roller bearing which has side flanges on the inner ring. Insert the prongs of the puller behind the flange and pull the bearing off the shaft.

Hydraulic Method

Remove lock plate or lock nut. Provide a means for pulling the bearing off the shaft. A bearing puller applied to the outer ring may be used without fear of brinelling the races since only a nominal force is required for removal by this system. Or, the shaft may be placed in a vertical position, simply allowing gravity and slight manual force to remove the bearing. Pull the bearing all the way off the shaft quickly to prevent freezing after the loss of oil pressure as the bearing clears the oil groove.

Mounting Bearings in Housings

To facilitate installing the bearing in the housing and to minimize fretting corrosion during operation, coat the housing with a light-grade machine oil. Make sure the outer ring is square with the housing bore before inserting it. If the outer ring becomes misaligned and sticks, do not attempt to force it farther into the housing. Use a soft brass or steel bar and tap the outer ring until it becomes free and is realigned. An assembly plate for holding the outer ring in alignment with the inner ring is often useful for applications in which conditions tend to result in misalignment.

In cases of outer-ring rotation, where the outer ring is a tight fit in the housing, the housing member can be expanded by heating.

Cast housings should be annealed. Split housings present no alignment problems during installation, but the housings should be checked carefully for size and roundness.

Internal Clearance

Internal clearance is the amount of radial play within a bearing. This clearance accommodates the effects of tight shaft and housing fits, radial thermal expansion, speed, and other operating conditions.

The internal clearance in radial bearings other than tapered roller bearings is defined as *diametral clearance.* The diametral clearance is the amount of radial play built into the bearings (see Figs. 2-26 and 2-27).

Straight-bore or tapered-bore bearings installed with interference fits will result in a reduction of internal or diametral clearance. Several factors influence the reduction of internal clearance. When the bearing inner ring is pressed onto a solid steel shaft, reduction of internal clearance is approximately 80 percent of the shaft interference fit. For an outer ring pressed into a steel or cast-iron housing, the reduction of internal clearance is about 60 percent of the housing-interference fit. If the shaft is hollow, the housing walls are thin, or materials other than steel are used, clearance reduction may vary considerably from the above percentages.

Tapered-bore spherical roller bearings require a slightly greater interference fit with the shaft than straight-bore bearings of corresponding size. The resultant fit cannot conveniently be determined by direct shaft measurement. It can be checked by the distance the bearing is pressed onto a carefully gauged tapered shaft, or more easily by the effects of radial expansion of the inner ring as it is pushed onto the taper.

Lubrication

Since the lubricant affects bearing life and operation, selecting the proper lubricant is an important design function. The purpose of lubrication in bearing applications is to:

1. Minimize friction at points of contact within the bearings
2. Protect the highly finished bearing surfaces from corrosion
3. Dissipate heat generated within the bearings
4. Remove foreign matter or prevent its entry into the bearings

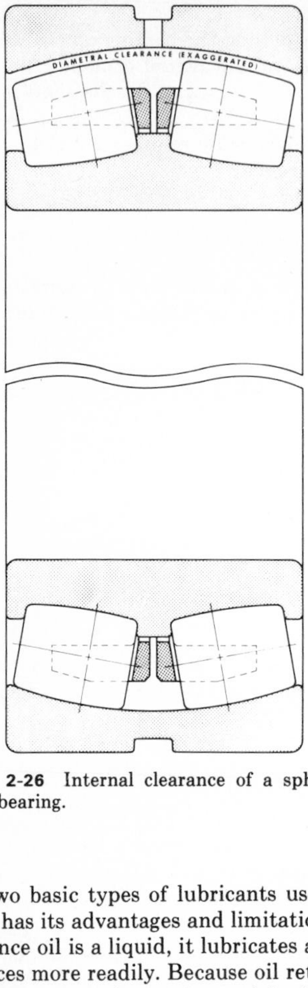

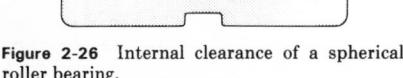

Figure 2-26 Internal clearance of a spherical roller bearing.

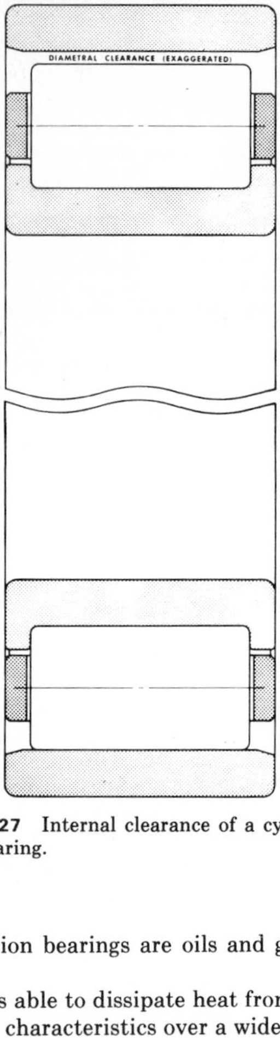

Figure 2-27 Internal clearance of a cylindrical roller bearing.

Two basic types of lubricants used with antifriction bearings are oils and greases. Each has its advantages and limitations.

Since oil is a liquid, it lubricates all surfaces and is able to dissipate heat from these surfaces more readily. Because oil retains its physical characteristics over a wider range of temperatures, it may be used for high-speed and high-temperature applications. The quantity of oil supplied to the bearing may be accurately controlled. Oil lubricants can be circulated, cleaned, and cooled for more effective lubrication.

Grease, which is easier to retain in the bearing housing, aids as a sealant against foreign matter and corrosive fumes.

Oil Lubrication

1. Oil is a better lubricant for high speeds or high temperatures. It can be cooled to help reduce bearing temperature.
2. Oil is easier to handle, and with oil it is easier to control the amount of lubricant reaching the bearing. It is harder to retain in the bearing. Lubricant losses may be higher than with grease.
3. As a liquid, oil can be introduced into the bearing in many ways, such as drip feed, wick feed, pressurized circulating systems, oil bath, or air-oil mist. Each is suited to certain types of applications.
4. Oil is easier to keep clean for recirculating systems.

Grease Lubrication

1. Restricted to lower-speed applications within operating-temperature limits of the grease.
2. Easily confined in the housing. This is important in the food, textile, and chemical industries.
3. Bearing enclosure and seal design simplified.
4. Improves the efficiency of mechanical seals to give better protection to the bearing.

For all new applications, a competent lubrication engineer should study the operating conditions and recommend the specific lubricant required.

BIBLIOGRAPHY

For additional or more detailed information on a specific bearing type or application, reference should be made to the bearing manufacturers' catalogs and or one of the following:

AFBMA Standards, The Anti-Friction Bearing Manufacturers Association, Inc., 2341 Jefferson Davis Highway, Arlington, VA 22202.
SKF Industries, Inc., Front Street and Erie Avenue, P. O. Box 6731, Philadelphia, PA 19132.
The Timken Company, 1835 Dueber Avenue S.W., Canton, OH 44706.
The Torrington Company, 3702 West Sample Street, P. O. Box 1684, South Bend, IN 46634.

PART 2

Plain Bearings

adapted with permission from

"Mechanical Drives" issue
Machine Design Magazine
A Penton/IPC Publication

INTRODUCTION

A *plain bearing* is any bearing that works by sliding action (with or without lubricant). This group encompasses essentially all types other than rolling-element bearings. Many of the basic principles dealing with plain-bearing loading and lubrication are explained in the prior section, and the reader is advised to review these basics before continuing with the more specific and detailed treatment here.

Plain bearings are often referred to as *sleeve bearings* or *thrust bearings,* terms that designate whether the bearing is loaded axially or radially.

Lubrication is critical to the operation of plain bearings, so their application and function is also often referred to according to the type of lubrication principle used. Thus, terms such as hydrodynamic, fluid-film, hydrostatic, boundary-lubricated, and self-lubricated are designations for particular plain bearings.

BEARING TYPES

Journal or Sleeve Bearings

These bearings (Fig. 2-28) are cylindrical or ring-shaped bearings designed to carry radial loads. The terms *sleeve* and *journal* are used more or less synonymously since sleeve refers to the general configuration while journal pertains to any portion of a shaft supported by a bearing. In another sense, however, the term journal may be reserved for two-piece bearings used to support the journals of an engine crankshaft. Sleeve bearings may be lubricated by either hydrostatic, hydrodynamic, or boundary means. In practice, most of them are designed to operate hydrodynamically, and hydrodynamic lubrication is normally implied when the terms sleeve or journal bearing are used.

The simplest and most widely used types of sleeve bearings are cast-bronze and porous-bronze (powdered-metal) cylindrical bearings. Cast-bronze bearings are oil- or grease-lubricated. Porous bearings are impregnated with oil and often have an oil reservoir in the housing.

Cast bearing bores are held within a few thousandths of an inch and are often finish-machined after installation to provide extreme accuracy of shaft position. Porous bearings are sized more accurately and are mounted with a special sizing tool so that no finish machining is needed. Machining poses the danger of smearing or closing the all-important pores.

Oscillating or slow-moving sleeve bearings cannot develop hydrodynamic lubrication, so their design criteria are different from those of other sleeve bearings. Oscillating bearings are generally designed on the basis of allowable wear or upon the load that produces "pounding out."

Engine bearings are designed to operate under hydrodynamic conditions. Their surfaces are smooth, and clearances are optimized to maximize load capacity while allowing enough oil flow to aid cooling.

Plastic bearings are being used increasingly in place of metal. Originally, plastic was used only in small, lightly loaded bearings where cost saving was the primary objective. More recently plastics with high load-carrying capacities have been developed for direct replacement of metals. Also, plastics are used because of functional advantages, including resistance to abrasion, and are being made in large sizes.

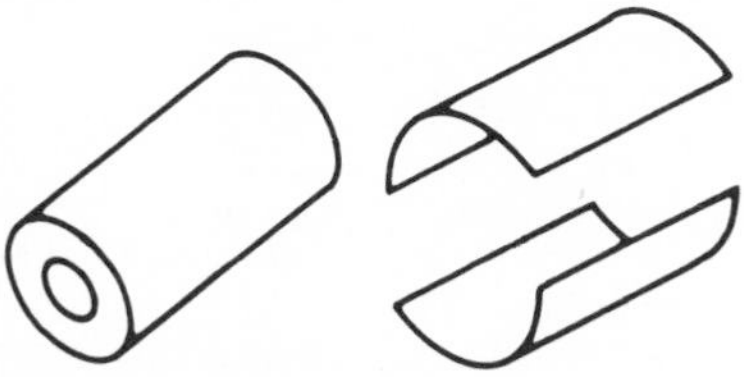

Figure 2-28 Sleeve or journal bearings.

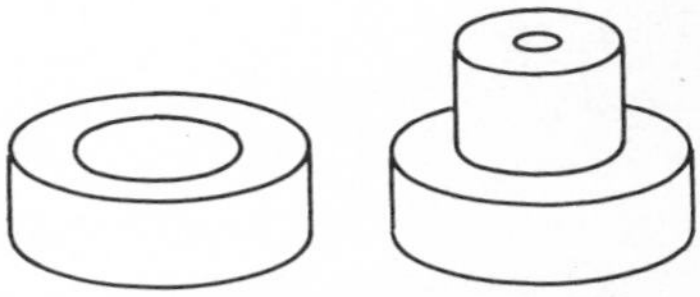

Figure 2-29 Thrust bearings.

Thrust Bearings

This type of bearing (Fig. 2-29) differs from a sleeve bearing in that loads are supported axially rather than radially. Thin, disklike thrust bearings are called thrust washers.

Hydrodynamic action in a thrust bearing can be improved by putting grooves in the surface, rather than having it flat. Small surface irregularities—either etched-in deliberately or resulting as a normal consequence of manufacturing—aid hydrodynamic action. Even assembly misalignment can help produce hydrodynamic behavior. Geometry of the grooves can be optimized for particular loads, speeds, and oil viscosities. For full hydrodynamic action, a thrust bearing must have tapered lands on tilting pads.

Spherical Bearings

The outside diameter of this type of bearing is spherical (or at least curved) to permit "wobble" of the bearing axis. This ball-and-socket action compensates for shaft and mount misalignment; thus these bearings (Fig. 2-30) are called *self-aligning*. Once the spherical tilt accommodates misalignment, subsequent bearing relative motion is normally between the shaft and the cylindrical bore in the ball.

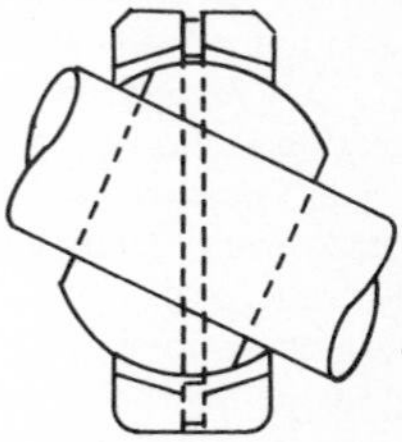

Figure 2-30 Spherical bearings.

When the ball is enclosed at the end of a link, the bearing is called a *rod-end bearing*. These bearings are normally used in oscillating linkages, and most of the relative motion is between the ball and the socket.

Hydrostatic Bearings

Hydrostatic bearings (Fig. 2-31) can be either of the thrust or sleeve (journal) type. Bearing characteristics are highly predictable, including such characteristics as stiffness, friction, and oil flow. Advantages include nearly zero starting friction, the ability to carry heavy loads at low speeds, and negligible wear. But hydrostatic bearings are expensive and require bulky equipment to provide pressurized oil. However, machines with hydrostatic bearings maintain their accuracy, so savings in depreciation can be significant. This type of lubrication is used most frequently in machine tools, but it is also employed in gyroscopes and other precision instruments where low friction is important.

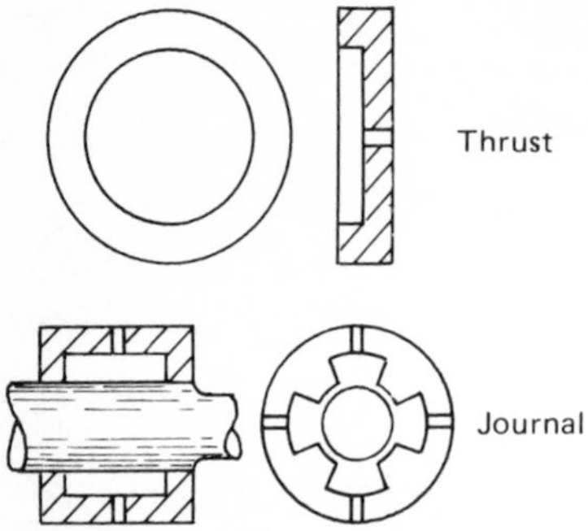

Figure 2-31 Hydrostatic bearings.

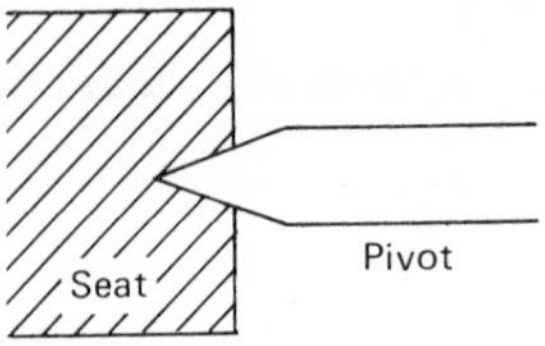

Figure 2-32 Pivot bearing.

Jewel and Pivot Bearings

These bearings (Fig. 2-32) are suitable for tiny, lightly loaded shafts in such applications as watches, meters, and instruments. The shaft does not pass through the bearing but rests in a cone or cup on each end. The pivots (pointed shaft ends) are usually hardened steel, and seats are usually sapphire, glass, beryllium copper, carbide, or carbon. The bearing normally is lubricated with a light nongumming oil.

These bearings operate in the boundary regime, but because of the small radius of contact, they have lower friction than rolling-element bearings. Jewel bearings come in a variety of standard designs.

Gas Bearings

Gas bearings are lubricated with a gas (usually air) instead of with an oil. They can either be externally pressurized (hydrostatic) or self-acting (hydrodynamic). A special type of self-acting gas bearing uses metal-foil interleaved to form a raceway and is thus called a *foil bearing.*

The operating principles and relative merits of hydrodynamic and hydrostatic gas bearings are similar to those of their oil-lubricated counterparts. The hydrostatic type can support loads at rest or low speed but requires an expensive and complex pumping system. The hydrodynamic type is simpler but must have significant relative motion to function and must be built to very close tolerances.

Gas bearings are used mainly for applications demanding extremely high rotational speeds or uniformly low frictional drag. Ambient atmospheres can be used as the working lubricant, so gas bearings also are used where possible contamination from oil lubricants must be avoided. The design of gas bearings is usually complex and requires sophisticated engineering. They usually are an integral part of the equipment in which they are used. However, some gas bearings are available as off-the-shelf components.

Most gas bearings demand small operating clearances, so they require precision manufacturing methods. They also are susceptible to overload, vibration, and dust contamination. Another problem is that they are subject to instabilities caused by the dynamics of a gas film or by unbalanced rotating masses.

HOW BEARINGS FAIL

The failure mode anticipated in a bearing influences decisions on the selection of a bearing type, bearing material, lubricant, and bearing surface finish. Typical failure modes include:

- Loss of shaft positioning from wear; seizure by overheating and loss of clearance
- Seizure by failure of lubricant and galling of contacting surfaces
- Destruction of journal by hard particles imbedded in the bearing surface
- Edge loading caused by dimensional instability of bearings and housings, or by shaft deflection
- Fatigue of bearing surface; and high friction

Friction

The frictional properties of any plain bearing depend on the lubrication system. Either hydrodynamic or hydrostatic lubrication can provide low friction. The lowest friction levels are attained in gas bearings or with magnetically supported bearings.

Friction in hydrodynamic and hydrostatic bearings is a function of lubricant viscosity and shear rate. Shear rate increases with increasing rotational speed and decreasing film thickness. The friction coefficient is generally below 0.001.

The friction in boundary and self-lubricated bearings varies widely and is difficult to predict for a given bearing-lubricant system. The range of coefficients of friction is 0.01 to 0.10 for boundary lubrication and 0.01 to 0.3 for self-lubrication.

Caution must be used when applying friction-coefficient handbook data. Conditions under which the values were measured should be known, and these conditions should be duplicated in the application. Coefficient of friction tends to increase with increasing surface roughness with dryness and cleanliness of surfaces, and with decreasing temperature.

Wear

Wear of plain bearings is strongly influenced by the state of lubrication and, conversely, wear characteristics influence the various lubrication states.

Hydrostatic bearings do not wear if operating properly because the bearing surfaces are separated by a film of oil, even when the bearing is stationary. Erosion of flow restrictors in hydrostatic bearings with high flow rates can and eventually does cause failure. Plugging of flow restrictors by wear debris can cause catastrophic failure. In gas-lubricated, externally pressurized bearings, occasional rubs and resulting wear can occur under impact and vibration loads.

Hydrodynamic bearings wear very slowly. Wear occurs during start-up and slowdown when the speed is too low to produce sufficient fluid pressure to support the bearing surfaces on a lubricant film. If hard debris (harder than the journal surface) imbeds in a babbitt or plastic bearing and protrudes above the bearing surface, the journal can wear seriously during start-up.

In a severe and catastrophic type of journal wear—known as *wire wooling*—the journal surface is machined by hard scabs of wear debris that pack into the babbit surface. In this failure the journals are deeply grooved and can no longer generate a hydrodynamic film.

Boundary-lubricated and self-lubricating bearings wear much faster than fluid-film bearings. Self-lubricating plastic bearings wear at higher rates than boundary-lubricated, metal-alloy bearings.

Babbitts are subject to fatigue damage in which pieces of the babbitt metal spall out of the surface. This might be considered as a catastrophic type of failure after a prolonged (but undetectable) period of wear buildup. Babbitt fatigue is most likely in bearings subjected to reciprocating or vibrating loads.

Babbitts and leaded bronzes are also susceptible to cavitation erosion. This is a form of wear in which shock waves, from gas bubbles collapsing in the lubricant, produce surface pits. Eventually, the removal of surface material impairs performance. Cavitation erosion has a characteristic lacy appearance and often is found around lubrication feed holes and slots in bearings subjected to vibratory or impact loading—such as in internal combustion engines.

Heat

Heat is generated either by shearing of the oil film or by rubbing contact. In hydrostatic and hydrodynamic bearings, heat generation at running speeds is the result of oil shear, and the amount of temperature rise can be estimated if oil viscosity and shear rates are known. Temperature can be regulated by controlling the oil flow through the bearing or by using external cooling.

High-speed and close-clearance fluid-film bearings are difficult to keep cool. The flow rate through a journal varies with the cube of oil-film thickness and linearly with supply pressure.

Boundary-lubricated and self-lubricating bearings are more sensitive to sliding velocity than fluid-film types because the coefficient of friction is as much as 10 times greater in the first two. Frictional heating is a function of bearing pressure, sliding velocity, and coefficient of friction. Therefore, if the coefficient of friction remains constant for a range of loads and speeds, a rough indication of heat load is provided by the *pressure-velocity* (PV) factor.

Plastic bearing materials are sensitive to PV because of their low thermal conductivities and high thermal-expansion rates.

Lubrication

Although some materials have an inherent lubricity or can be lubricated by virtue of a film of slippery solid, most bearings operate with a fluid film—generally oil but sometimes a gas.

By far the largest number of bearings are oil-lubricated. The oil film can be maintained through pumping by a pressurization system—in which case the lubrication is termed *hydrostatic*. Or it can be maintained by a squeezing or wedging of lubricant produced by the rolling action of the bearing itself—termed *hydrodynamic* lubrication. If loads are too high or speeds too slow, the hydrodynamic action begins to break down (allowing, in some cases, metal-to-metal contact), a condition referred to as *boundary* lubrication.

Hydrostatic Lubrication

The main virtue of hydrostatic lubrication is that it can accommodate heavy loads at low speeds because it does not depend upon relative motion to maintain the lubrication film. Instead, lubricant is supplied from a special pump and feed lines to the bearing or bearings. The oil is fed through flow restrictors, which generally are in the stationary part of the bearing. The flow restrictors automatically adjust the oil flow for the applied load. Another advantage of this lubrication system is low deflection in certain load ranges, making it preferred for many high-precision machine tools. The disadvantage of hydrostatic lubrication is its high cost and complexity.

Hydrodynamic Lubrication

This form of lubrication occurs more or less naturally in properly finished, sized, and lubricated holes and shafts. Essentially, rotation of the journal causes it to "roll up" one side of the bearing, thus creating a wedge-shaped channel into which lubricant is forced. The forcing of the lubricant into this wedge creates sufficient pressure to keep the journal riding on the oil film. This form of the lubricant is generally preferred because it is simple and dependable. Also, the lubrication action improves as speed increases, which in most applications goes hand-in-hand with an increase in loads experienced as speed increases. Its main drawbacks are an inability to carry heavy loads at low speed and appreciable wear under frequent stops or starts, or motion reversals.

The oil for hydrodynamic lubrication can be fed from an oil reservoir, or the bearing can be made of a porous metal that is impregnated with oil that "bleeds" to the bearing surface as the shaft rotates. (Most porous-metal bearings, however, operate under boundary or mixed-film conditions.)

Boundary Lubrication

This form of lubrication is essentially a breakdown of hydrodynamic action. At high loads or low speeds, the pressure of the hydrodynamic film cannot prevent metal-to-metal contact. So the opposing surfaces partially ride on an oil film and partially rub together as surface high points contact.

Lubrication is provided by lubricant decomposition products or surface-active additives which form a thin, soft, solid film on the metal surfaces and prevent metal-junction adhesion.

Boundary lubrication is not the most desirable operating mode, yet at times it is completely unavoidable. It is found mainly with slow-moving loads where the cost and expense of a hydrostatic system is not warranted. Hinge bearings in aircraft landing gears, for example, do not move fast enough to develop hydrodynamic films, yet hydrostatic systems would be too heavy, costly, and cumbersome.

Boundary-lubricated bearings using grease require special consideration to ensure adequate distribution of the lubricant. Grease travels only where is is carried on a moving surface, so the moving shaft carries a gob of grease at the center of the bearing, in a band only somewhat wider than the original gob. For this reason, sleeve bearings are grooved to distribute the grease over the entire load-bearing area, ensuring an adequate grease supply for boundary-film maintenance. In addition, the grooves collect wear debris, preventing it from damaging the bearing surface.

Many different groove designs are used (see Fig. 2-33). Some were developed for special applications; however, all are designed for easy manufacture. Basic principles to follow when selecting a groove design are:

- Grooves must carry the grease over the entire bearing width.
- Grooves must be placed so that the shaft contacts the grease within them. (A simple axial slot opposite the point of maximum load does not feed to the load zone because of internal clearances.)
- Groove edges should be rounded so that they do not act as scrapers that clean grease from the shaft.

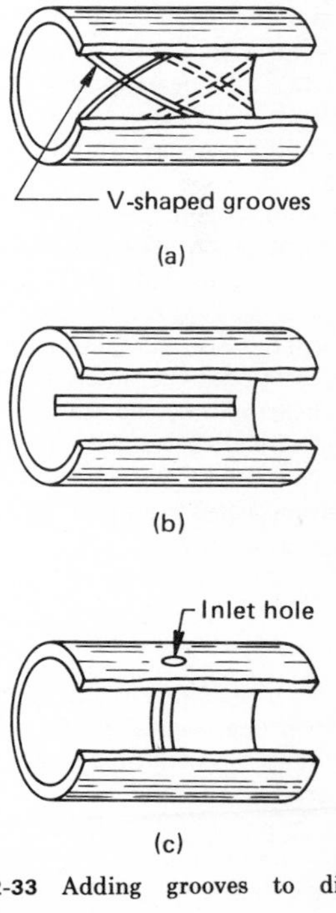

Figure 2-33 Adding grooves to distribute grease.

Groove cross section is usually V-shaped, but it can be rectangular or semicircular. A flat-bottom V groove provides ample reservoir capacity and reduces edge scraping. Groove width and depth are not critical as long as they do not weaken the bearing structurally or carve up too much of the bearing area. As a rule, groove depth should not exceed one-third the bearing wall thickness.

The 00 design is appropriate for continuous rotation under heavy loads. Conventional design calls for the grooves to be machined within about 0.125 in of the ends to retain the lubricant within the bearing area. However, under heavy loads where bearing temperature is high and debris becomes a problem, the grooves should exit through the ends, so that periodic flushing with fresh grease removes debris and lubricant decomposition products. Experiments with these *nonrecirculating* grooves indicate lower wear than with recirculating grooves.

The *straight axial* groove is a simple arrangement for moderate, unidirectional loads and continuous rotation. The axial slot should be located about 20° from the load line, counter to the direction of shaft rotation. This design spreads grease over the bearing width and still provides maximum bearing area. However, its lubricating effectiveness is not as good as that of the 00 design.

A circumferential slot in the center of the bearing is effective for oscillating bearings, especially in applications where load direction varies.

COMPARING PLAIN BEARINGS AND ROLLER-ELEMENT BEARINGS

Friction

The torque required to put a bearing into motion from rest is usually higher than the torque required to keep it running once it is in motion. *Starting friction* thus has an important influence on the power required in a drive system.

Externally pressurized bearings have very low starting torque. Roller bearings have a low starting torque. Unpressurized sleeve (fluid-film) bearings have substantially higher starting torque. The coefficient of friction at start-up for self-lubricated bearings is highly variable. It may range from 0.04 to 0.16.

The fluid-film bearing has a high starting torque because it must pass through boundary lubrication stages as it comes up to speed. Once running under a hydrodynamic film, the fluid-film bearing exhibits friction characteristics comparable to a rolling-element bearing.

At running speed, the friction characteristics of various bearing types cannot be summarized in a few words. The externally pressurized bearing runs with low friction. Friction in a self-lubricating sleeve bearing is quite variable, depending upon the application.

Running friction for a rolling-element bearing is lower than starting friction. If torque characteristics are critical to a design, recommended practice is to measure starting and running frictional characteristics experimentally.

If a bearing must be started repeatedly under heavy load, rolling-element bearings are a better choice than sleeve bearings. If the increased complexity is acceptable, externally pressurized (hydrostatic) bearings are the best choice of all. If starting load is light and load increases gradually with speed, the conventional hydrodynamic sleeve bearing usually is preferred.

Noise

Whenever a machine is subject to a noise-reduction program, bearings are normally involved in one way or another. Even if bearings are not generating noise, they often have much to do with noise transfer.

Generally, rolling-element bearings are noisier than fluid-film bearings. Minor inaccuracies in rollers or raceways can generate sounds that are amplified by the machine structure. Improving bearing quality can reduce this effect.

Fluid-film bearings under steady radial load generally do not produce any noise whatsoever. However, if this type of bearing is reversed frequently, it can generate considerable noise if it doesn't have enough lubricant to fill the bearing. Fluid-film bearings running unstable in a whirling mode can also produce noise.

Size

A bearing requiring a separate pressurized lubrication system requires more total space than a self-lubricating type. The relative space required at the actual load-support point is not a clear-cut matter. A pressurized bearing can conceivably be more compact than self-contained bearings at the load-support point.

For self-contained bearings, the sleeve bearing requires less radial space than a rolling-element bearing; however, the sleeve bearing requires slightly more axial space. Needle bearings require about as much space as journal bearings.

Cost

Hardware Costs

In high-volume lots, sleeve bearings and bushings are considerably less costly than rolling-element bearings. In mid-range volumes, prices for the two types are comparable. For special designs in small quantities, sliding bearings are usually more costly than rolling-element bearings. Dry-film and boundary-lubricated bearings usually employ proprietary materials that are expensive. Powdered-metal bearings, however, are inexpensive.

Design Costs

Rolling-element and dry-lubricated bearings normally require the least engineering cost to the end user. Manufacturers of rolling-element bearings, in particular, can provide considerable cost-saving assistance by virtue of well-documented design manuals.

Self-acting sleeve bearings, in contrast, may require considerable end-user design effort except for light-duty applications or where there is considerable application experience.

The behavior of externally pressurized bearings usually can be predicted easily by simple calculations, but considerable design effort may be required to verify the design completely.

Shop Costs

Rolling-element bearings normally require precise housings and shafts and thus require fairly costly machining for products in which they are used. Sleeve-bearings, in contrast, generally operate well with less finely prepared machine finishes. Many plain bearings operate satisfactorily with lathe-turned journals.

Maintenance Costs

If the bearing lubrication is self-contained, maintenance costs are normally determined by sealing requirements. If there is full-pressure lubrication, costs may be determined by the amount of filtration needed to keep out contaminants. Generally, rolling-element bearings have the lowest maintenance costs because of the lower lubrication requirements. The very minimum maintenance cost is associated with self-lubricating bearings—provided that they deliver sufficient service life.

Replacement Costs

These costs depend more on the specific design than on the type of bearing used. In general, however, sliding bearings normally are replaced more easily than rolling-element types. Both types can be damaged during installation if not handled properly. Sliding bearings can often be replaced quickly by machining from bar stock or by altering available stock sizes.

Cost of Failure

Rolling-element bearings usually give ample warning that they are approaching failure (by virtue of increasingly noisy operation). They usually fail from fatigue. Sliding bearings, on the other hand, usually perform well until just moments before violent failure.

If a rolling-element bearing fails at high speed, the failure is usually total and catastrophic. With a journal bearing, the effect is normally less drastic. Often, only a bit of polishing can put it back into service. However, sliding bearings are capable of total catastrophic failure.

chapter 7-3

Shaft Drives and Couplings

PART 1

Belt Drives

by

Wally Erickson
Chief Engineer, Product Application Department
The Gates Rubber Company
Denver, Colorado

BELT DRIVES

Belt drives are the most widely used method of power transmission. Improvements in the design and manufacture of belts have broadened their application and utility.

Belt drives employ V, flat, synchronous, or ribbed (poly V) belts. Although the different types of belt drives are considered interchangeable, this is really not the case. Each type is restricted to a well-defined application area.

For example, flat belts are satisfactory for drives operating at high speed and low horsepower. On the other hand, flat-belt drives become overly large when the primary consideration is high-power transmission, and they must give away to V-belt drives in such applications.

V belts provide the best overall power transmission and capability per dollar cost and unit of space. V belts also are the only type of belt that can be used on variable-speed drives.

Synchronous belts are used in applications requiring precise speed ratios and synchronization. These belts also offer lower static tensions, but operating tensions are about the same as on V-belt drives. The ability to operate over small sheave diameters often makes them attractive alternatives to V belts even when precise speed control is not important.

V Belts

The problems of high tension and instability with flat belts led to the development of V belts. These belts have deep, V-shaped cross sections that wedge into sheaves to provide the required traction. Because of this wedging action, V belts are highly stable and can operate at tensions considerably lower than those needed by flat belts. Thus, V-belt drives are more compact, allowing shafts and bearings to be smaller.

The load in a V belt is carried by a fiber tensile section located near the top of the belt. This section can contain one or several layers of cord, depending on the method of manufacture (Fig. 3-1).

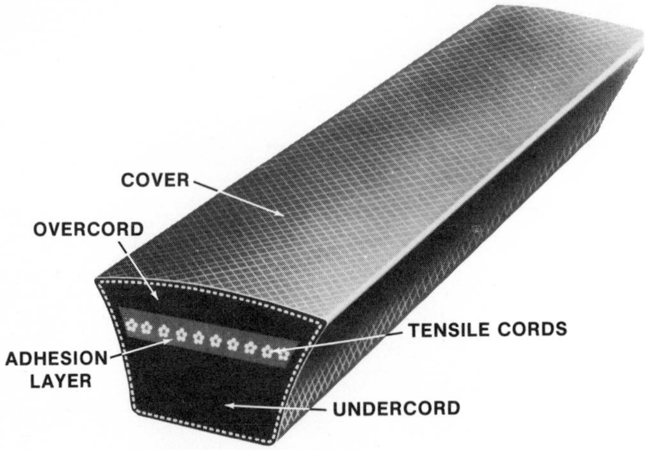

Figure 3-1 Typical V-belt construction.

V belts operate through a wide range of belt speeds. Standard sheaves are limited to 6500 ft/min (33 m/s) upper limit. Speed or peak capacity varies with the type of belt and section.

There are four basic types of V belts: industrial, light-duty, agricultural, and automotive.

Industrial V Belts

Classical. This line of belts includes five standard cross sections—A, B, C, D, E, (13C, 16C, 22C, 32C, 39C)*—ranging in width from ½ in (13 mm) for the A section to 1½ in (39

*Metric sizes are shown in parentheses.

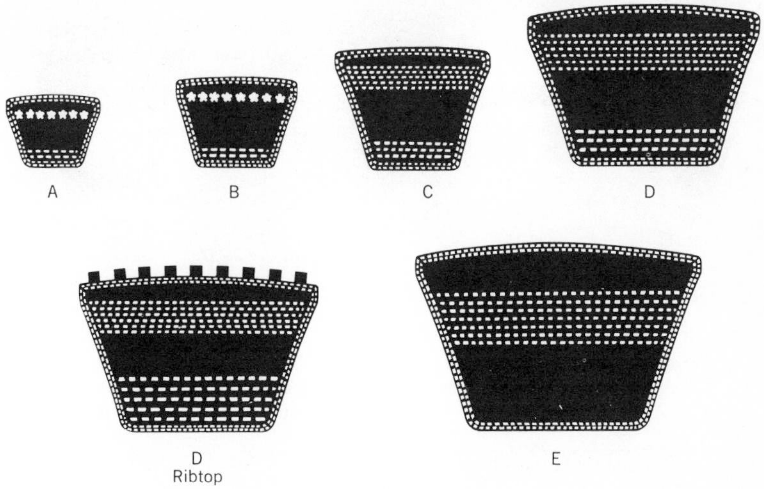

Figure 3-2 Classical V-belt cross sections.

mm) for the E section (Fig. 3-2). Classical belts can be teamed up in multiples of two or more belts, and multiple-belt drives can deliver up to several hundred horsepower continuously and can absorb reasonable shock loads. Temperature limits range from −30 to +140°F (−34 to +60°C).

Classical belts are available in the joined configuration, where several belt strands are connected by a tie band across the top. The tie band improves lateral stability and solves the problems of belts turning over and jumping off sheaves. The band rides above the sheave and does not interfere with the wedging action of any of the individual belt strands (Fig. 3-3).

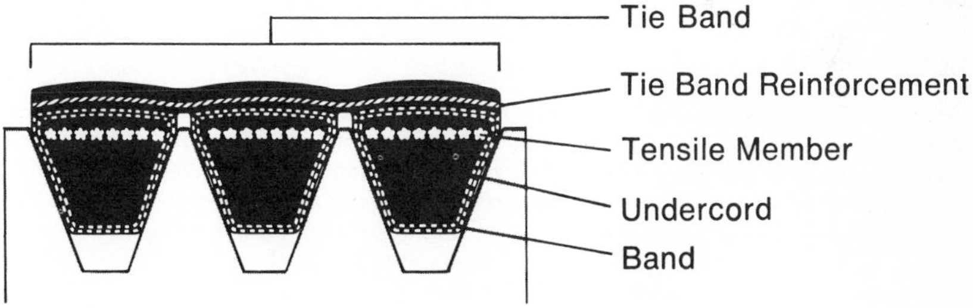

Figure 3-3 Joined V-belt tie-band configuration.

A widely used variation of the classical belt is the molded-notch or molded-cog belt. Being more flexible, the cog belt can operate on smaller-than-normal sheave diameters on drives with limited space for sheaves.

Narrow. These belts serve the same application area as multiple, classical V belts but with a lighter, more compact drive. Three cross sections—3V, 5V, 8V (9N, 15N, 25N)—replace the five classical cross sections (Fig. 3-4).

Narrow V belts usually provide substantial space and weight savings over classical belts. For instance, narrow belts can transmit up to three times the horsepower of conventional belts in the same drive space—or the same horsepower in one-half to two-thirds the space. In addition, the use of narrow belts results in a more aesthetically pleasing machine.

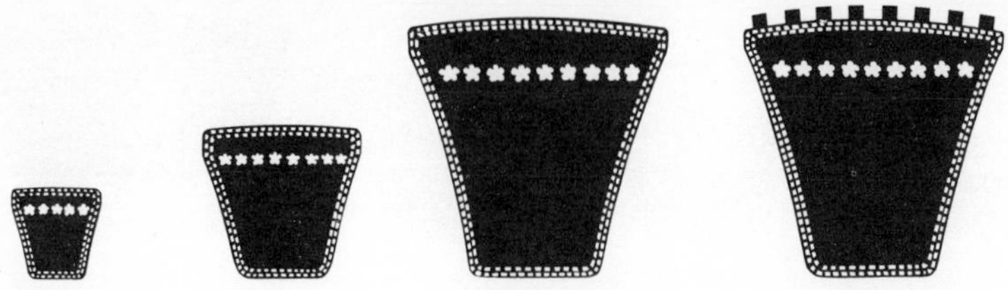

Figure 3-4 Narrow V-belt cross sections.

Variable-Speed Belts. Within certain limits, V belts are well suited to drives that must run at varying input and output speeds. Speed ratio on these drives is controlled by moving one sheave sidewall relative to the other so the belt rides at different pitch diameters.

Unlike gear transmissions where speed variation is limited to finite steps, variable-speed V-belt drives offer an infinite selection of speed ratios within a certain range. In simple applications, speed is varied manually by adjusting the sheave sidewalls while the drive is stopped. On other drives, speed ratio is varied while the drive is operating, either by manual control or by an automatic control sensitive to the speed and torque requirements of the machine.

But adjustable-speed operation generates high frictional heat and sidewall stresses. Until recently, belt fibers and molding compounds could not handle the strains of high-torque applications, and variable-speed belts were largely limited to industrial applications. In the past few years, manufacturers have found belt materials that withstand the heat better, and they have devised ways to optimize the belt configuration. As a result, variable-speed belts are moving into more demanding applications.

These belts are used in industrial, agricultural, automotive, and recreational vehicle applications where speed variation is a prerequisite.

Light-Duty V Belts

These belts are similar to classical belts, except they have slightly thinner cross sections. They are best suited for smaller sheaves, found on fractional-horsepower drives. Most drives designed for light-duty belts are single, rather than multiple drives. Usually these drives are operated intermittently. Service requirements may vary widely from 2 to 3 h/week for power lawn mowers to 40 or more hours per week for office machines.

Standard cross sections for light-duty V belts are 2L, 3L, 4L, 5L (6R, 9R, 12R, 16R) and range in width from ¼ to ⅝ in (6 to 16 mm).

Agricultural V Belts

While agricultural applications require heavy-duty drives, the requirements are significantly different. For example, industrial drives run at fairly constant loads, whereas agricultural drives are subjected to intermittent shock loads. In addition, most agricultural drives require the belt to bend in reverse over small-diameter sheaves; therefore, agricultural belts have more flexible jackets.

Agricultural V belts are available in joined, hexagonal, or double-V styles. Double-V belts are used on drives with more than one driven shaft where shaft location and direction of rotation make it necessary to drive from both sides of the shaft.

Standard cross sections for fixed-speed, classical agricultural V belts are HA, HB, HC, HD, HE (13F, 16F, 22F, 32F, 39F). Cross sections for the double-V belts are HAA, HBB, HCC and HDD (13FD, 16FD, 22FD, 32FD).

Automotive V Belts

Automotive accessory drives require special, narrow V belts that can fit in the limited space available in engine compartments. These belts must not only be able to transmit high horsepower, they must also do so over relatively small-diameter pulleys.

Flat Belts

The flat belt is still a widely used method of power transmission. These belts are flexible over small diameters, and some types adapt to almost any drive configuration. They provide good shock resistance, need no lubrication, are quiet, and offer good design flexibility.

One disadvantage of flat belts is they require higher belt tensions than V belts; this results in higher shaft and bearing loads. The need for higher tensions may cause more stretch, causing the belt to slip and require retensioning.

Synchronous Belts

V belts and flat belts "creep" somewhat in use, so they are not suitable for drives requiring synchronized input and output. Synchronous belts were designed to overcome this limitation. These belts eliminate slip by transmitting power through the positive engagement of belt teeth against pulley teeth. Thus, precise speed ratios and synchronization are possible in such applications as machine-tool indexing heads and automotive camshafts (Fig. 3-5).

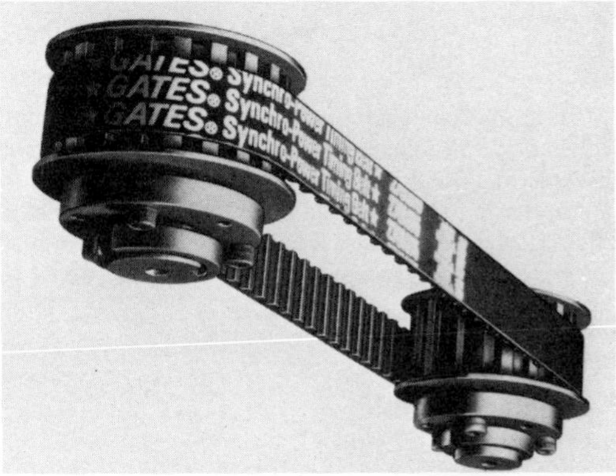

Figure 3-5 Synchronous V belt.

Synchronous belt drives have an advantage over gears and chains in that they transmit reasonably high loads with low noise and without lubrication. The shock-absorbing characteristics of the rubber teeth against the metal pulley can be helpful. However, synchronous belts wear rapidly and fail permanently if pulleys are not aligned to close tolerances.

There is one situation (when synchronization is not important) in which synchronous belts have an advantage over V belts. These belts have an extremely high modulus, low-stretch tensile cord to maintain uniform spacing between teeth. This low growth under load results in minimal need for center-distance adjustment. Therefore, synchronous belts are often used on drives with limited space for center distance movement.

There are two basic configurations to synchronous belt teeth: the standard trapezoidal shape and a recently developed deeper, rounded tooth which increases belt torque capacity.

V-Ribbed Belts

Ribs on the bottom side of V-ribbed belts (Fig. 3-6) mate with corresponding grooves in the pulley. This tracking guides the belt and makes it more stable than a flat belt. The ribs fill the pulley grooves completely; therefore, V-ribbed belts do not have the wedging action of V belts, so they must operate at higher tensions.

Figure 3-6 V-ribbed belt.

V-ribbed belts have high lateral stability, much the same as joined V belts. However, the tensile section rides above the grooves and the ribs can be deflected.

While V-ribbed belts are not as flexible as flat belts, they still perform fairly well on small-diameter pulleys, provided that the shafts and bearings can handle high belt tensions. V-ribbed belts are typically used in light-duty applications requiring high speed ratios, e.g., in clothes dryers.

The belts come in five standard sections, ranging in rib width from approximately $\frac{1}{16}$ in. to $\frac{3}{8}$ in. (1.6 to 9.4 mm). The sections are H, J, K, L, M (PH, PJ, PK, PL, PM).

BELT SELECTION

To aid in selection of the proper belt for the proper application, manufacturers put technical and design information on their belts. In addition, the Rubber Manufacturers' Association and Mechanical Power Transmission Association have worked together to publish engineering standards for most types of belts (see references at end of the chapter). These standards contain information from which a drive can be designed.

In basic terms there are four questions that need to be answered in drive design:

1. What horsepower is required of the drive?
2. What is the speed of the *driver* shaft?
3. What is the speed of the *driven* shaft?
4. What is the approximate center distance?

Also, while selecting or evaluating the drive, consider the following points:

1. If you need to keep the width of the sheave face at a minimum, select the largest-diameter drive from the group.
2. Larger-diameter sheaves will keep drive tensions (and shaft pull) at a minimum.
3. Larger-diameter sheaves will generally give a more economical drive, but not so large that multiple-belt capability is sacrificed.
4. If space is limited, consider using the smallest-diameter drive from the group. However, sheaves on electric drives must be at least as large as the National Electrical Manufacturers' Association minimum standards.
5. When the belt is midway between two belt cross-sectional sizes, the larger cross section will usually be a more economical drive.

Selecting an optimum belt involves too many factors and belt types to be discussed here, but it can be readily accomplished using the reference material and manufacturers' literature.

MAINTENANCE AND SAFETY

Once belts have been selected, they require a minimum of maintenance, but certain procedures can help reduce equipment downtime and increase safety.

Installing Belts

When installing belts on any multiple-belt drive, always replace all the belts because older belts naturally become stretched or worn from use. If old and new belts are mixed, the new belts will be tighter, will carry more than their share of the load, and will probably fail before their time.

Be sure to use a set of belts from a single manufacturer. If brands are mixed, the belts may have different characteristics and they could work against each other. This would result in unusual strain and would reduce the life of the belts.

There are two ways to assure a well-matched drive system depending on the method of matching used by the individual belt manufacturer:

- If the belt manufacturer uses the older conventional matching system, each belt is measured under tension on V sheaves and marked with a match number designating the small increment of length within the overall belt length tolerance. Then the match numbers are grouped within the limits given in Table 3-1.
- Because of improved production methods, some manufacturers, like Gates, can build belts to overall length tolerances that are within the RMA standard matching limits. This means any belts of a specified length can be grouped together to form a matched set.

After the belts have been properly selected, it is time to put them on the drive.

The most important rule is never to pry or roll the belts onto the sheave groove. Prying the belt onto the sheave shortens the belt's life, even though damage may not be visible. Safety also dictates that belts not be rolled onto sheaves; this is one of the most dangerous acts that can be done around V-belt drives. If the sheaves turn, clothing or fingers can be caught and painfully damaged. Rolling is not the easiest way to install belts. The best way to install a belt is to use the drive adjustment to move the sheaves closer together and then drop the belt into the sheave groove.

A sturdy pry bar will help move the motor. Keep the take-up rails free of dirt, rust, and grit; lubricate them from time to time. This will make belt change easier and safer.

The final V-belt installation procedure involves properly tensioning the drive for trouble-free service. Here are three helpful tensioning tips:

1. The best tension for a V belt is the lowest tension at which the belt will not slip under a full load. Too much tension shortens belt life; too little causes slippage, which also leads to premature belt failure.
2. Take up the drive until the belts are snug in their grooves. Run the drive for 15 min to seat the belts; then impose the peak load. If the belts slip, tighten them until they no longer slip at peak loads.

TABLE 3-1 Belt Matching Limits

Belt length		
Inches	Millimeters	Matching limit
up to 100	up to 2540	Use only one number
100–200	2540–5080	Within two consecutive match numbers only
200–300	5080–7620	Within three consecutive match numbers only
300–400	7620–10,160	Within four consecutive match numbers only
400–500	10,160–12,700	Within five consecutive match numbers only
over 500	over 12,700	Within six consecutive match numbers only

3. Check the V-belt tension frequently during the first day of operation. Check periodically thereafter, and make necessary adjustments.

Most V-belt manufacturers publish tensioning procedures for stock drives which give more accurate values.

V-Belt Safety

Safety is a critical factor in the efficient operation of belt drives. The maintenance crews can take several positive steps to help keep drives running smoothly and safely:

1. Keep belted drives properly guarded. Virtually all regulatory agencies (particularly the Occupational Safety and Health Administration), insurance companies, and safety authorities require that drives be completely guarded. The guard should allow proper ventilation of the drive, but should have no gap where workers can reach inside and become caught in the drive.
   The guard is a positive safety feature, but also enhances maintenance by protecting the drive from environmental conditions, debris, and tampering.
2. Always turn the equipment off for maintenance, lock the controls, and place a "down for maintenance" sign on the machine whenever work is performed on the drive.
3. Never roll a belt onto a sheave. As mentioned earlier, use the drive take-up and simply drop the belts into the sheave grooves.

TABLE 3-2 Why V Belts Fail

Trouble area and observation	Cause	Remedy
Worn side patterns	Constant slip	Retension drive until belt stops slipping
	Misalignment	Realign sheaves
	Worn sheaves	Replace with new sheaves
	Incorrect belt	Replace with new belts
Bottom of belt cracking	Belt slipping, causing heat buildup and gradual hardening of undercord	Install new belt, tension to prevent slip
	Idler installed on wrong side of belt	Refer to a V-belt installation manual
	Improper storage	Refer to belt storage information available from manufacturer
Bottom and sides burned	Belt slipping under starting or stalling load	Replace belt and tighten drive until slipping stops
	Worn sheaves	Replace sheaves
Belt turnover	Foreign material in grooves	Remove material, shield drive
	Misaligned sheaves	Realign the drive
	Worn sheave grooves	Replace sheave
	Tensile member broken through improper installation	Replace with new belt(s)
	Incorrectly aligned idler pulley	Carefully align idler, checking alignment with drive loaded and unloaded
Belt pulled apart	Extreme shock load	Remove cause of shock load
	Belt came off drive	Check drive alignment, foreign material in drive; ensure proper tension and drive alignment

V-Belt Inspection

If proper V-belt inspection procedures are followed, 90 percent of all maintenance problems are eliminated. This is because a properly installed V belt is a remarkably trouble-free piece of equipment. But, to assure continued trouble-free service, a quick inspection of the belt drive should be made a part of the routine maintenance schedule.

Listen to the sounds that could mean trouble: *slapping* of belts against the drive guard could be caused by an improperly placed guard, loose belts, or excessive vibration; *squealing* can be caused by poorly tensioned belts or foreign material (grease, dirt, or paint) in the sheave grooves. (Never use a belt dressing that can cause belts to slip or collect dirt and grit.)

Watch the drive while it runs, removing the guard *temporarily* if necessary. With multiple-belt drives, there will be some variation in tension, but there should be a tight side and a slack side. If one or more belts are loose or tight, check one of the following problems:

1. **Improper Tension** Drive may be poorly positioned. Measure tension and adjust movable sheaves as necessary.
2. **Damaged Belt** Remove the belt and carefully inspect to see if the belt is broken internally.
3. **Improper Matching** Replace belts.

When the drive is turned off, inspect the belts for wear. If sidewall wear is excessive, check drive alignment.

V-Belt Failure

The five most common symptoms, causes, and remedies of belt failure are listed in Table 3-2.

BIBLIOGRAPHY

Publications of the Rubber Manufacturers Association, Washington, D.C.:

Number	Title
IP-3	*Power Transmission Belt Technical Information (PTBTI).* A complete set of IP-3-1 thru 3-13, listed below (also available separately).
IP-3-1	*V-Belt Heat Resistance* (1972).
IP-3-2	*V-Belt Oil Resistance* (1972).
IP-3-3	*Static Conductive V-Belts* (1972).
IP-3-4	*Storage of V-Belts (1972).*
IP-3-5	*A Drive Design Procedure for Double V-Belts* (1972).
IP-3-6	*Effect of Idlers on V-Belt Performance* (1972).
IP-3-7	*V-Flat Drives* (1972).
IP-3-8	*Steel Cable V-Belts* (1972).
IP-3-9	*Joined V-Belts* (1972).
IP-3-10	*V-Belt Drives with Twist* (1975).
IP-3-11	*Inspection Guide for Automotive V-Belt Drives* (1973).
IP-3-12	*Belt Tension Calculation Method for Automotive V-Belt Drives* (1975).
IP-3-13	*Mechanical Efficiency of Power Transmission Belt Drives* (1978).
IP-20	Specification: Joint MPTA/RMA/RAC *Classical Multiple V-Belts* (1977). A, B, C, D, and E Cross Sections. (An American National Standard.)
IP-21	Specification: Joint RMA/MPTA *Double V-Belts* (1965). AA, BB, CC, DD Cross Sections.
IP-22	Specification: Joint MPTA/RMA/RAC *Narrow Multiple V-Belts* (1978). 3V, 5V, and 8V Cross Sections.
IP-23	Specification: Joint RMA/MPTA *Single V-Belts* (1968). 2L, 3L, 4L, and 5L Cross Sections.
IP-24	Specification: Joint MPTA/RMA/RAC *Synchronous Belts* (1978). MXL, XL, L, H, XH, and XXH Belt Sections.
IP-25	Specification: Joint MPTA/RMA/RAC *Variable Speed Belts* (1979). Twelve Cross Sections.
IP-26	Specification: Joint MPTA/RMA/RAC *V-Ribbed Belts* (1977). H, J, K, L, and M Cross Sections

PART 2

Flexible Couplings

by

Michael M. Calistrat
Manager, Mechanical Development
Metal Products Division
Koppers Company, Inc.
Baltimore, Maryland

INTRODUCTION

If one could perfectly align two shafts, they could be connected with a muff or two flanged hubs bolted together. Once this is done, it must be ensured that neither of the machines will move on the foundation and that the foundation will not settle. Actually, some misalignment between a driving and driven shaft is a fact of life and this is why *flexible coupling must* be used. The American Gear Manufacturers Association (AGMA) defines flexible couplings as machine elements which transmit torque without slip and accommodate misalignment between the driving and driven shafts.

Flexible couplings are perhaps the most mistreated parts of any machinery, both at selection time and at installation time. Through proper coupling selection and good alignment procedure, high maintenance costs and lost production time can be avoided.

Different types of couplings can accommodate various misalignments; selecting the one that accommodates the largest misalignment is not always the wisest choice. Sometimes a larger misalignment is made possible by a reduction in the power transmitted or a reduction in the useful life of the couplings.

Coupling catalogs usually list the maximum misalignment for each coupling. At installation, one should strive for the best alignment possible, so that the coupling has enough reserve misalignment capacity to accommodate future worsening of the operating conditions.

The misalignment can change for many reasons such as foundation settling, bearing wear, and distortions caused by vibration and temperature changes as well as by the movement of the connected machines under external forces induced through pipes, belts, and chains.

SHAFT MISALIGNMENT

In order to perform a good alignment, one should understand the various relative positions two connected shafts can have.

Parallel Misalignment

This type of misalignment (Fig. 3-7) is the easiest to understand, measure, and correct. Assuming that the motor shaft is higher above the base than the pump shaft, how do we

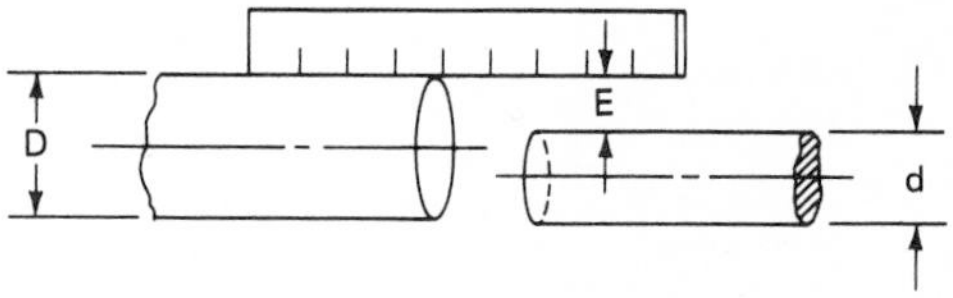

Figure 3-7 Parallel misalignment.

measure this offset? If the two shafts have the same diameter we can measure the offset directly using a straightedge and feeler gauges, as shown. If the two shafts do not have the same diameter, which is often the case, we must first measure dimension E and then calculate the offset:

$$\text{Offset} = E - \frac{D - d}{2}$$

If the result is positive, then the large shaft is higher; if the result is negative then the small shaft is higher. The offset misalignment can be corrected easily by adding shims under the machine with the lower shaft.

Angular Misalignment

This is more difficult to measure and correct. Seldom can one see pure angular misalignment.

Combination Misalignment

This type (see Fig. 3-8) is what is most likely to be found between two shafts; it cannot be corrected in one step. When combined misalignment is discovered, the first step should be an attempt to correct the angular misalignment. This can be observed by using a straightedge and feeler gauges. It can be corrected by shimming either the front or the rear of the motor (or pump) base. Once the angular misalignment is corrected, the parallel misalignment can be treated as shown above.

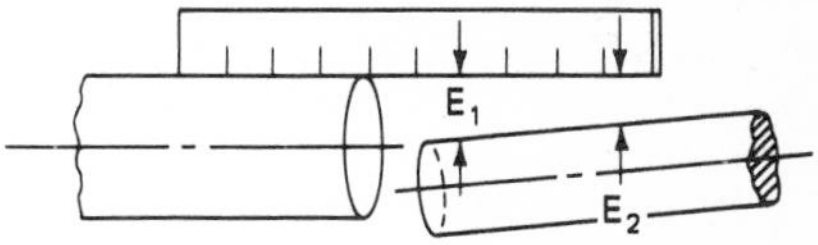

Figure 3-8 Combination misalignment.

In reality, aligning the two shafts is much more complicated, because in most cases the couplings cannot be installed on the shafts after the machines are in place. Sometimes the distance between the shafts is too large to be spanned with a straightedge. For proper alignment, the following steps are to be followed:

1. With a dial indicator point resting on the shaft, verify if the shafts are running true or if they are out of round. A machine with a bent shaft cannot be aligned!
2. The couplings should be installed on the shafts; they should then be checked for running true, both on the diameter and on the face. An eccentrically bored coupling will always operate in the misaligned condition and will have a short life.
3. For closely spaced shafts and moderate speeds, the method of aligning two machines using a straightedge and feeler gauges is satisfactory (Fig. 3-9).
4. The best alignment procedure, known as the *reverse-indicator* method, requires two readings: one with a dial indicator attached to one shaft and the point on the other,

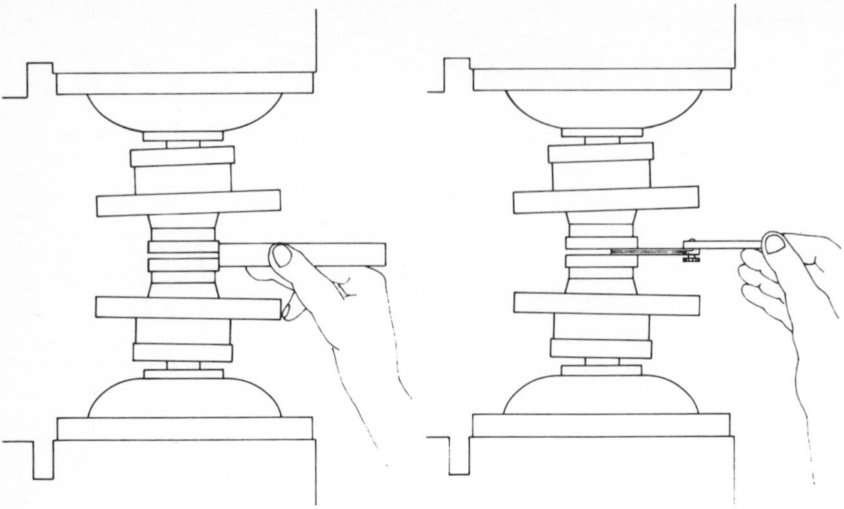

Figure 3-9 Coupling alignment.

and the second reading with the dial indicator reversed. For this method only small, light dial indicators and a sturdy support arm (rather than a magnetic base) should be used.

The importance of good alignment cannot be overemphasized. Fortunately many good articles on the subject are available, and some companies even specialize in selling courses in alignment as well as alignment tools.[1–8]

COUPLING INSTALLATION

Couplings are installed in two steps: first, each half-coupling is installed on its shaft; second, after the machines are aligned, the two coupling halves are bolted together either directly or through a spacer.[9]

The torque must be transmitted from the driving shaft to the first coupling hub, then through the coupling, and finally from the second hub to the driven shaft. In fractional horsepower applications the hubs are attached to the shafts through set screws. When the power is ½ hp (370 W) or more, then one or even two keys are used. Again, for small shafts a set screw over the key prevents the coupling from moving axially, but larger couplings are pressed on the shafts. Tapered shafts are often used, and they have a threaded extension. A nut and locking washer are used to retain the coupling on the shaft. The keys must fit perfectly in the keyway at the sides, but there *must* be a slight radial clearance.

It is important to have interference between the coupling hubs and the shafts because flexible couplings tend to resist misalignment, particularly under torque. The *restoring* forces generated by the couplings tend to rock the hubs on the shafts. Without interference, the movement between the hubs and shafts will wear the hub bore (fretting) and the coupling must be replaced. A safe rule is to use 0.0005 in per inch of diameter for interference.

Tapered shafts are often used because of the ease of installing and removing the hubs. First one should check to make sure there is good contact between the hub bore and the shaft. This is done by painting the bore with Prussian blue and then slightly rotating the hub on the shaft, without the key.

A minimum of 50 percent contact should be obtained. In no case should one attempt to obtain a better contact through lapping between the hub and the shaft. Lapping would generate a ridge on the shaft which in turn could cause the shaft to fail. A plug and ring should be used if lapping is necessary.

To calculate the *draw* of the hub on the shaft, use the formula

$$a = D \times i \times \frac{12}{t}$$

where

D = diameter, in
i = interference, in/in
t = taper, in/ft

For instance, for a 2-in shaft, 0.0005-in/in interference, and 1 ½-in/ft taper, the draw is

$$a = 2 \times 0.0005 \times \frac{12}{1.5} = 0.008 \quad \text{in}$$

For metric units:

$$a = D \times \frac{i}{t}$$

where

D = diameter, mm
i = interference, mm/mm
t = taper, mm/mm

For instance, for a 50-mm shaft, 0.00005-mm/mm interference, and 0.125 mm/mm taper, the draw is

$$a = 50 \times \frac{0.00005}{0.125} = 0.02 \quad \text{mm}$$

TYPES OF COUPLINGS

Depending on the method used to accommodate the misalignment, flexible couplings can be divided into: (1) sliding-element couplings, (2) flexing-element couplings, and (3) combination sliding and flexing couplings.

Until recently all sliding-element couplings required lubrication. With the advent of synthetic materials, some of these couplings can operate dry. For this reason, some people prefer to divide couplings into lubricated and dry categories; but for a better understanding of the mechanism of torque transmission and coupling dynamics, the division into the aforementioned three categories is preferable.

Whether one should use lubricated or dry couplings depends a lot on the type of maintenance available. A properly serviced lubricated coupling can last as long as the connected equipment. A dry coupling does not require periodic servicing, but the flexing elements, whether they are metal or elastomer, have to be replaced sooner or later. An interesting article on this subject can be found in Ref. 10.

Sliding-Element Couplings

These types of couplings accommodate misalignment by sliding between two or more of their components. This sliding, and the forces generated by the transmitted torque, generate wear. In order to provide adequate life, these couplings are either lubricated or use elements made of low-friction plastics. Sliding-element couplings have two half-couplings, because each sliding pair of elements can accommodate only angular misalignment; it takes two such pairs, properly spaced, to accommodate parallel misalignment. One can better understand this fact if each pair of sliding elements is assumed to be a hinge joint (Fig. 3-10). It can be seen that the maximum offset (parallel misalignment) between the two shafts is a function of the angle of misalignment α the coupling can accommodate *and* the distance L between the two joints. In analyzing Fig. 3-10, one can understand the advantage of using a coupling with a large shaft separation: it is easier to align, and changes in the positions of the connected machines have a small influence on the coupling alignment.

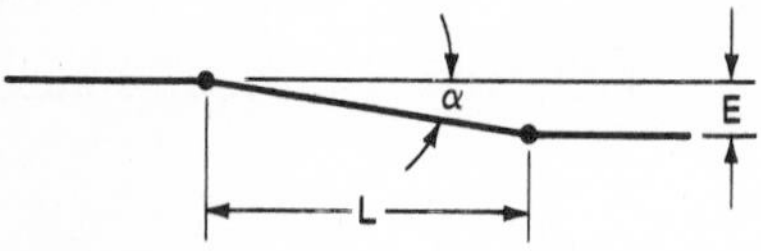

Figure 3-10 The effect of shaft separation.

Gear-Type Couplings

The gear-type couplings are probably the most versatile coupling design. They can be manufactured for almost any application, from a few horsepower to thousands of horsepower, from less than 1 r/min to more than 20,000 r/min. For a given application, a gear coupling is generally smaller and lighter than any other type. Gear couplings can be used on machines with closed coupled shafts or for long separations between the connected shafts. On the other hand, they require periodic relubrication (about every 6 months), they are rigid torsionally, and they are more expensive than other types of couplings.

A gear coupling for closed coupled shafts has two halves bolted together (Fig. 3-11). Each half-coupling has only three components: a hub, a sleeve, and a seal. The hub has a set of external teeth and is very similar to a pinion. The sleeve has a matching set of internal teeth cut in such a way that when the sleeve is slid over the hub there is play (backlash) between the meshing teeth. The seal is installed in a groove machined in the end plate of the sleeve and serves the double purpose of retaining the lubricant and preventing dirt or water from entering the coupling. The sleeves are also provided with one or two grease fittings or plugs. For long shaft separations, a spacer is introduced between the two sleeves. The flanges are connected with eight or more bolts, and a paper gasket, or an O ring, is installed between the flanges for sealing the joint.

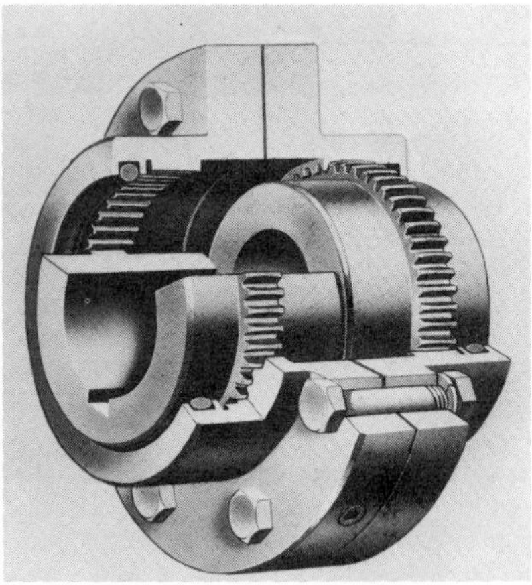
Figure 3-11 Gear-type coupling.

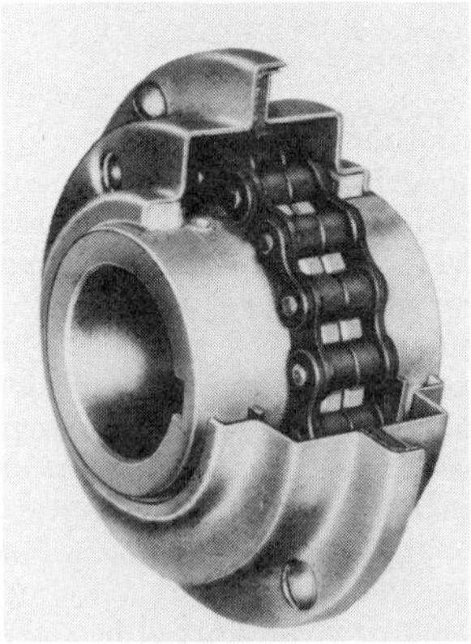
Figure 3-12 Chain coupling. *(Reproduced by permission of FMC Corporation.)*

Chain Couplings

The chain couplings excel through their simplicity; two sprockets and a length of double-row chain is all that is needed. They are generally used at low speeds, except when a special metal or plastic cover is used to contain the lubricant which otherwise will be thrown off by the centrifugal forces (Fig. 3-12). Chain couplings are used in closed coupled applications.

Steel-Grid Couplings

This type of coupling is in many respects similar to the gear coupling. It has two hubs with external teeth (Fig. 3-13) but of a special profile. Instead of the sleeves with internal teeth, it has a steel grid that loops through the spaces between the teeth. Because the steel grid flexes to some extent under torque, the steel-grid coupling is torsionally less rigid than the gear-type coupling.

Coupling Lubrication

Couplings that incorporate sliding elements require lubrication to minimize wear and thus increase their useful life. With few exceptions this type of coupling is grease-lubricated. Using the proper lubricants and procedures rewards the user with long, trouble-free service from the coupling. Not every grease is suitable for coupling lubrication. Manufacturers' catalogs list only a few greases. If these recommendations, or the greases listed in them, are not available, the following guidelines should be used:

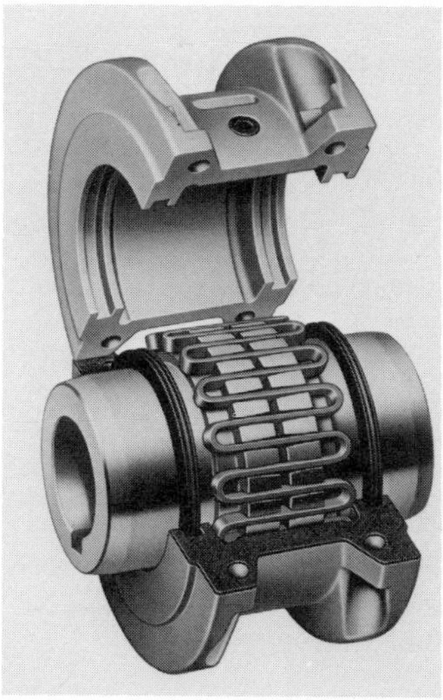

Figure 3-13 Steel-grid coupling. *(Reproduced by permission of The Falk Corporation.)*

1. Because couplings rely on the centrifugal effect to force the lubricant between the sliding surfaces, heavy greases are not good choices. NLGI No. 1 greases are the best compromise between good lubrication and adequate sealing.
2. Because the wear of the coupling decreases as the viscosity of the base oil of a grease increases, one should select a grease blended with an oil having a viscosity of no less than 900 SSU (Saybolt second(s) universal) at 100°F. This information can be obtained from the grease manufacturer.
3. Because greases will separate into oil and soap when subjected to centrifugal forces for a long time and because the soap used in greases is not a lubricant, one should select a grease having very little soap—preferably less than 8 percent of the total weight.

Couplings should be lubricated every 6 months, and before new grease is pumped the coupling should be opened and cleansed of the old lubricant.

Flexing-Element Couplings

These types of couplings accommodate misalignment through flexing of one or more of their components. This flexing can in time cause the failure of the element which must then be replaced. It is obvious that the less misalignment the coupling must accommodate, the less flexing the elements must suffer, and the coupling can provide longer trouble-free service.

Depending on the material used for the flexing element, couplings can be divided into two types: (1) metallic-element and (2) elastomer-element. Metallic-element couplings can accommodate only angular misalignment at each flexing point. To accommodate parallel (offset) misalignment, a coupling needs two flexing elements. As shown in Fig. 3-10, the greater the distance between the elements, the more offset the coupling can accommodate. Elastomer-element couplings can accommodate offset with only one element. They are designed for close-coupled machines; however, when used with a special centering bushing, they can be used for long shaft separations.

Metallic-Element Couplings

The flexible element is not a single piece; rather it is a pack of many thin stamped disks, usually made of stainless steel (Fig. 3-14). Coupling sizes vary from miniature to large. With few exceptions, they are not used at high speeds. The multiple-disk pack offers the advantage of a redundant system, and the coupling can operate even after one or more of the disks fail. However when the disks are replaced, the pack should be replaced as a whole rather than only the broken disks. One drawback of the metal-disk couplings is that they tolerate very little error in the axial spacing of the machines. On the other hand, this disadvantage becomes an advantage when a limited end-float coupling is required, as is the case with sleeve-bearing motors that rely on their magnetic centering and have no thrust bearings.

Figure 3-14 Disk-pack coupling. *(Reproduced by permission of Rexnord Corp.)*

Elastomer-Element Couplings

There are quite a few designs using elastomer elements: some using rubber with or without plies and some using plastics. Each model has its own advantages and disadvantages. Many times, availability in one particular area determines which couplings will be used. Only the most popular types will be discussed here.

The Rubber Tire. The rubber-tire element (Fig. 3-15) is clamped at each hub, and it is split axially to permit the element's replacement without moving the connected machines.

The Rubber Doughnut. The rubber-doughnut element (Fig. 3-16) is bolted with radial fasteners to the hubs, and in the process it is also precompressed so that it never

Figure 3-15 Rubber-tire coupling. *(Reproduced by permission of Reliance Electric Co.)*

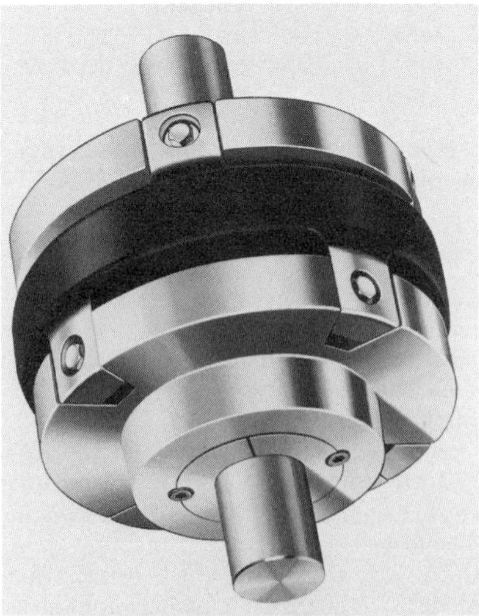

Figure 3-16 Rubber-doughnut coupling.

works in tension. It is split axially at one of the inserts to facilitate installation without the need to disturb the connected machines.

The Splined Element. The splined element (Fig. 3-17) slides inside the hubs axially, and it is either rubber or plastic. To replace the element, one of the hubs must be pushed back axially. For close-coupled shafts, the element is split axially so that the machines do not have to be moved at installation of the element.

The Jaw. This coupling (Fig. 3-18) is also known as the spider coupling because of the shape of the elastomer element. This type of coupling is perhaps the simplest, but it has the following disadvantages: (1) it can accommodate little misalignment and (2) it can usually transmit less than 100 hp (74.6 kW) and, similarly to the splined-element coupling, one of the hubs has to be moved axially in order to replace the element.

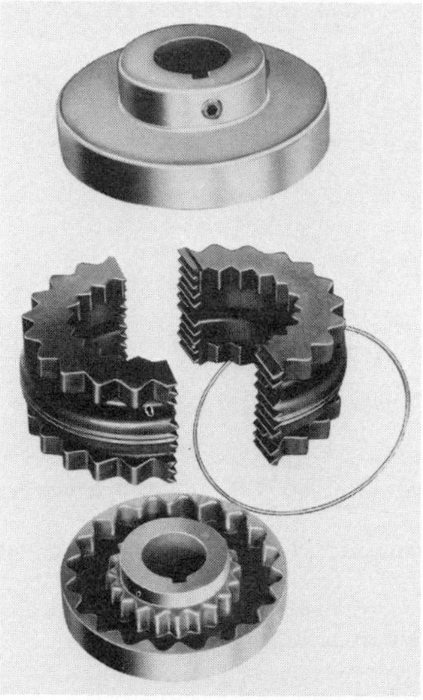

Figure 3-17 Splined-element coupling. *(Reproduced by permission of T. B. Wood's Sons Co.).*

Figure 3-18 Jaw coupling. *(Reproduced by permission of Lovejoy, Inc.)*

COUPLING SELECTION

Usually couplings are supplied as part of any new equipment. Instead of having to select a new coupling, one is faced only with the need to replace an old one, or part of an old one. Assuming that the equipment manufacturer selected the right coupling type and size, couplings generate few problems. There are cases, however, when either the coupling does not live up to expectations or when a new piece of equipment is purchased without a driver and a coupling must be selected. The process is not simple because there is no application for which only one type of coupling would work. An excellent article on this subject can be found in Ref. 11. The best approach is to let an application engineer from a coupling manufacturer make the selection. Today most manufacturers make more than one type of coupling and can objectively recommend the best one for the application.

Choosing a coupling of the correct size is very important. To do this one must know not only the required power and speed, but also the severity of service the coupling must accommodate. A correction factor, or service factor, must be applied.

Coupling manufacturers rate their couplings in horsepower per 100 r/min. For instance, if a pump requires 50 hp (37.3 kW) at 1750 r/min, it needs a coupling that can handle 2.86 hp (2.13 kW) at 100 r/min. This is correct only if the pump is centrifugal and is driven by an electric motor. In this case the service factor is 1. If we have a double-acting reciprocating pump driven by an internal-combustion engine, we have to use a service factor of 2.0 + 1.0 = 3.0 for a gear coupling and 2.0 + 0.5 = 2.5 for an elastomer coupling, according to one manufacturer. As a result, we must choose a gear coupling that can handle 8.58 hp (6.4 kW) at 100 r/min or an elastomer coupling that can handle

7.15 hp (5.3 kW) at 100 r/min. It seems that we can choose a smaller coupling if we choose the elastomer type. However, the elastomer coupling will be about 8¾ in (222 mm) in diameter, while the gear coupling will be only half that size! If size is not important, price can be the next selection criterion. But the price of the coupling alone is not a good guide; one should consider the total cost including maintenance, replacement parts, lost production, etc.

Although couplings represent a small percentage of the total cost of a piece of machinery, they can cause as much, if not more, trouble than the rest of the equipment if they are not properly selected. Buying an inadequate size or type of coupling will never be economical in the long run.

REFERENCES

1. Dreymala, J.: "Try Dial Indicators for Close Alignment of Coupling Connected Machinery," *Power,* June 1971, pp. 96–98.
2. Jackson, C.: "Techniques for Alignment of Rotating Equipment," *Hydrocarbon Processing,* January 1976, pp. 81–85.
3. *Compressor Handbook,* Gulf Publishing Co., Houston, Texas, 1979.
4. Campbell, A. J.: "Optical Alignment Saves Equipment Downtime," *Oil and Gas Journal,* November 1975, pp. 54–56.
5. Murray, M. G.: "Minimum Movement Machinery Alignment," *Hydrocarbon Processing,* January 1979, pp. 112–114.
6. "Reverse Indicator Method of Alignment," Hughes & Associates, Houston, Texas, 1974.
7. "M905 Series, Dyn Align Bars," Dymac, San Diego, California, 1974.
8. "Acculign Gauges," Boyce Engineering, Houston, Texas.
9. Calistrat, M. M.: "Flexible Coupling Installation," *Proceedings of the National Conference on Power Transmission, 1981,* Illinois Institute of Technology, Chicago.
10. Wright, J.: "Which Shaft Coupling is Best—Lubricated or Non-lubricated?," *Hydrocarbon Processing,* April 1975, pp. 191–193.
11. Wright, J.: "Which Flexible Coupling?," *Power Transmission & Bearing Handbook,* 1971/1972, Industrial Publishing Co., Cleveland.

PART 3

Chain Drives

by
Thomas J. Rost, P.E.
Senior Product Engineer
Industrial Chain Division
PT Components, Inc.
Indianapolis, Indiana

INTRODUCTION

Chain drives are one of the most efficient methods used to transmit mechanical power between two or more rotating shafts that cannot be directly coupled. A chain drive consists of a series of links assembled together and two or more sprockets (Fig. 3-19). The positive-engagement sprockets are keyed to the rotating shafts between which power is transmitted. Roller chain, engineering steel chain, and silent chain are the three types of chain used in industrial drive applications. The benefits of a chain drive compared to a belt drive or gear drive are:

1. Shaft center distances are relatively unrestricted.
2. Chains are easily installed.
3. Chain drives do not slip or creep, resulting in overall high efficiency.
4. The load in a chain drive is distributed over a number of sprocket teeth simultaneously.
5. Chain drives operate in adverse environmental conditions.

This presentation will deal only with chain-drive applications. Additional information on this subject and other chain applications, along with sprocket-selection procedures, can be obtained from manufacturers' catalogs and the sources listed in the reference section. The reader is cautioned that all calculations are examples to aid in understanding concepts. All design work should be done with the cooperation of a manufacturer.

DRIVE-SELECTION PROCEDURE

The proper steps for selecting a chain drive are listed below:

1. Determine and record all applicable operating factors, such as
   a. Source and type of power

Figure 3-19 A roller chain drive operates multiple rolls in a heat-treating furnace.

b. Size and speed of driving shaft
c. Size and speed of driven shaft
d. Approximate center distances between shafts
e. Relative position of shafts
f. Type of driven equipment
g. Operating conditions
h. space limitations

2. Establish the service factor from the actual operating conditions. The service factor is a factor by which the transmitted horsepower is multiplied to compensate for drive conditions. The composite or final service factor is the product of the separate service factors multiplied together (see Table 3-3).
3. Calculate the factored horsepower value by multiplying the horsepower to be transmitted by the final service factor.
4. Make a trial chain selection based on the factored horsepower and revolutions per minute of the small sprocket (Fig. 3-20).
5. Determine the number of teeth in the small sprocket from the speed-horsepower charts.
6. Check the small sprocket for bore capacity, number of teeth, and availability.
7. Divide the speed of the faster-turning shaft by the speed of the slower shaft to determine the drive ratio.
8. Calculate the chain length and exact sprocket centers.
9. Determine the chain sag. The chain sag should be approximately 2 percent of the distance between shaft centers at the initial installation.
10. Determine the method of lubrication.

DRIVE-SELECTION EXAMPLE

Application

Select a roller chain drive for the following conditions:

Source of power	Electric motor
Horsepower to be transmitted	10 hp (7.5 kW)
Size of driving shaft	2.438 in (62 mm) diameter

Speed of driving shaft	100 r/min
Drive equipment	Uniformly fed coal elevator for power plant
Size of driven shaft	2.938 in (75 mm) diameter
Speed of driven shaft	42 r/min
Approximate center distance	24.00 in (610 mm)
Relative position of shafts	On same horizontal plane
Space limitations	None

Solution

Service Factor. The service factor listed in Table 3-3 for a uniformly fed elevator driven by a gear motor is 1.0.

Equivalent Horsepower. The equivalent horsepower is $10 \times 1.0 = 10$ hp.

Trial Chain. From Fig. 3-20 note that the intersection of the 100-r/min vertical line and the 10-hp single-strand horizontal line falls in the area for No. 100 chain. Thus, the trial chain is No. 100 single strand.

Small Sprocket. In Table 3-4 for No. 100 roller chain, the 100-r/min column lists 10.3 hp which corresponds closely to the equivalent horsepower of 10 required for this application. This rating is for single-strand chain when used with a 17-tooth sprocket.

Check the Small Sprocket. As shown in the rating table, the maximum bore of a 17-tooth No. 100 sprocket is larger than the 2.438 in (62 mm) bore required; therefore, the selection is satisfactory.

Drive Ratio. The drive ratio equals

$$\frac{100 \text{ r/min}}{42 \text{ r/min}} = 2.38{:}1$$

Number of Teeth in Large Sprocket. The number of teeth in the large sprocket equals $2.38 \times 17 = 40.4$ teeth. Use a 40-tooth sprocket.

Center-Distance and Chain-Length Computations. To calculate the sprocket centers and chain length for a given drive, these symbols are used for the formulas in the following text:

e = desired sprocket centers, in (given as 24.00 in)
E = exact sprocket centers, in
g = pitch diameter of small sprocket, in (for a 17-tooth, No. 100 sprocket the pitch diameter = 6.803 in)
G = pitch diameter of large sprocket, in (for a 40-tooth, No. 100 sprocket the pitch diameter = 15.932 in)
N = actual length of chain, pitches
P = chain pitch, in (the No. 100 = 1.250 in pitch)
t = number of teeth in small sprocket (17 teeth)
T = number of teeth in large sprocket (40 teeth)

Calculate factor A using the formula

$$A = \frac{G - g}{2e} = \frac{15.932 - 6.803}{(2)(24.00)} = 0.19019$$

Refer to Table 3-5 and select factors B, C, and D corresponding to value A or the next higher value. Since $A = 0.19019$, select the next higher listed value of 0.19081. Corresponding factors for B, C, and D are 1.9633, 0.4389, and 0.5611, respectively. The number of pitches in the chain equals the sum of the pitches between sprockets and the pitches around the sprockets, or

$$\mathrm{N} = \frac{Be}{P} + Ct + Dt = \frac{(1.9633)(24.00)}{1.250} + (0.4389)(17) + (0.5611)(40)$$
$$= 67.601 = 68 \text{ pitches}$$

TABLE 3-3 Service Factors for Various Operating Conditions

Driven equipment	Service Factors		
	Input power		
	Internal combustion engine with hydraulic drive	Electric motor or turbine	Internal combustion engine with mechanical drive
Agitators, liquid stock	1.0	1.0	1.2
Beaters	1.2	1.3	1.4
Blowers, centrifugal	1.0	1.0	1.2
Boat propellers	1.4	1.5	1.7
Compressors			
centrifugal	1.2	1.3	1.4
reciprocating, 3 or more cylinders	1.2	1.3	1.4
reciprocating, singular, 2 cylinders	1.4	1.5	1.7
Conveyors			
uniformly loaded or fed	1.0	1.0	1.2
not uniformly loaded or fed	1.2	1.3	1.4
reciprocating	1.4	1.5	1.7
Cookers, cereal	1.0	1.0	1.2
Crushers	1.4	1.5	1.7
Elevators, bucket			
uniformly loaded or fed	1.0	1.0	1.2
not uniformly loaded or fed	1.2	1.3	1.4
Fans, centrifugal	1.0	1.0	1.2

Feeders			
rotary table	1.0	1.0	1.2
apron, belt, screw, rotary vane	1.2	1.3	1.4
reciprocating	1.4	1.5	1.7
Generators	1.0	1.0	1.2
Grinders	1.2	1.3	1.4
Hoists	1.2	1.3	1.4
Kettles, brew	1.0	1.0	1.2
Kilns and dryers, rotary	1.2	1.3	1.4
Lineshafts			
light or normal service	1.0	1.0	1.2
heavy service	1.2	1.3	1.4
Machinery			
uniform load, nonreversing	1.0	1.0	1.2
moderate pulsating load, nonreversing	1.2	1.3	1.4
severe impact or variable load, reversing	1.4	1.5	1.7
Mills			
ball, pebble and tube	1.2	1.3	1.4
hammer, rolling	1.4	1.5	1.7
Pumps			
centrifugal	1.0	1.0	1.2
reciprocating, 3 or more cylinders	1.2	1.3	1.4
Screens, rotary, uniformly fed	1.2	1.3	1.4
Basis for service factors: Uniform load	1.0	1.0	1.2
Moderate shock load	1.2	1.3	1.4
Heavy shock load	1.4	1.5	1.7

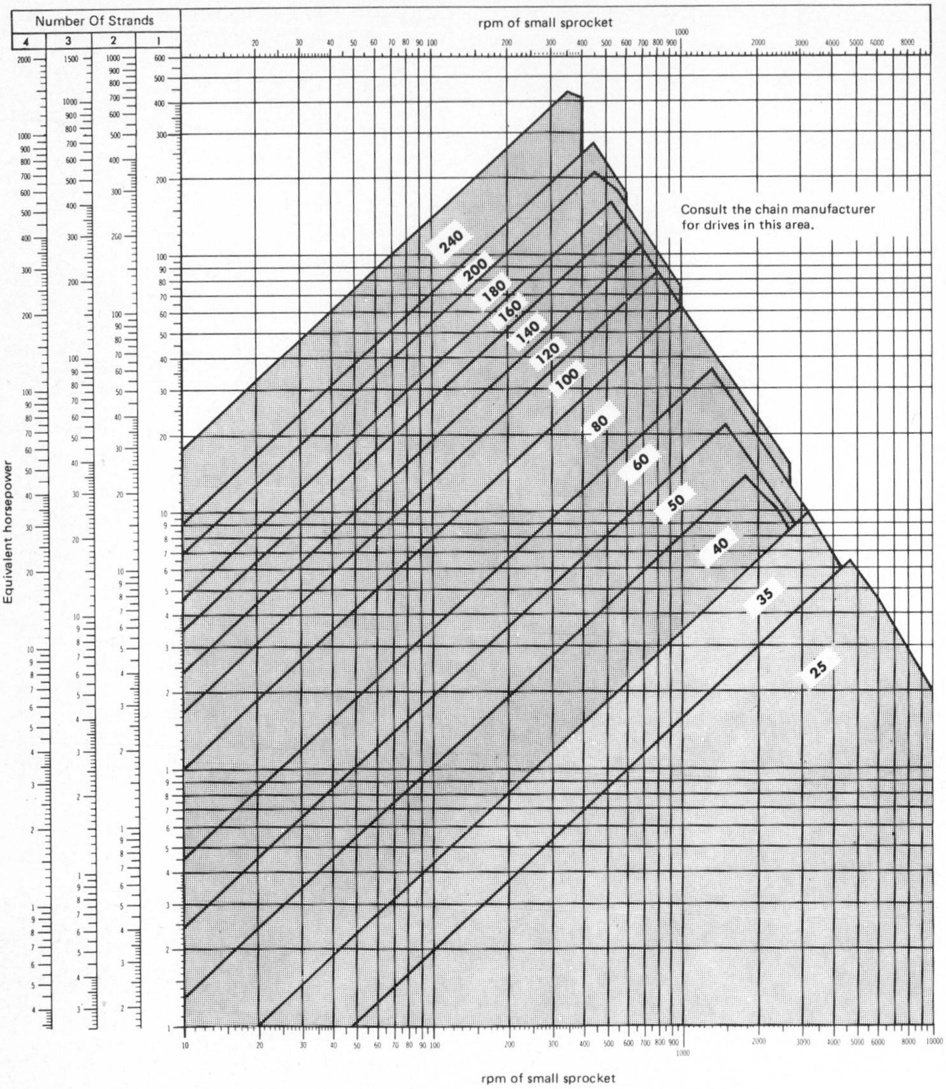

Figure 3-20 Trial selection chart for ANSI standard roller chains.[1]

Select an even whole number nearest to the calculated number of pitches. The exact sprocket center is found by the following formula:

$$E = \frac{(N - Ct - Dt)P}{B} = \frac{68 - (0.4389)(17) - (0.5611)(40)}{1.9633} 1.250$$

$$= 24.254 \text{ in}$$

Lubrication. The No. 100 rating table specified type B bath or disk lubrication. The drive selected for this application consists of:

17-tooth No. 100 driving sprocket

40-tooth No. 100 driven sprocket

Oil-retaining casing for oil-bath lubrication

TABLE 3-4 Speed-Horsepower Ratings for No. 100 Roller Chain

Number of teeth in small sprocket	Maximum bore, inches	Horsepower for single strand chain▲																			
		RPM of small sprocket																			
		100	500	900	1200	1800	2500	3000	3500	4000	4500	5000	5500	6000	6500	7000	7500	8000	8500	9000	10000
11	.313	0.05	0.23	0.39	0.50	0.73	0.98	1.15	1.32	1.38	1.16	0.99	0.86	0.75	0.67	0.60	0.54	0.49	0.45	0.41	0.35
12	.375	0.06	0.25	0.43	0.55	0.80	1.07	1.26	1.45	1.57	1.32	1.12	0.97	0.86	0.76	0.68	0.61	0.56	0.51	0.47	0.40
13	.438	0.06	0.27	0.47	0.60	0.87	1.17	1.38	1.58	1.77	1.49	1.27	1.10	0.96	0.86	0.77	0.69	0.63	0.57	0.53	0.45
14	.563	0.07	0.30	0.50	0.65	0.94	1.27	1.49	1.71	1.93	1.66	1.42	1.23	1.08	0.96	0.86	0.77	0.70	0.64	0.59	0.50
15	.563	0.08	0.32	0.54	0.70	1.01	1.36	1.61	1.85	2.08	1.84	1.57	1.36	1.20	1.06	0.95	0.86	0.78	0.71	0.65	0.56
16	.563	0.08	0.34	0.58	0.76	1.09	1.46	1.72	1.98	2.23	2.03	1.73	1.50	1.32	1.17	1.05	0.94	0.86	0.78	0.72	0.61
17	.625	0.09	0.37	0.62	0.81	1.16	1.56	1.84	2.11	2.38	2.22	1.90	1.64	1.44	1.28	1.14	1.03	0.94	0.86	0.79	0.67
18	.750	0.09	0.39	0.66	0.86	1.24	1.66	1.96	2.25	2.53	2.42	2.07	1.79	1.57	1.39	1.25	1.12	1.02	0.93	0.86	0.73
19	.813	0.10	0.41	0.70	0.91	1.31	1.76	2.07	2.38	2.69	2.62	2.24	1.94	1.70	1.51	1.35	1.22	1.11	1.01	0.93	0.79
20	.875	0.10	0.44	0.74	0.96	1.38	1.86	2.19	2.52	2.84	2.83	2.42	2.10	1.84	1.63	1.46	1.32	1.20	1.09	1.00	0.86
21	.875	0.11	0.46	0.78	1.01	1.46	1.96	2.31	2.66	2.99	3.05	2.60	2.26	1.98	1.76	1.57	1.42	1.29	1.17	1.08	0.92
22	.938	0.11	0.48	0.82	1.07	1.53	2.06	2.43	2.79	3.15	3.27	2.79	2.42	2.12	1.88	1.69	1.52	1.38	1.26	1.16	0.99
23	1.000	0.12	0.51	0.86	1.12	1.61	2.16	2.55	2.93	3.30	3.50	2.98	2.59	2.27	2.01	1.80	1.62	1.47	1.35	1.24	1.06
24	1.063	0.13	0.53	0.90	1.17	1.69	2.27	2.67	3.07	3.46	3.73	3.18	2.76	2.42	2.15	1.92	1.73	1.57	1.44	1.32	1.12
25	1.188	0.13	0.56	0.94	1.22	1.76	2.37	2.79	3.21	3.61	3.96	3.38	2.93	2.57	2.28	2.04	1.84	1.67	1.53	1.40	1.20
28	1.250	0.15	0.63	1.07	1.38	1.99	2.68	3.15	3.62	4.09	4.54	4.01	3.47	3.05	2.70	2.42	2.18	1.98	1.81	1.66	1.42
30	1.313	0.16	0.68	1.15	1.49	2.15	2.88	3.40	3.90	4.40	4.89	4.45	3.85	3.38	3.00	2.68	2.42	2.20	2.01	1.84	1.57
32	1.500	0.17	0.73	1.23	1.60	2.30	3.09	3.64	4.18	4.72	5.25	4.90	4.25	3.73	3.30	2.96	2.67	2.42	2.21	2.03	1.73
35	1.688	0.19	0.80	1.36	1.76	2.53	3.41	4.01	4.61	5.20	5.78	5.60	4.86	4.26	3.78	3.38	3.05	2.77	2.53	2.32	1.98
40	1.875	0.22	0.92	1.57	2.03	2.93	3.93	4.64	5.32	6.00	6.68	6.85	5.93	5.21	4.62	4.13	3.73	3.38	3.09	2.83	2.42
Lubrication type■		A		B							C										

▲ Ratings are based on a service factor of 1. For a complete list of service factors, refer to Table 3-3.
The ratings tabled above apply directly to lubricated, single strand, standard roller chains.

■ Type A: Manual or drip (Maximum chain speed 500 FPM)
Type B: Bath or disk (Maximum chain speed 3500 FPM)
Type C: Forced (pump)

TABLE 3-5 Factors for Chain-Length and Center-Distance Computations

A	B	C	D	A	B	C	D
.00000	2.0000	.5000	.5000	.19937	1.9598	.4361	.5639
.00436	2.0000	.4986	.5014	.20364	1.9581	.4347	.5653
.00873	1.9999	.4972	.5028	.20791	1.9563	.4333	.5667
.01309	1.9998	.4958	.5042	.21218	1.9545	.4319	.5681
.01745	1.9997	.4944	.5056	.21644	1.9526	.4306	.5694
.02181	1.9995	.4931	.5069	.22070	1.9507	.4292	.5708
.02618	1.9993	.4917	.5083	.22495	1.9487	.4278	.5722
.03054	1.9991	.4903	.5097	.22920	1.9468	.4264	.5736
.03490	1.9988	.4889	.5111	.23345	1.9447	.4250	.5750
.03926	1.9985	.4875	.5125	.23769	1.9427	.4236	.5764
.04362	1.9981	.4861	.5139	.24192	1.9406	.4222	.5778
.04798	1.9977	.4847	.5153	.24615	1.9385	.4208	.5792
.05234	1.9973	.4833	.5167	.25038	1.9363	.4194	.5806
.05669	1.9968	.4819	.5181	.25460	1.9341	.4181	.5819
.06105	1.9963	.4806	.5194	.25882	1.9319	.4167	.5833
.06540	1.9957	.4792	.5208	.26303	1.9296	.4153	.5847
.06976	1.9951	.4778	.5222	.26724	1.9273	.4139	.5861
.07411	1.9945	.4764	.5236	.27144	1.9249	.4125	.5875
.07846	1.9938	.4750	.5250	.27564	1.9225	.4111	.5889
.08281	1.9931	.4736	.5264	.27983	1.9201	.4097	.5903
.08716	1.9924	.4722	.5278	.28402	1.9176	.4083	.5917
.09150	1.9916	.4708	.5292	.28820	1.9151	.4069	.5931
.09585	1.9908	.4694	.5306	.29237	1.9126	.4056	.5944
.10019	1.9899	.4681	.5319	.29654	1.9100	.4042	.5958
.10453	1.9890	.4667	.5333	.30071	1.9074	.4028	.5972
.10887	1.9881	.4653	.5347	.30486	1.9048	.4014	.5986
.11320	1.9871	.4639	.5361	.30902	1.9021	.4000	.6000
.11754	1.9861	.4625	.5375	.31316	1.8994	.3986	.6014
.12187	1.9851	.4611	.5389	.31730	1.8966	.3972	.6028
.12620	1.9840	.4597	.5403	.32144	1.8939	.3958	.6042
.13053	1.9829	.4583	.5417	.32557	1.8910	.3944	.6056
.13485	1.9817	.4569	.5431	.32969	1.8882	.3931	.6069
.13917	1.9805	.4556	.5444	.33381	1.8853	.3917	.6083
.14349	1.9793	.4542	.5458	.33792	1.8824	.3903	.6097
.14781	1.9780	.4528	.5472	.34202	1.8794	.3889	.6111
.15212	1.9767	.4514	.5486	.34612	1.8764	.3875	.6125
.15643	1.9754	.4500	.5500	.35021	1.8733	.3861	.6139
.16074	1.9740	.4486	.5514	.35429	1.8703	.3847	.6153
.16505	1.9726	.4472	.5528	.35837	1.8672	.3833	.6167
.16935	1.9711	.4458	.5542	.36244	1.8640	.3819	.6181
.17365	1.9696	.4444	.5556	.36650	1.8608	.3806	.6194
.17794	1.9681	.4431	.5569	.37056	1.8576	.3792	.6208
.18224	1.9665	.4417	.5583	.37461	1.8544	.3778	.6222
.18652	1.9649	.4403	.5597	.37865	1.8511	.3764	.6236
.19081	1.9633	.4389	.5611	.38268	1.8478	.3750	.6250
.19509	1.9616	.4375	.5625	.38671	1.8444	.3736	.6264

TABLE 3-5 (*continued*)

A	B	C	D
.39073	1.8410	.3722	.6278
.39474	1.8376	.3708	.6292
.39875	1.8341	.3694	.6306
.40275	1.8306	.3681	.6319
.40674	1.8271	.3667	.6333
.41072	1.8235	.3653	.6347
.41469	1.8199	.3639	.6361
.41866	1.8163	.3625	.6375
.42262	1.8126	.3611	.6389
.42657	1.8089	.3597	.6403
.43051	1.8052	.3583	.6417
.43445	1.8014	.3569	.6431
.43837	1.7976	.3556	.6444
.44229	1.7937	.3542	.6458
.44620	1.7899	.3528	.6472
.45010	1.7860	.3514	.6486
.45399	1.7820	.3500	.6500
.45787	1.7780	.3486	.6514
.46175	1.7740	.3472	.6528
.46561	1.7700	.3458	.6542
.46947	1.7659	.3444	.6556
.47332	1.7618	.3431	.6569
.47716	1.7576	.3417	.6583
.48099	1.7535	.3403	.6597
.48481	1.7492	.3389	.6611
.48862	1.7450	.3375	.6625
.49242	1.7407	.3361	.6639
.49622	1.7364	.3347	.6653
.50000	1.7321	.3333	.6667
.50377	1.7277	.3319	.6681
.50754	1.7233	.3306	.6694
.51129	1.7188	.3292	.6708
.51504	1.7143	.3278	.6722
.51877	1.7098	.3264	.6736
.52250	1.7053	.3250	.6750
.52621	1.7007	.3236	.6764
.52992	1.6961	.3222	.6778
.53361	1.6915	.3208	.6792
.53730	1.6868	.3194	.6806
.54097	1.6821	.3181	.6819
.54464	1.6773	.3167	.6833
.54829	1.6726	.3153	.6847
.55194	1.6678	.3139	.6861
.55557	1.6629	.3125	.6875
.55919	1.6581	.3111	.6889
.56280	1.6532	.3097	.6903
.56641	1.6483	.3083	.6917
.57000	1.6433	.3069	.6931
.57358	1.6383	.3056	.6944
.57715	1.6333	.3042	.6958
.58070	1.6282	.3028	.6972
.58425	1.6231	.3014	.6986
.58779	1.6180	.3000	.7000
.59131	1.6129	.2986	.7014
.59482	1.6077	.2972	.7028
.59832	1.6025	.2958	.7042
.60182	1.5973	.2944	.7056
.60529	1.5920	.2931	.7069
.60876	1.5867	.2917	.7083
.61222	1.5814	.2903	.7097
.61566	1.5760	.2889	.7111
.61909	1.5706	.2875	.7125
.62251	1.5652	.2861	.7139
.62592	1.5598	.2847	.7153
.62932	1.5543	.2833	.7167
.63271	1.5488	.2819	.7181
.63608	1.5432	.2806	.7194
.63944	1.5377	.2792	.7208
.64279	1.5321	.2778	.7222
.64612	.15265	.2764	.7236
.64945	1.5208	.2750	.7250
.65276	1.5151	.2736	.7264
.65606	1.5094	.2722	.7278
.65935	1.5037	.2708	.7292
.66262	1.4979	.2694	.7306
.66588	1.4921	.2681	.7319
.66913	1.4863	.2667	.7333
.67237	1.4804	.2653	.7347
.67559	1.4746	.2639	.7361
.67880	1.4686	.2625	.7375
.68200	1.4627	.2611	.7389
.68518	1.4567	.2597	.7403
.68835	1.4507	.2583	.7417
.69151	1.4447	.2569	.7431
.69466	1.4387	.2556	.7444
.69779	1.4326	.2542	.7458
.70091	1.4265	.2528	.7472
.70401	1.4204	.2514	.7486
.70711	1.4142	.2500	.7500

ROLLER CHAIN DRIVES

Roller chain drives are used in a wide range of power-transmission applications for all basic industries such as food processing, materials handling, textiles, and machine tools.

Fourteen standard sizes of single- and multiple-width roller chain listed in ANSI B29.1[1] are available for service (Figs. 3-21 and 3-22). Table 3-6 shows the chain number,

Figure 3-21 Single-width roller chain.

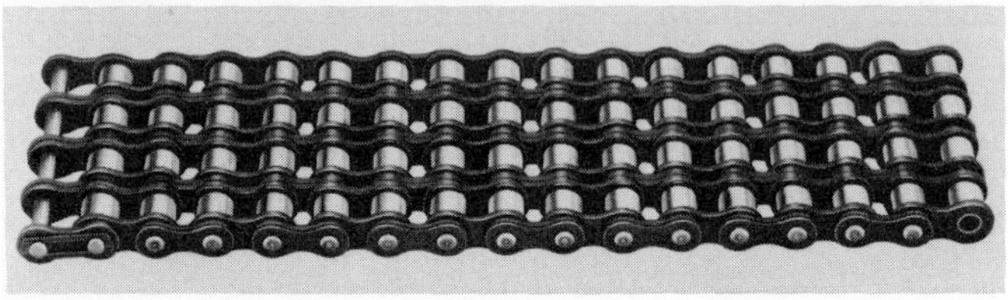

Figure 3-22 Multiple-width roller chain.

the pitch size, and the average ultimate strength for single-width chain. The average ultimate strength for multiple-width roller chain is equal to the average ultimate strength of the single-width chain times the number of strands in the multiple-width chain. Speed and horsepower ratings are the prime considerations in selecting a chain drive. The ratings are normally listed for the smaller sprocket, regardless of whether it is the drive or driven member. Chain manufacturers should be consulted when special conditions such as composite duty cycles, idlers, or more than two sprockets are involved in the drive cycle.

The speed and horsepower ratings are based upon approximately 15,000 h of service life at full-load operation and a service factor of 1.0. Operating conditions which establish the service factor are shown at the bottom of Table 3-3. Strand factors for multiple-width chains are given in Table 3-7. Note that the strand factors are not equal to the number of widths in a multiple-width chain.

Specialty roller chains have been developed for particular applications. For example,

TABLE 3-6 Standard Single-Width Roller Chain Size and Strength

	Chain pitch		Average ultimate strength	
Chain number	in	mm	lb	kg
25	0.250	6.35	875	400
35	0.375	9.52	2,100	950
40	0.500	12.70	3,700	1,680
41	0.500	12.70	2,000	900
50	0.625	15.88	6,100	2,770
60	0.750	19.05	8,500	3,850
80	1.000	25.40	14,500	6,580
100	1.250	31.75	24,000	10,900
120	1.500	38.10	34,000	15,400
140	1.750	44.45	46,000	20,900
160	2.000	50.80	58,000	26,300
180	2.250	57.15	80,000	36,300
200	2.500	63.50	95,000	43,100
240	3.000	76.20	130,000	59,000

TABLE 3-7 Multiple-Strand Factors for Multiple-Width Chains

Number of strands	Multiple-strand factor
2	1.7
3	2.5
4	3.3
5	4.1
6	5.0
7 or more	Consult manufacturer

double-pitch chain is an economical choice for slower-speed drives on relatively long centers. Heavy-series roller chain is used when conditions demand additional capacity to withstand occasional shock loads. Flexible-joint-type chain has been designed for smaller-horsepower drives where shafts must operate out of normal alignment. Lubricated-joint chains have oil-impregnated bushings to provide cleaner and longer operating life where restricted lubrication or absence of external lubrication is essential. Standard roller chain made of stainless steel is recommended for applications where high resistance to corrosive attack is required.

ENGINEERING STEEL CHAIN DRIVES

Engineering steel chain drives are especially suited for heavy-duty applications. The normally offset sidebar chain (Fig. 3-23) can handle speeds up to 1000 ft/min (305 m/min) and power requirements as high as 500 hp. These chains are commonly used in elevator drives, conveyor drives, drum drives, and applications with poor operating conditions.

The eight sizes of engineering steel chain available are listed in ANSI Standard B29.10[1]. Table 3-8 shows the chain number, the pitch size, and the average ultimate strength for the eight chains. Speed and horsepower ratings are the prime considerations in selecting a chain drive. Normally, the ratings are listed for the smaller sprocket, regardless of whether it is the drive or driven member. Chain manufacturers should be consulted for proper drive selections when special conditions are encountered.[2]

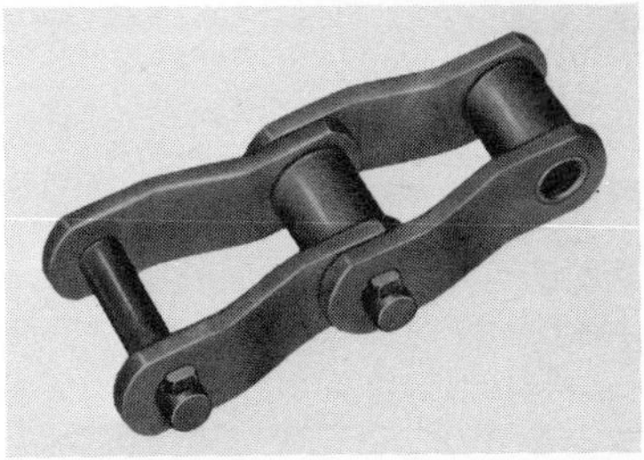

Figure 3-23 Engineering steel drive chain.

TABLE 3-8 Engineering Steel–Chain Size and Strength

Chain number	Chain pitch		Average ultimate strength	
	in	mm	lb	kg
2010	2.500	63.5	62,500	28,300
2512	3.067	77.9	90,000	40,800
2814	3.500	88.9	122,000	55,3000
3315	4.073	103.4	141,000	63,900
3618	4.500	114.3	193,000	87,500
4020	5.000	127.0	250,000	113,400
4824	6.000	152.4	360,000	163,300
5628	7.000	117.8	490,000	222,200

The speed and horsepower ratings are based upon approximately 15,000 h of service life at full-load operation and a service factor of 1.0. Operating conditions which establish the service factor are shown at the bottom of Table 3-3.

Two additional types of engineering chain are also available for use in special applications. Welded steel chain is used in some slow-speed drives subject to shock loads and other difficult operating conditions. Cast pintle chain is used for drives handling light loads at slow speeds.

SILENT-CHAIN DRIVES

Silent chain is a drive chain constructed of sidebars, pins, and bushings with no rollers (Fig. 3-24). The sidebars are designed to mesh with sprocket teeth in a gear-type engagement.

Figure 3-24 Silent drive chain.

ment. Silent drive chain is selected for high-speed–high-load applications and smooth, quiet operations for many industrial services such as electric generating plants, automotive test stands, machine tools, and ventilating systems.

There is a wide range of silent chain-link configurations available from various manufacturers.[2] For this reason, silent chains are not normally interchangeable on different manufacturers' sprockets. However, the "SC" series of silent chains shown in ANSI Standard B29.2[1] is interchangeable on sprockets between manufacturers.

DRIVE ARRANGEMENTS

Illustrated in Fig. 3-25 are the drive arrangements recommended for optimum drive life. The preferred direction of rotation is indicated, although arrangements *A*, *B*, and *C* will

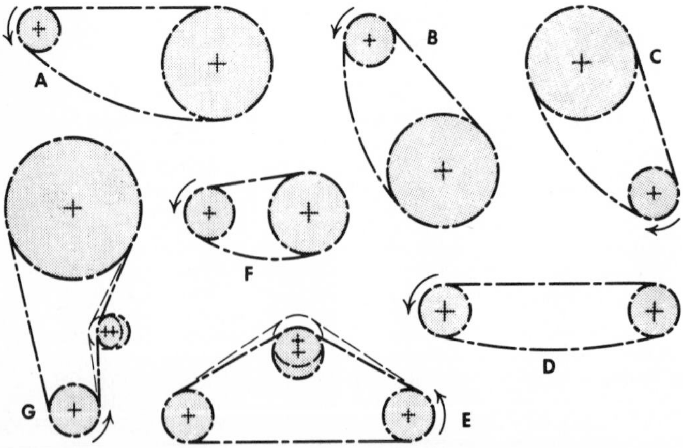

Figure 3-25 Preferred drive arrangements.

operate satisfactorily in either direction. Consult the chain manufacturer for approval on other drive arrangements.[3]

CHAIN TIGHTENERS

Chain tighteners are used to obtain a desired tension between shaft centers. Where centers approach the vertical relative position or there is pulsating operation, it is important to provide a means for adjusting the chain tension in order to prolong chain life. A chain tightener, sometimes referred to as an idler sprocket, should be located so as to operate against the slack or return side of the chain. Pictured below are illustrations of chain tighteners employed in chain drives (Fig. 3-26).

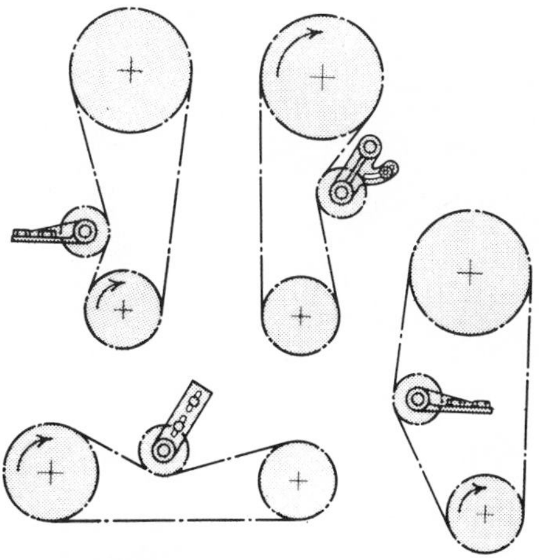

Figure 3-26 Chain tighteners used to maintain tension.

LUBRICATION

Lubrication is the most important factor in maintaining high chain efficiency and providing a long service life.

The primary purpose of chain lubrication is to provide a clean film of oil at all load-carrying points where relative motion occurs. The lubrication method recommended in the speed-horsepower selection charts should be used to ensure proper lubrication at all times for a drive chain selection.

Several methods of lubrication have been developed to fit a particular range of horsepower, chain speed, and relative position of shafts. Manual or drip lubrication is used for open running drives which operate in a nonabrasive atmosphere. These methods should be confined to low-horsepower drives with a chain speed under 600 ft/min (183 m/min). Bath lubrication is the simplest automatic method of lubricating encased chain drives and is highly satisfactory for low or moderate speeds. Disk lubrication is used for moderately high-speed drive arrangements unsuitable for oil-bath lubrication. Forced lubrication or oil-pump lubrication is recommended for large-horsepower drives, heavily loaded drives, high-speed drives, or where oil-bath or oil-disk lubrication cannot be used (Fig. 3-27).

Chains can be lubricated with any neutral grade of straight mineral oil in the 20 to 50 viscosity range, depending on the temperature. For difficult operating conditions, such as high temperature or an abrasive atmosphere, consult a lubricant manufacturer for proper lubricant selection.

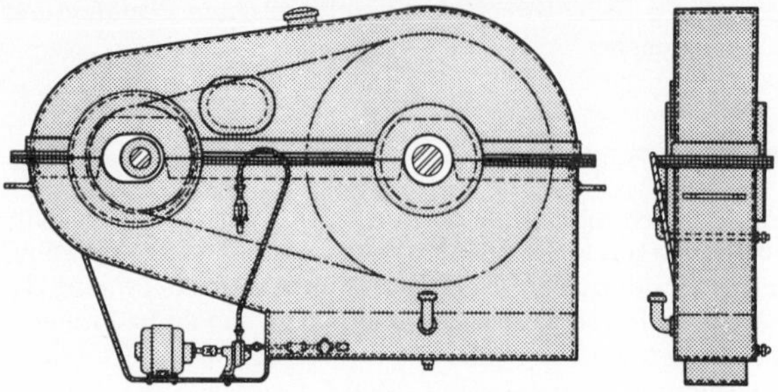

Figure 3-27 Forced lubrication. An oil pump is used to provide a continuous spray of oil to the inside of the lower span of chain.

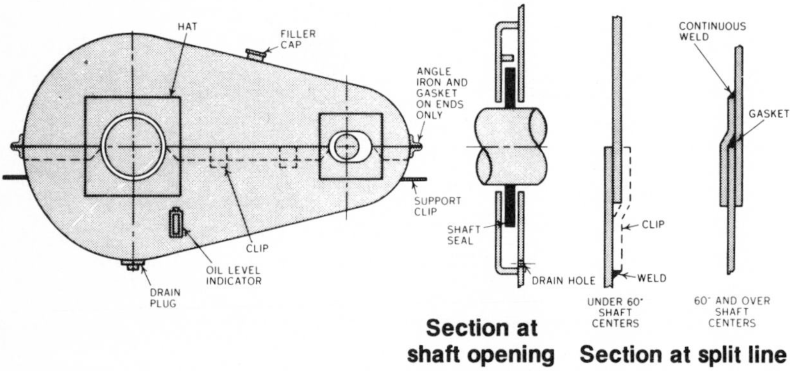

Figure 3-28 Oil-bath or oil-disk casing.

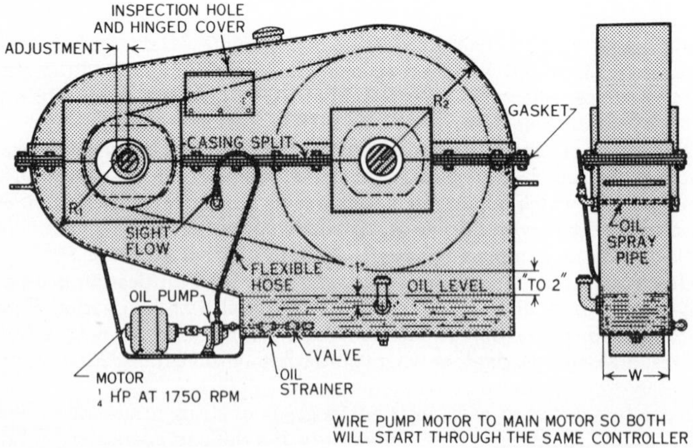

Figure 3-29 Casing for forced lubrication.

CASINGS

Casings are important considerations in providing adequate lubrication to prolong the efficiency and service life of a chain drive. Casings contain the lubricant, exclude foreign materials from the lubricant, and function as a safety guard. There are four basic types of casings used with chain drives.

The oil-bath or oil-disk casing is used with the bath and disk methods of lubrication, with variations from the standard casing available depending on the chain speed and shaft center distances (Fig. 3-28). The second type of casing is used in outdoor applications when water, dust, and other contaminants are present, when extra protection against oil leakage is needed, or when forced lubrication is used (Fig. 3-29). The weathertight casing is the third type of construction, and it is used when additional precaution against contamination is desired. Guard-type casings are used primarily as a safety device rather than a lubrication accessory and are employed when other types of casings are not compulsory.

MAINTENANCE

A good maintenance program for a chain drive should include proper installation, alignment, and periodic inspections. After an installation inspection, an initial inspection should come after 100 h of operation; a second inspection should come after 500 h, followed by regular intervals thereafter. The following is recommended as part of the maintenance routine:

1. **Check Sprocket and Shaft Alignment** Wear on the inner surface of the roller link sidebars and/or wear on both sides of the sprocket teeth indicate drive misalignment.
2. **Check Sprocket Tooth Wear** Hook-shaped teeth are an indication of a worn sprocket.
3. **Check Chain Tension** Incorrect chain tension can cause improper sprocket action.
4. **Check the Oil Level** The oil level should be inspected when the drive is idle, allowing sufficient time for the lubricant to accumulate in the sump.
5. **Check Oil Flow** Forced lubrication should be frequently inspected. Sight flow gauges must be mounted above the spray line to allow visual examination to ensure proper lubrication.
6. **Change Oil** Oil should be changed initially after the first 500 h of operation. Thereafter, the oil should be changed after 2500 h of operation.
7. **Inspect the Chain Parts** Look for visual indications of any improper operation such as unusual wear.

REFERENCES

1. American National Standards Institute, Standards (use most recent issue).
   ANSI B29.1, "Precision Power Transmission Roller Chains, Attachments and Sprockets."
   ANSI B29.2, "Inverted Tooth (Silent) Chains and Sprocket Teeth."
   ANSI B29.10, "Heavy Duty Offset Sidebar Power Transmission Roller Chains and Sprocket Teeth."
2. American Chain Association Publications
   Design Manual for Roller and Silent Chain Drives; Applications Handbook for Engineering Steel Chains.
3. Chain Manufacturers' Catalogs.

PART 4

Flexible Shafts for the Transmission of Rotary Motion

prepared by

Thomas C. Roberts
Product Engineering Manager
S. S. White Industrial Products Division
Pennwalt Corporation

INTRODUCTION

A rotary-motion flexible shaft can be a useful alternative to other types of power transmission when the engineer is faced with (1) hard-to-align shafts, (2) relative motions between shafts, (3) one or more angles in the power-transmission path, (4) power transmission at angles other than 90°, or (5) the need to absorb the shock of sudden stops and starts and reduce vibration.

Shaft Alignment

When mounting a motor to drive a pump, a gearbox, or any device with self-contained bearings, a conventional coupling can absorb only small amounts of misalignment (typically less than 2° angularity and 0.005 in parallel misalignment). The large tolerance for misalignment inherent in the flexible shaft allows the production engineer to do away with time-consuming and expensive realignment, requiring dial indicators, every time a motor is removed and replaced.

Relative Motion Between Shafts

Use in printing machinery and paper-coating equipment, or whenever power must be transmitted from a stationary motor to a moving carriage, are natural applications for flexible shafting. The shafting will absorb the relative motion between the driving and the driven member without resorting to constant-velocity universal joints.

More than One Angle in Power-Transmission Path

When the power must be transmitted through several angles or a complex path to the driven member, a flexible shaft is much easier to apply than a series of universal gearboxes. When using gears, an angle other than 90° will require two gearboxes with their attendant alignment and expense problems.

Power Transmission at Angles Other than 90°

Flexible shafting can be used for small angles and angles other than the standard 90° available with gearboxes. As a side benefit, the flexible shafting does not have to be mounted as a gearbox needs to be.

Shock and Vibration Absorption

The flexible shaft will act as a heavily damped spring in the system to absorb the shock of sudden machinery stops and starts. It will also tend to smooth out vibration in the drive system. This heavy internal damping can cause a temperature rise in the shaft if it is used with a continuously pulsating driver, such as a single-cylinder engine. The shaft should be oversized in applications like this to reduce the heating.

The engineer can apply flexible shafting in many applications where solid shafting and/or gearboxes would provide an expensive system that is difficult to align and maintain. It is suggested that the shafting manufacturer be consulted as early as possible in the design process to allow use of predesigned shafting whenever possible. Consultation in the early design stage will eliminate use of nonstandard sizes, which, in turn, will result in lower final cost and shorter delivery cycles.

WHAT IS A ROTARY-MOTION FLEXIBLE SHAFT?

A rotary-motion flexible shaft (RMFS) is a relatively simple and trouble-free device. The manufacturer will take one central wire mandrel, wrap several successive layers of wire concentrically around this mandrel, and square the ends or attach suitable fittings to make a flexible shaft. Depending on the number of wires in each layer and their diameter, the manufacturer can design either (1) a stiff, flexible shaft which exhibits very little backlash in either direction, making it ideal as a remote-control shaft, or (2) a more flexible shaft which is ideal for power transmission in one direction, sometimes at speeds as high as 20,000 r/min. Available flexible-shaft diameters range from 0.050 to 1.0 in (12.5 to 25 mm) and power shafts are capable of transmitting up to 50 hp (Fig. 3-30).

Critical Properties

Typically, the plant engineer is not expected to design a flexible shaft; that can be left to the engineering department of the flexible-shaft manufacturer. However, the engineer must know what is expected of the flexible shaft and must provide the manufacturer with certain basic data so that a suitable selection or design can be provided. Among the factors that are critical are: (1) type of shaft, remote control or power; (2) manual or dynamic operation; (3) maximum torque to be transmitted; (4) horsepower of driving unit; (5) peak allowable revolutions per minute; (6) direction(s) of rotation; (7) total permissible angular deflection; (8) cycling time or duty-cycle rest periods available; (9) minimum bend radius; (10) length of shaft; (11) ambient temperature; (12) unusual service conditions. Note that not all these data will apply in all instances, but the more background the manufacturer can be given, the more suitable and economical will be the flexible-shaft design.

In many instances, a predesigned shaft, or RMFS coupling, will solve the plant engineer's power transmission problem. However, if the requirements are unusual, a special shaft may have to be designed. Obviously, whenever possible, predesigned shafts should be selected because they are the less costly solution.

Casings and Fittings

In many applications, a bare flexible shaft as described above, equipped with suitable fittings, is perfectly adequate for the task at hand. In other cases, however, the flexible shaft should be provided with special fittings and a casing.

End fittings are installed on both the casing and the shaft core itself. The casing fittings keep the shaft assembly fixed in place; neither these fittings nor the casing rotate.

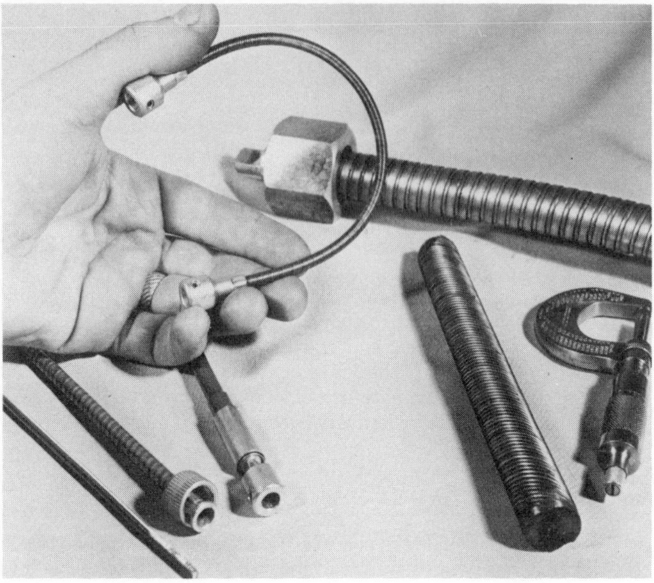

Figure 3-30 Standard flexible-shaft assemblies come in a wide range of sizes and configurations, with several types of end fittings; some of these are shown here.

Casing fittings may be of several types, among them a loose male (or female) threaded coupling nut, or quick disconnect. With *shaft*-end fittings, even more choices are available: integrally formed drive square, fork or tang fitting, male or female spline, hollow square, panel mounting, or set screw. The choice of fittings depends to a large extent on the application, how accessible the shaft is, and how often it is expected to be serviced. If a shaft is to rotate at high speeds for prolonged periods and has to be lubricated regularly to compensate for this severe service, it is recommended that the assembly be designed for easy withdrawal of the shaft from the casing. In that event, one end of the shaft is usually swaged square, because this offers a drive that does not permanently lock the shaft into the casing at both ends.

A casing provides the following benefits: (1) It keeps the flexible shaft properly lubricated for longer periods of time. (2) It offers intermediate points of attachment, when shaft length exceeds 18 in. (3) It keeps plant personnel from coming into contact with a high-speed rotating member. (4) It protects the flexible shaft against hostile environments.

Casings can be classified into two types: metallic and covered. Metallic casings are of either two- or four-wire construction. They are neat, strong, durable, and bendable. Although they will retain grease, they are not oil- or watertight.

Covered casings are more expensive because their construction is more elaborate. Following an inner liner that may be either metallic or plastic, the casing is reinforced with layers of steel braid, and finished off with a final outer sheath of rubber or plastic. This type of casing permits the flexible shaft to operate under water, hydraulic fluid, or other liquid, so long as the fluid does not attack the elastomeric outer covering.

Shafts Thrive on Speed

Power shafts function most efficiently when operated at high speed. The reason stems from basic engineering principles: Horsepower transmitted is directly proportional to speed.

$$hp = kTN,$$

where T is torque, N is speed, and k is a constant to reconcile the units of torque and peripheral speed being used in the equation

Typically, if torque is expressed in foot pounds and speed in revolutions per minute, hp = 0.00019 × ft·lb × r/min. For torque expressed in inch pounds, hp = 0.000016 × in·lb × r/min. Therefore, for a given horsepower requirement, the higher the shaft speed, the less the torque on the shaft, therefore, the smaller the shaft needed to accomplish the task. This factor should be kept in mind during the early stages of design, because quite often the insertion of a suitable speed-reduction device at the correct location in the drive train may serve to reduce the overall cost of the assembly.

A power shaft is to rotate at the highest permissible speed; it is, therefore, important to consider this in the design. If a speed reducer is necessary, insert it *at the correct end of the shaft.* Thus, if motor speed is 5000 r/min and final operating speed is to be 500 r/min, the speed reducer should be inserted at the output, *not* at the motor end. Conversely, if a speed*up* is desired, the speed increaser should be inserted at the motor end (Fig. 3-31).

Shaft Radius of Curvature

Although rotary-motion flexible shafts are almost snakelike in their ability to reach into inaccessible areas, there is a limit to how much they may be bent without damaging them. This limit is called the *n*on*d*amaging *r*adius (NDR). In addition, there is a larger radius which is the *m*inimum *o*perating *r*adius (MOR). Each shaft construction (diameter, number of wires per layer, and number of layers) leads to a different MOR, a figure that is always available from the manufacturer.

NDR is that minimum radius beyond which the shaft would be permanently damaged. Such a condition generally would only occur during handling or installation. The NDR is a smaller radius than the MOR. It is determined by test.

MOR is the zero torque point on the torque capacity curve for any particular shaft. (It is a kind of misnomer because in normal circumstances a shaft would not be run in the MOR because technically it could not carry any load.) The MOR is established by calculations based on the diameter and material of the mandrel wire.

It should come as no surprise that the torque load a flexible shaft can comfortably manage is an inverse function of the MOR. The more a flexible shaft is curved, the higher the internal friction, the greater the heat generated, and the smaller the permissible continuous load. Here again, these figures are known for each construction and should be rigidly adhered to if overload or early failure of the shaft is to be avoided. Table 3-9 shows typical values for a range of high-tensile-steel power shafts, including such basic parameters as diameter, radius of curvature vs. torque capacity, torsional breaking load, and weight per foot.

In using Table 3-9, it must be remembered that the torque-carrying capacities are shown for shafts rotating in the direction that tightens up the outer layer. If the same shaft is rotated so as to unwind the outer layer—clockwise for a right-lay flexible shaft or counterclockwise for a left-lay shaft—its torque capability is reduced by up to 50 percent. See Fig. 3-32 for an example of right- and left-hand shaft application.

Shaft Life

As a rule of thumb, a power shaft running under no load at the MOR can be expected to have a useful life in excess of 10^8 revolutions. Such factors as service temperature, number of shaft bends, torque output, speed, shock loading, and operating radius can reduce shaft life.

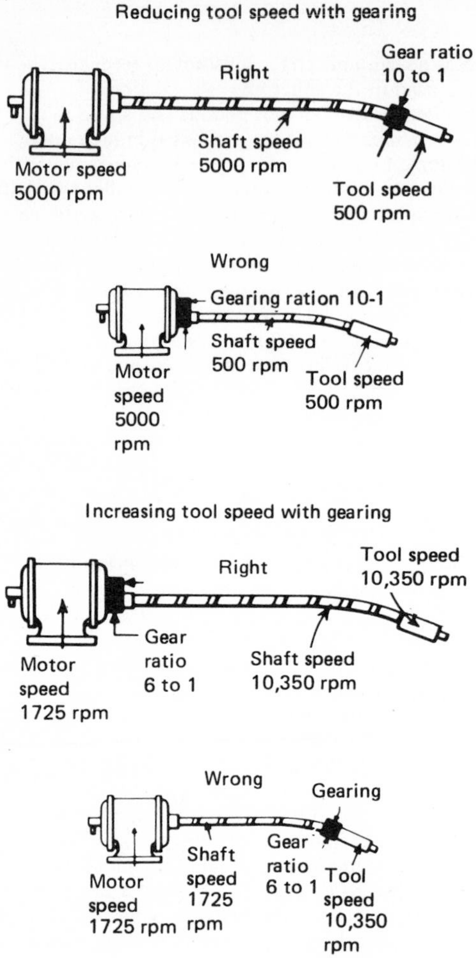

Figure 3-31 Changing tool speed with gearing.

A remote-control shaft, under manual operation, will last many years. When remote-control shafts are used for dynamic operations, it is very important that the duty cycle not exceed 5 min ON and 2 min OFF.

SHAFT SELECTION

To select the proper shaft, we need to determine the following items:

1. Torque requirement
2. Shaft radius of curvature
3. Length of path between driving and driven systems
4. Operating speed
5. Acceptable torsional deflection or backlash

TABLE 3-9 Power-Drive Flexible Shafts*,†

Shaft number‡	Shaft diameter, in	Min operating radius, in	Dynamic torque capacity winding direction, lb·in input								Torsional breaking load for straight shafts winding dir, lb·in§	Approx weight of shafting, lbs. per 100 ft
			Radius of curvature, in									
			25	20	15	12	10	8	6	4		
050–9	0.050	2.0					0.26	0.24	0.22	0.16	0.9	0.5
068–9	0.068	2.5					0.60	0.55	0.46	0.30	2.6	0.9
098–9	0.098	2.5				1.9	1.8	1.7	1.4	0.91	8.5	2.0
130–9	0.130	3.0			3.8	3.6	3.4	3.1	2.4	1.7	15.0	3.3
150–9	0.150	4.0			5.0	4.7	4.4	3.9	3.1	1.4	24	4.5
187–9	0.187	4.0		13.5	12.6	11.8	11.0	9.8	7.8	4.0	55	7.2
250–9	0.250	4.0	25	24	22	21	19	16	12		100	12.4
312–9	0.312	4.5	49	46	43	40	37	32	24		200	19.6
375–9	0.375	4.5	67	63	58	53	48	40	26		280	28.5
437–9	0.437	5.0	100	95	85	77	68	54	30		440	38.1
500–9	0.500	5.5	133	125	113	102	90	70	40		550	51.8
625–9	0.625	6.5	230	210	190	160	130	95			1100	79.0
750–9	0.750	9.0	405	365	300	235	170				2000	114.0

*The data shown in the shaft selection tables are suitable for most applications.

†The most popular sizes of shafts are listed. Other sizes can be made on special order if the volume warrants.

‡Shafts can be supplied in either "left-lay" or "right-lay." When ordering, the letter "L" or "R" should be inserted in the shaft number to indicate the lay (i.e., 150L9 is a 0.150 shaft, "left-lay").

§Shaft will break or helix under these loads. For routine tests, do not specify more than approximately 50 percent of these figures, otherwise the flexible shaft may be seriously damaged.

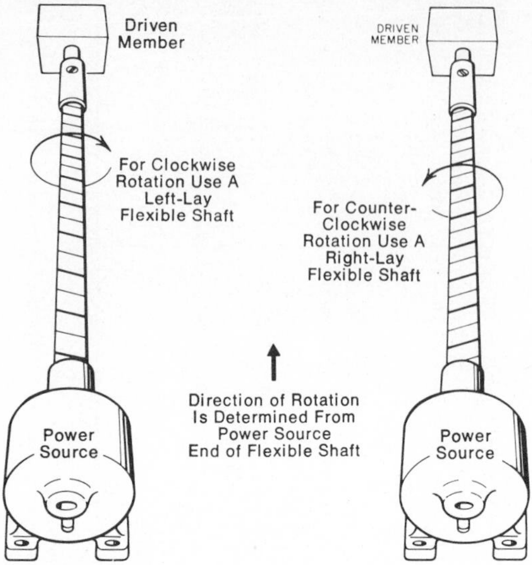

Figure 3-32 How to determine proper shaft lay.

Determining the torque requirement on a shaft may not be easy. Remote-control shaft requirements can be determined by attaching a lever of known length to the device to be turned and pulling the end of the lever with a spring scale. Use the highest torque as the required torque. Measuring torque on a power shaft is more difficult because the requirement is dynamic. The best way is to instrument load with a torque cell and measure the torque under operating conditions. Component manufacturers usually have these data available. A shaft efficiency of about 90 percent should be attainable for most applications; therefore, increasing the output torque by about 10 percent leads to the input torque.

The shaft path should then be examined. A drawing should be made which shows the path in a true view. The MOR should be determined from this drawing. A full-scale prototype can be used. This length of the shaft can also be determined at this time.

For power shafts, the operating speed is generally given. For remote-control shafts, in which the speed is essentially zero, the amount of torsional deflection or dead band acceptable between the turning device and the turned device is the important consideration.

Sample Problem

A power shaft is driven by a motor to drive a fan. The fan requires 2 hp at 1728 r/min. A layout of the installation shows that a 6-ft shaft with one bend of 12-in radius and one bend of 8-in radius are required. How big a shaft is required and what will the approximate life be in hours?

Solution

$$\text{Torque} = \frac{\text{hp} \times 63{,}025}{\text{r/min}}$$
$$= \frac{2.0 \times 63{,}025}{1728} = 72.9 \text{ in}\cdot\text{lb}$$

When assumed efficiency e is 90 percent,

$$e = \frac{\text{torque out}}{\text{torque in}}$$

so

$$\text{Torque in} = \frac{\text{torque out}}{e} = \frac{72.9}{0.90} = 81 \text{ in} \cdot \text{lb}$$

Entering Table 3-9 at the smallest radius of curvature (8 in in this case) we determine that a ⅝-in shaft is required.

The shaft selected would be a 6-ft-long shaft of ⅝-in diam. See Fig. 3-31 to determine if a left- or right-hand-lay shaft is required.

This diameter shaft is more expensive than a smaller shaft. For this reason, consider increasing each bend to at least a 25-in radius, which will allow use of a 7/16-in shaft at considerable cost saving.

$$\text{Shaft life} = \frac{10^8 \text{ r}}{N \text{ r/min} \times 60 \text{ min/h}}$$

$$= \frac{10^8}{1728 \times 60} = 964.5 \text{ h}$$

Exact life is dependent on the service and environment of the shaft.

DOS AND DON'TS

Used intelligently, flexible shafts are the product designer's friend. They are more flexible than universal joints; they are more versatile than gear systems because they are totally unaffected by the exact angle or offset necessary; finally, flexible shafts offer an inherent shock-absorption capability and ease of installation and maintenance unmatched by other forms of rotary-motion transmission.

However, flexible shafts do have to be treated carefully if they are to provide the service life built into them. They cannot be bent completely out of shape and must be used at radii equal to or greater than the MOR specified by the manufacturer. They cannot serve as steps of a ladder. They should be secured approximately every 18 in to prevent the possibility of helixing. They must be lightly lubricated for preventive maintenance at regular intervals. Replacement must match the original design because a control shaft is not interchangeable with a power shaft. The type of service expected must be clearly specified. Two differently built shafts, even of the same diameter and length, are not interchangeable. A remote-control shaft is intended for low-speed continuous or high-speed intermittent operation in either direction with minimum backlash. On the other hand, a power shaft is intended for high-speed continuous operation, but in only one direction. Above all, flexible shafts must be designed for the task they are to perform. For this reason, early consultation between the user's engineering department and the shaft manufacturer is highly recommended.

Follow these simple, basic suggestions, and the design flexibility of a rotary-motion flexible shaft will be applied to utmost advantage.

TYPICAL APPLICATIONS

Flexible-shaft couplings, in lengths of 6 in or less, are ideal for connecting a driven member to its power source when a small amount of angular, parallel, or combined misalignment must be accommodated. Not only does such a motor-type coupling readily compensate for the misalignment, but it will also handle a certain amount of shock loading and attenuate vibration.

Longer segments of flexible shaft, from 1 to 100 ft (0.3 to 30.5 m), are ideal for transmitting rotary motion to very distant locations, either for remote control of valves and other apparatus or for power transmission.

Flexible-shaft applications in industry range far and wide, so that it would be quite a task to try to list them all. However, a few typical cases may help to demonstrate how rotary-motion problems have been avoided or solved with a flexible shaft:

An operating device on a swinging oven door has to rotate freely, no matter how much the angle changes as the door opens and closes.

Two rotating devices have to be synchronized through a central gearbox, even though one of the devices moves laterally while the other stays fixed in place.

A crane operator has to know which way the cable is unreeling, even when neither reel nor load is visible from the cab.

Condenser or other heat-exchange tubing has to be cleaned out, 35 ft (10.7 m) from the header end.

Several door-actuating devices have to be moved simultaneously, all by one driving motor and each one at a different angle to it.

A small, buried control has to be readily turnable from "outside," a valve has to be closed from a remote location, a potentiometer has to be adjusted in an "unreachable" spot, and a switch has to be operated from far away.

The above examples serve to illustrate the wide variety of problems that flexible shafts can cope with. This type of drive should always be investigated by the plant engineer as a possible alternative to conventional drives in many applications.

chapter 7-4

Fluid Seals

prepared by

Parker-Hannifin Corp.
Seal Group
Lexington, Kentucky

Authors

Chris S. Louskos
Seal Group Staff

John B. Painter
Gasket Division

Ralph E. Peterson
Packing Division

Richard G. Ramsdell
Seal Group Staff

John B. Scannell
Packing Division

INTRODUCTION

Packings and *seals* are devices or materials designed to create and/or maintain a fluid-pressure differential across the interface or gap between two relatively movable and/or separable components of a fluid system. (*Fluids*, in this context, include liquids and gases, with or without entrained solids.)

Included within this category of packings and gaskets are static seals, reciprocating dynamic seals, rotary dynamic seals, and flexural sealing devices, involving an almost unlimited variety of sizes and configurations and a broad range of common and exotic materials and material combinations.

Because of the complexity of the subject, only a brief overview is possible in this handbook. Although the information provided may help in solving relatively simple problems, a competent supplier should be consulted for assistance with most sealing applications.

MATERIALS

Table 4-1 lists the most commonly used seal and packing materials and gives a few pertinent facts about each. All of the facts given for any material do not necessarily apply to all members of the group. They merely suggest the range of uses of materials within that group. For instance, elastomers are represented as having a temperature range of −178 to +500°F (−115 to 260°C), but only a few specific compounds will serve very long at either of these extremes and no compound will function at both. Similarly, asbestos is said to have good resistance to strong acids and bases, but only the blue asbestos has good resistance to strong basic solutions.

STATIC SEALS

Gaskets

Gaskets are seals placed between two static faces (Fig. 4-1). They are made of deformable materials which, when clamped, will flow into surface imperfections of the mating surfaces to prevent fluids from escaping through the joint. For relatively low pressures and temperatures, nonmetallic gaskets may be used. For more severe applications, metallic gaskets are required.

As a rough guide to this selection, it is common practice to multiply the operating pressure, in pounds per square inch, by the operating temperature, in degrees Fahrenheit. If this value exceeds 250,000, the use of metallics is indicated.* In any case, however, nonmetallic gaskets should not be used at temperatures above 850°F (450°C) or at pressures above 1200 lb/in^2 (8.3 MPa). Generally pressure, temperature, and fluid determine the gasket material, while dimensional and mechanical features of the joint determine the gasket type.

*In SI units, find the value of $[P(1.8C + 32)]$, where P is the pressure in megapascals and C is temperature in degrees Celsius. If the result exceeds 1720, the use of metallics is indicated.

TABLE 4-1 Materials Commonly Used in Seals

Materials and notes	Temperature range, °F (°C)	Commonly sealed fluids	Types of seals in which the material is commonly used
Aluminum	−300 to +800 (−185 to +430)	Water, weak acids except acetic, steam, air, oxygen, dry bromine and chlorine, aliphatic and aromatic fluids, acetone, alcohols, petroleum oils, hot sulfur-bearing gases	Compression packing (as foil), gaskets (solid or as jacketing over asbestos, rubber, or other filler)
Asbestos, white (chrysotile) and blue (crocidolite); fire-resistant fibrous minerals: May be woven into fabrics, braided, or pressed into solid form; often impregnated with elastomer or TFE (tetrafluoroethylene), or reinforced with metal	−300 to +1000 (−185 to +540)	Water, strong acids and bases, steam, air, chlorine, alcohols, petroleum fluids	Compression packing, gaskets (as sheet or with metals in corrugated, jacketed, or spiral-wound types)
Brass	−300 to +500) (−185 to +260)	Water, mild acids and bases, oxygen, steam, dry bromine and chlorine, aliphatic and aromatic fluids, acetone, alcohols, petroleum oils except sulfur-containing oils	Diaphragms, gaskets (as sheet, corrugated, or spiral-wound types)
Bronze	−300 to +500 (−185 to +260)	Similar to brass	Gaskets (with asbestos or TFE in spiral-wound types)
Cast iron	−50 to +1000 (−45 to +540)	Similar to iron	Mechanical seals, piston rings; not usually a gasket material
Ceramics: Excellent resistance to oxidation at elevated temperatures, but quite abrasive unless given a very smooth finish	−300 to >1500 (−185 to >815)	Water, strong acids and bases, steam, air, chlorine, alcohols, petroleum fluids; chemical resistance similar to glass	Mechanical seals, filler material for metallic types of gaskets
Copper	−300 to +600 (−185 to +315)	Water, mild acids and bases, oxygen, steam, dry bromine and chlorine, aliphatic and aromatic fluids, acetone, alcohols, petroleum oils except sulfur-containing oils; not recommended for use above 600°F (315°C) unless oxygen-free	Compression packing (foil), gaskets (solid, as jacketing over filler or with filler in spiral-wound-type)
Cork: Cork particles bonded together in sheet form with resin or an elastomer	−22 to +300 (−30 to +150)	Mild acids and bases, water, coolants, petroleum oils	Gaskets
Elastomers: See Table 4-2 for details	−178 to +500 (−115 to +260)	Wide range of fluids; see Table 4-2	Cup and hat packings, diaphragms, gaskets, O rings, oil seals, PolyPak, T seals, U cups, V packings, backup rings, wipers; reinforced with fabric in some of these products

TABLE 4-1 Materials Commonly Used in Seals (*Continued*)

Materials and notes	Temperature range, °F (°C)	Commonly sealed fluids	Types of seals in which the material is commonly used
Glass fibers	——	Added to nylon, TFE, etc., to improve resistance to wear, extrusion, creep, etc.	——
Graphite: Excellent chemical resistance, but oxidizes at elevated temperatures; good lubricant; porosity and oxidation effect can be modified by impregnating	−450 to +1500 (−270 to +815)	Water, all concentrations of acids and bases if not strongly oxidizing, nitric acid to only 100°F if concentration over 25%, steam, air to 960°F, aliphatic and aromatic fluids, acetone, alcohols, petroleum oils	Compression packing (in fiber form), gaskets, mechanical seals; in powdered form, added to elastomers and plastics to reduce friction
Hastelloy®: Union Carbide trademark for a series of high-strength nickel-base corrosion-resistant alloys	−300 to +2000 (−185 to +1100)	Noted for their superior corrosion resistance; individual applications should be investigated; boiling acids and bases, salts, chlorine, hypochlorites, and sea-water	Gaskets (solid, as jacketing over filler or with filler in spiral-wound type), mechanical seals
Inconel®: International Nickel trademark for nickel-chromium alloys	−300 to +2000 (−185 to +1100)	Noted for high-temperature strength and corrosion resistance; individual applications should be investigated; resists chloride-ion-stress corrosion cracking, organic acids in food products, alkaline sulfur compounds, ammonia, dry gases, steam, air, carbon dioxide; resists progressive oxidation to 2000°F (1100°C), sulfur atmospheres	Gaskets (solid, as jacketing over filler or with filler in spiral-wound type), mechanical seals
Iron: Soft and low-carbon steel	−50 to +1000 (−45 to +540)	Widely used on an economical heavy-cross-section gasket; sulfuric acid at high concentrations, hydrochloric not satisfactory at any concentration, most alkalies, air, water, steam, oxygen, acetone, acetylene	Gaskets (solid as jacketing over filler or with filler in spiral-wound type) mechanical seals
Lead: Soft metal, melts above 500°F (260°C)	−300 to +212 (−185 to +100)	Water, dry bromine and chlorine, aliphatic and aromatic fluids, acetone, alcohols, petroleum oils	Compression packing (as foil or insert in channel-type packing); gaskets (solid or jacketing over filler)
Leather: Porosity can be controlled or virtually eliminated with waxes, oils, etc., that also extend its temperature range and the types of fluids it can withstand	−70 to +212 (−55 to +100)	Water, weak acids and bases, air, aliphatic and aromatic fluids, alcohols, petroleum oils	Cup and hat packings, gaskets, oil seals, U cups, V packings

TABLE 4-1 Materials Commonly Used in Seals (*Continued*)

Materials and notes	Temperature range, °F (°C)	Commonly sealed fluids	Types of seals in which the material is commonly used
Molybdenum disulfide (MoS_2) in a silvery powder form	——	——	May be mixed with most seal materials to reduce friction without causing corrosion
Monel®: International Nickel trademark for group of alloys primarily of nickel and copper plus small amounts of other ingredients	−300 to +1500 (−185 to +815)	Noted for high-temperature properties and corrosion resistance; resists chloride-ion-stress corrosion cracking; most acids (including hydrofluoric) and alkalies; not satisfactory with strong oxidizing acids; fresh and seawater, air, dry gases, neutral and alkaline salts	Gaskets (solid, as jacketing over asbestos, or with asbestos in spiral-wound type), mechanical seals
Nickel	−300 to +1400 (−185 to +760)	Not as all-around-resistant as Monel; resists chloride-ion-stress corrosion cracking; fresh and seawater, alkalies, natural and alkaline salts; not satisfactory with strong, hot, sulfurous and oxidizing acids	Gaskets, as jacketing over asbestos
Nylon (polyamide)	−65 to +300 (−55 to +150)	Air, hydraulic fluids	Fabric, as reinforcement for U cups, V rings, diaphragms; solid, for backup rings
Paper: Due to porosity it is impregnated except where needed merely to exclude dust and dirt	to +300 (to +150)	Aliphatic and aromatic fluids, petroleum oils	Gaskets
Plastics: See nylon, polyurethane, TFE			
Polyamides: See nylon			
Polyurethane: A very tough, wear-, extrusion-, and abrasion-resistant group of materials spanning the range from elastomers to plastics; good resistance to petroleum fluids, air, and aging	−65 to +200 (−54 to +130)	Petroleum-base hydraulic fluids, especially in high-pressure heavy-duty systems; not satisfactory with hot water	Cup and hat packings, O rings, PolyPak, U cups, V packings, adapters, backup rings, scrapers
Rubber: See elastomers (Table 4-2)			
Silver	−300 to +1200 (−185 to +650)	Food and drug industry; acetic acid and acetic anhydride, carbon tetrachloride, wet chlorine, formaldehyde, formic acid, hydrofluoric acid over 65%, magnesium chloride, oxalic acid	Gaskets, plating for spring-type metal seals

TABLE 4-1 Materials Commonly Used in Seals (*Continued*)

Materials and notes	Temperature range, °F (°C)	Commonly sealed fluids	Types of seals in which the material is commonly used
Stainless steel	−300 to +1600 (−185 to +870)	Resistant to multitude of corrosive media depending on operating conditions and alloy selection; water, air, acids, bases, gases, alcohols, petroleum fluids,	Mechanical seals, gaskets (solid, as jacketing over asbestos in spiral-wound type)
Steel: Low-carbon	−50 to +1000 (−45 to +540)	Same as iron	Gaskets, mechanical seals
TFE (also PTFE): (poly) tetrafluoroethylene or Teflon® Du Pont; a plastic having excellent chemical resistance and low friction; softens above 500°F (260°C); creeps under stress	−300 to +500 (−185 to +260)	Water, all concentrations of acids and bases, steam, air, oxygen, bromine, chlorine, aliphatic and aromatic fluids, acetone, alcohols, petroleum oils	Compression packing, cup and hat packings, diaphragms, O rings, piston rings, spring-actuated U cups, tape, backup rings, cap strips, coating for spring-type metal seals, impregnant for asbestos, rubber, etc., mechanical seals
Titanium	−300 to +2000 (−185 to +1100)	Nitric acid, except fuming, other oxidizing acids, mixed acids, wet chlorine, phosphoric acid to 30%, chlorine compounds, hydrogen sulfide, fresh and seawater, salt solutions, alkaline solutions, organic acids	Gaskets
Vegetable fibers, such as cotton, flax, hemp, jute, and ramie: Usually impregnated with neoprene or other rubber to reduce porosity, while the fibers reinforce the rubber and reduce swelling in some fluids; exposed fibers hold fluid, aiding lubrication	−20 to +200 (−30 to +95)	Water, ammonia, aliphatic and aromatic fluids, petroleum oils	Compression packing, gaskets
Wool felt	−100 to +160 (−75 to +70)	Water, ammonia, aliphatic and aromatic fluids, petroleum oils	Gaskets, impregnated with an elastomer or plain, for dust seals

Nonmetallic Gaskets

Nonmetallic gaskets in general are more economical than metallic. They are also softer, thereby sealing at a lower seating stress. These gaskets are made from rubber, both synthetic and natural, paper, plant fibers, cork, cork combined with rubber, asbestos with rubber or other binders, compressed asbestos sheet packing, PTFE (polytetrafluoroethylene or Teflon®), and carbon sheet.

An almost limitless variety of nonmetallic gasket materials is produced by combinations of the above materials. This allows the designer to select the most economical combination to provide the sealing capability, strength, and durability required to fit the operating conditions of a given system.

Metallic and Combination Gaskets

Corrugated Gaskets. Corrugated gaskets are made from thin metal which is corrugated with concentric waves (Fig. 4-2). These are essentially line-contact seals with multiple corrugations providing a labyrinth effect. They are used in special shapes, with complicated hole and corrugation patterns, as engine head and manifold gaskets, and also used in lightweight fuel and hydraulic systems. This type of gasket, with asbestos cord or other filler, is used extensively in large, lightly bolted hot-gas duct systems. With this construction, large, oddly-shaped gaskets can be fabricated in pieces and assembled on the flange.

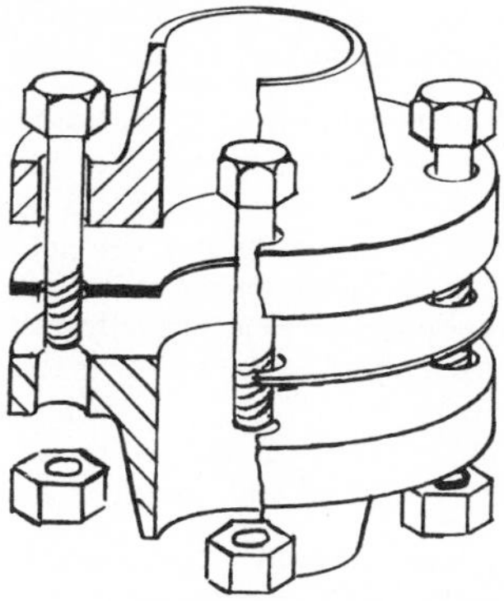

Figure 4-1 Flat gasket between two flanges.

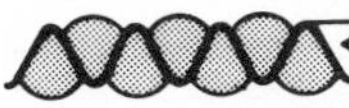

Figure 4-2 Corrugated gasket with filler. *(Reproduced by permission of Parker Gasket Division, North Brunswick, N.J.)*

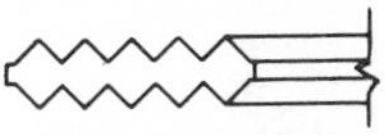

Figure 4-3 Flat-metal gasket with grooves. *(Reproduced by permission of Parker Gasket Division, North Brunswick, N.J.)*

Flat Metal Gaskets. Flat metal gaskets are washer-shaped and are relatively thin. The face width is at least 1.5 times the thickness (Fig. 4-3). They can be used with flat surfaces as cut, or grooves may be machined in the surfaces to reduce the contact area. These reduced-area types have less friction, and are therefore useful in screwed attrition joints.

All types seal by brute compressive force flowing the gasket into the flange contact surfaces, and finishes are therefore important. Nevertheless, this can be an economical gasket. Some uses include valve bonnets, ammonia fittings, heat exchangers, and tongue-and-groove joints.

Spiral-Wound Gaskets. Spiral-wound gaskets are made by spirally winding a V-shaped metal strip with a soft filler such as asbestos (Fig. 4-4). These gaskets have good resilience and sealability. This type is suited to assemblies subject to extremes in joint relaxation, temperature or pressure cycling, and shock or vibration. They are available in a wide variety of metals and filler materials and are produced in circular or moderately noncircular shapes. The inner and outer metal plies of the gasket must be under compression. Preferred flange surface finish is 125 to 250 rms.

Metal-Jacketed Gaskets. Metal-jacketed gaskets are made with a soft compressible filler partially or wholly enclosed in a metal jacket (Fig. 4-5). The entire inner lap must be under compression since this is the primary seal. They are used for circular and noncircular applications including heat exchangers, valves, pumps, compressors, and boilers. In some instances the double-jacketed type is used as rib work in a spirally wound outer

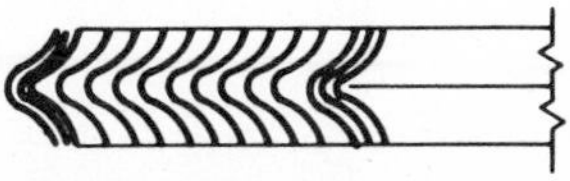

Figure 4-4 Spiral-wound gasket with asbestos filler. *(Reproduced by permission of Parker Gasket Division, North Brunswick, N.J.)*

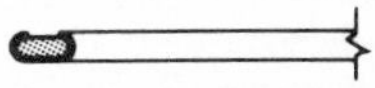

Figure 4-5 Metal-jacketed gasket with filler partially enclosed. *(Reproduced by permission of Parker Gasket Division, North Brunswick, N.J.)*

gasket for heat exchangers. They require 20 to 30 percent compression and are not normally used for joints requiring close maintenance of compressed thickness. When temperatures exceed 900°F (480°C), metallic fillers can be used.

Heavy-Cross-Section Gaskets. The heavy-cross-section gaskets are widely used in the petroleum and processing industries. These gaskets are designed for use in flanges that are specially machined to accept them. Among this group are specialized cross sections including oval, octagonal, lens, Bridgeman, and delta (Fig. 4-6). They are used in high-pressure and high-temperature services including oil-field drilling and production equipment, pressure vessels, valve bonnets, and piping systems. Some of these gaskets are pressure-actuated.

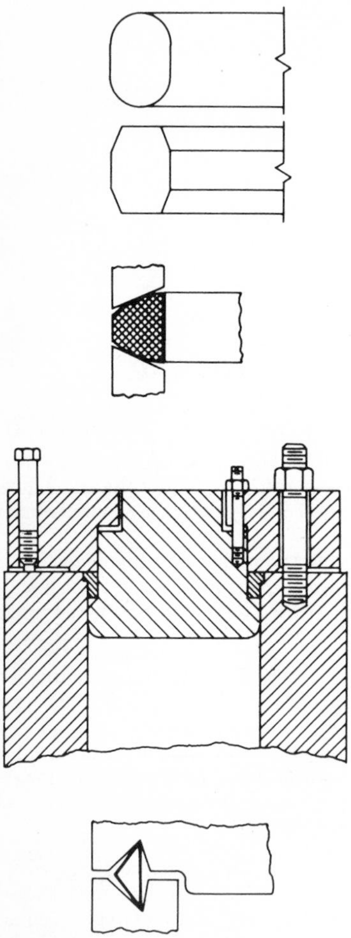

Figure 4-6 Heavy-cross-section gaskets; from the top down: oval gasket *(Reproduced by permission of Parker Gasket Division, North Brunswick, N.J.)*; octagonal gasket *(Reproduced by permission of Parker Gasket Division, North Brunswick, N.J.)*; lens gasket *(Reproduced by permission from Machine Design, September 13, 1973)*; Bridgeman joint *(Reproduced by permission from Petroleum Refinery, May 1956)*; delta gasket *(Reproduced by permission from Machine Design, September 13, 1973)*.

Round-Cross-Section Solid-Metal Gaskets. These gaskets are usually made from wire formed to size and welded (Fig. 4-7). They seal by line contact with high local gasket stresses at low flange loadings. Flange faces are usually grooved or otherwise faced to accurately locate the gasket during assembly. This type is useful in vacuum or other light bolted flange systems where efficient seals are needed.

Light-Cross-Section Gaskets. This type can be used in elastomeric O-ring-type installations sealing extremes in vacuum, high-temperature or -pressure, cryogenic, or nuclear applications (Fig. 4-8). They require very little load for initial

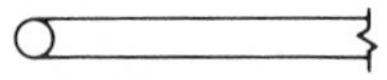

Figure 4-7 Round-cross-section metal gasket. *(Reproduced by permission of Parker Gasket Division, North Brunswick, N.J.)*

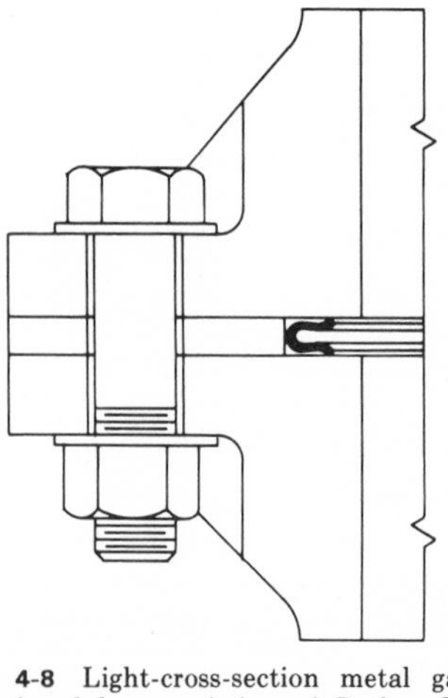

Figure 4-8 Light-cross-section metal gasket. *(Reproduced by permission of Parker O-Seal Division, Culver City, California)*

sealing. Pressure of the system acts on the specially designed seal bellows area, increasing tightness. Flange finish can be 32 to 100 rms for coated seals and 4 to 32 for uncoated.

Installation

Installation of any gasket is a very important factor in achieving a sealed joint. Flange surfaces must be clean and free from nicks, scratches, weld splatter, and any other foreign material. They must be flat and free from warpage. The clamping force or bolt load must be applied in an even stepwise manner to assure uncocked flanges and a uniformly applied load. *It is the joint that leaks. It is the gasket that seals.*

O Rings

The O ring is a versatile, compact, inexpensive sealing device (Fig. 4-9). It is made in a variety of elastomers and may be used as a face seal, a radial squeeze seal, a tube-fitting seal, or occasionally in other configurations.

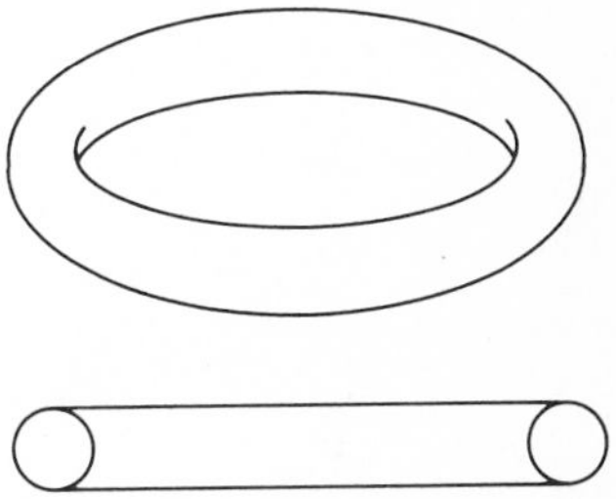

Figure 4-9 O ring. *(Reproduced by permission from Parker O-Ring Handbook No. ORD5700, page A1-1, Parker O-Ring Division, Lexington, Kentucky.)*

Materials

O-ring elastomers are selected for their ability to withstand the conditions of particular sealing applications. The most commonly used O-ring-seal elastomers, together with some other elastomers that are occasionally discussed, are described in Table 4-2.

Applications

O rings in use are confined in a cavity, or "gland," in which the cross section is squeezed in one direction to provide the lines of contact with mating surfaces that create the sealing function. The combination of an O ring and its gland makes up the O-ring seal.

In using an O ring as a seal, the amount of squeeze on the cross section may be critical. With too much squeeze, there may be assembly problems, or the O ring may be damaged at assembly. With too little, the seal is likely to leak at low temperatures.

Though the O ring is squeezed in one direction, it must be free to bulge at right angles to the direction of squeeze, and have extra space to allow for both normal volume swell in contact with the system fluid and for thermal expansion (Fig. 4-10). In some applications, it is acceptable to provide little excess void if the fluid causes little or no swelling, if the surrounding structure is so rugged that it can contain the high pressures caused by a confined rubber that would normally swell, or if the O ring is so completely confined that the fluid can contact only a minuscule portion of its surface, thus effectively preventing swell.

Standard gland dimensions established by O-ring manufacturers generally allow for a moderate amount of swelling.

1. Radial squeeze static O-ring seals, whether of the rod (male gland, Fig. 4-11) or piston (female gland, Fig. 4-12) type, usually have a moderate amount of squeeze that is sufficient to assure good low-temperature performance but not enough to cause undue assembly problems.
2. Face-type rectangular grooves for O-ring seals may provide a heavier squeeze, which makes for a more reliable sealing function (Fig. 4-13). The extra squeeze is acceptable in this design because the cross section is generally deformed by a straight axial pull exerted by flange or cover bolts, so that assembly is not a problem. A variation is the dovetail groove, designed to capture the O ring in a face-type assembly so that the ring cannot drop out when the cover is removed (Fig. 4-14). These are particularly

TABLE 4-2 Elastomers Commonly Used for Seals

No. and designations*	Temperature range, °F (°C)	Properties and uses
1. Butadiene (P, C) BR (A)	−80 to +225 (−62 to +110)	Very good low-temperature flexibility, good resistance to automotive-brake-fluid, water, low-molecular-weight alcohols; preferred because of better high-temperature resistance; often blended with other polymers; primary uses are tires and vibration mounts for low temperature
2. Butyl (P) isobutene-isobutene-isoprene (C) IIR (A)	−75 to +225 (−60 to +110)	Used for vacuum and gases because of its low permeability rate; in other sealing applications, generally superseded by ethylene propylene because of the superior heat resistance and recovery of ethylene propylene; types of seal: diaphragms, gaskets, O rings
3. Epichlorohydrin (P) Hydrin® (Goodrich) 3a. Polychloromethyl Oxirane (C) CO (A)	−40 to +325 (−40 to +163)	Low permeability rate, good resistance to weathering and to oils and other hydrocarbons; usefulness limited because it tends to corrode metals and has poor recovery; types of seal: diaphragms, gaskets
3b. Ethylene oxide (oxirane) + chloromethyl oxirane (C) ECO (A)	−50 to +325 (−45 to +163)	
4. Ethylene propylene (P) 4a. Ethylene propylene copolymer (C) EPM (A) 4b. Ethylene propylene terpolymer (C) EPDM (A)	−70/60 to +250/400 (−57/−50 to +120/205)	Very good resistance to water, steam, weak acids and bases, Skydrols® (Monsanto), and other phosphate esters; swells severely in petroleum products and other hydrocarbons; types of seal: diaphragms, gaskets, O rings, oil seals, T seals, U cups, V packings
5. Fluorocarbon elastomer (P, C) Fluorel (P) (® 3M) Viton (P) (® Du Pont) FKM (A) formerly FPM (A)	−40/0 to +400 (−40/−20 to +200)	Good resistance to wide range of fluids including hydrocarbons and air; hardens in high-temperature water; special, very expensive type (Du Pont Viton GLT) good down to −40°F, otherwise low-temperature limit for seals is 0 to −15°F; excellent high-temperature resistance; types of seal: diaphragms, gaskets, O rings, oil seals, T seals, U cups
6. Fluorosilicone (P, C) FVMQ (A)	−100 to +350/400 (−73 to +175/205)	Has resistance to a wide range of fluids, including hydrocarbons; an expensive elastomer generally restricted to static seals owing to lack of toughness; types of seal: gaskets, O rings
7. Natural Rubber (P) natural polyisoprene (C) NR (A)	−30 to +225 (−34 to +110)	Has excellent recovery, but seldom used for seals because of poor resistance to aging and to most fluids
8. Neoprene (P) chloroprene (P, C) CR (A)	−70/−40 to +250/300 (−55/40 to +120/150)	An "in-between" elastomer, having fair resistance to both hydrocarbons and to oxidation and aging; one of the earliest synthetics; primary seal usage is in refrigerants and silicate esters; types of seal: braided packing, diaphragms, gaskets, O rings, U cups

TABLE 4-2 Elastomers Commonly Used for Seals (*Continued*)

No. and designations*	Temperature range, °F (°C)	Properties and uses
9. Nitrile (P) Buna-N (P) Nitrile butadiene (C) NBR (A)	−70/−20 to +180/275 (−55/−30 to +80/135)	The most commonly used elastomer for seals owing to its resistance to petroleum fluids (fuels, hydraulic and lubricating oils), its useful temperature range, and its low cost; types of seal: braided packing, cup and hat packings, diaphragms, gaskets, O rings, oil seals, sealants, T seals, U cups, V packings
10. Perfluoroelastomer (P, C) Kalrez® (Du Pont) (P)	−30 to +500 (−34 to +260)	Resists a very wide range of fluids and extra-high temperatures; use limited owing to extremely high cost.
11. Phosphazene (P) PNF ® (Firestone) (P) Phosphonitrilic fluoroelastomer (C)	−85 to +350 (−65 to +177)	Similar to fluorosilicone in its fluid resistance and temperature range but tougher, making it suitable for dynamic use; as this is written, it is too new and too expensive to have gained wide acceptance as a seal material, but this may change in the next few years
12. Polyacrylate (P, C) ACM (A)	0 to +350 (−18 to +177)	Resists hydrocarbons, ozone, sunlight, but water resistance, mechanical properties, and recovery are poor; often used to seal automatic transmission and power-steering fluids; types of seal: gaskets, O rings
13. Polysulfide (P, C) Thiokol (P) EOT (A)	−75 to +225 (−60 to +107)	Resists many solvents that no other elastomer can handle; has good resistance to oxygen and ozone and good low-temperature flexibility; heat resistance, strength, and recovery poor; types of seal: cup and hat packing, diaphragms, gaskets, O rings
14. Polyurethane (P) 14a. Polyester polyurethane (C) AU (A) 14b. Polyether polyurethane (C) EU (A)	−65 to +200 (−54 to +93)	Very tough, high-tensile-strength elastomer with excellent resistance to abrasion and wear; good resistance to petroleum-type hydraulic fluids and other hydrocarbons and to ozone and aging; poor resistance to water, acids, high temperatures; poor recovery; types of seal: cup and hat packings, O rings, U cups, V seals; also used for backup rings and wipers
15. SBR (P, A) GR-S (P) Buna-S (P) styrene butadiene (C)	−70 to +225 (−57 to 107)	Most commonly used elastomer today because most automobile tires are made from it; very limited use in seals owing to its poor resistance to hydrocarbons, ozone, sunlight, and aging
16. Silicone (P) Silastic® (Dow Corning) (P) 16a. Phenyl vinyl methyl silicone (C) PVMQ (A)	−75/60 to 400/500 (−60/−50 to +205/260)	Has wide temperature range with good resistance to oxygen, ozone, high-temperature-air aging; excellent recovery, but is not very tough; has poor resistance to most fluids, and is quite permeable to gases; in seal use, is confined almost entirely to static applications; types of seal: diaphragms, gaskets, O rings, sealants
16b. Vinyl methyl silicone (C) VMQ (A)	−175 to +400 (−115 to +205)	

*P = popular term, T = trade name, C = Chemical name, A = designation per ASTM D1418.

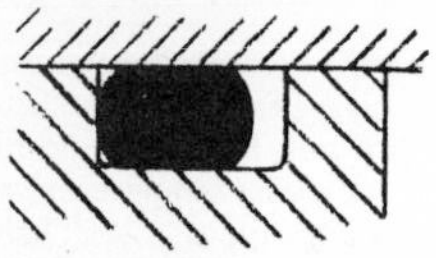

Figure 4-10 Assembled O ring showing squeezed section and extra void.

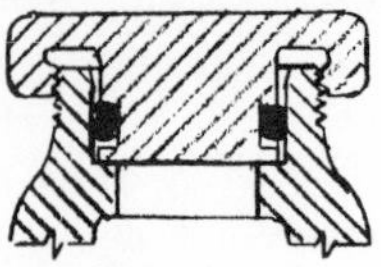

Figure 4-11 Static O-ring seal—male gland. *(Modified with permission from Parker O-Ring Handbook ORD5700, Figure A5-1, Parker O-Ring Division, Lexington, Kentucky.)*

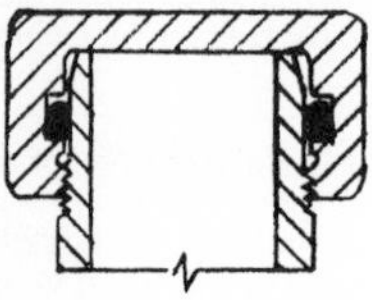

Figure 4-12 Static O-ring seal—female gland. *(Modified with permission from Parker O-Ring Handbook ORD5700, Figure A5-1, Parker O-Ring Division, Lexington, Kentucky.)*

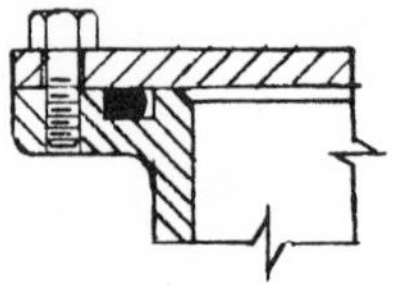

Figure 4-13 Static-face-type O-ring seal.

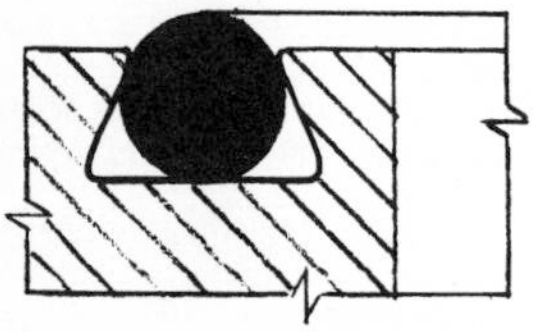

Figure 4-14 O ring in a dovetail groove. *(Modified with permission from Parker O-Ring Handbook ORD5700, Parker O-Ring Division, Lexington, Kentucky.)*

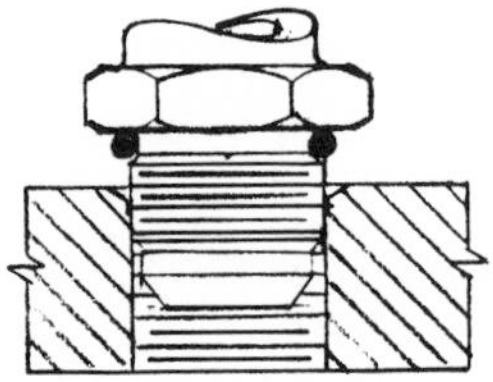

Figure 4-15 MS33656-tube-fitting end with O ring and MS33649 boss.

vulnerable to modification. If deviation from the standard is desired, it must be studied carefully to assure that a maximum-tolerance O ring will not overfill it and a minimum-tolerance ring will still be retained.

3. Tube fittings are sealed with a separate series of O-ring sizes established for the purpose. In this type of application, the gland is essentially triangular. There are two standards in the United States for the O-ring cavity, or "boss." The older is described in Military Standard MS33649 (Fig. 4-15). The new improved style is per MS16142 (Fig. 4-16).

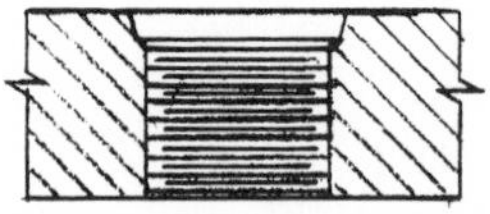

Figure 4-16 Improved tube-fitting boss, SAE J514 and MS16142.

Sealants and Tapes

Sealants are liquids or pastes that are applied between mating surfaces to prevent leakage. Some are formulated to harden after being applied to a joint so that they can contain high pressures. Others remain semiliquid for long periods of time so that joints can be opened easily after prolonged use. Sealants of this latter type are well known for threaded pipe joints. Tapes are also used for this and similar applications. In selecting a sealant, as for seals of other types, a material must be found that will function through the full anticipated temperature range and will not be adversely affected by the fluids and pressures that will be encountered.

RECIPROCATING SEALS

Introduction

The effectiveness of a reciprocating piston or rod seal made of an elastomer or a deformable thermoplastic material depends on the three variables of seal design and the effect of many factors on these variables. The primary factors influencing reciprocating seal design are the fluid to be sealed, the temperatures and pressures to be encountered, the length and speed of stroke, the surface finishes, the amount of clearance, the type of bearings, the space available for the seal, and the desired performance.

Seal Design

The three variables of seal design are the overall shape, the lip configuration, and the properties of the material.

Seal Shape

If permeability is disregarded, the only requirement for an effective seal is an unbroken line of contact between the sealing element and the mating surfaces. In a dynamic seal, this ideal situation can only be approached. Preferably, the seal surface rides on a thin film of the system fluid that provides lubrication and retards wear (Fig. 4-17).

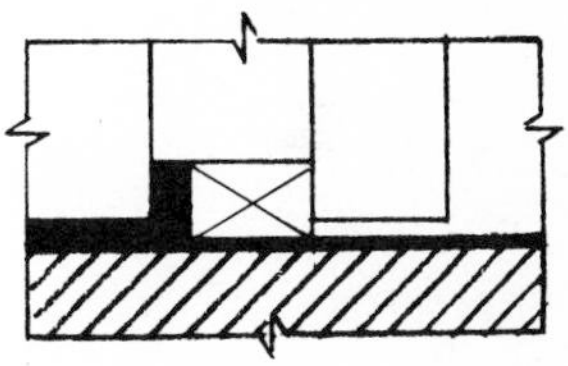

Figure 4-17 Seal riding on thin film of fluid.

Figure 4-18 shows a number of typical seal shapes ranging from pure lip seals on the left to pure compression or pure squeeze types on the right.

When there is little or no system pressure, and assuming that all the seal designs are made in the same material, the only force available to cause a pure lip-type seal to conform with the mating surface is the light load required to bend the lip. A pure compression type, however, generates a much higher wiping or sealing load because a relatively incompressible material must be deformed.

Examining these seal shapes, it will be seen that with low fluid pressure, a pure lip-type seal should have little friction and wear, but also it cannot be expected to seal as positively as the pure compression types. The pure compression types, however, will produce more friction and will wear more rapidly. Similarly, the intermediate shapes can be expected to produce intermediate results.

Lip Configuration

A refinement of the basic seal shape analysis takes into account the shape of the sealing lip. Figure 4-19 shows the five common lip shapes as they appear in the free state and as they appear in the installed state. Also shown is a force vector distribution curve associated with each lip shape. High unit loading improves sealability and film breaking. Therefore seals, especially squeeze seals, with the lip shape shown in No. 4, a back-beveled lip, would be extremely dry seals and would be excellent seals for wiping oil film. If lower friction (and lower sealability) is desired, a lip shape such as No. 2 would be selected because of the low force vectors associated with this particular shape.

Properties of Seal Material

Stability. Long strokes, too slow or too rapid speed, poor lubrication, eccentricity due to inaccurate machining or eccentric loads, and a number of other factors tend to cause a seal to twist in its gland, resulting in premature seal failure. The ability of a seal to resist this twisting action is called *stability*.

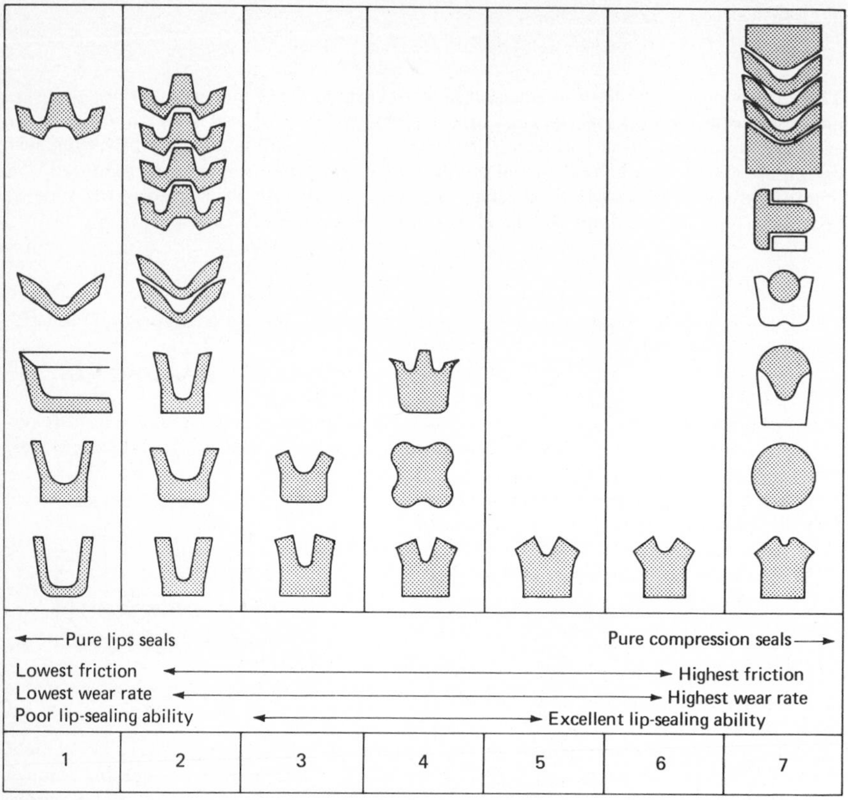

Figure 4-18 Basic shape chart. *(Reproduced by permission of Parker Packing Division, Salt Lake City.)*

Seal stability generally depends on two factors: the basic shape of the seal and the amount of gland fill (Fig. 4-20). A square seal is more stable than a round seal, and a rectangular seal is more stable than a square seal. Usually, the more completely a seal fills the gland, the greater will be its stability. For most seals, however, it is important that the gland not become overfilled since this will cause excessive friction, wear, and extrusion of the seal. When a seal becomes unstable, leakage usually results and the seal is often damaged as well.

In addition to the shape factors mentioned above, a stiffer, higher-modulus material will resist twisting much better than a softer, more flexible material.

Fluid Compatibility. When the basic seal shape has been determined and the lip shape has been established, the seal material must be selected. One of the first considerations should be fluid compatibility. A fluid is considered incompatible with a compound if it causes enough physical-property changes to reduce the sealing function or shorten the working life of the compound. Many fluids will cause seals to swell enough to produce excessive friction and wear. Other fluids may cause physical breakdown of the basic polymer used in the seal, whereas another class of fluids may harden the material so that it no longer has sufficient resilience to conform and establish a sealing line.

Memory and Resilience. The "memory" of a seal material is another important property. It is defined as the ability to return to the original shape on removal of a deforming force, and it is measured by use of the compression set test. Resilience is sim-

Contact type number	Free shape of contact	Low-pressure sealing ability	Ability to clean surface without trapping particles	Hydroplaning tendency	Deformed lip shape and force vector distribution
1		High	Low	Medium	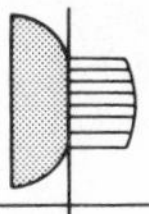
2		Low	Very high	High	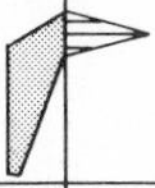
3		High	High	Medium	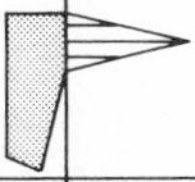
4		Very high	Low	Low	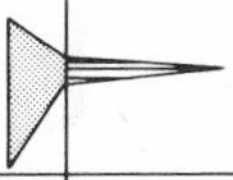
5		Medium/high	High	Medium/high	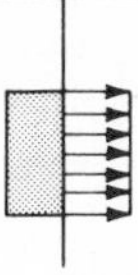

Notes:

(a) Highest unit loading will give highest sealability.
(b) Concentration of vector forces best for wiping.
(c) Lubrication will pass low unit loading patterns.

Figure 4-19 Lip shapes and unit loading chart. *(Reproduced by permission of Parker Packing Division, Salt Lake City.)*

ilar, but implies quick recovery, enabling the seal material to follow rapid variation in the surface passing under the sealing line. With rapid response, very little fluid will be lost on each cycle. The resilience of elastomers is determined with a Bashore resiliometer.

Temperature Range. Closely associated with fluid compatibility is the heat resistance of seal compounds. Heat resistance is a time-temperature function. The higher the temperature experienced by a seal, the shorter its life will be. Some problems stem from the fact that most seal materials expand about 10 times as much as do the metallic materials that contain them. The more serious problems, however, are related to resilience and hardness. High temperatures tend to reduce the resilience and the hardness of seal materials, though many of these materials will eventually harden if the high temperature persists for an extended time. The loss of memory at elevated temperatures, however, means that the seal will assume the shape of the gland. As the temperature falls, the deformed seal then shrinks more than its surroundings and pulls away from the mating surfaces, resulting in leakage.

Seal materials also lose resilience at low temperatures, though the resilience is regained on warming. Nevertheless, as the temperature drops below the point where resilience is lost, the seal shrinks away from the contacting surfaces and leaks until it is warmed again.

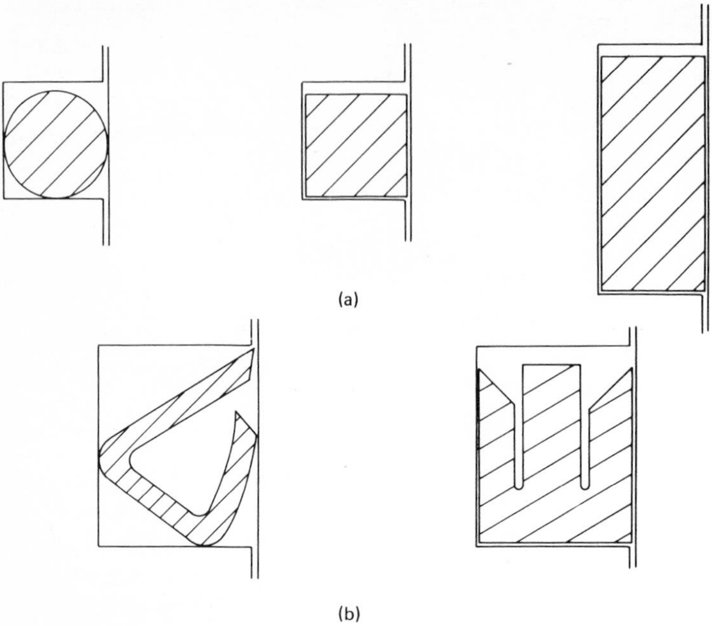

Figure 4-20 Stability guide: *(a)* basic shape; *(b)* gland fill. The square shape is better than the round and the rectangular is better than the square. The higher the gland fill, the more stable the seal. Overfill can cause high friction and heat, and subsequent seal damage. Stiffer seal materials also improve sealability *(Reproduced by permission of Parker Packing Division, Salt Lake City.)*

The temperatures at which these two effects become a problem vary with the seal material, and in many cases the fluid being sealed poses more restrictive limits on the temperature range than does the seal material.

Abrasion Resistance. The life of a seal can be directly related to its abrasion resistance. This property can be measured by various types of abrader wheels or pads. However, these tests generally give only relative abrasion resistance between polymers and do not give good results for predicting seal life in a particular application. Polyurethane and carboxylated nitrile are known for their excellent abrasion resistance, with polyurethanes being the leader.

Extrusion Resistance. Extrusion resistance is another property which must be considered in maximizing seal life. Extrusion resistance is a function of the seal compound and the size of the extrusion gap between the piston bore and the piston, or between the rod and the rod throat. It is also inversely proportional to temperature, pressure, and the frictional drag on the seal.

Extrusion of the seal is one primary cause of seal leakage. This type of leakage can be massive in that a seal can be progressively torn away until a major gap develops at the lip.

Extrusion can be controlled by following one or more of the directions listed below.

1. Use a backup ring.
2. Select an extrusion-resistant compound for the seal itself. (High-tensile-strength modulus and, sometimes, hardness will correlate with increased extrusion resistance.)
3. Reduce the clearance gap between the two dynamic surfaces.

4. Maintain concentricity between the cylinder and bore and between the rod and its housing. (This requires the use of bearings within the cylinder or nonmetallic wear bands on the piston or rod to avoid side loading the seals. Seals are not designed to carry side loads.)

Installation

The elongation of a seal compound is important if stretch-in applications are being considered. The forces required to install a seal should also be taken into account if large cross sections are being considered, even in rod applications, where the seal is folded, rather than stretched, in assembly. The 100 percent modulus of a seal compound is a good guide to ease of installation. Figures 4-21 and 4-22 are also helpful.

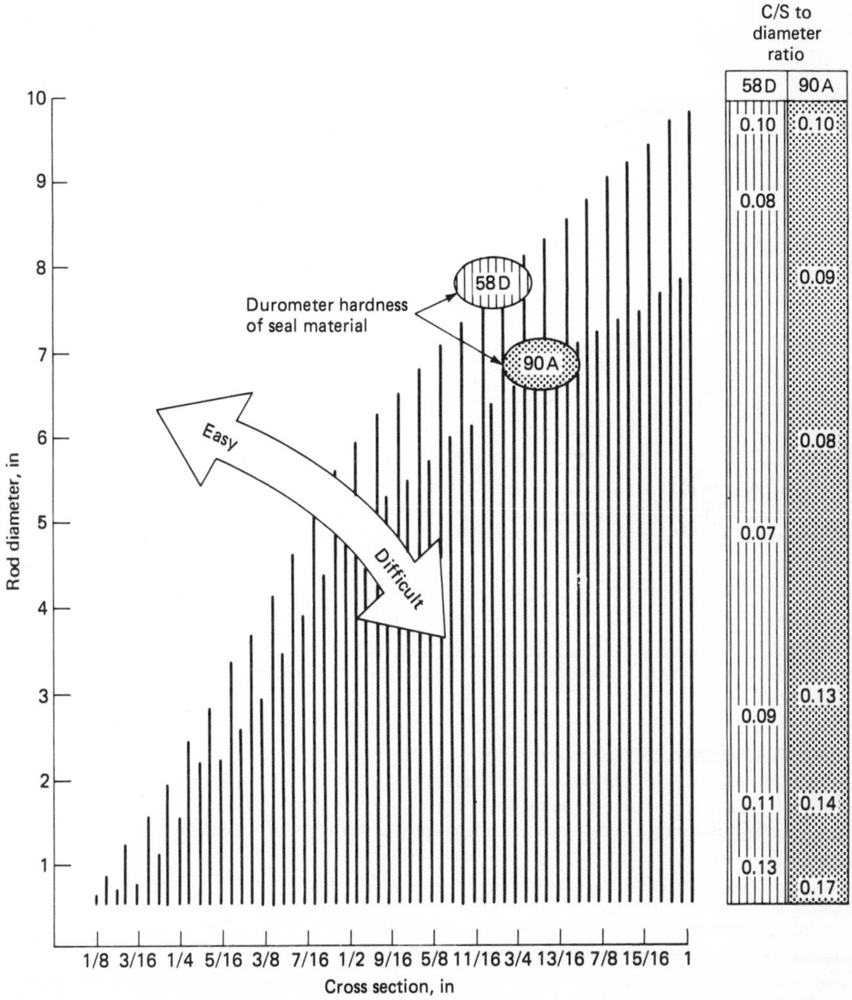

Figure 4-21 Rod-seal installation guide. For thermoplastic materials only: Rigid materials require split seals or separable gland. *(Reproduced by permission of Parker Packing Division, Salt Lake City.)*

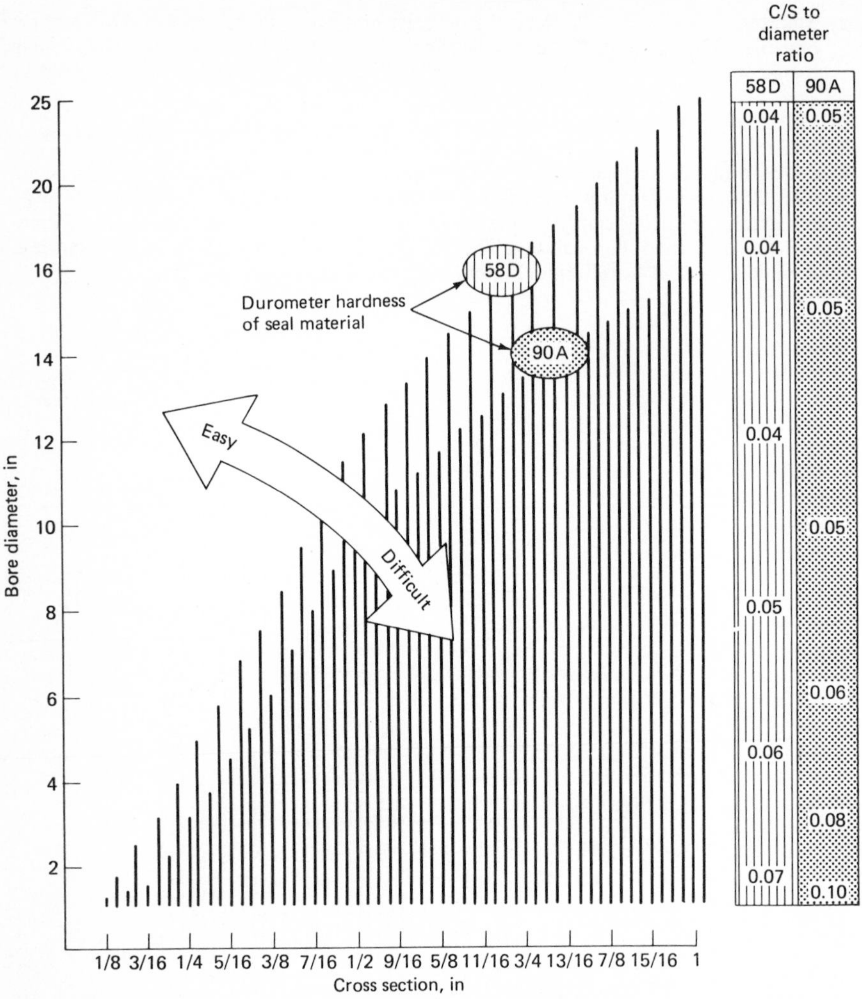

Figure 4-22 Piston-seal installation guide. For thermoplastic materials only: Rigid materials require split seals or split piston. *(Reproduced by permission from Parker Packing Division, Salt Lake City.)*

Backups (Auxiliary Devices)

The most common technique for preventing extrusion is to use auxiliary devices or backups. Backups are added to seals to improve performance. Not only will they prevent extrusion, but many times they will improve the stability of a seal package and its seal ability.

There are two primary types of backups in use, the positively actuated backup and the non-positively actuated backup (Fig. 4-23).

Non-positively actuated backups depend on axial forces which cause lateral deformation of the backup material to close the extrusion gap. Because lateral deformation of the material is required for this type of backup, the materials are limited to those having a fairly high degree of plastic or elastic deformation.

Positively actuated backups use radial forces to close the gap. Since this type of

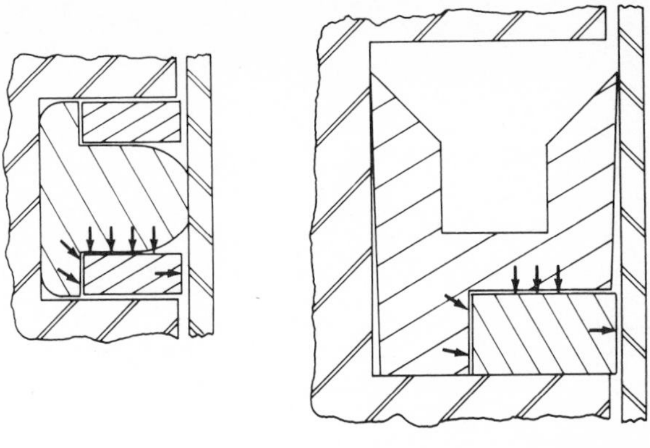

(a)

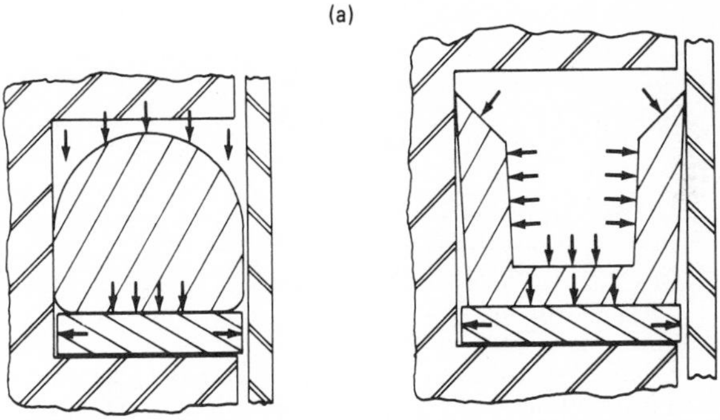

(b)

Figure 4-23 Extrusion resistance: (*a*) Positively actuated backups. (*b*) non-positively actuated backups. *(Reproduced by permission of Parker Packing Division, Salt Lake City.)*

backup does not require lateral deformation, a wide variety of hard, stiff materials, including metals, can be used. This type of backup is usually split to allow for radial movement.

Exclusion Devices (Wipers and Scrapers)

Although exclusion devices are not seals, they play an important part in extending seal life and minimizing leakage. Whenever a reciprocating rod is exposed to dust, ice, snow, mud, or other abrasive materials, it is important that a wiper be included in the system to clean the rod before it can carry these foreign materials in through the seal, abrading it and contaminating the system.

There are many types and styles of exclusion devices available, and one can be found that is appropriate for almost any environment.

Metal Considerations

Seal leakage can also be a function of face-roughness values. Tests and experience have shown that, for dynamic surfaces, surface roughness should be in the range of 10 to 20 rms. With rougher finishes, friction and wear increase, instability increases, and the potential for leakage increases. Tests and experience also show that finishes can become *too smooth*! Dynamic surfaces below 10 rms have also been shown to increase friction and wear and may shorten seal life (Fig. 4-24). It can be seen that the optimum dynamic

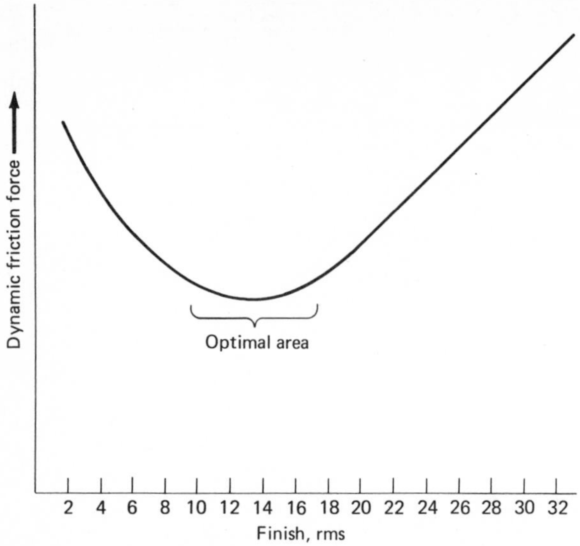

Figure 4-24 Friction-vs.-surface-roughness curve. *(Reproduced by permission of Parker Packing Division, Salt Lake City.)*

surface finish lies between 10 and 20 rms. (Static surface finish can range from 16 to 32 rms and still maintain a good seal line.)

It is important that all sharp corners be removed from the gland area or any area the seal passes during installation. If the seal is damaged by the metal prior to reaching its designed location, leakage can be an immediate result. Seal installation should be designed as carefully as the seal gland. Lead-in tapers should be used, avoiding threaded areas during installation. Other precautions like this will help assure reduced leakage and better sealing applications. Installation tools are also of great value. Gland dimensions should be adhered to carefully, as recommended by the seal supplier. If deviations must be used, the seal supplier should be consulted and the modification tested extensively prior to production.

Notes

1. When in doubt contact the seal supplier for guidance.
2. Test all new seal designs under conditions as close as possible to the conditions of use. (The vendor can provide guidance but cannot actually test all seals under all the conditions encountered in the field.)
3. If a premature failure occurs, save the failed seal and the mating parts to assist in a failure analysis.

Rigid Packings

Rigid packings, generally of cast iron or reinforced TFE (tetrafluoroethylene), are most commonly used on gas compressors. These range in form from the simple bevel or step-joint piston ring to rather complex spring-assisted segmental styles (Fig. 4-25). The need for "clear" (unlubricated) compressed gas or air systems has accentuated the use of the reinforced TFE style.

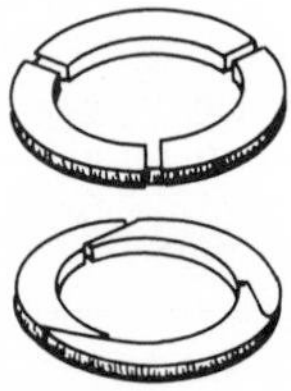

Figure 4-25 Segmental metal rings. *(From: W. Staniar, "Plant Engineering Handbook," 2d ed., Fig. 27-23d, p. 27-22. Copyright © 1959, McGraw-Hill Book Co., Inc. Reproduced by permission of the publisher.)*

ROTARY SEALS

Introduction

Rotary sealing devices may be subdivided into four categories:

1. Clearance seals (controlled clearance, labyrinth, wind-back)
2. Mechanical face seals (mechanical seals, face seals)
3. Compression packings
4. Lip seals (oil seals)

Although most types are commonly used alone, severe conditions often require two or more types in a single application.

Rotary packings and seals present a specific group of problems to the seal designer. Typically rubbing speeds are high, and the seal-rubbing interface is confined to a very limited area. The problem of dissipating the resultant heat differentiates rotary applications from reciprocating applications where the heat is incurred over a relatively large area of a rod or a cylinder bore.

Factors such as fluid, temperature, pressure, interface velocity, allowable leakage, necessary life, space, and cost will determine which type of rotary seal is best suited for any individual application. No attempt can be made in this limited discussion to provide all the information needed to make such a selection. Rather, the following section is intended to acquaint the reader with the basic types, principles, and terminology.

Clearance Seals

Clearance seals, as a group, are intended to reduce leakage while avoiding actual contact between the moving parts.

Labyrinth Seals

A simple labyrinth is shown in Fig. 4-26. Numerous variations of this type are feasible, involving stepped diameters, barrier fluid areas, drains, etc. Since clearances must be kept small and contact generally avoided, substantial design problems result when wobble, end play, and thermal expansion are taken into account.

Labyrinths have, essentially, no pressure or speed limitations because of their noncontacting nature. Similarly, no temperature limits are imposed other than those of the materials of construction. Since these materials are not required to be deformable, and no rubbing contact is involved, material choices for high temperatures normally present no problem.

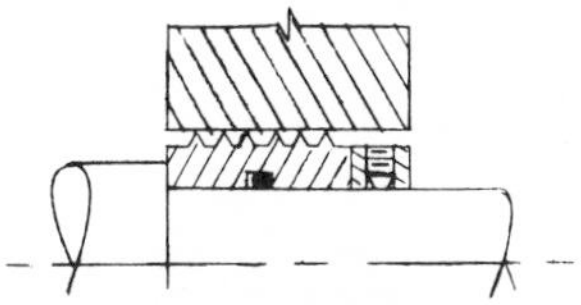

Figure 4-26 Labyrinth seal.

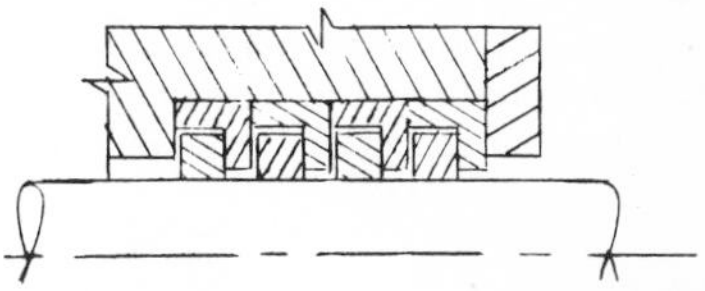

Figure 4-27 Multiple-unit floating bushing seal.

Bushing Seals

Bushings, most simply, are sleeves with a small clearance around a shaft, leakage being limited by laminar- and turbulent-flow considerations in the small annulus thus created. A noncontact situation is desired, and the need for small clearance, when viewed in the light of shaft runout, angularity, vibration, thermal expansion, etc., generally results in a floating-sleeve design. Figure 4-27 represents a multiple-unit floating-sleeve or bushing design, each bushing being free to center itself diametrally with reference to the shaft. As in other rotary-seal types, designs will often allow for drains, buffer or barrier fluids, etc.

Wind-Back Seals

The tendency of fluid in close proximity to a moving surface (as in a small annulus) to move with the surface is made use of in a variety of self-pumping or *wind-back* applications. Figure 4-28 shows a sample representational wind-back, where the tendency of the fluid to flow to the left is balanced by the pumping force of the helical land.

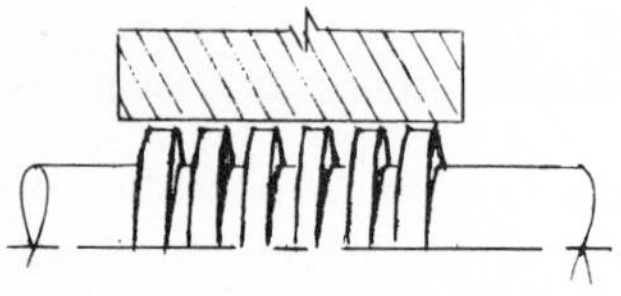

Figure 4-28 Wind-back seal.

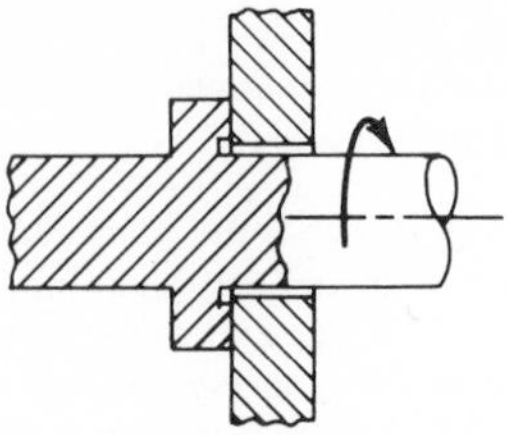

Figure 4-29 Simplified face seal. *(Reproduced by permission from "Sealing Is Our Business," Fig. 5-15. Copyright © January 1975, Parker Seal Group, Lexington, Kentucky.)*

Mechanical Face Seals

A face seal in its simplest conceptual form would be as illustrated in Fig. 4-29. As may be observed, the sealing interface is on a plane at right angles to the rotational axis. Thus a certain amount of shaft runout could be accommodated without distorting the relationship between the two parts at the sealing interface.

This conceptual design, however, could not compensate for wear at the seal interface, nor could it accommodate axial motion.

In Fig. 4-30 the design is modified to provide replaceable rubbing portions, and to provide axial spring force to accommodate wear and small amounts of axial motion. A replaceable seat *A* has been inserted in the housing, a replaceable head *B* replaces the rotating shoulder, a spring *C* bearing against a collar *D* keeps the seal interface in contact, while secondary seals *E* and *F* seal the leak paths thus created. Secondary seal *E* is essentially static, whereas *F* will experience motion as end play and wear are accommodated. Figure 4-30 is thus a simple mechanical face seal. Seal types discussed here may be considered to be modifications or expansions of this basic configuration.

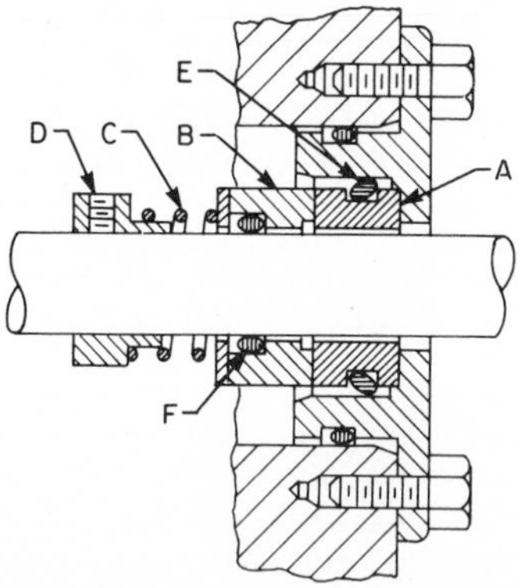

Figure 4-30 Typical mechanical face seal. *(Reproduced by permission from "Sealing Is Our Business," Fig. 5-18. Copyright © January 1975, Parker Seal Group, Lexington, Kentucky.)*

Balance Seals

Figure 4-30 shows the contained fluid pressure against the left radial surface of the head *B* as it tends to force the head to the right, thus increasing the interface pressure established by the spring in proportion to the fluid pressure contained. This is shown in simplified form in Fig. 4-31. If the design is modified by use of a stepped shaft, as in Fig. 4-32, one will observe that the annular area on the left end of the head is reduced, proportionately reducing the fluid-pressure induced component of the interface pressure. In industry terminology, Fig. 4-32 is referred to as "balanced" and Fig. 4-31 as "unbalanced." Balanced seals are normally specified for higher pressures or for the sealing of "light" liquids.

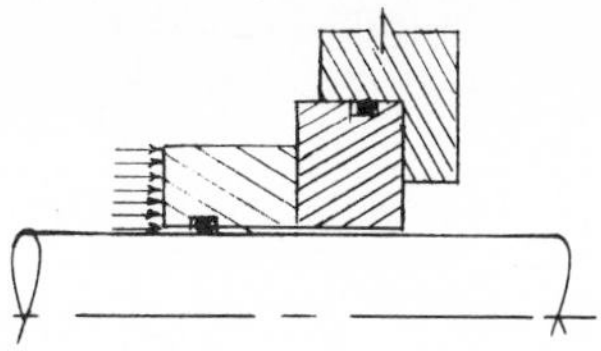

Figure 4-31 Unbalanced face-seal head.

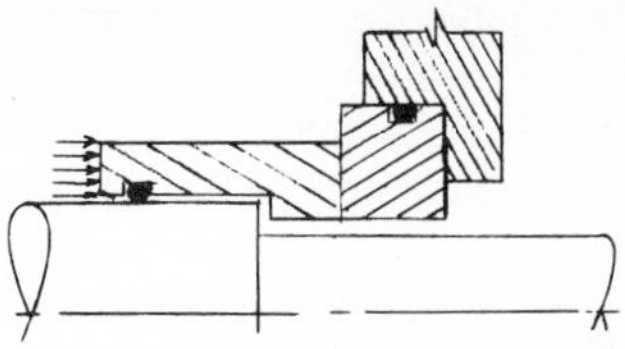

Figure 4-32 Balanced face-seal head.

Rotating Seat. Seals may be designed and installed with the seat stationary (as in the preceding examples) or rotating (Fig. 4-33).

Outside Installation. A further variation is shown later in Fig. 4-35. Figures 4-30 and 4-33 show the seal head on the fluid side of the seat (an inside installation), and Fig. 4-34 shows an outside installation.

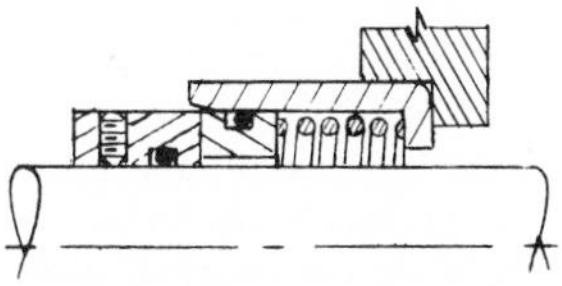

Figure 4-33 Mechanical seal with rotating seat.

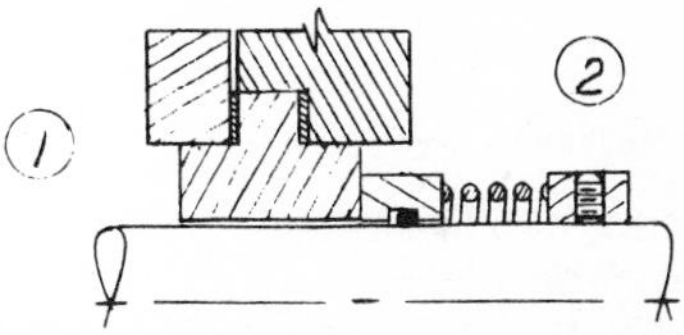

Figure 4-34 Seal head on outside of seat: (1) fluid; (2) atmosphere.

Clamped Seat. Figure 4-34 shows a further variation with the seat clamped, as opposed to the relatively free O-ring mounting shown in the preceding sketches. The secondary seal at the seat is a flat gasket.

Double Seals

It is often necessary or desirable to circulate a secondary fluid in a chamber formed by a double (Fig. 4-35) or tandem (Fig. 4-36) seal arrangement. This secondary fluid could

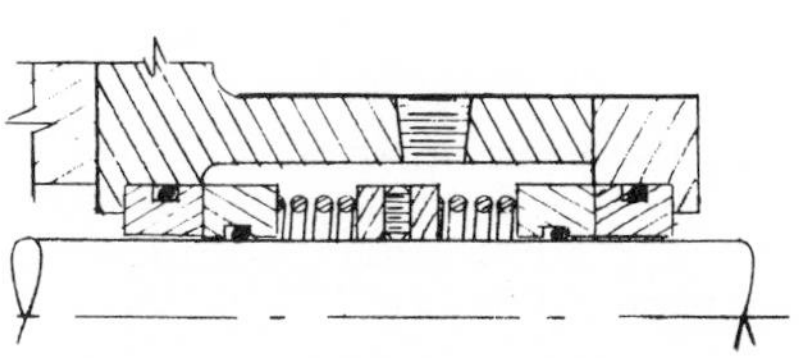

Figure 4-35 Double mechanical seal.

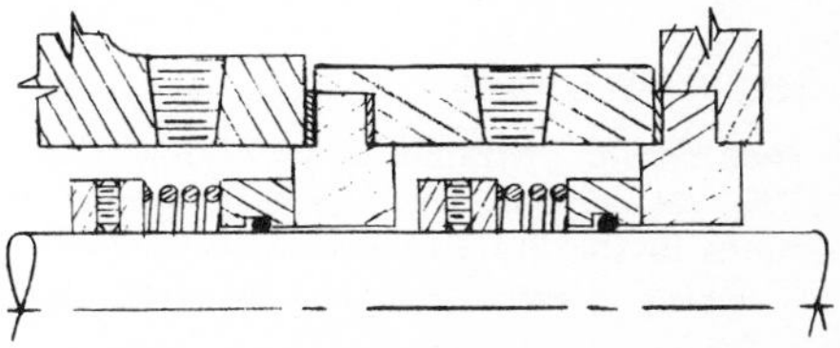

Figure 4-36 Tandem-type mechanical seal.

be at a higher or lower pressure than the primary fluid. Such an arrangement may be used for a wide variety of purposes, ranging from safety (in the case of a toxic or flammable primary fluid) to simple redundancy. Problems with fluids containing abrasives, or those which tend to crystallize on cooling or on contact with the atmosphere, are commonly handled in this fashion. This method may also be utilized to provide interface lubrication when sealing gases.

An alternative form of rotary seal (a throttle bushing, rings of soft packing, an oil seal, etc.) may be employed either inboard or outboard of the primary mechanical face seal for similar purposes.

For simplicity all of the examples shown depict a single spring. However, a number of small, evenly spaced springs are often employed to provide a more uniform load around the circumference of the head.

Seals with Bellows. The motion of the secondary seal associated with the head tends, in most designs, to "fret" or wear the adjacent sealed surface. In order to avoid this, recourse is often had to bellows designs, of elastomer or TFE, in which the motion in question is absorbed by deformation of the bellows (Fig. 4-37).

A further modification (Fig. 4-38) uses a welded metal bellows which performs a dual function as a motion-absorbing seal and as the actuating spring.

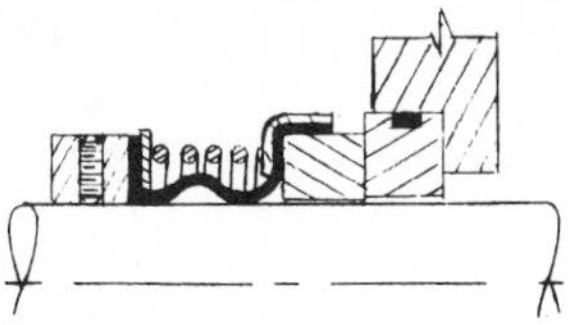

Figure 4-37 Mechanical seal with elastomeric or TFE bellows.

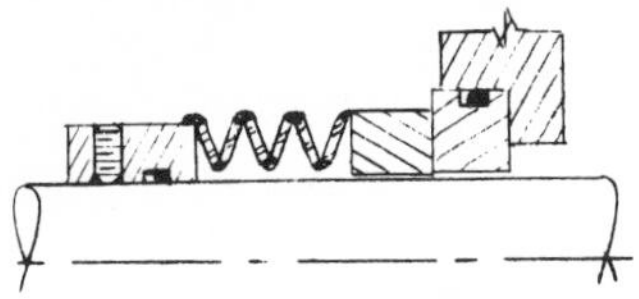

Figure 4-38 Mechanical seal with metal bellows.

The Secondary Seal.

In the preceding paragraphs, O rings, gaskets, and bellows have been mentioned as secondary seals. Actually, a wide variety of secondary seals are used in mechanical face-seal assemblies. The secondary seal at the seat is generally static, leading most often to use of a simple O-ring gasket. The secondary seal associated with the head, however, is subjected to various degrees of small-magnitude motion as the head moves in response to axial shaft motion and face wear. Thus, a wider variety of wedges, X rings, V rings, coated O rings, and other, more sophisticated, seal forms will be noted.

Compression Packings

Compression packings, also called soft packings, braided packings, and rope packings, are among the oldest packings known to industry. Such packings characteristically depend on axial pressure provided mechanically by a gland follower to radially distort the rings into intimate contact with the sealed surfaces. A broad range of materials and combinations of materials and impregnants or saturants, ranging from flax, cotton, and asbestos to the latest exotics, are used in compression-packing manufacture.

Types of Construction

Cross Braid. Strands of yarn cross the surface diagonally, interlocking to form a dense, yet flexible, structure that cannot unravel in service. Lubrication of individual strands in the braiding process provides a more uniform distribution of antifriction materials and yarn density for increased life of the packing.

Square Braid. Strands of yarn cross over and under other strands in the same running direction and produce square cross sections. Other names for this construction are square, plaited, or flax braid. This type of construction is normally specified for high-speed rotary service at low pressure.

Braid over Braid. This type is formed by one or more layers of braided yarns covering a core of braided, twisted, or homogeneous materials and producing an initially round construction. It may be calendered square or rectangular into a dense, highly lubricated packing for low-speed and high-pressure applications.

Twisted. Strands of yarn are twisted in the same running direction to the required round size. This type is recommended only for low-speed, low-pressure utility use when the packing space is small or when the available cross section is larger than the required packing space. The individual strands are easily removed to fit emergency requirements.

Laminated. Coils, spirals, or rings are cut from molded slabs of fabric and rubber materials. This construction is normally used for the liquid end of reciprocating rams or pump pistons.

Wrapped. Rubber-impregnated fabric materials are rolled or calendered into a square or rectangular cross section alone or over a homogeneous core for high resilience. Typical use is in applications with high lateral movement in the equipment.

Plastic. Lubricated plastic packings contain oil, fiber materials, graphite, grease, or mica. They are usually square and within a cotton jacket for handling or within an asbestos jacket to prevent extrusion. They are normally used to seal gases or in applications in which the packing must supply part or all of the lubrication.

Metallic. Metal packings of lead, copper, aluminum foil, or ribbon are spirally wrapped, folded, or twisted into a square or rectangular cross section. Normal use is in high-pressure, high-temperature service and corrosive fluids.

Cutting and Installing Packing Rings

Pumps and Agitators.

1. Remove old packing from the stuffing box. Clean the box and shaft thoroughly and examine the shaft for wear or scoring. Replace the shaft or sleeve if wear is excessive. Check the bearing by moving the shaft up and down.
2. Use the right size of coil packing to be sure the packing will fit and can compensate for shaft wear, if any. To determine the correct packing size, measure the diameter of the shaft and then the inside diameter of the stuffing box. Subtract the shaft diameter from the inside-diameter measurement and divide by 2 for the required size.
3. Cut the packing into individual rings. *Never wind packing in a continuous length into a stuffing box.* For pumps and agitators cut rings with a butt (square) joint (Fig. 4-39). The best way to cut packing rings is to cut them on a mandrel of the same diameter as the shaft in the stuffing-box area. Hold the coiled packing tightly and firmly on the mandrel but do not stretch it excessively. Cut the ring and try it in the stuffing box to make certain that it fills the packing space properly, with no gap in the joint at the outside diameter of the ring.

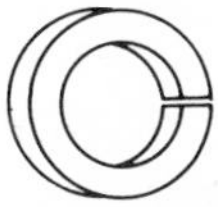

Figure 4-39 Butt joint. *(Reproduced by permission of Parker Packing Division, Carson City, Nevada.)*

4. Install one ring at a time. Make sure it is clean and has not picked up any dirt in handling before installing it. If clean oil is available, lubricate each ring thoroughly; the shaft and the inside of the stuffing box would also benefit from lubrication. Joints of successive rings should be staggered and be kept at least 90° apart. Each ring should be firmly seated with a tamping tool. When enough rings have been individually seated to allow the nose of the follower to reach them, the individual tamping should be supplemented by the follower. Never depend entirely on the follower to seat a set of rings properly; this practice will jam the last rings installed but leave the front rings loose in the box. The result is excessive and rapid wear of rear rings, erratic packing performance, or sometimes, twisting and tearing of the front rings which are loose in the stuffing box.
5. After last ring is installed, take up bolts finger tight. Start the pump, and take up bolts until leakage is decreased to no more than 10 drops per minute. Stopping leakage entirely at this point will cause the packing to burn up.
6. Allow the packing to leak freely when starting up a newly packed pump. It will take about one working day to break in a set of packing to a point where the leakage is stabilized at a uniform acceptable rate.
7. If at all possible, provide, through a lantern ring (Fig. 4-40), means of lubricating the

shaft and packing by supplying grease, oil, water, or the liquid handled in the pump. Make sure the lantern ring, as installed, is slightly behind the lubricant fitting so it will move under the fitting as follower pressure is applied.

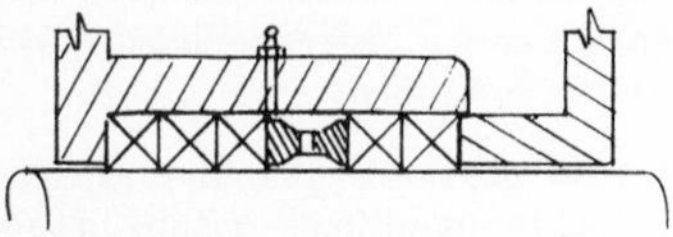

Figure 4-40 Packing set with lantern ring in a stuffing box. *(Modified from Parker Compression Packing Catalog, No. PPD3901-A, by permission of Parker Packing Division, Carson City, Nev.)*

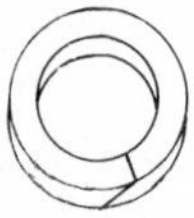

Figure 4-41 Skive joint. *(Reproduced by permission of Parker Packing Division, Carson City, Nev.)*

Valves and Expansion Joints.

1. Carefully perform all the operations outlined under steps 1, 2, 3, and 4 as given for pumps and agitators. Rings used on valves and expansion joints should be cut with a skive joint (45°). See Fig. 4-41.
2. Bring the follower down on the packing to the point where heavy resistance to wrenching is felt. During this time turn the valve stem back and forth to determine ease of turning. Do not wrench to the point at which the stem will not turn.
3. After the valve has been on the line a day or so, even if no leakage exists, the follower should be tightened slightly. If it is leaking at all, tighten the follower until there is no leakage.

Lip-Type Rotary Seals

Lip-type rotary seals are often called *oil seals*, though they are by no means restricted to sealing oils, or *rotary-shaft seals*. Elements common to this class of seals are a flexible sealing lip that rides on the surface of the rotating shaft and a formed metal cup that holds the sealing element and is pressed into the seal housing. Generally there is also a garter spring that helps maintain contact of the lip (Fig. 4-42). The sealing element may be elastomeric, leather, or plastic.

The primary use of oil seals is in low-pressure applications, generally below 8 lb/in^2. Surface speeds may be quite high, ranging up to 4000 ft/min. Occasionally there are special designs that can tolerate much higher pressures, but at high pressure the surface speed must be slow.

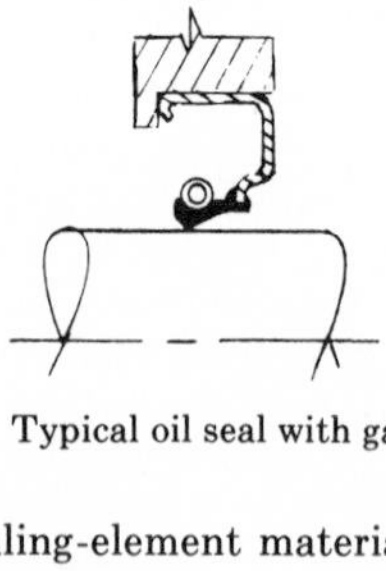

Figure 4-42 Typical oil seal with garter spring.

Oil seals are well suited to low-pressure applications because the narrow contact band of the sealing lip keeps friction and its associated heat low while permitting ready dissipation of the heat that is generated. The sealing-element material may be varied to resist special fluids and unusual temperature extremes.

Dirt-Lip Design

One of the most common variations on the basic design is a second lip pointing outward for use where abrasive dust could damage the primary sealing lip or where external fluid splash could enter and contaminate the internal medium (Fig. 4-43). The secondary or *dirt lip* is not spring-loaded, but relies on the resilience of the material to maintain contact with the shaft.

Nose-Seal Gasket

Another modification includes a nose-seal gasket to prevent static leakage around the drawn cup (Fig. 4-44).

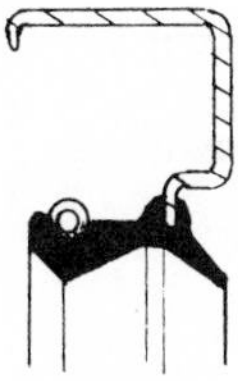

Figure 4-43 Oil seal with dirt lip. *(Reproduced by permission of Parker O-Seal Division, Culver City, California.)*

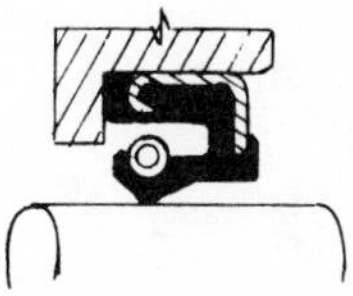

Figure 4-44 Oil seal with nose-seal gasket. *(Modified with permission from "Sealing Is Our Business," Copyright © 1975, Parker Seal Group, Lexington, Kentucky.)*

Cup Designs

Besides variations in the sealing element, the metal cup may be made in any number of styles. For instance, a pry-out flange may be incorporated to make the seal accessible from either the outside (Fig. 4-45) or from the fluid side. Often an inner case, which protects the spring and the sealing lip and makes the assembly more rigid, is provided (Fig. 4-46).

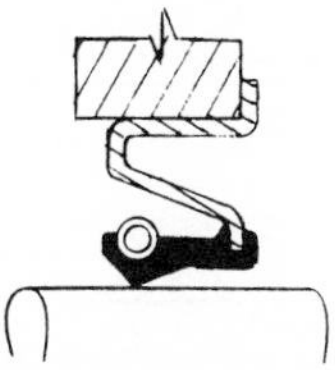

Figure 4-45 Oil seal with pry-out flange. *(Modified with permission from "Sealing Is Our Business," Copyright © 1975, Parker Seal Group, Lexington, Kentucky.)*

Figure 4-46 Oil seal with inner case. *(Modified with permission from "Sealing Is Our Business," Copyright © 1975, Parker Seal Group, Lexington, Kentucky.)*

FLEXURAL SEALING DEVICES

Diaphragms, bellows, and expansion joints are variations of a class of sealing devices that are attached to one or more components of a system, absorbing the relative motion between such components or absorbing fluid displacement by deformation of the sealing device itself.

The simplest of these devices is the flat diaphragm (Fig. 4-47). These are generally cut from reinforced or unreinforced elastomeric sheet, although metal, leather, and plastics find use in some situations. The range of motion which can be accommodated by the

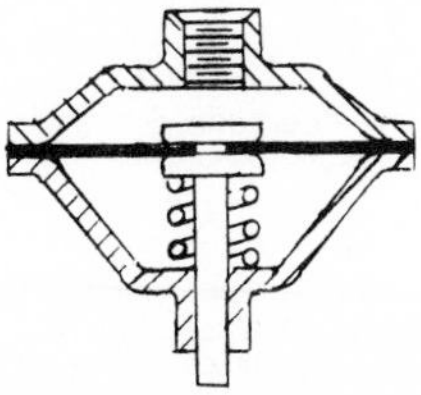

Figure 4-47 Flat diaphragm. *(From sketch by J. B. Scannell.)*

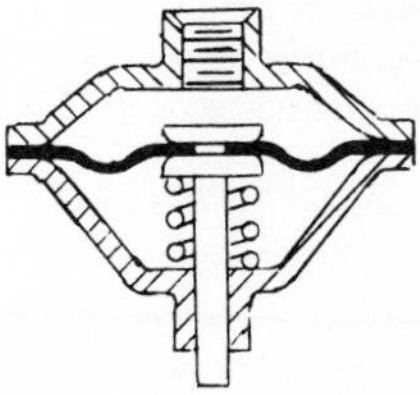

Figure 4-48 Diaphragm with convolution. *(From sketch by J. B. Scannell.)*

flat diaphragm is limited, and a molded or formed convolution is often added to accommodate additional travel (Fig. 4-48).

The *tubular diaphragm*, or bellows, accommodates large axial motion for a given diameter (Fig. 4-49), as does the rolling diaphragm (Fig. 4-50).

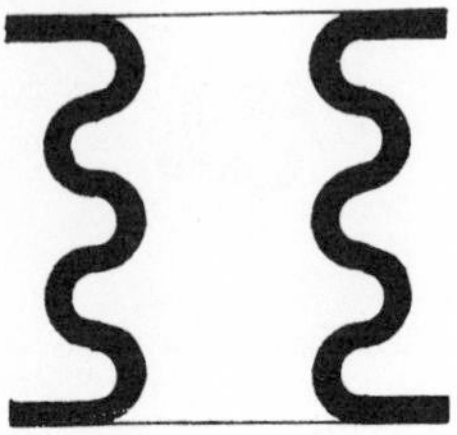

Figure 4-49 Bellows. *(From sketch by J. B. Scannell.)*

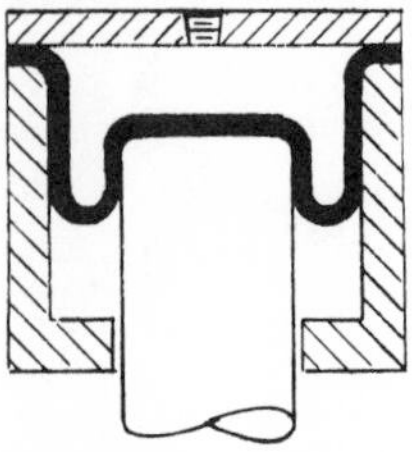

Figure 4-50 Rolling diaphragm. *(From sketch by J. B. Scannell.)*

A heavy-duty form of bellows is the *expansion joint* (Fig. 4-51), most commonly fabricated of rubber with fabric and metal reinforcement, but also of TFE or with a TFE liner. These devices absorb vibration and relative movement between sections of piping or duct systems.

Although a degree of standardization exists among these products, most are produced to order for specific applications, and competent suppliers should be consulted for detailed information.

INFORMATION FOR SEAL DESIGN

When assistance is needed in seal design, the consultant will need specific details about the application. The following comments may be helpful in collecting the necessary pieces of information.

Data Needed for All Seal Designs

1. **Temperature**
   a. Maximum operating temperature
   b. Time at maximum temperature. The time at maximum temperature is particularly important when the maximum is too high for normal seal materials. If the time is sufficiently short and will be experienced once or only a few times, an inexpensive material may be suitable since degradation due to high temperature, though generally irreversible, is time-dependent.
   c. Minimum operating temperature. If the minimum operating temperature is very low, it will have a great influence on the selection of a seal material. In general,

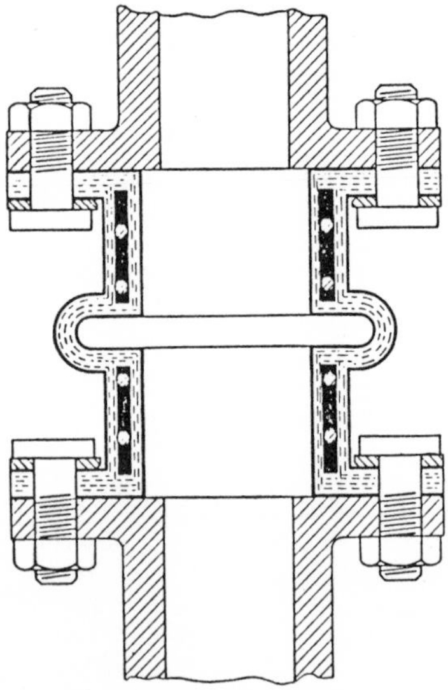

Figure 4-51 Expansion joint. *(From: W. Staniar, "Plant Engineering Handbook," 2d ed., Fig. 27-43, p. 27-52. Copyright © 1959, McGraw-Hill Book Co., Inc. Reproduced by permission of the publisher.)*

however, low-temperature effects are reversible. If a seal does not operate at low temperature, it will generally function normally when warmed up. A seal is considered to be operating, in this sense, if it must contain a fluid, even though the device in which it is installed may not operate at the low temperature.

   d. Normal operating temperature

2. **Pressure**
   a. Maximum pressure
   b. Operating pressure
   c. Minimum pressure (or vacuum level for vacuum seals).
3. **Medium** (to be sealed) Length of time of seal contact with the medium.
4. **Configuration** The seal consultant needs to know the arrangement, materials, and sizes of surfaces in the vicinity of the seal. In some cases, a few dimensions will suffice. Often it is necessary to provide dimensional drawings or sketches of the pertinent parts. Surface roughness values of mating parts are also important to seal design.

Additional Data for Reciprocating Seals

1. Length of stroke
2. Stroke rate
3. Mode of operation (i.e. actuator, operated by the fluid pressure, or pump, generating fluid pressure).
4. Duty cycle
5. Eccentricity (There should be bearings to minimize the effects of side loading. How much eccentricity do they permit?)

section 8

Materials Handling

Prepared by

K. W. Tunnell Company, Inc.
King of Prussia, Pennsylvania

chapter 8-1

Planning Materials Handling

MATERIALS-HANDLING DEFINITIONS

Materials-handling technology includes hardware and systems which can be categorized as follows:

- Containerization
- Fixed-path handling
- Mobile handling
- Warehousing

Containerization. This classification covers the broad spectrum of confinement methods that are used for storage through all phases of the manufacturing, or process, cycle. The materials-handling engineer employs the unit-size principle to optimize the quantity, size, and weight of the load to be handled or moved and is able to specify the best container after considering material and other production-system parameters. Pallets, skids, tote boxes, wire-mesh containers, covering a wide range of sizes and materials, are included within the category.

Fixed-Path Handling. This classification applies to movement and storage of unit loads of material with an intermittent or a continuous flow over a fixed path from one point to another. Fixed-path-handling equipment is secured to, and is considered part of, the facility. Once installed it is more difficult to modify or replace; therefore a considerable amount of planning and interfacing with other functions has to be considered. This equipment, if installed above the floor surface, can effectively utilize what would otherwise be dead space. Chutes, conveyors, elevators, bridge cranes, palletizing equipment, and robots are examples of fixed-path-handling equipment.

Mobile Handling. This classification inclues all handling systems that move material over various paths within a manufacturing or processing cycle. Handling equipment allows a considerable degree of flexibility in moving material in an intermittent flow but requires special facility requirements, such as aisle sizes, clearances, door openings, and running and maneuvering surfaces. Equipment in this category consumes more energy per unit load moved than most other systems and generally requires trained personnel for operation. Equipment in this category ranges from simple two-wheeled hand trucks to specially designed over-the-road vehicles and also includes skid trucks, floor trucks, powered walkie lift trucks, powered lift trucks, and mobile hydraulic cranes.

Warehousing. This classification of materials handling considers the systems, equipment practices, and requirements dedicated to the following operations within the manufacturing, processing, or distribution cycle:

- Receiving
- Storage of raw, in-process, and finished materials
- Movement in and out of storage
- Order picking and accumulation
- Containerization for shipping
- Loading and shipping

This area of materials handling involves a wider range of planning and analysis. Consider some of the following factors:

1. Location of activity
2. Sizing and physical characteristics relating to product size, type, and volume
3. Number of stockkeeping units
4. Storage equipment
5. Selection of materials-movement methods
6. Packaging methods for shipping

The range of solutions covers the full spectrum, from a single warehouse that shares movement equipment with other parts of the operation to a self-contained, specially equipped, fully automated warehouse.

INTRODUCTION

Materials handling deals with the movement of materials from receiving, through fabrication, to the shipment of the finished product. More broadly, handling and distribution are considered to be one overall system. This viewpoint gives consideration to all handling activities involved in movement of materials from all sources of supply through the various plant and central warehouse operations to the customer distribution network.

The activities related to the flow of the materials are either viewed as separate activities or treated as one element in an integrated handling system. Not everyone agrees that materials-handling activities should be viewed as an overall system. Progressive plant engineering personnel, however, recognize that materials handling represents the efficient integration of workers, materials-control systems, and equipment, as well as the movement of materials. Materials-handling applications must take into account operating costs and time-phased material flow.

Inefficient materials handling and storage increase product cost, delay product delivery, and consume excessive square feet of plant and warehouse space. Studies have indicated that actual materials-handling cost runs between 20 and 50 percent of the total product cost *even though it does not add any value to the finished product.* In addition, from 80 to 95 percent of the total overall time devoted to processing a customer order from fabrication to shipment is devoted to materials handling and storage. Product manufacturing time, therefore, is only a small percentage of the overall process time.

A properly designed and integrated materials-handling system provides tremendous cost-saving opportunities and customer-service improvement potential. The correct selection of a handling method can reduce handling costs per unit by as much as 200 percent. Improvement in storage, such as high-rise storage applications, can reduce unit storage space cost by 20 to 40 percent. Work-in-process inventory can be reduced 30 to 50 percent through compressed cycle times. The reduced cycle times will also result in shorter customer delivery cycles.

The significant impact of materials-handling costs on total product cost has resulted in a great deal of attention and substantial resources being directed toward discovering more efficient methods to reduce handling costs. This effort is expected to receive even greater concentration in the future. The following trends are beginning:

- Most manufacturing managers now recognize that materials handling is a prime area of cost-reduction opportunities.
- Most companies will either establish a full department or assign an individual to the responsibilities of analyzing and solving materials-handling problems.
- Materials handling and storage will increasingly be treated as an integral part of the manufacturing and processing cycle.
- Techniques which analyze problems and evaluate alternatives for materials flow and storage requirements in quantitative terms will increasingly replace intuitive solutions.
- Computer-based technology will employ more powerful and sophisticated techniques such as queueing theory, simulation facilities, and flow-planning techniques to consider and select optimum solutions from a wide range of variables and materials-handling options.
- Handling systems will become more automated by employing computer-controlled systems, robot loading and unloading, driverless vehicles, and automatic storage and retrieval systems. These automation principles will be applied in receiving, manufacturing operations, warehousing, and shipping.

It should be recognized that materials handling is an extremely broad-based subject which more often deals with the application of equipment and mechanical devices than fundamental engineering principles or basic physical laws. At least in part, it requires the application of subjective and experienced judgment and has even been described (with some justification) as more of an art than a science. Based on this fact and the trends in materials handling already discussed, this chapter outlines the methodology for solving materials-handling problems as well as the classification of hardware.

However, it is suggested that plant engineers involved continually with materials-handling projects should be familiar with additional sources of current information not only from reference books but also from specialized trade periodicals, professional societies, and trade associations. (See the reference section at the end of this chapter.)

SOLVING MATERIALS-HANDLING PROBLEMS

Materials-handling activities and installations vary in complexity from operation to operation. The plant engineer can solve most materials-handling problems by keeping two points in mind: one is to thoroughly understand the materials-handling principles; the other is to recognize that a materials-handling system is composed of a series of interrelated handling activities. The plant engineer should therefore apply materials-handling principles to improve each separate materials-handling activity and then interrelate the handling activities by applying flow-planning principles.

Principles of Materials Handling

The collective experience and knowledge of many materials-handling experts has been organized into a framework of generalized principles. The principles are basic and can be used universally. They include:

1. Integrate as many handling activities, such as receiving, storage, production, and inspection, as is practical into a coordinated system.
2. Arrange operation sequence and equipment layout so as to optimize materials flow.
3. Simplify handling by reducing, eliminating, or combining unnecessary movement and/or equipment.
4. Use gravity to move material wherever practical.
5. Make optimum use of the building cube.
6. Increase the quantity, size, and weight of the load handled.
7. Use mechanized or automated handling equipment whenever it can be economically or safely justified.
8. Select handling equipment on the basis of lowest overall cost when considering the material to be handled, the move to be made, and the methods to be utilized.
9. Standardize methods as well as types and sizes of handling equipment.
10. Use methods and equipment that perform a variety of tasks and applications.
11. Plan preventive maintenance, and schedule regular repairs on all handling equipment.
12. Determine the effectiveness of handling performance in terms of expense per unit handled.
13. Move materials in as direct a path as possible, minimizing backtracking.
14. Deliver materials directly to work areas whenever practical, and plan the minimum of material in the area.
15. Move the greatest weight or bulk the shortest distance.
16. Provide alternative plans in case of a breakdown.

Steps in Solving Handling Problems

The general methods that are used for solving other operational problems are applicable in the materials-handling area. The factors that must be considered relate to how to most efficiently move certain volumes and types of materials by a particular method. The steps involved in systematically solving these problems consist of:

- Problem identification
- Problem analysis and quantification
- Selection and evaluation of alternatives
- Project justification

Problem Identification

Identification of materials-handling problems includes determining the impact of interfacing activities such as production control, manufacturing, vendors, shipping, and receiving. The buildup of material in front of a machine or a truck dock may not be a problem of too little storage but rather one of lot sizing or inefficient truck-loading systems. Most importantly, a costly handling route between two distant machines may not be caused by the handling mechanism but by the location of the equipment itself.

Problem Analysis and Quantification

Qualitative and quantitative answers are obtained through use of industrial engineering techniques such as flow diagrams, flowcharts, from-to charts, and activity-relationship charts. For the detailed application of these manual techniques see Refs. 1 to 4.

Computer-aided techniques are useful when large amounts of data are involved.[5]

Selection and Evaluation of Alternatives

In the selection of alternatives three general types of criteria are involved.

Movement. Movement involves the study of routes in terms of the combination of handling equipment and containers jointly rather than on an individual product basis. Under this criterion, distances from and to locations, outside travel, and frequencies would be minimized.

Criteria Which Cannot be Directly Costed. These involve criteria such as:

1. **Performance** The potentials for relocation of equipment at future time periods, as related to the handling equipment, and for material design changes, as they relate to the containers themselves.
2. **Delicacy** The nature of the part and its requirements for special handling, dunnage, or containers—particularly to avoid damage.
3. **Interfaces** Production control and manufacturing departments, vendors, shipping, receiving, and intersite movement requirements.
4. **Uniformity** The need to standardize or at least unitize containers to provide uniform handling characteristics. This provides for the use of idealized container sizing and packing techniques, including proper dunnage for irregular loads. On the other hand, it requires the special handling and designs for special irregular-sized parts.

Cost-Effectiveness. This involves the analysis in concept of all standard operating costs, equipment life, maintenance and spare usage for equipment, and intermediate storage.

Generally, the analysis will be dominated by the cost-effectiveness of the alternatives involving cost components as follows.

- **Capital Investment Costs.** One-time charges incurred at the time of equipment procurement that include:
  1. Equipment cost, including freight
  2. Installation costs
  3. Special maintenance requirements
  4. Special power and fuel facilities
  5. Rearrangement and alteration of facilities to accommodate equipment
  6. Engineering support
  7. Supplies
- **Fixed Costs.** Determined or assigned to a system, a piece of equipment, or an activity on a time-period basis; include:
  1. Depreciation
  2. Taxes
  3. Insurance
  4. Supervision
- **Variable Costs.** Can be considered the cost of performing an operation or activity. In the case of equipment, it is the cost of using the equipment. The following items are included within this component.
  1. Equipment-operating personnel or personnel manually performing a materials-handling task

TABLE 1-1 Annual Operating Cost

Cost component	Present manual method, $	Proposed method, $		
		Conveyor	Fork lift, electric	Fork lift, propane
1. Capital equip. investment				
Equipment cost		6,000	25,000	27,000
Freight		500	800	800
Installation		8,000	—	—
Fuel-power facilities		—	4,000	—
Alterations to facility		10,000	—	—
Engineering support		1,000	—	—
Supplies		—	—	—
Total capital investment		25,500	29,800	27,800
2. Fixed cost				
Depreciation		5,100	6,560	6,160
Taxes		750	3,500	4,000
Insurance		200	1,000	1,000
Supervision		—	1,400	1,400
Total fixed cost		6,050	12,460	12,560
3. Variable cost				
Operators-loaders	(5)* 50,000	(2)* 20,000	(1)* 12,000	(1)* 12,000
Power-fuel	—	1,200	2,300	800
Lubrication	—	20	150	150
Maintenance	—	—	1,500	2,000
Total variable cost	50,000	21,220	15,950	14,950
4. Indirect cost				
Space occupied	—	1,500	—	—
Effect on taxes	—	1,000	—	—
Changes in prod. rate	—	—	—	—
Downtime	—	200	—	—
Total indirect	—	2,700	—	—
Total operating cost (2 + 3 + 4)	50,000	29,970	28,410	27,510
Annual cost savings		20,030	21,590	22,490

*Number of units employed.

2. Power and fuel costs
3. Lubricants
4. Maintenance labor supplies

- **Indirect Costs.** Affected in other areas of company operation as a result of changing a method, adding new equipment, or changing the materials-handling system, and may consist of:
  1. Space occupied
  2. Effect on taxes
  3. Values of repair parts
  4. Changes in production rate and quality of product
  5. Downtime

A summary of such an analysis is contained in Table 1-1.

JUSTIFICATION OF MATERIALS-HANDLING PROJECTS

The most common methods of determining the profitability of materials-handling investment are payoff period, return on investment (ROI), and discounted cash flow.

The *payoff-period* method indicates the amount of time that new equipment or a system will take to produce the savings to recover the capital investment. The investment is divided by the annual savings to give the time (in years) needed to break even. In Table 1-1:

$$\text{Payoff period} = \frac{\text{total capital investment}}{\text{annual cost savings}}$$

This method is a good risk indicator and measure which can be useful to indicate the projects that would be worth considering for closer study, but the actual profitability of new equipment depends on how much useful life is left after the payoff period. Some caution is therefore advised if the payoff period is to be the sole determinant for equipment justification, because cheaper equipment having a low useful life will always appear to be the best investment opportunity.

Simple ROI is another gross indicator that can be used to set priorities for capital investments. Here again, the effect of useful equipment life is not considered, so this method should not be used for determining the profitability of the proposed equipment:

$$\text{ROI} = \frac{\text{annual cost savings}}{\text{total capital investment}}$$

The ROI method that considers the effect of useful equipment life is

$$\text{ROI} = \frac{\text{annual savings} - \text{capital investment/useful equipment life}}{\text{capital investment}}$$

The discounted-cash-flow method of determining ROI indicates in a more realistic manner the equipment cost and return on investment by considering:

1. Savings and cost over equipment life period.
2. Net cash flow of the savings and depreciation.
3. Present worth of each year's cash flow. A factor is used to reduce the cash flow for each year to the amount of cash that would be required today to earn a desired rate of interest.
4. The effect of taxes on the rate of return.

The ROI is calculated as follows (Table 1-2):

1. Determine the cost savings for each year of the equipment useful life.

TABLE 1-2 Example of Calculating ROI* by Cash-Flow Analysis

Factors	Amounts
Total capital investment from annual operating cost	$30,800
Equipment life	5 years
Depreciation straight line	5 years, 20%/year
Savings per year	$22,490

Cash flow						Trial 1 @ 45%		Trial 2 @ 50%	
Yr	Cost savings	Taxes	Savings after taxes	Depreciation	Net cash flow	Factor	Present worth	Factor	Present worth
1	22,490	11,245	11,245	6,160	17,405	0.690	12,009.45	0.667	11,609.14
2	22,490	11,245	11,245	6,160	17,405	0.476	8,284.78	0.444	7,727.82
3	22,490	11,245	11,245	6,160	17,405	0.328	5,708.84	0.296	5,151.88
4	22,490	11,245	11,245	6,160	17,405	0.226	3,933.53	0.198	3,446.19
5	22,490	11,245	11,245	6,160	17,405	0.156	2,715.18	0.132	2,297.46
					78,715		32,651.78		30,232.49

*Interpolating

$$\frac{32{,}651.78 - 30{,}800}{32{,}651.78 - 30{,}232.49} \times 5 = \frac{1{,}851.78}{2{,}419.29} \times 5 = 3.83$$

Add 3.83% + 45% = 48.83% Return on Investment

TABLE 1-3 Present-Worth Values

	Interest, percent											
Years	6	8	10	12	15	20	25	30	35	40	45	50
1	0.943	0.926	0.909	0.893	0.870	0.833	0.800	0.769	0.741	0.714	0.690	0.667
2	0.890	0.857	0.826	0.797	0.756	0.694	0.640	0.592	0.549	0.510	0.476	0.444
3	0.840	0.794	0.751	0.712	0.658	0.579	0.512	0.455	0.406	0.364	0.328	0.296
4	0.792	0.735	0.683	0.636	0.572	0.482	0.410	0.350	0.301	0.260	0.226	0.198
5	0.747	0.681	0.621	0.568	0.497	0.402	0.328	0.269	0.223	0.186	0.156	0.132
6	0.705	0.630	0.564	0.507	0.432	0.335	0.262	0.207	0.165	0.133	0.108	0.088
7	0.665	0.583	0.513	0.452	0.376	0.279	0.210	0.159	0.122	0.095	0.074	0.058
8	0.627	0.540	0.466	0.404	0.327	0.323	0.168	0.123	0.091	0.068	0.051	0.039
9	0.592	0.500	0.424	0.361	0.284	0.194	0.134	0.094	0.067	0.048	0.035	0.026
10	0.558	0.463	0.386	0.322	0.247	0.162	0.107	0.072	0.050	0.035	0.024	0.017
11	0.527	0.429	0.350	0.288	0.215	0.135	0.086	0.056	0.037	0.025	0.017	0.012
12	0.497	0.397	0.319	0.257	0.187	0.112	0.069	0.043	0.027	0.018	0.012	0.008
13	0.469	0.368	0.290	0.229	0.162	0.094	0.055	0.033	0.020	0.013	0.008	0.005
14	0.442	0.340	0.263	0.205	0.141	0.078	0.044	0.025	0.015	0.009	0.006	0.003
15	0.417	0.315	0.239	0.183	0.123	0.065	0.035	0.020	0.011	0.006	0.004	0.002
16	0.394	0.292	0.218	0.163	0.107	0.054	0.028	0.015	0.008	0.005	0.003	0.002
17	0.371	0.270	0.198	0.146	0.093	0.045	0.022	0.012	0.006	0.003	0.002	0.001
18	0.350	0.250	0.180	0.130	0.081	0.038	0.018	0.009	0.004	0.002	0.001	0.001
19	0.330	0.232	0.164	0.116	0.070	0.031	0.014	0.007	0.003	0.002	0.001	0.000
20	0.312	0.214	0.149	0.104	0.061	0.026	0.012	0.005	0.002	0.001	0.001	0.000
	1.030	1.039	1.049	1.059	1.073	1.097	1.120	1.143	1.166	1.189	1.211	1.233

2. Deduct the estimated percentage for taxes for each year of equipment life.
3. Add depreciation for each year of the depreciation period.
4. Determine net cash flow, which is the algebraic total of items 1, 2, and 3 above.
5. Consult present-worth value table (Table 1-3) and select an interest value for the first trial.
6. Multiply the net cash flow by the factor selected in step 5.
7. If the present-worth cash flow is higher than the capital investment, select the present-worth factor for the higher percentage; if lower, select present-worth factor for the lower percentage.
8. Continue the discounted-cash-flow trials until the total discounted cash flow equals the capital investment cost. Interpolation will generally be necessary to determine the exact percent of ROI.
9. The trial present-worth calculations that equal the net cash flow total are those that are used to determine the ROI.

REFERENCES AND BIBLIOGRAPHY

1. Sims, E. Ralph, Jr.: *Planning and Managing Material Flow,* Industrial Education Institute, Boston, 1968.
2. Apple, James M.: *Material Handling Systems Design,* Ronald, New York, 1971.
3. Muther, Richard, and Knut Haganas: *Systematic Handling Analysis,* Management & Industrial Research Publications, 1969.
4. Muther, Richard: *Systematic Layout Planning,* Industrial Education Institute, Boston, 1961.
5. Tompkins, J. A.: "Computer-Aided Plant Layout," *Modern Material Handling,* 7 part series, May–September 1978.
6. Merkle, W.: "Dock Planning Guide," *Material Handling Engineering,* August 1980.
7. Bolz, Harold A., et al (ed.): *Materials Handling Handbook,* Wiley-Interscience, New York, 1958.

chapter 8-2

Containerization

INTRODUCTION

One of the basic principles of materials handling is that materials should be converted wherever possible to unit loads to avoid manual handling. A unit load is defined as a standard container package containing one or more items that can be handled in a standard way. The *unit-load principle* suggests that the larger the load to be handled or moved, the lower the overall handling cost. To meet this objective, materials-handling systems must be designed to handle the materials-handling volume within the constraints imposed by load size as well as the material properties involved in the production or process cycle. The decisions regarding size, shape, and configuration of the unit load should also take into account compatibility.

Some *guidelines for the specification of unit load sizes* leading to the design of containerization methods and hardware to transport and store materials include:

1. Use the same pallet or container throughout the system, or at least standardize on a limited number of containers wherever possible.
2. Plan to use raw material or parts directly out of the original container.
3. Use stackable containers to permit stacking without racks.
4. Consider collapsible containers to save space and freight costs, if they are to be used also as returnable shipping containers.
5. Use nesting.
6. Be sure that the size selected fits efficiently into standard trailers and/or railcars if containers are to be used for shipping.
7. Design or select containers suitable for mechanical handling.
8. Plan containers to accommodate a wide range of products and parts.

9. Design containers to fit into building geometry.
10. Design containers that do not require special orientation to accomplish movement.
11. Use the lightest-weight material possible.
12. Consider the use of expendable materials
13. Use containers through which contents can be identified.
14. Keep the design simple and inexpensive.

CONTAINERIZATION HARDWARE

Containerization hardware can generally be grouped into five main categories:

- Pallets
- Containers
- Tote boxes and bins
- Dunnage
- Outer securement

Standard Pallets

Pallets are used mainly as supports, carrier surfaces, or storing structures for unit loads.

The most commonly used material is *wood,* and pallets are available in a number of different hardwood and softwood varieties (Table 2-1). The type of wood, like any other material that is specified, should depend on load capacity, load requirements, durability, and the handling and storage environment. In general, softwood pallets are lighter and suitable for shipping pallets, while hardwood pallets are stronger, have a longer life, and are less susceptible to the wear and tear associated with interplant movement. Local, indigenous woods should be specified wherever possible to minimize costs.

Principles of Pallet Construction

Nomenclature, Design, Style, and Size. The principal pallet parts and the most commonly used construction features are indicated in Fig. 2-1. By convention, the length of the pallet is the first-stated dimension, the dimensions are always stated in inches, and the width is the dimension that is parallel to the top of the deck boards.

Types. Wood pallets fall into three general groups:

1. **Expendable** (one-way pallets) Cost is the major factor and the design and construction must meet the requirements for this purpose.
2. **Special-Purpose** Design and construction must meet the special requirements for the product or material to be moved or stored.
3. **General-Purpose** Uses standard design and features which enable the pallet to be used in a wide range of applications and also to be replaced and exchanged easily.

Pallet Configuration. This is specified by a combination of design, style, and construction features. The National Wooden Pallet and Container Association has established the descriptions of each parameter.

Typical pallet configurations are shown in Fig. 2-2. Pallets are available in a wide range of sizes; however, the most popular size is the 48 × 40 in pallet which accounts for over 27 percent of all pallets produced.

There is movement within some industries, particularly the food and grocery industries, to standardize pallet sizes. It has been determined that size standardization, among other obvious benefits, could also increase the use of pallet pools or exchange programs, which would have cost advantages throughout the distribution cycle.

TABLE 2-1 Strength Properties of Commercial Woods Employed for Pallets (Figures shown are for 12 percent moisture content.)

Species	Static bending fiber stress at proportional limit, lb/in²*	Compression perpendicular to grain, lb/in²*	General properties
Group IV			
Oak, red	8,400	1,260	Heaviest hardwood species; greatest nail-holding power and beam strength; best shock-resisting capacity; greatest tendency to split at nails; difficult to dry
Oak, white	7,900	1,410	
Maple, sugar	9,500	1,810	
Beech	8,700	1,250	
Birch	10,100	1,250	
Hickory, true	10,900	2,310	
Ash, white	8,900	1,510	
Pecan	9,100	2,040	
Group III			
Ash, black	7,200	940	More inclined to split when nailed; greater nail-holding and shock-resisting power, beam strength, and easier to dry than group IV
Gum, black	7,300	1,150	
Maple, silver	6,200	910	
Gum, red	8,100	860	
Sycamore	6,400	860	
Tupelo	7,200	1,070	
Elm, white	7,600	850	
Group II			
Douglas fir	7,400	950	Relatively free from splitting when nailed; moderate nail-holding power and shock-resisting capacity; lightweight, easy to work, holds shape well, and easy to dry
Hemlock (W)	6,800	680	
Larch (Tamarack)	8,000	990	
N.C. pine	7,700	1,000	
Southern yellow pine (longleaf)	9,300	1,190	
Group I			
Aspen	5,600	460	
Cottonwood	5,700	470	
Redwood	6,900	860	
Spruce	6,700	710	
Sugar pine	5,700	590	
Ponderosa pine	6,300	740	
White fir	6,300	610	
White pine (N)	6,300	550	
White pine (W)	6,200	540	
Yellow poplar	6,100	580	

*Multiply by 6900 for newtons per square meter.

Design. The most common designs of wood pallets are:

1. **Two-Way Pallets** Permit the entry of forklift or hand pallet trucks from two sides only and in opposite directions
2. **Four-way pallets** Permit entry on all four sides
   a. **Notched Stringer Design** Has four-way entry *only* with forklift trucks, and two-way entry with hand pallet trucks
   b. **Block Design** Has four-way entry with both forklift and hand pallet trucks.

Style. There are two styles of wood pallets, and they are (Fig. 2-2):

1. **Single-Face Pallet** Has only one deck as the top surface;
2. **Double-Face Pallet** Has both top and bottom decks and comes in two different designs, viz.:
   a. **Reversible** Has identical top and bottom decks, and goods may be stacked on either deck

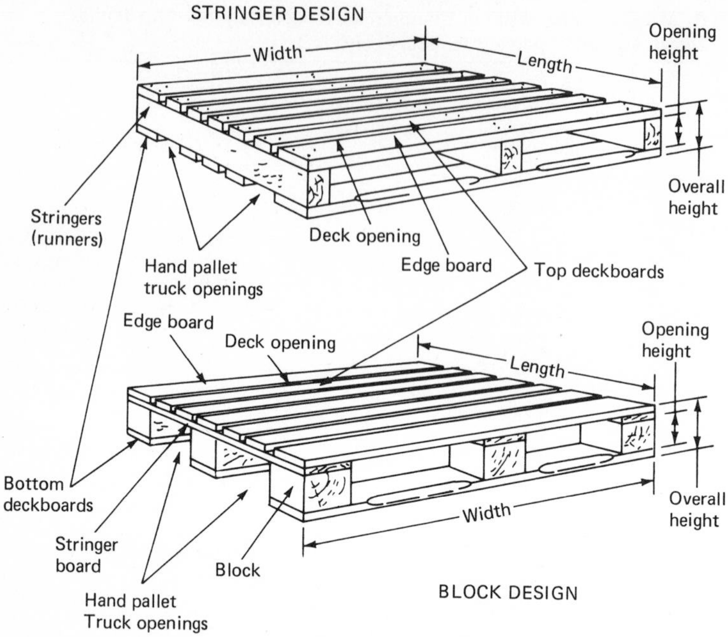

Figure 2-1 Principal parts of wooden pallets. *(National Wooden Pallet and Container Association.)*

b. Nonreversible Top and bottom decks have different configurations, and substitute goods may be stacked only on the top deck

Constructions. Wood pallet constructions are as follows:

1. **Flush Stringer** A pallet in which the outside stringers or blocks are flush with the ends of the deckboards
2. **Single Wing** A pallet in which the outside stringers are set inboard of the top deck, while the stringers are flush with the ends of the bottom deckboards
3. **Double Wing.** A pallet in which the outside stringers are set inboard of both top and bottom deckboards to accommodate bar slings or other devices for handling pallets

Maintenance and Repair

Procedures should also be established within the system to identify worn pallets that require repair or disposal. To accomplish this effectively, the acquisition date should be marked on the pallet and older pallets should be inspected periodically to detect wear. The following are guidelines for repair operations:

1. Never repair a pallet a second time.
2. Never repair more than three deckboards or one stringer on a given pallet. If the *average* replacement is more than 1½ boards per pallet, repair is uneconomical.
3. Productivity should average 100 repaired pallets per worker per 8-h shift for those on the repair line—forklift support and supervisory personnel excluded.
4. Cost of repair should not exceed half the price of a new, similar pallet.

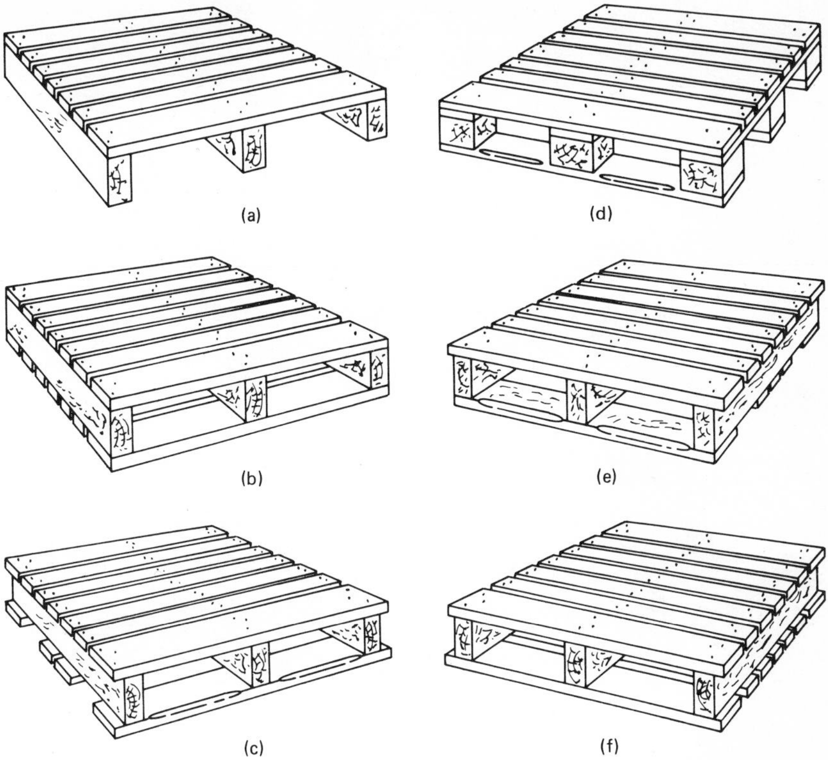

Figure 2-2 Typical pallet configurations. (*a*) Single-face; (*b*) double-face, reversible; (*c*) double-wing, double-face, nonreversible; (*d*) double-face, nonreversible; (*e*) single-wing, double-face, nonreversible; (*f*) double-wing, double-face, reversible. *(National Wooden Pallet Association.)*

Pallets for Use with Forklifts

Expendable Wood Pallets. These pallets are used to support a unit load for one-way and one-time use. Pallets of this type must be specified with the capacity to carry unit load but do not require the durability of reusable types. The single-face style (Fig. 2-2) is primarily used for this purpose. Plywood deck surfaces are frequently used for expendable pallets.

Metal Pallets. These pallets can be made of corrugated steel, expanded metal, steel wire, aluminum, and combinations of metal and wood. Metal pallets are more expensive than wood pallets and are used mainly for movement of materials inside the plant where additional strength and life is required.

Corrugated-Metal Pallet Bases. These pallets (Fig. 2-3) are often integrated into the design of corrugated-steel containers with a number of other features. This permits wide versatility in parts handling and storage in the plant. The style variations available are similar to those of their wood pallet counterparts to permit movement in both two- and four-way entry bases by forklift trucks and pallet hand trucks.

All-Steel, Single-Face Pallets. Supported on three runners, this type is designed to handle heavy loads and containers. Recessed side channels bound in flanges can be incorporated in the design to permit safe movement by hand. Their double-faced, reversible design eliminates sharp edges, and thus prevents damage to bagged materials.

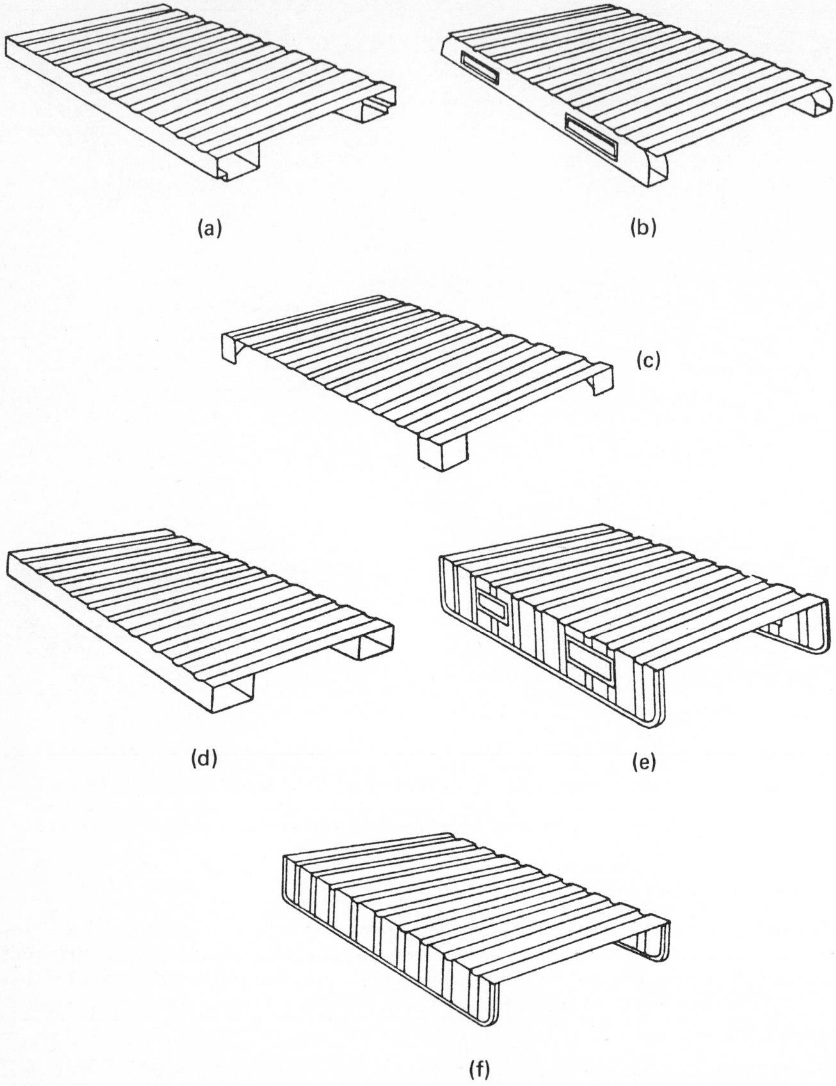

Figure 2-3 Corrugated-steel pallet container bases. (*a*) Two-way entry pallet style, (*b*) four-way entry pallet style, (*c*) box or angle-style pallet, (*d*) two-way box runner, (*e*) skid or pallet truck, (*f*) standard pallet with runners.

One-Piece, Formed-Metal Pallets. These pallets have a built-in nesting feature that permits a number of empty pallets to be stored conveniently. They are useful where pallet storage space is scarce.

Wire-Mesh Pallets. These pallets use galvanized or painted steel or aluminum deck sections with formed, corrugated support structures and are used where durability and light weight are required. The wire-mesh pallet, like the corrugated-metal pallets, are often incorporated as bases in wire-mesh containers.

Cardboard Pallets. These pallets (Fig. 2-4) are useful for light unit loads that are less than 1500 lb (700 kg) and for stacked loads that are less than 1000 lb (450 kg) per pallet leg. Because of their low cost, they are ideal as expendable pallets and can be expanded on a modular basis to become shipping containers.

Plastic Pallets. These pallets are more expensive than wood ones and in some cases are more expensive than metal ones. The main uses of plastic pallets are in the food or pharmaceutical industries where a high standard of cleanliness is required.

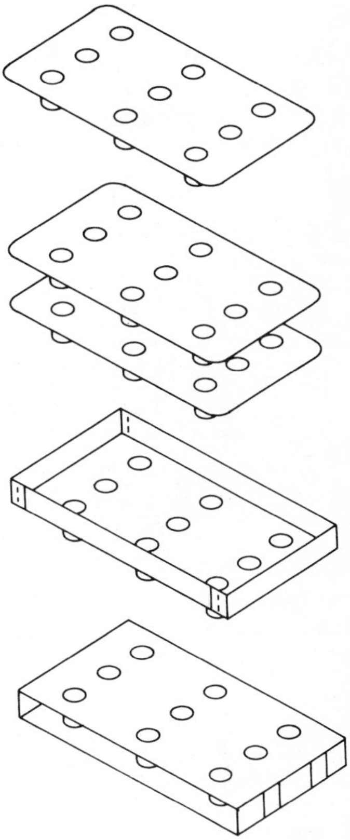

Figure 2-4 Cardboard expendable pallet. *(Menosha Corporation.)*

Other Types of Pallets

In addition to using forklift trucks or hand-operated equipment for major movements in the plant, there are other handling requirements for movement of materials within and between manufacturing operations. Generally there is very little standardization in this area, since the carrier surfaces have to be specified to be compatible with equipment or product.

Slip-Sheet Systems. Such systems enable a unit load to be handled and moved without being supported on a pallet type of platform. Slip sheets (Fig. 2-5) are made of heavy corrugated paperboard, plastic, or kraft fiber composition and function as the base surface for the unit load. Special equipment or *push-pull* forklift truck attachments are required to move and handle loads unitized by this method. The cost benefits of slip-sheet systems are obvious: in addition to lower initial cost, storage space requirements are $\frac{1}{100}$ of the cube required for empty pallets and shipping costs are less than for comparable loads using wood pallets.

Slave Pallets. These are used for assembling unit loads before transfer to other containers, for moving odd-shaped loads on conveyors, for serving as accumulating and transfer platforms for automated computer-controlled storage and retrieval systems, and for supporting unit-load containers that are not designed for use on conveyors.

Slave pallets are normally plywood-sheet surfaces, but if interim storage is required in racks where the pallet-edge surfaces are supported by shelf angles, pallet specifications become more critical from the standpoint of supporting loads and safety. As a general rule, to achieve maximum strength and stiffness, face grain should be across supports. Design criteria include pallet size, total uniform load, permissible deflection, clear span, and uniform load in pounds per square foot (newtons per square meter). Selection of the proper grade and thickness of plywood can then be determined. Information regarding recommended maximum uniform loads and deflections is available from the American Plywood Association.

Air Pallets. This type uses a bed of air to support a unit load and enables large loads to be moved and maneuvered. Air pallets or air-film equipment can be used to convey

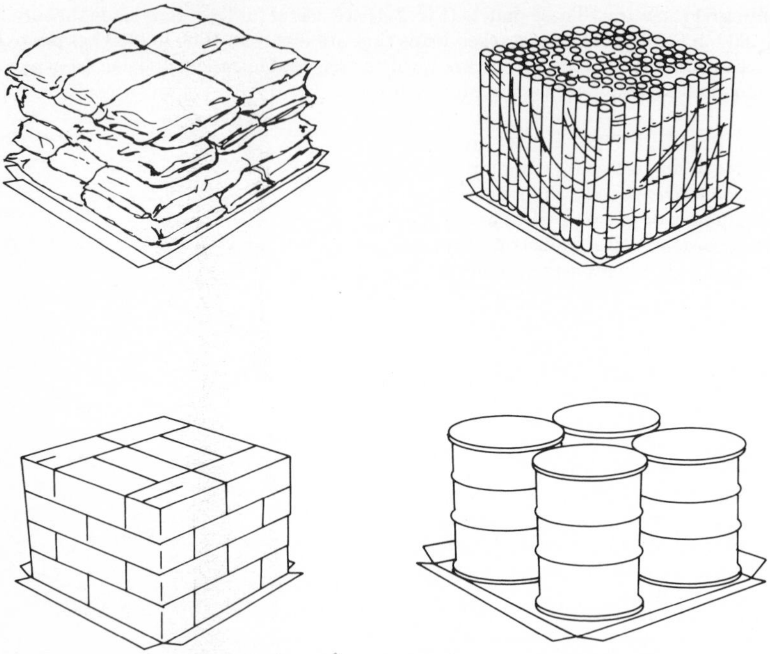

Figure 2-5 Typical uses of slip-sheet system. *(Little Giant Products.)*

parts, rotate work stock, and move palletized loads in and out of buildings as well as trucks, railcars, and other conveyances.

Portable Stacking Racks

These racks are used for storage of palletized loads that cannot be stacked on each other. Pallet stacking frames (Fig. 2-6) are used to confine and protect irregularly shaped, fragile, or nonuniform loads during in-process or temporary storage. The pallet itself is the base unit and rests on the frame of the pallet beneath it. The second kind of portable stacking rack (Fig. 2-7) consists of a base unit and removable post and end frames. Pallets are stored on the base units and the base units, when stacked, nest in the end frame.

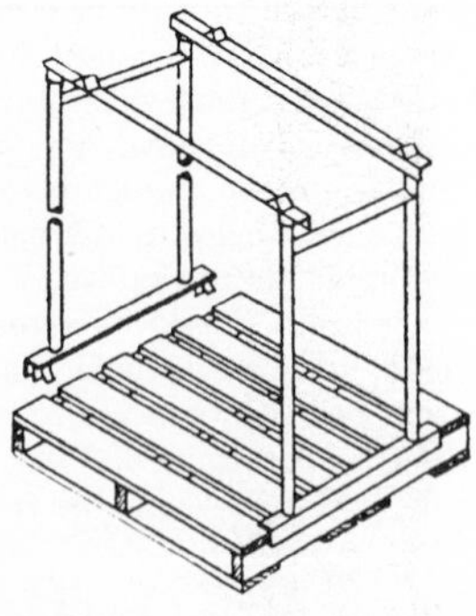

Figure 2-6 Pallet stacking frame using pallet as base unit.

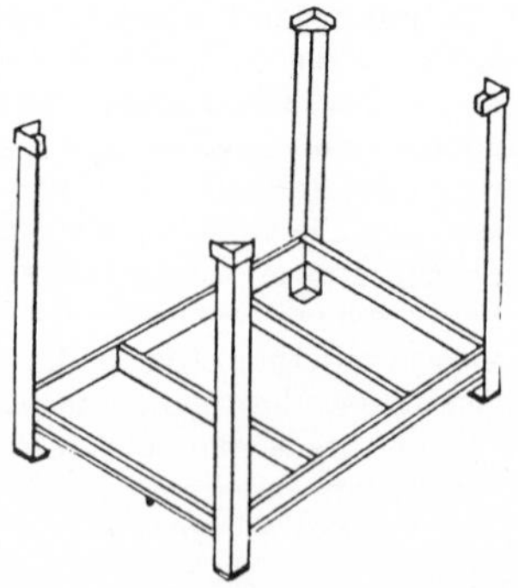

Figure 2-7 Pallet stacking frame with base unit.

Pallet Loading Patterns

A pallet pattern is an arrangement of units on a pallet and ideally is the most effective way of loading a pallet with the least loss of cube. There are a number of ways to select the optimum pattern, ranging from trial and error to the use of computer models. No matter what technique is used, the following factors must be taken into consideration:

1. **Size of Material** There may be several ways, one way, or no way to place a given size material onto a given pallet.
2. **Weight of Material** In the case of very heavy material, fewer layers will be stacked on a pallet. To a certain extent the number of layers will depend on the strength of the containers, if any are used.
3. **Size of Unit Load** Taken as a whole, the length, width, and especially the height of the load must be considered.
4. **Loss of Space within Unit Load** Some patterns have too many large gaps between units. This kind of piling is particularly bad when paper pallets are used because the weight should be distributed evenly and the units should brace each other.
5. **Compactness** Some patterns do not tie together well; they will not interlock.
6. **Methods of Binding Products in Patterns** If the units of a load are glued together, one kind of pattern may be ideal; with strapping, another type may be the best; and if no fastening at all is used, some combination of stacking may be the most suitable method to interlock and hold the load together.

Some general rules that should be followed in establishing pallet patterns are:

1. Interlocking unit loads should be used when possible to make the most effective use of the cube and to provide load stability.
2. Overhang, where unit loads extend beyond the edge of a pallet, should be avoided or minimized to a point where container damage or load stability is not affected. The added dimensions caused by this condition should not exceed the width or length of the shipping conveyance or reduce the utilization of the conveyance.
3. Underhang, where unit loads do not fill the deck surface and where there are large voids, should be avoided.
4. Utilize the basic pallet patterns (Fig. 2-8) effectively. Use block patterns for containers of equal width and length. This type of pattern is the least stable and may require bonding and fastening if considerable movement is involved.
5. Brick, row, and pinwheel patterns are used for containers of unequal length or width. All three patterns result in the interlocking that stabilizes a load.

Pallet patterns have been developed empirically for materials that have rectangular dimensions. The U.S. Navy Research and Development Facility has developed such a pattern.[1]

Container capacities range from 500 to 6000 lb (230 to 2700 kg), and standard base sizes cover the range of sizes of wood pallets, including 40 × 48 in (1.4 × 1.7 m). The 44 × 54 × 40 in (1.6 × 2 × 1.4 m) size is ideal for use when making shipments by rail or truck because the container dimensions are exact multiples of trailer and railcar dimensions and therefore can use the cube of these conveyances fully.

Metal Containers

Three types of metal containers are in general use: wire mesh, noncorrugated steel, and corrugated steel. Current development trends tend toward increasing versatility by incorporating features such as stacking and dumping capability and pallet-type bases to allow and facilitate movement.

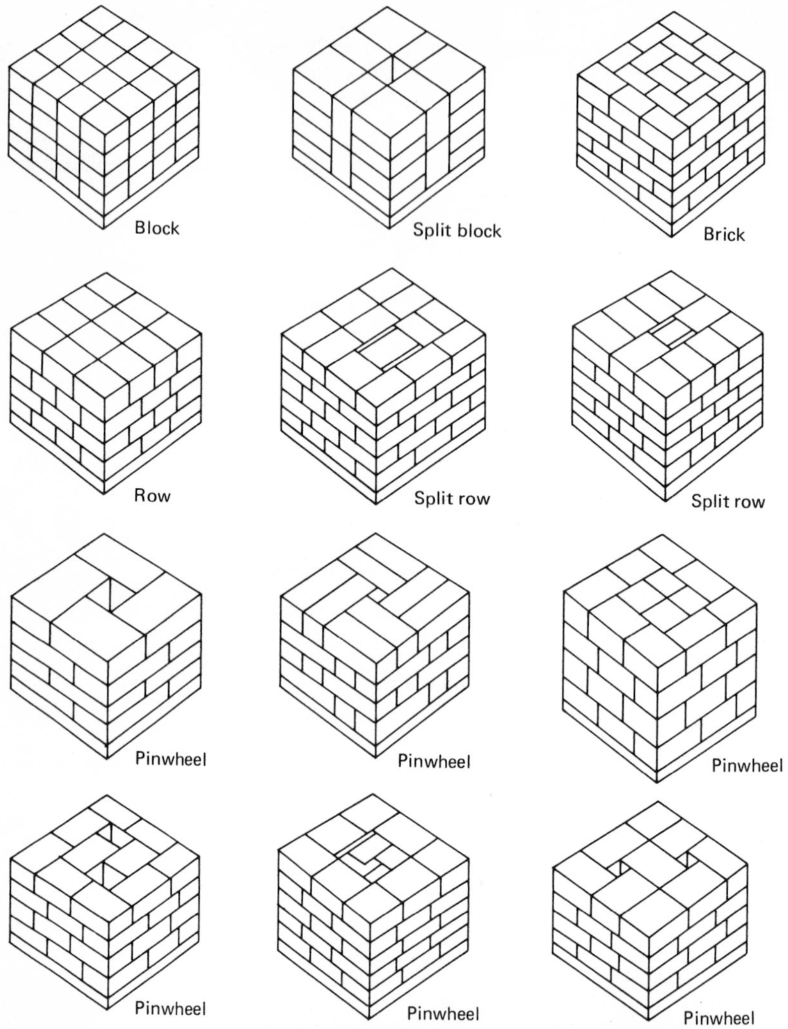

Figure 2-8 Typical pallet patterns. Paper or fiberboard binders are used between layers if necessary.

Welded-Wire-Mesh Containers

These containers (Fig. 2-9) are fabricated from welded wire for containment of materials. Additional structural sections are added for additional strength, and optional features can be included for specific uses and applications. The kinds of material and product that can be handled is limited only by size that would fall through the wire mesh and total container volume. Mesh openings from ½ × ½ in (1.3 × 1.3 cm) to 4 × 4 in (10 × 10 cm) are available to accommodate a wide range of product or material sizes.

Advantages that are generally cited in using this type of container are:

- Lightweight as compared with other metal containers
- Allows visibility of product for easy and quick identification

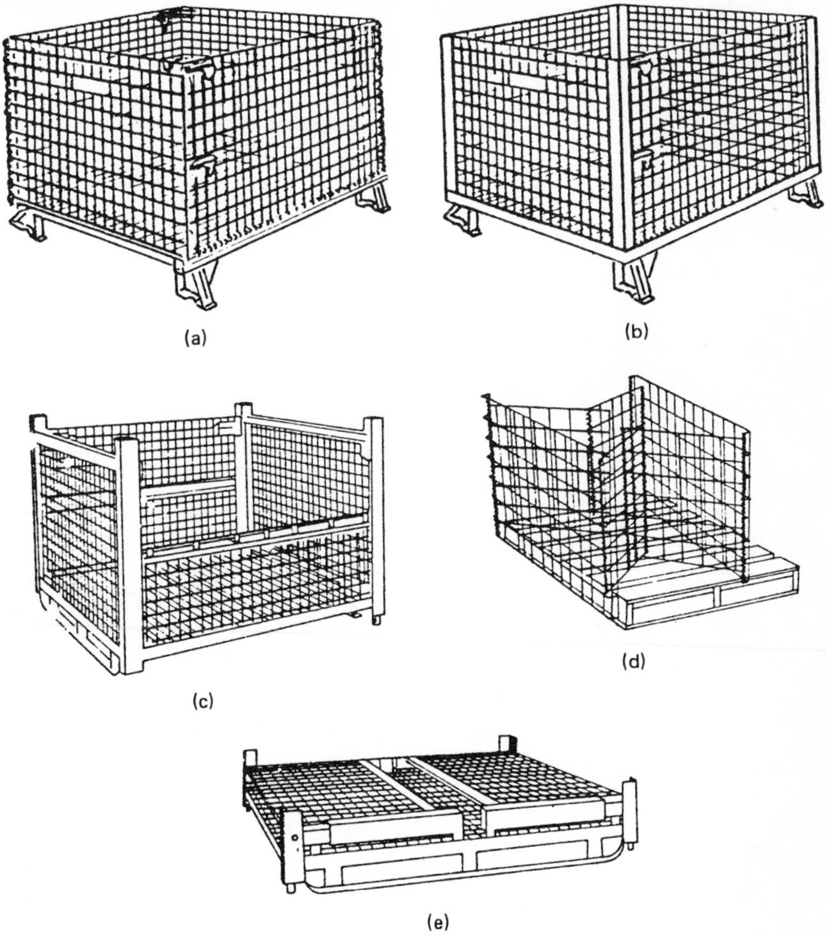

Figure 2-9 Kinds of welded-wire-mesh containers. (*a*) Collapsible wire container, (*b*) rigid wire container, (*c*) heavy-duty rigid, (*d*) wrap-around, (*e*) folded-down collapsible rigid.

- Self-cleaning by shedding debris
- Material can sometimes be processed in the container in such processes as degreasing, cleaning, and air drying.

Wire-mesh container stacking is accomplished in two ways: (1) interlocking through the corner posts and (2) the arch (saddle) of the upper container rests on the rim of the container below.

Wrap-Around Collapsible Frame. This forms the most basic type of wire-mesh container by using a standard pallet as a base.

Collapsible Container. This type of container is constructed to permit the sides to fold down when not in use. The obvious advantage of this design, saving space, must be weighed against its shorter life and smaller payload capacity.

Rigid Wire Container. This type is designed with additional vertical structural members that increase the strength of the containers, resulting in longer life and greater capacity than the collapsible types can provide. A heavy-duty rigid container has been developed to handle heavier loads in more strenuous environments. Additional horizontal bracing has been added to achieve the capability.

Collapsible Rigid Container. This is a heavy-duty version of the collapsible container. It is nearly equal in strength to the heavy-duty rigid type, but the higher original cost must be compared with savings on return freight costs.

Corrugated-Steel Containers

These are probably the strongest type of metal container available, since corrugation permits a longer surface of material to be used for a given size of a container than any other method of construction, and it is fabricated from hot-welded steels 0.105 to 0.179 in (2.7 to 4.6 mm) thick. This type of container has progressed from the one-type-only gondola bin to configurations that are almost virtually materials handling systems in themselves. They can be considered as consisting of three basic *modules:* base, lifting and stacking aids, and container box, with options for specific applications.

Container Bases

Container bases are used to facilitate surface movement and can be essentially categorized as metal pallets since there are provisions for two- and four-way entry with forklift trucks and other materials handling equipment. The general features of these bases were covered in the pallet section of this chapter.

Container Sections

Container sections can include a number of options (Fig. 2-10) to provide hopper dumping or accessibility, drop-bottom dumping, and end gates. Air- or hydraulic-actuated equipment and tilt stands are used in conjunction with, and as means to extend the versatility of, containers in assembly or machining stations. Container sections come in a wide range of heights; the deeper containers are used for dumping applications, and the shallower sizes are more suitable for manual handling of parts.

Several lifting and stacking aids can be designed or are available as part of the container system. Overhead movement can be accomplished by the addition of overhead crane lugs, chain-sling lugs, and bar-lift or sling-lift notches in the base. Stacking is accommodated by corner stacking angles, self-aligning hairpin brackets, and stacking lugs.

Industry Specifications

Specifications issued by the Industrial Metal Containers Product section of the Material Handling Institute specify the materials, design, construction, testing procedures, environmental considerations, utilization, and safety practices regarding the use of metal containers. The safety guidelines do not recommend the use of wire-mesh containers for applications where parts and material protrude or for dumping purposes. A load and capacity rating plate is used to indicate the load capacity of each container and the number of containers that can be safely stacked on top of each other.

Wood Containers

Wood containers are of three basic types.

Types

1. Bins with bases and closed sides and ends
2. Boxes with bases, closed sides and ends, and a top
3. Crates with open or slatted sides and ends

Wood containers with standard pallet bases are finding increasing use in applications where mechanized handling and storage are required for products ranging from solid

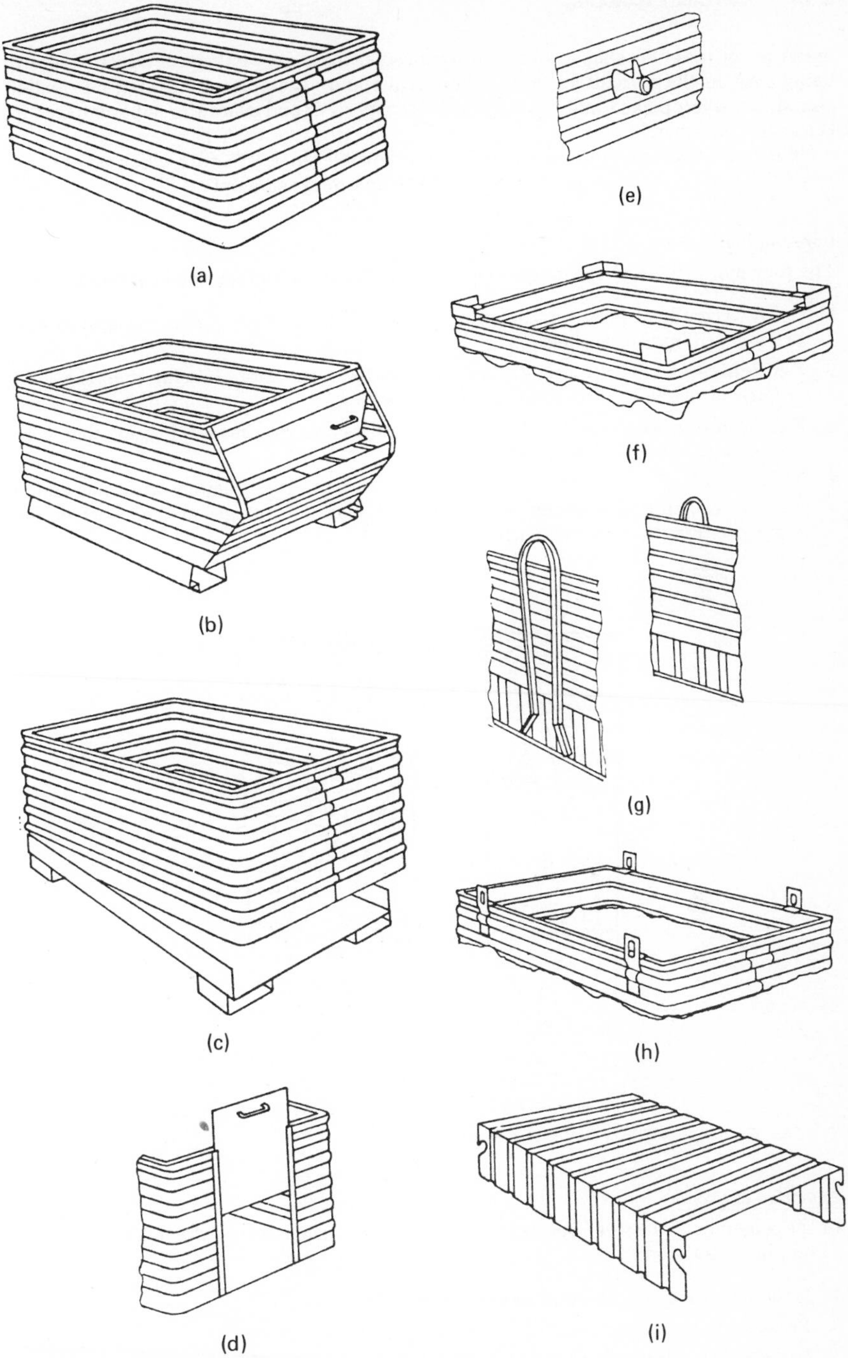

Figure 2-10 Corrugated-metal-container styles and lifting and stacking aids. (*a*) Basic corrugated box, (*b*) hopper-front pallet base container, (*c*) drop-bottom container, (*d*) end gate option, (*e*) overhead crane lugs, (*f*) corner-stacking lugs, (*g*) chain sling and stacking lugs, (*h*) bar lift or sling sling lift notches.

materials of irregular shapes and sizes to granular materials. Pallet containers are now being used in the agricultural area, where fruits and vegetables are loaded into pallet containers when picked and then washed, transported, and placed in supermarkets in the same container.

The pallet container design may include collapsible sides, a feature which saves space when the container is not in use or when it is being returned empty.

Construction

The four major types of end-panel and side-panel construction in wood containers are:

1. **Solid** Usually of plywood, this type of construction provides great strength as well as a smooth interior surface.
2. **Vertical-Slat** Used for deep containers and requires more bracing than the solid or horizontal-slat construction to prevent rocking.
3. **Horizontal-Slat** Used for shallower containers and requires less bracing than containers with vertical-slat construction.
4. **Wirebound** Provides the economy of lightweight construction with the added strength of being wire-bound (Fig. 2-11). The main advantages of this construction is a high-strength container with low tare weight.

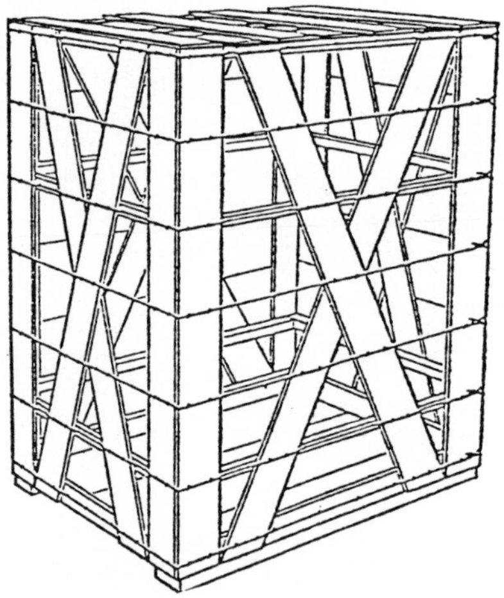

Figure 2-11 Typical wirebound container.

Selection of Wood

In the selection of the wood species to be used for wood containers, the following factors should be considered:

1. Intended usage and life requirements of the container
2. Pressure to be exerted on the bottom and sides by the goods
3. Permissible bulging of sides and edges
4. Degree of impact resistance required
5. Type of handling equipment used

6. Degree of weather and water resistance required for exposure to weather or cleaning operations

Corrugated-Cardboard Containers

Corrugated-cardboard containers offer a wide range of economical solutions for packaging or materials handling problems. There are literally hundreds of unique designs of corrugated containers. Because of the relatively low cost, it is feasible to "tailor-make" a configuration that is an optimum design for a particular product and situation. This container may be one of, or a modification of five common types.

Types

1. **Regular Slotted Container** The most commonly used style. All flaps are of equal length, and other flaps meet when closed. Contents of the box are protected by one thickness of corrugation on the side and two thicknesses on top. If additional top and bottom protection is required, both outer top and bottom flaps can be designed to overlap.
2. **Telescope Box** A two-piece box designed so one part fits into the other.
3. **Five-Panel Folder** A corrugated flat sheet, scored into five panels, which folds into a four-sided tube-type container that is closed by end flaps.
4. **One-Piece Folder or Book Fold** Used for shipping books, catalogs, wearing apparel.
5. **Gaylord** A corrugated container that has a base, generally consisting of two or more wood runners that allow the container to be moved by fork lift equipment.

Corrugated Material and Construction

The basic corrugated material is referred to as "double-face" or single-wall, consisting of outer facing, corrugated medium, and inner facing, joined by adhesive. Corrugated material is also available in single-face, double-wall, and triple-wall designs.

Types of Board. These include Fourdrinier kraft linerboard, the highest-quality and -cost material made from virgin pulpwood, and cylinder linerboard which is generally made from a combination of reclaimed fibers and virgin pulp. The weight of a linerboard required for the contents of a corrugated container is specified in the Uniform Freight Classification and Motor Truck Classification.

Corrugating Medium. Straw, reclaimed fibers, or woods can be used as a corrugating medium. The combination of corrugated-medium thickness, weight, and flute configuration determines the strength and moisture-lockout properties of the container. Corrugating mediums are specified by thickness and weight per 1000 ft^2 (MSF).

Flute Configuration. This specifies the number of corrugations per lineal foot. The three most common flutes used are:

1. **A Flute** The highest flute with the least number of corrugations (36 per linear foot, 0.1875-in high) (12 percm, 4.2 mm). When used in an upright position, A flute has the best stacking strength and greater capacity to absorb shock in the direction of the thickness.
2. **B Flute** Has the greatest number of corrugations per foot with the lowest flute height and is stiffer and less shock-absorbent than A flutes, but it has greater crush resistance to loads placed in the direction of thickness.
3. **C Flute** A compromise between A and B flutes.

Methods of Fastening Joints. The accepted methods are taping, stitching, or gluing. Common carriers publish detailed regulations governing the method to be used in regard to content type, weight, and other factors.

Corrugated boards may be specially treated to provide additional properties to the container, such as a coating and lamination, to:

- Retard slippage
- Inhibit mold
- Retain temperature
- Increase water or moisture resistance

Tote Boxes and Bins

Tote boxes are used for unit loads of smaller parts that can be moved manually through the operation or can be stacked in a larger container to become part of a unit load. Tote boxes are available mainly in metal and plastic and are also fabricated from other materials such as wood, cardboard, fiberboard, and Plexiglas.

Plastic Tote Boxes

Plastic tote boxes have many applications where small and light parts are handled and in environments where protection from corrosive chemicals or a high degree of cleanliness is required. Plastic containers are easily cleaned without harm or deterioration to the material. Molding capabilities permit desirable features to be incorporated as an

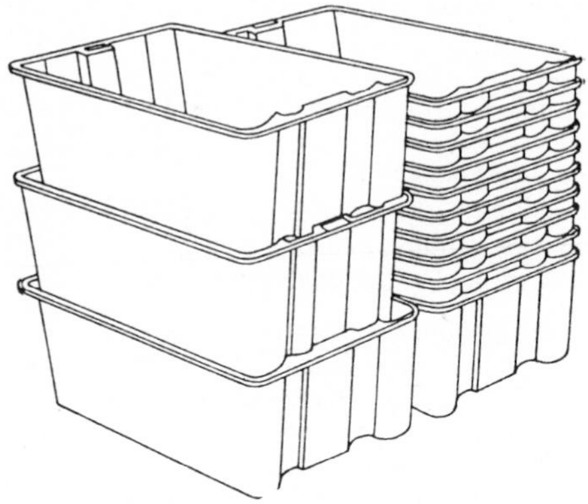

Figure 2-12 Typical tote boxes.

integral part of the container, permitting a number of provisions for nesting and stacking. A typical tote box is shown in Fig. 2-12. There are three major types:

Straight-Nesting. This refers to tote boxes that can be nested when not used. This type requires the use of lids or covers if stacking is required but results in maximum product protection, since one box will not fall into another. The design of straight-nesting boxes is characterized by tapered sides which reduce cube utilization and should not be used for storing on shelves.

Straight Stacking. This method of stacking boxes is ideal, because of minimum tapered sides, for shelf storage and use as an inner container of maximum cube utilization. Since the sides and ends support the loads, greater weights can be stacked than when using the straight-nesting variety.

Combination of Stack and Nest. This feature is available in some boxes and can be altered by the orientation of boxes in relation to each other.

The most common plastic materials and their major properties are indicated in Table 2-2.

TABLE 2-2 Plastic Materials Used in Tote Boxes

Material	General properties
ABS (acrylonitrile-butadiene-styrene)	High impact absorption, good compressive strength; more expensive than other thermoplastics
High-density polyethylene	Good to excellent stiffness, excellent temperature range, −40 to 150°F (−71 to 51°C); commonly used for food applications with USDA and FDA approval
High-impact polypropylene	More durable than polyethylene, not as stiff; tendency to crack at temperatures below 0°F (−32°C)
High-impact polystyrene	Extremely stiff, excellent compressive load strength, good temperature range; tendency to crack easily under high impact; readily attacked by solvents and oils
FRP (fiberglass-reinforced polyester)	Exceptional compressive load strength; can be heat-, fire-, and wear-resistant

DUNNAGE AND OUTER SECUREMENT OF CONTAINERS AND UNIT LOADS

Dunnage

Dunnage refers to inner package containment methods or material that is used to protect the contents of a container from damage. This is done in one of two ways, either by preventing the movement of the contents or by providing a cushioning medium to absorb shocks.

Plastic and other petroleum-base materials are used for dunnage because of their light weight and low bulk density.

Dunnage Materials

Polystyrene is used to cushion package contents in three general forms:

1. Loose foam strands are used to fill the air space in the package and provide a cushioning barrier around the contents.
2. Polystyrene can be molded to the general form of the part in the container and actually becomes an inner case for the part.
3. Polystyrene is also used for corner forms to strengthen corrugated containers.

Bubblepack is two thin sheets of polyethylene with air entrapped within the "bubble" sections when laminated together. The contents of a container may be wrapped in bubblepack for protection during shipment or movement.

Corrugated board can be easily formed into many shapes to protect, support, and cushion products. The trend is to reduce the number of inserts by combining features into a single interior form, reducing the cost and inventory expense of packaging materials. Typical inner packings, portions, and sheets used for this purpose are shown in Fig. 2-13.

Outer Securements

Container closure can be achieved by sealing flaps with glue and tape by stapling, and by strapping with plastic or steel bands.

The glue and tape sealing method lends itself to automation by use of case-sealing equipment which automatically dispenses glue or tape close to the flaps. Stapling can also be automated by passing containers through, under, and between staple heads.

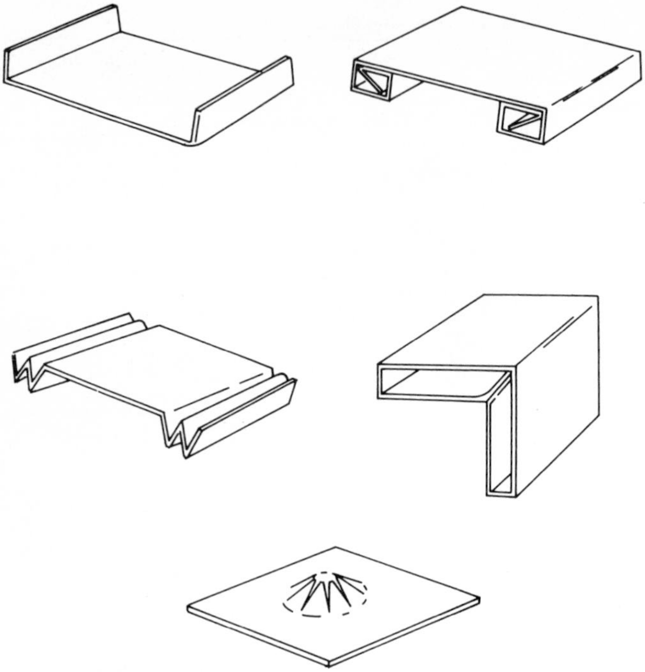

Figure 2-13 Typical inner packings, portions, and sheets.

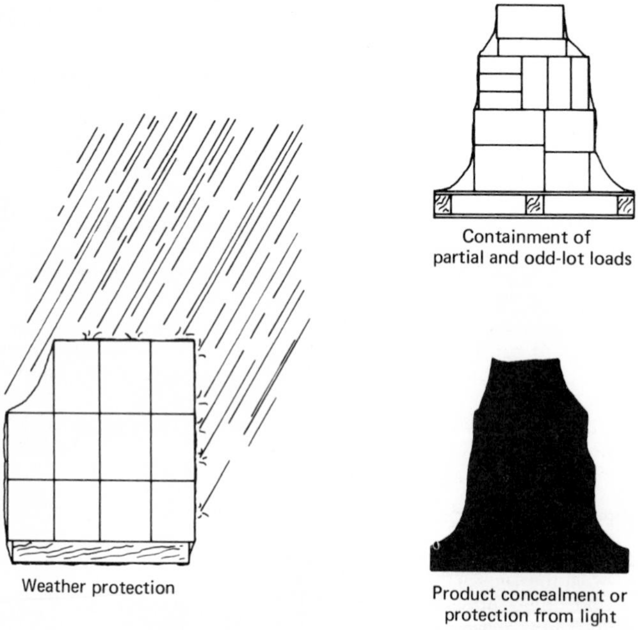

Figure 2-14 Other benefits from use of shrink wrap. *(Black Body Corp.)*

Strapping

Strapping can be used to secure not only containers but stand-alone unitized loads and palletized unit loads as well. There are two major strapping classifications: steel and plastic. Steel strapping is dimensionally stable under all but extreme conditions and plastic strapping is more resilient.

Common cold-rolled-steel strapping, the least stretchable of all steel strapping, is good for strapping lightweight packages or pallet loads that are not subject to high impact or shock. Heavy-duty steel strapping, both hot- and cold-rolled steel, absorbs high impacts without breaking; however, it stretches under great stress and requires staples to keep it in place.

Plastic strapping continues to stretch under tensioned loads and therefore should not be used where continuous loads are present. However, their resilience keeps them tight on a package or load that shrinks or settles.

Steel strapping may be applied by manually operated tensioning tools which notch or crimp steels to retain the strapping. Plastic strapping can also be applied both manually and automatically in the same manner as steel strapping. Friction-welding and heat-sealing systems are also available.

Stretch Wrap

Stretch wrap is one method of confining unitized or pallet loads by wrapping a film of polyethylene material around the load. The stretch film is wound under light tension by rotating the load on a platform on horizontal- and spiral-type stretch film equipment. Horizontal-type equipment uses a full sheet wrap, while spiral-type equipment bands the load by dispensing several overlapping layers. Vertical wrapping equipment rotates the film around the load and is generally used when the load is not on a pallet.

Shrink Wrap

Shrink wrap uses a polystyrene film that can be a bag or flat sheet that is put over or around the unit load and then placed in an oven. Bag design should be selected to achieve the desired protection required or to optimize the holding power for the load configuration. While in the oven, the film reaches a temperature of 240°F (116°C). During the cooling cycle, the film molecules try to return to their original compact orientation, shrinking the film tightly around the load.

Three types of ovens are used for this purpose, depending on the volume handled. Closet types can normally handle up to 20 loads per hour; bell-type ovens are for higher volume operations, up to 20 loads per hour; and tunnel or feedthrough ovens are generally specified for automated lines and handle more than 75 loads per hour.

In addition to securing products in a unit load, there are extra benefits in using shrink wrap (Fig. 2-14). Irregular, hard-to-pack loads (like those in order-picking warehouses) can be secured with shrink wrap. Products can be protected from damage caused by weather, dirt, or moisture. Clear film allows for easy viewing for identification and inventorying of product. Opaque film can be used to conceal or protect product from light.

REFERENCES AND BIBLIOGRAPHY

1. "Storage and Materials Handling," Departments of the Army, Navy, Air Force, and U.S. Marine Corps, Washington, D.C., 1955.
2. "Unit Load Stretch Wrapping," *Modern Materials Handling,* October 1979.
3. Schumf, G., "Slip Sheets," *Material Handling Engineering,* July 1980.

chapter 8-3

Fixed-Path Equipment

INTRODUCTION

Conveyors, cranes, and hoists are generally considered fixed-path materials-handling equipment, since they often become a fixed part of the physical plant. Once in place, a considerable amount of time, disruption, and cost is needed to change the arrangement of the equipment. It is therefore very important to plan the installation of these pieces of equipment very carefully.

A complete materials-handling system can include a wide variety of fixed-path equipment for unit loads and bulk materials handling, as well as mobile handling equipment and storage racks. This further complicates the planning process, since the fixed-path equipment not only has to satisfy the requirements of the fixed-path handling but must also be compatible with the overall flow of the total handling system.

There are many considerations involved in planning fixed-path equipment installations; many of these considerations are unique to a specific type or class of equipment, but general areas that must be addressed in the planning and exploration stage are:

- **Flexibility of the system.** Must a wide range of unit-load sizes or bulk material be handled or conveyed?
- **State of Materials To Be Handled.** Is it in a unit load or bulk form?
- **Weight, Dimensions, and Physical Properties of the Material Being Handled or Moved.** Is it fragile, light, firm, or does it have other properties that require special attention?
- **Loading and Unloading Methods.** Is it handled manually or received from or delivered to other equipment such as lift trucks, palletizers, or packaging equipment?
- **Capacity of Equipment.** Does the conveying speed match the speed or capacities of the equipment it is being interfaced with? Is there sufficient capacity or length to accumulate material when required?

- **Supporting-System Requirements.** Is the material to be sorted, accumulated, weighed, or further processed while being handled or conveyed?
- **Environmental Conditions.** Must provisions be made for dust, high or low temperature, high humidity, or other ambient conditions in the plant or outside?
- **Safety.** What special precautions must be taken to protect operating personnel or personnel working near the equipment? What provisions must be made to comply with regulatory requirements?
- **Maintenance.**
- **Facility Restrictions.** Are overhead heights or floor loading capacities adequate for supporting and accommodating equipment? Is there sufficient plant area? Will the fixed-path system impede access to equipment and the flow of personnel and other materials within the plant?
- **Horizontal or Vertical Distances To Be Covered.** What hardware is required to negotiate inclines and declines throughout the system?
- **Power and Energy Requirements.**

There are many types and varieties of fixed-path equipment. Each of the major classifications of this type of equipment are discussed and described here, but no effort will be made to list all of the items that are contained in each classification. The classifications that are covered include:

- Conveyors
- Sorting, consolidating, and diverting devices
- Hoists and cranes
- Guided vehicles
- Robots

CONVEYORS

Conveyors are gravity or power devices commonly used to move uniform loads continuously from point to point over fixed paths. The primary function of the conveyor is to move materials when the loads are uniform, and the routes do not vary. The movement rate and direction is usually fixed although the system can be designed to bypass cross traffic. The major types of conveyor and related devices are chutes and wheel and roller conveyors.

Chutes

Chutes are the simplest fixed-path devices that use gravity to convey bulk or unit loads down declines. Straight and spiral types are available. The spiral chute (Fig. 3-1) is a continuous trough over which bulk materials or discrete objects are guided in a helical path.

Wheel and Roller Conveyors

These depend on both gravity and power to move materials. Objects of various shapes can be handled by changing the cross section of the rolling surface or by aligning the objects in the conveyor framework. These conveyors are generally used to move materials horizontally.

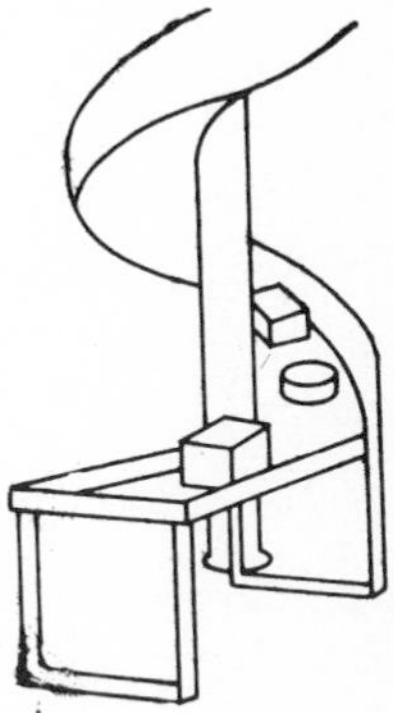

Figure 3-1 Spiral chute.

Considerations for Chutes and Wheel and Roller Conveyors

The following sections discuss points that must be considered in specifying and designing both of these classes of conveyors.

Load Characteristics. These include maximum and minimum sizes of loads and the shapes and carrying surfaces of all units. The suitability of a load configuration to be handled on roller or wheel conveyor is important. Unsupported packages such as bags (Fig. 3-2) are not recommended for this type of equipment.

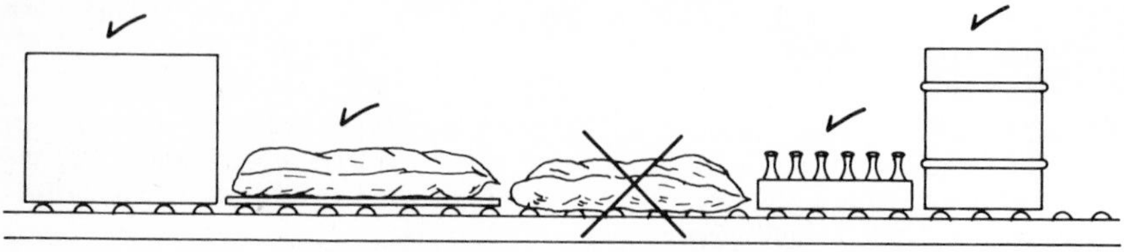

Figure 3-2 Can it be handled on a roller conveyor? *(Litton UHS.)*

Operating Conditions. These include the size and weight of the conveying surfaces, environmental conditions, and loading and unloading methods. These considerations determine the type and capacity of frame, roller, or wheel material and sizes, and the type of bearings that should be used.

Roller or Wheel Spacing and Pattern. This is determined by the size of the minimum package or unit load. (See Fig. 3-3.) To determine roller centers, divide the minimum load length by three. Wheel pattern should be specified to provide a minimum of five wheels under the package. Other guidelines include:

1. A minimum of three rollers under a hard bottom surface
2. A minimum of four rollers under a flexible bottom surface

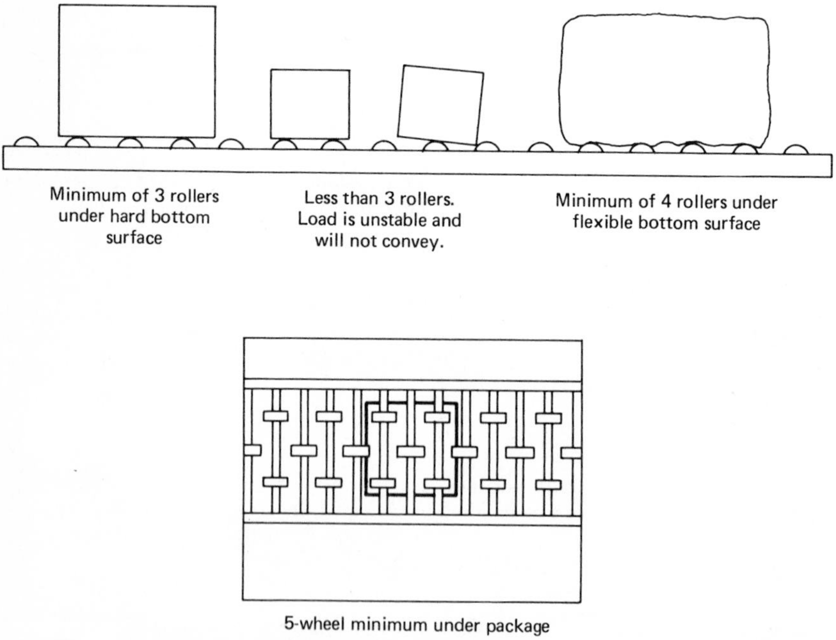
Minimum of 3 rollers under hard bottom surface
Less than 3 rollers. Load is unstable and will not convey.
Minimum of 4 rollers under flexible bottom surface
5-wheel minimum under package

Figure 3-3 Roller and wheel spacing guidelines. *(Litton UHS.)*

Roller and wheel capacity is determined by dividing the weight of the heaviest load to be handled by the minimum number of rollers or wheels under the carrying surface of the load. If special requirements must be considered, such as drop, shock, or side loads, then a roller with a higher load rating will have to be considered.

Conveyor Width; Wheel and Roller Setting. The width of conveyor is determined by the back-to-back frame dimension required to ensure sufficient clearance to carry the load around a 90° curve. The minimum clearance is dependent on the roller setting. If rollers are set high, the conveyor can handle loads up to 1.25 times the width of the conveyor. If the rollers are set low, a minimum of 1 in (2.5 cm) between frame and load must be allowed on each side. Package skew should also be considered in determining this dimension. Specification of curve sections can be calculated graphically from the chart in Fig. 3-4. The design of curve sections depends on the size and shapes of loads. Tracking of packages is important, especially when there are a number of turns involved and the skewing effect becomes cumulative. This effect can be minimized by using tapered rollers (Fig. 3-5) or a double-roller differential section.

Bearing Selection. This is dependent upon conveyor operating conditions. Plain ball bearings are used for indoor use where severe environmental conditions are not present. Dust-tight bearings, which are designed to run dry, are ideal in dusty atmospheres. Greased bearings require more force to turn and their use should be kept to a minimum on gravity-type conveyors.

Conveyor Pitch. The pitch that is required cannot be easily determined because it is dependent on a combination of many factors such as:

- Weight and stability of unit load
- Smoothness of bottom of load
- Firmness of load surface
- Length of rollers
- Number of rollers under load
- Length of runs
- Types of bearings

The suggested pitch for a number of types of unit loads is shown in Table 3-1 and should be used as a general guide for determining pitch.

Conveyor Support and Frame Capacity. Conveyor supports can be one of three types: permanent-floor, ceiling-hung, or portable. Supporting points (Fig. 3-6) should be located to handle the load equally. The design load is the weight of the conveyor section plus the maximum unit load for that section of the conveyor.

Special Considerations for Wheel Conveyors. Wheel conveyors are used for light-duty applications and have several advantages over roller conveyors for lightweight unit loads. Gravity wheel conveyors consist of a series of wheels which can be one of many different styles and materials, mounted on common axles and supported between two frames. They are generally less expensive and lighter in weight, making them ideal for portable applications. Light unit loads travel better on wheel conveyors because less pitch is needed and less force is required to start the wheels in motion (see Table 3-1). Another advantage inherent in the use of multicontact conveying surfaces is that the wheels provide a turning action which enables the package to maintain its original position.

Frames that support the wheel axles are either steel or aluminum. Low weight and corrosion resistance are properties that make aluminum attractive, but steel should be used where conditions require the conveyor to be more rugged.

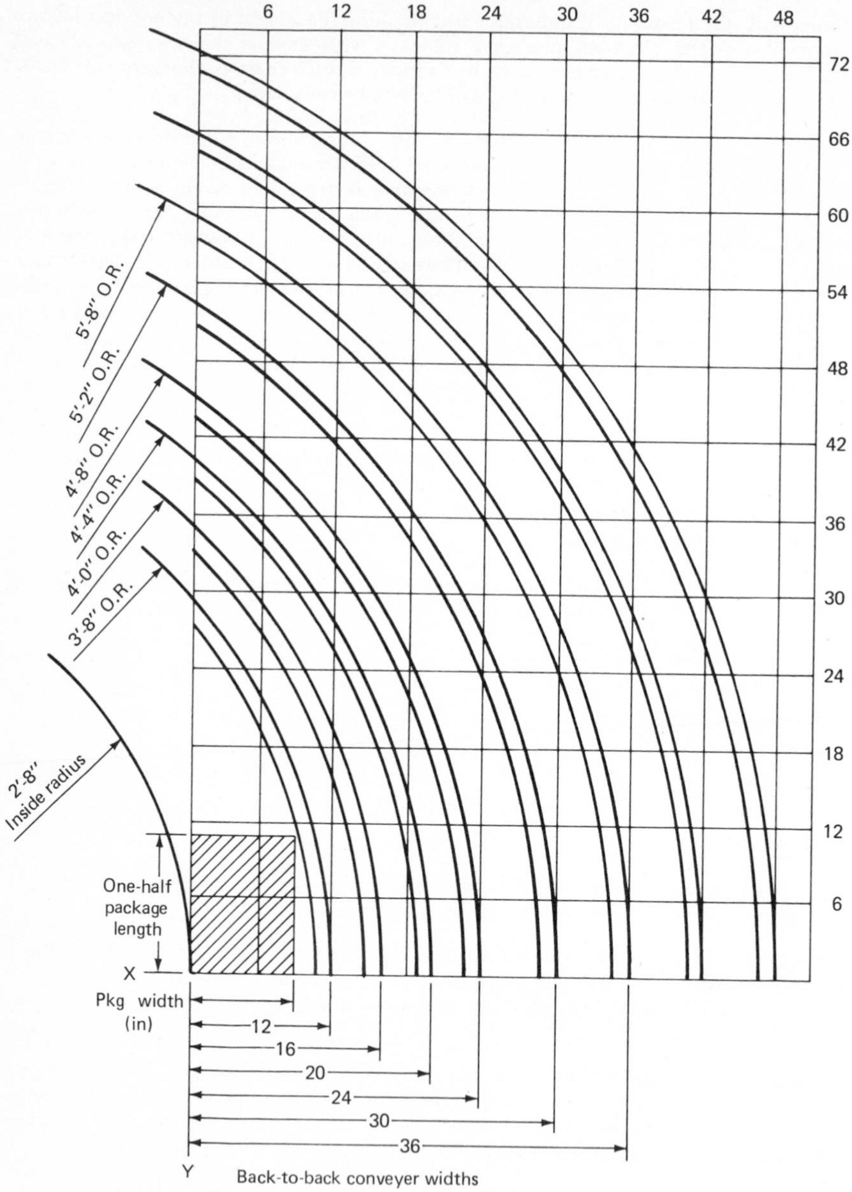

Figure 3-4 Curve-radius selection.

Types of Wheels. There is a wide variety of both metal and plastic wheels now available, including:

- **Steel Wheels with Ball Bearings.** The most rugged and commonly used wheels, these are used where long life is a requirement. Life potential of this wheel is 10 times that of aluminum. Steel wheels can be covered with *neoprene tires* and are used to

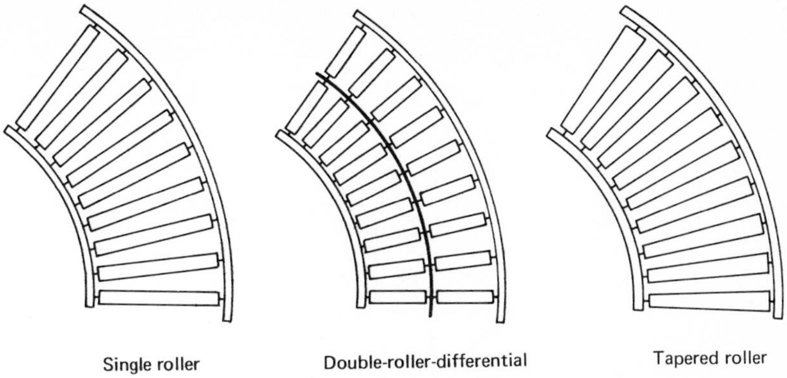

Figure 3-5 Roller-conveyor curve sections. *(Litton UHS.)*

reduce shock, prevent slippage, increase traction, prevent marring fragile surfaces, and reduce noise.

- **Aluminum Wheels with Steel or Plastic Ball Bearings.** Used in applications where weight is a factor, particularly for portable conveyors. Plastic ball bearings should be used in corrosive atmospheres.
- **Nylon Wheels.** Used where resistance to salt, water, and chemicals is required, as well as in applications where conveyors are cleaned frequently. Nylon wheels also do not mar or mark containers.
- **Polypropylene Plastic Wheels.** Have many properties which make them ideal for a wide range of applications. This material is highly impervious to a wide range of corrosive materials and is temperature-resistant from 230 to −30°F (110 to −34°C). The wheels do not absorb moisture and can be steam-cleaned repeatedly.
- **Hysteretic Wheels.** Metal wheels surrounded by a tire formed by elastomeric material, these are used for line storage of heavy loads. The purpose is to absorb the energy of the initial load impact and to control the movement of the load to a safe speed.

Lubrication. Metal wheels can be greased, oiled, or dry. Nylon and plastic wheels are always furnished dry. Oiled or dry bearings should be used where high temperatures can cause thinning and leakage of grease. Dry bearings are recommended where temperatures are below 0°F (−18°C).

Special Types of Wheel Conveyors. These have been designed for handling specific products or for special industries. Samples of these are shown in Fig. 3-7.

Powered Conveyors. These are designed for continuous control of products on level surfaces, through inclines and declines, and around curves. Many of the same points that must be considered for gravity conveyors are applicable to powered conveyors. Powered roller and belt conveyors are the major powered types used to move unit loads.

Powered Roller Conveyors. These are used mainly to accumulate loads because power-drive disengagement can be accomplished very simply when the forward motion of the unit load is stopped. Generally, power-drive disengagement is triggered when the unit load meets an obstruction, creating an opposite reaction which causes the carrying roller bushing to move up an angular slot, thus relieving the pressure and contact between the drive belt and rollers.

Powered roller conveyors are either chain- or belt-driven. Chain-driven units are used in heavy-duty applications and where oil or contaminants would have an adverse effect on the belting. The belt-drive-powered trains are designed for accumulation where the

TABLE 3-1 Suggested Pitch for Gravity Conveyor

Container being conveyed	Approx. cont. wt., lb	Gravity roller conv. plain and dust-tight brgs.			Gravity roller conv. pres. lubricated brgs.			Pi gravity wheel conv. or live rail conv.		
		Pitch per ft, pitch per 10′ 0″ sec, and pitch per 90° curve								
		1″/ft	10′ 0″	90°	1″/ft	10′0″	90°	1″/ft	10′0″	90°
Cartons (fiber smooth bottom)	10	5/8	6 1/4	5	—	—	—	3/8	3 3/4	4
Cartons (fiber smooth bottom)	20	1/2	5	5	—	—	—	1/4	2 1/2	3 1/2
Cartons (fiber smooth bottom)	45	3/8	3 3/4	3 1/2	3/4	7 1/2	6	3/16	1 7/8	3
Cartons (fiber smooth bottom)	100	1/4	2 1/2	3	5/8	6 1/4	5	1/4–3/16	—	3
Fiber beverage carton empty	5	—	—	—	—	—	—	1 1/2	15	10
Fiber beverage carton empty bottles	35	1/2	5	4–5	3/4	7 1/2	6	3/8	3 3/4	4
Fiber beverage carton filled bottles	45	3/8	3 3/4	3 1/2	5/8	6 1/4	5	1/4	2 1/2	3 1/2
Wood cases or boxes	5	5/8	6 1/4	5	—	—	—	3/8	3 3/4	5
Wood cases or boxes	10	3/8	3 3/4	4	3/4	7 1/2	6	1/4	2 1/2	5
Wood cases or boxes	25	1/4	2 1/2	4	5/8	6 1/4	5	1/4	2 1/2	4
Wood cases or boxes	50	3/16	1 7/8	3	1/2	5	4	1/4	2 1/2	3 1/2
Wood cases or boxes	100	3/16	1 7/8	3	3/8	3 3/4	3 1/2	1/8	1 1/4	2 1/2
Steel drums	20	5/8	6 1/4	5 1/2	3/4	7 1/2	6	—	—	—
Steel drums*	55	7/16–1/2	4 1/4–5	5	5/8	6 1/4	5	—	—	—
Steel drums	120	3/8	3 3/4	4	1/2	6	4 1/2	—	—	—
Steel drums	250	1/4	2 1/2	3	3/8	3 3/4	3 1/2	—	—	—
Steel drums*	550	1/4–3/16	2 1/2–1 7/8	3	1/4	2 1/2	3	—	—	—

Wood barrels	100	½	5	5	⅝	6¼	6	—	—	—
Wood barrels	400	⅜	3¾	4	½	5	5	—	—	—
Metal and fiberglass tote boxes	10	½	5	5	—	—	—	⅜	3¾	4
Metal and fiberglass tote boxes	25	⅜	3¾	4	¾	7½	6	¼	2½	3½
Metal and fiberglass tote boxes	50	¼	2½	3	⅝	6¼	5	3/16	1⅞	3
Milk cans empty	25	⅝	6¼	6	¾	7½	6½	—	—	—
Milk cans full	105	⅜	3¾	4	½	5	5	—	—	—
Milk crates empty	10	¾	7½	6	1	10	7½	—	—	—
Milk crates empty bottles	45	½	5	5	⅝	6¼	5	—	—	—
Milk crates full bottles	60	⅜	3¾	4	½	5	5	—	—	—
Wood pallets smooth runner†	350	⅜	3¾	—	7/16	4¼	—	—	—	—
Wood pallets smooth runner†	750	5/16	3⅛	—	⅜	3¾	—	—	—	—
Wood pallets formica base	350	¼	2½	—	5/16	3½	—	—	—	—
Wood pallets formica base†	750	⅛	1¼	—	3/16	2½	—	—	—	—
Wood pallets pine plywood†	350	⅜	3¾	—	7/16	4¼	—	—	—	—
Wood pallets pine plywood†	750	5/16	3⅛	—	⅜	3¾	—	—	—	—
Wood beverage case empty	6	13/16	8⅜	6¾	—	—	—	¼	2½	3½
Wood beverage case empty bottles	30	½	5	5	¾	7½	6	⅛	1¼	2½
Wood beverage case full bottles	40	7/16	4¼	4	½	5	5	⅛	1¼	2½
Multiwall bags, firm	50	—	—	—	—	—	—	⅝	6¼	6
Multiwall bags, firm	100	—	—	—	—	—	—	½	5	5

*Varies with new or reconditioned drums

†Depends on the way it is nailed and if banded or not. Bad nailing or banding will considerably increase the pitch and make a planned slope impractical.

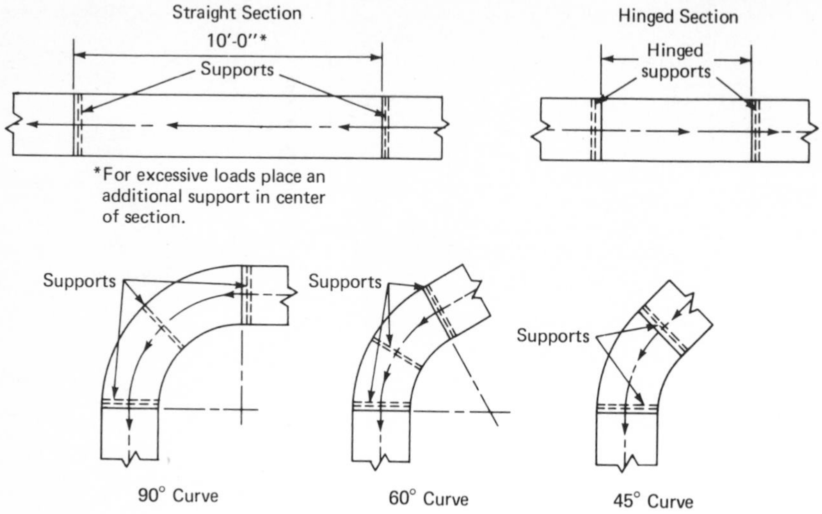

Figure 3-6 Suggested support locations.

belt-to-roller pressure is very light or for transportation sections where belt-to-roller pressure is increased by center takeup rollers and the use of high friction drive belts.

Powered roller conveyors are not used for inclines greater than 5° since the contact force between the unit load and roller surface is not sufficient to overcome the gravity force due to a low coefficient of friction. This type of conveyor is not normally used over straight runs because of the higher cost as compared with belt conveyors.

Belt Conveyors

Belt conveyors consist of an endless moving belt which carries materials within a supporting frame. The belt can be made from a variety of materials and may or may not be

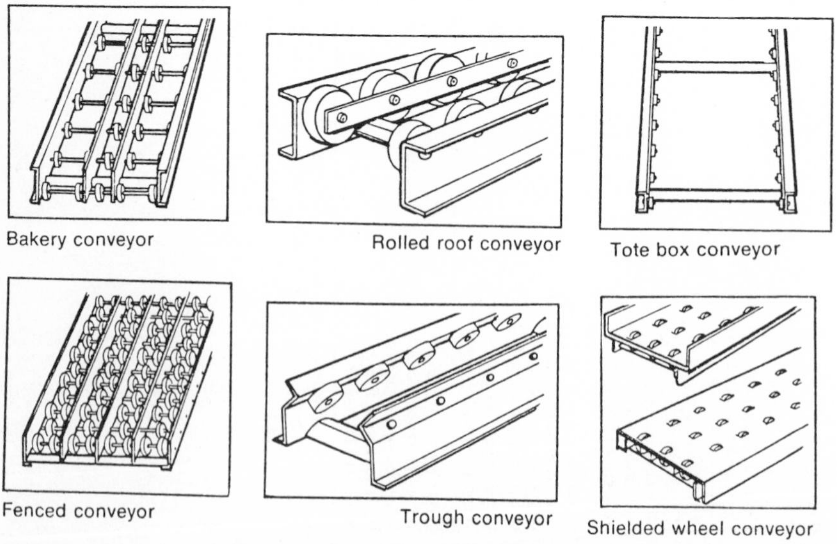

Figure 3-7 Special types of wheel conveyors.

equipped with cleats or other grabbing devices. The belt may be supported by a solid-slider-type bed of wood or metal or by rollers.

Conveyor manufacturers suggest friction surface belts on inclines up to 13°, and rough-top rubber belts should be used for inclines up to 25°. In applications where a steeper incline is required, heavily textured, nubbed or cleated surface belts can be used. Also, special requirements for belt surface material should be considered in applications where chemical or oil resistance is needed or for mandated cleanliness requirements.

Belt Conveyor Parameters. The parameters that must be defined before equipment is specified are belt speed, length, maximum load on belt at one time, tension loads, power requirements, and support and mounting hardware. Belt speed should be specified to be compatible with process equipment and other materials-handling hardware. Belt length should be adequate to accumulate the maximum expected product capacity. The maximum load on the belt at any one time can be calculated from the following formula:

$$P = \frac{K}{S \times 60 \text{ min}} \times C \text{ (ft)}$$

where

P = maximum product weight
K = load per hour, lb
S = speed of belt, ft/min
C = center-to-center distance

For example, if

Load per hour in pounds = 10,000 lb
Speed of belt = 50 ft/min
Center-to-center distance = 36 ft

Then, the maximum product weight is

$$P = \frac{10{,}000}{50 \times 60} \times 36$$
$$= 1200 \text{ lb}$$

Considerations for Bulk-Materials Belt Conveyors. These considerations are similar to those for all conveyors; however, the properties of the materials to be moved affect the parameters of and the specification of the conveyor. The use of belt conveyors for bulk materials is limited by the characteristics of materials, some of which are:

- Stickiness, which may prevent materials from completely discharging from the conveyor or interfering with the power-train components.
- Temperatures that exceed 150°F (71°C) would cause deterioration or damage to most belt materials.
- Chemical reactions of conveyed materials with belt material. Some oils, chemicals, fats, and acids can damage belts.
- Large lump size becomes a factor and generally requires the system to be oversized for the amount of weight being moved.

Weight and friction are the common factors that determine the amount of incline that is possible for unit-load and bulk-materials-handling belt conveyors. Bulk-materials belt conveyors must also include materials characteristics such as size consistency, shape of lumps, moisture content, angle of repose, and flowability. The maximum angle of incline for various bulk materials is shown in Table 3-2. The ideal combination of belt width and speed (Table 3-3) is determined by the characteristics of the materials handled.

Metal-Belt Conveyors. Similar in design to standard belt conveyors, these differ in that their surface is a belt of woven or solid metal. The materials include carbon steel,

TABLE 3-2 Maximum Angle of Incline

Material carried	Maximum angle of incline, deg*	Material carried	Maximum angle of incline, deg*
Alumina, dry, free-flowing, ⅛″ lumps	10 to 12	Grain	8–16
Beans, whole	8	Ore (see stone)	15–20
Coal, anthracite	16	Packages	15–25
Coal, bituminous, sized, lumps over 4 in	15	Pellets, depending on size, bed of material and concentricity (taconite, fertilizer, etc.)	5–15
Coal, bituminous, sized,† lumps 4 in and under	16	Rock (see stone)	15–20
Coal, bituminous, unsized†	18	Sand, very free-flowing¶	15
Coal, bituminous, fines, free-flowing‡	20	Sand, sluggish (moist)§	20
Coal, bituminous, fines, sluggish§	22	Sand, tempered foundry	24
Coke, sized	17	Stone, sized, lumps over 4 in	15
Coke, unsized	18	Stone, sized, lumps 4 in and under, over ⅜ in	16
Coke, fines and breeze	20	Stone, unsized, lumps over 4 in	16
Earth, free-flowing‡	20	Stone unsized, lumps 4 in and under, over ⅜ in	18
Earth, sluggish§	22	Stone, fines ⅜ in and under	20
Gravel, sized, washed	12	Wood chips	27
Gravel, sized, unwashed	15		
Gravel, unsized	18		

*For ascending conveyors when uniformly loaded and with constant feed.

‡Angle of repose 30 to 45°.

†See second footnote (†) to Table 3-3 for definitions of "sized" and "unsized" as used in Materials Carried columns of Table 3-2.

§Angle of repose greater than 45°.

¶Very wet or very dry, with angle of repose less than 30°.

TABLE 3-3 Recommended Belt Speed as Determined by Material Handled

Material		Recommended belt speed, ft/min*												
		Belt width, in												
Characteristics	Example	14	16	18	20	24	30	36	42	48	54	60	72	84
Maximum size lumps, sized or unsized														
Mildly abrasive	Coal, earth	300	300	400	400	450	500	550	600	600	650	650	650	650
Very abrasive, not sharp	Bank gravel	300	300	400	400	450	500	550	550	600	600	600	600	600
Very abrasive, sharp and jagged	Stone, ore	250	250	300	350	400	450	500	500	550	550	550	550	550
Half max. lumps, sized or unsized														
Mildly abrasive	Coal, earth	300	300	400	400	500	600	650	700	700	700	700	700	700
Very abrasive	Slag, coke, ore, stone, cullet	300	300	400	400	500	600	650	650	650	650	650	650	650
Flakes	Wood chips, bark, pulp	400	450	450	500	600	700	800	800	800	800	800	800	800
Granular ⅛″ to ½″ lumps	Grain, coal, cottonseed, sand	400	450	450	500	600	700	800	800	800	800	800	800	800
Fines														
Light, fluffy, dry, dusty	Soda ash, pulverized coal						220–250 ft/min							
Heavy	Cement, flue dust						250–300 ft/min							
Fragile, where degradation is harmful	Coke, coal											200–250 ft/min		
	Soap chips						150–200 ft/min							

*Normal for belts traveling horizontally on ball- or roller-bearing idlers. For picking belts, speed is usually 50 to 100 ft/min. Belts with discharge plows should not travel faster than 200 ft/min. For tripers, recommended speed is 300 to 400 ft/min. Trippers for higher-speed applications can be furnished. A speed of at least 300 ft/min should also be maintained for proper discharge when using 35 to 45° idlers, and also for materials tending to cling to belt.

†*Unsized* means a uniform mixture of material in which not more than 10 percent are lumps ranging from maximum size to ½ maximum size, at least 15 percent are fines or lumps smaller than ⅒ maximum, and the remaining 75 percent are lumps of any size smaller than ½ maximum. *Sized* means a uniform mixture in which not more than 20 percent are lumps ranging from maximum size to ½ maximum size, and the remaining 80 percent are lumps no larger than ½ maximum size and no smaller than ⅒ maximum size.

galvanized steel, chromium stainless steels, and other metals or special alloys that are required for a specific application and environment. Wire belts are also available for use where processing temperatures vary from 320 to 2500°F (160 to 1416°C). Wire-belt conveyors are used primarily to move product or unit loads through processes that include liquid or chemical treatment, heat treatment, or kiln firing operations. Wire belts can be cleaned or sterilized while in motion. Mesh openings in the belt permit circulation of water, gases, heat, and cooling air. Typical uses for metal-belt conveyors include operations such as spray-washing glass containers, moving baked goods to ovens, conveying cathode-ray tubes through various processes, and moving hot forgings from automatic die casting equipment.

Tracking of a wire-mesh belt is a problem since the belt is formed by a number of different sections joined together and a wide range of temperatures used in processing operations causes expansion and contraction of the belt material. The conveyor specification must frequently include one of the following features to compensate for these conditions and assure straight belt tracking.

- Multitooth sprocket belt drive
- Belt aligners, consisting of pulleys or rollers mounted on the supporting frame
- Self-tracking belt that has V-shaped wires on the underside that run through grooved driver drums

Surface Chain Conveyors

Surface chain conveyors (Fig. 3-8) include sliding chain, pusher bar, slat, and tow types and car-type trolley conveyors.

Sliding Chain Conveyors. These are the simplest type since they use the chain itself to convey packages down two sliding tracks. The conveyors are used to handle heavier loads than can belt conveyors, such as loaded pallets or unitized loads, but they have the same incline limitations as belt and powered-roller conveyors.

Pusher Bar Conveyors. These are able to convey loads up steeper inclines (to 45°) because the load is pushed by a car connected to the chain drives and this arrangement moves the load along a metal bed or trough. Pusher bar conveyors are generally used for floor-to-floor movement in multistory warehouses or facilities.

Slat Conveyors. These employ an endless chain to drive a conveyor surface of nonoverlapping, noninterlocking slats made of wood or metal. Slat conveyors can be used as moving work tables and to move heavy unit loads and are ideal for applications where the conveyor surface must be flush with a work station or with a floor surface. In the latter application, the installation will permit industrial trucks to cross over or be carried on the slat surface. Slat conveyors can be operated on inclines or declines, the angle of which is limited by the friction between slat surface and the load. Cleats may be added to support loads where steeper inclines are required.

Tow Conveyor. A tow conveyor uses an endless chain either supported from an overhead track or running in a track below the floor surface to tow trucks, dollies, or carts. The under-the-floor tow conveyor system is the type most commonly found in warehouses, and it is versatile because it can be looped around storage areas and moved down aisles and can be moved on spur sections for loading and unloading operations and empty car storage. The track recessed in the floor allows the use of the floor surface for other equipment, but the track and chain drive system cannot easily be relocated once it is installed. Carts and trucks used in a tow conveyor system can range from the ordinary pallet truck fitted with tow pins to engage with the chain drive to special carts or trucks designed for a specific application.

Car-Type Trolley Conveyors. These employ an endless chain to pull a series of small cars or trolleys which carry the material to be moved. These often have fixtures to be used in assembly lines or contain molds for use in foundry processing operations.

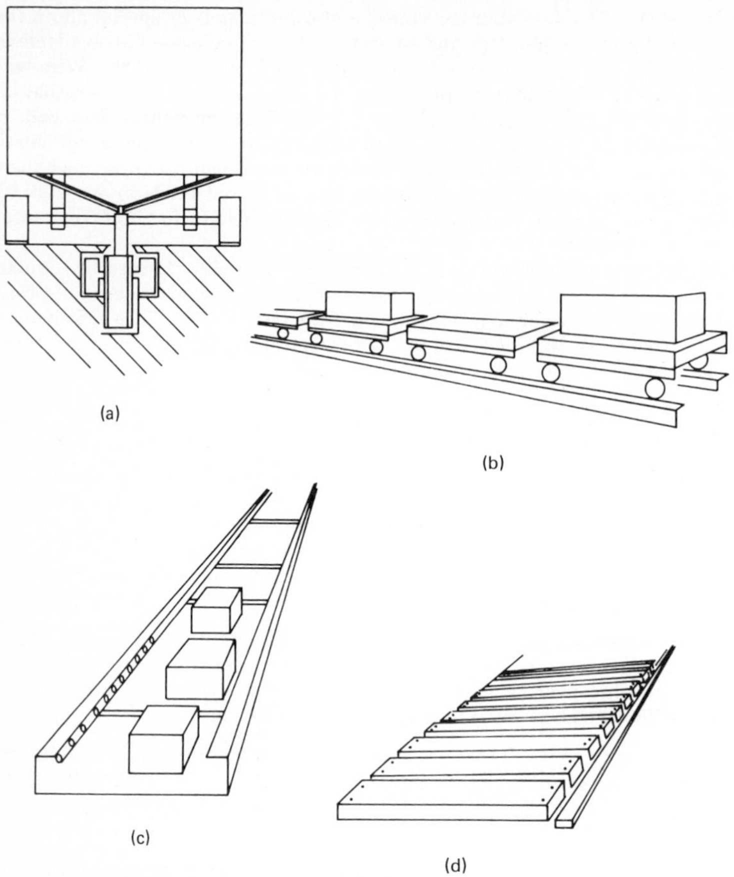

Figure 3-8 Types of conveyors. (*a*) In-floor towline, (*b*) trolley conveyor, (*c*) pusher bar conveyor, (*d*) slat conveyor.

Overhead Conveyors

Overhead conveyors include both trolley conveyors and power and free types of equipment. These conveyors are supported and function within a trolley track driven by a chain power drive to move parts or product. The path of the conveyor can be straight, inclined, declined, and around corners; it can make optimum use of building geometry and follow the general work flowpath within the limitations of building constraints and equipment design parameters. Conveyors can be supported independently or attached to existing beams and trusses, depending on the load factors involved.

In order to determine equipment design parameters, the following procedure should be followed.

1. From process flowcharts, determine all operations to be serviced by the conveyor.
2. Determine the path of the conveyor on a scaled plant layout (Fig. 3-9), showing all obstructions such as columns, walls, machinery, and work aisles.
3. Develop a vertical elevation view to determine incline and decline dimensions (Fig. 3-10). At this point a three-dimensional view could be prepared to give a multiplan view of the proposed installation.

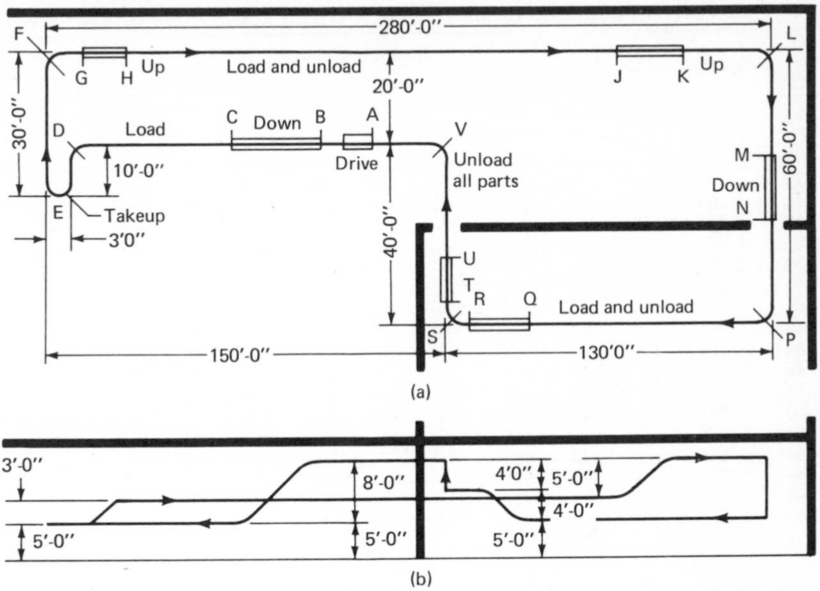

Figure 3-9 Plan and vertical elevation views of trolley conveyor system.

4. Determine the movement rate, unit load size, spacing, and carrier design.
5. Modify turn radii to provide clearances that allow for desired clearance (Fig. 3-10) on turns.
6. Modify loading spacing to provide clearance on inclines and declines. As inclines and declines get steeper, load spacing has to be increased to provide a constant clearance or separation between loads. Table 3-4 shows the load spacing required for a given separation at various incline angles.
7. Redraw the conveyor path and vertical elevation views using new radii and incline information.
8. Compute the chain pull, which is the total weight of the chain, trolleys, and other

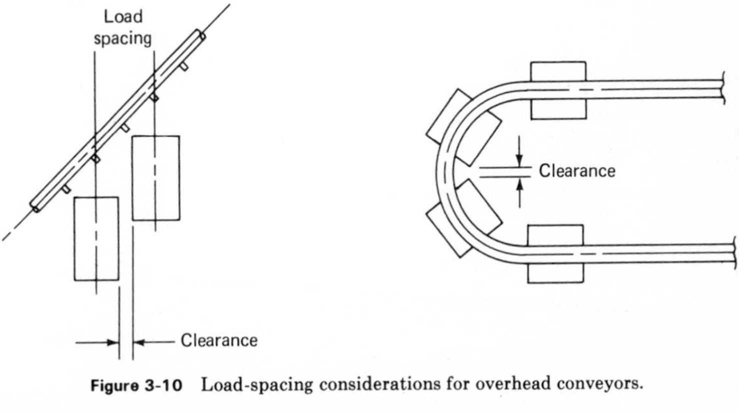

Figure 3-10 Load-spacing considerations for overhead conveyors.

TABLE 3-4 Load Clearance on Inclined Track for Overhead Conveyors

Load spacing, in	Incline angle, deg											
	5	10	15	20	25	30	35	40	45	50	55	60
	Horizontal centers, in											
12	12	11⅞	11⅝	11¼	10⅞	10⅜	9⅞	9¼	8½	7¾	6⅞	6
16	15⅞	15¾	15½	15⅛	14½	13⅞	13⅛	12¼	11⅜	10⅜	9¼	8
18	18	17¾	17⅜	17	16⅜	15⅝	14¾	13⅞	12¾	11⅝	10⅜	9
24	24	23⅝	23¼	22⅝	21¾	20⅞	19¾	18⅜	17	15½	13¾	12
30	29⅞	29⅝	29	28¼	27¼	26	24⅝	23	21¼	19⅜	17¼	15
32	31⅞	31½	31	30⅛	29	27¾	26¼	24½	22⅝	20⅝	18⅜	16
36	35⅞	35½	34¾	33⅞	32⅝	31¼	29½	27⅝	25½	23⅛	20⅝	18
40	39⅞	39⅜	38⅝	37⅝	36¼	34⅝	32¾	30⅝	28¼	25¾	23	20
42	41⅞	41⅜	40⅝	39½	38⅛	36⅜	34⅜	32¼	29¾	27	24⅛	21
48	47⅞	47¼	46⅜	45⅛	43½	41⅝	39⅜	36¾	34	30⅞	27⅝	24
54	53⅞	53¼	52¼	50¾	49	46¾	44¼	41⅜	38¼	34¾	31	27
56	55⅞	55⅛	54⅛	52⅝	50¾	48½	45⅞	42⅞	39⅝	36	32⅛	28
60	59¾	59⅛	58	56⅜	54⅜	52	49⅛	46	42½	38⅝	34½	30
64	63¾	63	61⅞	60⅛	58	55½	52½	49	45¼	41⅛	36¾	32
72	71¾	70⅞	69⅝	67¾	65¼	62⅜	59	55¼	51	46¼	41⅜	36
80	79¾	78⅞	77¼	75¼	72½	69⅜	65½	61⅜	56⅝	51½	45⅞	40

components, as well as the weight of the carriers and load. For example, for a given system the tentative chain pull is calculated as follows:

$$\text{Total tentative chain pull} = 700 \times 60.0 \times 0.03 = 1260$$

where 700 = conveyor length, ft
0.03 = coefficient of friction, %
60.0 = 10.0 lb/ft (chain and trolleys) + 12.5 lb/ft (carriers) + 37.5 lb/ft (line load)

For this initial calculation, inclines and declines are assumed to be level sections if the number of declines balances out the number of inclines; however, for each additional incline, the weight has to be added to determine the total chain pull. If, in our example, a vertical incline that raises a load 8 ft is required, then the additional chain pull is

$$37.5 \text{ lb} \times 8\text{-ft lift} = 300 \text{ lb}$$

The total chain pull then becomes 1260 + 300 lb = 1560 lb.

9. Select tentative conveyor size based on trolley load and chain pull.
10. Select vertical curve radii.
11. Determine power requirements and drive locations. This requires a point-to-point calculation of chain pull around the complete path of the conveyor, which is shown in Fig. 3-9. The following three formulas are used to compute point-to-point chain pull.
    a. Pull for each straight horizontal run:

$$P_H = XWL$$

where
X = 0.02 for standard ball-bearing trolleys
W = total moving weight, lb/ft (empty or loaded, as the case may be)
L = length of straight run, ft

b. Pull for each traction wheel or roller turn:

$$P_T = YP$$

where
Y = 0.02 for traction wheel or roller turn
P = pull at turn, lb

c. Pull for each vertical curve:

$$P_V = XWS + ZP + HW(1 + Z)$$

where
X = 0.02 for standard ball-bearing trolleys
W = total moving weight, lb/ft
S = horizontal span of vertical curve, ft
H = total change of level of conveyor, ft (plus, when conveyor is traveling up the curve; minus, when conveyor is traveling down the curve).
Z = 0.03 for 30° incline; 0.045 for 45° incline; 0.06 for 60° incline; 0.09 for 90° incline
P = pull at start of curve, lb

Drive horsepower may be calculated from the following formula:

$$\text{Drive hp} = \frac{\text{drive capacity (lb)} \times \text{maximum speed}}{33{,}000 \times 0.6}$$

12. Design conveyor supports and superstructures.

13. Design guards which are required by federal, state, and other codes under high trolley runs, particularly over aisles and work areas. Guard panels are normally made from woven or welded wire mesh with structural angles and channels to suit the size and weight of the material being handled.

Power and Free Conveyors. These consist of two separate trolley systems: one moves and is powered by a chain drive; the other has a track under the powered track that accommodates a free-moving trolley containing a carrier or fixture from which a load is suspended (Fig. 3-11). In the powered mode, the powered trolley system is engaged with the free trolley through contact of a pusher dog on the powered system to a retractable dog on the free system. Disengagement is accomplished by contact with another load or by actuating the dog acutator. The main advantage of this system is that one carrier can be stopped whenever and wherever desired without interrupting the whole system. The versatility of the power and free conveyor can be utilized in a process or production line where operations do not take the same amount of time to complete, or where units have to be accumulated for off-line operations, such as for repair. This type of conveyor is used in many diverse industries in applications which include engine and transmission assembly, butchering operations in meat-packing plants, and nonindustrial environments such as hospital distribution of medical supplies and patient meals.

The same design criteria apply to power and free conveyors as to other chain-driven trolley systems.

Bulk-Materials Vertical Conveyors

Bulk-materials vertical conveyors (Fig. 3-12) are generally used to lift bulk materials up to silos, hoppers, or other storage containers from which the material may be dispensed into a mixing, packing, truck-loading operation, or directly to a process. Some of the industries that use this equipment include glass, agricultural fertilizer, and powdered chemicals.

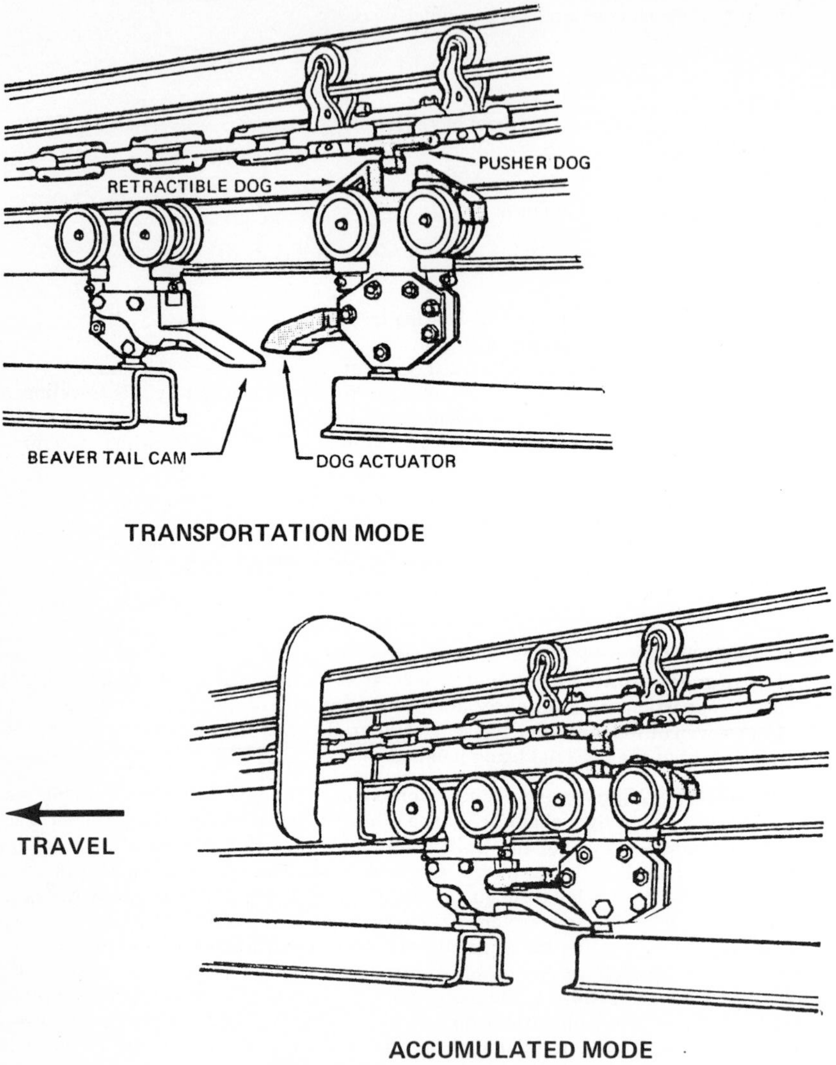

Figure 3-11 Power and free trolley.

Skip Hoists. These are used to lift bulk materials handled in batches to very high points. A bucket which carries the material moves vertically in guides and is raised and lowered by a hoist-operated cable.

Gravity Discharge Conveyor-Elevators. These carry material in both horizontal and vertical paths. The buckets are rigidly mounted on two strands of chain running in tracks. Material is loaded into a bucket at the base of the equipment by feeding material into a lower trough, and discharge is effected when the bucket position changes in the horizontal run.

Bulk-Flos. These lift material by the use of flights attached to a chain drive which is

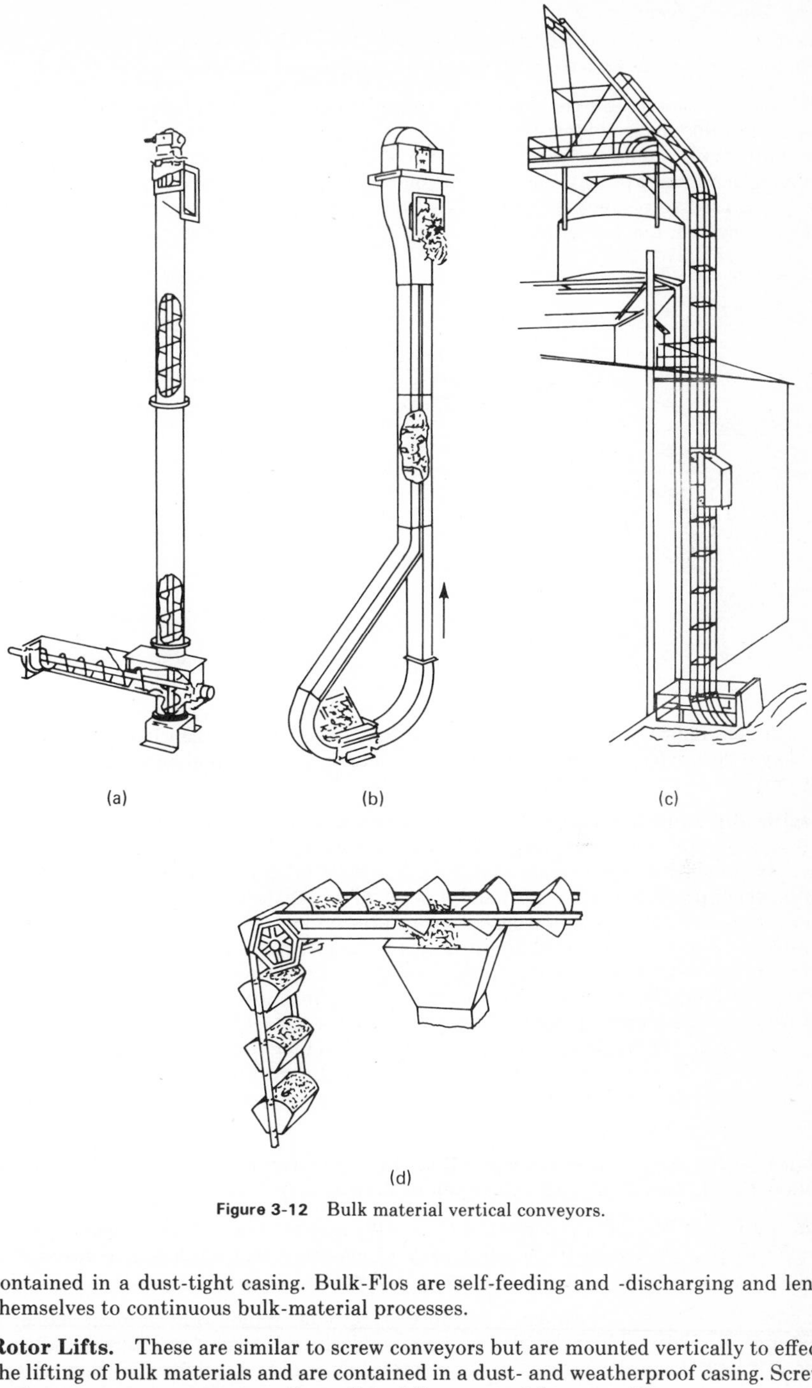

Figure 3-12 Bulk material vertical conveyors.

contained in a dust-tight casing. Bulk-Flos are self-feeding and -discharging and lend themselves to continuous bulk-material processes.

Rotor Lifts. These are similar to screw conveyors but are mounted vertically to effect the lifting of bulk materials and are contained in a dust- and weatherproof casing. Screw feeders or conveyors are generally used to deliver material to rotor lifts.

Other Specialty Conveyors

There are innumerable variations on standard conveying systems, some of which are unique to individual industries. Six common examples are described below.

Screw Conveyor. This conveyor (Fig. 3-13) consists of a screw rotating in a stationary trough and the material moving along its length by rotation of the screw. This type of conveyor serves a dual purpose since it can also be used to perform processes such as blending and mixing of material while the material is being moved. The conveyor is generally enclosed to prevent dust or fumes from escaping and allow the conveyor to be cooled or heated. Loading or discharging can be located at any point along a conveyor.

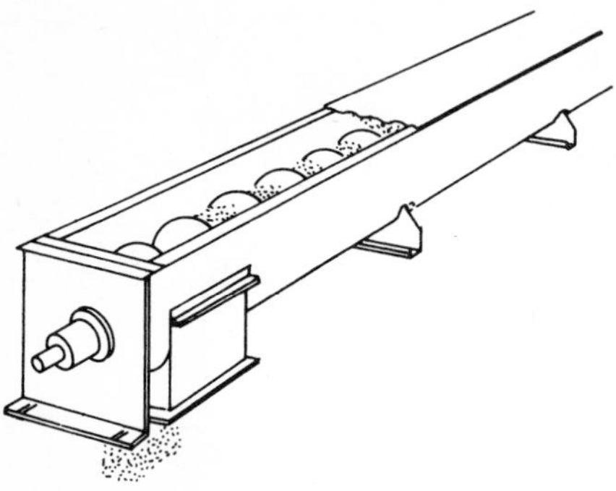

Figure 3-13 Screw conveyor.

Spiral Track Conveyors. These conveyors (Fig. 3-14) consist of a continuous spiral track with a power drive which turns the track, moving anything which is hung on it. It has wide application in the garment industry. It is generally used for items weighing less than 10 lb (5 kg). Interlocking nylon wafers can permit turns to be made in any direction on a radius of 18 in (46 cm).

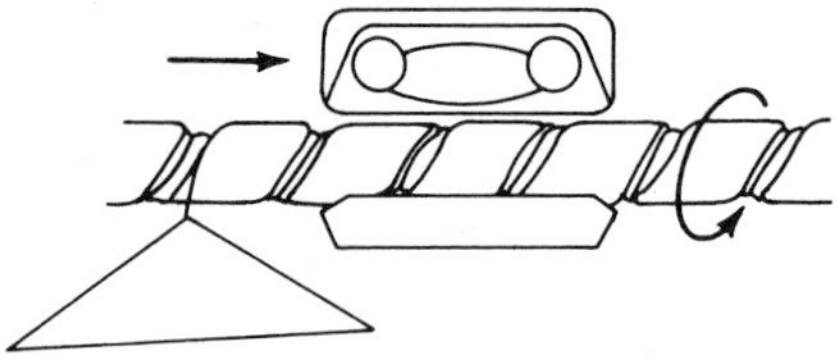

Figure 3-14 Spiral track conveyor.

Oscillating and Vibrating Conveyors. These use the natural frequency vibration of a trough to provide a conveying action to move material. Oscillating conveyors use a mechanically driven power train to move a trough carrying material against spring supports which provide a fast return and downward stroke, causing the trough to vibrate and convey the material. Vibrating conveyors utilize some form of magnetic pulsation to create this vibration motion. Wider variations of frequency are possible by simple control for vibrating conveyors, enabling speed changes compensating for material differences.

Application of both types of conveyors is growing in a number of different industries for uses such as: conveying light food products such as cereal in the food industry; moving, cooling, and breaking up lumps of casting sand in foundries; quenching and removal of glass cullet in water-filled troughs in the glass industry; removing ferrous from nonferrous materials in separation systems; and feeding small parts into automatic packaging or assembly equipment.

Flight Conveyors. These conveyors (Fig. 3-15) use scraper plates to push nonabrasive bulk material through a trough which can be horizontal or inclined.

Apron Conveyors. These conveyors (Fig. 3-16) use a series of interlocking apron pans supported in a stationary frame for conveying materials that are heavy, abrasive and lumpy, such as ore, stone, industrial refuse, and waste materials.

Pneumatic Tubes. These use a pressure or vacuum system to move materials or a container at relatively high speed. The major application is that of an internal mail carrier, although it can also be used to move certain types of high-volume fine particulate.

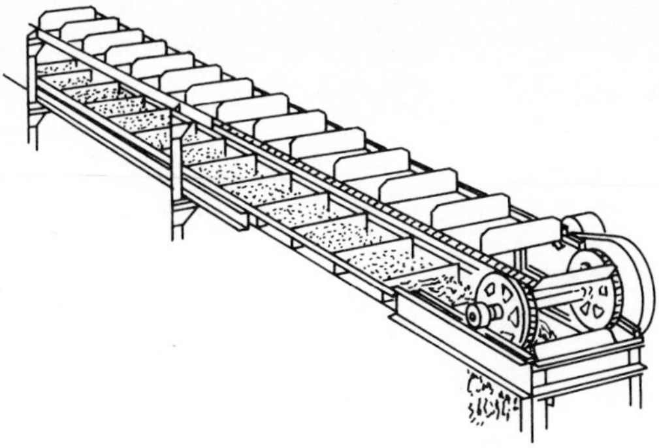

Figure 3-15 Flight conveyor.

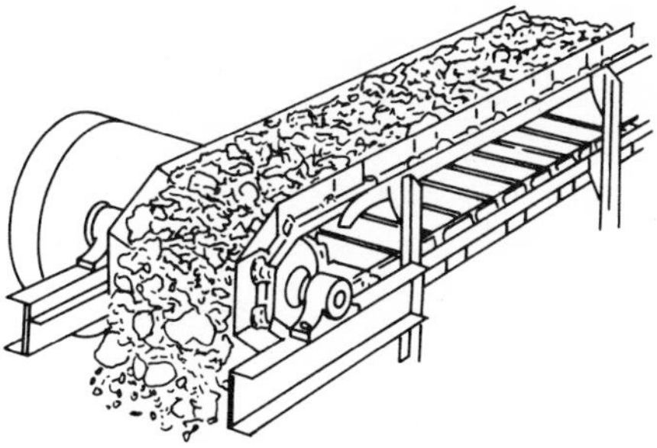

Figure 3-16 Apron conveyor.

SORTING, CONSOLIDATING, AND DIVERTING DEVICES

A materials handling system must frequently have the ability at some point to identify, sort, and divert parts, products, or unit loads. Peripheral accessories and equipment do this, ranging from simple mechanical diverters to sophisticated optical recognition reading devices, which can actually read and identify alphanumeric characters and sort 20,000 items per hour and which are used mainly for check and mail handling. Whatever the complexity of the system, three basic elements must be considered: identification of the item to be sorted or consolidated, recognition of the item, and the command to activate the mechanisms to divert the item.

Simple Mechanical Sorting

Simple mechanical sorting utilizes inherent differences such as size, shape, weight, or other physical differences to identify or recognize items; it generally is contact sorting in which an item must make contact with a channel or feeler guides or discerns physical differences and contacts a cam or other simple mechanism to activate a diverter chute or other diverting device.

Diverting Mechanisms

Diverting mechanisms can be grouped into devices that deflect, push off, drive off, or tilt; many variations are included within each group (Fig. 3-17).

Electromechanical Sorting

Electromechanical sorting uses noncontacting identification devices that can sense both inherent differences and applied differences. These are identified on the load or package

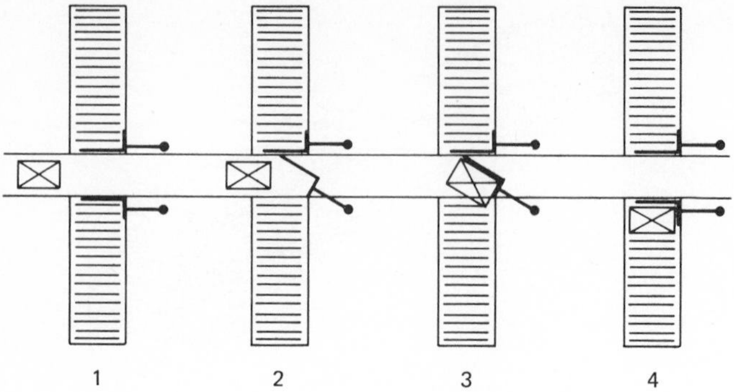

(a)

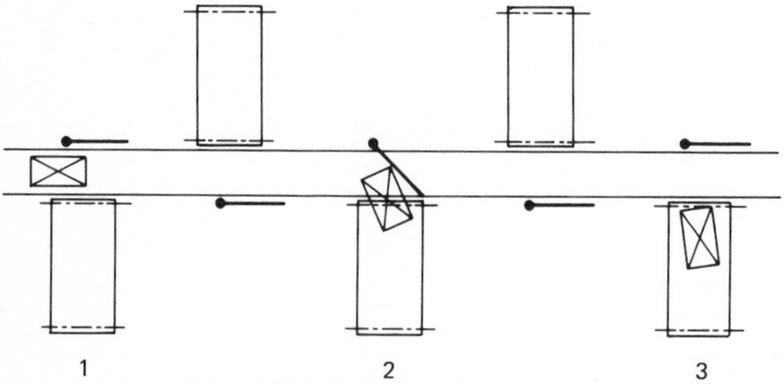

(b)

Figure 3-17 Diverting mechanisms.

by a code that can be discriminated by a sensing or scanning device and that triggers a diverting mechanism.

Photosensors. These are the most commonly used sensing and scanning devices. A photoelectric control consists of a light source, photoreceiver, amplifier, and output. A beam of light from the light source activates the photosensitive elements of the photoreceiver which produces an electric signal, which drives a relay to activate the diverting mechanism. When used as a sensor, photoelectric controls can be used to:

- Sense the presence or absence of containers or products on a conveyor
- Detect over- and undersize products
- Sort products by size
- Count items

Photoelectric controls can also be arranged in an array or ganged in a manner that a code on a container can be read. Each sensor that goes into the scanning device will detect a specific mark or blank space from the code which when scanned produces binary information into a logic function of a controller. This in turn supplies a signal for the diverting mechanism. Typical codes are ladderlike in format, and this allows a scanning device to read the code vertically.

The use of code scanning in a materials-handling system must consider the following criteria:

- Required information content
- Method of code application
- Range and depth of field requirements
- Nature and speed of product flow
- Value and volume of products

The state of the art in scanning technology now includes laser scanners and fiber optics. These are used as optical recognition devices which can actually identify alphanumeric characters. The charge-coupled device (CCD) is a self-contained device that scans a package and is able to detect many different levels of light intensity. It can produce an accurate representation of the object being scanned, enabling the CCD to be used as an observation or inspection device.

Palletizers

Palletizers receive individual packages, cases, or bags from a conveyor and automatically arrange them on a pallet in a predetermined pattern with the required number of tiers. Each tier need not have the same predetermined pattern. Generally, units to be stacked are received on a control belt at the entrance to the machine. At this point the unit will be counted and oriented depending on the patterns required. As each row is completed, a pusher moves the cases onto an apron. When the tier is completed, the apron is withdrawn, depositing the tier on the pallet or tier below. The operation is repeated until the pallet load is completed, when it is discharged and replaced with an empty pallet.

Large volumes of standard units are required, and it is estimated that palletizers become economical when approximately 900 units per hour require palletizing. The larger palletizers can handle in excess of 6000 cases per hour of certain products.

Depalletizers

Depalletizers are highly specialized pieces of equipment which automatically depalletize cartons and cases. Automatic squaring mechanisms permit the handling of loose pallet loads. Depalletizers generally operate in the range of 3500 cases per hour and are seen primarily in beverage distribution.

HOISTS AND CRANES

Hoists and cranes are materials-handling equipment used to move varying loads intermittently within a fixed area. The loads vary in size and weight and are not uniform. Most of the materials movement is devoted to raising and lowering loads, although some units are so constructed as to permit them to travel laterally over a specific area. The types of hoists, cranes, and attachments are listed below.

Hand and Powered Hoists

Hand and powered hoists (Fig. 3-18) are the most basic and economical lifting equipment which enables an operator to move a large load, up to 50 tons, vertically by using some kind of mechanical advantage.

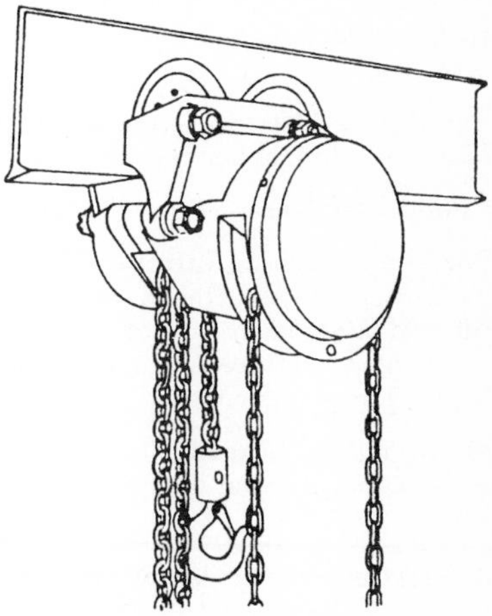

Figure 3-18 Hoist.

Jib Cranes

Jib cranes (Fig. 3-19) consist of a hoist that is mounted on a boom track. The hoist mechanism can be moved laterally in the track and the boom can be turned in an arc limited by the building restrictions or the mounting arrangement of the boom. Jib cranes are classified into basic groups of bracket jib, cantilever jib, and pillar jib. Load capacities range from small manually operated cranes to loading towers that exceed 300 tons.

Bridge Cranes

Bridge cranes consist of a hoist mounted on a guider bridge which is supported by two trucks on each end and rides on runways supported by building members. Top-running bridges, where end trucks ride on top of runway tracks are able to support a total bridge and load weight of hundreds of tons, but underhung or bottom-running bridges, where the trucks are suspended from the lower flanges of the runway track, normally are used for loads less than 20 tons. Bridge cranes can be operated manually or powered or in the cases of very large cranes can be operated by remote control (Fig. 3-19).

Gantry Cranes

The gantry crane is very similar to a bridge crane except it is supported by self-contained vertical support members that travel in tracks on the floor surface and it is generally used where overhead runways are not feasible due to building restrictions. The gantry crane system also has the advantage of being usable in outdoor operations without the construction of an expensive supporting structure (Fig. 3-19).

Stacker Cranes

The stacker crane consists of a rigid mast suspended from an overhead bridge that travels laterally. A platform or a set of forks moves up and down on slider bars to lift and lower loads. The stacker crane is most commonly used to place or retrieve loads to and from racks from both sides of an aisle. In automatic storage and retrieval systems, the stacker crane is computer-controlled. The computer has the rack location of each item stored in the memory and is able to command the load-carrying platform to a specific location for storage and retrieval of a load.

Lifters

A lifter (Fig. 3-20) is an attachment suspended from the load hook of a hoist or crane that permits a load to be handled more easily or quickly than possible with a hook, and many load configurations cannot be handled with a hook. In many cases, lifters are designed for a specific application, but there are many standard types that are available for a wide range of applications. Lifters are categorized by the method in which the load is carried.

Supporting Lifters. These carry the load on the surface of the lifter, on bearing surfaces of cradles, or hooks and slings attached to lifters.

Clamping Lifters. These hold the load by surface friction or by squeezing load.

Surface-Attaching Lifters. These consist of both magnetic or vacuum types. Magnetic lifters can use either a permanent magnet that requires a strip-off device to release the load or an *on-off magnet* that can be activated by applying a voltage. Vacuum pads can be used to lift loads with nonporous and smooth surfaces and are commonly used to handle glass and aluminum.

Manipulating Lifters. These move the load through one or more axes for operations such as positioning or dumping.

GUIDED (DRIVERLESS) VEHICLES

Guided (driverless) vehicles move material over fixed paths but do not require the use of an operator or a mechanical drive train located below the floor surface or an overhead

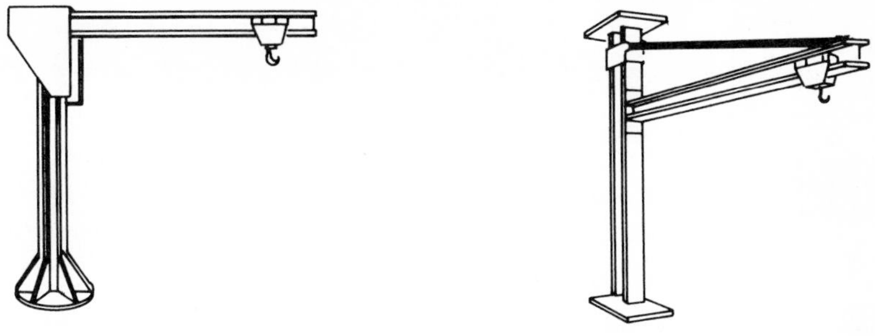

(a)

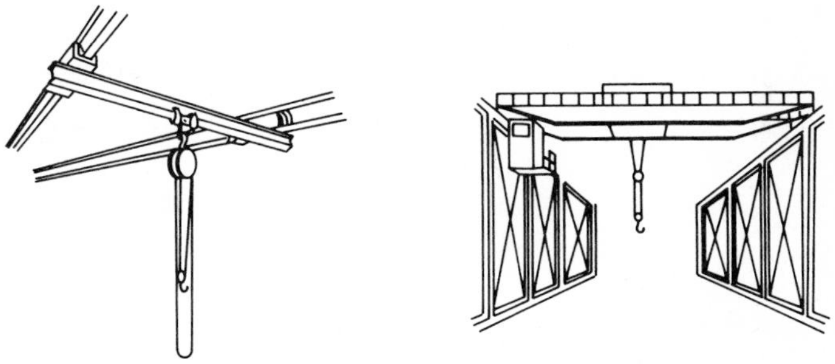

(b)

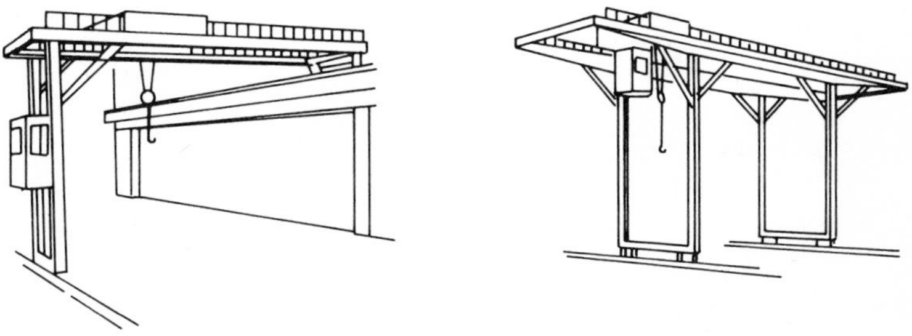

(c)

Figure 3-19 Types of cranes. (*a*) Jib crane, (*b*) bridge crane, (*c*) gantry crane.

towline. They are useful when a variety of materials must be moved over long distances to and from a variety of fixed destinations. There are three identifiable types of vehicles: first, the driverless tractor (Fig. 3-21) which hauls trailers or cartloads of material; second, the individual unit-load or pallet mover (Fig. 3-22); and third, the multishelved self-contained vehicle. The last type is used primarily to move mail in office buildings or for food and supply deliveries in hospitals.

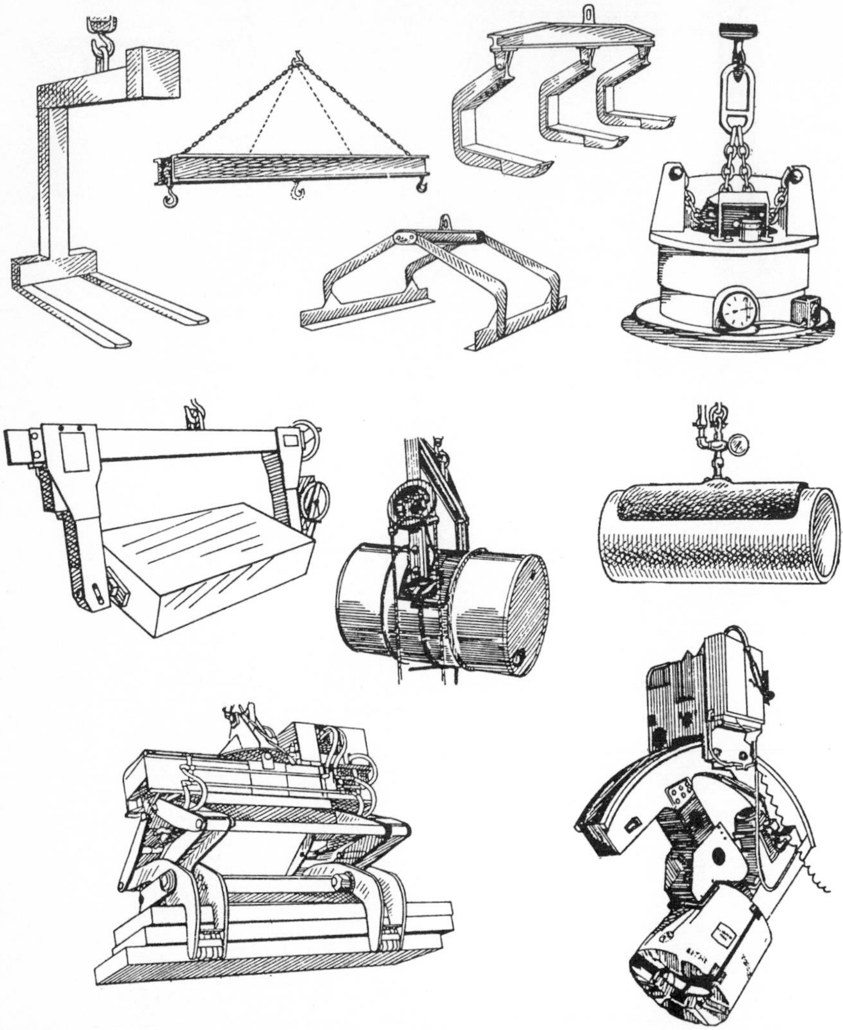

Figure 3-20 Types of lifters. *(Reproduced with permission from Material Handling Engineering Handbook and Directory, 1977/1978,* published by *Material Handling Engineering,* Cleveland.)

Guidance and Control Systems

Guidance and control systems are similar for all three systems. Two systems are used: optical, where the unit follows a line taped or painted on the floor surface; or magnetic, where a thin wire is set in a shallow channel sealed over in the floor. This latter system is less flexible and more costly to control but is not subject to obliteration or wear, which can be a problem in certain factory environments.

The driverless tractor, being unable to reverse on its own trailers, generally requires a closed-loop system. However, multiple-loop systems can be used. Unit-load movers are generally reversible and can operate on a spur.

The programming information which determines the paths and stops can be preset on the tractor programmer or can be controlled from a central dispatching point. These systems generally have the logic to allow the tractor to take the shortest route to the destination without traveling through the entire loop. Radio-control-transmitters are often used to reposition the train within a loading station, eliminating unnecessary walking in operations such as order picking or loading the train at the receiving dock.

Loading and Unloading

Although all vehicles can be loaded and unloaded with operator assistance, both tractors and unit-load movers can have automatic load and unload features. The tractor-trailer arrangement can have an automatic uncoupling option. More common are options whereby the trailers have rollers on the carrying surface and the loading-unloading stations where a pusher can be used to move the load. Similar systems can be used for unit-load movers, sometimes using powered roller systems. More common is the lifting device established in Fig. 3-22. This has particular potential in manufacturing operations where materials can be brought directly into the work station.

Routes are dependent on surface conditions. Cracked and broken slab can cause discontinuity in the tape or wire guides. Inclines and declines within a plant must be considered, in which case an acceleration or deceleration feature must be specified for the equipment. External routes linked with automatic door control, internal traffic lights, and automatic ramps to cover rail lines have been used. However, external use of this equipment is not widespread, and external surfaces must be prepared very carefully, especially in regions where snow and ice are involved.

Safety

Driverless tractors are available with many more safety options than any other automatic conveying system and include such features as encounter detection, sonic detectors, and optical detectors, which will all shut the tractor down if an object is detected in the path. Additional safety devices include a strobe light, siren, and panic buttons which can override all other controls. Using warning signs and placing mirrors at corners and blind spots are good preventive measures and so is keeping the tractor speed below 5 mi/h.

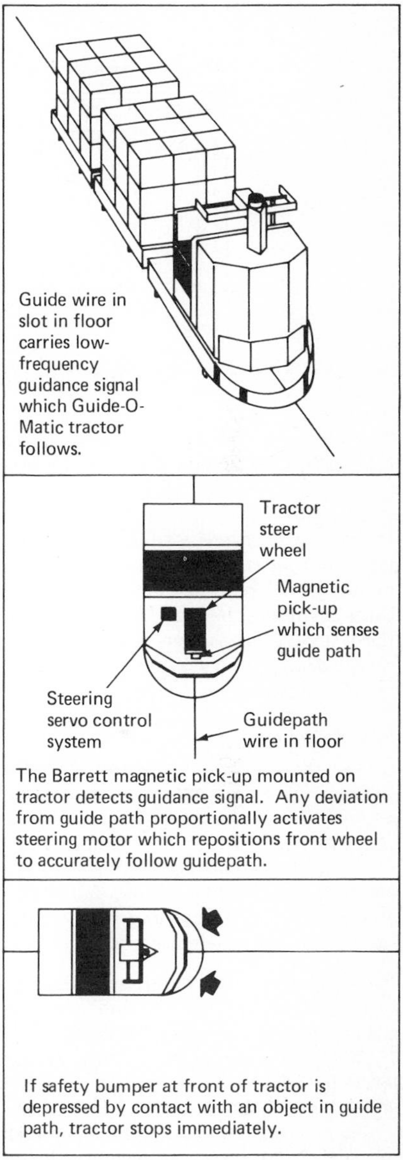

Guide wire in slot in floor carries low-frequency guidance signal which Guide-O-Matic tractor follows.
Tractor steer wheel
Magnetic pick-up which senses guide path
Steering servo control system
Guidepath wire in floor
The Barrett magnetic pick-up mounted on tractor detects guidance signal. Any deviation from guide path proportionally activates steering motor which repositions front wheel to accurately follow guidepath.
If safety bumper at front of tractor is depressed by contact with an object in guide path, tractor stops immediately.

Figure 3-21 Typical features of a driverless tractor system.

ROBOTS

Robots are programmable machines capable of automatically moving individual parts or objects over precise paths in space.[1] A robot can also be programmable so that it is able

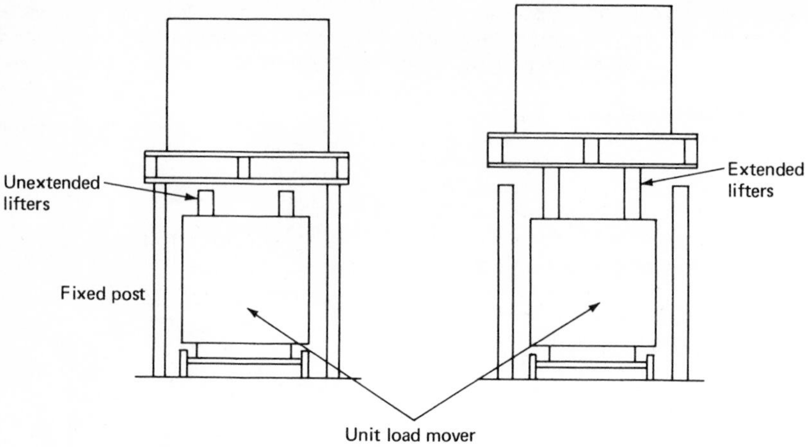

Figure 3-22 Individual unit load or pallet mover.

to move parts through different paths, capable of performing repetitive motions, able to duplicate the movements of the human arm by moving parts through four axes in space.

Applications

Present applications related to materials handling include machine loading and unloading, conveyor transfer, and pallet loading. The most practical applications for materials handling will be those areas that require repetitious manual operations, particularly those involving the interface between workers and machines. Robots are also ideal for these types of operations in poor working environments, such as those where heat, cold, fumes, or radiation exposure is present. Painting and welding are typical major potential application areas.

Design Components

Robots (Fig. 3-23) are available with a wide range of capabilities and in various design configurations. The major components include a manipulator which actually performs an operation and moves parts, a controller that stores data and directs the movements of the manipulator, and the energy source to power the robot.

A sophisticated robot with six axes of motion can perform many of the same movements as the shoulder, elbow, and wrist. Simpler, less expensive units with two degrees of freedom, called *put-and-place units,* are typically used for machine loading and should become widely used in the materials handling field in the next decade.

Manipulator. The handling of objects by the manipulator is facilitated by the use of tools that give the robot "hand" capability. The general categories for this purpose are either grippers or surface-lift devices.

Mechanical Grippers. These grippers (Fig. 3-24) are usually movable fingerlike levers paired to work in opposition to each other. They can be thought of as mechanical equivalents of the thumb and forefinger.

Surface-Lift Devices. These can include simple forklift attachments, vacuum pickups (Fig. 3-25), hooks, or magnetic devices.

Controller. The controller initiates the motions of the manipulator through a sequence at the desired points and stops the motion when required. The controller can be programmed by adjustment of mechanical cams, stops, and limit switches on the simpler types of put-and-place robots. The more sophisticated robots can be "taught" a sequence of movements by an operator. In the teaching mode, the programmer manually moves the manipulator through the motions of the operation, and the coordinates of the path are stored in the controller memory.

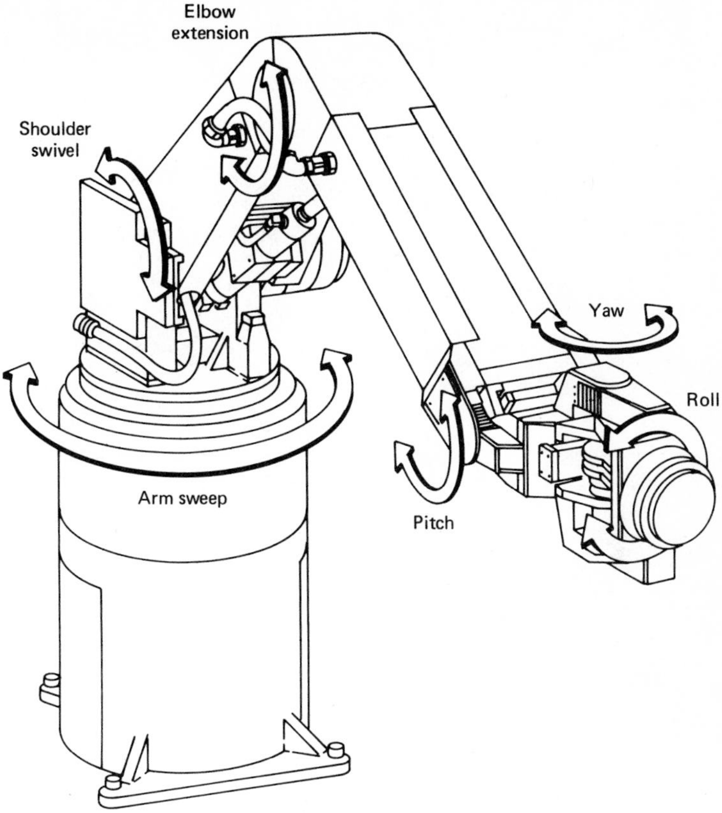

Figure 3-23 Robot with six axes of motion.

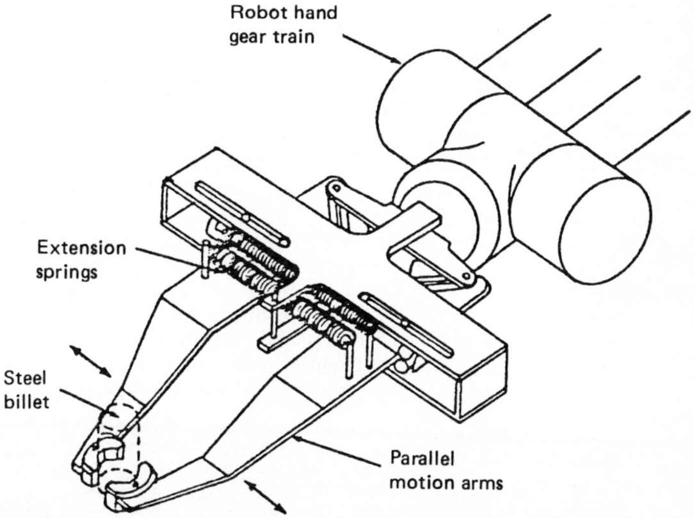

Figure 3-24 Robot grippers equipped with spring-loaded fingers.

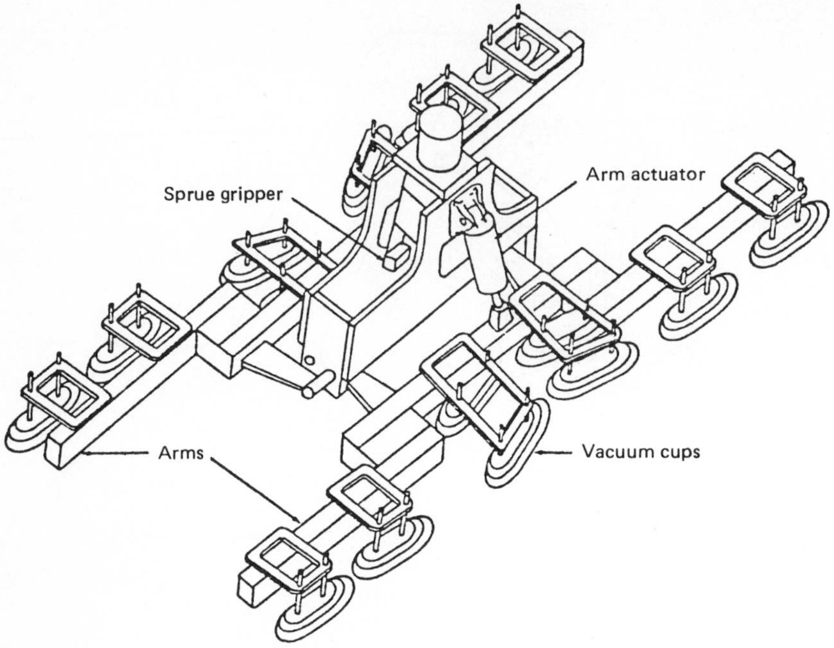

Figure 3-25 Vacuum pickup device for robot.

Energy Sources. Nonservo, or *pick-and-place* robots operate through activation of a hydraulic or pneumatic system and are the simplest, lowest-cost units. They have limited flexibility in terms of program capability and positioning capability but are highly reliable. In the operation of this type of robot, as the sequence is indexed, the manipulator members move until the present limit of travel is reached. Since there are only two positions for each axis to assume, programming can be done by adjusting the end stops for each axis to establish the operation sequence.

Servo-type robots use servo motors or valves to move the manipulator members and can be further classified into either point-to-point or continuous-path types. *Point-to-point* servo robots are programmed or taught by feeding them manipulator-position data at discrete points and, in performing a task, they will internally select a path to that point. *Continuous-path* servo robots are programmed or taught to follow a precise path and are used for operations where movement is important, particularly in spray painting.

Future Developments

The technology of robots will be expanded in the future to include the capability to discriminate differences in objects by optical- or mechanical-sensing devices which would send a feedback signal to the controller which will make a decision to initiate a movement command to the manipulator. Further future developments include speech recognition for robot programming and three-dimensional optical-sensing devices. Also, while robots now in operation are generally large, floor-mounted units, future robots will also include table-mounted units able to assist in small subassembly and final assembly operations.

Planning Considerations for the Use of Robots

There are four points that must be considered when evaluating the feasibility of using a robot in materials handling. They are rate of handling, weight of the object, orientation of the object, and number of different items to be handled.

Rate of Handling. Robots are not high-speed handling equipment. If the handling rate is greater than 15 items per minute, another approach should be considered.

Weight of Object. The weight-handling capacity of robots is presently 500 to 2000 lb, depending on the type of robot. The heavier the load, the lower the handling rate.

Orientation of the Object. Position of the object is important and should be consistent. A primary limitation of current robots is the precise orientation required of parts to be picked up by the robot and, hence, a feeding or positioning mechanism to the robot itself is often required.

Number of Items to be Handled. Setup time for product changes can be reduced by quick changeover grippers and automatic program selecting capability. In cases where dissimilar parts are handled in the same operation, a multipurpose gripper or "hand" should be used, along with a sensing device that can command the robot to switch to a preset program.

REFERENCE AND BIBLIOGRAPHY

1. Tanner, W. R.: *Industrial Robots,* Vol. 1 and 2, Society of Manufacturing Engineers, Dearborn, Mich., 1979.
2. *Automated Storage/Retrieval Systems Planning Guide,* Clark Handling Systems, 525 W. 26th St., Battle Creek, MI 49016
3. *Automated Storage/Retrieval Systems Justifications,* Clark Handling Systems 525 W. 26th St., Battle Creek, MI 49016.
4. Industrial Robots, *Modern Material Handling,* April 1980.

chapter 8-4

Mobile Materials-Handling Equipment

INTRODUCTION

The group of equipment that is described as mobile materials-handling equipment is made up of machines that essentially depend on a self-contained power source for movement and are independent in their movement route. The equipment, being self-contained material movers, provides a flexible, relatively inexpensive transportation link between plant activities. This broadly classified group of equipment includes devices and equipment from the simplest two-wheeled hand truck to highly sophisticated movers controlled by computer-based systems.

Within the mobile materials-handling equipment group there is a wide array of general-purpose and specialized material movers. Basically, there are two broad categories of mobile equipment. The powered equipment depends on a built-in power source for its operation. The unpowered device relies on a detachable prime mover, either a piece of powered equipment, or in many cases, the equipment operator. The least complex equipment provides transportation between two points without positioning or lifting capabilities. Other units lift or roughly position the load being transported as well as move the material. The multiple-axis movers transport the load; they also have a position capability along two or more axes to accomplish loading and unloading.

Generically, mobile materials-handling equipment falls into five groups, each of which will be discussed in this chapter:

1. Floor trucks and operator-powered movers
2. Powered lift trucks

3. Burden carriers
4. Tractors and tractor trains
5. Mobile industrial cranes

APPLICATION CONSIDERATIONS

Equipment Utilization and Selection

From available records, it appears that mobile equipment often has a low level of utilization. Powered equipment is often employed well beyond its economic life, generating penalty costs in spare-parts inventories, maintenance, and productivity.* Eleven thousand (11,000) engine hours, or approximately 5 years, has been calculated to be the average economic life of a powered vehicle. Other general considerations in establishing equipment requirements include:

- Unit-load condition and size and center of load
- Terrain, environment, and aisle width in the movement area
- Length, type, and frequency of moves
- Positioning requirements of load(s)
- Hazards inherent in movement area
- Operating economies and maintenance ease
- Maintenance and spares
- Standardization of equipment
- Critical nature of operation(s) serviced

Factors in Wheel Selection and Use

Solid Wheels. These are made in semi-steel, forged steel, or molded plastic, hard rubber, and composite materials. They should be limited to small diameters and low-speed movement and should not be used to transmit power. They have low resistance to roll, but a short life span when overloaded or subjected to rough floor conditions. They will cause load vibration because of a lack of cushioning.

Rubber-Cushioned Tired Wheels. These consist of a metal wheel having a machined diameter onto which a rubber tire is pressed or molded. It has the lightest load-carrying capacity of those used on mobile equipment. Minimal power is required to move material, since rolling friction is minimized.

Oil-Resistant Tired Wheels. The tires are made of special oil-resistant rubber compounds which will resist the degrading effects of oil on rubber.

High-Traction Tired Wheels. The tires are made of rubber impregnated with abrasive or other materials to give additional traction on ice or in wet conditions.

Low-Power Tired Wheels. The tires are fabricated from rubber compounds that offer minimum roll resistance and have lower power requirements, causing less drain on battery-operated equipment.

Nonmarking Tired Wheels. The tires use a rubber compound filler other than carbon to avoid floor marking and contamination.

Conductive Tired Wheels. The tires avoid the chance of static sparking in hazardous or explosive environments by maintaining vehicle-to-floor conductivity.

*"When to Replace Your Lift Truck," *Material Handling Engineering,* July 1980.

Laminated Tired Wheels. The tires for these wheels are made up of sections of pneumatic tire carcasses threaded onto a steel band. Such tires are extremely tough, with a harsh ride. They are well suited to littered environments, such as scrap yards, and trash handling.

Polyurethane Tired Wheels. Though more expensive than rubber, these wheels have a significantly higher load-carrying capacity and are less susceptible to cuts than most rubber and rubber-compound wheels. Wheel hardness of polyurethane tires results in a harsher ride and increased plant floor damage.

Inflatable Tired Wheels. These wheels have vulcanized, reinforced rubber tires similar to automotive tires. The tires are both tube and tubeless. They generally carry a lower load rating for their size than solid-tire wheels. Their use will provide greater load cushioning, higher speed capability, easier maintenance, and less floor damage.

Factors in Internal-Combustion-Engine Selection and Use

Internal-Combustion Engines. These are used in outdoor applications, in well-vented interiors, in nonhazardous environments, and where noise is not a factor.

Industrial Engine. Typically, this heavier engine is designed to operate in a lower rpm range than an automobile engine. It can be expected to give about 10,000 h of useful life before overhaul. At an equivalent operating speed of 20 mi/h in an automobile, this would equate to 200,000 mi.

Automotive Engine. This is of lighter construction than the industrial engine and, because of the quantities in which it is produced, is of relatively lower cost. It generally operates most efficiently in a higher rpm range than the industrial engine and can be expected to give about 7000 h of useful life prior to overhaul. This life is equivalent to about 140,000 mi of automobile travel. An advantage of this type of engine is the availability of replacement parts through automotive supply firms.

Air-Cooled Engine. This is restricted to lighter-duty applications where weight, size, and initial cost are the prime concerns. The absence of a separate cooling system is a distinct advantage, although this engines life expectancy is a relatively short 1500 to 2000 h of operation.

Diesel Engine. Typically, this type is installed in large pieces of equipment where the additional size and cost is not significant. However, because of recent improvements in engine design, diesel engines are becoming more prominent in smaller trucks. This is largely due to the reduced need for periodic maintenance, greater fuel economy per hour of operation, and longer expected life—up to 20,000 h.

Factors in Battery-Powered-Vehicle Selection and Use

Battery-Electric Equipment. This is mechanically simpler in design than engine-driven equipment. Typically, the high-torque dc electric-drive motor is coupled directly to the drive axle through a constant-mesh drive train. An electronic SCR speed-control device regulates the motor's revolutions per minute through operator foot control. Direction is reversed electrically with a delay interlock to avoid reversing motor direction while in motion.

Storage Battery. These must be replenished frequently either by recharging or by exchanging them for fully charged batteries. Batteries used in a given piece of equipment should provide ample power to operate effectively for an 8-h day as determined by their ampere-hour (Ah) rates. The Ah rating, to some degree, limits the effective operating range of battery-operated equipment and requires that routine schedules for replenishment are followed. Also, because of the weight of a large storage battery, equipment application is sometimes adversely limited.

Advantages of Battery Vehicles. The advantages are low fume emission and heat contamination, quietness and cleanliness, and generally lower maintenance requirements.

Types of Batteries. The two primary types of batteries used are lead-acid and nickel-iron-alkaline. A lead-acid battery will provide 2.0 to 2.3 V per cell, while the nickel-iron-alkaline battery will provide 1.2 V per cell. Voltages used for modern battery-powered mobile equipment are 12, 24, 36, 48, and 72, with some higher voltages used in larger equipment.

Advantages. The advantages of the lead-acid battery are a lower initial cost, high ampere-hour capacity, and low resistance to self-discharge. The nickel-iron-alkaline battery is desirable because of its longer life expectancy, resistance to physical damage, noncorrosive electrolyte (KOH), and more rapid and less critical recharge rates.

Recharging Times. These are adjusted for different batteries by dividing the Ah rating of the battery by the 8-h Ah rating of the charger and multiplying by 8. For example, a battery having a 600-Ah rating and a 450-Ah charger will require

$$(600 \div 450) \times 8 = 10.64 \text{ h}$$

FLOOR TRUCKS AND OPERATOR-POWERED MOVERS

This type of equipment is the most fundamental materials-handling aid available. The basic simplicity permits easy adaptation for single-purpose application. Standard catalogs indicate the wide variety available, often designed for specific industries. However, custom design may be specified with very little, if any, cost penalty.

Generally, floor trucks are described as follows.

Two-Wheeled Hand Trucks

Two-wheeled hand trucks (Fig. 4-1) are essentially levers on two wheels. The axle connecting the wheels serves as the fulcrum of the lever and carries up to 80 percent of the total load moved. The two-wheeled cart is normally used for short nonrepetitive moves of smaller loads over smooth floors. Carts are generally 48 to 64 in (1.2 to 1.6 m) high,

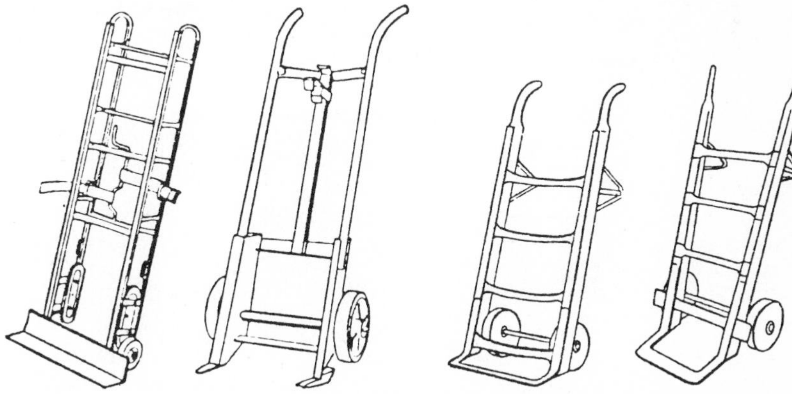

(a) (b) (c) (d)

Figure 4-1 Two-wheeled hand trucks. (*a*) Appliance type, (*b*) drum and barrel mover, (*c*) general type with Western handle, (*d*) general type with Eastern handle. (Reproduced with permission from *Material Handling Engineering Handbook and Directory, 1979–1980*, published by *Material Handling Engineering* magazine, Cleveland, Ohio.)

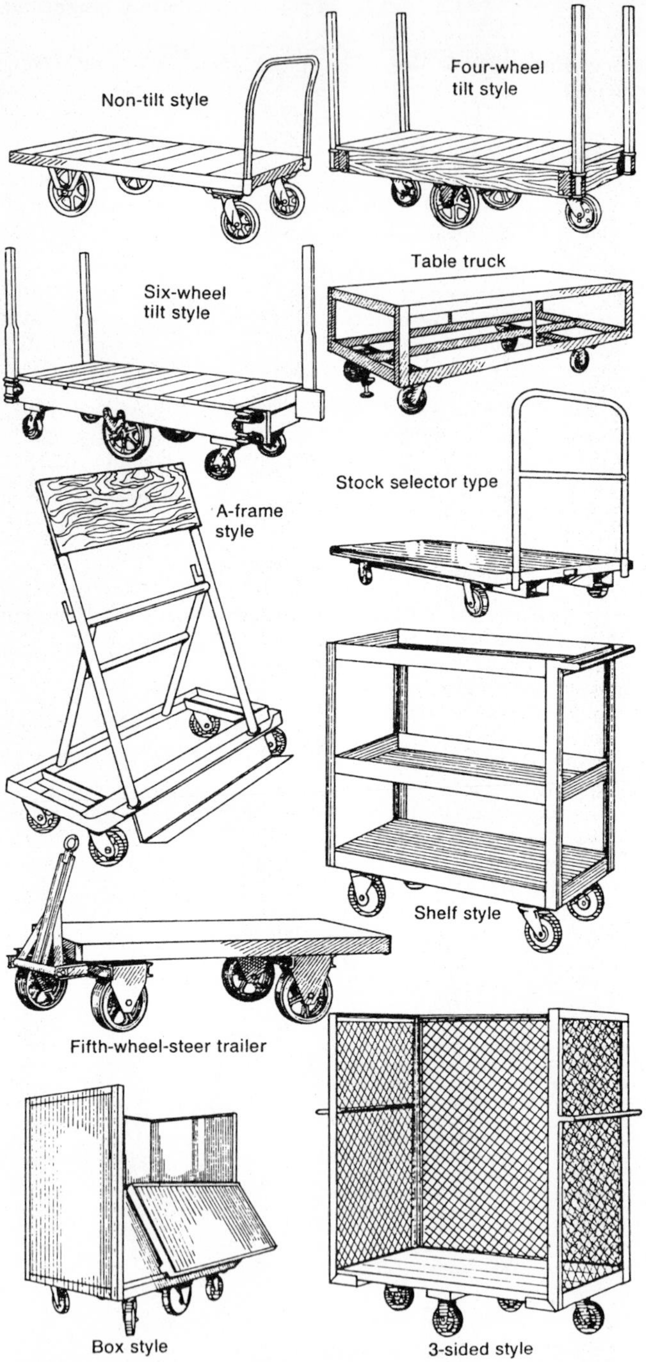

Figure 4-2 Factory trucks and wheel arrangement patterns. (Reproduced with permission from *Material Handling Engineering Handbook and Directory, 1979–1980*, published by *Material Handling Engineering* magazine, Cleveland, Ohio.)

and are designed to carry a variety of materials in bags, barrels, bales, boxes, and bins. Typical accessories include height extension, stair climbers, safety brake, spread clamps, and straps.

Dollies

Dollies are smaller-wheeled platforms upon which a load is placed for short distance and intermittent moves. Typically, dollies are fitted with caster-type wheels and are either pulled or pushed by an operator.

Factory Trucks

Factory trucks (Fig. 4-2) are wheeled platforms or containers either moved by an operator or towed by detachable power units. There is a wide variety of devices in this group and an even wider variety of uses for materials movement and as mobile storage.

The hand factory truck is hand-powered, guided by the direction of the moving force, and closely related to the dolly. Several wheel-arrangement patterns are available with tradeoffs between maneuverability and stability.

The towed factory truck is connected to the prime mover by a tow bar which provides the steering direction. Both two-wheel and four-wheel steering are available on towed factory trucks. Two-wheeled steering is generally the least expensive and most commonplace. Because of the steering geometry involved, each truck will follow a turn of shorter radius than the preceding vehicle. As several of these units are connected in trains, the continual tightening of turns requires more space for maneuvering.

The four-wheel-steered truck, with properly adjusted steering, is capable of following the same path as the vehicle in front of it. Where long trains are economically justified and desirable, the four-wheel-steered devices may be used to minimize commitment of valuable manufacturing space to aisles.

The Semilive Skid

The semilive skid is a rectangular platform or box having two wheels on one end and two fixed supports on the other. The end having the fixed supports is also fitted with a heavy pickup pin to which a two-wheeled jack is attached. The jack and handle are used as the lifting device and tiller, allowing the skid to be maneuvered by the operator.

Hydraulic-Lift Trucks

Hydraulic-lift trucks (Fig. 4-3) are used for moves at the workplace and occasional moves over short distances. They generally range in capacity from 2500 to 4500 lb. The operator uses a jacklike manually operated hydraulic system to elevate a loaded pallet sufficiently off the floor to move it. Some units use an electrically driven hydraulic system to lift the load, often above the maximum 5 in of the manually operated system. Hydraulic-lift trucks generally use forks for lifting pallets or platforms for special containers and for moving and positioning heavy loads such as dies.

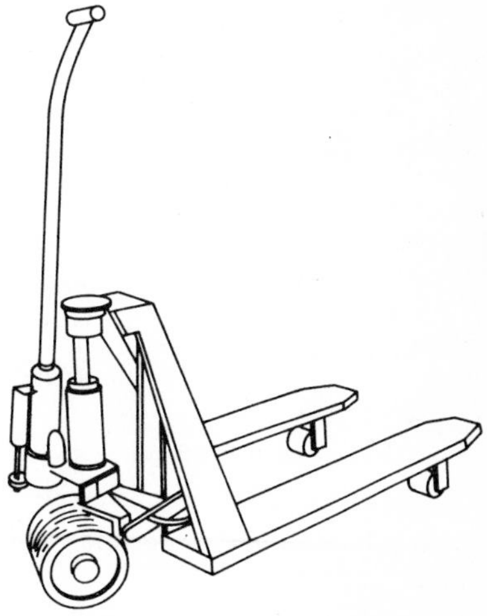

Figure 4-3 Hydraulic-lift truck.

POWERED-LIFT TRUCKS

This equipment group represents what is probably the largest and most varied of equipment for materials handling. The powered-lift truck owes its popularity to its versatility, being able to easily pick up a unit load, transport it quickly in a variety of environments, and then position the load vertically at almost any point within the capability of the equipment. Depending on the volumes involved, they become less economical for moves over 300 ft (90 m) since rated speeds are generally between 5 and 10 mi/h (7 and 14 km/h). Powered lift trucks are usually fitted with lifting forks to carry a unit load, although a wide variety of special load-carrying attachments can be used in place of forks. Power for lift trucks is either by internal-combustion engine or battery electricity.

The various pieces of equipment in this group can be operated over a variety of terrains, depending upon the design and, specifically, the wheel and tire combination used. Load-carrying capacities from 1000 to over 40,000 lb (450 to 18,000 kg) are common. Large vehicles are available with capacities in excess of 100,000 lb (45,000 kg). The very large vehicles are generally used outside, particularly for the moving and stacking of shipping containers.

Establishing aisle widths and their relation to fork-truck selection are critical when significant storage areas are involved. Clearly, the narrower the aisles, the more rows of storage. Equipment manufacturers have been ingenious in designing specialty trucks to operate in narrow aisles. It should be noted that manufacturers specify equipment turning circles and, thus, aisles will require space to aid in fork-truck maneuverability. Specialty trucks designed to operate in narrow aisles permit better space utilization, but tend to trade off some aspect of performance, a factor to be considered in specifying specialty as opposed to general-purpose equipment.

Truck capacity is generally calculated as follows (see Fig. 4-4):

A = distance, in, from center of front axle to heel of fork
B = distance, in, from heel of fork to center of load
C = distance ($A + B$) from center of front axle to center of load
D = length, in, of load on fork
W = weight of load, lb

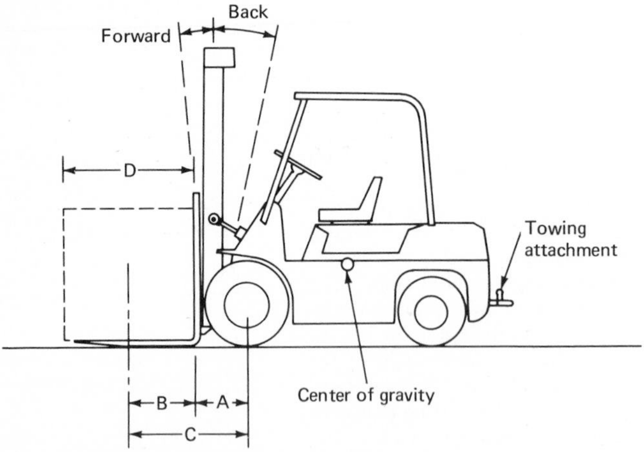

Figure 4-4 Rated truck capacity and counterbalanced truck.

1. Inch·Pound Rating

$$\text{Inch·pound rating} = W \times C$$

2. Maximum Load Length for Given Load

$$C = \frac{\text{inch·pound rating}}{W}$$

3. Maximum Load for Given Load Length

$$W = \frac{\text{inch}\cdot\text{pound rating}}{C}$$

A specific example is given to illustrate the actual calculations.

1. Truck is rated 3000 lb (W) at 20 in [3000-lb load which has a center 20 in (B) from heel of fork].
2. Distance from center of axle to heel of fork is 10 in (A).
3. Pallet load to be handled is 2000 lb:

$$C = A + B = 10 + 20 = 30 \text{ in}$$
$$\text{Inch-pound rating} = W \times C = 3000 \times 30 = 90{,}000 \text{ in}\cdot\text{lb}$$
$$C = \frac{\text{inch}\cdot\text{pound rating}}{W} = \frac{90{,}000}{2000} = 45 \text{ in}$$
$$B = C - A = 45 - 10 = 35 \text{ in}$$
$$D = 2 \times B = 2 \times 35 = 70 \text{ in allowable load length}$$

4. When selecting attachments, refer to the truck manufacturer to determine the amount of negative effect the attachment has on the truck's useful load-carrying capacity.

Aisle widths are generally established as follows:

A = aisle width
TR = turning radius of truck
L = load length
C = aisle clearance (total on both sides)
AX = distance from rear corner of load to centerline of axle:

$$A = TR + L + C + AX$$

The several varieties of powered-lift trucks are described below.

Counterbalanced Trucks

The counterbalanced trucks (Fig. 4-4) use their large, carefully positioned weight mass to offset (counterbalance) the moved load mass. These trucks are generally equipped with a tilting mast which will "tilt" the lifting mechanism rearward from the vertical lifting position and further counterbalance the load during movement. The load is positioned fully in front of the truck so that the truck structure does not interfere with adjacent stacks of material. This minimizes the aisle widths that are required.

Straddle Trucks

The straddle trucks (Fig. 4-5) differ from the counterbalanced type in that they do not depend on weight mass to counteract the weight of the load being handled. Instead, the straddle forklift positions the two main load-carrying wheels at or forward of the material load center. The truck is extremely stable as a result of this arrangement.

The straddle design is more compact and of lighter weight than the counterbalanced type. It is necessary, when negotiating loads into or out of racks, that either the straddle truck be equipped with an extending fork mechanism (pantograph) or the racks be positioned or constructed to allow the forward wheels of the truck to enter them.

Side-Loading Trucks

Side-loading trucks (Fig. 4-6) are a unique combination of a straddle-lift truck and a narrow-aisle truck. They are used where there are narrow aisles, where rapid transportation is called for, and where long narrow loads such as pipe and bar stock are handled. Side-loading trucks do not have to be turned to engage or place loads.

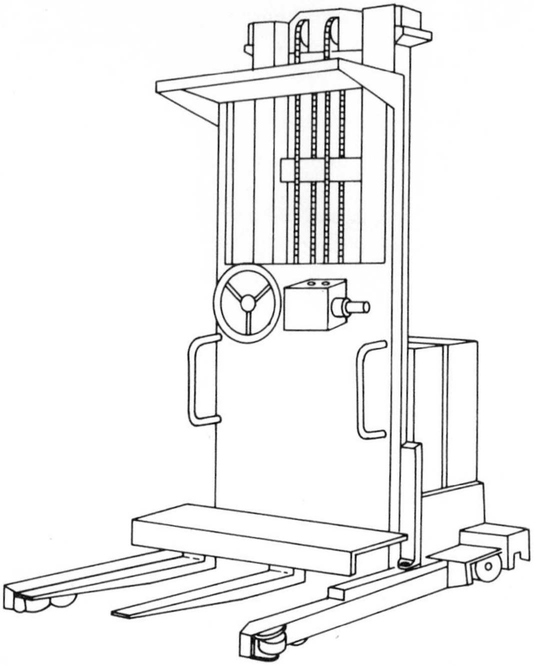

Figure 4-5 Straddle truck.

Nonrider Lift Trucks

Nonrider lift trucks (Fig. 4-7) are those where the operator walks along with the truck, directing the operation through a control unit attached to the truck. These units have basically the same features found in larger counterbalanced and straddle trucks. They are used for lifting and stacking light loads and moving these loads short distances.

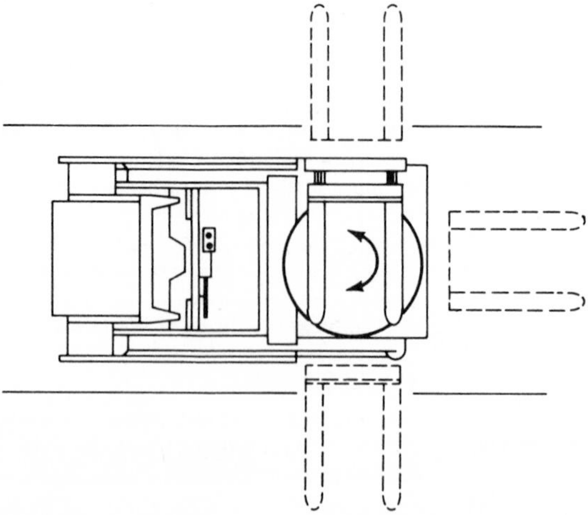

Figure 4-6 Side-loading truck.

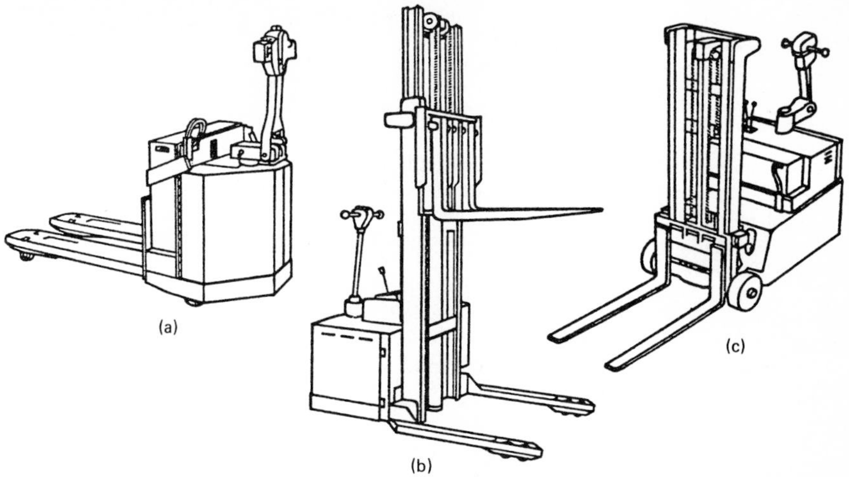

Figure 4-7 Nonrider lift trucks. (Reproduced with permission from *Material Handling Engineering Handbook and Directory, 1979–1980*, published by *Material Handling Engineering* magazine, Cleveland, Ohio.)

Straddle Carriers

Straddle carriers (Fig. 4-8) are large-capacity, highly maneuverable, powered lift trucks. To load and unload, the vehicle is driven over the unit load(s). The actual loading and unloading is extremely fast, although precise positioning of loads requires other methods. Unit loads can be transported at rates approaching highway speeds.

Order-Picker Trucks

Order-picker trucks have an elevated platform forward of the mast from which the truck and the platform can be operated. Typically, the trucks are used for picking partial loads in narrow aisles to heights of 24 ft, allowing for significant labor and space saving.

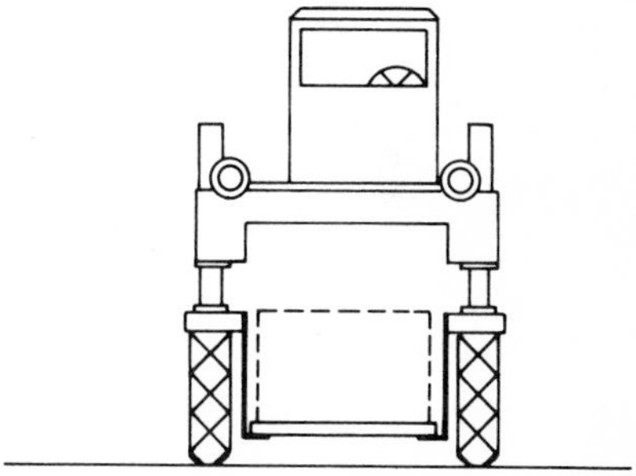

Figure 4-8 Straddle carrier.

Materials-Handling Attachments

The most widely used attachments are the forks themselves. They can be set at various widths and generally range between 30 and 60 in (80 and 160 cm) in length. The forks should be at least two-thirds the length of the maximum load to be lifted.

Standard two-stage uprights provide a lift height of approximately 18 ft (5.5 m), and three- and four-stage uprights provide heights to 20 ft (6 m). Certain specialty vehicles are designed to operate above 20 ft (6 m). The difference in fork height and total extended height is generally 4 ft (1.2 m), reflecting the height of the backrest. For low buildings, *free-lift* trucks should be specified. This feature permits the forks to be raised to lift loads to nearly half the total lift height without extending the uprights.

Frequently, a forklift truck will be fitted with an attachment or combination of attachments which allows the vehicle to perform special handling functions or simply allows it to operate more efficiently in a given situation. In some cases, these attachments replace the conventional forks for handling products which the forks cannot. In other instances, the attachments are used to augment the original fork function by giving the load-carrying forks additional motions.

When selecting attachments, it is always wise to consult with the truck manufacturer since attachments have a negative effect on a truck's useful load-carrying capability. When attachments are installed, the truck's information plate must be restamped, indicating the new effective truck capacity as required by OSHA 1910.178(4).

Attachments usually limit a forklift truck to a specialized function and, to some extent, limit its overall in-plant versatility. Some of the more simply designed attachments mount on the fork attachment rails and require only a few minutes to install or remove. The more complicated attachments, particularly those requiring hydraulic connections, should be considered as permanent conversions.

The following is a list and a brief description of some of the more common attachments (Fig. 4-9):

1. **Ram** A single projection, mounted in place of the forks for carrying coiled materials which can be easily entered horizontally. Rams have a variety of lengths and diameters to handle a variety of products, from steel coils to rolled carpet.

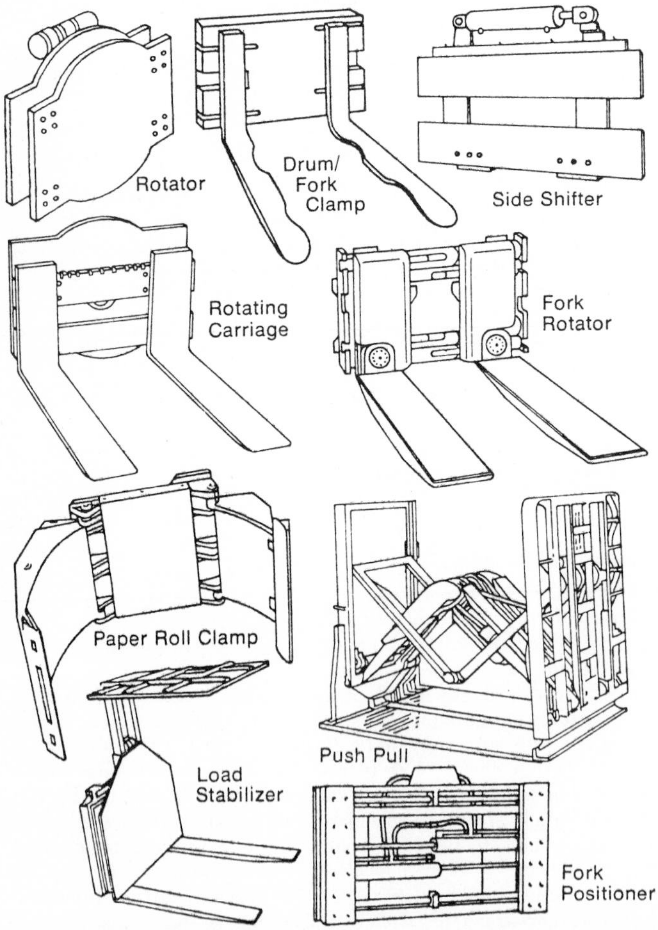

Figure 4-9 Common materials handling attachments. (Reproduced with permission from *Material Handling Engineering* 1979–1980, published by *Materials Handling Engineering* magazine, Cleveland, Ohio.)

2. **Barrel Attachment** Used to grasp the top seam of a steel drum and transport it in the vertical position.
3. **Concrete-Block Fork** This and the similar brick fork are designed specifically for handling stacks of masonry products without pallets.
4. **Paper-Roll Clamp** Specifically designed to carefully grasp and transport rolled materials in the vertical position. It is frequently combined with a rotator which allows the roll to be carried horizontally in the case of loosely rolled or easily damaged materials.
5. **Push-Pull** Uses a polished platen instead of forks to carry the load. Its purpose is to position loads in dense environments without the use of a pallet. In place of a pallet, a thin slip sheet is used under the load. This sheet is grasped by hydraulic clamps and pulled into the truck platen for loading and pushed off again into its next position.
6. **Bale Clamp** Used to grasp and carry baled materials and depends on hydraulic pressure to grasp the bales from the sides.
7. **Scoop** Used to handle loose or granular bulk material and consists of a metal bucket mounted in place of the fork, with a dumping capability usually provided. Tilting the bucket for loading and transport is accomplished by tilting the lift truck's mast forward and back.
8. **Squeeze Clamp** Used to grasp the sides of boxed products in a manner similar to the bale clamp, except that the grasping arms are smooth and deliver an even pressure to the carton to avoid damaging its contents. This device eliminates the need for pallets. It requires, however, additional side clearance on each side of the material moved to accommodate the clamps.
9. **Top-Handling Lift** Used to handle folded cartons by hooking into the folded lip of the top carton. The most important advantage is that extremely high storage density can be accomplished since only minimum side clearances are required without the use of either pallets or slip sheets.
10. **Side Shifter** Used with almost any type of attachment as well as forks, the side shifter allows loads to be positioned accurately from right to left without relocating the truck. Its major function is to speed the positioning of loads and to minimize rack space between loads. The side shifter will also reduce wear on the truck itself by reducing repositioning.
11. **Adjustable Forks** Where a variety of pallet and load sizes are encountered, adjustable forks are used. While most fork arrangements are manually adjustable, the mechanically adjustable forks allow the operator to accomplish the operation while remaining in the driver's seat.
12. **Load Stabilizers** To assure that loosely arranged and unstable loads are firmly contained during transit, various load stabilizers are available. Such a device is essentially a vertical clamp which exerts a downward pressure on a load and thus holds it in position while it is being moved.
13. **Clamping Forks** Similar in design to the fork positioner, clamping forks can be used to pick up loads in a conventional manner or may be used to clamp loads between the forks. This device is quite commonly used with special notched forks for transporting drums.
14. **Rotator** The use of a rotator allows a load to be rotated through 360 degrees, generally for dumping. The rotator is used with unit-load devices that fully enclose the forks and thus remain attached to the fork during rotation. They are also used with various clamping devices when rotation is required.
15. **Extended-Reach Forks** Commonly used with straddle forklifts to enable the truck to reach a load in the racks while the forward wheels remain outside of the rack space. The attachment also allows racked materials to be reached when two-deep storage is used. The reaching mechanism consists of a hydraulically operated pantograph system between the truck mast and the forks.

BURDEN CARRIERS

In the manufacturing process where sufficient volumes are involved, conveyor systems are often used to move materials from point to point. When smaller volumes or several moves of varying density are involved, however, a fixed-platform device is often used. These fixed-platform vehicles depend on an auxiliary loading and unloading method and are not tied to a specific unit-load module. Such devices are called burden carriers.

Burden carriers come in a wide variety of sizes and shapes. They are available in two basic types (Fig. 4-10). One is the walkie (nonriding) type and the other is the riding type. Both are available with battery-electric and internal-combustion power sources. They are usually limited in load-carrying capacity. High loads are generally handled by other types of handling equipment.

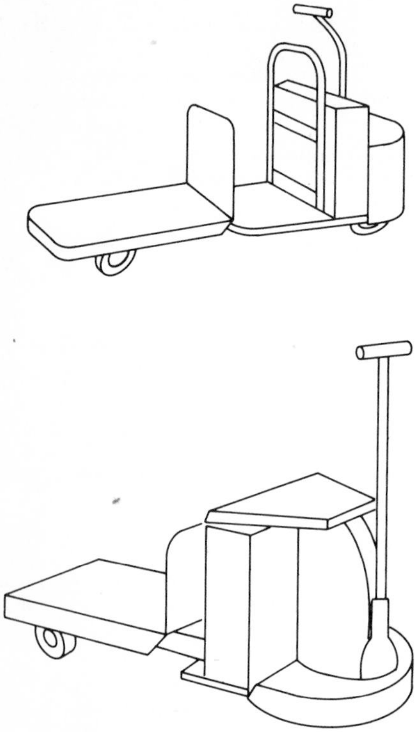

Figure 4-10 Types of burden carriers.

Walkie Burden Carrier

The walkie burden carrier is typically a three-point suspension hauler using battery-electric power, although some units are available that are powered by a small air-cooled engine. They are similar in design to the previously discussed walkie lift trucks, except that they have a fixed platform. Load ranges of 1000 to 3000 lb are available and application is limited to noncontained loads. Loading is generally done by hand or, in the case of heavier loads, by hoists and cranes.

Rider Burden Carrier

The rider-type burden carrier is often tailored to a variety of special applications such as personnel carriers, fire trucks, and portable maintenance shops. In its simplest form, the truck provides a driver's seat and a flat load-carrying bed. In this configuration it serves most commonly as a miscellaneous hauler to deliver supplies and materials in-house for distances of more than 300 ft (90 m). The power source for the rider-type truck is fairly evenly divided between air-cooled gasoline engines and battery-electricity. Both three-point and four-point suspensions are common, with an operating suspension system being incorporated in many larger units, along with pneumatic tires. These vehicles are able to negotiate rougher terrains and may attain speeds of up to 20 mi/h (30 km/h).

TRACTORS AND TRACTOR TRAINS

The term tractor (Fig. 4-11) refers to a detachable power source supplying locomotion to one or a group of load-bearing vehicles not having on-board power. The tractor is a steerable mover which is directed by an operator. They are generally classified according to their drawbar pulling rating (DPR) into small, medium, and large sizes.

On all grades above 5 percent, the individual manufacturer should be consulted since a variety of other factors which must be considered vary with individual tractor designs.

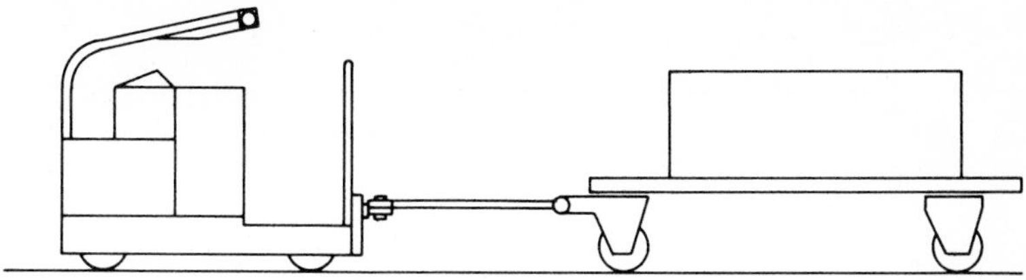

Figure 4-11 Tractor used in industrial applications.

Minimum safety criteria for industrial tractors are covered in OSHA Standard Section 1910.178 (Powered Industrial Trucks) and should be referred to when equipment is being selected.

The main application for these vehicles is the movement of goods in volume over distances too long to be economically moved by fork trucks—approximately 300 ft (90 m). Since the tractor trains are not self-loading, a system of tractor loading stations and surplus tractors and trailers is required, involving constant hitching and unhitching of tractors and trailers. An alternative is to use a forklift as a tractor with the operator of the fork truck loading the trailers.

Apart from fork trucks, five types of tractors are used for most industrial applications.

Highway Tractors

Highway tractors are typically used in over-the-road applications and are relatively specialized to serve this purpose. They do, however, find application in large manufacturing complexes for the movement of materials between remote locations where warranted by the speed and density of materials flow. This type of tractor is also frequently used in factory shipping yards for the positioning of both loaded and unloaded semitrailers.

Walkie Tractors

Walkie tractors are the smaller variety of industrial tractors. These tractors are battery-electric with motive power, braking, and steering being provided by a single wheel or a close-coupled pair of wheels. The drive mechanism is tiller-controlled through hand controls, as in other walkie equipment, and dead-man controls are provided. Two other wheels are provided for stability at the rear of the unit. A variety of coupling devices are available for attaching the tractor to trailers and semilive skids.

Walkie-Rider Tractors

The walkie-rider tractor is essentially a larger version of the walkie tractor. The major differences are that in a walkie-rider tractor a platform is provided for the operator to stand on during operation and two travel speeds are provided. A slow speed comparable to the operator's walking pace, on the order of 3 mi/h (5 km/h) and a higher speed of roughly 7 mi/h (11 km/h) is common in this type of equipment. Owing to higher operating speeds, these tractors have wider operating ranges. Because of the longer range of these tractors, larger-capacity batteries are used and, therefore, the units are heavier and larger than pure walkie tractors.

Rider Tractors

Rider tractors are available in both stand-up and sit-down configurations. The stand-up variety is more compact and generally applicable to more congested situations. The sit-down tractor is generally larger and is used where higher speed and longer distances, up to ½ mi (0.8 km) or more, are encountered. Battery-electric and internal-combustion engines are used as power sources in both versions; however, battery-electric power is more prevalent in the stand-up models, and the internal-combusion engine is the frequent choice in sit-down tractors.

Specialty Tractors

Specialty tractors are usually confined to very heavy load applications and are often built as an integral part of the load carrier itself. Two more common applications of these specialty tractors are large bulk-handling carriers for molten metals and granular materials and for spotting of railway cars.

MOBILE INDUSTRIAL CRANES

Mobile industrial cranes (Fig. 4-12) serve a variety of plant and production-related materials-handling functions. They are especially adaptable to loads of large or unusual

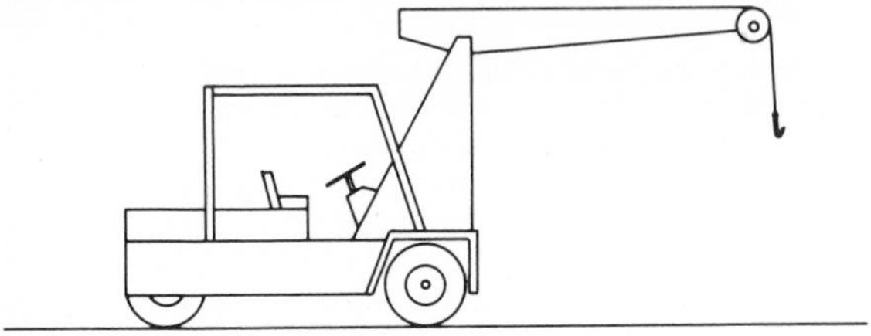

Figure 4-12 Mobile industrial crane.

size and where careful placement is required. In some applications, they are used only to position a given load, while in other applications they are used as both prime mover and positioner.

Mobile cranes differ from other plant hoisting equipment in that they operate independently of any supporting structure. The primary advantage of a crane is its ability to reach into places not normally accessible by other types of materials-handling equipment. With the exception of straddle cranes, the industrial crane depends on a boom for its reach and lift capability. It is the positioning of the boom that ultimately determines where a load will be placed and how large a load can be safely lifted.

The following text discusses the types of mobile cranes in use.

Portable Hand-Powered Crane

The portable hand-powered crane is similar in design to a small manual-lift truck, except that the load-carrying forks have been replaced by a boom and hook. This equipment is commonly used to move and position work pieces into and out of process equipment where volumes do not warrant a permanently installed hoisting system. It is also frequently found in maintenance and repair shops to assist in the disassembly and reassembly of in-plant equipment. Lifting is accomplished either through a hand winch and cable system or a manually operated hydraulic system. Typical lifting capacities are limited to 2000 lb (900 kg) or less.

Stevedore Crane

The stevedore crane is a nonswinging crane which requires that the hook be positioned by maneuvering the entire vehicle. This limits its use to relatively unobstructed areas. The boom may be extended outward by the operator to reach the load and returned back to a position closer to the vehicle for transport. The crane is a relatively fast vehicle which is used to pick up a load and transport it to a final destination.

The front, load-carrying wheels are also the powered wheels, with steering being achieved by the trailing wheels. Both three- and four-point suspensions are used, and the crane is often used to tow factory trucks while also loading and unloading them. Typical load capacities range from 2 to 4 tons (1800 to 3600 kg).

Swing-Boom Crane

The swing-boom crane is a larger-capacity crane than the stevedore crane and is used more for positioning loads than for transportation. The boom structure is constructed so that it can be rotated by the operator through 180°. Outriggers are provided for stability. They are usually powered by diesel or spark-ignition engines, with battery-electric power also being available.

Full Revolving Cranes

Full revolving cranes are capable of swinging a load through a full 360° and are generally the largest of mobile cranes. Their use is normally one of positioning loads as opposed to transportation. This type of crane will often be mounted on a truck-type chassis for rapid movement between jobsites. Power is provided by diesel or spark-ignition engines through direct hydraulic torque converters.

Load-lifting capacities at the boom's most upright position can be as high as 100 tons (90,000 kg) with reaches in excess of 100 ft (30 m) possible. Power is provided by diesel or spark-ignition engines through direct hydraulic torque converters. Industrial applications are almost totally limited to construction activities and maintenance of large structures.

Straddle Crane

The straddle crane has no boom but has a wheel-mounted framework on which are mounted two hoists. These hoists are capable of moving within the limits of the framework for precise load positioning. The straddle crane is related to the straddle carrier. It is a highly versatile crane, finding application as both positioner and mover of materials. In addition, its design is such that it is extremely stable and can move at relatively high speeds.

The framework consists of four vertical columns mounted above the vehicle's high flotation wheels, supporting two horizontal crane rails which carry the traveling hoists. Load-carrying capacity ranges from approximately 10 to 60 tons (9000 to 55,000 kg) per hoist for an aggregate capacity of approximately 20 to 120 tons (18,000 to 110,000 kg).

A variety of power systems are used, all of which are engine-driven. Hydraulic systems are those most commonly used for transport, hoisting, and positioning. However, one manufacturer employs an engine-driven electrical power plant to operate the crane's functions through electric motors.

The straddle crane is capable of operating in high-density areas and is highly maneuverable since all four of its wheels can be turned and powered independently. Common applications are in steel storage yards, loading and handling of shipping containers, commercial concrete castings, truck and car loading, and boatyards. Special load-handling devices are easily adapted to this crane, increasing its versatility.

chapter 8-5

Warehousing and Storage

INTRODUCTION

In the overall materials-handling system, warehousing provides the facilities, equipment, personnel, and techniques required to receive, store, and ship raw materials, goods in process, and finished goods. Storage facilities, equipment, and techniques vary widely depending on the nature of the material to be handled. Characteristics of materials, such as size, weight, durability, shelf life, and order lot size, are factors in designing a warehousing system and in solving warehousing problems.

Economics is also of great importance in the design of warehousing systems. Storage and retrieval costs are incurred, but add no value to the product. Thus, the investment in storage and handling equipment and in floor space must be based on minimizing unit storage and handling costs.

Other factors to be considered in designing warehousing systems include control of inventory size and location, provisions for quality inspection, provisions for order picking and packing, staging for receiving and shipping, appropriate numbers of shipping and receiving docks, and maintenance of records:

WAREHOUSING ACTIVITIES

Warehousing activities vary according to material amounts and characteristics. However, the activities associated with warehousing generally include the following procedures:

1. Unload inbound vehicles.
2. Accumulate received material in a staging area.
3. Examine the quantity and quality and assign a storage location.

4. Transport the material to the storage area.
5. Place the material in the assigned storage location.
6. Retrieve the material from storage and place it in an order-picking line, if a picking line is used.
7. Fill orders, if applicable.
8. Sort and pack, if applicable.
9. Accumulate for shipping.
10. Load and check outbound vehicles.

WAREHOUSING ADMINISTRATIVE CONTROL

Associated with the physical handling and storage of materials is an administrative control system. The administrative control system provides for:

1. Acknowledging receipt of goods for accounting purposes
2. Verifying the quality and quantity of received goods
3. Updating the inventory records to reflect receipts
4. Locating all goods in storage
5. Updating the inventory records to reflect shipments
6. Notification to the accounting function of shipments for billing purposes

Many administrative control systems are computerized and/or automated. The cost-effectiveness of such systems over manual systems depends on such factors as:

1. The number of line items in storage
2. The number of customers served
3. The volume of goods shipped

Generally, computerization and automation are cost-effective for industries and distribution centers having many line items in storage, many customers, and a large volume of goods shipped. Distributors of grocery, health, and beauty-aid products often have computerized and automated systems.

TYPES OF MATERIALS

Materials to be stored may be broadly classified as bulk materials or packaged goods. Bulk materials such as fuels, chemicals, minerals, and grain are stored in specialized storage facilities and transported in pipes, screw conveyors, power shovels, etc. In the many industries which handle and store bulk goods, each accomplishes these tasks with very specialized equipment and techniques. This discussion will be limited to warehousing packaged goods. The reader should consult specific publications that apply to bulk materials-handling industries for particulars in bulk handling.

Within the packaged-goods classification, materials are subdivided into categories according to their state of completion in the manufacturing process. Categories include raw materials, goods in process, and finished goods.

Raw Materials. These vary widely in characteristics, depending on the industry. A few examples are raw foods and ingredients for food processors, thousands of small parts for electronics assemblers, engines and motors for manufacturers of vehicles, and wood and finishes for furniture manufacturers. Raw materials are the goods on which the manufacturing process will operate to produce salable products. Indeed, the finished goods of one manufacturer often become the raw materials of another.

Goods in Process. This refers to goods which have completed some but not all of the manufacturing process. Typically, a manufacturing process involves several operations utilizing different equipment, skills, and materials. Goods in process are stored while awaiting the next manufacturing operation. They are often stored along the manufacturing process rather than in the warehouse proper.

Finished Goods. These goods are those which have completed the manufacturing process and are stored in inventory to fill customer orders. Finished goods may be further subdivided into reserve and order-picking stock. Customer orders are filled from order-picking stock while the picking stock is replenished from reserve stock.

The amount of raw materials, goods in process, and finished goods to be handled and stored varies considerably from industry to industry. Industries having large inventories of raw materials usually are converters of bulk materials such as paper and steel. Manufacturers of highly complex equipment such as computers and automobiles require a significant amount of raw-materials storage for parts as well.

Industries having significant needs for goods in process handling and storage are those whose manufacturing process is not automated. Machine-shop and electronic-assembly operations are examples.

Finished-goods handling and storage capacity are a function of manufacturing volume and product bulk. Industries having high-volume and high-bulk output generally require a considerable handling capacity for finished goods. The paper conversion and bottling industries are examples.

CONSIDERATIONS IN WAREHOUSE PLANNING

The objective of warehouse planning is to provide space and equipment to hold and preserve goods until they are used or shipped in the most cost-effective manner. The efficient accomplishment of warehousing activities listed in Chap. 8-1 is dependent on thorough planning. The following sections discuss these considerations as a guide to the warehouse planner.

Type and Number of Materials

The type and number of materials to be stored and handled form the basis for warehouse planning. The physical characteristics of the material, to a great extent, determine materials-storage and -handling methods. Physical factors include dimensions, weight, shape, and durability. As a first step in warehouse planning, all materials to be stored must be identified and their physical characteristics listed.

The quantity of each material item to be stored must be established. The planner may require assistance from sales management for finished-goods inventory levels and manufacturing management for establishing levels of raw materials and goods in process. In establishing inventory levels, seasonality, changes in product mix, and expected turnover rate become factors.

With the inventory level of each item of stored material established, a storage unit is selected. A storage unit is the least number of an item which is stored as one unit. Examples include a single crated refrigerator, a pallet containing 20 cases of canned goods, and a bundle of pipe. The storage unit is usually selected according to the physical characteristics of the material, the available handling and storage equipment, the quantity, and the manner in which the material is received or shipped.

A storage unit may be larger than a shipping unit or a manufacturing unit. In this case, order-picking facilities are provided for items used or shipped in lots smaller than a storage unit. The service level of storage in an order-picking operation must be established as well.

Factors affecting order-picking stock levels include minimum order quantity, volume, and the physical characteristics discussed earlier. Sheet-metal screws, for example, might be in 3-months supply, while cased canned goods might be in only 8-h supply.

Storage Equipment

Storage-equipment selection follows the establishment of the reserve and order-filling inventory storage units and levels.

In the case of selecting equipment for an existing building, the constraints of the building itself must be taken into consideration. Storage equipment must be compatible with floor loading capacity, clear height beneath sprinklers and structural steel, column spacing, and location of shipping and receiving docks, etc.

The characteristics of the storage unit, pallet, drum, bundle, etc., largely determine the type of storage equipment required. The inventory levels to be maintained determine the number of pieces of storage equipment. Materials characteristics and the volume of materials movement generally are deciding factors in selecting materials-handling equipment. Materials-handling equipment is discussed in Chaps. 8-3 and 8-4.

Storage equipment usually consists of general-purpose or specialty storage racks of varying height, depth, and load capacity. However, the warehouse floor may serve as all or part of the required equipment. Storage units such as pallets of cased canned goods, which have the rigidity and stability to support loads placed on top of them, are normally stored on the floor in stacks. Rolls of paper and coils of steel are frequently stacked on end. Storage units which have rigidity and are many in number lend themselves to floor-stacking techniques.

Heavy or bulky storage units which lack rigidity or which are few in number are generally better stored in storage racks. Storage units which are small, such as wristwatches or thumbtacks, are suitable for storage in shelving and bins. Containers used in conjunction with shelving or by themselves are discussed in Chap. 8-2.

Some types of available storage equipment are described below. Custom-designed special equipment is offered by many storage-equipment manufacturers.

Pallet Frames

Pallet frames (Fig. 5-1) are useful where materials lack the rigidity or stability to be stacked on the floor and where there are a large number of storage units in inventory. The pallet frame attaches to the pallet and extends above the material. The frame acts as a structure on which another pallet is stacked. Pallets so stacked are often placed several stacks deep and thus conserve floor space as compared with pallet racks which require aisle access. The frames are removable for pallet loads not requiring support.

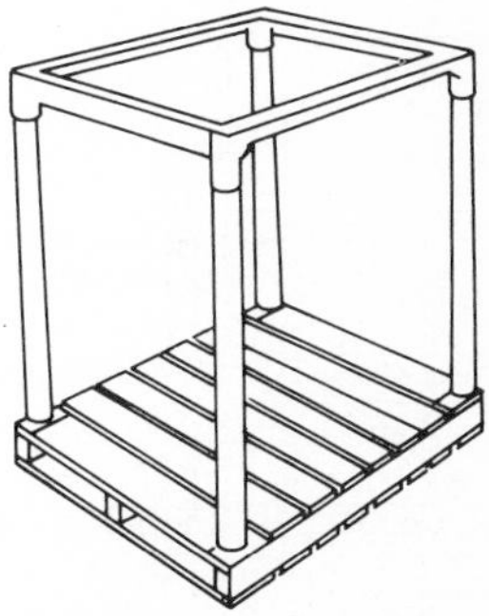

Figure 5-1 Pallet frame.

Pallet Racks

Pallet racks are the most commonly used storage aids and are available in many configurations adapted to particular materials characteristics and turnover rates. Pallet racks, for the purpose of this discussion, are classified into five groups.

One-Deep Pallet Racks. These are used when many items with small inventory quantities must be stored for ready accessibility. They may be configured to accommodate containers and other unit loads in addition to pallets. They are also used for order picking when it is most economical to pick directly from storage units.

One-deep pallets racks consist of vertical upright frames connected by horizontal crossbeams on which pallets and containers are placed one deep. Uprights are available in various heights and depths, and crossbeams are available in various lengths to accommodate most storage unit sizes. The load-carrying capacity for the upright and beam combinations is established by the manufacturers.

The normal storage height for this type of rack is from 20 to 24 ft (6 to 7 m) from the floor to the top of the top load. Lifting operations tend to be inefficient at greater heights because it becomes too difficult for the lift operator to accurately place the storage unit. Specialized equipment for heights greater than 24 ft (7 m) is available, however.

The horizontal crossbeams are adjustable so that the vertical height of the rack may be divided into as many storage levels as desired. The individual storage-opening height is tailored to suit the height of the storage unit. Clearance of 4 to 6 in (10 to 15 cm) from the top of the load to the bottom of the crossbeam above it is usually provided. In establishing the maximum height of the top load, the height of the fire-protection sprinklers must be considered. Most fire-protection codes and fire-insurance underwriters require a minimum clearance of 18 in (45 cm) between the top storage unit and the sprinklers. The warehouse planner should consult the local applicable fire code and the insurance underwriter.

The horizontal width of the storage opening is determined by two factors. These are the maximum weight and the maximum width of the loads to be stored in the opening. It should be noted that the load width may be larger than the pallet width because of load overhang. Normally, 4 in (10 cm) is provided horizontally between loads and between loads and uprights. Typically, two pallet loads are placed side by side in one opening. When the horizontal dimension of the opening has been determined, the rack manufacturer's catalog is consulted to select a compatible crossbeam length and weight capacity.

At this point in warehouse planning, the planner should calculate the floor load resulting from the fully loaded pallet rack. The floor loading will become a design parameter for new construction. In the case of an existing facility, the floor will be confirmed as adequate or the rack arrangement will be shown to be unfeasible.

The number of pallet racks required is determined by dividing the maximum number of storage units by the number of those units contained in one rack.

Two-Deep Pallet Racks. These racks (Fig. 5-2) are similar in design to the one-deep pallet rack except that two pallets, one behind the other, are stored in each position.

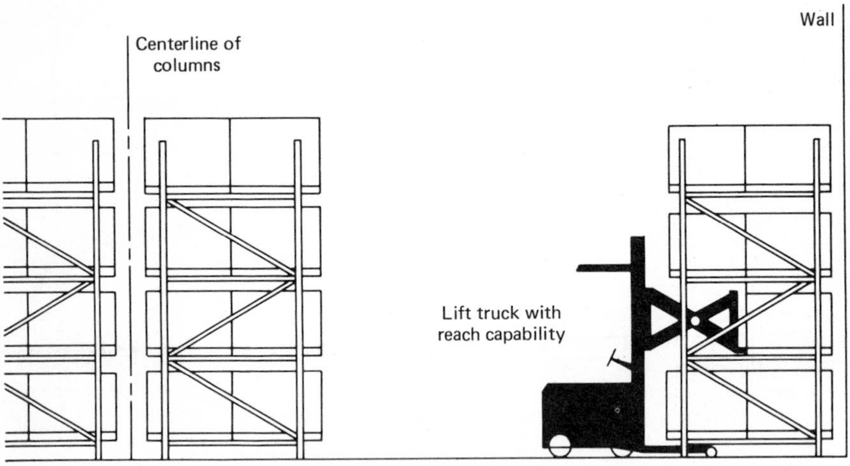

Figure 5-2 Two-deep pallet racks.

Two-deep racks are used when there is insufficient floor space to accommodate the required number of one-deep racks. Two one-deep racks are normally placed side by side and require aisle access from each side of the two-rack combination. The two-deep rack requires access from only one side but stores the same amount of material as the two one-deep racks placed side by side. The ratio of aisle to storage is thus reduced by using two-deep racks.

Two-deep racks have some costs associated with them, however. Lift equipment must be fitted with extended reach capability in order to position loads in the rear storage position. The efficiency of storing and retrieving the load in the rear storage position is less than with single-position loading and unloading. Two-deep racks are often more expensive than their two, one-deep, side-by-side counterparts. Storage units are sometimes damaged when positioned in or retrieved from the rear position.

The manner of selecting the height, depth, width, and number of two-deep racks is similar to that for one-deep racks.

Drive-In or Drive-Through Pallet Racks. These racks (Fig. 5-3) are designed to provide storage several pallets deep. The racks consist of vertical uprights which are

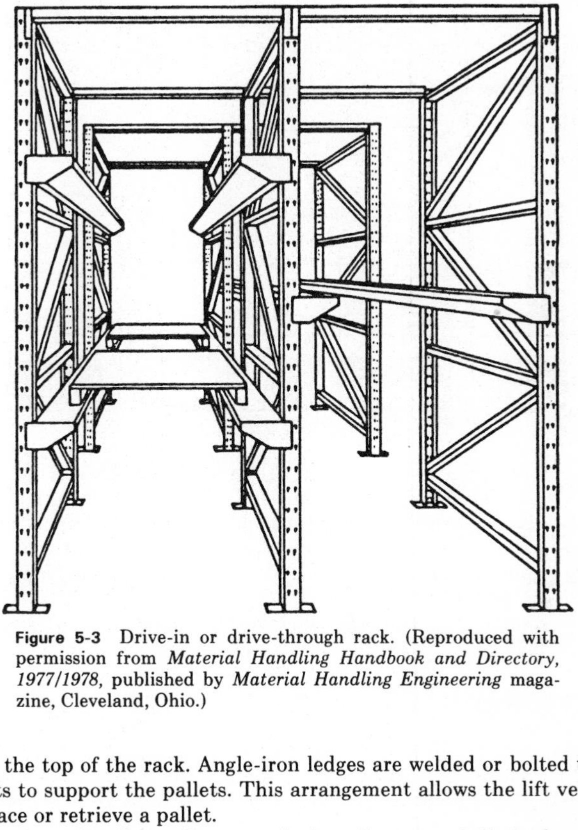

Figure 5-3 Drive-in or drive-through rack. (Reproduced with permission from *Material Handling Handbook and Directory, 1977/1978,* published by *Material Handling Engineering* magazine, Cleveland, Ohio.)

braced across the top of the rack. Angle-iron ledges are welded or bolted to the insides of the uprights to support the pallets. This arrangement allows the lift vehicle to enter the rack to place or retrieve a pallet.

Drive-in or -through pallet racks are used where floor space is limited and where there are many storage units of a particular item to be accommodated. Palletized items shipped or delivered by the entire truckload would be candidates for storage in this type of rack. The total number of positions of storage in one storage aisle in the rack could be designed to contain one truckload.

There are several limitations of drive-in or -through racks. Pallets stored in the rear or middle of a storage aisle cannot be retrieved until those in front are removed. This feature limits first-in–first-out inventory control except by loading or unloading entire storage-aisle lots one at a time. When entire storage-aisle lots are so treated, empty pallet positions are created if less than the entire aisle is filled or emptied. Storage efficiency is thereby reduced.

The lift operator must move and lift the storage unit in very confined spaces. Damage to the goods as a result of close tolerances is more frequent in this rack than in others.

It is also clear that the lift operator's efficiency is reduced by the requirement of driving into the rack in confined spaces.

Since the ledges that support the pallets are fixed, storage units of uniform dimensions are required. Storage units having overhang on the side of the pallet generally do not lend themselves to this type of storage. Finally, the lift vehicle may be no wider than the distance between the ledges. In selecting drive-in or -through racks, the planner should keep in mind that the usual maximum height is again 20 to 24 ft (6 to 7 m) and that drive-in racks usually are no more than six pallets deep. Due to their depth and the relative lack of bracing of their own, higher drive-in and -through racks generally require bracing to the building structure.

Gravity-Flow Racks. These racks (Fig. 5-4) are constructed to contain several pallets in depth and to support the pallets on inclined roller conveyors. Pallets are loaded on

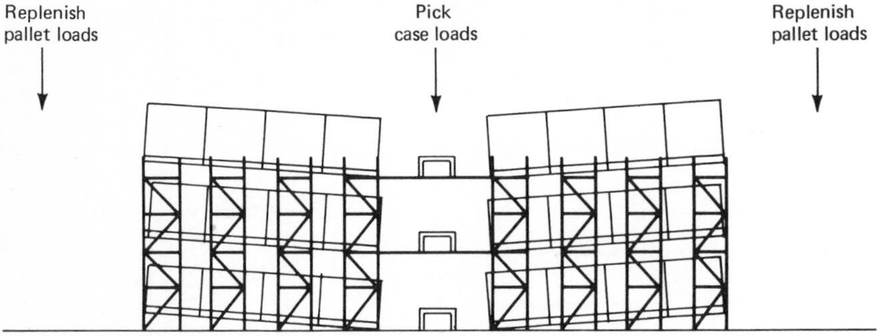

Figure 5-4 Gravity-flow racks.

the high side of the roller conveyor and removed from the low side. As a pallet is removed from storage, the pallets behind it roll down toward the retrieval opening.

Gravity-flow pallet racks and smaller versions for individual cartons and containers are commonly employed in order-picking operations. A continuous supply of an item is presented to the order picker without replenishment interference. Gravity-flow racks are also useful in maintaining first-in–first-out inventory control.

Gravity-flow pallet racks generally do not exceed six pallets in depth because of the high cost. However, the depth of the rack may be designed to contain a particular time period's supply. This configuration is appropriate when continuous replenishment is not employed.

The height of gravity-flow racks seldom exceeds 24 ft (7 m). The height of the rack is often limited to that conveniently reached by the order picker. In some cases two-level picking on the inside by personnel on foot is replenished by lift vehicles from the back side. See Fig. 5-4.

The height and width of storage or picking openings in a gravity-flow system are usually fixed with little or no adjustment conveniently possible. Instead, storage units are arranged to fit the gravity-flow configuration.

Gravity-flow racks may not be suited to very unstable storage units due to the shock of impact of movement and sudden stops on the inclined roller conveyor. Such difficulties are overcome by placing the unstable items in suitable containers.

Logic-Flow Racks. These are, in principle, designed to accomplish the same functions as gravity-flow racks. Instead of gravity providing the motive force to move full storage units to the picking opening, a powered conveyor does so. In most applications the order picker operates the powered conveyor with start-stop control. This arrangement eliminates the shock of impact experienced in gravity conveyors. Unstable and very delicate storage units are handled and stored in this manner.

In general, due to its high cost, the logic-flow rack is employed only for very specific, small storage situations. For the most part, these systems are manufactured from custom designs.

Bins and Shelving

Bins and shelving are widely used for the storage of goods in small lot sizes as raw materials, goods in process, and as finished goods, particularly in order-picking applications. They are available in many sizes, strengths, and degrees of closure. Indeed, pallet racks, previously discussed, may easily be converted to shelving.

In selecting shelf and bin storage, the planner determines for each storage item an appropriate shelf opening and depth or bin-drawer size. Shelves, bins, and drawers may be fitted with dividers to contain more than one item. The degree of protection from dust, light, theft, etc., determines the degree of closure required. Shelving and bin closures are available from totally open to totally enclosed and individually locked. Where many items require the same degree of protection, the shelving-bin system may be enclosed in a protective enclosure such as a clean room or refrigerated room.

Shelf, bin, and drawer arrangements can be obtained as separate units or in customized combinations. Customized combinations are more costly but may be justified in situations where stock may be advantageously stored in some picking order. Order-picking efficiency is maximized by reducing search and travel time on the part of the order picker.

Automatic Storage and Retrieval Systems

Automatic storage and retrieval systems (Fig. 5-5) are employed to achieve highly dense storage and very efficient placement and retrieval of materials. Many of the other activities listed under "Warehousing Activities" at the beginning of this chapter may also be mechanized and partially or fully automated as well.

The mechanization and automation of warehousing activities require a high capital investment and a very comprehensive feasibility study to justify the investment. The success of mechanized and automated warehousing also requires the *complete commitment by management* to support the planning, design, procurement, installation, and *especially* debugging. The time period from planning to start-up is often in excess of 3 years.

Mechanized and automated warehousing systems may be considered by the planner if some or all of the following conditions exist:

- Many varieties of items in storage
- High-volume storage items
- High turnover in general
- Highly seasonal storage items
- High cost of land and floor space
- High labor costs
- Need for rapid customer service
- Random storage desirable
- Storage units uniform in size

Automatic storage and retrieval systems, whether automated or not, achieve their high density by storing goods at greater heights than in conventional racks. *High cube* warehousing from 20 to 100 ft (6 to 30 m) is in use. At heights above 20 ft (6 m) the system may become the structure of the building to which walls and the roof are attached.

The materials-handling equipment, normally stacker cranes, travels on rails between the storage units and is guided by rails at the top of the storage units. The stacker cranes are capable of servicing the storage units on either side. Very narrow aisles result, further increasing density.

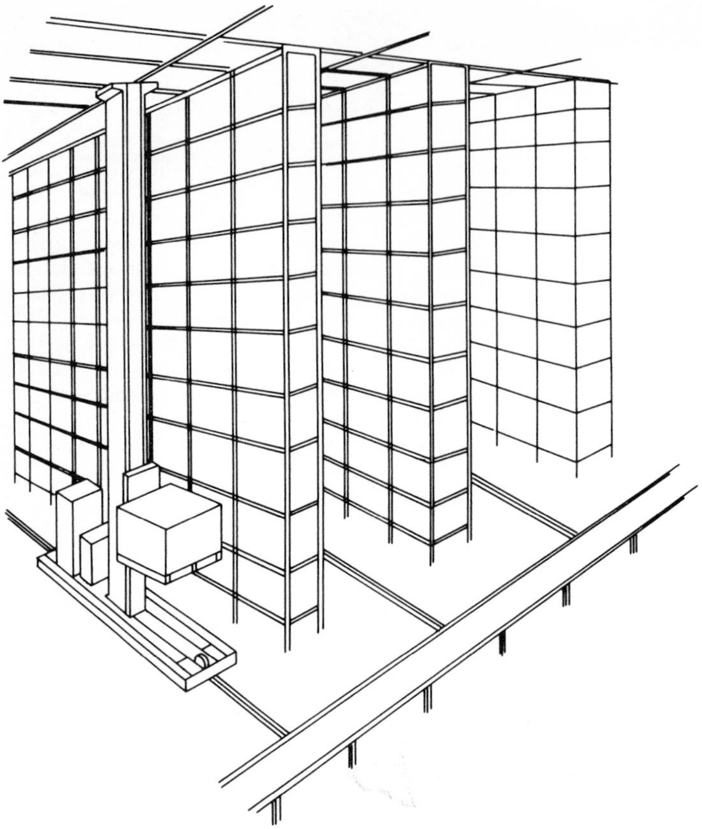

Figure 5-5 Automated storage and retrieval system.

The operator of the stacker crane travels with it both horizontally and vertically. In semiautomatic installations, the operator selects the bay and level on a keyboard. The stacker crane positions itself and the operator at the storage opening for placement or retrieval of a load. In a fully automatic system, a single operator controls the movements of several stacker cranes from a console, usually with the aid of a computer.

Goods to be placed in storage are often delivered to the stacker crane by conveyor; goods to be removed are delivered to the user by conveyor, too. Conveyors for these functions may be semi- or fully automatic as well.

The degree of mechanization and automatic control of warehousing varies from user to user and from manufacturer to manufacturer. The planner should consider engaging consultants and equipment manufacturers in the planning process. Most manufacturers provide planning guides to assist in identifying requirements. See Refs. 2 and 3, page 8-59, for more information. Prior to or during the identification of requirements for a mechanized-automated system, the planner should also determine requirements for a comparable conventional warehouse system. The capital investment and the operating cost of each system are than analyzed for economic justification.

Typically the automated system will require a higher initial capital investment but incur lower annual operating costs than a conventional system. The automated system would be economically justified if its payback period and return on investment are satisfactory to management. There also may be tax implications in choosing an automated storage and retrieval system. Where the racking structure supports the building walls and roof, the structure may be considered equipment. Equipment may be depreciated at a faster rate than buildings. Other factors influencing the decision to mechanize and automate include:

- Competitive advantage in servicing customers
- Image
- Reliability and the need for backup systems
- Degree of sophistication of current warehousing personnel
- Degree to which the market will change
- Time to become operational
- Availability of capital

Storage and retrieval for small lots can be accomplished with mechanized and automated arrangements. Mechanization usually provides for transporting bins of material to a stationary order picker. Carousels are frequently used for this purpose. See Fig. 5-6.

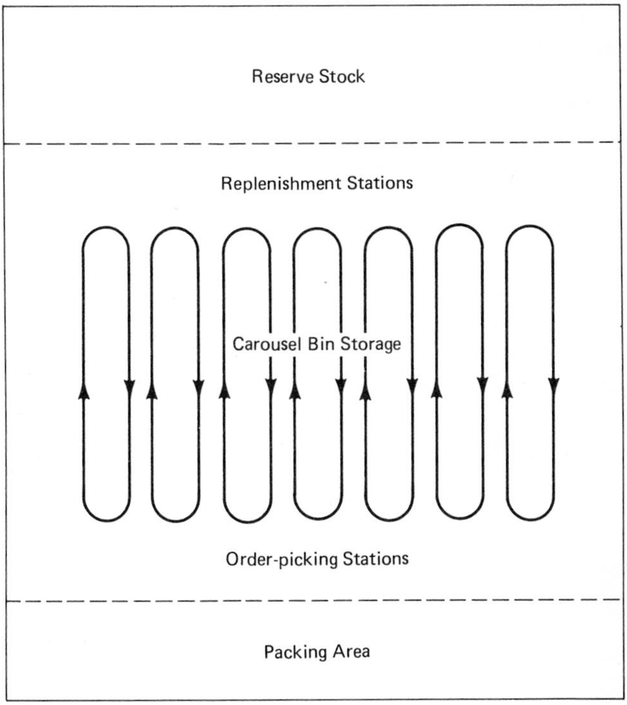

Figure 5-6 Schematic plan of automated bin storage.

The operator may select the picking face to be transported by an on-off control panel in a semiautomatic configuration. Frequently, the picking list is prepared by computer in the order of storage. In an automated system, the computer automatically positions the next picking face as the order picker indicates completion of the last pick or that the item is out of stock. Automation may also include provisions for automatic packing-list preparation, billing, and reordering.

Whether manual or mechanized, shelving and bin storage may be arranged in multi-levels by the use of mezzanines. See Fig. 5-7. This configuration is appropriate when high bay space is available and high storage density is required. In general, high-volume and high-weight items are stored in lower levels, and lighter, slow-moving items in upper levels.

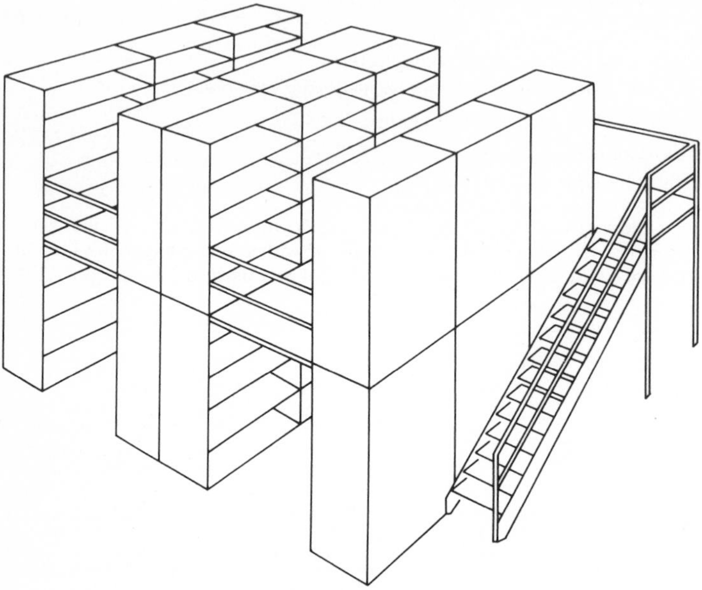

Figure 5-7 Multilevel shelving.

Shipping and Receiving Docks

Shipping and receiving docks are an important factor in warehouse planning. Of prime importance is the number of docks to be provided and how the docks are to be equipped.

In warehousing situations involving very high inbound and outbound volume, the principles of queuing theory are applied to assess peak dock activity. The peak dock activity, using alternative numbers of docks, may then be simulated using computer methods to determine the best combination of receiving and shipping docks. Suppliers of dock equipment will often assist in the planning.

In practice, warehouses having moderate to low inbound-outbound traffic are analyzed from historic or expected data. The peak rate of deliveries and shipments is determined and expressed in vehicles per hour. The rate at which vehicles are loaded and unloaded is also determined and expressed in vehicles per hour per dock. Dividing the vehicle rate of arrival by the rate at which the vehicles are loaded or unloaded results in the number of docks required.

Example

$$\text{Peak rate of truck deliveries} = 10 \text{ trucks per hour}$$

$$\text{Rate at which trucks are unloaded} = 2 \text{ trucks per hour per dock}$$

$$\text{No. receiving docks required} = \frac{\text{rate of truck arrivals}}{\text{rate trucks are unloaded}} = \frac{10 \text{ trucks per hour}}{2 \text{ trucks per hour per dock}} = 5 \text{ docks}$$

The type of dock and the equipment provided at the dock are dependent upon the type of material to be handled, the need for specialized loading/unloading equipment, and the need for security or weather protection.

Types of Docks

Types of docks to be considered include the following.

Indoor Docks. These are designed to accommodate the delivery vehicle under the roof of the warehouse building. Accommodation for entire tractor-trailers, trailers only, small delivery trucks, and railcars are common. These arrangements are appropriate where there is need to contain heated or cooled air, where security is important, and where materials-handling systems such as bridge cranes load or unload the material.

Flush Docks. These are constructed flush with the warehouse floor and with the outer wall of the building. This type of dock is generally built at a height above the outside grade level to accommodate the height of the vehicles to be serviced. Where the terrain is flat, ramps down to the docks are usually provided. This type of dock is normally provided with dock seals for weather protection. Also typical are dock leveler installations to accommodate minor differences in the heights of vehicle load beds.

Open Docks. These are less expensive than those previously discussed. However, they offer the least protection from weather and pilferage. Open docks are appropriate in warmer climates when handling goods which are not weather-sensitive or pilferable. When installed, a canopy over the dock area is usually provided.

Sawtooth Docks. These are arranged at an angle to the face of the building wall. This configuration is applicable where maneuvering room for shipping and delivery vehicles is limited. The sawtooth arrangement may be fully enclosed, covered, or open, as described above.

Dock Equipment

This must accommodate the materials to be handled, the vehicle to be serviced, the loading and unloading equipment, and the need for weather and security protection. Some standard equipment includes the following.

Dock Doors. These should be of the overhead type, counterbalanced for ease of operation. The doors should ride vertically along the wall of the building. Doors with tracks curving inward, such as a household garage door, could be damaged by materials-handling equipment. Doors are sized to accommodate the loaded materials-handling equipment passing through them and/or the size and configuration of the delivery-shipping vehicle. Options available for dock doors include the material from which the door is made, insulation, windows, and mechanized door-opening or -closing operators.

Dock Levelers and Dock Boards. These provide a bridge between the delivery or shipping vehicle and the dock (Fig. 5-8). They also serve to accommodate differences in height between vehicles and the dock. Dock levelers are typically installed to be flush with the floor of the dock when retracted. Models which attach to the outside of the dock as a retrofit are also available. The levelers are activated or retracted by spring pressure or hydraulics. Dock boards are reinforced steel plates which are manually lifted into place to form the bridge.

Factors used in selecting the appropriate dock leveler or board include: the weight of the heaviest load and materials-handling vehicle combination to cross it, the distance which the unit must span for bridging, and the combined width of the load and materials-handling vehicle. Manufacturers of this equipment offer a wide variety of weight capacities and sizes.

Weather-Protection Equipment. This equipment (Fig. 5-9) consists of devices to seal or cover the loading-dock area. Truck-dock seals used in conjunction with flush docks are popular. They have flexible construction and are often inflated. The arriving truck backs into the seal surrounding the door to effect the seal. Also available are hood-type seals which are mechanically activated. Made of flexible material, the hood extends out from the dock and conforms to the shape of the truck or railcar.

Other devices in common use are inside docks (discussed above) and canopies extend-

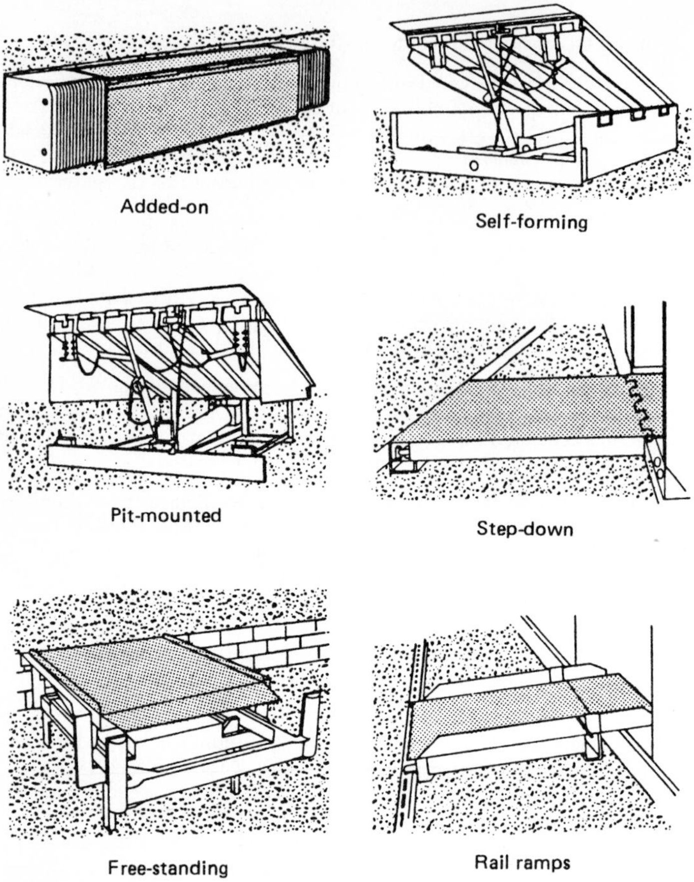

Figure 5-8 Types of dock levelers and dock boards. (Reproduced with permission from *Material Handling Engineering Handbook and Directory, 1977/1978,* published by *Material Handling Engineering* magazine, Cleveland, Ohio.)

ing over the outside dock area for rain protection. Where high activity means that dock doors are seldom closed, weather curtains are indicated. The curtains consist of strips of clear flexible material covering the door opening. Materials-handling equipment can drive through the curtain. When there is no traffic through the curtain, the strips act, to some extent, to seal the door from weather.

Dock Lighting

Dock lighting is required for nighttime operations. Floodlights are typical for lighting outside driveways, rails, and maneuvering areas to facilitate spotting delivery and shipping vehicles. Lighting for open docks is required to facilitate loading- and unloading-vehicle movement. When dock seals or hoods are used, lighting may be required for the inside of the truck or railcar. Portable or adjustable fixed lighting for these purposes is available. See Chap. 3-5, "Lighting."

Warehouse Layout

The warehouse layout is the final and perhaps the most important step in the planning process. Prior to undertaking the layout, the planner establishes the activities to be com-

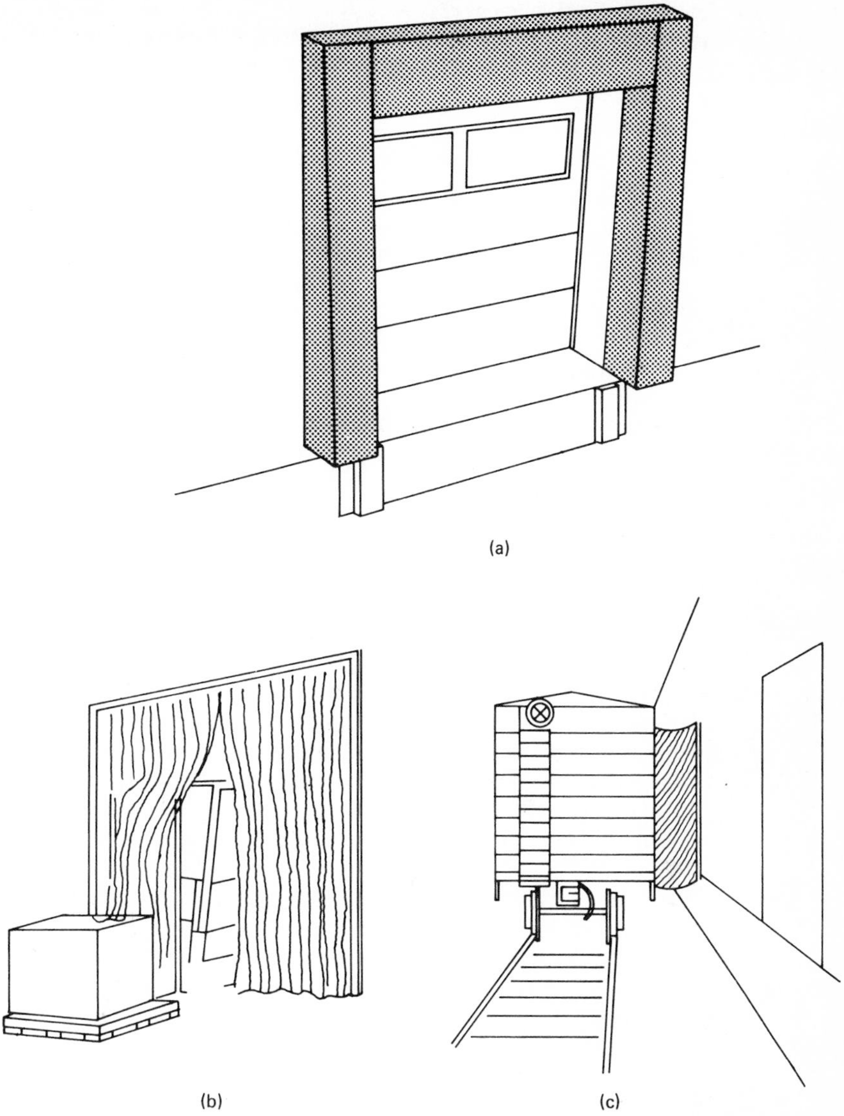

Figure 5-9 Types of dock weather protection. (*a*) One-deep pallet rack, (*b*) storage floor plan.

pleted and the type and number of materials to be stored and handled, storage and handling equipment, and docks. (See "Warehousing Activities" at the beginning of this chapter.) The warehouse layout should be planned to provide the space and arrangement that makes the best use of

- Storage cubes
- Efficiency of the flow of materials from activity to activity
- Effective communications between activities

Because thousands of combinations of types, sizes, weights, and volumes of materials have been observed, specific warehouse layout characteristics cannot be described in this book. However, general principles for warehouse design are discussed below.

Location in Storage

The location in storage (Fig. 5-10) of particular items is of importance. The following points should be considered.

- Items having a high turnover should be located near the user. The user may be a manufacturing operation, the shipping docks, or a quality-inspection area.
- Items having a high turnover should be stored and retrieved in the most convenient level vertically—slow movers high and fast movers low.
- Heavy and/or difficult-to-move items should be stored low.
- Where few items but large volumes of commodities are characteristic, individual loads of an item should be stored together in semidedicated areas.
- Where many items, but few of each, are encountered, random storage should be considered. A locator system, perhaps computerized, may be necessary.

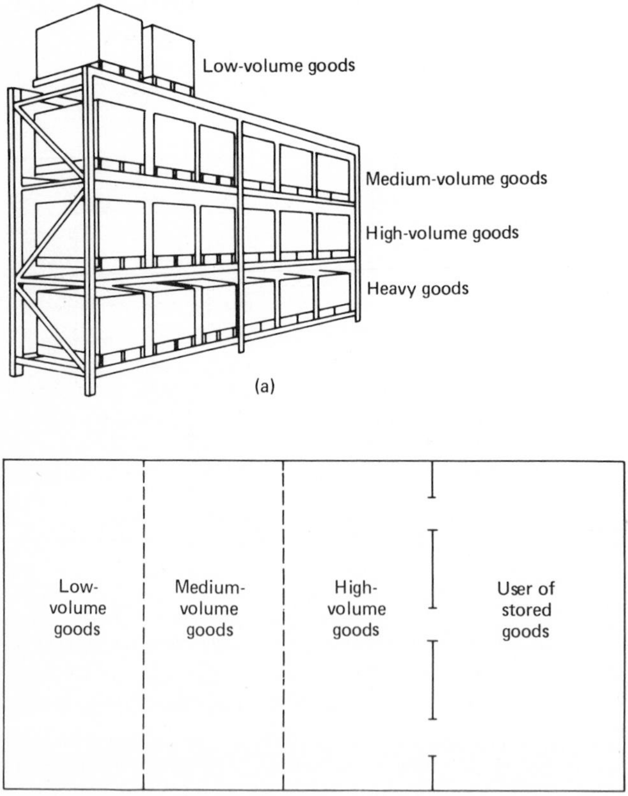

Figure 5-10 Location in storage according to volume.

- The nature of some storage items may require them to be stored in dedicated space. Some examples include hazardous materials, items of high value, and perishable goods.

Aisles

- Minimum aisle width is determined by the loaded maneuvering characteristics of materials-handling equipment. Determination of minimum aisle width is discussed in Chap. 8-4.
- Aisle width may be reduced by imposing one-way traffic.
- Aisles should open from the supplying area and open to the user area for maximum efficiency.
- Aisles should not be located next to walls as only one storage face is presented.

Storage Equipment Location and Arrangement

- The arrangement may be effected by column spacing in existing buildings or may determine the column spacing in new facilities. Normally, one-deep, two-deep, and drive-in racks, as well as shelving, are placed end on end, back-side along column lines. This arrangement eliminates column interference with aisles. See Fig. 5-2.
- One-deep, two-deep, and drive-in racks, as well as shelving, are most effectively placed back-to-back in open floor areas. This arrangement minimizes access aisle requirements. In the case of racks, space between them must be provided for pallet overhang. Frequently, the width of a line of columns provides this space.
- Storage racks, except gravity-flow and logic-flow, in addition to shelving, are efficiently placed along walls with openings for doors and fire-protection equipment. See Fig. 5-2.
- The height of the storage equipment is limited to that which provides no less than 18 in (45 cm) of clearance beneath fire-protection sprinklers. Local codes or fire underwriters may require a different clearance.

Docks

- Receiving and shipping docks are usually located to accommodate the flow of materials in the manufacturing process. The most common manufacturing flow patterns are straight-through and U-shaped. See Fig. 5-11.
- In the straight-through processes, raw materials are received and stored at the beginning of the manufacturing process. Finished goods appear at the end of the process and are stored and shipped from that location. Receiving and shipping areas which are so separated generally require more personnel and docks than an equivalent U-shaped arrangement.
- In U-shaped process flow, raw materials arrive at the same side of the building as that from which the finished goods are shipped. Shipping and receiving docks may be separated by no more than an imaginary line. This arrangement may offer economies in lower personnel requirements and in the number of docks since receiving and shipping personnel and equipment may be interchanged when necessary. The U-shaped process flow is also advantageous if high bay storage for both raw materials and finished goods is required.
- Spacing between docks is established to minimize interference between the docks during operations.
- Areas adjacent to docks for staging off-loaded material or material awaiting shipment are normally required.
- Provisions for enclosed space near docks may include administrative offices, personnel comfort facilities, and quality-inspection areas.

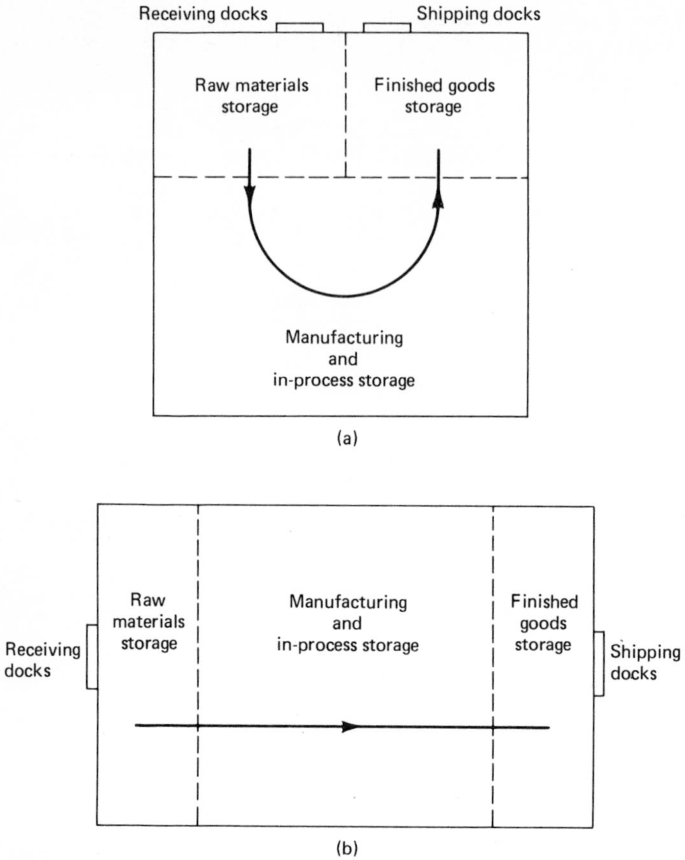

Figure 5-11 Materials flow patterns. (*a*) U-shaped materials flow, (*b*) straight-through materials flow.

Building Characteristics

- Storage height may be limited not only by fire codes but also by local zoning restrictions.
- Building services such as piping and space heaters should be placed in aisles to avoid interfering with storage equipment and to be accessible for maintenance.
- Lighting is normally designed to aid materials-handling equipment operators and order-picking personnel in locating and identifying stored items. Architectural and engineering firms normally assist the planner in determining the number and location of lighting fixtures.
- The type and number of units of fire-protection equipment are governed by fire code and fire underwriter requirements. Generally, the flammability and the amount of material stored determines the requirements.
- Floors may be enhanced to increase durability and housekeeping qualities. Coatings are available to increase surface hardness and wearability and to reduce dust. Typically, 3- to 4-in (8- to 10-cm) lines are painted on the floor to mark traffic and storage aisles.

section 9

Hydraulic and Pneumatic Systems

chapter 9-1

Hydraulic Systems

by
Ken S. Satija, Ph.D., P.E.
United Engineers & Constructors, Inc.
Philadelphia, Pennsylvania

INTRODUCTION

A system comprises a group of devices forming a network. A hydraulic system involves the devices and networks that operate by the pressure of liquids, whereas a pneumatic system works with gases. The most predominant liquid in hydraulic systems is water, the second is oil. In pneumatic systems, air is the most common fluid with natural gas and steam also in frequent use. Water, oil, and/or air systems and components will be mainly discussed in these chapters.

A typical system includes a pressurizer source (a high elevation tank or a pump), a conveyance system (channels, pipes, valves, etc.), and a sink. Thus, in a fluid (liquid or gas) system, a fluid is conveyed from one point with a certain energy to another point. Because of the resistance in the system, some pressure energy is lost as heat energy in the network.

It is necessary to know the fundamentals of fluid mechanics in order to understand, design, and maintain a system. The first part of this chapter is devoted to just that; the remainder discusses typical system components and some examples of actual plant systems.

BASIC FLUID MECHANICS

Pressure

Pressure is the most important term in fluid mechanics. The movement of the fluid and the forces on the equipment due to the fluid are basically due to the pressure in the fluid. By definition, pressure is the force per unit area of a surface with which the fluid is in contact, and always acts normal to the surface considered:

$$p = \frac{F}{A} = wH \tag{1}$$

where

p = pressure intensity, lb/ft^2 (kg/m^2)
F = total force acting on a surface, lb (kg)
A = area of the surface, ft^2 (m^2)
w = weight density of the fluid, lb/ft^3 (kg/m^3)
H = head = height of free surface of the fluid above a certain point on the surface, ft (m)

TABLE 1-1 Density of Common Fluids (at Atmospheric Pressure)

Fluid	Temperature, °F	°C	Specific density, lb/ft^3
Air	0	−18	0.0862
	60	16	0.0763
	100	38	0.0709
	200	93	0.0601
Ammonia	60	16	0.0455
Carbon dioxide	60	16	0.117
Gasoline	60	16	42.0
Helium	60	16	0.0106
Hydrogen	60	16	0.00531
Mercury	60	16	847.0
Methane (natural gas)	60	16	0.0424
Oxygen	60	16	0.0844
Nitrogen	60	16	0.0739
Sulfur dioxide	60	16	0.173
Water (fresh)	32	0	62.4
	70	21	62.3
	180	83	60.6
	212	100	59.8
Water (salt, sea)	60	16	64.0

The last part of Eq. (1) really emanates from the principles of hydrostatics, in which the only energy available comes from the static head of the column of fluid. Thus the hydrostatic pressure at a point is equal to the pressure created by the height of the fluid column above that point. Since the density of gases is usually very small compared with liquids (see Table 1-1), the hydrostatic pressure of gases is usually neglected in practical engineering unless a very high level of precision is desired.

A standard atmosphere (the atmospheric pressure) is taken as 14.7 psia (pounds per square inch absolute), or 34 ft of a column of water, or 760 mmHg (29.92 inHg). *Absolute* is used with reference to zero pressure. A pressure gauge normally measures and displays pressure *relative to the atmosphere*. Thus a gauge reading of zero is actually a pressure exactly equal to prevailing atmospheric pressure. In fluid mechanics, unless specifically mentioned otherwise, gauge pressures are understood. Thus a pressure given as 30 lb/in^2 automatically means 30 psig (pounds per square inch gauge). However, it must be kept in mind that consistent units must be used while working out problems with the help of the equations such as Eq. (1).

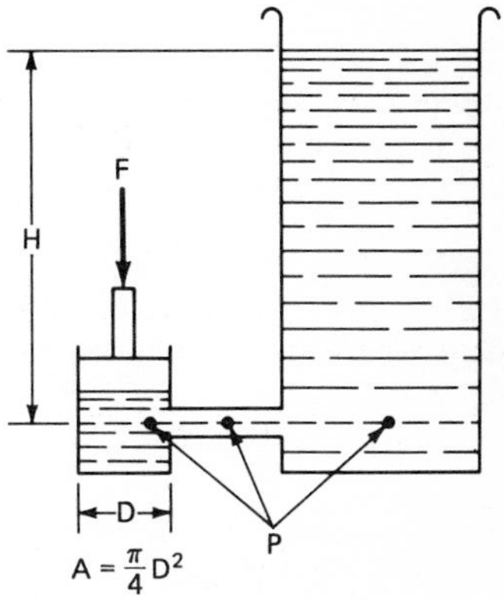

Figure 1-1 Relation of pressure to head.

The term *pressure head* with reference to Eq. (1) can easily be understood by referring to Fig. 1-1, which also depicts that hydrostatic pressure is transmitted equally in all directions at a point, the so-called *Pascal's law.* If a point P is surrounded by fluid on all sides, the hydrostatic forces are balanced, and the net force at that point is zero; that is, the point stays in equilibrium. Otherwise, there will be a net force trying to move that point.

Density

The density or the specific weight w of a substance is the weight of a unit volume of the substance. For water, w = 62.4 lb/ft^3 (999.6 kg/m^3) is commonly used (1 g/cm^3 = 1000 kg/m^3). Table 1-1 lists specific weight of several common fluids.

$$\text{Mass density} = \rho = \frac{\text{mass}}{\text{unit volume}} = \frac{w}{\mathbf{g}} \tag{2}$$

where

$\mathbf{g}$ = acceleration due to gravity, = 32.2 ft/s^2 (9.81 m/s^2)

G_f = relative density

= specific gravity

$$= \frac{\text{weight of the substance in air}}{\text{weight of an equal volume of water}}$$

$$= \frac{\text{weight of the substance in air}}{\text{loss of weight of the substance in water}}$$

$$= \frac{\text{weight of the substance in air}}{\text{buoyant force on the substance if immersed in water}}$$

$$= \frac{\text{density of the substance}}{\text{density of water}}$$

This is Archimedes' principle.

Viscosity

The resistance of a fluid to a shearing force is determined by viscosity. If the force tends to move a fluid particle with a velocity of V ft/s (m/s), then dynamic or absolute viscosity is

$$\mu = \frac{\tau}{dV/dy} \quad \text{lb·s/ft}^2 \quad (\text{poise} = \text{P} = \text{dyn·s/cm}^2) \tag{3}$$

where dV/dy is the velocity gradient and distance y is measured normal to V.

The kinematic viscosity is

$$\nu = \frac{\mu}{\rho} \quad \text{ft}^2\text{/s} \quad (\text{stokes} = \text{cm}^2\text{/s} = \text{St}) \tag{4}$$

and τ is the shear stress in fluid, lb/ft² (kg/cm²). Viscosity values for common fluids are shown in Table 1-2.

TABLE 1-2 Viscosity of Common Fluids

Fluid	Temperature °F	(°C)	Kinematic viscosity, ft²/s
Air	0	(−18)	1.26×10^{-4}
	60	(16)	1.46×10^{-4}
	100	(38)	1.80×10^{-4}
	200	(94)	2.40×10^{-4}
Water (fresh)	32	(0)	1.93×10^{-5}
	70	(21)	1.05×10^{-5}
	180	(82)	0.385×10^{-5}
	212	(100)	0.319×10^{-5}
Water (salt, sea)	60	(16)	0.319×10^{-5}

Hydrostatic Forces

Consider an inclined plane surface (with area A) of a tank containing a liquid as shown in Fig. 1-2. The center of gravity of the plane is at point cg. Point O is the point of intersection of the plane (extended if necessary) and the surface of liquid. Other nomen-

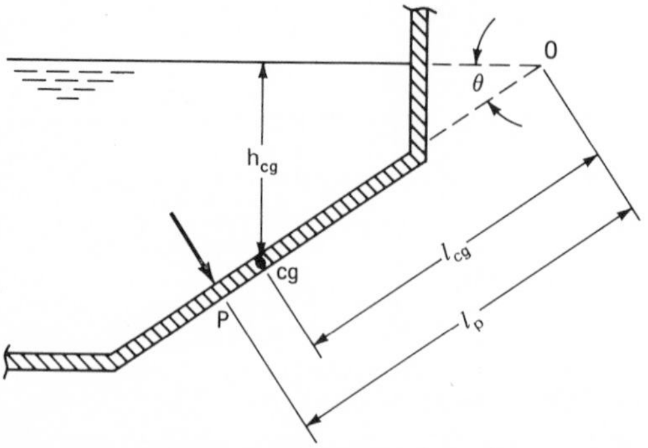

Figure 1-2 Hydrostatic forces.

clature is clear from the figure. Then the total hydrostatic force F (always normal to a surface) on the inclined surface and its point of action P are given by

$$F = wh_{cg}A \qquad \text{lb (kg)} \tag{5}$$

$$l_p = l_{cg} + \frac{I_{cg}}{Al_{cg}} = \frac{I_0}{Al_{cg}} \qquad \text{ft (m)} \tag{6}$$

where

I_{cg} = second moment of the area (moment of inertia) about the center of gravity of the plane (as usually given in mechanics books)

I_0 = moment of inertia about point O

Continuity Equation

For no other inflow or outflow from a subsystem, the mass flow rate at upstream and downstream end are equal; i.e.,

$$M_1 = M_2 \qquad (Q_1 = Q_2 \text{ if } \rho \text{ is constant}) \tag{7}$$

or

$$\rho_1 A_1 V_1 = \rho_2 A_2 V_2 \tag{8}$$

If a branch comes into a point on the subsystem, the total outflow from the point is equal to the total inflow to the point, i.e.,

$$M_1 + M_2 = M_3 \tag{9}$$

Energy Equation

Consider flow in a closed conduit of nonuniform circular cross section (Fig. 1-3). Also consider two cross sections 1 and 2 of this conduit with the following quantities applicable to them:

p_1 = pressure at cross section 1, lb/ft^2 (kg/m^2)
p_2 = pressure at cross section 2, lb/ft (kg/m^2)
v_1 = velocity at cross section 1, ft/s (m/s)
v_2 = velocity at cross section 2, ft/s (m/s)
z_1 = height above datum of cross section 1, ft (m)
z_2 = height above datum of cross section 2, ft (m)

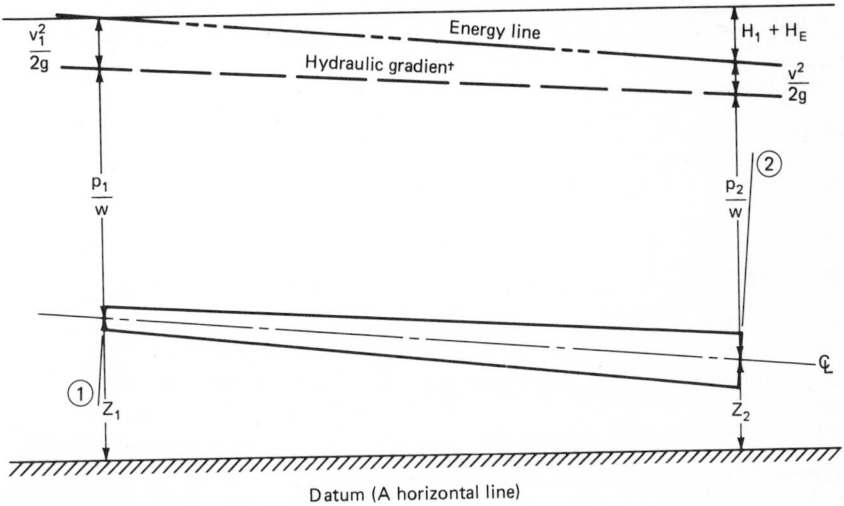

Figure 1-3 Energy equation.

All these quantities are referenced to the centerline of the conduit. Then, in terms of head, the energy equation is

$$\frac{p_1}{w} + \frac{V_1^2}{2g} + z_1 = \frac{p_2}{w} + \frac{V_2^2}{2g} + z_2 + H_1 + H_e \tag{10}$$

where

H_1 = head losses in the flow, such as friction, ft (m)

H_e = head energy extracted from the system (negative if energy was added), ft (m)

If H_1, and H_e are zero, as for an ideal flow, the remaining equation (10) is also called the Bernoulli theorem.

Each side of Eq. (10) is equal to the total energy in the system. The total energy must be constant, so the plot of both sides of Eq. (10) falls on a horizontal line, representing the total energy in the system. If H_1 and H_e are excluded from the equation, then we get a net *energy line.* If the velocity head is also excluded, then we get the *hydraulic gradient line,* representing the gradient of the hydrostatic pressure.

Momentum Principles

According to Newton's second law

$$\begin{aligned} \text{Momentum force} &= \text{rate of change of momentum in a direction} \\ &= \text{initial momentum} - \text{final momentum in that direction} \\ &= \frac{wq}{g} V_1 - \frac{wq}{g} V_2 \quad \text{lb (kg)} \end{aligned} \tag{11}$$

Therefore, the total force on an element will be the sum of the pressure force, pA, defined in Eq. (1) and the momentum force given by Eq. (11), both considered in the same direction. For example, on a 90° elbow (Fig. 1-4), the forces must be considered separately along the x and y axes, and then the resultant computed:

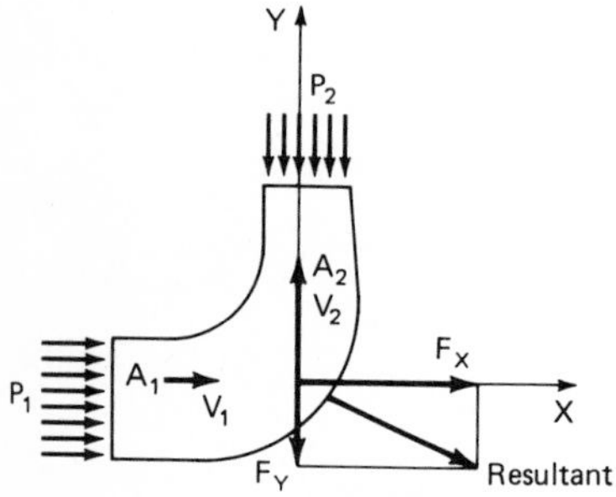

Figure 1-4 Forces on a bend.

$$F_x = p_1A_1 + \frac{wq}{g} V_l - \frac{wq}{g} 0 \tag{12}$$

(Final velocity in the x direction is zero.)

$$F_y = -p_2A_2 + \frac{wq}{g} 0 - \frac{wq}{g} V_2 \tag{13}$$

(Positive y is upward; initial momentum is zero; pressure always acts normal to the surface and toward the surface of the element considered.)

Head Losses

Frictional Flow in Closed Pipes

A pipe flow is considered closed and pressurized if the water surface is not exposed to atmospheric pressure. Under actual conditions, there is always a resistance to flow which causes a pressure head loss. (Basically, this loss of energy is converted to heat and is usually dissipated.)

The most commonly used equation for head loss due to friction, $H_{L,f}$, is the Darcy-Weisbach equation

$$H_{L,f} = \frac{fL}{D} \frac{V^2}{2g} \quad \text{ft (m)} \tag{14}$$

where

f = friction coefficient
D = diameter of pipe, ft (m)
L = length of pipe, ft (m)

The friction coefficient f depends upon the material and roughness of the pipe and the Reynolds number (N_R) of flow:

$$N_R = \frac{VD}{\nu} = \frac{V\,4R}{\nu} \tag{15}$$

where

R = hydraulic mean radius of flow = A/P = $D/4$ for pipe, ft (m)
P = wetted perimeter of flow boundary, ft (m)

Figure 1-5 is a chart for the determination of f for a certain value of N_R and relative roughness. The lowest curve is for any pipe called "smooth," i.e., with negligible roughness.

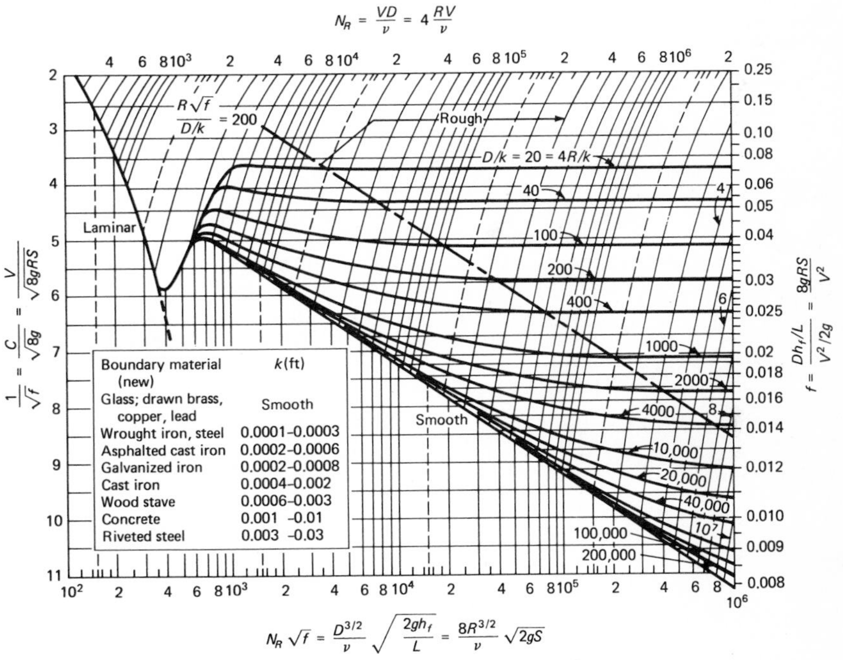

Figure 1-5 Friction in pipes and conduits.

Sometimes the Hazen-Williams formula is used for computing frictional flow and is given by

$$V = 1.318 C_1 R^{0.63} S^{0.54} \qquad \text{feet per second} \tag{16}$$

where

C_1 = Hazen-Williams coefficient (see Table 1-3)
S = slope of the hydraulic gradient

$$= \frac{H_{L_f}}{L} \tag{17}$$

TABLE 1-3 Some Values of the Hazen-Williams Coefficient C_1 (U.S. Customary Units)

Extremely smooth and straight pipes	140
New, smooth cast-iron pipes	130
Average cast-iron, new riveted steel pipes	110
Vitrified sewer pipes	110
Cast-iron pipes, some years in service	100
Cast-iron pipes, in bad condition	80

Other Losses

There are also other losses in a flow besides the frictional pressure drop; losses occur at every transition, control, bend, valve, etc. These are usually given as a function of the velocity head $V^2/2\mathbf{g}$. Table 1-4 provides the values of the coefficient K with which $V^2/2\mathbf{g}$ (or its variations, as indicated) is multiplied to get the head losses due to fittings on a piping system. (Sometimes head loss is computed in terms of equivalent length of pipe.)

Flow in Open Channels

If the water surface in a conduit (as in a channel) is open to atmospheric pressure, it is called *open-channel flow*. In this instance, the surface pressure is the same all along the flow, i.e., atmospheric. That is why, instead of total energy, *specific energy E* is employed in open-channel flow:

$$E = y + \frac{V^2}{2g} \qquad \text{ft (m)} \tag{18}$$

where y = depth of water in open-channel flow, ft (m)

The plot of Eq. (18) of E vs. y (for a constant Q) will yield a curve of the type shown in Fig. 1-6.

For any particular energy value, there are two depths y_A and y_B, except at one particular point where there is only one depth (the *critical depth* y_c), corresponding to a

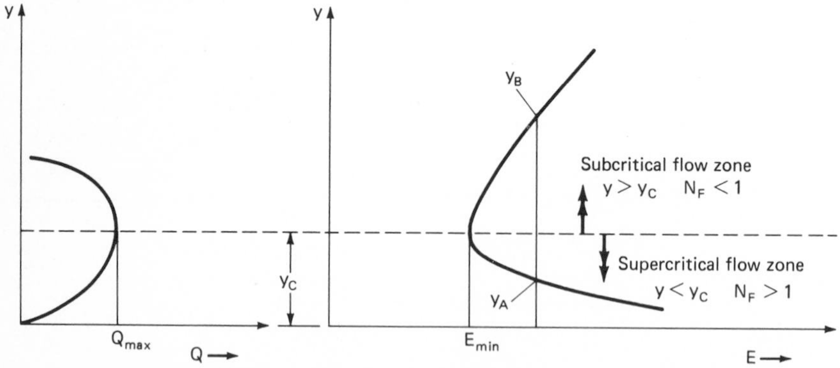

Figure 1-6 Critical depth.

minimum specific energy E_{min}. It is also known that if discharge vs. depth is plotted, the maximum discharge point on this curve corresponds to the critical depth. This is indicated at the left end of Fig. 1-6. It is also known that the Froude number N_F of flow is equal to one for a flow with critical depth $N_{F,c}$ = 1), where

$$N_F = \frac{V}{\sqrt{\mathbf{g}y}} \tag{19}$$

TABLE 1-4A Typical Loss of Head Coefficients (Subscript 1 = Upstream and Subscript 2 = Downstream)

Item		Average lost head
1. From tank to pipe (entrance loss)		
Flush connection		$0.50 \frac{V_2^2}{2g}$
Projecting connection		$1.00 \frac{V_2^2}{2g}$
Rounded connection		$0.05 \frac{V_2^2}{2g}$
2. From pipe to tank (exit loss)		$1.00 \frac{V_1^2}{2g}$
3. Sudden enlargement		$\frac{(V_1 - V_2)^2}{2g}$
4. Gradual enlargement		$K \frac{(V_1 - V_2)^2}{2g}$
5. Sudden contraction		$K_c \frac{V_2^2}{2g}$
6. Elbows, fittings, valves*		$K \frac{V_2^2}{2g}$
Some typical values of K are:		
45° bend	0.35 to 0.45	
90° bend	0.50 to 0.75	
Tees	1.50 to 2.00	
Gate valves (full open)	about 0.25	
Check valves (open)	about 3.0	
Ball valve (open)	about 0.1	
Butterfly valve (open)	about 1.0	
Globe valve (open)	about 2.0	
Plug valve (open)	about 0.5	

*In the one-quarter-open position, a valve will have a K value several times that in the full-open position.

TABLE 1-4B Values of K*—Contractions and Enlargements

Sudden contraction*		Gradual enlargement for total angles of cone*						
d_1/d_2	K_c	4°	10°	15°	20°	30°	50°	60°
1.2	0.08	0.02	0.04	0.09	0.16	0.25	0.35	0.37
1.4	0.17	0.03	0.06	0.12	0.23	0.36	0.50	0.53
1.6	0.26	0.03	0.07	0.14	0.26	0.42	0.57	0.61
1.8	0.34	0.04	0.07	0.15	0.28	0.44	0.61	0.65
2.0	0.37	0.04	0.07	0.16	0.29	0.46	0.63	0.68
2.5	0.41	0.04	0.08	0.16	0.30	0.48	0.65	0.70
3.0	0.43	0.04	0.08	0.16	0.31	0.48	0.66	0.71
4.0	0.45	0.04	0.08	0.16	0.31	0.49	0.67	0.72
5.0	0.46	0.04	0.08	0.16	0.31	0.50	0.67	0.72

*Values from H. W. King and E. F. Brater, *Handbook of Hydraulics*, McGraw-Hill, New York, 1976.

The critical depth for a rectangular channel can be calculated by

$$y_c = \left(\frac{q^2}{\mathbf{g}}\right)^{1/3} \tag{20}$$

where

$$q = \frac{Q}{B} = \text{discharge per unit width} \tag{21}$$

Also

$$y_c = \frac{2}{3} E_{\min} = \frac{V_c^2}{\mathbf{g}} \tag{22}$$

where

$$V_c = \text{velocity corresponding to critical depth} = \frac{q}{y_c} \tag{23}$$

Equation (20) can also be rewritten as

$$q_{\max} = \sqrt{\mathbf{g} y_c^3} \tag{24}$$

For a *trapezoidal* channel, the equations are

$$\frac{Q^2}{g} = \frac{A_c^3}{\text{B}} \quad \text{or} \quad \frac{V_c^2}{\mathbf{g}} = \frac{A_c}{B}$$

where

$$B = \text{top width of the channel} \qquad \text{ft (m)} \tag{25}$$

A channel flowing with a depth greater than the critical depth has a *subcritical* flow. If normal depth of flow is smaller than the critical depth, it is called *supercritical* flow.

The *normal flow* (uniform flow, where depth is not changing with distance, and the water surface is parallel to the bed) is calculated by using the Manning or the Chezy formula with the hydraulic gradient slope S equal to the bed slope S_0 (see Table 1-5).

TABLE 1-5 A Few Average Values of n for Use in Kutter's and Manning's Formulas

Type of open channel	n
Smooth cement lining, best planed timber	0.010
Planed timber, new wood-stave flumes, lined cast iron	0.012
Good vitrified sewer pipe, good brickwork, average concrete pipe, unplaned timber, smooth metal flumes	0.013
Average clay sewer pipe and cast-iron pipe, average cement lining	0.015
Earth canals, straight and well maintained	0.023
Dredged earth canals, average condition	0.027
Canals cut in rock	0.040
Rivers in good condition	0.030

The Manning formula is

$$V = \frac{1.486}{n} R^{2/3} S^{1/2} \qquad \text{(feet per second)} \tag{26}$$

where

S = hydraulic slope
S_0 = bed slope = vertical drop per length of channel
n = Manning roughness coefficient (Table 1-5)

The Chezy formula is

$$V = C\sqrt{RS} \tag{27}$$

where

$$C = \sqrt{\frac{8\mathbf{g}}{f}} \qquad \text{(Darcy-Weisbach } f\text{)} \tag{28}$$

Also

$$C = \frac{1.486}{n} R^{1/6} \qquad \text{(Manning } n\text{)} \tag{29}$$

Kutter gave a more complex value:

$$C = \frac{41.65 + (0.00281/S) + (1.811/n)}{1 + (n/\sqrt{R})[41.65 + (0.00281/S)]} \tag{30}$$

If the depth of flow at a particular cross section of a channel is supercritical ($y < y_c$) and if the conditions downstream of that section can allow subcritical flow ($y > y_c$), for example, a normal flow depth downstream, then the flow will jump from the smaller depth to the larger depth. This jump is called the *hydraulic jump* and is associated with

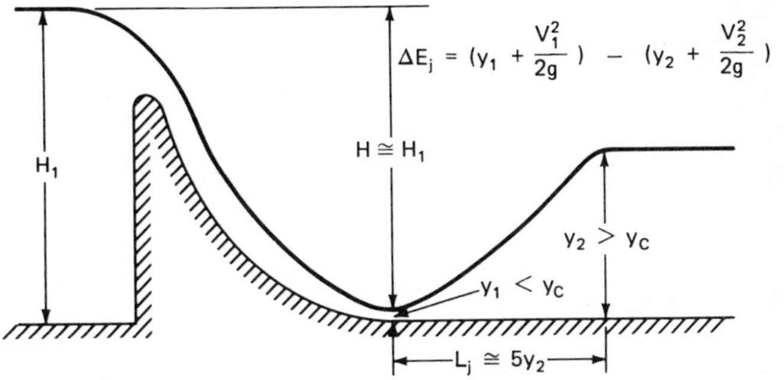

Figure 1-7 Hydraulic jump.

a lot of turbulence, and therefore loss of energy. Consider the flow downstream of a fall (say from a dam). The velocity downstream of the fall (Fig. 1-7) is such that almost all the potential energy has changed to kinetic energy; i.e.,

$$V_1 = \sqrt{2\mathbf{g}H} \tag{31}$$

Then

$$y_1 = \frac{q}{v_1} \tag{31a}$$

and the depth after the jump y_2 is given by

$$y_2 = \frac{y_1}{2}\left(\sqrt{1 + 8N_{f1}^2} - 1\right) \tag{32}$$

The most economical section of a rectangular channel is obtained when

$$R = \frac{y}{2} \tag{33}$$

This means that there will be the least-wetted perimeter and therefore the least frictional energy losses and hence the maximum flow.

Nonuniform Flow

When the water surface is not parallel to the bed, the depth of flow varies with the distance (depth is not equal to the normal depth mentioned before), and this flow is termed nonuniform flow. The water flowing at all the flow restrictions, transitions, falls, controls, spillways, etc., takes on a curved surface profile. The calculations of such surface profiles are made in small steps for better accuracy, or may be done in a single step for less accuracy with the equation

$$\frac{dy}{dL} = \frac{S_0 - S}{1 - \dfrac{V^2}{gy}} \tag{34}$$

Another form of this equation is

$$\Delta L = \frac{E_1 - E_2}{S - S_0} \tag{35}$$

Here, S is the normal slope computed from the Chezy or the Manning formula. E_1 and E_2 are the specific energies at the two ends of a reach of length ΔL (under consideration).

Measurements

The static pressure, velocity, and mass rate of flow are the usual quantities required to be measured in fluid mechanics. The *pressure* is usually measured by a U-tube manometer containing a liquid of known density (Fig. 1-8). One leg of the U tube is connected to the point where the pressure is to be measured. The other leg is open to the atmosphere. The vertical distance between the liquid surfaces in the two legs is a measure of the pressure head in feet of that liquid. It can be converted into psig by multiplying by the density of the liquid [see Eq. (1)]. As an alternative, use the pressure of a liquid immiscible with the first one. The pressure of a liquid can also be measured by just inserting a standpipe (piezometer) at the point where pressure is to be measured. The

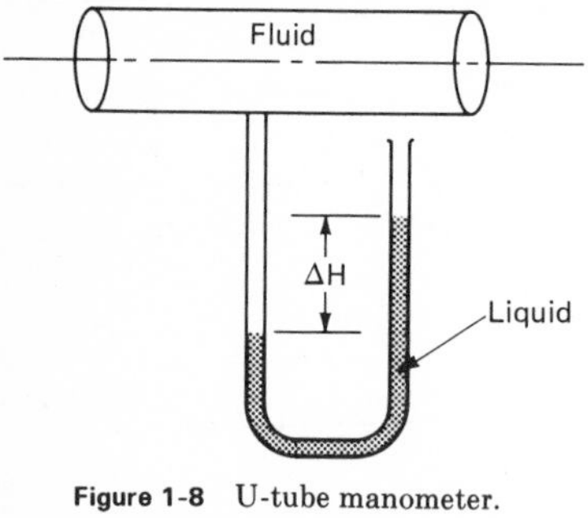

Figure 1-8 U-tube manometer.

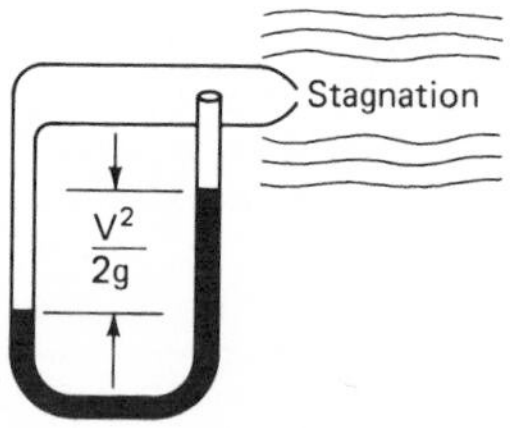

Figure 1-9 Pitot tube.

pressure in a fluid is also measured by using a diaphragm pressure gauge in which the scale is calibrated such that the displacement of the diaphragm within the system transmitted through mechanical, pneumatic, or hydraulic linkages directly gives a pressure reading on the dial.

The velocity of flow is usually obtained by measuring the differential pressure head ΔH between two points of flow (usually at a transition) and using proper formulation, such as

$$V = C_v\sqrt{2g\,\Delta H} \tag{36}$$

where

ΔH = differential head, ft (m), of fluid flowing (not gauge fluid)
C_v = coefficient of velocity

The Pitot tube (Fig. 1-9) was specifically developed to measure velocity and is a special form of U tube in which the stagnation leg measures $p_0/w + V^2/2\mathbf{g}$ and the other leg gets signal $p_0/2w$. Therefore, the differential head is equal to $V^2/2\mathbf{g}$, thus giving velocity. The quantity or mass *rate of flow* is most accurately measured by collecting in a weighing or graduated tank for a definite time. However, it is usually calculated by measuring the velocity of flow and multiplying it by a suitable area of cross section.

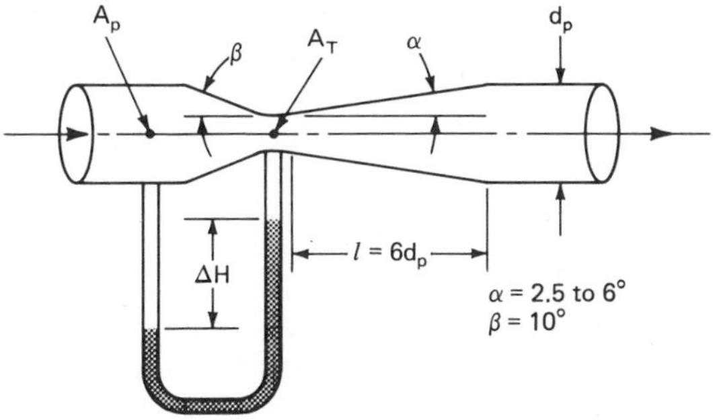

Figure 1-10 Venturimeter.

This principle has been further extended by using special transition elements or fittings for which the coefficient of discharge is already known from laboratory measurements. For example, a venturimeter (Fig. 1-10) has the following formula:

$$Q = A_T C_d \frac{\sqrt{2\mathbf{g}\,\Delta H}}{1 - (A_T/A_p)^2} \tag{37}$$

where

Q = quantity of flow, ft³/s (m³/s) (multiply by density to get mass rate)
C_d = coefficient of discharge of the meter, 0.95 to 0.99
A_T = area of cross section at throat, ft² (m²)
A_p = area of cross section of pipe, ft² (m²)
ΔH = differential head, ft (m) of fluid flowing (not the gauge liquid)

Orifices and nozzles have also been used with the same basic principle of acquiring velocity and then converting to discharge. However, the jet of flow in such a case is contracted at the discharge, "vena contracta" (Fig. 1-11), giving

$$C_c = \text{coefficient of contraction} = 0.65 \text{ usually} = \frac{A_{\text{jet}}}{A_o} \tag{38}$$

where

A_{jet} = area of vena contracta, ft² (m²)
A_o = area of orifice nozzle, ft² (m²)

Then

$$Q = VA_{\text{jet}} \tag{39}$$

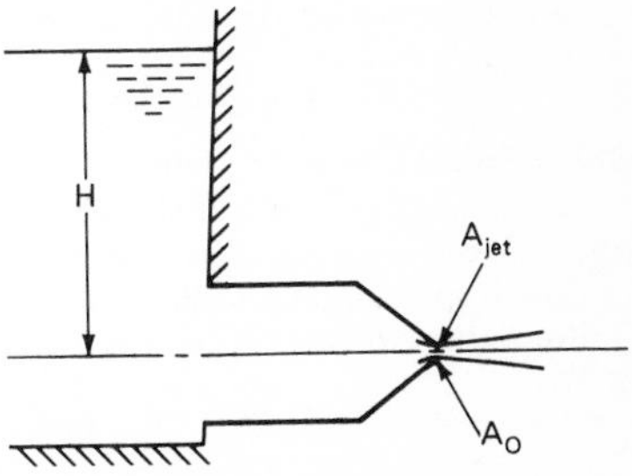

Figure 1-11 Vena contracta.

V is calculated from the static head H on the nozzle. Direct flow-reading, magnetic, ultrasonic, and other meters have also been used recently.

The flow in open channels is usually measured by utilizing *weirs* (Fig. 1-12). For a rectangular, sharp-crested weir, Francis gave the following relation:

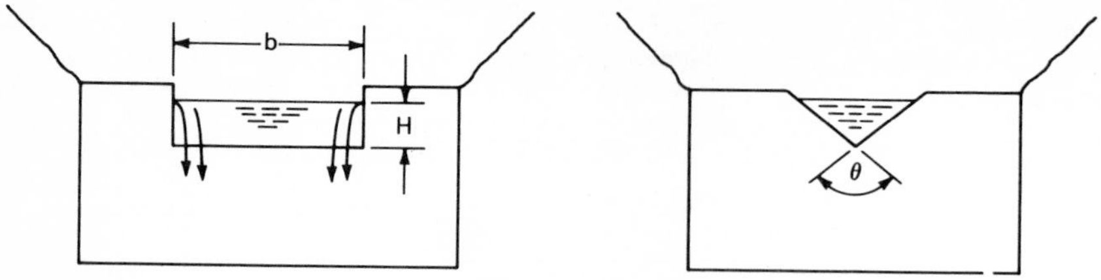

Figure 1-12 Weirs.

$$Q = C\left(b\frac{NH}{10}\right)\left[\left(H + \frac{V_0^2}{2\mathbf{g}}\right)^{3/2} - \left(\frac{V_0^2}{2\mathbf{g}}\right)^{3/2}\right] \tag{40}$$

where

Q = quantity of liquid flow, ft³/s (m³/s)
b = length of rectangular weir, ft (m)
H = head above weir crest, taken far upstream, ft (m)
V_0 = velocity of approach flow, taken far upstream, ft/s (m/s)
N = number of end contractions
= 2, if weir length < channel width at both ends
= 1, if weir length < channel width at one end
= 0, if weir length = channel width
C = 3.33 (U.S. customary units) = 1.84 (SI units)

However, Bazin gave the following relation for rectangular weir:

$$Q = \left(0.405 + \frac{c}{H}\right)\sqrt{2\mathbf{g}}\, bH^{3/2} \tag{41}$$

where c = 0.00984 (U.S. customary units) = 0.03 (SI units)

If the end contractions and approach velocity are neglected, a simplified formula can be used. For a triangular weir (V notch), the formula is

$$Q = \begin{cases} \frac{8}{15}\, c \tan\frac{\theta}{2}\sqrt{2g}\, H^{5/2} & \text{(U.S. customary units)} \\ 2.36c \tan\frac{\theta}{2}\, H^{5/2} & \text{(SI units)} \end{cases} \tag{41a}$$

where

c = a coefficient of contraction = 0.6
H = head above the lowest point on the notch, taken far upstream, ft (m)
θ = total angle subtended by the notch

Measurements by Using Electrical Data

Motor Input-Output. The flow-rate pump head can also be determined by measurement of voltage-current at the pump motor (or turbine-generator). If the current, voltage, and power factor of the pump motor are measured in the control room, the bus-voltage drop should be properly accounted for.[14]*

For a three-phase motor, the motor output brake horsepower (bhp) is given by

$$\text{bhp} = \sqrt{3}\,\frac{VI\cos\theta}{746}\,\eta_m \tag{42a}$$

where

V = terminal voltage, V
I = current, A

*References for Chapters 9-1 and 9-2 are combined following Chapter 9-2.

$\cos\theta$ = power factor
η_m = motor efficiency
746 W = 1 hp

For getting flow rate, the following equation for pump water horsepower (whp) is used:

$$\text{whp} = \begin{cases} \dfrac{\gamma QH}{550} & \text{(U.S. customary units)} \\ \dfrac{\gamma QH}{75} & \text{(SI units)} \end{cases} \tag{42b}$$

where

γ = density of water being pumped, lb/ft³ (kg/m³)
Q = rate of flow, ft³/s (m³/s)
H = TDH of pump, ft (m) of water
550 ft·lb/s = 1 hp(U.S. customary units)
75 kg·m/s = 1 hp (SI units)

Another relation required is for pump bhp:

$$\text{bhp} = \frac{\text{whp}}{\eta_p} \tag{42c}$$

where η_p is the efficiency of the pump. Using Eqs. (3), (4), and (5),

$$Q = \begin{cases} \dfrac{\sqrt{3}VI\cos\theta\,\eta_m\eta_p}{746}\,\dfrac{550}{\gamma H} & \text{(U.S. customary units)} \\ C\,\dfrac{VI\cos\theta\,\eta_m\eta_p}{\gamma H} & \text{(SI units)} \end{cases} \tag{42d}$$

where $C = 1.275$ for Q, ft³/s.

η_m and η_p are involved as unknowns. They have to be obtained from manufacturers' curves by trial and error, by assuming a Q, finding efficiencies from those curves, and using those efficiencies to get a new Q.

Note—For a turbine, the only difference is that the inverse of $\eta_m\eta_p$ is used.

1-3. PRINCIPAL COMPONENTS OF FLUID SYSTEMS

Following is a list of the components most frequently used in fluid systems:

- Accumulators
- Actuators
- Compressors, vacuum pumps, and blowers
- Couplings, quick-connect, hydraulic
- Couplings, quick-connect, pneumatic
- Cylinders
- Demineralizers
- Dryers
- Expansion joints
- Filters, air-line
- Filters, hydraulic
- Fittings, tube and port, hydraulic
- Fittings, tube and port, pneumatic
- Fluids
- Gauges, pressure and flow, hydraulic
- Gauges, pressure and flow, pneumatic
- Heat exchangers
- Hose, hose fittings, and hose assemblies, hydraulic
- Hose, hose fittings, and hose assemblies, pneumatic
- Hydrostatic drives
- Intensifiers
- Interface devices
- Joints, rotating and swivel, hydraulic and pneumatic
- Logic devices, air
- Lubricators
- Manifolds
- Measuring devices and meters
- Motors, air rotary

Motors, electrohydraulic stepping
Motors, hydraulic rotary
Motors, low-speed–high-torque
Pipe
Pulsation dampers
Pumps
Reservoirs and accessories
Restrictive orifices
Seals, scrapers, backup rings, and boots
Shock absorbers
Steam traps
Strainers
Switches, flow
Switches, level
Switches, limit and proximity
Switches, pressure and temperature, hydraulic
Switches, pressure and temperature, pneumatic
Transducers, pressure
Tubing
Valves, directional control
Valves, electrohydraulic servo
Valves, flow control
Valves, pressure control
Valves, level control

Various materials are used in these components. It is important to be aware of their compatibility with fluids used in industry. Table 1-6 provides such data.

Some of the major components of fluid systems will be discussed here in detail. Consult other sections of this handbook as well.

PUMPS

A pump is the most essential component of a hydraulic system in the absence of another source of pressure. The most common other source of pressure is a reservoir kept at high elevation (or potential energy). However, water at high elevation automatically exerts a static pressure and does not require much more explanation. Since pumps have diverse characteristics according to their speed and size and require extended discussion for a thorough understanding, only the most essential details will be discussed here. The Hydraulic Institute[7] classification of pumps given in Fig. 1-13 is useful.

Types of Pumps

Positive-Displacement-Type Pumps

These pumps (reciprocating plunger, rotary-gear, screw, piston, vane, etc.) create pressure by a direct-contact push to the liquid. They are employed generally for medium-pressure, low-flow applications, such as a metering pump to displace a known volume of fluid in a specified time. Large-size positive-displacement pumps usually become uneconomical because of space requirements.

Rotary-Gear Pumps

A rotary-gear pump, shown in Fig. 1-14, depends for its action on a pair of gears closely fitted in a housing. Fluid enters the tooth spaces usually on the inlet side and is carried along the casing to the outlet side.

Centrifugal Pumps

The pumps more commonly used in plant applications are of the centrifugal type, in which the rotary motion of the impeller imparts centrifugal force to the liquid, thereby raising the pressure. The selection of the impeller of a particular centrifugal pump (Fig. 1-15) is based upon the specific speed N_s as defined below:

$$N_s = \frac{N\sqrt{Q}}{H^{3/4}} \qquad (43)$$

Text continues on p. 9-24

TABLE 1-6 Suitability of Various Metals and Alloys for Various Fluids*,†

Corrosive	Steel‡ C.I. D.I.	Brz.	316SS	CA-20	CD4MCu	Mon	Ni	H-B	H-C	Ti	Zi
Acetaldehyde, 70°F	B	A	A	A	A	A	A		A	A	A
Acetic acid, 70°F	X	A	A	A	A	B	B	A	A	A	A
Acetic acid, <50%, to boiling	X	B	A	A	B	B	B	C	A	A	A
Acetic acid, >50%, to boiling	X	X	B	A	C	B	B	X	A	A	A
Acetone, to boiling	A	A	A	A	A	A	A	A	A	A	A
Aluminum chloride, <10%, 70°F	X	B	C	B	C	B	C	A		B	A
Aluminum chloride, >10%, 70°F	X	X	C	B	C	C	X	A		B	A
Aluminum chloride, <10%, to boiling	X	X	X	C	X	X	X	A		X	A
Aluminum chloride, >10%, to boiling	X	X	X	X	X	X	X	A	X	X	A
Aluminum sulfate, 70°F.	X	B	A	A	A	B	B	B	B	A	A
Aluminum sulfate, <10%, to boiling	X	B	B	A	B	X	X	A	A	A	A
Aluminum sulfate, >10%, to boiling	X	C	C	B	C	X	X	B	B	C	B
Ammonium chloride, 70°F	X	X	B	B	B	B	B		A	A	A
Ammonium chloride, <10%, to boiling	X	X	B	B	C	B	B	B	A	A	A
Ammonium chloride, >10%, to boiling	X	X	X	C	X	C	C		C	C	C
Ammonium fluosilicate, 70°F	X	X	C	B	C	X	X		C	X	X
Ammonium sulfate, <40%, to boiling	X	X	B	B	C	B	B	X	B	A	A
Arsenic acid, to 225°F	X	X	C	B	C	X	X				
Barium chloride, 70°F, <30%	X	B	C	B	C	B	B	B	B	B	B
Barium chloride, <5%, to boiling	X	B	C	B	C	B	B	B	B	A	A
Barium chloride, >5%, to boiling	X	C	X	C	X	C	C	C	C	C	C
Barium hydroxide, 70°F	B	X	A	A	A	B	A	B	B	A	A
Barium nitrate, to boiling	C	X	B	B	B		B	B		B	B
Barium sulfide, 70°F	C	X	B	B	B	X	X			A	A
Benzoic acid	X	C	B	B	B	B	B	A	A	A	A
Boric acid, to boiling	X	C	B	B	B	C	C	A	A	B	B
Boron trichloride, 70°F, dry	B	B	B	B	B	B	B	B	B		
Boron trifluoride, 70°F, 10%, dry	B	B	B	A	B	A	A		A		
Brine (acid), 70°F	X	X	X	X	X				B	B	
Bromine (dry), 70°F	X	X	X	X	X	X	C	B	B	X	X
Bromine (wet), 70°F	X	X	X	X	X	X	C		B	X	X
Calcium bisulfite, 70°F	X	X	B	B	B	X	X		B	A	A
Calcium bisulfite, to hot	X	X	C	B	C	X	X		C	A	A

TABLE 1-6 Suitability of Various Metals and Alloys for Various Fluids*,† (*Continued*)

Corrosive	Steel‡ C.I. D.I.	Brz.	316SS	CA-20	CD4MCu	Mon	Ni	H-B	H-C	Ti	Zi
Calcium chloride, 70°F	B	C	B	B	B	B	B	A	A	A	A
Calcium chloride, <5%, to boiling	C	C	B	B	B	A	A	A	A	A	A
Calcium chloride, >5%, to boiling	X	C	C	B	C	C	C	A	A	B	B
Calcium hydroxide, 70°F	B	B	B	B	B	B	B		A	A	
Calcium hydroxide, <30%, to boiling	C	B	B	B	B	B	B		A	A	
Calcium hydroxide, >30%, to boiling	X	X	C	C	C	C	C		B	A	
Calcium hypochlorite, <2%, 70°F	X	X	X	C	X	X	X		A	A	A
Calcium hypochlorite, >2%, 70°F	X	X	X	C	X	X	X		B	A	B
Carbolic acid, 70°F (phenol)	C	B	A	A	A	A	A	A	A	A	A
Carbon bisulfide, 70°F	B	B	A	A	A	B	B			A	
Carbonic acid, 70°F	B	C	A	A	A	C	B	A	A	A	A
Carbon tetrachloride, dry to boiling	B	B	A	A	A	A	A	B	B	A	A
Chloric acid, 70°F	X	X	X	B	C	X	X	X	C		
Chlorinated water, 70°F	C	C	B	B	B				A	A	A
Chloroacetic acid, 70°F	X		X	X						A	B
Chlorosulfonic acid, 70°F	X	X	X	C	X	X	X	A	A	B	X
Chromic acid, <30%	X	X	C	B	C	X	X		B	A	A
Citric acid	X	C	A	A	A	C	C	A	A	A	A
Copper nitrate, to 175°F	X	X	B	B	B	X	X	X	X	B	
Copper sulfate, to boiling	X	C	C	B	C	X	X		A	A	A
Cresylic acid	C	C	B	B	B	C	C	B	B		
Cupric chloride	X	C	X	X	X	C	X		C	B	X
Cyanohydrin, 70°F	C		B	B	B						
Dichloroethane	C	B	B	B	B	C	B	B	B	A	B
Diethylene glycol, 70°F	A	B	A	A	A	B	B	B	B	A	A
Dinitrochlorobenzene, 70°F, dry	C	B	A	A	A	A	A	A	A	A	A
Ethanolamine, 70°F	B	X	B	B	B	C	X			A	A
Ethers, 70°F	B	B	B	A	A	B	B	B	B	A	A
Ethyl alcohol, to boiling	A	A	A	A	A	A	A	A	A	A	A
Ethyl cellulose, 70°F	A	B	B	B	B	B	B	B	B	A	A
Ethyl chloride, 70°F	C	B	B	A	B	B	B	B	B	A	A
Ethyl mercaptan, 70°F	C	X	B	A	B			B	B		
Ethyl sulfate, 70°F	C	B	B	A	B	B					
Ethylene chlorohydrin, 70°F	C	B	B	B	B	B	B	B	B	A	A
Ethylene dichloride, 70°F	C	B	B	B	B	B	B	B	C	A	A

Ethylene glycol, 70°F	B	B	B	B	B	B	B	A	A	A	A
Ethylene oxide, 70°F	C	X	B	B	B	B	B	A	A	A	A
Ferric chloride, <5%, 70°F	X	X	X	X	X	X	X	X	A	A	B
Ferric chloride, >5%, 70°F	X	X	X	X	X	X	X	X	B	B	X
Ferric nitrate, 70°F	X	X	B	A	B	X	X		B		
Ferric sulfate, 70°F	X	X	C	B	C	C	C		B	B	B
Ferrous sulfate, 70°F	X	C	C	B	C	C	C	B	B	A	A
Formaldehyde, to boiling	B	B	A	A	A	B	B	B	B	A	A
Formic acid, to 212°F	X	C	X	A	B	C	C	A	A	C	A
Freon, 70°F	A	A	A	A	A	A	A	A	A	A	A
Hydrochloric acid, <1%, 70°F	X	X	C	B	C	B	B	B	A	B	A
Hydrochloric acid, 1–20%, 70°F	X	X	X	X	X	X	X	C	B	X	A
Hydrochloric acid, >20%, 70°F	X	X	X	X		X	X	C	B	X	B
Hydrochloric acid, <½%, 175°F	X	X	C	C	C	X	X	B	A	X	A
Hydrochloric acid, ½–2%, 175°F	X	X	X		X	X	X	B	B	X	A
Hydrocyanic acid, 70°F	X	X	C	B	C	C	C	C	C		
Hydrogen peroxide, <30% <150°F	C	X	B	B	B	B	B	B	B	A	A
Hydrofluoric acid, <20%, 70°F	X	B	X	B	C	C	C	C	B	X	X
Hydrofluoric acid, >20%, 50°F	X	C	X	C	X	C	C	C	B	X	X
Hydrofluoric acid, to boiling	X	X	X	X	X	C	X		C	X	X
Hydrofluosilicic acid, 70°F	X		C	B	C				B		
Lactic acid, <50%, 70°F	X	B	A	A	A	X	C	B	B	A	A
Lactic acid, >50%, 70°F	X	B	B	B	B	C	C	B	B	A	A
Lactic acid, <5%, to boiling	X	X	C	B	C	X	X	B	B	A	A
Lime slurries, 70°F	B	B	B	B	A	B	B	B	B	B	B
Magnesium chloride, 70°F	C	C	B	A	B	C	C	A	A	A	A
Magnesium chloride, <5%, to boiling	X	C	C	B	C	C	C	A	A	A	A
Magnesium chloride, >5%, to boiling	X	C	X	C	X	C	C	B	B	B	B
Magnesium hydroxide, 70°F	B	A	B	B	A	B	A	B	B	A	
Magnesium sulfate	C	C	B	A	B	B	B	C	C	B	B
Maleic acid	C	C	B	B	B	C	C	B	B	A	
Mercaptans	A	X	A	A	A	X	X				
Mercuric chloride, <2%, 70°F	X	X	X	X	X	X	C		B	A	A
Mercurous nitrate, 70°F	C	X	B	B	B	C			C		
Methyl alcohol, 70°F	A	A	A	A	A	A	A	A	A	A	A
Naphthalene sulfonic acid, 70°F	X	C	B	B	B	C	C	B	B		
Naphthalenic acid, to hot	C	C	B	B	B	C	C	B	B		
Nickel chloride, 70°F	X	X	C	B	C	C	X	A		B	B
Nickel sulfate	X	C	B	B	B	C	C		B		A

TABLE 1-6 Suitability of Various Metals and Alloys for Various Fluids*,† (*Continued*)

Corrosive	Steel‡ C.I. D.I.	Brz.	316SS	CA-20	CD4MCu	Mon	Ni	H-B	H-C	Ti	Zi
Nitric acid	X	X	B	B	B	X	X			B	B
Nitrobenzene, 70°F	A	C	A	A	A	B	B	B	B	A	
Nitroethane, 70°F	A	A	A	A	A	A	A	A	A	A	A
Nitropropane, 70°F	A	A	A	A	A	A	A	A	A	A	A
Nitrous acid, 70°F	X	X	X	C	X	X	X				
Nitrous oxide, 70°F	C	C	C	C	C	X	X		C		
Oleic acid	C	C	B	B	B	C	C	C	C	C	C
Oleum, 70°F	B	X	B	B	B	X	X	B	B	B	
Oxalic acid	X	C	C	B	C	C	C	B	B	X	A
Palmitic acid	B	B	B	A	B	B	B				
Phenol (see carbolic acid)											
Phosgene, 70°F	C	C	B	B	B	C	C	B	B		
Phosphoric acid, <10%, 70°F	X	C	A	A	A	C	C	A	A	A	A
Phosphoric acid, >10–70%, 70°F	X	C	A	A	A	C	C	B	C	B	B
Phosphoric acid, <20%, 175°F	X	C	B	B	B	C	C	A	A	C	B
Phosphoric acid, >20%, 175°F <85%	X	C	C	B	C	C	C	B	C	C	C
Phosphoric acid, >10%, boil, <85%	X	C	X	C	C	C	C	C	C	C	C
Phthalic acid, 70°F	C	B	B	A	B	B	B	B	B	A	A
Phthalic anhydride, 70°F	B	C	A	A	A	A	A	A	A		
Picric acid, 70°F	X	X	C	B	C	C	X		B		
Potassium carbonate	B	B	A	A	A	B	B	B	B	A	A
Potassium chlorate	B	C	A	A	A	C	C		B	A	A
Potassium chloride, 70°F	C	C	B	A	B	B	B	B	B	A	A
Potassium cyanide, 70°F	B	X	B	B	B	C	C	B	B		
Potassium dichromate	B	B	A	A	A	B	B		B	A	A
Potassium ferricyanide	C	B	B	B	B	B	B	B	B	A	A
Potassium ferrocyanide, 70°F	X	B	B	B	B	B	B	B	B		B
Potassium hydroxide, 70°F	C	C	B	A	B	A	A	B	C	B	A
Potassium hypochlorite	X	C	C	B	C	X	X		B	A	
Potassium iodide, 70°F	C	B	B	B	B	B	B	B	B	A	A
Potassium permanganate	B	B	B	B	B	C	B		B		
Potassium phosphate	C	C	B	B	B					B	B
Seawater, 70°F	C	B	B	A	B	A	A	A	A	A	A
Sodium bisulfate, 70°F	X	C	C	B	C	C	C	B	B	B	A
Sodium bromide, 70°F	B	C	B	B	B	B	B	B	B		

Sodium carbonate	B	B	B	A	B	B	B	B	B	A	A
Sodium chloride, 70°F	C	B	B	B	B	A	A	B	B	A	A
Sodium cyanide	B	X	B	B	B	X	X			B	
Sodium dichromate	B	X	B	B	B					B	
Sodium ethylate	B	A	A	A	A	A	A				
Sodium fluoride	C	C	B	B	B	B	B	C	C	B	B
Sodium hydroxide, 70°F	B	B	B	A	B	A	A	A	A	A	A
Sodium hypochlorite	X	X	C	C	C	X	X		B	A	B
Sodium lactate, 70°F	B	C	C	C	C	C		C	C		
Stannic chloride, <5%, 70°F	X	C	X	C	X	C	C	B	B	A	A
Stannic chloride, >5%, 70°F	X	X	X	X	X	X	X	B	C	B	B
Sulfite liquors, to 175°F	X	C	B	B	B	C	C		B	A	
Sulfur (molten)	B	X	A	A	A	C	C	C	A	A	
Sulfur dioxide (spray), 70°F	C	C	B	B	B	C	C		B	C	
Sulfuric acid, <2%, 70°F	X	C	B	A	B	C	C	A	A	B	A
Sulfuric acid, 2–40%, 70°F	X	C	C	B	C	C	C	A	A	X	A
Sulfuric acid, 40%, <90%, 70°F	X	X	X	B	X	X	X	A	A	X	C
Sulfuric acid, 93–98%, 70°F	B	X	B	B	B	X	X	B	B	X	C
Sulfuric acid, <10%, 175°F	X	C	X	B	X	X	X	A	C	X	B
Sulfuric acid, 10–60% & >80%, 175°F	X	X	X	B	X	X	X	B	C	X	C
Sulfuric acid, 60–80%, 175°F	X	X	X	X	X	X	X	B	C	X	C
Sulfuric acid, <¾%, boiling	X	X	C	B	C	X	X	B	B	X	B
Sulfuric acid, ¾–40%, boiling	X	X	X	C	X	X	X	B	C	X	B
Sulfuric acid, 40–65% & >85%, boil	X	X	X	X	X	X	X	X	X	X	X
Sulfuric acid, 65–85%, boiling	X	X	X	X	X	X	X	X	X	X	X
Sulfurous acid, 70°F	X	C	C	B	C	X	X	B	B	A	B
Titanium tetrachloride, 70°F	C		C	B	C	C			C		
Trichlorethylene, to boiling	B	C	B	B	B	B	B	B	B	A	A
Urea, 70°F	C	C	B	B	B	C	C	C	C	B	B
Vinyl acetate	B	B	B	B	B				B		
Vinyl chloride	B	C	B	B	B	C	C	C	B	A	
Water, to boiling	B	A	A	A	A	A	A	A	A	A	A
Zinc chloride	C	C	B	A	B	B	B	B		A	A
Zinc cyanide, 70°F	X	B	B	B	B	B	B	B	B	B	B
Zinc sulfate	X	C	A	A	A	C	C	C	C	A	

* *Source:* Goulds Pumps, Inc., Seneca Falls, N.Y.

†Ni, Nickel, ASTM A 296 Gr. CZ-100; H-Bm Hastelloy alloy-B, ASTM A 494; H-C, Hastelloy alloy-C, ASTM A 494; Ti, Titanium unalloyed, ASTM B 367 Gr. C-1; Zi, Zirconium.

‡Code: A, fully satisfactory; B, useful resistance; C, limited use; X, unsuitable.

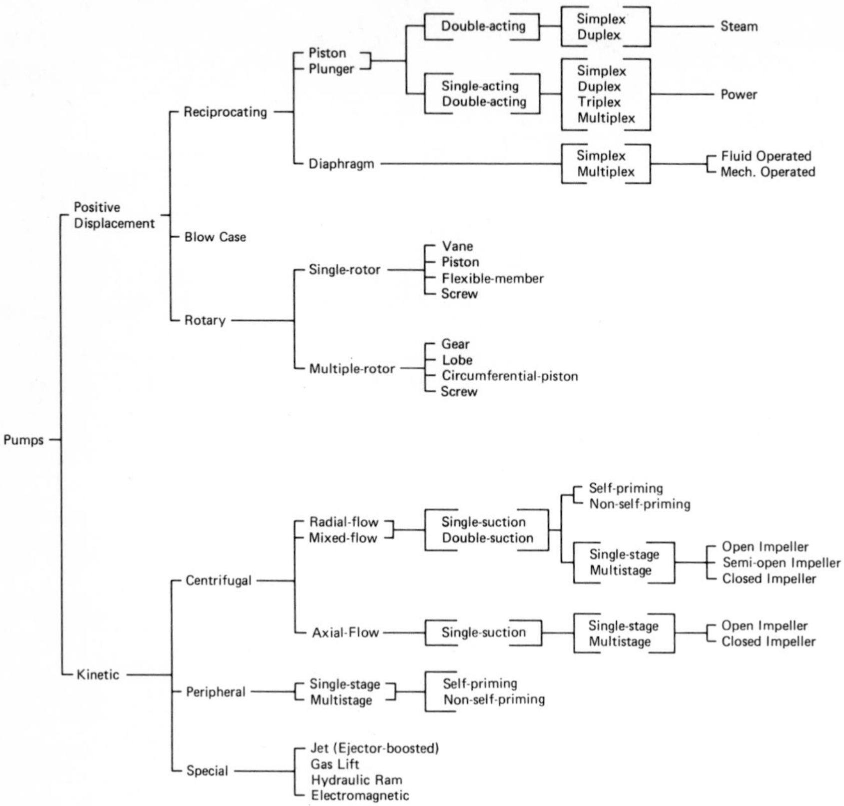

Figure 1-13 Classification of pumps.

where

N = pump rotation speed, r/min
Q = nominal rate of flow to be pumped, gal/min (m^3/s) (should be halved for double suction pump)
H = total pressure head to be created by pump, ft (m)

In the specific speed range of 1000 to 6000 r/min double-suction impellers can also be used if axial thrust needs to be balanced.

Pump Terms

Total Dynamic Head

The total dynamic head (TDH) of the pump needs to be determined. With reference to Fig. 1-16,

$$H = \text{TDH} = h_s^* + h_d + h_L \tag{44}$$

Also

$$\text{H} = h_{g,1} - h_{g,2} + \frac{V_2^2}{2\mathbf{g}} - \frac{V_1^2}{2\mathbf{g}} \tag{45}$$

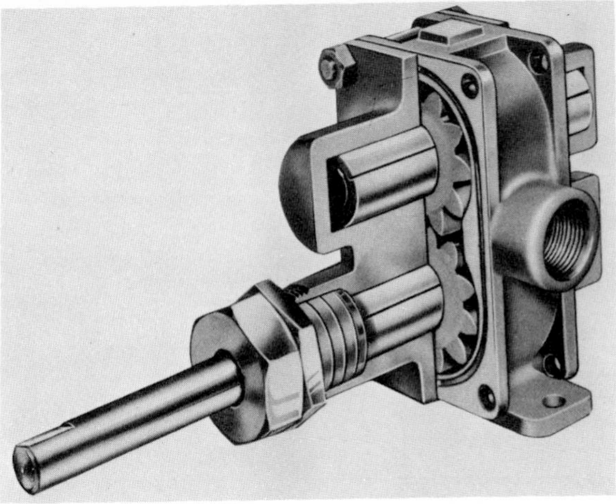

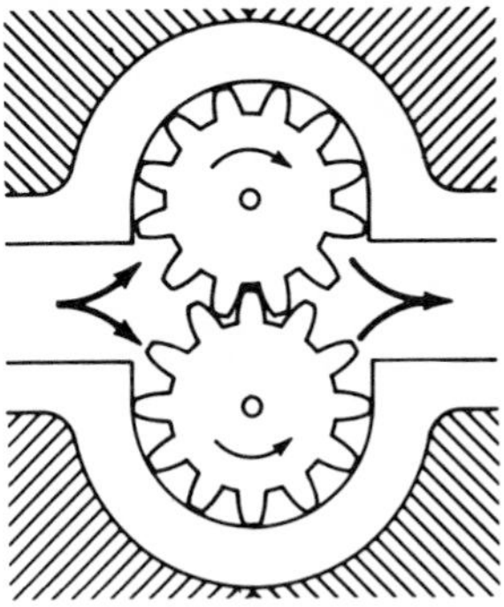

Figure 1-14 Rotary gear pump. *(ECO Pump Co.)*

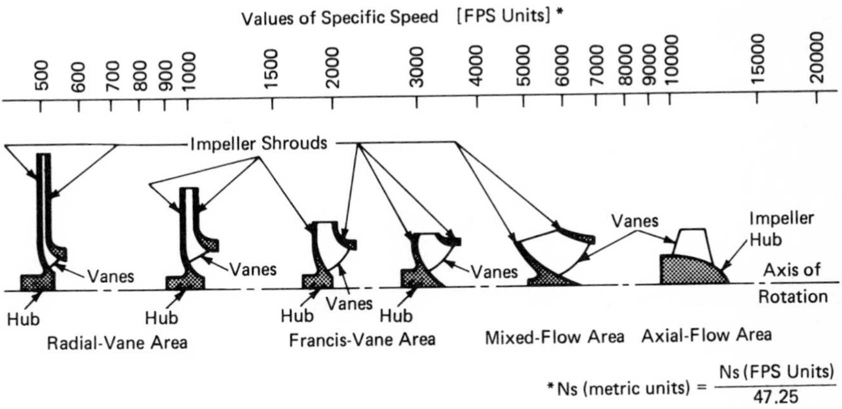

Figure 1-15 Selection of impeller design based on specific speed.

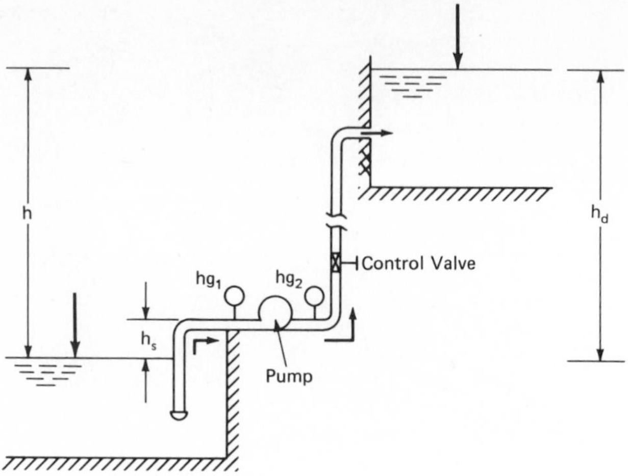

Figure 1-16 Definition sketch for pump heads.

TABLE 1-7 Centrifugal Pump Steam Turbine Specification Sheet

Client				Job no.		
Plant location				Date		
Inquiry no.				Equip. no.		
Pump service				No. required		
Operating Conditions				**Specifications***		
Fluid pumped				Pump type	Cat. no.	
				Stages	r/min	
Pumping temp., °F				Suct. size, in.	Location	
Spec. grav. @ P.T.				Noz. series	Flange facing	
Visc. @ P. T., cP				Disch. size, in.	Location	
Vap. press. @ P.T.				Noz. series	Flange facing	
Solids % & comp.				Hydraulic hp	Efficiency, %	
Suct. head, psia			ft	bhp	Max. hp	
Disch. head, psia			ft	NPSH required by pump, ft fluid		
TDH, ft, normal			Design	Pump materials & details*		
Pump cap., gal/min, normal			Design	Case		Max. imp.
NPSH available, ft, fluid				Case Th'kn.	Corr. all.	
Nonoverloading				Shaft	Dia.	Keyway
Rotation (facing coupling end)				Shaft sleeves		Brinell
Motor by				Impeller	Type	O.D.
hp	V	Ph.	Freq.	Impeller wear ring		
Frame	St. torque			Case wear ring		
Mfr. & type				Radial brg.	Mfg. no.	
Enclosure	Class	Group		Thrust brg.	Mfg. no.	

where
h_s^* = static suction head of the pump, ft (m)
h_d = static delivery discharge head of the pump, ft (m)
h_L = head losses in the piping, including entrance, friction, bends, fittings, exit, etc., ft (m)
$h_{g,1}^*$ = pressure (converted to head) indicated by suction gauge (vacuum gauge), ft (m). The asterisk indicates $+Ve$ if pump is higher than suction water surface and $-Ve$ if pump is lower than suction water surface.
$h_{g,2}$ = pressure (converted to head) indicated by discharge gauge, ft (m)
H = total static head of the pump, ft (m)

Net Positive Suction Head

The net positive suction head (NPSH) is another term that needs to be determined for the cavitation-free design and performance of a pump. It is the net absolute static pressure above the vapor pressure of the liquid in the pump:

$$\text{NPSH} = P_b - h_s^* - h_{f,s} - P_v \tag{46}$$

where
P_b = barometric pressure (absolute), ft (m)
P_v = vapor pressure of the liquid being pumped, ft (m)
$h_{f,s}$ = head losses in suction line up to pump, ft (m)

Operating Conditions			Specifications*	
Turbine by			Packing	No. rings
Mfr.	Model		Seal cage	Seal fluid
Type			Packing gland	
Rated hp			Mech. seal	Mfr.
Inlet steam, psia	temp., °F		Coupling	
Exh. steam, psia	Temp., °F	Quality	Base plate	
No. hr steam fl 3/4 L	1/2 L		Drip (lip) (pan)	
Turbine performance curves, details, cuts, and			Wt. pump, lb.	
dimension sheets, etc., must be submitted by bidder			Wt. base plate, lb.	
Driver mounted by			Wt. driver, lb.	
Remarks			Total wt., lb.	
			Case design temp. °F	
			Case design press., psig	
			Case hydrostatic test, psig	
			Pump performance curves, details, cuts, dimension	
			sheets, etc., must be submitted by bidder	

Bidder in proposal to complete these spaces and others				
left vacant. Successful bidder shall supply:				
Copies of certified dimension drawings				
Copies of performance curves for fluid pumped				
Copies of spare parts list				
Copies of operating instructions	No.	Date	Revision	Appr.

The available NPSH (NPSHA) at a pump application (in a pump house) should be greater than the NPSH required (NPSHR) by the pump for an efficient (cavitation-free) operation. If NPSHA is less than NPSHR, it means that vapor is likely to form inside the pump. This vapor can accumulate and form bubbles which, when burst, can produce cavities (nonsmooth surfaces on impeller vanes).

Suction Specific Speed

Here another term comes into the picture, the *suction specific speed* S_s:

$$S_s = \frac{N\sqrt{Q}}{(\text{NPSHR})^{3/4}} \tag{47}$$

The Hydraulic Institute[7] recommends S_s between 7480 and 10,690 r/min. See p. 9-22 for definitions of N and Q.

Data Specifications

A useful blank form for a specification sheet of pump data is given in Table 1-7. All data should be filled in and sent to vendors for further completion and bid preparation.

Pump Characteristic Curves

Pump characteristic parameters that are a measure of the pump performance are discharge, head, horsepower, efficiency, and NPSH. These characteristics vary with the pump type, size, speed, etc. Some typical characteristics are shown in Fig. 1-17.

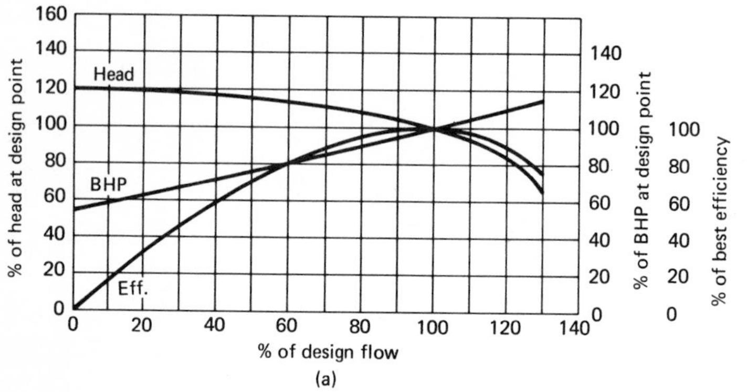

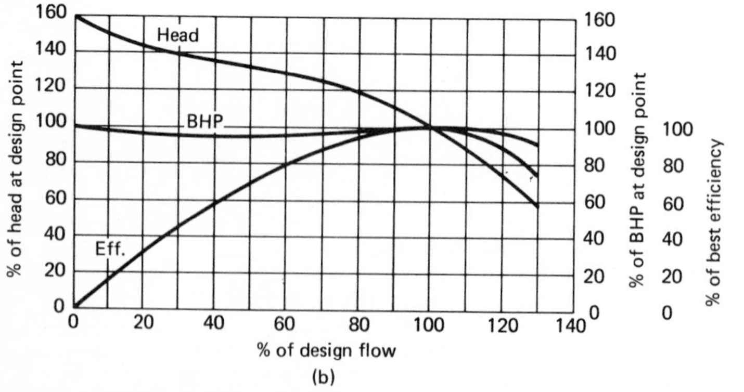

Figure 1-17 Pump characteristic curves: (*a*) radial-flow pump, (*b*) mixed-flow pump. (*Goulds Pumps, Inc.*)

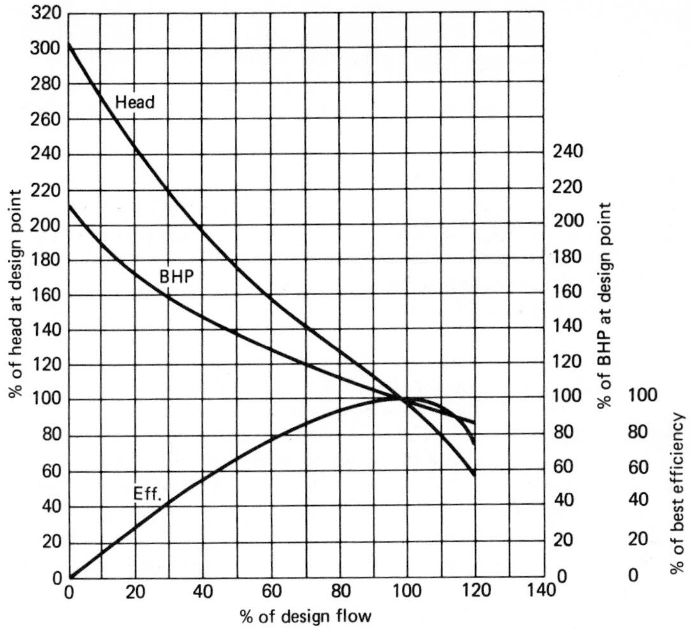

(c)

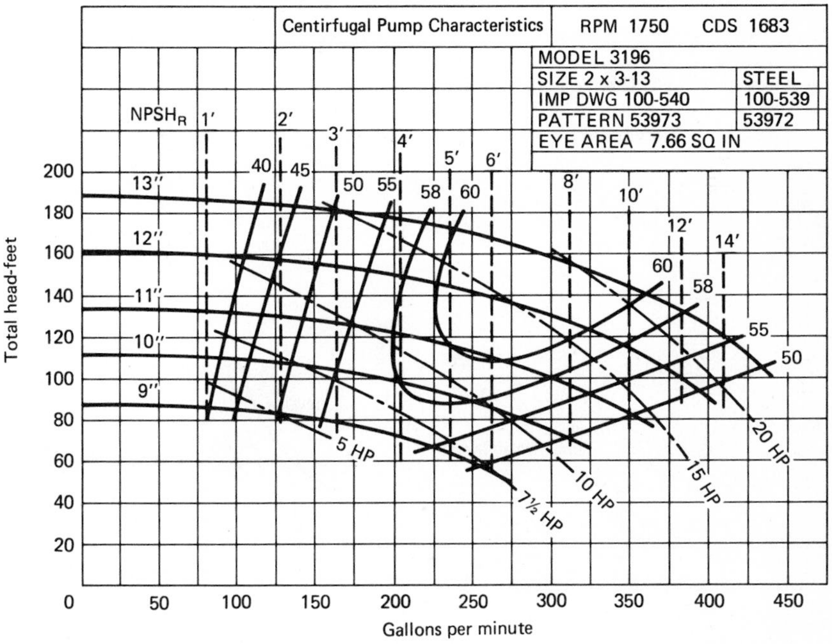

(d)

Figure 1-17 *(Continued)* (*c*) axial-flow pump, (*d*) composite performance curve. *(Goulds Pumps, Inc.)*

The normal operating point of the pump should be selected somewhere on the sloping portion of the curve, not too close to shutoff or run-out conditions. The manufacturer usually recommends a pump that supplies within +10 percent of design flow or +5 percent of design TDH.

Figures 1-18 and 1-19 are typical curves developed by manufacturers to aid in pump selection.

Pump Construction and Maintenance

Figures 1-20 and 1-21 show cross sections through some typical pumps. To control leakage, mechanical seals (discussed elsewhere in this book) serve best. The American National Standards Institute is often referred to for a proper design of pumps. ANSI B73.1 is a standard for horizontal centrifugal pumps.

Proper design of the pumphouse and pump-suction bays, inlet conditions, etc., is very important for proper performance of the pumps.[8]

Basic dimensions of the pumps required are first evolved on the basis of the Hydraulic Institute publication, "Standards for Centrifugal Pumps"; however, the final dimensions on the construction drawings have to be based on the pump manufacturers' recommendations, as their requirements are evolved on the basis of their own model tests. A matter of great importance is to determine the pump TDH and the NPSH which in turn reflect the overall dimensions of the suction intakes. Shape of the intake, depth, and clearances are among the important critical factors that affect the performance of large circulating-water pumps.

Hydraulic performance of a large-capacity pump is also affected by the formation of submerged or surface vortices in the sump area, which in practice are not visually observable because of the cover on the sump necessary to mount the pumping equipment and motors. Therefore, it is imperative to give careful consideration to the sump dimensions, preferably by the way of model experiments (discussed later). Moreover, the size of the pump house (including intake canal, cooling-tower basin for closed system, etc.) should be enough to store (or bypass) any transient quantities caused by sudden pump failures or operation of valves. Another important aspect of the pump house is to provide a device for measuring the actual flow and discharge pressure of each pump.

Preliminary dimensions of a pump house can be arrived at by using the guidelines given in Ref. 7 or a vendor's engineering data book.

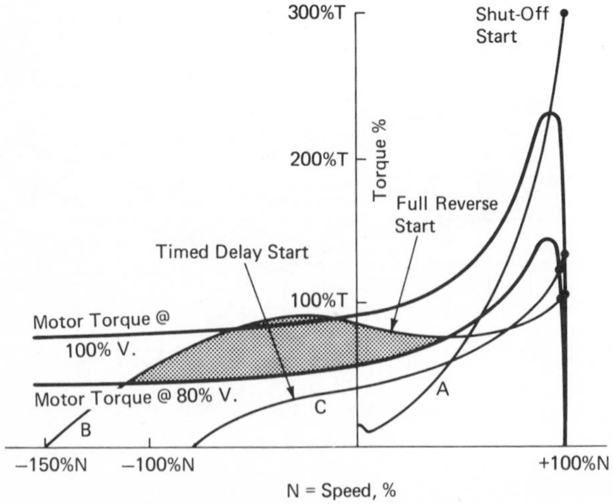

Figure 1-18 Start-up and shutdown of pumps. *(Allis-Chalmers.)*

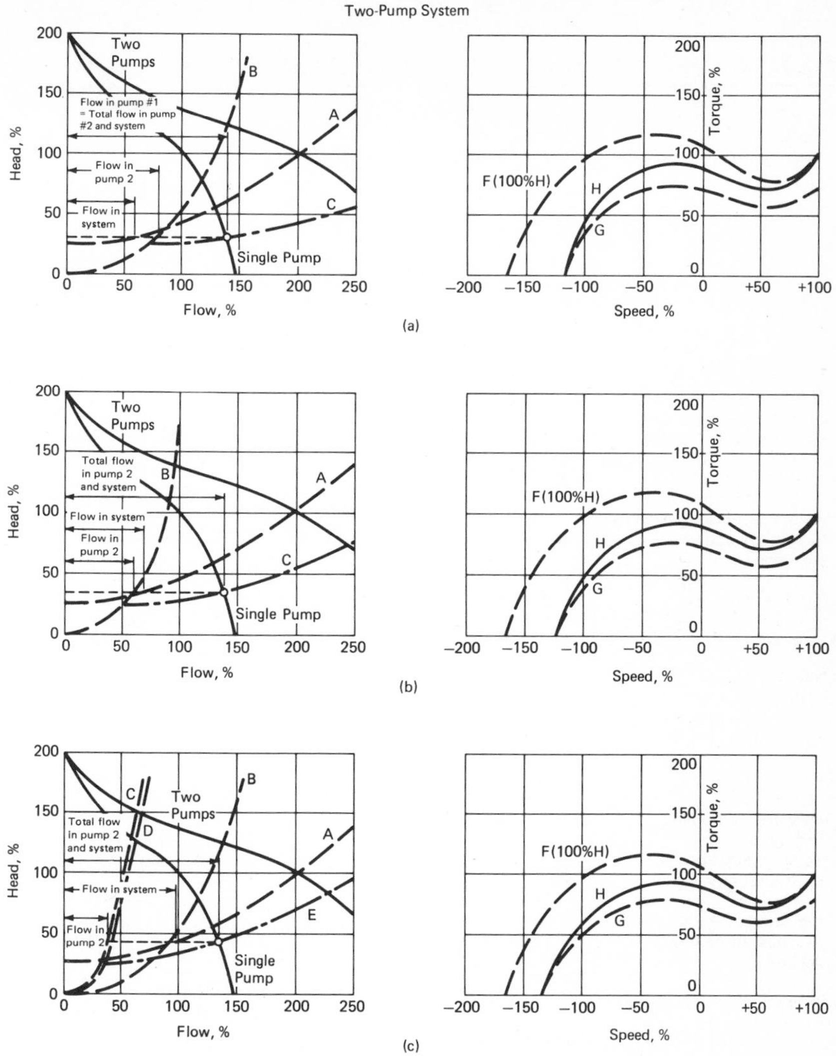

Figure 1-19 Starting large custom pumps against reverse flow. (*a*) Friction curves with pump characteristics are used to determine complete speed-torque curve. Pump 1 is operating and pump 2 is being started against reverse flow with valve open. (*b*) Starting at zero speed with valve open and reverse flow in pump, the speed-torque curve is identical. Because anti-reverse device prevents reverse rotation, maximum torque requirements are less. (*c*) Interlocking of valve operator and motor starter requires consideration of friction-head characteristic of the valve. The figure gives head vs. flow curves for single-pump operation and for two pumps in parallel operation, when both are running in the forward direction. The transient phase of starting is shown by dashed/dotted lines. The corresponding torque requirements are shown on the right. *(Allis-Chalmers.)*

Maintenance

Basically, pump maintenance requires proper lubrication according to the manufacturer's recommendations. Stuffing boxes and packings must be inspected and maintained as necessary. If proper setup, operation, and maintenance are not performed, the defects shown in Table 1-8 will be evident.[9] They should be corrected as soon as discovered.

TABLE 1-8 Causes and Symptoms of Malfunctions of Pumps

Causes	Symptoms: The pump fails to deliver liquid	Capacity is inadequate	Head is inadequate	The pump cuts out after starting	Power required by pump is too high	Excessive leakage of stuffing box	Packing must be renewed too frequently	The pump vibrates or is noisy	Bearings heat up	Pump runs heavily or seizes
Pump and suction line are not sufficiently primed with the liquid handled	●	●		●				●		●
Excessive suction lift	●	●		●				●		
Insufficient margin between suction lift and vapor pressure	●	●						●		●
Liquid contains gas		●	●	●						
Air pocket in suction line	●	●		●						
Air leak in suction line		●		●						
Air leaks into pump along stuffing box		●		●						
Foot valve too small		●								
Foot valve partly clogged		●						●		
Foot valve and suction line not fully submerged	●	●		●				●		
Connection of water seal on suction stuffing box blocked				●			●			
Lantern ring in stuffing box incorrectly fitted				●		●	●			
Speed too low	●	●	●							
Speed too high					●					
Incorrect direction of rotation	●		●							
Total manometric head of system greater than manometric head of the pump	●	●	●		●					
Total manometric head of system lower than manometric head of the pump					●					
Specific gravity of liquid handled not what it was originally supposed to be					●					

TABLE 1-8 Causes and Symptoms of Malfunctions of Pumps (*Continued*)

Causes	Symptoms: The pump fails to deliver liquid	Capacity is inadequate	Head is inadequate	The pump cuts out after starting	Power required by pump is too high	Excessive leakage of stuffing box	Packing must be renewed too frequently	The pump vibrates or is noisy	Bearings heat up	Pump runs heavily or seizes
Viscosity of liquid handled not what it was originally supposed to be		●	●		●					
Pump operating at too low a capacity								●		●
Parallel connection unsuitable for the specific operating conditions	●	●	●							●
Line, impeller, or pump casing clogged	●	●			●			●		
Pump set incorrectly aligned					●	●	●	●	●	●
Foundation not level								●		
Pump shaft warped					●	●	●	●	●	
A rotating part runs against a stationary part, e.g., impeller runs against wearing rings					●			●	●	●
Bearing(s) defective								●	●	●
Wearing rings worn		●	●							
Impeller damaged		●	●					●		
Pump shaft or shaft sleeve worn locally at the stuffing box						●	●			
Stuffing box incorrectly packed					●	●	●			
Type of packing used unsuitable for liquid handled						●	●			
Impeller out of balance						●	●	●	●	●
Failure to apply water cooling when handling hot liquids						●	●			
Clearance between pump shaft and bore of pump casing at the bottom of the stuffing box too great							●			

TABLE 1-8 Causes and Symptoms of Malfunctions of Pumps (*Continued*)

Causes	Symptoms: The pump fails to deliver liquid	Capacity is inadequate	Head is inadequate	The pump cuts out after starting	Power required by pump is too high	Excessive leakage of stuffing box	Packing must be renewed too frequently	The pump vibrates or is noisy	Bearings heat up	Pump runs heavily or seizes
Liquid for water seal contains impurities							●			
Gland overtightened							●			
Axial fit of complete pump shaft with impeller incorrect								●	●	●
Insufficient or excessive lubrication									●	
Lubricant contains impurities									●	●
Bearings incorrectly fitted									●	●

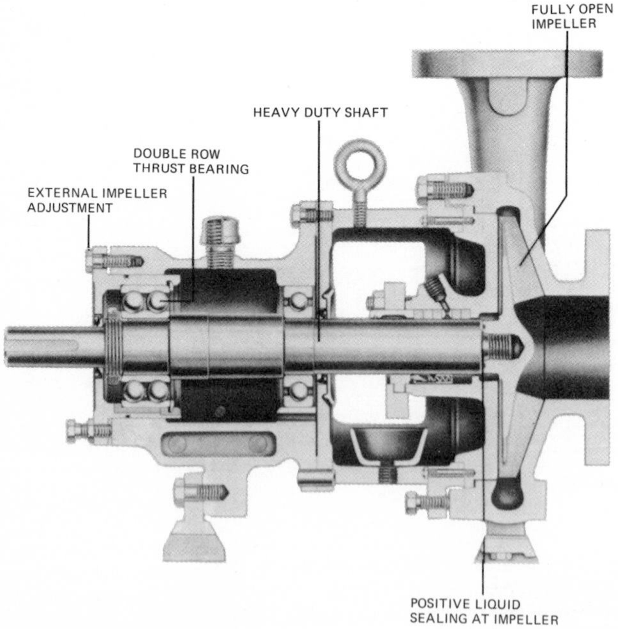

Figure 1-20 Cross-sectional view of a horizontal centrifugal pump. *(Goulds Pumps, Inc.)*

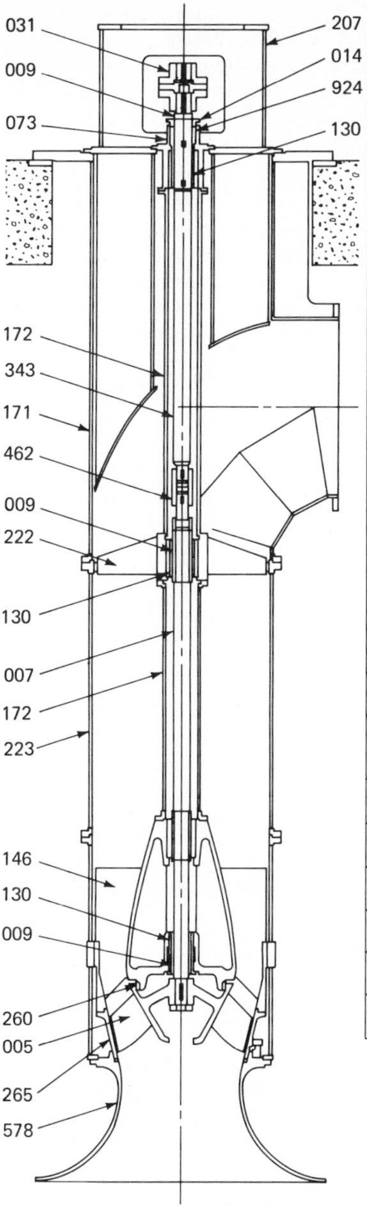

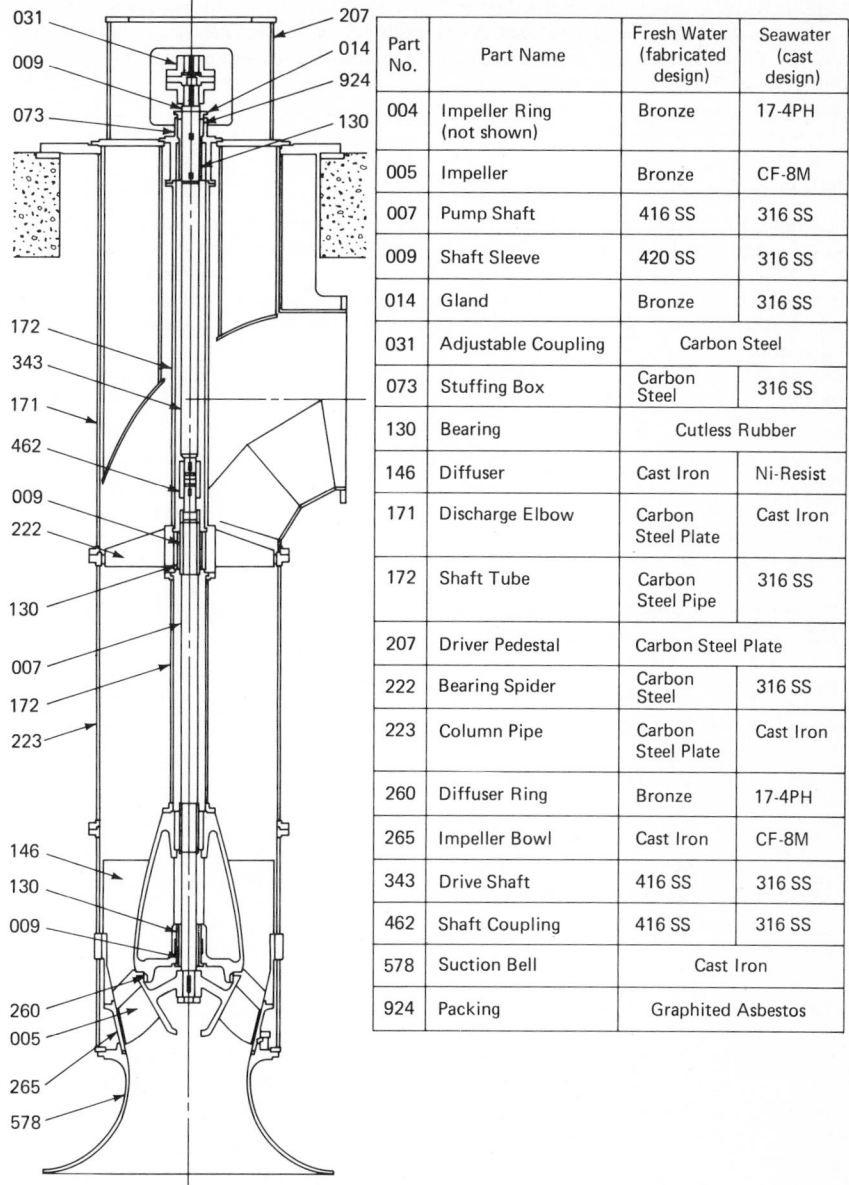

Part No.	Part Name	Fresh Water (fabricated design)	Seawater (cast design)
004	Impeller Ring (not shown)	Bronze	17-4PH
005	Impeller	Bronze	CF-8M
007	Pump Shaft	416 SS	316 SS
009	Shaft Sleeve	420 SS	316 SS
014	Gland	Bronze	316 SS
031	Adjustable Coupling	Carbon Steel	
073	Stuffing Box	Carbon Steel	316 SS
130	Bearing	Cutless Rubber	
146	Diffuser	Cast Iron	Ni-Resist
171	Discharge Elbow	Carbon Steel Plate	Cast Iron
172	Shaft Tube	Carbon Steel Pipe	316 SS
207	Driver Pedestal	Carbon Steel Plate	
222	Bearing Spider	Carbon Steel	316 SS
223	Column Pipe	Carbon Steel Plate	Cast Iron
260	Diffuser Ring	Bronze	17-4PH
265	Impeller Bowl	Cast Iron	CF-8M
343	Drive Shaft	416 SS	316 SS
462	Shaft Coupling	416 SS	316 SS
578	Suction Bell	Cast Iron	
924	Packing	Graphited Asbestos	

Figure 1-21 Vertical mixed-flow column pump. *(Allis-Chalmers.)*

1-5. VALVES

A valve is used to regulate the flow in a pipe. A hydraulic system may contain several valves, some of which may be just for full open-close positions, while other valves regulate the quantity of flow as desired. The latter are usually categorized as control valves. Control valves can be either manually or automatically operated depending upon the desired amount of flow, according to pressure, temperature, level, etc. (See Chap. 11-2,

"Automatic Controls.") Instrumentation is usually provided in the system to measure and sense the quantities. In automatic operation of the valve, the sensors send signals directly to the air-electric or other operating mediums that supply force to the automatic-controlling mechanisms. The valves are available as flanged, screwed, or welded inlet-outlet ports. The standards normally referenced for valves are ANSI-B16.10 and ASME Boiler and Pressure Vessel Code VIII, III. See also Chap. 10-7, "Valves."

Valve Pressure Drop

The valve loss coefficient K was mentioned previously (p. 9-10). Another way to represent the pressure drop through the valves (especially for control valves) is called the *flow coefficient* C_v. It is defined as the number of gallons per minute of water which will pass through a given flow restriction with a pressure drop of 1 lb/in². (The actual fluid does not necessarily have to be water. There are formulas available for calculating C_v for other fluids also.)

The relationship between C_v and K is easy to derive from definitions. For liquid service,

$$C_v = Q\sqrt{\frac{G_f}{\Delta P}} \tag{48}$$

$$\Delta P = K\frac{V^2}{2} \tag{48a}$$

where

Q = liquid flow rate, gal/min = $448.8AV\ldots$
G_f = specific gravity of liquid, for water = 1
ΔP = pressure drop at the valve for flow Q, lb/in²
V = flow velocity 1 ft/sp

For critical and flashing flows, when the pressure drop cannot be obtained from Eq. (48a), the equation is

$$\Delta P = C_f^2\,\Delta P_s \tag{49}$$

$$\Delta P_s = P_l - P_v \tag{50}$$

where

C_f = critical flow factor = 0.6 to 0.9 depending upon valve type
P_1 = pressure upstream of the valve
P_v = vapor pressure of liquid

For gas and vapor flow, the above formulas are modified to suit the flow units.[10] Chapter 10-7, "Valves," describes the more important valves in plant use.

OTHER COMPONENTS OF A SYSTEM

Besides pumps and valves, system piping must be considered. This includes associated fittings such as bends, reducers, branches, elbows, and T connections. (See Chaps. 10-2 and 10-3.) Standard pipe dimensions, i.e., nominal diameter, inside diameter, thickness, etc., for various schedule pipes, are available from literature.[6,15] Table 1-9 gives the important dimensions and burst pressures for some of the sizes and classes. The velocity of flow in a pipe should normally range from 8 to 15 ft/s (2.44 to 4.57 m/s) in order to keep the pressure drop reasonably low for a given pump size and also to deliver an optimum amount of fluid.

Other components are often a part of hydraulic systems. These include heat exchangers, filters, strainers, demineralizers, spray nozzles, steam traps, expansion joints, restrictive orifices, and measuring devices (already discussed).

Graphic Symbols for Fluid Power Systems

According to the American National Standards Institute definition, fluid power systems are those that transmit and control power through the use of a pressurized fluid within

TABLE 1-9 Important Pipe Dimensions and Burst Pressures (U.S. Customary Units)

		Schedule 40 (standard)		Schedule 80 (extra strong)		Schedule 160		Double (extra strong)	
Nominal pipe size, in	Outside diameter of pipe, in	Inside diameter of pipe, in	Burst pressure, lb/in^2	Inside diameter of pipe, in	Burst pressure, lb/in^2	Inside diameter of pipe, in	Burst pressure, lb/in^2	Inside diameter of pipe, in	Burst pressure, lb/in^2
¼	0.540	0.364	16,000	0.302	22,000	—	—	—	—
⅜	0.675	0.493	13,500	0.423	19,000	—	—	—	—
½	0.840	0.622	13,200	0.546	17,500	0.466	21,000	0.252	35,000
¾	1.050	0.824	11,000	0.742	15,000	0.614	21,000	0.434	30,000
1	1.315	1.049	10,000	0.957	13,600	0.815	19,000	0.599	27,000
1¼	1.660	1.380	8,400	1.278	11,500	1.160	15,000	0.896	23,000
1½	1.900	1.610	7,600	1.500	10,500	1.338	14,800	1.100	21,000
2	2.375	2.067	6,500	1.939	9,100	1.689	14,500	1.503	19,000
2½	2.875	2.469	7,000	2.323	9,600	2.125	13,000	1.771	18,000

an enclosed circuit. ANSI Y32.10 gives graphic symbols commonly used for drawing circuit diagrams of fluid power systems. They show connections, flow paths, and functions of components represented. Some of the important symbols are given here in Fig. 1-22.

Figure 1-23 is an illustration of a power system and its symbolic representation. This example illustrates a pump that transfers liquid from reservoir through a check valve and a four-way control valve to hydraulic cylinders, the other ends of which are connected to another open reservoir containing liquid.

A TYPICAL HYDRAULIC SYSTEM

In order to illustrate the use of hydraulic components, one hydraulic system will be discussed here in detail. It is the circulating-condenser, cooling-water system, one of the very important systems for a thermal power plant.[8] It is important because of the large quantity of water required which in turn necessitates large open channels, pipes, valves, etc., in the system.

Hydraulics

A typical circulating-water system for a power plant is shown schematically in the plan view in Fig. 1-24. Large quantities of cooling water are withdrawn from a river by pumps provided in the intake canal. Trash racks and traveling screens are usually provided in the intake canal in order to keep the unwanted debris from entering the pumps and condensers. The discharge from the condensers can be thrown off-site into a large body of water (e.g., the ocean) for final heat dissipation. A weir is usually provided downstream of the condenser to avoid the impact on the condensers from the downstream water-level fluctuations (e.g., due to tides). In order to dissipate the extra energy available from the fall over the weir and to accommodate a hydraulic jump, a stilling basin is usually provided downstream of the weir. If there is another waterway (say river) to be crossed downstream before approaching the ocean, pipes can pass under that waterway. Finally, under modern environmental practice, the hot water cannot be discharged into the ocean too near the shore. Therefore, water is pumped from the discharge canal to give offshore final discharge submerged for good heat dissipation. Valves are provided in the piping to control flow. No further discussion of heat dissipation is required here because the purpose of this chapter is not just to understand one system, but how to engineer a fluid system in general. A comprehensive knowledge of circulating water systems, if required, can be obtained from Ref. 8.

The first step in design is to plot a developed view of the system to scale, with all the hydraulic components (mentioned previously) indicated at least schematically. All the horizontal and vertical distances to these components should be plotted and the piping drawn properly. *Developed view* indicates plotting in two dimensions only. If there is a third dimension, it should be rotated horizontally to bring it to the second dimension at the point where it occurs. Thus all the dimensions get accounted for. The vertical scale can be distorted (different from horizontal scale) in order to be able to see the elevations of each point clearly. All the components, transitions, bends, etc., should be properly accounted for.

For the circulating water system mentioned above, a developed picture is shown in Fig. 1-25, with some assumed dimensions. The next step is to plot a hydraulic gradient line for this system. In order to do this, all the head losses in the open channels, piping, and appurtenances have to be calculated and pump TDH established for the known flow rate, as mentioned in previous sections. In Fig. 1-25, the losses for valves, bends, etc., and the friction loss in the pipeline drop from the upstream end to the downstream end of the pipe as shown on the hydraulic gradient line.

The calculations of head losses in pipe fittings, valves, etc., and determination of pump TDH can best be accomplished by using a tabular form, e.g., Table 1-10. This keeps the number of computation sheets to a minimum, and the important factors are not omitted. It should be noted that the head-loss data for certain components, e.g., heat

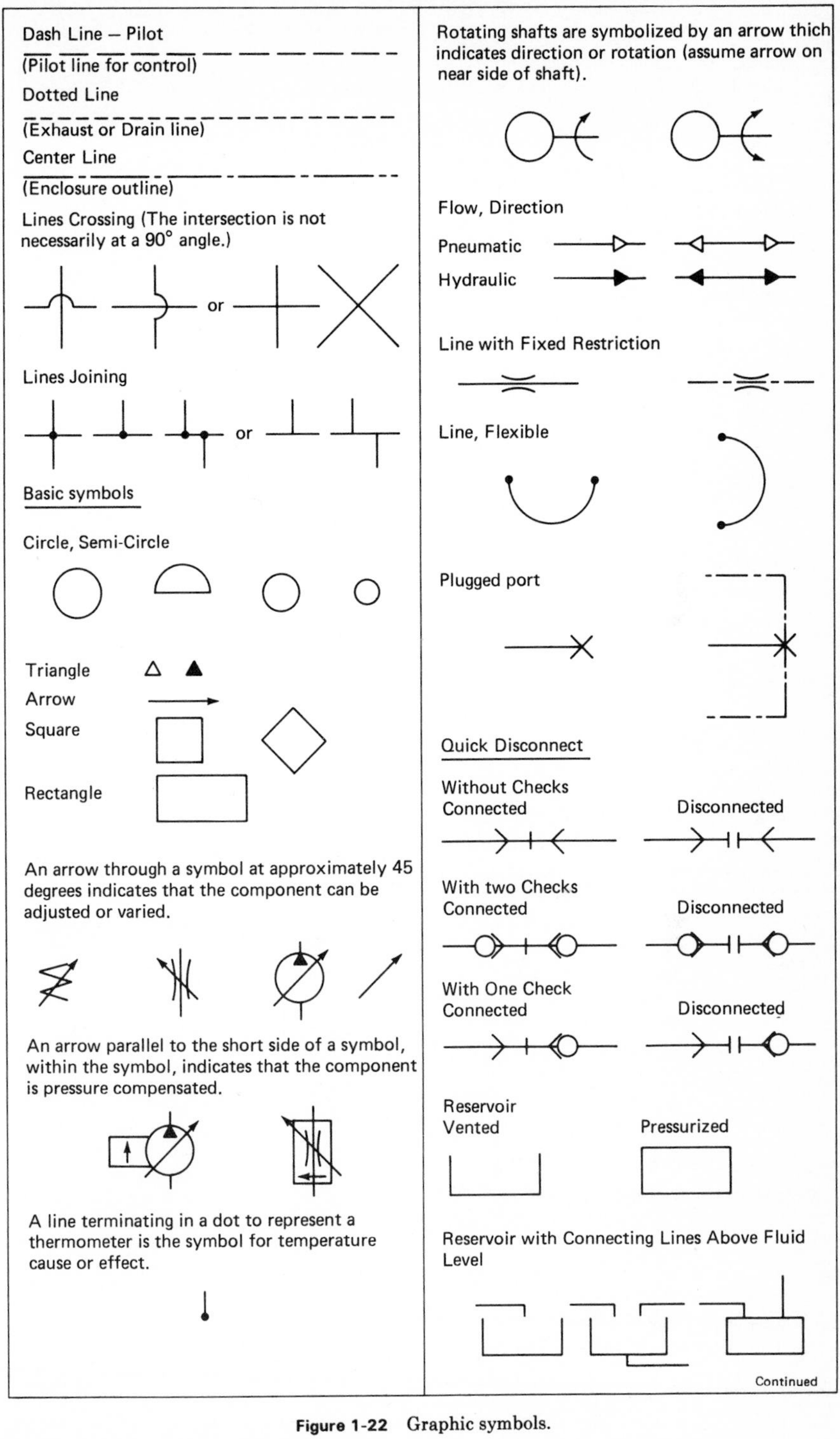
Dash Line — Pilot
(Pilot line for control)
Dotted Line
(Exhaust or Drain line)
Center Line
(Enclosure outline)
Lines Crossing (The intersection is not necessarily at a 90° angle.)
or
Lines Joining
or
Basic symbols
Circle, Semi-Circle
Triangle
Arrow
Square
Rectangle
An arrow through a symbol at approximately 45 degrees indicates that the component can be adjusted or varied.
An arrow parallel to the short side of a symbol, within the symbol, indicates that the component is pressure compensated.
A line terminating in a dot to represent a thermometer is the symbol for temperature cause or effect.
Rotating shafts are symbolized by an arrow thich indicates direction or rotation (assume arrow on near side of shaft).
Flow, Direction
Pneumatic
Hydraulic
Line with Fixed Restriction
Line, Flexible
Plugged port
Quick Disconnect
Without Checks
Connected
Disconnected
With two Checks
Connected
Disconnected
With One Check
Connected
Disconnected
Reservoir
Vented
Pressurized
Reservoir with Connecting Lines Above Fluid Level
Continued

Figure 1-22 Graphic symbols.

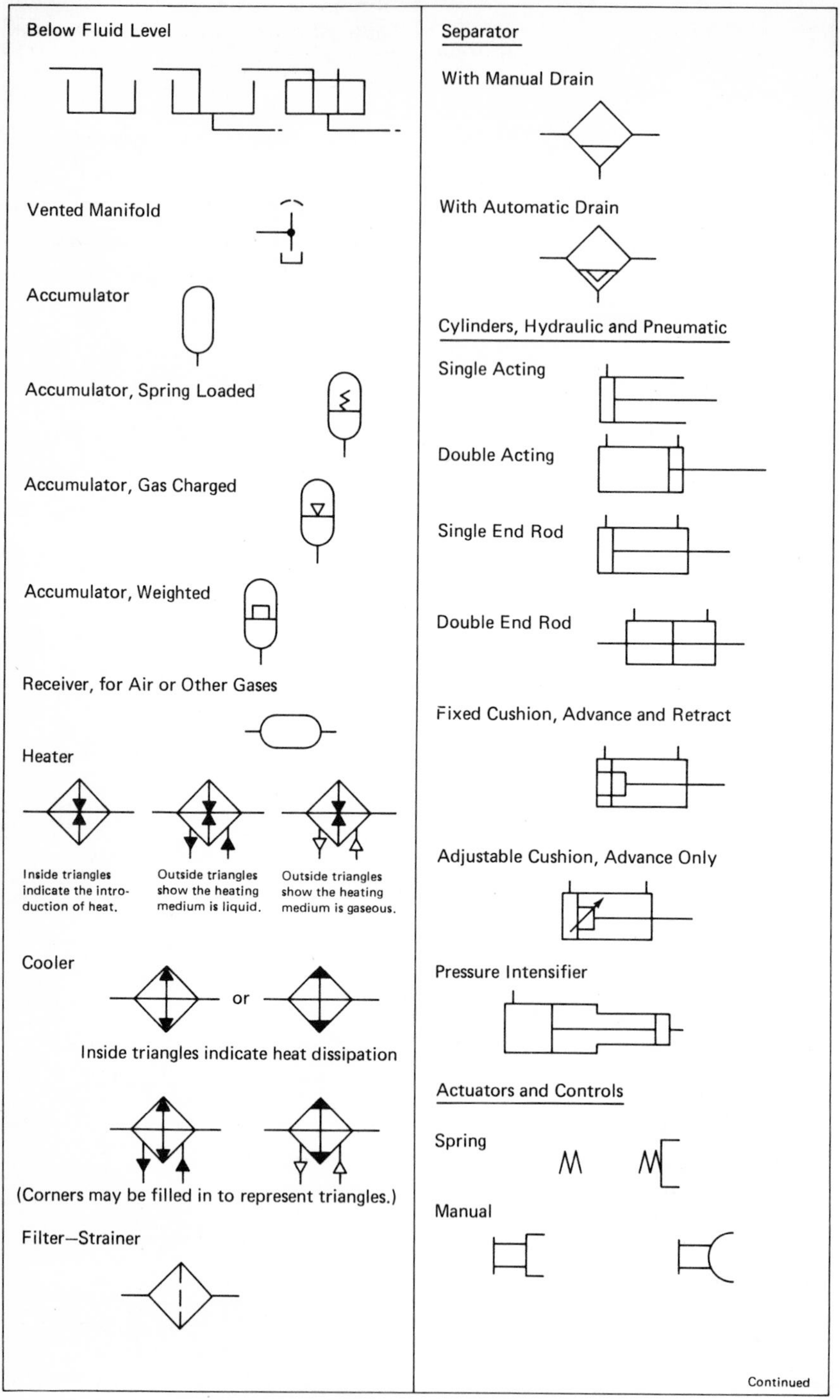
Below Fluid Level
Vented Manifold
Accumulator
Accumulator, Spring Loaded
Accumulator, Gas Charged
Accumulator, Weighted
Receiver, for Air or Other Gases
Heater
Inside triangles indicate the introduction of heat.
Outside triangles show the heating medium is liquid.
Outside triangles show the heating medium is gaseous.
Cooler
or
Inside triangles indicate heat dissipation
(Corners may be filled in to represent triangles.)
Filter—Strainer
Separator
With Manual Drain
With Automatic Drain
Cylinders, Hydraulic and Pneumatic
Single Acting
Double Acting
Single End Rod
Double End Rod
Fixed Cushion, Advance and Retract
Adjustable Cushion, Advance Only
Pressure Intensifier
Actuators and Controls
Spring
Manual
Continued

Figure 1-22 *(continued)* Graphic symbols.

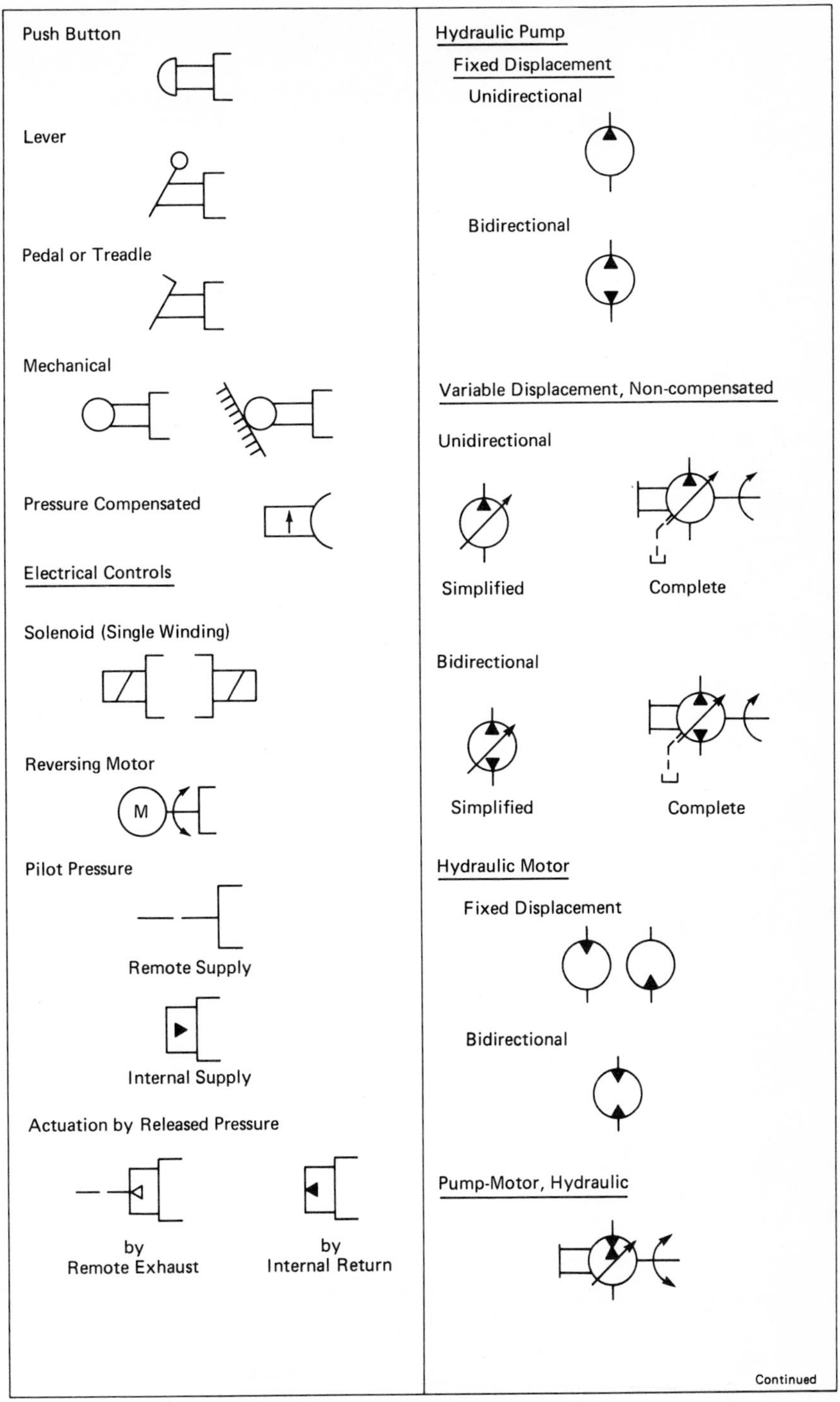
Push Button
Lever
Pedal or Treadle
Mechanical
Pressure Compensated
Electrical Controls
Solenoid (Single Winding)
Reversing Motor
M
Pilot Pressure
Remote Supply
Internal Supply
Actuation by Released Pressure
by
Remote Exhaust
by
Internal Return
Hydraulic Pump
Fixed Displacement
Unidirectional
Bidirectional
Variable Displacement, Non-compensated
Unidirectional
Simplified
Complete
Bidirectional
Simplified
Complete
Hydraulic Motor
Fixed Displacement
Bidirectional
Pump-Motor, Hydraulic
Continued

Figure 1-22 *(Continued)* Graphic symbols.

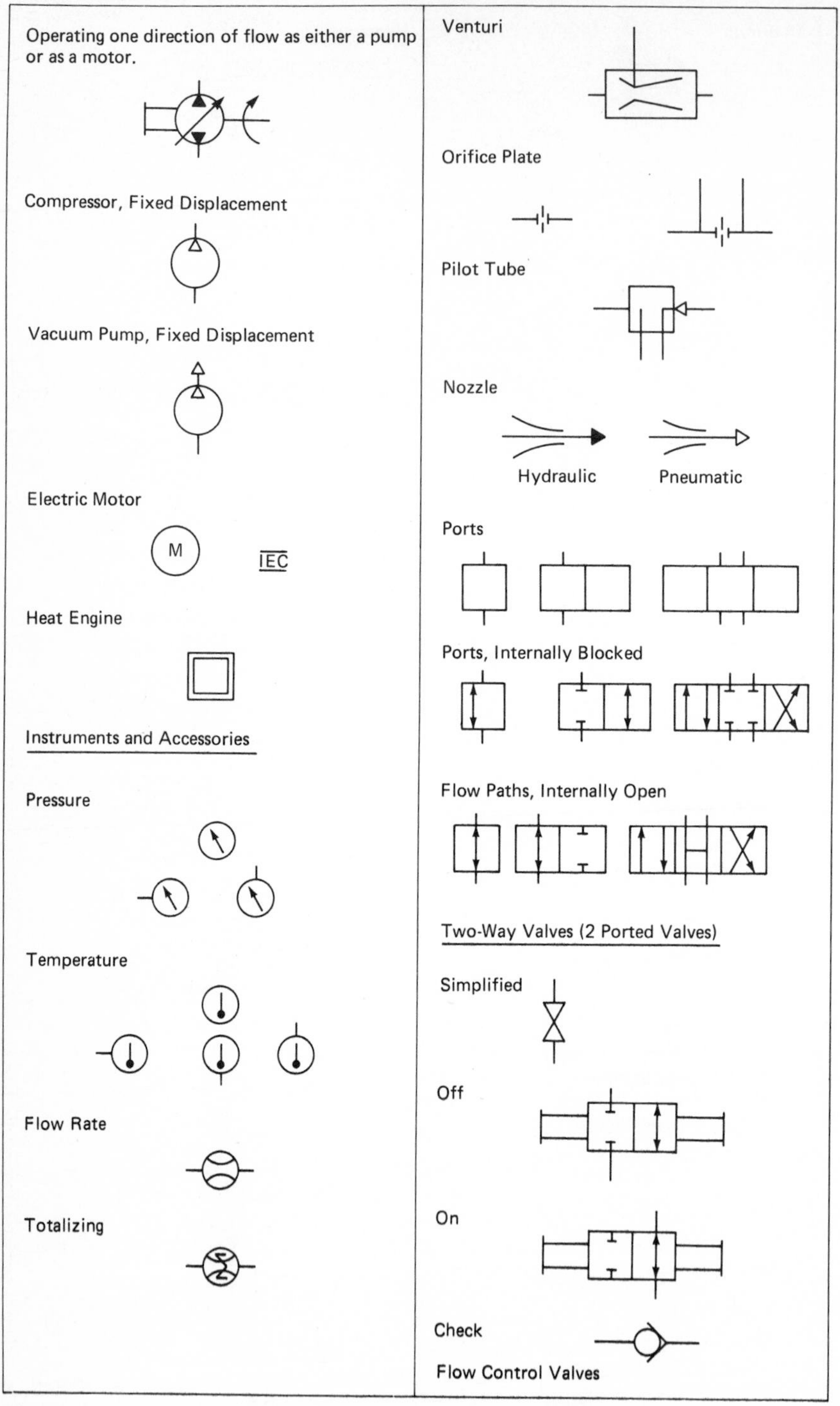
Operating one direction of flow as either a pump or as a motor.
Compressor, Fixed Displacement
Vacuum Pump, Fixed Displacement
Electric Motor
M
IEC
Heat Engine
Instruments and Accessories
Pressure
Temperature
Flow Rate
Totalizing
Venturi
Orifice Plate
Pilot Tube
Nozzle
Hydraulic
Pneumatic
Ports
Ports, Internally Blocked
Flow Paths, Internally Open
Two-Way Valves (2 Ported Valves)
Simplified
Off
On
Check
Flow Control Valves

Figure 1-22 *(Continued)* Graphic symbols.

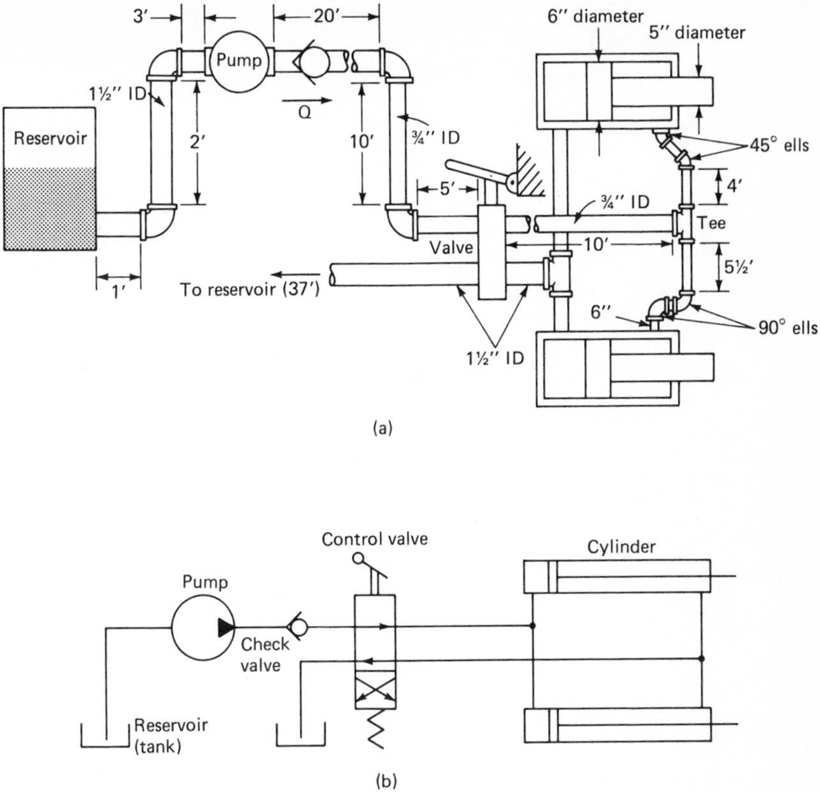

Figure 1-23 Illustration of power systems symbols.

exchanger (condenser in this typical example), orifice plates, control valves, filters, etc., should be obtained from the catalogs of the manufacturers of specific components.

The TDH of the pump should be conservatively estimated so as not to fall short of pressure in the system, especially as the system gets older and scale deposits form, thereby increasing the pressure drops. If a little extra pressure is available from the pump in the beginning, the runout of the pump (excessive flow) can be controlled by throttling the flow at the control valves in the system. Therefore, it is usually good to have some throttling-type valves in the system.

The TDH of the pump and performance of a system can be determined or verified in the field on the large-size systems because laboratory facilities are usually not large enough to handle full-scale flows. Field testing can be achieved by following one of the several measurement techniques outlined in a previous section. More sophisticated instrumentation and techniques are also available these days, such as, annubar, current meter, etc., which are discussed in Ref. 14. More sophisticated flowmeters, e.g., magnetic-acoustic, are also available in the market.

Hydraulic transients and water-hammer aspects for varied system-pump-valve operations should be studied.

Process Instrumentation

The engineering of systems, such as the circulating water system discussed above, of course involves other aspects besides hydraulics. The most important, from a systems

TABLE 1-10 Pump Calculation Sheet (U.S. Customary Units)

Project ______ Design gal/min ____ Design TDH ____ Pump no. ______
Pump service ______ Date ______ Sch. & pipe matl. ______ Viscosity @ min P.T. ______ cP(μ)
Pump no. ______ Fluid pumped ______ Spec. grav. @ min P.T. ______ (SG)
Lb/h ______ Temp., max ______ Min ______ F Kinematic viscosity ______ cS(z)
Normal gal/min ______ From ______ To ______ Vapor press. @ max P.T. ______ mmHg(VP)
Design gal/min ______
Reference drawings

		Suction line				Discharge line		
Line no. & schedule		___	___	___		___	___	___
Nom. line size, inches I.D. in; ft D/d	d	___	___	___	d	___	___	___
Velocity, ft/s	V	___	___	___	V	___	___	___
Frict. loss, ft fluid/100 ft	or f	___	___	___	F	___	___	___
Reynolds number $V^2/2g$	Re	___	___	___	Re	___	___	___
Lineal feet of pipe, or k$		___	___	___		___	___	___
Gate valves	___@	___ = ___	___ = ___	___ = ___	___@	___ = ___	___ = ___	___ = ___
Globe valves	___@	___ = ___	___ = ___	___ = ___	___@	___ = ___	___ = ___	___ = ___
Ells	___@	___ = ___	___ = ___	___ = ___	___@	___ = ___	___ = ___	___ = ___
Tees, thru branch	___@	___ = ___	___ = ___	___ = ___	___@	___ = ___	___ = ___	___ = ___
Tees, thru run	___@	___ = ___	___ = ___	___ = ___	___@	___ = ___	___ = ___	___ = ___
Ells 45° or other L	___@	___ = ___	___ = ___	___ = ___	___@	___ = ___	___ = ___	___ = ___
Butterfly/diaphragm	___@	___ = ___	___ = ___	___ = ___	___@	___ = ___	___ = ___	___ = ___
Check valves	___@	___ = ___	___ = ___	___ = ___	___@	___ = ___	___ = ___	___ = ___
Total equivalent length	EL				EL			
Line frict. (F × EL/100)1.15	F_s	− ___	− ___	− ___	F_d	+ ___	+ ___	+ ___
Vessel pres. psig × 2.31/SG	P_s**	+ ___	+ ___	+ ___	P_d**	+ ___	+ ___	+ ___
Liquid elev. above pump, ft	h_s*	+ ___	+ ___	+ ___	h_d	+ ___	+ ___	+ ___
Tank contr. or enl. loss	h_c	− ___	− ___	− ___	h_e	+ ___	+ ___	+ ___
Orifice pl., lb/in² × 2.31/SG						+ ___	+ ___	+ ___
Contr. valve lb/in² × 2.31/SG						+ ___	+ ___	+ ___
Exchanger, lb/in² × 2.31/SG						+ ___	+ ___	+ ___
Filter (dirty)						+ ___	+ ___	+ ___
						+ ___	+ ___	+ ___
Alg. sum = suct. or disch. head	Hs				Hd			
Total dynamic head, Hd–Hs	TDH	___ Ft(	″ Suct.,	″ Disch.)				
NPSH available:								
Abs pressure head	P_s† + 34/SG	+ ___	+ ___	+ ___				
Static head	h_s*	+ ___	+ ___	+ ___				
Friction head	$h_c + F_s$	− ___	− ___	− ___				
Vap. pres. head	0.045 × VP/SG	− ___	− ___	− ___				
Alg. sum = NPSH ft fluid		___	___	___				

Formulas

1 lb/in² = (2.31/SG) = ______ ft. fluid

$V = 0.408 \text{ gal/min} \cdot d^2$

$d = \sqrt{\dfrac{\text{gal/min}}{2.45v}}$

$z = \mu/\text{SG}$

$h_c = V^2/128.8$

$h_e = V^2/64.4$

$Re = \dfrac{3162 \times \text{gal/min} \times \text{SG}}{d \times \mu}$

$-$K$ can be given in terms of f. K for pipe $= f1/d$

Notes typ. calc., etc.

*Minus for suction lift.

†Minus for pressure less than atmospheric. For elevations above sea level, decrease 34 by 1.1 ft for each 1000 ft of altitude above sea level.

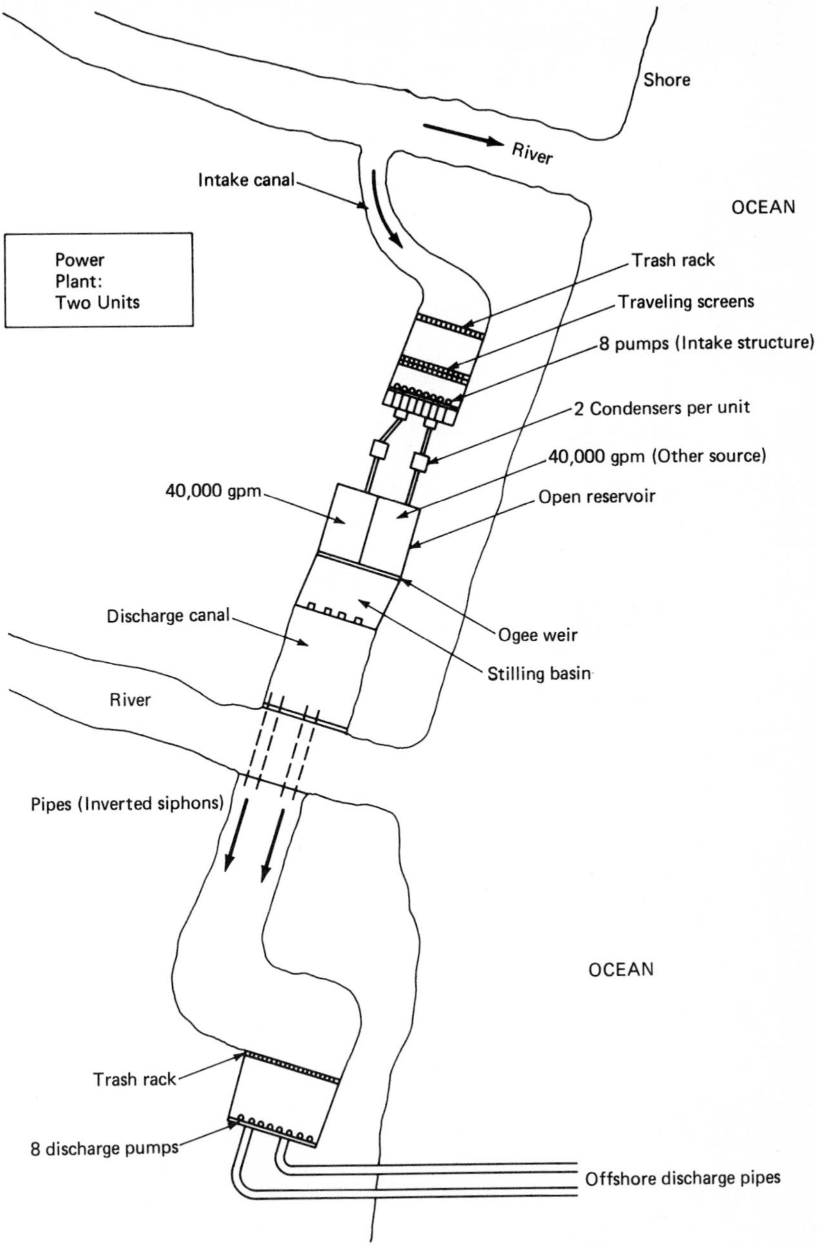

Figure 1-24 Cooling water system for a power plant.

point of view, is to establish the various modes of operation and logic of the system components, e.g., pumps, control valves, etc. For this purpose, process instrumentation and control and logic diagrams are developed with reference to the system descriptions. More details on these are shown in Chapter 9-2, "Pneumatic Power Systems," but is applicable to both.

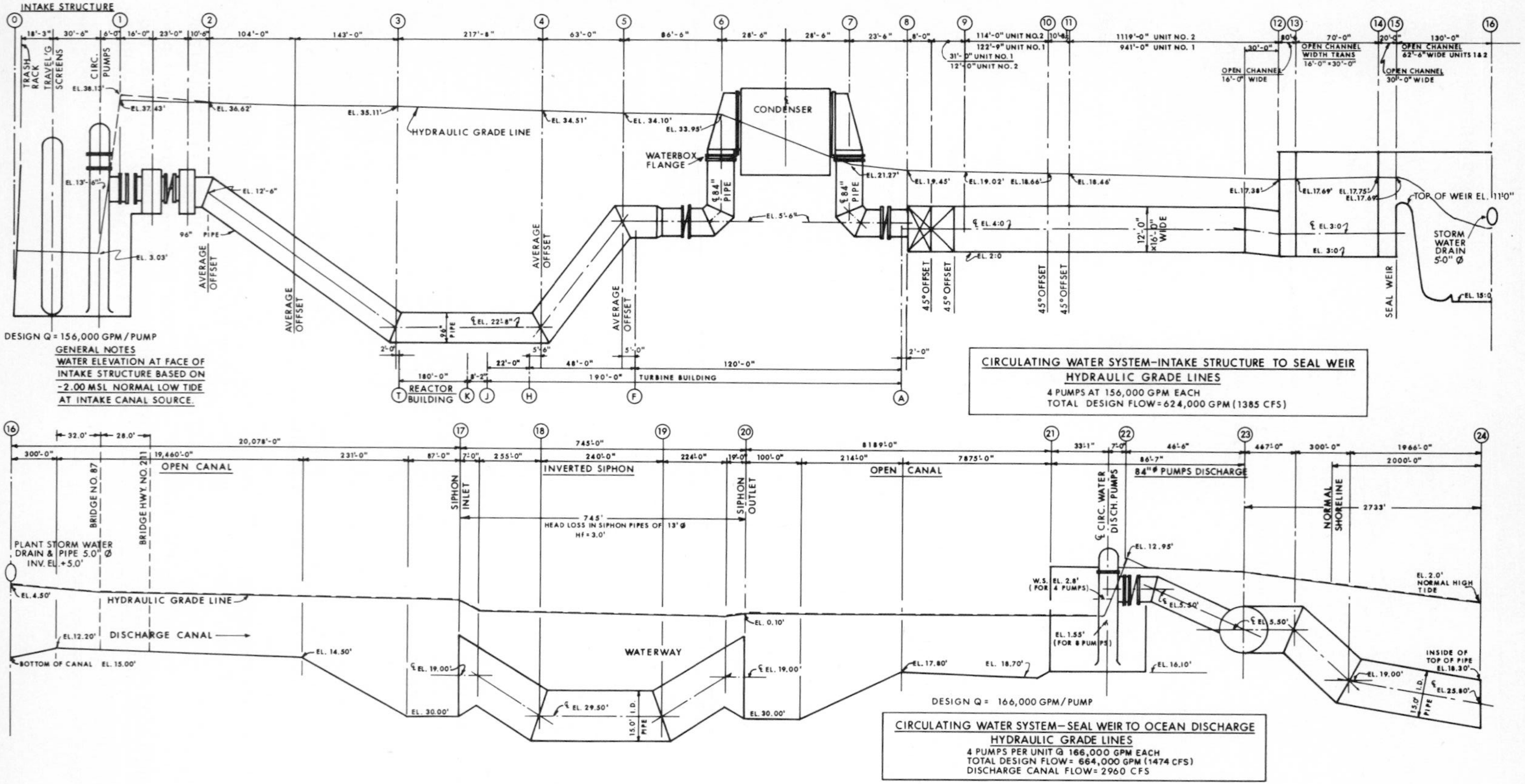

Figure 1-25 Hydraulic gradient of a circulating water system.

Other Aspects

Other important aspects include preparation of detailed technical specifications for the various types of equipment in the system. Besides the process aspects of the equipment, the materials and safety aspects should also be accounted for. A table of recommended materials and metals for various applications has already been discussed. It is recommended that the manufacture of the equipment should be closely monitored for quality assurance reasons and proper shop and field testing should be conducted *before* commercial operation.

Of course, maintenance of the systems and components is also a very important aspect. Only equipment properly inspected and maintained through lubrication and replacement of parts at periodic intervals will fulfill its life expectancy. Some checkpoints similar to those given in the discussion on pumps should be developed for all systems and components.

MODEL EXPERIMENTS

The science of hydraulic engineering depends on practical experience obtained from prototypes, or models. For the design of any hydraulic system, e.g., circulating water system, it is advisable to conduct model studies in order to determine the flow, pressure, and velocity distributions. In recent years, hydraulic-model studies have been conducted on the pump houses, channels, pipes, intake structures, diffusers, surrounding fluid fields, etc.

Intake, Discharge, and Pump House

Pump-house (and channels) hydraulics is unusual in that the flow pattern in the approach to the pump suction must be uniform, tranquil, and nonvortexing; it has a flow velocity of about 1 ft/s (0.305 m/s). Failing to conform to these criteria can result in inefficient performance of the system. Model studies on pump houses are conducted in order to ensure these criteria by providing enough length, width, transition, splitters, baffles, rounding of corners, etc., in the approach channel. Such model studies are usually conducted at a geometric scale of 1:8 to 1:15. Kinematic and dynamic similarities are achieved by further scaling according to the laws of Froude, Reynolds, etc. One important consideration is to see that supercritical flow and hydraulic jump (limiting the amount of flow and creating turbulence) do not exist in the approach channel. Sometimes flags or strings are attached at the bottom of the sump below the suction bell in order to see that all the flags are pointing radially and uniformly all around the bell. Lighted transparent (glass) window sections are provided at the important points, and dye injection tests are conducted in order to visualize the flow patterns.

Velocities are usually measured by miniature current meters, and pressures (or depth of water) are generally measured by piezometers. The pump bells are made geometrically similar, but the pumps themselves are usually not modeled in these tests. Instead, a siphon arrangement usually serves for the correct amount of simulated flow. The models for pumps and their internal hydraulics of vanes are generally tested by pump manufacturers only in order to determine their pump characteristics for the prototype.

A general question comes up, "Why it should be necessary to test a sump pump and not just utilize the existing experience or literature for design?" The answer is that most layouts and flow quantities are typical. If a pump house has been working efficiently for a long time and another similar one (geometrically and dynamically) is required, there is no need to test it. However, if a design is provided based on existing literature, e.g., Hydraulic Institute standards, usually the dimensions are very large and, therefore, result in expensive construction. Usually, a lot of construction quantities and costs can be saved by reducing the size of pump house to its optimum by a model test, the cost of which is usually only a fraction of the savings. Moreover, a design based on literature or experience may not guarantee a certain desirable flow pattern without conducting a test.

Hydrothermal Diffusion Problems

Another field in which extensive test work on models has been conducted recently is hydrothermal diffusion at the point of hot-water discharge into a natural body of water. The importance of this field has been increased due to the recent public concern for environmental protection. It is generally considered that hot-water discharge is harmful to ecological balance. Therefore, a careful study in the area of hot-water discharge should be conducted in order to determine the extent of potential harm and to reduce it to allowable limits. Physical model studies have proved useful to determine the isotherm patterns in the vicinity of discharges. However, there is a general concern that some of the hot water may be returning from the far field. This type of model study will need very large models and has been tried in some cases with only a limited success.

For the near field studies, physical models have generally been constructed to a scale of 1:80 to 1:130. The densimetric Froude number is generally used for similarity. This is also done for the bottom topography of the body of water. Most of the time, the purpose of testing the model is to determine what dimensions, shape, and size of discharge structure (diffuser with ports) will give the most efficient mixing in the least surface area resulting in maximum temperature isotherms. Temperature distributions in the surrounding area are measured by a series of thermostats. Data accumulation and processing is done side by side with the experiment by using computers. Several arrangements are tried, and the best ones are selected with their corresponding merits and demerits. Dye tests are also conducted to observe the flow patterns. It is important to model enough area around the diffuser. The temperature in the model room has to be properly controlled. The tidal effects, currents, and surface tension should be properly controlled or modeled.

Other Model Tests

Another field in which model studies are helpful is determining flow patterns (velocity distributions) near intake and discharge structures in order to determine biota intake and sedimentation-erosion quantities. Also, some model studies have been conducted for flow into and out of the individual intakes and diffuser ports, respectively. Some model studies have also been conducted to determine the hydrodynamic forces on immersed structures.

HYDRAULIC FORCE AND TORQUE

What we have discussed so far in hydraulic systems is largely related to what is involved in transmitting the fluid from one point to another. However, sometimes the purpose is not just to deliver a bulk quantity but also to deliver pressure to activate other components. The pressure is ultimately utilized either as a force or a torque.

Some simple uses of hydraulic pressure as a force become evident in applications such as a hydraulic press (which operates basically on Pascal's law, discussed previously), a hydraulic-brake system on an automobile, or a hydraulic crane for lifting objects. Some indirect uses, by passing through actuators and valves, become evident on applications such as the hydraulic milling machine, shaper or surface grinder, or lathe. For rotary action, hydraulic couplings and torque converters are used.

Pressure Accumulator

This is a device to accumulate or store liquid under pressure delivered by the pump when it is not required by the machine. The pressure can be later supplied to the machine when needed.

Various industrial presses require separate pumping units to furnish liquid at the desired pressure. Such pumps are known as *press pumps*. Normally, the pressure generated by these pumps ranges from 50 to 150 kg/cm^2 and is uniform throughout the supply period. However, the demand for liquid and its required pressure is variable. At some intervals the machine may not be doing any work at all, and to deal with such operating conditions an arrangement to receive and store the pressurized liquid being constantly supplied by the pump is necessary. The device must be able to deliver the liquid back to the machine on demand. In some cases it may be even desirable to retrieve the stored liquid at a pressure higher than that provided by the pump itself. All this is done by the pressure or hydraulic accumulator.

Hydraulic Accumulator

The hydraulic accumulator (refer Fig. 1-26) consists of a cylinder and a plunger generally known as a *ram*. One side of the cylinder is connected to the press pump and other to

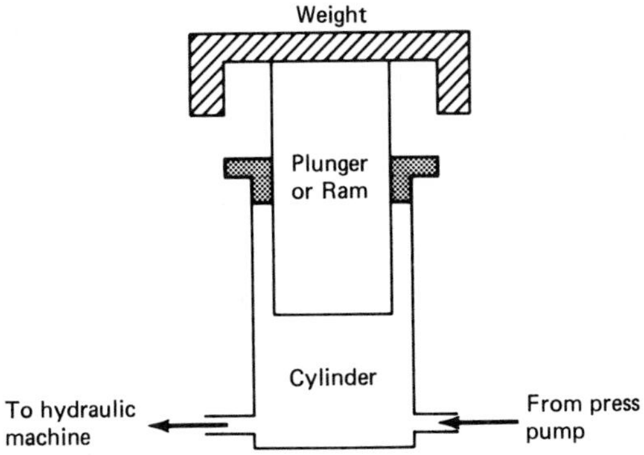

Figure 1-26 Hydraulic accumulator.

the hydraulic machine. Either the cylinder or the ram may be fixed. Generally the cylinder is fixed and the ram moves up and down to accommodate a variable quantity of liquid inside by the cylinder.

Accumulators may be *dead-load* or *variable-load* type. In the former, dead weights are employed to press the plunger in, while the latter employs steam pressure. The main advantage of the steam-pressure type is that the pressure may be varied at will, but it is handicapped in many applications by the need of a boiler to supply steam. However, it can be used on ships if steam is readily available.

The accumulator also serves the purpose of a pressure regulator. A suitable arrangement can be easily designed to switch on a pump motor after a predetermined travel of the ram.

Capacity of Accumulator

This is the maximum amount of energy stored by an accumulator. The storage capacity is equal to the potential energy of the lifted ram together with its weight.

Let

d = diameter of ram
s = stroke or lift of ram
p = intensity of pressure of water supplied

The total moving weight or weight of the ram is

$$W = \frac{\pi}{4} d^2 p \tag{51}$$

The work done in lifting the ram, or capacity of the accumulator, is

$$Ws = \frac{\pi}{4} d^2 ps \tag{52}$$

The volume of the accumulator is $(\pi/4)d^2s$, and

$$\text{Capacity of ram} = p \times \text{volume} \tag{53}$$

The capacity of the accumulator and that of the ram are same.

Pressure Intensifier

The pressure intensifier, sometimes known as *differential accumulator,* is a device to multiply the pressure supplied by the pump to suit the requirements of a high-pressure machine. Often a fluid pressure machine requires a high pressure at a particular stage in its operation. It can be easily provided by the intensifier.

Normally, a simple intensifier (refer Fig. 1-27) consists of two coaxial rams or pistons moving in cylinders as shown in Fig. 1-24. Low-pressure liquid is admitted to the ram or piston of large cross-sectional area, which then transmits force to a small ram or piston

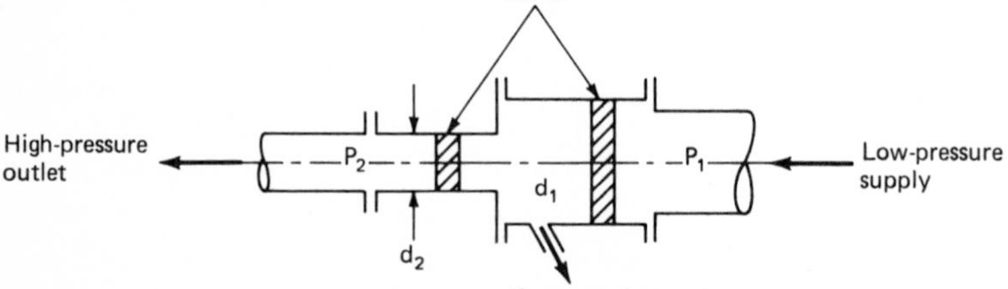

Figure 1-27 Pressure intensifier.

by a rod connecting the two. Since the piston on the left-hand side has a smaller cross-sectional area, the pressure of the liquid coming out will be high. The volume between the two pistons must be vented.

Let d_1 and d_2 be the diameters of the two rams and p_1 and p_2 the respective pressure of the liquid inside them. Then, if the ram moves slowly, forces

$$p_1 = p_2$$

or

$$p_1 \frac{\pi}{4} d_1^2 = p_2 \frac{\pi}{4} d_2^2$$

or

$$p_2 = p_1 \frac{d_1^2}{d_2^2} \tag{54}$$

Figure 1-28 shows a modified form of intensifier which consists of two coaxial rams inside a cylinder. The cylinder and the outer ram are fixed. *A*, *B*, and *C* are the valves provided for fixed cylinder, sliding ram, and fixed ram, respectively. To start with, the hollow sliding ram is filled with liquid; valves *A* and *C* are open. A liquid having a low pressure p_1 enters from the mains through *A* and forces the sliding ram upwards, so that the liquid inside the sliding ram goes out through valve *C* at a greater pressure p_2. When the sliding ram has reached its top position, valve *B* opens while valves *A* and *C* close. The liquid, now trapped between the fixed cylinder and sliding ram, enters inside the hollow sliding ram which then comes back to its starting position. This completes one cycle of the intensifier. Sometimes, compressed air is supplied to the larger cylinder in place of low-pressure hydraulic supply, in which case the intensifier is known as a hydropneumatic accumulator or intensifier.

Steam can also be supplied to the larger cylinder in place of low-pressure hydraulic supply or compressed air. It is then called a *steam intensifier.*

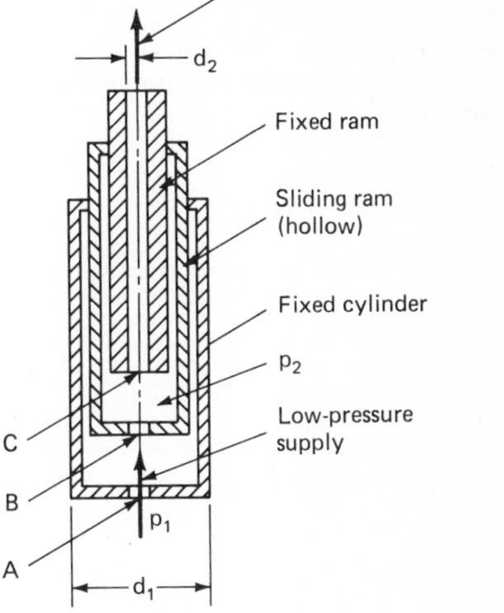

Figure 1-28 Modified form of intensifier.

Fluid or Hydraulic Coupling

The fluid coupling consists of a radial pump impeller keyed to a driving shaft *A* (refer to Fig. 1-29) and a reaction (radial-type) turbine runner keyed to driven shaft *B*. There is no mechanical connection between the driving and driven shafts. The impeller and turbine runner together form a casing completely filled with oil, the fluid with which shafts *A* and *B* are to be coupled. The fluid is usually conventional lubricating oil. If the shaft *A* is made to revolve slowly, the oil, due to a forced vortex, will flow out from the impeller and will strike the turbine runner

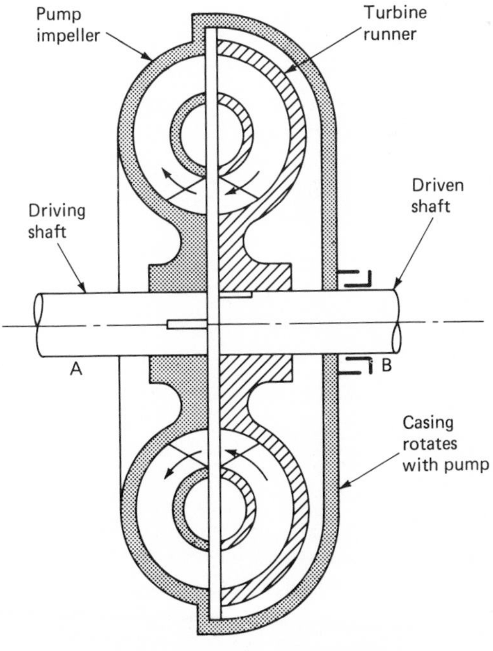

Figure 1-29 Fluid or hydraulic coupling.

blades. After sufficient head has been built up by increasing the speed of A, the fluid will drive the turbine runner and thus set the shaft B in motion. When at full speed, the two shafts rotate at almost the same rate; in practice, owing to slip, the driven shaft speed is typically about 2 percent less. Thus, the efficiency of the typical coupling would be 98 percent. If both driver and follower were to rotate at the same speed, no circulation of oil could take place. It is the difference between centrifugal forces set up in the driver and the follower which causes oil circulation. The necessary reduction in the speed of the driven shaft thus maintains the continuous flow of oil from impeller to the turbine runner. The blades of impeller and runner are generally a straight radial type.

REFERENCES

References are listed at the end of Chap. 9-2.

chapter 9-2

Pneumatic Systems

by
Ken S. Satija, Ph.D., P.E.
United Engineers & Constructors, Inc.
Philadelphia, Pennsylvania

INTRODUCTION

Pneumatic systems involve gases (usually compressed air, including negative gauge pressures for a vacuum system). The most important component of a pneumatic system, therefore, is a compressor (or blower, or fan). Other components include filters, aftercoolers, moisture separators, air tanks, air dryers, piping, valves, etc., the basic functions of which can be deduced just from their names. Some of these components have already been discussed as parts of hydraulic systems. A compressor usually supplies pressurized gas into a line. A pressure regulator is provided on every branch where gas is to be used at a specific pressure.

The fundamentals of airflow are similar to those of the flow of water which was discussed previously in great detail. The methods of calculating the pressure drops in order to compute the pressure requirements of a compressor are similar to those with water, with a few minor exceptions due to the compressibility of air. First, the density, viscosity, etc., of air at the operating temperature must be considered.

Second, the flow computations with air are made in terms of cubic feet per minute. In dealing with equations for gases, absolute temperature and pressure are used. Absolute temperature is the temperature above absolute zero, which is equivalent to $-460°F$ ($-273°C$). When the Fahrenheit temperature is converted to absolute, it is sometimes referred to as degrees Rankine (°R), and °C converted to absolute is kelvins (K). Thus

$$T(\mathrm{K}) = 273 + T(°\mathrm{C})$$

$$T(°R) = 460 + T(°F)$$

Similarly,

$$P \text{ absolute (lb/in}^2) = P \text{ gauge (lb/in}^2) + 14.7$$
$$P \text{ absolute (kg/m}^2) = P \text{ gauge (kg/m}^2) + 10{,}335$$

The pressure drop for flow through pipes is calculated usually to start with airflow conditions at 100 psig and 60°F. Then the pressure drop is converted to the actual flow temperature t and pressure p conditions by the following equation:

$$\frac{460 + t}{520} \frac{100 + 14.7}{p + 14.7} = \frac{\Delta p}{\Delta p_{100}} \quad \text{(U.S. customary units)} \tag{1}$$

where

Δp_{100} = pressure drop for flow of air at 100 psi, 60°F
Δp = pressure drop for actual air

It should be kept in mind that this pressure drop is for a flow volume, actual cubic feet per minute (acfm), at p, t conditions. The corresponding standard cubic feet per minute (scfm) of air, that is, at 14.7 psia and 60°F, can be computed from the following equation:

$$\frac{520}{460 + t} \frac{14.7 + p}{14.7} = \frac{\text{scfm}}{\text{acfm}} \quad \text{(general gas)} \tag{2}$$

Reference 6 gives tables for specific numbers corresponding to the above equations, in order to simplify calculations.

Tables[6] are also available for computing the pressure drop at different flow rates. These tables are usually specific p and t values only, but one can easily proceed with calculating pressure drop Δp using the Harris formula:

$$p = \frac{cL}{R} \frac{Q^2}{D^5} \tag{3}$$

where

c = a coefficient of pressure drop = $(0.1025/D^{0.33})$ (U.S. customary units)
L = length of pipe, ft
R = compression ratio = $(p + 14.7)/14.7$
Q = flow rate, ft^3/s of free air
D = pipe diameter, in

FANS

In plant applications, fans are used for circulating air in rooms, as well as delivering air for plant systems.

Forward-curved-fan characteristics are given in Fig. 2-1. The horsepower curve is continuously increasing with flow rate. Such fans are used in low-pressure applications.[17]

Backward-curved fans have a horsepower curve that rises slightly in the beginning and then drops. This is the type most commonly used in large air-handling systems.

Vane-axial fans that are mostly used in air-conditioning systems have a dip in the horsepower curve.

All types of fans mentioned above have an unstable surge zone in the low-flow zone, as the fan is started or stopped. This zone is considered unsuitable for normal fan operation. Similarly, a steep portion of the pressure curve towards the end, where pressure drops fast with a small change in flow, is unsuitable (and not very predictable) for normal operation. Therefore the system-demand curve should fall within the selection area. The best zone of selection on the curve is where there is a uniform gradual well-defined variation of pressure with flow (one pressure for one flow and not varying too fast).

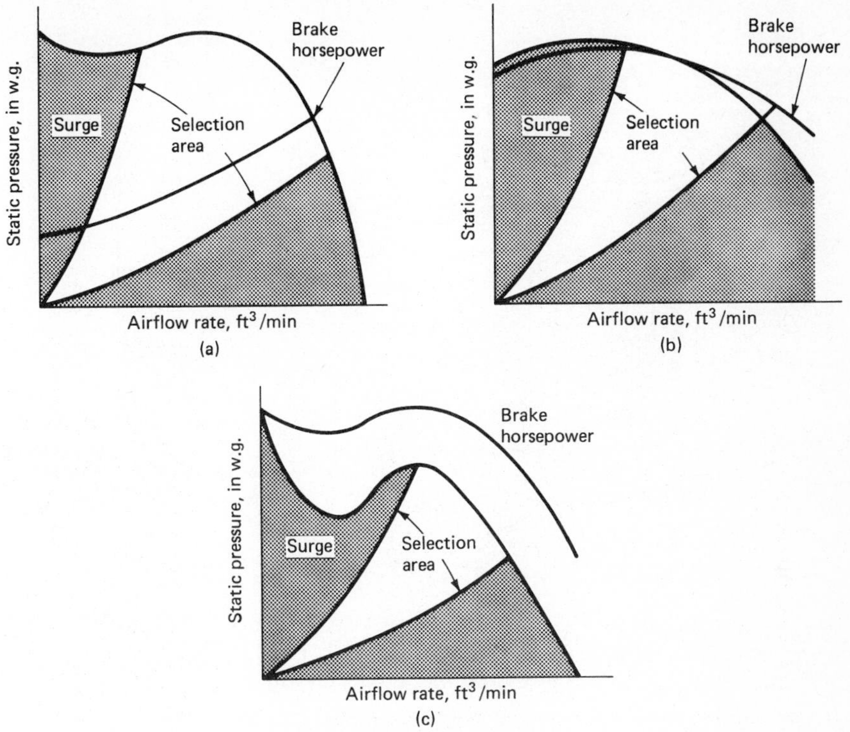

Figure 2-1 Fan characteristics: (*a*) forward-curved vanes, (*b*) backward-curved vanes, (*c*) axial vanes.

Small fans usually have a bigger zone to modulate airflow but are usually less efficient.

COMPRESSORS

Compressors are just like pumps in basic construction as well as function. The basic function of a compressor is to pressurize gas just as a pump pressurizes liquid. Like pumps, compressors have positive displacement (reciprocating, rotary screw, sliding vane) as well as turbo types. Metal diaphragm compressors[16] are used for compressing gases where leakage to the environment could be hazardous or wasteful. Reciprocating compressors operate on the principle of a piston sliding in a cylinder. A turbocompressor is shown in Fig. 2-2.

The horsepower of a centrifugal compressor is given by the following equation:

$$\text{hp} = \begin{cases} \dfrac{WH_{\text{ad}}}{550\eta_{\text{ad}}} & \text{(U.S. customary units)} \\ \dfrac{WH_{\text{ad}}}{75\eta_{\text{ad}}} & \text{(SI units)} \end{cases} \tag{4}$$

where

W = weight of flow

η_{ad} = adiabatic efficiency of the compressible flow and all other efficiencies, including mechanical, disk friction, motor, etc.

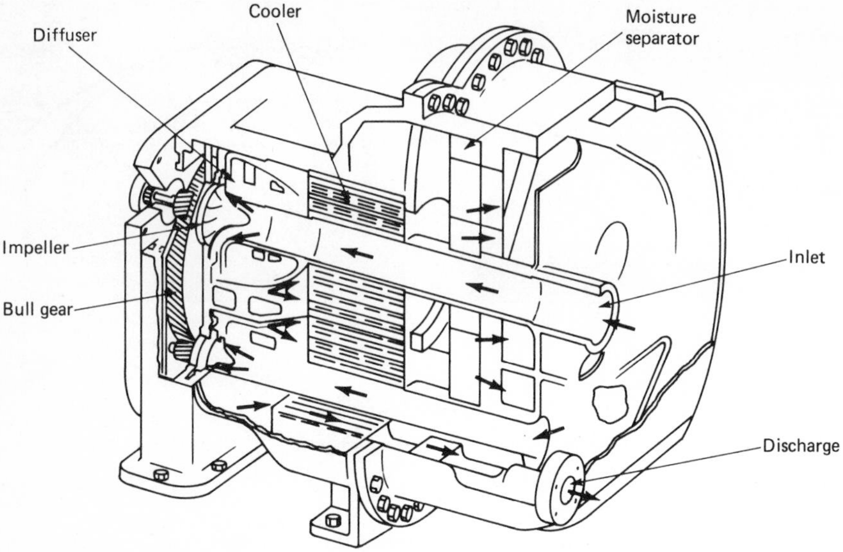

Figure 2-2 Centrifugal compressor. *(Ingersoll-Rand Corp.)*

H_{ad} = adiabatic head of the compressible flow including all the pressure losses in the system and the final pressure required for use.

H_{ad} can also be calculated from the following equation:

$$H_{ad} = \frac{1545\ T_1}{\overline{m}\sigma}(R_c - 1) \tag{5}$$

where

$\sigma = k_1/k$

k = coefficient of adiabatic expansion

= 1.4 for air

$\overline{m}$ = molar weight

R_c = compression ratio $= \dfrac{\text{after compression abs. press.}}{\text{before compression abs. press.}}$

T_1 = absolute temperature at suction

There are two basic types of turbocompressors, axial and centrifugal. Input energy is of course provided by an engine or motor. In the axial compressor the air is forced to move parallel to the centerline of the propeller. Stationary vanes divert the flow to the succeeding row of rotating vanes.

Centrifugal compressors are more common. A high-speed impeller forces airflow in a radial direction inside the compressor, thereby causing a pressure rise. The air at the periphery is then ducted to the successive stages.

Pulseless, high volumetric flow rates with oil-free discharge are the basic attractions of centrifugal compressors. Typical performance charts of compressors are shown in Fig. 2-3, for backward-leaning vs. radial vanes. Notice the general loss of capacity and higher power requirements of the radial vanes. The heat is rejected by the compressor to the air (therefore, there are cooling water requirements). The horsepower required by the compressor for a certain scfm capacity and the number of stages are important parameters for a final selection.

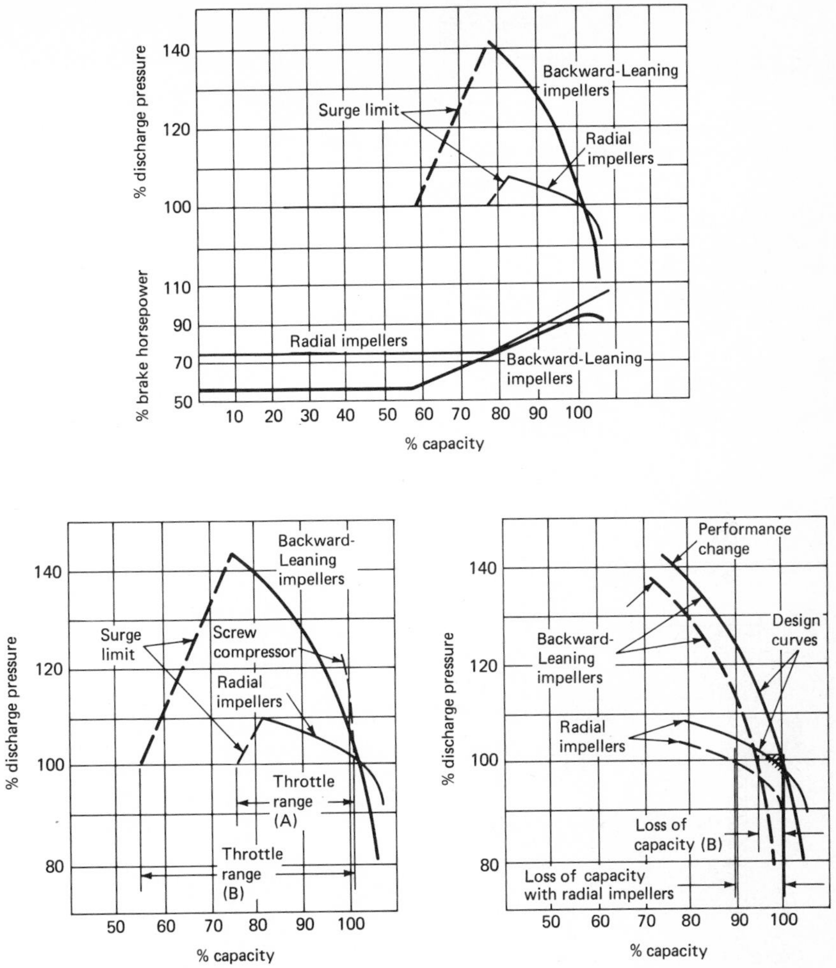

Figure 2-3 Typical performance curves of centrifugal compressors with backward-leaning impellers (A) vs. compressors with radial impellers (B). *(Ingersoll-Rand Corp.)*

More stages are often desirable. API Standards 617 and 618 are commonly used in connection with compressors.

TYPICAL PNEUMATIC SYSTEMS

A typical air-supply system diagram along with the primary controls and final uses is shown in Fig. 2-4. Free air is sucked through a filter into the compressor and discharged into a service air receiver (after passing through an aftercooler to remove the heat from the compressor and moisture separator). The air receiver is kept under pressure and also serves as a surge chamber. The relief valve protects against overpressures. From the receiver, the air can either be directly used or dried and chilled before use, as in instrument-air applications.

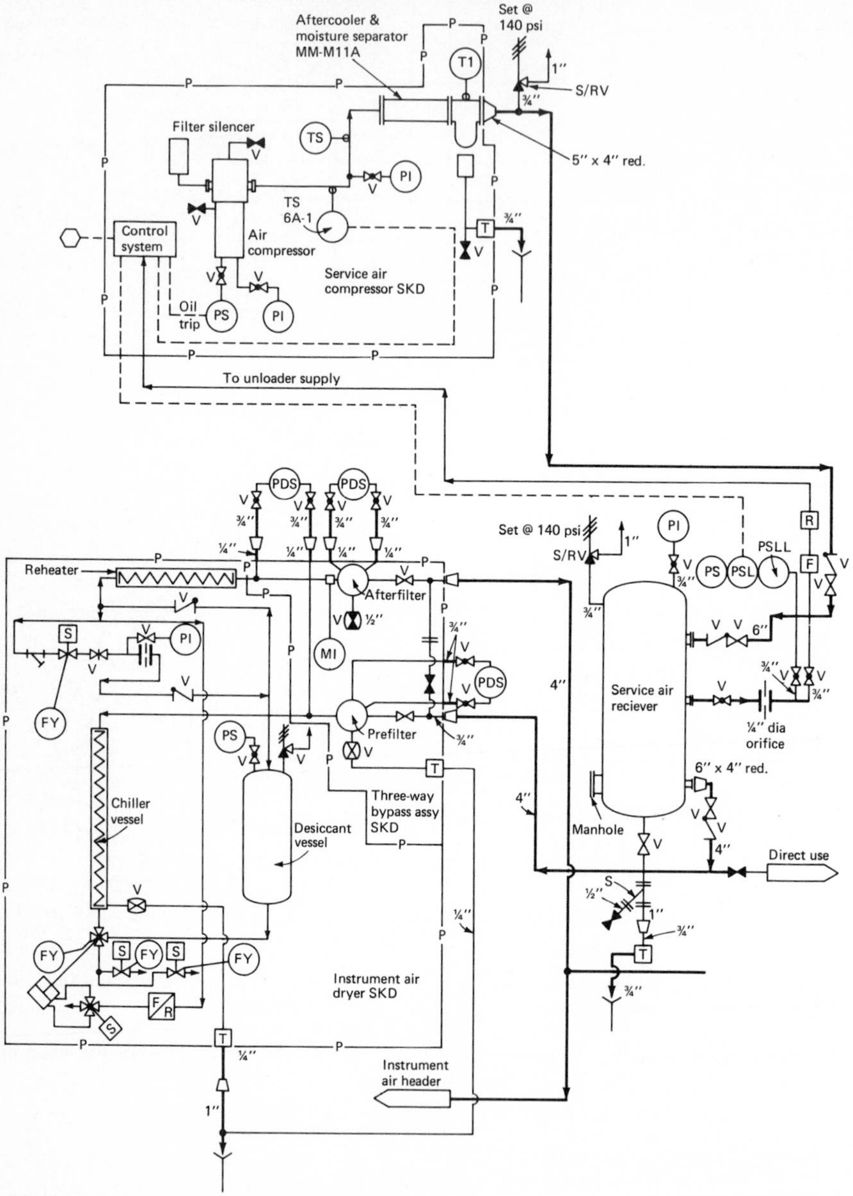

Figure 2-4 Typical pneumatic system. Key: PI = pressure indicator, T1 = temperature indicator, PS = pressure switch, TS = temperature switch, T = moisture trap, S/RV = safety/relief valve, V = valve, PDS = pressure differential switch, S = solenoid, FI = flow indicator, PSL = pressure switch low, PSLL = pressure switch low-low.

Another typical pneumatic system would include a vacuum pump (really a "negative" air compressor) to pump out air *from* a system as in applications to a condenser that functions best under vacuum, or in a negative-pressure air-vent system. The vacuum pumps function in a way similar to vacuum cleaners. One important component in such a system is a valve that shuts itself and the vacuum system upon sensing a particular water level in a container at the top of the water box of a condenser. The valve reopens and the vacuum system starts when air is again sensed inside the water box. There are other arrangements that require air to enter the system when a vacuum is sensed in order to avoid either pipe collapse or system cavitation due to vapor pressures forming inside. One such air and vacuum valve is shown in Fig. 2-5, with the corresponding performance graph shown in Fig. 2-6. Such a valve has a large orifice through which a great amount of air escapes when the system is being filled with liquid. Once the system is filled, the fluid lifts the float in the valve, closes the orifice, and stays closed until the system is drained.

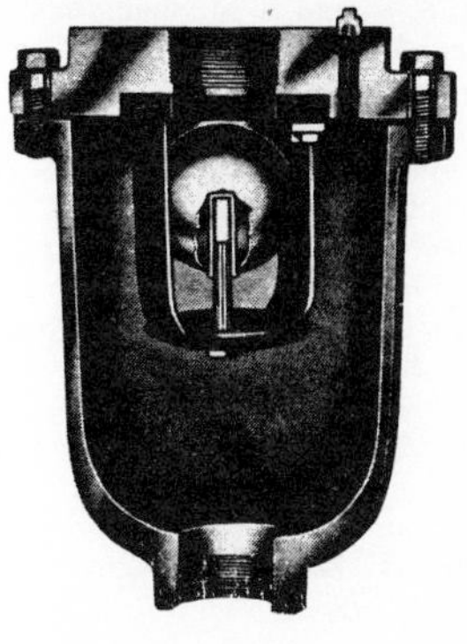

Figure 2-5 Air and vacuum valve. *(APCO Valve & Primer Corp.)*

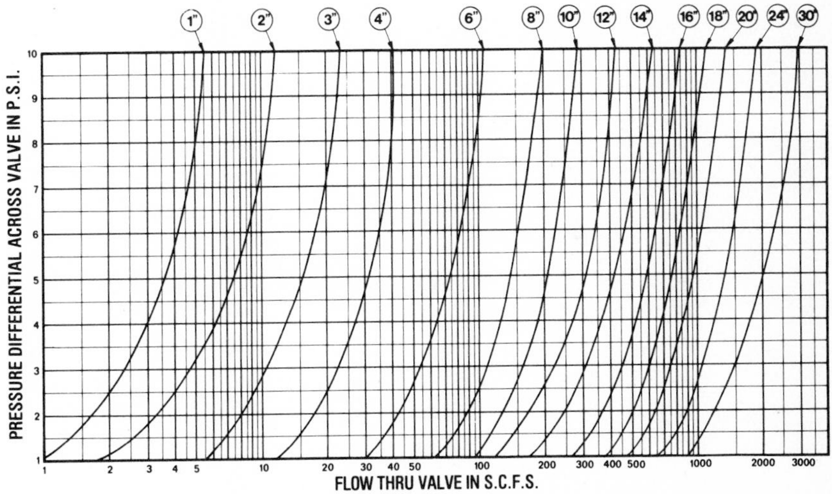

Figure 2-6 Performance graph for air and vacuum valve. *(APCO Value & Primer Corp.)*

REFERENCES AND BIBLIOGRAPHY

1. Miller, Donald S.: *Internal Flow: A Guide to Losses in Pipe and Duct Systems,* British Hydrodynamic Research Association, Cranfield-Bedford, England, 1971.
2. Giles, Donald V.: *Fluid Mechanics & Hydraulics,* Schaum Outline Series, McGraw-Hill, New York, 1962.
3. King, H. W., and E. F. Brater: *Handbook of Hydraulics,* McGraw-Hill, New York, 1976.
4. Rouse, Hunter: *Elementary Mechanics of Fluids,* Wiley, New York, 1946.
5. Simon, Andrew L.: *Practical Hydraulics,* Wiley, New York, 1976.

6. *Flow of Fluids Through Valves, Fittings, and Pipe,* Crane Company, New York, 1976.
7. "Standard for Centrifugal, Rotary & Reciprocating Pumps," Hydraulic Institute, Cleveland, Ohio, 13th ed., 1975.
8. Satija, K. S., and N. M. Shah: *On Circulating Water for Power Plants,* Proceedings Second World Congress, International Water Resources Association, India, 1975.
9. Lal, Jagdish: *Hydraulic Machines,* Metropolitan Book Company Pvt. Ltd., Delhi, India, 1975.
10. *Masoneilan Handbook for Control Valve Sizing,* 6th ed., Masoneilan International, Inc., Norwood, Mass., 1977.
11. *Power Engineers' Valve Manual,* Technical Publishing Co., Burlington, Ill., 1962.
12. *Standard of Tubular Exchanger Manufacturers Association (TEMA),* 6th ed., White Plains, N. Y., 1978.
13. Henke, R. W.: *Introduction to Fluid Mechanics,* Addison-Wesley, Reading, Mass., 1972.
14. Satija, K. S.: "Prototype Testing of Circulating Water Systems," paper presented at the *ASCE Hydraulics Division Conference,* Seattle, Wash., 1975.
15. Wylie, E. B., and V. L. Streeter: *Hydraulic Transients,* McGraw-Hill, New York, 1969.
16. Lingston, E. H.: "Metal Diaphragm Compressors," *Plant Engineering,* April 19, 1979.
17. Patterson, N. R.: "Variable Air Volume Systems", *Plant Engineering,* February 7, 1980.
18. Stewart, H. L.: "Plant Engineer's Fluid Power Handbook," *Plant Engineering,* June 1977 to December 1979.

section 10

Piping and Valving

chapter 10-1

Piping System Design

by

D. G. Wilson

Vice President and General Manager,
Facility Plans, Engineering and Construction
United States Steel Corp
Pittsburgh, Pennsylvania

INTRODUCTION

In industrial plants where fluids are transferred from one place to another, it has been found that the application of piping is extremely important. Therefore, a knowledge of how this piping is designed is critical to the plant engineer, for both the engineer's understanding of plant functions and his or her ability to make plant modifications to improve efficiency of operations and optimize costs.

Fundamental Considerations

Many elements make up a successful piping installation, and a conscientious effort should be made by the designer and the installer to achieve a trouble-free, maintainable system.

Proper and adequate system planning is considered the first step. With pipeline diameters established and equipment and end-point locations determined, an accurate sketch or drawing should be prepared. Lines should be routed to minimize length of pipe runs, yet allow for adequate flexibility for thermal expansion. Piping subject to water hammer from sudden valve closures must be provided with suitable anchors or pulsation-damping devices.

Lines for conveying condensable vapor or gases, such as steam, should be sloped for drainage, with provision for liquid removal by manual or automatic blowdown. Field investigations to determine existing site conditions may be necessary to confirm proposed line routings. Piping materials should be selected to be compatible with the fluid. Maximum pressures and temperatures of fluid within the line should be used to determine thicknesses of pipe and fitting wall. Bare carbon steel piping and supports should be painted with a suitable enamel or epoxy paint to minimize atmospheric corrosion. Consideration should be given to cathodic protection and/or coating for underground lines.

The method of fabrication of a piping system should be selected on the basis of maintainability, initial cost, service life, and overall end use. Smaller-diameter, low-pressure piping is frequently installed with threaded joints. Hazardous or flammable fluids may necessitate the use of welded fittings to minimize the potential for leakage. Periodic inspection of piping materials, weld joints, fitting makeup, etc., should be undertaken during the installation phase. Welding qualifications, tests, and procedures may be obtained from the American Welding Society or the American Society of Mechanical Engineers and used as guidelines for fabrication.

After fabrication and erection, the piping system should be pressure tested. Pressure testing should be performed with water and limited to a hydrostatic test pressure of 1.5 times the design pressure. When greater sensitivity for detecting leakage is desired, halide leak testers or helium mass spectrometers may be employed.

PRESSURE DROP AND SIZE DETERMINATION

System Definition

Before starting calculations, the system being studied must be defined. The following information should be gathered or developed.

1. A flow diagram of the process or plant section
2. Physical and mechanical properties of the system and fluids
3. Pumping requirements or power input to the fluids

Fluid Flow Equation

The basic equation for fluid flow can be derived from a consideration of thermodynamics and the interrelation of various forms of energy.

For 1 lb (kg) of an incompressible fluid, flowing between two points in a horizontal circular pipe under isothermal steady conditions and turbulent flow, with no work done on the system, the expression takes the form of the familiar Fanning or Darcy equation:

$$F = \left(2f\frac{L}{D}\right)\frac{V^2}{g_c} = \left(2f\frac{L}{D}\right)\frac{G^2}{g_c P^2} \tag{1}$$

or

$$p = \left(2f\frac{L}{D}\right)\frac{V^2 P}{g_c} = \left(2f\frac{L}{D}\right)\frac{G^2}{g_c P}$$

where

F = friction loss, ft·lbf/lb of fluid flowing = (N·m/kg)*
V = average linear velocity, ft/s (m/s)
L = length of pipe, ft (0.305 m)
f = friction factor, dimensionless
= factor, 32.17 ft·lb/(s^2)(lbf) [kg/m/(s^2)(N)]
D = pipe diameter, ft (m)
G = VP = mass velocity, lb/(s)(ft^2) (cross section) [kg/(s)(m^2)]
P = fluid density, lb/ft^3 (kg/m^3)
p = absolute pressure, lbf/ft^2 (Pa)

Equation (1) can be used to find values for pressure drop or head loss, size of the pipeline, and the rate of fluid flowing. Given any two unknowns, the third can be found. The friction factor f has been shown to be dependent on the dimensionless Reynolds number, $N_{\mathrm{Re}} = DVP/u = DG/u$, where u = fluid viscosity, lb/(ft)(h) [kg/(m)(h)].

A correlation developed by Moody,[1] relating the friction factor, pipe roughness, and N_{Re} is shown in Fig. 1-1 where E/D is the ratio of the surface roughness and pipe diameter and is dimensionless.

A useful chart for solving Eq. (1) has been developed by Generaux[2] and is shown in Fig. 1-2. Convenient nomographs and calculation methods for fluid flow parameters have also been developed by Crane Company[3] for both liquids and gases.

The Fanning equation (1) can also be used for determining the unknown parameters for gases, provided that the overall system pressure drop does not exceed 10 percent of the absolute internal pressure.

Miscellaneous Friction Losses

Friction losses due to sudden expansion and contraction and losses in valves and fittings can be included in Eq. (1). The friction head F can be split into a summation of terms covering the different losses.

The equations presented below for evaluating these frictional effects utilize the "velocity head" method.[4]

Friction Loss from Sudden Expansion of Cross Section

$$F_e = \frac{K_e(V_1 - V_2)^2}{2g_c} \qquad (2a)$$

and

$$K_e = 1 - \frac{S_1}{S_2} \qquad (2b)$$

Friction Loss from Sudden Contraction of Cross Section

$$F_c = K_c \frac{V_1^2}{2g_c} \qquad (3a)$$

and

$$K_c = 0.4\left(1 - \frac{S_1}{S_2}\right) \qquad (3b)$$

*The second set of units for each symbol denotes corresponding International System (SI) units.

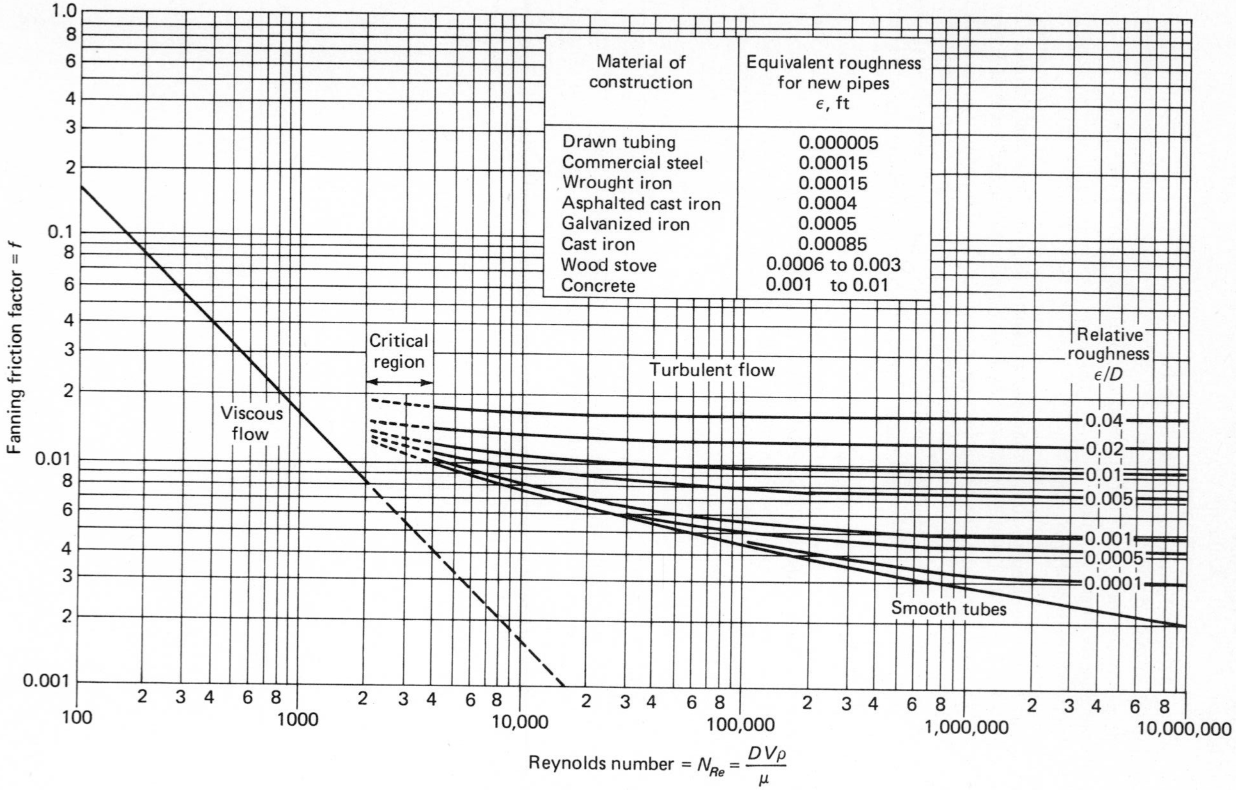

Figure 1-1 Fanning friction factors for long straight pipes.

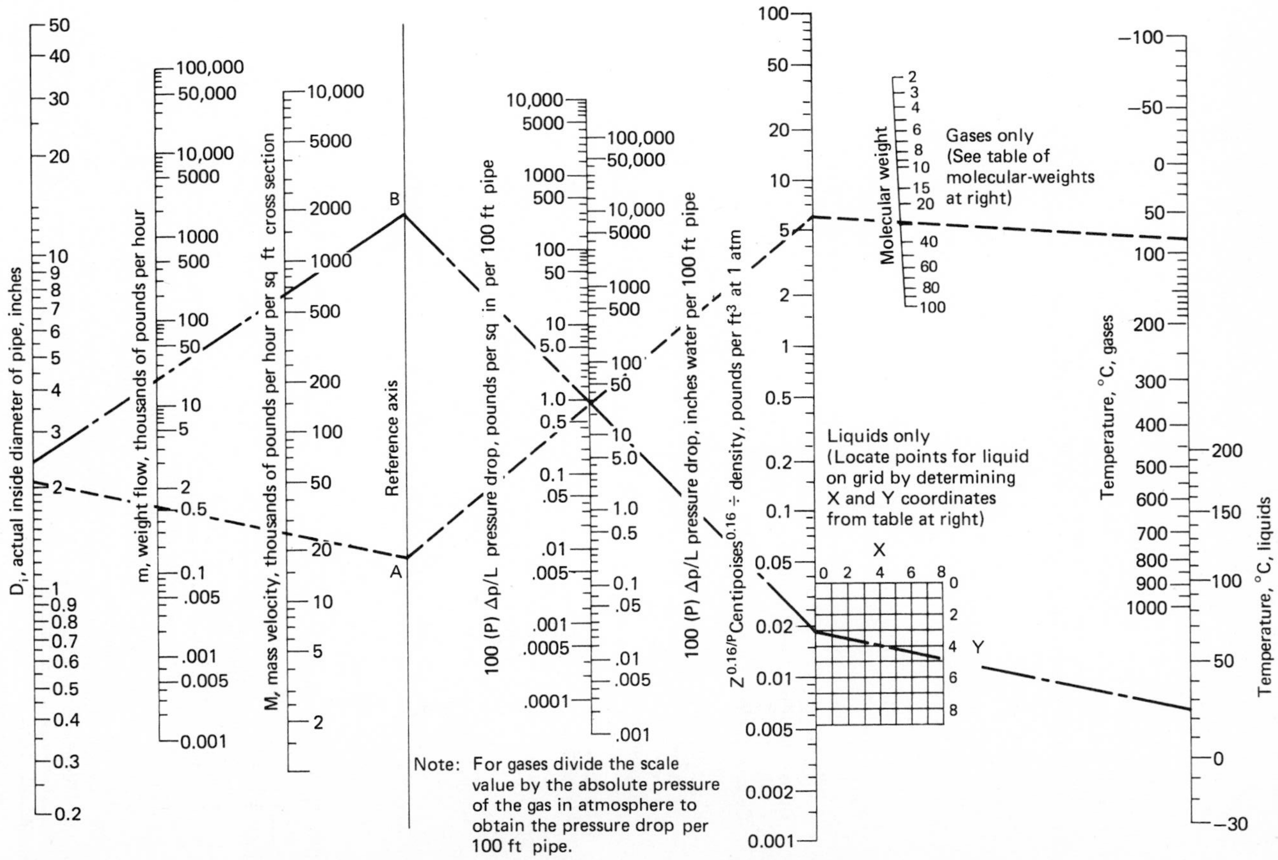

Figure 1-2 Pipe flowchart for turbulent flow.

TABLE 1-1 Additional Frictional Loss for Turbulent Flow Through Fittings and Valves[k]

Type of fitting or valve	Additional friction loss, equivalent no. of velocity heads, K_f
45°ell, standard[a,b,e,g,i]	0.35
45°ell, long radius[b]	0.2
90°ell, standard[a,b,d,g,i,m]	0.75
Long radius[a,b,e,g]	0.45
Square or miter[m]	1.3
180° bend close return[a,b,g]	1.5
Tee, std., along run, branch blanked off[g]	0.4
Used as ell, entering run[d,h]	1.0
Used as ell, entering branch[b,d,h]	1.0
Branching flow[f,h,l]	1[o]
Coupling[b,g]	0.04
Union[g]	0.04
Gate valve,[a,g,j] open	0.17
¾ open[p]	0.9
½ open[p]	4.5
¼ open[p]	24.0
Diaphragm valve,[n] open	2.3
¾ open[p]	2.6
½ open[p]	4.3
¼ open[p]	21.0
Globe valve,[g,j] bevel seat, open	6.0
½ open[p]	9.5
Composition seat, open	6.0
½ open[p]	8.5
Plug disk, open	9.0
¾ open[p]	13.0
½ open[p]	36.0
¼ open[p]	112.0
Angle valve,[a,g] open	2.0
Y or blowoff valve,[a,j] open	3.0
Plug cock[c] 0 = 5°	0.05
10°	0.29
20°	1.56
40°	17.3
60°	206.0
Butterfly valve,[c] 0 = 5°	0.24
10°	0.52
20°	1.54
40°	10.8
60°	118.0
Check valve,[a,g,j] swing	2.0[q]
Disk	10.0[q]
Ball	70.0[q]
Foot valve[g]	15.0
Water meter[m] disk	7.0[r]
Piston	15.0[r]
Rotary (star-shaped disk)	10.0[r]
Turbine-wheel	6.0[r]

[a]"Flow of Fluids through Valves, Fittings, and Pipe," Tech. Paper 410, Crane Co., 1969.

[b]Freeman, *Experiments upon the Flow of Water in Pipes and Pipe Fittings*, American Society of Mechanical Engineers, New York, 1941.

[c]Gibson, *Hydraulics and Its Applications*, 5th ed., Constable, London, 1952, p. 250.

[d]Giesecke and Badgett, *Heating/Piping/Air Conditioning*, **4**(6):443–447 (1932).

[e]Giesecke, *Journal of the American Society of Heating and Ventilation Engineers*, **32**:461 (1926).

[f]Gilman, *Heating/Piping/Air Conditioning*, **27**(4):141–147, (1955).

[g]*Pipe Friction Manual*, 3d ed., Hydraulic Institute, New York, 1961.

[h]Hoopes, Isakoff, Clarke, and Drew, *Chemical Engineering Progr.*, **44**:691–696 (1948).

[i]Ito, *Journal of Basic Engineering*, **82**:131–143 (1960).

Friction Loss from Valves and Fittings

$$F_f = K_f \frac{V_1^2}{2g_c} \tag{4}$$

where

F_e, F_c, F_f = friction loss, ft·lbf/lb [(m)(N)/kg]
$V_1, V_2,$ = volumetric average velocity for the upstream and downstream points, ft/s (m/s)
S_1, S_2 = upstream and downstream pipe cross-sectional areas, ft^2 (m^2)
K_f = valve and fitting component factor, dimensionless (see Table 1-1)

Economics

While the equations discussed above permit determination of the pipe diameter, the selection of a particular pipe size will depend upon a balance between the cost of materials, power consumption, and maintenance. Hence, when selecting pipe sizes, the overall costs of the system must be considered. One method of approach is that taken by Peters and Timmerhaus[5] for establishing optimum pipe diameters.

Small Computers

Desk-type computers are fast and reduce the need for engineers to access large computers for solving problems.

One such unit, the Hewlett-Packard HP 67/97, is capable of storing programs on magnetic strips for reuse without having to reprogram it each time a program is used. The availability of a large library reduces most engineering calculations to routine operations which can be completed quickly.

Benenati[6] has developed a program for pressure-drop calculations that uses the Hewlett-Packard 67/97 computer for solving the previously discussed Fanning equation (1).

PIPE STRESSES

To determine pressure-retaining ability and mechanical strength in piping components, a thorough knowledge of strength of materials is required. All standard piping materials in common use today have been developed with this theoretical background and have

---

[j]Lansford, *Loss of Heat in Flow of Fluids through Various Types of 1½ M. Valves,* Univ. Illinois Eng. Expt. Sta. Bull., ser. 340, 1943.

[k]Lapple, *Chemical Engineering,* **56**(5):96–104 (1949), general survey reference.

[l]McNown, *Proceedings of the American Society of Civil Engineers,* **79**(separate 258):1–22 (1953) discussion, *ibid.* **80**(separate 396):19–45 (1954).

[m]Schoder and Dawson, *Hydraulics,* 2d ed., McGraw-Hill, New York, 1934, p. 2130

[n]Streeter, *Product Engineering,* **18**(7):89–91 (1947).

[o]This is pressure drop (including friction loss) between run and branch, based on velocity in the main stream before branching. Actual value depends on the flow split, ranging from 0.5 to 1.3 if main stream enters run and from 0.7 to 1.5 if main stream enters branch.

[p]The fraction open is directly proportional to stem travel or turns of hand wheel. Flow direction through some types of valves has a small effect on pressure drop (see Freeman, op. cit.). For practical purposes this effect may be neglected.

[q]Values apply only when check valve is fully open, which is generally the case for velocities more than 3 ft/s for water.

[r]Values should be regarded as approximate because there is much variation in equipment of the same type from different manufacturers.

Source Perry, R. H. and C. H. Chilton (eds), *Chemical Engineers Handbook,* 5th ed., McGraw-Hill, New York, 1973, p.5–36. Reproduced with permission.

been service-tested to achieve the predicted performance. In addition, all piping systems or replacement piping components are now being specified to conform with an applicable code requirement. Most municipalities require compliance with building codes such as those issued by the American National Standards Institute for building permits to be issued.

Circumferential Stress

Neglecting external loading effects, thermal expansion, or corrosion allowance, the expression used to determine the pipe wall thickness from the internal pressure takes the form

$$t = \frac{pD}{2S}$$

where
- t = wall thickness, in (m)
- D = pipe diameter, in (m)
- S = hoop stress, lb/in^2 (Pa)
- p = internal pressure, lb/in^2 (Pa)

Usually the pipe diameter and internal pressure have been determined by flow requirements, and it is necessary to find the pipe wall thickness. The allowable hoop stress S is a function of the pipe material selected and the temperature to which the material will be exposed (i.e., normally the pipe contents' temperature).

Problem

A 12-in-diam carbon steel pipe is to be used to convey water at 60°F (15.4°C) and a pressure of 400 psig (2758 kPa). The pipe wall thickness must be determined.

Solution

By reference to handbook data it is found that the minimum yield strength corresponds to 30,000 lb/in^2 (20 700 MPa) for carbon steel and a 3:1 safety factor is considered adequate; therefore

$$t = \frac{400 \text{ lb/in}^2 \times 12 \text{ in}}{2 \times 10{,}000 \text{ lb/in}^2}$$
$$= 0.24 \text{ in } (0.61 \text{ m})$$

Some pressure piping codes require the use of a more rigorous determination of wall thickness using empirical factors for manufacturing tolerances and finite allowable stress-temperature ranges. As presented in ANSI B31.1, "Power Piping Code," the equation takes the form

$$t_{in} = \frac{PD_o}{2(S_E + Py)} + A$$

where
- t_{in} = required wall thickness, in (m)
- P = internal pressure, lb/in^2 (Pa)
- D_o = outside diameter of pipe, in (m)
- S_E* = allowable stress of material, lb/in^2 (Pa), temperature-dependent
- A* = additional thickness, in (m) to compensate for corrosion, threading, etc.
- y* = temperature coefficient

*Values determined by reference to code appendixes.

Combined Stresses

In multiple-plane systems, it often becomes necessary to examine the torsional stresses resulting from thermal expansion. When these and bending stresses are combined, a resultant fiber stress of higher intensity may develop. Piping codes often require the calculated combined stress to fall within an allowable stress range for hot and cold conditions.

Combined stress may be calculated by use of the formula

$$S = \tfrac{1}{2}\left[S_L + S_C + \sqrt{4S_t^2 + (S_L - S_C)^2}\right]$$

where

S = total bending stress
S_L = maximum combined stress
S_C = hoop stress
S_t = torsional or shear stress

Bending Stresses

Bending stresses are developed in piping systems by loads resulting from thermal expansion or weight of the pipe.

In high-temperature piping systems, it is required that a thorough analysis be made of maximum bending stresses resulting from thermal expansion. Often, it is necessary to reconfigure pipeline geometry and to provide specially designed spring hanger supports to maintain allowable pipe stresses.

The bending stress resulting from piping weight effects may be calculated by use of the following equation:

$$S = \frac{WL^2}{\mathrm{SM}}$$

where

S = bending stress, lb/in^2 (Pa)
W = weight, lb/in (kg/m)
L = distance between supports, in (m)
SM = section modulus of the pipe cross section, in^3 (m^3)

For a circular cross section the section modulus can be determined by

$$\mathrm{SM} = \frac{\pi}{32}\,\frac{D_o^4 - D_2^4}{D_o}$$

where

D_o = outside diameter, in (m)
D_2 = inside diameter, in (m)

Table 1-2 gives the recommended pressure-temperature ratings for ASTM A 53 grade A carbon steel pipe.

Pipe Deflection

Consideration should also be given to pipe deflection between supports; this is often the governing parameter in determining supporting spans. Deflections in horizontal pipelines should be maintained between 0.1 and 0.5 in (2.5 and 12.5 mm) at their midpoint. Additional supports should be provided at concentrated load areas such as valves, flanges, or direction changes. The following equation may be used to determine the midspan deflection of a horizontal pipeline with free ends:

$$\Delta = \frac{5WL^4}{384EI}$$

TABLE 1-2 Pressure-Temperature Ratings of Pipe, Seamless Carbon Steel to ASTM A 53 Grade A

			Temperature and stress	
		Wall	to 650°F 12,000 lb/in² stress	750°F 10,700 lb/in² stress
Pipe size, in	Schedule number	thickness, in	Working pressure, lb/in²	
½	40	0.109	882	787
	80	0.147	1947	1736
1	40	0.133	960	856
	80	0.179	1777	1585
2	40	0.154	724	646
	80	0.218	1329	1185
3	40	0.216	875	780
	80	0.220	1422	1218
4	40	0.237	1146	1022
	80	0.337	1660	1481
6	40	0.280	914	815
	80	0.432	1435	1279
8	20	0.250	622	555
	40	0.322	806	718
	80	0.500	1270	1133
10	20	0.250	497	443
	40	0.365	729	650
	80	0.594	1208	1077
12	20	0.250	418	373
	40	0.406	683	609
	80	0.688	1178	1050

where

Δ = deflection, in (m)
W = weight, lb/in (kg/m)
L = length of span, in (m)
E = modulus of elasticity
I = moment of inertia

When calculations are not performed, Table 1-3 may be used to determine recommended spans. Recommendations are based upon deflections of 0.1 in and combined bending and shear stress of 1500 lb/in² using Schedule 40 pipe.

TABLE 1-3

Pipe Size, in	1	2	3	4	6	8	12	16	20	24
Water service, ft	7	10	12	14	17	19	23	27	30	32
Air service, ft	9	13	15	7	21	24	30	35	39	42

STANDARDS AND SPECIFICATIONS

"Power Piping," ANSI B31.1. = 1977
"Fuel Gas Piping," ANSI B31.2. = 1968
"Chemical Plant and Petroleum Refinery Piping," ANSI B31.3. = 1976
"Refrigeration Piping," ANSI B31.5. = 1974
"Gas Transmission and Distribution Piping systems," ANSI B31.8. = 1975
"Building Services Piping," ANSI B31.9.
"Thickness Design of Cast-Iron Pipe," ANSI/AWWA C101-67 (1977) [A21.1]
"Thickness Design of Ductile-Iron Pipe," ANSI/AWWA C150-76 [A21.50.]
"Pipe Hangers and Supports, Materials and Design," MSS Standard SP-58.
"Pipe Hangers and Supports—Selection and Application," MSS Standard SP-69.
"Steel Water Pipe 6″ and Larger," AWWA Standard C-200-75.
"Recommended Practice for the Design and Installation of Pressure Relieving Systems in Refineries," API Standard RP 520; "Guide for Pressure Relief and Depressuring Systems," API Standard RP 521.
Hydraulic Institute Standards.

BIBLIOGRAPHY

Perry, R. H. and C. H. Chilton (eds.): *Chemical Engineers' Handbook,* 5th ed. McGraw-Hill, New York, 1973.

King, R. C.: *Piping Handbook,* 5th ed., McGraw-Hill, New York, 1967.

Clark, L., and R. Davidson: *Manual for Process Engineering Calculations,* 2d ed., McGraw-Hill, New York, 1962.

"Flow of Fluids through Valves, Fittings and Pipe", Tech. Paper 410, Crane Company, New York, N.Y., 1976.

Shanley, F. R.: *Strength of Materials,* McGraw-Hill, New York, 1957.

Piping Design & Engineering, Grinell Co., (DIV.ITT) Providence R.I. 3d ed., 1971.

Baumeister, Theodore, Eugene A. Avallone, and Theodore Baumeister III (eds.): *Marks' Standard Handbook for Mechanical Engineers,* 8th ed. McGraw-Hill, New York, 1978.

REFERENCES

1. *Transactions of the American Society of Mechanical Engineers,* **66**:671–684 (1944).
2. Generaux: *Chemical and Metallurgical Engineering,* **44**:241 (1937).
3. Tech. Paper 410, Crane Company, 1976.
4. McCabe and Smith: *Unit Operations of Chemical Engineering,* 3d ed., McGraw-Hill, New York, 1967, p. 108.
5. Peters and Timmerhaus: *Plant Design and Economics,* McGraw-Hill, New York, 1968.
6. Benenati, R. F.: "Solving Engineering Problems on Programmable Pocket Calculators," *Chemical Engineering,* February 1977.

chapter 10-2

Metallic Piping and Fittings

by

D. G. Wilson

Vice President and General Manager,
Facility Plans, Engineering and Construction
United States Steel Corp.
Pittsburgh, Pennsylvania

INTRODUCTION

Piping and tubing are the most commonly used construction materials today. The primary purpose of conveying liquid, gas, air, and slurry is augmented by piping's use as a support member as well as in the manufacture of products such as rollers, cylinders, conduit, recreation equipment, partitions, etc.

Orders for piping and tubing are, in addition to specifying standard weight designations and/or wall thickness, accompanied with a specification such as those published by ASTM, API, WWP, ASME, AWWA, or ANSI, which establishes the minimum standards required of the product. Standard pipe is generally covered under specifications published by ASTM.

Classifications for pipes are standard weight (Std), extra strong (XS), and double extra strong (XXS). In pipe sizes ⅛ to 10 in nominal, ANSI Schedule 40 thicknesses are identical to standard-weight pipe. Schedule 80 (⅛ to 8 in nominal) is identical to extra-strong pipe, and Schedule 160 falls between extra-strong and double extra-strong pipe. It should be noted that wall thickness does not apply to the outside diameter (OD) of the pipe, but only affects the inside diameter (ID), i.e., the thicker the wall the smaller the ID. Refer to Table 2-1.

DESCRIPTION, CLASSIFICATION, AND APPLICATIONS

Standard-weight pipe is used for low-pressure applications for gas and water or general plumbing. Extra-strong pipe with its heavier wall is for medium-pressure applications, whereas double extra-strong pipe is for high-pressure applications.

Pipe is manufactured from steel, cast iron, wrought iron, brass, and copper. Tubing is manufactured from steel, copper, stainless steel, and aluminum.

Pipe, as manufactured, is generally classified as seamless, continuous weld, electric weld, and double submerged-arc weld pipe. Tubing is classified as seamless or welded. The type of pipe to be used will depend upon the service condition, internal pressure, temperature, life expectancy, and corrosion. When specifying pipe, always remember to observe the applicable federal, state, local, and industrial codes.

Several Applications

Seamless Pipe

Seamless types are manufactured by piercing a solid billet and are available in sizes ⅛ to 26 in. Because of the seamless method of manufacturing, this type is widely used in industry for air, gas, steam, water, and oil piping.

Continuous Weld Pipe

This type is manufactured by butt welding a continuous metal sheet which passes through forming rolls, is welded, and then passes through a stretch reduction mill; it is available in sizes to 4 in and is used for gas, water, air, steam, structural applications, sprinkler systems, electric raceways, fencing, etc. It generally is less expensive than seamless pipe.

Electric Weld Pipe

This is manufactured by forming a strip from flat steel using high-frequency welding and sizing methods. It is available in sizes 4 to 20 in and is used in the steel, oil, and natural gas industries as well as for pipe piling and for slurry lines.

Double Submerged-Arc Weld Pipe

This type is manufactured from a prepared flat steel plate, is formed on presses, and is either mechanically or hydrostatically expanded: It is welded inside and outside and generally comes in sizes ranging from 20 to 48 in. It is used in construction, oil, natural gas, and water applications.

Pipe sizes larger than 48 in are true fabrication, rolled, and welded.

Tubing

There exists in the industry an intermingling of the terms "tubing" and "pipe." Round tubing is used for conveying fluids, tubing with other shapes for fabrication purposes.

TABLE 2-1 Pipe Dimensions and Properties

Nom. pipe size, in	Wall thickness: Iron pipe size	Wall thickness: Sch. no.	Dimensions: Outside diam, in	Dimensions: Inside diam, in	Dimensions: Wall thkn., in	Weights: Plain end pipe, lb/ft	Weights: Water in pipe, lb/ft	Areas, Surface: Outside, ft²/ft	Areas, Surface: Inside, ft²/ft	Areas, Cross-sectional: Flow, in²	Areas, Cross-sectional: Metal, in²	Properties: Moment of inertia, in⁴	Properties: Section modulus, in³	Properties: Radius of gyration, in
½	Std	40	0.840	0.622	0.109	0.850	0.132	0.220	0.1637	0.3040	0.2503	0.0171	0.0407	0.2613
	XS	80	0.840	0.546	0.147	1.087	0.101	0.220	0.1433	0.2340	0.3200	0.0201	0.0478	0.2505
	XXS		0.840	0.252	0.294	1.714	0.022	0.220	0.0660	0.0499	0.5043	0.0242	0.0577	0.2192
¾	Std	40	1.050	0.824	0.113	1.130	0.230	0.275	0.2168	0.5330	0.3326	0.0370	0.0705	0.3337
	XS	80	1.050	0.742	0.154	1.473	0.187	0.275	0.1948	0.4330	0.4335	0.0448	0.0853	0.3214
	XXS		1.050	0.434	0.308	2.440	0.063	0.275	0.1137	0.1479	0.7180	0.0579	0.1103	0.2840
1	Std	40	1.315	1.049	0.133	1.678	0.374	0.344	0.2740	0.8640	0.4939	0.0873	0.1328	0.4205
	XS	80	1.315	0.957	0.179	2.171	0.311	0.344	0.2520	0.7190	0.6388	0.1056	0.1606	0.4066
	XXS		1.315	0.599	0.358	3.659	0.122	0.344	0.1570	0.2818	1.0760	0.1405	0.2136	0.3613
1½	Std	40	1.900	1.610	0.145	2.717	0.882	0.497	0.4213	2.0361	0.8001	0.3099	0.3262	0.6226
	XS	80	1.900	1.500	0.200	3.631	0.765	0.497	0.3927	1.7672	1.0689	0.3912	0.4118	0.6052
	XXS		1.900	1.100	0.400	6.408	0.412	0.497	0.2903	0.9502	0.8859	0.5678	0.5977	0.5489
2	Std	40	2.375	2.067	0.154	3.65	1.45	0.622	0.540	3.355	1.075	0.666	0.561	0.787
	XS	80	2.375	1.939	0.218	5.02	1.28	0.622	0.507	2.953	1.477	0.868	0.731	0.766
	XXS		2.375	1.503	0.436	9.03	0.77	0.622	0.393	1.774	2.656	1.312	1.104	0.703
2½	Std	40	2.875	2.469	0.203	5.79	2.07	0.753	0.646	4.788	1.704	1.530	1.064	0.947
	XS	80	2.875	2.323	0.276	7.66	1.83	0.753	0.610	4.238	2.254	1.924	1.339	0.924
	XXS		2.875	1.771	0.552	13.70	1.07	0.753	0.463	2.464	4.028	2.871	1.997	0.844
3	Std	40	3.500	3.068	0.216	7.58	3.20	0.916	0.802	7.393	2.228	3.017	1.724	1.164
	XS	80	3.500	2.900	0.300	10.25	2.86	0.916	0.761	6.605	3.016	3.892	2.225	1.136
	XXS		3.500	2.300	0.600	18.58	1.80	0.916	0.601	4.155	5.466	5.993	3.424	1.047

4	Std	40	4.500	4.026	0.237	10.79	5.51	1.178	1.055	12.730	3.174	7.231	3.214	1.510
	XS	80	4.500	3.826	0.337	14.98	4.98	1.178	1.002	11.497	4.407	9.610	4.271	1.477
	XXS		4.500	3.152	0.674	27.54	3.38	1.178	0.826	7.803	8.101	15.284	6.793	1.374
6	Std	40	6.625	6.065	0.280	18.97	12.5	1.73	1.59	28.90	5.58	28.14	8.50	2.24
	XS	80	6.625	5.761	0.432	28.57	11.3	1.73	1.51	26.07	8.40	40.49	12.22	2.19
	XXS		6.625	4.897	0.864	53.16	8.1	1.73	1.28	18.83	15.64	66.33	20.02	2.06
8	Std	40	8.625	7.981	0.322	28.55	21.6	2.26	2.09	50.03	8.40	72.49	16.81	2.94
	XS	80	8.625	7.625	0.500	43.39	19.8	2.26	2.01	45.67	12.76	105.70	24.51	2.88
	XXS		8.625	6.875	0.875	72.42	16.1	2.26	1.80	37.13	21.30	161.98	37.56	2.76
10	Std	40	10.750	10.020	0.365	40.48	34.1	2.81	2.62	78.85	11.91	160.71	29.90	3.67
	XS	60,80S	10.750	9.750	0.500	54.74	32.3	2.81	2.55	74.66	16.10	211.94	39.43	3.63
	XXS	140	10.750	8.750	1.000	104.13	26.1	2.81	2.29	60.13	30.63	367.81	68.43	3.46
12	Std	40	12.750	11.938	0.406	53.6	48.5	3.34	3.13	111.9	15.74	300.3	47.1	4.37
	XS	80	12.750	11.374	0.688	88.6	44.0	3.34	2.98	101.6	26.07	475.7	74.6	4.27
	XXS	120	12.750	10.750	1.000	125.5	39.3	3.34	2.81	90.8	36.91	641.7	100.7	4.17
14	Std	30	14.000	13.250	0.375	54.6	59.7	3.67	3.47	137.9	16.05	372.8	53.2	4.82
	XS		14.000	13.000	0.500	72.1	57.4	3.67	3.40	132.7	21.21	483.8	69.1	4.78
16	Std	30	16.000	15.250	0.375	63	79.1	4.19	4.00	182.6	18.41	562	70.3	5.53
	XS	40	16.000	15.000	0.500	83	76.5	4.19	3.93	176.7	24.35	732	91.5	5.48
18	Std		18.000	17.250	0.375	71	101.2	4.71	4.51	233.7	20.76	807	89.6	6.23
	XS		18.000	17.000	0.500	93	98.2	4.71	4.45	227.0	27.49	1053	117.0	6.19
20	Std	20	20.000	19.250	0.375	79	126.0	5.24	5.04	291.1	23.12	1113	111.3	6.94
	XS	30	20.000	19.000	0.500	104	122.8	5.24	4.97	283.5	30.63	1457	145.7	6.90
22	Std	20	22.000	21.250	0.375	87	153.7	5.76	5.56	354.7	25.48	1490	135.4	7.65
	XS	30	22.000	21.000	0.500	115	150.2	5.76	5.50	346.4	33.77	1953	177.5	7.61
24	Std	20	24.000	23.250	0.375	95	183.9	6.28	6.09	424.6	27.83	1942	161.9	8.35
	XS		24.000	23.000	0.500	125	180.0	6.28	6.02	416.0	36.90	2550	213.0	8.31

Both shapes are seamless and welded, ferrous and nonferrous, and come in sizes ranging to 10 in. However, for this discussion tubing can be differentiated from piping when used with compression, bite, swaged, or flared-tube fittings to convey liquids, gases, or air. Tubing can be further defined as that product which can be easily bent or formed, generally because it has a thinner wall. It is beyond the scope of this chapter to adequately convey all the information available. It is therefore suggested that when the end use is known (boiler tubes, heat-exchanger tubes, hydraulic, refining tubes, etc.), the ASME, ANSI, and SAE standards be used.

FITTINGS AND COUPLINGS

Standard fittings are available in the form of 45 and 90° elbows, tees, reducers, unions, crosses, Y branches, couplings, and caps that have been screwed together, socket-welded, flanged, butt-welded, or soldered.

Materials used in fabrication are forged steel, cast steel, cast iron, malleable iron, galvanized steel, and brass.

Because of the difficulty of assembly, screwed fittings are limited to smaller sizes and are generally classified as 125, 150, and 250 lb; they are suitable for service with low-pressure steam, gas, air, and water, and are supplied in brass and malleable and cast iron. Forged-steel fittings are for higher pressure ratings to 6000 lb/in^2. Although available in small sizes, flanged and butt-welded fittings are generally for larger pipes.

Tubing is most frequently supplied in sizes 2 in and smaller for use with flared, bite-type, soldered, and swaged tube fittings.

AVAILABLE SIZES AND FORMS

The ANSI (formerly ASA) standards cover the dimensional and physical properties of fittings. Because there are large numbers of fittings, information about type and material should be sought in standards and manufacturers' handbooks, such as those published by Crane, Grinnell, Tube Turns Division of National Cylinder Gas, and Midwest Piping. Tables 2-3 to 2-10 (pages 10-22 t0 10-29) are a sampling of those available.

INSTALLATION AND MAINTENANCE

The joining of metallic pipe and fittings to form a complete piping system may be accomplished in several ways; the following methods are most popular.

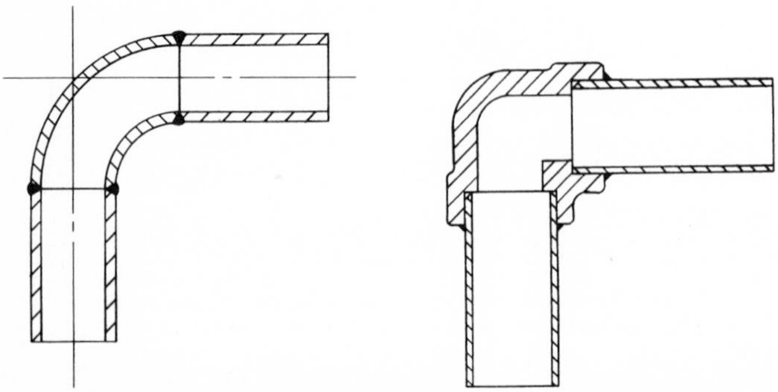

Figure 2-1 (*a*) Butt-welded elbow; (*b*) socket-welded elbow.

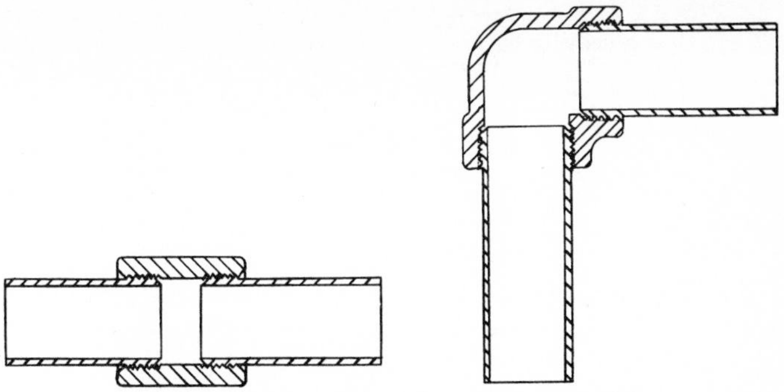

Figure 2-2 (*a*) Threaded coupling; (*b*) threaded elbow.

Welding

This technique requires the fusion of two metal surfaces to form one. Arc welding uses heat produced by an electric arc to coalesce the metals. Filler metals may also be used as fluxing agents. Gas welding uses a gas flame to produce the required fusion temperatures. See Fig. 2-1.

Brazing

Most commonly used in nonferrous domestic piping systems, brazing is a process whereby a filler metal, having a melting point above 800°F (425°C) but lower than that of the base metal, is heated to accomplish the joining. A capillary action occurs, drawing the filler metal between the closely fitted base-metal surfaces.

Threading

This method is customarily used in smaller-diameter, lower-pressure piping systems. Straight- and tapered-thread designs are available. Threaded fittings are made in a variety of material, including malleable iron, forged steel, bronze, etc. Pressure ratings are provided by the American National Standards Institute (ANSI) for threaded fittings and appear stamped on the body of each fitting produced. Fabrication of pipe is performed by die-cutting threads onto the pipe ends to mate with those of the fitting. See Fig. 2-2.

Flanging

Flanged joints are used in systems requiring frequent disassembly or to facilitate maintenance at equipment connections. Successful application of flanges relies upon achieving the correct mating surfaces of the flange faces and the seating element of the gasket into a leak-tight joint. Pipe mating the ends of flanges may be threaded or suitable for butt-welding, slip-on attachment, or socket-welding.

Utmost importance is placed upon gasket selection and uniformity of bolt tension. See Table 2-2. Gaskets must be selected for fluid compatibility, temperature, and pressure conditions. Bearing surfaces of nuts must have a smooth machine finish, and threads should be properly lubricated at assembly. Final tensioning of bolts or studs should be performed with a torque wrench. See Fig. 2-3.

Tubing and Tube Connectors

Tubing is in widespread use throughout the industrial market; its applications include fluid power, instrumentation, and general-service piping.

TABLE 2-2 Recommended Bolt Torques (Based upon Carbon Steel with 30×10^6 lb/in² Electric Modules)

Stud diameter, in	Threads per in	Torque, lb·ft
⅜	10	107
⅝	11	89
⅞	9	162
1	8	244
1⅛	8	322
1⅛	8	410
1⅜	8	510
1½	8	615

The use of smaller diameters and thinner wall sections allows tubing to be easily bent into a variety of configurations. This increased flexibility and workability often results in a cost savings by minimizing fitting requirements.

Tube fittings are primarily designed for use with extruded tubing (see Fig. 2-4) and are not used for standard pipe joining. Fittings are designed so that the interlocking threaded unions and ferrules maintain a pressure-tight seal. It is necessary to properly tension locking nuts, debur cut tube ends, and avoid crimped or scored tube surfaces.

STANDARDS AND SPECIFICATIONS

ASTM Ferrous Material Specifications

Pipe—Seamless for High-Temperature Service

A 106 Carbon Steel
A 335 Ferritic Alloy Steel
A 376 Austenitic Central Station Service
A 405 Ferritic Alloy, Special Heat-Treated

Pipe—Seamless and Welded

A 53 Carbon Steel
A 120 Black and Zinc-Coated (Ordinary Use)
A 312 Austenitic Stainless Steel
A 333 Carbon Steel (Low-Temperature Service)
A 530 General Requirements for Specialized Carbon- and Alloy-Steel Pipe

Pipe—Welded

A 134 Arc-Welded Steel Plate 16 in and over
A 135 Electric-Resistance-Welded Steel
A 139 Arc-Welded Steel 4 in and over
A 155 Arc-Welded Steel for High-Temperature Service
A 211 Spiral-Welded Steel or Iron
A 252 Specification for Welded and Seamless Pipe Piles
A 358 Arc-Welded Chrome-Nickel Alloy High-Temperature Service
A 523 Specification for Plain-End Seamless and Electric-Resistance-Welded Steel Pipe for High-Pressure Pipe-Type Cable Circuits
A 589 Specifications for Seamless and Welded Carbon Steel Water Well Pipe

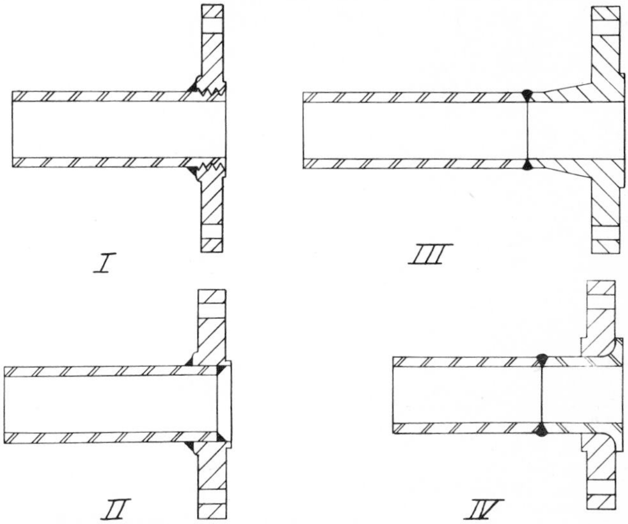

Figure 2-3 Standard flange-to-pipe configurations: I, screwed and back welded; II, slip-on welding flange; III, welding neck, butt-welded to pipe; IV, lap joint, stub-end butt-welded to pipe.

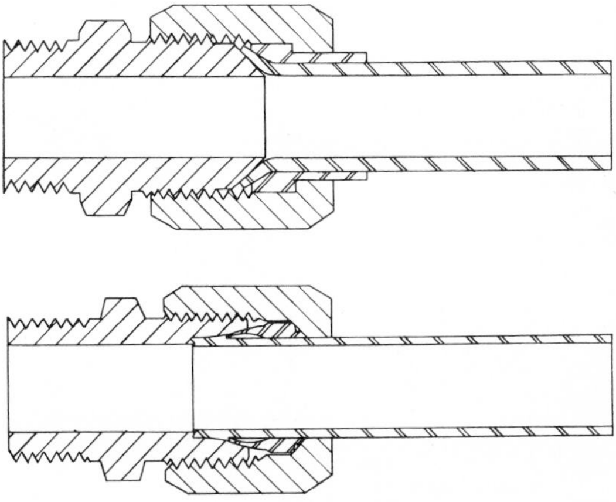

Figure 2-4 Compression fittings to join tubing to threaded unions.

TABLE 2-3 Dimensions of Long Radius, 90° Butt-Welding Elbows (Standard Weight: ANSI B16.9-1978, A 234)

Nominal pipe size*	OD	ID	Wall thickness	Center to face	Pipe sched. numbers	Approx wt, lb
2½	2.875	2.469	0.203	3¾	40	2.92
3	3.500	3.068	0.216	4½	40	4.58
3½	4.000	3.548	0.226	5¼	40	6.43
4	4.500	4.026	0.237	6	40	8.70
5	5.563	5.047	0.258	7½	40	14.7
6	6.625	6.065	0.280	9	40	22.9
8	8.625	7.981	0.322	12	40	46.0
10	10.750	10.020	0.365	15	40	81.5
12	12.750	12.000	0.375	18	St†	119
14	14.000	13.250	0.375	21	30	154
16	16.000	15.250	0.375	24	30	201
18	18.000	17.250	0.375	27	St†	256
20	20.000	19.250	0.375	30	20	317
22	22.000	21.250	0.375	33	St†	385
24	24.000	23.250	0.375	36	20	458
26	26.000	25.250	0.375	39	St†	539
30	30.000	29.250	0.375	45	St†	720
34	34.000	33.250	0.375	51	St†	926
36	36.000	35.250	0.375	54	St†	1040
42	42.000	41.250	0.375	63	St†	1420

*All dimensions are in inches except the last column.
†Standard weight.
Source: Tube Turns Division, National Cylinder Gas Co.

TABLE 2-4 Dimensions of Concentric and Eccentric Butt-Welding Reducers (Standard Weight: ANSI B16.9-1978, ASTM A 234)*

Nominal pipe size	Length	Approx wt, lb
2½ × 1	3½	1.30
2½ × 1¼	3½	1.47
2½ × 1½	3½	1.51
2½ × 2	3½	1.60
3 × 1¼	3½	1.70
3 × 1½	3½	1.89
3 × 2	3½	2.00
3 × 2½	3½	2.16
3½ × 1¼	4	2.35
3½ × 1½	4	2.52
3½ × 2	4	2.71
3½ × 2½	4	2.96
3½ × 3	4	3.05
4 × 1½	4	2.73
4 × 2	4	3.17
4 × 2½	4	3.34
4 × 3	4	3.50
4 × 3½	4	3.61
5 × 2	5	5.05
5 × 2½	5	5.52
5 × 3	5	5.73
5 × 3½	5	5.86
5 × 4	5	5.99
6 × 2½	5½	7.61
6 × 3	5½	8.00
6 × 3½	5½	8.14
6 × 4	5½	8.19
6 × 5	5½	8.65
8 × 3½	6	12.8
8 × 4	6	13.1

Nominal pipe size	Length	Approx wt, lb
8 × 6	6	13.4
8 × 6	6	13.9
10 × 4	7	21.1
10 × 5	7	21.8
10 × 6	7	22.3
10 × 8	7	23.2
12 × 5	8	30.5
12 × 6	8	31.1
12 × 8	8	32.1
12 × 10	8	33.4
14 × 6	13	55.8
14 × 8	13	57.2
14 × 10	13	60.4
14 × 12	13	63.4
16 × 8	14	70.2
16 × 10	14	72.9
16 × 12	14	75.6
16 × 14	14	77.5
18 × 10	15	86.9
18 × 12	15	89.2
18 × 14	15	90.9
18 × 16	15	94.0
20 × 12	20	134
20 × 14	20	135
20 × 16	20	138
20 × 18	20	142
22 × 14	20	148
22 × 16	20	151
22 × 18	20	154

Nominal pipe size	Length	Approx wt, lb	
22 × 20	20	157	
24 × 16	20	160	
24 × 18	20	163	
24 × 20	20	167	
26 × 18	24	200	
26 × 20	24	200	
26 × 22	24	200	
26 × 24	24	200	
30 × 20	24	220	
30 × 24	24	220	
30 × 26	24	220	
30 × 28	24	220	
		Conc.	Ecc.
34 × 24	24	270	229
34 × 26	24	270	237
34 × 30	24	270	253
34 × 32	24	270	261
36 × 24	24	340	237
36 × 26	24	340	245
36 × 30	24	340	261
36 × 32	24	340	269
36 × 34	24	340	277
42 × 24	24	260	
42 × 26	24	270	
42 × 30	24	285	
42 × 32	24	295	
42 × 34	24	300	
42 × 36	24	310	

*All pipe sizes and lengths are in inches
Source: Tube Turns Division, National Cylinder Gas Co.

TABLE 2-5 Dimensions and Weights of Steel Pipe—Plain End—by Iron Pipe Size Weight Class

Sizes, in		Standard (Std)		Extra strong (XS)		Double extra strong (XXS)	
Nom. ID*	OD	Wall, in	Wt per ft, lb	Wall, in	Wt per ft, lb	Wall, in	Wt per ft, lb
⅛	0.405	0.068	0.24	0.095	0.31		
¼	0.540	0.088	0.42	0.119	0.54		
⅜	0.675	0.091	0.57	0.126	0.74		
½	0.840	0.109	0.85	0.147	1.09	0.294	1.71
¾	1.050	0.113	1.13	0.154	1.47	0.308	2.44
1	1.315	0.133	1.68	0.179	2.17	0.358	3.66
1¼	1.660	0.140	2.27	0.191	3.00	0.382	5.21
1½	1.900	0.145	2.72	0.200	3.63	0.400	6.41
2	2.375	0.154	3.65	0.218	5.02	0.436	9.03
2½	2.875	0.203	5.79	0.276	7.66	0.552	13.69
3	3.500	0.216	7.58	0.300	10.25	0.600	18.58
3½	4.000	0.226	9.11	0.318	12.50	0.636	22.85
4	4.500	0.237	10.79	0.337	14.98	0.674	27.54
5	5.563	0.258	14.62	0.375	20.78	0.750	38.55
6	6.625	0.280	18.97	0.432	28.57	0.864	53.16
8	8.625	0.322	28.55	0.500	43.39	0.875	72.42
10	10.750	0.365	40.48	0.500	54.74	1.000	104.13
12	12.750	0.375	49.56	0.500	65.42	1.000	125.49
14	14.000	0.375	54.57	0.500	72.09		
16	16.000	0.375	62.58	0.500	82.77		
18	18.000	0.375	70.59	0.500	93.45		
20	20.000	0.375	78.60	0.500	104.13		
22	22.000	0.375	86.61	0.500	114.81		
24	24.000	0.375	94.62	0.500	125.49		
26	26.000	0.375	102.63	0.500	136.17		
28	28.000	0.375	110.64	0.500	146.85		
30	30.000	0.375	118.65	0.500	157.53		
32	32.000	0.375	126.66	0.500	168.21		
34	34.000	0.375	134.67	0.500	178.89		
36	36.000	0.375	142.68	0.500	189.57		

*Although pipe 2 in and larger will be identified by OD former nominal sizes are shown.
Source: "Wrought Steel and Wrought Iron Pipe," ANSI B36.10-1970.

TABLE 2-6 Dimensions of Long Radius 45° Butt-Welding Elbows (Standard Weight: ANSI B16.9-1978, ASTM A 234)

Nominal pipe size*	OD	ID	Wall thickness	Center to face	Radius	Pipe schedule numbers	Approx wt, lb
2½	2.875	2.469	0.203	1¾	3¾	40	1.64
3	3.500	3.068	0.216	2	4½	40	2.43
3½	4.000	3.548	0.226	2¼	5¼	40	3.29
4	4.500	4.026	0.237	2½	6	40	4.31
5	5.563	5.047	0.258	3⅛	7½	40	7.30
6	6.625	6.065	0.280	3¾	9	40	11.3
8	8.625	7.981	0.322	5	12	40	22.8
10	10.750	10.020	0.365	6¼	15	40	40.4
12	12.750	12.000	0.375	7½	18	St†	59.5
14	14.000	13.250	0.375	8¾	21	30	76.5
16	16.000	15.250	0.375	10	24	30	100
18	18.000	17.250	0.375	11¼	27	St†	128
20	20.000	19.250	0.375	12½	30	20	158
22	22.000	21.250	0.375	13½	33	St†	192
24	24.000	23.250	0.375	15	36	20	229
26	26.000	25.250	0.375	16	39	St†	269
30	30.000	29.250	0.375	18½	45	St†	358
34	34.000	33.250	0.375	21	51	St†	463
36	36.000	35.250	0.375	22¼	54	St†	518
42	42.000	41.250	0.375	26	63	St†	707

*All dimensions are in inches except the last column.
†Standard weight
Source: Tube Turns Division, National Cylinder Gas Co.

TABLE 2-7 ANSI Schedule Number Wall Thicknesses for Plain End Steel Pipe*

Sizes, in		ANSI Schedule number wall thicknesses									
Nom. ID	OD	10	20	30	40	60	80	100	120	140	160
⅛	0.405				0.68		0.095				
¼	0.540				0.88		0.119				
⅜	0.675				0.091		0.126				
½	0.840				0.109		0.147				0.188
¾	1.050				0.113		0.154				0.219
1	1.315				0.133		0.179				0.250
1¼	1.660				0.140		0.191				0.250
1½	1.900				0.145		0.200				0.281
2	2.375				0.154		0.218				0.344
2½	2.875				0.203		0.276				0.375
3	3.500				0.216		0.300				0.438
3½	4.000				0.226		0.318				
4	4.500				0.237		0.337		0.438		0.531
5	5.563				0.258		0.375		0.500		0.625
6	6.625				0.280		0.432		0.562		0.719
8	8.625		0.250	0.277	0.322	0.406	0.500	0.594	0.719	0.812	0.906
10	10.750		0.250	0.307	0.365	0.500	0.594	0.719	0.844	1.000	1.125
12	12.750		0.250	0.330	0.406	0.562	0.688	0.844	1.000	1.125	1.312
14	14.000	0.210	0.312	0.375		0.594	0.750	0.938	1.094	1.250	1.406
16	16.000	0.250	0.312	0.375	0.500	0.656	0.44	1.031	1.219	1.438	1.594
18	18.000	0.250	0.312	0.438	0.562	0.750	0.938	1.156	1.375	1.562	1.781
20	20.000	0.250	0.375	0.500	0.594	0.812	1.031	1.281	1.500	1.750	1.969
22	22.000	0.250	0.375	0.500		0.875	1.125	1.375	1.625	1.875	2.125
24	24.000	0.250	0.375	0.562	0.688	0.969	1.219	1.531	1.812	2.062	2.344
26	26.00	0.312	0.500								
28	28.000	0.312	0.500	0.625							
30	30.000	0.312	0.500	0.625							
32	32.000	0.312	0.500	0.625	0.688						
34	34.000	0.344	0.500	0.625	0.688						
36	36.000	0.312	0.500	0.625	0.750						

*The weight of steel pipe is expressed in pounds per foot. All weights of carbon and alloy steel pipe are calculated on the basis of a cubic inch of steel weighing 0.2833 lb. Two useful formulas relating to pipe weights are

$$\text{Outside diameter} \times \text{wall thickness} \times 10.68 = \text{pounds per foot}$$

$$\text{Pounds per foot} \times 2.64 = \text{tons per linear mile}$$

Source: "Wrought Steel and Wrought Iron Pipe," ANSI B36.10-1970.

TABLE 2-8 Pressure-Temperature Ratings of Pipe (Seamless Carbon Steel to ASTM A 53 Grade A)

Pipe size, in	Schedule number	Wall thickness, in	Working pressure, lb/in²	
			−20 to 650°F 12,000 lb/in² stress	750° 10,700 lb/in² stress
½	40	0.109	882	787
	80	0.147	1947	1736
1	40	0.133	960	856
	80	0.179	1777	1585
2	40	0.154	724	646
	80	0.218	1329	1185
3	40	0.216	875	780
	80	0.220	1422	1218
4	40	0.237	1146	1022
	80	0.337	1660	1481
6	40	0.280	914	815
	80	0.432	1435	1279
8	20	0.250	622	555
	40	0.322	806	718
	80	0.500	1270	1133
10	20	0.250	497	443
	40	0.365	729	650
	80	0.594	1208	1077
12	20	0.250	418	373
	40	0.406	683	609
	80	0.688	1178	1050

TABLE 2-9 Dimensions of American 150-lb Standard Malleable-Iron Threaded Fittings (Straight Sizes)*

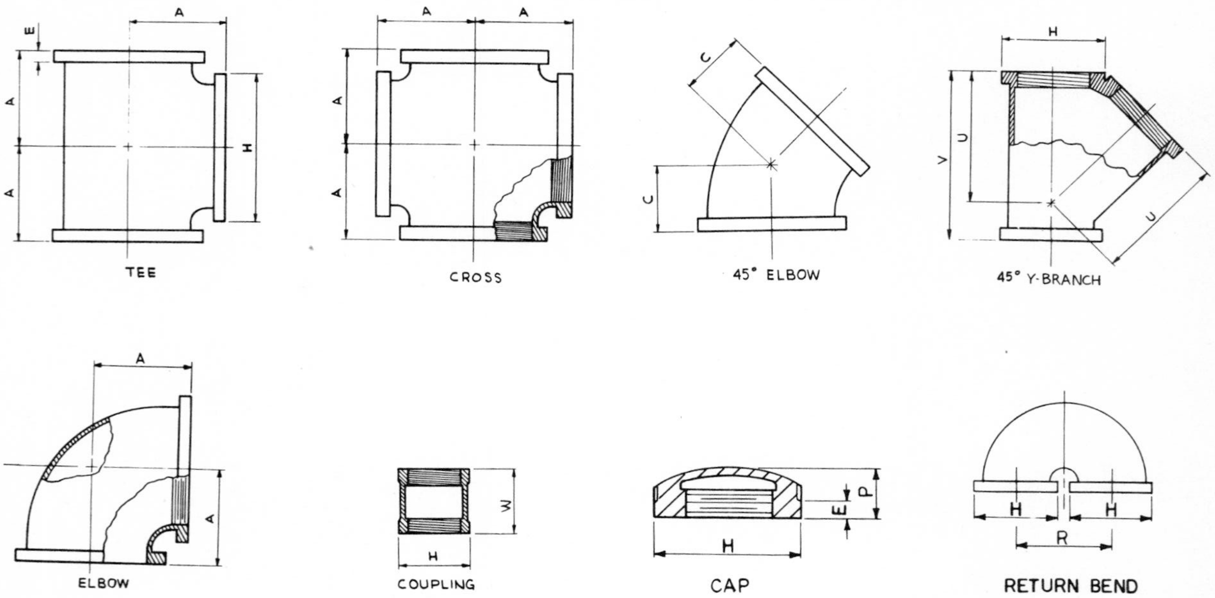

Size	Dimension								R		
	A	H	E	C	V	U	W	P	Closed	Medium	Open
⅛	0.69	0.693	0.200	—	—	—	—	0.96			
¼	0.81	0.844	0.215	0.73	—	—	1.06				
⅜	0.95	1.015	0.230	0.80	1.93	1.43	1.16				
½	1.12	1.197	0.249	0.88	2.32	1.71	1.34	0.87	1.000	1.25	1.50
¾	1.31	1.458	0.273	0.98	2.77	2.05	1.52	0.97	1.250	1.50	2.00
1	1.50	1.771	0.302	1.12	3.28	2.43	1.67	1.16	1.500	1.875	2.50
1¼	1.75	2.153	0.341	1.29	3.94	2.92	1.93	1.28	1.750	2.25	3.00
1½	1.94	2.427	0.368	1.43	4.38	3.28	2.15	1.33	2.188	2.50	3.50
2	2.25	2.963	0.422	1.68	5.17	3.93	2.53	1.45	2.625	3.00	4.00
2½	2.70	3.589	0.478	1.95	6.25	4.73	2.88	1.70	—	—	4.50
3	3.08	4.285	0.548	2.17	7.26	5.55	3.18	1.80	—	—	5.00
3½	3.42	4.843	0.604	2.39	—	—	3.43	1.90			
4	3.79	5.401	0.661	2.61	8.98	6.97	3.69	2.08			
5	4.50	6.583	0.780	3.05	—	—	—	2.32			
6	5.13	7.767	0.900	3.46	—	—	—	2.55			

*The complete standard (ANSI B16.3-1977) covers also reducing couplings, elbows, tees, crosses, and service or street elbows and tees. All dimensions in inches.

TABLE 2-10 Dimensions of Americn 125-and 250-lb Standard* Cast-Iron Threaded Fittings (Straight Sizes)(ANSI B16.4-1977)

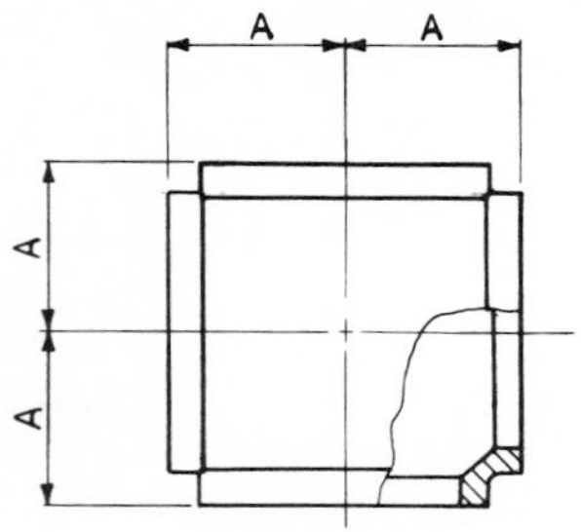

CROSS

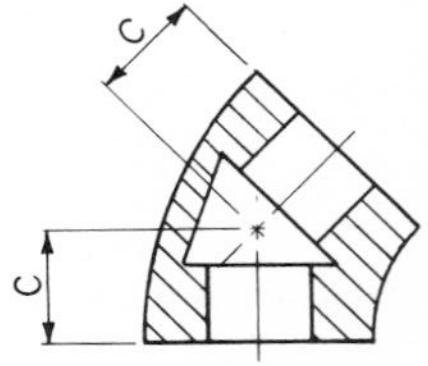

45° ELBOW

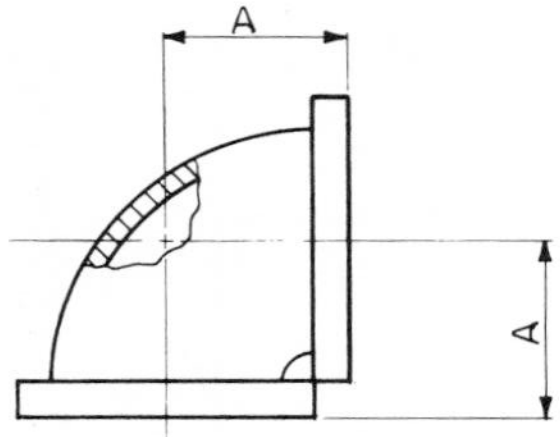

ELBOW

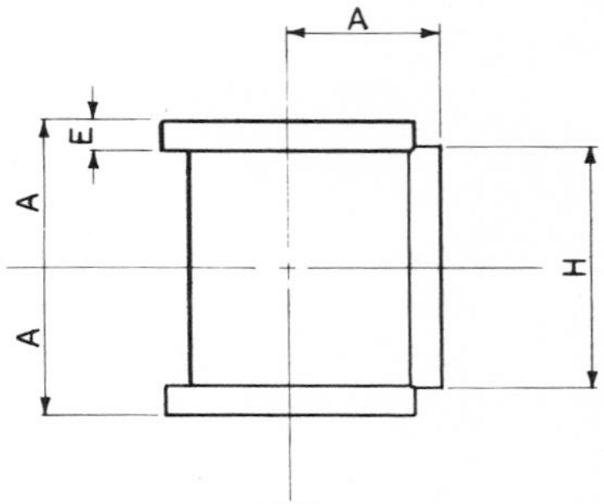

TEE

Size	125-lb fittings				250-lb fittings			
	A	H	E	C	A	H	E	C
¼	0.81	0.93	0.38	0.73	0.94	1.17	0.49	0.81
⅜	0.95	1.12	0.44	0.80	1.06	1.36	0.55	0.88
½	1.12	1.34	0.50	0.88	1.25	1.59	0.60	1.00
¾	1.13	1.63	0.56	0.98	1.44	1.88	0.68	1.13
1	1.50	1.95	0.62	1.12	1.63	2.24	0.76	1.31
1¼	1.75	2.39	0.69	1.29	1.94	2.73	0.88	1.50
1½	1.94	2.68	0.75	1.43	2.13	3.07	0.97	1.69
2	2.25	3.28	0.84	1.68	2.50	3.74	1.12	2.00
2½	2.70	3.86	0.94	1.95	2.94	4.60	1.30	2.25
3	3.08	4.62	1.00	2.17	3.38	5.36	1.40	2.50
3½	3.42	5.20	1.06	2.39	3.75	5.98	1.40	2.63
4	3.79	5.79	1.12	2.61	4.13	6.61	1.57	2.81
5	4.50	7.05	1.18	3.05	4.88	7.92	1.74	3.19
6	5.13	8.28	1.28	3.46	5.63	9.24	1.91	3.50
8	6.56	10.63	1.47	4.28	7.00	11.73	2.24	4.31
10	8.08	13.12	1.68	5.16	8.63	14.37	2.58	5.19
12	9.50	15.47	1.88	5.97	10.00	16.84	2.91	6.00

*The 125-lb standard covers also reducing elbows and tees. The 250-lb standard covers only the straight sizes.

All dimensions in inches.

†This applies to elbows and tee only.

Pipe—Forged and Bored High-Temperature Service

A 369	Ferritic Alloy Steel
A 430	Austenitic Stainless Steel

Pipe—Centrifugally Cast High-Temperature Service

A 426	Ferritic Alloys
A 451	Austenitic Alloys
A 452	Austenitic Cold-Wrought Alloys

Tube—Seamless

A 179	Low-Carbon Steel for Heat Exchanger and Condensers
A 192	Carbon Steel for Boilers at High-Temperature Service
A 199	Cold-Drawn Intermediate Alloy-Steel
A 210	Medium-Carbon Steel for Boilers
A 213	Ferritic and Austenitic Alloys for Boilers

Tube—Seamless and Welded

A 268	Ferritic Stainless Steel for General Use
A 269	Austenitic Stainless Steel Tubing for General Service
A 450	General Requirements for Carbon, Ferritic, and Austenitic Alloy

Tube—Welded

A 178	Arc-Welded Carbon Steel Boiler Tubes
A 214	Electric-Resistance-Welded Carbon Steel
A 226	Carbon Steel, High-Pressure Service
A 249	Austenitic Stainless Steel
A 254	Copper-Brazed Steel

Plates for Pressure Vessels

A 240	Stainless, Chromium and Chrome-Nickel
A 285	Low- and Intermediate-Temperature-Service Carbon Steel
A 299	Carbon-Manganese-Silicon Steel
A 387	Chromium-Molybdenum Alloy Steel
A 515	Carbon Steel for Intermediate- and High-Temperature Service
A 516	Carbon Steel for Moderate- and Low-Temperature Service

Welding Fittings—Factory-Made

A 234	Wrought Carbon and Ferritic Alloy Steel
A 403	Austenitic Steel

Castings

A 47	Malleable Iron
A 48	Gray Iron
A 126	Gray Iron for Valves, Flanges, and Pipe Fittings

A 197 Cupola, Malleable-Iron
A 389 Ferritic Alloy, Special Heat-Treated
A 395 Ductile Iron for Pressure Parts

ASTM Nonferrous Material Specifications

Pipe—Seamless

B 42 Copper (Standard Sizes)
B 43 Red Brass (Standard Sizes)
B 302 Threadless Copper

Pipe and Tube—Seamless

B 161 Nickel
B 165 Nickel-Copper Alloy
B 167 Nickel-Chromium-Iron Alloy
B 241 Aluminum Alloy Extruded
B 251 General Requirements for Wrought Copper and Copper Alloy
B 315 Copper Silicon Alloy
B 423 Nickel-Iron-Chromium-Molybdenum-Copper Alloy

Tube—Copper Seamless

B 68 Bright Annealed
B 75 For General Engineering Purpose
B 88 For General Plumbing Purpose
B 111 Condenser Tubes and Ferrule Stock
B 280 For Air-Conditioning and Refrigeration Field Service

Tube—Aluminum Seamless

B 210 Drawn
B 234 Drawn, for Condenser and Heat Exchangers

Tube—Welded

B 547 Aluminum-Alloy, Formed and Arc-Welded

Welding Fittings—Factory-Made

B 361 Wrought Aluminum and Aluminum-Alloy

American National Standards (ANSI)

A21.1-1977, Thickness Design of—Cast-Iron Pipe
A21.6-1975, Cast-Iron Pipe Centrifugally Cast in Metal Molds, for Water or Other Liquids
A21.10-1977, Gray-Iron and Ductile-Iron Fittings, 3 in through 48 in, for Water and other Liquids
A21.50-1976, Thickness Design of Ductile-Iron Pipe

A21.51-1976	Ductile-Iron Pipe, Centrifugally Cast in Metal Molds or Sand-Lined Molds for Water or Other Liquids
B1.1-1974	Unified Inch Screw Threads
B2.1-1968	Pipe Threads (except Dryseal)
B1.20.3-1976	Dryseal Pipe Threads (Inch)
B16.1-1975	Cast Iron Pipe Flanges and Flanged Fittings, Class 25, 125, 250, and 800
B16.3-1977	Malleable-Iron Threaded Fittings, Class 150 and 300
B16.4-1977	Cast Iron Threaded Fittings, Class 125 and 250
B16.5-1977	Steel Pipe Flanges and Flanged Fittings, Including Ratings for Class 150, 300, 400, 600, 900, 1500, and 2500
B16.9-1978	Factory-Made Wrought Steel Buttwelding Fittings
B16.11-1973	Forged Steel Fittings, Socket-Welding and Threaded
B16.14-1976	Ferrous Pipe Plugs, Bushings, and Locknuts with Pipe Threads
B16.15-1978	Cast Bronze Threaded Fittings, Class 125 and 150
B16.18-1978	Cast Bronze Alloy Solder-Joint Pressure Fittings
B16.22-1973	Wrought Copper and Bronze Solder-Joint Pressure Fittings
B16.24-1979	Bronze Pipe Flanges and Flanged Fittings, 150 and 300
B16.25-1979	Buttwelding Ends
B16.28-1978	Wrought Steel Buttwelding Short Radius Elbows and Returns
B16.34-1977	Steel Valves , Flanged and Butt-Welding End
B31.3-1976	Chemical Plant and Petroleum Refinery Piping
B31.4-1974	Liquid Petroleum Transportation Piping Systems
B31.8-1975	Gas Transmission and Distribution Piping Systems
B36.10-1979	Welded and Seamless Wrought Steel Pipe
B36.19-1976	Stainless Steel Pipe

MSS Standard Practices

SP-6	Finishes—On Flanges, Valves, and Fittings
SP-9	Spot-Facing for Bronze, Iron, and Steel Flanges
SP-25	Marking for Valves, Fittings, Flanges, and Unions
SP-43	Wrought Stainless Steel Butt-Welding Fittings
SP-45	Bypass and Drain Connection
SP-48	Steel Butt-Welding Fittings, 26 in and larger
SP-51	Corrosion-Resistant Cast Flanges and Flanged Fittings—150 lb
SP-58	Pipe Hangers and Supports—Materials and Design
SP-69	Pipe Hangers and Supports—Selection and Application
SP-75	High-Strength Wrought Welding Fittings
SP-79	Socket-Welding Reducer Inserts

API Specifications

5L	Line Pipe

ASME Codes

ASME Boiler and Pressure Vessel Code

AWWA Standards

C-101	Computation of Strength and Thickness of Cast Iron Pipe
C-106	Cast Iron Pipe Centrifugally Cast in Metal Molds
C-108	Cast Iron Pipe Centrifugally Cast in Sand-Lined Molds
C-110	Cast Iron Fittings, 2 through 48 in
C-111	Rubber-Gasketed Joint for Cast Iron Pipe and Fittings
C-112	2 and 2¼-in Cast Iron Pipe, Centrifugally Cast
C-150	Thickness Design of Ductile-Iron Pipe
C-151	Ductile-Iron Pipe, Centrifugally Cast in Metal or Sand-Lined Molds
C-201	Fabricated Electrically Welded Steel Pipe
C-208	Dimensions for Steel Water Pipe Fittings
C-600	Installation of Cast Iron Water Mains

Federal Specifications

WW-P-421	Pipe, Cast, Gray, and Ductile Iron, Pressure (for Water and Other Liquids)

SOURCES OF LISTED SPECIFICATIONS AND STANDARDS

API	American Petroleum Institute 1801 K Street, N.W. Washington, DC 20006
ANSI	American National Standards Institute 1430 Broadway New York, NY 10018
ASME	American Society of Mechanical Engineers, Inc. 345 East 47th Street New York, NY 10017
ASTM	American Society for Testing and Materials 1916 Race Street Philadelphia, PA 19103
AWS	American Welding Society, Inc. 2501 N.W. 7th Street Miami, FL 33125
AWWA	American Water Works Association 6666 W. Quincy Avenue Denver, CO 80235
MSS	Manufacturers Standardization Society of the Valve and Fittings Industry 1815 North Fort Myer Drive Arlington, VA 22209
PFI	Pipe Fabrication Institute 1326 Freeport Road Pittsburgh, PA 15238
SAE	Society of Automotive Engineers 400 Commonwealth Drive Warrendale, PA 15096

Federal Specifications

Superintendent of Documents
United States Government Printing Office
Washington, DC 20402

BIBLIOGRAPHY

King, R. C.: *Piping Handbook,* 5th ed., McGraw-Hill, New York, 1967.
Piping Design and Engineering, 3d ed., ITT Grinnell Co., Providence, R.I. 1971.
Rase, H. F.: *Piping Design for Process Plants,* Wiley, New York, 1963.
Welding Fittings and Forged Flanges, Crane Co., New York, 1967.

chapter 10-3

Plastics Piping

by
Stanley A. Mruk
Technical Director, Plastics Pipe Institute
A Divison of the Society of the Plastics Industry
New York, New York

INTRODUCTION

Outstanding chemical resistance is a principal reason for the growing use of plastics piping in practically every phase of U.S. industry including the pharmaceutical, chemical, food, paper and pulp, electronics, oil and gas production, water and waste treatment, mining, power generation, steel production and metal-refining industries. With the general recognition of its other features (e.g., it is easy to work with, durable, and economically advantageous), plastics piping has also become widely accepted for a broad range of applications for other reasons than just its chemical inertness. Current major uses include water mains and services, gas distribution, storm and sanitary sewers, plumbing (drain, waste, and vent piping and hot- and cold-water piping), electrical conduit, power and communications ducts, chilled-water piping, and well casing.

The diversity of plastic pipe—reflecting the many available materials, wall constructions, diameters, and techniques for joining—is so great and the attendant technology is evolving at such a rate that it is difficult to present in a concise reference all the pertinent information available. This chapter, therefore, offers only basic and general information. In applying plastic pipe, the reader is advised to ensure that all appropriate code requirements and government safety regulations are complied with. Sources and references for additional information are provided for this purpose at the end of the chapter.

DESCRIPTION, CLASSIFICATION, AND TYPICAL USES

"Plastic" pipe is as indefinite a term as "metal" pipe. As with metal products, plastic pipes are made from a variety of materials. Plastics used for pipe exhibit a wide range of properties and characteristics. The variabilities in properties of plastics are derived not only from the chemical composition of the basic synthetic resin, or polymer, but are also largely determined by the kind and amount of additives, the nature of reinforcement, and the process of manufacture. For example, it is possible to formulate mixtures of polyvinyl chloride (PVC) resins plus appropriate additives that range from a clear, soft, and pliable product (such as that used for laboratory tubing and upholstery) to a rigid and strong product (such as for pressure pipe). Another example is the construction of reinforced thermosetting resin pipe (RTRP) by using glass-fiber winding techniques that can adjust to the desired value the ratio of the circumferential (or hoop) strength to the axial strength.

Plastics are divided into two basic groups, thermosetting and thermoplastic, both of which are used in the manufacture of plastic pipe. Thermoplastics, as the name implies, soften upon the application of heat and reharden upon cooling: they can be formed and reformed repeatedly. This characteristic permits them to be easily extruded or molded into a wide variety of useful shapes, including pipe and fittings. Because thermoplastics are shaped, by a die or mold, while in a "molten" state their properties are essentially isotropic (i.e., independent of direction). In some processes and under some conditions, some anisotropy (i.e., direction dependence) may result.

Thermosetting plastics, on the other hand, form permanent shapes when cured by the application of heat or a "curing" chemical. Once shaped and cured during the manufacturing process, they cannot be reformed by heating. The excellent adhesion properties of thermosetting resins permits their utilization in composite structures by which strength and stiffness can be greatly enhanced through the use of reinforcements and fillers. The greater strength and higher temperature limits of these composite structures permit RTRP to handle fluids at temperatures and pressures beyond the limits for thermoplastic pipe. Piping systems now available are capable of operating at temperatures in excess of 300°F (148°C). All important commercial constructions of thermosetting pipe utilize some form of reinforcement and/or filler. By orienting the reinforcement, RTR pipes can be given directionally dependent properties.

Thermoplastic Pipe

Most thermoplastic pipes and fittings are made from materials containing no reinforcements, although fillers are occasionally used. Pipe is manufactured by the extrusion process, whereby molten material is continuously forced through a die that shapes the product. After being formed by the die, the soft pipe is simultaneously sized and hardened by cooling it with water. Fittings and valves are usually produced by the injection molding process, in which molten plastic is forced under pressure into a closed metal mold.

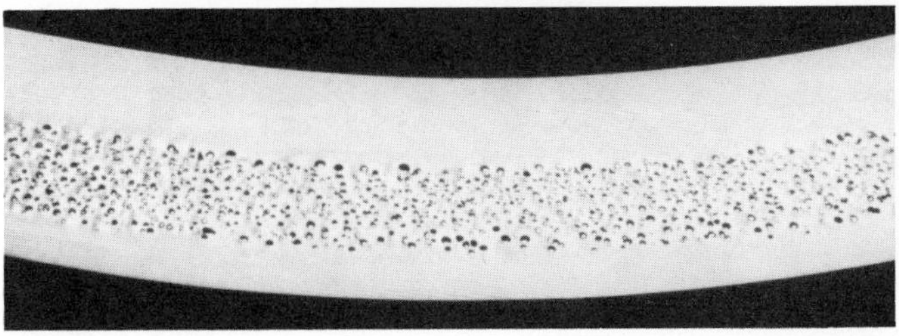

Figure 3-1 ABS foam-core construction. (Borg-Warner Chemicals.)

After cooling, the mold is opened and the finished part is removed. Some items, especially larger-sized fittings for which there is insufficient demand to justify construction of injection-molding tooling, are fabricated from pipe sections, or sheets, by utilizing thermal or solvent cementing fusion techniques. To compensate for the lower strength, the fitting may either be made from a heavier wall stock or reinforced with a fiberglass-resin overwrap. The engineer designing a pressure-rated system should make sure that the pressure ratings of the selected fittings are adequate.

There is some thermoplastic pipe made of a foam-core construction (for example, ASTM* F 628) in which the pipe wall consists of thin inner and outer solid skins sandwiching a high-density foam (Fig. 3-1). The primary benefit of such construction is improved ring and longitudinal (beam) stiffness in relation to the material used. Because the foam-wall structure results in some loss of strength, applications for foam-core pipe are in nonpressure uses, such as for above- and below-ground drainage piping, which can take advantage of the more material-efficient ring and beam stiffness.

For buried nonpressure applications, a composite pipe (ASTM D 2680) is produced that consists of two concentric tubes that are integrally braced with a truss webbing. The resultant openings between the concentric tubes are filled with a lightweight concrete. This construction increases both the ring and the beam stiffness. Composite pipe is used only for nonpressure buried applications such as sewerage and drainage.

Several other processes for improving the radial (i.e., ring) stiffness of thermoplastic pipe for buried applications have in common the formation of some type of rib reinforcement. A well-established technique is forming corrugations in the pipe wall. Corrugated polyethylene pipe (ASTM F 405) in sizes from 2 to 12 in (5 to 30 cm) is widely used for building foundations, land, highway, and agricultural drainage, and communications ducts. Ribbed pipe also is commercially made by the continuous spiral winding of the plastic over a mandrel of a specially shaped profile. Adjacent layers of this profile are fused to each other to form a cylinder that is smooth on the inside and has ribbed reinforcements on the outside (Fig. 3-2). The smooth inside diameter is preferable for many applications, such as sewerage, because it creates no flow disturbances. Pipes with ribbed construction are available in PVC and polyethylene (PE). PE pipes, which are made with hollow ribs to minimize material usage, are available in sizes from 18 to 96 in (45 cm to 1.8 m) in diameter.

The distinctive characteristics of the principal thermoplastic piping materials are discussed in the following pages.

Polyvinyl Chloride

Polyvinyl chloride (PVC) piping is made only from compounds containing no plasticizers and minimal quantities of other ingredients. To differentiate these materials from flexible, or plasticized PVCs (from which are made such items as upholstery, luggage, and laboratory tubing) they have been labeled rigid PVCs in the United States and unplasticized PVC (RuPVC) in Europe. Rigid PVCs used in piping range from type I to type III, as identified by an older classification system that is still much in use. In this system,

*American Society for Testing and Materials.

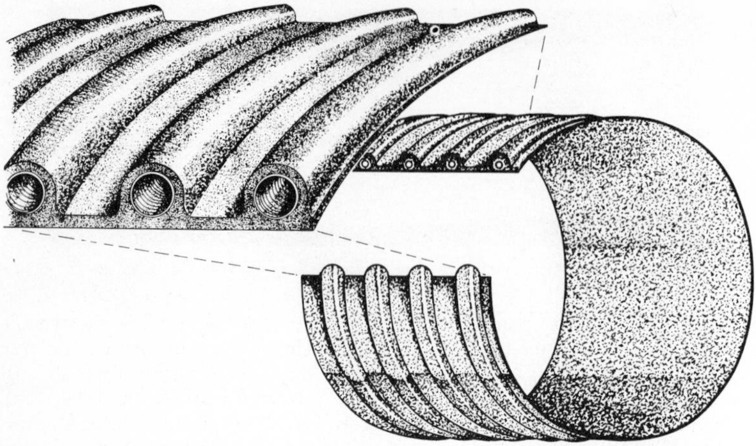

Figure 3-2 Typical profile of hollow ribbed polyethylene pipe. (*Spiral Engineered Systems.*)

the type designations are supplemented by grade designations (e.g., grade 1 or 2) which further define the material's properties. Type I materials, from which most pressure and nonpressure pipe is made, have been formulated to provide optimum strength as well as chemical and temperature resistance. Type II materials are those formulated with modifiers that improve impact strength but that also somewhat reduce, depending on modifier type and quantity, the aforementioned properties of Type I materials. There is little call for Type II pipe, as the impact strength of the stronger Type I pipe is more than adequate for most uses. Type III materials contain some inert fillers which tend to increase stiffness concomitant with some lowering of both tensile and impact strength and chemical resistance. Some nonpressure PVC piping, such as that used for conduit, sewerage, and drainage, is made from Type III PVCs.

The currently used classification system for rigid PVC materials for piping and other applications is described in ASTM D 1784, "Standard Specification for Rigid Polyvinyl Chloride and Chlorinated Polyvinyl Chloride Compounds." This specification categorizes rigid PVC materials by numbered cells that designate value ranges for the following properties: impact resistance (toughness), tensile strength, modulus of elasticity (rigidity), deflection temperature (temperature resistance), and chemical resistance. The following table cross-references the designations of the principal PVC materials from the older to the newer classification system.

By cell classification system of ASTM D1784 minimum cell class	*By older system*	
	Type and grade	*Designation*
12454-B	Type I, Grade 1	PVC 11
12454-C	Type I, Grade 2	PVC 12
14333-D	Type II, Grade 1	PVC 21
13233	Type III, Grade 1	PVC 31

Because (as expanded in the discussion on properties) short-term properties of plastic materials are not a reliable predictor of long-term capabilities, those PVC materials that have been formulated for long-term pressure applications are also designated by their categorized maximum recommended hydrostatic design stress (RHDS) for water at 73.4°F (23°C) as determined from long-term pressure testing. The most commonly used designation system for PVC pressure-piping materials is based on the above older designation system with two added digits that identify, in hundreds of pounds per square inch, the maximum recommended design stress.* For example: PVC 1120 is a type I, grade 1 PVC (minimum cell class 12454-B) with a maximum recommended HDS of 2000 lb/in^2 (13.8 MPa) for water at 73.4°F (23°C); PVC 2110 is a type 2, grade 1 PVC (minimum cell class 14333-D) with an RHDS of 1000 lb/in^2 (6.9 MPa). Other PVC material designations available for pressure pipe are listed later in Table 3-4. Most pressure-rated PVC pipe is made from PVC 1120 materials.

The combination of good long-term strength with higher stiffness explains why PVC has become the principal plastic pipe material for both pressure and nonpressure applications. Major uses include: water mains; water services; irrigation; drain, waste, and vent (DWV) pipes; sewerage and drainage; well casing; electric conduit; and power and communications ducts. A much broader range of fittings, valves, and appurtenances of all types is available in PVC than in any other plastic.

Chlorinated Polyvinyl Chloride

As implied by its name, chlorinated polyvinyl chloride (CPVC) is a chemical modification of PVC. CPVC has properties very similar to PVC but the extra chlorine in its structure extends its temperature limitation by about 50°F (28°C), to nearly 200°F (93°C) for pressure uses and about 210°F (99°C) for nonpressure applications. ASTM D 1784, the rigid PVC materials specification, also covers CPVC which it classifies as Class 23477-B. By the older designation system, it is known as type IV, grade 1 PVC. CPVCs for pressure pipe are designated CPVC 4120 (i.e., type IV, grade 1 CPVC with a maximum recommended hydrostatic design stress of 2000 lb/in^2 (13.8 MPa) for water at 73.4°F (23°C). At 180°F (82°C) the maximum recommended hydrostatic design stress* for CPVC is 500 lb/in^2 (3.4 MPa).

Principal applications for CPVC are for hot or cold water piping and for many industrial uses which take advantage of its higher temperature capabilities and superior chemical resistance.

Polyethylene

Polyethylene (PE) is the best-known member of the polyolefin group—plastics that are formed by the polymerization of straight-chain hydrocarbons—known as olefins—that include ethylene, propylene, and butylene. Polyethylene plastics are tough and flexible even at subfreezing temperatures. They are generally formulated with only an antioxidant (for protection during processing) and some pigment (usually carbon black) or other agent designed to screen out ultraviolet radiation in sunlight which over long-term exposure could be damaging to the natural-color polymer.

ASTM D 1248, the PE molding and extrusion materials specification, classifies these materials into three types depending on the density of the natural resin. Type I consists of lower-density materials which are relatively soft and flexible and have low heat resistance. Type II PEs are of medium density, slightly harder, more rigid and more resistant to elevated temperatures; they also have better tensile strength. Type III materials show maximum hardness, rigidity, tensile strength, and resistance to the effects of increasing temperature. Pipe is made almost exclusively from type II and type III PEs. ASTM D 1248 also provides for grade designations to further classify PEs according to other physical characteristics. PE piping materials for pressure piping are classified by a designation system that combines the type and grade coding with that for the maximum RHDS* for water at 73.4°F (23°C). PEs utilized for pressure piping, and so designated, are listed in Table 3-4 (p. 10-49).

The more recently issued standard, ASTM D 3350, classifies PE piping materials according to broader physical property criteria, including long-term strength, and it is expected to become the primary PE piping material standard.

*Since the maximum recommended design stress is for continuous water pressure at 73°F, it is up to the designer to determine the extent, if any, by which this stress should be reduced to account for any departure from these conditions and the need for a suitable margin of safety against other considerations. See the discussion on design.

Outstanding toughness and relatively low flexural modulus, which permits coiling of smaller-diameter pipe, are large factors in PE's prominence in gas-distribution and water-service piping. Other features, such as heat fusibility, good abrasion resistance, and availability in large diameters [up to 96 in (2.5 m)], account for the use of PE piping for chemical transfer lines, slurry transport, sewage force mains, intake and outfall lines, power ducts, and renewal (by insertion into the old pipe) of deteriorated sewer, gas, water, and other pipes.

Polybutylene

Polybutylene (PB) is a unique polyolefin. Its stiffness resembles that of low-density PE, yet its strength is higher than that of high-density PE. However, its most significant feature is that it retains its strength better with increasing temperature. Its upper temperature limit is higher than that for any PE: nearly 200°F (93°C) for pressure uses and somewhat higher for nonpressure applications.

PB piping materials are covered by ASTM D 2581. Materials for pressure applications are designated PB 2110; the last two digits signify a maximum recommended hydrostatic design stress* of 1000 lb/in^2 (6.9 MPa) for water at 73.4°F (23°C). At 180°F (83°C) this design stress is 500 lb/in^2 (3.4 MPa).

Major applications of PB pipe tend to take advantage of its improved temperature resistance and its toughness. They include hot or cold water piping and industrial uses such as hot effluent lines. Because of its excellent abrasion resistance, PB pipe is also used for slurry lines.

Polypropylene

Polypropylene (PP) is a polyolefin similar in properties to Type III PE but is slightly lighter in weight, more rigid, and more temperature-resistant. PPs are classified by ASTM D 2146 into two types: Type I covers homopolymers that generally have the greatest rigidity and strength but offer only moderate impact resistance; Type II covers copolymers of propylene and ethylene, or other olefins, which are less rigid and strong but have much improved toughness, particularly at lower temperatures. As indicated by Table 3-4 there are some PPs classified for use in pressure pipe.

Although on the basis of its short-term properties PP shows better resistance to temperature, this material is inherently somewhat more sensitive to thermal aging than PE. To overcome this sensitivity, specially formulated grades containing appropriate heat-stabilizer systems have been developed. For pressure uses, the adequacy of long-term heat stability is evaluated by means of long-term pressure tests of the PP formulation, conducted at various higher temperatures. A PP material containing flame-retardant additives is available for drainage-piping applications.

Polypropylene piping is used for chemical waste and drainage and various other industrial uses that take advantage of excellent chemical resistance, good rigidity and strength, and higher-temperature operating limits. One feature that accounts for its use in laboratory and industrial drainage is its superior resistance to many organic solvents.

Acrylonitrile-Butadiene-Styrene

Acrylonitrile-butadiene-styrene (ABS) is a family of materials formed from three different monomers (chemical building blocks), acrylonitrile, butadiene, and styrene. The proportions of the components and the way in which they are combined can be varied to produce a wide range of properties. Acrylonitrile contributes rigidity, strength, hardness, chemical resistance, and heat resistance; butadiene contributes toughness; and styrene contributes gloss, rigidity, and ease of processing.

*Since the maximum recommended design stress is for continuous water pressure at 73°F, it is up to the designer to determine the extent, if any, by which this stress should be reduced to account for any departure from these conditions and the need for a suitable margin of safety against other considerations. See the discussion on design.

ASTM D 1788 classifies ABS plastics into numbered cells that designate value ranges for each of three properties: impact strength (toughness), tensile stress at yield (strength), and deflection temperature under load. ABS pipe materials are categorized into types and grades in accordance with established minimum cell requirements for each type and grade. Like the other major thermoplastics, ABS materials for pressure pipe are designated by a coding that identifies both short-term properties and long-term strength. For example, ABS 1316 is a type I, grade 3 (minimum cell classification 3-5-5 per D 1788) material with a maximum RHDS* of 1600 lb/in^2 (11.2 MPa) for water at 73.4°F (23°C). Other ABS pressure-pipe materials are listed in Table 3-4 (p. 10-49).

An advantageous combination of toughness with good strength and stiffness largely accounts for the most common uses for ABS piping, i.e., for drain, waste, and vent (DWV) applications as well as for sewers, well casings, and communications ducts. One especially tough formulation of ABS is utilized to manufacture piping for compressed-air service. Most other thermoplastic materials are not recommended for above-ground compressed-air service because, should they fail, the pipe failure mechanism would sometimes produce flying fragments which could be injurious.

Polyvinylidine Fluoride

Polyvinylidine fluoride (PVDF) is a fluoroplastic, i.e., its chemical composition includes the element fluorine. Fluoroplastics are distinguished by their exceptional chemical and solvent resistance, excellent durability, and broad working-temperature range. To these features PVDF adds good strength and toughness and radiation resistance (which explains its use for conveying radioactive materials). The ASTM Specification covering PVDF molding and extrusion materials is D 3222. A standard for PVDF pressure piping is under development at ASTM. Standard recommended hydrostatic design stresses for PVDF have not yet been established, and an effort to develop these values is underway. Manufacturers of PVDF piping systems should be consulted for recommended design-stress values.

PVDF piping is utilized primarily for chemical processing and other industrial applications that are beyond the reach of the more common thermoplastics. These include the handling of active materials such as chlorine and bromine and piping materials that subject it to higher temperatures. Because of its immunity to radiation, PVDF piping is also used in reprocessing nuclear wastes.

Other Thermoplastics

Some thermoplastics of lesser current commercial importance are either used for special reasons or were more popular some time ago. Among these is *chlorinated polyether,* a tough rigid material of outstanding chemical resistance and useful to about 220°F (104°C). PVDF has largely displaced it as a specialty material. *Cellulose acetate butyrate* (CAB) was extensively used in petroleum production for conveying saltwater and crude oil as well as for natural gas distribution. However, because of its relatively low strength and moderate resistance to temperature and chemicals, it has been largely displaced by other materials. Nylon, a tough, strong, and heat-resistant material that is also very resistant to aromatic and chlorinated solvents, has some special uses. One of these is for coiled small-diameter tubing for compressed-air service. Nylon insert (with barbs) fittings are also made for use with polyethylene pipe.

Polyacetal, a strong, hard plastic with relatively good temperature resistance, is being increasingly used for various hot- or cold-water plumbing components such as fittings, valves, and faucets.

*Since the maximum recommended design stress is for continuous water pressure at 73°F, it is up to the designer to determine the extent, if any, by which this stress should be reduced to account for any departure from these conditions and the need for a suitable margin of safety against other considerations. See the discussion on design.

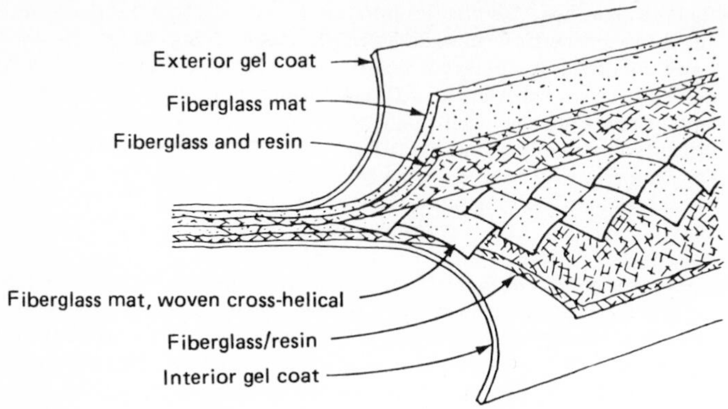

Figure 3-3 Fiberglass-reinforced pipe.

Reinforced Thermosetting Resin Pipe

Reinforced thermosetting resin pipe (RTRP) is a composite largely consisting of a reinforcement imbedded in, or surrounded by, cured thermosetting resin. Included in its composition may be granular or platelet fillers, thixotropic agents, pigments, or dyes. The most frequently used reinforcement is fiberglass, in any one or a combination of the following forms: continuous filament, chopped fibers, and mats (Fig. 3-3). While reinforcements such as asbestos or other mineral fibers are sometimes used, fiberglass-reinforced pipe (FRP) is by far the most popular. One form of FRP, called reinforced plastic mortar pipe (RPMP), consists of a composite of layers of thermosetting resin–sand aggregate mixtures that are sandwiched by layers of resin-fiberglass reinforcements. In another construction, the sand is replaced by glass microspheres. The high content of reinforcements in RTRP, which may run from 25 to 75 percent of the total pipe weight, and the specific design of the composite wall construction are the major determinants of the ultimate mechanical properties of the pipe. The resin, although also influencing these properties somewhat, is the binder that holds the composite structure together, and it supplies the basic source of temperature and chemical resistance. Glass fibers, as well as many other reinforcements, do not have high resistance to chemical attack. For enhanced chemical and/or abrasion resistance, RTRP construction may include a liner consisting of plastic (thermosetting or thermoplastic), ceramic, or other material. The outer surface of the pipe—especially that of the larger diameter sizes—may also be made "resin rich" to better resist weathering, handling, and spills. Reinforced thermosetting resin pipe is available in a variety of resins, wall constructions, and liners with diameters ranging from 1 in (2.5 cm) to more than 16 ft (5 m). Stock and specially fabricated fittings are readily available.

Filament Winding

Pipe is produced by machine-winding, under controlled tension and in predetermined patterns, of glass reinforcement—which may consist of continuous-filament strands, woven roving, or roving tape—onto the outside of a mandrel. The reinforcement may be saturated with a liquid resin or pre-impregnated with partially cured resin. After the pipe has been formed and cured, the mandrel is removed. The dimension of the inside diameter is set by the mandrel; that of the outside diameter by the thickness of the wall.

A high glass content and a precise machine-controlled fiber orientation which permits control of the ratio of circumferential to axial strength makes this pipe more economically suited for certain uses including higher-pressure applications. Since for a given pressure rating the axial strength of filament-wound pipe tends to be lower than that for pressure-molded pipe, it may have to be offset by more frequent support spacing. The use of automatic machines results in pipes of closer dimensional tolerance and in mechanical properties that are very consistent. To provide a stronger barrier against corrosion of the fiberglass filaments, a resin-rich liner, 0.02 to 0.1 in (0.5 to 2.5 mm) thick, is usually deposited on the inside of the pipe. In one process a thick, highly chemically resistant liner is produced by overwrapping a PVC, CPVC, or PVDF pipe (the only materials used at this time) with resins that can bond tightly to the thermoplastic.

The matched bell-and-spigot socket and the butt-and-strap methods are those primarily used to join filament-wound pipe.

Centrifugal Casting

In this process, resin and reinforcements (such as chopped fibers) are applied to the inside of a rotating mold. After the pipe has cured by action of heat or a catalyst, it is removed from the mold. The outside diameter of the pipe is set by the mold dimensions: the inside diameter is determined by the amount of material introduced into the mold. This type of pipe is almost fully machine-made and it provides very consistent mechanical properties and the closest tolerances. Because its glass fiber content is lower than that of filament-wound pipe it offers generally better chemical resistance. For additional resistance it may be made with a corrosion-resistant liner. The strength properties of centrifugally cast pipe are less direction-dependent than those of filament-wound pipe. Fittings are normally of the socket type for bell and spigot joining. The joint may be overwrapped if added strength is desired.

The smaller-diameter fittings for RTRP are usually produced by a molding (compression or injection) process that is similar to that used with thermoplastics. Fittings may also be made by the filament-winding or contact-molding process whereby the part is first fabricated on a steel or fiberglass form, then cured and removed. Centrifugal casting is also sometimes employed. By all these procedures (except filament winding) random orientation of the reinforcement is obtained.

Resins Used for Piping

Brief descriptions of the more important resins used for piping are given in the following text.

Epoxies

Epoxy resins are strong and have good resistance to solvents, salts, caustics, and dilute acids. Epoxies are cross-linked by curing agents which become an integral part of the polymer and affect the thermal, chemical, and physical properties of the polymer. For instance, the maximum service temperature of epoxy pressure pipe cured with anhydrides is 180°F (83°C), and it has little resistance to caustics; that cured with aromatic amines can be used at temperatures above 225°F (107°C), and it has good caustic resistance.

The major use of epoxy pipe is in oil fields where its resistance to corrosion and paraffin buildup makes it preferable to steel pipe for crude collection and saltwater injection lines. Other uses are in the chemical process industry, in heating and air conditioning, in food processing, for gasoline and solvents, and in mining applications (including abrasive slurry transport, communications ducts, and power conduits).

Polyesters

Although typically not as strong as the epoxies, polyesters offer good resistance to mineral acids, bleaching solutions, and salts. The most commonly used polyester resins for pipe are isophthalic polyesters and bisphenol A fumarate polyesters. Isophthalics have poorer resistance to caustics and oxidizers. Bisphenol A fumarates have improved resistance to these materials and are widely used in paper mills for bleach lines.

Isophthalic resin pipe is used in waste-treatment and power plants in services where corrosive conditions are not severe. Maximum operating-temperature limits for pressure pipe vary, depending upon the specific material, but are generally below 200°F (93°C).

Vinyl Esters

These resins include chemical features of both epoxies and polyesters. Vinyl ester resins offer better chemical resistance, somewhat higher temperature limits, and better solvent resistance than ordinary polyesters but generally do not compare to epoxies in these properties. Vinyl ester resins are preferred over polyesters because they are more chemical-resistant than the isophthalics and less brittle than the bisphenol A fumarates. Typical services are in fertilizer plants (acid lines), chlorine plants (chlorine-saturated brine lines), and paper mills (caustic and black-liquor lines).

Furans

Furan resins offer very good chemical, solvent, and temperature resistance—up to about 300°F (150°C). Because they extend the limitations of the other resins, they are often selected for use in the processing industries in place of exotic metal piping.

Desirable Qualities of FRP Resins

The performance criteria that usually determine the choice of one or more FRP resins over another include chemical resistance, mechanical strength, heat resistance, and (for the manufacturer) processability. A qualitative summary of FRP resin performance is presented in Table 3-1. For the final choice of the appropriate FRP resin(s) and liner combination of a given service (chemical environment, weathering exposure, abrasion resistance, etc.) more detailed information, including case history data, should be obtained.

PROPERTIES

Compared with other materials, pipes made of plastic have less strength and rigidity and are more temperature-sensitive. However, they offer these essential properties sufficiently to satisfy the performance requirements of most industrial piping applications. Moreover, they have good to excellent strength-to-weight ratios; are durable and easy to install and maintain; and have outstanding chemical resistance. The thermoplastics are lower than thermosets in strength, rigidity, and maximum operating temperature. However, their chemical resistance tends to be superior. Thermosets, in contrast, are capable of handling corrosive fluids at pressures and temperatures well beyond the service limits of most thermoplastic pipe.

Physical and Mechanical

Typical physical, mechanical, and thermal properties of the major thermoplastic piping materials are presented in Table 3-2. Those for thermosets are given in Table 3-3. The actual values for any pipe will vary according to the specific material(s) used and, in the case of composite products, such as the thermosets, will also depend on the specific wall construction.

Because the properties of plastics are influenced by time of loading, temperature, and environment, data-sheet values for mechanical properties such as those presented in Tables 3-2 and 3-3 are not satisfactory for design purposes. The stress-strain, and strain-stress, responses of plastics reflect their viscoelastic nature. The viscous, or fluidlike, component tends to damp or slow down the response between strain and stress. For example, if a load is continuously applied on a plastic material, it creates an instantaneous initial deformation that then increases at a decreasing rate. This further deformation response is known as *creep.* If the load is removed at any time, there is a partial immediate recovery of the deformation followed by a gradual creep recovery. If on the other hand the plastic is deformed (i.e., strained) to a given value that is then maintained, the initial load (stress) created by the deformation slowly decreases at a decreasing rate. This is known as the *stress-relaxation response.* The ratio of the actual values

TABLE 3-1 Qualitative Summary of FRP Resin Performance*

Resin	Chemical resistance			Other properties		
	Acids	Bases	Solvents	Processability	Strength	Heat resistance
Polyester resins	Fair–good	Poor	Fair	Good	Fair–good	Fair–good
Isophthalic acid based	——	O	+	+	+	——
Het acid based	——	O	+	+	+	+
BPA/fumarates	+	——	——	+	+	——
Vinyl ester terminated polyesters	——	——	——	+	+	+
Vinyl ester resins	Good	Fair–good	Fair–good	Good	V. good	Good–v. good
BPA/ECH epoxy derived						
$n = 0$	+	——	+	+	++	+
$n = 2$	+	+	——	+	++	——
Phenolic-Novolac epoxy derived	+	——	+	——	+	+
Epoxy resins	Fair	Good	V. good	Fair	V. good	Good–v. good
Aliphatic amine cured	——	——	+	——	+	+
Aromatic amine cured	——	+	++	——	++	++
Anhydride cured	——	O	+	+	+	+
Lewis acid cured	+	+	+	——	+	++
Furan resins	Fair	Good	V. good	Poor	Good	V. good
Furfuryl alcohol derived	——	+	++	O	+	++

*++ = Very Good, + = Good, —— = Fair, and O = Poor.

Source: Based on table from M. B. Launikitis, "Chemically Resistant FRP Resins," *Proceedings of 1977 Plastics Seminar*, National Association of Corrosion Engineers, Houston, Texas.

TABLE 3-2 Typical Physical Properties of Major Thermoplastic Piping Materials*†

	ASTM	ABS		PVC			PE				
	Test no.	I	II	I	II	CPVC	II	III	PB	PP	PVDF
Specific gravity	D 792	1.04	1.08	1.40	1.36	1.54	0.94	0.95	0.92	0.92	1.76
Tensile strength, lb/in² (× 10³)	D 638	4.5	7.0	8.0	7.0	8.0	2.4	3.2	4.2	5.0	7.0
Tensile modulus, lb/in² (× 10⁶)	D 638	0.3	0.3	0.41	0.36	0.42	0.12	0.13	0.06	0.2	0.22
Impact strength, Izod, ft·lb/in notch	D 256	6	4	1	6	1.5	>10	>10	>10	2	3.8
Coeff. of linear expansion, in/(in)(°F)(× 10⁻⁵)	D 696	5.5	6.0	3.0	5.0	3.5	9.0	9.0	7.2	4.3	7.0
Thermal conductivity, (Btu)(in)/(h)(ft²)(°F)	C 177	1.35	1.35	1.1	1.3	1.0	2.9	3.2	1.5	1.2	1.5
Specific heat, Btu/(lb)(°F)	——	0.32	0.34	0.25	0.23	0.20	0.54	0.55	0.45	0.45	0.29
Approx operating limit‡											
°F, nonpressure	——	180	180	150	130	210	130	160	210	200	300
°F, pressure	——	160	160	140	120	200	120	140	200	150	280

*The properties of a piping material may vary from one commercial material to another. The pipe manufacturer should be consulted for specific properties.

†Consult Table 3-4 for values of long-term strength at 73°F.

‡Exact operating limit may vary from each particular commercial plastic material (consult manufacturer). Effects of environment should also be considered.

TABLE 3-3 Typical Physical Properties of Glass-Fiber-Reinforced Thermosetting Resin Pipe*†

Property at 75°F	Test no.	Polyester	Filament wound epoxy	Filament wound vinyl ester	Filament wound reinforced plastics mortar
Specific gravity	D 792	1.6	1.9	1.9	1.7 to 2.2
Tensile strength, lb/in^2 ($\times 10^3$)	D 2105				
Hoop direction		35	20 to 50	20 to 50	15 to 60
Axial direction		25	2 to 10	2 to 10	3 to 20
Tensile modulus, lb/in^2 ($\times 10^6$)	D 2105				
Hoop direction		1.4	1.5 to 4	1.5 to 4	1.4 to 3.6
Axial direction		1.4	0.8 to 2.0	0.8 to 2.0	0.8 to 1.8
Coeff. of linear expansion, in/(in)(°F)($\times 10^{-5}$)	D 696				
Hoop direction		1.3	0.5 to 0.7	0.5 to 0.8	1.0
Axial direction		1.3	1.0 to 1.8	1.0 to 1.8	1.5
Thermal conductivity, (Btu)(in)/(h)(ft^2)(°F)	C 177	0.9	1.5 to 4.2	1.5 to 4.2	1.3
Approximate temperature limit‡					
°F, nonpressure		240	300	250	200
°F, pressure		200	240	220	140

*The properties of a piping material may vary from one commercial material to another. The pipe manufacturer should be consulted for specific properties.

†Consult Table 3-4 for values of long-term strength at 73°F.

‡Exact operating limit may vary from each particular commercial plastic material (consult manufacturer). Effects of environment should also be considered.

of stress to strain for a specific time under continuous stressing, or straining, is commonly referred to as the *effective creep modulus,* or the *effective stress-relaxation modulus.* In the case of thermoplastics this effective modulus is significantly influenced by time. For continuous loading of 20 years' duration, it can be from one-quarter to one-third the value of the short-term modulus. In the case of reinforced thermosets the viscous response is of lower order and the long-term effective modulus tends to be at least three-quarters of the short-term values. Most pipe manufacturers are prepared to provide values of effective moduli for specific materials and loading conditions.

The effective strength of plastics is influenced by time, temperature, and environment. For example, the breaking point for a thermoplastic material under short-term tensile testing is reached only after considerable material deformation has taken place, at least 10 percent, and in some cases over 100 percent (the ultimate elongation for reinforced thermosets is lower than for thermoplastics). Under long-term continuous loading material failure (or an unacceptable level of material damage) will occur at much lower deformations than in tensile testing. With thermoplastics the strain levels at failure can be as low as 3 percent, and with some reinforced thermosets they may be below 0.5 percent. Material damage, and not creep or excessive deformation, represents the durability limit for plastics subject to long-term loading. These durability limits, are time-, temperature-, and environment-dependent.

Durability under Continuous Loading

To establish its longer-term hydrostatic strength, plastics pipe is tested in accordance with ASTM D 1598, "Time to Failure of Plastic Pipe Under Constant Internal Pressure." After obtaining a sufficient number of stress vs. time-to-fail points that must span a testing time from 10 to 10,000 h, the data are extrapolated to determine the estimated average 100,000 h strength. The extrapolating procedures are those of ASTM method D 2837, "Obtaining Hydrostatic Design Basis for Thermoplastic Materials," or procedure B of ASTM D 2992, "Obtaining Hydrostatic Design Basis for Reinforced Thermosetting Resin Pipe and Fittings." The extrapolated long-term hydrostatic strength (LTHS) so determined is then rounded off into the appropriate hydrostatic design basis (HDB). The HDBs are design-stress categories represented by a series of preferred numbers (i.e., 2000, 2500, 3150, 4000, 5000, etc.) that ascend in steps of 25 percent. The following relationship, known as the ISO (International Standards Organization) plastics pipe hoop stress equation, is used to relate the test pressure and pipe dimensions to the resultant circumferential stress in the pipe wall:

$$S = \frac{p}{2}\frac{D_m}{t} \tag{1a}$$

where

S = hoop stress, lb/in² (N/m²)
p = internal hydraulic pressure, lb/in² (N/m²)
D_m = mean pipe diameter, in (cm)
t = minimum pipe wall thickness, in (cm)

Once the HDB is established and then reduced to a hydrostatic design stress (HDS) by the application of an appropriate service (or design) factor, the same formula, rewritten as follows, may be used to compute the pipe pressure rating:

$$p = 2\text{HDS}\,\frac{t}{D_m} \tag{1b}$$

where

HDS = hydrostatic design stress, lb/in² (N/m²)
= HDB × SF
SF = service factor

The selected service factor considers two general groups of conditions: The first encompasses the normal variations in material and pipe manufacture and the second those of the pipe's application and use (environment, hazards, life expectancy). The general practice has been to use service factors of not more than 0.5 for static water pressure at 73°F (23°C). Smaller factors are used for more demanding conditions.

The HDBs for static water pressure at 73°F (23°C) for the major thermoplastic pipe materials are presented in Table 3-4. As indicated by this table, the ASTM material designations for pressure thermoplastics code the material according to its short-term

TABLE 3-4 Hydrostatic Design Basis* (Strength Categories) for Thermoplastic Pipe Compounds Determined with Water at 73.4°F (23°C)

Material designation†	HDB lb/in^2 (MPa), 73.4°F (23°C)‡
PE 1404	800 (5.5)
PE 2305	1000 (6.9)
PE 2306	1250 (8.6)
PE 3306	1250 (8.6)
PE 3406	1250 (8.6)
PE 3408	1600 (11.2)
PVC 1120	4000 (27.6)
PVC 1220	4000 (27.6)
PVC 2110	2000 (13.8)
PVC 2112	2500 (17.2)
PVC 2116	3150 (21.7)
PVC 2120	4000 (27.6)
CPVC 4120	4000 (27.6)
ABS 1208	1600 (11.2)
ABS 1210	2000 (13.8)
ABS 1316	3150 (21.7)
ABS 2112	2500 (17.2)
CAB MH08	1600 (11.2)
CAB SO04	800 (5.5)
PB 2110	2000 (13.8)
PP 1110	2000 (13.8)
PP 1208	1600 (11.2)
PP 2105	1000 (6.9)

*Per ASTM D 2837.

†The last two digits code the maximum recommended hydrostatic design stress (RHDS), expressed in hundreds of pounds per square inch. RHDS = HDB × 0.5, where 0.5 is the generally accepted maximum value for the design factor. Lower values than 0.5 may be justified by certain operating and safety considerations. The first two digits code the material according to short-term properties.

‡Since thermoplastics, even though of the same ASTM designation, may be affected differently by increasing temperature, HDBs at higher temperatures must be established for each specific commercial product. A number of such products have HDBs for temperatures as high as 180°F (82°C). Consult the most current Plastics Pipe Institute (355 Lexington Ave, New York, NY 10017) Technical Report TR-4 for latest listing of HDBs of commercial pipe compounds.

properties and its maximum RHDS. The RHDS is determined by applying a factor of 0.5 to the material's HDB for water at 73°F (23°C). Values of HDB for other environments and temperatures will be different, even for materials of the same ASTM designation. For example, not all PE 3306s will show the same long-term strength at 120°F (49°C). For other than the standard HDB rating conditions the pipe manufacturer should be consulted for data or recommendations on the specific pipe material.

In the case of thermosets, because of their composite construction, their HDB is determined not only by the properties of the materials used but also by the specific wall construction and manufacturing process. Typical HDBs for some machine-made RTRPs for water service at 73°F (23°C) are given in Table 3-5. The pipe manufacturer should be consulted for HDBs for the specific pipe construction under consideration.

Durability under Cyclic Loading

The higher shorter-term strength of plastics permits them to easily tolerate relatively infrequent short-lived excursions in pressure beyond those established for static pressure conditions. However, under repeated cyclic stressing, as in pressure surging, plastics tend to fatigue and their long-term strength is reduced. The fatigue sensitivity varies from plastic to plastic, and for each material it is dependent on various factors including the amplitude and the frequency of the cyclic stressing. The reduction in limiting strength of thermoplastics may range from slight to as much as one-half the value under static conditions. For services with only moderate and infrequent surging, as is the case in many water piping systems, selecting the pipe pressure rating solely on the basis of the maximum static pressure has been demonstrated to be effective. The American Water Works Association Standard for PVC Water Piping (AWWA C-900) establishes pipe pressure classes with a built-in consideration of cyclic pressure.

TABLE 3-5 Hydrostatic Design Basis for Machine-Made Reinforced Thermosetting Resin Pipe

ASTM pipe specification*	Type	Classification per ASTM D 2310†			HDB, lb/in²‡	
		Grade	Class	Designation	Static	Cyclic
D 2517 (gas pressure pipe)	Filament-wound	Epoxy, glass-fiber-reinforced	No liner	RTRP-11AD RTRP-11AW	— 16,000	5,000 —
D 2996	Filament-wound	Epoxy, glass-fiber-reinforced	No liner	RTRP-11AD RTRP-11AW	— 16,000	5,000 —
	Filament-wound	Epoxy, glass-fiber-reinforced	Reinforced epoxy liner	RTRP-11FE RTRP-11FD	— —	6,300 5,000
	Filament-wound	Polyester, glass-fiber-reinforced	Reinforced polyester liner	RTRP-12EC RTRP-12ED RTRP-12EU	— — 12,500	4,000 5,000 —
	Filament-wound	Polyester, glass-fiber-reinforced	No liner	RTRP-12AD RTRP-12AU	— 12,500	5,000 —
D 2997	Centrifugally cast	Polyester, glass-fiber-reinforced	Polyester liner	RTRP-22BT RTRP-22BU	10,000 12,500	— —
	Centrifugally cast	Epoxy, glass-fiber-reinforced	Epoxy liner	RTRP-21CT RTRP-21CU	10,000 12,500	— —

*See Table 3-8 (p. 10-57) for the abbreviated title of the pipe standard.

†"Standard Classification for Machine-Made Reinforced Thermosetting Resin Pipe," ASTM D 2310. (The first three symbols following RTRP designate type, grade, and class; the last symbol codes the HDB.)

‡Multiply by 6894 to convert to pasculs.

Reinforced thermosetting plastic pipes are somewhat more sensitive to fatigue than thermoplastic pipes. Accordingly, they are often pressure-rated for surge service on the basis of their HDB as determined by Procedure A of the aforementioned ASTM D 2992 from test data obtained in accordance with ASTM D 2143, "Test for Cyclic Pressure Strength of Reinforced Thermosetting Plastic Pipe." Typical HDB values for cyclic pressure service for certain RTRPs are also shown in Table 3-5.

Durability under Continuous Straining

Stress relaxation gradually reduces the initial stress level that is first generated when a plastic material is strained to a given level and then maintained there. As a consequence of this reduction in stress, plastics can tolerate somewhat larger strains under constant-strain (i.e., stress-relaxation) conditions than the ultimate strain that ensues in a constant-load condition (under which there is no stress relaxation). At present there is no unanimously preferred method for establishing limiting strain values for conditions of stress relaxation. Appendix X2 to ASTM D 3262, "Standard Specification for Reinforced Plastic Mortar Sewer Pipe," includes a method, not officially part of the standard, by which one may evaluate the strain limits of the subject pipe in an acid environment. The strain limits for RTRPs tend to be significantly lower than those for thermoplastics piping. In fact, under conditions of continuous straining, limiting strain is not a design constraint for most thermoplastics. When dealing with situations such as in pipe bending or in deflection of pipe under earth loading (in which limiting strain is a possible consideration), recommended design values for maximum permissible strain should be obtained for the specific material and end-use conditions.

Durability under Cyclic Straining

Both ultimate strength and allowable strain may be reduced by fatigue. Guidance should be sought from the pipe manufacturer.

Other Mechanical Properties

In addition to the test for long-term strength, a number of special tests have also been established for evaluating other relevant properties of plastic pipe:

ASTM D 2444	"Impact Resistance of Thermoplastic Pipe and Fittings by Means of a Tup (Falling Weight)"
ASTM D 2105	"Longitudinal Tensile Properties of Reinforced Thermosetting Pipe and Tube"
ASTM D 2925	"Measuring Beam Deflection of Reinforced Thermosetting Pipe Under Full Bore Flow"
ASTM D 2412	"External Loading Properties of Plastic Pipe by Parallel-Plate Loading"
ASTM D 2924	"External Pressure Resistance of Reinforced Thermosetting Resin Pipe"

Flow Properties

Plastic pipes offer minimal resistance to fluid flow. Tests indicate that they may be characterized as hydraulically smooth conduits according to the well-established rational flow formulas. The Hazen-Williams equation, a frequently used engineering approximation formula, yields good correlations for water flow when a C_H factor of 150 to 155 is used. For calculating water flows in open channels, or nonfull flow in conduits, Manning's equation is often used. The Manning n factor for plastics for clean water is approximately 0.0085 to 0.0095. With sewerage, values of $n = 0.010$ to 0.012 are used since they provide margin for flow-disturbing influences such as sedimentation and slime growth.

Since plastics do not corrode, their original flows do not deteriorate with time. Plastics are also easier to clean than other materials.

Chemical Resistance

Plastics do not corrode in the sense that metals do. Being nonconductors they are immune to galvanic or electrochemical effects. If they are affected by an environment, it is generally through direct chemical attack, solvation, or strain corrosion. In addition to differences in chemical resistance due to the nature of the plastic and pipe wall construction, the extent of resistance may also depend on time and temperature of contact and, in some cases, on the presence of an externally applied stress.

In direct chemical attack, the molecules of either the polymer or the reinforcement are altered, and the alteration leads to a gradual deterioration of properties. Such attack can be brought about by strong oxidizing and reducing agents and by ultraviolet and other radiation. Thermoplastics as a group tend to be inherently more resistant than thermosets to chemical attack. Good protection against ultraviolet radiation (which might be required with the more ultraviolet-sensitive pipes that are to be used in continuous exposure to the weather) is afforded by incorporation into the formulation of an opacifier such as a finely divided carbon black, titanium dioxide, or some other opaque pigment.

Solvation is the absorption of an organic solvent by the plastic. Its effect may range from a slight swelling and softening, with minor effects on properties, to a complete solution. The solvent cementing of ABS and PVC pipe is based on solvation. By the use of selective solvents that evaporate after their task is completed, solvent cementing makes it possible to create a monolithic joint that retains the properties of the base material. Thermosets, because of their cross-linked chemical structure, tend to have superior resistance to solvation.

In strain corrosion, damage will occur only under the combined action of strain (i.e., stress) and environment. In the case of thermoplastics this form of attack is called environmental stress cracking. The mechanism, although not fully understood, is essentially the development and ensuing slow growth and propagation of cracks by the combined action of stress and a sensitizing agent. Stress-cracking agents tend to be materials such as detergents and alcohols that have a surface-wetting tendency. Stress cracking may be controlled by selecting stress-crack-resistant grades of material or by creating designs that ensure that the stress, or strain, is below the threshold value necessary to set this mechanism in action.

In the case of thermosetting materials, the formation of crazes or microcracks exposes the glass fiber, or other reinforcement, to possible chemical attack. To protect against this, a tough resilient liner is used, and the stress, or strain, level is limited to established safe values which will preclude the formation of crazing under the specific exposure conditions.

Table 3-6 presents a broad guide to the chemical resistance of the major thermoplastic piping materials. Table 3-1 includes a general guide to FRP resin performance. Final selection of material and pipe wall construction for chemical or corrosive service should follow consultation with more detailed chemical-resistance information. Because ultimate resistance is affected by stress, time, and temperature, it may not be reliably predicted by shorter-term "soak" tests. Successful previews in similar service are the best guide. Lacking these, new applications would best be evaluated by actual service testing. An advantage of service testing is that it is sure to include some minor (but often overlooked) contaminant which could influence the final result.

In the case of RTRP piping, the liner is considered the first line of defense and therefore a critical factor in corrosive service selection. Special resin liners, which may be deposited in extra thicknesses, are available for extra protection. Pipes with thermoplastic liners are also available.

Inertness to Potable Water and Other Fluids

Nearly all of the base materials from which plastic pipes are made are inert to potable water. It is possible, however, that through the addition of ingredients such as stabilizers,

TABLE 3-6 Thermoplastic Piping Materials: Chemical-Resistance Guide for Ambient Temperatures*

Attacking chemicals	ABS	PVC I	PVC II	CPVC	PE	PB	PP	PVDF
Inorganic compounds								
Acids, dilute	G	G	L	G	G	G	G	G
Acids, concentrated 80%	L	L	L	G	L	L	L	G
Acids, oxidizing	L	P	P	L	P	P	P	G
Alkalies, dilute	G	G	G	G	G	G	G	G
Alkalies, concentrated 80%	L	G	L	G	G	G	G	G
Gases, acid (HCl and HF), dry	L	L	L	L	G	G	G	G
Gases, acid (HCl and HF), wet	L	G	L	G	G	G	G	G
Gases, ammonia, dry	L	G	L	G	G	G	G	G
Gases, halogens, dry	L	L	L	L	L	L	P	G
Gases, sulfur gases, dry	P	G	L	G	G	L	P	G
Salts, acidic	G	G	G	G	G	G	G	G
Salts, basic	G	G	G	G	G	G	G	G
Salts, neutral	G	G	G	G	G	G	G	G
Salts, oxidizing	L	L	L	L	G	G	G	G
Organic compounds								
Acids	G	G	G	L	G	G	G	G
Acid anhydrides	L	L	L	P	L	L	L	L
Alcohols, glycols	L	G	L	G	L†	G	G	G
Esters, ethers, ketones	P	P	P	P	L	L	L	L
Hydrocarbons, aliphatic	L	L	L	G	L	L	L	G
Hydrocarbons, aromatic	P	P	P	L	P	P	P	G
Hydrocarbons, halogenated	L	L	L	L	P	P	P	L
Natural gas (fuel)	G	G	G	G	G	G	G	G
Mineral oil	G†	G	G	G	L†	G	G	G
Oils, animal and vegetable	G†	G	G	G	L†	G	G	G
Synthetic gas (fuel)	L	L	L	L	L	L	L	G

*G, good; P, poor, L, limited knowledge: determination requires precise knowledge of individual conditions.

†Stress-crack-resistant grade should be used.

catalysts, modifiers, and pigments, the final formulation may render a pipe inadequate for potable water in terms of its effect on the toxicological safety and the taste and odor quality of the water. To safeguard against this, most potable water piping standards require that the pipe be evaluated for this purpose by a laboratory recognized by the public health profession. A very commonly used laboratory is the National Sanitation Foundation (NSF), Ann Arbor, Michigan. NSF evaluates, and lists as acceptable, those plastic pipes that satisfy the requirements of their standards for potable water service.

Because they do not contaminate fluids with metallic ions, plastic pipes are often utilized in the transport of pure materials, including de-ionized water. For food service there are pipes available that have been made from materials approved by the Food and Drug Administration.

COMMON METHODS FOR JOINING PLASTIC PIPES

Plastic pipe may be joined by a variety of methods (see Table 3-7), the choice of which is influenced in some cases by the properties of the basic material. For example: the polyolefins (PE, PB, and PP) may not be solvent-cemented; the vinyls (PVC and CPVC) and ABS heat-fuse with relative difficulty; and the thermosets may not be either heat-fused or solvent-cemented. Of the available choices, the one that is selected will depend upon pipe performance, installation, and maintenance requirements as well as availability of fitting and joining equipment. Heat fusion (of PE, PP, PB, and PVDF), solvent-

TABLE 3-7 Techniques for Joining Plastic Pipe

Method of joining	Thermoplastic pipe ABS	PVC	CPVC	PE	PP	PB	PVDF	RTR pipe
Adhesive	—	—	—	—	—	—	—	o
Solvent cements	o	o	o	—	—	—	—	—
Heat fusion	—	—	—	o	o	o	o	—
Threading*	o	o	o	o	o	—	o	o
Flanged connectors†	o	o	o	o	o	o	o	o
Grooved joints‡	o	o	o	o	o	—	o	o
Mechanical compression§,¶	o	o	o	o	o	o	o	o
Elastomeric seal¶	o	o	o	o	o	o	o	o
Flaring	—	—	—	o	—	o	—	—
Insert	—	—	—	—	—	o	—	—

*Molded thread adapters are available for attachment on the pipe by another technique. Threads may not be cut in thermosetting pipe. Some thermosetting threaded connections may be adhesive-bonded for extra strength. For threading, the wall thickness of thermoplastic pipe should be not less than Schedule 80.

†Flanged adapters are applied on pipe by heat fusion, solvent-cementing, or threading.

‡Minimum wall thicknesses are prescribed depending on the pipe material.

§With thinner-walled pipe, stiffening inserts must be used.

¶Many designs of elastomeric seal and compression fittings provide no thrust restraint and therefore may be used only in situations, such as buried pipe, in which the pipe is restrained from pullout. Elastomeric seal and compression fittings are available in special designs that incorporate end restraint.

cementing (of ABS, PVC, CPVC), and adhesive joining (of reinforced thermosetting resin pipe), all methods which produce monolithic joints, are preferred for applications requiring maximum strength and optimum chemical resistance (Fig. 3-4). Bell (sometimes referred to as socket) and spigot connections are utilized to join pipe by these three techniques. RTRP may also be joined by the butt-and-strap method whereby two pieces of pipe are butted together and overlays of a laminate are then applied over the butted section and allowed to cure (Fig. 3-5). Larger-diameter polyolefin pipe is joined by the heat fusion of the pipe butt ends.

Flanged connections are often used in industrial applications, particularly when making transitions to other materials, such as when connecting to a metal valve or to a tank outlet, and when it is advantageous to provide for easy removal of a pipe section or other component from the system for cleaning, maintenance, or other purpose.

Threading is also used with plastic pipe. However, molded threads are preferred for thermoplastics and are required for most thermosetting pipe. Molded threaded adapters are available and may be applied by solvent-cementing or with adhesive, whichever is applicable. Threads may not be cut on most thermosetting pipe for they may damage the structural integrity of the pipe wall. Thermoplastic pipe may be threaded, provided its wall thickness is not less than a prescribed minimum, normally at least that of Schedule 80 pipe.

For installations not excluding the use of elastomeric sealants such as neoprene or red rubber, there are mechanical-compression as well as bell and spigot connectors which incorporate such sealants into their design. Much thermoplastic and thermosetting piping specifically made for buried water and sewer lines is available with integral elastomeric-seal bell and spigot connectors. Such connectors greatly facilitate pipe construction, partly because of the ease of making the connection (a stab fit) and partly because the connection may be made under almost any weather or field condition.

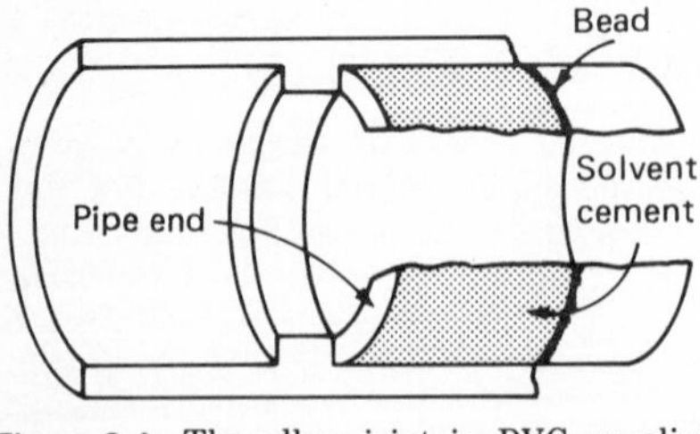

Figure 3-4 Threadless joint in PVC coupling (inside view).

Sometimes connectors utilizing grooved pipe are used with standard grooved-end

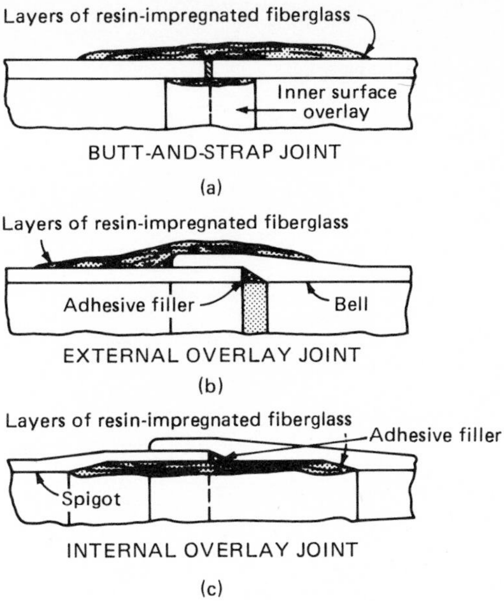

Figure 3-5 Butt and strap joints.

systems such as Victaulic or Gustin-Bacon. With thermoplastic pipe of sufficient wall thickness, the grooves may be cut or rolled in some cases. Cutting is not permitted with thermosetting pipe. Grooved adapters are available for both thermoplastic and thermosetting pipes.

COMMERCIALLY AVAILABLE PRODUCTS

Plastics piping is manufactured in an imposing array of materials, constructions, diameters, wall thicknesses, lengths, and fitting types. The more important products are listed in Table 3-8 which also reports for each product the available size range and its important end uses. Whenever the product is covered by a major standard, the applicable document is identified. Because of the dynamic rate at which new plastic piping standards are currently being written (and older ones revised), the reader is advised to check with the major standards issuing organizations for the most current listing.*

Fittings for larger-diameter pipes are not listed in Table 3-8 because they are often custom-fabricated rather than available from stock. Also not listed in Table 3-8 are piping components such as valves and flanges. In diameters ranging from ⅜ to 4 in, valves made from PVC, CPVC, PP, and PVDF are available in a great variety of different styles including check, ball, diaphragm, globe, gate, and needle. They are available with socket (for solvent-cementing or heat-fusion joining), butt, flanged, or threaded ends. Valves are also available in "Tru" union style and with multiports. A number of models may be obtained with pneumatic or electric actuators for automatic valve positioning. In the ⅜- to 4-in (0.95- to 10-cm) sizes a large array of other piping components, such as strainers, expansion joints, roof and floor drains, and line tapping fittings, is also available. Many

*The Plastics Pipe Institute, a Division of the Society of the Plastics Industry, 355 Lexington Avenue, New York, NY 10017, regularly updates its TR-5, "Standards for Plastics Piping," which lists standards issued by all major United States, Canadian, and international standards organizations (ISO) on both thermoplastic and thermosetting piping.

TABLE 3-8 Principal Commercially Available Plastic Piping Products

				End use																
				Water supply and distribution							Sewer and drain					Industrial			Duct	
Pipe material	Product standard*	Title (abbreviated) of standard or brief product description	Diameter range, in	Mains	Services	Drop pipe, wells	Well casing	Various: industrial, coml.,	Distributing: cold only	Distributing: hot & cold	Collecting system	Building connections	Drainage	Drain, waste, & vent	Natural gas distribution	Corrosives and abrasives	Compressed gases†	Liquid fuels	Conduit (above ground)	Duct (below ground)
ABS, PVC PE & PB	ASTM D 2513	Thermoplastic Gas Pressure Pipe & Fittings	¼–12												XX		X			
ABS, PVC & SR	ASTM F 480	Thermoplastic Water Well Casing	2–12				X													
ABS, PVC & PP	ASTM D 3311	DWV Plastic Fittings Patterns	1¼–6											X						
ABS & PVC	ASTM F 409	ABS & PVC Accessible & Replaceable Tube and Fittings	3–12											X						
ABS	ASTM D 1527	ABS Plastic Pipe, Sch. 40 & 80	⅛–12	X	X	X		X	X			X				X	X			
	D 2282	ABS Plastic Pipe, SDR-PR	⅛–12	X	X	X		X	X			X				X	X			
ABS	D 2465	ABS Plastic Pipe Fittings, threaded Sch. 80	¼–6	X	X	X		X	X			X				X	X			
	D 2468	ABS Plastic Pipe Fittings, socket, Sch. 40	¼–8	X	X	X		X	X			X				X	X			
	D 2469	ABS Plastic Pipe Fittings, socket, Sch. 80	⅛–8	X	X	X		X	X			X				X	X			
	D 2661	ABS DWV Pipe and Fittings	1¼–6										X	X						
	D 2680	ABS Composite Sewer Pipe & Fittings	8–15									X	X							
	D 2750	ABS Utility Conduit & Fittings	1–6																	X
	D 2751	ABS Sewer Pipe and Fittings	1–6									X	X							
	F 628	ABS Foam Core DWV	1¼–6										X	X						

PVC	ASTM																			
	D 1785	PVC Plastic Pipe, Sch. 40–80 & 120	⅛–12	X	X	X		X	X			X				X	X			
	D 2241	PVC Plastic Pipe, SDR-PR	⅛–24	X	X	X		X	X			X				X	X			
	D 2464	PVC Plastic Pipe Fittings, threaded sch. 80	⅛–6	X	X	X		X	X			X				X	X			
	D 2466	PVC Plastic Pipe Fittings, socket, sch. 40	⅛–8	X	X	X		X	X			X				X	X			
	D 2467	PVC Plastic Pipe Fittings, socket, Sch. 80	⅛–8	X	X	X		X	X			X				X	X			
	D 2665	PVC DWV Pipe & Fittings	1¼–6											X						
	D 2672	PVC Plastic Pipe, bell end	⅛–8	X	X	X		X	X			X				X	X			
	D 2729	PVC Drain Pipe & Fittings	2–6										X							
	D 2740	PVC Plastic Tubing	½–1¼		X	X		X												
	D 2949	3-in PVC Thin Wall DWV Piping	3											X						
	D 3033	PVC Sewer Pipe & Fittings, type PSP	4–15								X	X	X							
	D 3034	PVC Sewer Pipe & Fittings, type PSM	4–15								X	X	X							
	D 3036	PVC Line Couplings, socket type	1–8	X	X		X									X	X			
	ASTM F 512	PVC Conduit for Buried Installation	2–6																	
		PVC Sewer Pipe	8–27								X		X							
	AWWA C900	PVC Pressure Pipe for Water	4–12	X				X												
	UL 514	Electrical Outlet Boxes & Fittings	½–6																X	
	UL 651	Rigid Non-metallic Conduit	½–6																X	
	NEMA TC-2	Electrical Plastic Tubing & Conduit	½–6																X	
	TC-3	PVC Fittings for Conduit & Tubing	½–6																X	
PVC & CPVC	API 5LP	PVC and CPVC Line Pipe	½–12					X								X	X	X		
CPVC	ASTM																			
	D 2846	CPVC Hot Water Distribution Systems	⅜–2							X										
	F 437	CPVC Plastic Pipe Fitting, threaded, Sch. 80	¼–6					X		X						X	X	X		

TABLE 3-8 Principal Commercially Available Plastic Piping Products (*Continued*)

				End use																
				Water supply and distribution							Sewer and drain					Industrial			Duct	
Pipe material	Product standard*	Title (abbreviated) of standard or brief product description	Diameter range, in	Mains	Services	Drop pipe, wells	Well casing	Various: industrial, coml.,	Distributing: cold only	Distributing: hot & cold	Collecting system	Building connections	Drainage	Drain, waste, & vent	Natural gas distribution	Corrosives and abrasives	Compressed gases†	Liquid fuels	Conduit (above ground)	Duct (below ground)
	F 438	CPVC Plastic Pipe Fittings, socket, Sch. 40	¼–6					X		X						X	X	X		
	F 439	CPVC Plastic Pipe Fittings, socket, Sch. 80	¼–6					X		X						X	X	X		
	F 441	CPVC Plastic Pipe, Sch. 40 & 80	¼–12					X		X						X	X	X		
	F 442	CPVC Plastic Pipe, SDR-PR	¼–12					X		X						X	X	X		
	F 443	CPVC Bell End Pipe	⅛–8					X		X						X	X	X		
PE	ASTM																			
	D 2104	PE Plastic Pipe, Sch. 40	½–6	X	X	X		X	X		X	X				X	X	X		
	D 2239	PE Plastic Pipe, SDR-PR	½–6	X	X	X		X	X		X	X				X	X	X		
	D 2447	PE Plastic Pipe, OD-based, Sch. 40 & 80	½–12	X	X	X		X	X		X	X				X	X	X		
	D 2609	Plastic Insert Fittings for PE Pipe	½–4		X	X		X									X			
	D 2683	PE Fittings, socket-fusion type for OD-based pipe		X	X	X		X	X		X	X				X	X	X		
	D 2737	PE Plastic Tubing	½–2		X	X		X	X								X	X		
	D 3261	PE Fittings, butt-fusion type	½–10	X	X	X		X	X		X	X				X	X	X		
	F 405	PE Corrugated Tubing & Fittings	3–8										X							
	F 714	PE Plastic Pipe, SDR-PR, larger diam	3–63	X				X			X	X	X			X	X	X		

	AWWA C901	PE Pipe, tubing & fittings for water	½–3		X															
	‡	PE Pipe Spirally Wound	18–96								X	X	X							
	API 5LE	PE Line Pipe	½–12					X	X							X	X	X		
PB	ASTM-																			
	D 2662	PB Plastic Pipe, SDR-PR	½–6	X	X	X		X		X						X	X	X		
	D 2666	PB Plastic Tubing	½–2		X	X		X		X						X	X	X		
	D 3000	PB Plastic Pipe, SDR-PR, OD-controlled	½–6	X	X	X		X		X						X	X	X		
	D-3309	PB Plastic Hot-Water Distributing Systems	¼–2		X					X										
	‡	PB Plastic Pipe, SDR-PR, larger diam	3–42	X	X						X	X	X			X	X	X		
	AWWA C902	PB Pipe, tubing & fitting for water			X															
PP	‡	PP Plastic Pipe, Sch. 40 and 80	½–6									X	X	X		X	X	X		
	‡	PP Plastic Pipe Fittings, socket, Sch. 80	½–6									X	X	X		X	X	X		
	‡	PP Plastic Pipe Fittings, threaded, Sch. 80	½–6													X	X	X		
	‡	PP Chemical Drainage Pipe & Fittings	1½–4										X	X						
PVDF	‡	PVDF Plastic Pipe, Sch. 80	½–4													X	X	X		
	‡	PVDF Plastic Pipe, SDR-PR	½–4													X	X	X		
	‡	PVDF Plastic Pipe Fittings, socket, Sch. 80	½–4													X	X	X		
	‡	PVDF Plastic Pipe Fittings, threaded, Sch. 80	½–4													X	X	X		
RTRP	ASTM																			
	D 2517	RTR Pipe & Fittings for Gas	2–12												X		X			
	D 2996	Filament Wound RTR Pipe	1–12	X	X	X		X		X						X	X	X		
	‡	Filament-Wound Large Diameter RTR Pipe	12–160					X								X	X	X		
	D 2997	Centrifugally Cast RTR Pipe	1½–12	X	X	X		X	X							X	X	X		
	‡	RTR Pipe and Fittings	12–72					X								X	X	X		
	API 5AR	RTR Pipe Casing & Tubing	1½–10			X	X													
	API 5LR	RTR Line Pipe	1–12													X	X	X		
	‡	RTR Pipe Flanges	1–72	X				X									X			

TABLE 3-8 Principal Commercially Available Plastic Piping Products (*Continued*)

Pipe material	Product standard*	Title (abbreviated) of standard or brief product description	Diameter range, in	End use: Water supply and distribution: Mains	Services	Drop pipe, wells	Well casing	Various: industrial, coml.,	Distributing: cold only	Distributing: hot & cold	Sewer and drain: Collecting system	Building connections	Drainage	Drain, waste, & vent	Natural gas distribution	Industrial: Corrosives and abrasives	Compressed gases†	Liquid fuels	Duct: Conduit (above ground)	Duct (below ground)
TRP/RPMP	AWWA C950	RTR and RPM Pipe for Water	1–144	X	X			X		X						X	X	X		
RPMP	ASTM																			
	D 3262	RPM Sewer Pipe	8–108								X	X	X							
	D 3517	RPM Pressure Pipe	8–108					X			X					X	X	X		
	D 3754	RPM Sewer and Industrial Pipe	8–144					X			X	X	X			X	X	X		
	D 3840	RPM Non-Pressure Pipe Fittings	8–144								X	X	X							
	MIL-P-28584	Steam Condensate Lines, RTRP Filament Wound	2–6					X												
	MIL-P-22245	Pipe and Pipe Fittings, glass-fiber-reinforced plastic	2–16	X	X	X	X	X		X						X	X	X		

*Issuing agency identified in discussion on standards.

†Thermoplastic piping is not normally recommended for above-ground service because of safety considerations should the pipe fail. A special grade is available (see ABS piping).

‡Standard currently under development by ASTM.

of these items may be obtained up through about 8-in (20-cm) diam. Larger-sized components are sometimes specially fabricated, or metallic products are used which are connected to the plastic pipe by means of flanges or some other suitable connector.

Standard and special design manholes and access holes are available in both thermoplastic and thermosetting materials for sewerage and drainage applications. ASTM D 3753, "Specification for Glass Fiber Reinforced Manholes," is, to date, the only adopted standard for such products.

PRODUCT STANDARDS, CODES, AND APPROVALS

Standards

The primary source of standards on plastics piping is the American Society for Testing and Materials (ASTM), 1916 Race Street, Philadelphia, PA 19103. (See Table 3-8.) There are a number of other organizations that also develop such standards, generally on products related to their particular interest or activities. These include the following: American Water Works Association (AWWA), 6666 West Quincy Avenue, Denver, CO 80235, which issues standards for water distribution; the American Petroleum Institute (API), 300 Corrigan Tower Building, Dallas, TX 75201, on piping for oil and gas production; the National Electrical Manufacturers Association (NEMA), 2101 L Street, N.W., Washington, DC 20037, on conduit and ducting; and Underwriters Laboratories (UL), 333 Pfingston Road, Northbrook, IL 60062, on conduit and ducting. In addition, the federal government and some state agencies have also issued standards on plastic pipe, generally for projects in which they exercise regulatory or financial functions, or for purchases on their own behalf. Typical of these are the standards issued by the U.S. Department of Agriculture (USDA), the U.S. Department of Defense (DOD), the Federal Housing Administration (FHA), and the General Services Administration (GSA). Most of these documents parallel the basic requirements of ASTM and other listed standards.

The most frequently used dimensioning scheme for setting the outside diameter (OD) of plastic pipe is the traditional iron pipe size (IPS) system of commercial wrought steel pipe (ANSI B36.10). Most thermoplastic pipes are available with standard IPS outside diameters. Some PE and PB pipes which are designed to be joined with insert-type fittings are sized to the same inside diameters as Schedule 40 wrought steel pipe. In the case of thermoset pipes their outside diameters may be exactly, or approximately equal to the reference dimension depending on whether the pipe is sized from the outside in (as in centrifugally casting) or from the inside out (as when filament-winding over a mandrel). Other diameter systems that are utilized include:

- **Copper Tubing Size (CTS).** Based on standard outside diameters of copper tubes. CTS pipe is used for water and gas services and for hot- or cold-water plumbing.
- **Cast-Iron (CI) Pipe Size.** PVC and RTR water main piping is made for this outside-diameter basis and also conforms to the IPS system.
- **International Standards Organization (ISO) Sizes.** Some of the larger polyolefin pipes are made with these internationally set outside diameters.

Much pipe, especially in the larger sizes (for which compatibility with traditionally sized plastic piping components such as valves and fittings is not necessary), is made to fit into special diameter-sizing systems determined by the specific product standard or by the pipe manufacturer. Most of these systems have established diameter dimensions of such proportions that the resultant size of the inside bore is close to the pipe nominal diameter.

The wall thicknesses of most thermoplastic pipe of solid and homogeneous wall construction are defined in accordance with the standard dimension ratio (SDR) concept whereby the ratio of *average* outside diameter to *minimum* wall thickness is a constant value for each SDR pipe series over the entire range of pipe diameters. The standard diameter ratios that have been adopted by ASTM and other standard-writing organizations represent a series of preferred numbers that increase in steps of 25 percent, as follows: 11, 13.5, 17, 21, 26, 32.5, 41, etc. The advantage of establishing wall-thickness categories for thermoplastic pipe according to a constant ratio of diameter to wall thickness is evident from inspection of Eq. (1*b*): within each SDR category the pressure rating is the same for all pipe sizes. There is some pipe made to other than the established SDRs; for this diameter the actual wall thicknes ratio is referred to as diameter ratio (DR).

Thermoplastic pipes specifically intended for industrial uses are often made to the IPS Schedule 40, 80, and 120 system which sets not only the outside diameter but also the wall thickness for each nominal size in each schedule. Since in a given pipe schedule the ratio of diameter to wall thickness tends to decrease with increasing diameter, so too does the pipe pressure rating. Schedule 80 wall thickness, or greater, is required whenever pipe is threaded. The pressure rating of thermoplastic threaded pipe is reduced to half that for unthreaded pipe.

The wall thicknesses of thermoset pipe are not defined in the same way as for thermoplastic pipe because key properties such as strength and stiffness depend not only on material but also on exact wall construction, which can vary not only among manufacturers but even with pipe diameter. The resultant pipe wall thickness will generally be set by the performance requirements of the pipe. Some standards set minimum values for wall thickness. Thermosetting pipe may not be field-threaded. If the pipe is to be joined by threading, it is available with factory-applied molded threads. Threaded adapters are also available.

Codes

The use of piping for plumbing, fire protection, and for the transport of hazardous materials may be subject to the provisions of a code and/or to those of local, state, federal, or other regulations. All the major model plumbing codes which have become adopted, or referenced, by state and local jurisdictions permit and prescribe to a varying but fairly extensive degree the use of plastics piping for hot-cold water lines; water services; drain, waste, and vents (DWV); sewerage; and drainage. Plastics piping is also covered by other codes, such as the following which are of interest to industrial users:

American National Standards Institute Codes

ANSI B31.3-1980	Chemical Plant and Petroleum Refinery Piping
ANSI B31.8-1975	Gas Transmission and Distribution Piping Systems
ANSI Z223.1-1977	National Fuel Gas Code

Department of Transportation, Hazardous Materials Board, Office of Pipeline Safety Operations

Code of Federal Regulations (CFR), Title 49, Part 192, Transportation of Natural Gas and Other Gas by Pipeline: Minimum Federal Safety Standards

Code of Federal Regulations (CFR), Title 49, Part 195, Transportation of Liquids by Pipeline, Minimum Federal Safety Standards.

The National Fire Protection Association (Quincy, Mass.) Model Codes

NFPA 30	Flammable and Combustible Liquids Code
NFPA 54	National Fuel Gas Code
NFPA 70	***National Electrical Code***®*
NFPA 70A	Electrical Code for One and Two Family Dwellings
NFPA 34	Outdoor Piping

**National Electrical Code*® is a Registered Trademark of The National Fire Protection Association, Quincy, MA 02269.

Approvals

Some standards and various jurisdictions and authorities require that before a pipe may be used for certain applications it first must be approved for that use by a recognized, or specifically designated, organization. Organizations with listing and approval programs for plastic pipe include the following:

For Potable Water

The National Sanitation Foundation, NSF Bldg., P.O. Box 1468, Ann Arbor, MI 48105

Canadian Standards Association, 178 Rexdale Boulevard, Rexdale, Ontario, Canada, M9W 1R3

For Drain, Waste, and Vent

The National Sanitation Foundation and Canadian Standards Association (See above.)

For Meat- and Food-Processing Plants

U.S. Department of Agriculture, 14th and Independence S.W., Room 0717 South, Washington, DC 20250

For Underground Fire Protection Systems

Underwriters Laboratories, Inc., 333 Pfingston Road, Northbrook, IL 60062

Factory Mutual Research Corporation, 1151 Boston-Providence Turnpike, P.O. Box 688, Norwood, MA 02062

For Underground Gasoline and Petroleum Lines

Underwriters Laboratories Inc. (See above.)

DESIGN AND INSTALLATION

Standard piping products offered for specific uses such as cold water; hot or cold water; drain, waste, and vent; sewerage; and drainage are largely predesigned. For example, CPVC and PB hot- or cold-water tubing systems made in accordance with their respective standards ASTM D 2846 and D 3309 are pressure-rated at 100 lb/in^2 (690 kPa) for water at 180°F (83°C). Design and installation recommendations are included in an appendix to these documents. Since most standards for products dedicated to a specific application contain design and installation recommendations, such documents should be consulted by designers and installers. The installation of plastic plumbing piping products is often regulated by the applicable plumbing code. More detailed information on design and installation may be obtained from most manufacturers and plastic pipe trade associations (see "Additional Information").

When design is conducted to meet special requirements of a given application, particular attention must be given to the effects of solvents and corrosives on piping properties. Temperature and unusual loadings (such as cyclic pressure and vibration) must also be considered. Most manufacturers of industrial piping products can provide performance data and design and installation recommendations for special as well as ordinary conditions. Various references that provide design and installation information are listed under "Additional Information."

Fundamentally, the principles of the design and installation of plastic piping are the same as those applying to steel piping. However, because of differences in their properties, certain aspects of the design and installation of plastics may require different emphasis and solutions. The following paragraphs briefly discuss these more important aspects. Detailed design and installation recommendations for each specific product should be followed.

Pressure Rating

Continuous Pressure

The pipe pressure rating should be based on the design stress [see Eq. (1*b*)] that is established by taking into account the anticipated effect of time, temperature, and environment on the pipe strength properties. It should be recognized that because of material and fabrication differences, plastic fittings and joints may have a pressure rating lower than that for the pipes. The design stress should include adequate margin for safety considerations. With this in mind it should be recognized that most thermoplastics, excepting those specially formulated for this purpose, are not suitable for conveying compressed gases in above-ground service. In the event of accidental pipe failure, the large potential energy stored in compressed gases could precipitate a catastrophic-type failure mechanism that sometimes produces dangerous flying debris. Thermosets and certain specially formulated thermoplastics resist such failure and are suitable for this application.

Vacuum or External Pressure

The service capabilities of thinner-walled pipes made of less rigid materials may be determined in some cases not by internal pressure but by vacuum conditions created by transients (surges) or by the external pressure loading present inside a large tank. The buckling resistance of plastic pipes may be estimated using Timoshenkos's classic elastic buckling equation including the following adaptation which gives consideration to the effect of pipe ovality:

$$P_c = \frac{2E}{1 - \mu^2}\left(\frac{t}{D_m}\right)^3 C \tag{2}$$

where

P_c = collapsing pressure of unconstrained pipe, lb/in^2
E = effective modulus of elasticity of pipe material, lb/in^2
μ = Poisson ratio (approximately from 0.35 to 0.45 for thermoplastics for short-term loading)
t = pipe-wall thickness, in
D_m = pipe mean diameter, in
C = factor correcting for pipe ovality = $(r_o/r_i)^3$, where r_i is the major radius of curvature of the ovalized pipe, and r_o is the radius assuming no ovalization

For short-term loading conditions, the values of E and μ as obtained from short-term tensile tests yield reasonable correlations. For long-term loading, appropriate values as determined from long-term loading tests should be employed.

Cyclic Pressure

The shock load or high-pressure surge created by sudden closure of valves could exceed the pressure capabilities of a pipe and, if the pipe is not properly anchored, could result in fitting failure by overstraining of joints. Excessive surging should be eliminated by control of the rate of valve closure or the installation of accumulators. All plastic pipe can tolerate some surging in excess of working pressure, and because of its greater flexibility and viscoelastic properties the pressures generated by surging are of lower order than those for metal piping and are more quickly damped. However, under frequent and continuous surging the long-term strength of plastic pipe tends to be reduced by fatigue. Under these conditions an appropriately lowered value of hydrostatic design stress should be used.

Considerations for Above-Ground Uses

Supports, Anchors, and Guides

Most manufacturers supply information on support spacings. Typical recommendations are presented in Tables 3-9 and 3-10. Values are usually given for either single or con-

TABLE 3-9 Typical Recommended Maximum Support Spacing, in Feet, for Thermoplastic Pipe for Continuous Spans and for Uninsulated Lines Conveying Fluids of Specific Gravity up to 1.35

Pipe dimension		PVC			CPVC				PVDF				PP			
Nominal diam, in	Wall schedule	60°F	100°F	140°F	60°F	100°F	140°F	180°F	80°F	100°F	140°F	160°F	60°F	100°F	140°F	180°F
½	Schedule 40	4½	4	2½	5	4½	4	2½	3¾	3½	2	Continuous support recommended	1¾	1¾	1½	1¼
¾		5	4	2½	5½	5	4	2½	4	3¾	2½		2	2	1¾	1¾
1		5½	4½	2½	6	5½	4½	2½	4¼	4	2½		2	2	2	1¾
1¼		5½	5	3	6	5½	5	3	—	—	—		2½	2¼	2	2
1½		6	5	3	6½	6	5	3	4½	4¼	2½		2½	2½	2¼	2
2		6	5	3	6½	6	5	3	4½	4½	2¾		3	2¾	2½	2¼
3		7	6	3½	8	7	6	3½					3½	2¾	3	2¾
4		7½	6½	4	8½	7½	6½	4					4	3¾	3½	3
6		8½	7½	4½	9½	8½	7½	4½								
8		9	8	4½												
½	Schedule 80	5	4½	2½	5½	5	4½	2½	4½	4½	2½	Continuous support recommended	2	2	2	1½
¾		5½	4½	2½	6	5½	4½	2½	4½	4½	3		2½	2½	2¼	2
1		6	5	3	6½	6	5	3	5	4¾	3		2½	2½	2¼	2
1¼		—	—	—	—	—	—	—	—	—	—		3	2¾	2½	2½
1½		6½	5½	3½	7	6½	5½	3½	5½	5	3		3	3	2¾	2½
2		7	6	3½	7½	7	6	3½	5½	5¼	3		3½	3¼	3	2¾
3		8	7	4	9	8	7	4					4	4	3½	3½
4		9	7½	4½	10	9	7½	4½					4½	4½	4	3½
6		10	9	5	11	10	9	5								
8		11	9½	5½												

TABLE 3-10 Typical Recommended Maximum Support Spacing, in Feet, for Fiberglass-Reinforced Pipe for Temperatures up to 150°F (65°C) for Uninsulated Lines Conveying Fluids of up to 1.25 Specific Gravity

Nominal pipe size, in	Continuous span	Single span
1	7.5	6.3
1½	8.5	7.2
2	9.9	8.4
3	11.2	9.4
4	11.9	10.0
6	14.2	11.9
8	15.7	13.2
10	17.4	14.6
12	18.9	15.9

tinuous spans and for a given liquid specific gravity. A most important consideration is to ensure that the span distance between supports is based on the maximum system temperature. Depending on the piping material, pipe dimensions, and application, the support span may be dictated by the permissible deflection (generally about ½ in (1.3 cm) at midspan) or maximum allowable stress. The allowable stress should provide allowance for stresses generated by fluid pressure and by thermal and other loadings. Vertical pipe runs can be supported either in compression or tension. Long runs should be checked to ensure that the tensile or compressive load does not exceed the permissible design value. Standard strap, sling, clamp, clevis, and saddle supports providing at least 120° of contact are generally recommended (Fig. 3-6). Supports offering narrow or point contact should be avoided. Valves and other heavy piping components should be individually supported.

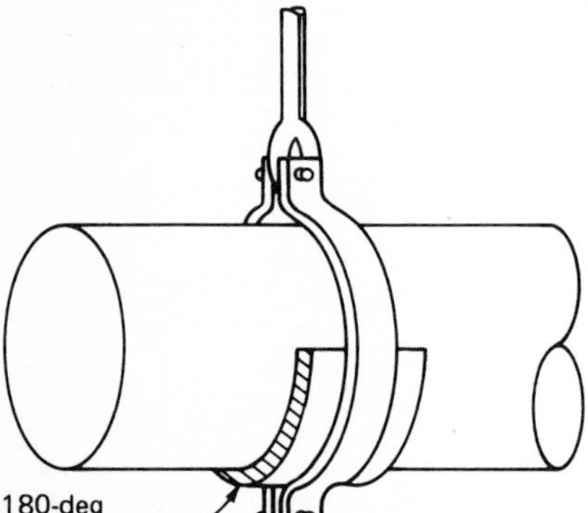

Figure 3-6 Wear-saddle prevents damage from standard hanger.

Anchors, which divide a pipe system into sections, must positively restrain the movement of pipe against all applied and developed forces, particularly dynamic loading. Anchors should generally be employed near changes of direction when transitioning to another piping material or when there is a change in line size. In long, straight runs, anchor spacing is generally recommended at about 200 to 300 ft (60 to 90 m). Anchor spacing and location will also be determined by the selection of anchoring for the control of expansion and contraction.

When the pipe is restrained against expansion and contraction it should be guided to prevent buckling. The guides should encircle the pipe but be loose enough to allow it to move freely in its axial direction.

Control of Expansion and Contraction

Because of the inherent flexibility of plastic piping, it is generally possible to design a pipe system so that no expansion joints are necessary. Their use should be avoided where possible, since they are expensive, and they remove the ability of the pipe to carry longitudinal loads. Offsetting this load with anchors can be an added problem with larger pipes. Preferred techniques for dealing with expansion and contraction are: (1) anchoring and guide spacing, (2) changing direction (offset legs), and (3) using expansion loops.

Although plastics expand more than steel, their relatively low modulus of elasticity results in significantly less end load for the same temperature change. The smaller thermal forces can generally be readily relieved by changes in direction. However, when using directional change to absorb thermal forces, neither the pipe nor any of its components should be subjected to a bending stress (or strain) in excess of that recommended by the manufacturer. The stress may be controlled by placement of anchors not closer than a calculated distance from the point of change of direction. The length of the legs of expansion loops should similarly be determined.

Oscillations and Vibrations

Because plastic pipe is so much more flexible than steel pipe, oscillations due to changes in velocity of fluid set up more easily and tend to be of greater amplitude. In long runs this is generally no problem but they could damage connected piping by subjecting it to excessive stresses or strains. The solution is to restrain the pipe by the use of anchors. High-amplitude vibrations from connected equipment, such as pumps, should be isolated from the piping by the use of flexible connectors.

Considerations for Below-Ground Uses

In terms of their underground performance all plastic pipes are classified as flexible, which signifies that when they are properly installed they are capable of developing sufficient diametrical deformation, without incurring material failure to fully activate soil support forces. Soil-assisted flexible pipes can easily support earth loads that would crush stronger rigid pipes for which total load bearing ability is almost entirely dependent on their own strength. To activate soil support, flexible pipes must be embedded in soils that are stable and that have been properly placed and densified around the pipe. ASTM documents D 2321, "Underground Installation of Flexible Thermoplastic Sewer Pipe," and D 3839, "Underground Installation of Flexible Reinforced Thermosetting Pipe and Reinforced Plastic Mortar Pipe," present detailed recommendations on the proper installation of flexible plastic pipe designed for nonpressure uses. Nonpressure pipe, which is generally of thinner wall construction and not rounded by internal pressure, requires somewhat more care in installation than heavier-wall pressure pipe. ASTM D 2774, "Underground Installation of Thermoplastics Pressure Piping," presents recommendations for solving this problem.

The basic principle of the installation of buried plastic piping is to embed it in a soil of such quality that the resultant ultimate pipe deflection is controlled to an acceptable value that is limited by either the pipe performance requirements or the pipe material capabilities. The former generally permits deflection up to about 10 percent (some engineers may set conservatively lower values) while the latter is determined by the maximum allowable stress, or strain, in the pipe wall for the given pipe material and construction. The pipe supplier can provide the limiting deflection values for a given pipe material and construction. These values may depend on the fluids being handled. With thermoplastic pipe of solid wall construction deflection will seldom be limited by material performance constraints.

The extent to which a flexible pipe will deflect when embedded in a given quality of soil may be estimated by a variety of methods. One of the better-known relationships, sometimes called the Iowa equation, was developed for flexible metal conduits at Iowa State University. A modification of this equation is

$$\frac{\Delta x}{D_i} = \frac{L_D K P}{EI/r^3 + 0.061E'}$$

where

Δx = horizontal deflection of the pipe, in (For relatively small deflections, the change Δy in vertical diameter of a circular section deforming elliptically is equal to 1.10 Δx. As an approximation, it is often assumed $\Delta y = \Delta x$.

D_i = pipe inside diameter prior to loading, in

L_D = deflection lag factor compensating for the time dependence of soil deformation, dimensionless

TABLE 3-11 Bureau of Reclamation Values of E' for Iowa Formula for Initial Average Deflection of Flexible Pipe

Soil type for pipe embedment material per ASTM D 2321	Soil type description (United Classification System, ASTM D 2487)	E', lb/in² for degree of compaction of embedment (proctor density, %)*			
		Dumped	Slight (>85%)	Moderate (85–95%)	High (>95%)
I	Manufactured angular, granular materials (crushed stone or rock, broken coral, cinders, etc.)	1000 (+4%)	3000 (+4%)	3000 (+3%)	3000 (+2%)
II	Coarse-grained soils with little or no fines	N.R.†	1000 (+4%)	2000 (+3%)	3000 (+2%)
III	Coarse-grained soils with fines	N.R.	N.R.	100 (+3%)	2000 (+2%)
IV	Fine-grained soils	N.R.	N.R.	N.R.	N.R.
V	Organic soils (peats, mulches, clays, etc.)	N.R.	N.R.	N.R.	N.R.

*Values in parentheses give the approximate limit of deflection beyond the average deflection that is computed by using the given E' values. These limits are for pipe of relatively low stiffness. As pipe stiffness increases, the limit is narrowed.

†N.R. indicates use not recommended by ASTM D 2321.

K = bedding constant which varies with the angle of bedding (i.e., bedding support), dimensionless (The bedding constant ranges from 0.110 for a point support on the bottom of a pipe to 0.083 for full support. For plastic pipe, the typical value is taken as 0.10.)
P = total vertical pressure acting on the pipe, lb/in²
r = pipe radius, in
E = modulus of elasticity of pipe material, lb/in²
I = moment of inertia of pipe wall per unit of length, in⁴/in (For round pipe $I = t_a^3/12$, in which t_a is the average wall thickness.)
E' = modulus of passive soil resistance, lb/in²

As a result of extensive field investigations of the load vs. deflection characteristics of various flexible pipes the U.S. Bureau of Reclamation* has developed a series of soil reaction E' values for use in the Iowa equation under the assumption that $K = 0.1$ and $D_L = 1.0$. These E' values may be used to estimate a pipe's initial average deflection. To assist in estimating the initial maximum (i.e., acceptance) deflection as a consequence of both soil loading and installation factors, the Bureau of Reclamation has also reported the observed upper limits of deflection values. Both these limits, which primarily apply to pipes of lower ring stiffness, and the values of E' are shown in Table 3-11 as a function of the embedment materials recommended by D 2321.

In actual practice, it is seldom necessary to go through the Iowa equation calculation, for if the recommended installation practices in D 2321 are followed, initial installed deflections can quite readily be held to approximately 5 percent and less. Burial of the thinner-walled, more flexible pipes may also require consideration of the adequacy of the pipe's wall compressive strength as well as its buckling stability (Fig. 3-7).

Recommendations for the design and installation of buried plastic pipe may be obtained from pipe manufacturers', trade associations, or from the listed information given under "Additional Information."

*Amster K. Howard, "Modulus of Soil Reaction Values for Buried Flexible Pipe," *Journal of the Geotechnical Division of the American Society of Civil Engineers,* vol. 103, No GTI, January 1977, pp. 33–43.

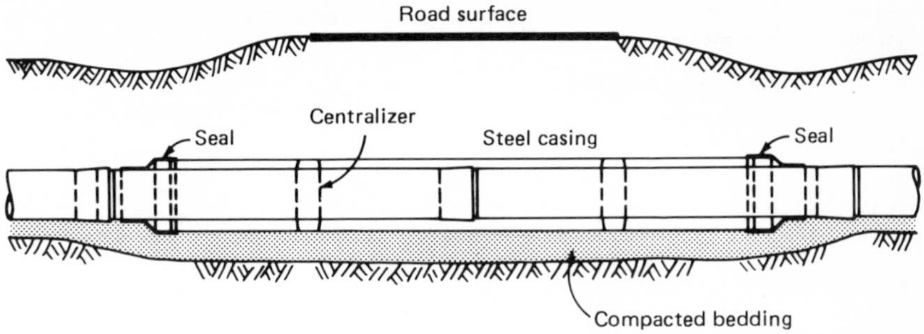

Figure 3-7 Steel casing protects pipe from concentrated loadings.

ADDITIONAL INFORMATION

The various plastic pipe trade associations issue reports, manuals, and lists of references on design and installation of their members' products. A list of current reports may be obtained by contacting each organization as follows:

Reinforced Thermosetting Piping

Reinforced Plastics/Composites Institute, a Division of the Society of the Plastics Industry, 355 Lexington Avenue, New York, NY 10017

The Materials Technology Institute of the Chemical Process Industries, Inc., 1380 Dublin Road, Columbus, OH 43215

Thermoplastic Pipe (Industrial, Gas Distribution, Sewerage, Water, and General Uses)

The Plastics Pipe Institute, a Division of the Society of the Plastics Industry, Inc., 355 Lexington Avenue, New York, NY 10017

Thermoplastics Pipe (Plumbing Applications)

Plastics Pipe & Fittings Association, 999 North Main Street, Glen Ellyn, IL 60137

PVC Piping (Water Distribution, Sewerage, and Irrigation)

Uni-Bell Plastics Pipe Association, 2655 Villa Creek Drive, Suite 164, Dallas, TX 75234

BIBLIOGRAPHY

The following references include useful information on plastics piping:

"PVC Pipe, Design and Installation," AWWA Manual No. M23, American Water Works Association, Denver, 1980.

"Standard for Reinforced Thermosetting Resin Pipe," AWWA C-950, American Water Works Association, Denver. (Appendix to standards includes much design and installation information.)

Britt, William F., Jr.: "Design Considerations for FRP Piping Systems," *Proceedings of the 1979 Conference on Managing Corrosion with Plastics,* National Association of Corrosion Engineers, Houston.

Cheremisinoff, Nicholas P., and N. Paul: *The Fiberglass-Reinforced Plastics Deskbook,* Ann Arbor Science, Ann Arbor, Michigan, 1978

Cooney, J. L.: "Guidelines for the Inspection and Maintenance of FRP Equipment and Piping," *Proceedings of the 1979 Conference on Managing Corrosion with Plastics,* National Association of Corrosion Engineers, Houston.

Escher, G. A.: "Transition to FRP, Basic Guidelines for Piping Designers and Users," *Proceedings of the 1975 Conference on Managing Corrosion with Plastics,* National Association of Corrosion Engineers, Houston.

Escher, G. A., and W. B. MacDonald: "Chemical and Mechanical Properties of Butt and Strap Joints," *Proceedings of the 1975 Conference on Managing Corrosion with Plastics,* National Association of Corrosion Engineers, Houston.

Greenwood: "Buried FRP Pipe—Performance Through Proper Installation," *Proceedings of the 1975 Conference on Managing Corrosion with Plastics,* National Association of Corrosion Engineers, Houston.

Kutschke, C. T.: "Use of Plastic Pipe for Industrial Applications," *Proceedings of the 1975 Conference on Managing Corrosion with Plastics,* National Association of Corrosion Engineers, Houston.

Launikitis, M. B.: "Chemically Resistant FRP Resins," *Proceedings of the 1977 Conference on Managing Corrosion with Plastics,* National Association of Corrosion Engineers, Houston.

Mallison, John H.: *Chemical Plant Design with Reinforced Plastics,* McGraw-Hill, New York, 1969.

Mruk, Stanley: "Thermoplastics Piping: A Review," *Proceedings of the 1979 Conference on Managing Corrosion with Plastics,* National Association of Corrosion Engineers, Houston.

Petroff, Larry J., and Luckenbill, Michael: "Flexibility of the Design of Fiberglass Pipe," *Proceedings of the 1981 International Conference on Plastic Pipe,* American Society of Civil Engineers, New York.

Plastics Piping Manual, The Plastics Pipe Institute, 355 Lexington Avenue, New York, 1976.

Rolston, Albert: "Fiberglass Composite and Fabrication," *Chemical Engineering,* January 28, 1980, pp. 96–110.

Rubens, A. C.: "Designing RTRP Systems Utilizing Published Engineering Data," *Proceedings of the 1979 Conference on Managing Corrosion with Plastics,* National Association of Corrosion Engineers, Houston.

Schrock, B. J.: "Thermosetting Resin Pipe," Preprint 3088, American Society of Civil Engineers, New York, 1977.

Proceedings of the International Conference on Underground Plastic Pipe, March 30–April 1, 1981, New Orleans, La., American Society of Civil Engineers, New York.

chapter 10-4

Asbestos-Cement Pipe and Fittings

by

Leo J. Horvath

Director, Technical Affairs

Association of Asbestos Cement Pipe Producers, Arlington, Virginia

DESCRIPTION AND APPLICATIONS

Asbestos-cement (A/C) pipe, composed of a mixture of portland cement and asbestos fiber, with or without silica, is completely inorganic and free from metallic substances. In the manufacturing process, these ingredients are combined, mixed with water, and formed as pipe on a rotating steel mandrel, creating a dense wall with a smooth interior surface. The pipe is cured in autoclave ovens for dimensional and chemical stability; during the curing time the silica reacts with the free lime, present in normally cured cement products, to form relatively insoluble calcium silicate compounds. These give A/C pipe added resistance to corrosive soils and fluids; as a nonconductor, it is immune to electrolysis.

The resulting A/C pipe is light in weight, resists corrosion, maintains good flow characteristics, and possesses high crush, flexural. and hydrostatic strength.

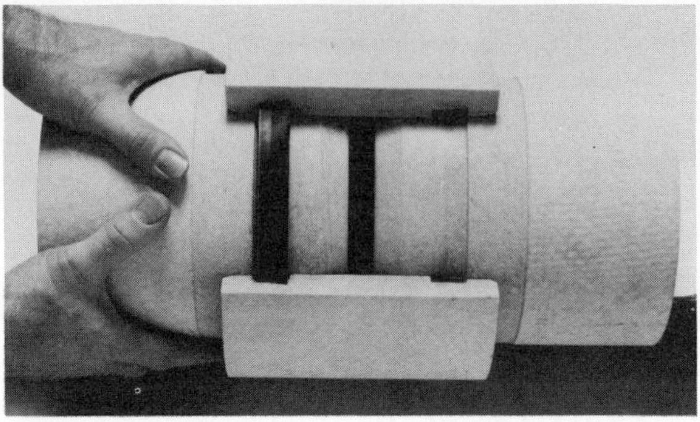

Figure 4-1 Asbestos-cement pipe coupling. *(CertainTeed Corporation.)*

The most common applications for A/C pipe are for water mains, sewage force mains, and gravity sewer systems. Other uses for A/C pipe are storm drains, perforated underdrains, electric and telephone conduits, irrigation systems, air and vent ducts, and building sewers. Because of its high corrosion resistance to many chemicals and freedom from rust and metallic oxides, A/C pipe is also used in many industrial services and, on occasion, for overhead process piping services. In addition, pre-insulated pressure pipe utilizes an outer asbestos-cement casing to contain an inner insulated core capable of handling underground installations of steam, chilled liquids, and high-temperature water.

FITTINGS AND COUPLINGS

Standard asbestos-cement or plastic fittings such as elbows, tees, wyes, adapters, and couplings are available for most sewer-pipe sizes and for building sewer lines. Metal fittings other than the couplings are normally used for pressure-pipe connections. A/C pipe products are designed for quick compatibility with cast-iron and ductile-iron fittings and easy interfacing with accessories made of these materials.

The joining of one A/C pipe to another is achieved easily with a push-together coupling, consisting of an asbestos-cement, plastic, or fiberglass sleeve with solid rubber rings as shown in Fig. 4-1. Compressing the rubber rings between the sleeve and the factory-machined pipe ends provides a tight seal that resists shock, vibration, and earth movement while compensating for the expansion and contraction of the pipe lengths. These joints, when used on pipe under pressure, withstand pressure equal to the rated pipe pressures with adequate safety factors. A coupling is assembled at the factory on one end of each standard section of A/C pipe.

AVAILABLE SIZES AND FORMS

Asbestos-cement pipe is manufactured in standard lengths of either 10 or 13 ft (3 or 4 m) and is also available in half or quarter lengths.

Sewer Pipe

Gravity Sewer Systems

A/C pipe is manufactured in five strength classifications which are designated as Classes 1500, 2400, 3300, 4000, and 5000. The class designation represents minimal crushing strength in pounds per linear foot of pipe, regardless of size. The pipe is manufactured with nominal inside diameters from 4 to 36 in (100 to 900 mm) and in some areas up to

TABLE 4-1 Asbestos-Cement Pressure Pipe, Minimum Crushing Load

Nominal pipe size, in	lb per linear ft		
	Class 100	Class 150	Class 200
4	4,100	5,400	8,700
6	4,000	5,400	9,000
8	4,000	5,500	9,300
10	4,400	7,000	11,000
12	5,200	7,600	11,800
14	5,200	8,600	13,500
16	5,800	9,200	15,400

42 in (1050 mm). Larger diameters and higher strength classifications may also be obtained on special order.

Sewage Force Mains

A/C pipe is manufactured in three classes of operating pressures: 100, 150, and 200. The class designation represents the operating pressure in pounds per square inch. These three classes of pressure pipe, available in sizes from 4 to 16 in (100 to 400 mm), are capable of withstanding the crushing loads shown in Table 4-1.

Water Pipe

Distribution Systems

For distribution systems which have relatively unpredictable flows and many appurtenances, A/C pressure pipe is available in sizes from 4 to 16 in (100 to 400 mm) for operating pressures of 100, 150, and 200 lb/in^2 designated as Classes 100, 150 and 200, respectively. The crushing strengths of these three classes which are also used for pressure sewer systems are given in Table 4-1.

Transmission Systems

For transmission systems which have relatively steady flow and few appurtenances, A/C pipe is available in sizes from 18 to 42 in (450 to 1050 mm). The strength classifications of 30, 35, 40, 45, 50, 60, 70, 80, and 90 represent one-tenth of minimum allowable hydrostatic bursting strength in pounds per square inch for A/C transmission pipe. A/C transmission-piping systems are designed on the combined loading theory in which the design engineer takes into consideration and determines the operating and surge pressures, the earth and superimposed external loads, and a safety factor. The relationship of design internal pressures and external loads for A/C transmission pipe is shown in Table 4-2.

TABLE 4-2 Asbestos-Cement Transmission Pipe, Design Internal Pressure (P, lb/in^2) and Design External Loads (W, lb/linear ft)

Pipe size, in	Strength Classification								
	30 P = 300 W	35 P = 350 W	40 P = 400 W	45 P = 450 W	50 P = 500 W	60 P = 600 W	70 P = 700 W	80 P = 800 W	90 P = 900 W
18	2,500	3,000	4,000	5,000	6,500	8,500	11,000	14,000	18,000
20	2,500	3,500	4,500	5,500	7,100	9,500	12,000	15,000	20,000
21	2,500	3,500	4,500	5,800	7,300	9,700	12,500	16,000	21,000
24	2,800	3,800	5,000	6,200	8,100	11,000	15,000	19,000	24,000
27	3,500	4,200	5,500	7,000	8,800	12,500	16,500	20,500	27,000
30	3,500	4,500	6,000	7,500	9,700	13,500	18,000	22,500	30,000
33	3,500	5,000	6,500	8,000	10,500	14,500	19,500	24,500	33,000
36	4,000	5,000	7,000	9,000	11,200	16,000	21,000	26,000	36,000
39	4,200	5,300	7,500	9,700	12,000	17,200	22,500	28,000	39,000
42	4,300	5,700	8,000	10,500	13,000	18,500	24,000	30,000	42,000

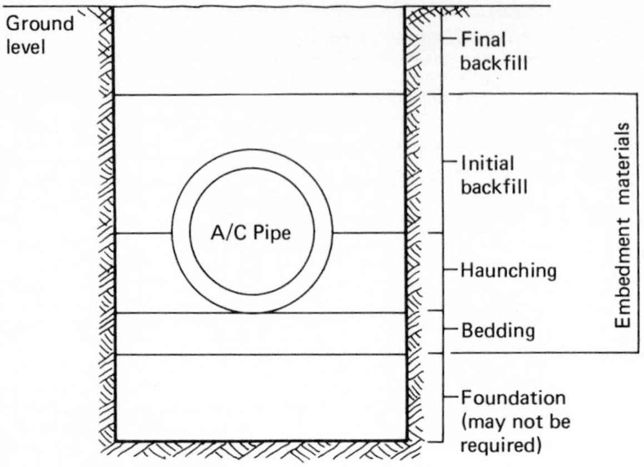

Figure 4-2 Trench cross section and terminology used in embedment and backfilling asbestos-cement pipe.

INSTALLATION AND MAINTENANCE

Asbestos fibers with a tensile strength four to five times greater than that of steel provide significant reinforcement in A/C pipe products to withstand internal hydrostatic pressure, external loads, or combinations of these forces. Even so, good construction procedures applicable for any underground piping systems are equally appropriate for A/C piping systems. Recommended practices for receiving, storage and handling, joint assembly, installation, and inspection and testing are covered in detail in A/C pipe manufacturers' literature, in American Society of Testing and Materials (ASTM) and American Water Works Association (AWWA) standards, and in the plans and specifications of the piping-system's design engineer. Terminology commonly used in a trench installation of A/C pipe is shown in Fig. 4-2. Following are highlights on installation of A/C pipe from these standards and specifications.

Receiving, Storage, and Handling

When receiving A/C pipe, each shipment should be inspected and inventoried. See Fig. 4-3. The pipe is inspected and carefully loaded at the factory using methods acceptable to the carrier whose responsibility it is to deliver the pipe in good condition. It is the responsibility of the receiver to ensure that there has been no loss or damage. Pipes should be stored, if possible, at the work site in the package units provided by the manufacturer. Rubber gaskets should be protected from oil and grease, direct sunlight, excessive exposure to heat, and electric motors which produce ozone. At all times, A/C pipe should be handled with care to avoid damage. Whether moved by hand, skidways, hoists, forklifts, or other handling equipment, A/C pipe should not be thrown, dragged, or bumped.

Trench Excavation

As a general rule, do not open the trench too far ahead of laying the pipe. In preparation for pipe installation, place (string) pipe as near the trench as possible but leave adequate space to perform necessary excavating and other related functions. If the trench is open, place the pipe on the opposite side from the excavated earth. When necessary to prevent caving, trench excavation should be sheeted and braced or sloped in accordance with

Figure 4-3 Asbestos-cement pipe shipment. *(CertainTeed Corporation.)*

applicable laws and ordinances. The trench width at the ground surface should be ample to permit the pipe to be laid and the backfill to be placed and throughly compacted. The trench should be kept free from water at all times. See Fig. 4-4.

Pipe should not be lowered into the trench until the pipe bed has been brought to correct grade. Any part of the trench bottom below grade should be backfilled with thoroughly compacted material. When an unstable subgrade condition is encountered, additional trench depth should be excavated and refilled with suitable, thoroughly compacted foundation material. All rocks, boulders, and large stones should be removed to provide a clearance of 6 to 9 in (150 to 225 mm) below and on all sides of pipe and fittings. When excavation is completed, a bed of sand, crushed stone, or earth that is free from rocks, frozen earth, or clods larger than 1 in (25.4 mm) should be placed and thoroughly compacted to properly regrade the trench bottom. A minimum clearance of 2 in (50 mm) below the coupling should be provided at each joint to permit proper joint assembly. The pipe should be provided continuous support between coupling holes.

Figure 4-4 Asbestos-cement pipe installation. *(CertainTeed Corporation.)*

Joint Assembly

Pipe and accessories should be lowered carefully into the trench by hand or with suitable equipment to avoid damaging the pipe and fittings or injuring the installers. Pipe ends, the coupling interior, and especially the exposed coupling groove and rubber gasket should be thoroughly cleaned prior to assembly. Joint assembly should be performed as recommended by the manufacturer. After alignment, insertion of the rubber gasket, and thorough lubrication of the pipe end as specified by the pipe manufacturer, the assembly of the joint is completed by a sliding action during which the lubricated pipe end slides under the rubber gasket located in the coupling groove and into the coupling to an automatic stop point. Each pipe joint is sealed with a coupling consisting of a sleeve and compressed rubber rings which keep the pipe ends separate, automatically providing for expansion, contraction, and joint flexibility. When pipelaying is not in progress, the open ends of installed pipe should be kept closed to prevent entrance of trench water, dirt, and other foreign matter into the pipeline.

Pipe Embedment and Backfilling

All pipe embedment material should be selected carefully, free from organic debris. The embedment material and its placement and compaction are important considerations in assuring that the pipe will satisfactorily resist trench loading conditions during construction and after the pipeline is completely installed.

Pipe-loading carrying capability is also greatly influenced by the degree of soil compaction around the lower half of the pipe section to provide satisfactory haunching. The initial backfill material should be placed to a minimum depth of 1 ft (30 cm) over the top of the pipe. After placement and compaction of pipe-embedment materials, the final backfill, which is usually placed by machine, need not be as carefully selected as the initial material but should contain no large stones or rocks, frozen soil, or debris.

Maintenance

Pipeline projects should be tested upon completion of installation to assure functional water and sewer systems construction. Before testing, all parts of the pipeline must be backfilled and braced to prevent movement under pressure. Since asbestos-cement will absorb some water, the line must be filled with water for a minimum of 24 h before being subjected to a hydrostatic pressure test. Approved methods of testing asbestos-cement pressure pipe (pressure-strength test and leakage testing) and nonpressure sewer pipe (infiltration, exfiltration, or low-pressure air-loss tests) are described in specifications listed in Table 4-3).

Implementation of proper installation, inspection, and testing procedures should assure maintenance-free, long-term performance of buried asbestos-cement piping systems.

In turn, good work practices complement proper construction procedures. Airborne asbestos fiber has been identified as a possible health hazard. The asbestos fiber in asbestos-cement pipe are not free, but are encapsulated, or locked into, the cement binder. Based on the best scientific data available to date, the proper use of A/C pipe does not pose a health hazard by reason of ingestion of asbestos fibers. Experience has shown,

TABLE 4-3 Specifications

	AWWA
C 400	Asbestos-Cement Distribution Pipe, 4 to 16 in
C 401	Selection of Asbestos-Cement Distribution Pipe, 4 to 16 in
C 402	Asbestos-Cement Transmission Pipe, 18 to 42 in
C 403	Selection of Asbestos-Cement Transmission Pipe, 18 to 42 in
C 603	Installation of Asbestos-Cement Pressure Pipe
	ASTM
C 296	Asbestos-Cement Pressure Pipe
C 668	Asbestos-Cement Transmission Pipe
C 428	Asbestos-Cement Nonpressure Sewer Pipe
C 644	Asbestos-Cement Nonpressure Small Diameter Sewer Pipe
C 663	Asbestos-Cement Storm Drain Pipe
C 508	Asbestos-Cement Underdrain Pipe
C 875	Asbestos-Cement Conduit and Fittings
D 1869	Rubber Rings for Asbestos-Cement Pipe

however, that minimizing exposure to airborne dust is the only effective method of preventing asbestos-related diseases. In 1978, AWWA published Manual No. 16, "Work Practices for Asbestos-Cement Pipe," which, based on results of field testing and study, recommends general guides to achieve safe and clean jobsite conditions when working with A/C pipe products.

STANDARDS AND SPECIFICATIONS

Asbestos-cement pipe is specified by pipe diameter, class or strength, and type of joint. Codes and specifications applicable to A/C pipe are issued by a number of organizations including various federal agencies, fire protection associations, and national and regional plumbing associations. A/C pipe should conform, as appropriate, to the standard specifications (Table 4-3) published by the American Water Works Association (AWWA) and the American Society for Testing and Materials (ASTM).

BIBLIOGRAPHY

American Water Works Association: "Work Practices for Asbestos-Cement Pipe," AWWA No. M16, 1978, Denver, Colorado.

Johns-Manville: "Pressure Pipe Installation Guide," 1979, Denver, Colorado.

ACKNOWLEDGMENT

For the preparation of this chapter, the author has drawn on standards published by the American Water Works Association and the American Society for Testing and Materials, as well as on source material supplied by asbestos-cement pipe manufacturers.

chapter 10-5

Pipe Insulation

by
Michael R. Harrison
Manager, Engineering and Technical Services
Johns-Manville Sales Corp., Denver, Colo.

HEAT-TRANSFER FUNDAMENTALS

Heat energy is transferred from one location to another by three different mechanisms: *conduction, convection,* and *radiation.* In insulation design theory, the objective is to minimize the contribution of each mode in the most efficient and economical manner. As temperatures vary, the relative importance of each transfer mechanism also varies, making different insulation designs appropriate for various applications.

Conduction

Energy transfer by conduction is a result of atomic or molecular motion. As molecules become heated, their vibration increases and energy is transferred to surrounding molecules. Conduction occurs in all three forms of matter: gas, liquid, and solid. Within most insulation, solid conduction is minimized by using an open-pore structure and a minimum amount of solid material. Gas conduction is more difficult to control, but to achieve much greater insulation efficiency, this mode must be limited. One method employs a vacuum, thus eliminating the gas from the system. This is very effective but costly, since the vacuum seal must be maintained in order to assure adequate performance. The second method of controlling gas conduction is to replace the air in the insulation by a heavier gas such as Freon®. Again, the seal must be maintained to avoid eventual air and moisture migration back into the cell structure.

Convection

Convective currents are established when a hot fluid (gas or liquid) rises from a heat source and is replaced by a cooler fluid which in turn is heated and rises, carrying the energy with it. In an insulation structure with many small cells, convection is minimized since the gas cannot freely pass through the structure. Most insulations are of sufficient density and formation to eliminate this mode within them, but convection plays a very important part in transferring energy from the insulation surface to the surrounding environment.

Radiation

As temperature increases, electromagnetic radiation gains in significance with regard to the total amount of energy transferred. Radiation occurs in a vacuum as well as in a gaseous environment, and its magnitude is dependent on the emittance of the radiating and receiving surfaces as well as the temperature difference between them. To control radiant flow, low-emittance surfaces are used in conjunction with absorbers and reflectors within the insulation itself. The mass density of the insulation is very important with a higher density reducing the level of radiation transfer.

There are obviously trade-offs that must be made in controlling the various heat-transfer mechanisms. Figure 5-1 illustrates the contribution to total conductivity of each mechanism at three different temperatures. The most efficient insulation design, both thermally and economically, will vary depending on the application conditions. References 1 and 2 are basic texts on heat transfer for further study.

Heat Flow

The level of heat flow to or from a system is directly proportional to the difference between the system and ambient temperatures and inversely proportional to the thermal resistance placed in the heat flow path:

$$\text{Heat flow} = \frac{\text{temperature difference}}{\text{resistance to heat flow}}$$

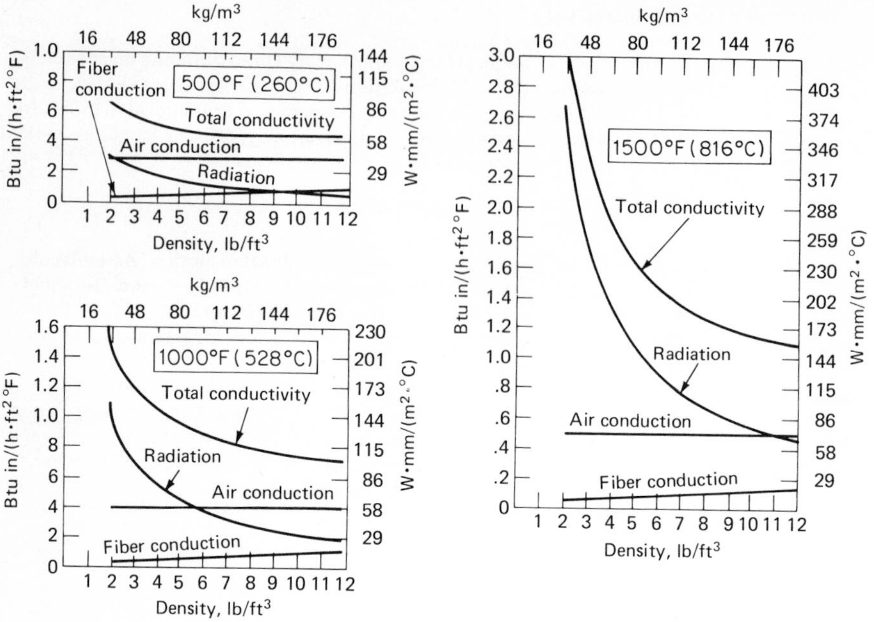

Figure 5-1 Contribution of each mode of heat transfer.[3]

In this light, the temperature difference is the forcing function, and as long as there is a differential, energy will flow. No amount of thermal resistance can completely stop the heat transfer; it can only slow the rate at which it occurs.

Total thermal resistance is generally composed of two distinct types of resistances: insulation and surface. Insulation resistance for a homogeneous material is determined by dividing the insulation thickness *tk* by its thermal conductivity

$$R_I = tk/k$$

The thermal conductivity, or *k* value, is an experimentally measured property of the insulation indicating the amount of heat transferred in 1 h through 1 ft^2 of 1-in-thick insulation with a temperature difference of 1°F. The units of *k* are (Btu)(in)/(h)(ft^2)(°F) [W·mm/m^2·°C]. The accurate measurement of thermal conductivity is very important since materials are often compared on this basis. The American Society for Testing and Materials (ASTM) has developed standardized test methods for measuring thermal conductivity; and for pipe insulation, the standard test is C 335, found in ASTM Part 18.[4] Since thermal conductivity increases with temperature, it is important to use the *k* value at the insulation mean or average temperature rather than the value at either the operating temperature (too high) or the ambient temperature (too low). For insulations of nonhomogeneous structure, a thermal conductivity based on 1-in (2.5-cm) thickness is not appropriate. Here, the *C* value on thermal conductance is used to represent the heat transfer through the actual thickness of insulation. Therefore,

$$R_I = 1/C$$

with *C* expressed in Btu/(h)(ft^2)(°F) [W/(m^2)(°C)].

The other type of thermal resistance is surface resistance

$$R_s = 1/f$$

where *f* represents the surface film coefficient. Surface emittance, air velocity across the surface, and temperature difference all influence the value of R_s. Table 5-1 lists various

TABLE 5-1 Values for Surface Resistance, R_s, (h)(ft²)(°F)/Btu [m²·°C/W]

A. Values for Still Air

$t_s - t_a$		Plain, fabric dull metal	Aluminum	Stainless steel
°F	°C	$\epsilon = 0.95$	$\epsilon = 0.2$	$\epsilon = 0.4$
10	5	0.53 (0.093)	0.90 (0.158)	0.81 (0.142)
25	14	0.52 (0.091)	0.88 (0.155)	0.79 (0.139)
50	28	0.50 (0.088)	0.86 (0.151)	0.76 (0.133)
75	42	0.48 (0.084)	0.84 (0.147)	0.75 (0.132)
100	55	0.46 (0.081)	0.80 (0.140)	0.72 (0.126)

B. R_s Values with Wind Velocities

Wind velocity		Plain, fabric dull metal		Aluminum		Stainless steel	
mi/h	km/h	mi/h	km/h	mi/h	km/h	mi/h	km/h
5	8	0.35	0.06	0.41	0.07	0.40	0.07
10	16	0.30	0.05	0.35	0.06	0.34	0.06
20	32	0.24	0.04	0.28	0.05	0.27	0.05

*For heat-loss calculations, the effect of R_s is small compared with R_I, so the accuracy of R_s is not critical. For surface-temperature calculations, R_s is the controlling factor and is, therefore, quite critical. The values presented in Table 5-1 are commonly used values for piping and flat surfaces. More precise values based on surface emittance and wind velocity can be found in the referenced texts.

Source: Johns-Manville, Ref. 5

R_s values for three common surface types, temperature differentials, and wind velocities. These values will be used in subsequent heat-transfer calculations.

Since thermal resistances are additive, they are very convenient to work with in calculating heat transfer. From the basic definition,

$$Q = \frac{\Delta t}{R_I + R_s} = \frac{\Delta t}{tk/k + R_s} \tag{1}$$

becomes the basic equation for heat transfer through insulation. Use of this equation is illustrated later.

Insulation Effectiveness

Before leaving the fundamentals, the importance of insulation should be illustrated. Table 5-2 shows the amount of heat transfer from a bare surface at a given temperature differential. Essentially, these values are calculated from the fundamental heat-transfer equation [Eq. (1)] with the insulation thickness being zero. Listed below are three different sets of operating conditions for an 8-in (20-cm) pipe. To show the effectiveness of insulation, heat losses are shown for the bare pipe and for the pipe with 1 in of fiberglass insulation applied, even though a greater thickness would normally be used.

Operating temp., °F	*Ambient temp., °F*	*Bare heat loss Btu/(h)(ft)*	*Heat loss with 1 in fiberglass, Btu/(h)(ft)*	*Reduction in heat loss, %*
200	80	617	70	88.7
350	80	1882	188	90.1
500	80	3998	353	91.2

It is obvious that insulation should be used in all cases where the energy being lost is costly, useful energy that needs to be conserved.

TABLE 5-2 Heat Loss From Bare Surfaces*

Nominal pipe size, inches	Temperature difference, °F															
	50	100	150	200	250	300	350	400	450	500	550	600	700	800	900	1000
½	22	47	79	117	162	215	279	355	442	541	650	772	1047	1364	1723	2123
¾	27	59	99	147	203	269	349	444	552	677	812	965	1309	1705	2153	2654
1	34	75	124	183	254	336	437	555	691	846	1016	1207	1637	2133	2694	3320
1¼	42	94	157	232	321	425	552	702	873	1070	1285	1527	2071	2697	3406	4198
1½	49	107	179	265	367	487	632	804	1000	1225	1471	1748	2371	3088	3899	4806
2	61	134	224	332	459	608	790	1004	1249	1530	1837	2183	2961	3856	4870	6002
2½	74	162	271	401	556	736	956	1215	1512	1852	2224	2643	3584	4669	5896	7267
3	89	197	330	489	677	897	1164	1480	1841	2256	2708	3219	4365	5685	7180	8849
3½	102	225	377	558	773	1024	1329	1690	2102	2576	3092	3675	4984	6491	8198	10100
4	115	254	424	628	869	1152	1496	1901	2365	2898	3479	4135	5607	7304	9224	11370
4½	128	282	471	698	965	1280	1662	2113	2628	3220	3866	4595	6231	8116	10250	12630
5	142	313	524	776	1074	1424	1848	2350	2923	3582	4300	5111	6931	9027	11400	14050
6	169	373	624	924	1279	1696	2201	2799	3481	4266	5121	6086	8254	10750	13580	16730
7	195	430	719	1064	1473	1952	2534	3222	4007	4910	5894	7006	9501	12380	15630	19260
8	220	486	813	1203	1665	2207	2865	3643	4531	5552	6666	7922	10740	13990	17670	21780
9	246	542	907	1343	1859	2464	3198	4066	5057	6197	7440	8842	11990	15620	19720	24310
10	275	606	1014	1502	2078	2755	3576	4547	5655	6930	8320	9888	13410	17470	22060	27180
11	300	661	1106	1638	2267	3005	3901	4960	6169	7560	9076	10790	14630	19050	24060	29660
12	326	718	1202	1779	2463	3265	4238	5338	6701	8212	9859	11720	15890	20700	26140	32210
14	357	783	1319	1952	2703	3582	4650	5912	7354	9011	10820	12860	17440	22710	28680	35350
16	408	901	1508	2232	3090	4096	5317	6759	8407	10300	12370	14700	19940	25970	32790	40410
18	460	1015	1698	2514	3480	4612	5987	7612	9467	11600	13930	16550	22450	29240	36930	45510
20	510	1127	1885	2790	3862	5120	6646	8449	10510	12880	15460	18380	24920	32460	40990	50520
24	613	1353	2263	3350	4638	6148	7980	10150	12620	15460	18570	22060	29920	38970	49220	60660
30	766	1690	2827	4186	5795	7681	9971	12680	15770	19320	23200	27570	37390	48700	61500	75790
Flat	98	215	360	533	738	978	1270	1614	2008	2460	2954	3510	4760	6200	7830	9650

*Losses given in Btu per hour per linear foot of bare pipe at various temperature differences and Btu per hour per square foot for flat surfaces.
Source: Reference 3.

PIPE INSULATION MATERIALS

Table 5-3 lists the principal insulations being used along with their important properties. To assure accurate information, current data sheets from the manufacturer should always be consulted before specifying a particular product. Table 5-3 provides a brief description of each type of material together with the benefits and drawbacks of each.

Calcium Silicate

Lime, silica, and reinforcing fibers are used to form these rigid insulations. Since no organic binders are used, the products are noncombustible and maintain their physical integrity to very high temperatures. These materials are known for their exceptional durability and strength and are the high-quality standard in industrial plant environments where physical abuse is always a problem. The calcium silicate insulations are more costly than fibrous materials but are more thermally efficient at higher temperatures.

Cellular Glass

This product is also rigid and completely inorganic, being composed of millions of completely sealed glass cells. It is unique among the insulation materials in that it is a totally closed cell and will not absorb any liquids or vapors. Although its thermal conductivity is higher than most of the other insulations', the product is widely used in below-ambient-temperatures and buried applications where moisture is often a problem. Similarly, lines carrying volatile liquids often use cellular glass around valves and fittings to minimize the hazard of a saturated insulation. The product is load-bearing, but it is also brittle and subject to thermal shock at high temperatures.

Expanded Perlite

A naturally occurring material, perlite is expanded at high temperature to form a structure of tiny air cells within a vitrified product. Organic and inorganic binders are used in conjunction with reinforcing fibers to hold the perlite structure together. At temperatures below the organic oxidation point, the products have very low moisture absorption, but this increases at elevated temperatures. The products are rigid and load-bearing, but are more brittle and less thermally efficient than the calcium silicate materials.

Mineral Fiber and Rock Wool

Formed from molten rock or slag, these products are more refractory than fiberglass, but they utilize similar organic binders for structural integrity. The products are not load-bearing and contain varying amounts of unfiberized material. The greatest drawback of the mineral wool materials is the short fiber length which allows vibration and physical abuse to cause severe damage, particularly after the organic binder is oxidized. However, the products are less costly than most of the rigid insulations.

Glass Fiber

Fiberglass pipe insulations are formed from molten glass and bonded with organic resins. The principal products for commercial piping systems are the one-piece molded insulations which install rapidly by hinging open and then closing around the pipe. Flexible wraparound products are also available for large-diameter pipes and vessels. The fiberglass products are very thermally efficient and easy to work with. The only issue surrounding their use is for applications above the binder oxidation temperature of 400 to 500°F. Most manufacturers rate their products above the binder temperature and are confident of its performance due to the fiber matrix composed of long glass fibers. How-

TABLE 5-3 Properties of Various Industrial Pipe Insulation Materials

Insulation type	Temp. range, °F (°C)	Thermal Conductivity (Btu)(in)/(h)/(ft^2)/(°F) at T mean, °F(°C) (W·mm/m^2·°C)			Compresive strength, lb/in^2 (kPa) at % deformation	Classification or flame-spread smoke developed*	Cell structure (permeability and moisture absorption*
		75°F (42°C)	200°F (93°C)	500°F (260°C)			
Calcium silicate	to 1500 (815)	(41.6) 0.37 (52)	(93.3) 0.41 (59)	(260) 0.53 (76)	100 to 250 @ 5% (689 to 1722)	Noncombustible	Open cell
Cellular glass	−450 to 900 (268 to 482)	0.38 (54)	0.45 (65)	0.72 (104)	100 @ 5% (689)	Noncombustible	Closed cell
Expanded perlite	to 1500 (815)	—	0.46 (66)	0.63 (91)	90 @ 5% (620)	Noncombustible	Open cell
Mineral fiber	to 1900 (1038)	0.23 to 0.34 (33 to 49)	0.28 to 0.39 (40 to 56)	0.45 to 0.82 (65 to 118)	1 to 18 @ 10% (6.9 to 124)	Noncombustible to 25/50	Open cell
Glass fiber	to 850 (454)	0.23 (33)	0.30 (43)	0.62 (89)	.02 to 3.5 @ 10% (.13 to 24)	Noncombustible to 25/50	Open cell
Urethane foam	−100 to −450 to 225 (−73 to −268 to 107)	0.16 to 0.18 (23 to 26)	—	—	16 to 75 @ 10% (110 to 516)	25 to 75 140 to 400	95% closed cell
Isocyanurate foam	to 350 (177)	0.15 (22)	—	—	17 to 25 @ 10% (117 to 172)	25 55 to 100	93% closed cell
Phenolic foam	−40 to 250 (−40 to 121)	0.23 (33)	—	—	13 to 22 @ 10% (89 to 151)	25/50	Open cell
Elastomeric closed cell	−40 to 220 (−40 to 104)	0.25 to 0.27 (36 to 39)	—	—	40 @ 10% (275)	25 to 75 115 to 490	Closed cell

*FHC from ASTM E 84 and UL 723 which indexes flame and smoke development to that generated by Red Oak which has 100/100 FHC.

ever, there are certain applications which combine severe vibration with high temperature. In such cases, the fiberglass products, like the mineral wools, may tend to sag on the pipe and lose some of their efficiency.

Foams

Four products in the general foam category are now being used, primarily for cold service and plumbing applications. Polyurethane and isocyanurate foamed plastics offer the lowest thermal conductivity since the cells are filled with fluorocarbon blowing agents which are heavier than air. However, the products still need to be sealed to prevent the migration of air and water vapor back into the cells. There have been problems with urethanes regarding both dimensional stability and fire safety; the isocyanurates were developed, in part, to deal with these problems. However, a 25/50 FHC (fire hazard classification) is still being sought for these materials.

The phenolic foams have the required level of fire safety but do not offer thermal efficiencies much different from that of fiberglass. They are limited in temperature range, but their rigid structure allows them to be used without special pipe saddle-supports on small lines.

Elastomeric closed-cell materials are the fourth type and are used primarily in plumbing and refrigeration work. These products are both flexible and closed-cell, permitting rapid installation without an additional vapor barrier for moderate design conditions. The major problem is smoke generation in excess of the 25/50 FHC rating; this restricts their use in certain areas.

MATERIALS SELECTION AND APPLICATION

There are many factors involved in selecting the best insulation for a specific application. First, the service requirements and location must be analyzed to determine which products will, for example, meet the pipe temperature requirements and also be able to withstand the abuse anticipated. Also, special considerations such as fire protection, removability, and chemical resistance must all be reviewed in the selection process.

Finally, three insulation cost factors should be carefully analyzed. The *initial cost* of material and installation is straightforward; competitive products can be readily compared. *Maintenance costs* are not as clear-cut and vary greatly from one material to another, with the softer materials usually requiring more maintenance in industrial environments. The last cost element is that of the *heat lost* through the insulation system. If the same thickness is specified for several different materials, the product with the lowest thermal conductivity will provide the lowest lost-heat cost. Similarly, a material that requires less maintenance may well retain its original performance longer than a material which is easily damaged and degraded. All of these costs should be reviewed before choosing the lowest bid package alternative which only deals with initial costs.

The following paragraphs deal with insulation selection based on operating temperature, design considerations, and common industry usage.

Cryogenic

Cryogenic insulation systems [−455 to −150°F (−271 to −101°C)] are usually custom-engineered due to the critical nature of such service. The two problems encountered are moisture migration and insulation efficiency. Unequal vapor pressure serves to drive moisture to the cold pipe which results ultimately in ice formation and the subsequent deterioration of the insulation. Multiple vapor barriers are used to prevent this, and cellular glass adds additional security since it is totally closed-cell. The system must also be thermally efficient so as to keep the outside surface temperature above the ambient dew point. This may require outlandish thicknesses of conventional insulations, although the more efficient plastic foams are frequently used for this reason. The most severe applications employ vacuum products with multiple layers of reflective foil or vacuum cavities with powder fill.

Low-Temperature

Refrigeration, plumbing, and HVAC systems operate in the range of −150 to 212°F (−101 to 100°C). Cellular glass, fiberglass, and foam products are the most frequently used products. Vapor-barrier requirements are still important for all applications below ambient temperatures, but the level of protection needed becomes less as ambient conditions are approached. Applications above ambient require little special attention with the exception that the plastic foams begin to reach their temperature limits around 200°F (93°C).

Much of this temperature service is in residential and commercial construction. A variety of fire codes are in force for insulation—ranging from a requirement for noncombustibility to allowing smoke development up to 400. Many codes call for a flame spread of 25 or less, and products that carry a composite 25/50 FHC are suitable for virtually all applications. In general, foam products and cellular glass are used for most of the below-freezing applications while fiberglass is the primary product used for chilled water and above-ambient work. Also, premolded PVC fitting covers with fiberglass inserts are frequently used as a quick and efficient way of insulating the wide variety of fittings encountered.

Intermediate-Temperature

Almost all steam and hot-process piping systems operate in this temperature range: 212 to 1000°F (100 to 538°C). The products most frequently used are calcium silicate, fiberglass, mineral wool, and expanded perlite. Choices are usually based on thermal conductivity and resistance to physical abuse where applicable. The fiberglass products are the most thermally efficient in the lower temperatures, with calcium silicate being the best at higher temperatures. *It is important to use the insulation mean or average temperature when comparing thermal conductivities* in order to make the proper comparison.

The fiberglass pipe insulations reach their temperature limits within this range, usually at 500, 650, or 850°F (260, 343, or 454°C). However, they all have organic binders (as do the mineral wools) that begin to oxidize from 400 to 500°F (204 to 260°C). This should be recognized when applying the products to piping above this temperature, but should not be used as an arbitrary cutoff point for fibrous products unless the service conditions are severe enough to warrant it. Vibration and physical abuse conditions should be analyzed for each particular application.

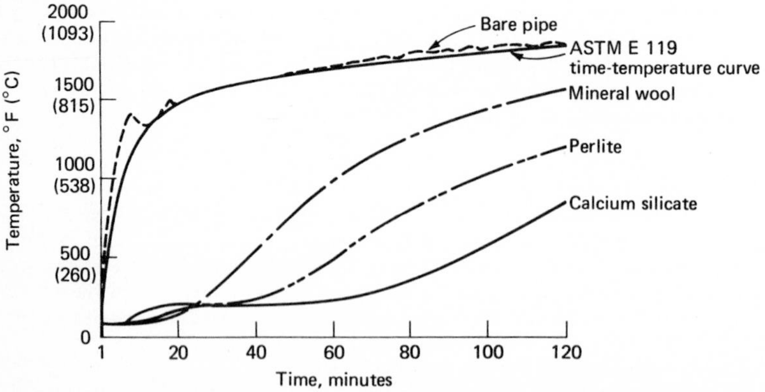

Figure 5-2 Fire-resistance test data for pipe insulation. *(Journal of Thermal Insulation.[6])*

Fire safety is another concern, in this instance from the standpoint of fire protection rather than fire hazard. Figure 5-2 shows the results of fire tests run with three high-temperature materials. The water of hydration in the calcium silicate prolonged the time required for the steel pipe to reach an unacceptable temperature.

Valves and fittings are most often insulated with field-mitered sections of the standard pipe insulation or shop-fabricated fitting covers. There is limited usage of mesh-enclosed blankets which are laced around large valves. Also, cellular glass valve and flange covers are often used in conjunction with other insulations when the fluid being transported is volatile or has a low flash point.

Reviewing the entire range, fiberglass insulation is most often used in lower-temperature applications where physical abuse is not a problem, because of its thermal efficiency and rapid installation. Calcium silicate is the standard type for higher-temperature work and in severe service. Expanded perlite provides the benefit of moisture resistance at lower temperatures, and mineral wool offers initial cost economies for higher-temperature work.

High-Temperature

Superheated steam, exhaust ducting, and some process operations are the few piping applications in the temperature range from 1000 to 1600°F (538 to 871°C). Again, calcium silicate, expanded perlite, and mineral wool are the usual materials employed. These products reach their temperature limits within this range, and the higher temperature requirements are usually met by ceramic fiber blanket materials wrapped around the piping.

Installation Techniques and Specifications

The proper installation of pipe insulation is critical to the in-place performance of the product. References 7 and 8 provide detailed schematics and procedures for such work and should be consulted by plant maintenance workers involved in insulation application. Similarly, the proper selection of protective coatings and jackets will greatly influence the long-term product performance. These references, along with technical data from the coating and jacketing manufacturers, provide appropriate information for proper selection.

INSULATION THICKNESS AND HEAT-LOSS CALCULATIONS

After the material is selected, an appropriate thickness of the material as well as the heat loss or gain with that thickness must be determined. The proper amount of insulation to use depends on the *thermal design objective* of the system: What is the insulation supposed to accomplish? There are four broad categories which encompass most insulation objectives:

1. Personnel protection
2. Condensation control
3. Process control
4. Economics

Terminology

The following symbols, definitions, and units will be used throughout the calculations.

t_a = ambient temperature, °F (°C)

t_s = surface temperature of insulation next to ambient, °F (°C)

t_h = hot-surface temperature, normally the operating temperature (cold-surface temperature in cold applications), °F (°C)
k = thermal conductivity of insulation, always determined at mean temperature, (Btu)(in)/(h)(ft^2)(°F) [W·mm/m^2·°C)]
t_m = $(t_h + t_s)/2$, mean temperature of insulation, °F (°C)
tk = thickness of insulation, in (mm)
r_1 = actual outer radius of steel pipe or tubing, in (mm)
r_2 = $r_1 + tk$, radius to outside of insulation on piping, in (mm)
eq tk = $r_2 \ln (r_2/r_1)$, equivalent thickness of insulation on a pipe, in (mm)
f = surface air film coefficient, Btu/(h)(ft^2)(°F) [W/m^2·°C]
$1/f$ – surface resistance, (h)(ft^2)(°F)/Btu [m^2·°C/W]
R_I = tk/k, thermal resistance of insulation, (h)(ft^2)(°F)/Btu [m^2·°C/W]
Q_F = heat flux through a flat surface, Btu/(h)(ft^2) [W/m^2]
Q_p = $Q_F \times 2\pi r_2/12$, heat flux through a pipe, Btu/(h)(ft) (For SI, Q_p and $Q_F \times 2\pi r_2$, W/m)
Δ = difference by subtraction, unitless
RH = relative humidity, percent
DP = dew-point temperature, °F (°C)

SI Conversions from USCS Units

$$k_{\text{Eng}} \times 144.2279 = k_{\text{SI}}$$
$$R_{\text{Eng}} \times 0.17611 = R_{\text{SI}}$$
$$Q_{F,\,\text{Eng}} \times 3.155 = Q_{F,\,\text{SI}}$$
$$Q_{p,\,\text{Eng}} \times 0.962 = Q_{p,\,\text{SI}}$$

Specific SI units use kelvins rather than degrees Celsius, but in Δ calculations, there is no difference between the two. Degrees Celsius is used here for convenience.

Calculation Fundamentals

There are two concepts that must be understood before proceeding with the calculations. First, in a system in steady-state equilibrium, the heat transfer through each portion of the system is the same. Specifically, the heat transfer through the insulation is equal to the heat transfer from the insulation surface to the surrounding air.

Since the heat loss is proportional to both the temperature difference and thermal resistance of any portion of the system, the following equations result:

$$Q = \frac{t_h - t_a}{R_I + R_s} = \frac{t_h - t_s}{R_I} = \frac{t_s - t_a}{R_s} \tag{2}$$

The second principle relates to the equivalent-thickness concept of pipe insulation. Since the insulation surface area is greater than the bare pipe surface, the pipe surface "sees" a greater insulation thickness than is actually there. This equivalent thickness (*eq tk*) is used as the insulation thickness for the thermal calculation, but the actual thickness is used for calculating the physical number of square feet of surface area per linear foot. The equation for equivalent thickness is

$$eq\ tk = r_2 \ln \frac{r_2}{r_1} \tag{3}$$

where r_1 and r_2 represent the inner and outer radii of the insulation system, respectively. Figure 5-3 provides for easy conversion to equivalent thickness from any specific thickness and pipe size. Some materials are manufactured in true even thicknesses, whereas others follow the ASTM schedule of simplified thicknesses[9] to accommodate proper nesting of double-layer construction. In any event, the proper determination of insulation resistance for pipe insulation is

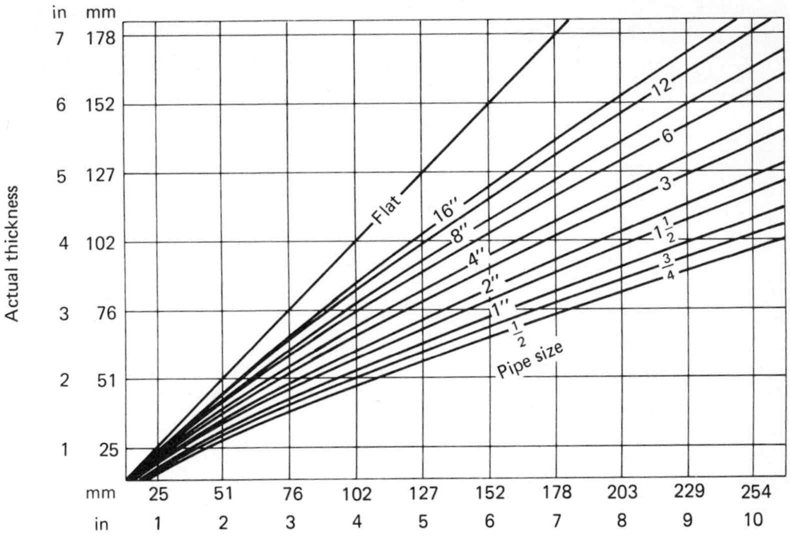

Figure 5-3 Equivalent thickness chart.[5]

$$R_I = \frac{eq\ tk}{k}$$

Figures 5-4 and 5-5 provide typical thermal conductivity values for sample calculations.

Personnel Protection

To protect workers from getting burned on hot piping, the insulation surface temperature should be within the safe touch range of 130 to 150°F; 140°F (60°C) is the temperature most often specified. Both ambient temperature and surface resistance R_s are important in this calculation. As noted in Table 5-1, the R_s values for aluminum are

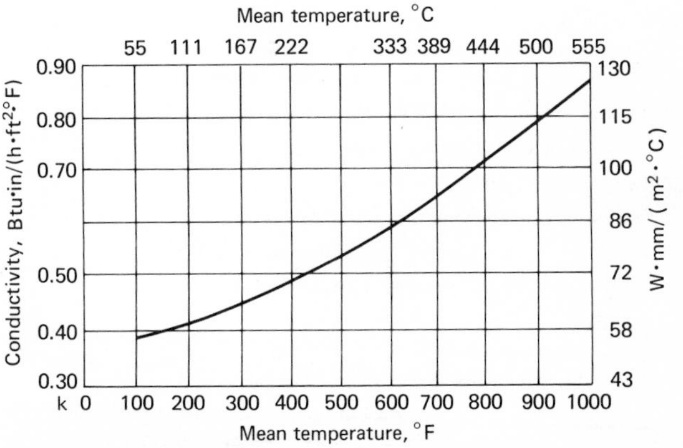

Figure 5-4 Thermal conductivity for typical calcium silicate pipe insulation.

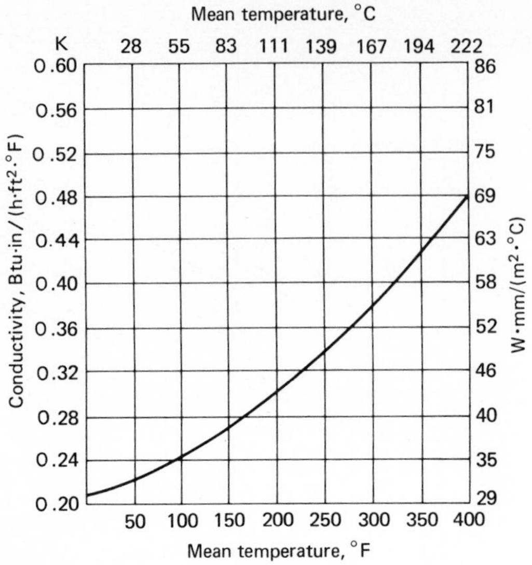

Figure 5-5 Thermal conductivity for typical fiberglass pipe insulation.

greater than for dull surfaces and will result in higher surface temperatures for the aluminum over the same thickness of insulation.

To calculate the thickness required to achieve a specific surface temperature, the basic heat loss equation is manipulated as follows:

$$Q = \frac{t_h - t_s}{R_I} = \frac{t_s - t_a}{R_s}$$

$$R_I = R_s \frac{t_h - t_s}{t_s - t_a} \qquad \text{and} \qquad R_I = \frac{eq\ tk}{k}$$

$$\therefore\ eq\ tk = kR_s \frac{t_h - t_s}{t_s - t_a}$$

EXAMPLE An 8-in pipe operating at 700°F (371°C) in an 85°F (29°C) ambient with aluminum jacketing. Determine the thickness of calcium silicate required to keep the surface at 140°F (60°C).

Step 1. Determine k at $t_m = (740 + 140)/2 = 420°\text{F}$:

$$k = 0.49\ (\text{Btu})(\text{in})/(\text{h})(\text{ft}^2)(°\text{F}) \qquad [71\ \text{W}\cdot\text{mm}/(\text{m}^2)(°\text{C})]$$

from Fig. 5-4 for calcium silicate.

Step 2. Determine R_s from Table 5-1 for aluminum:

$$t_s - t_a = 140 - 85 = 55°\text{F}$$

So, $R_s = 0.85\ (\text{h})(\text{ft}^2)(°\text{F})/\text{Btu}\ [0.1497\ (\text{m}^2)(°\text{C})/\text{W}]$

Step 3. Calculate

$$eq\ tk = (0.49)(0.85)\frac{700 - 140}{140 - 85}$$

$$= 4.24\ \text{in}$$

$$\text{In SI} = (71)(0.1497)\frac{371.11 - 60}{60 - 29.44}$$

$$= 10.8\ \text{mm}$$

Step 4. Determine from Fig. 5-3 that 4.24 in *eq tk* on an 8-in pipe can be accomplished with 3.25-in (83 mm) insulation. Always rounding off to the next half-inch increment, the recommendation would be 3.5-in (89 mm) calcium silicate.

Condensation Control

On cold systems, the insulation thickness must be sufficient to keep the insulation surface above the dew point of the ambient air. Table 5-4 gives the dew point for various temperature and relative-humidity combinations. The calculation is the same as for personnel protection with the surface temperature taking on the value of the dew-point temperature:

$$eq\ tk = kR_s \frac{(t_h - \mathrm{DP})}{(\mathrm{DP} - t_a)}$$

EXAMPLE A 4-in diam chilled water line is operating at 40°F (4.4°C) in an ambient temperature of 90°F (32.2°C) and 90 percent RH. Determine the amount of fiberglass pipe insulation with a kraft jacket required to prevent condensation.

Step 1. The dew point from Table 5-4 for 90°F, 90 percent RH is DP = 87°F (30.6°C)

Step 2. Determine k at $t_m = (40 + 87)/2 = 63.5$°F (17.5°C); $k = 0.23$ from Fig. 5-5 for fiberglass

Step 3. Determine R_s from Table 5-1 for fabric jacket and $t_a - \mathrm{DP} = 90 - 87 = 3$°F (−16.1°C). ∴ $R_s = 0.54$

Step 4. Calculate $eq\ tk = (0.23)(0.54)\ (40 - 87)/(87 - 90) = 1.95$ in (49.5 mm)

Step 5. Actual thickness required from Fig. 5-3 for a 4-in (101-mm) pipe is 1.5 in (38.1-mm). It would not be unreasonable to specify 2-in (50.8-mm) thickness for this application since 1.5 in is borderline.

Table 5-5 gives the thickness of fiberglass pipe insulation required to prevent condensation for various operating temperatures and three different ambient conditions.

Process Control

There are many complex flow calculations required to determine the amount of insulation required to maintain process temperature or allow a specific temperature drop. The simple calculation below is for determining heat loss only and is basic to all of the process control calculations. The heat loss from a pipe is

$$Q_p = Q_F \frac{2\pi r_2}{12} = \frac{t_h - t_a}{R_I + R_s} \frac{2\pi r_2}{12} \qquad \text{Btu/(h)(ft)} \tag{4}$$

with

$$R_I = \frac{eq\ tk}{k}$$

In this calculation, a surface temperature t_s must first be estimated to arrive at a t_m, k, and R_s. Then, when the t_s estimate is checked, the calculation can be redone with the new estimate if necessary.

EXAMPLE A 16-in diam steam line operating at 850°F (454°C) in an 80°F (27°C) ambient temperature is insulated with 3.5 in (89mm) of calcium silicate with an aluminum jacket. Determine the heat loss per linear foot and the surface temperature.

Step 1. Assume $t_s = 140$°F, $t_m = \dfrac{850 + 140}{2} = 495$°F; k from Fig. 5-4 for calcium silicate is 0.53

Step 2. Determine R_s for aluminum from Table 5-1 for $t_s - t_a = 60$°F; $R_s = 0.85$.

Step 3. Determine *eq tk* from Fig. 5-3 for 3.5 in on a 16-in pipe, $eq\ tk = 4.2$ in:

$$R_I = \frac{\text{eq}\ tk}{k} = \frac{4.2}{0.53} = 7.92$$

TABLE 5-4 Dew-Point Temperature

Dry-bulb temp, °F	Percent Relative Humidity 10	15	20	25	30	35	40	45	50
5	−35	−30	−25	−21	−17	−14	−12	−10	−8
10	−31	−25	−20	−16	−13	−10	−7	−5	−3
15	−28	−21	−16	−12	−8	−5	−3	−1	1
20	−24	−16	−11	−8	−4	−2	2	4	6
25	−20	−15	−8	−4	0	3	6	8	10
30	−15	−9	−3	2	5	8	11	13	15
35	−12	−5	1	5	9	12	15	18	20
40	−7	0	5	9	14	16	19	22	24
45	−4	3	9	13	17	20	23	25	28
50	−1	7	13	17	21	24	27	30	32
55	3	11	16	21	25	28	32	34	37
60	6	14	20	25	29	32	35	39	42
65	10	18	24	28	33	38	40	43	46
70	13	21	28	33	37	41	45	48	50
75	17	25	32	37	42	46	49	52	55
80	20	29	35	41	46	50	54	57	60
85	23	32	40	45	50	54	58	61	64
90	27	36	44	49	54	58	62	66	69
95	30	40	48	54	59	63	67	70	73
100	34	44	52	58	63	68	71	75	78
105	38	48	56	62	67	72	76	79	82
110	41	52	60	66	71	77	80	84	87
115	45	56	64	70	75	80	84	88	91
120	48	60	68	74	79	85	88	92	96
125	52	63	72	78	84	89	93	97	100
−15.0	−37.2	−34.4	−31.7	−29.4	−27.2	−25.6	−24.4	−23.3	−22.2
−12.2	−35.0	−31.7	−28.9	−26.7	−25.0	−23.3	−21.7	−20.6	−19.4
−9.4	−33.3	−29.4	−26.7	−24.4	−22.2	−20.6	−19.4	−18.3	−17.2
−6.7	−31.1	−26.7	−23.9	−22.2	−20.0	−18.9	−16.7	−15.6	−14.4
−3.9	−28.9	−26.1	−22.2	−20.0	−17.8	−16.1	−14.4	−13.3	−12.2
−1.1	−26.1	−22.8	−19.4	−16.7	−15.0	−13.3	−11.7	−10.6	−9.4
1.7	−24.4	−20.6	−17.2	3.0	−12.8	−11.1	−9.4	−7.8	−6.7
4.4	−21.7	−17.8	−15.0	−12.8	−10.0	−8.9	−7.2	−5.6	−4.4
7.2	−20.0	−16.1	−12.8	−10.6	−8.3	−6.7	−5.0	−3.9	−2.2
10.0	−18.3	1.0	−10.6	−8.3	−6.1	−4.4	−2.8	−1.1	0.0
12.8	−16.1	−11.7	−8.9	−6.1	−3.9	−2.2	0.0	1.1	2.8
15.6	−14.4	−10.0	−6.7	−3.9	−1.7	0.0	1.7	3.9	5.6
18.3	−12.2	−7.8	−4.4	−2.2	0.6	3.3	4.4	6.1	7.8
21.1	−10.6	−6.1	−2.2	0.6	2.8	5.0	7.2	8.9	10.0
23.9	−8.3	−3.9	0.0	2.8	5.6	7.8	9.4	11.1	12.8
26.7	−6.7	−1.7	1.7	5.0	7.8	10.0	12.2	13.9	15.6
29.4	−5.0	0.0	4.4	7.2	10.0	12.2	14.4	16.1	17.8
32.2	−2.8	2.2	6.7	9.4	12.2	14.4	16.7	18.9	20.6
35.0	−1.1	4.4	8.9	12.2	15.0	17.2	19.4	21.1	22.8
37.8	1.1	6.7	11.1	14.4	17.2	20.0	21.7	23.9	25.6
40.6	3.3	8.9	13.3	16.7	19.4	22.2	24.4	26.1	27.8
43.3	5.0	11.1	15.6	18.9	21.7	25.0	26.7	28.9	30.6
46.1	7.2	13.3	17.8	21.1	23.9	26.7	28.9	31.1	32.8
48.9	8.9	15.6	20.0	23.3	26.1	29.4	31.1	33.3	35.6
51.7	11.1	17.2	22.2	25.6	28.9	31.7	33.9	36.1	37.8

Percent relative humidity									
55	60	65	70	75	80	85	90	95	100
−6	−5	−4	−2	−1	1	2	3	4	5
−2	0	2	3	4	5	7	8	9	10
3	5	6	8	9	10	12	13	14	15
8	10	11	13	14	15	16	18	19	20
12	15	16	18	19	20	21	23	24	25
17	20	22	23	24	25	27	28	29	30
22	24	26	27	28	30	32	33	34	35
26	28	29	31	33	35	36	38	39	40
30	32	34	36	38	39	41	43	44	45
34	37	39	41	42	44	45	47	49	50
39	41	43	45	47	49	50	52	53	55
44	46	48	50	52	54	55	57	59	60
49	51	53	55	57	59	60	62	63	65
53	55	57	60	62	64	65	67	68	70
57	60	62	64	66	69	70	72	74	75
62	65	67	69	72	74	75	77	78	80
67	69	72	74	76	78	80	82	83	85
72	74	77	79	81	83	85	87	89	90
76	79	82	84	86	88	90	91	93	95
81	84	86	88	91	92	94	96	98	100
85	88	90	93	95	97	99	101	103	105
90	92	95	98	100	102	104	106	108	110
94	97	100	102	105	107	109	111	113	115
99	102	105	107	109	112	114	116	118	120
104	107	109	111	114	117	119	121	123	125
−21.1	−20.6	−20.0	−18.9	−18.3	−17.2	−16.7	−16.1	−15.6	−15.0
−18.9	−17.8	−16.7	−16.1	−15.6	−15.0	−13.9	−13.3	−12.8	−12.2
−16.1	−15.0	−14.4	−13.3	−12.8	−12.2	−11.1	−10.6	−10.0	−9.4
−13.3	−12.2	−11.7	−10.6	−10.0	−9.4	−8.9	−7.8	−7.2	−6.7
−11.1	−9.4	−8.9	−7.8	−7.2	−6.7	−6.1	−5.0	−4.4	−3.9
−8.3	−6.7	−5.6	−5.0	−4.4	−3.9	−2.8	−2.2	−1.7	−1.1
−5.6	−4.4	−3.3	−2.8	−2.2	−1.1	0.0	0.6	1.1	1.7
−3.3	−2.2	−1.7	−0.6	0.6	1.7	2.2	3.3	3.9	4.4
−1.1	0.0	1.1	2.2	3.3	3.9	5.0	6.1	6.7	7.2
1.1	2.8	3.9	5.0	5.6	6.7	7.2	8.3	9.4	10.0
3.9	5.0	6.1	7.2	8.3	9.4	10.0	11.1	11.7	12.8
6.7	7.8	8.9	10.0	11.1	12.2	12.8	13.9	15.0	15.6
9.4	10.6	11.7	12.8	13.9	15.0	15.6	16.7	17.2	18.3
11.7	12.8	13.9	15.6	16.7	17.8	18.3	19.4	20.0	21.1
13.9	15.6	16.7	17.8	18.9	20.6	21.1	22.2	23.3	23.9
16.7	18.3	19.4	20.6	22.2	23.3	23.9	25.0	25.6	26.7
19.4	20.6	22.2	23.3	24.4	25.6	26.7	27.8	28.3	29.4
22.2	23.3	25.0	26.1	27.2	28.3	29.4	30.6	31.7	32.2
24.4	26.1	27.8	28.9	30.0	31.1	32.2	32.8	33.9	35.0
27.2	28.9	30.0	31.1	32.8	33.3	34.4	35.6	36.7	37.8
29.4	31.1	32.2	33.9	35.0	36.1	37.2	38.3	39.4	40.6
32.2	33.3	35.0	36.7	37.8	38.9	40.0	41.1	42.2	43.3
34.4	36.1	37.8	38.9	40.6	41.7	42.8	43.9	45.0	46.1
37.2	38.9	40.6	41.7	42.8	44.4	45.6	46.7	47.8	48.9
40.0	41.7	42.8	43.9	45.6	47.2	48.3	49.4	50.6	51.7

TABLE 5-5 Minimum Insulation Thickness Of Fiberglass Pipe Insulation Needed To Prevent Condensation (Based on Still Air and AP Jacket)

Operating pipe temperature	80°F (26.6°C) & 90% RH			80°F (26.6°C) & 70% RH			80°F (26.6°C) & 50% RH		
		Thickness			Thickness			Thickness	
	Pipe size, in	in	mm	Pipe size, in	in	mm	Pipe size, in	in	mm
0–34°F (−18–1°C)	Up to 1	2	51	Up to 8	1	25	Up to 8	½	13
	1¼ to 2	2½	63	9 to 30	1½	38	9 to 30	1	25
	2½ to 8	3	76						
	9 to 30	3½	89						
35–49°F (1–9°C)	Up to 1½	1½	38	Up to 4	½	13	Up to 30	½	13
	2 to 8	2	51	4½ to 30	1	25			
	9 to 30	3	76						
50–70°F (10–21°C)	Up to 3	1½	38	Up to 30	½	13	Up to 30	½	13
	3½ to 20	2	51						
	21 to 30	2½	63						

Source: Johns-Manville, Ref 5.

Step 4. Calculate heat loss per square foot:

$$Q_F = \frac{850 - 80}{7.92 + 0.85} = 87.8 \text{ Btu/(h)(ft}^2) \qquad [(227 \text{ W/m}^2)]$$

Step 5. Check surface temperature assumption by

$$t_s = t_a + (R_s \times Q_F) = 80 + (0.85 \times 87.8)$$
$$= 155°\text{F}$$

This leads to a new $t_m = 502.5$ which is close enough to assumed $t_m = 495$ that k will not change.

Step 6. Calculate heat loss per linear foot:

$$[\text{nt}Q_p = Q_F \frac{2\pi r_2}{12} = 87.8 \frac{2\pi(8 + 3.5)}{12}$$
$$= 529 \text{ Btu/(h)ft} \qquad (508 \text{ W/m})$$

Freeze Protection

Water lines must either be heat-traced or have a controlled flow to prevent freezing. These calculations can be performed for various combinations of water temperature, line length, and ambient conditions. Table 5-6 is presented as an estimating tool for freeze protection. It shows the number of hours until freezing occurs along with the gallons per minute flow required in a 100-ft (30.3 m) pipe run to prevent freezing. The calculations are based on fiberglass pipe insulation with $k = 0.23$, initial water temperature of 42°F (6°C), and an ambient air temperature of −10°F (−23°C). The flow rate for the 100-ft (30.3-m) length may be prorated for longer or shorter pipes.

Economics

With energy costs on the rise, insulating for the purpose of energy conservation and economics is becoming the standard for most insulation-thickness calculations. The first economic thickness equations were presented in a 1926 paper by L. B. McMillan.[8] Since then, many nomographs and computer programs have been developed to perform these complex calculations.[10] Discounted cash-flow techniques, fuel escalation, maintenance costs, installed insulation costs, and tax effects are all involved in the computations. However, the economic thickness of insulation (ETI) concept is still quite simple. Economic thickness is defined as that thickness of insulation at which the cost of the next

TABLE 5-6 Freeze Protection Data, Hours to Freeze and Flow Rate (gal/min) Required to Prevent Freezing*

	Insulation thickness					
	1 in		2 in		3 in	
Nominal pipe size IPS	Hours to freeze	Flow rate, gal/min per 100 ft	Hours to freeze	Flow rate, gal/min per 100 ft	Hours to freeze	Flow rate, gal/min per 100 ft
½	0.30	0.087	0.42	0.282	0.50	0.053
¾	0.47	0.098	0.66	0.070	0.79	0.058
1	0.66	0.113	0.96	0.078	1.16	0.065
1½	0.90	0.144	1.35	0.096	1.67	0.078
2	1.72	0.169	2.64	0.110	3.31	0.088
2½	2.13	0.195	3.33	0.124	4.24	0.098
3	2.81	0.228	4.50	0.142	5.80	0.110
4	3.95	0.279	6.49	0.170	8.49	0.130
5	5.21	0.332	8.69	0.199	11.54	0.150
6	6.48	0.386	10.98	0.228	14.71	0.170
7	7.66	0.437	13.14	0.255	17.75	0.189
8	8.89	0.487	15.37	0.282	20.89	0.207

*Based upon 42°F (6°C) initial water temperature and −10°F (−23°C) ambient temperature.
Source: Johns-Manville, Ref. 5.

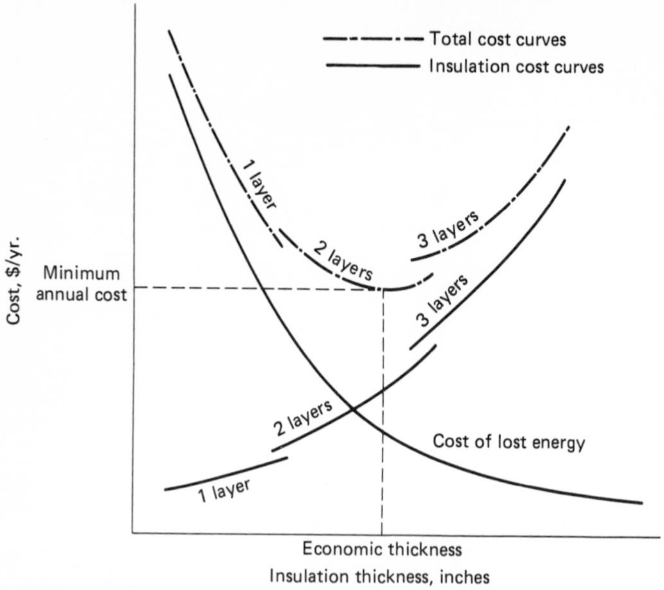

Figure 5-6 Economic thickness of insulation (ETI).

increment is just offset by the energy savings due to that increment over the life of the project.

Figure 5-6 graphically illustrates the ETI procedure which adds the installed insulation costs to the fuel costs to arrive at the point of minimum annual cost.

The latest version of the ETI program is sponsored by the Thermal Insulation Manufacturers Association (TIMA) and the National Insulation Contractors Association (NICA). The computer program is available from TIMA[12] and several insulation manufacturers offer to run the program for clients. Figure 5-7 shows the ETI program output. The first sections simply reproduce the input data used in the calculations. The output data begin with the seven columns of information. Column 1 is the insulation thickness which produces the other results. Column 2 is the annual cost associated with each thickness; the minimum value here corresponds with the economic thickness. Column 3 is the discounted payback period for the investment. Column 4 is the discounted present value of the energy saved per foot over the life of the project. Columns 5 and 6 are the heat loss and surface temperature, respectively, for each thickness. Column 7 is the cost of the installed insulation per linear foot.

The ETI program is quite sophisticated and allows for almost all the variables encountered in insulation design. Many corporate staff and consulting engineers have developed similar programs particularly suited to their specific needs. This is a very positive trend and establishes thermal insulation in its proper role, that of being a primary method of reducing energy consumption and improving overall plant efficiencies.

Economic Thickness of Insulation for Hot Surfaces (ETIHOT)
New and Retrofit Jobs

Project :	Project no:
System :	Location :
Date :	Engineer :
Contact:	

Fuel type: Steam
Present price: 6.00 $/MMBTU
Efficiency: 100.00%

Annual fuel inflation rate	=	10.0%
After-tax value of money	=	10.0%
Incremental heating equipment investment	=	2.00 $/1000 Btu/h
Effective income tax rate	=	46.0%
Heating-plant depreciation period	=	20.0 yr
Economic life of new insulation	=	20.0 yr
Emissivity of jacketing	=	0.90
Percent of new insulation cost for annual insulation maintenance	=	1.0%/yr
Percent of annual fuel bill for heating plant maintenance	=	5.0%/yr
Annual average wind speed	=	0.00 mi/h
Emissivity of present surface	=	0.80

Calculations are performed using linear cost and thickness data

Pipe description		=	8.00-in IPS-horizontal pipe
Operating temperature		=	250.0°F
Ambient temperature		=	80.0°F
Previous Insulation	Code 0	=	None
New insulation	Code 1	=	Fiberglass
Previous thickness of insulation		=	0.0 in
Hours of operation		=	5400.0 per yr
Reference thickness of insulation for payback calculation		=	0.00 in
Piping complexity factor		=	1.00
Conductivities supplied by user	New	=	None supplied
	Old	=	None supplied

Thick. of new insul., in	Annual cost, $/ft	Payback period, yr	Pres. value of heat saved, $/ft	Heat loss, Btu/(h)(ft)	Surf. temp., °F	Insul. cost, $/ft
0.00	35.99	——	0.00	830.40	249.6	0.00
1.50	4.22	0.61	28.03	72.60	96.6	10.04
2.00	3.81	0.74	28.57	58.00	92.7	12.09
2.50	3.64	0.87	28.90	49.09	90.1	14.15
3.00 S.L.	3.60	1.00	29.13	43.05	88.4	16.20
3.00 D.L.	3.83	1.14	29.13	42.92	88.4	18.34

ETI corresponds to the thickness with the lowest annual cost, 3.00 in.

Figure 5-7 TIMA ETI computer program output. Heat savings and present values are projected based on given input variables. Actual values may vary depending on the actual service conditions.

REFERENCES

1. McAdams, W. H.: *Heat Transmission*, McGraw-Hill, New York, 1954.
2. Sparrow, E. M., and R. D. Cess: *Radiation Heat Transfer*, McGraw-Hill, New York 1978.
3. Greebler, P.: "Thermal Properties and Applications of High Temperature Aircraft Insulation," American Rocket Society, 1954. Reprinted in *Jet Propulsion*, November-December 1954.
4. American Society for Testing and Materials: *Annual Book of ASTM Standards*, part 18, Thermal and Cryogenic Insulating Materials Building Seals and Sealants; Fire Tests; Building Constructions; Environmental Acoustics.
5. Johns-Manville Sales Corporation, Industrial Products Division, Technical Data Sheets, Denver, Colo.
6. Kanakia, M., W. Herrera, and F. Hutto, Jr.: "Fire Resistance Tests for Thermal Insulation," *Journal of Thermal Insulation*, vol. 1, no. 4, April 1978, Technomic Pub. Co., Westport, Conn.
7. *Commercial & Industrial Insulation Standards*, Midwest Insulation Contractors Assn., Inc., Omaha, 1979.
8. Malloy, J. F.: *Thermal Insulation*, Reinhold, New York, 1969.
9. ASTM part 18, op. cit., STD C 585.
10. McMillan, L. B.: "Heat Transfer Through Insulation in the Moderate and High Temperature Fields: A Statement of Existing Data," No. 2034, The American Society of Mechanical Engineers, New York, 1934.
11. *Economic Thickness of Industrial Insulation*, Conservation Paper No. 46, Federal Energy Administration, Washington, D.C., 1976. Available from Superintendent of Documents, U.S. Government Printing Office, Washington, DC 20402, Stock #041-018-00115-8.
12. TIMA ETI Computer Program, Thermal Insulation Manufacturers Association, 7 Kirby Plaza, Mt. Kisco, NY 10549.

chapter 10-6

Industrial Hose

by
Charles H. Artus
Product Application Engineer
Gates Rubber Company
Denver, Colorado

INTRODUCTION

There are many varieties of industrial hose and associated couplings. Each type is designed to give satisfactory and safe service life for a particular application where needed for flexible conveyance of liquids, gases, and certain solids in powder or granular form.

To obtain maximum service for any type of industrial hose, the user must: consider the application, select hose and couplings to match the application, take reasonable care of the hose assembly, and observe all applicable safety regulations.

Even regular care and maintenance would help very little to prolong the life of an industrial hose assembly *that was not originally selected to suit the application.* Hoses will soon fail regardless of the care given them in situations that are beyond the design of the hose.

For example: A hose with a natural rubber tube will fail in gasoline service. The tube stock swells, the hose loses its tensile strength and discolors the fluid. High pressures will burst a hose designed for low pressure; coupled assemblies will fail when low-pressure couplings are used on high-pressure hoses. And a hose with a thin tube not resistant to abrasion will soon wear through when handling abrasive products.

So that safest and most reliable service can be obtained, the plant engineer needs to know a few basic things about industrial hose—such as the kinds of hose available, limitations and applications, care and maintenance. We look at these factors one by one.

TYPES OF INDUSTRIAL HOSE

Following are brief descriptions of, and recommendations for, the basic kinds of industrial hose.

Air Hose

Air hose is used for efficient handling of air and compressed air in industrial applications. These hoses can be used for air tools, air drills, air and vapor ducts, agriculture, paint spraying, and welding.

Water Hose

Water hoses are designed for transport of water (but not recommended for use as vibration dampers or in closed water systems). Applications include water suction, water discharge, cleanup, and general-purpose water usage.

Steam Hose

This is a very specialized hose used to convey wet saturated steam, dry saturated steam, and superheated steam.

Materials-Handling Hose

These are specially designed to handle a wide variety of bulk commodities. Applications include: beverage and food, sand suction, dredge sleeves, cement, plaster, vacuums, sand blasting, fish handling, grains and flour.

Acid and Chemical Hose

These hoses are designed to withstand the corrosive effects of caustic, acidic, oxidizing, or chlorinated liquids as well as toxic substances such as anhydrous ammonia.

Petroleum-Transfer Hose

This is manufactured particularly to handle fuel oil, liquid petroleum gases, hot asphalt, gasoline, and diesel fuel.

HOSE CONSTRUCTION

Industrial hose generally consists of three parts—tube, reinforcement, and cover—plus all the necessary cements. See Fig. 6-1.

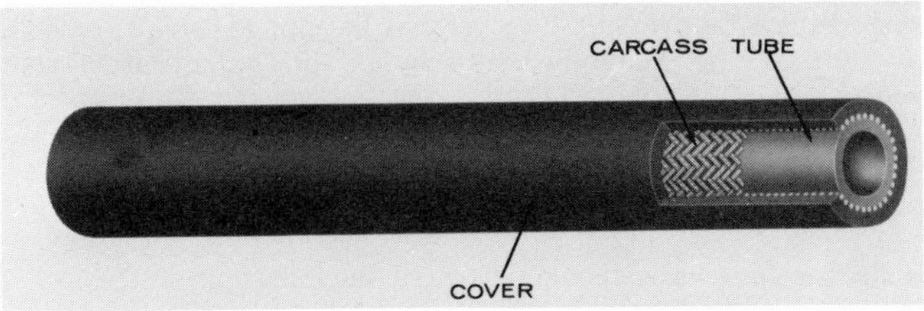

Figure 6-1 Reinforced-hose construction. (*Gates Rubber Company.*)

Tube

The *tube* is the inner layer and its function is to contain the substance being transported and evenly transmit forces, due to pressure, to the reinforcing element. Made of rubber or plastic, it is either extruded or made from sheet stock by spiraling or laminating.

Reinforcement

Outside the tube is the *reinforcement.* This is the material which provides the overall strength, pressure resistance, and sometimes the collapse resistance.

For some industrial hose applications, the service pressure is low enough that the strength of the tube stock is sufficient to meet the required service. This hose is called tubing, or nonreinforced, hose. All other hoses require some type of textile or wire reinforcement. The usual reinforcing materials are textile fibers or filaments or metal wires. Common textile reinforcements are made of cotton, rayon, nylon, and polyester; and they may take the form of a yarn- or fabric-like material. Wire reinforcements vary as to size and composition of the material used to make the wire. Identification of the hose usually relates to the method of reinforcement. For example, a hose with a braid reinforcement would be called a braided hose. Following are the principal forms of reinforcement.

- **Knit.** The knit reinforcement is applied around the tube by a knitting machine. A lock-stitch pattern provides greater resistance to diametrical hose growth. Knit hose is used in low-pressure ranges, seldom exceeding 100 to 120 lb/in^2.
- **Braid**. Braided hose has the reinforcement—either yarn or wire—applied by using horizontal or vertical braiders. The terms vertical and horizontal refer to the axis of the hose as it is being braided. Hose with a braid reinforcement finds applications in many areas, including air, water, hydraulic, and steam.
- **Spiral**. Spiral-reinforced hose has all the wire or textile strands in one layer laid parallel on the hose in one direction. At least two layers, or multiples of two layers, of reinforcement are normally required; with the layers spiraled in alternating directions to form a balanced construction. Additional layers of reinforcement may be applied to increase the service-pressure level, with a usual maximum of six layers. Typical applications of spiral hose range from low-pressure general-purpose to high-pressure hydraulic.
- **Wrapped**. The greatest variety of hoses are those in which woven reinforcement is wrapped onto the tube. These hoses range from ¼-in-ID air hose up to 36 to 48-in-diam suction and discharge hose. Wrapped hoses have the reinforcement applied by either spiraling or laminating the material onto the tube.

Cover

The *cover* is usually a rubber or plastic layer placed over the reinforcement. It protects the reinforcement and serves to keep elements such as weather, abrasion, and chemicals from weakening it.

The application of a cover over the reinforcement is completed in much the same manner as the tube is fabricated. The cover may be applied using an extruder to form a seamless, cylindrical tube around the hose or by spiraling or laminating calendered cover material onto a mandrel-supported hose.

SELECTION OF INDUSTRIAL HOSE

The industrial hose purchaser should determine the following:

1. Inside diameter of hose required
2. Length required
3. Material to be conveyed, including chemical makeup and temperature range
4. Suction and/or discharge rated working pressure
5. Hose ends or fittings required
6. External environment (temperature range, corrosive fluids, complete weather conditions)
7. External stresses (kinking, crushing, pulling, excessive bending)
8. Special requirements, i.e., government and safety regulations, special tests, etc.

When analyzing the cost differences among industrial hoses to be selected for in-plant piping, several factors must be considered.

First consider the method of construction. A vertical braided hose is the least expensive because it is built in long, continuous lengths and its reinforcement consists of yarn. Vertical braided hoses have small inside diameters, up to 1½ in. The horizontal braided hose construction provides larger hose diameters and greater working pressures than the vertical braided construction. Reinforcements consist of one or more layers of braided fibers or wire. Horizontal braided-wire reinforcement is similar to the standard horizontal braided hose with the addition of a reinforcing wire spiralled between the fiber braid reinforcement. Wrapped and wire-reinforced hoses are the most expensive, primarily because they are individually hand-built. Wire spiraled onto the hose between multiple plies of wrapped fabrics also adds to the cost of making these hoses.

Realizing that the mass-produced vertical braided hose is less expensive than wrapped hose and that yarn reinforcement usually costs less than wire, the next cost factor to be considered is the stock type used to construct the tube and cover of the hose. Table 6-1 shows the elastomers most commonly used in industrial hose in order of relative cost, from top to bottom.

The general properties listed in Table 6-1 provide a reliable guideline. However, the user should always follow the manufacturer's recommendations as to the use of any particular rubber composition. This is especially true with respect to the resistance of the rubber composition to the materials it is exposed to, from within and without.

Although the construction methods and materials used to make the hose provide a cost comparison, a true cost index must take into account certain variables. These include the dimensions of the hose, the quantity of materials used, and the changing price of many petroleum-based raw materials.

Table 6-2 is a section of hose types most commonly found in industrial plants. The hoses are grouped by general application, description, and cost index.

In applying any product, the more the plant engineer knows about the details of application, the better the product that can be selected to those needs. *This is particularly true with industrial hose because of the many variables involved.*

The process of selecting a hose usually involves one of two situations:

1. Making the optimum choice from several suitable selections for general application.
2. Making the one choice which satisfies all the requirements of a special application such as transferring anhydrous ammonia or LP gas.

TABLE 6-1 Characteristics of Hose Stock Types and Relative Cost Comparisons*

Common name	Usage in hose	Typical hose application	General properties
EPDM (Ethylene propylene diene)	Tube and cover	Steam (dry or wet), hot water, engine coolant, air (hot or cold), mild chemicals, and generally nonoily products	Excellent ozone, chemical, and aging characteristics; poor resistance to petroleum-based fluids
Natural rubber or synthetic (styrene-butadiene) rubber (SBR)	Tube and cover	Materials handling, water, air, chemicals, and generally nonoily products	Excellent abrasion and good low-temperature resistance; poor resistance to petroleum-based fluids
Buna-N	Tube	Fuel oils, diesel oils, aromatics, gasolines, and other petroleum products special applications	Excellent resistance to petroleum-based fluids; good physical properties; good resistance to heat and chemicals
Neoprene	Tube and cover	General-purpose air and water, saturated steam, materials handling, acid-chemical, and petroleum	Good to excellent abrasion, flame, petroleum, weathering, ozone, and heat resistance
Butyl	Tube and cover	Dry steam, engine coolant, hot air, chemicals, and generally nonoily products	Very good weathering resistance; low permeability to air; good physical properties
Hypalon	Tube and cover	Acid-chemical transfer	Excellent ozone, chemical, and aging resistance; good abrasion and heat resistance; fair resistance to petroleum-based fluids
Gatron	Tube	Acid-chemical transfer	Excellent general chemical, ozone, and weathering resistance; wide temperature flexibility range
FPM	Tube	Acid-chemical transfer	Excellent resistance to aromatic fluids and chlorinated hydrocarbons; outstanding heat resistance; excellent weathering resistance

*Chemical and physical properties are subject to a considerable amount of control through compounding. Characteristics given below for each stock type are therefore generalized to some degree. *These stocks are listed in order of relative costs,* with EPDM being the least expensive and FPM being the most expensive. Neoprene and Buna-N are in the same general cost range.

TABLE 6-2 Hose Types in Industrial Use

Application	Method of construction	Tube stock	Cover stock	Relative* cost index
General-purpose hose				
Low-pressure air or water	Spiral or vertical braided	EPDM	EPDM	1
High-pressure air or water (improved resistance to heat, oil, and abrasion)	Vertical braided	Buna-N	Neoprene	2
Petroleum-product transfer				
Aromatic fuels at full suction	Horizontal braided, wire reinforced	Buna-N	Neoprene	3
Butane-propane	Horizontal braided	Neoprene	Neoprene	4
Steam hose				
Saturated and superheated steam up to 250 lb/in^2 (1.72 MPa) and 450°F (232°C)	Wire braided (one or two layers)	EPDM	EPDM	5
Ssaturated steam up to 200 lb/in^2 (1.38 MPa) and 450°F (232°C)	Wire braided (one or two layers)	EPDM	EPDM	4
Materials-handling hose				
Bulk commodity, highly abrasion resistant	Wire reinforced	SBR	SBR	5
Food handling, primarily oils up to 150°F (65°C)	Wire reinforced	Food-grade neoprene	Neoprene	6
Dusty materials or abrasives in water suspension	Wire reinforced	Gum rubber (to absorb impact)	SBR	8
Acid-chemical transfer hose				
Strong, oxidizing solutions	Horizontal braided, wire reinforced	Hypalon	Neoprene	6
Highly aromatic fluids and chlorinated hydrocarbons	Horizontal braided, wire reinforced	FPM	Neoprene	12

*Using general-purpose hose as least expensive. Relative cost will vary, depending on size.

In the first case, final selection is usually made on the basis of cost. In the second, it is usually a matter of satisfactorily meeting the most important or critical conditions *first* and then the conditions of lesser importance.

SELECTION OF INDUSTRIAL COUPLINGS

In most hose catalogs, couplings are listed by name, and each is described as to construction, application, thread, and method of attaching it to the hose. Adapters and fittings are available for special applications. A complete up-to-date catalog should be obtained from the manufacturer before buying decisions are made.

Couplings can be grouped by a pressure rating of low, medium, or high. These terms have a general meaning as follows:

Low pressure	100 lb/in^2 (680. kPa) maximum
Medium pressure	315 lb/in^2 (2135 kPa) maximum
High pressure	800 lb/in^2 (5520 kPa) maximum

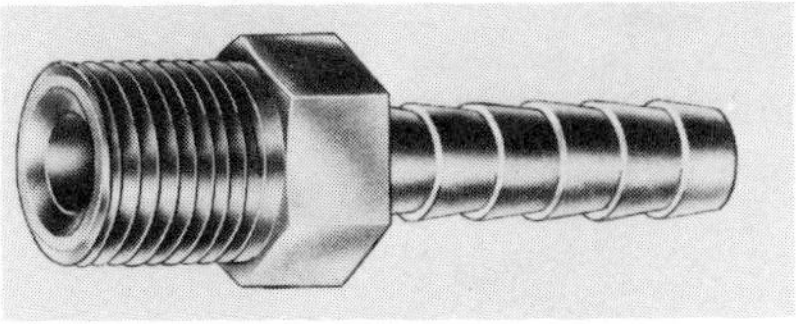

Figure 6-2 Air-hose coupling.

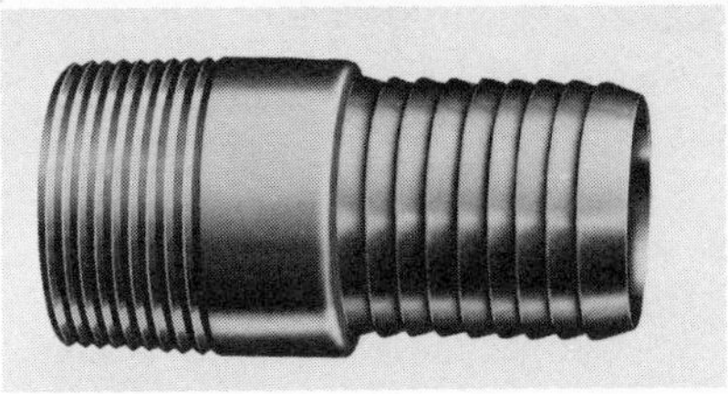

Figure 6-3 Combination nipple. (*Gates Rubber Company.*)

Coupling ratings apply only when the recommended clamps or bands are used. Care should be taken that the proper size of clamp is selected, since many clamps will fit only one size of hose or a narrow range of sizes.

GENERAL APPLICATIONS

Five different coupling types can be used to handle approximately 90 percent of all hose applications found in industrial plants. The standard air-hose coupling or insert (Fig. 6-2) is a machine brass, low- or medium-pressure coupling for use on air hose, general-purpose hose or food-handling hoses (except meat products).

The combination nipple (Fig. 6-3) is made of swaged steel pipe with scored or serrated shank and NPT (American Standard Taper Pipe Threads) threads on the male end. This nipple can be used with low- or medium-pressure water hose or materials handling hose.

Quick-connecting couplings (Fig. 6-4) are used primarily in medium-pressure applications where the hose is frequently connected and disconnected such as at tank farms, bulk-liquid storage receptacles, and tank trucks. The two parts of this coupling fit snugly together and are held in place by two cams on the female shank coupler which rotate against a groove in the male adapter.

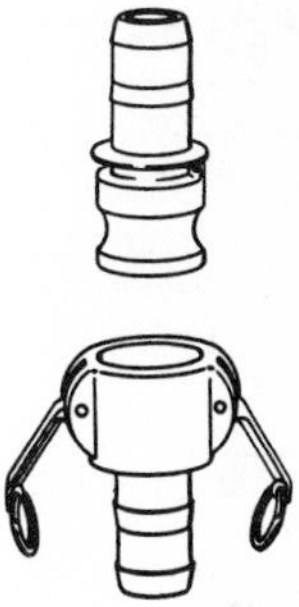

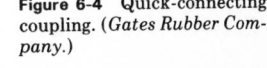

Figure 6-4 Quick-connecting coupling. (*Gates Rubber Company.*)

For maximum safety when handling high-pressure air, water, or steam, special high-pressure couplings must be employed. These are interlocking couplings, with a ground joint seal or with a washer joint seal (Fig. 6-5).

A coupling especially recommended for use on air hose is the universal quick-acting coupling. Several types of heads are available, but all have the same size attaching heads regardless of the hose size (Figs. 6-6 to 6-9).

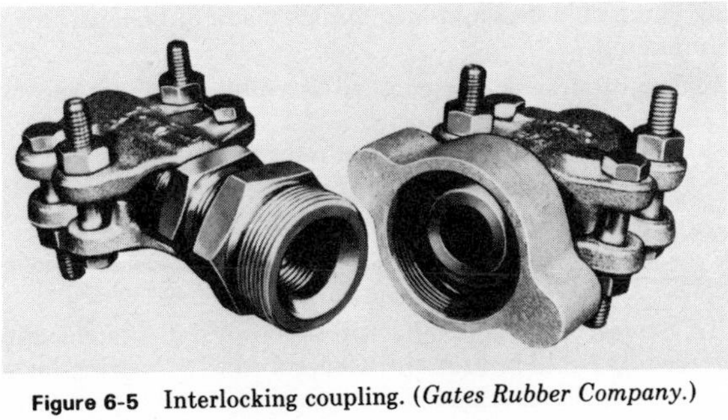

Figure 6-5 Interlocking coupling. (*Gates Rubber Company.*)

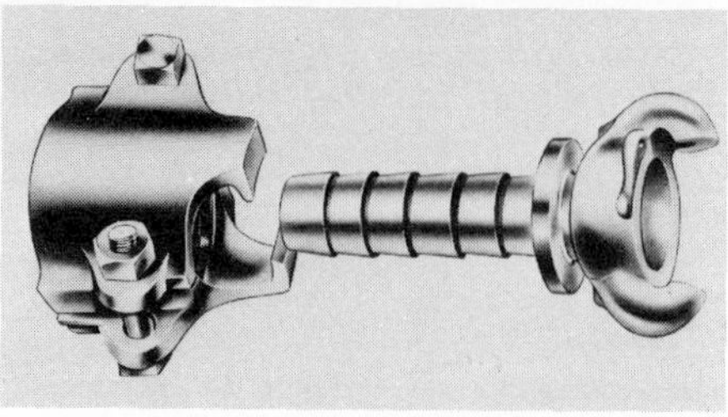

Figure 6-6 Quick-acting coupling, hose end (right), and clamp. (*Gates Rubber Company.*)

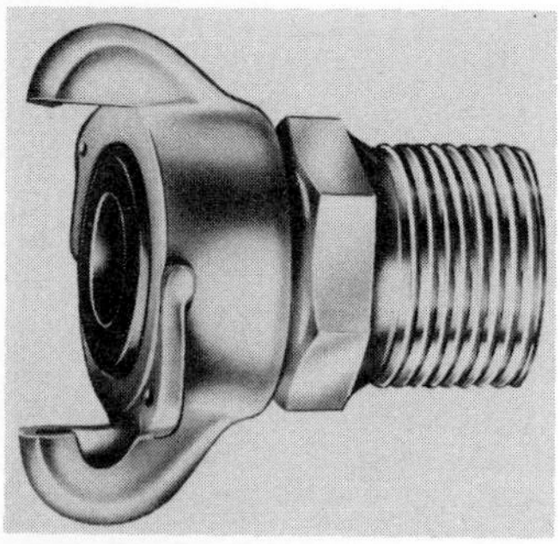

Figure 6-7 Quick-acting coupling, male end. (*Gates Rubber Company.*)

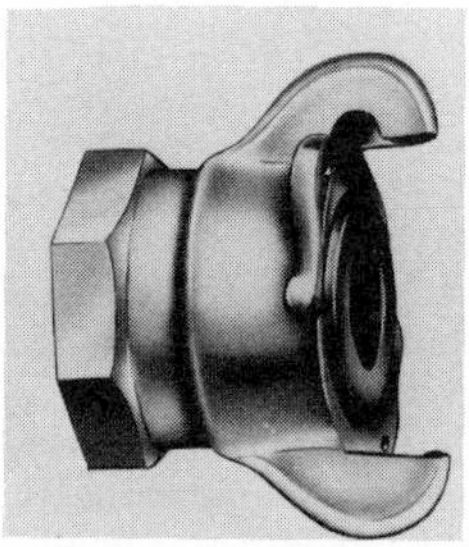

Figure 6-8 Quick-acting coupling, female end. (*Gates Rubber Company.*)

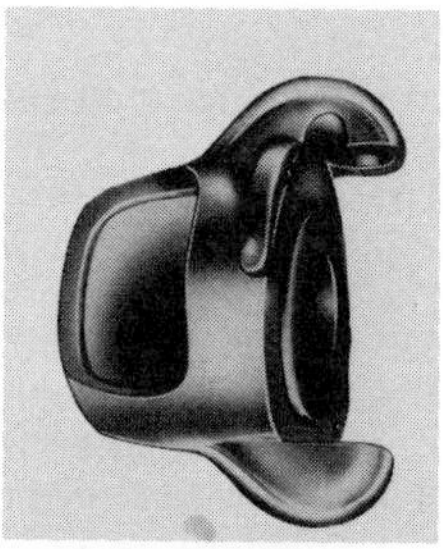

Figure 6-9 Quick-acting coupling, blank. (*Gates Rubber Company.*)

CRITICAL APPLICATIONS

For certain critical fluids, specific couplings are used, e.g., nonsparking for inflammable fluids. Below are guides to critical applications:

1. Use only the couplings recommended by the manufacturer for conveying:
   Steam
   LP gas
   Anhydrous ammonia
   Corrosive chemicals
   Petroleum products
2. For any high-temperature application, use only interlocking-type couplings.
3. For conveying flammable fluids, use couplings made of nonsparking materials, such as brass or aluminum.
4. For ground fueling of aircraft, use coupled assemblies only, as recommended by the supplier.

MAINTENANCE

Hose Care

Hose has definite service limitations and will certainly fail prematurely *if care is not taken to avoid exceeding these limitations.* General rules for caring for a hose are quite elementary but can be easily overlooked. They are especially important *because of the limited amount of maintenance or repair that can be made to extend the life of the hose assembly.*

Storage is important. If new hose is stored carefully, hose shelf life will be about 5 years before gradual deterioration begins. The hose should be stored in its original packing container or crate, out of direct sunlight. Avoid extremes of temperatures and exposure to ozone or direct heat. If the hose is shipped coiled, lay coils flat on the shelf; hose shipped straight should be stored straight.

Considerations for Hose Service

Environment

The general rule is to avoid those conditions which will accelerate hose aging. If conditions are unusually severe, a hose designed for that application must be used.

First, avoid extreme heat or cold unless the hose has been designed and built to withstand such extremes. The typical industrial hose will give satisfactory life in a temperature range of 0 to 150°F (−18 to 65°C). At −20°F (−7°C) the hose will lose some flexibility. At −30 to −40°F (−14 to −40°C) normal hose may crack if flexed sharply. Special hoses are available that will be serviceable down to −60°F (−50°C).

The limitations at about 150°F (65°C) vary with the type of elastomer and service, so general rules cannot be given. Always follow the manufacturer's recommendations for hose to be used in high ambient temperatures.

In addition, exposure to high concentrations of ozone will cause hose covers to crack. This is not a problem with routine applications, but the cracking can be severe in high smog areas, or near electric generating machinery.

Finally, consider weathering. Hoses in continuous service outdoors should have weather-resistant covers and, in all cases, should not be continuously exposed to oil or corrosive chemicals.

External Abuse

Do not over-bend the hose to the point of kinking. Always observe minimum-bend-radius recommendations. It is true that wire-reinforced hose may have greater rigidity, but it can be crushed or deformed by external weight or forces. Couplings and hose can also be damaged by too much end pull. A hard pull at any angle may kink the hose next to the coupling, especially in subzero temperatures.

Large-diameter hoses [4-in (10 cm) ID and greater] have some special considerations. One problem may be overstressing the hose carcass. Handle heavy hoses with slings every 6 to 10 ft (2 to 3 cm) and do not lift a long section from the middle with the ends hanging down.

Another concern—sometimes the hose cover is exposed to wear in one particular spot. In this case, the user should add a protective outside cover to avoid wearing through the cover and exposing the reinforcing.

Maintenance and Repair

Good maintenance programs including inspection, testing, and repair will ensure that maximum, safe service life is obtained from the hose and that damaged hose will be removed from service before it becomes a hazard.

All hoses should be inspected periodically. Hoses used in hazardous applications should be inspected at more frequent intervals. Basically, the user should be looking for evidence of stress or external abuse such as abrasion, cuts, and exposure to chemicals and oils. Obviously, the sources of these abuses should be corrected or eliminated.

The user should also look for evidence of failure that will necessitate taking the hose out of service or require recoupling or other repairs.

Inspection procedures for industrial hoses in nonhazardous applications are:

1. Lay hose out straight in a dry, light area.
2. Visually inspect hose for kinks, bulges, or soft spots in the cover, and excessive cover

wear that exposes reinforcing. Evidence of this kind usually means the hose should be removed from service.

3. Inspect couplings for signs of slippage. Examine hose adjacent to couplings for breakage. If necessary, hose can be cut off behind the couplings and recoupled. Retighten bolted clamps if necessary.
4. Excessive amounts of oil or chemicals on the cover should be wiped off. Periodic or specialized inspection and testing are necessary for hoses used in hazardous applications. These hoses should be inspected at definite intervals following established procedures. The *Hose Handbook* and individual bulletins published by the Rubber Manufacturers Association outline daily inspection procedures for hoses used for hard-to-handle anhydrous ammonia, liquid petroleum gas, oil suction and discharge, and aviation ground fueling, as well as for motor-vehicle hose.

Hose Repair

Hoses can be recoupled to replace damaged couplings or to rejoin the ends of a failed section of hose that is cut out of the original length. The extent of recoupling or coupling repair that can be done depends on the type of couplings and the size of the hose.

Hoses that have plain ends can be recoupled by cutting off the old coupling and applying a new one, provied the overall length is not too small. All braided hoses and some wrapped hoses up through 4-in (10 cm) ID have plain ends. Hoses having special ends or built-in couplings cannot normally be recoupled.

Large-diameter built-in nipples that have been worn thin by abrasives can be built up with plates. However, the rubber stock must be protected from any welding heat.

Beyond recoupling, there are very few ways in which a damaged or badly worn hose can be safely reclaimed, but the hose manufacturer should be consulted for recommendations.

STANDARDS

Table 6-3 lists applicable OSHA standards and their applications.

TABLE 6-3 OSHA Standards

Standards paragraph	Application
1910.106 Flammable and Combustible Liquids	General Transfer Curb pump, UL listed Flesible connectors, UL listed
1910.107 Spray Finishing Using Flammable and Combustible Materials	a. General, low, or medium pressure b. High pressure, airless, with static wire
1910.109 Explosives and Blasting Agents	As specified; special electrical properties
1910.110 Storage and Handling of LP Gas	As specified; UL listed
1910.111 Storage and Handling of Anhydrous Ammonia	Transfer service
1910.134 Respiratory Protection	Hoses for face masks
1910.158 Stand Pipe and Hose System for Fire Equipment	Various
1910.165a Sources of Standards	Fire extinguishers
1910.177 Indoor General Storage	Sprinkler systems
1910.243 Guarding of Portable Power Tools	Various air hoses
1910.1252 Welding, Cutting & Brazing	General welding and related

BIBLIOGRAPHY

OSHA standards are contained in *Federal Register,* October 18, 1972, vol. 37, no. 22, part II.

Hose Handbook, Rubber Manufacturer's Association, 1901 Pennsylvania Ave., N.W., Washington, DC 20006.

"Industrial Hose Finder," No. 39995. The Gates Rubber Company, P. O. Box 5887, Denver, CO 80217.

chapter 10-7

Valves

prepared by

Communications Committee
The Valve Manufacturers Association
McLean, Virginia

INTRODUCTION

A valve may be defined as a mechanical device by which the flow of liquid or gas may be started, stopped, or regulated by a movable part that opens, shuts, or partially obstructs one or more ports or passageways.

Plant engineers describe a valve as one of the most essential control instruments used in industry.

By the nature of their design and materials, valves can open and close, turn on and turn off, regulate, modulate, or isolate an extremely large array of liquids and gases, from the most basic to the most corrosive or toxic. They range in size from a fraction of an inch to 30 ft (9 m) in diameter. They can handle pressures ranging from vacuum to more than 20,000 lb/in^2 (140 MPa/m^2) and temperatures from the cryogenic region to 1500°F (815°C). Some applications require absolute sealing; in others leakage is not a factor.

CATEGORIES OF VALVES

Because of all these variables, there can be no universal valve; therefore, to meet the changing requirements of industry, innumerable designs and variations have evolved over the years as new materials have been developed. All these designs fall into nine major categories: gate valves, globe valves, ball valves, butterfly valves, pinch valves, diaphragm valves, plug valves, check valves, and relief valves.

These basic categories are described in the following paragraphs. It would be impossible to mention every feature of every valve manufactured, and we have not attempted to do this. Instead, a general overview of each type is presented in outline format, giving service recommendations, applications, advantages, disadvantages, and other information helpful to the reader. In many cases, a disadvantage inherent in a type of valve has been overcome or corrected by a particular manufacturer. Therefore, for specific applications, manufacturers' recommendations should be sought.

Gate Valves

A gate valve is a multiturn valve in which the port is closed by a flat-faced, vertical disk that slides at right angles over the seat (Fig. 7-1).

Recommended

- For fully opened or fully closed, nonthrottling service
- For infrequent operation
- For minimum resistance to flow
- For minimum amounts of fluid trapped in line

Applications

- General service, oil, gas, air, slurries, heavy liquids, steam, noncondensing gases and liquids, corrosive liquids

Advantages

- High capacity
- Tight shutoff
- Low cost
- Simple design and operation
- Little resistance to flow

Disadvantages

- Poor flow control
- High operating force
- Cavitates at low pressure drop
- Must be kept in full open or full closed position
- Throttling position will erode seat and disk

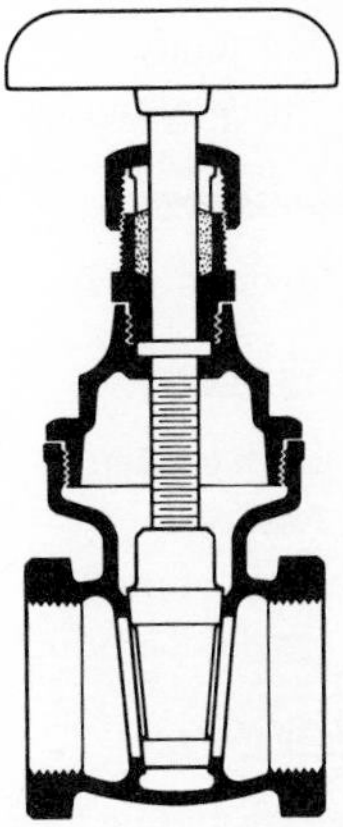

Figure 7-1 Gate valve.

Variations

- Solid wedge, flexible wedge, split wedge, double disk

Materials

- Body: bronze, cast iron, iron, forged steel, Monel, cast steel, stainless steel, PVC plastic
- Trim: various

Special Installation and Maintenance Instructions

- Lubricate on regular schedule
- Correct packing leaks immediately
- Always cool system when closing down a "hot" line and checking closed valves
- Never force valves closed with wrench or pry
- Open valves slowly to prevent hydraulic shock in line
- Close valves slowly to help flush trapped sediment and dirt

Ordering Specifications

- Type of end connections
- Type of wedge
- Type of seat
- Type of stem assembly
- Type of bonnet assembly
- Type of stem packing
- Pressure rating: operating and design
- Temperature rating: operating and design

Plug Valves

A plug valve is a quarter-turn valve that controls flow by means of a cylindrical or tapered plug with a hole through the center, which can be positioned from open to closed by a 90° turn (Fig. 7-2).

Recommended

- For full-open or full-closed service
- For frequent operation
- For low pressure drop across the valve
- For minimum resistance to flow
- For minimum amount of fluid trapped in line

Applications

- General service, slurries, liquids, vapors, gases, corrosives

Advantages

- High capacity
- Low cost
- Tight shutoff
- Quick operation

Disadvantages

- High torque for actuation
- Seat wear
- Cavitation at low pressure drop

Variations

- Lubricated, nonlubricated, multiport

Materials

- Iron, ductile iron, carbon steel, stainless steel, Alloy 20, Monel, nickel, Hastelloy, plastic-lined

Special Installation and Maintenance Instructions

- Allow space for operation of handle on wrench-operated valves
- For lubricated plug valves, lubricate before putting into service
- For lubricated plug valves, lubricate on regular schedule

Ordering Specifications

- Body material
- Plug material
- Temperature rating
- Pressure rating
- Port arrangement, if multiport valve
- Lubricant, if lubricated valve

Globe Valves

A globe valve is multiturn valve in which closure is achieved by means of a disk or plug that seals or stops the fluid on a seat generally parallel to the line flow (Fig. 7-3).

Recommended

- For throttling service or flow regulation
- For frequent operation
- For positive shutoff of gases or air
- Where some resistance to flow is acceptable

Applications

- General service, liquids, vapors, gases, corrosives, slurries

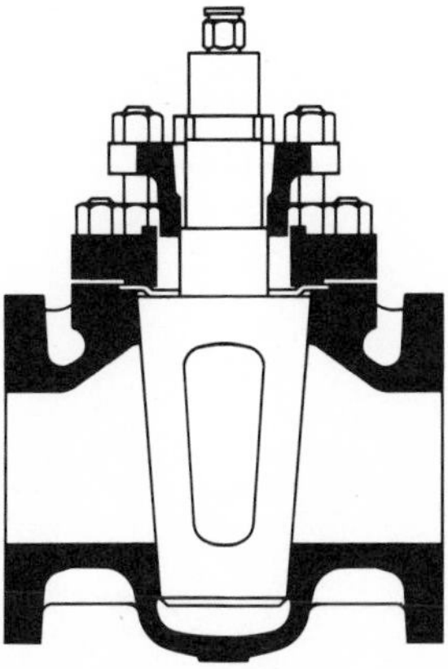

Figure 7-2 Plug valve.

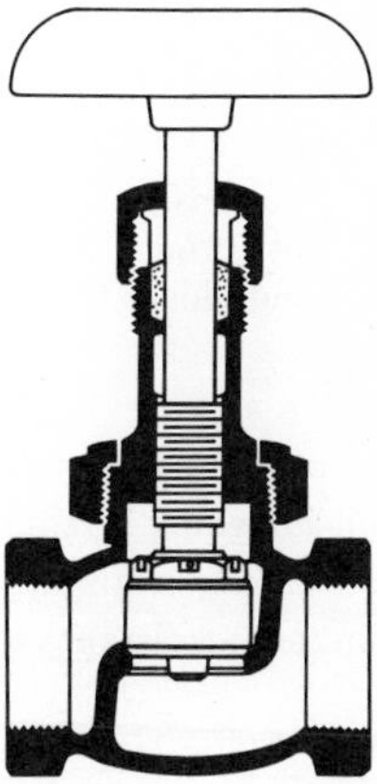

Figure 7-3 Globe valve.

Advantages

- Efficient throttling with minimum wire drawing or disk or seat erosion
- Short disk travel and fewer turns to operate, saving time and wear on stem and bonnet
- Accurate flow control
- Available in multiports

Disadvantages

- High pressure drop
- Relatively high cost

Variations

- Standard, Y pattern, angle, three-way

Materials

- Body: bronze, all iron, cast iron, forged steel, Monel, cast steel, stainless steel, plastics
- Trim: various

Special Installation and Maintenance Instructions

- Install so pressure is under disk, except in high-temperature steam service
- Lubricate on strict schedule
- Flush foreign matter off seat by opening valve slightly
- Correct packing leaks immediately by tightening the packing nut

Ordering Specifications

- Type of end connection
- Type of disk
- Type of seat
- Type of stem assembly
- Type of stem seal
- Type of bonnet assembly
- Pressure rating
- Temperature rating

Ball Valves

A ball valve is a quarter-turn valve in which a drilled ball rotates between resilient seats, allowing straight-through flow in the open position and shutting off flow when the ball is rotated 90° and blocks the flow passage (Fig. 7-4).

Recommended

- For on-off, nonthrottling service
- Where quick opening is required
- For moderate temperature requirements
- Where minimum resistance to flow is needed

Applications

- General service, high temperatures, slurries

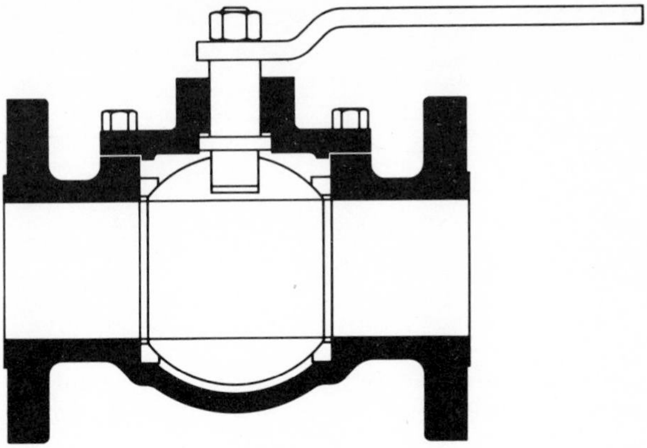

Figure 7-4 Ball valve.

Advantages

- Low cost
- High capacity
- Bidirectional shutoff
- Straight-through pattern
- Low leakage
- Self-cleaning
- Low maintenance
- No lubrication requirement
- Compact
- Tight sealing with low torque

Disadvantages

- Poor throttling characteristics
- High torque for actuation
- Susceptible to seal wear
- Prone to cavitation

Variations

- Top entry, split body or end entry, three-way, venturi, full-ported, reduced port

Materials

- Body: cast iron, ductile iron, bronze, brass, aluminum, carbon steels, stainless steels, titanium, tantalum, zirconium, and polypropylene and PVC plastics
- Seat: TFE, filled TFE, nylon, Buna-N, neoprene

Special Installation and Maintenance Instructions

- Allow sufficient space for operation of long handle

Ordering Specifications

- Operating temperature
- Type of port in ball
- Seat material
- Body material
- Operating pressure
- Full or reduced port
- Top entry or side entry

Butterfly Valves

A butterfly valve is a quarter-turn valve that controls flow by means of a circular disk with its port axis at right angles to the direction of flow (Fig. 7-5).

Recommended

- For full-open or full-closed service
- For throttling service
- For frequent operation
- Where positive shutoff is required for gases or liquids
- Where minimum amount of fluid trapped in line is allowed
- For low pressure drop across valve

Applications

- General service, liquids, gases, slurries, liquids with suspended solids

Advantages

- Compact, lightweight, low-cost
- Low maintenance
- Minimum number of moving parts
- No pockets
- High capacity
- Straight-through flow
- Self-cleaning

Disadvantages

- High torque for actuation
- Limited pressure-drop capability
- Prone to cavitation

Variations

- Wafer, lug wafer, flanged, screwed, fully lined, high-performance

Materials

- Body: iron, ductile iron, carbon steels, forged steel, stainless steels, Alloy 20, bronze, Monel
- Disk: all metals, elastomer coatings such as TFE, Kynar, Buna-N, neoprene, Hypalon
- Seat: Buna-N, Viton, neoprene, rubber, butyl, polyurethane, Hypalon, Hycar, TFE

Special Installation and Maintenance Instructions

- May be operated by lever, handwheel, or chainwheel
- Allow sufficient space for operation of handle if lever-operated
- Valves should remain in closed position during all handling and installation operations

Ordering Specifications

- Type of body
- Type of seat
- Body material
- Disk material
- Seat material
- Type of actuation
- Operating pressure
- Operating temperature

Diaphragm Valves

A diaphragm valve is a multiturn valve that effects closure by means of a flexible diaphragm attached to a compressor. When the compressor is lowered by the valve stem, the diaphragm seals and cuts off flow (Fig. 7-6).

Recommended

- For full-open or full-closed service
- For throttling service
- For service with low operating pressures

Applications

- Corrosive fluids, sticky and/or viscous materials, fibrous slurries, sludges, foods, pharmaceuticals

Advantages

- Low cost
- No packing glands

Figure 7-5 Butterfly valve.

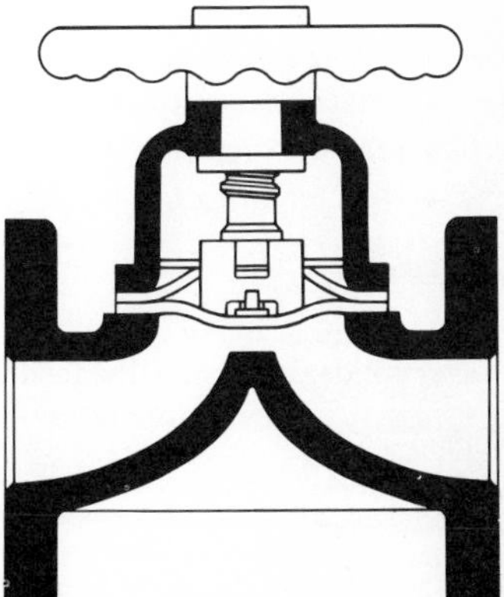

Figure 7-6 Diaphragm valve.

- No possibility of stem leakage
- Immune to problems of clogging, corroding, or gumming of media

Disadvantages

- Diaphragm subject to wear
- High torque under live-line closure

Variations

- Weir type and straight-through type

Materials

- Metallic, solid plastic, lined—wide variety of each

Special Installation and Maintenance Instructions

- Lubricate on a regular schedule
- Do not use bars, wrenches, or cheaters to close

Ordering Specifications

- Body material
- Diaphragm material
- End connections
- Type of stem assembly
- Type of bonnet assembly
- Type of operation
- Operating pressure
- Operating temperature

Pinch Valves

A pinch valve is a multiturn valve that effects closure by means of one or more flexible elements, such as diaphragms or rubber tubes, that can be pressed together to cut off flow (Fig. 7-7).

Recommended

- For on-off service
- For throttling service
- For moderate temperatures
- Where pressure drop through valve is low
- For services requiring low maintenance

Applications

- Slurries, mining slurries, liquids with large amounts of suspended solids, systems that convey solids pneumatically, food service

Advantages

- Low cost
- Low maintenance
- No internal obstruction or pockets to cause clogging

- Simple design
- Noncorrosive and abrasion-resistant

Disadvantages

- Limited vacuum application
- Difficult to size

Variations

- Exposed sleeve or body, encased metallic sleeve or body

Materials

- Rubber, white rubber, Hypalon, polyurethane, neoprene, white neoprene, Buna-N, Buna-S, Viton-A, butyl rubber, silicone, TFE

Special Installation and Maintenance Instructions

- Large sizes may require supports above or below the line if pipe supports are inadequate

Ordering Specifications

- Operating pressure
- Operating temperature
- Sleeve material
- Exposed or encased sleeve

Check Valves and Relief Valves

Two categories of valves are specific-purpose rather than general-service valves. These are check valves and relief valves. Unlike the other types described in this section, they are self-actuated valves and operate without outside control, depending for their operation on flow direction or pressures within the piping system. Since both types are nor-

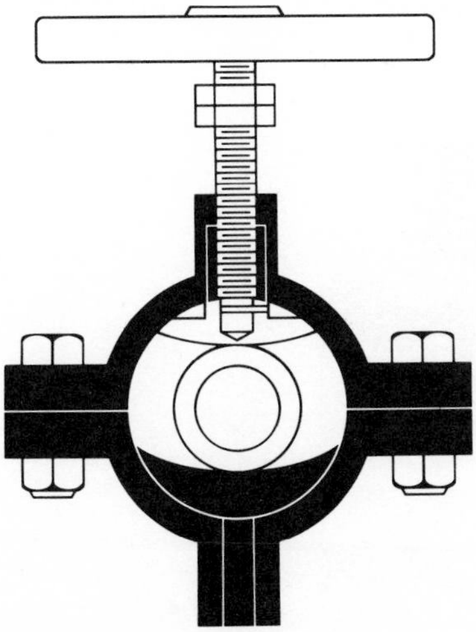

Figure 7-7 Pinch valve.

mally used in conjunction with flow-control valves, the choice of valve is often determined by the same conditions that determine the selection of the flow-control valve.

Check Valves

A check valve (Fig. 7-8) is designed to check reversal of flow. Fluid flow in the desired direction opens the valve; reversal of flow closes it. There are three basic styles of check valves: (1) swing check, (2) life check, and (3) butterfly check.

Swing Check Valve. A swing check valve has a hinged disk designed to open completely with line pressure and close when line pressure ceases and backflow begins. There are two designs: a Y pattern, which has an access opening in the body for easy regrinding of the disk without removing the valve from the line, and a straight-through pattern that has replaceable seat rings.

Recommended

- Where minimum resistance to flow is needed
- Where there is infrequent change of direction in the line
- For service in lines using gate valves
- For vertical lines having upward flow

Applications

- For low-velocity liquid service

Advantages

- Unobstructed view
- Turbulence and pressure within valve are very low
- Y-pattern disk can be reground without removing valve from line

Variations

- Tilting-disk check valve

Materials

- Body: bronze, all iron, cast iron, forged steel, Monel, cast steel, stainless steel, carbon steel
- Trim: various

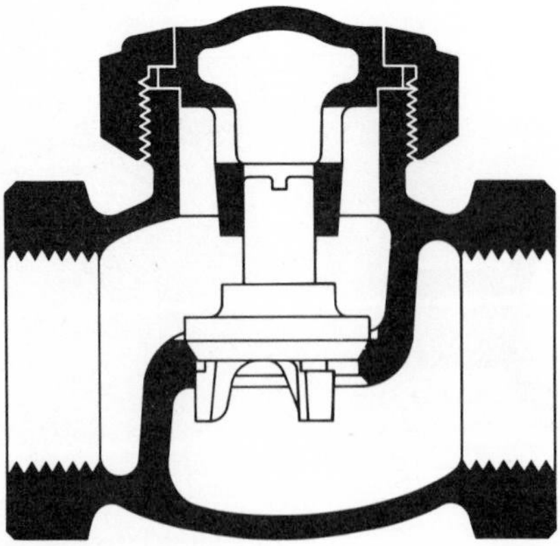

Figure 7-8 Check valve (lift type).

Special Installation and Maintenance Instructions

- In vertical lines, pressure should always be under seat
- If valve fails to seal, check seating surfaces
- If seat is damaged or scored, regrind or replace
- Before reassembling, clean internal portions thoroughly

Lift Check Valve. A lift check valve is similar in design to a globe valve except that the disk is lifted by the forward line pressure and closed by gravity and backflow.

Recommended

- Where there are frequent changes of direction in the line
- For use with globe and angle valves
- For use where pressure drop across valve is not a problem

Applications

- Steam, air, gas, water, and vapor lines with high flow velocities

Advantages

- Minimum travel of disk for full-open position
- Quick-acting

Variations

- Three body patterns: horizontal, angle, vertical
- Ball check, piston check, spring-loaded check, stop check

Materials

- Body: bronze, all iron, cast iron, forged steel, Monel, stainless steel, PVC, Penton, impervious graphite, TFE-lined
- Trim: various

Special Installation and Maintenance Instructions

- Line pressure should be under seat
- Horizontal pattern should be installed in horizontal lines
- Vertical pattern is used for vertical pipes with flow upward, from beneath seat
- If backflow leaks, check disk and seat

Butterfly Check Valve. A butterfly check valve has a split disk hinged on a shaft in the center of the disk so that a flexible sealing member attached to the disk is at a 45° angle to the valve body when the valve is closed. The disk then has to move only a short distance away from the body toward the center of the valve to open fully.

Recommended

- Where minimum resistance to flow in the line is needed
- Where there is frequent change of direction in the line
- For use with butterfly, plug, ball, diaphragm, or pinch valves

Applications

- For liquid or gas service

Advantages

- Body design lends itself to installation of various types of seat liners
- Less expensive for corrosion resistance
- Quiet operation
- Simplicity of design permits construction in large diameters
- May be installed in virtually any position

Variations

- Fully lined
- Soft-seated

Materials

- Body: steel, stainless steel, titanium, aluminum, PVC, CPVC, polyethylene, polypropylene, cast iron, Monel, bronze
- Flexible sealing members: Buna-N, Viton, butyl rubber, TFE, neoprene, Hypalon, urethane, Nordel, Tygon, silicone

Special Installation and Maintenance Instructions

- On lined valves, liner should be protected from damage during handling
- Make sure valve is installed so that forward flow opens valve

Relief Valves

A relief valve (Fig. 7-9) is a self-actuated valve designed to provide accurate automatic pressure regulation. The valve is used primarily for noncompressible fluid service and opens slowly as pressure increases, regulating the operating pressure.

Closely related to the relief valve is the safety valve, which opens quickly with a pop action to relieve excessive pressure caused by gases or compressible fluids.

Sizing is very important in relief valves and is determined by specific formulas.

Recommended

- In systems where a predetermined pressure range is required

Applications

- Hot water, steam, gases, vapor

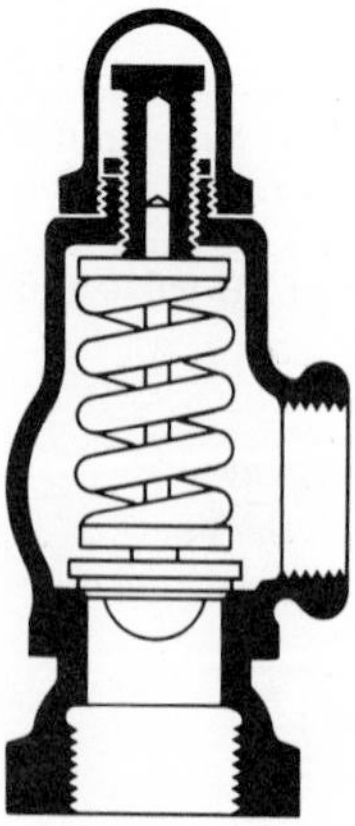

Figure 7-9 Relief valve.

Advantages

- Inexpensive
- No auxiliary power required for operation

Variations

- Safety, safety relief
- Diaphragm construction for valves used in corrosive service

Materials

- Body: cast iron, carbon steel, glass-TFE, bronze, brass, TFE-lined, stainless steel, Hastelloy, Monel
- Trim: various

Special Installation and Maintenance Instructions

- Should be installed in accordance with provisions of ASME Unfired Pressure Vessel Code
- Should be installed in readily accessible areas for inspection and maintenance

CODES AND STANDARDS

Various professional and industrial organizations have created codes and standards that are applicable to the design, selection, and use of industrial products. Certain of these standards are peculiar to specific industries, such as those set up by the American Water Works Association (AWWA), American Gas Association (AGA), American National Standards Institute (ANSI), American Society of Mechanical Engineers (ASME), and Manufacturers Standardization Society of the Valve and Fitting Industry (MSS). These are but a few of the existing organizations.

The valve specifier should be aware of those codes that apply to a specific industry and the application in question. For further information on the codes and standards available, these organizations may be contacted at the addresses listed below:

AWWA, 6666 W. Quincy Avenue, Denver, CO 80235

ANSI, 1430 Broadway, New York, NY 10018

AGA, 1515 Wilson Boulevard, Arlington, VA 22209

API, 2101 L Street N.W., Washington, DC 20037

ASME, 345 East 47th Street, New York, NY 10017

MSS, 1815 N. Fort Myer Drive, Arlington, VA 22209

BIBLIOGRAPHY

Beard, Chester S.: *Final Control Elements,* Rimback Publications Div., Chilton Co., Radnor, Pa., 1969.

Hutchison, J. W.: *ISA Handbook of Control Valves,* Instrument Society of America, Pittsburgh, 1971.

O'Keef, William, (Ed.): "Valves," *Power Magazine,* March 1971, pp. S-1 to S-16.

Schweitzer, Philip A.: *Handbook of Valves,* Industrial Press, New York, 1972.

section 11

Instrumentation and Automatic Controls

by

Ranjit S. Randhawa
The Foxboro Company,
Foxboro, Massachusetts

chapter 11-1

Instrumentation

TERMINOLOGY

Accuracy An ideal instrument would be one whose input-to-output relationship has a defined linear or nonlinear equation. Instruments, however, suffer from errors such as hysteresis, deadband, conformity, and repeatability. *Hysteresis* and *deadband* errors are illustrated in Fig. 1-1.

Hysteresis error Suppose, as input increases, the reading follows along curve *OBA*. If input returns to *O*, the reading may follow *AB′O*, a different path. The difference *BB′* at any input is the *hysteresis* error, a repeatable error due to energy absorption by the instrument element (typically a spring, diaphragm, or magnetic core).

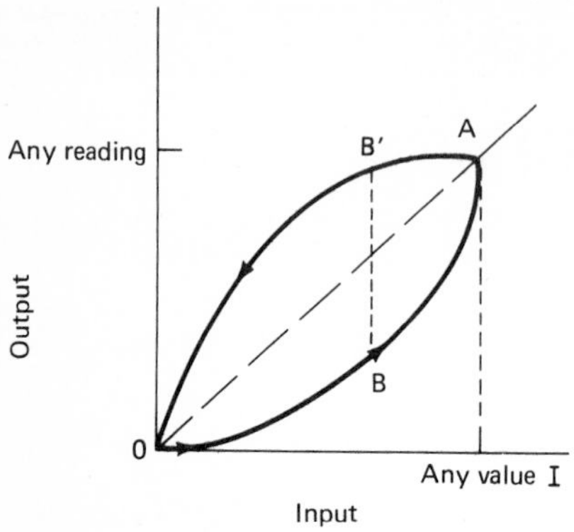

Figure 1-1 Hysteresis and deadband errors.

Deadband error Suppose input increases from *O* to any value *I*, and the output reading does not change (due to friction, for example) until the input is at *I*, whereupon the output reading jumps to *A*. If the input reverses, returning to *O*, the output may remain at *A* until the input returns to *O*; then the output jumps to *O*. The value of *I* is the *dead band*, the amount by which the input changes for a change in output.

Conformity The maximum deviation of an instrument's actual calibration curve as compared to its specified characteristic curve is called its *conformity.* There are two methods by which a numerical value for conformity is derived. First is the terminal-based method whereby the calibration curve is forced to coincide at the end points with the specified characteristic curve. With the second method, the calibration curve is forced to concide only with the lower-range value of the characteristic curve of the instrument.

Repeatability The closeness of agreement among a number of consecutive measurements of the output for the same value of the input under the same operating conditions, i.e., approaching from the same direction for full-range traverses, is called repeatability. It is usually measured as nonrepeatability and expressed as repeatability in percent of span. It does *not* include hysteresis since input measurements are varied in only one direction. See Fig. 1-2.

Accuracy rating This is a number or quantity that defines a limit which errors will not exceed when a device is used under specified operating conditions. When operating conditions are not specified, reference conditions are assumed. For example, a primary flow device may lose accuracy at higher fluid temperatures. Unless specified otherwise, however, a reference or standard test temperature is assumed.

As a performance specification, accuracy (or reference accuracy) is assumed to mean the accuracy rating of the device when used at reference operating conditions. The units being used must be stated explicitly. Preferably, a ± sign should precede the number or quantity, but absence of a sign is taken to mean ±.

Accuracy rating can be expressed in a number of ways. The following examples are typical:

1. Accuracy rating expressed in terms of the *measured variable itself.* Typical expression: The accuracy rating for a certain temperature recorder-indicator is ±2°F or ±1°C.
2. Accuracy rating expressed in terms of *span.* Typical expression: The accuracy rating is ±0.5 percent of span. (This percentage is calculated using scale units such as degrees Fahrenheit, pounds per square inch gauge, etc.). With that accuracy rating, a pressure transmitter with a span of 200 psig (1400 kPa) would be inaccurate by ± 1 psig (6.89 kPa).

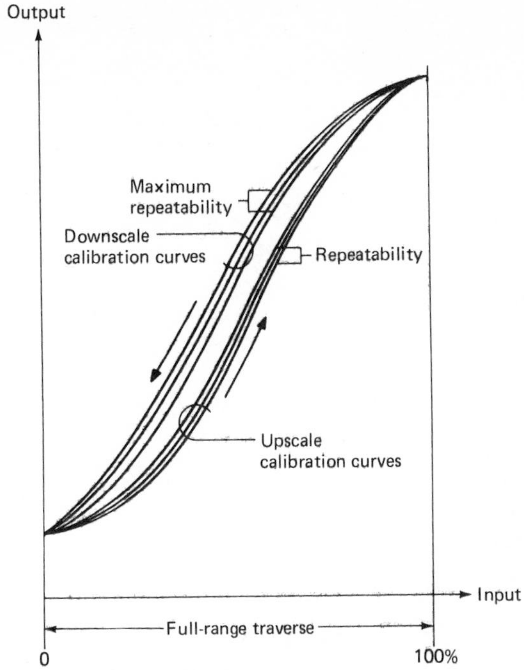

Figure 1-2 Repeatability.

3. Accuracy rating expressed in percent of the *upper-range value*. Typical expression: The accuracy rating is ±0.5 percent of the upper-range value. (This percentage is calculated using scale units such as kilopascals, degrees Fahrenheit, etc.) Thus, a temperature recorder with an upper range of 50-mV thermocouple input would be inaccurate by ±2.5 mV.
4. Accuracy rating expressed in percent of *scale length*. Typical expression: The accuracy rating is ±0.5 percent of scale length. A manometer with a scale length of 20 in (0.5 m) would have an error of 0.1 in (±0.25 cm).
5. Accuracy rating expressed in percent of *actual output reading*. Typical expression: The accuracy rating is ±1 percent of actual output reading. Thus, a manometer reading of 0.2 m with a ±1 percent reading error would have an error of ±0.2 cm *at that reading*.

Span The algebraic difference between the upper and lower range values. For example:

1. Range 0 to 150°F—span 150°F
2. Range −20 to 200°F—span 220°F
3. Range 20 to 150°C—span 130°C

For multirange devices, this definition applies to the particular range that the device is set to measure.

Span adjustment Means provided in an instrument to change the slope of the input-output curve. See Fig. 1-3.

Zero adjustment Means provided in an instrument to produce a parallel shift of an input-ouput curve. See Fig. 1-3.

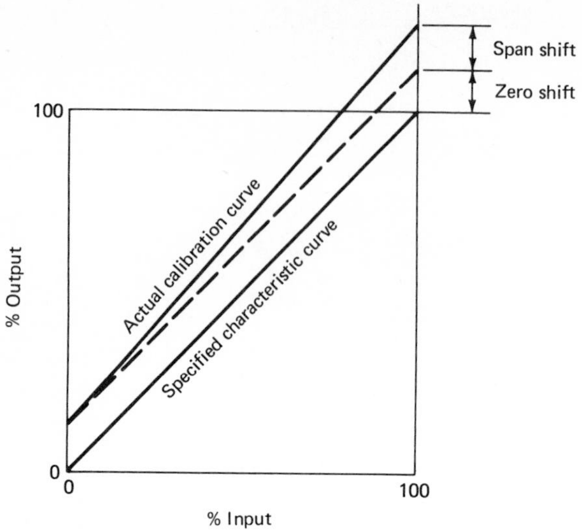

Figure 1-3 Accuracy rating span shift and zero shift.

Elevated zero range A range in which the zero value of the measured variable, measured signal, etc., is greater than the lower range value. The zero may be between the lower- and upper-range values, at the upper-range value, or above the upper-range value. Examples: range −20 to 200°F, −100 to −10°C. The terms *suppression, suppressed range,* or *suppressed span* are also frequently used to denote elevated zero range.

Suppressed zero range A range in which the zero value of the measured variable is less than the lower range value. For example: 20 to 100 scale range. The terms *elevation, elevated range,* and *elevated span* are also used to denote suppressed zero range.

Reproducibility The closeness of agreement among repeated measurements of the output for the same value of input made under the same operating conditions over a period of time, approaching from both directions. It is usually measured as a nonreproducibility and expressed as reproducibility in percent of span for a specified time interval.

INTRODUCTION

Measurement of physical phenomena is the basis of all technical activity. Obtaining, displaying, and conveying the quantitative facts of physical processes are the prime functions of instrumentation. This chapter presents the practical fundamentals of this technology.

LEVEL-MEASUREMENT METHODS

Level, as a process variable, is a common measurement both for control and indication. Various methods are used, and the selection of any one is based on many factors. Refer to Table 1-1.

Float Method

This is the simplest of level-measuring methods and makes use of a float which essentially follows the level in a closed or open vessel. The position of the float can be used to

TABLE 1-1 Level-Measurement Methods

Method	Type of liquid: Clean	Hard to handle	Solid	Range	Relative cost	Output type
Float	Good	Fair	—	75 mm–15 m (3 in–50 ft)	Low	Contact or tape
Displacement	Fair	Poor	—	150 mm–4 m (6 in–12 ft)	Med	Contact and/or signal
Head (pressure)	Good	Excellent	—	50 mm on (2 in on)	Med	Signal
Differential pressure	Good	Excellent	—	130 mm on (5 in on)	Med	Signal
Air (gas) bubbler	Fair	Fair	—	250 mm–75 m (10 in–250 ft)	Low	Signal
Capacitance	Good	Fair	Good	Wide	Med	Signal
Conductivity	Fair	—	Poor	Point	Med	Contact
Thermal	Good	—	—	Point	Med	Contact
Radiation	Good	Excellent	Good	Wide	High	Signal
Weighing	Good	Excellent	Good	Wide	High	Signal
Ultrasonic	Good	Excellent	Good	Wide	High	Signal

sense the level at a predetermined point by magnetically coupling the float to a mercury switch or miniature-type switch. This method is normally used for sensing high and/or low levels in a vessel. In large storage tanks, where a local indication or recording of the level is required, the float is coupled with a tape or cable which then, through a pulley mechanism, positions a local pointer for indication or applies a torque in a recorder. In both applications turbulence is kept to a minimum.

Displacement Method

The method employs a displacer which is located so that it is totally immersed when the level is at its predetermined maximum point (Fig. 1-4). The amount of force acting on the displacer is equal to the weight of the liquid displaced. The displacer weighs more than the maximum amount of liquid it can displace and is normally cylindrical in shape so that the relationship of buoyancy to submersion is linear. The displacer is linked to a torque tube which twists linearly with the buoyance of the displacer.

The buoyant force can be determined by

$$F = V\frac{L_w}{L}\mathrm{D} \tag{1}$$

where

V = total displacer volume
L_w = working length of displacer
L = total length of displacer
D = density of fluid

The level measurement is independent of the pressure in the vessel but may be subject to problems where extreme turbulence is encountered. Special precautions have to be taken under these circumstances.

The same principle of measurement can be used for measuring density. Here the displacer is kept fully immersed; the buoyant force is then a function of the span of density to be measured. A further application of this type of measurement is the interface level between two liquids. The interface level is allowed to vary over the length of the displacer while it is fully immersed. The force now depends upon the difference in densities of the two liquids. Standard displacers are designed for a range of buoyancies from 1.47 lb (0.67 kg) to about 12 lb (5.45 kg), including the weight of the hanger assembly.

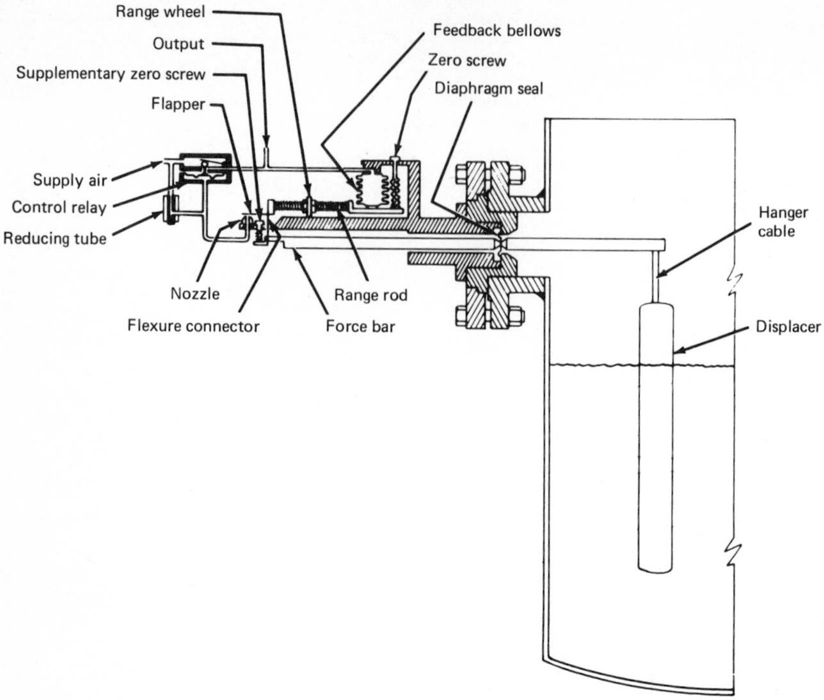

Figure 1-4 Schematic diagram of buoyancy transmitter with displacer.

Head-Pressure Method

This is a useful means of measuring level when a pressure transmitter is used to convert the head pressure in an open tank to an equivalent level. The transmitter output can be used for remote indication or control. The difficulties in this type of measurement usually arise because most tanks are closed vessels that are also pressurized, thus causing variations in pressure to affect level measurements. Furthermore, temperature variations

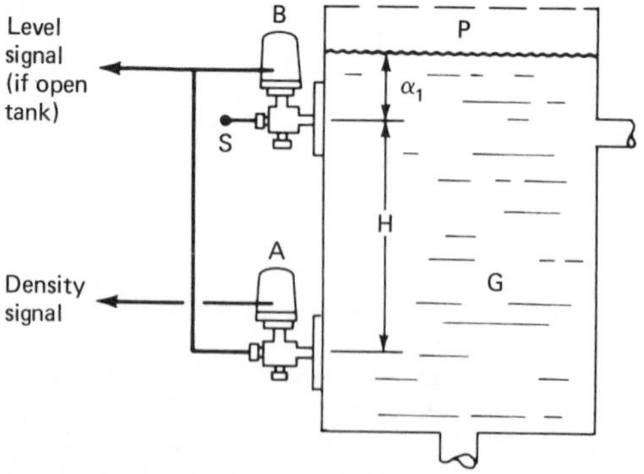

Figure 1-5 Differential pressure method of density measurement. *Key: A*, differential pressure transmitter; *B*, 1:1 repeater; *G*, specific gravity of liquid; *P*, pressure.

may cause density, too, to vary, which will then affect the level measurement. These problems limit the usefulness of this method for level measurements, but it is an excellent method for determining the density of liquids by keeping the level of the liquid constant in either a vessel with constant pressure or an open vessel with no pressure. An installation where pressure variations in the vessel can be taken into account will use a differential pressure measurement where the tank pressure is subtracted from the measured head. The differential pressure transmitter should be calibrated for a range of $H(G_2 - G_1)$, and its elevation is HG_1. Refer to Fig. 1-5.

Differential-Pressure Method

This is the most common method of measurement for both control and indication. Refer to Fig. 1-6.

A typical force-balance type of differential pressure (d/p) pneumatic transmitter (the d/p Cell*) is shown in Fig. 1-7. An electric transmitter would work essentially on the

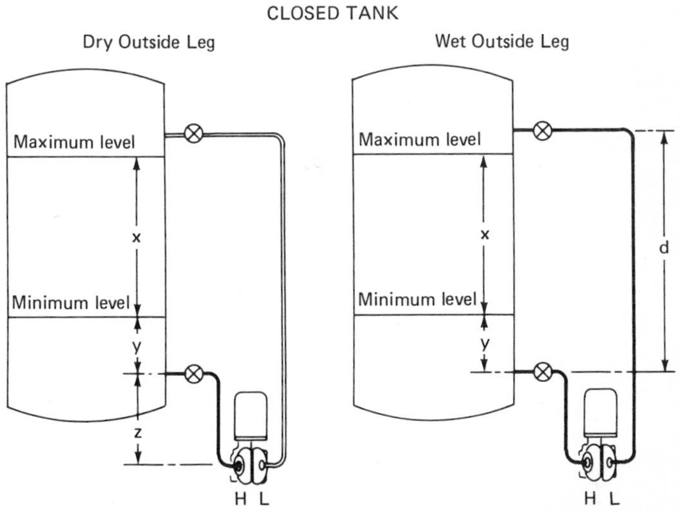

Figure 1-6 Differential-pressure level measurement. For dry outside leg, Span = xG_L and Elevation = $yG_L + zG_S$; for wet outside leg, Span = xG_L and Suppression = $dG_S - yG_L$. In both cases, G_L is the specific gravity of the liquid in the tank and G_S is the specific gravity of the liquid in the outside filled line. If the transmitter is at the level of the lower tank tap or if an air purge is used, $z = 0$.

same principle (force-balance) except that the detection, output, and feedback-balancing force would be accomplished with electric components, e.g., a differential transformer and a feedback motor. The output is normally a current signal, the standard being 4 to 20 mA.

Air Bubbler Method

This is a common measuring method for large, open-storage vessels. The level is measured by determining the pressure required to force air or gas into the liquid at a point beneath the surface. Figures 1-8 to 1-10 indicate various types of installations. The bubble pipe can be of any material and is notched at the end, as shown in Fig. 1-11. This prevents large bubbles from forming.

*d/p Cell is a trademark of The Foxboro Company, Foxboro, Massachusetts.

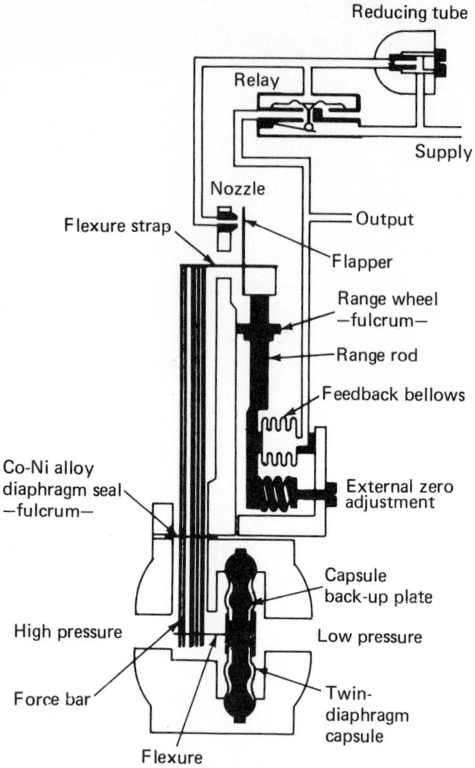

Figure 1-7 Differential-pressure transmitter. (*The Foxboro Company.*)

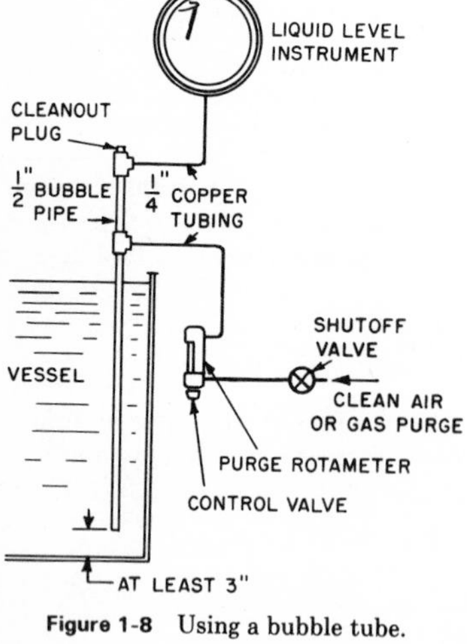

Figure 1-8 Using a bubble tube.

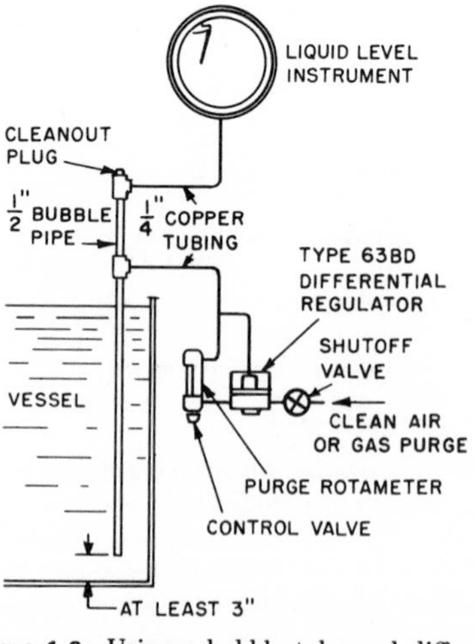

Figure 1-9 Using a bubble tube and differential-pressure regulator.

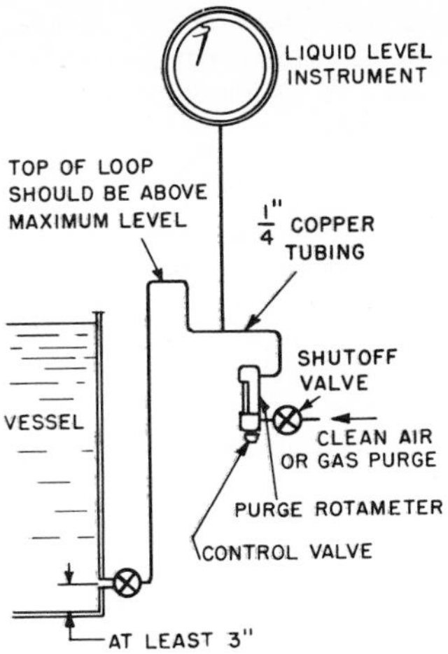

Figure 1-10 Purging directly into the side of a vessel.

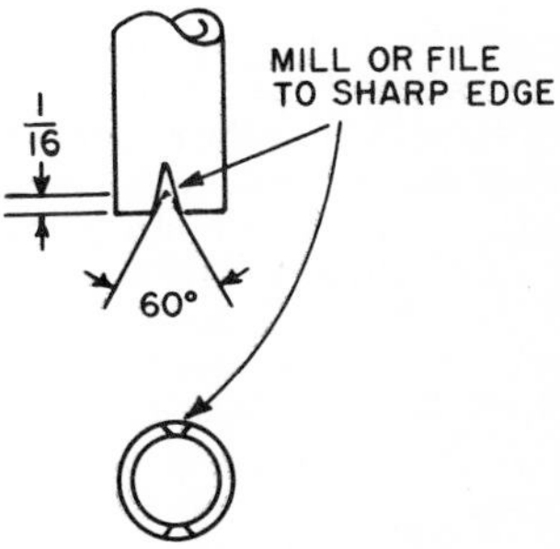

Figure 1-11 Detail of the notch in a bubble pipe.

Capacitance Method

The capacitance between two concentric cylinders is a direct function of the dielectric material between the two cylinders. Level measurements are accomplished by using a probe (one plate of the capacitor) and the tank, which acts as the second plate. As the level varies, the capacitance varies linearly, and this change in capacitance can be detected by using a bridge excited by a high-frequency oscillator. For nonconductive materials, an uninsulated probe can be used. Conductive materials require that the probe be coated with an insulator.

A number of possible errors can occur. For example, changes in the composition and in the temperature of a material will cause changes in its dielectric constant. The amount of water in a material has a profound effect on its dielectric constant. These changes will directly affect the level measurement. If the liquid tends to wet or adhere to the probe, then the level is measured only as far as the the probe is wetted, especially if the material is conductive. This is one of the serious drawbacks of this measuring method. The accuracy depends upon the differential capacitance, which is usually designed to be greater than 10 pF. The narrower this capacitive span, the more difficult the measurement.

Conductivity Method

This method is applicable for single-point measurement with a conductive process material. Two probes are used and placed such that an electric path is provided through the conductive material when both probes are covered. This is used to close the circuit of a relay, thereby providing a contact output for a certain level. Multipoint measurements can be made to provide point measurements for different levels. The method is inexpensive and the equipment fairly rugged, but it is not intrinsically safe unless special precautions are taken.

Thermal Element Method

The thermal element method is again used for point measurements and depends upon the fact the thermal conductivity of most process liquids is much higher than its associated vapor. When the thermal element comes in contact with liquid, the rate of heat transfer increases, causing its resistance to increase. This increased resistance is detected by an increased voltage drop which can then be used to activate a relay. The equipment used in this method is rugged and not affected by process changes, but the method is not intrinsically safe, and the heat input may cause changes in the quality of the measured product. This method, in general, has limited use in practice.

Radiation-Difference Method

This method is very useful in those applications in which the dangerous or extremely corrosive nature of the process fluid requires a noncontact type of measurement. The method depends upon the absorption of radiation by the process material. As the level of this material increases or decreases, the amount of radiation absorbed increases or decreases. This can then be determined by using a low-level detector, such as the Geiger-Müller tube. This method is applicable both for point measurement or a range of measurements. Safety considerations have to be kept in mind when considering installation, especially in terms of radiation exposure. The normal sources of radiation used are radium, cesium 13, and cobalt.

Weighing Method

This method, like radiation, is well suited for measuring process materials which do not allow contact-type measurement methods. Basically, the vessel weight is continuously measured by using either strain-gauge-type weigh cells or hydraulic pressure. Knowing the weight of the empty vessel, the weight of the process material in the vessel can be related to the depth of the material in the tank. Special precautions have to be taken to balance the vessel so that associated piping connected to the vessel does not affect the actual weight. The method has limited use in actual process control compared with some of the other methods, e.g., differential-pressure measurement using isolating seals for corrosive process materials.

Ultrasonic Method

This method uses an ultrasonic oscillator (20,000 Hz and higher) to excite a sensor. For point measurements, the signal is picked up by another sensor as long as a transmission path is available. When the level rises, the path is interrupted, thereby indicating the level. Another method is to allow the process material to damp the vibration of the sensor. This damping can be detected to provide a point measurement. Continuous measurements can be accomplished by using intermittent transmission and measuring electronically the time taken for the reflected signal to return to the sensor. This method is applicable to both liquids and solids; measurement of the level of solids in silos or storage bins has been the most popular application.

PRESSURE-MEASUREMENT INSTRUMENTS

Overview

Pressure is a fundamental process variable and its measurement can be used either directly for control or to infer other measurements, for example, level, flow, and temperature. There are many types of transducers that can be used. The most common are listed in Table 1-2.

The transducers can be linked to either a pneumatic or an electronic transmitter to develop a signal 3 to 15 psig (0.02 to 0.1 MPa) or 4 to 20 mA. The heart of the pneumatic transmitter is the flapper-nozzle assembly, including the pneumatic relay. The output from the relay is controlled by the back pressure developed in the nozzle which, in turn, is developed by the relationship of the flapper to the nozzle opening. Figures 1-12 and 1-13 indicate this relationship and a typical relay.

TABLE 1-2 Pressure-Measurement Elements*

Element	Local	Remote	Range	Contact with process	Relative cost	Accuracy, % of span
Bourdon tube	✓	✓	0–12,000 psig (0–83 MPa)	Yes*	Low	±0.25–5
Bellows		✓	0–3000 psig (0–20 MPa)	Yes*	Med	±0.5–1
Manometer	✓		0.05 in Hg–1 atm (3–100 kPa)	Yes	Low	0.1–1
Diaphragm	✓	✓	0.05 in Hg–1 atm (3–100 kPa)	Yes	Med	±0.5
Strain gauge		✓	1 atm on (0.1 MPa on)	No	High	±1
Dead weight	✓		1 atm on (0.1 MPa on)	No	High	High

*A chemical seal can be used for isolation.

There are two arrangements for making use of the flapper-nozzle relationship. These arrangements lead to either *motion-balance* or *force-balance* instruments. Figures 1-14 and 1-15 indicate schematics of the two types. Force-balance instruments tend to be more widely used because of their ruggedness, high degree of insensitivity to vibration, and method of mounting.

Electronic transmitters generally tend to be of the force-balance type. Figure 1-16 indicates a typical layout of a differential transformer type.

The pressure applied to the bellows capsule is applied to the force bar through a flexure and is transmitted to the detector. The lever system moves the ferrite disk (part of the detector), causing a change in the output of the differential transformer. This change in output is sensed by the amplifier oscillator circuit, and its output is rectified to provide the transmission signal. The feedback coil, in series with the output signal, supplies the balancing feedback force.

There are other types of detectors that have also been used: the inductance type, where the movement of the ferrite piece changes the air gap, which then changes the inductance of an oscillator circuit and thereby changes its output; and the variable reluctance type, where the movement of an armature causes the inductance ratio between two coils to change. The two coils are part of a bridge circuit which is rebalanced by a feedback amplifier, thereby changing the capacitance in the bridge. These devices can be considered to be motion-balance instruments.

Bourdon Tube

Simplest of the pressure-sensing elements, the Bourdon tube consists of a curved flattened tube with one end sealed and either connected to an indicator through linkage and gears or applied through a flexure mechanism to the force bar of a transmitter. The other end is connected to the process. Figure 1-17 shows a typical layout of an indicator. As pressure is applied, the tube tends to straighten, causing a small movement of the tip.

There are a number of other types of tube designs that have been developed. Figure 1-18 illustrates some of them.

The material of the tube can vary from bronze for some low-pressure ranges [up to 400 lb/in^2 (2.76 MPa)] to Ni-Span C* alloy which allows a C Bourdon tube to range up to 12,000 lb/in^2 (83 MPa). Absolute pressure measurements are limited to a maximum of 100 lb/in^2 (0.7 MPa).

In applications where the process fluid is extremely corrosive, it is normal to use pressure seals, connected directly or through capillary tubing. The system is solidly filled

*Ni-Span C is a trademark of Huntington Alloys, Inc., Huntington, West Virginia.

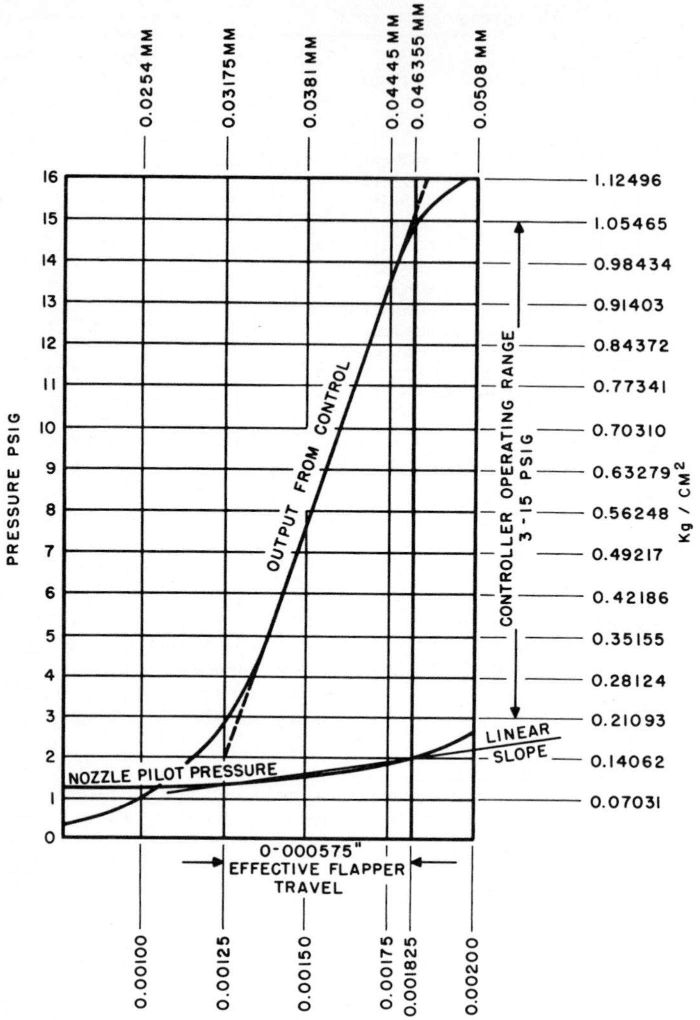

Figure 1-12 Curves for flapper-nozzle relation vs. operating pressure.

with a suitable liquid transmission medium. Process pressure is applied to the flexible member of the seal cavity in the measuring element, causing element movement in proportion to applied pressure. The spring rate of the flexible member must be less than that of the measuring element to allow full-range operation. A number of filling fluids have been used, the standard being DC-704, a silicone-base (Dow Corning Co.) product. This fluid has a low thermal coefficient of expansion with temperature limitation of 0 to 700°F (−18 to 370°C). The fluid must completely fill the system, as pockets of gas or air can cause large errors. There are many different types of seals available. Figure 1-19 indicates some of them.

Bellows

Refer to Fig. 1-20. Widely used both for measurement and control, bellows can be arranged as motion-balance or force-balance instruments. Figure 1-20 indicates the use

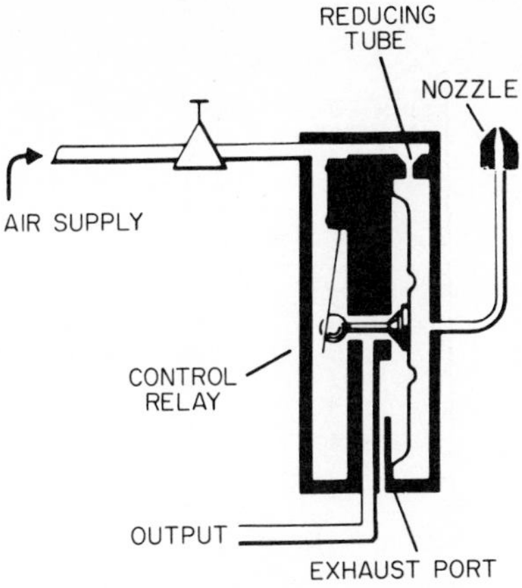

Figure 1-13 Pneumatic relay G.

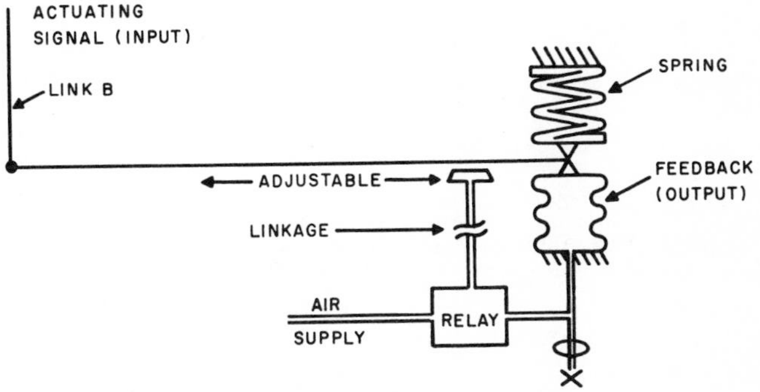

Figure 1-14 Motion-balance schematic.

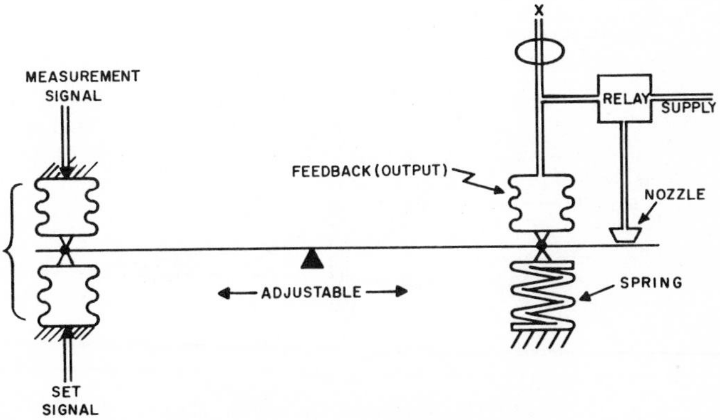

Figure 1-15 Force-balance schematic.

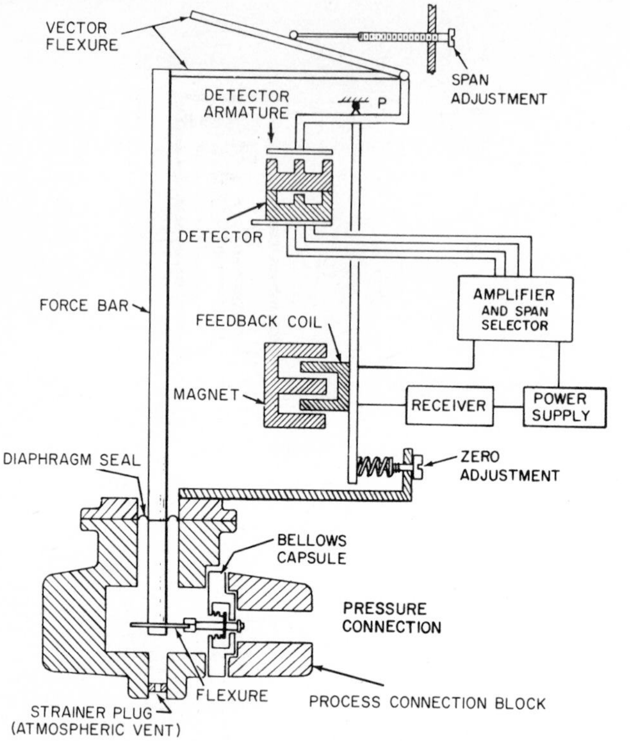

Figure 1-16 Differential transformer-type pressure transmitter.

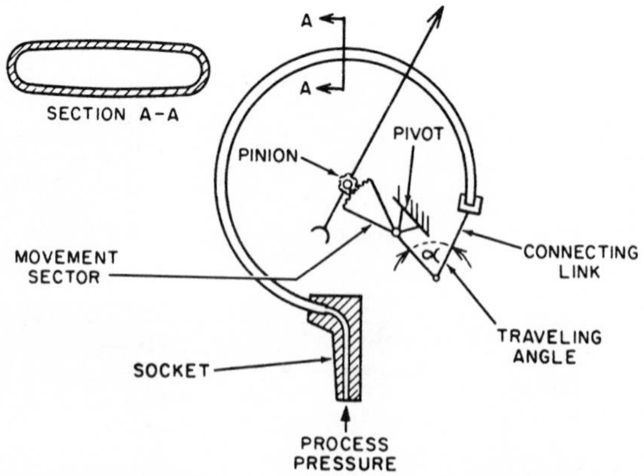

Figure 1-17 C Bourdon pressure element.

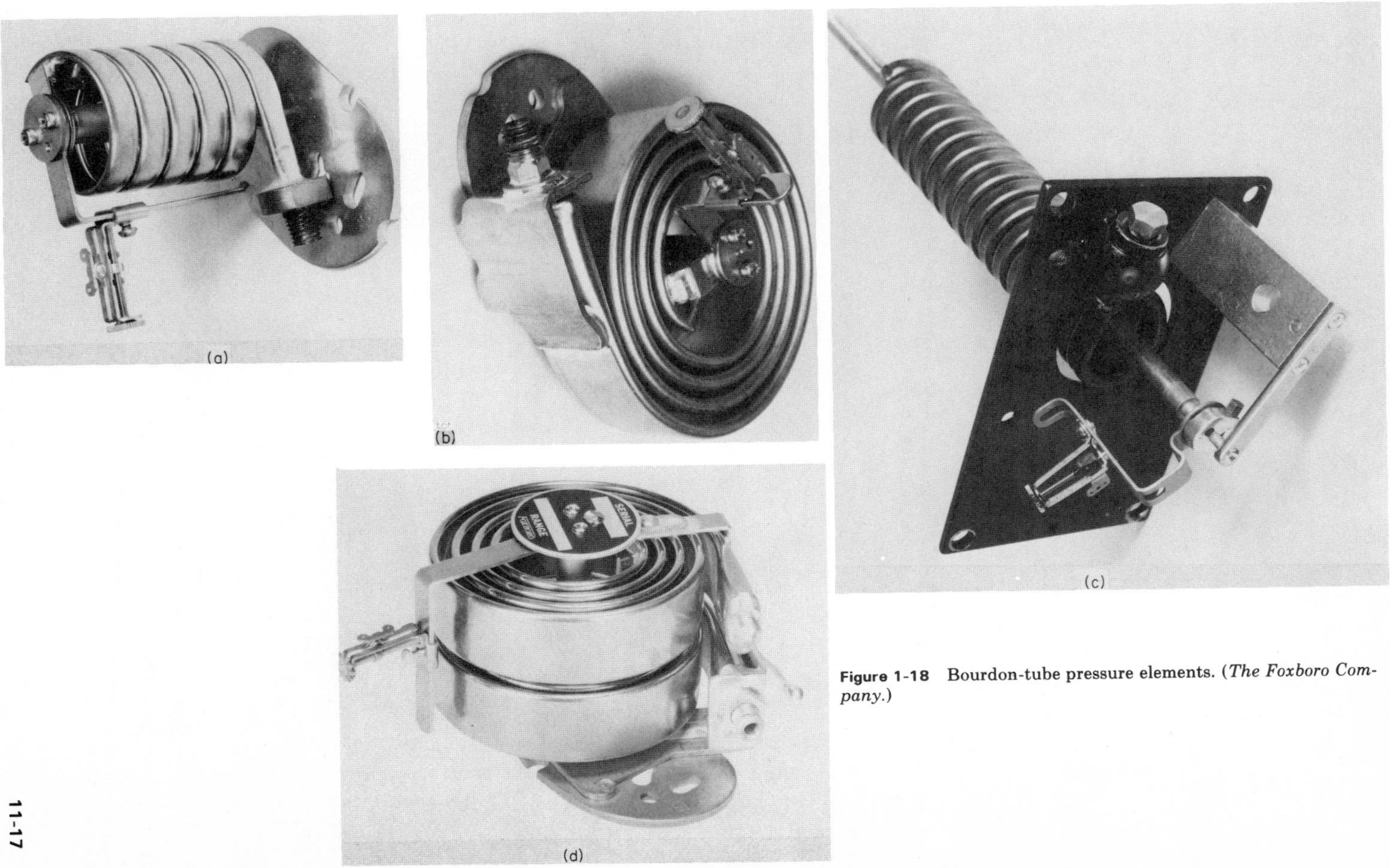

Figure 1-18 Bourdon-tube pressure elements. (*The Foxboro Company.*)

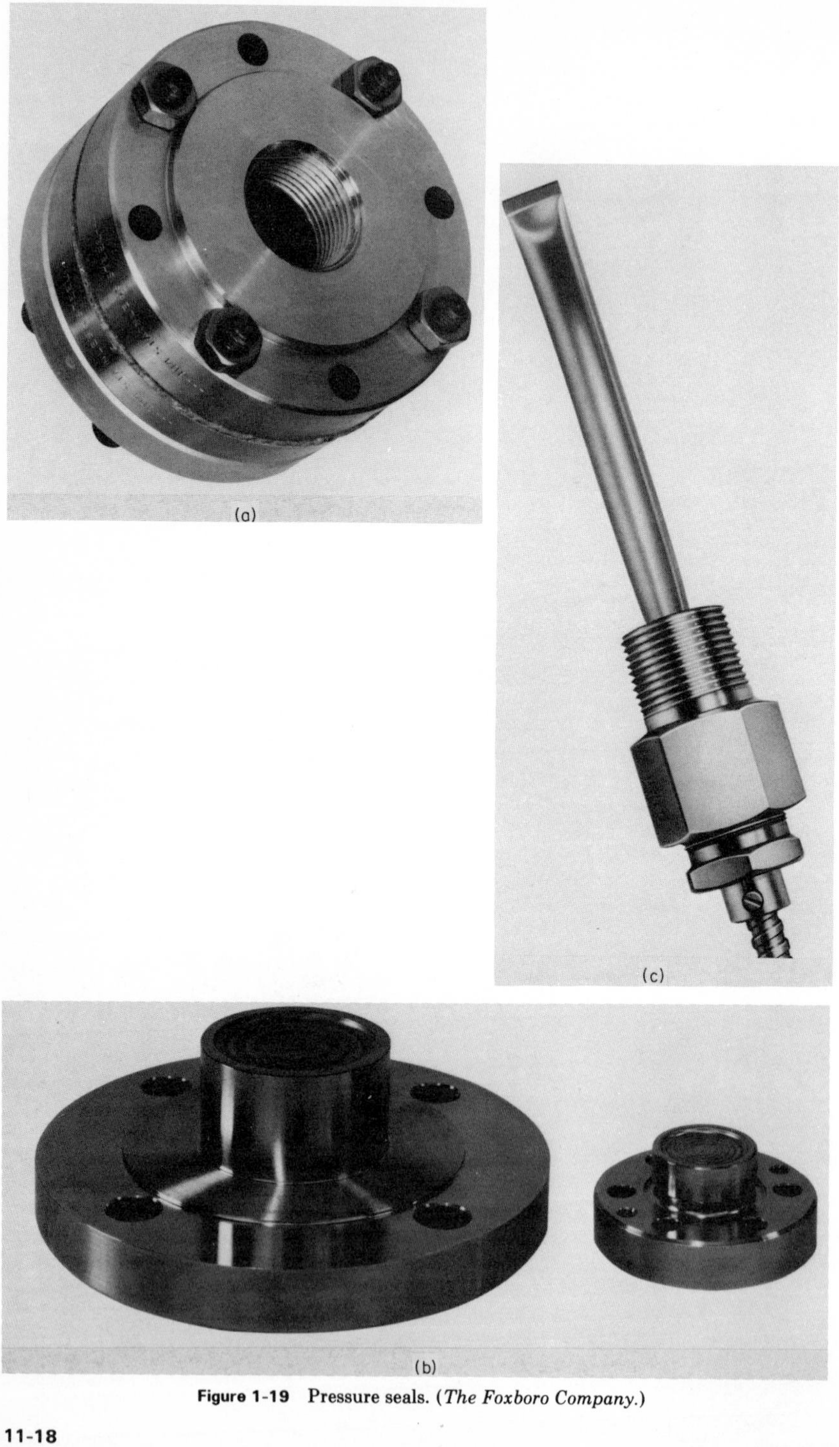

Figure 1-19 Pressure seals. (*The Foxboro Company.*)

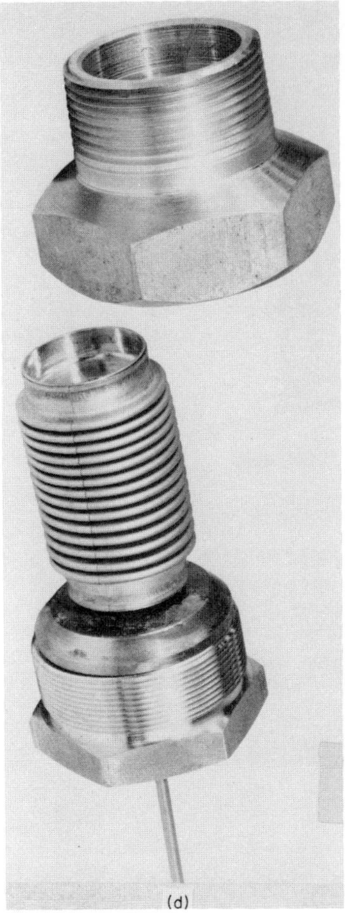

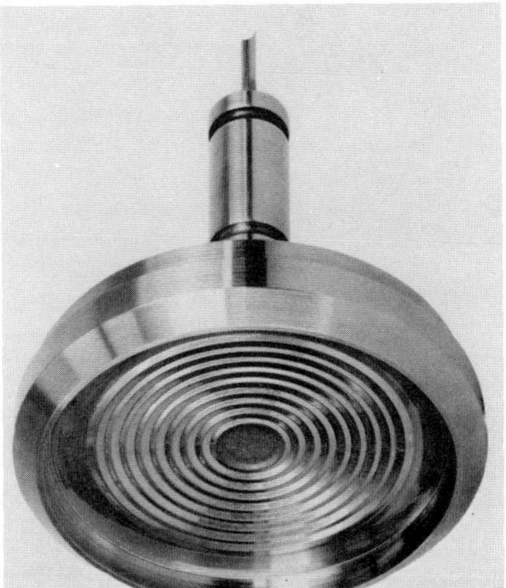

Figure 1-19 (*continued*) Pressure seals. (*The Foxboro Company.*)

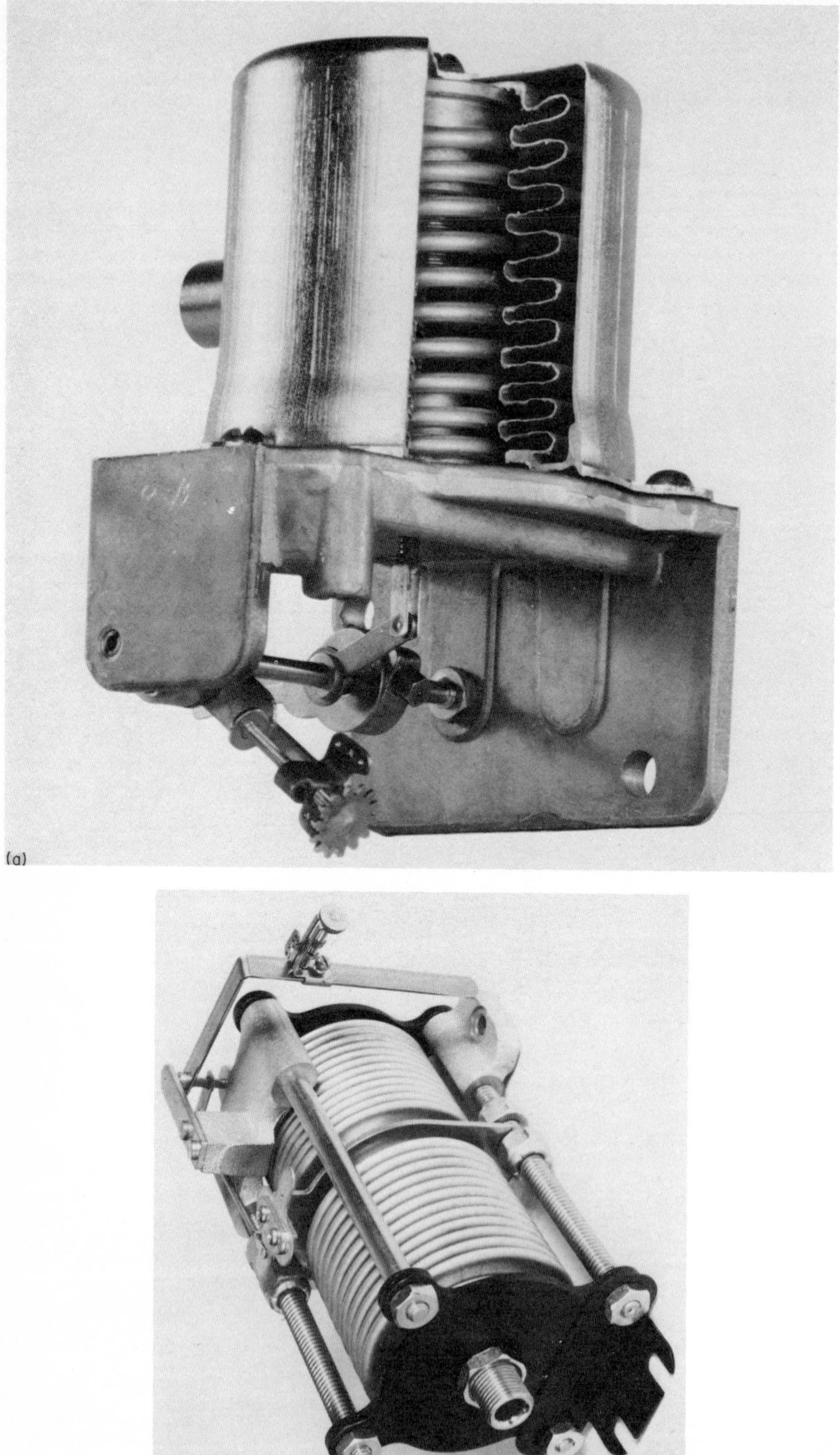

Figure 1-20 Bellows-type elements. (*The Foxboro Company.*)

of a bellows capsule for pressure measurement. Two-bellows systems are used when compensation is required, e.g., in absolute pressure measurement. One of the bellows is completely evacuated and acts in opposition to the measurement bellows. As atmospheric pressure varies, the evacuated bellows expands or contracts, thereby adding or subtracting from the measured absolute pressure. The size of the bellows can be varied for different pressure ranges from 0 to 7 lb/in^2 (0 to 50 kPa) to 0 to 2000 lb/in^2 (0 to 14 MPa).

Diaphragm

Refer to Fig. 1-21. The pressure-measuring element consists of a series of diaphragms which are connected together. Application of pressure causes the elements to expand,

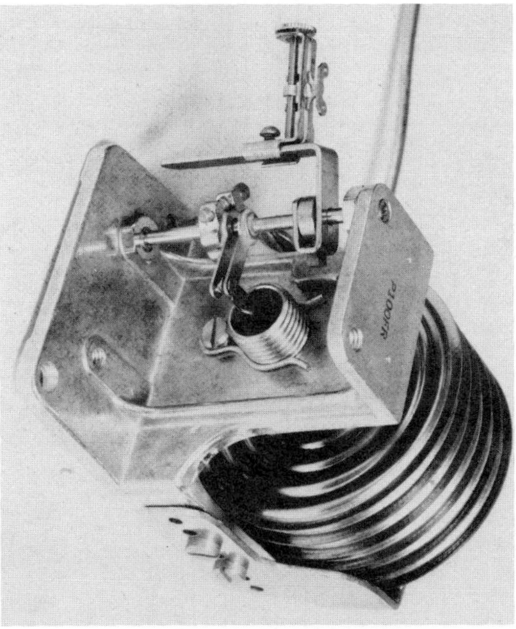

Figure 1-21 Diaphragm element. (*The Foxboro Company.*)

and the movement is then used for indication. These elements are typically used for low pressure applications of less than 5 lb/in^2 (1 kPa).

Figure 1-22 indicates a differential-pressure transmitter using a diaphragm capsule. The size of the capsule determines the range. Differential-pressure ranges from as low as 0 to 5 in H_2O (0 to 20 mm H_2O) to 0 to 850 in H_2O (0 to 330 cm H_2O) under static pressures up to 6000 lb/in^2 (41 MPa) are available.

Special Instruments

There are other types of pressure-measuring instruments that are available but have limited application in process industries. A manometer and a deadweight type are still quite popular, primarily for calibration of other pressure elements. The deadweight tester converts a known weight into a given liquid pressure which is then used in the calibration of an instrument.

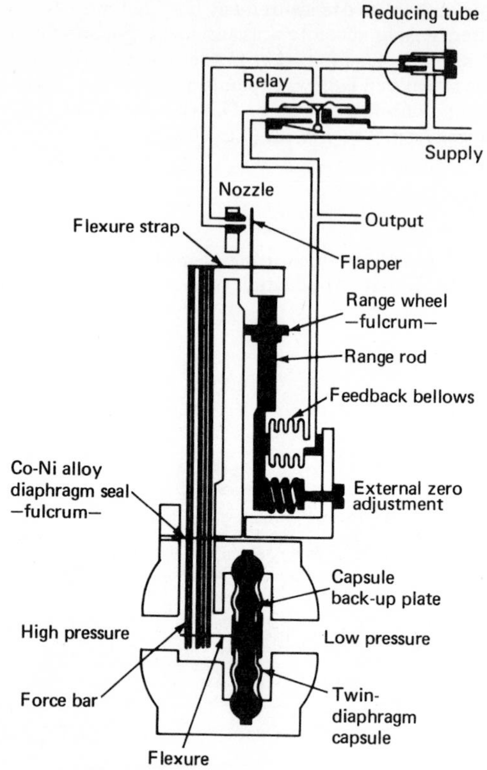

(a)

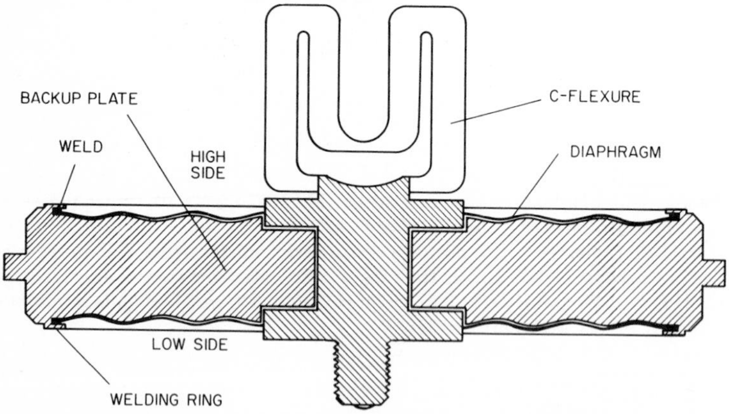

(b)

Figure 1-22 Diaphragm capsule for differential pressure measurement: (*a*) in situ in transmitter; (*b*) detailed structure.

TEMPERATURE-MEASURING METHODS

Refer to Table 1-3 for the more common methods used for measuring temperature in an industrial process environment. Each of these is discussed in detail in the following text.

TABLE 1-3 Temperature-Measuring Methods

	Local	Remote	Range, °F (°C)	Accuracy ± % span	Cost	Ease in replacement of components
Filled thermal system	✓	✓	−450 to 1400 (−260 to 760)	0.5 to 2	Low–med	Intermediate
Thermocouples		✓	−420 to 2000 (−250 to 1100)	0.3 to 1	Med	Easy
Resistance measurement		✓	−450 to 1625 (−260 to 900)	0.1 to 0.5	High	Difficult
Thermistors		✓	−200 to 500 (−130 to 260)	0.1	Med	Difficult
Bimetallic	✓		−80 to 800 (−60 to 420)	1 to 2	Low	Easy
Pyrometers	✓		High temperature	1 to 2	High	—
Paints	✓		100 to 2000 (40 to 1100)	—	Low	—

Filled Thermal Systems

The basis of measurement consists of a bulb connected by a capillary to a helical or C Bourdon tube element. The system is filled under pressure so that an increase in pressure causes a movement of the helical or Bourdon tube element. This movement can then be linked for local indication or through a transmitter for an electric or pneumatic signal.

Filled systems are classifed by SAMA (Scientific Apparatus Makers Association) into four classes with subclassifications in each class.

Classes IA and IB

These are liquid-filled systems. Class IA refers to fully compensated systems while Class IB refers to case-compensated systems. Fully compensated refers to compensation of ambient temperature variation along the length of the connecting capillary tubing. It consists of (in addition to the normal measuring bulb) a connecting capillary and pressure element, with a second capillary tube and pressure element. The two capillary tubes are run together while the pressure elements are connected such that equal amounts of motion nullify each other. Thus ambient-temperature variation along the length of the capillary tubes and at the case produce equal movements of the pressure elements and hence cancel the effect on the output of the instrument. Class IB, *case compensation,* nullifies ambient temperature variations at the case in a similar fashion but may only use a bimetallic strip instead of a pressure element. These instruments are generally mounted very close to the bulb.

The initial pressure in this class of instruments is very high, making it a volumetric instrument, free of static errors from bulb position with respect to the instrument case. Furthermore, being volumetric devices, the amount of available volume for expansion is limited, also limiting the amount of available overrange.

These instruments can also be used for measuring temperature differences by using two bulbs with associated capillary tubing and pressure elements. The pressure elements are arranged in a manner similar to Class IA fully compensated systems. Ambient-temperature variations are minimized by running the capillary tubes as close to each other as possible. Temperature-difference spans greater than 36°F (20°C) are normally required.

Classes IIA, IIB, IIC, and IID

In these systems, the bulb, capillary tube, and pressure-measuring element are partially filled with a volatile fluid such that the vapor-liquid interface level occurs in the bulb. As the temperature rises, the vapor pressure increases and this increase can then be sensed via the pressure element. The system is stable and unaffected by ambient-temperature variations. They are the simplest to manufacture and therefore least expensive, both initially and in use.

The difference characterizing the four subclassifications is as follows: In a Class IIA system the capillary tube and the pressure element are filled with liquid while the bulb has both vapor and liquid. These systems are used when the bulb temperature is higher than the ambient temperature. Class IIB systems are the reverse of Class IIA, with the bulb used for temperatures less than the ambient temperature. In Class IIA systems, the effect of the liquid head in the capillary has to be considered if the bulb is installed above or below the instrument case. Class IIC systems are systems where the bulb temperature crosses the ambient temperature over its operating range. This causes the capillary to be filled with liquid or vapor, resulting in static head errors unless the bulb is mounted at the same level as the case. Class IIC instruments are normally used as local indicators. Class IID systems are like Class IIC systems, except that the capillary tube and the pressure element are filled with a nonvolatile fluid, while the bulb contains liquid and vapor.

In all these classes the output is nonuniform and depends upon the vapor-pressure-vs.-temperature curve of the volatile fluid. Applications are geared such that normal operating temperatures are at the higher end of the scale where the scales are much wider, allowing greater readability.

Class III

These are gas-filled systems which are, like Class II systems, simple to construct, with wide applicable range limits. They are normally used with large bulbs which override the pressure variations in the capillary tube due to ambient-temperature changes. Gas pressure allows less power to be developed in the pressure-sensing element, making this class suited for wide temperature spans. The large bulbs are also well suited for average temperature measurements. Case compensation using a bimetallic strip is used in Class IIIB systems. Class IIIA systems use the normal parallel thermal system without a bulb.

Thermocouples

These sensors have in recent years achieved greater popularity with increasing use of electronic instruments. Basically, a thermocouple consists of two dissimilar metal wires, such as iron and constantan, joined to produce a thermal electromotive force (emf) when the junctions are at different temperatures. The measuring or hot junction is the end inserted in the medium where the temperature is to be measured. The reference or cold junction is the open end that is normally connected to the measuring-instrument terminals. The emf generated is a function of the difference in junction temperatures. Refer to Fig. 1-23.

The thermal emf developed by a thermocouple of two homogeneous metals is independent of the temperature gradient and distribution along the wires. Inhomogeneity in a thermocouple passing through a temperature gradient will produce undesirable emf's with resultant errors in temperature readings.

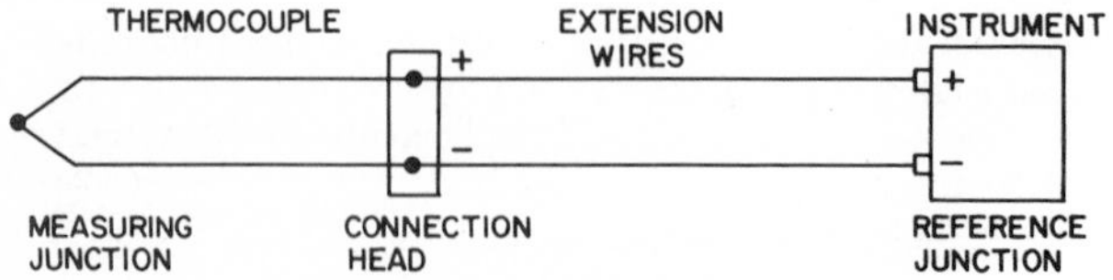

Figure 1-23 Thermocouple connection with extension wires.

TABLE 1-4 Thermocouple Comparison

ISA type	Positive wire	Negative wire	Recommended temp °F (°C)	Atmospheric conditions	Standard limits of error
T	Copper	Constantan	−450 to +750 (−270 to 400)	Oxidizing reducing	±2% of reading at low temperatures to ±¾% of reading at high temperatures
J	Iron	Constantan	0 to +1650 (−18 to +900)	Reducing	±¾% of reading at high temperatures
K	Chromel	Alumel	0 to 2300 (−18 to 1260)	Oxidizing or neutral	±¾% of reading at high temperatures
E	Chromel	Constantan	−300 to +1600 (−180 to 870)	Oxidizing	±¾% of reading at high temperatures
R, S	Platinum-rhodium	Platinum	0 to 2800 (−18 to 1530)	Oxidizing	0.5% of reading
B	Pt70–Rh30	Pt94–Rh6	0 to 3200 (−18 to 1760°C)	Inert or slow oxidizing	0.5% of reading

Introduction of intermediate metals into a thermocouple circuit will not affect the emf of the circuit, provided that the new junctions remain at the same temperature as the original junction. This fact is used in applying extension wires to cut costs, since some of the thermocouples are made of expensive metals. These extension wires must be made of materials having the same thermal emf characteristics as the thermocouple. If copper wires are used, temperature fluctuations at the reference junction (in this case the connection head) will introduce errors in proportion to the magnitude of the temperature changes. Use of proper extension wires moves the reference junction to the measuring instrument where temperature compensation can be applied. Figure 1-24 indicates the relationship of five standard thermocouples.

Table 1-4 highlights the different features of some standard thermocouples.

Figure 1-25 illustrates a simplified circuit of a Foxboro Series 33B transmitter. TC is

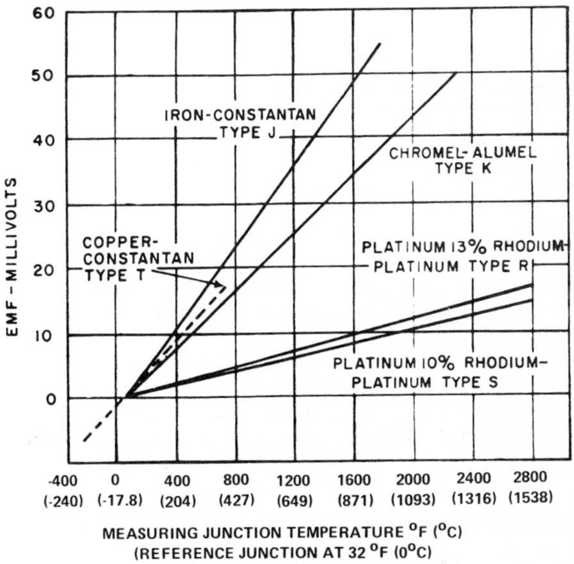

Figure 1-24 Thermocouple relationships.

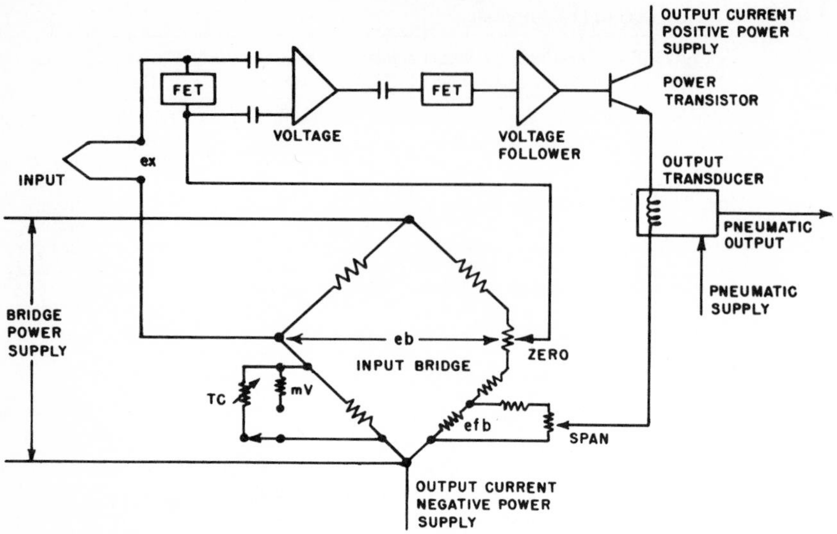

Figure 1-25 Simplified circuit diagram of Foxboro 33B series transmitter. (*The Foxboro Company.*)

a temperature-sensitive resistor which varies the bridge output as the actual reference-junction temperature varies from the design reference-junction temperature.

It should be noted that thermocouples are point-measuring devices. To get average temperature one should connect a number of thermocouples in parallel as shown in Fig. 1-26. Figure 1-27 shows the connection of two thermocouples for differential pressure measurement. The following points should be considered:

1. Swamping resistors are required to compensate for varying resistance of the thermocouple and extension wires.
2. In order to prevent ground loop currents the thermocouples must *not* be grounded.
3. The thermocouples and extension wires *must* be of the same type.
4. In differential measurement no reference-temperature compensation is required. Copper leads can be used between the connection box and the instrument.

Resistance Temperature Detectors

Based on the change of electric conductivity with temperature, the resistance temperature detector (RTD) consists of a coil of wire made of materials such as nickel and platinum. Spans as low as 5°F (3°C) can be achieved with a nickel RTD with a certified accuracy of 0.1°F (0.06°C). The sensor can be constructed with two, three, or four leads. For most industrial applications, two- or three-lead sensors are used in a Wheatstone bridge as shown in Fig. 1-28.

Differential temperature differences can also be measured by using two equal RTDs in adjacent sides of a Wheatstone bridge. The linearity of measurement is to an extent a function of the span of measurement. For wide-span instruments and extreme accuracy, nonlinear scales and charts are recommended. A certain amount of linearity can be achieved by connecting large-value equal resistors in series with the leads of the sensor.

Thermistors

These are metal oxide resistors having high temperature coefficients (usually negative), with resistance being a function of absolute temperature. They are used much like the

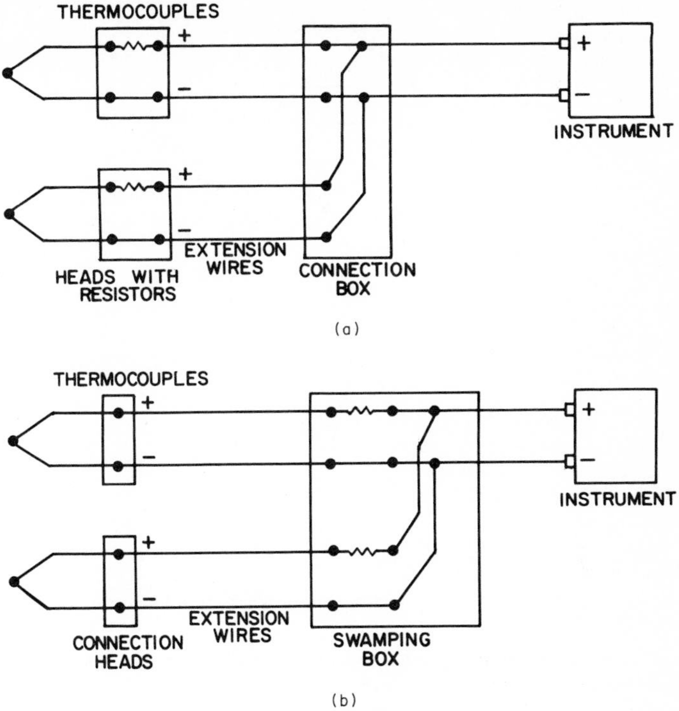

Figure 1-26 Average temperature circuits.

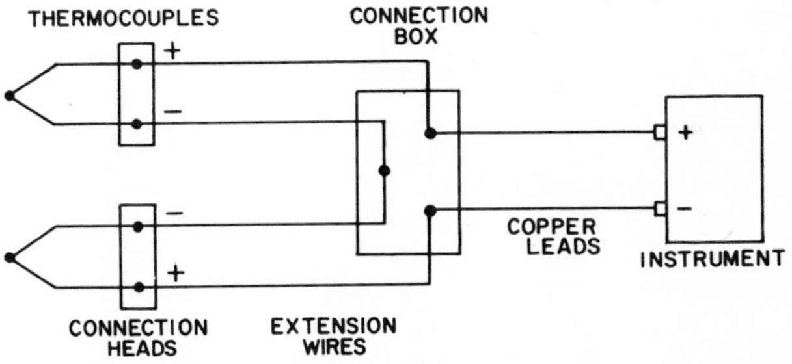

Figure 1-27 Differential temperature circuit. (*The Foxboro Company.*)

RTDs. The large temperature coefficient makes them useful for very narrow-span temperature measurement. Self-heating is a problem to be considered when current flow through the sensor causes sensor temperature to be higher than ambient temperature.

Though not widely used in process environment, the application of thermistors in various types of electronic circuitry for temperature compensation has been widespread. The temperature-resistance characteristic curve tends to be fairly nonlinear, making interchangeability of sensors a problem.

Radiation Pyrometers

These instruments are a special class used in those applications where very high temperatures are to be measured and normal contact-type measurement is not possible. The

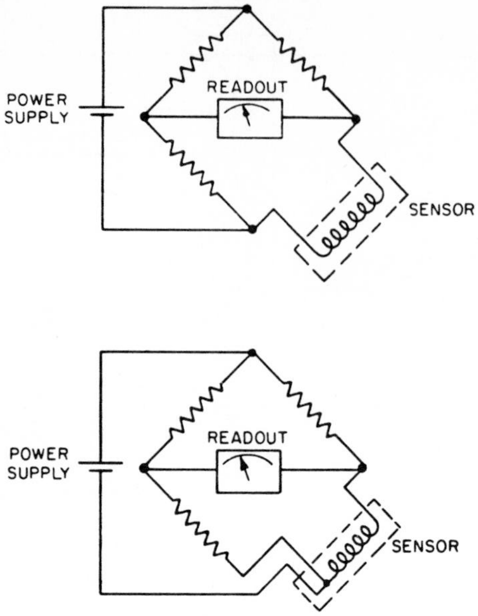

Figure 1-28 Two- and three-wire Wheatstone bridge.

basic principle is the use of an optical system to focus energy radiated to a detector. The system can be manual where this focused energy is compared with a calibrated optical filament, or automatic by using thermopiles or photomultiplier tubes. The accuracy depends on many factors, such as the emissivity of the source, reflections from other sources, and the line of sight. The temperature scale is nonlinear and the system tends to be expensive.

FLOW MEASUREMENT

The primary purpose of industrial control systems is to balance the material and energy flows in a process. Flow is the most common of the process variables. Accurage measurement and control are the two most important instrumentation functions. Table 1-5 lists some of the more common measurement methods and their characteristics.

Head-Type Devices

These are the most common types of measurement devices. Basically they depend upon a constriction in the fluid flow, thereby causing a pressure drop which can then be measured by a differential-pressure type of instrument.

Orifice Plates

Figure 1-29 indicates the pressure profile for an orifice plate.

The *vena contracta* is the location where the downstream flow has the maximum velocity and the minimum cross-sectional area. Figure 1-30 indicates this location with respect to the diameter ratio of orifice opening to pipe inside diameter. Figure 1-31 illustrates the various types of orifice plates.

The location of the orifice plate depends upon the type of taps used for the measurement of differential pressure. A certain section of straight pipe is required both upstream and downstream. Figure 1-32 is an example of the dimensions required using straight-

TABLE 1-5 Flow Measurements*

Head type	Liquids	Viscous liquid	Slurry	Gas	Solids	Linear	Rangeability	Cost	% full-scale accuracy	Indirect totalizer	Pressure loss
1. Orifice plates	✓	L		✓		SR	4:1	Low	¼–2	✓	High
2. Rotameters	✓	L	L	✓		✓	10:1	Med	½–2	—	F
3. Venturi tubes, nozzles	✓	L	✓	✓		SR	4:1	High	¼–3	✓	Med
4. Pitot tubes	✓			✓		SR	3:1	Low	2–5	—	L
5. Elbow	✓	L	L	✓		SR	3:1	Low	5–10	—	No
6. Target meters	✓	L	L			SR	4:1	Med	½–2	✓	High
7. Weirs, flumes	✓	L	L			NL	100:1	Low	2–5	—	Med
Velocity type											
1. Magnetic	✓	✓	✓			✓	20:1	High	½–1	✓	No
2. Vortex	✓	L				✓	10:1	Med	½–2	✓	Med
Displacement											
1. Positive displacement	✓	L				✓	20:1	Med	¼–1	✓	Med
2. Turbine	✓	L	L	✓		✓	20:1	Med	¼–1	✓	Med
Mass flow											
1. Weight types					✓	✓	20:1	Med	½–3	—	—
2. Solids flowmeters					✓	✓	20:1	Med	½–3	—	—

*L, limited; NL, nonlinear; SR, square root; F, fixed.

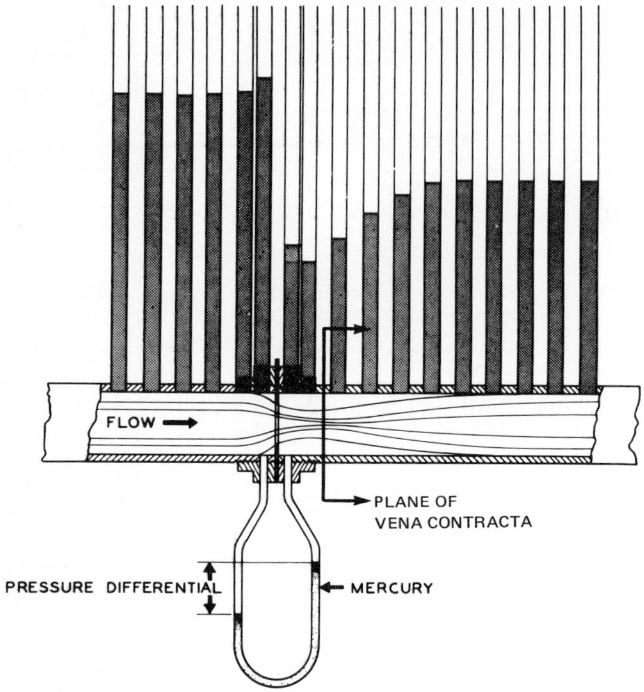

Figure 1-29 Pressure profile using an orifice plate.

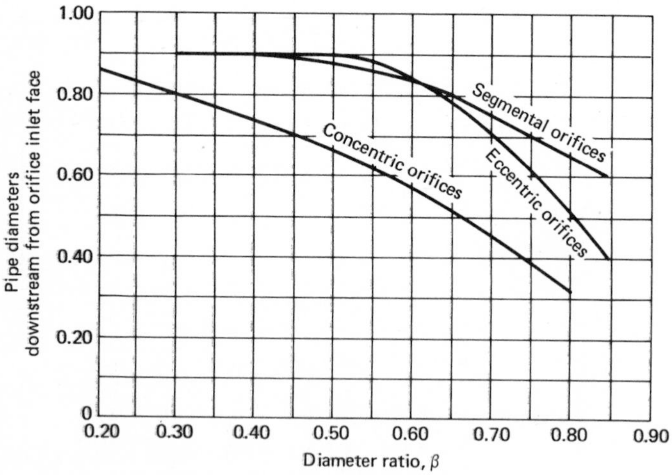

Figure 1-30 Location of the vena contracta.

ening vanes, which are devices that eliminate swirls, crosscurrents, and eddies set up by pipe fillings and valves in the upstream run.

Other types of primary elements that are used are shown in Fig. 1-33.

Figure 1-29 indicates that the pressure downstream of the primary element does not recover to its full value. Figure 1-34 indicates the permanent head loss as a percent of measured differential.

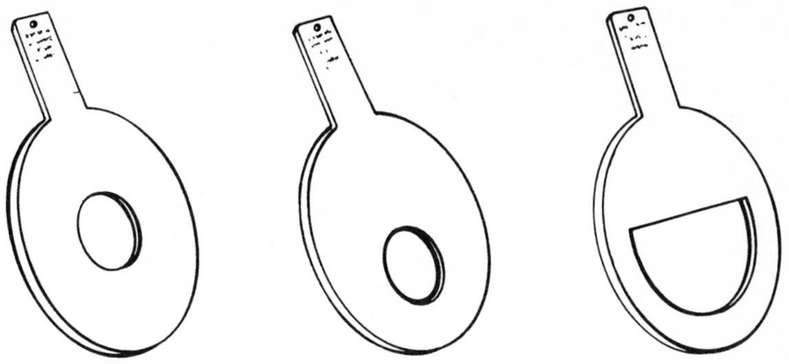

Figure 1-31 Types of orifice plates.

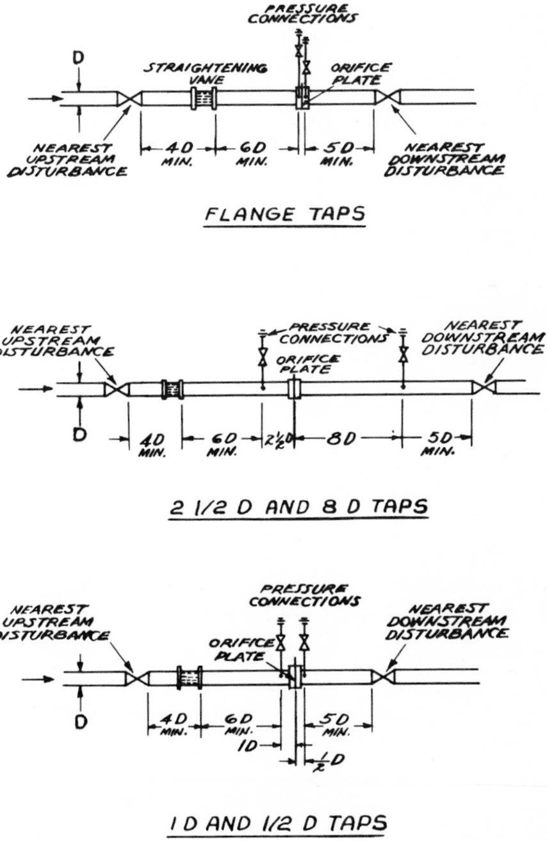

Figure 1-32 Liquid flow installation.

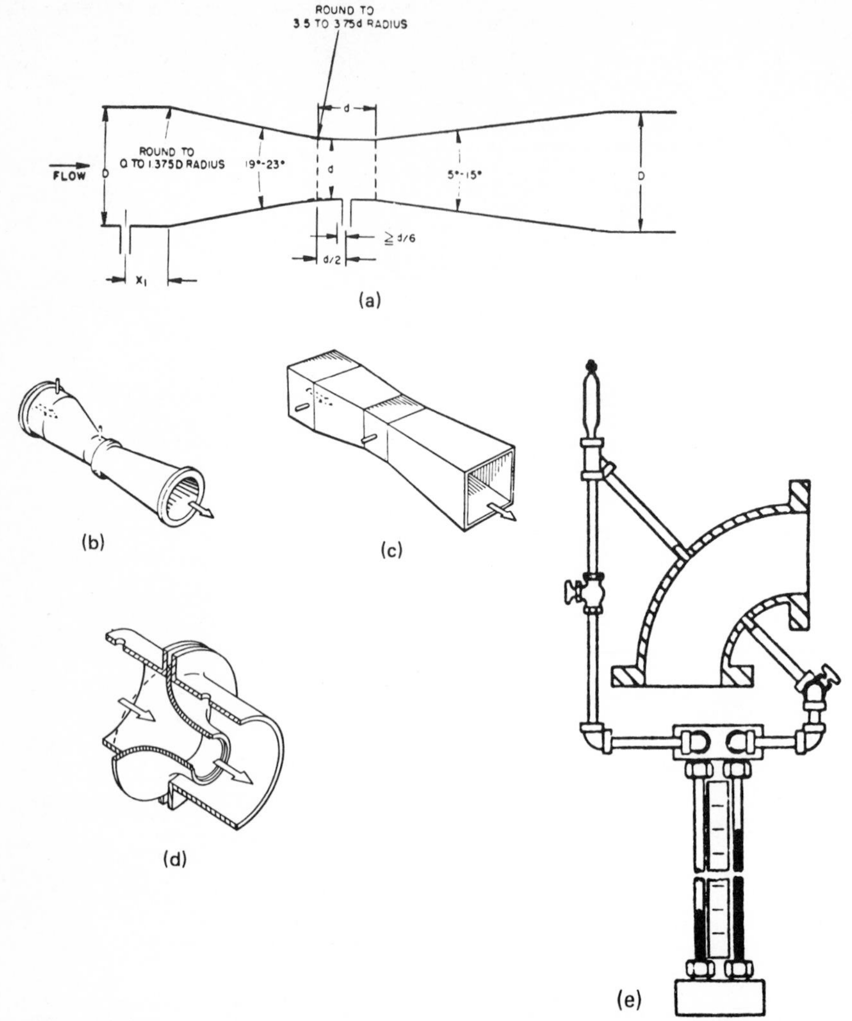

Figure 1-33 Flow nozzles, venturi tubes, and elbows. (*a*) The critical dimensions of the classical venturi tube. (*b*) An eccentric venturi tube. (*c*) A rectangular venturi tube. (*d*) Flow nozzle. (*e*) Elbow (used as primary device).

The following equations can be used for calculating flow rates for different services. Representative values for the S factor are given in Fig. 1-35.

Liquids:

$$\text{gal/min} = \frac{5.67SD^2\sqrt{h_w}}{G} \qquad \text{(to get L/min use 0.0066 instead of 5.67)}$$

Steam or other vapors:

$$\text{lb/h} = 359SD^2\sqrt{wh_w} \qquad \text{(to get kg/h use 0.01251 instead of 359)}$$

Gas:

$$\text{scfh} = 338.17SD^2F_gF_{tf}\sqrt{ph_w} \qquad \text{(to get Nm}^3\text{/h use 1 instead of 338.17)}$$

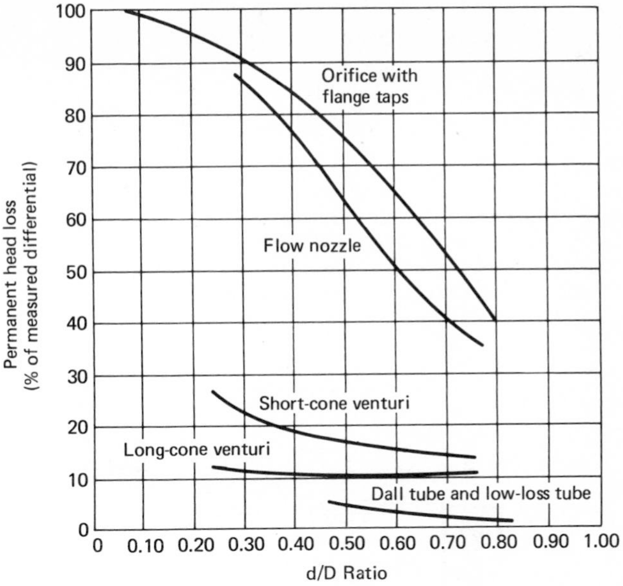

Figure 1-34 Permanent head loss for various devices.

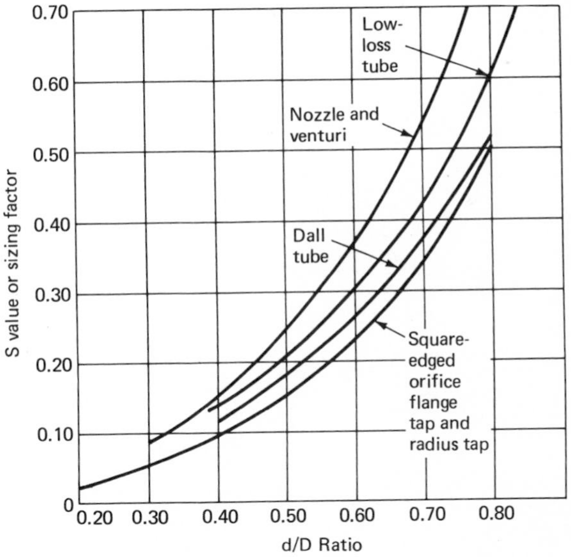

Figure 1-35 *S* values for various differential devices against beta ratios (d/D).

The standard condition is defined to be at 60°F (16°C) and 14.73 psia (760 mmHg), and the symbols in the equations are defined as follows:

S = flow coefficient (see Fig. 1-35)
G = specific gravity of the process fluid
h_w = differential pressure, mm H_20
w = steam or vapor density, lb/ft^3 (kg/m^3)

$F_g = \sqrt{1/G}$
F_{tf} = factor for flowing gas temperature, $\sqrt{520/(460 + °F)}$ or $\sqrt{228.7/(273.1 + °C)}$
P = static pressure psia (kg/cm^2 abs) [either p_1 (upstream) or p_2 (downstream) since the differential h_w is ordinarily chosen to be less than 4 percent of p]

Target Flowmeter

Another constriction-type device that has become popular recently is the target flowmeter. Figure 1-36 indicates a typical assembly with a pneumatic transmitter.

Steam- and water-flow measurements in outdoor installations are common applications. The complications of condensate pots, heat tracing, seal fluids, and antifreeze compounds to prevent freezing in cold weather are avoided.

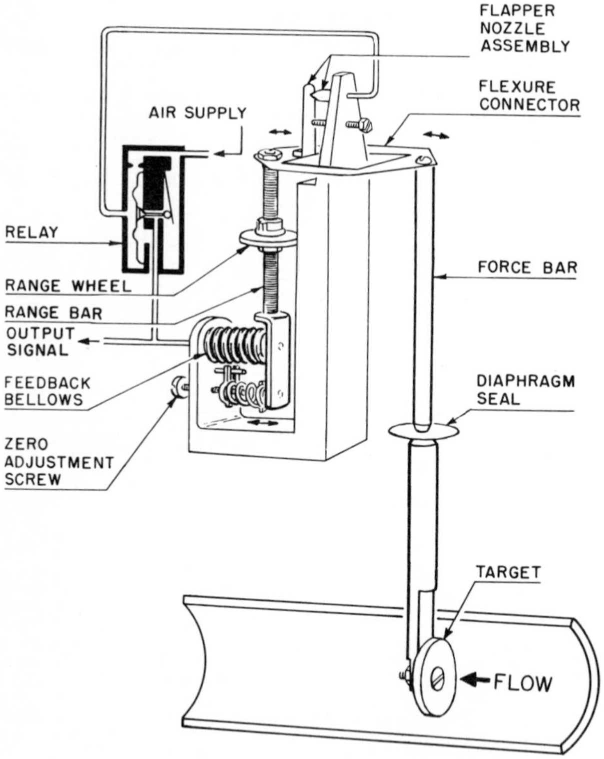

Figure 1-36 Target flowmeter. (*The Foxboro Company.*)

The flow in volume or mass units is proportional to the square root of the force on the target.

Measurement accuracy of the head-type meters discussed so far depends on the Reynolds number R_D, which can be defined by

$$R_D = \rho \frac{VD}{\mu}$$

where

ρ = density

V = velocity
D = pipe inside diameter (I.D.)
μ = viscosity

The higher the Reynolds number, the flatter the velocity profile of the flow across the pipe inside diameter. If we define a flow coefficient K_a as a factor that when multiplied

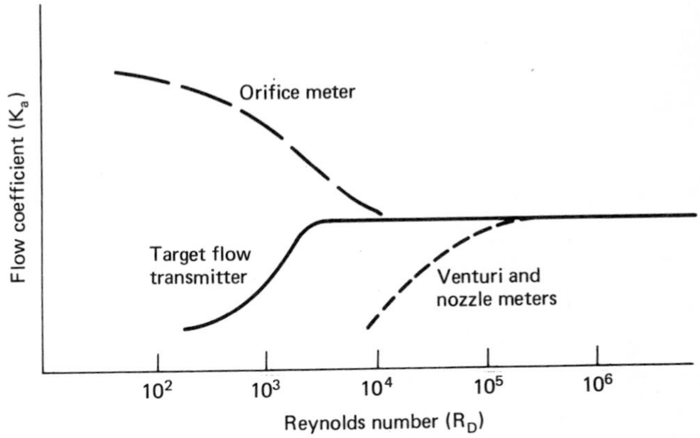

Figure 1-37 Typical flow-coefficient curves.

by the measured phenomena gives the required flow, then the plot of K_a vs. R_D for head-type meters is given in Fig. 1-37.

Weirs and Flumes

Open-channel measurements are important, especially in the waste- and water-treatment fields. Head H_a is developed by placing a weir in the flowpath as shown in Fig. 1-38. Aeration under the nappe is required for accurate flow measurement.

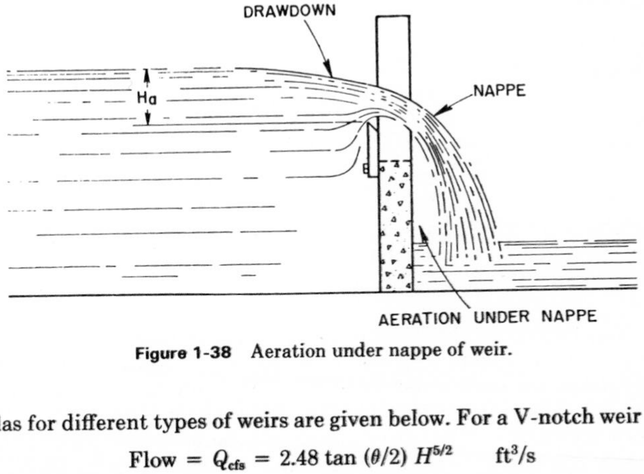

Figure 1-38 Aeration under nappe of weir.

Formulas for different types of weirs are given below. For a V-notch weir (Fig. 1-39*a*),

$$\text{Flow} = Q_{\text{cfs}} = 2.48 \tan(\theta/2)\, H^{5/2} \qquad \text{ft}^3/\text{s}$$

For a rectangular weir,

$$Q_{\text{cfs}} = 3.33\,(L - 0.2H)\, H^{3/2}$$

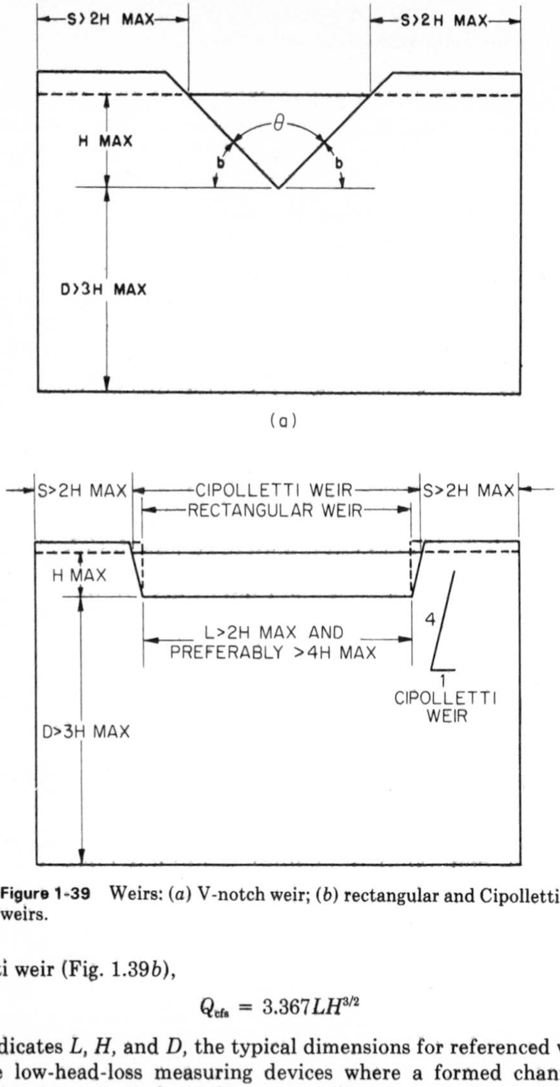

Figure 1-39 Weirs: (*a*) V-notch weir; (*b*) rectangular and Cipolletti weirs.

For a Cipolletti weir (Fig. 1.39*b*),

$$Q_{cfs} = 3.367LH^{3/2}$$

Figure 1-39 indicates *L, H,* and *D,* the typical dimensions for referenced weirs.

Flumes are low-head-loss measuring devices where a formed channel restriction changes static head to velocity head. Figure 1-40 illustrates the Parshall flume.

The three basic methods for measuring the head in a weir or flume are the float-and-cable, the in-flume float device, and the bubble tube.

Rotameters

Our discussion so far has emphasized meters having a variable differential head and a constant restriction area. Rotameters, however, are devices using a constant differential and a variable restriction area. These instruments are typically used in measurement of small liquid or gas flows with local indication. Transmitter-type instruments for measuring large amounts of liquid flow (e.g., oil) are commercially available.

This device consists of a vertical tapered tube through which the flow passes in an upward direction. A float moves up until the upward force acting on the float is balanced by the downward gravitational force. If the tube is made out of glass, the position of the

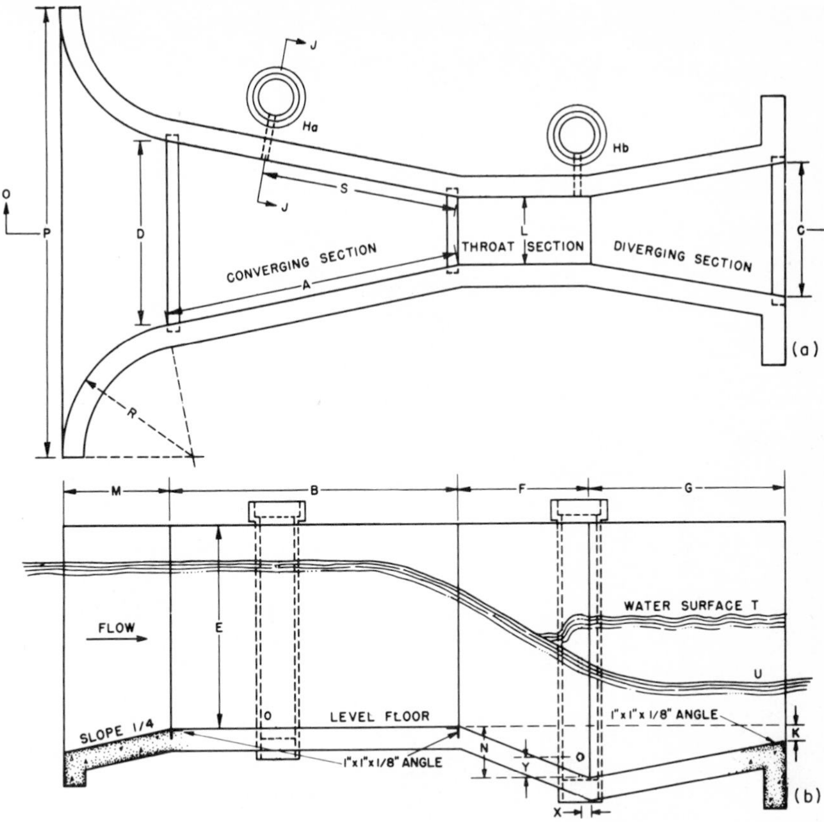

Figure 1-40 (*a*) Diagram and (*b*) dimensions of Parshall flume.

float is a direct and linear measurement of the flow. Transmitter-type instruments make use of a metallic tube in which the float is mechanically linked to the transmitter mechanism. Its accuracy is comparable to that of other head-type meters, with lower accuracy for indicating-type glass-tube meters. An advantage in transmitter-type instruments is the area available around the float for entrained fluid particles. The disadvantage is the mechanical linkage and its associated maintenance problems.

Velocity-Type Devices

The most common velocity device is the magnetic flowmeter. Its advantages are that no head loss occurs, it handles solids in suspension, no liquid connections are required, and an electronic output suitable for in-plant transmission is produced. Figure 1-41 shows a schematic drawing (*a*) and a cross section (*b*) of a typical instrument. Symbols used in Fig. 1-41 are listed:

E = generated voltage
C = meter constant
H = magnetic field
D = distance between conductors (pipe I.D.)
V = velocity of flow

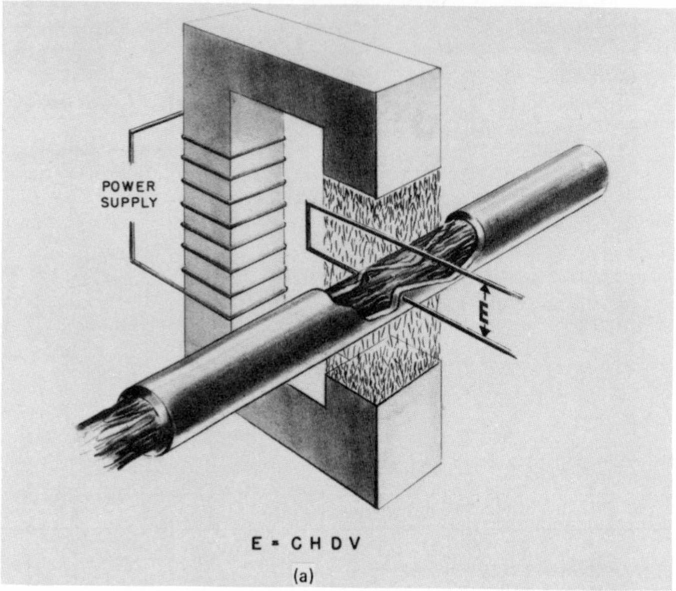

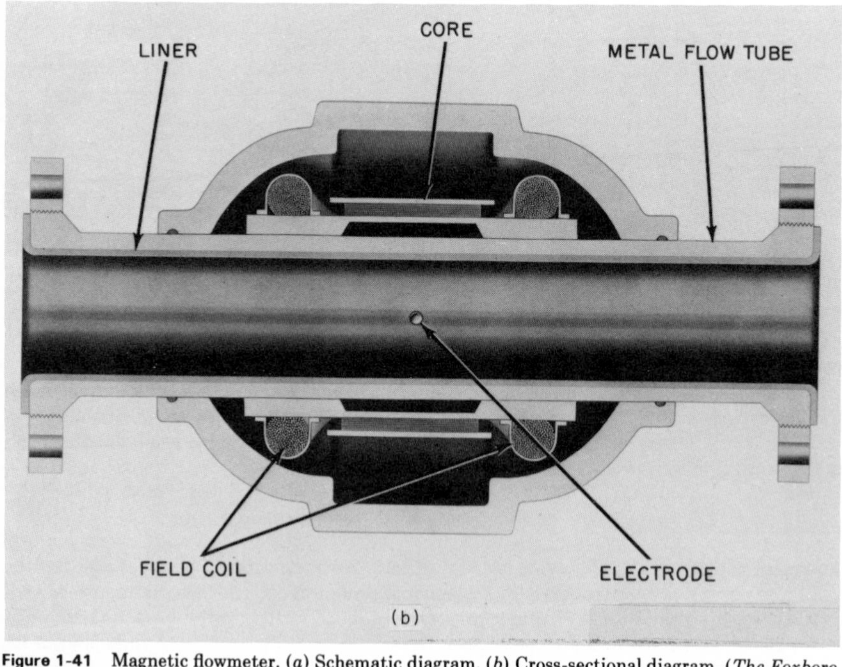

Figure 1-41 Magnetic flowmeter. (*a*) Schematic diagram. (*b*) Cross-sectional diagram. (*The Foxboro Company.*)

The output is linear with velocity for a constant magnetic field. The instrument is fairly rugged and can measure wide ranges of flow. To minimize current flow in the measuring circuit, a high-input impedance amplifier is used, with special precautions for shielding the input circuit.

Another velocity meter of recent development is the vortex-shedding meter. Liquid flowing through the meter housing passes a specially shaped vortex element which causes vortices to form and shed (separate) from alternate sides of the element at a rate proportional to the flow rate of the liquid. These vortices create an alternating differential pressure which is sensed by a detector located at the "tail" of the vortex generator. An ac voltage signal is produced in the flowmeter with a frequency synchronous with the vortex-shedding frequency.

The meter offers accuracies on a par with turbine and positive displacement meters but has the advantage of requiring no moving parts in the liquid stream. Therefore, the problems arising from overspeeding the turbine with two-phase flows, or from damage when slugs of liquid impinge upon it, do not exist. The meter equation is

$$f = kQ$$

where

f = frequency, pulses per minute
k = meter constant, pulses per unit volume
Q = flow rate

Figure 1-42 indicates the variation of k with respect to Reynolds number (signature curve) for a vortex flowmeter.

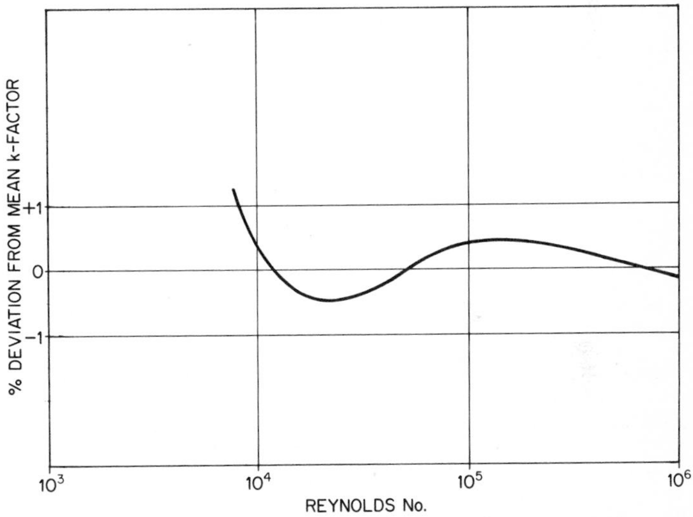

Figure 1-42 Typical signature curve of E83 vortex flowmeter.

Displacement Meters

There are many configurations used for these types of meters. A rotor is placed in the flowpath and turns as a function of the force imparted to it by the flowing fluid. This motion can either be mechanically linked to a totalizer indicator or magnetically coupled so that each rotation produces a pulse. The meter output is linear with flow and is capable of being used over a wide range. The advantages of this meter lie in its accuracy and ruggedness in clean fluid application. The two disadvantages are the susceptibility of the rotor bearing to dirt and its limitation under high-velocity gas flows. These may occur due to flashing of the liquid under certain conditions of process operation. *Turbine meters* have to be designed such that pressure drop across the meter does not cause the flowing fluid to flash (Fig. 1-43a). The meter accuracy is a function of viscosity, with the smaller-sized meters being affected more. Variation of k (equation similar to that for a vortex-shedding meter) is approximately ± 3 percent for a 1-in (2.5 cm) meter while it

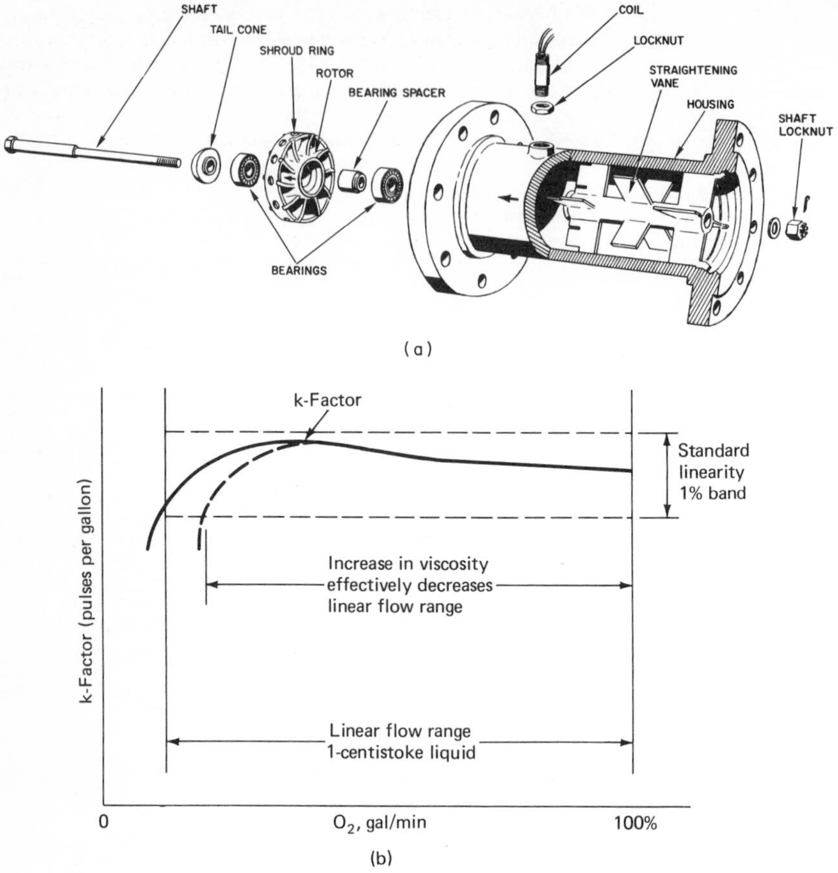

Figure 1-43 (*a*) Schematic of a turbine flowmeter. (*b*) A typical signature curve.

is approximately ±0.75 percent for a 6-in (10.5 cm) meter. Figure 1-43*b* is a typical signature curve of a turbine meter showing the effect of viscosity.

ANALYTICAL MEASUREMENTS

The successful operation of some complicated chemical processes is partially dependent upon analytical measurements and their use in process control. This discussion is limited to those measurements that can be made directly using a sensing electrode or other detectors. The accuracy and repeatability, to a certain extent, is a function of the "known" sample used for calibration. The application of the sensor is based to a great extent on the background chemistry of the process and, therefore, careful study of the alternatives is required before a particular choice can be made.

Chromatographs

These are general-purpose instruments used both for on-line process control and laboratory composition analysis. It would be difficult to run a modern chemical process without these instruments.

Components

The three basic components of a chromatograph are (1) the sample injection mechanism, (2) the separation column, and (3) the detector. Figure 1-44 shows the basic operation. As long as conditions within the analyzer remain the same, the three components indicated will appear at the same instant of time as measured from the start of the analysis. The height of the peak identifies the percent of that component present in the stream. The chart record at the end of the cycle is called a chromatogram.

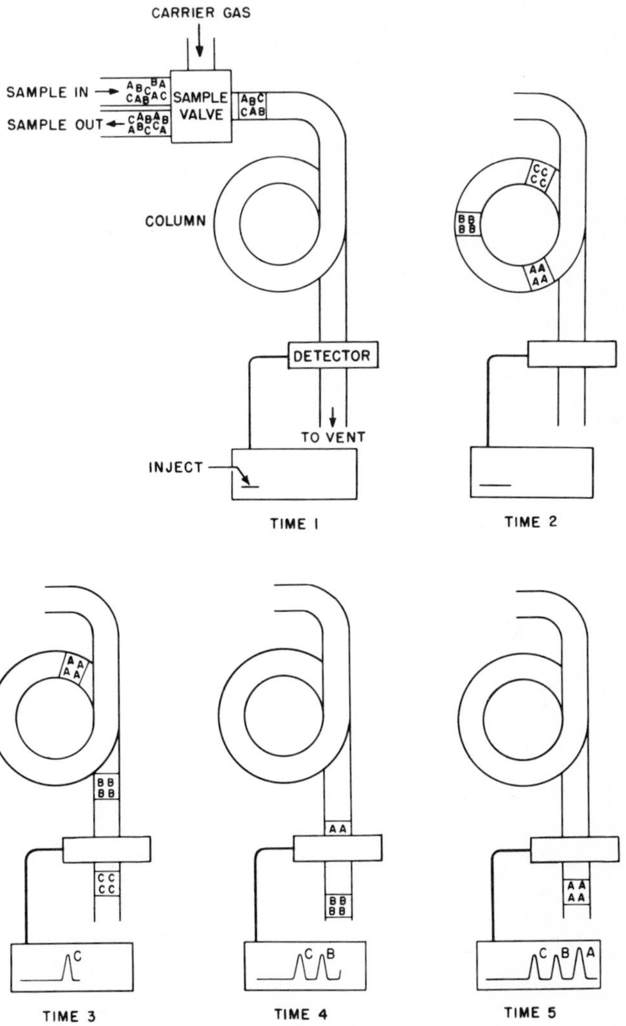

Figure 1-44 Chromatographic separation.

Analytical Conditions

The four major conditions that the analysis depends upon are (1) sample size, (2) carrier-gas flow rate, (3) analyzer temperature, and (4) carrier-gas pressure. Sample size must remain constant because detectors "see" only the actual amount of a given component, not a percent of the whole sample. If a sample twice as large as an earlier sample were used, the detector would show twice as much of a given component, assuming the sample was taken from the same source. Therefore, to get a true percent analysis the sample size must remain constant.

Carrier-gas flow rate and carrier-gas pressure are interrelated. A more rapid carrier-gas flow rate will "carry" the sample and components through the column faster, causing the time of elution to be shorter. Analyzer temperature must be held constant, since a higher temperature will cause the column elution rate to increase.

Columns

The column is used for separation on a time basis. It consists of a small-diameter tube made of type 316 stainless steel, varying in length from several inches (centimeters) to a few yards (meters). Three basic types of columns are used.

Partition Columns. The partition type uses a solid support, such as crushed firebrick coated with a high-boiling-point liquid (the liquid phase). Separation is achieved by the relative solubility of each component of the sample in the film of liquid (usually on oil) coating the support. A component of low solubility passes through the column much more quickly than one with high solubility. If the liquid phase had a high vapor pressure at the operating temperature, the flow of carrier gas would strip this liquid, leaving the column bare and thereby affecting the separating ability of the column. This phenomenon is known as *column bleed.*

Adsorption Columns. A second type of column is the adsorption column, where separation is based upon relative difference in adherence to an adsorbent material used to pack the column. These packing materials are surface-active solids such as charcoal, silica gel, and activated alumina. Components which adhere the least are eluted first.

Molecular Sieve. The third type of column includes Molecular Sieves, where the variation in molecular size of the components is used for the separation. The larger molecules

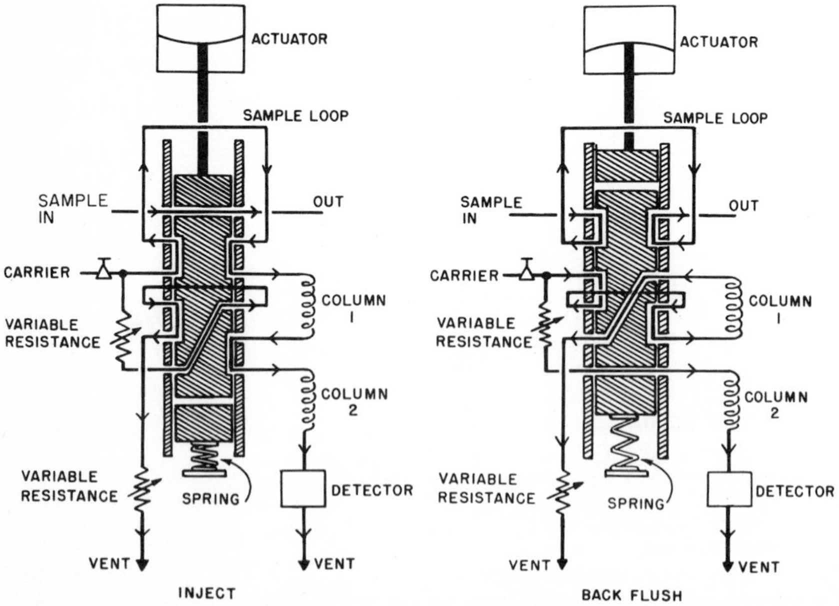

Figure 1-45 Schematic diagram of sample and back-flush valves.

are slowed down more than the smaller ones. These columns are most often used for hydrogen, oxygen, argon, nitrogen, carbon monoxide, and methane.

Backflushing. Figure 1-45 shows a schematic diagram of a sample and a backflush valve. Backflushing is used to preserve the life of a column and/or decrease the total cycle time for the particular application.

Detectors

There are many types of detectors that have been used. The thermal-conductivity type is based upon the difference in conductivity of the binary mixture flowing across the element. The flame-ionization type is based upon the ion current generated when the binary mixture of carrier gas and component is burned upon mixing with fuel, e.g., hydrogen. It does not work for inorganic compounds.

The gas-density type senses the change in density between the carrier gas and the binary mixture by using the change in differential density across an orifice or a "pneumatic" Wheatstone bridge.

The output of the detector is the chromatogram. Though this type of output is useful for laboratory-type analysis, a continuous-trend output is more useful for control. This is accomplished by incorporating extra circuitry to detect the peak for a given component and a memory circuit to hold this peak value between column cycles. The output can be a continuous signal which is used in the control loop.

Ion-Based Measurements

The most common of these is the measurement of pH. There are many applications for this measurement, with the field getting wider in scope. Some examples are:

1. Those chemical processes where the pH determines the reaction rate
2. Uniformity in product quality based upon maintaining correct pH valve
3. Neutralization of waste effluent for discharge into sewers and rivers (controlled by law)
4. Corrosion control in high-pressure boilers

The principle of measurement is based upon dissociation of chemical compounds into positively and negatively charged particles (known as cations and anions) in an aqueous solution. For a hypothetical chemical compound MA the chemical reaction may be written as

$$\mathrm{MA} \rightleftharpoons \mathrm{M^+} + \mathrm{A^-}$$

The amount of dissociation depends upon the compound and the temperature of the solution. At any fixed temperature, a fixed relationship exists between the ions and the undissociated compound. This relationship is based on the *activity* of the ions, a term which indicates the ability of an ion to take part in a reaction and is related to the *concentration* of the ions by

$$a = vc$$

where

a = activity of the ion
v = activity coefficient
c = concentration of the ion

The dissociation is related to the activity of the ions by the dissociation constant k as follows:

$$k = \frac{[a_{\mathrm{M^+}}][a_{\mathrm{A^-}}]}{a_{\mathrm{MA}}}$$

where $a_{\mathrm{M^+}}$, $a_{\mathrm{A^-}}$, and a_{MA} are the activity of the positive and negative ions and the undissociated compound, respectively.

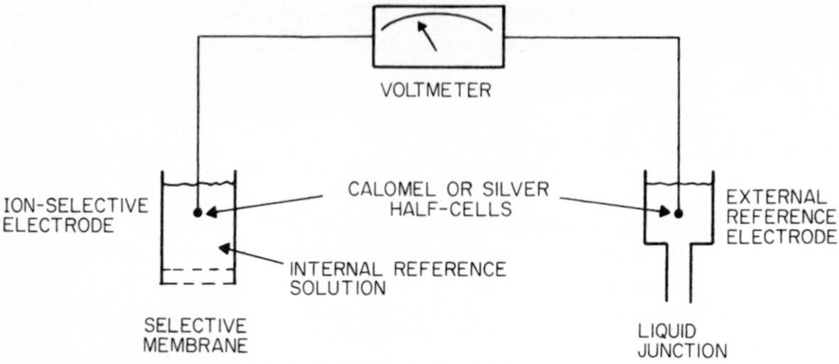

Figure 1-46 Basic arrangement for ion-selective measurement.

Therefore, if one could measure the activity of a particular ion, by knowing its activity coefficient the concentration of the particular ion in solution can be determined. If the ion of interest happens to be hydrogen $[H^+]$, the measurement is known as pH, which is a short form for $1/a_{H^+}$.

Recent research has been fruitful in developing sensors which can measure other types of ions, such as fluoride, silver, sulfide, chloride, and cyanide. Figure 1-46 shows a basic setup for measurement using two half-cells, one formed by the measuring electrode and the other by the reference electrode.

The potential generated by a measuring electrode is related to the ionic activity, and is expressed by the Nernst equation,

$$E = E^1 + \frac{2.3RT}{nF} \log \frac{a_1}{a_2}$$

where

E = electrode potential
E^1 = constant for a given electrode at a fixed temperature
R = gas law constant, 1.986 Btu/(lb)(mol)(°R) [1.986 cal/(g)(mol)(K)]
T = absolute temperature, K or °R
F = Faraday's constant, 96,490 C/g-ion
n = charge of an ion, including sign
a_1 = activity of measured ion in process solution
a_2 = activity of measured ion in the internal solution

If the internal solution is made up such that it has constant activity, the equation can be simplified to

$$E = E^\circ + \frac{2.3RT}{nF} \log a_1$$

where E° is the standard emf of the hydrogen cell.

For measurement of hydrogen activity the equation can be simplified to

$$E = 414.12 - [59.16 \text{ pH at } 77°\text{F } (25°\text{C})] \qquad \text{in mV}$$

Since the simplified Nernst equation has a temperature term, errors can exist for pH measurements at any other point than pH = 7.0, which is the isopotential point for this ion measurement. Figure 1-47 indicates the errors that can be expected.

Figure 1-48 shows the construction details of a glass pH measuring electrode with the internal solution buffered to have constant ph of 7.0. Figure 1-49 shows a flowing reference electrode with the ceramic tip acting as the liquid junction. The silver electrode with the silver chloride coating forms a silver–silver chloride half-cell in the reference electrode. To eliminate this potential in the overall measuring circuit (Figure 1-46), a

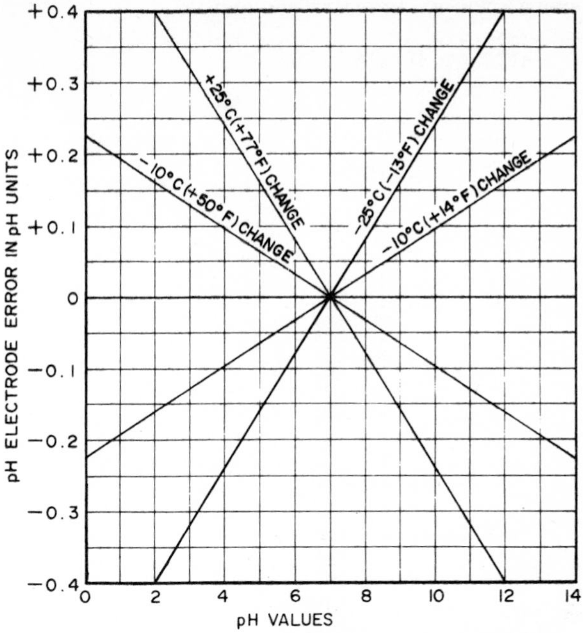

Figure 1-47 Graph of pH values at various solution temperatures vs. pH measurement errors due to temperature differences across the measurement electrode tip. No temperature error and no compensation required at pH 7.

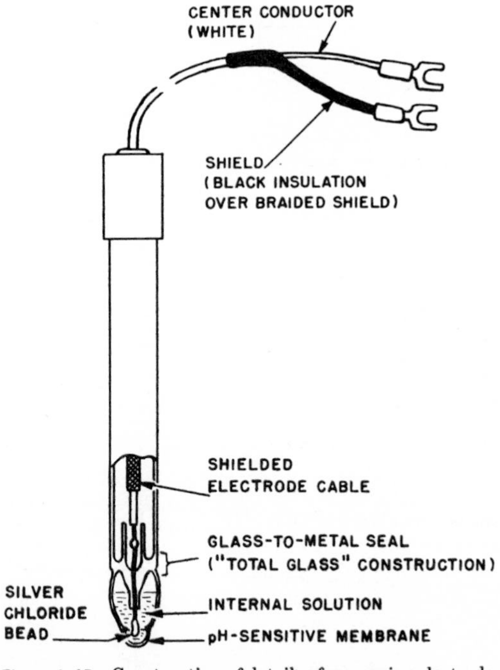

Figure 1-48 Construction of details of measuring electrode.

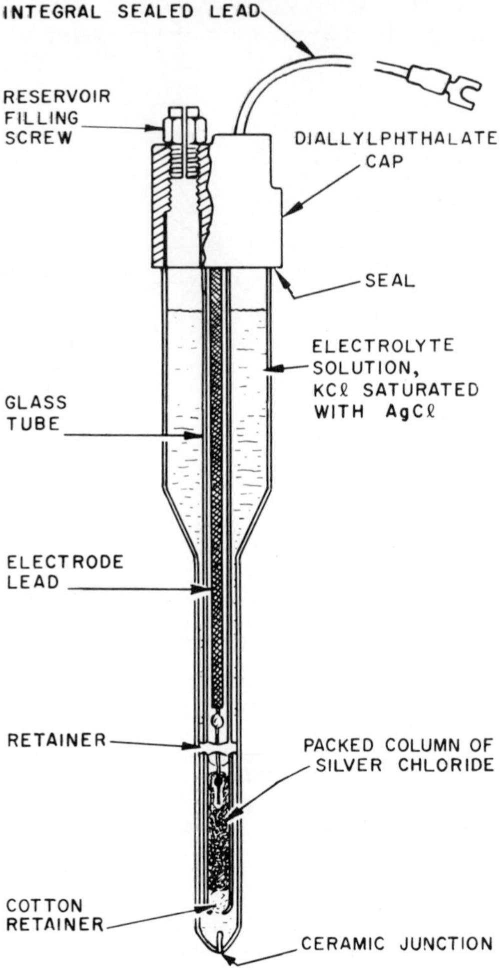

Figure 1-49 Construction details of flowing reference electrode.

similar half-cell is created in the measuring electrode, thereby having the meter read only the potential across the measuring electrode's ion-sensitive membrane. These electrodes can also come in the solid-state form, where the flow at the liquid junction is decreased to very small amounts.

It should be noted that the reference electrode does not create a measurement potential at the liquid junction since the flow of liquid prevents that. However, a small potential is still created by the ionic flow, and this may be compensated for in the measuring instrument. The measuring instrument must have a high input impedance to prevent current flows which can cause polarization problems in the electrodes. This high impedance requirement and millivolt measurement require special care in shielding of the measurement leads and grounding of the system.

While the Nernst equation was applied to the activity of specific ions, a more general form would be to consider it as follows:

$$E = E_0 + \frac{2.3RT}{nF} \log \frac{[a_{ox}]}{[a_{red}]}$$

where we consider the ratio of the activity of all the oxidized ions in solution to that of all the reduced ions. Oxidation here means the loss of an electron; reduction means a gain. The measuring electrode is a metallic plate (could be platinum), and the reference electrode is similar as for ion-selective measurement. E_0 is a constant which varies for differing types of reactions. These types of measurements are called oxidation-reduction-potential (or redox) measurements and are used when an oxidation or reduction type of chemical reaction is taking place, e.g., oxidation of cyanide or reduction of chromate in effluent treatment. The installations tend to be unique and depend greatly upon the background chemistry.

Conductivity

Aqueous solutions are electrically conductive, the amount being a function of temperature and concentration of ions in solution. It is expressed as specific conductance or conductivity in siemens. The conductance between two electrodes can be given by

$$C = \frac{k}{(1/A)} \quad \text{or} \quad \frac{K}{F}$$

where

F = cell factor in cm^{-1}, the ratio of the distance in cm between the two electrodes
A = area of the electrodes, cm^2
C = conductivity, normally given in microsiemens

The cell factor allows the spanning for the measuring instrument, which is basically a Wheatstone bridge. It should be noted that the measurement is non-ion-specific but is an indication of all ions present in solution. For calibration, known samples are used, thereby allowing the cell to be calibrated to read directly in concentration units.

The conductivity of a solution is a function of its temperature. Compensation can be provided if the solution's conductivity temperature curve is known. This limits the cell's application only to the solution for which the calibration was provided. When an electric current is passed through a solution, polarization can occur. One of the effects is electrolysis, whereby a gaseous layer can form on the electrode surface, thus increasing the relative resistance of the cell. For this reason, alternating currents are normally used.

Figure 1-50 illustrates the cell factor for various configurations of electrodes.

There are many applications where conductivity cells can provide a relatively inexpensive method for controlling processes like detecting impurities in boiler feedwater, concentration of black liquor, and other applications where the concentration of a known compound in solution has to be determined.

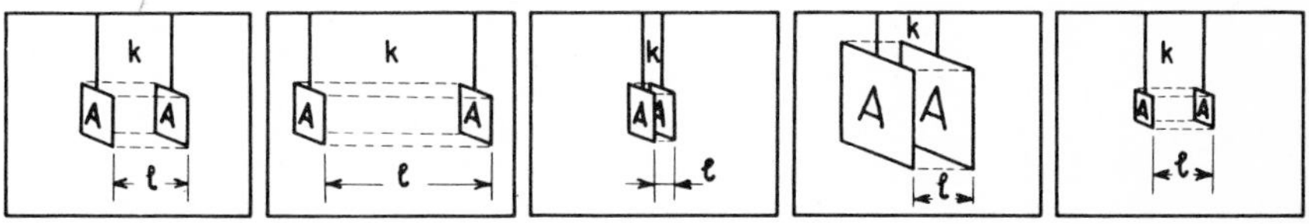

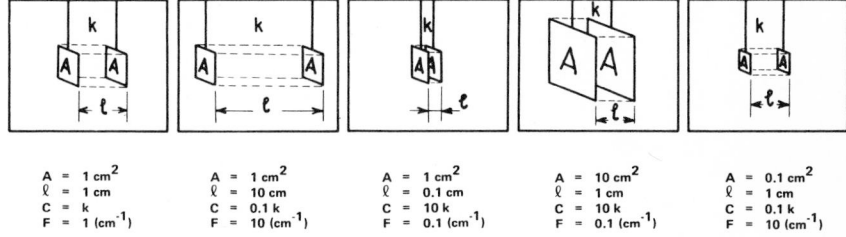

$A = 1\ cm^2$	$A = 1\ cm^2$	$A = 1\ cm^2$	$A = 10\ cm^2$	$A = 0.1\ cm^2$
$\ell = 1\ cm$	$\ell = 10\ cm$	$\ell = 0.1\ cm$	$\ell = 1\ cm$	$\ell = 1\ cm$
$C = k$	$C = 0.1\ k$	$C = 10\ k$	$C = 10\ k$	$C = 0.1\ k$
$F = 1\ (cm^{-1})$	$F = 10\ (cm^{-1})$	$F = 0.1\ (cm^{-1})$	$F = 0.1\ (cm^{-1})$	$F = 10\ (cm^{-1})$

Figure 1-50 Electrode cell factors.

Oxygen and Dissolved Oxygen

These measurements have become increasingly important in the combustion and water-treatment fields. The most popular method for oxygen measurement is the electrochemical detector, which works in principle by using the Nernst equation. Figure 1-51 indicates a schematic drawing of the basic cell, with P_1 the partial pressure of oxygen in the reference gas and P_2 the partial pressure of oxygen in the measured gas. The equation is

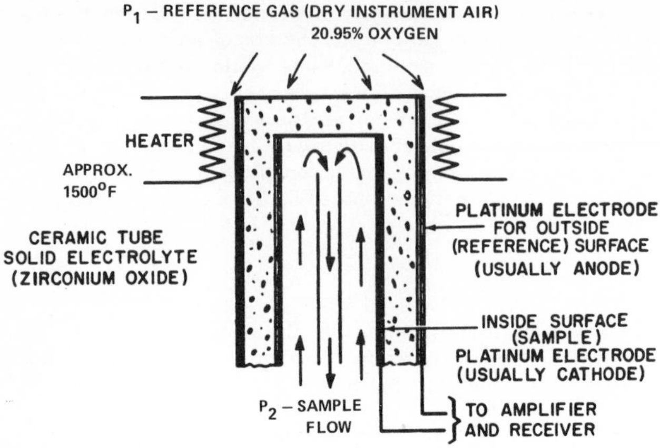

Figure 1-51 High-temperature electromechanical oxygen detector.

$$E = E^\circ + \frac{2.3RT}{nF} \log \frac{P_1}{P_2}$$

and the resulting logarithmic output voltage is approximately 53 mV per decade change in oxygen content. The output decreases as the concentration (actually activity) of measured oxygen increases. The measurement is a percent measurement as related to all components in the sample gas, including water vapor. The probe can be installed on-line with no sample preparation system required.

There are many devices available for measuring dissolved oxygen (DO). In general, the construction requires two electrodes in an electrolyte enclosed by a permeable membrane. Depending upon the electrodes, a galvanic reaction or polarization takes place when a small voltage is applied to the electrodes. The dissolved oxygen diffuses through the membrane and is dissolved in the electrolyte. An oxidation-reduction reaction causes

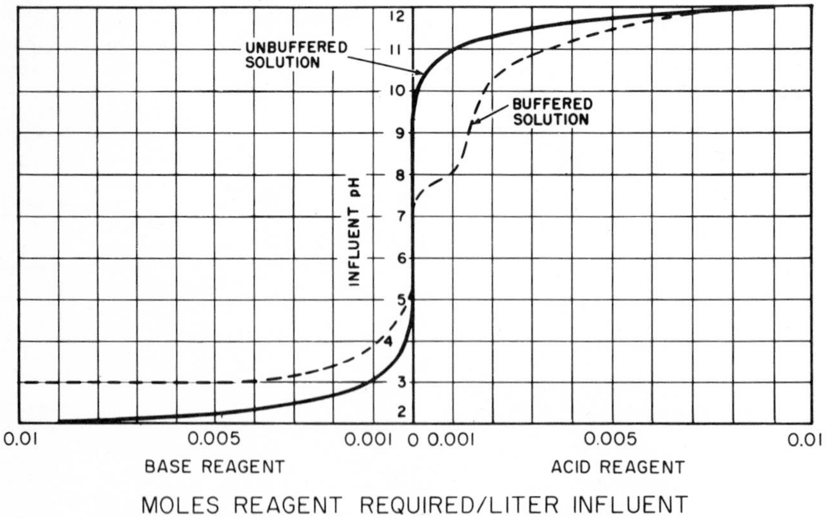

Figure 1-52 Typical neutralization curves for unbuffered solutions (strong acid or strong base) and buffered solutions.

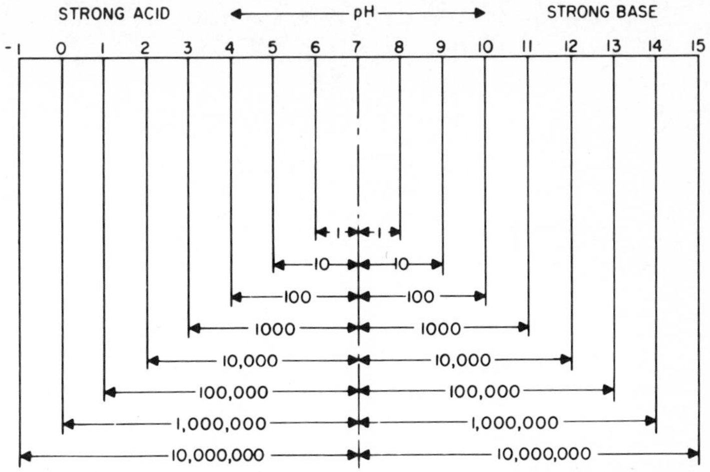

Figure 1-53 Graph of reagent demand. Reagent addition units are 10^{-6} mol/Li.

a current flow which can then be measured as a percentage of the maximum amount that could be present at the existing temperature. Since temperature determines the saturation capability of the solution to dissolve oxygen, automatic compensation is usually required.

General

From a control-application viewpoint, all instruments based upon the Nernst equation have to take into account the nonlinear nature of the measurement. The logarithmic nature of the measurement requires special consideration of the rangeability requirements of the system. Figure 1-52 illustrates the nonlinear nature of the neutralization curve for strong acid or strong base solutions and the effect of buffering in the solution.

The rangeability requirements can be noted in Fig. 1-53. It can also be noted that while neutralization of an acid from pH 6.0 to 7.0 requires one unit of base, to go from pH 2.0 to 7.0 requires 10,000 units. Therefore, if the influent stream varies from pH 2.10 to 7.10, the control system must have a rangeability of 10,000:1.

chapter 11-2

Automatic Controls

INTRODUCTION

Industrial processes are generally characterized by mass and energy flows. Application of automatic controls is concerned with the behavior of the process under dynamic or unsteady-state conditions where the accumulation of mass or energy cannot be tolerated. Hence the control problem is one in which the designer uses various engineering tools to match the supply against the demand over a period of time. The three terms, supply, demand, and time, can vary and depend to a great extent upon the type of processes considered. As processes get more and more complex and the availability of raw materials, including energy, becomes limited, the application of automatic control theory takes on greater significance.

Types of Controls

There are two basic types of control: open-loop and closed-loop.

Open-Loop Control

This is the simplest type of control that can be applied. It involves making an estimate of the amount of control action required based upon achieving a desired objective without regard to the actual conditions of the process. For example, a washing machine operates without regard to the actual condition of the clothes. The amount of detergent used and the settings on the machine are an estimate of the control action required in achieving the objective (clean clothes), they are not based on the actual state of our objective. This type of control is generally inadequate and is seldom encountered in industrial process work; however, this type of control cannot be completely ignored. With the advent of computers and their memory capability, control sequences based upon historical data may be considered in the future, especially for the kinds of processes where a measurement of the final objective may be difficult, if not impossible.

Closed-Loop Feedback

This is the most common type of control mechanism used. Any process in which the process variable under control is measurable allows the use of such control strategy. The importance of such a loop can be judged by the fact that most bio-socioeconomic processes incorporate some kind of closed-loop feedback mechanism, also known as a *servomechanism* (Fig. 2-1).

The controlled variable is the process variable which we are trying to maintain at some desired value (called the set point). Industrial processes are characterized by the

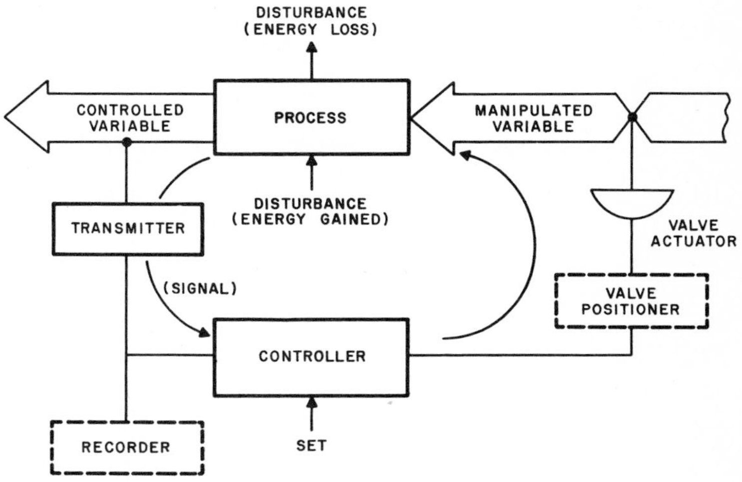

Figure 2-1 Closed-loop control. (*The Foxboro Company.*)

many types of control variables that are encountered, e.g., temperatures, flows, and levels. The function of the transmitter is to quantify this variable in terms of signals, which could be pneumatic, electric, hydraulic, or just a mechanical output like the position of a lever. It should be remembered that not all these measurements are linear, i.e., the output signal is not necessarily linear with respect to the controlled variable or, even if it is linear, is not a direct reading of the controlled variable, e.g., the controlling flow based directly on the pressure drop ΔP across an orifice meter. In all these cases, linearization can be achieved, thereby making the transmitter output a direct indication of the controlled variable.

The manipulated variable is that which the controller varies in its efforts to maintain the controlled variable at set point. The controller output is a signal to, say, a valve actuator, causing the valve to move to a position which would depend upon the value of the signal, type of valve, and the process conditions under which it is operating. The valve positioner causes a fixed relationship between the controller signal and valve position. Even though this relationship could be linear, the relationship between valve position and flow-through is generally complex and nonlinear. The nonlinearities in measurement and in the control of the manipulated variable have a significant influence upon the performance of the control loop.

The feedback control loop operates in an environment where constant disturbances are taking place. These disturbances affect the controlled variable and could be due to changes in the manipulated variable other than those instituted by the controller; e.g., changes in mass or energy flows to the input of the process, or changes of the same variable on the output side of the process. Changes can also be initiated by the operator when changing the set point of the loop. The controller loop must therefore perform both as a regulator and as a servomechanism. The strategy used is straightforward and makes use of the error (the difference between set point and the measured controlled variable) to develop a control signal which will drive the error to zero and thereby achieve the objective. The performance of the loop can be judged as some integral function of error and time, the controller and its action then being set to minimize this function.

The existence of this error means a certain economic loss in the production process in terms of either the production of off-specification product, the use of excessive amounts of energy, or both. As these considerations become more and more important in the future, control loops will tend to get complex, and input disturbances will be measured and used in strategies called *feed-forward.* Here the measured inputs are used to compute the required positioning of the manipulated variable based upon some mathematical model of the process. The amount of correction required is then based on the variation of the inputs to the process. As the models tend to be simplified representations of the process, feed-forward strategies are not accurate over the entire operating range of the process. Feed-forward control strategies are normally used in conjunction with feedback, as shown in Fig. 2-2.

It should now be noted that the feedback controller does not have any direct control of the process but acts to take minor trim action on the feed-forward model. The major control function is accomplished by the feed-forward scheme. The limited action by the feedback controller allows a much more stable operation of the process, allowing set

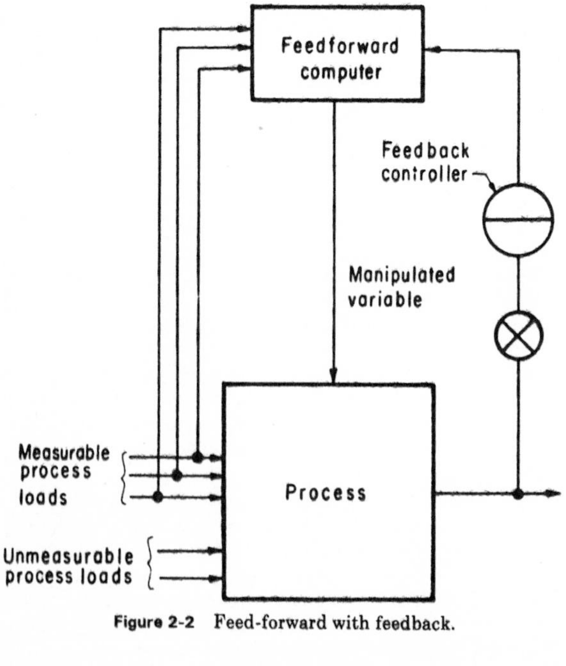

Figure 2-2 Feed-forward with feedback.

points to be set closer to the required specification values and also allowing increased throughput.

The time response of the entire control loop is made up of the sum of the responses of the primary element, the transmitter, all receivers in series with the controller, the controller, the final operator, and the process itself. There are two types of time elements which occur in industrial processes: first, dead time (or transportation lag) and, second, the resistance-capacitance (*RC*) time constant. To determine how much of each exists, a simple open-loop test can be performed as follows. With the loop in steady-state operation, the controller output is manually changed by a step amount. The response of the controlled variable is plotted as a function of time, as shown in Fig. 2-3.

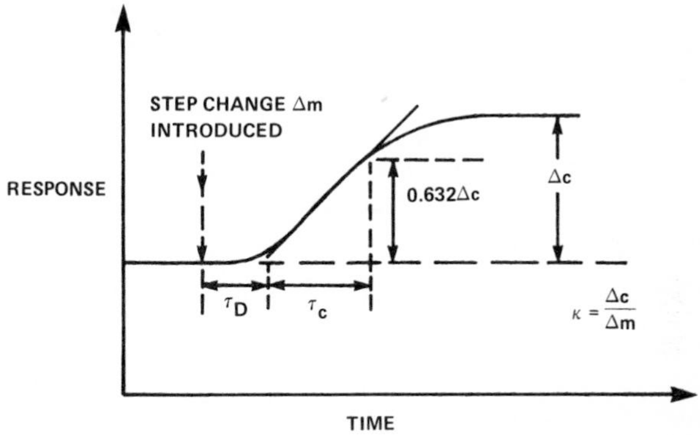

Figure 2-3 Open-loop response.

The graphical construction is designed to approximate a process consisting of a dead time τ_D and a capacitance time τ_c. Most processes will have some combination of these two elements, with the control problem becoming easier as the ratio of τ_D/τ_c becomes smaller. While this example is a simple model of the process, more complex models are available which approxmiate the response curve more closely. These, however, become cumbersome for practical use. Time elements introduce a phase lag; i.e., the output is delayed with respect to the input. The phase lag varies as the frequency of the input signal and becomes larger as the frequency increases.

If a loop is to oscillate with a sustained cycle, the phase shift of an upset, after going through all the elements in the loop, must be exactly 360 degrees.

All feedback controllers contain an element called *negative feedback,* which introduces a phase shift of 180 degrees. The remainder of the control loop must create an additional 180 degrees for sustained oscillations to exist. The point at which this occurs is the natural period of the loop as long as the controller contributes no phase shift.

In addition to the phase shift in a loop, one must consider the gains of the various elements in a control loop. There are two types of gains, static and dynamic. Figure 2-3 introduces the static gain κ as the change in output divided by the change in input. Dynamic gain would be the same ratio but would be a function of the frequency of the input signal. The gain at any given frequency would be the product of the static and dynamic gains. A loop gain could then be defined as the product of the gains of the individual elements of a loop. It is a dimensionless number which is an indication of the stability of the loop. As the loop gain becomes greater than 1, the loop tends to oscillate with larger and larger amplitude, while a value less than 1 would cause a certain amount of damping of the amplitude of oscillation. Normal industry practice is to achieve a loop gain of 0.5, which is also called ¼ amplitude damping, where every amplitude is half the previous one. These criteria do not necessarily prevail in all industrial control loops, but they are typical. Particular process requirements may mandate other dynamic responses.

CONTROLLERS

The introduction to this chapter described a feedback loop. Figure 2-1 indicated the four basic elements, one of which is the controller. The actual hardware used could be pneumatic, electronic (analog or digital), and hydraulic. Any controller can be considered to be composed of two parts, as shown in Fig. 2-4.

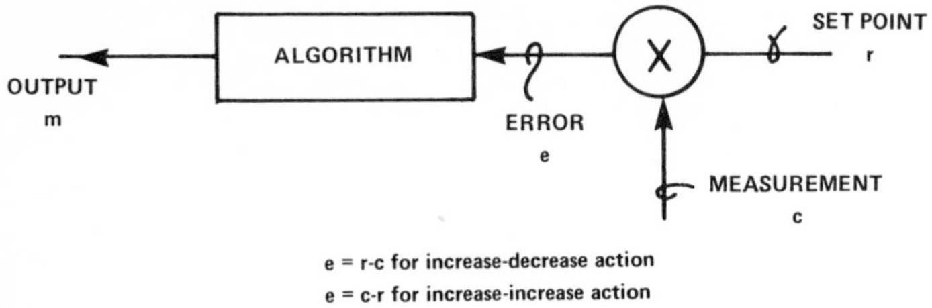

Figure 2-4 Basic controller elements.

The error detector determines the magnitude and, more important, the direction of the error; the algorithm defines the specific action of any one particular type of controller. The type of error signal implemented defines the direction of the output as a function of the measurement. For example, if the measurement *c* exceeds the set point *r*, the error would be negative, which then causes the algorithm to decrease the output *m*. If an "air-to-open" type of valve had been used, the decrease in output would tend to close the valve, which would decrease the flow to the process and thereby decrease the measurement back toward the set point. There are occasions in which an "air-to-close" valve may have been used, in which case the action of the controller would have to be changed. This option to change the action of the controller is a normal feature of the controller.

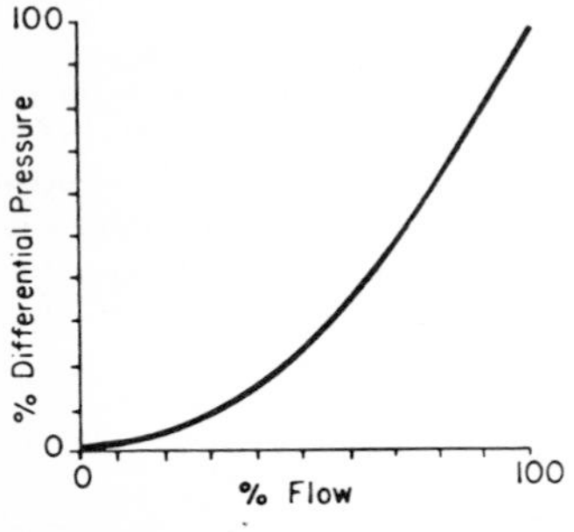

Figure 2-5 Flow measurement by differential pressure means.

While most control theory is based upon the linearity of the gain of various elements which make up a loop, practical design and operation must take into account the nonlinearities that occur due to variations in set point and/or variations in process load. For example, we have noted in Chapter 11-1, under "Flow Measurement," that the flow has a square-root relationship with the differential pressure measured. Figure 2-5 indicates this relationship.

The effect of this type of nonlinearity (it could occur in any of the four elements of a control loop) is normally to give good control at one set of operating conditions only. This could cause damaging results in those cases in which the controller had been set during low system gain because, at high system gain, the loop would tend toward oscillation. Elimination of these nonlinearities, by using special hardware or by selecting loop elements having equal and opposite characteristics, goes a long way in the overall stability of the loop, e.g., selecting a quick-opening valve whose characteristics would be opposite to a head-type flow measurement.

Selection of a controller is based to a great extent on the process characteristics and the precision needed to control it at an exact set point. The four basic control actions generally used are discussed in the text that follows.

Two-Position (ON-OFF)

This is the simplest of all control actions available. The output of the controller is at either 100 or 0 percent and could be either an analog signal or a contact actuation. The result in either case is that the final actuator is completely open or completely closed. The effect on the measured variable depends upon the type of process, specifically the capacity time constant. For a large-capacity process, the measured variable would change slowly to allow reasonable final actuator action. The measurement oscillations would be correspondingly small. Compare this to a process having very little capacity; the measured variable would tend to change very quickly, which would cause rapid actuator action, and this in turn would cause large and rapid oscillations in the measurement. Under these circumstances and where precise control is required, a throttling type of control action would be preferred.

There are three basic methods of achieving two-position control. Figure 2-6 illustrates the control action.

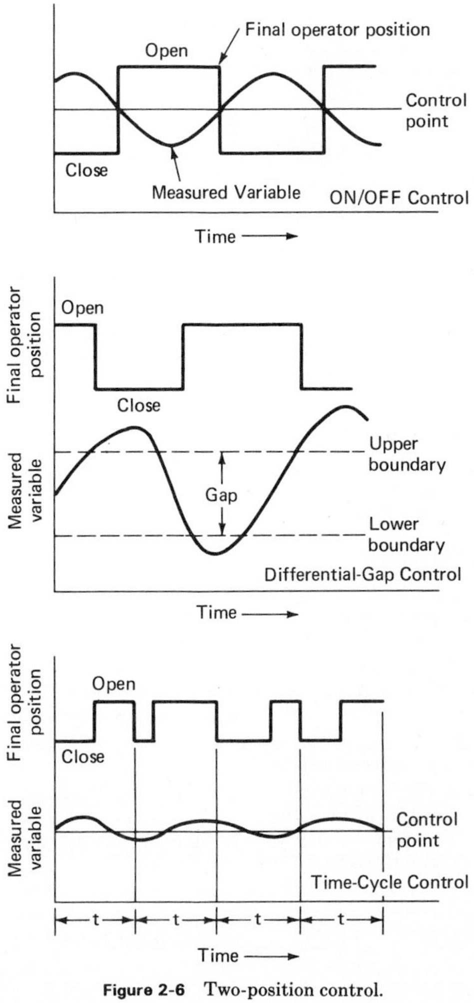

Figure 2-6 Two-position control.

A thermostat for an electric heater is an example of an ON-OFF control system. Here, the controller operates continuously to control the temperature of the process. Another application is for safety shutdown of a process when safe operating conditions have been exceeded.

To avoid frequent cycling and maybe damaging results to the final actuator, a differential-gap controller is used. A band or gap exists around the control point. When the measured variable exceeds the upper boundary of the gap, the final actuator is closed and remains closed until measurement drops below the lower boundary of the gap, at which point the final actuator opens. The control is not precise, but it does prevent excessive wear on the final actuator. In industry, this type of control is often found in noncritical-level control applications where the level could be anywhere between two limits.

In time-cycle control, a time base t is established. During this time period, the final operator is closed for a certain percentage of the time and open for the remainder. The ratio of closed time to open time is determined by the relationship between the measured variable and the control point.

A time-cycle controller is normally set up so that, when the measured variable equals the desired control point, the final operator will be open for half the time cycle and closed for the other half. As the measured variable drops below the control point, the final operator will remain open longer than it is closed. This type of control is often found on electrically operated heaters and on dry-solid control gates where a throttling gate position would cause buildup and clogging.

Proportional

To better understand this and the control actions described later, an open-loop response will be considered first. The open loop means that the controller is considered by itself with an artificially generated measurement signal. A change in input can be generated by changing either the set point or the measurement signal.

Figure 2-7 is a schematic diagram of a force-balance-type pneumatic proportional control; Fig. 2-8 is a schematic of an equivalent electronic controller using an operational amplifier.

The basic equation implemented is

$$m = (100/\%\text{PB})(r - c) + \text{bias}$$

where

r = set point
c = measurement
$\%$PB = proportional band, in percent

The factor 100/%PB is also termed the *gain* of the controller. The *bias* term allows the output to be approximately 50 percent of the signal range of the controller when measurement equals set point. This condition can occur only under one set of operating conditions, i.e., when the output from the controller positions the final actuator such that the manipulated process variable exactly matches the process requirements for the given load. For different operating conditions, the output from the controller is a function of the error (the difference between the set point and measurement) and the set proportional band (PB). Figure 2-9 shows the effect of PB on valve travel for various settings.

Line A indicates that for a PB of 100 percent, the measurement has to change from 0 to 100 percent of its span for a 0 to 100 percent throttling action. As the PB is decreased, lines B, C, and D indicate that smaller changes in measurement allow full throttling action. When the PB is greater than 100 percent, the throttling action is reduced to less than 0 to 100 percent.

It should be noted that a limited type of proportional control—with correspondingly limited possibilities of application—can be handled by a transmitter. From the preceding discussion, it can be deduced that a penumatic transmitter can perform as a 100

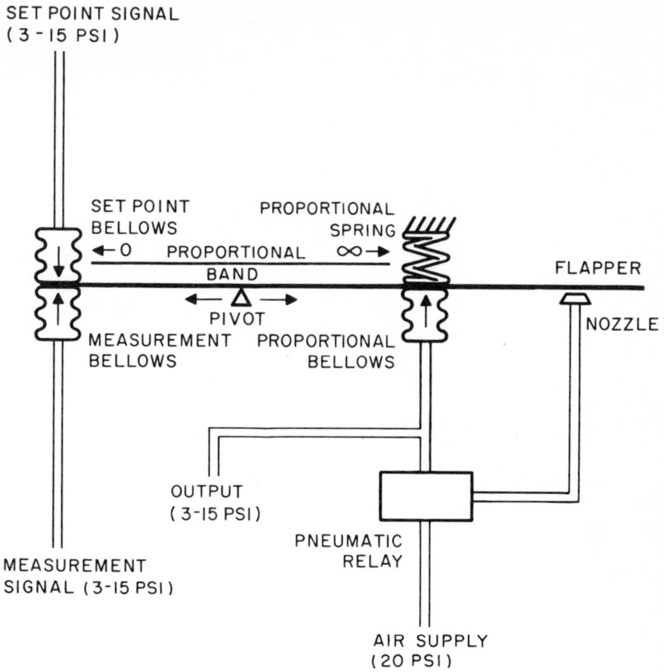

Figure 2-7 Pneumatic force-balance proportional controller.

percent PB *controller* when connected directly to a pneumatic final actuator. Some tank-level controls are implemented this way, especially when strict level control is not required.

The controller algorithm indicates that, for an output other than 50 percent, an error must exist; the size of this error is dependent upon the PB (gain setting in the controller). There is nothing in the algorithm which would cause this error to be eliminated. This error is called the *offset*, and its size dependent upon the particular PB (gain) for a given set of conditions. Figure 2-10 illustrates the types of responses that can be achieved.

As the proportional band is decreased, the amount of offset decreases, but the response tends to become more oscillatory. This should be obvious when the loop gain is considered. As the PB is decreased, the controller gain increases, which causes the loop

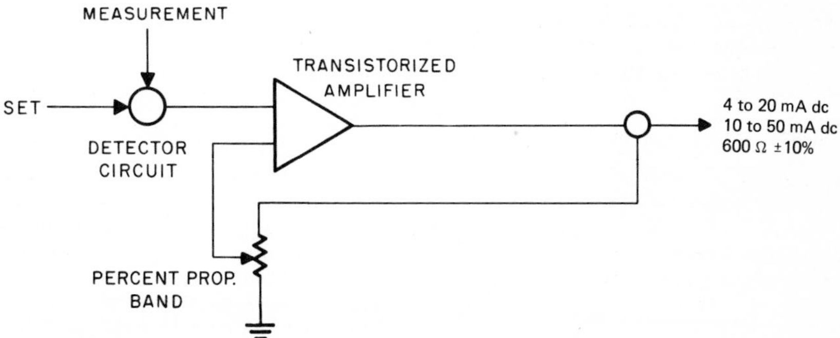

Figure 2-8 Electronic proportional controller.

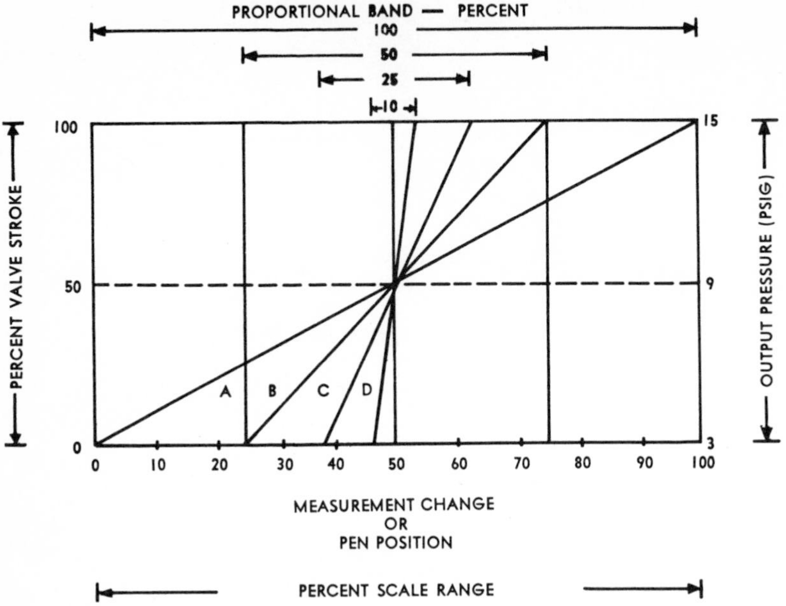

Figure 2-9 Proportional response to changes in proportional band (gain).

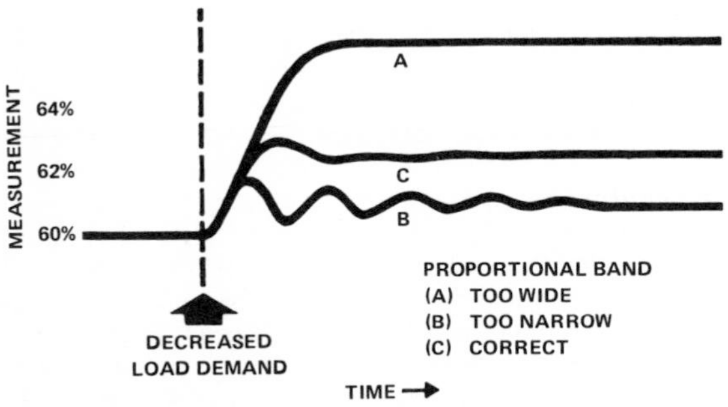

Figure 2-10 Proportional control response curves.

gain to increase thereby causing larger oscillations in the response. Large-capacity processes in general require narrow proportional bands, while processes having fast reaction times can accommodate wide proportional band settings only.

Proportional-Plus-Integral

These types of controllers are also known as *two-mode* controllers. In the discussion of proportional control, it was shown that the wide proportional bands required for some processes lead, in turn, to large offset errors. This is an intolerable condition in the majority of process applications. The application of the integral (also known as reset action) allows the final actuator-to-measurement relationship to change. This relationship changes when the measurement is not precisely at its set point. The effect is to cause the final actuator position to change in the direction, and at a predetermined rate, that will allow the measurement to equal the set point. Figures 2-11 and 2-12 are schematics of pneumatic and electronic proportional-plus-integral controllers.

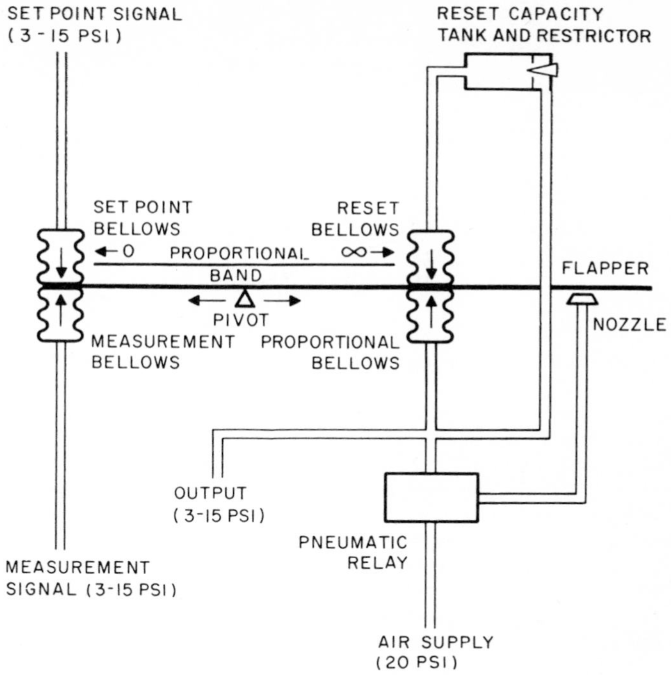

Figure 2-11 Pneumatic force-balance proportional-plus-integral controller.

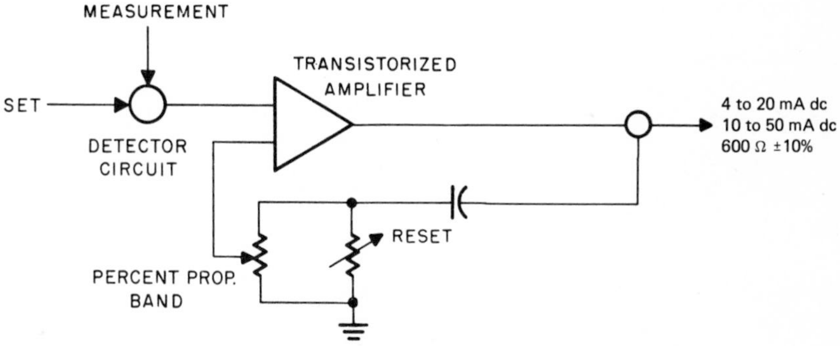

Figure 2-12 Electronic proportional-plus-integral controller.

The basic equation implemented in these controllers is

$$m = (100/\%\text{PB})\left(e + \frac{1}{R_T}\int e\, dt\right)$$

where

e = error $(r - c)$
R_T = the integral time constant

We see from the above equation that the output of the controller is the sum of both the proportional action and the integral (reset) action. Figure 2-13 illustrates open-loop response of a proportional-plus-integral controller.

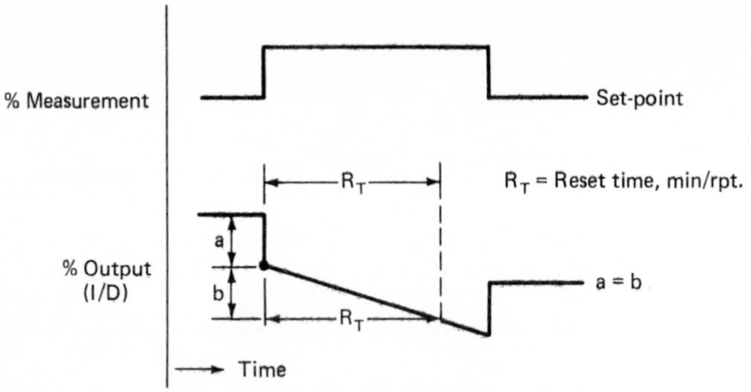

Figure 2-13 Open-loop response of proportional-plus-integral controller.

It should be noted that the integral action R_T can be defined as the time (in minutes) taken by the controller to change its output by an amount equal to the proportional action. This definition allows R_T to be calibrated in terms of minutes per repeat. An alternative way to set this calibration is the reciprocal of minutes per repeat, i.e., repeats per minute. Both these terms are equally popular among the various manufacturers. Though the introduction of integral action eliminates the offset error, it introduces a phase lag in the controller. The amount of gain is no longer the simple (100/%PB) term of the proportional controller, but the vector sum of the proportional part and integral part which gives a resultant larger than either. The phase lag introduced is the amount by which this resultant lags the proportional action. Two results can be deduced from this discussion. First, the addition of integral action for a given proportional action will cause the control-loop oscillations to increase; second, this oscillation will have a longer period than the original oscillations. Figure 2-14 shows response curves for a temperature loop where the integral is measured in minutes per repeat.

If the controller is unable to return the measurement to the set point, the integral action continues to increase or decrease the output until the saturation limit of the controller is reached. If at this time the measurement starts to change, the controller output does not change from its saturation limit until the measurement crosses the set point. If the system has any capacity, the measurement will overshoot the set point by an amount which depends upon the capacity and integral setting. This limitation is called *integral windup* and has to be seriously considered on discontinuous or batch-type processes.

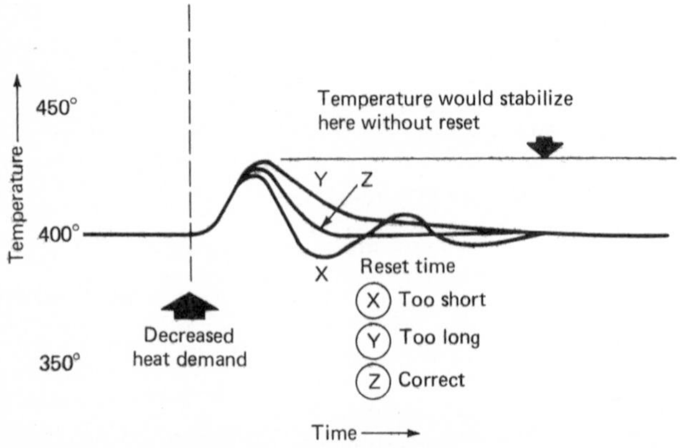

Figure 2-14 Proportional-plus-integral control response curves.

There are special batch-type controllers available which prevent the output of the controller from reaching saturation limits, thereby causing less or no overshoot in the measurement for the above conditions.

Proportional-Plus-Derivative

The addition of derivative action to the proportional controller can improve the response of the system, especially in those batch-type processes where integral windup can cause problems. The amount of derivative action is proportional to the rate-of-change of the measurement. Figure 2-15 illustrates the open-loop response and defines the derivative time constant.

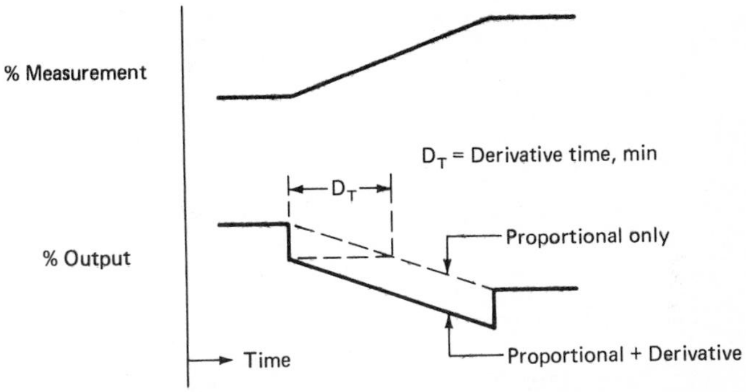

Figure 2-15 Open-loop response of proportional-plus-derivative controller.

Since the derivative control is a function of the rate of change of measurement, introduction of this mode in applications where the measurement can change very rapidly, or if the measurement is noisy, can degrade the controllability of the loop. The most useful applications are those where the system has large capacities with little dead time.

Proportional-Plus-Integral-Plus-Derivative

This is also called the three-mode controller. The open-loop response indicates the sum of all three modes in its output. Figures 2-16 and 2-17 are schematics of a pneumatic and an electronic controller, respectively.

In our previous discussion, integral action was considered to cause an additional lag in the control loop. The derivative has an opposite effect, i.e., it causes a phase *lead* and is, therefore, sometimes also called *preact.* The total gain of the controller is the vector sum of the individual mode gains and the phase lag or lead introduced in the controller is the angle the resultant-gain vector makes with respect to the proportional-gain vector. If the resultant-gain vector has a phase lead, the resultant response of the system will have a period of oscillation smaller than the natural period. In general, if the integral time in minutes per repeat is made equal to the derivative time in minutes, the response is quite satisfactory. For large-capacity processes, the amount of derivative action may be increased.

Table 2-1 gives a brief summary of some standard process loops that are common in industrial process plants. Indicated are some major characteristics of the processes and the control systems.

Multivariable Control

The four common types of multivariable control loops are *ratio, cascade, feed-forward,* and *override.* The importance of these types of loops can be appreciated by considering

TABLE 2-1 Summary of Common Loops

Variable	Process	Control System
Flow	Very fast Most lags are in the control system Nonlinear (square) measurement common Noisy	Proportional-plus-reset controllers Low gain, fast reset Derivative hurts Linear valves for differential pressure measurement Equal percentage valves for linear measurement Valve is the major dynamic element
Pressure, liquid	Fast Most lags are in the control system Nonlinear (square) Noisy	Proportional-plus-reset controllers Gain near 1, fast reset rate Derivative of no value Linear valve
Pressure, gas	Single capacity No dead time Linear, no noise Simple process	Self-acting or high gain proportional controllers Reset seldom necessary Derivative unnecessary Valve characteristic relatively unimportant
Pressure vapor	Dynamics vary Dead time possible Slow compared to other pressure processes Linear, no noise	Three-response controllers Settings vary Equal percentage valves
Level	Single capacity (integrating) No dead time Linear Infrequent noise	Precise control: High gain or proportional-plus-reset controllers Averaging control: Low gain proportional plus reset or specialized controllers Valve characteristic unimportant
Temperature	Multiple-capacity system Dead time possible (especially in heat exchangers) Nonlinear No noise	Three-response controllers Settings vary, but gain usually above 1 Derivative of limited value if dead time is large Equal percentage valves Measurement dynamics are important
Composition	Dynamics vary Dead time usually present Usually linear Sometimes noisy due to poor mixing	Proportional-plus-reset controller Low gain, variable reset rate Derivative sometimes useful On-line analyzers fast, often noisy, pH nonlinear Sampling systems complicate both measurement and control, add dead time Linear valves

that the response of any controlled variable in the industrial process environment will, in general, be a function of a number of other variables that could be either on the load side or on the input side of the unit. To account for these variables, multivariable control loops have to be considered.

Ratio

Here the controlled variable is the ratio of two measured variables, where one of these is the controlled variable and the other is the "wild" variable. The wild variable is not necessarily uncontrolled, it could be a controlled variable of another loop. These systems need not be limited to two components but can have the wild variable set the ratio, through separate relays, to several controlled variables. An alternative to this method

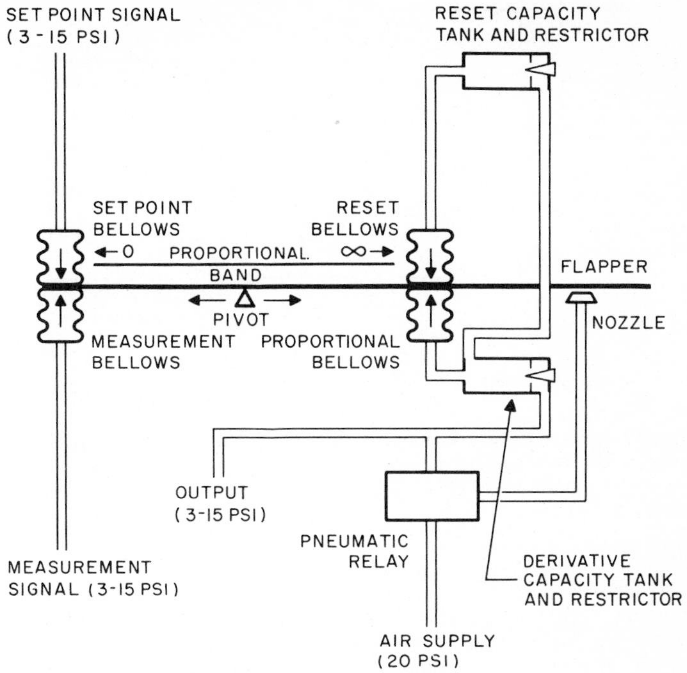

Figure 2-16 Pneumatic force-balance proportional-plus-integral-plus-derivative controller.

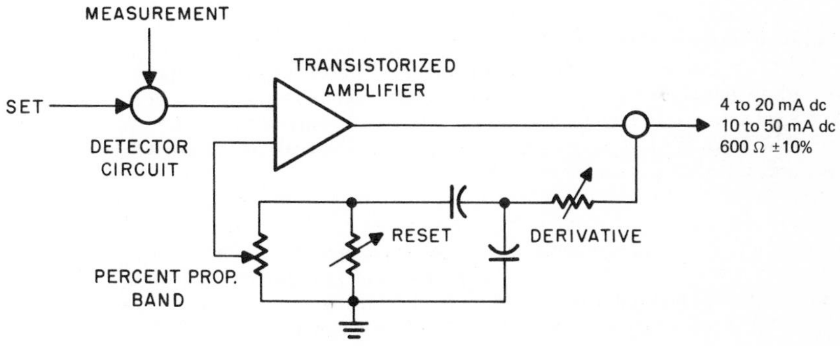

Figure 2-17 Electronic proportional-plus-integral-plus-derivative controller.

would be to set the ratio as a percentage of the required total flow, which itself is set by a master demand signal. The two types of systems are called *series* and *parallel*, respectively. The series approach has one basic advantage in having an inherent interlock between the wild variable and controlled variable.

There are a number of practical applications where these control schemes have been found useful. Examples are blending, control of air-to-fuel ratio in boilers, and ratioing of reactant flow in chemical processes. In steady-state operations, the system operates like an ordinary flow loop, i.e., dynamically quite fast. However, under changing loads, nonrecoverable errors can exist that can be removed in memory-type blending systems. Figure 2-18 indicates an example of the required setup.

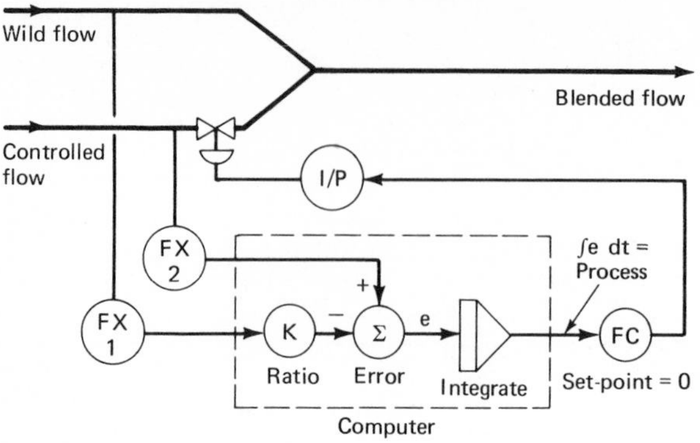

Figure 2-18 Memory-blending system controls accumulated flow ratio.

These systems can be either analog or digital, and frequently employ pulse-type measuring instruments for precise measurement of total flow. It should be noted that a certain amount of downstream capacity is required so as to be able to eliminate the accumulated error.

Cascade

In a single feedback loop, the output of the controller is directly used to manipulate the final actuator. Under these conditions, fluctuations on the input side of the manipulated variable will be noticed by the controller when the measurement changes. An example would be the header pressure change on a steam supply being used to control the outlet temperature from a heat exchanger. This after-the-fact control can be avoided by setting a secondary control loop which gets its set point from the primary controller. Figure 2-19 is a typical setup.

To properly apply these systems, the response of the secondary loop must be considerably faster than that of the primary loop. If this condition is not met, the resulting instability is the result of nearly simultaneous set-point and measurement changes in the secondary loop. It is for this reason that it is not recommended that a positioner be used on a final actuator in flow loops. The valve actuator contributes most of the capacity lag in a flow loop. With a positioner, this capacity lag exists in the secondary loop, with resultant instability of the entire flow loop.

A note of caution in actual operation of cascade loops is the integral windup that can occur in the primary controller, if the secondary controller were to be manual or if it were to fail. Special precautions have to be taken, and these could be either to interlock the stations or to use external integral feedback of the primary controller. The latter is the more useful method and is implemented by using the manipulated variable (steam flow in Fig. 2-19) as the external integral feedback signal.

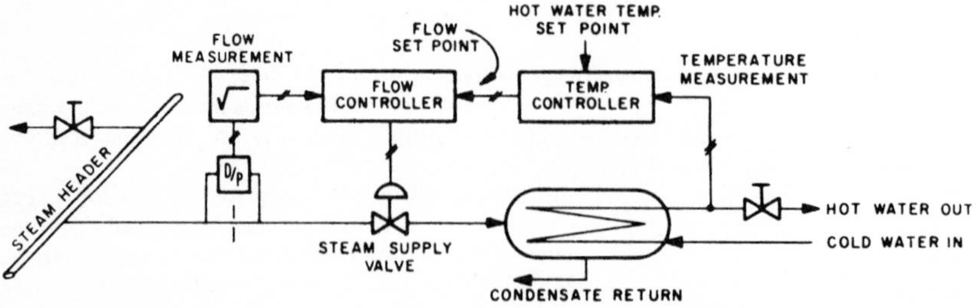

Figure 2-19 Cascade control of a heat exchanger.

Feed-Forward

The introduction to this chapter included the basic strategy and philosophy of feed-forward control systems. With the availability of both analog and digital computation elements such as multipliers, adders, etc., it becomes relatively easy to physically put together the hardware to perform the feed-forward calculation. The difficulty arises in setting up the required process model based on energy- or mass-balance equations. These equations must also take into consideration the dynamics involved for each disturbance variable considered in the model. For example, in the previous discussion on cascade control of a heat exchanger, it is observed that the load disturbances due to changes in both the temperature of the incoming cold water and the amount of flow would affect the controlled variable (the outlet temperature).

A simple model based upon *energy in* equalling *energy out* could be set up with dynamic compensation for the rate of change of product coming in. This model could then be tuned for a given set of conditions by adjusting the value K as shown in Fig. 2-20.

If our process model had been perfect, a perfect control would have been possible. But this perfection is difficult to achieve, because of first the large number of variables that can affect the control variable and second the difficulty of achieving exact mathematical representation of the process. To take into account these inaccuracies, a feedback loop is normally used and is called the *feedback trim.* As the name suggests, the action of this loop is not to take over complete control of the manipulated variable but

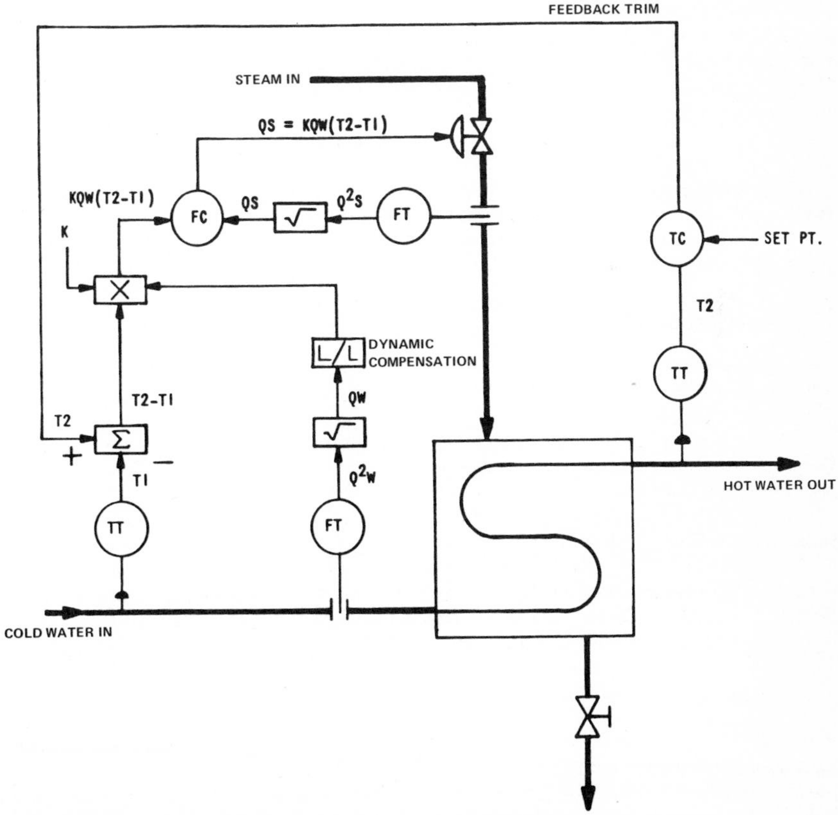

Figure 2-20 Feed-forward control with dynamic and feedback trim compensation.

to adjust the feed-forward model for varying operating conditions. Since this action can be slow, the "tuning" constants of the feedback trim controller are normally set much wider than if it had been the main control mechanism.

There are numerous examples of the use of feed-forward control in varying industrial processes. The oldest practical application is the setting of feedwater flow via steam flow for the control of drum level in a boiler. There is a certain amount of inherent safety in most feed-forward control systems. For example, in drum-level control the loss of steam flow automatically shuts off the feedwater flow. This action is the opposite of what a feedback controller would do—open the feedwater actuator if drum level were below set point. This inherent safety feature was also described in our discussion of series-type valve-control systems, which can be considered to be feed-forward control systems. Distillation towers, evaporators, and waste-steam neutralization are some examples of processes where feed-forward systems have been successfully applied.

Override

These systems are based on relays which allow the selection of the lowest or the highest input signals. The systems are used in the protection of equipment, in auctioneering, and in areas where redundant instrumentation has been used.

In the protection of equipment, it may be necessary to limit one variable to maintain safe operation. For example, a pump may have to be protected from both high discharge pressure and motor overload. If discharge pressure is the primary control, it will maintain control until an overload condition is detected. At that instant, the overload controller is selected for control by a low-select relay. Another example is the interlocking of fuel and air in boiler combustion control. Here, fuel is interlocked by air through a low-select relay, thereby preventing fuel flow from exceeding airflow; airflow is interlocked by fuel flow through a high-select relay, thereby preventing airflow from decreasing below fuel flow.

Auctioneering simply describes the selection of the highest or lowest of a set of signals. For example, if it is necessary to know the hottest spot in a reactor bed, a selection through a high-select relay for a number of inputs measuring temperature across the reactor bed could be used. In redundant instrumentation, the controls are set up to select a signal from a number of transmitters measuring the same variable. These are used in situations where loss of signal could produce an awkward control-system response.

Multiple-Loop Systems

Most industrial processes are composed of many individual control loops where one variable controls one manipulated variable or many input variables but still only one manipulated variable.

The first general problem is selection of the input variable that should control a selected manipulated variable. This is not obvious in some complex situations. However, a general rule is to select a manipulated variable having the greatest influence on the controlled variable. Other variables of interest would have to be taken into account by setting up cascaded systems or by operating them at fixed set points.

The second problem, the interaction between control loops, is a little more complex. An example of severe interaction is controlling both flow and pressure of gas in a pipe. For the smallest of upstream or downstream load upsets, the two control loops would interact because of the approximately equal natural period of oscillation of the two loops. The solutions are complex and use various decoupling circuits or detuning of controllers to prevent interaction. A physical solution is to structure the plant such that the dynamics of one variable could be changed by introducing capacity in one loop. In most cases the solutions are expensive and require a thorough knowledge of the process under consideration. The type of control hardware used also affects the problem. Although analog, pneumatic, and electronic components are available for most computing functions, a digital computer system provides the greatest ease and flexibility.

ELECTRONIC DIGITAL COMPUTERS

The present revolution in very large scale integrated (VLSI) circuit design is leading to a host of applications in the process-control field. These applications fall into three broad categories—smart measuring instruments, communications links, and controllers. These can be considered either in terms of dedicated application or shared application, where one controller can perform many functions on a time-shared basis.

Hardware

The foundation for these applications is the microprocessor, a basic building block that allows the application designer flexibility in configuring electronic circuitry to particular requirements. This flexibility is achieved because the microprocessor works somewhat similarly to a computer; it has a memory, an arithmetic and logic unit, a controller, and input-output circuitry. Refer to Fig. 2-21.

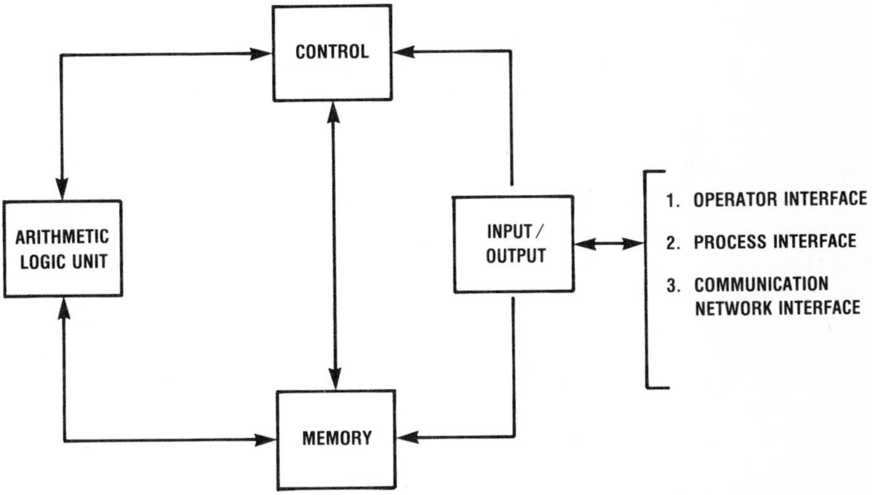

Figure 2-21 Basic processor.

The memory could be dedicated and part of the microprocessor or it could be external but under the processor's control. In either case, this fixed piece of memory (usually an ROM—read-only memory) contains words where the bit pattern for a particular set of words contains the requisite logic for the controller in the microprocessor to do a particular function. For example, this could be to read a word from the input and place it in the arithmetic logic unit (ALU). Another instruction could get another word from the input circuitry and also take it to the ALU. A third set of instructions could then add the two words together to give the required answer. Finally, one more instruction could take this answer and place it in the output circuitry. Given this basic instruction set, one could then develop the necessary logic for the microprocessor to accomplish the required functions.

This logic can now be implemented in two ways: The first is to design the microcode in such a manner that the microprocessor cycles through this code continuously, thereby dedicating the circuitry to performance of a predetermined function. The second is to define the microcode in such a manner that various sections of it could be used independently, and a particular function could then be externally selected. This external selection is accomplished by having an external memory (a RAM—random access memory) which contains words that, when decoded by the microprocessor, will point to the internal microcode instruction set that is required to be run. This allows the user to define requirements in a higher-level language for a given processor, rather than to work in terms of the machine-level microcode. Further, it now allows the processor to possibly do multiple tasks, where each task could be defined as a unique set of a higher-language code, basically a program. Thus, many programs could exist in the external memory and the processor would then run through each program, thereby performing many functions. The processor-memory combination looks basically like a regular computer, at least from the hardware description presented so far. It should be noted that there are many identifying characteristics of the various microprocessors now available. The first is word size, i.e., how many bits define a word? This could range from 4 bits to 16 bits. The 4-bit processors can be combined together to form larger word sizes. This is known as *bit slicing*. The size of the instruction set determines the power and flexibility in programming. Associated with the increased flexibility of the instruction set is the number of independently addressable (and therefore accessible) registers in the processors. This allows more complex programming. Another feature is the speed at which the microprocessor can perform the required function defined for each instruction in its higher-language repertoire.

In using a microprocessor as a computer, while considerations of the characteristics of the hardware are necessary, the major component that defines the uniqueness of each microprocessor is the operating system. In process control, the operating system complexity increases manyfold due to the real-time operation of the process. The operating system, also called the real-time executive, is a block of program logic that has five major functions to solve:

1. Scheduling the time of the processor among the many control programs
2. Managing the external main memory
3. Handling input and output for the various programs
4. Maintaining a data base for use by the tasks (programs) and also for reporting to management control
5. Allowing some mechanism whereby tasks can communicate with each other

As hardware costs of the microcomputer have steadily fallen, the software (executive programs) costs have taken on the major share in the overall cost of the computer installation. These needs lead to a large section of the main memory being taken by the real-time executive and up to 30 percent of the computer time. The rest of the memory space and processor time is available for actual application software.

These constraints are not necessarily a drawback. With the price of semiconductor memories falling rapidly, the expense of a large main memory is of minor consideration. Moreover, a limited memory space can be shared among many tasks or control loops, thus making even small microprocessor-based computers cost-effective compared with analog systems. With this price effectiveness is the increased reliability of the semiconductor components, allowing for an increased faith in the computer and thereby decreasing the requirements for backup. This backup could be either a duplicate computer or an analog system. In either case, the backup takes over control of the process on failure of the main system.

These computer-based systems are available in many sizes and varying capabilities. They can vary from being rather large and capable of controlling many hundreds of loops plus data-base management for report generation to being small dedicated units controlling a few loops. These smaller dedicated units can be linked together through a communication network, thereby leading into the beginning of application, i.e., distributed control. Each dedicated unit will perform its specialized function, which could be either measurement or control, or both. Normally, one large computer in the network would exist to act as the supervisor or master station.

In the application of computers for process control, some additional requirements

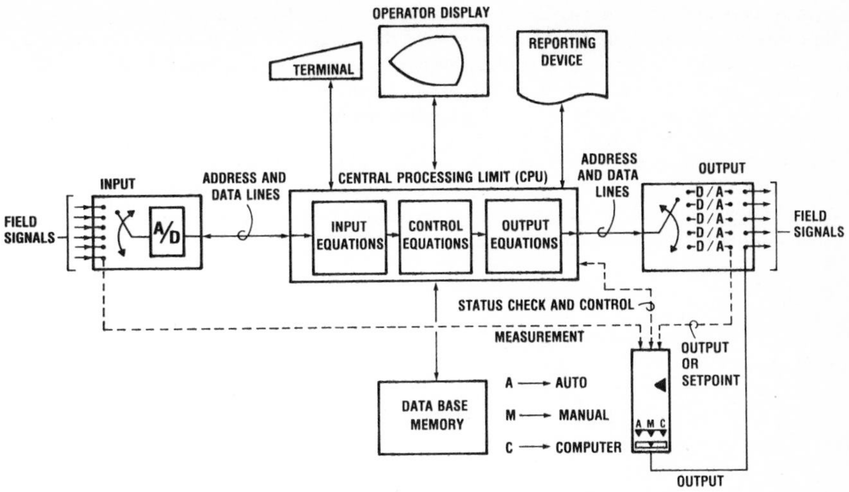

Figure 2-22 Basic elements of a computer for process control.

have to be considered. Primary is the additional hardware required to multiplex a large number of inputs and outputs, which are analog in nature. A conversion process must be performed on these analog signals to present them in a digital word recognizable by the computer and then to convert the digital value back to an analog form useful to the final actuator. Refer to Fig. 2-22.

Software

The software, the second requirement, should be capable of operating in a real-time mode. For example, if one were to schedule a given control program to run every 5 s, the operating system would have to be smart enough to recognize the priority of this program with respect to others and ensure that this schedule is met. This priority positioning of the real-time programs allows many control loops to be set up, each working at a certain given scan cycle. This flexibility allows certain loops to be scheduled as a function of the process characteristics. Thus, a process which is inherently fast (as given by its natural period) should also have its control program scheduled fast. A general rule is to have this schedule 4 to 10 times faster than the natural period of the process. An additional feature that the application software should have is the easy configuration of the various control loops required by a particular plant. This feature distinguishes the use of a computer from an analog system. No physical wiring or tubing changes are necessary. Hence, it is possible to change the control configuration as plant conditions change or as more sophisticated control strategies are implemented. The number of control loops and the ease of configuration or modification are functions of a particular manufacturer's software package and should be considered in any evaluation.

Operator Interface

The third important point for process control computers is the operator interface with the process. It is necessary for the operator of the process to have a clear, up-to-date picture of the conditions in the plant. Conditions requiring immediate attention—various alarms, for example—should be clearly and immediately brought to the operator's attention. Next, the software should be able to recognize any action taken and update outputs to the plant. Graphic displays on CRTs (cathode-ray tubes) are now the most common method by which the operator interfaces with the plant. Three major factors have to be considered. First, the ease with which these displays can be created and if there are any limitations to the number of different displays a system allows. Second, the speed with which (1) various displays can be accessed by the operator and (2) the programs that update and read actions taken by the operator from a given display can be obtained. Third, the ease with which various levels of information can be accessed or changed. For example, one may want to go from a plant unit display to accessing a particular loop of the unit, to a particular section of the loop, to ultimately change some parameter value which could then reflect the alarm value. This could possibly affect the alarm limits, tuning constants, or some calibration or characterization curves. Again, levels of sophistication vary among manufacturers.

Loop Control

Control of loops on a sample basis affects the controllability of the loop. As the sample time is increased, the loop control degrades. This is related directly to the dollar cost of the operation. However, a decrease in sample rate increases the load on the computer which decreases its ability to do other computations. This can be related to the dollar cost of the computing effort. Figure 2-23 shows the relationship between the two costs.

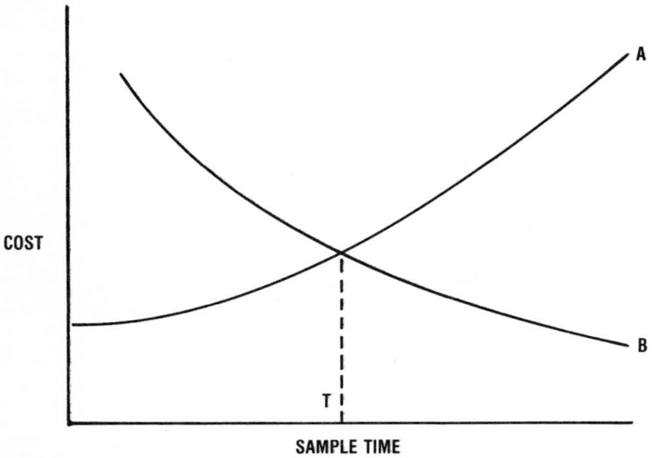

Figure 2-23 Sample rate costs.

Loop control is accomplished by basically solving equations for each loop. Normally there are three sections to these equations:

1. Input equation which reads the digital value from the input hardware (A/D), checks for validity, characterizes for some preset or calculated calibrations, and converts these into numbers meaningful to both the computer and the operator. They can be used to check for alarm limits and also can do some filtering of noisy measurement signals.
2. Control equation which basically gets its measurement from the input section, a set-point value either from the operator or from some other program which may be calculating the required set point, and using the two basic values, calculating an output value for the final actuator.
3. Output value calculated could be checked for limit or alarm conditions, characterized for the final actuator and output hardware, and finally sent to the D/A converter and a sample-and-hold device which is required to retain the output value between sample times.

It should be appreciated that the inherent computational ability of the computer allows the use of complex control algorithms. This enhances the controllability of the plant. The control software should offer some standard algorithms (e.g., PID) and the capability for the user to create and use his or her own.

Backup

An important consideration in any computer application for continuous process control is safety. This involves computer backup. There are basically two options. One is to back up the main system with an identical system. The second is to have an analog backup for each loop. The analog backup would have the capability of either taking over full automatic control or being simply a manual backup in case of computer failure. This requires that the interface hardware recognize the computer failure status and transfer the analog loop to a predefined mode, auto or manual. On the other hand, the computer should be able to recognize the status of the backup station and any action the operator takes with it. Thus, when the operator does transfer to computer control, a smooth bumpless transfer should take place, again a function of the control software. The backup station need not be analog, but could be another computer (mostly dedicated), leading to the concept of distributed control and network hierarchies. Each of these dedicated computers is a particular plant unit control and a number of them are capable of talking to the main computer.

Whatever backup method is implemented, there are two basic methods of operating. First, the main computer does the full control with the backup simply tracking. This is called the DDC (direct digital control) mode. In the second, the main computer does not actually control the final actuator, but provides only a set point to the backup, which becomes the main controller. This is called SPC (supervisory control or set-point control). There are certain advantages in this mode. Failure of the supervisory computer does not affect the plant operation. The computation load decreases quite a bit since the dedicated controller provides the main control action. This decrease could be effectively utilized to run management reporting programs, build an historical data base, and institute both complex feed-forward control strategies and adaptive tuning techniques. In this mode of operation, the computer's advantage of computational power and its memory capability will be fully utilized.

Historical Data Base

The historical data base is the ability to store past plant operating data on a real-time basis, i.e., the data must relate to the actual time of plant operations to serve any useful purpose in either reporting for management control, for cost and efficiency calculations, or in using the data to plot or display trends of the various plant variables. Flexibility and memory size play an important part in the software that performs these functions. Flexibility in data-base development is the ability to add or delete variables to the data base. Furthermore, the data base should allow easy access for the reporting or trending software. When trying to determine the number of variables that can be used, one must take into account size considerations, how often the data collection is performed, and for how long a period the data are to be stored on-line for a given system. These factors have to be considered to determine the amount of on-line memory that is required. The reporting function may have three modes of operation: first, reports scheduled on a regular basis; second, reports called for on demand by operating personnel; and third, exception reports that are printed under given plant conditions.

Tuning Techniques

Feed-forward and adaptive tuning techniques lend themselves to easier treatment since complex equations and algorithms can easily be solved. The availability of memory allows past loop responses to be used to update terms of a given process model. These updated terms can be used to calculate tuning constants of a given control algorithm (for example, proportional gain, integral, and derivative of a PID control equation).

Whereas our discussion has been concentrated on the control of continuous loops, the use of a computer for logical operations can also be easily accomplished. For example, it can be used to control a batch process where very little continuous control may be needed but a large number of logical operations have to be performed. The computer inputs and outputs will be primarily digital (i.e., contacts) with some analog signals. The computer algorithms have to be set up to sequence through the required batch cycle while continuously monitoring the plant conditions and ensuring the completion of each batch cycle. Again, flexibility should be available to stop the sequence at any step, to continue the sequence, and to allow the operator manual sequencing of the process. Further, the operator should be allowed to change operations in a given sequence, for example, change the temperature ramp profile of a reactor or change the given product recipe. The software should also be able to perform relay-type logic, thus eliminating the need for hardware relay logic. Specialized computers specifically designed for these functions are called programmable logic controllers (PLC). These devices have a limited operating system which allows the running of one preprogrammed application program. Operator interface, if provided, is limited to some predefined function keys. The present trend is to make these devices more flexible, bringing them into the domain of minicomputers.

Because the technology is changing so rapidly now, the plant engineer should make a careful comparative study of available program controller units before making any key purchase decisions.

section 12

Noise and Vibration Control

chapter 12-1

Noise Control

by

Paul Jensen

Manager, Industrial Services
Bolt Beranek and Newman Inc.
Cambridge, Massachusetts

INTRODUCTION

Industrial plants are inherently noisy. Not until the 1960s and 1970s, however, did concern for public health and welfare lead to the intensive study and analysis of industrial noise and promulgation of regulations to control it. Most important of the regulations was the Occupational Safety and Health Act of 1970, whose Standard 1910.95 established limits for occupational noise exposure and thus gave impetus to the development of industrial noise control. Permissible noise exposures for employees, according to the act, cannot exceed those given in Table 1-1.

The act also covers combined exposures to noise and means of controlling noise:

> When the daily noise exposure is of two or more periods of noise exposure at different levels, their combined effect should be considered, rather than the individual effect of each. If the sum of the following fractions

TABLE 1-1 Permissible Noise Exposure Levels

Duration per day, h	Sound level, dBA,* slow response
8	90
6	92
4	95
3	97
2	100
1½	102
1	105
½	110
¼ or less	115

*See definition of dBA later in this chapter.

$$E = \frac{C_1}{T_1} + \frac{C_2}{T_2} + \cdots + \frac{C_n}{T_n}$$

exceeds unity, then, the mixed exposure should be considered to exceed the limit value. C_n indicates the total time of exposure at a specified sound level, and T_n indicates the exposure limit at that level.

Exposure to impulsive or impact sound should not exceed 140 dB peak sound pressure level.

When employees are subjected to sound levels exceeding those listed in the table, feasible administrative or engineering controls shall be utilized. If such controls fail to reduce sound levels within the levels of the table, personal protective equipment shall be provided and used to reduce sound levels within the levels of the table.

If the variations in sound level involve maxima at intervals of 1 second or less, it is to be considered continuous.

In all cases where the sound level exceeds the values shown herein, a continuing, effective hearing conservation program shall be administered.

To enforce the standard, OSHA inspectors are empowered to enter and inspect any workplace in the United States, after presenting their credentials to the person in charge. Representatives of the employer and of the employees may accompany the inspector on the tour of the work place.

If a violation is found during an inspection, an attempt will be made to settle the issues without going to court. However, employers who fail to comply with the standard may be taken to court by OSHA. Penalties range from civil fines up to $10,000 through criminal penalties up to $20,000 plus a year of imprisonment.

The OSHA standard has been adopted by other federal agencies, notably the Bureau of Mines. In 1978, the Department of Defense issued its own noise standard, Instruction No. 6055.3, which limits exposure of members of the armed forces to hazardous noise.

The present OSHA standard described above is being considered for revision now. In 1974, the Department of Labor published proposed rules that, if accepted, will keep the permissible sound level at 90 dBA for an 8-h exposure but will incorporate in the dosage calculation the allowance of 85 dBA for 16 h.

In addition, the proposed rule spells out very specifically the adoption of a stringent hearing conservation program for all workers who are exposed to an equivalent sound level of 85 dBA for 8 h (50 percent of allowable noise exposure).

OSHA proposes to limit exposure to impulses at 140 dB to 100 per day and to permit a tenfold increase in the number of impulses for each 10-dB decrease in the peak pressure of the impulse. For example, the number of impulses allowed at 130 dB would be 1000 per day, and the number of impulses allowed at 120 dB would be 10,000 per day.

As with the existing standard, OSHA states in the proposed regulation the methods that must be used to ensure compliance. Engineering and administrative controls, to the extent that they are technically and economically feasible, are to be used to reduce employee noise exposure to permissible limits. When these controls are unable to reduce noise to permissible limits, they nonetheless are to be used to reduce noise to the lowest level feasible and are to be supplemented by personal protective equipment. One exception is noted: The proposed rule would allow hearing protectors to be the sole means of noise control if an employee is exposed to noise beyond permissible limits for only one day each week.

CONCEPTS AND VOCABULARY OF NOISE CONTROL

Sound Noise is simply unwanted sound; it is a series of vibrations in the air. Human ears are sensitive to these vibrations; they sense them and pass them on to the brain. Acoustics—the branch of physics that deals with the production, transmission, and control of sound—has its own concepts and vocabulary. The following items are the key terms and concepts needed by an engineer working in noise control.

Frequency The frequency f of a sound is the number of its vibrations. Units of frequency may be expressed in cycles per second (cps) or in hertz (Hz). Since the frequency is a description of a periodic phenomenon, it is reciprocal of the time period, or the time necessary for the phenomenon to repeat.

The audible frequency range (16 to 20,000 Hz) has been divided into a series of octave bands and one-third-octave bands. Just as with an octave on a piano keyboard, an octave in sound analysis represents the frequency interval between a given frequency and twice that frequency. The interval is identified by the center frequency, representing the geometric mean of the bounds of that interval. The internationally agreed upon 1000-Hz center frequency determines the center frequencies of the remaining bands. The center frequencies and approximate cutoff frequencies are listed in Table 1-2.

Velocity of sound A sound wave travels at a velocity that depends primarily on the elasticity and the density of the medium. In air at normal temperature, the velocity of sound c is approximately 340 m/s (1100 ft/s).

Wavelength Under normal conditions, sound waves always travel through air at essentially the same velocity. Therefore, at any frequency, the wavelength λ of the sound (i.e., the distance that the sound travels in one cycle) can be calculated by

$$\lambda = \frac{c}{f} m \qquad \text{(ft)}$$

Sound power Sound power in watts (W) describes the energy of the sound source. This sound power may be

1. The total power radiated by the sound source over its entire frequency range
2. The power radiated in each of a series of frequency bands
3. The power radiated in a limited frequency range

Sound intensity Sound intensity I is equal to the sound power radiated in a specified direction through a unit area normal to this direction. If a steady-state emission of sound energy (power) is assumed from the sound source, in free space the energy will spread itself thinner as it travels further from the sound source.

Intensity, though useful when discussing wave motion and energy radiation and spreading, has the practical disadvantage of employing very small numbers and failing to relate directly to human hearing experience. A more convenient form of expressing people's hearing sensation is *sound pressure.*

Sound pressure Sound waves produce changes in the density of air as they travel through it. These changes in the air density cause pressure fluctuations around the ambient static pressure, or sound pressure. In most cases, these disturbances created in the air cannot be easily expressed mathematically in time and space. The reason is that,

TABLE 1-2 Center and Approximate Frequency Limits for the Standard Set of Contiguous Octave and One-Third Octave Bands Covering the Audio Frequency Range

	Frequency, Hz					
	Octave			One-third octave		
Band	Lower band limit	Center	Upper band limit	Lower band limit	Center	Upper band limit
12	11	16	22	14.1	16	17.8
13				17.8	20	22.4
14				22.4	25	28.2
15	22	31.5	44	28.2	31.5	35.5
16				35.5	40	44.7
17				44.7	50	56.2
18	44	63	88	56.2	63	70.8
19				70.8	80	89.1
20				89.1	100	112
21	88	125	177	112	125	141
22				141	160	178
23				178	200	224
24	177	250	355	224	250	282
25				282	315	355
26				355	400	447
27	355	500	710	447	500	562
28				562	630	708
29				708	800	891
30	710	1,000	1,420	891	1,000	1,122
31				1,122	1,250	1,413
32				1,413	1,600	1,778
33	1,420	2,000	2,840	1,778	2,000	2,239
34				2,239	2,500	2,818
35				2,818	3,150	3,548
36	2,840	4,000	5,680	3,548	4,000	4,467
37				4,467	5,000	5,623
38				5,623	6,300	7,079
39	5,680	8,000	11,360	7,079	8,000	8,913
40				8,913	10,000	11,220
41				11,220	12,500	14,130
42	11,360	16,000	22,720	14,130	16,000	17,780
43				17,780	20,000	22,390

over an extended portion of the vibrating surface that is creating the airborne disturbance, some portions are compressing the surrounding air, while other portions are causing rarefactions.

The unit approved by the International Standards Organization (ISO) for measuring sound pressure is the pascal (Pa), though the terms microbars (μbar), dynes per square centimeter (dyn/cm²), and newtons per square meter (N/m²) have all been used.

Conversion factors are

$$1 \text{ bar} = 10^5 \text{ Pa}$$
$$1\ \mu\text{bar} = 1 \text{ dyn/cm}^2$$
$$1 \text{ dyn/cm}^2 = 10^{-1} \text{ Pa}$$
$$1 \text{ N/m}^2 = 1 \text{ Pa}$$

Sound power level The sound power level (PWL) is the designation in decibels of the ratio of two sound powers. The PWL is independent of the environment and the distance from the sound source.

When the PWL of a sound source is being given, it is customary to use a reference level of 10^{-12} W:*

$$\text{PWL} = 10 \log \frac{W}{10^{-12}} \quad \text{dB}$$

Sound intensity level The sound intensity level (IL) is the designation in decibels of the ratio of two intensities. The IL is influenced by the environment and the distance from the sound source.

When indicating IL, it is customary to use a reference level of 10^{-12} W/m^2:

$$\text{IL} = 10 \log \frac{I}{10^{-12}} \quad \text{dB}$$

Sound pressure level The sound pressure level (SPL) is the designation in decibels of the ratio of two pressures squared. The reason for using pressures squared is that all sounds consist of positive pressure disturbances (compressions) and negative pressure disturbances (rarefactions) measured from the static pressure. The mean value of the sound pressure disturbances would not be a meaningful value (it would be zero, since there are as many compressions as rarefactions). When indicating SPL, it is customary to use a reference level p_0 of 2×10^{-5} Pa. This reference value is roughly equivalent to the threshold of hearing at a frequency of 1000 Hz:

$$\text{SPL} = 10 \log \left(\frac{p}{2 \times 10^{-5}} \right)^2 = 20 \log \frac{p}{2 \times 10^{-5}} \quad \text{dB}$$

Some relationships between sound pressure and sound pressure level ($p_0 = 2 \times 10^{-5}$ Pa) are shown in Table 1-3.

TABLE 1-3 Sound Pressure and Sound Pressure Level ($p_0 = 2 \times 10^{-5}$ Pa)

Sound pressure, p	Sound pressure level, L, dB
p_0	0
$1.12p_0$	1
$1.26p_0$	2
$2p_0$	6
$3.16p_0$	10
$10p_0$	20
10^2p_0	40
10^4p_0	80
10^6p_0	120

Sound level There are many single-number schemes for evaluating sounds according to people's response to noise. The simplest of these single-number schemes is the sound level in dBA. The "A" means that the frequency spectrum has been weighted by an electrical network in the sound-measuring equipment before the single number was derived. It is simple to measure the A-weighted sound level and, fortunately, this single number correlates as well as or better than other commonly known noise ratings. The frequency response of the A-weighting is shown in Table 1-4.

Table 1-5 shows some typical sound pressure levels in industrial environments.

Decibel The decibel (dB) is a dimensionless unit for expressing the ratio of two numerical values on a logarithmic scale. It is convenient to use decibels in dealing with sound power, sound intensity, or sound pressure because of the tremendous range of values of these quantities that can be perceived by the ear. Audible intensities range from 10^{-12} to 10 W/m^3.

*Originally, 10^{-13} was used as a reference power. When this is the case, the power level must be reduced by 10 dB for use in the formulas of this chapter.

TABLE 1-4 A Scale Weighting

Octave band center frequency, Hz	One-third octave band center frequency, Hz	A-weighting relative response, dB
	20	−50.5
	25	−44.7
31.5	31.5	−39.4
	40	−34.6
	50	−30.2
63	63	−26.2
	80	−27.5
	100	−19.1
125	125	−16.1
	160	−13.4
	200	−10.9
250	250	− 8.6
	315	− 6.6
	400	− 4.8
500	500	− 3.2
	630	− 1.9
	800	− 0.8
1000	1000	0
	1250	+ 0.6
	1600	+ 1.0
2000	2000	+ 1.2
	2500	+ 1.3
	3150	+ 1.2
4000	4000	+ 1.0
	5000	+ 0.5
	6300	− 0.1
8000	8000	− 1.5
	10000	− 2.5
	12500	− 4.3
16000	16000	− 6.6
	20000	− 9.3

TABLE 1-5 Typical Sound Pressure Levels in Industrial Plants

	Octave band center frequency, Hz							
	A	125	250	500	1000	2000	4000	8000
Synthetic spinning machine	95	76	79	84	85	88	90	89
Rock crusher	101	102	100	99	96	93	85	77
Letterpress	96	93	94	94	93	89	84	80
Hammer mill	96	94	92	92	89	87	85	79
Hand-held sand blaster	95	85	82	82	87	88	90	96
Plastic extruder	97	103	99	94	89	88	85	82
Candy wrapper	90	78	82	83	84	84	80	75
Fly shuttle loom	102	86	87	91	95	98	95	89
Can filling/seamer	98	86	87	92	92	93	89	87
Ring twister	94	88	86	88	90	89	82	78
Chipper	105	92	98	100	100	99	99	106
Wood chipper	109	104	110	110	106	96	88	80
Billet heater	102	96	107	101	91	84	78	75
Buffing machine	101	88	85	89	98	94	89	83
Punch press	102	94	96	96	96	97	95	87
Wire-drawing machine	96	91	93	92	92	90	82	76
Molding machine	98	80	82	98	90	91	89	81
Concrete-block machine	106	100	102	101	101	101	96	90

Addition of decibels Since decibels are logarithmic units, they are not added arithmetically. Decibels are added by converting them to power, intensity, or pressure, adding these quantities, and then converting them back to decibels. In other words, 80 dB and 80 dB do not add up to 160 dB, but to 83 dB. Figure 1-1 gives the most convenient way to combine decibels.

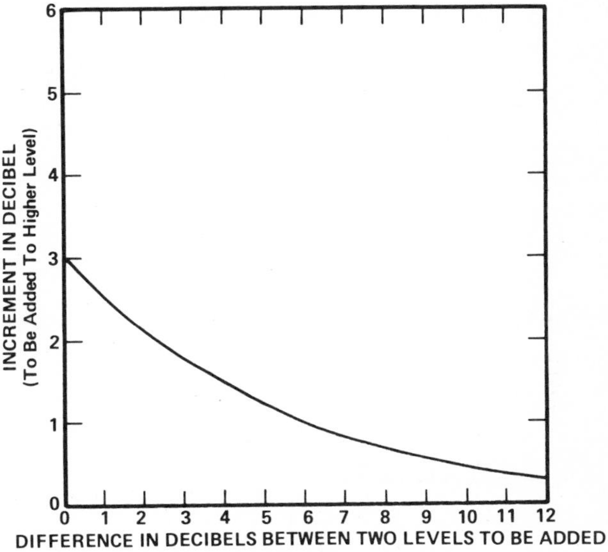

Figure 1-1 Chart for combining sound levels.

The inverse square law Under free-field conditions of sound radiation, the sound intensity is proportional to the square of the sound pressure:

$$I = \frac{p^2}{c\rho_0} \qquad \text{W/m}^2$$

where

p = sound pressure, Pa
c = velocity of sound, m/s
ρ_0 = density of air, kg/m^3

When the sound intensity is related to the sound power of a source, the result is

$$I = \frac{p^2}{c\rho_0} = \frac{W}{4\pi r^2} \qquad \text{W/m}^2$$

$$p^2 = c\rho_0 \frac{W}{4\pi r^2} \qquad \text{Pa}^2$$

The relationship between the sound pressures squared at two distances from the source becomes

$$\frac{p_1^2}{p_2^2} = \frac{r_2^2}{r_1^2}$$

This classic relationship is called the *inverse square law*.

Reverberation and reverberation time Reverberation is the persistance of sound after the source has stopped.

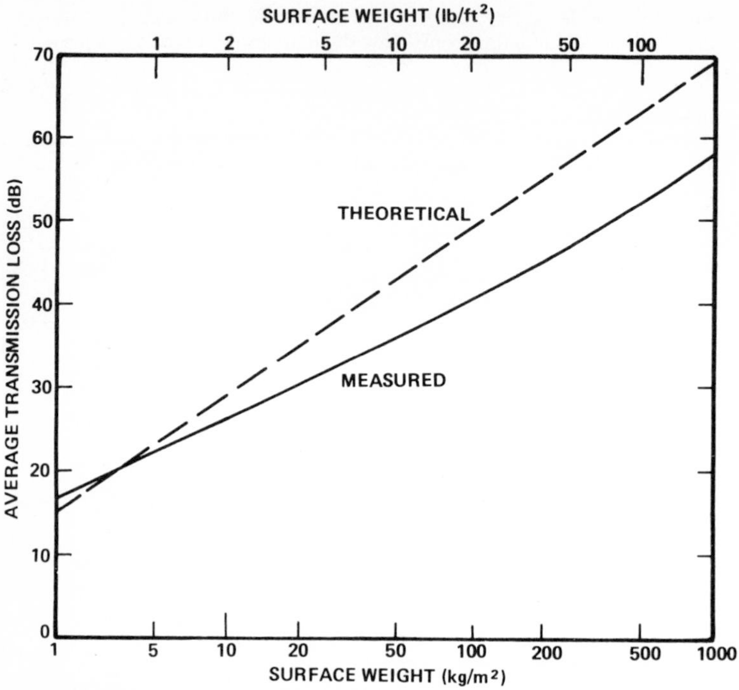

Figure 1-2 Average transmission loss of single homogeneous walls in the 100- to 3150-Hz frequency range.

The reverberation time t in a room has been defined as the time required for the sound level to decrease 60 dB after the source has stopped.

Transmission coefficient The transmission coefficient τ is the fraction of incident sound energy that is transmitted through a barrier. Thus

$$\tau = \frac{W_2}{W_1}$$

where W_1 is the incident sound energy in watts and W_2 is the sound energy transmitted in watts.

Transmission loss The transmission loss TL of a barrier is defined as the ratio of transmitted sound energy to the incident sound energy.

In logarithmic form, this ratio is expressed as

$$\mathrm{TL} = 10 \log \frac{W_1}{W_2} \qquad \mathrm{dB}$$

This can be written as

$$\mathrm{TL} = 10 \log \frac{1}{\tau} \qquad \mathrm{dB}$$

To calculate in detail the transmission loss for even simple constructions is difficult. As a result, engineers rely in most cases on laboratory data obtained in tests between specially constructed rooms.

Mass law The mass law says that for each doubling of the weight of a single-thickness wall, the average transmission loss increases by 6 dB. Figure 1-2 shows a graphic presentation of the transmission loss of a single wall. The empirically determined curve (solid) is lower than the theoretical curve (dotted). Walls with surface weight less than 200 kg/m^2 (40 lb/ft^2) gain only 5 dB per doubling, while for walls whose surface weight is above 200 kg/m^2, the mass law is valid again.

Composite transmission loss The composite transmission loss of a barrier is determined when the transmission coefficients for each part are known. The sound power transmitted by the two elements with a common incident sound power is

$$W_{\text{trans}} = \tau_1 S_1 + \tau_2 S_2 \qquad \text{W}$$

where τ_1 and τ_2 are the transmission coefficients for the individual parts, S_1 and S_2 are the areas of the individual parts in square meters (square feet), m^2 (ft^2), and W is watts.

The composite transmission loss becomes

$$\text{TL} = 10 \log \frac{S_1 + S_2}{\tau_1 S_1 + \tau_2 S_2} \qquad \text{dB}$$

Noise reduction The difference in sound pressure levels between two rooms is called the noise reduction (NR). The NR accounts for all sound paths and is thus more inclusive than the transmission loss.

The noise reduction can be expressed as

$$\text{NR} = \text{TL} - 10 \log \frac{S}{A_2} - C_1 + C_2 \qquad \text{dB}$$

where

S = area of the common wall, m^2 (ft^2)
A_2 = total absorption in the receiving room, m^2-sabin (ft^2-sabin)
C_1 = correction which depends on air leaks*
C_2 = correction which depends on flanking transmission.*

THEORY AND PHYSIOLOGY OF NOISE

Human Response to Noise

The psychophysical characteristics of a noise—those that determine how people react to it—cannot be measured with any currently available instrumentation. The only method that can be used is to expose a sufficiently large group of people *(test population)* to the noise in question in physical, psychological, and social situations similar to that of the people influenced by the noise—obviously a cumbersome and time-consuming method.

Because of such difficulties in developing data, acousticians, sociologists, and others have tried to develop objective measurement procedures for noise. But people obviously do not respond the same way to the same noise. Their response depends on their previous noise exposure, psychological attitude, socioeconomic status, the activity they are engaged in when they hear the noise, and other factors.

However, rating scales derived from physical measurements of the noise have been developed. These correlate reasonably accurately with the *average* human response to noise, particularly when the tests are confined to the laboratory environment. These scales rate human response in terms of loudness, perceived noise level (related to annoyance), articulation index (related to the ability to converse with ease), and other factors.

Numerous single-number rating schemes have been proposed for evaluating different noises according to one aspect or another of people's subjective response to the noise. Some of the ratings are quite simple; others are very complicated. Among the better known are

- Overall sound pressure level
- A-weighted sound level (dBA)

*C_1 and C_2 are zero for no air leaks and no flanking transmission.

- Equal energy sound level (L_{eq})
- Loudness level (LL)
- Articulation index (AI in percent)
- Speech interference level (SIL)
- Noise criterion curves (NC)
- Perceived noise level (PNdB)
- Noise and number index (NNI)
- Traffic noise index (TNI)
- Noise pollution level (NPL).

Acousticians and engineers have responded to the demand for noise control by developing the *systems approach*. Each problem of noise control is viewed as having three components:

- A noise *source*
- A noise *path*
- A noise *receiver*

Each component is considered in the solution of the problem. Engineers must

- Determine the characteristics of the noise sources
- Determine the tolerance of the receiver for intruding noise
- Determine how much and what kind of control must be used at the source and/or along the path to achieve the desired level of sound

Acoustical Treatments

Sound that originates in an enclosed space, such as a room or a factory, will spread until it reaches the surfaces, where it will either be absorbed or reflected. If room surfaces are hard, sound will reverberate, intermittent sounds will be mixed together, and steady sounds will add up. The result will be a relatively noisy space. If room surfaces are soft, the space will be relatively quiet.

It is important, then, that the amount of sound either absorbed or reflected can be quantified. The sound absorption efficiency of a material is determined as the fraction of incident sound energy that is absorbed by the surface. This is called the sound absorption coefficient α:

$$\alpha = \frac{I_a}{I_i}$$

where I_i is the sound intensity impinging on the material in watts per square meter and I_a is the sound intensity absorbed by the material, in watts per square meter. If $\alpha = 1.0$, all impinging sound energy is absorbed. If $\alpha = 0.0$, all impinging sound energy is reflected.

If 1 m^2 (ft^2) of material absorbs 20 percent of the impinging sound energy, 5 m^2 (ft^2) will absorb as much as 1 m^2 (ft^2) having complete efficiency. The sound absorption A provided by a material can be determined by

$$A = \alpha S \qquad \text{m}^2\text{-sabin}$$

where S is the surface area in square meters.

But rooms are constructed of several materials, each having different absorption coefficients. The total sound absorption then becomes

$$A = \alpha_1 S_1 + \alpha_2 S_2 + \cdots + \alpha_n S_n = \Sigma\, a_n S_n$$

Table 1-6 lists absorption coefficients of many typical building materials, and Table 1-7, the absorption coefficients of commonly used acoustic—or sound-absorbing—materials.

TABLE 1-6 Sound Absorption Coefficients of General Building Materials and Furnishings*

Materials	Coefficients, Hz 125	250	500	1000	2000	4000
Brick, unglazed	0.03	0.03	0.03	0.04	0.05	0.07
Brick, unglazed, painted	0.01	0.01	0.02	0.02	0.02	0.03
Carpet, heavy, on concrete	0.02	0.06	0.14	0.37	0.60	0.65
Same, on 40-oz hair felt or foam rubber	0.08	0.24	0.57	0.69	0.71	0.73
Same, with impermeable latex backing on 40-oz hair felt or foam rubber	0.08	0.27	0.39	0.34	0.48	0.63
Concrete block, coarse	0.36	0.44	0.31	0.29	0.39	0.25
Concrete block, painted	0.10	0.05	0.06	0.07	0.09	0.08
Fabrics						
Light velour, 10 oz/yd², hung straight, in contact with wall	0.03	0.04	0.11	0.17	0.24	0.35
Medium velour, 14 oz/yd², draped to half area	0.07	0.31	0.49	0.75	0.70	0.60
Heavy velour, 18 oz/yd², draped to half area	0.14	0.35	0.55	0.72	0.70	0.65
Floors						
Concrete or terrazzo	0.01	0.01	0.015	0.02	0.02	0.02
Linoleum, asphalt, rubber, or cork time on concrete	0.02	0.03	0.03	0.03	0.03	0.02
Wood	0.15	0.11	0.10	0.07	0.06	0.07
Wood parquet in asphalt on concrete	0.04	0.04	0.07	0.06	0.06	0.07
Glass						
Large panes of heavy plate glass	0.18	0.06	0.04	0.03	0.02	0.02
Ordinary window glass	0.35	0.25	0.18	0.12	0.07	0.04
Gypsum board, ½ in nailed to 2 × 4's 16 in. o.c.	0.29	0.10	0.05	0.04	0.07	0.09
Marble or glazed tile	0.01	0.01	0.01	0.01	0.02	0.02
Openings						
Stage, depending on furnishings			0.25–0.75			
Deep balcony, upholstered seats			0.50–1.00			
Grills, ventilating			0.15–0.50			
Plaster, gypsum or lime, smooth finish on tile or brick	0.013	0.015	0.02	0.03	0.04	0.05
Plaster, gypsum or lime, rough finish on tile or brick	0.14	0.10	0.06	0.05	0.04	0.03
Same, with smooth finish	0.14	0.10	0.06	0.04	0.04	0.03
Plywood paneling, 3/8 in thick	0.28	0.22	0.17	0.09	0.10	0.11
Water surface, as in a swimming pool	0.008	0.008	0.013	0.015	0.020	0.025
Air, sabins per 1000 ft³ at 50% RH				.9	2.3	7.2

*Complete tables of coefficients of the various materials that normally constitute the interior finish of rooms may be found in the various books on architectural acoustics. The following short list will be useful in making simple calculations of the reverberation in rooms.

Materials Selection

The most commonly used materials for control of noise in industry are absorbers and transmission-loss materials for airborne sound and vibration isolators and dampers for solid-borne sound. Selection of materials is governed by factors other than acoustical. These factors may be broadly classified as environmental and regulatory. Environmental factors include:

- Moisture, water spray, water immersion
- Oil, grease, dirt
- Vibration

TABLE 1-7 Sound Absorption Coefficients of Common Acoustic Materials

Materials*	Frequency, Hz 125	250	500	1000	2000	4000
Fibrous glass (typically 4 lb/ft³) hard backing						
1 in thick	0.07	0.23	0.48	0.83	0.88	0.80
2 in thick	0.20	0.55	0.89	0.97	0.83	0.79
4 in thick	0.39	0.91	0.99	0.97	0.94	0.89
Polyurethane foam (open cell)						
¼ in thick	0.05	0.07	0.10	0.20	0.45	0.81
½ in thick	0.05	0.12	0.25	0.57	0.89	0.98
1 in thick	0.14	0.30	0.63	0.91	0.98	0.91
2 in thick	0.35	0.51	0.82	0.98	0.97	0.95
Hair felt						
½ in thick	0.05	0.07	0.29	0.63	0.83	0.87
1 in thick	0.06	0.31	0.80	0.88	0.87	0.87

*For specific grades, see manufacturer's data.

- Temperature
- Erosion by fluid flow

Regulatory factors include:

- Lead-bearing material forbidden near food processing lines
- Restrictions on materials that may be in contact with foods being processed—glass, Monel, or stainless steel permitted
- Requirements for material not to be damaged by disinfecting
- Firebreak requirements on ducts, pipe runs, shafts
- Flame-spread rate limits on acoustically absorbing materials
- Fire-endurance limits on acoustically absorbing materials
- Restrictions on shedding of fibers in air by acoustically absorbing materials
- Elimination of uninspectable spaces in which vermin may hide
- Requirements for secure anchoring of heavy equipment
- Restrictions on hole sizes in machine guards (holes can reduce radiated noise of vibrating sheets)

NOISE-CONTROL TECHNIQUES

As previously noted, noise can be controlled at its *source,* along the *paths* it travels through air or through structures, and at the ears of the *receiver.* Industry today uses techniques that include both noise-source treatments and transmission path treatments. Personal protective equipment, such as earmuffs and earplugs, is often useful and efficient in reducing a worker's daily noise dose, although it is no substitute for engineered noise control. Often overlooked in general discussions of controlling industrial noise are methods that require no machine modifications or additions—good maintenance and operating procedures and planned replacement of machines or machine parts. These methods are described at the end of this section.

The primary techniques used in controlling noise in industry are the following. Noise-source treatments:

- Elimination or modification of specific equipment elements

- Substitution or redesign of entire machines or processes
- Surface damping

Transmission-path treatments:

- Mufflers
- Addition of room absorption
- Construction of noise shelters or other forms of personnel enclosures
- Vibration isolation
- Barriers
- Lagging

Receiver treatments:

- Earplugs
- Earmufflers

Noise-Source Treatments

Modifications or Substitutions

Quiet components are available to replace the noise-generating parts of certain machines. The equipment or components may need some engineering design and analysis to ensure that the machines are compatible with the manufacturing process. However, in many instances, off-the-shelf replacement items can be purchased for direct application.

The use of alternative materials for selected components can also reduce the noise generated by a machine or part of a machine. For example, the noise produced by the impact and subsequent vibration of metal parts can be significantly reduced by the addition of soft buffers or by the replacement of some non-load-bearing parts by plastic or other heavily damped materials. (*Damped* materials do not respond greatly or "ring" when struck.)

Surface Damping

Machinery housings often consist of large areas of flexible metal plate; if they are set into vibration, such plate areas may become significant acoustical radiators of airborne sound. Vibration is caused by some forcing mechanism such as the internal oscillating, rotating, or reciprocating components of the machine. The action of these components will tend to excite the surfaces at their resonant (or natural ring) frequency. The adding of damping (vibration-energy-absorbing) material to the surface will cause the amplitude of vibration at the resonant frequency to be reduced. Adding more material can stiffen and change the natural frequency so that it is no longer so easily excited.

Damping (viscoelastic) materials can be applied either as a free layer or as a constrained layer. (See Fig. 1-3*a* and *b*.) The material can be either troweled on, glued and baked on, or (for intermittent use) attached by a magnetic layer. Practical concerns regarding the specific material are ease of cleaning, toxicity, durability, and temperature-independent efficiency in the required frequency ranges. As a rule of thumb, for a damping material to have any effectiveness, it must be at least as thick as the material to be damped.

Transmission-Path Treatments

Mufflers

Mufflers are used to reduce noise associated with air- or gas flow. Examples are noise at the intake or exhaust of engines, at the inlet and outlet of fans, or generated by a high-velocity air or stream jet. Mufflers are also used in passageways to control the escape of

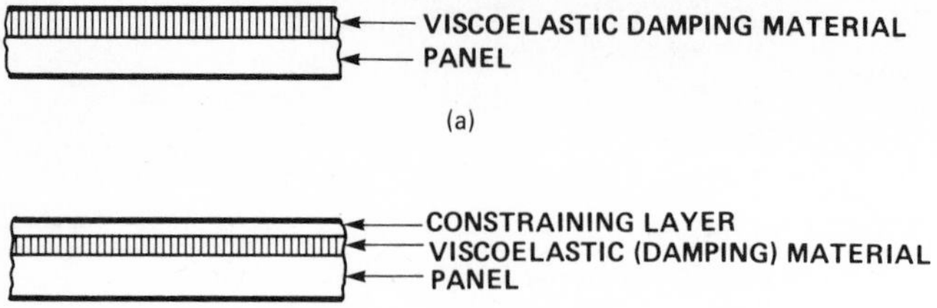

Figure 1-3 (*a*) Panel with free layer of viscoelastic (damping) material. (*b*) Panel with constrained layer of viscoelastic (damping) material.

noise in paths that must carry materials or personnel. A piece of acoustically lined duct, through which punch press punchings may pass, will also serve as a simple muffler to reduce noise escaping from the impacting region of the press.

There are two basic types of muffler: one is a dissipator of the acoustic energy; the other relies on reflection to confine the acoustic energy.

Figure 1-4 shows the acoustical performance of standard dissipative mufflers. The pressure drop through a muffler (the added drag the muffler puts on the flow) can be minimized by maintaining large openings and low velocities of flow through the muffler. (Indeed, if the velocity is not sufficiently reduced, the muffler can become, in turn, an

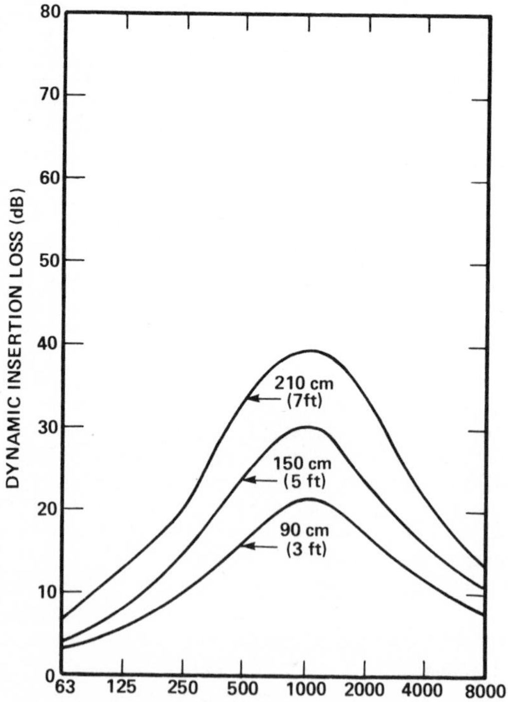

Figure 1-4 Acoustical performance of mufflers of different lengths.

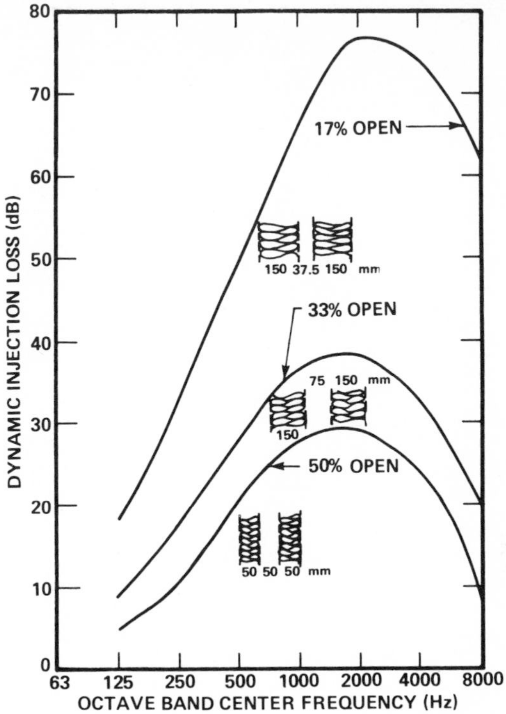

Figure 1-5 Acoustical performance of 90-cm (3-ft) long mufflers with different spacing and thickness of baffles.

additional noise source.) However, the distance between absorptive surfaces of a dissipative muffler is critical to the band of frequencies of sound absorbed. Sets of mufflers in series may sometimes be necessary to attenuate sound in different frequency bands.

Figure 1-5 shows the acoustical performance of dissipative mufflers with different spacing and thickness of baffles. The problem of contamination of the porous surfaces of the sound-absorbing material can be solved by covering the surface with a very thin, nonporous, limp, but durable skin, which is designed to allow sound to pass through it, yet still contain the flow.

The practical effectiveness of any design is likely to be limited by the available space, since it is theoretically possible to lengthen a muffler to produce any required attenuation (fading of sound). Practical design may range from an array of 15-ft- (5-m-) long absorptive splitters for the large duct of an induced-draft fan on a boiler to the 3-ft- (1-m-) long device (reactive muffler) presently fitted to an automobile engine.

Addition of Room Absorption

One commonly used method for noise reduction in a room is to clad the interior surfaces (ceiling and walls) with highly absorptive materials. The acoustical materials will have a minor noise reduction effect close-in to the noise sources, while at distances several meters from the noise sources, significant noise reduction (5 to 10 dB) can be obtained. Figure 1-6 shows the sound pressure level drop-off as a function of distance from a sound source for rooms with various amounts of acoustical materials. As Fig. 1-6 shows, a 3-dB drop in sound occurs for each doubling of the total amount of absorption at distances away from the noise sources (in the reverberant field).

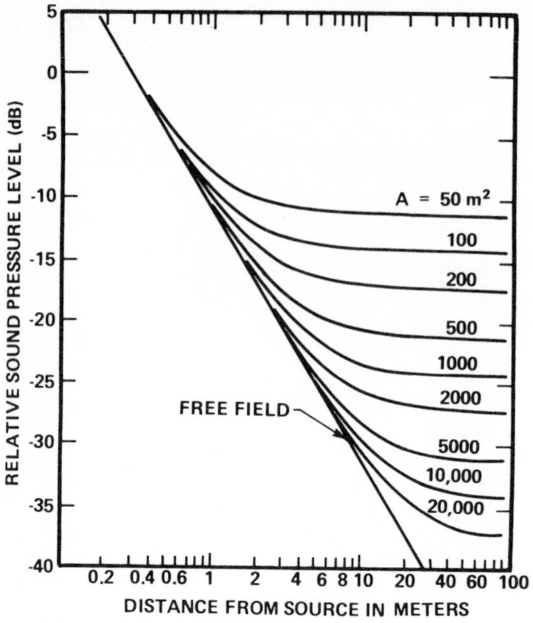

Figure 1-6 Drop-off of sound pressure level as a function of distance from sound source.

Enclosures

Enclosures, another useful means of controlling noise, are structures that completely surround the noise source and thus contain the sound it generates. However, enclosures can cause a buildup of high-level acoustic energy within it. Enclosures, therefore, usually consist of a wall with mass chosen to provide the required attenuation with a inner lining of porous material to dissipate the buildup of acoustic energy produced. In some cases where complete enclosures are built, the machines inside may require placement on vibration isolators—devices that prevent the transmission of structure-borne noise to the outer surfaces of the enclosure.

Employees exposed to high-level noise from a number of sources can also be protected by acoustic booths. These can range from small open-fronted telephone-booth-sized cabinets (into which the operator steps while observing the operation of a semiautomatic machine) to completely enclosed control consoles. In many cases, acoustic booths for personnel can provide an island of protection; an employee exposed to high-level noise during part of the workday can be protected well enough to reduce total exposure during normal working hours to less than the permissible exposure.

The design of worker enclosures requires suitable walls to produce the necessary sound reduction and also some internal sound absorption to prevent any reverberant buildup of transmitted sound. The use of fixed windows allows the necessary orientation of the booth, which can be important if an open entrance is to be used. The design and location of such booths can require a careful review and measurement program of the acoustical situation and may entail some local room acoustical treatment.

If there are gaps in the enclosure, sound can escape. The greater the percentage of open area of an enclosure, the smaller the reduction in radiated sound. Table 1-8 gives an indication of the effects of openings in otherwise well-designed enclosures.

In a practical sense, enclosure problems are not difficult to overcome but do require some accommodation. For ease of maintanance, enclosures can be constructed to lift vertically upward and off the machinery by use of overhead cranes. Access openings can be

TABLE 1-8 Noise Level Effects of Openings

Percentage of open area in enclosure	Maximum average noise reduction, dB
50	3
25	6
10	10
1	20

provided by use of tunnels lined with acoustically absorbent material (in effect, mufflers). Access for controls can be designed with hinged covers that lift easily, or the controls can be relocated. Some enclosure panels can be lifted automatically at the correct point in the machine cycle to provide access. Ducts can be provided with small ventilating fans to produce a controlled cooling airstream and, if combined with filters, can allow a controlled environment for some operations.

Vibration Isolation

Large vibrating areas (machinery housings or large sections of framing) may radiate large amounts of sound, even when the vibration is almost imperceptible. For noise control, the surfaces can be acoustically "isolated" from the vibrating drive mechanism by use of vibration isolation mounts, breaks, or pads installed between the vibrating source and the radiating surface. (See also Chap. 12-2, "Vibration Control.")

Barriers

A barrier is a solid wall used to shield a receiver from the direct radiated sound waves of a machine. To be effective, it must be sufficiently massive to provide the required reduction in the sound transmitted through the barrier, and it must be sufficiently wide and high to prevent the sound refracted around the edges from becoming significant. A barrier positioned halfway between the source and the receiver will require the greatest dimensions to produce the same acoustic shielding. Barriers, therefore, should be placed close to either the source or the receiver.

Barriers can also be used to shield a group of workers not working on noisy machines from an area containing noisy machines. Of course, the effectiveness of such barriers can be significantly reduced by reflected sound, from a ceiling or nearby sidewalls, for example. Therefore, barriers should be combined with ceiling and possibly wall treatment to reduce this alternative sound path. Figure 1-7 shows the acoustical performance of a barrier.

Barriers can be provided with acoustically absorptive material on the side facing the source to avoid a reflected buildup of sound near the machine.

Simple small barriers can be located to provide individual operator protection on some machinery. In this case, a transparent material may allow visual monitoring. On small machines the controls may still be reached under or around the barrier. Such arrangements can also act as safety shields.

Although a barrier is designed to stop the sound from reaching a given receiver, the barrier does not necessarily have to impede the passage of materials and products. The use of overlapping entrances to produce a visual and acoustic blockage can allow easy access for forklift trucks and the location of conveyors.

Lagging

Lagging of pipes, ducts, and other radiating surfaces requires the application of a double-layer construction, consisting of a resilient inner layer, for example, 1 to 4 in (2.5 to 10 cm) of glass fiber material and an airtight heavy outer layer. In many cases, lagging can be combined with or can replace the thermal insulation on a pipe and, as such, presents no installation or maintenance problem. The outer coating of sheet aluminum or mild steel can often be retained, as long as this outer layer is isolated everywhere from the inner vibrating structure. This separation is essential since the wrapping effectively pro-

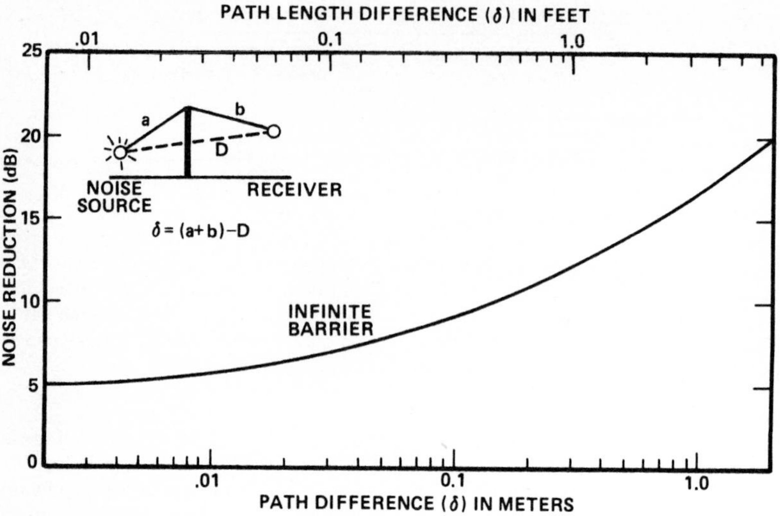

Figure 1-7 Average noise reduction of acoustical barrier of infinite length.

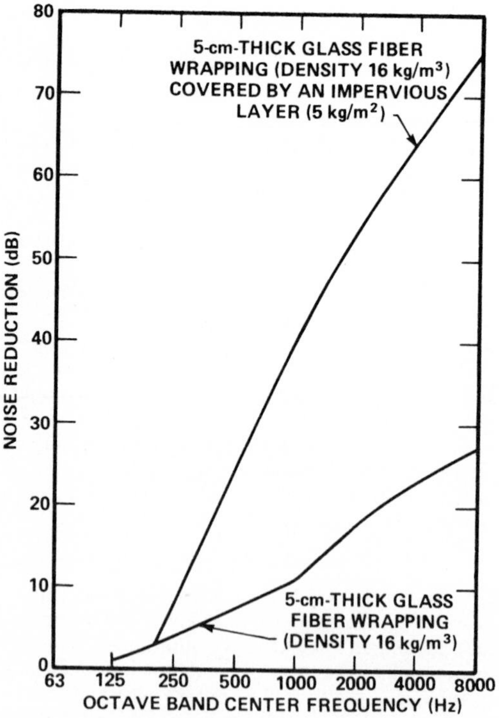

Figure 1-8 Acoustical performance of pipe wrapping.

vides an outer nonvibrating surface, separated by an absorbent material to dissipate sound energy produced by the pipe. An acoustic "short-circuit," such as at a valve where the outer jacket touches the pipe work, can allow the vibration to be transmitted directly and, hence, sound to be radiated.

Acoustic lagging requirements can be specified as part of the normal design supply and procurement process.

Figure 1-8 shows the acoustical performance of an acoustic lagging of a pipe.

Receiver Treatments

Earplugs

Earplugs are the simplest and least-expensive form of ear protection against excessive sound levels. All plugs share the feature of being inserted into the ear canal; they fall into three general categories:

1. Prefabricated earplugs of either rubber or plastic attempt to provide complete occlusion of sound from the inner ear by making an airtight seal over the ear canal.
2. Individually molded earplugs, made of silicone or synthetic rubber, are inserted in liquid form ½ in into the user's ear. The liquid can cure into a soft, customized earplug in 10 min. The greater weight of these plugs (11 to 13 g) is distributed over a larger area, so they are still comfortable.
3. Temporary or disposable ear plugs—made from materials such as wax-impregnated cotton, glass fiber, down, or foam (which are better than just cotton balls)—constitute the simplest form of plug.

Earplugs are not a permanent investment. They are easily lost, and the shape of the plastics is often deformed over time.

Earmuffs

Earmuffs are designed to isolate excessive sound levels from the ear by placing an airtight shell over the outer ear canal and attenuating the sound by the cover and cavity. Most earmuffs share a similar form of construction, differing mainly in materials and details.

An outer dome, made of hard shatterproof thermoplastic or the like, provides an initial shield. The heavier this dome (more mass) and the deeper the cavity around the ear (more air space separation), the more low-frequency sounds are absorbed.

Within this dome, an open-cell polyurethane filler attenuates high-frequency sounds (absorption occurs in the pores) and also eliminates the sea-shell effect of a hollow cover (high percent of absorption in a reflective space).

The seal with the ear must be airtight and yet flexible enough to prevent direct transmission of vibration from the rigid dome. For this, earmuff manufacturers use a vinyl-covered fluid-filled sack or a flexible semisoft plastic foam. The liquid seal is more effective for low-frequency attenuation.

The force of the headband must be enough to secure the seal yet be loose enough not to cause discomfort. This force ranges from 1 to 2.4 lb; an average is 2 lb (1 kg). The force also varies with head size. Values above 2.4 lb have been found to be uncomfortable and unacceptable.

The weight of earmuffs ranges from 6.75 to 16.0 oz (200 to 500 g); the average is 10 to 11 oz (300 to 330 g). The heavier muffs are generally more efficient. The weight should be broadly distributed on the skull by a wide headband. Earmuffs can also be held by a strap across the nape of the neck or by a safety helmet, which may offer some shielding of direct sound from the skull in certain frequencies. This reduction of bone-conducted sound can be a factor in extremely loud situations.

All surfaces in contact with the skin must be nontoxic and immune to degradation or irritation caused by perspiration, ear wax, humidity, skin oils, or dirt.

With any protector placed in or over the ear, acceptance by the user is an important factor. At sound levels that are painful to the ear, workmen seldom need prodding to wear protection. However, earmuff comfort becomes a determinant as sound levels decrease because workers can readily adapt to noise that—while sufficient to cause hearing deterioration over time—does not itself cause physical pain.

Since individual components of a set of earmuffs may become damaged or worn out, manufacturers usually provide individual replacement parts. The initial cost investment, durability, maintenance, and availability of replacement parts are factors that, along with noise-reduction characteristics and user comfort, will aid in the choice of a hearing protector.

At moderate and high noise levels, communication is enhanced when either earplugs or earmuffs are worn. This is a valuable safety aid so workmen can hear warnings.

Techniques That Require No Modifications or Additions

No discussion on noise-control techniques should omit a brief discussion of ways to reduce noise that do not involve equipment modification or addition. Examples are: proper maintenance, good operating procedures, and equipment replacement.

Proper Maintenance

Malfunctioning or poorly maintained equipment makes more noise then properly maintained equipment. Steam leaks, for example, generate high sound levels (and also waste money). Bad bearings, worn gears, slapping belts, improperly balanced rotating parts, or insufficiently lubricated parts can also cause unnecessary noise. Similarly, improperly adjusted linkages or cams or improperly positioned machine guards often make unnecessary contact with other parts and result in noise. Missing machine guards can also allow noise to escape. These types of noise sources share one characteristic. Their noise emissions can be readily controlled, though there is no simple way to predict how much noise reduction can be achieved through proper maintenance.

Operating Procedures

The way an operation is performed can cause workers to be overexposed to noise. Some operations are monitored by workers stationed near a noise source. At times, the distance is more critical in terms of noise exposure than the operation necessitates. In other words, the operator can be stationed at some other, quieter location without degrading work performance. Some operations can be monitored or performed from inside an operator "refuge," a booth or a room. Relocation of machine control systems can often augment this type of noise control.

Noise reduction obtained by relocating operators can be estimated by measuring sound levels at the existing station and the planned new station. If an operator booth is employed, noise reductions can be expected to range from 10 to 30 dB, with the higher value for booths with good windows and doors and the lower value for booths that are open to the environment on one or two sides.

Equipment Replacement

In some cases, the modification most readily available is quieter equipment that can be used to perform the same task. For example, several major manufacturers now sell quieted electric motors or quieted compressors. Quieted versions of equipment typically are more expensive than unquieted ones. Certainly, situations will arise when the purchase of different or newer equipment may be appropriate for production purposes, and these situations may be effectively combined with noise considerations. *Be aware that new equipment may not necessarily be quieter just because it is new.* Noise specifications can play a significant role in quieting an environment when an upgrading or expansion program is undertaken, and these specifications will be more important as pressure increases on equipment manufacturers to produce quieter equipment.

INSTRUMENTATION

Before a noise measurement program is started, its objective should be clearly understood. The objective could be to determine the approximate value of the noise during a short period of time or to find out whether the noise environment is hazardous to the health and welfare of the worker. The objective might also be to see if the noise exceeds some locally adopted sound level limits. Depending on the purpose and the desired accuracy, a wide range of descriptors and instrumentation is available.

A basic research program, that is, one that would assess noise throughout an entire plant, requires a large amount of detailed data. A noise-control investigation of a particular area or machine, on the other hand, normally requires much less detailed information, and many monitoring systems are set up to detect only relatively simple changes. For example, a system might be used to set off a flashing light whenever a sound level exceeds 90 dBA for periods longer than ½ s.

The simplest instrumentation available to determine sound pressure levels and sound levels is a sound-level meter. There are four different classifications of portable sound-level meters:

1. Precision
2. General purpose
3. Survey
4. Special purpose

The precision and tolerances of the indicating meter and the weighting networks vary significantly for the various types.

A sound-level meter must be kept calibrated if it is to provide meaningful data. Most equipment is battery-operated; the batteries must be fresh and capable of supplying the instrument with sufficient power. The manufacturer's instructions on how to check batteries appear on every instrument. A battery check is followed by a calibrator-pistonphone check. The calibrator-pistonphone is placed over the microphone and turned on. It will supply a pure tone at a known sound pressure level, which will allow appropriate adjustments to the meter reading, in accordance with the manufacturer's procedure.

Sound-level meters measure noise only at a given point at the time of observation. If the noise being measured is constant in both space and time, meters will give an accurate representation of the situation. However, if the sound level changes with time and location (for instance, as an operator moves around), it will be necessary either to record the sound level manually with short time intervals (5 to 10 s) or tape record the noise data, and later analyze the time history of the noise. The second approach is preferable when a worker's noise exposure is related to duty cycles or product flow. In this case, extrapolations can be made, on the basis of total day production, to determine the noise exposure of an employee over a full day.

In industry, there are situations in which time pressure is great, more than one person must be monitored, and duty cycles are not easily definable. In such cases, audiodosimeters can be used to measure an employee's noise exposure. The *audiodosimeters,* devices about the size of a cigarette pack, are worn by employees to record the noise exposure of the wearers wherever they go. The microphone can be fixed to the unit or detached from the unit and placed in the hearing zone of the wearer. The audiodosimeters can be obtained with an internal circuit that integrates the sound level and time in accordance with the OSHA regulation (halving of the exposure time for each 5-dB rise in sound level), the Department of Defense instruction (halving of the exposure time for each 4-dB rise in sound level), or the International Standard Organization standard (halving of the exposure time for each 3-dB increase in sound level).

STANDARDS

Two internationally known organizations have produced standards that acousticians use almost daily. Table 1-9 lists the most important of these standards and indicates the agencies that created them—the American National Standards Institute (ANSI) and the American Society for Testing and Materials (ASTM).

TABLE 1-9 Sound-Level Standards

American National Standards Institute (ANSI)	
S1.1-1960 (R1976)	Acoustical Terminology (including mechanical shock and vibration)
S1.2-1962 (R1976)	Method for the Physical Measurement of Sound (partially revised by S1.13-1971 and S1.21-1972)
S1.4-1971 (R1976)	Specification for Sound Level Meters (IEC 123)
S1.6-1976 (R1976)	Preferred Frequencies and Band Numbers for Acoustical Measurements (ISO 454; agrees with ISO 266)
S1.7-1970 (R1975)	Standard Test Method for Sound Absorption Coefficients by the Reverberation Method (See ASTM C 423-77)
S1.8-1969 (R1974)	Preferred Reference Quantities for Acoustical Levels
S1.10-1966 (R1976)	Method for the Calibration of Microphones
S1.11-1966 (R1976)	Specifications for Octave, Half-Octave, and Third-Octave Band Filter Sets
S1.13-1971 (R1976)	Methods for the Measurement of Sound Pressure Levels
S1.21-1972	Methods for the Determination of Sound Power Levels of Small Sources in Reverberation Rooms
S1.23	Method for the Designation of Sound Power Emitted by Machinery and Equipment (See ASA Standard Catalog No. 5-1976)
S2.9-1976	Nomenclature for Specifying Damping Properties of Materials
S1.26-1978	Method for the Calculation of the Absorption of Sound by the Atmosphere
S1.36-1979	Survey Methods for the Determination of Sound Power Levels of Noise Sources
S3.17-1975	Method for Rating the Sound Power Spectra of Small Stationary Noise Sources (See ASA CAT. NO. 4-1975)
S3.19-1974	Method for the Measurement of Real-Ear Protection of Hearing Protectors and Physical Attenuation of Earmuffs (See ASA CAT. NO. 1-1975)
American Society for Testing and Materials (ASTM)	
C 423-77	Standard Test Method for Sound Absorption Coefficients by the Reverberation Room Method
C 634-77	Standard Definitions of Terms Relating to Environmental Acoustics
E 90-75	Standard Method for Laboratory Measurement of Airborne Sound Transmission Loss of Building Partitions
E 336-77	Standard Test Method for Measurement of Airborne Sound Insulation in Buildings
E 413-73	Standard Classification for Determination of Sound Transmission Class
E 477-73	Standard Method of Testing Duct Liner Materials and Prefabricated Silencers for Acoustical and Airflow Performance
E 596-78	Standard Method for Laboratory Measurement of the Noise Reduction of Sound-Isolating Enclosures

BIBLIOGRAPHY

American Industrial Hygiene Association: *Industrial Noise Manual,* AIHA, Akron, Ohio, 1975.

American Institute of Physics: *Glossary of Terms Frequently Used in Acoustics*, AIP, New York, 1960.

Bell, L. H.: *Fundamentals of Industrial Noise Control,* Harmony, Trumbull, Conn., 1973.

Beranek, L. L.: *Noise and Vibration Control,* McGraw-Hill, New York, 1971.

Cheremisinoff, P. N., and P. P. Cheremisinoff: *Industrial Noise Control Handbook,* Ann Arbor Science, Ann Arbor, Mich., 1977.

Diehl, G. M.: *Machinery Acoustics,* Wiley, New York, 1973.

Jensen, P., C. R. Jokel, and L. N. Miller: *Industrial Noise Control Manual,* U.S. Department of Health, Education, and Welfare, NIOSH, Cincinnati, Ohio, 1978.

Lipscomb, D. M., and A. C. Taylor: *Noise Control Handbook of Principles and Practices,* Van Nostrand Reinhold, New York, 1978.

Morse, P. M., and K. U. Ingard: *Theoretical Acoustics,* McGraw-Hill, New York, 1968.

chapter 12-2

Vibration Control

by
Eric E. Ungar
Principal Engineer
Bolt Beranek and Newman Inc.
Cambridge, Mass.

CHARACTERIZATION OF VIBRATIONS

Vibration refers to oscillatory (back and forth) motions of structures, mechanical systems, or components of these. A vibration generally is characterized by the displacement, velocity, or acceleration measured at one or more points on the item of interest in specific directions of interest (e.g., perpendicular to a floor or wall).

An observer of the time variation of a vibration—for example, one who watches the displayed signal obtained from a sensor on an oscilloscope or chart recorder—often obtains a record that approximates the one shown in Fig. 2-1*a*. Such a regular curve, which corresponds mathematically to a sine or cosine, is called *sinusoidal* or *simple harmonic*. Note that it deviates from zero (the middle position) equally in both directions; the maximum excursion from zero in one direction is called the *amplitude A*, the total excursion in both directions is the *double amplitude 2A* (or sometimes the peak-to-peak value). Amplitudes may be given in units of displacement, velocity or acceleration, depending on how the vibration is measured.

The time interval T between successive peaks is called the *period* and usually is measured in seconds. The number of vibration cycles (i.e., the number of periods) that occur per second is called the *frequency f* and is generally measured in hertz (Hz), which is the internationally standardized name that has replaced cycles per second (cps).

In practice one rarely obtains a simple record like that of Fig. 2-1*a*. One is more likely to obtain a record that looks like Fig. 2-1*b*. This record may consist of a basic sinusoid like that of Fig. 2-1*a*, to which there are added one or more sinusoids with higher frequencies (shorter periods) and generally smaller amplitudes. Fig. 2-1*b* is said to represent a *multifrequency* or *complex* vibration. The component with the lowest frequency (greatest period) is called the *fundamental* component. Components that occur at frequencies that are integer multiples of that of the fundamental one are called *harmonics*.

Vibrations that have no well-defined period or amplitude—i.e., in essence, when the record never repeats itself—are called *nonperiodic* or *irregular*. A sample of a nonperiodic vibration record is given in Fig. 2-1*c*.

A vibration that has essentially the same amplitude over an extended time period is called *steady*, whereas a vibration with time-varying amplitude is called *transient*. Figures 2-1*a*, *b*, and *c* illustrate steady vibrations, whereas Fig. 2-1*d* illustrates a typical decaying transient containing a single-frequency component; similar transients with multiple-frequency components or with irregular behavior may readily be visualized.

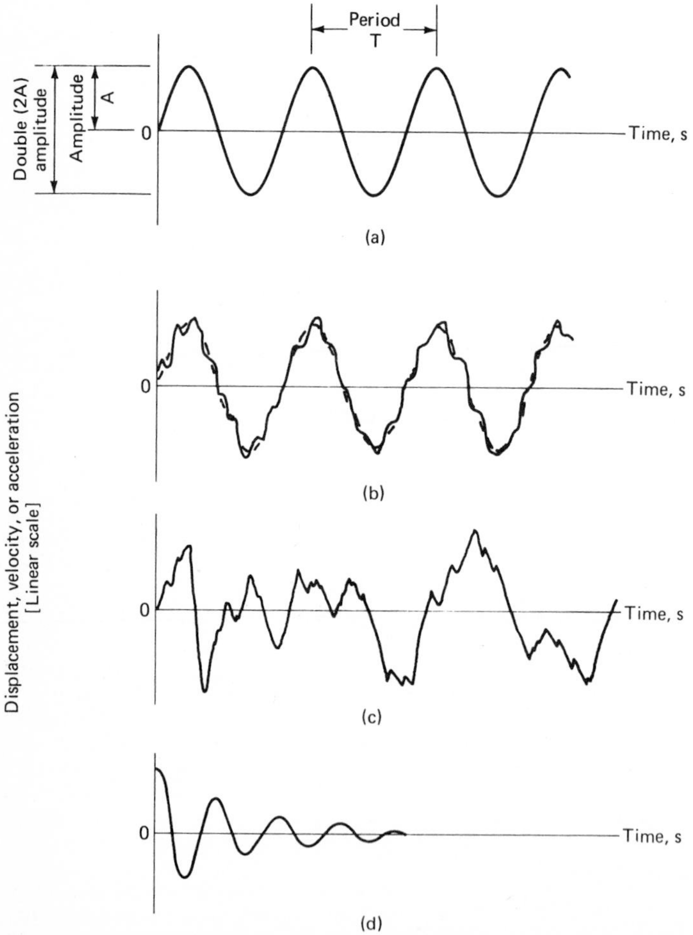

Figure 2-1 Typical vibration records: (*a*) steady sinusoidal or simple harmonic vibration; (*b*) steady multifrequency vibration; (*c*) irregular (nonperiodic) vibration; (*d*) decaying transient single-frequency vibration.

It is often useful to characterize a vibration in terms of a plot of amplitude vs. frequency. Such a plot is called a *spectrum.* One may obtain such a plot by passing the (electric) vibration signal through a variable filter to a readout device. The filter permits only selected frequency components to reach the readout, which then indicates the amplitudes of these selected components. There are also available instruments called *spectrum analyzers* that automatically provide amplitude-frequency displays for vibration signals fed into them.

Knowing the amplitude at a given frequency in terms of any one of the three motion quantities (displacement d, velocity v, acceleration a), one can calculate the amplitude in terms of the other two from:

$$a = 2\pi f v = (2\pi)^2 f^2 d$$

$$v = \frac{a}{2\pi f} = 2\pi f d \qquad (1)$$

$$d = \frac{a}{(2\pi)^2 f^2} = \frac{v}{2\pi f}$$

where a,v, and d always contain the same length units and seconds and f is in hertz. For example, to d given in mils there corresponds v in mils/s and a in mils/s^2. For $v = 10$ ft/s and 20 Hz, $a = 2\pi(20)(10) = 1256$ ft/s^2 and also $d = 10/2\pi(20) = 0.080$ ft. This can readily be converted to other units, e.g., to g's for acceleration (where $1g = 32.2$ ft/s^2 = 386 in/s^2) or to metric units.

Figure 2-2, which is based on Eq. (1), is a convenient chart for the approximate conversion between motion quantities.

CAUSES OF VIBRATIONS

Vibrations are always caused by unsteady forces, that is, by forces that may be oscillatory in magnitude or direction or by forces that are suddenly applied or released. These forces need not be due to mechanical causes; electromagnetic, aerodynamic or fluid-related forces also are often encountered in practice.

Unbalances in rotating machines produce net centrifugal forces that change direction in space as the machine rotates. For a machine with a horizontal shaft, such a force acts upward at one instant and downward a half-rotation later, thus producing a force that acts on the floor sinusoidally at a frequency that corresponds to the *shaft rotation frequency* f_r (hertz) = $N/60$, where N denotes the shaft rotation speed in revolutions per minute. As such a machine comes up to speed, it produces transient vibrations that increase in frequency and amplitude until steady operating conditions are reached; when such a machine is turned off and coasts toward a stop, it produces decaying transient vibrations with ever-decreasing amplitude and frequency.

Reciprocating machines also produce unbalanced inertia forces which are transmitted to the machine housing and supports. The primary components of these forces occur at the crankshaft's rotational frequency and at the first few integer multiples of that frequency. These forces generate vibrations along the direction of piston travel, as well as perpendicular to that direction, in the plane of the crank, and also produce vibratory moments or couples; the relative magnitudes depend on the cylinder arrangement and the degree of dynamic balance.

Fans, blowers, and pumps tend to generate steady vibrations, due to both unbalances and repetitive fluid pulses. The latter occur primarily at the *blade passage frequency,* which is the frequency with which blades pass a fixed point. For a rotor with n blades, rotating at N r/min, the blade-passage frequency is given by f_b (hertz) = $nN/60$, where N is in revolutions per minute.

Turbulent flows of water or air in a duct, or flows from a blower or airjet impinging on a surface, typically produce irregular forces on the structural surfaces. Similarly, irregularly repeated impacts, e.g., due to footfalls produced by many people walking on a floor, tend to produce irregular vibrations.

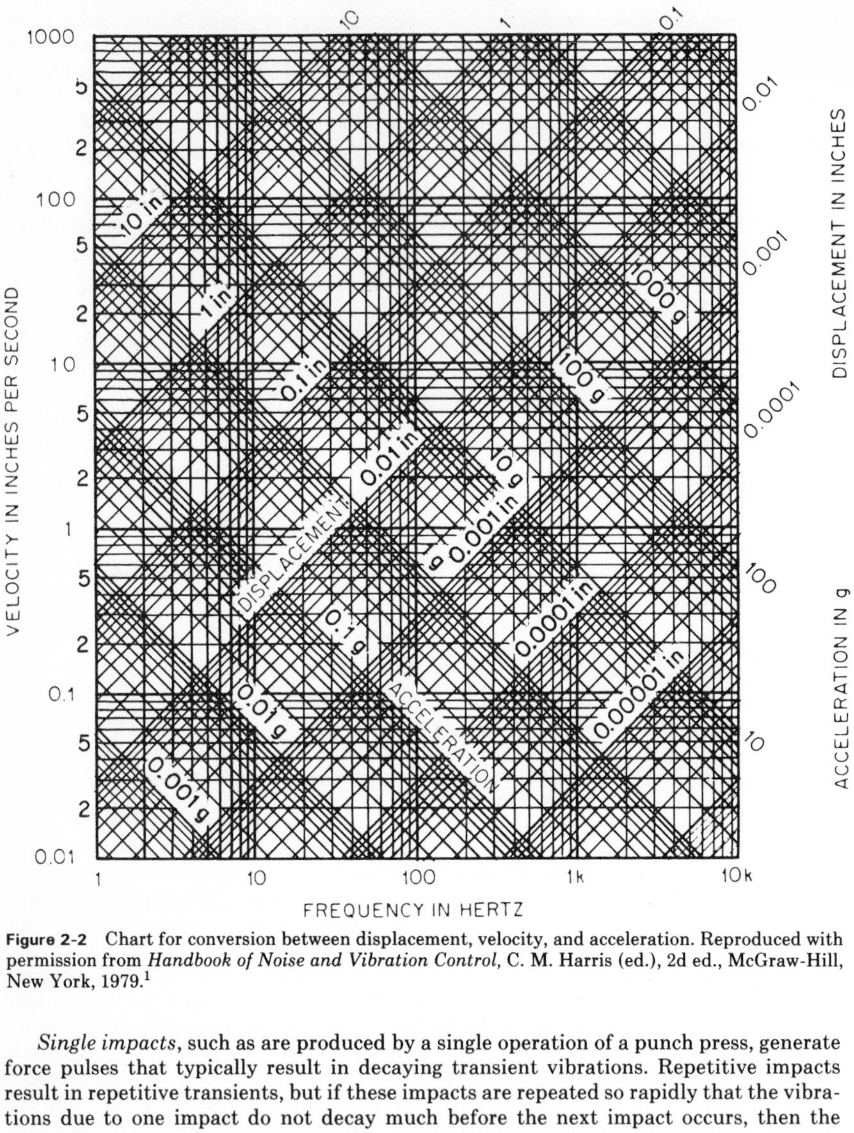

Figure 2-2 Chart for conversion between displacement, velocity, and acceleration. Reproduced with permission from *Handbook of Noise and Vibration Control,* C. M. Harris (ed.), 2d ed., McGraw-Hill, New York, 1979.[1]

Single impacts, such as are produced by a single operation of a punch press, generate force pulses that typically result in decaying transient vibrations. Repetitive impacts result in repetitive transients, but if these impacts are repeated so rapidly that the vibrations due to one impact do not decay much before the next impact occurs, then the vibrations tend to have more of a steady irregular character. Thus, impacts of materials in a gravity chute against the chute bottoms and sides tend to produce essentially nonperiodic (irregular) vibrations.

Rattling, slippage, and nonlinearities (deviations from proportionality between forces and displacements) associated with large excursions generally introduce higher-frequency components than those associated with the basic forces and motions.

EFFECTS OF VIBRATIONS; THE NEED FOR VIBRATION CONTROL

Excessive vibrations can have adverse effects on personnel, equipment, and structures. Vibrations can annoy people, can interfere with their ability to perform or concentrate on mental tasks, can make it difficult for people to carry out precise movements or to make precise readings of instruments, and in extreme cases can lead to physical disabilities and injuries. Vibrating surfaces act somewhat like loudspeaker membranes and radiate sound, which again may range from annoying to painful to injurious, depending on its intensity. Vibrations may also produce such secondary effects as rattling of windows or motion of lights, which also tend to be annoying or distracting.

Vibration of a machine may reduce the life of its components, particularly those most highly loaded. Oscillatory stresses induced in machine parts, supports, building structures, and also in connections (hold-down bolts, pipes, cables) tend to produce failures of these items due to structural fatigue. Machine tools subjected to excessive vibrations may produce poor finishes; some precision equipment (optical systems, microscopes, gauges, microassembly equipment) cannot be effectively used at all in the presence of vibrations.

The need for vibration control occurs wherever there exist adverse effects due to vibrations. The amount of reduction that is required depends on the existing vibration and on what level of vibration is acceptable; zero vibration is as much an impossibility as an immovable object or an irresistible force. For many items of sensitive equipment, the manufacturers indicate what vibrations should not be exceeded. The vibration limits acceptable to people are available in handbooks.[2] In many other cases, unfortunately, no solid vibration criteria are available, so that one is forced to proceed by trial and error.

Control of vibration is a highly developed specialized branch of engineering, which has been the subject of numerous papers, many texts, and several handbooks. Many of these publications deal primarily with analyses or specialized problems and thus require study and interpretation before they can be put to practical use. The list of references included at the end of this chapter is limited to items believed to be most directly applicable to plant engineering.

DIAGNOSING A VIBRATION PROBLEM

It is usually convenient to consider any vibration problem in terms of: (1) the *source* of the undesirable vibrations, (2) the *receiver*, i.e., whatever is adversely affected by vibrations and requires protection, and (3) the *path* along which vibrations from the source reach the receiver. In practical situations, many sources often contribute to the vibrations experienced by a single receiver, and vibrations from a given source often reach a single receiver via several paths.

The most convincing approach to identify the vibration source responsible for a given problem consists of turning off all possible sources, then turning them on one at a time while observing the resulting effects at the receiver of interest. Similarly, one can best identify the predominant paths by interrupting one at a time or by disabling all and then reestablishing one at a time. These procedures can only rarely be carried out in practice to their full extent, but they can often be carried out sufficiently to provide valuable partial, if not full, insight into the problem.

It is often useful to supplement the on-off procedures discussed in the foregoing paragraph by comparison of the vibration spectra measured on or near various sources with the spectrum measured at the receiver. If the receiver, for example, experiences adverse vibrations only at 50 and 80 Hz, then a source that generates vibrations at only 30 Hz cannot be responsible for these receiver vibrations; one would look for sources that produce vibrations near 50 and 80 Hz. Similarly, measurements made along the important contributing paths would reveal 50- and 80-Hz components, whereas these components would be present to a lesser extent along the less significant paths.

There are also available *correlation methods* for source and path identification. These methods, however, require specialized equipment and expertise.

In dealing with any vibration problem, one must keep in mind the phenomenon called *resonance.* Any mechanical system or structure has a number of frequencies at which it can be set into vibration very easily; these are called *natural frequencies.* (The lowest of these, called the *fundamental natural frequency,* is often most easily excited and of greatest importance.) Resonance occurs if a system is subjected to a vibratory force or motion at one of its natural frequencies; large vibrations can then occur *even with small inputs.* The natural frequencies of a system can be determined readily; if a system is deflected and released, or if it is struck, it will vibrate at one or more of its natural frequencies. In order to be able to identify these, however, it is usually necessary to have the items of concern turned off, so that the natural frequencies will not be masked by the excitation frequencies.

A *stroboscope* is often useful for diagnostic purposes. This consists of a light that is made to flash at precisely timed intervals. The light is aimed at a vibrating part and the flashing frequency is adjusted manually until the vibrating part appears to stand still; the frequency of the vibration can then be read from the instrument. Then, by changing the flashing frequency slightly, one can observe the vibration in apparent slow motion to see where the largest excursions occur.

VIBRATION CONTROL STRATEGY[3]

It is generally best to control a vibration at its source, because this approach avoids problems at all potential receivers. However, in cases where only a limited number of receivers are of concern and where control at the source(s) is not feasible, control at the receiver(s) may be preferable. Reduction or elimination of vibrations at the source typically involves improving the dynamic balance of rotating or reciprocating equipment, substituting items with lesser vibrations for those with more (e.g., centrifugal pumps for reciprocating pumps), or changing operating speeds to eliminate resonance conditions. Reduction of the adverse effects of vibrations at the receiver generally involves substitution of less-vibration-sensitive items or processes, or adding stiffening or mass judiciously in order to eliminate resonances, if any are present.

Vibration *isolation* generally turns out to be the most cost-effective means for vibration control. Isolation involves insertion of soft flexible elements in the propagation path so as to reduce the transmitted forces and motions. Because of the multitude of paths that can begin at any source or terminate at any receiver, isolation is best accomplished near a source or receiver.

Some other vibration control methods that are useful only under certain specific circumstances are described at the end of this chapter.

Vibration Isolation at the Source

Basic Principles

The basic concepts of vibration isolation can be understood with the aid of the schematic sketch of Fig. 2-3, which shows a machine that generates a vertical oscillatory force of amplitude F (e.g., due to an imbalance) rigidly attached to a base, which in reality may be a machine, a foundation, an inertia block, or the floated floor of a building. This base, in turn, is mounted atop a supporting structure via a series of springs.

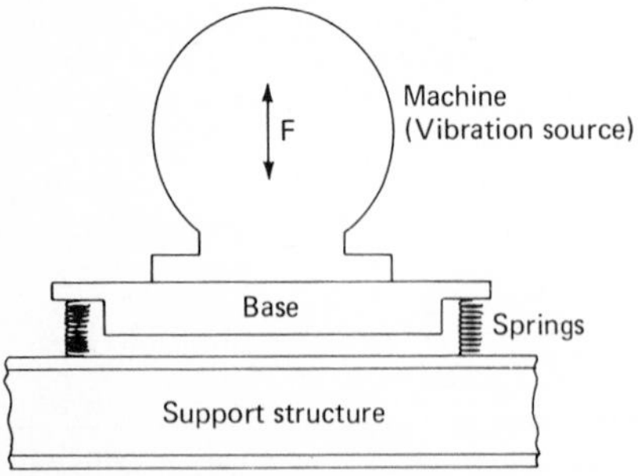

Figure 2-3 Conceptual sketch of isolation of vibration source.

The oscillatory force produces motion by accelerating the combined mass m of the machine and the base. This motion of the base produces oscillatory compression (and extension) of the springs (superposed on the static compression due to the weight they support), which in turn gives rise to oscillatory forces on the support.

If the force varies slowly, i.e., at low frequencies, then the inertia of the mass offers little opposition to the motion, and the force essentially acts directly to compress the springs. The machine-base mass here moves just enough for the total spring force to match the externally applied force, as it would if the force were applied statically, and thus the entire applied force is transmitted to the support structure. On the other hand, if the force F varies rapidly, i.e., at high frequencies, then the inertia of the mass opposes the motion to such an extent that the inertia's effect is much greater than that of the springs. The springs then compress and extend very little, and only the spring forces resulting from these small spring deflections are transmitted to the support structure.

The ratio of the amplitude F_S of the total force that the springs exert on the support (assumed rigid) to the amplitude F of the exciting force is called the *transmissibility T*. The transmissibility also is equal to F_S/F_R,* where F_S denotes the amplitude of the oscillatory force that acts on the support if the base is resiliently supported and F_R represents the corresponding oscillatory force that acts on the support structure if the base is rigidly fastened to it. Thus, the transmissibility also indicates the relative reduction of the force transmitted to the support structure (and of its motion) that results from the insertion of the isolation springs. The *isolation efficiency E*, which is defined by $E = 1 - T$ indicates what fraction of the exciting force is prevented from acting on the support, or, equivalently, by what factor the motion of the support is reduced due to use of the springs or springlike elements. For example, if a given isolation system results in a transmissibility of 0.05, then its isolation efficiency is 0.95, indicating that the use of the isolators reduces the vibrations of the support structure by 95 percent.

In order to have a reasonable beneficial effect, the springs (or other resilient elements) must be soft enough so that their total stiffness k is less than

$$k_{max}\ (\text{lb/in}) = \frac{m\ (\text{lb}) \times f^2(\text{Hz})}{40} \tag{2}$$

where m represents the total mass of the machine and base (in pounds) and f denotes the disturbing frequency (in hertz) of concern. For springs that have a straight-line force-deflection characteristic (the slope of which corresponds to the stiffness k)—for example, for steel coil compression springs that do not become solid at the greatest expected deflection — another way of specifying a soft enough system is to require that the *static deflection s* of the springs (that is, the deflection under the static load composed of the machine and base)

$$s\ (\text{in}) = \frac{m\ (\text{lb})}{k\ (\text{lb/in})}$$

be greater than

$$s_{min}\ (\text{in}) = \frac{40}{f^2\ (\text{Hz})} \tag{3}$$

If the resilient elements satisfy the maximum-stiffness (or minimum-static-deflection) requirement of the foregoing paragraph, then the transmissibility T and isolation efficiency E may be found from

$$T = 1 - E \approx \frac{10k\ (\text{lb/in})}{m\ (\text{lb}) \times f^2\ (\text{Hz})} = \frac{k}{4k_{max}} = \frac{s_{min}}{4s} \tag{4}$$

Practical Considerations

As evident from the foregoing expression, the lowest disturbing frequency f corresponds to the greatest vibration transmission. Therefore, an isolation system must be designed for the lowest disturbing frequency of concern.

In general, the vibration transmission can be reduced in two ways: (1) by using softer resilient elements (reducing the total stiffness k) or (2) by increasing the supported mass

*This equality holds only if the force produced by the source is unaffected by the degree to which the motion is restrained.

m. It should be noted that the use of softer springs (beyond the $k < k_{max}$ requirement) leads to greater static deflection, but produces practically no change in the vibratory motion of the base. Because the static deflection is proportional to the supported mass and the motion of that mass is inversely proportional to the mass, an increase in that mass results in increase of static deflection and in a reduction of the vibratory motion of the machine and base.

It is important to note that useful isolation can be achieved only if the resilient elements (springs) are considerably softer than the supporting structure, i.e., only if the resilient elements deflect considerably more under a given load than does the support structure. Otherwise, the support structure provides the predominant resilience and the springs merely serve to transmit the forces to the support essentially without attenuating them.

Selection of an isolation system must account for the fact that an excitation frequency *f* that is produced by a machine usually varies with the machine's rotational speed. As a machine is brought up to speed or slowed to a stop, a speed at which the exciting frequency matches the natural frequency of the machine on its resilient supports may be encountered. At this speed, which may be estimated from

$$N_r \approx 188 \sqrt{\frac{k\ (\text{lb/in})}{m\ (\text{lb})}} \approx \frac{188}{\sqrt{s\ (\text{in})}} \tag{5}$$

there may occur intense vibrations. Their magnitude depends on how fast the machine passes through this resonance speed and on the damping characteristics of the isolation system. If the machine accelerates or decelerates rapidly, vibrations do not have time to buildup. For machines that accelerate or decelerate slowly, the magnitude of the vibration produced at resonance is inversely proportional to the damping in the system.

Figure 2-4 is a convenient chart for estimating the major isolation system parameters one needs in order to achieve a desired vibration reduction or transmissibility. The dashed line in the figure illustrates the case where there exists a disturbing frequency of 3000 r/min (corresponding to 50 Hz), at which it is desired to reduce the vibration by 99.9 percent, i.e., to 0.01 times its original value. The chart shows that here a value for k/m of about 2.5 (lb/in)/lb is needed; thus, for a 2000-lb mass, a total stiffness of $2.5 \times 2000 = 5000$ lb/in is required. (Of course, a lesser stiffness would provide better isolation than that prescribed, whereas a greater stiffness would result in poorer isolation.) One may also read from the chart that to $k/m = 2.5$ (lb/in)/lb there corresponds a static deflection of about 0.4 in and a resonance frequency of about 300 r/min or 5 Hz.

Damping refers to a spring's or a structure's capability for dissipating oscillatory energy; a bell or a steel spring rings (i.e., vibrates) for a long time after it is struck; that is, it takes a relatively long time to dissipate the vibratory energy imparted to it. On the other hand, a rubber or cork rod vibrates only briefly after an impact; it dissipates vibratory energy rapidly and thus is highly damped. For machines that accelerate or coast down slowly, isolation elements of highly damped materials (e.g., rubber or cork) should be used, or energy-dissipation devices (e.g., friction pads or dashpots) should be added in parallel with the resilient members. In some cases, the incorporation of snubbers in the isolation systems may suffice to limit the excursions of the base as the machine passes through resonance. Such snubbers may, for example, be in the form of rubber cones that are mounted so that the base bumps against them when its excursion exceeds a given amount.

If slow acceleration or coast-down is no problem, then the type of isolator material and the isolator configuration essentially make no difference; the only parameter that counts is the total stiffness *k* under the expected load conditions and operating frequencies. For many resilient elements, particularly metal springs, the stiffness is practically independent of frequency and also of load within the design load range. Whenever vibrations in more than one direction (e.g., in the horizontal, as well as the vertical direction) are of concern, however, the proper resilience for isolation in all directions must be pro-

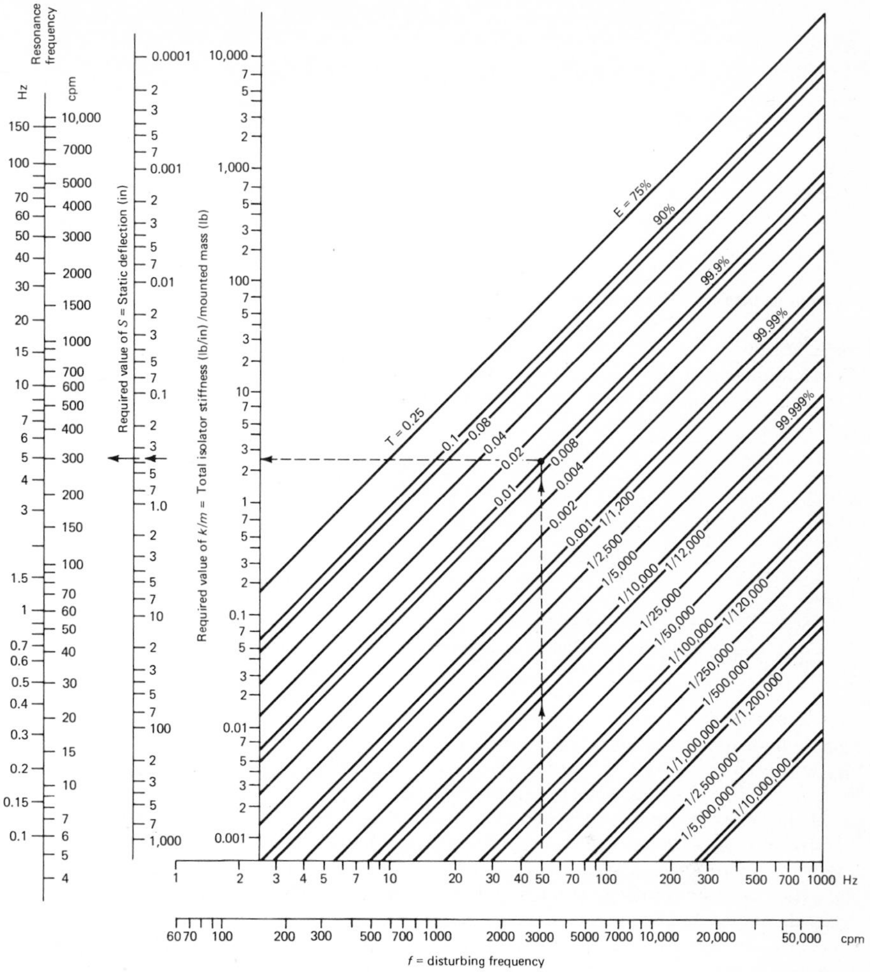

Figure 2-4 Chart for estimation of isolation-system requirements.

vided. In this case, appropriately selected and aligned springs, specially configured rubber pads, or suitably chosen commercial isolators should be employed.

Because of the beneficial effect of increased mass of the machine base, it usually is desirable to mount several machines on the same base. In this way, the vibration transmitted from each machine is attenuated by the mass of the base and also by the masses of the other machines. If such a common base is to be effective, it must not have any resonances of its own at or near the operating speeds of all machines mounted on it; ideally, its fundamental resonance should occur above all operating frequencies. This implies that such bases should be as stiff as possible.

Attention must also be given to avoidance of excessive rocking vibrations of isolated bases. For this purpose it is usually useful to employ wide bases, so that the springs act with large moment arms. It is also beneficial, particularly for bases supporting several items, to have the springs distributed so that they "pick up" the loads locally, e.g., so as to have stiffer or more closely spaced springs near the heavier items and less stiff or more widely spaced springs near the lighter items.

In practice one usually needs to make provisions to avoid the transmission of vibrations via paths that bypass the isolated base. For this purpose it usually is desirable to provide flexible sections of piping, electrical conduit, cable, and ducts between the isolated machines and the unisolated surroundings. (Appropriate devices especially designed for vibration isolation are commercially available; devices such as loops and bellows that are primarily designed to accomodate thermal expansion rarely are adequate for vibration isolation.) Similarly, hold-down bolts through isolators, and similar arrangements that in effect serve as vibration "short-circuit" paths around the isolators, must be removed or themselves isolated, e.g., with soft rubber sleeves and washers.

Isolation of Vibration-Sensitive Items

Basic Principles

The fundamental concepts pertaining to isolation of items to be protected from vibrations may be visualized with the aid of the schematic sketch of Fig. 2-5. This shows a

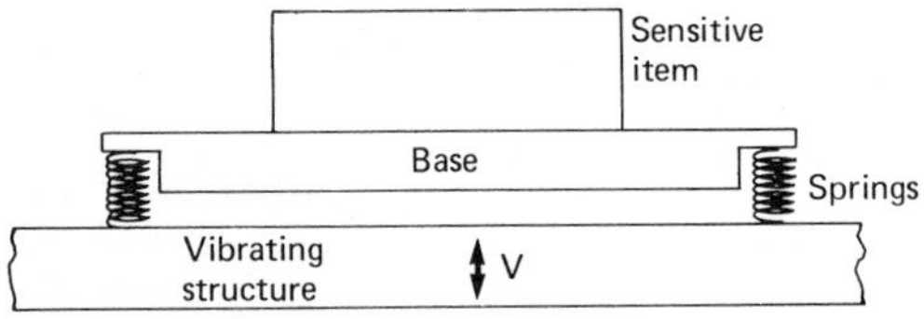

FGigure 2-5 Conceptual sketch of isolation of a sensitive item.

vibration-sensitive item attached to a base, which is fastened to a vibrating structure (e.g., a floor of a building housing vibration sources) via resilient elements, represented in the sketch by springs. In reality, the base may consist of a machine frame or foundation, an inertia block, or a floated (isolated) slab or floor.

If the sensitive item is attached to the vibrating structure directly, without springs, then it vibrates with the same amplitude as the vibrating structure. The same is true if the attachment springs are stiff. However, if the springs are soft, then the oscillatory deflection of the vibrating structure leads to only relatively small forces acting on the base; the inertia of the mass of the base resists acceleration and thus keeps the resulting motion small. This inertia effect is more pronounced for larger accelerations, i.e., higher frequencies, and therefore better isolation is obtained at higher frequencies, all other things being equal.

The ratio of the amplitude of the motion V_I (displacement, velocity, or acceleration) of the sensitive item to that of the vibrating structure, V (measured in terms of the same quantity as is V_I), is called the *motion transmissibility* T_v, or just *transmissibility*.* This transmissibility also turns out to be equal to V_R/V_I, where V_R represents the amplitude of the motion that the sensitive item experiences if it is rigidly fastened to the vibrating structure and V_I represents the corresponding amplitude that the sensitive item experiences if the isolators (springs) are used.† Thus, the motion transmissibility indicates the relative reduction in the motion of the sensitive item that results from use of the isolation elements (e.g., springs).

*It is somewhat unfortunate that the same word, "transmissibility," is used to refer to the force transmissibility T associated with source isolation as is used to refer to the motion transmissibility T_v associated with receiver (sensitive-item) isolation. However, because T and T_v obey the same relation, this usage tends to cause little confusion.

†This equality holds only if the motion of the vibrating structure is unaffected by the magnitude of the spring forces that act on that structure.

The *isolation efficiency** E_v is defined by $E_v = 1 - T_v$ and thus indicates what fraction of the motion of the vibrating structure is kept from the sensitive item. For example, if $T_v = 0.01$, then the isolation efficiency is 99 percent; the sensitive item vibrates only 1 percent as much as it would if it were rigidly attached to the vibrating structure.

In order for the isolation system to be significantly effective, the resilient elements must be soft enough so that their total stiffness k is less than k_{max} as given by Eq. (2), with m here representing the total mass of the sensitive item and its base and f denoting the lowest frequency of concern. For resilient elements, such as steel coil springs, whose force-deflection characteristic plots as a straight line, the "soft enough spring" requirement corresponds to the requirement that the static deflection s of the springs under the total static weight of the sensitive item and its base be greater than s_{min} as given by Eq. (3). If the resilient elements satisfy these requirements, then the motion transmissibility T_v is given by the same expression as is the force transmissibility T, namely, Eq. (4).

Practical Considerations

Because the lowest disturbing frequency f corresponds to the greatest transmissibility T, an isolation system must be designed with respect to the lowest disturbing frequency of interest.

Figure 2-4 is a convenient chart for estimating the major isolation system parameters needed in order to achieve a desired vibration reduction or transmissibility. The dashed line in the figure illustrates the approach for a case where there exists a disturbing frequency of 3000 r/min (corresponding to 50 Hz). At this frequency it is desired to reduce the vibration by 99.9 percent, i.e., to 0.01 of its original value. The chart shows that here a value of k/m of about 2.5 (lb/in)/lb is needed; thus, for a 2000-lb mass, a total stiffness of $2.5 \times 2000 = 5000$ lb/in is required. (Of course, a lesser stiffness would provide better isolation than that prescribed, whereas a greater stiffness would result in poorer isolation.) One may also read from the chart that to $k/m = 2.5$ (lb/in)/lb there corresponds a static deflection of about 0.4 in and a resonance frequency of about 300 r/min or 5 Hz.

Once the resilient elements are soft enough to satisfy the requirements indicated in the foregoing paragraph, inprovement in the isolation (reduction in the transmissibility) can be obtained by (1) using softer springs and/or (2) increasing the mass of the base. Reducing the spring stiffness by a given factor has the same effect on reducing the motion of the isolated item as increasing the combined mass of the item and its base by the same factor. Use of softer springs is usually less difficult and less costly than obtaining significantly increased base mass; thus, use of softer springs is generally preferable. However, softer springs are subject to greater static deflection and also tend to permit the isolated item to vibrate or to rock more due to motions of its own parts (e.g., of pistons, slides, spindles). In order to minimize this rocking and secondary vibration, it is often useful to balance the isolated items or to distribute the springs so as to oppose the dynamic loads directly (by placing more or stiffer springs where the greater loads act and having the spring force axes aligned with the direction of the loads).

In all cases, in order to be able to provide useful isolation, the resilient elements (springs) must be more flexible than the vibrating structure to which they are attached and the base supporting the sensitive item. That is, for a given force applied at the center of gravity of the isolated item, the resilient elements must deflect more than either one of the structures to which they are attached. The base of the sensitive item, in addition to being as heavy as possible, should be as stiff as possible.

It often is advantageous to have several sensitive items mounted on the same base; then the total mass of all items and the base act together to assist in the isolation of each item. In this case it is important that the base be as rigid as possible, so that all masses are well interconnected dynamically. In selecting items to be mounted on a common base, it must be kept in mind that items rigidly fastened to the same base can transmit vibra-

*Again, "isolation efficiency" is common usage, although *motion isolation efficiency* is more appropriate.

tions and rocking motions to each other; thus it is generally inadvisable to mount extremely sensitive items together with items that may cause dynamic disturbances.

Any connections between the isolated item or its base and the vibrating structure must be less stiff than the resilient (spring) elements if these connections are not to transmit more vibration than the springs. Thus, care must be taken to use flexible bellows, tubing, hoses, conduits, cables, etc., with coils and loops where appropriate, to reduce the vibration transmission via these connections.

Some Specialized Control Techniques

Two-Stage Isolation

In the cases illustrated by Figs. 2-2 and 2-3, a rigid connection is implied between the base and the equipment it supports. If resilient connections (e.g., rubber pads or springs) are used instead, then there are two stages of isolation: one below the base and one above it. Two-stage isolation systems can provide increased isolation at high frequencies, but they introduce additional system resonances at which there may occur large motions and dynamic forces. Careful isolation-system design is important to avoid these potential adverse effects.

Structural Damping

The large vibrations that occur if any system or structure vibrates at a resonance may best be reduced by *detuning*—changing the excitation or the system so as to avoid operation at resonance. If such changes cannot be made—either for operating reasons or because a system is subject to multifrequency excitation or because it may have a multitude of resonances at closely spaced frequencies—then increasing the damping constitutes essentially the only means for obtaining useful vibration reductions.

For thin metal panels, increased damping can usually be obtained most conveniently by means of layers of viscoelastic (tacky, rubbery) damping material that may be sprayed, troweled, or glued onto the metal panels. A variety of such materials is commercially available, and suppliers of these materials have experience and data pertaining to desirable material thicknesses.

For thick plates of metal or concrete, or for beams, it is often useful to employ a viscoelastic damping material as the middle layer of a sandwich configuration, with the outer layers consisting of the structural material. Careful design of such damped sandwich configurations, based on well-known but somewhat intricate specialized procedures, is required if these are to work properly.

In cases where large structures vibrate severely, it is sometimes useful to fasten boxes filled with sand or some other granular material at the locations where the greatest motions occur. Agitation of grains—and their "rattling" against each other and against the sides of the box—extracts energy from the vibration, thus reducing it.

Vibration Absorbers

Vibration absorbers, often also called *dynamic absorbers* or *dampers,* are useful primarily where a single fixed-frequency component is of concern. Such an absorber consists of a mass that is attached to a vibrating system via a spring, with the mass and spring values chosen so that the resonance frequency of this absorber system coincides with the frequency of concern. When this mass-spring system is attached to the base or housing of the vibrating machine or to the vibrating structure, it acts to reduce the vibratory motion of the item to which it is attached. The motion of the added mass is opposite to that of the structure to which it is attached; thus, the mass makes the spring push down on the vibrating structure when the latter moves upward, and vice versa.

For a vibrating component or structure with an attached absorber, the product of the absorber mass and its excursion amplitude needs to be equal to the product of the mass of the component and its excursion. Thus, absorbers with small masses need to operate at large excursions; trading off between absorber masses and excursions is part of the design process.

REFERENCES AND BIBLIOGRAPHY

1. Harris, C. M. (ed.) *Handbook of Noise and Vibration Control,* 2d ed., McGraw-Hill, New York, 1979.
2. Harris, C. M., and C. E. Crede (eds.): *Shock and Vibration Handbook,* 2d ed., McGraw-Hill, New York, 1976.
3. Harris, C. M.: "Vibration Control Principles," Chap. 19 of Ref. 3.

Den Hartog, J. P.: *Mechanical Vibrations,* 4th ed., McGraw-Hill, New York, 1956.

Peterson, A. P. G., and E. E. Gross, Jr.: *Handbook of Noise Measurement,* 7th ed., General Radio Co., Concord, Mass., 1972. Explains noise and vibration, their effects and their measurement and outlines vibration control approaches.

Tonndorf, J., H. E. von Gierke, and W. D. Ward: "Criteria for Noise and Vibration Exposure," Chap. 18 of Ref. 3.

Ungar, E. E., and R. Cohen: "Vibration Control Techniques," Chap. 20 of Ref. 3.

section 13

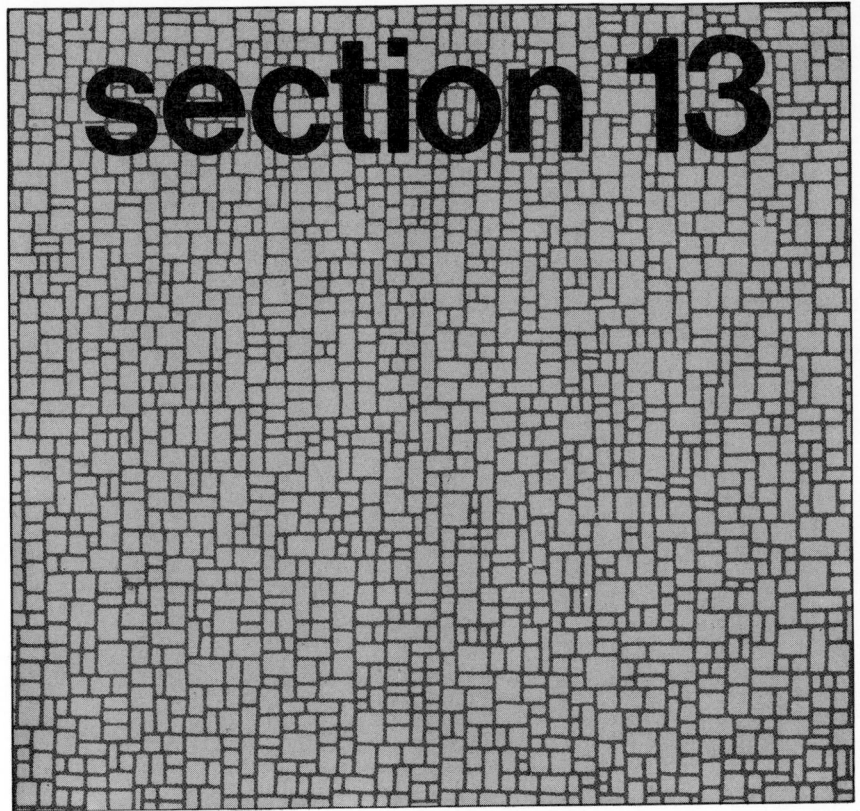

Pollution Control and Waste Disposal

chapter 13-1

Air-Pollution Control

by
Karl J. Danz, P.E.
Principal Engineer

Richard B. Ruch, Jr.
Project Manager

Benjamin A. Schranze, P.E.
Roy F. Weston, Inc.
West Chester, Pennsylvania

REGULATORY REQUIREMENTS

Although considerable federal and state legislation concerning air quality was developed during the 1960s, the primary statutory framework now in place was established by the Clean Air Act of 1970. This legislation was amended in 1974 to incorporate a program to ensure the prevention of significant air quality deterioration (PSD) in existing "clean" air regions; it was amended again in 1977 to include provisions which formalize the permitting of new (or to be modified) facilities. The Act was up for reauthorization in 1981, but will probably not be amended in final form until 1982.

TABLE 1-1 National Ambient-Air Quality Standards: Primary Standards, $\mu g/m^3$

Pollutant	Annual standard	Short-term standard	Short-term time, h
Sulfur dioxide (SO_x)	80	365	24
Particulates	75	260	24
Carbon monoxide (CO)	—	10 000	8
		40 000	1
Ozone (O_3)	—	240	1
Hydrocarbons (HC)	—	160	3 (6–9 A.M.)
Nitrogen dioxide (NO_x)	100	—	
Lead (Pb)	1.5 (3 mo)	—	

Regulations and Standards

National Ambient Air Quality Standards

The National Ambient Air Quality Standards (NAAQS) define the quality of air which must be achieved to prevent adverse effects. The primary air quality standards specify levels of pollution which cannot be exceeded without threatening adverse effects on human health. The secondary air quality standards set limits not to be exceeded without adverse effects on public welfare, including property, vegetation, and other influences. For each criterion pollutant there is a set of these standards, each with its own regulatory program to limit atmospheric concentration levels at or below the levels set by its standards. Refer to Table 1-1.

The secondary standards are identical to these primary standards, except for sulfur dioxide (for which there is a 3-h, short-term secondary standard of 1300 $\mu g/m^3$) and particulates (for which there is an annual secondary standard of 60 $\mu g/m^3$ as a guide to assessing achievement of the 24-h, short-term secondary standard of 150 $\mu g/m^3$).

State Implementation Plan

The State Implementation Plan (SIP) provides a scheme under which the National Air Quality Standards are expected to be achieved and maintained. These plans, which must be approved by federal EPA, provide comprehensive programs to reduce pollution through a variety of specific abatement measures, some of which are delineated in company-specific compliance programs. Many deficiencies in the SIPs are apparent, and modifications can be made, using specific legal procedures set up by EPA.

Air Quality Control Region

The Air Quality Control Region (AQCR) is a basic geographic delineation of areas having specific pollution control programs. There are 247 AQCRs in the United States in which air quality is defined by ambient-air monitoring data, estimates made from mathematical dispersion analyses, or both. Each AQCR has its own program, specified in the SIP, which has been developed to maintain acceptable air quality.

New Source Performance Standards

The New Source Performance Standards (NSPS) are an indirect way of limiting emissions from new sources and improving air quality.These standards now cover a number of basic industrial and process categories. A new plant is subject to the NSPS only if these standards are proposed by the authorities before the plant is under construction. Thus, before proceeding with a new plant, a company should determine which categories are covered by the NSPS. Industries affected by the NSPS are shown in Table 1-2. Industrial categories scheduled for NSPS development are listed in Table 1-3.

Prevention of Significant (Air Quality) Deterioration

Prevention of significant (air quality) deterioration (PSD) is a regulatory program requiring preconstruction approval of new plants which may have significant emissions levels. In some states that have not adopted their own PSD-permitting program, the federal EPA Regional Office has the reviewing and approval authority. The major aspects of this program include:

- **Area Classification System.** All areas in the United States undergo an area classification scheme. This permits moderate amounts of industrial development routinely, but it does not allow air quality degradation to the point where air quality standards are approached or exceeded. The *Class I* category involves pristine areas subject to the tightest control. *Class II* covers areas of moderate growth. *Class III* covers areas of major industrialization.
- **Increments of Air Quality.** Numerical limitations on the additional pollution which may be allowed above existing baseline concentrations of sulfur dioxide and total suspended particulates.
- **Use of Best Available Control Technology (BACT).** Each major new plant must install a type of pollution control device which has been deemed acceptable for the particular discharge involved. This is determined on a case-by-case basis, but EPA provides guidelines concerning acceptable control methods.
- **Preconstruction Review.** A company is prohibited from commencing construction on a new source until a review by the authorities has been completed and an opportunity has been provided for public response on any disputed item.

It is likely that the PSD regulations contained in the Clean Air Act will be significantly amended in 1982.

Major Emitting Facilities

Plants subject to PSD review include all major emitting facilities. A major emitting facility is defined as any one of 28 source categories (Table 1-4) that have the potential to emit 100 tons/year or more of any pollutant regulated under the Clean Air Act. Any other source not on the list of 28 source categories, but having the potential to emit 250 tons/year or more of any regulated pollutant, is included in the definition of major emitting facilities. These emission cutoffs apply to both new sources and modifications to existing sources. Modified major stationary sources that increase the potential to emit greater than the significant limits shown in Table 1-6 are subject to PSD review.

Sources subject to PSD require:

- Monitoring of ambient air quality or the use of available representative data
- Demonstration that the plant will not violate the applicable PSD increment or any other air quality standard, based on ambient air quality modeling analysis
- Installation of BACT
- Commitment to conduct postconstruction monitoring

TABLE 1-2 Industries Affected By New Source Performance Standards Issued Under the Clean Air Act

Steam generator	Primary zinc smelter
Municipal incinerator	Primary lead smelter
Portland cement plant	Primary aluminum reduction plant
Nitric acid plant	Coal cleaning plant
Asphalt concrete plant	Lime plants
Petroleum refinery	Grain elevators
Petroleum storage	Kraft pulp mills
Secondary lead smelter	Lignite-fired steam generators
Secondary brass and bronze smelter	Sulfur-recovery plants and refineries
Iron and steel mill	Electric steam generators
Sewage treatment plant	Primary aluminum reduction plants*
Ferroalloy production	Stationary gas turbines*
Phosphate fertilizer	Internal combustion engines*
Primary copper smelter	Glass manufacturing

*Proposed

TABLE 1-3 Industrial Categories for Which New Source Performance Standards Are To Be Developed*

Stationary fuel combustion
14. Stationary internal combustion engines

Metallurgical processes
10. By-product coke ovens
23. Foundries: gray iron
41. Foundries: steel
42. Secondary aluminum
20. Secondary copper
66. Secondary zinc
67. Uranium refining

Mineral products
57. Asphalt roofing
49. Brick and related clay products
60. Castable refractories
58. Ceramic clay
48. Fiberglass
38. Glass
45. Gypsum
19. Metallic mineral processing
13. Mineral wool
18. Nonmetallic mineral processing
64. Perlite
21. Phosphate rock preparation
43. Sintering: clay and fly ash

Polymers and resins
54. ABS-SAN resins
12. Acrylic resins
50. Phenolic resins
62. Polyester resins
30. Polyethylene
55. Polypropylene
53. Polystyrene
51. Urea-melamine resins

Food and agricultural
68. Alfalfa dehydrating
44. Ammonium sulfate
59. Ammonium nitrate fertilizer
69. Animal feed defluorination
63. Starch
70. Urea (for fertilizer and polymers)
27. Vegetable oil

Waste incineration
11. Incineration: industrial-commercial

Basic chemical manufacture
1. Synthetic organic chemical manufacturing
61. Borax and boric acid
47. Hydrofluoric acid
65. Phosphoric acid: thermal process
40. Potash
46. Sodium carbonate

Chemical products manufacture
52. Ammonia
2. Carbon black
31. Charcoal
71. Detergent
17. Explosives
7. Fuel conversion
34. Printing ink
35. Synthetic fibers
28. Synthetic rubber
29. Varnish

Evaporative Loss Sources
6. Dry cleaning
9. Graphic arts
15. Industrial surface coating: autos
3. Industrial surface coating: cans
8. Industrial surface coating: fabric
37. Industrial surface coating: large appliances
32. Industrial surface coating: metal coils
5. Industrial surface coating: paper

Petroleum industry
25. Crude oil and natural gas production
72. Gasoline additives
4. Petroleum refinery: fugitive sources
33. Transportation and marketing

Wood processing
24. Chemical wood pulping: acid sulfite
22. Chemical wood pulping: neutral sulfite (NSSC)
36. Plywood manufacture

Consumer products
56. Textile processing

Minor source categories
Lead acid battery manufacture
Solvent metal cleaning (degreasing)
Industrial surface coating: metal furniture

*This list was issued under Section 311 of the Clean Air Act by EPA on 31 August 1978 (43 *Federal Register* 38872-77). The numbers were assigned by EPA to reflect priorities, the lowest numbers indicating highest priority.

- Public hearing
- Issuance of construction permit

Source applicability under PSD review must be determined for each new or modified facility with a potential increase of emissions of any regulated pollutants. Because of the complexity of these regulations, it is suggested that a qualified engineer familiar with the regulations be consulted.

TABLE 1-4 Major Emitting Facilities

1. Fossil-fuel-fired steam electric plants of more than 250,000,000 Btu per hour of heat input	15. Phosphate-rock processing plants
2. Coal-cleaning plants (with thermal dryers)	16. Coke-oven batteries
3. Kraft pulp mills	17. Sulfur-recovery plants
4. Portland cement plants	18. Carbon black plants (furnace process)
5. Primary zinc smelters	19. Primary lead smelters
6. Iron and steel mills	20. Fuel-conversion plants
7. Primary aluminum ore reduction plants	21. Sintering plants
8. Primary copper smelters	22. Secondary metal production facilities
9. Municipal incinerators capable of charging more than 250 tons of refuse per day	23. Chemical process plants
10. Hydrofluoric acid plants	24. Fossil-fuel boilers of more than 250,000,000 Btu per hour of heat input
11. Sulfuric acid plants	25. Petroleum storage and transfer facilities with a capacity exceeding 300,000 barrels
12. Nitric acid plants	26. Taconite ore processing facilities
13. Petroleum refineries	27. Glass fiber processing plants
14. Lime plants	28. Charcoal production facilities

Nonattainment Area

A nonattainment area is an area where any ambient-air quality standard is being violated. The 1977 Amendments to the Clean Air Act require each state to tighten abatement requirements to assure compliance of all primary standards by 1982. Extensions to 1987 are allowed for photochemical oxidants (ozone) and carbon monoxide. Unless SIP revisions have been approved by EPA which reflect the method to eventually attain air quality standards, no new sources may be constructed in such areas. In areas covered by SIP revisions, any new plant (or modification to an existing plant) which can emit more than a specified emission level will require, for permitting purposes:

- **Air Quality Impact.** Sources located in attainment area cannot significantly impact a nonattainment area above established threshold concentrations (Table 1-5).
- **Emission Offset.** Reductions from existing sources which exceed the anticipated emissions from the new source are required.
- **Use of Lowest Achievable Emission Rate (LAER) Control Technology.** The most stringent emissions limitation scheme will be required for the new facility. The LAER requirement is determined individually, but unlike BACT does not need to take other factors into account (such as capital and outside maintenance costs).
- **Other Sources in Compliance.** All other plants in the state owned by the same company must be on approved emissions compliance schedules.
- **Assurance.** State assurance that the applicable SIP is being carried out.

Offsets and LAER are required for those pollutants whose potential emissions will equal or exceed 50 tons/year for all pollutants except carbon monoxide, which is 100 tons per year.

TABLE 1-5 Threshold Increases in Ambient-Air Concentrations for Nonattainment Areas

	Averaging time				
Pollutant	Annual	24 hours	8 hours	3 hours	1 hour
SO_2	1.0 μg/m^3	5 μg/m^3		25 μg/m^3	
TSP	1.0 μg/m^3	5 μg/m^3			
NO_2	1.0 μg/m^3				
CO			0.5 mg/m^3		2 mg/m^3

Air Quality Models

Air quality dispersion models are analytical tools used by industry and regulatory authorities to simulate the air quality impact of individual or multiple emission sources. It is recognized that, in the absolute sense, the performance of these air quality simulation models is generally poor. In a relative sense, the models provide a creditable indication of air quality impact over various averaging periods of concern. However, these models, which must be approved by the appropriate regulatory agencies prior to use, constitute one of the few available means of evaluating a source's compliance with PSD increments or NAAQS in the absence of measured air quality data in the vicinity of an emission source.

The U.S. EPA has specified certain basic modeling procedures in order to develop consistency in the application and use of the various EPA-approved models. The actual model selected and used will depend on a variety of factors including the source type and emission characteristics. These procedures are described and a list of the EPA-approved models is given in the EPA publication, *Guideline on Air Quality Models*, issued in April 1978 and revised in December 1980. Because each case may have some special or unique considerations such as proximity to complex terrain, stack downwash, etc., it may be in your best interest to consult an individual with more intimate familiarity with details of such modeling procedures.

Preliminary Screening Analysis

In order to reduce the number of cases in which more sophisticated and expensive modeling analysis is required, EPA allows the use of preliminary screening tests to isolate those cases where there appears to be little prospect of exceeding applicable the terms of NAAQS. The screening tests use conservative estimates of emission(s) characteristics and diffusion phenomenology in simple algorithms. If the screening computation shows that emissions from the proposed new source would not exceed the significant impact levels or one-half the PSD increments, the ambient-air quality analysis is considered satisfied and no further modeling is required.

Procedure in Modeling

If more sophisticated modeling is required, the following steps are required.

- Define *baseline* ambient air quality which corresponds to conditions existing on August 7, 1977. In some cases, a hypothetical baseline air quality must be projected for the specific locality of a proposed new plant, using representative data which may be available from ambient-air monitoring stations in the same general area. Data from the most recent 3-year period should be used.
- Determine to what extent the *increment* above the baseline available for new growth may have been consumed by prior PSD applicants. Major source construction commencing after August 7, 1977 cannot be included in the baseline, but must be counted against the increment. Additional emissions caused by fuel switches (by coal conversion or gas curtailment) can also be counted against the PSD increments.
- Determine the incremental impact of all significantly emitted pollutants. In this case, consideration will be limited not only to the computation of the worst short-term and annual concentrations but also to estimating the effect on nearby nonattainment or Class I areas. If concentrations exceed the significant impact levels (Table 1-6), then an emission offset/LAER analysis must be conducted. If it is predicted that concentrations in the attainment area will be in excess of "de minimis" levels, ambient-air monitoring may be required. (See Table 1-7.)

Recommended Models

EPA has recommended various models which it feels have had acceptable validation and which will provide replicate results. For convenience, EPA has made numerous air quality simulation model codes available to anyone wanting to make modeling estimates; these codes are collectively referred to as the *Users' Network for Applied Modeling of Air Pollution (UNAMAP)*. They are available on nine-track magnetic tape from the

TABLE 1-6 Significant Net Emissions Increase

Pollutant	Increase, ton/yr
CO	100
NO_x (as NO_2)	40
SO_2	40
Particulate matter	25
Ozone	40 of volatile organic compounds
Lead	0.6
Asbestos	0.007
Beryllium	0.0004
Mercury	0.1
Vinyl chloride	1
Fluorides	3
Sulfuric acid mist	7
Hydrogen sulfide	10
Total reduced sulfur	10
Reduced sulfur compounds	10
Other CAA pollutants	> 0

TABLE 1-7 Concentration Impact Values Above Which Ambient-Air Monitoring May Be Required

Pollutant	Concentration, $\mu g/m^3$	Avg. time (max), h
CO	575	8
NO_2	14	24
TSP	10	24
SO_2	13	24
Lead	0.1	24
Mercury	0.25	24
Beryllium	0.0005	24
Fluorides	0.25	24
Vinyl chloride	15	24
Total reduced sulfur	10	1
H_2S	0.04	1
Reduced sulfur compounds	10	1

National Technical Information Service (NTIS), Springfield, Va. Some of the available models include:

- **CDM (climatological dispersion model).** This determines long-term (seasonal or annual) quasistable pollutant concentrations at any ground-level receptor using average emission rates from point and area sources and a joint frequency distribution of wind direction, wind speed, and stability for the same period.
- **PTPLU (point-plume).** An interactive program which performs an analysis of the maximum short-term concentrations from a single point source as a function of stability and wind speed. The final plume height is used for each computation.
- **PTDIS (point-distance).** An interactive program which estimates short-term concentrations directly downwind of a point source at distances specified by the user. The effect of limiting vertical dispersion by a mixing height can be included, and gradual plume rise to the point of final rise is also considered. An option allows the calculation of isopleth half-widths for specific concentrations at each downwind distance.
- **PTMTP (point-multipoint).** An interactive program involving multiple point sources for a number of arbitrarily located receptor points at or above ground level. Plume rise is determined for each source. Downwind and crosswind distances are determined for each source-receptor pair. Concentrations at a receptor from various sources are assumed additive. Hourly meteorological data are used; both hourly concentrations and averages over any averaging time from 1 to 24 hours can be obtained.

Other Useful Models

Other models which have been found useful and are acceptable to the regulatory authorities include:

- **Valley Model.** A point-source model which is applicable to some complex terrain situations
- **CRSTER.** Provides a refined analysis of a point source where there are no significant meteorological or terrain complexities
- **Multiple-point Terrain (MPTER).** Similar to CRSTER, except that it is capable of simulating the air quality impact of multiple point sources at more than one location.
- **Texas Climatological Model (TCM).** A multisource model used to evaluate the long-term impacts of complex source configurations
- **RAM.** A multisource model used to evaluate the short-term impacts of complex sources on air quality
- **Texas Episodal Model (TEM).** Similar in capability to RAM
- **Industrial Source Complex (ISC).** A multisource model which is capable of dealing with the aerodynamic downwash of effluent from a stack that does not meet "good engineering practice" (GEP) requirements as well as the impact from a variety of point, area, and line sources.

Good Engineering Practice (GEP) Stack Height Requirements

In October 1981, EPA proposed revisions to stack height regulations which were originally proposed in January 1979 (Federal Register, October 7, 1981). The newly proposed revisions redefined a variety of terms including "stack" and "establishing a de minimis stack height."

Under the revised regulations, a stack is any point in a source designed to emit solids, liquids, or gases into the air, including a vent or duct. Flares have been omitted from this definition.

The previously proposed regulations defined a minimum stack height (i.e., the distance from the ground-level elevation of a plant to the elevation of the stack outlet) as 100 ft (30 m). Otherwise, for a source to get credit for that protion of its stack height built higher than this level in order to avoid excessive concentrations caused by nearby structures, a GEP stack formula was specified. The formula equals the stack height (H_g) to the height of the structure (H) plus one and one half times the height or width of the structure (L), whichever is less ($H_g = H + 1.5L$). The formula was derived from the traditional engineering practice of building a stack to approximately two and one-half times the height of a nearby structure.

Like the previously proposed regulations, the newly revised regulations do not limit the physical stack height of any source. See also Chap. 4-7. Instead, they set limits on the stack height to be used in ambient-air quality modeling for the purpose of setting an emission limitation.

Under the newly proposed regulations, EPA has defined a GEP stack height as the greater of (1) 65 m; or (2) $H_g = H + 1.5L$, where H_g = GEP stack height, H = height of nearby building structure(s), and L = lesser dimension (height or projected width) of nearby building structure(s).

A stack height less than GEP should be evaluated to ensure that excessive concentrations will not routinely occur offsite and cause the potential of exceeding any NAAQS or PSD increment.

Air Quality Monitoring

Before beginning any monitoring program, the applicant must obtain approval from federal EPA or the state. This prior approval is necessary to prevent possible delays in the application review caused by collection of unacceptable data.

Consideration should be given to the effects of existing sources, terrain, meteorological conditions, existence of fugitive or re-entrained dusts, etc. The number of sites is directly related to the expected spatial variability of the pollutant in the area(s) of study. The suggested approach is first to use appropriate dispersion modeling techniques to estimate the air quality impact of the proposed source and any existing sources within the impact range of the proposed source for each pollutant. The modeled pollutant contribution of the proposed source should be analyzed in conjunction with the model results for existing sources to determine the general location of maximum concentration. Monitoring should then be conducted in or as close to these areas as possible. Generally, one to four sites would be sufficient in most situations with multisource settings. Preconstruction monitoring may be required for periods ranging from 4 to 12 months.

Air Quality Monitoring Equipment

As specified by EPA, all ambient-air quality monitoring must be done with continuous reference or equivalent methods, with the exception of total suspended particulate (TSP) matter for which continuous reference or equivalent methods do not exist. For TSP, samples must be taken in accordance with the reference method which is normal technique.

The TSP reference method is described in 40 CFR 50. The identification of continuous reference or equivalent methods for SO_2, CO, O_3, and NO_2 which have been designated in accordance with 40 CFR 53, can be obtained by writing the Environmental Monitoring and Support Laboratory, Department E (MD-77), U.S. Environmental Protection Agency, Research Triangle Park, NC 27711.

For SO_2, CO, NO_2, O_3, and meteorological parameters, continuous analyzers must be used. Thus, continuous sampling (over the time period determined necessary) is required. For TSP, daily sampling (i.e., one sample every 24 h) is required except in areas where the applicant can demonstrate that significant pollutant variability is not expected. In these situations, a sampling schedule less frequent than every day would be permitted. A minimum of one sample every 6 days will be required for these areas, however.

At least one year of meteorological data should be available for input to dispersion models used in analyzing the impact of the proposed new source on ambient-air quality and for the analysis of effects on soils, vegetation, and visibility in the vicinity of the proposed source. In some cases, representative data are available from sources such as the National Weather Service. In many situations, however, on-site data collection will be required.

A rigorous quality assurance program is required as an integral part of any ambient-air monitoring plan. The specific quality assurance procedures to be followed, including the siting of ambient monitoring stations, are described in an EPA document titled, "Ambient Air Monitoring Guidlines for Prevention of Significant Deterioration (PSD)," November 1980.

Reporting

Air quality and meteorological data must be reported to the permit-granting authority at the time of the permit application, but this could be done every 3 months. The actual reporting frequency and data format, however, should be based on an agreement between the applicant and the permit-granting authority.

The quality assurance data, including precision checks and accuracy audits, should be submitted for each site along with the summarized air quality data.

BACT/LAER Determination

In the preparation of a federal PSD permit application as well in as filing most applications for state permits for attainment pollutant emissions from any new or modified source, a control technology review in terms of BACT determination is required. The rationale for the selection of a particular control technology for a certain emission source has to be based on economic, environmental, and engineering considerations in the BACT review process. Generally, the control technology specified for a particular pollutant which an NSPS emits is considered BACT.

For example, BACT for the control of particulates from a municipal-size incinerator is a wet venturi scrubber system. A venturi scrubber system to limit particulate emissions from an industrial boiler capable of being cofired on fuels including low-sulfur coal, wood wastes, and/or sludge is also considered BACT. However, if this incinerator or industrial boiler were located within a particulate nonattainment area or BACT-controlled emissions would have the potential result of a significant impact on a nearby particulate nonattainment area, an LAER technology evaluation would be required. For the two sources cited above, LAER technology is considered the use of a fabric filter (baghouse) system. A summary of previously determined BACT/LAER control technology for various source categories is contained in an EPA document titled, "A Compilation of BACT/LAER Determinations Revised", May 1980.

EMISSIONS SURVEY

The initial step in defining the emissions levels and determining the BACT or LAER technology for a projected facility, or to control atmospheric emissions from an existing plant, is an emissions survey. All applicable pollutant sources and quantities of emissions must be identified. The results of this survey will provide management with a representative and comprehensive overview and with sufficient detail from which to formulate abatement plans and design programs and to file permits. The basic steps in such a survey include:

- Cataloging emission sources
- Quantifying emissions
- Preparing a source identification file

Cataloging Emission Sources

General Plant Information

The first step in such cataloging activities is to provide general plant information. This information is required to provide a general background and overview of the proposed or existing plant and/or its modifications. The following data should be provided:

- A schematic block flow diagram of the process(es) for the plant showing the flow of raw materials into and out of the process(es), and the design of the air pollution control equipment. This diagram should allow for release of all emissions, process and fugitive, to the atmosphere.
- A materials balance across the process(es) and across the air pollution control equipment.
- Airflows, either shown on a schematic block flow diagram for the process(es) and the control equipment, or provided separately.
- A floor plan of all of the plant building and exit stacks, showing width and height of buildings and stacks, *present* and *proposed.*
- Details of proposed stack(s): height, diameter, exit gas flow, and temperature.
- Details of the design characteristics of the air pollution control equipment.
- Details of the dust generated by the raw and finished materials for assessment of fugitive emissions from the new operation.

Process Flowsheets

The second step is to prepare process flowsheets in sufficient detail to indicate the flow of all raw materials, additives, by-products, and waste streams. The flowsheet should identify all points of feed input, and all points at which atmospheric, liquid, and solid wastes are discharged. The engineer or supervisor responsible for each process should

POWER PLANT SURVEY FORM

Type of Heat Exchanger	Primary	Standby
Coal-fired	☐	☐
Oil-fired	☐	☐
Gas-fired	☐	☐

If multiple-fired, check appropriate boxes

Rated Input Capacity________Btu/hr
Maximum Operating Rate________Btu/hr
Rated Steam Output________lb/hr____(a)____Btu/lb steam
Maximum Steam Output________lb/hr____(a)____Btu/lb steam
Furnace Volume width____ft x depth____ft x height____ft = ____cu ft
Operating Schedule________hr/day____day/wk____wk/yr

Coal Firing

Type of Firing ☐ Grate ☐ Spreader stoker ☐ Pulverized coal ☐ Cyclone

Type ________ ☐ Dry bottom ☐ Wet bottom

Fly Ash Reinjection ☐ Yes ☐ No

Soot Blowing

☐ Continuous
☐ Intermittent

Time Interval Between Blowing____minutes

Duration____minutes

Outside Coal Storage ☐ Yes ☐ No

Maximum Amount Stored Outside____tons

Figure 1-1 Sample presurvey data sheet for fossil-fuel-fired steam generators.

verify that the flowsheets identify all sources. Many plants have numerous sources that must be identified correctly for identification purposes.

Survey Data Sheets. Analysis of process flowsheets can be further verified by review of prior permit applications, process blueprints, photographs, and inspection manuals. With the aid of these and any other resources available, the emissions surveyor can then develop checklists in the form of survey data sheets that will be used in the plant tour to ensure complete and efficient gathering of pertinent data. These survey data sheets will pertain chiefly to process and feed data, and control equipment and emissions data. Figure 1-1 shows an example of a process survey data sheet, representing presurvey eval-

TABLE 1-8 General Control Equipment Data

A. At maximum continuous production rate (MCPR)
   1. Inlet and outlet absolute cubic feet per minute (ACFM).*
   2. Inlet and outlet gas temperatures.
   3. a. Inlet and outlet percent of H_2O.
      b. Dew point.
   4. Inlet pollutant levels for
      a. TSP, lb/h, gr/scf†, etc., also provide particulate size distribution.
      b. SO_2, lb/h, gr/scf, etc
      c. NO_x, lb/h, gr/scf, etc.
      d. HC, lb/h, gr/scf, etc.
      e. CO, lb/h, gr/scf, etc.
   5. Expected and guaranteed efficiencies for each of the above pollutants.
   6. Expected and guaranteed pollutant levels at outlet for each of the above pollutants.
   7. If available, make, model, and type of control system(s).
   8. Explain basis for ACFM and °F selection for design. **CAUTION**—*These may be different from data used in dispersion modeling.*
   9. Material balance across control equipment.
   10. Block flow diagram of control equipment.
   11. Inlet particulate size distribution.

B. For system(s) which may have a lower efficiency at normal operating rates as compared to MCPR, provide all data outlined above at normal operating rate.

*Actual cubic feet per minute.
†Standard cubic feet.
Note: If available, provide specification sheets. Provide drawings of internals.

uation of a fuel-fired combustion source. Such an example can be modified to apply to most types of control equipment now in operation.

The process survey data sheet should include the following:

- Detailed information on operating conditions for the process as designed
- Identification of normal operation as continuous, batch, or intermittent, with frequency of emission discharges for each operation
- Description of raw materials, products, and wastes
- Values for normal operating temperature, equipment performance ratings, flows, pressures, and similar data that are routinely monitored and/or recorded

Control Equipment Survey Information. The information given in a control equipment survey sheet is outlined in Tables 1-8 to 1-12. Table 1-8 shows information for a general survey evaluation sheet, while Tables 1-9 to 1-12 outline data requirements for control methods.

Identical information and data must be developed and evaluated for any new or modified facility. This information and data are needed to select the control device or system that represents BACT or LAER control technology.

TABLE 1-9 Multiclone and Cyclone Data

1. Number of tubes for multiclones (multiple cyclones)
2. Length of tubes
3. Fractional size efficiency vs. *P* curves
4. Diameter of tubes for multiclone or diameter of cyclone
5. Pressure drop, in H_2O column
6. Particulate size distribution into cyclone and/or multiclone
7. Grain loading (gr/dscf)* at inlet and/or outlet
8. ACFM at °F
9. Design efficiency
10. Disposal and handling of dry, collected dust
11. Preventive maintenance program

*Dry standard cubic feet.

TABLE 1-10 Scrubber for SO_2 and/or Other Pollutants

1. Design inlet and outlet volume flows, temperatures, percent moisture, and SO_2 loadings.
2. Type of scrubber. Wet or dry. If wet, operational mode.
3. Is reheat necessary? Describe.
4. Percentage of exhaust gases? Describe.
5. Any prequench section? Describe.
6. Type of demister.
7. Type of internal construction.
8. Minimum number of isolatable modules.
9. Minimum liquid-to-gas ratio.
10. Minimum-maximum pH of scrubbing liquor.
11. Maximum gas face velocity.
12. Minimum ratio of reagent to SO_2 and chemical composition solution.
13. Pressure drop, in H_2O column.
14. Is any bypass to be provided?
15. Outline any measures to be taken to produce more uniform SO_2 loading to the scrubber (e.g., precleaning, coal blending, etc.).
16. Design efficiency for PM and/or SO_2 and/or other pollutants.
17. Redundancy of control equipment.
18. Parameters for packed-bed scrubbers.
    a. Type of packing
    b. Packing size and/or shape
    c. Packing height
19. Parameters for spray scrubbers:
    a. Number of nozzles
    b. Nozzle droplet size
    c. Nozzle design and/or shape
    d. Nozzle pressure psig
20. Materials of construction.
21. Disposal of sludge or wet dust.
22. Effluent to streams from sludge pond.
23. Percolation from sludge ponds to aquifers.
24. Preventive maintenance program.

Tour of Plant Facilities

The third step is a tour of the plant facilities, if an existing plant is involved. The tour should include discussions with the process engineer or supervisor in order to ensure identification of all sources, verify the process and control-equipment flowsheets, and account for any equipment modifications already made. Sufficient data should be gathered to allow computation of material balances in order to have a basis for quantifying each emission source.

The plant tour starts at the files. There the survey will gather design specifications

TABLE 1-11 Fabric Filter or Baghouse

1. Type of filter media.
    a. Chemical composition
    b. Woven cloth or felt
    c. Porosity of new cloth
    d. Weight of cloth in ounces per square yard
    e. Napped or nonnapped fibers in woven bags
    f. Life of bags
2. Air-to-cloth ratio.
3. Cleaning mechanism.
4. Number of compartments. Can one compartment be shut down for maintenance while rest of the baghouse continues operation? How is overall collection efficiency affected?
5. How is damage from gas dew point circumvented? What is dew point?
6. Overall collection efficiency.
7. Materials of construction.
8. Disposal of collected materials.
9. Preventive maintenance program.

TABLE 1-12 Electrostatic Precipitator (ESP)

1. Collection area, square feet per 1000 ACFM.
2. Number of fields.
3. Number of compartments. Can one compartment be shut down for maintenance while rest of ESP continues operation? How is the overall collection efficiency affected during maintenance?
4. Gas velocity through ESP, feet per second.
5. Description of gas conditioning techniques.
6. Design of charging area: weighed wires, rigid frames, etc.
7. Temperature of operation.
8. Resistivity of fly ash or particulate matter at temperature of operation.
9. Inlet gas details: How is turbulence minimized? Any aerodynamic gas flow studies?
10. Rapping or cleaning details: provisions to minimize reentrainment when cleaning.
11. Design efficiency for preventive maintenance.
12. Design of collecting rappers to prevent bridging, plugging, or improper operation.
13. Materials of construction.
14. Disposal of collected materials.
15. Preventive maintenance program.

for each process and control device. Correspondence may also yield pertinent information relating to operation and maintenance of process and control equipment, current status of compliance, comments of control agencies, public complaints, and the like. This kind of background information can enhance the understanding needed for a meaningful on-site inspection of each process and control device. If the particular (existing) emission source requires an air permit, most of the pertinent information on the operating and emission characteristics are contained in the permit. In most states, permits are renewed every 2 or 5 years.

Each air contaminant source has a duct that vents from the process to an outside chimney or stack. The exhaust gas is moved by a fan or, in some instances where heat is applied, by natural draft. For each operation, the ducting should be followed from the process to the point of entry to the atmosphere. In some instances, the exhaust gas stream is difficult to follow. Introduction of makeup air, splitting of gas streams, and ducting of several operations to a common stack complicate the overall exhaust system and require careful tracing to ensure that exhaust-gas paths are properly defined. Placement of fans and control devices must be noted. Air-conditioning, heating, and makeup vents must not be mistaken for process stacks. After defining all process systems, the surveyor should check the roof to identify any emission points that are left over. The check ensures that all egress points are accounted for.

Quantifying Emissions

All of the data obtained thus far from the process flowsheets, process survey forms, control equipment survey forms, stack survey forms, photographs, correspondence, discussions, and the plant tour can now be organized for development of an emission survey plan. This plan will indicate the quantity of emissions to be expected from each source, with possible variations due to season, time of day, feed materials, and similar variables. The emissions characterization should identify all important parameters affecting control of the pollutants and possible sampling techniques. These data will be used to establish a compliance program for each source. These programs will describe the plans that will be implemented by the company to achieve compliance and should contain the following increments of progress or milestones:

- Date of submittal of the final control plan or PSD permit to the appropriate air pollution control agency.
- Date of award of accepted control plan or PSD permit.
- Date by which contracts for emission control systems or process modifications will be awarded, or date by which orders will be issued for purchase of component parts to accomplish emission control or process modification.

- Date of initiation of on-site construction or installation of emission control equipment or process change.
- Date by which on-site construction or installation of emission control equipment or process modification is to be completed.
- Date by which final compliance or full operation is to be achieved.

Figure 1-2 presents an example of the activities that must be completed before compliance can be achieved. Depending on the nature of the emission source and the complexity and size of the modifications required, the time requirements for compliance can range from a few months to several years.

Quantification techniques can then be applied in order to develop and utilize the emission survey information. Mass balances usually can be established around each process, particularly when the throughputs and the composition of raw materials are known. The materials balance will indicate the extent of solid and liquid, as well as gaseous wastes.

A search of applicable air pollution control regulations will provide the basis for the mass balance. The control regulations state what pollutants are regulated and define each pollutant. The definition of each pollutant determines the conditions under which the pollutant is sampled and its chemical or physical makeup. For instance, because water vapor is not considered an air pollutant, it is not necessary to accurately account for it in the mass balance. Emissions of SO_2 or organic substances are usually regulated and must be estimated in the materials balance.

A materials balance for gaseous pollutants can be determined by analysis of raw materials, fuels, and products to give the gaseous pollutant potential of many of the compounds liberated during a combustion or chemical process. Knowledge of fuel composition is especially useful in estimating emissions, since many gaseous compounds in the fuel become airborne after combustion (e.g., sulfur in fuel-oil exhausts as sulfur dioxide). Other constituents, such as ash and volatile matter, directly affect the quantity of particulate emissions. The results of stack sampling tests can also be used to quantify pollutant emissions.

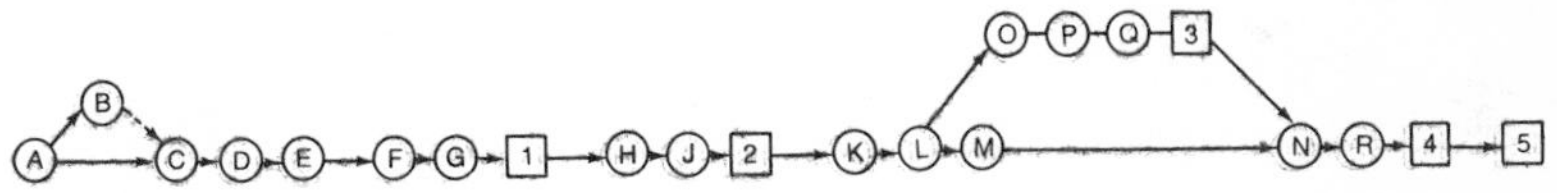

□ = Milestones
2 = Activity and duration in weeks

Milestones	Elapsed Time (weeks)
1 Date of submittal of final control plan to appropriate agency.	______
2 Date of award of control device contract.	______
3 Date of initiation of on-site construction or installation of emission-control equipment.	______
4 Date by which on-site construction or installation of emission-control equipment is completed.	______
5 Date by which final compliance is achieved.	______

Activities

Designation	
A-C	Preliminary investigation
A-B	Source Tests
C-D	Evaluate control alternatives
D-E	Commit funds for total program
E-F	Prepare preliminary control plan and compliance schedule for agency
F-G	Agency review and approval
G-I	Finalize plans and specifications
I-H	Procure control device bids
H-J	Evaluate control device bids
J-2	Award control device contract
2-K	Vendor prepares assembly drawings

Designation	
K-L	Review and approval of assembly drawings
L-M	Vendor prepares fabrication drawings
M-N	Fabricate control device
L-O	Prepare engineering drawings
O-P	Procure construction bids
P-Q	Evaluate construction bids
Q-3	Award construction contract
3-N	On-site construction
N-R	Install control device
R-4	Complete construction (system tie-in)
4-5	Start up, shakedown, preliminary source test

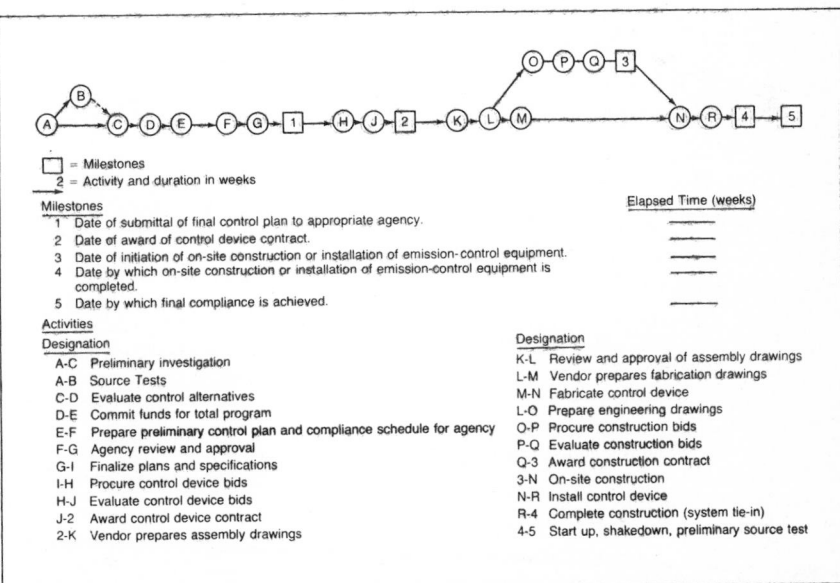

Figure 1-2 Sample compliance schedule chart.

In reviewing any permit application submitted, the regulatory authorities typically use an EPA publication titled, "A Compilation of Emission Factors," (AP-42) in order to estimate the potential uncontrolled and controlled emissions of a particular pollutant. This document and all its supplements should be obtained and maintained as reference material by the plant engineer.

Preparing a Source Identification File

A source identification file provides a means of standardizing data for the emission survey. For each pollutant source, a standard identification form gives a description of the process, a summary of emission data, the current compliance status, and proposed actions, if any are intended. A basic source identification form is shown in Fig. 1-3. Some sources involve several emission points with more than one pollutant at each point.

The source identification file should be indexed to provide easy access by any concerned party. The index should list all sources and identify each emission point for each source. Assignment of a number for each emission point will facilitate an alphanumeric search for emission points in the source identification file.

EMISSION CONTROL METHODS

The control equipment alternatives might be broadly grouped under the category of process or raw material changes to reduce the quantity of pollutant, eliminate production of a particularly undesirable pollutant, or collection and removal of the pollutant from the gas stream. Current designs employ some combination of these alternatives, with interaction between the alternatives. For example, using a scrubber to selectively remove a pollutant may lower an exhaust-gas temperature which could significantly reduce the thermal updraft effect, thereby reducing the dispersion of a conventional stack and necessitating the installation of a much taller stack. The problem of disposal of an undesirable liquid or solid waste removed from a gas stream may be greater than the problem of handling the gaseous waste itself. Clearly, in these situations, there must be a compromise. The control system must be considered to be a whole rather than individual parts.

The most important process parameters that must be collected for selection of control equipment are the following:

Flue gas characteristics
- Total flue gas flow rate
- Flue gas temperature
- Control efficiency required
- Particle size distribution
- Particle resistivity
- Composition of emissions
- Corrosiveness of flue gas over operating range
- Moisture content
- Stack pressure

Process or site characteristics (field survey)
- Reuse or recycling of collected emissions
- Availability of space
- Availability of additional electrical power
- Availability of water
- Availability of wastewater treatment facilities
- Frequency of start-up and shutdown

Emission Point No. ____________________

Emission Point Name ____________________

Date of Record ____________________

Source Name ____________________

Description of Source ____________________

Type of Permit ____________________

Date of Permit ____________________

Applicable Regulation(s) ____________________

Particulate Emissions ____________ units ____________

Allowable Emissions ____________ units ____________

Method of Determination ____________________

Gaseous Emissions ____________________

type ____________ units ____________

Allowable Emissions ____________ units ____________

Method of Determination ____________________

Compliance Status ____________________

Date Contract Awarded ____________________

Date Construction Began ____________________

Monitoring ____________________

ambient ____________________

stack ____________________

Figure 1-3 Source identification form.

Mechanical Collectors

The most familiar and widely used mechanical collector is the cyclone separator.(Fig. 1-4). Gas enters tangentially at the top of a cylindrical shell and is forced downward in a spiral of decreasing diameter in a conical section. Particles are centrifugally thrust outward and forced to spiral downward to the bottom, which is closed by an air lock. Since gas cannot escape at the bottom, it is forced to turn and travel, still whirling, back up the center of the vortex and out the top. Particles are discharged from the bottom through the air lock.

The tighter the spiral in which gas must flow, the greater the centrifugal force acting on a particle of given mass, and thus the more efficient a cyclone can be. Top diameters of cyclones range from more than 120 in to as small as 24 in, and capabilities reach 85 percent efficiency with particles as small as 10 μm at pressure drops from 0.5- to 3-in water gauge.

Cyclones alone are seldom adequate for pollution control, except where the load consists almost entirely of coarse particles, as in woodworking shops, or where particle density is unusually high. Cyclones are often used when there is a special reason to separately collect reusable coarse particles from useless fines (in fluid-bed catalyst regenerators, for example, where the fines pass through for subsequent separation by more efficient means). Cyclones are sometimes applied where gas cooling is necessary, where the cyclone is followed by a fabric filter; the cyclone scalps the coarse fractions of the dust load and at the same time provides cooling.

Cyclones are either applied singly, or manifolded to divide large gas flows between several individual units. They are also applied as two or more units of decreasing diameter in series for large loadings with a substantial percentage of small particles. The simple cyclone design, however, is not practical in very small diameters for capture of very small particles. Instead, the cyclone principle is employed in a different way in what is called a *multitube mechanical collector.*

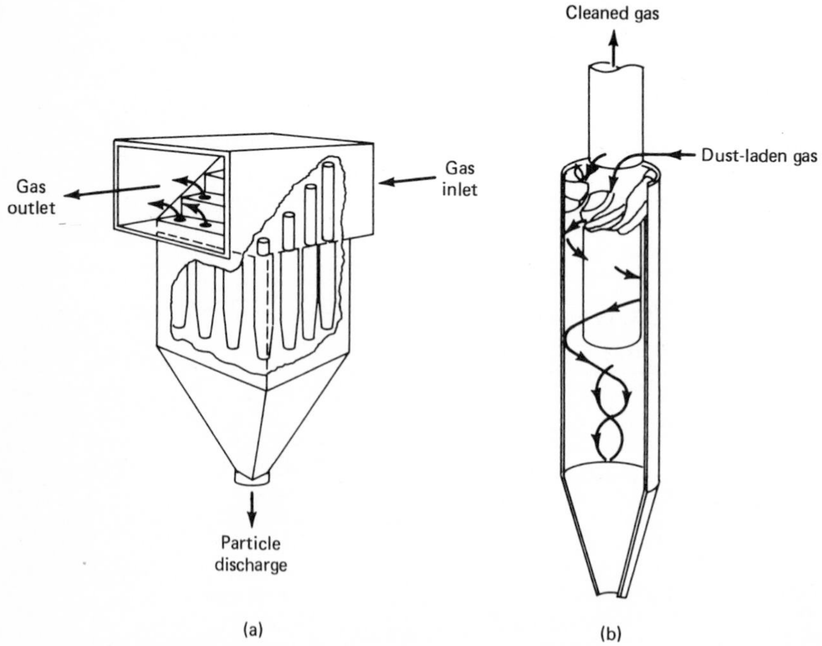

Figure 1-4 Mechanical collector-cyclone.

As the name implies, the multitube collector shown in Fig. 1-4 involves a large number of individual cyclone types in parallel within a single collector housing or shell. Dividing the gas stream among many tubes permits small tube diameters without requiring excessive gas velocity or pressure drop through the individual cyclone tubes. Tube diameters range from 4 to 10 or 12 in, with the larger sizes more common. Particles as small as 5 μm can be collected; efficiencies range up to 95 percent.

In a tube of a typical multitube collector, gas flows from the inlet section of the housing directly downward through an annular space between the outer tube wall and a smaller, concentric, exit tube. Fixed vanes in the annulus impart helical motion to the gas stream, thus applying centrifugal force for particle separation. As the gas stream slows down, reverses, and flows up to the exit tube, exit vanes may straighten out the cleaned-gas flow (which accomplishes a measure of gas pressure recovery to minimize overall pressure drop across the collector).

The smaller the tube diameter, the greater the tendency for large particles to bounce back off walls into the rising exit-gas stream. Tube diameter, inlet velocity to the collector, and the number of tubes in parallel (which determines the velocity through the individual tube) must be carefully selected for each application. It is vital that careful design ensure equal distribution of the inlet gas among all tubes to avoid excessive velocity in some and inadequate velocity in others, a special concern where angled or vertical gas flues are used to or from a multitube collector. In applications where particles have a tendency to be sticky, the smallest tube diameters cannot be used, and it may be necessary to eliminate exit recovery vanes.

Multitube mechanical collectors can be used alone for pollution control where the load fraction represented by particles smaller than 5 μm is small and the strictest emission codes do not apply. Increasingly, however, multitube collectors are used to scalp coarse particles from a gas stream ahead of a more efficient type of collector. Application to boilers that have been converted from oil to gas to pulverized coal is an example. Here, multitube collectors separate, for return to the boiler, coarse fuel particles with their content of unburned carbon, while ash fines pass through to a more efficient particulate collector.

Scrubbers

Scrubbers are compact inertial collectors capable of separating solid or liquid particles from a gas stream. They are also used to separate a chemically reactive or soluble gas constituent from other gas constituents in a flue gas stream. The most common use for gas separation involves its use in *flue gas desulfurization* (FGD).

In the simplest application of scrubbing, liquid is sprayed in at the top of a column and collision for particulate wetting or capture occurs as drops fall through a rising gas stream. Pressure drop is low, but application is limited to situations where 50 percent efficiency or less is acceptable and the percentage of particles smaller than 10 μm is low, or to scalping coarse particles ahead of a precipitator or a more efficient scrubber. A spray column is often used primarily for quenching hot gas, with coarse particulate removal a useful but incidental effect.

Centrifugal Scrubbers

In a centrifugal scrubber, column design and directed sprays cause drops and gas to mix in a rising vortex so that centrifugal force increases the momentum of collisions between particles and drops. Thus, smaller particles can be captured, and efficiencies as high as 90 percent can be achieved with particles as small as 5 μm, at pressure drops from 2- to 6-in water gauge.

Packed Scrubbers

A column may be fitted with impingement plates, wetted mesh, or fibrous packing, or packed with saddles, rings, or other solid shapes. In such scrubbers, at typically 1- to 10-in water-gauge pressure drop, efficiencies can range up to 95 percent with particles as small as 5 μm. Packed beds designed for gas absorption, however, are subject to fouling if the gas stream contains a significant fraction of solid particulates. In designs that use sprays to wash the packing, or that are packed with small spheres agitated by the gas flow, the fouling problem is reduced. A recent development employs a moist chemical-foam packing, which drains slowly from the scrubber with captured particulates and is replaced with fresh material.

Venturi Scrubbers

A venturi tube operating on the eductor or ejector principle, with scrubbing liquid as the motive fluid, can collect particles down to submicrometer size with efficiencies as high as 90 percent, if grain loading is low. Gas-pressure drop is not depended on for power input, and there can even be a gain in gas pressure across such a scrubber. The disadvantages are the requirement for substantial scrubbing-liquid flow at high pressure and the scrubber's inability to remove large particles because the induced gas-stream velocity is low.

A scrubber sometimes referred to as a *submerged-jet* type actually discharges gas through an orifice at, rather than under, the surface of a pool of scrubbing liquid. As it skims the surface, the gas atomizes some of the liquid into droplets for particle capture. A baffled flowpath causes the heavier particle-droplet agglomerations to drop back into the pool. Efficiencies can reach 90 percent, with particles down to 2 μm in size, at pressure drops from 2- to 6-in or more, water gauge. The special characteristic of this type of scrubber is its almost negligible power requirement for liquid pumping.

In what is usually called the *venturi scrubber*, Fig. 1-5, the drop in gas pressure across a venturi tube or orifice accelerates the gas stream. Scrubbing liquid is sprayed in at the throat or orifice (or is flowed in to be atomized by the passing gas stream) and mixes turbulently with the gas. Turbulence is ordinarily undesirable in fluid flow because it represents a useless consumption of power. But, in these scrubbers, turbulence is the cause of collisions between liquid droplets and particulates at high relative velocity, permitting capture of very small particles.

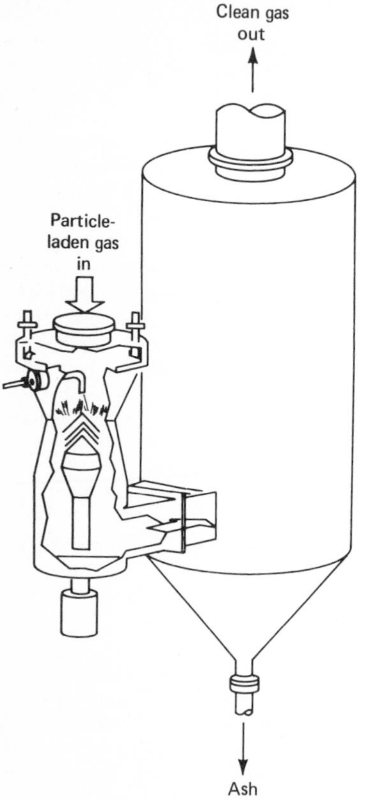

Figure 1-5 Scrubber.

If the venturi throat or orifice area of a venturi scrubber is fixed, pressure drop (and, therefore particulate capture capability) will vary with gas-flow rate. If gas flow drops significantly below the design value, pressure drop will be inadequate to achieve design efficiency. Therefore, venturi-type scrubbers have been designed to permit variation of the throat or orifice area, manually or under control of pressure-differential sensors, to maintain optimum pressure drop for stable operation and constant collection efficiency.

When a venturi scrubber is employed, mechanical means can be provided for adjustment of throat area while maintaining the proper throat configuration. In the flooded-disk scrubber, the orifice through which gas is accelerated is the annular space between a horizontal disk positioned in a tapered duct and the duct wall. Pressure drop is maintained, despite flow variations, by raising or lowering the disk in the tapered duct to increase or decrease the orifice area.

Scrubbing liquid is supplied at the center of the upper face of the disk, where gas pressure forces it outward. At the disk periphery, liquid is sheared off and atomized by the passing high-velocity gas stream. Since this type of unit involves no spray nozzles or other constricted liquid passages, it is particularly suitable for recirculating liquid with a high solids content; high liquid pressure is not required. Because the orifice is a narrow annulus, its area can be large, where a high rate of gas flow is involved, without requiring distribution of scrubbing liquid over an excessive area.

Fabric Filters

Fabric filters collect solid particulates by passing gas through cloth bags that most particles cannot penetrate. As the layer of collected material builds, the pressure differential required for continued gas flow increases; consequently, the accumulated dust must be removed at frequent intervals. Cotton, wool, glass, and various synthetic fibers are used in fabric filters, which are shaped in the form of cylindrical bags or envelopes of roughly elliptical cross section.

Fabric filters are capable of 99+ percent collection efficiency with particles down to submicrometer size. High efficiency is attained with moderate pressure drops, typically in the range from 2- to 4-in water gauge. Power input is thus comparable to that of multitube mechanical collectors, while capability in collection of fine particle sizes is much greater. Operating cost, including maintenance, is somewhat higher, however, because some moving parts are involved and bags must be periodically replaced. Unlike wet scrubbers, the performance of fabric-filters is relatively unaffected by variations in gas-flow rate. Fabric filters are sometimes preceded by mechanical dust collectors, or settling chambers in the baghouse, when excessive grain loading or abrasive, coarse particles, or both, are involved.

The simplest type of fabric filter is the mechanical, or more properly, mechanically cleaned type. Near the bottom of the filter housing is a horizontal tube sheet separating the inlet dusty-air plenum from the clean-air plenum into which the bags discharge. Bags, closed at the top, are suspended with their open bottom ends tightly connected to openings in the tube sheet. Gas flow is upward, into the bags, through the fabric, into the clean-air plenum, and out. Particles accumulate inside the bags and, when dislodged, fall down into hoppers for disposal.

The traditional method of surface filter cleaning is mechanical shaking of the structure from which the bags are suspended. Design and operation are simple, but maintenance requires access to the shaking mechanism within the housing. The same gas-flow and bag arrangement is employed in the *shakerless*, or *reverse-air*, fabric filter. A fan, ducting, and dampers are provided to supply a reverse flow of air from the clean side of bags, through to their interiors, to dislodge accumulated particles. Bags are usually stiffened by wire rings to prevent their complete collapse when the airflow reverses.

The mechanical-shake action is violent enough to accelerate bag wear if the fabric is not resistant to abrasion or if unusually abrasive particles are being collected. In fact, the action can be positive enough for cleaning felted fabrics in some applications, although most mechanical fabric filters are surface filters. The gentler reverse-airflow cleaning action is preferred where abrasion will be a problem and is usually required when high gas temperature dictates use of glass-fiber bags; glass fabrics cannot be expected to withstand the mechanical shaking action employed in most applications, even when the fibers have been lubricated with graphite or silicone.

If air jets (Fig. 1-6) are used for cleaning instead of low-velocity reverse airflow, felted fabrics can be used and bags can be cleaned while contaminated gas continues to flow to them. A reverse-jet-type fabric filter employs a ring of nozzles around each bag and a mechanism to traverse the rings up and down the bag length. It does not require the elaborate inlet duct-and-damper arrangement necessary for continuous duty with mechanical fabric filters. But this advantage is substantially offset by the cost of providing and maintaining the complex cleaning mechanism, all of which is inside the filter housing, and by the possibility of misalignment that will allow abrasion of bags by the moving rings.

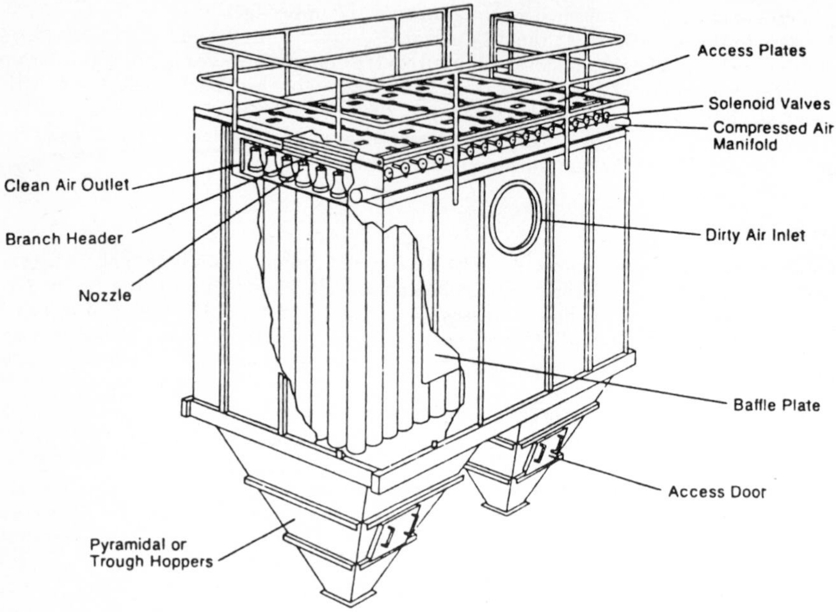

Figure 1-6 Fabric filter.

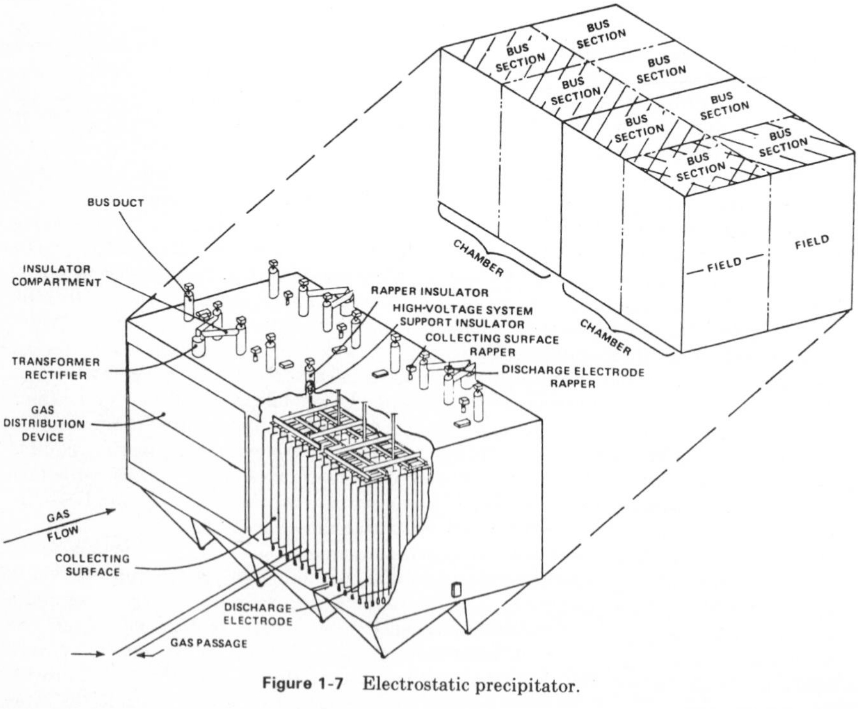

Figure 1-7 Electrostatic precipitator.

Electrostatic Precipitators

The electrostatic precipitator, Fig. 1-7, an extremely efficient air pollution control device, can remove more than 99 percent of the undesirable particulates in a gas stream. This high efficiency is possible because, unlike other pollution control devices, the precipitator applies the collecting force only to the particles to be collected, not to the entire gas stream. Thus, an extremely low, energy-saving power input, about 200 W/1000 ft^3/min (0.5 m^3/s), is required.

In operation, a voltage source creates a negatively charged area, usually by means of wires suspended in the gas-flow path. On either side of this charged area are grounded collecting plates. The high potential difference between these plates and the discharge wires creates a powerful electric field. As the polluted gas passes through this field, particles suspended in the gas become electrically charged and are drawn out of the gas flow by the collecting plates. They adhere to these plates until removed for storage or disposal. Removal is accomplished mechanically by periodic vibration, rapping, or rinsing. The gas stream, now free of particulate pollution, continues on for release to the atmosphere.

Precipitators may be designated as either plate or pipe varieties. The plate type, just described, is widely used for dry dust. The pipe type, for removal of liquid or sludge particles and particulate fumes, uses the same principle, but with a different mechanical setup; the discharge electrodes are suspended within a series of pipes (collecting plates) contained in a cylindrical shell, under a header plate.

As noted earlier, the precipitator applies the separating force directly to the particles, regardless of gas-stream velocity, thereby requiring less power than other control devices. And, because they can operate completely dry, they are ideal for use in the following situations:

- Where water availability and disposal are problems.
- Where very high efficiency on fine materials is required; for example, electrostatic precipitators are removing particulate matter as small as 0.05 μm from zinc oxide fumes with 97 to 98 percent efficiency.
- Where large volumes of gas must be treated. In nonferrous metals production, gas flows of 600 000 ft^3/min (250 m^3/s) at temperatures up to 800°F (425°C) are treated with up to 99.6 percent efficiency.
- Where valuable dry materials must be recovered, as in rotary-kiln or spray-drying operations, and in calcining.

Process Modifications

Pollutant generation can frequently be limited by using different processes of raw materials to produce the same end product. This is by far the best method of control if it can be used. In many cases, it is the most cost-effective way to reduce the emission of pollutants.

As an example, modifications to combustion equipment are used to regulate the combustion process to reduce emissions of nitrogen oxides. The amount of nitrogen oxides emitted from combustion is largely a function of the control of temperature and residence time in the primary flame zone. Accordingly, burners are being designed which produce significantly lower quantities of nitrogen oxides or are arranged in the furnace to minimize nitrogen oxides formation.

Low excess-air firing can reduce nitrogen oxide emissions by 10 to 30 percent. Essentially, oxygen input to the burners is reduced, thus reducing the chance of nitrogen oxides formation.

Nitrogen oxides may also be reduced 30 to 70 percent by using a two-stage combustion process. Substoichiometric quantities of primary air are supplied at the burner. Complete burnout is accomplished by using secondary air supplied at a lower temperature.

Flue gas recirculation is another technique used to reduce nitrogen oxides by 20 to 60 percent. The maximum flame temperature is reduced by dilution of the primary flame zone with the recirculated combustion flue gases. The oxygen concentration is also reduced, making nitrogen oxides formation less likely.

Not all of these processes can be used simultaneously, but by designing a combustion system using a judicious combination of the alternatives, nitrogen oxide emissions may be reduced by as much as 70 to 90 percent.

BIBLIOGRAPHY

1. EPA: "Compilation of BACT/LAER Determinations—Revised," EPA 450/2-80-70, May 1980.
2. EPA: "Compilation of Emission Factors," AP-42 and Supplements, Research Triangle Park, North Carolina, April 1981.
3. EPA, OAQPS: "Ambient Monitoring Guidelines for Prevention of Significant Deterioration (PDS)," EPA 450/4-80-012, November 1980.
4. EPA, OAQPS: *Guidelines for Air Quality Maintenance Planning and Analysis,* vol. 10, "Procedures for Evaluating Air Quality Impact of Near Stationary Sources," EPA-450/4-77-001, Research Triangle Park, North Carolina, October 1977.
5. EPA, OAQPS: "Guideline on Air Quality Models," EPA-450/2-78-027, Research Triangle Park, N.C., April 1978.
6. EPA, OAQPS, Region III: "Guidelines for Determining BACT," Philadelphia, December 1978.
7. EPA, Region III: "Permit Application Kit, Prevention of Significant Air Quality Deterioration," Air Programs Branch, Philadelphia, November 1978.
8. EPA, Technology Transfer: "Industrial Guide for Air Pollution Control," Contract 68-01-4147, by PED Co. Environmental, Inc., June 1978.
9. *Federal Register,* 1977 Clean Air Act, PSD, SIP Requirements, Part 52, June 19, 1978.
10. *Federal Register,* 40 CFR Parts 51 and 52, September 5, 1979.
11. *Federal Register,* 40 CFR Parts 51 and 52, February 5, 1980.
12. Quarles, J., Jr.: "Federal Regulation of New Industrial Plants," P.O. Box 998, Ben Franklin Station, Washington, DC 20044, January 1979.
13. Rymarz, T. M., and D. H. Klipstein: "Removing Particulates from Gases," *Chemical Engineering Deskbook,* vol. 82, no. 21, October 6, 1975, pp. 113–120.

chapter 13-2

Liquid-Waste Disposal

by
William M. Throop, P.E.
Manager, Market Development
Envirex Inc.
Waukesha, Wisconsin

GLOSSARY

Alkalinity Ability to neutralize acids—determined by the water's content of carbonates, bicarbonates, hydroxides, and borates, silicates, and phosphates, if present. Expressed in milligrams per liter of calcium carbonate ($CaCo_3$).

BOD_5 (biochemical oxygen demand) A measure of oxygen metabolized, in milligrams per liter, in five days by microorganisms that consume biodegradable organics in wastewater under aerobic (with air) conditions.

COD (chemical oxygen demand) The amount of oxygen, in milligrams per liter, needed to oxidize both organic and oxidizable inorganic compounds.

Effluent The liquid end product discharging from a process.

Floating matter Matter that passes through a 2000-μm sieve and separates by flotation in an hour.

Settleable solids Solids larger than 0.01 mm in diameter settling in two hours under quiescence.

Suspended solids Small filterable particles of solid pollutants in wastewater. The examination of suspended solids and the BOD_5 test constitute the two main determinations for water quality.

Total solids All dissolved, suspended, and settleable solids contained in a liquid.

Turbidity The amount of suspended matter in wastewater; quantity obtained by measuring its light-scattering ability.

INTRODUCTION

Generally, pollution is the addition of harmful or objectionable material to water in concentrations of sufficient quantity to result in measurable degradation of water quality.

Industrial wastewater pollution is not only a liability; it is a total systems problem. Various interactions between air, water, and solid wastes must be evaluated on an overall plant basis. The treatment of a water pollutant, for example, may create an air pollutant. Pollution is pollution de facto; as long as the deleterious material does not meet the standards of the regulatory agency, it is pollution. Harm or injury no longer has to be proved.

DESCRIPTION OF THE PROBLEM

The passage of the Federal Water Pollution Control Act as amended in 1977 under Public Law 92-217 forced the plant engineer to become familiar with its many ramifications. Among the provisions of this act that are of direct interest to industry is the establishment of water quality standards typified by Table 2-1, column A, for maintaining aquatic life.[1] These restrictions are enforced by the requirement that a permit be issued before discharges are permitted under the National Pollutant Discharge Elimination System (NPDES). Every holder of a NPDES permit is required to comply with monitoring sampling, recording, and reporting requirements.

Many local ordinances have pretreatment requirements limiting high effluent concentrations of wastes and toxic materials which might adversely affect treatment processes of publicly owned treatment works (POTW). A typical list of effluent limitations is shown in Table 2-1, column B.[2] When discharging to POTWs, industry is expected to pay its proportionate share of capital cost of the POTWs collection and treatment equipment. These user fees are usually based on a multiplier of BOD_5, suspended solids, and liquid volume, and vary with each municipality.

For specific pollutants, effluent guidelines for specific industrial categories are published in the *Code of Federal Regulations* (40 CFR 401).

Different effluent levels are allowable depending on the following:

1. Industrial subcategory
2. Control technology required (e.g., best available technology economically achievable)
3. Existing or new source (new sources are more severely regulated than existing sources)

TABLE 2-1 Maximum Discharge Limits

Constituent	A Direct discharge or recycle, mg/L	B* To POTW
Ammonia nitrogen (as N)	1.5	—
Arsenic (total)	1.0	—
Barium (total)	5.0	—
Boron (total)	1.0	1.0
Cadmium (total)	0.05	2.0
Chromium (total)	—	25.0
Chromium (total hexavalent)	0.05	10.0
Chromium (total trivalent)	1.0	—
Copper (total)	0.02	3.0
Cyanide (total)	0.025	@150°F & pH 4.5 = 2.0
Fluoride (total)	1.4	—
Iron (total)	1.0	50.0
Iron (dissolved)	0.5	—
Lead (total)	0.1	0.5
Manganese (total)	1.0	—
Mercury (total)	0.0005	0.0005
Nickel (total)	1.0	10.0
Fats, oils, and greases (FOG)	15.0	100.0 total
pH	5.0–10.0	4.5–10.0
Phenols	0.1	—
Phosphorus (as P)	1.0	—
Selenium (total)	1.0	—
Silver	0.005	—
Sulfate	500.0	—
Temperature	—	150°F (65°C)
Zinc (total)	1.0	—
Total dissolved solids	1000.0	No limit

*Units are milligrams per liter unless otherwise specified.

4. Where the effluents are discharged (effluent levels discharged into POTWs are different from direct discharges into navigable water)

For details pertaining to emissions by specific industry, the *Code of Federal Regulations* should be consulted. Because the promulgation of effluent guidelines is an ongoing process, the EPA should be contacted for the latest information.

All pollutants are classified as either conventional, toxic, or nonconventional. Conventional pollutants include BOD_5, TSS (total suspended solids), and pH. There are 129 priority pollutants that appear on the toxics list in the *Federal Register* 43(164)4108 (February 1978). Nonconventional pollutants are those that are neither toxic nor conventional, such as nitrogen, oil, and grease. Best conventional pollutant control technology will be required for conventional pollutants by July 1, 1984. Best available technology economically achievable will be required for toxic and nonconventional pollutants by the same date.

PLANT SURVEY

The accurate measurement of flow volume and pollutants in the waste flow are essential in assessing any wastewater problem, and in designing a wastewater treatment system. Limitations on effluent (see Table 2-1) make it imperative to analyze flows and impurities quickly, accurately, and at reasonable cost.

The best approach is to make a comprehensive wastewater survey that will (1) determine the quantity of wastewater discharge, (2) locate the major sources of waste within the plant, (3) determine wastewater composition, (4) explore in-plant or process changes

TABLE 2-2 Typical Process Discharge Volumes, BOD_5, and Suspended Solids for Industrial Wastewater Before Treatment

	Unit processed	Discharge per unit		BOD_5, mg/L	Suspended solids, mg/L
		Gallons	Liters		
Aluminum & copper	lb (kg)	12–13	45–50	N/A	300–500
Automotive	Car	10800	40900	190	215
Beverage, malt	bbl (L)	330	1250	390–1800	70–100
Canning					
Fruit	Case	20–40	75–150	300–1600	200–500
Vegetable	Case	50–100	190–380	700–2000	300–2000
Coal washing	ton (tonne)	125	138	N/A	2000–3000
Cooking	ton (tonne)	1500–2800	1650–3090	50–200	90
Dairy, milk	gal (L)	4–12	15–45	1800	560–4000
Electrical	kWh (kJ)	80	110	N/A	50–2000
Laundry					
Commerical wash	ton (tonne)	8600	36000	600–1860	400–2200
Industrial	ton (tonne)	5000	20000	650–1300	4900–8600
Manufacturing, gen. fabr.	ton (tonne)	700	3000	50–1500	200–15 000
Meat packing					
Cattle	Animal	400–2000	1515–7575	400–900	400–800
Chicken	Bird	8–9	30–34	150–2400	100–1500
Hogs	Animal	300–600	1136–2273	1000	650
Office building	Person	30–45	114–170	117	176
Paint, latex	gal (L)	3	11	2000–3000	15 000–60 000
Paper making	lb. (kg)	65	108	200–800	500–1200
Pharmaceutical	——	——	——	600–2500	500–1000
Phenolic resins	ton (tonne)	75 000	313 000	11500	40
Railway maintenance	Locomotive	3000	11 360	500–800	200–600
Refining	bbl crude	770	2900	100–500	300–700
Rubber, synthetic	Car tire	500	1890	25–1600	60–2200
Steel					
Cold-rolled	lb (kg)	9	15	150	100–300
Hot-rolled	lb (kg)	18	30	80	500–2000
Tanning, hide	lb (kg)	8–12	30–45	900	6000
Textiles					
Synthetic	lb (kg)	12–25	45–95	1500–6000	500
Wool	lb (kg)	70	265	900	100
Vegetable oil	gal (L)	22	83	3050	900

to minimize the waste problem, (5) establish the basis for wastewater treatment, and (6) evaluate effect of wastes on the receiving stream.

Composition of wastewater varies with the amount of impurities initially present in water and the chemical analysis of any pollutants that are added. While domestic sewage has a fairly uniform composition, industrial wastes have an almost infinite variety of characteristics, as shown in Table 2-2. Wastewaters should be analyzed for at least BOD_5, COD, color, total solids (suspended and dissolved), pH, and turbidity. Other impurities of interest will vary with the source and type of wastewater.

Concentration of pollutants must be correlated with average, minimum, and maximum flows encountered. The analysis program must also take effluent water quality standards into account (see Table 2-1) and the BOD_5 reduction required to meet them. Any toxic impurities in the wastewater that adversely affect water quality must be determined.

Sampling and Flow Measurement

The starting point in any wastewater survey is an effective program of sampling and flow measurement. To be useful, a sample of wastewater must accurately represent the source from which it is taken and be large enough to run all the laboratory tests required. This means the method of sampling must be tailored to the type and kind of wastewater flow.

A close check of each waste source will reveal whether flow is continuous or intermittent and any wide swings in flow rate.

It is also important to know if the concentration of pollutants changes drastically or is fairly constant. The presence of oil or excessive suspended solids may also cause problems.

An integral part of this program is the need to obtain flow information on various in-plant streams as well as the plant outfalls. Wastewater flows are measured for the following reasons:

1. To determine the quantity of water being discharged, as well as variations in the flow rate
2. To determine the number of pounds of constituents being discharged on the basis of the analytical data and the determined flow rate
3. To evaluate segregation possibilities
4. To determine the effect of the wastewater discharge on the receiving stream, if applicable

Measuring Rate of Flow

Rates of flow can be approximated by the methods discussed in the following paragraphs.

Water Meters on Influent Lines. Water consumption in the plant should be determined during a wastewater survey to check on wastewater-flow measurements and to compute a water balance for the plant. Meters can also be installed at particular water-using operations to obtain flow data.

Container and Stopwatch. The time required to obtain a given volume of water in a container is measured. Volume can be determined either by weight added or by a calibrated collection container. The weight of water added is divided by 8.34 lb/gal to determine the number of gallons collected. The flow is then determined by the formula

$$\frac{\text{Gal in container} \times 60}{\text{Time, s, to fill container}} = \frac{\text{gal/min}}{15.85} = \text{L s}^{-1}$$

If the container fills in less than 10 s, the accuracy of this method is questionable.

Weirs. A weir acts like a dam or obstruction, with the water flowing through the notch, which is usually rectangular or V-shaped.

To ensure accurate weir measurements:

1. The weir crest must be sharp or at least square-edged. Steel is the best construction material, but tempered wood is also used.
2. The weir must be ventilated. There must be air on the underside of the falling water.
3. Leaks around the weir plate must be sealed.
4. The weir must be exactly level.
5. Weirs should be kept clean.
6. The head on the weir should be measured at a distance of 2.5 times the head upstream from the weir.
7. The channel upstream from the weir should be straight, level, and free from disturbing influences. A stilling box may be used to quiet the water flow.
8. The weir should be sized after the flow is estimated by other methods. The head on any weir should be greater than 3 in (7.6 cm) but not more than 2 ft (61 cm).

The flow over V-notched (triangular) weirs and rectangular weirs can be taken from the nomographs shown in Fig. 2-1.

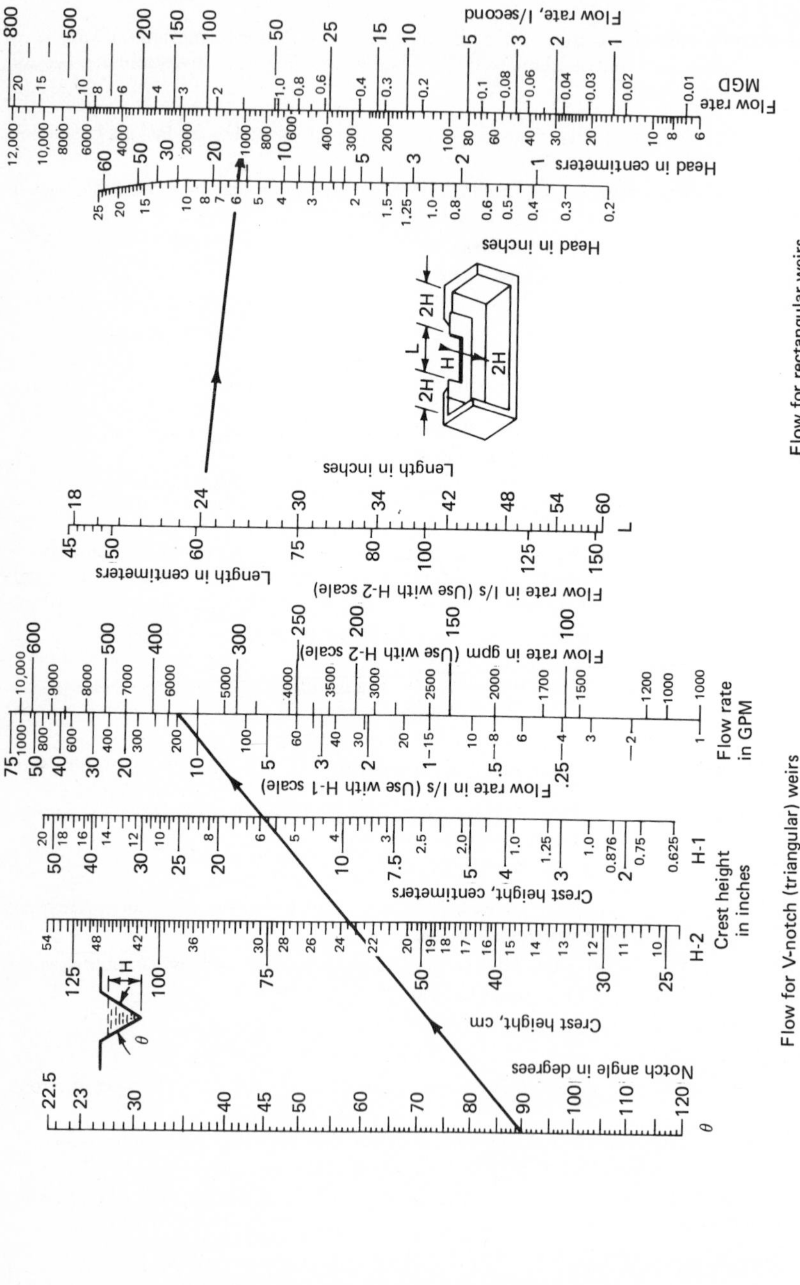

Figure 2-1 Nomographs for measuring flow over weirs.

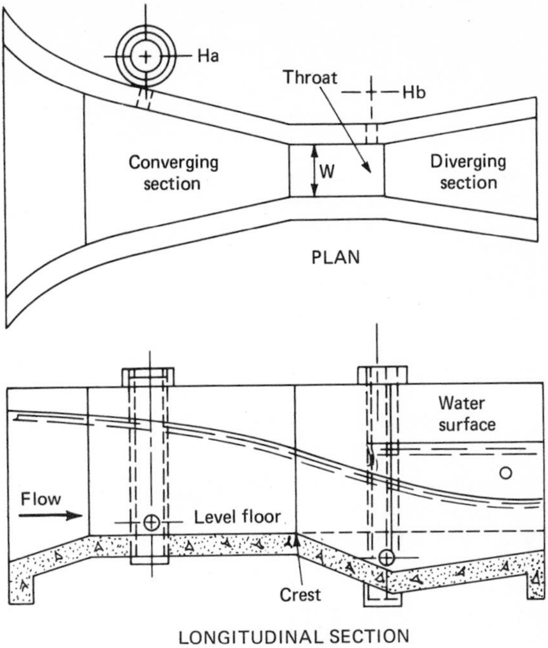

Figure 2-2 Parshall flume.

Parshall Flume. A Parshall flume (Fig. 2-2) can be used to measure flows in open channels at or near ground surface. This device is valuable when it is not possible to dam the water. It is also advisable for a permanent installation because it is self-cleaning.

Flow under submerged conditions can be calculated from readings taken at gauges. If the water surface downstream from the flume is high enough to retard the rate of discharge, submerged flow exists. When there is no backwater effect, water passing through the throat and diverging section assumes a level which corresponds to the floor of the channel. This pattern demonstrates free flow.

The flow of a free discharge from a Parshall flume is calculated by

$$Q = 4WH_an$$

where

Q = flow, ft³/s
W = throat width, ft
H_a = head of water above level floor, ft
$n = 1.522W^{0.026}$

or in metric units

$$Q = 8.52 \times 10^3(11.4 \log W)H_a(1.57 + 0.09 \log W)$$

where

Q = flow, m³/h
W = throat width, m
H_a = head of water above level floor, m

Flow under submerged conditions can be calculated from readings taken at gauges, one located at a point two-thirds the length of the converging section measured back from the crest of the flume H_a and one located near the downstream end of the throat section H_b. Degree of submergence is given by the ratio H_b/H_a.

Sample Collection and Analysis

Wastewaters are sampled and analyzed to identify those pollutants that require treatment and to select the proper treatment process.

Collecting Samples. Since a wastewater's characteristics can vary considerably, composite samples are collected to obtain a truer representation of the waste. Small samples are collected at frequent intervals during the sampling period. They are mixed together to form the composite sample.

Compositing Samples. Depending on plant operation, 8-, 16-, or 24-h composites can be collected. Daily sampling for 3 days generally constitutes a sampling program. Samples can be composited on the basis of the following criteria.

Flow. The amount of sample collected at any time during the sampling period is proportional to the flow of wastewater at that time.

Time. The same amount of sample is collected at every interval during the sampling period regardless of variations in wastewater flows.

Sample Size. The sample size collected at any one time should be at least 200 mL. Composite samples can be collected either manually or with automatic samplers. Automatic, battery-operated samplers are available for collecting composite samples on the basis of flow or time.

Amount of sample to be collected depends upon the laboratory tests to be run. The amount of sample required for each test to be performed should be determined before the sampling program is begun to ensure that sufficient sample is collected.

Analytical Determinations. Determinations that may be conducted on a wastewater sample are:

pH	Copper
Alkalinity or acidity	Nickel
Total hardness	Zinc
Chloride	Chromium, hexavalent
Sulfate	Chromium, total
Suspended solids	Iron
Volatile suspended solids	Manganese
Settleable solids	Solvent soluble (oil)
Total nitrogen	Phenol
Ammonia nitrogen	Biochemical oxygen demand (BOD_5)
Total phosphate	Chemical oxygen demand (COD)
Total solids	Total organic carbon (TOC)
Volatile total solids	Cyanide

Standard methods are available for conducting these determinations. Phases involved in a full pollution control program are:

I. Definition of the problem and development of an action plan (survey or feasibility study)

II. Detailed engineering

III. Construction and start-up

An outline of the steps involved in each of these phases is shown in Table 2-3.

TABLE 2-3 Pollution Control Program

I. Survey or feasibility study
  A. Fact finding:
    1. Develop a plant water balance for average and peak operating conditions.
    2. Inventory all industrial processes using water.
    3. Determine characteristics of the receiving waterway both upstream and downstream from plant's discharge.
    4. Determine chemical characteristics of waste streams.
    5. Study all operations using water and producing wastes.
    6. Determine local requirements with respect to pollution.
  B. Analyze data to determine:
    1. Sources of offending contaminants.
    2. Feasibility of segregating contaminated wastes requiring treatment from dilute wastes which would be acceptable without treatment.
    3. Availability of "natural" dilution waters, that is, waters employed for useful purposes but not contaminated.
    4. Quality of effluent required for compliance with discharge standards.
    5. Whether treatment is necessary.
  C. Exploit in-plant and/or process changes to minimize the problems by:
    1. Reducing wastes or waste volume at sources.
    2. Exploring the possibilities for reuse of process materials without treatment.
    3. Investigating recovery of valuable process materials.
    4. Reexamining the degrees of treatment required to meet standards.
    5. Reevaluating to decide whether treatment is necessary.
  D. Detailed report on the engineering survey:
    1. Recommend a preliminary course of action.
    2. Advise management whether a waste-treatment plant is necessary.
    3. Describe the general type of plant required.
    4. Provide preliminary estimate of construction cost.
    5. Prepare preliminary estimate of operating costs.

II. Detailed engineering
  A. Process design and evaluation:
    1. Assign liaison and engineering personnel as required.
    2. Evaluate bench scale or pilot plant data.
    3. Translate the total evaluated data into process flow diagrams and functional specifications for the treatment plant.
    4. Prepare plot plan showing layout on plant site.
    5. Assemble an engineering report for review and approval.
    6. Obtain preliminary approval of regulatory agency.
  B. Definitive engineering:
    1. Prepare detailed engineering flow diagrams which form the basis of final plant design.
    2. Obtain approval of overall plant design.
    3. Complete the definitive design.
    4. Obtain final approval and permit of regulatory agency.

III. Construction and start-up
  A. Procurement and scheduling:
    1. Prepare complete equipment specifications, bills of material, and preliminary timetable.
    2. Prepare item delivery and installation schedule.
    3. Use critical-path scheduling when warranted.
    4. Coordinate and inspect all phases of the work performed by fabricators.
  B. Facilities erection and testing:
    1. Plan, supervise, and coordinate erection of the complete wastewater-treatment plant.
    2. Conduct unit tests, after assembly, to assure proper functioning of all related facilities.
    3. Inspect, adjust, and calibrate instruments and controls to conform to high accuracy standards, with engineers performing the work.
  C. Operator training:
    1. Prepare detailed operating manuals for all unit operations in the plant.
    2. Assemble vendors' manuals for use by plant personnel in maintaining, repairing, and replacing mechanical, instrument, and electric equipment parts.
    3. Assist with training of operating crews while construction work is in the final stages.
  D. Start-up of treatment facilities:
    1. Initiate a control testing program.
      a. Operational
      b. Quality of effluent
    2. Initiate an efficiency testing program.
    3. Establish conditions for operations.
    4. Initiate a development program.
    5. Establish record-keeping procedures.
  E. Supervise operation.

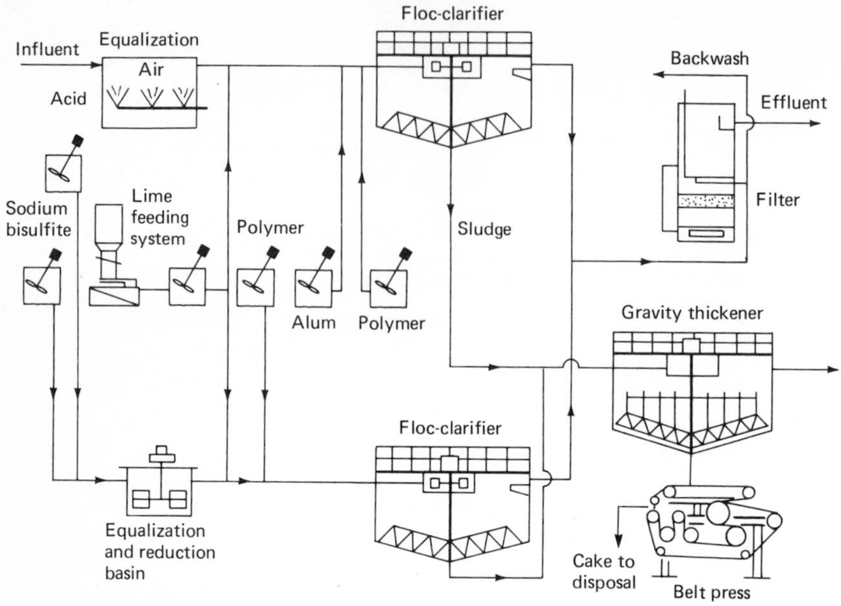

Figure 2-3 General manufacturing wastewater treatment.

TREATMENT

The necessity of effective industrial wastewater treatment must be considered an integral part of the manufacturing process, and the cost of treatment must be charged against the product.

The method of treatment depends on economic considerations and the degree of treatment required. The best alternative system for pollutant removal must be selected on the basis of a case-by-case study of efficiency and actual costs. It must be recognized that a complete system may involve several unit components and that pretreatment is required before tertiary treatment.

Figure 2-3 illustrates various treatment units combined to form a treatment system. Figure 2-4 shows the various treatment processes classified according to type, and illustrates how they can be combined to give the desired effluent quality.

Primary Treatment

The physical removal or combined chemical coagulation and physical removal of solids from wastewater is classified as primary treatment, especially if these processes are followed by biological treatment. Gravity or flotation units are used to remove suspended or coagulated colloidal material. Solids thus concentrated can be treated more economically by disposal, incineration, or biological degradation.

Oily waters are usually treated separately to remove the oil prior to mixing them with other waste streams. Chemicals can be used to enhance gravity separation of oil when emulsions are present.

Figure 2-5 presents a general guide for the design of a gravity separator based on parameters set forth by the American Petroleum Institute. It is desirable to have a minimum depth of 5 ft (1.5 m), a maximum horizontal flow-through velocity not exceeding 3 ft^3/min (0.015 L/s), and a minimum length-to-width ratio of 3:1 with a depth-to-width ratio of 0.3 to 0.5.

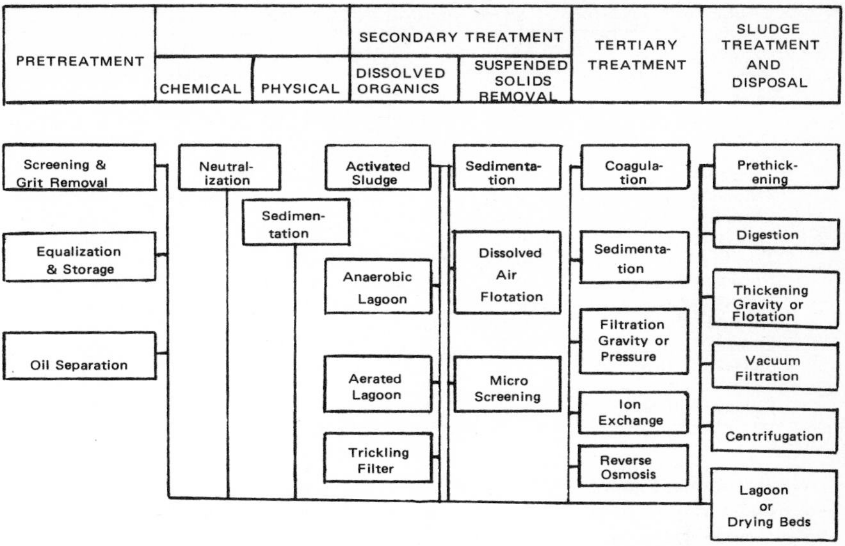

Figure 2-4 Alternatives for wastewater pollutant removal processes and how they may be combined in treatment programs.

Alkaline or acidic waste streams must be neutralized before secondary treatment or discharge.

Secondary Treatment

Secondary treatment is used to reduce soluble organic pollutants that are degradable to certain levels of organic materials remaining (the degradation products). If the effluent quality required is higher than that which can be obtained by biological treatment, tertiary processes are needed to remove the degradable with the undegradable fractions.

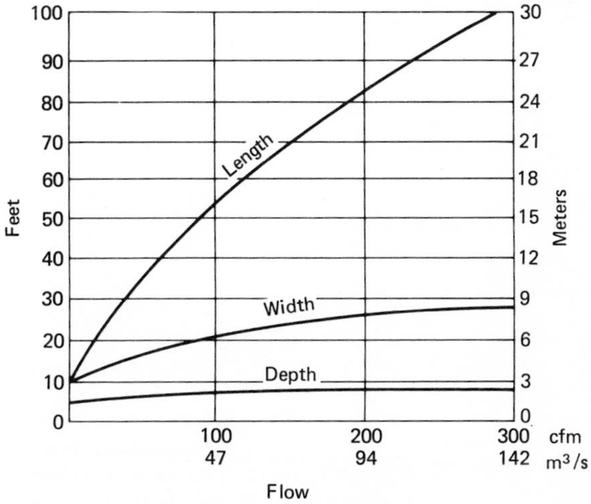

Figure 2-5 Design of gravity separator.

TABLE 2-4 Treatment-Process Removal Efficiencies

Treatment method	Removal efficiency, %				
	BOD_5	Suspended solids	Total dissolved solids	Fats, oils, & greases	Alkalinity
Screening	0–5	5–20	0	0	0
Sedimentation	5–15	15–6	0	5–15	0
Chemical precipitation	25–60	30–90	0–50	10–40	80–95
Dissolved-air flotation	10–30	70–85	10–20	80–95	10–20
Trickling filter	40–85	80–90	0–30	10–20	10–25
Nonaerated lagoon	30–70	30–70	30–80	0–40	10–20
Aerated lagoon	50–80	50–90	0–40	5–15	10–20
Activated-sludge	70–90	85–95	0–40	0–15	15–30
Filtration	80–85	30–70	10–60	0–10	10–20
Activated carbon	95–99	95–99	10	N/A	85–90
Ion-exchange	N/A	N/A	95–99	N/A	99
Reverse osmosis	N/A	N/A	99	N/A	99

Commonly used secondary-treatment processes include the completely mixed activated-sludge process, extended aeration, aerobic and aerated lagoons, trickling filters, and anaerobic and facultative waste stabilization ponds.

Table 2-4 is indicative of removal efficiencies for unit treatment processes.

EQUIPMENT

Various unit items of equipment are combined to provide the degree of treatment required.

Bar Screens

A mechanically cleaned bar screen, like the one shown in Fig. 2-6, is the simplest tool for removing debris or suspended matter which could damage equipment or disrupt the treatment process. All solids with larger than ¾- to 2-in (2- to 5-cm) bar rack openings are trapped on the upstream side and removed by moving rakes.

Clarifiers

Clarifiers used for removal of settleable solids and readily floating oils and greases are either rectangular or circular basins (Figs. 2-7 and 2-8). Clarifiers are sized on the basis of settling rate (area) and detention time (volume).

Typical overflow rates vary from 250 to 1400 gal/(day)(ft^2) [10 to 50 m^3/(day)(m^2)]. Detention time is in the range of 1 to 4 h.

Flocculation Systems

Flocculation is the agglomeration of finely divided suspended matter and floc caused by gently stirring or agitating the wastewater. The resulting increase in particle size increases the settling rate and improves suspended solids removal by providing more efficient contact between suspended solids, dissolved impurities, and chemical coagulants.

Mechanical flocculation uses paddles slowly rotating on a horizontal or vertical axis (Fig. 2-9). Peripheral speed at the paddle tip is about 1 ft/s. Various other mechanical devices are used to achieve the same result. An air flocculation system has diffusers along

Figure 2-6 Mechanically cleaned bar screen. (*Envirex.*)

one side of the basin near the bottom to produce a gentle rollover action perpendicular to flow.

The size of the required basin is determined by the detention time, which is normally in the range of 20 to 30 min at rated flow. In some cases involving industrial wastes, detention may be reduced to as little as 10 min.

Combined Equipment

Many clarifier designs, such as the solids contact unit (Fig. 2-10), combine mixing, flocculation, and coagulation in one basin. This may have economic advantages and may

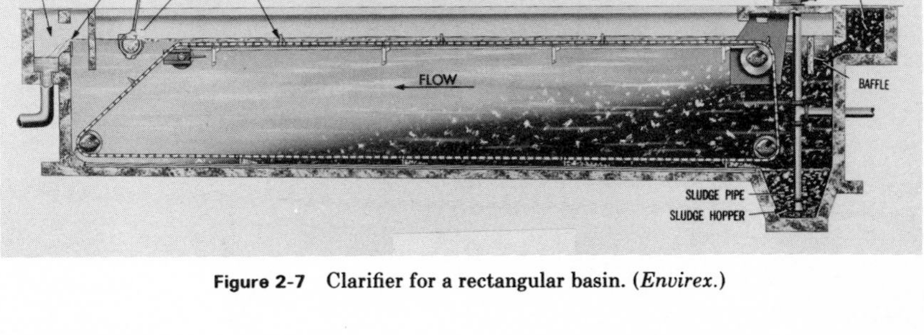

Figure 2-7 Clarifier for a rectangular basin. (*Envirex.*)

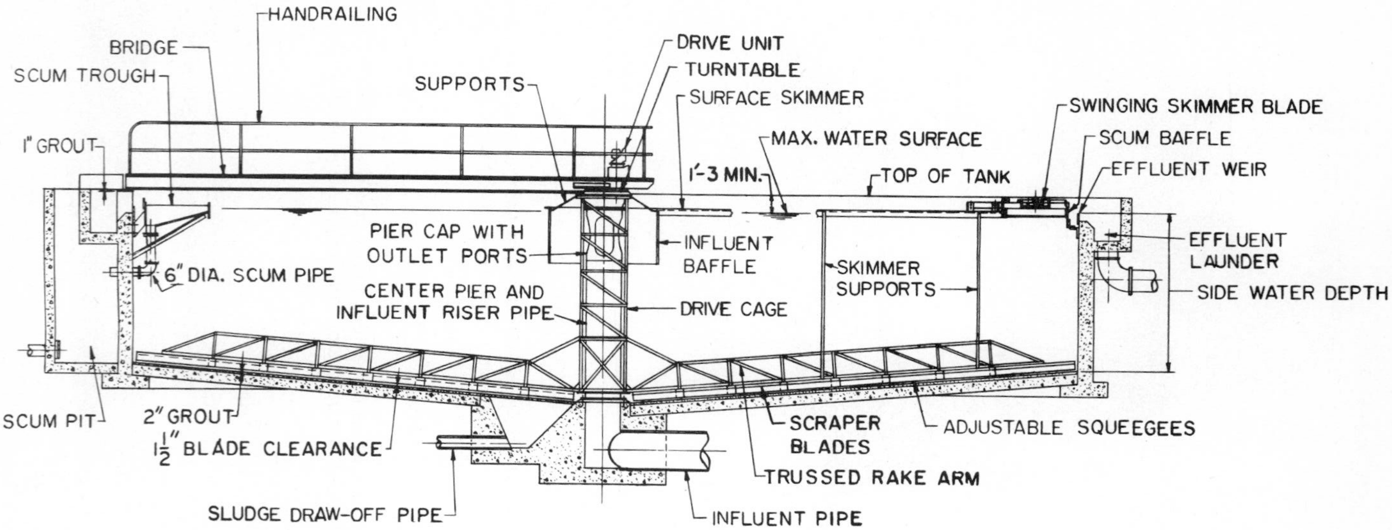
HANDRAILING
BRIDGE
SCUM TROUGH
1" GROUT
SUPPORTS
DRIVE UNIT
TURNTABLE
SURFACE SKIMMER
MAX. WATER SURFACE
1'-3 MIN.
TOP OF TANK
SWINGING SKIMMER BLADE
SCUM BAFFLE
EFFLUENT WEIR
EFFLUENT LAUNDER
SIDE WATER DEPTH
PIER CAP WITH OUTLET PORTS
INFLUENT BAFFLE
6" DIA. SCUM PIPE
CENTER PIER AND INFLUENT RISER PIPE
DRIVE CAGE
SKIMMER SUPPORTS
SCUM PIT
2" GROUT
1 1/2" BLADE CLEARANCE
SCRAPER BLADES
ADJUSTABLE SQUEEGEES
TRUSSED RAKE ARM
SLUDGE DRAW-OFF PIPE
INFLUENT PIPE

Figure 2-8 Clarifier for a circular basin.

produce better-quality effluent with shorter overall detention time than the approach using separate treatment units.

Flotation Systems

Flotation is sedimentation in reverse to remove floatable materials and solids with a specific gravity so close to that of water that they settle very slowly or not at all. The principle of air flotation is based on the fact that when the pressure on a liquid is reduced, dissolved gases are released as extremely fine bubbles. These bubbles attach themselves to any suspended matter present and rapidly float them to the surface, where they concentrate and can be removed by skimming.

Pressure-flotation units dissolve air in the water under pressure and then release it to the atmosphere in the flotation tank.

Flotation equipment may be circular or rectangular.

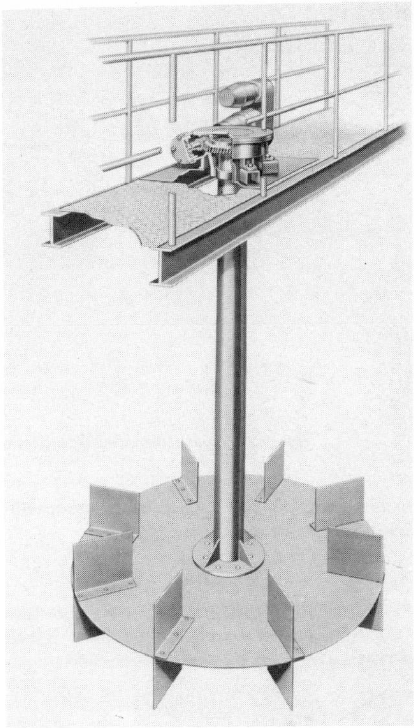

Figure 2-9 Mechanical flocculation device. (*Envirex.*)

Recycle Pressurization

The rectangular unit in Fig. 2-11 illustrates the use of recycle pressurization. Air is injected into the effluent recycle stream before it discharges into the inlet compartment of the flotation unit. There it is mixed with the incoming raw waste and releases the required air for flotation. The amount of effluent recycled varies from 25 to 50 percent of the forward flow. This approach has advantages when the raw waste is highly variable in composition or contains large amounts of solids.

Flotation units are normally designed for a feed-flow detention period of 15 min and an overflow rate of 1.5 to 4.0 gal/(min)(ft^2) [760 to 2025 L/(day)(m^2)]. Increased deten-

Figure 2-10 Solids contact clarifier. (*Envirex.*)

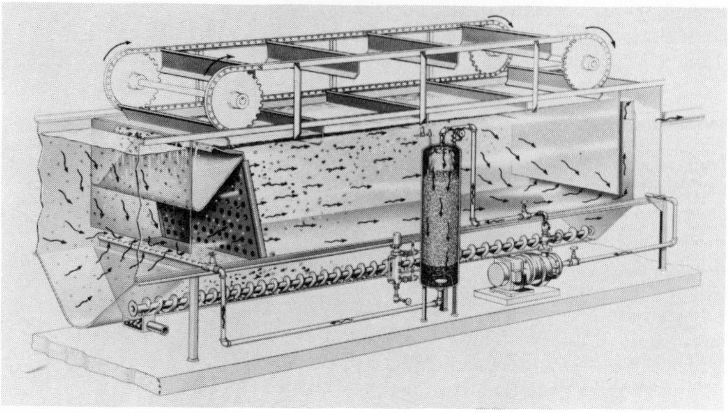

Figure 2-11 Rectangular flotation unit with recycle pressurization. (*Envirex.*)

tion time is needed if floated sludge must be thickened since surface area is based on the loading rate of solids.

High-Rate Gravity Filters

High-rate gravity filters remove suspended solids and operate in the range of 5 to 10 gal/(min)(ft^2) [2532 to 5063 L/(day)(m^2)]. They may be vertical or horizontal and filled with a variety of media such as anthracite coal, sand, and gravel.

PRODUCT RECOVERY

In many cases, industrial wastes contain valuable products such as high-value metals, acids, and other substances which can be used for manufacturing by-products, and these, when recovered, will yield high economic returns. Also obtainable are solvents, recovered with activated-carbon adsorption used for removal and recycling of solvents contained in the waste as vapors; these solvents include hydrocarbons, esters, alcohols, freons, ketones, and chlorinated or fluorinated organic compounds.

In the plating industry, rinse waters contain many of the following contaminants, usually in intolerable amounts: hexavalent chromium; sodium cyanide; complex cyanides of the heavy metals, such as cadmium, copper, zinc, and sometimes silver and gold; soluble nickel salts, strong mineral acids, and strong alkalis. Of these, the most toxic are the hexavalent chromium and cyanide ions.

Batch Treatment Method

There are several basic methods of treating such rinse waters for disposal or recovery of products. The oldest is the batch treatment method whereby the rinse waters are collected for treatment and disposal. The destruction of toxic chemicals is accomplished by oxidation of the cyanides and reduction of chomium. Chlorine (Cl_2) is used to destroy cyanides (CN^-), and the chlorinated rinse water is pumped into a reduction tank containing sulfuric acid and a reducing agent such as ferrous sulfate ($FeSO_4$), sodium bisulfite ($NaHSO_4$), or sulfur dioxide (SO_2). Chromium is reduced from its hexavalent state (Cr^{6+}) to the trivalent state (Cr^{3+}) and precipitated as a hydroxide, along with other metal hydroxides, by the addition of lime. The clear water from the clarifier is then ready for discharge into a sewer, and the sludge can be vacuum-filtered or disposed of in an acceptable manner.

Continuous-Flow Method

Another treatment of rinse water is the continuous-flow method. To be effective, this requires good instrumentation to monitor the flow and deliver the proper amount of reactant.

Ion Exchange

When the volume of chromic acid and sodium or potassium cyanide used is great enough, the recovery of the cyanides and chromic acid by ion exchange and evaporation has obvious economic advantages.

An integrated system for ion-exchange treatment of cyanide and chromium wastes has been found to be suitable for a plating operation that is split up into several parts. The system involves countercurrent rinsing, and the basic advantage of this system is that the toxic contaminants are destroyed before they can enter the rinse water, rather than having to be removed later.

Ion exchange plays an important part in chemical reclamation. The rinse waters are kept separate in the following categories:

1. Hard chrome rinse waters
2. Cyanide rinse waters containing silver, gold, or any other valuable metal
3. Miscellaneous wastes such as alkaline cleaners and acids

The hard chrome rinse waters are passed through a cation exchange and a strong-base anion exchanger. The contaminating metals such as copper, nickel, and trivalent chromium are exchanged on the cation exchanger, while the hexavalent chromium is exchanged on the anion exchanger. Upon exhaustion of the system, the cation exchanger is regenerated with an acid and the anion exchanger regenerated with caustic soda.

The regenerant effluent from the cation exchanger can be discharged to a collection chamber for neutralization. The regenerant effluent from the anion exchanger will contain sodium chromate and some caustic soda. This material can then be converted to chromic acid by passing it back through the cation exchanger so that the two exchangers may be alternated between service and conversion.

Table 2-5 shows recovery value of by-products from plating wastes.

An ion-exchange demineralization unit with a cation-anion mixed bed results in an effluent of less than 1 mg/L of total dissolved solids at a cost of roughly 5 cents per 1000 gal of waste treated. Membrane reverse-osmosis units can be used, but at substantially increased costs of five times or more than ion exchange.

Closed-Loop System

In the area of metal fabrication, parts are frequently washed or rinsed to clean away oils and other wastes. With a little care, improved wash operations cut downtime, reduce energy requirements, and cut soap costs. This is accomplished by a closed-loop wash-water treatment system.

TABLE 2-5 Economic Value of By-Product Recovery

Process	By-product	Concentration, mg/L	1980 value, \$/1000 gal
Rinse water	Copper	100–500	0.85–4.26
	Nickel	150–900	1.14–9.94
	Zinc	70–350	0.57–3.40
	Cadmium	50–250	1.70–8.50
	Chromium	400–2000	4.50–21.30
	Tin	100–600	1.14–6.80
Aluminum, bright dip	Phosphoric acid	10%	630

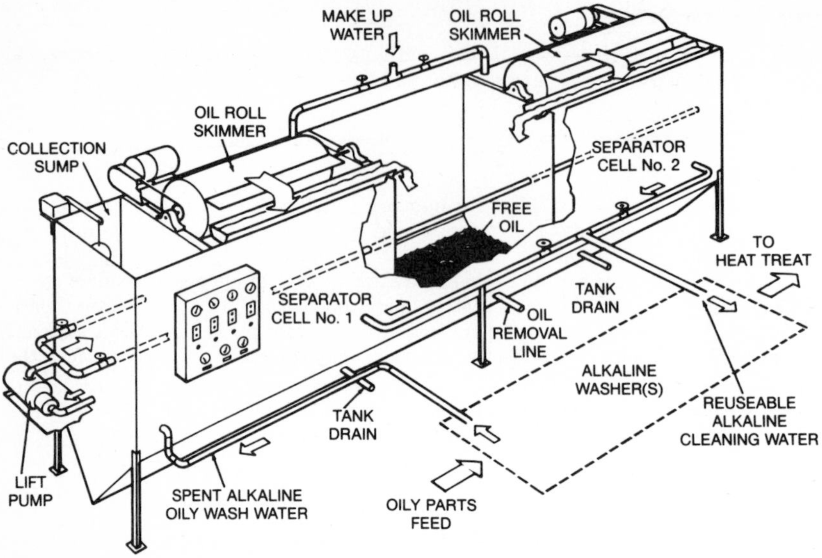

Figure 2-12 Alkaline wash system.

A typical closed-loop system (Fig. 2-12) is best described as a continuous-batch oil separator. It has dual compartments holding caustic wash solution, each equipped with an oil roll skimmer and separated by a waste tank. Piping leads from each compartment to a series of washers and back to a pump. Automated valves control flow from the pump to one of the two compartments.

One compartment continuously supplies caustic solution to a group of washers as the other stands for 24 h, allowing heavy materials to settle as oils float to the surface. Then, surface oils containing less than 0.1 percent water are skimmed off and drained into a waste tank; these may then be sold to an oil reclamation firm. While one wash solution in the first compartment undergoes treatment, the clean solution in the other compartment is circulated through the washers.

Treatment of Spent Coolants

Spent coolants can be treated for recovery of oil by acidification and centrifugation or by heat treatment at 223°F (106°C) for 22 h.

In the heat-treatment process, three distinct layers develop after there is water loss of approximately 12 percent through evaporation. The top layer consists of reusable oil, amounting to approximately 24 percent of the original scum volume; the middle layer, containing approximately 16 percent of the scum, is called the *rag layer* and consists of flocculated particulate matter; the bottom layer contains approximately 48 percent of the total scum and appears as relatively clear water. This bottom water layer is returned to the spent-coolant holding tanks for retreatment, and the rag, or intermediate, layer is combined with the swarfing dust and other heavy solids for removal by haulage.

A portion of the recovered oil can be used as the fuel for heating the floated scum, thereby making the system self-sustaining. The balance can be reprocessed for reuse as coolant. A simple product balance would show that for every 100 L of coolant treated, from 0.3 L to as much as 5 L of reusable oil can be recovered. A minimum flow of about 30,000 L of coolant per day might be selected as an economic break-even point.

At this minimum flow, taking, on an average, 3 L of recoverable oil for each 100 L of feed, 750 L per day of coolant would be recovered. Of this, approximately 50 L would be required to heat the scum to 223°F (106°C) for the 22-h break period, leaving a net total of 700 L of recovered oil.

Cost recovery can vary, depending upon the type of coolant used. Caution should be exercised when this system is used where plated parts are machined, as there is the hazard of cyanide buildup in the oil after prolonged usage.

Water Recovery

Recycled water may ultimately be the major valuable product because of increasing water-supply costs, increasing water-treatment costs, and mounting charges for using municipal wastewater facilities. The recovery of product fines, usable water, and thermal energy are important methods of reducing overall waste-disposal costs and should be seriously considered in every case.

Frequently, waste streams can be eliminated or reduced by process modifications or improvements. A notable example of this is the use of save-rinse and spray-rinse tanks in plating lines. This measure brings about a substantial reduction in waste volume and frequently a net reduction in metal dragout.

Segregation of waste streams is a necessity at times, not from the product-recovery point of view, but from the operational point of view. An example of this is segregation of acidic metal rinses from cyanide streams to avoid the production of toxic hydrogen cyanide (HCN) and thus eliminate potential safety hazards.

The prime requirement of waste treatment, by-product recovery, and water reuse is that the principal product or products of the plant be satisfactory to the consumer, and the secondary requirement is that the operation of the plants be efficient and economical.

By-Product Recovery and Use

The urgent problems facing industries are how to recover by-products from the waste materials inherent in every industrial operation and what to do with the by-products. Confronted with the growing dangers of pollution, the anticipation of government regulation, and the loss of valuable materials through unprofitable waste-disposal methods, industries are forced to develop sophisticated refining methods for processing chemical and industrial by-products and even to develop markets for by-products.

Industry is becoming increasingly aware of the necessity for pollution abatement and product recovery, not only because of its effect upon the general welfare of the public, but also because of its own dependence upon rivers and streams for suitable water for manufacturing processes. Industries are also increasingly aware of the fact that the benefits accruing from pollution abatement through product recovery may be quite significant.

SOURCE CONTROL AND WATER REUSE AND RECYCLING

The plant layout and arrangement of process and manufacturing sequences must be considered with regard to wastewater pollution control. This means undertaking a complete engineering survey of water use to develop an accurate water balance for peak and average operating conditions. In effect, what is needed is a complete inventory of all plant operations that use water and produce wastes.

With these data in hand, a fresh look should be taken at plant and process operations to see if any changes can be made that will reduce the amount of water used or decrease the flow of wastewater produced. A simple process adjustment is often all that is required to lower the concentration of pollutants. Perhaps there are valuable chemicals that can be recovered; or there may be alternative approach that might change the nature of the waste to make it easier to handle. Sometimes, segregating a contaminated process water from the rest of the waste discharge can reduce the size of the wastewater-treatment system.

The survey will uncover applications for water that can be recycled for repeated reuse. Some wastewater, now discharged to the sewer, may be well suited for cooling or boiler feed. The economics of using cooling towers to replace once-through cooling systems should be reconsidered. In some cases it may even pay to switch to air-cooled heat exchangers.

Machinery or operations having a common waste product can be connected to a centralized system of contaminant collection and treatment. With early involvement, the plant engineer can anticipate potential pollution problems and advise preventive measures. Where possible, the plant engineer should strive to prevent pollution at the source, thus minimizing the need to incorporate pollution control facilities. Typical methods of preventing pollution at the source include:

1. Substituting process materials
2. Modifying manufacturing procedures
3. Changing production equipment
4. Recycling process water

Examples of materials substitution are the use of chlorinated solvents of less toxicity for carbon tetrachloride and using a paint with reduced lead or zinc content.

Cascading or countercurrent water-use systems (where an operation that requires relatively low quality water uses wastewater from another operation) or recycling systems (where water is treated and returned to the same operation) can greatly reduce intake water requirements.

The quantities listed in column B, Table 2-1, should be considered the maximum pollutant values for reuse of process water.

REFERENCES AND BIBLIOGRAPHY

1. "Environmental Register," Illinois Environmental Protection Agency, Springfield, Ill., March 12, 1979.
2. "Industrial Waste Ordinance," Metropolitan Sanitary District of Greater Chicago, Chicago, Ill., January 19, 1978.

Beslievre, Edmund, and Schwartz, Max: *The Treatment of Industrial Wastes,* Industrial Water Engineering Bookshelf, Darien, Conn., 1978.

Eckenfelder, W. Wesley, Jr.: *Industrial Water Pollution Control,* McGraw-Hill, New York, 1966.

Lund, Herbert F.: *Industrial Pollution Control Handbook,* McGraw-Hill, New York, 1971.

Nemerow, Nelson L.: *Theories and Practices of Industrial Waste Treatment,* Addison-Wesley, Reading, Mass., 1963.

chapter 13-3

Solid-Waste Disposal

by
Charles Albert Johnson, Ph.D.
Technical Director
National Solid Wastes Management Association
Washington, D.C.

INTRODUCTION

One of the most significant factors in the successful operation of any industrial plant is the proper disposal of solid waste.

Obviously, such waste can be simply hauled away as generated by an outside contractor, or by plant personnel, to an appropriate disposal site. This chapter, however, discusses the alternative: employing an in-plant system to reduce the volume and weight of solid waste as generated so as to reduce haulage costs. However, hazardous material must be disposed of (usually without treatment) by specialists; this subject is discussed in the last portion of this chapter.

SYSTEM DESIGN

Due to large variations in the types of waste found in an industrial plant, each refuse-handling system should be custom-designed to fit the needs of each plant. The key to a successful design is understanding current operations—that of the plant engineer as well as the operations of the outside solid-waste managers (refuse collectors, equipment distributors, and manufacturers) servicing the plant. With this combination of knowledge, the most economical and efficient waste-handling system for a plant can be designed.

First, five essential factors must be analyzed:

1. The volume of waste produced
2. Its composition and characteristics

3. Special handling requirements
4. Location and other physical constraints
5. Requirements for safety and security

DISPOSAL METHODS

Having evaluated the five criteria, the plant engineer must determine the best method of disposal to be used. Two common methods employed in many industrial plants are (1) *compaction* and (2) *incineration.*

Compaction

Compaction is the method whereby a large volume of solid waste is reduced (squeezed) under high pressure, minimizing both container space and the number of times the container must be emptied. This reduces collection costs and permits a safe, clean method of handling in-plant refuse. The installation of an efficient compaction system, including chutes, conveyors, and large refuse-compactor containers, can reduce in-plant trash-handling costs by as much as 90 percent. Also, with an appropriate installation, "rear-door" pilfering can be practically eliminated.

Purchasers often want to specify *compaction ratio* or the density that can be produced by a compactor. Most manufacturers decline to guarantee specific compaction performance because it depends upon both the specific machine employed and the material being compacted. Typically, however, compactors will reduce the volume of wastes by factors ranging from 2 to 5.

Stationary Compactors

If stationary compaction is chosen, several factors must be considered. The parameters of stationary compactors include charging-chamber volume, pressure of the packer ram, cycle time, penetration of the packing ram into the compaction container, and compactor base size.

To assist purchasers of stationary compactors, the Waste Equipment Manufacturers Institute (WEMI) of the National Solid Wastes Management Association has developed a standardized method of rating compactors. Periodically WEMI publishes a book of standardized ratings for the equipment sold by its members.

Figure 3–1 shows a sketch of the most widely used style of stationary compactor, the horizontal detachable unit. Ratings are also provided for compactors with pivoting rams and self-contained compactors in which the compaction ram is integrally mounted within a refuse container.

Charging Chamber Volume. This must be large enough to accommodate, without any difficulty, the largest piece of refuse generated by the user.

Additional considerations must be taken into account when determining charging-chamber size. For example, the industrial plant may utilize in-plant trash carts of a specific size, so the charging chamber should be able to receive the entire contents of a full trash cart or a full container load. In many instances, this could mean that the charging area should be substantially greater than the largest single piece of refuse generated at the location.

Physical Dimensions of the Compactor. It may seem obvious that a compactor machine must fit properly into the space where it is intended to operate. However, errors made in this area are common, and utmost care is advised in planning. Conversely, a number of installation possibilities exist which may enable the utilization of a machine which, at first glance, would not appear to fit into the available space.

As a general rule, the placing of a stationary compactor should be calculated not only on the basis of the actual operating machine, but must include: (1) space for access to

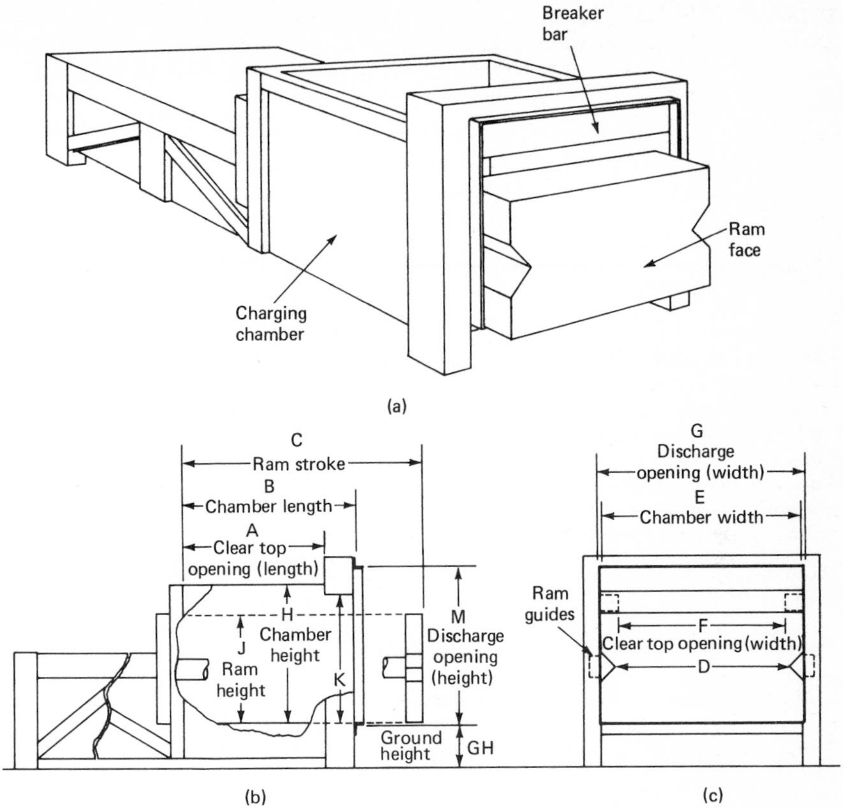

Figure 3-1 Horizontal commercial-industrial stationary compactor.

the charging chamber, (2) space for the container for the compacted wastes, and (3) space required for the pickup vehicle to maneuver, pull away the full container, replace an empty one, and then haul the full box away. Finally, sufficient room must be provided at the loading area to accommodate trash carts, containers, or even a conveyor chute leading to the hopper.

Dock space may be saved and plant security increased by installing the machine through the wall so that the charging area and the working mechanism remain inside the building, but the container is attached to the machine through an opening in the wall. Also, chutes from upper floors may run to the charging area, decreasing internal transportation costs and reducing the need for additional dock space.

Cycle Time. Cycle time is an important parameter of a container specification. It refers to the time required for a fully retracted ram face to pack the refuse from the charging box into the container and return to its original fully retracted position ready for another load of refuse.

Cycle times run from as short as 20 s to more than 1 min. Short cycle times are important if the application under consideration requires the capability of accepting refuse very quickly. A purchaser should carefully consider whether a compactor with a short cycle time is needed. Often compactors built with a rapid cycle utilize a small-diameter hydraulic cylinder which consequently exerts a relatively low force on the ram face, thereby sacrificing compaction pressure and material density.

Pressure of the Ram Face. A third important parameter of a stationary-compactor specification is the pressure of the ram face. Pressure, in pounds per square inch, is more important than total force in determining compaction density.

The ram pressure is the total force exerted on the ram, divided by the area of the ram. For a hydraulically operated compactor, the total ram force is the product of the hydraulic pressure and the hydraulic cylinder area.

For example, a compactor with a ram face measuring 30 x 60 in (0.8 x 1.6 m) has an area of 1800 in^2 (1.3 m^2). If it is acted upon by a 5-in-(12-cm) diameter hydraulic cylinder with a fluid pressure of 2000 lb/ft (14 x 10^6 N/m^2), the total ram force will be 39 200 lb (175 000 N), and the pressure exerted by the ram will be 21.7 lb/in^2 (140 000 N/m^2).

Most compactors have both a normal and maximum pressure rating. The higher pressure is used when finally packing out a container to make it easy to detach the container and ensure that the waste remains within it.

Penetration of the Ram into the Container. The penetration of the ram face into the container is an important factor in the operation of a compactor. Essentially, it represents the position of the forward ram stroke available for final compaction of the refuse load into the container. As the container begins to fill up with compacted waste, there is a tendency for the portion of refuse closest to the ram to fall back into the charging area. This is of most concern when the container is detached because waste falling from the container will litter the environment. The further a ram penetrates into the compaction container, the easier it is to load and detach a container cleanly.

Base Size. Compactor specifications are summarized as a single parameter, the base size. This is defined as the volume of waste theoretically moved through the compactor in a single stroke. Purchasers commonly use the base size as the primary specification, adding the parameters, listed above as necessary. Table 3-1 gives typical uses of compactors of various base sizes.

Rated stationary compactors manufactured by members of the WEMI carry the NSWMA Rating Seal. This seal assures performance conforming to established standards, certified by a registered professional engineer. Table 3-2 gives a hypothetical sample listing of rated compactors.

In addition to obtaining a compactor, the plant engineer will have to specify a container, which is usually provided by the refuse hauler. Before purchasing any equipment, compactor or container, the engineer should always consult the hauler to determine what type, size, and weight of containers can be handled. Some points to watch for are:

1. The container must be able to withstand the pressure of the waste without damage or distortion.
2. Roll-off and drop-off hoists, which are used to remove containers, have weight limits. The hoist will not be able to lift containers which exceed these limits.
3. State highway laws limit the amount of weight which can be carried on each chassis.
4. The length of the container affects the compaction ratio. The larger the container, generally the less the compaction.

There is no single formula for use in selecting a compactor for a plant. However, careful consideration of the tangible and intangible factors that have been outlined here will aid in avoiding costly mistakes.

TABLE 3-1 Typical Compactor Applications

Use	Base size	
	yd^3	m^3
High-rise apartment building	Up to 1	0.8
Commercial establishment	1 to 4	0.8 to 3
Industrial plants	1.5 to 7	1.2 to 5.5
Transfer stations	7 to 12	5.5 to 9

TABLE 3-2 NSWMA Commercial-Industrial Stationary Compactor Ratings

Manufacturer: Hypothetical Manufacturing, Inc. — Date: August 1977

Model number	NSWMA base size, yd^3	Clear top opening length (L) width (W), in	Chamber length, in	Ram stroke, in	Ram penetration, in	Force rating normal (N) maximum (M), (ram lb/in^2 per force, lb)	System pressures normal (N) maximum (M)	Ram face, in	Volume displacement rate, yd^3/h	Rated motor size, hp	Cycle time, s	Discharge opening width (W) height (H), in	Ground height, in	Base unit weight, lb
A	0.38	L = 24.0 W = 34.5	24.5	30.0	5.5	N = 25.3/18200 M = 27.8/20000	N = 1450 M = 1600	W = 36.0 H = 20.0	39	1	35	W = 37.0 H = 24.5	15.5	1800
B	1.06	L = 36.0 W = 51.5	39.0	50.0	11.0	N = 28.5/36200 M = 30.8/39100	N = 1850 M = 2000	W = 53.0 H = 24.0	106	10	36	W = 53.5 H = 32.5	16.0	4000
C	1.60	L = 40.0 W = 58.5	41.0	55.0	14.0	N = 21.6/38800 M = 24.3/43700	N = 2000 M = 2250	W = 60.0 H = 30.0	144	10	40	W = 61.0 H = 37.5	15.5	5100
D	1.91	L = 48.0 W = 58.5	49.0	65.0	16.0	N = 24.3/43700 M = 27.6/49600	N = 2250 M = 2550	W = 60.0 H = 30.0	146	10	47	W = 61.0 H = 37.5	15.5	5900
E	3.43	L = 84.5 W = 58.0	89.0	110.0	21.0	N = 55.8/100400 M = 61.4/110500	N = 2000 M = 2200	W = 60.0 H = 30.0	225	20	55	W = 61.5 H = 37.5	16.5	13000

Note: The National Solid Wastes Management Association presents this technical information on commercial-industrial stationary compaction equipment to assist users in the selection of equipment. The National Solid Wastes Management Association does not present these data as a warranty for equipment performance. The ratings represent data supplied by manufacturers that have been computed according to the NSWMA criteria and certified for accuracy by a registered professional engineer, selected by the manufacturer.

In-Plant Incineration

An alternative method of disposal is *incineration.* Although incinerators are more expensive to build and install than stationary compactors, their use can provide additional savings in refuse hauling costs. Energy recovery in the form of steam may also be considered as a source of additional savings.

When one is considering an incinerator, one should evaluate the following factors:

1. Any limitations imposed by air pollution emission limits in the area of the plant.
2. The physical constraints imposed by the dimensions of the plant.
3. The appropriateness in quantity and composition of the solid waste generated within the plant: a sufficiently high fired Btu value.
4. Fuel requirements. Most incinerators require supplementary fuel, either gas or oil, to control air pollution. Fuel costs can be high, and fuel availability may become uncertain. However, energy recovery can offset some of the costs.
5. By-product disposal: Adequate methods of disposal must be provided to handle the by-products of the incinerated waste.

Controlled-air incinerators (Fig. 3–2), the most frequently used incinerators today, were first commercially available in the United States in the early 1960s, but they were not really accepted until the late 1960s and early 1970s. This acceptance was mostly due to the increasing demands for high performance as measured by very low particulate emissions and a very high reduction ratio.

Most controlled-air incinerators employ two chambers. These chambers are desig-

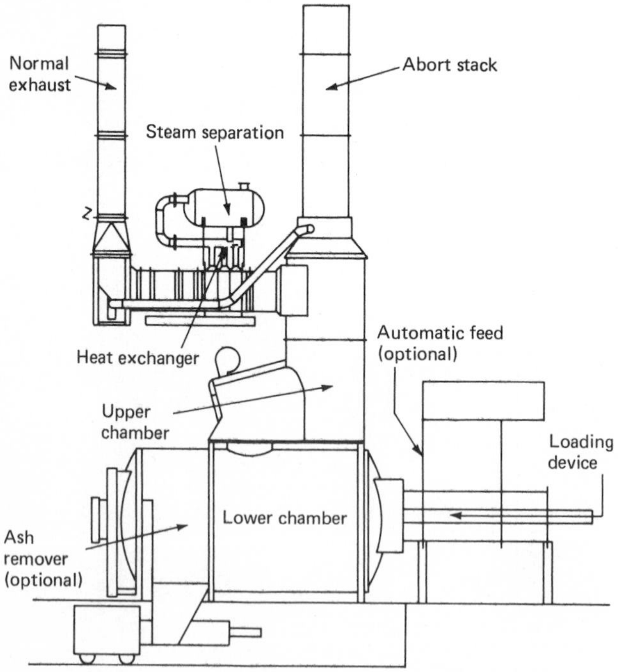

Figure 3-2 Controlled-air incinerator. (*Adapted with permission from the* American City & County Magazine.).

nated lower, or primary, and upper, or secondary. Performance of the antipollution functions of the system depends on controlling the conditions within these two chambers. The lower chamber is required to operate at low interior gas velocities and under controlled temperature conditions. This is done by limiting the air introduced into the primary chamber to less than the amount required for complete combustion (hence, the system is sometimes called "starved-air" incineration). This gives the lower chamber the operating characteristics of a partial oxidation system.

The heat released in the lower chamber is controlled by limiting the introduction of combustion air to an amount which will give partial oxidation of the waste in the chamber. The heat is sufficient to sustain the partial oxidation reactions. The gases from the lower chamber pass into the upper chamber through a turbulent zone, where additional air is added and ignition takes place to complete the oxidation reactions. The noncombustible portion of the waste and the carbonaceous residue from the reactions remain in the lower chamber. The noncombustibles are rendered sterile by the relatively high temperature while the carbonaceous material is further oxidized by the incoming air. The result is a high-quality sterile ash.

The gas velocity in the lower chamber is influenced by several factors. The gas which evolves from the chamber is a result of the interaction of the air, the auxiliary fuel, and the oxidation and volatilization products from the waste. The quantity of gas from the waste can vary substantially depending on chamber conditions of the waste and could therefore alter the gas velocity in the lower chamber significantly. The airflow controls of the upper and lower chambers are integrated in order to minimize cycling and provide a uniform flow of gases. This is important for controlling pollution performance and especially so for an efficient energy-recovery system.

When volatilization proceeds at an excessive rate as a result of the high temperatures, two distinct adverse effects ensue. First, the velocity in the lower chamber will exceed the design velocity, and particles which are too large to be oxidized properly will be carried into the upper chamber. Second, the gases will flow to the upper chamber at a rate which exceeds the capacity of the chamber and can result in excessive particulate emissions or smoking.

The function of the upper chamber is to complete the oxidation reactions of the combustible products as they are received from the lower chamber. In order to accomplish this, conditions in the chamber must be controlled within a rather narrow band, from inputs which vary rather widely. The control system is designed to maintain the required conditions by modulating both air and fuel to the system. This in effect controls the air input, auxiliary fuel input, and gas flow from the lower chamber.

The gases pass from the lower chamber into the upper chamber through a turbulent mixing region in a controlled manner. Additional air is introduced into the system and the gases are ignited, again under controlled conditions. The gas temperature at this point is somewhat higher than in the lower chamber, and the atmosphere is oxidizing (more than sufficient air for complete combustion). Temperatures in the upper chamber are limited to less than 2500°F (1400°C) in order to minimize production of nitrogen oxides and in the interest of equipment durability. On the lower-temperature side, it is recommended that at least 1800°F (1000°C) be maintained in order to stabilize an adequate reaction rate to complete the combustion process. The desired operating temperature point is adjustable but is factory preset for maximum performance. The primary means of controlling the temperature in the upper chamber is to control the quantity of combustion air. Air quantity is decreased when the temperature drops below the set point and increased when the temperature rises above the set point.

If heat is to be recovered, the gas temperature at the inlet to the heat exchanger is not allowed to exceed 1800°F (1000°C). This is done to protect the heat exchanger and is an addition to the normal safety controls. An over-temperature condition will automatically drive the hot-gas flow to the abort stack.

Compared with compaction, incineration is a more costly and more complex way of disposing of waste; however, the energy-recovery potential may in the long run prove to be a decisive factor for incineration in many plants.

Hazardous Waste Disposal

Disposal of hazardous wastes has become a serious problem for many industrial plant engineers. This problem is especially critical for chemical industries, and most such companies now have entire departments devoted to this concern. Many other industries also have hazardous wastes to dispose of, generally in lesser amounts.

Transportation, treatment, storage, and disposal of hazardous wastes are now very tightly regulated under provisions of Title C of the Solid Waste Disposal Act. Even though generators of small quantities of hazardous wastes are exempted from the full scope of the regulations, it is not a safe practice merely to intermingle hazardous wastes, even in small quantities, with other solid waste without taking special precautions.

To arrange for disposal of hazardous wastes, a plant engineer should contact a waste service company that is fully permitted to transport and dispose of those wastes. The company should provide a written statement certifying where the wastes are to be taken and how they are to be treated or disposed of. The disposal facilities should be inspected and the actual waste disposition verified.

The hazardous-waste generator will be requested to initiate a manifest for shipments of hazardous wastes going off-site for disposal and will also be required to designate where the wastes are to be taken.

Transportation of hazardous wastes, inter- or intra-state, is regulated by the Department of Transportation and the Environmental Protection Agency under both the Hazardous Materials Transportation Act and the Solid Waste Disposal Act. The plant engineer should be very sure that the transporter chosen is in full compliance with these regulations.

section 14

Plant Safety and Sanitation

chapter 14-1

Safety Considerations in Plant Operations

by

Arthur Spiegelman, P.E.
M & M Protection Consultants
A Technical Service of Marsh & McLennan Inc.
New York, New York

SAFETY

A well-organized plant safety program should encompass all phases of the plant environment and operations. From the standpoint of this handbook we are primarily concerned with safety controls and devices in the environment and only make reference to supervisory controls of the safety program.

Safeguarding industrial hazards has always been one of the principal tasks of the plant engineer. Production depends on the ability to maintain a continuous flow of materials without the interruptions caused by accidents. Engineering controls and safeguards serve to protect inexperienced workers or those workers who are distracted or who suffer from fatigue.

The basic elements of a good safety program include:

1. Management leadership
2. Assignment of responsibility
3. Maintenance of safe working environment
4. Training program
5. Record system
6. Medical follow-through

The plant engineer is basically concerned with items 1, 2, and 3, but items 4, 5, and 6 are also important.

LEGAL ASPECTS OF SAFETY

The early legal aspects of industrial safety were limited to laws which were developed in conjunction with workers' compensation acts. They were primarily aimed at providing for just compensation to the injured party, but they also established accident investigation procedures and some regulation of hazards.

The past two decades have sent a greater public demand for safety in the workplace. With it came a proliferation of federal laws and regulations which include the following.

Laws Applicable to Industrial Plants

1. Environmental Protection Agency
   - Toxic Substances Control Act of 1976
   - Marine, Protection, Research and Sanctuaries Act of 1972
   - Safe Drinking Water Act of 1974
   - Water Pollution Control Act of 1972
   - Clean Air Act of 1970
   - National Environment Policy Act of 1969
   - Atomic Energy Act of 1954
   - Clean Water Act
   - Endangered Species Act of 1973
   - Energy Supply and Environmental Coordination Act
   - Fish and Wildlife Coordination Act
2. Solid Waste Disposal Act of 1965
3. Occupational Safety and Health Act of 1971
4. Department of Transportation
   - Transportation Safety Act
   - Hazardous Materials Transportation Act of 1974
   - Ports and Waterways Safety Act of 1972
5. Resource Conservation and Recovery Act of 1976
6. Federal Insecticide, Fungicide, and Rodenticide Act of 1972
7. Consumer Product Safety Commission
   - Federal Hazardous Substances Act
   - Consumer Product Safety Act
   - Poison Prevention Packaging Act
8. Mine Safety and Health Act

The Occupational Safety and Health Act will probably have a far greater effect on business and industry than the other legislation. The law requires all employers in the private sector of business to furnish their employees with a workplace free from those recognized hazards likely to cause physical harm or death. Standards are published or referenced in the act. This 800-page document is entitled *OSHA Safety and Health Standards* (29 CFR 1910). It is available from the Superintendent of Documents, Washington, DC 20402, and is recommmended as a reference for plant engineers.

The adoption of specific federal legislation has also led to an increase in the number of civil suits. The violation of a federal standard may become prima facie evidence of negligence. Such cases have been very costly.

Recent developments at OSHA have placed greater emphasis on the health aspects of worker's compensation. What started in specialized industries, such as coal mining and asbestos, is spreading to the chemical industry and others. Much more of the social responsibility for the well being of the employee is being directed to the employer.

All employers must keep accurate records of work-related injuries, illnesses, and deaths. Any injury that involves medical treatment, loss of consciousness, restrictions of motion or work, or job transfer, must be recorded. Required information must be logged and specific information regarding each case detailed. At the end of the calendar year a summary of logged cases must be posted in each plant.

INSTRUMENTATION AND CONTROLS

Some of the most important advances in safety over the past two decades have been achieved in the area of instrumentation and controls. Standards of the Instrument Society of America (ISA) should be followed. The plant engineer is directly concerned with these devices since they are involved in controlling the operations, materials flow, and environment of the plant. Instrumentation and controls fall into the following categories:

1. Temperature systems
2. Pressure systems
3. Flow systems
4. Analytical and testing systems (i.e., gas analysis, solids and dust analysis, reaction product tests, oxygen analysis, pollution instrumentation, electronic components, and system checkout equipment)
5. Weighing, feeding, and batching systems
6. Force and power systems
7. Motion and geometric systems (i.e., speed, velocity, vibrations)
8. Humidity and moisture determinations
9. Radiation systems
10. Electrical determinations
11. Communication systems
12. Automatic process controllers (i.e., computer process control, safety in instrumentation and control systems, electronic, pneumatic and hydraulic control devices)
13. Controlling elements (i.e., valves, actuators, electric motor drives and controls, mercury switches, etc.)

The importance of instrumentation and controls cannot be overestimated. From the standpoint of safety and environmental control the following applications are dominant:

1. Detection of leaks in equipment
2. Survey of operating areas for escape of toxic materials
3. Detection of flammable or explosive mixtures in the atmosphere or process lines
4. Monitoring plant stacks and other areas for the accidental discharge of toxic gases, vapors, or smokes.
5. Analysis of waste streams for toxic or other objectionable material.
6. Control of waste-treatment or product-recovery facilities

Automatic process control, unlike manual control, provides continuous monitoring and corrective action. However, automatic controllers cannot react to new conditions, nor can they predict beyond the data programmed into them. Therefore, the ability to foresee an upset condition in a process and the capability of developing fail-safe actions may well determine whether a serious accident is averted.

Analytical and testing instrumentation provides the plant engineer with much needed information. For instance, decisions can be made in these areas:

1. Purity of raw materials entering a process. Contaminants may be hazardous.
2. Process control by automatization of sampling.
3. Process troubleshooting by continuous analysis to prevent process upsets.
4. Determination of product quality to provide an impetus to a product safety program.

PLANT ENGINEER'S FUNCTION WITHIN THE SAFETY COMMITTEE

A safety program is a management method to develop specific safety objectives, assign responsibility, and obtain desired results. The plant engineer should participate with the safety committee from the standpoint of developing adequate inspection and control procedures in order to maintain a safe work environment and control of safety devices. Safety devices may be anything from a pressure-relief valve to a machine guard.

Inspection and Maintenance

In many plants an inspection committee is appointed. A safety inspection may be done in conjunction with the regular plant inspection and maintenance program. This is an opportunity for the plant engineer to check all safety devices. The safety inspection should focus attention on those items directly concerned with accident prevention. The plant engineering staff should have a working knowledge of safety standards if their inspections are to cover safety as well as normal wear and tear of equipment. A general plant inspection should include:

1. All buildings and physical equipment
2. Inspection of new machinery before it is placed in operation
3. Inspection of walking and working surfaces and means of egress
4. Special equipment such as powered platforms, personnel lifts, and vehicle-mounted work platforms
5. Materials handling and storage facilities, elevators, cranes
6. Machinery, machinery guards, and electrical equipment.
7. Compressed gas and air equipment
8. Special equipment such as pressure vessels, drums and furnaces, and welding, cutting, and brazing equipment
9. Hand and portable powered tools and other hand-held equipment
10. Environmental control, equipment, ventilation, and pollution controls for toxic and hazardous substances

A systematic method of inspection and maintenance is preferred. Some companies require *preventive* maintenance programs which call for the replacement of *critical* equipment periodically regardless of inspection results. Others require general inspections of the entire premises, annually.

Areas of the plant that have the potential of developing into catastrophic hazards require special inspection procedures. Items that may develop into a catastrophe include: (1) structural failure, (2) fire or explosion, (3) release of hazardous gases or vapors.

Some elements of a plant may require inspection and maintenance on the basis of federal, state or local laws. Items such as elevators, boilers, and unfired pressure vessels are in this category. This equipment may require the use of specially trained and licensed inspectors either from a governmental agency or an insurance carrier.

A careful record should be kept of all inspections and recommendations. This is particularly important in the event that an accident occurs at this location which results in litigation. Some companies assign inspection tasks to maintenance personnel, electricians, and others who are charged with the job of repairing the equipment. Supervisors should continuously examine their own work areas to make sure that tools, machinery, and other types of equipment are safe to handle.

Inspection methods should be established for all new equipment and processes. Nothing should be placed into operation until all safeguards have been checked and operations evaluated by the plant engineer. Safe operating instructions should be given to all workers concerned.

Inspection Procedures

The following inspection criteria should be applied:

1. Inspectors must be familiar with the company's safety and health policies as well as the particular laws and regulations that pertain. Frequently, these regulations are

only minimum requirements and it may be necessary to exceed them to secure adequate safety.

2. Inspectors should have available an analysis of all accidents that have occurred in the plant within the past year.
3. The inspectors should utilize all aids available, including inspection checklists, report forms, and other pertinent information.

An inspection report should be divided into three areas of interest:

1. A report on imminent hazards which require immediate corrective action.
2. A routine report on unsatisfactory (nonemergency) conditions which need corrective action.
3. A general report on the overall safety conditions of the facility.

ACCIDENT PREVENTION

The four basic steps for preventing accidents are as follows:

1. Elimination of the hazard
2. Control of the hazard
3. Training of personnel to be aware of and avoid the hazard
4. Utilization of personal protective equipment

The plant engineer is primarily concerned with steps 1 and 2 by ensuring a safe design for plant, physical facilities, and machinery.

BUILDING STRUCTURE

The plant engineer is directly concerned with the inspection and maintenance of the building structures. Slippery floors, stairs, runways, ramps, and other means of access are involved in many serious plant accidents. About one-fifth of the industrial injuries result from falls. Many of these can be avoided by careful design, construction, and maintenance. Applicable standards and codes should be applied.

Basic Rules

Housekeeping is a prime consideration in minimizing the hazard of falls. Some of the basic rules in this area are as follows:

1. All places of employment should be kept clean and orderly. Floors shall be maintained in a clean and dry condition with adequate drainage.
2. All passageways including aisles, ramps, and stairways should be kept clear and in good repair.
3. Floor loading shall be kept well within prescribed limits.
4. All floor and wall openings should be guarded by standard railings or properly constructed closures.
5. Stairways should conform to acceptable standards.
6. Exits should be sufficient in number and properly located so that the building can be evacuated quickly in an emergency.
7. All ladders should conform to the applicable standards and codes and they should be properly maintained.
8. Ladders should not take the place of fixed stairways.
9. Workers should follow good practices in the use of ladders with regard to placement,

support, angle between horizontal base and vertical plane of support, and proximity to other hazards (i.e., electrical).

10. Scaffolds, which are in effect elevated working platforms, should be designed with an adequate factor of safety and protection for the workers.
11. Scaffolds must be maintained, inspected, and guarded on all exposed sides. Metal scaffolds should not be constructed near electrical equpment.

Standards

The following standards apply to building structures:

ANSI Z4.1-1979, "Requirements for Sanitation in Places of Employment"

41 CFR 50-204.2 *Code of Federal Regulations,* "Public Contracts and Property Management," General Safety and Health Standards.

ANSI A58.1-1972, Building Code Requirements for "Minimum Design Loads in Building and Other Structures"

ANSI A12.1-1973, "Safety Requirements for Floor and Wall Openings, Railings, and Toeboards"

ANSI A64.1-1968, "Requirements for Fixed Industrial Stairs"

ANSI A14.1-1975, "Safety Requirements for Portable Wood Ladders"

ANSI A14.2-1972, "Safety Requirements for "Portable Metal Ladders"

ANSI A14.3-1974, "Safety Requirements for Fixed Ladders"

ANSI A10.8-1977, "Safety Requirements for Scaffolding"

ANSI A92.1-1977, "Manually Propelled Mobile Ladder Stands and Scaffolds"

MEANS OF EGRESS FOR INDUSTRIAL OCCUPANCIES

A means of egress is a continuous and unobstructed way to exit from any point in a building or structure to a public way. It includes vertical and horizontal ways of travel and intervening room spaces and other areas. The number of exit facilities required is specified in standards and codes, and it depends on the structure, occupants, and hazard exposures.

Inspections of means of egress have shown the following items at fault in such facilities:

1. Improper number of exits and locations.
2. Poor illumination—lack of emergency lighting.
3. Lack of directional signs.
4. Poor housekeeping—obstructions.
5. Improper and hazardous floor surfaces.
6. Faulty operation of exit doors.

Standard

The following standard applies to egress for industrial occupancies:

NFPA 101 1970, "Life Safety Code"

POWERED PLATFORMS, PERSONNEL LIFTS, AND VEHICLE-MOUNTED WORK PLATFORMS

Powered platforms, personnel lifts, and vehicle-mounted work platforms are means of elevating workers on suspended operated work platforms or personnel lifts. The platforms are used for exterior building maintenance work, the personnel lifts are used to lift workers to an elevated jobsite, and the vehicle-mounted work platforms are used to position personnel to an elevated work site. All of these installations should be in conformance with the appropriate ANSI standards.

Every work platform should be tested and inspected frequently. New platforms should be tested before they are placed in service. Each installation should be inspected and tested at least every 12 months and should undergo a maintenance inspection and test every 30 days. Results of all inspections and tests should be logged, including the date, time, and inspector.

Items which require special inspections are as follows:

1. Governors
2. Initiating devices
3. Both independent braking systems
4. Interlocks and emergency electric devices
5. Electric systems
6. Emergency communication system

Maintenance is required specifically on all parts of the equipment related to safe operations. Broken or worn parts and electric devices should be replaced promptly. Gears, shafts, bearings, brakes, and hoisting drums should be maintained in proper alignment. Gears should be replaced when there is evidence of appreciable wear. All parts should be kept free from dirt. Wires or ropes should be replaced when they are damaged or in a deteriorated condition. Guardrails and toeboards are required for working platforms. Load-rating plates must be conspicuously posted and adhered to.

The personnel lifts require frequent inspections. Their use should be limited to *personnel.* They should not be used to lift construction materials. The inspections should include but are not limited to the following items:

Steps	Drive pulley
Rails and supports	Electrical systems
Belt and belt tension	Vibration—alignment
Handholds	Brake systems
Floor landing	Warning lights
Limit switches	Pulleys—clearance

Standards

The following standards apply to platforms, personnel lifts, and vehicle-mounted work platforms:

ANSI A120.1-1970, "Safety Code for Powered Platforms for Exterior Building Maintenance"

ANSI A9.2-1969, "American National Standard for Vehicle-Mounted Elevating and Rotating Work Platforms"

ANSI A90.1-1969, "Safety Code for Manlifts"

VENTILATION

Ventilation systems protect the health and environment of plant workers by removing objectionable dusts, fumes, vapors, or gases. They are essential safety devices and must be properly installed and maintained. Ventilation systems can be divided into two primary groups, local systems and general systems.

Local Systems

Local exhaust systems prevent the accumulation of toxic or flammable materials near the process unit. Due to the variety of work and the types of equipment, it is necessary to design hoods specifically developed to be as close to the operation as possible. Air-sampling devices are available to determine whether the atmosphere has been adequately cleared or constitutes a hazard. Safety standards may specify local exhaust ventilation for particular processes.

After contaminated air passes from the hood to exhaust ducts, it is sent through a cleaning system or to the outdoors. Such exhaust air must conform to the regulations of the EPA. Dusts may be collected by means of cyclone dust collectors (Fig. 1-1) or electrostatic precipitators. Many types of air-cleaning devices are used. The selection is determined by the degree of the hazard associated with the dust. Gases or vapors may be removed from the waste air by absorption or adsorption processes which dissolve or

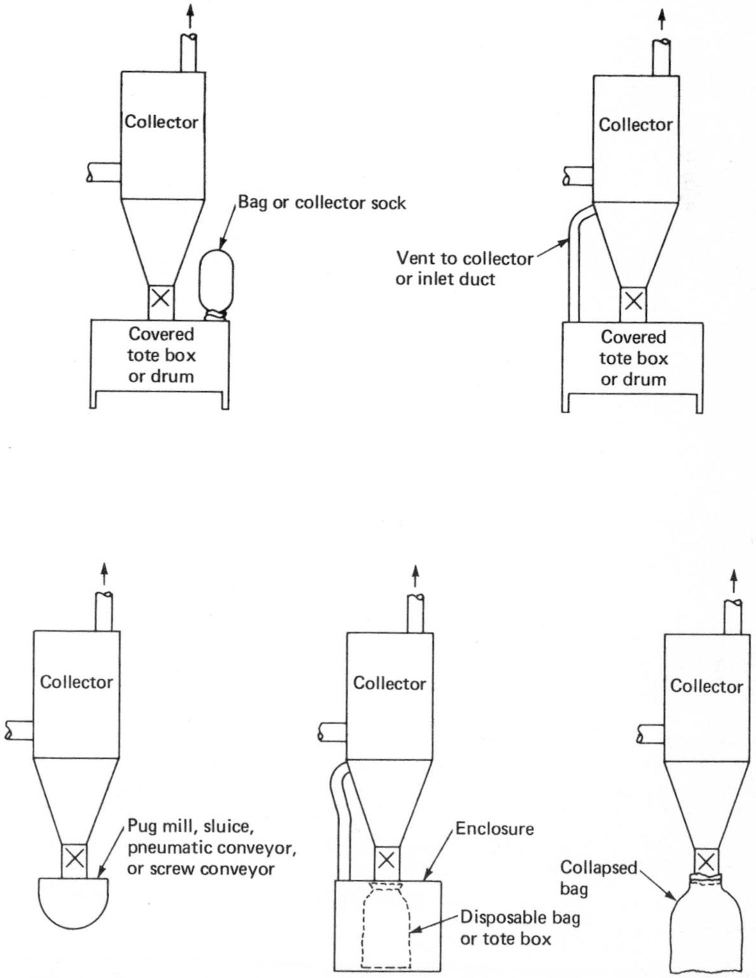

Figure 1-1 Dry-type dust collectors for dust disposal. *(American Conference of Governmental Industrial Hygienists.)*

react with the waste product chemically. Some gases and vapors are rendered harmless by passing them through a combustion chamber where they are changed to acceptable gases. In other cases condensers are utilized to liquefy toxic vapors for removal.

When ventilation is used to control potential exposure to workers it is adequate to reduce the concentration of contaminant so that the hazard is removed.

Local exhaust ventilation is used for a great many industrial operations such as anodizing, pickling, metal cleaning, and open-surface tank operations. There is a standard developed for this purpose. Other uses include spray booths, dip tanks, fume controls in electric welding grinding operations, cast-iron machining, dust control in foundries, ventilation of internal combustion engines, and kitchen range hoods.

One of the basic maintenance items in the operation of local exhaust ventilation systems is to ensure that the flow of air is unobstructed. A minimum maintained velocity must exist in order to meet the health and safety requirements. Where flammable gases or vapors are removed, the electric equipment must conform to code requirements.

At intervals of not more than 3 months the hood and duct system should be inspected for evidence of corrosion, damage, or obstruction. In any event if the airflow is found to be less than required, it should be increased to that required.

General Systems

General ventilation in the workplace contributes to the comfort and efficiency of the employees. It also serves to clear the air of hazardous contaminants and excessive heat or humidity. When local exhaust ventilation cannot be applied due to the many sources of vapor release, general ventilation is prescribed. The publication of the American Society of Heating, Refrigerating and Air Conditioning Engineers, *Handbook of Fundamentals*, defines the prescribed methods and requirements for the development of acceptable ventilation systems. The plant engineer should inspect this equipment frequently to make sure that it performs as required.

Standards

The following standards pertain to ventilation:

ANSI Z9.4-1979, "Ventilation and Safe Practices of Abrasive Blasting Operations"

ANSI Z43.1-1966, "Ventilation Control of Grinding, Polishing, and Buffing Operations"

ANSI Z9.3-1964 (R1971), "Safety Code for Design, Construction and Ventilation of Spray Finishing Operations"

ANSI Z9.1-1977, "Practices for Ventilation and Operation of Open-Surface Tanks"

41 CFR 50-204.10 *Code of Federal Regulations,* "Public Contracts and Property Management," General Safety and Health Standards: Occupational noise exposure.

41 CFR 50-204.22 *Code of Federal Regualtions,* "Public Contracts and Property Management," Radiation Standards: Exposure to airborne radioactive material.

COMPRESSED GASES

Plant engineers should be aware of the safety requirements for compressed-gas handling and storage. Pressure-relief devices for gas cylinders, portable tanks, and cargo tanks should be installed and maintained in accordance with Compressed Gas Association (CGA) pamphlets S1.1-1963 and S1.2-1963.

References

The following references pertain to compressed gases:

CGA G-1-1966, "Cylinders"

CGA G1.3-1959, "Piped Systems"
CGA G1.4-1966, "Generators and Filling Cylinders"

MATERIALS HANDLING AND STORAGE

Powered Industrial Trucks

Safety requirements for powered industrial trucks involve the following:

1. The plant engineer is directly concerned with the safety requirements relating to fire protection: design, maintenance, and use of fork trucks, tractors, platform lift trucks, motorized hand trucks, and other specialized industrial trucks.
2. All new industrial trucks should meet the requirements of ANSI B56.1-1969, "Powered Industrial Trucks."
3. In locations used for the storage of hazardous liquids or liquefied or compressed gases, only approved trucks designated as DS, ES, GS, or LPS may be used. In areas containing combustible dusts approved EX or ES trucks may be used.
4. Any power-operated industrial truck not in safe operating condition should be removed from service.

Overhead and Gantry Cranes

Safety requirements for overhead and gantry cranes include:

1. All new and existing overhead and gantry cranes should meet the design specifications of ANSI B30.2.0-1976, "Safety Standard for Overhead and Gantry Cranes (Top Running Bridge, Multiple Girder)."
2. Exposed moving parts such as gears, set screws, projecting keys, chains, chain sprockets, and reciprocating components which might constitute a hazard under normal operating conditions should be guarded.
3. Both holding brakes and control brakes should be tested to determine whether they are within required tolerances.
4. Brakes on trolleys and bridges should be examined and adjusted when necessary.
5. Inspections fall into two classes: *frequent inspections,* which fall into daily or monthly intervals, and *periodic inspections,* which fall into 1- to 12-month intervals. *Frequent inspections* should include (*a*) all functional operating mechanisms for maladjustments interfering with daily operations, (*b*) deterioration or leakage in air or hyraulic systems, (*c*) hooks and hoist chains and ropes including end connections, (*d*) all functional operating mechanisms. *Periodic inspections* should include (*a*) all items listed under frequent inspections, (*b*) deformed, cracked, or corroded members, (*c*) worn, cracked, or distorted parts, (*d*) excessive wear on brake system, (*e*) improper performance of power plants, (*f*) excessive wear of chain drive sprockets and excessive chain stretch, (*g*) defective electrical apparatus.
6. Preventive maintenance based on crane manufacturers' recommendations should be established.
7. Any unsafe conditions—including those in ropes—disclosed in the inspection requirements should be corrected before operation of the crane is allowed.

Crawler, Locomotive, and Truck Cranes

Safety requirements for crawler, locomotive, and truck cranes include:

1. All new and existing locomotive and truck cranes should meet the design specifications of ANSI B30.5-1968, "Safety Code for Crawler, Locomotive, and Truck Cranes."

2. Load ratings should not exceed the stipulated percentages for cranes with indicated types of mountings.
3. Inspections are frequent and periodic as cited under overhead gantry cranes.

Derricks

All new derricks constructed after August 31, 1971, should meet the design specifications of ANSI B30.6-1977, "Safety Code for Derricks." Those derricks constructed prior to that date should be modified to conform. The plant engineer should check the standard to assure compliance.

Inspection procedures are similar to those required for cranes under frequent or periodic inspections. Derricks not in regular use for a period of 6 months should be given a complete inspection before use.

Maintenance and repair is an important part of the plant engineer's assignment and should include the following steps:

1. Any unsafe conditions shown in the inspection should be corrected.
2. Adjustments should be made to assure correct functioning of all components.
3. Ropes and wires should be thoroughly inspected and replaced or repaired.
4. No derrick should be loaded beyond the rated load.

Helicopter Cranes

Helicopter cranes should comply with the applicable regulations of the Federal Aviation Administration. The weight of an external load should not exceed the helicopter manufacturer's rating. All equipment should be checked frequently. The cargo hooks should be tested prior to each day's operation to determine that the release functions properly, both electrically and mechanically.

Slings

All slings used in conjunction with materials handling equipment made from alloy-steel chains, wire rope, metal mesh, natural or synthetic fiber rope, and synthetic web (nylon and polypropylene) should conform to the applicable standards. They should be inspected daily for damage or defects.

Standards

The following standards pertain to materials handling and storage:

41 CFR 50-204.3 *Code of Federal Regulations*, "Public Contracts and Property Management," General Safety and Health Standards: Material Handling and Storage.

NFPA 231-1970, "General Indoor Storage"

NFPA 505-1969, "Powered Industrial Trucks"

ANSI B56.1-1975, "Safety Standard for Powered Industrial Trucks—Low Lift and High Lift Trucks"

ANSI B30.2.0-1976, "Safety Standard for Overhead and Gantry Cranes"

ANSI B30.5-1968, "Safety Code for Crawler, Locomotive, and Truck Cranes"

ANSI B30.6-1977, "Safety Code for Derricks"

MACHINERY AND MACHINE GUARDING

One or more methods of machine guarding should be provided to protect the operator and other employees in the machine area from hazards such as those created by point of operation, ingoing nip points, rotating parts, flying chips, and sparks. Examples are barrier guards, two-hand tripping devices, electronic safety devices, etc.

General

The following machines usually require point-of-operation guarding:

Guillotine cutters
Shears (Figs. 1-2 and 1-3)
Alligator shears
Power presses (Fig. 1-4)
Milling machines
Pointer saws
Jointers
Forming rolls and calenders
Revolving drums, barrels, and containers
Exposed blades

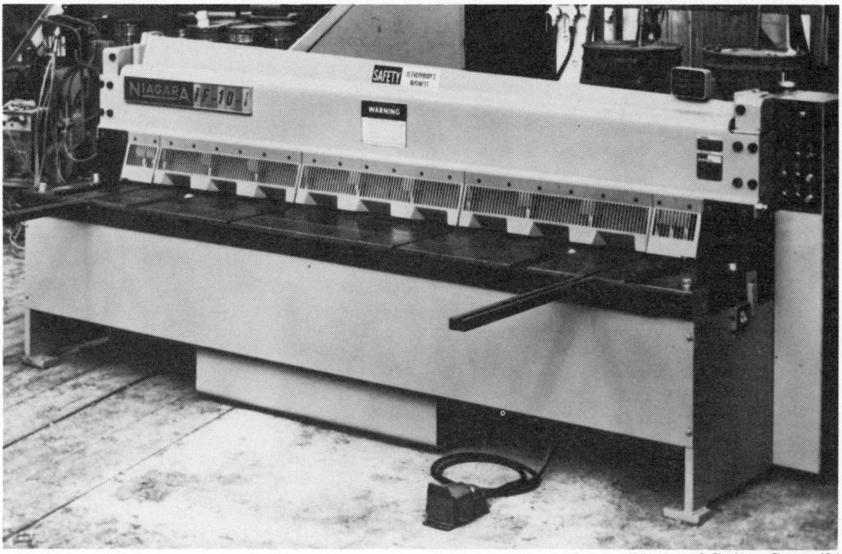

Figure 1-2 Metal shear protected by finger guards with plastic windows. *(National Safety Council.)*

Woodworking Machinery

Woodworking machinery includes automatic cutoff saws, circular saws, hinged saws, revolving double arbor saws, hand-fed ripsaws, hand-fed crosscut saws, circular resaws, swing-cut saws, radial saws, bandsaws, jointers, tenoning machines, boring and mortising machines, wood shapers, planers, lathes, sanding machines, guillotine veneer cutters, and miscellaneous woodworking machines. These machines should be properly guarded in accordance with appropriate ANSI standards. All belts, pulleys, gears, shafts, and moving parts should be guarded in accordance with requirements.

Abrasive Wheel Machinery

Abrasive wheel machinery includes cylindrical grinders, surface grinders, swing frame grinders, and automatic snagging machines. The guard design and specifications should be in accordance with appropriate ANSI standards.

Figure 1-3 Paper shear with two hand controls for clamping and shearing operations. *(National Safety Council.)*

MILLS AND CALENDERS IN THE RUBBER AND PLASTICS INDUSTRIES

All new and existing installations of mills and calenders should comply with the OSHA requirements and pertinent standards.

Mill safety controls consisting of safety trip controls, pressure-sensitive body bars, safety trip rods, safety trip wire cables, or wire center cords should be installed in all mills either singly or in combination. Fixed guards should be used where applicable.

Calender safety controls should be provided on all calenders within reach of the operator and the bite. Calenders and mills should be installed so that persons cannot normally reach over or under, or come in contact with, the roll bite. All trip and emergency switches should require manual resetting. A suitable alarm should be provided in conjunction with the safety devices.

MECHANICAL POWER PRESSES

All mechanical power presses (excluding pneumatic power presses, bulldozers, hot-banding and hot-metal presses, forging presses, hammers, riveting machines, and similar types of fastener applicators) should conform to the OSHA regulations and applicable standards. Mechanical power presses require particular emphasis as they are involved in many accidents. The machine components should be designed, secured, and covered to minimize hazards. Safeguards should include:

Figure 1-4 Swing-in safety prop for air hammer press. *(National Safety Council.)*

1. Brakes or blocks capable of stopping the motion of the slide (Fig. 1-4)
2. Foot pedals, protected to prevent unintended operation
3. Hand-operated levers to prevent premature or accidental tripping
4. Two-hand trips to have individual operator hand controls arranged to require the use of both hands to trip the press (Fig. 1-3).
5. Air-controlling equipment to be protected against foreign material
6. Brake monitoring systems installed on the press to indicate when the performance of the braking system has deteriorated
7. Point-of-operation guards installed and used in accordance with the standard's requirements
8. A regular program of periodic and regular inspections on a weekly basis of power presses
9. Reports of injuries to employees operating mechanical power presses required by OSHA, with an analysis of each accident on a 30-day basis

FORGING MACHINES

The safety requirements apply to use of lead casts and other uses of lead in the forge or die shop. All equipment should comply with OSHA requirements and the applicable standards. This should include:

1. Thermostatic control of heating elements required for melting lead

2. Industrial hygiene and personal protective equipment required due to the toxicity and heat condition of the lead
3. All presses and hammers controlled by operators who are protected by required guards
4. Devices included to lock out the power when dies are being changed

MECHANICAL POWER TRANSMISSION

This section pertains to mechanical power transmission equipment (see "Standards"), with the exception of small belts operating at a reduced speed and reduced requirements for the textile industry to prevent accumulation of combustible lint. Standard guards are usually secured with the following materials: expanded metal, perforated or solid sheet metal on a frame of angle iron, or an iron pipe securely fastened to the floor or frame of machine. The standard requirements include guards for the following:

1. Flywheels located 7 ft or less from the floor
2. Cranks and connecting rods when exposed to contact
3. Tail rods or extension piston rods
4. Shafting–horizontal, vertical, and inclined
5. Power transmission apparatus and pulleys
6. Belts, ropes, chain drums, gears, sprockets, and chains
7. Shafting, collars, couplings, belt shifters, clutches, perches, or fasteners

Periodic inspection is required of all power transmission equipment. Intervals not exceeding 60 days are recommended.

Standards

The following standards apply to mechanical power transmission:

41 CFR 50-204.5 *Code of Federal Regulations,* "Public Contracts and Property Management, General Safety and Health Standards: Machine Guarding.

ANSI O1.1-1975, "Safety Requirements for Woodworking Machinery"

ANSI B7.1-1978, "Safety Requirements for the Use, Care, and Protection of Abrasive Wheels."

ANSI B28.1-1967 (R1972), "Safety Specifications for Mills and Calenders in the Rubber and Plastic Industries"

ANSI B11.1-1971, "Safety Requirements for Construction, Care, and Use of Mechanical Power Presses"

ANSI B24.1-1971, "Safety Requirements for Forging"

ANSI B15.1-1972, "Safety Standard for Mechanical Power Transmission Apparatus"

HAND AND PORTABLE POWERED TOOLS AND OTHER HAND-HELD EQUIPMENT

Portable tools must be equipped with adequate guards and should comply with OSHA and the appropriate standards. This includes:

1. Portable circular saws, saber, scroll, and jigsaws, portable belt sanding machines, portable abrasive wheels, explosive actuated fastening tools, power lawnmowers, jacks.

2. Warning instructions should be supplied with the equipment. All equipment should be inspected periodically.

Standards

The following standards pertain to hand-held equipment:

ANSI A10.3-1977, "Safety Requirements for Power Actuated Fastening Systems"

ANSI B7.1-1978, "Safety Requirements for the Use, Care, and Protection of Abrasive Wheels"

ANSI B71.1-1972, "Safety Specifications for Power Lawn Mowers, Lawn and Garden Tractors, and Lawn Tractors"

41 CFR 50-204.4 and 50-204.8 *Code of Federal Regulations,* "Public Contracts and Property Management," General-Safety and Health Standards: Tools and equipment; Use of compressed air.

ANSI O1.1-1975, "Safety Requirements for Woodworking Machines"

ANSI B30.1-1975, "Safety Standard for Jacks"

ANSI Z9.4-1979, "Ventilation and Safe Practices of Abrasive Blasting Operations"

WELDING, CUTTING, AND BRAZING

Most welding and cutting operations are mobile and generally used for construction, demolition, maintenance, and repair. Production line welding and cutting equipment is permanently installed; the hazards can be controlled through proper design and operation. The oxygen and acetylene used for welding and cutting must be handled and stored in accordance with the pertinent standards.

Compressed-gas cylinders should be handled with care. The fusible safety plugs in acetylene cylinders melt at about 212°F (100°C). This is an obvious fire hazard if the cylinder is subjected to heat. Oxygen, on the other hand, will react strongly with oils or other hydrocarbons upon contact.

Regulators and pressure gauges should be used on the appropriate cylinders. Leaking cylinders must be removed from the building and taken away from sources of ignition. All equipment should be kept free from oily or greasy substances.

Piping systems should be tested and proved gas-tight at 1½ times the maximum operating pressure. Service piping systems should be protected by pressure-relief devices discharging upward to a safe location. Backflow protection should be provided by an approved device that will prevent oxygen from flowing into the fuel system.

Acetylene generators should be of approved construction and clearly marked with the maximum rate of acetylene production and the pressure limitations. Relief valves should be regularly operated to ensure proper functioning. Storage of all chemicals should conform to the requirements of the standards.

ARC-WELDING AND CUTTING EQUIPMENT

Arc-welding apparatus should comply with the requirements of the National Electrical Manufacturers "Standard for Electric Arc-Welding Apparatus," NEMA-EW-1 1962, and for ANSI/UL551-1976, "Safety Standards for Transformer-Type-Arc-Welding Machines." The design requirements include the following items of concern to the plant engineer:

1. Input power terminals, tip change devices, and live metal parts should be completely enclosed.
2. Terminals for welding leads should be protected from accidental electrical contact by

personnel or by metal objects. The frame or case of the welding machine should be grounded as specified.

3. Printed rules and instructions concerning operation of the equipment supplied by the manufacturer should be strictly followed.
4. All parts of the equipment should be frequently inspected. Cables with damaged insulation or exposed bare conductors should be replaced.

RESISTANCE-WELDING EQUIPMENT

All resistance-welding equipment should be installed in accordance with article 630D of the ***NEC®***.* The following items of particular interest to the plant engineer are cited:

1. Controls on all automatic or air and hydraulic clamps should be arranged or guarded to prevent accidental actuation.
2. All doors and access and control panels should be kept locked and interlocked to prevent access by unauthorized persons to live portions of the equipment.
3. All press-welding machine operations, where there is a possibility of the operator's fingers being under the point of operation, should be effectively guarded in a manner similar to that prescribed for punch-press operations.
4. Flash-welding equipment should be equipped with hoods to control flying flash and ventilation of fumes.
5. Combustible material must be removed from the welding area and the basic fire-prevention requirements in accordance with ANSI/NFPA 51B-1977. Fire watchers are required wherever a fire problem exists. A welding permit system should be instituted to control hazardous exposures.
6. Welding operators should use all of the protective equipment specified in the standard, and ventilation should be used where required.

Standards

The following standards are for resistance-welding equipment:

ANSI Z49.1-1973, "Safety in Welding and Cutting"

ANSI/NFPA-51-1978, "Oxygen–Fuel Gas Systems for Welding and Cutting"

ANSI/NFPA 51B 1977, "Cutting and Welding Processes"

41 CFR 50-204.7 *Code of Federal Regulations,* "Public Contracts and Property Management," General Safety and Health Standards: Personal protective equipment

SPECIAL INDUSTRIES

Plant engineers involved in the following special industries should consult the OSHA regulations for specific requirements pertinent to their industry.

1. Pulp, paper, and paperboard mills
2. Textiles
3. Bakery equipment
4. Laundry machinery and operations
5. Sawmills
6. Pulpwood logging
7. Agricultural operations
8. Telecommunications

****NEC®*** is a Registered Trademark of the National Fire Protection Association, Quincy, MA 02269

Standards

Following are some special-industry standards:

ANSI P1.1-1969, "Safety Requirements for Pulp, Paper, and Paperboard Mills"

ANSI L1.1-1972, "Safety Requirements for the Textile Industry"

ANSI Z50.1-1977, "Safety Requirements for Bakery Equipment"

ANSI Z8.1-1972, "Safety Requirements for Commercial Laundry and Drycleaning Equipment and Operations"

ANSI O2.1-1969, "Safety Requirements for Sawmills"

ANSI O3.1-1978, "Safety Requirements for Pulpwood Logging"

ELECTRIC EQUIPMENT

All electric equipment in the plant should comply with OSHA regulations and the **National Electrical Code®** (NFPA 70-1981)*. It is particularly important for the plant engineer to be knowledgeable in the following sections:

1. Article 250, Grounding
2. Article 500, Hazardous(Classified) Locations

TOXIC AND HAZARDOUS SUBSTANCES

The exposure of employees to any of approximately 600 chemicals (see "Standards" at end of this section) should at no time exceed the ceiling value given for that material. To obtain compliance with this section of the OSHA regulations, it is necessary to develop administrative and engineering controls. When such controls are not feasible, protective equipment or other protective measures should be used.

Controls can be developed by environmental monitoring, personal monitoring, and employee observation. The plant engineer has an important part to play in the inspection and maintenance of the environmental monitoring equipment.

One method for controlling exposure is by local exhaust-ventilation and dust-collection systems. They should be constructed, installed, and maintained in accordance with ANSI Z9.2-1979, "Fundamentals Governing the Design and Operation of Local Exhaust Systems," which is incorporated by reference in this section of the handbook.

Twenty-three materials require special care in handling since they are in the category of regulated substances which present a cancer hazard. These are listed below:

Asbestos	beta-Propiolactone
Coal-tar pitch volatiles	2-Acetylaminofluorene
4-Nitrobiphenyl	4-Dimethylaminoayobenzene
alpha-Naphthylamine	*N*-Nitrosodimethylamine
Methylchloromethyl ether	Vinyl chloride
3,3-Dichlorobenzidene	Inorganic arsenic
bis(Chloromethyl) ether	Benzene
beta-Naphthylamine	Coke-oven emissions
Benzidine	Cotton dust
4-Amino diphenyl	1,2-dibromo-3-chloropropane
Ethylenimine	Acrylonitrile

*__National Electrical Code®__ is a Registered Trademark of the National Fire Protection Association, Quincy, MA 02269.

Controls on these materials are very detailed since the materials must be contained in a closed-system operation. The operating area must be restricted to authorized personnel only. Warning signs and instructions should be posted.

Standard

Refer to the following for standards on toxic and hazardous substances:

Tables 2-1, 2-2 and 2-3, "Threshold Limit Valves for Chemical Substances in Workroom Air, with Intended Changes," 1978 Conference of American Governmental Industrial Hygienists PO Box 1937, Cincinnati, OH 45201.

chapter 14-2

Fire Protection and Prevention*

by

Robert William Ryan
Assistant Director
Department of Environmental Safety
University of Maryland
College Park, Maryland

Richard Best
Senior Fire Analysis Specialist
Research and Fire Information Services
National Fire Protection Association
Quincy, Massachusetts

GLOSSARY

The following list contains the terms most often used in this chapter or in the fire protection field.

Boiling point The temperature at which the liquid boils when under normal atmospheric pressure (14.7 psia). The boiling point increases as pressure increases and is dependent on the total pressure.

Combustible A material or structure which can burn is labeled combustible. Combustible is a relative term; many materials which will burn under one set of conditions will not burn under others, e.g., structural steel is noncombustible, but fine steel wool is combustible. The term *combustible* does not usually indicate ease of ignition, burning intensity, or rate of burning, except when modified, as in *highly combustible interior finish.*

Fire prevention Measures directed toward avoiding the inception of fire.

Fireproof A misnomer for fire-resistive.

Fire load The amount of combustibles present in a given situation, usually expressed in terms of weight of combustible material per square foot. This measure is employed frequently to calculate the degree of fire resistance required to withstand a fire or to judge the rate of application and quantity of extinguishing agent needed to control or extinguish a fire.

Fire point The lowest temperature of a liquid in an open container at which vapors are evolved fast enough to support continuous combustion.

Fire resistance A relative term, used with a numerical rating or modifying adjective to indicate the extent to which a material or structure resists the effect of fire, e.g., "fire resistance of 2 h."

Fire-resistive Properties or designs to resist the effects of any fire to which a material or structure may be expected to be subjected. *Fire-resistive materials* or assemblies of materials are noncombustible, but noncombustible materials are not necessarily fire-resistive; fire-resistive implies a higher degree of fire resistance than noncombustible. *Fire-resistive construction* is defined in terms of specified fire resistance as measured by the standard time–temperature curve.

Fire-retardant Usually denotes a substantially lower degree of fire resistance than fire-resistive and is often used to refer to materials or structures which are combustible in whole or in part, but have been subjected to treatments or have surface coverings to prevent or retard ignition or the spread of fire under the conditions for which they are designed.

Flameproof, flameproofing Misleading terms and their use is discouraged in favor of *flame-retardant* or *flame-resistant.*

Flame-resistant A term that may be used more or less interchangeably with flame-retardant.

Flame-retardant Materials, usually decorative, which due to chemical treatment or inherent properties, do not ignite readily or propagate flaming under small to moderate exposure.

Flammable A combustible material that ignites very easily, burns intensely, or has a rapid rate of flamespread is called flammable. Flammable is used in a general sense without reference to specific limits of ignition temperature, rate of burning, or other property. *Flammable* and *inflammable* are identical in meaning. Flammable is used in preference to inflammable.

Flammable limits The extreme concentration limits of a combustible in an oxidant through which a flame will continue to propagate at the specified temperature and pressure. For example, hydrogen-air mixtures will propagate flame between 4.0 and 75 percent by volume of hydrogen at 21°C and atmospheric pressure. The smaller value is the lower (lean) limit, and the larger value is the upper (rich) limit of flammability. For liquid fuels in equilibrium with their vapors in air, a minimum temperature exists for each fuel above which sufficient vapor is released to form a flammable vapor-air mixture. There is also a maximum temperature above which the vapor concentration is too high to propagate flame. These minimum and maximum temperatures are referred to respectively as the lower and upper *flash points* in air. The flash-point temperatures for a combustible liquid vary directly with environmental pressure.

Flashover The phenomenon of a slowly developing fire (or radiant heat source) producing radiant energy at wall and ceiling surfaces. The radiant feedback from those surfaces gradually heats the contents of the fire area, and when all the combustibles in the space have become heated to their ignition temperature, simultaneous ignition occurs.

Flash point The lowest temperature at which the vapor pressure of a liquid will produce a flammable mixture and resultant flame. The flame will not continue to burn at this temperature if the source of ignition is removed.

Glowing combustion and flame Combustion is the process of exothermic, self-catalyzed reactions involving either a condensed-phase or a gas-phase fuel, or both. The process is usually (but not necessarily) associated with oxidation of a fuel by atmospheric oxygen. Condensed-phase combustion is usually referred to as glowing combustion, while gas-phase combustion is referred to as a flame.

Ignition temperature (autoignition temperature, autogenous ignition temperature The minimum temperature to which the substance in air must be heated in order to initiate, or cause, self-sustained combustion independently of the heating or heated element. The ignition temperature of a combustible solid is influenced by rate of airflow, rate of heating, and size and shape of the solid. Small sample tests have shown that as the rate of airflow and the rate of heating are increased, the ignition temperature of a solid drops to a minimum and then increases.

Latent heat Heat is absorbed by a substance when converted from a solid to a liquid and from a liquid to a gas. Conversely, heat is released during conversion of a gas to a liquid, or a liquid to a solid. Latent heat is the quantity of heat absorbed or given off by a substance in passing between liquid and gaseous phases (latent heat of vaporization) or between solid and liquid phases (latent heat of fusion). The high heat of vaporization of water is a reason for the effectiveness of water as an extinguishing agent.

Noncombustible Not combustible.

Nonflammable Not flammable.

Specific heat The heat, or thermal capacity, of a substance is the number of calories required to raise 1 g 1°C. The specific heats of various substances vary over a considerable range; for all common substances, except water, they are less than unity. Specific-heat figures are significant in fire protection as they indicate the relative quantity of heat needed to raise the temperature to a point of danger, or the quantity of heat that must be removed to cool a hot substance to a safe temperature.

Vapor density The weight of a volume of pure gas compared with the weight of an equal volume of dry air at the same temperature and pressure. A figure less than 1 indicates that a gas is lighter than air, and a figure greater than 1 that a gas is heavier than air. If a flammable gas with a vapor density greater than 1 escapes from its container, it may travel at a low level to a source of ignition.

Vapor pressure Because molecules of a liquid are always in motion (with the amount of motion depending on the temperature of the liquid), the molecules are continually escaping from the free surface of the liquid to the space above. Some molecules remain in space while others, due to random motion, collide with the liquid. If the liquid is in an open container, molecules (collectively called *vapor*) escape from the surface, and the liquid is said to evaporate. If, on the other hand, the liquid is in a closed container, the motion of the escaping molecules is confined to the vapor space above the surface of the liquid. As an increasing number strike and reenter the liquid, a point of equilibrium is eventually reached when the rate of escape of molecules from the liquid equals the rate of return to the liquid. The pressure exerted by the escaping vapor at the point of equilibrium is called *vapor pressure.* Vapor pressure is measured in millimeters of mercury (mm), or torr.

INTRODUCTION

Fire protection is an area in which most plant engineers can make a significant contribution to plant operations. At many facilities the chief engineer may also serve as the *fire marshal* or *fire chief.* Even at larger plants, where there is a full-time safety or fire-protection engineer responsible for plant protection and loss prevention, the plant engineer should be familiar with the fire problem, fire-prevention methods, and fire protection systems.

Fires in private industry have a great potential for significant economic and financial loss to both the industrial plant and the community. Recovery from an industrial fire includes not only replacement of equipment and facilities at higher costs, but also temporary and permanent lost business income, loss of skilled employees during the time the plant is closed, loss of profits on damaged finished goods, and extra expenses to restore operations. Many plants destroyed by fire do not reopen, contributing to local unemployment and disrupting the personal lives of employees.

This chapter is intended to give the plant engineer a basic understanding of fire protection and prevention, to provide some basic information about fire, and to identify other resources for the incorporation of fire safety in every aspect of plant operations.

In addition to the National Fire Protection Association in Quincy, Mass., and the Federal Emergency Management Agency in Washington, D.C., the following fire protection services are also available in most cities in the United States: the Society of Fire Protection Engineers, fire protection consultants, equipment manufacturers, testing laboratories, and companies specializing in special-hazard protection and control, fire investigations, fire-protection equipment testing, installation, and servicing. There are also state and regional training facilities, and the resources of municipal fire departments should not be overlooked.

This chapter does not attempt to discuss management programs which may have a significant impact on plant fire protection. Information regarding risk management, personnel training, plant fire brigades, plant emergency organization, and fire-prevention programs must be obtained from other sources. A list of sources is presented in the final section of this chapter.

THE NATURE OF FIRE

A simple method of visualizing what fire is and how burning takes place is to use a four-sided object, or tetrahedron. Each surface of the tetrahedron is used to represent one of the conditions necessary for fire to occur (see Figure 2-1).

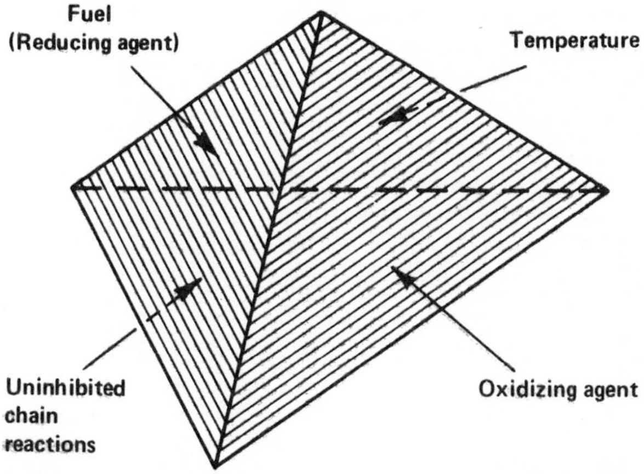

Figure 2-1 Tetrahedron fire model. *(From R. Tuve,* Principles of Fire Protection Chemistry, *NFPA, Quincy, Massachusetts, 1976.)*

The four components of this simple fire model are fuel, heat, oxidant, and the chemical chain reaction. While this model does not provide a complete scientific description of fire, it is sufficient for explaining most fire-protection concepts. In chemical terminology, the *fuel* component may also be referred to as a reducing agent. During the fire reaction the reducing agent loses electrons. The *heat* component includes both the heat which causes the fire and the heat emitted by the fire which causes it to be self-sustaining. The *oxidant* required for fire is most often provided by the oxygen in the ambient air, approximately 21 percent. Although oxygen is the most common oxidizing agent and is usually necessary for fire to occur, there are some chemicals that release oxygen and some that can burn in an oxygen-free atmosphere.

The *chemical chain reaction* of fire is a self-sustaining reaction yielding energy or products that cause further reactions of the same kind, or burning.

Vapor State of Fuel

Before a fuel can be burned it must be in a vapor state. Therefore, flammable gases are most easily ignited. Even solids and liquids, such as wood and gasoline, must be vaporized before they will burn. The decomposition of matter due to heat which generates flammable vapors is called *pyrolysis.*

The initial vaporization of the fuel may be caused by heat from a source of ignition such as a chemical reaction, electric energy, or mechanical heat energy. Ambient heat may also be sufficient to vaporize fuels which are normally liquids.

Once sufficient vaporization has occurred, the combustible vapors can be ignited by an open flame or spark or, at a sufficiently high temperature, the vapor and oxygen mixture will ignite spontaneously.

After ignition of the vaporized fuel has occurred, the heat generated by the fire will cause further vaporization of the fuel and the intensity of the fire will increase. This process is known as *radiation feedback*.

Heat Transfer

The heat which causes the fuel to ignite and the heat generated by the resulting fire can be transmitted by one or all of the following three methods: conduction, radiation, or convection.

- **Conduction.** Heat transferred by direct contact from one body to another.
- **Radiation.** Energy travel through space or materials as waves.
- **Convection.** Heat transferred by a circulating medium—either a gas or a liquid.

Products of Combustion

There are four categories of products of combustion: (1) fire gases, (2) flame, (3) heat, and (4) smoke. All of these products are produced in varying degrees by each fire. The material or materials that are involved in the fire and the resulting chemical reactions produced by the fire determine the products of combustion.

Fire Gases

The primary cause of loss of life in fires is the inhalation of heated, toxic, and oxygen-deficient gases and smoke. The amount and kind of fire gases present during and after a fire vary widely with the chemical composition of the material burning, the amount of available oxygen, and the temperature. The effect of toxic gases and smoke on people will depend on the time of exposure, the concentration of the gases in air, and the physical condition of the individual.

There are usually several gases present during a fire. Those that are most lethal are carbon monoxide, carbon dioxide, hydrogen sulfide, sulfur dioxide, ammonia, hydrogen cyanide, hydrogen chloride, nitrogen dioxide, acrolein, and phosgene.

Flame

The burning of materials in a normal oxygen-rich atmosphere is generally accompanied by flame. For this reason, flame is considered a distinct product of combustion. Burns can be caused by direct contact with flames or heat radiated from flames. Flame is rarely separated from the burning materials by any appreciable distance.

Heat

Heat is the combustion product most responsible for fire spread.

Exposure to heat from a fire will affect persons in proportion to the length of exposure and the temperature of the heat. The dangers of exposure to heat from fire range from minor injury to death. Exposure to heated air increases the heart rate and causes dehydration, exhaustion, blockage of the respiratory tract, and burns. Fire fighters should not enter atmospheres exceeding 120°F (48°C) to 130°F (54°C) without special protective clothing and masks. The maximum survivable breathing level of heat from fire in a dry atmosphere for a short period has been estimated at 300°F (148°C). Any moisture present in the air greatly increases the danger and sharply reduces the time of survival.

Smoke

Smoke is matter consisting of very fine solid particles and condensed vapor. Fire gases from common combustibles (such as wood) contain water vapor, carbon dioxide, and carbon monoxide. Under the usual conditions of insufficient oxygen for complete combustion, methane, methanol, formaldehyde, and formic and acetic acids are also present. These gases are usually evolved from the combustible with sufficient velocity to carry droplets of flammable tars which appear as smoke. Particles of carbon develop from the decomposition of these tars; they are also present in the fire gases from the burning of petroleum products, particularly from the heavier oils and distillates.

These small particles of carbon and tarlike particles that are visible and the phenomenon of fire gases rendered visible by the particles are generally defined as *smoke*.

Modes of Combustion

The combustion process may occur in two modes, *flaming* (including explosions) and *flameless* (including glow and deep-seated glowing embers). The flaming mode is characterized by relatively high burning rates. Intense and high levels of heat are usually associated with the flaming mode.

The flaming and flameless modes are not mutually exclusive; combustion may involve one or both modes. Often combustion may occur in the flaming mode and gradually make a transition to the flameless mode. At one point in this process both modes occur simultaneously.

Fire Control

Fire prevention and fire extinguishment can be described in terms of the fire model previously discussed. Fire prevention is generally a matter of keeping heat and fuel separated; or, in some processes, keeping heated fuel from combining with oxygen.

Fire extinguishment can be summarized by four methods:

1. Removal or dilution of air or oxygen to a point where combustion ceases.
2. Removal of fuel to a point where there is nothing remaining to oxidize.
3. Cooling of the fuel to a point where combustible vapors are no longer evolved or where activation energy is lowered to the extent that no activated atoms or free radicals are produced.
4. Interruption of the flame chemistry of the chain reaction of combustion by injection of compounds capable of quenching free-radical production.

Removal of Oxygen

The amount of dilution of oxygen necessary to stop the combustion varies greatly with the kind of material that is burning. Ordinary hydrocarbon gases and vapors will not burn when the oxygen level is below 15 percent. Acetylene will continue to burn unless the oxygen concentration is lowered, but will continue to glow on the surface even if the oxygen level is as low as 4 to 5 percent.

A fire in a closed space can extinguish itself by consuming the oxygen. However, incomplete combustion, which takes place when the oxygen is consumed, usually results in considerable generation of flammable gases.

A commonly used method of putting a fire out by removing or diluting the oxygen is by flooding the entire fire area with carbon dioxide or with some other inert gas.

Some burning materials react violently with water and can best be extinguished by covering them with a suitable inert material.

Fuel Removal

Fuel removal can be accomplished in a variety of ways. One of the most common examples is the practice of bulldozing a firebreak across the path of an advancing forest fire.

Fires in large coal or wood pulp piles can usually be controlled only by moving the pile out of the fire zone. Fires in large oil storage tanks have been controlled by pumping the oil out of the burning tank into an empty tank. If a gas line is ruptured and the gas ignited, shutting off the supply of gas is the only way to stop the fire.

If it is not practical to remove the fuel, extinguishment can be accomplished by shutting off the fuel vapors or by covering the burning or glowing fuel. The use of fire-fighting foams and dry powder extinguishers are effective procedures for covering or coating a fire.

Cooling

For most common combustibles such as wood, paper, and cloth, the simplest and most effective means of removing the heat of a fire is through the application of water. Water application can be varied and will depend on the fire.

Applying water to the burning fuel cools the fuel until the rate of release of combustible vapors and gases is reduced and ultimately stopped. Heat developed by a fire is carried away by radiation, conduction, and convection. This helps reduce the amount of heat and makes the use of water more effective. Only a relatively small proportion of the heat evolved needs to be cooled by the water in order to extinguish the fire.

Effective use of water cannot be accomplished if the water cannot reach the burning fuel directly. For this reason, areas where fire fighters cannot readily reach the fire with water streams, such as high-rise buildings and high-piled storage areas, must be provided with automatic sprinklers or other automatic fire-protection systems.

Interruption of Chemical Reaction

Extinguishment by cooling, by oxygen dilution, and by fuel removal is applicable to all classes of flaming and glowing fires. Extinguishment by chemical flame inhibition applies to the flaming mode only.

Note that in the fourth method of extinguishment the action occurs only during contact of the chemical agents with activated groups or with atoms being produced by the combustion process. In a sense this could be seen as a temporary extinguishment process, operating only when the agent particles are present in the flame. If activation energy continues to exist (as with ignition points for vapor or gas reignition) after withdrawal of the agent, the flame reaction will reestablish and will continue.

Summary

Combustion occurs under the following conditions:

1. An oxidizing agent, a combustible material, and an ignition source are essential for combustion.
2. The combustible material must be heated to its ignition temperature before it will burn.
3. Combustion will continue until:
   a. The combustible material is consumed or removed.
   b. The oxidizing agent concentration is lowered to below the concentration necessary to support combustion.
   c. The combustible material is cooled to below its ignition temperature.
   d. Flames are chemically inhibited.

THE PLANT FIRE PROBLEM: CAUSES AND PREVENTION

In general, the cause of most plant fires is the *exposure of a fuel to a source of heat.* Where the fuel, such as accumulations of trash or debris, is not necessary to plant operation, fires can be prevented by removal of the fuel. Where the exposed fuel, such as raw materials or finished products, is essential, the source of heat must be protected or controlled.

Sources of Heat of Ignition and Fuel

Some of the most common sources of heat and fuel that cause plant fires are heating and cooking equipment, smoking, electric equipment, burning, flammable liquids, open flames and sparks, incendiary (arson), spontaneous ignition, gas fires, and explosions. These sources of heat are summarized below:

Heating and Cooking Equipment

Defective or Overheated Equipment. This includes improperly maintained or operated furnaces, smoke pipes, vents, portable and stationary heaters, industrial and commercial furnaces, and incinerators.

Chimneys and Flues. Fire can arise from ignition of accumulated soot or inadequate separation from combustible material.

Hot Ashes and Coals. These can cause problems when improper disposal or disposal in combustible containers or with combustible debris occurs.

Improper Location. This can mean installation too close to combustibles or accumulation of combustibles near an appliance.

Smoking

Smoking in Flammable or Explosive Atmospheres

Discarding Smoking Materials in Combustible Debris

Electric Equipment

Wiring and Distribution Equipment. These include short-circuit faults, arcs, and sparks from damaged, defective, or improperly installed components.

Motors and Appliances. These include careless use, improper installation, and poor maintenance.

Burning

Trash and Rubbish. Burning trash and rubbish can furnish the fuel for accidental ignition; careless burning ignites other material.

Warming Fires. Careless burning ignites other material.

Flammable Liquids

Storage and Handling. These hazards include careless spills, leaking fuel, and overturned tanks.

Inadequate Safeguards. Fires can be started by improper storage containers or facilities, improper electrical equipment near open processes, or improper bonding and grounding of transfer processes.

Open Flames and Sparks

Sparks and Embers. Problems include ignition of roof coverings by sparks from chimneys, incinerators, rubbish fires, locomotives, etc.

Welding and Cutting. Hazards include ignition of combustibles by the arc or flame itself, heat conduction through the metals being welded or cut, molten slag and metal from the cut, or sparks.

Friction, Sparks from Machinery. Friction heat or sparks resulting from impact between two hard surfaces are a hazard.

Thawing Pipes. Open-flame devices are a hazard when used in the dangerous practice of thawing pipes.

Other Open Flames. These include ignition sources such as candles, locomotive sparks, incinerator sparks, and chimney sparks.

Lightning. This includes building fires caused by the effects of lightning.

Exposure. Exposure fires are those originating in places other than buildings, but which ignite buildings.

Incendiary, Suspicious. These are fires that are known to be or thought to have been set: fires set to defraud insurance companies, fires set by mentally disturbed persons, and fires set by malicious persons.

Spontaneous Ignition. This means fires resulting from the uncontrolled spontaneous heating of materials.

Gas Fires and Explosions. These are fires and explosions that involve gas that has escaped from piping, storage tanks, equipment, or appliances and fires caused by misuse or faulty operation of gas appliances.

PLANT FIRE HAZARDS

Texts and handbooks which describe the hazard and control technique for specific risks are listed in the last section of this chapter.

Tables 2-1 and 2-2 present a statistical profile of large-loss fire experience in the United States for 1978 and 1979. A large-loss fire is defined by the NFPA as one that caused $500,000 or more in direct fire damage. Indirect losses, such as business interruption losses, are not included.

Fire Hazards of Materials

Virtually all matter can be changed by exposure to sufficient quantities of energy. Energy in the form of heat was previously discussed. The heat energy which causes or is produced by a fire usually causes undesired changes to the material involved. The relative fire risks and products of burning of different types of materials are presented in this section.

Wood

When in contact with sufficient heat, all wood or wood-based products will ignite, the time of ignition depending on the ignition source and the length of exposure.

Wood or wood-based products can be treated with fire-retardant chemicals. When so treated, the flammability of wood and wood-based products is reduced. The flammability of these products can also be reduced when they are used in combination with other materials, such as insulation.

Moisture Content. Wood is made up primarily of carbon, hydrogen, and oxygen. Live wood cells retain considerable moisture. When the wood is dead, air replaces most of the water in the cellular structure of wood.

The fire behavior of wood and other combustible solids of the same size and shape varies greatly with the moisture content. Wet wood is harder to ignite and will not burn as fast as dry wood. Burning rate is also influenced by the moisture content in materials. See Table 2-3 for heat of combustion for various wood-based products composed with petroleum-based materials.

Even when exposed to a relatively high heat source for a prolonged period of time, ignition is generally difficult when the moisture content of wood (and similar fuels) is

TABLE 2-1 Property Uses Where Large-Loss Fire Occurred, 1980 (United States)

Property use	No. of large-loss fires	Loss	Total no. of large-loss fires for major property uses	Total loss for major property uses
Public Assembly			26	$ 60,210,000
Bowling establishments	3	$ 3,500,000		
Racetrack	1	20,000,000		
Churches	7	12,780,000		
Country clubs	3	5,800,000		
Historic buildings	2	2,750,000		
Restaurants, drinking places	7	10,650,000		
Theaters	3	4,730,000		
Educational			27	52,801,800
Public schools	22	40,113,000		
Residential schools	2	7,071,600		
Colleges	2	4,050,000		
Other educational	1	1,567,200		
Institutional			1	1,800,000
Prison	1	1,800,000		
Residential			25	93,594,500
Private dwellings	3	8,275,000		
Apartments	15	24,600,000		
Hotels	5	57,084,000		
Dormitories	2	3,635,500		
Stores and Offices			53	97,692,200
Food, beverage sales	4	4,325,000		
Clothing, fabric sales	3	3,000,000		
Furniture, appliance sales	12	18,500,000		
Drugstore	1	1,000,000		
Recreation, hobby, home repair	5	7,634,000		
Linen supply	1	1,000,000		
Vehicle sales/repair	3	4,000,000		
General-item stores	13	29,038,200		
Offices	10	27,515,000		
Other stores	1	1,680,000		
Basic Industry, Utility, Defense			17	49,570,000
Non-nuclear energy production	5	28,720,000		
Communication facility	1	2,000,000		
Energy distribution systems	2	3,000,000		
Fruit packing plants	2	4,500,000		
Forests	3	6,000,000		
Coal mine	1	1,000,000		
Mineral product manufacture	3	4,350,000		
Manufacturing			81	339,510,700
Food	6	11,635,000		
Textile	3	5,835,000		
Clothing, rubber manufacture	4	6,440,000		
Wood, furniture, paper, printing	18	47,199,900		
Chemical, plastic, petroleum	16	61,962,800		
Metal, metal products	9	33,911,500		
Other manufacturing	7	35,061,500		
Storage			72	172,950,000
Agricultural products	4	6,720,000		
Textile, textile products	4	12,083,500		
Processed food, beverage	3	8,710,000		
Petroleum, petroleum products	10	20,301,000		
Wood, paper products	23	56,316,000		
Chemical, plastic products	5	10,100,000		
Metal, metal products	3	15,023,500		
Vehicles	6	14,666,000		
General-item storage	14	29,030,000		

TABLE 2-1. Property Uses Where Large-Loss Fire Occured, 1980 (United States) (*Continued*)

Property use	No. of large-loss fires	Loss	Total no. of large-loss fires for major property uses	Total loss for major property uses
Special Properties			21	80,674,900
Under construction, unoccupied	15	35,707,500		
Outdoor properties (brush) including structures	2	40,233,900		
Rail transport	3	3,733,500		
Rail transport	1	1,000,000		
TOTAL:			323	$948,804,100

Source: Fire Journal, September 1980, National Fire Protection Quincy, Massachusetts.

TABLE 2-2 Percentage of Large-Loss Fires by Property Use Classification, 1979 (United States)

Property use	Percent of fires	Percent of loss
Public assembly	11.0	7.7
Educational	4.8	3.9
Institutional	0.5	0.4
Residential	6.6	7.0
Stores and offices	17.1	13.3
Basic industry, utility, and defense	6.0	4.6
Manufacturing	22.8	18.5
Storage	22.3	21.8
Special properties	8.9	22.8
	100.0	100.0

Source: Fire Journal, September, 1980, National Fire Protection Association, Quincy, Massachusetts.

above 15 percent. Once ignition and resultant fire have begun, heat radiation and the rate of pyrolysis reduce the importance of the moisture factor.

Plastics

There are thousands of plastic product formulations that are produced in a variety of shapes and sizes, such as solid shapes, films and sheets, foams, molded forms, synthetic fibers, pellets, and powders. They are classified into 30 major groups of plastics and polymers. In addition, most finished products contain additives such as colorants, reinforcing agents, fillers, stabilizers, and lubricants. These additives vary the chemical nature of the product still more.

Most plastics are combustible, but the degree of combustibility varies widely because of the range of chemical compositions and combinations. As a result, it is virtually impossible to assign a fire hazard or flammability limit to any general plastic group. The only method of determining the fire hazard of a particular plastic is to fire test the plastic under exact end-use conditions. See Tables 2-4 and 2-5.

Storage. The chemical composition, the physical form, and the manner and arrangement in which plastics are stored greatly affect the degree of fire hazard that is present. Large quantities of smoke are usually generated when stored plastics are involved in fire, a condition made more or less difficult by the amount of ventilation present or available in a given storage area.

Thermoplastics, such as polyethylene and plasticized polyvinylchloride, and ther-

TABLE 2-3 Heat of Combustion of Various Wood and Wood-Based Products and Comparative Substances*

Substance	Heating value, Btu/lb
Wood sawdust (oak)	8,493†
Wood sawdust (pine)	9,676†
Wood shavings	8,248†
Wood bark (fir)	9,496†
Corrugated fiber carton	5,970†
Newspaper	7,883†
Wrapping paper	7,106†
Petroleum coke	15,800
Asphalt	17,158
Oil (cottonseed)	17,100
Oil (paraffin)	17,640

*Extracted from *Kent's Mechanical Engineers' Handbook*, 12th ed., H. B. Carmichael and J. K. Salisbury, eds., Wiley-Interscience, New York, 1950.

†Dry.

mosets, such as polyesters, present severe fire hazards. Plastics in foamed material form present the most severe hazard of all. In a fire, thermoplastics will melt and break down and behave and burn like flammable liquids. Automatic sprinkler systems with high sprinkler-discharge densities are necessary for adequate fire protection.

Dusts

When some combustible solids are ground or rubbed into minute particles, the particles tend to mix with the air in much the same way that vapor or gas mixes with the air. The finer the dust particle, the more completely it will mix with the air and remain suspended in the air. Although dust particles from all combustible solids do not result in potentially explosive dust particles, a large number of combustible solids can yield explosive dust particles. See Table 2-6.

Metals

Nearly all metals will burn in air under certain conditions. See Table 2-7. Some oxidize rapidly in the presence of air or moisture, generating sufficient heat to reach their ignition temperatures. Others oxidize so slowly that heat generated during oxidation is dissipated before they become hot enough to ignite. Certain metals, notably magnesium, titanium, sodium, potassium, calcium, lithium, hafnium, zirconium, zinc, thorium, uranium, and plutonium, are referred to as combustible metals because of the ease of ignition of thin sections, fine particles, or molten metal. The same metals in massive solid form are comparatively difficult to ignite.

Hot, burning metals may react violently with the extinguishants used on fires involving ordinary combustibles or flammable liquids. A few metals, such as uranium, thorium, and plutonium, emit ionizing radiations that can complicate fire fighting and introduce a contamination problem.

Temperatures in burning metals are generally much higher than the temperature in burning flammable liquids. Some hot metals can continue burning in nitrogen, carbon dioxide, or steam atmospheres in which ordinary combustibles or flammable liquids would be incapable of burning.

Flammable and Combustible Liquids

The improper storage, handling, and use of flammable and combustible liquids has been the cause of many deaths, injuries, and disastrous fires.

TABLE 2-4 Small-Scale Tests for Combustibility of Plastics

Test method	Sample size, in	Position of sample	Ignition source, flame, in	Time and limit of exposure, s	Value reported, in/min	Usual material application
ASTM D 635	⅛ × ½ × 5	Horizontal	1	2–30	Burning rate	Rigid plastic
ASTM D 568	0.05 × 1 × 18	Horizontal	1	15	Burning rate	Films
ASTM D 229	1/32 × ½ × 5	Horizontal	1	30	Burning time for 4 samples	Elec. insulation
ASTM D 1692	½ × 2 × 6	Horizontal	1	60	Burning rate	Foam
UL 94*	⅛ × ½ × 5 ⅛ × ½ × 5	Horizontal Vertical	¾ ¾	30 2–10	Burning rate Extinguishment time	Rigid plastics Rigid plastics
NFPA 701	2¾ × 10	Vertical	1½	12	Length of char	Films
UL 214	2¾ × 10	Vertical	1½	12	Length of char	Films

*UL stands for Underwriters Laboratories Inc.
Source: NFPA, Quincy, Massachusetts.

TABLE 2-5 Medium- and Large-Scale Tests for Combustibility of Plastics

Test for	Number
Surface burning of building materials	NFPA 255 UL 723 ASTM E 84
Fire tests of building construction and materials	NFPA 251 UL 263 ASTM E 119
Fire tests of roof coverings	NFPA 256
Radiant panel test for flame spread	ASTM E 162
Factory mutual calorimeter test	
UL and FM corner wall tests	
Full-room burnouts—FM, UL*	
Flame-retardant films	NFPA 701 UL 214

*UL stands for Underwriters Laboratories Inc. FM for Factory Mutual Systems. See p. 14-69.

Source: NFPA, Quincy, Massachusetts.

TABLE 2-6 Common Combustible Solid Dusts Generating Severe Explosions*

	Maximum explosion pressure		Maximum rate of pressure rise	
Type of dust	psig	bar	psig/s	bar/s
Corn (processing)	95	6.55	6,000	413.7
Cornstarch	115	7.93	9,000	620.5
Potato starch	97	6.89	8,000	551.6
Sugar (processing)	91	6.27	5,000	344.7
Wheat starch	105	7.24	8,500	586.0
Ethyl cellulose plastic molding compound	102	7.03	6,000	413.7
Wood flour filler	110	7.58	5,500	379.2
Natural resin	87	6.0	10,000	689.5
Aluminum	100	6.9	10,000	689.5
Magnesium (powder)	94	6.48	10,000	689.5
Silicon (powder)	106	7.31	10,000	689.5
Titanium (powder)	80	5.52	10,000	689.5
Aluminum magnesium alloy (powder)	90	6.20	10,000	689.5

*Extracted from U.S. Bureau of Mines Investigations and Reports, Nos. 5753, RI 5971, RI 6516

It is the vapor from the evaporation of a flammable or combustible liquid when exposed to air or under the influence of heat, rather than the liquid itself, which burns or explodes when mixed with air in certain proportions in the presence of some source of ignition. There is a flammable range below which the vapor mixture is too lean to burn or explode, or above which the vapor mixture is too rich to burn or explode. (See Table 2-8.) For gasoline, the most common and widely used flammable liquid, the flammable range is 1.4 and 7.6 percent by volume. When the vapor-air mixture is near either the lower flammable limit (LFL) or upper flammable limit (UFL), the explosion is less intense than when the mixture is in the intermediate range. The violence of the explosion depends on the concentration of the vapor as well as the quantity of vapor-air mixture and the type of container. Thus, it is important in controlling the fire hazard, to store a flammable liquid in the proper type of closed container and minimize the exposure to air. When exposed to heat from a fire, a tank or other container may rupture with dangerous results if properly designed vents are not provided or if the exposed tank or con-

TABLE 2-7 Melting, Boiling, and Ignition Temperatures of Pure Metals in Solid Form*

	Temperature, °F		
Pure metal	Melting point	Boiling point	Solid metal ignition
Aluminum	1220	4445	above 1832†
Barium	1337	2084	347†
Calcium	1548	2625	1300
Hafnium	4032	9750	——
Iron	2795	5432	1706†
Lithium	367	2437	356
Magnesium	1202	2030	1153
Plutonium	1184	6000	1112
Potassium	144	1400	156†S‡
Sodium	208	1616	above 239
Strontium	1425	2102	1328†
Thorium	3353	8132	932†
Titanium	3140	5900	2900
Uranium	2070	6900	below 6900
Zinc	786	1665	1652*
Zirconium	3326	6470	2552*

*Because of the variations in available information, the values given must be considered as being approximate and for guidance only. Generally, if a metal melts and flows before burning, an ignition temperature can be determined.

†Ignition in oxygen

‡S Spontaneous ignition in moist air

Source: NFPA, Quincy, Massachusetts.

TABLE 2-8 Flash Points and Flammable Limits of Some Common Liquids and Gases

	Flash point		
Liquid (or gas at ordinary temps.)	°F	°C	Flammable limits, percent by volume
Acetylene	(Gas)		2.5–81.0+
Benzene	12	−11	1.3–7.1
Ether (ethyl ether)	−49	−45	1.9–36.0
Fuel oil			
Domestic, No. 2	100 (min.)	38	None at ordinary temps.
Heavy, No. 5	130 (min.)	54	None at ordinary temps.
Gasoline (high test)	−36	−38	1.4–7.4
Hydrogen	(Gas)		4.07–75.0
Jet fuel (A & A-1)	110 to 150	43 to 65	None at ordinary temps.
Kerosene (Fuel oil, No. 1)	100 (min.)	38	0.7–5.0
LPG (propane-butane)	(Gas)		1.9–9.5
Lacquer solvent (butyl acet.)	72	22	1.7–7.6
Methane (natural gas)	(Gas)		5.0–15.0
Methyl alcohol	52	11	6.7–36.0
Turpentine	95	35	0.8–(undetermined)
Varsol (standard solv.)	110	43	0.7–5.0
Vegetable oil (cooking, peanut)	540	282	(Ignition temp. = 833°F)

Source: NFPA, Quincy, Massachusetts.

tainer is not cooled by hose streams. The principal fire and explosion prevention measures under such circumstances are: (1) exclusion of sources of ignition, (2) exclusion of air, (3) keeping the liquid in a closed container, (4) ventilation to prevent the accumulation of vapor in the flammable range, and (5) use of an atmosphere of inert gas instead of air.

An arbitrary system of classifying liquids has generally been adopted.

Flammable Liquids. Flammable liquids are any liquid having a flash point below 100°F (38°C) and having a vapor pressure not exceeding 2068.6 mm at 100°F (38°C). Class I liquids include those having flash points below 100°F (38°C) and may be subdivided as follows:

Class IA includes those having flash points below 73°F (23°C) and a boiling point below 100°F (38°C).

Class IB includes those having flash points below 73°F (23°C) and a boiling point at or above 100°F (38°C).

Class IC includes those having flash points at or above 73°F (23°C) and below 100°F (38°C).

Combustible Liquids. Liquids with a flash point at or above 100°F (38°C) are referred to as combustible liquids. They are:

Class II liquids include those having flash points at or above 100°F (38°C) and below 140°F (60°C).

Class IIIA liquids include those having flash points at or above 140°F (60°C) and below 200°F (93°C).

Class IIIB liquids include those having flash points at or above 200°F (93°C).

Some typical liquids would be classed as follows:

Denatured Alcohol	Class IB
Fuel oil	Class II
Gasoline	Class IB
Kerosene	Class II
Peanut oil	Class IIIB
Turpentine	Class IC
Paraffin wax	Class IIIB

Storage and Handling. Proper storage and handling of flammable and combustible liquids are necessary to prevent fire or explosion. Ventilation to prevent accumulations of flammable vapors is of primary importance because there is the possibility of breaks or leaks in the storage and handling in a closed system. It is important to eliminate possible sources of ignition in an area where flammable liquids are stored, handled, or used.

Ventilation of an area where flammable liquids are manufactured or used can be accomplished by natural or mechanical means. Wherever possible, equipment such as compressors, stills, and pumps should be located in a spacious, open area. Most flammable liquids produce heavier-than-air vapors that flow along the ground or floor and settle in depressions. These can travel long distances and be ignited and flash back from a point remote from the origin of the vapors. NFPA 30, "Flammable and Combustible Liquids Code," is the accepted national standard for fire protection of such liquids.

NFPA standards specify the construction, installation, spacing, venting, and diking of above-ground and underground storage tanks, container storage in buildings, loading and unloading practices, safeguards for dispensing the liquids, and standards for transporting the liquids in trucks, ships, or pipelines (Tables 2-9 and 2-10).

Gases

There are many kinds of materials that exist in the form of gas. In general, gases are thought of and described when the substance exists in a gaseous state at normal temperature and pressure 70°F (21°C) and 14.7 psia (101430 N/m^2).

Classification by Chemical Properties. Gases can be broadly classified according to chemical properties, physical properties, or usage. Classification by chemical properties helps to define the hazards of gases to people and in fires.

TABLE 2-9 Storage Limitations for Inside Storage Rooms

Fire protection* provided	Fire resistance, h	Maximum size, ft^2	Allowable loading, gal/ft^2 floor area
Yes	2	500	10
No	2	500	4
Yes	1	150	5
No	1	150	2

*Fire-protection system of sprinkler, water spray, carbon dioxide, dry chemical, halon, or other acceptable type of system.

Source: NFPA, Quincy, Massachusetts.

TABLE 2-10 Storage Limitations for Warehouses or Storage Buildings

		Protected storage* maximum per pile height		Unprotected storage maximum per pile height	
Class liquid	Storage level	gal	ft	gal	ft
IA	Ground & upper floors	2,750	3	660	3
	Basement	(50)	(1)	(12)	(1)
		Not permitted		Not permitted	
IB	Ground & upper floors	5,500	6	1,375	3
	Basement	(100)	(2)	(25)	(1)
		Not permitted		Not permitted	
IC	Ground & upper floors	16,500	6	4,125	3
	Basement	(300)	(2)	(75)	(1)
		Not permitted		Not permitted	
II	Ground & upper floors	16,500	9	4,125	9
	Basement	(300)	(3)	(75)	(3)
		5,500	9	Not permitted	
		(100)	(3)		
Combustible	Ground & upper floors	55,000	15	13,750	12
	Basement	(1,000)	(5)	(250)	(4)
		8,250	9	Not permitted	
		(150)	(3)		

*A sprinkler or equivalent fire-protection system installed in accordance with the applicable NFPA Standard. (Numbers in parentheses indicate corresponding number of 55-gal drums.) *Note 1:* When two or more classes of materials are stored in a single pile, the maximum gallonage permitted in that pile is the smallest of the two or more separate maximum gallonages. *Note 2:* Aisles are provided so that no container is more than 12 ft from an aisle. Main aisles shall be at least 8 ft wide and side aisles at least 4 ft wide. *Note 3:* Each pile is separated from each other pile by at least 4 ft. When stored on suitably protected racks or when the storage is suitably protected, containers may be piled up but no closer than 3 ft to the nearest beam, chord, girder, or other obstructions. Good practice is to maintain 3 ft clearance below sprinkler deflectors or discharge orifices or other overhead fire protection systems.

Source: NFPA, Quincy, Massachusetts.

Flammable Gases. Any gas that will burn in the normal concentrations of oxygen in the air is a flammable gas. Like flammable liquid vapors, the burning of this gas in air is in a range of gas-air mixture (the flammable range).

Nonflammable Gases. Nonflammable gases will not burn in air or in any concentration of oxygen. A number of nonflammable gases, however, will support combustion. Such gases are often referred to as "oxidizers" or oxidizing gases. Common oxidizers are oxygen or oxygen in a mixture with other gases.

Nonflammable gases that will not support combustion are usually called *inert gases.* Common inert gases are nitrogen, carbon dioxide, and sulfur dioxide.

Toxic Gases. Toxic gases endanger life when inhaled. Gases such as chlorine, hydrogen sulfide, sulfur dioxide, ammonia, and carbon monoxide are poisonous or irritating when inhaled.

Reactive Gases. Reactive gases react with other materials or within themselves by a reaction other than burning. When exposed to heat and shock, some reactive gases rearrange themselves chemically. Such gases can produce hazardous quantities of heat or reaction products. Fluorine is a highly reactive gas. At normal temperatures and pressures it will react with most organic and inorganic substances, often fast enough to result in flaming. Other examples of reactive gases are acetylene and vinyl chloride.

Classification by Physical Properties. Gases can also be classified by their physical properties. They can be compressed, liquefied, or cryogenic.

Compressed Gases. A compressed gas is at normal temperature inside a gas container and exists solely in the gaseous state under pressure; common compressed gases are hydrogen, oxygen, acetylene, and ethylene.

Liquefied Gases. Liquefied gases can be liquefied relatively easily and stored at ordinary temperatures at relatively high pressure. Liquefied gas exists in both liquid and gaseous states; at storage pressure, both the liquid and gas in the liquefied-gas container are in equilibrium and will remain so as long as any liquid remains in the container. Liquefied gas is more concentrated than compressed gas.

Cryogenic Gases. Cryogenic gases are stored in a completely liquid state. They must be maintained in their containers as low-temperature liquids at relatively low pressure. Cryogenic gases must be stored in special containers that allow the gas from the liquid to escape in order to prevent a pressure buildup caused by the production of the gaseous state within the container, which would result in container failure.

Classification by Usage. An understanding of gases as they are classified by usage is important to those involved in fire protection because the terms of these classifications are used in codes, standards, and general industrial and medical terminology (Table 2-11).

Fuel Gases. These gases are customarily used for burning with air to produce heat which in turn is used as a source of heat (comfort and process), power, or light; the principal and most widely used fuel gases are natural gas and the liquefied petroleum gases, butane and propane.

Industrial Gases. These are classified by chemical properties customarily used in industrial processes, for welding and cutting, heat treating, chemical processing, refrigeration, water treatment, etc.

Medicinal Gases. These are for medical purposes such as anesthesia and respiratory therapy; cyclopropane, oxygen, and nitrous oxide gases are common medical gases.

Hazardous Materials

Corrosive Chemicals. Corrosive chemicals are usually strong oxidizing agents that can increase fire hazards. Caustics, which are classified as water- and air-reactive chemicals, are also corrosive.

Inorganic Acids. Concentrated aqueous solutions of inorganic acids are not in themselves combustible; however, in addition to their corrosive and destructive effect on living tissue, their chief fire hazard results from the possibility of their mixing with combustible materials or other chemicals, which could result in fire or explosion; almost all corrosive chemicals are strong oxidizing agents.

Halogens. These salt-producing chemicals are very active. They are noncombustible but will support combustion; presence of halogens cause turpentine, phosphorus, and finely divided metals to ignite spontaneously. The fumes are poisonous and corrosive.

Storage and Fire Protection for Corrosive Chemicals. Storage of corrosive chemicals should be provided with two considerations in mind: (1) protection against the dam-

TABLE 2-11 Combustion Properties of Common Flammable Gases

Gas	Btu/ft³ (Gross)	Limits of flammability percent by volume in air: Lower	Upper	Specific gravity (air = 1.0)	Air needed to burn 1 ft³ of gas, ft³	Ignition temp., °F
Natural gas						
High inert type*	958–1051	4.5	14.0	0.660–0.708	9.2	——
High methane type†	1008–1071	4.7	15.0	0.590–0.614	10.2	900–1170
High Btu type‡	1071–1124	4.7	14.5	0.620–0.719	9.4	——
Blast furnace gas	81– 111	33.2	71.3	1.04–1.00	0.8	——
Coke oven gas	575	4.4	34.0	.38	4.7	——
Propane (commercial)	2516	2.15	9.6	1.52	24.0	920–1120
Butane (commercial)	3300	1.9	8.5	2.0	31.0	900–1000
Sewage gas	670	6.0	17.0	.79	6.5	——
Acetylene	1499	2.5	81.0	.91	11.9	581
Hydrogen	325	4.0	75.0	.07	2.4	752
Anhydrous ammonia	386	16.0	25.0	.60	8.3	1204
Carbon monoxide	314	12.5	74.0	.97	2.4	1128
Ethylene	1600	2.7	36.0	.98	14.3	914
Methyl acetylene, propadiene, stabilized§	2450	3.4	10.8	1.48	——	850

*Typical composition CH_4, 71.9–83.2%; N_2, 6.3–16.20%.
†Typical composition CH_4, 87.6–95.7; N_2, 0.1–2.39.
‡Typical composition CH_4, 85.0–90.1; N_2, 1.2–7.5.
§MAPP® gas.
Source: NFPA, Quincy, Massachusetts.

aging effect of corrosive chemicals on living tissue and (2) guarding against any fire and explosion hazard that might be associated with the corrosive chemical.

Inorganic Acids. These should be stored in cool, well-ventilated areas that are not exposed to the sun or other chemical and waste materials; they should be protected from freezing temperatures. Water in spray form is the recommended procedure for fighting fires in inorganic acid storage areas; in fires involving perchloric acid, extra care should be taken since this may mix with other organic materials and result in an explosion.

Halogens. Fluorine and chlorine should be stored in special containers; fluorine may be safely stored in nickel or Monel cylinders. Impurities or moisture in the cylinder may cause an explosive reaction; chlorine, a serious inhalation hazard, should be stored in areas where ventilation is a prime consideration. Where chlorine leakage is suspected, use a self-contained breathing apparatus; fluorine requires protection of the self-contained breathing apparatus and special protective clothing.

Corrosive Vapors. Very often, corrosive vapors are a by-product of an industrial or chemical process. Ducts must be used to carry these vapors safely from the area. The type of duct used is determined by the vapor. Heavier gauge metal may be sufficient, although a protective coating or special lining may be required in the ducts. Stainless steel, asbestos cement, and plastic linings have been used with success depending on the corrosive vapor.

Radioactive Materials

Fire Protection for Radioactive Materials. The main concern in fire protection of radioactive materials is to prevent the release (or control the release) of these materials during fire extinguishment. Although fire protection operations are similar to those used when nonradioactive materials are involved, fire involving buildings or areas containing radioactive materials presents two additional considerations: (1) the presence of harmful radioactive materials might necessitate that normal fire-fighting procedures be changed and (2) because of the presence of radioactive matter, delay in salvage and resumption of normal operations may occur.

DESIGN AND CONSTRUCTION FOR FIRE SAFETY

The architectural design and building methods and materials used for a structure will often determine how a fire will be confined or will spread. Firesafety objectives must be determined before a facility is designed. Often *code compliance* is not a sufficient standard to meet the level of risk acceptable to a particular organization.

Fire safety design decisions are necessary in at least three objective areas: life safety, property protection, and continuity of operations. Systems-analysis techniques are used by fire protection engineers to determine the level of protection to meet design objectives.

Building and Site Planning for Fire Safety

Two categories of decisions should be made in the design process to provide effective firesafe design: interior building functions and exterior site planning. Building fire defenses, active and passive, should be designed so that the building assists in the suppression of fire.

Firesafety Planning for Buildings

Interior layout, circulation patterns, finish material, and building services are all important fire-safety considerations in building design. Building design also has a significant influence on the efficiency of fire-department operations. Manual fire-suppression activities should be considered during architectural design.

Fire Fighting Accessibility to Building's Interior. This includes access to the building itself as well as access to the interior of the building. Spaces in which fire fighting access and operations are restricted because of architectural, engineering, or functional requirements should be provided with effective protection. A complete automatic sprinkler system is often the best solution. Other methods which may be used include access panels in interior walls and floors, fixed nozzles in floors with fire department connections, and roof vents and access openings.

Ventilation. This is of vital importance in removing smoke, gases, and heat. Appropriate skylights, roof hatches, emergency escape exits, and similar devices should be provided when the building is constructed.

Ventilation of building spaces performs the following important functions:

1. Protection of life by removing or diverting toxic gases and smoke
2. Improvement of the environment in the vicinity of the fire by removal of smoke and heat; enables fire fighters to advance close to the fire
3. Control of the spread or direction of fire by setting up air currents that cause the fire to move in a desired direction
4. Provision of a release for unburned, combustible gases before they acquire a flammable mixture, thus avoiding a backdraft explosion

NFPA 204, "Smoke and Heat Venting Guide" recommends automatic smoke venting of large industrial buildings.

Curtain Boards as Venting Aids. In large-area buildings, unless vented areas are subdivided by means of walls or partitions, curtain boards are essential. The function of curtain boards is to delay and limit the horizontal spread of heat by providing the horizontal confinement needed to obtain the desired *stack* action. The depth of such curtain boards largely determines the height of the stack which affects the capacity of the vent. If an area is protected by automatic sprinklers, curtain boards have added values: confinement of heat to speed up operation of sprinklers over the fire and obstructed lateral spread of heat to minimize the operation of an excessive number of sprinklers. See Fig. 2-2.

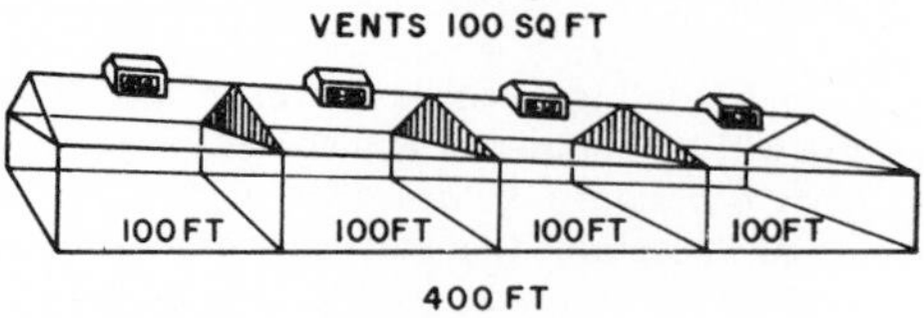

Figure 2-2 Curtain boards and roof vents. *(From* Fire Protection Handbook, *14th ed., G. P. McKinnon (ed.), NFPA, Quincy, Massachusetts, 1976.)*

Firesafety Planning for Sites

Proper building design for fire protection should include a number of factors outside the building itself. The site on which the building is located will influence the design. Among the more significant features are traffic and transportation conditions, fire department accessibility, and water supply. Inadequate water mains and poor spacing of hydrants have contributed to the loss of many buildings.

Traffic and Transportation. Fire department response time is a vital factor in building design considerations; traffic access routes, traffic congestion at certain times of the day, traffic congestion from highway entrances and exits, and limited-access highways have significant effects on fire department response distances and response time.

Fire Department Access to the Site. Is the building easily accessible to fire apparatus? Ideal accessibility occurs where a building can be approached from all sides by fire-department apparatus; congested areas, topography, or buildings and structures located appreciable distances away from the street can cause difficulty and prevent effective use of fire apparatus. Inadequate attention to site details can place the building in an unnecessarily vulnerable position; if fire defenses are compromised by preventing adequate fire department access, the building itself must make up the difference in more complete internal protection.

Water Supply to the Site. Are the water mains adequate, and are the hydrants properly located? The number, location, and spacing of hydrants and the size of the water mains are vital considerations; consult local standards and insurance requirements.

Interior Finish

Interior finish is defined as those materials that make up the exposed interior surface of wall, ceiling, and floor constructions. The common interior-finish materials are wood, plywood, plaster, wallboards, acoustical tile, insulating and decorative finishes, plastics, and various wall coverings.

Some building codes include floor coverings under their definition of interior finishes.

Fire Tests

It is possible to estimate the damage that fire can cause to a building by studying: (1) the amount and kind of combustible materials in the buildings and (2) the way they are distributed throughout the building. These two factors indicate the rate of combustion, the duration of the fire, and the degree of difficulty to extinguish the fire.

The effects of fire on the components of a building (such as the columns, floors, walls, partitions, and ceiling or roof assemblies) are tested against both time and temperature.

Results of the tests are recorded in hours or minutes and indicate the duration of fire resistance.

Ratings for flame spread of interior finish materials have been established with the use of the 25 ft (7.6 m) tunnel developed by A. J. Steiner at Underwriters Laboratories Inc. These ratings are used in the NFPA **Life Safety Code®** and in other codes to indicate the areas in which finishes of varying flame spread characteristics may be used (Table 2-12). The five classifications used in the **Life Safety Code®** are:

Class	*Flame-Spread Range*
A	0–25
B	26–75
C	76–200
D	201–500
E	over 500

The higher the flame spread, the greater the hazard. For example, in a new hospital, Class A materials would be required for most areas, and Class D or E materials would not be permitted at all.

Materials are measured on a relative scale with cement asbestos board rated 0 and red oak flooring rated 100. Some highly combustible wallboards have received ratings as high as 1500.

One of the best sources of information showing the wide variety of building assemblies and giving the fire-resistance ratings of beams, columns, floors, walls, and partitions is the Underwriters Laboratories Inc. *Fire Resistance Index.*

Confinement of Smoke and Fire

Design criteria for plant facilities are generally based on estimated fire severity (Table 2-13). Specific industrial-hazard fire-severity data may be obtained from insurance organizations (Table 2-14 and Fig. 2-3).

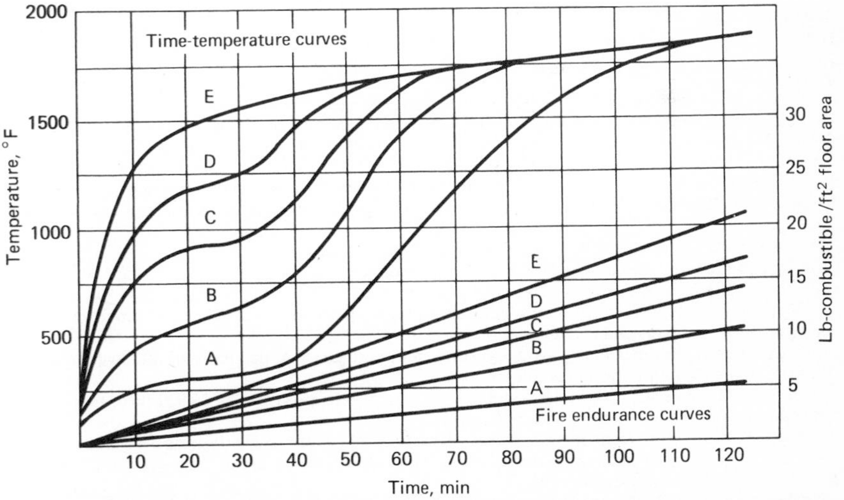

Figure 2-3 Possible classification of building contents for fire severity and duration. The straight lines indicate the length of fire endurance based upon amounts of combustibles involved. The curved lines indicate the severity expected for the various occupancies (see Table 2-14). There is no direct relationship between the straight and curved lines, but, for example, 10 lb of combustibles per square foot will produce a 90-min fire in a C occupancy, and a fire severity following the time-temperature curve C might be expected. *(From R. Tuve,* Principles of Fire Protection Chemistry, *NFPA, Quincy, Massachusetts, 1976.)*

TABLE 2-12 Summary of Life Safety Code Requirements for Interior Finish

Occupancy	Class of interior finish[a]		
	Exits	Access to exits	Other spaces
Places of assembly—class A[b]	A	A	A or B
Places of assembly—class B[c]	A	A	A or B
Places of assembly—class C[d]	A	A	A, B, or C
Educational	A	A	A, B, or C
Educational—unsprinklered open-plan buildings[e]	A	A or B	A or B
Flexible-plan buildings[f]	A	A	C or low height partitions
Institutional, existing—hospitals, nursing homes, residential-custodial care	A or B	A or B	A or B
Institutional, new—hospitals, nursing homes, residential-custodial care	A	A	A B in individual room with capacity not more than 4 persons
Residential, new—apartment houses	A or B	A or B	A, B, or C
Residential, existing—apartment houses	A or B	A, B, or C	A, B, or C
Residential—dormitories	A or B	A, B, or C	A, B, or C
Residential, new—1- and 2-family, lodging or rooming houses			A, B, or C
Residential, existing—1- and 2-family, lodging or rooming houses			A, B, C, or D
Residential, new—hotels	A or B	A or B	A, B, or C
Residential, existing—hotels	A or B	(1) A or B if required path of exit travel; (2) A, B, or C if not used as required path of exit travel	A, B, or C
Mercantile—class A[g]	A or B		Ceilings—A or B Walls—A, B, or C
Mercantile—class B[h]	A or B		Ceilings—A or B Walls—A, B, or C
Mercantile—class C[i]	A or B		A, B, or C
Office	A or B	A or B	A, B, or C
Industrial	A, B, or C	A, B, or C	A, B, or C
Towers	A or B		A or B

[a]There are five classes of interior finish: Class A, flame spread 0–25; Class B, flame spread 25–75; Class C, flame spread 75–200; Class D, flame spread 200–500; and Class E, over 500. Where a standard system of automatic sprinklers is installed, an interior finish with a flame spread rating not over Class C may be used in any location where Class B is normally specified, and with a rating of Class B in any location where Class A is normally specified, unless specifically prohibited elsewhere in the *Life Safety Code*.

[b]Class A Places of Assembly—1000 persons or more.

[c]Class B Places of Assembly—300 to 1000 persons

[d]Class C Places of Assembly—100 to 300 persons

[e]Open plan buildings—includes all buildings where no permanent partitions are provided between rooms or between rooms and corridors.

[f]Flexible plan buildings have movable corridor walls and movable partitions of full height construction with doors leading from rooms to corridors.

[g]Class A Mercantile Occupancies—stores having aggregate gross area of 30,000 ft^2 or more, or utilizing more than 3 floor levels for sales purposes.

[h]Class B Mercantile Occupancies—stores of less than 30,000 ft^2 aggregate gross area, but over 3,000 sq ft,, or utilizing any floors above or below street floor level for sales purposes, except that if more than 3 floors are utilized, store shall be Class A.

[i]Class C Mercantile Occupancies—stores of 3000 ft^2 or less gross area, used for sales purposes on street level only. (A single balcony or mezzanine floor with less than half the area of the street level floor and which is used for sales purposes is not counted as another floor.)

Source: NFPA, Quincy, Massachusetts.

TABLE 2-13 Estimated Fire Severity for Offices and Light Commercial Occupancies*

Combustible content, total, including finish, floor and trim, lb/ft²	Heat potential assumed Btu/ft²†	Equivalent fire severity, approximately equivalent to that of test under standard curve for the following periods
5	40,000	30 min
10	80,000	1 h
15	120,000	1½ h
20	160,000	2 h
30	240,000	3 h
40	320,000	4½ h
50	380,000	7 h
60	432,000	8 h
70	500,000	9 h

*From Gordon P. McKinnon (ED.) *Fire Protection Handbook,* NFPA, Quincy, 1976. Data applying to fire-resistive buildings with combustible furniture and shelving.

†Heat of combustion of contents taken at 8000 Btu/lb up to 40 lb/ft²; 7600 Btu/lb for 50 lb, and 7200 Btu for 60 lb and more to allow for relatively greater proportion of paper. The weights contemplated by the tables are those of ordinary combustible materials, such as wood, paper, or textiles.

TABLE 2-14 Fire Severity Expected by Occupancy (See Fig. 2-3)

Temperature curve A (slight)
Well-arranged office, metal furniture, noncombustible building
Welding areas containing slight combustibles
Noncombustible power house
Noncombustible buildings, slight amount of combustible occupancy
Temperature curve B (moderate)
Cotton and waste-paper storage (baled) and well-arranged, noncombustible building
Paper-making processes, noncombustible building
Noncombustible institutional buildings with combustible occupancy
Temperature curve C (moderately severe)
Well-arranged combustible storage, e.g., wooden patterns, noncombustible buildings
Machine shop having noncombustible floors
Temperature curve D (severe)
Manufacturing areas, combustible products, noncombustible building
Congested combustible storage areas, noncombustible building
Temperature curve E (standard fire exposure—severe)
Flammable liquids
Woodworking areas
Office, combustible furniture and buildings
Paper working, printing, etc.
Furniture manufacturing and finishing
Machine shop having combustible floors

Source: NFPA, Quincy, Massachusetts.

Fire Doors

Fire doors are the most widely used and accepted means of protecting vertical and horizontal openings. Suitability of fire doors is determined by nationally recognized testing laboratories; doors that have not been tested cannot be relied upon for effective protection. The doors are tested as they are installed in the field, that is, with the frame, hardware, wired-glass panels, and other accessories necessary to complete the installation.

Nearly all building codes use NFPA 80, "Standard for Fire Doors and Windows." This standard establishes the minimum ratings for the five most commonly encountered types of openings in walls. They are as follows:

1. Class A openings are in walls separating buildings or dividing a single building into fire areas. Doors for the protection of these openings have a fire protection rating of 3 h.

2. Class B openings are in enclosures of vertical communication through buildings (stairs, elevators, etc.). Doors for the protection of these openings have a fire protection rating of 1 or 1½ h.
3. Class C openings are in corridor and room partitions. Doors for the protection of these openings have a fire protection rating of ¾ h.
4. Class D openings are in exterior walls which are subject to severe exposure from the outside of the building. Doors and shutters for the protection of these openings have a fire protection rating of 1½ h.
5. Class E openings are in exterior walls which are subject to moderate or light fire exposure from outside of the building. Doors, shutters, or windows for the protection of these openings have a fire-protection rating of ¾ h.

It is important to note that this classification applies to the various types of openings and not to the fire door itself. A fire door is not a Class A fire door. It ia a door that is suitable for a Class A opening.

The following excerpt from the NFPA *Fire Protection Handbook** is a description of the various types of construction for fire doors:

> *Composite Doors.* These are of the flush design and consist of a manufactured core material with chemically impregnated wood-edge banding and untreated wood-face veneers, or laminated plastic faces, or surrounded by and encased in steel.
>
> *Hollow-Metal Doors.* These are of formed steel of the flush and paneled designs of No. 20 gage or heavier steel.
>
> *Metal-Clad (Kalamein) Doors.* These are of flush and paneled design consisting of metal covered with steel of 24 gage or lighter.
>
> *Sheet-Metal Doors.* These are of formed No. 22 gage or lighter steel and of the corrugated, flush, and paneled designs.
>
> *Rolling Steel Doors.* These are of the interlocking steel slat design or plate steel construction.
>
> *Tin-Clad Doors.* These are of two- or three-ply wood-core construction, covered with No. 30 gage galvanized steel or terneplate (maximum size 14 by 20 in.) or No. 24 gage galvanized steel sheets not more than 48 in. wide.
>
> *Curtain-Type Doors.* These consist of interlocking steel blades or a continuous formed-spring-steel curtain in a steel frame.
>
> The suitability of a fire door should be judged on the class of opening in which it is to be installed, not on the fire-resistance rating of the wall in which it is to be installed. . . .
>
> If the opening is in a wall dividing a building into separate fire areas, the door should be suitable for installation in a Class A opening (3-h fire protection rating). The same door can be used whatever the fire resistance rating of the wall. If a wall encloses a vertical communication, the door should be suitable for a Class B opening.

The NFPA fire doors and windows standard† gives recommendations on the installation of suitable approved doors, windows, and shutters, and it also specifies how the opening should be constructed and how the door or window should be mounted, equipped, and operated.

FIRE-DETECTION AND-ALARM SYSTEMS

There are several general systems and many devices which can be effectively used to detect fire and transmit a warning. This section briefly describes this equipment.

Heat Detectors

Heat-detection devices are categorized in two ways: (1) those that respond when the detection element reaches a predetermined temperature (fixed-temperature-types), and

**Fire Protection Handbook,* 14th ed., NFPA, Quincy, Massachusetts, 1976, p. 6-84.
†NFPA 80, "Standard for Fire Doors and Windows," NFPA, Quincy, Massachusetts, 1979.

(2) those that respond to an increase in heat at a rate greater than some predetermined value (rate-of-rise-types). Some devices combine both principles. The same principles apply whether the devices are of the spot-pattern type, in which the thermally sensitive element is a unit, or the line-pattern type, in which the element is continuous along a line or a circuit.

Fixed-Temperature Detectors

Thermostats are the most widely used fixed-temperature heat detectors in signaling systems.

Bimetallic Thermostats. The common form of thermostat is the bimetallic type that utilizes the different coefficients of expansion of two metals under heat to cause a movement resulting in closing of electrical contacts (Fig. 2-4).

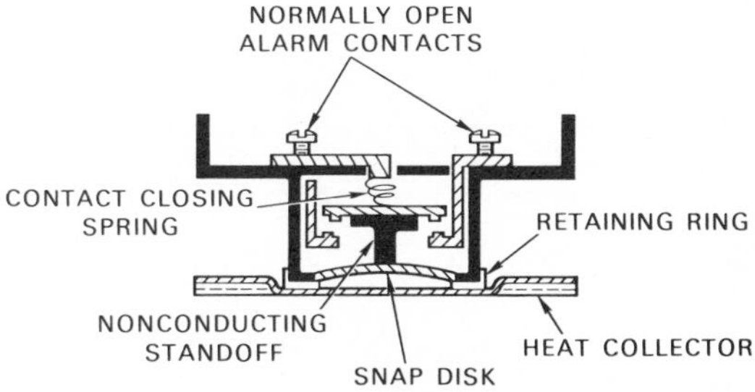

Figure 2-4 Spot-type, fixed-temperature snap-disk detector.

Snap-Action Disk Thermostats. A metal disk goes from concave to convex when the temperature rating of the thermostat is reached. One special advantage of these thermostats is that when the temperature goes down, they are restored to their original condition.

Line Thermostats. The *thermostat cable* is a line type of thermostat. The cable is made up of two metals separated from each other by a heat-sensitive covering applied directly to the wires. When the rated temperature is reached, the covering melts and the two wires come in contact to initiate an alarm. The section of wire affected must be replaced after operation.

Other Types. Other forms of fixed-temperature heat detectors are the *fusible link,* occasionally employed to restrain operation of an electrical switch until the point of fusion is reached, and the *quartzoid bulb thermostat,* which depends on removal of the restriction by breaking the bulb. Both of these units require replacement after operation.

Rate-of-Rise Detectors

Fire detectors that operate on the rate-of-rise principle function when the rate of temperature increase exceeds a predetermined rate. Detectors of this type combine two functioning elements, one of which initiates an alarm on a rapid rise of temperature, while the other acts to delay or prevent an alarm on a slow temperature rise. Advantages to rate-of-rise devices are: (1) they can be set to operate more rapidly under most conditions than can fixed-point devices; (2) they are effective across a wide range of ambient temperatures; (3) they recycle rapidly and are usually readily available for continued service; (4) they tolerate slow increases in ambient temperature without giving an alarm. The disadvantages of rate-of-rise detectors for some applications are their susceptibility to false alarms where there is a rapidly increasing temperature and their possible failure to respond to a fire that propagates very slowly.

Pneumatic-Tube Detectors. These operate on the rate-of-rise principle. When the temperature increases at a certain rate, the air in the tube expands and causes a diaphragm to move and close a cirucit, thus causing an alarm. The device will not cause an alarm if the temperature rise is too slow.

Combined Rate-of-Rise and Fixed-Temperature Detectors

Thermostats have been developed to take advantage of the rate-of-rise feature to sense a fast-developing fire; the fixed-temperature part takes care of a fire whose growth is slow. The typical form of the rate-of-rise thermostat is a vented air chamber that heats up in a flexible diaphragm carrying electric contacts (Fig. 2-5). Heat outside the chamber causes air within the chamber to expand. When such expansion exceeds the capacity of the vent to relieve pressure, the diaphragm is flexed, thus closing the electric contacts. Slow changes in ambient temperature near the chamber allow it to "breathe" through its vent, and the diaphragm is not moved sufficiently to cause an alarm.

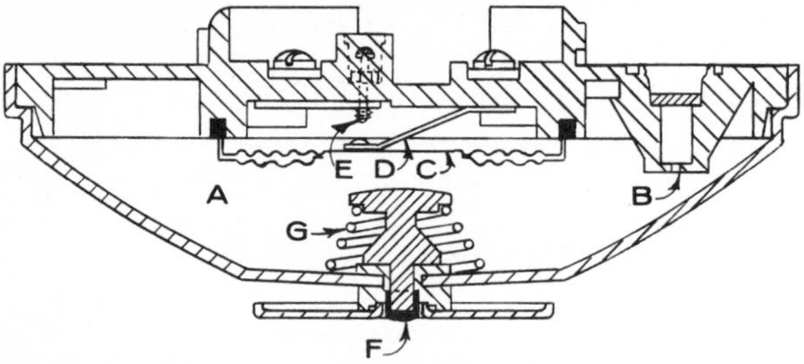

Figure 2-5 A spot-type combination rate-of-rise, fixed-temperature device. The air in chamber A expands more rapidly than it can escape from vent B. This causes pressure to close electrical contact D between diagram C and insulated screw E. Fixed-temperature operation occurs when fusible alloy F melts releasing spring G which depresses the diaphrahm closing the contact points.

Rate-Compensation (Anticipation and Differentiation) Devices

These provide an assured actuation at some predetermined maximum temperature and compensate for changes in rates of temperature rise (Fig. 2-6).

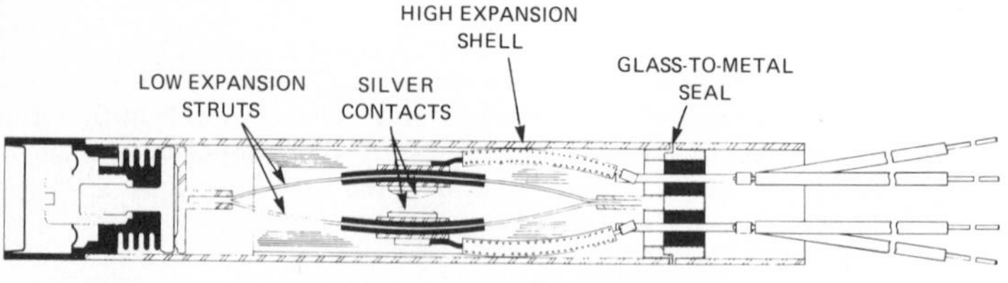

Figure 2-6 Rate-compensation heat detector. *(From* Fire Protection Handbook, *14th ed., G. P. McKinnon (ed.), NFPA, Quincy, Massachusetts, 1976.)*

Smoke Detectors

There are four types of smoke detectors: (1) photoelectric, (2) beam-type, (3) ionization, and (4) sampling detectors.

Photoelectric Detectors

These detectors operate on a beam of light. The smoke either obscures a beam of light directly, or enters a refraction chamber where the smoke reflects the light into the photocell. The change in electric current resulting from either partial obscuring of a photoelectric beam by smoke particles, or the scattering of light onto a photosensitive device, causes an alarm sound when the smoke reaches a sufficient density. (Fig. 2-7 *a* and *b*, respectively)

Beam-Type Detectors

These employ a light beam that is carried between elements at extreme ends or sides of the protected area and crosses the area to be protected. The beam is projected into a photosensing cell. Smoke between the light source and the receiving photocell reduces the light that reaches the cell, activating the alarm (Fig. 2-8).

Ionization Detectors

These detectors consist of one or more ionization chambers and the necessary related amplification circuits. The ionization detector has as a sensing element, the ionization chamber, in which air is made electrically conductive (ionized) by a minute source of radioactive material. A voltage applied across the ionization chamber causes a very small electric current to flow as the ions travel to the electrode of opposite polarity. When smoke particles enter the chamber, they attach themselves to the ions and cause a reduction in mobility and thus a reduction in current flow. The reduced current flow increases the voltage on the electrodes which, when reaching a predetermined level, results in an alarm (Fig. 2-9).

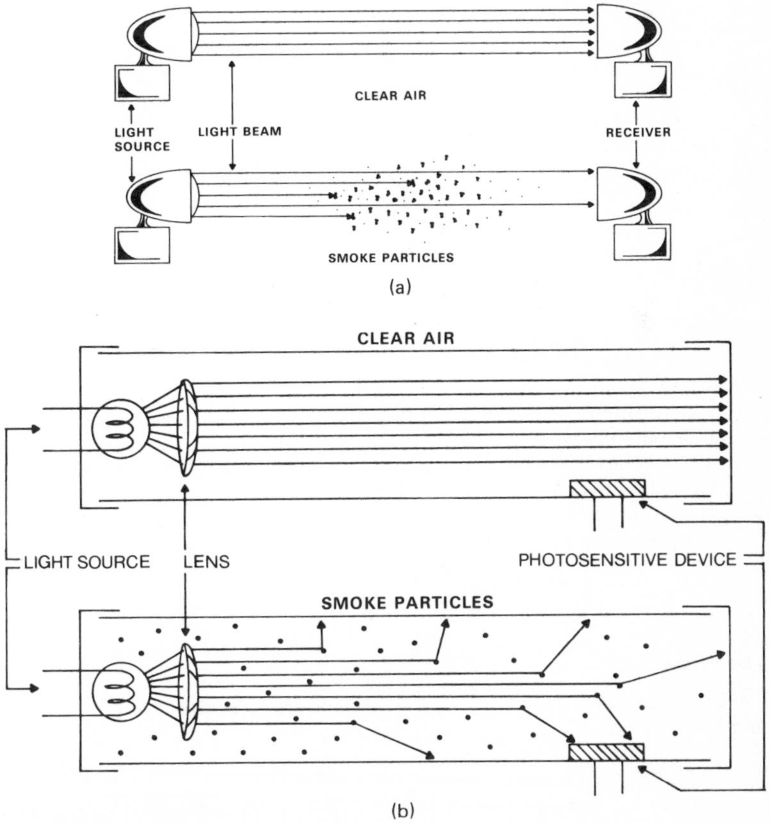

Figure 2-7 Principle of operation for (*a*) a photoelectric obscuration smoke detector, and (*b*) a photoelectric scattering smoke detector.

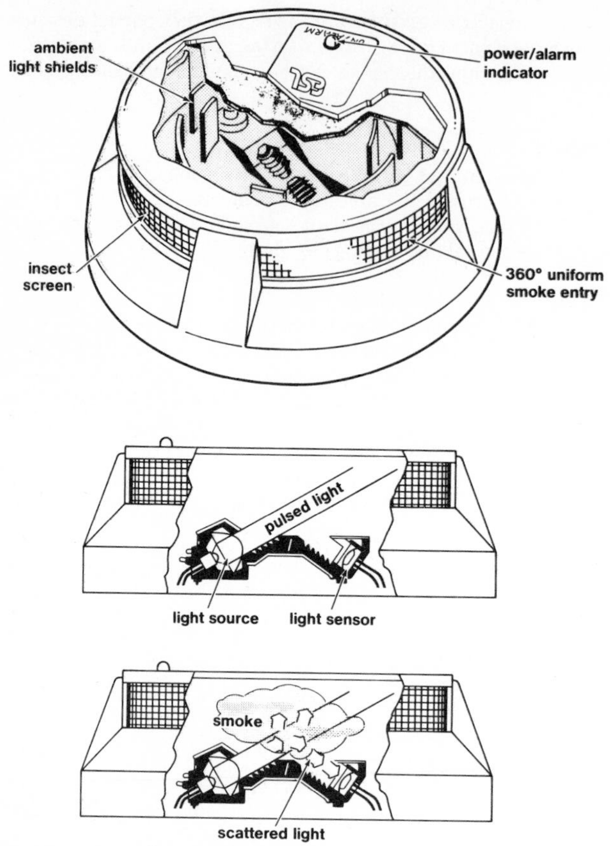

Figure 2-8 Cross-sectional view of a photoelectric light-scattering smoke detector. *(Electro Signal Lab., Inc.)*

Sampling Detectors

These consist of tubing distributed from the detector unit into the area(s) to be protected. An air pump draws air from the protected area back to the detector. A cloud-chamber smoke detector is a form of sampling detector. The air pump draws a sample of air into a high-humidity chamber; the pressure is lowered slightly. If smoke particles are present, the moisture in the air condenses on them forming a cloud in the chamber. The density of this cloud is measured by the photoelectric principle. When the density is greater than a predetermined level, the detector responds to the smoke and the alarm is activated.

Flame Detectors

There are four basic types of flame detectors: (1) infrared, (2) ultraviolet, (3) photoelectric, and (4) flame flicker.

Infrared and Ultraviolet

These detectors have sensing elements responsive to radiant energy outside the range of human vision.

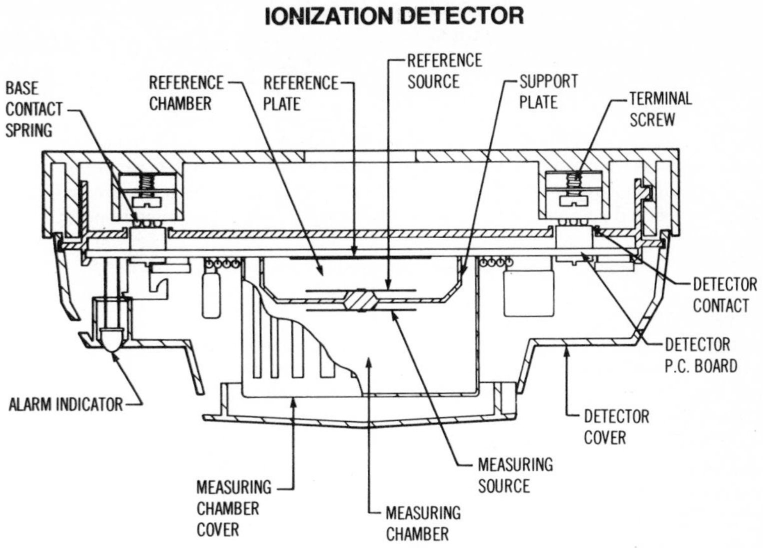

Figure 2-9 Cross-sectional view of an ionization smoke detector. *(Pyrotronics, Inc.)*

Photoelectric

This type employs a photocell that either changes its electric conductivity or produces an electric potential when it is exposed to radiant energy.

Flame Flicker

This is a photoelectric type of detector that includes means to prevent response to visible light unless the observed light is modulated at a frequency characteristic of the flicker of a flame.

Gas Detectors

Various detection devices will monitor the amount of flammable gases or vapors in an area. Portable gas detectors are used to detect the presence of combustible gas or vapor in basements, sewers, manholes, etc. Other devices will analyze the air samples brought into the device from various points. Gas and vapor testing equipment is valuable for preventing fires and explosions in petroleum and chemical plants and in industries where combustible vapors may be generated.

Fire-detection devices are usually installed in systems which combine manually activated fire-alarm stations and audible and visual warning devices. They may also be connected to fire-suppression systems in some hazardous areas.

Alarm Systems

The detection and alarm systems in a plant building should be connected to a constantly supervised monitoring system. The most common systems are described below.

Local Systems

These systems produce a signal manually or automatically at the protective premises for an alarm of fire and for required supervisory services, including supervision of a security guard's rounds, supervision of sprinkler water-flow alarm service and of sprinkler systems, etc. Local systems are used for the protection of property and for the protection of life by indicating the necessity for evacuation of the building.

Proprietary Systems

Proprietary systems are used for individual properties where the system is under constant supervision by competent and experienced personnel in a central supervisory station at the property protected. Such systems are usually found in large industrial plants; signals are received at a central supervisory station where experienced operators are on duty at all times. The central supervisory station is under the control of the owner or occupant of the protected property and is usually on or near that property.

Remote-Station Systems

These are usually used to protect premises on which there is frequently no one present. The signal is received at fire-alarm headquarters or at the office of a communications agency, usually located at a distance from the protected property. Signals are transmitted and received on privately owned equipment; the agency receiving the signals may be a municipal fire department or a communications agency capable of receiving the signals and acting upon them.

Auxiliary Systems

This type connects devices in the protected plant with the municipal fire alarm system. Alarms are received at fire-alarm headquarters on the same equipment and by the same alerting methods as alarms transmitted from municipal street boxes. Signals are recorded at a municipal fire department; connecting facilities between the protected property and the fire department are part of the municipal fire alarm system. Devices in the protected plant are customarily owned and maintained by the property owner. Equipment that connects the devices to the city's circuits is owned and maintained by the municipality, or leased by it, as part of the municipal alarm system and limited to alarm service only.

Central Station Systems

Central station systems are operated by firms whose principal business is the furnishing and maintaining of supervised signaling service. The central station services properties subscribing to the service; alarm and signaling devices on the subscribers' property are connected to the central station where operators are on hand to receive the signal and take the appropriate action. Central station operators retransmit alarms to the fire department.

Standards for the installation and maintenance of fire detection and alarm equipment are listed in the last section of this chapter.

FIRE-SUPPRESSION SYSTEMS

This section describes systems which extinguish or control fires.

Water Supply

The most common type of fire-suppression systems rely on water. Therefore it is essential that adequate supplies of water be provided and maintained. A thorough discussion of water-supply systems is presented in Sec. 6 of this handbook.

The plant water-supply system or nearby public water supply will be the primary fire-suppression system used by the plant fire brigade or public fire department. Water must be provided in quantity and pressure sufficient for supplying automatic sprinkler systems and fire hoses, in addition to normal plant requirements. When the public water supply is inadequate for plant protection, supplemental private supplies are necessary.

Pipe Networks

The minimum recommended pipe size for fire protection is 6 in; the pipe network is looped in a grid pattern, and no leg is more than 600 ft long.

It is usually cost-effective to use larger pipe sizes since the installation costs are relatively the same (Table 2-15).

TABLE 2-15 Comparison of Pipe Capacity

Size of pipe, in	Relative capacity
6	1.0
8	2.1
10	3.8
12	6.2
14	9.3
16	13.2

Source: NFPA, Quincy, Massachusetts.

Wherever possible pipe networks should be arranged in loops to provide water flow from two directions when necessary (Fig. 2-10).

A pipe 8 in or larger is necessary to supply:

- More than one hydrant on a dead-end main
- A dead-end main exceeding 500 ft
- Two hydrants on a loop exceeding 1500 ft
- Three hydrants on a loop exceeding 1000 ft
- Four or more hydrants
- Hydrants with three or four outlets
- Water at low pressure

Fire Hydrants

Fire hydrants are provided on public mains to allow the fire department to draw water with mobile pumpers to supply sprinkler and standpipe systems, or hose streams. Fire hydrants are provided on private mains to allow the fire brigade or fire department to supply hose streams and where the supply is adequate to support sprinkler and standpipe systems with mobile pumpers.

Hydrants are available in wet-barrel (California, Fig. 2-11) and dry-barrel (base valve, Fig. 2-12) types. Base-valve hydrants are necessary where there is any chance of freezing.

Hydrants on plant pipe networks should be located every 250 ft and about 50 ft from the buildings protected. They must be protected from damage by vehicles or machinery.

The available water flow for fire suppression is determined by flow testing the hydrant system. The water flow at 20 psig is calculated since this is the minimum pressure required by fire department pumpers.

Valves in pipe lines supplying fire-protection water are generally required to be indicating valves. These include underground gate valves with indicator post, underground butterfly valves with indicator post, and outside screw and yoke (OS&Y) gate valves.

Incorrectly shut valves have been the primary cause of sprinkler systems failing to control fires.

Fire Pumps

Fire pumps are essentially the same as water-supply pumps discussed in Sec. 6. Additional considerations required for fire pumps include:

- Use of equipment approved for fire pumps
- Use of approved accessories

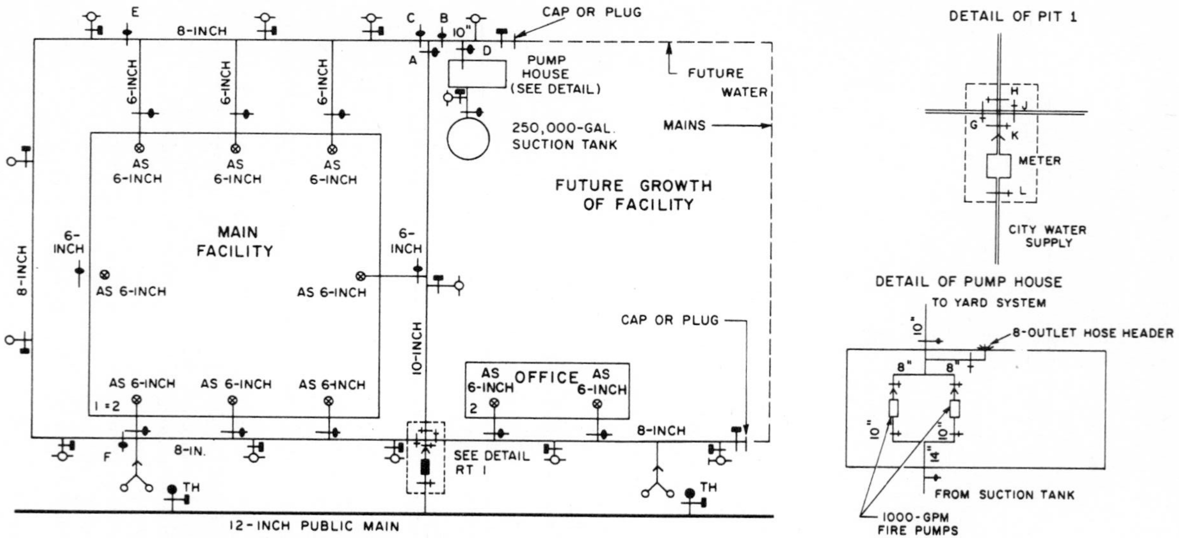

Figure 2-10 Water-pipe network. *(From* Fire Protection Handbook, *14th ed., G. P. McKinnon (ed.), NFPA, Quincy, Massachusetts, 1976.)*

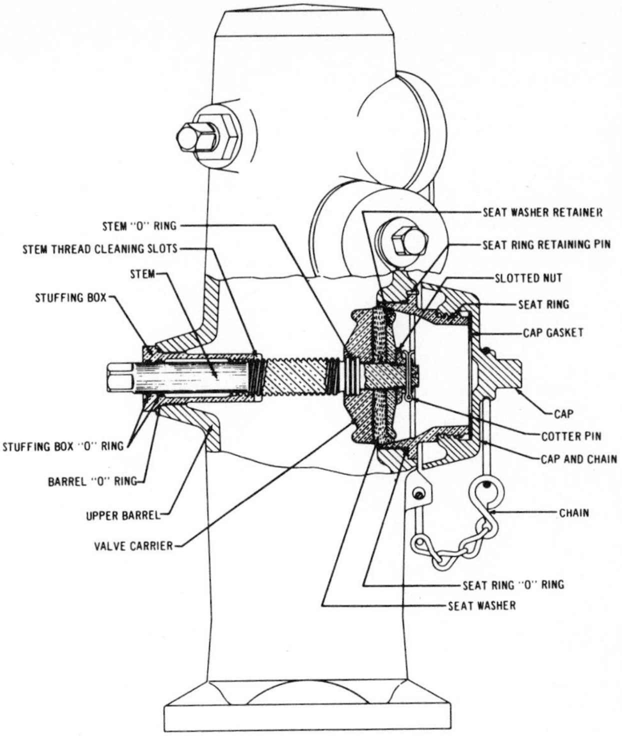

Figure 2-11 Wet-barrel hydrant. *(Mueller Co.)*

- Adequate capacity to meet fire-flow demands
- Selection of fire pump driver based on reliability, adequacy, economy, and safety of power source
- Automatic operation
- Safe location for uninterrupted service
- Annual testing
- Maintenance

Sprinkler Systems

A sprinkler system, for fire-protection purposes, is an integrated system of underground and overhead piping designed in accordance with fire-protection engineering standards. The installation includes a water supply, such as a gravity tank, fire pump, reservoir or pressure tank and/or connection by underground piping to a city main. The portion of the sprinkler system above ground is a network of specially sized or hydraulically designed piping installed in a building, structure, or area, generally overhead, and to which sprinklers are connected in a systematic pattern. The system includes a controlling valve and a device for actuating an alarm when the system is in operation. The system is usually activated by heat from a fire and discharges water over the fire area.

Wet-Pipe Systems

These are under water pressure at all times; water will be discharged immediately when an automatic sprinkler operates (Fig. 2-13).

Water flowing from the wet-pipe sprinkler system actuates an alarm valve that gives off a signal. Figure 2-13 illustrates the total concept of the wet-pipe automatic sprinkler system.

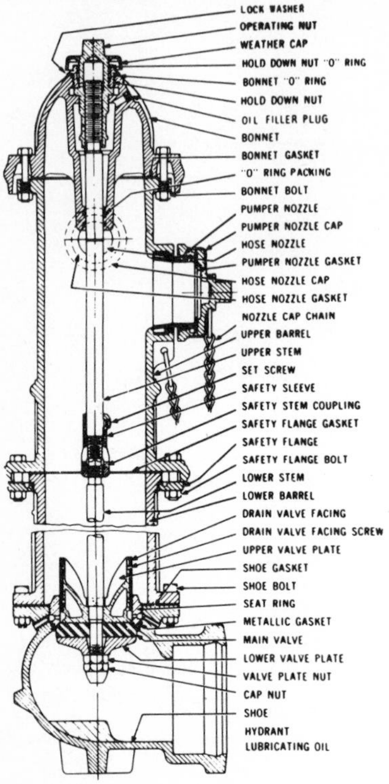

Figure 2-12 Dry-barrel hydrant. *(Mueller Co.)*

The essential features of wet-pipe sprinkler systems, which represent about 75 percent of sprinkler installations, include provisions for water supplies, piping, and location and spacing of sprinklers. This system is generally used wherever there is no danger that the water in the pipes will freeze and wherever there are no special conditions requiring one of the other systems. Inspection of the wet-pipe sprinkler system at regular intervals is essential, and weekly inspection of all water-control valves and alarm-control valves is recommended.

Where subject to temperatures below freezing, the ordinary wet-pipe system cannot be used. There are two recognized methods of maintaining automatic sprinkler protection in such locations: (1) through the use of systems where water enters the sprinkler piping only after operation of a control valve (dry-pipe, pre-action, etc.) and (2) by the use of antifreeze solution in a portion of the wet-pipe system.

Regular Dry-Pipe Systems

In locations where there is danger of freezing, it is the usual practice to install a dry-pipe system. In regular dry-pipe systems, the sprinkler piping contains air or nitrogen under pressure instead of water, and admission of the water is controlled by a dry-pipe valve. When a sprinkler is opened by heat from a fire, the pressure is reduced, a dry-pipe valve is opened by water pressure, and water flows out of any opened sprinklers.

Pre-Action Systems

Systems in which the air in the piping may or may not be under pressure are called pre-action systems. These systems are designed primarily to protect properties on which the danger of water damage from broken sprinklers or piping could be serious. The water-supply valve is actuated independently of the opening of the sprinkler heads by an automatic fire-detection system; the valve is opened sooner than with the dry-pipe system, and the alarm is given when the valve is opened.

The pre-action system has several advantages over a dry-pipe system. The valve is opened sooner because the fire detectors have less thermal lag than sprinklers. The detection system also automatically rings an alarm. Fire and water damage is decreased because water is on the fire sooner, and the alarm is given when the valve is opened. Sprinkler piping is normally dry; thus, pre-action systems are nonfreezing and applicable to dry-pipe service.

Deluge Systems

Deluge systems are used for areas of extra occupancies. All sprinkler heads are open at all times so that when the water comes on, the entire area is flooded; when heat from a

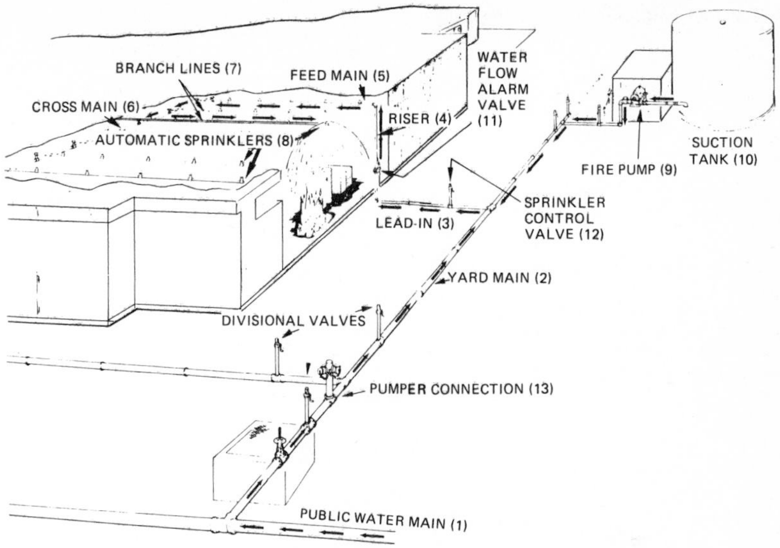

Figure 2-13 Sprinkler system. *(Factory Mutual System.)*

fire actuates the fire-detecting device, water flows to and is discharged from all sprinklers on the piping system, thus "deluging" the protected areas. These systems are often used in airplane hangars and in areas where flammable liquids are handled or stored.

By using sensitive thermostatic controls operating on the rate-of-rise or fixed-temperature principle, or controls designed for individual hazards, it is possible to apply water to a fire more quickly than with systems in which operation depends on opening of sprinklers only as the fire spreads.

Sprinkler Heads

Sprinkler heads are designed with temperature ratings ranging from 135°F (57°C) to as high as 500°F (260°C). Ratings of 165°F (74°C) are usual for use in buildings that are maintained at normal, constant temperatures.

The location and spacing of sprinkler heads depends on the degree of hazard and type of construction. The "Standard for the Installation of Sprinkler Systems," NFPA 13, provides detailed design and maintenance specifications (Table 2-16).

Standpipes

The four generally recognized standpipe system concepts are described in (Table 2-17):

1. A wet-standpipe system, having supply valve open and water pressure maintained at all times. This is the most desirable type of system.
2. A dry-standpipe system arranged to admit water to the system through manual operation of approved remote-control devices located at each hose station. The water-supply control mechanism introduces an inherent reliability factor that must be considered.
3. A dry-standpipe system in an unheated building. The system should be arranged to admit water automatically by means of a dry-pipe valve or other approved device. The depletion of system air at the time of use introduces a delay in the application of water to the fire and increases the level of competency required to control the pressurized hose and nozzle assembly during the charging period.

TABLE 2-16 Summary of Spacing Rules

Type of construction	Maximum distance of deflectors below ceiling, in			
	In bays*		Under beams	
	Comb.	Noncomb.	Comb.	Noncomb.
Smooth ceiling	10	12	14	16
Beam and girder	16	16	20	20
Panel up to 300 ft²	18	18	22	22
Bar joists	10	12	—	—
Open wood joists—center 3 ft or less	6	—	—	—

Minimum below ceiling is 1 in.
Minimum below beams 1 in, maximum 4 in. Do not exceed maximum below ceiling.

Maximum coverage per sprinkler

Light hazard	200 ft² smooth ceiling and beam and girder construction
	130 ft² open wood joist
	168 ft² all other types of construction
Ordinary hazard	130 ft² all types of construction except
	100 ft² high piled storage
Extra hazard	90 ft² all types of construction

Direction of lines: Either direction to facilitate hanging except: across beams for beams on girders 3 ft to 7½ ft on centers and across joists for wood joists (open or sheathed) and bar joists (through or under)

Maximum spacing between lines and sprinklers

Light and ordinary hazard	15 ft except 12 ft for high piled storage
Extra hazard	12 ft

*Not more than 4 in below beams where lines run across beams.
Source: NFPA 13, "Installation of Sprinkler Systems," Quincy, Massachusetts.

4. A dry-standpipe system having no permanent water supply. This type would be used for reducing the time required for fire departments to put hose lines into action on upper floors of tall buildings. This type of system might also be used in buildings during construction, where allowed in lieu of the wet standpipe in unheated areas.

Water-Spray Fixed Systems

Water-spray fixed systems are generally used to protect flammable liquid and gas tankage; piping and equipment; electrical equipment such as transformers, oil switches, and rotating electrical machinery; and openings in firewalls and floors through which conveyors pass. The type of water spray required for any particular hazard depends, of course, upon the nature of the hazard and the purpose for which the protection is provided.

NFPA 15, "Standard for Water Spray Fixed Systems for Fire Protection," calls for piping, valves, pressure gauges, and detection systems of an approved type. The spray nozzles generally used in these systems are open, and the pipes, especially outdoor ones that are subject to freezing temperatures, are usually dry.

Foam Extinguishing Systems

Foam extinguishing systems have been used extensively for many years, especially in the petrochemical industry, for the extinguishment of flammable liquid fires. The principal

TABLE 2-17 Summary of National Fire Protection Association Standpipe Standards*

Type	Intended use	Size hose and distribution	Minimum size pipe	Minimum water supply
Class I	Heavy streams Fire department Trained personnel Advanced stages of fire	2½-in connections All portions of each story or section within 30 ft of nozzle with 100 ft of hose	4 in up to 100 ft 6 in above 100 ft (275 ft maximum unless pressure regulated)	500 gal/min 1st standpipe 250 gal/min each additional 2500 gal/min maximum) 30-min duration 65 lb/in² at top outlet with 500 gal/min flow
Class II	Small streams Building occupants Incipient fire	1½-in connections (distribution same as class I)	2 in up to 50 ft 2½ in above 50 ft	100 gal/min per building 30-min duration 65 lb/in² at top outlet with 100 gal/min flowing
Class III	Both of above	Same as class I with added 1½-in outlets or 1½-in adapters and 1½-in hose.	Same as class I	Same as class I

*From NFPA 14, "Standard for the Installation of Standpipe and Hose Systems," Quincy, Massachusetts.

kinds of foam are chemical and mechanical (determined by how they are generated), and these classes are further subdivided.

Special compatible foam concentrates result in the generation of a foam that does not break down as readily as ordinary foam when mixed with dry chemical. Other special foams are available for application on fires in alcohols, esters, ketones, and ethers (called water-soluble or polar liquids). This concentrate produces a foam that does not deteriorate like ordinary foam when in contact with water-miscible solvents.

Carbon Dioxide Systems

Carbon dioxide is a noncombustible gas that has been effectively used to extinguish certain types of fires. It acts to reduce the oxygen in the fire area to a point where it will no longer support combustion (Table 2-18). Because carbon dioxide is stored under pressure, it can readily be discharged from its cylinder or extinguisher. Carbon dioxide is inert and will not conduct electricity. It can be used safely on energized electric equipment fires without causing damage to the equipment.

Because carbon dioxide does little or no damage to equipment or materials with which it comes in contact, it is very useful for protection of rooms with contents of high value and contents subject to water damage. Typical of such occupancies are fur vaults, record storage rooms, computer rooms, and rooms housing live electric equipment. Carbon dioxide is also widely used for extinguishing flammable liquid fires because the carbon dioxide gas rapidly spreads above the surface of the burning liquid and shuts off the oxygen.

Halogenated Agents and Systems

One of the earliest halons used for fire extinguishing was carbon tetrachloride. Concern about the toxic properties of carbon tetrachloride began to manifest itself about 1960, and the *NFPA and the extinguisher-testing laboratories discontinued any recommendation of carbon tetrachloride as a fire extinguishing agent.*

TABLE 2-18 Minimum Carbon Dioxide Concentrations for Extinguishment*

Material	Theoretical min. CO_2 concentration, (%)
Acetylene	55
Acetone	26†
Benzol, benzene	31
Butadiene	34
Butane	28
Carbon disulfide	55
Carbon monoxide	53
Coal gas or natural gas	31†
Cyclopropane	31
Dowtherm	38†
Ethane	33
Ethyl ether	38†
Ethyl alcohol	36
Ethylene	41
Ethylene dichloride	21
Ethylene oxide	44
Gasoline	28
Hexane	29
Hydrogen	62
Isobutane	30†
Kerosene	28
Methane	25
Methyl alcohol	26
Pentane	29
Propane	30
Propylene	30
Quench, lubricating oils	28

*From Gordon P. McKinnon (ed.) *Fire Protection Handbook,* NFPA, Quincy, Massachusetts, 1976.

†The theoretical minimum extinguishing concentrations in air for the above materials were obtained from Bulletin 503, U.S. Bureau of Mines, Washington, 1952. Those marked † were calculated from accepted residual oxygen values.

A halon is a hydrocarbon (hydrogen and carbon) in which some of the hydrogen atoms have been replaced by such elements as bromine, chlorine, or fluorine, or by combinations of these (Table 2-19). A number of halons are toxic, thus making them undesirable for general use; two of them, Halon 1301 and Halon 1211, have acceptable levels of toxicity and excellent flame extinguishment properties.

Halon 1211 and Halon 1301 are the only two agents recognized by the NFPA Technical Committee on Halogenated Fire Extinguishing Agent Systems. Both Halon 1211 and 1301 are widely used for protection of electric equipment (both are nonconductors of electricity), airplane engines, and computer rooms. As both of these halons rapidly vaporize, they leave little corrosive or abrasive residue to clean up and do not interfere as much with visibility during fire fighting as foam or carbon dioxide. These halons are used for hand extinguishers and fixed systems.

Dry-Chemical Extinguishing Systems

Dry-chemical extinguishing agents consist of finely divided powders that effectively extinguish a fire when applied to the fire by portable extinguishers, hose lines, or fixed systems. The original dry powder was sodium bicarbonate (ordinary baking soda). Potassium bicarbonate and other chemical powders, with additives to make the powders free flowing and more moisture resistant, are now in use. Dry chemical has been found to be

TABLE 2-19 Some Physical Properties of the Common Halogenated Fire Extinguishing Agents

Agent	Chemical formula	Halon no.	Type of agent	Approx. boiling point, °F	Approx. freezing point, °F	Specific gravity of liquid at 68°F (water = 1)	Approx. critical temp., °F	Estimated pressure, psig		Latent heat of vaporization, cal/g water = 540 cal/g CO_2 = 138 cal/g
								At 130°F	At critical temp.	
Carbon tetrachloride	CCl_4	104	Liquid	170	−8	1.595	—	—	—	46
Methyl bromide	CH_3Br	1001	Liquid	40	−135	1.73	—	—	—	62
Bromochloromethane	$BrCH_2Cl$	1011	Liquid	151	−124	1.93	—	—	—	—
Dibromodifluoromethane	Br_2CF_2	1202	Liquid	76	−223	2.28	389	23	585	29
Bromochlorodifluoromethane	$BrCClF_2$	1211	Liquefied gas*	25	−257	1.83	309	75	580	32
Bromotrifluoromethane	$BrCF_3$	1301	Liquefied gas	−72	−270	1.57	153	435	560	28
Dibromotetrafluoroethane	BrF_2CCBrF_2	2402	Liquid	117	−167	2.17	—	3.8	—	25

*May be kept as a liquid at reduced temperatures.

Source: NFPA, Quincy, Massachusetts.

an effective extinguishing agent for fires in flammable liquids and in certain types of ordinary combustibles and electric equipment, depending upon the type of dry chemical used.

Dry-chemical extinguishing systems are used to protect flammable-liquid storage rooms, dip tanks, kitchen range hoods, deep-fat fryers, and similar hazardous areas and appliances. Because dry chemical is nonconductive, these systems are useful in the protection of oil-filled transformers and circuit breakers. Dry-chemical systems are not recommended for telephone-switchboard or computer protection. Dry-chemical hose-line systems are used in crash trucks and for the protection of aircraft hangars. Dry chemicals are used in portable fire extinguishers.

Combustible-Metal Extinguishing Systems

A number of metals and metal powders found in industrial situations and in transport will burn. Some metals burn when heated to high temperatures by friction or exposure to external heat. Others burn from contact with moisture or in reaction with other materials. These metals and metal powders require special extinguishing agents and special fire-fighting techniques. Some result in explosions and very high temperatures, and some react violently with water. Still others give off toxic fumes when burning.

Some combustible metal extinguishing agents' success in handling metal fires has led to the terms *approved extinguishing powder* and *dry powder*. Such terms have been accepted in describing extinguishing agents for metal fires, and should not be confused with the name *dry chemical*, which normally applies to an agent suitable for use on flammable-liquid and live electric equipment fires. Graphite powder, talc, and sand have all been used to smother metal fires.

Portable Fire Extinguishers

Portable fire extinguishers are required in most plants by local, state, and federal regulations and insurance requirements. Where there are trained personnel available to use the proper extinguisher on a small incipient fire, extinguishers may prove useful in preventing a larger, more devastating fire.

The limitations of extinguishers, personal exposure to fire and smoke, capacity, range, selectivity, and availability necessitate that training be provided if they are expected to be effective. Use of extinguishers should be simultaneous with notification of the fire brigade or department.

Types of Portable Fire Extinguishers

The kind and number of extinguishers needed for particular types of fires are specified in NFPA 10, "Standard for Portable Fire Extinguishers." The most common types of extinguishers in use are the pressurized water, carbon dioxide, and multipurpose dry chemical. Other extinguishers commonly used are water pump tanks, liquefied gases such as Halon 1211, and combustible-metal-type dry powder.

Application of Portable Fire Extinguishers

NFPA 10, "Standard for Portable Fire Extinguishers," classifies fires in four ways:

Class A. Fires involving ordinary combustible materials (wood, cloth, paper, rubber, and many plastics) requiring the heat-absorbing (cooling) effects of water, water solutions, or the coating effects of certain dry chemicals that retard combustion (Table 2-20).

Class B. Fires involving flammable or combustible liquids, flammable gases, greases, and similar materials where extinguishment is most readily secured by excluding air (oxygen), inhibiting the release of combustible vapors, or interrupting the combustion chain reaction (Table 2-21).

Class C. Fires involving live electric equipment where safety to the operator requires the use of electrically nonconductive extinguishing agents. (*Note:* When electric equipment is de-energized, the use of Class A or B extinguishers may be indicated.)

TABLE 2-20 Fire Extinguisher Size and Placement for Class A Hazards

		Areas to be protected per extinguisher		
Basic minimum extinguisher rating for area specified	Maximum travel distances to extinguishers, ft	Light hazard occupancy, ft^2	Ordinary hazard occupancy, ft^2	Extra hazard occupancy, ft^2
1-A	75	3,000	†	†
2-A	75	6,000	3,000	†
3-A	75	9,000	4,500	3,000
4-A	75	11,250	6,000	4,000
6-A	75	11,250	9,000	6,000
10-A	75	11,250*	11,250*	9,000
20-A	75	11,250*	11,250*	11,250*
40-A	75	11,250*	11,250*	11,250*

*11,250 ft^2 is considered a practical limit.

†Protection requirements may be fulfilled by several extinguishers of the minimum specified rating with the approval of the authority having jurisdiction.

Source: NFPA, Quincy, Massachusetts.

TABLE 2-21 Fire Extinguisher Size and Placement for Class B Hazard Excluding Protection of Deep Layer Flammable Liquid Tanks (For Extinguishers Labeled After June 1, 1969)

Type of hazard	Basic minimum extinguisher rating	Maximum travel distance to extinguishers, ft
Light	5-B	30
	10-B	50
Ordinary	10-B	30
	20-B	50
Extra	20-B	30
	40-B	50

Source: NFPA, Quincy, Massachusetts.

Class D. Fires involving certain combustible metals (such as magnesium, titanium, zirconium, sodium, potassium, etc.) requiring a heat-absorbing extinguishing medium not reactive with the burning metals.

Figures 2-14 through 2-16 illustrate fire extinguishing agents, classifications, and symbols.

CODES AND STANDARDS

Model Fire Prevention Codes

Currently, there are five "model" fire prevention codes available in the United States. Each of these five model codes utilizes NFPA standards as the basis for the technical details of its fire-prevention and fire-control measures. These five codes are: (1) the NFPA's *Fire Prevention Code,* which was developed by a committee representing state and city fire marshals and other concerned interests, and which was adopted after going through the usual NFPA standards-making process and submission procedure at an annual meeting of the association; (2) the American Insurance Association's *Fire Prevention Code,* which was the first model fire prevention code issued, and which has been adopted by many communities; (3) the *Uniform Fire Code,* published by the International Conference of Building Officials in cooperation with the Western Fire Chiefs

1. Extinguishers suitable for "Class A" fires should be identified by a triangle containing the letter "A." If colored, the triangle shall be colored green.*

2. Extinguishers suitable for "Class B" fires should be identified by a square containing the letter "B." If colored, the square shall be colored red.*

3. Extinguishers suitable for "Class C" fires should be identified by a circle containing the letter "C." If colored, the circle shall be colored blue.*

4. Extinguishers suitable for fires involving metals should be identified by a five-pointed star containing the letter "D." If colored, the star shall be colored yellow.*

Figure 2-14 Fire extinguisher identification.* Recommended colors per PMS (Pantone Matching System): GREEN—*Basic Green*, RED—*192*, BLUE—*Process Blue*, YELLOW—*Basic Yellow*. *(From NFPA 10, "Standard for Portable Fire Extinguishers," 1978.)*

Association; (4) the *BOCA Basic Fire Prevention Code,* published by the Building Officials and Code Administrators International, Inc.; and (5) the *Southern Standard Fire Prevention Code* of the Southern Building Code Congress.

Model Building Codes

When a code commission or building official is revising an outdated building code, there are four model codes that are usually used as guides. These four model codes are: (1) the *National Building Code,* (2) the *Uniform Building Code,* (3) the *Standard Building Code,* (4) the *BOCA Basic Building Code.*

Fire Safety Standards-Making Organizations

American National Standards Institute (ANSI)

ANSI sets public requirements for national standards and develops and publishes them on a wide range of subjects. In order to achieve uniformity in voluntary and mandatory state and federal standards, it coordinates voluntary standardization activities of concerned organizations.

ANSI standards cover a variety of products, materials, and equipment that is used both in highly specialized fields and in nearly all other areas of modern life. The ANSI publishes standards on ceramic tiles, chemical process equipment, home appliances, electronics equipment, motion picture film and equipment, acids, refractory materials, oil burners, office machines and supplies, hospital supplies, and combustion engines.

Fire Extinguisher/Agent Characteristics

SUITABLE FOR USE ON TYPE OF FIRE	AGENT CHARACTERISTICS	Available Sizes	Horizontal Range	Discharge Time
DRY CHEMICAL B C	Sodium Bicarbonate or Potassium Bicarbonate or Potassium Chloride. Discharges a white or bluish cloud. Leaves residue. Non-freezing.	1 to 30 lbs.	5 to 20 ft.	8 to 25 Sec.
MULTIPURPOSE DRY CHEMICAL A B C OR B C A CAPABILITY	Basically Ammonium Phosphate. Discharges a yellow cloud. Leaves residue. Non-freezing. Some extinguishers utilizing this agent do not have an "A" rating – however, they are designated as having "A" capability.	2 to 30 lbs.	5 to 20 ft.	8 to 25 Sec.
FOAM B	Basically water and detergent. Discharges a foamy solution After evaporation, leaves a powder residue. Protect from freezing.	21 oz.	4 to 6 ft.	24 Sec.
CARBON DIOXIDE B C	Basically an inert gas that discharges a cold white cloud. Leaves no residue. Non-freezing.	2½ to 20 lbs.	3 to 8 ft.	8 to 30 Sec.
HALON 1211 A B C	Basically halogenated hydrocarbons. Discharges a white vapor Leaves no residue Non-freezing.	2 to 9 lbs.	8 to 15 ft.	8 to 15 Sec.
WATER ▲ A	Basically tap water. Discharges in a solid or spray stream. Protect from freezing!	2½ Gal.	30 to 40 ft.	1 Minute

▲ NOTE: Pump tanks available.

NOTE: A garden hose connected to a suitable weather protected hose connection is advisable for use in fighting Class A fires. This should not be considered as a replacement for extinguishers.

NOTE: 1 ft. = 0.305 m: 1 lb. = 0.454 kg: 1 gal. = 3.785*l*.

Figure 2-15 Fire extinguisher classification. *(From NFPA 10, "Standard for Portable Fire Extinguishers," 1978.)*

American Society for Testing and Materials (ASTM)

ASTM develops and publishes standards on finished products and on materials used in manufacturing and construction. Because some products and materials are used only within certain companies, industries, and government agencies, not all ASTM standards are developed by the full-consensus system. However, standards that deal with commod-

Typical Pictorial Extinguisher Marking Labels

*NOTE: Recommended colors, per PMS (Pantone Matching System):
(BLUE–*299*)
(RED–*Warm Red*)

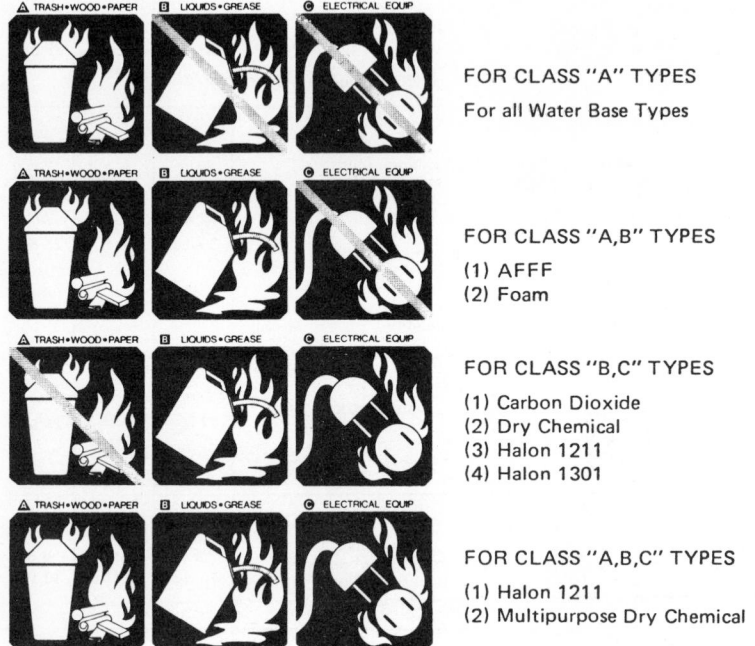

FOR CLASS "A" TYPES
For all Water Base Types

FOR CLASS "A,B" TYPES
(1) AFFF
(2) Foam

FOR CLASS "B,C" TYPES
(1) Carbon Dioxide
(2) Dry Chemical
(3) Halon 1211
(4) Halon 1301

FOR CLASS "A,B,C" TYPES
(1) Halon 1211
(2) Multipurpose Dry Chemical

Color Separation Identification (picture symbol objects are white; background borders are white)

BLUE *	–	background for "YES" symbols
BLACK	–	background for symbols with slash mark ("NO")
RED *	–	slash mark for black background symbols

Figure 2-16 Fire extinguisher symbols. *(From NFPA 10, "Standard for Portable Fire Extinguishers," 1978.)*

ities used by the general public are developed by a full-consensus procedure, wherein all interested parties are fairly represented in the committee writing the standard. The standard committee is made up of anyone technically qualified or knowledgeable in the area of the committee's scope.

National Fire Protection Association (NFPA)

NFPA codes and standards encompass the entire scope of fire prevention, fire protection, fire fighting, and fire hazards, ranging from the **National Electrical Code,*** believed to be the most widely adopted set of safety requirements in the world, to codes or standards of specific limited areas that are nevertheless important in controlling a life or fire hazard.

****National Electrical Code***® is a Registered Trademark of the National Fire Protection Association, Quincy, MA 02269.

Once a code or standard has been adopted by the NFPA, it becomes available for adoption by any organization or jurisdiction having enforcement authority. A number of NFPA standards are widely used and commonly referenced in fire legislation.

Fire Testing and Research Laboratories

There are many laboratories in the United States capable of performing, in varying degrees, fire tests of materials and/or equipment; many of these same laboratories, as well as other laboratories, have facilities for conducting fire-related research work. Generally, these laboratories can be classified into three categories: (1) private and industrial laboratories, (2) university laboratories, and (3) government laboratories.

Private and Industrial Laboratories. In the United States there are approximately 65 private and industrial laboratories that perform a wide range of fire tests. Space does not permit that each be described in detail. However, there are two, Underwriters Laboratories Inc. and Factory Mutual Laboratory Facilities, whose work warrants particular emphasis.

Underwriters Laboratories Inc. (UL). Annually UL publishes lists of manufacturers whose products, when tested, have proved acceptable under appropriate standards and which are subjected to one of the follow-up services provided by the laboratories as a counter-check. The word "listed" appears on UL labels attached to these products as authorized evidence that these products have been found to be in compliance with the laboratories' requirements.

Factory Mutual Systems (FM). Factory Mutual maintains testing facilities in Norwood, Mass., and also conducts large-scale applied research in its 1-acre, 60-ft-high FM test center in West Gloucester, R.I. Factory Mutual Laboratory facilities are available on a contract basis through Factory Mutual Research.

University and Federal Government Laboratories. More than forty American colleges and universities are equipped with laboratories for fire testing and research. In addition to the colleges and universities that serve primarily as institutions for fire science training and educations, there are others, both private and state-supported, whose engineering, physics, or science departments engage in such activities.

Several departments of the federal government—Agriculture, Air Force, Army, Commerce, Navy, and Transportation—as well as independent agencies also have research laboratories located throughout the country. These facilities are a direct result of an increasing national interest in fire safety as well as other safety- and health-related issues.

Insurance Organizations

Many important groups perform varied fire protection and inspection services on behalf of the insurance industry and its insureds. For example, the Association of Mill and Elevator Mutual Insurance Companies serves the mill and elevator industry's needs; the American Institute of Marine Underwriters is organized to serve the marine underwriters and to promote, advance, and protect their interests.

There are five large insurance organizations, however, that serve a wide range of casualty and property insurers and contribute to fire protection in many ways. They are: (1) American Insurance Association, (2) American Mutual Insurance Alliance, (3) Factory Mutual System, (4) Industrial Risk Insurers, and (5) Insurance Services Office.

BIBLIOGRAPHY

Bryan, John L.: *Automatic Sprinkler & Standpipe Systems,* NFPA, Quincy, Massachusetts, 1976.

Bryan, John L.: *Fire Suppression and Detection Systems,* Glencoe Press, Beverly Hills, 1974.

Bugbee, Percy: *Principles of Fire Protection,* NFPA, Quincy, Massachusetts, 1978.

Factory Mutual: *The Handbook of Property Conservation,* Factory Mutual System, Norwood, Massachusetts.

Factory Mutual: *Loss Prevention Data Books,* Factory Mutual System, Norwood, Massachusetts.

Factory Mutual: *Property Conservation Workbook,* Factory Mutual System, Norwood, Massachusetts 1979.

Kimball, Warren Y.: *Fire Department Terminology,* NFPA, Quincy, Massachusetts, 1970.

Magison, Ernest C.: *Electrical Instruments in Hazardous Locations.* 3d ed., Instrument Society of America, Triangle Park, North Carolina, 1978.

McKinnon, Gordon P. (ed.): *Fire Protection Handbook,* 15th ed., NFPA, Quincy, Massachusetts, 1981.

McKinnon, Gordon P. (ed.): *Industrial Fire Hazards Handbook,* NFPA, Quincy, Massachusetts, 1979.

NFPA: *Guide to OSHA Fire Protection Regulations,* NFPA, Quincy, Massachusetts.

NFPA: *Industrial Fire Brigades,* NFPA, Quincy, Massachusetts, 1978.

NFPA: *Introduction to Fire Protection,* NFPA, Quincy, Massachusetts, 1982.

NFPA: *National Fire Codes,* NFPA, Quincy, Massachusetts, annual.

Planer, Robert G.: *Fire Loss Control, A Management Guide,* Marcel Dekker, New York, 1979.

Roytman, M. Ya.: *Principles of Fire Safety Standards for Building Construction,* Amerind, New Delhi, India, 1975.

Tuck, Charles A. (ed.): *NFPA Inspection Manual,* NFPA, Quincy, Massachusetts, 1976.

Tuve, Richard C.: *Principles of Fire Protection Chemistry,* NFPA, Quincy, Massachusetts 1976.

U.S. Department of Labor: *General Industry Standards,* part 1910, title 29, Code of Federal Regulations, Occupational Safety and Health Administration.

Williams, C. A., Jr., and Heins, R. M.: *Risk Management and Insurance,* McGraw-Hill, New York, 1976.

Zajic, J. E., and Himmelmann, W. A., *Highly Hazardous Materials Spills and Emergency Planning,* Marcel Dekker, New York, 1978.

chapter 14-3

Electrical Hazard Protection and Prevention

by

Robert L. Smith, Jr.
George W. Walsh
General Electric Company
Schenectady, New York

GLOSSARY

Basic impulse insulation level (BIL) A reference level expressed in impulse crest of a standard 1.2 × 50-μs wave. The withstand of apparatus insulation, as demonstrated by suitable tests, shall be equal to or greater than the basic impulse insulation level.

Chopped wave An impulse voltage wave that is suddenly reduced substantially to zero value by the sparkover of an air gap.

Crest value of a wave The maximum value (voltage or current) attained by a wave.

Discharge current The surge current that flows through an arrester.

Discharge voltage The voltage that appears across the terminals of an arrester during the passage of discharge current.

Follow current The power-frequency current that flows through an arrester during and following the passage of discharge current.

Front of wave sparkover Voltage at which arrester-gap sparkover occurs on the front of a wave rising at the rate of 100 kV/μs for each 12 kV of arrester rating for arresters rated from 3 to 240 kV and 2000 kV/μs for arresters rated above 240 kV. For arresters rated less than 3 kV, 10 kV/μs; and all rotating machine arresters, 10 ± 3 kV/μs.

Full wave An impulse voltage wave that rises to crest value and then decays in a normal manner without modification by the sparkover of an air gap or similar occurrence (contrast with "chopped wave").

Impulse A surge of unidirectional polarity.

Impulse sparkover voltage The highest value of voltage attained prior to the flow of discharge current when an impulse of a given wave shape and polarity is applied across the line and ground terminals of an arrester.

Microsecond One-millionth of a second (abbreviated μs).

Sparkover A disruptive discharge between electrodes of a measuring gap, voltage-control gap, or protective device.

Surge A transient variation in the current or potential at a point in a circuit.

Surge arrester A protective device for limiting surge voltages on equipment by discharging a surge current. It prevents continued flow of follow current and is capable of repeating these functions.

Switching surge (slow-front) Impulse of front time of 30 to 2000 μs, generally, with time to half-crest on the tail appreciably longer than twice the time to crest.

Traveling wave The resulting wave when the electric variation in a circuit takes the form of energy translation along a conductor, such energy being always equally divided between current and potential forms.

Valve-type lightning arrester An arrester having an element consisting of a resistor with a nonlinear voltampere characteristic which limits the follow current to a value the series gap can interrupt. If the arrester has no series gap, the characteristic element limits the follow current to a magnitude which does not interfere with normal operation of the system.

Voltage rating of an arrester The highest permissible rms alternating voltage that may be present between the line and ground terminals of the arrester while it is performing its operating duty cycle.

Wave The variation of current and/or voltage in an electric circuit with respect to time or distance.

Wave front That part of a surge or impulse which occurs prior to reaching the crest value.

Wave shape The variation of voltage or current of an impulse with time. For test purposes it is expressed as a combination of two numbers. The first, an index of wave-front steepness, is the time in microseconds from a virtual zero to the instant at which the crest value is reached. The second, an index of duration, is the time in microseconds from the virtual zero to the instant at which one-half of the crest value is reached on the wave tail. Examples are 1.2×50 and 8×20 waves.

Wave tail That part between the crest value and the end of an impulse.

OVERLOAD AND FAULT PROTECTION

Medium- and low-voltage systems require both *overload* and *fault* protection. An overload is circuit operation in excess of its capability for a time long enough to cause damage or dangerous overheating. A fault is either a short circuit or an open circuit. A bolted short circuit is a solid connection between two phase conductors, a phase conductor and ground, or all three phase conductors. Protective devices activated by overloads or faults include fuses which melt as well as relays and direct-acting trip devices which open circuit breakers, open contactors, or sound alarms.

OVERLOAD PROTECTION

Overloads may cause damaging elevated temperatures to persist for extended periods of time. Resistance temperature detectors are overload-protective devices which sense temperature directly. They are sometimes used to protect the most vulnerable parts of major equipment. Other overload-protective devices measure current in a circuit element and operate after a specific overcurrent persists for a time inversely related to current magnitude. Some current-detecting devices are ambient-compensated to make the protection sensitive to total hot-spot temperature. Since current unbalances resulting from unbalanced voltages cause rotating machine overheating, some protective devices sense current unbalance; others sense voltage unbalance. See Table 3-1 for a list of common overload-protective devices.

FAULT PROTECTION

Short-Circuit Fault Protection

Short-circuit currents of large magnitude flow when circuit insulation suddenly fails. These currents usually are much greater than normal load or overload currents. Sudden

TABLE 3-1 Overload Protective Devices

Device	Device function number	Measured quantity	Sensor	Usual application
Fuse	—	Current	Fusible element	Low- or medium-voltage circuits
Direct-acting trip device	—	Current	Trip device sensor	Low-voltage circuits
Thermal overcurrent relay	49	Current	Relay element	Low-voltage motor controller size 4 and smaller
			Current transformer	Low-voltage motor controller size 5 and larger
			Current transformer	Medium-voltage motor controller or circuit breaker
Time-delay magnetic or solid-state overcurrent relay	51	Current	Current transformer	Medium-voltage circuits
Temperature relay	49	Temperature	Resistance temperature detectors	Motors 1500 hp and over Transformers over 10 MVA
Combination solid-state relay	49	Current and temperature	Current transformers and resistance temperature detector	Motors 1500 hp and over
Solid-state or magnetic voltage balance relay	60	Voltages	Potential transformers	Motor buses
Solid-state or magnetic current balance relay	46	Currents	Current transformers	Motors 1500 hp and larger

failure of insulation results from either gradual deterioration or sudden overvoltage. Short-circuit current flow usually damages adjacent circuit components or equipment by arcing, explosive heating, or fire. Under short-circuit conditions, currents much greater than load currents flow in all circuit components between the source and the fault. A system short-circuit study can be conducted to reveal the magnitudes of short-circuit current throughout a system for important fault locations and can guide selection of system components rated for the calculated duties.

Short-circuit protective devices such as fuses, overcurrent trip devices, or relays operate in durations that can be preset or preselected depending on the magnitude of fault current. Series overcurrent devices between a source and a fault all experience the same short-circuit current, changed in magnitude only by the presence of transformers. This fact can be used to provide a completely selective protective device system in which the system device closest to the point of fault operates first and other series devices operate later. When several devices exist between the point of fault and the source, device characteristics other than progressively longer times are appropriate, because delays to attain complete time interval selectivity tend to get too long for source end devices.

If more than one source supplies a system, each source circuit breaker usually requires directional as well as nondirectional relays for best continuity of service. See References 8 and 9 for description of this type of relaying.

Important system components require rapid removal of a short circuit in order to minimize fault damage. Differential relaying accomplishes this without compromising system selectivity. Differential relaying compares the currents entering and leaving a circuit element and operates if the difference exceeds the setting or sensitivity of the relay. This relaying requires separate current transformers for both the incoming and outgoing conductors to a circuit element in order to obtain the proper sensitivity and accuracy. A special exception for motor circuits uses three current transformers, one for *both ends* of *each* motor phase winding.

Fault pressure relays provide excellent protection for internal transformer tank faults by detecting the sudden change in internal tank pressure which accompanies the initial insulation breakdown when a fault occurs. These completely selective relays usually are applied to main stepdown liquid-filled transformers. They can also be applied to smaller liquid-filled transformers.

Occasionally, distance relays instead of overcurrent relays are used to provide selective operation for large industrial systems. These relays measure the impedance from the relay to some system location. They operate only for faults within the protected zone.

Providing a reasonably selective system with appropriate system component protection involves compromises between overcurrent selectivity and protection. A system coordination study performed by a qualified power-system protection engineer incorporates the appropriate compromises to assure optimum system selectivity and protection. See Table 3-2 for a list of common short-circuit fault-protective devices.

Open-Circuit Fault Protection

An open-circuit fault condition results from the opening of any or all phase conductors at any point between the source and the load. Single-phase or three-phase undervoltage devices or relays easily detect this condition instantaneously on systems with no motors or generators connected. Protective devices consist of instantaneous or time-delay circuit-breaker undervoltage trip devices, instantaneous undervoltage relays with or without a timer, or time-delay undervoltage relays. These devices may be connected directly to a circuit or connected through potential transformers. They may operate to open a source or feeder circuit breaker, initiate an automatic throwover to an alternative source, or start an automatic sequence to energize the system from a standby generator.

Local generation can maintain both voltage and frequency at normal system values after a system disconnection, provided the generation is not overloaded. Overloading causes both of these quantities to decay.

TABLE 3-2 Short-Circuit Fault Protective Devices

Device	Device function number	Measured quantity	Detection devices	Usual application
Fuse	—	Current	Fuse element	Low- or medium-voltage circuits
Direct-acting trip device	—	Current	Trip device sensor	Low-voltage circuits
Overcurrent relay		Current	Current transformer	Medium-voltage circuits:
	50/51			Feeder-phase-time and instantaneous
	50/51N			Feeder residual ground time and instantaneous
	50GS			Feeder-ground sensor, instantaneous
	51GS			Feeder-ground sensor, time
	51N			Incoming or tie-residual ground, time
	51G			Neutral ground, time
	51			Incoming or tie phase, time
Differential relay		Current	Current transformer	Medium-voltage circuits:
	87T			Transformer over 10 MVA
	87T-G			Transformers, resistance grounded, overcurrent 10 MVA
	87M			Motors 1500 hp and over
	87B			Buses with 750 MVA or over, short circuit
	87L			Lines, ties to subbuses
	87G			Generators, any rating
Fault pressure relay	63FP	Pressure	Tank-mounted relay	Medium-voltage liquid-filled transformers over 10 MVA
Directional overcurrent relays	67 67N	Current and voltage	Current and potential transformers	Medium-voltage sources connected to a common bus
Distance relays	21	Current and voltage	Current and potential transformers	Large industrial medium- or high-voltage circuits

Motors connected to an electric system maintain the system voltage for some short time after the source system is disconnected. The length of time this voltage is maintained and its frequency depend on motor size, the amount of connected motor load, and motor plus load inertia.

Out-of-phase reclosing may ensue if power is lost for systems without local generation or if the loss overloads local generation and system voltage decay is not rapid enough to cause undervoltage relays to operate and disconnect the local system from the source before the source is automatically reclosed. This can result in extensive motor and associated equipment damage. To mitigate this possible event, various arrangements of sensitive reverse power, under- or overfrequency, or synchronizing check relays are available.

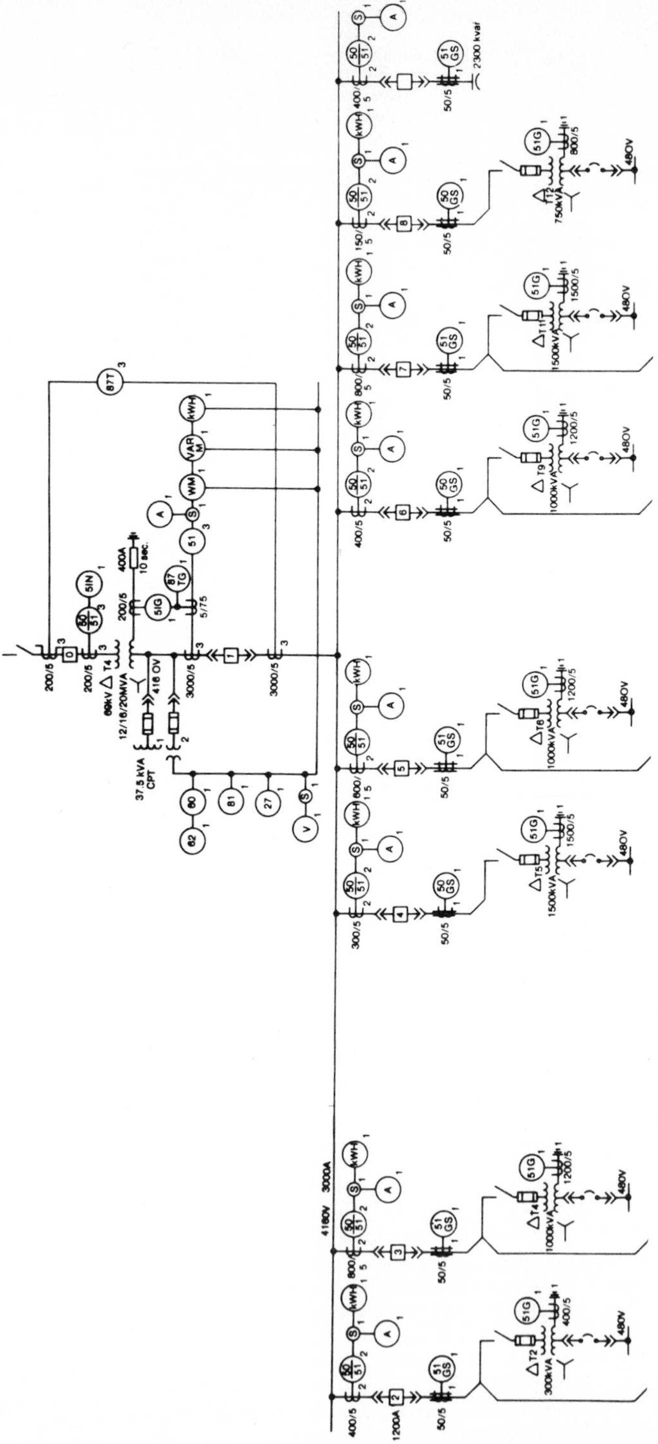
4160V 3000A
2300 kvar
37.5 kVA CPT
12/16/20MVA
87T
480V

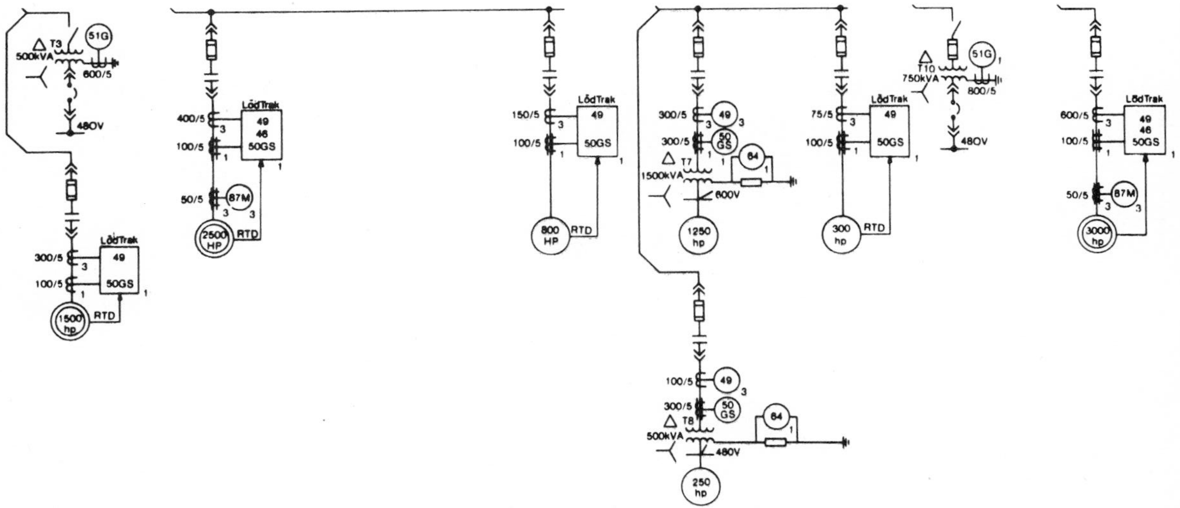

Figure 3-1 Typical industrial-plant one-line diagram. *(From* Industrial Power Systems Magazine, *General Electric Co., Schenectady, N.Y., March 1980, pp. 8–9.)*

Over-undervoltage relays connected to a single line-to-ground potential transformer detect line-to-ground faults on an incoming line that is disconnected from a grounded system at the source end. Such relays also can detect the return of normal voltage to a de-energized source and initiate automatic reclosing of the incoming-line circuit breaker.

Selection of appropriate relay settings for other than overcurrent relays is an important part of any system coordination study. Such selection may depend on the results of load flow, load shedding, or stability studies. The selection of the type of study necessary is at the discretion of the power-system protection engineer. See Table 3-3 for a list of common open-circuit fault-protective devices.

Figure 3-1 shows a typical industrial plant one-line diagram. Device function numbers identify the protective devices included. Figure 3-2 compares the time-current characteristics of some protective devices.

Figure 3-3 is a time-current curve for a portion of a system similar to Fig. 3-1. It is of utmost importance to identify each device represented by manufacturer and model number or other manufacturer's nomenclature when making a corodination study for selecting device settings.

Additional discussion of fault protection will be found in Sec. 3, Chap. 1, "Power Distribution Systems."

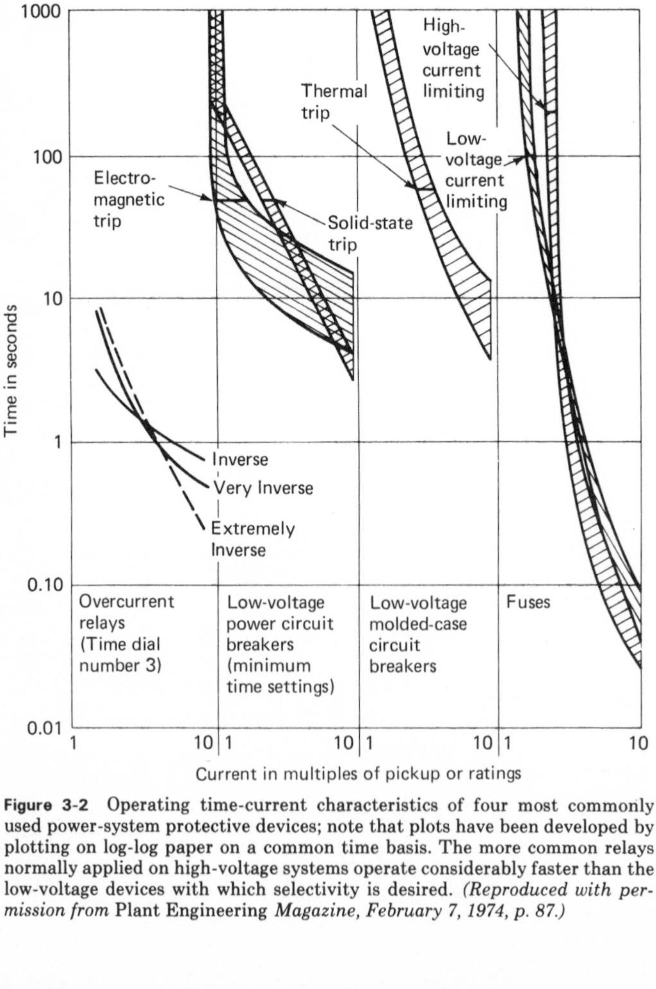

Figure 3-2 Operating time-current characteristics of four most commonly used power-system protective devices; note that plots have been developed by plotting on log-log paper on a common time basis. The more common relays normally applied on high-voltage systems operate considerably faster than the low-voltage devices with which selectivity is desired. *(Reproduced with permission from* Plant Engineering *Magazine, February 7, 1974, p. 87.)*

TABLE 3-3 Open-Circuit Fault Protective Devices

Device	Device function number	Measured quantity	Sensor	Usual application
Undervoltage device	——	Voltage	Potential transformer	Low-voltage motor circuits
Single-phase undervoltage relay	27	Single-phase voltage	Potential transformer	Low or medium voltage or incoming line
Three-phase undervoltage and phase-sequence relay	47	Three-phase voltage	Potential transformers	Low or medium voltage or incoming line
Directional power relay	32	Current and voltage	Current and potential transformers	Low- or medium-voltage incoming line or generator circuits
Frequency relay	81	Frequency	Potential transformer	Generator bus
Under-overvoltage relay	27/59	Voltage	Potential transformer	Incoming line circuits
Synchronizing check relay	25	Voltage and phase angle	Potential transformer	Nonsynchronous source tie circuits

SURGE-VOLTAGE PROTECTION OF ELECTRIC-POWER-SYSTEM EQUIPMENT

APPARATUS-INSULATION CAPABILITY

Surge protection is intended to avoid surge stresses beyond the fairly well standardized surge capabilities of power-system-apparatus insulation. Apparatus-insulation capability and surge-arrester protective ability are both established by tests with various elevated voltage and/or current waveshapes.

Tables 3-4 to 3-7 list surge capabilities of basic power-system equipment and ac rotat-

TABLE 3-4 Impulse Test Levels for Liquid-Filled Transformers*

Insulation class and nominal bushing rating, kV	Windings				Bushing withstand voltages		
	High-pot. tests, kV	Chopped wave: Minimum time to flashover, kV	Chopped wave: Minimum time to flashover, μs	BIL full wave (½ × 50), kV	60-cycle 1-min dry, kV (rms)	60-cycle 10-s, kV (rms)	BIL impulse full wave (1.5 × 40) kV (crest)
1.2	10	54(36)	1.5(1)	45(30)	15(10)	13(6)	45(30)
2.5	15	69(54)	1.5(1.25)	60(45)	21(15)	20(13)	60(45)
5.0	19	88(69)	1.6(1.5)	75(60)	27(21)	24(20)	75(60)
8.7	26	110(88)	1.8(1.6)	95(75)	35(27)	30(24)	95(75)
15.0	34	130(110)	2.0(1.8)	110(95)	50(35)	45(30)	110(95)
25.0	50	175	3.0	150	70	70(60)	150
34.5	70	230	3.0	200	95	95	200
46.0	95	290	3.0	250	120	120	250
69.0	140	400	3.0	350	175	175	350
92.0	185	520	3.0	450	225	190	450
115.0	230	630	3.0	550	280	230	550
138.0	275	750	3.0	650	335	275	650
161.0	325	865	3.0	750	385	315	750

*Values in parentheses are for distribution transformers, instrument transformers, constant-current transformers, step- and induction-voltage regulators, and cable potheads for distribution cables. Data taken from industry standards.

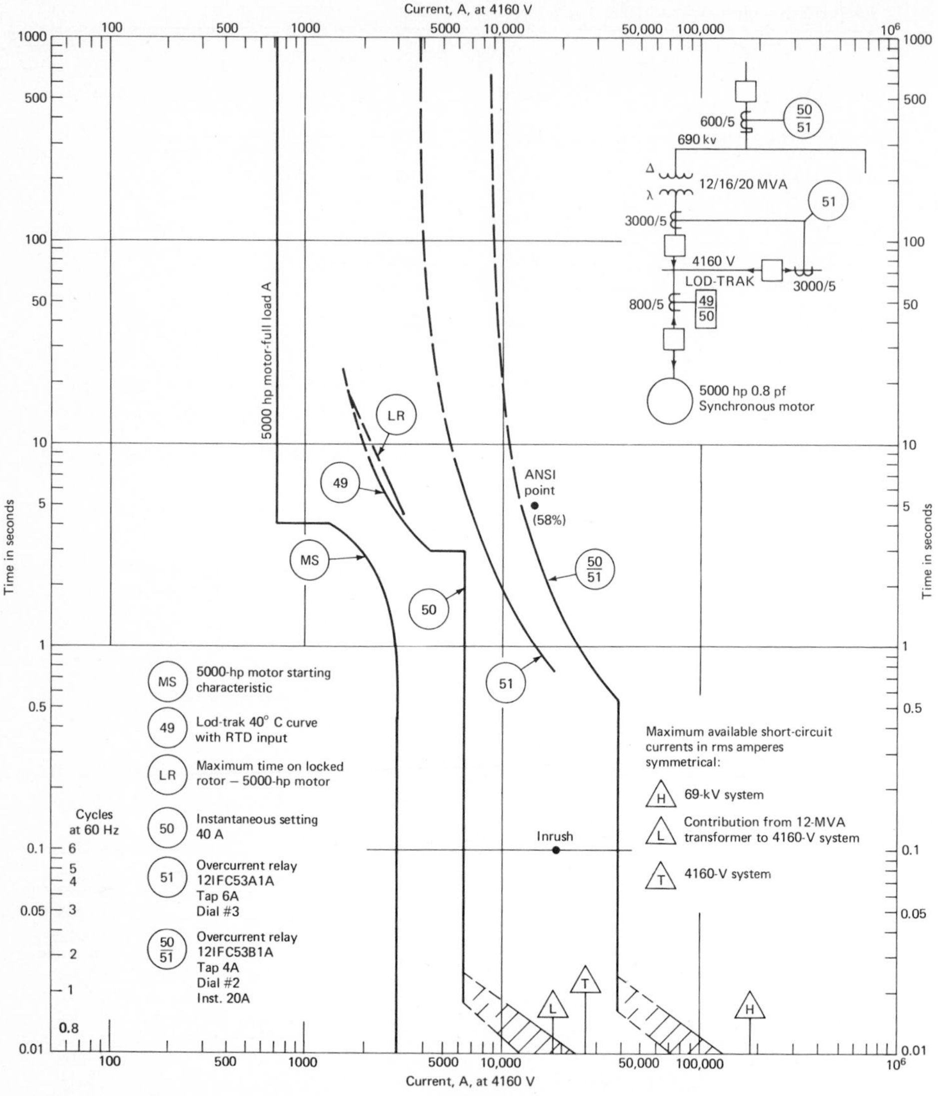

Figure 3-3 Time-current, phase-overcurrent protection, for a portion of a system similar to Fig. 3-1. *(From "Cement Plant Power Distribution," GET-2672C, General Electric Co., Schenectady, N.Y., December 1979.)*

ing machines. Impulse withstand of open-wire lines varies somewhat in specific value, depending upon construction, maintenance, weather, etc., but is generally considered well above that of associated transformers. An open-wire 13.8-kV distribution circuit, for example, is typically considered to have a 400-kV BIL. While cables do not have assigned BILs (basic impulse insulation levels), they too have impulse capability significantly higher than associated liquid-filled transformers.

TABLE 3-5 Impulse Test Levels for Power Circuit Breakers, Switchgear Assemblies, and Metal-Enclosed Buses

Voltage rating, kV	BIL full wave (1.2 × 50), kV, crest	Voltage rating, kV	BIL full wave (1.2 × 50), kV, crest	Voltage rating, kV	BIL full wave (1.2 × 50), kV, crest
2.4	45	23	150	115	550
4.6	60	34.5	200	138	650
7.2	75*	46	250	161	750
13.8	95	69	350	230	900
14.4	110	92	450	345	1300

*95 kV for metal-clad switchgear with power circuit breakers.

TABLE 3-6 Impulse Test Levels for Dry-Type Transformers*

Nominal equipment voltage		High pot. test, kV (rms)	Standard BIL (1.2 × 50), kV, crest
Delta or ungrounded wye	Grounded wye		
120–1,200		4	10
	1,200/693	4	10
2,520		10	20
	4,360/2,520	10	20
4,160–7,200		12	30
	8,720/5,040	10	30
8,320		19	45
12,000–13,800		31	60
	13,800/7,970	10	60
18,000		34	95
	22,860/13,200	10	95
2,300		37	110
	24,940/14,400	10	110
27,600		40	125
	34,500/19,920	10	125
34,500		50	150

*Data from Ref. 11, ANSI C57.12.01, "General Requirements for Dry-Type Distribution and Power Transformers."

TABLE 3-7 AC Rotating Impulse Levels

Motors			Generators		
Voltage rating, kV	High pot. test, kV (rms)	Impulse capability,* kV, crest	Voltage rating, kV	High pot. test, kV (rms)	Impulse capability,* kV, crest
2.3	5.6	9.9	2.4	5.8	10.2
4.0	9.0	15.9	4.16	9.3	16.5
6.6	14.2	25.1	6.9	14.8	26.2
13.2	27.4	48.4	13.8	28.6	50.6

*Equivalent BIL, commonly accepted values. Not standardized.

SURGE-PROTECTION TECHNIQUES

Electric-power-system components are protected against lightning and switching-produced surges by *interception* and *diversion* of the surges to ground. Shielding via overhead static wire and by conducting structures is used to greatly reduce the probability of significant lightning penetration of the system. Surge arresters are applied between

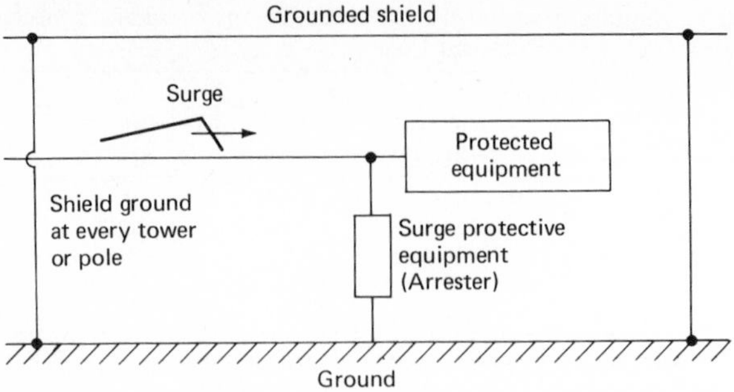

Figure 3-4 Schematic diagram of basic lightning-protection philosophy. Grounded shield intercepts a very large percentage of direct strokes. Surges that appear on incoming circuits are diverted to ground by surge arresters located at or near terminals of equipment to be protected.

line and ground at selected locations to divert surges from important and sensitive power-system and utilization equipment. Additionally, surge capacitors are sometimes applied at the terminals of motors and generators to reduce the surge voltage gradient. See Fig. 3-4.

SHIELDING

Lightning exposure to plant power systems is principally through outdoor supply substations and associated overhead lines. Effectively shielded installations are desirable. These are described in standards[13] as having shielding against direct strokes for the station and for all connected lines, at least for ½ mi from the station (line end protection).

Shielding of outdoor substation areas is usually provided by masts, or equivalent, which are designed to form a protective zone (not likely to be entered by lightning) within which all vulnerable parts will lie. With a single mast the protective zone is usually considered to be a cone having its apex at the top of the mast and whose sides make an angle with the vertical of 30 to 45°. With two or more masts the protective zone of each is increased somewhat in the area between them. With the usual spacings between masts, this shielding angle may increase to 60°. See Fig. 3-5.

Incoming line shielding is provided by overhead ground wire(s) which should be grounded at each pole or tower, through as low a ground resistance as it is practicable to

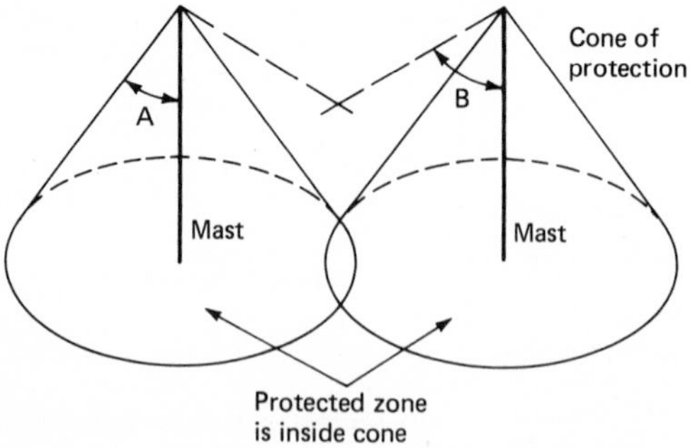

Figure 3-5 Shielding by masts. Protected zone with single mast is cone as illustrated with angle A of 30° to 45°. Cone of protection between masts extends up to 60°, angle B.

obtain, and should be connected to the ground bus at the substation. Low ground resistance is particularly important for the ground connection at the first few poles or towers adjacent to the substation. The associated protective zone is quite commonly considered to be of triangular cross section perpendicular to the shield wire with apex at the shield wire and with base at ground level extending a distance of twice the shield wire height on *each* side of the pole or tower. See Fig. 3-6.

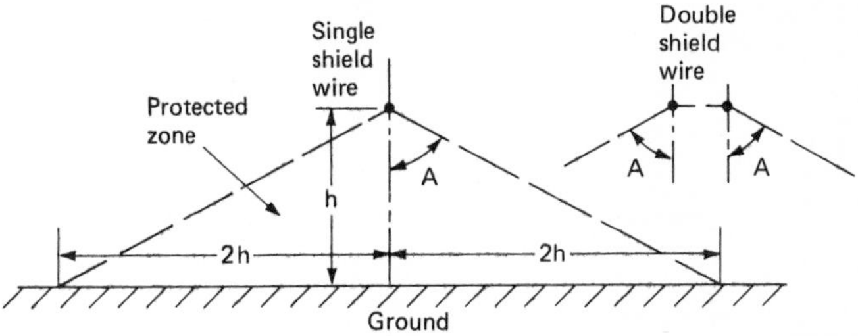

Figure 3-6 Commonly considered zone of protection provided by overhead shield wires for transmission (230 kV and below), subtransmission, and distribution lines. Specific shielding practices vary widely.

Outdoor substations not meeting the above criteria are generally classified as noneffectively shielded. Since noneffectively shielded applications entail a much higher surge exposure, it is important that the lowest practical ground resistance be obtained to minimize voltage gradients at the substation caused by surge current discharge to ground and to enhance arrester-protective performance.

Proper connection to earth is vital to the satisfactory performance of shielding and surge arresters. Resistance to ground should not exceed 1 Ω for large substations and generating stations, and 5 Ω for small substations and industrial plants. The ***National Electrical Code***[5] sets maximum resistance to earth at 25 Ω and requires arrester ground lead size be not less than No. 6 AWG.

SURGE ARRESTERS

Valve-type arresters are used for surge protection of plant power systems. Two basic valve material designs are employed to provide the necessary nonlinearity: (1) the zinc oxide arrester and (2) the gapped silicon carbide arrester. Zinc oxide–based valve elements have sufficient nonlinearity to accommodate both surge and nonsurge (normal operating) conditions. However, the silicon carbide–based valve elements require a series gap which effectively isolates them from continuous power frequency voltage. The gap sparks over at a predetermined elevated voltage which engages the valve element.

Arrester-Protective Characteristics

Three classes of arresters are recognized by standards[12] for medium-voltage and high-voltage systems—station class, intermediate class, and distribution class. Low-voltage-system (below 1000 V) arresters are classified as secondary arresters. Arrester manufacturers list arrester-surge-protective performance characteristics by arrester class. These performance characteristics are based on surge-voltage and surge-current test waveshapes also specified by standards. See Fig. 3-7. The two surge-voltage tests most frequently used for such listing are the front-of-wave sparkover and the 1.2 × 50-μs wave test. Equivalent front-of-wave test data is also listed for the nongapped zinc oxide arrest-

****National Electrical Code***® is a Registered Trademark of the National Fire Protection Association, Quincy, MA 02269.

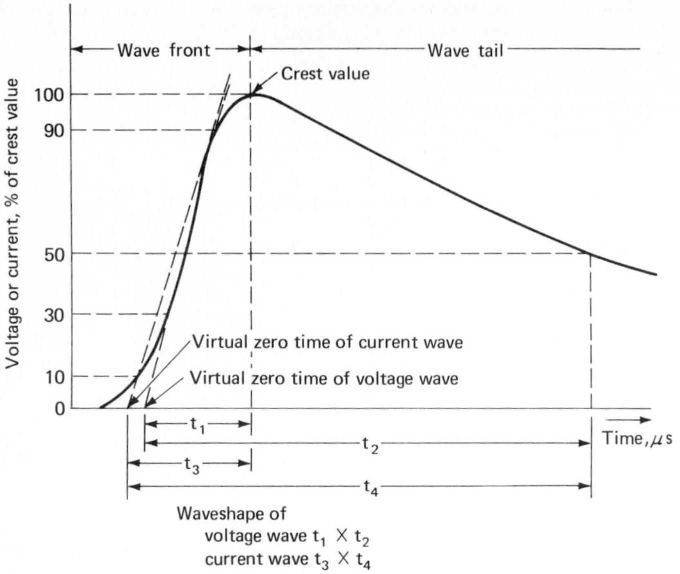

Figure 3-7 Wave geometry used to designate waveshape for purposes of standardizing surge-arrester protective characteristics and specifying insulation surge capability.

ers. The standardized 8 × 20-μs current wave is the basis for listing arrester-discharge-voltage levels. See Tables 3-8 to 3-11. As such, the listed characteristics usually provide the *maximum* surge voltage permitted (between arrester terminals) at the arrester location for the specified voltage and current surge duties.

Zinc oxide valve element technology is relatively new to power-system surge-arrester designs. It promises to affect future improvements in intermediate-class and distribution-class arresters in similar fashion to that already initiated in station-class designs. Accordingly, it is recommended that the most up-to-date vendor data and standards be used in selection of arresters.

Application of Arresters

The application of arresters depends upon three basic selections: (1) selection of arrester rating, (2) selection of arrester class, and (3) selection of arrester location.

Selection of Arrester Rating

A surge arrester should be selected with the *minimum* rating that will have a satisfactory service life on the power system. This will provide the greatest practical margin of safety for the protected insulation and at lowest cost. Expressed as a percentage this protective margin is defined as

$$100 \times \left(\frac{\text{apparatus insulation withstand}}{\text{surge-arrester protective level}} - 1 \right)$$

The generally recommended protective margin is 20 percent for impulse coordination (front-of-wave, full wave) and 15 percent for switching surge coordination.

Although protective margins are increased by reducing arrester ratings, there is a practical limit to such reduction because arresters are rated in relation to their ability to withstand power-frequency voltage during and following surge current discharge. The

TABLE 3-8 Surge-Protective Characteristics of Station-Class Arresters* for Zinc Oxide Valve Elements

Arrester rating, kV, rms	Equivalent front-of-wave protective level, kV, crest	Maximum switching surge protective level, kV, crest	Maximum discharge voltage, kV, crest, at indicated impulse current using an 8 × 20-μs current wave						
			1.5 kA	3.0 kA	5 kA	10 kA	15 kA	20 kA	40 kA
2.7	8.0	5.5	5.9	6.2	6.5	6.9	7.3	7.6	8.8
3.0	9.4	6.4	6.9	7.3	7.6	8.1	8.5	8.9	10.4
4.5	13.6	9.3	10.0	10.5	11.0	11.7	12.3	12.9	15.0
5.1	15.2	10.4	11.2	11.8	12.3	13.1	13.8	14.5	16.8
6.0	18.6	12.7	13.7	14.4	15.0	16.0	16.9	17.7	20.5
7.5	22.0	15.0	16.3	17.1	17.8	19.0	20.0	21.0	24.3
8.5	25.0	17.1	18.5	19.4	20.3	21.6	22.8	23.9	27.7
9.0	27.6	18.8	20.4	21.4	22.4	23.8	25.1	26.3	30.5
10	30.6	19.4	22.6	23.8	24.8	26.4	27.9	29.2	33.8
12	36.8	25.1	27.1	28.5	29.8	31.7	33.4	35.1	40.6
15	45.8	31.2	33.8	35.6	37.1	39.5	41.6	43.7	50.6
18	55.0	37.5	40.6	42.7	44.5	47.4	50.0	52.4	60.7
21	61.7	42.1	45.6	47.9	50.0	53.2	56.1	58.8	68.1
24	70.5	48.1	52.1	54.7	57.1	60.8	64.1	67.2	77.8
27	79.1	53.9	58.4	61.4	64.0	68.2	72.0	75.3	87.3
30	88.2	60.1	65.1	68.4	71.4	76.0	80.2	84.0	97.3
36	106	72.1	78.1	82.1	85.6	91.2	96.2	101	117
39	115	78.0	84.5	88.8	92.7	98.7	104	109	126
45	132	90.2	97.6	103	107	114	120	126	145
48	140	95.7	103.6	109	114	121	128	134	155
54	142	106	103	111	115	122	128	135	157
60	158	118	114	123	128	136	143	150	174
66	174	130	125	135	141	150	157	165	191
72	189	142	137	147	153	163	172	180	209
90	237	177	171	184	191	204	215	225	261
96	253	189	182	196	204	218	229	240	278
108	285	213	205	220	230	245	258	270	313
120	316	236	228	245	255	272	286	300	348
132	347	260	251	269	280	299	315	330	383
144	379	283	273	294	306	326	344	360	417
168	442	331	319	343	357	381	401	420	487
172	453	339	326	350	365	390	410	430	499
180	474	354	341	367	382	408	430	450	522
192	505	378	364	391	408	435	458	480	556
228	599	449	432	465	484	516	544	570	661
240	630	472	455	489	510	543	572	600	696

*Data in this table courtesy General Electric Co.
†Based on 500 A for 2.7–54-kV ratings and 3000 A for ratings 60 kV and above.

circumstance of a *line to ground* on the system and the associated *line to ground* voltage on the unfaulted phases is the prime criterion for selection of minimum arrester rating. This depends upon the degree of system grounding. Standards[12] define *coefficient of grounding* as "... The ratio E_{LG}/E_{LL}, expressed as a percentage, of the highest rms line-to-ground power frequency voltage E_{LG} on a sound phase, at a selected location, during a fault to ground affecting one or more phases to the line-to-line power-frequency voltage E_{LL} which would be obtained, at the selected location, with fault removed." Thus, the *minimum required arrester rating is the maximum line-to-line operating voltage times the coefficient of grounding.*

Standards[13] contain various aids to determine coefficient of grounding. The majority of industrial-plant systems in the 2.4- to 13.8-kV range are resistance-grounded with a 100 percent coefficient of grounding, as is also the case for ungrounded systems. Accord-

TABLE 3-9 Surge-Protective Characteristics of Station-Class Arresters* for Silicon Carbide Valve Elements

Arrester rating, kV, rms	Maximum ANSI std. front-of-wave sparkover, kV, crest †	Maximum 1.2 × 50 μs, sparkover, kV, crest	Maximum switching surge protective characteristic, kV, crest	Minimum 60 Hz sparkover, kV, rms	Maximum discharge voltage, crest, kV, at indicated impulse current, 8 × 20 μs					
					1.5 kA	5.0 kA	10.0 kA	15.0 kA	20.0 kA	40.0 kA
3	11	10	8.25	4.5	5.0	6.4	7.3	7.8	8.3	10.2
4.5	16.5	15	12.4	6.8	7.4	9.5	10.8	11.6	12.3	15.1
6	19	16	15.5	9.0	9.8	12.6	14.3	15.3	16.3	19.9
7.5	24	20	19.5	11.3	12.2	15.7	17.7	19.0	20.3	24.8
9	28.5	24	23.5	13.5	14.6	18.8	21.2	22.7	24.3	29.6
12	37	32	31	18	19.4	24.9	28.1	30.2	32.1	39.2
15	46.5	40	39	22.5	24.2	31.0	35.0	37.5	40.0	48.8
18	55.5	48	46.5	27	28.9	37.1	41.8	44.8	47.8	58.5
21	65	56	55.5	31.5	33.7	43.2	48.7	52.3	55.5	68.0
24	74	64	62	36	38.4	49.2	55.5	59.5	63.5	77.5
27	83	72	70	40.5	43.1	55.3	62.5	67.0	71.2	87.0

30	92	80	78	45	47.8	61.5	69.5	74.5	79.0	96.6
36	111	96	93	54	57.5	73.5	83.0	89.2	94.5	115.0
39	120	104	101	58.5	62.5	79.5	89.5	96.0	102.0	125.0
48	148	128	124	72	76.0	97.5	110	117	125	153
60	170	141	136	84	95.0	122.0	137	147	156	190
72	204	169	163	100	114	146	164	170	187	227
78	220	184	177	109	123	158	178	191	202	246
84	237	198	191	117.5	133	170	191	205	217	265
90	254	212	204	126	142	182	204	219	232	283
96	270	226	218	134.5	151	194	218	234	248	302
108	304	254	245	151	170	218	245	262	278	339
120	338	282	272	168	188	241	272	292	309	376
132	372	310	299	185	207	262	294	315	333	402
144	405	338	326	201.5	226	287	321	344	363	439
168	483	395	381	235	263	334	374	401	422	510
180	521	400	400	243	281	358	400	424	452	550
192	559	427	426	260	300	382	427	457	482	585
228	677	510	506	308	355	452	510	546	575	695
240	718	535	533	324	374	476	535	574	605	730
258	775	575	573	349	402	515	575	616	650	785

*Data in this table courtesy General Electric Co.

†ANSI/IEEE 28-1979, "Surge Arresters for Alternating-Current Power Circuits."

TABLE 3-10 Surge-Protective Characteristics of Intermediate-Class Arresters* for Silicon Carbide Valve Elements

Arrester rating, kV, rms	Maximum ANSI front-of-wave sparkover, kV crest	Maximum 1.2 × 50 μs sparkover, kV, crest	Maximum discharge voltage, kV, crest, at indicated impulse current 8 × 20 μs wave				
			1.5 kA	3 kA	5 kA	10 kA	20 kA
3	11	11	6.5	7	7.4	8.3	9.5
4.5	16	15	9.5	10.1	10.8	12.0	14.0
6	21	19	12.5	13.3	14.5	16.0	18.5
7.5	26	23.5	15.8	16.7	17.8	20.0	22.8
9	31	27.5	19.0	20.0	21.0	23.5	27.0
10	35	31	21.8	23.0	24.5	27.2	31.5
12	40	35.5	24.7	26.1	28.0	31.0	36.0
15	50	43.5	31.0	32.5	35.0	39.0	45.0
18	59	51.5	37	39	42	47	54
21	68	59	43.0	45.5	49.0	55.0	63.0
24	78	67	49.0	52.0	56.0	62.0	72.0
30	97	81	61.5	65.0	70.0	78.0	90.0
36	116	95	74.0	78.0	83.0	94.0	108
39	126	102	80.0	84.5	90.0	102	117
45	144	116	92.0	98.0	104	117	135
48	154	123	98.0	104	111	125	145
60	190	153	123	130	139	156	180
72	228	180	148	156	166	187	216
90	282	223	184	195	208	233	270
96	300	236	197	208	222	249	288
108	335	263	221	234	249	281	324
120	370	290	246	260	277	311	360

*Data in this table courtesy General Electric Co.

ingly, a large majority of plant medium-voltage systems require minimum arrester ratings of at least 100 percent of the maximum operating voltage of the system (100 percent arresters). Grounded systems typically require 75 to 80 percent arresters. See Table 3-12. Multigrounded (four-wire) distribution systems and some high-voltage transmission systems (69 kV and above) may safely use lower rated arresters. These systems require individual determination of their coefficients of grounding to ensure the most economical, secure arrester rating selection.

Selection of Arrester Class

In order of cost, protective efficiency, and durability, the three classes of arresters are: (1) station class, (2) intermediate class, and (3) distribution class. As a general guide to arrester-class usage vs. equipment size, the following may be considered typical practice:

Station Class. Component protection of 7.5 MVA and above substations and large or essential rotating machines

Intermediate Class. Component protection of 1 to 20 MVA substations, overhead lines, and rotating machines

Distribution Class. Distribution-class apparatus, dry-type transformers, and small rotating machines

These classes overlap considerably. There is a tendency to use higher-class arresters at higher voltages. In some cases of distribution-class arrester application, low sparkover models are recommended, and, in fact, are necessary. Limitations on the available range of ratings of distribution-class and intermediate-class arresters eliminate them as a choice in higher-voltage applications.

TABLE 3-11 Surge-Protective Characteristics of Distribution-Class Arresters* for Silicon Carbide Valve Elements

Arrester rating, kV, rms	Maximum ANSI front-of-wave sparkover, kV, crest	Maximum discharge voltage, kV, crest at indicated 8 × 20 μs impulse current wave				
		1500 A	2500 A	5000 A	10,000 A	20,000 A
3	15	9.5	10	11	12	13.5
	11†	9.5	10	11	12	13.5
4.5	17†	14	15	17	19	20.5
6	29	19	20	22	24	27
	21†	19	20	22	24	27
7.5	26.5†	23	25	28	31	34
9	40.5	28	30	33	36	40
	32†	28	30	33	36	40
10	44.5	28	30	33	36	40
	32†	33	35	39	43	47.5
12	56	37	40	44	48	54
	39.5†	37	40	44	48	54
15	63	42	45	50	54	61
	46†	46	49	55	60	67
18	75	50	55	61	66	74
21	89	60	64	72	78	88
27	98	73	79	87	96	107

*Data in this table courtesy General Electric Co.
†Low-sparkover model data.

TABLE 3-12 Voltage Ratings of Arresters Usually Selected for Three-Phase Systems

Nominal system voltage, kV	Voltage rating of arrester, kV	
	System neutral ungrounded or resistance-grounded	System neutral solidly grounded
0.120/0.208Y	0.65	0.175
0.240	0.65	0.65
0.480	0.65	0.65
0.600	0.65	0.65
2.4	3 (2.7)	3 (2.7)
4.16	4.5	3 (4.5)
4.8	6.0 (5.1)	4.5 (5.1)
6.9	7.5	6.0
12.47	15, 12	10, 9
13.8	15	12, 10
23	24	24, 21, 18
34.5	39, 36	30, 27
46	48	39
69	72	60, 54
115	120	90, 96, 108
138	144	108, 120
161	—	120, 132, 144
230	—	172, 180, 192

While actual lightning-protective practices may necessarily vary from one type of installation to the next, all installations are either effectively shielded or noneffectively shielded. There are different degrees of jeopardy for each of these two basic categories, and the degree may vary with a change in system arrangement or operating mode.

Excessive arrester discharge currents in noneffectively shielded installations are a prime cause of arrester failure, and inadvertent loss of protection may occur when arrester discharge voltages exceed anticipated levels. In effectively shielded systems, arrester discharge currents are unlikely to exceed 5000 A as a conservative maximum, but 20,000 A is a conservative maximum for noneffectively shielded systems. This encourages use of station-class and intermediate-class arresters in noneffectively shielded installations.

Location of Arresters

The ideal location for surge arresters, from the standpoint of protection, is directly at the terminals of equipment to be protected. Usually, low-BIL apparatus (certain dry-type transformers and rotating machines) requires surge-protective devices in direct shunt with associated insulation. Otherwise practical circumstances dictate often that arresters be remote, requiring that one set of arresters protect more than one piece of apparatus.

For remote arresters it is necessary to estimate the depreciation in protection caused by separation distances between the arresters and the protected equipment. These determinations require the use of traveling wave mechanics by a practiced surge protection engineer. The following are *general guides* for locating arresters relating to typical components and their arrangements.

Effectively Shielded Substations. Arresters are required on each overhead line as it enters the substation, for protection of disconnect switches, buses, etc. Separation distances of 75 to 200 ft, and sometimes more, between arresters and full-BIL equipment can be tolerated at 23 kV and above. At 15 kV and below, practice usually avoids appreciable separation distance. Usually arresters are applied at transformer incoming line terminals.

Noneffectively Shielded Substations. Arresters should be applied at or very close to terminals of transformers and breakers. A minimum separation distance between arresters and protected equipment for overcurrent protection equipment is permissible.

Metal-Clad Switchgear. Metal-clad switchgear installed in substations (as above) should be protected in the same fashion as transformers. In effectively shielded substations where continuous metallic-sheathed cable (or equivalent) intervenes between the switchgear and exposed line, an arrester at the junction of line and cable suffices if it is station class, intermediate class, or special distribution class. When exposure is through a power transformer protected on the exposed side, generally arresters are not necessary at the switchgear.

Dry-Type Transformers. Full-BIL dry-type transformers should be protected as described above for full-BIL equipment. Reduced-BIL dry-type transformers (with BILs below comparably rated liquid-filled distribution transformers) require arresters at their terminals when connected directly to overhead lines. When exposure is through continuous metallic-sheathed cable (or equivalent), a line-cable junction arrester may or may not protect the low-BIL dry-type transformer. In such cases, a special (low sparkover) distribution-class arrester at the transformer terminals will suffice. If exposure is through another transformer, usually arresters are not required at the dry-type transformer.

Cable. Cable should be surge-protected the same as full-BIL transformers.

Aerial Cable. Arresters are required at junctions of open-wire line and aerial cable. Messenger and sheath should be grounded through a low value of ground resistance at every pole. Aerial cable should be considered the same as open-wire line for protection of terminal equipment.

ROTATING-MACHINE PROTECTION

Medium-voltage (2.3- to 13.8-kV) rotating machines require: (1) a strictly effectively shielded environment, (2) arresters at the terminals of the machine, (3) surge capacitors at the terminals of the machine, and (4) strict adherence to good grounding practices. The surge capacitors are used to reduce the voltage gradient of surges that may be damaging to machine turn insulation. Such capacitors should be connected in the closest possible shunt relation to the machine line-to-ground insulation. Table 3-13 lists commonly available surge-protective capacitors. Note surge-protective capacitors are rated on a line-to-line (rms) voltage basis with associated designated capacitance per pole (terminal to case).

TABLE 3-13 Ratings and Sizes of Surge-Protective Capacitors

Voltage rating, rms, V, L-L	Maximum voltage, rms, V, L-L	Poles per unit	μF per pole
0–650	715	3	1.0
2,400	2,640	3	0.5
4,160	4,576	3	0.5
6,900	7,590	1	0.5
13,800	15,180	1	0.25
24,000	26,400	1	0.125

Switching events within the distribution system (e.g., motor switching, fault inception, fault removal, insulation flashover, capacitor switching, etc.) present surge exposure possibilities to motors and generators. In practice, a high percentage of machines above 4 kV have arresters and surge capacitors. At least half of the 4-kV motor installations are so protected, while at 2.3 kV only a minority are so protected. Low-voltage motors and generators (below 1000 V) seldom have such surge protection, except in exposed pumping applications.

REFERENCES AND BIBLIOGRAPHY

Standards (Use Latest Issue)

1. ANSI C42.100-(year), "IEEE Standard Dictionary of Electrical and Electronics Terms."

IEEE Recommended Practices

2. IEEE 141-(year), "IEEE Recommended Practice for Electric Power Distribution for Industrial Plants."
3. IEEE 142-(year), "IEEE Recommended Practice for Grounding of Industrial and Commercial Power Systems."
4. IEEE 242-(year), "IEEE Recommended Practice for Protection and Coordination of Industrial and Commercial Power Systems."
5. NFPA 70-(year), ***National Electrical Code.***®
6. NFPA 70E-(year), "Electrical Safety Requirements for Employee Workplaces."
7. NFPA 70B-(year), "Recommended Practice for Electric Equipment Maintenance."

ANSI Standards for Equipment

8. ANSI C19 Series, "Package Control Equipment."
9. ANSI C37.2-1970, "Switchgear."

10. ANSI C51:1-1978, "Safety Standards for Construction and Guide for Selectors, Installation and Use of Electric Motors and Generators."
11. ANSI C57 Series, "Transformers, Regulators, Reactors."
12. IEEE Std 28-(1974); also ANSI C62.1-(1974), "Surge Arresters for Alternating-Current Power Circuits."
13. ANSI C62.2-(1969), "Guide for Application of Valve Type Lightning Arresters for Alternating-Current Systems."

Books

14. Beeman, D. L. (ed.): *Industrial Power Systems Handbook*, McGraw-Hill, New York, 1955.
15. *Electrical Transmission and Distribution Reference Book*, Westinghouse Electric Corp., Pittsburgh, 1950.
16. *Applied Protective Relaying*, Westinghouse Electric Corp., Pittsburgh, 1976.
17. Smeaton, R. W. (ed.): *Switchgear and Control Handbook*, McGraw-Hill, New York, 1977.

chapter 14-4

Security Equipment

by

A. J. Grosso
Vice President, Engineering
American District Telegraph Co.
New York, New York

INTRODUCTION

The term *security* may be defined as the protection of life and property. Two of the basic hazards are fire (see Chap. 14-2) and crime. Crime hazards, which will be covered here, consist of intruders who enter a premise to steal physical property or secret information or to commit sabotage. The purpose of crime security is to deter the entrance of potential criminals or vandals, to detect intruders, and to alert a guard force as promptly as possible. A basic premise of good security is protection in depth—the provision of more than one line of defense to discourage incursions, to slow the advance of knowledgeable intruders, and to provide secondary defenses in case of penetration beyond the first line.

Security is much more than an alarm system and must include:

1. A complete plan, attitude, and purpose by the persons involved in "securing" life and property
2. Physical security

3. Detection and reporting methods
4. Signal transmission
5. Signal handling
6. Continued review of quality, value, and changing conditions and the intent to amend and upgrade as necessary

TYPES OF SYSTEMS

The alarm industry recognizes four types of alarm systems:

1. **Local Alarm** A system which produces an audible and/or visible signal at the protected premises only as a result of an alarm condition or fault (capability degradation).
2. **Direct-Connect (or Headquarters)** A system in which signals are transmitted to police or fire headquarters is sometimes combined with a local alarm system.
3. **Central Station** A system in which signals of various types are transmitted to an independent control center where trained personnel are maintained continuously to supervise the status of protected premises and take appropriate action upon the receipt of signals. The central station is generally considered to be the preferred form of protection.
4. **Proprietary** This system is similar to the central station system except that the control center is staffed and maintained by the owner of the protected properties.

In practice, fire-protection and crime-protection systems are closely related, and the equipment and services needed for both types of systems are usually provided by the same equipment manufacturers, installers, and service companies.

Modern practice, especially for proprietary systems, in large plants, allows for the integration of the fire- and burglar-alarm systems with the building automation services. Air conditioning, energy management, closed-circuit television, access control, and communications, together with the protection services, are controlled from a central location under the direction of a computer that has been programmed to initiate the proper actions and responses required under a wide variety of operating and emergency conditions.

SECURITY PLANNING

The planning of a plant-wide security system is a technical matter that may require the services of a specialist, either an independent consultant or the representative of a security company. Consultation should begin before the plans for a new building are completed as there are aspects of architecture and design which affect security.

With the aid of the expert, a security system can be designed to meet the sometimes conflicting requirements for:

1. **Likelihood of Attack** A precious-metals refinery is more likely to be attacked than a junk storage area. Criminals prefer readily portable articles of high value. The local crime rate is also an important factor.
2. **Police Response** A large, well-equipped police force located within a few minutes travel time establishes a smaller security risk than a remotely located plant in a rural setting.
3. **Building Construction** A modern masonry structure is more resistant to attack than an old wooden building or, in an extreme case, an air house.
4. **Insurance** The security expert can be a guide in the selection of equipment and techniques for both security and fire-alarm systems that will meet the requirements

of such approval agencies as the Underwriters Laboratories, Inc., Factory Mutual, and National Fire Protection Association which lead to insurance premium discounts.

5. **Technical Considerations** These include a variety of technical considerations leading to reliable security systems which produce a minimum of nuisance alarms and other trouble conditions and assure compliance with any local codes and ordinances.

PHYSICAL SECURITY

Physical security is passive in nature as opposed to the active methods for the detection and reporting of intruders. It is often the first line of defense and can be seen in the form of fences, walls, and moats. It includes visual definition of boundary lines as psychological barriers; for example, low fences, a line of gardens, well-kept lawns, and a change of walkway materials, e.g., from concrete to brick.

National records indicate that crime against property averages 90 percent of all reported crime and that, in the majority of all burglary cases, entry was gained through doors or windows. In spite of this knowledge, there is a consistent use of inferior, low-security hardware and other materials including locks.

In some localities in the United States there are codes and ordinances which set forth standards for door frames, doors, door locks, windows, window locks, elevators, and other openings such as skylights, hatchways, air ducts, vents and transoms. In the absence of local standards, or the presence of outdated codes, reference should be made to the Law Enforcement Assistance Administration, National Institute of Law Enforcement and Criminal Justice Standards (NILECJ; See Bibliography).

Physical security is dependent on good housekeeping. The best security equipment and alarm systems lose effectiveness when doors are not locked at night.

DETECTION AND REPORTING METHODS

Outdoor Perimeter Protection

The protection of an outdoor perimeter is the first line of defense in a protection-in-depth security plan. It can start with a masonry wall or a fence constructed of wood. A widely used construction is the familiar chain-link (wire mesh) fence six or more feet in height and topped with an outward slanting barbed-wire or barbed-tape-coils section to discourage fence climbers. While difficult to climb for the inexperienced, such a barrier is susceptible to attack with wire cutters, or by jacking. Thus, it may be necessary to provide a patrolling guard, with or without dogs.

An alternative to the expense of a guard force is the use of an electronic barrier consisting of fence-mounted transducers to detect and annunciate the mechanical vibrations caused by attempts to climb or cut the fence.

Another form of electronic barrier comprises a series of posts strung with a number of wires in telephone-line style. This type of fence presents no physical barrier to an intruder but the electromagnetic field set up by the current flowing through the wires is disturbed by the presence of an intruder. The field disturbance is detected electronically and annunciated at a guard station.

In an effort to reduce the costs of installing and maintaining a long line of posts and wires, projected energy beams are sometimes used. A beam of energy (which may be light in the invisible infrared frequency spectrum or microwave energy) is projected from a source to a receiver and, when interrupted by the passage of an intruder, an alarm is annunciated at the guard station. (See Fig. 4-1.) *Note:* This type of system is best adapted to flat terrain where the beam length can be relatively long. When the terrain is hilly, or very irregular in outline, the fence-post type of system is preferred to reduce equipment cost.

All types of electronic intrusion barriers require both means to keep animals at a safe distance and constant maintenance to control the growth of vegetation to prevent an

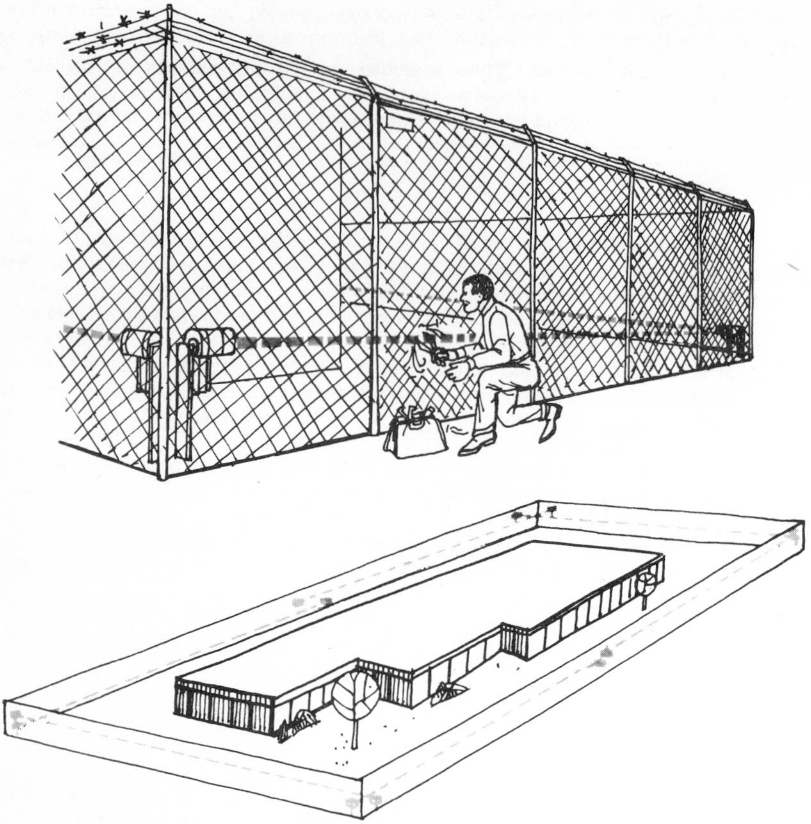

Figure 4-1 Use of light beams to protect restricted area.

excessive rate of nuisance alarms. Erosion of the earth under the barrier must be controlled to prevent access paths.

Portal Protection

Outdoor perimeter fences as well as building entrances must have portals that are guarded in such a manner that authorized persons may be admitted and others excluded. This may be done by means of a guard who provides other services such as directing visitors.

The cost of guard service may be reduced by an electrical lock on the door and a voice intercom connection as is commonly provided in apartment house lobbies. Where voice identification is not sufficient, a closed-circuit television system will permit visual inspection before admission. Obviously, a single guard at a central location can monitor many portals.

The simplest means of admission control is to provide authorized persons with a key to the door. Pushbutton locks avoid the problems of lost and stolen keys since it is easier and cheaper to change the combination when necessary than to provide a new lock and set of keys.

Card access systems avoid the problems associated with metallic keys and provide other advantages. In its simplest form, the card access system is the equivalent of a metallic door key. When a photograph plus other information is applied to the card, it becomes an identification card presented to building guards when needed. With additional coding, a zoned system may be established in which many persons may be allowed to enter a first door, a lesser number to enter another door, and only a select few to pass through a third door.

Computer-based systems provide such sophisticated functions as sounding an alarm when the wrong card is presented, keeping a record of who entered what door at what time, and establishing control according to the time of day.

Building Perimeter Protection

Experience has shown that the large majority of burglarious attacks are made on doors and windows. The basic device employed for their protection is known as the *burglar-alarm contact* and has evolved from clumsy mechanical contrivances to the modern magnetically operated switch. Areas of glass have been protected for almost a century by the familiar current-carrying foil strip whose rupture initiates an alarm signal. Foil is now being replaced by transducer devices cemented to the glass and tuned to respond to the characteristic sound frequencies produced when the glass is broken.

An older form of protection for windows and other openings is the *burglar-alarm screen* consisting of a small-diameter, current-carrying wire embedded in a frame of wooden dowels. To pass such a barrier, the intruder must break at least some of the dowels (and the embedded wire) and so actuate the alarm.

Since it is also possible to effect entry by penetration of the walls, floors, and ceilings, such surfaces are protected by *pads* of wire or foil mounted on building-material panels attached to the surface to be protected. Figure 4-2 illustrates these as well as other devices described below.

Interior Protection

In accordance with the principle of protection in depth, *traps* are employed to reveal the intruder who has penetrated the perimeter. A simple and effective trap is the pressure mat hidden under the carpet which produces a signal when stepped on by the intruder. The *floor trap* is a trip-wire device in which the wire need not necessarily be broken; its displacement alone will initiate the alarm. A photoelectric beam made invisible by the use of infrared light is a more advanced form of trip wire.

Discrete objects such as safes and file cabinets may be protected by contacts to give warning of their being opened but it is better to have an alarm prior to that event. For this purpose, capacitance alarm systems which produce an electric field around the protected object will give the alarm signal upon the approach of the intruder before he or she touches the object.

An earlier form of protection still in use is the "cabinet" made of wood lined with foil or fine wire surrounding the object to be protected. Penetration of the cabinet, as in the case of the burglar alarm screen, initiates an alarm signal.

Space Protection

About 1950, the concept of space protection was introduced in the form of the ultrasonic motion detector. The space to be protected is flooded with sound energy in the ultrasonic range, and the motions of the intruder cause a Doppler shift in the frequencies detected at the receiver which is employed to actuate the alarm signal. See Fig. 4-3. Problems in some situations caused by air turbulence led to the introduction of generally similar systems operating on energy in the microwave range (Fig. 4-4) and, most recently, to passive infrared systems which detect the warmth of the intruder's body.

Designated areas may be kept under constant surveillance by means of closed-circuit television cameras or a record of activities established by constantly or intermittently operated photographic cameras. Alternatively, surveillance may be provided by microphones which relay the sounds made by an intruder to a listening post.

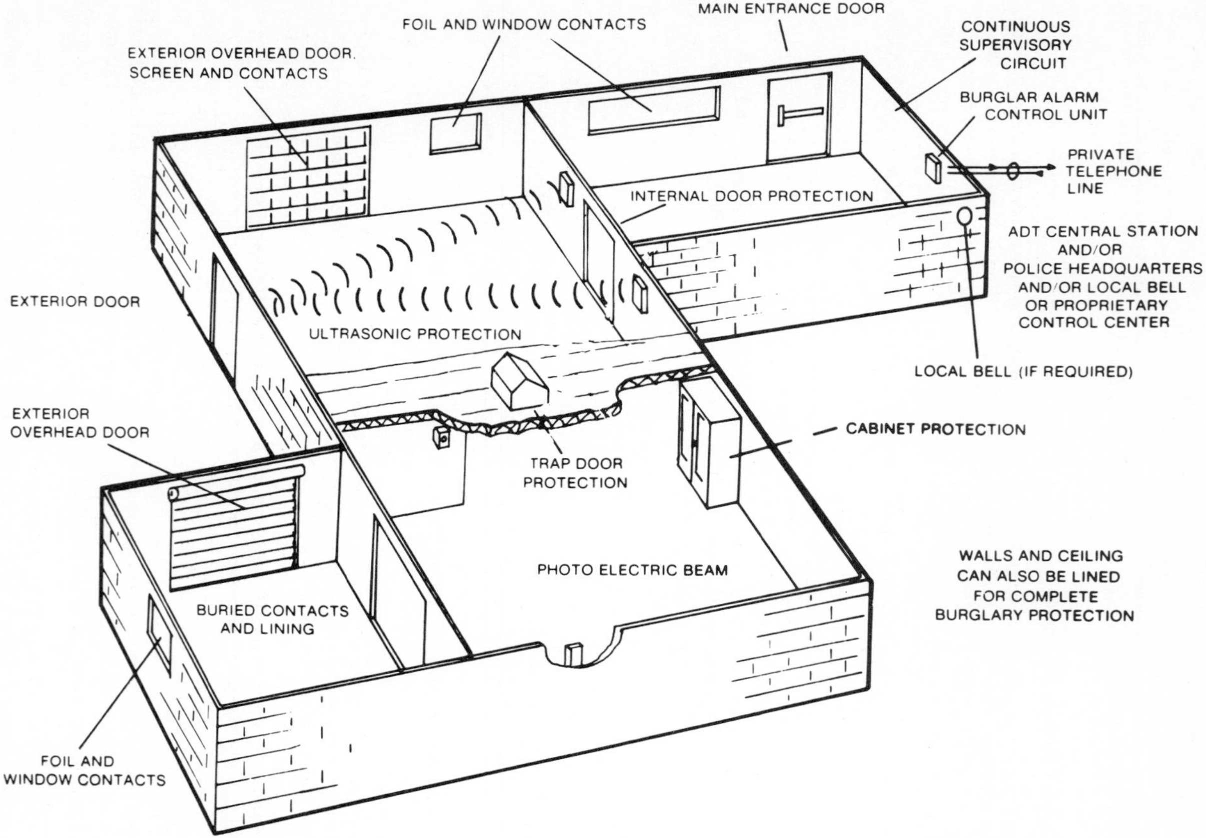

Figure 4-2 Typical perimeter and interior protection plan. *(American District Telegraph Co.)*

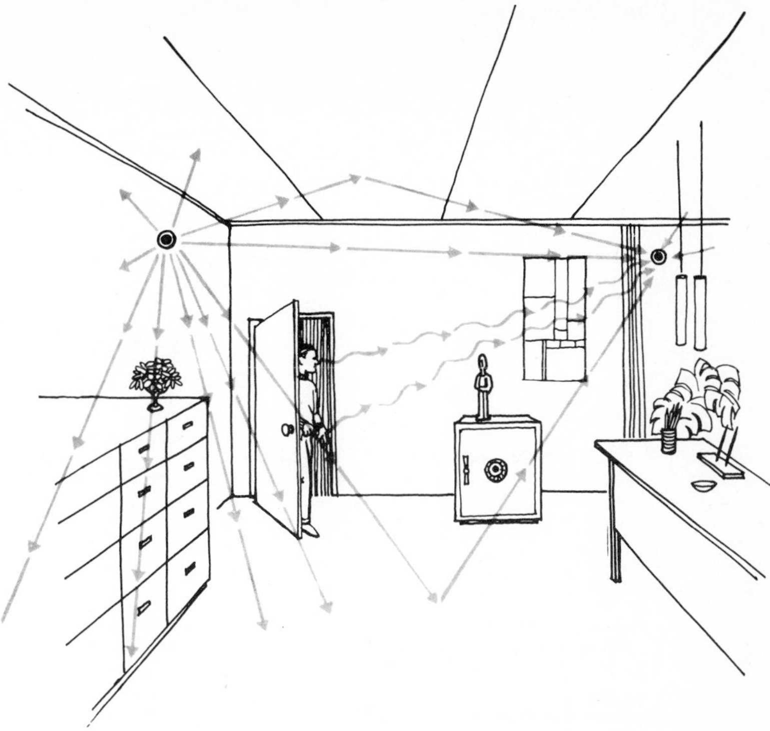

Figure 4-3 Diagram shows how system utilizes Doppler effect, a scientific phenomenon which causes sound waves to change tone when hitting a moving surface. The change triggers an alarm.

A relatively new device is the rail television system in which a mobile camera travels on a track-type rail to provide coverage of an area which would otherwise require multiple cameras.

Vaults like those used for safe deposit, fur, and narcotics are often protected by alarm systems responsive to the sounds created during an attack which range from the single, high-intensity burst of a dynamite blast to the low-level noise produced by the scraping away of mortar with a pointed tool such as a screwdriver. Two basic types of system are employed: one is responsive to airborne sounds and the other to sounds transmitted through the structure of the vault.

Attacks conducted with oxyacetylene torches or burning bars are detected by heat and/or smoke detectors adapted from fire-protection technology.

Night depositories can be protected by lead-sheathed, current-carrying cables embedded in the concrete at the time of construction. Rupture of the cable during the course of an attack results in an alarm.

Holdup Alarms

It is often advisable to provide a holdup alarm system for payroll departments and other high-risk areas so that assistance may be summoned rapidly either during or immediately after a raid. This system could also actuate automatic cameras, photographic or television. The initiating devices are concealed hand- or foot-operated switches and short-range radio transmitters which are carried on the belt or in a pocket.

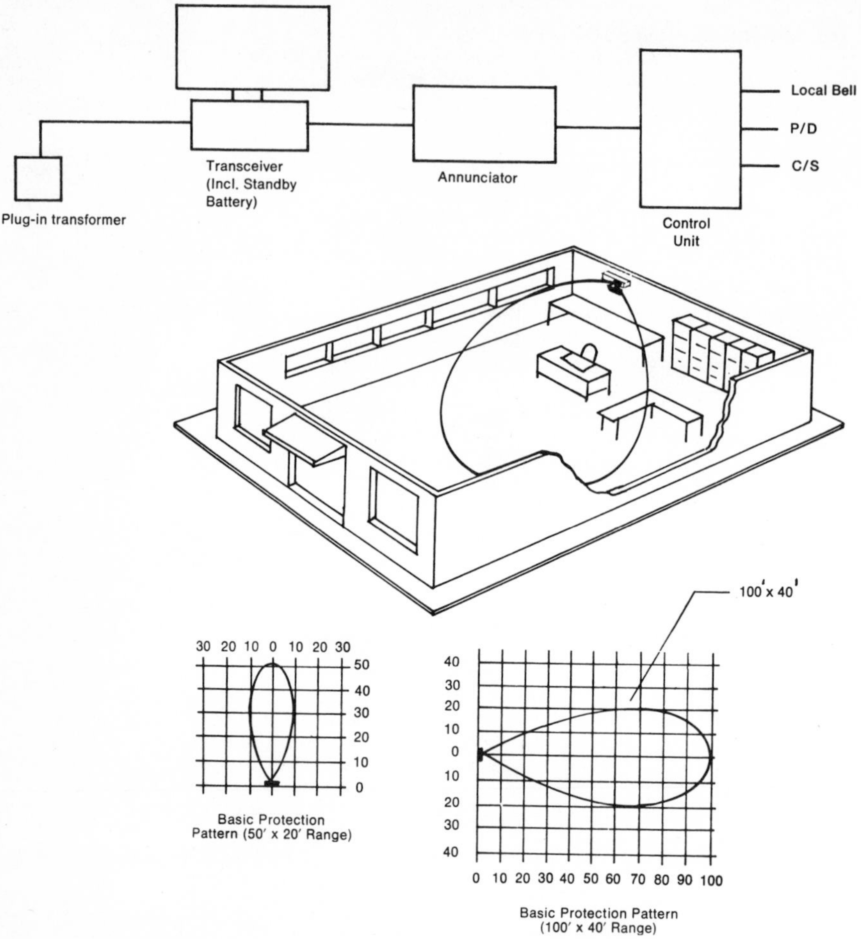

Figure 4-4 Typical protection patterns with microwave beams.

There are also a variety of *cash-drawer* devices intended to permit the surreptitious initiation of a signal: pressure pads which can be pressed by the cashier when removing money from the drawer; a money clip which is a switch whose contacts are held apart by one or more pieces of paper currency whose withdrawal allows the contacts to close and actuate a silent alarm.

A recent addition is the exploding-money package. A bundle of what appears to be currency contains tear gas and marker dye. It is actuated by an electric field at the exit door as the perpetrator departs. After a short delay, the package ruptures to scatter the tear gas and dye. This assists in the capture and identification of the bandit.

SIGNAL TRANSMISSION

The basic burglar-alarm circuit is that of the doorbell; i.e., operate a switch and the bell rings. In practice, such a simple, normally open circuit is not acceptable because loss of the power supply or a break in the line (accidental or intentional) would not be discovered until the system failed to operate when needed.

It is, therefore, the general practice to monitor a small current continually passing through the circuit and initiate alarm signals when its value rises above or falls below specified levels. Means are sometimes provided for the operator to distinguish between true alarms and circuit faults.

In central-station service, two basic types of circuit are employed. Direct-wire service employs a dedicated telephone connection from each subscriber to an individual readout device at the central station and is the equivalent of a private-wire telephone connection. Circuit burglar-alarm service connects a number of subscribers to a common circuit path, as in party-line telephone service, and the individuals are identified at the central station by means of coded signals. The McCulloh circuit arrangement is usually employed in this type of service to permit continued operation when faults appear on the line.

Where high-security risks are involved, some form of line security should be provided to detect attempts to compromise the transmission lines. The technical sophistication of the modern criminal element poses a substantial threat to security.

Traditionally, the central-station services have been provided by dc circuits which have distance limitations. As more and more subscribers leave the city for suburban locations, the central stations have turned to ac signal transmission to relieve the problem. Interrogate-response polling techniques enhance the reliability and security of the service.

An intermediate step is the use of automatic telephone dialers connected to the switched network which eliminates the need to wait for (and pay for) the availability of a dedicated telephone line. The best form of this service is provided by digital dialers which check for the completion of a connection before transmission of the signal. Tape-recorded voice message dialers have presented service problems which have led to their being prohibited by the police in certain locations. Dialers are sometimes used as backup for other forms of transmissions.

The radio transmission of alarm signals has been impeded by cost and federal regulations. Nonetheless, the transmission of signals by radio can be a valuable method when other facilities are unreliable or nonexistent and when used as a backup for the conventional transmission facilities.

SIGNAL HANDLING

A traditional problem in the monitoring of security systems has been the processing of the large number of signals that do not represent an emergency situation. These signals are caused by the opening and closing of protected premises at the proper time of day and the signals received from patrolling watchmen according to schedule.

Carefully programmed computers are now used to process routine signals automatically with the added advantage of automatically providing printouts of desired records. Computer-based operations relieve the pressure on the operators during busy periods and generally improve the efficiency and reliability of security services.

Modern microprocessor technology has been especially effective in proprietary systems where it provides such assistance to the operators as illuminated maps showing the exact location of the alarm source and simultaneously presenting detailed instructions as to the actions to be taken by the operator.

REVIEW AND UPGRADING

Security is a never-ending procedure because it must be constantly reviewed and upgraded to allow for changes in the physical structure of the protected premises as well as their content and the environment, as discussed under "Security Planning." In addition, consideration must be given to the replacement of aging equipment with new-technology devices offering superior performance.

BIBLIOGRAPHY

National Institute of Law Enforcement and Criminal Justice (NILECJ)-STD-0306.00, May 1976, "Physical Security of Door Assemblies and Components," National Institute of Law Enforcement and Criminal Justice, Law Enforcement Assistance Administration, U.S. Department of Justice.

NILECJ-STD-0316.00, March 1979, "Physical Security of Window Units," National Institute of Law Enforcement and Criminal Justice, Law Enforcement Assistance Administration, U.S. Department of Justice.

Security World Publishing Company, P.O. Box 272, Culver City, CA 90230, specializes in books and periodicals relating to security.

chapter 14-5

Toxic Substances and Radiation Hazards

by

Jaswant Singh, Ph.D., C.I.H.
Vice President/Technical Director
Clayton Environmental Consultants, Inc.
Southfield, Michigan

GLOSSARY

Aerosols Liquid droplets or solid particles dispersed in air that are of fine enough particle size (0.01 to 100 μm) to remain so dispersed for a period of time.

Alveoli Tiny air sacs of the lungs, formed by a dilation at the end of a bronchiole; through the thin walls of the alveoli, the blood takes in oxygen and gives up its carbon dioxide through respiration.

Anthrax A highly virulent bacterial infection picked up from infected animals and animal products.

Aplastic anemia A condition in which the bone marrow fails to produce an adequate number of red blood corpuscles.

Asbestos A hydrated magnesium silicate in fibrous form.

Asbestosis A disease of the lungs caused by the inhalation of fine airborne asbestos fibers.

Asphyxia Suffocation from lack of oxygen. *Chemical asphyxia* is produced by a substance, such as carbon monoxide, that combines with hemoglobin to reduce the blood's capacity to transport oxygen. *Simple asphyxia* is the result of exposure to a substance, such as carbon dioxide, that displaces oxygen.

Bronchi The two main branches of the trachea that go into the right and left lung.

Bronchiole The smallest of the many tubes that carry air into and out of the lungs.

Byssinosis Disease occurring to those who experience prolonged exposure to heavy air concentrations of cotton dust.

Carcinoma Malignant tumors derived from epithelial tissues, that is, the outer skin, the membranes lining the body cavities, and certain glands.

Cesium 137 An isotope of the element cesium having an atomic mass number of 137. One of the important fission products.

Chloracne A disease caused by chlorinated polyphenyls and naphthalenes acting on sebaceous glands and the liver.

Chronic bronchitis An inflammation of the bronchial tubes occurring over a long period of time and/or frequently.

Cilia Tiny hairlike "whips" in the bronchi and other respiratory passages that aid in the removal of dust trapped on these moist surfaces.

Conjunctivitis Inflammation of the delicate mucous membrane (conjunctiva) that lines the eyelids and covers the front of the eyeball.

Contact dermatitis Dermatitis caused by a primary irritant.

Cristobalite A crystalline form of free silica. Quartz in refractory bricks and amorphous silica in diatomaceous earth are altered to cristobalite when exposed to high temperatures (calcined).

Curie A measure of the activity, or the rate at which a radioactive material throws off particles. The radioactivity of one gram of radium is a curie. One curie corresponds to 37 billion disintegrations per second or 37 becquerels. The becquerel (abbreviated Bq) is the SI unit that supersedes the curie (abbreviated Ci).

Cutie-pie A portable instrument equipped with a direct-reading meter used to determine the level of radiation in an area.

Cyclone As used in industrial-hygiene monitoring, a particle-size selector whose operation is based on imparting sufficient tangential velocities to relatively larger (heavier) particles sufficient to cause impaction on the walls of a conical chamber, while permitting smaller (respirable) particles to remain entrained in the air system.

Diatomaceous earth A soft, gritty, amorphous silica composed of small aquatic plants. Used in filtration and decolorization of liquids. Calcined and flux-calcined diatomaceous earth contains appreciable amounts of cristobalite.

Dose The total amount of a substance taken into the body by all routes of exposure during some recognized time period.

Dyspnea Shortness of breath, difficult or labored breathing.

Edema A swelling of body tissues as a result of being waterlogged with fluid.

Electromagnetic radiation The propagation of varying electric and magnetic fields through space at the speed of light, exhibiting the characteristics of wave motion.

Emphysema A lung disease, in which the walls of the alveoli have been stretched too thin and broken down; frequently accompanied by impairment of the heart action.

Epidemiology The science of correlating incidence and distribution of disease with causative factors or agents.

Etiology The study or knowledge of the causes of disease.

Fibrosis A growth of fibrous tissue in an organ in excess of that naturally present. A condition marked by increase of interstitial fibrous tissue.

Forced vital capacity (FVC) The maximum volume of air that can be expelled with maximum effort after a full inspiration.

Free crystalline silica Silicon dioxide with the SiO_2 molecule oriented in a fixed tetrahedral (crystalline) pattern, the most prevalent forms being quartz, cristobalite, and tridymite.

Hemoglobin The red coloring matter of the blood which carries the oxygen.

Ionizing radiation Refers to *(a)* electrically charged or neutral particles, or *(b)* electromagnetic radiation which will interact with gases, liquids, or solids to produce ions.

Industrial hygiene The science of recognizing, evaluating, and controlling environmental stresses arising in or from the workplace.

Inertial impaction The forceful impingement on, or striking of, a particle on a surface with resulting adherence.

LD_{50} Abbreviation of lethal dose 50, the dose which is required to produce death in 50 percent of the exposed species. Death is usually reckoned as occurring within the first 30 days.

Leukemia A blood disease distinguished by overproduction of white blood cells. It may result from overexposure to radiation or it may generate spontaneously.

Lymph A clear, colorless fluid which circulates through the vessels of the lymphatic system.

Maser Microwave amplification by stimulated emission of radiation.

Metastasis Spread of malignancy from the site of primary cancer to secondary sites due to transfer through the lymphatic or blood system.

mg/m^3 Milligrams of substance per cubic meter of air; a common unit of exposure concentration.

Milliroentgen One one-thousandth of a roentgen. A roentgen is a unit of radioactive dose.

mppcf Millions of particles per cubic foot of air; a common unit of exposure concentration for mineral dusts.

Narcosis Stupor or unconsciousness produced by chemical substances.

Necrosis Destruction of body tissue.

Papilloma A small growth or tumor of the skin or mucous membrane.

Phagocyte A cell in the body that characteristically engulfs foreign material and consumes debris and foreign bodies, bacteria, and other cells.

Pharynx A part of the alimentary canal located between the mouth and the esophagus.

Pneumoconiosis A disease of the lungs caused by irritation of dusts and other particles.

Pulmonary function tests Measurement of ventilatory capacity of the lung by a series of tests, such as forced expiratory volume (FEV) and forced vital capacity (FVC).

Quartz The most prevalent, naturally occurring form of free silica, the basic raw material for the industrial sand industry.

Rad Standard unit of radioactive dose. It supersedes the roentgen.

Radioactivity Emission of energy in the form of alpha, beta, or gamma radiation from the nucleus of an atom.

Radiologist A specialist in the diagnostic and therapeutic use of x-rays and other forms of ionizing radiant energy.

Respirable Capable of penetrating into the lower respiratory tract, generally regarded as requiring a particle size of 10 μm or less.

Respirable mass That portion of total suspended particulate matter capable of penetrating into the lower respiratory tract.

Roentgen A unit of radioactive dose or exposure is called a roentgen (abbreviated R). A roentgen is that amount of x- or gamma radiation that will produce one electrostatic unit of charge, of either sign, in one cubic centimeter of dry air at standard temperature and pressure. It is equivalent to 2.58×10^{-4} C/kg in air.

Roentgenogram The shadow picture formed on a sensitized film or plate by x-rays passing through a body.

Sensitizer A chemical that, at first exposure, may or may not cause irritation. After extended or repeated exposure, some individuals develop an allergic type of skin irritation called sensitization dermatitis.

Siderosis Lung disease resulting from inhalation of iron oxide.

Silicosis A lung disease resulting from fibrosis of the lungs due to inhalation of silica dust.

Spirometry The measurement of air movement in or out of the lungs with the use of a spirometer.

Synergism Combined action of substances whose total effect is greater than the sum of their separate effects.

Talc A hydrous magnesium silicate used in ceramics, cosmetics, paint, pharmaceuticals, and soap.

TLV Threshold-limit value. An exposure level under which most people can work consistently for 8 h/day, day after day, with no harmful effects. A table of these values and accompanying precautions is published annually by the American Conference of Governmental Industrial Hygienists (ACGIH); TLV is a registered trademark of ACGIH.

Time-weighted average concentration A calculated average obtained by dividing the sum of products of concentration times time for all activities by the total time of exposure.

Trachea The cartilaginous and membranous tube (windpipe) by which air passes to and from the lung.

Tridymite A vitreous, colorless form of free silica formed when quartz is heated to 870°C (1598°F). A form of crystalline silica rarely found in naturally occurring deposits.

X-ray diffraction Since all crystals act as three-dimensional gratings for x-rays, the pattern of diffracted rays is characteristic for each crystalline material. This method is of particular value in determining the presence or absence of crystalline silica in industrial dusts.

INTRODUCTION

The practical considerations of occupational health programs in general, and control of workplace contaminants in particular, took on a new cast with the passage by Congress in 1970 of the Occupational Safety and Health Act, which requires all employers to comply with strict health and safety standards. Responsibility for enforcing the act was assigned by Congress to the Occupational Safety and Health Administration (OSHA), an agency established under the Department of Labor. In addition, a National Institute for Occupational Safety and Health (NIOSH) was created within the Department of Health, Education, and Welfare (now Health and Human Services) to conduct research, establish educational programs, and make recommendations to OSHA for health and safety standards.

Enactment of occupational health standards by OSHA has forced many plant engineers to become involved in the evaluation and control of industrial hygiene problems not only in existing facilities, but in the design of new facilities as well. For example, industrial hygiene considerations play a dominant role when considering ventilation systems (local and general), makeup air, enclosure and/or isolation of processes using toxic chemicals, handling and storage of toxic materials, and cleanup of exhaust emissions. Adequate disposal of toxic or hazardous wastes is also important to prevent environmental contamination and exposure of workers handling them. Plant engineers, therefore, need to be aware of toxic chemical and radiation hazards and how to effectively deal with them.

Industrial hygiene is defined by the American Industrial Hygiene Association (AIHA) as "that science and art devoted to the recognition, evaluation, and control of those environmental factors or stresses, arising in or from the workplace, which may cause sickness, impaired health, and well-being, or significant discomfort and inefficiency." An industrial hygienist is someone trained in "engineering, chemistry, physics, or medicine or related biological sciences (augmented by) special studies and training in all of the above cognate sciences" to practice industrial hygiene as defined above. A fully staffed industrial-health team also includes, at minimum, an *industrial physician,* an *industrial toxicologist,* a *lab-analysis technician,* and an *environmental* or *process engineer.*

TOXIC SUBSTANCES

Identifying Hazards: The Workplace Survey

Identifying potential health hazards is often relatively straightforward. For example, it is safe to assume that petrochemical workers will probably be exposed to one or more of a variety of organic vapors, that sandblasters risk overexposure to respirable dusts (crystalline silica, etc.), or that radiation is a potential hazard affecting workers in nuclear power plants.

In many other situations, the presence of a hazard may only be suspected, as when employee complaints of disease symptoms fall into an identifiable pattern or when strange odors are detected in the workplace. These cases call for a more rigorous type of exploratory survey, to both identify the hazard and measure its extent.

Whether the identity of the contaminant is known or not, a workplace survey should be conducted that includes an inventory of the processes taking place in the work area or entire plant, the chemicals used in these processes, the raw materials, by-products, etc. Generally, a comprehensive investigation requires the services of a highly experienced professional, such as a certified industrial hygienist. When the nature of the contaminant is indeterminate, the industrial hygienist should work with an industrial physician and a process engineer.

Many times in an investigation, visibility is a positive indicator of a hazard. The absence of visibility of dusts and fumes, however, does not mean that there are not dangerous levels of contaminants present, since many contaminants constitute a hazard at levels much too low to be visible.

In the same way, the investigator uses the sense of smell to pinpoint many contaminants. For example, it is possible to distinguish between the haylike odor of phosgene and the fishlike odor of trimethylamine. Here again, however, certain precautions must be observed. For example, although hydrogen sulfide has a very distinct rotten-egg-like odor, it can also dull the sense of smell after prolonged exposure, and thus effectively mask heavy concentrations of many other substances.

Tables 5-1 and 5-2 provide examples of typical occupational exposure to particulate and gaseous toxic agents.

TABLE 5-1 Examples of Occupational Exposure to Particulate Toxic Agents

Contaminant	Physical state	Occupation
Asbestos	Dust, fiber	Fireproofers, insulation strippers, asbestos-cement workers, auto-garage mechanics, construction workers, shipbuilding and repair workers, gasket manufacturing workers, rubber compounders, vinyl-tile workers
Silica	Dust	Abrasive blast cleaners, pottery makers, glass makers, cement workers, coal miners, construction workers, enamellers, foundry workers, smelters
Lead	Dust, fume	Babitters, glass makers, foundry workers, pottery makers, printers, paint manufacturers and sprayers, can and dye makers
Chromic acid	Mist	Electroplaters, picklers, colored glass, ink, and refractory makers
Arsenic	Dust, fume, or gas (arsine)	Copper smelters, brass manufacturing workers, ceramic and glass makers, insecticide manufacturing workers, electroplaters
Beryllium	Dust, fume	Beryllium metal and alloy manufacturing workers, alloy machinists, glass and neon tube makers, rocket fuel manufacturing workers
Coal-tar pitch volatiles	Dust, fume	Metallurgical operations workers, metal casters, petroleum refinery workers, coking operations workers

TABLE 5-2 Examples of Occupational Exposure to Gaseous Toxic Agents

Contaminant	Physical state	Occupation
Carbon monoxide	Gas	Cokers, smelters, metal casters, forklift drivers, garage operators, heat treaters, coal conversion workers, pottery makers
Nitrogen oxides	Gas	Welders, electroplaters, forklift drivers, fertilizer manufacturing workers, explosive manufacturers, dye workers
Fluorocarbons	Gas or vapor	Food processers, storage workers
Sulfur dioxide	Gas	Brewers, copper smelters, ore roasters, petroleum refiners, glass makers, powerplant operators, paper makers
Benzene	Vapor	Petroleum refiners, coke oven workers, organic chemical manufacturing workers, gasoline station (coke ovens) attendants, ink makers, insecticide makers, lithographers, paint makers, rubber makers
Toluene	Vapor	Core makers (foundry operation), polyurethane foam makers and users, spray painters, ship welders

Work and Process Inventory

The first step is to obtain descriptions of job functions within the company. This information may be provided by the personnel department and/or manufacturing staff, and should include the following:

- Nature of the job
- Description of the process
- Work duration
- Nature of potential exposure (if known)

This information allows the investigator to group various workers according to similarity of exposure to chemical and physical stresses. This grouping is essential later in selecting representative workers who may need to be sampled for exposure levels.

Chemical Inventory

It is important that the investigator conducting the industrial hygiene survey be knowledgeable about the chemicals that are used in the plant. In preparing a chemical inventory (preferably with the help of the purchasing department), all processes using chemicals are taken into consideration, such as waste treatment, boiler operations, air conditioning, and any other pertinent sources including those involving raw materials, intermediate products, finished products, and cleaning compounds.

A major problem in recognizing chemical hazards is the many chemical formulations of different suppliers and manufacturers. Industry uses a large variety of materials sold under various trade names, and obtaining sufficient or accurate information on the composition of each one is often difficult. Obtaining *material-safety data sheets* (MSDSs) from the supplier can be helpful in this regard. Updating MSDSs is recommended using various relevant publications and fact sheets, including the hygiene guides prepared by the American Industrial Hygiene Association and the chemical-safety data sheets supplied by various organizations, including the Chemical Manufacturers Association.

Consideration must also be given to toxic materials that may be present as impurities in relatively safe materials—diatomaceous earth, for example, which is supposedly 100 percent amorphous silica and thus generally considered a safe material. Diatomaceous earth has extensive applications, including use as a filtering aid and as a filler in cosmetics, detergents, and other household products. However, certain varieties of diatomaceous earth used mostly as filtration aids have been found to contain substantial quantities of crystalline silica, a widely recognized health hazard.

Moreover, some materials that apparently exist at subtoxic levels in the workplace can substantially appreciate in toxicity by undergoing chemical reaction within the body. For example, a metabolic reaction occurs in the case of benzidine dyes. The carcinogenic nature of benzidine has been recognized for many years. Benzidine-based dyes have been considered relatively safe on the assumption that they contain very little free or unreacted benzidine. Studies of analyses of urine samples collected from workers exposed to these dyes, however, indicate that in work areas where these dyes are in extremely small or even undetectable concentrations, metabolization of benzidine dye results in significantly elevated benzidine levels within the body.

Measurement of Worker Exposure

A variety of techniques are used in measuring exposure of workers to toxic chemicals. Techniques used for measuring particulates differ somewhat from those used to measure gases and vapors.

Sampling to determine worker exposure to particulates is most often done by filtration, which uses a battery-operated pump that draws air through a filter, usually 37 mm in diameter and made of paper, fiberglass, or synthetic materials such as mixed cellulose-ester, polyvinyl chloride, or polycarbonate. An impinger is recommended in sampling for some particulates. The solution within the impinger can also collect gases and vapors and, in fact, impingers are used more for this purpose than for sampling for particulates.

Route of Entry

Contaminants enter the body principally in three ways:

- Inhalation (through the respiratory tract)
- Skin absorption (through the skin)
- Ingestion (through the digestive tract)

Inhalation is by far the most common access for airborne contaminants to the body because of the continuous need to oxygenate the tissue cells and because of intimate contact with the body's circulatory system.

The effect of exposure to toxic agents is usually classified as acute or chronic.

Acute Exposure

Acute exposure is characterized by exposure to high concentrations of the toxic material over a short period. The exposure occurs quickly and can result in immediate damage to the body. For example, inhaling high concentrations of carbon monoxide gas or carbon tetrachloride vapors will produce acute poisoning.

Chronic Exposure

Chronic exposure occurs when there is continuous absorption of small amounts of contaminants over a long period. Each dose, taken independently, would have little toxic effect, but the quantity accumulated over a number of years can result in serious damage. Chronic poisoning can also be produced by exposure to small amounts of harmful material that produces irreversible damage to tissues and organs so that the injury rather than the poison accumulates. An example of such a chronic effect is silicosis, a disease produced by inhaling crystalline silica dust over a period of years.

Nature of Contaminants

Airborne contaminants can be present as liquids or solids, as gaseous material in the form of a true gas or vapor, or in combination of both gaseous and particulate matter. Most often, airborne contaminants are classified according to physical state and physiological effect on the human body. Knowledge of these classifications is necessary for proper evaluation of the work environment. One must also consider the route of entry and action of the contaminant.

Physiological Classification of Toxic Effects

Irritants

Irritants cause inflammation of the moist mucous surfaces of the body. Irritants are corrosive, but inflammation of tissues may result from concentrations well below those needed to produce corrosion. Examples of irritant materials include aldehydes, alkaline and acid mists, and ammonia. Materials that affect both the upper respiratory tract and lung tissues are chlorine and ozone. Irritants that affect primarily the terminal respiratory passages are nitrogen dioxide and phosgene.

Asphyxiants

Asphyxiants deprive the tissues of oxygen. They are generally divided into two classes—simple and chemical.

Simple Asphyxiants. These are physiologically inert gases that deprive the tissue of oxygen by diluting the available atmospheric oxygen. Examples include nitrogen, carbon dioxide, hydrogen, helium, and aliphatic hydrocarbons such as methane.

Chemical Asphyxiants. These prevent either oxygen transport in blood or normal oxygenation of the tissues. Chemical asphyxiants are active far below the level required for damage from simple asphyxiants. Examples include carbon monoxide, hydrogen cyanide, and nitrobenzene.

Primary Anesthetics

Anesthetics depress the central nervous system, particularly the brain. Examples include ethylene and ethyl ether.

Systemic Poisons

Systemic poisons cause injury to particular organs or body systems. The halogenated hydrocarbons (such as carbon tetrachloride) can damage the liver and kidneys, whereas benzene, aniline, and phenol may cause damage to the blood-forming system. Examples of materials classified as neurotoxic agents include carbon disulfide, methyl alcohol, tetraethyl lead, and organic phosphorus insecticides. Examples of metallic systemic poisons include cadmium, lead, manganese, and mercury.

Chemical Carcinogens and Teratogens

Chemical carcinogens can cause tumors in mammalian species. Carcinogens may induce a tumor type not usually observed, or induce an increased incidence of a tumor type normally seen, or induce such tumors at an earlier time than would otherwise be expected. In some instances, the worker's initial stages of exposure to the carcinogen and the tumor appearance are separated by a latent period of 20 to 30 years. Examples of chemical carcinogens include benzo(a)pyrene, beta-naphthylamine, vinyl chloride, and chromates.

Chemical teratogens are chemicals that produce malformation of developing cells, tissues, or fetal organs. These effects may result in growth retardation or in degenerative toxic effects. Examples of chemical teratogens include acetamide and methyl mercury.

Particulate Size

Airborne particulate matter varies in size from less than 0.01 to more than 25 μm (one micrometer equals approximately 1/25,000 inch). These particles are invisible to the naked eye.

Nonrespirable particulates consist of particles that are either too large to escape the respiratory tract's defenses before they can reach the lungs or that are too small to be retained in the lungs even if they get that far. Nonrespirable particulates present other types of health problems. Toxic fumes and nonrespirable dusts that are inhaled or ingested (or, less commonly, that come into contact with the skin) may be eventually absorbed into the bloodstream and cause systemic poisoning. Other nonrespirable dusts, inert "nuisance" substances like limestone and gypsum, can severely irritate the upper respiratory tract, endangering health as well as causing great discomfort.

Particulates that are small enough to find their way into the lungs and remain there, including dusts fine enough to be classified as *respirable dusts,* can produce serious chronic conditions such as the group of diseases known as *pneumoconiosis*—lung diseases like coal miners' "black lung disease" and silicosis.

In view of the above, it is clear that the measurement of dust exposure in many cases needs to be limited to that fraction of inhaled particles small enough to be deposited in alveolar spaces. Thus, a size-selective air-sampling method is required that will separate this "respirable fraction" from the coarser material that is deposited in the upper respiratory tract. The size selector most commonly used in respirable dust sampling has the following characteristics:

Aerodynamic diameter, μm (unit density sphere)	% passing selector
2.0	90
2.5	75
3.5	50
5.0	25
10.	0

Time-Weighted Average Exposures

Threshold-limit values usually refer to time-weighted concentrations for a 7- or 8-h workday and 40-h workweek, although recent efforts of NIOSH and OSHA have been aimed at defining permissible exposure limits for up to 10-h workdays.

Short-term exposures to concentrations above the threshold limit are permitted provided they are compensated for by equivalent excursions below the limit during the workday. For example, if the permissible exposure limit is 10, a worker could be permitted to work in concentrations as high as 15 for 4 h, provided that the remainder of this 8-h shift did not result in exposure to concentrations above 5, thus yielding an average 8-h weighted concentration of 10.

Stated mathematically, the time-weighted average concentration, based on an 8-h shift, is defined as

$$\text{TWA} = \frac{1}{8}\sum_{i=1}^{N} \overline{C}_i t_i$$

where $\overline{C}$ is the average concentration of the contaminant at location i, N is the total number of work locations, Σ refers to the summation of the products of concentration $\overline{C}$ in percentage and t is the time in house spent at location i. Expressed otherwise,

$$\text{TWA} = \frac{(\text{exposure time})(\text{con. } A) + (\text{exposure time})(\text{con. } B) + \cdots}{\text{total work time per shift}}$$

Obviously, this approach requires reliance upon personal-breathing-zone sampling or detailed job analyses for all classifications studied, and a comprehensive program of sampling sufficient to establish average airborne concentrations at all work sites.

Whom to Sample

Sampling programs are based upon statistical considerations, as it may be impractical and is often unnecessary to sample every worker's exposure. Thus, various statistical approaches have been devised to determine the minimum number of workers to be sampled to achieve adequate representation. A simple approach recommended by NIOSH is summarized as follows:

Number of employees exposed	Minimum number of employees whose individual exposures shall be determined
1–20	50% of the total number of exposed employees
21–100	10 plus 25% of the excess over 20 exposed employees
Over 100	30 plus 5% of the excess over 100 exposed employees

Occupational Safety and Health Standards

Measured results are compared to OSHA health standards or other widely accepted standards such as the American Conference of Governmental Industrial Hygienists (ACGIH) threshold-limit values.

Federal Standards (29 CFR 1910 Subpart Z)

The first compilation of health and safety standards promulgated by OSHA was based on the existing federal and national consensus standards.

Table 5-3 lists the standards in Sections 1910.1001 through 1910.1046 of Title 29, Chapter XVII, Section 1910.1000 (available through OSHA as Publication 2206). These standards are detailed sets of regulations for individual substances. In addition to setting exposure limits, they require exposure monitoring and medical surveillance of exposed employees. OSHA plans to promulgate many more complete standards.

Private Organizations

The first influential American organization in the field of occupational safety was the American Conference of Governmental Industrial Hygienists (ACGIH). The ACGIH publishes each year its revised and updated list of occupational exposure limits, commonly referred to as TLVs (threshold-limit values). This listing is more extensive than

TABLE 5-3 Individual Chemical Substance Standards

		Exposure limits		
29 CFR 1910	Chemical name	TWA	Ceiling	Action level
1000	Air contaminants	—		
1001	Asbestos	2 fibers/cm^3	10 fibers/cm^3	
1003	4-Nitrobiphenyl	*		
1004	α-Naphthylamine	*		
1005	Moca (deleted)			
1006	Methyl chloromethyl ether	*		
1007	3, 3′-Dichlorobenzidine	*		
1008	bis (Chloromethyl) ether	*		
1009	β-Naphthylamine	*		
1010	Benzidine	*		
1011	4-Aminodiphenyl	*		
1012	Ethylenimine	*		
1013	β-Propiolactone	*		
1014	2-Acetylaminofluorene	*		
1015	4-Dimethylaminoazobenzene	*		
1016	*N*-Nitrosodimethylamine	*		
1017	Vinyl chloride	1 ppm	5 ppm/15min	0.5 ppm
1018	Inorganic arsenic	10 μg/m^3		5 μg/m^3
1025	Lead	50 μg/m^3		30 μg/m^3
1028	Benzene	1 ppm	5 ppm/15 min	0.5 ppm
1029	Coke-oven emissions	150 μg/m^3		
1043	Cotton dust	200 μg/m^3 (yarn mfr.) 750 μg/m^3 (weaving) 500 μg/m^3 (nontextile)		
1044	1,2-Dibromo-3-chloropropane	1 ppb		
1045	Acrylonitrile	2 ppm	10 ppm/15 min	1 ppm
1046	Cotton dust (gins)	—		

*These carcinogen standards require that these chemicals be handled only in completely enclosed systems with exposures reduced to the lowest feasible level.

the OSHA listing of air contaminants in CFR 29 1910.1000. It should be noted that the term TLV is a registered trademark of the ACGIH and should not be used to refer to OSHA permissible exposure limits. Another organization involved in the promulgation of such standards, although its concerns extend into several other areas, is the American National Standards Institute (ANSI). Two younger organizations are the American Board of Industrial Hygienists (ABIH) and the American Industrial Hygiene Association (AIHA), both established in the 1950s; among the services they provide are the certification of industrial hygienists and of laboratories who analyze workplace samples.

Control Strategy—Reducing or Eliminating Hazards

A strategy for reducing or eliminating workplace hazards can include one of two approaches (or both): engineering or administrative controls.

Engineering Controls. Some examples are: substituting relatively safer chemicals for more toxic ones; altering a process in such a way as to reduce worker exposure to contaminants; changing or upgrading ventilation systems; isolating or enclosing contaminated areas.

Administrative Controls. An example is adjusting work schedules so that workers receive only a fraction of their present exposure to contaminants.

As an example of an engineering control, if asbestos is used as insulation, it may be possible to replace it with fiberglass or calcium silicate, materials which do not present the same degree of hazard. Here again, however, it should never be assumed that such a substitution completely eliminates the hazard. The hazardous potential for fiberglass is not completely known, for example, while certain varieties of calcium silicate, generally regarded as a relatively safe substance, have been found to contain measurable amounts of asbestos fibers and/or free crystalline silica, another highly hazardous material.

A control strategy must take into account work *practices,* which can be a major factor in exposure. Many poor work practices result from unawareness of the potential hazards of a toxic material. For example, in one paint factory, workers were observed dusting off their clothes with a compressed-air line, starting from the shoes and proceeding upward. The raw material in the plant's primary process included, among other substances, lead chromate. Sampling measurements in the plant indicated that worker exposures to lead and hexavalent chromium (both highly toxic materials) were not excessive under normal working conditions. It was determined that *this dusting procedure, however, exposed workers to higher concentrations of lead and chromium during a 2- to 4-min period than they were exposed to during the remainder of the day* when handling these materials frequently. Vacuuming of dusty clothing (rather than blowing the dust off) could easily prevent this exposure. In such cases, employee-awareness programs can be highly useful in preventing work practices that may result in unnecessary exposure.

The evaluation of existing and anticipated engineering controls is a major goal of an effective industrial-hygiene program. OSHA considers engineering controls to be the primary and most effective means of reducing or eliminating exposure to toxic substances in the workplace. The extent of engineering controls needed can be better established after sampling measurements have been taken, other aspects such as work practices, housekeeping, etc., have been evaluated, and corrective actions have been instituted. Examples of poor engineering controls are all too plentiful.

Exposure Surveillance Programs

A comprehensive ongoing occupational health or industrial hygiene program includes a monitoring program that ensures surveillance of worker health in addition to determining compliance with the applicable regulatory standards. The *surveillance program* consists not only of ongoing sampling of worker exposure to airborne concentrations of chemicals but also monitoring for exposure to such physical agents as noise, radiation, and heat stress. The frequency of sampling will depend upon the levels of various contaminants, the toxicity of the material, and the specific requirements under the applicable codes, and may vary from weekly to biannually.

Biological monitoring of worker exposure to chemical hazards through analysis of samples of blood, urine, expired air, etc., is required in those cases where measurement of airborne concentration alone is not a reliable indicator of hazard potential. In many instances, the metabolite (chemical produced within the body) of a toxic substance rather than the substance itself may be measured in the biological fluids.

Medical surveillance of exposed workers is also a highly useful and necessary tool for protection against toxic exposures. Pre-employment and periodic medical examinations can reveal the presence of toxic effects at an early stage, when a cure is often possible. Medical examinations should include specific organ functions to detect changes relative to the specific contaminants to which the worker is exposed.

RADIATION HAZARDS

Radiation is energy which is emitted, transmitted, or absorbed in wave or energetic particle form. The electromagnetic (EM) waves consist of electric and magnetic forces. When these forces are disturbed, EM radiation results. Figure 5-1 is an arrangement of known EM radiations according to their frequency and/or wavelength. It includes microwave, infrared, visible, and ultraviolet radiation. The range of biological effects of exposure within the electromagnetic spectrum is, therefore, extremely broad and diverse.

Radiation is generally divided into two categories: *ionizing* and *nonionizing*. Ionizing radiation includes x-rays, alpha particles, and beta, gamma, and neutron rays. Nonionizing radiation includes ultraviolet, visible, infrared, microwave, and radio-frequency. Laser radiations fall into most of these bands. Examples of occupational exposure to ionizing and nonionizing radiation are shown in Table 5-4.

Ionizing Radiation

Of the type of ionizing radioactivity, alpha particles are the least penetrating—paper and skin will ordinarily stop alpha particles.

Beta radiation has considerably more penetrating power than alpha particles. X-rays and gamma rays both have very good penetrating power and require the use of heavy

TABLE 5-4 Example of Occupational Exposure to Ionizing and Nonionizing Radiation

Type of radiation	Occupation
Ionizing radiation (radioactive isotopes, x-rays)	Nuclear powerplant workers, food preservers, electron microscopists, biologists, food sterilizers, high-voltage repairmen, ceramic workers, drug makers
Ultraviolet radiation	Meat curers, movie projectionists, pipeline workers, paint curers, nurses, bacteriologists, dentists, food preservers, lithographers, laboratory workers, welders, textile inspectors, plastic curers, printers
Infrared radiation	Bakers, electricians, furnace workers, glass blowers, heat treaters, solderers, welders, firemen, steel workers
Microwave, radio-frequency radiation	Automotive workers, paper product workers, plastic heat-sealing workers, rubber-product workers, textile workers, tobacco workers, electronics workers, advertising sign workers
Laser	Medical technicians, surgeons, aerospace workers, semiconductor workers

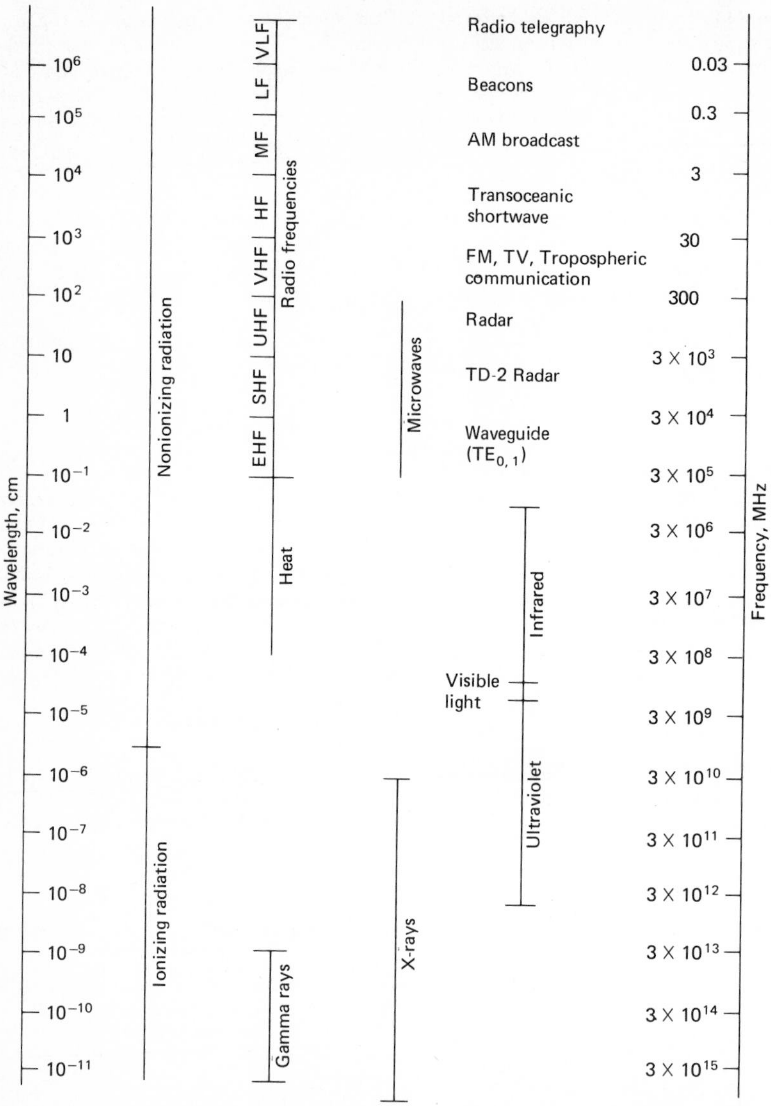

Figure 5-1 Electromagnetic spectrum.

shielding material (lead). Neutrons are very penetrating and require shielding with materials of higher hydrogen atom content rather than the use of mass alone.

Alpha emitters are an internal hazard because they do not have the ionizing ability to travel very long distances. One must take precaution against breathing or ingesting the alpha emitters. Beta emitters are also generally considered an internal hazard.

Regulations Regarding Ionizing Radiation

Jobs involving exposure to ionizing radiation fall under the provisions of Nuclear Regulatory Commission (NRC) standards (10 CFR Part 20). OSHA standards (Section

1910.96) on ionization radiation apply in cases where employees may not be protected under the NRC standards.

Table 5-5 summarizes the most frequent types of radiation encountered in health physics surveys and the type of detection most commonly used.

A photographic film badge can be used for determining beta and gamma doses. Such film badges give a sufficiently quantitative indication of the integrated weekly doses of individuals. These should be recorded along with other records of radiation levels.

TABLE 5-5 Types of Detectors Used for Various Types of Radiation

Type of detector	Type of radiation
Proportional or scintillation counter	Alpha
Geiger-Mueller tube or proportional counter	Beta
Ionization chamber	X and gamma
Proportional counter	Neutron

Nonionizing Radiations

Ultraviolet Radiation

For purposes of assessing the biological effects of ultraviolet radiation, the wavelengths of interest can be restricted to 0.1 to 0.4 μm. The major source of ultraviolet radiation is the sun; common constructed sources are mercury discharge lamps, welding and plasma torches, xenon discharge lamps, and lasers. The symptoms of overexposure to ultraviolet radiation are those characteristic of a severe sunburn.

Currently, there are no OSHA standards for exposure to ultraviolet radiation. OSHA Section 1910.97 concerning nonionizing radiation includes exposure levels and warning signs. However, this is an advisory and not a mandatory standard.

NIOSH has a recommended standard for occupational exposure to ultraviolet radiation. The American Conference of Governmental Industrial Hygienists (ACGIH) has developed TLVs (threshold-limit values) for ultraviolet radiation in the spectral region between 200 and 400 nm (see Table 5-6).

Infrared Radiation

Infrared radiation (ir) can be considered associated with the heat given off by all bodies that radiate heat and covers wavelengths from 0.75 μm to about 1 mm. Infrared rays are

TABLE 5-6 Permissible Ultraviolet Exposures

Duration of exposure per day	Effective irradiance, E_{eff}, $\mu W/cm^2$
8 h	0.1
4 h	0.2
2 h	0.4
1 h	0.8
30 min	1.7
15 min	3.3
10 min	5.
5 min	10.
1 min	50.
30 s	100.
10 s	300.
1 s	3,000.
0.5 s	6,000.
0.1 s	30,000.

mostly absorbed by the skin and burn the skin similar to the way the sun does. Like ultraviolet radiation, infrared radiation is invisible and can seriously damage the eyes. Although infrared radiation may not be felt, over the years it can cause permanent eye damage. Water can be used as a barrier, as it can absorb most ir waves. Regular clothing protects the skin against ir, but goggles should be worn when there is a potential for overexposure.

Except for thermal burns (below wavelengths of 1.5 μm), infrared radiation is insignificant as a health hazard. However, when highly intense and compacted sources of radiant energy are being used, as with lasers, injury can occur in fractions of a second, before pain is evident.

Microwave Radiation

There has been a great deal of exposure to microwaves among people in the armed forces because microwave radiation is used for radar. With the advent of microwave ovens, however, this type of radiation is now encountered in the home. Microwaves have wavelengths of 3 m to 3 mm and their effect is related to power intensity and time of exposure as well as to the wavelength. Their ability to heat the body allows them to be used for medical treatment. This ability, however, can also be a hazard to the overexposed, unprotected worker. Microwaves penetrate deeply into the body and cause body temperature to rise; if the body temperature rises high enough, the person can go into a coma and die.

Before personnel are assigned to work in or about radar equipment, they should be given a complete physical including eye and blood examinations. Workers should be instructed never to look directly at a radar beam and should be given physicals periodically and whenever exposure is indicated. Most microwave measuring devices are based on (1) bolometry, (2) colorimetry, (3) voltage and resistance changes in detectors, and (4) radiation pressure on a reflecting surface. The bolometry method is one of the most widely used in commercially available power meters.

One troublesome fact in the measurement of microwave radiation is that the near field (reactive field) of many sources may produce unpredictable radioactive patterns. Energy density rather than power density may be a more appropriate means of expressing hazard potential in the near field.

Radio-Frequency Radiation

Radio-frequency (rf) electromagnetic radiation obeys the general laws of electromagnetic radiation and is characterized by the following basic parameters: frequency f in hertz (1 Hz = 1 cycle per second), propagation time T, wavelength λ in meters, and velocity c, which in free space under normal conditions is equal to the velocity of light, i.e., 300,000 km/s. These quantities are interrelated according to the formula

$$\lambda = cT$$

Radio-frequency radiation covers wavelengths from 300 m to 1 mm. For measurement, either electric or magnetic field strength meters or power meters are used.

Using power meters, the rf energy causes a change of a temperature-sensitive element which is monitored by a bridge circuit and meter calibrated in $\mu W/m^2$ or mW/m^2.

All effects of rf radiation may be classified as either thermal or nonthermal, leading to heat stress or disturbance of the nervous system, respectively.

Nonionizing Radiation Instruments

Ultraviolet and Infrared Radiation Monitoring Devices

A variety of instruments are commonly used to measure ultraviolet and infrared radiation. They are classified according to the type of detector used, which is generally one of two types: thermal detectors or photoelectric detectors. Thermal detectors are those in which the absorbed radiation is degraded to heat and subsequently coverted to an electric signal by changing the electric resistance of a filament. Photoelectric detectors are based on the principle that the absorbed photons eject electrons from a photo-emissive material. Most of these instruments are precalibrated by the manufacturer, but should be routinely checked prior to field use.

Microwave Monitoring Devices

Most microwave radiation detectors consist of the following components: the test antenna, the attenuator, a bolometer or thermistor, and a power meter. These detectors are usually in the form of a single integrated unit. The test antenna is specific to certain wavelengths and, therefore, interchangeable. These units must be calibrated at frequencies throughout the band to ensure accuracy.

Laser Monitoring Devices

A wide variety of laser radiation detectors are commercially available to fulfill the diversified needs due to different wavelengths, pulse durations, and power and energy densities of various laser instruments. Laser radiation detectors are generally based upon two basic principles—photon and thermal detection.

Lasers

Laser technology has increased rapidly since its inception in 1960. Presently it involves virtually every scientific field in one way or another.

Laser is an acronym for light amplification by stimulated emission of radiation. A laser is a source of coherent energy which may appear in the near and far infrared, long and short ultraviolet, and visible regions. The properties of lasers are similar to those for various regions of the electromagnetic spectrum, except that the laser normally achieves great power densities. A laser beam does not rapidly diverge since a laser operates at specific frequencies in the electromagnetic spectrum. It travels in one direction in straight lines.

There are four types of laser beams, classified by generating medium:

- Solid-state (the ruby crystal is the most common)
- Gaseous-state (the helium-neon is the most common)
- Semiconductor or junction
- Liquid-state, using organic dyes as a medium

Some lasers operate continuously while others operate in pulses, sometimes as short as 10^{-11} s. Output levels range from milliwatts to kilowatts for continuous operation and up to gigawatts in pulse operations.

The two main areas of potential health effects from laser radiation are the eye and the skin. Associated electrical, chemical, and explosive hazards are also possible. Knowledge of and continued interest in laser safety are necessary in developing applications of lasers.

Ocular Hazards

Protective eyewear should be worn whenever hazardous conditions may result from operation of a laser product. Laser radiation in the infrared region is highly absorbed at the surface of the cornea and could induce opacities in the cornea and destruction of the protective epithelial layer. In the ultraviolet region, exposure may result in extreme discomfort, but a moderate exposure is not thought to produce permanent damage. In the region of the spectrum from 320 to 1500 nm (the visible range is normally 380 to 760 nm), eye hazards are confined primarily to the retina and choroid. The most critical area for vision is the fovea. Thus, safety guidelines for laser radiation are designed to protect against foveal area damage. There is an often undetected hazard from diffused laser radiation in the visible region due to the ability of the eye to focus such radiation to a very small spot on the retina. If the laser beam is incident upon a diffuse surface, the illuminated diffuse surface can serve as a secondary extended source. In this case, the retinal power density is unaltered regardless of viewing angle and how far away the surface is from the viewer.

Skin Hazards

The effects of laser radiation on the skin may vary from mild reddening (erythema), to blistering and charring, depending on the amount of energy absorbed, the wavelengths of the radiation, skin pigmentation, individual sensitivity, and duration of exposure.

Electrical and Explosion Hazards

Live parts of circuits and components with peak open-circuit potentials over 42.5 V are considered hazardous, unless limited to less than 0.5 mA circuit components of combustible materials. For example, transformers are potential fire hazards unless individual noncombustible enclosures are provided.

In the event of a tube or lamp failure, components such as electrolytic capacitors may explode if subjected to voltages higher than their ratings. A misdirected laser beam, in this case, could steam the coolant of a high output laser system to trigger an explosion.

Chemical Hazards

A misdirected laser beam could decompose nearby chemical compounds to give off possibly toxic components.

Recently liquid lasing mediums (mostly complex organic dyes) have been gaining popularity due to the advantage of tunable frequency. However, many of these organic dyes are toxic themselves and, therefore, pose a potential exposure problem for laser handlers.

Control Measures

OSHA currently has no occupational health standards for laser products. Threshold limits established by the American Conference of Governmental Industrial Hygienists (ACGIH) are the recommended values to use. All laser products should meet specifications of the ***National Electrical Code***® (NFPA 70-1981),* Articles 300 and 400. Adherence to control procedures required under this standard should eliminate any significant probability of detrimental health effects from laser product operations. Therefore, an extensive, medical surveillance protocol for laser operators is not required under this standard.

BIBLIOGRAPHY

American Industrial Hygiene Association Journal, 66 South Miller Road, Akron, OH 44313, 1940–

American Journal of Public Health, American Public Health Association, 1015 Eighteenth Street, N.W., Washington, DC 20036, 1911–

Archives of Environmental Health, American Medical Association, 535 North Dearborn Street, Chicago, IL 60610, 1950.

Cember, H.: *Introduction to Health Physics.* Pergamon, New York, 1969.

Clarke, A. M.: *Ocular Hazards from Lasers and Other Optical Sources,* CRC Press, Boca Raton, Florida, 1970.

Clayton, F. E., and G. D. Clayton (eds.): *Patty's Industrial Hygiene and Toxicology: General Principles,* vol. I, 3d ed., 1978. *Toxicology,* vol. IIA, 3d ed., Wiley-Interscience, New York, 1981.

Cleary, S. F.: *The Biological Effects of Microwave and Radiofrequency Radiations,* CRC Press, Boca Raton, Florida, 1970.

Cralley, L. J., and L. V. Cralley (eds.): *Patty's Industrial Hygiene and Toxicology: Theory and Rationale of Industrial Hygiene Practice,* vol. III, Wiley-Interscience, New York, 1979.

****National Electrical Code*** is a trademark of the National Fire Protection Association, Quincy, MA 02269.

Cralley, L. V., G. D. Clayton, and J. A. Jurgill (eds.): *Industrial Environmental Health: The Worker and the Community*. Academic, New York, 1972.

Documentation of the Threshold Limit Values for Substances in Workroom Air, 4th ed, American Conference of Governmental Industrial Hygienists, Cincinnati 1980.

Dreisbach, R. H.: *Handbook of Poisoning: Diagnosis and Treatment,* 8th ed., Lange Medical Publications, Palo Alto, California.

Drinker, P., and T. Hatch: *Industrial Dust: Hygienic Significance, Measurement and Control,* 2d ed. McGraw-Hill, New York, 1954.

Engineering Manual for Control of In-Plant Environment in Foundries, American Foundrymen's Society, Des Plaines, Illinois, 1956.

Environmental Health Monitoring Manual, United States Steel Corporation, 1973.

Hamilton, A., and H. L. Hardy: *Industrial Toxicology,* 3d ed. Publishing Sciences Group, Acton, Massachusetts, 1974.

Hunter, D.: *The Diseases of Occupations,* 5th ed., Little, Brown, Boston, 1974.

The Industrial Environment, Its Evaluation and Control, 3d ed., National Institute for Occupational Safety and Health, Rockville, Maryland, 1973.

Industrial Ventilation, a Manual of Recommended Practice, 13th ed., Committee on Industrial Ventilation, American Conference of Governmental Industrial Hygienists, Lansing, Michigan, 1974.

Intersociety Committee: *Methods of Air Sampling and Analysis,* American Public Health Association, Washington, D.C., 1972.

Kinsman, S. (ed.): *Radiological Health Handbook.* U.S. Bureau of Radiological Health, Washington, D.C., 1970.

NIOSH Manual of Analytical Methods, National Institute for Occupational Safety and Health, Cincinnati, vol. I to V, 1979.

NIOSH Manual of Sampling Data Sheets, National Institute for Occupational Safety and Health, Cincinnati, 1974.

Hemeon, W. (ed.): *Plant and Process Ventilation,* 2d ed., Industrial Press, New York, 1963.

Plunkett, E. R.: *Handbook of Industrial Toxicology,* 2d ed., Chemical Publishing, New York, 1976.

Sax, N. I.: *Dangerous Properties of Industrial Materials,* 5th ed., Van Nostrand Reinhold, New York, 1979.

Norwood, W. D.: *Health Protection of Radiation Workers.* Charles C. Thomas, Springfield, Illinois, 1975.

Schwartz, L., L. Tulipen, and D. J. Birmingham (eds.) *Occupational Diseases of the Skin,* 3d ed., Lea and Febiger, Philadelphia, 1957.

Olishifski, J. B., and F. E. McElroy: *Fundamentals of Industrial Hygiene,* National Safety Council, Chicago, 1971.

Shreve, R. N. and J. Brink: *Chemical Process Industries,* 4th ed. McGraw-Hill, New York, 1977.

Wendholz, M. (ed.): *The Merck Index of Chemicals and Drugs,* 9th ed. Merck & Co., Rahway, N. J., 1976.

chapter 14-6

Sanitation Control and Housekeeping

by

Don Williams
Technical Service Director
Huntington Laboratories, Inc.
Huntington, Indiana

GLOSSARY

Acid A compound that gives a pH below 7, produces hydrogen ions in water, and reacts with or neutralizes alkalis. Most soils including oils, greases, and waxes are acids.

Alkali A compound that gives a pH between 7 and 14, produces hydroxide ions in water, and reacts with or neutralizes acids. Hard-water films, carbonates, potash, and phosphates are alkaline.

Deodorize Destroying, masking, or modifying foul and unpleasant odors. May be accomplished by killing bacteria that make foul odors.

Detergent Any product that cleans. Usually a synthetic detergent but may be a soap or even an abrasive material.

Disinfectant Product used on inanimate surfaces, which destroys microorganisms but not necessarily spores. Also called germicide.

Finish A protective coating used as a top coat.

Germicide See Disinfectant.

Hard floors Concrete, terrazzo, ceramic, quarry, slate, etc.

Polymer Usually a synthetic plastic used as a floor seal or finish. Examples are acrylic, styrene, and urethane.

Resilient floors Vinyl, vinyl asbestos, asphalt, rubber, linoleum, etc.

Sanitary Relating to health or to the preservation of or restoration of health and hygiene.

Sanitation Use of sanitary measures to clean and maintain a building.

Sanitize Reduce bacterial counts to safe levels as determined by health requirements.

Sealer A product used to prevent excessive absorption of finish coats into porous surfaces. An undercoat.

INTRODUCTION

Proper housekeeping procedures play an important role in the total plant operation. Custodial employees work with a great variety of chemicals and equipment and perform unique procedures. They work in a building that has taken considerable money to construct, and improper procedures can shorten the normal life of various parts of it. In a relatively few years, the owners will have *spent more money on housekeeping than the initial cost of the building and its furnishings.* Custodians should be well-trained to perform their tasks. Table 6-1 gives average cleaning times for various duties so that an estimate of efficiency can be determined. In addition to maintaining the appearance and prolonging the life of the building, good housekeeping provides immeasurable safety, hygienic, and sanitary benefits.

Safety is a key reason for good custodial procedures. The Occupational Safety and Health Administration of the U.S. Department of Labor states in its Standards and Interpretations, 1910.22(a) Housekeeping:

> (1) All places of employment, passageways, storerooms, and service rooms shall be kept clean and orderly and in a sanitary condition.
>
> (2) The floor of every workroom shall be maintained in a clean, and so far as possible, a dry condition. Where wet processes are used, drainage shall be maintained, and false floors, platforms, mats, or other dry standing places should be provided where practicable.
>
> (3) To facilitate cleaning, every floor, working place, and passageway shall be kept free from protruding nails, splinters, holes, or loose boards.

Other benefits from the results of a well-trained custodial work force are:

1. A clean, sanitary environment contributes to the health of all the personnel.
2. Training improves the morale and mental attitude of each worker.
3. Each worker can have a feeling of pride in a job well done.
4. A well-maintained building is easier to keep in shape, thus saving time and money.
5. A well-maintained building and a crew of employees that enjoy their work contribute to goodwill and public relations.
6. The life of the various parts of the building and the building itself can be greatly extended by proper housekeeping procedures.

TRAINING TIPS

Some key tips to use in training are:

1. The longer a soil, stain, or coating is allowed to remain on the surface, the harder it will be to remove.
2. Quality products made by reputable manufacturers are usually best in the long run.

TABLE 6-1 Cleaning Operation Time Estimates

Floor operations	1000 ft², min
Sweeping	
Halls & corridors	10
General rooms	20
Mopping	
Dust mop (unobstructed)	5
Dust mop (obstructed)	10
Damp mop (unobstructed)	16
Damp mop (obstructed)	30
Wet mop and rinse (unobstructed)	30
Wet mop and rinse (obstructed)	50
Scrubbing	
Hand scrub 12″ brush	300
Deck scrub	100
Machine scrub	
Machine scrub 12″ diam	48
Machine scrub 14″ diam	40
Machine scrub 16″ diam	36
Machine scrub 18″ diam	31.5
Machine scrub 19″ diam	30
Machine scrub 21″ diam	27
Machine scrub 23″ diam	25
Machine scrub 24″ diam	24
Machine scrub 32″ diam	18
Machine scrub 36″ diam	16
Automatic scrub machine (24″)	6
Vacuum pickup	
Vacuum pickup (unobstructed)	20
Vacuum pickup (obstructed)	30
Wax	
Wax	30
Machine polish (19″ machine)	15
Rectangular machine (48″ plate)	3.5
Buff with steel wool	20
Strip and rewax (1 operator)	150
Dry strip and rewax (1 operator)	120
Spray buffing	
Spray buffing (unobstructed)	30
Spray buffing (obstructed)	45
Carpeting	
Vacuuming (unobstructed)	20
Vacuuming (obstructed)	30
Spot vacuuming	16
Shampoo (dry-foam)	60
Pile lift	30

Furniture and fixtures operations	per unit, min
Dusting	
Air conditioners	0.30
Ashtrays (desk)	0.25
Bookcases (3-tier sect.)	0.30
Chairs	0.30
Cigarette stands	0.40
Couch	0.25
Desks	0.80
Desk trays	0.15
File cabinets (4 drawer)	0.40
Lockers	0.20
Radiators	0.30
Tables (medium)	0.50
Telephones	0.15
Towel dispensers	0.12
Towel disposal cans	0.40
Typewriter & stand	0.50
Wash basin (office)	0.60
Wastebasket	0.50
Window sill	0.20
Venetian blinds std. size	3.50

Washrooms	1000 ft², min
Cleaning	
Cleaning commode	3.83
Door (spot wash both sides)	0.83
Mirrors	0.66
Sanitary napkin dispenser	0.16
Urinals	3.00
Washbasin—soap dispenser	3.00
General washroom cleaning	
General cleaning per 1000 ft²	120.00

Miscellaneous operations	1000 ft², min
Wall washing	
Painted walls (manual)	240
Painted walls (machine)	150
Marble walls (manual)	92
Ceiling washing	
Ceiling washing (manual)	300
Ceiling washing (machine)	180
Window washing	
Single pane	125
Multipane	170
Frosted single pane	190
Opaque glass	50
Plate glass	35
Office partitions (glass)	110

TABLE 6-1 Cleaning Operation Time Estimates (*Continued*)

Miscellaneous operations	per unit, min	Miscellaneous operations	per unit, min
Dusting lamps & light fixtures		Whisk or vacuum armchair	1
Wall fluorescent fixtures	0.13	Whisk or vacuum couch	2
Desk fluorescent lamp	0.30	Shampooing armless chair	4
Table lamp & shade	0.58	Shampooing armchair	7
Floor lamp & shade	0.58	Shampooing couch	20
Washing fluorescent light fixtures		Stairway cleaning	
Ceiling fixture (eggcrate) 4′ ea.	9	Sweep and dust 1 flight 15 steps	6
Ceiling fixture (eggcrate) 8′ ea.	12	Damp mop 1 flight 15 steps	5
		Scrubbing (hand)	20
Fabric upholstery cleaning			
Whisk or vacuum armless chair	0.50		

*Reprinted with permission from International Sanitary Supply Association. These cleaning time estimates represent average cleaning times. Layout, obstacles, maintenance level desired, environmental conditions, etc., will affect cleaning time.

3. Chemicals should be measured and applied properly when required.
4. Two thin coats of a finish are better than one thick coat.
5. Always follow directions on labels or given in training, for safety as well as effectiveness.
6. Do not depend upon perfumed products to cover up odors. Clean thoroughly and kill germs with disinfectants or sanitizers and there will not be putrefaction odors.
7. Surfaces vary considerably and aging often changes them.
8. Use products and procedures that will do the best job and that are made for that particular surface.
9. Clean and maintain equipment in proper working order.
10. Learn how to use equipment to the best advantage.
11. Use products with the best blend of properties to do the job intended.
12. Read supplier directions, literature, trade books, and magazines for generating ideas and becoming aware of new products.
13. Talk with fellow custodians about common problems and situations.
14. Attend local, regional, or state training programs where possible.
15. Practice cooperating with other departments, and they will cooperate with you.

PLANT DESIGN

The actual design of the plant can play a key role in contributing to simpler maintenance. Too often, those responsible for maintaining the building are never consulted during the early planning of the building design. If given a chance, many experienced custodians can provide some good ideas in planning or rearranging plant areas. Following are some basic suggestions:

1. Custodial service closets should be provided on every floor and be centrally located to save walking time.
2. The closet should have sufficient shelving, floor space, a floor-level sink, hard sur-

face floors and wall racks for mops and brooms, and other facilities that fit the needs.

3. Keys for doors, cabinets, machines, etc., should be planned and organized to make locking and unlocking by the custodians quick and easy as well as secure.
4. Access through one room, closet, or office to another should not be permitted.
5. Restrooms should be planned to make use and cleaning easy. (Wall-hung fixtures, placement of fixtures for sequential use, glazed walls, ceramic floors, and floor drains are good choices.)
6. Kick plates and push plates can help the appearance and cleanability of the custodial closet, restroom, and other doors.
7. Wall and furniture surfaces should be easily washable.
8. Entrance mats should be either recessed or integrated so that tracked-in soils are reduced.
9. Light-colored carpeting should not be laid where soiling will show quickly.
10. OSHA states in Standards and Interpretations 1910.22(b) aisles and passageways:

> (1) Where mechanical handling equipment is used, sufficient safe clearances shall be allowed for aisles, at loading docks, through doorways and wherever turns or passage must be made. Aisles and passageways shall be kept clear and in good repair, with no obstruction across or in aisles that could create a hazard.
>
> (2) Permanent aisles and passageways shall be appropriately marked.

EQUIPMENT

Cleaning equipment may vary somewhat according to type of surfaces, size of area, and other factors. Though it is tempting to order large versions of all equipment to apparently save time in covering large surface areas, there are many occasions when the larger equipment version will be too clumsy, heavy, and hard to adapt to the inevitable smaller areas. When it is impractical to own a variety of sizes to fit every need, consider a compromise size. Many units have attachments to broaden their usage in various areas; they may combine two or more functions and may be of the rider type. Details and demonstrations should always be obtained from equipment suppliers prior to commitment. All equipment should be cleaned after each use and kept in working order. Standardization of purchases can make training easier, requires stocking of fewer replacement parts, and makes possible larger quantity discounts. Following are some of the basic items:

1. Rotary *floor machines* for use in scrubbing, stripping, scouring, and polishing resilient and hard-surface floors as well as shampooing carpets. A 20-in model is most efficient.
2. Wet or dry *vacuums* are available in a number of different varieties. They may use a wand or squeegee, be lightweight backpack models, be extra quiet, contain filters, and be of various sizes or suction powers. Sweeper versions may be of the rider type.
3. *Automatic scrubber vacuums* to scrub soils and pick up the scrub solution.
4. Pressure sprayers for fast and effective cleaning.
5. Wall-washing equipment with many variations.
6. *Carpet extractors* for cleaning and rinsing on location.
7. *Compactors* for trash disposal.
8. Mops, buckets, wringers, carts, pads, sprayers, brushes, dispensers, and many other miscellaneous items.

CHEMICALS

Understanding some of the basic properties of the products used for sanitation should become one of the most important goals of maintenance personnel. Each product will

TABLE 6-2 Solving Floor-Finish Problems*

SOLUTIONS	Scratching	Black marking	Dirt pickup	Furniture sticking	Tackiness	Discoloring of light floor	Poor initial gloss	Slipperiness	Streaking, uneven, or dull film	Unpleasant odor	Powdering or dusting	Marring, smearing, or scuffing	Poor water resistance	Poor detergent resistance
	PROBLEMS													
Do not pour used coating back into container					*	*	*		*	*		*	*	*
Apply more coats or use sealer		*					*							
Apply thin, continuous films			*	*	*				*		*	*		
Remove all previous coating or factory finish and recoat	*	*	*	*	*	*	*	*	*		*	*	*	*
Check freezing damage	*	*	*				*		*		*	*	*	*
Use fine or medium scrubbing pads rather than coarse pads	*						*		*					
Allow adequate drying time between coats				*	*		*		*		*			
Wipe, do not rub while applying							*		*					
False impression—check natural color of flooring						*								
Use clean equipment and applicator						*			*				*	*
Remove excessive buildup on surface			*			*					*			
Remove dust and loose soil regularly and use carpet at entranceways	*							*			*			
Spray clean and buff or recoat	*	*					*							
Allow adequate drying time before resetting furniture—ventilate				*	*									
Avoid applying in hot and humid weather—ventilate well				*	*								*	
Avoid or remove soap or oil films on floor					*		*	*	*			*		
Floor too cold—apply closer to room temperature, 70° (21°C)	*								*		*			
Change to metal interlocked polymer														*
Buff dry stripped floor before applying coating							*		*					
Do not apply polymer finish over a wax	*							*			*			
Change to hard polymer finish		*	*									*		
Unstable plasticizer in vinyl tile—allow curing time		*	*		*									

Source Huntington Laboratories, Inc., Huntington, Indiana.

*Most waxes and polymer finishes are quality formulations, made of the finest raw materials, under most careful laboratory control. Yet, many outside factors may cause the finest product to do a poor job. This chart explains some of the problems encountered and their solutions.

perform certain tasks if used properly but have the potential to harm surfaces or personnel if improperly applied. Use only where and in the manner recommended. Observe the cautions on labels to preclude dangerous situations. Use the mildest abrasive, coupled with the mildest chemical possible, to complete the task and protect the surface.

A cleaning product may be either all-purpose for a variety of tasks or highly specialized. Mild *alkaline cleaners* will remove oily, greasy soils from any surface not harmed by water alone. Special alkaline cleaners are recommended for specific tasks such as strippping of floor coatings, window washing, carpet cleaning, wall washing, etc. Strong alkaline cleaners and degreasers are for heavy industrial soil buildups. *Acid cleaners* remove lime and hard-water deposits or rust stains from hard surfaces such as toilet bowls, metals, ceramics, and concrete. *Solvents* are generally used solely for spot removers in housekeeping.

Abrasive or mechanical action of some sort is usually necessary for the best cleaning procedures. Scouring powders, abrasive hand pads, abrasive floor pads, brushes, pressure sprayers, and simple "elbow grease" rubbing are supplements to the chemicals.

Germicides combined in certain cleaners can help control harmful microorganisms and unpleasant odors. This results in a healthful as well as a pleasant environment.

Floor coatings are of two types, sealer or finish. Sealers are generally designed for porous surfaces to smooth them out and make them easier to maintain. Concrete and terrazzo sealers stop dusting and may be used to provide a good base for a finish. Wood sealers bring out the natural beauty of wood, prevent penetration of stains, and provide gloss, nonslip, and other desirable properties. Resilient-tile sealers are mainly for the porous tile and provide a good base for a finish. Most sealers are solvent-based and very hard to remove, often requiring special removers. Note that some products may perform both as a sealer and finish.

Finishes generally are water-based and easily removed with a detergent stripping solution. The first finishes were made of natural waxes that required much dry buffing to retain a smooth, glossy appearance. The relatively recent chemical revolution has brought about many new coatings. At first synthetic *waxes* replaced the natural waxes, but more recently hard *polymers* have prevailed. These polymers dry hard and glossy without buffing, require less maintenance, and perform best on the light floors in use today. Table 6-2 outlines specific solutions to floor finish problems.

REFERENCES

Edwards, J. K. P.: *Floors and Their Maintenance*, Butterworths, London, 1972.

Feldman, Edwin B.: *Building Design for Maintainability*, McGraw-Hill, New York, 1975.

Feldman, Edwin B.: *Housekeeping Handbook for Institutions, Business and Industry*, Frederick Fell Publishers, New York, 1978.

General Industry Standards and Interpretations, U.S. Department of Labor, Occupational Safety and Health Administration, 1977.

Meyers, Earl M.: "Standardized Housekeeping Program Improves Utilization of Manpower and Materials," *Maintenance Engineering*, October 1972.

Ruhlin, Robert R.: "Work Control: A Sure Way to Improve Building Maintenance," *Buildings*, February 1974.

Sack, Thomas F.: *A Complete Guide to Building and Plant Maintenance*, Prentice-Hall, Englewood Cliffs, N.J., 1971.

Sipes, Sherrill F., Jr.: "Plan to Manage Your Maintenance Program," *Buildings*, May 1971.

section 15

Energy-Conservation Techniques

by

Richard Ryan, P.E.

Manager, Plant Engineering
Hamilton Standard Division
United Technologies Corp.
Windsor Locks, Connecticut

INTRODUCTION

Volumes have been written regarding energy-conservation techniques, and many key technical procedures are outlined in other, appropriate sections of this handbook. This section describes important elements which, when integrated with the other information, will provide a total energy conservation program. Assuming that all the simple and obvious steps have already been taken to eliminate waste, this section deals with the importance of planning, reducing distribution losses, and optimizing plant operations through an energy management system (EMS).

ENERGY-CONSERVATION STEPS

There are three logical steps in the industrial energy-conservation program. First, take out the "fat." That is, implement simple programs to reduce waste. This can yield 10 percent or more savings without significant investment. The second step is to engineer the plant for greater efficiency. This encompasses programs such as adding insulation to steam and hot-water pipes, improving maintenance on steam traps, and improving powerhouse efficiency. This can give us another 10 to 20 percent with nominal investment. When all this has been done, we logically come to the third step, optimizing the entire plant operation. In the majority of plants, optimization can best be accomplished with a computerized EMS.

THE MOST IMPORTANT STEP—THE ENERGY PLAN

After the easy steps toward energy conservation have been completed, the next steps are to re-engineer the facility and optimize its use. First, let us use the best management tool: developing and formalizing our action plan.

Scope

The plan should encompass all the facilities of the company and should segregate information logically by location, type of business, or other classification suitable to the business. The plan should cover at least a 5-year period to give it sufficient visibility and to ensure that correct judgments are made in determining major capital funding. The 5-year period is suitable for dealing with existing situations without looking to the technology of the 21st century to solve today's problems of energy supply, demand, and pricing.

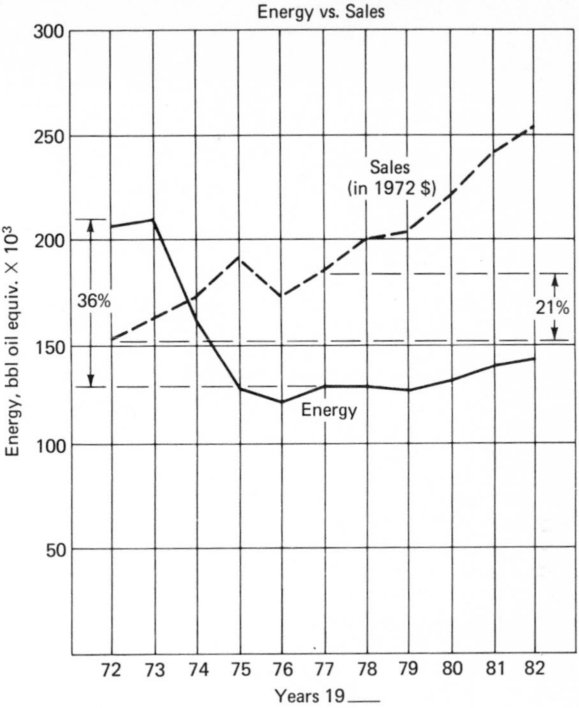

Figure 1-1 Oil use vs. sales in constant dollars.

Inputs necessary to develop this plan include forecasts of sales, production volume, staff, and facility usage. This information is needed to determine the driving forces behind energy consumption. Initially, it may be assumed that an increase in sales or production volume indicates an increase in energy use at the same percentage. When using dollars as an index, it is important to realize that the projected increase may be distorted by inflation. To avoid this, a base dollar value is established. One such base that might be used would be equivalent barrels of oil vs. sales held in consistent dollars such as shown in Fig. 1-1. For fuel-oil consumption, a similar index choice could be degree-days if only comfort heating were involved. Even many homes would find this inappropriate, however, since hot-water heating may be an integral part of the system. In the latter case, a dual index could be used, but the important thing is that some index appropriate to the business be chosen and monitored.

At Hamilton Standard, British thermal units-(Btu) per sales dollar (1972) is used as an overall index of energy-conservation performance. The base year 1972 is used because the former Federal Energy Administration and other generally accepted agencies use 1972 as a base year because of the oil embargo. Current and future sales projections are extended back to 1972 to be consistent; however, updated plans use 1977 as a new base year for improved accuracy of adjusting base indices.

Figure 1-2 shows our plan through 1982 compared with the results of the previous 5 years.

Analysis of Sources

Begin the survey of operations by starting with energy suppliers. Most manufacturing plants use fuel oil or natural gas, electricity, and other forms of energy. Each is, of course, subject to cost escalation, curtailment, or even a breakdown of normal supply. The anal-

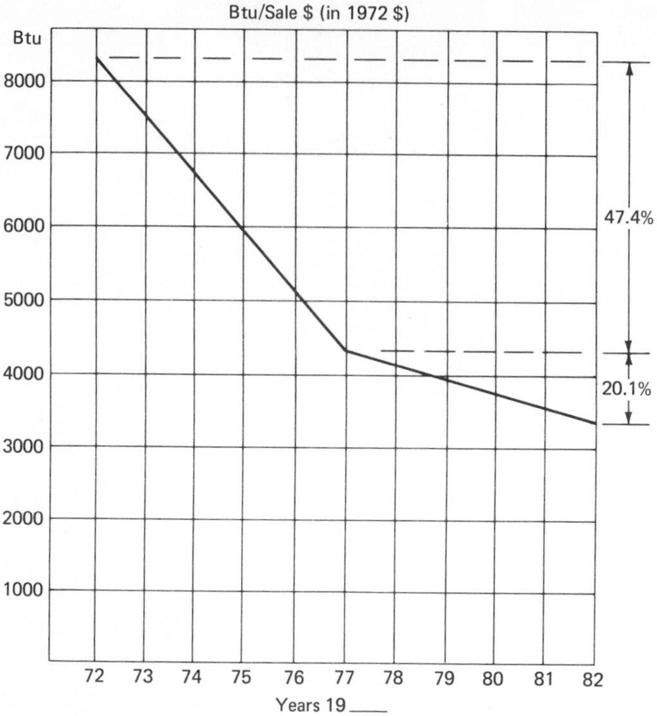

Figure 1-2 Projections vs. past results.

ysis should result in an understanding of the following aspects of the supply system: transportation, cost, pricing formula, and quantities used.

The information about energy sources should be available on a Btu basis and an energy-cost basis. Costs reflect current importance, while Btu's are important in planning the long-range strategy. At Hamilton Standard two-thirds of the British thermal units used are derived from oil, while two-thirds of the cost is related to electric power purchased. See Figs. 1-3 and 1-4.

The Btu-use information is important in that a sudden curtailment or increase in the price of fuel oil might immediately affect the strategy or require the use of alternative sources, e.g., electric heat. An important part of the analysis of the sources is the reliance on management procedures relative to storage and supply.

In an area where brownouts have occurred or are likely to occur, there should be a brownout procedure available for immediate use. If fuel oil is delivered by truck, it is necessary both to have an adequate supply on hand and to develop an alternative means of delivery should there be a trucker's strike.

For purposes of evaluation, it is important to calculate energy by means of a common denominator, i.e., British thermal units or barrels of oil equivalent (if oil is used). Table 1-1, "Energy-Conversion Factors," outlines the approximate Btu value per unit of conventional fuel.

Energy Uses

Step two in the development of an energy plan is to identify the energy used. Just as the comptroller establishes a budget for the amount of dollars available, the energy manager should take steps to budget the amount of energy that can be afforded. Determining

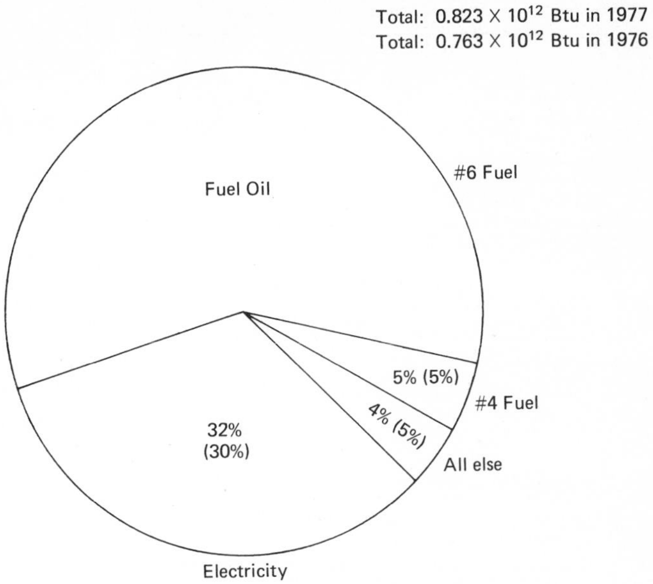

Figure 1-3 Energy sources, Btu basis.

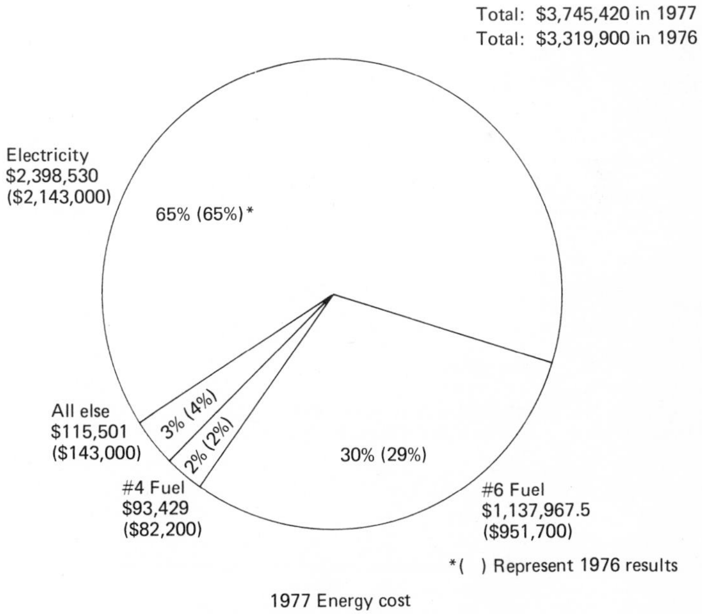

Figure 1-4 Energy sources, cost basis.

TABLE 1-1 Energy-Conversion Factors

Fuels	Btu per unit
Electricity	3,413/kWh
Natural gas	$1{,}100/ft^3$
Propane	91,600/gal
Gasoline	125,000/gal
Kerosene	135,000/gal
Distillate oil #2 (including diesel)	140,000/gal
Residual oil #5 and #6 (bunker C)	150,000/gal
Coal	25,000,000/ton
Coke	26,000,000/ton

where it is used is much more difficult than reviewing the bills to find out how much was used and where it was obtained, but it is much more important. It is necessary to break down the uses so that each segment can be thoroughly studied. The breakdown also determines which areas offer the best opportunities for savings, since it clearly outlines major uses and costs. Figures 1-5 and 1-6 show a sample analysis of energy usage. Categories may vary from plant to plant by location or type of energy, but it is important to identify them as closely as possible.

A simple approach such as reading the main meters at regular intervals could define an electric usage pattern in a particular plant. Follow that up with sampling current or wattage readings at specific branches. Calculating some known relatively constant quantity, such as the kilowatthours of lighting used, can lead to the usage of kilowatthours and power demand for a given area.

Analysis of Energy Use by Area

When the major functional areas of energy use have been established, the compilation will show where to put resources at the earliest possible time. When examining the use

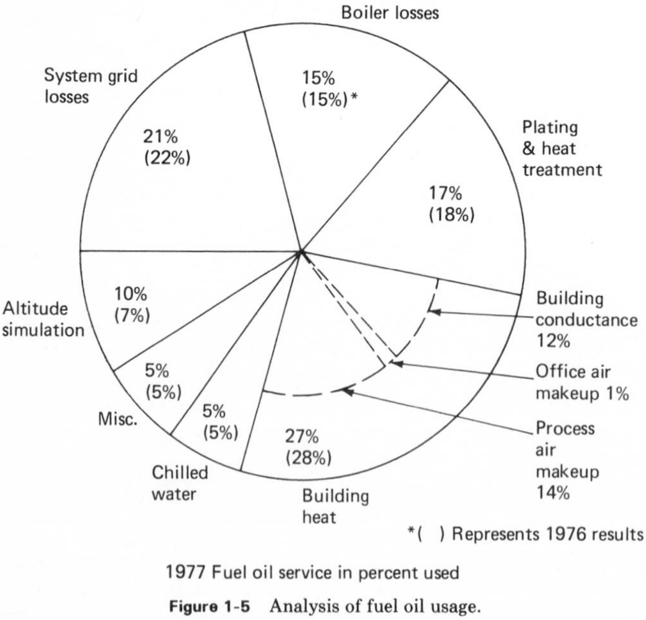

Figure 1-5 Analysis of fuel oil usage.

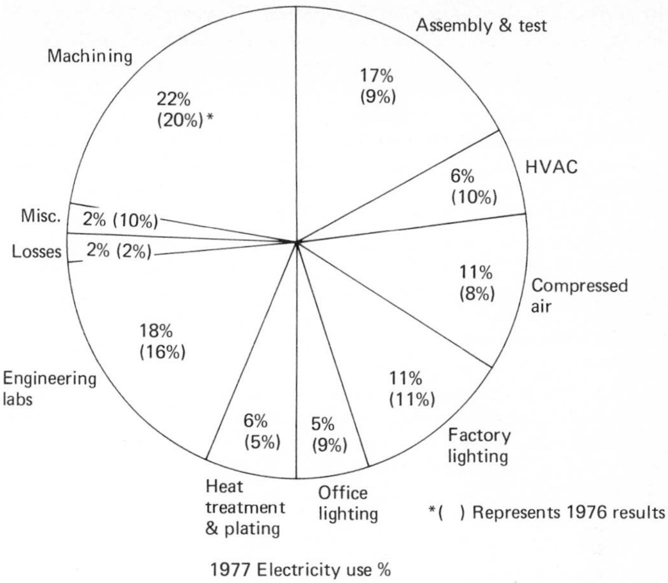

Figure 1-6 Analysis of electricity usage.

charts for fuel oil, many people are surprised to discover that a major use of fuel is actual fuel losses!

The primary losses in boilers are usually due to combustion inefficiency. Other typical losses include those in distribution pipes, both insulated and uninsulated, as well as heat loss within equipment to which steam or oil is supplied. Be sure to categorize all losses and correct them on a step-by-step basis. For example, an effective survey and modernization of the steam-trap system will probably show the need for changing many trap sizes, and for the establishment of a more effective steam-trap maintenance program.

REDUCING HEAT-DISTRIBUTION LOSSES

Distribution losses basically fall into two categories, losses through the piping insulation and steam loss to equipment. Piping losses can be reduced through more insulation or by eliminating losses needlessly occurring because of unnecessary piping in the overhead. At Hamilton Standard significant reductions were achieved in steam losses by examining distribution needs and closing off portions of the distribution where absolutely not needed. In some cases it required installation of small steam boilers to prevent steam running through miles of distribution piping to get to a needed process. Overlooked by many casual inspectors, elimination of unnecessary steam piping is a big savings that can have a large payback. If a steam line is indeed needed, examine the possibility of improved maintenance or of lowering operating pressure.

One of the major actions that can be taken to reduce losses through steam lines is to deliberately shut down the powerhouse on weekends. Traditional practice is to avoid deliberate shutdowns because of possible occurrence of leaks. At Hamilton Standard this did not occur, and the shutdown saved about 5000 gal of fuel oil each day of shutdown. Also, pressure systems were advantageously rearranged. In one case, where conservation significantly reduced needs, the low-pressure (12 lb/in^2) distribution system used for heating the building was removed and building heating was switched to another source. Process use was serviced by high-pressure steam (175 lb), and heating requirements were supplied by adding a low-pressure reducing station. This eliminated over 1000 ft of 8-in low-pressure steam piping and afforded a saving of over $6600 annually in fuel costs (at 1974 rates). Moreover, the old pressure line was salvaged to provide compressed air to a production test facility in the building. This conversion took place in 1980, and the salvage avoided the cost of a new compressed-air line which was estimated at $257,000. Again, a good energy-conservation program can result in significant cost savings in many categories if a company takes advantage of all the rearrangement opportunities that analysis can yield.

Boiler controls, in water boilers especially, still use technology developed when fuel oil cost $2.00 a barrel. Today's controls are much more sophisticated, and since the cost of a barrel of oil has almost doubled every year, it pays to examine automation with several vendors. The boiler blowdown cycle is a good candidate for automation. Before installation of boiler blowdown at Hamilton Standard, operators tested for dissolved solids and chemical residuals in the conventional way and blew down as they thought was required—dumping out excessive hot water and steam. The automatic blowdown device continuously monitors the conductivity of the blowdown and controls the blowdown valve automatically. The annual savings for automatic blowdown was calculated at about 100 barrels of oil, less than the cost of the device. Installing viscosity control equipment to keep the fuel at the correct atomization point and oxygen measurement equipment to keep excess air to a minimum also are useful investments to consider. Payback for most plants would easily be within one year.

ESTABLISHING A COMPLETE STEAM TRAP PROGRAM

Definitions of Terms Related to Steam Use

Saturated Steam

Saturated steam is steam at the temperature corresponding to the boiling temperature of water at the pressure in the boiler.

Absolute Pressure

Absolute pressure is the pressure, in pounds per square inch, above a perfect vacuum (psia).

Gage Pressure

Gage pressure is pressure, in pounds per square inch, above atmospheric pressure, which is 14.7 lb/in^2 (100 kN/m^2) absolute at sea level.

Heat of Vaporization

Heat of vaporization is the amount of heat required to change 1 lb (1 kg) of boiling water to 1 lb (1 kg) of steam; it is equal to the same amount of heat when the steam is condensed back into a pound of water.

Condensate

Condensate is the by-product of a steam system formed because of unavoidable radiation in heating and process equipment during heat transferal to the substance being heated. Steam upon condensing gives up its latent heat, and the hot condensate must be removed immediately; however, to conserve the heat still remaining, the condensate should be returned to the boiler.

Steam Traps

One of the biggest culprits causing loss in steam systems is the common steam trap found in every large steam system. A steam trap is an automatic device intended to hold the steam in a piece of equipment until it condenses and gives up its heat of vaporization.

The condensed steam, or condensate, is then passed out of the system using the pressure of the live steam. *Proper operation of the steam trap is extremely important in order to control cost and ensure production.* Primary concerns are long-term economies, through control of energy waste, and the maintenance of the steam traps. A good steam-trap maintenance program is essential to the life-cycle costs of the steam system. However, the selection of type is also important. Many plants will have several different types and brands of steam traps, depending upon the date of installation, engineer, and the availability of the items. Installed during low fuel cost, their selection was not as critical then as it is now in our energy-short society.

Three types of steam traps are discussed in this section: the inverted bucket trap, the thermostatic trap, and the float and thermostatic trap. Note that in many plants disk or impulse traps may also be found. These should be removed from service since they are considered unreliable and wasteful of steam.

Bucket Traps

Bucket traps have the longest service life—about 42 months—compared with a disk trap which averages fewer than 18 months. Traps need service when they start to pass live steam. At an operating condition of 150 psig (1000 kN/m^2 gage) inlet and 0 psig discharge, an average disk trap will lose about 80 lb (36 kg) of steam per hour after two years of service, where a bucket trap under the same conditions of service would lose only 2 lb (1 kg) steam per hour. At a cost of $1.00 for 100,000 Btu, this would mean a disk trap would waste almost $6000 in steam per year. This shows reasons why life-cycle costs must be considered in the design of an energy-using system. Bucket traps are designed to have good resistance to water hammer and are recommended for either high- or low-pressure steam applications. They have intermittent discharge and, because of the sudden surge of condensate, can handle dirt better than other types of traps. A high back pressure on a bucket trap only reduces its capacity.

Float and Thermostatic Traps

In comparison with other types, float and thermostatic traps offer superior air-handling ability because they provide a continuous discharge. Float and thermostatic traps are typically used for unit heaters handling outside air and are the first choice for applications with modulating control valves. Float and thermostatic traps are best suited for low-pressure applications, with high-pressure applications reverting to the bucket traps.

Thermostatic Traps

Thermostatic traps are used where potential freeze-ups are a problem, e.g., in tracer lines. Since they are controlled by bimetallic elements, these traps offer good resistance to water hammer and may be used for both high- or low-pressure applications. The main characteristic of these traps is that they back up condensate and therefore require a few feet of uninsulated piping upstream of the trap to minimize condensate backup. Note that a thermostatic trap is the only trap requiring an uninsulated cooling leg.

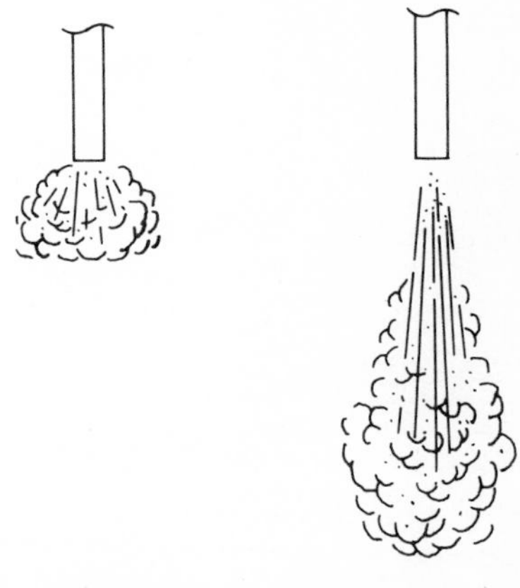

Figure 1-7 Flash and live steam.

Testing Steam Traps

Steam traps should be tested at least once each year. The test procedure varies for each type of trap. *All traps must be tested while in use.* For example, when testing a trap serving a plating-tank heating coil, be sure that steam is flowing at the time of the test. All disk traps should be replaced by appropriate bucket, float and thermostatic, or thermostatic traps.

Before traps can be effectively tested, the

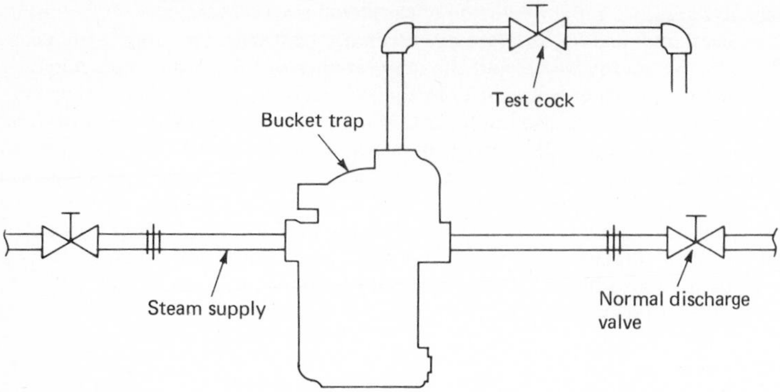

Figure 1-8 Bucket-trap test.

difference between live and flash steam must be understood. Flash steam is a slow, lazy vapor. Live steam has a bluish tinge and high velocity. Flash steam can be eliminated entirely by pouring a cup of water on the discharge pipe, while live steam cannot. See Fig. 1-7.

It is this observation for live-vs.-flash steam in the trap discharge that is the most effective test to see whether or not the trap is operating properly for our inverted bucket traps, thermostatic traps, and float and thermostatic traps. Tests are presented in outline form as follows.

A. Bucket Trap Tests Visual or sound tests are excellent.

1. **Visual Test** A test cock shall be installed as shown in Fig. 1-8. Open the test cock and close the formal discharge valve. Observe for flash or live steam.
2. **Sound Test** Use the ultrasonic tester. Shut off steam inlet valve and condensate return valve. Zero out background noises. Open both valves and listen for operation of the trap. The sounds of the bucket opening and closing should be distinct, as well as the sound of steam and air bubbling up through the trap vent hole.
3. **Temperature Test** Measuring the temperatures at the inlet to and discharge from the trap is unreliable. This test is not an acceptable indicator for bucket traps.

B. Float and Thermostatic Trap

1. **Visual Test** A visual test is the first choice. A test cock shall be installed as shown in Fig. 1-9. Open the test cock and close the normal discharge valve. Observe for flash or live steam.

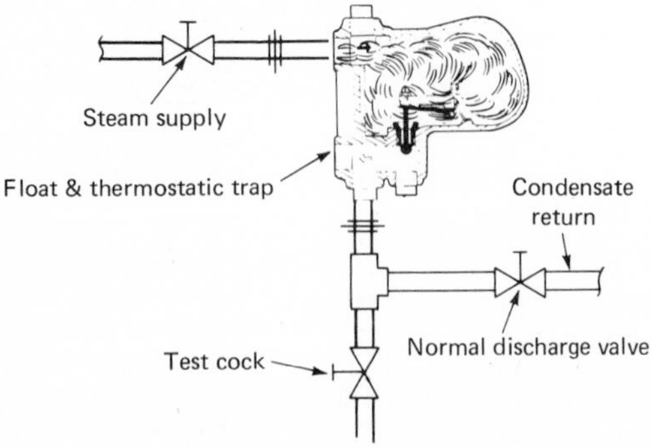

Figure 1-9 Float and thermostatic-trap test.

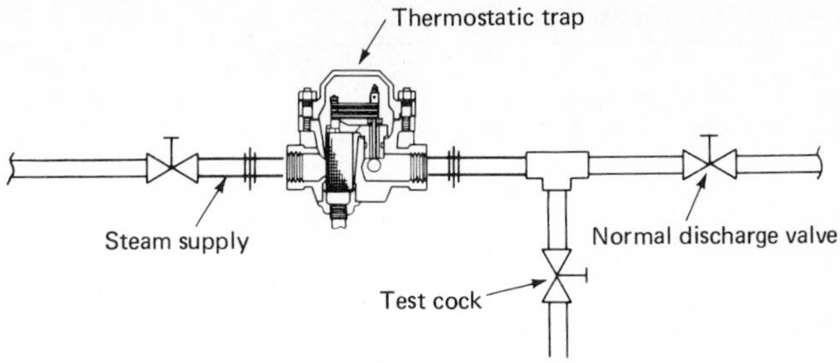

Figure 1-10 Thermostatic-trap test.

2. **Sound Test** Sound test is the second choice. Use the ultrasonic tester. Shut off the steam inlet valve and condensate discharge valve. Zero out background noises. Open both valves and listen for a continuous rumble.
3. **Temperature Test** Measuring the inlet and discharge temperatures is unreliable and is not acceptable.

C. **Thermostatic Trap**

1. **Visual Test** A visual test is the first choice. Open the test cock and close the normal discharge valve. Observe for flash or live steam (Fig. 1-10).
2. **Sound Test** Use the ultrasonic tester. Listen to the trap cycle open and shut as it discharges condensate.
3. **Temperature Test** The discharge side of the trap should be about 15°F (8°C) cooler than the steam side. The temperature test is reasonable for thermostatic traps, but a visual test is preferred.

UNDERSTANDING THE ELECTRIC BILL

The demand-factored electric bill (Fig. 1-11) is the most complex energy cost to understand. Since electric energy is available for everyone to use, generally without central control, we must understand the cost elements. The terms are:

Demand. The highest average kilowatt use measured over a short interval of time, usually ¼ or ½ h; measured in kilowatts.

013 919 00150 9 STATEMENT FEBRUARY 1, 1981

SPECIAL PROVISIONS:
NO. 1, 2

HIGH E-DEM LAST 11 MON 17414.1 JUN
$36,647.98 D $195,048.25 E $234.75 R

FUEL ADJUSTMENT

JAN

FROM	TO	READING	RATE	CON-STANT	DEMAND	REACTIVE DEMAND	KILOWATT HRS. 100 CUBIC FEET	
DEC 28	JAN 29	1703	35	2400	15877.6	8877.8	7500000–E	314,533.73

7350000 NET KWH X$0.005210 = $38,293.50 ELEC FUEL ADJ

Bills not paid in 28 days of statement date bear monthly interest of 1% on unpaid balance.

276,260.23

TOTAL $314,553.73

PLEASE KEEP THIS PORTION FOR YOUR RECORDS

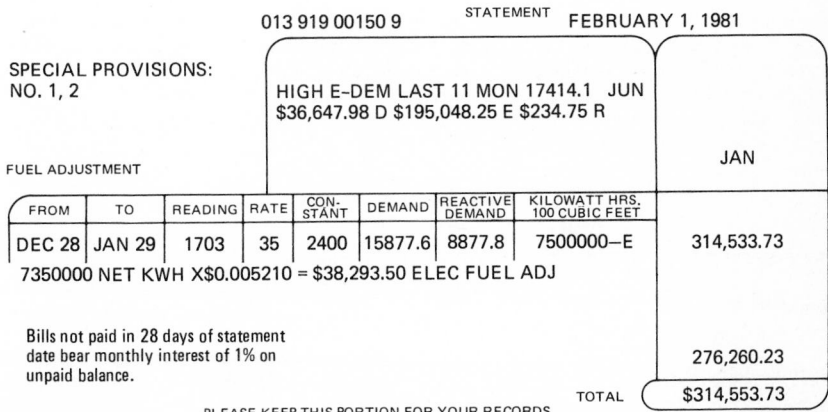

Figure 1-11 Demand-factored electric bill.

TABLE 1-2 Typical Monthly Demand Rate

Demand charge	
First 50 kW of demand or less	\$234.00
Next 150 kW of demand	3.17 per kW
All over 200 kW of demand	2.40 per kW
Reactive demand charge	
All reactive kVA in excess of 50% of the demand	25¢ per kVA
Energy charge	
First 200 times the demand	2.94¢ per kWh
Next 100 times the demand	2.55¢ per kWh
Next 100 times the demand	2.42¢ per kWh
All over 400 times the demand	2.28¢ per kWh

Energy. The total kilowatthour use for the billing period—usually 30 days; measured in kilowatthours.

Reactive Charge or Demand. A penalty imposed by some utilities when power factor is poor. Separate metering is used to measure reactive voltamperes.

Fuel-Oil Adjustment. A surcharge usually applied as a multiplier to the kilowatthour (energy) cost to reflect changes in fuel cost from a pre-established base.

Typically, large customers have a stepped billing schedule where the demand is factored into the energy charge, in such a way that a high demand causes a major penalty (Table 1-2).

Using the simple rate shown with a demand of 15,877 kW and a total consumption of 7,500,000 kWh as shown on the bill, the demand rate is calculated as follows:

$$\begin{aligned}\text{Demand charge} &= \$234.00 + 150(\$3.17) + (15{,}877 - 200)\$2.40 \\ &= \$38{,}334.30\end{aligned}$$

$$\begin{aligned}\text{Energy charge} &= 200(15{,}877)(0.0294) + 100(15{,}877)(0.0255) + 100(15{,}877) \\ &\quad + (7{,}500{,}000 - 400(15{,}877)(0.0228) \\ &= \$198{,}467.21.\end{aligned}$$

More specifically, the equation can be reduced to:

$$\text{Energy cost} = 1.73D + 0.022344E$$

where D = demand kilowatts and E = total kilowatthours

Thus, a high demand has a hidden effect on the energy portion of the electric bill. The fuel adjustment is typically a multiplier applied to the energy portion of the bill, and thus is also affected by the demand. When considering the impact of energy savings, it is also important to know if the demand is affected, e.g., the shifting of lighting, or whether energy only (off-peak use) is applicable. Then the appropriate billing element can be used. Since off-peak costs are considerably less than peak costs, steps should be taken to control peak demand. This can be done through a single control operating a single device or, in the case of a large, diversified plant, by means of an EMS described later in this section. Where known elements of the energy use cannot be shifted off-peak, such as some lighting requirements, an attempt should be made to improve their efficiency.

IMPROVING THE EFFICIENCY OF THE LIGHTING SYSTEM

Lighting is an item common in all building energy use. The impact of lighting has an effect not only on the energy charge, but also on the electric demand charge. Indirectly it also has an effect on the air-conditioning systems. Today it is possible to practice energy conservation—and maintain safety, security, and employee morale—by using more efficient light sources.

Lamp Terminology

An adequate description of a lighting system should begin with a discussion of general information pertinent to lighting design and evaluation. (A detailed treatment of this subject is found in Chap. 3-5.) The terms commonly used are defined as follows:

1. *Efficiency* is defined as the light output (lumens) per unit of electric input and the measured unit is lumens per watt.
2. *Life* refers to the average life of the lamp, based on the survival of at least 50 percent of the lamps.
3. *Maintenance factor* refers to the gradual reduction of light from the lighting system as a function of time. It depends on a variety of factors, but one to be primarily concerned about is efficiency reduction, or *lumen depreciation.*
4. Warm-up (start-up) time from a cold start refers to the time, in minutes, that it takes the bulb to reach the full lighting intensity when first energized.
5. *Restrike time* refers to the time, in minutes, that it takes to restart a lamp already warm (hot restart). This is important, of course, in a plant where there is no artificial light or insufficient light to see exits in case of a quick power interruption.

The chart of lamp output (Fig. 1-12) clearly shows why more light efficiency can be obtained by changing light sources.

HID Light Sources

Mercury Lamps

Incandescent and fluorescent systems are well known. Mercury, metal halide, and high-pressure sodium lamps are all in the high-intensity discharge (HID) family. To introduce HID lighting, various types can be compared with the most common (and the one that has been around the longest), mercury. A ballast is used to start the lamp. This establishes an electric field between the main electrodes and the starting electrodes, causing an emission of electrons and slowly ionizing the argon gas in the lamp and creating an argon arc between the main electrodes (Fig. 1-13). Heat from the argon arc gradually vaporizes mercury which joins the arc stream until the mercury takes over, increasing in intensity until full light output is reached in 5 to 7 min—sometimes longer in colder weather. The main advantages of the mercury lamp when it was introduced was its extremely long life, up to 24,000 h, and its ability to have good luminaire efficiency since it is a point source and is more easily focused and directed than other types.

In the past, the time for warm-up and restrike has been offset by starting mercury lamps earlier than necessary, not shutting them down during short periods of no light requirements, and supplementing certain fixtures with incandescent lamps to handle the

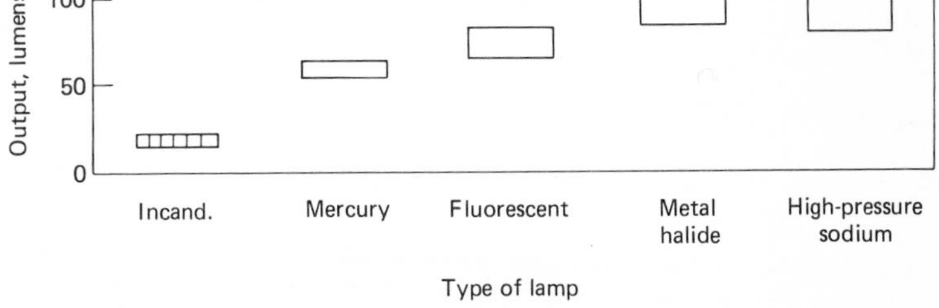

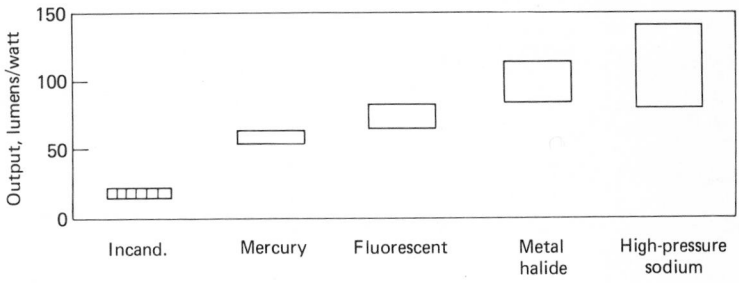

Figure 1-12 Lamp output vs. type.

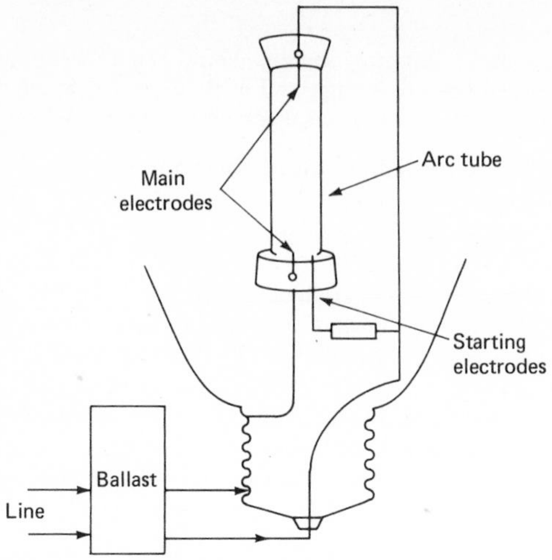

Figure 1-13 HID lighting.

restrike time during a power dip. Mercury lamps come in a variety of color corrections so that good color can be obtained at a slight premium cost (Fig. 1-14).

Metal Halide Lamps

A new light source with greater efficiency, great color, and wide applications is the metal halide lamp. The halide lamp fires in the same fashion as a mercury lamp, but the halide compounds are also vaporized as the lamp warms up. Several color steps may be observed during the final warm-up period, but the lamp has very good white color (Fig. 1-15).

There is a large installation of these lamps at Hamilton Standard. It has saved 7 $\times$ 10^6 kWh annually over the system it replaced. Metal halide lamps have good efficiency and excellent color characteristics, but shorter life than mercury lamps.

High-Pressure Sodium Lamps

If color is not important, using high-pressure sodium (HSP) lamps (Fig. 1-16) should be considered. These fire in a manner similar to that of mercury lamps, except there is no starting electrode, so initially a high-voltage spike from the ballast establishes a xenon gas arc between the main electrodes. Small amounts of mercury and sodium then vaporize rapidly to produce a warm, yellow-white light. The high-pressure sodium lamp can be easily identified by its characteristic bulb shape.

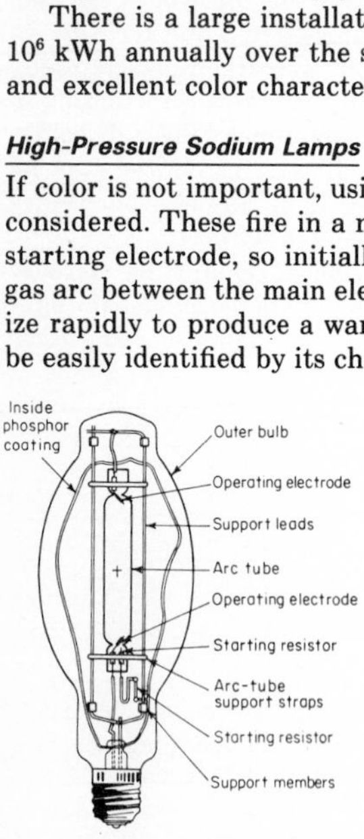

Figure 1-14 Mercury lamp.

Figure 1-15 Metal halide lamp.

Figure 1-16 High-pressure sodium lamp.

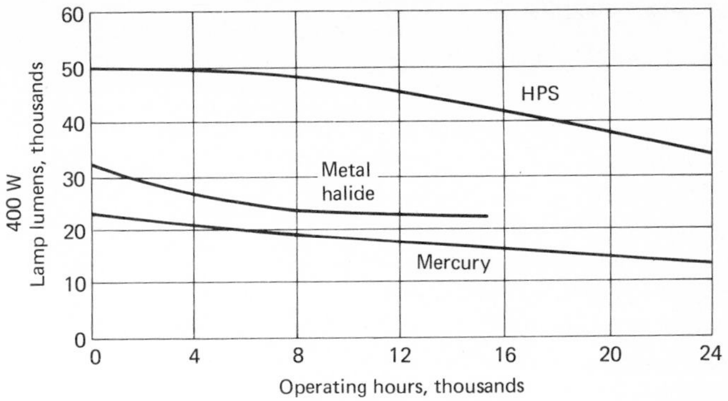

Figure 1-17 Lumen depreciation.

This is the most efficient light source made in the United States. It has good life, minimum starting and restrike timing problems, and good color distinguishability. Its drawback is that although the color is accepted outdoors, most workers are not familiar with its indoor use.

Application Considerations

High-pressure sodium lamps retain their advantages over time. Figure 1-17 shows why it is disadvantageous to try to get each last lumen out of old lamps since it means paying for almost the same wattage as when they were new, without the benefits.

In many cases, new lamps will be used to replace fluorescent lamps which have shorter life and less efficiency, but have none of the problems relative to starting time. Fluorescent lights, because of their long, slender shape are a diffuse rather than a point source; this property results in more uniform lighting. When using an HID light source, be sure to recognize it as a point source and to distribute the light uniformly through the proper mounting height, fixture use, and wattage selection.

HID Lamp Summary

It is apparent from Figs. 1-14 to 1-16 that the high-pressure sodium lamp has the greatest efficiency, longest life, shortest warm-up time, quickest restart time, and best luminaire efficiency because of its smaller point source.

AN ENERGY MANAGEMENT SYSTEM

An energy management system (EMS) offers:

- Conservation of *all forms* of energy through optimization techniques and reduced usage
- Increased staff efficiency through centralized monitoring and control of services
- Incipient failure recognition of plant engineering services and equipment
- Economic payback through increased energy and staff efficiency
- Simple system integration of all types of equipment and services under plant engineering control
- Future growth flexibility

The EMS consists of a remote data collection system, a centralized display system, and methods of remote control.

Typical software features are:

- Electric load shedding or smoothing
- Electric load cycling
- Optimized equipment start or stop
- Building precooling
- Zone-temperature night setback
- Enthalpy damper control
- Hot- or cold-deck temperature optimization
- Chilled-water temperature optimization
- Weather-data collection
- Equipment trending

Industrial-commercial patterns of energy use are shown (Fig. 1-18) with energy use peaking at midday and early afternoon. Smaller second- and third-shift use is typical. Graphically, the chart shows the anticipated savings if a rigorous control is provided. Manual control can be achieved by bringing the data back to an operator who controls use or, in the case of computer control, by bringing the information back to a computer having controls following preset directions. These directions can be very specific preset actions that the computer can perform without operator command, such as switching on and off electric heating to the plant or shutting off certain air-handling units.

Electric Load Shedding

To explain a typical EMS operation, let us explore control by using electric load shedding.

Load shedding is primarily aimed at demand control by shifting operations off-peak, but the fact that some loads never come on also saves energy in the form of kilowatt-

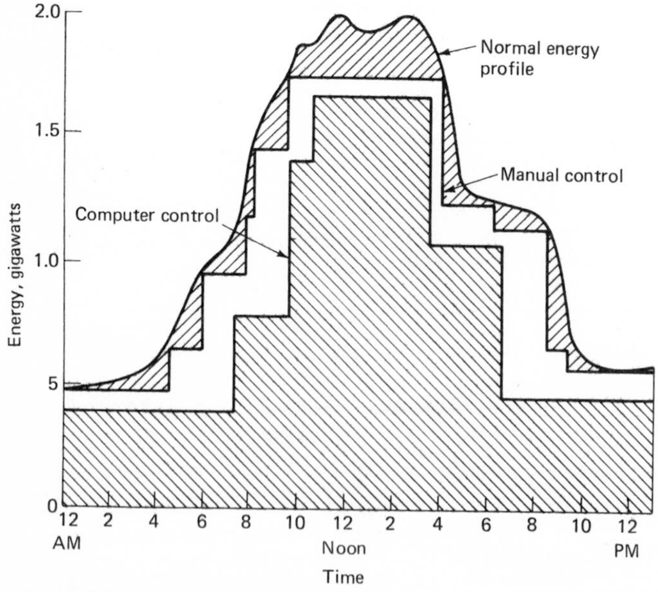

Figure 1-18 Lamp summary comparison—typical energy-use profiles.

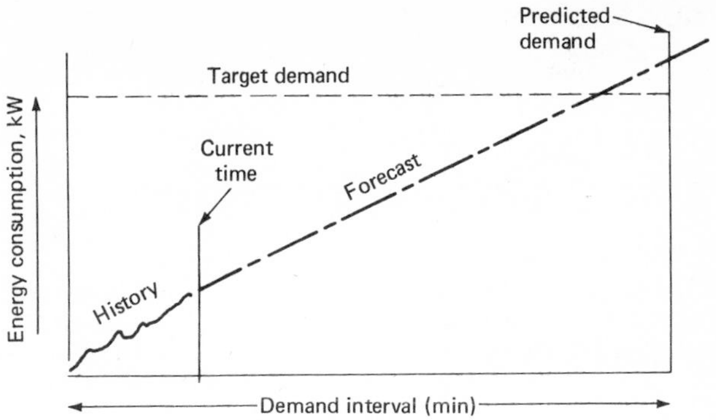

Figure 1-19 Demand forecasting.

hours. The techniques used here are targeting, forecasting, and load shedding (Fig. 1-19). Each day starts off the first demand interval (normally a 15-min interval) with a "not to exceed" target demand. The demand usage is constantly measured by parallel relays from the power company's metering devices. Each minute, a prediction of peak demand for duration of the interval is forecast. This forecast is based on the data for the previous 3 min and the time remaining in the demand interval. Based on this prediction, loads are automatically shed, on a predetermined priority basis, as needed to prevent exceeding the target demand. This maximum target demand can be easily varied by the system operator. By setting the target demand lower than the historical high peak, automatic peak shaving or limiting can be achieved. This results in a savings on monthly demand charges (Table 1-3) and on energy consumption charges if they are based on a formula using peak demand, as previously shown.

TABLE 1-3 Typical Monthly Load Shedding Savings

	Before		After	
	Demand, kW			
Highest demand	18,000		16,000	
First 50 kW	− 50	$ 225	− 50	$ 225
	17,950		15,950	
Next 150 kW @ $2.90/kW	− 150	$ 435	− 150	$ 435
Remainder @ 2.20/kW	17,800	$36,160	15,800	$34,760
		$39,820 total		$35,420 total
			$4,400 Demand savings	
	Consumption, kWh			
Total kWh	6,000,000		6,000,000	
demand × 200 @ $.028/kWh	−3,600,000	$100,800	3,200,000	$ 89,600
	2,400,000		2,800,000	
Demand × 100 @ $.028/kWh	−1,800,000	$ 43,200	−1,600,000	$ 38,400
Remainder @ $.022/kWh	600,000	$ 13,200	1,200,000	$ 26,400
		$157,200 total		$154,400 total
			$2,800 Consumption savings	
Monthly savings due to 2000-kW demand decrease				$7,200 total

Equipment typically used to control demand includes:

- Heating, ventilating, and air conditioning
- Hot-water heaters
- Process heating units
- Lighting (partial)

Electric Load Smoothing

An associated technique, using reverse strategy to load demand control, is "smoothing" after the demand level peak is reached. The target is gradually reduced. When a demand interval occurs without shedding loads, target demand is reduced by 10 percent during each succeeding interval until loads are shed. This decreasing-increasing, in conjunction with demand control, has the effect of smoothing out the daily demand curve, and, in effect, reducing base-load energy consumption; thus a two-part savings can be realized: electric consumption savings, and peak demand savings. See Table 1-3.

Electric Load Cycling

Electric load cycling is used to reduce base-load energy consumption. Many electric loads can be cycled *off* for periods of time each hour with no noticeable degradation of performance. A typical example of this is an air-handler motor. In most instances, these motors can be cycled *off* a minimum of 10 min each half-hour, with no discomfort to the affected zones. This totals over 2½ h of reduced electric consumption for each 8-h working shift. See Table 1-4.

TABLE 1-4 Typical Monthly Load Cycling Savings

Total hp cycled	2,680
Total kW cycled	(2,680 × 0.746) = 2,000 kW
Cycled kWh/day	(2,000 × 2.5) = 5,000 kWh
Monthly kWh cycled	(5,000 × 30) = 150,000 kWh
150,000 kWh @ $0.022/kWh = $3,300	
Monthly savings due to cycling 2,680 hp $3,300	

HVAC-Related Load Shedding

Recently, there has been much discussion and advertising about electric load shedding and the associated dollar savings. Very little is heard about load shedding for steam, chilled water, boilers, chillers, or cooling towers. Experience at Hamilton Standard (both on EMS and outside-site surveys) indicates that at least twice the savings are attainable through HVAC-related load shedding. For example, Hamilton Standard's potential annual savings in electricity are approximately $100,000 as opposed to potential HVAC related savings of $300,000 annually. This 3:1 ratio dictates a *total* energy optimization system over one that addresses only electric energy optimization.

To intelligently shed HVAC-type loads, the following information must be collected and assimilated by the EMS:

- Outside weather conditions, such as dry-bulb temperature, wind speed, wind direction, relative humidity
- Zone temperatures
- Return air temperature and relative humidities
- Time of day
- Occupancy schedules
- Management's set-point policy and control strategy.

In addition, certain key controls have to be under control of the EMS. Typical parameters are:

- Zone temperature set point
- Damper position
- Chilled-water temperature reset
- Hot- or cold-deck temperature reset
- On-off speed control (if available) of fan systems.

The common element in the HVAC system is the air-handler unit. A typical system is shown in Fig. 1-20. This allows optimization by maximizing outside air potential and permits alternative fan operation for load shedding or cycling with little or no change in basic equipment. A schematic diagram, seen on a terminal, through an associated slide projector, or via print retrieval is a valuable trouble-shooting tool for maintenance.

With this information and control, logical load shedding of steam and chilled water can be achieved. Resetting zone temperatures up or down according to occupancy, outside conditions, and near future predictions can result in steam or chilled-water reduction.

Resetting hot- or cold-deck temperatures down or up to *just* satisfy the current worst zone load can also result in steam or chilled-water savings.

Scheduling dampers according to outside enthalpy vs. zone enthalpy instead of dry-bulb comparison can save energy by utilizing the air source with the least coil load.

Resetting chilled-water temperature to *just* satisfy the current worst air-handler load can result in significant savings. This also has the advantage of overriding the zone thermostats which are far from tamper-proof.

These simple ways of optimizing plant operations by taking computer control of the fan systems will allow improved efficiency of major plant equipment such as central chillers, since they are no longer required to operate to an independent, worst-case set point, but can operate only to need.

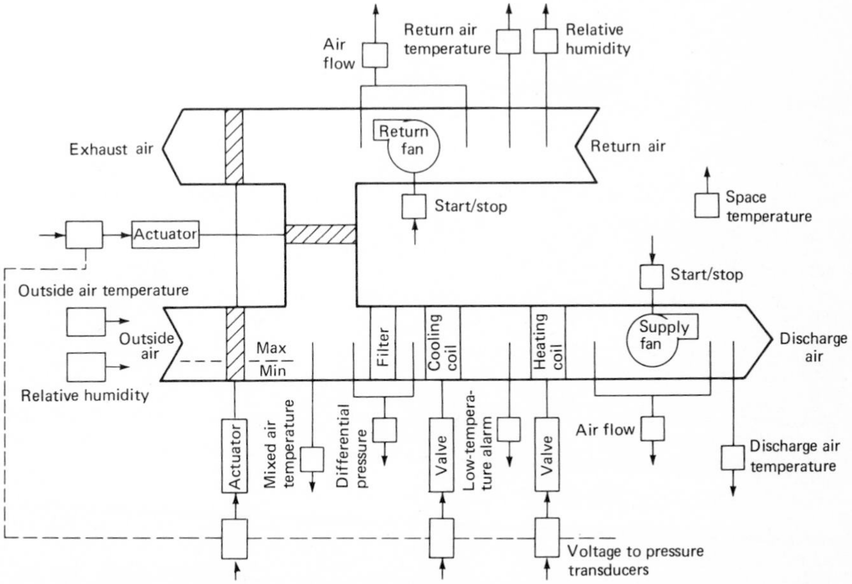

Figure 1-20 Air-handler system.

Maximizing Operations

The EMS computer must be tailored to fit each plant. Remember that you have key information on plant needs with the ability to forecast future needs. This means that, based on historical trends plus key indicators which are being monitored, you have the ability to determine equipment combinations that should be online to meet the forecast load. Your computer should have a backup plan so that, should the load be higher than anticipated, certain nonimportant things could be shed quickly; e.g., there could be a shift from steam air compressors to electrically driven air compressors should a higher steam load occur. These same sets of circumstances can be applied to the compressed-air, chilled-water, or electric distribution system. In short, there are unlimited opportunities to improve day-to-day operations through the use of a computer.

part C

The Maintenance Function: Basic Equipment and Supplies

section 16

Equipment for Maintenance and Repair

chapter 16-1

The Maintenance Facility

by
William S. Harrison, P.E.
Director of Technical Services
Engelhard Industries Division
Engelhard Corp.
Plainville, Massachusetts

INTRODUCTION

Every maintenance department needs a well-planned workplace. Though large machine repairs are usually done on site, many components may need to be returned to the shop for repair. When possible, removal of equipment from the production floor is more likely to yield a better, more complete repair or rebuild. The nonrepetitive nature of the tasks demands a location where skill and tools can be combined and applied without the distractions of production.

Plan your facility to suit the size and type of responsibilities of your firm. The shop in a textile mill with looms will be equipped differently than that in a chemical plant. Common factors to consider: basic work area; workbenches; mobile or permanent toolbox area; parts-cleaning facility; basic machine tools; access for hoists, cranes, or forktrucks; spare parts and raw-material storage; welding area; electrical area; ventilation-fume exhaust; utilities—air, gas, water, etc.

Specialized areas such as instrument repair, pipe fitting, plumbing, blacksmithing, and others are added to suit. It is seldom advisable to combine the functions of model shop, development shop, or gauge control in the maintenance shop because of the difference in work content.

There are many workable layout variations. Access, manipulation in disassembly, and temporary storage of components are important. Some operations may require ladders, stands, or scaffolding. Vehicle or forktruck repair requires pits or lifts and special ventilation.

If space allows, the following ideas should be considered:

1. An area separated from production is preferable.
2. A main aisle for access with workplaces on each side is efficient.
3. Benches can be used to delineate the individual stations at right angles to the aisle or placed parallel to the aisle at a distance equal to the depth of the workplace.
4. Electrical and cleaning stations should be out of the way or at the perimeter.
5. Parts need racks or bins; fencing may be needed for security; some items can be double-decked for space saving.
6. Pipe and steel racks should be accessible to handling equipment.
7. Offices should be accessible, but out of the way—at the perimeter, on the mezzanine, etc.

Think ahead! Items may come in piecemeal but may have to go out assembled. When planning the facility in a very large plant, such as automotive, aircraft, or appliance, employ the *satellite concept* using small shops equipped to handle the more routine demands of a particular area.

BASIC TOOLS

Basic tools are common hand tools, powered or manual and usually portable. The simple drill press, grinder, and small lathe are generally included as are common electrical tools like a voltage tester, multimeter, megger, etc.

Selection criteria for tools include:

1. **Cost** Always important.
2. **Quality** Remembered longer than cost.
3. **Design** Feel, ease of use, functionality.
4. **Size** Be conservative, do not overload any tool or instrument.
5. **Reliability-Durability** Works consistently and can stand hard use.
6. **Reputation of the maker** Can be meaningful.
7. **Repair facilities** Proximity; time is valuable.

In general, there are some quality and capability differences between hand tools used at home and those in the factory. However, power and machine tools are substantially different. It seldom pays to mix the two. The timely success of a department should not depend on weekend- or home-type tools.

LARGER FACILITIES

Beyond the basic hand tools necessary in every shop, the plant engineer must consider work demand and cost. The cost and complexity of use of some modern tools may dictate that you buy the service instead. Your decisions will depend on the following criteria:

1. Cost of the tool
2. Frequency of use
3. Space requirements
4. Skill requirements
5. Priority of need

The last is difficult to assess, given the demanding nature of the profession. Avoid the tendency to collect unique tools that are rarely used. Only the larger and/or specialized shops can justify diamond-sawing and core-drilling equipment for concrete work or the latest infrared scanning rigs for predictive plant-maintenance work.

After the basic considerations come the functional requirements of each plant which dictate specialized tool needs. A detailed list for each industry is beyond the scope of this book. Apply the basic criteria given above to each facility. Implications to be considered include:

1. Age of plant machinery
2. Reaction time expected by management
3. Value of lost production time
4. Availability of outside services
5. Training required to operate sophisticated equipment

Management may lean toward minimizing an investment in costly, specialized tools since it has a negative effect on return on invested capital, particularly if the service can be bought. Plan acquisitions that take into account prevailing ideas.

Other acquisitions to be considered include magnetic-base drill press, wire-feed welding equipment, cleaning stations with pumped solvent, steam-cleaning rigs, sand-blasting gear, drill press or radial drill 1 in (25 mm) up, Bridgeport ®* mill or K & T Universal, surface grinder, flame-cutting table and gear, impact drills to 1½- or 2-in (38- or 51-mm), pneumatic chisels (street busters), optical alignment equipment, electronic balancing equipment, vibration analysis equipment, portable breathing equipment, gear pulling and/or portapower rigs, industrial vacuum cleaners, ground detection testers, thermocouple calibration galvanometers, portable generators, portable hydraulic personnel lifts, sound measurement frequency analyzers, environmental gas-liquid test equipment, and power hack and band saws.

LIGHTING AND POWER†

Whether planning a new facility, revamping an old, or just relocating within a plant, the steps are the same. Adopt a methodical approach to avoid avoid costly errors. Electrical work can represent a sizable fraction of total project cost, so:

1. Analyze the power requirements for existing and projected equipment.
2. Allow for expansion.
3. Allow for temporary power to equipment being repaired.
4. Decide on a lighting requirement.
5. Carefully integrate the light and power design with the plant layout.

Medium to larger maintenance facilities in plants fewer than 30 years old should have, and most probably will have, 460-V-ac, three-phase power available. This is the *standard* industrial *low*-voltage distribution. A four-wire, Y configuration is good since wiring economy can be effected with 277-V fluorescents if they are the lighting choice.

Even the most modern plants need 230-V, three-phase and 115/230-V, single-phase power. These are easily obtained via single or three-phase stepdown transformers. This is generally termed *miscellaneous power* and is used for smaller machine tools or for power equipment being repaired, tested, or constructed.

There are no firm rules for sizing the main electric feeder to a maintenance facility. The task requires professional expertise. As a rule of thumb a *demand factor* (ratio of maximum demand to total connected load) for the average facility is probably in the

*Registered trademark, Bridgeport Machine Works.

†See also Sec. 3 of this handbook.

range of 0.2 to 0.3. It is important to consider again whether most of the work is within the facility or elsewhere.

Only extremely specialized shops would require electric service at higher voltages. Such installations should be left to specialists in the profession.

A typical installation will have:

1. Machine and equipment loads
2. 117-V single-phase receptacles
3. 208- or 230-V, single-phase receptacles
4. 208- or 230-V, three-phase receptacles
5. 460-V, three-phase receptacles
6. Special receptacles for welders, etc.
7. Lighting circuitry and switching

In a new plant, try to include special receptacles for welders and other equipment at multiple locations *in the work areas* to save time and money.

How do you electrify a facility? The scheme depends on the size, quantity, and location of the equipment. Machines can be fed from power distribution panels, busways, miscellaneous power panels, motor control centers, connection boxes, and multiple fused disconnects. When recommended practice is followed, all conform with the **National Electrical Code®**.

Provide more than adequate light in the work area; 100 footcandles (fc) at bench level is a good rule. This may be supplemented by fixtures installed directly over workbenches with local control. In a new facility, high-intensity sodium vapor fixtures are most effi-

***National Electrical Code®** is the trademark of the National Fire Prevention Association, Quincy, Mass.

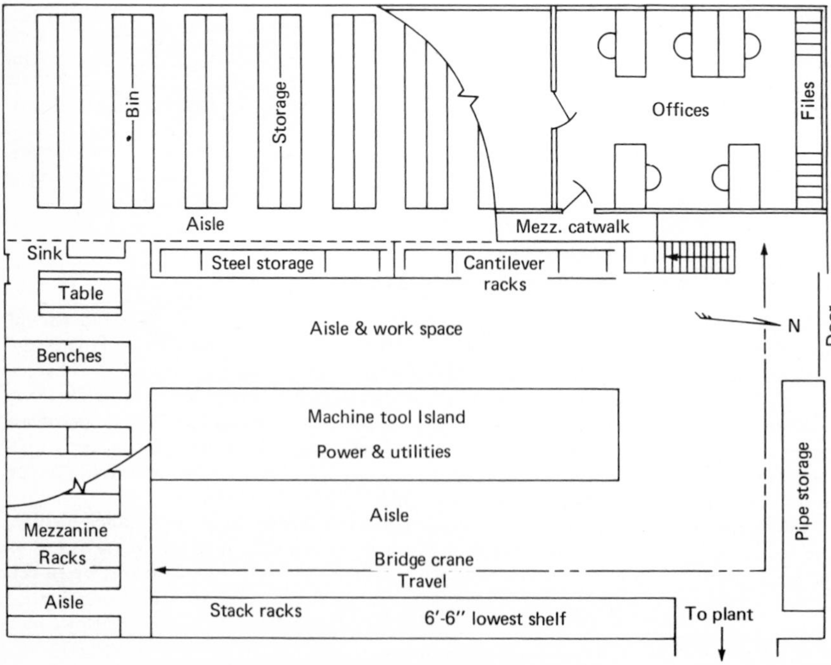

Figure 1-1 A compact contralized maintenance facility, 58 × 78 ft (17.5 × 24 m).

cient if color is not a factor. The time-proven fluorescent is available in a design using 25 percent less than normal energy for comparable lumen output.

Provide portable stands with a bank of floodlights to use in the facility or out on the job. If the normal mode of operation is not centered in the facility, consider switching. This can yield overall levels of illumination only as required. Efficient use of energy is a must.

A FACILITY IDEA

There are many ways of fitting together the necessary elements of an efficient facility. Figure 1-1 shows the layout of a shop serving a metalworking factory of over 250,000 ft^2 (23,200 m^2) and serviced by 30 to 35 technicians

Several features in the figure are:

1. One side is occupied by 7-ft (2.13-m)-high bins with a controlled-access aisle connecting all bin aisles.
2. Above the bins and supported by them are supervisory and plant engineering offices.
3. On the end is a workbench, lunch, and washup area.
4. Above No. 3 is a mezzanine with rack storage for more bulky supplies.
5. The second side has stacking racks of 4-ft (1.22-m) depth with the first shelf at 6 ft 6 in (2 m) to allow tool-chest storage and work space for grinders and sanders. On the wall, not shown, is a V-belt rack. Storage continues to an eave height of 14 ft 6 in (4.4 m).
6. The remaining end has a long pipe rack with multiple levels and bulk storage on top. A door occupies the remaining length.
7. Two aisles, one 9 ft (2.7 m), the other 10 ft (3 m), flank the machinery *island.* The wider aisle allows a truck to drive in and be unloaded by a 2-ton (1820-kg) bridge crane which sweeps the entire working area. The aisles provide forktruck access and working area. Only portable units or equipment under repair are allowed in the aisles.
8. Machine tools are installed back-to-back with electric power at three voltages and single- and three-phase in between. The wireway run at head height does not interfere with the sweep of the bridge crane. Power panels, transformers, etc., are located on the south wall. An air line also runs between the machines.

An arrangement to be workable and efficient should consider storage and work space. It should have maximum density for storage with easy access and a maximum open work area with hoisting and handling service.

BIBLIOGRAPHY

Muther, Richard: *Systematic Layout Planning,* Cahners, Boston, 1976.
Dergamo, E. Paul: *Materials and Processes in Manufacturing,* Macmillan, New York, 1957.
Ball, John E.: *Carpenter's & Builder's Library,* vol. 4, Audel, Indianapolis, 1965.

chapter 16-2

Selection, Use, and Care of Basic Tools

by

William S. Harrison, P.E.
Director of Technical Services
Engelhard Industries Division
Engelhard Corp.
Plainville, Massachusetts

INTRODUCTION

In earliest days, the use of tools distinguished the intelligence and evolution of the human race. Tools multiplied productivity, and this is still the objective in the acquisition and use of good tools.

Craftspeople must have the proper tools to accomplish their work safely, quickly, and correctly. In most cases involving machinery maintenance, it is impossible to complete a task without a substantial number of common hand tools and several special pullers or shaped wrenches.

When individuals and companies depend on tools for their existence, it pays to select good, even the finest, tools. It takes very little experience to perceive the difference in *feel, balance, fit,* and *design for the job* of a fine tool or tool assembly.

Virtually all fine hand tools and many of their component parts are manufactured from alloy steel forgings. It is the continuous-grain skeleton following every curve and shape intricacy that provides the strength within a relatively light section. One has only to try to fit the oversize gripping hook of a cheap puller in a tight space behind a gear to appreciate the design and strength of the forged, alloy steel version of the same tool. Or contemplate that delicate hooked probe that a dentist inserts between your teeth. As a professional yourself, select the best you can find. Price is not the only criterion, but fine tools are seldom sold as loss leaders.

Even a small shop will need an extensive assortment of hand tools, portable electric tools, and a few basic machines. If you must live within a tight budget, be even more selective. The fine tool can usually be *stretched* over a wider range of application than the poorer grade.

The best source for tools is an industrial-supply or mill-supply distributor. Look for one with a good reputation for service and repair since all power tools will eventually require maintenance. Several complete tool lines are sold directly rather than through local distributors, for example, those sold by Snap-On Tool Corp. and Sears Roebuck and Co.

Our purpose is to present a guide or checklist for selection. Rather than use space for pictures of the many types and sizes, we recommend that each shop keep on hand the catalogs of the many reputable manufacturers. These can be acquired either through their distributors or directly. Use them for more detailed information on the tools listed and briefly described here.

SCREWDRIVERS

Types

The types of screwdrivers are:

1. **Standard** Common slot, sized by blade width, ⅛ to about ¾ in (3 to 19 mm) (larger sizes less used as slots disappear in favor of socket heads).
2. **Phillips** Cross-slotted, tapered from tip, sized 0 to 4.
3. **Reed Prince** Variation of Phillips, shallow taper from tip, little or no vertical taper.
4. **Clutch Head** Two intersecting circles when viewed head on, sized by the wider dimension.
5. **Auto-Return, Ratcheting** Spiral-grooved shafts or gear-and-pawl mechanisms to enable semicontinuous turning without changing hand grip, sized by the blade. Many adapt to a range of interchangeable blades.
6. **Screw-Holding or Magnetic** Spring steel fingers grip under screw head; two-piece blade is offset to bear tightly against both slot sides; entire blade is permanently magnetized.

7. **Precision or *Jewelers'*** Slot type usual, Phillips more rare, blade width in decimal inches like 0.040, 0.070 in (1 to 2 mm) for tiny screws like those found in watches, instruments. Generally with free rotating finger rest on top and knurled steel body.

Selection and Use

Fine screwdrivers have forged steel blades or ends. Blade ends are ground to the proper size. Choose the size that fits snugly with little clearance. Any tendency to rock in the slot will ultimately wear the blade or round off the corners. Never use a screwdriver for a wedge, prybar, or punch. Should a blade tip break, it is a tedious task to grind it to the correct angle and size without using a vise and surface grinder. It is preferable to replace a broken screwdriver since hardness and temper of the blade would likely be affected by regrinding.

WRENCHES

The term *wrench* is generic, and there are literally hundreds of tools that hold, grip, and turn. Whether needed for gripping nuts, cap screws, pipe, shafts, or other components, choose a tool of proper gripping capacity and leverage. Never extend capacity with pipe extensions over a handle. If the jaw or grip breaks, accidents are quite likely to occur when an individual is exerting an undesigned pull or push on the tool. Observe manufacturers' specifications as to capacity and safe torque limits. Overloading can distort a jaw, grip, or socket, making the tool useless and in many cases *peening, rounding,* or *damaging* the fastener or device being loosened.

Types

Most common wrenches are available in both U.S. customary and metric sizes:

Open End. $\frac{3}{16}$ to 2 in (4 to 50 mm)

Box End. $\frac{3}{16}$ to 2 in (4 to 50 mm)

Combination. Open and box combined

Open End. One end grip, smooth tapered handle

Socket. $\frac{3}{16}$ to 4 in (5 to 200 mm) with drive $\frac{1}{4}$ to $1\frac{1}{2}$

Where access allows, choose a wrench that contacts all sides of a fastener or object. For a hex head, a 6 point is always preferable to 12 point.

Adjustable End. 4 to 24 in (100 mm to 0.6 m) length

Stillson. 8 in to 6 ft (0.2 to 1.8 m) length, adjustable, serrated jaws for pipe work

Monkey. 6 to 24 in (150 mm to 0.6 m) length, adjustable, smooth jaws, generally made obsolete by adjustable end wrenches

Box for Tubing Application. $\frac{3}{8}$ to $\frac{3}{4}''$ (9.5 to 19 mm); essentially a six-point box-end wrench with a section of the perimeter cut away to permit entry of a thin-wall tube and access to a tube fitting when it is connected. Prevents *rounding* of soft brass fittings.

HAMMERS

There are three basic types of hammers, and most of the many special types are adaptations of the basic types for the work requirement. The basic types are curved-claw (carpenter style), ball-peen (machinist style), and maul or sledge.

In addition to steel, modern hammers use materials such as rubber, filled plastics, and other nonmetallics for the heads; steel, wood, or reinforced plastic is used for the handles. Hickory is still the prime handle material for its shock deadening and *balance,* though fiberglass-reinforced plastic has similar qualities and does not readily chip from a missed blow.

Proper care begins with using a hammer of correct weight for the task; position yourself so that a direct, well-aimed blow of proper force can be struck. Handles should never be used for prying or as levers. Touch up minor nicks on the striking face and head with a file. Keep the handles clean to prevent slipping or turning in the hand. Sand chipped or splintered wood. Handles may be replaced, but this painstaking process is almost a lost art. If in doubt, do not take the chance of using a hammer with a loose head; this is an extremely dangerous condition. Always wear safety glasses.

Types

Curved-Claw (Carpenter's Hammer)

This type is made with heads varying in weight from 4 to 24 oz (113 to 680 g); 16 or 20 oz (450 or 570 g) is the preferred weight. Curved-claw hammers have a flat face for nail striking. Avoid extraction of nails beyond 16d; use a nail puller or wrecking bar which is better suited for the task.

Ball-Peen (Machinist's Hammer)

This type of hammer varies in weight from 4 to 40 oz (113 g to 1.1 kg) or more; 12 or 16 (340 or 450 g) is the preferred weight. Ball-peen hammers have a crowned face for striking and a peen end for upsetting pins, rivets, etc. Surfaces are rounded to minimize chipping and denting since these tools generally strike metals of hardness comparable to their own.

Maul or Sledge

The heads of mauls vary in weight of 4 to 20 lb (1.8 to 9.1 kg). Usually they are double-faced with a crown. They are generally used for striking or driving large keys or shafts and are used for masonry demolition only as a last resort.

Other Types

Special types include bricklayer's, mason's, upholsterer's (tack), shingle, cobbler's, special sheet-metal forming, and a variety of half-hammer, half-axe, or half-hatchet combinations.

DRILLS

Most drills used for maintenance are electric since hand drills are not practical in metalwork except as noted. Keyed Jacobs chucks are used on all but the lightest tools, and handle bits up to ¾ in (19 mm). With occasional cleaning and oiling, they are completely reliable. Tools should be kept clean and oiled as recommended and not overloaded. Stalling is a sign of overload and will cause burning of the commutator and wear on brushes. Industrial-quality drills usually have a simple system for brush replacement, and a yearly change is good preventive maintenance.

Common practice in the manufacture of industrial twist-drill bits is to supply a straight shank equivalent to the diameter of the bit up to ½ in (12.7 mm) diam. Beyond ½ in, a Morse taper shank is more common. The taper on all sizes from 0 to 7 is approximately ⅝ in/ft (15.9 mm per 305 mm). This taper develops so much frictional resistance that it is termed *self-holding*. The small end terminates in a flat-sided tang which fits in a matching female slot in the chuck or socket. This is used to remove the bit via a tapered key or wedge which is inserted in the open slot and tapped lightly, thus driving the bit out of the taper. Refer to a *machinists' handbook* to determine which size taper applies to a specific bit diameter.

While straight-shank twist drills *are* available in the larger sizes, another advantage of the Morse taper is its self-centering and positive alignment. It is advisable to specify drilling equipment, particularly drill presses, to include Morse taper chucks or sockets for diameters over ½ in (12.7 mm).

Hand Drills

The smallest hand drill is called a *pin vise* and has a straight knurled shank and tiny hand chuck to hold bits in *jewelers' sizes.* It is rotated by finger action. Geared hand drills are made with straight or pistol-grip handles with chucks of ⅛- or ¼-in (3.2- or 6.35-mm) capacity. Larger sizes are termed *breast drills* from the shaped plate which is borne on the chest while the crank handle is being turned. Most of these units are approaching antique status. A better bet is a *cordless* or battery-powered unit.

Electric Drills

All shops require several electric drills of different sizes and styles. Factors to consider in their selection are discussed in the following paragraphs.

Style

Pistol-grip drills are most common and can be used for most applications. Some heavier versions have a side handle to apply extra force; an end handle may be preferred in heavier, repetitive drilling; a spade handle is common from ½ in (12.7 mm) and up and accompanied by a companion side handle to allow two-hand control and force; extended spade and side handles are found on drills ¾ in (19 mm) and up to allow two or more individuals to hold and push the drill. For tight clearances there are right-angle styles that are particularly useful for electricians and plumbers.

Capacity

Capacity is defined as the largest-diameter drill bit that the machine will accept. Capacity ranges are usually ¼, ⅜, ½, ⅝, ¾, 1, 1¼ in (6.35, 9.5, 12.7, 15.9, 19, 25.4, 31.9 mm).

A practical minimum bit size is 1/16 in for the ¼-in chuck. Special drills and chucks are available which will handle numbered sizes to 97, but these would find little use in maintenance. Material to be drilled will govern speed and power requirements and must be considered with capacity.

Speed

It is important to know the material to be drilled and to match this using handbook formulas for the required speed within broad limits. In general, harder metals like nickel, stainless steel, and carbon steel require the lowest speeds. Softer metals such as aluminum, magnesium, and zinc require medium speeds, and most woods require high speeds. As an example, applying this to a drill of ¼-in (6.35-mm) capacity, speeds of about 1500, 3500, and 5000 r/min would represent reasonable compromises. The newer variable-speed or multirange tools may yield some flexibility, though it is very difficult to judge and control their speed.

Power and Construction

Industrial-quality drills are designed with sufficient power for the particular size. Close comparison will reveal some with slightly more power than others. Before choosing, examine the other features. Look for head-treated gears, ball or roller bearings, replaceable brushes, and a well-molded case. Expect these industrial-quality, heavy-duty portable drills to cost anywhere from 5 to 10 times what you might pay for a home-workshop type. They are built to last and are usually worth repairing when they do falter. Even the smallest shop would have use for two ¼-in (6.35-mm) drills, low and medium speeds, a ½-in (12.7-mm) low-speed unit, and a ¾-in (19-mm), 300- to 400-r/min unit.

Magnetic Drill Press

Medium to large shops should consider a magnetic-base portable drill press. These units make possible the drilling of clean, correctly aligned holes in locations where individuals could not hold and apply force to a hand drill. The geared feed handle yields point pressures comparable to those of a regular nonportable drill press.

Drill Stands

These are small, portable, column-and-stand, lever-feed units that hold portable drills. They are handy for repetitive drilling and provide better alignment where throat capacity allows their use.

Hammer Drills and Impact Drills

These drills combine rotational motion of the bit with repeated vibratory hammer blows. This type of drill is mandatory when drilling concrete, brick, or other refractory-like material and is used in combination with carbide-type drill bits. Do not use carbide bits with regular, nonimpact drills in concrete or similar materials. This is a quick way to ruin the drill or its gears and bearings.

ELECTRIC HAND TOOLS

Depending on the type and size of shop and the capability you are attempting to build you may consider some of the following electric portable hand tools:

Types

The types of electric hand tools are:

Screwdrivers-tappers
Impact wrenches (also air-driven)
Routers, wood or metal
Shears
Nibbers
Circular saws
Reciprocating hacksaws
Saber saws or jigsaws
Portable band saws
Sanders, belt-type
Sanders, disk or rotary
Sanders, reciprocating
Grinders, disk-type
Grinders, wheel-type
Grinders, die high-speed
Percussion hammers
Demolition hammers
Portable planers
Riveters
Chisels
Nail guns
Staple guns
Electrohydraulic pull-and-squeeze units

Selection and Use

The same rules of selection apply here. Examine capacities and match them to needs. Many distributors will demonstrate tools or will even lend tools for trial. All these specialty tools are potentially dangerous. Follow the manufacturer's instructions, and proceed slowly and carefully. If possible, learn from a worker who is already skilled in the use of the tool. Many of the more powerful units develop substantial force reactions in operation or if they bind or jam. *Make certain you have the proper footing and stance to resist these forces. Wear safety glasses.*

Clean tools after use. Impact or percussion tools usually require special lubrication procedures.

The electrohydraulic pull-and-squeeze units are extremely versatile. They employ compact hydraulic pumps which supply fluid under high pressure to cylinders of various strokes. The cylinders combine with puller jaws and grips to remove press fit bearings, sleeves, and gears. They generally come in sets with assorted special tools to handle a wide range of applications where pulling or pushing forces ranging up to 10 tons (9 metric tons) are required. A diverse set may cost half a year's craft wage, but it is a worthwhile investment.

DRILL PRESS

Perhaps the most widely used machine tool in the shop, the drill press is a versatile unit that allows proper matching of speed and feed for the drill bit and the material being penetrated. The types applicable to plant engineering and maintenance are bench, floor, upright, and radial. The essential size or capacity of a drill press is described by the following parameters.

Drill Press Parameters

Size

Size is measured by the distance from the centerline of the spindle-chuck-quill to the outside or nearest point of the column. In effect, it is the diameter of the largest disk in which a center hole could be drilled.

Capacity

The capacity is the maximum diameter of the bit that can be inserted in or driven by the chuck or bit holder.

Travel

Travel is defined as the vertical distance that the bit can be fed via movement of the quill and spindle assembly.

Speeds

Individual speeds may be selected through pulley or gear combinations to match bit diameter and material hardness.

Feeds

Different vertical feed rates, in inches per revolution, are available via the power train.

Types of Drill Presses

Bench Type

This is generally limited to a size of 15 in (380 mm) and a travel of 4 in (102 mm). It is limited in application by the distance from the bit to the pedestal.

Floor Type

This is essentially the same as the bench type except for the longer column and greater clearance for a higher or larger workpiece. The smallest shop should include a floor-type press with capacity to ¾ in. (19 mm) via a Morse taper chuck. Versatility of this unit would be augmented if it included power feed and a minimum size of 18 in.

Upright

Generally upright implies a floor-type press with heavy box or tube column and power feed.

Radial

A radial drill is one consisting of a large-diameter round column fixed to a base; an arm is attached to the column and may rotate about it; the arm may be raised or lowered; a drilling head moves radially on the arm via a lead screw; the head provides multiple spindle rotation (speeds) and feeds. It is very versatile for drilling multiple holes in large, difficult-to-move workpieces. Size is specified via column diameter and radial travel of the head.

A radial drill press affords the capability to machine larger machined components whose dimensions exceed the capacity of regular drill presses.

GRINDER

With the exception of portable or off-hand grinders, the most important function of grinders in the maintenance shop is to sharpen the cutting tools of other machines.

Types

The types most used in our applications are:

Surface Grinder with Reciprocating Table and Horizontal Spindle

This is the workhorse of the shop in terms of capability. It can be used to sharpen many types of cutters with the different-shaped grinding wheels that are available. The grinding wheel rotates on a horizontal spindle as the table reciprocates or travels right to left (and back) in a horizontal direction. It is excellent for establishing the thickness of small, critical parts.

Surface Grinder with Rotating Table and Vertical Spindle

This is commonly called a Blanchard grinder. It is used for clearance purposes to establish parallelism between two critical surfaces or to establish a bearing or mounting surface. This type of surface grinding is a prime restoration step in many maintenance operations.

Universal Tool and Cutter Grinder

This is an ingenious adaptation of the common horizontal-shaft shop grinder that employs manual table reciprocation as well as vertical and rotational adjustment of the horizontal wheelhead to enable proper access to virtually all types of cutters with the ability to hold or grind proper rake, lip, nose, and relief angles. It is a versatile but somewhat specialized tool.

Portable Grinders

Among the often-used types is the angle-head, hand-held grinder-sander which accepts a variety of abrasives and has provision to attach appropriate wheel guards. Typical abrasives are a thin, flexible disk on a rubber wheel; a thicker, rigid disk; a formed, or dished, disk; a common cup shop wheel; and a wire brush. Portable grinders handle many operations such as weld grinding, descaling, paint preparation, etc.

Handy for certain finer work is the small, high-speed hand grinder which accepts mandrel-mounted abrasive *stones*. These are often called *die grinders*.

Selection

Grinding speed is quite critical. The machine must closely match the grit and size of the wheel or stone; otherwise the abrasive will not cut properly or will tend to *load* up. More important, there is a danger of wheel disintegration, particularly at improper low speeds. Safety glasses, preferably with side shields, are a *must* in any grinding or sanding operation. Gloves and long sleeves are essential with the portable angle grinders to protect from sparks, grit, dust, or loosened particles.

SANDER

Sanders are useful in many phases of building maintenance, carpentry, or cabinet work; they are more necessary in larger shops than in smaller ones.

Types

Disk and Belt Sanding Machine

This is a nonportable bench- or stand-mounted machine specifically designed for finish woodworking. It generally combines a 10- or 12-in (254- or 305-mm)-diam disk wheel with a 6- or 8-in (152- or 203-mm)-wide belt in combination with appropriate tables to hold and feed the work.

Portable Belt Sander

This is available with a 3- or 4-in (76- or 102-mm) belt width. It is ideal for rough sanding of wood with high removal rates. Some models include self-contained dust-collection systems.

Reciprocating-Oscillating Sander

This is available in several pad sizes (effective area) from about 2 x 5 to 4½ x 9 in (51 x 127 to 114 x 228 mm). Generally, it is most effective for fine or finished woodwork.

Use

Most important with sanders is the rule, "Let the paper do the work." Little, if any, additional pressure is necessary. Choose *open-coat* abrasives that do not load. Synthetic aluminum oxide abrasives have a longer life than flint or garnet abrasive.

LATHE

The lathe is the basic machine for metal turning and is important to even the smallest shop. Size is specified by two figures, swing and distance between centers. Together they specify the diameter and length of the largest cylinder that may be turned. Swing is twice the radial distance or *clearance* between the line connecting the center of the headstock and tailstock and the nearest point of the movable carriage.

There are many varieties of lathes and the names may describe a special operational capability or a degree of precision. The lathe illustrated in Fig. 2-1 is a 14-in swing by South Bend ®, one of several manufacturers and a specialist in the small- to medium-size range of 9 to 14 in (228 to 356 mm). Available for lathes of this size and type are many attachments that broaden the capability of the machine. Milling attachments, internal and external grinders, and others may be mounted on the compound and yield added versatility. These, plus an endless variety of chucks, centers, collets, and tools, allow a good machinist to make and repair many precise and complex machinery components. The lathe may be considered the most useful machine tool.

Proper cleaning and lubrication will preserve new *fits* for many years. Careful gear changing is a must since most of the speed and feed selection is achieved via spur gearing which does not lend itself to moving meshes. Safe operation always includes wearing safety glasses and avoiding loose clothing.

Figure 2-1 Typical maintenance lathe, 14-in swing South Bend®. *(South Bend Lathe, Inc., South Bend, Indiana).*

MILLING MACHINE

The milling machine matches the lathe in utility, and a medium-sized milling machine is a basic tool in all but the smallest shops. When parts cannot be procured in a timely manner, they must be made. When this happens, a milling machine is indispensable. Choose equipment with power capability of 1 to 4 hp.

The size of a mill is specified by the power available to the milling spindle, quill (or spindle) travel, collet capacity (diameter of the tool shank that can be held), table size, table travel in three dimensions (longitudinal or side-to-side, cross or front-to-back, vertical, or up-and-down, travel of the knee), spindle speeds available, and table and/or quill feeds available. Table travel under power feed may be different from travel under manual feed.

Figure 2-2 shows the Bridgeport ® Turret Miller, an optimum-size machine for the average facility. Many attachments have been developed for this machine by its maker and other tool companies since so great a number are in use by industry. Power table feed, optical or electronic dimension readout, slotting heads, and many other versatile attachments are available. While this particular machine is in wide use, there are several other good mills which are available for consideration.

A mill gives the capability of reproducing many broken or worn machinery parts. Except in the case of very accurate or special-purpose equipment, shop tools, if well cared for, will duplicate factory tolerances when skillfully operated within their capacity. The mills described offer diverse capability; vertical or horizontal milling plus angles between; boring, slotting, reaming, slitting, straddle milling, etc., via a wide range of tooling and attachments.

To use a milling machine properly requires a certain level of skill. The operator must first understand the use of basic cutting tools, i.e., end mills, shell mills, keyway cutters, side-cutting mills, etc. It is imperative that handbooks or the machine manual be con-

Figure 2-2 Typical maintenance milling machine, 2-hp Bridgeport® turret mill. *(Bridgeport Machines Division of Textron, Inc., Bridgeport, Connecticut).*

sulted to determine the feed, speed, and depth of cut that can be used within the power and rigidity capability of the milling machine. Bearing on this is the sharpness of the cutting tool and the geometry of the grind. An operator who is not thoroughly familiar with the operating controls of a particular machine should carefully study the manual or be given thorough training.

It is foolish to risk injury to either the person or the machine. Know exactly what movement will occur when a lever or handle is moved or engaged. A milled and defaced table is a symptom of lack of knowledge and attention. All tools are potentially dangerous; yet they are vital, valuable, and versatile. Keep your mill clean, free of chips, properly lubricated, and within its load capability and it will perform for a long time.

GAUGES AND OTHER MEASURING TOOLS

Most maintenance operations will depend on an appraisal of condition or degree of wear. This appraisal is best made by comparing *original* or *new* dimensions with existing or *worn* dimensions. The smallest shop requires precision scales, levels, vernier calipers, and micrometers. These will enable a wide variety of comparative checks on shaft diameters, thicknesses, lengths, and anything within the range of the instrument. Generally, we would be limited to a dimension approximating 24 to 36 in (0.61 to 0.91 m).

Measurement of the position or coordinates of machine components in different planes or at distances beyond those above is extremely difficult, as is the similar operation of alignment verification, either during original setup or restorative operations.

Two excellent methods are available which do not depend on a *continuous* measurement device (one that extends from one surface to another). These are optical and laser measuring and alignment techniques. Both depend on the intrinsic ability to establish a reference or base line in space. This could be compared with placing widely separated parts of a machine or assembly on a very large surface plate, except that the surface plate reference can be defined in any plane. In effect, there is almost infinite flexibility in moving or placing this precise datum in space.

A combination example may illustrate the versatility of these techniques. Assume we are to place 400 ft (122 m) of 12-in (304-mm) drainage pipe in a ditch with a continuous slope of 1 in (25.4 mm) in 10 ft (3.04 m). (This is a minimum slope and hence, quite critical.) A crude technique would be to place a 5-ft (1.52-m) section of concrete pipe, establish an angle of slope by calculating the proportional rise to run, and then use a carpenter's level or a line level to maintain this. More accurate would be the older optical technique of the dumpy level or the more precise transit combined with the surveyor's calibrated rod. After the rod is placed on each section, a *sighting* or *shooting* is taken, a slope is calculated, and adjustments are made. This would establish a reasonably accurate slope compared to a perfectly level, horizontal plane. Alignment in a vertical plane would have been more difficult. All the measurements might have been to the outside of the pipe and subject to wall-thickness variation.

Imagine, instead, that you could slide each section of pipe over a perfectly straight 12-in (304-mm) diam rod, thereby realizing perfect alignment, and remove the rod when alignment is complete. Lasers give this capability. The *gun* is centered via a *spider* (centering frame) in the pipe; the beam is properly inclined. As each section of pipe is laid, a target spider is inserted in the end and the section is quickly adjusted till the beam passes through the exact center of the target and therefore the pipe is perfectly aligned. The technique is so simple as to be almost beyond belief! Each section is perfectly concentric to the desired axis, the rod, and the needlelike, nonscattered laser beam.

If your responsibility includes substantial erection and millwright work, investigate both the older optical and the newer laser techniques.

No scale, level, plumb line, stretched wire, or the like can come close to the precision and accuracy of either technique.

PORTABLE ELECTRIC INSTRUMENTS*

The variety of electric instruments grows daily. With the advent of solid-state circuitry, sophisticated instruments once found only in a laboratory or test shop are now made in hand-held varieties.

When selecting an electric instrument the most important features to look for are perhaps reliability—will it take a beating and continue to operate; accuracy—is the reading accurate within a reasonable deviation, say 2 percent; and functionality—does it do the job for me?

There are so many new manufacturers of electric gear that it is difficult to sort out the reputable brands. Be willing to consider some of the new equipment because it is quite good. However, you could waste a lot of money on your own trials. Therefore, talk to artisans, suppliers, and plant engineers.

Types of Portable Electric Instruments

The most common are discussed below.

*See also Chap. 3-7 of this Handbook.

Voltage Tester

A voltage tester is a handy pocket instrument with two probes. It is used to verify the presence of a voltage (a *live* circuit) and the approximate voltage level.

Multimeter

A multimeter is portable and battery operated; sometimes it is called a VOM, volt-ohm-milliammeter. It is used for more accurate measurements in diagnostic work, such as isolating high resistance, checking progressive voltage drop, etc., and is required in even the smallest shop.

Phase Tester

A phase tester is a portable unit which tells in which direction a three-phase motor will turn before it is connected to the line. It saves cut-and-try time and also identifies the phases for hookup or balancing purposes.

Clamp-On Ammeter

A clamp-on ammeter has loop jaws that separate to allow encircling of a current-carrying conductor without disconnecting it. The magnitude of the current is inductively measured and indicated on a meter. This tool is invaluable. It is now available in solid-state versions which also read voltage and resistance.

Megger Ohmmeter–Insulation Tester

This is another invaluable tool used to measure the breakdown resistance or resistance to ground of a current-carrying device. It is an excellent device for identifying grounded phase bars in a busway or for testing doubtful insulation on any electric device. It reads out in megohms. As a general rule the absolute minimum insulation resistance for voltages below 1000 is 1 MΩ or 1×10^6 Ω.

Others

There are numerous other instruments, but describing them all exceeds the scope of this book. Remember that by their nature these devices are somewhat fragile and should be treated with care. Meters of all types should be calibrated at least every 2 years. Some difficulties may be experienced with solid-state devices which do not have the inherent damping of the older moving-coil devices. They are often affected by spurious waveforms on a line and may yield erroneous readings. Users must be alert to this possibility in choosing an instrument.

BIBLIOGRAPHY

Scharff, Robert: *The Complete Book of Home Workshop Tools,* McGraw-Hill, New York, 1979.
Weygers, Alexander G.: *The Making of Tools,* Van Nostrand Reinhold, New York 1973.
Oberg, Erik and Franklin Jones: *Machinery's Handbook,* 18th ed. Industrial Press, New York 1969.
Carroll, Grady: *Industrial Instrument Servicing Handbook,* McGraw-Hill, New York, 1960.

chapter 16-3

Welding, Cutting, Brazing, and Soldering

by
Howard Cary
Vice President, Welding Systems
Hobart Brothers Co.
President, Hobart Brothers Technical Center
Troy, Ohio

GLOSSARY*

Arc blow Magnetic disturbance of the arc which causes it to waver from its intended path.

Arc length The distance from the end of the electrode to the point where the arc makes contact with the work surface.

Arc voltage The voltage across the welding arc. It is measured with a voltmeter.

As-welded The condition of the weld metal, welded joints, and weldments after welding and prior to any subsequent aging or thermal, mechanical, or chemical treatments.

Backhand welding A welding technique in which the welding torch or gun is directed opposite to the progress of welding. It is sometimes referred to as the *pull gun technique* in gas metal arc welding and flux-cored arc welding. See travel angle, work angle, and drag angle.

Backing Material (metal, asbestos, carbon, granulated flux, etc.) backing up the joint during welding.

Back-step welding A welding technique wherein the increments of welding are deposited opposite the direction of progression.

Base metal The metal to be welded, soldered, or cut.

Braze A weld in which coalescence is produced by heating to a suitable temperature and by using a filler metal, having a liquidus above 800°F (427°C) and below the solidus of the base metals. The filler metal is distributed between the closely fitted surfaces of the joint by capillary attraction.

Butt weld A weld made in the joint between two pieces of metal approximately in the same place. See Fig. 3-2.

Carbon steel Carbon steel is a term applied to a broad range of material containing carbon, 1.7 percent max.; manganese, 1.65 percent max.; and silicon, 0.60 percent max. See the following table for classifications.

*Refer to Fig. 3-3 for a useful illustration of definitions of common welding nomenclature.

Type of Steel	Carbon Content, %
Low-carbon	0.15 max.
Mild-carbon	0.15–0.29
Medium-carbon	0.30–0.59
High-carbon	0.60–1.70

Cast iron A wide variety of iron-base materials—containing carbon, 1.7 to 4.5 percent; silicon, 0.5 to 3 percent; manganese, 0.2 to 1.3 percent; phosphorus, 0.8 percent max.; sulfur, 0.2 percent max.; molybdenum, nickel, chromium, and copper—can be added to produce alloyed cast irons.

Covered electrode A filler metal electrode, used in arc welding, consisting of a metal core wire with a relatively thick covering which provides protection for the molten metal from the atmosphere, improves the properties of the weld metal and stabilizes the arc.

Crater A depression at the termination of a weld bead or in the weld pool beneath the electrode.

Depth of fusion The depth of fusion of a groove weld is the distance from the surface of the base metal to that point within the joints at which fusion ceases. See Fig. 3-3.

Direct current Electric current which flows only in one direction. In welding it is an arc-welding process in which the power supply at the arc is direct current. It is measured by an ammeter.

Downhill welding A pipe-welding term indicating that the weld progresses from the top of the pipe to the bottom of the pipe. The pipe is not rotated.

Drag (thermal cutting) The offset distance between the actual and the theoretical exit points of the cutting oxygen stream measured on the exit surface of the material.

Elongation Extension produced between two gauge marks during a tensile test. It is expressed as a percentage of the original gauge length which should also be given.

Face of a weld The exposed surface of a weld, on the side from which welding was done. See Fig. 3-3.

Fillet weld A weld of approximately triangular cross section joining two surfaces approximately at right angles to each other in a lap joint, T joint, or corner joint. See Fig. 3-3.

Flat position The position in which welding is performed from the upper side of the joint and the face of the weld is approximately horizontal—sometimes called *downhand welding*. See Fig. 3-4.

Flux A fusible material used to dissolve and/or prevent the formation of oxides, nitrides, or other undesirable inclusions formed in welding which might contaminate the weld.

Flashback A recession of the flame into or back of the mixing chamber of the torch.

Flux-cored arc welding (FCAW) An arc-welding process in which coalescence is produced by heating with an arc between a continuous filler-metal (consumable) electrode and the work. Shielding is obtained from a flux contained within the electrode. Additional shielding may or may not be obtained from an externally supplied gas or gas mixture.

Gas metal arc welding (GMAW) (MIG) An arc welding process in which coalescence is produced by heating with an arc between a continuous filler-metal (consumable) electrode and the work. Shielding is obtained entirely from an externally supplied gas or gas mixture.

Gas-shielded arc welding See MIG and TIG welding.

Gas tungsten arc welding (GTAW) (TIG) An arc welding process wherein coalescence is produced by heating with an arc between a single tungsten (nonconsumable) electrode and the work. Shielding is obtained from a gas (argon or helium or a mixture of them). Pressure may or may not be used and filler metal may or may not be used.

Groove weld A weld made in the groove between two members to be joined. See Fig. 3-3.

Heat-affected zone (HAZ) That portion of the base metal which has not been melted, but the mechanical or microstructure properties of which have been altered by the heat of welding or cutting.

Horizontal position See Fig. 3-4.

Impact resistance Energy absorbed during breakage by impact of a specially prepared notched specimen, the result being commonly expressed in foot-pounds.

Kerf The width of the cut produced during a cutting process.

Lap joint A joint between two overlapping members.

Leg of a fillet weld The distance from the root of the joint to the toe of the fillet weld. See Fig. 3-3.

Low-alloy steel Low-alloy steels are those containing a low percentage of alloying elements.

Melting rate The weight or length of electrode melted in a unit of time.

MIG welding See gas metal arc welding.

Open-circuit voltage The voltage between the output terminals of the welding machine when no current is flowing in the welding circuit. It is measured by a voltmeter.

Overhead position The position of welding wherein welding is performed from the underside of the joint. See Fig. 3-4.

Overlap Protrusion of weld metal beyond the bond at the toe of the weld.

Pass A single longitudinal progression of a welding operation along a joint of weld deposit. The result of a pass is a weld bead.

Peening Mechanical working of metal by means of impact blows with a hammer or power tool.

Penetration The distance the fusion zone extends below the surface of the part(s) being welded.

Porosity Gas pockets or voids in metal.

Post heating Heat applied to the work after a welding, brazing, soldering, or cutting operation.

Preheating The heat applied to the work prior to the welding, brazing, soldering, or cutting operation.

Pool That portion of a weld that is molten at the place the heat is applied.

Radiography The use of radiant energy in the form of x-rays or gamma rays for the nondestructive examination of metals.

Reduction of area The difference between the original cross-sectional area and that of the smallest area at the point of rupture; usually stated as a percentage of the original area.

Reversed polarity The arrangement of arc-welding leads in which the work is the negative pole and the electrode is the positive pole in the arc circuit. Abbreviated as DCEP.

Rheostat A variable resistor which has one fixed terminal and a movable contact (often erroneously referred to as a "two-terminal potentiometer"). Potentiometers may be used as rheostats, but a rheostat cannot be used as a potentiometer because connections cannot be made to both ends of the resistance element.

Root of a weld The points, as shown in cross section, at which the bottom of the weld intersects the base metal surfaces. See Fig. 3-3.

Root opening The separation between the members to be joined at the root of the joint.

Shielded-metal arc welding An arc-welding process wherein coalescence is produced by heating with an electric arc between a covered metal electrode and the work. Shielding is obtained from decomposition of the electrode covering. Pressure is not used and filler metal is obtained from the electrode.

Silver solder A term erroneously used to denote silver-base brazing filler metal.

Size of a weld *Groove weld:* The joint penetration (depth of chamfering plus the root penetration when specified). *Fillet weld:* For equal fillet welds, the leg length of the largest isosceles right triangle which can be inscribed within the fillet-weld cross section. For unequal fillet welds, the leg lengths of the largest right triangle which can be inscribed within the fillet-weld cross section. See Fig. 3-3.

Slag inclusion Nonmetallic solid material entrapped in weld metal or between the weld metal and base metal.

Spatter In arc and gas welding, the metal particles expelled during welding and which do not form a part of the weld.

Stick welding See shielded-metal arc welding.

Straight polarity The arrangement of arc-welding leads in which the work is the positive pole and the electrode is the negative pole of the arc circuit. Abbreviated as DCEN.

Stress relief, heat treatment The uniform heating of structures to a sufficient temperature below the critical range to relieve the major portion of the residual stresses followed by uniform cooling.

Stringer bead A type of weld bead made without appreciable transverse oscillation.

Tack weld A weld (generally short) made to hold parts of a weldment in proper alignment until the final welds are made. Used for assembly purposes only.

Tensile strength The maximum load per unit of original cross-sectional area obtained before rupture of a tensile specimen. Measured in pounds per square inch.

Throat of a fillet weld Shortest distance from the root of a fillet weld to the face. See Fig. 3-3.

TIG welding See gas tungsten arc welding.

Toe of a weld The junction between the face of the weld and the base metal. See Fig. 3-3.

Tungsten electrode A nonfiller metal electrode, used in arc welding, consisting of a tungsten wire.

Ultimate tensile strength The maximum tensile stress which will cause a material to break. Usually expressed in pounds per square inch.

Underbead crack A crack in the heat-affected zone not extending to the surface of the base metal.

Undercut A groove melted into the base metal adjacent to the toe of the weld and left unfilled by weld metal.

Uphill welding A pipe-welding term indicating that the welds are made from the bottom of the pipe to the top of the pipe. The pipe is not rotated.

Vertical position The position of welding in which the axis of the weld is approximately vertical. See Fig. 3-4.

Weaving A technique of depositing weld metal in which the electrode is oscillated.

Weld A localized coalescence of metal in which coalescence is produced by heating to suitable temperatures, with or without the application of pressure and with or without the use of filler metal. The filler metal either has a melting point approximately the same as the base metal or has a melting point below that of the base metal but above 800°F (427°C).

Weld metal That portion of a weld which has been melted during welding.

Welding procedure The detailed methods and practices including joint welding procedures involved in the production of a weldment using a specific process.

Welding rod Filler metal, in wire or rod form, used in gas welding and brazing processes, and those arc-welding processes in which the electrode does not furnish the filler metal.

Weldment An assembly, the component parts of which are joined by welding.

Whipping A term applied to an inward and upward movement of the electrode which is employed in vertical welding to avoid undercut.

BASIC PRINCIPLES

Welding is the most economical and efficient way to permanently join metal. It is the only way of joining two or more pieces of metal to make them act as a single piece. Welding is used to join almost all the commercial metals. However, some metals are more difficult to weld than others. There are more than 100 different welding processes and process variations which include brazing, soldering, and thermal cutting. These are broken into seven groups of welding processes and six groups of allied processes, as shown by the master chart of welding and allied processes, Fig. 3-1.

Many of the processes can be applied in different ways; i.e., they may be applied as manual, semiautomatic, machine, or fully automatic processes. Manual-process applications require a high degree of skill, while automatic welding requires a minimum amount of welding skill. This is important with respect to training and qualification of personnel.

There are certain fundamentals which must be grasped if one is to properly understand welding. Some of the most important are as follows:

The *weld joint* relates to the junction arrangement of the members to be joined. There are five basic types of joints, similar to those used by other crafts. These are: the butt joint, the corner joint, the edge joint, the lap joint, and the T joint. They are sometimes used in combination.

Another basic concept relates to the *type of weld.* See Fig. 3-2. The weld is the localized coalescence of metal at the specific junction of the parts. The many different types of welds are best described by the shape they show in cross section. Most popular are the *fillet* welds, followed by the *groove* weld. There are seven basic types of groove welds: square groove, bevel groove, V groove, J groove, U groove, flare V, and flare bevel. Many of these can be used in combination as double-groove welds. In order to completely describe a weld joint, the weld and the joint should both be defined (a "single V-welded butt joint," for example). See Fig. 3-3.

Some welds use *filler metals.* In the resistance-welding processes filler materials are normally not used. In the arc-welding processes filler metals are usually used. When filler metal is used, it must be properly specified in order to produce a weld joint of the specified strength.

The various *welding positions* (see Fig. 3-4) are particularly important when the welder's skill is involved. There are four basic positions: flat, horizontal, overhead, and vertical. These positions are rather obvious, but specific definitions are used. In identifying welding procedures, the position of the weld is highly important.

Another important factor is the *type of process:* electric or some other. The arc-welding processes are all electrically related, as are the resistance-welding processes. For gas welding, heat is obtained by chemical reactions of one type or another. In other cases, pressure is used, and it can be applied in various ways. In general, it is best to refer to the master chart of processes and then to the specific process that is to be employed.

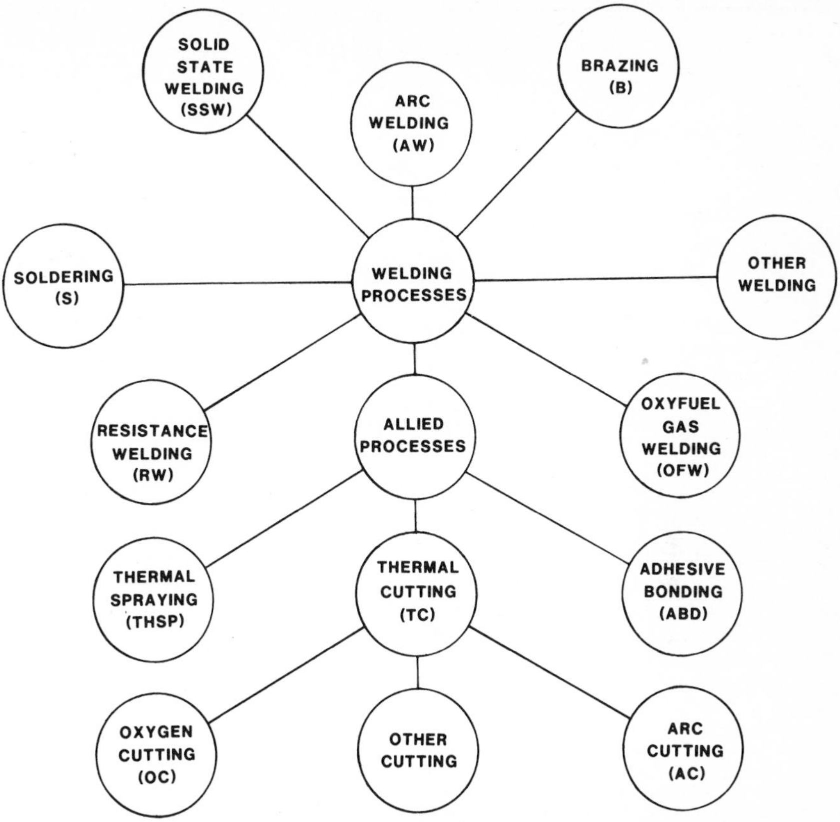

Figure 3-1 American Welding Society master chart of welding and allied processes.

Each process has its specific advantages and rationale for use. Many of them also have specific shortcomings and can not be used effectively in certain applications.

The safety and health of welders must always be considered. Welding is no more dangerous than other industrial occupations providing that the recommended safety precautions are followed. These precautions are given here and also appear on labels of filler metals and fluxes and on equipment for making welds.

This chapter describes the more popular welding processes in sufficient detail so that they can be properly understood and effectively applied. Included also are a glossary of the basic definitions of welding, a discussion of safety and health aspects, and guides to selecting filler material for welding various metals.

WELDING SAFETY AND HEALTH

These comments on safety are presented early in this chapter because it is important to understand the hazards involved *before* any welding operations are planned or begun.

Welding is no more hazardous than any other metalworking occupation providing that proper precautionary measures are followed. Hazards exist with any welding process. Welding is safe when safe practices are followed. The welder should follow safety precautions and supervisors must enforce safety regulations. The two most important

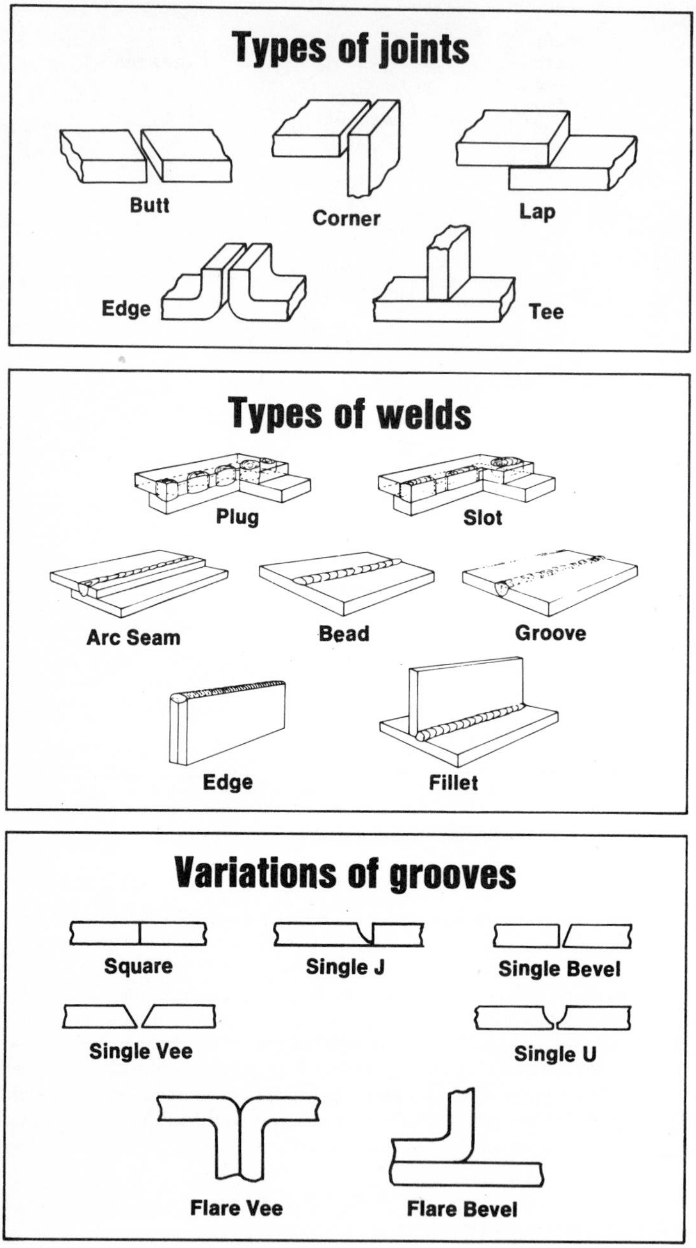

Figure 3-2 Types of joints, welds, and grooves. *(American Welding Society.)*

GROOVE WELD

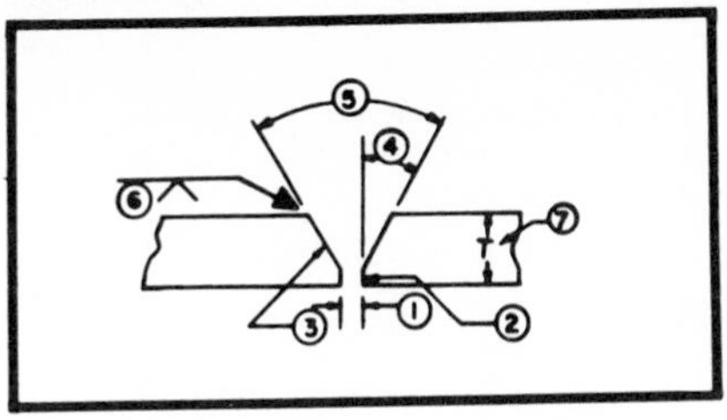

1. ROOT OPENING (RO): The separation between the members to be joined at the root of the joint.
2. ROOT FACE (RF): Groove face adjacent to the root of the joint.
3. GROOVE FACE: The surface of a member included in the groove.
4. BEVEL ANGLE (A): The angle formed between the prepared edge of a member and a plane perpendicular to the surface of the member.
5. GROOVE ANGLE (A): The total included angle of the groove between parts to be joined by a groove weld.
6. SIZE OF WELD(S): The joint penetration (depth of chamfering plus root penetration when specified).
7. PLATE THICKNESS (T): Thickness of plate welded.

FILLET WELD

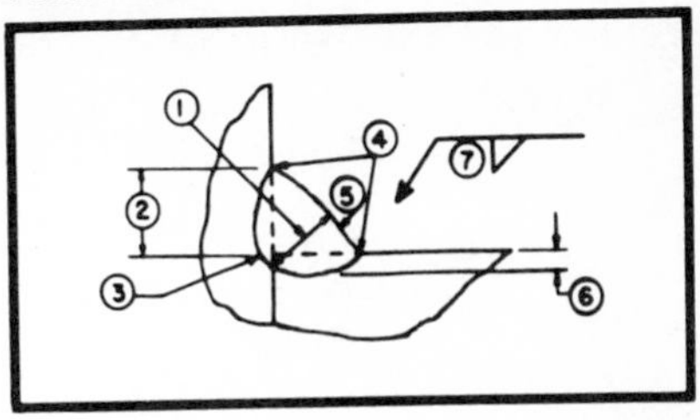

1. THROAT OF A FILLET WELD: The shortest distance from the root of the fillet weld to its face.
2. LEG OF A FILLET WELD: The distance from the root of the joint to the toe of the fillet weld.
3. ROOT OF WELD: Deepest point of useful penetration in a fillet weld.
4. TOE OF A WELD: The junction between the face of a weld and the base metal.
5. FACE OF WELD: The exposed surface of a weld on the side from which the welding was done.
6. DEPTH OF FUSION: The distance that fusion extends into the base metal.
7. SIZE OF WELD(S): Leg length of the fillet.

Figure 3-3 Common terms applied to a weld. *(American Welding Society.)*

regulations concerning the subject are the American National Standard Z49.1, "Safety in Welding & Cutting," available from the American Welding Society and OSHA, "Safety & Health Standard 22CFR1910," available from the U.S. Department of Labor.

General Safety Rules

To protect yourself and others, read and understand these rules.

1. Electric shock
- **a** Electric shock can kill.
- **b** Do not touch live electric parts.
- **c** Make sure that the welding equipment is properly installed, the case is grounded, and the equipment is in good working condition.
- **d** Avoid welding in a wet or damp area. If this is unavoidable wear rubber boots and stand on a dry, insulated platform. Stay dry.
- **e** Always use insulated electrode holders. When not in use, hang the holder on brackets provided. Never place it under your arm.
- **f** Make sure that all electric connections are tight, clean, dry, and insulated.
- **g** Never attempt to repair electric equipment inside the welding machine or inside control panels, etc.
- **h** Make sure that power cables are insulated. Make sure that welding cables are insulated. Do not wrap cables around your body.
- **i** Do not use cables with frayed, cracked, or bare spots in the insulation. If there is a splice in the welding cable, make sure it is tight and insulated.

2. Arc radiation
- **a** Arc rays can injure eyes and burn skin.
- **b** Protect your eyes from the rays of the arc. Wear a head shield with the proper filter shade when welding or cutting. See a lens shade selector chart.

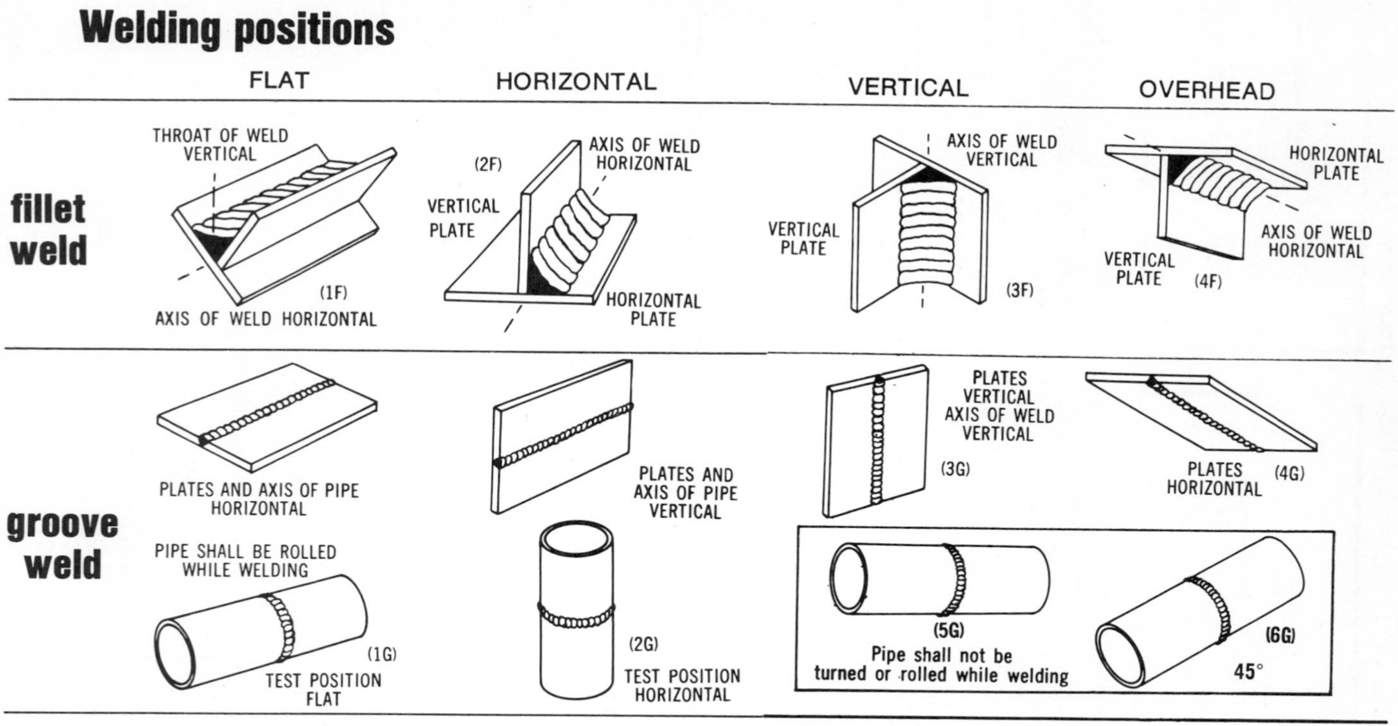

Figure 3-4 Welding positions. *(American Welding Society.)*

  - **c** Be sure protective equipment is in good condition. Wear safety glasses in the work area at all times.
  - **d** Wear protective clothing suitable for the welding work being done. Wear leather gloves and aprons with sleeves for heavy-duty welding. Protective clothing should shield the skin from arc rays.
  - **e** Do not weld near degreasing operations. Arc rays may turn vapors into dangerous fumes.
  - **f** Protect others from arc rays or flash with protective screens or barriers painted with nonreflecting paint.

3. Air contamination
  - **a** Fumes and gases can be dangerous to your health.
  - **b** Keep your head out of the fumes. Do not get too close to the arc.
  - **c** Use enough ventilation and/or exhaust at the arc to keep fumes and gases from your breathing zone. Use natural drafts or fans to keep fumes away from your face.
  - **d** Use mechanical exhaust when welding lead, cadmium, chromium, manganese, beryllium, bronze, zinc, or galvanized steel.
  - **e** Do not weld in confined spaces without extra precautions.
  - **f** Do not weld on plated materials or material covered with vinyl or heavy paint without mechanical exhaust. The coatings can release toxic fumes or gases.
  - **g** Read and obey the warning label that appears on all containers of welding materials.
4. Fire and explosion
  - **a** Arc welding and flame cutting involve high-temperature arcs and open flames which can create fires.
  - **b** Keep your work area neat, clean, dry, and free of hazards.
  - **c** Have fire-fighting equipment ready for immediate use and know how to use it.
  - **d** Do not weld near flammable, volatile, or explosive liquids or gases. Remove all potential fire hazards from welding area.
  - **e** Do not weld on or near fuel tanks of engine-driven equipment.
  - **f** Do not weld on containers such as drums, barrels, or tanks that may have held combustibles or hazardous materials without taking extra special precautions. See the AWS Bulletin, "Safe Practices for Welding and Cutting Containers That Have Held Combustibles."
  - **g** Do not weld on sealed containers or compartments without providing vents and taking extra precautions.
5. Compressed gases
  - **a** Handle all compressed-gas cylinders with extreme care. Keep cylinder caps on when cylinder is not in use.
  - **b** Make sure that all gas cylinders are secured to the wall or other structural support. Protect them from mechanical shocks.
  - **c** *Never* strike an arc on a compressed-gas cylinder. A gas cylinder should *not* be a part of an electric circuit.
  - **d** When compressed-gas cylinders are empty, close the valve and mark the cylinder "EMPTY."
  - **e** Store compressed-gas cylinders in a safe place with good ventilation. Acetylene cylinders and other fuel-gas cylinders should be stored separately from oxygen cylinders. Avoid excessive heat.
  - **f** Acetylene cylinders should be stored and used in the vertical position.
6. Cleaning and chipping welds and other hazards
  - **a** Wear protective chipping goggles when chipping weld slag. Chip away from your face.

b When you are grinding or using power tools, you should wear safety glasses with side shields under the welding helmet.

c Dispose of electrode stubs in containers; stubs on the floor are a safety hazard.

d When working above ground make sure that scaffolds, ladders, or work surfaces are substantial and solid.

e When welding in high places without railings, use a safety belt or lifeline.

f When working in noisy areas or using noisy processes, wear ear protection.

g When working in confined areas take special precautions because of fire and explosion problems with fuels. Guard against inert gas or fume buildup from welding. Provide lookouts and special ventilation.

ARC WELDING AND CUTTING PROCESSES

The Shielded-Metal–Arc-Welding Process

Process

Shielded-metal arc welding (SMAW), sometimes called stick welding or manual metal arc welding, is the most popular welding process in use today (Fig. 3-5). It is an electrical

Figure 3-5 Application of shielded-metal arc welding. *(Hobart Brothers Company.)*

arc-welding process which fuses the parts to be welded by heating them with an arc between a covered consumable metal electrode and the work. Shielding is obtained from the decomposition of the electrode covering. This process became very popular in the early 1930s when the different kinds of coatings were developed. SMAW is normally manually applied and can be used for welding thin and thick steels and some nonferrous metals in all positions. The process requires a relatively high degree of welder skill. The diagram of the SMAW process in Fig. 3-6, shows the covered electrode, the core wire, the arc, the shielding atmosphere, the weld, and the solidified slag. Deposited metal is obtained from the end of the electrode, which melts and crosses the arc.

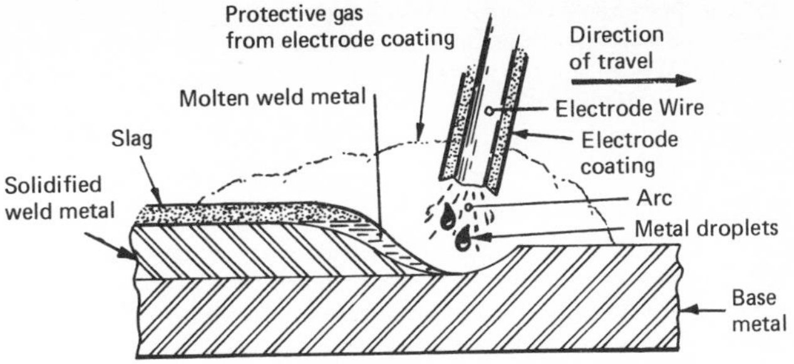

Figure 3-6 Process diagram for shielded-metal arc welding.

Application

This manually controlled process welds all nonferrous metals ranging in thickness from 18 gauge (0.048 in) to the maximum encountered. When material thicknesses are over ¼ in, a bevel edge preparation is used and a multipass welding technique is employed. The process allows for all-position welding. The arc is under the control of, and is visible to, the welder. Slag removal is required.

Equipment

The major parts needed for the SMAW process are: (1) the welding machine (power source), (2) the covered electrode, (3) electrode holder, and (4) welding leads or cables to complete the welding circuit. These are shown in Fig. 3-7.

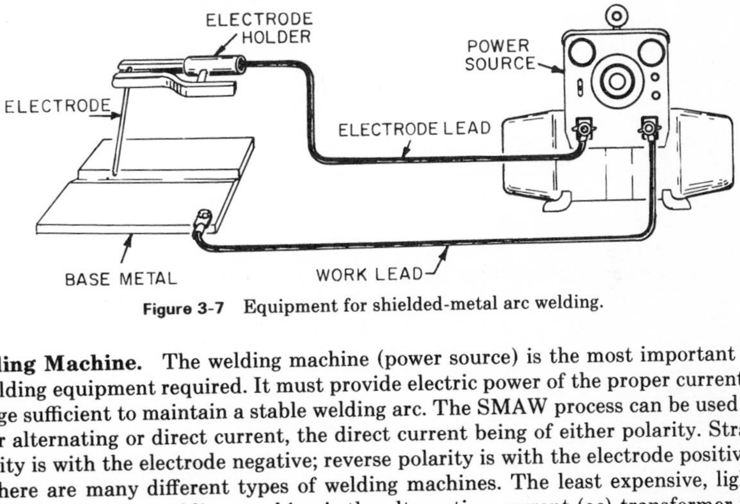

Figure 3-7 Equipment for shielded-metal arc welding.

Welding Machine. The welding machine (power source) is the most important item of welding equipment required. It must provide electric power of the proper current and voltage sufficient to maintain a stable welding arc. The SMAW process can be used with either alternating or direct current, the direct current being of either polarity. Straight polarity is with the electrode negative; reverse polarity is with the electrode positive.

There are many different types of welding machines. The least expensive, lightest weight, and smallest welding machine is the alternating-current (ac) transformer type. It provides alternating welding current at the arc. It is usually a single-control type of machine having one knob which is used to vary the current output. Other types have plug-in connectors or tap switches for this purpose. Transformer machines range from the smallest hobby type up to heavy industrial machines for automatic welding.

The rectifier-type welding machine converts ac power to dc power and provides direct current at the arc. This type of machine usually has a single control knob; however, range switches are sometimes used. These machines come in various sizes.

Another power source is the ac-dc transformer-rectifier type that is especially designed to allow either ac or dc welding. A selector switch allows either alternating current or direct current of either polarity.

Another type of power source for welding is the dual-control dc generator. This type of machine allows for adjustment of the open-circuit voltage and output slope as well as the welding current. Where electric power is available, the generator is driven by an electric motor. Away from the power line, the generator will be driven by an internal-combustion engine fueled by either gasoline or diesel oil. Belt-driven generators with a power takeoff are also available.

Electrode Holder. The electrode holder, which is held by the welder, is used to grip the electrode and carry the welding current to it. Only electrically insulated holders should be used. They come in different types, primarily the pincher type and the twist-collet type. They also come in different sizes rated according to the maximum current that can be used. Holders having larger current ratings are heavier and will also accommodate larger cables. Personal preference of the welder has much to do with the selection of electrode holders.

Welding Leads. The welding cables and connectors provide the electric circuit necessary to transmit power from the welding machines to the arc. The *electrode lead* forms one side of the circuit and runs from the electrode holder to the electrode terminal of the power source. The *work lead* (erroneously called ground lead) is the other side of the circuit and runs from the work clamp to the work terminal of the welding machine. Welding cables are made of many strands of copper wire; aluminum, however, is sometimes used. The cable is covered by a sheath of tough insulating material to protect it and avoid short circuits. The cable size is based on the welding current to be used. Cable sizes range from AWG No. 6 to AWG No. 4/10, which is the largest and is used for heavy-duty applications. The leads should be no longer than is required for the work to be done.

Covered Electrodes. Covered electrodes come in various diameters from $\frac{1}{16}$ to $\frac{1}{4}$ in, and their length is normally 14 or 18 in. The electrodes are available for welding different types and strengths of metals. This depends upon the composition of the core wire and the type of electrode coating. Other types have been designed to match most common metals and also to deposit hard surfaces.

The covering on the electrode is designed to provide (1) gas shielding—obtained by the decomposition of some of the ingredients in the coating to shield the arc from the atmosphere, (2) deoxidizers—for purifying the deposited weld metal, (3) slag formers—to protect the deposited weld metal from the atmosphere, (4) ionizing elements—to make the electrode operate more smoothly, especially on alternating current, (5) alloying elements—to provide deposited metal matched to the base metal, and (6) iron powder—to improve the productivity of the electrode.

Covered electrodes are specified by the American Welding Society Specification A5.1, "Carbon Steel Covered Arc Welding Electrodes," and A5.5, "Low Alloy Steel Covered Electrodes," as well as others for the different electrodes available.

Gas Tungsten Arc-Welding Process

Process

Gas tungsten arc welding (GTAW), also known as TIG welding, heliarc welding, heli welding and argon arc welding is one of the newer welding processes. It is illustrated in Fig. 3-8. It is an electric arc welding process which fuses the parts to be welded by heating them with an arc between a nonconsumable tungsten electrode and the work. Filler metal may or may not be used. Shielding is obtained from an inert gas or an inert gas mixture. The process is normally applied manually and is capable of welding steels and nonferrous metals in all positions. The process is commonly used to weld thin metals and for the root-pass welding on tubing and pipe. It requires a relatively high degree of welder skill and produces excellent quality welds.

GTAW was developed by the aircraft industry in the early 1940s to join hard-to-weld metals, particularly magnesium, aluminum, and stainless steels. Figure 3-9 is a diagram of the GTAW process. The tungsten electrode is fastened in a torch which also has a

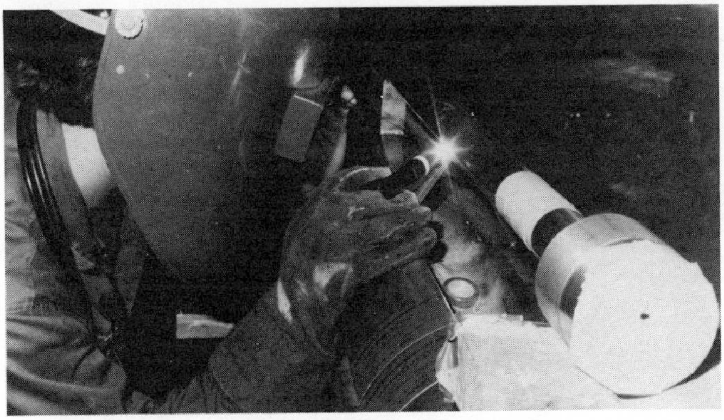

Figure 3-8 Application of gas tungsten arc welding. *(Hobart Brothers Company.)*

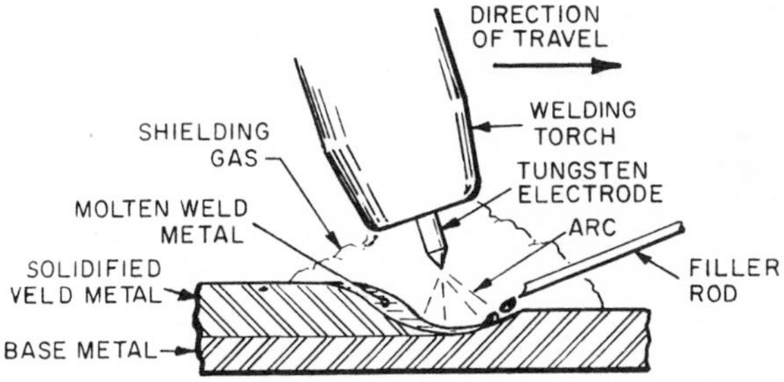

Figure 3-9 Process diagram for gas tungsten arc welding.

nozzle for directing the shielding gas around the arc area. The arc is between the tungsten electrode and the work. Filler metal in the form of a rod or wire is usually fed manually although it can be fed automatically.

Application

The outstanding features of the GTAW process are (1) the ability to produce high-quality welds on almost all metals and alloys; (2) little or no postweld cleaning is required; (3) the arc and weld pool are clearly visible to the welder; (4) there is no filler metal crossing the arc, hence little or no weld spatter; (5) welding is possible in all positions; (6) there is no production of slag which might be trapped in the weld. GTAW can be used for welding aluminum, magnesium, stainless steel, bronze, silver, copper and copper alloys, nickel and nickel alloys, cast iron, and steel. It will weld a wide range of metal thicknesses but is most popular on thinner gauges. Argon is usually used as the shielding gas, although helium or argon-helium mixtures are sometimes used.

Equipment

The major components required for GTAW are shown in Fig. 3-10. These items are (1) the welding machine or power source, (2) the GTAW torch, including the tungsten electrode, (3) the shielding gas and controls, and (4) the filler rod when required. There are several optional accessories available including a remote-controlled foot rheostat which permits the welder to control current while welding; others are arc timers and controllers, high-frequency units, water circulating systems, and specialized devices.

Welding Machine. A specially designed welding machine or power source is used for GTAW. In general, power sources for GTAW have drooping characteristics. Both alter-

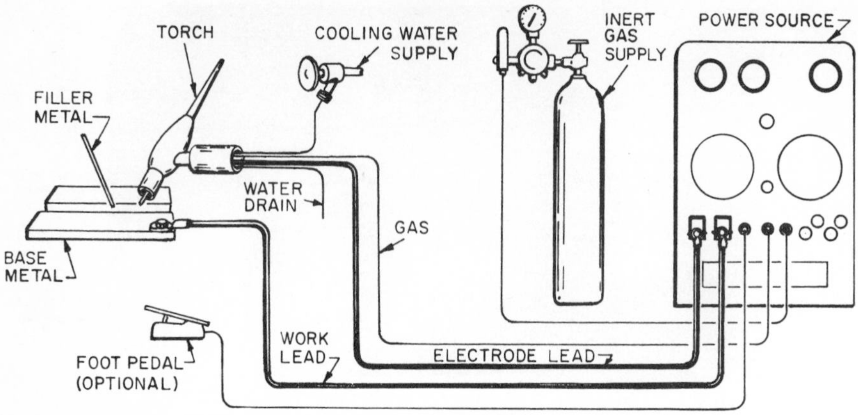

Figure 3-10 Equipment for gas tungsten arc welding.

nating and direct current are used. A transformer-type, transformer-rectifier-type, or generator-type power source can be employed. The power source usually contains a high-frequency generator which is used to aid arc starting when welding with direct current and it used continuously when welding with alternating current. The selection of alternating or direct current depends on the material being welded. Alternating current is recommended for welding aluminum and magnesium. Direct current is recommended for welding stainless steels, carbon steels, copper and its alloys, nickel and its alloys, and precious metals.

Most machines designed for GTAW include solenoid valves for controlling the shielding gas and cooling water, when used. The high-frequency spark gap oscillator is also included in the welding machine, as well as the special connectors for attaching the welding torch and cable assembly to the machine. The welding machine may also include meters and programmers. Some machines provide pulsed current capability.

Welding machines designed for GTAW can also be used for SMAW and several other processes.

It is possible to use conventional ac or dc power sources designed for SMAW. However, a high-frequency attachment is usually required, and the machine must be downrated when welding with alternating current. Best results are obtained with a welding machine specifically designed for GTAW.

Welding Torch. The GTAW torch holds the tungsten electrode and directs the shielding gas and welding power to the arc. Torches come in different sizes, and the larger sizes are usually water-cooled. The torches normally come equipped with a cable assembly which directs the gases, welding-power current, and cooling water (when used) from the machine to the torch.

Tungsten Electrodes. The electrodes used with the GTAW process are made of tungsten or tungsten alloys. Tungsten has the highest melting point of any metal (6170°F or 3405°C) and is considered nonconsumable. When properly used, the electrode does not touch the molten weld puddle. If the tungsten electrode accidentally touches the weld puddle, it becomes contaminated and must be cleaned immediately. If it is not cleaned, an erratic arc will result. Electrodes are available in three alloys as well as pure tungsten. Pure tungsten is the least expensive; however, alloys containing 1 or 2 percent thoriated tungsten are quite popular. This type of electrode is somewhat more expensive but is recommended for welding specific metals. Another alloy is the zirconiated tungsten which is often used for x-ray quality work. The different tungstens are identified by AWS Specification A5.12, "Specifications for Tungsten Arc Welding Electrodes," and are color-coded for ease of recognition. Tungsten electrodes come in diameters ranging from

0.020 in (.5 mm) up through ¼ in (6 mm). The electrode surfaces come in either a ground finish or a cleaned finish. The lengths of tungsten rods are normally 3 to 6 in (7.5 to 15 cm).

Shielding Gas. An inert shielding gas must be used. Argon is the most popular; however, helium is used in certain applications, and in some cases a mixture of argon and helium is used. Argon is more easily obtainable and is less expensive than helium. Also it is heavier than helium and provides better shielding at lower flow rates. The arc produced with helium for shielding is considered hotter and is used for obtaining deeper penetration. Helium is also used for welding in the overhead position. Helium is usually used at a higher flow rate than argon.

Filler Metal. Though filler metal may or may not be used, it is normally used except when one is welding very thin material. The composition of the filler metal should match that of the base metal. Filler metals may not be available in every alloy; therefore, filler-metal charts show the recommended type for use. The size of the filler-metal rod depends on the thickness of the base metal and the welding current. Filler metal is usually added manually to the weld puddle, but automatic feed is sometimes used. AWS specifications provide information about filler wires available.

Welding Safety

Welding safety for GTAW is essentially the same as for the other arc-welding processes; however, the process should not be used near chemical-cleaning tanks since the arc rays may change the gas vapors to poisonous vapors. Also, the filter glass shade in the helmet must be related to the proper welding current. Ventilation should be provided when one is welding in confined areas since argon, being relatively heavy, will tend to stay in the area and gradually displace the breathing air of the welder.

Gas Metal Arc-Welding Process

Process

The gas metal arc-welding process (GMAW) is an arc-welding process which fuses the parts to be welded by heating them with an arc between a continuous, consumable solid wire electrode and the work. Shielding is obtained from an externally supplied gas or gas mixture. The process is normally applied semiautomatically; however, it can be applied by machine or by automatic equipment. The process can be used to weld thin and fairly thick metals, both steel and nonferrous. The arc is visible to the welder, and it can be used in all positions. A lesser degree of welding skill is required; however, the equipment is more complex than that used for SMAW.

This process, shown in Fig. 3-11, is one of the newer arc welding processes. It was developed in the early 1950s and became extremely popular in the 1960s.

This process is sometimes called MIG welding (standing for metal–inert gas welding), or microwire welding, short-arc welding, dip transfer welding, CO_2 welding, etc. The electrode is melted in the heat of the arc, and the metal is transferred across the arc to become the deposited weld metal.

The GMAW process is also shown in Fig. 3-12. The illustration shows the electrode wire, the nozzle of the welding gun or torch, the shielding-gas envelope, and the arc between the end of the electrode and the base metal.

There are a number of variations of the GMAW process. These depend on the type of shielding gas which relates to the type of metal transfer across the arc as follows:

- MIG welding using inert-gas shielding on nonferrous metals.
- Short circuiting transfer (microwire) normally using CO_2 gas or CO_2 gas mixtures and small-diameter electrode wire allows welding in all positions and on thin metals.
- CO_2 welding using CO_2 shielding gas and larger electrode wires and is restricted to steels.

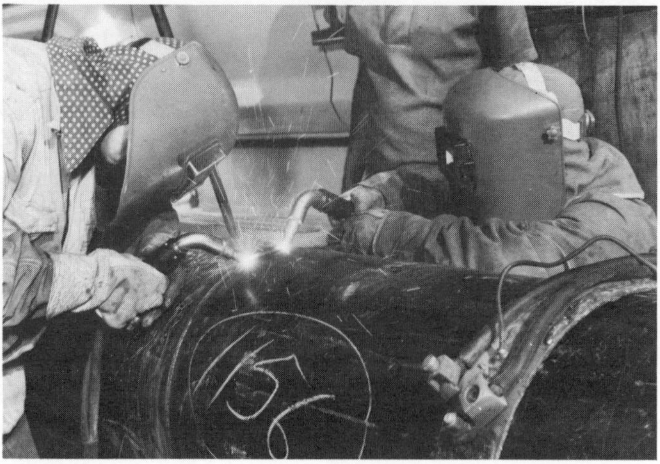

Figure 3-11 Application of gas metal arc welding. *(Hobart Brothers Company.)*

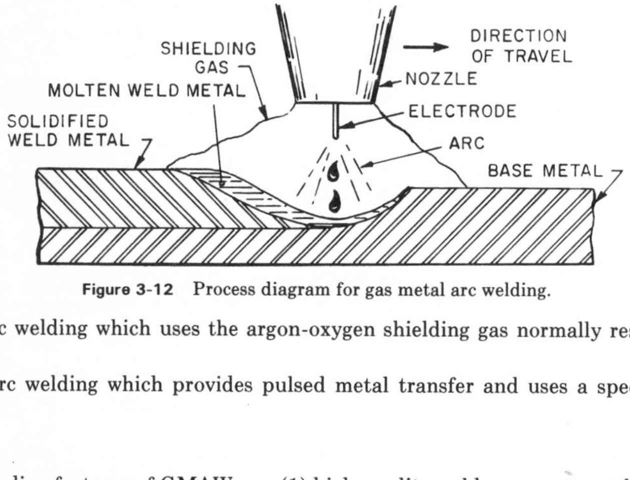

Figure 3-12 Process diagram for gas metal arc welding.

- Spray arc welding which uses the argon-oxygen shielding gas normally restricted to steels.
- Pulsed arc welding which provides pulsed metal transfer and uses a special power source.

Application

The outstanding features of GMAW are: (1) high-quality welds on most metals; (2) minimum postweld cleaning is required; (3) the arc and weld pool are visible to the welder; (4) welding is possible in all positions depending on electrode size; (5) relatively high-speed welding; (6) there is little or no slag produced; (7) it is considered a low-hydrogen-type welding process.

Variations of the process offer special advantages. The short-circuiting arc (microwire) will weld most steels in the thinner gauges. CO_2 welding allows for high-speed travel on steel. The spray arc variation produces high-speed welds with minimum spatter and cleanup, and the MIG process welds the nonferrous metals at a higher rate of speed than the GTAW process.

Equipment

Major components required for GMAW are shown in Fig. 3-13. These are (1) the welding machine or power source, (2) the electrode wire-feed system and control, (3) the welding gun and cable assembly (for semiautomatic welding or welding torch for automatic welding), (4) the shielding gas supply and controls, and (5) the consumable electrode.

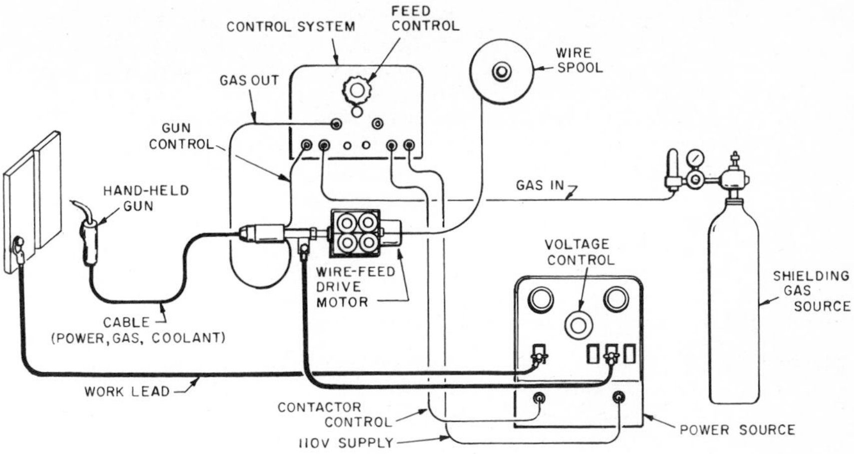

Figure 3-13 Equipment for gas metal arc welding.

Welding Machine. The power source for GMAW is normally a constant-voltage (CV) or constant-potential (CP) type. Its characteristic output volt-ampere curve is essentially flat with a small droop. Thus the output voltage is approximately the same even though the welding current changes. The output voltage is adjusted at the power source which can be a transformer-rectifier, a motor-generator, or an engine-driven generator. A CV or CP power source does not have a welding-current control and is not used for the shielded-metal arc process. The welding-current output is determined by the electric load on the machine which depends on the electrode wire-feed speed. A dc electrode positive (DCEP) arrangement is normally used. Machines for this process are available from 150 A up to as high as 1000 A and should be rated at 80 to 100 percent duty cycle. They should include a contactor and meters and should provide 115-V ac power for the electrode wire feeder.

Wire Feeder. The wire-feed system must be matched to the power supply. The CV system of welding relies on the relationship between the electrode wire burn-off rate and the welding current. This relationship is fairly constant for a given electrode wire size, composition, and shielding atmosphere. At a given wire-feed speed rate, the welding machine will supply the proper amount of current to maintain a steady arc. Thus the wire-feed speed control adjusts the welding current. The CV welding system is a self-regulating system and is recommended when using small-diameter electrode wires. A miniaturized wire feeder built into the welding gun is popular for welding with small-diameter aluminum wire. The wire-feed system and controls are essentially the same for semiautomatic, machine, or automatic welding.

Welding Gun. The welding gun and cable assembly are used to carry the electrode wire, the welding current, and the shielding gas to the welding arc. For higher-current applications, water-cooled guns are used and the water is also carried through the cable assembly. There are two general types of welding guns, the pistol-grip and curved-head (gooseneck). The gooseneck type is more popular for small-diameter-electrode wire. The pistol-grip type is usually used for welding with larger electrode wires and for welding with nonferrous electrodes. Guns used for heavy-duty work at high currents and guns using inert gas for shielding at medium to high currents are water-cooled. For machine or automatic welding, a welding torch is used. The automatic torches are either air- or water-cooled depending on the welding application, as mentioned above. For CO_2 welding a side-delivery gas nozzle is often used with automatic torches. The wire guides in all guns and torches must match the size of the electrode wire being used.

Shielding Gas. The shielding gas displaces the air around the arc to prevent contamination by the oxygen and nitrogen in the atmosphere. The gaseous shielding envelope must efficiently shield the arc area in order to obtain high-quality weld metal. Various shielding gases can be used depending on the process variation and the base metal being welded. Carbon dioxide is the least expensive and is very popular. Mixtures of CO_2 and argon and mixtures of argon and oxygen are also used. Shielding gas must be specified "welding grade." This means the gas has a high purity and low moisture content indicated by its dew-point temperature. The type of gas for shielding and the flow rate are given by welding procedure tables for welding various metals with the different process variations. The gas-flow rates depend on the type of gas, metal being welded, welding position, etc. When welding outside or when air currents disturb the gas shield, higher gas-flow rates are necessary. For high flow of CO_2 gas, two or more cylinders are manifolded together to avoid freezing of the CO_2 pressure regulators.

Electrode. The composition of the electrode for GMAW must be selected to match the metal being welded, the variation of the process, and the shielding atmosphere. The diameter or size of the electrode depends on the variation of the process and the welding position. All electrode wires are normally solid and bare, except for a thin, protective coating on carbon steel wires. The welding procedure tables indicate the proper electrode wire, type, and size for welding different metals. Electrode wires are available in a wide variety of diameters, spools, coils, and reels and are specified by AWS specifications.

Flux-Cored Arc-Welding Process

Process

The flux-cored arc-welding process (FCAW), also known as FabCO®*, Dualshield, Fabshield®*, self-shield, Innershield, etc., is an arc-welding process which fuses the parts to be welded by heating them with an arc between a continuous flux-filled electrode wire and the work. Shielding is obtained from gas generated by the decomposition of the flux within the tubular wire; however, additional shielding may be obtained from an externally supplied gas or gas mixture. The process, usually applied semiautomatically, is shown in Fig. 3-14. It also can be applied by machine or automatic equipment. It is normally used for welding medium-thick steels and stainless steel and for surfacing. It is not normally used for welding nonferrous metals. Small-diameter electrodes enable the use of all positions of welding. With larger-size electrodes, the welder is restricted to the flat and horizontal positions. The arc is visible to the welder and the skill level required is similar to that for GMAW.

The process diagram shown in Fig. 3-15 shows the two variations, with the optional items for the externally gas-shielded variation indicated by the dotted lines. The flux-cored electrode wire and the arc between it and the base metal are shown. The process normally produces a slag covering which must be removed after welding. The externally gas-shielded variation was the original process and employed CO_2 for external shielding. The other, or self-shielding variation, generates sufficient shielding gas from the decomposition of the ingredients in the core of the electrode wire. In either case, the gas shield prevents the atmospheric oxygen and nitrogen from reaching the arc area. This process was developed in the mid-1950s and became popular in the 1960s.

Application

The two variations of the process provide slightly different welding features. With external shielding gas, the features of the process are (1) extremely smooth, sound, high-quality welds, (2) deep penetration, and (3) good properties for x-ray–quality welds.

The gasless or self-shielding variation offers the following features: (1) elimination of gas supply, controls, and gas nozzle, (2) moderate penetration, and (3) ability to weld in drafts or breezes.

*® Hobart Brothers Co.

Figure 3-14 Application of flux-cored arc welding. (*Hobart Brothers Company.*)

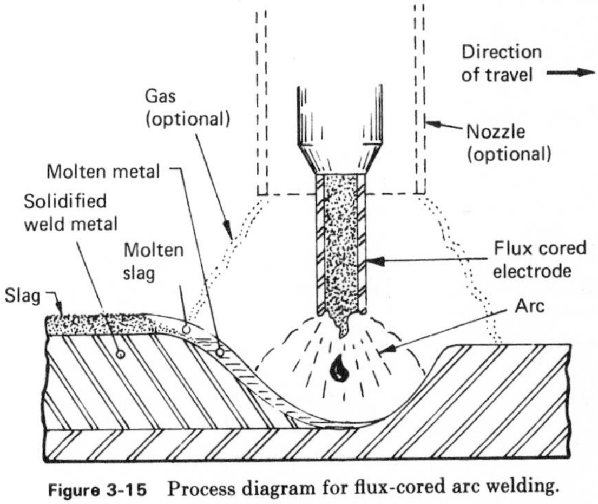

Figure 3-15 Process diagram for flux-cored arc welding.

Both variations have the following features: (1) high deposition rates, (2) visibility of the arc to the welder, (3) all position welding based on the size of the electrode, and (4) similarity of the weld-joint design to those used for the other arc-welding processes.

Both variations are normally restricted to the welding of carbon and stainless steels and for overlaying. The external gas-shielded version can be used for welding many low-alloy steels.

Equipment

The major components required for the FCAW process are shown in Fig. 3-16. Equipment is generally similar to that used for GMAW and is common for both variations

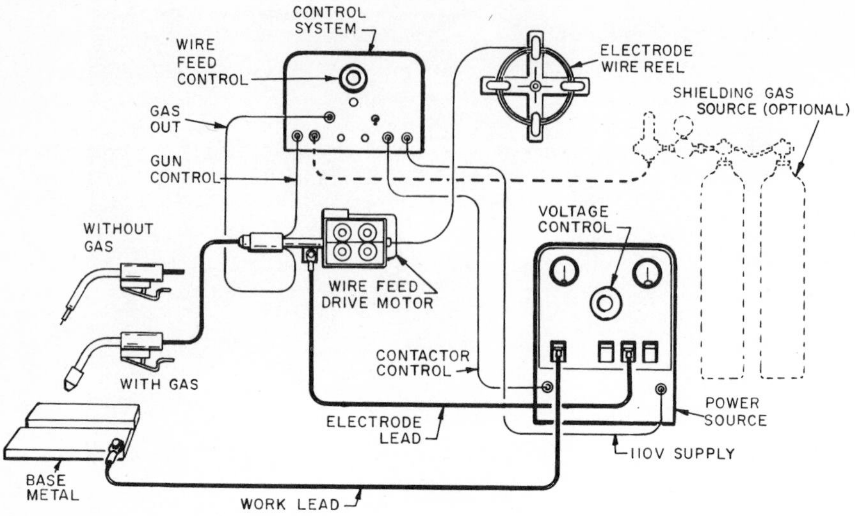

Figure 3-16 Equipment for flux-cored arc welding.

except for the gas shielding system. The items involved are (1) the welding machine or power source, (2) the wire-feed drive system and control, (3) the welding gun and cable assembly for semiautomatic welding or a welding torch for automatic welding, and (4) the flux-cored electrode wire. The external gas-shielded version requires the external shielding gas supply, flowmeter-regulator, gas valves and control, and the gas nozzle on the gun. The self-shielding type uses a lightweight gun; however, such guns often will include smoke-exhaust nozzles.

Welding Machine. The welding machine or power source for flux-cored arc welding is normally a CV or CP type. These types of welding machines have an output characteristic volt-ampere curve that is essentially flat with a minimum droop. The output voltage for the welding machine is adjusted by a control on the welding power source which can be either a transformer-rectifier or a generator driven by a motor or an engine. CV-type power sources do not have a current control and, therefore, cannot be used for welding with the SMAW process. The welding-current output is determined by the electric load on the power source. This is dependent upon the electrode wire-feed speed rate. Direct-current electrode positive (DCEP) is the arrangement normally used; however, some electrodes use direct current with the electrode negative. Alternating current is used rarely. Power sources are available for FCAW ranging from 150 to 1000 A and should be rated at 80 to 100 percent duty cycle. They should include a contactor and meters, and should provide 115-V-ac power for the electrode wire feeder.

Wire Feeder. The wire-feeding mechanism feeds the flux-cored electrode wire automatically from a coil or spool to the cable assembly and welding gun into the arc. The wire-feed system must match the type of power supply used. The CV-type power supply is normally used; therefore, a constant-speed wire-feed system with adjustable speed is used. The wire-feed speed rate controls the welding current. The CV welding system is a self-regulating system. Voltage-sensing wire-feed systems can be used when matched to a drooping-characteristic-type power source, but they are not too popular for FCAW. Basically the same type of wire feeder that is used for GMAW can be used for FCAW.

Welding Gun. The welding gun is used to deliver the electrode wire, the current, and the shielding gas (when used) to the arc area. Guns with shielding gas nozzles are water-cooled for high current, 500 A or more, heavy-duty cycle welding. Water cooling is not used for the gasless variation welding gun. Both pistol-grip and gooseneck guns are available. Sometimes with the gasless variation, a special insulated extension which adds to the electrical stickout is added to the gun to provide higher deposition rates.

Shielding Gas (External Gas-Shielded Variation). The shielding gas displaces the air around the arc area, preventing contamination by oxygen and nitrogen of the atmosphere. CO_2 is normally used as the shielding gas for steel; however, for stainless steel and certain alloy steels, a gas mixture is used. The type of shielding gas must be related to the electrode wire and base metal. Gas-flow rates depend on the type of gas being used, the metal being welded, welding position, welding current, etc. Procedure tables provide this information.

Electrode Wire. The electrode wire employed must be selected to match the composition and mechanical properties of the base metal. The selection must also be based on whether it is to be used with external shielding gas or not. Procedure tables usually indicate the type of electrode wire to be used. Various diameters are available for different applications. Electrode wires are packaged on spools, coils, and in payoff-type packs. The American Welding Society classifies flux-cored electrodes according to the strength level, properties, and deposited weld metal composition.

Submerged Arc-Welding Process

Process

Submerged arc welding (SAW), also known as welding under powder, hidden arc welding and union melt welding, is an arc-welding process which fuses the parts to be welded by heating them with an arc or arcs between a bare electrode or electrodes and the work. The arc is shielded by a blanket of granular flux on the work. The process is normally applied by machine or automatically but is used on a limited basis semiautomatically. It is used to weld medium to thick steels in a flat or horizontal position. Manual welding skill is not required; however, a technical understanding of the equipment and the welding procedure is necessary. SAW, shown in Fig. 3-17, was developed in 1930 by the

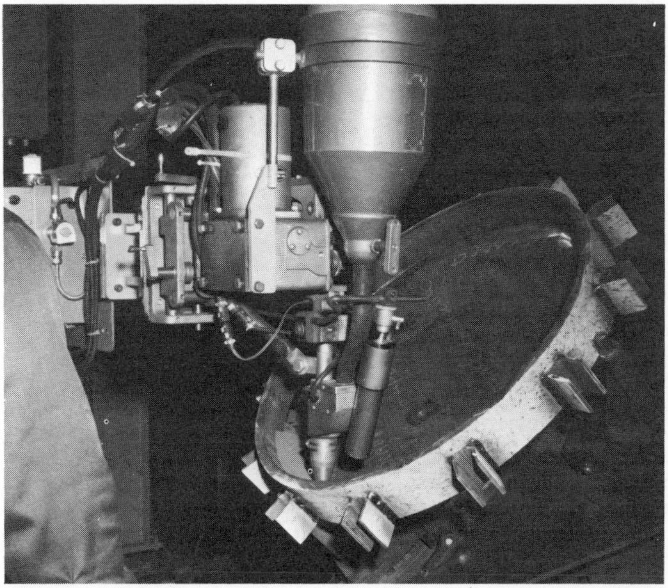

Figure 3-17 Application of submerged arc welding. (*Hobart Brothers Company.*)

National Tube Company to make longitudinal welds in pipe. It has become extremely popular for heavy plate welding because it produces high-quality weld metal at a minimum cost. Figure 3-18 shows the base metal, the consumable electrode wire, the granular

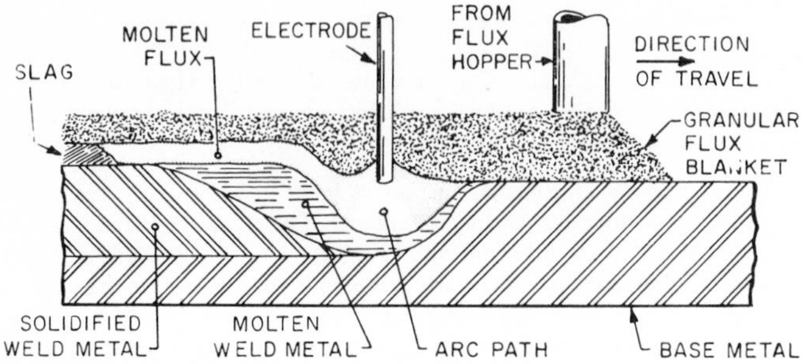

Figure 3-18 Process diagram for submerged arc welding.

flux covering, the slag cover, the arc area, and the molten metal. SAW is normally used for welding steels and is not used for welding nonferrous metals.

Application

The outstanding features of the SAW process are (1) high metal deposition rates, (2) high welding travel speed, (3) deep penetration, (4) good x-ray quality, (5) smooth weld appearance, (6) easily removed slag covering, and (7) a wide range of weldable metal thicknesses. The arc is not visible to the welder. The automatic or machine methods of application are most commonly used. The semiautomatic application method is less popular. SAW is used to weld low- and medium-carbon steels, low-alloy high-strength steels, quenched and tempered steels, and many stainless steels. It is also used for hardsurfacing, hardfacing, and buildup work. Metal thicknesses ranging from 16 gauge to ½ in are welded with no edge preparation. With edge preparation and multiple-pass welding, the maximum thickness welded is practically unlimited. SAW is restricted to the flat and horizontal positions.

Equipment

The major equipment components required for SAW are shown in Fig. 3-19. These are (1) the welding machine (power source), (2) the wire-feeding mechanism and control, (3) the welding torch for automatic welding or the welding gun and cable assembly for semiautomatic welding, (4) the flux hopper and flux feeding mechanism, and (5) the travel mechanism for automatic welding. A flux recovery system is usually included in an automatic installation.

Welding Machine. The welding machine or power source for SAW can be either an ac or dc power source. It must be rated at a 100 percent duty cycle since welding operations are continuous and the length of time in operation will normally exceed the 10-min base period used for rating duty cycle. For dc SAW, the CV-type or CC-type power source can be used. The CV type is more common for small-diameter electrode wires, usually ⅛ in and smaller in diameter. The CC type is more commonly used for larger-diameter electrode wires, usually 5/32 in and larger. The wire feeder must be matched to the type of power source used. When alternating current is employed, the machine is a CC type. Welding machines for SAW range in size from 200 to 1000 A. In some cases two or more electrode wires are employed in the same puddle, and in other cases one electrode may be on direct current and the other on alternating current.

Wire Feeder. The wire-feeding mechanism and its associated control feed the electrode wire into the welding arc. When a CC or drooping-type power source is employed,

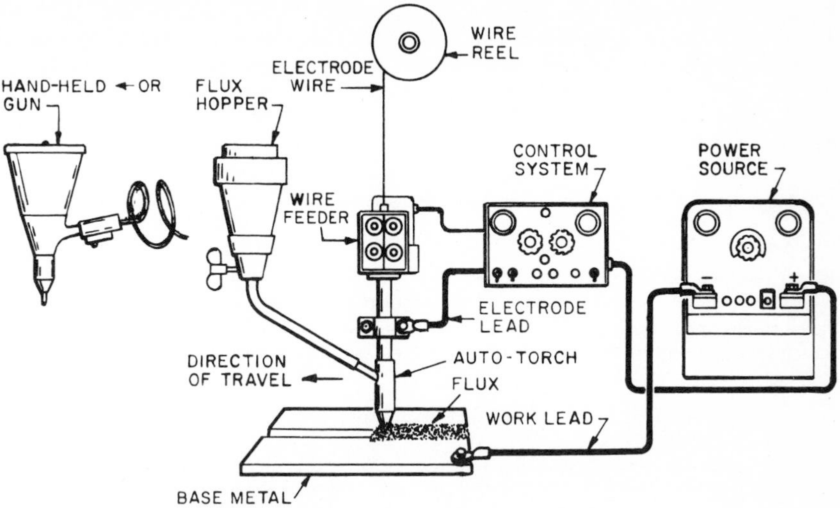

Figure 3-19 Equipment for submerged arc welding.

a voltage-sensing wire-feeder system must be used. This type of wire feeder maintains a specific arc voltage and feeds the electrode wire at the proper rate to maintain this value. If a CV or flat-characteristic power source is used, the constant-speed wire feeder and control should be employed. In this case, the wire feeder feeds the electrode wire at a constant but adjustable rate in order to draw the prescribed welding current from the power source. The arc voltage is adjusted by changing the output voltage of the power source. The control system initiates the arc, provides the proper electrode wire-feed speed and, in automatic operation, performs other necessary functions such as start and stop of fixture travel.

Welding Torch or Gun. For automatic welding the torch directs the electrode wire into the arc and transfers the welding current to the wire as it leaves the torch. For automatic welding, the torch is usually attached to the electrode wire-feeder and travel mechanism. A flux hopper is usually attached to or is adjacent to the torch. For semiautomatic operations a welding gun and cable assembly are used to transmit the electrode wire and the welding current to the arc and to provide the flux at the welding zone. A small flux hopper may be attached to the gun, and it dispenses flux over the weld area in accordance with the manipulation of the gun. In another system, the flux is fed through a conduit to the gun from a hopper and is dispensed at the welding zone. Semiautomatic guns usually have a trigger switch for initiating the arc.

Welding Flux. The SAW flux is a granular, fusible material which is poured over the arc area. This flux performs the same functions as the covering on a coated electrode. It protects the arc and molten metal from atmospheric contamination, acting as a scavenger to clean and purify the weld metal. Additionally, it may be used to add alloy elements to the deposited weld metal. A portion of the flux is melted by the heat of the welding arc. The molten flux then cools and solidifies, forming a slag on the surface of the weld. The portion of the flux which is not melted can be recovered and reused. There are different grades and types of submerged arc flux and it is important to select the proper flux-wire combination to match the chemistry and properties of the metal being welded. AWS Specification 5.17 provides the information necessary to match the properties of the metal being welded.

Electrode. The electrode wires used for SAW are usually solid and bare except for a thin, protective coating on the surface, usually copper. The electrode contains deoxidizers which help clean and scavenge the weld metal to produce a quality weld. Alloying elements may also be included in the composition of the electrode. The electrode composition and the type of flux must be matched to the requirements of the base metal in order to provide a quality weld. This is covered by AWS Specification 5.17. Electrode wires are available in diameter sizes of 1/16, 5/32, 5/64, 3/32, 1/8, 3/16, 7/32, and 1/4 in. Wire is usually available in coils ranging from 50 to 1000 lb.

The Electroslag Welding Process Consumable Guide System

Process

Electroslag welding (ESW), also known as Porta-Slag* or slag welding, is a welding process that fuses the parts to be welded with molten slag which melts the filler metal and the surfaces of the work to be welded. The molten weld pool is shielded by a slag covering which moves along the joint as welding progresses. The process, shown Fig. 3-20, is not

Figure 3-20 Application of electroslag welding.(*Hobart Brothers Company.*)

an arc-welding process, except that an arc is used to start the process. After stabilization the molten slag provides the necessary heat for welding. The process is applied automatically. It is a limited-application process used for making vertical welds on medium to heavy thicknesses of steel. Manual welding skill is not required, but a technical knowledge of the process is necessary to operate the equipment.

A diagram of the electroslag welding process is shown in Fig. 3-21. Molding shoes are used to form a cavity with the parts to be welded; this cavity contains the molten flux pool, the molten weld metal, and the solidified weld metal. Shielding from the atmosphere is provided by the pool of molten flux. The consumable guide variation is the simplified version of electroslag welding. This variation is shown in detail in Fig. 3-22. The electrode is directed to the bottom of the joint by a guide tube. The guide carries

*®Hobart Brothers Co.

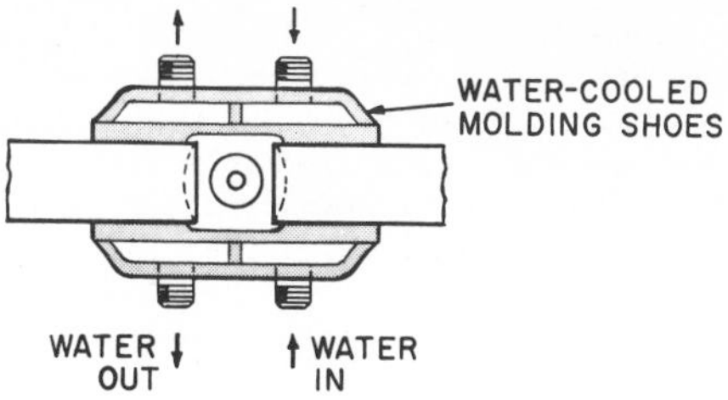

Figure 3-21 Process diagram of electroslag welding (top view).

the welding current and transfers it to the electrode which passes through its hollow core. The guide tube melts off just above the flux bath. The electrode wire protrudes into the molten flux bath and gradually melts as it is fed deeper into the molten pool. The melted metal from the guide tube, from the electrode wire, and from the edges of of the joint collect at the bottom of the flux pool and form the molten weld metal. The molten weld metal slowly solidifies and joins the parts being welded. There is no arc except at the start of the weld before the granulated flux melts from the heat of the arc to become the molten slag. Welding is done with the axis of the joint in the vertical position. The molding shoes are usually water-cooled, and their surface determines the contour of the finished weld. The shoes are fixed and nonsliding. The other version of ESW, not using the consumable guide, uses sliding shoes that move upward along the joint as the weld is made. The electrode feed head is usually mounted above the weld joint and does not move. For welding thicker materials, the head may be oscillated to provide a wider joint. Extra electrodes and guides may be employed for making extra-wide joints. The square-groove weld joint is normally used. The welding process is limited to a minimum thickness of ¾ in (20 mm) and a maximum thickness with one power source of 3 in (7.5 cm). With two wires and additional power sources, the thickness can be increased to 12 in (30 cm). Once the system is started it continues automatically until the weld is completed. Additional flux is added until the joint is completed. Starting tabs and runoff tabs are employed. They are removed when the weld is completed. A thin slag covering, which is easily removed, adheres to the surface of the welds.

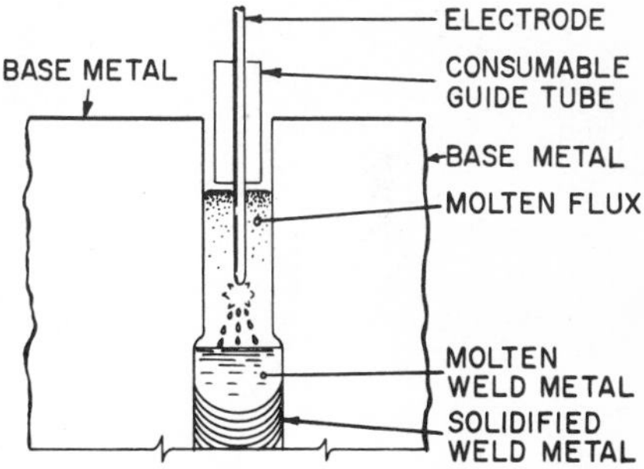

Figure 3-22 Process diagram of electroslag welding (side view: consumable guide tube).

Application

The electroslag process using the consumable guide version has the following features: (1) extremely high metal deposition rates, (2) ability to weld thick materials in one pass, (3) joint preparation and fit-up requirements more tolerant than those for other arc-welding processes, (4) little or no angular distortion, (5) electrode utilization approaching 100 percent, and (6) flux consumption much lower than with the submerged arc process. In addition, once the weld is started, the process is continuous until the weld is completed.

The electroslag process with the consumable guide will weld low-carbon steels, low-alloy high-strength steels, medium-carbon steels, alloy steels, and stainless steels. Quenched and tempered steels can be welded; however, subsequent heat treating is required to maintain weld-joint properties. This is because of the slow cooling cycle inherent in the process. The process can be used for welding joints from as short as 4 in (10 cm) to as long or as high as 12 ft (4 m). A single electrode and guide are used on materials ranging from ¾ to 2 in (20 mm to 5 cm) in thickness. For materials from 2 to 5 in (5 to 12 cm) in thickness, the electrode and guide tube are oscillated in the joint. With material from 5 to 12 in (12 to 30 cm), two electrodes and guide tubes are used and are oscillated in the joint. Oscillation ensures an even distribution of heat in the joint and maintains uniform penetration into the base metal. Use of oscillation reduces the number of electrodes required for welding thicker materials.

Equipment

The major components required for the consumable guide version of electroslag welding are shown in Fig. 3-23. These are (1) the welding machine or power source (one required

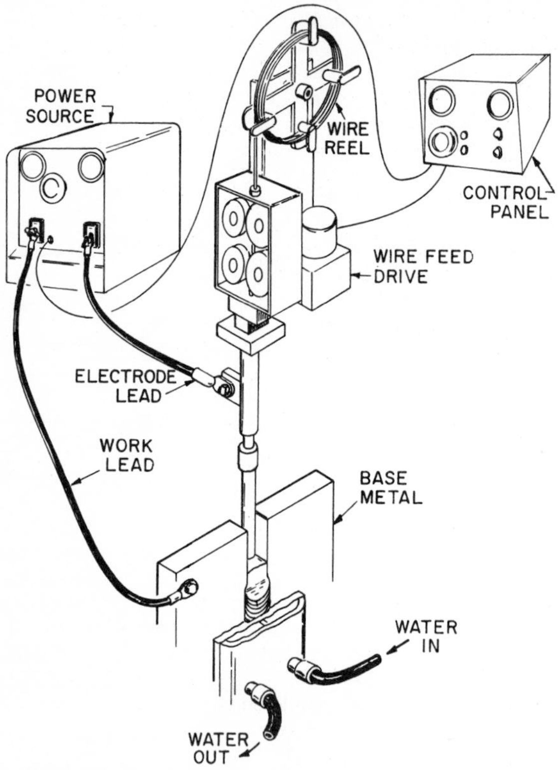

Figure 3-23 Equipment for electroslag welding.

per electrode), (2) the electrode feed head or wire feeder and control system, (3) the mounting device and oscillation mechanism, when required, and (4) the molten-metal-retaining shoes, usually water-cooled. The consumable guide version allows the welding equipment to be taken to the work and is normally mounted on the work itself.

Welding Machine. The welding machine or power source is normally a CV dc transformer-rectifier machine. The power source must be rated at 100 percent duty cycle and must include a contactor and provisions for remote adjustment. Power sources ranging in size from 500 to 1000 A, CV, are used. DCEP is normally employed. Some variations of the process use ac power sources. The CV-type power source used for electroslag welding can be used for SAW, GMAW, and FCAW.

Electroslag Flux. The flux used for ESW must be designed specifically for the process. It must have a balanced composition to provide the proper electric conductivity in the molten state as well as the proper viscosity and melting temperature. It should have satisfactory shielding characteristics and must have specific deoxiding properties. There is no standard specification for electroslag flux. The amount of flux consumed depends on the fit of the molding shoes to the face of the weld joint. It is normally a fairly constant amount and can be determined by the face width and length of the weld.

Electrode Wire. Electrode wire used for ESW must be matched to the base metal being welded. Normally, electrode wires can be of a composition similar to those used for CO_2 gas-shielded GMAW. Electrode wire must have a minimum covering of copper to minimize seizing to the inside of the guide tube. The electrode is normally solid wire, and the most popular size is 3/32 in diam. AWS Specification No. A5.25 is used for specifying electrode wires for electroslag welding.

Guide Tube. The consumable guide version of ESW requires a guide to transmit the electrode wire to the bottom of the weld-joint cavity. Normally, a heavy-wall seamless tube can be used. The size is usually 5/8 in OD and the composition must match the composition of the metal being welded. The guide tube melts during the process and contributes a small portion of the deposit welding metal. Guides of other shapes than round tubes can be employed. In some cases the guide tube is coated with a covering which matches the composition of the electroslag flux. When bare tubes are used and if the joint is extremely long, the consumable guide tube may be fitted with intermittent insulating materials to avoid short-circuiting against the side of the joint.

Electrode wire is supplied in coils or large reels or from payoff packs. Appropriate reel-dispensing equipment is required.

Plasma Arc-Welding Process

Process

Plasma arc welding (PAW), sometimes referred to as needle arc or microplasma, is an electric arc-welding process which fuses the parts to be welded by heating them with a constricted arc between the electrode and the work (transferred-arc mode). Shielding is obtained from the hot ionized gases issuing from the torch orifice. Auxiliary inert shielding gas or a mixture of inert gases supplements the shielding gas system. Pressure may or may not be used, and filler metal may or may not be used. *Plasma,* the fourth state of matter, is defined as a gas which has been heated to a high enough temperature to become ionized. When it is ionized, the gas, or plasma, becomes electrically conductive.

The process, shown in Fig. 3-24, is commonly applied manually but may be applied as a machine or fully automatic process. It can be used to weld almost all metals and can be used in all positions at lower currents. It is normally used on thinner materials. The process requires a fairly high degree of welder skill for manual application and a knowledge of the equipment. There are two ways of using the PAW process. One is known as the melt-in technique and the other, the keyhole technique. The melt-in technique is very similar to gas tungsten arc welding. The keyhole technique actually makes a hole which is then filled as welding progresses. Process details in the keyhole mode are shown in Fig. 3-25.

Application

PAW is similar to GTAW. There is one major difference in that the arc in PAW between the electrode and the work is constricted and forced to go through a small hole or orifice in the torch. A gas is also forced through the orifice; this creates the plasma. The temperature of the plasma is considerably higher than the temperature of the gas tungsten

Figure 3-24 Application of plasma arc welding. (*Hobart Brothers Company.*)

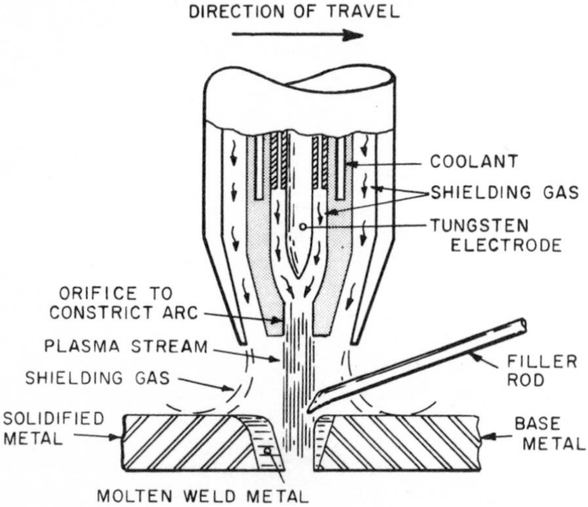

Figure 3-25 Process diagram for plasma arc welding.

arc. The process diagram (Fig. 3-25), shows the tungsten electrode inside the torch and the plasma extending through the torch orifice to the work. The plasma fed through the orifice is ionized and has a columnar form rather than the flare common with GTAW. The ionized gas, or plasma, travels at extremely high speeds and has a force action on the base material, which is important when using the keyhole technique.

One of the advantages of PAW over GTAW is the columnar structure of the plasma which reduces the effect of changes in torch-to-work distance. The high-velocity, high-temperature plasma causes deep penetration in the base metal and allows full penetration of keyhole, single-pass, butt-welding joints. The welds produced have unusually deep penetration with a relatively narrow bead width. The plasma process will weld all the metals that are welded with the GTAW process.

Equipment

The major components required for PAW are shown in Fig. 3-26 and include (1) a welding machine or power source, (2) a special plasma arc console which contains the control

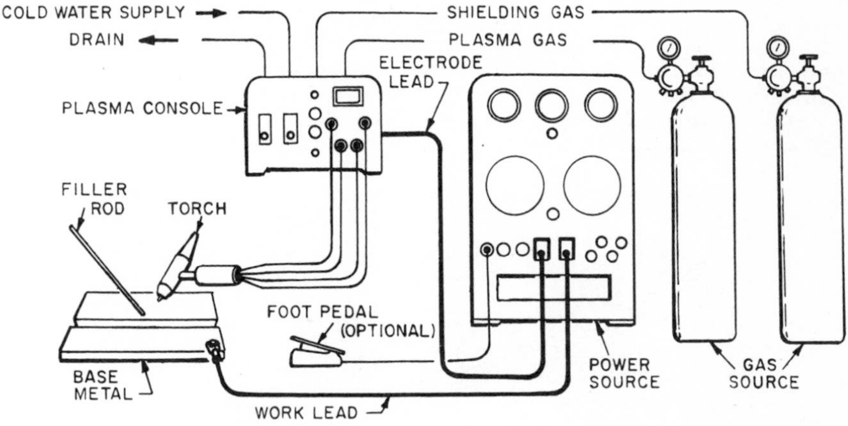

Figure 3-26 Equipment for plasma arc welding.

system, (3) the plasma welding torch, (4) the source of plasma and shielding gas, and (5) filler material when required.

Welding Machine. The power source for PAW is the CC type with a drooping output characteristic. A GTAW power source is normally used for plasma welding since it includes a contactor, remote current control, and provisions for shielding gas and cooling water. For more complex weldments, programmed current control including upslope and downslope and pulsing is sometimes used.

Plasma Console. This unit contains a high-frequency arc starter, a nontransferred pilot-arc current supply, torch protection devices, flow meters, and other meters. It may also include the water control system and protective interlocks to protect the plasma arc torch.

Plasma Torch. The torch contains a tungsten electrode (usually 2 percent thoriated type) and a nozzle having a constricting orifice. Since the arc is enclosed within the torch, all plasma torches are water-cooled. The torches are either manual or machine type with the smaller sizes restricted to the manual applications. Different sizes are available depending upon the current level to be employed.

Shielding Gas–Plasma Gas. Inert gas, often argon, is normally used as the plasma gas. In addition, argon, helium, or a mixture of the two is used as an auxiliary gas to shield the arc and arc area from the atmosphere. Argon is more commonly used because it is less expensive and easier to obtain. It also provides for better shielding because it is heavier than air.

Filler Metal. Filler metal may or may not be used. It is not used on very thin metals but is normally used for normal sheet-metal thickness and heavier material. The composition of the filler metal should match that of the base metal. Welding procedure charts will show the recommended filler material used for different base metals. The size of the filler metal (filler rod diameter) depends on the thickness of the base metal and the welding current. Filler metal is usually added to the puddle manually, but automatic feed can be used.

Stud Welding Process

Process

Stud welding (SW), also known as stud arc welding, is a special-purpose arc-welding process used to attach studs to base metal. Partial shielding is obtained by a ceramic ferrule surrounding the stud. It is a machine-welding process using a special gun that holds the stud and makes the weld. The process is normally used on steels in the flat and horizontal positions. A relatively low degree of welding skill is required.

The SW process, shown in Fig. 3-27, was developed in the mid-1930s to satisfy the

Figure 3-27 Application of stud welding. (*Ohio Edison Co.*)

need to weld brackets or retainers to steel plate, particularly in the shipbuilding industry. It became popular for securing wood decking to steel plate, for attaching bracket hangers, etc.

The operation of the SW process is shown in Fig. 3-28. This operation is as follows: (1) A stud gun holds the stud in contact with the workpiece, and the welding operator presses the gun trigger which causes the welding current to flow through the circuit to the stud, which is an electrode, to the work surface. (2) The welding current activates the solenoid within the gun which draws the stud away from the work surface and estab-

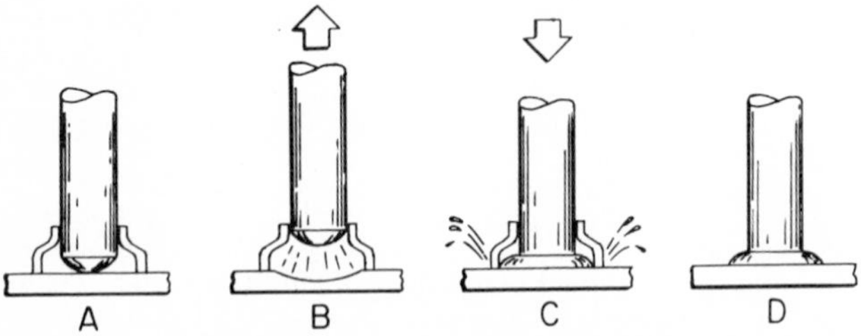

Figure 3-28 Process diagram for stud welding.

lishes an arc. The intense heat of the arc melts the surface of the workpiece and the end of the stud. The length of the arc time is controlled by a timer built into the control unit. (3) When the welding current is automatically shut off, the gun solenoid releases its pull on the stud and a spring action plunges the stud into the molten pool of the workpiece. (4) The molten stud end and the molten pool on the work surface solidify, and the stud weld is completed. The ferrule is broken off and discarded. The process is either a machine application or an automatic application, with machine application being the most popular. For automatic application, the studs are fed automatically into the gun. Welding can be done in all positions; however, flat and horizontal positions are those most commonly used.

Application

SW is widely used for attaching studs and other similar devices to plate or structural members. Studs are normally thought of as threaded, round fasteners; however, rectangular devices, hooks, pins, brackets, and other configurations can be stud-welded to appropriate backing materials. A popular application is the attachment of shear connectors in structural steelwork. Shear connectors are round, usually with a head attached to the upper flange of beams over which concrete is poured, most commonly for bridges and decking. The shear connectors ensure that the steel and concrete work together as a structural member. Another popular application is the use of studs for attaching wood decking over steel decking on ships, particularly aircraft carriers. SW is used for welding connectors used to attach pipe hangers, electric boxes, and other miscellaneous items in ship construction. It is also widely used for attaching insulation to the inside of steel structures. Other uses include attaching and holding insulation to pipe surfaces, attaching studs to hold inspection plates, etc. It is normally used for steels and stainless steels, but variations of the process can be used on nonferrous metals.

Equipment

The major components for SW are shown in Fig. 3-29. These include (1) the welding machine or power source, (2) the stud gun, (3) the control unit, (4) the studs, and (5) the disposable ferrules.

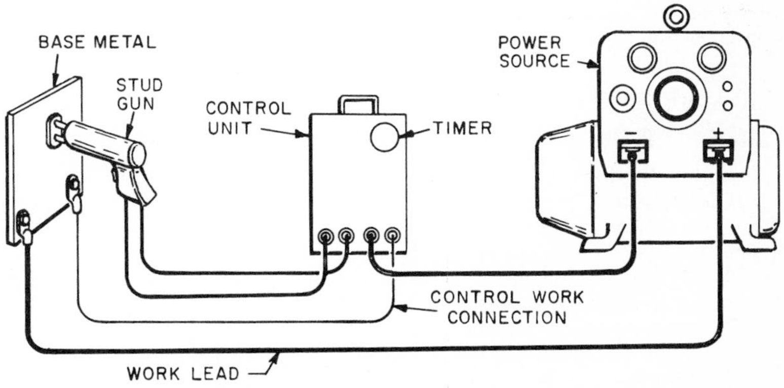

Figure 3-29 Equipment for stud welding.

Welding Machine. The welding machine or power source is a dc power source which can be a transformer-rectifier or a generator, either motor- or engine-driven. The welding current is dictated by the size or diameter of the stud. The electrode, or the stud, is the negative pole (straight polarity). Amperage required for smaller studs in the $\frac{5}{16}$-in-diam size ranges from 200 to 500 A. With larger size studs, the amperage can be as high as 2300 A. For high current requirements, two or more power sources are connected in parallel. Welding machines for SW should have high overload capacity and a relatively high open-circuit voltage of 95 to 100 V. A CC or drooping-characteristic-type power source is required.

Stud Gun. The stud gun holds the stud and has a switch which starts the control sequence. It also includes the solenoid which provides the withdrawal or lift action to establish the arc. A spring mechanism within the gun applies the pressure required to plunge or push the stud into the pool of the workpiece. The gun should be properly adjusted to accommodate the size of the stud that is being used and to provide the correct arc length during the arc period. The stud gun is normally hand-held and must be held perpendicular to the work. The process can be automated. The stud gun must match, or be of the same make as, the control unit.

Control Unit. The control unit consists of a welding-current contactor, a timing device, and the necessary interconnections. Some control units regulate the speed at which the stud is pushed into the molten base metal; this kind of regulation tends to eliminate spatter and provide more control over the weld shape and quality. The control unit must be of the same type or make as the stud gun.

The welding current passes through the stud gun to provide power for the solenoid.

Studs. Steel studs range in diameter from ⅛ to 1 in (3 to 25 mm) and vary in length; they can be threaded or plain. There are many other types of devices that can be welded with the stud system. These are shown in Fig. 3-30. Studs produced by different manufacturers contain somewhat different fluxing devices on the end of the stud. In most cases, the arcing end contains a charge of welding flux or some other device for shielding the arc area. The fluxes protect the weld and the arc from atmospheric contamination and contain scavengers which purify the melted metal. Essential welding-cycle data are shown in Table 3-1.

Ferrules. A ferrule is used with each stud. Ferrules are made of a ceramic material and are broken off and discarded after each weld is made. They shield the arc area, protect the welding operator, and eliminate the need for a helmet. The ferrule concentrates the heat during welding and confines the molten metal to the weld area. It helps prevent oxidation of the molten metal during the arcing cycle, but it must be made to fit the studs being used. There is no common specification for studs or ferrules, which are manufactured by stud-welding companies.

Air–Carbon Arc Cutting and Gouging Process

Process

The air–carbon arc (AAC) cutting and gouging process is also known as carbon-arc gouging. It is an arc-cutting process in which metals to be cut are melted by the heat of a carbon arc and the molten metal is removed by a blast of air. It is shown in Fig. 3-31. A high-velocity air jet traveling parallel to the electrode hits the molten puddle just behind the arc and blows the molten metal out of the puddle. It is usually a manually controlled operation and can be used in all positions. It can also be applied automatically. The process normally creates considerable noise, and ear protection is recommended. A special electrode holder includes the air-jet opening. The other features of the process are similar to carbon-arc welding. Figure 3-32 shows the details of the process.

Application

The AAC cutting and gouging process is used to cut metal, gouge out defective metal sections, remove old or inferior welds, back-gouge roots of welds, and to prepare grooves for welding. The AAC cutting process is used where slightly ragged edges are not objectionable. It is normally used on steels but can be used on other metals. It is popular for preparing scrap metal for remelting. The surface of some metals deteriorates when cut or gouged by this process. The area of the cut is relatively small since the molten metal

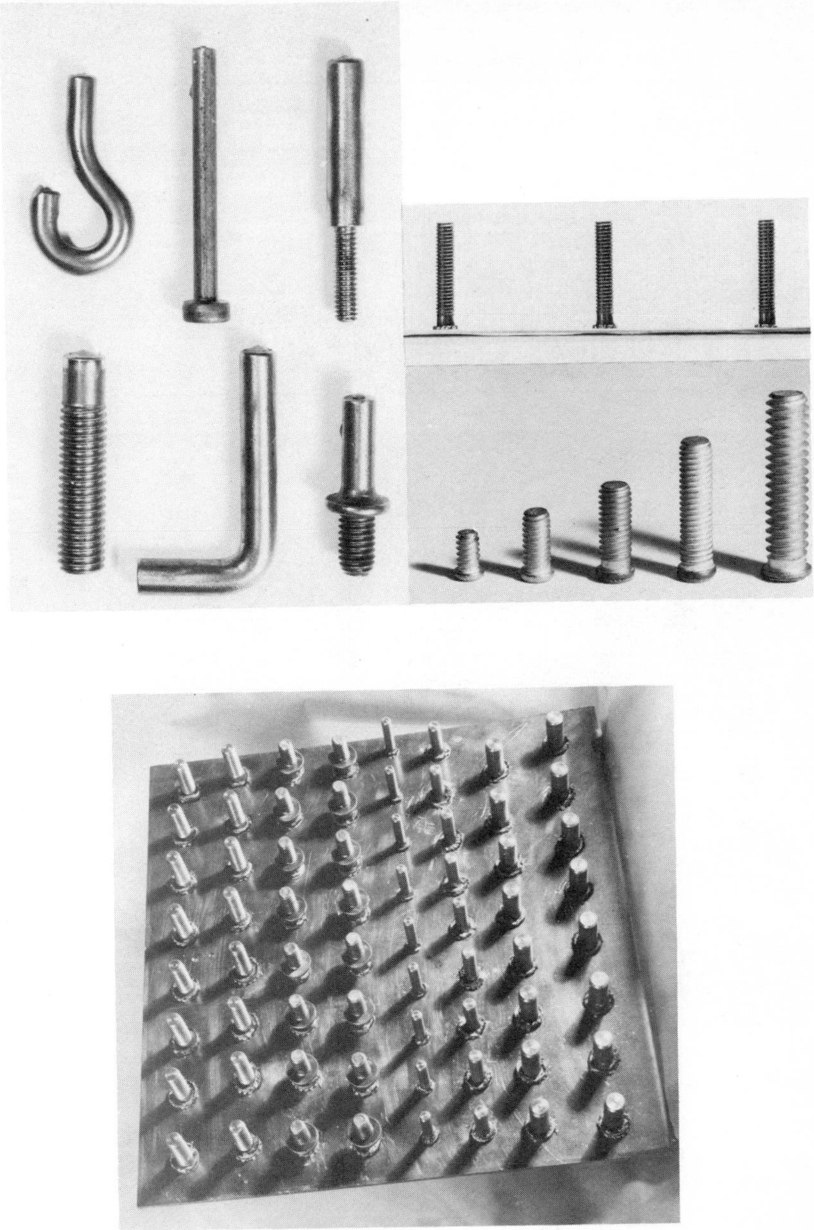

Figure 3-30 Uses for stud welding. (*Nelson Stud Welding Company.*)

TABLE 3-1 Stud-Welding Data

Stud diam, in	Current, A DCEP	Welding specifications			
		Voltage, V	Time, s	Lift, in	Plunge, in
3/16	300	30	7	1/16	1/8
1/4	400	30	10	1/16	1/8
5/16	500	30	15	1/16	1/8
3/8	600	28	20	1/16	1/8
7/16	700	28	25	1/16	1/8
1/2	900	28	30	3/32	5/32
5/8	1150	28	40	3/32	5/32
3/4	1600	26	50	1/8	3/16
7/8	1800	24	60	1/8	3/16
1	2000	24	70	1/8	3/16

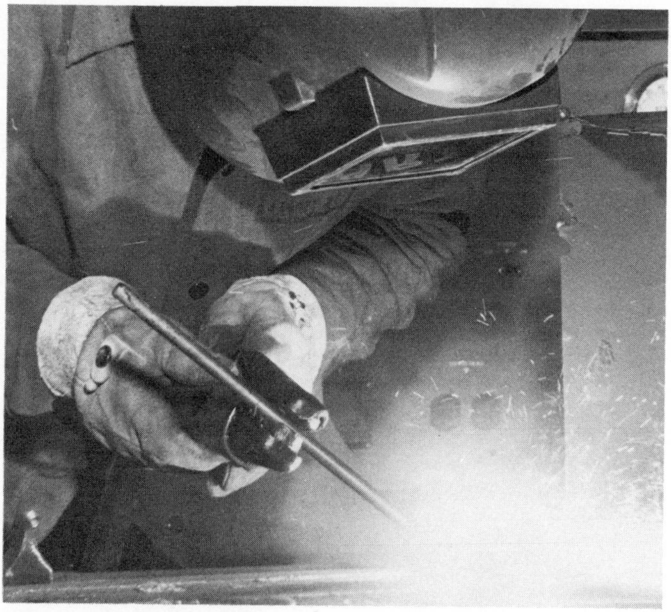

Figure 3-31 Application of air–carbon arc cutting. (*Hobart Brothers Company.*)

is quickly removed. The surrounding area does not reach high temperatures, thus reducing the tendency toward warpage and cracking. In some cases the surface must be ground to provide quality weld-joint preparation.

Equipment

Equipment for cutting and gouging is the same as for carbon-arc welding and for SMAW, with the exception of the special electrode holder and the required compressed-air supply. The necessary equipment is shown in Fig. 3-33. This consists of (1) the welding machine or power source, (2) the special electrode holder or torch, (3) a carbon electrode, and (4) the compressed-air supply.

Welding Machine. The welding machine or power source is normally a CC drooping-characteristic type either a transformer-rectifier or generator. CV machines with flat characteristics may be used, but precautions must be taken to operate them within their

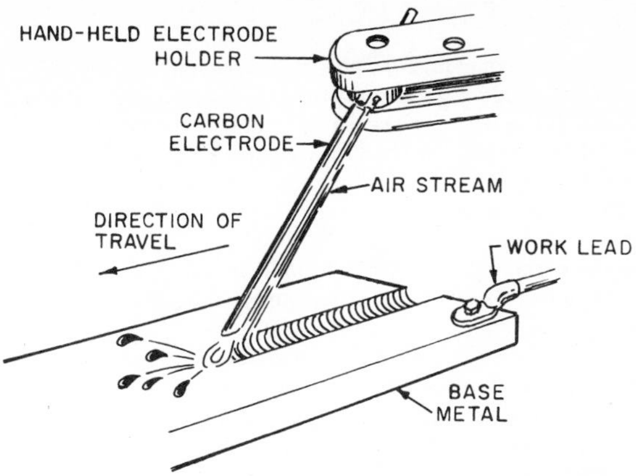

Figure 3-32 Process diagram for air–carbon arc cutting.

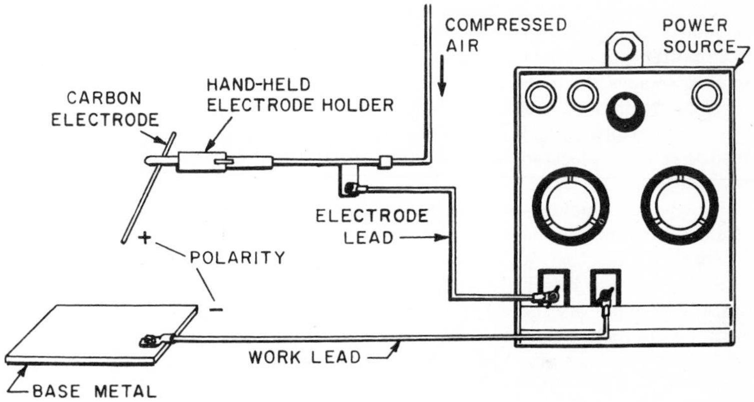

Figure 3-33 Equipment for air–carbon arc cutting.

rated output. Specially designed heavy-duty machines are used for AAC cutting or gouging with large electrodes. Machines of this type are available with a capacity up to 1000 A. The ac power source CC types can be used for special applications; however, ac-type carbons are required.

Electrode Holder. The electrode holder or torch is of a special design which includes the air-jet's stream nozzle and valve. In addition, it must clamp the carbon electrode tightly. Electrode holders come in several sizes depending on the size of the carbon electrode to be used. Larger holders may be water-cooled. The cable assembly includes the compressed-air hose which is connected to the supply.

Electrodes. Electrodes used for AAC cutting and gouging can be of pure carbon or the graphite type. Electrodes are also available with a copper coating which tends to make the electrodes last longer. Electrodes erode away rapidly during heavy-duty cutting. Electrode diameters may be $\frac{3}{16}$, $\frac{1}{4}$, $\frac{5}{16}$, $\frac{3}{8}$, $\frac{1}{2}$, or $\frac{5}{8}$ in. Larger sizes are also available. AWS Specification A5 provides specifications for electrodes for this process.

Air Supply. A supply of dry compressed air is required. The air pressure is not critical and ranges from 80 to 100 lb/in^2. It is normally obtained from shop lines or from an air compressor.

Other Arc Cutting Processes

The electric arc, a highly concentrated energy source, can be useful for cutting metals. The arc alone does not produce good-quality cuts, but, when assisted by a jet of oxygen or air, or by plasma, the quality of the cut is greatly improved. There are several arc-welding processes which can also be used for cutting. They are the plasma arc, the carbon arc, the arc between covered electrodes and the work, and a system that uses a special, hollow, covered electrode whereby oxygen can be introduced inside the electrode to produce a quality cut.

The principle of cutting with an arc is to melt metal. This is done by increasing the heat input faster than heat is extracted from the arc area. On thin materials, the molten metal will fall away by gravity, and a rather crude cut will result. In emergency situations, the arc alone can be used. This is not recommended for industrial applications.

For arc cutting the parts to be cut must be arranged so that the area beneath the parts will receive the molten metal without creating problems. For carbon-arc cutting, a single carbon electrode is used with a dc power source using the electrode on the negative pole. The carbon should be sharpened to a long taper approximately half its diameter at the end. It should be gripped close to the arc end to avoid overheating. Position the material to be cut so that it projects over a table edge with a container to catch the molten metal. The arc is struck on the edge of the plate, with a fairly long arc maintained until a puddle is melted at the edge. The arc should then be shortened; this will help force the molten puddle to fall away from the material. A sawing-type motion can be used to help remove the metal from the material being cut. The "icicles" which tend to form on the bottom of the cut can be removed by the arc. Holes may be pierced in steel plates up to ⅜-in thick by striking the arc and holding it in one spot until a large puddle is formed. Feed the electrode downward and force it through the plate as the metal becomes molten.

Cutting with the shielded-metal electrode can be accomplished in much the same way as the carbon-arc cutting described above. A smaller-size electrode is used, usually ⅛-, 5/32-, or 3/16-in diam, but the welding machine must have sufficient capacity for the size of the electrode selected. The E 6011 type is recommended. The coating on the electrode provides a little more arc force than is available with the carbon arc. If electrodes are quickly dipped in water prior to use for cutting, they tend to provide more of a cut before they are consumed in the arc. The shielded-metal arc and carbon-arc cuts are extremely rough and are normally used only for emergency situations when the proper cutting equipment may not be available.

Oxy arc cutting is a proprietary method using covered electrodes having a hollow core. A special electrode holder is required. This electrode allows oxygen to pass through the hole in the center of the electrode. A valve on the electrode holder allows the oxygen to be started or stopped. In this process, the arc is struck in the normal way and, as soon as the metal becomes molten, the oxygen is turned on, this provides a jet that oxidizes the molten metal and carries it away. This process is extremely useful for cutting materials such as cast iron, high-chromium stainless steels, and other hard-to-cut substances.

GAS WELDING AND CUTTING

Oxyacetylene Gas Welding Process

The oxyacetylene gas welding (OAW) process—sometimes called gas welding, oxyfuel gas welding, or torch welding—is an oxy-fuel gas process used to fuse the parts to be welded by heating with a gas flame or flames obtained from the combustion of the acetylene with oxygen. The process, shown in Fig. 3-34, may be used with or without filler metal added to fill gaps or grooves. It can be used on thin- to medium-thickness metals of many types. It is most commonly used on nonferrous metals and can be used in all positions. It is applied as a manual process and requires a relatively high degree of skill on the part of the welder.

Figure 3-34 Application of oxyacetylene welding. (*Hobart Brothers Company.*)

Oxyacetylene welding is the oldest of the modern welding processes. It came into popularity in the late 1800s and is used for repair and maintenance, overlaying, sheet-metal welding, and small-diameter-pipe welding.

The oxyacetylene welding process is diagrammed in Fig. 3-35. The oxyacetylene flame

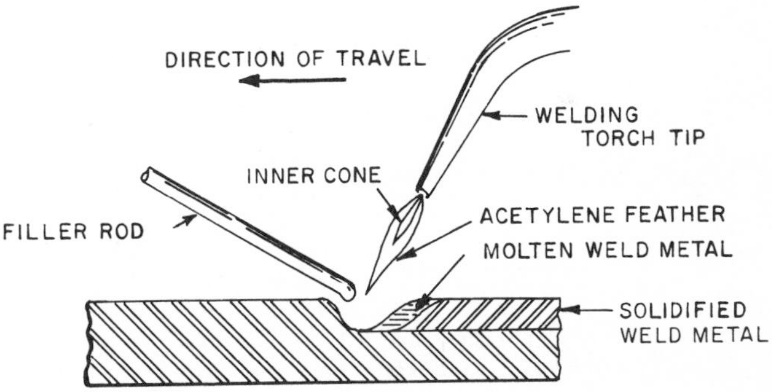

Figure 3-35 Process diagram for oxyacetylene welding.

is extremely hot, approaching 6300°F (3200°C). This hot flame melts the surface of the materials to be joined so that they flow together to produce a weld. Filler material in the form of a rod is added to fill gaps or grooves. The mixing of the oxygen and fuel gas takes place in the welding torch, and the flame is initiated by means of a spark lighter. An atmosphere provided by the burning of the gases shields the molten metal from the atmosphere.

Process

The oxygen and acetylene flow through the hoses from the supply source or from individual cylinders to the welding torch, where they are mixed and burned at the torch tip.

The reaction for this is

$$2C_2H_2 + O_2 \rightarrow 4CO + 2H_2$$

This is the primary reaction that occurs in the inner cone of the flame adjacent to the welding tip. The secondary reactions, as shown,

$$4CO + 2O_2 \rightarrow 4CO_2$$
$$2H_2 + O_2 \rightarrow 2H_2O$$

occur in the outer portion of the flame, and the extra oxygen is obtained from the atmosphere. Note that CO_2 and water vapor result from the secondary reactions. The CO_2 formed shields the molten metal from the atmosphere.

In the combustion of oxygen and acetylene the gases are mixed in the torch in about equal volume, but the remaining oxygen required is from the atmosphere. When the proportions of oxygen and acetylene from the cylinders are the same, the type of flame is referred to as a neutral flame (shown in Fig. 3-36). The exact adjustment is such that the inner cone is just defined, with no feather of acetylene appearing in the flame. This type of flame is used mostly for welding, brazing, and heating. When slightly more acetylene is applied to the neutral flame, a visible feather is seen extending from the inner cone, as shown in Fig. 3-36. This is known as a reducing flame. It has excess acetylene and is used for welding alloy steels, aluminum, and cast iron or for certain surfacing applications. It is slightly cooler than the neutral flame.

When additional oxygen is supplied, the inner cone becomes darker and shorter and the entire flame is smaller and hotter. This flame is called an oxidizing flame because of the excess oxygen. An oxidizing flame (Fig. 3-36) is normally not used except for oxygen flame cutting.

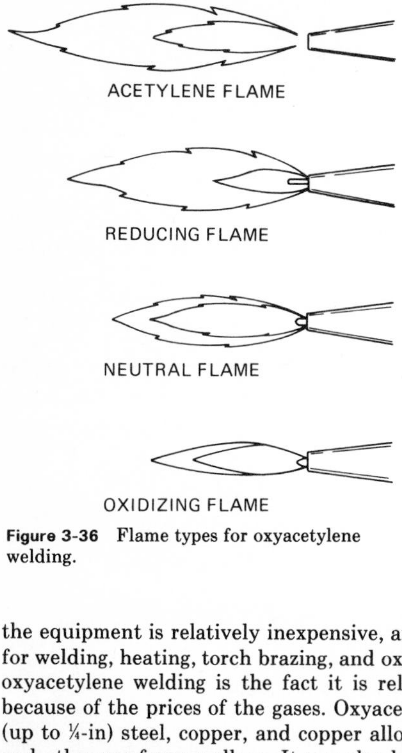

Figure 3-36 Flame types for oxyacetylene welding.

Application

The oxyacetylene welding process has certain advantages: (1) the equipment is very portable, (2) it is a highly versatile process, (3) the weld pool is visible to the welder, (4) welding is possible in all positions, (5) the equipment is relatively inexpensive, and (6) the same basic equipment can be used for welding, heating, torch brazing, and oxygen flame cutting. The main disadvantage of oxyacetylene welding is the fact it is relatively slow and relatively expensive to use because of the prices of the gases. Oxyacetylene welding is most useful for joining thin (up to ¼-in) steel, copper, and copper alloys, and it can be used for welding aluminum and other nonferrous alloys. It can also be used for overlaying, for surfacing, for wear resistance, etc. as well as for heating metals for bending, straightening, etc. Its industrial applications include maintenance and repair, auto-body repair, welding small-diameter piping, brazing, and light manufacturing.

Equipment

Equipment for oxyacetylene welding includes: (1) the welding torch and tips, (2) the hose for transporting the gas from the supply to the torch, (3) regulators for oxygen and acetylene (normally attached to cylinders or to the supply-pipe system), (4) a cylinder or

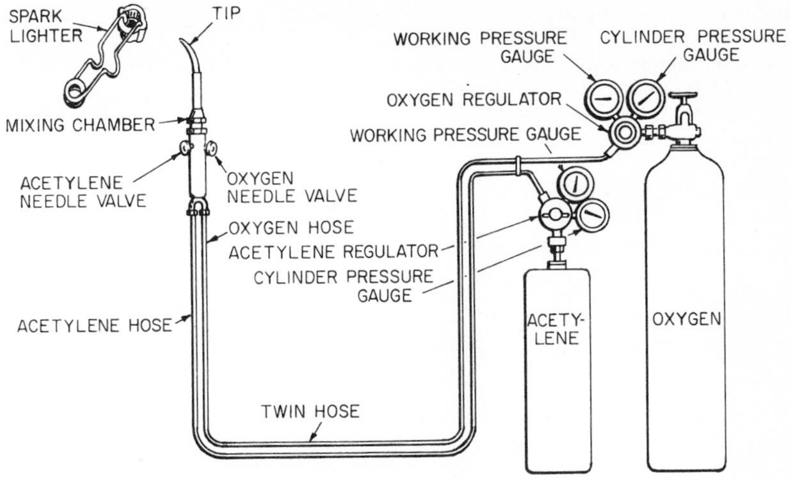

Figure 3-37 Equipment for oxyacetylene welding.

supply of oxygen, and (5) a cylinder or supply of acetylene. In addition, a spark lighter, torch, and cylinder wrench are required. Often, a cylinder cart for transporting the cylinders and apparatus is used. This arrangement is shown in Fig. 3-37.

Welding Torch. The welding torch, sometimes called a *blowpipe,* is the major piece of equipment for the process. It performs the function of mixing fuel gas with oxygen to produce the required type of flame, which is then directed manually as desired. The torch consists of a handle, or body, which contains the hose connections for oxygen and acetylene. It also contains the oxygen and acetylene valves (sometimes called *needle valves*) for regulating gas flow into the torch and a mixing chamber. A medium-pressure oxyacetylene torch is shown in Fig. 3-38. Different sizes of tips can be attached to the torch.

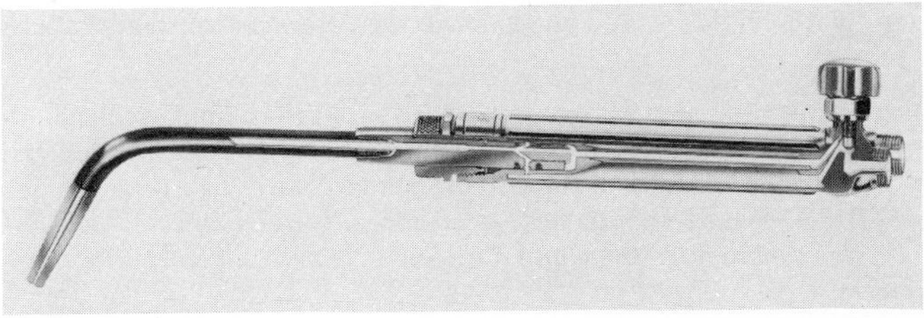

Figure 3-38 Medium-pressure torch for oxyacetylene welding. (*Smith Torch Company.*)

These are identified by manufacturers' numbers which indicate the hole or orifice in the end of the tip. Unfortunately, there is no standard system for identifying tip sizes, and each manufacturer has its own system; however, in every case the system relates to a drill size which identifies the diameter of the hole.

Gases. The gases used for oxyacetylene welding are oxygen and acetylene. Acetylene produces the highest-temperature flame and is considered the all-purpose fuel for this process. Acetylene is colorless but it has an easily detected odor somewhat like the odor of onions. When the torch is used for heating, other fuel gases may be used, such as natural gas, propane, and proprietary fuel gases. Different tips and mixers are usually needed when other gases are used.

Regulators. The pressure of the gas used for oxyacetylene welding is relatively low; however, the pressure of the gas in the supply system or in individual cylinders is relatively high. Therefore, a device known as a *gas regulator* is used to reduce the pressure from the high side to the correct working pressure for the torch. This is a complex unit, made of needle valves, springs, and diaphragms, for precisely producing the lower pressure used by the torch. Figure 3-39 shows a gas regulator. The regulators for oxygen and

Figure 3-39 Gas regulator for oxyacetylene welding. (*Hobart Brothers Company.*)

acetylene are different and *cannot be interchanged.* Oxygen connections have right-hand threads and acetylene and other fuel gas connections have left-hand threads. A gas regulator will keep the gas pressure constant and has gauges showing the pressure going to the torch and sometimes showing the pressure of the supply. Two-stage regulators are normally used with cylinders, and single-stage regulators are normally used for supply lines.

Gas Cylinders. Oxygen and acetylene are both supplied in individual cylinders. A pair of cylinders used for oxyacetylene welding is shown in Fig. 3-40. Oxygen cylinders are made of a high-strength steel and contain oxygen at a very high pressure: up to 24,000 lb/in^2. ***CAUTION****—Cylinders must be treated carefully and inspected periodically. Mistreating cylinders can damage them and may cause them to explode, creating a very dangerous situation.*

Acetylene, on the other hand, is stored at a relatively low pressure. Acetylene cannot be stored safely over 15 lb/in^2. It is dissolved in liquid acetone which is contained by a filler material in the cylinder. An acetylene cylinder will have a working pressure of 250 lb/in^2; however, most of the acetylene is dissolved in the acetone, which keeps it stable and eliminates the danger from high-pressure free acetylene. Acetylene cylinders should always be kept upright because of the liquid inside the cylinder. In addition, an acetylene cylinder should always be kept away from high temperatures and should be treated with the respect due *any* gas cylinder. There is no uniform national color code for gas cylinders. Each company supplying gases has its own color code. However, there is standardization of threads on the fittings of the cylinders; remember, oxygen cylinders have right-hand threads and acetylene cylinders have left-hand threads.

Cylinder Carts. For portable installations, cylinder carts are usually employed. This allows the cylinders to be affixed to a structure even though it is portable. It allows the storage of the hoses and torch and is useful for maintenance applications.

Safety Precautions

The safety precautions for oxyacetylene and gas welding are somewhat special for the process. For your own safety and the safety of those about you, it is important to follow these safety directions when you are using oxyacetylene welding equipment.

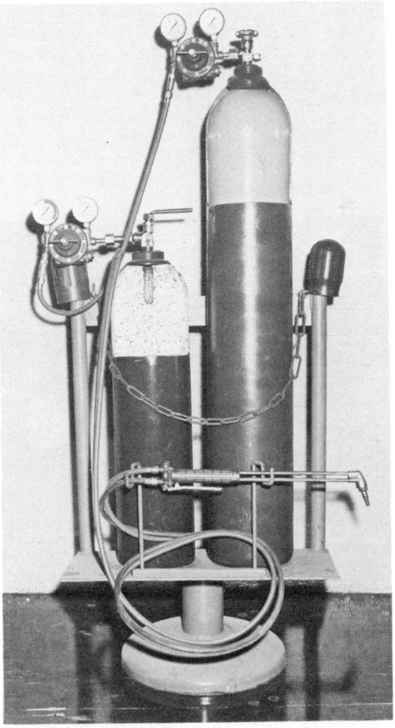

Figure 3-40 Gas cylinders for oxyacetylene welding. (*Hobart Brothers Company.*)

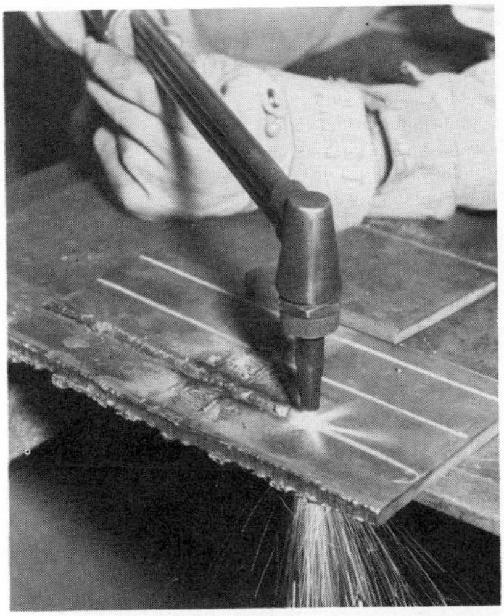

Figure 3-41 Application of oxy–fuel gas cutting. (*Hobart Brothers Company.*)

For additional information concerning the oxyacetylene welding process refer to Sec. 6-2 of Howard Cary's *Modern Welding Technology,* Prentice-Hall, Englewood Cliffs, N.J., 1979.

Oxy–Fuel Gas Cutting Process

Process

The oxy–fuel gas cutting (OFC) process, also known as oxygen cutting, gas cutting, burning, and so on, is a thermal process used to sever metals by heating the metal with a flame to an elevated temperature and then using pure oxygen to oxidize the metal and produce the cut. Different fuel gases can be used, including acetylene, natural gas, propane, and a variety of proprietary or trade-name fuel gases. The process shown in Fig. 3-41 can be applied manually or by machine. It can be used to cut ferrous materials in sections varying from thin to thick, and it can be used in all positions. Manual OFC requires a fairly high degree of skill to produce quality cuts.

Details of the process are diagrammed in Fig. 3-42. This diagram shows the torch and cutting tip, the preheating flames to bring the metal up to the kindling temperature, and the oxygen jet supplied to oxidize, or "burn," metal away to produce the cut.

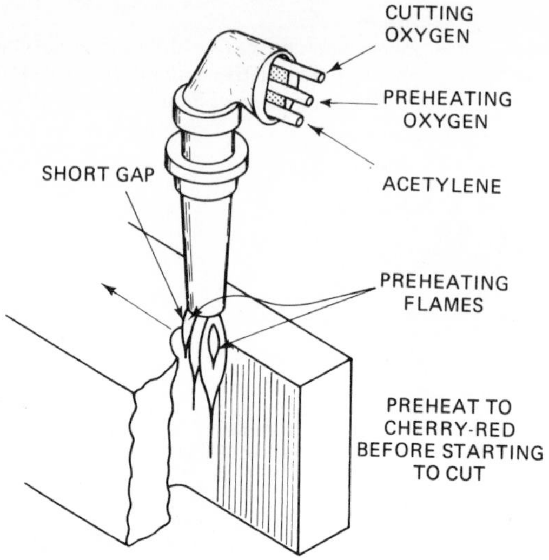

Figure 3-42 Process diagram for oxy–fuel gas cutting.

Application

This cutting process (1) is very portable, (2) is versatile, (3) allows cutting in all positions, (4) uses relatively inexpensive equipment, and (5) can be used to cut steels. The disadvantages of the process are: (1) it cannot be used to cut nonferrous materials, and (2) the cut surfaces are not as smooth as mechanically cut surfaces. It is widely used throughout industry as a manual process. It is also widely used as a machine-cutting process with automatic torch controls and seam-following devices. When it is used as an automatic process, extremely smooth surfaces can be obtained.

Equipment

Oxy–fuel gas welding equipment includes: (1) the cutting torch and tips, (2) oxygen and fuel gas hoses, (3) regulators for oxygen and fuel gas or acetylene, and (4) a supply of oxygen and fuel gas from cylinders or a piping system. The equipment is essentially the same as used for oxyacetylene welding.

The Cutting Torch. The cutting torch can be a combination cutting and welding torch or a torch especially designed for cutting only. The gases are mixed within the torch, and needle valves control the quantity of each gas flowing into the mixing chamber. A lever-type valve controls the oxygen flow for cutting. Various sizes and types of tips are used with the cutting torch for specific applications of cutting, gouging, beveling, etc. The cutting tips are sized by the oxygen orifice size in the cutting tip. There is no standard cutting-size designation, and each company has its own system; however, each cutting tip size relates to the standard drill size for the cutting orifice. In this way they can be related to the thickness of metal to be cut. Preheat flames are arranged around the central cutting orifice and are sufficient to bring the metal to the kindling temperature prior to cutting.

The rest of the equipment is the same as that used in the oxyacetylene welding process.

Gases. The gas used for oxygen cutting is normally pure oxygen, while the fuel gas is always a hydrocarbon gas, often acetylene. Other fuel gases used are natural gas, propane, and a variety of proprietary liquid-petroleum-base or propane-base gases. The selection of fuel gas is extremely complex; however, the fuel gas is used for the preheating flame that brings the material to be cut up to its kindling temperature. The basic cutting

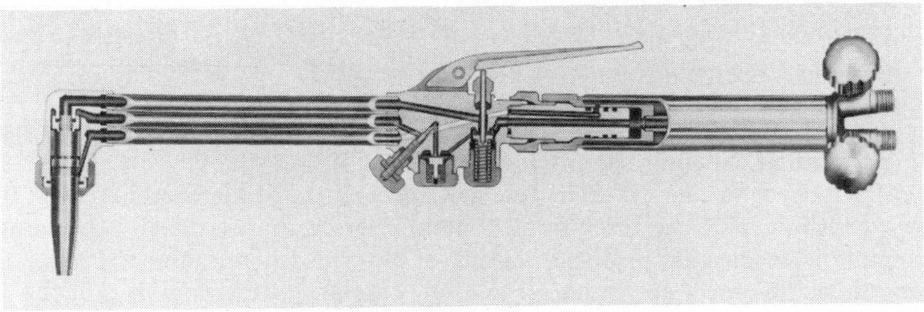

Figure 3-43 Cutting assembly with welding torch body. (*Smith Torch Company.*)

process using oxygen is not affected by the choice of preheat fuel gas that is used. The type of fuel gas relates primarily to the time period required to bring the material up to the kindling temperature. Figure 3-43 shows the flame-cutting assembly.

Safety Precautions

Safety precautions for oxygen–fuel gas cutting are extremely important because it is widely used in maintenance and construction work. The normal precautions involving gases under pressure should be observed. Additional precautions relate to the cutting of vessels and containers that may be sealed and/or may have contained combustible materials. This should not be done without taking extra-special precautions. Another problem relating to oxygen cutting is the fact that white-hot metal from the cut will travel many, many feet and will retain sufficient heat to set combustible materials on fire. Metallic or noncombustible material should be used to backstop the hot metal being ejected from the cut. Cutting should *never* be attempted in confined areas without first testing the atmosphere and providing a fire watch or observers to continually watch the cutter while it is in operation.

BRAZING AND SOLDERING

Distinction

The primary difference between brazing and soldering is the arbitrary temperature of 840°F (450°C). Both are a group of processes which join materials by heating them to a suitable temperature; both use a filler metal which is distributed between closely fitted surfaces of the joint by capillary attraction. Solder, the filler metal used in soldering, has a melting temperature below 840°F (450°C), while brazing alloy, the filler metal used for brazing, has a melting temperature above 840°F (450°C). Both solder and brazing alloy have a composition somewhat different from that of the base metal. Also, for both soldering and brazing, a fluxing material is normally used.

There is one other term that should be mentioned—*braze welding*. It refers to a process that is similar to, but different from, brazing in that capillary attraction is not used to distribute the filler metal. Braze welding is used quite often to join cast-iron sections.

Heating

The method of heating the materials to be joined is the method usually used to differentiate between the different soldering and brazing processes. The same methods of heating can be used for both soldering and brazing. These include using a gas torch, which is one of the most common methods of both brazing and soldering; dipping the materials in flux or molten filler metal; generating heat in the parts by means of a furnace; applying heat by means of induction or infrared radiation; generating heat by means of the resistance of the parts to current flow; and introducing heat by means of an iron—a method used only for soldering.

Torch brazing is the method discussed here for applying heat to the parts to be joined. In torch brazing, heat can be applied by using different fuel gases and oxygen or air combinations.

The torch for melting high-temperature brazing alloys is the same as that used for oxy–fuel gas welding, whereas the torch for soldering uses a fuel-gas–air system. Different torches are used for the different fuel gas and oxygen or air combinations. In each case, however, the use of the torch and its manipulation are essentially the same. The basic principle is to provide uniform heating of the parts being joined. Proper fluxing and proper fit of the parts are essential to allow capillary attraction to pull the molten filler metal into the joint.

Joints

Both brazed and soldered joints require close fit of the parts to be joined. This is necessary to provide the capillary attraction to pull the alloy filler metal into the joint to provide sufficient area of filler metal to ensure a sufficiently strong joint. The lap-type joint is most commonly used since it provides for sufficient faying surfaces to attract the filler material. Butt-type joints are rarely used for soldering or brazing. One of the most common types of joints is the socket joint used for pipe and tubing.

Fluxes

Flux is almost always used in torch brazing. The flux helps maintain cleanliness of the faying surfaces so that the filler metal will adhere properly. The joints should be properly cleaned before applying flux because cleaning the surface *is not* the function of the flux. It does, however, help by combining with, dissolving, or inhibiting the formation of chemical compounds which would interfere with the quality of the joint. The flux also protects the surface during the heating operation. The type of flux to be used is chosen on the basis of the process function and metal to be joined. The flux and the filler metal must also be matched. Manufacturer's information concerning fluxes must be followed since there are no established specifications covering fluxes. However, the American Welding Society provides flux type numbers with recommendations for use. AWS Specification A5.8 on filler metals for brazing and ASTM Specification B 32 on the composition and uses of solders provide further information. The particular alloy or type of filler metal to be used depends upon the process and the metals being joined.

Summary

Quality brazes or soldered joints can be made by following the basic principles of cleanliness, fluxing, joint detail, and matching the proper flux and filler metal alloy.

OTHER WELDING PROCESSES

The previous sections provided information concerning arc-welding processes and some of the other welding and cutting processes most commonly used by plant engineers. However, there are many other welding processes used in manufacturing that should be mentioned.

The American Welding Society's "Master Chart for Welding and Allied Processes" (Fig. 3-44) shows seven families of welding processes, two families of allied processes (thermal spraying and adhesive bonding), and three families of thermal cutting processes. Previously we have described arc welding, brazing, soldering, oxy–fuel gas welding, oxygen cutting, and arc cutting. Some other important processes are now briefly described.

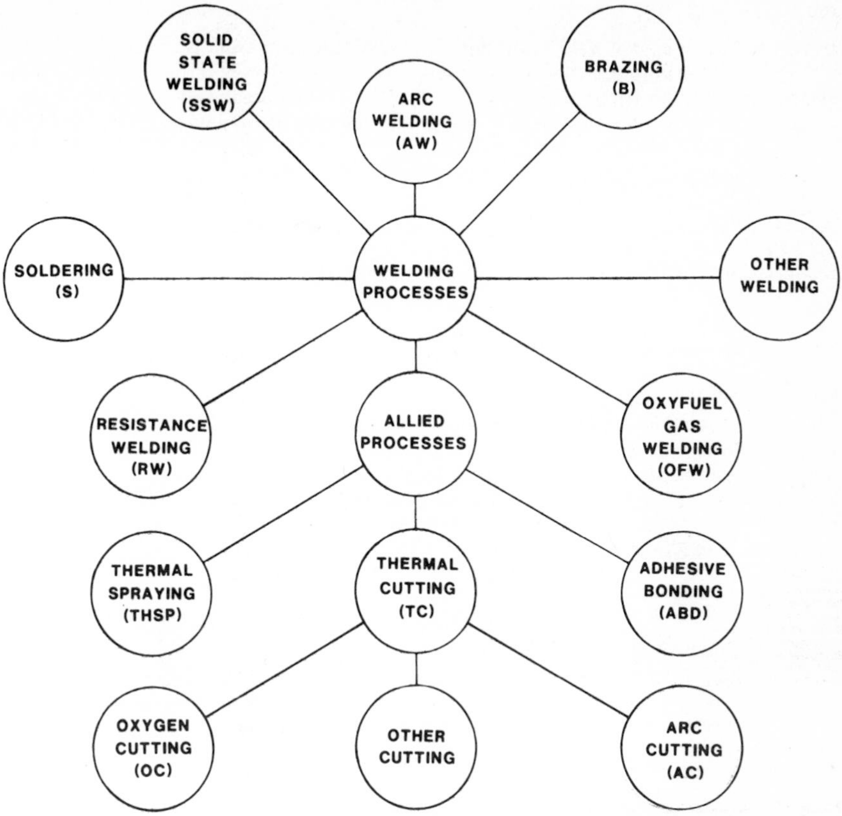

Figure 3-44 American Welding Society master chart of welding and allied processes.

Resistance Welding

Resistance welding is a group of welding processes that produce joints of metal by means of heat obtained from resistance and pressure. The resistance is that of the work to the electric current in a circuit of which the work is a part, and the pressure is applied externally. Spot welding is the most popular of the resistance-welding processes. Spot welding is accomplished with a machine which uses electrodes to carry the current to and through the joint being welded. The electrodes also apply the pressure which is necessary to force the parts together after the current has heated the metal to the welding temperature. Resistance welding is extremely fast, and filler metal is normally not required. It is very popular for welding automobile bodies and for making household appliances. Other resistance-welding processes are projection welding, seam welding, flash welding, and high-frequency resistance welding, with many variations of each. Most metals can be resistance-welded. Special precautions are required, however, for certain metals.

Solid-State Welding

The solid-state family of welding processes includes friction welding, cold welding, ultrasonic welding, and other less-important processes.

Friction Welding

In friction welding, the weld is produced by heat obtained from a mechanical sliding motion between rubbing surfaces. The process usually involves rotating one part against another to generate frictional heat at the junction. When a suitably high temperature has been reached, rotational motion ceases and pressure is applied to create the weld. Equipment is similar to a lathe: it is extremely fast, and no filler metal is required. It is restricted primarily to mass-production industries.

Cold Welding

Cold welding is a solid-state process whereby pressure at room temperature is used to produce the weld. The metals are substantially deformed, and extremely high pressures are required on extremely clean interfacing surfaces. This process is restricted to thinner materials, and it is often used for welding nonferrous materials such as aluminum and copper. It is also used to weld aluminum to copper.

Explosion Welding

Explosion welding is also a solid-state process. In this case the weld is obtained by high-velocity movement together of the parts to be joined. The movement is caused by an explosion, and heat is not applied. The interface between the parts welded shows a saw-tooth-type configuration. Heat is instantly produced from the shock wave associated with impact. This process is often used to weld dissimilar parts together and is used for overlaying or cladding materials.

Ultrasonic Welding

Ultrasonic welding is another of the solid-state processes. It produces the joint by local application of high-frequency energy to the parts being welded, while they are held together under pressure. Welding occurs when the electrode, which couples the energy to the work, is vibrating at ultrasonic frequencies. This, plus pressure, creates the weld. Ultrasonic welding is restricted to thinner materials and is quite often used in the packaging industries.

Electron-Beam Welding

Electron-beam (EB) welding is one of the most important non-arc welding processes. In EB welding, the heat for welding is obtained from a concentrated beam of high-velocity electrons impinging upon the surface of the work. Pressure is not used, but filler metal is sometimes added.

EB welding was initially done in a vacuum chamber. The work and work-moving devices, as well as the electron beam, were contained in the chamber. The electron beam is generated by an electron gun and is similar to that in an x-ray tube. The work had to be taken to the machine and it had to fit within the chamber. Evacuation of the chamber was a major part of the operation. Recently, however, specially designed chambers which allow continuous entrance and exit of parts have been used in mass-production industries. A lower vacuum in the chamber is sometimes used. EB welding in the atmosphere is now also possible. However it is restricted to operating close to the electron gun, which must be in a vacuum chamber. The capital expense for EB welding is quite high, and this type of welding, therefore, is restricted to specialty materials and special applications.

Laser-Beam Welding

Laser-beam welding is very similar to electron-beam welding except that the heat is obtained from the application of a concentrated coherent light beam impinging on the surface of the work; a vacuum chamber is not required. However, the generation of a laser beam is extremely complex and expensive and the electrical efficiency of the process is relatively low. This process is quite new, and additional developments are expected. The laser beam is used for cutting as well as for welding and will cut nonmetals as well as metals. At this time, there are more applications for laser cutting than for laser welding. However, developments in this field are accelerating and the reader is urged to investigate the state of the art when considering the various alternatives for any application.

Thermite Welding

One of the older welding processes still in use is thermite welding. In this process, the weld is produced by heating the parts to be joined with superheated liquid metal obtained from a chemical reaction between a metal oxide and aluminum. The filler metal is obtained from the superheated liquid metal. The heat is obtained from an exothermic reaction between iron oxide and aluminum. This reaction occurs immediately above the weld, and when it has gone to completion, the superheated liquid flows into the weld area and is retained by a mold. The process is used for joining rails, reinforcing bars, and other similar items. It is also used to join castings used in ship construction.

QUALITY CONTROL AND INSPECTION METHODS

The quality of welds can be determined by nondestructive testing methods. Welds made in most commercial metals normally equal the strength of the base metal. This result depends upon the proper selection of the process and procedure, including the filler metal. Welds in metals having special properties resulting from heat treatment or working may not equal the strength of the base metal because the heat of making the weld will cause these special properties to deteriorate in the area adjacent to the weld. For these types of metals special precautions are required. For all other welds, however, the quality of the weld can be determined and controlled. Adherence to procedures that are known to produce quality welds is recommended. After the weld has been made, it can be inspected by a number of nondestructive evaluation techniques. The most popular is visual inspection. Visual inspection (VT) is used by welders, supervisors, and inspectors for potential defects such as undersized welds which can be checked by gauges, rough or irregular surface, surface cracks, surface porosity, undercut, etc. In addition, weld quality can be determined by at least four other evaluation techniques. These are summarized in Table 3-2.

Visual Inspection

Process and Applications

Visual welding inspection is the most widely used and most valuable welding inspection technique. In particular, it is the most effective for noncritical welding production. Visual inspection requires less time than any other inspection method and is also the least expensive. In addition to being a weldment inspection technique, it allows inspection of the welding procedures themselves and thus is also a preventive tool. The inspector is able to watch and require procedure conformity during weldment production.

Visual inspection throughout the forming of a single weldment can catch errors in each step and items which might develop errors, such as faulty materials and procedures. Repairs can be made on an incompleted piece of work. Inspectors can check the basic materials, the joint preparation, process manipulation, and welding technique long before the weldment is completed. This prevention and early correction of the welds is particularly important on highly critical or expensive weldments.

Inspectors can note errors in weld preparation, dimensions, alignment, fit-up, cleanliness, welding procedure, warpage, finish, and mishandling in marking. They can detect scabs, seams, scale, surface slag, laminations, roughness, spatter, craters, surface porosity, undercuts, overlaps, cracks, and inadequate penetration. They can check for many of these at once and can note several defects simultaneously.

For any other welding inspection technique, inspectors need, primarily, to be able to interpret a series of indicators. With visual inspection, they must know welding more

TABLE 3-2 Guide to Welding Quality Control (NDT) Techniques

Technique	Equipment	Defects detected	Advantages	Disadvantages	Other considerations
Visual, VT	Pocket magnifier, welding viewer, flashlight, weld gauge, mirror	Weld preparation, fit-up; cleanliness, roughness, spatter, undercuts, overlaps, inadequate penetration and size; welding procedures	Easy to use; fast, inexpensive, usable at all stages of production	For surface conditions only; dependent on subjective opinion of inspector	Most universally used inspection technique
Magnetic particle, MT	Iron powder, wet, dry, or fluorescent; commercial power source; black light for the fluorescent type	Surface and near-surface discontinuities, cracks, etc.; subsurface porosity and slag on light materials	Indicates discontinuities not visible to the naked eye; useful in checking edges prior to welding, also, repairs; no size restriction	Used on magnetic materials only; surface roughness may distort magnetic field	Testing should be from two perpendicular directions to catch discontinuities which may be parallel to one set of magnetic lines
Liquid penetrant, PT	Fluorescent or visible commercial penetrating liquids and developers; black light for the fluorescent type	Defects open to the surface only	Very small, tight, surface imperfections show up. Easy to apply and to interpret; inexpensive; use on either magnetic or nonmagnetic materials	Somewhat time-consuming in the various steps of the processes	Often used on root pass of highly critical pipe welds; if material improperly cleaned, some indications may be misleading
Radiographic, RT	X-ray or gamma-ray equipment; film-processing equipment; film-viewing equipment; penetrometers	Most internal discontinuities and flaws; limited by direction of discontinuity	Provides permanent record; indicates both surface and internal flaws; applicable on all materials	Usually not suitable for fillet-weld inspection; film exposure and processing critical; slow and expensive	Most popular technique for subsurface inspection; required by many codes and specifications
Ultrasonic, UT	Commercial ultrasonic units and probes; reference and comparison patterns	Can locate all flaws located by other methods with the addition of other exceptionally small flaws	Extremely sensitive; use restricted to only very complex weldments; can be used on all materials	Time-consuming; demands highly developed interpretation skill; permanent record not normally obtained	For irregularly shaped parts, immersion testing often used; required by some codes

thoroughly and be able to inspect all areas of the weldment production. The technique depends upon the alertness, eyesight, welding knowledge, and subjective judgment of the inspectors.

Visual inspection is unreliable on subsurface conditions and discovery of these must result primarily on the inspectors' judgment of the welders' actual work. Tiny, fine flaws can be overlooked very easily and can be covered by peening and hammering while removing slag.

Because of the simplicity, absence of elaborate equipment, and low cost of visual inspection, it can be relied upon too heavily when used entirely by itself rather than in conjunction with more sensitive inspection methods.

Equipment

A pocket magnifier, flashlight, borescope, dentist's mirror, weld gauge, straightedge, T-square, and weld standards are all helpful pieces of equipment in visual inspection.

WELDING CODES AND QUALIFICATIONS OF WELDERS

Before a welder can begin work on any job covered by a welding code or specification he or she must become certified under the code that applies. Many different codes are in use, and it is exceedingly important that the specific code is referred to when one is taking qualifying tests. (Standard welding symbols as shown in Fig. 3-45 are used throughout the industry.) In general, the following types of work are covered by codes: pressure vessels and pressure piping, highway and railway bridges, public building, tanks and containers that will hold flammable or explosive materials, cross-country pipelines, aircraft, ordnance material, ships, and boats. A qualified welding procedure is normally required.

Certification is obtained differently under the various codes. Certification under one code will not necessarily qualify a welder to weld under a different code. In most cases certification for one employer will not allow the welder to work for another employer (except in cases where welders are qualified by an association of employers). Also, if the welder uses a different process or if the welding procedure is altered drastically, recertification is required. In most codes, if the welder is continually employed, welding recertification is not required, providing the work performed meets the quality requirement. An exception is the military aircraft code which requires requalification every 6 months.

Qualification tests may be given by responsible manufacturers or contractors. On pressure vessels, the welding procedure must be qualified before the welders can be qualified. Under some codes this is not necessary. To become qualified, the welder must make specified welds using the selected process, base metal, thickness, electrode type, position, and joint design. Standard test specimens must be made under the observation of a qualified person. In government specifications, a government inspector must witness the making of welding specimens. Specimens must be properly identified and prepared for testing.

The most common test is the guided-bend test. In some cases x-ray examinations, fracture tests, or other tests are employed. Satisfactory completion of test specimens, provided that they meet acceptability standards, qualifies the welder for specific types of welding. The welding that is allowed depends on the particular code. In general, the code indicates the range of thicknesses which may be welded, the positions which may be employed, and the alloys which may be welded.

Qualification of welders is an extremely technical subject and cannot be adequately covered here. The actual code must be obtained and studied prior to taking any test.

The most important codes are:

AWS D1.1, "Structural Welding Code"
"Welding Qualifications," Sec. IX of "ASME Boiler and Pressure Vessel Code"
API #1104, "Standard For Welding Pipelines and Related Facilities"

These codes can be obtained from the sponsoring associations.

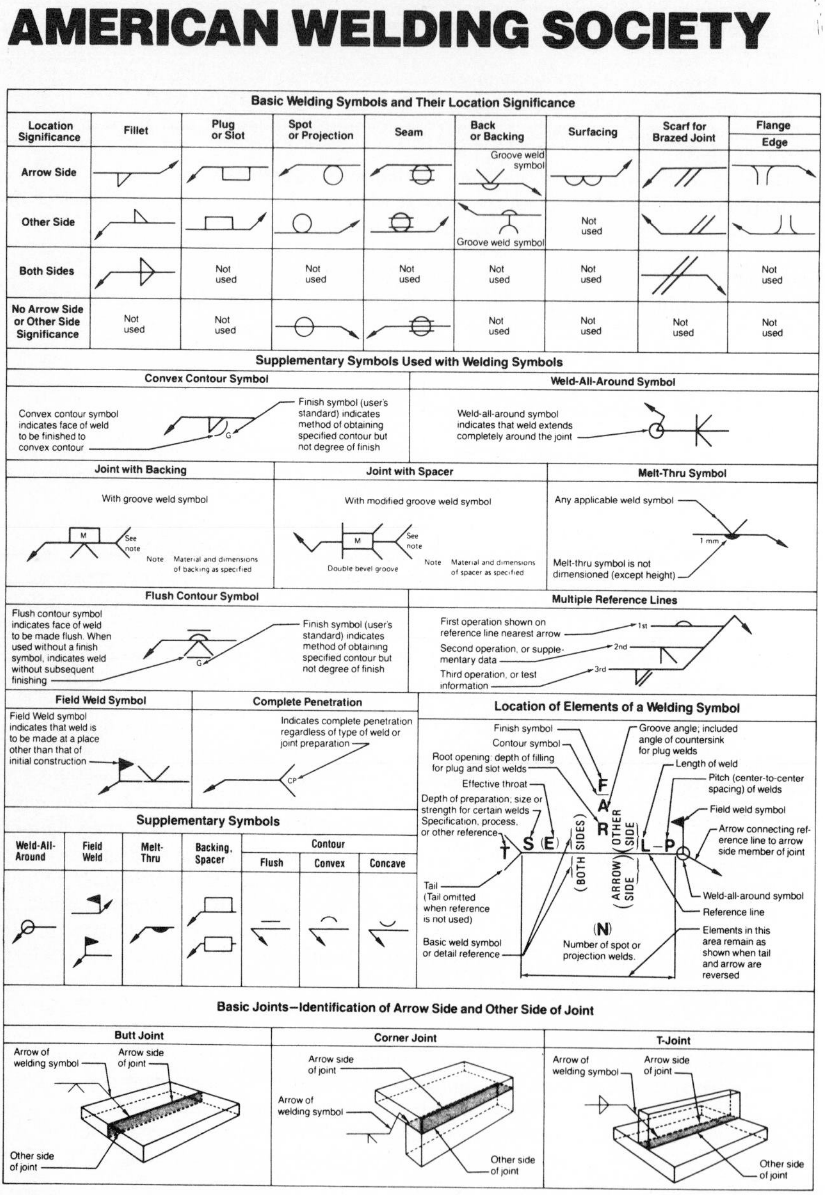

Figure 3-45 American Welding Society standard welding symbols. Continued on pages 16-73 to 16-75.

POWER SOURCES FOR ARC WELDING

Many different types and sizes of arc-welding machines are available. It is important to select the best machine and the one most suited for the particular work to be done. The following information describes the different types of machines available, thus allowing the selection of the one most ideally suited for each type of work.

STANDARD WELDING SYMBOLS

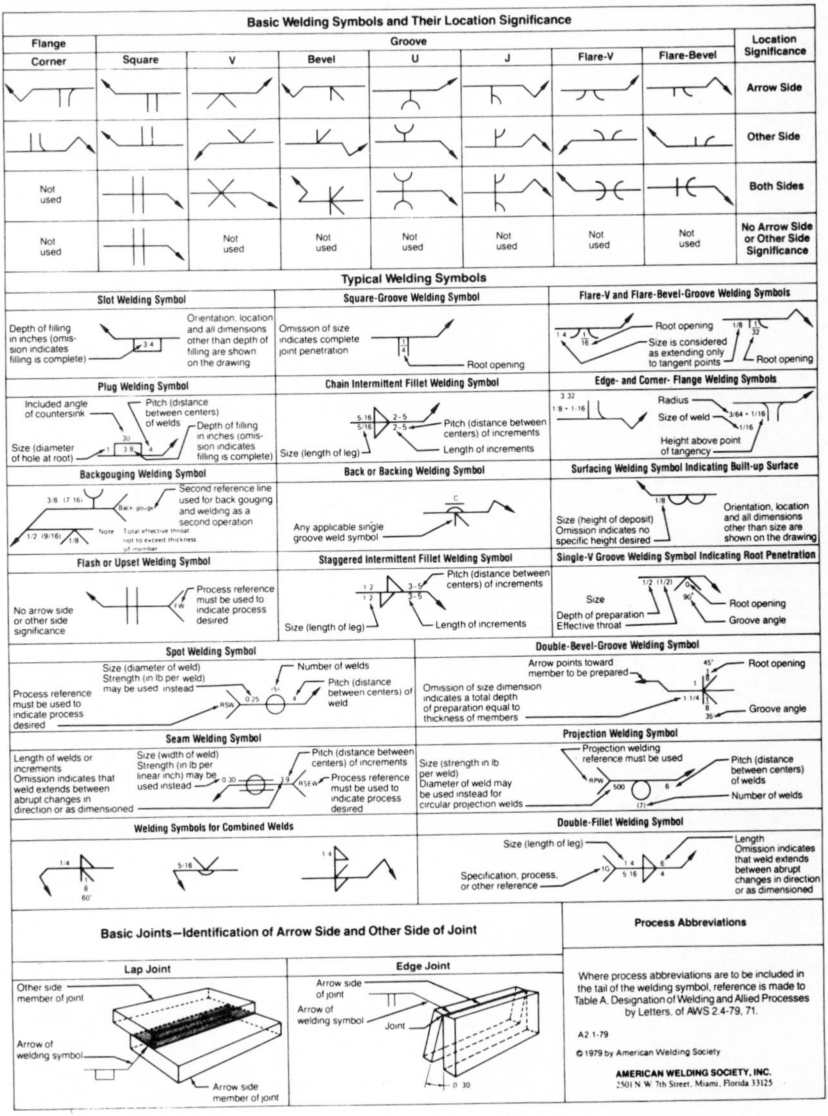

Arc-welding machines can be classified in many different ways, such as rotating machines, static machines, electric-motor–driven machines, internal-combustion-engine–driven machines, transformer-rectifiers, limited-input welding machines, conventional CV voltage welding machines, and single-operator machines or multiple-operator machines.

There are two basic categories of power sources: the conventional or CC or variable-voltage welding machine with the drooping volt-ampere curve and the CV or CP or mod-

AMERICAN WELDING SOCIETY STANDARD WELDING SYMBOLS

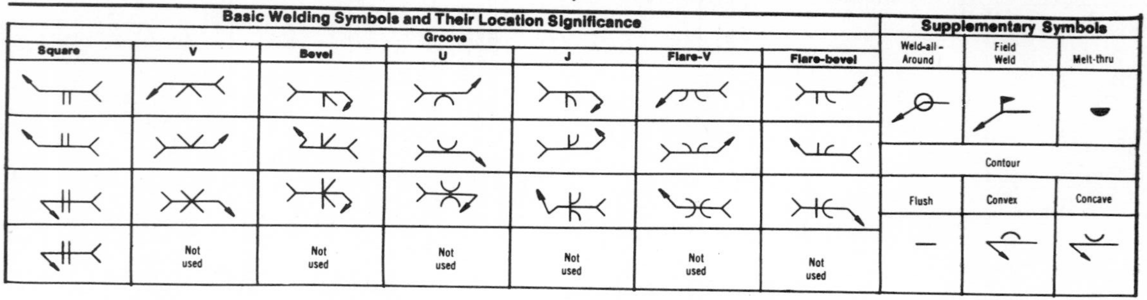

Basic Welding Symbols and Their Location Significance
Groove
Square
V
Bevel
U
J
Flare-V
Flare-bevel
Not used
Not used
Not used
Not used
Not used
Not used
Supplementary Symbols
Weld-all-Around
Field Weld
Melt-thru
Contour
Flush
Convex
Concave

Designation of Welding and Allied Processes by Letters

GMAW-P ... gas metal arc welding—pulsed arc
GMAW-S ... gas metal arc welding— short circuiting arc
GTAC ... gas tungsten arc cutting
GTAW ... gas tungsten arc welding
GTAW-P ... gas tungsten arc welding—pulsed arc
HFRW ... high frequency resistance welding
HPW ... hot pressure welding
IB ... induction brazing
INS ... iron soldering
IRB ... infrared brazing
IRS ... infrared soldering
IS ... induction soldering
IW ... induction welding
LBC ... laser bream cutting
LBW ... laser beam welding
LOC ... oxygen lance cutting
MAC ... metal arc cutting
OAW ... oxyacetylene welding
OC ... oxygen cutting
OFC ... oxyfuel gas cutting
OFC-A ... oxyacetylene cutting
OFC-H ... oxyhydrogen cutting
OFC-N ... oxynatural gas cutting
OFC-P ... oxypropane cutting
OFW ... oxyfuel gas welding
OHW ... oxyhydrogen welding
PAC ... plasma arc cutting
PAW ... plasma arc welding
PEW ... percussion welding
PGW ... pressure gas welding
POC ... metal powder cutting
PSP ... plasma spraying
RB ... resistance brazing
RPW ... projection welding
RS ... resistance soldering
RSEW ... resistance seam welding
RSW ... resistance spot welding
ROW ... roll welding
RW ... resistance welding
S ... soldering
SAW ... submerged arc welding
SAW-S ... series submerged arc welding
SMAC ... shielded metal arc cutting
SMAW ... shielded metal arc welding
SSW ... solid state welding
SW ... stud arc welding
TB ... torch brazing

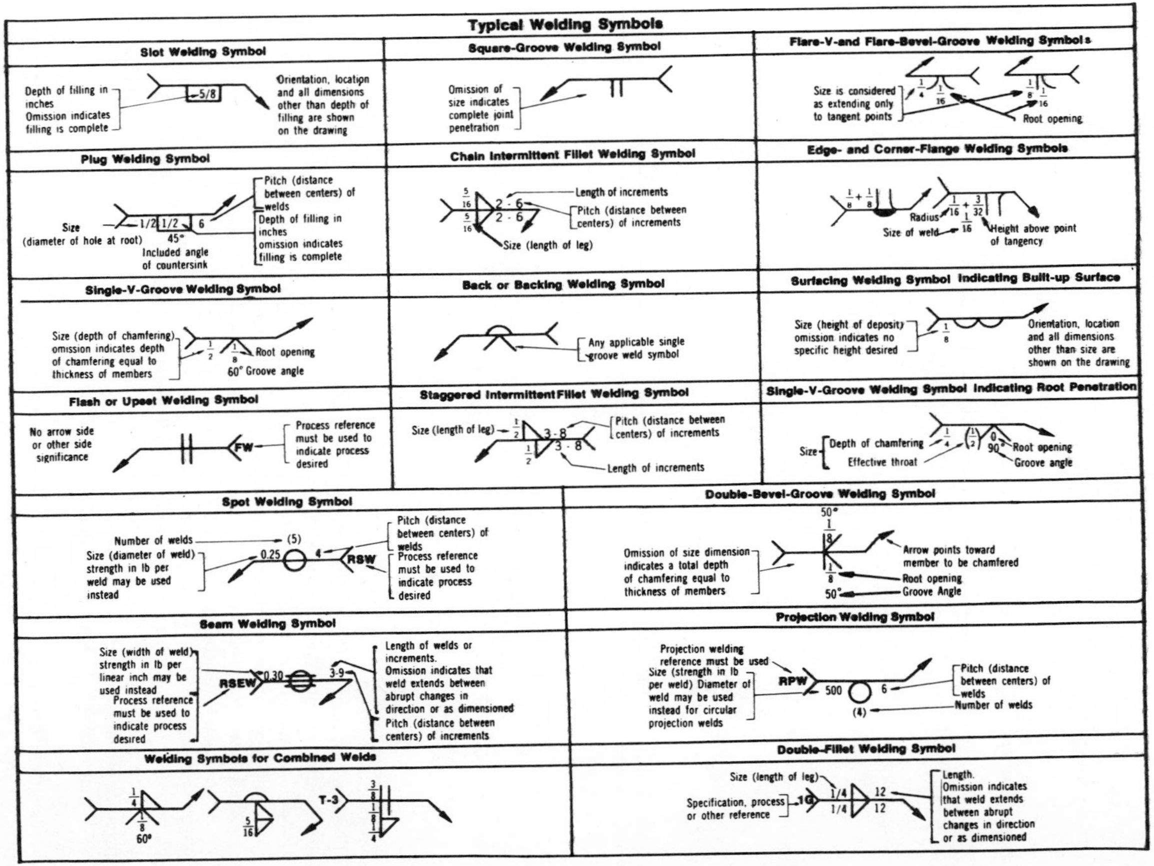
Typical Welding Symbols
Slot Welding Symbol
Depth of filling in inches Omission indicates filling is complete
5/8
Orientation, location and all dimensions other than depth of filling are shown on the drawing
Square-Groove Welding Symbol
Omission of size indicates complete joint penetration
Flare-V-and Flare-Bevel-Groove Welding Symbols
Size is considered as extending only to tangent points
Root opening
Plug Welding Symbol
Size (diameter of hole at root)
1/2 1/2 6
45°
Included angle of countersink
Pitch (distance between centers) of welds
Depth of filling in inches omission indicates filling is complete
Chain Intermittent Fillet Welding Symbol
2-6
Length of increments
Pitch (distance between centers) of increments
Size (length of leg)
Edge- and Corner-Flange Welding Symbols
Radius
Size of weld
Height above point of tangency
Single-V-Groove Welding Symbol
Size (depth of chamfering) omission indicates depth of chamfering equal to thickness of members
Root opening
60° Groove angle
Back or Backing Welding Symbol
Any applicable single groove weld symbol
Surfacing Welding Symbol Indicating Built-up Surface
Size (height of deposit) omission indicates no specific height desired
Orientation, location and all dimensions other than size are shown on the drawing
Flash or Upset Welding Symbol
No arrow side or other side significance
FW
Process reference must be used to indicate process desired
Staggered Intermittent Fillet Welding Symbol
Size (length of leg)
3-8
Pitch (distance between centers) of increments
Length of increments
Single-V-Groove Welding Symbol Indicating Root Penetration
Size
Depth of chamfering
Effective throat
90°
Root opening
Groove angle
Spot Welding Symbol
Number of welds
(5)
Size (diameter of weld) strength in lb per weld may be used instead
0.25
4
RSW
Pitch (distance between centers) of welds
Process reference must be used to indicate process desired
Double-Bevel-Groove Welding Symbol
50°
Omission of size dimension indicates a total depth of chamfering equal to thickness of members
Arrow points toward member to be chamfered
Root opening
Groove Angle
Seam Welding Symbol
Size (width of weld) strength in lb per linear inch may be used instead
Process reference must be used to indicate process desired
RSEW
0.30
3-9
Length of welds or increments. Omission indicates that weld extends between abrupt changes in direction or as dimensioned
Pitch (distance between centers) of increments
Projection Welding Symbol
Projection welding reference must be used
Size (strength in lb per weld) Diameter of weld may be used instead for circular projection welds
RPW
500
6
(4)
Pitch (distance between centers) of welds
Number of welds
Welding Symbols for Combined Welds
60°
T-3
Double-Fillet Welding Symbol
Size (length of leg)
Specification, process or other reference
1/4
12
Length. Omission indicates that weld extends between abrupt changes in direction or as dimensioned

ified CV machine with the fairly flat characteristic curve. The conventional CC machine can be used for manual welding and, under some conditions, for automatic welding. The CV machine is used *only* for continuous electrode wire arc-welding processes operated automatically or semiautomatically. These types of machines are best understood by comparing their respective volt-ampere characteristic output curves. This type of curve is obtained by loading the welding machine with variable resistance and plotting the voltage at the electrode and work terminals for each amperage output. Figure 3-46 shows an example of this curve.

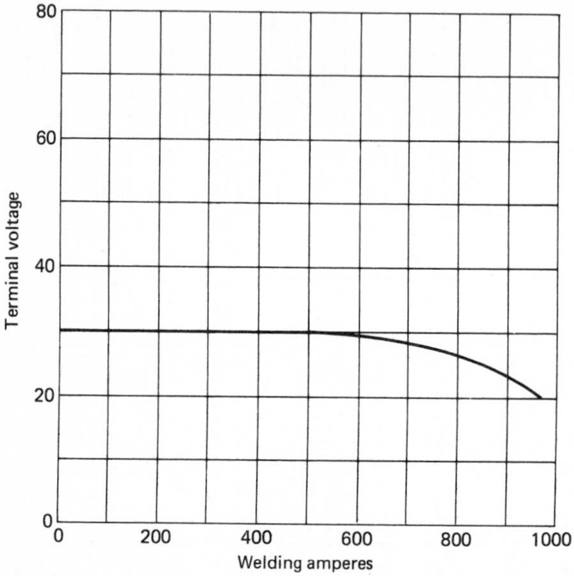

Figure 3-46 Voltage-current curve.

Conventional or Constant-Current Welding Machines

The conventional or CC welding machine is used for manual, covered (stick) electrode arc welding or SMAW, the gas tungsten (TIG) process or GTAW, carbon-arc welding (CAW), arc gouging, and stud welding (SW). It can be used for automatic welding with larger-sized electrode wire, but only with a *voltage-sensing* wire feeder.

The CC welder produces a volt-ampere output curve such as is shown in Fig. 3-47.

A brief study of the curve reveals that a machine of this type produces maximum output voltage with no load (zero current), and that as the load increases the output voltage decreases. Under normal welding conditions the output voltage is between 20 and 40 V. The open-circuit voltage is between 60 and 80 V; CC machines are available that produce either ac or dc welding power or both ac and dc.

When welding with covered electrodes on CC welding machines, the arc voltage is largely controlled by the welder and has a direct relationship to the arc length. As the arc length is increased (a long arc), the arc voltage increases. If the arc length is decreased (a short arc), the arc voltage decreases. The output curve shows that when the arc voltage increases (long arc), the welding current decreases, or when the arc voltage decreases (short arc), the welding current increases. Thus, without changing the machine setting, the welder can vary the current in the arc or welding heat to a limited extent by lengthening or shortening the arc.

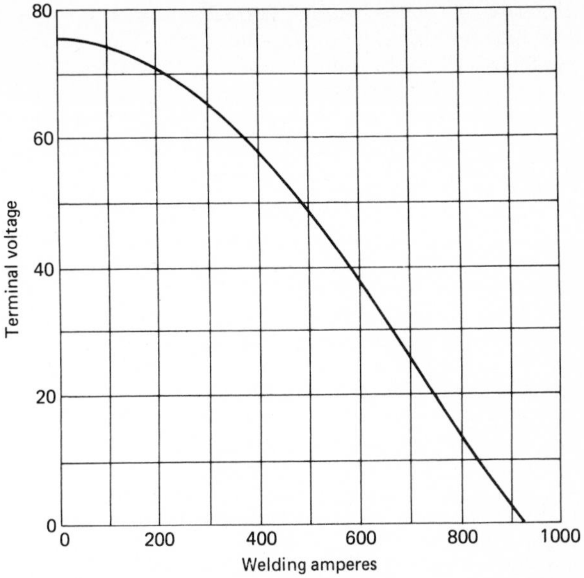

Figure 3-47 Voltage-current curve for constant-current machines.

CC machines can produce ac or dc welding power and can be rotating (generators) or static (transformers or transformer-rectifier) machines. The generator can be powered by a motor for shop use or by an internal-combustion engine (gasoline, LP gas, or diesel) for field use. Engine-driven welders can have either water- or air-cooled engines, and many of them provide auxiliary power for emergency lighting, power tools, etc.

Generator Welding Machines

On dual-control machines, normally generators, the slope of the output curve can be varied. The open-circuit, or no-load voltage is controlled by the fine-adjustment control knob. This control is also the fine welding-current adjustment during welding. The range switch provides coarse adjustment of the welding current. In this way, a soft or harsh arc can be obtained. With the flatter curve, and its low open-circuit voltage, a change in arc voltage will produce a greater change in output current. This produces a digging arc, preferred for pipe welding. With the steeper curve and its high open-circuit voltage, the same change in arc voltage will produce less of a change in output current. This is a soft, or quiet arc, useful for sheet-metal welding. In other words, the dual control, conventional, or CC welding generator allows the most flexibility for the welder. These machines can be driven by an electric motor or internal-combustion engine.

The CC, or drooping volt-ampere characteristic, machine can also be used for automatic welding processes. However, to use this type of welding machine, the automatic wire-feeding device must compensate for changes in arc length. This requires rather complex control circuits which involve feedback from the arc voltage or *voltage sensing.* This type of system is not used for the small-diameter electrode wire welding process.

Transformer Welding Machines

The transformer-type welding machine is the least expensive, lightest, and smallest of any of the different types. It produces alternating current for welding. The transformer takes power directly from the line, transforms it to the power required for welding and, by means of various magnetic circuits, inductors, etc., provides the volt-ampere characteristics proper for welding. The welding current output of a transformer may be adjusted in many different ways. The simplest method of adjusting output current is to use a tapped secondary coil on the transformer. This is a popular method employed by many of the limited-input, small welding transformers. The leads to the electrode holder and the work are connected to plugs, which may be inserted in sockets on the front of the machine in various locations to provide the required welding current. On some machines a tap switch is employed instead of the plug-in arrangement. In any case, exact current adjustment is not entirely possible.

On industrial types of transformer welding machines a continuous-output current control is usually employed. This can be a mechanical or electric control. The mechanical method involves moving the core of the transformer or moving the position of the coils within the transformer. The more advanced method of adjusting current output is by means of electric circuits. In this method the core of the transformer or reactor is saturated by an auxiliary electric circuit which controls the amount of current delivered to the output terminals. By adjusting a small knob, it is possible to provide continuous current adjustment from minimum to maximum output.

Although the transformer type of welder has many desirable characteristics, it also has some limitations. The power required for a transformer welder must be supplied by a single-phase system. This may create an unbalance in the power-supply lines, which is objectionable to most power companies. In addition, transformers have a rather low power-factor demand unless they are equipped with power-factor–correcting capacitors. The addition of capacitors corrects the power factor under load and produces a reasonable power factor which is not objectionable to electric power companies.

Transformer welding machines have the lowest initial cost, are the least expensive to operate, and require least space. In addition, ac welding power supplied by transformers reduces arc blow which can be troublesome on many welding applications.

Transformer-Rectifier Welding Machines

Some types of electrodes can be operated successfully only with dc power. A method of supplying dc power to the arc, other than using a rotating generator, is by adding a rectifier, an electric device which changes alternating current into direct current. Rectifier welding machines can be made to use a three-phase input. The three-phase input machine overcomes the line unbalance mentioned before.

In this type of machine the transformers feed into a rectifier bridge which then produces direct current for the arc. In other cases, where both alternating and direct current may be required, a single-phase ac transformer is connected to the rectifier. By means of a switch the welder can select either alternating or direct, straight- or reverse-polarity current for the welding requirement. In some types of ac-dc machines a high-frequency oscillator, plus water- and gas-control valves, are installed. This then makes the machine suited for gas tungsten arc welding as well as for manual coated-electrode welding.

The transformer-rectifier welding machines are available in different sizes and for single-phase or three-phase power supply. They may also be arranged for different primary voltages from the power line. The transformer-rectifier unit is more efficient electrically than the generator and provides quiet operation.

Multiple-Operator Welding System

This system uses a heavy-duty, high-current, and relatively high-voltage power source which feeds a number of individual-operator welding stations. At each welding station, a variable resistance is adjusted to drop the current to the proper welding range. Depending on the duty cycle of the welders, one welding machine can supply welding power simultaneously to a number of welders. The current supplied at the individual station has a drooping characteristic similar to the single-operator welding machines described above. The power source, however, has a CV output like those described next. The welding machine size and the number and size of the individual welding-current control stations must be carefully matched for an efficient multiple-operator system.

Constant-Voltage Welding Machines

A CV, sometimes called CP, or modified CV power source, is a welding machine that provides a nominally constant voltage to the arc regardless of the current in the arc. The characteristic curve of this type of machine is typified by the volt-ampere curve. *This type of machine can only be used for semiautomatic or automatic arc welding* using a continuously fed electrode wire. Furthermore, these machines are made to produce only direct current.

In continuous wire welding, the burn-off rate of a specific size and type of electrode wire is proportional to the welding current. As the welding current increases, the amount of wire burned off increases proportionally. This is graphically shown on the burn-off rate vs. current chart (Fig. 3-48). Thus, it can be seen that if wire were fed into an arc at

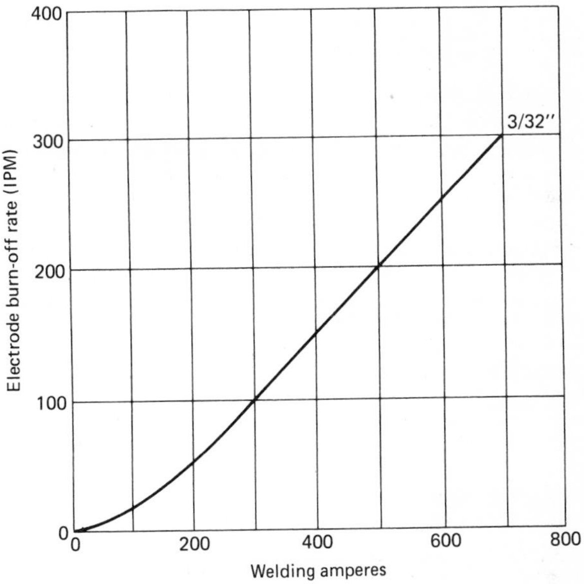

Figure 3-48 Chart showing burnoff rate vs. current.

a specific rate it would automatically draw a proportionate amount of current from a constant-voltage power source. The CV machine provides the amount of current required from it by the load imposed on it. The wire is fed into the arc by means of a constant-speed feed motor. This feed motor can be adjusted to increase or decrease the rate of wire feed. The system is inherently self-regulating. Thus, if the electrode wire were fed in faster, the current would increase. If it were fed in more slowly, the current would decrease automatically. The current output of the welding machine is thus set by the speed of the wire-feed motor. The voltage of the machine is regulated by an output control on the power source. Thus, only two controls maintain the proper welding current and voltage when the CV system is used.

The characteristic curves of CV machines have a slight inherent droop. This droop can be increased, or the slope made steeper, by various methods. Many machines have different tapes, or controls, for varying the slope of the characteristic curve. It is important to select the slope most appropriate to the process and the type of work being welded. CV machines can be either the generator type or the transformer-rectifier type.

Combination CV-CC Welding Machines

The most flexible type of welding machine is a combination type that can provide dc welding power with either a drooping or flat output characteristic volt-ampere curve by using different terminals and/or changing a switch. This type of welding machine is the most universal machine available. It allows the welder to use any of the arc-welding processes. The combination machine can be either a generator or a transformer-rectifier power source.

Specifying a Welding Machine

Selection of the welding machine is based on:

1. The process or processes to be used
2. The amount of current required for the work
3. The power available to the jobsite
4. Convenience and economic factors

These criteria plus information about each of the arc-welding processes indicate the type of machine required. The size of the machine is based on the welding current and duty cycle required. Welding current, duty cycle, and voltage are determined by analyzing the welding job and considering weld joints, weld sizes, etc., and by consulting welding procedure tables. The incoming power available dictates this fact. Finally, the job situation, personal preference, and economic considerations narrow the field to the final selection. The local welding equipment supplier should be consulted to help make the selection.

To order a welding machine properly, the following data should be given:

1. Manufacturer's type designation or catalog number
2. Manufacturer's identification or model number
3. Rated load voltage
4. Rated load amperes (current)
5. Duty cycle
6. Voltage of power supply (incoming)
7. Frequency of power supply (incoming)
8. Number of phases of power supply (incoming)

Welding-Machine Duty Cycle

Duty cycle is defined as the ratio of arc time to total time. For a welding machine, a 10-min time period is used. Thus for a 60 percent duty-cycle machine the welding load would be applied continuously for 6 min and would be off for 4 min. Most industrial-type CC (drooping) machines are rated at 60 percent duty cycle. Most CV (flat) machines used for automatic welding are rated at 100 percent duty cycle.

Figure 3-49, showing percent of working time vs. welding current, represents the ratio of the square of the rated current to the square of the load current multiplied by the rated duty cycle. Rather than work out the formula, use this chart. Draw a line parallel to the sloping lines through the intersection of the machine's rated current output and rated duty cycle. For example, a question might arise whether a 400-A, 60 percent duty cycle machine could be used for a fully automatic requirement of 300 A for a 10-min welding job. Line *A* shows this to be possible. It shows that the machine can be used at slightly over 300 A at a 100 percent duty cycle. Conversely, there may be a need to draw more than the rated current from a welding machine, but for a short period. Line *B*, for example, shows that the 200-A, 60 percent rated machine can be used at 250 A, providing the duty cycle does not exceed 40 percent (or 4 min out of each 10 min).

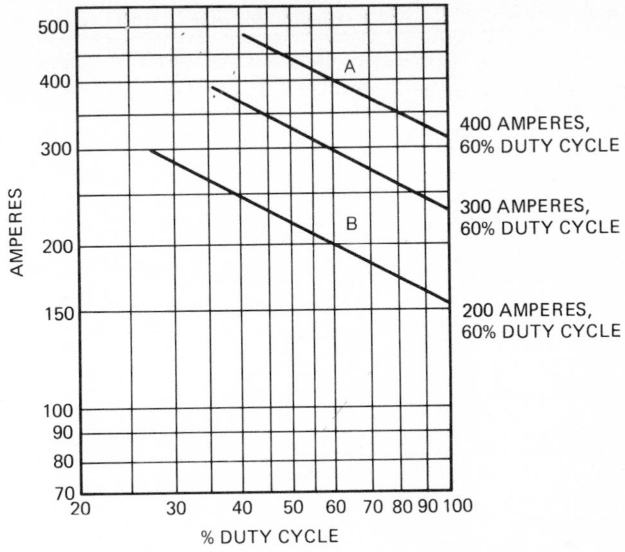

Figure 3-49 Percent of working time vs. welding current.

Use Fig. 3-49 to compare various machines. Relate all machines to the same duty cycle for a true comparison.

WELDING CABLE SELECTION

The size and length of the welding leads, welding cable and work cable, have a definite influence on the cost of welding. As the length of the leads is increased, their diameter should also be increased in order to avoid excessive voltage drop between the machine and the electrode and, particularly, to avoid wasted power as a result of the cables heating excessively.

To determine the power lost in the welding work leads, measure the voltage at the welding machine terminals. Then measure the voltage at the arc (meter connected between electrode holder and the work). Measure the welding current. The voltage loss in the leads equals the difference between the voltage at the terminals and that at the holder. Multiply this by the welding current, and the result is the power lost:

$$V \text{ (at terminal)} - V \text{ (at holder)} \times \text{welding current} = \text{power lost}$$

For example:

$$(35V - 32V) \times 250A = 750W \text{ lost}$$

Recommended Cable Sizes for Leads of Various Lengths

Table 3-3 shows cable sizes recommended for various lengths of leads. The footage shown includes the complete welding circuit—both welding lead and work lead combined. For example, the 60-ft column refers to two 30-ft leads.

DC Voltage Drop per 100 ft of Leads

Table 3-4 shows the voltage drop that will occur in a given length of cable of a given AWG size during welding with a given current value.

TABLE 3-3 Suggested Copper Welding Cable Size (AWG) Guide

Weld type	Weld current, A	Length of cable circuit, ft*—cable size 60	100	150	200	300	400
Manual (low-duty cycle)	100	4	4	4	2	1	1/0
	150	2	2	2	1	2/0	3/0
	200	2	2	1	1/0	3/0	4/0
	250	2	2	1/0	2/0		
	300	1	1	2/0	3/0		
	350	1/0	1/0	3/0	4/0		
	400	1/0	1/0	3/0			
	450	2/0	2/0	4/0			
	500	2/0	2/0	4/0			
Automatic (high-duty cycle)	400	4/0	4/0				
	800	4/0 (2)	4/0 (2)				
	1200	4/0 (3)	4/0 (3)				
	1600	4/0 (4)	4/0 (4)				

*Length of cable circuit equals total of electrode and work cables.

TABLE 3-4 DC Voltage Drop per 100 ft of Leads*

Welding current, A	Cable size (AWG) 2	1	1/0	2/0	3/0	4/0
50	1.0	0.7	0.5	0.4	0.3	0.3
75	1.3	1.0	0.8	0.7	0.5	0.4
100	1.8	1.4	1.2	0.9	0.7	0.6
125	2.3	1.7	1.4	1.1	1.0	0.7
150	2.8	2.1	1.7	1.4	1.1	0.9
175	3.3	2.6	2.0	1.7	1.3	1.0
200	3.7	3.0	2.4	2.0	1.5	1.2
250	4.7	3.6	3.0	2.4	1.8	1.5
300		4.4	3.4	2.8	2.2	1.7
350			4.0	3.2	2.5	2.0
400			4.6	3.7	2.9	2.3
450				4.2	3.2	2.6
500				4.7	3.6	2.8
550					3.9	3.1
600					4.3	3.4
650						3.7
700						4.0

*Figures in this table are for three-conductor cable. For four-conductor cable, reduce the ampere rating of each wire size by 20 percent.

When a cable is overheated, its life is shorter than when it is used without overheating. These figures assume that all connections are tight and that electrode holder and ground connections are in good condition.

For a higher amperage than given in the table, divide the load *equally* across two input cables of sufficient size to carry half the load.

To determine proper power cable wire size, consult the welding machine nameplate or data sheet for the amperage drawn at input line voltage. The data are based on NEC "Minimum Standards for Welding Equipment," Sec. 630.

WELDING DIFFERENT METALS

In order to produce a quality weld it is necessary to know the composition or analysis of each piece of metal that is to be welded. Once these properties or specifications are known, the proper filler metal can be selected so that the deposited weld metal will meet or exceed the mechanical properties of the base metal and have approximately the same composition and physical properties. The following two conditions must be fulfilled.

Base-Metal Properties

The base metal, or parts to be welded, must be known in order to determine their mechanical properties. The deposited weld metal must be selected to overmatch the mechanical properties of the base material. This can be done by selecting filler metals in accordance with some of the rules that follow.

Base-Metal Composition

The composition or analysis of the base metal, or metal to be welded, must be known. This can be determined if the specifications or trade name of the material is known. The filler metal to be used will then be selected in accordance with some of the rules that follow.

The exact selection procedure for filler metals varies somewhat depending upon the welding process that has been selected and, broadly, on the classification of metal that is to be welded.

Shielded-Metal Arc Welding

The shielded-metal–arc-welding process is the most popular and is commonly used for welding carbon steels, low-alloy steels, and stainless steels as well as for surfacing and other specialized applications. It is normally not used for welding aluminum, magnesium, titanium, and other "hard-to-weld" metals. Shielded-metal arc welding and the gas-shielded processes are used for welding the nickel alloys, copper alloys, high-strength steels, tool steels, and similar materials.

The following guidelines are to be used for selecting covered electrodes for welding carbon and low-alloy steels. These are related to the American Welding Society Filler Metal Specifications A5.1, "Carbon Steel Covered Arc Welding Electrodes," and A5.5, "Low Alloy Steel Covered Arc Welding Electrodes." The classification for these types of electrodes is shown in Table 3-5. The prefix letter "E" designates an electrode. The first two or three digits indicate tensile strength and other mechanical properties. The third (or fourth) digit indicates the welding position that can be used, and the last digit indicates usability of the electrode. The suffix letter, when used following the four- or five-digit classification, designates the composition of the deposit weld metal. This is normally used for low-alloy, high-strength electrodes and does not apply to the 60XX classification. The suffix letters and the nominal composition are shown in Table 3-6.

For exact data the filler metal specification should be consulted.

The operational factors relating to covered electrodes are as follows:

Welding Position

Electrodes are designed to be used in specific positions. The third (or fourth) digit of the electrode classification indicates the welding position that can be used. Match the electrode to the welding position that will be encountered.

Welding Current

Some electrodes are designed to operate best with direct current, others on alternating current. Some will operate on either alternating or direct current. The last digit indicates the welding current usability. Select the electrode to match the type of power source that will be used.

TABLE 3-5 AWS Classification System for Covered Mild and Low-Alloy Steel Electrodes

A. Prefix: E designates an electrode
B. First two or three digits: mechanical properties

Classification	Minimum tensile strength, lb/in² (MPa)	Minimum yield strength, lb/in² (MPa)	Minimum elongation, %
E60XX	62 000 (427)	50 000 (345)	22
E70XX	70 000 (483)	57 000 (393)	22
E80XX	80 000 (552)	67 000 (462)	19
E90XX	90 000 (621)	77 000 (531)	17
E100XX	100 000 (690)	87 000 (600)	16
E110XX*	110 000 (758)	97 000 (669)	15
E120XX*	120 000 (827)	107 000 (738)	14

C. Third (or fourth) digit: applicable welding positions
EXX1X: flat, horizontal, vertical, and overhead
EXX2X: flat and horizontal fillet
D. Last digit: electrode usability

Classification	Current†	Arc	Penetration	Covering slag	Iron powder, % ‡
EXX10	dcep	Digging	Deep	Cellulose-sodium	0–10
EXX11	ac, dcep	Digging	Deep	Cellulose-potassium	0
EXX12	ac, dcen	Medium	Medium	Rutile-sodium	0–10
EXX13	ac, dcen, dcep	Soft	Light	Rutile-potassium	0–10
EXX14	ac, dcen, dcep	Soft	Light	Rutile-iron powder	25–40
EXX15	dcep	Medium	Medium	Low-hydrogen–sodium	0
EXX16	ac, dcep	Medium	Medium	Low-hydrogen–potassium	0
EXX18	ac, dcep	Medium	Medium	Low-hydrogen–iron powder	25–40
EXX20 and EXX22 (single pass)	ac, dcen, dcep	Medium	Medium	Iron oxide-sodium	0
EXX24	ac, dcen, dcep	Soft	Light	Rutile–iron powder	50
EXX27	ac, dcen, dcep	Medium	Medium	Iron oxide–iron powder	50
EXX28	ac, dcep	Medium	Medium	Low-hydrogen–iron powder	50
EXX48 (vertical down)	ac, dcep	Medium	Medium	Low-hydrogen–iron powder	25–50

*Low-hydrogen-type coating only.
†dcep = electrode positive—reverse polarity; dcen = electrode negative—standard polarity.
‡Iron powder percentage based on weight of the covering.

Thickness and Shapes of Base Metal

Weldments may include thick and heavy material of complicated design. The electrode selected should have maximum ductility to avoid weld cracking. Select the low-hydrogen types, EXX15, 16, 18, or 28.

Weld Design and Fit-up

Welding electrodes are designed with a digging, a medium, a soft, or a light-penetrating arc. The last digit of the classification indicates this usability factor. Deep-penetrating electrodes with a digging arc should be used when edges are not beveled or fit-up is tight. At the other extreme, light-penetrating electrodes with a soft arc are required when welding on thin material or when root openings are too wide.

TABLE 3-6 Chemical Composition of Deposited Weld Metal

	Weld metal composition, %*						
Suffix	C	Mn	Si	Ni	Cr	Mo	V
A1	0.12	0.60 or 1.00†	0.40 or 0.80†	—	—	0.40–0.65	—
B1	0.12	0.90	0.60 or 0.80†	—	0.40–0.65	0.40–0.65	—
B2L	0.05	0.90	1.00	—	1.00–1.50	0.40–0.65	—
B2	0.12	0.90	0.60 or 0.80†	—	1.00–1.50	0.40–0.65	—
B3L	0.05	0.90	1.00	—	2.00–2.50	0.90–1.20	—
B3	0.12	0.90	0.60 or 0.80†	—	2.00–2.50	0.90–1.20	—
B4L	0.05	0.90	1.00	—	1.75–2.25	0.40–0.65	—
B5	0.07						
‡	0.15	0.40–0.70	0.30–0.60	—	0.40–0.60	1.00–1.25	0.05
C1	0.12	1.20	0.60 or 0.80†	2.00–2.75	—	—	—
C2	0.12	1.20	0.60 or 0.80†	3.00–3.75	—	—	—
C3	0.12	0.40–1.25	0.80	0.80–1.10	0.15	0.35	0.05
D1	0.12	1.25–1.75	0.60 or 0.80†	—	—	0.25–0.45	—
D2	0.15	1.65–2.00	0.60 or 0.80†	—	—	0.25–0.45	—
‡							0.10
G	—	1.00 min.	0.80 min.	0.50 min.	0.30 min.	0.20 min.	min.
M	0.10	0.60–2.25†	0.60 or 0.80†	1.40–2.50	0.15–1.50†	0.25–0.55†	0.05

*Compositions are maximum unless otherwise indicated.
†Amount depends on electrode classification.
‡A suffix is not aplied to E60XX classification.

Service Conditions and Specifications

Weldments subjected to severe service conditions such as low temperature, high temperature, and shock loading, need special consideration. Select the electrode to match the base-metal properties including composition, ductility, and toughness. This is indicated by the toughness requirement of the specification. Usually low-hydrogen-type electrodes are required.

Production Efficiencies and Job Conditions

Certain electrodes are designed for high deposition rates but may be used only under certain position requirements. Where they can be used, select the high-iron powder types, the EXX24, 27, or 28 types. Other conditions may require some experimentation to determine the best electrode for the job, allowing for the most efficient production.

Carbon and low-alloy steel electrodes may be classed into four general groups:

F-1	High-deposition group	E6020, E7024, E6027, E6028
F-2	Mild-penetrating group	E6012, E6013, E7014
F-3	Deep-penetrating group	E6010, E6011
F-4	Low-hydrogen group	E6015, E7016, E7018, E6028

Electrodes in the same grouping operate and are run in the same general manner.

Electrode Selection for Constructional Steels

Welding procedure information is provided for most trade-name steels. This information is presented with a section for the constructional steels of each of the following steel companies: Armco Steel Corporation and its Sheffield Division, Bethlehem Steel Corporation, Great Lakes Steel Corporation, Inland Steel Corporation, Jones & Laughlin Steel Corp., United States Steel Corporation, Republic Steel Corporation, and Youngstown Sheet and Tube Company. The information given conforms to recommendations of that particular steel company. The trade name of each steel is given. In cases where the proprietary steel is also known to conform to certain ASTM specifications, this information is also given. These listings are not intended to list all the products of the companies named, but an attempt has been made to include the steels commonly used in welded fabrication at the time of this writing.

General Recommendations for Preheating and Electrode Selection

The recommendations are:

1. No welding should be done when the ambient temperature is below 0°F (−18°C). When the base metal's temperature is below 32°F, preheat the base metal to at least 70°F (0°C) (21°C) and maintain this minimum temperature during welding. Light sections require only local preheating, but heavy sections require general preheating. For structures, the American Welding Society specifies that the preheat be maintained on all surfaces of the plate within 3 in of the point of welding.
2. Electrodes that are not of the low-hydrogen type can be used to weld thinner sections of mild carbon steel when proper preheat temperatures are maintained. Low-hydrogen electrodes are recommended for thicker sections of steel and for low-alloy steel in all thickness ranges. When low-hydrogen electrodes are used, they must be thoroughly dry. They may be kept dry by storing them in a heated box and removing them from it immediately prior to use. Specific recommendations are found in the discussion of the Republic Steel Company's products. See Tables 3-12 and 3-13 in that discussion.
3. Any preheating indicated should be done prior to any tack welding as well as prior to the principal welding, and the temperature should be maintained as a minimum interpass temperature as welding proceeds.
4. When low-alloy steels are welded to lower-strength grades, select electrodes to match the strength of the lower-strength steel, but use welding practice suitable for the higher-strength steel.

Popular ASTM Constructional Steels

Popular steels are listed by the ASTM as follows:

A 36, Structural Steel

A 131, Structural Steel for Ships

A 201, Carbon-Silicon Steel Plates of Intermediate Tensile Ranges for Fusion Welded Boilers and other Pressure Vessels

A 212, High Tensile Strength Carbon-Silicon Plates for Boilers and Other Pressure Vessels

A 242, (Weldable grade) High Strength Low Alloy Structural Steel

A 283, Low and Intermediate Tensile Strength Carbon Steel Plates of Structural Quality

A 441, High Strength Low Alloy Manganese Vanadium Steel

Armco Steel Company (Including Sheffield Steels)

SSS-100, SSS-100A, and SSS-100B. Use low-hydrogen rods only. For high restraint, use 500°F (260°C) maximum preheat (400°F or 205°C maximum for SSS-100A and SSS-100B). It is important to keep heat input low to obtain fast cooling. Do not weave the electrode more than 2½ times the electrode diameter. Multipass welding employing the stringer bead technique should be used. Allow beads to cool below 250°F (120°C) (200°F or 93°C for SSS-100A and SSS-100B) before making additional passes.

Use E110XX electrodes (E120XX in cases of high restraint). If SSS-100 is welded to a lower-strength material, low-hydrogen electrodes should be selected to match the strength of the weaker material. When postweld stress-relief heat treatment is applied to SSS-100 series steel weldments, the weld metal should not contain added vanadium. Stress-relief temperature is 1100°F (600°C). Fillet welds may be made with E90XX or E100XX. For joints where the weld metal is expected to provide yield and tensile strengths equal to that of the base metal, electrodes of the E110XX series are ordinarily employed.

Abrasion-Resistant SSS-100 Series Plates (321, 360 and 400 Bhn). The same procedures as outlined for the SSS-100 constructional alloys should be followed using techniques ordinarily employed on hardenable alloy steels. Weld metal with the lowest permissible strength often is selected to assure adequate ductility and toughness in the weld deposits. Where the hardness or toughness of a welded zone appears unsuited for service conditions, a postweld tempering may be done. However, the temperature should be limited to 800°F (430°C) to avoid lowering the overall hardness of the heat-treated plate.

Sheffield Hi-Strength A (ASTM A 242). Also called Armco High Strength #1. Use E60XX electrode and E70XX for multipass welding. Preheating or postheating is not necessary.

Sheffield Hi-Strength B (ASTM A 441). Also called Armco High Strength #5. Use E60XX or E70XX electrodes.

Sheffield Hi-Strength C (Grades 45, 50, 55, 60). Grade designations indicate minimum yield point. Grade 50 is called Armco High Strength #6; grade 45 is called Armco High Strength #7. No preheating is required for C steels. Use E70XX electrodes for grades 45 and 50. Use E80XX for grades 55 and 60. Use low-hydrogen electrodes for grades 55 and 60.

Sheffield Shef-Ten (ASTM A 440). This steel is primarily applicable to riveted and bolted structures. Steel under this specification can be satisfactorily welded under controlled conditions and proper procedures. No preheating is necessary in thicknesses up to and including ½ in. Preheat temperatures ranging from "hand warm" for thickness only slightly greater than ½ in to 350°F (180°C) for the greater thickness may be necessary. Preheating is required for cold metal.

Shef-Lo-Temp and Shef-Super-Low-Temp. Use low-hydrogen electrodes. Use E80XX-C1 or E80XX-C2. Weld deposits with high nickel content should be used for low-temperature applications.

Armco (Sheffield) Abrasion-Resisting Steel. Preheating to 300 to 400°F (150 to 205°C) and postheating or slow cooling after welding are recommended for this steel. Preheat also for tack welding. Use low-hydrogen electrodes of E100XX grade. If full-strength joints are not needed, lower-strength electrodes such as E70XX and E80XX may be used.

Aluminized Steel Type 1. Resistance welding is the best method for this steel. Corrosion resistance can be restored to weld areas by metallizing. Low-hydrogen E-6016 rods and 18-8 stainless steel rods can also be used. All slag and oxides must be removed to prevent corrosion. Weld areas should then be metallized.

Armco Hi-Strength #4R and 4S. No preheat required. Use E7018 or E7018-A1 electrode.

Armco QTC. Use low-hydrogen electrodes. E90XX and E100XX electrodes are suggested. Caution should be exercised to ensure dryness of both plate surfaces to be welded and flux coatings on electrodes.

Inland Steel Company

Hi-Steel (A 242). Use E-7018 or E-7018-A1 electrodes. Preheat of 200 to 500°F (93 to 260°C) may be required in the heavier sections or highly restrained weldments.

Tri-Steel (A 441). Use E-7018 or E-7018-A1 electrodes. Preheat may be required in the heavier sections or highly restrained weldments.

Hi-Man (A 440). Welding same as Tri-Steel. This steel is not recommended for welding. It can be welded when special care is taken, but is intended primarily for riveted or bolted construction.

INX Steels (45, 50 and 55). Use E-7018 or E-7018-A1 electrodes. Preheat may be required in heavier sections or highly restrained weldments.

INX Steels (60, 65 and 70). Use E-8018 or E-9018 electrodes. Preheat may be required when there is restraint.

Bethlehem Steel Company

V Steels. Low-hydrogen electrodes are recommended for most applications, particularly for thicknesses over ¾ in and for all thicknesses in V-55, V-60, and V-65. If the temperature of the shop or steel falls below 50°F (10°C) for conditions where no preheat is shown, preheat the steel to 100°F (38°C) before welding. Where V steels are to be welded to lower-strength grades, select electrodes to match the strength of the lower-strength steel, but employ welding practice for the higher-strength steel. See Table 3-7.

TABLE 3-7A Preheat Schedule for V Steels

Thickness, in	Type of electrode	Preheat schedule V-45	V-50	V-55	V-60	V-65
<⅜	Conventional	None	None	None	100°F	150°F
	Low-hydrogen	None	None	None	None	None
>⅜ to ¾	Conventional	None	100°F	150°F	200°F	250°F
	Low-hydrogen	None	None	None	100°F	150°F
>¾ to 1½	Conventional	150°F	150°F	200°F	250°F	—
	Low-hydrogen	None	None	100°F	150°F	
>1½ to 2	Conventional	200°F	250°F	300°F	—	—
	Low-hydrogen	150°F	150°F	200°F		
>2 to 3	Conventional	300°F	300°F	350°F	—	—
	Low-hydrogen	200°F	250°F	300°F		

TABLE 3-7B Recommended Electrodes for Manual Arc Welding of V Steels

Steel	Electrode
V-45	E60XX
	E70XX
V-50	E60XX
	E70XX
V-55	E70XX
V-60	E70XX
	E80XX
V-65	E70XX
	E80XX

Bethlehem Abrasion-Resisting Steel, Grade 235. A preheat and interpass temperature of 300 to 400°F (150 to 205°C) is recommended. Slow cooling is necessary after welding to prevent cracking. If the weld will be subject to impact, normalize at 1650°F (900°C). Use E10018-D2 or E10018-M rods. If preheat is impractical, use 309 or 312 stainless steel rods.

Mayari R(A242). Use E7018 or E7018-A1 rods. For greater corrosion resistance, use E8018-B2. For thicknesses from 1½ to 2 in (3.8 to 5 cm), preheat at 100°F (38°C); for thicknesses from 2 to 4 in (5 to 10 cm), preheat at 200°F (93°C).

Medium Manganese. Use E7018 or E7018-A1 rods.

Preheat temperature, °F (°C)

½″ to 1″	1 to 1½″	1½″ to 2″	2 to 4	over 4″
100 (38)	200 (93)	200 (93)	300 (150)	350+ (177+)

Manganese Vanadium. Use E7018 or E7018-A1 rods.

Preheat temperature, °F (°C)

To 1″	1 to 1½″	1½ to 2″	2 to 4	over 4″
50 (10)	100 (38)	200 (93	300 (150)	350+ (177+)

Great Lakes Steel Corporation

GLX-45-W, GLX-55-W, GLX-50-W, GLX-60-W. These steels are classified according to their yield strength. Thus, GLX-45-W indicates a mild carbon steel with a yield strength of 45,000 lb/in^2 minimum. Electrodes recommended are those that would be used at the same carbon grade, except that the class should overmatch the parent metal in strength. For multipass welds, electrodes of the AWS E70XX class are satisfactory.

Steel	Electode
N-A-XTRA 80	Use E9015 or E9018
N-A-XTRA 90	Use E10015 or E10018
N-A-XTRA 100	Use E11015 or E11018
N-A-XTRA 110	Use E12015 or E12018

Equivalent electrodes for alternating current or iron-bearing coatings are satisfactory.

X-A-R 15 and X-A-R 30. The high hardenability of these steels requires care in welding. Preheat and interpass temperature may be required for highly restrained weldments and when welding material more than ¾ in (19 mm) thick. E10018-M, E10018-D2, or E12018-M electrodes should be used.

Jones & Laughlin Steel Corp.

J & L Cor-Ten. Use AWS E7018 or AWS E7018-A1 electrodes. Where greater corrosion resistance is required use AWS E8018-B2. No preheat is required.

Jalloy-S-90. Preheat is not normally required unless a unique stress-distribution condition arises. Use AWS E10XX low-hydrogen series.

Jalloy-S-100. Preheat same as for Jalloy-S-90. Use AWS E110XX low-hydrogen series.

Jalloy-S-110. Preheat same as for Jalloy-S-90. Use AWS E120XX low-hydrogen series.

Jalloy-AR-280, Jalloy-AR-320, Jalloy-AR-360, and Jalloy-AR-400. Must use a low-hydrogen rod. Preheat is not normally required unless a unique stress-distribution condition arises. Use AWS E100XX, E110XX, or E120XX. If preheat is required, 200 to 400°F (93 to 205°C) should be used.

JLX-45-W, JLX-50-W, JLX-55-W, JLX-60-W. Use E7018 or E7018-A1 electrodes. No preheat is required.

J & L Nickel-Copper-Titanium High-Strength Forming Steel. Preheat is required only if a code must be met. Use an E60XX electrode (E6010 or E6012). For highly stressed conditions, an E70XX rod may be required.

Jalten No. 1, No. 2, and No. 3. Use AWS E60XX electrode. An E70XX rod may be required for highly stressed conditions. Use 400 to 600°F (205 to 316°C) preheat in critical cases.

Speed Case and Speed Treat. Must use low-hydrogen rods, normally E70XX, but may have to go higher depending on design. A short arc is desirable with proper current an important factor. Preheat may be required.

Speed Alloy. Must use low-hydrogen electrode. Preheat to 500°F (260°C) before welding. Normalizing or full annealing is recommended after welding. A short arc with proper current is essential. Use AWS E100XX rods of low-hydrogen series.

United States Steel Corporation

USS Spec	ASTM Spec
Man-Ten	A440
Tri-Ten	A441
Cor-Ten	A242

The general rules on preheat and electrode selection apply unless superseded by a specific note for that steel.

Cor-Ten Steel (A 242). When the welded area in multiple-pass welds is to simulate or approach the color of Cor-Ten steel after atmospheric exposure, electrodes containing 2½ percent nickel (AWS E8016-C3) or 3½ percent nickel (AWS E8016-C2) are suggested for shield-metal arc welding. For single-pass welding, carbon steel electrodes of the E60XX and E70XX group are satisfactory. Low-hydrogen electrodes are preferred for thickness greater than ½ in (12 mm). A preheat temperature above the minimum given may be required for highly restrained welds.

For shielded-metal arc welding the practices in Table 3-8 are suggested. The use of an austenitic stainless steel electrode of the E309, E310, or E312 class is recommended for joining USS Cor-Ten to austenitic stainless steel.

Tri-Ten Steel (ASTM A 441 and A 242). Low-hydrogen electrodes are preferred for manual arc welding. USS Tri-Ten can also be welded to austenitic stainless steel by using electrodes of the E309, E310, or E312 class. For suggested welding practices see Table 3-9.

Man-Ten and Man-Ten A440 Steels. Man-Ten A440 is intended for riveted or bolted structures. Man-Ten is considered weldable under carefully controlled conditions (see Table 3-10). Low-hydrogen electrodes are preferred for USS Man-Ten Steel. Man-Ten A440 is intended for riveted or bolted structures. However, when attachments are made by welding, the minimum preheat temperatures in Table 3-10 are necessary.

TABLE 3-8 SMAW Welding Practices for Cor-Ten Steel

Electrode	Thickness, in	Suggested min. preheat or interpass temp, °F (°C)
Low-hydrogen type (E60 or E70, 16, 18 or 28) of ASTM A-233	<1 incl*	50 (10)
	>1 to 2 incl	100 (38)
	>2 to 5 incl	200 (93)
Other than low-hydrogen type (E60 or E70 group) of ASTM A-233	to ½	50 (10)
	>½ to 2 incl	200 (93)
	>2 to 5 incl	300 (150)

*incl = inclusive.

TABLE 3-9 Welding Practices for Tri-Ten Steel

Electrode	Thickness, in	Min preheat or interpass temp, °F (°C)
Low-hydrogen type (E60 or E70, 15, 16, 18 or 28) of ASTM A-233	to 1 incl*	50 (10)
	> 1 to 2 incl	100 (38)
	>2	200 (93)
Other than low-hydrogen type (E60 or E70 group) of ASTM A-233	to ½ incl	50 (10)
	>½ to 1 incl	100 (38)
	>1 to 1½ incl	200 (93)
	>1½ to 2 incl	250 (120)
	>2	300 (150)

*incl = inclusive.

TABLE 3-10 Suggested Welding Practices for Man-Ten and Man-Ten A440 Steels

Electrodes	Thickness, in	Man-Ten A440 min preheat or interpass temp, °F (°C)	Man-Ten min preheat or interpass temp, °F (°C)
Low-hydrogen type (E60 or E70 15, 16, 18 or 28) of ASTM A-233	to ⅜ incl	50 (10)	50 (10)
	>⅜ to ½ incl*	100 (38)	50 (10)
	>½ to 1 incl	200 (93)	100 (38)
	>1 to 2 incl	300 (150)	200 (93)
	>2	300 (150)	300 (150)
Other than low-hydrogen type (E60 or E70 group) of ASTM A-233	to ½ incl	NR*	200 (93)
	>½ to 1 incl	NR	300 (150)
	>1 to 2 incl	NR	300 (150)
	>2	NR	NR

*incl = inclusive; NR = not recommended.

A-R Steel. Low-hydrogen electrodes of the ASTM-AWS E9015, E9016, E9018, E10016, and E10015 grades should be used for welding USS A-R steel. If the weldment is to be postweld heat-treated, the E10016 or other electrodes containing more than 0.05 percent vanadium should not be used for welding. Recommended preheating for A-R steel is

Thickness, in	Minimum preheat or interpass temperature, †F (°C)
<½ inclusive	300 (150)
>½ to 2 inclusive	400 (205)

Postweld heat treatment for 1 h of thickness is recommended (1025°F or 555°C).

Ex-Ten Steel. Welding operations with Ex-Ten occasionally result in residual stresses or increased hardness of sufficient magnitude to require postweld heat treatment, particularly in the higher strength grades. See Table 3-11. USS Ex-Ten steel can be welded to structural carbon steel or to other USS high-strength steels using an ordinary mild or low-alloy steel electrode as desired, subject to the practices suggested for the grade. Ex-Ten steel can be welded to austenitic stainless steel by using electrodes of the E309, E310, or E312 class.

TABLE 3-11 Suggested Welding Practice for Ex-Ten Steel

		Low-hydrogen electrode		Non-low-hydrogen electrode	
Grade	Thickness, in	Min preheat or interpass temp, °F (°C)	Electrode	Min preheat or interpass temp, °F (°C)	Electrode
70	to ⅜ incl*	50 (10)	E8015, 16, 18	NR*	NR
60	to ⅜ incl	50 (10)	E70, 15, 16, 18 or 28	50 (10)	E70XX
	>⅜ to ¾ incl	50 (10)	E70, 15, 16, 18 or 28	200 (93)	E70XX
	>¾ to 1¼ incl	100 (38)	E70, 15, 16	NR	NR
50	to ⅜ incl	50 (10)	E60 or E70, 15, 16, 18 or 28	50 (10)	E60XX or E70XX
	⅜ to ¾ incl	50 (10)	E60 or E70, 15, 16, 18 or 28	100 (38)	E60XX or E70XX
	¾ to 1 incl	50 (10)	E60 or E70	200 (93)	E60XX or E70XX
	1 to 1½ incl	100 (38)	E60 or E70, 15, 16, 18 or 28	200 (93)	E60XX or E70XX
	1½ to 2 incl	100 (38)	E60 or E70, 15, 16, 18 or 28	NR	NR
	>2	200 (93)	E60 or E70, 15, 16, 18 or 28	NR	NR
42	to ¾ incl	50 (10)	E60 or E70, 15, 16, 18 or 28	50 (10)	E60XX or E70XX
	¾ to 1 incl	50 (10)	E60 or E70, 15, 16, 18 or 28	100 (38)	E60XX or E70XX
	1 to 1½ incl	50 (10)	E60 or E70, 15, 16, 18 or 28	200 (93)	E60XX or E70XX
	1½ to 2 incl	50 (10)	E60 or E70, 15, 16, 18 or 28	NR	NR
	>2	150 (66)	E60 or E70, 15, 16, 18 or 28	NR	NR

*incl, inclusive; NR, not recommended.

Republic Steel Company

Steel		Electrode
Republic	70	E80XX, E70XX, E10018, E10016
Republic	65	E70XX, E80XX, E10016, E10018
Republic	50	E60XX, E70XX, E7016-E7018
Republic	M-1	E60XX, E7016 or 18
Republic	M-2	Same as M-1
Republic	A-441	E60XX, E70XX
X-45-W		E60XX, E70XX
X-50-W		E60XX, E70XX
X-55-W		E60XX, E70XX
X-60-W		E60XX, E70XX

See Tables 3-12 and 3-13 for preheat temperatures.

Welding Stainless Steel

Selection of Covered Electrodes for Welding Stainless Steels

In order to properly select the electrode for welding stainless steels or, more correctly, corrosion-resisting steels, it is necessary to know and understand the numbering system used. This numbering system, established by the American Iron and Steel Institute

TABLE 3-12 Preheat for Non-Low-Hydrogen Rods, °F

Rod	To ½ in	½ to 1 in	1 to 1½ in	1½ to 2 in	2 to 4 in	Over 4 in
70	NR	NR	NR	NR	NR	NR
65	NR	NR	NR	NR	NR	NR
50	50	50	50			
M-1	50	50	200	200		
M-2	50	50	200	200		
A-441	NR	NR	NR	NR	NR	NR
X-45-W	50	150	150	200	NR	NR
X-50-W	50	150	150	200	NR	NR
X-55-W	150	250	250	250	NR	NR
X-60-W	200	250	250	300	NR	NR

*NR, not recommended.

TABLE 3-13 Preheat for Low-Hydrogen Rods, °F

Rod	To ½ in	½ to 1 in	1 to 1½ in	1½ to 2 in	2 to 4 in	Over 4 in
70	50	100	150			
65	50	50	100			
50	50	50	50	150	200	
M-1	50	50	50	50	150	
M-2	50	50	50	50	150	
A-441	50	200	300	300	300	350+
X-45-W	50	50	50	150	200	
X-50-W	50	50	50	150	250	
X-55-W	50	100	100	200	300	
X-60-W	50	100	150			

(AISI), is based on the composition of the stainless steel, i.e., type 308, type 312, etc. These are three-digit numbers which classify the steel according to its metallurgical structure. Stainless steels are sometimes known and identified according to their principal alloying element, such as 18/8, 25/20, and so on.

Iron is the main element of all stainless steels. However, to make it corrosion-resistant, chromium must be present in an amount of 11.5 percent or more. The addition of chromium to iron provides a fine film of chromium oxide which forms on the surface and acts as a barrier to further oxidation, rust, or corrosion. The addition of nickel in the proper ratio results in a stainless steel series referred to as chrome-nickel types. They all contain a percentage of nickel and are nonmagnetic. The addition of nickel increases the corrosion resistance, ductility, electric resistance, impact properties, and fatigue resistance.

There are three basic classes of stainless steel which are grouped according to their metallurgical microstructure. They are known as the austenitic, martensitic, and ferritic types. The properties of these three classes of stainless steel differ and require different welding electrodes and procedures.

Electrodes for welding stainless steels are identified by the AISI three-digit number following the prefix letter "E" and followed by the usability classification, normally 15 or 16. The usability classification can also be followed by a letter such as "L" indicating "low carbon." All stainless steel electrodes have a low-hydrogen-type coating. The EXXX-15 type indicates a lime coating which is normally used with direct current. The EXXX-16 type uses titanium-base coating and can be used with alternating or direct current. Type 16 is a smoother-running electrode. However, type 15 may be the best for out-of-position welding. The mechanical properties are not specified for stainless steel electrodes since the deposit weld metal will be nearly identical to the base metal and will have physical properties normal to the composition of the metal deposit. Table 3-14 gives the proper electrodes for welding the different types of stainless steel.

TABLE 3-14 Electrode Selection for Welding Stainless Steels

AISI No.	Chemical analyses of stainless steels, % Carbon	Manganese	Silicon	Chromium	Nickel	Other elements	Hobart Electrode No.
				Austenitic			
201	0.15 max.	5.5–7.5	1.0	16.0–18.0	3.5–5.5	N_2 0.25 max.	308
202	0.15 max.	7.5–10.	1.0	17.0–19.0	4.0–6.0	N_2 0.25 max.	308
301	0.15 max.	2.0	1.0	16.0–18.0	6.0–8.0	——	308
302	0.15 max.	2.0	1.0	17.0–19.0	8.0–10.0	——	308
302B	0.15 max.	2.0	2.0–3.0	17.0–19.0	8.0–10.0	——	308
303	0.15 max.	2.0	1.0	17.0–19.0	8.0–10.0	S 0.15 min.	308DC
303Se	0.15 max.	2.0	1.0	17.0–19.0	8.0–10.0	Se 0.15 min.	308DC
304	0.08 max.	2.0	1.0	18.0–20.0	8.0–12.0	——	308
304L	0.03 max.	2.0	1.0	18.0–20.0	8.0–12.0	——	308L
305	0.12 max.	2.0	1.0	17.0–19.0	10.0–13.0	——	308
308	0.08 max.	2.0	1.0	19.0–21.0	10.0–12.0	——	308
309	0.20 max.	2.0	1.0	22.0–24.0	12.0–15.0	——	309
309S	0.08 max.	2.0	1.0	22.0–24.0	12.0–15.0	——	309
310	0.25 max.	2.0	1.50	24.0–26.0	19.0–22.0	——	310
310S	0.08 max.	2.0	1.50	24.0–26.0	19.0–22.0	——	310
314	0.25 max.	2.0	1.5–3.0	23.0–26.0	19.0–22.0	——	310DC
316	0.08 max.	2.0	1.0	16.0–18.0	10.0–14.0	Mo 2.0/3.0	316
316L	0.03 max.	2.0	1.0	16.0–18.0	10.0–14.0	Mo 2.0/3.0	316L
317	0.08 max.	2.0	1.0	18.0–20.0	11.0–15.0	Mo 3.0/4.0	317
321	0.08 max.	2.0	1.0	17.0–19.0	9.0–12.0	Ti 5 × C min.	347
347	0.08 max.	2.0	1.0	17.0–19.0	9.0–13.0	Cb + Ta 10 C min.	347
348	0.08 max.	2.0	1.0	17.0–19.0	9.0–13.0	Ta 0.10 max.	347
				Martensitic			
403	0.15 max.	1.0	0.5	11.5–13.0	——	——	410
410	0.15 max.	1.0	1.0	11.5–13.5	——	——	410
414	0.15 max.	1.0	1.0	11.5–13.5	1.25–2.5	——	410
416	0.15 max.	1.25	1.0	12.0–14.0	——	S 0.15 min.	410DC
416Se	0.15 max.	1.25	1.0	12.0–4.0	——	Se 0.15 min.	410DC
420	Over 0.15	1.0	1.0	12.0–14.0	——	——	410
431	0.20 max.	1.0	1.0	15.0–17.0	1.25–2.5	——	430
440A	0.60–0.75	1.0	1.0	16.0–18.0	——	Mo 0.75 max.	——
440B	0.75–0.95	1.0	1.0	16.0–18.0	——	Mo 0.75 max.	——
440C	0.95–1.2	1.0	1.0	16.0–18.0	——	Mo 0.75 max.	——
				Ferritic			
405	0.08 max.	1.0	1.0	11.5–14.5	——	Al 0.1/0.3	410
430	0.12 max.	1.0	1.0	14.0–18.0	——	——	430
430F	0.12 max.	1.25	1.0	14.0–18.0	——	S 0.15 min.	430DC
430FSe	0.12 max.	1.25	1.0	14.0–18.0	——	Se 0.15 min.	430DC
446	0.20 max.	1.50	1.0	23.0–27.0	——	N 0.25 max.	309

Welding Nonferrous Metals

The Gas Tungsten Arc Process and Gas Metal Arc Process

The gas-shielded arc-welding processes are more popular for welding nonferrous metals. When using these processes it is best to select a welding electrode having a composition similar to the metal being welded. Unfortunately, electrodes are not available in every conceivable composition and, therefore, charts showing recommended electrodes for different types of metals are available. A guide for the choice of filler metals for welding aluminum is given in Table 3-15. This chart can be used for either GMAW or GTAW. Similar charts are available for welding magnesium. However, for welding with nickel-base or copper-base electrodes, the manufacturer of the base metal or manufacturers of nonferrous electrodes should be consulted.

TABLE 3-15 Guide to the Choice of Filler Metal for Aluminum

	Base metal	Casting alloys		6061, 6062, 6063, 6151	5456	5454	5154, 5254 (1)	5086, 5356	5083	5052, 5652 (1)	5005, 5050	3004, CLAD 3004	1100, 3003, CLAD 3003	1060
		43, 355, 356	214, A214 B214, F214											
A	1060	ER4043	ER4043	ER4043	ER5356 (3), (5)	ER4043	ER4043	ER5356 (3), (5)	ER5356 (3), (5)	ER4043	ER1100 (3)	ER4043	ER1100 (3)	ER1260
B	1100, 3003 CLAD 3003	ER4043	ER4043 (5)	ER4043 (5)	ER5356 (3), (5)	ER4043 (5)	ER4043 (5)	ER5356 (3), (5)	ER5356 (3), (5)	ER4043 (5)	ER4043 (5)	ER4043 (5)	ER1100 (3)	
C	3004, CLAD 3004,	ER4043	ER4043 (5)	ER4043 (2)	ER5356 (5)	ER5356 (3), (5)	ER5356 (5)	ER5356 (5)	ER5356 (5)	ER5356 (3), (5)	ER4043 (5)	ER4043 (4), (5)		
D	5005, 5050	ER4043	ER4043 (5)	ER4043 (2)	ER5356 (5)	ER4043 (5)	ER5356 (3), (5)	ER5356 (5)	ER5356 (5)	ER4043 (5)	ER4043 (4), (5)			
E	5052, 5652 (1)	ER4043 (5)	ER4043 (2)	ER4043 (2)	ER5356 (5)	ER5356 (2), (3)	ER5356 (2)	ER5356 (5)	ER5356 (5)	ER5652 (2), (3)				
F	5083	ER5356 (3), (5)	ER5356 (5)	ER5356 (5)	ER5183 (5), (6)	ER5356 (5)	ER5356 (5)	ER5356 (5)	ER5183 (5), (6)					
G	5086, 5356	ER5356 (3), (5)	ER5356 (5)	ER5356 (5)	ER5356 (5)	ER5356 (5)	ER5356 (5)	ER5356 (5)						
H	5154 5254 (1)	ER5356 (3), (5)	ER5356 (2)	ER5356 (2), (3)	ER5356 (5)	ER5356 (2)	ER5254 (2)							
I	5454	ER5356 (3), (5)	ER5356 (2)	ER5356 (2), (3)	ER5356 (5)	ER5554 (5)								
J	5456	ER5356 (3), (5)	ER5356 (5)	ER5356 (5)	ER5556 (5), (6)									
K	6061, 6062, 6063, 6151	ER4043 (5)	ER5356 (2), (3)	ER5356 (2), (3)										

L	214, A214 B214, F214	ER4043 (5)	ER5356 (2)
M	43,355, 356	ER4043 (4)	

Source American Welding Society.

(1) Base-metal alloys 5652 and 5254 are used for hydrogen peroxide service. ER5254 filler metal is used for welding both alloys for low-temperature service (150F and below). ER5652 filler metal is used for welding 5652 for high-temperature service (150F and above).

(2) ER5154, ER5254, ER5183, ER5356, ER5554, and ER5556 may be used. In some cases, they provide: (1) improved color match after anodizing treatment, (2) highest weld ductility, and (3) higher weld strength. ER5554 is suitable for elevated temperature service.

(3) ER4043 may be used for some applications.

(4) Filler metal with the same analysis as the base metal is sometimes used.

(5) ER5356, ER5183, or ER5556 may be used.

(6) ER5356 is the third choice.

chapter 16-4

Metal Resurfacing by Thermal Spraying

by
Richard J. DuMola
Materials Engineer
METCO Inc.
Westbury, New York

GLOSSARY

Listed below are the terms most commonly used in the thermal spray industry together with their definitions.

Abrasive A hard material such as sand, aluminum oxide, steel grit, or silicon carbide used to clean and roughen a surface.

Base metal The part or substrate to be resurfaced.

Blasting The process in which an abrasive material is propelled, usually by air pressure, onto a surface to effect both cleaning and roughening.

Coating The spray material which is applied to the base metal.

Deposition rate The amount of material which adheres to the base metal per unit of time.

Bond or bond strength A measure of how well a sprayed coating has adhered to the base metal.

Thermal spraying A group of processes which involve the melting and accelerating of an atomized spray of particles onto a surface forming a solid coating.

INTRODUCTION

The application of a coating by the thermal spray process is an established industrial method for resurfacing metal parts. The process is characterized by the simultaneous melting and transporting of the spray material, usually a metal or ceramic, onto the surface of the part to be coated. The spray material is propelled in the form of fine molten droplets which, upon striking the part, flatten, solidify, and adhere by a mechanical and metallurgical interaction. Each applied layer of spray material bonds tenaciously to the previously deposited layer. The process is continued until the desired coating thickness is achieved.

Thermal spraying can be used to apply a coating to machine element or structural parts to satisfy any one of the following broad requirements:

1. To repair worn areas on parts damaged in service
2. To restore dimension to mismachined parts
3. To increase the service life of a part by optimizing the surface physical properties

In addition to satisfying any one of these broad requirements, thermal spraying can be a cost-effective repair procedure when compared to the high cost of replacing worn or mismachined parts and the economic losses incurred as a result of machine downtime.

The primary advantages of thermal spraying over other methods of metal resurfacing are the wide range of chemically different materials which can be sprayed, the high coating deposition rate which allows thick coatings to be applied economically, and the portability of the spray equipment.

PHYSICAL PROPERTIES OF COATINGS

Thermally sprayed coatings are composed of individual particles of the spray material alloyed and mechanically interlocked together to form a solid coating. In general, there is only limited metallurgical bonding between the coating and the base metal. The coating adheres primarily by a mechanical anchoring mechanism. To ensure adequate bonding of the coating, the base metal must be free from oil or dirt contamination and should be roughened by machining or blasting.

Sprayed coatings are harder and more wear-resistant than cast or wrought alloys of the same material. The increased properties are due to fine oxides and a combination of work-hardening and rapid quenching of the spray particles upon impact with the base metal. Rapid quenching causes hard metastable phases to form.

Some degree of porosity is present in all sprayed coatings and results from the presence of air gaps between the spray particles. Typically, thermally sprayed coatings are 80 to 95 percent as dense as cast or wrought alloys of the same material. In applications where the coating is used as a bearing surface, the porosity helps to retain lubricating oil and gives the coating a degree of self-lubricity. In corrosion applications where it is necessary to protect the base metal, the coating should be sealed with an epoxy or aluminum vinyl paint to close off the pores.

The surface texture or roughness of thermally sprayed coatings is coarser than cast or wrought surfaces, and a subsequent finishing operation such as sanding, machining, or grinding is often required before the resurfaced part can be placed into service. Thermally sprayed coatings are generally not as machinable as wrought or cast alloys of the same chemistry due to the presence of oxides in the coating. Because tool wear is greater, sprayed coatings should be machined with the most abrasion-resistant carbide cutting tools available.

MATERIAL SELECTION

The first step in the selection of a thermally sprayed coating for a specific in-plant application is to define the coating function. For example, if a badly worn shaft is to be repaired by thermal spraying, then the desired coating function is increased wear resistance. Table 4-1 lists the most common coating functions encountered, one or two typical application areas, and the appropriate thermal spray materials which satisfy each coating function. It should be noted that for each coating function a number of spray materials are indicated. In order to pinpoint the best material for the application, secondary considerations such as equipment available, coating thickness required, material available, and final method of finishing should be evaluated.

PROCESS DESCRIPTION

Thermal spraying consists of four basic processes: wire flame spraying, powder flame spraying, arc wire spraying, and plasma arc spraying. Of these processes wire flame spraying and powder flame spraying are the most widely used in industry. (See Fig. 4-1.)

Both wire and powder flame spraying utilize the heat generated by the combustion of an oxygen fuel flame (typically oxyacetylene, oxypropane, or oxyhydrogen) to melt the spray material. A wire flame spray gun pulls the wire into the combustion flame by means of either a self-contained, variable-speed, air-driven turbine or an electric motor. A high-pressure stream of air both constricts the combustion flame and atomizes the molten tip of the wire, forming a spray of metal. A powder flame spray gun operates by feeding a fine powder into the combustion flame by a combination of suction and gravity. The powder is both melted and propelled by the combustion flame. The equipment typically required for either wire or powder flame spraying is listed in Table 4-2.

Both arc wire spraying and plasma arc spraying utilize an electric arc rather than a

TABLE 4-1 Coating Function and Material Selection*

Coating function	Applications	Material selection
Adhesive wear	Bearings Piston guides	Aluminum bronze Phosphor bronze Tin-base babbitts
Abrasive wear	Shafts Couplings Cutting blades	Hardfacing alloys Molybdenum Carbon steel Stainless steel Tungsten carbide
Atmospheric and saltwater corrosion	Exposed steel structures	Aluminum Zinc
High-temperature oxidation	Exhaust mufflers Annealing pans	Nickel chromium Aluminum
Restoration of dimensions	Mismachined parts and castings	Carbon steels Stainless steel Nickel alloys Aluminum

*Adapted from *Handbook of Coating Recommendations,* METCO Inc., New York, 1970.

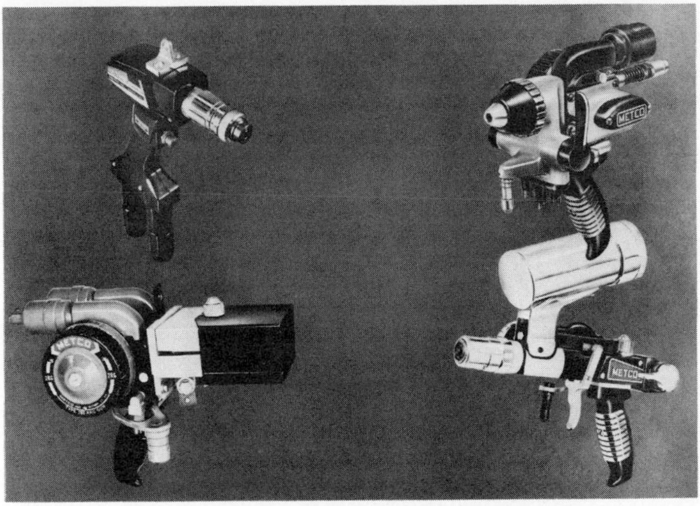

Figure 4-1 Thermal spray equipment. (*METCO Inc.*)

combustion flame to melt the spray material. Arc wire spraying is becoming increasingly important because it yields lower operating costs and higher deposition rates than wire flame spraying. Plasma arc spraying is utilized primarily where optimum coating density and overall quality are critical requirements as in aircraft applications.

Thermal spraying will generate airborne dust and metal fumes to varying degrees, depending on the material being sprayed. As a result, always provide adequate ventilation in the spray area to ensure operator safety. In cases where sufficient ventilation is not possible, spray operators should be equipped with dust masks or auxiliary ventilation equipment such as an exhaust hood or exhaust booth.

Each thermal spray process produces an intense bright light generated by either an electric arc or combustion flame. Eye protection as provided by dark glasses or an approved welding helmet is required.

In cases where the spray material is known to be toxic or suspected of containing potentially toxic elements, the recommendations of the thermal spray equipment manufacturer should be followed.

TABLE 4-2 Equipment Requirements

Wire or powder flame spray gun
Oxygen and fuel gas cylinders
Two-stage oxygen regulator
Two-stage acetylene regulator
Gas flowmeters (oxygen, fuel gas, and air)
Compressed air (30 ft^3/min at 75 lb/in^2)*
Gas hoses
Wire straightener (optional)

*85 m^3/min at 500,000 N/m^2

BIBLIOGRAPHY

Ballard, W. E.: *Metal Spraying and the Spray Deposition of Ceramics and Plastics*, 4th ed., Charles Griffin, London, 1963.

Burns, R. M., and W. W. Bradley: *Protective Coatings for Metals,* 3d ed., Reinhold, New York, 1967.
Ehrhardt, R. A., and A. Mendizza: "Sprayed Metal Coatings" in *Metals Handbook,* 1948 ed., American Society for Metals, Metals Park, Ohio, 1948.
Ingham, H. S., and A. P. Shepart: *METCO Flame Spray Handbook,* vols. I and II, METCO Inc., New York, 1964.
Longo, F. N.: *Handbook of Coating Recommendations,* METCO Inc., New York, 1972.
"Thermal Spray Terms and Their Definitions," Booklet AWS C2.9-70, American Welding Society, New York, 1970.

chapter 16-5

Structural Adhesives

by

Weldon M. Scardino, P.E.
Consulting Engineer
Centerville, Ohio

GLOSSARY*

Adherend A body which is held to another body by an adhesive.

Adhesive, contact An adhesive that is apparently dry to the touch and which will adhere to itself instantaneously upon contact.

Adhesive, heat-setting An adhesive that requires a temperature above 87°F (31°C) to set it.

*Reprinted with permission from the *Annual Book of ASTM Standards,* Part 22. Copyright American Society for Testing and Materials, 1916 Race Street, Philadelphia, PA 19103.

Adhesive, pressure-sensitive A viscoelastic material which in solvent-free form remains permanently tacky. Such material will adhere instantaneously to most solid surfaces with the application of very slight pressure.

Adhesive, room-temperature setting An adhesive that sets in the temperature range of 68° to 86°F (20° to 30°C).

Bond strength The unit load, applied in tension, compression, flexure, peel, impact, cleavage, or shear, required to break an adhesive assembly with failure occurring in or near the plane of the bond.

Catalyst A substance that markedly speeds up the cure of an adhesive when added in minor quantity compared with the amounts of the primary reactants.

Creep The dimensional change with time of a material under load, following the initial instantaneous elastic or rapid deformation.

Cure To change the physical properties of an adhesive by chemical reaction.

Failure, adhesive Rupture of an adhesive bond such that the separation appears to be at the adhesive-adherend interface.

Failure, cohesive Rupture of an adhesive bond such that the separation appears to be within the adhesive.

Fillet Adhesive which fills the corner or angle where two adherends are joined.

Flow Movement of adhesive during the bonding process, before the adhesive is set.

Hardener A reacting substance or mixture that promotes or controls the curing reaction.

Joint, lap Joint made by overlapping and bonding two adherends.

Polymerization A chemical reaction whereby monomer molecules link together to form larger ones.

Post cure To expose an adhesive to additional cure (usually heat), following initial cure.

Primer A coating applied to a surface, prior to the adhesive, to improve the bond performance.

Storage life Time during which adhesive can be stored and still be suitable for use (also called *shelf life*).

Structural adhesive Adhesive used for transferring required loads between adherends exposed to service environments typical for the structure involved.

Surface preparation Physical and/or chemical preparation of an adherend to make it suitable for adhesive bonding.

Thermoplastic A material that will repeatedly soften when heated and harden when cooled.

Thermoset A material that will undergo or has undergone a chemical reaction by the action of heat, catalyst, etc., leading to a relatively infusible state.

Time, assembly The time interval between the spreading of the adhesive on the adherend and the application of pressure or heat, or both, to the assembly.

Working life The period of time during which an adhesive, after mixing with catalyst, solvent, or other compounding ingredients, remains suitable for use.

INTRODUCTION

The use of adhesives, both structural and nonstructural, is increasing rapidly. The advent of the epoxies made much of this possible, although they are not the only types of structural adhesives. Adhesives are used for simple repairs to equipment having low operational stresses and also in many applications requiring strength equal to or greater than that of the parent material or part. For instance, epoxy repairs to concrete and many types of reinforced plastics as well as metals can be expected to reinforce these base materials, so that later breakage will be outside the adhesive joint. Adhesives can now be used to replace rivets, screws, and welding.

The modern plant engineer is often faced with a variety of fastening problems to be solved with a minimum of cost and time. The many adhesives available today can often meet all of these requirements best; i.e., they solve the problem, and are quick and inexpensive to use. Frequently, adhesives furnish the *only* solution to problems where mechanical fasteners simply cannot be used.

In addition to frequently furnishing the best or only solution to a fastening problem, adhesives usually provide structural joints with the best fatigue resistance. This is true for two reasons: the ability of the adhesive to damp, absorb, and distribute stresses, and the fact that no holes are needed for fastening. Most failures of fastened parts originate at fastener holes in the form of fatigue or stress cracks which get progressively larger. A bonded joint eliminates this problem.

However, adhesives are not a panacea for all fastening problems. The successful use of adhesives requires a careful choice of adhesive properties to suit the adherends, proper joint design, proper surface preparation, proper cure temperature and pressure, and an adhesive system that can resist the environmental exposure to which the bonded joint will be subjected.

CLASSIFICATIONS

Hot-Melt Adhesives

These are thermoplastic polyvinyl acetate and other chemicals either melted at point of application or melted in a container or gun and dispensed as a liquid at the point of application. The hot melts have long been used in shoe construction and packaging and are now being used in carpet splicing and many other applications.

Advantages

Rapid application; fast setting; low cost; indefinite shelf life; moisture and solvent resistance; ability to bond to many surfaces; nontoxic; one part, no mixing.

Disadvantages

Poor heat resistance [softening point of typical materials is 104 to 212°F (40 to 100°C)]; special equipment needed for application; low shear strength [500 to 1000 lb/in^2 (3450 to 6900 kPa) although some can go as high as 3500 lb/in^2 (24 129 kPa)]; poor creep resistance.

Cyanoacrylate Adhesives

The cyanoacrylate adhesives, or *miracle glues,* are well known for their advertising aimed at the home market. They cure by the action of the minute amount of moisture present on most surfaces.

Advantages

Very fast setting and curing (seconds or minutes); ease of application; special equipment not needed; good tensile and shear strength; one part, no mixing.

Disadvantages

Work best on nonpermeable surfaces; closely mated surfaces needed for most types; poor impact strength; poor solvent, moisture, and heat resistance; limited to small bonding areas because of fast set; careful handling needed (can bond skin almost instantly).

Anaerobic Adhesives

These one-part materials cure when confined in a joint that excludes oxygen. Long used for thread sealing and locking, they are available in many strength variations and viscosities. Some are true structural adhesives. Primer is required on some surfaces to counteract contamination or to "activate" the surface.

Advantages

One component, no mixing; ease of application; can be somewhat gap-filling as compared to cyanoacrylates; tough, good impact resistance; temperature operating limit as high as 300 to 450°F (149 to 232°C); fast cure; long shelf life, easy to store.

Disadvantages

Sometimes require primer; not for all substrates; for nonpermeable surfaces only.

Acrylic Adhesives

Acrylic adhesives are thermosetting liquids and pastes. Early-generation adhesives have been improved and have advantages in a variety of operations. They are fast-setting, easy to mix and use, and they can tolerate a broad range of bond-joint gaps. Their strength is close to that of the epoxies, and some have very good moisture resistance when cured. For special applications, the curing agent can be applied to one side in advance as a primer.

Advantages

Fast cure at room temperature; high-strength peel and shear; good moisture resistance obtainable; good gap filling; tolerant of surface contamination; with primer, can be used on permeable surfaces.

Disadvantages

Strong, objectionable odor; limited container open time.

Epoxies

Probably the most widely used industrial structural adhesives, these 100 percent solid thermosetting materials have been used for several decades and are constantly being improved. The peel strength and impact resistance have been improved by modification with elastomers. Recently, improved durability versions for *primary* structural bonding of aircraft parts have been developed and "qualified." Nylon-epoxy and epoxy-polyamide adhesives have good impact strength and elongation, but poor moisture and heat resistance. The phenolic epoxies have excellent heat resistance, but poor elongation and impact resistance.

Epoxy adhesives can be obtained as one-part liquids, pastes, or films and as two-part pastes. The one-part adhesives have a much shorter shelf life than the two-part adhesives and usually must be stored at 0 to 40°F (−18 to 4°C). The two-part adhesives generally cure at room temperature and require only contact pressure. However, they have limited heat and moisture resistance. The one-part epoxy adhesives fall in two general classes—those that cure at 250°F (121°C) and are usable at 180°F (82°C) and those that cure at 350°F (177°C) and are usable to 300 to 350°F (149 to 177°C). An adhesive that cures at an intermediate temperature [300°F (149°C)] is also available. The film versions are particularly suitable for large and intricate surface areas and are easy to apply. The pastes and liquids can be applied by spatula, roller, or spray. Some epoxies can also be obtained in powdered form for electrostatic and hot-dip application. For best results, prefitting and proper pressure to obtain bond lines of approximately 4 to 10 mils (0.1 to 0.25 mm) or 6 mils (0.15 mm) are necessary for most adhesives. Bond-line "shimming" with glass beads or filaments is sometimes used with paste-type adhesives to obtain the optimum bond thickness. The film adhesives are easier to control in this respect, as most are on a "carrier" which helps to perform this task. As a general rule, the higher the cure temperature, the greater the heat and humidity resistance and the lower the peel and impact strength of an epoxy. Only a few room-temperature adhesives can match the properties of the heat-cured epoxies.

Frequently, room-temperature-curing epoxies that set rapidly use an amine curing agent, which can be toxic. These fast-setting epoxies are also quite brittle. The room-temperature-curing adhesives that use polyamide curing agents have good impact resistance.

Epoxies can be modified by fillers to give various properties, including electric conductance. Some typical uses of epoxies involve everything from attaching jewels to jewelry to bonding aircraft primary structures. Almost all helicopter rotor blades are epoxy-bonded laminated metal or composite structures. Most aircraft leading edges, trailing edges, spoilers, rudders, and elevators are epoxy-bonded assemblies. These can be metal or composite (either graphite-epoxy or fiberglass-epoxy).

Advantages

Obtainable in many different forms: liquids, paste, films, or powder—one-part, or two-part; variability of cure time, cure temperature, and pot life; inert to most solvents and moisture (depending on type); wide variety of shear, peel, and stiffness properties available; strongest and most dependable for most structural bonding applications.

Disadvantages

Require careful surface preparation—usually chemical treatment; some equipment needed for mixing and dispensing paste types; require proper attention to cure process; limited shelf life for one-component types—require refrigerated storage; require primer for best environmental results.

Phenolics

Phenolic adhesives cover a wide range of types: water dispersions for bonding wood, one-component heat-curable liquids; solutions, films, or powders and liquid solutions plus catalyst for lower-temperature cures. Because of their heat-resistant properties, various combination adhesives are formulated, such as nitrile phenolics, vinyl phenolics, and epoxy phenolics. Nitrile-phenolic adhesives have been very successfully used to *bond-seal* aircraft integral fuel tanks.

Advantages

Heat resistance; wide range of forms available; humidity and solvent resistance.

Disadvantages

Brittle; emission of volatiles upon curing; not good gap fillers; lower strength than epoxies; high pressure required during cure, 100 to 200 lb/in^2 (690 to 1380 kPa).

Polyurethanes

Polyurethane adhesives are 100 percent solid materials in a liquid form. They may be one or two components and can be room-temperature or heat cured. Their use is limited.

Advantages

Nonbrittle; good for joining dissimilar materials; good for low-temperature (cryogenic) applications—strengths to 5000 lb/in^2 (34 500 kPa) at 100°F (−73°C); high peel susceptibility.

Disadvantages

Moisture-sensitive, before and after cure; poor lap shear at room and elevated temperature.

Silicones

The silicones are 100 percent solid pastes or liquids that usually cure at room temperature. Most one-part materials cure by reacting with moisture in the air. Because of their high cost and low shear strength, their use is confined to specialty applications where their unique properties are necessary.

Advantages

One- or two-part; wide range of temperature use [−80 to 500°F (−62 to 260°C)]; high peel and impact strength; moisture resistance; good for dissimilar adherend joining; one part, easy to apply; solvent resistance.

Disadvantages

High cost; poor lap shear strength–usually not more than 500 to 600 lb/in^2 (3.45 to 4.14 MPa); poor creep resistance; for best adhesion usually requires primer; some can be reverted by heat and moisture.

Polyimides

Polyimide adhesives may be obtained as thermoplastic liquids and as one- or two-part thermosetting films or pastes. Their use is limited mostly to applications requiring their high-temperature resistance. Curing takes a long time at 500°F (260°C) and usually requires a high-temperature postcure.

Advantages

High-temperature resistance.

Disadvantages

High cost; limited production availability; high cure temperatures; emission of volatiles during cure by some types; brittle (low impact strength).

JOINT DESIGN

Most adhesives are strong in tension or shear strength but very weak in cleavage and/or peel strength (see Fig. 5-1). All cleavage and peel forces should be eliminated from adhesive joints to the greatest extent possible. Good joint design should also allow for the maximum possible bond area and mechanical locking as well as adhesive bonding. The simple lap joint of Fig. 5-2 can be improved in many ways: by increasing the thickness of each adherend or by making a type of double-lap shear joint. Both joints reduce the peel and cleavage forces at the edge of the bond joint caused by the eccentricity of the adherends. Other typical bonded joints are shown in Fig. 5-2. Figure 5-3 shows both good and bad ways to bond flexible adherends.

PREPARATION FOR BONDING

Surface Preparation

Surface preparation is the most critical step in the adhesive bonding process. Unless a satisfactory surface preparation is accomplished, the bond will fail adhesively and unpredictably at the adherend-primer interface. With proper surface preparation, bonds can be accomplished that will allow any failure to be cohesive in nature, thus realizing the predicted strength of the adhesive, and/or primer combination. The proper surface prep-

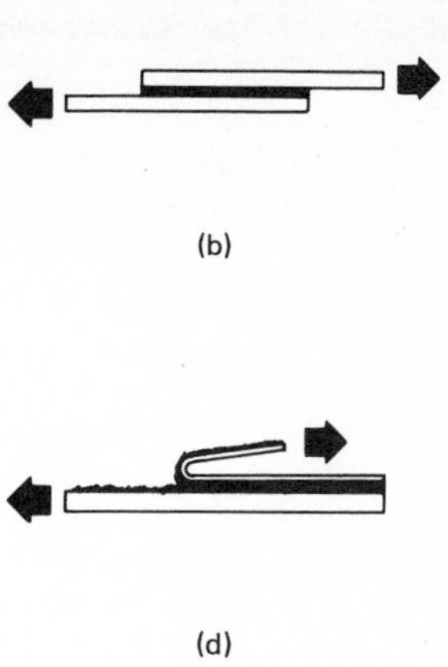

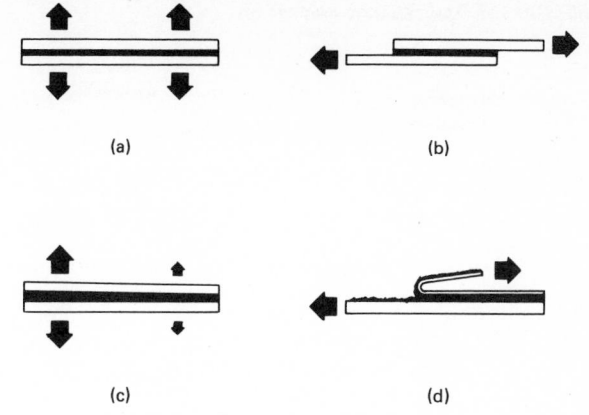

Figure 5-1 Stresses on bonded joints.

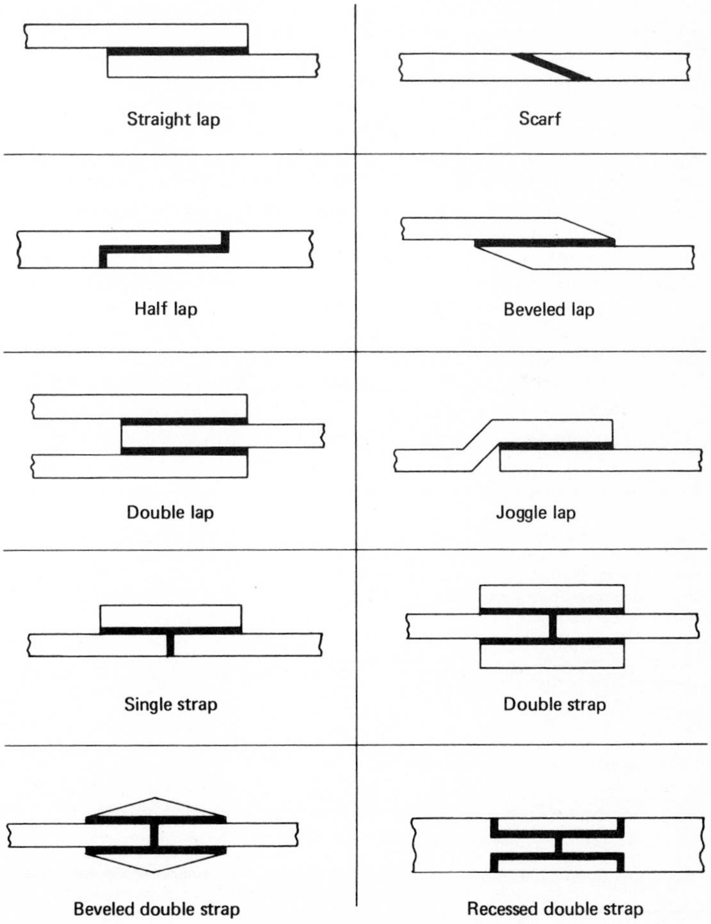

Figure 5-2 Types of bonded joints.

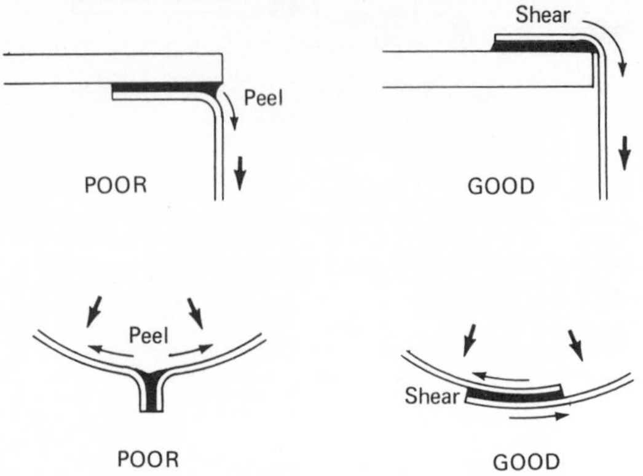

Figure 5-3 Joints for flexible adherends.

aration is a key factor, not only in the initial strength of a bonded joint but, even more importantly, *in its long-term environmental resistance.*

Surface-preparation methods must, as a minimum, remove oil, grease, or any coating whose bond strength to the adherend is apt to be less than that of the adhesive bond. Simple abrasion and/or solvent wiping are used for some metallic and plastic adherends. However, on most metals these simple surface preparations are usually not sufficient to obtain good adhesion or long-term environmental resistance. For metals, usually the preferred surface treatment is one that chemically removes organic contamination and the original oxide coating. This is followed by immediate priming or bonding or else, as for aluminum, the building up of a controlled oxide layer. ANSI/ASTM D 2093[5] and ANSI/ASTM D 2651-79[5] contain detailed recommended methods for preparation of various plastic and metal surfaces. (See Tables 5-1 and 5-2.) When using epoxy adhesives that cure at 250°F (121°C) or lower, the preferred treatment for most aluminum alloys is a patented phosphoric acid anodizing process described in ARP 1524[6] or the nonpatented, hand-applied version described in ARP 1575.[7]

The most widely used aluminum surface preparation for bonding is the FPL (Forest Products Laboratory) etch. This is described in ANSI/ASTM D 2651-79.[5] Recent

TABLE 5-1 Surface Treatments for Plastics*

Material	Procedure
Cellulose acetate, cellulose acetate butyrate, cellulose nitrate, cellulose propionate, ethyl cellulose, methyl styrene, polycarbonate, polystyrene, vinyl chloride, polymethylmethacrylate	1. Methanol wipe 2. Sand 3. Methanol wipe 4. Dry
Epoxy, polyester, phenolic, urea-formaldehyde, diallyl phthalate, melamine, nylon, polyurethane	1. Acetone wipe 2. Sand 3. Dry wipe 4. Acetone wipe
Polyethylene, polypropylene, chlorinated polyether, polyformaldehyde	Must be treated with sulfuric acid–dichromate solution (see ASTM D 2093)[5]
Teflon	Must be treated with sodium-naphthalene-complex, a strong oxidizer (see ASTM D 2093)[5]

*Abstracted from ANSI/ASTM D 2093 (Ref. 5)

TABLE 5-2 Surface Treatments for Metals*

Material	Procedure
Aluminum	1. Phosphoric acid anodizing per ARP 1524[6] or ARP 1575[7] or 2. FPL (Forest Products Lab.) etch:* Degrease, hot alkaline cleaner, hot sulfuric acid–sodium dichromate etch, rinse, and dry.
Carbon steel and stainless steel	Degrease, followed by mechanical abrasion–vapor blasting, sand blasting, etc.
Titanium, magnesium, copper	Require chemical treatment.

*Abstracted from ANSI/ASTM D 2093 (Ref. 5)

changes in the process have been incorporated in the document; these changes improve the environmental resistance of bonds prepared using this process.

Environmental Considerations

In selecting an adhesive for a particular application, one of the most important considerations is the environment or surroundings to which the adhesive joint will be subjected. Of course, the force acting on the joint is of prime consideration, and the adhesive joint must be capable of carrying the maximum expected load (without excessive creep) and amount of fatigue or cyclic stresses. Cyclic stresses, particularly slow ones, are much more damaging to an adhesive joint than a steady stress. The adhesive selected for a particular application must be able to resist these loads and stresses not only initially but also after exposure to the most severe environmental factors to be encountered during the life of the adhesive joint. Heat and humidity are usually the most damaging environmental factors for most bonded joints. Thermal-expansion stresses created between dissimilar materials having widely different coefficients of thermal expansion, e.g., a plastic-to-metal bond joint, require low-modulus (nonbrittle) adhesives for best performance. Other deleterious factors are solvents and ultraviolet or other energy. Always choose an adhesive that is resistant to these factors; do not plan on coating the adhesive joint with some "protective" coating which can possibly crack or eventually become permeable to solvents or moisture.

Primers

For many adhesives, a primer is not merely desirable but absolutely essential in order to obtain maximum bond strength and environmental durability. Usually, the adhesive manufacturer recommends a compatible adhesive primer. The primer performs several functions. The primer, being less viscous than the adhesive, wets the adherend surface and adheres to it better than does the adhesive. Corrosion-resistant primers used particularly with epoxies and aluminum adherends contain chromates that leach out and protect the adherend from corrosion. Corrosion-inhibited adhesive primers (CIAPs) are essential to environmentally durable aluminum bonding.

SPECIFICATIONS AND STANDARDS

The principal sources of specifications and standards are those issued by the government and by industry "voluntary" associations. The government documents are free. The industry documents may be purchased both individually or (for ASTM) in related groups or "books." Consult their indexes for the ASTM and AMS specifications, and the DoDISS (Department of Defense Index of Specifications and Standards) for all Government specifications and standards. The DoDISS also includes many industry specifications and standards that have been officially accepted for use by the Government.

Federal and Military Specifications Standards and Handbooks, Naval Publications and Forms Center, 5801 Tabor Avenue, Philadelphia, PA 19120, (215)697-2179.

SAE (AMS) Specifications, Society of Automotive Engineers, 400 Commonwealth Drive, Warrendale, PA 15096, (412)776-4841.

ASTM Specifications, American Society for Testing and Materials, 1916 Race Street, Philadelphia, PA 19103, (215)299-5400.

REFERENCES AND SOURCES OF INFORMATION

1. *Adhesives Desk-Top Data Bank,* 3d ed., The International Plastics Selector, Inc., San Diego, 1980–1981.
2. Thrall, E. W., and R. W. Shannon: *Adhesive Bonding of Aluminum Alloys for Aircraft,* Marcel Dekker, New York, 1982.
3. *Adhesives Red Book, Directory of the Adhesives Industry.* Communication Channels, Inc., Atlanta, 1982.
4. *Adhesives in Modern Manufacturing,* Society of Manufacturing Engineers (SAE), Dearborn, 1970.
5. *ASTM Annual Book of Standards,* part 22, American Society for Testing and Materials, 1982.
6. ARP 1524, "Aerospace Recommended Practice, Surface Preparation and Priming of Aluminum Alloy Parts for High Durability Structural Bonding, Phosphoric Acid Anodizing," 1978.
7. ARP 1575, "Aerospace Recommended Practice, Surface Preparation and Priming of Aluminum Alloy Parts for High Durability Structural Adhesive Bonding, Hand Applied Phosphoric Acid Anodizing," 1979.

chapter 16-6

Diagnostic Instrumentation for Machinery Maintenance

by

J. Mark Gilstrap
Technical Manager, Mechanical Engineering Services
Bently Nevada Corp.
Minden, Nevada

INTRODUCTION

In recent years, the level of sophistication of rotating machinery has indeed increased. Modern machines are operating at higher speeds, temperatures, pressures, and flow rates. It is common practice to continuously monitor the *process variables* (pressure, temperature, flow) in a machine in order to assess total plant operation. The techniques used to monitor these variables are described elsewhere in this Handbook.* When

*See Chap. 11-1.

machinery reliability is considered, it is also necessary to monitor the onstream mechanical condition of that machinery. Of the available parameters to accurately determine mechanical integrity, the measurement of machine *vibration and position* characteristics has consistently proved to be a powerful tool.[1] Furthermore, when a monitor system indicates a problem with a machine, the early diagnosis of the malfunction is not only desirable, but often mandatory.

Successful vibration monitoring and analysis require an intimate familiarity with the types of measurements, transducer characteristics and applications, plus the capabilities and limitations of the diagnostic instrumentation. Ultimately, the vibration signals must be reduced to a hard-copy data format for engineering evaluation and presentation to plant maintenance, operations, and/or management personnel. This chapter will deal with the types of monitors and the various instruments used for machinery diagnosis and will describe the data presentations available.

MONITORING SYSTEMS

Monitoring systems provide the function known as *preventive maintenance*. The objective is to prevent, or at least limit, machinery damage due to mechanical or process malfunctions. It is economically favorable to monitor critical machinery.[2] The benefits are increased plant and personnel safety, decreased maintenance costs, reduced spare parts, lower insurance rates, and those factors associated with the highest dollar values, reduced downtime and increased plant availability. Thus, those machines which operate in a dangerous process, involve high capital expenditures, have high maintenance costs, or are critical to a major part of a plant's operation should have the most complete monitoring systems available. Semicritical machines would justify a less-sophisticated monitoring system, and noncritical machines may justify only a scanning-type microprocessor system or periodic measurements with portable instrumentation.

Standard monitor systems usually provide two alarm levels, an *alert* or first level of warning and a *danger* or shutdown alarm. Monitors are also equipped with a circuit and indicating light which ensures the proper operation of the transducer and field wiring. The American Petroleum Institute has published the most complete specification on monitoring systems for rotating machinery.[3] Recently, computer systems have been used increasingly in plants with many major machine trains. Computers can provide the absolute-limit alarms found in conventional monitors, but can also monitor rate of change and significant bandwidths (increases or decreases) and can compare one measured variable with others on the same machine. Computers are excellent for routine data storage, trending, alarm sequencing, and data comparison in a machine-upset condition.

Careful attention must be given to the selection of the appropriate measurement transducers for a monitor system.[4] With respect to machine vibration, transducers are naturally divided into two groups: (1) velocity transducers and accelerometers which measure machine housing motion, and (2) proximity probes which measure shaft motion. The objective of monitoring is to protect the machine against the most likely potential malfunctions. The most common malfunctions should then be evaluated in terms of shaft-related or machine-housing-related mechanisms. Then the transducer selected should be the type that will be the most reliable indicator of a change in the mechanical condition of the machine. Of course, on some machines, shaft vibrations may be transmitted to some extent to the housing and vice versa. However, the extent of transmissibility is a function of the *mechanical impedance* of a machine and is highly variable. Transducer selection should not be compromised for the sake of ease of installation; a monitor system can only be as reliable as the initial measurement.

It must be recognized that monitoring is not analyzing. A monitor may indicate when a machine is in trouble and may even indicate the severity of the problem. But the identification of the malfunction requires the use of additional instrumentation for machinery analysis.[5] Also, monitors are great averagers of events. For routine data acquisition on machines in *normal* operation, monitor amplitude levels should be recorded periodically or even continuously. But monitor readings can be enhanced greatly by the use of specialized instruments which measure more than amplitude alone. In some cases where transducer selection for monitoring is somewhat compromised, it will be beneficial to have permanently installed unmonitored transducers on the same machine. These transducers can then be measured periodically with portable monitors or specialized instrumentation.

INSTRUMENTATION FOR DIAGNOSIS

If a monitor system indicates a change in the running condition of a machine, then additional instruments are usually necessary to qualify and quantify the significance of that change. Some of these instruments are misnamed "analyzers." An instrument does not *analyze* data, it merely reduces it into a format for subsequent analysis by the cognizant technician or engineer. These so-called analyzers cannot compete with the human brain. Sound engineering judgment is still an absolute necessity.

Instruments used for machinery diagnostics aid the engineer in the function of *predictive maintenance.* Once a monitor or a periodic measurement of amplitude indicates a machine is in trouble, then these instruments must be used to predict: (1) what the mechanical or process malfunction is, (2) how long the machine can operate safely before a shutdown becomes necessary, (3) what operational changes may be made on-line to reduce the severity of the trouble and allow the machine to operate longer until an orderly shutdown can be planned, and (4) upon shutdown, what corrective action is necessary to restore the mechanical integrity of the machine.

It should be obvious that before any machinery analysis begins, the transducers providing the measurement signals should be verified for accuracy. Assuming that the basic measurement system has been installed and maintained properly, a complete recalibration should not be necessary. A few fundamental checks with a voltmeter and an oscilloscope will suffice. Besides, an unnecessary system recalibration will steal valuable time from the analysis program. Once proper transducer operation is verified, it is generally recommended that the more basic instruments be used in the initial analysis. Many of the most frequently occurring machine problems can be detected through the use of very basic instruments. On the other hand, the very sophisticated instruments may overlook some of the basic machinery data and yield misleading conclusions. For example, rotor unbalance, misalignment, internal rubs, and oil whirl can usually be recognized on an oscilloscope display.

Keyphasor

A Keyphasor is a function which provides a once-per-turn reference signal to shaft rotation.[6] The Keyphasor signal can be from a proximity probe observing a shaft discontinuity (such as a keyway) or from an optical transducer observing a light (or dark) spot on the shaft. When the Keyphasor signal pulse is generated, the shaft is at a known rotational position; the shaft event (discontinuity or bright spot) is directly under the Keyphasor probe. This signal can then be compared with respect to time to the shaft vibration signal. With this information, the rotational position of the shaft is related to the angular position of the shaft in its vibration cycle at any point in time. The Keyphasor function is not used alone with any instrument, except for a tachometer to measure shaft rotative speed. But, in conjunction with vibration signals, it is used with such instruments as oscilloscopes and tracking filters. Specific use of the Keyphasor will be explained in following sections.

Voltmeter

The voltmeter is a basic tool for transducer and monitor-system calibration and can also be a valuable machinery diagnostic aid when proximity probes are used. The dc output signal from a proximity transducer represents the *average* distance between the probe and the shaft. This information is processed directly by axial-thrust position monitors. However, for radially mounted probes, the standard vibration monitor usually only processes the ac or dynamic signal. The dc output of radial probes represents the average shaft centerline position within the bearing clearance. This information can be used to detect such parameters as: (1) shaft lift-off from the bottom of the bearing during machine start-up, (2) shaft attitude angle, (3) eccentricity ratio, (4) oil film thickness, (5) internal or external misalignment, (6) other steady-state unidirectional shaft preloads, (7) bearing wear, and (8) electrostatic discharge.[7] A plot of the average shaft centerline position relative to the bearing clearance during a machine start-up can be made from hand-logged probe gap voltages or from a magnetic tape recording with an *XY* plotter. A typical plot is shown in Fig. 6-1.

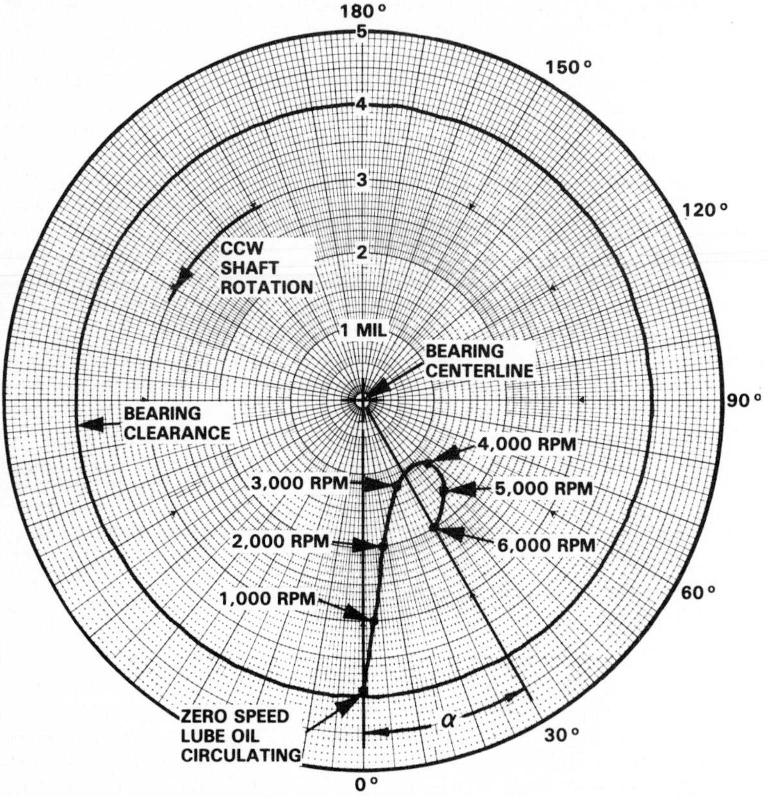

Figure 6-1 Change in average radial shaft centerline position during start-up. Diametral bearing clearance = 8 mils.

Oscilloscope

The oscilloscope is another basic tool for ensuring proper transducer and monitor-system calibration. It is also *the most fundamental* vibration-analysis instrument. Although other instruments are called *real-time analyzers,* the scope comes the closest to meeting the definition. The scope can display the raw, composite vibration signal from any type of transducer and can also show the dc signal of a proximity probe. It is most useful when used with a Keyphasor signal, and an oscilloscope camera can be used for hard-copy documentation of the displays.

The oscilloscope has two basic presentation forms: (1) time-base or waveform, and (2) orbit (Lissajous) or *XY* displays. The time-base display uses the sweep function of the scope to observe vibration amplitude with respect to time. Usually the sweep time is very short—three or four shaft revolutions—and the Keyphasor pulse can be superimposed on the time-base waveform to indicate the period of one shaft revolution. In the time-base mode, several vibration characteristics can be measured directly on or determined from the display: (1) vibration amplitude, (2) vibration phase angle, (3) vibration frequency, and (4) shaft rotative speed. When the Keyphasor signal is superimposed on the vibration signal(s), then it is easy to determine synchronous and nonsynchronous vibrations. It is particularly informative to use a scope with the output of a filter instrument and compare it with the unfiltered waveform.

The second form of presentation is the orbit or Lissajous. This is possible on scopes which have an *XY* or left-vs.-right channel function. The orbit is the dynamic motion of the shaft centerline as measured in two perpendicular planes. Usually an orbit is from two shaft-observing proximity probes at 90°, but it can be a casing orbit from *XY* seismic transducers. Essentially, all of the information (including the Keyphasor) available in the time-base display is also available, or can be confirmed, in the orbit display. In addition, shaft orbit *shape* has become a fundamental form of vibration analysis. Very unique orbit shapes indicate problems such as misalignment, oil whirl, and rubs, including the actual location of the rub. Unbalance simply shows an orbit which grows from a small circle or ellipse to a larger one. A Keyphasor with an orbit can be used very effectively for balancing.[8]

The oscilloscope can be twice as useful an instrument when it is used with a scope camera. The camera provides documentation for historical purposes. Most scope manufacturers have a camera and mounting bezel available with their scopes. It is also easy to use the camera in a multiple-exposure mode to obtain more information on one photograph than would normally be provided in a single oscilloscope display. A photo that looks like Fig. 6-2 can be obtained two ways: (1) a double exposure on a four-trace scope, or (2) a quadruple exposure on a dual-trace scope.

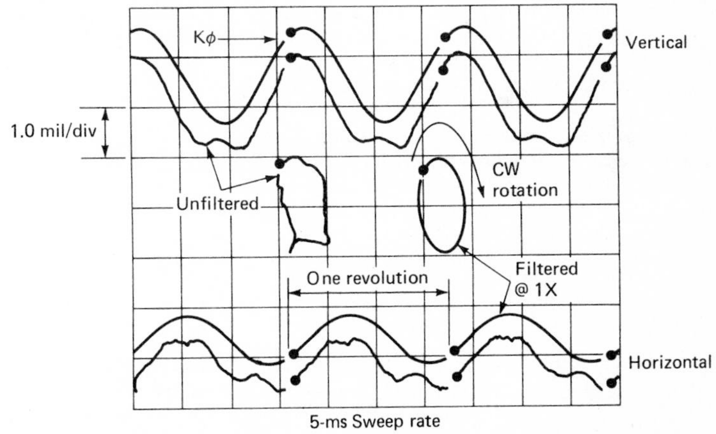

Figure 6-2 Oscilloscope tracing showing filtered and unfiltered waveforms and orbits.

Filters

Almost every instrument with the exceptions of the voltmeter, oscilloscope, and recording devices used for vibration analysis employs some sort of filter. This means that a certain vibration frequency or band of frequencies will be isolated and either (*a*) discarded or (*b*) retained for analysis. So, by definition, a filter will throw away some information. The general rule to follow is to be sure the filter is not discarding any information which may be important.

The most common type of filter is the bandpass filter. This filter will pass (or retain) a certain frequency (or very narrow band of frequencies) and discard all others. The bandpass filter is used to identify specific frequencies present in a composite vibration signal and to measure the amplitudes at each frequency. High-pass and low-pass filters are opposites. A high-pass filter discards all information below a certain cutoff frequency while a low-pass filter discards all information above a certain frequency. These filters are used to eliminate high- or low-frequency noise or unrelated vibrations at the ends of the frequency spectrum. A band-reject or notch filter is the opposite of a bandpass filter. It is used to discard a certain frequency and retain all higher and lower frequencies. Often a machine will exhibit one predominant vibration frequency; a notch filter can eliminate that frequency to observe the total machine action at all other frequencies.

Many common machine malfunctions have specific associated vibration frequencies. The use of filters to identify these frequencies would seemingly lead directly to the identification of particular machine malfunctions. There are even several charts published relating vibration frequency to different machine malfunctions.[5] It should be noted that frequency analysis charts alone are not the final answer to vibration analysis. The overall shape of the waveform and orbit as viewed on an oscilloscope and the measurement of phase angle are two very important vibration characteristics which are unavailable through frequency analysis. Determination of vibration frequency may even be the *first* step in machine problem solving, but it should not be the last.

Tunable Filter

A filter instrument which typically has several of the above functions and also an adjustment to control the tuned and cutoff frequencies is called a tunable filter. These instruments have a vibration readout meter so they can also be used for periodic overall measurements on unmonitored machines. A tunable filter may have two bandpass filters—broad and narrow. The broad filter will have a wide bandwidth or low Q, and vice versa for the narrow filter. A narrow filter may pinpoint a certain vibration frequency and associated amplitude more accurately, but it suffers from a lower response time. The narrow filter may be essential for definition of low-frequency components, but the broad filter will be more useful for higher frequencies and provides a better response time when tracking.

Most tunable filters have analog dc outputs for driving an XY plotter in order to generate amplitude-vs.-frequency spectrum information. Some even have an auto-sweep feature where the tuning section is swept automatically through a certain frequency range. Any filtered output is especially useful when viewed on an oscilloscope, particularly with a Keyphasor. The filtered signals of Fig. 6-2 were made by means of a tunable filter with the bandpass filter adjusted to running-speed frequency.

Vector Filter Phase Meter

The vector filter is a unique type of filter instrument since it displays vibration at a specific frequency as a true vector quantity. A given vibration frequency has both magnitude (amplitude) and direction (phase angle). The most common use of a vector filter is for observation of the unbalance response of a machine. The instrument requires inputs from two transducers, a Keyphasor and a vibration transducer. The bandpass filter in the instrument is automatically tuned to running-speed frequency by the Keyphasor; the vibration probe measures amplitude, and the two transducers together provide a measurement of vibration phase angle. These three measurements—shaft revolutions per minute, amplitude, and phase angle—have associated dc analog outputs for use with an XY plotter. As a machine runs from rest to operating speed, this information will produce a Bodé plot, Fig. 6-3. Since vibration due to rotor unbalance occurs at running-speed frequency, the Bodé plot represents the fundamental unbalance response of a

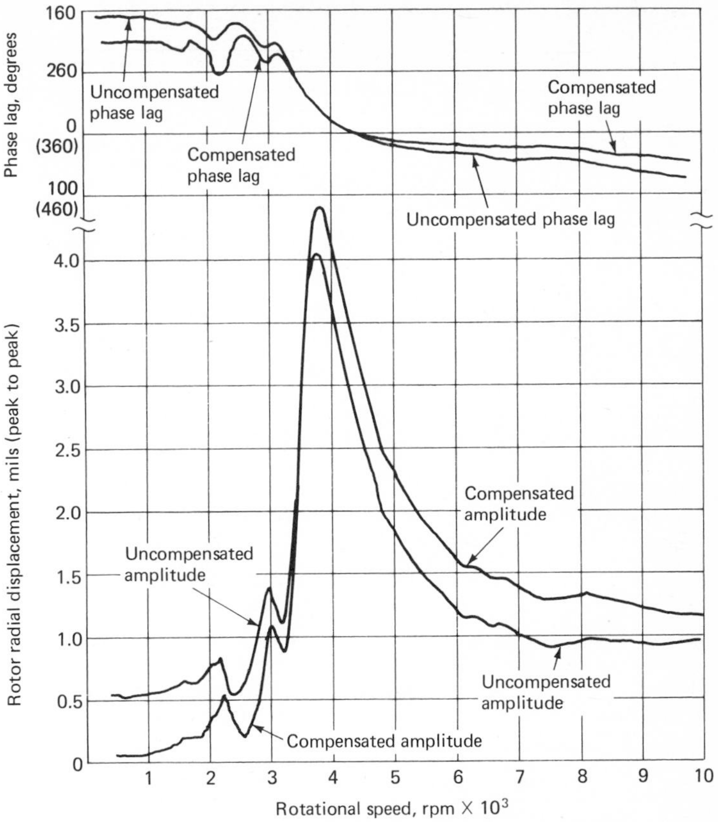

Figure 6-3 Bodé diagrams of compensated and uncompensated rotor displacement during start-up.

machine during start-up or shutdown. The Bodé plot is used to measure balance resonance frequencies (critical speeds), initial bow vectors, and amplification factors.[8]

Since an unbalance vector has magnitude and direction, it can be represented as a polar quantity. The vector filter will also generate two dc analog voltages for use with a plotter to represent these vector quantities on an XY basis. This information results in the polar plot, Fig. 6-4. The same information available from a Bodé plot is also provided by the polar plot. In addition, the polar representation shows the direction (phase angle) of the initial bow vector, the direction of initial rotor deflection due to unbalance, and the direction of rotor deflection in the self-balance speed region. Both the Bodé and polar plots are extremely useful in rotor balancing.

Spectrum Display

This instrument, often called a *spectrum analyzer,* consists of many bandpass filters, each fixed at a different frequency. The instrument usually provides a CRT display of vibration amplitudes (on the vertical scale) at specific frequencies (on the horizontal scale). The amplitude and frequency measurements are displayed continuously on the CRT and any changes in the spectrum content can be identified while the machine is running. These units usually incorporate an averaging function whereby transient data can be captured over several spectrum samples. A peak hold function allows the display

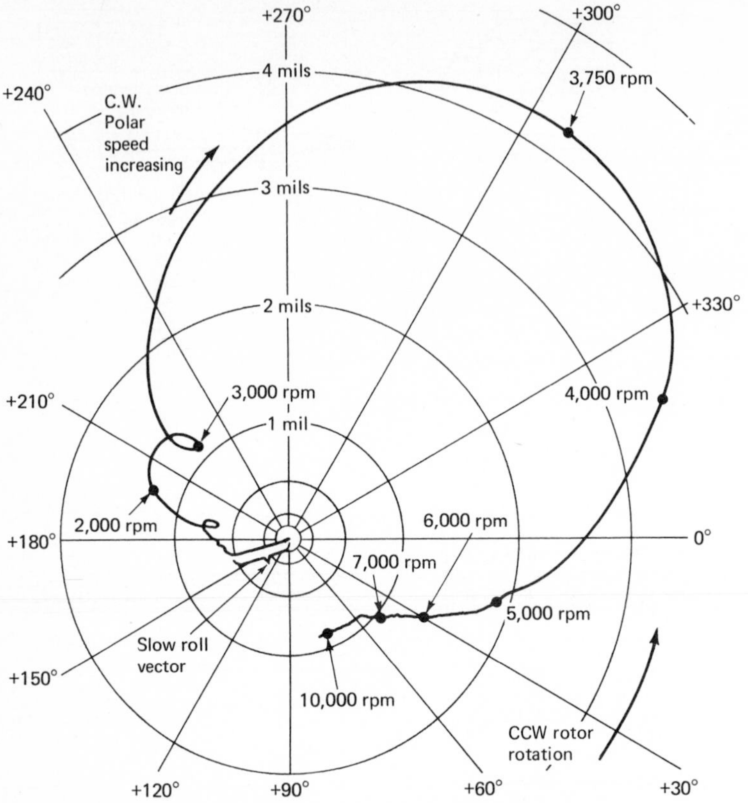

Figure 6-4 Polar plot of rotor displacement during start-up.

of the maximum amplitude at any frequency band over a period of time. Both amplitude and frequency measurements have associated dc analog outputs for use with an *XY* plotter.

Since many common machine malfunctions produce unique vibration frequencies, the spectrum display may be used to indicate the presence of these malfunctions and the relative magnitudes of each. A common use of this spectrum information is called *spectrum analysis* or *signature analysis.* This involves the comparison of vibration spectra taken at different points on a machine or the comparison of spectra taken at the same measurement location at different points in time. Figure 6-5 shows such a spectrum comparison. If spectra are acquired at different rotative speeds during a start-up or shutdown, the resulting presentation is the cascade or waterfall plot, Fig. 6-6.

Data Documentation

Previous sections have described the use of the oscilloscope camera and the plotter for hard-copy documentation of data presentations from various instruments. Recently, microprocessor and computer systems have been developed to display some of this same information on the computer CRT—oscilloscope waveforms and orbits, Bodé and polar plots, spectra, etc. Most computer CRTs are equipped with functions to provide direct hard-copy documentation via a graphics printer or outputs to a conventional plotter. One additional data-documentation instrument is used universally with all the above-men-

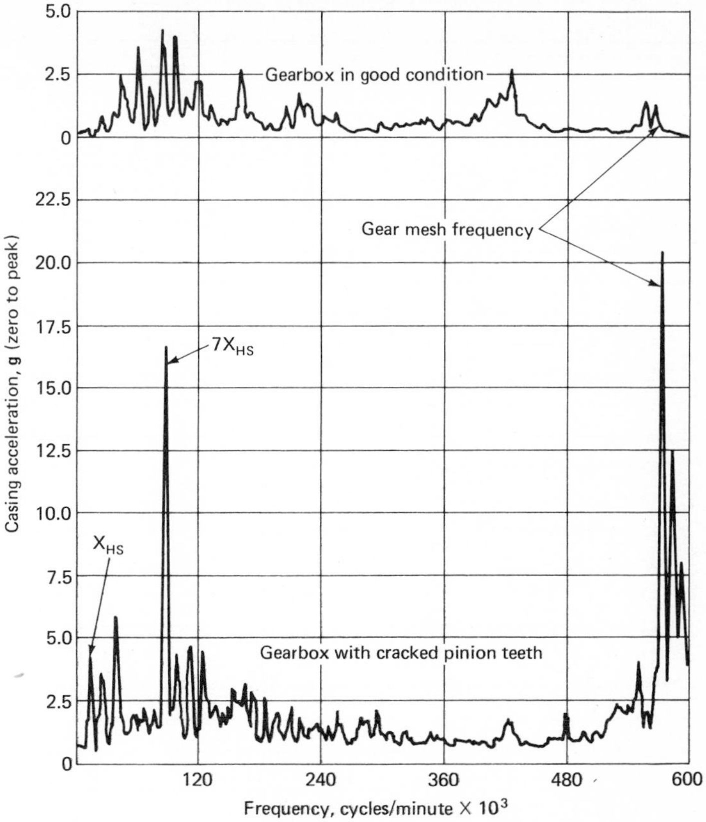

Figure 6-5 A comparison between gearbox casing acceleration characteristics.

tioned instruments—the magnetic-tape recorder. The tape recorder can capture dynamic and steady-state (in the case of FM recorders) transducer signals directly. Then in the playback mode, the signals can be sent from the recorder to these various data reduction instruments, exactly as they would have come from the transducers directly.

Tape recorders are especially useful when many measurement points need to be documented simultaneously, such as during a machine train start-up. Oscilloscopes, digital vector filters, and spectrum displays can of course be used directly with the transducers during the start-up, but most of these instruments are only one- or two-channel devices. However, after the start-up, the transducer signals can be reproduced from the magnetic tape and displayed and documented with each of these instruments independently.

INFORMATION SOURCES

Technical data and specification sheets for the various instruments used in machinery analysis are available from various instrument manufacturers. Most manufacturers also publish technical articles describing the specific uses of these products. Articles authored by persons in industry regarding machinery malfunction diagnosis are periodically published in industrial trade journals. Two of the best publications are *Hydrocarbon Processing* for the petrochemical industry and *Power* for the power generation industry. To

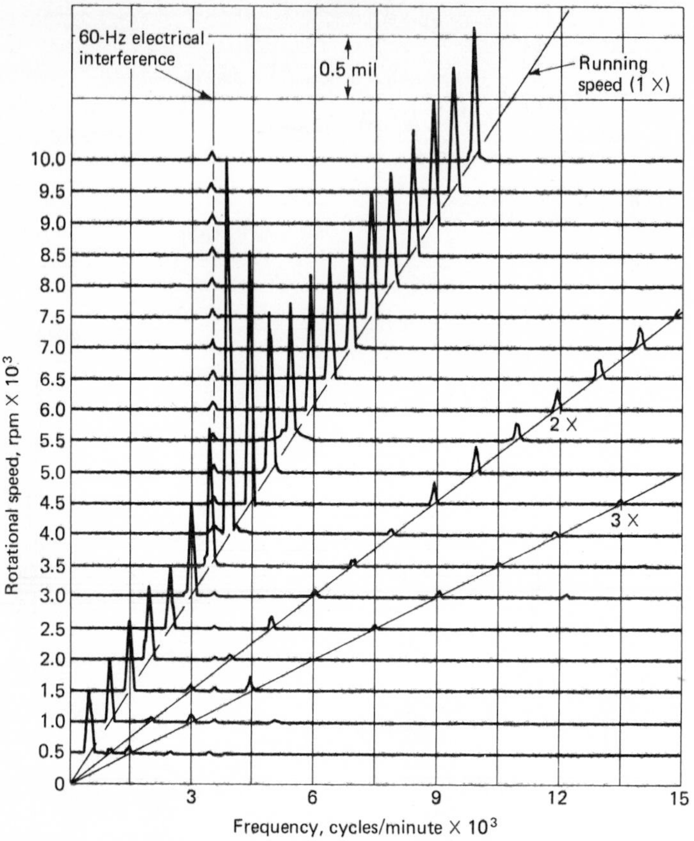

Figure 6-6 Cascade spectrum analysis of rotor displacement during start-up.

date, the most comprehensive book published on the subjects of machinery measurements, monitoring systems, data acquisition and reduction instruments, and machinery malfunction identification is by Charles Jackson, of Monsanto Company. This book, *The Practical Vibration Primer,* is published by Gulf Publishing Company, Houston, Texas. Other specific references are in the following list.

REFERENCES

1. Applications Note, "Vibration Measurement—Basic Parameters for Predictive Maintenance on Rotating Machinery," Bently Nevada Corporation, Minden, Nevada.
2. Dodd, V. Ray: "Machinery Monitoring Update," presented at the Texas A & M Sixth Turbomachinery Symposium, Houston, Texas., December 1977.
3. Standard 670, "Noncontacting Vibration and Axial Position Monitoring System," 1st ed., American Petroleum Institute, Washington, D.C., June 1976.
4. Applications Note, "Machinery Protection Systems for Various Types of Rotating Machinery," Bently Nevada Corporation, Minden, Nevada.
5. Jackson, Charles: *The Practical Vibration Primer,* 1979, Gulf Publishing Company, Houston, Texas.
6. Applications Note, "The Keyphasor—A Necessity for Machinery Diagnosis," Bently Nevada Corporation, Minden, Nevada.

7. Applications Notes, (1) "Alignment Loading of Gear Type Couplings," (2) "Preloads on Rotating Shafts," (3) "Shaft Position Changes Reveal Machinery Behavior/Malfunctions," (4) "Mechanical Degradation Due to Electrostatic Shaft Voltage Discharge," (5) "Attitude Angle and the Newton Dogleg Law of Rotating Machinery," and (6) "Plotting Average Shaft Centerline Position," Bently Nevada Corporation, Minden, Nevada.
8. Jackson, Charles: "Balance Rotors by Orbit Analysis," *Hydrocarbon Processing,* January 1971.
9. Standard 612, "Special-Purpose Steam Turbines for Refinery Services," 2d ed., American Petroleum Institute, Washington, D.C., June 1979.

section 17

Lubricants and Lubrication Systems

chapter 17-1

Lubricants: General Theory and Practice

by

Anne Bernhardt,
Staff Engineer
Gulf Research and Development Company
Petroleum Products Department
Pittsburgh, Pennsylvania

GENERAL THEORY

Functions of Lubricants

Lubricants perform a variety of functions. The primary, and most obvious, function is to reduce friction and wear in moving machinery. In addition, lubricants can

- Protect metal surfaces against rust and corrosion

- Control temperature and act as heat-transfer agents
- Flush out contaminants
- Transmit hydraulic power
- Absorb or damp shocks
- Form seals

Because reducing friction is such an important function of lubricants, it is necessary to understand how they perform.

Friction

Friction is the resistance to motion between two bodies in intimate contact. Two types of friction can be identified: solid (or dry) friction and fluid friction.

Solid Friction. Solid friction occurs when there is physical contact between two solid bodies moving relative to each other. The type of motion divides solid friction into two categories, sliding and rolling friction.

Sliding Friction. This is the resistance to movement as one body slides over another. Solid surfaces which appear smooth to the eye will in fact consist of many peaks and valleys. The resistance to motion is due primarily to the interlocking of these asperities. Under conditions of extreme pressure, the heat generated by sliding friction can result in welding of the points of contact.

Rolling Friction. This is the resistance to motion as one solid body rolls over another. It is caused primarily by the deformation of the rolling elements and support surfaces under load. For a given load, rolling friction is significantly less than sliding friction.

Fluid Friction. Fluid friction occurs when two solid bodies in relative motion are completely separated by a fluid. It is caused by the resistance to motion between the molecules in the fluid. For a given load, fluid friction usually is significantly less than solid friction. The film thickness, relative to the height of the surface asperities, distinguishes three types of lubrication:

- Full or thick-film lubrication
- Mixed-film lubrication
- Boundary lubrication

Full or Thick-Film Lubrication. This exists when the lubricant film between two surfaces is of sufficient thickness to completely separate the asperities on the two surfaces. In this case, true fluid friction exists between the moving surfaces and no metal-to-metal contact will occur (Fig. 1-1*a*).

Mixed-Film Lubrication. This exists when the lubricant film between the two surfaces is of sufficient thickness to separate most of the surface asperities but some metal-to-metal contact may occur (Fig. 1-1*b*).

Boundary Lubrication. This exists when the film thickness is equal to the asperity heights and extensive metal-to-metal contact occurs (Fig. 1-1*c*).

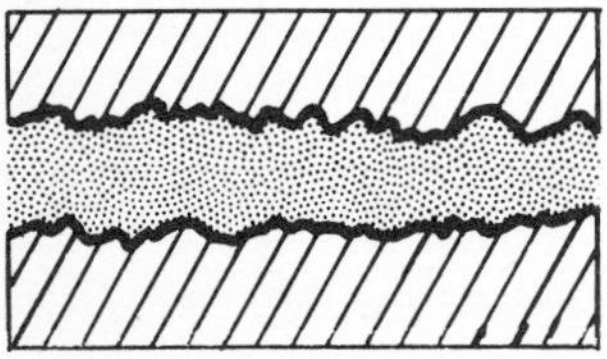

(a) FULL-FILM LUBRICATION

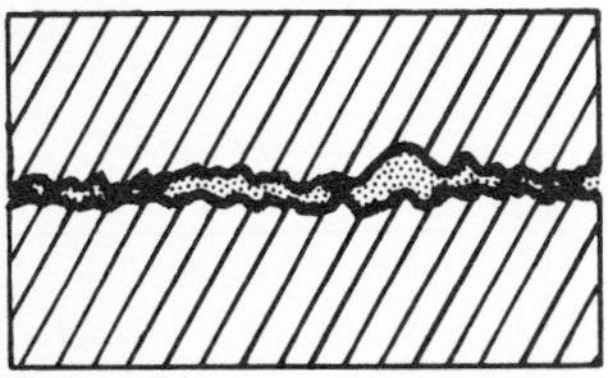

(b) MIXED-FILM LUBRICATION

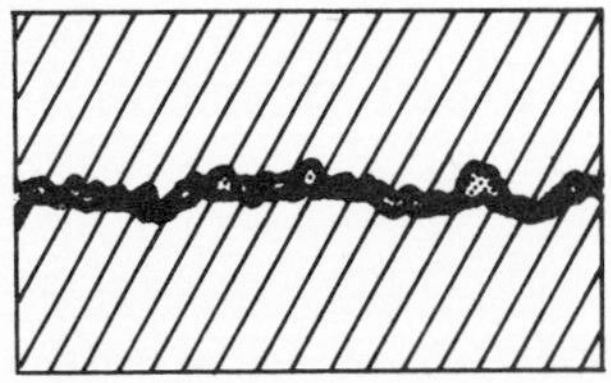

(c) BOUNDARY LUBRICATION

Figure 1-1 (*a*) Full-film lubrication. (*b*) Mixed-film lubrication. (*c*) Boundary lubrication.

Formation of the Lubricant Film

The lubricant film may be formed and maintained in one of two ways:

- Hydrostatically
- Hydrodynamically

Hydrostatic Lubrication

Hydrostatic lubrication occurs when the film is formed by pumping the lubricant under pressure between the bearing surfaces. The surfaces may or may not be moving with respect to each other. The hydrostatic pressure acts to completely separate the surfaces, and full-film lubrication is established.

Hydrodynamic Lubrication

Hydrodynamic lubrication depends on motion between the two solid surfaces to generate and maintain the lubricating film. In a plain bearing that is not rotating, the shaft will rest on the bottom of the bearing and will tend to squeeze any lubricant out from between the surfaces. When the shaft begins to rotate, a very thin film of lubricant will tend to adhere to the shaft surface and will be drawn between the shaft and the bearing. A film that will ultimately separate the load-bearing surfaces is established. A lubricant film generated in this manner is called a *hydrodynamic film.*

The thickness of the hydrodynamic oil film developed in a properly designed plain bearing is dependent on the oil viscosity, the bearing load, speed, metallurgy, and quality of the bearing surfaces. The dimensionless bearing parameter ZN/P conveniently describes the combined effect of viscosity Z, speed N, and load P.

The thickness of the hydrodynamic film and the amount of friction developed in the bearing can be predicted by means of the bearing parameter ZN/P. Plotting the bearing coefficient of friction vs. the bearing parameter for a particular bearing and lubricant gives a characteristic curve similar to the one in Fig. 1-2. Experience has shown that the

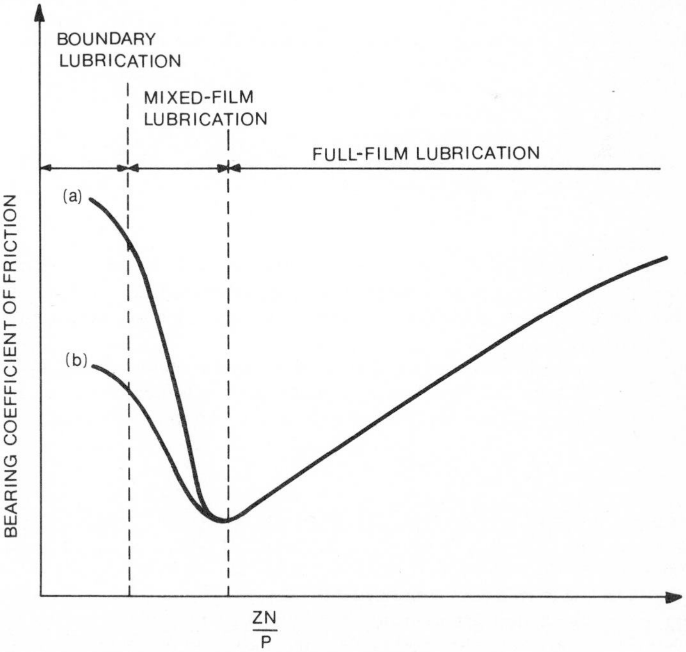

Figure 1-2 Typical ZN/P vs. coefficient of friction curve.

thickness of the lubricant film developed in a bearing can be determined by estimating where on the curve a bearing is operating (see Fig. 1-2).

Changes in the quality of the bearing's metallurgy or surface finish, or of the lubricant's "oiliness" or film strength, will cause shifts in coefficient of friction under boundary or mixed-film lubrication conditions. For example, when holding everything else constant, adding an oiliness additive to the lubricant will shift the bearing performance curve from curve *a* to curve *b*. As this indicates, it is possible to reduce the amount of friction generated in a particular bearing under boundary or mixed-film lubrication conditions through the use of certain additives.

TYPES OF LUBRICANTS

There are three major categories of lubricants:

- Fluid lubricants
- Greases (semisolid lubricants)
- Solid lubricants

Each lubricant will have its own physical properties which will affect its performance in different applications. A knowledge of the various types of lubricants on the market today and a basic understanding of their advantages and limitations is most helpful in the selection of the optimum lubricant for a particular application.

Fluid Lubricants

Fluid lubricants are the most widely used. The most common are petroleum oils, synthetic fluids, and animal or vegetable oils. Many other fluids can fulfill a lubrication function under special conditions when the use of oils may be precluded.

Petroleum or Mineral Oils

Petroleum or mineral oils, refined from petroleum crude, are sometimes referred to as *conventional oils* due to their wide acceptance as lubricants.

Synthetic Fluids

Synthetic fluids include all artificially made fluids used for lubricating purposes. Included in this category are synthesized hydrocarbons, esters, silicones, polyglycols, and phosphate esters. These are discussed in more detail in Chap. 17-2.

Animal and Vegetable Oils

Animal and vegetable oils, as the terms indicate, are oils made from either animal fat or vegetables. They are used primarily where food contact is likely to occur and the lubricant must be edible. Their main disadvantage is that most of them tend to deteriorate rapidly in the presence of heat.

In the past, oils made from animal fats, such as sperm whale oil and lard oil, were frequently used for their "oiliness" properties. Today, however, these are frequently being replaced with synthesized fatty oils which perform the same function.

Greases

Greases are fluid lubricants with thickeners dispersed in them to give them a solid or semisolid consistency. The fluid lubricant content of a grease performs the actual lubricating function. The thickener acts solely to hold the lubricant in place, to prevent leakage, and to block the entrance of contaminants.

Many types of thickeners are used in the manufacture of modern greases. Each type imparts certain properties to the finished product. Table 1-3 describes some of the typical properties and applications of greases manufactured with certain common thickeners.

Solid Lubricants

Solid lubricants, such as graphite, molybdenum disulfide (moly), and PFTE (*poly**t**etra**f**luoro**e**thylene), are not only used by themselves but are also frequently added to oils and greases to improve their performance under boundary lubrication conditions. Chapter 17-3 discusses these in more detail.

IMPORTANT LUBRICANT CHARACTERISTICS

Various physical and chemical properties of lubricants are measured and used to determine a lubricant's suitability for different applications.

Oil Properties

Viscosity

Of the various lubricant properties and specifications, viscosity (also referred to as the "body" or "weight") normally is considered the most important. It is a measure of the force required to overcome fluid friction and allow an oil to flow.

Industry uses several different systems to express the viscosity of an oil. Table 1-1 gives a comparison of some of the most common. Lubricant specifications usually express viscosity in Saybolt Universal Seconds (SUS or SSU) at 100 and 210°F (37.8 and 98.9°C) and/or in centistokes (cSt) at 40 and 100°C (104 and 212°F). Viscosity expressed in centistokes is called the *kinematic viscosity.*

With the current move toward metrication and the establishment of the International Organization for Standardization (ISO) viscosity grade identification system, the centistoke has become the preferred unit of measure. The ISO viscosity grade system contains 18 grades covering a viscosity range from 2 to 1500 cSt at 40°C. Each grade is approximately 50 percent more viscous than the next lower grade.

Laboratories determine oil viscosity experimentally using a viscometer (Fig. 1-3). The viscometer measures an oil's kinematic viscosity by the time (in seconds) it takes a specified volume of lubricant to pass through a capillary of a specified size, at a specified temperature. The kinematic viscosity is then derived by calculations based on constants for the viscometer and the time it took the sample to pass through the instrument.

Viscosity Index

The viscosity index (VI) is an empirical measure of an oil's change in viscosity with temperature. The greater the value of the viscosity index, the less the oil viscosity will change with temperature. Originally ranging from 0 to 100, viscosity indexes greater than 100 are now achieved with certain synthetic oils or through the use of additives.

Oxidation Stability

When a lubricant is exposed to heat and air, a chemical reaction called *oxidation* takes place. Products of this reaction include carbonaceous deposits, sludge, varnish, resins, and corrosive and noncorrosive acids. Oxidation usually brings with it an increase in the viscosity of the oil.

The rate of oxidation is dependent on the chemical composition of the oil, the ambient temperature, the amount of surface area exposed to air, the length of time the lubricant has been in service, and the presence of contaminants which can act as catalysts to the oxidation reaction.

Depending on the intended end use of the oil, the oxidation stability will be measured or expressed in different ways. All of the oxidation stability tests are based on placing a

TABLE 1-1 Commonly Used Industrial Lubricant Viscosity Classifications

ISO/ASTM viscosity grade system						
ISO viscosity grade no. (ISO VG)	Kinematic viscosity, cSt at 40°C	Saybolt viscosity, SUS at 100°F (approximate)*	Former ASTM-ASLE grade no.†	AGMA lubricant no.‡	SAE viscosity no. (approximate)§	SAE gear viscosity no. (approximate)¶
2	1.98–2.42	32.8–34.4	32	—	—	—
3	2.88–3.52	36.0–38.2	36	—	—	—
5	4.14–5.06	40.4–43.5	40	—	—	—
7	6.12–7.48	47.2–52.0	50	—	—	—
10	9.00–11.0	57.6–65.3	60	—	—	—
15	13.5–16.5	75.8–89.1	75	—	—	—
22	19.8–24.2	104.6–126.0	105	—	5W	—
32	28.8–35.2	149.1–181.7	150	—	10W	75W
46	41.4–50.6	214–262	215	1 (R & O)	20W	—
68	61.2–74.8	317–389	315	2 (R & O, EP)	20	80W
100	90.0–110	468–574	465	3 (R & O, EP)	30	—
150	135–165	709–866	700	4 (R & O, EP)	40	85W
220	198–242	1047–1283	1000	5 (R & O, EP)	50	90
320	288–352	1533–1876	1500	6 (R & O, EP)	—	—
460	414–506	2214–2719	2150	7 (EP, comp)	—	140
680	612–748	3298–4044	3150	8 (EP, comp)	—	—
1000	900–1100	4882–5994	4650	8A (EP, comp)	—	250
1500	1350–1650	7383–9060	7000	—	—	—

*SUS viscosities are approximate and are based on typical 95 VI single-grade oils.

†Numbers are equivalent to the *Plant Engineering* magazine designation. The American Society of Lubrication Engineers (ASLE) and the American Society for Testing and Materials (ASTM) used a viscosity grade system based on SUS at 100°F. This system is now obsolete.

‡American Gear Manufacturers Association (AGMA) proposed revisions will equate AGMA lubricant no. to the ISO VG.

§Engine Oil Viscosity Classification—SAE J 300d.

¶Axle and Manual Transmission Lubricant Viscosity Classification—SAE J 306c.

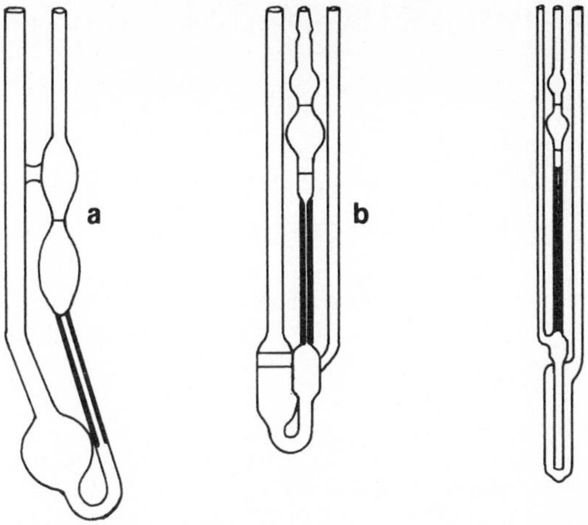

Figure 1-3 Capillary tube viscosimeters used to measure kinematic viscosity: (*a*) modified Ostwald, (*b*) Ubbelohde, (*c*) Fitz Simmons. (*Gulf Oil Corporation.*)

sample of oil under conditions which will greatly increase the rate of oxidation. Buildups of reaction products are then measured. The American Society for Testing and Materials (ASTM) D-943 test is the most widely used. Conducted under prescribed conditions, it measures the time (in hours) for the acidity of a sample of oil to increase a specified amount. The more stable the oil, the longer it will take for the change in acidity to occur.

Used-oil analysis to determine if the oil is suitable for further service is based on a comparison between the used oil and the new oil. Increases in viscosity, acidity, and buildups of insoluble contaminants are usually indicators that oxidation has occurred.

Thermal Stability

Thermal stability is a measure of an oil's ability to resist chemical change due to temperature. Since oxygen is present in most lubricant applications, the term *thermal stability* is frequently used in reference to the oxidation resistance of an oil.

Chemical Stability

Chemical stability defines an oil's ability to resist chemical change. Usually it, too, is used to refer to the oxidation stability of an oil. Chemical stability, other than resistance to oxidation, sometimes can refer to an oil's inertness in the presence of various metals and outside contaminants.

Carbon Residue

The carbon-forming tendencies of an oil can be determined with a test in which the weight percent of the carbon residue of a sample is measured after evaporation and pyrolysis.

Neutralization Number

The neutralization number (neut. no.) is a measure of the acidity or alkalinity of an oil. Usually reported as the total acid number (TAN) or total base number (TBN) it is expressed as the equivalent milligrams of potassium hydroxide required to neutralize the acidic or basic content of a 1-g sample of oil. Increases in the TAN or decreases in the TBN are usually indicators that oxidation has occurred.

Lubricity

Lubricity is the term used to describe an oil's "oiliness" or "slipperiness." If two oils of the same viscosity are used in the same application and one causes a greater reduction in friction than the other, it is said to have better lubricity than the first. This is strictly a descriptive term.

Saponification Number

The saponification number (SAP no.) is an indicator of the amount of fatty material present in an oil. The SAP no. will vary from 0, for an oil containing no fatty material, to 200 for 100 percent fatty material.

Demulsibility

Demulsibility is the term used to describe an oil's ability to shed water. The better the oil's demulsibility, the more rapidly the oil will separate from water after the two have been mixed together.

API Gravity

API gravity is a relative measure of the unit weight of a petroleum product. It is related to the specific gravity in the following manner:

$$\text{API gravity} = \frac{141.5}{\text{specific gravity}} - 131.5$$

Pour Point

Pour point is the lowest temperature at which an oil will flow in a certain test procedure. It is usually not advisable to use an oil at temperatures lower than 15°F (8°C) above its pour point.

Flash Point

Flash point is the oil temperature at which vapors from the oil ignite when an open flame is passed over a test sample.

Fire Point

Fire point is the oil temperature at which vapors from the oil will sustain a continuous flame. The fire point is usually approximately 60°F (33°C) above the flash point.

Grease Properties

Penetration

Penetration is an indicator of a grease's relative hardness or softness and not a criterion of quality. Measured on a penetrometer at 77°F (25°C), it is the depth of penetration (in tenths of millimeters) into the grease of a standard 150-g cone. The softer the grease, the greater the penetration number will be.

If the penetration test is performed on an "undisturbed" sample, the results are reported as unworked penetration. If the sample has been subjected to extrusion by a reciprocating perforated piston for a number of strokes (most commonly 60 strokes) prior to the penetration test, the results are reported as worked penetration. It is normally desirable to have as little difference between the worked and unworked penetrations as possible.

NLGI Consistency Numbers

The National Lubricating Grease Institute (NLGI) has developed a number system ranging from 000 (triple zero) to 6 to identify various grease consistencies. This system is used by most of industry. Table 1-2 gives the NLGI numbers, their corresponding worked penetration ranges, and their descriptions (their corresponding consistencies). Most multipurpose greases are of either a no. 1 or no. 2 consistency.

TABLE 1-2 NLGI Classification of Greases

NLGI consistency grade	Worked penetration ASTM D 217-60T	Description
000	445–475	Very fluid
00	400–430	Fluid
0	355–385	Semifluid
1	310–340	Very soft
2	265–295	Soft
3	220–250	Semistiff
4	175–205	Stiff
5	130–160	Very stiff
6	85–115	Hard

Dropping Point

Dropping point is the temperature at which a grease liquefies and will flow. Generally it is not advisable to use a grease at temperatures higher than 50°F (28°C) below its dropping point.

Soap

The thickener used to manufacture greases can be called "soap." Many greases use metallic soaps as thickeners. Table 1-3 shows a comparison of some of the key properties of greases manufactured with different soaps and their typical applications.

Fillers

Solid lubricants, particularly molybdenum disulfide, are frequently added to greases to enhance their performance. These solid lubricants are called *fillers.*

ADDITIVES USED IN LUBRICATING OILS

It is possible, through the use of chemical additives, to improve a lubricant's natural ability to protect metal surfaces, to resist chemical changes, and to drop out contaminants.

Since industrial lubricating oils are frequently described by the additives they contain, it is helpful to understand the function of the major types of additives. Following are general definitions of some of the most common, listed in alphabetical order:

Air release agents Assist the oil in the release of entrapped air.

Antifoam agents Promote the rapid breakup of foam bubbles.

Antiseptic agents or bactericides Prevent the growth of microorganisms and bacteria. These are found primarily in water-soluble oils.

Antiwear agents Decrease the coefficient of friction and reduce wear under boundary or mixed-film lubrication conditions.

Demulsifiers Assist the natural ability of an oil to separate rapidly from water. These agents can be helpful in preventing rust since they help to keep water out of the oil and thus away from the metal surfaces.

Detergent-dispersant agents Prevent the formation of varnish and sludge. They are most commonly found in engine oils.

Emulsifiers Permit the mixing of oil and water to form stable emulsions. They are used primarily in the manufacture of water-soluble oils.

TABLE 1-3 Grease Application Guide

	Thickener						
	Lithium	Calcium	Sodium	Calcium complex	Aluminum complex	Polyurea	Clay or "Bentone"
Properties							
Dropping point, °F	350–375	200–225	325–350	500+	500+	550	
Average max usable temp, °F	275	175	250	325	300	350	275
High-temperature characteristics	Good	Poor	Fair-Good	Good	Good	Excellent	Good
Thermal stability	Good	Poor	Fair-Good	Good	Good	Excellent	Good
Low-temperature characteristics	Good	Fair	Fair	Fair–good	Fair–good	Good	Good
Pumpability	Excellent	Fair–good	Poor	Fair	Fair–good	Good–excellent	Good
Mechanical stability	Excellent	Good	Poor	Fair–good	Excellent	Excellent	Good
Oil separation	Good	Poor	Fair–good	Excellent	Good	Excellent	Excellent
Water resistance	Good	Excellent	Poor	Good	Excellent	Excellent	Good
Texture	Smooth and buttery	Smooth and buttery	Buttery to fibrous	Smooth and buttery	Smooth and buttery	Smooth and buttery	Smooth and buttery
Rust protection	Fair to good	Poor	Excellent	Excellent	Good	Excellent	Poor
Oxidation stability	Fair to good	Poor	Poor	Good	Good	Excellent	Good
Other properties			Good adhesive and cohesive properties	Good inherent EP properties			
Applications							
	Multipurpose All applications except extra high temperatures	Where water is dominant factor Wet, moderate low temperature conditions Plain and roller bearings, water pumps, slides	Antifriction and plain bearings Electric motors, fans Must be used in dry conditions	High temperatures Corrosive conditions Do not use in centralized lubrication systems	Multipurpose Moderately high temperatures	Multipurpose High temperatures Antifriction and plain bearings Electric motors, fans Wet conditions Corrosive conditions	Multipurpose High temperatures

Extreme-pressure agents Protect against metal-to-metal contact and welding after the oil film has been ruptured by high loads or sliding velocities. The majority of the extreme-pressure oils on the market today are of the sulfur-phosphorus type and are noncorrosive to most metals including brass. This was not true of some of the earlier formulations, and many misconceptions still exist in this regard.

"Oiliness" or fatty compounds Improve the lubricity or slipperiness of an oil. These compounds are also helpful in resisting water wash-off.

Oxidation inhibitors Prevent or retard the oxidation of an oil, thereby reducing the formation of deposits and acids.

Pour-point depressants Lower the pour point of paraffinic petroleum oils.

Rust and corrosion inhibitors Improve an oil's ability to protect metal surfaces from rust and corrosion.

Tackiness agents Improve the adhesive qualities of an oil.

Viscosity index improvers Increase the viscosity index of an oil by increasing an oil's viscosity at high temperatures. These additives are most widely used in motor oils to create multigrade oils.

LUBRICANT SELECTION

Practical lubrication is more an art than an exact science. Proper lubricant selection depends on the equipment design, the operating conditions, and the method of application.

Most equipment manufacturers provide lubrication recommendations based on design, normal operating conditions, and past experience. Whenever possible, these recommendations should be followed. In addition, most reputable oil suppliers keep in close contact with major equipment builders and are available to consult with users on lubricant selection.

The recommendations included in this chapter are based on standard practices and are intended solely as guidelines.

General Selection Guidelines

The design of the equipment and the expected operating conditions will determine which functions the lubricant is expected to perform and will dictate the type of lubricant and additives that will be best suited.

The oil of proper viscosity for an application is a function of speed, load, and ambient temperature. Conditions of high loads and slow speeds will require a high-viscosity oil. Similarly, a low-viscosity oil is best suited to conditions of low loads and high speeds. Ideally, one would like to select the oil of lowest possible viscosity that is capable of maintaining a lubricant film between the moving surfaces. Selection of a higher-viscosity oil than is needed can result in power losses and temperature buildups due to the higher internal fluid friction of the lubricant.

The effect of operating temperatures on the selection of the lubricant should not be overlooked. Since oil viscosity decreases as temperatures increase, it is necessary to select higher-viscosity fluids for high-temperature applications and lower-viscosity fluids for low-temperature applications in order to ensure adequate lubricant film thickness and minimal fluid friction. Fluids with high viscosity indexes (high VI) should be used for applications where wide temperature ranges are anticipated.

Operating Limits of Petroleum Oils

As a result of additive technology, a suitable petroleum-base lubricating oil can be found for most applications. Exceptions can exist where fire-resistant fluids are required or extreme temperature conditions exist.

Where fire-resistant fluids are required, petroleum oils are not suitable. In some instances water-oil emulsions are acceptable if operating temperatures are below 150°F (65°C) (to avoid excessive water evaporation) and if the equipment is designed to handle these fluids. Two types of water-oil emulsions are currently in use. The high-water-base fluids, (95/5 fluids as they are sometimes called). are an oil-in-water emulsion containing 95% water and 5% oil. The other is the invert emulsion, a water-in-oil emulsion, which contains approximately 40% water.

Unsatisfactory performance of petroleum oils can occur under three types of extreme temperature conditions:

- Excessively high temperatures
- Excessively low temperatures
- Wide temperature variations

Petroleum oils will adequately withstand very high temperatures for very short periods of time. Problems will occur when the oil is subjected to high temperatures for extended periods of time. The rate of oxidation of petroleum oils subjected to constant temperatures above 115°F (45°C) will approximately double for every 15 to 20°F (8 to 10°C) rise in temperature. Temperatures above 200°F (95°C) will almost always result in excessive sludge and deposit formation and should be avoided. In circulating systems, the reservoir should always be cool enough to comfortably hold a hand on it. The oxidation rate is usually negligible at temperatures below 115°F (45°C).

Petroleum oils should not be used at temperatures less than 10 to 15°F (5 to 8°C) above their pour point. For applications subjected to large temperature variation only high-viscosity-index fluids should be used. The viscosity index should be high enough to ensure that the oil viscosity remains within the recommended limits at both the high- and low-temperature extremes to which the equipment is subjected.

Plain-Bearing Lubrication

Plain bearings, also called journal or sleeve bearings, comprise one of the simplest machine components. The type of motion between the bearing and the shaft is pure sliding.

In plain bearings, the lubricant must reduce sliding friction, carry away any heat generated in the bearing, prevent rust and corrosion, and serve as a seal to prevent the entry of foreign material.

Barring any unusual operating conditions, plain bearings will operate satisfactorily with any lubricant of the correct viscosity. Special operating conditions may require the use of oils containing additives. Antiwear and extreme-pressure oils may be desirable for plain bearings operating intermittently or under very high loads. Rust- and corrosion-inhibited oils are generally preferred for humid operating environments.

Most plain bearings are designed to operate under full-film hydrodynamic lubrication. Referring to Fig. 1-2, and assuming the bearing load and oil viscosity to be constant, the lubricant film development would be expected to follow the ZN/P curve as the shaft speed increases. If oil of the proper viscosity is selected for the load and speed conditions, full-film hydrodynamic lubrication will prevail during continuous operation.

Numerous mathematical models of plain-bearing lubrication have been used in attempts to accurately select the best oil viscosity for a plain bearing. Unfortunately, these models are complicated and expensive to develop. For this reason, except in special cases, lubricant viscosity selection is usually based on standard practices established through experience. Table 1-4 presents a general guide for viscosity selection for plain bearings subjected to average loading.

Plain bearings may be grease-lubricated if their operating speed does not exceed approximately 6 ft/s (2 m/s). At higher speeds, excessive temperature buildup could result.

In general, relatively soft greases are used for centralized systems and harder greases

TABLE 1-4 Oil Viscosity Selection for Plain Bearings

Bearing speed factor, r/min × shaft diameter		Viscosity at operating temperature	
in	mm	cSt	SUS
Below 750	Below 1,900	130–325	600–1500
750–2,000	1,900–50,800	65–130	300–600
2,000–4,000	50,800–101,600	32–65	150–300
4,000–10,000	101,600–254,000	14–32	75–150
Above 10,000	Above 254,000	5–14	40–75

for compression cups and open journals. Each application should be considered on its own merits, taking into consideration the operating conditions. Temperature and water contamination require particular attention.

Plain bearings are frequently grooved (Fig. 1-4) to improve the distribution and flow of the lubricant. Normally, two important rules should be followed when grooving a plain bearing:

- Grooves should not extend into the load-carrying area of the bearing because this would increase unit pressures.
- Groove edges should be rounded to prevent scraping the lubricant off the journal.

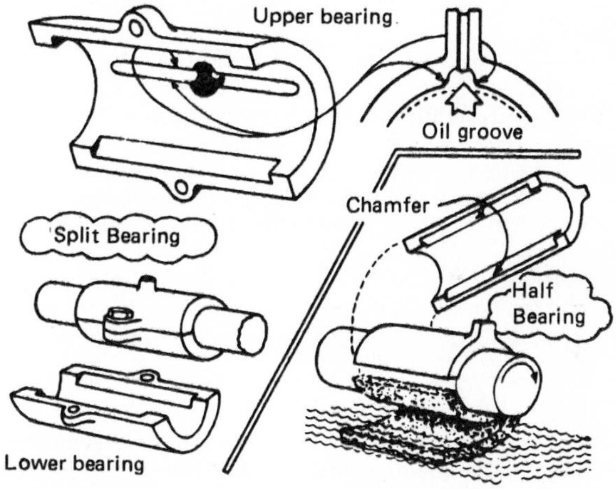

Figure 1-4 Oil groove. *(U.S. Steel Corporation, Reprinted from* The Lubrication Engineers Manual, *Copyright 1971.)*

Antifriction-Bearing Lubrication

Antifriction or roller bearings use balls or rollers to substitute rolling friction for sliding friction. This type of bearing has closer tolerances than do plain bearings and is used where precision and higher speeds are required.

In antifriction bearings a lubricant facilitates easy rolling, reduces the friction generated between the rolling elements and the cages or retainers, prevents rust and corrosion, and serves as a seal to prevent the entry of foreign material.

High-quality rust- and oxidation-inhibited (R & O) oils are generally recommended, especially where high-temperature conditions may oxidize the oil and so lead to the formation of deposits which could interfere with the free action of the rolling elements.

TABLE 1-5 Oil Viscosity Selection for Antifriction Bearings

Speed factor bearing bore, mm × r/min	Operating temperatures		Viscosity	
	°F	°C	ISO viscosity grade	SUS at 100°F
Up to 75,000	−40–32	−40–0	15–32	70–150
	32–150	0–65	32–100	150–600
	150–200	65–93	100–220	600–1200
	200–250	93–21	220–680	1100–3000
75,000–200,000	−40–32	−40–0	7–22	50–100
	32–150	0–65	22–68	100–300
	150–200	65–93	68–100	300–600
	200–250	95–121	150–320	700–2100
200,000–400,000	−40–32	−40–0	7–15	50–70
	32–150	0–65	15–46	70–200
	150–200	65–93	32–68	150–300
	200–250	93–121	68–150	400–900
Above 400,000	−40–32	−40–0	5–10	40–60
	32–150	0–65	10–32	60–150
	150–200	65–93	22–46	100–200
	200–250	93–121	68–100	300–600

Extreme pressure and antiwear additives may also be desirable under conditions of heavy or high shock loads.

Because of its better cooling ability, oil is generally preferred to grease. Table 1-5 gives general guidelines to the proper viscosity selection of oils for antifriction bearings.

Grease may be used to lubricate antifriction bearings running at low speeds and located in areas where they are likely to receive infrequent attention.

The selection of the proper type and grade of grease depends on the operating conditions and the method of application. Generally, soft greases (i.e., NLGI no. 1 consistency) with low base oil viscosity are preferred for use at low temperatures and in central systems. Harder greases (i.e., NLGI no. 2 consistency) with low base oil viscosity perform better at high speeds.

Care should be taken not to overgrease antifriction bearings because this can lead to excessive temperature buildup. Generally the bearing housing should be ⅓ to ½ filled.

Gear Lubrication

The motion between gear teeth as they go through mesh is a combination of sliding and rolling. The type of gear, the operating load, speed, temperatures, method of application of the lubricant, and metallurgy of the gears, are all important considerations in the selection of a lubricant.

Industrial gearing may either be enclosed, in which case the gears and the bearings which support them are operated off the same lubricant system; or open, in which case the mountings are lubricated separately from the gears themselves.

Due to the high sliding forces encountered in enclosed worm and hypoid gears, lubricant selection for these should be considered separately from lubrication of other types of enclosed gears.

As with all equipment, the first rule in selecting a gear lubricant is to follow the manufacturer's recommendation, if at all possible. In general, one of the following types of oils is used:

Rust- and Oxidation-Inhibited Oils. R & O oils are good-quality petroleum-based oils containing rust and oxidation inhibitors. These oils provide satisfactory protection for most lightly to moderately loaded enclosed gears.

Extreme-Pressure Oils. EP oils are usually high-quality petroleum-based oils containing extreme-pressure additives. These products are especially helpful when high-load conditions exist and are a must in the lubrication of enclosed hypoid gears.

Compounded Oils. These are usually petroleum-based oils containing 3 to 5 percent fatty or synthetic fatty oils (usually animal fat or acidless tallow). They are usually used for worm-gear lubrication where the fatty content helps reduce the friction generated under high sliding conditions.

Heavy Open-Gear Compounds. These are very-heavy-bodied tarlike substances designed to stick to the metal surfaces. Some are so thick they must be heated or diluted with a solvent to soften them for application. These products are used in cases where the lubricant application is intermittent.

A large number of gear lubrication models and viscosity selection guides exist. In the United States, the most widely used selection method employs the American Gear Manufacturers Association (AGMA) standards. Under its specifications for enclosed industrial gear drives the AGMA has defined lubricant numbers which designate viscosity grades for gear oils. These grades are currently being revised to correspond to the ISO viscosity grades. Table 1-1 identifies the newly proposed AGMA viscosity numbers with their corresponding ISO viscosity grades.

As a rule, low speeds and high pressures require high-viscosity oils. Intermediate speeds and pressures require medium-viscosity oils, and high speeds and low pressures require low-viscosity oils. Table 1-6 gives some very broad guidelines for viscosity and type of lubricant for industrial gearing.

Open gears operate under conditions of boundary lubrication. The lubricant can be applied by hand or via drop-feed cups, mechanical force-feed lubricators, or sprays.

Heavy-bodied oils with good adhesive and film-strength properties are required because centrifugal forces tend to throw the lubricant off the gear teeth.

TABLE 1-6 Oil Selection for Enclosed Gear Drives

Service	ISO viscosity grade	Oil type
Helical, herringbone, straight-bevel, spiral-bevel, and spur-gear drives		
Operating at normal speeds and loads	220	EP or R & O
Operating at normal speeds and high loads	220	EP
Operating at high speeds (above 3600 r/min)	68	EP or R & O
Worm-gear drives	460	Compounded or EP
Hypoid-gear drives		
Normal speeds (1200–2000 r/min)	220	EP
High speeds (above 2000 r/min)	150	EP
Low speeds (below 1200 r/min)	460	EP

Compressor Lubrication

The compressor model and type, the loading, the gas being compressed, and other environmental conditions dictate the type and viscosity of the oil to be used. Most compressors are lubricated with petroleum oils; however, there has been considerable interest in synthetic lubricants for compressor lubrication in recent years.

Compressing gases other than air creates problems which require special lubrication consideration because of possible chemical reactions between the gas being compressed and the lubricant. Since no two cases are alike, it is recommended that the compressor manufacturer and lubricant supplier be consulted for recommendations for a particular operation.

Oils for use in compressors should have the following characteristics:

Good Stability

A good compressor oil must have high oxidation stability to minimize the formation of gum and carbon deposits. Such deposits can cause valve sticking. This can lead to very-high-temperature conditions and compressor malfunction.

Good Demulsibility

A good compressor oil must be able to shed water readily to prevent formation of emulsions which could interfere with proper lubrication.

Anticorrosion and Antirust Properties

A compressor lubricant must protect the valves, pistons, rings, and bearings against rust and corrosion. This is especially important in humid atmospheres or in compressors that operate intermittently.

Good Antiwear Properties

Good compressor oils must form and maintain a strong lubricant film at relatively high temperatures; therefore, good antiwear properties are required.

Nonfoaming Properties

This requirement is especially important in crankcases where air-oil mixtures could impair good lubrication.

Low Pour Point

This property is necessary only for low-temperature start-up. Usually it is a factor only in portable air compressors which will frequently be used outdoors.

Proper Viscosity

Table 1-7 summarizes the oil viscosity requirements for various types of compressors. The operator's manual should be consulted for the manufacturer's viscosity recommendations for the prevailing operating temperatures and conditions.

TABLE 1-7 Lubricant Selection for Compressors

Type of compressor	Type of service	Recommended ISO viscosity grade
Reciprocating		
Crankcase	All	68–100
Cylinders		
Under 300 lb/in²	Dry air	68–100
	Wet air	100
Over 300 lb/in²	Dry air	100–150
	Wet air	220
Rotary		
Sliding-vane type		
Air- or water-cooled	Dry air	32–68
	Wet air	46–68
Oil-flooded	Dry air	150
	Wet air	150
Lobe or impeller type	Air	32–46
Liquid-piston type	Air	32
Dynamic compressors		
Centrifugal	Air	32
Axial-flow	Air	32

SPECIFICATIONS AND STANDARDS

"Lubrication of Industrial Enclosed Gear Drives," AGMA Standard 250.04 American Gear Manufacturers Association, Arlington, Va., November, 1974.

"Lubrication of Industrial Open Gearing," AGMA Standard 251.02 American Gear Manufacturers Association, Arlington, Va., November 1974.

"ASLE Standards for Machine Tool Petroleum Fluids," American Society of Lubrication Engineers, Park Ridge, Ill., January 1972.

Annual Book of ASTM Standards, Parts 23, 24, 25, and 47; American Society for Testing and Materials, Philadelphia.

BIBLIOGRAPHY

Bailey, Charles A., and Joseph S. Aarons (eds.): *The Lubrication Engineers Manual,* United States Steel Corp., Pittsburgh, 1971.

Bearings and Their Lubrication, Gulf Oil Corp., Pittsburgh, 1952.

Billett, Michael: *Industrial Lubrication: A Practical Handbook for Lubrication and Production Engineers,* Pergamon, Oxford, England, 1979.

Brewer, Allen F.: *Basic Lubrication Practice,* Reinhold, New York, 1955.

Brewer, Allen F.: *Effective Lubrication,* Robert E. Krieger, Huntington, N.Y., 1974.

Ellis, E. G.: *Fundamentals of Lubrication,* 2d ed., Scientific Publications, Broseley, England, 1970.

Fuller, Dudley D.: *Theory and Practice of Lubrication for Engineers,* Wiley, New York, 1956.

Gunther, Raymond C.: *Lubrication,* Chilton, Philadelphia, 1971.

Neale, M. J. (ed.): *Tribology Handbook,* Wiley, New York, 1973.

O'Connor, J. J. and J. Boyd (ed.): *Standard Handbook of Lubrication Engineers,* McGraw-Hill, New York, 1968.

Shigley, Joseph E.: *Mechanical Engineering Design,* 2d ed., McGraw-Hill, New York, 1972.

chapter 17-2

Synthetic Lubricants

by

Marvin Campen,
Product Manager—Synthetic Fluids
Gulf Oil Chemicals Company
Houston, Texas

DEFINITION AND CLASSIFICATION

Synthetic lubricants, or synlubes, are made by compounding (or blending) chemically synthesized base fluids with conventional lubricant additives. (For some applications the synthesized base fluids are used neat, without additives.) Synthesized base fluids are formed by combining low-molecular-weight (MW) components via chemical reactions into higher-MW compounds. Thus each base fluid's molecular structure is planned and controlled and its properties are predictable. Synthetic base fluids do not occur in nature, as does the crude petroleum oil from which conventional mineral-oil-base stocks are derived by refining processes. Only by *super-refining,* which has never been economically feasible, could petroleum-base stocks be obtained with properties comparable to those of synthesized base fluids. Yet most synthetic base fluids come from petroleum via synthesis using components derived from petroleum hydrocarbons.

Principal reasons for the increasing uses of synlubes are their abilities to:

- Work where conventional lubricants will not
- Comply with certain specifications or regulations (as military, OSHA, safety, pollution)
- Provide enhanced cost effectiveness, including energy savings.

The distinguishing feature of all synthetics is superiority in one or more respects to the mineral oils. Advantageous characteristics of the most widely used synlubes, in varying degrees, are:

- Low-temperature fluidity
- High-temperature oxidation stability and fire resistance (high flash, fire, and autoignition points)
- Low volatility in relation to viscosity
- High viscosity index (Less change of viscosity with temperature)

Some synthetics are chemically inert; some are fire-resistant or nonflammable.

Thus synlubes are high-performance, problem-solving lubricants that provide operating benefits such as: (1) less frequent lubrication ("sealed for life" in some machines); (2) less maintenance; (3) higher productivity; (4) lower parts rejection (due to more uniform tolerances and quality); (5) longer machine life; (6) reduced fire hazard; (7) greater resistance to acids, bases, and solvents; (8) more economical metalworking-oil or coolant applications; and (9) easier reclamation or disposal.

Although synthetic base fluids make possible these benefits, proper compounding of base fluids with performance additives is essential to achieving finished synlubes with the desired advantages. Getting one's money's worth from a more-expensive, high-quality synlube is best assured by buying specification-grade, approved synlubes from reputable suppliers who know how to maximize the advantages of each base fluid's properties which include (1) *rheology,* or flow characteristics; (2) *volatility;* (3) *stability,* thermal, oxidative, hydrolytic, chemical; (4) *additives' compatibility* and *response;* and (5) *effects on elastomers, paint, and other softwear.*

The *classification* of synthetic base fluids was proposed by an ad hoc ASTM committee and subsequently adopted by SAE in Standard J 357a and by RCCC, as follows:

1. **Synthetic Hydrocarbons**
   Alkylated aromatics (dialkylbenzenes, DAB) and cycloaliphatics
   Polyalphaolefins (PAO)
   Polybatenes
2. **Organic Esters**
   Dibasic acid esters (diesters)
   Polyol esters
   Polyesters
3. **Other**
   Halogenated hydrocarbons
   Phosphate esters
   Polyglycols (polalkylene glycols, PAG)
   Polyphenyl ethers
   Silicate esters
   Silicones
4. **Blends** Synthetic base stock may be mixtures of the above which are compatible.

By specific omission from this classification, other compounds and mixtures such as the following are *not classified* as synthetic lubricants: (1) molybdenum disulfide (MoS_2), graphite, and other solids; (2) plastics (PE, PTFE, nylon, et al.); (3) inert and liquefied gases; (4) liquid metals (as Na, K); (5) exotic natural products (as jojoba oil); and (6) surfactant-chemical-concentrate, high-water-content, or "water-additive" coolants and other metalworking fluids (which are nevertheless called "synthetic" by some people in the metalworking industry).

Unless a lubricant consists of at least 90 percent synthetic base fluid, it should not be designated a synlube; rather it should be labeled "partial synthetic," "fortified synthetic," "semisynthetic," or some similar term.

PROPERTIES AND USES OF SYNTHETIC LUBRICANTS

The most widely used industrial synlube classes are: (1) polyalphaolefins, (2) organic esters, (3) phosphate esters, (4) polyalkylene glycols, and (5) silicones. Their operating-temperature ranges are shown in Fig. 2-1, their performance vs. mineral oil, in Table

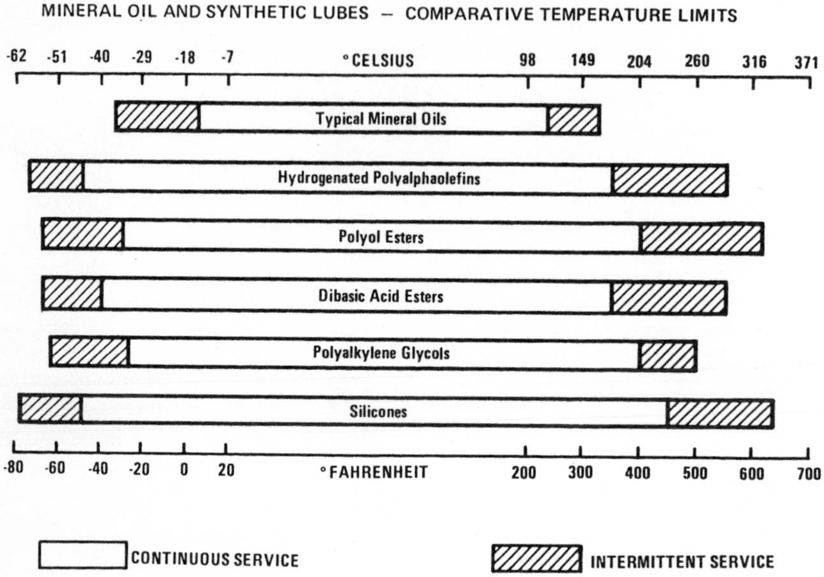

Figure 2-1 Mineral oil and synlubes—comparative temperature limits.

2-1. Most are supplied in several viscosity grades and can be formulated with proper additives to make industrial lubricants—circulating oils, gear and bearing lubes, and greases and other functional fluids for use with gears, traction drives, compressors, pumps, turbines, calenders, motors, hydraulic systems, valves, instrumentation, and other machinery and equipment, including metalworking applications (see Table 2-2). Many companies compound such lubricants from their own or from purchased synthetic base fluids and additives. Viscosity grades range from 1.5 cSt at 212°F (100°C) to firm greases.

Polyalphaolefins

Polyalphaolefins (PAO), or olefin alpha oligomers, are derived from linear alphaolefins which are made from ethylene, a basic building block from petroleum or natural gas liquids. Manufacturers are Bray, Ethyl/Cooper, Gulf, Lubrizol, Mobil, Emery, and Uniroyal. Shell, Conoco, Texaco, Arco, and others are potential manufacturers.

The largest volume use for PAO is in crankcase oils for internal combustion engines—gasoline, diesel, and natural gas. Principal industrial PAO lubricants are for gears, calenders, textile machinery, conveyors, gas turbines, hydraulic systems, instrumentation, rotary-screw and reciprocating compressors, pumps, equipment, and many grease applications—particularly in harsh environments (chlorine, HCl, and hot process gases). PAO products serve well as lubricants and sealant fluids for rotary mechanical seals of chemical process pumps and agitated kettles. The longer functional life of a PAO grease in a

TABLE 2-1 Relative Performance of Synlubes vs. Mineral Oil*

Properties	Mineral oil	Synthetics						
		Synthesized hydrocarbons		Organic esters				
		Polyalpha-olefin (PAO)	Dialkylated (C_{12}) benzene (DAB)	Dibasic	Hindered polyol	Polyalkylene glycol (PAG)	Phosphate ester	Silicone fluid
Viscosity-temperature properties (VI)	F	G	F	VG	G	G	P	E
Low-temperature fluidity, low pour point	P	G	G	G	G	G	F	G
High-temperature oxidation resistance with inhibitors	F	VG	G	G	E	F	F	G
Compatibility with mineral oils	E	E	E	G	F	P	F	P
Low volatility	F	E	G	E	E	G	G	G
Effect on most paints and finishes	N	N	N	S	M	M	C	S
Stability in presence of water (hydrolytic stability	E	E	E	F	F	VG	F	G
Antirust properties, with inhibitor	E	E	E	F	F	G	F	G
Additive solubility	E	G	E	G	G	F	G	P
Elastomer swelling tendency—buna rubber	L	N	L	M	H	L	H	L

*Letter signifies performance level: P = poor, F = fair, G = good, VG = very good, E = excellent, M = moderate, H = high, C = considerable, N = none or nil, S = slight, L = light.

TABLE 2-2 Principle Uses of Synthetic Lubricants*

	Int. comb. engines	Gears and bearings	Greases	Compressors turbines†	Hydraulic fluids†	Water emulsions
Polyalphaolefins	E	E	E	VG	E	F
Organic esters	G	VG	VG	VG	VG	P
Polyalkylene glycols	—	VG	G	G	E(FR)	E
Phosphate esters	—	—	—	VG(FR)	E(FR)	—
Silicones	—	VG	E	E	E	—
Mineral oils	VG	VG	VG	VG	VG	F

*Letter signifies performance level: P = poor, F = fair, G = good, VG = very good, E = excellent
†FR = Fire-resistant.

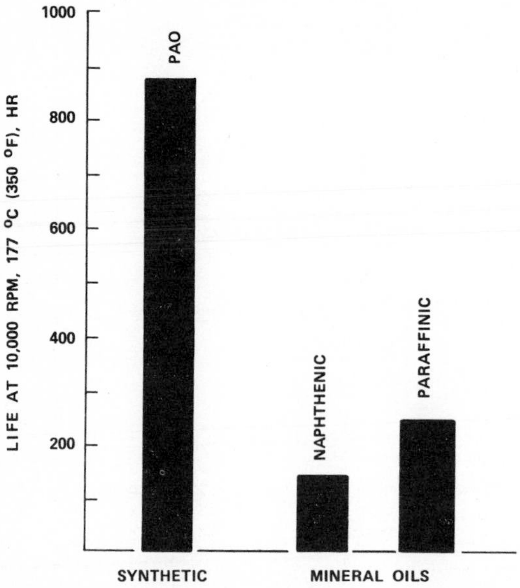

Figure 2-2 Functional-life ball-bearing test for grease, Federal Standard 791, No. 333.

high-speed ball-bearing test is shown in Fig. 2-2. Food-grade synlubes, compounded from FDA-approved PAO white oils, also are available.

Because PAO lubes are compatible with mineral oils and with existing systems designed for mineral oil products, they can be used in such systems without changing seals, hoses, or materials of pumps and other equipment. PAOs are often incorporated into organic ester lubes to assure seal compatibility.

Organic Esters

Diesters

Diester, or dibasic acid ester, base fluids are made by reacting short-chain C_{8-13} oxo alcohols with dibasic acids, such as adipic and phthalic (from petroleum), azelaic (from tallow), and dimer acids (from tall oil). Diesters are a subclass of polyesters, which include tri- and tetra-esters. More than 50 United States companies have the capacity for manufacturing at least 2 × 10^8 gal/year of these and many other diesters, primarily for plasticizers for PVC (vinyl) products. Manufacturers specializing in lubricant-grade diesters are Emery, Exxon, Hatco, Inolex, Mobil, Röhm and Haas, Tenneco, and Witco.

Diester lubricants are used mainly in oil-flooded air and natural-gas compressors. Their polar nature, due to the carbonyl group in their chemical structure, facilitates additive acceptance and provides solvation properties that help maintain clean intake and exhaust valves, thereby extending lubricant drain periods and improving ignition safety. Several makers of reciprocating compressors specify diester lubricants for some models. Diester lubricants are used in oven chain lubricants, in some gas-turbine engine oils, and as greases in certain low-temperature applications.

Polyol Esters

Polyol, or neopentyl, esters are formed by combining, linear short-chain (C_{5-10}) fatty acids with polyols, usually pentaerythritol (PE) or trimethylolpropane (TMP). Such "hindered esters" exhibit very high thermal stability, inherently higher than diesters. Emery, Hatco, Hercules, Witco/Humko, Mobil, and Stauffer are the current manufacturers of lubricant-grade polyol esters.

Polyol esters are chiefly used in military and commercial jet aircraft and in surface gas-turbine engines supplied by the jet-engine makers (Pratt & Whitney—TP & M, Detroit Diesel-Allison, GE aircraft-type models). Excellent gear oils, greases, and other industrial lubricants can be formulated from polyol ester fluids. However, their use in industry has been constrained by comparably performing, but lower-cost, diester and PAO lubricants. Polyol esters are sometimes combined with PAO to enhance solubility of certain additives and to optimize elastomer seal swell characteristics.

Polyalkylene Glycols

Polyalkylene glycols (PAG or polyglycols) are synthesized by combining ethylene oxide and propylene oxide under a variety of processing conditions. Consequently PAGs are available in the widest range of viscosities and hydrophobic/hydrophilic balances of any synthetic functional fluid (Table 2-3). The fluids are noncarbonizing and possess high viscosity indices. A limiting characteristic is their incompatibility with mineral oils and several conventional lubricant additives. Union Carbide is the dominant supplier. BASF, Dow, Olin, and Texaco also supply many PAG grades.

A large use for PAG is in metalworking fluids, e.g., coolants. The different degrees of water solubility of the different grades enable metal heat-treating quenchants to be selected for the desired rate of heat removal. Other large-volume applications of compounded PAG include heat-transfer oils, fire-resistant hydraulic fluids, rubber and textile processing aids, and gas compressor lubricants. PAG lubricants are unsuitable for air and refrigeration compressors. There are some gear lubricant and grease applications.

TABLE 2-3 Polyalkylene Glycol Fluids

Viscosity, cSt				Pour point			Flash—C.O.C.		Auto-ignition	
212°F (100°C)	100°F (38°C)	0°F (−18°C)	Viscosity index	°F	°C	Sp. gr. 20/20°C	°F	°C	°F	°C
2.75	11.7	270	83	−70	−57	0.960	325	163	410	210
6.80	35.3	1,400	169	−50	−46	0.983	460	238	572	300
18.5	112	7,100	196	−30	−34	0.997	505	263	671	335
53	365	31,000	219	−10	−23	1.002	515	268	752	400
165	1,100	71,000	281	−20	−29	1.063	545	285	779	415
255	1,970	—	282	40	4.4	1.094	490	254	797	425
2600	19,400	—	414	40	4.4	1.097	620	327	833	445

Phosphate Esters

Phosphate esters derive from the reaction of phosphorus oxychloride with cresylic acids, synthetic alkyl phenols, or certain alcohols. FMC Houghton, Monsanto, and Stauffer are the largest suppliers, with IMC (Sobin/Montrose) expanding its position.

Fire resistance is unquestionably the most notable property of inorganic phosphate esters. Where combustibility is a hazard, OSHA, Factory Mutual, and other standards increasingly require phosphate ester lubricants. Low volatility and chemical stability are other advantageous properties, but their incompatibility with mineral oil systems is often a disadvantage. Phosphate esters craze and soften certain plastics, coatings, neoprene and nitrile elastomers, and pipe-joint compounds.

Hydraulic fluids and compressor lubricants are the principal uses. Some greases also are used where safety from fire or from very high temperatures is important.

Silicones

Silicones are the reaction products of silicon (from sand or quartz) and different halocarbons, such as alkyl or aryl chlorides. Fluorosilicones are also available. Dow Corning is the major supplier. GE, Union Carbide, and Stauffer's SWS affiliate are becoming more active.

Chemical inertness is the major advantage of silicones. Low flammability and self-extinguishing properties make silicones desirable for many uses. Widespread applicability is limited by their incompatibility with mineral oils and certain additives.

Many specialty applications exist for silicones: moisture-proof seals and lubricants for ignition and electronic equipment and greases for valves and swivel joints exposed to chlorine gas and strong oxidizing or corrosive chemicals. Silicones are used as hydraulic fluids and compressor lubricants. Silicone brake fluids are being used in new systems designed to benefit from their unique properties.

Other Synlubes

Alkylated Aromatics

Alkylated aromatics (dialkylbenzenes, or DAB) were originally co-products of the manufacture of linear alkylbenzene (LAB) "soft" detergent alkylate. Conoco is the major manufacturer. DAB is compounded into low-temperature engine oils, gear lubricants, hydraulic fluid, and grease. Chevron recovers some bottoms from their dodecylbenzene (DDB) "hard" detergent alkylate plant. This highly branched-chain DAB is the base for a high-quality refrigerator-compressor oil and a dielectric fluid.

Cycloaliphatics

Cycloaliphatics are obtained by dimerizing and then hydrogenating the by-product α-methyl styrene from synthetic phenol plants. Monsanto is the only manufacturer today, but Sun also has the technology.

The unique property of these synthesized hydrocarbons is their high traction coefficient. Lubricants compounded from such fluids become shear-resistant semisolids at the high momentary contact pressure in adjustable-speed traction drives. This enables power to be transmitted from one smooth-rolling element to another—without gears, chains, belts, etc. The driving member drives the driven member with essentially no slipping (see Fig. 2-3).

Applications for traction drives include wire-drawing machines, injection molders, filament-payout stands, boring machines, press-roll drives, spring coilers, and gun mills.

Polybutenes

Polybutenes are oligomers of C_4 olefins, principally isobutene. Amoco, Chevron, and Cosden sell some of what they produce, but Exxon and Lubrizol use most of their output as intermediates for lubricant additives.

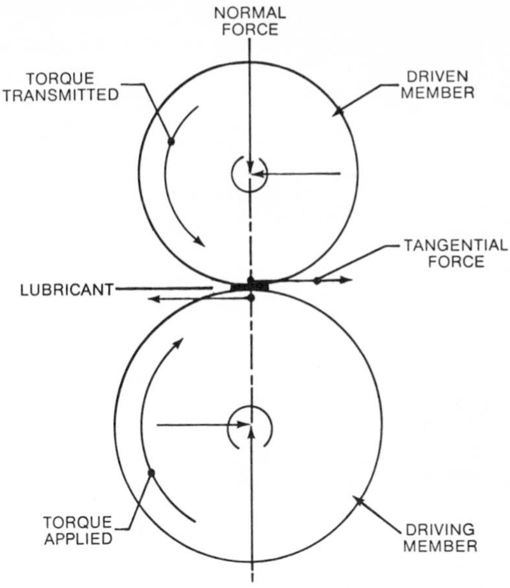

Figure 2-3 Traction coefficient = tangential force/normal force.

Polybutenes combine lubricity with complete burn-off at relatively mild elevated temperatures [600°F (315°C)]. Thus no deposits or stains are left on the hot surfaces. Because of these properties, polybutenes are used in rolling and wire drawing of nonferrous metals. They are also used as lubricants for compressors of nonoxidizing gases, including ethylene and propylene compressors in polyethylene and polypropylene plants.

Halogenated Hydrocarbons

Halogenated hydrocarbons are primarily classified as (1) chlorofluorocarbons, made by Halocarbon, and (2) perfluorinated polyethers, made by duPont and Montedison, and purified further by Bray (distilled fractions).

The extremely inert chlorofluorocarbons are essentially fireproof. They are used in liquid, grease, and plastic form in oxygen plants and in vacuum pumps in gaseous diffusion plants separating isotopes of uranium hexafluoride, an extremely reactive compound. Also they lubricate diaphragm compressors of radioactive gases in nuclear power plants and serve as special fire-resistant hydraulic fluids and suppressants. Because of their high cost, systems to recover the fluid from the gases are generally installed.

Perfluorinated polyethers combine nonflammability with good lubricity. They perform well at temperature extremes and in contact with gases such as oxygen, ozone, fluorine, BF_3, UF_6, etc. Other industrial lubricant applications include furnaces, ovens, chains, rollers, conveyors, and plastic film orienters.

Polyphenyl Ethers

Polyphenyl ethers are best characterized as bis(pheno-oxyphenoxy) benzene, made by Monsanto. This compound possesses extraordinary resistance to degradation by heat, oxygen, radiation, hydrolysis, and chemical attack. It is particularly effective in lubricating electric contacts, especially noble metals. The oil also is used as a very-high-temperature bearing lubricant, in critical heat-transfer systems and in cases where radiation resistance is important.

Silicate Esters

Silicate esters are made from silica sand and phenols or polyols by Chevron, Monsanto, Olin, and Union Carbide. The principal applications are in fire-resistant hydraulic fluids, dielectric coolants, and heat-transfer mediums. Excess moisture causes hydrolysis of silicate esters.

New Developmental Synthetic Functional Fluids

New developmental synthetic functional fluids include perfluorinated triazine and other *s*-triazines, perfluoro alkylated ethers and substituted polyphenyl ethers, as well as heterocyclic derivatives.

SELECTING THE RIGHT SYNLUBE

Some original equipment makers (OEMs) and operating agencies specify certain synthetic lubricants. In other instances, the OEMs or agencies provide a list of approved synlubes or a list of the types and properties of synlubes that have been found satisfactory. In an increasing number of situations, use of the proper synlube will markedly improve operations and result in lower overall costs. Two examples (Tables 2-4 and 2-5) illustrate simple cost-benefits analyses. They show that synlubes, when properly employed in certain applications, are well worth their higher prices.

TABLE 2-4 Cost Benefits of Synthetic Lubricant: Reciprocating Sodium Hydroxide Metering Pump, Teflon-Impregnated Asbestos Packing

Per year	Packing only (no lubrication)	Packing with fluorosilicone grease
Packing changes	10	1
Labor costs (for changes)	$500*	$ 50*
Packing costs	50	5
Grease cost	——	90
Total maintenance costs	$550*	$145*

Benefits: • $405* annual savings • Less production downtime (worth $$$)

*In 1980 dollars.

TABLE 2-5 Annual Costs (in 1980 Dollars) for 20 Reactor Drives

	Mineral oil	PAO synlube
Lube price, $ -gal	2	10
Lube, changes	Monthly	4 Months*
Requirements, gal	600	200
Cost,	1,200	2,000
Labor, 1 h/change/drive at $30/h;	7,200	2,400
Average repair costs		*
Labor at 16 h/repair,	2,880	960
Materials at $100/repair,	1,200	400
Lost productivity	?	——
Reduced equipment service	?	——
Total tangible costs,	12,480	5,760
Savings,	——	6,720

*Lube change and failure rate is ⅓ of that with mineral oil.

The high costs of new equipment, parts, and labor necessitate keeping machines running better and longer. Such increased efficiency and extended life can be obtained by the judicious use of synthetic lubricants. As costs soar and downtime becomes more expensive, the economics will swing more and more in favor of longer-life, maintenance-saving synlubes—even though their price may be higher than mineral oil's. Where safety is the overriding consideration, cost of the safer synlube will of course remain secondary.

There is no perfect synlube for all applications. Tradeoffs always exist. For some uses mixed-base synlubes are best. Each synthetic has its own individual characteristics. The most appropriate synlube must be carefully selected on the basis of characteristics most suited for the specific conditions and equipment. Endorsements and recommendations of equipment builders and of synlube suppliers should be considered seriously. Only proven products from reliable suppliers should be used. It is wise to review experiences of others in similar situations before trying a new synlube. Synlube papers published by ASLE, ASME, NLGI, and other technical associations usually contain worthwhile information. Brochures and bulletins of synlube suppliers, of course, present much detailed data. More and more independent lubricant suppliers are expanding their lines of synlubes, and are increasingly able to provide high-performance products supported by competent technical service.

Over the longer term, as petroleum supplies continue to decline, more and more lubricants will be synthetic. Starting materials for synlubes may increasingly come from shale oil and coal; also from vegetable and animal oils and other renewable sources, such as biomass. Someday, in the not too distant future, civilization may well run primarily on synthetic, long-lasting, high-performance lubricants.

BIBLIOGRAPHY

Campen, M.: "PAO-based Lubricants in Power Transmission Applications," *6th Annual Conference on Power Transmission,* Chicago, November 1979.

Green, R. L., and Langenfeld, F. L.: "Lubricants for Traction Drives," *Machine Design,* May 2, 1974.

Gunderson, R. W., and A. W. Hart: *Synthetic Lubricants,* Reinhold, New York, 1962.

Manley, L. W.: "New Developments in Synthetic Lubricants," *World Petroleum Congress,* Bucharest, September 1979.

Mueller, E. R.: "Polyalkylene Glycol Fluids and Lubricants," *ASLE Seminar on Synthetic Lubes,* Pittsburgh, February 10, 1977.

O'Connor, J. J., and Boyd, J.: *Standard Handbook of Lubrication Engineering,* McGraw-Hill, New York, 1968, chap. 11, "Synthetic Liquid Lubricants."

Reid, H. F.: *A New Report on Ester Lubricants,* Hatco Chemical Co., Fords, N.J., June 1977.

Smith, R. E.: "Silicone Lubricants for the Chemical Processing Industry," *30th Annual Meeting ASLE,* Atlanta, May 5–8, 1975.

Wolfe, G. F., Cohen, M., and Dimitroff, V. T.: "Ten Years Experience with Fire Resistant Fluids in Steam Turbine Electrohydraulic Controls," *24th Annual ASLE Meeting,* Philadelphia, May 5–9, 1969.

Suppliers' technical bulletins and brochures.

SOURCES OF ADDITIONAL INFORMATION

BASF Wyandotte Corporation, Parsippany, N.J.
Bray Oil Company, Irvine, Calif.
Chevron USA, San Francisco, Calif.
Conoco Oil Company, Houston, Tex.
Dow Corning Corporation, Midland, Mich.
Emery Industries Incorporated, Cincinnati, Ohio.
Ethyl/Edwin Cooper, St. Louis, Mo.
Exxon USA, Houston, Tex.
FMC Corporation, Chicago, Ill.

General Electric Company, Waterford, N.Y.
Gulf Oil Company, Houston, Tex.
Gulf Oil Chemicals Company, Houston, Tex.
Hatco Chemical Corporation, Fords, N.J.
Hercules, Incorporated, Wilmington, Del.
Humko Sheffield Chemical Company, Memphis, Tenn.
Mobil Corporation,
Monsanto Company, St. Louis, Mo.
Olin Corporation, Stamford, Conn.
Stauffer Chemical, Westport, Conn.
Tenneco Chemicals, Incorporated, Piscataway, N.J.
Union Carbide Corporation, New York, N.Y.
Uniroyal Chemical, Naugatuck, Conn.
Witco, Fairfax, Va.

chapter 17-3

Solid Lubricants

by

Rodger C. Dishington
Senior Engineer

Walter D. Janssens
Director, Solid Research

Jack R. Waite
Manager, Engineering Field Services
Imperial Oil and Grease Co.
Los Angeles, California

INTRODUCTION

Solid lubricants are selected solid materials with friction-reducing, low-shear properties. Typical of these materials are graphite, MoS_2 (molybdenum disulfide), PTFE (polytetrafluoroethylene), and mica. They may be used as powders in their natural form, or dispersed in fluids (oils, water) and greases; or they may be added to binders, as pigments to paint, and used as *dry-film* bonded lubricants.

USES

There are many exotic uses for solid lubricants in aerospace, electronics, and instrumentation industries, and machine designers should consider their use wherever conventional lubrication is impracticable. However, solid lubricants play an important role in the practice of industrial "preventive maintenance" as an antiwear, load-bearing component in plant lubricants. Their major benefits are: longer service life of machinery, extended lubrication intervals, and the ability to function in environments hostile to conventional lubricants. They are also used extensively by equipment manufacturers in "lubricated-for-life" components; they are the basis for the 30,000-mile chassis lubrication interval in automobiles.

All machined surfaces, even the ground and polished ones, can be defined in terms of surface roughness, or the dimensions of microscopic high spots called *asperities*. Between the rubbing surfaces of bearings, the lowest coefficient of friction is achieved by maintaining separation of these surfaces by a full *fluid film* (hydrodynamic lubrication). As the oil-film thickness decreases with higher loads, shock loading, or diminishing surface speeds, asperities on opposing surfaces come into contact in *mixed-film* lubrication with a resultant increase in friction.

With increasing severity, a much higher coefficient of friction is experienced during *boundary* lubrication when asperity contact is sufficient to cause microscopic welding and shearing between the two rubbing surfaces. Under the pressures and heat of contact, the solid lubricants may or may not react with the metal substrates, but they do interpose themselves as a barrier coating in which shearing may take place more easily and with less friction than if the bearing metals were shearing in pure contact. As the wear phenomenon is related to the shearing of contacting asperities, wear also may be reduced by the interposing of a solid lubricant barrier.

FORMS AND APPLICATIONS

The useful forms in which solid lubricants are prepared include powders, pastes (compounds), bonded coatings, greases, and dispersions. Powders, burnished into rubbing metal surfaces, have a limited wear endurance and offer virtually no protection from the atmosphere. Pastes are heavy concentrates (up to 65 percent) of solid lubricants in a fluid or grease base; they offer longer, but limited, service life and may include corrosion protection. Bonded coatings may be applied by spray or brush (like paint), by plasma techniques, or by an impingement process if tolerances are critical.

Greases used in antifriction bearings usually contain less than 3 to 5 percent solids, and their particle size is closely monitored, especially for precision and high-speed bearings. Heavy-duty and special-purpose greases may contain up to 25 percent solids. Finished lubricating oils may contain dispersed solid lubricants for improved load-carrying and antiwear characteristics; concentrated dispersions in oil are available as lubricating-oil additives.

The carrier fluids in pastes, greases, and dispersions are not always petroleum-base, but may be water-base or any number of synthetic fluids including glycols, esters, synthetic hydrocarbons, or silicones. Most solid lubricants can tolerate hazardous environments better than most carrier fluids can; thus the carrier is usually selected on the basis of the expected environment. For example: some solids, primarily graphite and MoS_2, are excellent oven-bearing and high-temperature chain lubricants in their powdered form; certain petroleum, glycol, and ester fluids, selected for their favorable volatility are used as media to carry the solids into the rubbing areas to reduce friction and wear even after the fluid has evaporated.

Rubbing metal surfaces are most subject to high friction and wear when they are new or resurfaced, and their ultimate condition and service life depend upon whether they succeed or fail in "running in," or seating properly. Early destruction of contacting asperities or their orderly distortion will determine the true load-bearing area. The less destructive the run-in, the lower will be the magnitude of instantaneous loading. Thus among the most common industrial applications are *wear-in, press-fit,* and *threaded connections.* An expansion on industrial uses appears in Table 3-1.

In addition to the aforementioned materials, hundreds of solid lubricants have been described in technical literature. Those described include metallic oxides and sulfides, soft metals, calcium fluoride, zinc pyrophosphate, talc, and vermiculite. The low-shear characteristics of solid lubricants may be the result of crystal structure, interstitial mat-

TABLE 3-1 Some Industrial Uses of Solid Lubricants

Industrial application	Product form
Wear-in—protection against galling and seizure of newly machined surfaces at start-up and early running-in; examples: gears, slides and ways, cams, valve sleeves, splines, bearings,	Preassembly: powder, paste, bonded coating Initial fill: finished oil dispersion or concentrate
Press-fit—to reduce pressure required, prevent galling, seizure, and possible misalignment; at times clearances are negative; examples: antifriction bearings, splines, keyed shafts, sleeves	Preassembly: powder, paste
Threaded connections, fasteners—to reduce torque loss due to friction, galling, and seizure; promotes optimum uniformity in the tension of assembly bolts and facilitates nondestructive disassembly	Preassembly: pastes (thread compounds), bonded coatings
Life-of-part prelubrication—where maintenance lubrication is impossible, or improbable; examples: enclosed mechanisms, hinges, locks, linkages, instruments, appliances	Preassembly: powder, paste, bonded coating
Lubrication of machine in operation—applied by all conventional systems: drop-feed, circulation, reservoir, air mist; grease gun, automatic grease dispensing systems; for heavy-duty installations and/or extending lubrication intervals and machine service life	Ongoing plant lubrication: finished lubricating oils and greases containing dispersions of solid lubricants
Antiwear, load-bearing additive in lubricants—to fortify conventional lubricating oils and greases used in plant lubrication, initial fills for wear-in, and in units experiencing progressive wear rates	Pastes added to greases, and concentrated dispersions added to oils
Additive to metalworking fluids—to reduce friction, lengthen life, and reduce metal pickup on punching and forming dies and all types of cutting tools	Concentrated dispersions added to metalworking fluids and pastes added to compounds
Reduce fretting (friction oxidation)—to protect against fretting corrosion of metal surfaces under static loads (vibrating) as on bearings, bearing housings, splines, and various press-fit components	Preassembly: pastes and bonded coatings
Antiwear, load-bearing additive in self-lubricated components—to extend life of parts and reduce friction and distortion when blended into rubber, plastics, elastomers, and sintered metals	Powder is added as a component in the raw material for the fabrication of bushings, O rings, etc.
Dry-film lubrication—for use in dusty atmospheres to minimize the adherence of abrasive particles to metal surfaces of open gears, bearings, cams, and slides	Bonded coatings, pastes, greases
High-temperature applications—oven conveyor chains and bearings, kiln-car wheelbearings, mechanical devices operating at temperatures above the capability of fluids	Dispersions in fluids and greases designed to volatilize leaving mainly the solid lubes; some bonded coatings
Equipment exposed to destructive environments—acids, alkali, solvents, detergent, steam, etc. For load-carrying, antiwear characteristics.	Greases, constructed from materials also resistant to the environment

ter, bond strength, or chemical interaction of the surface and the solid. The effectiveness of solid lubricants stems from the almost fail-proof film which they form on moving surfaces, usually beyond the yield strength of the metal asperities. The mechanism of solid lubrication is not dependent on any single property of the lubricant; it is an interdependence of the surface, the solid lubricant composition, the geometry of the particles and the metal surface, and the nature of the processes that occur on or near the bearing surfaces.

CHARACTERISTICS

Solid lubricants composed of two or more materials often combine their most favorable characteristics to provide synergistic lubricating properties that are superior to any single lubricating solid. A brief review of the characteristics of the most commonly used solids can give insight into the lubrication mechanism of solids alone or in combination. The most widely used inorganic and metal base materials are graphite and MoS_2; the most widely used film-forming plastic is PTFE.

Inorganic Lubricants

Graphite and MoS_2 differ in composition, general properties, and type of chemical bonding, but they do have in common their layer-lattice structure. Their characteristic crystal structures are layers of sheets within which the atoms are tightly packed and strongly bonded; but these sheets (laminae) are separated by relatively large distances and held together by weak residual forces. In graphite, a crystalline form of carbon, the distances of atoms within the sheets is 1.4×10^{-8} cm, but between sheets it is as much as 3.4×10^{-8} cm. When under shear, there are strong forces within the graphite sheets, but the forces holding the sheets together are much weaker, allowing them to slip over one another.

A theoretical explanation for the lubricity of molybdenum disulfide is similarly found in its molecular structure. Each lamina of this compound is composed of a layer of molybdenum atoms with a layer of sulfur atoms on each side. The sulfur and molybdenum layers bond tightly but the adjacent laminae interface at their layers of sulfur atoms which form weak bonds between the laminae. The weak sulfur bonds between the laminae form the slippage planes of low shear resistance as between the carbon layers in graphite. Both graphite and MoS_2 display an affinity for metal substrates, and under high loads both have been found to alloy with ferrous metal, forming even greater bonds at the surface.

Under heavy forces (loads) perpendicular to bearing surfaces, these laminae are compressed and oriented parallel to the bearing surfaces and have the strength to resist rupture. The low friction reflects the low resistance of the laminae sliding on one another. Cohesive forces within graphite and MoS_2 are sufficient to allow for self-healing of interruptions in the solid lubricant film.

Plastics

Of the plastic materials used as solid lubricants, PTFE has the greatest affinity for metal surfaces, the lowest internal shear resistance, and the greatest ability to be self-healing. As no plastics have a tight molecular structure or laminar nature, they are unable to take the heavy loads carried by graphite or MoS_2. At loads up to 25,000 lb/in^2 (1760 kg/cm^2) however, PTFE has the lowest coefficient of friction of all solid lubricants.

Common Characteristics

A primary characteristic of these three leading solid lubricants is their ability to lubricate as dry powders. Almost all other solid lubricants require some addition of fluid or grease to lubricate for any period without noise and increasing friction. Nor do other solids have the excellent film-forming and self-healing characteristics of these three which often form the base to which other solids are added. Some of the following are the characteristics which are sought by the use or addition of other solids: particle geometry suited to antiseize or wear-in of coarse surfaces, improved protection against fretting wear, improved bulk-film load-carrying capability, and the thickening of some fluids for greases. Some white solid lubricants, e.g., zinc sulfide, are used alone where a white or colorless lubricant is required for appearance. Solids, even though not laminar, still become interposed between rubbing surfaces when dispersed in fluids or greases.

It is again the superior properties of graphite, MoS_2, and PTFE that allow for their use in very adverse environments. In high-temperature operations graphite is commonly used to 800°F (425°C) and intermittently to 1200°F (650°C). In vacuum at high temperature, or in high vacuum, graphite loses a water-vapor layer which is found naturally adsorbed at the surfaces of the laminae and is felt to be largely responsible for the weak attraction or bond between them. Loss of the water vapor results in a significant increase in friction and a tendency toward abrasion. On the other hand, MoS_2 has a lower coefficient of friction in the absence of water vapor and it can withstand very high vacuum, but it is not as resistant to oxidation as is graphite. MoS_2 performs well up to 650°F (345°C). Above this temperature, in air, its oxidation rate increases until at 750°F (400°C) it oxidizes quite rapidly. PTFE will perform well in vacuum and at temperatures up to 600°F (315°C) at which it begins to decompose.

Solvents and chemicals provide some of the environments most destructive to conventional lubricants. Again, the leaders are the most impervious of all solid lubricants, being virtually unaffected by direct contact. Graphite and PTFE are nearly indestructable by any chemicals, and MoS_2 is relatively unaffected except by hot strong oxidizing acids like aqua regia and chromic acid. At times, the removal of MoS_2 from metal surfaces requires strong caustic cleaners and abrasive action. Special solvent-, chemical-, and detergent-resistant greases containing solid lubricants require also the careful selection of fluid carriers and thickeners.

SUMMARY

There is a need for a solid lubricant wherever a hydrodynamic oil film cannot be sustained. Whether the film becomes insufficient because of high pressure or shock loading, elevated temperatures, or exposure to solvents and chemicals or is depleted by the extension of lubrication cycles, solid lubricants can extend service life by inhibiting asperity welding (wear), galling, and the ultimate seizure of bearing metals. All bearings are in boundary lubricating conditions during stop-start operations since full fluid-film lubrication requires motion to propagate.

BIBLIOGRAPHY

Acheson, E. G.: U.S. Patent 813,426, February 5, 1907.

ASLE Proceedings, International Conference on Solid Lubrication, Chicago, 1971.

ASLE Proceedings, 2nd International Conference on Solid Lubrication, Chicago, 1978.

Bowden, F. P., and D. Tabor: *The Friction and Lubrication of Solids,* Oxford, New York, 1950.

Braithwaite, E. R.: *Solid Lubricants and Surfaces,* Pergamon, New York, 1964.

Braithwaite, E. R.: *Lubrication and Lubricants,* Elsevier, New York, 1967.

Campbell, M. E.: *Solid Lubricants, A Survey,* NASA SP-5059 (01), 1972.

Deoine, N. J., E. R. Tourson, J. P. Cerini, and R. J. McCartney: "Solids and Solid Lubrication," *Lubrication Engineering,* January 1965.

E. Hall: U.S. Patent 2,700,623, April 26, 1950, and U.S. Patent 2,703,768, April 21, 1954.

McCabe, J. T.: "Molybdenum Disulfide—Its Role in Lubrication," reprint of a paper presented at International Industrial Lubrication Exhibition, New Hall, Westminster, London, March 8–11, 1965.

Neale, M. J. (ed.): *Tribology Handbook,* Wiley, New York, 1973.
Notes on Solid Lubricants, Seminar Proceedings at Rensselaur Polytechnic Institute, 1966.
Solid Lubricants, NTIS/PS 75/715, 78/0816, 78/0817, 78/0818, 78/0819, NTIS Bulletins, U.S. Dept. of Commerce, Washington, D.C.
Stock, A. J.: "Graphite, Molybdenum Disulfide and PTFE—A Comparison" *Lubrication Engineering,* August 1963.
Ubbelohde, Q. R., and F. A. Lewis: *Graphite and its Crystal Compounds,* Oxford/Clarendon, 1960.
Waite, J. R., and W. D. Janssens: "Use of Inorganic Solids in Lubricating Oils," Presented at ASLE Annual Meeting, Cleveland, Ohio, May 6–9, 1968.
Youse, E. L., NLGI Spokesperson: *Characteristics and Selection of Graphite as a Lubricant,* January 1962, Kansas City, Mo.

chapter 17-4

Lubrication Systems

by

Anne Bernhardt
Staff Engineer
Gulf Research and Development Company
Petroleum Products Department
Pittsburgh, Pennsylvania

INTRODUCTION

An effective lubrication system is any system or device which dispenses the correct lubricant, at the correct point, in the correct amount, at the correct time. Systems may vary from hand oiling to a complicated centralized system. Escalating costs and the development of high-speed precision machinery are necessitating changes in plant lubrication practices.

ALL-LOST SYSTEMS

All-lost or once-through lubrication (Fig. 4-1) systems are those in which the lubricant is used only once. Manual all-lost systems such as hand oiling, individual grease fittings, wick lubricators, oil cups, and drop-feed oilers are rapidly becoming a part of history. These systems are inexpensive to install but require close attention on the part of the operator to ensure that each point is relubricated on a regular basis for adequate lubrication.

The most common all-lost systems in use today are automatic. Mist systems and mechanical force-feed lubricators are common examples. The reason for their popularity

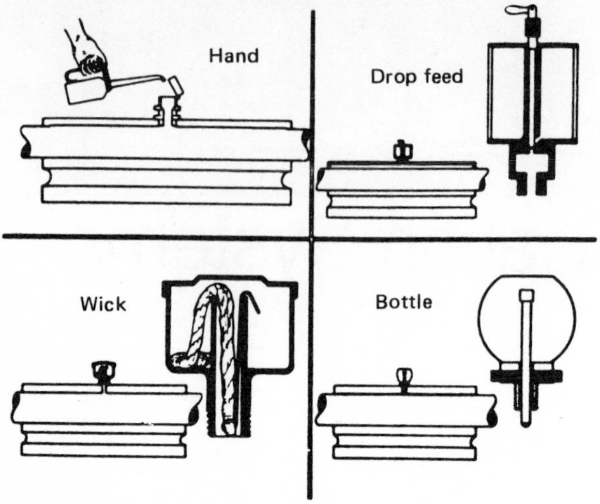

Figure 4-1 Once-through oiling *(U.S. Steel Corporation, reprinted with permission from "The Lubrication Engineers Manual," Copyright 1971).*

is their ability to lubricate more than one point on a machine from a central reservoir and to automatically dispense the lubricant in metered amounts at the point of application.

Mist systems rely on compressed air to atomize the oil into fine droplets and deliver it through pipes to the point of application. Frequently used to lubricate bearings and gears, the air passing over the part assists the oil in carrying off heat and preventing the entry of dirt.

Mechanical force-feed lubricators were originally designed to deliver oil to cylinders but are no longer limited to that application. They consist of small pumping units mounted on a common shaft which take oil from a reservoir through pipes to the point of application. The driving mechanism may be an electric motor or some moving part of the machine being lubricated. Each pumping unit may be set to deliver the precise amount of oil which is required at a particular application point. Both the mist system and the mechanical force-feed lubricator require little maintenance beyond ensuring that no lines are plugged.

OIL-RESERVOIR SYSTEMS

Unlike all-lost systems, oil-reservoir or self-contained systems reuse the same oil over and over again. These methods depend on a common housing containing the oil and the parts to be lubricated (Fig. 4-2).

Gears and cylinders lubricated by these methods usually depend on the splashing action of one or more moving parts dipping into the pool of oil at the bottom of the housing.

Bearings lubricated by self-contained systems may be splash-lubricated or rely on a ring, chain, or collar to dip into the oil and carry the oil to the top of the journal. Collar bearings are used at higher speeds than rings or chains since the latter will tend to slip excessively at high speeds, precluding adequate lubrication.

To ensure adequate lubrication it is important that the oil be maintained at the proper level. Insufficient oil could result in a lack of lubrication, while overfilling can cause foaming and temperature buildups due to excessive churning.

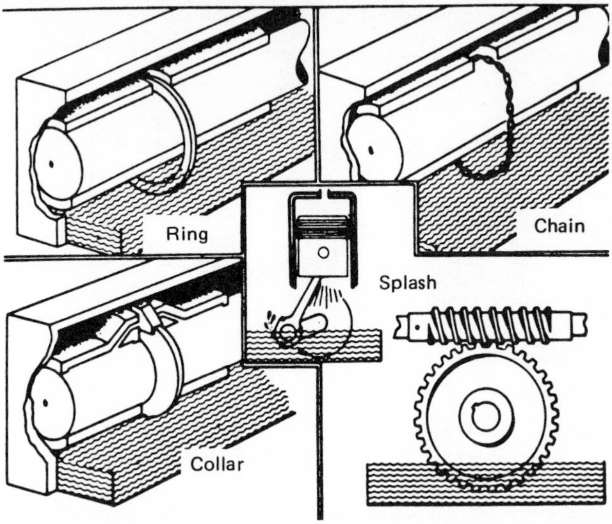

Figure 4-2 Oil reservoirs *(U.S. Steel Corporation, Reprinted with permission from "The Lubrication Engineers Manual," Copyright 1971).*

CENTRALIZED SYSTEMS

Like oil-reservoir systems, centralized systems use the oil over and over again. They can range from a simple reservoir, pump, and return-line setup to complex systems with electronic controls, servo valves, heat exchangers, and filters.

Depending on the complexity of the system, costs vary greatly. The cost-effectiveness of centralized systems depends heavily on the length of time the fluid can remain in circulation before it needs to be changed.

To ensure maximum fluid life, oil-reservoir temperatures should be controlled as well as the amount of contaminants in the oil. If petroleum oils are used, reservoir temperatures should be maintained between 110 and 130°F (43 and 54°C) for optimum fluid life. Synthetics can be operated at somewhat higher temperatures. Reservoir temperatures can be controlled through the use of heat exchangers and proper reservoir design.

The oil reservoir should be large enough to allow the oil to rest for a minimum of 15 min before being recirculated. It should be baffled to ensure that returning oil is not immediately pumped back into circulation. This rest time in the reservoir allows the oil to drop out contaminants and dissipate the heat it has picked up while in circulation.

The oil-reservoir's fluid level is also very important to the trouble-free operation of circulating systems. If the suction line is not completely submerged in oil at all times, pump cavitation could result. Also, it is important that the return line be submerged in the oil to reduce air entrainment and thus prevent foaming problems which could occur if the returning oil is allowed to splash into the reservoir. Figure 4-3 is an example of a properly designed reservoir.

Monitoring and Warning Devices

A properly designed centralized or automatic lubrication system can provide effective equipment lubrication with very little human intervention. The main drawback is the risk of catastrophic equipment failure which may result when a malfunction occurs. To prevent this, monitoring and warning devices are installed to alert operators to lubrication malfunctions. These may be bells, sirens, flashing lights, automatic equipment shut-

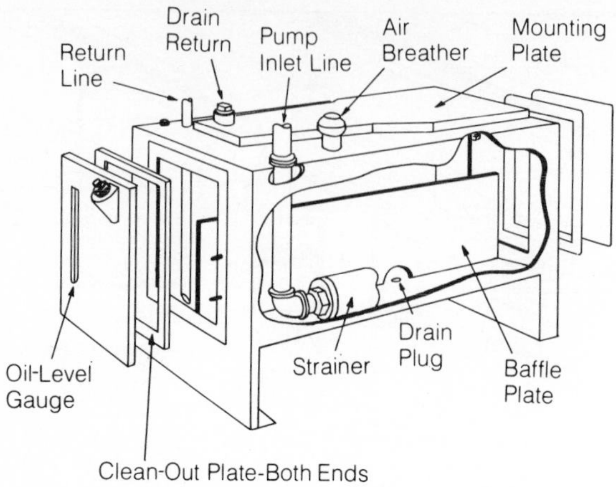

Figure 4-3 Typical circulating system reservoir.

down, or telltale indicators. All of these can be effective if maintained in proper operating condition.

CRITERIA FOR SELECTING A LUBRICATION SYSTEM OR DEVICE

A lubrication system should be selected with one purpose in mind: *to get the proper quantity of the correct lubricant where and when it is needed.* Before purchasing or installing a new lubrication system it is recommended that the application be studied to aid in the selection of the most economical and effective system for a particular application. Factors to be considered include:

Equipment Considerations

- The components to be lubricated
- The lubricant to be applied
- The lubrication-point accessibility
- The number of lubrication points the system is expected to service

Operation Condition Considerations

- Equipment speeds
- Operating temperatures
- Expected relubrication intervals

Economic and Plant Practice Considerations

- Plant's past experience with various types of lubrication systems
- Available capital
- Staff available to maintain and monitor systems
- Downtime costs of the equipment

Oil companies, equipment builders, and lubricant-dispensing-equipment suppliers are available to assist in the design and selection of lubrication systems. A list of some of the better-known manufacturers of lubricant-dispensing equipment follows.

LUBRICANT-DISPENSING EQUIPMENT MANUFACTURERS

Alemite Division, Stewart-Warner Corp., Chicago, Ill.
Bijur Lubricating Corp., Rochelle Park, N.J.
Farval Division, Fluid Control Division, Eaton Corp., Cleveland, Ohio
Lincoln St. Louis, A Division of McNeil Corp., St. Louis, Mo.
Lubriquip, Cleveland, Ohio
Madison-Kipp Corp., Madison, Wis.
C. A. Norgren Co., Littleton, Colo.
Oil-Rite Corp., Manitowoc, Wis.

BIBLIOGRAPHY

Bailey, Charles A., and Joseph S. Aarons (eds.): *The Lubrication Engineers Manual,* United States Steel Corp., Pittsburgh, 1971.
Brewer, Allen F.: *Basic Lubrication Practice,* Reinhold, New York, 1955.
Brewer, Allen F.: *Effective Lubrication,* Robert E. Krieger, Huntington, N.Y., 1974.

section 18

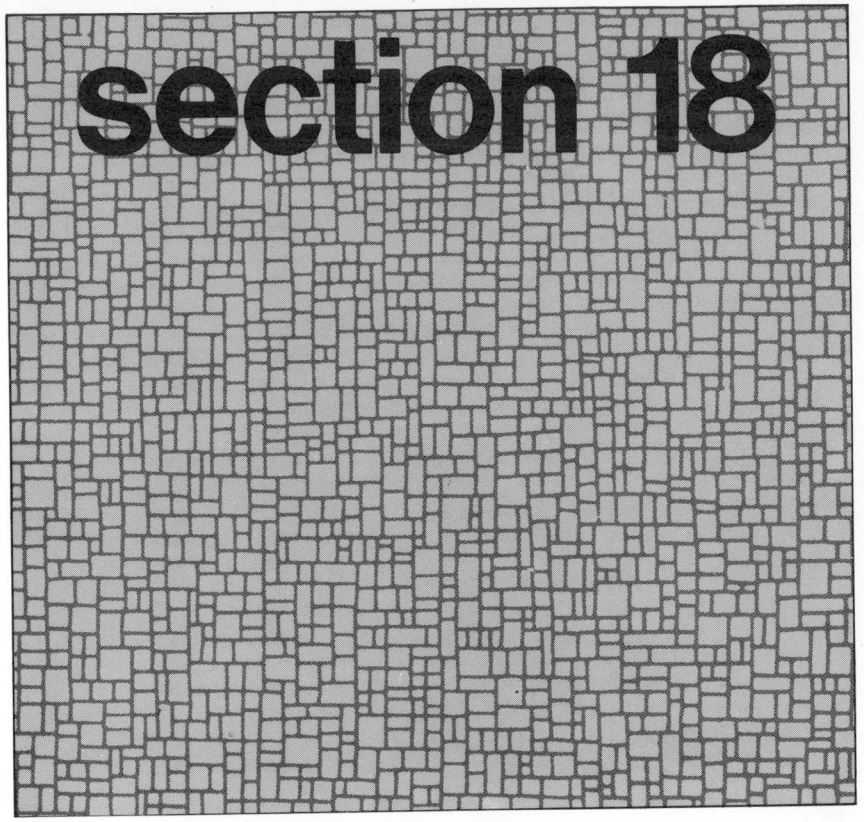

Corrosion and Deterioration of Materials

chapter 18-1

Causes and Control of Deterioration

by
Philip H. Maslow, P.E., FCSI
Consultant
Chemical Materials for Construction
Brooklyn, NY

INTRODUCTION

All materials deteriorate. External building materials and assemblies are no exception to this law. Scientific observation shows that some materials deteriorate at a faster rate than others, depending on a set of controlling conditions, or causal factors. Often, these conditions can be modified to alter the rate of deterioration to a preferred or acceptable level.

The *total performance* of a building refers to the ability of both external and internal materials to fulfill their intended function over the useful life of the building. The performance of the external materials has special significance here in terms of maintenance costs and replacement costs. The differences between external and internal performance are tied into the differences between the effect of wear caused by use and that of wear by the environment. Experience with new building types has shown the need for considerable caution because one of the problems in the use of new materials is their breakdown or failure, particularly on the exterior.

Deterioration is almost the antonym of durability. Durability is essentially a *state* whereas deterioration is a *rate*. Deterioration is the result of corrosion, chemical action, atmospheric pollution, structural movement, and user wear and tear. The environment comprises a set of elements such as air-temperature, rainfall rate, wind speed, pollution level, and smoke concentration. These elements have special significance because they are the ones over which the least control can be exercised and have the widest range of effects on exterior building materials.

Standard test results indicate that the type of atmospheric pollutants normally found in urban areas accelerate the breakdown of concrete, brick, mortar, paints, and various metals and plastics. Some of the pollutants most commonly encountered in urban and industrial areas are carbon dioxide, carbon monoxide, sulfur dioxide, sulfur trioxide, and nitrous oxide. These pollutants react with moisture in the atmosphere to form carbonic, sulfurous, sulfuric, and nitric acids, respectivly. And these corrosive acids act upon all building materials.

Because of new zoning regulations and current design trends, many new commercial buildings, and even industrial buildings, have plazas, walks, and landscaped areas containing reflecting pools, fountains, and, in some cases, huge sculptures. These areas can be paved with materials varying from concrete, paving brick, bituminous concrete shapes, and terrazzo to natural minerals such as slate, bluestone, granite, and Travertine marble.

For walls there is a choice of brick, poured-in-place concrete, precast concrete, cast stone, ceramic veneers, marble, limestone, granite, etc. In addition, there are window walls combined with masonry and curtain walls, which are all glass with metal frames.

Plastics have entered the construction field in recent years to such an extent that millions of pounds of plastics are now used annually in a variety of installations. Dozens of test methods for evaluating plastics indicate how extensively they are utilized in construction: these include tests for abrasion resistance, bearing strength, brittleness, crazing resistance, deformation under load, flame resistance, flexural strength, indentation hardness, mildew resistance, shatterproofness, shear strength, tensile strength, thermal expansion, and water absorption.

The history of metals in architectural applications extends over hundreds of years: similarly, large-scale use of metals on building surfaces has a long tradition. Today, the demands of technology have resulted in the development and production of many alloys covering an extensive range of properties. Fundamental to the deterioration of metals is the phenomenon of corrosion. Ferrous metals are particularly subject to corrosion. Copper, lead, and zinc corrode, initially, on exposure to the atmosphere, but the product of their corrosion becomes a protective layer which serves to isolate the metal from further attack. Stainless steel and aluminum become passive upon exposure to the atmosphere. These metals acquire a surface film which isolates the metal and effectively protects it from corrosive attack. To protect aluminum even further, it is subject to modification by alloying with various metals and by anodizing. This is a process in which the thickness of the surface film is increased by electrolytic oxidation. By including various organic dyes in the electrolytic bath, the aluminum may be anodized in a range of colors. Coatings to protect carbon steel and other sensitive metals have also been developed.

Water is one of the prime elements in the composition of the Earth. Although it is basic to the sustenance of life, it is also a highly destructive material and has a great effect on the durability of the exterior of a building. The sources of water problems and the resultant damage to a structure extend from the roof, through the fabric of the building, to the foundation below grade.

The causes of water infiltration at the roof area can be traced to several sources. The parapet is a common source of water infiltration because of the lack of proper flashing details, poor mortar joints, poor coping stones, and poor sealants. Roof expansion joints, if not properly designed or installed, can lead to splitting of the roofing membrane and to subsequent water penetration. The detailing and construction of the roofing membrane, beginning with the vapor barrier, the insulation, the plies, the flood coat, and the gravel ballast, all have a bearing on the durability of the entire roofing system.

The envelope of a building is also subject to penetration by water. Rain does not affect the roof area exclusively; it can wreak havoc on all sides of a building, especially, when there is a wind factor behind it. The sources of infiltration vary with the kind of wall fabric. There are absorption factors to consider, and these factors vary with the construction material. Joints and sealants, flashing and flashing details, and weep systems—all have a bearing on waterproofing and/or water infiltration.

The effects of water damage to the exterior are both unsightly and damaging structurally. Masonry can spall, especially in a freeze-thaw cycle or in a continuous saturation process. Rusting of structural steel members can result in serious deterioration of beams, relieving angles, and lintels. Of course, the damage from water entering the interior affects plaster, paint, drywall, ceilings, floors, carpets, curtains, and furniture.

Modern curtain walls and window walls are equally vulnerable to water infiltration. Water may come through stack or expansion joints in column covers, mullions, or tracks. Water may enter because of improper metal-to-metal joinery, improper joint design, poor sealant details, poor shop seals and/or field seals, and poor gutter or weep systems.

Glazing systems in every kind of window detail, including skylights, are also subject to water infiltration. One source of infiltration can be a faulty glazing system. Poor workmanship, inadequate materials, and poor design of the glass surrounds are other major reasons for water infiltration.

This chapter explains the causes of deterioration of concrete, masonry, metals, wood, and plastic and outlines methods of maintenance, repair, and rehabilitation.

CONCRETE*

CAUSES FOR DETERIORATION OF CONCRETE

The visual symptoms of concrete deterioration are cracking, spalling, and disintegration. Each is obvious and may occur individually or in any combination. Most of the basic causes for the deterioration of concrete are noted in the following text.

Inadequate Design Details

Deterioration is often caused by design deficiencies such as nonworking joints between precast members; poorly sealed joints; inadequate drainage at foundations, at horizontal surfaces such as roofs and tops of walls and sills, and at inadequately placed or incomplete weep holes; unanticipated shear stresses in piers, columns, or abutments; incompatibility of materials (such as concrete in contact with aluminum or concrete containing calcium chloride in contact with reinforcing steel); neglect of the cold flow factor in the deformation of concrete under stress; and inadequate provision for expansion joints, control joints, and contraction joints.

Construction Operations

Problems during construction include settlement of the subgrade, movement of formwork that is inadequately designed or built, vibration of concrete during set, segregation of the concrete with heavy aggregates settling, and removal of forms before concrete has achieved minimal strength. Segregation can be minimized by proper water-cement ratios, which will produce concrete of proper consistency that is "placed" rather than "poured." Proper compaction by vibration is mandatory.

Drying Shrinkage

Contraction caused by loss of moisture from green concrete may be minimized by using proper water-cement ratios, proper vibration, adequate reinforcement, and effective curing membranes or other curing methods.

Temperature Stress

Variations in atmospheric and internal temperatures can cause enough stress in the concrete to cause cracking. Daily and seasonal variations in atmospheric temperature and the coefficient of thermal expansion of concrete must be considered when designing and locating joints.

Moisture Absorption

Premature spalling can be caused by moisture absorption. De-icing salts accelerate this phenomenon. The use of proper sealers and air-entrained concrete minimize this type of disintegration.

Corrosion of Reinforcing Steel

Corrosion of reinforcement may be attributed to the types of cements, sands, and aggregates used in the concrete mix, as well as to the type and amount of water. Under some conditions, admixtures, such as accelerators and water-reducing agents, can contribute to corrosion of reinforcing steel.

*This material adapted with permission from *Plant Engineering,* October 13,1977, pp. 215–222.

Chemical Reaction

Many types of chemicals are deleterious to concrete. Salts in soil, as well as bacteria, can affect concrete. Saltwater is highly corrosive to steel in concrete. Carbon dioxide and other products developed by the exhaust systems of internal-combustion engines add acids to the atmosphere that attack concrete.

Wear

Erosion or abrasion are common factors that affect concrete, particularly in traffic areas. Hydraulic structures, such as dams, are subject to cavitation erosion because of rapidly flowing water.

Impact

Floors in industrial plants are particularly subject to damage by heavy impact. Heavy reinforcement, combined with a concrete of high compressive strength are the usual methods for providing impact resistance.

Weathering

All concrete exposed to the atmosphere is subject to attack from the elements, from ultraviolet light, and from the myriad chemicals found in the atmosphere and in industrial plants.

DIAGNOSING DURABILITY PROBLEMS

Deterioration may be the result of corrosion, chemical action, atmospheric pollution, structural movement, or wear and tear. Diagnosing the causes of deterioration in a concrete structure may be a matter of eliminating possibilities. The process for identifying possible causes should include:

- Checking for errors in basic design. The knowledge and experience of an architect or a structural engineer will usually be required.
- Relating basic symptoms (cracking, spalling, and disintegration) to possible causes.
- Checking first for readily identifiable causes, such as corrosion of the reinforcing steel where the concrete has broken away to reveal the rusty bars and concrete that has been damaged by impact or shock, revealing reinforcing steel that is not yet corroded.
- Making a specific investigation of the concrete mix design, methods of construction, curing methods, time of year of installation, reputation of the contractor, etc.
- Analyzing clues. Each of the principal symptoms should be examined individually. If the concrete is disintegrating, check for chemical attack, erosion, or general weathering. If the concrete is spalling, check for variations in internal temperature, chemical and electrolytic corrosion of the reinforcing steel, and poor design details as well. If the concrete is cracking, check all the foregoing factors and for variations in atmospheric temperature and accidents during construction.
- Using the process of elimination. If initial construction or design details are responsible, redesign and corrective measures may be necessary. If external causes are responsible, protective measures must be exercised. Protection and maintenance are necessary to maintain the structure in a sound condition.

There are several accepted procedures and methods for repairing cracks in concrete structures, spalled concrete floors and pavements, and disintegrated concrete members. Table 1-1 outlines several types of materials for such repairs. The materials described can also be used to repair masonry members.

TABLE 1-1 Latex and Epoxy Adhesives and Bonding Agents for Concrete

	Latices			Epoxies				
						Epoxy-polyamide		
	Acrylic	Polyvinyl-acetate (nonre-emulsifiable)	Butadiene-styrene	Epoxy-polysulfide binder only	Binder with sand	Binder only	Binder with sand	Epoxy–coal tar binder only
Appearance	Milky white	Milky white	Milky white	Light straw to amber	Light straw to amber	Light straw to amber	Light straw to amber	Black
Solids content, %	45	55	48	100	100	95 to 100	95 to 100	100
Reference specifications	MIL-B-19235	MIL-B-19235	MIL-B-19235	MMM B-350A; AASHTO M-200	MMM G-650A; AASHTO M-200	AASHTO M-200	AASHTO M-200	AASHTO M-200
Chemical resistance								
Acids	Fair	Fair	Fair	Excellent	Excellent	Excellent	Excellent	Excellent
Alkalis	Very good	Very good	Very good	Excellent	Excellent	Excellent	Excellent	Excellent
Salts	Very good	Very good	Very good	Excellent	Excellent	Excellent	Excellent	Excellent
Solvents	Fair to good	Fair to good	Fair to good	Excellent	Excellent	Excellent	Excellent	Excellent
Compressive strength, lb/in^2 (2-in cubes; ASTM C 109)	3200 to 4100	3400 to 3600	3300 to 4000	8000 to 10,000	12,000 to 15,000	6000 to 8000	10,000 to 13,000	3000 to 4000
Tensile strength, lb/in^2 (1-in briquettes; ASTM C 190)	580 to 615	340 to 450	450 to 580	—	—	—	—	—
Tensile strength, lb/in^2 (ASTM D 638)	—	—	—	3000 to 3500	—	—	3500 to 4000	400 to 800
Tensile elongation, % (ASTM D 368)	—	—	—	2.5 to 15	—	6 to 25	—	35 to 40

Flexural strength, lb/in² (bar; ASTM C 348)	950 to 1400	1000 to 1250	1250 to 1650	—	—	—	—	—
Compressive double-shear strength, lb/in² (MMM G-650A)	—	—	—	900 to 1000	700 to 1000	400 to 500	500 to 650	300 to 400
Application notes	Suitable for indoor and outdoor exposure on concrete, steel, wood, thin section toppings; shotcrete, plaster bond within 45 to 60 min; not suitable for extreme chemical exposure			Suitable for filling cracks in concrete to bond both sides of crack into an integral member; preparation of epoxy mortars by adding sand	Suitable for bonding hardened concrete and other materials to hardened concrete; setting dowels; bonding plastic concrete to hardened concrete; bonding skid-resistant materials to hardened concrete	Suitable for filling cracks in concrete to bond both sides of crack into an integral member; preparation of epoxy mortars by adding sand	Suitable for bonding hardened concrete and other materials to hardened concrete; setting dowels; bonding plastic concrete to hardened concrete; bonding skid-resistant materials to hardened concrete	Suitable for preparation of epoxy mortars by adding sand; bonding skid-resistant materials to hardened concrete; membrane between asphalt and concrete
	Not for conditions of high hydrostatic head		Not for constant water immersion					
	Do not use with air entrainers	Can be used with accelerators, retarders, and water-reducing agents, but not with air entrainers	Not for use with air entrainers or accelerators	For maximum chemical and physical properties; highest cost; not for use on surfaces treated with rubber or resin curing membranes, dirty surfaces, weak concrete, or bituminous surfaces	For maximum chemical and physical properties; highest cost; not for use on surfaces treated with rubber or resin curing membranes, dirty surfaces, weak concrete, or bituminous surfaces	For maximum chemical and physical properties; not for use on surfaces treated with rubber or resin curing membranes, dirty surfaces, weak concrete, or bituminous surfaces		For resistance to grease, oil, gasoline, and traffic; use on bituminous concrete; lower cost applications of nonskid membranes; not to be used for bonding new wet concrete to old or where black color will be undesirable

REPAIRING CRACKS

It has often been said that concrete is destined to crack. The purpose of proper design and the objective of the design engineer are to minimize cracking without the expectation of eliminating it. When cracking does develop, it is important to determine the basic causes, as well as its extent, before deciding on the method of repair.

The surface of concrete will often exhibit shrinkage cracks—these are not necessarily defects, but they are aesthetically undesirable. They are usually attributed to the use of a high-slump concrete that contains excessive water. Cracking occurs when excess moisture leaves the concrete too soon—before the concrete has sufficient tensile strength. Such cracks can be minimized by using a sheet membrane or curing compounds. Although this type of cracking may not necessarily lead to problems, and it may often be ignored, applying a sealer based on a synthetic-rubber compound to protect the concrete from further damage is prudent.

Cracks can be described as active or dormant. An active crack will open and close with changes in temperature and with cyclic movement of the structure. Dormant cracks may not go through such movement, but they may still leak, collect dirt, interfere with traffic, etc. A structural crack may usually be attributed to inadequate structural design, insufficient strength (material composition), or poorly designed joints. If joints are not provided in concrete slabs, the concrete will create its own joints by cracking, and these cracks will continue to develop until the concrete member comes to equilibrium.

Active cracks may be sealed with an elastomeric sealant. Dormant cracks may be sealed with a fluid epoxy sealer that can be pumped into the crack or allowed to flow in by gravity. An epoxy sealer of 100 percent solids content will seal the crack without shrinkage and will join the crack faces to re-form a monolithic structure. This seal will be strong enough to resist further cracking. However, should stress still occur, cracking will take place somewhere else in the structure.

Horizontal cracks may be filled by simply pouring in a liquid epoxy sealer until it overflows, indicating that the crack has been filled.

Vertical cracks may be filled with a liquid epoxy sealer. First, the face of the crack is sealed with a fast setting epoxy compound, which is allowed to cure thoroughly. Small holes are then drilled into the crack through the epoxy seal, and nipples are installed in these holes and bonded with the same fast-setting epoxy. The low-viscosity epoxy sealer is then injected into the lowest nipple. Pressure is maintained until liquid begins to seep out of the next higher nipple. These two nipples are then plugged and the same procedure is resumed with the third nipple. This operation is continued until the entire crack is filled. After the sealer has cured, the nipples may be cut off flush with the concrete surface.

The tensile strength of a cracked concrete member may also be restored by stitching. U-shaped iron rods known as stitching dogs are inserted in drilled holes to transfer stress across the crack. Holes are drilled on both sides of a structural crack, far enough away not to cause additional breaks but not in parallel position, which would produce a plane of weakness. The legs of the stitching dogs are designed to be long enough to provide adequate pull strength. After the legs of the stitching dogs are inserted into the holes, the holes may be grouted with a nonshrinking grout. The crack itself should be sealed with an elastomeric or an epoxy sealer to prevent water from entering.

The elastomeric sealants should be based on polysulfide or urethane rubbers (preferably a two-component formulation in traffic grade) and comply with Federal Specification TT-S-00227E. The epoxy sealers may be epoxy-polysulfides (complying with Corps of Engineers Specification MMM B-350A) or epoxy-polyamides (complying with Specification M-200-65 of the American Association of State Highway and Transportation Officials). The epoxy compounds are described in Table 1-1.

In crack repair, certain procedures are detrimental to the successful repair of the crack.

- Filling cracks with new mortar will result in further cracking.

- Placing a topping over a crack to seal it, unless the topping is elastomeric, will inevitably result in the crack passing through the topping itself.
- Repairing a crack without relieving the restraints that caused it will cause cracking elsewhere.
- Repairing a crack with exposed reinforcing that has begun to corrode should not be done until the steel has been cleaned and protected with a rust-inhibitive paint.
- Burying a joint that has been repaired prevents frequent inspection to determine whether further failure has occurred.

REPAIRING SURFACES OF PAVEMENT AND SLABS

A spalled concrete surface is the beginning of continued disintegration of the concrete. If the cost of replacement is at all comparable with the cost of repair, replacement is recommended. Replacement is also recommended when changing the level of the final surface is impractical. Otherwise, the concrete can be resurfaced with new concrete, epoxy topping, latex mortar, or iron topping.

Resurface with New Concrete

When there is no problem in changing the level of the surface, it may be resurfaced with new concrete. It may be laid with or without bonding, although bonding is always recommended. The old surface should be prepared by removing all loose material and contaminants.

A simple bonding can be made by scrubbing a neat cement slurry (a mixture of straight portland cement and water) into the surface. The new concrete may then be placed with the expectation of a good bond.

Concrete placement should follow recommended practice. The concrete mix should have a water-cement ratio of 0.50 or less and a slump of 3 to 4 in. It is also recommended that reinforcement be embedded in the middle of the topping slab, rather than allowed to lie at the bottom of the slab.

Finishing procedures will depend on the required surface: use a wood float finish for a regular surface and steel troweling for a hard, smooth, sealed finish. The slab should be cured by covering it with sheet materials or liquid curing membranes. Concrete overlays of this type should be at least 2½ to 3 in thick.

A more positive bond may be achieved by using an epoxy bonding agent, either a filled epoxy-polysulfide or an epoxy-polyamide bonding agent (see Table 1-1). The epoxy bonding agent may be a proprietary formulation or one meeting a specification of a governmental agency. The bonding agent must be thoroughly mixed and applied by brush, broom, or spray. Do not apply more bonding agent than can be covered with new concrete while the bonding agent is still tacky. If the film of bonding agent has set, apply fresh epoxy adhesive before applying new concrete.

Shearing of the new concrete course from the base slab at the bond line is unlikely if the bonding-agent film is still tacky when the topping is placed. However, should a crack develop in the base slab, there is a definite possibility that this crack will transfer through the epoxy bonding-agent glue line and through the new wearing course.

New urethane bonding agents, which are often used between slab waterproofing membranes, can inhibit crack transfer from the base slab to the topping. The urethane membrane is also a two-component system that is applied in the same way as an epoxy bonding agent. Being elastomeric, the urethane membrane can absorb more of the stresses that are set up as the concrete overlay cures and contracts. It is also flexible enough to stretch and bridge cracks that develop in the substrate, as well as in the wearing course, thereby preventing transfer of cracks.

A relatively inexpensive bonding agent that may be used to ensure a positive bond of a new concrete overlay is a latex-reinforced cement slurry grout. Portland cement and sand (in a ratio of 1:3 by volume) are combined with a gauging liquid that is a mixture of equal volumes of water and an acrylic or polyvinyl acetate latex emulsion, as described in the table. This emulsion must be nonre-emulsifiable. The gauging liquid, based on the latex blend, is added to the cement-sand powder until a creamy paste is developed. This paste is scrubbed into the surface of the base slab, covering the surface thoroughly. New concrete is then placed on this bond line, which is approximately ⅟₁₆ in thick.

It is recommended that the latex slurry not be applied too far ahead of the placement of the concrete.

The latex bonding agent should not be used by itself because too much could be absorbed by a porous substrate, leaving a minimal thickness in the glue line. It may also dry too quickly to form an effective bond by the time new concrete is applied. It is best used in the form of a slurry grout.

Resurface with an Epoxy Topping

When the level of the surface must be kept close to the original elevation and a chemically resistant and tougher wearing surface must be provided, an epoxy topping may be the answer. After the base slab has been thoroughly cleaned by sandblasting, grinding, or acid etching, it is rinsed and allowed to dry. The epoxy mortar is supplied as a three-package proprietary system consisting of a base epoxy resin, a catalyst, and a measured quantity of dry, salt-free sand. The ingredients are mixed together to form a mortar which is applied to a thickness ranging from ⅛ to ⅜ in, screeded, and troweled smooth (Fig. 1-1). Most systems also include a primer, based on an epoxy system, to be applied before the epoxy mortar.

Figure 1-1 Surface of concrete slab or pavement can be repaired with thin trowel-applied overlay of sand-filled epoxy resin.

Resurface with a Latex Mortar

Although less resistant to chemicals than epoxies, latex mortar toppings also allow resurfacing without significantly changing surface elevation. The mortar may be made by blending 1 part (by volume) portland cement with 3 parts mason's sand. Latex is then added at the rate of 10 percent solids based on the cement content. Enough water is added to make a trowellable mixture. A typical formulation would be:

Portland cement (Type I)	1 bag (94 lb)
Mason's sand	3 bags (300 lb)

Latex emulsion (50 percent solids)	2 gal
Water	4–5 gal

Before the latex mortar is applied, a latex cement slurry should be applied as a bonding agent. The slurry may have the same latex as that used in the mortar topping. Latex mortar is best spread and leveled with a wood float rather than a steel trowel, because the latex tends to rise to the surface and cause a drag on the steel trowel.

After the latex mortar is finished, a sealer should be applied. The sealer, which functions as a curing membrane and protects the latex mortar against contamination from grease, oil, and deicing salts, may be based on a chlorinated rubber, a butadiene styrene rubber, or a methyl methacrylate resin. This type of sealer may be applied to all new concrete.

Resurface with an Iron Topping

Areas subject to very heavy traffic, particularly to steel-wheeled vehicles, require an extra durable surface such as that produced by an iron topping. It is usually applied in thicknesses of 1 in. Specially graded iron particles are substituted for a major percentage of the sand aggregate in a mortar or concrete mix. This topping may be bonded to a substrate, using a neat cement slurry, a latex cement slurry, or a urethane or epoxy bonding agent. This iron topping is usually available in a ready mixed form or can be blended at the jobsite following the recommendations of the supplier of the iron particles.

REPAIRING DISINTEGRATED CONCRETE MEMBERS

Methods and materials for repairing columns, beams, piers, and precast concrete panels will generally be similar to those used in repairing concrete pavements. However, because such concrete members are either load-bearing or an integral part of a structure, it is not always possible to remove and replace them. Therefore, repairs must be made to the existing structure using the best means available.

Before any repair or replacement of concrete sections is attempted, steel that is exposed by spalled or disintegrated concrete must be treated to prevent further corrosion, which may have been the initial cause of the disintegration. Rusty steel is best cleaned by sandblasting—a process that will also prepare the concrete surface by removing disintegrated and loose material. The steel should be treated with a rust-inhibitive chemical or coating, which may be a zinc-rich primer or another quality metal primer, and allowed to dry before new concrete is placed.

If the disintegration is deep, it may be necessary to use jackets or forms to hold the fresh concrete in place until it hardens. A bonding agent is recommended to ensure proper adhesion of the new concrete to the base substrate. A latex-cement slurry is the simplest and most economical bonding agent. However, for maximum adhesion, an epoxy bonding agent is recommended. This epoxy compound, painted on the reinforcing steel as well as on the surrounding concrete before placing the concrete, will also serve as a rust-inhibitive primer. A urethane bonding agent may not be suitable since this repair work is not on a deck over occupied areas requiring waterproofing properties.

Repair concrete should have a low water-cement ratio and a low slump. It should also be consolidated in the forms by direct vibration or by vibrating the form. After the concrete has been set for the minimum period of time, the forms may be removed and a liquid sealer may be applied. The sealer will continue to assist in the cure of the concrete and will protect the concrete against weathering.

If the disintegration is shallow, it may be possible to use a latex-fortified or an epoxy mortar to patch the area. The mixtures of latex and cement and sand described under "Repairing Surfaces of Pavement and Slabs" may be used. An epoxy mortar may also be used, but it may not blend readily with the surrounding concrete in color or appearance.

Proper preparation of the surface and proper priming are still necessary. If latex mortar is used, a sealer should be applied. It is not necessary to apply a sealer to an epoxy mortar since this compound is dense and resistant enough to require no additional protection.

Shotcrete or gunned mortar can be used on spalled areas by an experienced contractor who has the necessary equipment. The mortar or cement plaster is usually formulated to a 1:4.5 ratio of cement to sand, although richer 1:3 mixtures are often used. If, as in some cases, the shotcrete does not have the necessary adhesive qualities, a bonding agent (a latex-cement slurry or an epoxy compound) is used as a prime coat. Shotcrete applied in very thin layers may have to be reinforced with a latex emulsion admixture. When the necessary equipment can be obtained, this application can be made by plant maintenance people.

MASONRY*

Masonry structures, like all plant structures, are susceptible to deterioration caused by natural weathering and deleterious effects of the industrial environment. Steps can be taken to modify the rate of attack when the basic principles involved in weathering deterioration are understood.

BRICK MASONRY CONSTRUCTION

Although bricks may be made of many materials, the term *brick masonry* is normally applied only to that type of construction employing comparatively small units of burned clay or shale. Ordinary brick is economical, and, when hard burned and laid in good mortar, it is one of the most durable construction materials available for buildings.

Clay is produced naturally by the weathering of rocks. Shale, produced in much the same way but with compression and perhaps heating, is denser than clay and more difficult to mine. Various brick colors and textures result from different chemical compositions and methods of firing.

Clay is ground, mixed with water, and molded into bricks by several methods: (1) stiff-mud process in which stiff, plastic clay is pushed through a die and cut into desired lengths, (2) soft-mud process in which clay is pressed into forms, and (3) dry process in which relatively dry clay is put into molds and compressed at pressures from 550 to 1500 lb/in^2.

After some drying, green bricks are fired in large kilns. The total firing process takes between 75 and 100 h. The brick must be gradually brought to the vitrification point (the temperature at which clays begin to fuse) and then must be gradually cooled.

Common brick, also known as hard or kiln-run, is made from ordinary clay or shale and is fired in the usual manner. Overburned bricks, called "clinkers," are unusually hard and durable.

A standard brick is 8 in long, 2¼ in deep, and 3¾ to 3⅞ in wide and weighs approximately 4½ lb. It should be rough enough to assure good bonding with mortar and should not absorb more than 10 to 15 percent of its weight in water during a 24-h soaking.

Types of Brick

Common bricks are often classified according to their position in the kiln, as follows:

Arch and *clinker bricks* are close to the fire in the kiln and are overburned and extremely hard and durable. They are often irregular in shape and size.

Red, well-burned and *straight-hard* are well-burned, hard, and durable bricks.

*This material adapted with permission from *Plant Engineering,* November 10, 1977, pp. 203–207.

Stretcher bricks come from these classifications and are selected for uniformity of hardness, size, and durability.

Rough-hard bricks are in the clinker class.

Soft and *salmon* bricks are farthest from the fire in the kiln and are underburned, soft, and less durable.

Special kinds of brick include the following:

Face brick is made from specially selected materials to control color, texture, hardness, uniformity, and strength. It is used in veneering and exterior tiers, chimneys, etc.

Pressed brick is made by the dry process and has regular smooth faces, sharp edges, and perfectly square corners. This type is generally used as face brick.

Glazed brick has the front surface glazed in white or other colors. It is used in dairies, hospitals, and other buildings where cleanliness and ease of cleaning are important.

Fire-brick is made from a special type of fire clay to resist high temperatures.

Imitation brick is usually made from portland cement and sand rather than from clay. It is not burned, but has qualities similar to good mortar.

Mortar

Mortar serves several functions in brick construction. It holds bricks together, compensates for brick irregularities, and distributes load or pressure among the units.

Properties of mortar depend, to a large extent, on the type and quantity of sand used. Good mortar is made from sharp, clean, and well-screened sand. When sand is too fine, the mortar has less "give" and water works out of it, making it stiff and difficult to trowel. It may also set before the bricks can be placed. Too much sand robs the mortar of its cohesive consistency and makes it difficult to work with.

Mortar may be classified into five general types, on the basis of composition: straight lime, straight cement, cement lime, masonry cement, or lime pozzolan.

Natural cement is an important constituent of masonry cement because of its gradual strength-gaining properties, high plasticity, excellent water retention, and good adherence to aggregates. Combining natural cement with portland cement, which has early-strength properties, provides a hydraulic cement ideally suited for masonry mortar. Cement-lime mortars are usually classified in accordance with the ratios of cement, lime, and sand, by volume, in Table 1-2.

TABLE 1-2 Cement Lime Mortars for Masonry Construction

Mortar	Use	Proportions*			Minimum compressive strength, lb/in²
		Portland cement	Hydrated lime	Damp, loose sand	
Type M	Maximum compressive strength	1	1/4	2 7/8–3 3/4	2500
Type S	Maximum bond	1	1/2	3 3/8–4 1/2	1800
Type N	General purpose	1	1	4 1/2–5	750
Type O	Non-load-bearing interior construction	1	2	6 3/4–9	350

*Parts by volume.

†Sand quantity is 2 1/4 to 3 times combined volume of cement and lime.

Causes of Deterioration of Brick Masonry

Repeated natural destructive forces, mild though they may be, break down hard rock into clay. These same forces act on clay bricks, fired tile, and fired terra cotta to cause deterioration. Natural stones and the binding mortars of masonry construction are similarly affected. In addition, airborne chemicals in industrial atmospheres and pollutants from internal-combustion engines contribute greatly to rapid soiling and chemical destruction of these binding materials and masonry units.

Frost Damage

One of the more destructive agents of weathering is frost. Water expands 9 percent as it freezes. Under certain conditions, such expansion may produce stresses that disrupt the bricks and cause spalling.

To take up water, a brick must be porous. There are two measures of water absorption: one obtained after soaking the brick for 24 h in cold water and another, larger one obtained after boiling the brick for 5 h. The difference between the two values represents the so-called sealed pores (pores that are not accessible to water under normal conditions, such as wetting by rain). If all open pores are filled with water, the unfilled sealed pores provide space into which water can expand on freezing with little or no development of stress.

Efflorescence

Pitting and spalling of clay products and natural stones is associated with efflorescence, a phenomenon in which salts percolate through the member and crystallize on the surface of a brick, stone, or mortar joint. These salts may be the sulfates of calcium, magnesium, sodium, potassium, and, in some cases, iron. Soluble salts may be present in the clay, or they may be formed by the firing process (oxidation of pyrites, or reaction of sulfurous fuel gases with carbonates in the clay).

Other sources of soluble salts are portland cement and hydraulic lime mortars, which contain soluble sulfates and carbonates of sodium and potassium; mortar containing magnesium lime, which can produce destructive magnesium sulfate; and gypsum plaster or dry wall. Salts may also be drawn by capillary action from the ground soil and limestone or concrete copings.

For efflorescence to occur, salts must be present, water must be available to take them into solution, and a drying surface on which evaporation can proceed to deposit crystals at the surface must exist. Water is always available and drying is periodic. The potential for efflorescence can be reduced by specifying well-fired brick that, generally, contains less soluble salts, or by specifying special-quality brick formulated with minimum soluble salts. The place where efflorescence appears is no indication of its source because solutions of salts may migrate considerable distances. Efflorescence on a particular surface merely indicates that it provided a convenient drying area. When a very dense impermeable mortar touches a more permeable brick, efflorescent salts will often appear on the brick, although the salts may have migrated from the mortar.

Glazed brick veneers do not always eliminate efflorescence. Although the glazing is impermeable to water entry from the exterior, moisture can still move slowly from the interior of the building toward the exterior face. Salts transferred to the exterior glazing create enough pressure to produce shaling of the glazed face. This phenomenon is common, especially when the brick has been inadequately fired. The result is unattractive, and the brick is exposed to further degradation.

Dimensional Changes

Expansion or contraction of building units may not in itself be harmful. However, the continuation of differential movements of dissimilar materials may give rise to difficulties. Live-load changes and foundation settlement can cause whole buildings to move. Building joints must be properly designed and located to accommodate these movements. Lime-sand mortar, used in older structures, is able to accommodate large movements without distress. Modern, higher-strength cement mortars, on the other hand, are less flexible and may shrink excessively on setting to cause cracking.

The movement of moisture in porous building materials can cause expansion and contraction. Expansion generally takes place with wetting; shrinkage occurs with drying.

Water taken up by new brick when it is laid in fresh mortar causes the unit to expand Meanwhile, the mortar is shrinking. Such action can be minimized by wetting kiln-fresh bricks before they are used, avoiding excessively rigid mortars, and providing adequate joints. Reinforced concrete frames may also lead to failure of brick cladding when there is vertical shrinkage of the frame and when no movement joints are provided.

Steel columns clad with brick, common in structures built two decades or more ago, also can cause brick failure. As water enters this cladding, through either the brick or the mortar, the steel column rusts. Expansion of the rust pushes the brick cladding away from the column.

Mortar Deterioration

Mortar may decay from the formation of calcium sulfo-aluminate (which causes expansion and loss of mortar strength) and by the attack of pollutants in the atmosphere. Portland cement contains tricalcium aluminate, which reacts with sulfates in solution to form calcium sulfo-aluminate. Exhaust gases from automobiles contain sulfur dioxide, sulfur trioxide, and nitrous oxides. These oxides react with moisture in the atmosphere to form sulfurous acid, sulfuric acid, and nitric acid, which are the attacking agents. As attack continues over the years, the mortar joints may crack, the surface of the joint may spall off, and the mortar may become softer and more crumbly.

Sulfate-resistant cements used in mortar will inhibit this disintegration.

REPOINTING MORTAR JOINTS

Repointing, or tuckpointing, is the process of removing deteriorated mortar from masonry joints and replacing it with new mortar to correct some perceptible problem, such as falling mortar, loose bricks, or damp walls. All contributing factors to the problem should be thoroughly investigated and corrected before repointing because the great amount of hand work and special materials required make repointing expensive and time consuming. Matching bricks may have to be obtained or specially made. Existing mortar may have to be analyzed before a repointing mortar can be formulated to match its color, texture, and physical properties.

It is a common error to assume that hardness or high strength is a measure of durability. A mortar that is stronger and harder than the masonry units will not "give," causing stress concentrations in the masonry units. These stresses are usually relieved by cracking and spalling. Mortar should contain as much sand as possible (consistent with workability) to help reduce shrinkage while drying. It should have good cohesive and adhesive qualities, be easy to handle on the pointing tool, and have good water retention (to resist rapid loss of water through absorption by the brick). It should not be sticky.

There is some controversy as to whether high-lime mortar is preferable to portland cement mortar for tuck pointing. High-lime mortar is suggested for use on old buildings because it is soft and porous, has low volume change, and is slightly soluble in water.

A slight amount of high-lime mortar will dissolve in rain and precipitate in small cracks; during drying, these small cracks and voids will seal. A small amount of white portland cement will accelerate setting of this normally slow-setting mortar. Even if the building was originally constructed with cement mortar, high-lime mortar may be recommended to reduce shrinkage and potential stresses at the edges of the masonry.

Mortar Mixes

The mixes outlined in Table 1-3 provide a starting point for developing a visually and physically acceptable mortar.

Sometimes, small amounts (5 to 10 percent) of finely divided iron are added to the repointing mixes to provide slight expansion rather than normal shrinkage. Too much iron may produce excessive expansion and may cause iron-rust stain. Repointing mortars

TABLE 1-3 Trial Mortar Mixes for Repointing Masonry

Mortar	Portland cement	Hydrated lime	Sand	Acrylic latex (50% solids)
Formula A	¼ bag	1 bag	3 ft³	—
Formula B	1 bag	1–1½ bags	5–6½ ft³	—
Formula C	1 bag	—	3 ft³	2 gal

can be further modified by adding (1) water-reducing agents to keep water content or water-cement ratio low, (2) waterproofing admixtures such as stearate soaps to minimize the absorption of water, (3) air-entraining agents to increase resistance to freeze-thaw weathering in areas of extreme exposure, and (4) mineral oxide colors to blend the mortar color with the brick.

Requirements for Mortar Materials

Materials used in preparing pointing mortars should comply with the following specifications and requirements:

Lime. "Standard Specification for Hydrated Lime for Masonry Purposes (ASTM C 207)," Type S; Federal Specification SS-L-351B.

Cement. "Standard Specification for Portland Cement (ASTM C 150)," Type I or II; Federal Specification SS-C-192G(3). The cement should not have more than 0.60 percent alkali (sodium oxide), or not more than 0.15 percent water-soluble alkali by weight.

Sand. "Standard Specification for Aggregate for Masonry Mortar (ASTM C 144)," Federal Specification SS-A-281B(1), paragraph 3.1

Water. Potable water free from acids, alkalis, and organic materials.

Execution of the Work

Generally, old mortar should be cut out to a depth of ¾ to 1 in to ensure an adequate bond between old and new mortar and to prevent popouts. For joints that are less than ⅜ in wide, cutting back ½ in is usually sufficient. Using power tools is risky, unless they are handled by a skilled mason, because bricks can be easily damaged by such equipment. The use of hand chisels is still the best procedure.

Dry mortar ingredients should be mixed first. They should be prehydrated with only enough water to make a damp, stiff mortar to help prevent drying shrinkage. After an hour or two, the mortar is mixed with additional water to provide trowelability.

The joints should be thoroughly cleaned and moistened before the mortar is placed. A chemical bonding agent may be used; however, care must be exercised in painting it into the joint so that it is not smeared on the brick face. Mortar must not be placed before the bonding agent has set.

The mortar is best placed in ¼-in layers and then packed until the void is filled. When the final layer of mortar is thumbprint hard, the joint should be tooled to the desired shape with the correct size of pointing tool. If old bricks have worn, rounded corners, the final mortar surface should be slightly recessed to avoid leaving uneven joints that may be damaged easily.

The small amount of excess mortar left on the wall from a careful repointing job can be removed with a bristle brush before it hardens. Hardened mortar can be removed with a wooden paddle or chisel. Care should be exercised in using any chemicals, especially acids.

Grouting techniques can be used to repoint joints that show only minor defects such as very shallow deterioration, hairline cracks, or slight loss of adhesion to the brick. One method, often referred to as a bagging operation, can be used on glazed brick. A grouting mortar such as formula C in Table 1-3 is mixed to the consistency of a creamy paste and brushed onto the wall surface. After a short period (15 to 30 min, depending on temperature and degree of set) burlap bags or rags are used to bag or wipe the grout from the surface of the nonabsorptive glazed brick; grout is left only in the mortar joint. After an entire section is done, the area may be washed to remove any grout remaining on the glazed brick faces.

Another method involves the same bagging or grouting operation, but it may be done on unglazed or any standard face brick. This method involves masking each brick with tape cut to the same dimensions as the brick. The same type of latex grout is applied by brush over a wider area. The masking tape is removed after the grout sets, leaving neat, clean, and sharp mortar joints. This method is commonly called mask and grout.

Some mortar joints, especially very narrow ones, may be sealed with an elastomeric sealant applied by a caulking gun. Cracked bricks may also be repaired by widening the crack and sealing it with an elastomeric sealant.

CLEANING PLANT BUILDINGS

There are a number of valid reasons for cleaning building exteriors. Before building units are repaired or replaced, the original colors and textures of these units must be known so they can be properly matched. And, when new materials are installed, uniform appearance and weathering of the entire structure are desirable.

Preventive maintenance is an often-overlooked reason for cleaning. Dirt provides a much greater surface area than clean building materials; and, the more surface area that is exposed to atmospheric pollutants, the greater are the possibilities for destructive chemical reactions to be started. Dirty areas remain wet longer, resulting in more severe freeze-thaw cycling. And wet, dirty areas can support microorganisms that can cause disintegration, dissolution, and staining.

Selecting an appropriate cleaning method can be challenging, because dirt composition is so complex. Dirt is a surface deposit of finely divided solids held together by various organic materials. The solids are primarily carbon soot, siliceous dust, and inorganic sulfates. The organic binders consist largely of hydrocarbons from incomplete combustion products of various fuels. A combination of adsorption and electrostatic attractive forces hold the dirt to the masonry. Other adherent factors include efflorescent salts, leached cementitious materials, and recrystalized carbonates.

Acidic cleaners can be very damaging, particularly to marble and limestone, and alkaline cleaners can also be harmful. It is recommended that cleaners be tested on small areas to determine their effectiveness and reaction to the substrates.

One of the most versatile techniques for cleaning building exteriors is water washing. Although it requires minimal expenditure for materials and equipment, it can be time consuming, particularly if hand scrubbing is involved. There are three types of water-washing procedures: low pressure, high pressure, and steam cleaning.

The low-pressure wash is carried out over an extended period. The prolonged spraying loosens heavy dirt deposits; then, moderate pressure (200 to 600 lb/in^2) can be used to flush away loosened dirt.

High-pressure water cleaning involves equipment capable of supplying water to a special high-pressure gun that jets the water at pressures of 1000 to 1800 lb/in^2. The gun can deliver up to 1700 gal of water per hour. A special nozzle can aerate the water, thereby minimizing physical damage to the masonry.

Steam cleaning involves the use of low-pressure (10 to 30 lb/in^2), large-diameter (½ in) nozzles; steam is generated from a flash boiler. The equipment is relatively more expensive and presents some safety hazards to the operators. It is also possible to add detergents or surfactants to the water in the flash boiler. Adequate dirt removal requires an average working time of 1 min/ft^2.

Chemical cleaners can be acidic (low pH) or alkaline (high pH). Both types are used with surfactants (1 to 2 percent) to promote detergency and wetting. Acidic cleaners are often based on hydrofluoric acid or phosphoric acid in concentrations of less than 5 percent in water. Alkaline cleaners are often based on sodium hydroxide, ammonium hydroxide, or ammonia. Sometimes, all purpose cleaners, such as ammonium bifluoride, are used.

Abrasive blasting, both dry and wet processes, are effective on all substrates, but they require experienced mechanics to minimize damage to surfaces. Pressures used are usually between 20 and 110 lb/in^2, and working distances are from 3 to 12 in. The abrasives are usually silica sand, but crushed slags and coal wastes are often used. The mesh sizes are very fine, either 0 or 00. Round particles are less abrasive and damaging than crushed grains.

Cleaning may precede or follow replacement of masonry units and repointing, depending on conditions. Cleaning before repair helps reveal original colors and textures and prepares substrates for receiving new mortars and bonding agents. Cleaning after repairs helps remove any excess droppings, splashings, or other accidental contamination. Whether done before or after, cleaning is a very important aspect of masonry and concrete restoration.

SEALING THE SURFACE

A final step in the entire restoration process involves sealing all porous surfaces on the exterior facade. This procedure waterproofs the masonry to minimize the ingress of water, protects against attack by pollutants and other chemicals, minimizes the collection of dirt, and protects against graffiti damage.

One of the best sealers is a methyl methacrylate in organic solvent, containing 15 to 20 percent solids and, preferably, having a matte finish. The sealer may be applied in one or two coats, depending on the porosity of the substrate, by brush, spray, or roller. The sealer is water white, will not yellow or embrittle, and may be effective for 5 to 10 years or longer.

Silicone compounds are not particularly recommended, primarily because they are effective only for relatively short periods and are water-repellent rather than waterproofing.

METALS*

GLOSSARY OF COMMON METAL CORROSION TERMS

Active A state in which a metal tends to corrode (opposite of passive, noble).

Anode Electrode in a cell at which surface oxidation or corrosion occurs (opposite of cathode).

Bimetallic corrosion Galvanic corrosion.

Cathode Electrode in a cell at which reduction occurs (opposite of anode). Practically no corrosion occurs here.

Cell An electrochemical circuit consisting of an anode, a cathode, and electrolyte. The anode and cathode may be different metals or dissimilar areas of same metal. Corrosion generally occurs only at anodic areas.

Corrosion Direct, chemical or electrochemical reaction of a metal with its environment, and general destruction of any material resulting from reaction with environment.

Corrosion fatigue Damage to metal from combination of corrosion and fatigue (cyclic stresses).

Couple A cell formed by two dissimilar metals in an electrolyte.

Electrolyte A chemical mixture capable of carrying ions. It is usually a liquid (often an aqueous solution).

Erosion Deterioration of a surface by the abrasive action of moving fluids.

*This material adapted with permission from *Plant Engineering,* November 24, 1977, pp. 103–106.

Erosion-corrosion Combined effects of erosion and corrosion on a metal surface.

Fatigue Cracking failure of a material resulting from repeated cyclic stress below the normal tensile strength.

Galvanic cell An electrolytic cell, consisting of two dissimilar electrodes (anode and cathode) in an electrolyte, capable of producing electric energy by electrochemical action. The anode and cathode may be dissimilar metals or dissimilar areas of the same metal. Corrosion occurs at anodic areas.

Galvanic corrosion Corrosion associated with the current in a galvanic cell (also called couple action).

Galvanic series Listing of metals arranged according to their relative corrosion potentials in some specific environment (often seawater).

Ion An electrically charged atom or group of atoms (radical).

Local cell Galvanic cell produced by differences in composition in the metal or the electrolyte.

Noble metal A metal, or alloy such as silver, gold, copper, having high resistance to corrosion or oxidation.

Oxidation Loss of electrons by a chemical reaction such as corrosion (opposite of reduction).

Passive A state of a metal that tends to slow corrosion or anodic reactions (opposite of active). It is a surface phenomenon.

Patina A green coating that slowly forms on copper and copper alloys exposed to the atmosphere. It consists mainly of copper sulfates, carbonates, and chlorides.

Reduction Gain of electrons (opposite of oxidation).

Rust The reddish-brown corrosion product of iron and ferrous alloys. It is primarily hydrated ferric oxide.

Corrosion is the critical performance factor for metals in structures exposed to weather and industrial atmospheres. Corrosion can be completely destructive, or it can form a protective film over the surface, stopping the corrosion process. Methods of controlling the destructive effects of corrosion vary according to the type of metal and the environment.

Corrosion may be defined as the direct chemical or electrochemical reaction of a metal with its environment. Chemical attack is simple dissolution of the metal. Electrochemical attack requires an electrolyte, anode, cathode, and return circuit (see Fig. 1-2). The electrolyte of a simple galvanic corrosion cell is a substance capable of carrying ions. It is typically an aqueous solution of small amounts of dirt, salts, acids, or alkalis from the atmosphere and rain or condensed moisture. The anode and cathode of the corrosion cell may be dissimilar metals or two areas of the same metal in dissimilar electrolytes.

The corrosion potential of a galvanic couple (galvanic corrosion cell of dissimilar metals) is indicated by the position of the metals in the galvanic series—a list of metals and alloys arranged according to their relative corrosion potentials—for the given environment (seawater in the example shown in Fig. 1-3).

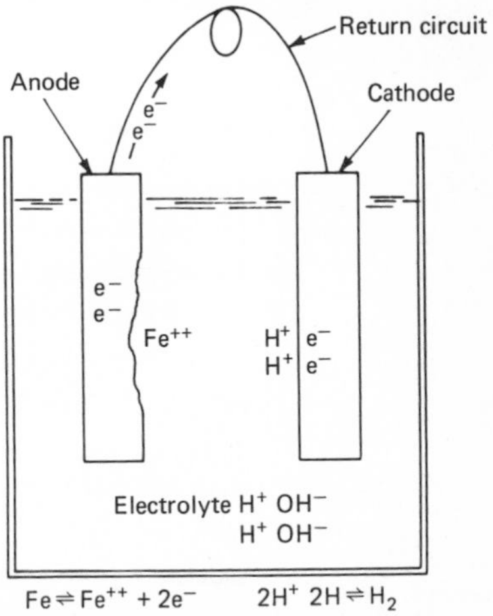

Figure 1-2 Four essential elements of the basic corrosion cell are electrolyte, anode, cathode, and return circuit.

Most metals used in building construc-

Active

Magnesium
Zinc
Beryllium
Aluminum Alloys
Cadmium
Mild Steel, Cast Iron
Low Alloy Steel
Stainless Steel (Active) Types 410, 416, 430, 302, 304, 321, 347
Austenitic Nickel Cast Iron
Stainless Steel (Active) Types 316, 317
Aluminum Bronze
Naval Brass, Yellow Brass, Red Brass
Tin
Copper
Pb-Sn Solder (50/50)
Admiralty Brass, Aluminum Brass
Manganese Bronze
Silicon Bronze
Tin Bronzes (G & M)
Stainless Steel (Passive) Types 410, 416
Nickel Silver
90-10 Copper-Nickel
80-20 Copper Nickel
Stainless Steel (Passive) Type 430
Lead
70-30 Copper-Nickel
Nickel-Aluminum Bronze
Nickel-Chromium alloy 600
Silver Braze Alloys
Nickel 200
Silver
Stainless Steel (Passive) Types 302, 304, 321, 347
Nickel-Copper alloys 400, K-500
Stainless Steel (Passive) Types 316, 317
Nickel-Iron-Chromium alloy 825
Ni-Cr-Mo-Cu-Si alloy B
Titanium
Ni-Cr-Mo alloy C
Platinum
Graphite

Noble or Passive

tion are alloys: single-phase alloys in which the atoms of the different metals are combined in a single crystal structure, or two-phase or multiphase alloys in which the metals are physically distinct. Because separate phases can form anodic and cathodic regions, a multiphase alloy is more susceptible to certain types of galvanic corrosion than a single-phase alloy or a pure metal. Deformation of stress of a metal member can also produce galvanic cells that lead to corrosion.

Many metals used in buildings become passive when exposed to the atmosphere. Such metals include aluminum, stainless steel, copper, lead, and zinc. They corrode on initial exposure to the atmosphere and then acquire a protective layer of corrosion products that isolates the metal from further attack. The green patina of copper is one of the most visible examples of this phenomenon.

Corrosion is promoted by a variety of conditions—contact of metals with other building materials (such as uncured concrete, mortar, or plaster), salts in the soil, contaminated rainwater, or acids from certain types of timber. Calcium chloride added to concrete during cold-weather construction can intensify galvanic corrosion of embedded metals.

ALUMINUM

The four principal categories of wrought and cast aluminum for construction are various grades of pure aluminum, heat-treatable alloys, non-heat-treatable alloys, and casting alloys. Pure aluminum, which is about as ductile as lead, is used for supported roofing and flashing. Heat-treatable alloys, based on blends of aluminum, magnesium, silicon, and copper, are used for fastenings and structural purposes because of their high strengths. Non-heat-treatable alloys containing manganese or magnesium are used in sheet roofing and cladding. Casting alloys usually contain up to 12 percent silicon and perhaps some copper and magnesium.

Aluminum may be used as manufactured, or it may be furnished with a wide range of treatments. The best-known treatment is anodizing, in which the thickness of the surface film is increased by electrolytic oxidation. Because the film is a suitable base for dyeing, organic dyes are included in the electrolytic bath to produce a wide range of colors in the finished aluminum. The anodic film should be 1 to 2.5 mils thick if it is to be exposed to the weather.

Aluminum may also be surface-treated with chromates, phosphates, or fluorides. Vitreous enameling and various types of paints are also common treatments. Manufacturer-applied finishes, usually baked polymeric systems, are popular for roofing and siding applications.

Prolonged exposure to the atmosphere, particularly in areas of high pollution, causes spots of white crystalline corrosion products to form on the surface of aluminum, including the anodized type. In a sooty atmosphere, the metal may acquire a gray color. Pitting corrosion can affect the useful life of aluminum members in industrial atmospheres.

Generally, maintenance of aluminum building members consists of washing the surface fairly regularly to remove foreign matter that can damage anodized finishes. Washing is particularly recommended for areas that are not effectively washed by rainfall. Other finishes also benefit from washing.

Aluminum members are often shop-coated with a 1- to 2-mil-thick film of an acrylic sealer. This film will ultimately weather to a point at which it must be renewed if protection is to continue. A methyl methacrylate sealer, similar to those used in sealing con-

Figure 1-3 (facing page) Galvanic series of metals and alloys in seawater. Metals and alloys are listed in the order of their corrosion potential in seawater. When two metals are coupled, those close together in the list are less susceptible to galvanic corrosion than those widely separated in list.

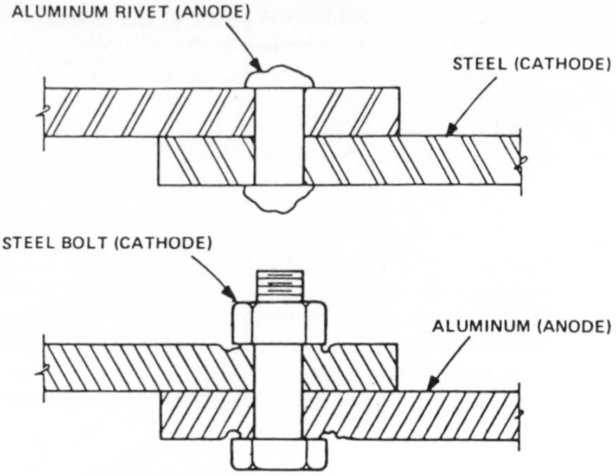

Figure 1-4 In a galvanic couple of aluminum and steel, anodic aluminum corrodes while cathodic steel is relatively unaffected. It is important that fasteners be compatible with building materials.

crete and masonry structures, may be used. Painted aluminum may be touched up or repainted.

For structural applications, it is suggested that:

- Aluminum roofs be pitched to prevent water ponding and promote washing of the surface by rain
- Vapor barriers and proper ventilation be used to minimize condensation on the underside of an aluminum member
- Aluminum fittings (or nonmagnetic stainless steel and galvanized fittings) be used with aluminum building components to prevent galvanic corrosion (see Fig. 1-4)
- Aluminum and steel be separated by insulating washers, bituminous or plastic materials, or paint
- Water from copper roofs or pipes be prevented from discharging onto aluminum
- Aluminum embedded in concrete, masonry, or stonework be protected with a bituminous paint
- Aluminum that will contact certain timbers be protected by paint or bituminous coating to reduce the chance of attack from acidic wood (unseasoned oak, western red cedar, and redwood) and wood preservatives that contain water-soluble salts or copper salts that can corrode aluminum.

ZINC

Although zinc can be used in the form of rolled sheet and strip, its widest use in building is for protective plating. When exposed to weather, zinc on the surface of galvanized steel develops a passive film that protects the underlying zinc and, thus, the steel. When the surface is scratched or scored, the anodic zinc corrodes, protecting the cathodic steel from attack.

Zinc can be applied to metal surfaces with hot-dip galvanizing, electrolytic processes, metal spraying, sherardizing, and zinc-rich paints. Hot-dip galvanizing provides excellent coverage and bond, leaving a 3- to 5-mil-thick film. The interface is actually a zinc-iron alloy. Electrolytic processes provide a thin coating up to 1 mil thick. Metal spraying provides a 4- to 20-mil-thick zinc film. Sherardizing, the heating of small components in a container of zinc dust, produces films 0.5 to 1.5 mils thick. The dry film of zinc-rich paint should be at least 90 percent zinc to provide adequate galvanic protection to a steel substrate.

Atmospheric exposure gives the zinc surface a coating of zinc carbonate that reacts with sulfurous and sulfuric acids in the atmosphere to form zinc sulfate, which is water-soluble and can be washed off by rain.

Zinc sheeting is susceptible to attack from condensation on its underside. Proper ventilation of roof spaces and using vapor barriers and suitable underlays can minimize such conditions. Zinc surfaces should be designed to allow rapid drying.

Local air pollution should be considered when designing zinc roofs, and proper fasteners should be specified. Zinc should be prevented from contacting copper and the chlorides and sulfates in concrete and mortar. Acids from certain timbers and some wood preservatives can also attack zinc. Bituminous coatings are recommended for protection.

Weathered galvanized members can be painted after being properly cleaned. Freshly galvanized members can also be painted. The Zinc Institute now recommends solvent cleaning or detergent washing, rather than acid etching, before painting. Suitable primers include zinc dust, zinc oxide primers, vinyl wash primers, and latex emulsion primers.

COPPER

One of the major uses for copper in buildings is in roofing—gutters, flashing, and cladding. It is also used as an alloying metal in brass, bronze, and weathering steel.

Characteristically, copper weathers by the formation of a protective green patina, which, chemically, consists of copper hydroxide salts of sulfate, chloride, nitrate, and carbonate. This process may take 5 to 10 years, depending on air pollution, humidity, and temperature. The patina can be produced artificially, so new copper can be made to match old copper.

Protective lacquers are available that preserve the polished finish of copper or brass. One such coating is Incralac A (a trademark of the International Copper Research Association). It is a methyl methacrylate resin dissolved in a blend of xylol and toluol with a thixotropic agent to leave thicker films, and benzo-triazole, an inhibitor that prevents discoloration of the metal.

Copper shows good compatibility with other building materials. It has long been used with concrete as a water-stop material. Its use in roofing is traditional because of its outstanding durability.

LEAD

In its pure form, lead has been used in plumbing, roofing, cladding, and waterproofing applications. However, lead is now typically alloyed with 6 to 7 percent antimony to increase stiffness and strength. It can be used to coat iron, steel, and copper.

Lead's shiny metallic appearance dulls rapidly in the atmosphere through the formation of an oxide; with prolonged exposure, carbonates and sulfates form, leaving a whitish-gray appearance. Because lead is a relatively soft sheet material and has little rigidity, it requires strong supporting structures.

Lead's galvanic action is relatively insignificant, but contact between lead and aluminum should be prevented. Lead can corrode in contact with damp, uncured, cementitious materials and damp acidic timbers. The corrosion is minor and can be prevented if the lead is coated with bituminous paint. However, most paints do not adhere well to lead.

STEEL

Although there are many uses for wrought and cast iron in building construction, steel is the major type of ferrous metal used in plant construction. Steel and iron deteriorate primarily from four basic causes.

- Corrosion (rusting) is the chief cause of damage to steel and creates a major maintenance problem. The corrosion products of steel are reddish brown scales or flakes of iron oxides, hydroxides, and other compounds loosely bonded to the parent metal. Rusting is an electrochemical process that requires an electrolyte (usually water).
- Abrasion (erosion) may be distinguished by the worn, smooth appearance of the surface. It usually results from the working of moving parts, members subjected to wave action by water (such as piling), flue gases with a high ash content, and windborne dust, sand, and debris.
- Fatigue is failure of a structural member resulting from repetitive, fluctuating loads at or below allowable design values. Symptoms are small, often undetectable, fractures which may result in sudden collapse. Steel members subjected to repetitive loadings should be inspected frequently, particularly at riveted and welded points.
- Impact of moving objects can lead to distortion. Sometimes, bowing or buckling can result from overstress, weakening of a steel member from corrosion, or both. Corrosion of steel lintels over windows can reduce effective thickness, resulting in bowing from the weight of bricks above the lintel.

Corrosion and abrasion problems can be solved by proper maintenance. Impact and fatigue are engineering considerations.

Corrosion of iron and of carbon and low-alloy steels can be controlled by protective coatings. Corroded members must be properly cleaned before appropriate primers and finish paint systems are applied. Abrasive blasting is the most effective method of preparing corroded steel for painting. Cleaning to commercial grade is usually adequate, but blasting to white metal may be necessary when high-performance coatings such as epoxies or urethanes are to be used, or when the steel is to be exposed to a highly corrosive atmosphere. Other cleaning methods include power wire brushing, hand wire brushing and scraping, and flame cleaning.

STAINLESS STEEL

High strength and outstanding resistance to atmospheric corrosion are major advantages of stainless steels (alloys containing from 17 to 20 percent chromium). Various modifications of these alloys incorporate up to 8 percent nickel and 3 percent molybdenum. The straight chromium alloys are magnetic; those containing nickel and molybdenum are nonmagnetic. Stainless steel is used in window framing, curtain walls, cladding, and fasteners.

A stainless steel surface develops a thin oxide film that is highly corrosion-resistant and self-repairing. Regular washing to remove dirt and other deposits is normally the only maintenance required. Stainless steel causes very minimal galvanic corrosion of other metals.

Weathering steel, a high-strength, low-alloy steel for architectural applications, also develops a passive surface. These alloys, containing small percentages of carbon, manganese, phosphorus, sulfur, silicon, copper, chromium, and nickel, develop a deeply colored, reddish-brown coating. It develops with time and only with complete exposure to the elements. Until the weathered surface comes to equilibrium and the coating develops, adjacent concrete, unglazed brick, stucco, granite, marble, unpainted galvanized steel, flat and semigloss paints, and some types of glass may develop a rust stain. Unexposed areas of this steel, like ordinary carbon steel, must be painted.

WOOD AND PLASTIC*

Wood and plastic building materials resist many of the natural and industrial environmental factors that affect concrete, masonry, and metals. However, wood and plastic are

*This material adapted with permission from *Plant Engineering*, February 16, 1978, pp. 149–152.

susceptible to other types of attack—fungi and insects can destroy wood, and ultraviolet light and temperature can affect some plastics.

WOOD

When exposed to the elements, wood is affected by water and light. Pollutants have little effect on wood except to dirty it. It has good resistance to chemical corrosion, and its thermal expansion is slight.

Wood normally contains 12 to 18% water, but it can absorb moisture and swell up to 5 percent. Ultraviolet light can cause chemical and color change (accelerated by the extraction of water-soluble materials). The stresses set up by fluctuating moisture content and ultraviolet light cause splitting and checking (defects in the wood surface that allow more water to enter to produce gross dimensional changes). Weathering effects of moisture and sunlight can be reduced by applying a film-forming or penetrating finish.

Biological Attack

Wood is subject to decay caused by certain types of fungi. In extreme cases, such as in tropical climates, unprotected timber may be destroyed in a few months. Keeping wood adequately dry (less than 20 percent moisture) minimizes fungal damage. Structural timber should be protected from the weather by a well-ventilated shelter, and wood members should not be framed close to the ground.

Deprivation of air causes fungi to become dormant and, finally, die. Encasement in concrete or heavy bituminous coatings can shut off the air supply; ordinary painting will not. Embedment of timbers in soil can also shut off access to air, but it can cause other problems.

Temperatures between 50 and 90°F are optimum for the growth of fungi. They become dormant at temperatures over 110°F and near 32°F, but can reactivate when temperatures moderate.

Protection against fungal decay can be provided by various chemical treatments.

In wood exposed to marine atmospheres, attack by marine borers may lead to destruction. Aggressive organisms such as teredo or shipworms, various mollusks, and wood lice attack wood by burrowing or boring and gnawing.

Their terrestrial counterparts—termites, beetles, caterpillars, bees, and ants—eat the wood. Such attack reduces the strength of the timbers and weakens and ultimately destroys the structure.

Termites are one of the most destructive organisms to timber. They are found in almost all areas of the country, with greater concentrations in warmer and more humid regions. Their diet is based on cellulose. There are wood-dwelling and earth-dwelling types. Visible symptoms of termite attack are subtle and difficult to recognize.

Protection of most wood against termites is provided by chemical treatments or by poisoning the ground with chemicals such as copper sulfate, sodium fluosilicate, borax, paradichlorobenzene, and various commercial poisons.

Chemical Preservatives

A number of species of wood have natural decay resistance. The heartwood of several species native to North America—especially redwood, cedar, cypress, and juniper—has good decay resistance; however, the sapwood of substantially all common species has poor resistance. Regardless of the species, chemical protection of woods is recommended. The preservative treatment makes the wood poisonous to fungi, insects, and marine borers. There are three classes of preservatives: waterborne, oilborne, and creosote.

Waterborne Salts

Waterborne salts for wood preservation include zinc chloride, chromated zinc chloride, copperized chromated zinc chloride, zinc meta arsenite, chromated zinc arsenate, chromated copper arsenate, ammoniacal copper arsenite, acid copper chromate, and fluor chrome arsenate phenol. These chemicals leave little odor and have little effect on the appearance of the wood, which may also be painted. The zinc chloride salts, at high penetrations, also provide fire retardance. Wood should be reseasoned before use because this type of chemical treatment injects a large amount of water into the wood.

Oilborne Preservatives

These include penta (pentachlorophenol) and copper naphthenate. Penta does not change the color of the wood, but copper naphthenate gives wood a green shade. Paintability after treatment depends on the type of oil or solvent used as a vehicle.

Creosote

Creosote is excellent for preventing decay, especially in exterior use or in contact with water. Its advantages include relative insolubility in water (giving it a high degree of permanence), good penetration, good availability, and low cost; in addition, it causes little dimensional change in the wood. Disadvantages include its potential fire hazard; a distinctive, sometimes unpleasant odor that can affect foods; volatile vapors that can affect plant life; black color; and staining of adjacent woods or porous materials. Treated wood cannot be painted, and on hot days it may sweat and become wet and tacky.

Treatment Processes

Methods for treating timber with chemicals include pressure, coating, dipping and steeping, thermal, and diffusion.

Pressure Process

These processes produce relatively deep penetration of the preservative into the wood. Although various processes differ in detail, the basic principle of all is the same—wood is placed in a pressure vessel that is filled with preservative. Pressurization then drives preservative into the wood to meet penetration and retention specifications.

Coating

Coating by brush or spray may be done at the site. At least two, preferably three, applications are made; each coat is applied after the previous one has been absorbed. Brush and spray treatments are generally used only when more effective treatments are not practical.

Dipping and Steeping

This involves immersing the wooden member in the preservative liquid. From a few minutes to as long as several days of immersion may be required. It provides greater penetration of the preservative than brush and spray treatments, but it generally less effective than pressure processes.

Thermal Treatment

Thermal treatment, or hot and cold dipping, is similar to dipping, except the member is first heated in the preservative in an open tank and is then submerged in cold preservative. This procedure provides deeper penetration of preservative than dipping and steeping.

Diffusion Processes

Diffusion processes can be used while the timber is in place. Water-soluble preservatives, carried in bandages or pastes or in retaining rings applied to the member, diffuse into the water present in the wood.

All of these methods should meet the standards and specifications of the American Wood-Preservers' Association (AWPA); see Tables 1-4 to 1-6.

TABLE 1-4 Selected AWPA Standards for Preservatives

Standard numbers	Preservative	Symbol	Trade names†
P1 or P13	Coal tar creosote	—	—
P2 or P12	Creosote–coal tar solution	—	—
P5	Acid copper chromate	ACC	Celcure*
	Ammoniacal copper arsenite	ACA	Chemonite*
	Chromated copper arsenate	CCA	Type A: Greensalt* Langwood* Type B: Boliden* CCA Koppers CCA-B Osmose K-33* Type C: Chrom-Ar-Cu(CAC)* Osmose K-33C* Wolman* CCA Wolmanac* CCA
	Chromated zinc chloride	CZC	—
	Fluor chromate arsenate phenol	FCAP	Osmosalts* (Osmosar*) Tanalith* Wolman* FCAP Wolman* FMP
P8 and P9	Pentachlorophenol	Penta	—

*Reg. U.S. Pat. Off.
†*Source:* American Wood-Preservers' Association.

TABLE 1-5 Selected AWPA Standards for Pressure-Treatment Processes

AWPA* standard	Product
C1	All timber products (general)
C2	Lumber, timber, bridge ties, and mine ties
C3	Piles
C4	Poles
C9	Plywood
C11	Wood blocks for floors and platforms
C14	Wood for highway construction
C18	Piles and timbers for marine construction
C23	Round poles and posts used in building construction
C29	Lumber to be used for the harvesting, storage, and transportation of foodstuffs

*American Wood-Preservers' Association.

Repairing Termite-Damaged Wood

Wood that has been weakened or damaged by termite attack may be repaired without removing the wooden member from the structure. The procedure is similar to pressure-grouting cracks in concrete. For termite-damaged wood, a form sleeve must be built around the damaged wooden member; then, a low-viscosity epoxy-resin compound is injected until all voids have been filled. The form sleeve may be treated with oil or wax to prevent the epoxy resin from sticking to the form. After the epoxy resin has hardened, the wood member's structural strength may be even greater than originally. If, as so often happens, the termite damage is inside the wooden member and the outside shell is intact, the epoxy-resin sealer may be invisible. However, should it be exposed, its light amber color may blend with the wood's color.

Sometimes, epoxy-resin compounds used for repairing wood may be manufactured with protective chemicals such as pentachlorophenol to help prevent further attack by termites. Termites cannot damage the epoxy compound. These special epoxy-resin compounds are generally available in boat yards or marinas, where they are widely used to repair wood damaged by rot or marine borers.

TABLE 1-6 Preservatives and Minimum Retentions for Various Wood Products[a]

Product and service condition	AWPA product[b] standard	Recommended minimum net retention, lb/ft³[c]						
		Waterborne preservatives[d]					Oilborne[d,e]	
		CCA	ACA	ACC	CZC	FCAP	Penta[f]	Creosote
LUMBER AND TIMBER								
Above ground	C2	0.23	0.23	0.25	0.46	0.22	0.40	8
Soil or water contact								
Nonstructural	C2	0.40	0.40	0.50	NR	NR	0.50	10
Structural	C14	0.60	0.60	NR	NR	NR	NR	12
In salt water	C14	2.5	2.5	NR	NR	NR	NR	25
PLYWOOD								
Above ground	C9	0.23	0.23	0.25	0.46	0.22	0.40	8
Soil or water contact	C9	0.40	0.40	0.50	NR	NR	0.50	10
PILING								
Soil or fresh water	C3	0.80	0.80	NR	NR	NR	0.60	12
In salt water								
Severe borer hazard (Limnoria)	C18	2.5 & 1.5[g]	2.5 & 1.5[g]	NR	NR	NR	NR	NR
Moderate borer hazard (Pholads)	C18	NR	NR	NR	NR	NR	NR	20
Dual treatment (Limnoria and pholads)								
First treatment	C18	1.0	1.0	NR	NR	NR	NR	—
Second treatment	C18	—	—	NR	NR	NR	NR	20
POLES								
Utility in normal service	C4	0.60	0.60	NR	NR	NR	0.38	7.5
Utility in severe decay & termite areas	C4	0.60	0.60	NR	NR	NR	0.45	9.0
Building poles (structural)	C23	0.60	0.60	NR	NR	NR	0.45	9.0
POSTS								
Fence								
Round, half-round, quarter-round	C14	0.40	0.40	0.50	NR	NR	0.40	8
Sawn four sides	C14	0.50	0.50	0.62	NR	NR	0.50	10
Guardrail and sign								
Round	C14	0.50	0.50	NR	NR	NR	0.50	10
Sawn four sides	C14	0.60	0.60	NR	NR	NR	0.60	12

[a] Key to symbols: ACA, ammoniacal copper arsenate; ACC, Acid copper chromate; CCA, chromated copper arsenate; CZC, chromated zinc chloride; FCAP, Fluor chrome arsenate phenol; NR, not recommended; Penta, pentachlorophenol.

[b] See Table 1-5.

[c] Minimum net retentions conforming to AWPA standards for softwood lumber and plywood. Retentions for piles, poles, and posts are for sourthern pine. AWPA Standard C1 applies to all processes.

[d] See Table 1-4.

[e] Creosote, creosote–coal tar solution, and oilborne penta are not recommended for applications that require clean, paintable, or odor-free wood.

[f] Penta can be applied in liquid petroleum gas or light solvents to provide a clean, paintable surface

[g] Two assay zones: 0 to 0.5 in and 0.5 to 2.0 in.

PLASTICS

Use of reinforced plastics in all construction is growing by millions of pounds yearly. These materials are finding extensive use in ceiling and floor systems, piping, skylights, translucent panels, structural shapes, grating, etc.

Plastics are usually strong, durable, lightweight, and resilient; they are easy to manufacture and install and have low maintenance costs. Color can be built in over a wide range, enhancing their use for exterior decoration.

Plastics, a term used for synthetic or modified natural polymers, are classified into two major groups: thermoplastics, which can be softened by heat and lend themselves readily to molding, and thermosetting plastics, which do not soften under heat once they are cured. Plastics used for exterior applications in the construction industry can be formulated to almost any set of properties, including light stability, rigidity, opacity, water absorption, and abrasion and fire resistance.

Weathering of plastics involves ultraviolet radiation, infrared radiation, water, temperature, microorganisms, industrial gases, and stresses from wind and snow loadings. Some plastics discolor after weathering, but discoloration can usually be minimized by selecting materials containing ultraviolet absorbers. Certain pigments can also increase stability for exterior exposure.

Thermoplastic polymers should not be used in areas where high temperatures are present to cause distortion. Distortion temperatures of commercial plastic materials should be a guide in this respect. Low temperatures can cause embrittlement. Oxidation can cause changes in the molecular structure of a plastic similar to those in paint films. In most cases, plastics are reasonably resistant to industrial pollution and microorganisms, but they will discolor as dirt collects. However, frequent simple washings can maintain plastics adequately.

The most important generic types of plastics for exterior application include polyvinyl chloride, glass-reinforced polyester, acrylics, phenolics, and amino resins.

Polyvinyl Chloride (PVC)

One of the most extensively used plastics in construction, PVC is often used for roofing panels, gutters and downspouts, pipes, cladding, wall and floor coverings, and window frames. PVC film has also been used extensively on metal sheeting as a protective and decorative finish. Heavy use is made of PVC in hidden items, such as water stops, vapor barriers, and waterproofing membranes.

PVC is a thermoplastic material with a wide range of properties determined by stabilizers, plasticizers, ultraviolet absorbers, lubricants, and other additives. For fire-resistant properties, antimony trioxide is often used. Methods of manufacture include extrusion, injection molding, blow molding, and calendering.

When properly formulated and fabricated under controlled conditions, a PVC product can have a life of 20 to 30 years. Certainly, a PVC waterstop, buried in a concrete foundation wall, must last for the life of the structure.

Translucent or transparent PVC sheet does not weather as well as the opaque form because stabilizers and ultraviolet absorbers affect light-transfer properties.

If weathering causes color change or differential fading, the PVC member can be painted after proper preparation of the surface by washing and light sanding.

Glass-Reinforced Polyester

Laminates of glass-reinforced polyester find wide use in automobile bodies, aircraft, boats, swimming pools, tanks, prefabricated housing systems, curtain walls, and lightweight building panels.

Polyesters are thermosetting resins produced by the reaction of mixtures of glycols and dibasic acids. The compound is comparable to an alkyd resin used in paints. However, it is further modified by dissolving it in styrene, and it is cured into a thermosetting plastic by the addition of catalysts and accelerators. The plastic can be modified with additives similar to those used in the PVC compounds.

The polyester resin, by itself, is not strong enough for industrial use, so it must be reinforced with glass fibers. A laminate is manufactured by spreading the catalyzed polyester resin onto a form or mold. While it is still uncured, woven glass cloth, swirled mat, or chopped strands are laid up into the film. Then, more coats of catalyzed polyester and glass fibers are added until the necessary number of layers is installed. The final coat of polyester is normally heavy enough to cover the glass fibers thoroughly. The last layer of resin will be exposed to weathering and can have color and all the necessary additives built in.

The polyester resin may be modified with methyl methacrylate resins for better clarity and transparency. Recent developments involve the use of acrylates or polyvinyl fluoride as surface coatings bonded to the laminate for longer gloss retention. The product is cured at ambient temperatures, although heat will accelerate the cure and ensure a more satisfactory laminate. The same procedures are also followed in producing epoxy-resin laminates, although these are not as widely used in the construction industry.

A glass-reinforced polyester laminate may fail if the surface resin is not thick enough to cover the glass strands, or if weathering wears away the surface color. Change of color, fading, and loss of gloss may also develop. Good maintenance depends on regular and thorough washing to remove dirt. When the surface must be refurbished, steel wool may be used to remove dirt and loose resin and fibers. A surface layer of polyester resin or an acrylic sealer may be applied. Frequent washing to remove dirt is usually the best method of maintenance.

Acrylics

Methyl methacrylate is the basic monomer for acrylic resins, materials that found their first extensive uses in cockpit covers of airplanes. Today, methyl methacrylate and its modifications are used in making window panes, fascia panels, skylights, sunshades, bath and shower enclosures, roof lights, etc. These thermoplastic materials are water-white and have almost the same light transmission properties as glass. They have good impact resistance, can be easily formed and machined, are easy to handle and install, and possess outstanding weathering resistance and durability. However, acrylics have low abrasion resistance and a very high coefficient of thermal expansion.

Periodic cleaning and washing of acrylic members is recommended as the best method of maintenance. However, they are easily scratched. Polishing with a soft rouge can remove scratches without affecting transparency. Colored acrylic members may be produced, and there is little or no fading or change in color. Painting is seldom required.

Phenolics and Amino Resins

Reacting aldehydes with phenol and amino compounds (such as urea and melamine) can produce thermosetting plastics with good chemical resistance. One of the first synthetic resins of this type was Bakelite, a phenolic resin. This group of plastics is used in making laminates, usually with reinforcement, for curtain wall paneling, wall linings, corrugated roofing, etc. Melamine formaldehyde is used in making the laminate known as Formica. These plastics, particularly the phenolics, can change color and weathering and develop very fine crazing patterns with aging. Painting can restore surface appearance and color after proper washing and light sanding. Frequent washing is recommended for maintenance.

Other plastics are used in plant structures, although less extensively. Among these are the acrylonitrile butadiene styrene (ABS) resins, the polyvinyl fluoride resins, the polycarbonate resins, and the polyurethanes. The epoxy resins have been used extensively in structural applications (such as flooring) and adhesives. Specialized applications include use in chemically resistant coatings and in plasters for exposed aggregate wall finishes.

As research and development proceed, improved formulations of available plastics and completely new resins will find their way into plant structures.

chapter 18-2

Paints and Protective Coatings

by

Robert J. Klepser
Senior Development Chemist
Heavy Duty Maintenance Coatings
PPG Industries, Inc.
Coatings and Resins Division
Houston, Texas

BASICS OF PROTECTIVE COATINGS

The Corrosion Process

This chapter covers the use of coatings to protect a variety of surfaces. A discussion of corrosion and other surface deterioration can be found in Chap. 18-1.

Because steel is most susceptible to corrosion attack, it is, in most situations, the surface of primary concern. A clear understanding of the processes by which a useful steel structure is reduced to a collection of rusty scrap is essential to controlling corrosion and making intelligent recommendations for protective coatings systems.

The reddish rust on a piece of steel is created electrochemically by the iron in it combining with atmospheric oxygen. Most of the iron ores found in nature are oxides, or combinations of iron and oxygen. The most common form of iron ore is hematite (Fe_2O_3), an oxide of iron which is equivalent to the form of common rust. These natural oxides are converted into usable iron and steel products by exposing the ores to a very vigorous reduction reaction which separates the iron and oxygen. Because iron has a strong affinity for oxygen it is necessary to deal with the ever-present tendency of iron to recombine with oxygen and form the stable natural oxides once again. This change is an electrochemical process accompanied by the production of minute, measurable electric currents—a process very similar to that creating electricity in a battery. Therefore, to create rust or iron oxide on a piece of steel, there must be an anode, a cathode, and an electrolyte. In a corrosion cell, the anode is the negative electrode where corrosion occurs (oxidation), the cathode is the positive electrode where no corrosion occurs (reduction), and the electrolyte is an ionic conductor, usually an aqueous solution.

This tiny corrosion cell is duplicated many times over a steel surface so that the eye sees what appears to be a uniform layer of rust which in reality is a series of multiple corrosion sites. As corrosion products build up over an anode, the surface can at this point become less active electrically and anodes and cathodes can reverse roles. The process will continue until all of the steel is converted to rust, or something happens to break the external circuit.

Coatings in Corrosion Control

This electrochemical rust-forming circuit can be broken by using a barrier. Not only steel may be protected in this manner but other substrates such as concrete and wood may also be shielded from the environment by a barrier. Protective coatings, which serve as barriers, are undoubtedly the engineers' principal materials for easily creating the barrier and giving the desired insulation.

A coating may be defined as a material which is applied to a surface as a fluid and which forms, by chemical and physical processes, a solid continuous film bonded to the surface.

Composition of Coatings

Most coatings consist of three principal parts: pigment, binder, and solvent which dictate the ultimate protection and performance of the coating. The components of a coating all interact to accomplish the purpose for which the coating was designed. The pigments and binder make up the solids of the coating that remain after the coating has dried. The pigments obscure the coated surface, contribute color, prevent premature degradation of the binder by ultraviolet light, and inhibit corrosion. The solvent reduces the viscosity of the coating to permit application—and then evaporates. The binder proceeds from the liquid to the solid state, and in so doing, isolates the surface being coated from the elements.

The interrelationship of coating components is illustrated in Fig. 2-1

Pigments are included in coatings to perform any one or a combination of the following functions:

1. Adding opacity or hiding power
2. Adding color
3. Giving corrosion inhibition
4. Providing resistance to light, heat, water, and chemicals
5. Adjusting the flow properties of the wet coatings
6. Contributing strength

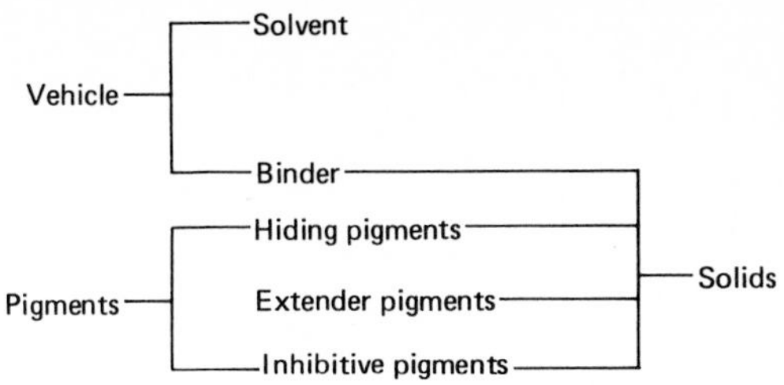

Figure 2-1 Coating components.

Contribution of Each Coating Component

Pigments

Pigments whose principal function is to contribute opacity to the coating are called "hiding" or "prime" pigments. The principal white hiding pigment is titanium dioxide. There are hundreds of colored hiding pigments which, when used singly and in groups, give coatings a variety of colors.

Pigments used to reduce or prevent the corrosion of the coated surface are called inhibitive pigments. The use of inhibitive pigments in modern industrial coatings brings to the war against corrosion all of the principles of corrosion control, including cathodic protection, passivation of the coated surface, and creation of barrier films.

Pigments protect the binder in the coating from the degrading effects of solar radiation. All pigments do this to some degree, but the hiding pigments do it best. Pigments in the dry binder film also reduce the permeability to corrosive elements. Furthermore, pigments add strength and abrasion resistance to coatings. Some pigments are also used to adjust the consistency of coatings so that they may be applied more effectively.

Binders

The binder portion of the vehicle is the component responsible for the curing process as well as for defining basic performance properties. Traditionally these binders have been natural oils derived from plant sources. Some of the more familiar ones are linseed oil and tung oil. Although oil-based coatings are still used in architectural paints, they have been almost totally replaced by synthetic-resin-based coatings for use in heavy-duty applications. These newer coatings dry faster, show better chemical resistance, and give longer service life.

Synthetic-resin binders evolve from the liquid to the solid state by several different drying or cure mechanisms:

1. Solvent evaporation, generally occurring in coatings known as lacquers
2. Drying by a chemical reaction with atmospheric oxygen which is known as oxidation
3. Drying of two-component coatings by a variety of chemical reactions involving each of the components

The simplest drying mechanism is the physical process of evaporation. As solvent evaporates, polymer molecules become more compacted and coil and intertwine. These polymer molecules eventually become bonded together and adhere to the substrate. As an example, vinyl and chlorinated rubber-based coatings dry by this mechanism and demonstrate excellent moisture and chemical resistance. Because this drying or cure process is only physical, not chemical, this type of coating will soften at elevated temperatures and redissolve in strong solvents, the classical definition of a lacquer.

Oil-based coatings, or synthetic binders modified with drying oils, dry by a chemical reaction with atmospheric oxygen. Atmospheric oxygen creates active cross-linking sites of the drying-oil component; these sites connect to form a three-dimensional chemically bonded network. Alkyd and epoxy ester–type binders are examples of systems that cure by oxidation.

The third type of drying or cure mechanism is characteristic of two-component coatings. Two reactive materials, a resin and a polyfunctional cross-linking agent, are mixed prior to use. The functional cross-linking agent reacts with the resin molecules, forming a chemically bonded matrix. Because this type of coating is more chemical- and heat-stable, two-component coatings can be formulated to give heat and solvent resistance, as well as chemical resistance. Two-component epoxy and urethane coatings are representative of this type of drying and cure mechanism.

Solvents

The other portion of the vehicle is the solvent. A solvent dissolves the essential substances in the formulation. It breaks them down and holds the molecular units more or less evenly distributed throughout the solution. The purpose of the solvent is to adjust the flow and viscosity properties of the coating so that it may be successfully brushed, rolled, or sprayed on a surface. In every case, the solvent is designed only to be a carrier for the coating during application and then to evaporate from the coating prior to the completion of the drying cycle.

Volatility or evaporation of solvents is an important property to be considered in the selection of solvents for coatings. The volatility of the solvent determines the drying time of lacquer-type coatings and the tack-time—a time after which flow no longer occurs—of chemically cured coatings. A highly volatile solvent often renders a coating unsprayable by causing the coating to dry before it can touch the surface to be painted—resulting in a phenomenon known as dry spray. A solvent of low volatility often limits the practical applied film thickness of the coating by extending the tack time excessively and causing the coating to sag or run.

SURFACE PREPARATION

Reasons for Surface Preparation

No matter how carefully a coating is formulated and manufactured, or how sound the research on which it was developed, or how sophisticated its chemical technology, the coating will fail prematurely in service if the surface to which it is applied is inadequately prepared. No coating can form a strong bond with a surface if there is something beneath the coating that is weakly bonded to that surface. Dirt, rust, scale, oil, wax, moisture, or other foreign matter provide a weak foundation to hold a paint or coating and consequently cause loss of adhesion. Impurities such as oil or water can prevent adhesion even when present in such small quantities as to be invisible. *Proper surface preparation is vital for best service-life results.*

Types of Substrates

Steel

Probably the most widely used structural material in industrial plants is steel, and it is the surface most frequencly encountered in maintenance operations involving paint. Beginning with new steel construction, let us examine some of the surface-preparation problems which are usually encountered when applying a protective paint coating to its surface. Mill scale is one of the most prevalent contaminants found on new, rolled-steel plates, bars, beams, etc.

Mill scale is a hard, blue-black layer formed on steel during the hot-rolling process. It often adheres tightly to new steel, but usually loosens because it is brittle. Mill scale must be removed for best painting results.

Red rust (iron oxide) is another contaminant that is familiar to anyone who has seen an unprotected piece of steel outdoors. It varies in color from bright red to deep brown and may be loose and powdery or hard and brittle. In any case, it also provides a weak foundation for paint, contributes to the breakdown of a coating applied directly over it, and promotes further corrosion if it is painted over rather than removed. For proper coating performance, *all rust* must be removed.

Before proceeding with the surface cleaning of steel, the surface must be inspected for imperfections and troublesome areas, then corrected as follows:

1. Rough welds and other sharp edges must be ground smooth.
2. Weld spatter must be knocked off with a scraper, chisel, or other means.
3. Seams joined with skip or tack welds must be rewelded with a continuous, smooth bead.
4. Rivets and bolts must be set firm and tight, or caulked if they are angled or not flush.
5. Crevices and pits should be caulked.

Welding should be done before the cleaning operation, and caulking after the cleaning operation.

Choose a surface preparation method that will clean the steel and be within the limits of cost, accessibility, contamination of the manufacturing or cleaning process, damage to machinery and equipment, and physical harm to personnel.

Abrasive Blast Cleaning. Blast cleaning is the best method for cleaning steel because it is effective in removing all scale and corrosion products, as well as chemical contaminants. It also provides an anchor pattern in the steel for the coating due to the roughness of the blast pattern and the increase in surface area. The various types of blasting abrasives which are used perform differently, and produce characteristically different types of surface pattern. The pattern, or profile, resulting from blasting is extremely important in its effect on a coating's performance.

As an example, if the blasted surface is too smooth, there will not be adequate "anchor" for the coating. On the other hand, if the surface is too rough, sharp pinnacles of metal are likely to project through the coating and the steel will remain unprotected. Sand is the most widely used abrasive because of its availability and low cost. It is important to use hard sand that does not produce excessive dust. It is also important to use a clean sand free of clay and other foreign matter. In other words, a washed, graded silica sand should be used.

Conventional sandblasting procedure is to use dry sand and dry air. However, "wet" blasting is sometimes used because dusting is thereby reduced. In wet blasting, water is injected at the nozzle or is mixed with sand in the pot. In this process, very little dust is produced, but a surface covered with wet sand remains, and this must be washed off. Inhibitors are used in the water to prevent the steel from flash rusting.

Steel grit is sometimes used in place of sand. These grit particles are hard, angular-shaped pieces of cast iron, malleable iron, or steel with sharp edges. Steel grit has several advantages over sand; it produces less dust, cuts faster, and can be reclaimed for future use. Its disadvantages are: (1) higher cost and (2) a tendency to roughen the metal excessively. Again, like sand, the proper size of steel grit must be selected. Too fine a grit may not give a good pattern; conversely, coarse grit will cut deeply into the surface, leaving points of metal sticking up. In addition, steel grit used over an extended period under humid conditions will eventually rust—contaminating rather than cleaning the surface.

Steel or iron shot may also be used as an abrasive. But shot blasting is relatively inefficient because the particles are round and smooth and tend to hammer foreign matter into the surface without removing it.

Table 2-1 shows the maximum height of profile produced by various types of abrasives in blast-cleaning steel.

Surface Preparation Procedures. There are accepted standards that cover the cleaning of steel by several methods including abrasive blasting. These standards have been established by the Steel Structures Painting Council (SSPC) and the National Association of Corrosion Engineers (NACE). The following paragraphs are summaries of these standards.

TABLE 2-1 Characteristics of Various Blasting Media

Abrasive	Maximum particle size	Maximum height of profile	
		mils	μm
Sand, very fine	Through 80 mesh*	1.5	38.1
Sand, fine	Through 40 mesh	1.9	48.3
Sand, medium	Through 18 mesh	2.5	63.5
Sand, large	Through 12 mesh	2.8	71.1
Crushed iron grit #G-50	Through 25 mesh	3.3	83.8
Crushed iron grit #G-40	Through 18 mesh	3.6	91.4
Crushed iron grit #G-25	Through 16 mesh	4.0	101.6
Crushed iron grit #G-16	Through 12 mesh	8.0	203.2
Iron shot #S-230	Through 18 mesh	3.0	76.2
Iron shot #S-330	Through 16 mesh	3.3	83.8
Iron shot #S-390	Through 14 mesh	3.6	91.4

*U.S. Sieve series.

Solvent Cleaning: SSPC-SP 1-63. Solvents, emulsions, cleaning compounds, steam cleaning, or similar materials and methods are used to remove oil, grease, dirt, drawing compounds, and other similar foreign matter from surfaces prior to painting. Certain corrosive salts, such as chlorides and sulfates, must be removed with water before cleaning the surface with hydrocarbon-type solvents.

When rags or waste are used with solvents for cleaning, they should be replaced frequently with clean ones, since they absorb grease and then act as a transportation agent for the grease or other contaminants. Use plenty of clean rags, and as an added safety measure, wipe the surface with clean saturated rags.

Solvent cleaning should be undertaken only with adequate ventilation. Caution should always be observed, because of the possibility of fire.

Hand-Tool Cleaning: SSPC-SP 2-63. Hand-tool cleaning is generally confined to wire-brushing, scraping, chipping, and sanding. Surface preparation of this quality is not recommended if finishing systems with poor wetting characteristics such as vinyl coatings are to be applied. This type of cleaning is normally used to prepare the surface in areas where corrosion is a minor problem. On weld seams or spots, always remove the surrounding weld flux spatter, since it promotes premature paint failure.

Power-Tool Cleaning: SSPC-SP 3-63. Power tools of various kinds and shapes, either electric or pneumatic, are used for this type of surface preparation. Rotary wire brushes, chipping hammers, abrasive grinders, descalers, needle guns, and sanding machines are included in the equipment required for power-tool surface preparation.

This method of surface preparation is often used to remove rust and scale left on the surface after other methods have been used to remove the bulk of the contamination. Care must always be exercised; inexperienced workers will sometimes mistake burnished mill scale for bright clean metal. The primer will not adhere to burnished areas of this kind, and premature paint film failure results.

Power-tool cleaning is considered to be somewhat more efficient than hand cleaning if care and good judgment are used by the mechanics.

Flame Cleaning of New Steel: SSPC-SP 4-63. One traverse of the flame-cleaning head, followed by wire-brushing, is normally sufficient to remove rust and scale from new or old steel that has been partially weathered. Badly rusted or heavily scaled steel should be treated as previously painted or "old work." At best, flame cleaning will give results somewhere between that obtained with power tools and that from commercial blast cleaning. The individual job determines the types of gas and the speed of traverse required for acceptable cleaning.

White-Metal Blast Cleaning: SSPC-SP 5-63 (NACE No. 1). This system provides the maximum surface preparation and should result in the best performance possible from the painting system chosen. Surface preparation of this magnitude is often done in the field after the equipment or structure is in place.

This system of cleaning is mandatory for preparing the interior of tanks prior to the application of a lining. The cost is comparatively high, and it is usually used for work only when such costs are warranted. In some cases, accessibility after erection is not possible; therefore, under such conditions, the use of blasting to white metal before erection is considered economical.

The use of white-metal blast cleaning in areas of high humidity without having the steel rust back necessitates the choice of proper blasting conditions so that no rust-back can occur—and painting must be completed, at least, on the same day. A white-metal blast-cleaned surface finish is defined by the SSPC as a surface with a gray-white, uniform metallic color, slightly roughened to form a suitable anchor pattern for coatings. When viewed without magnification, the surface should be free of all oil, grease, dirt, visible mill scale, rust, corrosion products, oxides, paint, or any other foreign matter.

Commercial Blast Cleaning: SSPC-SP 6-63 (NACE No. 3). This type of blast cleaning is generally considered adequate for most surfaces requiring a normally clean surface for painting. It is also economical and generally recommended for surfaces requiring fast-drying protective coatings.

A commercial blast-cleaned surface finish is defined as one in which all oil, grease, dirt, rust scale, and foreign matter have been completely removed from the surface, and all rust, mill scale, and old paint have been completely removed except for slight shadows, streaks, or discolorations caused by rust stain, mill scale, oxides, or slight, tight residues of paint or coating that may remain. If the surface is pitted, slight residues, rust, or paint may be found in the bottom of pits. At least two-thirds of each square inch of surface area should be free of all visible residues, and the remainder should be limited to the light discoloration, slight staining, or light residues mentioned above.

Brush-Off Blast Cleaning: SSPC-SP 7-63 (NACE No. 4). This is a relatively low-cost method of cleaning and is often used at the jobsite to clean materials that are shop- and field-coated before installation. This type of cleaning is not usually recommended for severe environments. It is often used to remove temporary coatings applied for the protection of equipment in transit or storage. It is also used to remove old finishes in bad condition.

A brush-off blast-cleaned surface finish is defined by the SSPC as one from which all oil, grease, dirt, rust-scale, loose mill scale, loose rust, and loose paint or coatings are removed completely. However, tight mill scale and tightly adhering rust, paint, and coatings are permitted to remain—provided that all mill scale and rust have been exposed to the abrasive blast pattern sufficiently to expose numerous flecks of the underlying metal fairly uniformly distributed over the entire surface.

Pickling: SSPC-SP 8-63. Pickling is a method of preparing metal surfaces for painting by completely removing all mill scale, rust, and rust scale by chemical reaction, electrolysis, or both. It is intended that the pickled surface should be completely free of all scale, rust, and foreign matter. Furthermore, the surface should be free of unreacted or harmful acid, alkali, or smut.

Near-White Blast Cleaning: SSPC-SP 10-63T (NACE No. 2). Near-white blast cleaning is a method of preparing metal surfaces for painting or coating by removing nearly all mill scale, rust, rust scale, paint, or foreign matter by the use of abrasives propelled through nozzles or by centrifugal wheels.

A near-white blast-cleaned surface finish is defined as one in which all oil, grease, dirt, mill scale, rust corrosion products, oxides, paint, or other foreign matter have been completely removed from the surface except for very light shadows, very slight streaks, or slight discolorations caused by rust stain, mill scale oxides, or slight, tight residues of paint or coating that may remain. At least 95 percent of each square inch of surface area should be free of all visible residues, and the remainder should be limited to the light discolorations mentioned above.

The complete specification for these procedures may be found in the Steel Structures

Painting Council publication, *Good Painting Practice.*[1] Pictorial standards for these procedures, SSPC-VIS-1, are also available from the Steel Structures Painting Council (see "Specifications and Standards.")

Other Substrates

In addition to steel, there are other substrates that may be encountered occasionally in maintenance operations. These surfaces also must be prepared properly to receive a coating system.

Cast Iron. Cast iron is a porous material and is likely to absorb moisture or other liquids with which it comes in contact. To drive absorbed material out of its pores, cast iron should be heated before sand or grit blasting. This can be done by placing it in an oven for 8 to 12 h at 300°F (149°C) or by heating it with torches until this temperature is reached.

Zinc. New zinc surfaces, galvanized or metal-sprayed, should first be wiped with solvent. The clean surface is then treated with a specially formulated acid-treatment solution. Such solutions are designed to etch zinc, thereby providing sufficient "tooth" for the application of coatings. However, if zinc-coated surfaces are allowed to properly weather, sufficient roughness is obtained to permit direct coating without prior acid treating or sand blasting.

Copper or Brass. These metals should be blasted lightly to remove oxides and provide "tooth" for the coating.

Aluminum. Aluminum surfaces must first be degreased by solvent cleaning. The clean surface may then be cleaned with either a chromate-type conversion treatment, a phosphate chemical treatment, or a wash primer treatment.

Concrete. Concrete is coated to protect it from chemical attack or physical damage by spalling and cracking. There are several factors which must be considered when preparing concrete to receive a coating system:

1. Laitance is a thin layer of incompletely hydrated cement which floats to the surface. Since this layer has poor strength and adheres loosely to the concrete, it must be removed. This may be accomplished by either sand blasting or acid etching.
2. Efflorescence is a deposition of salts on the concrete surface, caused by moisture released during the curing process. These deposits are alkaline and must be removed by acid etching.
3. Form oils, used for easy removal of concrete pouring forms, present problems because a coating will not wet the oils. These oils must be removed by detergent cleaning before acid etching or sand blasting the surface.
4. Concrete hardeners are sometimes used to increase surface hardness and decrease the permeability of concrete. They migrate to the surface and cannot be removed by acid etching. They must be removed by sand blasting.

How to Etch Concrete. Carefully prepare a solution consisting of one part full-strength muriatic acid, mixed with three parts water. Use 1 gal of solution to 100 ft^2 (9.29 m^2) of floor, and scrub well while applying. Be sure to use enamel, plastic, or wooden pails. The acid *must be added* to the water in order to prevent adverse effects. Allow the solution to remain on the floor until it stops bubbling. Scrub the surface well, using a stiff bristle brush. Flush off thoroughly with clean water. After the surface has dried (less than 15 percent moisture), it is ready for painting. *Use appropriate eye- and skin-protective devices.*

USE OF PROTECTIVE COATINGS

Systems Concept

There is a right way and a wrong way to select coatings. For instance, a coat of paint can be "slopped" on every couple of years, or a system can be engineered that will last for a much longer time.

Coatings are very selective with respect to the types of environment and corrosion they are capable of protecting against. They are equally selective concerning some of their other properties such as impact resistance, abrasion resistance, flexibility, bondability to a substrate, and their appearance. *No one generic class of coatings is universally acceptable.* Sometimes there can be a dilemma—such as the one in which the coating provides the protection required for use on catwalks but does not give adequate abrasion resistance.

In most cases, the best result can be obtained by combining two, and sometimes more, coatings into a coating system. The typical coating system usually consists of a primer applied to the metal surface and a top coat. Primers are selected for these characteristics:

1. Bond to metal surface
2. Rust-inhibitive pigment content

Topcoats are usually characterized by:

1. Attractive appearance
2. Color retention and resistance to uv radiation from the sun
3. Low permeability to moisture, chemicals, etc.
4. Abrasion and impact resistance
5. Chemical resistance

With the variety of coatings available today, and the many painting systems possible with these coatings, the concept of the coating system adds versatility to paint technology. With the greater selection available in the coating-system concept, there is now an excellent chance that the user can zero in on the exact protection required without paying a premium for overprotection. See Ref. 2.

In order to arrive at the best coating system for the purpose, certain information must be available, including:

1. Type of surface to be coated
2. Type of surface preparation that can be achieved
3. Type of chemical exposure (if any) expected
4. Operating temperature of surface to be coated
5. Type of structure being coated
6. Degree of abrasion and impact resistance desired

Coating Characteristics

It would be difficult to thoroughly discuss all of the characteristics of the many coating types which are available. Table 2-2 summarizes the important properties of some of the more common coating types. These properties would be considered the optimum to be expected from the coatings. Should more detailed or specific coating information be required, it would be wise to contact a reputable coatings supplier.

TABLE 2-2 Coating Characteristics

Coating type	Type of drying mechanism	Resistance*							Remarks
		Mineral acid	Alkali	Solvent	Water	Weather	Max. drying temperature		
							°F	°C	
Epoxy	Catalyzed (two-component)	Good	Excellent	Fair	Good	Good (chalks)	300	149	
Vinyl	Solvent evaporation	Excellent	Good	Poor	Very good	Very good	120	49	
Chlorinated rubber	Solvent evaporation	Good	Good	Poor	Good	Good	120	49	
Urethane	Catalyzed (two-component)	Excellent	Excellent	Fair	Good	Excellent	300	149	
Silicone	Catalyzed (heat)	Very good	Fair	Poor	Excellent	Excellent	1000	538	
Alkyd	Oxidation	Poor	Poor	Poor	Poor	Good	180	82	
Silicone alkyd	Oxidation	Poor	Poor	Poor	Fair	Very good	300	149	Min 25% silicone
Epoxy ester	Oxidation	Fair	Fair	Poor	Fair	Good (chalks)	250	121	Min 45% epoxy
Coal tar–epoxy	Catalyzed (two-component)	Very good	Excellent	Fair (bleeds)	Excellent	Fair (chalks and bronzes)	325	163	
Inorganic zinc	Hydration	Poor	Poor	Excellent	Excellent	Good (6–8 pH)	750	399	
Bitumastic	Solvent evaporation	Good	Good	Poor	Very good	Poor	150	65	Weatherable grades available

*Splash and spillage

Economics

Cost of Material

Up to this point, the systems concept or coating systems engineered for corrosion control has been discussed; but the economics of painting must also be considered. All too often persons purchasing coatings emphasize the cost per gallon. While the cost per gallon or per liter is a consideration, the decision could change when the costs are based on the volume solids and spreading rate. The following example will illustrate this.

	Cost, $/Gal	Volume solids, %	Dry-film thickness	
			mils	μm
Coating A	4.50	30	4	101.6
Coating B	4.10	25	4	101.6

Coating A

$$\frac{1604 \times 0.3 \text{ (volume solids)}}{4 \text{ mils}} = 122 \text{ ft}^2/\text{gal } (2.9 \text{ m}^2/\text{L})$$

where 1604 is a constant and

$$\frac{\$4.50/\text{gal}}{122 \text{ ft}^2/\text{gal}} = \$0.037/\text{ft}^2 \ (\$0.398/\text{m}^2)$$

Coating B

$$\frac{1604 \times 0.25 \text{ (volume solids)}}{4 \text{ mils}} = 101 \text{ ft}^2/\text{gal } (2.5 \text{ m}^2/\text{L})$$

where 1604 = a constant

$$\frac{\$4.10/\text{gal}}{101 \text{ ft}^2/\text{gal}} = \$0.04/\text{ft}^2 \ (\$0.431/\text{m}^2)$$

We can see from this example that coating A is the best bargain even though it has a greater cost per gallon. The nomograph shown in Fig. 2-2 will allow you to determine material costs and make comparisons quickly and simply.

System Life

Although these economics are impressive, there is still a tendency to look at "low-cost" systems because of modest painting budgets. A look at the cost breakdown of a typical coating job will show the materials cost.

Application labor	30 to 50%
Surface preparation labor	15 to 40%
Coating material	15 to 20%
Clean-up labor	5 to 10%
Tools and equipment	2 to 5%

In this frame of reference, the difference between a high-cost system and a low-cost system is not really significant. With the cost that goes into labor, regardless of materials costs, it would seem that the high-cost system should be the choice. Figure 2-3 further illustrates this.

Each step in the chart lines is either a recoat or touch-up job. The high-cost or properly engineered system does cost more initially, but the longer periods between touch-up

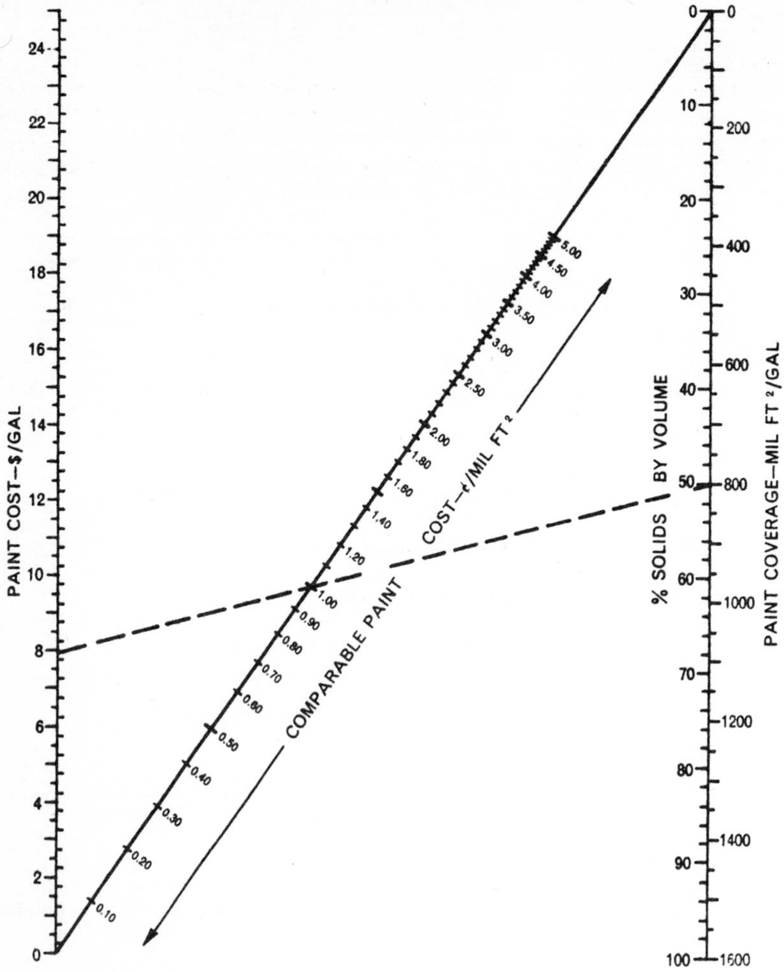

Figure 2-2 Nomograph for calculating paint material costs.

or recoat, as compared to the low-cost system, result in a lower cost of maintaining a structure over a period of years. Quality truly pays.

Inspection

To obtain the planned economies and to realize the maximum potential of a coating system, it is essential that periodic inspections be made before, during, and following the application. The inspection begins with the writing of the specification. Suitable standards for surface preparation, application, and inspection must be established before the job begins. A good specification alone will not guarantee a good coating system. It is necessary for a qualified inspector to see that all specifications are followed and that all defects are promptly remedied. The specification should include the duties and authority of the inspector along with all the quality control or measuring equipment to be used.

A successful coating system must have:

1. Proper film thickness for long-term durability
2. Coating continuity (freedom from holes)

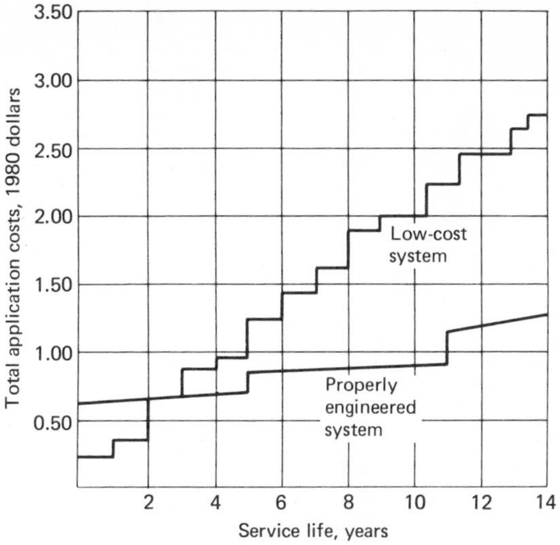

Figure 2-3 Service life vs. application costs.

3. Good adherence, or bonding, to the substrate
4. The ability to cure under proper conditions

Surface Preparation. Before the cleaning operation begins, all parties concerned with a particular job should have a clear understanding of the specified surface preparation. It is recommended that Steel Structures Painting Council specifications (see "Specifications") be used. These specifications may be found in SSPC's publication *Good Painting Practice.*[1]

All cleaning equipment should be inspected before the job begins. The surface cleaning operation should be inspected periodically to ensure compliance with health, safety, and quality of work. To aid in this inspection, there are tools available such as:

1. Steel Structures Painting Council Visual Standards, SSPS-VIS-1
2. CAPS (Clemtex Anchor Pattern Standards)

All cleaned surfaces should be coated before the end of the workday.

Application. Inspection of the application process actually begins with the materials. Records should be made of the code number and batch number of the coatings and of the area where they were applied. Material should be inspected for such deficiencies as skinning, thickening, gassing, gelling, and excessive settling. The painting materials should be mixed and thinned according to manufacturer's directions using mechanical mixers. Special attention should be given to two-component materials to ensure proper mixing, and the painter should have a thorough knowledge of the products' digestion time and pot life. In addition, application equipment should be inspected to ensure that it is in good working order and meets the requirements for the coating being applied.

The application of all coatings should be followed closely by an inspector making sure the equipment and material are behaving properly and the correct amount is applied. Particular attention should be given to adequate coverage on sharp edges, rivets, corners, and crevices. The amount of material applied can be checked with an instrument such as a Nordson prong-type gauge which reads wet-film thickness.

Weather conditions form another important factor in the paint application process. In general, coatings should not be applied at metal temperatures below 40°F (4°C) nor above 125°F (52°C) and at a temperature at least 5°F (2°C) above the dew point. In the case of catalyzed materials, the minimum application temperature is 60°F (15°C) for the metal being coated.

Dry Film. The dry film also should be inspected for defects such as peeling, blistering, pinholing, fish-eyeing, sagging, blushing, and failure to dry. In some cases, the defect may be serious enough that it should be corrected before applying the next coat. Each coat, as well as the total system, should be measured for dry-film thickness. This may be done with a magnetic type gauge such as the Mikrotest (Nordson Corp., Amherst, Ohio).

While the absence of pinholes is important for all coatings, it is critical for coatings being applied for tank or pipe linings. Coatings for these applications should be inspected for pinholes after drying, using a wet-sponge-type detector such as the Tinker-Razor Holiday Detector (Tinker-Razor, San Gabriel, Calif.).

APPLICATION OF COATINGS

Estimating Coverage

In the application of coatings, it is important to know how much material is needed for a given job. To make this determination, the total area to be coated must first be known. This information may be obtained from several sources.

A. Estimating from Weight of Steel:

Steel Plate. 98 ft^2 (3.6 m^2) per inch (centimeter) of thickness, both sides; for example, ½-in-(1.3-cm-) thick steel plate = 196 ft^2/ton (20.1 m^2/t), both sides

Angles. 3 × 3 × ½ in @ 9.4 lb/ft = 213 ft^2/ton (7.6 × 7.6 × 1.3 cm @ 14.1 kg/m = 21.8 m^2/t)

Channels. 6 in @ 8.2 lb/ft = 390 ft^2/ton (15.2 cm. @ 12.2 kg/m = 39.9 m^2/t)

I Beams. 12 in @ 35 lb/ft = 211 ft^2/ton (30.5 cm @ 52.1 kg/m = 21.6 m^2/t)

Piping. 4 in Std. @ 10.8 lb/ft = 219 ft^2/ton (10.2 cm Std. @ 16.1 kg/m = 22.4 m^2/t)

These are examples of the type if information found in steel company manuals or in the AISC (American Institute of Steel Construction) manual.

B. Estimating by Weight and Type of Structure

Structural steel, after erection, generally falls into classifications which may be readily identified:

1. Light structural steel, for example, would be represented by electric transmission towers and installations of that type.
2. Heavy structural steel is the equivalent of built-up bridge girders or designs which utilize built-up and fabricated heavy structural members.
3. Extra-heavy structural steel is used in structures which utilize large rolled structural shapes.
4. Medium or mixed structural steel will be the combination normally seen in trusses, etc., where a mixture of heavy to very heavy compression members are combined with substantial amounts of much lighter cross bracing.
   Using these descriptions, the average surface area per ton of steel for different types of construction is shown in Table 2-3.

C. Geometric and Equipment-Surface Relationships

Certain geometric relationships are repeated for convenience.

1. The area of a triangle is its base times one-half its altitude ($a = \frac{1}{2} bh$).
2. The area of a square or a rectangle is the product of two adjacent sides ($a = hw$).
3. The inside or outside of a cylinder (straight sides) is the product of its height, its diameter, and π (3.1416) ($a = h\pi d$).

TABLE 2-3 Average Surface Area of Steel

Type of construction	Area, ft^2 per ton	Average wt, U.S. tons	Area, mt^2 per ton	Average wt, metric tons
Light	300–500	400	30.7–51.2	41.0
Medium	150–300	225	15.4–30.7	23.0
Heavy	100–150	125	10.2–15.4	12.8
Extra heavy	50–100	75	5.1–10.2	7.7
Average industrial plants	200–250	225	20.5–25.6	23.0

4. The area of a dished head, such as a pressure vessel end, is usually 1.58 times the area of a flat circle ($a = 1.58\pi r^2$).

In addition, certain other relationships are helpful:

1. The area of a standard corrugated sheet with ribs 2½ or 1¼ in (6.4 or 3.2 cm) in size is 8 or 11 percent greater, respectively, than the area it covers.

2. Open wet steel joists should be considered as solid rather than open; double the measurement to account for both sides and add 10 percent.

3. In structural steel, add 20 percent for rivets, plates, and flanges. A 5 percent allowance for plates and flanges is adequate for welded structures.

4. Gratings may be approximated by multiplying the area covered by 4 to 6 (varying according to depth). Use an average of 5 for most gratings.

5. For window sash, allow 3 ft^2 (0.3 m^2) per pane.

6. For large valves, allow 20 ft^2 (1.8 m^2) per valve [above 4 in (10.2 cm)].

Another factor that must be considered in estimating coverage is the amount of material loss during the application process. The following are loss ranges for different methods of application:

Application	Loss, %
Brush	4–8
Roller	4–8
Conventional spray	20–40
Airless spray	10–20

The amount of loss will vary with the size and shape of the surface and the environmental conditions. Under adverse conditions of wind and small surfaces, spray loss can be 50 percent or more.

Now that we have several means of estimating area to be coated, we must determine the amount of paint needed. The following example will illustrate the steps to accomplish this:

Calculate the amount of paint needed to apply the finish coat, at a thickness of 2 mils dry, on the exterior of storage tank. The storage tank is constructed with 50 tons (45.5 t) of ½-in (1.3-cm) steel plate. The finish coat is an epoxy with 50 percent volume solids and is applied by airless spray.

Exterior surface of tank:

½-in (1.3-cm) steel plate, 98 ft^2/ton (10 m^2/t) per side

50 tons (45.5 t) × 98 ft^2/ton (10 m^2/t) = 4900 ft^2 (455 m^2)

Coverage rate of paint:

$$\frac{1604 \times \text{volume solids}}{\text{dry film thickness}} = \text{theoretical coverage}$$

$$\frac{1604 \times 0.5}{2} = \frac{802}{2} = 401 \text{ ft}^2\text{/gal (9.8 m}^2\text{/L)}$$

Allowance for spray loss (20 percent):

$$401\ ft^2/gal\ (9.8\ m^2/L) \times 80\% = 320.8\ ft^2/gal\ (7.8\ m^2/L)$$

Amount of paint needed:

$$\frac{4900\ ft^2 (455\ m^2)}{320.8\ ft^2/gal\ (7.8\ m^2/L)} = 15.27\ gal\ (58.3\ L)$$

Application

Application of coating materials not only involves application methods and equipment but also includes other factors of environmental conditions, materials preparation, film characteristics, curing, and recoating. These factors are generally covered in the coating manufacturers' literature. This literature should be read and understood before the job begins. Do not deviate from these instructions or take short cuts. This can only lead to problems that will be difficult and costly to resolve.

Application Equipment

Conventional Air Sprays. In conventional air spraying, paint is pushed through a nozzle in a spray gun by air pressure. At the nozzle, the paint is atomized by jets of air. The volume output of paint by air spray is relatively small. A conventional air spray setup consists of a pressure pot, air and fluid hose, spray gun, and air supply. The pressure pot should be equipped with a double regulator, to regulate both fluid pressure and atomizing air, and an agitator. All parts of the spray gun should be tightly connected, as should all hose connections. Any air leak will result in inefficient operation.

Airless Sprays. In airless spraying, paint is pumped mechanically under high pressure through a spray-gun nozzle. The paint is pumped at such a high velocity and pressure that it atomizes as it emerges from the small orifice in the nozzle. The volume of paint moved by airless spray is much greater than by air spray. Airless spray permits the application of higher viscosity materials, which will allow greater film build per coat. An airless spray setup consists of a hydraulic pump, high-pressure fluid hose, spray gun, and air supply.

Brushes and Rollers. Although heavy-duty maintenance coatings are normally applied by spray because it is the best and most efficient method, there are situations when it is impossible to spray-apply coating materials, e.g., in difficult-to-reach areas or areas where spray would be hazardous or damaging to adjacent areas. Brush and roller application is slower, less efficient, and more costly.

Brushes and roller equipment must be selected according to the type of material being applied. When strong solvent materials like epoxies are used, brushes and roller materials should be selected for their resistance to strong solvents.

SPECIFICATIONS AND STANDARDS

Surface Preparation

1. Surface preparation specifications (written).
2. SSPC-VIS, "1 Pictorial Surface Preparation Standards," Steel Structures Painting Council, 4400 Fifth Avenue, Pittsburgh, PA 15213.
3. CAPS—Clemtex Anchor Pattern Standards.
4. "Visual Comparator," Clemtex, Ltd., P.O. Box 15214, Houston, TX 77020.

Performance

1. SSPC-VIS 2, "Standard Methods of Evaluating Degree of Rusting on Painted Steel Surfaces," Steel Structures Painting Council, 4400 Fifth Avenue, Pittsburgh, PA 15213.
2. ASTM D 714, "Blister Standards," American Society for Testing and Materials, 1916 Race St., Philadelphia, PA 19103.

REFERENCES AND BIBLIOGRAPHY

1. *Steel Structures Painting Manual,* vol. 1, *Good Painting Practice,* vol. 2, *System and Specifications,* Steel Structures Painting Council, 4400 Fifth Avenue, Pittsburgh, PA 15213, 1966.
2. *Industrial Maintenance Painting,* National Association of Corrosion Engineers, P.O. Box 986, Katy, TX 77450, 1973.

NACE Standards and Technical Committee Reports, National Association of Corrosion Engineers, P.O. Box 986, Katy, TX 77450, in the following subject areas:
Surface Preparation for Protective Coatings
Protective Coatings for Atmospheric Service
Coatings and Linings for Immersion Service
Specialty Coatings
Painting and Decorating Craftsman's Manual and Textbook, Painting and Decorating Contractors of America, 7223 Lee Highway, Falls Church, VA 22046, 1978.

Uhlig H. H. (ed.): *The Corrosion Handbook,* Wiley, New York, 1948.

NACE Basic Corrosion Course Text, National Association of Corrosion Engineers, P.O. Box 986, Katy, TX 77450, 1977.

part D

Supplementary Technical Data

section 19

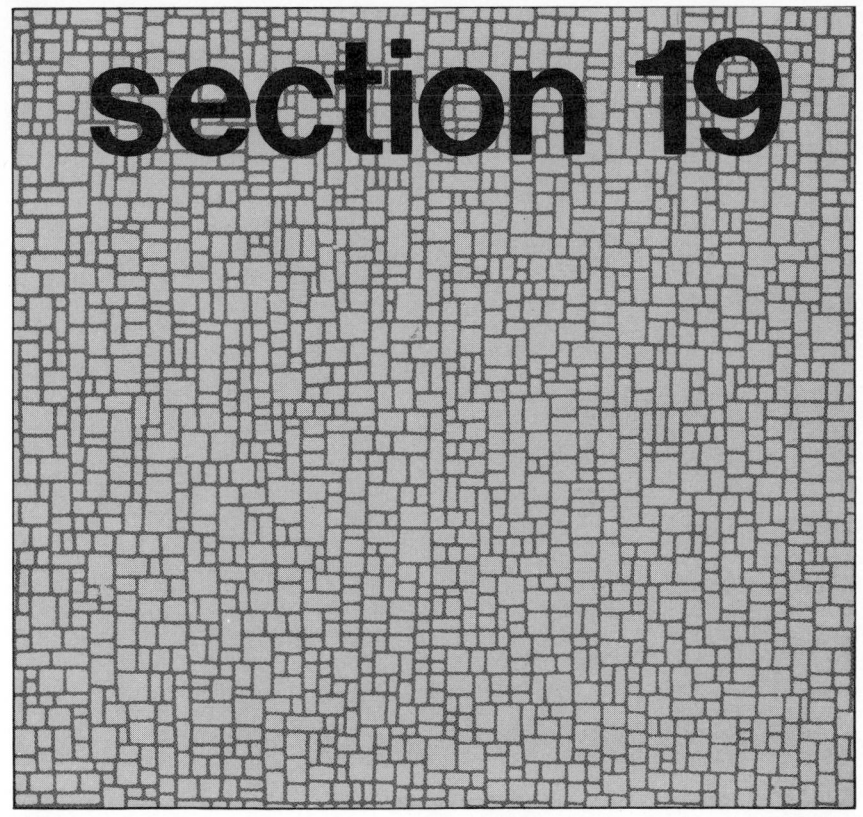

Properties of Structural Metals and Wood

chapter 19-1

Iron and Steel

prepared by

American Iron and Steel Institute
Washington, D.C.

GLOSSARY

Air-hardening steel A steel containing sufficient carbon and other alloying elements to harden fully during cooling in air or other gaseous mediums from a temperature above its transformation range. Same as self-hardening steel.

Alloy steel Steel containing significant quantities of alloying elements (other than carbon and the commonly accepted amounts of Mn, Si, S, P) added to effect changes in the mechanical or physical properties.

Annealing Heating iron or steel to a suitable temperature, holding it there for a suitable time, and then cooling it at a suitable rate, for such purposes as reducing hardness, improving machinability, facilitating cold working, producing a desired microstructure, or obtaining desired mechanical, physical, or other properties.

Arc cutting Cutting metal with an arc between an electrode and the metal itself.

Austenite A solid solution of one or more elements in face-centered cubic iron. Unless otherwise designated (such as nickel austenite), the solute is generally assumed to be carbon.

Bainite A decomposition product of austenite consisting of an aggregate of ferrite and carbide. In general, it forms at temperatures lower than those at which very fine perlite forms and higher than that at which martensite begins to form on cooling.

Braze welding Welding in which a groove, fillet, plug, or slot weld is made using a nonferrous filler metal having a melting point lower than that of the base metal but higher than 800°F.

Brinell hardness test A test for determining the hardness of a material by forcing a hard steel or carbide ball of specified diameter into it under a specified load. Hardness values are prefixed by the letters Bhn, for Brinell hardness number.

Brittleness The quality of a material that leads to crack propagation without appreciable plastic deformation.

Carbide A compound of carbon with one or more metallic elements.

Carbon equivalent An empirical value, determined by various carbon-equivalent formulas that represent the sum of the effects of various elements in steel on its hardenability, or in cast iron on its solidification characteristics.

Carbon steel Steel which owes its properties chiefly to various percentages of carbon without substantial amounts of other alloying elements.

Charpy test A pendulum type of impact test in which a specimen, supported at both ends as a simple beam, is broken by the impact of a falling pendulum. The energy absorbed in breaking the specimen is a measure of the impact strength of the metal.

Fatigue The progressive localized permanent structural change that occurs in a material subjected to repeated or fluctuating strains at stresses having a maximum value less than the tensile strength of the material. Fatigue may culminate in cracks or fracture after a sufficient number of fluctuations.

Ferrite A solid solution in which alpha iron is the solvent and which is characterized by a body centered cubic crystal structure.

Flakes Short discontinuous internal fissures in ferrous metals attributed to stresses produced by localized transformation and decreased solubility of hydrogen during cooling after hot working.

Flake graphite Excess carbon in gray cast iron which has precipitated in flake form during solidification.

Forging Plastically deforming metal, usually hot, into desired shapes with compressive force, with or without dies.

Grain An individual crystal in a polycrystalline metal or alloy.

Graphitic carbon Free carbon in steel or cast iron.

Hardenability The property that determines the depth and distribution of hardness induced by quenching.

Heat-affected zone The portion of the base metal which was not melted during brazing, cutting, or welding, but whose microstructure and physical properties were altered by the heat.

Martensite An interstitial supersaturated solid solution of carbon in iron having a body-centered tetragonal lattice. Its microstructure is characterized by an acicular, or needlelike pattern.

Martensitic steel Steel which has martensite as its chief constituent after cooling in air.

McQuaid-Ehn test The most commonly used test for the determination of austenitic grain size of metals. Through microscopic examination, the grain size of a sample can be compared to eight standard grain sizes ranging from the finest, #8, to the coarsest, #1.

Mechanical properties Those properties that reveal the reaction, elastic or inelastic, of a material to an applied force, or that involve the relationship between stress and strain.

Microstructure The structure of metals as revealed by examination of polished and etched samples with the microscope.

Normalizing Heating a ferrous alloy to a suitable temperature above the transformation range and subsequently cooling in still air at room temperature.

Notch sensitivity A measure of the reduction in strength of a metal caused by the presence of stress concentration. Values can be obtained from static, impact, or fatigue tests.

Overheating Heating a metal or alloy to such a high temperature that its properties are impaired. When the original properties cannot be restored by further heat treating, by mechanical work, or by a combination of heat treating and working, the overheating is known as burning.

Quench hardening Rapid cooling by immersion in liquids or gases or by contact with cold metal.

Rockwell hardness test A test for determining the hardness of a material based upon the depth of penetration of a specified penetrator into the specimen. Hardness values are prefixed by the letters HR and the hardness scale used, e.g., HRC, HRB, HR30T, etc.

Skelp A piece or strip of metal, manufactured to a suitable thickness, width, and edge configuration, from which pipe or tubing is made.

Stress relief A low-temperature heat treatment for the purpose of relieving residual stress and thereby improving mechanical stability of steels which have been placed in a state of strain as a result of either hardening, cold working, or welding.

Tempering A process of reheating quench-hardened or normalized ferrous alloy to a temperature below the transformation range and then cooling at any rate desired.

Temper brittleness Brittleness that results when certain steels are held within or are slowly cooled through a certain range of temperature below the transformation range.

Temper carbon This is a form of graphite developed in producing malleable iron. The iron is cast as white iron, and in an extended heat treatment starting at a temperature greater than 1600°F, graphite is precipitated within the iron. It is called temper carbon because it is formed in the solid state during heat treatment.

Tensile strength In tensile testing, the ratio of maximum load to original cross-sectional area. Also called ultimate strength, or ultimate tensile strength (UTS).

Toughness Ability of a metal to absorb energy and deform plastically before fracturing.

Transformation temperature The temperature at which a change in phase occurs, e.g., the transformation of ferrite to austenite during heating.

Yield point The load per unit of original cross section at which a marked increase in deformation occurs without increase in load.

IRON AND STEEL CLASSIFICATIONS

The terms iron and steel represent two broad classes of iron-based materials. The irons are primarily castings consisting of iron alloys which contain more than 2 percent carbon and from 1 to 3 percent silicon. The steels are iron-based alloys, usually containing less than 2 percent carbon and significant amounts of manganese and other alloying elements.

Iron

Gray Cast Iron

This most widely used form of cast iron is so named because of the dull gray surface of fractured pieces. This color is due mainly to the flakes of free graphite in the microstructure. Carbon is present at about 2.8 to 4.0 percent and silicon at 1 to 3 percent. Tensile strengths of 20,000 to 60,000 lb/in^2 are normal. Higher-strength levels usually require alloy additions. Gray iron is readily machinable and has outstanding properties for applications involving vibrational damping or moderate thermal shock.

White Cast Iron

White cast iron is used for applications requiring wear resistance, e.g., grinding balls and mill liners. By control of chemical composition, principally a low silicon content, and/or by accelerated cooling, the flake graphite common to gray iron is replaced with massive iron carbides which impart high abrasion-resistance characteristics.

Malleable Cast Iron

The composition limits for malleable iron are generally given as 2.0 to 3.0 percent carbon, 1.0 to 1.8 percent silicon, and 0.2 to 0.5 percent manganese. Malleable iron is produced by heat-treating white cast iron of a suitable composition. Malleable iron exhibits a higher tensile strength and better ductility than gray cast iron. Malleable iron is used for structural and decorative parts when the combination of moderate strength, good toughness, and machinability is desired.

Ductile Cast Iron

Ductile iron is similar in composition to gray cast iron. The addition of a small amount of magnesium or cerium, called nodulizing, results in the formation of graphite as spherical particles rather than the flakes found in gray iron. This nodular shape of the graphite results in superior ductility and tensile properties as compared to gray or malleable cast iron. Like gray iron, ductile iron can be heat-treated or alloyed to enhance certain properties, especially wear resistance.

Steel

Composition ranges for carbon and alloy steels have been established jointly by the Society of Automotive Engineers (SAE) and the American Iron and Steel Institute (AISI). The numbers assigned to the individual steels have a significance based on the system outlined in Table 1-1. Typical mechanical properties and specifications are listed in Tables 1-2, 1-3, and 1-4. Tables 1-2 and 1-3* are a guide to show the potential user what to expect of a given grade of steel in the indicated condition. Data were obtained from specimens of 0.505-in diameter which were machined from 1-in rounds; gage lengths were 2 in. Average properties of hot-rolled, normalized, and annealed material are listed in Table 1-2, while properties of quenched and tempered grades are for single heats (Table 1-3). Sources of the data are Bethlehem Steel Corp. and Republic Steel Corp.

Because of the many variables that affect a steel's properties, however, these listed properties should not be considered either as average or typical. Both strengths and ductilities may range up and down from the values given, depending on the compositions of individual heats of the same grade, section sizes, and internal structures. Properties of carbon steels and many alloy steels are also affected by residual elements (particularly nickel, chromium, and molybdenum), even though their amounts are limited to maximums by AISI and SAE specifications.

Fine-grained steels normally have better impact strength than coarse-grained types, a factor which should be considered when reviewing the results of Izod tests. Hardness values are not always related to corresponding tensile strengths. In particular, this effect

*From *1977 Metal Progress Databook*, ASM, with permission of the American Society for Metals.

TABLE 1-1 Classification and Standardization of the SAE and AISI Steels*

Numerals and digits	Type of steel and nominal alloy content	Numerals and digits	Type of steel and nominal alloy content
Carbon steels		Nickel-chromium-molybdenum steels	
10XX(a)	Plain carbon (Mn 1.00% max)	93XX	Ni 3.25; Cr 1.20; Mo 0.12
11XX	Resulfurized	94XX	Ni 0.45; Cr 0.40; Mo 0.12
12XX	Resulfurized and rephosphorized	97XX	Ni 0.55; Cr 0.20; Mo 0.20
15XX	Plain carbon (max Mn range—1.00 to 1.65%)	98XX	Ni 1.00; Cr 0.80; Mo 0.25
Manganese steels		Nickel-molybdenum steels	
13XX	Mn 1.75	46XX	Ni 0.85 and 1.82; Mo 0.20 and 0.25
Nickel steels		48XX	Ni 3.50; Mo 0.25
23XX	Ni 3.50	Chromium steels	
25XX	Ni 5.00	50XX	Cr 0.27, 0.40, 0.50 and 0.65
Nickel-chromium steels		51XX	Cr 0.80, 0.87, 0.92, 0.95, 1.00 and 1.05
31XX	Ni 1.25; Cr 0.65 and 0.80	Chromium steels	
32XX	Ni 1.75; Cr 1.07	50XXX	Cr 0.50 } C 1.00 min
33XX	Ni 3.50; Cr 1.50 and 1.57	51XXX	Cr 1.02 } C 1.00 min
34XX	Ni 3.00; Cr 0.77	52XXX	Cr 1.45 } C 1.00 min
Molybdenum steels		Chromium-vanadium steels	
40XX	Mo 0.20 and 0.25	61XX	Cr 0.60, 0.80 and 0.95; V 0.10 and 0.15 min
44XX	Mo 0.40 and 0.52	Tungsten-chromium steel	
Chromium-molybdenum steels		72XX	W 1.75; Cr 0.75
41XX	Cr 0.50, 0.80 and 0.95; Mo 0.12, 0.20, 0.25 and 0.30	Silicon-manganese steels	
Nickel-chromium-molybdenum steels		92XX	Si 1.40 and 2.00; Mn 0.65, 0.82 and 0.85; Cr 0.00 and 0.65
43XX	Ni 1.82; Cr 0.50 and 0.80; Mo 0.25	High-strength low-alloy steels	
43BVXX	Ni 1.82; Cr 0.50; Mo 0.12 and 0.25; V 0.03 min	9XX	Various SAE grades
47XX	Ni 1.05; Cr 0.45; Mo 0.20 and 0.35	Boron steels	
81XX	Ni 0.30; Cr 0.40; Mo 0.12	XXBXX	B denotes boron steel
86XX	Ni 0.55; Cr 0.50; Mo 0.20	Leaded steels	
87XX	Ni 0.55; Cr 0.50; Mo 0.25	XXLXX	L denotes leaded steel
88XX	Ni 0.55; Cr 0.50; Mo 0.35	(a) XX in the last two digits of these designations indicates that the carbon content (in hundredths of a percent) is to be inserted.	

*These steels have been identified by a numerical index system that is partially descriptive of their composition. The first digit indicates the type to which the steel belongs; thus "1" indicates a carbon steel; "2" indicates a nickel steel; "3" indicates a nickel-chromium steel. In the case of the simple alloy steels, the second number usually indicates the percentage of the predominating alloying element. Usually the last two or three digits indicate the average carbon content in "points" or hundredths of a percent. Thus "2340" indicates a nickel steel containing approximately 3 percent nickel (3.25 to 3.75) and 0.40 percent carbon (0.35 to 0.45).

Source: ASM Metal Handbook, 9th ed., 1978, vol. 1, p. 124. Reprinted with permission of the American Society for Metals.

occurs with carbon steels because they are shallow-hardening. Hardness tests are made on surfaces, and these hardnesses will not reflect the tensile strengths obtained with specimens representing bar centers. (Center hardnesses are usually lower than surface hardnesses.)

Hot-rolled properties for alloy steels are not given because these grades are customarily heat-treated. Because the samples were small enough to assure full quenching, values indicate strengths and ductilities which may be obtained with hardened, fine-grained steels of a similar section size at room temperatures.

Carbon Steel

Carbon steel owes its properties chiefly to various percentages of carbon without substantial amounts of other alloying elements. Carbon steel is also referred to as ordinary steel or plain carbon steel. Commercially, steel is classed as carbon steel under the fol-

TABLE 1-2 Properties of Selected Carbon and Alloy Steels—Hot-Rolled, Normalized, and Annealed

AISI No.*	Treatment	Yield strength, lb/in²	Tensile strength, lb/in²	Elongation, %	Reduction in area, %	Hardness, Bhn	Impact strength (Izod), ft·lb
1015	As rolled	45,500	61,000	39.0	61.0	126	81.5
	Normalized (1700°F)	47,000	61,500	37.0	69.6	121	85.2
	Annealed (1600°F)	41,250	56,000	37.0	69.7	111	84.8
1020	As rolled	48,000	65,000	36.0	59.0	143	64.0
	Normalized (1600°F)	50,250	64,000	35.8	67.9	131	86.8
	Annealed (1600°F)	42,750	57,250	36.5	66.0	111	91.0
1022	As rolled	52,000	73,000	35.0	67.0	149	60.0
	Normalized (1700°F)	52,000	70,000	34.0	67.5	143	86.5
	Annealed (1600°F)	46,000	65,250	35.0	63.6	137	89.0
1030	As rolled	50,000	80,000	32.0	57.0	179	55.0
	Normalized (1700°F)	50,000	75,500	32.0	60.8	149	69.0
	Annealed (1550°F)	49,500	67,250	31.2	57.9	126	51.2
1040	As rolled	60,000	90,000	25.0	50.0	201	36.0
	Normalized (1650°F)	54,250	85,500	28.0	54.9	170	48.0
	Annealed (1450°F)	51,250	75,250	30.2	57.2	149	32.7
1050	As rolled	60,000	105,000	20.0	40.0	229	23.0
	Normalized (1650°F)	62,000	108,500	20.0	39.4	217	20.0
	Annealed (1450°F)	53,000	92,250	23.7	39.9	187	12.5
1060	As rolled	70,000	118,000	17.0	34.0	241	13.0
	Normalized (1650°F)	61,000	112,500	18.0	37.2	229	9.7
	Annealed (1450°F)	54,000	90,750	22.5	38.2	179	8.3
1144	As rolled	61,000	102,000	21.0	41.0	212	39.0
	Normalized (1650°F)	58,000	96,750	21.0	40.4	197	32.0
	Annealed (1450°F)	50,250	84,750	24.8	41.3	167	48.0
1340	Normalized (1600°F)	81,000	121,250	22.0	62.9	248	68.2
	Annealed (1475°F)	63,250	102,000	25.5	57.3	207	52.0
3140	Normalized (1600°F)	87,000	129,250	19.7	57.3	262	39.5
	Annealed (1500°F)	61,250	100,000	24.5	50.8	197	34.2
4130	Normalized (1600°F)	63,250	97,000	25.5	59.5	197	63.7
	Annealed (1585°F)	52,250	81,250	28.2	55.6	156	45.5
4140	Normalized (1600°F)	95,000	148,000	17.7	46.8	302	16.7
	Annealed (1500°F)	60,500	95,000	25.7	56.9	197	40.2
4150	Normalized (1600°F)	106,500	167,500	11.7	30.8	321	8.5
	Annealed (1500°F)	55,000	105,750	20.2	40.2	197	18.2
4320	Normalized (1640°F)	67,250	115,000	20.8	50.7	235	53.8
	Annealed (1560°F)	61,625	84,000	29.0	58.4	163	81.0
4340	Normalized 1600°F)	125,000	185,500	12.2	36.3	363	11.7
	Annealed (1490°F)	68,500	108,000	22.0	49.9	217	37.7
4620	Normalized (1650°F)	53,125	83,250	29.0	66.7	174	98.0
	Annealed (1575°F)	54,000	74,250	31.3	60.3	149	69.0
4820	Normalized (1580°F)	70,250	109,500	24.0	59.2	229	81.0
	Annealed (1500°F)	67,250	98,750	22.3	58.8	197	68.5
5140	Normalized (1600°F)	68,500	115,000	22.7	59.2	229	28.0
	Annealed (1525°F)	42,500	83,000	28.6	57.3	167	30.0

1080	As rolled	85,000	140,000	12.0	17.0	293	5.0	5150	Normalized (1600°F)	76,750	126,250	20.7	58.7	255	23.2
	Normalized (1650°F)	76,000	146,500	11.0	20.6	293	5.0		Annealed (1520°F)	51,750	98,000	22.0	43.7	197	18.5
	Annealed (1450°F)	54,500	89,250	24.7	45.0	174	4.5	5160	Normalized (1575°F)	77,000	138,750	17.5	44.8	269	8.0
1095	As rolled	83,000	140,000	9.0	18.0	293	3.0		Annealed (1495°F)	40,000	104,750	17.2	30.6	197	7.4
	Normalized (1650°F)	72,500	147,000	9.5	13.5	293	4.0	6150	Normalized (1600°F)	89,250	136,250	21.8	61.0	269	26.2
	Annealed (1450°F)	55,000	95,250	13.0	20.6	192	2.0		Annealed (1500°F)	59,750	96,750	23.0	48.4	197	20.2
1117	As rolled	44,300	70,600	33.0	63.0	143	60.0	8620	Normalized (1675°F)	51,750	91,750	26.3	59.7	183	73.5
	Normalized (1650°F)	44,000	67,750	33.5	63.8	137	62.8		Annealed (1600°F)	55,875	77,750	31.3	62.1	149	82.8
	Annealed (1575°F)	40,500	62,250	32.8	58.0	121	69.0	8630	Normalized (1600°F)	62,250	94,250	23.5	53.5	187	69.8
1118	As rolled	45,900	75,600	32.0	70.0	149	80.0		Annealed (1550°F)	54,000	81,750	29.0	58.9	156	70.2
	Normalized (1700°F)	46,250	69,250	33.5	65.9	143	76.3	8650	Normalized (1600°F)	99,750	148,500	14.0	40.4	302	10.0
	Annealed (1450°F)	41,250	65,250	34.5	66.8	131	78.5		Annealed (1465°F)	56,000	103,750	22.5	46.4	212	21.7
1137	As rolled	55,000	91,000	28.0	61.0	192	61.0	8740	Normalized (1600°F)	88,000	134,750	16.0	47.9	269	13.0
	Normalized (1650°F)	57,500	97,000	22.5	48.5	197	47.0		Annealed (1500°F)	60,250	100,750	22.2	46.4	201	29.5
	Annealed (1450°F)	50,000	84,750	26.8	53.9	174	36.8	9255	Normalized (1650°F)	84,000	135,250	19.7	43.4	269	10.0
1141	As rolled	52,000	98,000	22.0	38.0	192	8.2		Annealed (1550°F)	70,500	112,250	21.7	41.1	229	6.5
	Normalized (1650°F)	58,750	102,500	22.7	55.5	201	38.8	9310	Normalized (1630°F)	82,750	131,500	18.8	58.1	269	88.0
	Annealed (1500°F)	51,200	86,800	25.5	49.3	163	25.3		Annealed (1550°F)	63,750	119,000	17.3	42.1	241	58.0

*All grades are fine-grained except for those in the 1100 series which are coarse-grained. Normalizing and annealing temperatures are given in parentheses.

Source: 1977 Metal Progress Databook, ASM. Reproduced with permission of the American Society for Metals.

TABLE 1-3 Properties of Selected Carbon and Alloy Steels—Quenched and Tempered

AISI No.*	Tempering temperature, °F	Tensile strength, lb/in²	Yield strength, lb/in²	Elongation, %	Reduction in area, %	Hardness, Bhn
1030†	400	123,000	94,000	17	47	495
	600	116,000	90,000	19	53	401
	800	106,000	84,000	23	60	302
	1000	97,000	75,000	28	65	255
	1200	85,000	64,000	32	70	207
1040†	400	130,000	96,000	16	45	514
	600	129,000	94,000	18	52	444
	800	122,000	92,000	21	57	352
	1000	113,000	86,000	23	61	269
	1200	97,000	72,000	28	68	201
1040	400	113,000	86,000	19	48	262
	600	113,000	86,000	20	53	255
	800	110,000	80,000	21	54	241
	1000	104,000	71,000	26	57	212
	1200	92,000	63,000	29	65	192
1050†	400	163,000	117,000	9	27	514
	600	158,000	115,000	13	36	444
	800	145,000	110,000	19	48	375
	1000	125,000	95,000	23	58	293
	1200	104,000	78,000	28	65	235
1050	400	—	—	—	—	—
	600	142,000	105,000	14	47	321
	800	136,000	95,000	20	50	277
	1000	127,000	84,000	23	53	262
	1200	107,000	68,000	29	60	223
1060	400	160,000	113,000	13	40	321
	600	160,000	113,000	13	40	321
	800	156,000	111,000	14	41	311
	1000	140,000	97,000	17	45	277
	1200	116,000	76,000	23	54	229
1080	400	190,000	142,000	12	35	388
	600	189,000	142,000	12	35	388
	800	187,000	138,000	13	36	375
	1000	164,000	117,000	16	40	321
	1200	129,000	87,000	21	50	255
1095†	400	216,000	152,000	10	31	601
	600	212,000	150,000	11	33	534
	800	199,000	139,000	13	35	388
	1000	165,000	110,000	15	40	293
	1200	122,000	85,000	20	47	235

AISI No.*	Tempering temperature, °F	Tensile strength, lb/in²	Yield strength, lb/in²	Elongation, %	Reduction in area, %	Hardness, Bhn
4130†	400	236,000	212,000	10	41	467
	600	217,000	200,000	11	43	435
	800	186,000	173,000	13	49	380
	1000	150,000	132,000	17	57	315
	1200	118,000	102,000	22	64	245
4140	400	257,000	238,000	8	38	510
	600	225,000	208,000	9	43	445
	800	181,000	165,000	13	49	370
	1000	138,000	121,000	18	58	285
	1200	110,000	95,000	22	63	230
4150	400	280,000	250,000	10	39	530
	600	256,000	231,000	10	40	495
	800	220,000	200,000	12	45	440
	1000	175,000	160,000	15	52	370
	1200	139,000	122,000	19	60	290
4340	400	272,000	243,000	10	38	520
	600	250,000	230,000	10	40	486
	800	213,000	198,000	10	44	430
	1000	170,000	156,000	13	51	360
	1200	140,000	124,000	19	60	280
5046	400	253,000	204,000	9	25	482
	600	205,000	168,000	10	37	401
	800	165,000	135,000	13	50	336
	1000	136,000	111,000	18	61	282
	1200	114,000	95,000	24	66	235
50B46	400	—	—	—	—	560
	600	258,000	235,000	10	37	505
	800	202,000	181,000	13	47	405
	1000	157,000	142,000	17	51	322
	1200	128,000	115,000	22	60	273
50B60	400	—	—	—	—	600
	600	273,000	257,000	8	32	525
	800	219,000	201,000	11	34	435
	1000	163,000	145,000	15	38	350
	1200	130,000	113,000	19	50	290
5130	400	234,000	220,000	10	40	475
	600	217,000	204,000	10	46	440
	800	185,000	175,000	12	51	379
	1000	150,000	136,000	15	56	305
	1200	115,000	100,000	20	63	245

1095	400	187,000	120,000	10	30	401	5140	400	260,000	238,000	9	38	490
	600	183,000	118,000	10	30	375		600	229,000	210,000	10	43	450
	800	176,000	112,000	12	32	363		800	190,000	170,000	13	50	365
	1000	158,000	98,000	15	37	321		1000	145,000	125,000	17	58	280
	1200	130,000	80,000	21	47	269		1200	110,000	96,000	25	66	235
1137	400	157,000	136,000	5	22	352	5150	400	282,000	251,000	5	37	525
	600	143,000	122,000	10	33	285		600	252,000	230,000	6	40	475
	800	127,000	106,000	15	48	262		800	210,000	190,000	9	47	410
	1000	110,000	88,000	24	62	229		1000	163,000	150,000	15	54	340
	1200	95,000	70,000	28	69	197		1200	117,000	118,000	20	60	270
1137†	400	217,000	169,000	5	17	415	5160	400	322,000	260,000	4	10	627
	600	199,000	163,000	9	25	375		600	290,000	257,000	9	30	555
	800	160,000	143,000	14	40	311		800	233,000	212,000	10	37	461
	1000	120,000	105,000	19	60	262		1000	169,000	151,000	12	47	341
	1200	94,000	77,000	25	69	187		1200	130,000	116,000	20	56	269
1141	400	237,000	176,000	6	17	461	51B60	400	—	—	—	—	600
	600	212,000	186,000	9	32	415		600	—	—	—	—	540
	800	169,000	150,000	12	47	331		800	237,000	216,000	11	36	460
	1000	130,000	111,000	18	57	262		1000	175,000	160,000	15	44	355
	1200	103,000	86,000	23	62	217		1200	140,000	126,000	20	47	290
1144	400	127,000	91,000	17	36	277	6150	400	280,000	245,000	8	38	538
	600	126,000	90,000	17	40	262		600	250,000	228,000	8	39	483
	800	123,000	88,000	18	42	248		800	208,000	193,000	10	43	420
	1000	117,000	83,000	20	46	235		1000	168,000	155,000	13	50	345
	1200	105,000	73,000	23	55	217		1200	137,000	122,000	17	58	282
1330†	400	232,000	211,000	9	39	459	81B45	400	295,000	250,000	10	33	550
	600	207,000	186,000	9	44	402		600	256,000	228,000	8	42	475
	800	168,000	150,000	15	53	335		800	204,000	190,000	11	48	405
	1000	127,000	112,000	18	60	263		1000	160,000	149,000	16	53	338
	1200	106,000	83,000	23	63	216		1200	130,000	115,000	20	55	280
1340	400	262,000	231,000	11	35	505	8630	400	238,000	218,000	9	38	465
	600	230,000	206,000	12	43	453		600	215,000	202,000	10	42	430
	800	183,000	167,000	14	51	375		800	185,000	170,000	13	47	375
	1000	140,000	120,000	17	58	295		1000	150,000	130,000	17	54	310
	1200	116,000	90,000	22	66	252		1200	112,000	100,000	23	63	240
4037	400	149,000	110,000	6	38	310	8640	400	270,000	242,000	10	40	505
	600	138,000	111,000	14	53	295		600	240,000	220,000	10	41	460
	800	127,000	106,000	20	60	270		800	200,000	188,000	12	45	400
	1000	115,000	95,000	23	63	247		1000	160,000	150,000	16	54	340
	1200	101,000	61,000	29	60	220		1200	130,000	116,000	20	62	280
4042	400	261,000	241,000	12	37	516	86B45	400	287,000	238,000	9	31	525
	600	234,000	211,000	13	42	455		600	246,000	225,000	9	40	475
	800	187,000	170,000	15	51	380		800	200,000	191,000	11	41	395
	1000	143,000	128,000	20	59	300		1000	160,000	150,000	15	49	335
	1200	115,000	100,000	28	66	238		1200	131,000	127,000	19	58	280

TABLE 1-3 Properties of Selected Carbon and Alloy Steels—Quenched and Tempered (*Continued*)

AISI No.*	Tempering tempera-ture, °F	Tensile strength, lb/in^2	Yield strength, lb/in^2	Elonga-tion, %	Reduc-tion in area, %	Hard-ness, Bhn	AISI No.*	Tempering tempera-ture, °F	Tensile strength, lb/in^2	Yield strength, lb/in^2	Elonga-tion, %	Reduc-tion in area, %	Hard-ness, Bhn
8650	400	281,000	243,000	10	38	525	9255	400	305,000	297,000	1	3	601
	600	250,000	225,000	10	40	490		600	281,000	260,000	4	10	578
	800	210,000	192,000	12	45	420		800	233,000	216,000	8	22	477
	1000	170,000	153,000	15	51	340		1000	182,000	160,000	15	32	352
	1200	140,000	120,000	20	58	280		1200	144,000	118,000	20	42	285
8660	400	—	—	—	—	580	9260	400	—	—	—	—	600
	600	—	—	—	—	535		600	—	—	—	—	540
	800	237,000	225,000	13	37	460		800	255,000	218,000	8	24	470
	1000	190,000	176,000	17	46	370		1000	192,000	164,000	12	30	390
	1200	155,000	138,000	20	53	315		1200	142,000	118,000	20	43	295
8740	400	290,000	240,000	10	41	578	94B30	400	250,000	225,000	12	46	475
	600	249,000	225,000	11	46	495		600	232,000	206,000	12	49	445
	800	208,000	197,000	13	50	415		800	195,000	175,000	13	57	382
	1000	175,000	165,000	15	55	363		1000	145,000	135,000	16	65	307
	1200	143,000	131,000	20	60	302		1200	120,000	105,000	21	69	250

*All grades are fine-grained except for those in the 1100 series which are coarse-grained. Normalizing and annealing temperatures are given in parentheses.
†Water quenched.

Source: 1977 Metal Progress Databook, ASM. Reproduced with permission of the American Society for Metals.

lowing conditions: when no minimum content is specified or guaranteed for aluminum, chromium, cobalt, columbium, molybdenum, nickel, titanium, tungsten, vanadium, zirconium, or any other alloying element added to obtain a desired alloying effect; when the specified or guaranteed minimum content for copper does not exceed 0.40 percent; when the maximum content specified or guaranteed for any of the following elements does not exceed the percentages noted: manganese—1.65 percent, silicon—0.60 percent, and copper—0.60 percent.

Alloy Steels

Steel is classified as alloy steel when the content of alloying elements exceeds certain limits. Alternatively, it is a steel in which a definite range of alloying elements is specified within the limits of the recognized commercial field of alloy steels: Al and Cr up to 3.99 percent; Co, Cb, Mo, Ni, Ti, W, V, Zr, or any other alloying elements added to obtain a desired alloying effect.

High-Strength Low-Alloy (HSLA) Steels

High-strength low-alloy steel represents a specific group of steels in which mechanical properties and, in some cases, resistance to atmospheric corrosion are obtained by the addition of moderate amounts of one or more alloying elements other than carbon.

Applications. These steels, because of their high strength/weight ratio, abrasion resistance and, in the case of certain compositions, improved atmospheric corrosion resistance, are adapted particularly for use in mobile equipment and other structures in which substantial weight savings are generally desirable. Typical applications are truck bodies, frames, structural members, scrapers, truck wheels, cranes, shovels, booms, chutes, and conveyors.

Stainless and Heat-Resisting Steel

These steels contain at least 4 percent chromium, and stainless steels contain at least 10 percent chromium with or without additions of other elements. Their principal use is in applications requiring resistance to oxidation and/or corrosion.

These steels may be subdivided into the following groups:

1. Martensitic stainless steels are iron-chromium alloys which are hardenable by heat treatment.
2. Ferritic stainless steels are iron-chromium alloys which cannot be hardened significantly by heat treatment.
3. Austenitic stainless steels are iron-chromium-nickel and iron-chromium-nickel-manganese alloys which are hardenable only by cold working.
4. Precipitation-hardening stainless steels are iron-chromium-nickel alloys with additional elements which are hardenable by solution-treating and aging.

All four groups of stainless steels achieve their "stainless" characteristics through their ability to form a tightly adhering film of iron chromium oxide which strongly resists attack by both the atmosphere and a wide variety of industrial gases and chemicals. This effect, plus the superior high-temperature strength characteristics exhibited by many of these alloys, accounts for their wide use at ordinary and elevated temperatures with a wide choice of mechanical properties and several distinct levels of corrosion resistance.

Hardenable Chromium Steels (Martensitic and Magnetic)

Metallurgically the martensitic steels may be considered to be high-chromium counterparts of the quenched and tempered carbon and alloy steels. Because of the effect of chromium on increasing hardenability and resistance to tempering, the martensitic stainless steels are capable of being heat-treated to a variety of desirable mechanical properties.

TABLE 1-4 Properties of Selected Carbon and Alloy Steels—Cold-Drawn Bars*

AISI no. size, in	As cold-drawn					Cold-drawn followed by low-temperature stress relief					Cold-drawn followed by high-temperature stress relief				
	Strength					Strength					Strength				
	Tensile, 1000 lb/in^2	Yield, 1000 lb/in^2	Elongation in 2 in, %	Reduction in area, %	Hardness, Bhn	Tensile, 1000 lb/in^2	Yield, 1000 lb/in^2	Elongation in 2 in, %	Reduction in area, %	Hardness, Bhn	Tensile, 1000 lb/in^2	Yield, 1000 lb/in^2	Elongation in 2 in, %	Reduction in area, %	Hardness, Bhn
1018, 1025															
⅝ to ⅞ incl.	70	60	18	40	143						65	45	20	45	131
Over ⅞ to 1¼ incl.	65	55	16	40	131						60	45	20	45	121
Over 1¼ to 2 incl.	60	50	15	35	121						55	45	16	40	111
Over 2 to 3 incl.	55	45	15	35	111						50	40	15	40	101
1117, 1118															
⅝ to ⅞ incl.	75	65	15	40	149	80	70	15	40	163	70	50	18	45	143
Over ⅞ to 1¼ incl.	70	60	15	40	143	75	65	15	40	149	65	50	16	45	131
Over 1¼ to 2 incl.	65	55	13	35	131	70	60	13	35	143	60	50	15	40	121
Over 2 to 3 incl.	60	50	12	30	121	65	55	12	35	131	55	45	15	40	111
1035															
⅝ to ⅞ incl.	85	75	13	35	170	90	80	13	35	179	80	60	16	45	163
Over ⅞ to 1¼ incl.	80	70	12	35	163	85	75	12	35	170	75	60	15	45	149
Over 1¼ to 2 incl.	75	65	12	35	149	80	70	12	35	163	70	60	15	40	143
Over 2 to 3 incl.	70	60	10	30	143	75	65	10	30	149	65	55	12	35	131
1040, 1140															
⅝ to ⅞ incl.	90	80	12	35	179	95	85	12	35	187	85	65	15	45	170
Over ⅞ to 1¼ incl.	85	75	12	35	170	90	80	12	35	179	80	65	15	45	163
Over 1¼ to 2 incl.	80	70	10	30	163	85	75	10	30	170	75	60	15	40	149
Over 2 to 3 incl.	75	65	10	30	149	80	70	10	30	163	70	55	12	35	143

1045, 1146, 1145															
⅝ to ⅞ incl.	95	85	12	35	187	100	90	12	35	197	90	70	15	45	179
Over ⅞ to 1¼ incl.	90	80	11	30	179	95	85	11	30	187	85	70	15	45	170
Over 1¼ to 2 incl.	85	75	10	30	170	90	80	10	30	179	80	65	15	40	163
Over 2 to 3 incl.	80	70	10	30	163	85	75	10	25	170	75	60	12	35	149
1050, 1137, 1151															
⅝ to ⅞ incl.	100	90	11	35	197	105	95	11	35	212	95	75	15	45	187
Over ⅞ to 1¼ incl.	95	85	11	30	187	100	90	11	30	197	90	75	15	40	179
Over 1¼ to 2 incl.	90	80	10	30	179	95	85	10	30	187	85	70	15	40	170
Over 2 to 3 incl.	85	75	10	30	170	90	80	10	25	179	80	65	12	35	163
1141															
⅝ to ⅞ incl.	105	95	11	30	212	110	100	11	30	223	100	80	15	40	197
Over ⅞ to 1¼ incl.	100	90	10	30	197	105	95	10	30	212	95	80	15	40	187
Over 1¼ to 2 incl.	95	85	10	30	187	100	90	10	25	197	90	75	15	40	179
Over 2 to 3 incl.	90	80	10	20	179	95	85	10	20	187	85	70	12	30	170
1144															
⅝ to ⅞ incl.	110	100	10	30	223	115	105	10	30	229	105	85	15	40	212
Over ⅞ to 1¼ incl.	105	95	10	30	212	110	100	10	30	223	100	85	15	40	197
Over 1¼ to 2 incl.	100	90	10	25	197	105	95	10	25	212	95	80	15	35	187
Over 2 to 3 incl.	95	85	10	20	187	100	90	10	20	197	90	75	12	30	179

*The tensile and yield strengths of carbon-steel bars are improved by cold-drawing. By comparison, the tensile strength of hot-rolled bars is about 10 percent less, and their yield strength is some 40 percent less. For example, a low-carbon steel, with a yield-tensile strength ratio of about 0.55 in the form of hot-rolled bars, will have a ratio of about 0.85 after cold-drawing. While there is some sacrifice in elongation, reduction in area, and impact strength, these changes are relatively insignificant in most structural applications or engineering components.

This improvement is of interest to the design engineer seeking a better strength-weight ratio or a reduction in costs by the elimination of alloy contents and heat treatment. The enhanced properties may also be useful in applications involving threads, notches, cut-outs, and in other design requirements that might affect strength adversely.

Turned and polished and turned, ground, and polished bars have the mechanical properties of hot-rolled bars.

Source: AISI Committee of Hot Rolled and Cold Finished Bar Producers.

Source: 1977 Metal Progress Databook, ASM, p. 31. Produced with permission of the American Society for Metals.

Type 410 is the basic general-purpose steel, used for steel valves, pump shafts, bolts, and miscellaneous parts requiring corrosion resistance and moderate strength up to 1000°F (537°C). Where high strength and abrasion resistance are required, grades 420 and 440C are recommended. For products such as quality bushings, cutlery, valves, and ball bearings, which require the highest hardness values obtainable, type 440C is always considered.

All of the martensitic stainless steels can be forged, pierced, rolled, and machined. Sulfur and selenium are added to some compositions for easier machining.

Representative mechanical properties and applications are shown in Table 1-5.

Nonhardenable Chromium Steels (Ferritic and Magnetic)

These types are essentially nonhardenable by heat treatment as compared to the martensitic steels. They develop their maximum softness, ductility, and corrosion resistance in the annealed condition.

Their chromium content ranges from 11 to 26 percent, but is usually between 17 and 26 percent, and these steels are essentially free from nickel. Unlike the martensitic steels, they are ferritic and are hardened only by cold-working.

The 17 percent chromium steels, types 430, 434, and 436, are the most popular ferritic stainless steels. These are widely used in the field of automotive trim, and appliances, and architectural hardware. Types 405 and 409 are low-cost stainless steels designed for use in the as-welded condition. These alloys will resist corrosion from soap, sugar solutions, ammonia, alcohol, crude oil, gasoline, and other mild reagents.

Nonhardenable Chromium-Nickel and Chromium-Nickel-Manganese Steels (Austenitic and Nonmagnetic)

These types are austenitic, are essentially nonmagnetic in the annealed condition, and do not harden by heat treatment. They are annealed by rapidly cooling from high temperatures to develop maximum softness and ductility as well as corrosion resistance. Cold-working develops a wide range of mechanical properties.

Owing to their high ductility, particularly at low temperatures, their all-around resistance to corrosion and general oxidation, and their creep strength, the austenitic stainless steels are used over a temperature range from −320 to 1650°F (−196 to 900°C). Since they are generally readily formed and welded, they are particularly versatile and are used over a wide range of applications.

Type 304, also known as 18-8, is the most common grade of austenitic chromium-nickel steel. It has excellent resistance to corrosion and oxidation, has high creep strength, and is frequently used at temperatures up to 1500°F. It is used in high-temperature service in such applications as high-pressure steam-pipe as well as boiler tubes, radiant superheaters, and oil-refinery tubes. Types 321 and 347 stainless steel are similar to type 304 except that titanium and columbium, respectively, have been added to these steels. The titanium and columbium additions combine with carbon and minimize intergranular corrosion that may occur in certain media after welding.

Type 316 stainless steel, which contains molybdenum, is able to withstand attack by many industrial chemicals and solvents and in particular inhibits pitting caused by chlorides. Because of the molybdenum addition, type 316 can withstand corrosive attack by sodium and calcium brines, hydrochlorite solutions, phosphoric acid, and the sulfite and sulfurous acids used in the pulp and paper industry.

Precipitation-Hardenable Stainless Steels (Martensitic and Magnetic)

These are iron-chromium-nickel alloys with additional elements which are hardenable by solution treating and aging. Products are fabricated in the solution-annealed condition then aged to develop optimum properties. These steels have been developed over the past 20 years and may be difficult to hot-process. They are used in high strength/weight ratio applications, including elevated temperatures. However, some grades, for example, S17400, are also used for ambient-temperature applications where high strength and corrosion resistance are required.

Representative mechanical properties and applications are shown in Table 1-5.

TABLE 1-5 Classification, Composition, and General Properties of Stainless Steel (*Continued*)

AISI type (UNS)	Typical composition, %[a]	Form[b]	Mechanical properties of annealed material at room temperature: Tensile strength, 10^3 lb/in^2	Yield strength, 10^3 lb/in^2	Elongation, %	Hardness	Characteristics and typical applications
			Austenitic[c]				
201 (S20100)	16·18 Cr, 3.5·5.5 Ni, 0.15 C, 3.5·7.5 Mn, 1.0 Si, 0.060 P, 0.030 S, 0.25 N	Sheets	115	55	55	Rb 90	High work-hardening rate low-nickel equivalent of type 301; flatware, automobile wheel covers, trim
		Strips	115	55	55	Rb 90	
		Tubing	115	55	55	Rb 90	
202 (S20200)	17·19 Cr, 4·6 Ni, 0.15 C, 7.5·10.0 Mn, 1.0 Si, 0.060 P, 0.030 S, 0.25 N	Sheets	105	55	55	Rb 90	General-purpose low-nickel equivalent of type 302; kitchen equipment; hub caps, milk handling
		Strips	105	55	55	Rb 90	
		Tubing	105	55	55	Rb 90	
205 (S20500)	16.5·18 Cr, 1·1.75 Ni, 0.12 0.25 C, 14·15.5 Mn, 1.0 Si, 0.060 P, 0.030 S, 1·1.75 Mo, 0.32 0.40 N	Plates	120	69	58	Rb 98	Lower work-hardening rate than type 202; used for spinning and special drawing operations, nonmagnetic and cryogenic parts
301 (S30100)	16·18 Cr, 6·8 Ni, 0.15 C, 2.0 Mn, 1.0 Si, 0.045 P, 0.030 S	Plates	105	40	55	Bhn 165	High work-hardening rate: used for structural applications where high strength plus high ductility is required. Railroad cars, trailer bodies, aircraft structurals, fasteners: automobile wheel covers, trim, pole line hardware
		Sheets	110	40	60	Rb 85	
		Strips	110	40	60	Rb 85	
		Tubing	105	40	50	Rb 85	
302 (S30200)	17·19 Cr, 8·10 Ni, 0.15 C, 2.0 Mn, 1.0 Si, 0.045 P, 0.030 S	Bars	85	35	60	Bhn 150	General purpose austenitic stainless steel; trim, food-handling equipment, aircraft: cowlings; antennas, springs, cookware: building exteriors; tanks, hospital household appliances, jewelry, oil refining equipment; signs
		Plates	90	35	60	Rb 80	
		Sheets	90	40	50	Rb 85	
		Strips	90	40	50	Rb 85	
		Tubing	85	35	50	Rb 85	
		Wire	90	35	60	Rb 83	
302B (S30215)	17·19 Cr, 8·10 Ni, 0.15 C, 2.0 Mn, 2.0·30 Si, 0.045 P, 0.030 S	Bars	90	40	50	Rb 85	More resistant to scale than type 302; furnace parts, still liners, heating elements, annealing covers, burner sections
		Plates	90	40	50	Rb 85	
		Sheets	95	40	55	Rb 85	
		Strips	95	40	55	Rb 85	
		Tubing	85	35	50	Rb 85	

TABLE 1-5 Classification, Composition, and General Properties of Stainless Steel (*Continued*)

AISI type (UNS)	Typical composition, %[a]	Form[b]	Mechanical properties of annealed material at room temperature: Tensile strength, 10^3 lb/in^2	Yield strength, 10^3 lb/in^2	Elongation, %	Hardness	Characteristics and typical applications
303 (S30300)	17·19 Cr, 8·10 Ni, 0.15 C, 2.0 Mn, 1.0 Si, 0.20 P, 0.15 S min, 0.60 Mo (optional)	Bars Tubing Wire	90 80 90	35 38 35	50 53 50	Bhn 160 Rb 76	Free-machining modification of type 302, for heavier cuts. Screw machine products, shafts, valves, bolts, bushings, nuts
303Se (S30323)	17·19 Cr, 8 10 Ni, 0.15 C, 2.0 Mn, 1.0 Si, 0.20 P, 0.060 S, 0.15 Se min						Free-machining modification of type 302, for lighter cuts; used where hot working or cold heading may be involved; aircraft fittings, bolts, nuts, rivets, screws, studs
304 (S30400)	18·20 Cr, 8·10.50 Ni, 0.08 C, 2.0 Mn, 1.0 Si, 0.045 P, 0.030 S	Bars Plates Sheets Strips Tubing Wire	85 82 84 84 85 90	35 35 42 42 35 35	60 60 55 55 50 60	Bhn 149 Bhn 149 Rb 80 Rb 80 Rb 80 Rb 83	Low-carbon modification of type 302 for restriction of carbide precipitation during welding. Chemical- and food-processing equipment, brewing equipment, cryogenic vessels, gutters, downspouts, flashings
304L (S30403)	18·20 Cr, 8·12 Ni, 0.03 C, 2.0 Mn, 1.0 Si, 0.045 P, 0.030 S	Plates Sheets Strips Tubing	79 81 81 78	33 39 39 34	60 55 55 55	Bhn 143 Rb 79 Rb 79 Rb 75	Extra-low-carbon modification of type 304 for further restriction of carbide precipitation during welding; coal hopper linings; tanks for liquid fertilizer and tomato paste
(S30430)	17·19 Cr, 8·10 Ni, 0.08 C, 2.0 Mn, 1.0 Si, 0.045 P, 0.030 S, 3·4 Cu	Wire	73	31	70	Rb 70	Lower work-hardening rate than type 305; severe cold-heading applications
304N (S30451)	18·20 Cr, 8·10.5 Ni, 0.08 C, 2.0 Mn, 1.0 Si, 0.045 P, 0.030 S, 1.10·0.16 N	Bars Sheets	90 90	42 48	55 50	Bhn 180 Rb 85	Higher nitrogen than type 304 to increase strength with minimum effect on ductility and corrosion resistance, more resistant to increased magnetic permeability; type 304 applications requiring higher strength

305 (S30500)	17·19 Cr, 10.50 13 Ni, 0.12 C, 2.0 Mn, 1.0 Si, 0.045 P, 0.030 S	Plates	85	35	55	——	Low work-hardening rate; used for spin forming, severe drawing, cold heading and forming; coffee urn tops; mixing bowls; reflectors
		Sheets	85	38	50	Rb 80	
		Strips	85	38	50	Rb 80	
		Tubing	80	36	56	Rb 80	
		Wire	85	34	60	Rb 77	
308 (S30800)	19·21 Cr, 10·12 Ni, 0.08 C, 2.0 Mn, 1.0 Si, 0.045 P, 0.030 S	Bars	85	30	55	Rb 80	Higher alloy steel having high corrosion and heat resistance; welding filler metals to compensate for alloy loss in welding; industrial furnaces
		Plates	85	30	55	Bhn 150	
		Sheets	85	35	50	Rb 80	
		Strips	85	35	50	Rb 80	
		Tubing	85	35	50	Rb 80	
		Wire	95[d]	60[d]	50[d]		
309 (S30900)	22·24 Cr, 12·15 Ni, 0.20 C, 2.0 Mn, 1.0 Si, 0.045 P, 0.030 S	Bars	95	40	45	Rb 83	High-temperature strength and scale resistance. Aircraft heaters, heat treating equipment, annealing covers, furnace parts, heat exchangers, heat-treating-trays, oven linings, pump parts
		Plates	95	40	45	Bhn 170	
		Sheets	90	45	45	Rb 85	
		Strips	90	45	45	Rb 85	
		Tubing	90	45	45	Rb 85	
		Wire	105	70	35	Rb 98	
309S (S30908)	22·24 Cr, 12·15 Ni, 0.08 C, 2.0 Mn, 1.0 Si, 0.045 P, 0.030 S						Low-carbon modification of type 309 welded constructions; assemblies subject to moist corrosion conditions
310 (S31000)	24·26 Cr, 19·22 Ni, 0.25 C, 2.0 Mn, 1.5 Si, 0.045 P, 0.030 S	Bars	95	45	50	Rb 89	Higher elevated-temperature strength and scale resistance than type 309; heat exchangers, furnace parts, combustion chambers, welding filler metals, gas turbine parts, incinerators, recuperators, rolls for roller hearth furnaces
		Plates	95	45	50	Bhn 170	
		Sheets	95	45	45	Rb 85	
		Strips	95	45	45	Rb 85	
		Tubine [e]	95	45	45	Rb 85	
		Wire	105	75	30	Rb 98	
310 S (S31008)	24·26 Cr, 19·22 Ni, 0.08 C, 2.0 Mn, 1.5 Si, 0.045 P, 0.30 S						Low carbon modification of type 310. Welded constructions; jet engine rings
314 (S31400)	23·26 Cr, 19·22 Ni, 0.25 C, 2.0 Mn, 1.5·3.0 Si, 0.045 P, 0.030 S	Bars	100	50	45	Bhn 180	More resistant to scale than type 310 Severe cold heading or forming applications. Annealing and carburizing boxes, heat treating fixtures, radiant tubes
		Plates	100	50	45	Bhn 180	
		Sheets	100	50	40	Rb 85	

TABLE 1-5 Classification, Composition, and General Properties of Stainless Steel (*Continued*)

AISI type (UNS)	Typical composition, %[a]	Form[b]	Mechanical properties of annealed material at room temperature: Tensile strength, 10^3 lb/in^2	Yield strength, 10^3 lb/in^2	Elongation, %	Hardness	Characteristics and typical applications
316 (S31600)	16·18 Cr, 10·14 Ni, 0.08 C, 2.0 Mn, 1.0 Si, 0.045 P, 0.030 S, 2.0·3.0 Mo	Bars	80	30	60	Rb 78	Higher corrosion resistance than types 302 and 304; high creep strength. Chemical- and pulp-handling equipment, photographic equipment, brandy vats, fertilizer parts, ketchup cooking kettles, yeast tubs
		Plates	82	36	55	Bhn 149	
		Sheets	84	42	50	Rb 79	
		Strips	84	42	50	Rb 79	
		Tubing	85	35	50	Rb 79	
		Wire	80	30	60	Rb 78	
316L (S31603)	16·18 Cr, 10·14 Ni, 0.08 C, 2.0 Mn, 1.0 Si, 0.20 P, 0.10 0.030 S, 2.0·3.0 Mo	Plates	81	34	55	Bhn 146	Extra-low-carbon modification of type 316; welded construction where intergranular carbide precipitation must be avoided. Type 316 applications requiring extensive welding
		Sheets	81	42	50	Rb 79	
		Strips	81	42	50	Rb 79	
		Tubing	80	35	55	Rb 78	
316F (S31620)	16·18 Cr, 10·14 Ni, 0 08 C, 2.0 Mn, 1.0 Si, 0.045 P, S min, 1.75·2.50 Mo	Bars	82	35	57	Bhn 143	Higher phosphorus and sulfur than type 316 to improve machining and nonseizing characteristics; automatic screw machine parts
		Sheets	85	38	60	Rb 85	
316H (S31651)	16·18 Cr, 10·14 Ni, 0.08 C, 2.0 Mn, 1.0 Si, 0.045 P, 0.030 S, 2·3 Mo, 0.10·0.16 N	Bars	90	42	55	Bhn 180	Higher nitrogen than type 316 to increase strength with minimum effect on ductility and corrosion resistance; type 316 applications requiring extra strength
		Sheets	90	48	48	Rb 85	
317 (S31700)	18·20 Cr, 11.15 Ni, 0.08 C, 2.0 Mn, 1.0 Si, 0.045 P, 0.030 S, 3.0·4.0 Mo	Bars	85	40	50	Bhn 160	Higher corrosion and creep resistance than type 316; dyeing and ink-manufacturing equipment
		Plates	85	40	50	Bhn 160	
		Sheets	90	40	45	Rb 85	
		Strips	90	40	45	Rb 85	
		Tubing	85	35	40	Rb 85	

317L (S31703)	18 20 Cr, 11·15 Ni, 0.03 C, 2.0 Mn, 1.0 Si, 0.045 P, 0.030 S, 3·4 Mo	Plates	85	35	55	Rb 80	Extra-low-carbon modification of type 317 for restriction of carbide precipitation during welding; welded assemblies
		Sheets	86	38	55	Rb 85	
		Tubing	86	50	55		
321 (S32100)	17·19 Cr, 9·12 Ni, 0.08 C, 2.0 Mn, 1 0 Si, 0.045 P, 0.030 S (Ti, 5 × C min)	Bars	85	35	55	Bhn 150	Stabilized for weldments subject to severe corrosive conditions and for service from 800 to 1600°F; aircraft exhaust manifolds, boiler shells, process equipment, expansion joints; cabin heaters; fire walls; flexible couplings; pressure vessels
		Plates	85	30	55	Bhn 160	
		Sheets	90	35	45	Rb 80	
		Strips	90	35	45	Rb 80	
		Tubing	85	35	50	Rb 80	
		Wire	95	65	40	Rb 89	
329 (S32900)	25·30 Cr, 3·6 Ni, 0.10 C, 2.0 Mn, 1.0 Si, 0.040 P, 0.030 S, 1·2 Mo	Bars	105	80	25	Bhn 230	Austenitic ferritic type with general corrosion resistance similar to type 316 but with better resistance to stress-corrosion cracking; capable of age-hardening; valves, valve fittings, piping, pump parts
		Strips	105	80	25	Bhn 230	
330 (N08330)	17·20 Cr, 34·37 Ni, 0.08 C, 2.0 Mn, 0.75·1.50 Si, 0.040 P, 0.030 S	Bars	85	42	45	Rb 80	Good resistance to carburization and to heat and thermal shock; heat-treating fixtures
		Plates	90	38	45	Rb 80	
		Sheets	80	38	40		
		Strips	80	38	40		
347 (S34700)	17·19 Cr, 9·13 Ni, 0 08 C, 2.0 Mn, 1.0 Si, 0.045 P, 0.030 S (Cb—Ta 10 × C)	Bars	90	35	50	Bhn 160	Similar to type 321 with higher creep strength; airplane exhaust stacks; welded tank cars for chemicals; jet engine parts
		Plates	90	35	50	Bhn 160	
		Sheets	95	40	45	Rb 85	
		Strips	95	40	45	Rb 85	
		Tubing	85	35	45	Rb 85	
		Wire	100^d	70^d	40^d	Rb 95^d	
348 (S34800)	17·19 Cr, 9·13 Ni, 0.08 C, 2.0 Mn, 1.0 Si, 0.045 P, 0.030 S, (Cb–Ta 10 × C min but 0.10 Ta max), 0.20 Co						Similar to type 321; low retentivity; tubes and pipes for radioactive systems; nuclear energy uses
384 (S38400)	15·17 Cr, 17·19 Ni, 0.08 C, 2.0 Mn, 1.0 Si, 0.045 P, 0.030 S	Wire	75	35	55	Rb 70	Suitable for severe cold-heading or cold-forming; lower cold-work-hardening rate than type 304; bolts, rivets, screws, instrument parts

TABLE 1-5 Classification, Composition, and General Properties of Stainless Steel (*Continued*)

AISI type (UNS)	Typical composition, %[a]	Form[b]	Mechanical properties of annealed material at room temperature: Tensile strength, 10^3 lb/in^2	Yield strength, 10^3 lb/in^2	Elongation, %	Hardness	Characteristics and typical applications
Ferritic[a] 405 (S40500)	11.5·14.5 Cr, 0.08 C, 1.0 Mn, 1.0 Si, 0.040 P, 0.030 S, 0.1·0.3 Al	Bars	70	40	30	Bhn 150	Nonhardenable grade for assemblies where air-hardening types such as 410 or 403 are objectionable; annealing boxes, quenching racks, oxidation-resistant partitions
		Plates	65	40	30	Bhn 150	
		Sheets	65	40	25	Rb 75	
		Tubing	65	40	25	Rb 80	
		Wire	90[d]	75[d]	15[d]		
409 (S40900)	10.5·11 75 Cr, 0.08 C, 1.0 Mn, 1.0 Si, 0.045 P, 0.045 S (Ti 6 × C, but with 0.75 max)	Bars	65	35	25	Rb 75	General-purpose construction stainless; automotive exhaust systems, transformer and capacitor cases, dry fertilizer spreaders, tanks for agricultural sprays
		Plates	65	35	25	Rb 75	
		Sheets	65	35	25	Rb 75	
		Strips	65	35	25	Rb 75	
429 (S42900)	14·16 Cr, 0.12 C, 1.0 Mn, 1.0 Si, 0.040 P, 0.030 S	Bars	71	45	30	Bhn 156	Improved weldability as compared to type 430; nitric acid and nitrogen fixation equipment
		Plates	70	40	30	Bhn 163	
430 (S43000)	16·18 Cr, 0.12 C, 1.0 Mn, 1.0 Si, 0.040 P, 0.030 S	Bars	75	45	30	Bhn 155	General-purpose nonhardenable chromium type; decorative trim, nitric acid tanks; annealing baskets; combustion chambers, dishwashers; heaters; mufflers; range hoods; recuperators; restaurant equipment
		Plates	75	40	30	Bhn 160	
		Sheets	75	50	25	Rb 85	
		Strips	75	50	25	Rb 85	
		Tubing	75	40	25	Rb 80	
		Wire	70	40	35	Rb 82	
430F (S43020)	16·18 Cr, 0.12 C, 1.25 Mn, 1.0 Si, 0.060 P, 0.15 S min, 0.60 Mo (optional)	Bars	80	55	25	Bhn 170	Free-machining modification of type 430, for heavier cuts; screw machine parts
		Wire	95[d]	85[d]	10[d]	Rb 92[d]	
430FSe (S43023)	16·18 Cr, 0.12 C, 1.25 Mn, 1.0 Si, 0.060 P, 0.060 S, 0.15 Se min						Free-machining modification of type 430, for lighter cuts; machined parts requiring light cold-heading or forming

434 (S43400)	16·18 Cr, 0.12 C, 1.0 Mn, 1.0 Si, 0.040 P, 0.030 S, 0.75·1.25 Mo	Sheets Strips Wire	77 77 79	53 53 60	23 23 33	Rb 83 Rb 83 Rb 90	Modification of type 430 designed to resist atmospheric corrosion in the presence of winter road-conditioning and dust-laying compounds; automotive trim and fasteners
436 (S43600)	16·18 Cr, 0.12 C, 1.0 Mn, 1.0 Si, 0.040 P, 0.030 S, 0.75·1.25 Mo (Cb—Ta 5 × C min, 0.70 max)	Sheets Strips	77 77	53 53	23 23	Rb 83 Rb 83	Similar to types 430 and 434; used where low "roping" or "ridging" required; general corrosion and heat-resistant applications such as automobile trim
442 (S44200)	18·23 Cr, 0.20 C, 1.0 Mn, 1.0 Si, 0.040 P, 0.030 S	Bars	80	45	20	Rb 90	High-chromium steel, principally for parts which must resist high service temperatures without scaling; furnace parts, nozzles, combustion chambers
446 (S44600)	23·27 Cr, 0.20 C, 1.5 Mn, 1.0 Si, 0.040 P, 0.030 S, 0.25 N	Bars Plates Sheets Strips Tubing Wire	80 85 80 80 80 95[d]	50 55 50 50 50 80[d]	25 25 20 20 25 15[d]	Rb 86 Rb 84 Rb 83 Rb 83 Rb 84 Rb 92[d]	High-resistance to corrosion and scaling at high temperatures especially for intermittent service; often used in sulfur-bearing atmosphere; annealing boxes, combustion chambers, glass molds, heaters, pyrometer tubes, recuperators, stirring rods, valves
Martensitic[c] 403 (S40300)	11.5·13.0 Cr, 0.15 C, 1.0 Mn, 0.5 Si, 0.040 P, 0.030 S	Bars Sheets Strips Tubing Wire	75 70 70 75 95[d]	40 45 45 40 80[d]	35 25 25 35 15[d]	Rb 82 Rb 80 Rb 80 Rb 80 Rb 92[d]	"Turbine quality" grade; steam turbine blading and other highly stressed parts including jet engine rings
410 (S41000)	11.5·13.5 Cr, 0.15 C, 1 0 Mn, 1.0 Si, 0.040 P, 0.030 S	Bars Plates Sheets Strips Tubing Wire	75 70 70 70 75 75	40 35 45 45 40 40	35 30 25 25 30 30	Rb 82 Bhn 150 Rb 80 Rb 80 Rb 82 Rb 82	General-purpose heat-treatable type; machine parts, pump shafts, bolts, bushings, coal chutes, cutlery, finishing tackle, hardware, jet engine parts, mining machinery, rifle barrels, screws, valves

TABLE 1-5 Classification, Composition, and General Properties of Stainless Steel (*Continued*)

AISI type (UNS)	Typical composition, %[a]	Form[b]	Mechanical properties of annealed material at room temperature: Tensile strength, 10^3 lb/in²	Yield strength, 10^3 lb/in²	Elongation, %	Hardness	Characteristics and typical applications
414 (S41400)	11.5-13.5 Cr, 1.25-2.50 Ni, 0.15 C, 1.0 Mn, 1.0 Si, 0.040 P, 0.030 S	Bars Plates Sheets Strips Wire	115 115 120 120 135[d]	90 90 105 105 115[d]	20 20 15 15 10[d]	Bhn 235 Bhn 235 Rb 98 Rb 98 Rc 29[d]	High hardenability steel; springs, tempered rules, machine parts, bolts, mining machinery, scissors, ships' bells, spindles, valve seats
416 (S41600)	12-14 Cr, 0.15 C, 1.25 Mn, 1.0 Si, 0.060 P, 0.15 S min, 0.60 Mo (optional)	Bars Tubing Wire	75 75 75	40 40 40	30 30 20	Rb 82 Rb 82 Rb 82	Free-machining modification of type 410, for heavier cuts; aircraft fittings, bolts, nuts, fire extinguisher inserts, rivets, screws
416Se (S41623)	12-14 Cr, 0.15 C, 1.25 Mn, 1.0 Si, 0.060 P, 0.060 S, 0.15 Se min						Free-machining modification of type 410, for lighter cuts; machined parts requiring hot working or cold heading
420 (S42000)	12-14 Cr, 0.15 C min, 1.0 Mn, 1.0 Si, 0.040 P, 0.030 S	Bars Wire	95 95	50 50	25 20	Rb 92 Rb 92	Higher-carbon modification of type 410; cutlery, surgical instruments, valves, wear-resisting parts, glass molds, hand tools, vegetable choppers
420F (S42020)	12-14 Cr, over 0.15 C, 1.25 Mn, 1.0 Si, 0.060 P, 0.15 S min, 0.60 Mo max (optional)	Bars Wire	95 100[d]	55 80[d]	22 15[d]	Bhn 220 Rb 99[d]	Free-machining modification of type 420; applications similar to those for type 420 requiring better machinability
422 (S42200)	11-13 Cr, 0.50-1.0 Ni, 0.02-0.25 C, 1.0 Mn, 0.75 Si, 0.025 P, 0.025 S, 0.75-1.25 Mo, 0.15-0.30 V, 0.75-1.25 W	Bars	145	125	18	Bhn 370	High strength and toughness at service temperatures up to 1200°F; steam turbine blades; fasteners

431 (S43100)	15·17 Cr, 1.25·2.50 Ni, 0.20 C, 1 0 Mn, 1 0 Si, 0.040 P, 0.030 S	Bars Wire	125 135[d]	95 115[d]	20 10[d]	Bhn 260 Rc 29[d]	Special-purpose hardenable steel used where particularly high mechanical properties are required; aircraft fittings, beater bars, paper machinery, bolts
440A (S44002)	16·18 Cr, 0.60·0.75 C, 1.0 Mn, 1.0 Si, 0.040 P, 0.030 S, 0.75 Mo	Bars Wire	105 105	60 60	20 18	Rb 95 Rb 95	Hardenable to higher hardness than type 420 with good corrosion resistance; cutlery, bearings, surgical tools
440B (S44003)	16·18 Cr, 0.75·0.95 C, 1.0 Mn, 1.0 Si, 0.040 P, 0.030 S, 0.75 Mo	Bars Wire	107 107	62 62	18 16	Rb 96 Rb 96	Cutlery grade; cutlery, valve parts, instrument bearings
440C (S44004)	16·18 Cr, 0.95·1.20 C, 1.0 Mn, 1.0 Si, 0.040 P, 0.030 S, 0.75 Mo	Bars Wire	110 110	65 65	14 13	Rb 97 Rb 97	Yields highest hardnesses of hardenable stainless steels; balls, bearings, races, nozzles, balls and seats for oil-well pumps, valve parts
501 (S50100)	4·6 Cr, 0.10 C min, 1.0 Mn, 1.0 Si, 0.040 P, 0.030 S, 0.40 0.65 Mo	Bars Plates	70 70	30 30	28 28	Bhn 160 Bhn 160	Heat-resistant; good mechanical properties at moderately elevated temperatures; heat exchangers, petroleum refining equipment
502 (S50200)	4·6 Cr, 0.10 C, 1.0 Mn, 1.0 Si, 0.040 P, 0.030 S, 0.40·0.65 Mo	Bars Plates Sheets Strips Wire	65 65 70 70 75	25 25 — — 30	30 30 30 30 30	Bhn 150 Bhn 150 Rb 75 Rb 75 Rb 72	More ductility and less strength than type 501; heat exchangers, petroleum refining equipment, gaskets
Precipitation Hardening[f] (S13800)	12 25·13.25 Cr, 7.5·8.5 Ni, 0.05 C, 0.10 Mn, 0.10 Si, 0.010 P, 0.008 S, 0.90·1.35 A1, 2.0·2.5 Mo, 0.010 N	Bars Plates	160 160	120 120	17 17	Rc 33 Rc 33	Martensitic (maraging) stainless that can be hardened by a single low-temperature heat treatment; aircraft parts, forged
(S15500)	14·15.5 Cr, 3.5·5.5 Ni, 0.07 C, 1.0 Mn, 1.0 Si, 0.040 P, 0.030 S, 2.5·4.5 Cu, (Cb—Ta 0.15·0.45)	Bars Plates Sheets Strips	160 160 160 160	145 145 145 145	15 15 15 15	Rc 35 Rc 35 Rc 35 Rc 35	Martensitic (maraging) stainless with high strength, hardness, and corrosion resistance; gears, cams, cutlery, shafting, aircraft parts

TABLE 1-5 Classification, Composition, and General Properties of Stainless Steel (*Continued*)

AISI type (UNS)	Typical composition, %[a]	Form[b]	Mechanical properties of annealed material at room temperature: Tensile strength, 10^3 lb/in^2	Yield strength, 10^3 lb/in^2	Elongation, %	Hardness	Characteristics and typical applications
(S17400)	15.5·17.5 Cr, 3·5 Ni, 0.07 C, 1.0 Mn, 1.0 Si, 0.040 P, 0.030 S, 3·5 Cu, (0.15·0.45 Cb—Ta)	Bars Plates Sheets	160 160 160	145 145 145	15 15 5	Rc 35 Rc 35 Rc 35	Similar to S15500, but with slightly higher chromium content; gears, springs, cutlery, fasteners, aircraft and turbine parts
(S17700)	16·18 Cr, 6.5·7.75 Ni, 0.09 C, 1.0 Mn, 1.0 Si, 0.040 P, 0.040 S, 0.75·1.50 Al	Bars Plates Sheets	130 130 130	40 40 40	10 10 35	Rb 90 Rb 90 Rb 85	Semiaustenitic stainless; can be cold-drawn and then hardened by a low-temperature heat treatment; springs, knives, pressure vessels

[a]Single values are maximums, except as noted.

[b]Forms listed are only those for which mechanical properties are given. Most types are available in many forms.

[c]Austenitic hardenable by cold working, not hardenable by heat treatment. Ferritic not hardenable by heat treatment or cold working. Martensitic hardenable by heat treatment.

[d]Soft temper.

[e]Composition for type 310 tubing varies slightly from AISI values. For standard compositon refer to ASIM A213.

[f]Mechanical properties are for a solution-treated condition.

Source: 1978 Metal Progress Databook, ASM. Reproduced with permission of the American Society for Metals.

Tool Steels

Tool steels are either carbon or alloy steels capable of being hardened and tempered. They may be utilized for application in certain hand tools; in mechanical fixtures for cutting, shaping, forming, and blanking of materials at either ordinary or elevated temperatures; and also in other applications where wear resistance is important. Tool steels are available in the form of hot- and cold-finished bars, special shapes, forgings, hollow bars, wire, drill rod, plate, sheets, strip, tool bits, and castings. Many tool steels are sold under trade names and are the result of experimental development for specific purposes.

ALLOYING ELEMENTS AND THEIR EFFECTS

General Information

The highest mechanical properties of the constructional alloy steels are developed principally by quenching and tempering. If an alloy steel has been well quenched, maximum surface hardness will be developed. It is important to remember that carbon is responsible for the maximum attainable surface hardness in the section being quenched, but practical considerations necessitate the use of alloys to get maximum surface hardness and depth of hardening as section sizes increase. Alloying elements such as nickel or chromium do not, in the constructional alloy steels, make the steel surface any harder, but rather improve the hardenability and the mechanical properties of the heat-treated alloy steel by influencing the rate of austenite transformation so that deep-hardening characteristics are obtained.

In the practical application of the constructional alloy steels, it has been desirable, in order to obtain the best combination of strength and ductility in the heat-treated condition, to have a steel containing both a ferrite strengthener, such as nickel, and the carbide-forming elements chromium and molybdenum. The latter elements allow the use of higher temperatures for tempering while maintaining high tensile strength and a good degree of ductility in addition to their function of promoting full hardening at slower cooling rates.

Specific Effects of Alloying Elements in the Constructional Alloy Steels

Carbon

Carbon is the principal hardening element in steel, with each additional increment of carbon increasing the hardness and tensile strength of the steel in the as-rolled or normalized condition. As the carbon content increases above approximately 0.85 percent, the resulting increase in strength and hardness is proportionately less than it is for the lower carbon ranges.

Manganese

Manganese is present in all commercial steels and contributes significantly to a steel's strength and hardness in much the same manner as does carbon, but to a lesser extent. Its effectiveness depends largely upon, and is directly proportional to, the carbon content of the steel.

Silicon

Silicon is one of the principal deoxidizers used in the manufacture of both carbon and alloy steels and, depending on the type of steel, can be present in varying amounts up to 0.35 percent for the purpose of deoxidation. It is used in greater amounts in some steels, such as the silicomanganese spring and tool steels, where its effects tend to complement those of manganese to produce unusually high strength combined with good ductility and shock resistance in the quenched and tempered condition.

Copper

Copper is added to steel primarily to improve the steel's resistance to corrosion. In the usual amounts of from 0.20 to 0.50 percent, the copper addition does not significantly affect the mechanical properties. In an industrial environment, copper-bearing steel (0.2 percent minimum) has approximately twice the corrosion resistance of plain carbon steel.

Nickel

Nickel is one of the fundamental steel-alloying elements. When present in appreciable amounts, it provides improved toughness (particularly at low temperatures), simplified and more economical thermal treatment, increased hardenability, less distortion in quenching, and improved corrosion resistance.

Chromium

Chromium is used in constructional alloy steels primarily to increase hardenability, to provide improved resistance to abrasion, and to promote carburization. Of the common alloying elements, chromium is surpassed only by manganese and molybdenum in its effect on hardenability. It forms the most stable carbide of any of the more common alloying elements, giving to high-carbon chromium steels exceptional wear resistance. Because its carbide is relatively stable at elevated temperatures, chromium is frequently added to steels used for high-temperature applications.

Molybdenum

Molybdenum exhibits a greater effect on hardenability per unit added than any other commonly specified alloying element except manganese. It is a nonoxidizing element and therefore is highly useful in the melting of steels where close hardenability control is desired. Molybdenum is unique in the degree to which it increases the high-temperature tensile and creep strengths of steel.

Vanadium

Vanadium is one of the strong carbide-forming elements. It dissolves to some degree in ferrite, imparting strength and toughness.

Aluminum

Aluminum is widely used as a deoxidizer and for control of inherent grain size.

Boron

Boron has the unique ability to increase the hardenability of steel when added in amounts as small as 0.0005 percent.

HEAT TREATMENT

Most soft carbon steels are used as-rolled, and the properties desired in the finished product are so predicated. However, in most cases involving higher-carbon or alloy steels, the ultimate property capability of a steel may be defined as an operation, or series of operations, involving the heating and cooling of steel in the solid state to develop the required properties. There are several different forms of heat treatment employed, and these modify the mechanical properties of the steel to suit the end use. Typical mechanical properties are listed in Tables 1-2, 1-3, and 1-4.

Normalizing

Normalizing involves heating the steel to a temperature of approximately 100 to 150°F above the upper critical temperature and following the heating by cooling in still air. It is often employed to remove undesirable or nonuniform structures produced by rolling, forging, welding, and brazing.

Annealing

Annealing in general is a combination of a heating cycle, a holding period, and a controlled cooling cycle. It is used to obtain a variety of results, among which are: softening or alteration of the grain structure, development of formability, machinability, and required mechanical properties, to the relief of residual stresses. The most common annealing procedure is the *full anneal.* In this process, the steel is heated to above the critical range, held for sufficient time to allow solution of carbon and alloying elements to take place, and then cooled very slowly, either by definitely controlling the cooling rate of the furnace, or by shutting down to allow the furnace and load to cool together.

Quenching and Tempering

Quenching and tempering usually consist of three successive operations: (1) heating the steel above the critical range so that it approaches a uniform solid solution; (2) hardening the steel by quenching it in oil, water, brine, or a fused-salt bath; and (3) tempering the steel by reheating it to a point below the critical range, to get the desired combination of strength and ductility.

There are two types of steel which do not lend themselves to this kind of heat treatment: most soft steels in which the carbon content is insufficient to form any appreciable amount of martensite (the hardest microconstituent of steel) on quenching and austenitic stainless, or heat-resisting, steels where the critical range is depressed below room temperature because of the presence of large amounts of certain alloying constituents.

PROPERTIES AFFECTED BY GRAIN SIZE

During heat treatment, time and temperature must be controlled in order to control the austenite grain size which has an important influence on steel properties.

Fine-grain steels (grain sizes 5 and finer) do not harden as deeply as coarse-grain steels, and they have less tendency to crack during heat treatment. Fine-grain steels exhibit greater toughness and shock-resistance properties that make them suitable for applications involving moving loads and high impact. Some carbon steels and practically all alloy steels are produced as fine-grain steel.

Coarse-grain steels exhibit definite machining superiority. For this reason, some parts which are intricately machined are made from coarse-grain steel.

Notch Toughness

Beginning around 1940, the number of large monolithic structures, such as welded ships, pipelines, storage vessels, etc., increased very rapidly. The sudden and complete failure of several of these structures at stresses well below their yield strengths indicated that other considerations besides the conventional tensile properties must be included when designing such structures. Analysis of these failures and re-analysis of previous failures indicated that the fracture usually initiated at notches. These notches were caused by (1) design features, e.g., two rigid members rigidly attached at a sharp angle (90° or less) to one another; (2) fabrication procedures, e.g., weld-arc strikes, tool gouges, etc.; or (3) flaws in the material, e.g., flakes, seams, weld porosity, etc. Almost all fabricated structures have notches of one type or another in them.

The Charpy impact test has become widely used in both assessing the behavior of steel in the presence of a notch and in quality control during steel-producing operations, but it does not provide a number that can be used directly in engineering design calculations.

TABLE 1-6 Available Forms of Carbon and Alloy Steels

Carbon steels

- Semifinished for forging
  - Forging quality
    - Special hardenability
    - Special internal soundness
    - Nonmetallic inclusion requirement
    - Special surface
- Carbon steel structural sections
  - Structural quality
- Carbon steel plates
  - Regular quality
  - Structural quality
  - Cold-drawing quality
  - Cold-pressing quality
  - Cold-flanging quality
  - Forging quality
  - Pressure vessel quality
  - Marine quality
- Hot-rolled carbon steel bars
  - Merchant quality
  - Special quality
    - Special hardenability
    - Special internal soundness
    - Nonmetallic inclusion requirement
    - Special surface
  - Scrapless nut quality
  - Axle shaft quality
  - Cold-extrusion quality
  - Cold-heading and cold-forging quality
- Cold-finished carbon steel bars
  - Standard quality
    - Special hardenability
    - Special internal soundness
    - Nonmetallic inclusion requirement
    - Special surface
  - Cold-heading and cold-forging quality
  - Cold-extrusion quality
- Hot-rolled sheets
  - Commercial quality
  - Drawing quality
  - Drawing quality special killed
  - Physical quality
- Cold-rolled sheets
  - Commercial quality
  - Drawing quality
  - Drawing quality special killed
  - Physical quality
- Porcelain enameling sheets
  - Commercial quality
  - Drawing quality
- Long terne sheets
  - Commercial quality
  - Drawing quality
  - Drawing quality special killed
  - Physical quality
- Galvanized sheets
  - Commercial quality
  - Drawing quality
  - Drawing quality special killed
  - Physical quality
  - Lock forming quality
- Electrolytic zinc-coated sheets
  - Commercial quality
  - Drawing quality
  - Drawing quality special killed
  - Physical quality
- Hot-rolled strip
  - Commercial quality
  - Drawing quality
  - Drawing quality special killed
  - Physical quality
- Cold-rolled strip
  - Specific quality descriptions not provided in cold-rolled strip since this product is largely produced for specific end use
- Tin mill products
  - Specific quality descriptions not applicable to tin mill products
- Carbon steel wire
  - Industrial quality wire
  - Cold-extrusion wires
  - Heading, forging, and roll threading
  - Wires
  - Mechanical spring wires
  - Upholstery spring construction wires
  - Welding wire
- Carbon steel flat wire
  - Stitching wire
  - Stapling wire
- Carbon steel pipe
- Structural tubing
- Line pipe
- Oil country tubular goods
- Steel specialty tubular products
  - Pressure tubing
  - Mechanical tubing
  - Aircraft tubing
- Hot-rolled carbon steel wire rods
  - Industrial quality
  - Rods for manufacture of wire intended for electric welded chain
  - Rods for heading, forging, and roll threading wire
  - Rods for lock-washer wire
  - Rods for scrapless nut wire
  - Rods for upholstery spring wire
  - Rods for welding wire

Alloy steels

- Alloy steel plates
  - Regular quality or structural quality
  - Drawing quality
  - Pressure vessel quality
  - Structural quality
  - Aircraft quality
  - Aircraft physical quality
- Hot-rolled alloy steel bars
  - Regular quality
  - Aircraft quality or steel subject to magnetic particle inspection
  - Axle shaft quality
  - Bearing quality
  - Cold heading quality

TABLE 1-6 *(Continued)*

Alloy steels	
Special cold-heading quality	Cold-heading quality
Rifle barrel quality, gun quality, shell or A.P. shot quality	Special cold-heading quality
Alloy steel wire	Rifle barrel quality, gun quality, shell or A.P. shot quality
Aircraft quality	Line pipe
Bearing quality	Oil country tubular goods
Special surface quality	Steel specialty tubular goods
Cold-finished alloy steel bars	Pressure tubing
Regular quality	Mechanical tubing
Aircraft quality or steel subject to magnetic particle inspection	Stainless and heat-resisting pipe, pressure tubing, and mechanical tubing
Axle shaft quality	Aircraft tubing
Bearing shaft quality	Pipe

Source: 1979 SAE Handbook, p. 3:02.

The notch toughness of most ferritic steels decreases with decreasing temperatures. Grain refinement, inclusion shape control, increased manganese/carbon ratio, additions of nickel, and heat treatment can improve the notch toughness of ferritic steels. In the case of quench and temper treatments, tempered martensite gives higher impact values when compared with other structures such as bainite.

AVAILABLE FORMS

Table 1-6 lists the available forms and shapes of carbon and alloy steels.

PUBLISHERS OF STANDARDS AND SPECIFICATIONS

AAR, Association of American Railroads.
ABS, American Bureau of Shipping.
ACI, Alloy Casting Institute.
AFS, American Foundrymans Society.
AISI, American Iron and Steel Institute.
AMS, Aerospace Materials Specifications.
ANSI, American National Standards Institute.
API, American Petroleum Institute.
ASM, American Society for Metals.
ASME, American Society of Mechanical Engineers.
ASTM, American Society for Testing and Materials.
CSA, Canadian Standards Association.
ICS, Iron Casting Society.
ISO, International Standards Organization.
MIL, U.S. Military Specifications.
SAE, Society of Automotive Engineers.
AWS, American Welding Society.

BIBLIOGRAPHY

Annual Standards, American Society for Testing and Materials, 1916 Race Street, Philadelphia, PA 19103, 1979.*

*Use the most recent edition available.

Metals Handbook, 9th ed., American Society for Metals, Metals Park, OH 44073.

Republic Alloy Steels, Republic Steel Corporation, Republic Building, Cleveland, OH 44101.

SAE Handbook, Society of Automotive Engineers, 400 Commonwealth Drive, Warrendale, PA 15096, 1979.*

Source Book, American Society for Metals, Metals Park, OH 44073.

Steel Castings Handbook, Steel Founders Society of America, 20611 Center Ridge Road, Rocky River, OH 44116, 1979.*

Steel Products Manual, American Iron and Steel Institute, 1000 16th Street, N. W., Washington, DC 20036

The Making, Shaping and Treating of Steel, 9th ed., United States Steel Corporation, 600 Grant Street, Pittsburgh, PA 15230.

Practical Metallurgy for Engineers, E. F. Houghton and Company, 303 West Lehigh Avenue, Philadelphia, PA 19133, 1952.

Design Guidelines for Selection and Use of Stainless Steel, American Iron and Steel Institute, 1000 16th Street, N.W., Washington, DC 20036.

*Use the most recent edition available.

chapter 19-2

Aluminum

prepared by

The Aluminum Association, Inc.
Washington, D.C.

GLOSSARY

Age hardening A process occurring at room temperature or through heating at elevated temperatures which for some alloys results in increased strength and hardness.

Aging Precipitation of alloying elements and compounds from solid solution resulting in a change in properties of an alloy, usually occurring slowly at room temperature (natural aging) and more rapidly at elevated temperatures (artificial aging).

Alloying The addition of one or more metals dissolved in molten aluminum (the base metal) to produce an alloy, with the purpose of changing or improving specific mechanical or physical characteristics of the base aluminum.

Annealing Thermal treatment to soften metal by removal of stress resulting from cold working or by coalescing precipitates from solid solution.

Cold working Plastic deformation of metal at such temperature and rate that strain hardening occurs.

Corrosion Deterioration of a metal by chemical or electrochemical reaction with its environment.

Electric conductivity Capacity of a material to conduct electric current.

Electric resistivity Electrical resistance of a body of unit length and unit cross-sectional area or unit weight.

Elongation Percentage increase in distance between two gouge marks that results from stressing the specimen in tension to fracture.

Forging Metal part worked to a predetermined shape by one or more processes such as hammering, upsetting, pressing, rolling, etc.

Hardness Resistance to plastic deformation, usually by indentation.

Heat treating Heating and cooling a solid metal or alloy to obtain desired conditions or properties.

Heat-treatable alloy Alloy which may be strengthened by a suitable thermal treatment.

Mechanical properties Properties of a material associated with elastic and inelastic reaction when force is applied or that involve the relation between stress and strain (tensile strength, endurance limit).

Non-heat-treatable alloy Alloy which can be strengthened by cold working.

Physical properties Properties other than mechanical properties that pertain to the physics of material (density, thermal expansion).

Strain hardening Modification of a metal structure by cold working resulting in an increase in strength and hardness with loss of ductility.

Workability Relative ease with which various alloys may be formed by rolling, extruding, forging, etc.

Wrought product Product which has been subjected to mechanical working by rolling, extruding, forging, etc. See Table 2-1.

TABLE 2-1 Forms Available

Wrought alloys
Sheet and plate
Foil
Rod, bar,wire, and tube
Extrusions
Forgings and impacts
Electric conductor
Cast alloys
Can be cast by any known foundry process
Aluminum powder and paste
Finely divided aluminum used in pigments and explosives, powdered metal

CHARACTERISTICS OF ALUMINUM

Properties of Pure Aluminum

Specific gravity is 2.7 times that of water and about one-third that of steel. One cubic inch weighs 0.10 lb (0.045 kg); 1 ft^3 (0.028 m^3) weighs 170 lb (77 kg). Pure aluminum melts at 1220°F (660°C). Its electric conductivity on a volume basis is about 65 percent of the International Annealed Copper Standard. However, pound-for-pound aluminum is a better conductor than copper, surpassed only by sodium.

Pure aluminum is relatively weak but is highly ductile. Its modulus of elasticity is approximately 10×10^6 lb/in^2 (6.895×10^4 MPa), compared with 3.0×10^7 lb/in^2 (2.07×10^5 MPa) for steel. Its elasticity is not substantially improved by alloying. Aluminum and its alloys are tough and neither lose ductility nor become brittle at cryogenic temperature.

It can be significantly strengthened by strain hardening, or cold working. Alloying with a variety of elements, together with heat treatment, results in even greater strengthening.

Oxide Layer

A protective coating—normally 50 to 100 Å (stoionm) thick—forms when fresh aluminum is exposed to oxygen. Anodizing can increase the oxide film's thickness by more than 1 mil to improve resistance to corrosion and abrasion. Aluminum oxide is transparent and hard and tenaciously adheres to the metal. The oxide is usually stable over a pH range of approximately 4.5 to 8.5 but is dissolved by most strong acids and alkalis. Some chemicals attack the oxide near the neutral point (pH 7). The oxide resists concentrated nitric acid at pH 1 and ammonium hydroxide at pH 13.

Classification of Alloys

Aluminum alloys are classified as either wrought or cast, depending on how they are formed. Aluminum can be formed by virtually every known process. Wrought alloys are available in sheet and plate, foil, extrusions, bar, rod and wire, forgings, etc. Cast alloys are formulated to flow into a sand or permanent mold, to be die-cast, or to be cast by any other process in which the casting is the final form.

ALLOYING AND HEAT TREATMENT

Alloy Designations

Copper, manganese, silicon, magnesium, and zinc are the principal alloying elements of aluminum. Others are added in smaller amounts for grain refinement and development of special properties.

Wrought (Table 2-2) and cast (Table 2-3) alloys are designated by numbers to identify them and broadly describe their alloy content. Wrought alloys have four-digit numbers. The first classifies the alloy by series or principal element. The second, other than zero, indicates a modification in the basic alloy. The third and fourth digits are arbitrary and specify the alloy in the series (see Table 2-4).

Cast alloys are given four-digit numbers, the last of which is separated by a decimal point. The first indicates alloy series or principal addition. The second and third specify alloy. The number following the decimal point tells whether the alloy composition is for final casting or in ingot form. A capital letter prefix indicates modification of the basic alloy.

TABLE 2-2 Wrought Alloy Designations

Alloy series	Description
1xxx	99.00% minimum aluminum
2xxx	Copper
3xxx	Manganese
4xxx	Silicon
5xxx	Magnesium
6xxx	Magnesium and silicon
7xxx	Zinc
8xxx	Other element
9xxx	Unused series

TABLE 2-3 Cast Alloy Designations

Alloy series	Description
1xx.x	99.00% minimum aluminum
2xx.x	Copper
3xx.x	Silicon + copper and/or magnesium
4xx.x	Silicon
5xx.x	Magnesium
6xx.x	Unused series
7xx.x	Zinc
8xx.x	Tin
9xx.x	Other element

TABLE 2-4 Classifications of Aluminum Alloys

Alloy	Typical uses
Wrought alloys	
1060	Chemical equipment, tank cars
1100	Sheet-metal work, cooking utensils, decorative
1350	Electric conductors
2011	Screw machine products
2017	Screw machine products fittings
2018	Aircraft-engine cylinders, heads, pistons
2014, 2024, 2618	Aircraft structures, engines; truck frames, wheels
2025	Forgings
2036	Auto panel sheet
2117	Rivets
3003, 3004	Sheet-metal work, chemical equipment, storage tanks
4032	Pistons
4043	Welding electrodes
4343	Brazing alloy
5005, 5050, 5062, 5657	Decorative, automotive trim; architectural, anodized; sheet-metal work, appliances
5083, 5086, 5454, 5456	Marine, welded structures, storage tanks, pressure vessels, armor plate, cryogenics
5154	Welded structures, storage tanks, pressure vessels
5652	Hydrogen peroxide, chemical storage vessels
6053	Wire, rod for rivets
6061, 6063	Marine; truck frames, bodies; structures, architectural, furniture
6066	Forgings, extrusions for welded structures
6070	Heavy-duty welded structures, pipelines
6262	Screw machine products
7001, 7075, 7178	High-strength structures, aircraft
7004	Structural, cryogenic, missile
Cast alloys	
308.0	General-purpose casting alloy
319.0	Engine components, valve bodies
356.0	High-strength, corrosion-resistant machinery components
850.0	Bearings

Temper Designations

An alloy's metallurgical condition is referred to as temper. Temper is identified by a letter following alloy designation.

"F" refers to *fabricated* and applies to formed products where there is no special control over thermal or work-hardening conditions.

"O" designates *annealed* and is used with wrought products which have been heated to effect recrystallization and produce the lowest strength condition. It is also used with castings which are annealed to uniform ductility and dimensional stability.

"H" identifies *strain-hardened* and applies to wrought products strengthened by strain hardening through cold working.

"W" designates *solution heat-treated.* This is an unstable temper applicable only to alloys which age spontaneously at room temperature after solution heat treatment. Solution heat treatment involves heating the alloy to approximately 1000°F (537°C).

"T" signifies *thermally treated* and applies to products which are heat-treated, sometimes with supplementary strain hardening, to produce a stable temper other than F or O.

Strengthening Processes

Since high-purity aluminum is soft and ductile, it requires strengthening for most commercial applications. This is accomplished by the addition of alloy-producing elements. Further strengthening is possible via methods that classify the alloys into non-heat-treatable and heat-treatable.

Non-Heat-Treatable Alloys

Their initial strength is enhanced by the addition of such alloying elements as copper, magnesium, zinc, and silicon. As the temperature is increased, their solid solubility in aluminum rises.

Heat treatment or solution heat treatment is an elevated-temperature process designed to put the soluble element or elements in solid solution. Rapid quenching, usually in water, follows. This momentarily "freezes" the structure and for a short time renders the alloy very workable. After several days of exposure to room-temperature aging (or room-temperature precipitation), the alloy is considerably stronger.

For an even stronger alloy, the metal can be heated for several hours at about 400°F (204°C). This treatment also stabilizes its properties and is called *artificial aging* or *precipitation hardening.*

By proper combination of solution heat treatment, quenching, cold working, and artificial aging, it is possible to obtain the highest strengths.

Heat-Treatable Alloys

To increase corrosion resistance of heat-treatable sheet and plate alloys with copper or zinc as the major element, the metal is clad with a high-purity aluminum, a low magnesium-silicon alloy, or an alloy with 1 percent zinc (clad alloys).

Annealing

All aluminum alloys can be annealed.

Effect of Alloying Elements

The 1xxx series has many uses, particularly electrical and chemical. These alloys feature excellent corrosion resistance, high thermal and electric conductivity, low mechanical properties, and very good workability. Strain hardening produces moderate strength increases. See Tables 2-5 and 2-6.

Copper

Copper is the major alloying element in the 2xxx alloys. The 2xxx alloys are solution heat-treated for optimum properties. In the heat-treated condition, mechanical properties are similar to and sometimes superior to mild steel. At times, artificially aging further increases the mechanical properties. Corrosion resistance is not as good as in most other aluminum alloys. Therefore, these alloys in sheet form are usually clad with a high-purity alloy or magnesium-silicon alloy to improve corrosion resistance. Alloy 2024 is widely used in aircraft.

Manganese

Although it is a major alloying element in the 3xxx series, only up to 1.5 percent of manganese can be added effectively to aluminum. Manganese is a major factor in 3003, a general-purpose alloy for moderate-strength applications requiring good workability.

Silicon

Enough silicon can be added to 4xxx series alloys to substantially lower the melting point without making them brittle. Resulting alloys are used in welding wire and as brazing

TABLE 2-5 Typical Mechanical Properties of Aluminum Alloys

			Ultimate tensile strength		Yield strength			
Alloy	Nominal composition	Temper	psi × 10^3	MN/m^2	psi × 10^3	MN/m^2	Elongation, % in 2 in	Hardness, Bhn
1100	99+% Al	O	13	90	5	34	35	23
		H14	18	124	17	117	9	32
		H18	24	165	22	152	5	44
2024	4.4% Cu	O	27	186	11	76	20	47
	1.5% Mg	T4	68	469	47	324	20	120
	0.6% Mn	T6	69	476	57	393	10	125
		T86	75		71	490	6	135
2219	6.3% Cu	O	25	172	11	76	20	—
	0.3% Mn	T37	57	393	46	317	11	117
		T87	69	476	57	393	10	130
3003	1.2% Mn	O	16	110	6	41	30	28
		H14	22	152	21	145	8	40
3004	1.2% Mn	O	26	179	10	69	20	45
	1.0% Mg	H34	35	241	29	200	9	63
		H38	41	283	36	248	5	77
4032	12.2% Si	T6	55	379	46	317	9	120
		H18	29	200	27	186	4	55
5005	0.8% Mg	O	18	124	6	41	30	30
		H14	23	159	22	152	6	41
		H18	29	200	28	193	4	51
5052	2.5% Mg	O	28	193	13	90	25	47
		H34	38	262	31	214	10	68
		H38	42	290	37	255	7	77
5456	5.1% Mg	O	45	310	23	159	24	70
	0.8% Mn	H343	56	386	43	296	8	94
6061	1.0% Mg	O	18	124	8	55	25	30
	0.6% Si	T4	35	241	21	145	22	65
		T6	45	310	40	276	12	95
7001	7.4% Zn	O	37	255	22	152	14	60
	3.0% Mg	T6	98	676	91	627	9	160
	2.1% Cu							
7075	5.6% Zn	O	33	228	15	103	17	60
	2.5% Mg	T6	83	572	73	503	11	150
	1.6% Cu	T73	73	503	63	434	13	—
308.0		F	28	193	16	110	2	70
319.0		T6	40	276	27	186	3	95
356.0		T6	37	255	27	186	5	80
850.0		T5	23	159	11	76	12	45

alloys. If large amounts of silicon are added, the alloys become dark gray when anodic oxide finishes are applied. Thus, they are sought for architectural applications.

Magnesium

Magnesium, the principal element in the 5xxx series, is one of the most effective and widely used alloying elements. When used as the major alloying element or with manganese, the result is a moderate- to high-strength non-heat-treatable alloy. 5xxx series alloys possess good welding characteristics and resistance to corrosion in marine environments. Certain limitations should be placed on the amount of cold working and safe operating temperatures permissible for the higher magnesium content alloys [more than about 3.5 percent for operating temperatures about 65°C (149°F)] to avoid possible stress corrosion.

TABLE 2-6 Typical Physical Properties of Aluminum Alloys

Alloy	Density, (68°F) (20°C) lb/in³	Specific gravity	Average coefficient thermal expansion		Melting range approx,		Thermal* conductivity		Electric† conductivity (68°F) (20°C) equal volume
			68–212°F	20–100°C	°F	°C	77°F	25°C	
1100	0.098	2.71	13.1	23.6	1190–1215	643–657	1600	0.55	62
2024	0.100	2.77	12.9	23.2	935–1180	502–638	1340	0.46	50
2219	0.103	2.84	12.4	22.3	1010–1190	543–643	1190	0.41	44
3003	0.099	2.73	12.9	23.2	1190–1210	643–654	1340	0.46	50
3004	0.098	2.72	13.3	23.9	1165–1210	629–654	1130	0.39	42
4032	0.097	2.69	10.8	19.4	990–1060	532–571	1070	0.37	40
5005	0.098	2.70	13.2	23.8	1170–1210	632–654	1390	0.48	52
5052	0.097	2.68	13.2	23.8	1125–1200	607–649	960	0.33	35
5456	0.096	2.65	13.3	23.9	1060–1180	571–638	810	0.28	29
6061	0.098	2.70	13.1	23.6	1080–1205	638–652	1250	0.43	47
7001	0.102	2.81	13.0	23.4	890–1160	477–627	870	0.30	31
7075	0.101	2.80	13.1	23.6	890–1175	477–635	900	0.31	33
308.0	0.101	2.79	11.9		970–1135	521–613			37
319.0	0.101	2.79	11.9		960–1120	516–604			27
356.0	0.097	2.68	11.9		1035–1135	557–613			41
850.0	0.104	2.88	12.6	22.7	435–1200	224–649			47

*Thermal conductivity: English units for "O" temper except alloys 7001 and 7075 which are T6.
†Electric conductivity: for annealed material in "O" temper except alloys 7001 and 7075 which are T6

TABLE 2-7 Comparative Characteristics of Various Aluminum Alloys*

Alloy	Corrosion resistance: General	Corrosion resistance: S.C.C.	Workability	Machineability	Brazability	Weldability (arc)
1100	A	A	A–C	E–D	A	A
2024	D	C–B	C–D	B	D	B–C
2219	D	C–B	C–D	B	D	A
3003	A	A	A–C	E–D	A	A
3004	A	A	A–C	D–C	B	B
4032	C	B	——	B	D	B
5005	A	A	A–C	E–D	B	A
5052	A	A	A–C	D–C	C	A
5456	A	B	B–C	D–C	D	A
6061	B	B–A	A–C	D–C	A	A
7075	C	C	D	B	D	C

*Key for resistance to stress corrosion cracking: A, no known service or laboratory failure; B, no known service failure; limited failure in laboratory tests of short transverse specimens; C, service failures with sustained tension stress acting in short transverse direction relative to grain structure; limited failures in laboratory tests of long transverse specimens; D, limited service failures with sustained longitudinal or long transverse stress. Workability and machineability: A through D for cold workability and A through E for machineability are relative ratings in decreasing order of merit. Brazability and weldability: A, generally weldable by all commercial procedures and methods; B, weldable with special techniques or for specific applications which justify preliminary trials or testing to develop welding prodedures and weld performance; C, limited weldability because of crack sensitivity or loss in resistance to corrosion and mechanical properties; D, no commonly used welding methods developed.

Magnesium Silicide

Alloys in the 6xxx series contain silicon and magnesium in approximate proportions to form magnesium silicide and make them heat-treatable. These alloys have good formability and corrosion resistance and medium strength. The series' principal alloy is 6061, the most versatile of the heat-treatable alloys.

Zinc

The major alloying element in the 7xxx series is zinc, which when joined with a smaller percentage of magnesium produces heat-treatable alloys of very high strength. Usually, other elements such as copper and chromium also are added in small quantities. Prominent in the series is 7075, one of the highest-strength alloys and found in air-frame structures and highly stressed parts.

Corrosion Resistance

In general, aluminum alloys have good corrosion resistance when exposed to the atmosphere, freshwater, seawater, most soils and foods, and many chemicals. That means they have a long service life without surface protection.

The higher the aluminum purity the greater its resistance to corrosion. But certain elements can be alloyed with aluminum without reducing corrosion resistance. In some instances, an improvement results. Sb, Bi, Pb, Si, Ti, and Zn have little or no effect on aluminum's corrosion resistance. Cu, Fe, and Ni have adverse effects. Their influence depends on the amount present, metallurgical condition of the alloy, and state of the addition in the alloy structure and environment. See Table 2-7.

STANDARDS AND SPECIFICATIONS

Aluminum Standards and Data, 6th ed., Aluminum Association, Washington, D.C., March 1979.

Aluminum Standards and Data, 1st ed., Aluminum Association, metric SI, Washington, D.C., March 1978.
ASTM B 209, "Aluminum—Alloy Sheet and Plate."
ASTM B 210, "Aluminum—Alloy Drawn Tubes."
ASTM B 211, "Aluminum—Alloy Bars, Rods and Wire."
ASTM B 221, "Aluminum—Alloy Extruded Bars, Rods, Wire, Shapes and Tubes."
All above ASTM specifications appear in part 7, *ASTM Book of Standards,* American Society for Testing and Materials, current edition, Philadelphia, Pa.

chapter 19-3

Wood

by

Duane E. Lyon
Associate Professor
Mississippi State University
and
Chairman, Wood Engineering and
Mechanical Properties Technical Committee
Forest Products Research Society

GLOSSARY

Board foot A unit of measurement of lumber represented by a board one foot long, twelve inches wide, and one inch thick or its cubic equivalent, based on nominal dimensions.

Fiberboard A generic term describing a family of sheet materials manufactured of refined or partially refined wood fibers.

Grain The direction, size, arrangement, appearance, or quality of the fibers in wood or lumber.

Green wood Freshly hewed or unseasoned wood.

Hardboard A generic term for a panel product manufactured from refined or partially refined wood fibers, consolidated under heat and pressure to a density of 31 or more pounds per cubic foot.

Hardwoods Generally one of the botanical groups of trees that have broad leaves in contrast to softwoods. The term has no reference to the actual hardness of the wood.

Heartwood The wood extending from the center of the log to the sapwood. Compounds in the heartwood make it more resistant to decay than sapwood.

Laminated wood An assembly made by bonding layers of veneer or lumber with an adhesive so that the grain of all laminations is essentially parallel.

Lumber The product of the saw and planing mill.

Boards Lumber nominally less than two inches thick and two or more inches wide.

Dimension Lumber with a nominal thickness of from two to four inches and a nominal width of two inches or more.

Dressed size The dimensions of lumber after being surfaced on a planing machine.

Nominal size The size by which lumber is sold in the market.

Rough lumber Lumber which has not been dressed (surfaced).

Structural lumber Lumber that is intended for use where its allowable properties are required. The grading of structural lumber is based on the strength of the piece as related to anticipated uses.

Timbers Lumber that is nominally five inches or more in least dimension.

Moisture content The amount of water contained in wood, usually expressed as a percentage of the weight of the oven-dry wood.

Particleboard A generic term for a panel manufactured from wood—essentially in the form of particles (as distinct from fibers).

Plywood A composite panel or board made up of cross-banded layers of veneer only or veneer in combination with a core of lumber or particleboard.

Sapwood The wood of pale color near the outside of the log.

Seasoning Removing moisture from green wood to improve its serviceability.

Softwoods Generally, one of the botanical groups of trees that in most cases have needlelike or scalelike leaves. The term has no reference to the actual hardness of the wood.

CLASSIFICATION OF WOOD PRODUCTS

The properties of wood are quite variable because of differences in structure among species. Woods presently considered to have major commercial value in North America are obtained from more than 60 types of native trees and more than 30 imported species.

The properties of wood can often be modified during processing. An example is the seasoning of lumber to increase strength. Properties can also be modified by comminution and reconstituting wood to make new products. Examples of structural products made this way are plywood, fiberboard, and particleboard. A wide variety of wood products are available for structural applications, and new products are constantly being developed.

PHYSICAL AND STRENGTH CHARACTERISTICS OF WOOD

The efficiency with which wood serves as a material of construction is determined by its strength and the factors which affect strength. The main factors that affect strength and related properties are discussed here.

Specific Gravity

Specific gravity is a measure of the amount of wood substance in a unit volume of wood. It is closely correlated with mechanical properties. Strength S and specific gravity G are related by the equation

$$S = KG^n$$

where K and n are constants depending on the property. For between-species variations, n ranges from 1.00 to 2.25, and for within-species variation from 1.25 to 2.5. Values of K and n may be found in Ref. 7. The average specific gravities of several species of wood are listed in Table 3-2.

It is interesting to note the extreme range of specific gravity of wood that can be found. The heaviest known wood is black ironwood (found in South Africa) with a specific gravity of 1.49; the lightest is *Aeschynomene hispida* (found in Cuba) having a specific gravity of 0.044.

Grain Angle

Because of its structure, wood has different strength properties parallel and perpendicular to the grain. Strength in tension, compression, and bending is greatest parallel to the grain and least across the grain. Shear strength is least parallel to the grain and greatest across the grain. The strength of wood at an angle to the grain may be estimated from the following formula:

$$N = \frac{PQ}{P \sin^n \theta + Q \cos^n \theta}$$

where N is the strength property at an angle θ from the fiber direction, Q is the strength across the grain, P is the strength parallel to the grain, and n is an empirically determined constant. Values of n for several properties are:

Property	n
Tensile strength	1.5–2.0
Compression strength	2.0–2.5
Bending strength	1.5–2.0
Modulus of elasticity	2.0

Moisture Content

Wood free of defects increases in strength as it dries below approximately 30 percent moisture content. The change in strength resulting from a 1 percent change in moisture content, expressed in percent, is approximately 4 for modulus of rupture, 2 for modulus of elasticity, and 5 for compression parallel to the grain. More exact relationships may be found in Ref. 7.

Volumetric Changes

Below the fiber-saturation point, water that evaporates from wood results in a reduction in wood volume. The amount of volumetric change is positively related to the changes in moisture content and density. The anisotropic nature of wood results in unequal shrinkage in the three principal grain directions. Shrinkage is greatest in the transverse grain direction parallel to the growth rings (tangential), and the total shrinkage from green to the oven-dry condition ranges from 4 to 13 percent, depending upon the species and density. In general, shrinkage increases with density. The transverse shrinkage perpendicular to the growth rings (radial) is usually about one-half that parallel to the growth rings and ranges from 2 to 8 percent. Total shrinkage along the grain ranges from 0.1 to 0.3 percent. Some representative shrinkage values are given in Table 3-1.

WOOD SEASONING

Most wood products are seasoned (dried) prior to use to remove the large amount of moisture present in freshly cut wood. Wood that has been dried offers a number of advantages, including reduced weight and shrinkage and increased strength and durability. Drying may be accomplished by one of several procedures. The two most common are air drying and kiln drying. A number of defects may develop during drying if the process is not carefully controlled. These defects are the result of drying stresses due to unequal shrinkage. Most kiln-drying procedures include moisture-equalizing and conditioning treatments to improve moisture uniformity throughout the thickness of the wood product and to relieve residual stresses. Improper drying may result in warping, checking, or more severe defects.

TABLE 3-1 Average Moisture Content of Green Wood and Shrinkage of Wood by Species*

	Moisture content		Shrinkage	
Species	Heartwood, %	Sapwood, %	Radial, %	Tangential, %
Ash, black	95	—	5.0	7.8
Ash, white	46	44	4.9	7.8
Aspen	95	113	3.5	6.7
Basswood, American	81	133	6.6	9.3
Beech, American	55	72	5.5	11.9
Birch, yellow	74	72	7.3	9.5
Cottonwood, eastern	162	146	3.9	9.2
Elm, American	95	92	4.2	7.2
Hickory, shagbark	70	52	7.0	10.5
Maple, sugar	65	72	4.8	9.9
Oak, red	80	69	4.0	8.6
Oak, white	64	78	5.6	10.5
Sweetgum	79	137	5.3	10.2
Sycamore, American	114	130	5.0	8.4
Poplar, yellow	83	106	4.6	8.2
Bald cypress	121	171	3.8	6.2
Cedar, eastern red	33	—	3.1	4.7
Cedar, Port Orford	50	98	4.6	6.9
Cedar, western red	58	249	2.4	5.0
Douglas-fir, coast	37	115	4.8	7.6
Fir, white	98	160	3.3	7.0
Hemlock, western	85	170	4.2	7.8
Larch, western	54	110	4.5	9.1
Pine, lodgepole	41	120	4.3	6.7
Pine, loblolly	33	110	4.8	7.4
Pine, longleaf	31	106	5.1	7.5
Pine, ponderosa	40	148	3.9	6.2
Pine, western white	62	148	4.1	7.4
Redwood, old growth	86	210	2.6	4.4
Spruce, Engelmann	51	173	3.8	7.1
Spruce, Sitka	41	142	4.3	7.5

*Expressed as a percentage of the green dimension.

Temperature Effects

In general, heating reduces and cooling increases the mechanical strength of wood. The change is immediate and irreversible for temperatures remaining above 200°F (93°C) for any appreciable period of time. The adverse effect of high temperature is more pronounced at high moisture contents. For elevated temperatures below 200°F (93°C), the immediate loss of strength is recovered when the wood is cooled to ambient conditions. When wood is repeatedly exposed to high temperature, the adverse effect on properties is cumulative.

The mechanical properties of the commercially important woods of the United States have been evaluated by the U.S. Forest Products Laboratory. These tests were conducted in accordance with ASTM Standard D 143,[2] which specifies small, clear specimens to eliminate the influence of naturally occurring physical defects in the wood.

Table 3-2 shows some mechanical properties for wood at 12 percent moisture content. These data have been abstracted to include several of the more important hardwood and softwood species and the mechanical properties of each which are likely to be uniquely important for specific uses encountered in structural applications. For additional data on other strength properties and other species, see Ref. 7.

Thermal Properties of Wood

Temperature affects several properties of wood. As wood is heated, it expands. The coefficient of thermoexpansion for wood averages near 0.6×10^{-6}/°F (1.1×10^{-6}/°C) for most native species. Wood is a good insulator and does not respond very fast to a change in environmental temperature. The coefficient of thermoconductivity for wood ranges from 0.4 to 0.7 Btu/(h) (°F) for a 1 ft^2 area of 1-in thickness at a moisture content of 12 percent. The thermoconductivity of wood increases with increasing specific gravity and moisture content.

PROTECTING WOOD

Protection from Biological Attack

At ordinary temperatures, wood is very stable and unless attacked by living organisms remains the same for centuries, either in air or under water. When wood products are not properly used or protected, they are susceptible to attack by decay fungi and several animal organisms including termites, carpenter ants, and some species of beetles.

Decay of wood can be prevented by keeping it dry and well-vented or by keeping it continually submerged in water to remove the air supply.

Most termites in North America live below ground level and will attack only wood that is in contact with earth or in close proximity to the soil. One species of termite, found along the West Coast and in the extreme southern areas of the country, does not require soil contact.

The resistance of heartwood to biological attack varies with wood species. The resistance of some North American woods to decay is shown in Table 3-3. The sapwood of all species is vulnerable to biological attack if not properly protected.

Wood Preservation

Wood preservatives fall into two main classes: (1) oil-type preservatives and (2) waterborne metallic salts. The former may be further subdivided into (*a*) coal-tar creosote with and without the mixture of cheaper materials such as petroleum or coal tar and (*b*) solutions of toxic organic chemicals such as pentachlorophenol dissolved in petroleum oils. Oil-type preservatives are used extensively for products that are exposed to ground contact where resistance to leaching is an important requirement of the preservative. Waterborne preservatives are used mainly for the treatment of lumber. Wood treated with a waterborne preservative is clean, paintable, and odorless.

TABLE 3-2 Mechanical Properties of Various Woods Grown in the United States Adjusted to 12 Percent Moisture Content*

Species	Specific gravity†	Static bending: Modulus of rupture, lb/in²‡	Static bending: Modulus of elasticity, 1,000 lb/in²	Compression parallel to grain maximum crushing strength, lb/in²	Compression perpendicular to grain stress at proportional limit, lb/in²	Tension perpendicular to grain maximum tensile strength, lb/in²	Side hardness§ lb	Maximum shearing strength parallel to grain, lb/in²
Ash, black	0.49	12,600	1,600	5,970	760	700	850	1,570
Ash, white	0.60	15,400	1,770	7,410	1,160	940	1,320	1,160
Aspen	0.38	8,400	1,180	4,250	370	260	350	850
Basswood	0.37	8,700	1,460	4,730	370	350	410	990
Beech	0.64	14,900	1,720	7,300	1,010	1,010	1,300	2,010
Birch, yellow	0.62	16,600	2,010	8,170	970	920	1,260	1,880
Cottonwood, eastern	0.40	8,500	1,370	4,910	380	580	430	930
Elm, American	0.50	11,800	1,340	5,520	690	660	830	1,510
Hickory, shagbark	0.72	20,200	2,160	9,210	1,760	—	—	2,430
Maple, sugar	0.63	15,800	1,830	7,830	1,470	—	1,450	2,330
Oak, red	0.63	14,300	1,820	6,760	1,010	800	1,290	1,780
Oak, white	0.68	15,200	1,780	7,440	1,070	800	1,360	2,000
Sweetgum	0.52	12,500	1,640	6,320	620	760	850	1,600
Sycamore	0.49	10,000	1,420	5,380	700	720	770	1,470
Yellow poplar	0.42	10,100	1,580	5,540	500	540	540	1,190
Bald cypress	0.46	10,600	1,440	6,360	730	270	510	1,000
Cedar, northern white	0.31	6,500	800	3,960	310	240	320	850
Cedar, Port Orford	0.42	11,300	1,730	6,470	620	400	560	1,080
Cedar, western red	0.33	7,700	1,120	5,020	490	220	350	860
Douglas fir, coast	0.48	12,400	1,950	7,240	800	340	710	1,130
Fir, white	0.39	9,800	1,490	5,810	530	300	480	1,100
Hemlock, western	0.45	11,300	1,640	7,110	550	340	540	1,250
Larch, western	0.52	13,100	1,870	7,640	930	430	830	1,360
Pine, lodgepole	0.41	9,400	1,340	5,370	610	290	480	880
Pine, ponderosa	0.40	9,400	1,290	5,320	580	420	460	1,130
Pine, loblolly	0.51	12,800	1,800	7,080	800	470	690	1,370
Pine, longleaf	0.58	14,700	1,990	8,440	960	470	870	1,500
Pine, western white	0.38	9,500	1,510	5,620	440	—	370	850
Redwood, old growth	0.40	10,000	1,340	6,150	700	240	480	940
Spruce, Engelmann	0.34	8,700	1,280	4,770	470	350	350	1,030
Spruce, Sitka	0.40	10,200	1,570	5,610	580	370	510	1,150

*Data compiled from Ref. 7.
†Specific gravity based on green volume and oven-dry weight.
‡To convert to SI units, multiply pounds per square inch by 0.1465 to obtain kilopascals.
§Load required to embed a 0.444-in ball to half its diameter.

TABLE 3-3 Grouping of Some Domestic Woods According to Heartwood Decay*

Resistant or very resistant	Moderately resistant	Slightly or nonresistant
Bald cypress (old growth)	Bald cypress (young growth)	Alder
Catalpa	Douglas-fir	Ashes
Cedars	Honeylocust	Aspens
Cherry, black	Larch, western	Basswood
Chestnut	Oak, swamp chestnut	Beech
Cypress, Arizona	Pine, eastern white	Birches
Junipers	Southern pine	Buckeye
Locust, black	Longleaf	Butternut
Mesquite	Slash	Cottonwood
Mulberry, red	Tamarack	Elms
Oak		Hackberry
Bur		Hemlocks
Chestnut		Hickories
Gambel		Magnolia
Oregon white		Maples
Post		Oaks (red and black species)
White		Pines (other than longleaf, slash, and eastern white)
Osage orange		Poplars
Redwood		Spruces
Sassafras		Sweetgum
Walnut, black		True firs (western and eastern)
Yew, Pacific		Willows
		Yellow poplar

*Compiled from Ref. 7.

The methods of applying the preservative may be divided into two classes: pressure and nonpressure. In pressure methods the wood is enclosed in a vessel, and the liquid preservative is forced into the wood under considerable hydrostatic pressure. Nonpressure methods do not utilize externally applied pressure, the preservative being applied by dipping, soaking, brushing, or spraying. Only pressure methods enable the preservative to penetrate deeply into the side grain for effective protection against biological attack.

Protection from Fire

The primary source of fire in buildings is usually its contents. Nevertheless, wood and wood products used in both structural members and as interior finish can be contributors to fire destruction. The most effective way to protect wood from exposure to fire is by pressure-impregnating it with fire-retardant chemicals. Commercial chemicals are mono- and diammonium phosphate, ammonium sulfate, borax, boric acid, and zinc chloride. Wood, properly treated, will be self-extinguishing once the primary source of heat and fire is exhausted or extinguished.

Protection from Corrosive Environments

Wood is degraded through exposure to strong acids and bases. Resistance to chemical attack is improved by various treatments including impregnation with coal tar, creosote–coal tar mixtures, or monomeric furfuryl alcohol that has been polymerized *in situ.*

Protection from Weathering

When unprotected wood is exposed to weather, the surface changes color, checks, and erodes away at the rate of about 0.25 in (6 mm) a century. If the wood does not decay, no protection may be needed. Often, however, mildew or water stains give the surface an undesirable appearance. This can be controlled by brushing the wood with a water-repellent finish that contains a fungicide. Pigmented penetrating stains are recommended for rough-sawn siding or plywood. Paints provide the most protection for wood against surface erosion, but paint will not prevent decay if conditions are favorable for fungal growth. Clear coatings of conventional spar or marine varnish are not recommended for exterior use.

STRESS GRADES AND WORKING STRESSES FOR LUMBER

The mechanical property values listed in Table 3-1 are ultimate values for wood free from defects. Lumber usually contains knots and cross grain. Therefore, structural lumber is graded by estimating strength and appearance. Working stresses for the structural grades of lumber are found in Ref. 6. Values for a few typical grades are shown in Table 3-4. Standard sizes for lumber are listed in Table 3-5. Working stresses vary according to the grade and size of lumber and their condition with respect to moisture content. Stresses are adjusted also for duration of load and for special conditions such as extreme temperature. Additional information may be found in references 1, 4, 5, and 7.

LAMINATED WOOD

Glued-laminated timbers, made with two or more layers of wood glued together with the grain of all layers approximately parallel, are widely used for structural timbers in buildings. Glued-laminated timbers may be straight or curved and may be designed to span more than 300 ft. Information on the design and use of such timbers may be found in references 1, 4, 5, and 7.

Wood-Base Panel Materials

Included in this category are plywood, insulating board, hardboard, particleboard, and the medium-density building fiberboards. Plywood is more dimensionally stable and more uniform in strength in the plane of the sheet than wood. Qualities of glue line and veneer permitted are set by the various commercial standards for plywood and determine the grades under which plywood is sold. In general, glue-line quality determines whether plywood is classed as being suitable for interior or exterior use.

U.S. Product Standard PS1(5.1) covers the basic specifications for the manufacture of construction plywood. Decorative hardwood plywood is described by U.S. Product Standard PS51(5.2). Plywood manufactured according to this standard will carry a grade trademark of a qualified testing agency.

Insulation board is of either interior or water-resistant quality and is usually manufactured for use where combinations of thermal and sound-insulating properties and stiffness and strength are desired. Hardboard with a density of 50 lb/ft^3 (0.8 g/cm^3) or more is used in many applications where a relatively thin, hard, uniform panel material is required.

Particleboard comes in many forms, including waferboards and flakeboards for exterior use. Thermosetting resins, usually urea- or phenolformaldehyde, are used to provide bonds of either interior or water-resistant quality. The important physical and mechanical properties of the various board products are summarized in Table 3-6.

TABLE 3-4 Typical Stress Grades and Working Stress for Structural Lumber (Normal Duration of Load and Dry Conditons of Use)*

Species	Grade	Size, in	Allowable Working Stress, lb/in²†			
			Bending (single member uses)	Horizontal shear	Compression parallel	Modulus of elasticity
California redwood	No. 1	2–4 thick	1400	100	1400	1,400,000
	No. 3	5 wide and wider	700	80	725	1,100,000
	Select decking	2 thick	1850	—	—	1,000,000
Douglas fir—larch	Select structural	2–4 thick, 5 wide and wider	1800	95	1400	1,800,000
	No. 3		725	95	675	1,500,000
	Dense No. 1	Beams and stringers	1550	85	1100	1,700,000
	Dense No. 1	Posts and timbers	1400	85	1200	1,700,000
Southern pine	Select structural	2–4 thick, 5 wide and wider	1350	75	1000	1,500,000
	No. 3		525	75	500	1,200,000
	No. 1	Beams and stringers	1000	70	675	1,300,000
	No. 1	Posts and timbers	925	70	750	1,300,000
	Commercial Dex	Decking	1100	—	—	1,300,000

*Compiled from "National Design Specification for Wood Construction," National Forest Products Association, Washington, D.C., 1977.

†To convert to SI units, multiply value in pounds per square inch by 6.895 to get kilopascals.

TABLE 3-5 Nominal and Minimum-Dressed Sizes of Boards, Dimension, and Timbers* (Thicknesses Apply to All Widths and All Widths to All Thicknesses)†

	Thicknesses			Face Widths		
		Minimum dressed			Minimum dressed	
Item	Nominal	Dry, in	Green, in	Nominal	Dry, in	Green, in
Boards	1	¾	25/32	2	1½	1 9/16
	1¼	1	1 1/32	3	2½	2 9/16
	1½	1¼	1 9/32	4	3½	3 9/16
				5	4½	4⅝
				6	5½	5⅝
				7	6½	6⅝
				8	7¼	7½
				9	8¼	8½
				10	9¼	9½
				11	10¼	10½
				12	11¼	11½
				14	13¼	13½
				16	15¼	15½
Dimension	2	1½	1 9/16	2	1½	1 9/16
	2½	2	2 1/16	3	2½	2 9/16
	3	2½	2 9/16	4	3½	3 9/16
	3½	3	3 1/16	5	4½	4⅝
				6	5½	5⅝
				8	7¼	7½
				10	9¼	9½
				12	11¼	11½
				14	13¼	13½
				16	15¼	15½
Dimension	4	3½	3 9/16	2	1½	1 9/16
	4½	4	4 1/16	3	2½	2 9/16
				4	3½	3 9/16
				5	4½	4⅝
				6	5½	5⅝
				8	7¼	7½
				10	9¼	9½
				12	11¼	11½
				14	—	13½
				16	—	15½
Timbers	5 and thicker	—	½ off	5 and wider	—	½ off

*Compiled from Ref. 6.
†To convert to SI units, divide dimension in inches by 2.54 to get centimeters.

SOURCES OF ADDITIONAL INFORMATION

Standards, specifications, and technical information on wood and wood products can be obtained from the various associations of the forest products industry. A few of these are listed here. A complete listing may be found in Ref. 3.

American Plywood Association, 1119 A. St., Tacoma, WA 98401.
American Wood-Preservers' Association, 1625 I St., Washington, DC 20036.
Forest Products Research Society, 2801 Marshall Court, Madison, WI 53705.
National Forest Products Association, 1619 Massachusetts Avenue, Washington, DC 20036.
National Particleboard Association, 2606 Perkins Place, Silver Spring, MD 20910.
U.S. Forest Products Laboratory, P.O. Box 5130, Madison, WI 53705.

TABLE 3-6 Strength and Mechanical Properties of Wood-Base Fiber and Particle Panel Materials*

Material	Specific gravity	Modulus of rupture, lb/in²†	Modulus of elasticity (bending), 1,000 lb/in²	Tensile strength parallel to surface, lb/in²	Tensile strength perpendicular to surface, lb/in²	Compression strength parallel to surface, lb/in²	24-h water absorption by weight, %	Thickness swelling, 24-h soak, %	Maximum linear expansion,‡ %	Thermal conductivity, (Btu) (in)/(h) (ft²) (°F)§
Building fiberboards										
1. Structural insulating board	0.16–0.42	200–800	25–125	200–500	10–25	—	—	—	0.2–0.5	0.27–0.45
2. Medium-density building fiberboard	0.53–0.80	1,900–6,000	325–700	1,000–4,000	40–200	1,000–3,500	—	2–10	0.2–1.3	0.50–0.60
3. Hardboard										
a. Untempered	0.80–1.28	3,000–7,000	400–800	3,000–6,000	—	1,800–6,000	3–30	10–25	0.2–0.4	0.54–0.75
b. Tempered	0.93–1.28	5,600–10,000	650–1,100	3,600–7,800	160–450	3,700–6,000	3–20	8–15	0.15–0.45	0.75–1.50
Particleboards										
1. Insulating type	0.40–0.59	800–1,400	150–250	—	20–30	—	—	—	0.30	0.55–0.75
2. Medium-density type	0.59–0.80	1,600–8,000	250–700	500–4,000	40–200	1,400–3,000	10–50	5–50	0.2–0.6	0.75–1.00
3. High-density type	0.80–1.12	2,400–7,500	350–1,000	1,000–5,000	125–450	3,500–5,200	15–40	15–40	0.2–0.85	1.00–1.25

*The data presented are general round-figure values accumulated from numerous sources. For more exact figures on a specific product, individual manufacturers may be consulted or actual tests made. Values are for general laboratory conditions of temperature and relative humidity.

†To convert to SI units, multiply by 6.895 to obtain kilopascals.

‡Expansion resulting from a change in moisture content from equilibrium at 50 percent RH to equilibrium at 90 percent RH.

§To convert to SI units, multiply by 1.602×10^4 to obtain values cm/(h)(m²)(°C).

REFERENCES

1. American Institute of Timber Construction: *Timber Construction Manual,* Wiley-Interscience, New York. 1974.
2. American Society for Testing and Materials: *Annual Book of ASTM Standards,* part 22, *Wood; Adhesives.* current edition, ASTM, Philadelphia.
3. *Directory of the Forest Products Industry,* Miller Freeman Publications, San Francisco, current ed.
4. Gurfinkel, G.: *Wood Engineering,* Southern Forest Products Association, New Orleans, 1973.
5. Hoyle, R. J., Jr.: *Wood Technology in the Design of Structures,* Mountain Press, Missoula, Mont., 1978.
6. National Forest Products Association: *National Design Specification for Wood Construction and Its Supplement,* current edition, NFPA, Washington, D.C.
7. U.S. Forest Products Laboratory: *Wood Handbook: Wood as an Engineering Material,* Agriculture Handbook No. 72, current edition, U.S. Dept. of Agriculture, Washington, D.C.

printers, and communication controllers. Quite often, minicomputers are linked into communication networks with a host system.

The rest of this chapter concentrates on minicomputers since they are the most universally applicable to plant engineering. Unless otherwise noted, the term "computers" refers to minicomputers.

Equipment Costs

Equipment and software costs vary greatly. For example, the chips used for microcomputers can be purchased for less than $100, whereas a mainframe computer costs millions of dollars. Software can be purchased for anywhere from a few hundred dollars to a few hundred thousand dollars. (All price citations are in 1981 dollars.)

Minicomputers can be purchased or rented for almost any application. The cost of the equipment must be broken down into its components (CPU, printer, display, storage). $10,000 would be the minimum purchase price of a midrange CPU in a minicomputer. This price can easily be brought up to close to $100,000 for some of the more complex and sophisticated CPUs. Desk-top printers (print speed measured in characters per second) start at $2500, while line printers (speed measured in lines per minute) start at $10,000. Simple displays start at $2000; with some of the more complex display units, with definable program keys are priced at $5000 and up. Graphic display terminals are also available. These terminals carry prices of $3000 and up, depending upon size and color capability. Disk storage systems start at $10,000 for disks of 50 to 100 megabytes. Diskette storage units are priced at $500 plus the price of the diskettes, which are $3 to 5 each.

As can be seen by the prices shown above, the minicomputer systems for use in plant engineering can start at $15,000 and go up to $100,000 with the addition of the peripherals usually needed. The rental fee for a machine of this type is typically 2 to 10 percent of the purchase price. Many systems of interest to the plant engineer are sold on a *turnkey basis*. This means that a vendor supplies both the computer hardware and application software to meet the user's needs. The price of such packages varies greatly depending upon their size and complexity.

APPLICATION AREAS

The applications of concern to the plant engineer can include systems that handle departmental functions, energy management, process control, computer-aided design, and maintenance. This section presents a brief list of typical applications, the functions and requirements of each, and the benefits to be derived from the system.

Departmental Functions

Budget System

This system could track planned expenditures for a given project. The expenses can be broken out by time period, type of expenditure (capital or expense), requesting department, budgeted or approved item, and actual charges against that budgeted item. The system would assist in projecting expenses. The main benefit of the system is for performing the "number crunching" involved in generating the various totals that are part of budgets. A system of this nature is ideal for placing on a minicomputer and, if extended over many departments, could justify placement on a mainframe computer.

Return on Investment (ROI)

This package assists the engineer in preparing the economic justification for a project. By inputting costs associated with various alternative solutions to a project, tax rates, and the time value of money, the program calculates the present worth of each alternative and the associated rate of return for any incremental investment made. This then allows selection of the most attractive alternative based on economic terms.

Space Planning

A computer can be used to track space occupants. The system could break space into various types (production, office, storage, lab, service, rearrangement, etc.), the occupants (purchasing, personnel, engineering, etc.), and where the space is located (building, floor, zone). The system should be in two data bases: one showing current space allocation by building, type, and occupant; the other showing projected space for the current year, the following year, and several successive years depending on the extent of the user's planning cycle. The system shows how space could change hands over time (which is useful in planning for rearrangements) and when new space is needed for expansion. The main benefit from this system is to enable the plant engineer to keep accurate records of space allocation and to permit future planning so that space needed to accomplish company objectives is available on a timely basis.

Project Status and Tracking

This system could consist of the assignments for the individual plant engineer with an assigment number, description, user department, budget number, planned hours, total hours, hours by quarters for large projects, total planned dollars, and expended dollars. Reports could be generated for the engineer to show a list of assignments, for the department manager to show jobs completed last month, jobs open by engineer, all jobs related by budget number, etc. The benefits of this system accrue to both engineer and department manager. For the engineer, it both keeps track of the projects with all the needed information and helps plan work for the coming months. For the department manager, it keeps track of all department work; for personnel planning, it helps accommodate future work.

Energy Management

Energy management software that runs on minicomputers can be purchased from a number of firms. For energy management, the minicomputer has to be connected to the boiler, fan, thermostat, and other controls via a sensor input/output and a multiplexing network (for large systems) in order to effectively manage the plant energy (heating, cooling, lighting) systems. Most of the software can be tailored to fit an individual user's requirements through user changes. The main benefit of this software is cost savings through energy conservation. See also Sec. 15.

Process Controls

Process controls, or machine control, occurs when the microcomputer, usually in the form of a programmable controller, performs one of its most useful functions. The microcomputer can replace banks of relay control panels at less cost and with greater ease of maintenance than mainframes. These applications are cost-justified on the basis of reduced levels of manual intervention to operate a system. See also Chapter 11-2.

Computer-Aided Design (CAD)

This can be a very useful tool for the plant engineer. There are several software packages capable of providing the needed drawing capability unique to plant engineering. Most packages require a mainframe computer but there is a rapid trend toward increased use of minicomputers. CAD has many benefits. It permits engineers to design without having to sit at a drawing board for extended periods. The engineer can make rough sketches and give them to the CAD operator for actual execution. The need for originals, brown lines, etc., is removed, because all drawings are stored in the computer. There is a large cost element for most CAD systems in that they usually must run on a mainframe computer with all its associated costs, but such justification is normally not difficult for a plant engineering section comprising 25 or more engineers. Smaller plants may justify minicomputer programs to the extent of their availability.

Maintenance

Inventory Control

The computer can keep track of all spare-parts numbers, quantity on hand, order quantity, and location of parts. This, coupled with a requisition in/out system for parts withdrawal or entry into the stocking areas, can help reduce the total size of inventory, as well as keep strict accountability of the spare-parts stock.

Preventive Maintenance (PM) Management

The system would keep track of all PM activity. The system could issue PM sheets daily for the machines that needed PM that day, based on the date of the machines' most recent PM. The machine could also keep track of the specifics of the PM and issue reports to management based on each user's need.

PLANNING CONSIDERATIONS

This section discusses planning for a computer system, initial sizing and justifying of the system, physical planning, and system security.

Sizing

To size a computer system, the user (if experienced) must rely on a vendor's assistance or on a consultant to ensure that the system will accomplish what the user wishes it to do. A starter system for the user who wishes to develop an individual application program could be a 128K CPU, with at least a 20-megabyte disk unit. The reason for the 20-megabyte minimum is that some operating systems need 7 to 8 megabytes on disk for operating-system use. Most vendors are willing to aid in this sizing by translating the user's requirements (response time, number of records, number of terminals, etc.) into the appropriate system requirements.

Justification

Most engineers, when first confronted with having to try to justify a computer system, cannot understand how such a system can be economically feasible. "But it costs $25,000." and "It takes up 400 square feet." are typical comments. *A computer system is justified by the worker-hours that will be saved by having the computer do the work compared with the cost of having a person do the work.* In order to justify a computer system, the user must know the projected savings (i.e., worker-hours saved times cost per worker-hour, including benefits) and the system and support costs (i.e., rental compared with purchase price, construction costs to house the machine if needed, monthly maintenance contract, electricity, etc.). The user may then run an ROI computation to determine if the labor saved will be sufficient to overcome the initial and continuing costs of owning or renting a computer system.

Physical Planning

Once the computer equipment has been selected, a layout or physical plan should be made. This will ensure that enough room is allowed for operating and servicing the equipment. To prepare a physical plan, carry out the following necessary steps:

1. Mark off on a grid the dimensions of the site, room, and floor where the equipment is to be located. A scale of ¼ in to 1 ft (1cm to 50 cm) is typical.
2. Make templates of the machines to be installed. Include on the template all required service and operating clearances as specified by the manufacturer. See Fig. 1-1 for typical equipment arrangements to observe.

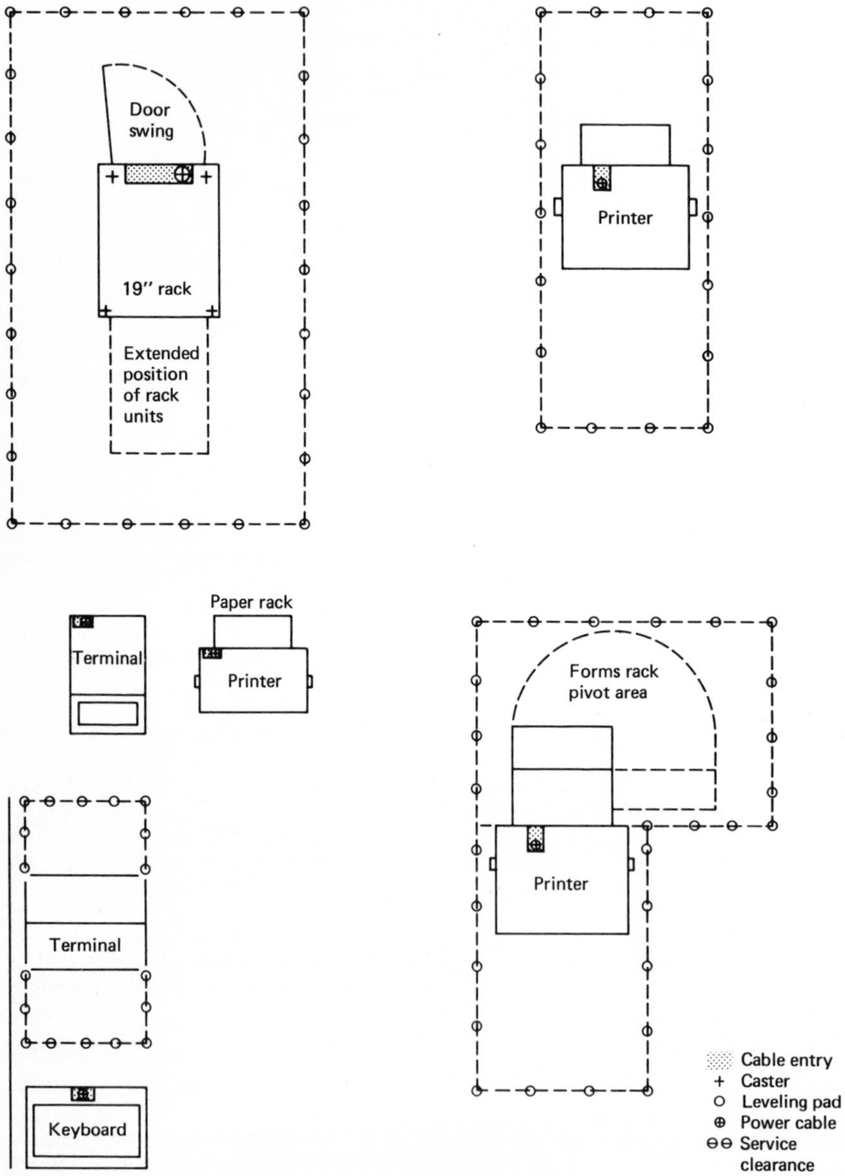

Figure 1-1 Typical equipment clearances.

3. Position the templates on the grid. You will then be able to observe if there is enough space and determine the best arrangement for the equipment.
4. Include all noncomputer equipment such as desks, tables, chairs, storage files, etc. Be sure to locate all columns and walls. See Fig. 1-2 for typical layouts.
5. Check cabling between pieces of equipment.
6. Check lengths of electric power cord for all equipment.

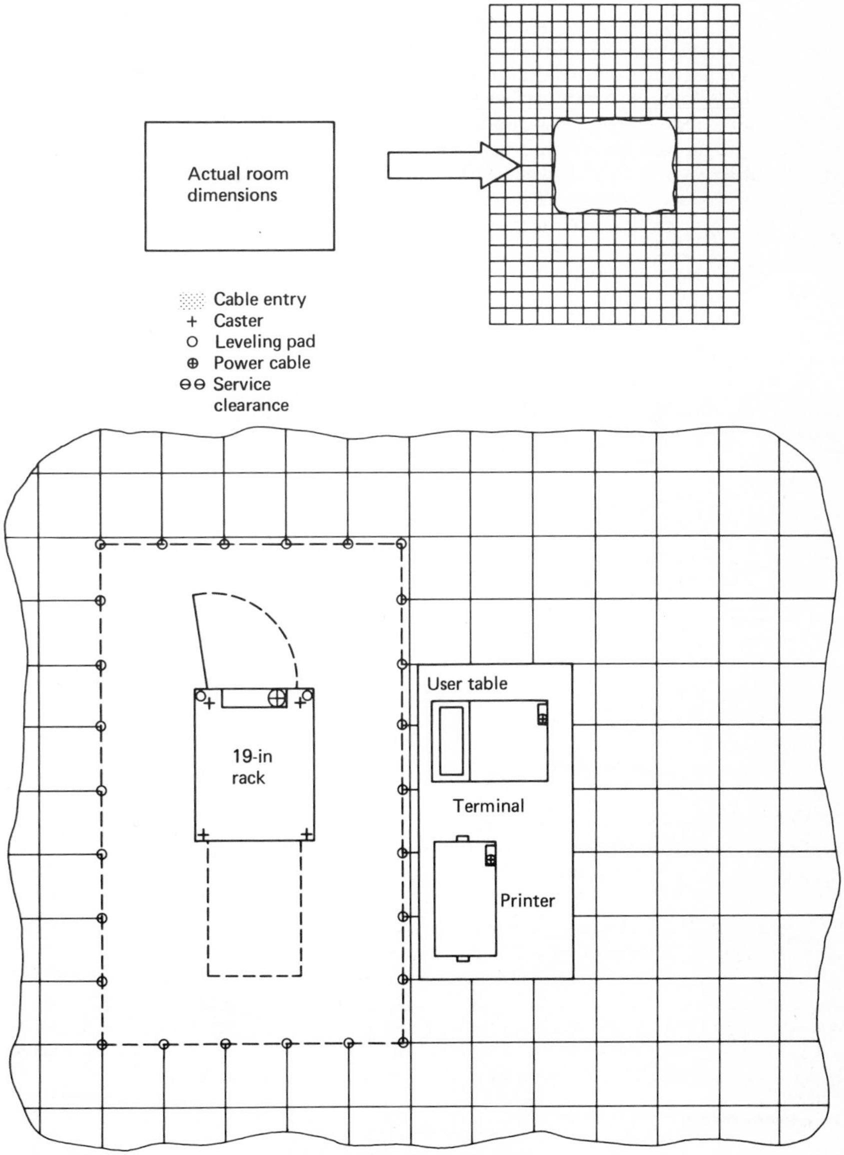

Figure 1-2 Typical layout.

7. Consider raised flooring if there are many cables that could present tripping hazards.
8. Investigate electrical, heating, and cooling requirements.

If the foregoing steps are followed, the layout produced should permit the installation of the equipment without any last-minute surprises, such as insufficient space planned or too little electric power provided.

Security

The purpose of system security, whether physical machine security or data-software security, is to prevent unauthorized access to the machine or to data within the machine. However, because no security system is perfect, a security system must also be capable of determining when security has been breached.

Physical security may consist simply of a desk-top computer that is stored in a locked room at night, or may be as complex as a computer room with door locks, on the top floor of a building that is kept locked, with security personnel patrolling the area. For a departmental system, physical security usually consists of one or more of the following methods of securing the machine:

1. For a desk-top unit, simply locking it up nights
2. For larger units, putting the CPU, printers, disks, and system terminal behind a locked door
3. A lockable ON-OFF power switch

For the minicomputer systems that may have the CPU behind locked doors, there are usually terminals that are spread out among the engineering area, and these terminals are usually *not* physically secure. This brings the discussion to *software security,* which is the key to preventing unauthorized use of computers. Software security will normally consist of sign-on numbers, passwords, user classes, and logging of all unauthorized attempts to access the system or data. The following description is the minimum software security system recommended for a minicomputer that consists of

1. CPU, printers, disk storage, and a system terminal in a locked room
2. User terminals in an unsecured area
3. Multiple users and programs running concurrently, using the same or various data bases (i.e., budget system, project tracking, maintenance tracking).

Because there are users from various areas using different data and because the user terminals are unsecured, a software security package must accomplish the following:

1. Prevent unauthorized sign-on
2. Prevent unauthorized users from accessing data that they do not have a need to know
3. Record attempted violations

To accomplish this, several system-wide programs must be written and installed. The first is a sign-on program that accesses a sign-on file to determine if the user number entered is in the file, if the correct password is entered, and if the password has not expired. If all three conditions are met, then the user is permitted to have access to the desired program. At this point, a validation routine is entered that accesses the sign-on file for the user classification and determines if the user is authorized to have access to the program and data. There should also be an error-logging routine and data set that will log all attempts at unauthorized access.

MAINTENANCE

System maintenance is broken down into hardware and software maintenance. *Hardware maintenance* involves the physical equipment, chips, disks, cabling, printers, and all other equipment attached to the CPU. If the user purchases a machine, whether a microcomputer or minicomputer, it is recommended that the user also purchase hardware maintenance services from the hardware manufacturer. *It is not recommended that hardware maintenance be attempted by the average user.* It is also recommended that the user purchase the programming language (APL, Basic, Fortran, PL/I) from a vendor and *have the vendor maintain the basic programming language.* The user can also purchase many different and varied "canned" software packages that may serve well enough to prevent the need to develop individual programs, thereby eliminating the necessity of ongoing software support. A word of caution: If the user changes a vendor's software package, any warranty is normally voided.

At the present time, numerous software programs are being developed for many plant engineering applications. References 2 and 3 provide up-to-date listings and sources for such programs.

REFERENCES

1. *American National Dictionary for Information Processing,* American National Standards Institute, Inc., New York, NY 10018.
2. *Engineering Computer Applications Newsletter,* Engineering Computer Applications, Inc., P.O. Box 3109, Englewood, CO 80111.
3. *The APEC Journal,* Automated Procedures for Engineering Consultants, Inc., Miami Valley Tower, Dayton, OH 45402.

appendix A

Sources of Up-to-Date Technical Information

prepared by

Robert C. Rosaler, P.E.
Vice President
James O. Rice Associates, Inc.
New York, New York

INTRODUCTION

Technical information is proliferating at ever-increasing rates, and the modern plant engineer *must* keep abreast of new developments to remain effective at the job. Reading trade magazines is helpful, but it should be only a starting point for the active practicing plant engineer. Consistent attendance at professional society meetings, industrial conferences, and trade shows is necessary to keep completely up to date.

Appendix A is a guide to such information sources.

TRADE PUBLICATIONS

Table A-1 lists current trade publications prominent in plant engineering. (Publications of professional societies are not included.) Although there is some inevitable duplication in coverage, these publications should be read on a regular basis for information pertinent to the engineer's area of activity. Most of these publications are available to plant engineers at no charge.

PROFESSIONAL SOCIETIES

These organizations provide opportunities for the plant engineer to associate with peers at regular intervals to exchange experiences and obtain new technical information. The kind of information received from these sources usually has more detail and depth than

TABLE A-1 Trade Publications

Name	Frequency	Published by
Energy User News	Biweekly	Technical Publishing Co., 1301 So. Grove Ave., Barrington, IL 60010
Industrial Maintenance and Plant Operation	Monthly	Ames Publishing Co., 1 W. Olney Ave., Philadelphia, PA 19120
Instruments & Control Systems	Monthly	Chilton Company, Chilton Way Radnor, PA 19089
Engineer's Digest	Monthly	Walker Davis Publications, Inc., P.O. Box 482 2500 Office Center, Willow Grove, PA 19090
Plant Energy Management	Monthly	
Plant Engineering	Biweekly	Technical Publishing Co., 1301 So. Grove Ave., Barrington, IL 60010
Pollution Engineering	Monthly	
Plant Engineering & Maintenance	Monthly	Clifford/Elliot Ltd., 1289 Marlborough Court Oakville, Ontario, Canada

commercial publications offer. All societies publish magazines for members, and most of these periodicals are published monthly. The societies listed below have large numbers of plant engineer members:

American Institute of Plant Engineers (AIPE), 3975 Erie Ave., Cincinnati, OH 45208
American Society of Heating, Refrigerating and Air Conditioning Engineers (ASHRAE), 345 E. 47th St., New York, NY 10017
American Society of Mechanical Engineers (ASME), 345 E. 47th St., New York, NY 10017
Association of Energy Engineers, Suite 340, 4025 Pleasantdale Rd., Atlanta, GA 30340
National Association of Corrosion Engineers (NACE), 1440 South Creek, Houston, TX 77084
National Society of Professional Engineers (NSPE), 2029 K St. N.W., Washington, DC 20006

TRADE ASSOCIATIONS

Practically every product manufacturer is represented by a trade association. Table A-2 includes some of the larger groups. For a complete listing of trade associations (and professional societies as well) see Ref. 1.

Most trade associations publish high-quality objective and educational information relating to their trade. They are an excellent first stop when researching an unfamiliar technical area.

TABLE A-2 Trade Associations

Association of Asbestos Cement Pipe Producers, 1600 Wilson Blvd., Arlington, VA 22209
Aluminum Association, 818 Connecticut Ave. N.W., Washington, DC 20006
American Boiler Manufacturers Association, 1500 Wilson Blvd., Arlington, VA 22209
American Concrete Institute, P.O. Box 1950, Redford Sta., Detroit, MI 48219
American Gear Manufacturers Association, 1901 N. Fort Myer Dr., Arlington, VA 22209
American Institute of Steel Construction, 400 N. Michigan Ave., Chicago, IL 60611
American Iron and Steel Institute, 1000 16th St. N.W., Washington, DC 20036
American Welding Society, Inc., 2501 N.W. 7th St., Miami, FL 33125
Forest Products Research Society, 2801 Marshall Ct., Madison, WI 53705
National Electrical Contractors Association, 7315 Wisconsin Ave., Washington, DC 20014
National Electrical Manufacturers Association, 2101 L St. N.W., Washington, DC 20037
National Fire Protection Association, Batterymarch Park, Quincy, MA 02269
National Roofing Contractors Association, 1515 N. Harlem Ave., Oak Park, IL 60302
National Solid Wastes Management Association, 1120 Connecticut Ave. N.W., Washington, DC 20036
Plastics Pipe Institute, 369 Lexington Ave., New York, NY 10017
Portland Cement Association, 5420 Old Orchard Rd., Skokie, IL 60077
Valve Manufacturers Association, P.O. Drawer II, McLean, VA 22101

INDUSTRIAL TRADE SHOWS AND CONFERENCES

Shows offer a convenient opportunity to observe new products first-hand and to discuss applications of such products with company representatives. Many shows have concurrent conferences and seminars presented by leading experts in all fields. Table A-3 lists pertinent data on leading United States of America trade shows. An annual list of all trade shows is given in Ref. 2.

TABLE A-3 Industrial Trade Shows

Name	Frequency	Managers
National Plant Engineering & Maintenance Show	Annual	Clapp & Poliak, Inc., 708 Third Ave. New York, NY 10017
Plant Engineering & Maintenance Show/East	Biennial	Clapp & Poliak, Inc., 708 Third Ave. New York, NY 10017
Plant Engineering & Maintenance Show/West	Biennial	Clapp & Poliak, Inc., 708 Third Ave. New York, NY 10017
Southern Plant Engineering & Maintenance Show	Biennial	Clapp & Poliak, Inc., 708 Third Ave. New York, NY 10017
Western Plant Engineering & Maintenance Show	Annual	American Institute of Plant Engineers, 3975 Erie Ave., Cincinnati, OH 45208

CONSULTING ENGINEERS

By accumulating varied experience in their specialized areas, firms of consulting engineers constitute an in-depth resource of up-to-date technical information readily available. Use of these organizations to supplement in-house capability in specific situations will usually pay off in time and expenditures.

Before retaining such organizations, their experience and background should be thoroughly explored, discussions held with principal engineers, and proposals solicited. Although price should, of course, be a factor in selection among competitors, the reputation and accomplishments of the consultant *in the field involved* is the prime consideration. See Ref. 3 for a directory listing consulting engineers.

MANUFACTURERS' CATALOGS

All engineers should keep the latest catalogs of frequently used products in their library at all times. In addition to excellent applications information, such data are necessary to make certain that current models are ordered.

It is always good policy to keep at least two or three competitive manufacturers' catalogs in each pertinent product line. This avoids excessive dependence on one design or method that may not be optimum for all applications.

REFERENCES

1. *National Trade and Professional Associations of the United States & Canada,* Columbia Books, Washington, D.C., published annually.
2. *Directory of Conventions,* Bill Communications, Philadelphia, Pa., published annually.
3. *Directory of Consulting Engineers,* the Association of Consulting Management Engineers, New York, N.Y., published annually.

appendix B

Metric Conversion Tables

Editor's Note:
Metric conversions, in terms of SI (International System) units are given in most of the text, graphs, and tables in this handbook. However, practical considerations precluded direct conversion in every instance.

Appendix B presents a table of convenient conversion factors into SI units from U.S. Customary and non-SI metric units for frequently used physical quantities involved in plant engineering.

CONVERSION TO SI UNITS: LISTING BY PHYSICAL QUANTITY

The first two digits of each numerical entry represent a power of 10. For example, the entry "−02 2.54" expresses the fact that 1 inch = 2.54×10^{-2} meter.

To Convert from	to	Multiply by
	Acceleration	
foot/second^2	meter/second^2	−01 3.048
inch/second^2	meter/second^2	−02 2.54
	Area	
acre	meter^2	+03 4.046
circular mil	meter^2	−10 5.067
foot^2	meter^2	−02 9.290
hectare	meter^2	+04 1.00
inch^2	meter^2	−04 6.4512
mile^2 (U.S. statute)	meter^2	+06 2.589
yard^2	meter^2	−01 8.361

To Convert from	to	Multiply by
	Density	
gram/centimeter³	kilogram/meter³	+03 1.00
lbm/inch³	kilogram/meter³	+04 2.768
lbm/foot³	kilogram/meter³	+01 1.602
slug/foot³	kilogram/meter³	+02 5.154
	Energy	
British thermal unit (ISO/TC 12)	joule	+03 1.055
British thermal unit (International Steam Table)	joule	+03 1.055
British thermal unit (mean)	joule	+03 1.056
British thermal unit (thermochemical)	joule	+03 1.054
British thermal unit (39° F)	joule	+03 1.060
British thermal unit (60° F)	joule	+03 1.055
calorie (International Steam Table)	joule	+00 4.187
calorie (mean)	joule	+00 4.190
calorie (thermochemical)	joule	+00 4.184
calorie (15°C)	joule	+00 4.186
calorie (20°C)	joule	+00 4.182
calorie (kilogram, (International Steam Table)	joule	+03 4.187
calorie (kilogram, mean)	joule	+03 4.190
calorie (kilogram, thermochemical)	joule	+03 4.184
foot lbf	joule	+00 1.356
killowatthour	joule	+06 3.60
ton (nuclear equivalent of TNT)	joule	+09 4.20
watt hour	joule	+03 3.60
	Energy/Area Time	
Btu (thermochemical)/foot² second	watt/meter²	+04 1.135
Btu (thermochemical)/foot² minute	watt/meter²	+02 1.891
Btu (thermochemical)/foot² hour	watt/meter²	+00 3.152
Btu (thermochemical)/inch² second	watt/meter²	+06 1.634
calorie (thermochemical)/cm² minute	watt/meter²	+02 6.973
watt/centimeter²	watt/meter²	+04 1.00
	Force	
dyne	newton	−05 1.00
kilogram force (kgf)	newton	+00 9.807
lbf (pound force, avoirdupois)	newton	+00 4.448

To convert from	to	Multiply by
ounce force (avoirdupois)	newton	−01 2.780
pound force, lbf (avoirdupois)	newton	+00 4.448
	Length	
caliber	meter	−04 2.54
chain (surveyor or gunter)	meter	+01 2.012
chain (engineer or ramden)	meter	+01 3.048
cubit	meter	−01 4.572
fathom	meter	+00 1.829
foot	meter	−01 3.048
foot (U.S. survey)	meter	−01 3.048
furlong	meter	+02 2.012
inch	meter	−02 2.54
link (engineer's or ramsden)	meter	−01 3.048
link (surveyor's or gunter)	meter	−01 2.012
micron	meter	−06 1.00
mil	meter	−05 2.54
mile (U.S. statute)	meter	+03 1.609
mile (international nautical)	meter	+03 1.852
mile (U.S. nautical)	meter	+03 1.852
yard	meter	−01 9.144
	Mass	
carat (metric)	kilogram	−04 2.00
ounce mass (avoirdupois)	kilogram	−02 2.835
ounce mass (troy or apothecary)	kilogram	−02 3.110
pound mass, lbm (avoirdupois)	kilogram	−01 4.536
pound mass (troy or apothecary)	kilogram	−01 3.732
ton (long)	kilogram	+03 1.016
ton (metric)	kilogram	+03 1.00
ton (short, 2000 pound)	kilogram	+02 9.072
	Power	
Btu (thermochemical)/second	watt	+03 1.054
Btu (thermochemical)/minute	watt	+01 1.757
calorie (thermochemical)/second	watt	+00 4.184
calorie (thermochemical)/minute	watt	−02 6.973
foot lbf/hour	watt	−04 3.766

To convert from	to	Multiply by
foot lbf/minute	watt	−02 2.260
foot lbf/second	watt	+00 1.356
horsepower (550 foot lbf/second)	watt	+02 7.457
horsepower (boiler)	watt	+03 9.809
horsepower (electric)	watt	+02 7.46
horsepower (metric)	watt	+02 7.355
horsepower (U.K.)	watt	+02 7.457
horsepower (water)	watt	+02 7.460
kilocalorie (thermochemical)/minute	watt	+01 6.973
kilocalorie (thermochemical)/second	watt	+03 4.184
	Pressure*	
atmosphere	newton/meter^2	+05 1.013
bar	newton/meter^2	+05 1.00
centimeter of mercury (0°C)	newton/meter^2	+03 1.333
centimeter of water (4°C)	newton/meter^2	+01 9.806
dyne/centimeter^2	newton/meter^2	−01 1.00
foot of water (39.2°F)	newton/meter^2	+03 2.989
inch of mercury (32°F)	newton/meter^2	+03 3.386
inch of mercury (60°F)	newton/meter^2	+03 3.377
inch of water (39.2°F)	newton/meter^2	+02 2.490
inch of water (60°F)	newton/meter^2	+02 2.488
kgf/centimeter^2	newton/meter^2	+04 9.807
kgf/meter^2	newton/meter^2	+00 9.807
lbf/foot^2	newton/meter^2	+01 4.788
lbf/inch^2 (psi)	newton/meter^2	+03 6.895
millibar	newton/meter^2	+02 1.00
millimeter of mercury (0°C)	newton/meter^2	+02 1.333
pascal	newton/meter^2	+00 1.00
psi (lbf/inch^2)	newton/meter^2	+03 6.895
	Speed	
foot/hour	meter/second	−05 8.467
foot/minute	meter/second	−03 5.08
foot/second	meter/second	−01 3.048
inch/second	meter/second	−02 2.54
kilometer/hour	meter/second	−01 2.778
knot (international)	meter/second	−01 5.144
mile/hour (U.S. statute)	meter/second	−01 4.470
mile/minute (U.S. statute)	meter/second	+01 2.682
mile/second (U.S. statute)	meter/second	+03 1.609

*The pascal is preferred here for most purposes. It is less cumbersome to use.

To convert from	to	Multiply by
	Temperature	
degree Celsius	kelvin (K)	K = °C + 273.15
degree Fahrenheit	kelvin	K = 5/9 (°F + 459.67)
degree Fahrenheit	degree Celsius	°C = 5/9 (°F − 32)
degree Rankine	kelvin	K = 5/9 (°R)
	Viscosity	
centistoke	meter²/second	−06 1.00
stoke	meter²/second	−04 1.00
foot²/second	meter²/second	−02 9.290
centipoise	newton second/meter²	−03 1.00
lbm/foot second	newton second/meter²	+00 1.488
lbf second/foot²	newton second/meter²	+01 4.788
poise	newton second/meter²	−01 1.00
poundal second/foot²	newton second/meter²	+00 1.488
slug/foot second	newton second/meter²	+01 4.788
	Volume	
acre foot	meter³	+03 1.233
barrel (petroleum, 42 gallons)	meter³	−01 1.590
board foot	meter³	−03 2.360
cord	meter³	+00 3.625
dram (U.S. fluid)	meter³	−06 3.697
fluid ounce (U.S.)	meter³	−05 2.957
foot³	meter³	−02 2.832
gallon (U.S. dry)	meter³	−03 4.405
gallon (U.S. liquid)	meter³	−03 3.785
inch³	meter³	−05 1.639
liter	meter³	−03 1.00
ounce (U.S. fluid)	meter³	−05 2.957
pint (U.S. dry)	meter³	−04 5.506
pint (U.S. liquid)	meter³	−04 4.732
quart (U.S. dry)	meter³	−03 1.101
quart (U.S. liquid)	meter³	−04 9.466
stere	meter³	+00 1.00
yard³	meter³	−01 7.646

index